18. $$\int \frac{u^2\,du}{\sqrt{a+bu}} = \frac{2}{15b^3}(3b^2u^2 - 4abu + 8a^2)\sqrt{a+bu} + C$$

19. $$\int \frac{u^n\,du}{\sqrt{a+bu}} = \frac{2u^n\sqrt{a+bu}}{b(2n+1)} - \frac{2an}{b(2n+1)}\int \frac{u^{n-1}\,du}{\sqrt{a+bu}}$$

20. $$\int \frac{du}{u\sqrt{a+bu}} = \begin{cases} \dfrac{1}{\sqrt{a}}\ln\left|\dfrac{\sqrt{a+bu}-\sqrt{a}}{\sqrt{a+bu}+\sqrt{a}}\right| + C & \text{if } a > 0 \\ \dfrac{2}{\sqrt{-a}}\tan^{-1}\sqrt{\dfrac{a+bu}{-a}} + C & \text{if } a < 0 \end{cases}$$

21. $$\int \frac{du}{u^n\sqrt{a+bu}} = -\frac{\sqrt{a+bu}}{a(n-1)u^{n-1}} - \frac{b(2n-3)}{2a(n-1)}\int \frac{du}{u^{n-1}\sqrt{a+bu}}$$

22. $$\int \frac{\sqrt{a+bu}\,du}{u} = 2\sqrt{a+bu} + a\int \frac{du}{u\sqrt{a+bu}}$$

23. $$\int \frac{\sqrt{a+bu}\,du}{u^n} = -\frac{(a+bu)^{3/2}}{a(n-1)u^{n-1}} - \frac{b(2n-5)}{2a(n-1)}\int \frac{\sqrt{a+bu}\,du}{u^{n-1}}$$

Forms Containing $a^2 \pm u^2$

24. $$\int \frac{du}{a^2+u^2} = \frac{1}{a}\tan^{-1}\frac{u}{a} + C$$

25. $$\int \frac{du}{a^2-u^2} = \frac{1}{2a}\ln\left|\frac{u+a}{u-a}\right| + C = \begin{cases} \dfrac{1}{a}\tanh^{-1}\dfrac{u}{a} + C & \text{if } |u| < a \\ \dfrac{1}{a}\coth^{-1}\dfrac{u}{a} + C & \text{if } |u| > a \end{cases}$$

26. $$\int \frac{du}{u^2-a^2} = \frac{1}{2a}\ln\left|\frac{u-a}{u+a}\right| + C = \begin{cases} -\dfrac{1}{a}\tanh^{-1}\dfrac{u}{a} + C & \text{if } |u| < a \\ -\dfrac{1}{a}\coth^{-1}\dfrac{u}{a} + C & \text{if } |u| > a \end{cases}$$

Forms Containing $\sqrt{u^2 \pm a^2}$

In formulas 27 through 38, we may replace

$\ln(u + \sqrt{u^2 + a^2})$ by $\sinh^{-1}\dfrac{u}{a}$

$\ln|u + \sqrt{u^2 - a^2}|$ by $\cosh^{-1}\dfrac{u}{a}$

$\ln\left|\dfrac{a + \sqrt{u^2 + a^2}}{u}\right|$ by $\sinh^{-1}\dfrac{a}{u}$

27. $$\int \frac{du}{\sqrt{u^2 \pm a^2}} = \ln|u + \sqrt{u^2 \pm a^2}| + C$$

28. $$\int \sqrt{u^2 \pm a^2}\,du = \frac{u}{2}\sqrt{u^2 \pm a^2} \pm \frac{a^2}{2}\ln|u + \sqrt{u^2 \pm a^2}| + C$$

29. $$\int u^2\sqrt{u^2 \pm a^2}\,du = \frac{u}{8}(2u^2 \pm a^2)\sqrt{u^2 \pm a^2} - \frac{a^4}{8}\ln|u + \sqrt{u^2 \pm a^2}| + C$$

30. $$\int \frac{\sqrt{u^2+a^2}\,du}{u} = \sqrt{u^2+a^2} - a\ln\left|\frac{a + \sqrt{u^2+a^2}}{u}\right| + C$$

31. $$\int \frac{\sqrt{u^2-a^2}\,du}{u} = \sqrt{u^2-a^2} - a\sec^{-1}\left|\frac{u}{a}\right| + C$$

32. $$\int \frac{\sqrt{u^2 \pm a^2}\,du}{u^2} = -\frac{\sqrt{u^2 \pm a^2}}{u} + \ln|u + \sqrt{u^2 \pm a^2}| + C$$

33. $$\int \frac{u^2\,du}{\sqrt{u^2 \pm a^2}} = \frac{u}{2}\sqrt{u^2 \pm a^2} - \frac{\pm a^2}{2}\ln|u + \sqrt{u^2 \pm a^2}| + C$$

34. $$\int \frac{du}{u\sqrt{u^2+a^2}} = -\frac{1}{a}\ln\left|\frac{a + \sqrt{u^2+a^2}}{u}\right| + C$$

35. $$\int \frac{du}{u\sqrt{u^2-a^2}} = \frac{1}{a}\sec^{-1}\left|\frac{u}{a}\right| + C$$

36. $$\int \frac{du}{u^2\sqrt{u^2 \pm a^2}} = -\frac{\sqrt{u^2 \pm a^2}}{\pm a^2 u} + C$$

37. $$\int (u^2 \pm a^2)^{3/2}\,du = \frac{u}{8}(2u^2 \pm 5a^2)\sqrt{u^2 \pm a^2} + \frac{3a^4}{8}\ln|u + \sqrt{u^2 \pm a^2}| + C$$

38. $$\int \frac{du}{(u^2 \pm a^2)^{3/2}} = \frac{u}{\pm a^2\sqrt{u^2 \pm a^2}} + C$$

Forms Containing $\sqrt{a^2 - u^2}$

39. $$\int \frac{du}{\sqrt{a^2-u^2}} = \sin^{-1}\frac{u}{a} + C$$

40. $$\int \sqrt{a^2-u^2}\,du = \frac{u}{2}\sqrt{a^2-u^2} + \frac{a^2}{2}\sin^{-1}\frac{u}{a} + C$$

41. $$\int u^2\sqrt{a^2-u^2}\,du = \frac{u}{8}(2u^2-a^2)\sqrt{a^2-u^2} + \frac{a^4}{8}\sin^{-1}\frac{u}{a} + C$$

42. $$\int \frac{\sqrt{a^2-u^2}\,du}{u} = \sqrt{a^2-u^2} - a\ln\left|\frac{a + \sqrt{a^2-u^2}}{u}\right| + C = \sqrt{a^2-u^2} - a\cosh^{-1}\frac{a}{u} + C$$

43. $$\int \frac{\sqrt{a^2-u^2}\,du}{u^2} = -\frac{\sqrt{a^2-u^2}}{u} - \sin^{-1}\frac{u}{a} + C$$

(This table is continued on the back endpapers)

Louis Leithold

THE CALCULUS with Analytic Geometry

FOURTH EDITION

HARPER & ROW, PUBLISHERS
New York

Cambridge
Hagerstown
Philadelphia
San Francisco

London
Mexico City
São Paulo
Sydney

1817

THE COVER AND CHAPTER OPENING ARTIST

Roman Opalka was born in France in 1931 and spent most of his adult life in Poland. Internationally recognized as the leading European maker of Conceptual Art, Opalka has had museum retrospectives all over the world, is represented in the collections of the world's great museums, and has won numerous major international awards.

In 1965, in Warsaw, Roman Opalka embarked upon an elaborate ritual of counting and painting that he intends to continue for the rest of his life. Starting with a tiny white "1" in the upper left-hand corner of a black canvas, Opalka has in 15 years of steady work filled over 70 canvases with numbers and has by now passed the 2,500,000 mark. Each canvas, or "detail," as he calls it, is identical in size, but the background of each successive work becomes 1% whiter. Eventually Opalka will be painting white numbers on a white background. As he paints, Opalka speaks each number into a tape recorder. In the future, only the drone of numbers in Polish read over a loudspeaker will serve to document his ordinal progress. Numerically Opalka is progressing toward infinity; visually he is moving into whiteness; spiritually he is attempting to move closer to enlightenment. These three destinations, all related, are symbolic of humankind's most universal searches. Opalka's canvases produce an effect of rhythm and texture that seems subdued and serene from a distance, but increasingly rich and complex as one approaches.

Sponsoring Editor: Sharmon J. Hilfinger
Special Projects Editor and Copyeditor: Karen Judd
Designer: Janet Bollow
Production Manager: Laura Argento
Compositor: Typothetae Book Composition
Printer and Binder: Kingsport Press
Art Studio: J & R Technical Services

Portions of this work originally appeared in *Essentials of Calculus for Business and Economics.* Copyright © 1978 by Louis Leithold. Published by Harper & Row, Publishers, Inc.

THE CALCULUS WITH ANALYTIC GEOMETRY, Fourth Edition

Library of Congress Cataloging in Publication Data

Leithold, Louis.
The calculus with analytic geometry.

Includes index.
1. Calculus. 2. Geometry, Analytic. I. Title.
QA303.L428 1981 515′.15 80-23144
ISBN 0-06-043935-1

ACKNOWLEDGMENTS

Reviewers of the Fourth Edition of *The Calculus with Analytic Geometry* and of previous editions

William D. Bandes, San Diego Mesa College
John V. Baxley, Wake Forest University
Archie D. Brock, East Texas State University
Phillip Clarke, Los Angeles Valley College
Reuben W. Farley, Virginia Commonwealth University
Jacob Golightly, Jacksonville University
Robert K. Goodrich, University of Colorado
Albert Herr, Drexel University
James F. Hurley, University of Connecticut
John Impagliazzo, State University of New York at Farmingdale
Henry Korn, Westchester Community College
Bert Liberi, Westchester Community College
Gordon L. Miller, Wisconsin State University
W. W. Mitchell, Jr., Phoenix College
Roger B. Nelsen, Lewis and Clark College
Robert A. Nowlan, Southern Connecticut State College
Sister Madeleine Rose, Holy Names College
George W. Schultz, St. Petersburg Junior College
Donald R. Sherbert, University of Illinois
John Vadney, Fulton-Montgomery Community College
David Whitman, San Diego State College
J. Terry Wilson, San Jacinto College

Assistants for Exercises

Paul Cooke, Suffolk Community College
Jacqueline Dewar, Loyola Marymount University
Thomas Hasiotis, Suffolk Community College
Ken Kast, Logicon, Inc.
Jean Kilmer, West Covina Unified School District
Michael Sentlowitz, Rockland Community College
James M. Thelen, Rockland Community College
Margaret R. Trivisone, Rockland Community College

Cover and Chapter Opening Artist

Roman Opalka, Paris, France
Courtesy of John Weber Gallery, New York

To these people, to the staff of Harper & Row, and to all the users of the first, second, and third editions who have suggested changes, I express my deep appreciation.

L. L.

To my former students at

Phoenix College
California State University at Los Angeles
University of Southern California
Suffolk Community College
State University of New York at Farmingdale
Manhattan College
Open University of Great Britain

from whom I have learned.

Contents

Preface

This *fourth edition* of THE CALCULUS WITH ANALYTIC GEOMETRY, like the other three, is designed for prospective mathematics majors as well as for students whose primary interest is in engineering, the physical and social sciences, or nontechnical fields. A knowledge of high-school algebra and geometry is assumed.

The text is available either in onc volumc or in two parts: Part I consists of the first fifteen chapters, and Part II comprises Chapters 15 through 20. Chapter 15 on Infinite Series is included in both parts to make the use of the two-volume set more flexible. The material in Part I consists of the differential and integral calculus of functions of a single variable and plane analytic geometry, and it may be covered in a one-year course of nine or ten semester hours or twelve quarter hours. The second part is suitable for a course consisting of five or six semester hours or eight quarter hours. It includes the calculus of several variables and a treatment of vectors in the plane, as well as in three dimensions, with a vector approach to solid analytic geometry.

The objectives of the previous editions have been maintained. I have endeavored to achieve a healthy balance between the presentation of elementary calculus from a rigorous approach and that from the older, intuitive, and computational point of view. Bearing in mind that a textbook should be written for the student, I have attempted to keep the presentation geared to a beginner's experience and maturity and to leave no step unexplained or omitted. I desire that the reader be aware that proofs of theorems are necessary and that these proofs be well motivated and carefully explained so that they are understandable to the student who has achieved an average mastery of the preceding sections of the book. If a theorem is stated without proof, I have generally augmented the discussion by both figures and examples, and in such cases I have always stressed that what is presented is an illustration of the content of the theorem and is not a proof.

Innovations in this edition occur throughout the text and are discussed in the paragraphs that follow. There are now over 6800 exercises that have been revised and graded to provide a wide variety of types that range from computational to applied and theoretical problems. Among the more than 2000 new exercises are additional introductory ones, as well as some involving metric units of measurement.

Chapter 1 contains basic facts about the real-number system and is less detailed than in previous editions. The introduction to analytic geometry in this chapter includes the traditional material on straight lines as well as that of the circle. Functions are presented in Section 1.5. I have defined a function as a set of ordered pairs and have used this idea to point up the concept of a function as a correspondence between sets of real numbers. The definitions of the sine and cosine functions have been moved forward to Section 1.7. This early introduction allows their use in Chapters 3 and 5 in examples of differentiation and antidifferentiation for nonalgebraic functions.

The discussion of limits and continuity in Chapter 2 is now complete in this one chapter. These topics are at the heart of any first course in the calculus. The notion of a limit of a function is first given a step-by-step motivation, which brings the discussion from computing the value of a function near a number, through an intuitive treatment of the limiting process, up to a rigorous epsilon-delta definition. A sequence of examples progressively graded in difficulty is included. All the limit theorems are stated, and some proofs are presented in the text, while other proofs have been outlined in the exercises. In the discussion of continuity, I have used as examples and counterexamples common everyday functions and have avoided those that would have little intuitive meaning.

In Chapter 3, before giving the formal definition of a derivative, I have defined the tangent line to a curve in order to demonstrate in advance the geometrical application of the derivative. The physical application of instantaneous velocity in rectilinear motion is presented after theorems on differentiation are proved. The chain rule is stated and proved in Section 3.7, and here the sine and cosine functions are used as examples of the rule for other than algebraic functions. Chapter 4 gives the traditional applications of the derivative to problems involving related rates, maxima and minima, and curve sketching, as well as some to business and economics.

The antiderivative is treated in Chapter 5. I use the term "antidifferentiation" instead of indefinite integration, but the standard notation $\int f(x)\,dx$ is retained. This notation will suggest that some relation must exist between definite integrals, introduced in Chapter 6, and antiderivatives, but I see no harm in this as long as the presentation gives the theoretically proper view of the definite integral as the limit of sums. In Chapter 5 the discussion of differential equations with variables separable has been rewritten with new examples added.

Exercises involving the evaluation of definite integrals by finding limits of sums are given in Chapter 6 to stress that this is how they are calculated. The introduction of the definite integral follows the definition of the measure of the area under a curve as a limit of sums. Elementary properties of the definite integral are derived and the fundamental theorem of the calculus is proved. It is emphasized that this is a theorem, and an important one, because it provides us with an alternative to computing limits of sums. Approximate integration is now presented immediately after the fundamental theorem. This material includes the statement of theorems for computing the bounds of the error involved in these approximations.

Applications of the definite integral are split into two chapters in this edition. Those in Chapter 7 include area, volume, length of arc, and work. The section on work has been rewritten to include a discussion of metric units, and there are new examples and exercises that use these units.

In Chapter 8, the topic of inverse functions has been rewritten using the one-to-one function approach and is now contained in two sections instead of one. The treatment of logarithmic and exponential functions in this chapter is the modern approach. The natural logarithm is defined as an integral, and the exponential function is defined as the inverse of the natural logarithmic function. An irrational power of a real number is then defined.

The tangent, cotangent, secant, and cosecant functions are defined in Chapter 9. The important trigonometric identities are derived and used to obtain the formulas for the derivatives and integrals of these functions. The treatment of inverse trigonometric functions has been modified from previous editions, and the discussion of hyperbolic functions has been moved forward to this chapter.

Chapter 10, on techniques of integration, involves one of the most important computational aspects of the calculus. I have explained the theoretical backgrounds of each different method after an introductory motivation. The mastery of integration techniques depends upon the examples, and I have used as illustrations problems that the student will certainly meet in practice, those that require patience and persistence to solve.

The sections on applications of the definite integral in Chapter 11 have been rewritten and they include new examples and exercises involving the techniques of integration learned in Chapter 10. There is additional discussion of metric units of measurement.

Polar coordinates and some of their applications appear in Chapter 12. In Chapter 13, conics are treated as a unified subject to stress their natural and close relationship to each other. The parabola is discussed in the first two sections. Then equations of conics in polar coordinates are obtained, and the cartesian equations of the ellipse and the hyperbola are derived from the polar equations. Additional applications of the parabola, ellipse, and hyperbola are included in this edition.

The topics of indeterminate forms, improper integrals, and Taylor's formula, and the computational techniques involved, are presented in Chapter 14. I have attempted in Chapter 15 to give as complete a treatment of infinite series as is feasible in an elementary calculus text. In addition to the customary computational material, I have included the proof of the equivalence of convergence and boundedness of monotonic sequences based on the completeness property of the real numbers and the proofs of the computational processes involving differentiation and integration of power series.

The first five sections of Chapter 16 on vectors in the plane can be taken up after Chapter 4 if it is desired to introduce vectors earlier in the course. The approach to vectors is modern, and it serves both as an introduction to the viewpoint of linear algebra and to that of classical vector analysis. The applications are to physics and geometry. Chapter 17 treats vectors in three-dimensional space, and, if desired, the topics in the first three sections of this chapter may be studied concurrently with the corresponding topics in Chapter 16.

Limits, continuity, and differentiation of functions of several variables are considered in Chapter 18. The discussion and examples are applied mainly to functions of two and three variables; however, statements of most of the definitions and theorems are extended to functions of n variables.

In Chapter 19, a section on directional derivatives and gradients is followed by a section that shows the application of the gradient to finding an equation of the tangent plane to a surface. Applications of partial derivatives to the solution of extrema problems are presented, and there is an expanded treatment of Lagrange multipliers including new examples and a discussion of the theory of the method. Three sections are devoted to line integrals and related topics.

The double integral of a function of two variables and the triple integral of a function of three variables, along with some applications to physics, engineering, and geometry are given in Chapter 20. Section 20.6 is new to this edition. It contains coverage of Green's theorem and Stokes' theorem in the plane.

Louis Leithold

CHAPTER 1 Real Numbers, Introduction to Analytic Geometry, and Functions

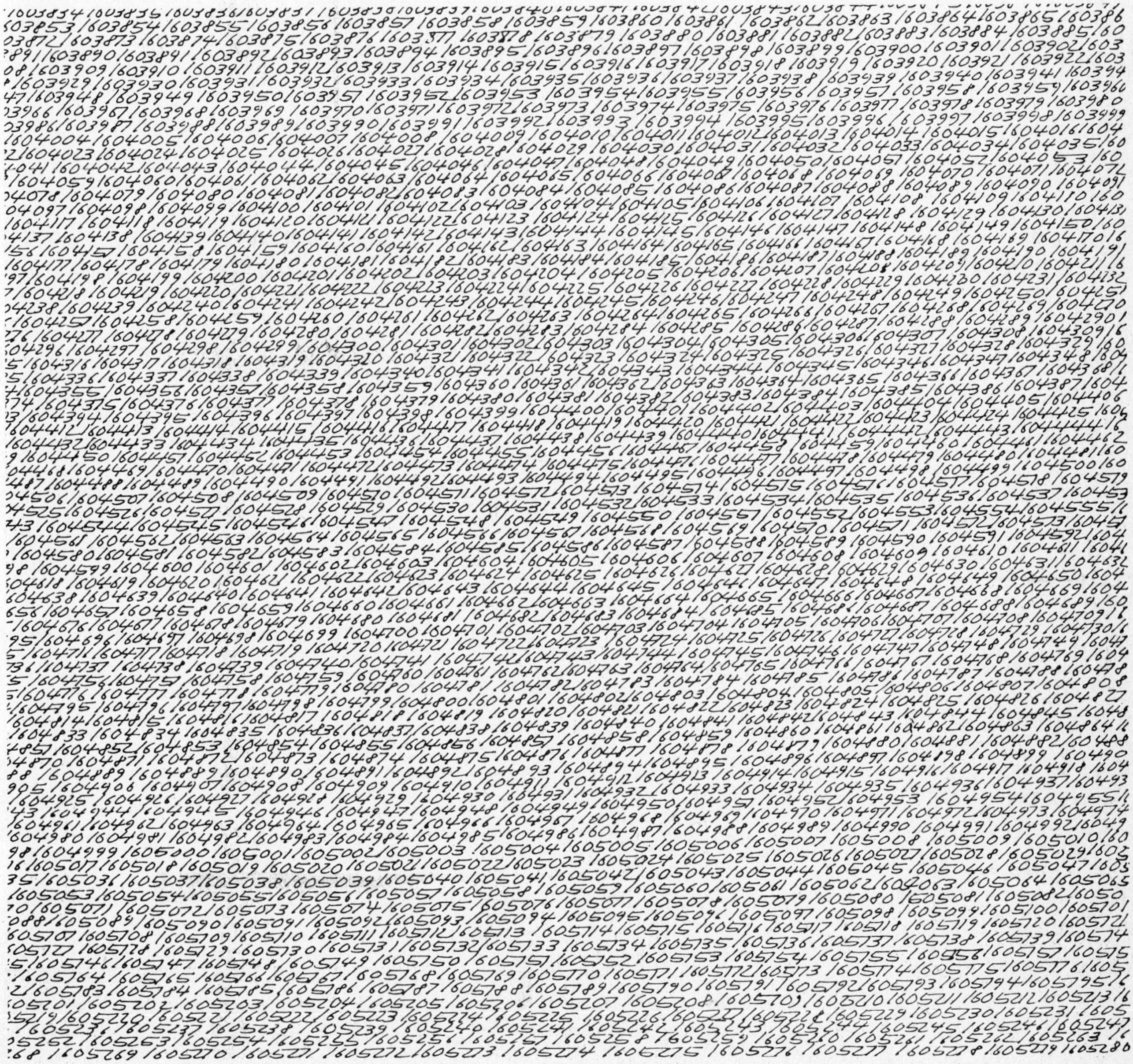

1.1 REAL NUMBERS AND INEQUALITIES

The **real number system** consists of a set of elements called **real numbers** and two operations called **addition** and **multiplication.** The set of real numbers is denoted by R^1. The operation of addition is denoted by the symbol $+$, and the operation of multiplication is denoted by the symbol $\cdot$. If a and b are elements of the set R^1, $a + b$ denotes the **sum** of a and b, and $a \cdot b$ (or ab) denotes their **product.** The operation of **subtraction** is defined by the equation

$$a - b = a + (-b)$$

where $-b$ denotes the **negative** of b such that $b + (-b) = 0$. The operation of **division** is defined by the equation

$$a \div b = a \cdot b^{-1}$$

where b^{-1} denotes the **reciprocal** of b such that $b \cdot b^{-1} = 1$.

The real number system can be completely described by a set of axioms (the word **axiom** is used to indicate a formal statement that is assumed to be true without proof). With these axioms we can derive the properties of the real numbers from which follow the familiar algebraic operations of addition, subtraction, multiplication, and division, as well as the algebraic concepts of solving equations, factoring, and so forth.

Properties that can be shown to be logical consequences of axioms are **theorems.** In the statement of most theorems there are two parts: the "if" part, called the **hypothesis,** and the "then" part, called the **conclusion.** The argument verifying a theorem is a **proof.** A proof consists in showing that the conclusion follows from the assumed truth of the hypothesis.

A real number is either a positive number, a negative number, or zero, and any real number can be classified as either **rational** or **irrational.** A rational number is any number that can be expressed as the ratio of two integers. That is, a rational number is a number of the form p/q, where p and q are integers and $q \neq 0$. The rational numbers consist of the following:

The **integers** (positive, negative, and zero)

$$\ldots, -5, -4, -3, -2, -1, 0, 1, 2, 3, 4, 5, \ldots$$

The positive and negative **fractions,** such as

$$\tfrac{2}{7} \qquad -\tfrac{4}{5} \qquad \tfrac{83}{5}$$

The positive and negative **terminating decimals,** such as

$$2.36 = \frac{236}{100} \qquad -0.003251 = -\frac{3{,}251}{1{,}000{,}000}$$

The positive and negative **nonterminating repeating decimals,** such as

$$0.333\ldots = \tfrac{1}{3} \qquad -0.549549549\ldots = -\tfrac{61}{111}$$

The real numbers that are not rational numbers are called **irrational numbers.** These are positive and negative **nonterminating nonrepeating** decimals, for example,

$$\sqrt{3} = 1.732\ldots \qquad \pi = 3.14159\ldots$$

We use some set notation and terminology in the discussions that follow. The idea of "set" is used extensively in mathematics and is such a basic concept that it is not given a formal definition. We can say that a **set** is a collection of objects, and the objects in a set are called the **elements** of a set. If every element of a set S is also an element of a set T, then S is a **subset** of T. In calculus we are concerned with the set R^1 of real numbers. Two subsets of R^1 are the set N of natural numbers and the set Z of integers.

We use the symbol $\in$ to indicate that a specific element belongs to a set. Hence we may write $8 \in N$, which is read "8 is an element of N." The notation $a, b \in S$ indicates that both a and b are elements of S. The symbol $\notin$ is read "is not an element of." Thus we read $\frac{1}{2} \notin N$ as "$\frac{1}{2}$ is not an element of N."

A pair of braces { } used with words or symbols can describe a set. If S is the set of natural numbers less than 6, we can write the set S as

$$\{1, 2, 3, 4, 5\}$$

We can also write the set S as

$$\{x, \text{ such that } x \text{ is a natural number less than } 6\}$$

where the symbol x is called a *variable*. A **variable** is a symbol used to represent any element of a given set. Another way of writing the above set S is to use what is called **set-builder notation,** where a vertical bar is used in place of the words "such that." Using set-builder notation to describe the set S we have

$$\{x \mid x \text{ is a natural number less than } 6\}$$

which is read "the set of all x such that x is a natural number less than 6."

Two sets A and B are said to be **equal,** written $A = B$, if A and B have identical elements. The **union** of two sets A and B, denoted by $A \cup B$ and read "A union B," is the set of all elements that are in A or in B or in both A and B. The **intersection** of A and B, denoted by $A \cap B$ and read "A intersection B," is the set of all elements that are in both A and B. The set that contains no elements is called the **empty set** and is denoted by $\varnothing$.

● ILLUSTRATION 1 Suppose $A = \{2, 4, 6, 8, 10, 12\}$, $B = \{1, 4, 9, 16\}$, and $C = \{2, 10\}$. Then

$$A \cup B = \{1, 2, 4, 6, 8, 9, 10, 12, 16\} \qquad A \cap B = \{4\}$$

$$B \cup C = \{1, 2, 4, 9, 10, 16\} \qquad B \cap C = \varnothing$$ ●

1.1.1 Definition If $a, b \in R^1$, then ***a* is less than *b*** (written $a < b$) if and only if $b - a$ is positive.

● ILLUSTRATION 2 (a) $3 < 8$ because $8 - 3 = 5$, and 5 is positive. (b) $-10 < -6$ because $-6 - (-10) = 4$, and 4 is positive. ●

1.1.2 Definition If $a, b \in R^1$, then ***a* is greater than *b*** (written $a > b$) if and only if b is less than a; with symbols we write

$$a > b \text{ if and only if } b < a$$

1.1.3 Definition The symbols $\leq$ ("is less than or equal to") $\geq$ ("is greater than or equal to") are defined as follows:

(i) $a \leq b$ if and only if either $a < b$ or $a = b$.
(ii) $a \geq b$ if and only if either $a > b$ or $a = b$.

The statements $a < b, a > b, a \leq b$, and $a \geq b$ are called **inequalities.** Some examples are $2 < 7$, $-5 < 6$, $-5 < -4$, $14 > 8$, $2 > -4$, $-3 > -6$, $5 \geq 3$, and $-10 \leq -7$. In particular, $a < b$ and $a > b$ are called **strict** inequalities, whereas $a \leq b$ and $a \geq b$ are called **nonstrict** inequalities.

The following theorems can be proved by using axioms for the set R^1 and 1.1.1 through 1.1.3.

1.1.4 Theorem

(i) $a > 0$ if and only if a is positive.
(ii) $a < 0$ if and only if a is negative.
(iii) $a > 0$ if and only if $-a < 0$.
(iv) $a < 0$ if and only if $-a > 0$.

1.1.5 Theorem If $a < b$ and $b < c$, then $a < c$.

● ILLUSTRATION 3 Because $-6 < 0$, then if $0 < x$, it follows from Theorem 1.1.5 that $-6 < x$. ●

1.1.6 Theorem If $a < b$, then $a + c < b + c$, and $a - c < b - c$ if c is any real number.

● ILLUSTRATION 4 $4 < 7$; so $4 + 5 < 7 + 5$; and $4 - 5 < 7 - 5$. ●

1.1.7 Theorem If $a < b$ and $c < d$, then $a + c < b + d$.

● ILLUSTRATION 5 $2 < 6$ and $-3 < 1$; so $2 + (-3) < 6 + 1$. ●

1.1.8 Theorem If $a < b$ and if c is any positive number, then $ac < bc$.

● ILLUSTRATION 6 $2 < 5$; so $2 \cdot 4 < 5 \cdot 4$. ●

1.1.9 Theorem If $a < b$ and if c is any negative number, then $ac > bc$.

● ILLUSTRATION 7 $2 < 5$; so $2(-4) > 5(-4)$. ●

1.1.10 Theorem If $0 < a < b$ and $0 < c < d$, then $ac < bd$.

● ILLUSTRATION 8 $0 < 4 < 7$ and $0 < 8 < 9$; so $4(8) < 7(9)$. ●

Theorem 1.1.8 states that if both members of an inequality are multiplied by a positive number, the direction of the inequality remains unchanged, whereas Theorem 1.1.9 states that if both members of an inequality are multiplied by a negative number, the direction of the inequality is reversed. Theorems 1.1.8 and 1.1.9 also hold for division, because dividing both members of an inequality by a number d is equivalent to multiplying both members by $1/d$.

The following theorems are identical to Theorems 1.1.5 to 1.1.10 except that the direction of the inequality is reversed.

1.1.11 Theorem If $a > b$ and $b > c$, then $a > c$.

● ILLUSTRATION 9 $8 > 4$ and $4 > -2$; so $8 > -2$. ●

1.1.12 Theorem If $a > b$, then $a + c > b + c$, and $a - c > b - c$ if c is any real number.

● ILLUSTRATION 10 $3 > -5$; so $3 - 4 > -5 - 4$. ●

1.1.13 Theorem If $a > b$ and $c > d$, then $a + c > b + d$.

● ILLUSTRATION 11 $7 > 2$ and $3 > -5$; so $7 + 3 > 2 + (-5)$. ●

1.1.14 Theorem If $a > b$ and if c is any positive number, then $ac > bc$.

● ILLUSTRATION 12 $-3 > -7$; so $(-3)4 > (-7)4$. ●

1.1.15 Theorem If $a > b$ and if c is any negative number, then $ac < bc$.

● ILLUSTRATION 13 $-3 > -7$; so $(-3)(-4) < (-7)(-4)$. ●

1.1.16 Theorem If $a > b > 0$ and $c > d > 0$, then $ac > bd$.

● ILLUSTRATION 14 $4 > 3 > 0$ and $7 > 6 > 0$; so $4(7) > 3(6)$. ●

We impose upon the set R^1 a condition called the **axiom of completeness** (Axiom 15.2.5). The statement of this axiom is deferred until Sec. 15.2 because it requires some terminology that is best introduced and discussed later. However, we now give a geometric interpretation to the set of real numbers by associating them with the points on a horizontal line, called an **axis.** The axiom of completeness guarantees that there is a one-to-one correspondence between the set R^1 and the set of points on an axis.

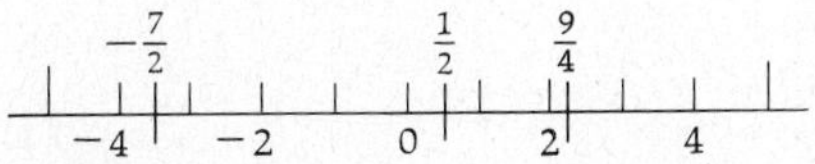

FIGURE 1.1.1

Refer to Fig. 1.1.1. A point on the axis is chosen to represent the number 0. This point is called the **origin.** A unit of distance is selected. Then each positive number x is represented by the point at a distance of x units to the right of the origin, and each negative number x is represented by the point at a distance of $-x$ units to the left of the origin (it should be noted that if x is negative, then $-x$ is positive). To each real number there corresponds a unique point on the axis, and with each point on the axis there is associated only one real number; hence there is a one-to-one correspondence between R^1 and the points on the axis. So the points on the axis are identified with the numbers they represent, and the same symbol is used for both the number and the point representing that number on the axis. We identify R^1 with the axis, and we call R^1 the **real number line.**

We see that $a < b$ if and only if the point representing the number a is to the left of the point representing the number b. Similarly, $a > b$ if and only if the point representing a is to the right of the point representing b. For instance, the number 2 is less than the number 5 and the point 2 is to the left of the point 5. We could also write $5 > 2$ and say that the point 5 is to the right of the point 2.

A number x is between a and b if $a < x$ and $x < b$. We can write this as a continued inequality as follows:

$$a < x < b \tag{1}$$

The set of all numbers x satisfying the continued inequality (1) is called an **open interval** and is denoted by (a, b). Therefore

$$(a, b) = \{x \mid a < x < b\}$$

The **closed interval** from a to b is the open interval (a, b) together with the two endpoints a and b and is denoted by $[a, b]$. Thus

$$[a, b] = \{x \mid a \leq x \leq b\}$$

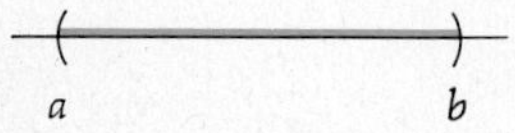

FIGURE 1.1.2

Figure 1.1.2 illustrates the open interval (a, b), and Fig. 1.1.3 illustrates the closed interval $[a, b]$.

The **interval half-open on the left** is the open interval (a, b) together with the right endpoint b. It is denoted by $(a, b]$; so

$$(a, b] = \{x \mid a < x \leq b\}$$

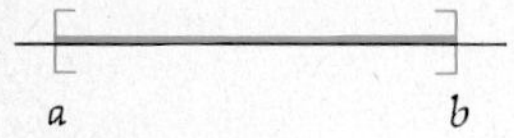

FIGURE 1.1.3

We define an **interval half-open on the right** in a similar way and denote it by $[a, b)$. Thus

$$[a, b) = \{x \mid a \leq x < b\}$$

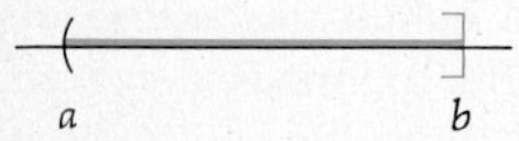

FIGURE 1.1.4

Figure 1.1.4 illustrates the interval $(a, b]$, and Fig. 1.1.5 illustrates the interval $[a, b)$.

We shall use the symbol $+\infty$ (positive infinity) and the symbol $-\infty$ (negative infinity); however, take care not to confuse these symbols with real numbers, for they do not obey the properties of the real numbers. We have the following intervals:

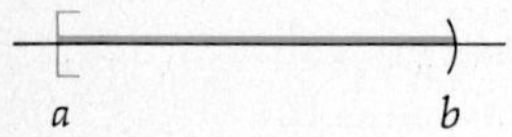

FIGURE 1.1.5

$$(a, +\infty) = \{x \mid x > a\}$$

$$(-\infty, b) = \{x \mid x < b\}$$

$$[a, +\infty) = \{x \mid x \geq a\}$$

$$(-\infty, b] = \{x \mid x \leq b\}$$

$$(-\infty, +\infty) = R^1$$

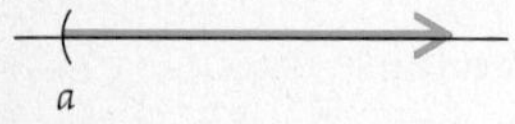

FIGURE 1.1.6

Figure 1.1.6 illustrates the interval $(a, +\infty)$, and Fig. 1.1.7 illustrates the interval $(-\infty, b)$. Note that $(-\infty, +\infty)$ denotes the set of all real numbers.

For each of the intervals (a, b), $[a, b]$, $[a, b)$, and $(a, b]$ the numbers a and b are called the **endpoints** of the interval. The closed interval $[a, b]$ contains both its endpoints, whereas the open interval (a, b) contains neither endpoint. The interval $[a, b)$ contains its left endpoint but not its right one, and the interval $(a, b]$ contains its right endpoint but not its left one. An open interval can be thought of as one that contains none of its endpoints, and a closed interval can be regarded as one that contains all of its endpoints. Consequently, the interval $[a, +\infty)$ is considered to be a

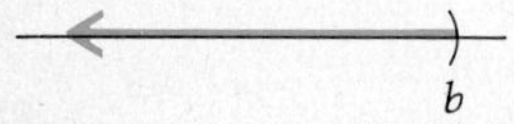

FIGURE 1.1.7

closed interval because it contains its only endpoint a. Similarly, $(-\infty, b]$ is a closed interval, whereas $(a, +\infty)$ and $(-\infty, b)$ are open. The intervals $[a, b)$ and $(a, b]$ are neither open nor closed. The interval $(-\infty, +\infty)$ has no endpoints, and it is considered both open and closed.

Intervals are used to represent "solution sets" of inequalities in one variable. The **solution set** of such an inequality is the set of all numbers that satisfy the inequality.

Example 1 Find the solution set of the inequality $2 + 3x < 5x + 8$. Illustrate it on the real number line.

Solution If x is a number such that

$$2 + 3x < 5x + 8$$

then

$$2 + 3x - 2 < 5x + 8 - 2$$
$$3x < 5x + 6$$
$$-2x < 6$$
$$x > -3$$

What we have proved is that if $2 + 3x < 5x + 8$, then $x > -3$. Each of the steps is reversible; that is, if we start with

$$x > -3$$

then

$$-2x < 6$$
$$2 + 3x < 5x + 8$$

Therefore we can conclude that

$$2 + 3x < 5x + 8 \quad \text{if and only if} \quad x > -3$$

So the solution set of the given inequality is the interval $(-3, +\infty)$, which is illustrated in Fig. 1.1.8.

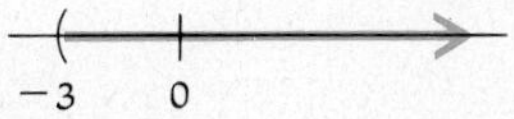

FIGURE 1.1.8

Example 2 Find the solution set of the inequality $4 < 3x - 2 \leq 10$. Illustrate it on the real number line.

Solution Add 2 to each member of the inequality to obtain

$$6 < 3x \leq 12$$
$$2 < x \leq 4$$

Each step is reversible; so the solution set is the interval $(2, 4]$, as illustrated in Fig. 1.1.9.

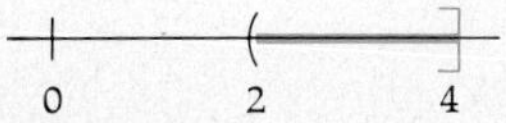

FIGURE 1.1.9

Example 3 Find the solution set of the inequality

$$\frac{7}{x} > 2$$

Illustrate it on the real number line.

Solution We wish to multiply both members of the inequality by x. However, the direction of the inequality that results will depend on whether x is positive or negative.

Observe that if $x < 0$, then

$$\frac{7}{x} < 0$$

which contradicts the given inequality. Therefore we need consider only $x > 0$. Multiplying on both sides of the given inequality by x we obtain

$$7 > 2x$$

$$\tfrac{7}{2} > x$$

$$x < \tfrac{7}{2}$$

Because the above steps are reversible, the solution set of the inequality is $\{x \mid x > 0\} \cap \{x \mid x < \frac{7}{2}\}$ or, equivalently, $\{x \mid 0 < x < \frac{7}{2}\}$, which is the interval $(0, \frac{7}{2})$, as illustrated in Fig. 1.1.10.

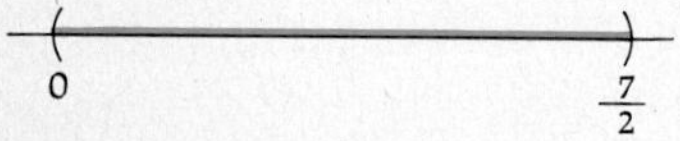

FIGURE 1.1.10

Example 4 Find the solution set of the inequality

$$\frac{x}{x-3} < 4$$

Illustrate it on the real number line.

Solution To multiply both members of the inequality by $x - 3$, we must consider two cases.

Case 1: $x - 3 > 0$; that is, $x > 3$.

Multiplying on both sides of the inequality by $x - 3$ we get

$$x < 4x - 12$$

$$-3x < -12$$

$$x > 4$$

Thus the solution set of Case 1 is $\{x \mid x > 3\} \cap \{x \mid x > 4\}$ or, equivalently, $\{x \mid x > 4\}$, which is the interval $(4, +\infty)$.

Case 2: $x - 3 < 0$; that is, $x < 3$.

Multiplying on both sides by $x - 3$ and reversing the direction of the inequality, we have

$$x > 4x - 12$$

$$-3x > -12$$

$$x < 4$$

Therefore x must be less than 4 and also less than 3. Thus the solution set of Case 2 is the interval $(-\infty, 3)$.

If the solution sets for Cases 1 and 2 are combined, we obtain $(-\infty, 3) \cup (4, +\infty)$, which is illustrated in Fig. 1.1.11.

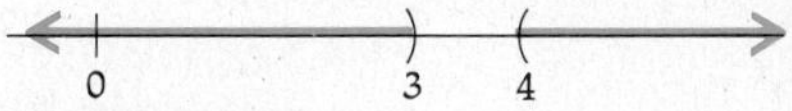

FIGURE 1.1.11

The concept of the *absolute value* of a number is used in some important definitions in the study of calculus. Furthermore, you will need to work with inequalities involving absolute value.

1.1.17 Definition The **absolute value** of x, denoted by $|x|$, is defined by

$$|x| = \begin{cases} x & \text{if } x \geq 0 \\ -x & \text{if } x < 0 \end{cases}$$

● ILLUSTRATION 15

$$|3| = 3 \qquad |-5| = -(-5) = 5 \qquad |8 - 14| = |-6| = -(-6) = 6$$ ●

From the definition, the absolute value of a number is either a positive number or zero; that is, it is nonnegative.

In terms of geometry, the absolute value of a number x is its distance from 0, without regard to direction. In general, $|a - b|$ is the distance between a and b without regard to direction, that is, without regard to which is the larger number. Refer to Fig. 1.1.12.

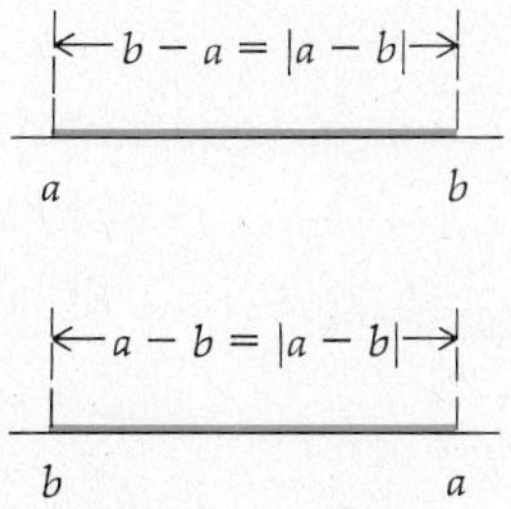

FIGURE 1.1.12

We have the following properties of absolute values.

1.1.18 Theorem $|x| < a$ if and only if $-a < x < a$, where $a > 0$.

Proof Because the theorem has an *if and only if* qualification, the proof requires two parts.

Part 1: Prove that $|x| < a$ if $-a < x < a$, where $a > 0$. Here, two cases must be considered: $x \geq 0$ and $x < 0$.

Case 1: $x \geq 0$.

Then $|x| = x$. Because $x < a$, we conclude that $|x| < a$.

Case 2: $x < 0$.

Then $|x| = -x$. Because $-a < x$, we apply Theorem 1.1.9 and obtain $a > -x$ or, equivalently, $-x < a$. But because $-x = |x|$, we have $|x| < a$.

In both cases, then,

$$|x| < a \quad \text{if } -a < x < a, \qquad \text{where } a > 0$$

Part 2: Prove that $|x| < a$ only if $-a < x < a$, where $a > 0$. Here we must show that whenever the inequality $|x| < a$ holds, the inequality $-a < x < a$ also holds. Assume that $|x| < a$ and consider the two cases $x \geq 0$ and $x < 0$.

Case 1: $x \geq 0$.

Then $|x| = x$. Because $|x| < a$, we conclude that $x < a$. Also, because $a > 0$, it follows from Theorem 1.1.15 that $-a < 0$. Thus $-a < 0 \leq x < a$, or $-a < x < a$.

Case 2: $x < 0$.

Then $|x| = -x$. Because $|x| < a$, we conclude that $-x < a$. Also, because $x < 0$, it follows from Theorem 1.1.9 that $0 < -x$. Therefore $-a < 0 < -x < a$, or $-a < -x < a$, which by applying Theorem 1.1.9 yields $-a < x < a$.

In both cases,

$$|x| < a \quad \text{only if } -a < x < a, \qquad \text{where } a > 0$$

■

1.1.19 Corollary $|x| \leq a$ if and only if $-a \leq x \leq a$, where $a > 0$.

1.1.20 Theorem $|x| > a$ if and only if $x > a$ or $x < -a$, where $a > 0$.

The proof of Theorem 1.1.20 is left as an exercise (see Exercise 57).

1.1.21 Corollary $|x| \geq a$ if and only if $x \geq a$ or $x \leq -a$, where $a > 0$.

The following examples illustrate the solution of equations and inequalities involving absolute values.

Example 5 Solve each of the equations for x: (a) $|3x + 2| = 5$; (b) $|2x - 1| = |4x + 3|$; (c) $|5x + 4| = -3$.

Solution

(a) $|3x + 2| = 5$

This equation will be satisfied if either

$$3x + 2 = 5 \quad \text{or} \quad -(3x + 2) = 5$$

Considering each equation separately we have

$$x = 1 \quad \text{and} \quad x = -\tfrac{7}{3}$$

which are the two solutions to the given equation.

(b) $|2x - 1| = |4x + 3|$

This equation will be satisfied if either

$$2x - 1 = 4x + 3 \quad \text{or} \quad 2x - 1 = -(4x + 3)$$

Solving the first equation we have $x = -2$; solving the second we get $x = -\frac{1}{3}$, thus giving two solutions to the original equation.

(c) $|5x + 4| = -3$

Because the absolute value of a number may never be negative, this equation has no solution.

Example 6 Find the solution set of the inequality $|x - 5| < 4$. Illustrate it on the real number line.

Solution If x is a number such that

$$|x - 5| < 4$$

then

$$-4 < x - 5 < 4 \quad \text{(by Theorem 1.1.18)}$$

$$1 < x < 9$$

Because each step is reversible, we can conclude that

$$|x - 5| < 4 \quad \text{if and only if} \quad 1 < x < 9$$

So the solution set of the given inequality is (1, 9), which is illustrated in Fig. 1.1.13.

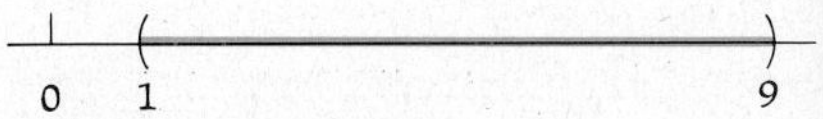

FIGURE 1.1.13

Example 7 Find the solution set of the inequality $|3x + 2| > 5$.

Solution By Theorem 1.1.20, the given inequality is equivalent to

$$3x + 2 > 5 \quad \text{or} \quad 3x + 2 < -5 \tag{2}$$

That is, the given inequality will be satisfied if either of the inequalities in (2) is satisfied.

Considering the first inequality, we have

$$3x + 2 > 5$$

$$x > 1$$

Therefore every number in the interval $(1, +\infty)$ is a solution.

From the second inequality,

$$3x + 2 < -5$$

$$x < -\tfrac{7}{3}$$

Hence every number in the interval $(-\infty, -\frac{7}{3})$ is a solution.

The solution set of the given inequality is therefore $(-\infty, -\frac{7}{3}) \cup (1, +\infty)$.

You may recall from algebra that the symbol $\sqrt{a}$, where $a \geq 0$, is defined as the unique *nonnegative* number x such that $x^2 = a$. We read $\sqrt{a}$ as "the principal square root of a." For example,

$$\sqrt{4} = 2 \qquad \sqrt{0} = 0 \qquad \sqrt{\tfrac{9}{25}} = \tfrac{3}{5}$$

Note: $\sqrt{4} \neq -2$ even though $(-2)^2 = 4$ because $\sqrt{4}$ denotes only the *positive* square root of 4.

Because we are concerned only with real numbers in this book, $\sqrt{a}$ is not defined if $a < 0$.

Example 8 Find all values of x for which $\sqrt{x^2 + 7x + 12}$ is real.

Solution $x^2 + 7x + 12 = (x + 3)(x + 4)$.

$$\sqrt{(x + 3)(x + 4)} \quad \text{is real when} \quad (x + 3)(x + 4) \geq 0$$

We find the solution set of this inequality. The inequality will be satisfied when both factors are nonnegative or when both factors are nonpositive, that is, if $x + 3 \geq 0$ and $x + 4 \geq 0$, or if $x + 3 \leq 0$ and $x + 4 \leq 0$. We consider two cases

Case 1: $x + 3 \geq 0$ and $x + 4 \geq 0$. That is,

$$x \geq -3 \quad \text{and} \quad x \geq -4$$

Both inequalities hold if $x \geq -3$, which is the interval $[-3, +\infty)$.

Case 2: $x + 3 \leq 0$ and $x + 4 \leq 0$. That is,

$$x \leq -3 \quad \text{and} \quad x \leq -4$$

Both inequalities hold if $x \leq -4$, which is the interval $(-\infty, -4]$.

If we combine the solution sets of Cases 1 and 2, then we have $(-\infty, -4] \cup [-3, +\infty)$.

From the definition of $\sqrt{a}$ it follows that

$$\sqrt{x^2} = |x|$$

For example, $\sqrt{5^2} = |5| = 5$, and $\sqrt{(-3)^2} = |-3| = 3$.

The following theorems about absolute value will be useful later.

1.1.22 Theorem If a and b are any numbers, then

$$|ab| = |a| \cdot |b|$$

Proof

$$\begin{aligned} |ab| &= \sqrt{(ab)^2} \\ &= \sqrt{a^2b^2} \\ &= \sqrt{a^2} \cdot \sqrt{b^2} \\ &= |a| \cdot |b| \end{aligned}$$

■

1.1.23 Theorem If a is any number and b is any number except 0,

$$\left|\frac{a}{b}\right| = \frac{|a|}{|b|}$$

The proof of Theorem 1.1.23 is left as an exercise (see Exercise 58).

1.1.24 Theorem
The Triangle Inequality

If a and b are any numbers, then

$$|a + b| \leq |a| + |b|$$

Proof By Definition 1.1.17, either $a = |a|$ or $a = -|a|$; thus

$$-|a| \leq a \leq |a| \tag{3}$$

Furthermore,

$$-|b| \leq b \leq |b| \tag{4}$$

From inequalities (3) and (4) and Theorem 1.1.7,

$$-(|a| + |b|) \leq a + b \leq |a| + |b|$$

Hence, from Corollary 1.1.19 it follows that

$$|a + b| \leq |a| + |b|$$ ■

Theorem 1.1.24 has two important corollaries that we now state and prove.

1.1.25 Corollary If a and b are any numbers, then

$$|a - b| \leq |a| + |b|$$

Proof

$$|a - b| = |a + (-b)| \leq |a| + |(-b)| = |a| + |b|$$ ■

1.1.26 Corollary If a and b are any numbers, then

$$|a| - |b| \leq |a - b|$$

Proof

$$|a| = |(a - b) + b| \leq |a - b| + |b|$$

thus, subtracting $|b|$ from both members of the inequality we have

$$|a| - |b| \leq |a - b|$$ ■

EXERCISES 1.1

In Exercises 1 through 22, find the solution set of the given inequality, and illustrate the solution set on the real number line.

1. $5x + 2 > x - 6$
2. $3 - x < 5 + 3x$
3. $\frac{2}{3}x - \frac{1}{2} \leq 0$
4. $3 - 2x \geq 9 + 4x$
5. $13 \geq 2x - 3 \geq 5$
6. $-2 < 6 - 4x \leq 8$
7. $2 > -3 - 3x \geq -7$
8. $2 \leq 5 - 3x < 11$
9. $\frac{4}{x} - 3 > \frac{2}{x} - 7$
10. $\frac{5}{x} < \frac{3}{4}$
11. $\frac{1}{x+1} < \frac{2}{3x-1}$
12. $\frac{x+1}{2-x} < \frac{x}{3+x}$
13. $x^2 > 4$
14. $x^2 \leq 9$
15. $(x - 3)(x + 5) > 0$
16. $x^2 - 3x + 2 > 0$
17. $1 - x - 2x^2 \geq 0$
18. $x^2 + 3x + 1 > 0$
19. $4x^2 + 9x < 9$
20. $2x^2 - 6x + 3 < 0$
21. $\frac{1}{3x-7} \geq \frac{4}{3-2x}$
22. $x^3 + 1 > x^2 + x$

In Exercises 23 through 30, solve for x.

23. $|4x + 3| = 7$ **24.** $|3x - 8| = 4$ **25.** $|5x - 3| = |3x + 5|$ **26.** $|x - 2| = |3 - 2x|$

27. $|7x| = 4 - x$ **28.** $2x + 3 = |4x + 5|$ **29.** $\left|\frac{x + 2}{x - 2}\right| = 5$ **30.** $\left|\frac{3x + 8}{2x - 3}\right| = 4$

In Exercises 31 through 36, find all the values of x for which the number is real.

31. $\sqrt{8x - 5}$ **32.** $\sqrt{x^2 - 16}$ **33.** $\sqrt{x^2 - 3x - 10}$

34. $\sqrt{2x^2 + 5x - 3}$ **35.** $\sqrt{x^2 - 5x + 4}$ **36.** $\sqrt{x^2 + 2x - 1}$

In Exercises 37 through 48, find the solution set of the given inequality, and illustrate the solution set on the real number line.

37. $|x + 4| < 7$ **38.** $|2x - 5| < 3$ **39.** $|3x - 4| \leq 2$ **40.** $|6 - 2x| \geq 7$

41. $|2x - 5| > 3$ **42.** $|x + 4| \leq |2x - 6|$ **43.** $|3x| > |6 - 3x|$ **44.** $|3 + 2x| < |4 - x|$

45. $|9 - 2x| \geq |4x|$ **46.** $|5 - 2x| \geq 7$ **47.** $\left|\frac{x + 2}{2x - 3}\right| < 4$ **48.** $\left|\frac{6 - 5x}{3 + x}\right| \leq \frac{1}{2}$

In Exercises 49 through 52, solve for x, and write the answer with absolute value notation.

49. $\frac{x - a}{x + a} > 0$ **50.** $\frac{a - x}{a + x} \geq 0$ **51.** $\frac{x - 2}{x - 4} > \frac{x + 2}{x}$ **52.** $\frac{x + 5}{x + 3} < \frac{x + 1}{x - 1}$

53. Prove Theorem 1.1.5. **54.** Prove Theorem 1.1.6.

55. Prove Theorems 1.1.8 and 1.1.9.

56. Prove that if $x < y$, then $x < \frac{1}{2}(x + y) < y$.

57. Prove Theorem 1.1.20. **58.** Prove Theorem 1.1.23.

1.2 THE NUMBER PLANE AND GRAPHS OF EQUATIONS

We now consider ordered pairs of real numbers. Any set of two real numbers form a **pair,** and when the order of the pair is designated, it is called an **ordered pair of real numbers.** If x is the first real number and y is the second, we denote this ordered pair by writing them in parentheses with a comma separating them as (x, y). Note that the ordered pair (3, 7) is different from the ordered pair (7, 3).

1.2.1 Definition The set of all ordered pairs of real numbers is called the **number plane,** and each ordered pair (x, y) is called a **point** in the number plane. The number plane is denoted by R^2.

Just as we can identify R^1 with points on an axis (a one-dimensional space), we can identify R^2 with points in a geometric plane (a two-dimensional space). The method used with R^2 is the one attributed to the French mathematician René Descartes (1596–1650), who is credited with the invention of analytic geometry in 1637.

A horizontal line is chosen in the geometric plane and is called the **x axis.** A vertical line is chosen and is called the **y axis.** The point of inter-

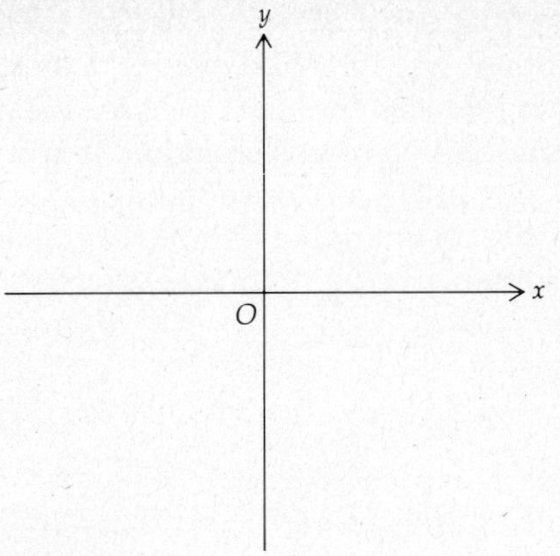

FIGURE 1.2.1

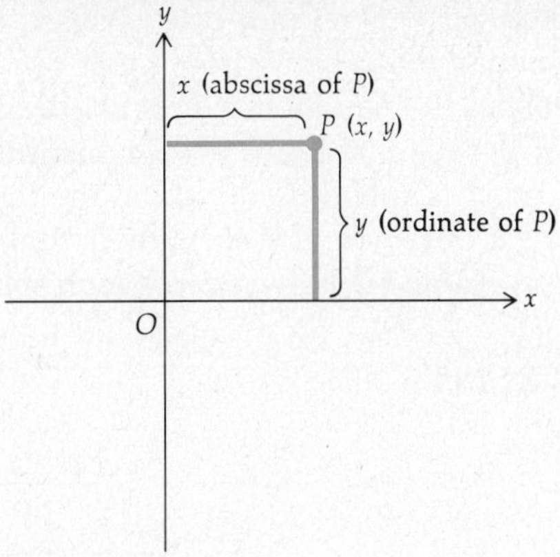

FIGURE 1.2.2

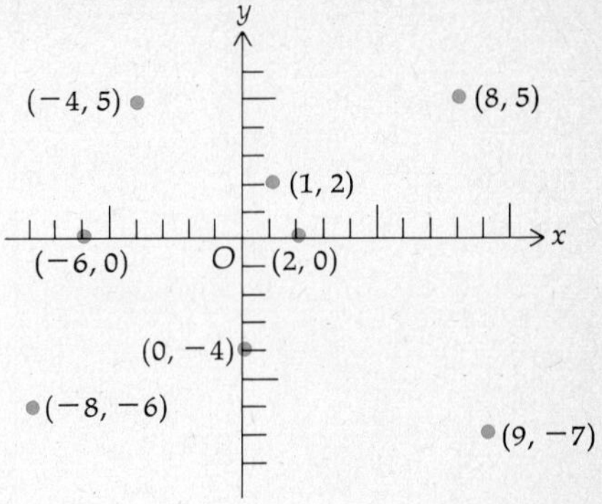

FIGURE 1.2.3

section of the x axis and the y axis is called the **origin** and is denoted by the letter O. A unit of length is chosen. Usually the unit length on each axis is the same. We establish the positive direction on the x axis to the right of the origin, and the positive direction on the y axis above the origin. See Fig. 1.2.1.

We now associate an ordered pair of real numbers (x, y) with a point P in the geometric plane. Refer to Fig. 1.2.2. The distance of P from the y axis, considered positive if P is to the right of the y axis and negative if P is to the left of the y axis, is called the **abscissa** (or **x coordinate**) of P and is denoted by x. The distance of P from the x axis, considered positive if P is above the x axis and negative if P is below the x axis, is called the **ordinate** (or **y coordinate**) of P and is denoted by y. The abscissa and the ordinate of a point are called the **rectangular cartesian coordinates** of the point. There is a one-to-one correspondence between the points in a geometric plane and R^2; that is, with each point there corresponds a unique ordered pair (x, y), and with each ordered pair (x, y) there is associated only one point. This one-to-one correspondence is called a **rectangular cartesian coordinate system.** Figure 1.2.3 illustrates a rectangular cartesian coordinate system with some points plotted.

The x and y axes are called the **coordinate axes.** They divide the plane into four parts, called **quadrants.** The first quadrant is the one in which the abscissa and the ordinate are both positive, that is, the upper right quadrant. The other quadrants are numbered in the counterclockwise direction, with the fourth being the lower right quadrant. See Fig. 1.2.4.

Because of the one-to-one correspondence, we identify R^2 with the geometric plane. For this reason we call an ordered pair (x, y) a *point.* Similarly, we refer to a *line* in R^2 as the set of all points corresponding to a line in the geometric plane, and we use other geometric terms for sets of points in R^2.

Consider the equation

$$y = x^2 - 2 \tag{1}$$

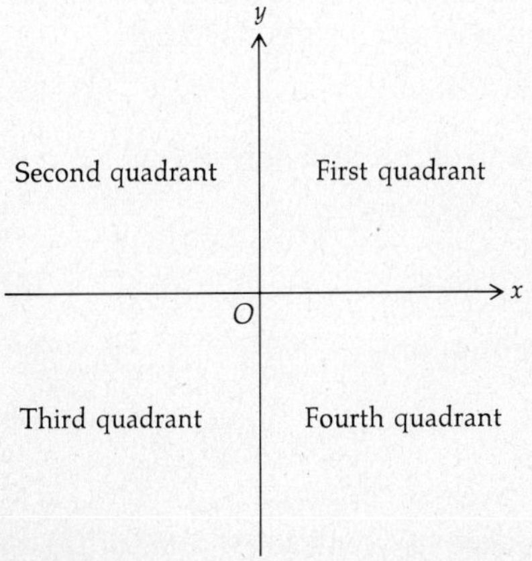

FIGURE 1.2.4

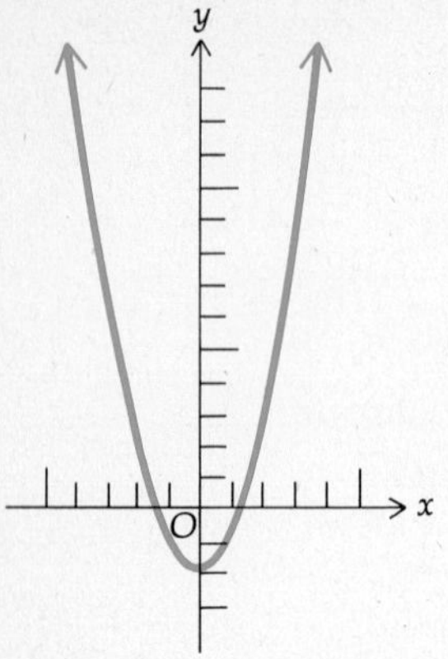

FIGURE 1.2.5

where (x, y) is a point in R^2. This is an equation in R^2. A solution of this equation means an ordered pair of numbers, one for x and one for y, that satisfies the equation. For example, if x is replaced by 3 in (1), we see that $y = 7$; thus $x = 3$ and $y = 7$ constitutes a solution of this equation. If any number is substituted for x in the right side of (1), a corresponding value for y is obtained. It is seen, then, that (1) has an unlimited number of solutions. Table 1.2.1 gives a few such solutions.

TABLE 1.2.1

x	0	1	2	3	4	-1	-2	-3	-4
$y = x^2 - 2$	-2	-1	2	7	14	-1	2	7	14

If we plot the points having as coordinates the number pairs (x, y) satisfying (1), we have a sketch of the graph of the equation. In Fig. 1.2.5 points are plotted whose coordinates are the number pairs obtained from Table 1.2.1. These points are connected by a smooth curve. Any point (x, y) on this curve has coordinates satisfying (1). Also, the coordinates of any point not on this curve do not satisfy the equation.

1.2.2 Definition The **graph of an equation** in R^2 is the set of all points (x, y) in R^2 whose coordinates are numbers satisfying the equation.

The graph of an equation is sometimes called the **locus** of the equation. The graph of an equation in R^2 is also called a **curve.** Unless otherwise stated, an equation with two unknowns, x and y, is considered an equation in R^2.

The graph in Fig. 1.2.5 is a **parabola.** The lowest point on the graph is $(0, -2)$ and is called the **vertex** of the parabola. This parabola opens upward. The y axis is called the **axis** of the parabola. A complete discussion of parabolas occurs in Chapter 13, where we show that the graph of an equation of the form $y = ax^2 + bx + c (a \neq 0)$ is a parabola having a vertical axis and opening upward if $a > 0$ and downward if $a < 0$. In the following example, the graph is a parabola whose axis is horizontal.

Example 1 Draw a sketch of the graph of the equation $y^2 - x - 2 = 0$.

Solution Solving the given equation for y we have

$$y = \pm\sqrt{x+2} \tag{2}$$

Equations (2) are equivalent to the two equations

$$y = \sqrt{x+2} \tag{3}$$

$$y = -\sqrt{x+2} \tag{4}$$

The coordinates of all points that satisfy (2) will satisfy either (3) or (4), and the coordinates of any point that satisfies either (3) or (4) will satisfy (2). Table 1.2.2 gives some of these values of x and y.

TABLE 1.2.2

x	0	0	1	1	2	2	3	3	-1	-1	-2
y	$\sqrt{2}$	$-\sqrt{2}$	$\sqrt{3}$	$-\sqrt{3}$	2	-2	$\sqrt{5}$	$-\sqrt{5}$	1	-1	0

Observe that for any value of $x < -2$ there is no real value for y. Also, for each value of $x > -2$ there are two values for y. A sketch of the graph of the given equation is shown in Fig. 1.2.6.

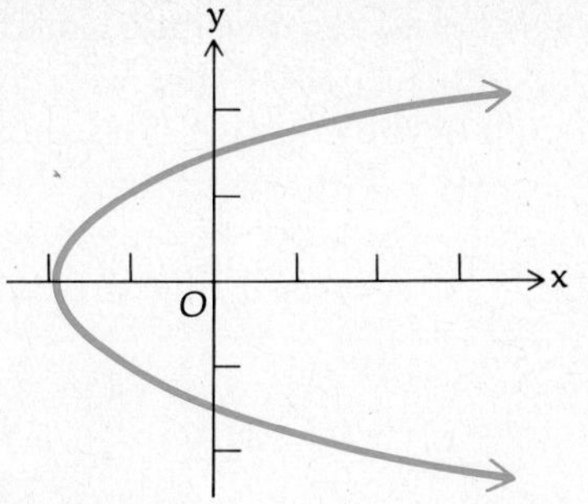

FIGURE 1.2.6

Example 2 Draw sketches of the graphs of the equations $y = \sqrt{x+2}$ and $y = -\sqrt{x+2}$.

Solution The equation $y = \sqrt{x+2}$ is the same as (3). The value of y is nonnegative; hence the graph of the equation is the upper half of the graph of (2). A sketch of this graph is shown in Fig. 1.2.7.

Similarly, the graph of the equation $y = -\sqrt{x+2}$, a sketch of which is shown in Fig. 1.2.8, is the lower half of the parabola of Fig. 1.2.6.

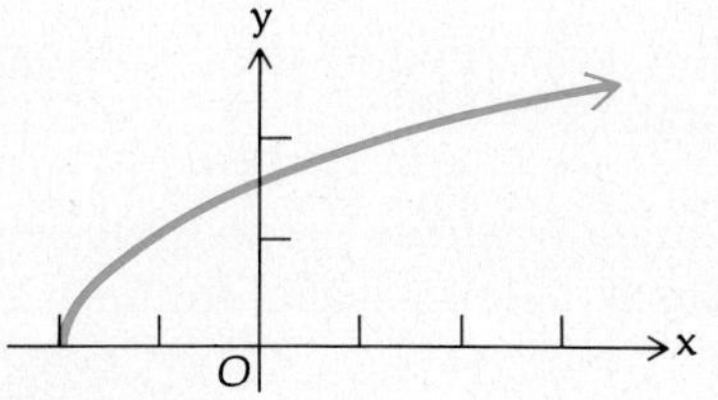

FIGURE 1.2.7

Example 3 Draw a sketch of the graph of the equation $y = |x + 3|$.

Solution From the definition of the absolute value of a number it follows that

$$y = \begin{cases} x + 3 & \text{if } x + 3 \geq 0 \\ -(x + 3) & \text{if } x + 3 < 0 \end{cases}$$

or, equivalently,

$$y = \begin{cases} x + 3 & \text{if } x \geq -3 \\ -(x + 3) & \text{if } x < -3 \end{cases}$$

Table 1.2.3 gives some values of x and y satisfying the given equation.

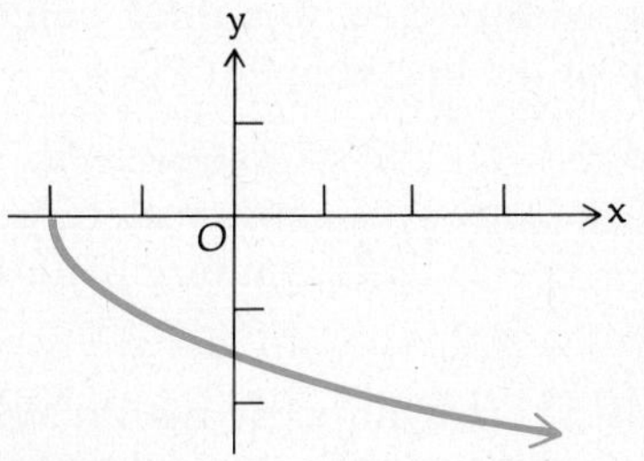

FIGURE 1.2.8

TABLE 1.2.3

x	0	1	2	3	-1	-2	-3	-4	-5	-6	-7	-8	-9
y	3	4	5	6	2	1	0	1	2	3	4	5	6

A sketch of the graph of the given equation is shown in Fig. 1.2.9.

Example 4 Draw a sketch of the graph of the equation

$$(x - 2y + 3)(y - x^2) = 0 \tag{5}$$

Solution By the property of real numbers that $ab = 0$ if and only if $a = 0$ or $b = 0$, we have from (5)

$$x - 2y + 3 = 0 \tag{6}$$

and

$$y - x^2 = 0 \tag{7}$$

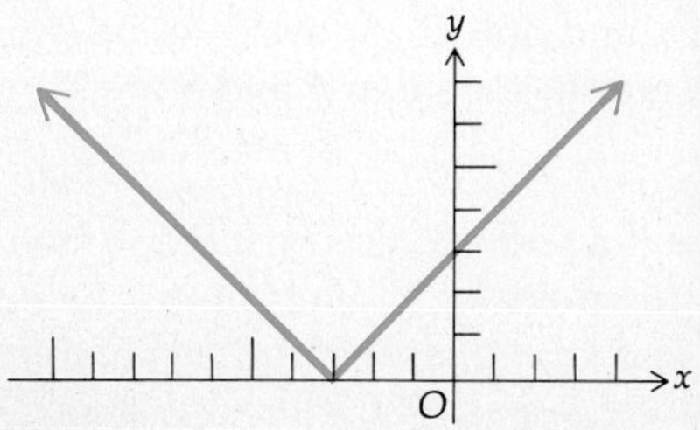

FIGURE 1.2.9

The coordinates of all points that satisfy (5) will satisfy either (6) or (7), and the coordinates of any point that satisfies either (6) or (7) will sat-

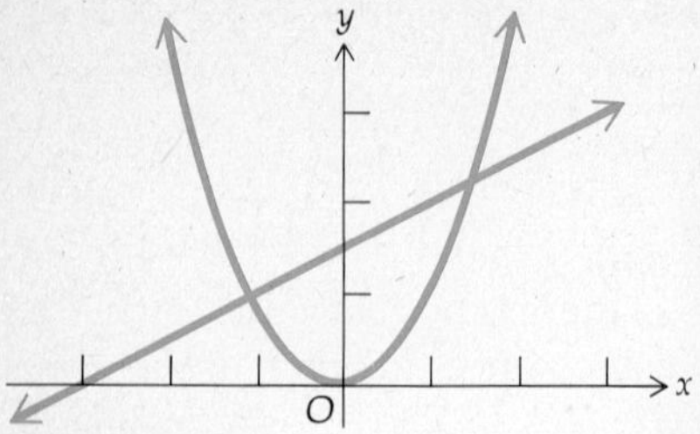

FIGURE 1.2.10

isfy (5). Therefore the graph of (5) will consist of the graphs of (6) and (7). Table 1.2.4 gives some values of x and y satisfying (6), and Table 1.2.5 gives some values of x and y satisfying (7). A sketch of the graph of (5) is shown in Fig. 1.2.10.

TABLE 1.2.4

x	0	1	2	3	-1	-2	-3	-4	-5
y	$\frac{3}{2}$	2	$\frac{5}{2}$	3	1	$\frac{1}{2}$	0	$-\frac{1}{2}$	-1

TABLE 1.2.5

x	0	1	2	3	-1	-2	-3
y	0	1	4	9	1	4	9

1.2.3 Definition An **equation of a graph** is an equation that is satisfied by the coordinates of those, and only those points on the graph.

• ILLUSTRATION 1 In R^2, $y = 8$ is an equation whose graph consists of those points having an ordinate of 8. This is a line parallel to the x axis and 8 units above it. •

In drawing a sketch of the graph of an equation, it is often helpful to consider properties of *symmetry* of a graph.

1.2.4 Definition Two points P and Q are said to be **symmetric with respect to a line** if and only if the line is the perpendicular bisector of the line segment PQ. Two points P and Q are said to be **symmetric with respect to a third point** if and only if the third point is the midpoint of the line segment PQ.

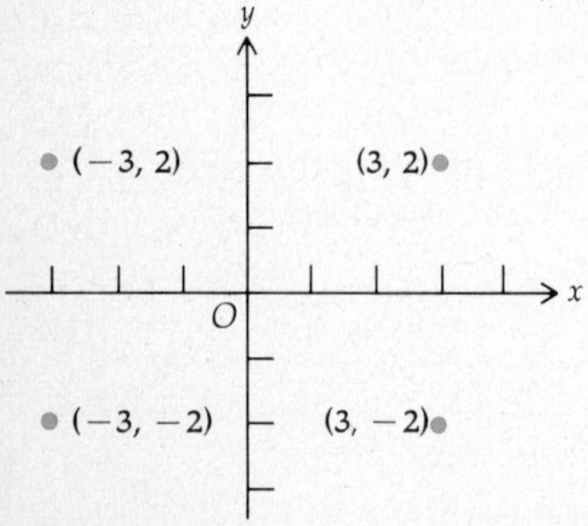

FIGURE 1.2.11

• ILLUSTRATION 2 The points (3, 2) and (3, -2) are symmetric with respect to the x axis, the points (3, 2) and (-3, 2) are symmetric with respect to the y axis, and the points (3, 2) and (-3, -2) are symmetric with respect to the origin (see Fig. 1.2.11). •

In general, the points (x, y) and $(x, -y)$ are symmetric with respect to the x axis, (x, y) and $(-x, y)$ are symmetric with respect to the y axis, and (x, y) and $(-x, -y)$ are symmetric with respect to the origin.

1.2.5 Definition The graph of an equation is symmetric with respect to a line l if and only if for every point P on the graph there is a point Q, also on the graph, such that P and Q are symmetric with respect to l. The graph of an equation is symmetric with respect to a point R if and only if for every point P on the graph there is a point S, also on the graph, such that P and S are symmetric with respect to R.

From Definition 1.2.5 it follows that if a point (x, y) is on a graph symmetric with respect to the x axis, then the point $(x, -y)$ also must be on the graph. And if both the points (x, y) and $(x, -y)$ are on the graph, then the graph is symmetric with respect to the x axis. Therefore the coordinates of the point $(x, -y)$ as well as (x, y) must satisfy an equation of the graph. Hence the graph of an equation in x and y is symmetric with respect to the x axis if and only if an equivalent equation is obtained when y is replaced by $-y$ in the equation. We have thus proved part (i) in the following theorem. The proofs of parts (ii) and (iii) are similar.

1.2.6 Theorem
Tests for Symmetry

The graph of an equation in x and y is

(i) symmetric with respect to the x axis if and only if an equivalent equation is obtained when y is replaced by $-y$ in the equation;
(ii) symmetric with respect to the y axis if and only if an equivalent equation is obtained when x is replaced by $-x$ in the equation;
(iii) symmetric with respect to the origin if and only if an equivalent equation is obtained when x is replaced by $-x$ and y is replaced by $-y$ in the equation.

The graph in Fig. 1.2.5 is symmetric with respect to the y axis, and for (1) an equivalent equation is obtained when x is replaced by $-x$. In Example 1 we have the equation $y^2 - x - 2 = 0$ for which an equivalent equation is obtained when y is replaced by $-y$, and its graph sketched in Fig. 1.2.6 is symmetric with respect to the x axis. The following example gives a graph that is symmetric with respect to the origin.

Example 5 Draw a sketch of the graph of the equation $xy = 1$.

Solution We see that if in the given equation x is replaced by $-x$ and y is replaced by $-y$, an equivalent equation is obtained; hence by Theorem 1.2.6(iii) the graph is symmetric with respect to the origin. Table 1.2.6 gives some values of x and y satisfying the given equation.

TABLE 1.2.6

x	1	2	3	4	$\frac{1}{2}$	$\frac{1}{3}$	$\frac{1}{4}$	-1	-2	-3	-4	$-\frac{1}{2}$	$-\frac{1}{3}$	$-\frac{1}{4}$
y	1	$\frac{1}{2}$	$\frac{1}{3}$	$\frac{1}{4}$	2	3	4	-1	$-\frac{1}{2}$	$-\frac{1}{3}$	$-\frac{1}{4}$	-2	-3	-4

From the given equation, $y = 1/x$. We see that as x increases through positive values, y decreases through positive values and gets closer and closer to zero. As x decreases through positive values, y increases through positive values and gets larger and larger. As x increases through negative values (i.e., x takes on the values $-4, -3, -2, -1, -\frac{1}{2}$, etc.), y takes on negative values having larger and larger absolute values. A sketch of the graph is shown in Fig. 1.2.12.

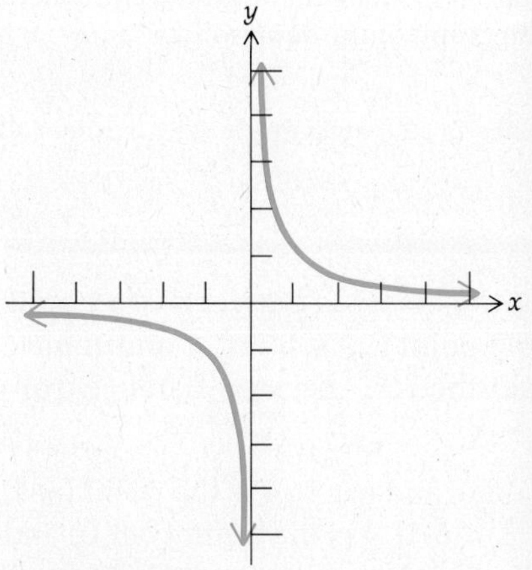

FIGURE 1.2.12

EXERCISES 1.2

In Exercises 1 through 6, plot the given point P and such of the following points as may apply:
(a) The point Q such that the line through Q and P is perpendicular to the x axis and bisected by it. Give the coordinates of Q.
(b) The point R such that the line through P and R is perpendicular to and bisected by the y axis. Give the coordinates of R.
(c) The point S such that the line through P and S is bisected by the origin. Give the coordinates of S.
(d) The point T such that the line through P and T is perpendicular to and bisected by the 45° line through the origin bisecting the first and third quadrants. Give the coordinates of T.

1. $P(1, -2)$ **2.** $P(-2, 2)$ **3.** $P(2, 2)$ **4.** $P(-2, -2)$ **5.** $P(-1, -3)$ **6.** $P(0, -3)$

In Exercises 7 through 36, draw a sketch of the graph of the equation.

7. $y = 2x + 5$ **8.** $y = 4x - 3$ **9.** $y = \sqrt{x+5}$ **10.** $y = \sqrt{x-3}$ **11.** $y = -\sqrt{x+5}$

12. $y = -\sqrt{x-3}$ **13.** $y^2 = x - 3$ **14.** $y = 5$ **15.** $x = -3$ **16.** $x = y^2 + 1$

17. $y = |x - 5|$ **18.** $y = -|x + 2|$ **19.** $y = |x| - 5$ **20.** $y = -|x| + 2$ **21.** $y = 5 - |x + 1|$

22. $y = 3 - x^2$ **23.** $4x^2 + 9y^2 = 36$ **24.** $4x^2 - 9y^2 = 36$ **25.** $y = 4 + x^2$ **26.** $2y = -x^3$

27. $y = 4x^3$ **28.** $y^2 = 4x^3$ **29.** $4x^2 - y^2 = 0$ **30.** $y = \sqrt[3]{x}$ **31.** $4x^2 + y^2 = 0$

32. $3x^2 - 13xy - 10y^2 = 0$ **33.** $x^4 - 5x^2y + 4y^2 = 0$ **34.** $(2x + y - 1)(4y + x^2) = 0$

35. $2y = \sqrt[3]{x-8}$ **36.** $(y^2 - x + 2)(y + \sqrt{x-4}) = 0$

In Exercises 37 through 42, draw a sketch of the graph of each of the equations.

37. (a) $x + 3y = 0$; (b) $x - 3y = 0$; (c) $x^2 - 9y^2 = 0$ **38.** (a) $2x - 5y = 0$; (b) $2x + 5y = 0$; (c) $4x^2 - 25y^2 = 0$

39. (a) $y = \sqrt{2x}$; (b) $y = -\sqrt{2x}$; (c) $y^2 = 2x$ **40.** (a) $y = \sqrt{-2x}$; (b) $y = -\sqrt{-2x}$; (c) $y^2 = -2x$

41. (a) $y = \sqrt{4 - x^2}$; (b) $y = -\sqrt{4 - x^2}$; (c) $x^2 + y^2 = 4$ **42.** (a) $y = \sqrt{x^2 - 4}$; (b) $y = -\sqrt{x^2 - 4}$; (c) $x^2 - y^2 = 4$

43. (a) Write an equation whose graph is the x axis. (b) Write an equation whose graph is the y axis. (c) Write an equation whose graph is the set of all points on either the x axis or the y axis.

44. (a) Write an equation whose graph consists of all points having an abscissa of 4. (b) Write an equation whose graph consists of all points having an ordinate of -3.

45. Prove that a graph that is symmetric with respect to both coordinate axes is also symmetric with respect to the origin.

46. The graph of an equation in x and y is symmetric with respect to the line with equation $y = x$ if and only if an equivalent equation is obtained when x is replaced by y and y is replaced by x. Show that the graph of the equation $ax^2 + by^2 = c$ is symmetric with respect to this line if and only if $a = b$.

47. Prove that a graph that is symmetric with respect to any two perpendicular lines is also symmetric with respect to their point of intersection.

1.3 THE DISTANCE FORMULA, THE CIRCLE, AND THE MIDPOINT FORMULAS

We now discuss the problem of finding the distance between two points in R^2. If A is the point (x_1, y_1) and B is the point (x_2, y_1) (i.e., A and B have the same ordinate but different abscissas), then the directed distance from A to B, denoted by $\overline{AB}$, is defined as $x_2 - x_1$.

● ILLUSTRATION 1 Refer to Fig. 1.3.1(a), (b), and (c). If A is the point $(3, 4)$ and B is the point $(9, 4)$, then $\overline{AB} = 9 - 3 = 6$. If A is the point $(-8, 0)$ and B is the point $(6, 0)$, then $\overline{AB} = 6 - (-8) = 14$. If A is the point $(4, 2)$ and B

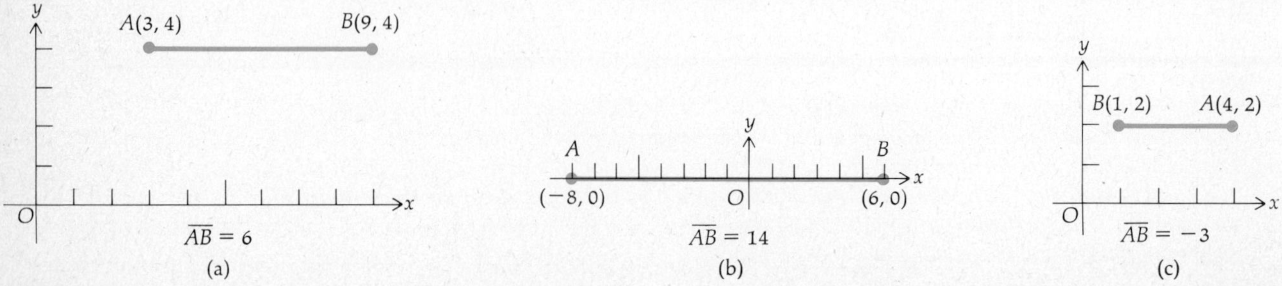

FIGURE 1.3.1

is the point (1, 2), then $\overline{AB} = 1 - 4 = -3$. We see that $\overline{AB}$ is positive if B is to the right of A, and $\overline{AB}$ is negative if B is to the left of A. ●

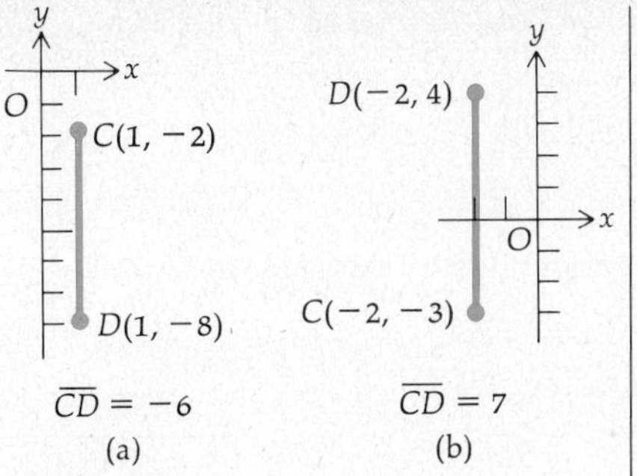

FIGURE 1.3.2

If C is the point (x_1, y_1) and D is the point (x_1, y_2), then the directed distance from C to D, denoted by $\overline{CD}$, is defined as $y_2 - y_1$.

● ILLUSTRATION 2 Refer to Fig. 1.3.2(a) and (b). If C is the point $(1, -2)$ and D is the point $(1, -8)$, then $\overline{CD} = -8 - (-2) = -6$. If C is the point $(-2, -3)$ and D is the point $(-2, 4)$, then $\overline{CD} = 4 - (-3) = 7$. The number $\overline{CD}$ is positive if D is above C, and $\overline{CD}$ is negative if D is below C. ●

We consider a directed distance $\overline{AB}$ as the signed distance traveled by a particle that starts at $A(x_1, y_1)$ and travels to $B(x_2, y_1)$. In such a case, the abscissa of the particle changes from x_1 to x_2, and the notation Δx ("delta x") is used to denote this change; that is,

$$\Delta x = x_2 - x_1$$

Therefore $\overline{AB} = \Delta x$.

It is important to note that the symbol Δx denotes the difference between the abscissa of B and the abscissa of A, and it *does not mean* "delta multiplied by x."

Similarly, if a particle is moving along a line parallel to the y axis from a point $C(x_1, y_1)$ to a point $D(x_1, y_2)$, then the ordinate of the particle changes from y_1 to y_2. This change is denoted by Δy or

$$\Delta y = y_2 - y_1$$

Thus $\overline{CD} = \Delta y$.

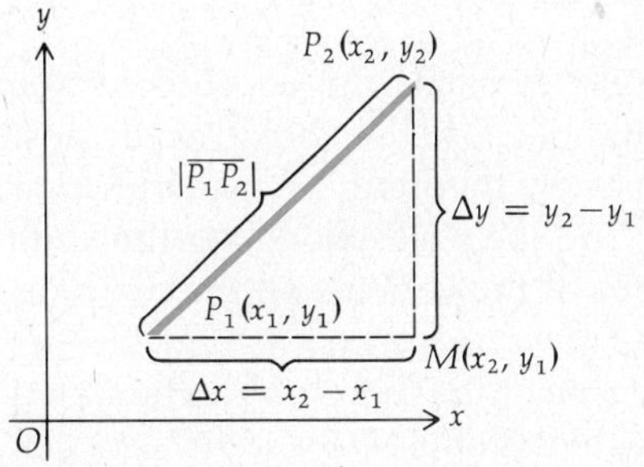

FIGURE 1.3.3

Now let $P_1(x_1, y_1)$ and $P_2(x_2, y_2)$ be any two points in the plane. We wish to obtain a formula for finding the nonnegative distance between these two points. Denote this distance by $|\overline{P_1P_2}|$. Absolute-value bars are used because we are concerned only with the length, which is a nonnegative number, of the line segment between the two points P_1 and P_2. To derive the formula, note that $|\overline{P_1P_2}|$ is the length of the hypotenuse of a right triangle P_1MP_2. This is illustrated in Fig. 1.3.3 for P_1 and P_2, both of which are in the first quadrant.

From the Pythagorean theorem,

$$|\overline{P_1P_2}|^2 = |\Delta x|^2 + |\Delta y|^2$$

So

$$|\overline{P_1P_2}| = \sqrt{|\Delta x|^2 + |\Delta y|^2}$$

That is,

$$|\overline{P_1P_2}| = \sqrt{(x_2 - x_1)^2 + (y_2 - y_1)^2} \tag{1}$$

Formula (1) holds for all possible positions of P_1 and P_2 in all four quadrants. The length of the hypotenuse will always be $|\overline{P_1P_2}|$, and the lengths of the two legs will always be $|\Delta x|$ and $|\Delta y|$ (see Exercises 33 and 34). This result is stated as a theorem.

1.3.1 Theorem The undirected distance between the two points $P_1(x_1, y_1)$ and $P_2(x_2, y_2)$ is given by

$$|\overline{P_1P_2}| = \sqrt{(x_2 - x_1)^2 + (y_2 - y_1)^2}$$

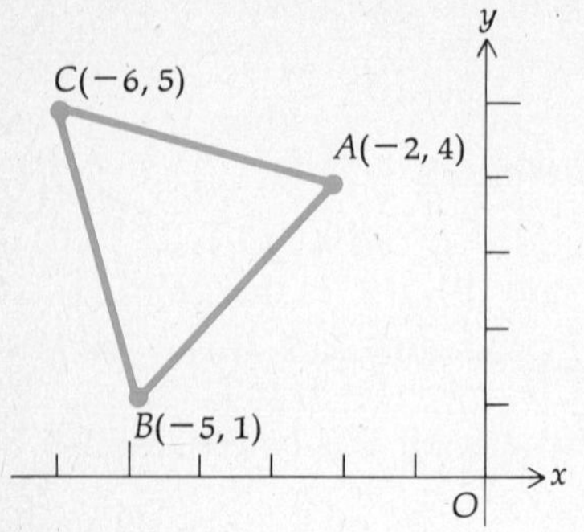

FIGURE 1.3.4

Example 1 Show that the triangle with vertices at $A(-2, 4)$, $B(-5, 1)$, and $C(-6, 5)$ is isosceles.

Solution The triangle is shown in Fig. 1.3.4.

$$|\overline{BC}| = \sqrt{(-6+5)^2 + (5-1)^2} = \sqrt{1+16} = \sqrt{17}$$

$$|\overline{BA}| = \sqrt{(-2+5)^2 + (4-1)^2} = \sqrt{9+9} = 3\sqrt{2}$$

$$|\overline{AC}| = \sqrt{(-6+2)^2 + (5-4)^2} = \sqrt{16+1} = \sqrt{17}$$

Therefore

$$|\overline{BC}| = |\overline{AC}|$$

Hence triangle ABC is isosceles.

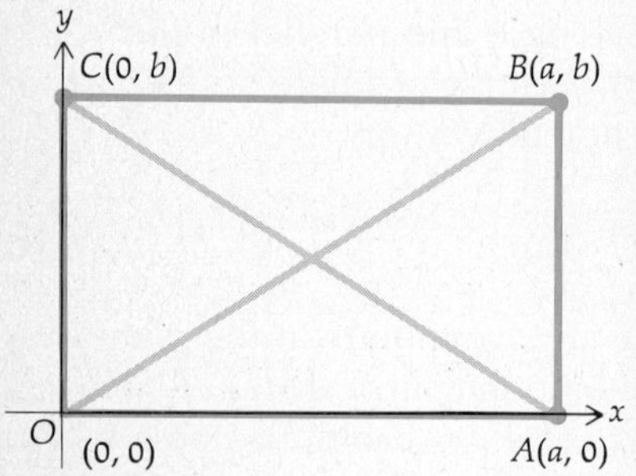

FIGURE 1.3.5

Example 2 Prove analytically that the lengths of the diagonals of a rectangle are equal.

Solution Draw a general rectangle. Because the coordinate axes can be chosen anywhere in the plane, and because the choice of the position of the axes does not affect the truth of the theorem, we take the origin at one vertex, the x axis along one side, and the y axis along another side. This procedure simplifies the coordinates of the vertices on the two axes. Refer to Fig. 1.3.5.

Now the hypothesis and the conclusion of the theorem can be stated.

Hypothesis: $OABC$ is a rectangle with diagonals OB and AC.

Conclusion: $|\overline{OB}| = |\overline{AC}|$.

Proof

$$|\overline{OB}| = \sqrt{(a-0)^2 + (b-0)^2} = \sqrt{a^2 + b^2}$$

$$|\overline{AC}| = \sqrt{(0-a)^2 + (b-0)^2} = \sqrt{a^2 + b^2}$$

Therefore

$$|\overline{OB}| = |\overline{AC}|$$

■

One of the simplest curves that is the graph of a second-degree equation in two variables is the *circle,* which is defined in Definition 1.3.2 below. The formula for obtaining an equation of a circle is derived by applying the distance formula.

1.3.2 Definition A **circle** is the set of all points in a plane equidistant from a fixed point. The fixed point is called the **center** of the circle, and the measure of the constant equal distance is called the **radius** of the circle.

1.3.3 Theorem The circle with center at the point $C(h, k)$ and radius r has as an equation

$$(x-h)^2+(y-k)^2=r^2 \tag{2}$$

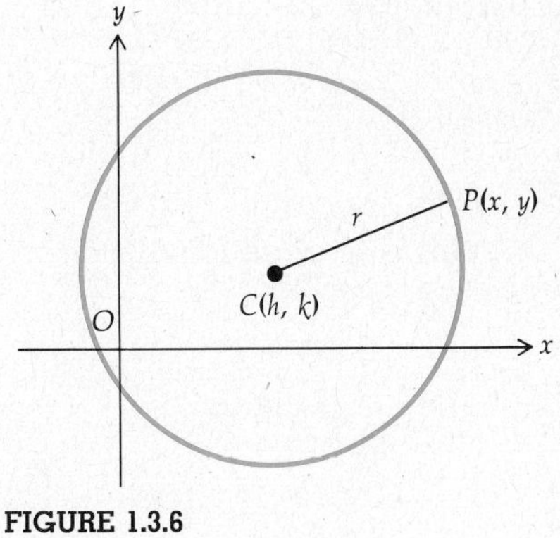

FIGURE 1.3.6

Proof The point $P(x, y)$ lies on the circle if and only if

$$|\overline{PC}|=r$$

that is, if and only if

$$\sqrt{(x-h)^2+(y-k)^2}=r$$

This is true if and only if

$$(x-h)^2+(y-k)^2=r^2$$

which is (2). Equation (2) is satisfied by the coordinates of those and only those points that lie on the given circle. Hence (2) is an equation of the circle. ■

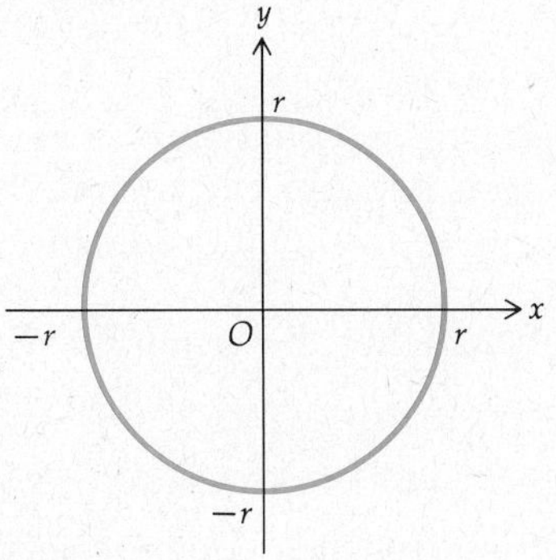

FIGURE 1.3.7

Figure 1.3.6 shows the circle with center at (h, k) and radius r. If the center of the circle is at the origin, then $h = k = 0$; therefore its equation is

$$x^2+y^2=r^2$$

Such a circle is shown in Fig. 1.3.7.

If the center and measure of the radius of a circle are known, the circle can be drawn by using a compass.

● ILLUSTRATION 3 The equation

$$x^2+y^2=9$$

is an equation of a circle having its center at the origin and radius 3. ●

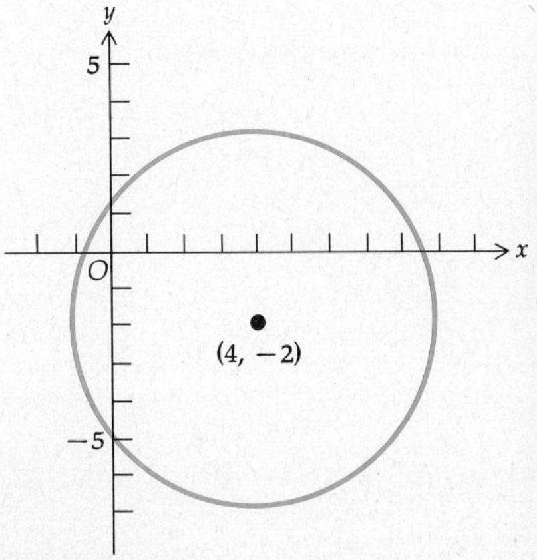

FIGURE 1.3.8

Example 3 Find an equation of the circle with center at $(4, -2)$ and radius equal to 5. Draw the circle.

Solution The circle is shown in Fig. 1.3.8. An equation of the circle is of the form of (2), where $h = 4$, $k = -2$, and $r = 5$. Substituting into (2) we have

$$(x-4)^2+[y-(-2)]^2=5^2$$
$$(x-4)^2+(y+2)^2=25$$

Squaring we have

$$x^2-8x+16+y^2+4y+4=25$$
$$x^2+y^2-8x+4y-5=0$$

Example 4 Show that the graph of the equation $x^2 + y^2 + 6x - 2y - 15 = 0$ is a circle, and find its center and radius.

Solution The given equation may be written as

$$(x^2 + 6x) + (y^2 - 2y) = 15$$

Completing the squares of the terms in parentheses by adding 9 and 1 on both sides of the equation, we have

$$(x^2 + 6x + 9) + (y^2 - 2y + 1) = 15 + 9 + 1$$

$$(x + 3)^2 + (y - 1)^2 = 25$$

Comparing this equation with (2) we see that this is an equation of a circle with its center at $(-3, 1)$ and a radius of 5.

In (2), removing parentheses and combining terms gives

$$x^2 + y^2 - 2hx - 2ky + (h^2 + k^2 - r^2) = 0 \tag{3}$$

Equation (3) is of the form

$$x^2 + y^2 + Dx + Ey + F = 0 \tag{4}$$

where $D = -2h$, $E = -2k$, and $F = h^2 + k^2 - r^2$.

Equation (4) is called the **general form** of an equation of a circle, whereas (2) is called the **center-radius form.** Because every circle has a center and a radius, its equation can be put in the center-radius form, and hence into the general form.

Now consider the question of whether or not the graph of every equation of the form

$$x^2 + y^2 + Dx + Ey + F = 0$$

is a circle. To determine this we attempt to write this equation in the center-radius form. The equation is rewritten as

$$x^2 + y^2 + Dx + Ey = -F$$

We complete the squares of the terms in parentheses by adding $\frac{1}{4}D^2$ and $\frac{1}{4}E^2$ on both sides, thus giving

$$(x^2 + Dx + \tfrac{1}{4}D^2) + (y^2 + Ey + \tfrac{1}{4}E^2) = -F + \tfrac{1}{4}D^2 + \tfrac{1}{4}E^2$$

or, equivalently,

$$(x + \tfrac{1}{2}D)^2 + (y + \tfrac{1}{2}E)^2 = \tfrac{1}{4}(D^2 + E^2 - 4F) \tag{5}$$

Equation (5) is in the form of (2) if and only if

$$\tfrac{1}{4}(D^2 + E^2 - 4F) = r^2$$

We now consider three cases, namely, $(D^2 + E^2 - 4F)$ as positive, zero, and negative.

Case 1: $(D^2 + E^2 - 4F) > 0$.

Then $r^2 = \frac{1}{4}(D^2 + E^2 - 4F)$, and so (5) is an equation of a circle having a radius equal to $\frac{1}{2}\sqrt{D^2 + E^2 - 4F}$ and its center at $(-\frac{1}{2}D, -\frac{1}{2}E)$.

Case 2: $D^2 + E^2 - 4F = 0$.

Equation (5) is then of the form

$$(x + \tfrac{1}{2}D)^2 + (y + \tfrac{1}{2}E)^2 = 0 \tag{6}$$

Because the only real values of x and y satisfying (6) are $x = -\frac{1}{2}D$ and $y = -\frac{1}{2}E$, the graph is the point $(-\frac{1}{2}D, -\frac{1}{2}E)$. Comparing (6) with (2) we see that $h = -\frac{1}{2}D$, $k = -\frac{1}{2}E$, and $r = 0$. Thus this point can be called a **point-circle.**

Case 3: $(D^2 + E^2 - 4F) < 0$.

Then (5) has a negative number on the right side and the sum of the squares of two real numbers on the left side. There are no real values of x and y that satisfy such an equation; consequently, the graph is the empty set.

Before the results of these three cases are stated as a theorem, observe that an equation of the form

$$Ax^2 + Ay^2 + Dx + Ey + F = 0, \qquad \text{where } A \neq 0 \tag{7}$$

can be written in the form of (4) by dividing by A, thereby obtaining

$$x^2 + y^2 + \frac{D}{A}x + \frac{E}{A}y + \frac{F}{A} = 0$$

Equation (7) is a special case of the general equation of the second degree,

$$Ax^2 + Bxy + Cy^2 + Dx + Ey + F = 0$$

in which the coefficients of x^2 and y^2 are equal and which has no xy term. The following theorem is a result of this discussion.

1.3.4 Theorem The graph of any second-degree equation in R^2 in x and y, in which the coefficients of x^2 and y^2 are equal and in which there is no xy term, is either a circle, a point-circle, or the empty set.

● ILLUSTRATION 4 The equation

$$2x^2 + 2y^2 + 12x - 8y + 31 = 0$$

is of the form (7), and therefore its graph is either a circle, a point-circle, or the empty set. The equation is put in the form of (2):

$$x^2 + y^2 + 6x - 4y + \tfrac{31}{2} = 0$$

$$(x^2 + 6x) + (y^2 - 4y) = -\tfrac{31}{2}$$

$$(x^2 + 6x + 9) + (y^2 - 4y + 4) = -\tfrac{31}{2} + 9 + 4$$

$$(x + 3)^2 + (y - 2)^2 = -\tfrac{5}{2}$$

Therefore the graph is the empty set. ●

Example 5 Find an equation of the circle through the three points $A(4, 5)$, $B(3, -2)$, and $C(1, -4)$.

Solution The general form of an equation of the circle is

$$x^2 + y^2 + Dx + Ey + F = 0$$

Because the three points A, B, and C must lie on the circle, the coordinates of these points must satisfy the equation. Thus

$$16 + 25 + 4D + 5E + F = 0$$
$$9 + 4 + 3D - 2E + F = 0$$
$$1 + 16 + D - 4E + F = 0$$

or, equivalently,

$$4D + 5E + F = -41$$
$$3D - 2E + F = -13$$
$$D - 4E + F = -17$$

Solving these three equations simultaneously we get

$$D = 7 \qquad E = -5 \qquad F = -44$$

Thus an equation of the circle is

$$x^2 + y^2 + 7x - 5y - 44 = 0$$

Let $P_1(x_1, y_1)$ and $P_2(x_2, y_2)$ be the endpoints of a line segment. We denote this line segment by P_1P_2. This is not to be confused with the notation $\overline{P_1P_2}$, which denotes the directed distance from P_1 to P_2. That is, $\overline{P_1P_2}$ denotes a number, whereas P_1P_2 is a line segment. Let $P(x, y)$ be the midpoint of the line segment P_1P_2. Refer to Fig. 1.3.9.

In Fig. 1.3.9, triangles P_1RP and PTP_2 are congruent. Therefore $|\overline{P_1R}| = |\overline{PT}|$, and so $x - x_1 = x_2 - x$, giving

$$x = \frac{x_1 + x_2}{2} \tag{8}$$

Similarly, $|\overline{RP}| = |\overline{TP_2}|$. Then $y - y_1 = y_2 - y$, and therefore

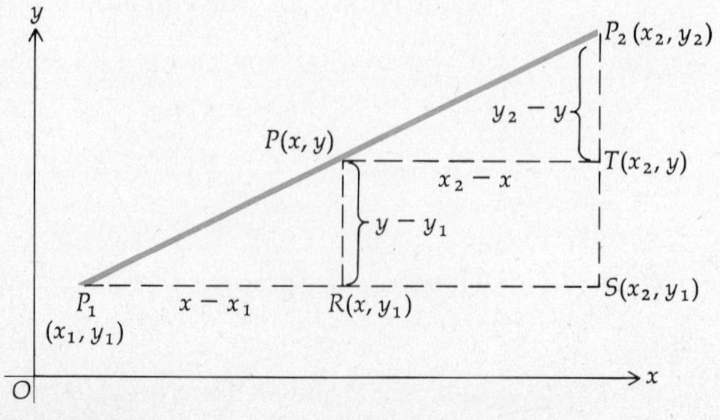

FIGURE 1.3.9

$$y = \frac{y_1 + y_2}{2} \tag{9}$$

Hence the coordinates of the midpoint of a line segment are, respectively, the average of the abscissas and the average of the ordinates of the endpoints of the line segment.

In the derivation of formulas (8) and (9) it was assumed that $x_2 > x_1$ and $y_2 > y_1$. The same formulas are obtained by using any orderings of these numbers (see Exercises 35 and 36).

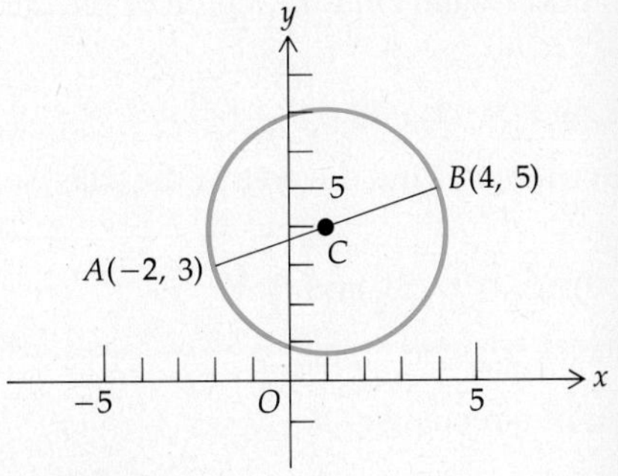

FIGURE 1.3.10

Example 6 Find an equation of the circle having a diameter with extremities at $A(-2, 3)$ and $B(4, 5)$.

Solution Because the extremities of a diameter are at the points A and B, the midpoint of the line segment AB is the center of the circle. See Fig. 1.3.10. If $C(h, k)$ is the center of the circle, then

$$h = \frac{-2+4}{2} \qquad k = \frac{3+5}{2}$$
$$= 1 \qquad\qquad = 4$$

The center is at $C(1, 4)$. The radius of the circle can be computed as either $|\overline{CA}|$ or $|\overline{CB}|$. If $r = |\overline{CA}|$, then

$$r = \sqrt{(1+2)^2 + (4-3)^2}$$
$$= \sqrt{10}$$

An equation of the circle is therefore

$$(x-1)^2 + (y-4)^2 = 10$$

$$x^2 + y^2 - 2x - 8y + 7 = 0$$

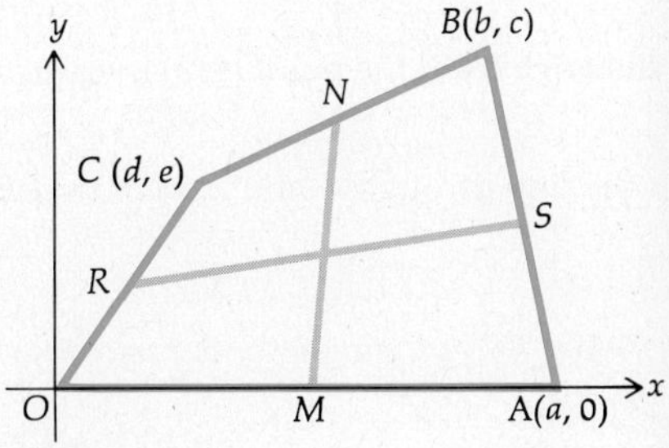

FIGURE 1.3.11

Example 7 Prove analytically that the line segments joining the midpoints of the opposite sides of any quadrilateral bisect each other.

Solution Draw a general quadrilateral. Take the origin at one vertex and the x axis along one side. This method simplifies the coordinates of the two vertices on the x axis. See Fig. 1.3.11.

Hypothesis: $OABC$ is a quadrilateral. M is the midpoint of OA, N is the midpoint of CB, R is the midpoint of OC, and S is the midpoint of AB.

Conclusion: MN and RS bisect each other.

Proof To prove that two line segments bisect each other, we show that they have the same midpoint. From formulas (8) and (9) we obtain the coordinates of M, N, R, and S. M is the point $(\frac{1}{2}a, 0)$, N is the point $(\frac{1}{2}(b+d), \frac{1}{2}(c+e))$, R is the point $(\frac{1}{2}d, \frac{1}{2}e)$, and S is the point $(\frac{1}{2}(a+b), \frac{1}{2}c)$.

The abscissa of the midpoint of MN is $\frac{1}{2}[\frac{1}{2}a + \frac{1}{2}(b+d)] = \frac{1}{4}(a+b+d)$.
The ordinate of the midpoint of MN is $\frac{1}{2}[0 + \frac{1}{2}(c+e)] = \frac{1}{4}(c+e)$.

Therefore, the midpoint of MN is the point $(\frac{1}{4}(a+b+d), \frac{1}{4}(c+e))$.
The abscissa of the midpoint of RS is $\frac{1}{2}[\frac{1}{2}d + \frac{1}{2}(a+b)] = \frac{1}{4}(a+b+d)$.
The ordinate of the midpoint of RS is $\frac{1}{2}[\frac{1}{2}e + \frac{1}{2}c] = \frac{1}{4}(c+e)$.
Therefore, the midpoint of RS is the point $(\frac{1}{4}(a+b+d), \frac{1}{4}(c+e))$.
Thus, the midpoint of MN is the same point as the midpoint of RS.
Therefore, MN and RS bisect each other. ■

EXERCISES 1.3

1. A median of a triangle is a line segment from a vertex to the midpoint of the opposite side. Find the length of the medians of the triangle having vertices $A(2, 3)$, $B(3, -3)$, and $C(-1, -1)$.
2. Find the length of the medians of the triangle having vertices $A(-3, 5)$, $B(2, 4)$, and $C(-1, -4)$.
3. Prove that the triangle with vertices $A(3, -6)$, $B(8, -2)$, and $C(-1, -1)$ is a right triangle. Find the area of the triangle. (*Hint:* Use the converse of the Pythagorean theorem.)
4. Find the midpoints of the diagonals of the quadrilateral whose vertices are (0, 0), (0, 4), (3, 5), and (3, 1).
5. Prove that the points $A(-7, 2)$, $B(3, -4)$, and $C(1, 4)$ are the vertices of an isosceles triangle.
6. Prove that the points $A(-4, -1)$, $B(-2, -3)$, $C(4, 3)$, and $D(2, 5)$ are the vertices of a rectangle.
7. By using distance formula (1), prove that the points $(-3, 2)$, $(1, -2)$, and $(9, -10)$ lie on a line.
8. Determine whether the points (14, 7), (2, 2), and $(-4, -1)$ lie on a line by using distance formula (1).
9. Prove that the points $A(6, -13)$, $B(-2, 2)$, $C(13, 10)$, and $D(21, -5)$ are the vertices of a square. Find the length of a diagonal.
10. If one end of a line segment is the point $(-4, 2)$ and the midpoint is $(3, -1)$, find the coordinates of the other end of the line segment.
11. If one end of a line segment is the point $(6, -2)$ and the midpoint is $(-1, 5)$, find the coordinates of the other end of the line segment.
12. The abscissa of a point is -6, and its distance from the point (1, 3) is $\sqrt{74}$. Find the ordinate of the point.
13. If two vertices of an equilateral triangle are $(-4, 3)$ and (0, 0), find the third vertex.
14. Find an equation that must be satisfied by the coordinates of any point that is equidistant from the two points $(-3, 2)$ and (4, 6).
15. Find an equation that must be satisfied by the coordinates of any point whose distance from the point (5, 3) is always two units greater than its distance from the point $(-4, -2)$.
16. Given the two points $A(-3, 4)$ and $B(2, 5)$, find the coordinates of a point P on the line through A and B such that P is (a) twice as far from A as from B, and (b) twice as far from B as from A.

In Exercises 17 through 20, find an equation of the circle with center at C and radius r. Write the equation in both the center-radius form and the general form.

17. $C(4, -3), r = 5$ **18.** $C(0, 0), r = 8$ **19.** $C(-5, -12), r = 3$ **20.** $C(-1, 1), r = 2$

In Exercises 21 and 22, find an equation of the circle satisfying the given conditions.

21. Center is at (1, 2) and through the point $(3, -1)$.
22. Through the three points (2, 8), (7, 3), and $(-2, 0)$.

In Exercises 23 through 26, find the center and radius of each circle, and draw a sketch of the graph.

23. $x^2 + y^2 - 6x - 8y + 9 = 0$ **24.** $2x^2 + 2y^2 - 2x + 2y - 7 = 0$

25. $3x^2 + 3y^2 + 4y - 7 = 0$ **26.** $x^2 + y^2 - 10x - 10y + 25 = 0$

In Exercises 27 through 32, determine whether the graph is a circle, a point-circle, or the empty set.

27. $x^2 + y^2 - 2x + 10y + 19 = 0$ **28.** $4x^2 + 4y^2 + 24x - 4y + 1 = 0$ **29.** $x^2 + y^2 - 10x + 6y + 36 = 0$

30. $x^2 + y^2 + 2x - 4y + 5 = 0$ **31.** $36x^2 + 36y^2 - 48x + 36y - 119 = 0$ **32.** $9x^2 + 9y^2 + 6x - 6y + 5 = 0$

33. Derive distance formula (1) if P_1 is in the third quadrant and P_2 is in the second quadrant. Draw a figure.

34. Derive distance formula (1) if P_1 is in the second quadrant and P_2 is in the fourth quadrant. Draw a figure.

35. Derive midpoint formulas (8) and (9) if P_1 is in the first quadrant and P_2 is in the third quadrant.

36. Derive midpoint formulas (8) and (9) if $P_1(x_1, y_1)$ and $P_2(x_2, y_2)$ are both in the second quadrant and $x_2 > x_1$ and $y_1 > y_2$.

37. Find the coordinates of the three points that divide the line segment from $A(-5, 3)$ to $B(6, 8)$ into four equal parts.

38. If the point $P(x, y)$ is on the line through $P_1(x_1, y_1)$ and $P_2(x_2, y_2)$ such that $|\overline{P_1P}|/|\overline{P_1P_2}| = k$, prove that the coordinates of P are given by

$$x = x_1 + k(x_2 - x_1) \quad \text{and} \quad y = y_1 + k(y_2 - y_1)$$

where

$0 < k < 1$ if P is between P_1 and P_2
$k > 1$ if P is not between P_1 and P_2 and closer to P_2
$k < 0$ if P is not between P_1 and P_2 and closer to P_1

In Exercises 39 through 43, use the formulas of Exercise 38 to find the coordinates of point P.

39. The point P is on the line segment between points $P_1(1, 3)$ and $P_2(6, 2)$ and is three times as far from P_2 as it is from P_1.

40. The point P is on the line segment between points $P_1(1, 3)$ and $P_2(6, 2)$ and is three times as far from P_1 as it is from P_2.

41. The point P is on the line through P_1 and P_2 and is three times as far from $P_2(6, 2)$ as it is from $P_1(1, 3)$ but is not between P_1 and P_2.

42. The point P is on the line through P_1 and P_2 and is three times as far from $P_1(1, 3)$ as it is from $P_2(6, 2)$ but is not between P_1 and P_2.

43. The point P is on the line through $P_1(-3, 5)$ and $P_2(-1, 2)$ so that $\overline{P_1P} = 4 \cdot \overline{P_1P_2}$.

44. From the origin, chords of the circle $x^2 + y^2 + 4x = 0$ are drawn. Prove that the set of midpoints of these chords is a circle.

45. Find an equation whose graph consists of all points equidistant from the points $(-1, 2)$ and $(3, 4)$. Draw a sketch of the graph of this equation.

46. Find an equation that must be satisfied by the coordinates of any point the sum of whose distances from $(0, -3)$ and $(0, 3)$ is 9. Draw a sketch of the graph of this equation.

47. Prove analytically that the sum of the squares of the distances of any point from two opposite vertices of any rectangle is equal to the sum of the squares of its distances from the other two vertices.

48. Prove analytically that the midpoint of the hypotenuse of any right triangle is equidistant from each of the three vertices.

49. Prove analytically that the line segment joining the midpoints of two opposite sides of any quadrilateral and the line segment joining the midpoints of the diagonals of the quadrilateral bisect each other.

50. Prove analytically that if the lengths of two of the medians of a triangle are equal, the triangle is isosceles.

1.4 EQUATIONS OF A LINE

Let l be a nonvertical line and $P_1(x_1, y_1)$ and $P_2(x_2, y_2)$ be any two distinct points on l. Figure 1.4.1 shows such a line. In the figure, R is the point (x_2, y_1), and the points P_1, P_2, and R are vertices of a right triangle; furthermore, $\overline{P_1R} = x_2 - x_1$ and $\overline{RP_2} = y_2 - y_1$. The number $y_2 - y_1$ gives the measure of the change in the ordinate from P_1 to P_2, and it may be positive, negative, or zero. The number $x_2 - x_1$ gives the measure of the change in the abscissa from P_1 to P_2, and it may be positive or negative. Because the line l is not vertical, $x_2 \neq x_1$, and therefore $x_2 - x_1$ may not be zero. Let

$$m = \frac{y_2 - y_1}{x_2 - x_1} \tag{1}$$

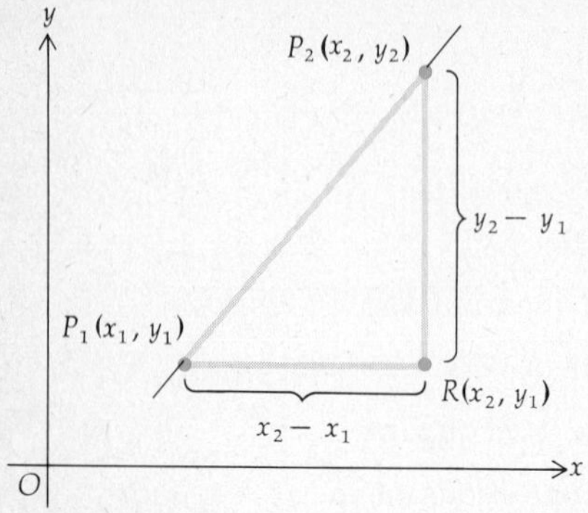

FIGURE 1.4.1

The value of m computed from this equation is independent of the choice of the two points P_1 and P_2 on l. To show this, suppose we choose two different points $\bar{P}_1(\bar{x}_1, \bar{y}_1)$ and $\bar{P}_2(\bar{x}_2, \bar{y}_2)$, and compute a number $\bar{m}$ from (1).

$$\bar{m} = \frac{\bar{y}_2 - \bar{y}_1}{\bar{x}_2 - \bar{x}_1}$$

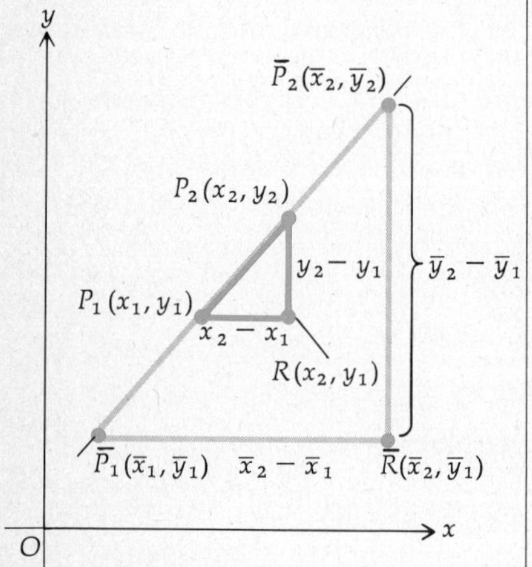

FIGURE 1.4.2

We shall show that $\bar{m} = m$. Refer to Fig. 1.4.2. Triangles $\bar{P}_1\bar{R}\bar{P}_2$ and P_1RP_2 are similar, so the lengths of corresponding sides are proportional. Therefore

$$\frac{\bar{y}_2 - \bar{y}_1}{\bar{x}_2 - \bar{x}_1} = \frac{y_2 - y_1}{x_2 - x_1}$$

or

$$\bar{m} = m$$

Thus the value of m computed from (1) is the same number no matter what two points on l are selected. This number m is called the *slope* of the line. Following is the formal definition.

1.4.1 Definition If $P_1(x_1, y_1)$ and $P_2(x_2, y_2)$ are any two distinct points on line l, which is not parallel to the y axis, then the **slope** of l, denoted by m, is given by

$$m = \frac{y_2 - y_1}{x_2 - x_1}$$

In Sec. 1.3 we defined $\Delta y = y_2 - y_1$ and $\Delta x = x_2 - x_1$. Substituting these values into (1) we have

$$m = \frac{\Delta y}{\Delta x}$$

$$\Delta y = m\,\Delta x \tag{2}$$

It is seen from (2) that if a particle is considered as moving along line l, the change in the ordinate of the particle is proportional to the change in the abscissa, and the constant of proportionality is the slope of the line.

If the slope of a line is positive, then as the abscissa of a point on the line increases, the ordinate increases. Such a line is shown in Fig. 1.4.3. In Fig. 1.4.4 there is a line whose slope is negative. For this line, as the ab-

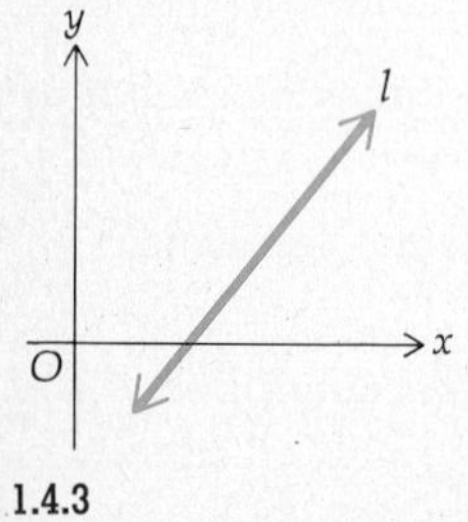

FIGURE 1.4.3

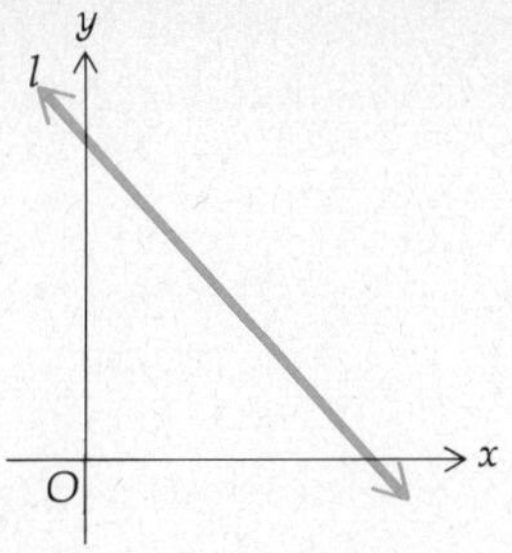

FIGURE 1.4.4

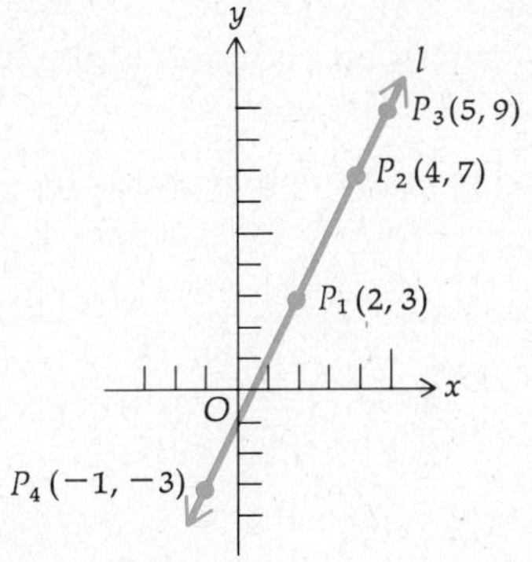

FIGURE 1.4.5

scissa of a point on the line increases, the ordinate decreases. Note that if the line is parallel to the x axis, then $y_2 = y_1$ and so $m = 0$.

If the line is parallel to the y axis, $x_2 = x_1$; thus (1) is meaningless because we cannot divide by zero. This is the reason that lines parallel to the y axis, or vertical lines, are excluded in Definition 1.4.1. Thus, a vertical line does not have a slope.

● ILLUSTRATION 1 Let l be the line through the points $P_1(2, 3)$ and $P_2(4, 7)$. The slope of l, by Definition 1.4.1, is given by

$$m = \frac{7 - 3}{4 - 2} = 2$$

Refer to Fig. 1.4.5. If $P(x, y)$ and $Q(x + \Delta x, y + \Delta y)$ are any two points on l, then

$$\frac{\Delta y}{\Delta x} = 2$$

$$\Delta y = 2\,\Delta x$$

Thus if a particle is moving along the line l, the change in the ordinate is twice the change in the abscissa. That is, if the particle is at $P_2(4, 7)$ and the abscissa is increased by one unit, then the ordinate is increased by two units, and the particle is at the point $P_3(5, 9)$. Similarly, if the particle is at $P_1(2, 3)$ and the abscissa is decreased by three units, then the ordinate is decreased by six units, and the particle is at $P_4(-1, -3)$. ●

Since two points $P_1(x_1, y_1)$ and $P_2(x_2, y_2)$ determine a unique line, we should be able to obtain an equation of the line through these two points. Consider $P(x, y)$ any point on the line. We want an equation that is satisfied by x and y if and only if $P(x, y)$ is on the line through $P_1(x_1, y_1)$ and $P_2(x_2, y_2)$. There are two cases.

Case 1: $x_2 = x_1$.

In this case the line through P_1 and P_2 is parallel to the y axis, and all points on this line have the same abscissa. So $P(x, y)$ is any point on the line if and only if

$$x = x_1 \tag{3}$$

Equation (3) is an equation of a line parallel to the y axis. Note that this equation is independent of y; that is, the ordinate may have any value whatsoever, and the point $P(x, y)$ is on the line whenever the abscissa is x_1.

Case 2: $x_2 \neq x_1$.

The slope of the line through P_1 and P_2 is given by

$$m = \frac{y_2 - y_1}{x_2 - x_1} \tag{4}$$

If $P(x, y)$ is any point on the line except (x_1, y_1), the slope is also given by

$$m = \frac{y - y_1}{x - x_1} \tag{5}$$

The point P will be on the line through P_1 and P_2 if and only if the value of m from (4) is the same as the value of m from (5), that is, if and only if

$$\frac{y - y_1}{x - x_1} = \frac{y_2 - y_1}{x_2 - x_1}$$

Multiplying on both sides of this equation by $(x - x_1)$ we obtain

$$y - y_1 = \frac{y_2 - y_1}{x_2 - x_1}(x - x_1) \tag{6}$$

Equation (6) is satisfied by the coordinates of P_1 as well as by the coordinates of any other point on the line through P_1 and P_2.

Equation (6) is called the **two-point** form of an equation of the line. It gives an equation of the line if two points on the line are known.

● ILLUSTRATION 2 An equation of the line through the two points $(6, -3)$ and $(-2, 3)$ is

$$y - (-3) = \frac{3 - (-3)}{-2 - 6}(x - 6)$$

$$y + 3 = -\tfrac{3}{4}(x - 6)$$

$$3x + 4y = 6$$ ●

If in (6) m is replaced by $(y_2 - y_1)/(x_2 - x_1)$, we get

$$y - y_1 = m(x - x_1) \tag{7}$$

Equation (7) is called the **point-slope** form of an equation of the line. It gives an equation of the line if a point $P_1(x_1, y_1)$ on the line and the slope m of the line are known.

● ILLUSTRATION 3 An equation of the line through the point $(-4, -5)$ and having a slope of 2 is

$$y - (-5) = 2[x - (-4)]$$

$$2x - y + 3 = 0$$ ●

If we choose the particular point $(0, b)$ (i.e., the point where the line intersects the y axis) for the point (x_1, y_1) in (7), we have

$$y - b = m(x - 0)$$

or, equivalently,

$$y = mx + b \tag{8}$$

The number b, which is the ordinate of the point where the line intersects the y axis, is called the y **intercept** of the line. Consequently, (8) is called the **slope-intercept** form of an equation of the line. This form is especially important because it enables us to find the slope of a line from its equation. It is also important because it expresses the y coordinate explicitly in terms of the x coordinate.

Example 1 Given the line having the equation $3x + 4y = 7$, find the slope of the line.

Solution The equation is solved for y.

$$y = -\tfrac{3}{4}x + \tfrac{7}{4}$$

Comparing this equation with (8), we see that $m = -\frac{3}{4}$ and $b = \frac{7}{4}$. Therefore the slope is $-\frac{3}{4}$.

Another form of an equation of a line is the one involving the intercepts of a line. The x **intercept** of a line is defined as the abscissa of the point at which the line intersects the x axis. The x intercept is denoted by a. If the x intercept a and the y intercept b are given, two points $(a, 0)$ and $(0, b)$ on the line are known. Applying (6), the two-point form, we have

$$y - 0 = \frac{b - 0}{0 - a}(x - a)$$

$$-ay = bx - ab$$

$$bx + ay = ab$$

Dividing by ab, if $a \neq 0$ and $b \neq 0$, we obtain

$$\frac{x}{a} + \frac{y}{b} = 1 \qquad (9)$$

Equation (9) is called the **intercept** form of an equation of the line. Obviously it does not apply to a line through the origin, because for such a line both a and b are zero.

Example 2 The point (2, 3) bisects that portion of a line which is cut off by the coordinate axes. Find an equation of the line.

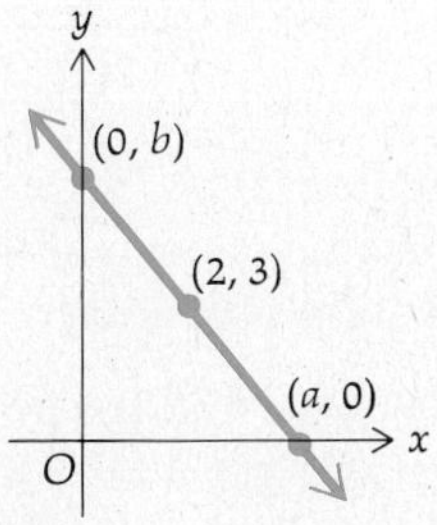

FIGURE 1.4.6

Solution Refer to Fig. 1.4.6. If a is the x intercept of the line and b is the y intercept of the line, then the point (2, 3) is the midpoint of the line segment joining $(a, 0)$ and $(0, b)$. By the midpoint formulas,

$$2 = \frac{a + 0}{2} \quad \text{and} \quad 3 = \frac{0 + b}{2}$$

Then

$$a = 4 \quad \text{and} \quad b = 6$$

The intercept form, (9), gives

$$\frac{x}{4} + \frac{y}{6} = 1$$

$$3x + 2y = 12$$

1.4.2 Theorem The graph of the equation

$$Ax + By + C = 0 \qquad (10)$$

where A, B, and C are constants and where not both A and B are zero, is a straight line.

Proof Consider the two cases $B \neq 0$ and $B = 0$.

Case 1: $B \neq 0$.

Because $B \neq 0$, we divide on both sides of (10) by B and obtain

$$y = -\frac{A}{B}x - \frac{C}{B} \tag{11}$$

Equation (11) is an equation of a straight line because it is in the slope-intercept form, where $m = -A/B$ and $b = -C/B$.

Case 2: $B = 0$.

Because $B = 0$, we may conclude that $A \neq 0$ and thus have

$$Ax + C = 0$$

$$x = -\frac{C}{A} \tag{12}$$

Equation (12) is in the form of (3), and so the graph is a straight line parallel to the y axis. This completes the proof. ■

Because the graph of (10) is a straight line, it is called a **linear equation.** Equation (10) is the general equation of the first degree in x and y.

Because two points determine a line, to draw a sketch of the graph of a straight line from its equation we need only determine the coordinates of two points on the line, plot the two points, and then draw the line. Any two points will suffice, but it is usually convenient to plot the two points where the line intersects the two axes. These points are given by the intercepts.

Example 3 Given line l_1, having the equation $2x - 3y = 12$, and line l_2, having the equation $4x + 3y = 6$, draw a sketch of each of the lines. Then find the coordinates of the point of intersection of l_1 and l_2.

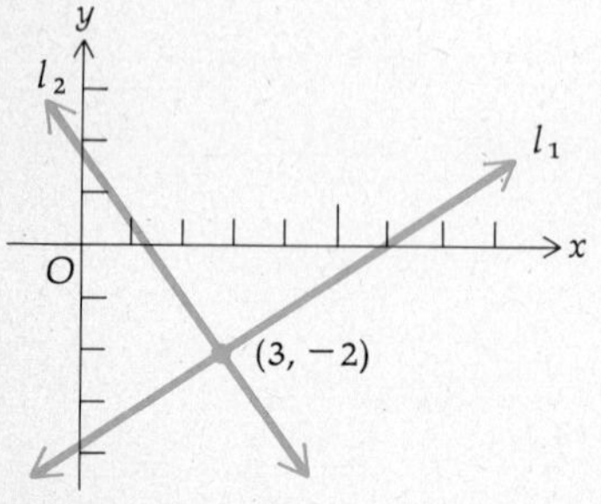

FIGURE 1.4.7

Solution To draw a sketch of the graph of l_1, we find the intercepts a and b. In the equation of l_1, substitute 0 for x and get $b = -4$. In the equation of l_1, substitute 0 for y and get $a = 6$. Similarly, we obtain the intercepts a and b for l_2, and for l_2 $a = \frac{3}{2}$ and $b = 2$. The two lines are plotted in Fig. 1.4.7.

To find the coordinates of the point of intersection of l_1 and l_2, we solve the two equations simultaneously. Because the point must lie on both lines, its coordinates must satisfy both equations. Both equations are put in the slope-intercept form.

$$y = \tfrac{2}{3}x - 4 \quad \text{and} \quad y = -\tfrac{4}{3}x + 2$$

Eliminating y gives

$$\tfrac{2}{3}x - 4 = -\tfrac{4}{3}x + 2$$
$$2x - 12 = -4x + 6$$
$$x = 3$$

So

$$y = \tfrac{2}{3}(3) - 4$$
$$= -2$$

Therefore the point of intersection is $(3, -2)$.

1.4.3 Theorem If l_1 and l_2 are two distinct nonvertical lines having slopes m_1 and m_2, respectively, then l_1 and l_2 are parallel if and only if $m_1 = m_2$.

Proof Let an equation of l_1 be $y = m_1x + b_1$, and let an equation of l_2 be $y = m_2x + b_2$. Because there is an "if and only if" qualification, the proof consists of two parts.

Part 1: Prove that l_1 and l_2 are parallel if $m_1 = m_2$.

Assume that l_1 and l_2 are not parallel. Let us show that this assumption leads to a contradiction. If l_1 and l_2 are not parallel, then they intersect. Call this point of intersection $P(x_0, y_0)$. The coordinates of P must satisfy the equations of l_1 and l_2, and so

$$y_0 = m_1x_0 + b_1 \quad \text{and} \quad y_0 = m_2x_0 + b_2$$

But $m_1 = m_2$, which gives

$$y_0 = m_1x_0 + b_1 \quad \text{and} \quad y_0 = m_1x_0 + b_2$$

from which it follows that $b_1 = b_2$. Thus, because $m_1 = m_2$ and $b_1 = b_2$, both lines l_1 and l_2 have the same equation, $y = m_1x + b_1$, and so the lines are the same. But this contradicts the hypothesis that l_1 and l_2 are distinct lines. Therefore our assumption is false. So we conclude that l_1 and l_2 are parallel.

Part 2: Prove that l_1 and l_2 are parallel only if $m_1 = m_2$.

Here we must show that if l_1 and l_2 are parallel, then $m_1 = m_2$. Assume that $m_1 \neq m_2$. Solving the equations for l_1 and l_2 simultaneously, we get, upon eliminating y,

$$m_1x + b_1 = m_2x + b_2$$

from which it follows that

$$(m_1 - m_2)x = b_2 - b_1$$

Because we have assumed that $m_1 \neq m_2$, this gives

$$x = \frac{b_2 - b_1}{m_1 - m_2}$$

Hence l_1 and l_2 have a point of intersection, which contradicts the hypothesis that l_1 and l_2 are parallel. So our assumption is false, and therefore $m_1 = m_2$. ■

In Fig. 1.4.8 the two lines l_1 and l_2 are perpendicular. We state and prove the following theorem on the slopes of two perpendicular lines.

1.4.4 Theorem If neither line l_1 nor line l_2 is vertical, then l_1 and l_2 are perpendicular if and only if the product of their slopes is -1. That is, if m_1 is the slope of l_1 and m_2 is the slope of l_2, then l_1 and l_2 are perpendicular if and only if $m_1 m_2 = -1$.

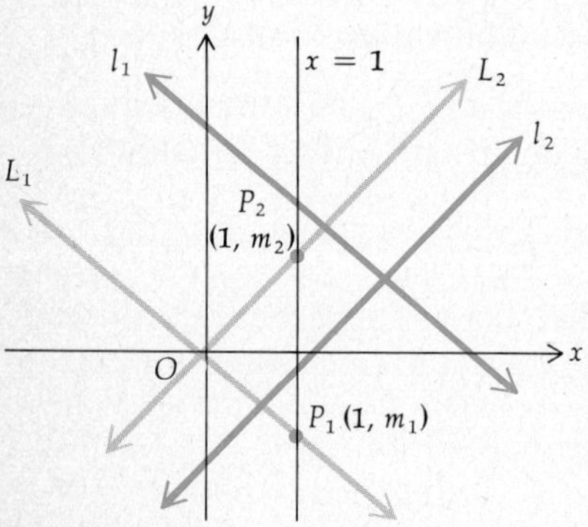

FIGURE 1.4.8

Proof

Part 1: Prove that l_1 and l_2 are perpendicular only if $m_1 m_2 = -1$.

Let L_1 be the line through the origin parallel to l_1, and let L_2 be the line through the origin parallel to l_2. See Fig. 1.4.8. Therefore, by Theorem 1.4.3, the slope of line L_1 is m_1 and the slope of line L_2 is m_2. Because neither L_1 nor L_2 is vertical, these two lines intersect the line $x = 1$ at points P_1 and P_2, respectively. The abscissa of both P_1 and P_2 is 1. Let $\bar{y}$ be the ordinate of P_1. Since L_1 contains the points (0, 0) and $(1, \bar{y})$ and its slope is m_1, it follows from Definition 1.4.1 that

$$m_1 = \frac{\bar{y} - 0}{1 - 0}$$

and so $\bar{y} = m_1$. Similarly, the ordinate of P_2 is shown to be m_2.

Applying the Pythagorean theorem to right triangle $P_1 O P_2$ we get

$$|\overline{OP_1}|^2 + |\overline{OP_2}|^2 = |\overline{P_1 P_2}|^2 \tag{13}$$

From the distance formula,

$$|\overline{OP_1}|^2 = (1 - 0)^2 + (m_1 - 0)^2 = 1 + m_1^2$$

$$|\overline{OP_2}|^2 = (1 - 0)^2 + (m_2 - 0)^2 = 1 + m_2^2$$

$$|\overline{P_1 P_2}|^2 = (1 - 1)^2 + (m_2 - m_1)^2 = m_2^2 - 2m_1 m_2 + m_1^2$$

Substituting into (13) gives

$$(1 + m_1^2) + (1 + m_2^2) = m_2^2 - 2m_1 m_2 + m_1^2$$

$$2 = -2m_1 m_2$$

$$m_1 m_2 = -1$$

Part 2: Prove that l_1 and l_2 are perpendicular if $m_1 m_2 = -1$.

Starting with $m_1 m_2 = -1$, we can reverse the steps of the proof of Part 1 in order to obtain

$$|\overline{OP_1}|^2 + |\overline{OP_2}|^2 = |\overline{P_1 P_2}|^2$$

from which it follows, from the converse of the Pythagorean theorem, that L_1 and L_2 are perpendicular; hence l_1 and l_2 are perpendicular. ■

Theorem 1.4.4 states that if two lines are perpendicular and neither one is vertical, the slope of one of the lines is the negative reciprocal of the slope of the other line.

● ILLUSTRATION 4 If line l_1 is perpendicular to line l_2 and the slope of l_1 is $\frac{2}{3}$, then the slope of l_2 must be $-\frac{3}{2}$. ●

Example 4 Prove by means of slopes that the four points $A(6, 2)$, $B(8, 6)$, $C(4, 8)$, and $D(2, 4)$ are the vertices of a rectangle.

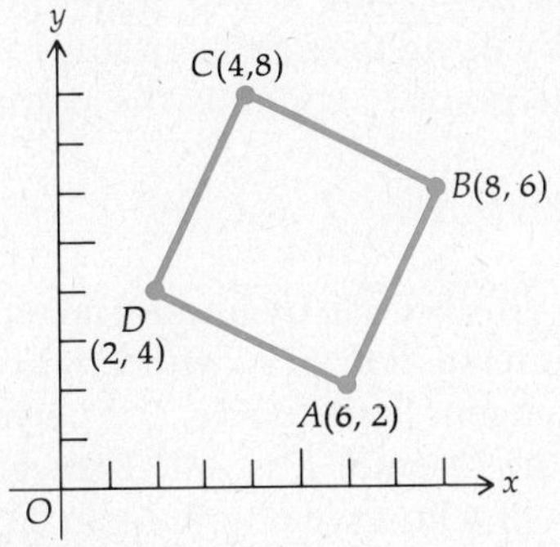

FIGURE 1.4.9

Solution See Fig. 1.4.9. Let

$$m_1 = \text{the slope of AB} = \frac{6-2}{8-6} = 2 \qquad m_2 = \text{the slope of } BC = \frac{8-6}{4-8} = -\frac{1}{2}$$

$$m_3 = \text{the slope of } DC = \frac{8-4}{4-2} = 2 \qquad m_4 = \text{the slope of } AD = \frac{4-2}{2-6} = -\frac{1}{2}$$

Because

$$m_1 = m_3, AB \parallel DC$$

$$m_2 = m_4, BC \parallel AD$$

$$m_1 m_2 = -1, AB \perp BC$$

Therefore quadrilateral $ABCD$ has opposite sides parallel and two adjacent sides perpendicular; thus $ABCD$ is a rectangle.

Example 5 Given the line l having the equation $2x + 3y - 5 = 0$, find an equation of the line perpendicular to line l and passing through the point $A(-1, 3)$.

Solution Because the required line is perpendicular to line l, its slope must be the negative reciprocal of the slope of l. We find the slope of l by putting its equation into the slope-intercept form. Solving the given equation for y we obtain

$$y = -\tfrac{2}{3}x + \tfrac{5}{3}$$

Therefore the slope of l is $-\frac{2}{3}$, and the slope of the required line is $\frac{3}{2}$. Because it is also known that the required line contains the point $(-1, 3)$, we use the point-slope form, which gives

$$y - 3 = \tfrac{3}{2}(x + 1)$$

$$2y - 6 = 3x + 3$$

$$3x - 2y + 9 = 0$$

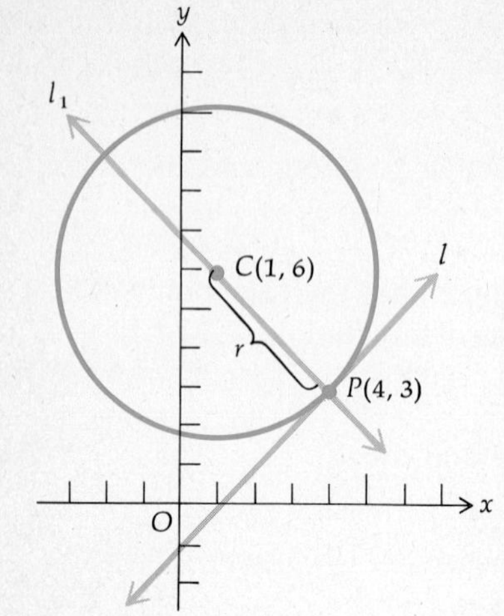

FIGURE 1.4.10

In the following example, a line is tangent to a circle. The definition of the tangent line to a general curve at a specific point is given in Sec. 3.1. However, for a circle we use the definition from plane geometry, which states that a tangent line at a point P on the circle is the line intersecting the circle at only the point P.

Example 6 Find an equation of the circle with its center at the point $C(1, 6)$ and tangent to the line l having the equation $x - y - 1 = 0$.

Solution See Fig. 1.4.10. Given that $h = 1$ and $k = 6$, if we find r, we can obtain an equation of the circle by using the center-radius form.

Let l_1 be the line through C and the point P, which is the point of tangency of line l with the circle.

$$r = |\overline{PC}|$$

Hence we must find the coordinates of P. This is done by finding an equation of l_1 and then finding the point of intersection of l_1 with l. Since l_1 is along a diameter of the circle, and l is tangent to the circle, l_1 is perpendicular to l. Because the slope of l is 1, the slope of l_1 is -1. Therefore, using the point-slope form of an equation of a line, we obtain as an equation of l_1

$$y - 6 = -1(x - 1)$$

$$x + y - 7 = 0$$

Solving this equation simultaneously with the given equation of l, namely,

$$x - y - 1 = 0$$

we get $x = 4$ and $y = 3$. Thus P is the point (4, 3). Because $r = |\overline{PC}|$,

$$r = \sqrt{(4 - 1)^2 + (3 - 6)^2}$$

$$= \sqrt{18}$$

So an equation of the circle is

$$(x - 1)^2 + (y - 6)^2 = (\sqrt{18})^2$$

$$x^2 + y^2 - 2x - 12y + 19 = 0$$

EXERCISES 1.4

In Exercises 1 through 4, find the slope of the line through the given points.

1. $(2, -3), (-4, 3)$ **2.** $(5, 2), (-2, -3)$ **3.** $(\frac{1}{3}, \frac{1}{2}), (-\frac{5}{6}, \frac{2}{3})$ **4.** $(-2.1, 0.3), (2.3, 1.4)$

In Exercises 5 through 15, find an equation of the line satisfying the given conditions.

5. The slope is 4 and through the point $(2, -3)$.

6. Through the two points (3, 1) and $(-5, 4)$.

7. The x intercept is -3 and the y intercept is 4.

8. Through the point (1, 4) and parallel to the line whose equation is $2x - 5y + 7 = 0$.

9. Through the point $(-2, 3)$ and perpendicular to the line whose equation is $2x - y - 2 = 0$.

10. Through the point $(-3, -4)$ and parallel to the y axis.

11. Through the point $(1, -7)$ and parallel to the x axis.

12. The slope is -2 and the x intercept is 4.

13. Through the point $(-2, -5)$ and having a slope of $\sqrt{3}$.

14. Through the origin and bisecting the angle between the axes in the first and third quadrants.

15. Through the origin and bisecting the angle between the axes in the second and fourth quadrants.

In Exercises 16 through 19, find the slope of the given line.

16. $x + 3y = 7$ 17. $4x - 6y = 5$ 18. $3x - 5 = 0$ 19. $2y + 9 = 0$

20. Find an equation of the line through the points $(3, -5)$ and $(1, -2)$, and write the equation in slope-intercept form.

21. Find an equation of the line through the points $(1, 3)$ and $(2, -2)$, and write the equation in intercept form.

22. Show that the lines having the equations $3x + 5y + 7 = 0$ and $6x + 10y - 5 = 0$ are parallel.

23. Show that the lines having the equations $3x + 5y + 7 = 0$ and $5x - 3y - 2 = 0$ are perpendicular.

24. Find the value of k such that the lines whose equations are $3kx + 8y = 5$ and $6y - 4kx = -1$ are perpendicular.

In Exercises 25, 26, and 27, for each set of points determine by means of slopes if the points are collinear.

25. (a) $(2, 3), (-4, -7), (5, 8)$; (b) $(2, -1), (1, 1), (3, 4)$

26. (a) $(4, 6), (1, 2), (-5, -4)$; (b) $(-3, 6), (3, 2), (9, -2)$

27. (a) $(2, 5), (-1, 4), (3, -2)$; (b) $(0, 2), (-3, -1), (4, 6)$

28. Three consecutive vertices of a parallelogram are $(-4, 1)$, $(2, 3)$, and $(8, 9)$. Find the coordinates of the fourth vertex.

29. Show by means of slopes that the points $(-4, -1)$, $(3, \frac{8}{3})$, $(8, -4)$, and $(2, -9)$ are the vertices of a trapezoid.

30. Prove by means of slopes that the three points $A(3, 1)$, $B(6, 0)$, and $C(4, 4)$ are the vertices of a right triangle, and find the area of the triangle.

31. Given the line l having the equation $2y - 3x = 4$ and the point $P(1, -3)$, find (a) an equation of the line through P and perpendicular to l; (b) the shortest distance from P to line l.

32. Find an equation of the line that has equal intercepts and that passes through the point $(8, -6)$.

33. Given the line l having the equation $Ax + By + C = 0$, $B \neq 0$, find (a) the slope; (b) the y intercept; (c) the x intercept; (d) an equation of the line through the origin perpendicular to l.

34. If A, B, C, and D are constants, show that (a) the lines $Ax + By + C = 0$ and $Ax + By + D = 0$ are parallel and (b) the lines $Ax + By + C = 0$ and $Bx - Ay + D = 0$ are perpendicular.

35. Find equations of the three medians of the triangle having vertices $A(3, -2)$, $B(3, 4)$, and $C(-1, 1)$, and prove that they meet in a point.

36. Find equations of the perpendicular bisectors of the sides of the triangle having vertices $A(-1, -3)$, $B(5, -3)$, and $C(5, 5)$, and prove that they meet in a point.

37. Find an equation of each of the lines through the point $(3, 2)$ that forms with the coordinate axes a triangle of area 12.

38. Let l_1 be the line having the equation $A_1x + B_1y + C_1 = 0$, and let l_2 be the line having the equation $A_2x + B_2y + C_2 = 0$. If l_1 is not parallel to l_2 and if k is any constant, the equation

$$A_1x + B_1y + C_1 + k(A_2x + B_2y + C_2) = 0$$

represents an unlimited number of lines. Prove that each of these lines contains the point of intersection of l_1 and l_2.

39. Given that an equation of l_1 is $2x + 3y - 5 = 0$ and that an equation of l_2 is $3x + 5y - 8 = 0$, by using Exercise 38 and without finding the coordinates of the point of intersection of l_1 and l_2, find an equation of the line through this point of intersection and (a) passing through the point (1, 3); (b) parallel to the x axis; (c) parallel to the y axis; (d) having slope -2; (e) perpendicular to the line having the equation $2x + y = 7$; (f) forming an isosceles triangle with the coordinate axes.

40. Find an equation of each straight line that is perpendicular to the line having the equation $5x - y = 1$ and that forms with the coordinate axes a triangle having an area of measure 5.

In Exercises 41 through 44, find an equation of the circle satisfying the given conditions.

41. Center is at $(-3, -5)$ and tangent to the line $12x + 5y = 4$.

42. Center is at $(-2, 5)$ and tangent to the line $x = 7$.

43. Tangent to the line $3x + y + 2 = 0$ at $(-1, 1)$ and through the point (3, 5).

44. Tangent to the line $3x + 4y - 16 = 0$ at (4, 1) and with a radius of 5 (two possible circles).

45. Find an equation of the line that is tangent to the circle $x^2 + y^2 - 4x + 6y - 12 = 0$ at the point (5, 1).

46. Find an equation of each of the two lines having slope $-\frac{4}{3}$ that are tangent to the circle $x^2 + y^2 + 2x - 8y - 8 = 0$.

47. Prove analytically that the diagonals of a rhombus are perpendicular.

48. Prove analytically that the line segments joining consecutive midpoints of the sides of any quadrilateral form a parallelogram.

49. Prove analytically that the diagonals of a parallelogram bisect each other.

50. Prove analytically that if the diagonals of a quadrilateral bisect each other, then the quadrilateral is a parallelogram.

51. Given the line $y = mx + b$ tangent to the circle $x^2 + y^2 = r^2$, find an equation involving m, b, and r.

52. Prove analytically that a line from the center of any circle bisecting any chord is perpendicular to the chord.

1.5 FUNCTIONS AND THEIR GRAPHS

We intuitively consider y to be a function of x if there is some rule by which a unique value is assigned to y by a corresponding value of x. Familiar examples of such relationships are given by equations such as

$$y = 2x^2 + 5 \tag{1}$$

and

$$y = \sqrt{x^2 - 9} \tag{2}$$

The formal definition makes the concept of a function precise.

1.5.1 Definition A **function** is a set of ordered pairs of numbers (x, y) in which no two distinct ordered pairs have the same first number. The set of all possible values of x is called the **domain** of the function, and the set of all possible values of y is called the **range** of the function.

In Definition 1.5.1, the restriction that no two distinct ordered pairs can have the same first number assures that y is unique for a specific value of x.

Equation (1) defines a function. Call this function f. The equation gives the rule by which a unique value of y can be determined whenever x is given; that is, multiply the number x by itself, then multiply that product by 2, and add 5. The function f is the set of all ordered pairs (x, y) such that x and y satisfy (1), that is,

$$f = \{(x, y) \mid y = 2x^2 + 5\}$$

The numbers x and y are **variables.** Because for the function f values are assigned to x, and because the value of y is dependent on the choice of x, x is the **independent variable** and y is the **dependent variable.** The domain of the function is the set of all possible values of the independent variable, and the range of the function is the set of all possible values of the dependent variable. For the function f under consideration, the domain is the set of all real numbers, and it can be denoted with interval notation as $(-\infty, +\infty)$. The smallest value that y can assume is 5 (when $x = 0$). The range of f is then the set of all positive numbers greater than or equal to 5, which is $[5, +\infty)$.

● ILLUSTRATION 1 Let g be the function which is the set of all ordered pairs (x, y) defined by (2); that is,

$$g = \{(x, y) \mid y = \sqrt{x^2 - 9}\}$$

Because the numbers are confined to real numbers, y is a function of x only for $x \geq 3$ or $x \leq -3$ (or simply $|x| \geq 3$), because for any x satisfying either of these inequalities a unique value of y is determined. However, if x is in the interval $(-3, 3)$, a square root of a negative number is obtained, and hence no real number y exists. Therefore we must restrict x, and so

$$g = \{(x, y) \mid y = \sqrt{x^2 - 9} \text{ and } |x| \geq 3\}$$

The domain of g is $(-\infty, -3] \cup [3, +\infty)$, and the range of g is $[0, +\infty)$. ●

It should be stressed that in order to have a function there must be *exactly one value* of the dependent variable for a value of the independent variable in the domain of the function.

1.5.2 Definition If f is a function, then the **graph** of f is the set of all points (x, y) in R^2 for which (x, y) is an ordered pair in f.

● ILLUSTRATION 2 Let $f = \{(x, y) \mid y = \sqrt{5 - x}\}$. A sketch of the graph of f is shown in Fig. 1.5.1. The domain of f is the set of all real numbers less than or equal to 5, which is $(-\infty, 5]$, and the range of f is the set of all nonnegative real numbers, which is $[0, +\infty)$. ●

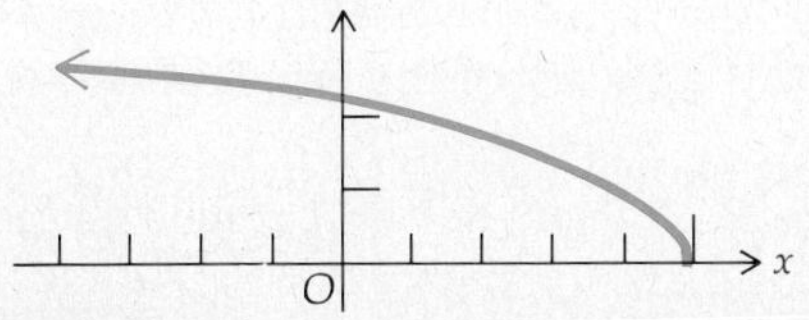

FIGURE 1.5.1

In the next illustration, a function is defined by more than one equation. Such a definition is permissible provided that for each number x in the domain there is a unique value for y in the range.

● ILLUSTRATION 3 Let g be the function which is the set of all ordered pairs (x, y) such that

$$y = \begin{cases} -3 & \text{if } x \le -1 \\ 1 & \text{if } -1 < x \le 2 \\ 4 & \text{if } 2 < x \end{cases}$$

The domain of g is $(-\infty, +\infty)$, while the range of g consists of the three numbers -3, 1, and 4. A sketch of the graph is shown in Fig. 1.5.2. ●

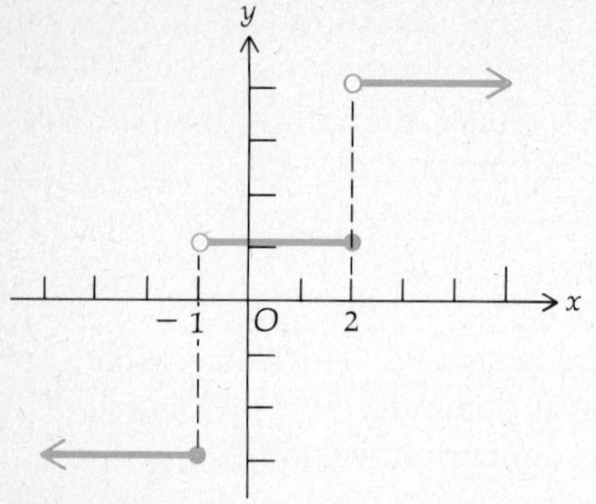

FIGURE 1.5.2

Observe in Fig. 1.5.2 that there is a break at $x = -1$ and another at $x = 2$. We say that g is discontinuous at -1 and 2. Continuous and discontinuous functions are considered in Sec. 2.6.

● ILLUSTRATION 4 Consider the set

$$\{(x, y) \mid x^2 + y^2 = 25\}$$

A sketch of the graph of this set is shown in Fig. 1.5.3. This set of ordered pairs is not a function because for any x in the interval $(-5, 5)$ there are two ordered pairs having that number as a first element. For example, both (3, 4) and (3, −4) are ordered pairs in the given set. Furthermore, observe that the graph of the given set is a circle with center at the origin and radius 5, and a vertical line having the equation $x = a$ (where $-5 < a < 5$) intersects the circle in two points. Refer to the figure. ●

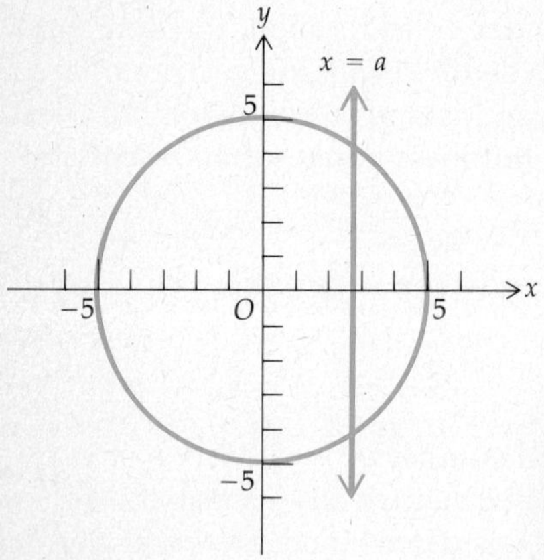

FIGURE 1.5.3

Example 1 Let $h = \{(x, y) \mid y = |x|\}$. Find the domain and range of h, and draw a sketch of the graph of h.

Solution The domain of h is $(-\infty, +\infty)$, and the range of h is $[0, +\infty)$. A sketch of the graph of h is shown in Fig. 1.5.4.

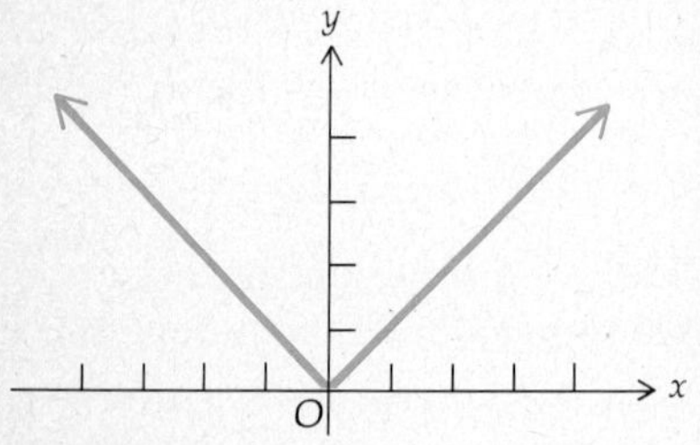

FIGURE 1.5.4

Example 2 Let F be the function which is the set of all ordered pairs (x, y) such that

$$y = \begin{cases} 3x - 2 & \text{if } x < 1 \\ x^2 & \text{if } 1 \le x \end{cases}$$

Find the domain and range of F, and draw a sketch of the graph of F.

Solution A sketch of the graph of F is shown in Fig. 1.5.5. The domain of F is $(-\infty, +\infty)$, and the range of F is $(-\infty, +\infty)$.

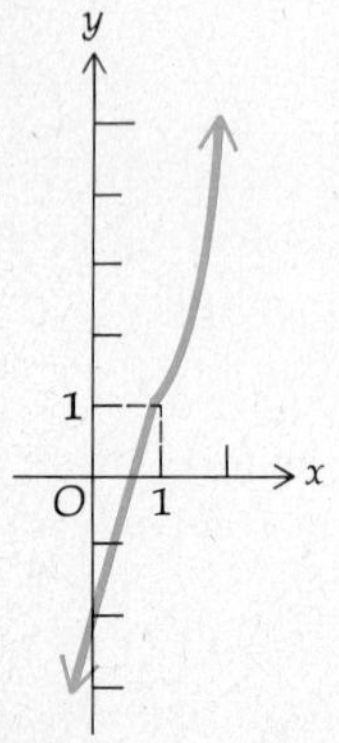

FIGURE 1.5.5

Example 3 Let G be the function which is the set of all ordered pairs (x, y) such that

$$y = \frac{x^2 - 9}{x - 3}$$

Find the domain and range of G, and draw a sketch of the graph of G.

Solution A sketch of the graph is shown in Fig. 1.5.6. Because a value for y is determined for each value of x except $x = 3$, the domain of G consists of all real numbers except 3. When $x = 3$, both the numerator and denominator are zero, and 0/0 is undefined.

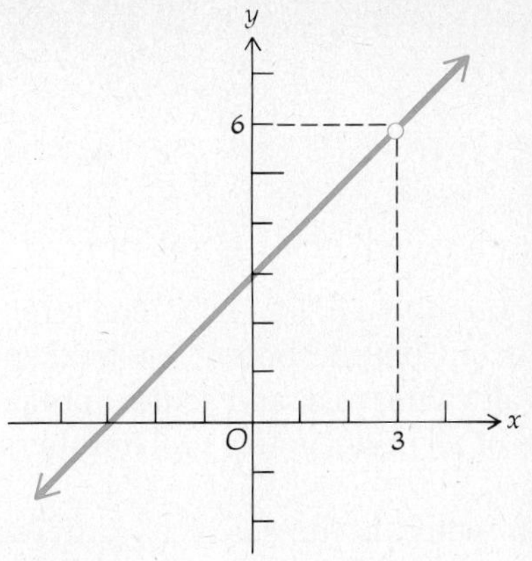

FIGURE 1.5.6

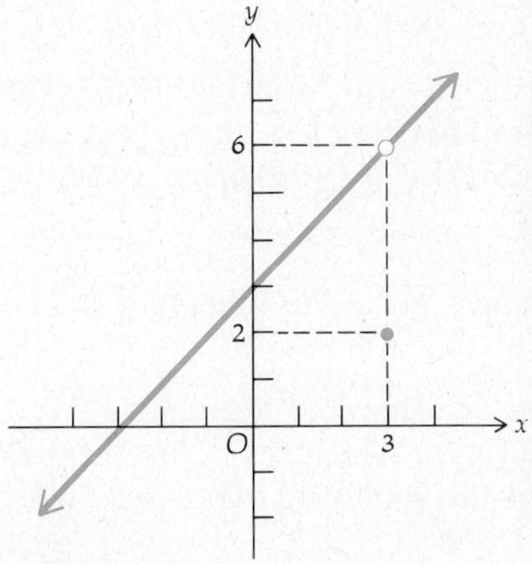

FIGURE 1.5.7

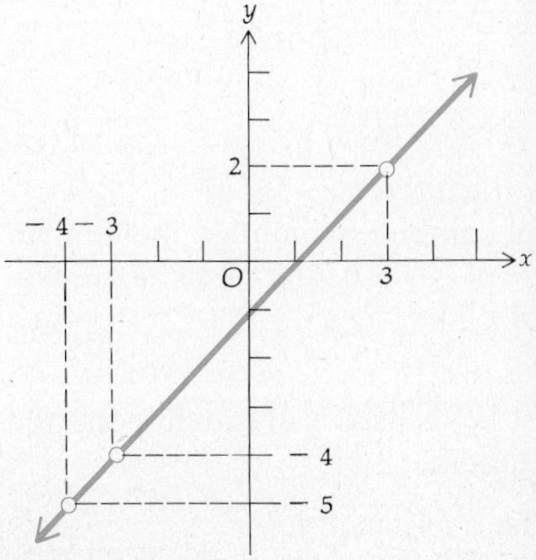

FIGURE 1.5.8

Factoring the numerator into $(x - 3)(x + 3)$ we obtain

$$y = \frac{(x-3)(x+3)}{(x-3)}$$

or $y = x + 3$, provided that $x \neq 3$. In other words, the function G consists of all ordered pairs (x, y) such that

$$y = x + 3 \quad \text{and} \quad x \neq 3$$

The range of G consists of all real numbers except 6. The graph consists of all points on the line $y = x + 3$ except the point $(3, 6)$.

Example 4 Let H be the function which is the set of all ordered pairs (x, y) such that

$$y = \begin{cases} x + 3 & \text{if } x \neq 3 \\ 2 & \text{if } x = 3 \end{cases}$$

Find the domain and range of H, and draw a sketch of the graph of H.

Solution A sketch of the graph of this function is shown in Fig. 1.5.7. The graph consists of the point $(3, 2)$ and all points on the line $y = x + 3$ except the point $(3, 6)$. Function H is defined for all values of x, and therefore the domain of H is $(-\infty, +\infty)$. The range of H consists of all real numbers except 6.

Example 5 Let ϕ be the function which is the set of all ordered pairs (x, y) such that

$$y = \frac{(x^2 + 3x - 4)(x^2 - 9)}{(x^2 + x - 12)(x + 3)}$$

Find the domain and range of ϕ, and draw a sketch of the graph of ϕ.

Solution A sketch of the graph of this function is shown in Fig. 1.5.8. Factoring the numerator and denominator we obtain

$$y = \frac{(x+4)(x-1)(x-3)(x+3)}{(x+4)(x-3)(x+3)}$$

The denominator is zero for $x = -4, -3$, and 3; therefore ϕ is undefined for these three values of x. For values of $x \neq -4, -3$, or 3, we may divide numerator and denominator by the common factors and obtain

$$y = x - 1 \quad \text{if } x \neq -4, -3, \text{ or } 3$$

Therefore the domain of ϕ is the set of all real numbers except -4, -3, and 3, and the range of ϕ is the set of all real numbers except those values of $x - 1$ obtained by replacing x by -4, -3, or 3, that is, all real numbers except -5, -4, and 2. The graph of this function is the straight line $y = x - 1$, with the points $(-4, -5)$, $(-3, -4)$, and $(3, 2)$ deleted.

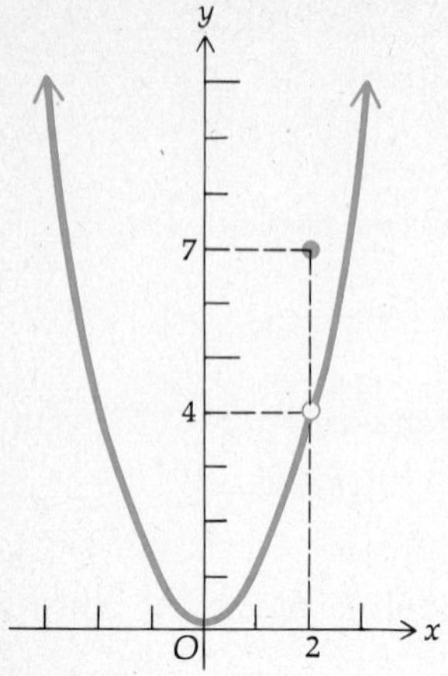

FIGURE 1.5.9

Example 6 Let f be the function which is the set of all ordered pairs (x, y) such that

$$y = \begin{cases} x^2 & \text{if } x \neq 2 \\ 7 & \text{if } x = 2 \end{cases}$$

Find the domain and range of f, and draw a sketch of the graph of f.

Solution A sketch of the graph of f is shown in Fig. 1.5.9. The graph consists of the point $(2, 7)$ and all points on the parabola $y = x^2$ except the point $(2, 4)$. Function f is defined for all values of x, and so the domain of f is $(-\infty, +\infty)$. The range of f consists of all nonnegative real numbers.

Example 7 Let h be the function which is the set of all ordered pairs (x, y) such that

$$y = \begin{cases} x - 1 & \text{if } x < 3 \\ 2x + 1 & \text{if } 3 \leq x \end{cases}$$

Find the domain and range of h, and draw a sketch of the graph of h.

Solution A sketch of the graph of h is shown in Fig. 1.5.10. The domain of h is $(-\infty, +\infty)$. The values of y are either less than 2 or greater than or equal to 7. So the range of h is $(-\infty, 2) \cup [7, +\infty)$ or, equivalently, all real numbers not in $[2, 7)$.

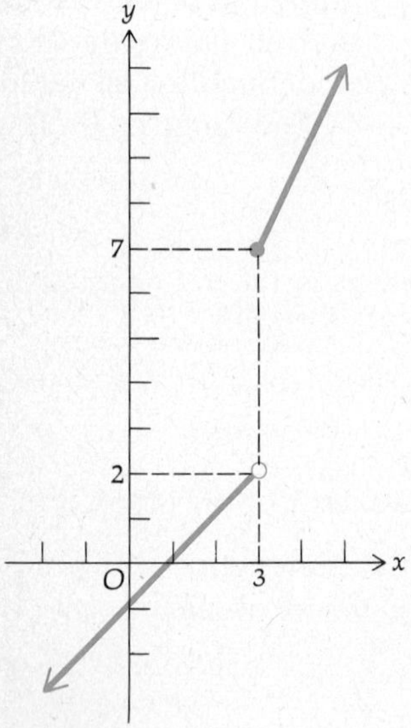

FIGURE 1.5.10

Example 8 Let $g = \{(x, y) \mid y = \sqrt{x(x-2)}\}$. Find the domain and range of g, and draw a sketch of the graph of g.

Solution Because $\sqrt{x(x-2)}$ is not a real number when $x(x-2) < 0$, the domain of g consists of the values of x for which $x(x-2) \geq 0$. This inequality will be satisfied when one of the following two cases holds: $x \geq 0$ and $x - 2 \geq 0$; or $x \leq 0$ and $x - 2 \leq 0$.

Case 1: $x \geq 0$ and $x - 2 \geq 0$. That is,

$$x \geq 0 \quad \text{and} \quad x \geq 2$$

Both inequalities hold if $x \geq 2$, which is the interval $[2, +\infty)$.

Case 2: $x \leq 0$ and $x - 2 \leq 0$. That is,

$$x \leq 0 \quad \text{and} \quad x \leq 2$$

Both inequalities hold if $x \leq 0$, which is the interval $(-\infty, 0]$.

The solutions for the two cases are combined to obtain the domain of g. It is $(-\infty, 0] \cup [2, +\infty)$. The range of g is the interval $[0, +\infty)$. Figure 1.5.11 shows a sketch of the graph of g.

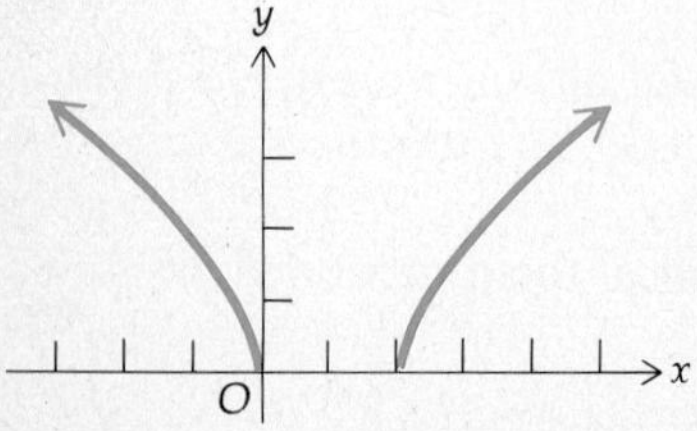

FIGURE 1.5.11

In the next illustration, the symbol $[\![x]\!]$ is used. $[\![x]\!]$ is defined as the greatest integer less than or equal to x; that is,

$$[\![x]\!] = n \quad \text{if } n \leq x < n + 1, \qquad \text{where } n \text{ is an integer}$$

This function is called the **greatest integer function.** It follows that $[\![1]\!] = 1$, $[\![1.3]\!] = 1$, $[\![\frac{1}{2}]\!] = 0$, $[\![-4.2]\!] = -5$, $[\![-8]\!] = -8$, $[\![9.8]\!] = 9$, and so on.

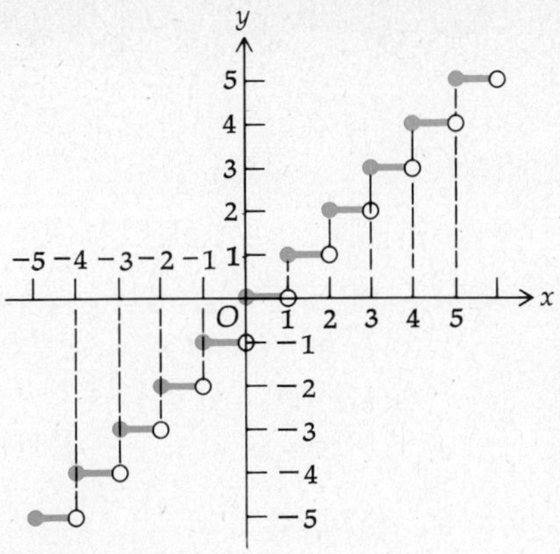

FIGURE 1.5.12

● ILLUSTRATION 5 Let F be the greatest integer function; that is, $F = \{(x, y) \mid y = [\![x]\!]\}$. A sketch of the graph of F is shown in Fig. 1.5.12. The following values are used to draw the sketch.

$-5 \le x < -4$	$[\![x]\!] = -5$
$-4 \le x < -3$	$[\![x]\!] = -4$
$-3 \le x < -2$	$[\![x]\!] = -3$
$-2 \le x < -1$	$[\![x]\!] = -2$
$-1 \le x < 0$	$[\![x]\!] = -1$
$0 \le x < 1$	$[\![x]\!] = 0$
$1 \le x < 2$	$[\![x]\!] = 1$
$2 \le x < 3$	$[\![x]\!] = 2$
$3 \le x < 4$	$[\![x]\!] = 3$
$4 \le x < 5$	$[\![x]\!] = 4$

The domain of F is the set of all real numbers. The range of F consists of all the integers. ●

Example 9 Let $G = \{(x, y) \mid y = [\![x]\!] - x\}$. Draw a sketch of the graph of G. State the domain and range of G.

Solution

If $0 \le x < 1$, $[\![x]\!] = 0$; so $G(x) = -x$.

If $1 \le x < 2$, $[\![x]\!] = 1$; so $G(x) = 1 - x$.

If $2 \le x < 3$, $[\![x]\!] = 2$; so $G(x) = 2 - x$.

If $-1 \le x < 0$, $[\![x]\!] = -1$; so $G(x) = -1 - x$.

and so on.

A sketch of the graph is shown in Fig. 1.5.13. The domain of G is thc set of all real numbers, and the range of G is $(-1, 0]$.

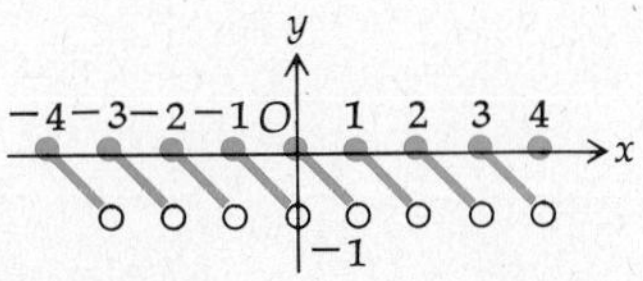

FIGURE 1.5.13

EXERCISES 1.5

In Exercises 1 through 14, find the domain and range of the given function, and draw a sketch of the graph of the function.

1. $f = \{(x, y) \mid y = 3x - 1\}$

2. $g = \{(x, y) \mid y = x^2 + 2\}$

3. $F = \{(x, y) \mid y = 3x^2 - 6\}$

4. $G = \{(x, y) \mid y = 5 - x^2\}$

5. $G = \{(x, y) \mid y = \sqrt{x + 1}\}$

6. $f = \{(x, y) \mid y = \sqrt{5 - 3x}\}$

7. $h = \{(x, y) \mid y = \sqrt{3x - 4}\}$

8. $f = \{(x, y) \mid y = \sqrt{4 - x^2}\}$

9. $g = \{(x, y) \mid y = \sqrt{x^2 - 4}\}$

10. $H = \{(x, y) \mid y = |x - 3|\}$

11. $\phi = \{(x, y) \mid y = |3x + 2|\}$

12. $F = \left\{(x, y) \mid y = \dfrac{4x^2 - 1}{2x + 1}\right\}$

13. $H = \left\{(x, y) \mid y = \dfrac{x^2 - 4x + 3}{x - 1}\right\}$

14. $g = \left\{(x, y) \mid y = \dfrac{x^3 - 3x^2 - 4x + 12}{x^2 - x - 6}\right\}$

In Exercises 15 through 40, the function is the set of all ordered pairs (x, y) satisfying the given equation. Find the domain and range of the function, and draw a sketch of the graph of the function.

15. $G: y = \begin{cases} -2 & \text{if } x \leq 3 \\ 2 & \text{if } 3 < x \end{cases}$

16. $h: y = \begin{cases} -4 & \text{if } x < -2 \\ -1 & \text{if } -2 \leq x \leq 2 \\ 3 & \text{if } 2 < x \end{cases}$

17. $f: y = \begin{cases} 2x - 1 & \text{if } x \neq 2 \\ 0 & \text{if } x = 2 \end{cases}$

18. $f: y = \begin{cases} x^2 - 4 & \text{if } x \neq -3 \\ -2 & \text{if } x = -3 \end{cases}$

19. $H: y = \begin{cases} x^2 - 4 & \text{if } x < 3 \\ 2x - 1 & \text{if } 3 \leq x \end{cases}$

20. $\phi: y = \begin{cases} x + 5 & \text{if } x < -5 \\ \sqrt{25 - x^2} & \text{if } -5 \leq x \leq 5 \\ x - 5 & \text{if } 5 < x \end{cases}$

21. $F: y = \begin{cases} x - 2 & \text{if } x < 0 \\ x^2 + 1 & \text{if } 0 \leq x \end{cases}$

22. $g: y = \begin{cases} 6x + 7 & \text{if } x \leq -2 \\ 4 - x & \text{if } -2 < x \end{cases}$

23. $F: y = \dfrac{(x + 1)(x^2 + 3x - 10)}{x^2 + 6x + 5}$

24. $G: y = \dfrac{(x^2 + 3x - 4)(x^2 - 5x + 6)}{(x^2 - 3x + 2)(x - 3)}$

25. $f: y = \sqrt{x^2 - 3x - 4}$

26. $h: y = \sqrt{6x^2 - 5x - 4}$

27. $g: y = \dfrac{x^3 - 2x^2}{x - 2}$

28. $f: y = \dfrac{x^3 + 3x^2 + x + 3}{x + 3}$

29. $h: y = \dfrac{x^3 + 5x^2 - 6x - 30}{x + 5}$

30. $F: y = \dfrac{x^4 + x^3 - 9x^2 - 3x + 18}{x^2 + x - 6}$

31. $f: y = |x| + |x - 1|$

32. $g: y = |x| \cdot |x - 1|$

33. $f: y = \sqrt[3]{x^2 - 1}$

34. $F: y = [\![x + 2]\!]$

35. $g: y = [\![x - 4]\!]$

36. $H: y = |x| + [\![x]\!]$

37. $G: y = x - [\![x]\!]$

38. $h: y = [\![x^2]\!]$

39. $g: y = \dfrac{[\![x]\!]}{|x|}$

40. $h: y = \dfrac{|x|}{[\![x]\!]}$

1.6 FUNCTION NOTATION, OPERATIONS ON FUNCTIONS, AND TYPES OF FUNCTIONS

If f is the function having as its domain variable x and as its range variable y, the symbol $f(x)$ (read "f of x") denotes the particular value of y that corresponds to the value of x.

● ILLUSTRATION 1 In Illustration 2 of Sec. 1.5,

$$f = \{(x, y) \mid y = \sqrt{5 - x}\}$$

Thus $f(x) = \sqrt{5 - x}$. Because when $x = 1$, $\sqrt{5 - x} = 2$, we have $f(1) = 2$. Similarly, $f(-6) = \sqrt{11}$, $f(0) = \sqrt{5}$, and so on. ●

When defining a function, the domain of the function must be given either explicitly or implicitly. For instance, if f is defined by

$$f(x) = 3x^2 - 5x + 2$$

it is implied that x can be any real number. However, if f is defined by

$$f(x) = 3x^2 - 5x + 2 \qquad 1 \leq x \leq 10$$

then the domain of f consists of all real numbers between and including 1 and 10.

Similarly, if g is defined by the equation

$$g(x) = \frac{5x - 2}{x + 4}$$

it is implied that $x \neq -4$, because the quotient is undefined for $x = -4$; hence, the domain of g is the set of all real numbers except -4.

If

$$h(x) = \sqrt{9 - x^2}$$

it is implied that x is in the closed interval $-3 \le x \le 3$, because $\sqrt{9 - x^2}$ is undefined (i.e., not a real number) for $x > 3$ or $x < -3$. So the domain of h is $[-3, 3]$, and the range of h is $[0, 3]$.

Example 1 Given that f is the function defined by $f(x) = x^2 + 3x - 4$, find: (a) $f(0)$; (b) $f(2)$; (c) $f(h)$; (d) $f(2h)$; (e) $f(2x)$; (f) $f(x + h)$; (g) $f(x) + f(h)$.

Solution

(a) $f(0) = 0^2 + 3 \cdot 0 - 4 = -4$

(b) $f(2) = 2^2 + 3 \cdot 2 - 4 = 6$

(c) $f(h) = h^2 + 3h - 4$

(d) $f(2h) = (2h)^2 + 3(2h) - 4 = 4h^2 + 6h - 4$

(e) $f(2x) = (2x)^2 + 3(2x) - 4 = 4x^2 + 6x - 4$

(f)
$$\begin{aligned} f(x + h) &= (x + h)^2 + 3(x + h) - 4 \\ &= x^2 + 2hx + h^2 + 3x + 3h - 4 \\ &= x^2 + (2h + 3)x + (h^2 + 3h - 4) \end{aligned}$$

(g)
$$\begin{aligned} f(x) + f(h) &= (x^2 + 3x - 4) + (h^2 + 3h - 4) \\ &= x^2 + 3x + (h^2 + 3h - 8) \end{aligned}$$

Example 2 Given $g(x) = \sqrt{3x - 1}$, find

$$\frac{g(x + h) - g(x)}{h} \qquad h \ne 0$$

Solution

$$\begin{aligned} \frac{g(x + h) - g(x)}{h} &= \frac{\sqrt{3(x + h) - 1} - \sqrt{3x - 1}}{h} \\ &= \frac{(\sqrt{3x + 3h - 1} - \sqrt{3x - 1})(\sqrt{3x + 3h - 1} + \sqrt{3x - 1})}{h(\sqrt{3x + 3h - 1} + \sqrt{3x - 1})} \\ &= \frac{(3x + 3h - 1) - (3x - 1)}{h(\sqrt{3x + 3h - 1} + \sqrt{3x - 1})} \\ &= \frac{3h}{h(\sqrt{3x + 3h - 1} + \sqrt{3x - 1})} \\ &= \frac{3}{\sqrt{3x + 3h - 1} + \sqrt{3x - 1}} \end{aligned}$$

In the second step of this solution, the numerator and denominator were multiplied by the conjugate of the numerator in order to rationalize the numerator, and this gave a common factor of h in the numerator and the denominator.

We now define some operations on functions. In the definition new functions are formed from given functions by adding, subtracting, multiplying, and dividing function values. Accordingly, these new functions are known as the *sum, difference, product,* and *quotient* of the original functions.

1.6.1 Definition Given the two functions f and g:

(i) their **sum,** denoted by $f + g$, is the function defined by

$$(f + g)(x) = f(x) + g(x)$$

(ii) their **difference,** denoted by $f - g$, is the function defined by

$$(f - g)(x) = f(x) - g(x)$$

(iii) their **product** denoted by $f \cdot g$, is the function defined by

$$(f \cdot g)(x) = f(x) \cdot g(x)$$

(iv) their **quotient,** denoted by f/g, is the function defined by

$$(f/g)(x) = f(x)/g(x)$$

In each case the *domain* of the resulting function consists of those values of x common to the domains of f and g, with the additional requirement in case (iv) that the values of x for which $g(x) = 0$ are excluded.

Example 3 Given that f is the function defined by $f(x) = \sqrt{x + 1}$ and g is the function defined by $g(x) = \sqrt{x - 4}$, find: (a) $(f + g)(x)$; (b) $(f - g)(x)$; (c) $(f \cdot g)(x)$; (d) $(f/g)(x)$. In each case, determine the domain of the resulting function.

Solution

(a) $(f + g)(x) = \sqrt{x + 1} + \sqrt{x - 4}$

(b) $(f - g)(x) = \sqrt{x + 1} - \sqrt{x - 4}$

(c) $(f \cdot g)(x) = \sqrt{x + 1} \cdot \sqrt{x - 4}$

(d) $(f/g)(x) = \dfrac{\sqrt{x + 1}}{\sqrt{x - 4}}$

The domain of f is $[-1, +\infty)$, and the domain of g is $[4, +\infty)$. So in parts (a), (b), and (c) the domain of the resulting function is $[4, +\infty)$. In part (d) the denominator is zero when $x = 4$; thus 4 is excluded from the domain, and the domain is therefore $(4, +\infty)$.

To indicate the product of a function f multiplied by itself, or $f \cdot f$, we write f^2. For example, if f is defined by $f(x) = 3x$, then f^2 is the function defined by $f^2(x) = (3x)(3x) = 9x^2$.

In addition to combining two functions by the operations given in Definition 1.6.1 we consider the *composite function* of two given functions.

1.6.2 Definition Given the two functions f and g, the **composite function,** denoted by $f \circ g$, is defined by

$$(f \circ g)(x) = f(g(x))$$

and the domain of $f \circ g$ is the set of all numbers x in the domain of g such that $g(x)$ is in the domain of f.

Example 4 Given that f is defined by $f(x) = \sqrt{x}$ and g is defined by $g(x) = 2x - 3$, find $F(x)$ if $F = f \circ g$, and find the domain of F.

Solution

$$\begin{aligned} F(x) = (f \circ g)(x) &= f(g(x)) \\ &= f(2x - 3) \\ &= \sqrt{2x - 3} \end{aligned}$$

The domain of g is $(-\infty, +\infty)$, and the domain of f is $[0, +\infty)$. So the domain of F is the set of real numbers for which $2x - 3 \geq 0$ or; equivalently, $[\frac{3}{2}, +\infty)$.

Example 5 Given that f is defined by $f(x) = \sqrt{x}$ and g is defined by $g(x) = x^2 - 1$, find: (a) $f \circ f$; (b) $g \circ g$; (c) $f \circ g$; (d) $g \circ f$. Also find the domain of the composite function in each part.

Solution The domain of f is $[0, +\infty)$, and the domain of g is $(-\infty, +\infty)$.

(a) $(f \circ f)(x) = f(f(x)) = f(\sqrt{x}) = \sqrt{\sqrt{x}} = \sqrt[4]{x}$

The domain of $f \circ f$ is $[0, +\infty)$.

(b) $(g \circ g)(x) = g(g(x)) = g(x^2 - 1) = (x^2 - 1)^2 - 1 = x^4 - 2x^2$

The domain of $g \circ g$ is $(-\infty, +\infty)$.

(c) $(f \circ g)(x) = f(g(x)) = f(x^2 - 1) = \sqrt{x^2 - 1}$

The domain of $f \circ g$ is $(-\infty, -1] \cup [1, +\infty)$ or, equivalently, all x not in $(-1, 1)$.

(d) $(g \circ f)(x) = g(f(x)) = g(\sqrt{x}) = (\sqrt{x})^2 - 1 = x - 1$

The domain of $g \circ f$ is $[0, +\infty)$. Note that even though $x - 1$ is defined for all values of x, the domain of $g \circ f$, by Definition 1.6.2, is the set of all numbers x in the domain of f such that $f(x)$ is in the domain of g.

1.6.3 Definition

(i) A function f is said to be an **even** function if for every x in the domain of f, $f(-x) = f(x)$.

(ii) A function f is said to be an **odd** function if for every x in the domain of f, $f(-x) = -f(x)$.

In both parts (i) and (ii) it is understood that $-x$ is in the domain of f whenever x is.

● ILLUSTRATION 2

(a) If $f(x) = 3x^4 - 2x^2 + 7$, then $f(-x) = 3(-x)^4 - 2(-x)^2 + 7 = 3x^4 - 2x^2 + 7 = f(x)$. Therefore f is an even function.

(b) If $g(x) = 2x^5 + 5x^3 - 8x$, then $g(-x) = 2(-x)^5 + 5(-x)^3 - 8(-x) = -2x^5 - 5x^3 + 8x = -(2x^5 + 5x^3 - 8x) = -g(x)$. Thus g is an odd function.

(c) If $h(x) = 2x^4 + 5x^3 - x^2 + 8$, $h(-x) = 2(-x)^4 + 5(-x)^3 - (-x)^2 + 8 = 2x^4 - 5x^3 - x^2 + 8$. The function h is neither even nor odd. ●

From the definition of an even function and Theorem 1.2.6(ii) it follows that the graph of an even function is symmetric with respect to the y axis. From Theorem 1.2.6(iii) and the definition of an odd function, the graph of an odd function is symmetric with respect to the origin.

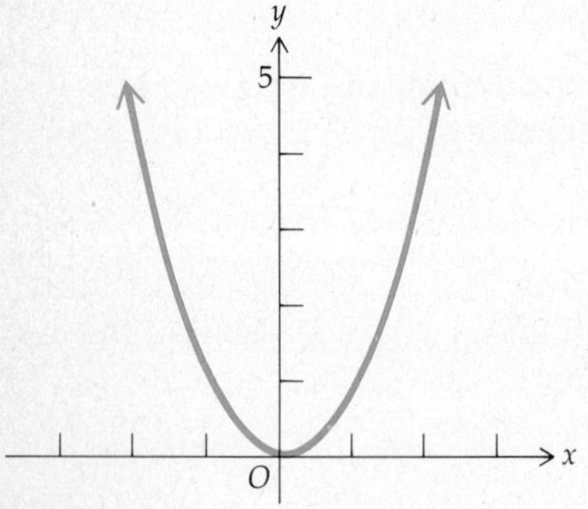

FIGURE 1.6.1

● ILLUSTRATION 3

(a) If $f(x) = x^2$, f is an even function, and its graph is a parabola that is symmetric with respect to the y axis. See Fig. 1.6.1.

(b) If $g(x) = x^3$, g is an odd function. The graph of g, shown in Fig. 1.6.2, is symmetric with respect to the origin. ●

If the range of a function f consists of only one number, then f is called a **constant function.** So if $f(x) = c$, and if c is any real number, then f is a constant function and its graph is a straight line parallel to the x axis at a directed distance of c units from the x axis.

If a function f is defined by

$$f(x) = a_0x^n + a_1x^{n-1} + a_2x^{n-2} + \cdots + a_{n-1}x + a_n$$

where n is a nonnegative integer, and $a_0, a_1, \ldots, a_n$ are real numbers ($a_0 \neq 0$), then f is called a **polynomial function** of degree n. Thus the function f defined by

$$f(x) = 3x^5 - x^2 + 7x - 1$$

is a polynomial function of degree 5.

If the degree of a polynomial function is 1, then the function is called a **linear function;** if the degree is 2, the function is called a **quadratic function;** and if the degree is 3, the function is called a **cubic function.** The particular linear function defined by $f(x) = x$ is called the **identity function.**

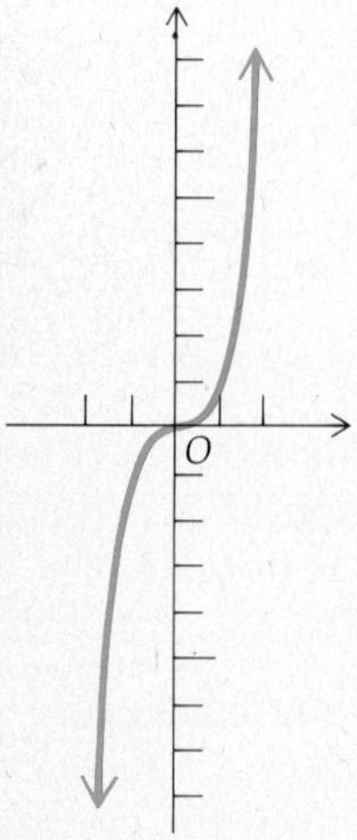

FIGURE 1.6.2

● ILLUSTRATION 4

The function f defined by $f(x) = 3x + 4$ is a linear function.

The function g defined by $g(x) = 5x^2 - 8x + 1$ is a quadratic function.

The function h defined by $h(x) = 8x^3 - x + 4$ is a cubic function. ●

If a function can be expressed as the quotient of two polynomial functions, it is called a **rational function.**

An **algebraic function** is a function formed by a finite number of algebraic operations on the identity function and the constant function. These algebraic operations include addition, subtraction, multiplication, division, raising to powers, and extracting roots. An example of an algebraic function is the function f defined by

$$f(x) = \frac{(x^2 - 3x + 1)^3}{\sqrt{x^4 + 1}}$$

In addition to algebraic functions, **transcendental functions** are considered in calculus. Examples of transcendental functions are trigonometric functions, logarithmic functions, and exponential functions, which are discussed in later chapters.

EXERCISES 1.6

1. Given $f(x) = 2x - 1$, find: (a) $f(3)$; (b) $f(-2)$; (c) $f(0)$; (d) $f(a + 1)$; (e) $f(x + 1)$; (f) $f(2x)$; (g) $2f(x)$; (h) $f(x + h)$; (i) $f(x) + f(h)$; (j) $\dfrac{f(x + h) - f(x)}{h}$, $h \neq 0$.

2. Given $f(x) = \dfrac{3}{x}$, find: (a) $f(1)$; (b) $f(-3)$; (c) $f(6)$; (d) $f(\frac{1}{3})$; (e) $f\left(\dfrac{3}{a}\right)$; (f) $f\left(\dfrac{3}{x}\right)$; (g) $\dfrac{f(3)}{f(x)}$; (h) $f(x - 3)$; (i) $f(x) - f(3)$; (j) $\dfrac{f(x + h) - f(x)}{h}$, $h \neq 0$.

3. Given $f(x) = 2x^2 + 5x - 3$, find: (a) $f(-2)$; (b) $f(-1)$; (c) $f(0)$; (d) $f(3)$; (e) $f(h + 1)$; (f) $f(2x^2)$; (g) $f(x^2 - 3)$; (h) $f(x + h)$; (i) $f(x) + f(h)$; (j) $\dfrac{f(x + h) - f(x)}{h}$, $h \neq 0$.

4. Given $g(x) = 3x^2 - 4$; find: (a) $g(-4)$; (b) $g(\frac{1}{2})$; (c) $g(x^2)$; (d) $g(3x^2 - 4)$; (e) $g(x - h)$; (f) $g(x) - g(h)$; (g) $\dfrac{g(x + h) - g(x)}{h}$, $h \neq 0$.

5. Given $F(x) = \sqrt{2x + 3}$, find: (a) $F(-1)$; (b) $F(4)$; (c) $F(\frac{1}{2})$; (d) $F(30)$; (e) $F(2x + 3)$; (f) $\dfrac{F(x + h) - F(x)}{h}$, $h \neq 0$.

6. Given $G(x) = \sqrt{2x^2 + 1}$, find: (a) $G(-2)$; (b) $G(0)$; (c) $G(\frac{1}{5})$; (d) $G(\frac{4}{7})$; (e) $G(2x^2 - 1)$; (f) $\dfrac{G(x + h) - G(x)}{h}$, $h \neq 0$.

7. Given $f(x) = \begin{cases} \dfrac{|x|}{x} & \text{if } x \neq 0 \\ 1 & \text{if } x = 0 \end{cases}$ find: (a) $f(1)$; (b) $f(-1)$; (c) $f(4)$; (d) $f(-4)$; (e) $f(-x)$; (f) $f(x + 1)$; (g) $f(x^2)$; (h) $f(-x^2)$.

8. Given $f(t) = \dfrac{|3 + t| - |t| - 3}{t}$, express $f(t)$ without absolute-value bars if (a) $t > 0$; (b) $-3 \leq t < 0$; (c) $t < -3$.

In Exercises 9 through 18, the functions f and g are defined. In each exercise define the following functions, and determine the domain of the resulting function: (a) $f + g$; (b) $f - g$; (c) $f \cdot g$; (d) f/g; (e) g/f; (f) $f \circ g$; (g) $g \circ f$.

9. $f(x) = x - 5$; $g(x) = x^2 - 1$

10. $f(x) = \sqrt{x}$; $g(x) = x^2 + 1$

11. $f(x) = \dfrac{x + 1}{x - 1}$; $g(x) = \dfrac{1}{x}$

12. $f(x) = \sqrt{x}$; $g(x) = 4 - x^2$

13. $f(x) = \sqrt{x}$; $g(x) = x^2 - 1$

14. $f(x) = |x|$; $g(x) = |x - 3|$

15. $f(x) = x^2 + 1$; $g(x) = 3x - 2$

16. $f(x) = \sqrt{x + 4}$; $g(x) = x^2 - 4$

17. $f(x) = \dfrac{1}{x + 1}$; $g(x) = \dfrac{x}{x - 2}$

18. $f(x) = x^2$; $g(x) = \dfrac{1}{\sqrt{x}}$

In Exercises 19 and 20, a function f is defined. In each exercise define the following functions, and determine the domain of the resulting function: (a) $f(x^2)$; (b) $f^2(x)$; (c) $(f \circ f)(x)$.

19. $f(x) = 2x - 3$ **20.** $f(x) = \dfrac{2}{x-1}$

21. Given $G(x) = |x - 2| - |x| + 2$, express $G(x)$ without absolute-value bars if x is in the given interval: (a) $[2, +\infty)$; (b) $(-\infty, 0)$; (c) $[0, 2)$.

In each part of Exercises 22 and 23, determine whether the given function is even, odd, or neither.

22. (a) $f(x) = 2x^4 - 3x^2 + 1$ (b) $f(x) = 5x^3 - 7x$ (c) $f(s) = s^2 + 2s + 2$ (d) $g(x) = x^6 - 1$
(e) $h(t) = 5t^7 + 1$ (f) $f(x) = |x|$ (g) $f(y) = \dfrac{y^3 - y}{y^2 + 1}$ (h) $g(z) = \dfrac{z-1}{z+1}$

23. (a) $g(x) = 5x^2 - 4$ (b) $f(x) = x^3 + 1$ (c) $f(t) = 4t^5 + 3t^3 - 2t$ (d) $g(r) = \dfrac{r^2 - 1}{r^2 + 1}$
(e) $f(x) = \begin{cases} 1 & \text{if } x \geq 0 \\ -1 & \text{if } x < 0 \end{cases}$ (f) $h(x) = \dfrac{4x^2 - 5}{2x^3 + x}$ (g) $f(z) = (z - 1)^2$ (h) $g(x) = \dfrac{|x|}{x^2 + 1}$
(i) $g(x) = \sqrt{x^2 - 1}$ (j) $f(x) = \sqrt[3]{x}$

24. Prove that if f and g are both odd functions, then $(f + g)$ and $(f - g)$ are also odd functions.

25. Prove that if f and g are both odd functions, then $f \cdot g$ and f/g are both even functions.

26. Prove that any function can be expressed as the sum of an even function and an odd function by writing

$$f(x) = \tfrac{1}{2}[f(x) + f(-x)] + \tfrac{1}{2}[f(x) - f(-x)]$$

and showing that the function having function values $\frac{1}{2}[f(x) + f(-x)]$ is an even function and the function having function values $\frac{1}{2}[f(x) - f(-x)]$ is an odd function.

In Exercises 27 and 28, use the result of Exercise 26 to express the functions defined by the given equations as the sum of an even function and an odd function.

27. (a) $f(x) = 2x - 7$; (b) $f(x) = x^2 + 2$; (c) $f(x) = x^3 - 1$; (d) $f(x) = x^4 + x^3 - x + 3$; (e) $f(x) = \dfrac{1}{x}$; (f) $f(x) = \dfrac{x-1}{x+1}$;
(g) $f(x) = |x| + |x - 1|$.

28. (a) $f(x) = 3x + 8$; (b) $f(x) = x^2 + 3x - 1$; (c) $f(x) = 8x^3 + 1$; (d) $f(x) = \sqrt[3]{x}$; (e) $f(x) = x^4 - 16$; (f) $f(x) = \dfrac{1}{x+1}$;
(g) $f(x) = |2x + 1|$.

29. Determine whether the composite function $f \circ g$ is odd or even in each of the following cases: (a) f and g are both even; (b) f and g are both odd; (c) f is even and g is odd; (d) f is odd and g is even.

30. There is one function that is both even and odd. What is it?

In Exercises 31 through 46, draw a sketch of the graph of the given function. In Exercises 33 through 46, the function U is the unit step function defined in Exercise 31, and the function sgn is the signum function defined in Exercise 32.

31. U is the function defined by $U(t) = \begin{cases} 0 & \text{if } t < 0 \\ 1 & \text{if } t \geq 0 \end{cases}$

This function U is called the **unit step function.**

32. The **signum function** (or **sign function**), denoted by sgn, is defined by

$$\operatorname{sgn} x = \begin{cases} -1 & \text{if } x < 0 \\ 0 & \text{if } x = 0 \\ 1 & \text{if } x > 0 \end{cases}$$

sgn x is read "signum of x."

33. $f(x) = U(x - 1)$ **34.** $g(x) = U(x) - U(x - 1)$ **35.** $g(x) = \operatorname{sgn}(x + 1) - \operatorname{sgn}(x - 1)$

36. $f(x) = \operatorname{sgn} x^2 - \operatorname{sgn} x$ **37.** $h(x) = x \cdot U(x)$ **38.** $g(x) = 2 - x \cdot \operatorname{sgn} x$

39. $f(x) = (x - 1) \cdot U(x)$ **40.** $F(x) = (x + 1) \cdot U(x + 1)$ **41.** $G(x) = (x + 1) \cdot U(x + 1) - x \cdot U(x)$

42. $F(x) = x - 2 \operatorname{sgn} x$ **43.** $h(x) = \operatorname{sgn} x - U(x)$ **44.** $f(x) = \operatorname{sgn} x \cdot U(x + 1)$

45. $g(x) = \operatorname{sgn} x + x \cdot U(x)$ **46.** $G(x) = \operatorname{sgn}(U(x))$

47. Find formulas for $(f \circ g)(x)$ if

$$f(x) = \begin{cases} 0 & \text{if } x < 0 \\ 2x & \text{if } 0 \le x \le 1 \\ 0 & \text{if } x > 1 \end{cases} \quad \text{and} \quad g(x) = \begin{cases} 1 & \text{if } x < 0 \\ \frac{1}{2}x & \text{if } 0 \le x \le 1 \\ 1 & \text{if } x > 1 \end{cases}$$

48. Find formulas for $(g \circ f)(x)$ for the functions of Exercise 47.

49. Find formulas for $(f \circ g)(x)$ if

$$f(x) = \begin{cases} 0 & \text{if } x < 0 \\ x^2 & \text{if } 0 \le x \le 1 \\ 0 & \text{if } x > 1 \end{cases} \quad \text{and} \quad g(x) = \begin{cases} 1 & \text{if } x < 0 \\ 2x & \text{if } 0 \le x \le 1 \\ 1 & \text{if } x > 1 \end{cases}$$

50. The function g is defined by $g(x) = x^2$. Define a function f such that $(f \circ g)(x) = x$ if (a) $x \ge 0$; (b) $x < 0$.

51. If $f(x) = x^2 + 2x + 2$, find two functions g for which $(f \circ g)(x) = x^2 - 4x + 5$.

52. If $f(x) = x^2$, find two functions g for which $(f \circ g)(x) = 4x^2 - 12x + 9$.

53. Prove that if f and g are both linear functions, then $f \circ g$ is a linear function.

1.7 THE SINE AND COSINE FUNCTIONS

In geometry an **angle** is defined as the union of two rays called the **sides,** having a common endpoint called the **vertex.** Any angle is congruent to some angle having its vertex at the origin and one side, called the **initial side,** lying on the positive side of the x axis. Such an angle is said to be in **standard position.** Figure 1.7.1 shows an angle AOB in standard position with OA as the initial side. The other side, OB, is called the **terminal side.** The angle AOB can be formed by rotating the side OA to the side OB, and under such a rotation the point A moves along the circumference of a circle, having its center at O and radius $|\overline{OA}|$ to the point B.

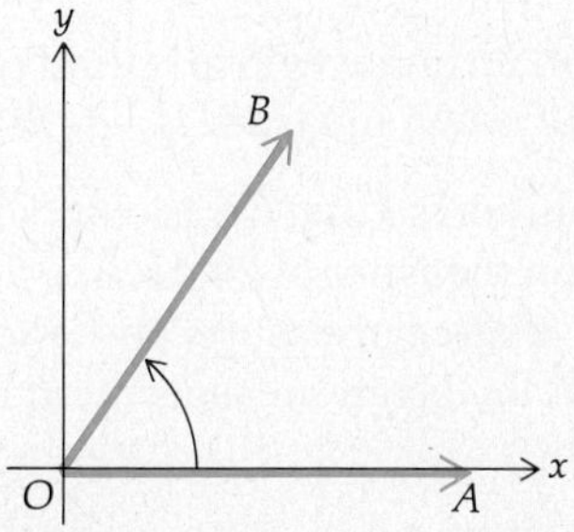

FIGURE 1.7.1

In dealing with problems involving angles of triangles, the measurement of an angle is usually given in degrees. However, in the calculus we are concerned with trigonometric functions of real numbers, and these functions are defined in terms of *radian measure.*

The length of an arc of a circle is used to define the radian measure of an angle.

1.7.1 Definition Let AOB be an angle in standard position and $|\overline{OA}| = 1$. If s units is the length of the arc of the circle traveled by point A as the initial side OA is rotated to the terminal side OB, the **radian measure, t,** of angle AOB is given by

$$t = s \quad \text{if the rotation is counterclockwise}$$

and

$$t = -s \quad \text{if the rotation is clockwise}$$

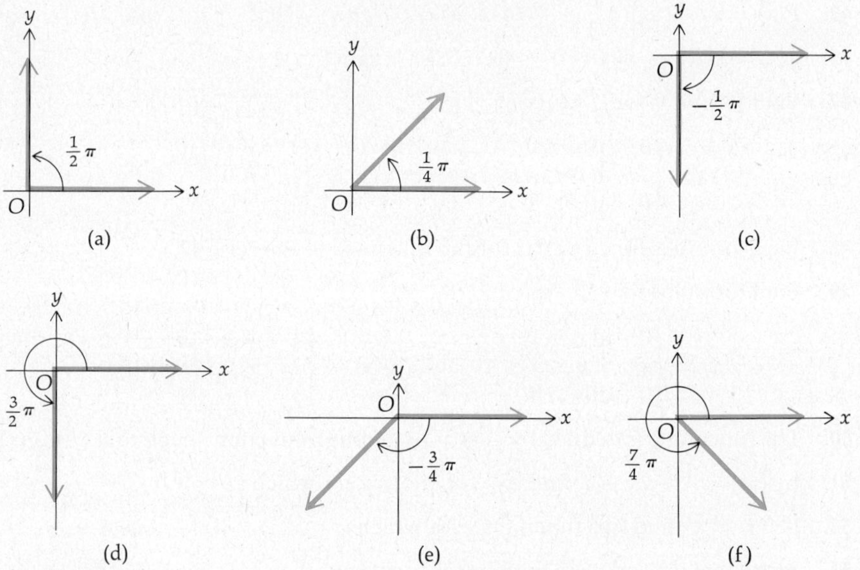

FIGURE 1.7.2

● ILLUSTRATION 1 From the fact that the measure of the length of the unit circle's circumference is 2π, the radian measures of the angles in Fig. 1.7.2(a), (b), (c), (d), (e), and (f) are determined. They are $\frac{1}{2}\pi$, $\frac{1}{4}\pi$, $-\frac{1}{2}\pi$, $\frac{3}{2}\pi$, $-\frac{3}{4}\pi$, and $\frac{7}{4}\pi$, respectively. ●

In Definition 1.7.1 it is possible that there may be more than one complete revolution in the rotation of OA.

● ILLUSTRATION 2 Figure 1.7.3(a) shows such an angle whose radian measure is $\frac{5}{2}\pi$, and Fig. 1.7.3(b) shows one whose radian measure is $-\frac{13}{4}\pi$. ●

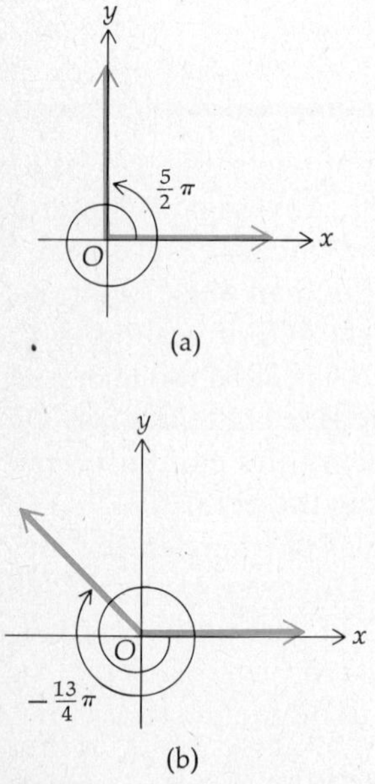

FIGURE 1.7.3

An angle formed by one complete revolution so that OA is coincident with OB has degree measure of 360 and radian measure of 2π. Hence there is the following correspondence between degree measure and radian measure (where the symbol $\sim$ indicates that the given measurements are for the same or congruent angles):

$$360° \sim 2\pi \text{ rad}$$

$$180° \sim \pi \text{ rad}$$

TABLE 1.7.1

degree measure	*radian measure*
30	$\frac{1}{6}\pi$
45	$\frac{1}{4}\pi$
60	$\frac{1}{3}\pi$
90	$\frac{1}{2}\pi$
120	$\frac{2}{3}\pi$
135	$\frac{3}{4}\pi$
150	$\frac{5}{6}\pi$
180	π
270	$\frac{3}{2}\pi$
360	2π

From this it follows that

$$1^\circ \sim \tfrac{1}{180}\pi \text{ rad}$$

and

$$1 \text{ rad} \sim \frac{180^\circ}{\pi} \approx 57^\circ 18'$$

From this correspondence the measurement of an angle can be converted from one system of units to the other.

● ILLUSTRATION 3

$$162^\circ \sim 162 \cdot \tfrac{1}{180}\pi \text{ rad} \qquad \tfrac{5}{12}\pi \text{ rad} \sim \tfrac{5}{12}\pi \cdot \frac{180^\circ}{\pi}$$

$$162^\circ \sim \tfrac{9}{10}\pi \text{ rad} \qquad \tfrac{5}{12}\pi \text{ rad} \sim 75^\circ$$

●

Table 1.7.1 gives the corresponding degree and radian measures of certain angles.

We now define the sine and cosine functions of any real number.

1.7.2 Definition

Suppose that t is a real number. Place an angle, having radian measure t, in standard position and let point P be at the intersection of the terminal side of the angle with the unit circle having its center at the origin. If P is the point (x, y), then the **cosine function** is defined by

$$\cos t = x$$

and the **sine function** is defined by

$$\sin t = y$$

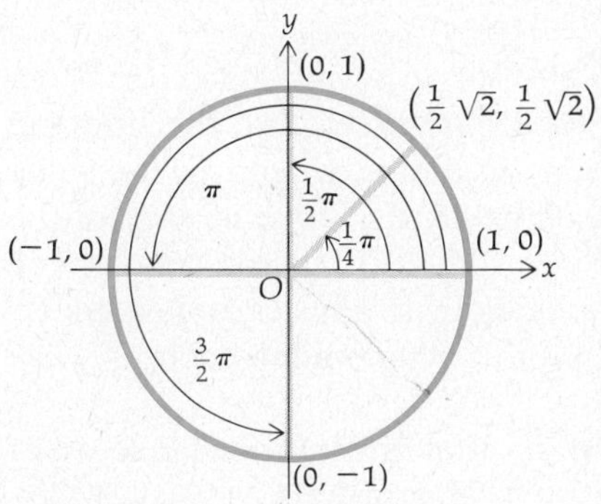

FIGURE 1.7.4

From the above definition it is seen that $\sin t$ and $\cos t$ are defined for any value of t. Hence the domain of the sine and cosine functions is the set of all real numbers. The largest value either function may have is 1, and the smallest value is -1. It will be shown later that the sine and cosine functions take on all values between -1 and 1, and from this it follows that the range of the two functions is $[-1, 1]$.

For certain values of t, the cosine and sine are easily obtained from a figure. From Fig. 1.7.4 we see that $\cos 0 = 1$ and $\sin 0 = 0$, $\cos \frac{1}{4}\pi = \frac{1}{2}\sqrt{2}$ and $\sin \frac{1}{4}\pi = \frac{1}{2}\sqrt{2}$, $\cos \frac{1}{2}\pi = 0$ and $\sin \frac{1}{2}\pi = 1$, $\cos \pi = -1$ and $\sin \pi = 0$, and $\cos \frac{3}{2}\pi = 0$ and $\sin \frac{3}{2}\pi = -1$. Table 1.7.2 gives these values and some others that are frequently used.

TABLE 1.7.2

x	$\sin x$	$\cos x$
0	0	1
$\frac{1}{6}\pi$	$\frac{1}{2}$	$\frac{1}{2}\sqrt{3}$
$\frac{1}{4}\pi$	$\frac{1}{2}\sqrt{2}$	$\frac{1}{2}\sqrt{2}$
$\frac{1}{3}\pi$	$\frac{1}{2}\sqrt{3}$	$\frac{1}{2}$
$\frac{1}{2}\pi$	1	0
$\frac{2}{3}\pi$	$\frac{1}{2}\sqrt{3}$	$-\frac{1}{2}$
$\frac{3}{4}\pi$	$\frac{1}{2}\sqrt{2}$	$-\frac{1}{2}\sqrt{2}$
$\frac{5}{6}\pi$	$\frac{1}{2}$	$-\frac{1}{2}\sqrt{3}$
π	0	-1
$\frac{3}{2}\pi$	-1	0
2π	0	1

An equation of the unit circle having its center at the origin is $x^2 + y^2 = 1$. Because $x = \cos t$ and $y = \sin t$, it follows that

$$\cos^2 t + \sin^2 t = 1 \tag{1}$$

Note that $\cos^2 t$ and $\sin^2 t$ stand for $(\cos t)^2$ and $(\sin t)^2$. Equation (1) is called an **identity** because it is valid for any real number t.

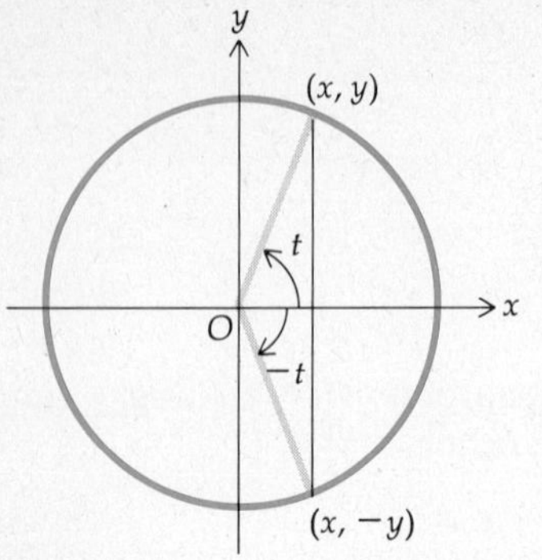

FIGURE 1.7.5

Figures 1.7.5 and 1.7.6 show angles having a negative radian measure of $-t$ and corresponding angles having a positive radian measure of t. From these figures observe that

$$\cos(-t) = \cos t \quad \text{and} \quad \sin(-t) = -\sin t \tag{2}$$

These equations hold for any real number t because the points where the terminal sides of the angles (having radian measures t and $-t$) intersect the unit circle have equal abscissas and ordinates that differ only in sign. Hence Eqs. (2) are identities. From these equations it follows that the cosine is an even function and the sine is an odd function.

From Definition 1.7.2 the following identities are obtained.

$$\cos(t + 2\pi) = \cos t \quad \text{and} \quad \sin(t + 2\pi) = \sin t \tag{3}$$

The property of cosine and sine stated by Eqs. (3) is called *periodicity,* which is now defined.

1.7.3 Definition A function f is said to be **periodic** with period $p \neq 0$ if whenever x is in the domain of f, then $x + p$ is also in the domain of f and

$$f(x + p) = f(x)$$

From the above definition and Eqs. (3) it is seen that the sine and cosine functions are periodic with period 2π; that is, whenever the value of the independent variable t is increased by 2π, the value of each of the functions is repeated. It is because of the periodicity of the sine and cosine that these functions have important applications in physics and engineering in connection with periodically repetitive phenomena such as wave motion and vibrations.

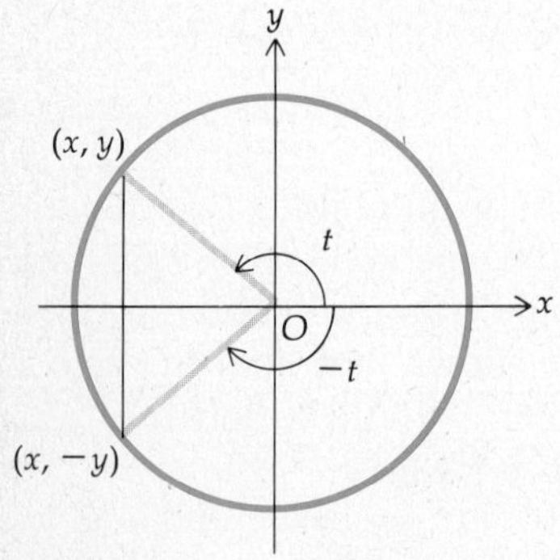

FIGURE 1.7.6

We now proceed to derive a formula that is useful because other important identities can be obtained from it. The formula is:

$$\cos(a + b) = \cos a \cos b - \sin a \sin b \tag{4}$$

where a and b are any real numbers.

Refer to Fig. 1.7.7 showing a unit circle with the points $Q(1, 0)$, $P_1(\cos a, \sin a)$, $P_2(\cos b, \sin b)$, $P_3(\cos(a + b), \sin(a + b))$, and $P_4(\cos a, -\sin a)$. Using the notation $\overparen{RS}$ to denote the measure of the length of arc from R to S we have $\overparen{QP_1} = a$, $\overparen{QP_2} = b$, $\overparen{P_2P_3} = a$, and $\overparen{P_4Q} = |-a| = a$. Because $\overparen{QP_3} = \overparen{QP_2} + \overparen{P_2P_3}$, then

$$\overparen{QP_3} = b + a \tag{5}$$

and because $\overparen{P_4P_2} = \overparen{P_4Q} + \overparen{QP_2}$, it follows that

$$\overparen{P_4P_2} = a + b \tag{6}$$

From (5) and (6), $\overparen{QP_3} = \overparen{P_4P_2}$; therefore the length of the chord joining the points Q and P_3 is the same as the length of the chord joining the points

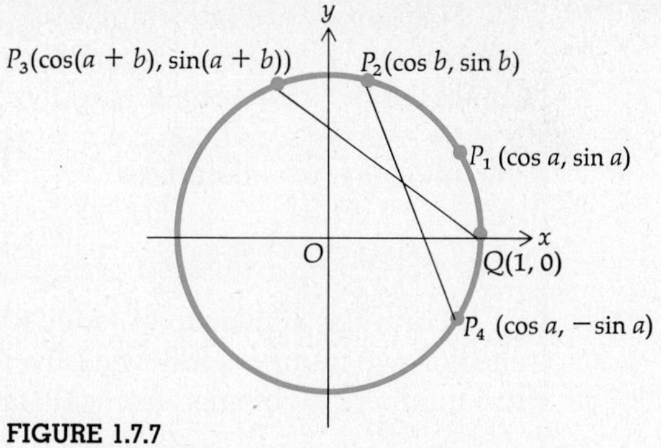

FIGURE 1.7.7

P_4 and P_2. By squaring the measures of these lengths,

$$|\overline{QP_3}|^2 = |\overline{P_4P_2}|^2 \tag{7}$$

From the distance formula,

$$|\overline{QP_3}|^2 = [\cos(a + b) - 1]^2 + [\sin(a + b) - 0]^2$$
$$= \cos^2(a + b) - 2\cos(a + b) + 1 + \sin^2(a + b)$$

and because $\cos^2(a + b) + \sin^2(a + b) = 1$, the above may be written as

$$|\overline{QP_3}|^2 = 2 - 2\cos(a + b) \tag{8}$$

Again from the distance formula,

$$|\overline{P_4P_2}|^2 = (\cos b - \cos a)^2 + (\sin b + \sin a)^2$$
$$= \cos^2 b - 2\cos a \cos b + \cos^2 a + \sin^2 b$$
$$+ 2\sin a \sin b + \sin^2 a$$

and because $\cos^2 b + \sin^2 b = 1$ and $\cos^2 a + \sin^2 a = 1$, the above becomes

$$|\overline{P_4P_2}|^2 = 2 - 2\cos a \cos b + 2\sin a \sin b \tag{9}$$

Substituting from (8) and (9) into (7) we obtain

$$2 - 2\cos(a + b) = 2 - 2\cos a \cos b + 2\sin a \sin b$$

from which follows

$$\cos(a + b) = \cos a \cos b - \sin a \sin b$$

which is formula (4).

A formula for $\cos(a - b)$ is obtained from formula (4) by substituting $-b$ for b:

$$\cos(a + (-b)) = \cos a \cos(-b) - \sin a \sin(-b)$$

Because $\cos(-b) = \cos b$ and $\sin(-b) = -\sin b$, then

$$\cos(a - b) = \cos a \cos b + \sin a \sin b \tag{10}$$

for all real numbers a and b.

Letting $a = \frac{1}{2}\pi$ in formula (10), we have the equation $\cos(\frac{1}{2}\pi - b) = \cos\frac{1}{2}\pi \cos b + \sin\frac{1}{2}\pi \sin b$; and because $\cos\frac{1}{2}\pi = 0$ and $\sin\frac{1}{2}\pi = 1$, it becomes $\cos(\frac{1}{2}\pi - b) = \sin b$. Now if in this equation we let $\frac{1}{2}\pi - b = c$, then $b = \frac{1}{2}\pi - c$ and we obtain $\cos c = \sin(\frac{1}{2}\pi - c)$. We have therefore proved the following two identities:

$$\cos(\tfrac{1}{2}\pi - t) = \sin t \quad \text{and} \quad \sin(\tfrac{1}{2}\pi - t) = \cos t \tag{11}$$

Formulas similar to (4) and (10) for the sine of the sum and difference of two numbers follow easily. A formula for the sine of the sum of two numbers is obtained from (10) and (11). From (11),

$$\sin(a + b) = \cos(\tfrac{1}{2}\pi - (a + b)) = \cos((\tfrac{1}{2}\pi - a) - b)$$

With (10) it follows that

$$\sin(a + b) = \cos(\tfrac{1}{2}\pi - a)\cos b + \sin(\tfrac{1}{2}\pi - a)\sin b$$

and from (11),

$$\sin(a + b) = \sin a \cos b + \cos a \sin b \tag{12}$$

for all real numbers a and b.

To obtain a formula for the sine of the difference of two numbers, write $a - b$ as $a + (-b)$ and apply formula (12). Thus

$$\sin(a + (-b)) = \sin a \cos(-b) + \cos a \sin(-b)$$

and because $\cos(-b) = \cos b$ and $\sin(-b) = -\sin b$,

$$\sin(a - b) = \sin a \cos b - \cos a \sin b \tag{13}$$

which is valid for all real numbers a and b.

By letting $a = b = t$ in formulas (12) and (4), respectively, the following formulas are obtained:

$$\sin 2t = 2 \sin t \cos t \tag{14}$$

and

$$\cos 2t = \cos^2 t - \sin^2 t \tag{15}$$

From the identity $\sin^2 t + \cos^2 t = 1$ we may rewrite formula (15) as

$$\cos 2t = 2\cos^2 t - 1 \tag{16}$$

or as

$$\cos 2t = 1 - 2\sin^2 t \tag{17}$$

Replacing t by $\frac{1}{2}t$ in formulas (16) and (17) and performing some algebraic manipulations, we get, respectively,

$$\cos^2 \frac{1}{2}t = \frac{1 + \cos t}{2} \tag{18}$$

and

$$\sin^2 \frac{1}{2} t = \frac{1 - \cos t}{2} \tag{19}$$

Formulas (14) through (19) are identities because they hold for all real numbers t.

By subtracting the terms of formula (4) from the corresponding terms of formula (10), the following is obtained:

$$\sin a \sin b = \tfrac{1}{2}[-\cos(a + b) + \cos(a - b)] \tag{20}$$

By adding corresponding terms of formulas (4) and (10),

$$\cos a \cos b = \tfrac{1}{2}[\cos(a + b) + \cos(a - b)] \tag{21}$$

and by adding corresponding terms of formulas (12) and (13),

$$\sin a \cos b = \tfrac{1}{2}[\sin(a + b) + \sin(a - b)] \tag{22}$$

By letting $c = a + b$ and $d = a - b$ in formulas (22), (21), and (20), we obtain, respectively,

$$\sin c + \sin d = 2 \sin \frac{c + d}{2} \cos \frac{c - d}{2} \tag{23}$$

$$\cos c + \cos d = 2 \cos \frac{c + d}{2} \cos \frac{c - d}{2} \tag{24}$$

and

$$\cos c - \cos d = -2 \sin \frac{c + d}{2} \sin \frac{c - d}{2} \tag{25}$$

If the terms of formula (13) are subtracted from the corresponding terms of formula (12) and $a + b$ and $a - b$ are replaced by c and d, respectively, then

$$\sin c - \sin d = 2 \cos \frac{c + d}{2} \sin \frac{c - d}{2} \tag{26}$$

You are asked to perform the indicated operations in the derivation of some of the above formulas in Exercises 17 to 22.

When referring to the trigonometric functions with a domain of angle measurements we use the notation $\theta°$ to denote the measurement of an angle if its degree measure is θ. For example, 45° is the measurement of an angle whose degree measure is 45 or, equivalently, whose radian measure is $\frac{1}{4}\pi$.

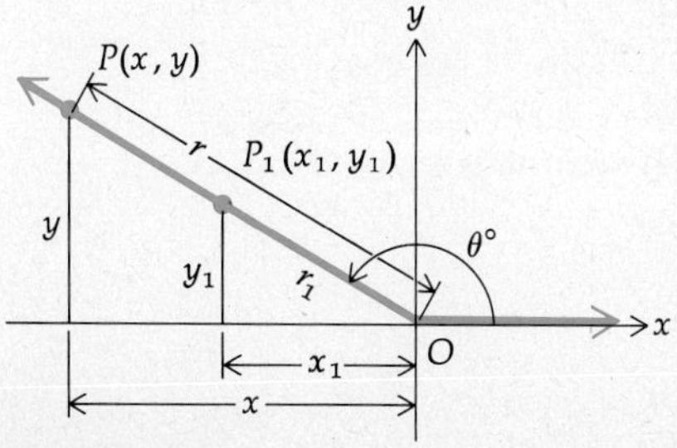

FIGURE 1.7.8

Consider an angle of $\theta°$ in standard position on a rectangular cartesian coordinate system. Choose any point P, excluding the vertex, on the terminal side of the angle and let its abscissa be x, its ordinate be y, and $|\overline{OP}| = r$. Refer to Fig. 1.7.8. The ratios x/y and y/r are independent of the choice of P because if the point P_1 is chosen instead of P, we see by Fig. 1.7.8 that $x/r = x_1/r_1$ and $y/r = y_1/r_1$. Because the position

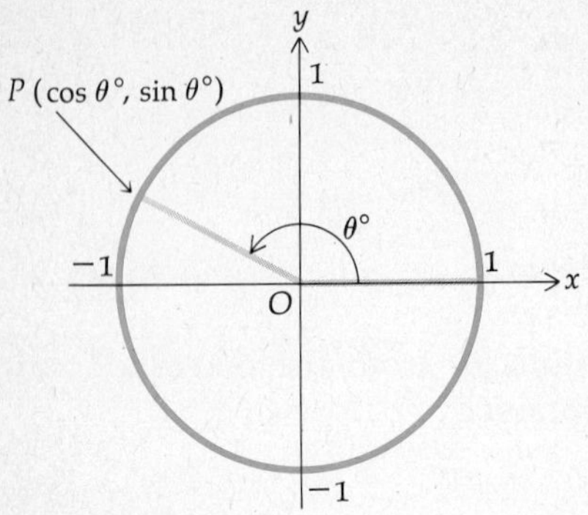

FIGURE 1.7.9

of the terminal side depends on the angle, these two ratios are functions of the measurement of the angle, and we define

$$\cos \theta° = \frac{x}{r} \quad \text{and} \quad \sin \theta° = \frac{y}{r} \tag{27}$$

Because any point P (other than the origin) may be chosen on the terminal side, we could choose the point for which $r = 1$; this is the point where the terminal side intersects the unit circle $x^2 + y^2 = 1$ (see Fig. 1.7.9). Then $\cos \theta°$ is the abscissa of the point and $\sin \theta°$ is the ordinate of the point. This gives the analogy between the sine and cosine of real numbers and those of angle measurements stated in the next definition.

1.7.4 Definition If α degrees and x radians are measurements for the same angle, then

$$\sin \alpha° = \sin x \quad \text{and} \quad \cos \alpha° = \cos x$$

The graphs of the sine and cosine functions are discussed in Sec. 3.6, where we can obtain more information about these graphs than here.

EXERCISES 1.7

In Exercises 1 and 2, find the equivalent radian measurement.

1. (a) 60°; (b) 135°; (c) 210°; (d) −150°; (e) 20°; (f) 450°; (g) −75°; (h) 100°

2. (a) 45°; (b) 120°; (c) 240°; (d) −225°; (e) 15°; (f) 540°; (g) −48°; (h) 2°

In Exercises 3 and 4, find the equivalent degree measurement.

3. (a) $\frac{1}{4}\pi$ rad; (b) $\frac{2}{3}\pi$ rad; (c) $\frac{11}{6}\pi$ rad; (d) $-\frac{1}{2}\pi$ rad; (e) $\frac{1}{2}$ rad; (f) 3π rad; (g) -2 rad; (h) $\frac{1}{12}\pi$ rad

4. (a) $\frac{1}{6}\pi$ rad; (b) $\frac{4}{3}\pi$ rad; (c) $\frac{3}{4}\pi$ rad; (d) -5π rad; (e) $\frac{1}{3}$ rad; (f) -5 rad; (g) $\frac{11}{12}\pi$ rad; (h) 0.2 rad

In Exercises 5 through 8, find the indicated value.

5. (a) $\sin \frac{1}{6}\pi$; (b) $\cos \frac{1}{4}\pi$; (c) $\sin(-\frac{3}{2}\pi)$; (d) $\cos(-\frac{2}{3}\pi)$

6. (a) $\cos \frac{1}{3}\pi$; (b) $\sin \frac{1}{4}\pi$; (c) $\cos(-\frac{1}{2}\pi)$; (d) $\sin(-2\pi)$

7. (a) $\cos \frac{5}{6}\pi$; (b) $\sin(-\frac{5}{4}\pi)$; (c) $\cos 3\pi$; (d) $\sin(-5\pi)$

8. (a) $\sin \frac{4}{3}\pi$; (b) $\cos(-\frac{1}{6}\pi)$; (c) $\sin 7\pi$; (d) $\cos(-\frac{5}{2}\pi)$

In Exercises 9 through 12, find all values of t in the interval [0, 2π) for which the given equation is satisfied.

9. (a) $\sin t = \frac{1}{2}\sqrt{3}$; (b) $\cos t = -\frac{1}{2}\sqrt{2}$

10. (a) $\cos t = -\frac{1}{2}$; (b) $\sin t = \frac{1}{2}\sqrt{2}$

11. (a) $\cos t = 0$; (b) $\sin t = -\frac{1}{2}$

12. (a) $\sin t = -1$; (b) $\cos t = -\frac{1}{2}\sqrt{3}$

In Exercises 13 and 14, find the indicated value by using formula (4) or (12).

13. (a) $\sin \frac{11}{12}\pi$ (*Hint:* $\frac{11}{12}\pi = \frac{2}{3}\pi + \frac{1}{4}\pi$); (b) $\cos \frac{7}{12}\pi$

14. (a) $\cos \frac{11}{12}\pi$; (b) $\sin \frac{7}{12}\pi$

In Exercises 15 and 16, find the indicated value by using formula (18) or (19).

15. (a) $\cos \frac{1}{12}\pi$; (b) $\sin \frac{1}{8}\pi$ **16.** (a) $\sin \frac{1}{12}\pi$; (b) $\cos \frac{1}{8}\pi$

17. Derive formula (20) by subtracting the terms of formula (4) from the corresponding terms of formula (10).

18. Derive formula (21) by a method similar to that suggested in Exercise 17.

19. Derive formula (22) by a method similar to that suggested in Exercise 17.

20. Derive formula (23) by using formula (22).

21. Derive formula (24) by using formula (21).

22. Derive formula (25) by using formula (20).

23. Derive a formula for $\sin 3t$ in terms of $\sin t$ by using formulas (12), (14), (15), and (1).

24. Derive a formula for $\cos 3t$ in terms of $\cos t$. Use a method similar to that suggested in Exercise 23.

In Exercises 25 through 27, find the indicated value without using tables.

25. (a) $\sin \frac{5}{12}\pi$; (b) $\cos(-\frac{11}{12}\pi)$ **26.** (a) $\cos \frac{5}{12}\pi$; (b) $\sin(-\frac{11}{12}\pi)$ **27.** (a) $\cos \frac{19}{12}\pi$; (b) $\sin(-\frac{5}{8}\pi)$

28. Express each of the following in terms of $\sin t$ or $\sin(\frac{1}{2}\pi - t)$: (a) $\sin(\frac{3}{2}\pi - t)$; (b) $\cos(\frac{3}{2}\pi - t)$; (c) $\sin(\frac{3}{2}\pi + t)$; (d) $\cos(\frac{3}{2}\pi + t)$.

29. Express the function values of Exercise 28 in terms of $\cos t$ or $\cos(\frac{1}{2}\pi - t)$.

30. Express each of the following in terms of $\sin t$ or $\sin(\frac{1}{2}\pi - t)$: (a) $\sin(\pi - t)$; (b) $\cos(\pi - t)$; (c) $\sin(\pi + t)$; (d) $\cos(\pi + t)$.

31. Express the function values of Exercise 30 in terms of $\cos t$ or $\cos(\frac{1}{2}\pi - t)$.

32. Find all values of t for which (a) $\sin t = 0$; (b) $\cos t = 0$.

33. Find all values of t for which (a) $\sin t = 1$; (b) $\cos t = 1$.

34. Find all values of t for which (a) $\sin t = -1$; (b) $\cos t = -1$.

35. Find all values of t for which (a) $\sin t = \frac{1}{2}$; (b) $\cos t = \frac{1}{2}$.

36. Find all values of t for which (a) $\sin t = -\frac{1}{2}\sqrt{2}$; (b) $\cos t = -\frac{1}{2}\sqrt{2}$.

37. Suppose f is a function which is periodic with period 2π, and whose domain is the set of all real numbers. Prove that f is also periodic with period -2π.

38. Prove that the function of Exercise 37 is periodic with period $2\pi n$ for every integer n. (*Hint:* Use mathematical induction.)

39. Prove that if f is defined by $f(x) = x - [\![x]\!]$, then f is periodic. What is the smallest positive period of f?

REVIEW EXERCISES FOR CHAPTER 1

In Exercises 1 through 8, find the solution set of the given inequality, and illustrate the solution on the real number line.

1. $3x - 7 \leq 5x - 17$ **2.** $8 < 5x + 4 \leq 10$ **3.** $\dfrac{x}{x-1} > \dfrac{1}{4}$ **4.** $\dfrac{3}{x+4} < \dfrac{2}{x-5}$

5. $2x^2 + x < 3$ **6.** $|2x - 5| > 7$ **7.** $|3 + 5x| \leq 9$ **8.** $\left|\dfrac{2-3x}{3+x}\right| \geq \dfrac{1}{4}$

In Exercises 9 through 11, solve for x.

9. $|3 - 4x| = 15$ **10.** $\left|\dfrac{2x - 1}{x + 3}\right| = 4$ **11.** $|3x - 4| = |6 - 2x|$

12. Find all the values of x for which $\sqrt{x^2 + 2x - 15}$ is real.

In Exercises 13 through 18, draw a sketch of the graph of the equation.

13. $y^2 = x - 4$ **14.** $y = |x - 4|$ **15.** $y = x^2 - 2$

16. $x^2 + 4y^2 = 16$ **17.** $9x^2 - 25y^2 = 0$ **18.** $4x^2 - y^2 = 16$

19. Find the abscissas of the points having ordinate 4 that are at a distance of $\sqrt{117}$ from the point $(5, -2)$.

20. Find an equation that must be satisfied by the coordinates of any point whose distance from $(-1, 2)$ is twice its distance from $(3, -4)$.

21. Find an equation that must be satisfied by the coordinates of any point whose distance from the point $(-3, 4)$ is $\sqrt{10}$.

22. Define the following sets of points by either an equation or an inequality: (a) the point-circle $(3, -5)$; (b) the set of all points whose distance from the point $(3, -5)$ is less than 4; (c) the set of all points whose distance from the point $(3, -5)$ is at least 5.

23. Prove that the quadrilateral having vertices at $(1, 2)$, $(5, -1)$, $(11, 7)$, and $(7, 10)$ is a rectangle.

24. Prove that the triangle with vertices at $(-8, 1)$, $(-1, -6)$, and $(2, 4)$ is isosceles, and find its area.

25. Prove that the points $(2, 4)$, $(1, -4)$, and $(5, -2)$ are the vertices of a right triangle, and find the area of the triangle.

26. Prove that the points $(1, -1)$, $(3, 2)$, and $(7, 8)$ are collinear in two ways: (a) by using the distance formula; (b) by using slopes.

27. Two vertices of a parallelogram are at $(-3, 4)$ and $(2, 3)$, and its center is at $(0, -1)$. Find the other two vertices.

28. Two opposite vertices of a square are at $(3, -4)$ and $(9, -4)$. Find the other two vertices.

29. Prove that the points $(2, 13)$, $(-2, 5)$, $(3, -1)$, and $(7, 7)$ are the vertices of a parallelogram.

30. Find an equation of the circle having its center at $(-2, 4)$ and radius $\sqrt{5}$. Write the equation in the general form.

31. Find an equation of the circle having the points $(-3, 2)$ and $(5, 6)$ as endpoints of a diameter.

32. Find an equation of the circle through the points $(3, -1)$, $(2, 2)$, and $(-4, 5)$.

33. Find the center and radius of the circle having the equation, $4x^2 + 4y^2 - 12x + 8y + 9 = 0$.

34. Find an equation of the line through the points $(2, -4)$ and $(7, 3)$, and write the equation in intercept form.

35. Find an equation of the line through the point $(5, -3)$ and perpendicular to the line whose equation is $2x - 5y = 1$.

36. Find an equation of the line that is the perpendicular bisector of the line segment from $(-1, 5)$ to $(3, 2)$.

37. Determine all values of k for which the graphs of the two equations $x^2 + y^2 = k$ and $x + y = k$ intersect.

38. Find an equation of the circle having as its diameter the common chord of the two circles $x^2 + y^2 + 2x - 2y - 14 = 0$ and $x^2 + y^2 - 4x + 4y - 2 = 0$.

39. Find an equation of the circle circumscribed about the triangle having sides on the lines $x - 3y + 2 = 0$, $3x - 2y + 6 = 0$, and $2x + y - 3 = 0$.

40. Find the shortest distance from the point $(2, -5)$ to the line having the equation $3x + y = 2$.

41. Find an equation of the line through the point of intersection of the lines $5x + 6y - 4 = 0$ and $x - 3y + 2 = 0$ and perpendicular to the line $x - 4y - 20 = 0$ without finding the point of intersection of the two lines. (*Hint:* See Exercise 38 in Exercises 1.4.)

42. The sides of a parallelogram are on the lines $x + 2y - 10 = 0$, $3x - y + 20 = 0$, $x + 2y - 15 = 0$, and $3x - y + 10 = 0$. Find equations of the diagonals without finding the vertices of the parallelogram. (*Hint:* See Exercise 38 in Exercises 1.4.)

43. Find equations of the lines through the origin that are tangent to the circle having its center at (2, 1) and radius 2.

44. Determine the values of k and h if $3x + ky + 2 = 0$ and $5x - y + h = 0$ are equations of the same line.

In Exercises 45 through 52, the function is the set of all ordered pairs (x, y) satisfying the given equation. Find the domain and range of the function, and draw a sketch of its graph.

45. $f: y = \sqrt{2x + 5}$

46. $g: y = \dfrac{x^2 - 16}{x + 4}$

47. $h: y = \dfrac{(x - 2)(2x^2 + 11x + 15)}{2x^2 + x - 10}$

48. $F: y = \begin{cases} x + 3 & \text{if } x < -3 \\ \sqrt{9 - x^2} & \text{if } -3 \le x \le 3 \\ 3 - x & \text{if } 3 < x \end{cases}$

49. $G: y = (x + 3) \cdot U(x + 3)$, where U is the unit step function defined in Exercise 31 in Exercises 1.6.

50. $H: y = 2x - \operatorname{sgn} x$, where sgn is the signum function defined in Exercise 32 in Exercises 1.6.

51. $g: y = \begin{cases} x^2 - 4 & \text{if } x \le 2 \\ 4 - x & \text{if } 2 < x \le 4 \\ x - 8 & \text{if } 4 < x \end{cases}$

52. $f: y = x + [\![x + 1]\!]$

53. Given $f(x) = 3x^2 - x + 5$, find: (a) $f(-3)$; (b) $f(-x^2)$; (c) $\dfrac{f(x + h) - f(x)}{h}$, $h \neq 0$.

54. Given $g(x) = \sqrt{x^2 - 1}$, find $\dfrac{g(x + h) - g(x)}{h}$, $h \neq 0$.

55. Determine if the function is even, odd, or neither: (a) $f(x) = 2x^3 - 3x$; (b) $g(x) = 5x^4 + 2x^2 - 1$; (c) $h(x) = 3x^5 - 2x^3 + x^2 - x$; (d) $F(x) = \dfrac{x^2 + 1}{x^3 - x}$.

In Exercises 56 through 60, for the given functions f and g, define the following functions and determine the domain of the resulting function: (a) $f + g$; (b) $f - g$; (c) $f \cdot g$; (d) f/g; (e) g/f; (f) $f \circ g$; (g) $g \circ f$.

56. $f(x) = \sqrt{x + 2}$ and $g(x) = x^2 + 4$

57. $f(x) = x^2 - 4$ and $g(x) = 4x - 3$

58. $f(x) = x^2 - 9$ and $g(x) = \sqrt{x + 5}$

59. $f(x) = \dfrac{1}{x - 3}$ and $g(x) = \dfrac{x}{x + 1}$

60. $f(x) = \sqrt{x}$ and $g(x) = \dfrac{1}{x^2}$

In Exercises 61 and 62, find all values of t in the interval $[0, 2\pi)$ for which the given equation is satisfied.

61. (a) $\sin t = \frac{1}{2}$; (b) $\cos t = -1$; (c) $\sin t = 0$

62. (a) $\cos t = 1$; (b) $\sin t = -\frac{1}{2}\sqrt{3}$; (c) $\cos t = \frac{1}{2}\sqrt{2}$

In Exercises 63 and 64, find the indicated value without using tables.

63. (a) $\sin \frac{2}{3}\pi$; (b) $\cos \frac{5}{4}\pi$; (c) $\sin(-\frac{5}{6}\pi)$; (d) $\cos \frac{13}{12}\pi$

64. (a) $\cos \frac{7}{6}\pi$; (b) $\sin(-\frac{7}{4}\pi)$; (c) $\cos \frac{5}{3}\pi$; (d) $\sin \frac{5}{8}\pi$

65. Prove that $x \operatorname{sgn} x = |x|$, where sgn is the signum function defined in Exercise 32 in Exercises 1.6.

66. Prove that if x is any real number, $|x| < x^2 + 1$.

67. Prove analytically that the three medians of any triangle meet in a point.

68. Prove that the two lines $A_1x + B_1y + C_1 = 0$ and $A_2x + B_2y + C_2 = 0$ are parallel if and only if $A_1B_2 - A_2B_1 = 0$.

69. Prove analytically that if the diagonals of a rectangle are perpendicular, then the rectangle is a square.

70. Prove analytically that the line segment joining the midpoints of any two sides of a triangle is parallel to the third side and that its length is one-half the length of the third side.

71. In a triangle the point of intersection of the medians, the point of intersection of the altitudes, and the center of the circumscribed circle are collinear. Find these three points and prove that they are collinear for the triangle having vertices at (2, 8), (5, −1), and (6, 6).

72. Prove analytically that the set of points equidistant from two given points is the perpendicular bisector of the line segment joining the two points.

CHAPTER 2 Limits and Continuity

2.1 THE LIMIT OF A FUNCTION

Consider the function f defined by the equation

$$f(x) = \frac{2x^2 + x - 3}{x - 1} \tag{1}$$

f is defined for all values of x except $x = 1$. We shall investigate the function values, $f(x)$, when x is close to 1 but not equal to 1. First let x take on the values 0, 0.25, 0.50, 0.75, 0.9, 0.99, 0.999, 0.9999, and so on. We are taking values of x closer and closer to 1 but less than 1; in other words, the variable x is approaching 1 through values that are less than 1. This is illustrated in Table 2.1.1. Now let the variable x approach 1 through values that are greater than 1; that is, let x take on the values 2, 1.75, 1.5, 1.25, 1.1, 1.01, 1.001, 1.0001, 1.00001, and so on. Refer to Table 2.1.2.

Observe from both tables that as x gets closer and closer to 1, $f(x)$ gets closer and closer to 5; and the closer x is to 1, the closer $f(x)$ is to 5. For instance, from Table 2.1.1, when $x = 0.9$, $f(x) = 4.8$; that is, when x is 0.1 less than 1, $f(x)$ is 0.2 less than 5. When $x = 0.999$, $f(x) = 4.998$; that is, when x is 0.001 less than 1, $f(x)$ is 0.002 less than 5. Furthermore, when $x = 0.9999$, $f(x) = 0.49998$; that is, when x is 0.0001 less than 1, $f(x)$ is 0.0002 less than 5.

Table 2.1.2 shows that when $x = 1.1$, $f(x) = 5.2$; that is, when x is 0.1 greater than 1, $f(x)$ is 0.2 greater than 5. When $x = 1.001$, $f(x) = 5.002$; that is, when x is 0.001 greater than 1, $f(x)$ is 0.002 greater than 5. When $x = 1.0001$, $f(x) = 5.0002$; that is, when x is 0.0001 greater than 1, $f(x)$ is 0.0002 greater than 5.

Therefore, from the two tables we see that when x differs from 1 by ± 0.001 (i.e., $x = 0.999$ or $x = 1.001$), $f(x)$ differs from 5 by ± 0.002 (i.e., $f(x) = 4.998$ or $f(x) = 5.002$). And when x differs from 1 by ± 0.0001, $f(x)$ differs from 5 by ± 0.0002.

Now, looking at the situation another way, we consider the values of $f(x)$ first. We see that we can make the value of $f(x)$ as close to 5 as we please by taking x close enough to 1. Another way of saying this is that we can make the absolute value of the difference between $f(x)$ and 5 as small as we please by making the absolute value of the difference between x and 1 small enough. That is, $|f(x) - 5|$ can be made as small as we please by making $|x - 1|$ small enough. But bear in mind that $f(x)$ never takes on the value 5.

TABLE 2.1.1

x	0	0.25	0.5	0.75	0.9	0.99	0.999	0.9999	0.99999
$f(x) = \dfrac{2x^2 + x - 3}{x - 1}$	3	3.5	4	4.5	4.8	4.98	4.998	4.9998	4.99998

TABLE 2.1.2

x	2	1.75	1.5	1.25	1.1	1.01	1.001	1.0001	1.00001
$f(x) = \dfrac{2x^2 + x - 3}{x - 1}$	7	6.5	6.0	5.5	5.2	5.02	5.002	5.0002	5.00002

A more precise way of noting this is by using two symbols for these small differences. The symbols usually used are ϵ (epsilon) and δ (delta). So we state that $|f(x) - 5|$ will be less than any given positive number ϵ whenever $|x - 1|$ is less than some appropriately chosen positive number δ and $|x - 1| \neq 0$ (since $x \neq 1$). It is important to realize that the size of δ depends on the size of ϵ. Still another way of phrasing this is: Given any positive number ϵ, we can make $|f(x) - 5| < \epsilon$ by taking $|x - 1|$ small enough; that is, there is some sufficiently small positive number δ such that

$$|f(x) - 5| < \epsilon \quad \text{whenever} \quad 0 < |x - 1| < \delta \tag{2}$$

From the two tables notice that $|f(x) - 5| = 0.2$ when $|x - 1| = 0.1$. So given $\epsilon = 0.2$, we take $\delta = 0.1$ and state that

$$|f(x) - 5| < 0.2 \quad \text{whenever} \quad 0 < |x - 1| < 0.1$$

This is statement (2) with $\epsilon = 0.2$ and $\delta = 0.1$.

Also, $|f(x) - 5| = 0.002$ when $|x - 1| = 0.001$. Hence if $\epsilon = 0.002$, we take $\delta = 0.001$, and then

$$|f(x) - 5| < 0.002 \quad \text{whenever} \quad 0 < |x - 1| < 0.001$$

This is statement (2) with $\epsilon = 0.002$ and $\delta = 0.001$.

Similarly, if $\epsilon = 0.0002$, we take $\delta = 0.0001$ and state that

$$|f(x) - 5| < 0.0002 \quad \text{whenever} \quad 0 < |x - 1| < 0.0001$$

This is statement (2) with $\epsilon = 0.0002$ and $\delta = 0.0001$.

We could go on and give ϵ any small positive value, and find a suitable value for δ such that $|f(x) - 5|$ will be less than ϵ whenever $|x - 1|$ is less than δ and $x \neq 1$ (or $|x - 1| > 0$). Now, because for any $\epsilon > 0$ we can find a $\delta > 0$ such that $|f(x) - 5| < \epsilon$ whenever $0 < |x - 1| < \delta$, we state that the limit of $f(x)$ as x approaches 1 is equal to 5 or, expressed in symbols,

$$\lim_{x \to 1} f(x) = 5$$

Observe that the numerator of the fraction in (1) can be factored so that

$$f(x) = \frac{(2x + 3)(x - 1)}{x - 1}$$

If $x \neq 1$, the numerator and denominator can be divided by $x - 1$ to obtain

$$f(x) = 2x + 3 \qquad x \neq 1 \tag{3}$$

Equation (3) with the stipulation that $x \neq 1$ is just as suitable as (1) for a definition of $f(x)$. From (3) it is apparent that $f(x)$ can be made as close to 5 as we please by taking x sufficiently close to 1, and this property of the function f does not depend on f being defined when $x = 1$. This fact gives the distinction between $\lim_{x \to 1} f(x)$ and the function value at 1; that is, $\lim_{x \to 1} f(x) = 5$, but $f(1)$ does not exist. Consequently, in statement (2),

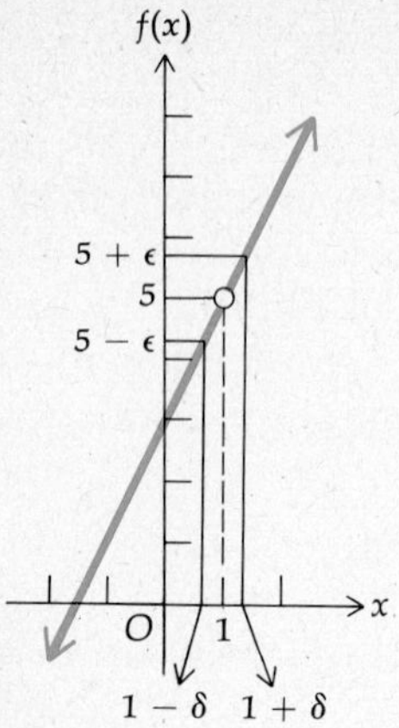

FIGURE 2.1.1

$0 < |x - 1|$ because we are concerned only with values of $f(x)$ for x close to 1, but not for $x = 1$.

Let us see what all this means geometrically for the particular function defined by either (1) or (3). Figure 2.1.1 illustrates the geometric significance of ϵ and δ. Observe that $f(x)$ on the vertical axis will lie between $5 - \epsilon$ and $5 + \epsilon$ whenever x on the horizontal axis lies between $1 - \delta$ and $1 + \delta$; or

$$|f(x) - 5| < \epsilon \quad \text{whenever} \quad 0 < |x - 1| < \delta$$

Another way of stating this is that $f(x)$ on the vertical axis can be restricted to lie between $5 - \epsilon$ and $5 + \epsilon$ by restricting x on the horizontal axis to lie between $1 - \delta$ and $1 + \delta$.

Note that the values of ϵ are chosen arbitrarily and can be as small as desired, and that the value of δ is dependent on the ϵ chosen. It should also be pointed out that the smaller the value of ϵ, the smaller will be the corresponding value of δ.

Summing up for this example, we state that $\lim_{x \to 1} f(x) = 5$ because for any $\epsilon > 0$, however small, there exists a $\delta > 0$ such that

$$|f(x) - 5| < \epsilon \quad \text{whenever} \quad 0 < |x - 1| < \delta$$

The limit of a function in general is now defined.

2.1.1 Definition Let f be a function which is defined at every number in some open interval I containing a, except possibly at the number a itself. The **limit of $f(x)$ as x approaches a is L**, written as

$$\lim_{x \to a} f(x) = L$$

if for any $\epsilon > 0$, however small, there exists a $\delta > 0$ such that

$$|f(x) - L| < \epsilon \quad \text{whenever} \quad 0 < |x - a| < \delta \tag{4}$$

In words, Definition 2.1.1 states that the function values $f(x)$ approach a limit L as x approaches a number a if the absolute value of the difference between $f(x)$ and L can be made as small as we please by taking x sufficiently near a, but not equal to a.

It is important to realize that in the above definition nothing is mentioned about the value of the function when $x = a$. That is, it is not necessary that the function be defined for $x = a$ in order for the $\lim_{x \to a} f(x)$ to exist.

A geometric interpretation of Definition 2.1.1 for a function f is shown in Fig. 2.1.2. A portion of the graph of f near the point where $x = a$ is shown in the figure. Because f is not necessarily defined at a, there is no point on the graph with abscissa a. Observe that $f(x)$ on the vertical axis will lie between $L - \epsilon_1$ and $L + \epsilon_1$ whenever x on the horizontal axis lies between $a - \delta_1$ and $a + \delta_1$. Stated another way, $f(x)$ on the vertical axis can be restricted to lie between $L - \epsilon_1$ and $L + \epsilon_1$ by restricting x on the horizontal axis to lie between $a - \delta_1$ and $a + \delta_1$. Thus

$$|f(x) - L| < \epsilon_1 \quad \text{whenever} \quad 0 < |x - a| < \delta_1$$

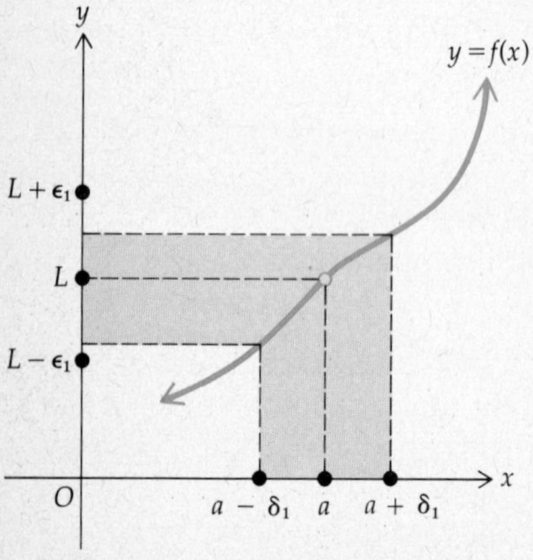

FIGURE 2.1.2

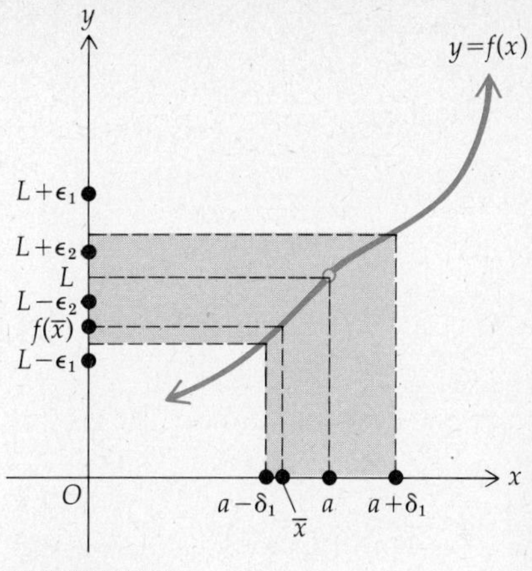

FIGURE 2.1.3

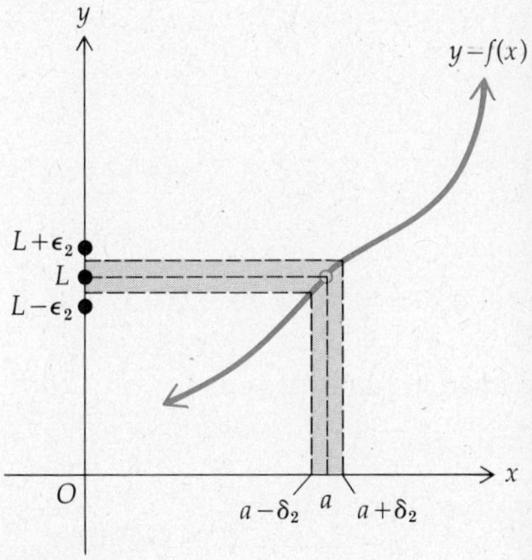

FIGURE 2.1.4

Figure 2.1.3 shows how a smaller value of ϵ can require a different choice for δ. In the figure it is seen that for $\epsilon_2 < \epsilon_1$, the δ_1 value is too large; that is, there are values of x for which $0 < |x - a| < \delta_1$, but $|f(x) - L|$ is not less than ϵ_2. For instance, $0 < |\bar{x} - a| < \delta_1$, but $|f(\bar{x}) - L| > \epsilon_2$. So we must choose a smaller value δ_2, shown in Fig. 2.1.4, such that

$$|f(x) - L| < \epsilon_2 \quad \text{whenever} \quad 0 < |x - a| < \delta_2$$

However, for any choice of $\epsilon > 0$, no matter how small, there exists a $\delta > 0$ such that statement (4) holds. Therefore $\lim_{x \to a} f(x) = L$.

Example 1 Let the function f be defined by the equation $f(x) = 4x - 1$. Given $\lim_{x \to 3} f(x) = 11$, find a δ for $\epsilon = 0.01$ such that

$$|f(x) - 11| < 0.01 \quad \text{whenever} \quad 0 < |x - 3| < \delta$$

Solution

$$\begin{aligned} |f(x) - 11| &= |(4x - 1) - 11| \\ &= |4x - 12| \\ &= 4|x - 3| \end{aligned}$$

Therefore we want

$$4|x - 3| < 0.01 \quad \text{whenever} \quad 0 < |x - 3| < \delta$$

or, equivalently,

$$|x - 3| < 0.0025 \quad \text{whenever} \quad 0 < |x - 3| < \delta$$

If we take $\delta = 0.0025$, we have

$$|(4x - 1) - 11| < 0.01 \quad \text{whenever} \quad 0 < |x - 3| < 0.0025 \tag{5}$$

In this example any positive number less than 0.0025 can be used in place of 0.0025 as the required δ. That is, if $0 < \gamma < 0.0025$ and statement (5) holds, then

$$|(4x - 1) - 11| < 0.01 \quad \text{whenever} \quad 0 < |x - 3| < \gamma$$

because every number x satisfying the inequality $0 < |x - 3| < \gamma$ also satisfies the inequality $0 < |x - 3| < 0.0025$.

The solution of Example 1 consisted of finding a δ for a specific ϵ. If for any ϵ we can find a δ that will work, we shall have established that the value of the limit is 11. This is done in the next example.

Example 2 Use Definition 2.1.1 to prove that $\lim_{x \to 3} (4x - 1) = 11$.

Solution The first requirement of Definition 2.1.1 is that $4x - 1$ be defined at every number in some open interval I containing 3 except possibly at 3. Because $4x - 1$ is defined for every value of x, any open interval containing 3 will satisfy this requirement.

Now we must show that for any $\epsilon > 0$ there exists a $\delta > 0$ such that

$$|(4x - 1) - 11| < \epsilon \quad \text{whenever} \quad 0 < |x - 3| < \delta$$

From Example 1 note that

$$\begin{aligned} |(4x - 1) - 11| &= |4x - 12| \\ &= 4|x - 3| \end{aligned}$$

Therefore we want

$$4|x - 3| < \epsilon \quad \text{whenever} \quad 0 < |x - 3| < \delta$$

$$|x - 3| < \tfrac{1}{4}\epsilon \quad \text{whenever} \quad 0 < |x - 3| < \delta$$

So if we choose $\delta = \frac{1}{4}\epsilon$, we have

$$4|x - 3| < 4\delta \quad \text{whenever} \quad 0 < |x - 3| < \delta$$

or, equivalently,

$$4|x - 3| < 4 \cdot \tfrac{1}{4}\epsilon \quad \text{whenever} \quad 0 < |x - 3| < \delta$$

$$4|x - 3| < \epsilon \quad \text{whenever} \quad 0 < |x - 3| < \delta$$

giving

$$|(4x - 1) - 11| < \epsilon \quad \text{whenever} \quad 0 < |x - 3| < \delta$$

if $\delta = \frac{1}{4}\epsilon$. This proves that $\lim_{x \to 3} (4x - 1) = 11$.

In particular, if $\epsilon = 0.01$, then we take $\delta = 0.01/4$, or 0.0025, which corresponds to the result in Example 1.

Any positive number $\delta' < \frac{1}{4}\epsilon$ can be used in place of $\frac{1}{4}\epsilon$ as the required δ in this example.

Example 3 Use Definition 2.1.1 to prove that $\lim_{x \to 2} x^2 = 4$.

Solution Because x^2 is defined for every value of x, any open interval containing 2 will satisfy the first requirement of Definition 2.1.1.

We must show that for any $\epsilon > 0$ there exists a $\delta > 0$ such that

$$|x^2 - 4| < \epsilon \quad \text{whenever} \quad 0 < |x - 2| < \delta \tag{6}$$

That is, we want

$$|(x - 2)(x + 2)| < \epsilon \quad \text{whenever} \quad 0 < |x - 2| < \delta$$

$$|x - 2||x + 2| < \epsilon \quad \text{whenever} \quad 0 < |x - 2| < \delta$$

$$|x - 2| < \frac{\epsilon}{|x + 2|} \quad \text{whenever} \quad 0 < |x - 2| < \delta \tag{7}$$

Because we want $|x - 2| < \epsilon/|x + 2|$ for all values of x in the open interval I required by Definition 2.1.1, we choose a particular interval for I and then determine a lower bound for $\epsilon/|x + 2|$. If the interval I is $(1, 3)$, which has 2 as its center and a radius of 1, then for all $x \in I$, $|x + 2| = x + 2$. Furthermore, for all $x \in I$,

$$\frac{\epsilon}{x + 2} > \frac{\epsilon}{5}$$

Thus $\epsilon/5$ is a lower bound for $\epsilon/|x+2|$ for values of x in I. Consequently, δ is taken to be the smaller of this lower bound and the radius of the interval I; that is, δ is the smaller of the two numbers $\epsilon/5$ and 1, which is written as $\delta = \min(\epsilon/5, 1)$. Then

$$|x-2| < \frac{\epsilon}{5} < \frac{\epsilon}{|x+2|} \quad \text{whenever} \quad 0 < |x-2| < \delta \tag{8}$$

if $\delta = \min(\epsilon/5, 1)$. From statement (8) it follows that for this value of δ, statement (7) and the equivalent statement (6) hold.

To show the significance of the δ, consider possible choices for ϵ. If $\epsilon/5 > 1$, that is, $\epsilon > 5$, take $\delta = 1$. Then

$$0 < |x-2| < 1 < \frac{\epsilon}{5} < \frac{\epsilon}{|x+2|} \quad \text{whenever} \quad 0 < |x-2| < \delta$$

If $0 < \epsilon/5 \le 1$, that is, $0 < \epsilon \le 5$, take $\delta = \epsilon/5$. Then

$$0 < |x-2| < \frac{\epsilon}{5} \le 1 \quad \text{and} \quad |x-2| < \frac{\epsilon}{5} < \frac{\epsilon}{|x+2|} \quad \text{whenever} \quad 0 < |x-2| < \delta$$

Thus for any $\epsilon > 0$,

$$0 < |x-2| < 1 \quad \text{and} \quad |x-2| < \frac{\epsilon}{|x+2|} \quad \text{whenever} \quad 0 < |x-2| < \delta$$

The statement $0 < |x-2| < 1$ implies that $x \ne 2$ and $x \in I$. The statement $|x-2| < \epsilon/|x+2|$ is equivalent to the statement $|x^2-4| < \epsilon$. Therefore Definition 2.1.1 holds, and $\lim_{x \to 2} x^2 = 4$.

Example 4 Use Definition 2.1.1 to prove that

$$\lim_{t \to 7} \frac{8}{t-3} = 2$$

Solution Because $8/(t-3)$ is defined for every value of t except 3, the open interval I required by Definition 2.1.1 can be any open interval containing 7 but not containing 3.

We must show that for any $\epsilon > 0$ there exists a $\delta > 0$ such that

$$\left|\frac{8}{t-3} - 2\right| < \epsilon \quad \text{whenever} \quad 0 < |t-7| < \delta \tag{9}$$

Because

$$\begin{aligned}\left|\frac{8}{t-3} - 2\right| &= \left|\frac{8-2(t-3)}{t-3}\right| \\ &= \left|\frac{14-2t}{t-3}\right| \\ &= \frac{2 \cdot |7-t|}{|t-3|} \\ &= |t-7| \cdot \frac{2}{|t-3|}\end{aligned}$$

statement (9) is equivalent to

$$|t-7| \cdot \frac{2}{|t-3|} < \epsilon \quad \text{whenever} \quad 0 < |t-7| < \delta$$

which is equivalent to

$$|t-7| < \frac{|t-3|\epsilon}{2} \quad \text{whenever} \quad 0 < |t-7| < \delta \tag{10}$$

Because we want $|t-7| < \frac{1}{2}|t-3|\epsilon$, we choose an open interval I required by Definition 2.1.1 that will determine a lower bound of $\frac{1}{2}|t-3|\epsilon$ for values of t in I. Let I be the open interval (6, 8), which has 7 as its center and a radius of 1. Then for all $t \in I$, $|t-3| = t-3$, and

$$\frac{(t-3)\epsilon}{2} > \frac{(6-3)\epsilon}{2} = \frac{3}{2}\epsilon$$

Therefore $\frac{3}{2}\epsilon$ is a lower bound of $\frac{1}{2}|t-3|\epsilon$ for values of t in I. Thus we take δ as the smaller of this lower bound and the radius of I; that is, $\delta = \min(\frac{3}{2}\epsilon, 1)$. Then

$$|t-7| < \frac{3}{2}\epsilon < \frac{|t-3|\epsilon}{2} \quad \text{whenever} \quad 0 < |t-7| < \delta$$

if $\delta = \min(\frac{3}{2}\epsilon, 1)$. For this value of δ, statement (10) and the equivalent statement (9) hold for all $t \in I$.

If $\frac{3}{2}\epsilon > 1$, that is, $\epsilon > \frac{2}{3}$, take $\delta = 1$. Then

$$0 < |t-7| < 1 < \frac{3}{2}\epsilon < \frac{|t-3|\epsilon}{2} \quad \text{whenever} \quad 0 < |t-7| < \delta$$

If $0 < \frac{3}{2}\epsilon \leq 1$, that is, $0 < \epsilon \leq \frac{2}{3}$, take $\delta = \frac{3}{2}\epsilon$. Then

$$0 < |t-7| < \frac{3}{2}\epsilon \leq 1 \text{ and } |t-7| < \frac{3}{2}\epsilon < \frac{|t-3|\epsilon}{2} \text{ whenever } 0 < |t-7| < \delta$$

Hence for any $\epsilon > 0$,

$$0 < |t-7| < 1 \quad \text{and} \quad |t-7| < \frac{|t-3|\epsilon}{2} \quad \text{whenever} \quad 0 < |t-7| < \delta$$

The statement $0 < |t-7| < 1$ implies that $t \neq 7$ and $t \in I$. The statement $|t-7| < \frac{1}{2}|t-3|\epsilon$ is equivalent to the statement

$$\left|\frac{8}{t-3} - 2\right| < \epsilon$$

Therefore Definition 2.1.1 holds, and $\lim_{t \to 7}[8/(t-3)] = 2$.

The following theorem states that a function cannot approach two different limits at the same time. It is called a **uniqueness theorem** because it guarantees that if the limit of a function exists, it is unique.

2.1.2 Theorem If $\lim_{x \to a} f(x) = L_1$ and $\lim_{x \to a} f(x) = L_2$, then $L_1 = L_2$.

Proof We shall assume that $L_1 \neq L_2$ and show that this assumption leads to a contradiction. Because $\lim_{x \to a} f(x) = L_1$, it follows from Definition 2.1.1 that for any $\epsilon > 0$ there exists a $\delta_1 > 0$ such that

$$|f(x) - L_1| < \epsilon \quad \text{whenever} \quad 0 < |x - a| < \delta_1 \tag{11}$$

Also, because $\lim_{x \to a} f(x) = L_2$, there exists a $\delta_2 > 0$ such that

$$|f(x) - L_2| < \epsilon \quad \text{whenever} \quad 0 < |x - a| < \delta_2 \tag{12}$$

Now, writing $L_1 - L_2$ as $L_1 - f(x) + f(x) - L_2$ and applying the triangle inequality, we have

$$\begin{aligned} |L_1 - L_2| &= |[L_1 - f(x)] + [f(x) - L_2]| \\ &\leq |L_1 - f(x)| + |f(x) - L_2| \end{aligned} \tag{13}$$

So from (11), (12), and (13) we may conclude that for any $\epsilon > 0$ there exists a $\delta_1 > 0$ and a $\delta_2 > 0$ such that

$$|L_1 - L_2| < \epsilon + \epsilon \text{ whenever } 0 < |x - a| < \delta_1 \text{ and } 0 < |x - a| < \delta_2 \tag{14}$$

If δ is the smaller of δ_1 and δ_2, then $\delta \leq \delta_1$ and $\delta \leq \delta_2$, and (14) states that for any $\epsilon > 0$ there exists a $\delta > 0$ such that

$$|L_1 - L_2| < 2\epsilon \quad \text{whenever} \quad 0 < |x - a| < \delta \tag{15}$$

However, if $\epsilon = \frac{1}{2}|L_1 - L_2|$, then (15) states that there exists a $\delta > 0$ such that

$$|L_1 - L_2| < |L_1 - L_2| \quad \text{whenever} \quad 0 < |x - a| < \delta \tag{16}$$

Because (16) is a contradiction, our assumption is false. So $L_1 = L_2$, and the theorem is proved. ■

EXERCISES 2.1

In Exercises 1 through 14, you are given $f(x)$, a, and L, as well as $\lim_{x \to a} f(x) = L$. Determine a number δ for the given ϵ such that $|f(x) - L| < \epsilon$ whenever $0 < |x - a| < \delta$.

1. $\lim_{x \to 3} (2x + 4) = 10$; $\epsilon = 0.01$

2. $\lim_{x \to 2} (4x - 5) = 3$; $\epsilon = 0.001$

3. $\lim_{x \to -1} (3 - 4x) = 7$; $\epsilon = 0.02$

4. $\lim_{x \to -2} (2 + 5x) = -8$; $\epsilon = 0.002$

5. $\lim_{x \to 3} x^2 = 9$; $\epsilon = 0.005$

6. $\lim_{x \to -4} x^2 = 16$; $\epsilon = 0.03$

7. $\lim_{x \to 2} (x^2 - 2x + 1) = 1$; $\epsilon = 0.001$

8. $\lim_{x \to 1} (x^2 - 5) = -4$; $\epsilon = 0.01$

9. $\lim_{x \to 1} \sqrt{x} = 1$; $\epsilon = 0.002$

10. $\lim_{x \to 4} \sqrt{x} = 2$; $\epsilon = 0.005$

11. $\lim_{x \to -2} \dfrac{x^2 - 4}{x + 2} = -4$; $\epsilon = 0.01$

12. $\lim_{x \to 1/3} \dfrac{9x^2 - 1}{3x - 1} = 2$; $\epsilon = 0.01$

13. $\lim_{x \to 2} \dfrac{1}{x} = \dfrac{1}{2}$; $\epsilon = 0.003$

14. $\lim_{x \to 2} \dfrac{1}{x + 1} = \dfrac{1}{3}$; $\epsilon = 0.02$

In Exercises 15 through 42, establish the limit by using Definition 2.1.1; that is, for any $\epsilon > 0$, find a $\delta > 0$ such that $|f(x) - L| < \epsilon$ whenever $0 < |x - a| < \delta$.

15. $\lim_{x \to 4} (2x + 1) = 9$

16. $\lim_{x \to 3} (3x - 5) = 4$

17. $\lim_{x \to 1} (5x - 3) = 2$

18. $\lim_{x\to -1}(4x+1) = -3$ **19.** $\lim_{x\to -2}(1+3x) = -5$ **20.** $\lim_{x\to -2}(7-2x) = 11$

21. $\lim_{x\to -1}\dfrac{x^2-1}{x+1} = -2$ **22.** $\lim_{x\to 3}\dfrac{x^2-9}{x-3} = 6$ **23.** $\lim_{x\to 1} x^2 = 1$

24. $\lim_{x\to -3} x^2 = 9$ **25.** $\lim_{x\to 5}(x^2-3x) = 10$ **26.** $\lim_{x\to 2}(x^2+2x-1) = 7$

27. $\lim_{x\to -3}(5-x-x^2) = -1$ **28.** $\lim_{x\to -1}(3+2x-x^2) = 0$ **29.** $\lim_{x\to -3}\dfrac{1}{x} = -\dfrac{1}{3}$

30. $\lim_{x\to 4}\dfrac{2}{x+2} = \dfrac{1}{3}$ **31.** $\lim_{x\to 3}\dfrac{4}{x-1} = 2$ **32.** $\lim_{x\to -4}\dfrac{1}{x+3} = -1$

33. $\lim_{x\to 5}\dfrac{2}{x-4} = 2$ **34.** $\lim_{x\to 6}\dfrac{x}{x-3} = 2$ **35.** $\lim_{x\to -2}\dfrac{3x}{3x+2} = \dfrac{3}{2}$

36. $\lim_{x\to 1/2}\dfrac{3+2x}{5-x} = \dfrac{8}{9}$ **37.** $\lim_{x\to 4}\sqrt{x+5} = 3$ **38.** $\lim_{x\to -1}\sqrt{1-3x} = 2$

39. $\lim_{x\to 1}\dfrac{\sqrt{x}-1}{x-1} = \dfrac{1}{2}$ **40.** $\lim_{x\to 1}\dfrac{1}{\sqrt{5-x}} = \dfrac{1}{2}$ **41.** $\lim_{x\to -3}\dfrac{1}{\sqrt{3-2x}} = \dfrac{1}{3}$

42. $\lim_{x\to 4}\dfrac{\sqrt{x}-2}{x-4} = \dfrac{1}{4}$

43. Prove that $\lim_{x\to a} x^2 = a^2$ if a is any positive number. **44.** Prove that $\lim_{x\to a} x^2 = a^2$ if a is any negative number.

45. Prove that $\lim_{x\to a}\sqrt{x} = \sqrt{a}$ if a is any positive number. **46.** Prove that $\lim_{x\to a} x^3 = a^3$ if a is any positive number.

47. Prove that $\lim_{x\to a} x^3 = a^3$ if a is any negative number.

48. Prove that $\lim_{x\to a}\sqrt[3]{x} = \sqrt[3]{a}$. (*Hint:* $a^3 - b^3 = (a-b)(a^2+ab+b^2)$.)

49. Prove that if $\lim_{x\to a} f(x)$ exists and is L, then $\lim_{x\to a}|f(x)|$ exists and is $|L|$.

50. Prove that if $f(x) = g(x)$ for all values of x except $x = a$, then $\lim_{x\to a} f(x) = \lim_{x\to a} g(x)$ if the limits exist.

51. Prove that if $f(x) = g(x)$ for all values of x except $x = a$, then if $\lim_{x\to a} g(x)$ does not exist, $\lim_{x\to a} f(x)$ does not exist. (*Hint:* Show that the assumption that $\lim_{x\to a} f(x)$ does exist leads to a contradiction.)

2.2 THEOREMS ON LIMITS OF FUNCTIONS

To find limits of functions in a straightforward manner, we need some theorems. The proofs of the theorems are based on Definition 2.1.1. These theorems, as well as other theorems on limits of functions appearing in later sections of this chapter, are labeled "limit theorems" and are so designated as they are presented.

2.2.1 Limit Theorem 1

If m and b are any constants,

$$\lim_{x\to a}(mx+b) = ma+b$$

Proof To prove this theorem we use Definition 2.1.1. For any $\epsilon > 0$ we must prove that there exists a $\delta > 0$ such that

$$|(mx+b)-(ma+b)] < \epsilon \quad \text{whenever} \quad 0 < |x-a| < \delta \tag{1}$$

Case 1: $m \neq 0$.

Because $|(mx + b) - (ma + b)| = |mx - ma| = |m| \cdot |x - a|$, we want to find a $\delta > 0$ for any $\epsilon > 0$ such that

$$|m| \cdot |x - a| < \epsilon \quad \text{whenever} \quad 0 < |x - a| < \delta$$

or, because $m \neq 0$,

$$|x - a| < \frac{\epsilon}{|m|} \quad \text{whenever} \quad 0 < |x - a| < \delta \tag{2}$$

Statement (2) will hold if $\delta = \epsilon/|m|$; so we conclude that

$$|(mx + b) - (ma + b)| < \epsilon \text{ whenever } 0 < |x - a| < \delta \text{ if } \delta = \frac{\epsilon}{|m|}$$

This proves the theorem for Case 1.

Case 2: $m = 0$.

If $m = 0$, then $|(mx + b) - (ma + b)| = 0$ for all values of x. So take δ to be any positive number, and statement (1) holds. This proves the theorem for Case 2. ■

● ILLUSTRATION 1 From Limit Theorem 1 it follows that

$$\begin{aligned}\lim_{x \to 2} (3x + 5) &= 3 \cdot 2 + 5 \\ &= 11\end{aligned}$$ ●

2.2.2 Limit Theorem 2 If c is a constant, then for any number a,

$$\lim_{x \to a} c = c$$

Proof This follows immediately from Limit Theorem 1 by taking $m = 0$ and $b = c$. ■

2.2.3 Limit Theorem 3

$$\lim_{x \to a} x = a$$

Proof This also follows immediately from Limit Theorem 1 by taking $m = 1$ and $b = 0$. ■

● ILLUSTRATION 2 From Limit Theorem 2,

$$\lim_{x \to 5} 7 = 7$$

and from Limit Theorem 3,

$$\lim_{x \to -6} x = -6$$ ●

2.2.4 Limit Theorem 4 If $\lim_{x \to a} f(x) = L$ and $\lim_{x \to a} g(x) = M$, then

$$\lim_{x \to a} [f(x) \pm g(x)] = L \pm M$$

Proof We shall prove this theorem with the plus sign. Given

$$\lim_{x \to a} f(x) = L \tag{3}$$

and

$$\lim_{x \to a} g(x) = M \tag{4}$$

we wish to prove that

$$\lim_{x \to a} [f(x) + g(x)] = L + M \tag{5}$$

To prove (5), use Definition 2.1.1; that is, for any $\epsilon > 0$ we must prove that there exists a $\delta > 0$ such that

$$|[f(x) + g(x)] - (L + M)| < \epsilon \quad \text{whenever} \quad 0 < |x - a| < \delta \tag{6}$$

Because (3) is given, it follows from the definition of a limit that for $\frac{1}{2}\epsilon > 0$ there exists a $\delta_1 > 0$ such that

$$|f(x) - L| < \tfrac{1}{2}\epsilon \quad \text{whenever} \quad 0 < |x - a| < \delta_1$$

Similarly, from (4), for $\frac{1}{2}\epsilon > 0$ there exists a $\delta_2 > 0$ such that

$$|g(x) - M| < \tfrac{1}{2}\epsilon \quad \text{whenever} \quad 0 < |x - a| < \delta_2$$

Now let δ be the smaller of the two numbers δ_1 and δ_2. Therefore $\delta \le \delta_1$ and $\delta \le \delta_2$. So

$$|f(x) - L| < \tfrac{1}{2}\epsilon \quad \text{whenever} \quad 0 < |x - a| < \delta$$

and

$$|g(x) - M| < \tfrac{1}{2}\epsilon \quad \text{whenever} \quad 0 < |x - a| < \delta$$

Hence

$$\begin{aligned} |[f(x) + g(x)] - (L + M)| &= |(f(x) - L) + (g(x) - M)| \\ &\le |f(x) - L| + |g(x) - M| \\ &< \tfrac{1}{2}\epsilon + \tfrac{1}{2}\epsilon = \epsilon \quad \text{whenever} \quad 0 < |x - a| < \delta \end{aligned}$$

In this way we have obtained statement (6), thereby proving that

$$\lim_{x \to a} [f(x) + g(x)] = L + M$$

The proof of Limit Theorem 4 using the minus sign is left as an exercise (see Exercise 42). ■

Limit Theorem 4 can be extended to any finite number of functions.

2.2.5 Limit Theorem 5

If $\lim_{x \to a} f_1(x) = L_1$, $\lim_{x \to a} f_2(x) = L_2$, . . . , and $\lim_{x \to a} f_n(x) = L_n$, then

$$\lim_{x \to a} [f_1(x) \pm f_2(x) \pm \cdots \pm f_n(x)] = L_1 \pm L_2 \pm \cdots \pm L_n$$

This theorem may be proved by applying Limit Theorem 4 and mathematical induction (see Exercise 43).

2.2.6 Limit Theorem 6

If $\lim_{x \to a} f(x) = L$ and $\lim_{x \to a} g(x) = M$, then

$$\lim_{x \to a} f(x) \cdot g(x) = L \cdot M$$

The proof of this theorem is more sophisticated than those of the preceding theorems, and it is often omitted from a beginning calculus text. The proof is outlined in Exercises 45 and 46.

● ILLUSTRATION 3 From Limit Theorem 3, $\lim_{x \to 3} x = 3$, and from Limit Theorem 1, $\lim_{x \to 3} (2x + 1) = 7$. Thus from Limit Theorem 6,

$$\begin{aligned} \lim_{x \to 3} x(2x + 1) &= \lim_{x \to 3} x \cdot \lim_{x \to 3} (2x + 1) \\ &= 3 \cdot 7 \\ &= 21 \end{aligned}$$ ●

Limit Theorem 6 also can be extended to any finite number of functions by applying mathematical induction.

2.2.7 Limit Theorem 7 If $\lim_{x \to a} f_1(x) = L_1$, $\lim_{x \to a} f_2(x) = L_2$, . . . , and $\lim_{x \to a} f_n(x) = L_n$, then

$$\lim_{x \to a} [f_1(x) f_2(x) \cdots f_n(x)] = L_1 L_2 \cdots L_n$$

The proof is left as an exercise (see Exercise 47).

2.2.8 Limit Theorem 8 If $\lim_{x \to a} f(x) = L$ and n is any positive integer, then

$$\lim_{x \to a} [f(x)]^n = L^n$$

The proof follows immediately from Limit Theorem 7 by taking

$$f(x) = f_1(x) = f_2(x) = \cdots = f_n(x) \quad \text{and} \quad L = L_1 = L_2 = \cdots = L_n$$

● ILLUSTRATION 4 From Limit Theorem 1, $\lim_{x \to -2} (5x + 7) = -3$. Therefore, from Limit Theorem 8 it follows that

$$\begin{aligned} \lim_{x \to -2} (5x + 7)^4 &= [\lim_{x \to -2} (5x + 7)]^4 \\ &= (-3)^4 \\ &= 81 \end{aligned}$$ ●

2.2.9 Limit Theorem 9 If $\lim_{x \to a} f(x) = L$ and $\lim_{x \to a} g(x) = M$, and $M \neq 0$, then

$$\lim_{x \to a} \frac{f(x)}{g(x)} = \frac{L}{M}$$

The proof, based on Definition 2.1.1, is frequently omitted from a beginning calculus text but is given in Sec. 2.7. Although the proof is postponed, this theorem is applied when necessary.

● ILLUSTRATION 5 From Limit Theorem 3, $\lim_{x \to 4} x = 4$, and from Limit Theorem 1, $\lim_{x \to 4} (-7x + 1) = -27$. Therefore from Limit Theorem 9,

$$\lim_{x \to 4} \frac{x}{-7x + 1} = \frac{\lim_{x \to 4} x}{\lim_{x \to 4} (-7x + 1)}$$

$$= \frac{4}{-27}$$

$$= -\frac{4}{27}$$ •

2.2.10 Limit Theorem 10 If $\lim_{x \to a} f(x) = L$, then

$$\lim_{x \to a} \sqrt[n]{f(x)} = \sqrt[n]{L}$$

if $L > 0$ and n is any positive integer, or if $L \leq 0$ and n is a positive odd integer.

A proof of this theorem is given in Sec. 2.7, and as with the preceding theorem, it is applied when necessary.

• ILLUSTRATION 6 From the result of Illustration 5 and Limit Theorem 10 it follows that

$$\lim_{x \to 4} \sqrt[3]{\frac{x}{-7x+1}} = \sqrt[3]{\lim_{x \to 4} \frac{x}{-7x+1}}$$

$$= \sqrt[3]{-\frac{4}{27}}$$

$$= -\frac{\sqrt[3]{4}}{3}$$ •

Following are some examples illustrating the application of the above theorems. To indicate the limit theorem being used we use the abbreviation "L.T." followed by the theorem number; for example, "L.T. 2" refers to Limit Theorem 2.

Example 1 Find $\lim_{x \to 3} (x^2 + 7x - 5)$, and, when applicable, indicate the limit theorems being used.

Solution

$$\lim_{x \to 3} (x^2 + 7x - 5) = \lim_{x \to 3} x^2 + \lim_{x \to 3} 7x - \lim_{x \to 3} 5 \qquad \text{(L.T. 5)}$$

$$= \lim_{x \to 3} x \cdot \lim_{x \to 3} x + \lim_{x \to 3} 7 \cdot \lim_{x \to 3} x - \lim_{x \to 3} 5 \qquad \text{(L.T. 6)}$$

$$= 3 \cdot 3 + 7 \cdot 3 - 5 \qquad \text{(L.T. 3 and L.T. 2)}$$

$$= 9 + 21 - 5$$

$$= 25$$

It is important at this point to realize that the limit in Example 1 was evaluated by direct application of the theorems on limits. For the function f defined by $f(x) = x^2 + 7x - 5$, note that $f(3) = 3^2 + 7 \cdot 3 - 5 = 25$, which is the same as $\lim_{x \to 3} (x^2 + 7x - 5)$. It is not always true that $\lim_{x \to a} f(x) = f(a)$ (see Example 4). In Example 1, $\lim_{x \to 3} f(x) = f(3)$ because the function f is continuous at $x = 3$. The meaning of continuous functions is discussed in Sec. 2.6.

Example 2 Find

$$\lim_{x \to 2} \sqrt{\frac{x^3 + 2x + 3}{x^2 + 5}}$$

and, when applicable, indicate the limit theorems being used.

Solution

$$\lim_{x \to 2} \sqrt{\frac{x^3 + 2x + 3}{x^2 + 5}} = \sqrt{\lim_{x \to 2} \frac{x^3 + 2x + 3}{x^2 + 5}} \qquad \text{(L.T. 10)}$$

$$= \sqrt{\frac{\lim_{x \to 2} (x^3 + 2x + 3)}{\lim_{x \to 2} (x^2 + 5)}} \qquad \text{(L.T. 9)}$$

$$= \sqrt{\frac{\lim_{x \to 2} x^3 + \lim_{x \to 2} 2x + \lim_{x \to 2} 3}{\lim_{x \to 2} x^2 + \lim_{x \to 2} 5}} \qquad \text{(L.T. 5)}$$

$$= \sqrt{\frac{(\lim_{x \to 2} x)^3 + \lim_{x \to 2} 2 \cdot \lim_{x \to 2} x + \lim_{x \to 2} 3}{(\lim_{x \to 2} x)^2 + \lim_{x \to 2} 5}} \qquad \text{(L.T. 6 and L.T. 8)}$$

$$= \sqrt{\frac{2^3 + 2 \cdot 2 + 3}{2^2 + 5}} \qquad \text{(L.T. 3 and L.T. 2)}$$

$$= \sqrt{\frac{8 + 4 + 3}{9}}$$

$$= \frac{\sqrt{15}}{3}$$

Example 3 Find

$$\lim_{x \to 3} \frac{x^3 - 27}{x - 3}$$

and, when applicable, indicate the limit theorems being used.

Solution Here there is a more difficult problem: Limit Theorem 9 cannot be applied to the quotient $(x^3 - 27)/(x - 3)$ because $\lim_{x \to 3} (x - 3) = 0$. However, factoring the numerator, we obtain

$$\frac{x^3 - 27}{x - 3} = \frac{(x - 3)(x^2 + 3x + 9)}{x - 3}$$

This quotient is $(x^2 + 3x + 9)$ if $x \neq 3$ (since if $x \neq 3$, the numerator and denominator can be divided by $x - 3$).

When evaluating $\lim_{x \to 3} [(x^3 - 27)/(x - 3)]$, we are considering values of x close to 3 but not equal to 3. Therefore it is possible to divide the numerator and denominator by $x - 3$. The solution to this problem takes the following form:

$$\lim_{x \to 3} \frac{x^3 - 27}{x - 3} = \lim_{x \to 3} \frac{(x - 3)(x^2 + 3x + 9)}{x - 3}$$

$= \lim_{x \to 3} (x^2 + 3x + 9)$ (dividing numerator and denominator by $x - 3$ because $x \neq 3$)

$= \lim_{x \to 3} x^2 + \lim_{x \to 3} (3x + 9)$ (L.T. 4)

$= (\lim_{x \to 3} x)^2 + 18$ (L.T. 8 and L.T. 1)

$= 3^2 + 18$ (L.T. 3)

$= 27$

Note that in Example 3 $(x^3 - 27)/(x - 3)$ is not defined when $x = 3$, but $\lim_{x \to 3} [(x^3 - 27)/(x - 3)]$ exists and is equal to 27.

Example 4 Given that f is the function defined by

$$f(x) = \begin{cases} x - 3 & \text{if } x \neq 4 \\ 5 & \text{if } x = 4 \end{cases}$$

find $\lim_{x \to 4} f(x)$.

Solution When evaluating $\lim_{x \to 4} f(x)$, we are considering values of x close to 4 but not equal to 4. Thus

$$\lim_{x \to 4} f(x) = \lim_{x \to 4} (x - 3)$$

$$= 1 \qquad \text{(L.T. 1)}$$

In Example 4 $\lim_{x \to 4} f(x) = 1$ but $f(4) = 5$; therefore $\lim_{x \to 4} f(x) \neq f(4)$. This is an example of a function that is discontinuous at $x = 4$. In terms of geometry, there is a break in the graph of the function at the point where $x = 4$ (see Fig. 2.2.1). The graph of the function consists of the isolated point (4, 5) and the straight line whose equation is $y = x - 3$, with the point (4, 1) deleted.

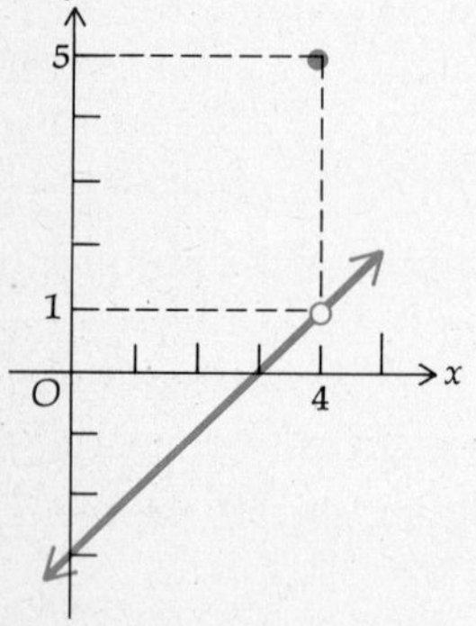

FIGURE 2.2.1

Example 5 Find

$$\lim_{x \to 4} \frac{\sqrt{x} - 2}{x - 4}$$

and, when applicable, indicate the limit theorems being used.

Solution As in Example 3, Limit Theorem 9 cannot be applied to the quotient $(\sqrt{x} - 2)/(x - 4)$ because $\lim_{x \to 4} (x - 4) = 0$. To simplify the quotient we rationalize the numerator by multiplying the numerator and denominator by $\sqrt{x} + 2$.

$$\frac{\sqrt{x} - 2}{x - 4} = \frac{(\sqrt{x} - 2)(\sqrt{x} + 2)}{(x - 4)(\sqrt{x} + 2)}$$

$$= \frac{x - 4}{(x - 4)(\sqrt{x} + 2)}$$

Because we are evaluating the limit as x approaches 4, we are considering values of x close to 4 but not equal to 4. Hence we can divide the numerator and denominator by $x - 4$. Therefore

$$\frac{\sqrt{x} - 2}{x - 4} = \frac{1}{\sqrt{x} + 2} \qquad \text{if } x \neq 4$$

The solution is as follows:

$$\lim_{x \to 4} \frac{\sqrt{x} - 2}{x - 4} = \lim_{x \to 4} \frac{(\sqrt{x} - 2)(\sqrt{x} + 2)}{(x - 4)(\sqrt{x} + 2)}$$

$$= \lim_{x \to 4} \frac{x - 4}{(x - 4)(\sqrt{x} + 2)}$$

$$= \lim_{x \to 4} \frac{1}{\sqrt{x} + 2} \quad \text{(dividing numerator and denominator by } x - 4 \text{ because } x \neq 4)$$

$$= \frac{\lim_{x \to 4} 1}{\lim_{x \to 4} (\sqrt{x} + 2)} \qquad \text{(L.T. 9)}$$

$$= \frac{1}{\lim_{x \to 4} \sqrt{x} + \lim_{x \to 4} 2} \qquad \text{(L.T. 2 and L.T. 4)}$$

$$= \frac{1}{\sqrt{\lim_{x \to 4} x} + 2} \qquad \text{(L.T. 10 and L.T. 2)}$$

$$= \frac{1}{\sqrt{4} + 2} \qquad \text{(L.T. 3)}$$

$$= \tfrac{1}{4}$$

To prove Limit Theorem 9 in Sec. 2.7 we need the following theorem. It is the special case of Limit Theorem 9 where $f(x) = 1$ and $g(x) = x$.

2.2.11 Theorem If a is any number except 0,

$$\lim_{x \to a} \frac{1}{x} = \frac{1}{a}$$

Proof Two cases need to be considered: $a > 0$ and $a < 0$. We prove the theorem if $a > 0$ and leave the proof for $a < 0$ as an exercise (see Exercise 48).

Because $1/x$ is defined for every x except zero, the open interval I required by Definition 2.1.1 can be any open interval containing a but not containing 0.

Considering $a > 0$, we must show that for any $\epsilon > 0$ there exists a $\delta > 0$ such that

$$\left|\frac{1}{x} - \frac{1}{a}\right| < \epsilon \quad \text{whenever} \quad 0 < |x - a| < \delta \tag{6}$$

Now

$$\left|\frac{1}{x} - \frac{1}{a}\right| = \left|\frac{a - x}{ax}\right| = \frac{|x - a|}{|a||x|}$$

and because $a > 0$,

$$\left|\frac{1}{x} - \frac{1}{a}\right| = |x - a| \cdot \frac{1}{a|x|}$$

Therefore statement (6) is equivalent to

$$|x - a| \cdot \frac{1}{a|x|} < \epsilon \quad \text{whenever} \quad 0 < |x - a| < \delta$$

$$|x - a| < a|x|\epsilon \quad \text{whenever} \quad 0 < |x - a| < \delta \tag{7}$$

Because we want $|x - a| < a|x|\epsilon$, we choose an open interval I required by Definition 2.1.1 that will determine a lower bound of $a|x|\epsilon$ for values of x in I. Because I must contain a and must not contain 0, choose I as the open interval $(\frac{1}{2}a, \frac{3}{2}a)$, which has a as its center and a radius of $\frac{1}{2}a$. For all $x \in I$, $|x| = x$, and

$$ax\epsilon > a \cdot \tfrac{1}{2}a \cdot \epsilon = \tfrac{1}{2}a^2\epsilon$$

Therefore $\frac{1}{2}a^2\epsilon$ is a lower bound of $a|x|\epsilon$ for values of x in I. Hence δ is taken as the smaller of this lower bound and the radius of I; that is, $\delta = \min(\frac{1}{2}a^2\epsilon, \frac{1}{2}a)$. Then

$$|x - a| < \tfrac{1}{2}a^2\epsilon < a|x|\epsilon \quad \text{whenever} \quad 0 < |x - a| < \delta$$

if $\delta = \min(\frac{1}{2}a^2\epsilon, \frac{1}{2}a)$. For this value of δ, statement (7) and the equivalent statement (6) hold for all $x \in I$. We have therefore proved that $\lim_{x \to a} (1/x) = 1/a$ if a is any positive number. ■

The proof of Limit Theorem 10 in Sec. 2.7 also utilizes a theorem that is a special case. This special case is given as Theorem 2.2.12, and its proof makes use of the following formula, where n is any positive integer:

$$a^n - b^n = (a - b)(a^{n-1} + a^{n-2}b + a^{n-3}b^2 + \cdots + a^2b^{n-3} + ab^{n-2} + b^{n-1}) \tag{8}$$

The proof of (8) follows.

$$a(a^{n-1} + a^{n-2}b + \cdots + ab^{n-2} + b^{n-1}) = a^n + a^{n-1}b + \cdots + a^2b^{n-2} + ab^{n-1} \tag{9}$$

$$b(a^{n-1} + a^{n-2}b + \cdots + ab^{n-2} + b^{n-1}) = a^{n-1}b + \cdots + a^2b^{n-2} + ab^{n-1} + b^n \tag{10}$$

Subtracting terms of (10) from those of (9) gives (8).

2.2.12 Theorem If n is a positive integer, then

$$\lim_{x \to a} \sqrt[n]{x} = \sqrt[n]{a}$$

if either

(i) a is any positive number,

or (ii) a is a negative number or zero and n is odd.

Proof We prove part (i) and leave the proof of part (ii) as an exercise (see Exercise 49).

Let a be any positive number. Because $\sqrt{x}$ is defined for every nonnegative number, the open interval I required by Definition 2.1.1 can be any open interval containing a and having a nonnegative number as its left endpoint.

We must show that for any $\epsilon > 0$ there exists a $\delta > 0$ such that

$$|\sqrt[n]{x} - \sqrt[n]{a}| < \epsilon \quad \text{whenever} \quad 0 < |x - a| < \delta \tag{11}$$

To express $|\sqrt[n]{x} - \sqrt[n]{a}|$ in terms of $|x - a|$ we use (8).

$$|\sqrt[n]{x} - \sqrt[n]{a}| = \left|\frac{(x^{1/n} - a^{1/n})[(x^{1/n})^{n-1} + (x^{1/n})^{n-2}a^{1/n} + \cdots + x^{1/n}(a^{1/n})^{n-2} + (a^{1/n})^{n-1}]}{(x^{1/n})^{n-1} + (x^{1/n})^{n-2}a^{1/n} + \cdots + x^{1/n}(a^{1/n})^{n-2} + (a^{1/n})^{n-1}}\right|$$

If (8) is applied to the numerator,

$$|\sqrt[n]{x} - \sqrt[n]{a}| = |x - a| \cdot \frac{1}{|x^{(n-1)/n} + x^{(n-2)/n}a^{1/n} + \cdots + x^{1/n}a^{(n-2)/n} + a^{(n-1)/n}|}$$

Therefore the inequality $|\sqrt[n]{x} - \sqrt[n]{a}| < \epsilon$ is equivalent to

$$|x - a| \cdot \frac{1}{|x^{(n-1)/n} + x^{(n-2)/n}a^{1/n} + \cdots + x^{1/n}a^{(n-2)/n} + a^{(n-1)/n}|} < \epsilon$$

$$|x - a| < |x^{(n-1)/n} + x^{(n-2)/n}a^{1/n} + \cdots + x^{1/n}a^{(n-2)/n} + a^{(n-1)/n}| \cdot \epsilon \tag{12}$$

Because we want inequality (12) to hold, we choose an open interval I required by Definition 2.1.1 that will determine a lower bound of

$$|x^{(n-1)/n} + x^{(n-2)/n}a^{1/n} + \cdots + x^{1/n}a^{(n-2)/n} + a^{(n-1)/n}|\epsilon$$

for values of x in I. Choose I as the open interval $(0, 2a)$, having a as its center and a as its radius. Then for all x in I,

$$|x^{(n-1)/n} + x^{(n-2)/n}a^{1/n} + \cdots + x^{1/n}a^{(n-2)/n} + a^{(n-1)/n}|\epsilon > a^{(n-1)/n}\epsilon$$

Therefore $a^{(n-1)/n}\epsilon$ is a lower bound for

$$|x^{(n-1)/n} + x^{(n-2)/n}a^{1/n} + \cdots + x^{1/n}a^{(n-2)/n} + a^{(n-1)/n}|\epsilon$$

for values of x in I. Hence δ is taken as the smaller of this lower bound and the radius of I; that is, $\delta = \min(a^{(n-1)/n}\epsilon, a)$. Then

$$|x - a| < a^{(n-1)/n}\epsilon < |x^{(n-1)n} + x^{(n-2)/n}a^{1/n} + \cdots + x^{1/n}a^{(n-2)/n} + a^{(n-1)/n}|\epsilon$$

whenever $0 < |x - a| < \delta$ if $\delta = \min(a^{(n-1)/n}\epsilon, a)$. For this value of δ, inequality (12) and thus statement (11) hold for all $x \in I$. Therefore we have proved that $\lim\limits_{x \to a} \sqrt[n]{x} = \sqrt[n]{a}$ if a is any positive number. ■

● ILLUSTRATION 7 From part (i) of Theorem 2.2.12 it follows that $\lim\limits_{x \to 5} \sqrt{x} = \sqrt{5}$ and $\lim\limits_{x \to 16} \sqrt[4]{x} = 2$; from part (ii) it follows that $\lim\limits_{x \to -27} \sqrt[3]{x} = -3$. A limit such as $\lim\limits_{x \to -4} \sqrt{x}$ is not considered because $\sqrt{x}$ is not defined on any open interval containing -4. ●

EXERCISES 2.2

In Exercises 1 through 35, find the value of the limit, and, when applicable, indicate the limit theorems being used.

1. $\lim_{x \to 5} (3x - 7)$

2. $\lim_{x \to -4} (5x + 2)$

3. $\lim_{x \to 2} (x^2 + 2x - 1)$

4. $\lim_{x \to 3} (2x^2 - 4x + 5)$

5. $\lim_{z \to -2} (z^3 + 8)$

6. $\lim_{y \to -1} (y^3 - 2y^2 + 3y - 4)$

7. $\lim_{x \to 3} \dfrac{4x - 5}{5x - 1}$

8. $\lim_{x \to 2} \dfrac{3x + 4}{8x - 1}$

9. $\lim_{t \to 2} \dfrac{t^2 - 5}{2t^3 + 6}$

10. $\lim_{x \to -1} \dfrac{2x + 1}{x^2 - 3x + 4}$

11. $\lim_{x \to 2} \dfrac{x - 2}{x^2 - 4}$

12. $\lim_{z \to -5} \dfrac{z^2 - 25}{z + 5}$

13. $\lim_{x \to -3/2} \dfrac{4x^2 - 9}{2x + 3}$

14. $\lim_{x \to 1/3} \dfrac{3x - 1}{9x^2 - 1}$

15. $\lim_{x \to -3} \dfrac{x^2 + 5x + 6}{x^2 - x - 12}$

16. $\lim_{t \to -2} \dfrac{2 - 3t - 2t^2}{16 + 6t - t^2}$

17. $\lim_{s \to 4} \dfrac{3s^2 - 8s - 16}{2s^2 - 9s + 4}$

18. $\lim_{x \to 4} \dfrac{3x^2 - 17x + 20}{4x^2 - 25x + 36}$

19. $\lim_{y \to -2} \dfrac{y^3 + 8}{y + 2}$

20. $\lim_{s \to 1} \dfrac{s^3 - 1}{s - 1}$

21. $\lim_{r \to 1} \sqrt{\dfrac{8r + 1}{r + 3}}$

22. $\lim_{x \to -3} \sqrt[3]{\dfrac{5 + 2x}{5 - x}}$

23. $\lim_{x \to 4} \sqrt[3]{\dfrac{x^2 - 3x + 4}{2x^2 - x - 1}}$

24. $\lim_{x \to 2} \sqrt{\dfrac{x^2 + 3x + 4}{x^3 + 1}}$

25. $\lim_{y \to -3} \sqrt{\dfrac{y^2 - 9}{2y^2 + 7y + 3}}$

26. $\lim_{t \to 3/2} \sqrt{\dfrac{8t^3 - 27}{4t^2 - 9}}$

27. $\lim_{x \to 9} \dfrac{3 - \sqrt{x}}{9 - x}$

28. $\lim_{t \to 0} \dfrac{2 - \sqrt{4 - t}}{t}$

29. $\lim_{x \to 0} \dfrac{\sqrt{x + 2} - \sqrt{2}}{x}$

30. $\lim_{x \to 1} \dfrac{\sqrt[3]{x} - 1}{x - 1}$

31. $\lim_{h \to 0} \dfrac{\sqrt[3]{h + 1} - 1}{h}$

32. $\lim_{x \to -2} \dfrac{x^3 - x^2 - x + 10}{x^2 + 3x + 2}$

33. $\lim_{x \to -1} \dfrac{2x^2 - x - 3}{x^3 + 2x^2 + 6x + 5}$

34. $\lim_{y \to 4} \dfrac{2y^3 - 11y^2 + 10y + 8}{3y^3 - 17y^2 + 16y + 16}$

35. $\lim_{x \to 3} \dfrac{2x^3 - 5x^2 - 2x - 3}{4x^3 - 13x^2 + 4x - 3}$

36. If $f(x) = x^2 + 5x - 3$, show that $\lim_{x \to 2} f(x) = f(2)$.

37. If $F(x) = 2x^3 + 7x - 1$, show that $\lim_{x \to -1} F(x) = F(-1)$.

38. If $g(x) = \dfrac{x^2 - 16}{x - 4}$, show that $\lim_{x \to 4} g(x) = 8$ but that $g(4)$ is not defined.

39. If $h(x) = \dfrac{\sqrt{x + 9} - 3}{x}$, show that $\lim_{x \to 0} h(x) = \frac{1}{6}$, but that $h(0)$ is not defined.

40. Given that f is the function defined by

$$f(x) = \begin{cases} 2x - 1 & \text{if } x \neq 2 \\ 1 & \text{if } x = 2 \end{cases}$$

(a) Find $\lim_{x \to 2} f(x)$, and show that $\lim_{x \to 2} f(x) \neq f(2)$. (b) Draw a sketch of the graph of f.

41. Given that f is the function defined by

$$f(x) = \begin{cases} x^2 - 9 & \text{if } x \neq -3 \\ 4 & \text{if } x = -3 \end{cases}$$

(a) Find $\lim_{x \to -3} f(x)$, and show that $\lim_{x \to -3} f(x) \neq f(-3)$. (b) Draw a sketch of the graph of f.

42. Use Definition 2.1.1 to prove that if

$$\lim_{x\to a} f(x) = L \quad \text{and} \quad \lim_{x\to a} g(x) = M$$

then

$$\lim_{x\to a} [f(x) - g(x)] = L - M$$

43. Prove Limit Theorem 5 by applying Limit Theorem 4 and mathematical induction.

44. Prove that if $\lim_{x\to a} f(x) = L$, then $\lim_{x\to a} (f(x) - L) = 0$.

45. Use Definition 2.1.1 to prove that if

$$\lim_{x\to a} f(x) = L \quad \text{and} \quad \lim_{x\to a} g(x) = 0$$

then

$$\lim_{x\to a} f(x) \cdot g(x) = 0$$

(*Hint:* In order to prove that $\lim_{x\to a} f(x) \cdot g(x) = 0$, we must show that for any $\epsilon > 0$ there exists a $\delta > 0$ such that $|f(x) \cdot g(x)| < \epsilon$ whenever $0 < |x - a| < \delta$. First show that there is a $\delta_1 > 0$ such that $|f(x)| < 1 + |L|$ whenever $0 < |x - a| < \delta_1$, by applying Definition 2.1.1 to $\lim_{x\to a} f(x) = L$, with $\epsilon = 1$ and $\delta = \delta_1$, and then use the triangle inequality. Then show that there is a $\delta_2 > 0$ such that $|g(x)| < \epsilon/(1 + |L|)$ whenever $0 < |x - a| < \delta_2$, by applying Definition 2.1.1 to $\lim_{x\to a} g(x) = 0$. By taking δ as the smaller of the two numbers δ_1 and δ_2, the theorem is proved.)

46. Prove Limit Theorem 6: If $\lim_{x\to a} f(x) = L$ and $\lim_{x\to a} g(x) = M$, then

$$\lim_{x\to a} [f(x) \cdot g(x)] = L \cdot M$$

(*Hint:* Write $f(x) \cdot g(x) = [f(x) - L]g(x) + L[g(x) - M] + L \cdot M$. Apply Limit Theorem 5 and the results of Exercises 44 and 45.)

47. Prove Limit Theorem 7 by applying Limit Theorem 6 and mathematical induction.

48. Prove Theorem 2.2.11 if $a < 0$.

49. Prove Theorem 2.2.12(ii).

2.3 ONE-SIDED LIMITS

When considering $\lim_{x\to a} f(x)$, we are concerned with values of x in an open interval containing a but not at a itself, that is, at values of x close to a and either greater than a or less than a. However, suppose that we have the function f for which, say, $f(x) = \sqrt{x - 4}$. Because $f(x)$ does not exist if $x < 4$, f is not defined on any open interval containing 4. Hence $\lim_{x\to 4} \sqrt{x - 4}$ has no meaning. However, if x is restricted to values greater than 4, the value of $\sqrt{x - 4}$ can be made as close to 0 as we please by taking x sufficiently close to 4 but greater than 4. In such a case, we let x approach 4 from the right and consider the **one-sided limit from the right,** or the **right-hand limit,** which is now defined.

2.3.1 Definition Let f be a function which is defined at every number in some open interval (a, c). Then the **limit of $f(x)$, as x approaches a from the right, is L,** written

$$\lim_{x\to a^+} f(x) = L$$

if for any $\epsilon > 0$, however small, there exists a $\delta > 0$ such that

$$|f(x) - L| < \epsilon \quad \text{whenever} \quad 0 < x - a < \delta \tag{1}$$

Note that in statement (1) there are no absolute-value bars around $x - a$ since $x - a > 0$, because $x > a$.

It follows from Definition 2.3.1 that

$$\lim_{x \to 4^+} \sqrt{x - 4} = 0$$

If when considering the limit of a function the independent variable x is restricted to values less than a number a, we say that x approaches a from the left; the limit is called the **one-sided limit from the left,** or the **left-hand limit.**

2.3.2 Definition Let f be a function which is defined at every number in some open interval (d, a). Then the **limit of $f(x)$, as x approaches a from the left, is L,** written

$$\lim_{x \to a^-} f(x) = L$$

if for any $\epsilon > 0$, however small, there exists a $\delta > 0$ such that

$$|f(x) - L| < \epsilon \quad \text{whenever} \quad 0 < a - x < \delta$$

We can refer to $\lim_{x \to a} f(x)$ as the **two-sided limit,** or the **undirected limit,** to distinguish it from the one-sided limits.

Limit Theorems 1–10 given in Sec. 2.2 remain unchanged when "$x \to a$" is replaced by "$x \to a^+$" or "$x \to a^-$."

Example 1 Let f be defined by

$$f(x) = \operatorname{sgn} x = \begin{cases} -1 & \text{if } x < 0 \\ 0 & \text{if } x = 0 \\ 1 & \text{if } 0 < x \end{cases}$$

(a) Draw a sketch of the graph of f. (b) Determine $\lim_{x \to 0^-} f(x)$ if it exists. (c) Determine $\lim_{x \to 0^+} f(x)$ if it exists.

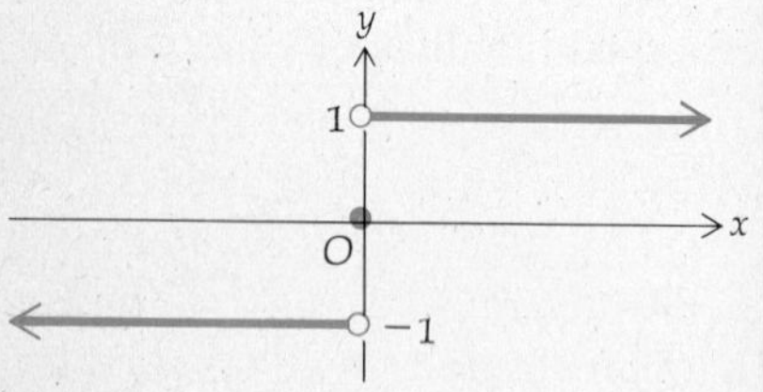

FIGURE 2.3.1

Solution A sketch of the graph is shown in Fig. 2.3.1. $\lim_{x \to 0^-} f(x) = -1$ since, if x is any number less than 0, $f(x)$ has the value -1. Similarly, $\lim_{x \to 0^+} f(x) = +1$.

In the above example, $\lim_{x \to 0^-} f(x) \neq \lim_{x \to 0^+} f(x)$. Because the left-hand limit and the right-hand limit are not equal, the two-sided limit, $\lim_{x \to 0} f(x)$, does not exist. The concept of the two-sided limit failing to exist because the two one-sided limits are unequal is stated in the following theorem.

2.3.3 Theorem $\lim_{x \to a} f(x)$ exists and is equal to L if and only if $\lim_{x \to a^-} f(x)$ and $\lim_{x \to a^+} f(x)$ both exist and both are equal to L.

The proof of this theorem is left as an exercise (see Exercise 34).

Example 2 Let g be defined by

$$g(x) = \begin{cases} |x| & \text{if } x \neq 0 \\ 2 & \text{if } x = 0 \end{cases}$$

(a) Draw a sketch of the graph of g. (b) Find $\lim_{x \to 0} g(x)$ if it exists.

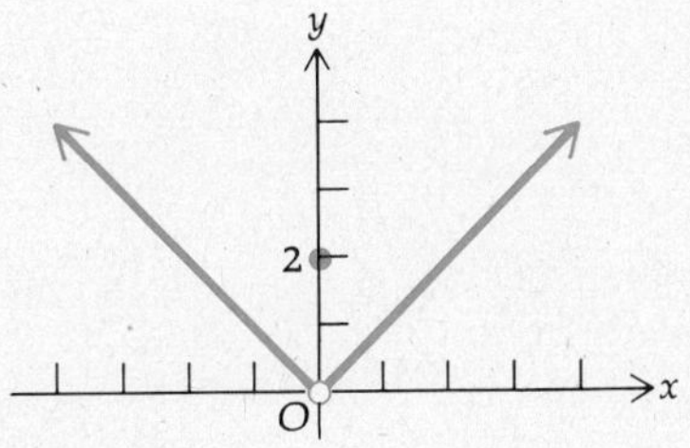

FIGURE 2.3.2

Solution A sketch of the graph is shown in Fig. 2.3.2.

$$\lim_{x \to 0^-} g(x) = \lim_{x \to 0^-} (-x) = 0 \quad \text{and} \quad \lim_{x \to 0^+} g(x) = \lim_{x \to 0^+} x = 0$$

Therefore by Theorem 2.3.3, $\lim_{x \to 0} g(x)$ exists and is equal to 0. Note that $g(0) = 2$, which has no effect on $\lim_{x \to 0} g(x)$.

Example 3 Let h be defined by

$$h(x) = \begin{cases} 4 - x^2 & \text{if } x \leq 1 \\ 2 + x^2 & \text{if } 1 < x \end{cases}$$

(a) Draw a sketch of the graph of h. (b) Find each of the following limits if they exist: $\lim_{x \to 1^-} h(x)$, $\lim_{x \to 1^+} h(x)$, $\lim_{x \to 1} h(x)$.

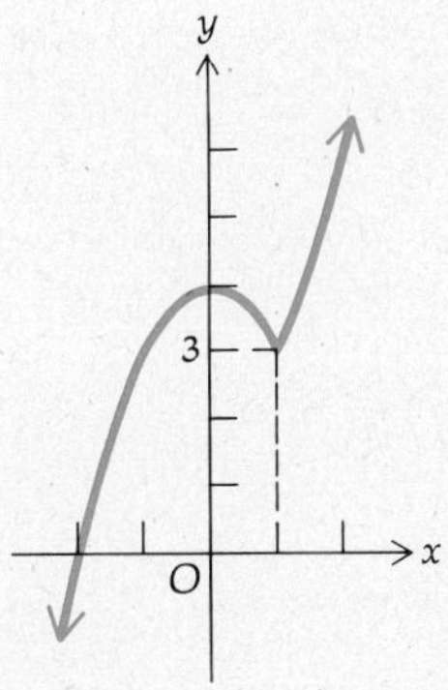

FIGURE 2.3.3

Solution A sketch of the graph is shown in Fig. 2.3.3.

$$\lim_{x \to 1^-} h(x) = \lim_{x \to 1^-} (4 - x^2) = 3$$

$$\lim_{x \to 1^+} h(x) = \lim_{x \to 1^+} (2 + x^2) = 3$$

Therefore by Theorem 2.3.3, $\lim_{x \to 1} h(x)$ exists and is equal to 3. Note that $h(1) = 3$.

Example 4 Let f be defined by

$$f(x) = \begin{cases} x + 5 & \text{if } x < -3 \\ \sqrt{9 - x^2} & \text{if } -3 \leq x \leq 3 \\ 5 - x & \text{if } 3 < x \end{cases}$$

(a) Draw a sketch of the graph of f. (b) Find, if they exist, each of the following limits: $\lim_{x \to -3^-} f(x)$, $\lim_{x \to -3^+} f(x)$, $\lim_{x \to 3^-} f(x)$, $\lim_{x \to 3^+} f(x)$.

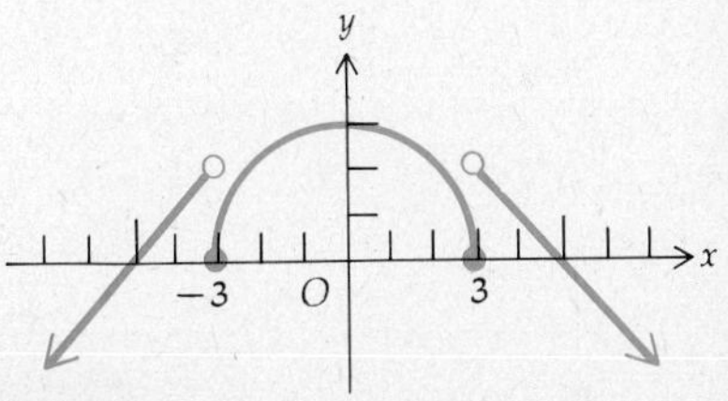

FIGURE 2.3.4

Solution A sketch of the graph of f is shown in Fig. 2.3.4.

$$\lim_{x \to -3^-} f(x) = \lim_{x \to -3^-} (x + 5) = 2$$

$$\lim_{x \to -3^+} f(x) = \lim_{x \to -3^+} \sqrt{9 - x^2} = 0$$

$$\lim_{x \to 3^-} f(x) = \lim_{x \to 3^-} \sqrt{9 - x^2} = 0$$

$$\lim_{x \to 3^+} f(x) = \lim_{x \to 3^+} (5 - x) = 2$$

EXERCISES 2.3

In Exercises 1 through 22, draw a sketch of the graph, and find the indicated limit if it exists; if the limit does not exist, give the reason.

1. $f(x) = \begin{cases} 2 & \text{if } x < 1 \\ -1 & \text{if } x = 1 \\ -3 & \text{if } 1 < x \end{cases}$; (a) $\lim_{x \to 1^+} f(x)$; (b) $\lim_{x \to 1^-} f(x)$; (c) $\lim_{x \to 1} f(x)$

2. $f(x) = \begin{cases} -2 & \text{if } x < 0 \\ 2 & \text{if } 0 \le x \end{cases}$; (a) $\lim_{x \to 0^+} f(x)$; (b) $\lim_{x \to 0^-} f(x)$; (c) $\lim_{x \to 0} f(x)$

3. $f(t) = \begin{cases} t + 4 & \text{if } t \le -4 \\ 4 - t & \text{if } -4 < t \end{cases}$; (a) $\lim_{t \to -4^+} f(t)$; (b) $\lim_{t \to -4^-} f(t)$; (c) $\lim_{t \to -4} f(t)$

4. $g(s) = \begin{cases} s + 3 & \text{if } s \le -2 \\ 3 - s & \text{if } -2 < s \end{cases}$; (a) $\lim_{s \to -2^+} g(s)$; (b) $\lim_{s \to -2^-} g(s)$; (c) $\lim_{s \to -2} g(s)$

5. $F(x) = \begin{cases} x^2 & \text{if } x \le 2 \\ 8 - 2x & \text{if } 2 < x \end{cases}$; (a) $\lim_{x \to 2^+} F(x)$; (b) $\lim_{x \to 2^-} F(x)$; (c) $\lim_{x \to 2} F(x)$

6. $h(x) = \begin{cases} 2x + 1 & \text{if } x < 3 \\ 10 - x & \text{if } 3 \le x \end{cases}$; (a) $\lim_{x \to 3^+} h(x)$; (b) $\lim_{x \to 3^-} h(x)$; (c) $\lim_{x \to 3} h(x)$

7. $g(r) = \begin{cases} 2r + 3 & \text{if } r < 1 \\ 2 & \text{if } r = 1 \\ 7 - 2r & \text{if } 1 < r \end{cases}$; (a) $\lim_{r \to 1^+} g(r)$; (b) $\lim_{r \to 1^-} g(r)$; (c) $\lim_{r \to 1} g(r)$

8. $g(t) = \begin{cases} 3 + t^2 & \text{if } t < -2 \\ 0 & \text{if } t = -2 \\ 11 - t^2 & \text{if } -2 < t \end{cases}$; (a) $\lim_{t \to -2^+} g(t)$; (b) $\lim_{t \to -2^-} g(t)$; (c) $\lim_{t \to -2} g(t)$

9. $f(x) = \begin{cases} x^2 - 4 & \text{if } x < 2 \\ 4 & \text{if } x = 2 \\ 4 - x^2 & \text{if } 2 < x \end{cases}$; (a) $\lim_{x \to 2^+} f(x)$; (b) $\lim_{x \to 2^-} f(x)$; (c) $\lim_{x \to 2} f(x)$

10. $f(x) = \begin{cases} 2x + 3 & \text{if } x < 1 \\ 4 & \text{if } x = 1 \\ x^2 + 2 & \text{if } 1 < x \end{cases}$; (a) $\lim_{x \to 1^+} f(x)$; (b) $\lim_{x \to 1^-} f(x)$; (c) $\lim_{x \to 1} f(x)$

11. $F(x) = |x - 5|$; (a) $\lim_{x \to 5^+} F(x)$; (b) $\lim_{x \to 5^-} F(x)$; (c) $\lim_{x \to 5} F(x)$

12. $f(x) = 3 + |2x - 4|$; (a) $\lim_{x \to 2^+} f(x)$; (b) $\lim_{x \to 2^-} f(x)$; (c) $\lim_{x \to 2} f(x)$

13. $G(x) = |2x - 3| - 4$; (a) $\lim_{x \to 3/2^+} G(x)$; (b) $\lim_{x \to 3/2^-} G(x)$; (c) $\lim_{x \to 3/2} G(x)$

14. $F(x) = \begin{cases} |x - 1| & \text{if } x < -1 \\ 0 & \text{if } x = -1 \\ |1 - x| & \text{if } -1 < x \end{cases}$; (a) $\lim_{x \to -1^+} F(x)$; (b) $\lim_{x \to -1^-} F(x)$; (c) $\lim_{x \to -1} F(x)$

15. $f(x) = \dfrac{|x|}{x}$; (a) $\lim_{x \to 0^+} f(x)$; (b) $\lim_{x \to 0^-} f(x)$; (c) $\lim_{x \to 0} f(x)$

16. $S(x) = |\text{sgn } x|$ (sgn x is defined in Exercise 32 of Exercises 1.6); (a) $\lim_{x \to 0^+} S(x)$; (b) $\lim_{x \to 0^-} S(x)$; (c) $\lim_{x \to 0} S(x)$

17. $f(x) = \begin{cases} 2 & \text{if } x < -2 \\ \sqrt{4 - x^2} & \text{if } -2 \le x \le 2 \\ -2 & \text{if } 2 < x \end{cases}$; (a) $\lim_{x \to -2^-} f(x)$; (b) $\lim_{x \to -2^+} f(x)$; (c) $\lim_{x \to 2^-} f(x)$; (d) $\lim_{x \to 2^+} f(x)$

18. $f(x) = \begin{cases} x+1 & \text{if } x < -1 \\ x^2 & \text{if } -1 \le x \le 1 \\ 1-x & \text{if } 1 < x \end{cases}$; (a) $\lim\limits_{x \to -1^-} f(x)$; (b) $\lim\limits_{x \to -1^+} f(x)$; (c) $\lim\limits_{x \to 1^-} f(x)$; (d) $\lim\limits_{x \to 1^+} f(x)$

19. $f(t) = \begin{cases} \sqrt[3]{t} & \text{if } t < 0 \\ \sqrt{t} & \text{if } 0 \le t \end{cases}$; (a) $\lim\limits_{t \to 0^+} f(t)$; (b) $\lim\limits_{t \to 0^-} f(t)$; (c) $\lim\limits_{t \to 0} f(t)$

20. $g(x) = \begin{cases} \sqrt[3]{-x} & \text{if } x \le 0 \\ \sqrt[3]{x} & \text{if } 0 < x \end{cases}$; (a) $\lim\limits_{x \to 0^+} g(x)$; (b) $\lim\limits_{x \to 0^-} g(x)$; (c) $\lim\limits_{x \to 0} g(x)$

21. $F(x) = \begin{cases} \sqrt{x^2-9} & \text{if } x \le -3 \\ \sqrt{9-x^2} & \text{if } -3 < x < 3 \\ \sqrt{x^2-9} & \text{if } 3 \le x \end{cases}$; (a) $\lim\limits_{x \to -3^-} F(x)$; (b) $\lim\limits_{x \to -3^+} F(x)$; (c) $\lim\limits_{x \to -3} F(x)$; (d) $\lim\limits_{x \to 3^-} F(x)$; (e) $\lim\limits_{x \to 3^+} F(x)$; (f) $\lim\limits_{x \to 3} F(x)$

22. $G(t) = \begin{cases} \sqrt[3]{t+1} & \text{if } t \le -1 \\ \sqrt{1-t^2} & \text{if } -1 < t < 1 \\ \sqrt[3]{t-1} & \text{if } 1 \le t \end{cases}$; (a) $\lim\limits_{t \to -1^-} G(t)$; (b) $\lim\limits_{t \to -1^+} G(t)$; (c) $\lim\limits_{t \to -1} G(t)$; (d) $\lim\limits_{t \to 1^-} G(t)$; (e) $\lim\limits_{t \to 1^+} G(t)$; (f) $\lim\limits_{t \to 1} G(t)$

23. F is the function of Exercise 42 in Exercises 1.6. Find, if they exist: (a) $\lim\limits_{x \to 0^+} F(x)$; (b) $\lim\limits_{x \to 0^-} F(x)$; (c) $\lim\limits_{x \to 0} F(x)$.

24. h is the function of Exercise 43 in Exercises 1.6. Find, if they exist: (a) $\lim\limits_{x \to 0^+} h(x)$; (b) $\lim\limits_{x \to 0^-} h(x)$; (c) $\lim\limits_{x \to 0} h(x)$.

25. Find, if they exist: (a) $\lim\limits_{x \to 2^+} [\![x]\!]$; (b) $\lim\limits_{x \to 2^-} [\![x]\!]$; (c) $\lim\limits_{x \to 2} [\![x]\!]$.

26. Find, if they exist: (a) $\lim\limits_{x \to 4^+} [\![x-3]\!]$; (b) $\lim\limits_{x \to 4^-} [\![x-3]\!]$; (c) $\lim\limits_{x \to 4} [\![x-3]\!]$.

27. Let $h(x) = (x-1)\operatorname{sgn} x$. Draw a sketch of the graph of h. Find, if they exist: (a) $\lim\limits_{x \to 0^+} h(x)$; (b) $\lim\limits_{x \to 0^-} h(x)$; (c) $\lim\limits_{x \to 0} h(x)$.

28. Let $G(x) = [\![x]\!] + [\![4-x]\!]$. Draw a sketch of the graph of G. Find, if they exist: (a) $\lim\limits_{x \to 3^+} G(x)$; (b) $\lim\limits_{x \to 3^-} G(x)$; (c) $\lim\limits_{x \to 3} G(x)$.

29. Given $f(x) = \begin{cases} 3x+2 & \text{if } x < 4 \\ 5x+k & \text{if } 4 \le x \end{cases}$. Find the value of k such that $\lim\limits_{x \to 4} f(x)$ exists.

30. Given $f(x) = \begin{cases} kx-3 & \text{if } x \le -1 \\ x^2+k & \text{if } -1 < x \end{cases}$. Find the value of k such that $\lim\limits_{x \to -1} f(x)$ exists.

31. Given $f(x) = \begin{cases} x^2 & \text{if } x \le -2 \\ ax+b & \text{if } -2 < x < 2 \\ 2x-6 & \text{if } 2 \le x \end{cases}$. Find the values of a and b such that $\lim\limits_{x \to -2} f(x)$ and $\lim\limits_{x \to 2} f(x)$ both exist.

32. Given $f(x) = \begin{cases} 2x-a & \text{if } x < -3 \\ ax+2b & \text{if } -3 \le x \le 3 \\ b-5x & \text{if } 3 < x \end{cases}$. Find the values of a and b such that $\lim\limits_{x \to -3} f(x)$ and $\lim\limits_{x \to 3} f(x)$ both exist.

33. Let $f(x) = \begin{cases} -1 & \text{if } x < 0 \\ 1 & \text{if } x > 0 \end{cases}$

Show that $\lim\limits_{x \to 0} f(x)$ does not exist but that $\lim\limits_{x \to 0} |f(x)|$ does exist.

34. Prove Theorem 2.3.3.

35. Given $f(x) = \begin{cases} x^2+3 & \text{if } x \le 1 \\ x+1 & \text{if } x > 1 \end{cases}$ and $g(x) = \begin{cases} x^2 & \text{if } x \le 1 \\ 2 & \text{if } x > 1 \end{cases}$

(a) Show that $\lim\limits_{x \to 1^-} f(x)$ and $\lim\limits_{x \to 1^+} f(x)$ both exist but are not equal, and hence $\lim\limits_{x \to 1} f(x)$ does not exist.
(b) Show that $\lim\limits_{x \to 1^-} g(x)$ and $\lim\limits_{x \to 1^+} g(x)$ both exist but are not equal, and hence $\lim\limits_{x \to 1} g(x)$ does not exist.
(c) Find formulas for $f(x) \cdot g(x)$.
(d) Prove that $\lim\limits_{x \to 1} f(x) \cdot g(x)$ exists by showing that $\lim\limits_{x \to 1^-} f(x) \cdot g(x) = \lim\limits_{x \to 1^+} f(x) \cdot g(x)$.

2.4 INFINITE LIMITS

Let f be the function defined by

$$f(x) = \frac{3}{(x-2)^2}$$

A sketch of the graph of this function is in Fig. 2.4.1.

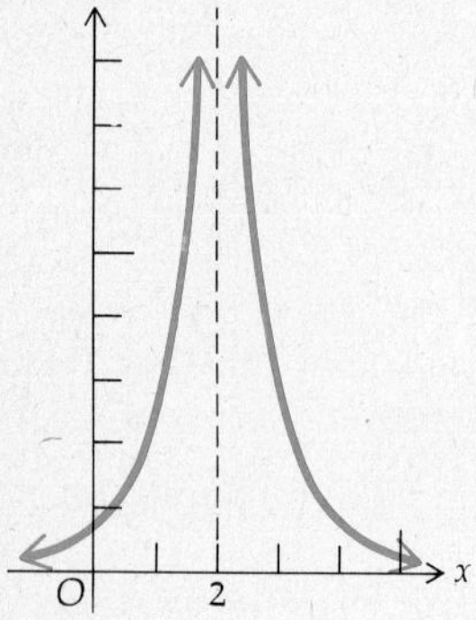

FIGURE 2.4.1

We investigate the function values of f when x is close to 2. Let x approach 2 from the right, and consider the values of $f(x)$ given in Table 2.4.1. From this table you intuitively see that as x gets closer and closer to 2 through values greater than 2, $f(x)$ increases without bound. In other words, we can make $f(x)$ greater than any preassigned positive number (i.e., $f(x)$ can be made as large as we please) for all values of x close enough to 2 and x greater than 2.

TABLE 2.4.1

x	3	$\frac{5}{2}$	$\frac{7}{3}$	$\frac{9}{4}$	$\frac{21}{10}$	$\frac{201}{100}$	$\frac{2001}{1000}$
$f(x) = \frac{3}{(x-2)^2}$	3	12	27	48	300	30,000	3,000,000

To indicate that $f(x)$ increases without bound as x approaches 2 through values greater than 2 we write

$$\lim_{x \to 2^+} \frac{3}{(x-2)^2} = +\infty$$

Now let x approach 2 from the left and consider the values of $f(x)$ given in Table 2.4.2. You intuitively see from this table that as x gets closer and closer to 2 through values less than 2, $f(x)$ increases without bound; so we write

$$\lim_{x \to 2^-} \frac{3}{(x-2)^2} = +\infty$$

Therefore, as x approaches 2 from either the right or the left, $f(x)$ increases without bound, and we write

$$\lim_{x \to 2} \frac{3}{(x-2)^2} = +\infty$$

TABLE 2.4.2

x	1	$\frac{3}{2}$	$\frac{5}{3}$	$\frac{7}{4}$	$\frac{19}{10}$	$\frac{199}{100}$	$\frac{1999}{1000}$
$f(x) = \frac{3}{(x-2)^2}$	3	12	27	48	300	30,000	3,000,000

2.4.1 Definition Let f be a function which is defined at every number in some open interval I containing a, except possibly at the number a itself. **As x approaches a, $f(x)$ increases without bound,** which is written

$$\lim_{x \to a} f(x) = +\infty \tag{1}$$

if for any number $N > 0$ there exists a $\delta > 0$ such that $f(x) > N$ whenever $0 < |x - a| < \delta$.

Another way of stating Definition 2.4.1 is as follows: "The function values $f(x)$ increase without bound as x approaches a number a if $f(x)$ can be made as large as we please (i.e., greater than any positive number N) for all values of x sufficiently close to a but not equal to a.

Note: It should be stressed again (as in Sec. 1.1) that $+\infty$ is not a real number; hence, when we write $\lim_{x \to a} f(x) = +\infty$, it does not have the same meaning as $\lim_{x \to a} f(x) = L$, where L is a real number. Equation (1) can be read as "the limit of $f(x)$ as x approaches a is positive infinity." In such a case the limit does not exist, but the symbol $+\infty$ indicates the behavior of the function values $f(x)$ as x gets closer and closer to a.

In an analogous manner we can indicate the behavior of a function whose function values decrease without bound. To lead up to this, consider the function g defined by the equation

$$g(x) = \frac{-3}{(x-2)^2}$$

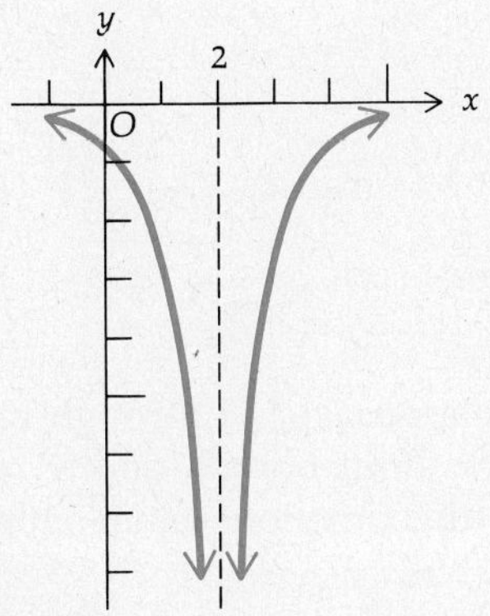

FIGURE 2.4.2

A sketch of the graph of this function is in Fig. 2.4.2.

The function values given by $g(x) = -3/(x-2)^2$ are the negatives of the function values given by $f(x) = 3/(x-2)^2$. So for the function g as x approaches 2, either from the right or the left, $g(x)$ decreases without bound, and we write

$$\lim_{x \to 2} \frac{-3}{(x-2)^2} = -\infty$$

2.4.2 Definition Let f be a function which is defined at every number in some open interval I containing a, except possibly at the number a itself. **As x approaches a, $f(x)$ decreases without bound,** which is written

$$\lim_{x \to a} f(x) = -\infty \tag{2}$$

if for any number $N < 0$ there exists a $\delta > 0$ such that $f(x) < N$ whenever $0 < |x - a| < \delta$.

Note: Equation (2) can be read as "the limit of $f(x)$ as x approaches a is negative infinity," observing again that the limit does not exist and the symbol $-\infty$ indicates only the behavior of the function values as x approaches a.

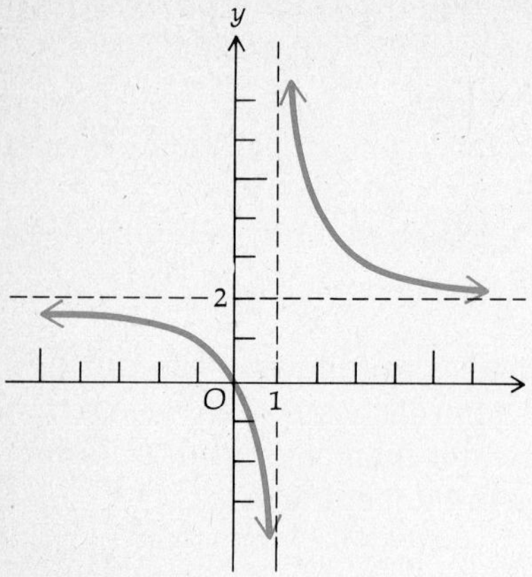

FIGURE 2.4.3

We can consider one-sided limits that are "infinite." In particular, $\lim_{x \to a^+} f(x) = +\infty$ if f is defined at every number in some open interval (a, c) and if for any number $N > 0$ there exists a $\delta > 0$ such that $f(x) > N$ whenever $0 < x - a < \delta$. Similar definitions can be given if $\lim_{x \to a^-} f(x) = +\infty$, $\lim_{x \to a^+} f(x) = -\infty$, and $\lim_{x \to a^-} f(x) = -\infty$.

Now suppose that h is the function defined by the equation

$$h(x) = \frac{2x}{x-1} \tag{3}$$

A sketch of the graph of this function is in Fig. 2.4.3. By referring to Figs. 2.4.1, 2.4.2, and 2.4.3, note the difference in the behavior of the function whose graph is sketched in Fig. 2.4.3 from the functions of the other two figures. Observe that

$$\lim_{x \to 1^-} \frac{2x}{x-1} = -\infty$$

and

$$\lim_{x \to 1^+} \frac{2x}{x-1} = +\infty$$

That is, for the function defined by (3), as x approaches 1 through values less than 1, the function values decrease without bound, and as x approaches 1 through values greater than 1, the function values increase without bound.

Before giving some examples, we need two limit theorems involving "infinite" limits.

2.4.3 Limit Theorem 11 If r is any positive integer, then

(i) $\lim_{x \to 0^+} \dfrac{1}{x^r} = +\infty$.

(ii) $\lim_{x \to 0^-} \dfrac{1}{x^r} = \begin{cases} -\infty & \text{if } r \text{ is odd} \\ +\infty & \text{if } r \text{ is even} \end{cases}$.

Proof We prove part (i). The proof of part (ii) is analogous and is left as an exercise (see Exercise 47). We must show that for any $N > 0$ there exists a $\delta > 0$ such that

$$\frac{1}{x^r} > N \quad \text{whenever} \quad 0 < x < \delta$$

or, equivalently, because $x > 0$ and $N > 0$,

$$x^r < \frac{1}{N} \quad \text{whenever} \quad 0 < x < \delta$$

or, equivalently, because $r > 0$,

$$x < \left(\frac{1}{N}\right)^{1/r} \quad \text{whenever} \quad 0 < x < \delta$$

The above statement holds if $\delta = (1/N)^{1/r}$. Therefore

$$\frac{1}{x^r} > N \quad \text{whenever} \quad 0 < x < \delta, \text{ if } \delta = \left(\frac{1}{N}\right)^{1/r}$$

■

● ILLUSTRATION 1 From Limit Theorem 11(i) it follows that

$$\lim_{x \to 0^+} \frac{1}{x^3} = +\infty \quad \text{and} \quad \lim_{x \to 0^+} \frac{1}{x^4} = +\infty$$

From Limit Theorem 11(ii)

$$\lim_{x \to 0^-} \frac{1}{x^3} = -\infty \quad \text{and} \quad \lim_{x \to 0^-} \frac{1}{x^4} = +\infty$$

●

2.4.4 Limit Theorem 12 If a is any real number, and if $\lim_{x \to a} f(x) = 0$ and $\lim_{x \to a} g(x) = c$, where c is a constant not equal to 0, then

(i) if $c > 0$ and if $f(x) \to 0$ through positive values of $f(x)$,

$$\lim_{x \to a} \frac{g(x)}{f(x)} = +\infty$$

(ii) if $c > 0$ and if $f(x) \to 0$ through negative values of $f(x)$,

$$\lim_{x \to a} \frac{g(x)}{f(x)} = -\infty$$

(iii) if $c < 0$ and if $f(x) \to 0$ through positive values of $f(x)$,

$$\lim_{x \to a} \frac{g(x)}{f(x)} = -\infty$$

(iv) if $c < 0$ and if $f(x) \to 0$ through negative values of $f(x)$,

$$\lim_{x \to a} \frac{g(x)}{f(x)} = +\infty$$

The theorem is also valid if "$x \to a$" is replaced by "$x \to a^+$" or "$x \to a^-$."

Proof We prove part (i) and leave the proofs of the other parts as exercises (see Exercises 48–50). To prove that

$$\lim_{x \to a} \frac{g(x)}{f(x)} = +\infty$$

we must show that for any $N > 0$ there exists a $\delta > 0$ such that

$$\frac{g(x)}{f(x)} > N \quad \text{whenever} \quad 0 < |x - a| < \delta \tag{4}$$

Since $\lim_{x \to a} g(x) = c > 0$, by taking $\epsilon = \frac{1}{2}c$ in Definition 2.1.1, it follows that there exists a $\delta_1 > 0$ such that

$$|g(x) - c| < \tfrac{1}{2}c \quad \text{whenever} \quad 0 < |x - a| < \delta_1$$

By applying Theorem 1.2.2 to the above inequality, it follows that there exists a $\delta_1 > 0$ such that

$$-\tfrac{1}{2}c < g(x) - c < \tfrac{1}{2}c \quad \text{whenever} \quad 0 < |x - a| < \delta_1$$

$$\tfrac{1}{2}c < g(x) < \tfrac{3}{2}c \quad \text{whenever} \quad 0 < |x - a| < \delta_1$$

So there exists a $\delta_1 > 0$ such that

$$g(x) > \tfrac{1}{2}c \quad \text{whenever} \quad 0 < |x - a| < \delta_1 \tag{5}$$

Now $\lim_{x \to a} f(x) = 0$. Thus, for any $\epsilon > 0$ there exists a $\delta_2 > 0$ such that

$$|f(x)| < \epsilon \quad \text{whenever} \quad 0 < |x - a| < \delta_2$$

Since $f(x)$ is approaching zero through positive values of $f(x)$, the absolute-value bars around $f(x)$ can be removed; hence, for any $\epsilon > 0$ there exists a $\delta_2 > 0$ such that

$$0 < f(x) < \epsilon \quad \text{whenever} \quad 0 < |x - a| < \delta_2 \tag{6}$$

From statements (5) and (6) we can conclude that for any $\epsilon > 0$ there exist a $\delta_1 > 0$ and a $\delta_2 > 0$ such that

$$\frac{g(x)}{f(x)} > \frac{\frac{1}{2}c}{\epsilon} \quad \text{whenever} \quad 0 < |x - a| < \delta_1 \quad \text{and} \quad 0 < |x - a| < \delta_2$$

Hence, if $\epsilon = c/(2N)$ and $\delta = \min(\delta_1, \delta_2)$, then

$$\frac{g(x)}{f(x)} > \frac{\frac{1}{2}c}{c/(2N)} = N \quad \text{whenever} \quad 0 < |x - a| < \delta$$

which is statement (4). Hence part (i) is proved. ■

The four parts of Limit Theorem 12 are demonstrated in the following illustration.

● ILLUSTRATION 2

(a) To evaluate

$$\lim_{x \to 4^+} \frac{2x - 1}{x - 4}$$

observe that $\lim_{x \to 4^+} (2x - 1) = 7 > 0$ and $\lim_{x \to 4^+} (x - 4) = 0$, where $x - 4$ is approaching 0 through positive values. Therefore we apply Limit Theorem 12(i) and have

$$\lim_{x \to 4^+} \frac{2x - 1}{x - 4} = +\infty$$

(b) To evaluate

$$\lim_{x \to 4^-} \frac{2x - 1}{x - 4}$$

note that $\lim_{x \to 4^-} (2x - 1) = 7 > 0$ and $\lim_{x \to 4^-} (x - 4) = 0$, where $x - 4$ is approaching 0 through negative values. Thus from Limit Theorem 12(ii),

$$\lim_{x \to 4^-} \frac{2x - 1}{x - 4} = -\infty$$

(c) To evaluate

$$\lim_{x \to 4^+} \frac{1 - 2x}{x - 4}$$

observe that $\lim_{x \to 4^+} (1 - 2x) = -7 < 0$ and $\lim_{x \to 4^+} (x - 4) = 0$, where $x - 4$ is approaching 0 through positive values. Hence from Limit Theorem 12(iii),

$$\lim_{x \to 4^+} \frac{1 - 2x}{x - 4} = -\infty$$

(d) To evaluate

$$\lim_{x \to 4^-} \frac{1 - 2x}{x - 4}$$

note that $\lim_{x \to 4^-} (1 - 2x) = -7 < 0$ and $\lim_{x \to 4^-} (x - 4) = 0$, where $x - 4$ is approaching 0 through negative values. Therefore, from Limit Theorem 12(iv) it follows that

$$\lim_{x \to 4^-} \frac{1 - 2x}{x - 4} = +\infty$$

Example 1 Find

(a) $\lim_{x \to 3^+} \dfrac{x^2 + x + 2}{x^2 - 2x - 3}$ (b) $\lim_{x \to 3^-} \dfrac{x^2 + x + 2}{x^2 - 2x - 3}$

Solution

(a) $$\lim_{x \to 3^+} \frac{x^2 + x + 2}{x^2 - 2x - 3} = \lim_{x \to 3^+} \frac{x^2 + x + 2}{(x - 3)(x + 1)}$$

The limit of the numerator is 14, which can be easily verified.

$$\begin{aligned} \lim_{x \to 3^+} (x - 3)(x + 1) &= \lim_{x \to 3^+} (x - 3) \cdot \lim_{x \to 3^+} (x + 1) \\ &= 0 \cdot 4 \\ &= 0 \end{aligned}$$

The limit of the denominator is 0, and the denominator is approaching 0 through positive values. Then from Limit Theorem 12(i),

$$\lim_{x \to 3^+} \frac{x^2 + x + 2}{x^2 - 2x - 3} = +\infty$$

(b) $$\lim_{x \to 3^-} \frac{x^2 + x + 2}{x^2 - 2x - 3} = \lim_{x \to 3^-} \frac{x^2 + x + 2}{(x - 3)(x + 1)}$$

As in part (a), the limit of the numerator is 14.

$$\lim_{x \to 3^-} (x-3)(x+1) = \lim_{x \to 3^-} (x-3) \cdot \lim_{x \to 3^-} (x+1)$$
$$= 0 \cdot 4$$
$$= 0$$

In this case, the limit of the denominator is zero, but the denominator is approaching zero through negative values. From Limit Theorem 12(ii),

$$\lim_{x \to 3^-} \frac{x^2+x+2}{x^2-2x-3} = -\infty$$

Example 2 Find

(a) $\lim_{x \to 2^+} \frac{\sqrt{x^2-4}}{x-2}$ (b) $\lim_{x \to 2^-} \frac{\sqrt{4-x^2}}{x-2}$

Solution

(a) Because $x \to 2^+$, $x-2>0$; so $x-2=\sqrt{(x-2)^2}$. Thus

$$\lim_{x \to 2^+} \frac{\sqrt{x^2-4}}{x-2} = \lim_{x \to 2^+} \frac{\sqrt{(x-2)(x+2)}}{\sqrt{(x-2)^2}}$$
$$= \lim_{x \to 2^+} \frac{\sqrt{x-2}\,\sqrt{x+2}}{\sqrt{x-2}\,\sqrt{x-2}}$$
$$= \lim_{x \to 2^+} \frac{\sqrt{x+2}}{\sqrt{x-2}}$$

The limit of the numerator is 2. The limit of the denominator is 0, and the denominator is approaching 0 through positive values. Therefore, by Limit Theorem 12(i) it follows that

$$\lim_{x \to 2^+} \frac{\sqrt{x^2-4}}{x-2} = +\infty$$

(b) Because $x \to 2^-$, $x-2<0$; so $x-2=-(2-x)=-\sqrt{(2-x)^2}$. Therefore

$$\lim_{x \to 2^-} \frac{\sqrt{4-x^2}}{x-2} = \lim_{x \to 2^-} \frac{\sqrt{2-x}\,\sqrt{2+x}}{-\sqrt{2-x}\,\sqrt{2-x}}$$
$$= \lim_{x \to 2^-} \frac{\sqrt{2+x}}{-\sqrt{2-x}}$$

The limit of the numerator is 2. The limit of the denominator is 0, and the denominator is approaching 0 through negative values. Hence by Limit Theorem 12(ii),

$$\lim_{x \to 2^-} \frac{\sqrt{4-x^2}}{x-2} = -\infty$$

Example 3 Find

$$\lim_{x \to 4^-} \frac{[\![x]\!] - 4}{x - 4}$$

Solution $\lim_{x \to 4^-} [\![x]\!] = 3$. Therefore $\lim_{x \to 4^-} ([\![x]\!] - 4) = -1$. Furthermore, $\lim_{x \to 4^-} (x - 4) = 0$, and $x - 4$ is approaching 0 through negative values. Hence from Limit Theorem 12(iv),

$$\lim_{x \to 4^-} \frac{[\![x]\!] - 4}{x - 4} = +\infty$$

Remember that because $+\infty$ and $-\infty$ are not numbers, the Limit Theorems 1–10 of Sec. 2.2 do not hold for "infinite" limits. However, there are the following properties regarding such limits. The proofs are left as exercises (see Exercises 51–53).

2.4.5 Theorem (i) If $\lim_{x \to a} f(x) = +\infty$, and $\lim_{x \to a} g(x) = c$, where c is any constant, then

$$\lim_{x \to a} [f(x) + g(x)] = +\infty$$

(ii) If $\lim_{x \to a} f(x) = -\infty$, and $\lim_{x \to a} g(x) = c$, where c is any constant, then

$$\lim_{x \to a} [f(x) + g(x)] = -\infty$$

The theorem is valid if "$x \to a$" is replaced by "$x \to a^+$" or "$x \to a^-$."

● ILLUSTRATION 3 Because $\lim_{x \to 2^+} \frac{1}{x - 2} = +\infty$ and $\lim_{x \to 2^+} \frac{1}{x + 2} = \frac{1}{4}$, it follows from Theorem 2.4.5(i) that

$$\lim_{x \to 2^+} \left[\frac{1}{x - 2} + \frac{1}{x + 2}\right] = +\infty$$ ●

2.4.6 Theorem If $\lim_{x \to a} f(x) = +\infty$ and $\lim_{x \to a} g(x) = c$, where c is any constant except 0, then

(i) if $c > 0$, $\lim_{x \to a} f(x) \cdot g(x) = +\infty$.

(ii) if $c < 0$, $\lim_{x \to a} f(x) \cdot g(x) = -\infty$.

The theorem is valid if "$x \to a$" is replaced by "$x \to a^+$" or "$x \to a^-$."

● ILLUSTRATION 4

$$\lim_{x \to 3} \frac{5}{(x - 3)^2} = +\infty \quad \text{and} \quad \lim_{x \to 3} \frac{x + 4}{x - 4} = -7$$

Therefore from Theorem 2.4.6(ii),

$$\lim_{x \to 3} \left[\frac{5}{(x - 3)^2} \cdot \frac{x + 4}{x - 4}\right] = -\infty$$ ●

2.4.7 Theorem If $\lim_{x \to a} f(x) = -\infty$ and $\lim_{x \to a} g(x) = c$, where c is any constant except 0, then

(i) if $c > 0$, $\lim_{x \to a} f(x) \cdot g(x) = -\infty$.

(ii) if $c < 0$, $\lim_{x \to a} f(x) \cdot g(x) = +\infty$.

The theorem is valid if "$x \to a$" is replaced by "$x \to a^+$" or "$x \to a^-$."

● ILLUSTRATION 5 In Example 2(b) we showed

$$\lim_{x \to 2^-} \frac{\sqrt{4 - x^2}}{x - 2} = -\infty$$

Furthermore,

$$\lim_{x \to 2^-} \frac{x - 3}{x + 2} = -\frac{1}{4}$$

Thus, from Theorem 2.4.7(ii) it follows that

$$\lim_{x \to 2^-} \left[\frac{\sqrt{4 - x^2}}{x - 2} \cdot \frac{x - 3}{x + 2}\right] = +\infty$$ ●

An aid in drawing the sketch of the graph of a function is to find, if there are any, the *asymptotes* of the graph. In this section we discuss *vertical asymptotes,* and in Sec. 2.5, *horizontal asymptotes.* Let the function f be defined by

$$f(x) = \frac{1}{(x - a)^2} \tag{7}$$

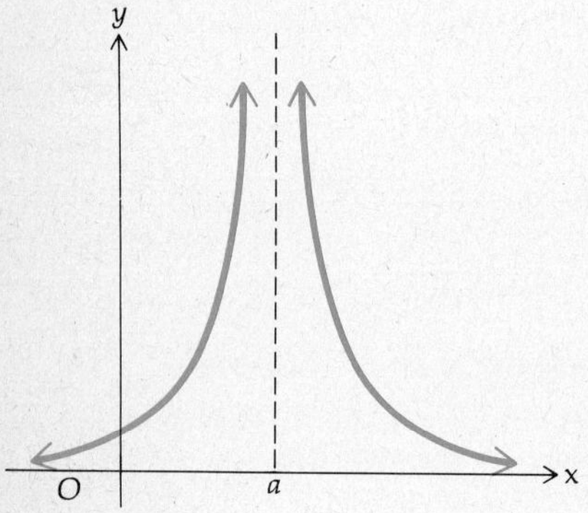

FIGURE 2.4.4

A sketch of the graph of f is in Fig. 2.4.4. Any line parallel to and above the x axis will intersect this graph in two points: one point to the left of the line $x = a$ and one point to the right of this line. Thus for any $k > 0$, no matter how large, the line $y = k$ will intersect the graph of f in two points; the distance of these two points from the line $x = a$ gets smaller and smaller as k gets larger and larger. The line $x = a$ is called a *vertical asymptote* of the graph of f.

2.4.8 Definition The line $x = a$ is said to be a **vertical asymptote** of the graph of the function f if at least one of the following statements is true:

(i) $\lim_{x \to a^+} f(x) = +\infty$.

(ii) $\lim_{x \to a^+} f(x) = -\infty$.

(iii) $\lim_{x \to a^-} f(x) = +\infty$.

(iv) $\lim_{x \to a^-} f(x) = -\infty$.

● ILLUSTRATION 6 Each of the figures 2.4.5 through 2.4.8 shows a portion of the graph of a function for which the line $x = a$ is a vertical asymptote. In Fig. 2.4.5, part (i) of Definition 2.4.8 applies; in Fig. 2.4.6, part (ii) applies; and in Figs. 2.4.7 and 2.4.8, parts (iii) and (iv), respectively, apply. ●

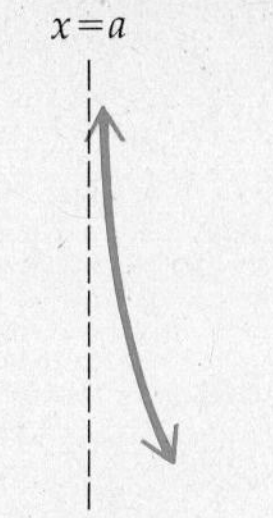

$\lim_{x \to a^+} f(x) = +\infty$

FIGURE 2.4.5

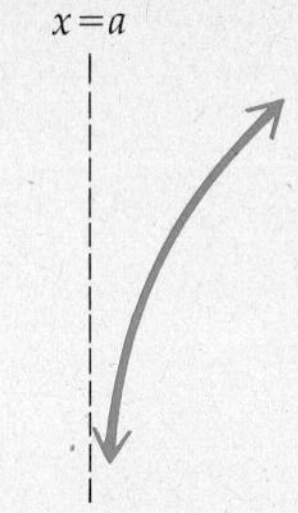

$\lim_{x \to a^+} f(x) = -\infty$

FIGURE 2.4.6

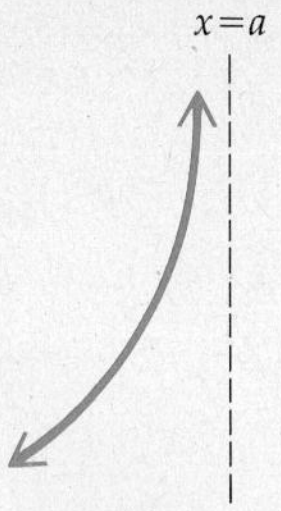

$\lim_{x \to a^-} f(x) = +\infty$

FIGURE 2.4.7

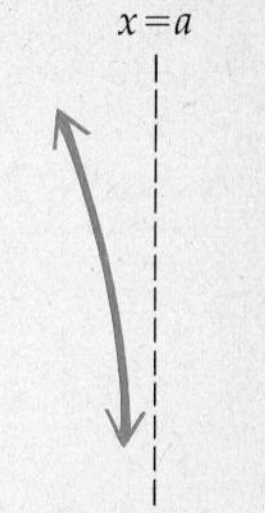

$\lim_{x \to a^-} f(x) = -\infty$

FIGURE 2.4.8

For the function defined by (7), both parts (i) and (iii) of the above definition are true. See Fig. 2.4.4. If g is the function defined by

$$g(x) = -\frac{1}{(x-a)^2}$$

then both parts (ii) and (iv) are true, and the line $x = a$ is a vertical asymptote of the graph of g. This is shown in Fig. 2.4.9.

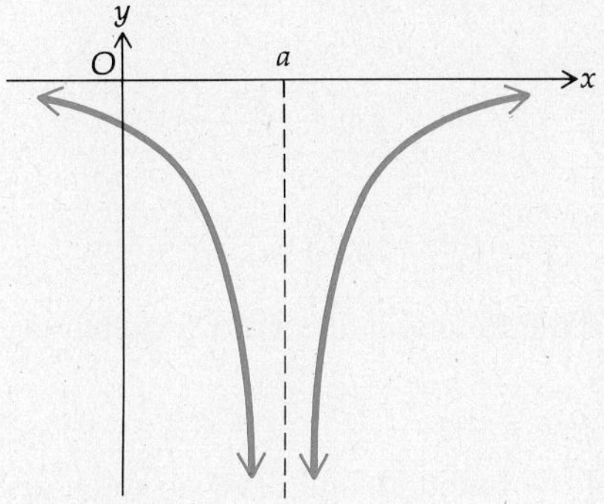

FIGURE 2.4.9

Example 4 Find the vertical asymptote of the graph of the function defined by the equation

$$f(x) = \frac{3}{x-3}$$

and draw a sketch of the graph of f.

Solution

$$\lim_{x \to 3^+} \frac{3}{x-3} = +\infty \quad \text{and} \quad \lim_{x \to 3^-} \frac{3}{x-3} = -\infty$$

It follows from Definition 2.4.8 that the line $x = 3$ is a vertical asymptote of the graph of f. A sketch of the graph of f is shown in Fig. 2.4.10.

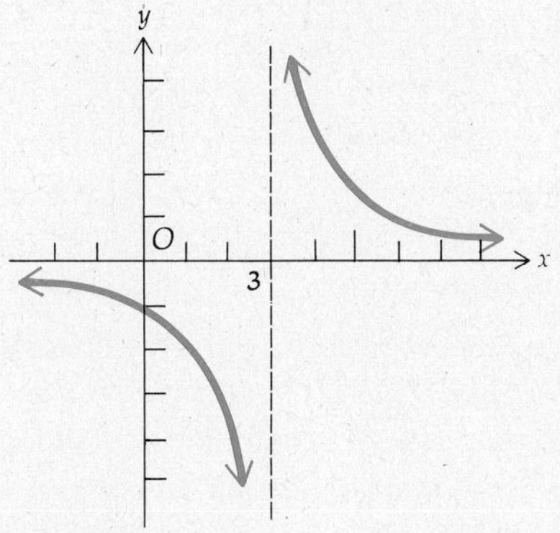

FIGURE 2.4.10

EXERCISES 2.4

In Exercises 1 through 32, evaluate the limit.

1. $\lim_{x \to 5^+} \frac{1}{x-5}$

2. $\lim_{x \to 5^-} \frac{1}{x-5}$

3. $\lim_{x \to 5} \frac{1}{(x-5)^2}$

4. $\lim_{x \to 1^-} \frac{x+2}{1-x}$

5. $\lim_{x \to 1^+} \frac{x+2}{1-x}$

6. $\lim_{x \to 1} \frac{x+2}{(x-1)^2}$

7. $\lim_{x \to -1^+} \frac{x-2}{x+1}$

8. $\lim_{x \to -1^-} \frac{x-2}{x+1}$

9. $\lim_{x \to -4^-} \frac{x}{x+4}$

10. $\lim_{x \to 4^+} \frac{x}{x-4}$

11. $\lim_{x \to -3^-} \frac{4x}{9-x^2}$

12. $\lim_{x \to 3^+} \frac{4x^2}{9-x^2}$

13. $\lim_{t\to 2^+} \dfrac{t+2}{t^2-4}$ **14.** $\lim_{t\to 2^-} \dfrac{-t+2}{(t-2)^2}$ **15.** $\lim_{t\to 2^-} \dfrac{t+2}{t^2-4}$ **16.** $\lim_{x\to 0^+} \dfrac{\sqrt{3+x^2}}{x}$

17. $\lim_{x\to 0^-} \dfrac{\sqrt{3+x^2}}{x}$ **18.** $\lim_{x\to 0} \dfrac{\sqrt{3+x^2}}{x^2}$ **19.** $\lim_{x\to 3^+} \dfrac{\sqrt{x^2-9}}{x-3}$ **20.** $\lim_{x\to 4^-} \dfrac{\sqrt{16-x^2}}{x-4}$

21. $\lim_{x\to 0^+} \left(\dfrac{1}{x} - \dfrac{1}{x^2}\right)$ **22.** $\lim_{x\to 0^+} \dfrac{x^2-3}{x^3+x^2}$ **23.** $\lim_{x\to 0^-} \dfrac{2-4x^3}{5x^2+3x^3}$ **24.** $\lim_{s\to 2^-} \left(\dfrac{1}{s-2} - \dfrac{3}{s^2-4}\right)$

25. $\lim_{t\to -4^-} \left(\dfrac{2}{t^2+3t-4} - \dfrac{3}{t+4}\right)$ **26.** $\lim_{x\to 0^-} \dfrac{2x^3-5x^2}{x^2-1}$ **27.** $\lim_{x\to 3^-} \dfrac{[\![x]\!]-x}{3-x}$ **28.** $\lim_{x\to 1^-} \dfrac{[\![x^2]\!]-1}{x^2-1}$

29. $\lim_{x\to 3^-} \dfrac{x^3+9x^2+20x}{x^2+x-12}$ **30.** $\lim_{x\to -2^+} \dfrac{6x^2+x-2}{2x^2+3x-2}$ **31.** $\lim_{x\to 1^+} \dfrac{x-1}{\sqrt{2x-x^2}-1}$ **32.** $\lim_{x\to 2^-} \dfrac{x-2}{2-\sqrt{4x-x^2}}$

33. For each of the following functions, find the vertical asymptote of the graph of the function, and draw a sketch of the graph: (a) $f(x) = \dfrac{1}{x}$, (b) $g(x) = \dfrac{1}{x^2}$, (c) $h(x) = \dfrac{1}{x^3}$, (d) $\phi(x) = \dfrac{1}{x^4}$.

34. For each of the following functions, find the vertical asymptote of the graph of the function, and draw a sketch of the graph: (a) $f(x) = -\dfrac{1}{x}$, (b) $g(x) = -\dfrac{1}{x^2}$, (c) $h(x) = -\dfrac{1}{x^3}$, (d) $\phi(x) = -\dfrac{1}{x^4}$.

In Exercises 35 through 43, find the vertical asymptote(s) of the graph of the function, and draw a sketch of the graph.

35. $f(x) = \dfrac{2}{x-4}$ **36.** $f(x) = \dfrac{3}{x+1}$ **37.** $f(x) = \dfrac{-2}{x+3}$ **38.** $f(x) = \dfrac{-4}{x-5}$

39. $f(x) = \dfrac{-2}{(x+3)^2}$ **40.** $f(x) = \dfrac{4}{(x-5)^2}$ **41.** $f(x) = \dfrac{5}{x^2+8x+15}$ **42.** $f(x) = \dfrac{1}{x^2+5x-6}$

43. $f(x) = \dfrac{2}{4x^2-4x-35}$

44. Use Definition 2.4.1 to prove that $\lim_{x\to -3} \left|\dfrac{5-x}{3+x}\right| = +\infty$.

45. Prove that $\lim_{x\to 2} \dfrac{3}{(x-2)^2} = +\infty$ by using Definition 2.4.1.

46. Prove that $\lim_{x\to 4} \dfrac{-2}{(x-4)^2} = -\infty$ by using Definition 2.4.2.

47. Prove Theorem 2.4.3(ii). **48.** Prove Theorem 2.4.4(ii). **49.** Prove Theorem 2.4.4(iii).

50. Prove Theorem 2.4.4(iv). **51.** Prove Theorem 2.4.5. **52.** Prove Theorem 2.4.6.

53. Prove Theorem 2.4.7.

2.5 LIMITS AT INFINITY

Consider the function f defined by the equation

$$f(x) = \frac{2x^2}{x^2+1}$$

A sketch of the graph of this function is shown in Fig. 2.5.1. Let x take on the values 0, 1, 2, 3, 4, 5, 10, 100, 1000, and so on, allowing x to increase with-

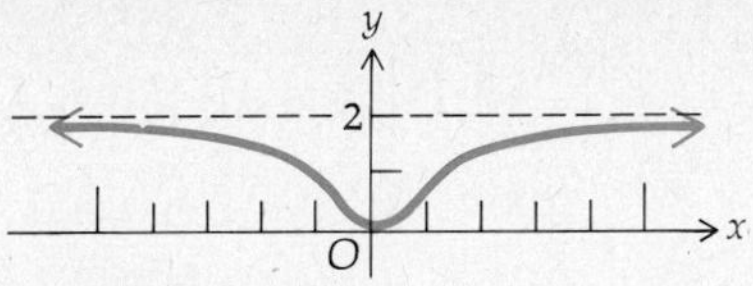

FIGURE 2.5.1

out bound. The corresponding function values are given in Table 2.5.1. Observe from Table 2.5.1 that as x increases through positive values, the function values $f(x)$ get closer and closer to 2. In particular, when $x = 4$,

$$2 - \frac{2x^2}{x^2 + 1} = 2 - \frac{32}{17} = \frac{2}{17}$$

TABLE 2.5.1

x	0	1	2	3	4	5	10	100	1000
$f(x) = \dfrac{2x^2}{x^2 + 1}$	0	1	$\frac{8}{5}$	$\frac{18}{10}$	$\frac{32}{17}$	$\frac{50}{26}$	$\frac{200}{101}$	$\frac{20{,}000}{10{,}001}$	$\frac{2{,}000{,}000}{1{,}000{,}001}$

Therefore the difference between 2 and $f(x)$ is $\frac{2}{17}$ when $x = 4$. When $x = 100$,

$$2 - \frac{2x^2}{x^2 + 1} = 2 - \frac{20{,}000}{10{,}001} = \frac{2}{10{,}001}$$

Hence the difference between 2 and $f(x)$ is $2/10{,}001$ when $x = 100$.

Continuing on, we see intuitively that the value of $f(x)$ can be made as close to 2 as we please by taking x large enough. In other words, the difference between 2 and $f(x)$ can be made as small as we please by taking x any number greater than some sufficiently large positive number. Or, going a step further, for any $\epsilon > 0$, however small, we can find a number $N > 0$ such that $|f(x) - 2| < \epsilon$ whenever $x > N$.

When an independent variable x is increasing without bound through positive values, we write "$x \to +\infty$." From the illustrative example above, then, we can say that

$$\lim_{x \to +\infty} \frac{2x^2}{x^2 + 1} = 2$$

2.5.1 Definition Let f be a function which is defined at every number in some interval $(a, +\infty)$. The **limit of $f(x)$, as x increases without bound, is L,** written

$$\lim_{x \to +\infty} f(x) = L \tag{1}$$

if for any $\epsilon > 0$, however small, there exists a number $N > 0$ such that

$$|f(x) - L| < \epsilon \quad \text{whenever} \quad x > N$$

Note: When $x \to +\infty$ is written, it does not have the same meaning as, for instance, $x \to 1000$. The symbol $x \to +\infty$ indicates the behavior of the variable x. However, (1) can be read as "the limit of $f(x)$ as x approaches positive infinity is L," bearing in mind this note.

Now consider the same function and let x take on the values 0, -1, -2, -3, -4, -5, -10, -100, -1000, and so on, allowing x to decrease through negative values without bound. Table 2.5.2 gives the corresponding function values of $f(x)$.

TABLE 2.5.2

x	0	-1	-2	-3	-4	-5	-10	-100	-1000
$f(x) = \dfrac{2x^2}{x^2+1}$	0	1	$\frac{8}{5}$	$\frac{18}{10}$	$\frac{32}{17}$	$\frac{50}{26}$	$\frac{200}{101}$	$\frac{20{,}000}{10{,}001}$	$\frac{2{,}000{,}000}{1{,}000{,}001}$

Observe that the function values are the same for the negative numbers as for the corresponding positive numbers. So we see intuitively that as x decreases without bound, $f(x)$ approaches 2; that is, $|f(x) - 2|$ can be made as small as we please by taking x any number less than some negative number having a sufficiently large absolute value. Formally we say that for any $\epsilon > 0$, however small, we can find a number $N < 0$ such that $|f(x) - 2| < \epsilon$ whenever $x < N$. Using the symbol $x \to -\infty$ to denote that the variable x is decreasing without bound, we write

$$\lim_{x \to -\infty} \frac{2x^2}{x^2 + 1} = 2$$

2.5.2 Definition Let f be a function which is defined at every number in some interval $(-\infty, a)$. The **limit of $f(x)$, as x decreases without bound, is L,** written

$$\lim_{x \to -\infty} f(x) = L \tag{2}$$

if for any $\epsilon > 0$, however small, there exists a number $N < 0$ such that

$$|f(x) - L| < \epsilon \quad \text{whenever} \quad x < N$$

Note: As in the note following Definition 2.5.1, the symbol $x \to -\infty$ indicates only the behavior of the variable x, but (2) can be read as "the limit of $f(x)$ as x approaches negative infinity is L."

Limit Theorems 2, 4, 5, 6, 7, 8, 9, and 10 in Sec. 2.2 and Limit Theorems 11 and 12 in Sec. 2.4 remain unchanged when "$x \to a$" is replaced by "$x \to +\infty$" or "$x \to -\infty$." We have the following additional limit theorem.

2.5.3 Limit Theorem 13 If r is any positive integer, then

(i) $\displaystyle\lim_{x \to +\infty} \frac{1}{x^r} = 0$

(ii) $\displaystyle\lim_{x \to -\infty} \frac{1}{x^r} = 0$

Proof of (i) To prove part (i) we must show that Definition 2.5.1 holds for $f(x) = 1/x^r$ and $L = 0$; that is, we must show that for any $\epsilon > 0$ there exists a number $N > 0$ such that

$$\left|\frac{1}{x^r} - 0\right| < \epsilon \quad \text{whenever} \quad x > N$$

or, equivalently,

$$|x|^r > \frac{1}{\epsilon} \quad \text{whenever} \quad x > N$$

or, equivalently, since $r > 0$,

$$|x| > \left(\frac{1}{\epsilon}\right)^{1/r} \quad \text{whenever} \quad x > N$$

In order for the above to hold, take $N = (1/\epsilon)^{1/r}$. Thus

$$\left|\frac{1}{x^r} - 0\right| < \epsilon \quad \text{whenever} \quad x > N, \text{ if } N = \left(\frac{1}{\epsilon}\right)^{1/r}$$

This proves part (i). ■

The proof of part (ii) is analogous and is left as an exercise (see Exercise 54).

Example 1 Find

$$\lim_{x \to +\infty} \frac{4x - 3}{2x + 5}$$

and, when applicable, indicate the limit theorems being used.

Solution To use Limit Theorem 13, divide the numerator and the denominator by x, thus giving

$$\lim_{x \to +\infty} \frac{4x - 3}{2x + 5} = \lim_{x \to +\infty} \frac{4 - 3/x}{2 + 5/x}$$

$$= \frac{\lim_{x \to +\infty} (4 - 3/x)}{\lim_{x \to +\infty} (2 + 5/x)} \qquad \text{(L.T. 9)}$$

$$= \frac{\lim_{x \to +\infty} 4 - \lim_{x \to +\infty} (3/x)}{\lim_{x \to +\infty} 2 + \lim_{x \to +\infty} (5/x)} \qquad \text{(L.T. 4)}$$

$$= \frac{\lim_{x \to +\infty} 4 - \lim_{x \to +\infty} 3 \cdot \lim_{x \to +\infty} (1/x)}{\lim_{x \to +\infty} 2 + \lim_{x \to +\infty} 5 \cdot \lim_{x \to +\infty} (1/x)} \qquad \text{(L.T. 6)}$$

$$= \frac{4 - 3 \cdot 0}{2 + 5 \cdot 0} \qquad \text{(L.T. 2 and L.T. 13)}$$

$$= 2$$

Example 2 Find

$$\lim_{x \to -\infty} \frac{2x^2 - x + 5}{4x^3 - 1}$$

and, when applicable, indicate the limit theorems being used.

Solution To use Limit Theorem 13, divide the numerator and the denominator by the highest power of x occurring in either the numerator or denominator, which in this case is x^3.

$$\lim_{x\to-\infty} \frac{2x^2 - x + 5}{4x^3 - 1} = \lim_{x\to-\infty} \frac{2/x - 1/x^2 + 5/x^3}{4 - 1/x^3}$$

$$= \frac{\lim_{x\to-\infty} (2/x - 1/x^2 + 5/x^3)}{\lim_{x\to-\infty} (4 - 1/x^3)} \qquad \text{(L.T. 9)}$$

$$= \frac{\lim_{x\to-\infty} (2/x) - \lim_{x\to-\infty} (1/x^2) + \lim_{x\to-\infty} (5/x^3)}{\lim_{x\to-\infty} 4 - \lim_{x\to-\infty} (1/x^3)} \qquad \text{(L.T. 5)}$$

$$= \frac{\lim_{x\to-\infty} 2 \cdot \lim_{x\to-\infty} (1/x) - \lim_{x\to-\infty} (1/x^2) + \lim_{x\to-\infty} 5 \cdot \lim_{x\to-\infty} (1/x^3)}{\lim_{x\to-\infty} 4 - \lim_{x\to-\infty} (1/x^3)} \qquad \text{(L.T. 6)}$$

$$= \frac{2 \cdot 0 - 0 + 5 \cdot 0}{4 - 0} \qquad \text{(L.T. 2 and L.T. 13)}$$

$$= 0$$

Example 3 Find

$$\lim_{x\to+\infty} \frac{3x + 4}{\sqrt{2x^2 - 5}}$$

and, when applicable, indicate the limit theorems being used.

Solution Divide the numerator and the denominator of the fraction by x. In the denominator let $x = \sqrt{x^2}$ since only positive values of x are being considered.

$$\lim_{x\to+\infty} \frac{3x + 4}{\sqrt{2x^2 - 5}} = \lim_{x\to+\infty} \frac{3 + 4/x}{\sqrt{2x^2 - 5}/\sqrt{x^2}}$$

$$= \frac{\lim_{x\to+\infty} (3 + 4/x)}{\lim_{x\to+\infty} \sqrt{2 - 5/x^2}} \qquad \text{(L.T. 9)}$$

$$= \frac{\lim_{x\to+\infty} (3 + 4/x)}{\sqrt{\lim_{x\to+\infty} (2 - 5/x^2)}} \qquad \text{(L.T 10)}$$

$$= \frac{\lim_{x\to+\infty} 3 + \lim_{x\to+\infty} (4/x)}{\sqrt{\lim_{x\to+\infty} 2 - \lim_{x\to+\infty} (5/x^2)}} \qquad \text{(L.T. 4)}$$

$$= \frac{\lim_{x\to+\infty} 3 + \lim_{x\to+\infty} 4 \cdot \lim_{x\to+\infty} (1/x)}{\sqrt{\lim_{x\to+\infty} 2 - \lim_{x\to+\infty} 5 \cdot \lim_{x\to+\infty} (1/x^2)}} \qquad \text{(L.T. 6)}$$

$$= \frac{3 + 4 \cdot 0}{\sqrt{2 - 5 \cdot 0}} \qquad \text{(L.T. 2 and L.T. 13)}$$

$$= \frac{3}{\sqrt{2}}$$

Example 4 Find

$$\lim_{x \to -\infty} \frac{3x + 4}{\sqrt{2x^2 - 5}}$$

Solution The function is the same as the one in Example 3; however, since we are considering negative values of x, in this case $\sqrt{x^2} = -x$ or, equivalently, $-\sqrt{x^2} = x$. Hence in the first step, when the numerator and denominator are divided by x, let $x = -\sqrt{x^2}$ in the denominator.

$$\lim_{x \to -\infty} \frac{3x + 4}{\sqrt{2x^2 - 5}} = \lim_{x \to -\infty} \frac{3 + 4/x}{\sqrt{2x^2 - 5}/(-\sqrt{x^2})}$$

$$= \frac{\lim_{x \to -\infty} (3 + 4/x)}{\lim_{x \to -\infty} (-\sqrt{2 - 5/x^2})}$$

$$= \frac{\lim_{x \to -\infty} 3 + \lim_{x \to -\infty} (4/x)}{-\sqrt{\lim_{x \to -\infty} (2 - 5/x^2)}}$$

The remaining steps are similar to those in the solution of Example 3, and in the denominator $-\sqrt{2}$ is obtained instead of $\sqrt{2}$. Hence the limit is $-3/\sqrt{2}$.

"Infinite" limits at infinity can be considered. There are formal definitions for each of the following.

$$\lim_{x \to +\infty} f(x) = +\infty \qquad \lim_{x \to -\infty} f(x) = +\infty$$

$$\lim_{x \to +\infty} f(x) = -\infty \qquad \lim_{x \to -\infty} f(x) = -\infty$$

For example, $\lim_{x \to +\infty} f(x) = +\infty$ if the function f is defined on some interval $(a, +\infty)$ and if for any number $N > 0$ there exists an $M > 0$ such that $f(x) > N$ whenever $x > M$. The other definitions are left as an exercise (see Exercise 51).

Example 5 Find

$$\lim_{x \to +\infty} \frac{x^2}{x + 1}$$

Solution Divide the numerator and the denominator by x^2.

$$\lim_{x \to +\infty} \frac{x^2}{x + 1} = \lim_{x \to +\infty} \frac{1}{1/x + 1/x^2}$$

Evaluating the limit of the denominator we have

$$\lim_{x \to +\infty} \left(\frac{1}{x} + \frac{1}{x^2}\right) = \lim_{x \to +\infty} \frac{1}{x} + \lim_{x \to +\infty} \frac{1}{x^2} = 0 + 0 = 0$$

Therefore the limit of the denominator is 0, and the denominator is approaching 0 through positive values.

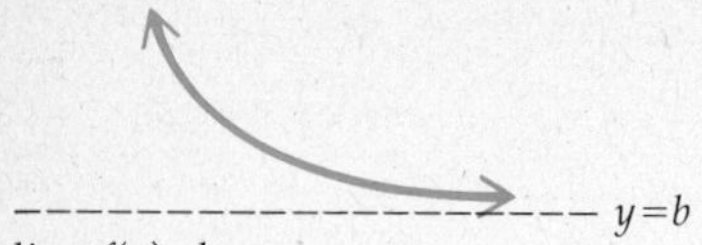

$\lim_{x \to +\infty} f(x) = b$

FIGURE 2.5.2

The limit of the numerator is 1, and so by Limit Theorem 12(i) (2.4.4) it follows that

$$\lim_{x \to +\infty} \frac{x^2}{x+1} = +\infty$$

Example 6 Find

$$\lim_{x \to +\infty} \frac{2x - x^2}{3x + 5}$$

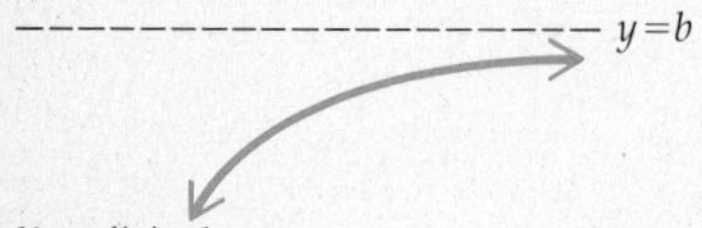

$\lim_{x \to +\infty} f(x) = b$

FIGURE 2.5.3

Solution

$$\lim_{x \to +\infty} \frac{2x - x^2}{3x + 5} = \lim_{x \to +\infty} \frac{2/x - 1}{3/x + 5/x^2}$$

The limits of the numerator and the denominator are considered separately.

$$\lim_{x \to +\infty} \left(\frac{2}{x} - 1\right) = \lim_{x \to +\infty} \frac{2}{x} - \lim_{x \to +\infty} 1 = 0 - 1 = -1$$

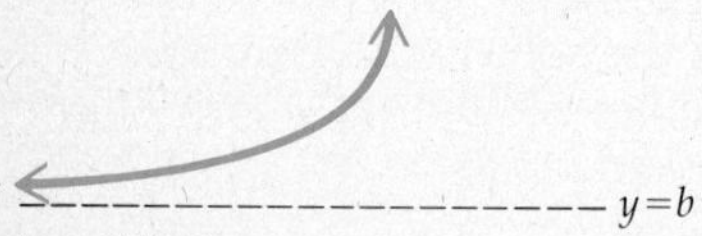

$\lim_{x \to -\infty} f(x) = b$

FIGURE 2.5.4

$$\lim_{x \to +\infty} \left(\frac{3}{x} + \frac{5}{x^2}\right) = \lim_{x \to +\infty} \frac{3}{x} + \lim_{x \to +\infty} \frac{5}{x^2} = 0 + 0 = 0$$

Therefore we have the limit of a quotient in which the limit of the numerator is -1 and the limit of the denominator is 0, where the denominator is approaching 0 through positive values. By Limit Theorem 12(iii) it follows that

$$\lim_{x \to +\infty} \frac{2x - x^2}{3x + 5} = -\infty$$

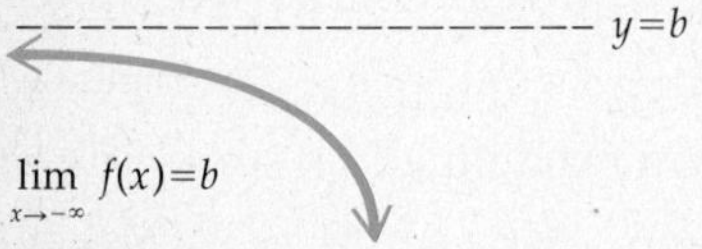

$\lim_{x \to -\infty} f(x) = b$

FIGURE 2.5.5

In Sec. 2.4 we gave the formal definition of a vertical asymptote of the graph of a function. A *horizontal asymptote* of a graph is a line parallel to the x axis, and the following is the formal definition.

2.5.4 Definition The line $y = b$ is said to be a **horizontal asymptote** of the graph of the function f if at least one of the following statements is true:

(i) $\lim_{x \to +\infty} f(x) = b$, and for some number N, $f(x) \neq b$ whenever $x > N$.

(ii) $\lim_{x \to -\infty} f(x) = b$, and for some number N, $f(x) \neq b$ whenever $x < N$.

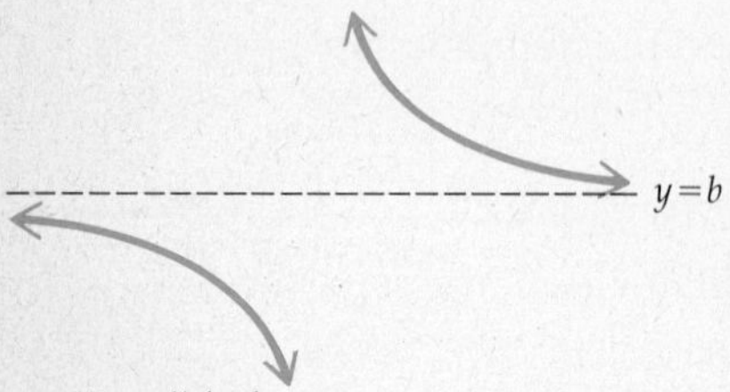

$\lim_{x \to +\infty} f(x) = b$

and

$\lim_{x \to -\infty} f(x) = b$

FIGURE 2.5.6

• ILLUSTRATION 1 In each of Figs. 2.5.2 through 2.5.6 is shown a portion of the graph of a function for which the line $y = b$ is a horizontal asymptote. In Figs. 2.5.2 and 2.5.3, part (i) of Definition 2.5.4 applies, and in Figs. 2.5.4 and 2.5.5, part (ii) is true. Both parts (i) and (ii) hold for the function whose graph appears in Fig. 2.5.6. •

In Fig. 2.5.7 there is a sketch of the graph of a function for which $\lim_{x \to +\infty} f(x) = b$ but there is no number N for which $f(x) \neq b$ whenever $x > N$. Consequently, the line $y = b$ is not a horizontal asymptote of the

FIGURE 2.5.7

graph of f. An example of such a function is given in Exercise 59 of Exercises 8.5.

Example 7 Find the horizontal asymptotes of the graph of the function f defined by

$$f(x) = \frac{x}{\sqrt{x^2+1}}$$

and draw a sketch of the graph.

Solution First consider $\lim_{x \to +\infty} f(x)$.

$$\lim_{x \to +\infty} f(x) = \lim_{x \to +\infty} \frac{x}{\sqrt{x^2+1}}$$

To evaluate this limit write $x = \sqrt{x^2}$ ($x > 0$, because $x \to +\infty$) and then divide the numerator and the denominator, under the radical sign, by x^2.

$$\begin{aligned}\lim_{x \to +\infty} \frac{x}{\sqrt{x^2+1}} &= \lim_{x \to +\infty} \frac{\sqrt{x^2}}{\sqrt{x^2+1}}\\ &= \lim_{x \to +\infty} \sqrt{\frac{1}{1+1/x^2}}\\ &= \sqrt{\frac{1}{1+\lim_{x \to +\infty}(1/x^2)}}\\ &= 1\end{aligned}$$

Therefore, by Definition 2.5.4(i) the line $y = 1$ is a horizontal asymptote.

Now consider $\lim_{x \to -\infty} f(x)$; in this case write $x = -\sqrt{x^2}$ because if $x \to -\infty$, $x < 0$. So

$$\begin{aligned}\lim_{x \to -\infty} f(x) &= \lim_{x \to -\infty} \frac{-\sqrt{x^2}}{\sqrt{x^2+1}}\\ &= \lim_{x \to -\infty} -\sqrt{\frac{1}{1+1/x^2}}\\ &= -\sqrt{\frac{1}{1+\lim_{x \to -\infty}(1/x^2)}}\\ &= -1\end{aligned}$$

Accordingly, by Definition 2.5.4(ii), the line $y = -1$ is a horizontal asymptote. A sketch of the graph appears in Fig. 2.5.8.

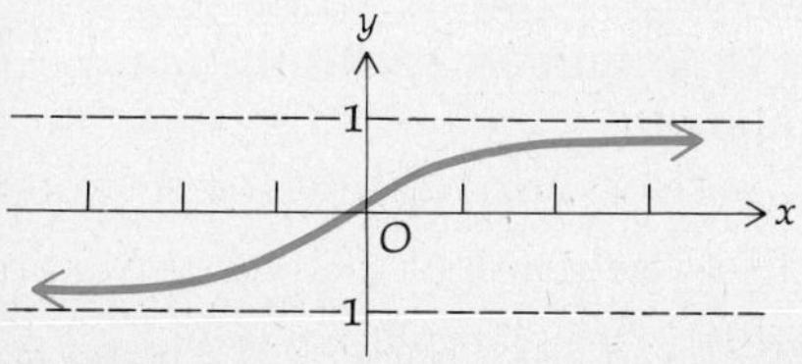

FIGURE 2.5.8

Example 8 Find the vertical and horizontal asymptotes of the graph of the equation $xy^2 - 2y^2 - 4x = 0$, and draw a sketch of the graph.

Solution Solve the given equation for y and obtain

$$y = \pm 2\sqrt{\frac{x}{x-2}} \tag{3}$$

Equation (3) defines two functions:

$$y = f_1(x) \quad \text{where } f_1 \text{ is defined by} \quad f_1(x) = +2\sqrt{\frac{x}{x-2}}$$

and

$$y = f_2(x) \quad \text{where } f_2 \text{ is defined by} \quad f_2(x) = -2\sqrt{\frac{x}{x-2}}$$

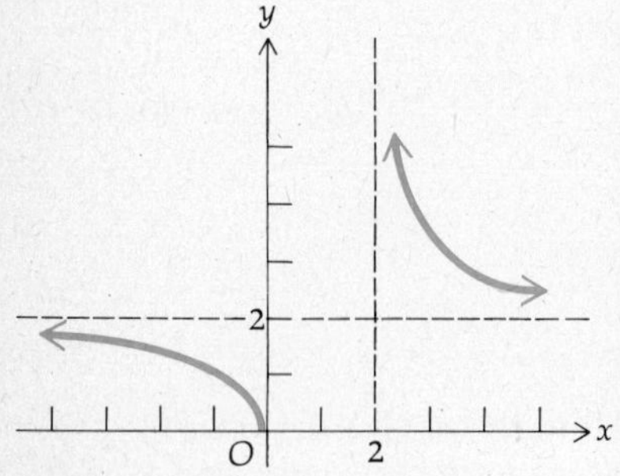

FIGURE 2.5.9

The graph of the given equation is composed of the graphs of the two functions f_1 and f_2. The domains of the two functions consist of those values of x for which $x/(x-2) \geq 0$. By using the result of Example 8 in Sec. 1.5 and excluding $x = 2$, we see that the domain of f_1 and f_2 is $(-\infty, 0] \cup (2, +\infty)$.

Now consider f_1. Because

$$\lim_{x \to 2^+} f_1(x) = \lim_{x \to 2^+} 2\sqrt{\frac{x}{x-2}} = +\infty$$

by Definition 2.4.8(i) the line $x = 2$ is a vertical asymptote of the graph of f_1.

$$\lim_{x \to +\infty} f_1(x) = \lim_{x \to +\infty} 2\sqrt{\frac{x}{x-2}} = \lim_{x \to +\infty} 2\sqrt{\frac{1}{1-2/x}} = 2$$

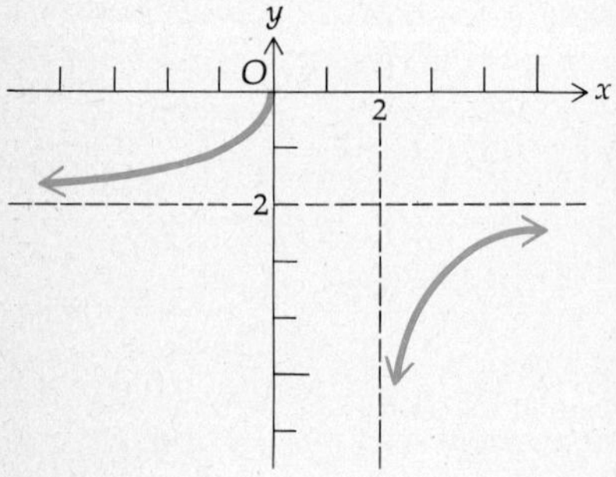

FIGURE 2.5.10

So by Definition 2.5.4(i) the line $y = 2$ is a horizontal asymptote of the graph of f_1.

Similarly, $\lim_{x \to -\infty} f_1(x) = 2$. A sketch of the graph of f_1 is shown in Fig. 2.5.9.

$$\lim_{x \to 2^+} f_2(x) = \lim_{x \to 2^+} \left[-2\sqrt{\frac{x}{x-2}}\right] = -\infty$$

Hence, by Definition 2.4.8(ii) the line $x = 2$ is a vertical asymptote of the graph of f_2.

$$\lim_{x \to +\infty} f_2(x) = \lim_{x \to +\infty} \left[-2\sqrt{\frac{x}{x-2}}\right] = \lim_{x \to +\infty} \left[-2\sqrt{\frac{1}{1-2/x}}\right] = -2$$

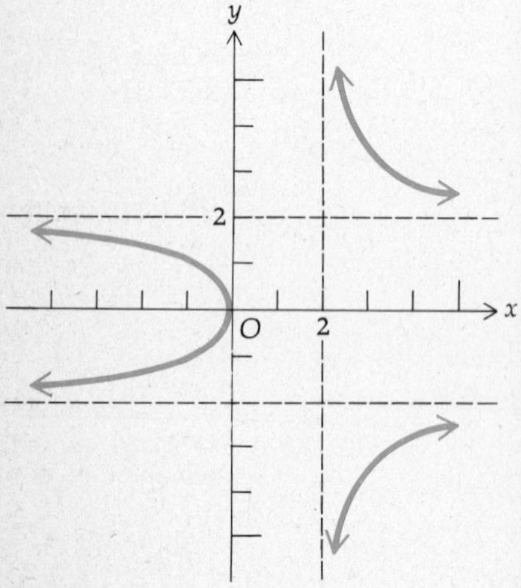

FIGURE 2.5.11

So by Definition 2.5.4(i) the line $y = -2$ is a horizontal asymptote of the graph of f_2.

Also, $\lim_{x \to -\infty} f_2(x) = -2$. A sketch of the graph of f_2 is shown in Fig. 2.5.10. The graph of the given equation is the union of the graphs of f_1 and f_2, and a sketch is shown in Fig. 2.5.11.

Example 9 Find the horizontal and vertical asymptotes of the graph of the function f if

$$f(x) = \frac{8x - 2x^2}{x^2 - 9}$$

Draw a sketch of the graph of f.

Solution

$$\lim_{x \to +\infty} f(x) = \lim_{x \to +\infty} \frac{\dfrac{8}{x} - 2}{1 - \dfrac{9}{x^2}}$$

$$= \frac{0 - 2}{1 - 0}$$

Thus

$$\lim_{x \to +\infty} f(x) = -2 \tag{4}$$

Also,

$$\lim_{x \to -\infty} f(x) = -2 \tag{5}$$

If the numerator and denominator are factored,

$$f(x) = \frac{2x(4 - x)}{(x - 3)(x + 3)}$$

Consider the four limits $\lim\limits_{x \to 3^+} f(x)$, $\lim\limits_{x \to 3^-} f(x)$ $\lim\limits_{x \to -3^+} f(x)$, and $\lim\limits_{x \to -3^-} f(x)$. To find $\lim\limits_{x \to 3^+} f(x)$, observe that $\lim\limits_{x \to 3^+} 2x(4 - x) = 6 > 0$ and $\lim\limits_{x \to 3^+} (x - 3)(x + 3) = 0$, where $(x - 3)(x + 3)$ is approaching 0 through positive values. Therefore by Limit Theorem 12(i) (2.4.4),

$$\lim_{x \to 3^+} f(x) = \lim_{x \to 3^+} \frac{2x(4 - x)}{(x - 3)(x + 3)} = +\infty \tag{6}$$

Since $\lim\limits_{x \to 3^-} 2x(4 - x) = 6 > 0$ and $\lim\limits_{x \to 3^-} (x - 3)(x + 3) = 0$, where $(x - 3)(x + 3)$ is approaching 0 through negative values, it follows from Limit Theorem 12(ii) that

$$\lim_{x \to 3^-} f(x) = \lim_{x \to 3^-} \frac{2x(4 - x)}{(x - 3)(x + 3)} = -\infty \tag{7}$$

Limit Theorem 12(iv) is used to find $\lim\limits_{x \to -3^+} f(x)$. Since $\lim\limits_{x \to -3^+} 2x(4 - x) = -42 < 0$ and $\lim\limits_{x \to -3^+} (x - 3)(x + 3) = 0$, where $(x - 3)(x + 3)$ is approaching 0 through negative values, then

$$\lim_{x \to -3^+} f(x) = \lim_{x \to -3^+} \frac{2x(4 - x)}{(x - 3)(x + 3)} = +\infty \tag{8}$$

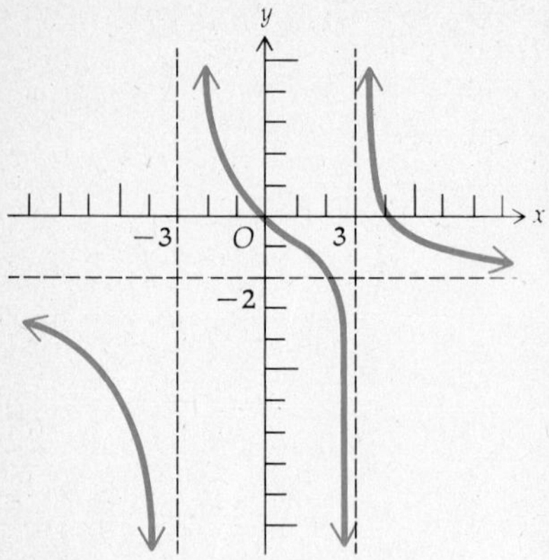

FIGURE 2.5.12

Because $\lim_{x \to -3^-} 2x(4 - x) = -42 < 0$ and $\lim_{x \to -3^-} (x - 3)(x + 3) = 0$, where $(x - 3)(x + 3)$ is approaching 0 through positive values, it follows from Limit Theorem 12(iii) that

$$\lim_{x \to -3^-} f(x) = \lim_{x \to -3^-} \frac{2x(4 - x)}{(x - 3)(x + 3)} = -\infty \tag{9}$$

From (4) and (5) and Definition 2.5.4 it follows that the line $y = -2$ is a horizontal asymptote of the graph of f. From (6)–(9) and Definition 2.4.8 we conclude that the lines $x = 3$ and $x = -3$ are vertical asymptotes of the graph.

From the given equation it follows that the graph of f contains the points (0, 0) and (4, 0). We plot a few other points and use the information given in (4)–(9) to obtain the sketch of the graph shown in Fig. 2.5.12.

EXERCISES 2.5

In Exercises 1 through 28, evaluate the limit.

1. $\lim_{x \to +\infty} \frac{1}{x^3}$

2. $\lim_{x \to -\infty} \frac{3}{x^4}$

3. $\lim_{t \to +\infty} \frac{2t + 1}{5t - 2}$

4. $\lim_{x \to -\infty} \frac{6x - 4}{3x + 1}$

5. $\lim_{x \to -\infty} \frac{2x + 7}{4 - 5x}$

6. $\lim_{x \to +\infty} \frac{1 + 5x}{2 - 3x}$

7. $\lim_{x \to +\infty} \frac{7x^2 - 2x + 1}{3x^2 + 8x + 5}$

8. $\lim_{s \to -\infty} \frac{4s^2 + 3}{2s^2 - 1}$

9. $\lim_{x \to +\infty} \frac{x + 4}{3x^2 - 5}$

10. $\lim_{x \to +\infty} \frac{x^2 + 5}{x^3}$

11. $\lim_{y \to +\infty} \frac{2y^2 - 3y}{y + 1}$

12. $\lim_{x \to +\infty} \frac{x^2 - 2x + 5}{7x^3 + x + 1}$

13. $\lim_{x \to -\infty} \frac{4x^3 + 2x^2 - 5}{8x^3 + x + 2}$

14. $\lim_{x \to +\infty} \frac{3x^4 - 7x^2 + 2}{2x^4 + 1}$

15. $\lim_{y \to +\infty} \frac{2y^3 - 4}{5y + 3}$

16. $\lim_{x \to -\infty} \frac{5x^3 - 12x + 7}{4x^2 - 1}$

17. $\lim_{x \to -\infty} \left(3x + \frac{1}{x^2}\right)$

18. $\lim_{t \to +\infty} \left(\frac{2}{t^2} - 4t\right)$

19. $\lim_{x \to +\infty} \frac{\sqrt{x^2 + 4}}{x + 4}$

20. $\lim_{x \to -\infty} \frac{\sqrt{x^2 + 4}}{x + 4}$

21. $\lim_{w \to -\infty} \frac{\sqrt{w^2 - 2w + 3}}{w + 5}$

22. $\lim_{y \to -\infty} \frac{\sqrt{y^4 + 1}}{2y^2 - 3}$

23. $\lim_{x \to +\infty} (\sqrt{x^2 + 1} - x)$

24. $\lim_{x \to +\infty} (\sqrt{x^2 + x} - x)$

25. $\lim_{r \to +\infty} (\sqrt{3r^2 + r} - 2r)$

26. $\lim_{x \to +\infty} (\sqrt[3]{x^3 + 1} - x)$

27. $\lim_{x \to -\infty} (\sqrt[3]{x^3 + x} - \sqrt[3]{x^3 + 1})$

28. $\lim_{t \to +\infty} \frac{\sqrt{t + \sqrt{t + \sqrt{t}}}}{\sqrt{t + 1}}$

In Exercises 29 through 40, find the horizontal and vertical asymptotes of the graph of the function defined by the given equation, and draw a sketch of the graph.

29. $f(x) = \frac{2x + 1}{x - 3}$

30. $f(x) = \frac{4 - 3x}{x + 1}$

31. $g(x) = 1 - \frac{1}{x}$

32. $h(x) = 1 + \frac{1}{x^2}$

33. $f(x) = \frac{2}{\sqrt{x^2 - 4}}$

34. $F(x) = \frac{-3x}{\sqrt{x^2 + 3}}$

35. $G(x) = \frac{4x^2}{x^2 - 9}$

36. $g(x) = \frac{x^2}{4 - x^2}$

37. $h(x) = \dfrac{2x}{6x^2 + 11x - 10}$ **38.** $f(x) = \dfrac{-1}{\sqrt{x^2 + 5x + 6}}$ **39.** $f(x) = \dfrac{4x^2}{\sqrt{x^2 - 2}}$ **40.** $h(x) = \dfrac{x}{\sqrt{x^2 - 9}}$

In Exercises 41 through 46, find the horizontal and vertical asymptotes of the graph of the given equation, and draw a sketch of the graph.

41. $3xy - 2x - 4y - 3 = 0$ **42.** $2xy + 4x - 3y + 6 = 0$ **43.** $x^2y^2 - x^2 + 4y^2 = 0$ **44.** $xy^2 + 3y^2 - 9x = 0$

45. $(y^2 - 1)(x - 3) = 6$ **46.** $2xy^2 + 4y^2 - 3x = 0$

For the functions defined in Exercises 47 and 48, prove that $\lim_{x \to +\infty} f(x) = 1$ *by applying Definition 2.5.1; that is, for any* $\epsilon > 0$, *show that there exists a number* $N > 0$ *such that* $|f(x) - 1| < \epsilon$ *whenever* $x > N$.

47. $f(x) = \dfrac{x}{x - 1}$ **48.** $f(x) = \dfrac{x^2 + 2x}{x^2 - 1}$

49. Prove that $\lim_{x \to -\infty} \dfrac{8x + 3}{2x - 1} = 4$ by showing that for any $\epsilon > 0$ there exists a number $N < 0$ such that $\left|\dfrac{8x + 3}{2x - 1} - 4\right| < \epsilon$ whenever $x < N$.

50. Prove part (i) of Limit Theorem 12 (2.4.4) if "$x \to a$" is replaced by "$x \to +\infty$."

51. Give a definition for each of the following: (a) $\lim_{x \to +\infty} f(x) = -\infty$; (b) $\lim_{x \to -\infty} f(x) = +\infty$; (c) $\lim_{x \to -\infty} f(x) = -\infty$.

52. Prove that $\lim_{x \to +\infty} (x^2 - 4) = +\infty$ by showing that for any $N > 0$ there exists an $M > 0$ such that $(x^2 - 4) > N$ whenever $x > M$.

53. Prove that $\lim_{x \to +\infty} (6 - x - x^2) = -\infty$ by applying the definition in Exercise 51(a).

54. Prove part (ii) of Limit Theorem 13 (2.5.3).

55. A rectangular field having an area of 2700 yd² is to be enclosed by a fence, and an additional fence is to be used to divide the field down the middle. The cost of the fence down the middle is \$4 per running yard, and the fence along the sides costs \$6 per running yard. Let x yd be the length of the dividing fence and $C(x)$ dollars be the cost of the fencing. (a) Write an equation defining $C(x)$, and state the domain of the function C. (b) Find $\lim_{x \to 0^+} C(x)$ and $\lim_{x \to +\infty} C(x)$, and explain these results in terms of the problem. (c) Draw a sketch of the graph of C.

56. A rectangular open tank is to have a square base, and its volume is to be 125 yd³. The cost per square yard for the bottom is \$8 and for the sides is \$4. Let x yd be the length of a side of the square base and $C(x)$ dollars be the cost of the material. (a) Write an equation defining $C(x)$, and state the domain of the function C. (b) Find $\lim_{x \to 0^+} C(x)$ and $\lim_{x \to +\infty} C(x)$, and explain these results in terms of the problem. (c) Draw a sketch of the graph of C.

57. A page of print is to contain 24 in.² of printed area, a margin of $1\frac{1}{2}$ in. at the top and bottom, and a margin of 1 in. at the sides. Let x in. be the width of the printed portion of the page and $A(x)$ in.² be the total area of the page. (a) Write an equation defining $A(x)$, and state the domain of the function A. (b) Find $\lim_{x \to 0^+} A(x)$ and $\lim_{x \to +\infty} A(x)$, and explain these results in terms of the problem. (c) Draw a sketch of the graph of A.

2.6 CONTINUITY OF A FUNCTION AT A NUMBER

In Sec. 2.1 we considered the function f defined by the equation

$$f(x) = \frac{(2x + 3)(x - 1)}{x - 1}$$

The function f is defined for all values of x except 1. A sketch of the graph consisting of all points on the line $y = 2x + 3$ except (1, 5) is shown in

Fig. 2.6.1. There is a break in the graph at the point (1, 5), and we state that the function f is *discontinuous* at the number 1.

If we define $f(1) = 2$, for instance, the function is defined for all values of x, but there is still a break in the graph, and the function is still discontinuous at 1. If we define $f(1) = 5$, however, there is no break in the graph, and the function f is said to be *continuous* at all values of x.

2.6.1 Definition The function f is said to be **continuous** at the number a if and only if the following three conditions are satisfied:

(i) $f(a)$ exists.
(ii) $\lim_{x \to a} f(x)$ exists.
(iii) $\lim_{x \to a} f(x) = f(a)$.

If one or more of these three conditions fails to hold at a, the function f is said to be **discontinuous** at a.

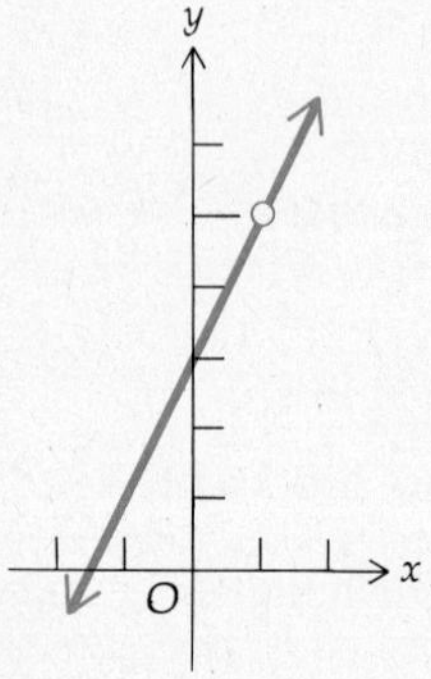

FIGURE 2.6.1

Some illustrations of discontinuous functions are now considered. In each illustration we draw a sketch of the graph, determine the points where there is a break in the graph, and show which of the three conditions in Definition 2.6.1 fails to hold at each discontinuity.

● ILLUSTRATION 1 Let f be defined as follows:

$$f(x) = \begin{cases} \dfrac{(2x + 3)(x - 1)}{x - 1} & \text{if } x \neq 1 \\ 2 & \text{if } x = 1 \end{cases}$$

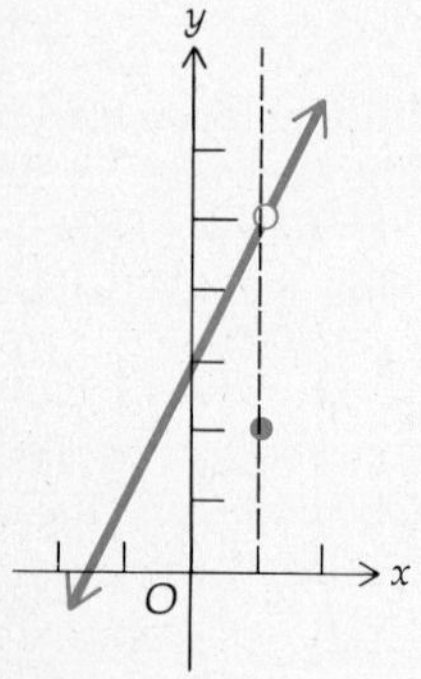

FIGURE 2.6.2

A sketch of the graph of this function is given in Fig. 2.6.2. Observe that there is a break in the graph at the point where $x = 1$. So we investigate there the conditions of Definition 2.6.1.

$f(1) = 2$; therefore, condition (i) is satisfied.

$\lim_{x \to 1} f(x) = 5$; therefore, condition (ii) is satisfied.

$\lim_{x \to 1} f(x) = 5$, but $f(1) = 2$; therefore, condition (iii) is not satisfied.

Thus f is discontinuous at 1. ●

Note that if in Illustration 1 $f(1)$ is defined to be 5, then $\lim_{x \to 1} f(x) = f(1)$ and f would be continuous at 1.

● ILLUSTRATION 2 Let f be defined by

$$f(x) = \frac{1}{x - 2}$$

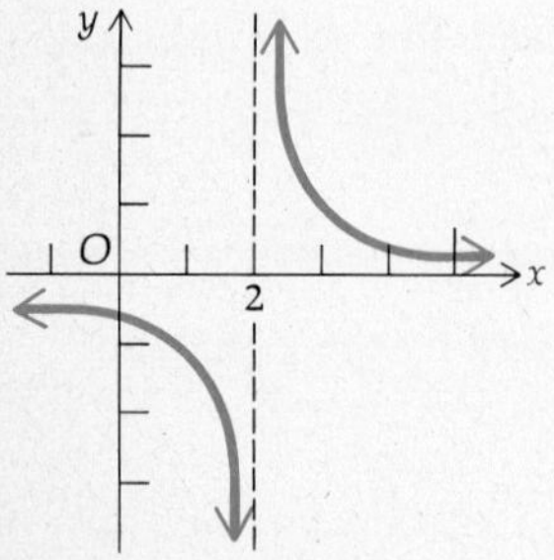

FIGURE 2.6.3

A sketch of the graph of f is given in Fig. 2.6.3. There is a break in the graph at the point where $x = 2$, and so we investigate there the conditions of Definition 2.6.1. $f(2)$ is not defined; therefore, condition (i) is not satisfied. Hence f is discontinuous at 2. ●

● ILLUSTRATION 3 Let g be defined by

$$g(x) = \begin{cases} \dfrac{1}{x-2} & \text{if } x \neq 2 \\ 3 & \text{if } x = 2 \end{cases}$$

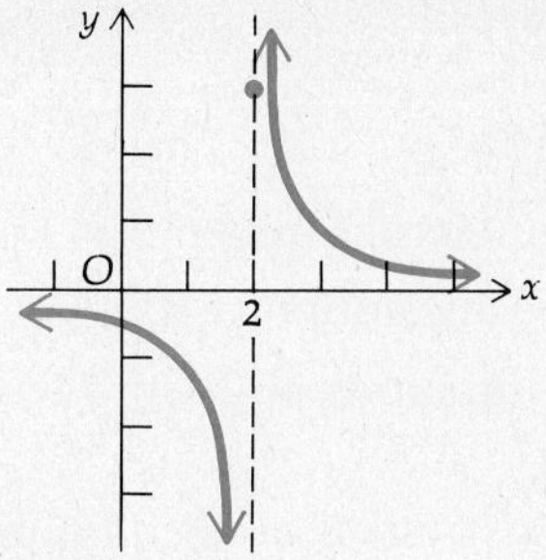

FIGURE 2.6.4

A sketch of the graph of g is shown in Fig. 2.6.4. The three conditions of Definition 2.6.1 are investigated at 2. $g(2) = 3$; therefore, condition (i) is satisfied. $\lim\limits_{x \to 2^-} g(x) = -\infty$, and $\lim\limits_{x \to 2^+} g(x) = +\infty$; therefore, condition (ii) is not satisfied. Thus g is discontinuous at 2. ●

● ILLUSTRATION 4 Let h be defined by

$$h(x) = \begin{cases} 3 + x & \text{if } x \leq 1 \\ 3 - x & \text{if } 1 < x \end{cases}$$

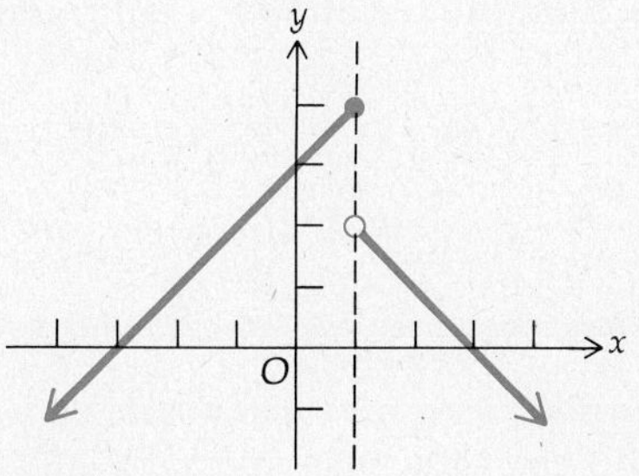

FIGURE 2.6.5

A sketch of the graph of h is shown in Fig. 2.6.5. Because there is a break in the graph at the point where $x = 1$, we investigate the conditions of Definition 2.6.1 at 1.

$h(1) = 4$; therefore, condition (i) is satisfied.

$$\lim_{x \to 1^-} h(x) = \lim_{x \to 1^-} (3 + x) = 4$$

$$\lim_{x \to 1^+} h(x) = \lim_{x \to 1^+} (3 - x) = 2$$

Because $\lim\limits_{x \to 1^-} h(x) \neq \lim\limits_{x \to 1^+} h(x)$, we conclude that $\lim\limits_{x \to 1} h(x)$ does not exist; therefore, condition (ii) fails to hold at 1. Hence h is discontinuous at 1. ●

● ILLUSTRATION 5 Let F be defined by

$$F(x) = \begin{cases} |x - 3| & \text{if } x \neq 3 \\ 2 & \text{if } x = 3 \end{cases}$$

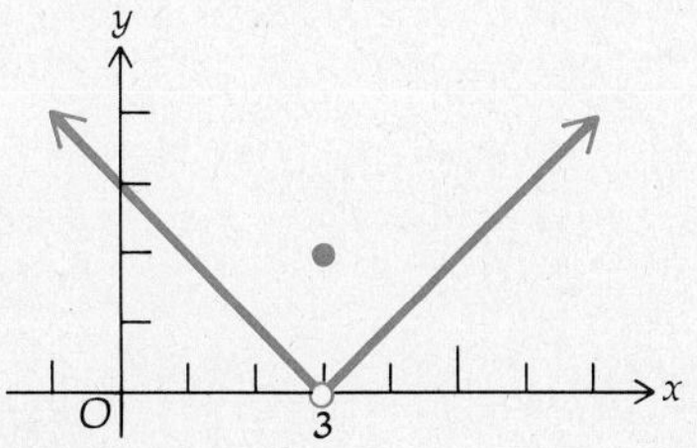

FIGURE 2.6.6

A sketch of the graph of F is shown in Fig. 2.6.6. We investigate the three conditions of Definition 2.6.1 at the point where $x = 3$. $F(3) = 2$; therefore, condition (i) is satisfied. $\lim\limits_{x \to 3^-} F(x) = 0$ and $\lim\limits_{x \to 3^+} F(x) = 0$. So $\lim\limits_{x \to 3} F(x)$ exists and is 0; therefore condition (ii) is satisfied. $\lim\limits_{x \to 3} F(x) = 0$ but $F(3) = 2$; therefore condition (iii) is not satisfied. Thus F is discontinuous at 3. ●

It should be apparent that the geometric notion of a break in the graph at a certain point is synonymous with the concept of a function being discontinuous at a certain value of the independent variable.

After Illustration 1 it was mentioned that if $f(1)$ had been defined to be 5, then f would be continuous at 1. This illustrates the concept of a *removable discontinuity.* In general, suppose that f is a function which is discontinuous at the number a, but for which $\lim\limits_{x \to a} f(x)$ exists. Then either $f(a) \neq \lim\limits_{x \to a} f(x)$ or else $f(a)$ does not exist. Such a discontinuity is called a **removable discontinuity** because if f were redefined at a so that $f(a) = \lim\limits_{x \to a} f(x)$, f becomes continuous at a. If the discontinuity is not removable, it is called an **essential discontinuity.**

Example 1 In each of the Illustrations 1 through 5, determine if the discontinuity is removable or essential.

Solution

In Illustration 1 the function is discontinuous at 1 but $\lim_{x \to 1} f(x) = 5$. By redefining $f(1) = 5$, then $\lim_{x \to 1} f(x) = f(1)$; so the discontinuity is removable.

In Illustration 2 the function f is discontinuous at 2. $\lim_{x \to 2} f(x)$ does not exist; hence the discontinuity is essential.

In Illustration 3, $\lim_{x \to 2} g(x)$ does not exist; so the discontinuity is essential.

In Illustration 4 the function h is discontinuous because $\lim_{x \to 1} h(x)$ does not exist; so again the discontinuity is essential.

In Illustration 5, $\lim_{x \to 3} F(x) = 0$, but $F(3) = 2$; so F is discontinuous at 3. However, if $F(3)$ is redefined to be 0, then the function is continuous at 3; so the discontinuity is removable.

EXERCISES 2.6

In Exercises 1 through 26, draw a sketch of the graph of the function; then by observing where there are breaks in the graph, determine the values of the independent variable at which the function is discontinuous and show why Definition 2.6.1 is not satisfied at each discontinuity.

1. $f(x) = \dfrac{x^2 + x - 6}{x + 3}$

2. $F(x) = \dfrac{x^2 - 3x - 4}{x - 4}$

3. $g(x) = \begin{cases} \dfrac{x^2 + x - 6}{x + 3} & \text{if } x \neq -3 \\ 1 & \text{if } x = -3 \end{cases}$

4. $G(x) = \begin{cases} \dfrac{x^2 - 3x - 4}{x - 4} & \text{if } x \neq 4 \\ 2 & \text{if } x = 4 \end{cases}$

5. $h(x) = \dfrac{x + 3}{x^2 + x - 6}$

6. $H(x) = \dfrac{x - 4}{x^2 - 3x - 4}$

7. $f(x) = \begin{cases} \dfrac{5}{x - 4} & \text{if } x \neq 4 \\ 1 & \text{if } x = 4 \end{cases}$

8. $g(x) = \begin{cases} \dfrac{1}{x + 2} & \text{if } x \neq -2 \\ 0 & \text{if } x = -2 \end{cases}$

9. $F(x) = \dfrac{x^4 - 16}{x^2 - 4}$

10. $h(x) = \dfrac{(x - 1)(x^2 - x - 12)}{x^2 - 5x + 4}$

11. $G(x) = \dfrac{x^2 - 4}{x^4 - 16}$

12. $H(x) = \dfrac{x^2 - 5x + 4}{(x - 1)(x^2 - x - 12)}$

13. $f(x) = \begin{cases} -1 & \text{if } x < 0 \\ 0 & \text{if } x = 0 \\ \sqrt{x} & \text{if } 0 < x \end{cases}$

14. $f(x) = \begin{cases} x - 1 & \text{if } x < 1 \\ 1 & \text{if } x = 1 \\ 1 - x & \text{if } 1 < x \end{cases}$

15. $g(t) = \begin{cases} t^2 - 4 & \text{if } t < 2 \\ 4 & \text{if } t = 2 \\ 4 - t^2 & \text{if } 2 < t \end{cases}$

16. $H(x) = \begin{cases} 1 + x & \text{if } x \leq -2 \\ 2 - x & \text{if } -2 < x \leq 2 \\ 2x - 1 & \text{if } 2 < x \end{cases}$

17. $g(x) = \begin{cases} \sqrt{-x} & \text{if } x < 0 \\ \sqrt[3]{x + 1} & \text{if } 0 \leq x \end{cases}$

18. $G(x) = \begin{cases} |2x + 5| & \text{if } x \neq -\frac{5}{2} \\ 3 & \text{if } x = -\frac{5}{2} \end{cases}$

19. $f(x) = \dfrac{x - 3}{|x - 3|}$

20. $F(x) = \begin{cases} \dfrac{x - 3}{|x - 3|} & \text{if } x \neq 3 \\ 0 & \text{if } x = 3 \end{cases}$

21. The greatest integer function (see Illustration 5 in Sec. 1.5).
22. The unit step function (see Exercise 31 in Exercises 1.6).
23. The signum function (see Exercise 32 in Exercises 1.6).
24. The function of Exercise 34 in Exercises 1.6.
25. The function of Exercise 45 in Exercises 1.6.
26. The function of Exercise 44 in Exercises 1.6.

In Exercises 27 through 38, prove that the function is discontinuous at the number a. Then determine if the discontinuity is removable or essential. If the discontinuity is removable, define f(a) so that the discontinuity is removed.

27. $f(x) = \dfrac{9x^2 - 4}{3x - 2}; a = \dfrac{2}{3}$

28. $f(s) = \begin{cases} \dfrac{1}{s+5} & \text{if } s \neq -5 \\ 0 & \text{if } s = -5 \end{cases}; a = -5$

29. $f(t) = \begin{cases} 9 - t^2 & \text{if } t \leq 2 \\ 3t + 2 & \text{if } 2 < t \end{cases}; a = 2$

30. $f(x) = \dfrac{x^2 - x - 12}{x^2 + 2x - 3}; a = -3$

31. $f(x) = \begin{cases} |x - 3| & \text{if } x \neq 3 \\ 2 & \text{if } x = 3 \end{cases}; a = 3$

32. $f(x) = \begin{cases} \dfrac{x^2 - 4x + 3}{x - 3} & \text{if } x \neq 3 \\ 5 & \text{if } x = 3 \end{cases}; a = 3$

33. $f(t) = \begin{cases} t^2 - 4 & \text{if } t \leq 2 \\ t & \text{if } t > 2 \end{cases}; a = 2$

34. $f(y) = \dfrac{\sqrt{y+5} - \sqrt{5}}{y}; a = 0$

35. $f(x) = \dfrac{3 - \sqrt{x+9}}{x}; a = 0$

36. $f(x) = \dfrac{\sqrt[3]{x+1} - 1}{x}; a = 0$

37. $f(x) = \dfrac{\sqrt{2 + \sqrt[3]{x}} - 2}{x - 8}; a = 8$

38. $f(x) = [\![x + 2]\!]; a = -1$

39. Let f be the function defined by

$$f(x) = \begin{cases} |x - [\![x]\!]| & \text{if } [\![x]\!] \text{ is even} \\ |x - [\![x + 1]\!]| & \text{if } [\![x]\!] \text{ is odd} \end{cases}$$

Draw a sketch of the graph of f. At what numbers is f discontinuous?

40. The function f is defined by

$$f(x) = \lim_{n \to +\infty} \frac{2nx}{1 - nx}$$

Draw a sketch of the graph of f. At what values of x is f discontinuous?

41. Given

$$f(x) = \begin{cases} \dfrac{\sqrt[3]{x + a^3} - a}{x} & \text{if } x \neq 0 \\ b & \text{if } x = 0 \end{cases}, a \neq 0$$

Find b such that f is continuous at $x = 0$.

42. Suppose the postage of a domestic first-class letter is computed as follows: \$0.15 for the first ounce or less, then \$0.13 for each ounce (or fractional part of an ounce) for the next 6 oz, and then \$1.06 if the weight is greater than 7 oz and less than or equal to 1 lb. If x oz is the weight of a letter, express the number of cents in the postage as a function of x. Draw a sketch of the graph of this function. Where is the function discontinuous? Show why Definition 2.6.1 is not satisfied at each discontinuity.

43. Shipping charges are usually based on a formula that offers a lower charge per pound as the size of the shipment is increased. Suppose x lb is the weight of a shipment, $C(x)$ dollars is the total cost of the shipment, and

$$C(x) = \begin{cases} 0.80x & \text{if } 0 < x \leq 50 \\ 0.70x & \text{if } 50 < x \leq 200 \\ 0.65x & \text{if } 200 < x \end{cases}$$

Draw a sketch of the graph of C. Determine where C is discontinuous, and show why Definition 2.6.1 is not satisfied at each discontinuity.

2.7 THEOREMS ON CONTINUITY

The following theorem about functions that are continuous at a number is obtained by applying Definition 2.6.1 and limit theorems.

2.7.1 Theorem If f and g are two functions that are continuous at the number a, then

(i) $f + g$ is continuous at a.
(ii) $f - g$ is continuous at a.
(iii) $f \cdot g$ is continuous at a.
(iv) f/g is continuous at a, provided that $g(a) \neq 0$.

To illustrate the kind of proof required for each part of this theorem, we prove part (i).

Because f and g are continuous at a, from Definition 2.6.1 it follows that

$$\lim_{x \to a} f(x) = f(a) \tag{1}$$

and

$$\lim_{x \to a} g(x) = g(a) \tag{2}$$

Therefore from (1) and (2) and Limit Theorem 4,

$$\lim_{x \to a} [f(x) + g(x)] = f(a) + g(a) \tag{3}$$

Equation (3) is the condition that $f + g$ is continuous at a, which furnishes the proof of (i).

The proofs of parts (ii), (iii), and (iv) are left as exercises (see Exercises 17–19).

2.7.2 Theorem A polynomial function is continuous at every number.

To prove this theorem, consider the polynomial function f defined by

$$f(x) = b_0x^n + b_1x^{n-1} + b_2x^{n-2} + \cdots + b_{n-1}x + b_n, \qquad b_0 \neq 0$$

where n is a nonnegative integer and $b_0, b_1, \ldots, b_n$ are real numbers. By successive applications of limit theorems it can be shown that if a is any number,

$$\lim_{x \to a} f(x) = b_0a^n + b_1a^{n-1} + b_2a^{n-2} + \cdots + b_{n-1}a + b_n$$

from which it follows that

$$\lim_{x \to a} f(x) = f(a)$$

thus establishing the theorem. The details of the proof are left as an exercise (see Exercise 20).

● ILLUSTRATION 1

(a) If $f(x) = x^3 - 2x^2 + 5x + 1$, then f is a polynomial function and therefore, by Theorem 2.7.2, is continuous at every number.

(b) If $g(x) = (x^2 - 5)^4$, then g is a polynomial function and thus is continuous everywhere.

(c) $f(x) \cdot g(x) = (x^3 - 2x^2 + 5x + 1)(x^2 - 5)^4$. From parts (a) and (b) and Theorem 2.7.1(iii) it follows that $f \cdot g$ is continuous everywhere. Note that because $(x^3 - 2x^2 + 5x + 1)(x^2 - 5)^4$ is a polynomial, it also follows directly from Theorem 2.7.2 that $f \cdot g$ is continuous everywhere. ●

2.7.3 Theorem A rational function is continuous at every number in its domain.

Proof If f is a rational function, it can be expressed as the quotient of two polynomial functions. So f can be defined by

$$f(x) = \frac{g(x)}{h(x)}$$

where g and h are two polynomial functions, and the domain of f consists of all numbers except those for which $h(x) = 0$.

If a is any number in the domain of f, then $h(a) \neq 0$; so by Limit Theorem 9,

$$\lim_{x \to a} f(x) = \frac{\lim_{x \to a} g(x)}{\lim_{x \to a} h(x)} \tag{4}$$

Because g and h are polynomial functions, by Theorem 2.7.2 they are continuous at a; so $\lim_{x \to a} g(x) = g(a)$ and $\lim_{x \to a} h(x) = h(a)$. Consequently, from (4).

$$\lim_{x \to a} f(x) = \frac{g(a)}{h(a)}$$

Thus f is continuous at every number in its domain. ■

Example 1 Given

$$f(x) = \frac{x^3 + 1}{x^2 - 9}$$

Determine all values of x for which f is continuous. Indicate the theorems applied.

Solution The domain of f is the set of all real numbers except those for which $x^2 - 9 = 0$. Because $x^2 - 9 = 0$ when $x = \pm 3$, it follows that the domain of f is the set of all real numbers except 3 and -3.

The function f is a rational function. Therefore by Theorem 2.7.3, f is continuous at all real numbers except 3 and -3.

The following theorem follows immediately from Theorem 2.2.12 and Definition 2.6.1.

2.7.4 Theorem If $g(x) = \sqrt[n]{x}$, where n is any positive integer, then g is continuous at a if either

(i) a is any positive number,

or (ii) a is a negative number or zero, and n is odd.

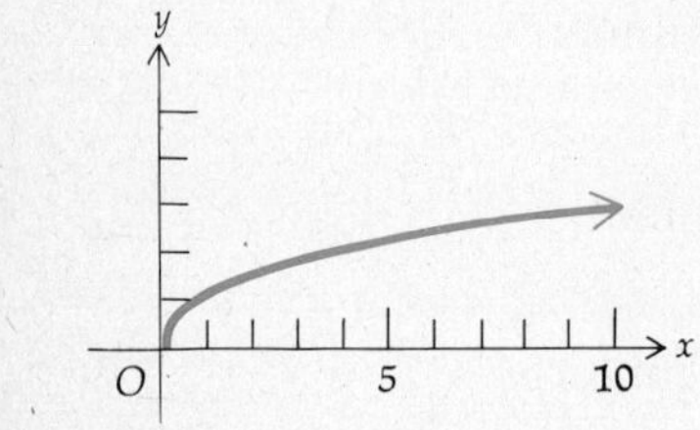

FIGURE 2.7.1

● ILLUSTRATION 2

(a) If $g(x) = \sqrt{x}$, then from Theorem 2.7.4(i) it follows that g is continuous at every positive number. In Fig. 2.7.1 there is a sketch of the graph of g.

(b) If $h(x) = \sqrt[3]{x}$, it follows from Theorem 2.7.4(i) and (ii) that h is continuous at every real number. A sketch of the graph of h is shown in Fig. 2.7.2. ●

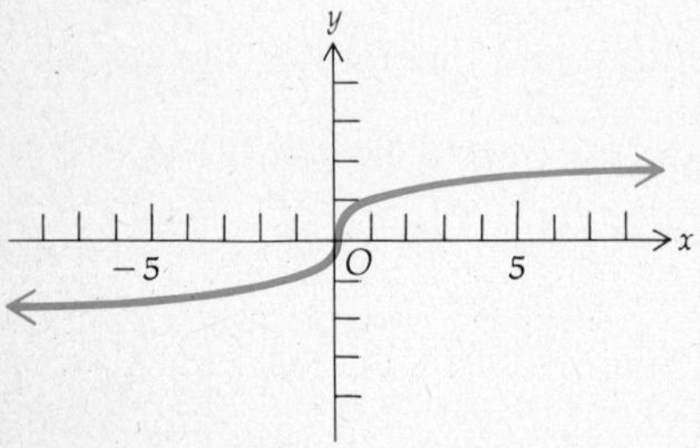

FIGURE 2.7.2

Definition 2.6.1 states that the function f is continuous at the number a if $f(a)$ exists, if $\lim_{x \to a} f(x)$ exists, and if

$$\lim_{x \to a} f(x) = f(a) \tag{5}$$

Definition 2.1.1 states that $\lim_{x \to a} f(x) = L$ if for any $\epsilon > 0$ there exists a $\delta > 0$ such that

$$|f(x) - L| < \epsilon \quad \text{whenever} \quad 0 < |x - a| < \delta$$

Applying this definition to (5) we have $\lim_{x \to a} f(x) = f(a)$ if for any $\epsilon > 0$ there exists a $\delta > 0$ such that

$$|f(x) - f(a)| < \epsilon \quad \text{whenever} \quad 0 < |x - a| < \delta \tag{6}$$

If f is continuous at a, $f(a)$ exists; thus in statement (6) it is not necessary that $|x - a| > 0$ because when $x = a$, (6) obviously holds. We have, then, the following theorem, which serves as a definition of continuity of a function that uses ϵ and δ notation.

2.7.5 Theorem The function f is said to be continuous at the number a if f is defined on some open interval containing a and if for any $\epsilon > 0$ there exists a $\delta > 0$ such that

$$|f(x) - f(a)| < \epsilon \quad \text{whenever} \quad |x - a| < \delta$$

Theorem 2.7.5 is used in proving the following important theorem regarding the limit of a composite function.

2.7.6 Theorem If $\lim_{x \to a} g(x) = b$ and if the function f is continuous at b,

$$\lim_{x \to a} (f \circ g)(x) = f(b)$$

or, equivalently,

$$\lim_{x \to a} f(g(x)) = f(\lim_{x \to a} g(x))$$

Proof Because f is continuous at b, we have the following statement from Theorem 2.7.5. For any $\epsilon_1 > 0$ there exists a $\delta_1 > 0$ such that

$$|f(y) - f(b)| < \epsilon_1 \quad \text{whenever} \quad |y - b| < \delta_1 \tag{7}$$

Because $\lim_{x \to a} g(x) = b$, for any $\delta_1 > 0$ there exists a $\delta_2 > 0$ such that

$$|g(x) - b| < \delta_1 \quad \text{whenever} \quad 0 < |x - a| < \delta_2 \tag{8}$$

Whenever $0 < |x - a| < \delta_2$, we replace y in statement (7) by $g(x)$ and obtain the following: For any $\epsilon_1 > 0$ there exists a $\delta_1 > 0$ such that

$$|f(g(x)) - f(b)| < \epsilon_1 \quad \text{whenever} \quad |g(x) - b| < \delta_1 \tag{9}$$

From statements (9) and (8) we conclude that for any $\epsilon_1 > 0$ there exists a $\delta_2 > 0$ such that

$$|f(g(x)) - f(b)| < \epsilon_1 \quad \text{whenever} \quad 0 < |x - a| < \delta_2$$

from which it follows that

$$\lim_{x \to a} f(g(x)) = f(b)$$

or, equivalently,

$$\lim_{x \to a} f(g(x)) = f(\lim_{x \to a} g(x))$$ ■

An immediate application of Theorem 2.7.6 is in proving Limit Theorems 9 and 10, the proofs of which were deferred until now. These theorems are restated and proved, but before the statement of each theorem a special case is given as an example.

Example 2 By using Theorem 2.7.6 and not Limit Theorem 9 (limit of a quotient), find

$$\lim_{x \to 2} \frac{4x - 3}{x^2 + 2x + 5}$$

Solution Let the functions f and g be defined by

$$f(x) = 4x - 3 \quad \text{and} \quad g(x) = x^2 + 2x + 5$$

Consider $f(x)/g(x)$ as the product of $f(x)$ and $1/g(x)$ and use Limit Theorem 6 (limit of a product). First, though, we must find $\lim_{x \to 2} 1/g(x)$, which is done by considering $1/g(x)$ as a composite function value.

If h is the function defined by $h(x) = 1/x$, then the composite function $h \circ g$ is the function defined by $h(g(x)) = 1/g(x)$. Now

$$\lim_{x \to 2} g(x) = \lim_{x \to 2} (x^2 + 2x + 5) = 13$$

To use Theorem 2.7.6, h must be continuous at 13, which follows from Theorem 2.2.11. Then

$$\begin{aligned}\lim_{x \to 2} \frac{1}{x^2 + 2x + 5} &= \lim_{x \to 2} \frac{1}{g(x)} \\ &= \lim_{x \to 2} h(g(x)) \\ &= h(\lim_{x \to 2} g(x)) \quad \text{(by Theorem 2.7.6)} \\ &= h(13) \\ &= \tfrac{1}{13}\end{aligned}$$

Therefore, by using Limit Theorem 6,

$$\begin{aligned}\lim_{x \to 2} \frac{4x - 3}{x^2 + 2x + 5} &= \lim_{x \to 2} (4x - 3) \cdot \lim_{x \to 2} \frac{1}{x^2 + 2x + 5} \\ &= 5 \cdot \tfrac{1}{13} \\ &= \tfrac{5}{13}\end{aligned}$$

Limit Theorem 9 If $\lim_{x \to a} f(x) = L$ and if $\lim_{x \to a} g(x) = M$ and $M \neq 0$, then

$$\lim_{x \to a} \frac{f(x)}{g(x)} = \frac{L}{M}$$

Proof Let h be the function defined by $h(x) = 1/x$. Then the composite function $h \circ g$ is defined by $h(g(x)) = 1/g(x)$. The function h is continuous everywhere except at 0, which follows from Theorem 2.2.11. Hence

$$\begin{aligned}\lim_{x \to a} \frac{1}{g(x)} &= \lim_{x \to a} h(g(x)) \\ &= h(\lim_{x \to a} g(x)) \quad \text{(by Theorem 2.7.6)} \\ &= h(M) \\ &= \frac{1}{M}\end{aligned}$$

From Limit Theorem 6 and the above result,

$$\begin{aligned}\lim_{x \to a} \frac{f(x)}{g(x)} &= \lim_{x \to a} f(x) \cdot \lim_{x \to a} \frac{1}{g(x)} \\ &= L \cdot \frac{1}{M} \\ &= \frac{L}{M}\end{aligned}$$

■

Example 3 By using Theorem 2.7.6 and not Limit Theorem 10, find

$$\lim_{x \to 7} \sqrt[4]{3x - 5}$$

Solution Let the functions f and h be defined by

$$f(x) = 3x - 5 \quad \text{and} \quad h(x) = \sqrt[4]{x}$$

The composite function $h \circ f$ is defined by $h(f(x)) = \sqrt[4]{3x - 5}$. $\lim_{x \to 7} f(x) = \lim_{x \to 7} (3x - 5) = 16$; so to use Theorem 2.7.6 for the composite function $h \circ f$, h must be continuous at 16, which follows from Theorem 2.7.4. The solution, then, is as follows.

$$\begin{aligned}\lim_{x \to 7} \sqrt[4]{3x - 5} &= \lim_{x \to 7} h(f(x)) \\ &= h(\lim_{x \to 7} f(x)) \quad \text{(by Theorem 2.7.6)} \\ &= h(16) \\ &= \sqrt[4]{16} \\ &= 2\end{aligned}$$

Limit Theorem 10 If $\lim_{x \to a} f(x) = L$, then

$$\lim_{x \to a} \sqrt[n]{f(x)} = \sqrt[n]{L}$$

if $L > 0$ and n is any positive integer, or if $L \le 0$ and n is a positive odd integer.

Proof Let h be the function defined by $h(x) = \sqrt[n]{x}$. Then the composite function $h \circ f$ is defined by $h(f(x)) = \sqrt[n]{f(x)}$. From Theorem 2.7.4 it follows that h is continuous at L if $L > 0$ and n is any positive integer or if $L \le 0$ and n is a positive odd integer. Therefore

$$\begin{aligned}\lim_{x \to a} \sqrt[n]{f(x)} &= \lim_{x \to a} h(f(x)) \\ &= h(\lim_{x \to a} f(x)) \quad \text{(by Theorem 2.7.6)} \\ &= h(L) \\ &= \sqrt[n]{L}\end{aligned}$$

■

Another application of Theorem 2.7.6 is in proving that *a continuous function of a continuous function is continuous.* This is now stated as a theorem.

2.7.7 Theorem If the function g is continuous at a and the function f is continuous at $g(a)$, then the composite function $f \circ g$ is continuous at a.

Proof Because g is continuous at a,

$$\lim_{x \to a} g(x) = g(a) \tag{10}$$

Now f is continuous at $g(a)$; thus we can apply Theorem 2.7.6 to the composite function $f \circ g$, thereby giving

$$\begin{aligned}\lim_{x \to a} (f \circ g)(x) &= \lim_{x \to a} f(g(x)) \\ &= f(\lim_{x \to a} g(x)) \\ &= f(g(a)) \quad \text{(by Eq. (10))} \\ &= (f \circ g)(a)\end{aligned}$$

which proves that $f \circ g$ is continuous at a. ■

The following example shows how Theorem 2.7.7 is used to determine the numbers for which a particular function is continuous.

Example 4 Given

$$h(x) = \sqrt{4 - x^2}$$

determine the values of x for which h is continuous.

Solution If $g(x) = 4 - x^2$ and $f(x) = \sqrt{x}$, then $h(x) = (f \circ g)(x)$. Because g is a polynomial function, g is continuous everywhere. Furthermore, f is continuous at every positive number (by Theorem 2.7.4). Therefore by Theorem 2.7.7, h is continuous at every number x for which $g(x) > 0$, that is, when $4 - x^2 > 0$. Hence h is continuous at every number in the open interval $(-2, 2)$.

In the next section ("Continuity on an Interval") the concept of continuity is extended to include continuity at an endpoint of a closed interval. *Right-hand continuity* and *left-hand continuity* are defined, and with these definitions we show that the function h of Example 4 is continuous on the closed interval $[-2, 2]$.

Example 5 Given

$$f(x) = \sqrt{\frac{2 - x}{3 + x}}$$

Determine the values of x for which f is continuous, and indicate the theorems you apply.

Solution Let g and h be the functions defined by

$$g(x) = \sqrt{x} \quad \text{and} \quad h(x) = \frac{2 - x}{3 + x}$$

The domain of g is the set of nonnegative numbers, and the domain of h is the set of all real numbers except -3.

$$(g \circ h)(x) = g(h(x))$$
$$= g\left(\frac{2-x}{3+x}\right)$$
$$= \sqrt{\frac{2-x}{3+x}}$$

Therefore

$$f(x) = (g \circ h)(x)$$

We wish to determine the domain of f. The domain of f is the set of all numbers for which $h(x)$ is nonnegative, that is, all numbers for which

$$\frac{2-x}{3+x} \geq 0 \tag{11}$$

To determine the values of x that satisfy (11), refer to Table 2.7.1. In the table we consider the intervals for which $x < -3$, $-3 < x < 2$, and $2 < x$, because the numerator changes sign when $x = 2$ and the denominator changes sign when $x = -3$.

TABLE 2.7.1

	$2-x$	$3+x$	$\frac{2-x}{3+x}$	$f(x)$
$x < -3$	+	−	−	does not exist
$x = -3$	5	0	undefined	does not exist
$-3 < x < 2$	+	+	+	+
$x = 2$	0	5	0	0
$2 < x$	−	+	−	does not exist

From the table observe that $f(x)$ exists if x is in the interval half-open on the left $(-3, 2]$; so this interval is the domain of f.

By Theorem 2.7.3, h is continuous at every number in its domain, that is, at all numbers except -3. The function g is continuous at all positive numbers; so from Theorem 2.7.4 it follows that f is continuous at all values of x for which $x \neq -3$ and

$$\frac{2-x}{3+x} > 0 \tag{12}$$

From Table 2.7.1 these are the numbers in the open interval $(-3, 2)$.

In Example 3 of Sec. 2.8, when our concept of continuity is extended to include continuity at an endpoint of a closed interval, it is shown that the function f of Example 5 is continuous on the interval $(-3, 2]$.

EXERCISES 2.7

In Exercises 1 through 16, determine all values of x for which the given function is continuous. Indicate which theorems you apply.

1. $f(x) = 3x^2 - 8x + 1$ **2.** $f(x) = (x^2 - 2)^4$ **3.** $f(x) = x^2(x + 3)^2$

4. $f(x) = (x - 5)^3(x^2 + 4)^5$ **5.** $g(x) = \dfrac{x}{x - 3}$ **6.** $h(x) = \dfrac{x + 1}{2x + 5}$

7. $F(x) = \dfrac{x + 1}{x^2 - 1}$ **8.** $G(x) = \dfrac{x - 2}{x^2 + 2x - 8}$ **9.** $f(x) = \dfrac{x^3 + 7}{x^2 - 4}$

10. $g(x) = \dfrac{x^2 - x - 2}{x^3 - 2x^2 + x - 2}$ **11.** $F(x) = \dfrac{x^2 + 2x - 8}{x^3 + 6x^2 + 5x - 12}$ **12.** $h(x) = \dfrac{x + 2}{x^3 - 7x - 6}$

13. $g(x) = \sqrt{x^2 - 16}$ **14.** $f(x) = \sqrt{16 - x^2}$ **15.** $h(x) = \sqrt[4]{x^2 + 1}$

16. $g(x) = \sqrt{x^2 + 4}$

17. Prove Theorem 2.7.1(ii). **18.** Prove Theorem 2.7.1(iii). **19.** Prove Theorem 2.7.1(iv).

20. Prove Theorem 2.7.2 and show step by step which limit theorems are used.

In Exercise 21 through 28, show the application of Theorem 2.7.6 to find the limit.

21. $\lim\limits_{x \to 3} \dfrac{1}{x^2 - 2x + 2}$ **22.** $\lim\limits_{t \to 3} \dfrac{2}{t^2 - 4}$ **23.** $\lim\limits_{t \to -1} \dfrac{t + 4}{t - 2}$ **24.** $\lim\limits_{x \to 2} \sqrt{x^3 + 1}$

25. $\lim\limits_{y \to -4} \sqrt[3]{3y + 4}$ **26.** $\lim\limits_{x \to 5} \dfrac{1}{\sqrt[3]{x^2 + 2}}$ **27.** $\lim\limits_{x \to 7} \dfrac{x^2 - 4}{\sqrt[4]{x^2 + 5x - 3}}$ **28.** $\lim\limits_{x \to 3} \dfrac{x + 3}{\sqrt[3]{1 - x^2}}$

In Exercises 29 through 36, functions f and g are defined. In each exercise define $f \circ g$, and determine all values of x for which $f \circ g$ is continuous.

29. $f(x) = x^3$; $g(x) = \sqrt{x}$ **30.** $f(x) = x^2$; $g(x) = x^2 - 3$ **31.** $f(x) = \sqrt{x}$; $g(x) = \dfrac{1}{x - 2}$

32. $f(x) = \sqrt{x}$; $g(x) = x + 1$ **33.** $f(x) = \dfrac{1}{x - 2}$; $g(x) = \sqrt{x}$ **34.** $f(x) = \sqrt[3]{x}$; $g(x) = \sqrt{x + 1}$

35. $f(x) = \dfrac{x + 1}{x - 1}$; $g(x) = \sqrt{x}$ **36.** $f(x) = \sqrt{x + 1}$; $g(x) = \sqrt[3]{x}$

In Exercises 37 through 54, determine all values of x for which the given function is continuous.

37. $f(x) = \sqrt{9 - x^2}$ **38.** $g(x) = \sqrt{x^2 - 9}$ **39.** $g(x) = \sqrt{\dfrac{x + 4}{x - 4}}$

40. $f(x) = \sqrt{\dfrac{4 - x}{4 + x}}$ **41.** $h(x) = \sqrt{\dfrac{2 - x}{5 + x}}$ **42.** $F(x) = \sqrt{6x^2 - x - 2}$

43. $f(x) = |x - 5|$ **44.** $f(x) = |9 - x^2|$ **45.** $g(x) = \dfrac{|x|}{x}$

46. $h(x) = \dfrac{x - 1}{|x - 1|}$ **47.** $f(x) = x^2(x^{-2} + x^{-1/2})^3$ **48.** $G(x) = \left(\dfrac{x^2}{x^2 - 4} - \dfrac{1}{x}\right)^{1/3}$

49. $g(x) = \left(\dfrac{2}{x-3} - \dfrac{x}{x^2-1}\right)^{1/3}$ **50.** $f(x) = (x^{-3} + (x-4)^{-1/2})^2$ **51.** $f(x) = [\![2x - 3]\!]$

52. $g(x) = \left[\!\!\left[\dfrac{2}{x}\right]\!\!\right]$ **53.** $g(x) = [\![\sqrt{1-x^2}]\!]$ **54.** $F(x) = 1 - x + [\![x]\!] - [\![1 - x]\!]$

55. Prove that if the function f is continuous at t, then $\lim_{h \to 0} f(t - h) = f(t)$.

56. Prove that if f is continuous at a and g is discontinuous at a, then $f + g$ is discontinuous at a.

57. If

$$f(x) = \begin{cases} -x & \text{if } x < 0 \\ 1 & \text{if } x \geq 0 \end{cases} \quad \text{and} \quad g(x) = \begin{cases} 1 & \text{if } x < 0 \\ x & \text{if } x \geq 0 \end{cases}$$

prove that f and g are both discontinuous at 0 but that the product $f \cdot g$ is continuous at 0.

58. Give an example of two functions that are both discontinuous at a number a, but whose sum is continuous at a.

59. Give an example to show that the product of two functions f and g may be continuous at a number a where f is continuous at a but g is discontinuous at a.

60. If the function g is continuous at a number a and the function f is discontinuous at a, is it possible for the quotient of the two functions, f/g, to be continuous at a? Prove your answer.

2.8 CONTINUITY ON AN INTERVAL

In Example 4 of Section 2.7 the function h, for which $h(x) = \sqrt{4 - x^2}$, was shown to be continuous at every number in the open interval $(-2, 2)$. Because of this fact we say that h is continuous on the open interval $(-2, 2)$. Following is the general definition of *continuity on an open interval.*

2.8.1 Definition A function is said to be **continuous on an open interval** if and only if it is continuous at every number in the open interval.

Example 1 If $f(x) = 1/(x - 3)$, on what open intervals is f continuous?

Solution The function f is continuous at every number except 3. Hence by Definition 2.8.1, f is continuous on every open interval that does not contain the number 3.

Refer again to the function h for which $h(x) = \sqrt{4 - x^2}$. We know that h is continuous on the open interval $(-2, 2)$. However, because h is not defined on any open interval containing either -2 or 2, we cannot consider $\lim_{x \to -2} h(x)$ or $\lim_{x \to 2} h(x)$. Hence, to discuss the question of the continuity of h on the closed interval $[-2, 2]$, we must extend the concept of continuity to include continuity at an endpoint of a closed interval. This is done by first defining *right-hand continuity* and *left-hand continuity.*

2.8.2 Definition The function f is said to be **continuous from the right at the number a** if and only if the following three conditions are satisfied:

(i) $f(a)$ exists.
(ii) $\lim_{x \to a^+} f(x)$ exists.
(iii) $\lim_{x \to a^+} f(x) = f(a)$.

2.8.3 Definition The function f is said to be **continuous from the left at the number** a if and only if the following three conditions are satisfied:

(i) $f(a)$ exists.
(ii) $\lim_{x \to a^-} f(x)$ exists.
(iii) $\lim_{x \to a^-} f(x) = f(a)$.

2.8.4 Definition A function whose domain includes the closed interval $[a, b]$ is said to be **continuous on** $[a, b]$ if and only if it is continuous on the open interval (a, b), as well as continuous from the right at a and continuous from the left at b.

Example 2 Prove that the function h, for which $h(x) = \sqrt{4 - x^2}$, is continuous on the closed interval $[-2, 2]$. Draw a sketch of the graph of h on $[-2, 2]$.

Solution The function h is continuous on the open interval $(-2, 2)$, and

$$\lim_{x \to -2^+} \sqrt{4 - x^2} = 0 = h(-2)$$

and

$$\lim_{x \to 2^-} \sqrt{4 - x^2} = 0 = h(2)$$

Thus by Definition 2.8.4, h is continuous on the closed interval $[-2, 2]$. A sketch of the graph of h on $[-2, 2]$ is shown in Fig. 2.8.1.

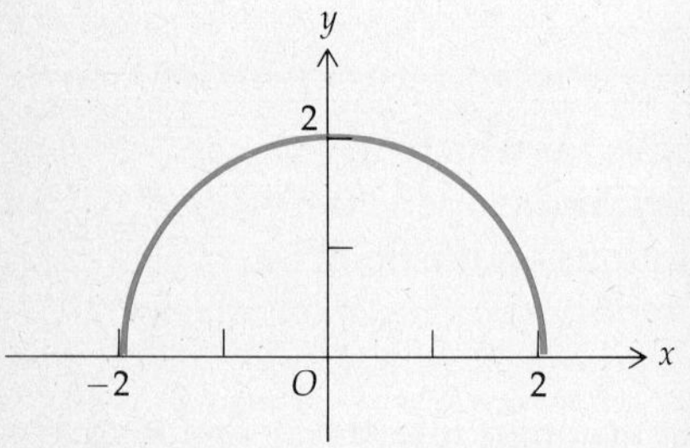

FIGURE 2.8.1

2.8.5 Definition

(i) A function whose domain includes the interval half-open on the right $[a, b)$ is said to be **continuous on** $[a, b)$ if and only if it is continuous on the open interval (a, b) and continuous from the right at a.
(ii) A function whose domain includes the interval half-open on the left $(a, b]$ is said to be **continuous on** $(a, b]$ if and only if it is continuous on the open interval (a, b) and continuous from the left at b.

Example 3 Given that f is the function defined by

$$f(x) = \sqrt{\frac{2 - x}{3 + x}}$$

determine whether f is continuous or discontinuous on each of the following intervals: $(-3, 2)$, $[-3, 2]$, $[-3, 2)$, $(-3, 2]$.

Solution In Example 5 of Section 2.7 it was determined that the domain of f is the interval half-open on the left $(-3, 2]$. It was also shown in that example that f is continuous at every number in the open interval $(-3, 2)$ and so is continuous on that interval. Because

$$\lim_{x \to 2^-} \sqrt{\frac{2-x}{3+x}} = 0$$
$$= f(2)$$

it follows that f is continuous from the left at 2. However, f is not continuous from the right at -3 because the function is not defined at -3. Furthermore,

$$\lim_{x \to -3^+} \sqrt{\frac{2-x}{3+x}} = +\infty$$

Therefore f is continuous on $(-3, 2]$ and discontinuous on $[-3, 2]$ and $[-3, 2)$.

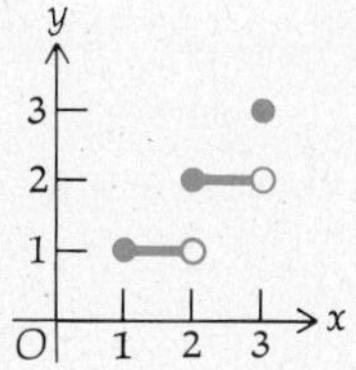

FIGURE 2.8.2

Example 4 Given $g(x) = [\![x]\!]$ and $1 \leq x \leq 3$. Discuss the continuity of g.

Solution Figure 2.8.2 shows a sketch of the graph of g. Suppose that u is any number in $(1, 2)$. Then $[\![u]\!] = 1$, and $\lim_{x \to u} [\![x]\!] = 1$. Hence if $1 < u < 2$, $\lim_{x \to u} g(u)$; so g is continuous on $(1, 2)$.

$$\lim_{x \to 1^+} g(x) = \lim_{x \to 1^+} [\![x]\!] = 1 = g(1)$$

Therefore g is continuous from the right at 1. But

$$\lim_{x \to 2^-} g(x) = \lim_{x \to 2^-} [\![x]\!] = 1$$

and $g(2) = 2$. So g is not continuous from the left at 2. The function g is therefore continuous on the interval half-open on the right $[1, 2)$.

Similarly, g is continuous at every number u for which $2 < u < 3$. At 2, g is continuous from the right since $\lim_{x \to 2^+} g(x) = g(2)$, but g is not continuous from the left at 3 since $\lim_{x \to 3^-} g(x) = 2$ and $g(3) = 3$. Hence g is also continuous on the interval $[2, 3)$.

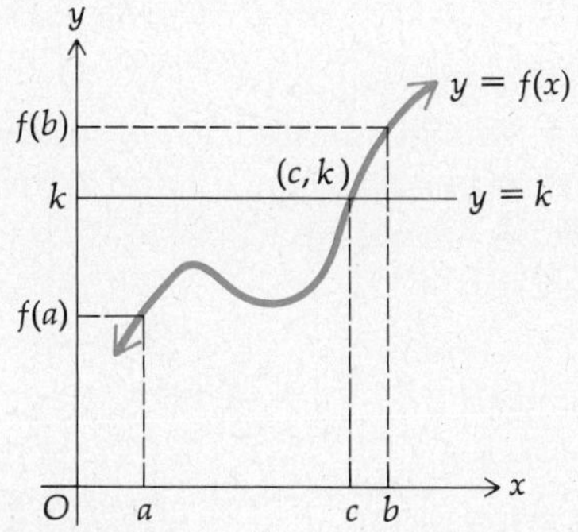

FIGURE 2.8.3

We now discuss an important theorem about a function that is continuous on a closed interval. It is called the **intermediate-value theorem,** and it is needed in later sections.

2.8.6 Theorem
Intermediate-Value Theorem

If the function f is continuous on the closed interval $[a, b]$, and if $f(a) \neq f(b)$, then for any number k between $f(a)$ and $f(b)$ there exists a number c between a and b such that $f(c) = k$.

The proof of this theorem is beyond the scope of this book; it can be found in an advanced calculus text. However, we discuss the geometric interpretation of the theorem. In Fig. 2.8.3, $(0, k)$ is any point on the y axis between the points $(0, f(a))$ and $(0, f(b))$. Theorem 2.8.6 states that the line $y = k$ must intersect the curve whose equation is $y = f(x)$ at the point (c, k), where c lies between a and b. Figure 2.8.3 shows this intersection.

Note that for some values of k there may be more than one possible value for c. The theorem states that there is always at least one value of c, but that it is not necessarily unique. Figure 2.8.4 shows three possible values of c (c_1, c_2, and c_3) for a particular k. Theorem 2.8.6 states that if the function f is continuous on a closed interval $[a, b]$, then f assumes every value between $f(a)$ and $f(b)$ as x assumes all values between a and b. The importance of the continuity of f on $[a, b]$ is demonstrated in the following illustration.

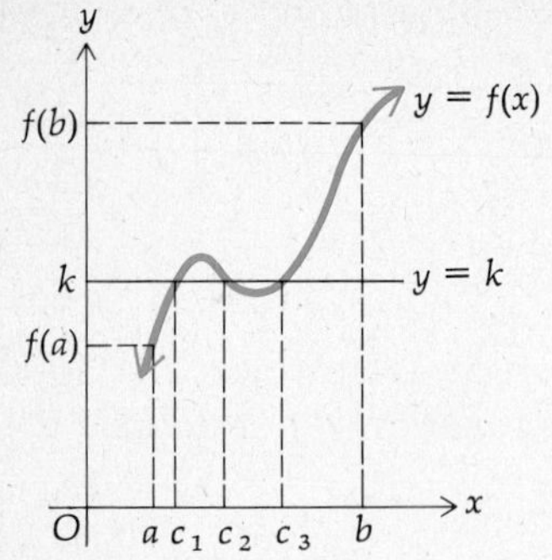

FIGURE 2.8.4

● ILLUSTRATION 1 Consider the function f defined by

$$f(x) = \begin{cases} x - 1 & \text{if } 0 \leq x \leq 2 \\ x^2 & \text{if } 2 < x \leq 3 \end{cases}$$

A sketch of the graph of this function is shown in Fig. 2.8.5.

The function f is discontinuous at 2, which is in the closed interval $[0, 3]$; $f(0) = -1$ and $f(3) = 9$. If k is any number between 1 and 4, there is no value of c such that $f(c) = k$ because there are no function values between 1 and 4. ●

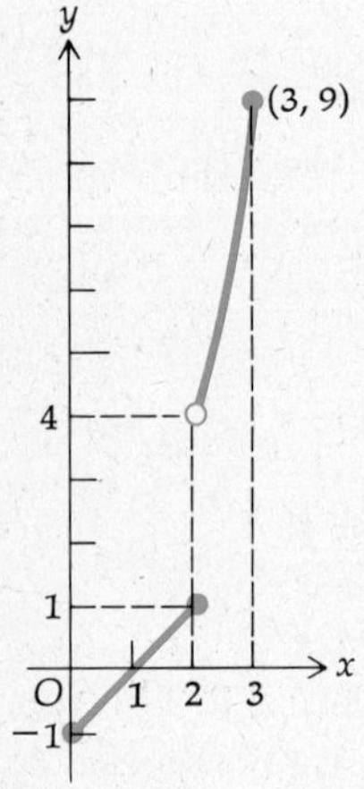

FIGURE 2.8.5

Another function for which Theorem 2.8.6 does not hold is given in the next illustration.

● ILLUSTRATION 2 Let the function g be defined by

$$g(x) = \frac{2}{x - 4}$$

A sketch of the graph of this function is shown in Fig. 2.8.6.

The function g is discontinuous at 4, which is in the closed interval $[2, 5]$; $g(2) = -1$ and $g(5) = 2$. If k is any number between -1 and 2, there is no value of c between 2 and 5 such that $g(c) = k$. In particular, if $k = 1$, then $g(6) = 1$, but 6 is not in the interval $[2, 5]$. ●

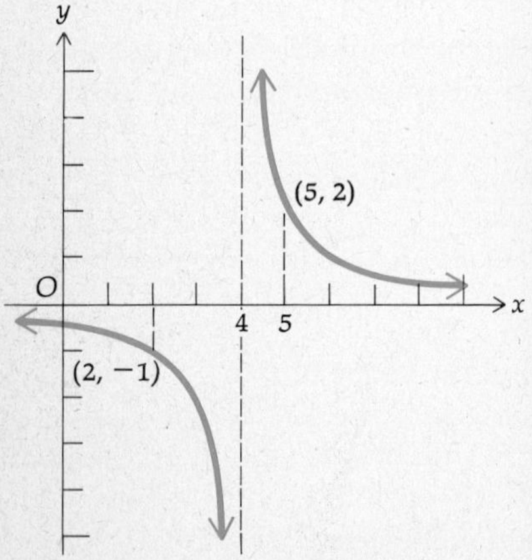

FIGURE 2.8.6

Example 5 Given the function f defined by

$$f(x) = 4 + 3x - x^2 \qquad 2 \leq x \leq 5$$

verify the intermediate-value theorem if $k = 1$.

Solution $f(2) = 6$; $f(5) = -6$. To find c, set $f(c) = k = 1$; that is,

$$4 + 3c - c^2 = 1$$

$$c^2 - 3c - 3 = 0$$

This gives

$$c = \frac{3 \pm \sqrt{21}}{2}$$

We reject $\frac{1}{2}(3 - \sqrt{21})$ because this number is outside the interval $[2, 5]$. The number $\frac{1}{2}(3 + \sqrt{21})$ is in the interval $[2, 5]$, and

$$f\left(\frac{3 + \sqrt{21}}{2}\right) = 1$$

EXERCISES 2.8

In Exercises 1 through 14, determine whether the function is continuous or discontinuous on each of the indicated intervals.

1. $f(x) = \dfrac{2}{x+5}$; $(3, 7), [-6, 4], (-\infty, 0), (-5, +\infty), [-5, +\infty), [-10, -5)$.

2. $g(x) = \dfrac{x}{x-2}$; $(-\infty, 0], [0, +\infty), (0, 2), (0, 2], [2, +\infty), (2, +\infty)$.

3. $f(t) = \dfrac{t}{t^2-1}$; $(0, 1), (-1, 1), [0, 1], (-1, 0], (-\infty, -1], (1, +\infty)$.

4. $f(r) = \dfrac{r+3}{r^2-4}$; $(0, 4], (-2, 2), (-\infty, -2], (2, +\infty), [-4, 4], (-2, 2]$.

5. $g(x) = \sqrt{x^2-9}$; $(-\infty, -3), (-\infty, -3], (3, +\infty), [3, +\infty), (-3, 3)$.

6. $f(x) = [\![x]\!]$; $(-\frac{1}{2}, \frac{1}{2}), (\frac{1}{4}, \frac{1}{2}), (1, 2), [1, 2), (1, 2]$.

7. $f(t) = \dfrac{|t-1|}{t-1}$; $(-\infty, 1), (-\infty, 1], [-1, 1], (-1, +\infty), (1, +\infty)$.

8. $h(x) = \begin{cases} 2x-3 & \text{if } x < -2 \\ x-5 & \text{if } -2 \le x \le 1 \\ 3-x & \text{if } 1 < x \end{cases}$; $(-\infty, 1), (-2, +\infty), (-2, 1), [-2, 1), [-2, 1]$

9. $f(x) = \sqrt{4-x^2}$; $(-2, 2), [-2, 2], [-2, 2), (-2, 2], (-\infty, -2], (2, +\infty)$.

10. $f(x) = \sqrt{\dfrac{2+x}{2-x}}$; $(-2, 2), [-2, 2], [-2, 2), (-2, 2], (-\infty, -2), [2, +\infty)$.

11. $F(y) = \dfrac{1}{\sqrt{3+2y-y^2}}$; $(-1, 3), [-1, 3], [-1, 3), (-1, 3]$.

12. $f(x) = \sqrt{3+2x-x^2}$; $(-1, 3), [-1, 3], [-1, 3), (-1, 3]$.

13. $G(x) = \sqrt{\dfrac{9-x^2}{4-x}}$; $(-\infty, -3), (-3, 3), [-3, 3], [-3, 3), [3, 4], (3, 4], [4, +\infty), (4, +\infty)$.

14. $g(x) = \sqrt{\dfrac{2+x}{25-x^2}}$; $(-\infty, -5), (-\infty, -5], [-5, -2], [-2, 5], [-2, 5), (-2, 5], (-2, 5), [5, +\infty), (5, +\infty)$.

In Exercises 15 through 22, determine the intervals on which the given function is continuous.

15. $f(x) = \dfrac{1}{x^2-4}$ 16. $g(y) = \dfrac{4}{y+1}$ 17. $h(x) = \sqrt{x^2-x-12}$ 18. $F(x) = \sqrt{\dfrac{x-5}{x+6}}$

19. The function of Exercise 35 in Exercises 1.6.

20. The function of Exercise 36 in Exercises 1.6.

21. The function of Exercise 39 in Exercises 1.5.

22. The function of Exercise 40 in Exercises 1.5.

23. Prove that the function g for which $g(x) = \sqrt{x}$ is continuous on the interval $[0, +\infty)$.

24. Prove that the function f of Example 4 in Section 2.3 is continuous on the closed interval $[-3, 3]$.

25. Prove that the function of Exercise 17 in Exercises 2.3 is continuous on the closed interval $[-2, 2]$.

26. Prove that the function of Exercise 18 in Exercises 2.3 is continuous on the closed interval $[-1, 1]$.

27. Prove that the function of Exercise 21 in Exercises 2.3 is continuous on the closed interval $[-3, 3]$.

28. Prove that the function of Exercise 22 in Exercises 2.3 is continuous on the closed interval $[-1, 1]$.

In Exercises 29 through 32, find the values of the constants c and k that make the function continuous on $(-\infty, +\infty)$, and draw a sketch of the graph of the resulting function.

29. $f(x) = \begin{cases} 3x + 7 & \text{if } x \leq 4 \\ kx - 1 & \text{if } 4 < x \end{cases}$

30. $f(x) = \begin{cases} kx - 1 & \text{if } x < 2 \\ kx^2 & \text{if } 2 \leq x \end{cases}$

31. $f(x) = \begin{cases} x & \text{if } x \leq 1 \\ cx + k & \text{if } 1 < x < 4 \\ -2x & \text{if } 4 \leq x \end{cases}$

32. $f(x) = \begin{cases} x + 2c & \text{if } x < -2 \\ 3cx + k & \text{if } -2 \leq x \leq 1 \\ 3x - 2k & \text{if } 1 < x \end{cases}$

33. Prove that the function of Exercise 55 in Exercises 2.5 is continuous on its domain.

34. Prove that the function of Exercise 56 in Exercises 2.5 is continuous on its domain.

35. Prove that the function of Exercise 57 in Exercises 2.5 is continuous on its domain.

36. If $f(x) = \sqrt{k^2 - x^2}$ and $k > 0$, prove that f is continuous on the closed interval $[-k, k]$.

In Exercises 37 through 44, a function f and a closed interval [a, b] are given. Determine if the intermediate-value theorem holds for the given value of k. If the theorem holds, find a number c such that f(c) = k. If the theorem does not hold, give the reason. Draw a sketch of the curve and the line y = k.

37. $f(x) = 2 + x - x^2$; $[a, b] = [0, 3]$; $k = 1$

38. $f(x) = x^2 + 5x - 6$; $[a, b] = [-1, 2]$; $k = 4$

39. $f(x) = \sqrt{25 - x^2}$; $[a, b] = [-4.5, 3]$; $k = 3$

40. $f(x) = -\sqrt{100 - x^2}$; $[a, b] = [0, 8]$; $k = -8$

41. $f(x) = \dfrac{4}{x + 2}$; $[a, b] = [-3, 1]$; $k = \frac{1}{2}$

42. $f(x) = \left\{\begin{matrix} 1 + x & \text{if } -4 \leq x \leq -2 \\ 2 - x & \text{if } -2 < x \leq 1 \end{matrix}\right\}$; $[a, b] = [-4, 1]$; $k = \frac{1}{2}$

43. $f(x) = \left\{\begin{matrix} x^2 - 4 & \text{if } -2 \leq x < 1 \\ x^2 - 1 & \text{if } 1 \leq x \leq 3 \end{matrix}\right\}$; $[a, b] = [-2, 3]$; $k = -1$

44. $f(x) = \dfrac{5}{2x - 1}$; $[a, b] = [0, 1]$; $k = 2$

45. Given that f is defined by

$$f(x) = \begin{cases} g(x) & \text{if } a \leq x < b \\ h(x) & \text{if } b \leq x \leq c \end{cases}$$

If g is continuous on $[a, b)$, and h is continuous on $[b, c]$, can we conclude that f is continuous on $[a, c]$? If your answer is yes, prove it. If your answer is no, what additional condition or conditions would assure continuity of f on $[a, c]$?

46. If f is the function defined by $f(x) = \sqrt[n]{x^m}$, where m and n are positive integers, prove that f is continuous on the interval $[0, +\infty)$.

47. Find the largest value of k for which the function defined by $f(x) = [\![x^2 - 2]\!]$ is continuous on the interval $[3, 3 + k)$.

48. Suppose f is a function for which $0 \leq f(x) \leq 1$ if $0 \leq x \leq 1$. Prove that if f is continuous on $[0, 1]$, there is at least one number c in $[0, 1]$ such that $f(c) = c$. (*Hint:* If neither 0 nor 1 qualifies as c, then $f(0) > 0$ and $f(1) < 1$. Consider the function g for which $g(x) = f(x) - x$ and apply the intermediate-value theorem to g on $[0, 1]$.)

49. Show that the intermediate-value theorem guarantees that the equation $x^3 - 4x^2 + x + 3 = 0$ has a root between 1 and 2.

50. Show that the intermediate-value theorem guarantees that the equation $x^3 + x + 3 = 0$ has a root between -2 and -1.

2.9 THE SQUEEZE THEOREM AND $\lim_{t \to 0} \frac{\sin t}{t}$

The following theorem, sometimes referred to as the **squeeze theorem,** is considered now because it is used to prove a significant limit in this section and some important theorems in later sections.

2.9.1 Theorem The Squeeze Theorem

Suppose that the functions f, g, and h are defined on some open interval I containing a except possibly at a itself, and that $f(x) \le g(x) \le h(x)$ for all x in I for which $x \ne a$. Also suppose that $\lim_{x \to a} f(x)$ and $\lim_{x \to a} h(x)$ both exist and are equal to L. Then $\lim_{x \to a} g(x)$ exists and is equal to L.

Before Theorem 2.9.1 is proved, consider the following illustration, which interprets the theorem geometrically.

● ILLUSTRATION 1 Let the functions f, g, and h be defined by

$$f(x) = -4(x-2)^2 + 3$$

$$g(x) = \frac{(x-2)(x^2-4x+7)}{x-2}$$

$$h(x) = 4(x-2)^2 + 3$$

The graphs of f and h are parabolas having their vertex at (2, 3). The graph of g is a parabola with its vertex (2, 3) deleted. Sketches of these graphs are shown in Fig. 2.9.1. The function g is not defined when $x = 2$; however, for all $x \ne 2$, $f(x) \le g(x) \le h(x)$. Furthermore, $\lim_{x \to 2} f(x) = 3$ and $\lim_{x \to 2} h(x) = 3$. The hypothesis of Theorem 2.9.1 is therefore satisfied, and it follows that $\lim_{x \to 2} g(x) = 3$. ●

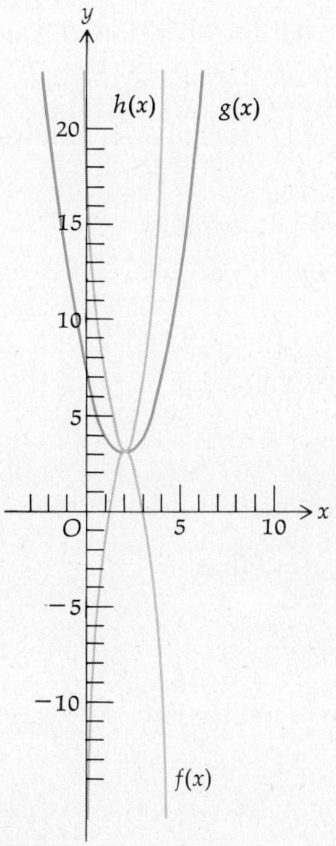

FIGURE 2.9.1

Proof of Theorem 2.9.1 To prove that $\lim_{x \to a} g(x) = L$ we must show that for any $\epsilon > 0$ there is a $\delta > 0$ such that

$$|g(x) - L| < \epsilon \quad \text{whenever} \quad 0 < |x - a| < \delta \tag{1}$$

It is given that

$$\lim_{x \to a} f(x) = L \quad \text{and} \quad \lim_{x \to a} h(x) = L$$

and so for any $\epsilon > 0$ there is a $\delta_1 > 0$ such that

$$|f(x) - L| < \epsilon \quad \text{whenever} \quad 0 < |x - a| < \delta_1$$

or, equivalently,

$$L - \epsilon < f(x) < L + \epsilon \quad \text{whenever} \quad 0 < |x - a| < \delta_2 \tag{2}$$

and a $\delta_2 > 0$ such that

$$|h(x) - L| < \epsilon \quad \text{whenever} \quad 0 < |x - a| < \delta_2$$

or, equivalently,

$$L - \epsilon < h(x) < L + \epsilon \quad \text{whenever} \quad 0 < |x - a| < \delta_2 \tag{3}$$

Let $\delta = \min(\delta_1, \delta_2)$, and so $\delta \leq \delta_1$ and $\delta \leq \delta_2$. Therefore it follows from statement (2) that

$$L - \epsilon < f(x) \quad \text{whenever} \quad 0 < |x - a| < \delta \tag{4}$$

and from statement (3) that

$$h(x) < L + \epsilon \quad \text{whenever} \quad 0 < |x - a| < \delta \tag{5}$$

We are given that

$$f(x) \leq g(x) \leq h(x) \tag{6}$$

From statements (4), (5), and (6),

$$L - \epsilon < f(x) \leq g(x) \leq h(x) < L + \epsilon \quad \text{whenever} \quad 0 < |x - a| < \delta$$

Therefore

$$L - \epsilon < g(x) < L + \epsilon \quad \text{whenever} \quad 0 < |x - a| < \delta$$

or, equivalently,

$$|g(x) - L| < \epsilon \quad \text{whenever} \quad 0 < |x - a| < \delta$$

which is statement (1). Hence

$$\lim_{x \to a} g(x) = L$$

■

Example 1 Given $|g(x) - 2| \leq 3(x - 1)^2$ for all x. Use the squeeze theorem to find $\lim_{x \to 1} g(x)$.

Solution Because $|g(x) - 2| \leq 3(x - 1)^2$ for all x, it follows from Theorem 1.1.18 that

$$-3(x - 1)^2 \leq g(x) - 2 \leq 3(x - 1)^2 \qquad \text{for all } x$$

$$-3(x - 1)^2 + 2 \leq g(x) \leq 3(x - 1)^2 + 2 \qquad \text{for all } x$$

Let $f(x) = -3(x - 1)^2 + 2$ and $h(x) = 3(x - 1)^2 + 2$. Then

$$\lim_{x \to 1} f(x) = 2 \quad \text{and} \quad \lim_{x \to 1} h(x) = 2 \tag{7}$$

Furthermore, for all x,

$$f(x) \leq g(x) \leq h(x) \tag{8}$$

It follows from (7), (8), and the squeeze theorem that

$$\lim_{x \to 1} g(x) = 2$$

Example 2 Use the squeeze theorem to prove that

$$\lim_{x \to 0} \left| x \sin \frac{1}{x} \right| = 0$$

Solution Because $-1 \leq \sin t \leq 1$ for all t, then

$$0 \leq \left| \sin \frac{1}{x} \right| \leq 1 \qquad \text{if } x \neq 0$$

Therefore, if $x \neq 0$,

$$\left|x \sin \frac{1}{x}\right| = |x|\left|\sin \frac{1}{x}\right|$$
$$\leq |x|$$

Hence

$$0 \leq \left|x \sin \frac{1}{x}\right| \leq |x| \qquad \text{if } x \neq 0 \tag{9}$$

Because $\lim_{x\to 0} 0 = 0$ and $\lim_{x\to 0} |x| = 0$, it follows from inequality (9) and the squeeze theorem that

$$\lim_{x\to 0}\left|x \sin \frac{1}{x}\right| = 0$$

In Sec. 3.5,

$$\lim_{t\to 0} \frac{\sin t}{t}$$

appears, and we need to know its value.

If $f(t) = (\sin t)/t$, $f(0)$ is not defined. However, the following theorem states that $\lim_{t\to 0} f(t)$ exists and is equal to 1. The squeeze theorem is used in its proof.

2.9.2 Theorem

$$\lim_{t\to 0} \frac{\sin t}{t} = 1$$

Proof First assume that $0 < t < \frac{1}{2}\pi$. Refer to Fig. 2.9.2, which shows the unit circle $x^2 + y^2 = 1$ and the shaded sector BOP, where B is the point $(1, 0)$ and P is the point $(\cos t, \sin t)$. The area of a circular sector of radius r and central angle of radian measure t is determined by $\frac{1}{2}r^2t$; so if S square units is the area of sector BOP,

$$S = \tfrac{1}{2}t \tag{10}$$

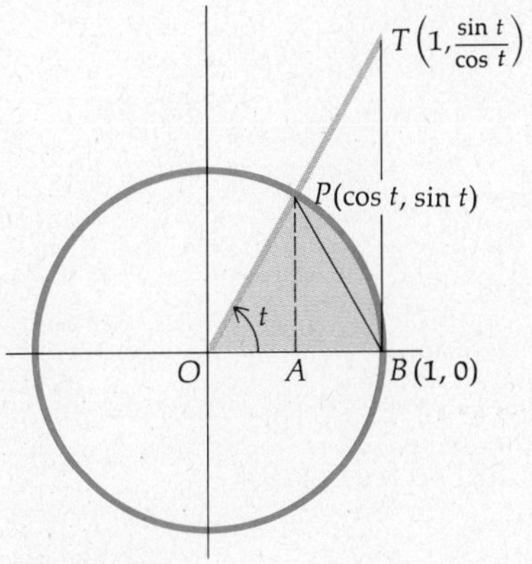

FIGURE 2.9.2

Consider now the triangle BOP, and let K_1 square units be the area of this triangle. Hence

$$K_1 = \tfrac{1}{2}|\overline{AP}| \cdot |\overline{OB}| = \tfrac{1}{2}(\sin t)\cdot(1) = \tfrac{1}{2}\sin t \tag{11}$$

The line through the points $O(0, 0)$ and $P(\cos t, \sin t)$ has slope $\sin t/\cos t$; therefore its equation is

$$y = \frac{\sin t}{\cos t}x$$

This line intersects the line $x = 1$ at the point $(1, \sin t/\cos t)$, which is the point T in the figure.

If K_2 square units is the area of right triangle BOT,

$$K_2 = \tfrac{1}{2}|\overline{BT}| \cdot |\overline{OB}| = \frac{1}{2}\cdot\frac{\sin t}{\cos t}\cdot 1 = \frac{1}{2}\cdot\frac{\sin t}{\cos t} \tag{12}$$

From Fig. 2.9.2 observe that

$$K_1 < S < K_2 \tag{13}$$

Substituting from (10), (11), and (12) into inequality (13),

$$\frac{1}{2}\sin t < \frac{1}{2}t < \frac{1}{2}\cdot\frac{\sin t}{\cos t}$$

Multiplying each member of the above inequality by $2/\sin t$, which is positive because $0 < t < \frac{1}{2}\pi$,

$$1 < \frac{t}{\sin t} < \frac{1}{\cos t}$$

Taking the reciprocal of each member of the above inequality and reversing the direction of the inequality signs,

$$\cos t < \frac{\sin t}{t} < 1 \tag{14}$$

From the right-hand inequality in the above,

$$\sin t < t \tag{15}$$

and from formula (19) in Sec. 1.7,

$$\frac{1-\cos t}{2} = \sin^2 \frac{1}{2}t \tag{16}$$

Replacing t by $\frac{1}{2}t$ in inequality (15) and squaring,

$$\sin^2 \tfrac{1}{2}t < \tfrac{1}{4}t^2 \tag{17}$$

Thus from (16) and (17) it follows that

$$\frac{1-\cos t}{2} < \frac{t^2}{4}$$

which is equivalent to

$$1 - \tfrac{1}{2}t^2 < \cos t \tag{18}$$

From (14) and (18) and because $0 < t < \frac{1}{2}\pi$,

$$1 - \frac{1}{2}t^2 < \frac{\sin t}{t} < 1 \qquad \text{if } 0 < t < \tfrac{1}{2}\pi \tag{19}$$

If $-\frac{1}{2}\pi < t < 0$, then $0 < -t < \frac{1}{2}\pi$; and so from (19),

$$1 - \frac{1}{2}(-t)^2 < \frac{\sin(-t)}{-t} < 1 \qquad \text{if } -\tfrac{1}{2}\pi < t < 0$$

But $\sin(-t) = -\sin t$; thus the above can be written as

$$1 - \frac{1}{2}t^2 < \frac{\sin t}{t} < 1 \qquad \text{if } -\tfrac{1}{2}\pi < t < 0 \tag{20}$$

From (19) and (20) we conclude that

$$1 - \frac{1}{2}t^2 < \frac{\sin t}{t} < 1 \qquad \text{if } -\tfrac{1}{2}\pi < t < \tfrac{1}{2}\pi \text{ and } t \neq 0 \tag{21}$$

Because $\lim_{t\to 0} (1 - \frac{1}{2}t^2) = 1$ and $\lim_{t\to 0} 1 = 1$, it follows from (21) and the squeeze theorem that

$$\lim_{t\to 0} \frac{\sin t}{t} = 1$$

■

Example 3 Find

$$\lim_{x\to 0} \frac{\sin 3x}{\sin 5x}$$

if it exists.

Solution We wish to write the quotient $\sin 3x/\sin 5x$ in such a way that Theorem 2.9.2 can be applied. If $x \neq 0$,

$$\frac{\sin 3x}{\sin 5x} = \frac{3\left(\dfrac{\sin 3x}{3x}\right)}{5\left(\dfrac{\sin 5x}{5x}\right)}$$

As x approaches zero, so do $3x$ and $5x$. Hence

$$\lim_{x\to 0} \frac{\sin 3x}{3x} = \lim_{3x\to 0} \frac{\sin 3x}{3x} = 1$$

and

$$\lim_{x\to 0} \frac{\sin 5x}{5x} = \lim_{5x\to 0} \frac{\sin 5x}{5x} = 1$$

Therefore,

$$\lim_{x\to 0} \frac{\sin 3x}{\sin 5x} = \frac{3 \lim_{x\to 0}\left(\dfrac{\sin 3x}{3x}\right)}{5 \lim_{x\to 0}\left(\dfrac{\sin 5x}{5x}\right)} = \frac{3 \cdot 1}{5 \cdot 1} = \frac{3}{5}$$

From Theorem 2.9.2 we can prove that the sine function and the cosine function are continuous at 0.

2.9.3 Theorem The sine function is continuous at 0.

Proof The three conditions necessary for continuity at a number are shown to be satisfied.

(i) $\sin 0 = 0$.

(ii) $\lim_{t \to 0} \sin t = \lim_{t \to 0} \frac{\sin t}{t} \cdot t = \lim_{t \to 0} \frac{\sin t}{t} \cdot \lim_{t \to 0} t = 1 \cdot 0 = 0.$

(iii) $\lim_{t \to 0} \sin t = \sin 0$.

Therefore the sine function is continuous at 0. ■

2.9.4 Theorem The cosine function is continuous at 0.

Proof

(i) $\cos 0 = 1$.

(ii)
$$\lim_{t \to 0} \cos t = \lim_{t \to 0} \sqrt{1 - \sin^2 t}$$
$$= \sqrt{\lim_{t \to 0} (1 - \sin^2 t)} = \sqrt{1 - 0} = 1.$$

Note: We can replace $\cos t$ by $\sqrt{1 - \sin^2 t}$ since $\cos t > 0$ when $0 < t < \frac{1}{2}\pi$ and when $-\frac{1}{2}\pi < t < 0$.

(iii) $\lim_{t \to 0} \cos t = \cos 0$.

Therefore the cosine function is continuous at 0. ■

The limit in the following theorem is also important. It follows from the previous three theorems and limit theorems.

2.9.5 Theorem
$$\lim_{t \to 0} \frac{1 - \cos t}{t} = 0$$

Proof

$$\lim_{t \to 0} \frac{1 - \cos t}{t} = \lim_{t \to 0} \frac{(1 - \cos t)(1 + \cos t)}{t(1 + \cos t)}$$
$$= \lim_{t \to 0} \frac{(1 - \cos^2 t)}{t(1 + \cos t)}$$
$$= \lim_{t \to 0} \frac{\sin^2 t}{t(1 + \cos t)}$$
$$= \lim_{t \to 0} \frac{\sin t}{t} \cdot \lim_{t \to 0} \frac{\sin t}{1 + \cos t}$$

By Theorem 2.9.2,

$$\lim_{t \to 0} \frac{\sin t}{t} = 1$$

and because the sine and cosine functions are continuous at 0 it follows that

$$\lim_{t\to 0} \frac{\sin t}{1+\cos t} = \frac{0}{1+0}$$
$$= 0$$

Therefore

$$\lim_{t\to 0} \frac{1-\cos t}{t} = 1 \cdot 0$$
$$= 0$$

■

Example 4 Find

$$\lim_{x\to 0} \frac{1-\cos x}{\sin x}$$

if it exists.

Solution Because $\lim_{x\to 0}(1-\cos x) = 0$ and $\lim_{x\to 0} \sin x = 0$, the limit theorems cannot be applied to the quotient $(1-\cos x)/\sin x$. However, if the numerator and denominator are divided by x, which is permissible because $x \neq 0$, we will be able to apply Theorems 2.9.2 and 2.9.5. Thus

$$\lim_{x\to 0} \frac{1-\cos x}{\sin x} = \lim_{x\to 0} \frac{\dfrac{1-\cos x}{x}}{\dfrac{\sin x}{x}}$$
$$= \frac{\lim_{x\to 0} \dfrac{1-\cos x}{x}}{\lim_{x\to 0} \dfrac{\sin x}{x}}$$
$$= \frac{0}{1}$$
$$= 0$$

EXERCISES 2.9

In Exercises 1 through 26, evaluate the limit if it exists.

1. $\lim_{x\to 0} \dfrac{\sin 4x}{x}$ **2.** $\lim_{x\to 0} \dfrac{2x}{\sin 3x}$ **3.** $\lim_{x\to 0} \dfrac{\sin 9x}{\sin 7x}$ **4.** $\lim_{t\to 0} \dfrac{\sin 3t}{\sin 6t}$ **5.** $\lim_{y\to 0} \dfrac{3y}{\sin 5y}$

6. $\lim_{x\to 0} \dfrac{\sin^3 x}{x^2}$ **7.** $\lim_{x\to 0} \dfrac{x^2}{\sin^2 3x}$ **8.** $\lim_{x\to 0} \dfrac{\sin^5 2x}{4x^5}$ **9.** $\lim_{x\to 0} \dfrac{x}{\cos x}$ **10.** $\lim_{x\to 0} \dfrac{1-\cos x}{1+\sin x}$

11. $\lim_{x \to 0} \frac{1 - \cos 4x}{x}$ **12.** $\lim_{z \to 0} \frac{1 - \cos 2z}{4z}$ **13.** $\lim_{x \to 0} \frac{3x^2}{1 - \cos^2 \frac{1}{2}x}$ **14.** $\lim_{x \to 0} \frac{1 - \cos^2 x}{2x^2}$ **15.** $\lim_{r \to 0} \frac{\sin^2 r + 1}{\sqrt{\cos^2 r + 3}}$

16. $\lim_{x \to 0} \frac{\sin x - 1}{2 \cos x + 1}$ **17.** $\lim_{x \to \pi/2} \frac{1 - \sin x}{\frac{1}{2}\pi - x}$ **18.** $\lim_{x \to \pi/2} \frac{\frac{1}{2}\pi - x}{\cos x}$

(*Hint for Exercises 17 and 18:* Let $t = \frac{1}{2}\pi - x$.)

19. $\lim_{t \to 0^+} \frac{\sin t}{t^2}$ **20.** $\lim_{x \to 0} \frac{1 - \cos x}{x^2}$ **21.** $\lim_{x \to 0} \frac{1 - \cos 2x}{\sin 3x}$

22. $\lim_{y \to 0^+} \frac{\sin 4y}{\cos 3y - 1}$ **23.** $\lim_{x \to \pi^+} \frac{\sin x}{x - \pi}$ **24.** $\lim_{x \to +\infty} x \sin \frac{1}{x}$

(*Hint:* Let $t = x - \pi$.) $\left(\textit{Hint: } \text{Let } t = \frac{1}{x}.\right)$

25. $\lim_{x \to 0} \frac{x^2 + 3x}{\sin x}$ **26.** $\lim_{x \to 0} \frac{\sin x}{3x^2 + 2x}$

In Exercises 27 through 30, use the squeeze theorem to find the limit.

27. $\lim_{x \to 0} x \cos \frac{1}{x}$ **28.** $\lim_{x \to 0} x^2 \sin \frac{1}{\sqrt[3]{x}}$

29. $\lim_{x \to 3} g(x)$, if $|g(x) + 4| < 2(3 - x)^4$ for all x. **30.** $\lim_{x \to -2} g(x)$, if $|g(x) - 3| < 5(x + 2)^2$ for all x.

In Exercises 31 and 32, find the limit, if it exists.

31. $\lim_{x \to 0} \frac{\sin (\sin x)}{x}$ **32.** $\lim_{x \to 0} \sin x \sin \frac{1}{x}$

33. Given: $1 - \cos^2 x \le f(x) \le x^2$ for all x in the open interval $(-\frac{1}{2}\pi, \frac{1}{2}\pi)$. Find $\lim_{x \to 0} f(x)$.

34. Given: $-\sin x \le f(x) \le 2 + \sin x$ for all x in the open interval $(-\pi, 0)$. Find $\lim_{x \to -\pi/2} f(x)$.

35. If $|f(x)| \le M$ for all x and M is a constant, use the squeeze theorem to prove that $\lim_{x \to 0} x^2 f(x) = 0$.

36. Suppose that $|f(x)| \le M$ for all x, where M is a constant. Furthermore, suppose that $\lim_{x \to a} |g(x)| = 0$. Use the squeeze theorem to prove that $\lim_{x \to a} f(x)g(x) = 0$.

37. If $|f(x)| \le k|x - a|$ for all $x \ne a$, where k is a constant, prove that $\lim_{x \to a} f(x) = 0$.

2.10 ADDITIONAL THEOREMS ON LIMITS OF FUNCTIONS

We now discuss four theorems that are needed to prove some important theorems in later sections. A graphical illustration is given after the statement of each theorem.

2.10.1 Theorem If $\lim_{x \to c} f(x)$ exists and is positive, then there is an open interval containing c such that $f(x) > 0$ for every $x \ne c$ in the interval.

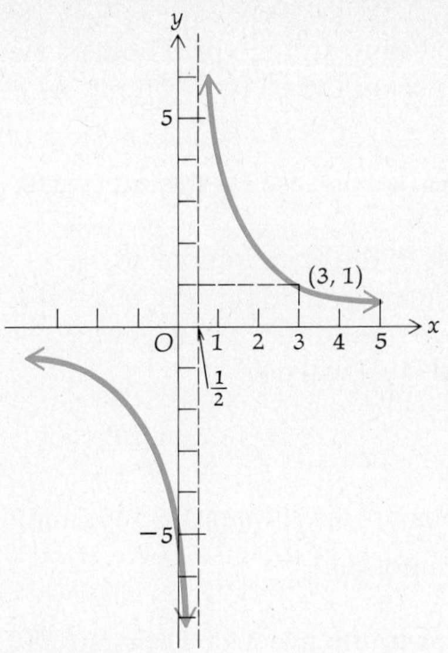

FIGURE 2.10.1

● ILLUSTRATION 1 Consider the function f defined by

$$f(x) = \frac{5}{2x - 1}$$

A sketch of the graph of f is in Fig. 2.10.1. Because $\lim\limits_{x \to 3} f(x) = 1$, and $1 > 0$, according to Theorem 2.10.1 there is an open interval containing 3 such that $f(x) > 0$ for every $x \neq 3$ in the interval. Such an interval is (2, 4). Actually, any open interval (a, b) for which $\frac{1}{2} \leq a < 3$ and $b > 3$ will do. ●

Proof of Theorem 2.10.1 Let $L = \lim\limits_{x \to c} f(x)$. By hypothesis, $L > 0$. Applying Definition 2.1.1 and taking $\epsilon = \frac{1}{2}L$, there is a $\delta > 0$ such that

$$|f(x) - L| < \tfrac{1}{2}L \quad \text{whenever} \quad 0 < |x - c| < \delta \tag{1}$$

Also, $|f(x) - L| < \frac{1}{2}L$ is equivalent to $-\frac{1}{2}L < f(x) - L < \frac{1}{2}L$ (refer to Theorem 1.1.18), which in turn is equivalent to

$$\tfrac{1}{2}L < f(x) < \tfrac{3}{2}L \tag{2}$$

Also, $0 < |x - c| < \delta$ is equivalent to $-\delta < x - c < \delta$ but $x \neq c$, which in turn is equivalent to

$$c - \delta < x < c + \delta \quad \text{but} \quad x \neq c$$

which is equivalent to stating that

$$x \text{ is in the open interval } (c - \delta, c + \delta) \quad \text{but} \quad x \neq c \tag{3}$$

From statements (2) and (3) we can replace (1) by the statement

$$\tfrac{1}{2}L < f(x) < \tfrac{2}{3}L \quad \text{when } x \text{ is in the open interval } (c - \delta, c + \delta) \text{ but } x \neq c$$

Since $L > 0$, it follows that $f(x) > 0$ for every $x \neq c$ in the open interval $(c - \delta, c + \delta)$. ■

Example 1 Given

$$f(x) = \frac{-3}{2x - 7}$$

(a) Show that $\lim\limits_{x \to 1} f(x) > 0$. (b) Verify Theorem 2.10.1 for this function by finding an open interval containing 1 such that $f(x) > 0$ for every $x \neq 1$ in the interval.

Solution

(a) $\displaystyle \lim_{x \to 1} \frac{-3x}{2x - 7} = \frac{-3}{-5} = \frac{3}{5} > 0$

(b) The values of x in the required open interval must be such that

$$\frac{-3x}{2x - 7} > 0 \tag{4}$$

Inequality (4) will be satisfied when both the numerator and denominator are positive or when both are negative. Two cases are considered.

Case 1: $-3x > 0$ and $2x - 7 > 0$.

That is, $x < 0$ and $x > \frac{7}{2}$. There is no value of x for which both of these inequalities hold. Therefore the solution set of Case 1 is the empty set.

Case 2: $-3x < 0$ and $2x - 7 < 0$.

That is, $x > 0$ and $x < \frac{7}{2}$. The solution set of Case 2 is the interval $(0, \frac{7}{2})$.

Thus inequality (4) is satisfied if x is in the open interval $(0, \frac{7}{2})$.

We conclude, then, that any open interval (a, b) for which $0 \le a < 1$ and $1 < b \le \frac{7}{2}$ will be such that it contains 1 and $f(x) > 0$ for every $x \ne 1$ in the interval. In particular, the interval $(\frac{1}{2}, \frac{3}{2})$ will do.

2.10.2 Theorem If $\lim\limits_{x \to c} f(x)$ exists and is negative, there is an open interval containing c such that $f(x) < 0$ for every $x \ne c$ in the interval.

The proof of this theorem is similar to the proof of Theorem 2.10.1 and is left as an exercise (See Exercise 13).

● ILLUSTRATION 2 Let

$$g(x) = \frac{6 - x}{3 - 2x}$$

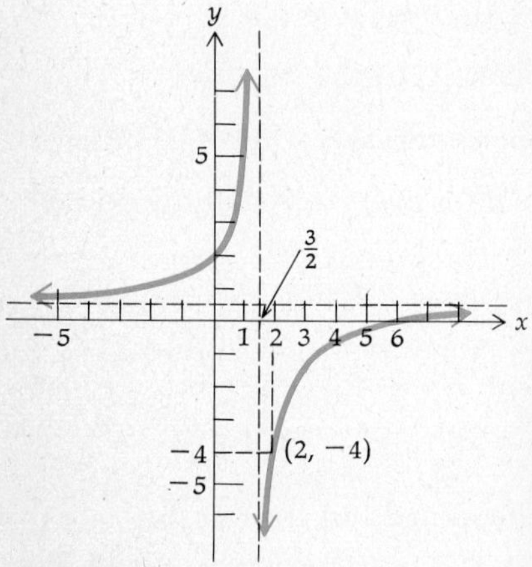

FIGURE 2.10.2

Figure 2.10.2 shows a sketch of the graph of g. $\lim\limits_{x \to 2} g(x) = -4 < 0$; hence, by Theorem 2.10.2 there is an open interval containing 2 such that $g(x) < 0$ for every $x \ne 2$ in the interval. Such an interval is $(\frac{3}{2}, 3)$. Any open interval (a, b) for which $\frac{3}{2} \le a < 2$ and $2 < b \le 6$ will suffice. ●

The following example also illustrates Theorem 2.10.2.

Example 2 Given

$$g(x) = \frac{1}{1 - 2x}$$

(a) Determine the values of k such that $\lim\limits_{x \to k} g(x) < 0$. (b) Find all open intervals (a, b) containing a particular value of k satisfying part (a) such that $g(x) < 0$ for all $x \ne k$ in (a, b). (c) Draw a sketch of the graph of g, and show on the graph the geometric interpretation of the results of parts (a) and (b).

Solution

(a) If $k \ne \frac{1}{2}$,

$$\lim_{x \to k} \frac{1}{1 - 2x} = \frac{1}{1 - 2k}$$

We wish to find the values of k for which

$$\frac{1}{1 - 2k} < 0 \qquad (5)$$

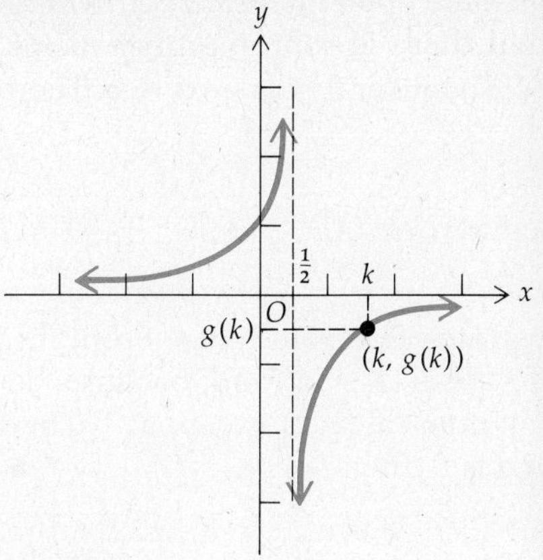

FIGURE 2.10.3

Inequality (5) will be satisfied if and only if the denominator is negative, that is, if and only if

$$1 - 2k < 0$$
$$-2k < -1$$
$$k > \tfrac{1}{2}$$

Therefore $\lim\limits_{x \to k} g(x) < 0$ whenever k is in the interval $(\frac{1}{2}, +\infty)$.

(b) Because $g(x) < 0$ if and only if $x > \frac{1}{2}$, the required open intervals (a, b) containing k such that $g(x) < 0$ are those for which $\frac{1}{2} \le a < k$ and $b > k$.

(c) A sketch of the graph of g is shown in Fig. 2.10.3. In the figure a value of $k > \frac{1}{2}$ is selected. Note that $\lim\limits_{x \to k} g(x) = g(k)$ and so g is continuous at k. The point $(k, g(k))$ is shown on the graph, and $g(k) < 0$. Observe that for all values of x in any open interval (a, b) where $\frac{1}{2} \le a < k$ and $b > k$, the graph is in the fourth quadrant, and so $g(x) < 0$.

2.10.3 Theorem Suppose that the function f is defined on some open interval I containing c, except possibly at c. Also suppose that there is some number M for which there is a $\delta > 0$ such that $f(x) \le M$ whenever $0 < |x - c| < \delta$. Then if $\lim\limits_{x \to c} f(x)$ exists and is equal to L, $L \le M$.

● ILLUSTRATION 3 Figure 2.10.4 shows a sketch of the graph of a function f satisfying the hypothesis of Theorem 2.10.3. From the figure note that $f(1)$ is not defined, but f is defined on the open interval $(\frac{1}{2}, \frac{3}{2})$ except at 1. Furthermore, $f(x) \le \frac{9}{4}$ whenever $0 < |x - 1| < \frac{1}{2}$. Thus it follows from Theorem 2.10.3 that if $\lim\limits_{x \to 1} f(x)$ exists and is L, then $L \le \frac{9}{4}$. From the figure, observe that there is an L and it is 2. ●

Proof of Theorem 2.10.3 We assume that $M < L$ and show that this assumption leads to a contradiction. If $M < L$, there is some $\epsilon > 0$ such that $M + \epsilon = L$. Because $\lim\limits_{x \to c} f(x) = L$, there exists a $\delta_1 > 0$ such that

$$|f(x) - L| < \epsilon \quad \text{whenever} \quad 0 < |x - c| < \delta_1$$

which is equivalent to

$$L - \epsilon < f(x) < L + \epsilon \quad \text{whenever} \quad 0 < |x - c| < \delta_1$$

Replace L by $M + \epsilon$; it follows that there exists a $\delta_1 > 0$ such that

$$(M + \epsilon) - \epsilon < f(x) \quad \text{whenever} \quad 0 < |x - c| < \delta_1$$

or, equivalently,

$$M < f(x) \quad \text{whenever} \quad 0 < |x - c| < \delta_1 \tag{6}$$

But, by hypothesis, there is a δ such that

$$f(x) \le M \quad \text{whenever} \quad 0 < |x - c| < \delta \tag{7}$$

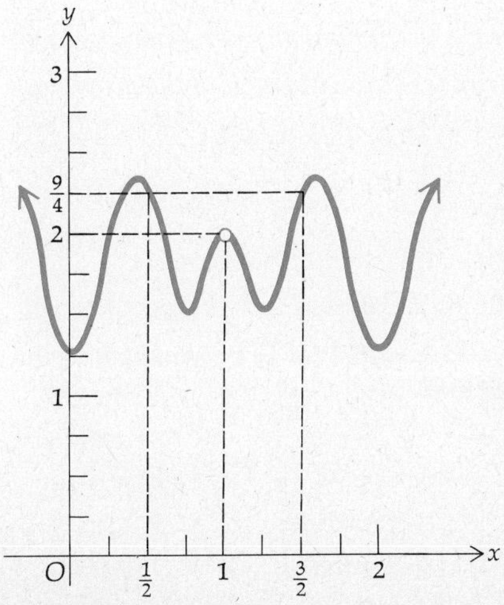

FIGURE 2.10.4

Statements (6) and (7) contradict each other. Hence our assumption that $M < L$ is false. Therefore $L \le M$. ■

2.10.4 Theorem Suppose that the function f is defined on some open interval I containing c, except possibly at c. Also suppose that there is some number M for which there is a $\delta > 0$ such that $f(x) \geq M$ whenever $0 < |x - c| < \delta$. Then if $\lim_{x \to c} f(x)$ exists and is equal to L, $L \geq M$.

The proof is left as an exercise (See Exercise 14).

● ILLUSTRATION 4 Figure 2.10.4 also illustrates Theorem 2.10.4. From the figure observe that $f(x) \geq \frac{3}{2}$ whenever $0 < |x - 1| < \frac{1}{2}$; and, because, as previously stated, f is defined on the open interval $(\frac{1}{2}, \frac{3}{2})$ except at 1, Theorem 2.10.4 states that if $\lim_{x \to 1} f(x)$ exists and is L, then $L \geq \frac{3}{2}$. ●

EXERCISES 2.10

In Exercises 1 through 4, a function f and a number c are given. (a) Show that $\lim_{x \to c} f(x) > 0$; and (b) verify Theorem 2.10.1 for the function f by finding an open interval containing c such that $f(x) > 0$ for every $x \neq c$ in the interval.

1. $f(x) = \dfrac{5}{2x + 4}$; $c = 3$ **2.** $f(x) = \dfrac{7}{2x - 1}$; $c = 4$ **3.** $f(x) = \dfrac{\sqrt{x} - 1}{x - 1}$; $c = 1$ **4.** $f(x) = \dfrac{x^4 - 1}{x^2 - 1}$; $c = -1$

In Exercises 5 through 8, a function g and a number c are given. (a) Show that $\lim_{x \to c} g(x) < 0$; and (b) verify Theorem 2.10.2 for the function g by finding an open interval containing c such that $g(x) < 0$ for every $x \neq c$ in the interval.

5. $g(x) = \dfrac{x^2 - 9}{x + 3}$; $c = -3$ **6.** $g(x) = \dfrac{2 - 7x + 3x^2}{2 - x}$; $c = 2$

7. $g(x) = \dfrac{2 - \sqrt{4 + x}}{x}$; $c = 0$ **8.** $g(x) = \dfrac{\sqrt{x + 1} - 2}{3 - x}$; $c = 3$

In Exercises 9 and 10, a function f is given. (a) Determine the values of k such that $\lim_{x \to k} f(x) > 0$. (b) Find all open intervals (a, b) containing a particular value of k satisfying part (a) such that $f(x) > 0$ for all $x \neq k$ in (a, b). (c) Draw a sketch of the graph of f, and show on the graph the geometric interpretation of the results of parts (a) and (b).

9. $f(x) = \dfrac{3x}{x + 1}$ **10.** $f(x) = \dfrac{x^2 - 25}{x^2 - 4x - 5}$

In Exercises 11 and 12, a function g is given. (a) Determine the values of k such that $\lim_{x \to k} g(x) < 0$. (b) Find all open intervals (a, b) containing a particular value of k satisfying part (a) such that $g(x) < 0$ for all $x \neq k$ in (a, b). (c) Draw a sketch of the graph of g, and show on the graph the geometric interpretation of the results of parts (a) and (b).

11. $g(x) = \dfrac{2x^2 + x - 6}{2x^2 + 11x + 14}$ **12.** $g(x) = 2 - \sqrt{x - 1}$

13. Prove Theorem 2.10.2. **14.** Prove Theorem 2.10.4.

REVIEW EXERCISES FOR CHAPTER 2

In Exercises 1 through 8, evaluate the limit and, when applicable, indicate the limit theorems used.

1. $\lim_{x\to 2}(3x^2 - 4x + 5)$ **2.** $\lim_{h\to 1}\dfrac{h^2-4}{3h^3+6}$ **3.** $\lim_{z\to -3}\dfrac{z^2-9}{z+3}$ **4.** $\lim_{x\to -2}\dfrac{x^2-x-6}{x^2-5x-14}$

5. $\lim_{x\to 1/2}\sqrt[3]{\dfrac{4x^2+4x-3}{4x^2-1}}$ **6.** $\lim_{y\to -4}\sqrt{\dfrac{5y+4}{y-5}}$ **7.** $\lim_{t\to 0}\dfrac{\sqrt{9-t}-3}{t}$ **8.** $\lim_{t\to 0}\dfrac{1-\sqrt[3]{1+t}}{t}$

In Exercises 9 through 18, establish the limit by using Definition 2.1.1; that is, for any $\epsilon > 0$ find a $\delta > 0$ such that $|f(x) - L| < \epsilon$ whenever $0 < |x - a| < \delta$.

9. $\lim_{x\to 3}(2x+1) = 7$ **10.** $\lim_{x\to -2}(8-3x) = 14$ **11.** $\lim_{x\to -4}x^2 = 16$ **12.** $\lim_{x\to 3}(2x^2-x-6) = 9$

13. $\lim_{r\to 5}\dfrac{2}{r-3} = 1$ **14.** $\lim_{y\to 5}\dfrac{3}{2y-1} = \dfrac{1}{3}$ **15.** $\lim_{t\to 4}\sqrt{t-3} = 1$ **16.** $\lim_{x\to -2}x^3 = -8$

17. $\lim_{x\to 3}\dfrac{5+3x}{2x+1} = 2$ **18.** $\lim_{t\to 4}\sqrt{2t+1} = 3$

In Exercises 19 through 22, draw a sketch of the graph of the function; then by observing where there are breaks in the graph, determine the values of the independent variable at which the function is discontinuous and show why Definition 2.6.1 is not satisfied at each discontinuity.

19. $f(x) = \dfrac{x+2}{x^2+x-2}$ **20.** $g(x) = \dfrac{x^4-1}{x^2-1}$ **21.** $g(x) = \begin{cases} 2x+1 & \text{if } x \le -2 \\ x-2 & \text{if } -2 < x \le 2 \\ 2-x & \text{if } 2 < x \end{cases}$ **22.** $F(x) = \begin{cases} |4-x| & \text{if } x \ne 4 \\ -2 & \text{if } x = 4 \end{cases}$

In Exercises 23 through 26, draw a sketch of the graph, and discuss the continuity of the function.

23. $f(x) = \sqrt{4x^2-9}$ **24.** $F(x) = \dfrac{|x^2-4|}{x+2}$ **25.** $g(x) = \sqrt{\dfrac{x-3}{4-x}}$ **26.** $h(x) = \sqrt{-x^2+7x-12}$

In Exercises 27 through 48, evaluate the limit if it exists.

27. $\lim_{x\to -1}\dfrac{2x^2-x-3}{3x^2+8x+5}$ **28.** $\lim_{y\to 3}\sqrt{\dfrac{y-3}{y^3-27}}$ **29.** $\lim_{x\to 3}f(x)$ if $f(x) = \begin{cases} x^2-1 & \text{if } x \le 3 \\ x+5 & \text{if } 3 < x \end{cases}$

30. $\lim_{x\to 1/3}(|3x-1| - 5)$ **31.** $\lim_{x\to 9}\dfrac{2\sqrt{x}-6}{x-9}$ **32.** $\lim_{y\to 5^-}\dfrac{\sqrt{25-y^2}}{y-5}$

33. $\lim_{x\to -4^+}\dfrac{2x}{16-x^2}$ **34.** $\lim_{s\to 7}\dfrac{5-\sqrt{4+3s}}{7-s}$ **35.** $\lim_{t\to 5}\dfrac{\sqrt{t-4}}{t^2-10t+25}$

36. $\lim_{x\to 0^-}\dfrac{x^2-5}{2x^3-3x^2}$ **37.** $\lim_{x\to 2^+}\dfrac{[\![x]\!]-1}{[\![x]\!]-x}$ **38.** $\lim_{x\to +\infty}\left(\dfrac{8x^3+7x-2}{7x^3+3x^2+5x}\right)^2$

39. $\lim_{x\to -\infty}\dfrac{\sqrt{x^2+5}}{2x-4}$ **40.** $\lim_{t\to +\infty}(\sqrt{t^2+t} - \sqrt{t^2+4})$ **41.** $\lim_{t\to 0}\dfrac{\sin 5t}{\sin 2t}$

42. $\lim_{x\to 0}\dfrac{1-\cos 3x}{\sin 3x}$ **43.** $\lim_{x\to 0}\dfrac{1-\cos^2 x}{x}$ **44.** $\lim_{x\to 8}\dfrac{\sqrt{7+\sqrt[3]{x}}-3}{x-8}$

45. $\lim_{t \to 0} \dfrac{\sqrt[3]{(t+a)^2} - \sqrt[3]{a^2}}{t}$ **46.** $\lim_{x \to 0} \dfrac{2x^2 - 3x}{2 \sin x}$

47. $\lim_{x \to 0} \dfrac{\sqrt[4]{x^4+1} - \sqrt{x^2+1}}{x^2}$ $\left(\textit{Hint:}\text{ Write } \dfrac{\sqrt[4]{x^4+1} - \sqrt{x^2+1}}{x^2} = \dfrac{1 - \sqrt{x^2+1}}{x^2} + \dfrac{\sqrt[4]{x^4+1} - 1}{x^2}.\right)$

48. $\lim_{x \to 1^+} \dfrac{[\![x^2]\!] - [\![x]\!]^2}{x^2 - 1}$

49. Draw a sketch of the graph of f if $f(x) = [\![1 - x^2]\!]$ and $-2 \le x \le 2$. (a) Does $\lim_{x \to 0} f(x)$ exist? (b) Is f continuous at 0?

50. Draw a sketch of the graph of g if $g(x) = (x - 1)[\![x]\!]$, and $0 \le x \le 2$. (a) Does $\lim_{x \to 1} g(x)$ exist? (b) Is g continuous at 1?

In Exercises 51 through 54, prove that the function is discontinuous at the number a. Then determine if the discontinuity is removable or essential. If the discontinuity is removable, define f(a) such that the discontinuity is removed.

51. $f(x) = \dfrac{x^2 + 2x - 8}{x^2 + 3x - 4}$; $a = -4$ **52.** $f(x) = \begin{cases} 4 - x^2 & \text{if } x < 1 \\ 2x + 3 & \text{if } 1 \le x \end{cases}$; $a = 1$

53. $f(x) = \dfrac{|2x - 1|}{2x - 1}$; $a = \dfrac{1}{2}$ **54.** $f(x) = \dfrac{2 - \sqrt{x + 4}}{x}$; $a = 0$

In Exercises 55 through 59, find the horizontal and vertical asymptotes of the graph of the function defined by the given equation, and draw a sketch of the graph.

55. $f(x) = \dfrac{5x}{7 + 4x}$ **56.** $f(x) = \dfrac{-2x}{\sqrt{x^2 - 5x + 6}}$ **57.** $f(x) = \dfrac{4x^2}{3x^2 + 1}$ **58.** $f(x) = \dfrac{-2}{x^2 - x - 6}$

59. $f(x) = \dfrac{5x^2}{x^2 - 4}$

60. Find the horizontal and vertical asymptotes of the graph of the equation $xy + 2 - 3y = 5x$. Draw a sketch of the graph.

In Exercises 61 through 64, functions f and g are defined. In each exercise define $f \circ g$, and determine all values of x for which $f \circ g$ is continuous.

61. $f(x) = \sqrt{x - 3}$; $g(x) = x + 2$ **62.** $f(x) = \operatorname{sgn} x$; $g(x) = x^2 - x$

63. $f(x) = \operatorname{sgn} x$; $g(x) = x^2 - 1$ **64.** $f(x) = \dfrac{4}{3x - 5}$; $g(x) = \sqrt{x}$

In Exercises 65 and 66, a function f and a closed interval [a, b] are given. Verify that the intermediate-value theorem holds for the given value of k, and find a number c such that f(c) = k. Draw a sketch of the curve and the line y = k.

65. $f(x) = x^2 + 3$; $[a, b] = [2, 4]$; $k = 10$ **66.** $f(x) = -\sqrt{16 - x^2}$; $[a, b] = [0, 4]$; $k = -1$

In Exercises 67 and 68, use the squeeze theorem to find the limit.

67. $\lim_{x \to 1} (x - 1)^2 \sin \dfrac{1}{\sqrt[3]{x - 1}}$ **68.** $\lim_{x \to 4} g(x)$ if $|g(x) + 5| < 3(4 - x)^2$ for all x

In Exercises 69 and 70, find the values of the constants a and b that make the function f continuous on $(-\infty, +\infty)$, and draw a sketch of the graph of f.

69. $f(x) = \begin{cases} 2x + 1 & \text{if } x \le 3 \\ ax + b & \text{if } 3 < x < 5 \\ x^2 + 2 & \text{if } 5 \le x \end{cases}$ **70.** $f(x) = \begin{cases} 3x + 6a & \text{if } x < -3 \\ 3ax - 7b & \text{if } -3 \le x \le 3 \\ x - 12b & \text{if } 3 < x \end{cases}$

71. Prove that $\lim_{t\to 2} \sqrt{t^2-4} = 0$ by showing that for any $\epsilon > 0$ there exists a $\delta > 0$ such that $\sqrt{t^2-4} < \epsilon$ whenever $0 < t - 2 < \delta$.

72. Prove that $\lim_{x\to 2} \left|\dfrac{x+3}{x-2}\right| = +\infty$ by showing that for any $N > 0$ there exists a $\delta > 0$ such that $\left|\dfrac{x+3}{x-2}\right| > N$ whenever $0 < |x-2| < \delta$.

73. Prove that $\lim_{x\to -\infty} \dfrac{3-2x}{x} = -2$ by showing that for any $\epsilon > 0$ there exists a number $N < 0$ such that $\left|\dfrac{3-2x}{x} + 2\right| < \epsilon$ whenever $x < N$.

74. Prove that $\lim_{x\to +\infty} \dfrac{9x+1}{3x-2} = 3$ by showing that for any $\epsilon > 0$ there exists a number $N > 0$ such that $\left|\dfrac{9x+1}{3x-2} - 3\right| < \epsilon$ whenever $x > N$.

75. Given $f(x) = \dfrac{4x^2-9}{2x^2+7x+6}$. (a) Show that $\lim_{x\to -3/2} f(x) < 0$. (b) Verify Theorem 2.10.2 for the function f by finding an open interval containing $-\frac{3}{2}$ such that $f(x) < 0$ for every $x \neq -\frac{3}{2}$ in the interval.

76. Given $f(x) = \dfrac{x-2}{x+5}$. (a) Determine the values of k such that $\lim_{x\to k} f(x) > 0$. (b) Find all open intervals (a, b) containing a particular value of k satisfying part (a) such that $f(x) > 0$ for all $x \neq k$ in (a, b). (c) Draw a sketch of the graph of f, and show on the graph the geometric interpretation of the results of parts (a) and (b).

77. Let f be the function defined by

$$f(x) = \begin{cases} 1 & \text{if } x \text{ is an integer} \\ 0 & \text{if } x \text{ is not an integer} \end{cases}$$

(a) Draw a sketch of the graph of f. (b) For what values of a does $\lim_{x\to a} f(x)$ exist? (c) At what numbers is f continuous?

78. If the function g is continuous at a and f is continuous at $g(a)$, is the composite function $f \circ g$ continuous at a? Why?

79. Give an example of a function for which $\lim_{x\to 0} |f(x)|$ exists but $\lim_{x\to 0} f(x)$ does not exist.

80. Give an example of a function f that is discontinuous at 1 for which (a) $\lim_{x\to 1} f(x)$ exists but $f(1)$ does not exist; (b) $f(1)$ exists but $\lim_{x\to 1} f(x)$ does not exist; (c) $\lim_{x\to 1} f(x)$ and $f(1)$ both exist but are not equal.

81. (a) Prove that if $\lim_{h\to 0} f(x+h) = f(x)$ then $\lim_{h\to 0} f(x+h) = \lim_{h\to 0} f(x-h)$. (b) Show that the converse of the theorem in (a) is not true by giving an example of a function for which $\lim_{h\to 0} f(x+h) = \lim_{h\to 0} f(x-h)$ but $\lim_{h\to 0} f(x+h) \neq f(x)$.

82. If the domain of f is the set of all real numbers, and f is continuous at 0, prove that if $f(a+b) = f(a)\cdot f(b)$ for all a and b, then f is continuous at every number.

83. If the domain of f is the set of all real numbers and f is continuous at 0, prove that if $f(a+b) = f(a) + f(b)$ for all a and b, then f is continuous at every number.

CHAPTER 3 The Derivative

3.1 THE TANGENT LINE

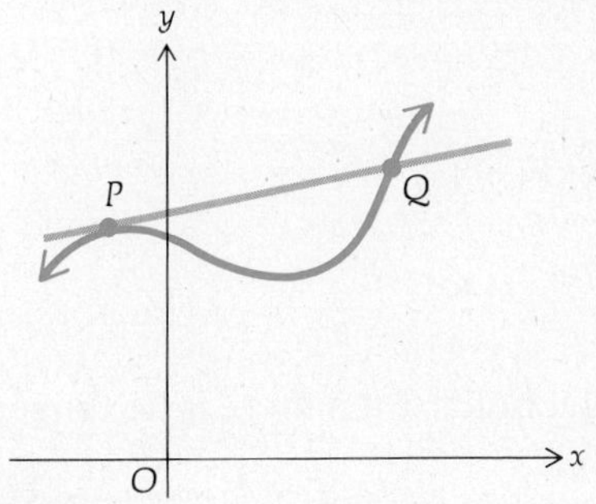

FIGURE 3.1.1

Many of the important problems in calculus depend on the problem of finding the tangent line to a given curve at a specific point on the curve. In plane geometry, if the curve is a circle, the tangent line at a point P on the circle is defined as the line intersecting the circle at only one point P. This definition does not suffice for a curve in general. For example, in Fig. 3.1.1 the line that should be the tangent line to the curve at point P intersects the curve at another point Q.

In this section we arrive at a suitable definition of the tangent line to the graph of a function at a point on the graph. We proceed by considering how to define the slope of the tangent line at a point, because if the slope of a line and a point on the line are known, the line is determined.

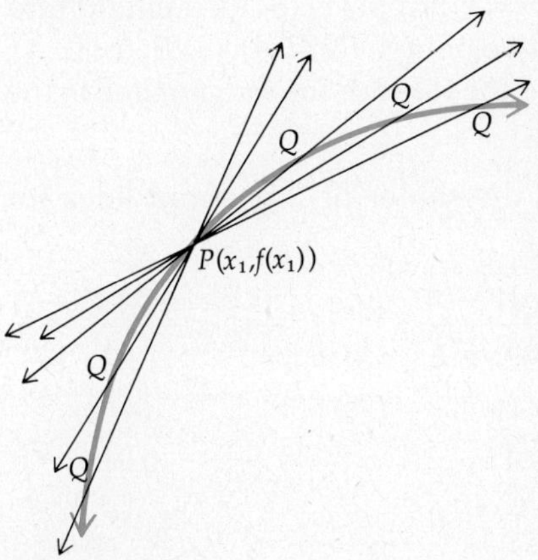

FIGURE 3.1.2

Consider the function f continuous at x_1. We wish to define the slope of the tangent line to the graph of f at $P(x_1, f(x_1))$. Let I be the open interval that contains x_1 and on which f is defined. Let $Q(x_2, f(x_2))$ be another point on the graph of f such that x_2 is also in I. Draw a line through P and Q. Any line through two points on a curve is called a **secant line;** therefore the line through P and Q is a secant line. In Fig. 3.1.2 the secant line is shown for various values of x_2. Figure 3.1.3 shows one particular secant line. In this figure Q is to the right of P. However, Q may be on either the right or the left side of P, as seen in Fig. 3.1.2.

Denote the difference of the abscissas of Q and P by Δx so that

$$\Delta x = x_2 - x_1$$

Δx may be either positive or negative. The slope of the secant line PQ then is given by

$$m_{PQ} = \frac{f(x_2) - f(x_1)}{\Delta x}$$

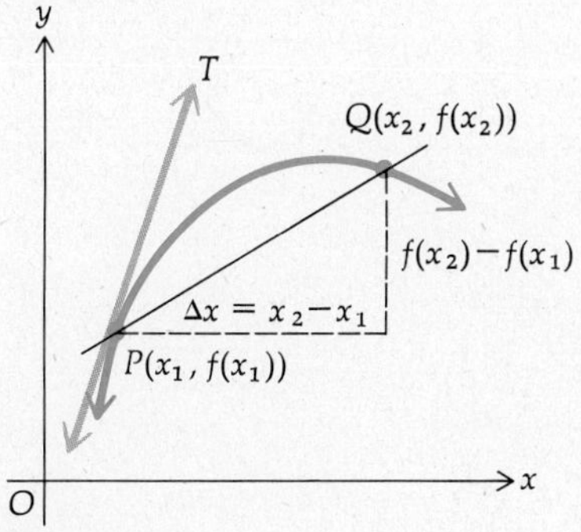

FIGURE 3.1.3

provided that line PQ is not vertical. Because $x_2 = x_1 + \Delta x$, the above equation can be written as

$$m_{PQ} = \frac{f(x_1 + \Delta x) - f(x_1)}{\Delta x}$$

Now think of point P as being fixed, and move point Q along the curve toward P; that is, Q approaches P. This is equivalent to stating that Δx approaches zero. As this occurs, the secant line turns about the fixed point P. If this secant line has a limiting position, it is this limiting position that we wish to be the tangent line to the graph at P. So we want the slope of the tangent line to the graph at P to be the limit of m_{PQ} as Δx approaches zero, if this limit exists. If $\lim\limits_{\Delta x \to 0} m_{PQ} = +\infty$ or $-\infty$, then as Δx approaches zero, the line PQ approaches the line through P that is parallel to the y axis. In this case we would want the tangent line to the graph at P to be the line $x = x_1$. The preceding discussion leads to the following definition.

3.1.1 Definition Suppose the function f is continuous at x_1. The **tangent line** to the graph of f at the point $P(x_1, f(x_1))$ is

(i) the line through P having slope $m(x_1)$, given by

$$m(x_1) = \lim_{\Delta x \to 0} \frac{f(x_1 + \Delta x) - f(x_1)}{\Delta x} \tag{1}$$

if this limit exists;

(ii) the line $x = x_1$ if

$$\lim_{\Delta x \to 0} \frac{f(x_1 + \Delta x) - f(x_1)}{\Delta x} = +\infty \text{ or } -\infty$$

If neither (i) nor (ii) of Definition 3.1.1 holds, then there is no tangent line to the graph of f at the point $P(x_1, f(x_1))$.

Example 1 Given the parabola $y = x^2$. In parts (a) through (c) find the slope of the secant line through the two points: (a) (2, 4), (3, 9); (b) (2, 4), (2.1, 4.41); (c) (2, 4), (2.01, 4.0401). (d) Find the slope of the tangent line to the parabola at the point (2, 4).

Solution Let m_a, m_b, and m_c be the slopes of the secant lines for (a), (b), and (c), respectively.

(a) $m_a = \dfrac{9-4}{3-2} = 5$

(b) $m_b = \dfrac{4.41-4}{2.1-2} = \dfrac{0.41}{0.1} = 4.1$

(c) $m_c = \dfrac{4.0401-4}{2.01-2} = \dfrac{0.0401}{0.01} = 4.01$

(d) $f(x) = x^2$. From (1) we have

$$\begin{aligned} m(2) &= \lim_{\Delta x \to 0} \frac{f(2+\Delta x) - f(2)}{\Delta x} \\ &= \lim_{\Delta x \to 0} \frac{(2+\Delta x)^2 - 4}{\Delta x} \\ &= \lim_{\Delta x \to 0} \frac{4 + 4\,\Delta x + (\Delta x)^2 - 4}{\Delta x} \\ &= \lim_{\Delta x \to 0} \frac{4\,\Delta x + (\Delta x)^2}{\Delta x} \\ &= \lim_{\Delta x \to 0} (4 + \Delta x) \\ &= 4 \end{aligned}$$

Example 2 Find the slope of the tangent line to the curve $y = x^2 - 4x + 3$ at the point (x_1, y_1).

Solution $f(x) = x^2 - 4x + 3$; therefore $f(x_1) = x_1{}^2 - 4x_1 + 3$, and $f(x_1 + \Delta x) = (x_1 + \Delta x)^2 - 4(x_1 + \Delta x) + 3$. From (1),

$$m(x_1) = \lim_{\Delta x \to 0} \frac{f(x_1 + \Delta x) - f(x_1)}{\Delta x}$$

$$= \lim_{\Delta x \to 0} \frac{[(x_1 + \Delta x)^2 - 4(x_1 + \Delta x) + 3] - [x_1{}^2 - 4x_1 + 3]}{\Delta x}$$

$$= \lim_{\Delta x \to 0} \frac{x_1{}^2 + 2x_1 \Delta x + (\Delta x)^2 - 4x_1 - 4\Delta x + 3 - x_1{}^2 + 4x_1 - 3}{\Delta x}$$

$$= \lim_{\Delta x \to 0} \frac{2x_1 \Delta x + (\Delta x)^2 - 4\Delta x}{\Delta x}$$

Because $\Delta x \neq 0$, the numerator and the denominator can be divided by Δx to obtain

$$m(x_1) = \lim_{\Delta x \to 0} (2x_1 + \Delta x - 4)$$

$$m(x_1) = 2x_1 - 4 \tag{2}$$

TABLE 3.1.1

x	y	m
2	−1	0
1	0	−2
0	3	−4
−1	8	−6
3	0	2
4	3	4
5	8	6

To draw a sketch of the graph of the equation in Example 2 we plot some points and a segment of the tangent line at some points. Values of x are taken arbitrarily and the corresponding value of y is computed from the given equation, as well as the value of m from (2). The results are given in Table 3.1.1, and a sketch of the graph is shown in Fig. 3.1.4. It is important to determine the points where the graph has a horizontal tangent. Because a horizontal line has a slope of zero, these points are found by setting $m(x_1) = 0$ and solving for x_1. Doing this calculation for this example we have $2x_1 - 4 = 0$, which gives $x_1 = 2$. Therefore, at the point having an abscissa of 2 the tangent line is parallel to the x axis.

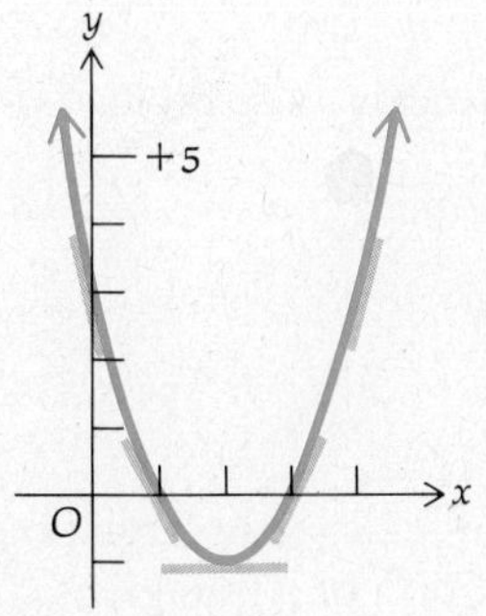

FIGURE 3.1.4

Example 3 Find an equation of the tangent line to the curve of Example 2 at the point (5, 8).

Solution Because the slope of the tangent line at any point (x_1, y_1) is given by

$$m(x_1) = 2x_1 - 4$$

the slope of the tangent line at the point (5, 8) is $m(5) = 2(5) - 4 = 6$. Therefore an equation of the desired line in the point-slope form is

$$y - 8 = 6(x - 5)$$

$$y = 6x - 22$$

3.1.2 Definition The **normal line** to a curve at a given point is the line perpendicular to the tangent line at that point.

Example 4 Find an equation of the normal line to the curve $y = \sqrt{x-3}$ that is parallel to the line $6x + 3y - 4 = 0$.

Solution Let l be the given line. To find the slope of l, write its equation in the slope-intercept form, which is

$$y = -2x + \tfrac{4}{3}$$

Therefore the slope of l is -2, and the slope of the desired normal line is also -2 because the two lines are parallel.

To find the slope of the tangent line to the given curve at any point (x_1, y_1), we apply Definition 3.1.1 with $f(x) = \sqrt{x-3}$.

$$m(x_1) = \lim_{\Delta x \to 0} \frac{\sqrt{x_1 + \Delta x - 3} - \sqrt{x_1 - 3}}{\Delta x}$$

To evaluate this limit we rationalize the numerator.

$$m(x_1) = \lim_{\Delta x \to 0} \frac{(\sqrt{x_1 + \Delta x - 3} - \sqrt{x_1 - 3})(\sqrt{x_1 + \Delta x - 3} + \sqrt{x_1 - 3})}{\Delta x(\sqrt{x_1 + \Delta x - 3} + \sqrt{x_1 - 3})}$$

$$= \lim_{\Delta x \to 0} \frac{\Delta x}{\Delta x(\sqrt{x_1 + \Delta x - 3} + \sqrt{x_1 - 3})}$$

Numerator and denominator are divided by Δx (since $\Delta x \neq 0$) to obtain

$$m(x_1) = \lim_{\Delta x \to 0} \frac{1}{\sqrt{x_1 + \Delta x - 3} + \sqrt{x_1 - 3}}$$

$$= \frac{1}{2\sqrt{x_1 - 3}}$$

Because the normal line at a point is perpendicular to the tangent line at that point, the product of their slopes is -1. Hence the slope of the normal line at (x_1, y_1) is given by

$$-2\sqrt{x_1 - 3}$$

As shown above, the slope of the desired line is -2. So we solve the equation

$$-2\sqrt{x_1 - 3} = -2$$

to give

$$x_1 = 4$$

Therefore the desired line is the line with slope -2 and through point $(4, 1)$ on the curve. The point-slope form of an equation of a line gives

$$y - 1 = -2(x - 4)$$

$$2x + y - 9 = 0$$

Refer to Fig. 3.1.5, which shows a sketch of the curve together with the line l, the normal line PN at $(4, 1)$, and the tangent line PT at $(4, 1)$.

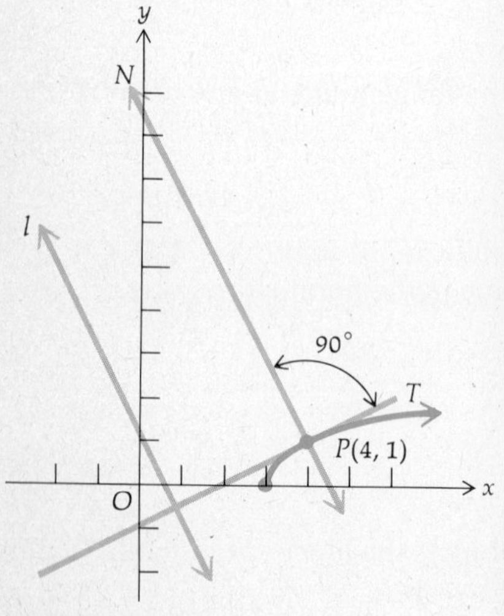

FIGURE 3.1.5

EXERCISES 3.1

In Exercises 1 through 6, find the slope of the tangent line to the graph at the point (x_1, y_1). Make a table of values of x, y, and m over the closed interval [a, b], and include in the table all points where the graph has a horizontal tangent. Draw a sketch of the graph, and show a segment of the tangent line at each of the plotted points.

1. $y = 9 - x^2$; $[a, b] = [-3, 3]$ **2.** $y = x^2 + 4$; $[a, b] = [-2, 2]$ **3.** $y = -2x^2 + 4x$; $[a, b] = [-1, 3]$

4. $y = x^2 - 6x + 9$; $[a, b] = [1, 5]$ **5.** $y = x^3 + 1$; $[a, b] = [-2, 2]$ **6.** $y = 1 - x^3$; $[a, b] = [-2, 2]$

In Exercises 7 through 12, find the slope of the tangent line to the graph at the point (x_1, y_1). Make a table of values of x, y, and m at various points on the graph, and include in the table all points where the graph has a horizontal tangent. Draw a sketch of the graph.

7. $y = x^3 - 3x$ **8.** $y = 7 - 6x - x^2$ **9.** $y = \sqrt{4 - x}$

10. $y = \sqrt{x + 1}$ **11.** $y = x^3 - 4x^2 + 4x - 2$ **12.** $y = x^3 - x^2 - x + 10$

In Exercises 13 through 24, find equations of the tangent line and normal line to the given curve at the indicated point. Draw a sketch of the curve together with the resulting tangent line and normal line.

13. $y = x^2 - 4x - 5$; $(-2, 7)$ **14.** $y = x^2 - x + 2$; $(2, 4)$ **15.** $y = \frac{1}{8}x^3$; $(4, 8)$ **16.** $y = x^2 + 2x + 1$; $(1, 4)$

17. $y = \sqrt{9 - 4x}$; $(-4, 5)$ **18.** $y = \sqrt{4x - 3}$; $(3, 3)$ **19.** $y = 6/x$; $(3, 2)$ **20.** $y = -8/\sqrt{x}$; $(4, -4)$

21. $y = x^4 - 4x$; $(0, 0)$ **22.** $y = 2x - x^3$; $(-2, 4)$ **23.** $y = \sqrt[3]{x}$; $(8, 2)$ **24.** $y = \sqrt[3]{5 - x}$; $(-3, 2)$

25. Find an equation of the tangent line to the curve $y = 2x^2 + 3$ that is parallel to the line $8x - y + 3 = 0$.

26. Find an equation of the tangent line to the curve $y = 3x^2 - 4$ that is parallel to the line $3x + y = 4$.

27. Find an equation of the normal line to the curve $y = 2 - \frac{1}{3}x^2$ that is parallel to the line $x - y = 0$.

28. Find an equation of each normal line to the curve $y = x^3 - 3x$ that is parallel to the line $2x + 18y - 9 = 0$.

29. Find an equation of the tangent line to the curve $y = \sqrt{4x - 3} - 1$ that is perpendicular to the line $x + 2y - 11 = 0$.

30. Find an equation of each of the normal lines to the curve $y = 1/x$ that is perpendicular to the line $x + 4y = 0$.

31. Find an equation of each line through the point $(2, -6)$ that is tangent to the curve $y = 3x^2 - 8$.

32. Find an equation of each line through the point $(3, -2)$ that is tangent to the curve $y = x^2 - 7$.

33. If $y = 4x^2$, then at any point (x_1, y_1) on the curve, $m(x_1) = 8x_1$. Prove analytically that there is no line through the point $(1, 5)$ that is tangent to the curve $y = 4x^2$.

34. If $y = 4 - x^2$, then at any point (x_1, y_1) on the curve, $m(x_1) = -2x_1$. Prove analytically that there is no line through the point $(1, 2)$ that is tangent to the curve $y = 4 - x^2$.

3.2 THE DERIVATIVE

In Sec. 3.1 the slope of the tangent line to the graph of $y = f(x)$ at the point $(x_1, f(x_1))$ is defined by

$$m(x_1) = \lim_{\Delta x \to 0} \frac{f(x_1 + \Delta x) - f(x_1)}{\Delta x} \tag{1}$$

if this limit exists.

This type of limit occurs in other problems too and has a specific name.

3.2.1 Definition The **derivative** of the function f is that function, denoted by f', such that its value at any number x in the domain of f is given by

$$f'(x) = \lim_{\Delta x \to 0} \frac{f(x + \Delta x) - f(x)}{\Delta x} \tag{2}$$

if this limit exists.

Another symbol that is used instead of $f'(x)$ is $D_x f(x)$, which is read "the derivative of f of x with respect to x."

If x_1 is a particular number in the domain of f, then

$$f'(x_1) = \lim_{\Delta x \to 0} \frac{f(x_1 + \Delta x) - f(x_1)}{\Delta x} \tag{3}$$

if this limit exists. Comparing formulas (1) and (3), note that the slope of the tangent line to the graph of $y = f(x)$ at the point $(x_1, f(x_1))$ is precisely the derivative of f evaluated at x_1.

Example 1 Given $f(x) = 3x^2 + 12$, find the derivative of f.

Solution If x is any number in the domain of f, then from (2),

$$\begin{aligned}
f'(x) &= \lim_{\Delta x \to 0} \frac{f(x + \Delta x) - f(x)}{\Delta x} \\
&= \lim_{\Delta x \to 0} \frac{[3(x + \Delta x)^2 + 12] - (3x^2 + 12)}{\Delta x} \\
&= \lim_{\Delta x \to 0} \frac{3x^2 + 6x\,\Delta x + 3(\Delta x)^2 + 12 - 3x^2 - 12}{\Delta x} \\
&= \lim_{\Delta x \to 0} \frac{6x\,\Delta x + 3(\Delta x)^2}{\Delta x} \\
&= \lim_{\Delta x \to 0} (6x + 3\,\Delta x) \\
&= 6x
\end{aligned}$$

Therefore the derivative of f is the function f' defined by $f'(x) = 6x$. The domain of f' is the set of all real numbers, which is the same as the domain of f.

Consider now formula (3), which is

$$f'(x_1) = \lim_{\Delta x \to 0} \frac{f(x_1 + \Delta x) - f(x_1)}{\Delta x}$$

In this formula let

$$x_1 + \Delta x = x \tag{4}$$

Then

$$\text{"}\Delta x \to 0\text{" is equivalent to "}x \to x_1\text{"} \tag{5}$$

From (3), (4), and (5) we obtain the following formula for $f'(x_1)$:

$$f'(x_1) = \lim_{x \to x_1} \frac{f(x) - f(x_1)}{x - x_1} \tag{6}$$

if this limit exists. Formula (6) is an alternative formula to (3) for computing $f'(x_1)$.

Example 2 For the function f of Example 1 find the derivative of f at 2 in three ways: (a) Apply formula (3); (b) apply formula (6); (c) substitute 2 for x in the expression for $f'(x)$ in Example 1.

Solution

(a) $f(x) = 3x^2 + 12$. From formula (3),

$$\begin{aligned}
f'(2) &= \lim_{\Delta x \to 0} \frac{f(2 + \Delta x) - f(2)}{\Delta x} \\
&= \lim_{\Delta x \to 0} \frac{[3(2 + \Delta x)^2 + 12] - [3(2)^2 + 12]}{\Delta x} \\
&= \lim_{\Delta x \to 0} \frac{12 + 12\,\Delta x + 3(\Delta x)^2 + 12 - 12 - 12}{\Delta x} \\
&= \lim_{\Delta x \to 0} \frac{12\,\Delta x + 3(\Delta x)^2}{\Delta x} \\
&= \lim_{\Delta x \to 0} (12 + 3\Delta x) \\
&= 12
\end{aligned}$$

(b) From formula (6),

$$\begin{aligned}
f'(2) &= \lim_{x \to 2} \frac{f(x) - f(2)}{x - 2} \\
&= \lim_{x \to 2} \frac{(3x^2 + 12) - 24}{x - 2} \\
&= \lim_{x \to 2} \frac{3x^2 - 12}{x - 2} \\
&= 3 \lim_{x \to 2} \frac{(x - 2)(x + 2)}{x - 2} \\
&= 3 \lim_{x \to 2} (x + 2) \\
&= 12
\end{aligned}$$

(c) Because, from Example 1, $f'(x) = 6x$, then $f'(2) = 12$.

If $y = f(x)$, then $f'(x)$ is the derivative of y with respect to x, and sometimes the notation $D_x y$ is used in place of $f'(x)$. The notation y' is also used for the derivative of y with respect to an independent variable if the independent variable is understood.

If the function f is given by the equation $y = f(x)$, let

$$\Delta y = f(x + \Delta x) - f(x)$$

so that from formula (2),

$$D_x y = \lim_{\Delta x \to 0} \frac{\Delta y}{\Delta x} \tag{7}$$

A derivative is sometimes indicated by the notation dy/dx. But this symbolism is avoided until dy and dx are defined.

Example 3 Given

$$y = \frac{2 + x}{3 - x}$$

find $D_x y$.

Solution

$$\begin{aligned}
D_x y &= \lim_{\Delta x \to 0} \frac{\Delta y}{\Delta x} \\
&= \lim_{\Delta x \to 0} \frac{f(x + \Delta x) - f(x)}{\Delta x} \\
&= \lim_{\Delta x \to 0} \frac{(2 + x + \Delta x)/(3 - x - \Delta x) - (2 + x)/(3 - x)}{\Delta x} \\
&= \lim_{\Delta x \to 0} \frac{(3 - x)(2 + x + \Delta x) - (2 + x)(3 - x - \Delta x)}{\Delta x(3 - x - \Delta x)(3 - x)} \\
&= \lim_{\Delta x \to 0} \frac{(6 + x - x^2 + 3\,\Delta x - x\,\Delta x) - (6 + x - x^2 - 2\,\Delta x - x\,\Delta x)}{\Delta x(3 - x - \Delta x)(3 - x)} \\
&= \lim_{\Delta x \to 0} \frac{5\,\Delta x}{\Delta x(3 - x - \Delta x)(3 - x)} \\
&= \lim_{\Delta x \to 0} \frac{5}{(3 - x - \Delta x)(3 - x)} \\
&= \frac{5}{(3 - x)^2}
\end{aligned}$$

Example 4 Given $f(x) = 1/\sqrt{x + 2}$, find $f'(x)$.

Solution

$$\begin{aligned}
f'(x) &= \lim_{\Delta x \to 0} \frac{f(x + \Delta x) - f(x)}{\Delta x} = \lim_{\Delta x \to 0} \frac{\dfrac{1}{\sqrt{(x + \Delta x) + 2}} - \dfrac{1}{\sqrt{x + 2}}}{\Delta x} \\
&= \lim_{\Delta x \to 0} \frac{\sqrt{x + 2} - \sqrt{(x + \Delta x) + 2}}{\Delta x \sqrt{(x + \Delta x) + 2}\,\sqrt{x + 2}}
\end{aligned}$$

$$= \lim_{\Delta x \to 0} \frac{(\sqrt{x+2} - \sqrt{(x+\Delta x)+2})(\sqrt{x+2} + \sqrt{(x+\Delta x)+2})}{\Delta x \sqrt{(x+\Delta x)+2}\,\sqrt{x+2}\,(\sqrt{x+2} + \sqrt{(x+\Delta x)+2})}$$

$$= \lim_{\Delta x \to 0} \frac{(x+2) - (x+\Delta x+2)}{\Delta x \sqrt{x+\Delta x+2}\,\sqrt{x+2}\,(\sqrt{x+2} + \sqrt{x+\Delta x+2})}$$

$$= \lim_{\Delta x \to 0} \frac{-\Delta x}{\Delta x \sqrt{x+\Delta x+2}\,\sqrt{x+2}\,(\sqrt{x+2} + \sqrt{x+\Delta x+2})}$$

$$= \lim_{\Delta x \to 0} \frac{-1}{\sqrt{x+\Delta x+2}\,\sqrt{x+2}\,(\sqrt{x+2} + \sqrt{x+\Delta x+2})}$$

$$= \frac{-1}{\sqrt{x+2}\,\sqrt{x+2}\,(\sqrt{x+2} + \sqrt{x+2})}$$

$$= \frac{-1}{2(x+2)^{3/2}}$$

Example 5 Given $f(x) = x^{1/3}$. (a) Find $f'(x)$. (b) Show that $f'(0)$ does not exist even though f is continuous at 0. Draw a sketch of the graph of f.

Solution

(a) $$f'(x) = \lim_{\Delta x \to 0} \frac{f(x+\Delta x) - f(x)}{\Delta x} = \lim_{\Delta x \to 0} \frac{(x+\Delta x)^{1/3} - x^{1/3}}{\Delta x}$$

We rationalize the numerator in order to obtain a common factor of Δx in the numerator and the denominator; this yields

$$f'(x) = \lim_{\Delta x \to 0} \frac{[(x+\Delta x)^{1/3} - x^{1/3}][(x+\Delta x)^{2/3} + (x+\Delta x)^{1/3}x^{1/3} + x^{2/3}]}{\Delta x[(x+\Delta x)^{2/3} + (x+\Delta x)^{1/3}x^{1/3} + x^{2/3}]}$$

$$= \lim_{\Delta x \to 0} \frac{(x+\Delta x) - x}{\Delta x[(x+\Delta x)^{2/3} + (x+\Delta x)^{1/3}x^{1/3} + x^{2/3}]}$$

$$= \lim_{\Delta x \to 0} \frac{1}{(x+\Delta x)^{2/3} + (x+\Delta x)^{1/3}x^{1/3} + x^{2/3}}$$

$$= \frac{1}{x^{2/3} + x^{1/3}x^{1/3} + x^{2/3}}$$

$$= \frac{1}{3x^{2/3}}$$

(b) $f'(0)$ does not exist because $\dfrac{1}{3x^{2/3}}$ is not defined when $x = 0$. However, the function f is continuous at 0 because

$$\lim_{x \to 0} f(x) = \lim_{x \to 0} x^{1/3}$$

$$= 0$$

$$= f(0)$$

A sketch of the graph of f is shown in Fig. 3.2.1.

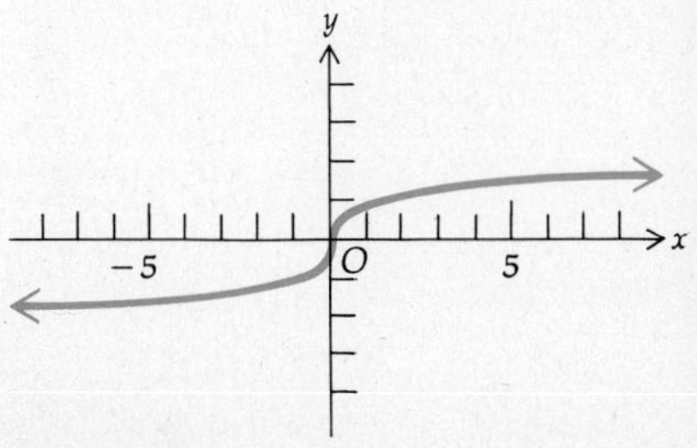

FIGURE 3.2.1

● ILLUSTRATION 1 For the function f of Example 5 consider

$$\lim_{\Delta x \to 0} \frac{f(x_1 + \Delta x) - f(x_1)}{\Delta x} \quad \text{at } x_1 = 0$$

We have

$$\lim_{\Delta x \to 0} \frac{f(0 + \Delta x) - f(0)}{\Delta x} = \lim_{\Delta x \to 0} \frac{(\Delta x)^{1/3} - 0}{\Delta x}$$

$$= \lim_{\Delta x \to 0} \frac{1}{(\Delta x)^{2/3}}$$

$$= +\infty$$

From Definition 3.1.1(ii) it follows that the line $x = 0$ is the tangent line to the graph of f at the origin. ●

Example 5 and Illustration 1 show that $f'(x)$ can exist for some values of x in the domain of f but fail to exist for other values of x in the domain of f.

3.2.2 Definition The function f is said to be **differentiable at x_1** if $f'(x_1)$ exists.

● ILLUSTRATION 2 From Definition 3.2.2 it follows that the function of Example 5 and Illustration 1 is differentiable at every number except 0. ●

3.2.3 Definition A function is said to be **differentiable on an interval** if it is differentiable at every number in the interval.

● ILLUSTRATION 3 In Example 1 $f(x) = 3x^2 + 12$, and the domain of f is the set of all real numbers. Because $f'(x) = 6x$, and $6x$ exists for all real numbers, it follows that f is differentiable everywhere. ●

EXERCISES 3.2

In Exercises 1 through 18, find $f'(x)$ for the given function by applying formula (2) of this section.

1. $f(x) = 7x + 3$
2. $f(x) = 8 - 5x$
3. $f(x) = -4$
4. $f(x) = 3x^2 + 4$
5. $f(x) = 4 - 2x^2$
6. $f(x) = 3x^2 - 2x + 1$
7. $f(x) = 4x^2 + 5x + 3$
8. $f(x) = 6 - 3x - x^2$
9. $f(x) = 8 - x^3$
10. $f(x) = x^3$
11. $f(x) = \sqrt{x}$
12. $f(x) = \sqrt{3x + 5}$
13. $f(x) = \dfrac{1}{x + 1}$
14. $f(x) = \dfrac{1}{\sqrt{3x}}$
15. $f(x) = \dfrac{1}{x^2} - x$
16. $f(x) = \dfrac{3}{1 + x^2}$
17. $f(x) = \dfrac{1}{\sqrt{x + 1}}$
18. $f(x) = \sqrt[3]{2x + 3}$

In Exercises 19 through 24, find $f'(a)$ for the given value of a by applying formula (3) of this section.

19. $f(x) = 1 - x^2$; $a = 3$
20. $f(x) = \dfrac{4}{5x}$; $a = 2$
21. $f(x) = \dfrac{2}{x^3}$; $a = 6$

22. $f(x) = \dfrac{2}{\sqrt{x}} - 1;\ a = 4$ **23.** $f(x) = \sqrt{x^2 - 9};\ a = 5$ **24.** $f(x) = \dfrac{1}{x} + x + x^2;\ a = -3$

In Exercises 25 through 30, find $f'(a)$ for the given value of a by applying formula (6) of this section.

25. $f(x) = 3x + 2;\ a = -3$ **26.** $f(x) = x^2 - x + 4;\ a = 4$ **27.** $f(x) = 2 - x^3;\ a = -2$

28. $f(x) = \sqrt{1 + 9x};\ a = 7$ **29.** $f(x) = \dfrac{1}{\sqrt{2x + 3}};\ a = 3$ **30.** $f(x) = \dfrac{1}{\sqrt[3]{x}} - x;\ a = -8$

In Exercises 31 through 36, find $D_x y$.

31. $y = \dfrac{4}{x^2} + 3x$ **32.** $y = \dfrac{4}{2x - 5}$ **33.** $y = \sqrt{2 - 7x}$

34. $y = \dfrac{1}{\sqrt{x - 1}}$ **35.** $y = \sqrt[3]{1 - 2x}$ **36.** $y = \sqrt{4 - x^2}$

37. If g is continuous at a and $f(x) = (x - a)g(x)$, find $f'(a)$. (*Hint:* Use formula (6).)

38. If g is continuous at a and $f(x) = (x^2 - a^2)g(x)$, find $f'(a)$. (*Hint:* Use formula (6).)

39. Given $f(x) = \sqrt[3]{x - 1}$, find $f'(x)$. Is f differentiable at 1? Draw a sketch of the graph of f.

40. Given $f(x) = \sqrt[3]{(4x - 3)^2}$, find $f'(x)$. Is f differentiable at $\frac{3}{4}$? Draw a sketch of the graph of f.

41. Given $f(x) = \sqrt[3]{1 - 3x}$, find $f'(x)$. For what value or values of x is f not differentiable?

42. If

$$f''(x) = \lim_{\Delta x \to 0} \frac{f'(x + \Delta x) - f'(x)}{\Delta x}$$

find $f''(x)$ if $f(x) = ax^2 + bx$.

43. Use the formula of Exercise 42 to find $f''(x)$ if $f(x) = a/x$.

44. Let f be a function whose domain is the set of all real numbers and $f(a + b) = f(a) \cdot f(b)$ for all a and b. Furthermore, suppose that $f(0) = 1$ and $f'(0)$ exists. Prove that $f'(x)$ exists for all x and that $f'(x) = f'(0) \cdot f(x)$.

45. If f is differentiable at a, prove that

$$f'(a) = \lim_{\Delta x \to 0} \frac{f(a + \Delta x) - f(a - \Delta x)}{2\,\Delta x}$$

(*Hint:* $f(a + \Delta x) - f(a - \Delta x) = f(a + \Delta x) - f(a) + f(a) - f(a - \Delta x)$.)

3.3 DIFFERENTIABILITY AND CONTINUITY

The function of Example 5 of Sec. 3.2 is continuous at the number zero but is not differentiable there. However, the graph of this function has a vertical tangent line at the origin, as shown in Illustration 1 of Sec. 3.2. In the following illustration there is a function that is continuous but not differentiable at zero and whose graph does not have a tangent line at the origin.

● ILLUSTRATION 1 Let f be the absolute-value function. Thus

$$f(x) = |x|$$

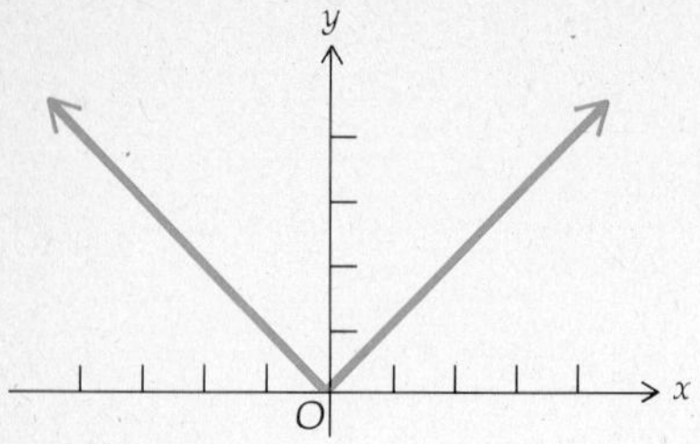

FIGURE 3.3.1

A sketch of the graph of this function was obtained in Example 1 of Sec. 1.5. It is shown here in Fig. 3.3.1. From formula (3) of Sec. 3.2,

$$f'(0) = \lim_{\Delta x \to 0} \frac{f(0 + \Delta x) - f(0)}{\Delta x}$$

if this limit exists. Because $f(0 + \Delta x) = |\Delta x|$ and $f(0) = 0$,

$$\lim_{\Delta x \to 0} \frac{f(0 + \Delta x) - f(0)}{\Delta x} = \lim_{\Delta x \to 0} \frac{|\Delta x|}{\Delta x}$$

Because $|\Delta x| = \Delta x$ if $\Delta x > 0$ and $|\Delta x| = -\Delta x$ if $\Delta x < 0$, one-sided limits at 0 must be considered.

$$\lim_{\Delta x \to 0^+} \frac{|\Delta x|}{\Delta x} = \lim_{\Delta x \to 0^+} \frac{\Delta x}{\Delta x} = \lim_{\Delta x \to 0^+} 1 = 1$$

and

$$\lim_{\Delta x \to 0^-} \frac{|\Delta x|}{\Delta x} = \lim_{\Delta x \to 0^-} \frac{-\Delta x}{\Delta x} = \lim_{\Delta x \to 0^-} (-1) = -1$$

Because $\lim_{\Delta x \to 0^+} \frac{|\Delta x|}{\Delta x} \neq \lim_{\Delta x \to 0^-} \frac{|\Delta x|}{\Delta x}$, it follows that the two-sided limit $\lim_{\Delta x \to 0} \frac{|\Delta x|}{\Delta x}$ does not exist. Therefore $f'(0)$ does not exist, and so f is not differentiable at 0.

Because $f'(0)$ does not exist and is neither $+\infty$ nor $-\infty$, there is no tangent line at the origin for the graph of the absolute value function. ●

Because the functions of the above illustration and Example 5 of Sec. 3.2 are continuous at a number but not differentiable there, it may be concluded that continuity of a function at a number does not imply differentiability of the function at that number. However, differentiability *does* imply continuity, which is given by the next theorem.

3.3.1 Theorem If a function f is differentiable at x_1, then f is continuous at x_1.

Proof To prove that f is continuous at x_1 we must show that the three conditions of Definition 2.6.1 hold there. That is, we must show that (i) $f(x_1)$ exists; (ii) $\lim_{x \to x_1} f(x)$ exists; and (iii) $\lim_{x \to x_1} f(x) = f(x_1)$.

By hypothesis, f is differentiable at x_1. Therefore $f'(x_1)$ exists. Because by formula (6) of Sec. 3.2

$$f'(x_1) = \lim_{x \to x_1} \frac{f(x) - f(x_1)}{x - x_1} \tag{1}$$

$f(x_1)$ must exist; otherwise the above limit has no meaning. Therefore condition (i) holds at x_1. Now consider

$$\lim_{x \to x_1} [f(x) - f(x_1)]$$

We can write

$$\lim_{x \to x_1} [f(x) - f(x_1)] = \lim_{x \to x_1} \left[(x - x_1) \cdot \frac{f(x) - f(x_1)}{x - x_1}\right] \tag{2}$$

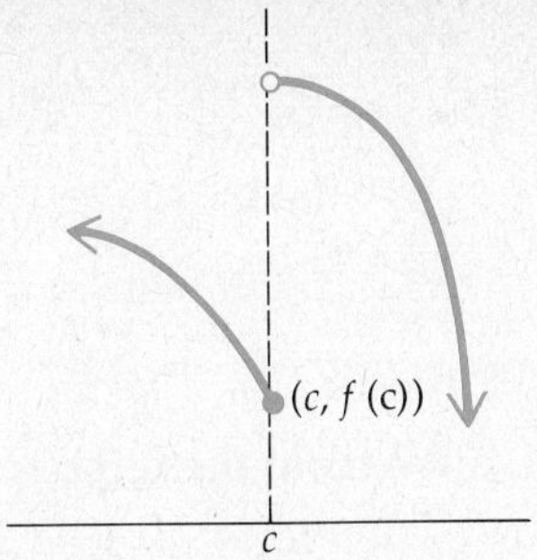

FIGURE 3.3.2

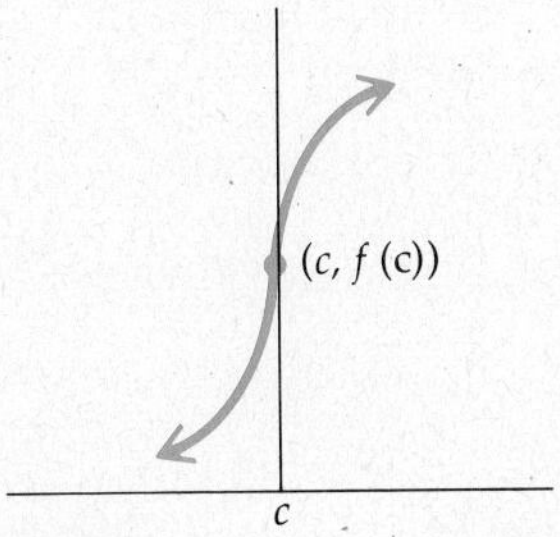

FIGURE 3.3.3

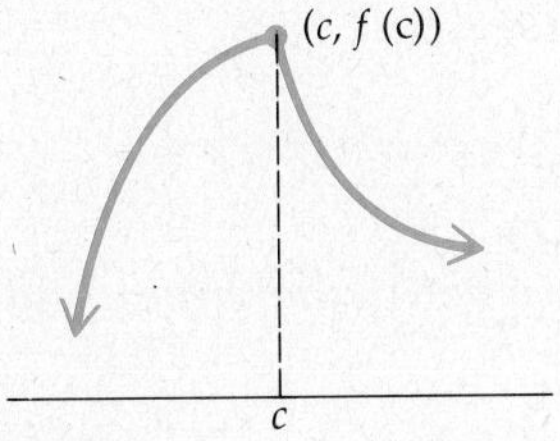

FIGURE 3.3.4

Because

$$\lim_{x \to x_1} (x - x_1) = 0 \quad \text{and} \quad \lim_{x \to x_1} \frac{f(x) - f(x_1)}{x - x_1} = f'(x_1)$$

we apply the theorem on the limit of a product (Theorem 2.2.6) to the right side of (2) and obtain

$$\lim_{x \to x_1} [f(x) - f(x_1)] = \lim_{x \to x_1} (x - x_1) \cdot \lim_{x \to x_1} \frac{f(x) - f(x_1)}{x - x_1} = 0 \cdot f'(x_1)$$

so that

$$\lim_{x \to x_1} [f(x) - f(x_1)] = 0$$

Then

$$\begin{aligned} \lim_{x \to x_1} f(x) &= \lim_{x \to x_1} [f(x) - f(x_1) + f(x_1)] \\ &= \lim_{x \to x_1} [f(x) - f(x_1)] + \lim_{x \to x_1} f(x_1) \\ &= 0 + f(x_1) \end{aligned}$$

which gives

$$\lim_{x \to x_1} f(x) = f(x_1) \tag{3}$$

From (3) it follows that conditions (ii) and (iii) for continuity of f at x_1 hold. Therefore the theorem is proved. ■

A function f can fail to be differentiable at a number c for one of the following reasons:

1. The function f is discontinuous at c. This follows from Theorem 3.3.1. Refer to Fig. 3.3.2 for a sketch of the graph of such a function.
2. The function f is continuous at c, and the graph of f has a vertical tangent line at the point where $x = c$. See Fig. 3.3.3 for a sketch of the graph of a function having this property. This situation also occurs in Example 5 of Sec. 3.2.
3. The function f is continuous at c, and the graph of f does not have a tangent line at the point where $x = c$. In Fig. 3.3.4 there is a sketch of the graph of a function satisfying this condition. Observe that there is a "sharp turn" in the graph at $x = c$. In Illustration 1 there is another such function.

Before giving an additional example of a function that is continuous at a number but not differentiable there, the concept of a **one-sided derivative** is introduced.

3.3.2 Definition If the function f is defined at x_1, then the **derivative from the right** of f at x_1, denoted by $f'_+(x_1)$, is defined by

$$f'_+(x_1) = \lim_{\Delta x \to 0^+} \frac{f(x_1 + \Delta x) - f(x_1)}{\Delta x} \tag{4}$$

or, equivalently,

$$f'_+(x_1) = \lim_{x \to x_1^+} \frac{f(x) - f(x_1)}{x - x_1} \tag{5}$$

if the limit exists.

3.3.3 Definition If the function f is defined at x_1, then the **derivative from the left** of f at x_1, denoted by $f'_-(x_1)$, is defined by

$$f'_-(x_1) = \lim_{\Delta x \to 0^-} \frac{f(x_1 + \Delta x) - f(x_1)}{\Delta x} \tag{6}$$

or, equivalently,

$$f'_-(x_1) = \lim_{x \to x_1^-} \frac{f(x) - f(x_1)}{x - x_1} \tag{7}$$

if the limit exists.

Example 1 Let f be the function defined by

$$f(x) = \begin{cases} 2x - 1 & \text{if } x < 3 \\ 8 - x & \text{if } 3 \leq x \end{cases}$$

(a) Draw a sketch of the graph of f. (b) Prove that f is continuous at 3. (c) Find $f'_-(3)$ and $f'_+(3)$. (d) Is f differentiable at 3?

Solution

(a) A sketch of the graph is shown in Fig. 3.3.5.

(b) To prove that f is continuous at 3, we verify the three conditions for continuity at a number.

(i) $f(3) = 5$

(ii) $\lim_{x \to 3^-} f(x) = \lim_{x \to 3^-} (2x - 1) = 5$

$\lim_{x \to 3^+} f(x) = \lim_{x \to 3^+} (8 - x) = 5$

Thus $\lim_{x \to 3} f(x) = 5$.

(iii) $\lim_{x \to 3} f(x) = f(3)$

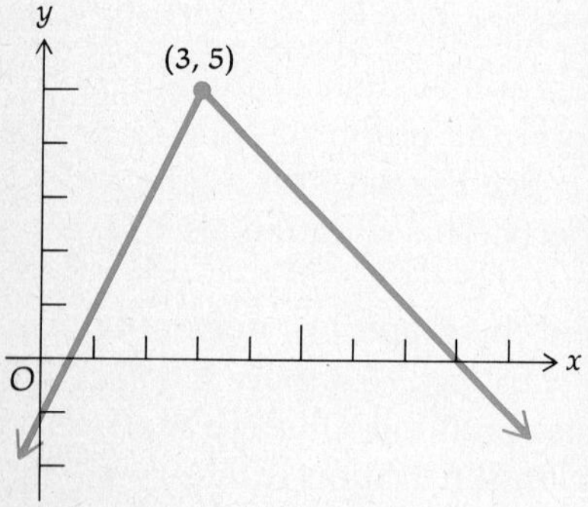

FIGURE 3.3.5

Because conditions (i), (ii), and (iii) all hold at 3, f is continuous at 3.

(c) $$f'_-(3) = \lim_{\Delta x \to 0^-} \frac{f(3 + \Delta x) - f(3)}{\Delta x}$$

$$= \lim_{\Delta x \to 0^-} \frac{[2(3 + \Delta x) - 1] - 5}{\Delta x}$$

$$= \lim_{\Delta x \to 0^-} \frac{6 + 2\,\Delta x - 6}{\Delta x}$$

$$= \lim_{\Delta x \to 0^-} \frac{2\,\Delta x}{\Delta x}$$

$$= \lim_{\Delta x \to 0^-} 2$$

$$= 2$$

$$f'_+(3) = \lim_{\Delta x \to 0^+} \frac{f(3 + \Delta x) - f(3)}{\Delta x}$$

$$= \lim_{\Delta x \to 0^+} \frac{[8 - (3 + \Delta x)] - 5}{\Delta x}$$

$$= \lim_{\Delta x \to 0^+} \frac{8 - 3 - \Delta x - 5}{\Delta x}$$

$$= \lim_{\Delta x \to 0^+} \frac{-\Delta x}{\Delta x}$$

$$= \lim_{\Delta x \to 0^+} (-1)$$

$$= -1$$

(d) Because

$$\lim_{\Delta x \to 0^-} \frac{f(3 + \Delta x) - f(3)}{\Delta x} \neq \lim_{\Delta x \to 0^+} \frac{f(3 + \Delta x) - f(3)}{\Delta x}$$

we conclude that

$$\lim_{\Delta x \to 0} \frac{f(3 + \Delta x) - f(3)}{\Delta x}$$

does not exist. Hence f is not differentiable at 3. However, the derivative from the left and the derivative from the right both exist at 3.

Example 2 Let f be defined by

$$f(x) = |1 - x^2|$$

(a) Draw a sketch of the graph of f. (b) Prove that f is continuous at 1. (c) Determine if f is differentiable at 1.

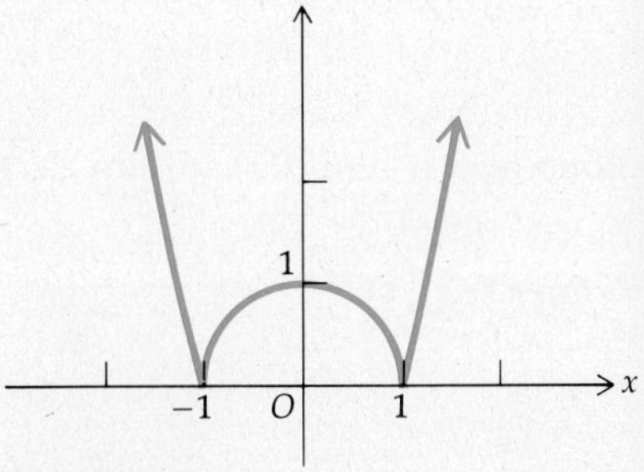

FIGURE 3.3.6

Solution By definition of absolute value, if $x < -1$ or $x > 1$, then $f(x) = -(1 - x^2)$, and if $-1 \leq x \leq 1$, $f(x) = 1 - x^2$. Therefore f can be defined as follows:

$$f(x) = \begin{cases} x^2 - 1 & \text{if } x < -1 \\ 1 - x^2 & \text{if } -1 \leq x \leq 1 \\ x^2 - 1 & \text{if } 1 < x \end{cases}$$

(a) A sketch of the graph of f is shown in Fig. 3.3.6.

(b) To prove that f is continuous at 1 we verify the three conditions for continuity.

(i) $f(1) = 0$

(ii) $\lim_{x \to 1^-} f(x) = \lim_{x \to 1^-} (1 - x^2) = 0$

$\lim_{x \to 1^+} f(x) = \lim_{x \to 1^+} (x^2 - 1) = 0$

Thus $\lim_{x \to 1} f(x) = 0$

(iii) $\lim_{x \to 1} f(x) = f(1)$

Because conditions (i)–(iii) all hold at 1, f is continuous at 1.

(c)
$$\begin{aligned} f'_-(1) &= \lim_{x \to 1^-} \frac{f(x) - f(1)}{x - 1} \\ &= \lim_{x \to 1^-} \frac{(1 - x^2) - 0}{x - 1} \\ &= \lim_{x \to 1^-} \frac{(1 - x)(1 + x)}{x - 1} \\ &= \lim_{x \to 1^-} [-(1 + x)] \\ &= -2 \end{aligned}$$

$$\begin{aligned} f'_+(1) &= \lim_{x \to 1^+} \frac{f(x) - f(1)}{x - 1} \\ &= \lim_{x \to 1^+} \frac{(x^2 - 1) - 0}{x - 1} \\ &= \lim_{x \to 1^+} \frac{(x - 1)(x + 1)}{x - 1} \\ &= \lim_{x \to 1^+} (x + 1) \\ &= 2 \end{aligned}$$

Because $f'_-(1) \neq f'_+(1)$, it follows that $f'(1)$ does not exist, and so f is not differentiable at 1.

The function of Example 2 is also not differentiable at $x = -1$, as can be shown by a method similar to that used for $x = 1$.

Example 3 Given

$$f(x) = \begin{cases} \dfrac{1}{x} & \text{if } x < b \\ 1 - \frac{1}{4}x & \text{if } b \leq x \end{cases}$$

(a) Determine a value of b for which f is continuous. (b) Is f differentiable at the value of b found in part (a)?

Solution

(a) The function f will be continuous at b if $\lim_{x \to b^-} f(x) = \lim_{x \to b^+} f(x) = f(b)$.

$$\begin{aligned} \lim_{x \to b^-} f(x) &= \lim_{x \to b^-} \frac{1}{x} \\ &= \frac{1}{b} \end{aligned}$$

$$\begin{aligned} \lim_{x \to b^+} f(x) &= \lim_{x \to b^+} (1 - \tfrac{1}{4}x) \\ &= 1 - \tfrac{1}{4}b \end{aligned}$$

$f(b) = 1 - \frac{1}{4}b$; therefore f will be continuous at b if

$$\frac{1}{b} = 1 - \frac{1}{4}b$$
$$4 = 4b - b^2$$
$$b^2 - 4b + 4 = 0$$
$$(b-2)^2 = 0$$
$$b = 2$$

Thus

$$f(x) = \begin{cases} \dfrac{1}{x} & \text{if } x < 2 \\ 1 - \frac{1}{4}x & \text{if } 2 \leq x \end{cases}$$

and f is continuous at 2.

(b) To determine if f is differentiable at 2 we compute $f'_-(2)$ and $f'_+(2)$.

$$f'_-(2) = \lim_{x \to 2^-} \frac{f(x) - f(2)}{x - 2} = \lim_{x \to 2^-} \frac{\frac{1}{x} - \frac{1}{2}}{x - 2} = \lim_{x \to 2^-} \frac{2 - x}{2x(x-2)} = \lim_{x \to 2^-} \frac{-1}{2x} = -\frac{1}{4}$$

$$f'_+(2) = \lim_{x \to 2^+} \frac{f(x) - f(2)}{x - 2} = \lim_{x \to 2^+} \frac{(1 - \frac{1}{4}x) - \frac{1}{2}}{x - 2} = \lim_{x \to 2^+} \frac{\frac{1}{2} - \frac{1}{4}x}{x - 2} = \lim_{x \to 2^+} \frac{2 - x}{4(x-2)} = \lim_{x \to 2^+} \frac{-1}{4} = -\frac{1}{4}$$

Because $f'_-(2) = f'_+(2)$, it follows that $f'(2)$ exists, and hence f is differentiable at 2.

EXERCISES 3.3

In Exercises 1 through 16, do each of the following: (a) Draw a sketch of the graph of the function; (b) determine if f is continuous at x_1; (c) find $f'_-(x_1)$ and $f'_+(x_1)$ if they exist; (d) determine if f is differentiable at x_1.

1. $f(x) = \begin{cases} x + 2 & \text{if } x \leq -4 \\ -x - 6 & \text{if } x > -4 \end{cases}$ $x_1 = -4$

2. $f(x) = \begin{cases} 3 - 2x & \text{if } x < 2 \\ 3x - 7 & \text{if } x \geq 2 \end{cases}$ $x_1 = 2$

3. $f(x) = |x - 3|$; $x_1 = 3$

4. $f(x) = 1 + |x + 2|;\ x_1 = -2$

5. $f(x) = \begin{cases} -1 & \text{if } x < 0 \\ x - 1 & \text{if } x \geq 0 \end{cases}$
$x_1 = 0$

6. $f(x) = \begin{cases} x & \text{if } x \leq 0 \\ x^2 & \text{if } x > 0 \end{cases}$
$x_1 = 0$

7. $f(x) = \begin{cases} x^2 & \text{if } x \leq 0 \\ -x^2 & \text{if } x > 0 \end{cases}$
$x_1 = 0$

8. $f(x) = \begin{cases} x^2 - 4 & \text{if } x < 2 \\ \sqrt{x - 2} & \text{if } x \geq 2 \end{cases}$
$x_1 = 2$

9. $f(x) = \begin{cases} \sqrt{1 - x} & \text{if } x < 1 \\ (1 - x)^2 & \text{if } x \geq 1 \end{cases}$
$x_1 = 1$

10. $f(x) = \begin{cases} x^2 & \text{if } x < -1 \\ -1 - 2x & \text{if } x \geq -1 \end{cases}$
$x_1 = -1$

11. $f(x) = \begin{cases} 2x^2 - 3 & \text{if } x \leq 2 \\ 8x - 11 & \text{if } 2 < x \end{cases}$
$x_1 = 2$

12. $f(x) = \begin{cases} x^2 - 9 & \text{if } x < 3 \\ 6x - 18 & \text{if } 3 \leq x \end{cases}$
$x_1 = 3$

13. $f(x) = \sqrt[3]{x + 1};\ x_1 = -1$

14. $f(x) = (x - 2)^{-2};\ x_1 = 2$

15. $f(x) = \begin{cases} 5 - 6x & \text{if } x \leq 3 \\ -4 - x^2 & \text{if } x > 3 \end{cases}$
$x_1 = 3$

16. $f(x) = \begin{cases} -x^{2/3} & \text{if } x \leq 0 \\ x^{2/3} & \text{if } x > 0 \end{cases}$
$x_1 = 0$

17. Given $f(x) = \sqrt{x - 4}$, prove that f is continuous from the right at 4. Prove that $f'_+(4)$ does not exist. Draw a sketch of the graph of f.

18. Given $f(x) = \sqrt{4 - x^2}$, prove that f is continuous on the closed interval $[-2, 2]$. Prove that neither $f'_+(-2)$ nor $f'_-(2)$ exists. Draw a sketch of the graph of f.

19. Given $f(x) = \sqrt{x^2 - 9}$, prove that f is continuous on $(-\infty, -3]$ and $[3, +\infty)$. Prove that neither $f'_-(-3)$ nor $f'_+(3)$ exists. Draw a sketch of the graph of f.

20. Given $f(x) = \sqrt{8 - x^3}$, prove that f is continuous from the left at 2. Prove that $f'_-(2)$ does not exist. Draw a sketch of the graph of f.

21. In Illustration 1 we have the absolute-value function, defined by $f(x) = |x|$, and this function is not differentiable at 0. Prove that $f'(x) = |x|/x$ for all $x \neq 0$. (*Hint:* Let $|x| = \sqrt{x^2}$.)

22. If $f(x) = |x|$ and $g(x) = -|x|$, find a formula for $(f + g)(x)$. Prove that $f + g$ is differentiable at 0 while neither f nor g is differentiable there.

23. Given $f(x) = x^{3/2}$, prove that f is continuous from the right at 0. Prove that $f'_+(0)$ exists, and find its value. Draw a sketch of the graph of f.

24. Given $f(x) = (1 - x^2)^{3/2}$, prove that f is continuous on the closed interval $[-1, 1]$. Prove that $f'(x)$ exists for all values of x in the open interval $(-1, 1)$, and that both $f'_+(-1)$ and $f'_-(1)$ exist. Draw a sketch of the graph of f.

25. Given

$$f(x) = \begin{cases} \dfrac{x + 4}{3} & \text{if } x \leq b \\ 2x - 1 & \text{if } b < x \end{cases}$$

(a) Determine a value of b for which f is continuous. (b) Is f differentiable at the value of b found in part (a)?

26. Find the values of a and b such that $f'(1)$ exists if

$$f(x) = \begin{cases} x^2 & \text{if } x < 1 \\ ax + b & \text{if } 1 \leq x \end{cases}$$

27. Find the values of a and b such that $f'(2)$ exists if

$$f(x) = \begin{cases} ax + b & \text{if } x < 2 \\ 2x^2 - 1 & \text{if } 2 \leq x \end{cases}$$

28. Given $f(x) = [\![x]\!]$, find $f'(x_1)$ if x_1 is not an integer. Prove by applying Theorem 3.3.1 that $f'(x_1)$ does not exist if x_1 is an integer. If x_1 is an integer, what can you say about $f'_-(x_1)$ and $f'_+(x_1)$?

29. Given $f(x) = (x - 1)[\![x]\!]$. Draw a sketch of the graph of f for x in $[0, 2]$. Find, if they exist: (a) $f'_-(1)$; (b) $f'_+(1)$; (c) $f'(1)$.

30. Given $f(x) = (5 - x)[\![x]\!]$. Draw a sketch of the graph of f for x in $[4, 6]$. Find, if they exist: (a) $f'_-(5)$; (b) $f'_+(5)$; (c) $f'(5)$.

31. Given $f(x) = (x - a)[\![x]\!]$, show that $f'_-(a) + 1 = f'_+(a)$.

32. Let the function f be defined by

$$f(x) = \begin{cases} \dfrac{g(x) - g(a)}{x - a} & \text{if } x \neq a \\ g'(a) & \text{if } x = a \end{cases}$$

Prove that if $g'(a)$ exists, f is continuous at a.

33. Given $f(x) = \operatorname{sgn} x$. (a) Prove that $f'_+(0) = +\infty$ and $f'_-(0) = +\infty$. (b) Prove that $\lim_{x \to 0^+} f'(x) = 0$ and $\lim_{x \to 0^-} f'(x) = 0$.

34. Given

$$f(x) = \begin{cases} 0 & \text{if } x \leq 0 \\ x^n & \text{if } x > 0 \end{cases}$$

where n is a positive integer. (a) For what values of n is f differentiable for all values of x? (b) For what values of n is f' continuous for all values of x?

35. A school-sponsored trip that can accommodate up to 250 students will cost each student \$15 if not more than 150 students make the trip; however, the cost per student will be reduced \$0.05 for each student in excess of 150 until the cost reaches \$10 per student. (a) If x students make the trip, express the number of dollars in the gross income as a function of x. (b) Because x is the number of students, x is a nonnegative integer. To have the continuity requirements to apply the calculus, x must be a nonnegative real number. With this assumption show that the function in part (a) is continuous at 150 but not differentiable there.

36. A manufacturer can make a profit of \$20 on each item if not more than 800 items are produced each week. The profit decreases \$0.02 per item over 800. (a) If x items are produced each week, express the number of dollars in the manufacturer's weekly profit as a function of x. (b) Because x is the number of items, x is a nonnegative integer. To have the continuity requirements to apply the calculus, x must be a nonnegative real number. With this assumption show that the function in part (a) is continuous at 800 but not differentiable there.

3.4 SOME THEOREMS ON DIFFERENTIATION OF ALGEBRAIC FUNCTIONS

The operation of finding the derivative of a function is called **differentiation,** which can be performed by applying Definition 3.2.1. However, because this process is usually rather lengthy, we state and prove some theorems that enable us to find the derivative of certain functions more easily. These theorems are proved by applying Definition 3.2.1. Following the proof of each theorem we state the corresponding formula of differentiation.

3.4.1 Theorem If c is a constant and if $f(x) = c$ for all x, then

$$f'(x) = 0$$

Proof

$$
\begin{aligned}
f'(x) &= \lim_{\Delta x \to 0} \frac{f(x + \Delta x) - f(x)}{\Delta x} \\
&= \lim_{\Delta x \to 0} \frac{c - c}{\Delta x} \\
&= \lim_{\Delta x \to 0} 0 \\
&= 0
\end{aligned}
$$

■

$$D_x(c) = 0 \tag{1}$$

The derivative of a constant is zero.

● ILLUSTRATION 1 If $f(x) = 5$, then

$$f'(x) = 0$$

●

3.4.2 Theorem If n is a positive integer and if $f(x) = x^n$, then

$$f'(x) = nx^{n-1}$$

Proof

$$
\begin{aligned}
f'(x) &= \lim_{\Delta x \to 0} \frac{f(x + \Delta x) - f(x)}{\Delta x} \\
&= \lim_{\Delta x \to 0} \frac{(x + \Delta x)^n - x^n}{\Delta x}
\end{aligned}
$$

Applying the binomial theorem to $(x + \Delta x)^n$ we have

$$
\begin{aligned}
f'(x) &= \lim_{\Delta x \to 0} \frac{\left[x^n + nx^{n-1}\Delta x + \dfrac{n(n-1)}{2!}x^{n-2}(\Delta x)^2 + \cdots + nx(\Delta x)^{n-1} + (\Delta x)^n\right] - x^n}{\Delta x} \\
&= \lim_{\Delta x \to 0} \frac{nx^{n-1}\Delta x + \dfrac{n(n-1)}{2!}x^{n-2}(\Delta x)^2 + \cdots + nx(\Delta x)^{n-1} + (\Delta x)^n}{\Delta x}
\end{aligned}
$$

The numerator and denominator are divided by Δx to obtain

$$f'(x) = \lim_{\Delta x \to 0} \left[nx^{n-1} + \frac{n(n-1)}{2!}x^{n-2}(\Delta x) + \cdots + nx(\Delta x)^{n-2} + (\Delta x)^{n-1}\right]$$

Every term except the first has a factor of Δx; therefore every term except the first approaches zero as Δx approaches zero. Therefore

$$f'(x) = nx^{n-1}$$

■

$$D_x(x^n) = nx^{n-1} \tag{2}$$

● ILLUSTRATION 2

(a) If $f(x) = x^8$, then

$$f'(x) = 8x^7$$

(b) If $f(x) = x$, then

$$\begin{aligned} f'(x) &= 1 \cdot x^0 \\ &= 1 \cdot 1 \\ &= 1 \end{aligned}$$

●

3.4.3 Theorem If f is a function, c is a constant, and g is the function defined by

$$g(x) = c \cdot f(x)$$

then if $f'(x)$ exists,

$$g'(x) = c \cdot f'(x)$$

Proof

$$\begin{aligned} g'(x) &= \lim_{\Delta x \to 0} \frac{g(x + \Delta x) - g(x)}{\Delta x} \\ &= \lim_{\Delta x \to 0} \frac{cf(x + \Delta x) - cf(x)}{\Delta x} \\ &= \lim_{\Delta x \to 0} c \cdot \left[\frac{f(x + \Delta x) - f(x)}{\Delta x}\right] \\ &= c \cdot \lim_{\Delta x \to 0} \frac{f(x + \Delta x) - f(x)}{\Delta x} \\ &= cf'(x) \end{aligned}$$

■

$$D_x[c \cdot f(x)] = c \cdot D_x f(x) \tag{3}$$

The derivative of a constant times a function is the constant times the derivative of the function if this derivative exists.

By combining Theorems 3.4.2 and 3.4.3 we obtain the following result: If $f(x) = cx^n$, where n is a positive integer and c is a constant,

$$f'(x) = cnx^{n-1}$$

$$D_x(cx^n) = cnx^{n-1} \tag{4}$$

● ILLUSTRATION 3 If $f(x) = 5x^7$, then

$$\begin{aligned} f'(x) &= 5 \cdot 7x^6 \\ &= 35x^6 \end{aligned}$$

●

3.4.4 Theorem If f and g are functions and if h is the function defined by

$$h(x) = f(x) + g(x)$$

then if $f'(x)$ and $g'(x)$ exist,

$$h'(x) = f'(x) + g'(x)$$

Proof

$$\begin{aligned} h'(x) &= \lim_{\Delta x \to 0} \frac{h(x + \Delta x) - h(x)}{\Delta x} \\ &= \lim_{\Delta x \to 0} \frac{[f(x + \Delta x) + g(x + \Delta x)] - [f(x) + g(x)]}{\Delta x} \\ &= \lim_{\Delta x \to 0} \frac{[f(x + \Delta x) - f(x)] + [g(x + \Delta x) - g(x)]}{\Delta x} \\ &= \lim_{\Delta x \to 0} \frac{f(x + \Delta x) - f(x)}{\Delta x} + \lim_{\Delta x \to 0} \frac{g(x + \Delta x) - g(x)}{\Delta x} \\ &= f'(x) + g'(x) \end{aligned}$$

■

$$D_x[f(x) + g(x)] = D_x f(x) + D_x g(x) \tag{5}$$

The derivative of the sum of two functions is the sum of their derivatives if these derivatives exist.

The result of the preceding theorem can be extended to any finite number of functions, by mathematical induction, and this is stated as another theorem.

3.4.5 Theorem The derivative of the sum of a finite number of functions is equal to the sum of their derivatives if these derivatives exist.

From the preceding theorems the derivative of any polynomial function can be found easily.

Example 1 Given $f(x) = 7x^4 - 2x^3 + 8x + 5$, find $f'(x)$.

Solution

$$\begin{aligned} f'(x) &= D_x(7x^4 - 2x^3 + 8x + 5) \\ &= D_x(7x^4) + D_x(-2x^3) + D_x(8x) + D_x(5) \\ &= 28x^3 - 6x^2 + 8 \end{aligned}$$

3.4.6 Theorem If f and g are functions and if h is the function defined by

$$h(x) = f(x)g(x)$$

then if $f'(x)$ and $g'(x)$ exist,

$$h'(x) = f(x)g'(x) + g(x)f'(x)$$

Proof

$$h'(x) = \lim_{\Delta x \to 0} \frac{h(x + \Delta x) - h(x)}{\Delta x}$$

$$= \lim_{\Delta x \to 0} \frac{f(x + \Delta x) \cdot g(x + \Delta x) - f(x) \cdot g(x)}{\Delta x}$$

If $f(x + \Delta x) \cdot g(x)$ is subtracted and added in the numerator, then

$$h'(x) = \lim_{\Delta x \to 0} \frac{f(x + \Delta x) \cdot g(x + \Delta x) - f(x + \Delta x) \cdot g(x) + f(x + \Delta x) \cdot g(x) - f(x) \cdot g(x)}{\Delta x}$$

$$= \lim_{\Delta x \to 0} \left[f(x + \Delta x) \cdot \frac{g(x + \Delta x) - g(x)}{\Delta x} + g(x) \cdot \frac{f(x + \Delta x) - f(x)}{\Delta x} \right]$$

$$= \lim_{\Delta x \to 0} \left[f(x + \Delta x) \cdot \frac{g(x + \Delta x) - g(x)}{\Delta x} \right] + \lim_{\Delta x \to 0} \left[g(x) \cdot \frac{f(x + \Delta x) - f(x)}{\Delta x} \right]$$

$$= \lim_{\Delta x \to 0} f(x + \Delta x) \cdot \lim_{\Delta x \to 0} \frac{g(x + \Delta x) - g(x)}{\Delta x} + \lim_{\Delta x \to 0} g(x) \cdot \lim_{\Delta x \to 0} \frac{f(x + \Delta x) - f(x)}{\Delta x}$$

Because f is differentiable at x, by Theorem 3.3.1 f is continuous at x; therefore $\lim_{\Delta x \to 0} f(x + \Delta x) = f(x)$. Also,

$$\lim_{\Delta x \to 0} \frac{g(x + \Delta x) - g(x)}{\Delta x} = g'(x)$$

$$\lim_{\Delta x \to 0} \frac{f(x + \Delta x) - f(x)}{\Delta x} = f'(x)$$

and

$$\lim_{\Delta x \to 0} g(x) = g(x)$$

thus giving

$$h'(x) = f(x)g'(x) + g(x)f'(x)$$ ■

$$D_x[f(x)g(x)] = f(x) \cdot D_x g(x) + g(x) \cdot D_x f(x) \tag{6}$$

The derivative of the product of two functions is the first function times the derivative of the second function plus the second function times the derivative of the first function if these derivatives exist.

Example 2 Given $h(x) = (2x^3 - 4x^2)(3x^5 + x^2)$, find $h'(x)$.

Solution

$$h'(x) = (2x^3 - 4x^2)(15x^4 + 2x) + (3x^5 + x^2)(6x^2 - 8x)$$

$$= (30x^7 - 60x^6 + 4x^4 - 8x^3) + (18x^7 - 24x^6 + 6x^4 - 8x^3)$$

$$= 48x^7 - 84x^6 + 10x^4 - 16x^3$$

In Example 2, note that if multiplication is performed before the differentiation, the same result is obtained. Doing this we have

$$h(x) = 6x^8 - 12x^7 + 2x^5 - 4x^4$$

Thus

$$h'(x) = 48x^7 - 84x^6 + 10x^4 - 16x^3$$

3.4.7 Theorem If f and g are functions and if h is the function defined by

$$h(x) = \frac{f(x)}{g(x)}, \qquad \text{where } g(x) \neq 0$$

then if $f'(x)$ and $g'(x)$ exist,

$$h'(x) = \frac{g(x)f'(x) - f(x)g'(x)}{[g(x)]^2}$$

Proof

$$h'(x) = \lim_{\Delta x \to 0} \frac{h(x + \Delta x) - h(x)}{\Delta x}$$

$$= \lim_{\Delta x \to 0} \frac{\dfrac{f(x + \Delta x)}{g(x + \Delta x)} - \dfrac{f(x)}{g(x)}}{\Delta x}$$

$$= \lim_{\Delta x \to 0} \frac{f(x + \Delta x) \cdot g(x) - f(x) \cdot g(x + \Delta x)}{\Delta x \cdot g(x) \cdot g(x + \Delta x)}$$

If $f(x) \cdot g(x)$ is subtracted and added in the numerator, then

$$h'(x) = \lim_{\Delta x \to 0} \frac{f(x + \Delta x) \cdot g(x) - f(x) \cdot g(x) - f(x) \cdot g(x + \Delta x) + f(x) \cdot g(x)}{\Delta x \cdot g(x) \cdot g(x + \Delta x)}$$

$$= \lim_{\Delta x \to 0} \frac{\left[g(x) \cdot \dfrac{f(x + \Delta x) - f(x)}{\Delta x}\right] - \left[f(x) \cdot \dfrac{g(x + \Delta x) - g(x)}{\Delta x}\right]}{g(x) \cdot g(x + \Delta x)}$$

$$= \frac{\lim\limits_{\Delta x \to 0} g(x) \cdot \lim\limits_{\Delta x \to 0} \dfrac{f(x + \Delta x) - f(x)}{\Delta x} - \lim\limits_{\Delta x \to 0} f(x) \cdot \lim\limits_{\Delta x \to 0} \dfrac{g(x + \Delta x) - g(x)}{\Delta x}}{\lim\limits_{\Delta x \to 0} g(x) \cdot \lim\limits_{\Delta x \to 0} g(x + \Delta x)}$$

$$= \frac{g(x) \cdot f'(x) - f(x) \cdot g'(x)}{g(x) \cdot g(x)}$$

$$= \frac{g(x)f'(x) - f(x)g'(x)}{[g(x)]^2}$$

■

$$D_x\left[\frac{f(x)}{g(x)}\right] = \frac{g(x)D_x f(x) - f(x)D_x g(x)}{[g(x)]^2} \tag{7}$$

The derivative of the quotient of two functions is the fraction having as its denominator the square of the original denominator, and as its numerator the denominator times the derivative of the numerator minus the numerator times the derivative of the denominator if these derivatives exist.

Example 3 Given

$$h(x) = \frac{2x^3 + 4}{x^2 - 4x + 1}$$

find $h'(x)$.

Solution

$$h'(x) = \frac{(x^2 - 4x + 1)(6x^2) - (2x^3 + 4)(2x - 4)}{(x^2 - 4x + 1)^2}$$

$$= \frac{6x^4 - 24x^3 + 6x^2 - 4x^4 + 8x^3 - 8x + 16}{(x^2 - 4x + 1)^2}$$

$$= \frac{2x^4 - 16x^3 + 6x^2 - 8x + 16}{(x^2 - 4x + 1)^2}$$

3.4.8 Theorem If $f(x) = x^{-n}$, where $-n$ is a negative integer and $x \neq 0$, then

$$f'(x) = -nx^{-n-1}$$

Proof If $-n$ is a negative integer, then n is a positive integer. We write

$$f(x) = \frac{1}{x^n}$$

From Theorem 3.4.7,

$$f'(x) = \frac{x^n \cdot 0 - 1 \cdot nx^{n-1}}{(x^n)^2} = \frac{-nx^{n-1}}{x^{2n}} = -nx^{n-1-2n}$$

$$= -nx^{-n-1}$$

■

Example 4 Given

$$f(x) = \frac{3}{x^5}$$

find $f'(x)$.

Solution $f(x) = 3x^{-5}$. Hence

$$f'(x) = 3(-5x^{-6}) = -15x^{-6} = -\frac{15}{x^6}$$

If r is any positive or negative integer, it follows from Theorems 3.4.2 and 3.4.8 that

$$D_x(x^r) = rx^{r-1} \tag{8}$$

and from Theorems 3.4.2, 3.4.3, and 3.4.8 we obtain

$$D_x(cx^r) = crx^{r-1} \tag{9}$$

EXERCISES 3.4

In Exercises 1 through 36, differentiate the given function by applying the theorems of this section.

1. $f(x) = 7x - 5$

2. $g(x) = 8 - 3x$

3. $g(x) = 1 - 2x - x^2$

4. $f(x) = 4x^2 + 4x + 1$

5. $f(x) = x^3 - 3x^2 + 5x - 2$

6. $f(x) = 3x^4 - 5x^2 + 1$

7. $f(x) = \frac{1}{8}x^8 - x^4$

8. $g(x) = x^7 - 2x^5 + 5x^3 - 7x$

9. $F(t) = \frac{1}{4}t^4 - \frac{1}{2}t^2$

10. $H(x) = \frac{1}{3}x^3 - x + 2$

11. $v(r) = \frac{4}{3}\pi r^3$

12. $G(y) = y^{10} + 7y^5 - y^3 + 1$

13. $F(x) = x^2 + 3x + \dfrac{1}{x^2}$

14. $f(x) = \dfrac{x^3}{3} + \dfrac{3}{x^3}$

15. $g(x) = 4x^4 - \dfrac{1}{4x^4}$

16. $f(x) = x^4 - 5 + x^{-2} + 4x^{-4}$

17. $g(x) = \dfrac{3}{x^2} + \dfrac{5}{x^4}$

18. $H(x) = \dfrac{5}{6x^5}$

19. $f(s) = \sqrt{3}(s^3 - s^2)$

20. $g(x) = (2x^2 + 5)(4x - 1)$

21. $f(x) = (2x^4 - 1)(5x^3 + 6x)$

22. $f(x) = (4x^2 + 3)^2$

23. $G(y) = (7 - 3y^3)^2$

24. $F(t) = (t^3 - 2t + 1)(2t^2 + 3t)$

25. $f(x) = (x^2 - 3x + 2)(2x^3 + 1)$

26. $g(x) = \dfrac{2x}{x + 3}$

27. $f(x) = \dfrac{x}{x - 1}$

28. $F(y) = \dfrac{2y + 1}{3y + 4}$

29. $H(x) = \dfrac{x^2 + 2x + 1}{x^2 - 2x + 1}$

30. $f(x) = \dfrac{4 - 3x - x^2}{x - 2}$

31. $h(x) = \dfrac{5x}{1 + 2x^2}$

32. $g(x) = \dfrac{x^4 - 2x^2 + 5x + 1}{x^4}$

33. $f(x) = \dfrac{x^3 - 8}{x^3 + 8}$

34. $f(x) = \dfrac{x^2 - a^2}{x^2 + a^2}$

35. $f(x) = \dfrac{2x + 1}{x + 5}(3x - 1)$

36. $g(x) = \dfrac{x^3 + 1}{x^2 + 3}(x^2 - 2x^{-1} + 1)$

37. Find an equation of the tangent line to the curve $y = x^3 - 4$ at the point $(2, 4)$.

38. Find an equation of the normal line to the curve $y = 4x^2 - 8x$ at the point $(1, -4)$.

39. Find an equation of the normal line to the curve $y = 10/(14 - x^2)$ at the point $(4, -5)$.

40. Find an equation of the tangent line to the curve $y = 8/(x^2 + 4)$ at the point $(2, 1)$.

41. Find an equation of the line tangent to the curve $y = 3x^2 - 4x$ and parallel to the line $2x - y + 3 = 0$.

42. Find an equation of the normal line to the curve $y = x^4 - 6x$ that is parallel to the line $x - 2y + 6 = 0$.

43. Find an equation of each of the normal lines to the curve $y = x^3 - 4x$ that is parallel to the line $x + 8y - 8 = 0$.

44. Find an equation of each of the tangent lines to the curve $3y = x^3 - 3x^2 + 6x + 4$ that is parallel to the line $2x - y + 3 = 0$.

45. Find an equation of each of the lines through the point (4, 13) that is tangent to the curve $y = 2x^2 - 1$.

46. Given $f(x) = \frac{1}{3}x^3 + 2x^2 + 5x + 5$. Show that $f'(x) \geq 0$ for all values of x.

47. If f, g, and h are functions and $\phi(x) = f(x) \cdot g(x) \cdot h(x)$, prove that if $f'(x)$, $g'(x)$, and $h'(x)$ exist, $\phi'(x) = f(x) \cdot g(x) \cdot h'(x) + f(x) \cdot g'(x) \cdot h(x) + f'(x) \cdot g(x) \cdot h(x)$. (*Hint:* Apply Theorem 3.4.6 twice.)

Use the result of Exercise 47 to differentiate the functions in Exercises 48 through 51.

48. $f(x) = (x^2 + 3)(2x - 5)(3x + 2)$

49. $h(x) = (3x + 2)^2(x^2 - 1)$

50. $g(x) = (3x^3 + x^{-3})(x + 3)(x^2 - 5)$

51. $\phi(x) = (2x^2 + x + 1)^3$

3.5 THE DERIVATIVE AS A RATE OF CHANGE

Consider a particle moving along a straight line. Such a motion is called **rectilinear motion.** One direction is chosen arbitrarily as positive, and the opposite direction is negative. For simplicity, assume that the motion of the particle is along a horizontal line, with distance to the right as positive and distance to the left as negative. Select some point on the line and denote it by the letter O. Let f be the function determining the directed distance of the particle from O at any particular time.

To be more specific, let s centimeters (cm) be the directed distance of the particle from O at t seconds of time. Then f is the function defined by the equation

$$s = f(t) \tag{1}$$

which gives the directed distance from the point O to the particle at a particular instant of time.

● ILLUSTRATION 1 Let

$$s = t^2 + 2t - 3$$

Then when $t = 0$, $s = -3$; therefore the particle is 3 cm to the left of point O when $t = 0$. When $t = 1$, $s = 0$; so the particle is at point O at 1 sec. When $t = 2$, $s = 5$; so the particle is 5 cm to the right of point O at 2 sec. When $t = 3$, $s = 12$; so the particle is 12 cm to the right of point O at 3 sec.

Figure 3.5.1 illustrates the various positions of the particle for specific values of t.

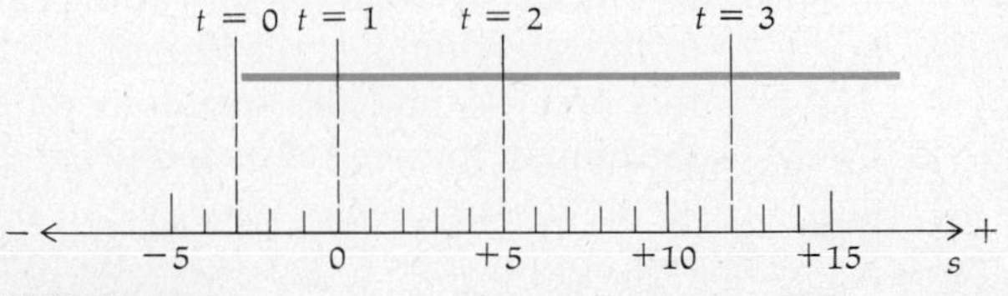

FIGURE 3.5.1

Between the time for $t = 1$ and $t = 3$, the particle moves from the point where $s = 0$ to the point where $s = 12$; thus in the 2-sec interval the change in the directed distance from O is 12 cm. The average velocity of the particle is the ratio of the change in the directed distance from a fixed point to the change in the time. So the number of centimeters per second in the average velocity of the particle from $t = 1$ to $t = 3$ is $\frac{12}{2} = 6$. From $t = 0$ to $t = 2$, the change in the directed distance from O of the particle is 8 cm, and so the number of centimeters per second in the average velocity of the particle in this 2-sec interval is $\frac{8}{2} = 4$. ●

In Illustration 1 the average velocity of the particle is obviously not constant; and the average velocity supplies no specific information about the motion of the particle at any particular instant. For example, if a car travels a distance of 100 kilometers (km) in the same direction and it takes 2 hr, we say that the average velocity in traveling that distance is 50 km/hr. However, from this information we cannot determine the speedometer reading of the car at any particular time in the 2-hr period. The speedometer reading at a specific time is referred to as the *instantaneous velocity.* The following discussion enables us to arrive at a definition of what is meant by "instantaneous velocity."

Let (1) define s (the number of centimeters in the directed distance of the particle from point O) as a function of t (the number of seconds in the time). When $t = t_1$, $s = s_1$. The change in the directed distance from O is $(s - s_1)$ cm over the interval of time $(t - t_1)$ sec, and the number of centimeters per second in the average velocity of the particle over this interval of time is given by

$$\frac{s - s_1}{t - t_1}$$

or, because $s = f(t)$ and $s_1 = f(t_1)$, the average velocity is found from

$$\frac{f(t) - f(t_1)}{t - t_1} \tag{2}$$

Now the shorter the interval is from t_1 to t, the closer the average velocity will be to what we would intuitively think of as the instantaneous velocity at t_1.

For example, if the speedometer reading of a car as it passes a point P_1 is 80 km/hr and if a point P is, for instance, 10 meters (m) from P_1, then the average velocity of the car as it travels this 10 m will very likely be close to 80 km/hr because the variation of the velocity of the car along this short stretch is probably slight. Now if the distance from P_1 to P were shortened to 5 m, the average velocity of the car in this interval would be even closer to the speedometer reading of the car as it passes P_1. We can continue this process, and the speedometer reading at P_1 can be represented as the limit of the average velocity between P_1 and P as P approaches P_1. That is, the *instantaneous velocity* can be defined as the limit of quotient (2) as t approaches t_1, provided the limit exists. This limit is the derivative of the function f at t_1. We have, then, the following definition.

3.5.1 Definition If f is a function given by the equation

$$s = f(t)$$

and a particle is moving along a straight line such that s is the number of units in the directed distance of the particle from a fixed point on the line at t units of time, then the **instantaneous velocity** of the particle at t units of time is $v(t)$ units of velocity where

$$v(t) = f'(t)$$

if it exists.

The instantaneous velocity may be either positive or negative, depending on whether the particle is moving along the line in the positive or the negative direction. When the instantaneous velocity is zero, the particle is at rest.

The **speed** of a particle at any time is defined as the absolute value of the instantaneous velocity. Hence the speed is a nonnegative number. The terms "speed" and "instantaneous velocity" are often confused. It should be noted that the speed indicates only how fast the particle is moving, whereas the instantaneous velocity also tells the direction of motion.

Example 1 A particle is moving along a horizontal line according to the equation

$$s = 2t^3 - 4t^2 + 2t - 1$$

Determine the intervals of time when the particle is moving to the right and when it is moving to the left. Also determine the instant when the particle reverses its direction.

Solution $f(t) = 2t^3 - 4t^2 + 2t - 1.$
Thus

$$\begin{aligned} f'(t) &= 6t^2 - 8t + 2 \\ &= 2(3t^2 - 4t + 1) \\ &= 2(3t - 1)(t - 1) \end{aligned}$$

Therefore

$$v(t) = 2(3t - 1)(t - 1)$$

The instantaneous velocity is zero when $t = \frac{1}{3}$ and $t = 1$. Therefore the particle is at rest at these two times. The particle is moving to the right when $v(t)$ is positive, and it is moving to the left when $v(t)$ is negative. We determine the sign of $v(t)$ for various intervals of t, and the results are given in Table 3.5.1.

TABLE 3.5.1

	$3t-1$	$t-1$	Conclusion
$t<\frac{1}{3}$	$-$	$-$	$v(t)$ is positive, and the particle is moving to the right
$t=\frac{1}{3}$	0	$-$	$v(t)$ is zero, and the particle is changing direction from right to left
$\frac{1}{3}<t<1$	$+$	$-$	$v(t)$ is negative, and the particle is moving to the left
$t=1$	$+$	0	$v(t)$ is zero, and the particle is changing direction from left to right
$1<t$	$+$	$+$	$v(t)$ is positive, and the particle is moving to the right

The motion of the particle, indicated in Fig. 3.5.2, is along the horizontal line; however, the behavior of the motion is indicated above the line. The accompanying Table 3.5.2 gives values of s and v for specific values of t.

TABLE 3.5.2

t	s	v
-1	-9	16
0	-1	2
$\frac{1}{3}$	$-\frac{19}{27}$	0
1	-1	0
2	3	10

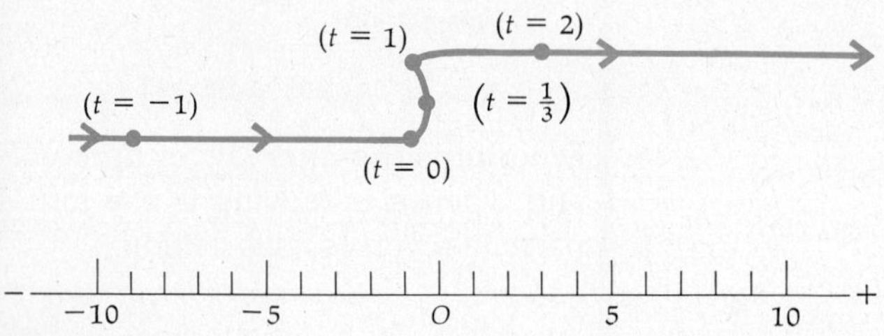

FIGURE 3.5.2

Example 2 A ball is thrown vertically upward from the ground with an initial velocity of 64 ft/sec. If the positive direction of the distance from the starting point is up, the equation of motion is

$$s = -16t^2 + 64t$$

If t is the number of seconds in the time that has elapsed since the ball was thrown, and s is the number of feet in the distance of the ball from the starting point at t sec, find (a) the instantaneous velocity of the ball at the end of 1 sec; (b) the instantaneous velocity of the ball at the end of 3 sec; (c) how many seconds it takes the ball to reach its highest point; (d) how high the ball will go; (e) the speed of the ball at the end of 1 sec and at the end of 3 sec; (f) how many seconds it takes the ball to reach the ground; (g) the instantaneous velocity of the ball when it reaches the ground. At the end of 1 sec is the ball rising or falling? At the end of 3 sec is the ball rising or falling?

TABLE 3.5.3

t	s	v
0	0	64
$\frac{1}{2}$	28	48
1	48	32
2	64	0
3	48	-32
$\frac{7}{2}$	28	-48
4	0	-64

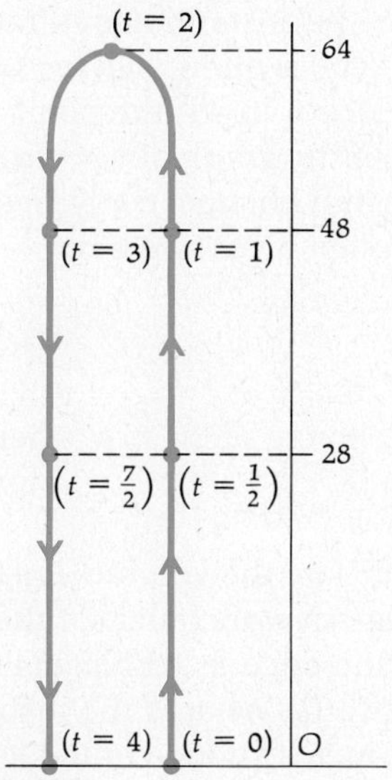

FIGURE 3.5.3

Solution $v(t)$ is the number of feet per second in the instantaneous velocity of the ball at t sec.

$$s = f(t), \qquad \text{where } f(t) = -16t^2 + 64t$$

$$f'(t) = -32t + 64$$

Therefore

$$v(t) = -32t + 64$$

(a) $v(1) = -32(1) + 64 = 32$; so at the end of 1 sec the ball is rising with an instantaneous velocity of 32 ft/sec.

(b) $v(3) = -32(3) + 64 = -32$; so at the end of 3 sec the ball is falling with an instantaneous velocity of -32 ft/sec.

(c) The ball reaches its highest point when the direction of motion changes, that is, when $v(t) = 0$. Setting $v(t) = 0$ we obtain $-32t + 64 = 0$. Thus $t = 2$.

(d) When $t = 2$, $s = 64$; therefore the ball reaches a highest point of 64 ft above the starting point.

(e) $|v(t)|$ is the number of feet per second in the speed of the ball at t sec.

$$|v(1)| = 32 \quad \text{and} \quad |v(3)| = 32$$

(f) The ball will reach the ground for $s = 0$. Setting $s = 0$ we have $-16t^2 + 64t = 0$, from which we obtain $t = 0$ and $t = 4$. Therefore the ball will reach the ground in 4 sec.

(g) $v(4) = -64$; when the ball reaches the ground, its instantaneous velocity is -64 ft/sec.

Table 3.5.3 gives values of s and v for some specific values of t. The motion of the ball is indicated in Fig. 3.5.3. The motion is assumed to be in a straight vertical line, and the behavior of the motion is indicated to the left of the line.

The concept of velocity in rectilinear motion corresponds to the more general concept of instantaneous rate of change. For example, if a particle is moving along a straight line according to the equation $s = f(t)$, the velocity of the particle at t units of time is determined by the derivative of s with respect to t. Because velocity can be interpreted as a rate of change of distance per unit change in time, the derivative of s with respect to t is the rate of change of s per unit change in t.

In a similar way, if a quantity y is a function of a quantity x, we may express the rate of change of y per unit change in x. The discussion is analogous to the discussions of the slope of a tangent line to a graph and the instantaneous velocity of a particle moving along a straight line.

If the functional relationship between y and x is given by

$$y = f(x)$$

and if x changes from the value x_1 to $x_1 + \Delta x$, then y changes from $f(x_1)$ to $f(x_1 + \Delta x)$. So the change in y, which may be denoted by Δy, is $f(x_1 + \Delta x) - f(x_1)$ when the change in x is Δx. The average rate of change of y per unit change in x, as x changes from x_1 to $x_1 + \Delta x$, is then

$$\frac{f(x_1 + \Delta x) - f(x_1)}{\Delta x} = \frac{\Delta y}{\Delta x} \qquad (3)$$

If the limit of this quotient exists as $\Delta x \to 0$, this limit is what we intuitively think of as the instantaneous rate of change of y per unit change in x at x_1. Accordingly, we have the following definition.

3.5.2 Definition If $y = f(x)$, the **instantaneous rate of change of y per unit change in x at x_1** is $f'(x_1)$ or, equivalently, the derivative of y with respect to x at x_1, if it exists there.

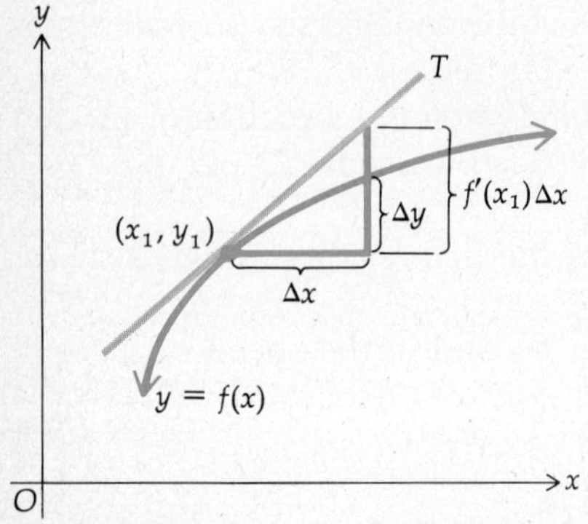

FIGURE 3.5.4

To illustrate geometrically, let $f'(x_1)$ be the instantaneous rate of change of y per unit change in x at x_1. Then if $f'(x_1)$ is multiplied by Δx (the change in x), we have the change that would occur in y if the point (x, y) were to move along the tangent line at (x_1, y_1) of the graph of $y = f(x)$. See Fig. 3.5.4. The average rate of change of y per unit change in x is given by the fraction in (3), and if this fraction is multiplied by Δx, we have

$$\frac{\Delta y}{\Delta x} \cdot \Delta x = \Delta y$$

which is the actual change in y caused by a change of Δx in x when the point (x, y) moves along the graph.

Example 3 Let V cubic centimeters (cm^3) be the volume of a cube having an edge of length e cm. Find the average rate of change of the volume per centimeter change in the length of the edge as e changes from (a) 3.00 to 3.20; (b) 3.00 to 3.10; (c) 3.00 to 3.01. (d) What is the instantaneous rate of change of the volume per centimeter change in the length of the edge when $e = 3$?

Solution Because the formula for finding the volume of a cube is $V = e^3$, let f be the function defined by $f(e) = e^3$. Then the average rate of change of V per unit change in e as e changes from e_1 to $e_1 + \Delta e$ is

$$\frac{f(e_1 + \Delta e) - f(e_1)}{\Delta e}$$

(a) $e_1 = 3$, $\Delta e = 0.2$, and $\dfrac{f(3.2) - f(3)}{0.2} = \dfrac{(3.2)^3 - 3^3}{0.2} = \dfrac{5.77}{0.2} = 28.8$

(b) $e_1 = 3$, $\Delta e = 0.1$, and $\dfrac{f(3.1) - f(3)}{0.1} = \dfrac{(3.1)^3 - 3^3}{0.1} = \dfrac{2.79}{0.1} = 27.9$

(c) $e_1 = 3$, $\Delta e = 0.01$, and $\dfrac{f(3.01) - f(3)}{0.01} = \dfrac{(3.01)^3 - 3^3}{0.01} = \dfrac{0.271}{0.01} = 27.1$

In part (a) we see that as the length of the edge of the cube changes from 3.00 cm to 3.20 cm, the change in the volume is 5.77 cm^3 and the average rate of change of the volume is 28.8 cm^3 per centimeter change in the length of the edge. There are similar interpretations of parts (b) and (c).

(d) The instantaneous rate of change of V per unit change in e at 3 is $f'(3)$.

$$f'(e) = 3e^2$$

Hence

$$f'(3) = 27$$

Therefore, when the length of the edge of the cube is 3 cm, the instantaneous rate of change of the volume is 27 cm³ per centimeter change in the length of the edge.

Example 4 A company estimates that if $1000x$ dollars are spent on advertising, it will sell y units of a commodity, where $y = 5 + 400x - 2x^2$. (a) Find the average rate of change of y with respect to x when the advertising budget is increased from \$10,000 to \$11,000. (b) Find the instantaneous (or marginal) rate of change of y with respect to x when the advertising budget is \$10,000.

Solution Let f be the function defined by

$$f(x) = 5 + 400x - 2x^2$$

Then the average rate of change of y with respect to x as x changes from x_1 to $x_1 + \Delta x$ is

$$\frac{f(x_1 + \Delta x) - f(x_1)}{\Delta x} \tag{4}$$

(a) To find the average rate of change of y with respect to x as x changes from 10 to 10 + 1 we use quotient (4) with $x_1 = 10$ and $\Delta x = 1$.

$$\begin{aligned}\frac{f(10 + 1) - f(10)}{1} &= \frac{f(11) - f(10)}{1}\\ &= [5 + 400(11) - 2(11)^2] - [5 + 400(10) - 2(10)^2]\\ &= 4163 - 3805\\ &= 358\end{aligned}$$

Therefore, when the advertising budget is increased from \$10,000 to \$11,000, the average rate of change of the number of units sold is 358 units per \$1000 increase in the advertising budget.

(b) The instantaneous rate of change of y with respect to x at 10 is $f'(10)$.

$$f'(x) = 400 - 4x$$

Therefore

$$f'(10) = 360$$

Hence when the advertising budget is \$10,000, the instantaneous rate of change of the number of units sold is 360 units per \$1000 increase in the budget.

Example 5 The annual gross earnings of a particular corporation t years after January 1, 1979, is p millions of dollars and $p = \frac{2}{5}t^2 + 2t + 10$. Find (a) the rate at which the gross earnings were growing January 1, 1981; (b) the rate at which the gross earnings should be growing January 1, 1985.

Solution

(a) On January 1, 1981, $t = 2$; hence we find $D_t p$ when $t = 2$.

$$D_t p = \tfrac{4}{5}t + 2 \qquad D_t p\Big]_{t=2} = \tfrac{8}{5} + 2 = 3.6$$

So on January 1, 1981, the gross earnings were growing at the rate of 3.6 million dollars per year.

(b) On January 1, 1985, $t = 6$ and

$$D_t p\Big]_{t=6} = \tfrac{24}{5} + 2 = 6.8$$

Therefore on January 1, 1985, the gross earnings should be growing at the rate of \$6.8 million per year.

The results of Example 5 are meaningful only if they are compared to the actual earnings of the corporation. For example, if on January 1, 1980, it was found that the earnings of the corporation for the year 1979 had been \$3 million, then the rate of growth on January 1, 1981, of \$3.6 million annually would have been excellent. However, if the earnings in 1979 had been \$300 million, then the growth rate on January 1, 1981, would have been very poor. The measure used to compare the rate of change with the amount of the quantity which is being changed is called the *relative rate.*

3.5.3 Definition If $y = f(x)$, the **relative rate of change of y per unit change in x at x_1** is given by $f'(x_1)/f(x_1)$ or, equivalently, $D_x y/y$ evaluated at $x = x_1$.

If the relative rate is multiplied by 100, the percent rate of change is obtained.

Example 6 Find the relative rate of growth of the gross earnings on January 1, 1981, and January 1, 1985, for the corporation of Example 5.

Solution

(a) When $t = 2$, $p = \frac{2}{5}(4) + 2(2) + 10 = 15.6$. Hence on January 1, 1981, the relative rate of growth of the corporation's annual gross earnings was

$$\frac{D_t p}{p}\Big]_{t=2} = \frac{3.6}{15.6} = 0.231 = 23.1\%$$

(b) When $t = 6$, $p = \frac{2}{5}(36) + 2(6) + 10 = 36.4$. Therefore on January 1, 1985, the relative rate of growth of the corporation's annual gross earnings should be

$$\left.\frac{D_t p}{p}\right]_{t=6} = \frac{6.8}{36.4} = 0.187 = 18.7\%$$

Observe that the growth rate of \$6.8 million for January 1, 1985, is greater than the \$3.6 million for January 1, 1981; however, the relative growth rate of 18.7% for January 1, 1985, is less than the relative growth rate of 23.1% for January 1, 1981.

EXERCISES 3.5

In Exercises 1 through 8, a particle is moving along a horizontal line according to the given equation, where s cm is the directed distance of the particle from a point O at t sec. Find the instantaneous velocity v(t) cm/sec at t sec; and then find $v(t_1)$ for the particular value of t_1 given.

1. $s = 3t^2 + 1;\ t_1 = 3$ **2.** $s = 8 - t^2;\ t_1 = 5$ **3.** $s = \frac{1}{4t};\ t_1 = \frac{1}{2}$ **4.** $s = \frac{3}{t^2};\ t_1 = -2$

5. $s = 2t^3 - t^2 + 5;\ t_1 = -1$ **6.** $s = 4t^3 + 2t - 1;\ t_1 = \frac{1}{2}$ **7.** $s = \frac{2t}{4 + t};\ t_1 = 0$ **8.** $s = \frac{1}{t} + \frac{3}{t^2};\ t_1 = 2$

In Exercises 9 through 14, the motion of a particle is along a horizontal line according to the given equation, where s cm is the directed distance of the particle from a point O at t sec. The positive direction is to the right. Determine the intervals of time when the particle is moving to the right and when it is moving to the left. Also determine when the particle reverses its direction. Show the behavior of the motion by a figure similar to Fig. 3.5.2, and choose values of t at random but include the values of t when the particle reverses its direction.

9. $s = t^3 + 3t^2 - 9t + 4$ **10.** $s = 2t^3 - 3t^2 - 12t + 8$ **11.** $s = \frac{2}{3}t^3 + \frac{3}{2}t^2 - 2t + 4$

12. $s = \frac{t}{1 + t^2}$ **13.** $s = \frac{t}{9 + t^2}$ **14.** $s = \frac{t + 1}{t^2 + 4}$

15. An object falls from rest, and $s = -16t^2$ where s ft is the distance of the object from the starting point at t sec, and the positive direction is upward. If a stone is dropped from a building 256 ft high, find (a) the instantaneous velocity of the stone 1 sec after it is dropped; (b) the instantaneous velocity of the stone 2 sec after it is dropped; (c) how long it takes the stone to reach the ground; (d) the instantaneous velocity of the stone when it reaches the ground.

16. A stone is dropped from a height of 64 ft. If s ft is the height of the stone t sec after being dropped, then $s = -16t^2 + 64$. (a) How long does it take the stone to reach the ground? (b) Find the instantaneous velocity of the stone when it reaches the ground.

17. A billiard ball is hit and travels in a straight line. If s cm is the distance of the ball from its initial position at t sec, then $s = 100t^2 + 100t$. If the ball hits a cushion that is 39 cm from its initial position, at what velocity does it hit the cushion?

18. If a stone is thrown vertically upward from the ground with an initial velocity of 32 ft/sec, then $s = -16t^2 + 32t$, where s ft is the distance of the stone from the starting point at t sec, and the positive direction is upward. Find (a) the average velocity of the stone during the time interval: $\frac{3}{4} \le t \le \frac{5}{4}$; (b) the instantaneous velocity of the stone at $\frac{3}{4}$ sec and at $\frac{5}{4}$ sec; (c) the speed of the stone at $\frac{3}{4}$ sec and at $\frac{5}{4}$ sec; (d) the average velocity of the stone during the time interval: $\frac{1}{2} \le t \le \frac{3}{4}$; (e) how many seconds it will take the stone to reach the highest point; (f) how high the stone will go; (g) how many seconds it will take the stone to reach the ground; (h) the instantaneous velocity of the stone when it reaches the ground. Show the behavior of the motion by a figure similar to Fig. 3.5.3.

19. If a ball is given a push so that it has an initial velocity of 24 ft/sec down a certain inclined plane, then $s = 24t + 10t^2$, where s ft is the distance of the ball from the starting point at t sec and the positive direction is down the inclined plane. (a) What is the instantaneous velocity of the ball at t_1 sec? (b) How long does it take for the velocity to increase to 48 ft/sec?

20. A rocket is fired vertically upward, and it is s ft above the ground t sec after being fired, where $s = 560t - 16t^2$ and the positive direction is upward. Find (a) the velocity of the rocket 2 sec after being fired, and (b) how long it takes for the rocket to reach its maximum height.

21. If A cm² is the area of a square and s cm is the length of a side of the square, find the average rate of change of A with respect to s as s changes from (a) 4.00 to 4.60; (b) 4.00 to 4.30; (c) 4.00 to 4.10. (d) What is the instantaneous rate of change of A with respect to s when s is 4.00?

22. The length of a rectangle is 4 in. more than its width, and the 4-in. difference is maintained as the rectangle increases in size. If A in.² is the area of the rectangle and w in. is the width of the rectangle, find the average rate of change of A with respect to w as w changes from (a) 3.00 to 3.20; (b) 3.00 to 3.10; (c) 3.00 to 3.01. (d) What is the instantaneous rate of change of A with respect to w when w is 3.00?

23. A right-circular cylinder has a constant height of 10.00 cm. If V cm³ is the volume of the cylinder and r cm is the radius of its base, find the average rate of change of V with respect to r as r changes from (a) 5.00 to 5.40; (b) 5.00 to 5.10; (c) 5.00 to 5.01. (d) What is the instantaneous rate of change of V with respect to r when r is 5.00?

24. A right-circular cone has a constant height of 6.00 in. If V in.³ is the volume of the cone and r in. is the radius of its base, find the average rate of change of V with respect to r as r changes from (a) 2.00 to 2.20; (b) 2.00 to 2.10; (c) 2.00 to 2.01. (d) What is the instantaneous rate of change of V with respect to r when r is 2.00?

25. Let r be the reciprocal of a number n. Find the instantaneous rate of change of r with respect to n and the relative rate of change of r with respect to n when n is (a) 4 and (b) 10.

26. Let s be the principal square root of a number x. Find the instantaneous rate of change of s with respect to x and the relative rate of change of s with respect to x when x is (a) 9 and (b) 4.

27. The demand equation for a particular kind of costume jewelry is $x = 100 - 3p - 2p^2$, where x units are demanded when p dollars is the price per unit. (a) Find the average rate of change of the demand with respect to the price when the price is increased from \$5 to \$5.50. (b) Find the instantaneous (or marginal) rate of change of the demand with respect to the price when the price is \$5.

28. The supply equation for a certain kind of pencil is $x = 3p^2 + 2p$, where $1000x$ pencils are supplied when p cents is the price per pencil. (a) Find the average rate of change of the supply with respect to the price when the price is increased from 10 cents to 11 cents. (b) Find the instantaneous (or marginal) rate of change of the supply with respect to the price when the price is 10 cents.

29. If water is being drained from a swimming pool and V liters is the volume of water in the pool t min after the draining starts, where $V = 250(1600 - 80t + t^2)$, find (a) the average rate at which the water leaves the pool during the first 5 min, and (b) how fast the water is flowing out of the pool 5 min after the draining starts.

30. Boyle's law for the expansion of a gas is $PV = C$, where P is the number of pounds per square unit of pressure, V is the number of cubic units in the volume of the gas, and C is a constant. Find the instantaneous rate of change of V with respect to P when $P = 4$ and $V = 8$.

31. In an electric circuit, if E volts is the electromotive force, R ohms is the resistance, and I amperes is the current, Ohm's law states that $IR = E$. Assuming that E is constant, show that R decreases at a rate proportional to the inverse square of I.

32. A balloon maintains the shape of a sphere as it is being inflated. Find the rate of change of the surface area with respect to the radius at the instant when the radius is 2 meters.

33. A cold front approaches the college campus. The temperature is T degrees t hours after midnight, and

$$T = 0.1(400 - 40t + t^2) \qquad 0 \le t \le 12$$

(a) Find the average rate of change of T with respect to t between 5 A.M. and 6 A.M. (b) Find the instantaneous rate of change of T with respect to t at 5 A.M.

34. It is estimated that a worker in a shop that makes picture frames can paint y frames x hours after starting work at 8 A.M., and

$$y = 3x + 8x^2 - x^3 \qquad 0 < x \leq 4$$

(a) Find the rate at which the worker is painting at 10 A.M. (b) Find the number of frames that the worker paints between 10 A.M. and 11 A.M.

35. Suppose that the number of people in the population of a particular city t years after January 1, 1981, will be $40t^2 + 200t + 10{,}000$. (a) Find the rate at which the population will be growing on January 1, 1990. (b) Find the rate at which the population will be growing on January 1, 1996. (c) Find the relative rate of growth of the population on January 1, 1990. (d) Find the relative rate of growth of the population on January 1, 1996.

36. Two particles, A and B, move to the right along a horizontal line. They start at a point O, s ft is the directed distance of the particle from O at t sec, and the equations of motion are

$$s = 4t^2 + 5t \quad \text{for particle } A$$

$$s = 7t^2 + 3t \quad \text{for particle } B$$

If $t = 0$ at the start, for what values of t will the velocity of particle A exceed the velocity of particle B?

37. The profit of a retail store is $100y$ dollars when x dollars are spent daily on advertising and $y = 2500 + 36x - 0.2x^2$. Use the derivative to determine if it would be profitable for the daily advertising budget to be increased if the daily advertising budget is (a) \$60 and (b) \$300. (c) What is the maximum value for x below which it is profitable to increase the advertising budget?

38. Show that for any linear function f the average rate of change of $f(x)$ as x changes from x_1 to $x_1 + k$ is the same as the instantaneous rate of change of $f(x)$ at x_1.

3.6 DERIVATIVES OF THE SINE AND COSINE FUNCTIONS

To show that the sine function has a derivative we apply (12) of Sec. 1.7, which is

$$\sin(a + b) = \sin a \cos b + \cos a \sin b \tag{1}$$

and make use of Theorems 2.9.2 and 2.9.5.

Let f be the function defined by

$$f(x) = \sin x$$

From the definition of a derivative,

$$f'(x) = \lim_{\Delta x \to 0} \frac{f(x + \Delta x) - f(x)}{\Delta x}$$

$$= \lim_{\Delta x \to 0} \frac{\sin(x + \Delta x) - \sin x}{\Delta x}$$

Formula (1) for $\sin(x + \Delta x)$ is used to obtain

$$f'(x) = \lim_{\Delta x \to 0} \frac{\sin x \cos(\Delta x) + \cos x \sin(\Delta x) - \sin x}{\Delta x}$$

$$= \lim_{\Delta x \to 0} \frac{\sin x[\cos(\Delta x) - 1]}{\Delta x} + \lim_{\Delta x \to 0} \frac{\cos x \sin(\Delta x)}{\Delta x}$$

$$= -\lim_{\Delta x \to 0} \frac{1 - \cos(\Delta x)}{\Delta x}\left(\lim_{\Delta x \to 0} \sin x\right) + \left(\lim_{\Delta x \to 0} \cos x\right) \lim_{\Delta x \to 0} \frac{\sin(\Delta x)}{\Delta x} \tag{2}$$

From Theorem 2.9.5,

$$\lim_{\Delta x \to 0} \frac{1 - \cos(\Delta x)}{\Delta x} = 0 \tag{3}$$

and from Theorem 2.9.2

$$\lim_{\Delta x \to 0} \frac{\sin(\Delta x)}{\Delta x} = 1 \tag{4}$$

Substituting from (3) and (4) into (2) we get

$$f'(x) = -0 \cdot \sin x + \cos x \cdot 1$$

$$= \cos x$$

We have proved the following theorem.

3.6.1 Theorem

$$D_x(\sin x) = \cos x \tag{5}$$

Example 1 Given, $f(x) = x^2 \sin x$, find $f'(x)$.

Solution We find the derivative of a product of two functions by applying Theorem 3.4.6. We have

$$f'(x) = x^2 D_x(\sin x) + D_x(x^2)\sin x$$

$$= x^2 \cos x + 2x \sin x$$

To find the derivative of the cosine function we proceed as with the sine function. Here we apply (4) of Sec. 1.7, which is

$$\cos(a + b) = \cos a \cos b - \sin a \sin b \tag{6}$$

Let g be the function defined by

$$g(x) = \cos x$$

Then

$$g'(x) = \lim_{\Delta x \to 0} \frac{g(x + \Delta x) - g(x)}{\Delta x}$$

$$= \lim_{\Delta x \to 0} \frac{\cos(x + \Delta x) - \cos x}{\Delta x}$$

Formula (6) for $\cos(x + \Delta x)$ is used to obtain

$$g'(x) = \lim_{\Delta x \to 0} \frac{\cos x \cos(\Delta x) - \sin x \sin(\Delta x) - \cos x}{\Delta x}$$

$$= \lim_{\Delta x \to 0} \frac{\cos x[\cos(\Delta x) - 1]}{\Delta x} - \lim_{\Delta x \to 0} \frac{\sin x \sin(\Delta x)}{\Delta x}$$

$$= -\lim_{\Delta x \to 0} \frac{1 - \cos(\Delta x)}{\Delta x}\left(\lim_{\Delta x \to 0} \cos x\right) - \left(\lim_{\Delta x \to 0} \sin x\right) \lim_{\Delta x \to 0} \frac{\sin(\Delta x)}{\Delta x} \quad (7)$$

We substitute from (3) and (4) into (7) and obtain

$$g'(x) = -0 \cdot \cos x - \sin x \cdot 1$$

$$= -\sin x$$

The following theorem has been proved.

3.6.2 Theorem

$$D_x(\cos x) = -\sin x \quad (8)$$

Example 2 Given

$$y = \frac{\sin x}{1 - 2\cos x}$$

find $D_x y$.

Solution We apply Theorem 3.4.7.

$$D_x y = \frac{(1 - 2\cos x)\, D_x(\sin x) - \sin x \cdot D_x(1 - 2\cos x)}{(1 - 2\cos x)^2}$$

$$= \frac{(1 - 2\cos x)(\cos x) - \sin x(2\sin x)}{(1 - 2\cos x)^2}$$

$$= \frac{\cos x - 2(\cos^2 x + \sin^2 x)}{(1 - 2\cos x)^2}$$

$$= \frac{\cos x - 2}{(1 - 2\cos x)^2}$$

Because $D_x(\sin x) = \cos x$ and $\cos x$ exists for all values of x, the sine function is differentiable everywhere and therefore continuous everywhere. Similarly, the cosine function is differentiable and continuous everywhere. Because the largest value either function may have is 1 and the smallest value is -1, it follows from the intermediate-value theorem that the range of each of these functions is $[-1, 1]$.

The graph of the sine function is now discussed. Let

$$f(x) = \sin x$$

Then

$$f'(x) = \cos x$$

To determine where the graph has a horizontal tangent, set $f'(x) = 0$ and get $x = \frac{1}{2}\pi + n\pi$, where $n = 0, \pm 1, \pm 2, \ldots$. At these values of x, $\sin x$ is either $+1$ or -1; and these are the largest and smallest values that $\sin x$ assumes. The graph intersects the x axis at those points where $\sin x = 0$, that is, at the points where $x = n\pi$ and n is any integer. From this information we draw a sketch of the graph of the sine function. It is shown in Fig. 3.6.1.

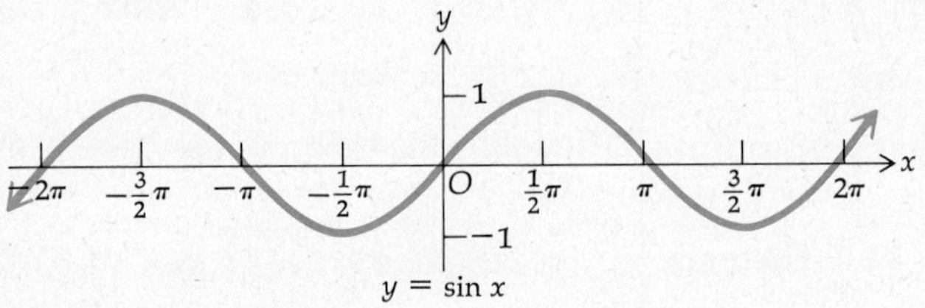

FIGURE 3.6.1

To obtain the graph of the cosine function, use the identity

$$\cos x = \sin(\tfrac{1}{2}\pi + x)$$

which follows from (12) of Sec. 1.7. Hence the graph of the cosine function is obtained from the graph of the sine function by translating the y axis $\frac{1}{2}\pi$ units to the right (see Fig. 3.6.2).

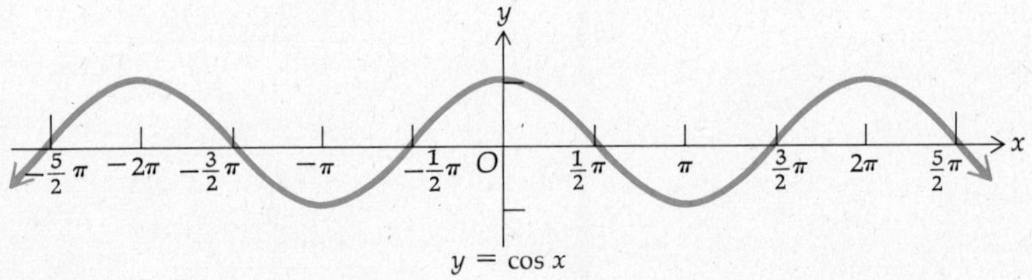

FIGURE 3.6.2

Example 3 Find an equation of the tangent line to the graph of the cosine function at the point $(\frac{3}{2}\pi, 0)$.

Solution If $f(x) = \cos x$, then $f'(x) = -\sin x$. Thus $f'(\frac{3}{2}\pi) = -\sin \frac{3}{2}\pi = -(-1) = 1$. From the point-slope form for an equation of the tangent line having slope 1 and containing the point $(\frac{3}{2}\pi, 0)$, we have

$$y - 0 = 1(x - \tfrac{3}{2}\pi)$$

$$y = x - \tfrac{3}{2}\pi$$

EXERCISES 3.6

In Exercises 1 through 18, find the derivative of the given function.

1. $f(x) = 3 \sin x$

2. $g(x) = \sin x + \cos x$

3. $f(t) = 2t \cos t$

4. $f(x) = 4x^2 \cos x$

5. $g(x) = x \sin x + \cos x$

6. $g(y) = 3 \sin y - y \cos y$

7. $h(x) = 4 \sin x \cos x$

8. $f(x) = x^2 \sin x + 2x \cos x$

9. $f(x) = x^2 \cos x - 2x \sin x - 2 \cos x$

10. $h(y) = y^3 - y^2 \cos y + 2y \sin y + 2 \cos y$

11. $g(z) = \dfrac{2 \cos z}{z + 1}$

12. $h(t) = \dfrac{\sin t}{t}$

13. $f(x) = \dfrac{\sin x}{1 - \cos x}$

14. $g(x) = \dfrac{x + 4}{\cos x}$

15. $f(y) = \dfrac{1 + \sin y}{1 - \sin y}$

16. $f(x) = \dfrac{\sin x - 1}{\cos x + 1}$

17. $g(x) = (x - \sin x)(x + \cos x)$

18. $f(z) = (z^2 + \cos z)(2z - \sin z)$

In Exercises 19 through 30, find $f'(a)$ for the given value of a.

19. $f(x) = x \cos x;\ a = 0$

20. $f(x) = x \sin x;\ a = \frac{3}{2}\pi$

21. $f(x) = \dfrac{\cos x}{x};\ a = \frac{1}{2}\pi$

22. $f(x) = \dfrac{1}{x^2 \cos x};\ a = \pi$

23. $f(x) = \dfrac{x^2}{\sin x};\ a = \frac{3}{2}\pi$

24. $f(x) = x^2 \cos x - \sin x;\ a = 0$

25. $f(x) = \sin x(\cos x - 1);\ a = \pi$

26. $f(x) = (\cos x + 1)(x \sin x - 1);\ a = \frac{1}{2}\pi$

27. $f(x) = x \cos x + x \sin x;\ a = \frac{1}{4}\pi$

28. $f(x) = \dfrac{\sin x + 1}{\cos x};\ a = \frac{1}{6}\pi$

29. $f(x) = \dfrac{2 \cos x - 1}{\sin x};\ a = \frac{2}{3}\pi$

30. $f(x) = \dfrac{\sin x}{\cos x - \sin x};\ a = \frac{3}{4}\pi$

In Exercises 31 through 34, derive the given differentiation formula. You will need to use the following identities:

$$\tan x = \frac{\sin x}{\cos x} \qquad \cot x = \frac{\cos x}{\sin x} \qquad \sec x = \frac{1}{\cos x} \qquad \csc x = \frac{1}{\sin x}$$

31. $D_x(\tan x) = \sec^2 x$

32. $D_x(\cot x) = -\csc^2 x$

33. $D_x(\sec x) = \sec x \tan x$

34. $D_x(\csc x) = -\csc x \cot x$

In Exercises 35 through 38, use the formulas of Exercises 31 through 34 to find the derivative of the given function.

35. $f(x) = 3 \sec x \tan x$

36. $g(x) = \cos x \cot x$

37. $g(t) = \dfrac{2 \csc t - 1}{\csc t + 2}$

38. $f(z) = \dfrac{\tan z + 1}{\tan z - 1}$

39. Find an equation of the tangent line to the graph of the sine function at the point $(\pi, 0)$.

40. Find an equation of the normal line to the curve $y = 2 \cos x$ at the point $(\frac{1}{6}\pi, \sqrt{3})$.

41. Find an equation of the normal line to the curve $y = 2 \sin x + 3 \cos x$ at the point $(0, 3)$.

42. Find an equation of the tangent line to the curve $y = x - 3 \sin x$ at the point (π, π).

43. If a body of weight W lb is dragged along a horizontal floor by means of a force of magnitude F lb and directed at an angle of θ radians with the plane of the floor, then F is given by the equation

$$F = \frac{kW}{k \sin \theta + \cos \theta}$$

where k is a constant and is called the coefficient of friction. If $k = 0.5$, find the instantaneous rate of change of F with respect to θ when (a) $\theta = \frac{1}{4}\pi$; (b) $\theta = \frac{1}{2}\pi$.

44. A projectile is shot from a gun at an angle of elevation having radian measure $\frac{1}{2}\alpha$ and an initial velocity of v_0 ft/sec. If R ft is the range of the projectile, then

$$R = \frac{v_0^{\,2}}{g} \sin \alpha \qquad 0 \leq \alpha \leq \pi$$

where g ft/sec² is the acceleration due to gravity. (a) If $v_0 = 480$, find the rate of change of R with respect to α when $\alpha = \frac{1}{2}\pi$ (i.e., the angle of elevation has radian measure $\frac{1}{4}\pi$). Take $g = 32$. (b) Find the values of α for which $D_\alpha R > 0$.

3.7 THE DERIVATIVE OF A COMPOSITE FUNCTION

Suppose that y is a function of u and u, in turn, is a function of x. For example, let

$$y = f(u) = u^{10} \tag{1}$$

and

$$u = g(x) = 2x^3 - 5x^2 + 4 \tag{2}$$

Equations (1) and (2) together define y as a function of x because if in (1) u is replaced by the right side of (2),

$$y = h(x) = f(g(x)) = (2x^3 - 5x^2 + 4)^{10} \tag{3}$$

where h is a composite function that was defined previously (refer to Definition 1.6.2).

An important theorem for finding the derivative of a composite function is the following one, known as the **chain rule.**

3.7.1 Theorem Chain Rule

If y is a function of u, defined by $y = f(u)$, and $D_u y$ exists, and if u is a function of x, defined by $u = g(x)$, and $D_x u$ exists, then y is a function of x and $D_x y$ exists and is given by

$$D_x y = D_u y \cdot D_x u$$

Before proving the chain rule, we give an illustration showing its application.

● ILLUSTRATION 1 The chain rule is applied to the function defined by (3), which is

$$y = (2x^3 - 5x^2 + 4)^{10}$$

To find $D_x y$ consider y as a function of u, where u is a function of x. Thus

$$y = u^{10}, \qquad \text{where } u = 2x^3 - 5x^2 + 4$$

Therefore, from the chain rule,

$$\begin{aligned} D_x y &= D_u y \cdot D_x u \\ &= 10u^9(6x^2 - 10x) \\ &= 10(2x^3 - 5x^2 + 4)^9(6x^2 - 10x) \end{aligned}$$

●

Proof of the Chain Rule Suppose that an increment of x, Δx, where $\Delta x \neq 0$, causes a change in u, Δu. That is,

$$u + \Delta u = g(x + \Delta x)$$

Thus, because $u = g(x)$,

$$\Delta u = g(x + \Delta x) - g(x)$$

Because Δu depends on Δx, it is possible that $\Delta u = 0$ for some values of Δx (see Exercises 49 and 50). We consider two cases, $\Delta u \neq 0$ for all values of Δx, or $\Delta u = 0$ for some values of Δx.

Case 1: $\Delta u \neq 0$ for all values of Δx.

Because $y = f(u)$, Δu causes a change in y, Δy. Then

$$\frac{\Delta y}{\Delta x} = \frac{\Delta y}{\Delta u} \cdot \frac{\Delta u}{\Delta x} \qquad \text{if } \Delta u \neq 0$$

Therefore

$$\lim_{\Delta x \to 0} \frac{\Delta y}{\Delta x} = \lim_{\Delta x \to 0} \frac{\Delta y}{\Delta u} \cdot \lim_{\Delta x \to 0} \frac{\Delta u}{\Delta x} \qquad \text{if } \Delta u \neq 0 \tag{4}$$

Because $u = g(x)$ and g is differentiable, then g is also continuous. Hence, as $\Delta x \to 0$, $\Delta u \to 0$. Thus, in the first limit on the right side of (4) we replace $\Delta x \to 0$ by $\Delta u \to 0$ and have

$$\lim_{\Delta x \to 0} \frac{\Delta y}{\Delta x} = \lim_{\Delta u \to 0} \frac{\Delta y}{\Delta u} \cdot \lim_{\Delta x \to 0} \frac{\Delta u}{\Delta x} \qquad \text{if } \Delta u \neq 0$$

Therefore

$$D_x y = D_u y \cdot D_x u$$

Case 2: $\Delta u = 0$ for some values of Δx.

From (7) of Sec. 3.2,

$$D_u y = \lim_{\Delta u \to 0} \frac{\Delta y}{\Delta u}$$

Hence when $|\Delta u|$ is small (close to zero but not equal to zero), the difference

$$\frac{\Delta y}{\Delta u} - D_u y$$

is numerically small. This difference is denoted by η. Thus

$$\eta = \frac{\Delta y}{\Delta u} - D_u y \qquad \text{if } \Delta u \neq 0$$

The above equation defines η as a function of Δu provided $\Delta u \neq 0$. If $\eta = F(\Delta u)$,

$$F(\Delta u) = \frac{\Delta y}{\Delta u} - D_u y \qquad \text{if } \Delta u \neq 0 \tag{5}$$

To define $F(\Delta u)$ when $\Delta u = 0$ we make use of the fact that in a later part of this proof we want the function F to be continuous at 0. Thus $F(0)$ is defined to be $\lim_{\Delta u \to 0} F(\Delta u)$. Using (5) for $\lim_{\Delta u \to 0} F(\Delta u)$ we have

$$\begin{aligned} F(0) &= \lim_{\Delta u \to 0} F(\Delta u) \\ &= \lim_{\Delta u \to 0} \frac{\Delta y}{\Delta u} - \lim_{\Delta u \to 0} D_u y \\ &= D_u y - D_u y \\ &= 0 \end{aligned}$$

Therefore

$$F(\Delta u) = \begin{cases} \dfrac{\Delta y}{\Delta u} - D_u y & \text{if } \Delta u \neq 0 \\ 0 & \text{if } \Delta u = 0 \end{cases}$$

Solving (5) for Δy we obtain

$$\Delta y = D_u y \cdot \Delta u + \Delta u \cdot F(\Delta u) \qquad \text{if } \Delta u \neq 0 \tag{6}$$

Note that (6) also holds if $\Delta u = 0$ because then $\Delta y = 0$ (remember that $\Delta y = f(u + \Delta u) - f(u)$).

If both sides of (6) are divided by Δx, where $\Delta x \neq 0$,

$$\frac{\Delta y}{\Delta x} = D_u y \cdot \frac{\Delta u}{\Delta x} + \frac{\Delta u}{\Delta x} \cdot F(\Delta u)$$

Taking the limit on both sides of the above equation as Δx approaches zero and applying limit theorems, we have

$$\lim_{\Delta x \to 0} \frac{\Delta y}{\Delta x} = \lim_{\Delta x \to 0} D_u y \cdot \lim_{\Delta x \to 0} \frac{\Delta u}{\Delta x} + \lim_{\Delta x \to 0} \frac{\Delta u}{\Delta x} \cdot \lim_{\Delta x \to 0} F(\Delta u)$$

Hence

$$D_x y = D_u y \cdot D_x u + D_x u \cdot \lim_{\Delta x \to 0} F(\Delta u) \tag{7}$$

We now show that $\lim_{\Delta x \to 0} F(\Delta u) = 0$ by making use of Theorem 2.7.6. We first express Δu as a function of Δx. Because

$$\lim_{\Delta x \to 0} \frac{\Delta u}{\Delta x} = D_x u$$

it follows that when $|\Delta x|$ is small and $\Delta x \neq 0$, $\Delta u / \Delta x$ differs from $D_x u$ by a small number that depends on Δx, which we call $\phi(\Delta x)$. Hence

$$\frac{\Delta u}{\Delta x} = D_x u + \phi(\Delta x) \qquad \text{if } \Delta x \neq 0$$

Multiplying by Δx we obtain

$$\Delta u = D_x u \cdot \Delta x + \phi(\Delta x) \cdot \Delta x \qquad \text{if } \Delta x \neq 0$$

The above equation expresses Δu as a function of Δx. Denote this function by G. Thus

$$G(\Delta x) = D_x u \cdot \Delta x + \phi(\Delta x) \cdot \Delta x \qquad \text{if } \Delta x \neq 0$$

Hence $F(\Delta u) = F(G(\Delta x))$, and

$$\lim_{\Delta x \to 0} F(\Delta u) = \lim_{\Delta x \to 0} F(G(\Delta x))$$

Because $\lim_{\Delta x \to 0} G(\Delta x) = 0$ and F is continuous at 0 (remember that we made it so), Theorem 2.7.6 can be applied to the right side of the above equation to obtain

$$\begin{aligned} \lim_{\Delta x \to 0} F(\Delta u) &= F(\lim_{\Delta x \to 0} G(\Delta x)) \\ &= F(0) \\ &= 0 \end{aligned}$$

So in (7) if $\lim_{\Delta x \to 0} F(\Delta u)$ is replaced by 0,

$$D_x y = D_u y \cdot D_x u + D_x u \cdot 0$$

Thus

$$D_x y = D_u y \cdot D_x u$$ ■

It should be noted that the proof of the chain rule for case 2 is also valid for case 1. The separate proof for case 1 is included to show you that the proof is simplified when we are dealing with functions for which Δu is never zero.

If composite function notation is used, the chain rule can be written as

$$D_x f(g(x)) = f'(g(x))g'(x) \tag{8}$$

● ILLUSTRATION 2 We compute the derivative of the function in Illustration 1 by the chain rule with composite function notation. If

$$h(x) = (2x^3 - 5x^2 + 4)^{10}$$

let $f(u) = u^{10}$ and $g(x) = 2x^3 - 5x^2 + 4$, and then

$$h(x) = f(g(x))$$

Therefore

$$\begin{aligned} h'(x) &= D_x f(g(x)) \\ &= f'(g(x))g'(x) \end{aligned}$$

Because $f'(u) = 10u^9$, then $f'(g(x)) = 10[g(x)]^9$ and $g'(x) = 6x^2 - 10x$. Thus

$$h'(x) = 10[g(x)]^9(6x^2 - 10x)$$
$$= 10(2x^3 - 5x^2 + 4)^9(6x^2 - 10x)$$

Example 1 Given

$$f(x) = \frac{1}{4x^3 + 5x^2 - 7x + 8}$$

find $f'(x)$ by the chain rule.

Solution Write $f(x) = (4x^3 + 5x^2 - 7x + 8)^{-1}$, and apply the chain rule to obtain

$$f'(x) = -1(4x^3 + 5x^2 - 7x + 8)^{-2} \cdot D_x(4x^3 + 5x^2 - 7x + 8)$$
$$= -1(4x^3 + 5x^2 - 7x + 8)^{-2}(12x^2 + 10x - 7)$$
$$= \frac{-12x^2 - 10x + 7}{(4x^3 + 5x^2 - 7x + 8)^2}$$

Example 2 Given

$$f(x) = \left(\frac{2x + 1}{3x - 1}\right)^4$$

find $f'(x)$.

Solution From the chain rule,

$$f'(x) = 4\left(\frac{2x + 1}{3x - 1}\right)^3 \cdot D_x\left(\frac{2x + 1}{3x - 1}\right)$$
$$= 4\left(\frac{2x + 1}{3x - 1}\right)^3 \frac{(3x - 1)(2) - (2x + 1)(3)}{(3x - 1)^2}$$
$$= \frac{4(2x + 1)^3(-5)}{(3x - 1)^5}$$
$$= -\frac{20(2x + 1)^3}{(3x - 1)^5}$$

● ILLUSTRATION 3 Let

$$h(x) = \sin 4x$$

To find $h'(x)$ by the chain rule, set $f(u) = \sin u$ and $g(x) = 4x$. Then

$$h(x) = f(g(x))$$

and

$$h'(x) = D_x f(g(x))$$
$$= f'(g(x))g'(x)$$

Because $f'(u) = \cos u$, then $f'(g(x)) = \cos[g(x)]$. Also, $g'(x) = 4$. Thus

$$h'(x) = (\cos 4x)(4)$$
$$= 4 \cos 4x$$

Example 3 Given $F(t) = \cos(3t^2 + 2t)$, find $F'(t)$.

Solution We use the chain rule and obtain

$$F'(t) = [-\sin(3t^2 + 2t)]D_t(3t^2 + 2t)$$
$$= [-\sin(3t^2 + 2t)](6t + 2)$$
$$= -2(3t + 1)\sin(3t^2 + 2t)$$

Example 4 Given $y = \sin(\cos x)$, find $D_x y$.

Solution We apply the chain rule.

$$D_x y = \cos(\cos x)[D_x(\cos x)]$$
$$= \cos(\cos x)[-\sin x]$$
$$= -\sin x[\cos(\cos x)]$$

Example 5 Given $f(x) = (3x^2 + 2)^2(x^2 - 5x)^3$, find $f'(x)$.

Solution Consider f as the product of the two functions g and h, where

$$g(x) = (3x^2 + 2)^2 \quad \text{and} \quad h(x) = (x^2 - 5x)^3$$

From Theorem 3.4.6 for the derivative of the product of two functions,

$$f'(x) = g(x)h'(x) + h(x)g'(x)$$

We find $h'(x)$ and $g'(x)$ by the chain rule.

$$f'(x) = (3x^2 + 2)^2[3(x^2 - 5x)^2(2x - 5)] + (x^2 - 5x)^3[2(3x^2 + 2)(6x)]$$
$$= 3(3x^2 + 2)(x^2 - 5x)^2[(3x^2 + 2)(2x - 5) + 4x(x^2 - 5x)]$$
$$= 3(3x^2 + 2)(x^2 - 5x)^2[6x^3 - 15x^2 + 4x - 10 + 4x^3 - 20x^2]$$
$$= 3(3x^2 + 2)(x^2 - 5x)^2(10x^3 - 35x^2 + 4x - 10)$$

Example 6 Given $\phi(x) = \sin^4 2x^2$, find $\phi'(x)$.

Solution

$$\phi'(x) = 4 \sin^3 2x^2[D_x(\sin 2x^2)]$$
$$= 4 \sin^3 2x^2[(\cos 2x^2)D_x(2x^2)]$$
$$= 4 \sin^3 2x^2[(\cos 2x^2)(4x)]$$
$$= 16x \sin^3 2x^2 \cos 2x^2$$

EXERCISES 3.7

In Exercises 1 through 36, find the derivative of the given function.

1. $f(x) = (2x + 1)^3$
2. $f(x) = (10 - 5x)^4$
3. $F(x) = (x^2 + 4x - 5)^4$
4. $g(r) = (2r^4 + 8r^2 + 1)^5$
5. $f(t) = (2t^4 - 7t^3 + 2t - 1)^2$
6. $H(z) = (z^3 - 3z^2 + 1)^{-3}$
7. $f(x) = (x^2 + 4)^{-2}$
8. $g(x) = \sin x^2$
9. $f(x) = 4 \cos 3x - 3 \sin 4x$
10. $G(x) = \sin^2 x$
11. $h(t) = \cos 2t^2 + 2 \cos^2 t$
12. $f(x) = \cos(3x^2 + 1)$
13. $g(x) = 12 \sin 3x \cos 4x$
14. $f(t) = 2 \sin^3 t \cos^2 t$
15. $h(u) = (3u^2 + 5)^3(3u - 1)^2$
16. $f(x) = (4x^2 + 7)^2(2x^3 + 1)^4$
17. $g(x) = (2x - 5)^{-1}(4x + 3)^{-2}$
18. $f(x) = (x^2 - 4x^{-2})^2(x^2 + 1)^{-1}$
19. $f(y) = \left(\dfrac{y - 7}{y + 2}\right)^2$
20. $g(t) = \left(\dfrac{2t^2 + 1}{3t^3 + 1}\right)^2$
21. $f(x) = \dfrac{2}{7x^2 + 3x - 1}$
22. $h(x) = \left(\dfrac{x + 4}{2x^2 - 5x + 6}\right)^3$
23. $f(r) = (r^2 + 1)^3(2r + 5)^2$
24. $f(y) = (y + 3)^3(5y + 1)^2(3y^2 - 4)$
25. $f(z) = \dfrac{(z^2 - 5)^3}{(z^2 + 4)^2}$
26. $g(x) = (2x - 9)^2(x^3 + 4x - 5)^3$
27. $G(x) = \dfrac{(4x - 1)^3(x^2 + 2)^4}{(3x^2 + 5)^2}$
28. $F(x) = \dfrac{(5x - 8)^{-2}}{(x^2 + 3)^{-3}}$
29. $g(t) = \sin^2(3t^2 - 1)$
30. $f(x) = \cos^2 x^2$
31. $f(x) = (\sin^2 x - x^2)^3$
32. $G(x) = (2 \sin x - 3 \cos x)^3$
33. $f(y) = \dfrac{3 \sin 2y}{\cos^2 2y + 1}$
34. $g(x) = \dfrac{1 + \sin 3x}{3x \cos 3x}$
35. $F(x) = 4 \cos(\sin 3x)$
36. $f(x) = \sin^2(\cos 2x)$

In Exercises 37 and 38, find an equation of the tangent line to the curve at each of the given points. Draw a sketch of the graph and segments of the tangent lines at the given points.

37. $y = (x^2 - 1)^2$; $(-2, 9), (-1, 0), (0, 1), (1, 0), (2, 9)$
38. $y = 2(4 - x)^{-2}$; $(0, \frac{1}{8}), (1, \frac{2}{9}), (2, \frac{1}{2}), (3, 2), (5, 2), (6, \frac{1}{2})$
39. Find an equation of the tangent line to the curve $y = (1 - x^3)^{-1}$ at the point where $x = -1$.
40. Find an equation of the normal line to the curve $y = (x + 3)^2(2x + 1)^3$ at the point where $x = -1$.
41. Find an equation of the normal line to the curve $y = 3 \cos 2x$ at the point where $x = \frac{1}{3}\pi$.
42. Find an equation of the tangent line to the curve $y = (x^2 - 4)^2(3x - 5)^{-2}$ at the point where $x = 1$.

In Exercises 43 through 46, a particle is moving along a straight line according to the given equation where s cm is the directed distance of the particle from an origin at t sec. (a) What is the instantaneous velocity of the particle at t sec? (b) Find the instantaneous velocity of the particle at t_1 sec for each of the given values of t_1.

43. $s = \dfrac{(t^2 - 1)^2}{(t^2 + 1)^2}$, $t \geq 0$; $t_1 = 1, 2$
44. $s = \left(\dfrac{3t}{2t + 1}\right)^4$; $t \geq 0$; $t_1 = \frac{1}{2}, 1$
45. $s = 5 \sin \pi t + 3 \cos \pi t$; $t_1 = \frac{1}{4}, \frac{3}{2}$
46. $s = 2 \cos \pi(t + 1)$; $t_1 = \frac{1}{4}, \frac{3}{2}$
47. Given $f(x) = x^3$ and $g(x) = f(x^2)$. Find (a) $f'(x^2)$; (b) $g'(x)$.
48. Given $f(u) = u^2 + 5u + 5$ and $g(x) = (x + 1)/(x - 1)$. Find the derivative of $f \circ g$ in two ways: (a) by first finding $(f \circ g)(x)$ and then finding $(f \circ g)'(x)$; (b) by using the chain rule.
49. Show that case 1 of the proof of the chain rule applies if $f(u) = u^2$ and $g(x) = x^3$, but that it does not apply if $f(u) = u^2$ and $g(x) = \sin 1/x$.
50. Show that case 1 of the proof of the chain rule does not apply if $f(u) = u^2$ and $g(x) = \operatorname{sgn} x$.

51. Suppose that f and g are functions such that $f'(x) = 1/x$ and $f(g(x)) = x$. Prove that if $g'(x)$ exists, then $g'(x) = g(x)$.

52. Use the chain rule to prove that (a) the derivative of an even function is an odd function, and (b) the derivative of an odd function is an even function, provided that these derivatives exist.

53. Use the result of Exercise 52(a) to prove that if g is an even function and $g'(x)$ exists, then if $h(x) = (f \circ g)(x)$ and f is differentiable everywhere, $h'(0) = 0$.

54. Derive the formula for the derivative of the cosine function by using the formula for the derivative of the sine function, the chain rule, and the identities

$$\cos x = \sin(\tfrac{1}{2}\pi - x) \quad \text{and} \quad \sin x = \cos(\tfrac{1}{2}\pi - x)$$

55. An airplane is flying parallel to the ground at an altitude of 2 km and at a speed of $4\frac{1}{2}$ km/min. If the plane flies directly over the Statue of Liberty, at what rate is the line-of-sight distance between the plane and the statue changing 20 sec later?

56. At 8 A.M. a ship sailing due north at 24 knots (nautical miles per hour) is at a point P. At 10 A.M. a second ship sailing due east at 32 knots is at P. At what rate is the distance between the two ships changing at (a) 9 A.M. and (b) 11 A.M.?

3.8 THE DERIVATIVE OF THE POWER FUNCTION FOR RATIONAL EXPONENTS

The function f defined by

$$f(x) = x^r \tag{1}$$

is called the **power function.** In Sec. 3.4 the following formula for the derivative of this function when r is a positive or negative integer was obtained.

$$f'(x) = rx^{r-1} \tag{2}$$

We now show that this formula holds when r is a rational number, with certain stipulations when $x = 0$.

First consider $x \neq 0$, and $r = 1/q$, where q is a positive integer. Equation (1) then can be written

$$f(x) = x^{1/q}$$

From Definition 3.2.1,

$$f'(x) = \lim_{\Delta x \to 0} \frac{(x + \Delta x)^{1/q} - x^{1/q}}{\Delta x} \tag{3}$$

To evaluate the limit in (3) we must rationalize the numerator. To do this we use formula (8) in Sec. 2.2, which is

$$a^n - b^n = (a - b)(a^{n-1} + a^{n-2}b + a^{n-3}b^2 + \cdots + a^2b^{n-3} + ab^{n-2} + b^{n-1}) \tag{4}$$

So the numerator of the fraction in (3) is rationalized by applying (4), where $a = (x + \Delta x)^{1/q}$, $b = x^{1/q}$, and $n = q$. So we multiply the numerator and denominator by

$$[(x + \Delta x)^{1/q}]^{(q-1)} + [(x + \Delta x)^{1/q}]^{(q-2)}x^{1/q} + \cdots + (x + \Delta x)^{1/q}(x^{1/q})^{(q-2)} + (x^{1/q})^{(q-1)}$$

Then, from (3), $f'(x)$ equals

$$\lim_{\Delta x \to 0} \frac{[(x + \Delta x)^{1/q} - x^{1/q}][(x + \Delta x)^{(q-1)/q} + (x + \Delta x)^{(q-2)/q}x^{1/q} + \cdots + x^{(q-1)/q}]}{\Delta x[(x + \Delta x)^{(q-1)/q} + (x + \Delta x)^{(q-2)/q}x^{1/q} + \cdots + x^{(q-1)/q}]} \tag{5}$$

Now if (4) is applied to the numerator, we get $(x + \Delta x)^{q/q} - x^{q/q}$, which is Δx. So from (5),

$$f'(x) = \lim_{\Delta x \to 0} \frac{\Delta x}{\Delta x[(x + \Delta x)^{(q-1)/q} + (x + \Delta x)^{(q-2)/q}x^{1/q} + \cdots + x^{(q-1)/q}]}$$

$$= \lim_{\Delta x \to 0} \frac{1}{(x + \Delta x)^{(q-1)/q} + (x + \Delta x)^{(q-2)/q}x^{1/q} + \cdots + x^{(q-1)/q}}$$

$$= \frac{1}{x^{(q-1)/q} + x^{(q-1)/q} + \cdots + x^{(q-1)/q}}$$

Because there are exactly q terms in the denominator of the above fraction

$$f'(x) = \frac{1}{qx^{1-(1/q)}} \quad \text{or, equivalently,} \quad f'(x) = \frac{1}{q}x^{1/q-1} \tag{6}$$

which is formula (2) with $r = 1/q$.

Now, in (1) with $x \neq 0$, let $r = p/q$, where p is any nonzero integer and q is any positive integer; that is, r is any rational number except zero. Then (1) is written as

$$f(x) = x^{p/q} \quad \text{or, equivalently,} \quad f(x) = (x^{1/q})^p$$

Because p is either a positive or negative integer, it follows from the chain rule and Theorems 3.4.2 and 3.4.8 that

$$f'(x) = p(x^{1/q})^{p-1} \cdot D_x(x^{1/q})$$

Applying formula (6) for $D_x(x^{1/q})$ we get

$$f'(x) = p(x^{1/q})^{p-1} \cdot \frac{1}{q}x^{1/q-1}$$

$$f'(x) = \frac{p}{q}x^{p/q-1/q+1/q-1}$$

$$f'(x) = \frac{p}{q}x^{p/q-1} \tag{7}$$

Formula (7) is the same as formula (2) with $r = p/q$.

If $r = 0$ and $x \neq 0$, (1) becomes $f(x) = x^0 = 1$. In this case $f'(x) = 0$, which can be written as $f'(x) = 0 \cdot x^{0-1}$. Therefore formula (2) holds if $r = 0$ with $x \neq 0$. We have therefore shown that formula (2) holds when r is any rational number and $x \neq 0$.

Now 0 is in the domain of the power function f if and only if r is a positive number, because when $r \leq 0$, $f(0)$ is not defined. Hence we wish to determine for what positive values of r, $f'(0)$ will be given by formula (2). We must exclude the values of r for which $0 < r \leq 1$ because for those values of r, x^{r-1} is not a real number when $x = 0$. Suppose, then, that $r > 1$. By the definition of a derivative,

$$f'(0) = \lim_{x \to 0} \frac{x^r - 0^r}{x - 0} = \lim_{x \to 0} x^{r-1}$$

When $r > 1$, $\lim_{x \to 0} x^{r-1}$ exists and equals 0 provided that r is a number such that x^{r-1} is defined on some open interval containing 0. For example, if $r = \frac{3}{2}$, $x^{r-1} = x^{3/2-1} = x^{1/2}$, which is not defined on any open interval containing 0 (since $x^{1/2}$ does not exist when $x < 0$). However, if $r = \frac{5}{3}$, $x^{r-1} = x^{5/3-1} = x^{2/3}$, which is defined on every open interval containing 0. Hence formula (2) gives the derivative of the power function when $x = 0$ provided that r is a number for which x^{r-1} is defined on some open interval containing 0. Thus we have proved the following theorem.

3.8.1 Theorem If f is the power function defined by $f(x) = x^r$, where r is any rational number, then

$$f'(x) = rx^{r-1}$$

For this formula to give $f'(0)$, r must be a number such that x^{r-1} is defined on some open interval containing 0.

An immediate consequence of Theorem 3.8.1 and the chain rule is the following theorem.

3.8.2 Theorem If f and g are functions such that $f(x) = [g(x)]^r$, where r is any rational number, and if $g'(x)$ exists, then

$$f'(x) = r[g(x)]^{r-1}g'(x)$$

Example 1 Given $f(x) = 4\sqrt[3]{x^2}$, find $f'(x)$.

Solution $f(x) = 4x^{2/3}$. From Theorem 3.8.1,

$$\begin{aligned} f'(x) &= 4 \cdot \frac{2}{3}[x^{2/3-1}] \\ &= \frac{8}{3}x^{-1/3} \\ &= \frac{8}{3x^{1/3}} \\ &= \frac{8}{3\sqrt[3]{x}} \end{aligned}$$

Example 2 Given $h(x) = \sqrt{2x^3 - 4x + 5}$, find $h'(x)$.

Solution $h(x) = (2x^3 - 4x + 5)^{1/2}$. From Theorem 3.8.2,

$$\begin{aligned} h'(x) &= \tfrac{1}{2}(2x^3 - 4x + 5)^{-1/2}(6x^2 - 4) \\ &= \frac{3x^2 - 2}{\sqrt{2x^3 - 4x + 5}} \end{aligned}$$

Example 3 Given

$$g(x) = \frac{x^3}{\sqrt[3]{3x^2 - 1}}$$

find $g'(x)$.

Solution The given fraction is written as a product.

$$g(x) = x^3(3x^2 - 1)^{-1/3}$$

From Theorems 3.4.6 and 3.8.2,

$$\begin{aligned} g'(x) &= 3x^2(3x^2 - 1)^{-1/3} - \tfrac{1}{3}(3x^2 - 1)^{-4/3}(6x)(x^3) \\ &= x^2(3x^2 - 1)^{-4/3}[3(3x^2 - 1) - 2x^2] \\ &= \frac{x^2(7x^2 - 3)}{(3x^2 - 1)^{4/3}} \end{aligned}$$

Example 4 Given $f(r) = \sqrt{4 \sin^2 r + 9 \cos^2 r}$, find $f'(r)$.

Solution $f(r) = (4 \sin^2 r + 9 \cos^2 r)^{1/2}$. We apply Theorem 3.8.2.

$$\begin{aligned} f'(r) &= \tfrac{1}{2}(4 \sin^2 r + 9 \cos^2 r)^{-1/2} \cdot D_r(4 \sin^2 r + 9 \cos^2 r) \\ &= \frac{8 \sin r \cdot D_r(\sin r) + 18 \cos r \cdot D_r(\cos r)}{2\sqrt{4 \sin^2 r + 9 \cos^2 r}} \\ &= \frac{8 \sin r \cos r + 18 \cos r(-\sin r)}{2\sqrt{4 \sin^2 r + 9 \cos^2 r}} \\ &= \frac{-10 \sin r \cos r}{2\sqrt{4 \sin^2 r + 9 \cos^2 r}} \\ &= -\frac{5 \sin r \cos r}{\sqrt{4 \sin^2 r + 9 \cos^2 r}} \end{aligned}$$

EXERCISES 3.8

In Exercises 1 through 32, find the derivative of the given function.

1. $f(x) = 4x^{1/2} + 5x^{-1/2}$

2. $f(x) = 3x^{2/3} - 6x^{1/3} + x^{-1/3}$

3. $g(x) = \sqrt{1 + 4x^2}$

4. $f(s) = \sqrt{2 - 3s^2}$

5. $f(x) = (5 - 3x)^{2/3}$

6. $g(x) = \sqrt[3]{4x^2 - 1}$

7. $g(y) = \dfrac{1}{\sqrt{25 - y^2}}$

8. $f(x) = (5 - 2x^2)^{-1/3}$

9. $h(t) = 2 \cos \sqrt{t}$

10. $g(x) = \sqrt{3 \sin x}$

11. $f(x) = (\sin 3x)^{-1/2}$

12. $f(y) = 3 \cos \sqrt[3]{2y^2}$

13. $g(x) = \sqrt{\dfrac{2x - 5}{3x + 1}}$

14. $h(t) = \dfrac{\sqrt{t} - 1}{\sqrt{t} + 1}$

15. $F(x) = \sqrt[3]{2x^3 - 5x^2 + x}$

16. $G(t) = \sqrt{\dfrac{5t + 6}{5t - 4}}$

17. $g(t) = \sqrt{2t} + \sqrt{\dfrac{2}{t}}$

18. $g(x) = \sqrt[3]{(3x^2 + 5x - 1)^2}$

19. $f(x) = (5 - x^2)^{1/2}(x^3 + 1)^{1/4}$

20. $g(y) = (y^2 + 3)^{1/3}(y^3 - 1)^{1/2}$

21. $F(x) = \dfrac{\sqrt{x^2 - 1}}{x}$

22. $f(x) = \sqrt{x^2 - 5}\,\sqrt[3]{x^2 + 3}$ **23.** $f(t) = \sqrt{\dfrac{\sin t + 1}{1 - \sin t}}$ **24.** $f(z) = \sin \sqrt[3]{z} \cos \sqrt[3]{z}$

25. $h(y) = \cos \sqrt{y^2 + 1}$ **26.** $f(x) = \sqrt{\dfrac{\cos x - 1}{\sin x}}$ **27.** $h(x) = \dfrac{\sqrt{x - 1}}{\sqrt[3]{x + 1}}$

28. $G(x) = \dfrac{4x + 6}{\sqrt{x^2 + 3x + 4}}$ **29.** $f(x) = \sqrt{9 + \sqrt{9 - x}}$ **30.** $g(y) = \sqrt[4]{\dfrac{y^3 + 1}{y^3 - 1}}$

31. $g(x) = \sqrt{1 + \cos^2 2x}$ **32.** $f(x) = \sqrt{x} \sin \sqrt{\dfrac{1}{x}}$

In Exercises 33 and 34, find an equation of the tangent line to the curve at each of the given points. Draw a sketch of the graph and segments of the tangent lines at the given points.

33. $y = (2x - 2)^{2/3}$; $(-3, 4)$, $(0, \sqrt[3]{4})$, $(1, 0)$, $(2, \sqrt[3]{4})$, $(5, 4)$

34. $y = (6 - 2x)^{1/3}$; $(-1, 2)$, $(1, \sqrt[3]{4})$, $(3, 0)$, $(5, -\sqrt[3]{4})$, $(7, -2)$

35. Find an equation of the tangent line to the curve $y = \sqrt{x^2 + 9}$ at the point $(4, 5)$.

36. Find an equation of the tangent line to the curve $y = (7x - 6)^{-1/3}$ that is perpendicular to the line $12x - 7y + 2 = 0$.

37. Find an equation of the normal line to the curve $y = x\sqrt{16 + x^2}$ at the origin.

38. Find an equation of the tangent line to the curve $y = \sqrt{\sin x + \cos x}$ at the point where $x = \frac{1}{4}\pi$.

39. An object is moving along a straight line according to the equation of motion $s = \sqrt{4t^2 + 3}$, with $t \geq 0$. Find the value of t for which the measure of the instantaneous velocity is (a) 0; (b) 1; (c) 2.

40. An object is moving along a straight line according to the equation of motion $s = \sqrt{5 + t^2}$. Find the value of t for which the measure of the instantaneous velocity is (a) 0; (b) 1.

41. A property development company rents each apartment at p dollars per month when x apartments are rented, and $p = 15\sqrt{300 - 2x}$. If R dollars is the revenue received from the rental of x apartments, then $R = px$. How many apartments must be rented before the rate of change of R with respect to x (the marginal revenue) is zero? (*Note:* Because x is the number of apartments rented, x is a nonnegative integer. However, to apply the calculus, assume that x is a nonnegative real number and then round off any noninteger values of x to the nearest integer.)

42. Given $f(u) = 1/u^2$ and $g(x) = \sqrt{x}/\sqrt{2x^3 - 6x + 1}$, find the derivative of $f \circ g$ in two ways: (a) by first finding $(f \circ g)(x)$ and then finding $(f \circ g)'(x)$; (b) by using the chain rule.

In Exercises 43 through 46, find the derivative of the given function. (Hint: $|a| = \sqrt{a^2}$.)

43. $f(x) = |x^2 - 4|$ **44.** $g(x) = x|x|$ **45.** $g(x) = |x|^3$ **46.** $h(x) = \sqrt[3]{|x|} + x$

47. Suppose $g(x) = |f(x)|$. Prove that if $f'(x)$ and $g'(x)$ exist, then $|g'(x)| = |f'(x)|$.

48. Suppose that $g(x) = \sqrt{9 - x^2}$ and $h(x) = f(g(x))$ where f is differentiable at 3. Prove that $h'(0) = 0$.

49. If g and h are functions and if f is the function defined by

$$f(x) = [g(x)]^r[h(x)]^s$$

where r and s are rational numbers, prove that if $g'(x)$ and $h'(x)$ exist,

$$f'(x) = [g(x)]^{r-1}[h(x)]^{s-1}[r \cdot h(x)g'(x) + s \cdot g(x)h'(x)]$$

In Exercises 50 through 53, use the result of Exercise 49 to find the derivative of the given function.

50. $g(x) = (4x + 3)^{1/2}(4 - x^2)^{1/3}$ **51.** $f(x) = (3x + 2)^4(x^2 - 1)^{2/3}$

52. $f(x) = \left(\dfrac{x + 1}{x^2 + 1}\right)^3 (x^3 + 4)^{1/3}$ **53.** $F(t) = (t^3 - 2t + 1)^{3/2}(t^2 + t + 5)^{1/3}$

3.9 IMPLICIT DIFFERENTIATION

If $f = \{(x, y) | y = 3x^2 + 5x + 1\}$, then the equation

$$y = 3x^2 + 5x + 1 \tag{1}$$

defines the function f explicitly. However, not all functions are defined explicitly. For example, for the equation

$$x^6 - 2x = 3y^6 + y^5 - y^2 \tag{2}$$

we cannot solve for y in terms of x; however, there may exist one or more functions f such that if $y = f(x)$, Eq. (2) is satisfied, that is, such that the equation

$$x^6 - 2x = 3[f(x)]^6 + [f(x)]^5 - [f(x)]^2$$

is true for all values of x in the domain of f. In this case the function f is defined *implicitly* by the given equation.

With the assumption that (2) defines y as at least one differentiable function of x, the derivative of y with respect to x can be found by the process called **implicit differentiation,** which we now do.

The left side of (2) is a function of x, and the right side is a function of y. Let F be the function defined by the left side of (2), and let G be the function defined by the right side of (2). Thus

$$F(x) = x^6 - 2x \tag{3}$$

and

$$G(y) = 3y^6 + y^5 - y^2 \tag{4}$$

where y is a function of x, say,

$$y = f(x)$$

So (2) can be written as

$$F(x) = G(f(x)) \tag{5}$$

Equation (5) is satisfied by all values of x in the domain of f for which $G[f(x)]$ exists.

Then for all values of x for which f is differentiable,

$$D_x[x^6 - 2x] = D_x[3y^6 + y^5 - y^2] \tag{6}$$

The derivative on the left side of (6) is easily found, and

$$D_x[x^6 - 2x] = 6x^5 - 2 \tag{7}$$

We find the derivative on the right side of (6) by the chain rule.

$$D_x[3y^6 + y^5 - y^2] = 18y^5 \cdot D_xy + 5y^4 \cdot D_xy - 2y \cdot D_xy \tag{8}$$

Substituting the values from (7) and (8) into (6) we obtain

$$6x^5 - 2 = (18y^5 + 5y^4 - 2y)D_xy$$

$$D_xy = \frac{6x^5 - 2}{18y^5 + 5y^4 - 2y}$$

Equation (2) is a special type of equation involving x and y because it can be written such that all the terms involving x are on one side of the equation and all the terms involving y are on the other side.

In the following illustration the method of implicit differentiation is used to find $D_x y$ from a more general type of equation.

● ILLUSTRATION 1 Consider the equation

$$3x^4y^2 - 7xy^3 = 4 - 8y \tag{9}$$

and assume that there exists at least one differentiable function f such that if $y = f(x)$, Eq. (9) is satisfied. Differentiating on both sides of (9) (bearing in mind that y is a differentiable function of x), and applying the theorems for the derivative of a product, the derivative of a power, and the chain rule, we obtain

$$12x^3y^2 + 3x^4(2yD_x y) - 7y^3 - 7x(3y^2D_x y) = 0 - 8D_x y$$

$$D_x y(6x^4y - 21xy^2 + 8) = 7y^3 - 12x^3y^2$$

$$D_x y = \frac{7y^3 - 12x^3y^2}{6x^4y - 21xy^2 + 8}$$ ●

Remember that we assumed that both (2) and (9) define y as at least one differentiable function of x. It may be that an equation in x and y does not imply the existence of any real-valued function, as is the case for the equation

$$x^2 + y^2 + 4 = 0$$

which is not satisfied by any real values of x and y. Furthermore, it is possible that an equation in x and y may be satisfied by many different functions, some of which are differentiable and some of which are not. A general discussion is beyond the scope of this book but can be found in an advanced calculus text. In subsequent discussions when we state that an equation in x and y defines y implicitly as a function of x, it is assumed that one or more of these functions is differentiable. Example 4, which follows, illustrates the fact that implicit differentiation gives the derivative of two differentiable functions defined by the given equation.

Example 1 Given $(x + y)^2 - (x - y)^2 = x^4 + y^4$, find $D_x y$.

Solution Differentiating implicitly with respect to x we have

$$2(x + y)(1 + D_x y) - 2(x - y)(1 - D_x y) = 4x^3 + 4y^3D_x y$$

$$2x + 2y + (2x + 2y)D_x y - 2x + 2y + (2x - 2y)D_x y = 4x^3 + 4y^3D_x y$$

$$D_x y(4x - 4y^3) = 4x^3 - 4y$$

$$D_x y = \frac{x^3 - y}{x - y^3}$$

Example 2 Find an equation of the tangent line to the curve $x^3 + y^3 = 9$ at the point (1, 2).

Solution We differentiate implicitly with respect to x.

$$3x^2 + 3y^2 D_x y = 0$$

Hence

$$D_x y = -\frac{x^2}{y^2}$$

Therefore, at the point (1, 2), $D_x y = -\frac{1}{4}$. An equation of the tangent line is then

$$y - 2 = -\tfrac{1}{4}(x - 1)$$

Example 3 Given $x \cos y + y \cos x = 1$, find $D_x y$.

Solution Differentiating implicitly with respect to x we get

$$1 \cdot \cos y + x(-\sin y)\, D_x y + D_x y(\cos x) + y(-\sin x) = 0$$

$$D_x y(\cos x - x \sin y) = y \sin x - \cos y$$

$$D_x y = \frac{y \sin x - \cos y}{\cos x - x \sin y}$$

Example 4 Given the equation $x^2 + y^2 = 9$, find (a) $D_x y$ by implicit differentiation; (b) two functions defined by the equation; (c) the derivative of each of the functions obtained in part (b) by explicit differentiation. (d) Verify that the result obtained in part (a) agrees with the results obtained in part (c).

Solution

(a) We differentiate implicitly.

$$2x + 2yD_x y = 0 \quad \text{and so} \quad D_x y = -\frac{x}{y}$$

(b) If the given equation is solved for y,

$$y = \sqrt{9 - x^2} \quad \text{and} \quad y = -\sqrt{9 - x^2}$$

Let f_1 be the function for which

$$f_1(x) = \sqrt{9 - x^2}$$

and f_2 be the function for which

$$f_2(x) = -\sqrt{9 - x^2}$$

(c) Because $f_1(x) = (9 - x^2)^{1/2}$, by using the chain rule we obtain

$$f_1'(x) = \tfrac{1}{2}(9 - x^2)^{-1/2}(-2x)$$

$$= -\frac{x}{\sqrt{9 - x^2}}$$

Similarly,

$$f_2'(x) = \frac{x}{\sqrt{9 - x^2}}$$

(d) For $y = f_1(x)$, where $f_1(x) = \sqrt{9 - x^2}$, it follows from part (c) that

$$f_1'(x) = -\frac{x}{\sqrt{9 - x^2}} = -\frac{x}{y}$$

which agrees with the answer in part (a). For $y = f_2(x)$, where $f_2(x) = -\sqrt{9 - x^2}$, we have from part (c)

$$\begin{aligned} f_2'(x) &= \frac{x}{\sqrt{9 - x^2}} \\ &= -\frac{x}{-\sqrt{9 - x^2}} \\ &= -\frac{x}{y} \end{aligned}$$

which also agrees with the answer in part (a).

EXERCISES 3.9

In Exercises 1 through 24, find $D_x y$ by implicit differentiation.

1. $x^2 + y^2 = 16$ **2.** $4x^2 - 9y^2 = 1$ **3.** $x^3 + y^3 = 8xy$ **4.** $x^2 + y^2 = 7xy$

5. $\frac{1}{x} + \frac{1}{y} = 1$ **6.** $\frac{3}{x} - \frac{3}{y} = 2x$ **7.** $\sqrt{x} + \sqrt{y} = 4$ **8.** $2x^3y + 3xy^3 = 5$

9. $x^2y^2 = x^2 + y^2$ **10.** $(2x + 3)^4 = 3y^4$ **11.** $x^2 = \frac{x + 2y}{x - 2y}$ **12.** $\frac{x}{\sqrt{y}} - 4y = x$

13. $\sqrt[3]{x} + \sqrt[3]{xy} = 4y^2$ **14.** $\sqrt{y} + \sqrt[3]{y} + \sqrt[4]{y} = x$ **15.** $\sqrt{xy} + 2x = \sqrt{y}$ **16.** $y + \sqrt{xy} = 3x^3$

17. $\frac{y}{\sqrt{x - y}} = 2 + x^2$ **18.** $x^2y^3 = x^4 - y^4$ **19.** $y = \cos(x - y)$ **20.** $x = \sin(x + y)$

21. $x \sin y + y \cos x = 1$ **22.** $\cos(x + y) = y \sin x$

23. $(x + y)^2 - (x - y)^2 = x^3 + y^3$ **24.** $y\sqrt{2 + 3x} + x\sqrt{1 + y} = x$

In Exercises 25 through 28, consider y as the independent variable and find $D_y x$.

25. $x^4 + y^4 = 12x^2y$ **26.** $y = 2x^3 - 5x$ **27.** $x^3y + 2y^4 - x^4 = 0$ **28.** $y\sqrt{x} - x\sqrt{y} = 9$

29. Find an equation of the tangent line to the curve $16x^4 + y^4 = 32$ at the point $(1, 2)$.

30. Find an equation of the normal line to the curve $9x^3 - y^3 = 1$ at the point $(1, 2)$.

31. Find an equation of the normal line to the curve $x^2 + xy + y^2 - 3y = 10$ at the point $(2, 3)$.

32. Find an equation of the tangent line to the curve $\sqrt[3]{xy} = 14x + y$ at the point $(2, -32)$.

33. Find the rate of change of y with respect to x at the point $(3, 2)$ if $7y^2 - xy^3 = 4$.

34. For the circle $x^2 + y^2 = r^2$, show that the tangent line at any point (x_1, y_1) on the curve is perpendicular to the line through (x_1, y_1) and the center of the circle.

35. At what point of the curve $x + \sqrt{xy} + y = 1$ is the tangent line parallel to the x axis?

36. There are two lines through the point $(-1, 3)$ that are tangent to the curve $x^2 + 4y^2 - 4x - 8y + 3 = 0$. Find an equation of each of these lines.

In Exercises 37 through 42, an equation is given. Do the following in each of these problems: (a) Find two functions defined by the equation, and state their domains. (b) Draw a sketch of the graph of each of the functions obtained in part (a). (c) Draw a sketch of the graph of the equation. (d) Find the derivative of each of the functions obtained in part (a), and state the domains of the derivatives. (e) Find $D_x y$ by implicit differentiation from the given equation, and verify that the result so obtained agrees with the results in part (d). (f) Find an equation of each tangent line at the given value of x_1.

37. $y^2 = 4x - 8$; $x_1 = 3$ **38.** $x^2 + y^2 = 25$; $x_1 = 4$ **39.** $x^2 - y^2 = 9$; $x_1 = -5$

40. $y^2 - x^2 = 16$; $x_1 = -3$ **41.** $x^2 + y^2 - 2x - 4y - 4 = 0$; $x_1 = 1$ **42.** $x^2 + 4y^2 + 6x - 40y + 93 = 0$; $x_1 = -2$

43. Prove that the sum of the x and y intercepts of any tangent line to the curve $x^{1/2} + y^{1/2} = k^{1/2}$ is constant and equal to k.

44. If $x^n y^m = (x + y)^{n+m}$, prove that $x \cdot D_x y = y$.

3.10 DERIVATIVES OF HIGHER ORDER

If f' is the derivative of the function f, then f' is also a function, and it is the **first derivative** of f. It is sometimes referred to as the **first derived function.** If the derivative of f' exists, it is called the **second derivative** of f, or the second derived function, and can be denoted by f'' (read as "f double prime"). Similarly, the **third derivative** of f, or the third derived function, is defined as the first derivative of f'' if it exists. The third derivative of f is denoted by f''' (read as "f triple prime").

The ***n*th derivative** of the function f, where n is a positive integer greater than 1, is the first derivative of the $(n - 1)$st derivative of f. We denote the nth derivative of f by $f^{(n)}$. Thus if $f^{(n)}$ is the nth derived function, we can write the function f itself as $f^{(0)}$. Another symbol for the nth derivative of f is $D_x{}^n f$. If the function f is defined by the equation $y = f(x)$, the nth derivative of f can be denoted by $D_x{}^n y$.

Example 1 Find all the derivatives of the function f defined by $f(x) = 8x^4 + 5x^3 - x^2 + 7$.

Solution

$$f'(x) = 32x^3 + 15x^2 - 2x$$

$$f''(x) = 96x^2 + 30x - 2$$

$$f'''(x) = 192x + 30$$

$$f^{(4)}(x) = 192$$

$$f^{(5)}(x) = 0$$

$$f^{(n)}(x) = 0 \qquad n \geq 5$$

Because $f'(x)$ gives the rate of change of $f(x)$ per unit change in x, $f''(x)$, being the derivative of $f'(x)$, gives the rate of change of $f'(x)$ per unit change in x.

If (x, y) is any point on the graph of $y = f(x)$, then $D_x y$ gives the slope of the tangent line to the graph at the point (x, y). Thus $D_x{}^2 y$ is the rate of change of the slope of the tangent line with respect to x at the point (x, y).

Example 2 Let $m(x)$ be the slope of the tangent line to the curve $y = x^3 - 2x^2 + x$ at the point (x, y). Find the instantaneous rate of change of m per unit change in x at the point $(2, 2)$.

Solution $m = D_x y = 3x^2 - 4x + 1$. The instantaneous rate of change of m per unit change in x is given by $D_x m$ or, equivalently, $D_x{}^2 y$.

$$D_x m = D_x{}^2 y = 6x - 4$$

At the point $(2, 2)$, $D_x{}^2 y = 8$. Hence, at the point $(2, 2)$ the change in m is 8 times the change in x.

The second derivative $f''(x)$ is expressed in units of $f'(x)$ per unit of x, which is units of $f(x)$ per unit of x, per unit of x. For example, in straight-line motion, if $f(t)$ centimeters is the distance of a particle from the origin at t seconds, then $f'(t)$ centimeters per second is the velocity of the particle at t seconds, and $f''(t)$ centimeters per second per second is the instantaneous rate of change of the velocity at t seconds. In physics, the instantaneous rate of change of the velocity is called the **instantaneous acceleration.** Therefore, if a particle is moving along a straight line according to the equation of motion $s = f(t)$, where the instantaneous velocity at t sec is given by v cm/sec and the instantaneous acceleration at t sec is given by a cm/sec^2, then a is the first derivative of v with respect to t or, equivalently, the second derivative of s with respect to t; that is,

$$v = D_t s \quad \text{and} \quad a = D_t v = D_t{}^2 s$$

When $a > 0$, v is increasing, and when $a < 0$, v is decreasing. When $a = 0$, then v is not changing. Because the speed of the particle at t sec is $|v|$ cm/sec, we have the following results:

(i) If $v \geq 0$ and $a > 0$, the speed is increasing.
(ii) If $v \geq 0$ and $a < 0$, the speed is decreasing.
(iii) If $v \leq 0$ and $a > 0$, the speed is decreasing.
(iv) If $v \leq 0$ and $a < 0$, the speed is increasing.

● ILLUSTRATION 1 A particle is moving along a horizontal line according to the equation

$$s = 3t^2 - t^3, \qquad t \geq 0 \tag{1}$$

where s cm is the distance of the particle from the origin at t sec. If v cm/sec is the instantaneous velocity at t sec, then $v = D_t s$. Therefore

$$v = 6t - 3t^2 \tag{2}$$

If a cm/sec^2 is the instantaneous acceleration at t sec, then $a = D_t{}^2 s$. So

$$a = 6 - 6t \tag{3}$$

Let us determine the values of t when any one of the quantities s, v, or a is 0. From (1),

$$s = 0 \quad \text{when} \quad t = 0 \text{ or } t = 3$$

From (2),

$$v = 0 \quad \text{when} \quad t = 0 \text{ or } t = 2$$

From (3),

$$a = 0 \quad \text{when} \quad t = 1$$

TABLE 3.10.1

	s	v	a	*Conclusion*
$t = 0$	0	0	6	Particle is at the origin. The velocity is 0 and is increasing. The speed is increasing.
$0 < t < 1$	+	+	+	Particle is at the right of the origin, and it is moving to the right. The velocity is increasing. The speed is increasing.
$t = 1$	2	3	0	Particle is 2 cm to the right of the origin, and it is moving to the right at 3 cm/sec. The velocity is not changing; so the speed is not changing.
$1 < t < 2$	+	+	−	Particle is at the right of the origin, and it is moving to the right. The velocity is decreasing. The speed is decreasing.
$t = 2$	4	0	−6	Particle is 4 cm to the right of the origin, and it is changing its direction of motion from right to left. The velocity is decreasing. The speed is increasing.
$2 < t < 3$	+	−	−	Particle is at the right of the origin, and it is moving to the left. The velocity is decreasing. The speed is increasing.
$t = 3$	0	−9	−12	Particle is at the origin, and it is moving to the left at 9 cm/sec. The velocity is decreasing. The speed is increasing.
$3 < t$	−	−	−	Particle is at the left of the origin, and it is moving to the left. The velocity is decreasing. The speed is increasing.

In Table 3.10.1 there are values of s, v, and a for these values of t: 0, 1, 2, and 3. Also indicated is the sign of the quantities s, v, and a in the intervals of t excluding 0, 1, 2, and 3. A conclusion is formed regarding the position and motion of the particle for the various values of t.

In Fig. 3.10.1 the motion of the particle is along the horizontal line and the behavior of the motion is indicated above the line. ●

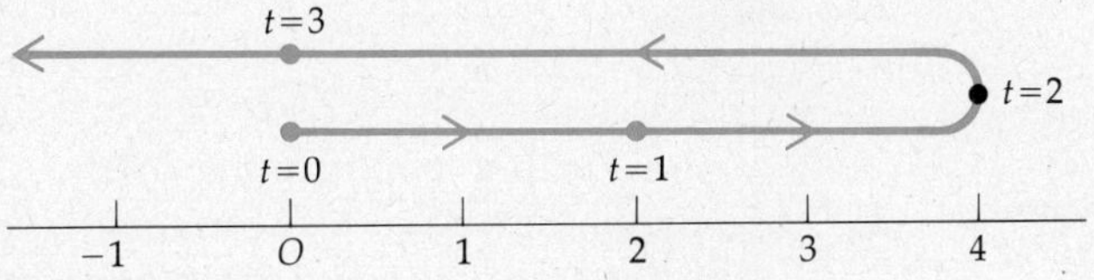

FIGURE 3.10.1

Example 3 A particle is moving along a straight line according to the equation of motion

$$s = \frac{1}{2}t^2 + \frac{4t}{t+1}$$

where s cm is the directed distance of the particle from the origin at t sec. If v cm/sec is the instantaneous velocity at t sec and a cm/sec² is the instantaneous acceleration at t sec, find t, s, and v when $a = 0$.

Solution

$$v = D_t s = t + \frac{4}{(t+1)^2}$$

$$a = D_t v = D_t^2 s = 1 - \frac{8}{(t+1)^3}$$

Setting $a = 0$ we have

$$\frac{(t+1)^3 - 8}{(t+1)^3} = 0$$

or

$$(t+1)^3 = 8$$

from which the only real value of t is obtained from the principal cube root of 8, so that

$$t + 1 = 2 \quad \text{or} \quad t = 1$$

When $t = 1$,

$$s = \frac{1}{2}(1)^2 + \frac{4 \cdot 1}{1 + 1} = \frac{5}{2}$$

and

$$v = 1 + \frac{4}{(1 + 1)^2} = 2$$

Therefore the acceleration is 0 at 1 sec when the particle is $2\frac{1}{2}$ cm from the origin and moving to the right at a velocity of 2 cm/sec.

Further applications of the second derivative are its uses in the second-derivative test for relative extrema (Sec. 4.2) and the sketching of the graph of a function (Secs. 4.8 and 4.9). An important application of other higher ordered derivatives is to determine infinite series as shown in Chapter 15.

The following example illustrates how the second derivative is found for functions defined implicitly.

Example 4 Given $4x^2 + 9y^2 = 36$, find $D_x^2 y$ by implicit differentiation.

Solution Differentiating implicitly with respect to x we have

$$8x + 18yD_x y = 0$$

$$D_x y = -\frac{4x}{9y} \tag{4}$$

To find $D_x^2 y$ we find the derivative of a quotient and keep in mind that y is a function of x. So

$$D_x^2 y = \frac{9y(-4) - (-4x)(9\,D_x y)}{81y^2} \tag{5}$$

If the value of $D_x y$ is substituted from (4) into (5),

$$D_x^2 y = \frac{-36y + (36x)\dfrac{-4x}{9y}}{81y^2} = \frac{-36y^2 - 16x^2}{81y^3}$$

Thus

$$D_x^2 y = \frac{-4(9y^2 + 4x^2)}{81y^3} \tag{6}$$

Because any values of x and y satisfying (6) must also satisfy the original equation, $(9y^2 + 4x^2)$ can be replaced by 36 to obtain

$$D_x^2 y = \frac{-4(36)}{81y^3} = -\frac{16}{9y^3}$$

EXERCISES 3.10

In Exercises 1 through 16, find the first and second derivative of the function defined by the given equation.

1. $f(x) = x^5 - 2x^3 + x$ **2.** $F(x) = 7x^3 - 8x^2$ **3.** $g(s) = 2s^4 - 4s^3 + 7s - 1$ **4.** $G(t) = t^3 - t^2 + t$

5. $F(x) = x^2\sqrt{x} - 5x$ **6.** $g(r) = \sqrt{r} + \dfrac{1}{\sqrt{r}}$ **7.** $f(x) = \sqrt{x^2 + 1}$ **8.** $h(y) = \sqrt[3]{2y^3 + 5}$

9. $f(t) = 4\cos t^2$ **10.** $g(t) = 2\sin^3 t$ **11.** $G(x) = \dfrac{1}{\sqrt{3 + 2x^2}}$ **12.** $f(x) = \dfrac{2 - \sqrt{x}}{2 + \sqrt{x}}$

13. $g(x) = \dfrac{x^2}{x^2 + 4}$ **14.** $g(x) = (2x - 3)^2(x + 4)^3$ **15.** $f(x) = \sqrt{\sin x + 1}$ **16.** $f(x) = \sin\sqrt{x} - \cos\sqrt{x}$

17. Find D_x^3y if $y = x^4 - 2x^2 + x - 5$.

18. Find D_t^3s if $s = \sqrt{4t + 1}$.

19. Find D_x^4y if $y = \dfrac{3}{2x - 1}$.

20. Find $f^{(4)}(x)$ if $f(x) = \dfrac{2}{x - 1}$.

21. Find $D_x^3f(x)$ if $f(x) = \dfrac{x}{(1 - x)^2}$.

22. Find $D_t^4f(t)$ if $f(t) = 3\sin 2t$.

23. Find $f^{(5)}(x)$ if $f(x) = \cos 2x - \sin 2x$.

24. Find D_v^3u if $u = v\sqrt{v - 2}$.

25. Given $x^2 + y^2 = 1$, show that $D_x^2y = -\dfrac{1}{y^3}$.

26. Given $x^2 + 25y^2 = 100$, show that $D_x^2y = -\dfrac{4}{25y^3}$.

27. Given $x^3 + y^3 = 1$, show that $D_x^2y = \dfrac{-2x}{y^5}$.

28. Given $x^{1/2} + y^{1/2} = 2$, show that $D_x^2y = \dfrac{1}{x^{3/2}}$.

29. Given $x^4 + y^4 = a^4$ (a is a constant), find D_x^2y in simplest form.

30. Given $b^2x^2 - a^2y^2 = a^2b^2$ (a and b are constants), find D_x^2y in simplest form.

31. Find the slope of the tangent line at each point of the graph of $y = x^4 + x^3 - 3x^2$ where the rate of change of the slope is zero.

32. Find the instantaneous rate of change of the slope of the tangent line to the graph of $y = 2x^3 - 6x^2 - x + 1$ at the point $(3, -2)$.

In Exercises 33 and 34, a particle is moving along a horizontal line according to the given equation, where s cm is the directed distance of the particle from the origin at t sec, v cm/sec is the velocity of the particle at t sec, and a cm/sec² is the acceleration of the particle at t sec. Find v and a in terms of t. Make a table similar to Table 3.10.1 that gives a description of the position and motion of the particle. Include in the table the intervals of time when the particle is moving to the left, when it is moving to the right, when the velocity is increasing, when the velocity is decreasing, when the speed is increasing, when the speed is decreasing, and the position of the particle with respect to the origin during these intervals of time. Show the behavior of the motion by a figure similar to Fig. 3.10.1.

33. $s = t^3 - 9t^2 + 15t,\ t \geq 0$ **34.** $s = \frac{1}{6}t^3 - 2t^2 + 6t - 2,\ t \geq 0$

In Exercises 35 through 39, a particle is moving along a straight line according to the given equation where s ft is the directed distance of the particle from the origin at t sec. Find the time when the instantaneous acceleration is zero, and then find the directed distance of the particle from the origin and the instantaneous velocity at this time.

35. $s = \frac{1}{3}t^3 - \frac{3}{2}t^2 + 2t + 1,\ t \geq 0$ **36.** $s = 2t^3 - 6t^2 + 3t - 4,\ t \geq 0$ **37.** $s = \dfrac{125}{16t + 32} - \dfrac{2}{5}t^5,\ t \geq 0$

38. $s = 9t^2 + 2\sqrt{2t + 1},\ t \geq 0$ **39.** $s = \frac{4}{9}t^{3/2} + 2t^{1/2},\ t \geq 0$

40. A particle moving in a straight line is said to have **simple harmonic motion** if the measure of its acceleration is always proportional to the measure of its displacement from a fixed point on the line and its acceleration and displacement are oppositely directed. Show that the straight-line motion of a particle described by $s = A \sin 2\pi kt + B \cos 2\pi kt$, where s cm is the directed distance of the particle from the origin at t sec, and A, B, and k are constants, is a simple harmonic motion.

41. Show that if a particle is moving along a straight line according to the equation of motion $s = a\cos(kt + \theta)$, where a, k, and θ are constants and s cm is the directed distance of the particle from the origin at t sec, then the motion is simple harmonic. (See Exercise 40.)

In Exercises 42 through 45, find formulas for $f'(x)$ and $f''(x)$ and state the domains of f' and f''.

42. $f(x) = \begin{cases} \dfrac{x^2}{|x|} & \text{if } x \neq 0 \\ 0 & \text{if } x = 0 \end{cases}$ **43.** $f(x) = \begin{cases} -x^2 & \text{if } x < 0 \\ x^2 & \text{if } x \geq 0 \end{cases}$ **44.** $f(x) = |x|^3$ **45.** $f(x) = \begin{cases} \dfrac{x^5}{|x|} & \text{if } x \neq 0 \\ 0 & \text{if } x = 0 \end{cases}$

46. For the function of Exercise 44, find $f'''(x)$ when it exists.

47. For the function of Exercise 45, find $f'''(x)$ when it exists.

48. Show that if $xy = 1$, then $D_x{}^2y \cdot D_y{}^2x = 4$.

49. If f', g', f'', and g'' exist and if $h = f \circ g$, express $h''(x)$ in terms of the derivatives of f and g.

50. If f and g are two functions such that their first and second derivatives exist and h is the function defined by $h(x) = f(x) \cdot g(x)$, prove that

$$h''(x) = f(x) \cdot g''(x) + 2f'(x) \cdot g'(x) + f''(x) \cdot g(x)$$

51. If $y = x^n$, where n is any positive integer, prove by mathematical induction that $D_x{}^ny = n!$

52. If

$$y = \frac{1}{1 - 2x}$$

prove by mathematical induction that

$$D_x{}^ny = \frac{2^n n!}{(1 - 2x)^{n+1}}$$

53. If k is any positive integer, prove by mathematical induction that

$$D_x{}^n(\sin x) = \begin{cases} \sin x & \text{if } n = 4k \\ \cos x & \text{if } n = 4k + 1 \\ -\sin x & \text{if } n = 4k + 2 \\ -\cos x & \text{if } n = 4k + 3 \end{cases}$$

54. Obtain a formula similar to that in Exercise 53 for $D_x{}^n(\cos x)$.

REVIEW EXERCISES FOR CHAPTER 3

In Exercises 1 through 16, find the derivative of the given function.

1. $f(x) = \dfrac{4}{x^2} - \dfrac{3}{x^4}$

2. $g(x) = 5(x^4 + 3x^7)$

3. $G(t) = (3t^2 - 4)(4t^3 + t - 1)$

4. $f(x) = (x^4 - 2x)(4x^2 + 2x + 5)$

5. $g(x) = \dfrac{x^3+1}{x^3-1}$

6. $h(y) = \dfrac{y^2}{y^3+8}$

7. $f(s) = (2s^3 - 3s + 7)^4$

8. $F(x) = (4x^4 - 4x^2 + 1)^{-1/3}$

9. $f(x) = \sqrt[3]{\dfrac{x}{x^3+1}}$

10. $g(x) = \left(\dfrac{3x^2+4}{x^7+1}\right)^{10}$

11. $F(x) = (x^2-1)^{3/2}(x^2-4)^{1/2}$

12. $g(x) = (x^4 - x)^{-3}(5 - x^2)^{-1}$

13. $f(x) = (x+1)\sin x - x\cos x$

14. $f(x) = \dfrac{1+\sin x}{x\cos x}$

15. $g(w) = \sin(\cos 3w) - 3\cos^2 2w$

16. $f(t) = \sin^2 3t\sqrt{\cos 2t}$

In Exercises 17 through 24, find $D_x y$.

17. $4x^2 + 4y^2 - y^3 = 0$

18. $y = \sqrt{1+x} + \sqrt{1-x}$

19. $y = \dfrac{1}{x - \sqrt{x^2-1}}$

20. $xy^2 + 2y^3 = x - 2y$

21. $\sin(x+y) + \sin(x-y) = 1$

22. $x^{2/3} + y^{2/3} = a^{2/3}$

23. $y = x^2 + [x^3 + (x^4+x)^2]^3$

24. $y = \dfrac{x\sqrt{3+2x}}{4x-1}$

25. A particle is moving in a horizontal line according to the equation $s = t^3 - 11t^2 + 24t + 100$, where s cm is the directed distance of the particle from the origin at t sec. (a) The particle is at the starting point when $t = 0$. For what other values of t is the particle at the starting point? (b) Determine the velocity of the particle at each instant that it is at the starting point, and interpret the sign of the velocity in each case.

26. A particle is moving in a horizontal line according to the equation $s = t^3 - 3t^2 - 9t + 2$, where s cm is the directed distance of the particle from a point O at t sec. The positive direction is to the right. Determine the intervals of time when the particle is moving to the right and when it is moving to the left. Also determine when the particle reverses its direction. Show the behavior of the motion by a figure similar to Fig. 3.5.2, and choose values of t at random, but include the values of t when the particle reverses its direction.

27. A particle is moving along a straight line according to the equation $s = 5 - 2\cos^2 t$, where s cm is the directed distance of the particle from the origin at t sec. If v cm/sec and a cm/sec² are, respectively, the velocity and acceleration of the particle at t sec, find v and a in terms of s.

28. An object is sliding down an inclined plane according to the equation $s = 12t^2 + 6t$, where s meters is the directed distance of the object from the top t sec after starting. (a) Find the velocity 3 sec after the start. (b) Find the initial velocity.

29. A ball is thrown vertically upward from the top of a house 112 ft high. Its equation of motion is $s = -16t^2 + 96t$, where s ft is the directed distance of the ball from the starting point at t sec. Find (a) the instantaneous velocity of the ball at 2 sec; (b) how high the ball will go; (c) how long it takes for the ball to reach the ground; (d) the instantaneous velocity of the ball when it reaches the ground.

30. Find equations of the tangent lines to the curve $y = 2x^3 + 4x^2 - x$ that have the slope $\frac{1}{2}$.

31. Find equations of the tangent line and normal line to the curve $2x^3 + 2y^3 - 9xy = 0$ at the point (2, 1).

32. Find an equation of the normal line to the curve $x - y = \sqrt{x+y}$ at the point (3, 1).

33. Find an equation of the normal line to the curve $y = \cos x$ at the point $(\frac{2}{3}\pi, -\frac{1}{2})$.

34. Find equations of the tangent and normal lines to the curve $y = 8\sin^3 2x$ at the point $(\frac{1}{12}\pi, 1)$.

35. Prove that the line tangent to the curve $y = -x^4 + 2x^2 + x$ at the point (1, 2) is also tangent to the curve at another point, and find this point.

36. Prove that the tangent lines to the curves $4y^3 - x^2y - x + 5y = 0$ and $x^4 - 4y^3 + 5x + y = 0$ at the origin are perpendicular.

37. Find $D_x{}^3y$ if $y = \sqrt{3 - 2x}$.

38. Given $D_xy = y^k$, where k is a constant and y is a function of x. Express $D_x{}^3y$ in terms of y and k.

39. Given $f(x) = \frac{1}{12}x^4 + \frac{2}{3}x^3 + \frac{3}{2}x^2 + 8x + 2$. For what values of x is $f''(x) > 0$?

40. Show that if $xy = k$, where k is a nonzero constant, then $D_x{}^2y \cdot D_y{}^2x = 4/k$.

41. If A square units is the area of an isosceles right triangle for which each leg has a length of x units, find (a) the average rate of change of A with respect to x as x changes from 8.00 to 8.01; (b) the instantaneous rate of change of A with respect to x when x is 8.00.

42. Stefan's law states that a body emits radiant energy according to the formula $R = kT^4$, where R is the measure of the rate of emission of the radiant energy per square unit of area, T is the measure of the Kelvin temperature of the surface, and k is a constant. Find (a) the average rate of change of R with respect to T as T increases from 200 to 300; (b) the instantaneous rate of change of R with respect to T when T is 200.

43. The supply equation for a calculator is $y = m^2 + \sqrt{m}$, where $100y$ calculators are supplied when m dollars is the price per calculator. Find (a) the average rate of change of the supply with respect to the price when the price is increased from \$16 to \$17; (b) the instantaneous (or marginal) rate of change of the supply with respect to the price when the price is \$16.

44. If $y = x^{2/3}$, find the relative rate of change of y with respect to x when (a) $x = 8$, and (b) $x = c$, where c is a constant.

45. Use only the definition of a derivative to find $f'(x)$ if $f(x) = \sqrt{4x - 3}$.

46. Use only the definition of a derivative to find $f'(x)$ if $f(x) = \dfrac{2}{\sqrt{3x + 1}}$.

47. Use only the definition of a derivative to find $f'(-5)$ if $f(x) = \dfrac{3}{x + 2}$.

48. Use only the definition of a derivative to find $f'(5)$ if $f(x) = \sqrt[3]{3x + 1}$.

49. Find $f'(x)$ if $f(x) = (|x + 1| - |x|)^2$.

50. Find $f''(x)$ if $f(x) = 3 \sin^2 x - 4 \cos^2 x$.

51. Find $f''(\pi)$ if $f(x) = \sqrt{2 + \cos x}$.

52. Find $f'(-3)$ if $f(x) = (|x| - x)\sqrt[3]{9x}$.

53. Given

$$f(x) = \begin{cases} x^2 + 2 & \text{if } x \leq 3 \\ 20 - x^2 & \text{if } 3 < x \end{cases}$$

(a) Draw a sketch of the graph of f. (b) Determine if f is continuous at 3. (c) Determine if f is differentiable at 3.

54. Given

$$f(x) = \begin{cases} x^2 - 16 & \text{if } x < 4 \\ 8x - 32 & \text{if } 4 \leq x \end{cases}$$

(a) Draw a sketch of the graph of f. (b) Determine if f is continuous at 4. (c) Determine if f is differentiable at 4.

55. Given

$$f(x) = \begin{cases} ax^2 + b & \text{if } x \le 1 \\ \dfrac{1}{|x|} & \text{if } x > 1 \end{cases}$$

Find the values of a and b such that $f'(1)$ exists.

56. Suppose

$$f(x) = \begin{cases} x^3 & \text{if } x < 1 \\ ax^2 + bx + c & \text{if } x \ge 1 \end{cases}$$

Find the values of a, b, and c such that $f''(1)$ exists.

57. Given $f(x) = |x|^3$. (a) Draw a sketch of the graph of f. (b) Find $\lim_{x \to 0} f(x)$ if it exists. (c) Find $f'(0)$ if it exists.

58. Given $f(x) = x^2 \operatorname{sgn} x$. (a) Discuss the differentiability of f. (b) Is f' continuous on its domain?

59. A particle is moving along a straight line, and

$$s = \sin(4t + \tfrac{1}{3}\pi) + \sin(4t + \tfrac{1}{6}\pi)$$

where s ft is the directed distance of the particle from the origin at t sec. Show that the motion is simple harmonic. (See Exercise 40 in Exercises 3.10.)

60. If a particle is moving along a straight line and $s = \cos 2t + \cos t$, prove that the motion is not simple harmonic.

61. A particle is moving in a straight line according to the equation $s = \sqrt{a + bt^2}$, where a and b are positive constants. Prove that the measure of the acceleration of the particle is inversely proportional to s^3 for any t.

62. If the two functions f and g are differentiable at the number x_1, is the composite function $f \circ g$ necessarily differentiable at x_1? If your answer is yes, prove it. If your answer is no, give a counterexample.

63. Suppose $f(x) = 3x + |x|$ and $g(x) = \frac{3}{4}x - \frac{1}{4}|x|$. Prove that neither $f'(0)$ nor $g'(0)$ exists but that $(f \circ g)'(0)$ does exist.

64. Give an example of two functions f and g for which f is differentiable at $g(0)$, g is not differentiable at 0, and $f \circ g$ is differentiable at 0.

65. Give an example of two functions f and g for which f is not differentiable at $g(0)$, g is differentiable at 0, and $f \circ g$ is differentiable at 0.

66. In Exercise 45 of Exercises 3.2, you are to prove that if f is differentiable at a, then

$$f'(a) = \lim_{\Delta x \to 0} \frac{f(a + \Delta x) - f(a - \Delta x)}{2\,\Delta x}$$

Show by using the absolute-value function that it is possible for the limit in the above equation to exist even though $f'(a)$ does not exist.

67. If $f'(x_1)$ exists, prove that

$$\lim_{x \to x_1} \frac{xf(x_1) - x_1 f(x)}{x - x_1} = f(x_1) - x_1 f'(x_1)$$

68. Let f and g be two functions whose domains are the set of all real numbers. Furthermore, suppose that (i) $g(x) = xf(x) + 1$; (ii) $g(a + b) = g(a) \cdot g(b)$ for all a and b; (iii) $\lim_{x \to 0} f(x) = 1$. Prove that $g'(x) = g(x)$.

69. The remainder theorem of elementary algebra states that if $P(x)$ is a polynomial in x and r is any real number, then there is a polynomial $Q(x)$ such that $P(x) = Q(x)(x - r) + P(r)$. What is $\lim_{x \to r} Q(x)$?

70. Suppose $g(x) = |f(x)|$. If $f^{(n)}(x)$ exists and $f(x) \neq 0$, prove that

$$g^{(n)}(x) = \frac{f(x)}{|f(x)|} f^{(n)}(x)$$

71. Prove that $D_x^{\,n}(\sin x) = \sin(x + \frac{1}{2}n\pi)$. (*Hint:* Use mathematical induction and the formulas $\sin(x + \frac{1}{2}\pi) = \cos x$ or $\cos(x + \frac{1}{2}\pi) = -\sin x$ after each differentiation.)

72. Suppose the function f is defined on the open interval $(0, 1)$ and

$$f(x) = \frac{\sin \pi x}{x(x-1)}$$

Define f at 0 and 1 so that f is continuous on the closed interval $[0, 1]$.

CHAPTER 4 Applications of the Derivative

4.1 RELATED RATES

There are many problems concerned with the rate of change of two or more related variables with respect to time in which it is not necessary to express each of these variables directly as functions of time. For example, suppose we have an equation involving the variables x and y, and both x and y are functions of a third variable t, where t units denotes time. Then, because the rate of change of x with respect to t and the rate of change of y with respect to t are given by $D_t x$ and $D_t y$, respectively, we differentiate on both sides of the given equation with respect to t by applying the chain rule and proceed as in the following example.

Example 1 In (a) and (b), x and y are functions of a third variable t. (a) If $5x + 3xy = 4$ and $D_t y = 3$, find $D_t x$ when $x = 2$. (b) If $x \cos y = 5$ and $D_t x = -4$, find $D_t y$ when $y = \frac{1}{3}\pi$.

Solution

(a) The given equation is

$$5x + 3xy = 4 \tag{1}$$

We differentiate on both sides of the equation with respect to t.

$$5D_t x + 3y\, D_t x + 3x\, D_t y = 0$$

$$D_t x = -\frac{3x}{3y + 5} \cdot D_t y \tag{2}$$

To find the value of y when $x = 2$ we replace x by 2 in (1) and obtain $y = -1$. From (2), with $x = 2$, $y = -1$, and $D_t y = 3$,

$$\begin{aligned} D_t x \Big]_{x=2} &= -\tfrac{6}{2}(3) \\ &= -9 \end{aligned}$$

(b) On both sides of the equation

$$x \cos y = 5 \tag{3}$$

we differentiate with respect to t.

$$(\cos y)D_t x - (x \sin y)D_t y = 0$$

$$D_t y = \frac{\cos y}{x \sin y} \cdot D_t x \tag{4}$$

From (3), when $y = \frac{1}{3}\pi$, $x = 10$. From (4), with $y = \frac{1}{3}\pi$, $x = 10$, and $D_t x = -4$,

$$\begin{aligned} D_t y \Big]_{y=\pi/3} &= \frac{\frac{1}{2}}{10(\frac{1}{2}\sqrt{3})}(-4) \\ &= -\tfrac{2}{15}\sqrt{3} \end{aligned}$$

Example 2 A ladder 25 ft long is leaning against a vertical wall. If the bottom of the ladder is pulled horizontally away from the wall at 3 ft/sec, how fast is the top of the ladder sliding down the wall when the bottom is 15 ft from the wall?

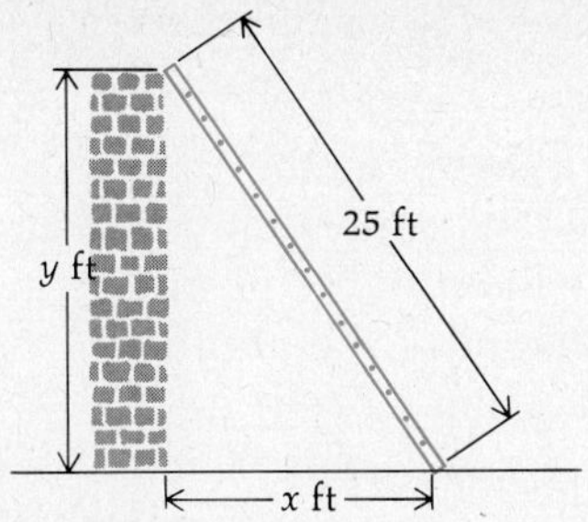

FIGURE 4.1.1

Solution Let t be the number of seconds in the time that has elapsed since the ladder started to slide down the wall, y be the number of feet in the distance from the ground to the top of the ladder at t sec, and x be the number of feet in the distance from the bottom of the ladder to the wall at t sec. See Fig. 4.1.1. Because the bottom of the ladder is pulled horizontally away from the wall at 3 ft/sec, $D_t x = 3$. We wish to find $D_t y$ when $x = 15$. From the Pythagorean theorem,

$$y^2 = 625 - x^2 \tag{5}$$

Because x and y are functions of t, we differentiate on both sides of (5) with respect to t and obtain

$$2y\, D_t y = -2x\, D_t x$$

$$D_t y = -\frac{x}{y} D_t x \tag{6}$$

When $x = 15$, it follows from (5) that $y = 20$. Because $D_t x = 3$, we get from (6)

$$D_t y \Big]_{y=20} = -\tfrac{15}{20} \cdot 3 = -\tfrac{9}{4}$$

Therefore the top of the ladder is sliding down the wall at the rate of $2\frac{1}{4}$ ft/sec when the bottom is 15 ft from the wall. The significance of the minus sign is that y is decreasing as t is increasing.

Example 2 involves a problem with rates of change of related variables. It is called a problem in related rates. In such problems the variables have a specific relationship for values of t, where t is a measure of time. This relationship is usually expressed in the form of an equation, as in Example 2 with Eq. (5). Values of the variables and rates of change of the variables with respect to t are often given at a particular instant. In Example 2, at the instant when $x = 15$, then $y = 20$ and $D_t x = 3$, and we wish to find $D_t y$.

The following steps represent a possible procedure for solving a problem involving related rates.

1. Draw a figure if it is feasible to do so.
2. Define the variables. Generally define t first, because the other variables usually depend on t.
3. Write down any numerical facts known about the variables and their derivatives with respect to t.
4. Obtain an equation involving the variables that depend on t.
5. Differentiate with respect to t on both sides of the equation found in step 4.
6. Substitute values of known quantities in the equation of step 5, and solve for the desired quantity.

Example 3 A tank is in the form of an inverted cone having an altitude of 16 m and a base radius of 4 m. Water is flowing into the tank

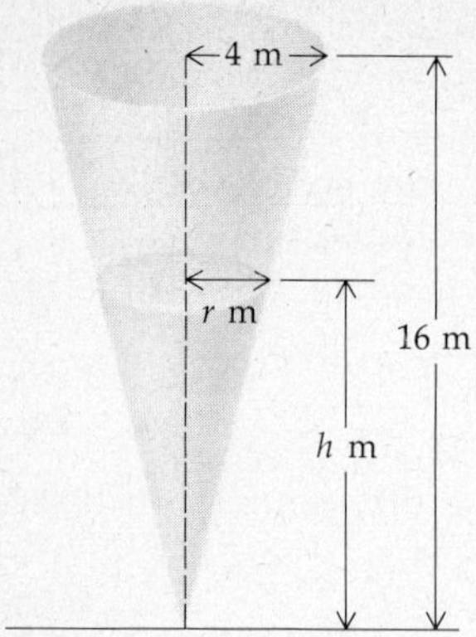

FIGURE 4.1.2

at the rate of 2 m^3/min. How fast is the water level rising when the water is 5 m deep?

Solution Let t be the number of minutes in the time that has elapsed since water started to flow into the tank, h the number of meters in the height of the water level at t min, r the number of meters in the radius of the surface of the water at t min, and V the number of cubic meters in the volume of water in the tank at t min.

At any time, the volume of water in the tank may be expressed in terms of the volume of a cone (see Fig. 4.1.2).

$$V = \tfrac{1}{3}\pi r^2 h \tag{7}$$

V, r, and h are all functions of t. Because water is flowing into the tank at the rate of 2 m^3/min, $D_tV = 2$. We wish to find D_th when $h = 5$. To express r in terms of h, we have from similar triangles

$$\frac{r}{h} = \frac{4}{16} \qquad r = \tfrac{1}{4}h$$

Substituting this value of r into (7) we obtain

$$V = \tfrac{1}{3}\pi(\tfrac{1}{4}h)^2(h) \quad \text{or} \quad V = \tfrac{1}{48}\pi h^3 \tag{8}$$

By differentiating on both sides of (8) with respect to t,

$$D_tV = \tfrac{1}{16}\pi h^2 \, D_th$$

Substituting 2 for D_tV and solving for D_th we get

$$D_th = \frac{32}{\pi h^2}$$

Therefore

$$D_th\Big]_{h=5} = \frac{32}{25\pi}$$

Therefore the water level is rising at the rate of $\frac{32}{25\pi}$ m/min when the water is 5 m deep.

Example 4 Two cars, one going due east at the rate of 90 km/hr and the other going due south at the rate of 60 km/hr, are traveling toward an intersection of the two roads. At what rate are the two cars approaching each other at the instant when the first car is 200 m and the second car is 150 m from the intersection?

Solution Refer to Fig. 4.1.3, where the point P is the intersection of the two roads.

Let t sec be the time elapsed since the cars started to approach P, x meters be the distance of the first car from P at t sec, y meters be the distance of the second car from P at t sec, and z meters be the distance between the two cars at t sec.

The first car is approaching P at the rate of 90 km/hr, which is 90,000/3600 m/sec or 25 m/sec. Because x is decreasing as t is increasing, it follows that $D_tx = -25$. Similarly, because 60 km/hr is 60,000/3600

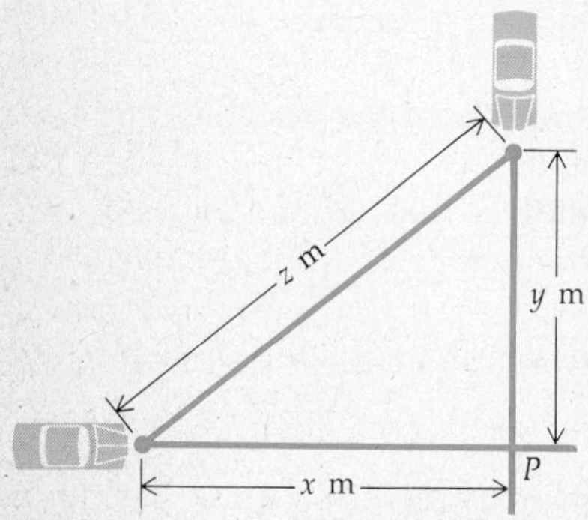

FIGURE 4.1.3

m/sec, which is $\frac{50}{3}$ m/sec, $D_t y = -\frac{50}{3}$. We wish to find $D_t z$ when $x = 200$ and $y = 150$. From the Pythagorean theorem,

$$z^2 = x^2 + y^2 \tag{9}$$

Differentiating on both sides of (9) with respect to t we obtain

$$2z\, D_t z = 2x\, D_t x + 2y\, D_t y$$

$$D_t z = \frac{x\, D_t x + y\, D_t y}{z} \tag{10}$$

When $x = 200$ and $y = 150$, it follows from (9) that $z = 250$. In (10) let $D_t x = -25$, $D_t y = -\frac{50}{3}$, $x = 200$, $y = 150$, and $z = 250$ to get

$$D_t z \bigg|_{z=250} = \frac{(200)(-25) + (150)(-\frac{50}{3})}{250}$$

$$= -30$$

Therefore, at the instant in question the cars are approaching each other at the rate of 30 m/sec.

Example 5 Suppose in a certain market that p dollars is the price of a crate of oranges, x is the number of thousands of crates of oranges supplied daily, and the supply equation is $px - 20p - 3x + 105 = 0$. If the daily supply is decreasing at the rate of 250 crates per day, at what rate is the price changing when the daily supply is 5000 crates?

Solution Let t days be the time that has elapsed since the daily supply of oranges started to decrease. Then p and x are both functions of t. Since the daily supply is decreasing at the rate of 250 crates per day, $D_t x = -\frac{250}{1000} = -\frac{1}{4}$. We wish to find $D_t p$ when $x = 5$. From the given supply equation we differentiate implicitly with respect to t and obtain

$$p\, D_t x + x\, D_t p - 20 D_t p - 3 D_t x = 0$$

$$D_t p = \frac{3 - p}{x - 20} D_t x \tag{11}$$

When $x = 5$, it follows from the supply equation that $p = 6$. Because $D_t x = -\frac{1}{4}$, from (11),

$$D_t p \bigg|_{p=6} = \frac{3 - 6}{5 - 20}\left(-\frac{1}{4}\right) = \frac{-3}{-15}\left(-\frac{1}{4}\right) = -\frac{1}{20}$$

Hence the price of a crate of oranges is decreasing at the rate of \$0.05 per day when the daily supply is 5000 crates.

EXERCISES 4.1

In Exercises 1 through 8, x and y are functions of a third variable t.

1. If $2x + 3y = 8$ and $D_t y = 2$, find $D_t x$.

2. If $x/y = 10$ and $D_t x = -5$, find $D_t y$.

3. If $xy = 20$ and $D_t y = 10$, find $D_t x$ when $x = 2$.

4. If $2 \sin x + 4 \cos y = 3$ and $D_t y = 3$, find $D_t x$ at $(\frac{1}{6}\pi, \frac{1}{3}\pi)$.

5. If $\sin^2 x + \cos^2 y = \frac{5}{4}$ and $D_t x = -1$, find $D_t y$ at $(\frac{2}{3}\pi, \frac{3}{4}\pi)$.

6. If $x^2 + y^2 = 25$ and $D_t x = 5$, find $D_t y$ when $y = 4$.

7. If $\sqrt{x} + \sqrt{y} = 5$ and $D_t y = 3$, find $D_t x$ when $x = 1$.

8. If $y(\sin x + 1) = 4$ and $D_t y = -4$, find $D_t x$ when $x = \pi$.

9. A kite is flying at a height of 40 ft. A child is flying it so that it is moving horizontally at a rate of 3 ft/sec. If the string is taut, at what rate is the string being paid out when the length of the string released is 50 ft?

10. A spherical balloon is being inflated so that its volume is increasing at the rate of 5 m^3/min. At what rate is the diameter increasing when the diameter is 12 m?

11. A spherical snowball is being made so that its volume is increasing at the rate of 8 ft^3/min. Find the rate at which the radius is increasing when the snowball is 4 ft in diameter.

12. Suppose that when the diameter is 6 ft, the snowball in Exercise 11 stopped growing and started to melt at the rate of $\frac{1}{4}$ ft^3/min. Find the rate at which the radius is changing when the radius is 2 ft.

13. Sand is being dropped at the rate of 10 m^3/min onto a conical pile. If the height of the pile is always twice the base radius, at what rate is the height increasing when the pile is 8 m high?

14. A light is hung 15 ft above a straight horizontal path. If a man 6 ft tall is walking away from the light at the rate of 5 ft/sec, how fast is his shadow lengthening?

15. In Exercise 14, at what rate is the tip of the man's shadow moving?

16. A man 6 ft tall is walking toward a building at the rate of 5 ft/sec. If there is a light on the ground 50 ft from the building, how fast is the man's shadow on the building growing shorter when he is 30 ft from the building?

17. A water tank in the form of an inverted cone is being emptied at the rate of 6 m^3/min. The altitude of the cone is 24 m, and the base radius is 12 m. Find how fast the water level is lowering when the water is 10 m deep.

18. A trough is 12 ft long and its ends are in the form of inverted isosceles triangles having an altitude of 3 ft and a base of 3 ft. Water is flowing into the trough at the rate of 2 ft^3/min. How fast is the water level rising when the water is 1 ft deep?

19. Boyle's law for the expansion of gas is $PV = C$, where P is the number of pounds per square unit of pressure, V is the number of cubic units of volume of the gas, and C is a constant. At a certain instant the pressure is 3000 lb/ft^2, the volume is 5 ft^3, and the volume is increasing at the rate of 3 ft^3/min. Find the rate of change of the pressure at this instant.

20. The adiabatic law (no gain or loss of heat) for the expansion of air is $PV^{1.4} = C$, where P is the number of pounds per square unit of pressure, V is the number of cubic units of volume, and C is a constant. At a specific instant the pressure is 40 lb./in.^{2} and is increasing at the rate of 8 lb/in.^{2} each second. What is the rate of change of volume at this instant?

21. A stone is dropped into a still pond. Concentric circular ripples spread out and the radius of the disturbed region increases at the rate of 16 cm/sec. At what rate does the area of the disturbed region increase when its radius is 4 cm?

22. Oil is running into an inverted conical tank at the rate of 3π m^3 per minute. If the tank has a radius of 2.5 m at the top and a depth of 10 m, how fast is the depth of the oil changing when it is 8 m?

23. An automobile traveling at a rate of 30 ft/sec is approaching an intersection. When the automobile is 120 ft from the intersection, a truck traveling at the rate of 40 ft/sec crosses the intersection. The automobile and the truck are on roads that are at right angles to each other. How fast are the automobile and the truck separating 2 sec after the truck leaves the intersection?

24. A woman on a dock is pulling in a boat at the rate of 50 ft/min by means of a rope attached to the boat at water level. If the woman's hands are 16 ft above the water level, how fast is the boat approaching the dock when the amount of rope out is 20 ft?

25. This week a factory is producing 50 units of a particular commodity and the amount being produced is increasing at the rate of 2 units per week. If C dollars is the cost of producing x units and $C = 0.08x^3 - x^2 + 10x + 48$, find the current rate at which the production cost is increasing.

26. The demand in a certain market for a particular kind of breakfast cereal is given by the demand equation $px + 25p = 4000$, where p cents is the price of one box and x thousands of boxes is the quantity demanded per week. If the current price of the cereal is 80 cents per box and the price per box is increasing at the rate of 0.2 cent each week, find the rate of change in the demand.

27. The supply equation for a certain commodity is $x = 1000\sqrt{3p^2 + 20p}$, where x units are supplied per month when p dollars is the price per unit. Find the rate of change in the supply if the current price is \$20 per unit and the price is increasing at the rate of \$0.50 per month.

28. Suppose that y is the number of workers in the labor force needed to produce x units of a certain commodity, and $x = 4y^2$. If the production of the commodity this year is 250,000 units and the production is increasing at the rate of 18,000 units per year, what is the current rate at which the labor force should be increased?

29. The demand equation for a particular shirt is $2px + 65p - 4950 = 0$, where x hundreds of shirts are demanded per week when p dollars is the price of a shirt. If the shirt is selling this week at \$30 and the price is increasing at the rate of \$0.20 per week, find the rate of change in the demand.

30. The measure of one of the acute angles of a right triangle is decreasing at the rate of $\frac{1}{36}\pi$ rad/sec. If the length of the hypotenuse is constant and 40 cm, find how fast the area is changing when the measure of the acute angle is $\frac{1}{6}\pi$.

31. Two trucks, one traveling west and the other traveling south, are approaching an intersection. If both trucks are traveling at the rate of k km/hr, show that they are approaching each other at the rate of $k\sqrt{2}$ km/hr when they are each m km from the intersection.

32. A horizontal trough is 16 m long, and its ends are isosceles trapezoids with an altitude of 4 m, a lower base of 4 m, and an upper base of 6 m. Water is being poured into the trough at the rate of 10 m^3/min. How fast is the water level rising when the water is 2 m deep?

33. In Exercise 32, if the water level is decreasing at the rate of 25 cm/min when the water is 3 m deep, at what rate is water being drawn from the trough?

34. A ladder 7 m long is leaning against a wall. If the bottom of the ladder is pushed horizontally toward the wall at 1.5 m/sec, how fast is the top of the ladder sliding up the wall when the bottom is 2 m from the wall?

35. A ladder 20 ft long is leaning against an embankment inclined 60° to the horizontal. If the bottom of the ladder is being moved horizontally toward the embankment at 1 ft/sec, how fast is the top of the ladder moving when the bottom is 4 ft from the embankment?

36. The volume of a balloon is decreasing at a rate proportional to its surface area. Show that the radius of the balloon shrinks at a constant rate.

37. If a ladder of length 30 ft that is leaning against a wall has its upper end sliding down the wall at the rate of $\frac{1}{2}$ ft/sec, what is the rate of change of the measure of the acute angle made by the ladder with the ground when the upper end is 18 ft above the ground?

38. Water is being poured at the rate of 8 ft^3/min into a tank in the form of a cone. The cone is 20 ft deep and 10 ft in diameter at the top. If there is a leak in the bottom and the water level is rising at the rate of 1 in./min, when the water is 16 ft deep, how fast is the water leaking?

4.2 MAXIMUM AND MINIMUM VALUES OF A FUNCTION

We have seen that the geometrical interpretation of the derivative of a function is the slope of the tangent line to the graph of the function at a point. This fact enables us to apply derivatives as an aid in sketching graphs. For example, the derivative may be used to determine at what points the tangent line is horizontal; these are the points where the derivative is zero. Also, the derivative may be used to find the intervals for which the graph of a function lies above the tangent line and the intervals for which the graph lies below the tangent line. Before applying the derivative to draw sketches of graphs, we need some definitions and theorems.

4.2.1 Definition The function f is said to have a **relative maximum value** at c if there exists an open interval containing c, on which f is defined, such that $f(c) \geq f(x)$ for all x in this interval.

Figures 4.2.1 and 4.2.2 each show a sketch of a portion of the graph of a function having a relative maximum value at c.

4.2.2 Definition The function f is said to have a **relative minimum value** at c if there exists an open interval containing c, on which f is defined, such that $f(c) \leq f(x)$ for all x in this interval.

Figures 4.2.3 and 4.2.4 each show a sketch of a portion of the graph of a function having a relative minimum value at c.

If the function f has either a relative maximum or a relative minimum value at c, then f is said to have a **relative extremum** at c.

The following theorem is used to locate the possible values of c for which there is a relative extremum.

4.2.3 Theorem If $f(x)$ exists for all values of x in the open interval (a, b) and if f has a relative extremum at c, where $a < c < b$, then if $f'(c)$ exists, $f'(c) = 0$.

Proof The proof will be given for the case when f has a relative minimum value at c.

If $f'(c)$ exists, then

$$f'(c) = \lim_{x \to c} \frac{f(x) - f(c)}{x - c} \tag{1}$$

Because f has a relative minimum value at c, by Definition 4.2.2 there exists a $\delta > 0$ such that

$$f(x) - f(c) \geq 0 \quad \text{whenever} \quad 0 < |x - c| < \delta$$

If x is approaching c from the right, $x - c > 0$, and therefore

$$\frac{f(x) - f(c)}{x - c} \geq 0 \quad \text{whenever} \quad 0 < x - c < \delta$$

By Theorem 2.10.4, if the limit exists,

$$\lim_{x \to c^+} \frac{f(x) - f(c)}{x - c} \geq 0 \tag{2}$$

Similarly, if x is approaching c from the left, $x - c < 0$, and therefore

$$\frac{f(x) - f(c)}{x - c} \leq 0 \quad \text{whenever} \quad -\delta < x - c < 0$$

so that by Theorem 2.10.3, if the limit exists,

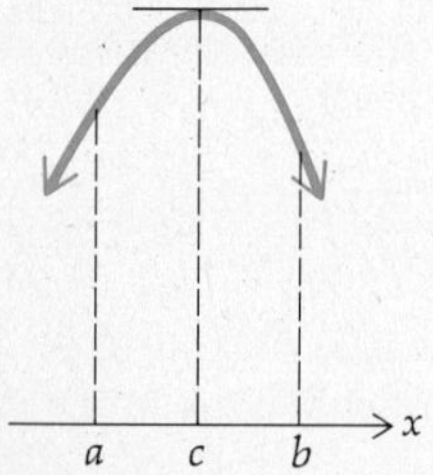

FIGURE 4.2.1

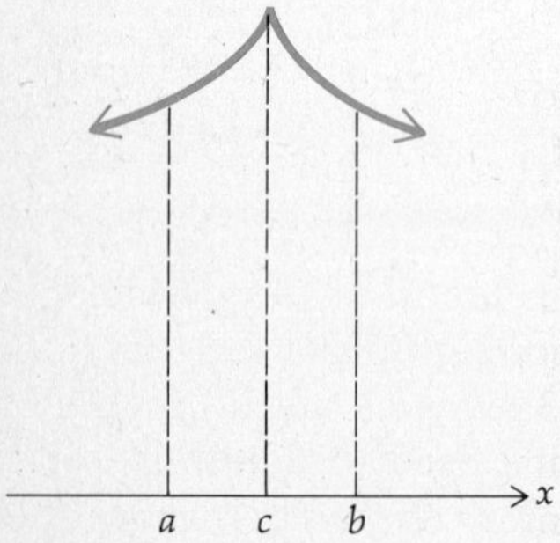

FIGURE 4.2.2

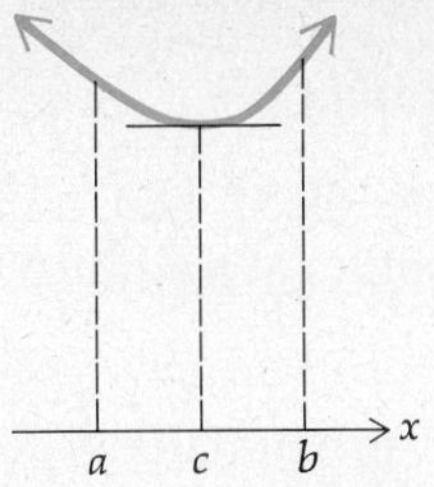

FIGURE 4.2.3

$$\lim_{x \to c^-} \frac{f(x) - f(c)}{x - c} \leq 0 \tag{3}$$

Because $f'(c)$ exists, the limits in inequalities (2) and (3) must be equal, and both must be equal to $f'(c)$. So from (2),

$$f'(c) \geq 0 \tag{4}$$

and from (3),

$$f'(c) \leq 0 \tag{5}$$

Because both (4) and (5) are taken to be true, we conclude that

$$f'(c) = 0$$

which was to be proved.

The proof for the case when f has a relative maximum value at c is similar and is left as an exercise (see Exercise 51). ■

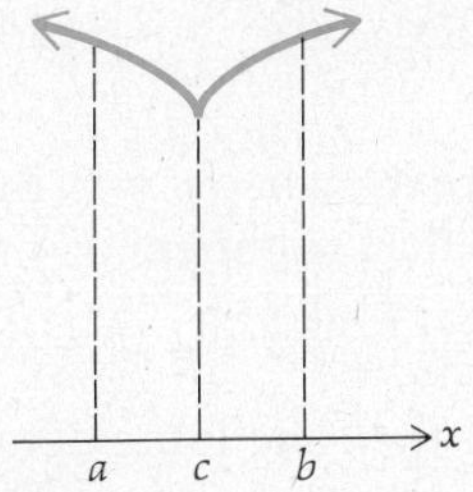

FIGURE 4.2.4

The geometrical interpretation of Theorem 4.2.3 is that if f has a relative extremum at c and if $f'(c)$ exists, then the graph of $y = f(x)$ must have a horizontal tangent line at the point where $x = c$.

If f is a differentiable function, then the only possible values of x for which f can have a relative extremum are those for which $f'(x) = 0$. However, $f'(x)$ can be equal to zero for a specific value of x, and yet f may not have a relative extremum there, as shown in the following illustration.

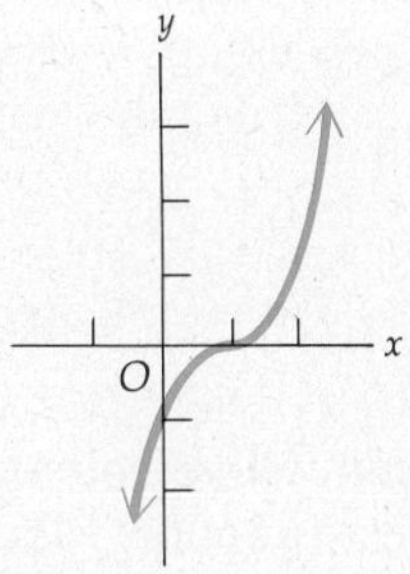

FIGURE 4.2.5

● ILLUSTRATION 1 Consider the function f defined by

$$f(x) = (x - 1)^3$$

A sketch of the graph of this function is shown in Fig. 4.2.5. $f'(x) = 3(x - 1)^2$, and so $f'(1) = 0$. However, $f(x) < 0$ if $x < 1$, and $f(x) > 0$ if $x > 1$. So f does not have a relative extremum at 1. ●

A function f may have a relative extremum at a number and f' may fail to exist there. This situation is shown in Illustration 2.

● ILLUSTRATION 2 Let the function f be defined as follows:

$$f(x) = \begin{cases} 2x - 1 & \text{if } x \leq 3 \\ 8 - x & \text{if } 3 < x \end{cases}$$

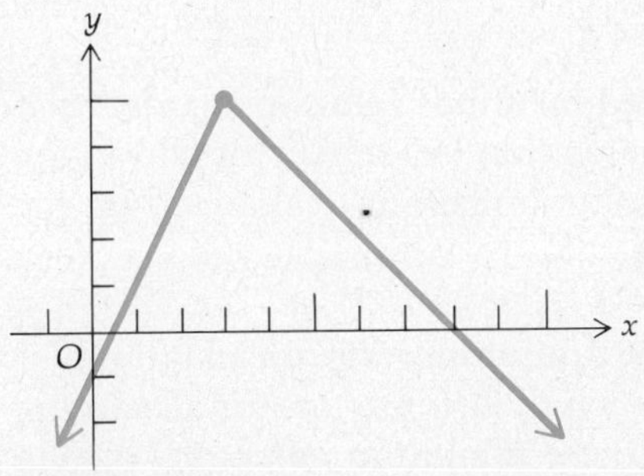

FIGURE 4.2.6

A sketch of the graph of this function is shown in Fig. 4.2.6. The function f has a relative maximum value at 3. The derivative from the left at 3 is given by $f'_-(3) = 2$, and the derivative from the right at 3 is given by $f'_+(3) = -1$. Therefore we conclude that $f'(3)$ does not exist. ●

Illustration 2 demonstrates why the condition "$f'(c)$ exists" must be included in the hypothesis of Theorem 4.2.3.

In summary, then, if a function f is defined at a number c, a necessary condition for f to have a relative extremum there is that either $f'(c) = 0$ or $f'(c)$ does not exist. But this condition is not sufficient.

4.2.4 Definition If c is a number in the domain of the function f and if either $f'(c) = 0$ or $f'(c)$ does not exist, then c is called a **critical number** of f.

Because of this definition and the previous discussion, a necessary condition for a function to have a relative extremum at a number c is for c to be a critical number.

Example 1 Find the critical numbers of the function f defined by $f(x) = x^{4/3} + 4x^{1/3}$.

Solution

$$f(x) = x^{4/3} + 4x^{1/3}$$

$$f'(x) = \tfrac{4}{3}x^{1/3} + \tfrac{4}{3}x^{-2/3} = \tfrac{4}{3}x^{-2/3}(x+1) = \frac{4(x+1)}{3x^{2/3}}$$

$f'(x) = 0$ when $x = -1$, and $f'(x)$ does not exist when $x = 0$. Both -1 and 0 are in the domain of f; therefore the critical numbers of f are -1 and 0.

Example 2 Find the critical numbers of the function g defined by $g(x) = \sin x \cos x$.

Solution Because $\sin 2x = 2 \sin x \cos x$,

$$g(x) = \tfrac{1}{2}\sin 2x$$

$$g'(x) = \tfrac{1}{2}\cos 2x(2) = \cos 2x$$

$g'(x) = 0$ when $\cos 2x = 0$, that is, when

$$2x = \tfrac{1}{2}\pi + n\pi, \qquad \text{where } n = 0, \pm 1, \pm 2, \ldots$$

Therefore the critical numbers of g are $\frac{1}{4}\pi + \frac{1}{2}n\pi$, where n is any integer.

We are frequently concerned with a function defined on a given interval, and we wish to find the largest or smallest function value on the interval. These intervals can be either closed, open, or closed at one end and open at the other. The greatest function value on an interval is called the *absolute maximum value*, and the smallest function value on an interval is called the *absolute minimum value*. Following are the precise definitions.

4.2.5 Definition The function f is said to have an **absolute maximum value on an interval** if there is some number c in the interval such that $f(c) \geq f(x)$ for all x in the interval. In such a case, $f(c)$ is the absolute maximum value of f on the interval.

4.2.6 Definition The function f is said to have an **absolute minimum value on an interval** if there is some number c in the interval such that $f(c) \leq f(x)$ for all x in the interval. In such a case, $f(c)$ is the absolute minimum value of f on the interval.

An **absolute extremum** of a function on an interval is either an absolute maximum value or an absolute minimum value of the function on the interval. A function may or may not have an absolute extremum on a given interval. In each of the following illustrations, a function and an interval are given, and we determine the absolute extrema of the function on the interval if there are any.

● ILLUSTRATION 3 Suppose f is the function defined by

$$f(x) = 2x$$

A sketch of the graph of f on $[1, 4)$ is shown in Fig. 4.2.7. The function f has an absolute minimum value of 2 on $[1, 4)$. There is no absolute maximum value of f on $[1, 4)$ because $\lim_{x \to 4^-} f(x) = 8$, but $f(x)$ is always less than 8 on the given interval. ●

● ILLUSTRATION 4 Consider the function f defined by

$$f(x) = -x^2$$

A sketch of the graph of f on $(-3, 2]$ is shown in Fig. 4.2.8. The function f has an absolute maximum value of 0 on $(-3, 2]$. There is no absolute minimum value of f on $(-3, 2]$ because $\lim_{x \to -3^+} f(x) = -9$, but $f(x)$ is always greater than -9 on the given interval. ●

● ILLUSTRATION 5 The function f defined by

$$f(x) = \frac{x}{1 - x^2}$$

has neither an absolute maximum value nor an absolute minimum value on $(-1, 1)$. A sketch of the graph of f on $(-1, 1)$ is shown in Fig. 4.2.9. Observe that

$$\lim_{x \to -1^+} f(x) = -\infty \quad \text{and} \quad \lim_{x \to 1^-} f(x) = +\infty$$ ●

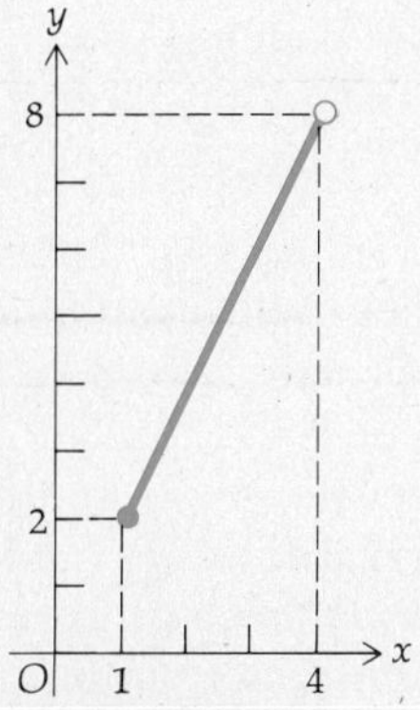

FIGURE 4.2.7

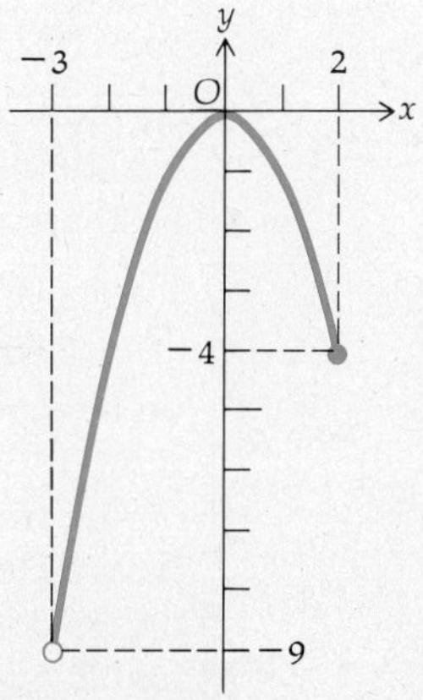

FIGURE 4.2.8

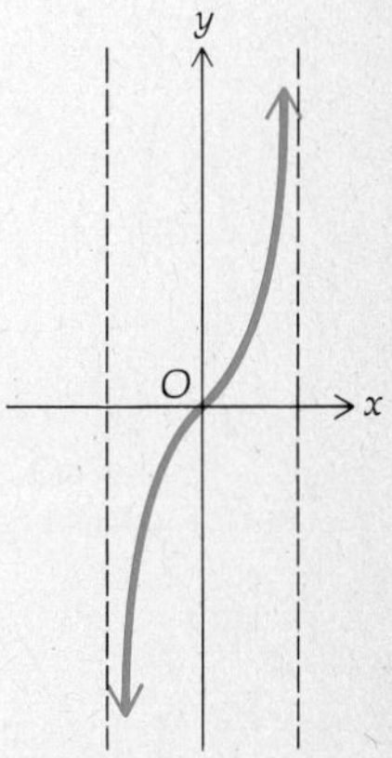

FIGURE 4.2.9

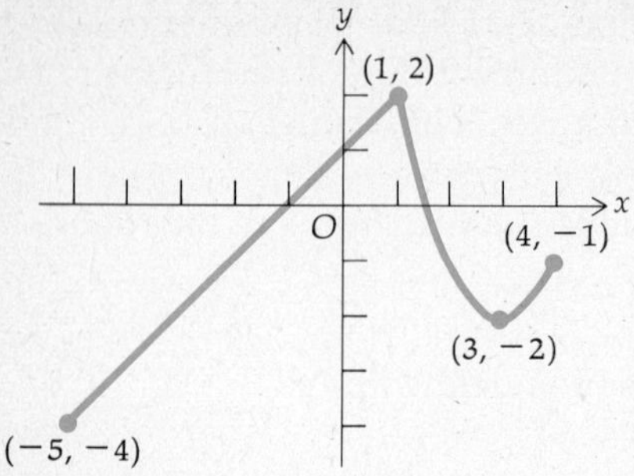

FIGURE 4.2.10

● ILLUSTRATION 6 Let f be the function defined by

$$f(x) = \begin{cases} x + 1 & \text{if } x < 1 \\ x^2 - 6x + 7 & \text{if } 1 \le x \end{cases}$$

A sketch of the graph of f on $[-5, 4]$ is shown in Fig. 4.2.10. The absolute maximum value of f on $[-5, 4]$ occurs at 1, and $f(1) = 2$; the absolute minimum value of f on $[-5, 4]$ occurs at -5, and $f(-5) = -4$. Note that f has a relative maximum value at 1 and a relative minimum value at 3. Also observe that 1 is a critical number of f because f' does not exist at 1, and 3 is a critical number of f because $f'(3) = 0$. ●

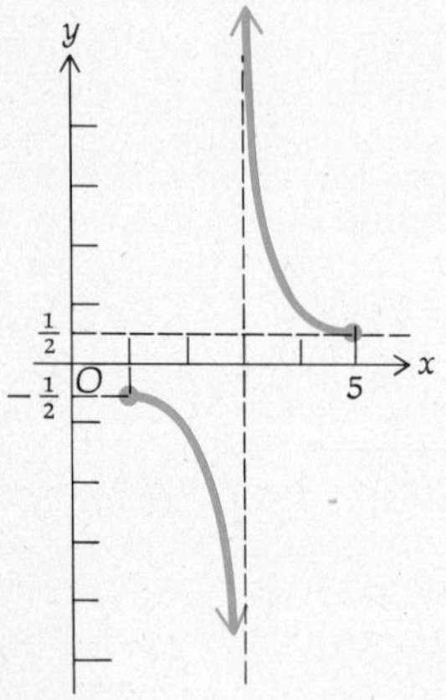

FIGURE 4.2.11

● ILLUSTRATION 7 The function f defined by

$$f(x) = \frac{1}{x - 3}$$

has neither an absolute maximum value nor an absolute minimum value on $[1, 5]$. See Fig. 4.2.11 for a sketch of the graph of f. $\lim\limits_{x \to 3^-} f(x) = -\infty$; so $f(x)$ can be made less than any negative number by taking $(3 - x) > 0$ and less than a suitable positive δ. Also, $\lim\limits_{x \to 3^+} f(x) = +\infty$; so $f(x)$ can be made greater than any positive number by taking $(x - 3) > 0$ and less than a suitable positive δ. ●

We may speak of an absolute extremum of a function when no interval is specified. In such a case we are referring to an absolute extremum of the function on the entire domain of the function.

4.2.7 Definition $f(c)$ is said to be the **absolute maximum value** of the function f if c is in the domain of f and if $f(c) \ge f(x)$ for all values of x in the domain of f.

4.2.8 Definition $f(c)$ is said to be the **absolute minimum value** of the function f if c is in the domain of f and if $f(c) \le f(x)$ for all values of x in the domain of f.

● ILLUSTRATION 8 The graph of the function f defined by

$$f(x) = x^2 - 4x + 8$$

is a parabola, and a sketch is shown in Fig. 4.2.12. The lowest point of the parabola is at $(2, 4)$, and the parabola opens upward. The function has an absolute minimum value of 4 at 2. There is no absolute maximum value of f. ●

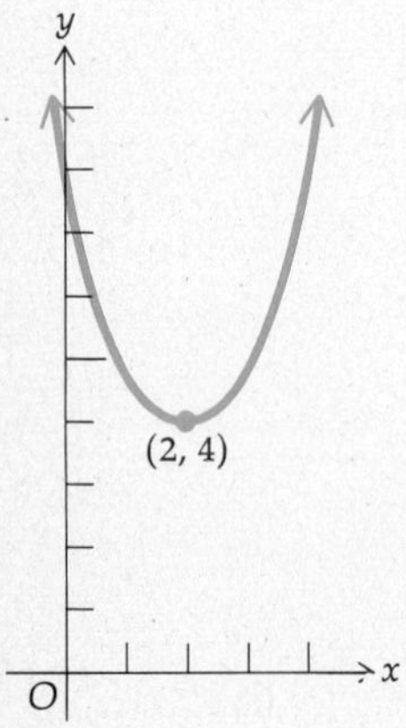

FIGURE 4.2.12

Referring back to Illustrations 3–8, we see that the only case in which there is both an absolute maximum function value and an absolute minimum function value is in Illustration 6, where the function is continuous on the closed interval $[-5, 4]$. In the other illustrations, either we do not have a closed interval or we do not have a continuous function. If a function is continuous on a closed interval, there is a theorem, called the *extreme-value theorem*, which assures that the function has both an absolute maximum value and an absolute minimum value on the interval. The proof of this theorem is beyond the scope of this book. You are referred to an advanced calculus text for the proof.

4.2.9 Theorem
Extreme-Value Theorem

If the function f is continuous on the closed interval $[a, b]$, then f has an absolute maximum value and an absolute minimum value on $[a, b]$.

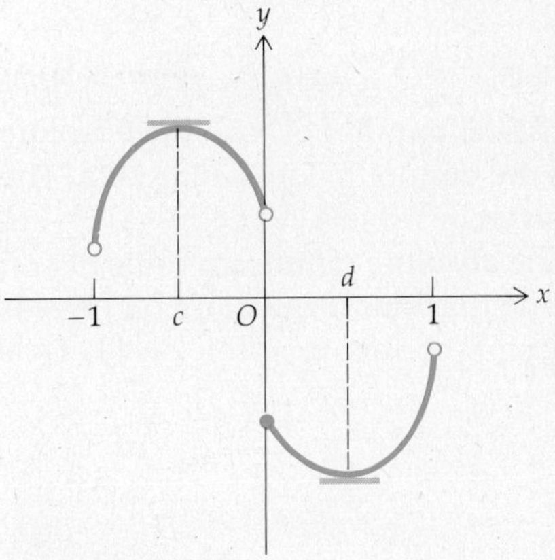

FIGURE 4.2.13

Theorem 4.2.9 states that continuity of a function on a closed interval is a sufficient condition to guarantee that the function has both an absolute maximum value and an absolute minimum value on the interval. However, it is not a necessary condition. For example, the function whose graph appears in Fig. 4.2.13 has an absolute maximum value at $x = c$ and an absolute minimum value at $x = d$, even though the function is discontinuous on the open interval $(-1, 1)$.

An absolute extremum of a function continuous on a closed interval must be either a relative extremum or a function value at an endpoint of the interval. Because a necessary condition for a function to have a relative extremum at a number c is for c to be a critical number, the absolute maximum value and the absolute minimum value of a continuous function f on a closed interval $[a, b]$ can be determined by the following procedure:

1. Find the function values at the critical numbers of f on $[a, b]$.
2. Find the values of $f(a)$ and $f(b)$.
3. The largest of the values from steps 1 and 2 is the absolute maximum value, and the smallest of the values from steps 1 and 2 is the absolute minimum value.

TABLE 4.2.1

x	-2	-1	$\frac{1}{3}$	$\frac{1}{2}$
$f(x)$	-1	2	$\frac{22}{27}$	$\frac{7}{8}$

Example 3 Given $f(x) = x^3 + x^2 - x + 1$, find the absolute extrema of f on $[-2, \frac{1}{2}]$.

Solution Because f is continuous on $[-2, \frac{1}{2}]$, the extreme-value theorem applies. To find the critical numbers of f, first find f':

$$f(x) = x^3 + x^2 - x + 1 \qquad f'(x) = 3x^2 + 2x - 1$$

$f'(x)$ exists for all real numbers, and so the only critical numbers of f will be the values of x for which $f'(x) = 0$. Set $f'(x) = 0$.

$$(3x - 1)(x + 1) = 0$$

Then

$$x = \tfrac{1}{3} \quad \text{and} \quad x = -1$$

The critical numbers of f are -1 and $\frac{1}{3}$, and each of these numbers is in the given closed interval $[-2, \frac{1}{2}]$. The function values at the critical numbers and at the endpoints of the interval are given in Table 4.2.1.

The absolute maximum value of f on $[-2, \frac{1}{2}]$ is therefore 2, which occurs at -1, and the absolute minimum value of f on $[-2, \frac{1}{2}]$ is -1, which occurs at the left endpoint -2. A sketch of the graph of this function on $[-2, \frac{1}{2}]$ is shown in Fig. 4.2.14.

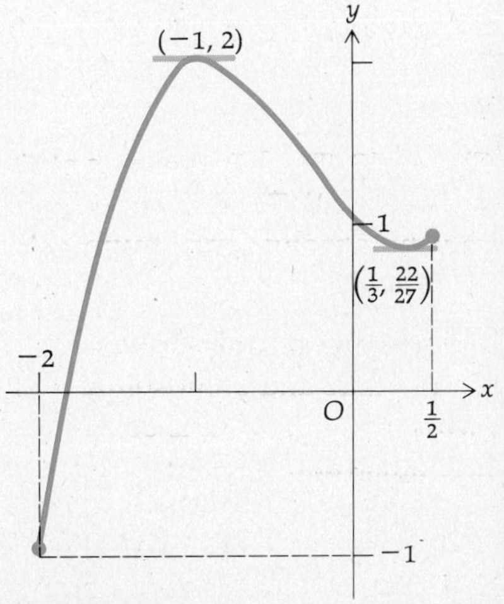

FIGURE 4.2.14

TABLE 4.2.2

x	1	2	5
$f(x)$	1	0	$\sqrt[3]{9}$

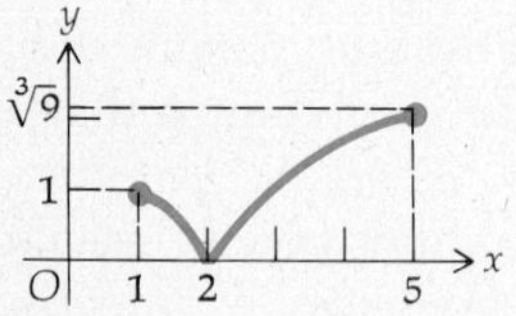

FIGURE 4.2.15

Example 4 Given $f(x) = (x - 2)^{2/3}$, find the absolute extrema of f on $[1, 5]$.

Solution Because f is continuous on $[1, 5]$, the extreme-value theorem applies.

$$f(x) = (x - 2)^{2/3} \qquad f'(x) = \frac{2}{3(x-2)^{1/3}}$$

There is no value of x for which $f'(x) = 0$. However, because $f'(x)$ does not exist at 2, we conclude that 2 is a critical number of f; so the absolute extrema occur either at 2 or at one of the endpoints of the interval. The function values at these numbers are given in Table 4.2.2.

From the table we conclude that the absolute minimum value of f on $[1, 5]$ is 0, occurring at 2, and the absolute maximum value of f on $[1, 5]$ is $\sqrt[3]{9}$, occurring at 5. A sketch of the graph of this function on $[1, 5]$ is shown in Fig. 4.2.15.

EXERCISES 4.2

In Exercises 1 through 16, find the critical numbers of the given function.

1. $f(x) = x^3 + 7x^2 - 5x$
2. $g(x) = 2x^3 - 2x^2 - 16x + 1$
3. $f(x) = x^4 + 4x^3 - 2x^2 - 12x$
4. $f(x) = x^{7/3} + x^{4/3} - 3x^{1/3}$
5. $g(x) = x^{6/5} - 12x^{1/5}$
6. $f(x) = x^4 + 11x^3 + 34x^2 + 15x - 2$
7. $f(t) = (t^2 - 4)^{2/3}$
8. $f(x) = (x^3 - 3x^2 + 4)^{1/3}$
9. $h(x) = \dfrac{x-3}{x+7}$
10. $f(t) = t^{5/3} - 3t^{2/3}$
11. $f(x) = \dfrac{x}{x^2 - 9}$
12. $f(x) = \dfrac{x+1}{x^2 - 5x + 4}$
13. $f(x) = \sin^2 3x$
14. $f(z) = \cos^2 4z$
15. $g(t) = \sin 2t \cos 2t$
16. $f(x) = \sin 2x + \cos 2x$

In Exercises 17 through 34, find the absolute extrema of the given function on the given interval, if there are any, and find the values of x at which the absolute extrema occur. Draw a sketch of the graph of the function on the interval.

17. $f(x) = 4 - 3x;\ (-1, 2]$
18. $f(x) = x^2 - 2x + 4;\ (-\infty, +\infty)$
19. $g(x) = \dfrac{1}{x};\ [-2, 3]$
20. $f(x) = \dfrac{1}{x};\ [2, 3)$
21. $f(x) = 2\cos x;\ [-\frac{2}{3}\pi, \frac{1}{3}\pi)$
22. $G(x) = -3\sin x;\ [0, \frac{3}{4}\pi)$
23. $f(x) = \sqrt{3 + x};\ [-3, +\infty)$
24. $f(x) = \sqrt{4 - x^2};\ (-2, 2)$
25. $h(x) = \dfrac{4}{(x-3)^2};\ [2, 5]$
26. $g(x) = \dfrac{3x}{9 - x^2};\ (-3, 2)$
27. $F(x) = |x - 4| + 1;\ (0, 6)$
28. $f(x) = |4 - x^2|;\ (-\infty, +\infty)$
29. $g(x) = \sqrt{4 + 7x};\ [0, 3)$
30. $F(x) = U(x) - U(x - 1);\ (-1, 1)$ (U is the unit-step function)
31. $f(x) = \begin{cases} \dfrac{2}{x-5} & \text{if } x \neq 5 \\ 2 & \text{if } x = 5 \end{cases};\ [3, 5]$
32. $f(x) = \begin{cases} |x + 1| & \text{if } x \neq -1 \\ 3 & \text{if } x = -1 \end{cases};\ [-2, 1]$
33. $f(x) = x - [\![x]\!];\ (1, 3)$
34. $h(x) = 2x + [\![2x - 1]\!];\ (1, 2)$

In Exercises 35 through 50, find the absolute maximum value and the absolute minimum value of the given function on the indicated interval by the method used in Examples 3 and 4 of this section. Draw a sketch of the graph of the function on the interval.

35. $g(x) = x^3 + 5x - 4; [-3, -1]$

36. $f(x) = x^3 + 3x^2 - 9x; [-4, 4]$

37. $f(x) = x^4 - 8x^2 + 16; [-4, 0]$

38. $f(x) = x^4 - 8x^2 + 16; [-3, 2]$

39. $f(x) = x^4 - 8x^2 + 16; [0, 3]$

40. $g(x) = x^4 - 8x^2 + 16; [-1, 4]$

41. $f(t) = 2 \sin t; [-\pi, \pi]$

42. $f(w) = 3 \cos 2w; [\frac{1}{6}\pi, \frac{3}{4}\pi]$

43. $f(x) = \dfrac{x}{x+2}; [-1, 2]$

44. $f(x) = \dfrac{x+5}{x-3}; [-5, 2]$

45. $f(x) = \dfrac{x+1}{2x-3}; [0, 1]$

46. $f(x) = \begin{cases} 2x - 7 & \text{if } -1 \le x \le 2 \\ 1 - x^2 & \text{if } 2 < x \le 4 \end{cases}; [-1, 4]$

47. $F(x) = \begin{cases} 3x - 4 & \text{if } -3 \le x < 1 \\ x^2 - 2 & \text{if } 1 \le x \le 3 \end{cases}; [-3, 3]$

48. $G(x) = \begin{cases} 4 - (x+5)^2 & \text{if } -6 \le x \le -4 \\ 12 - (x+1)^2 & \text{if } -4 < x \le 0 \end{cases}; [-6, 0]$

49. $f(z) = (z + 1)^{2/3}; [-2, 1]$

50. $f(x) = 1 - (x - 3)^{2/3}; [-5, 4]$

51. Prove Theorem 4.2.3 for the case when f has a relative maximum value at c.

4.3 APPLICATIONS INVOLVING AN ABSOLUTE EXTREMUM ON A CLOSED INTERVAL

We consider some problems in which the solution is an absolute extremum of a function on a closed interval. Use is made of the extreme-value theorem, which assures that both an absolute maximum value and an absolute minimum value of a function exist on a closed interval if the function is continuous on that closed interval. The procedure is illustrated by some examples.

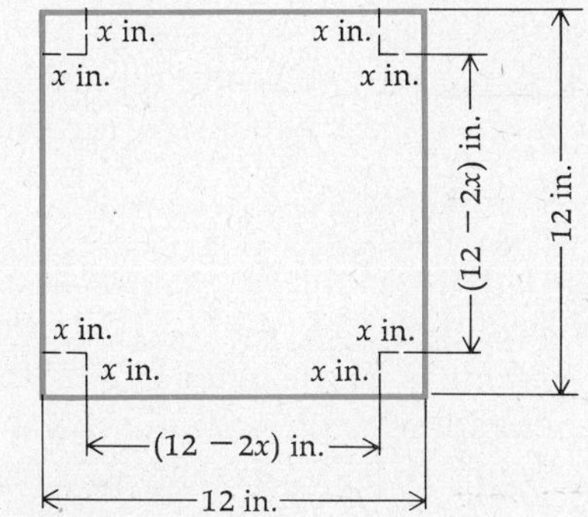

FIGURE 4.3.1

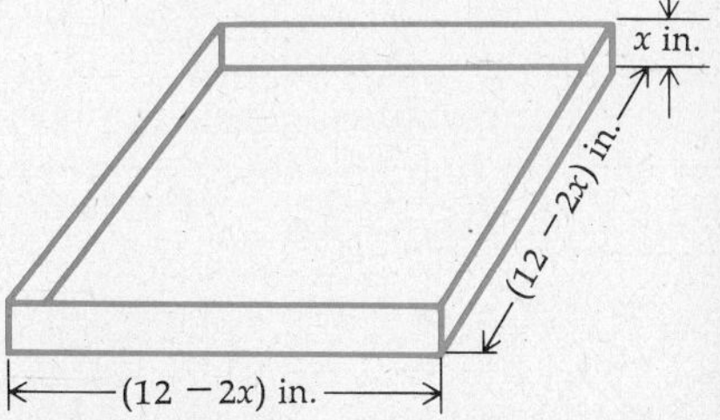

FIGURE 4.3.2

Example 1 A cardboard box manufacturer wishes to make open boxes from pieces of cardboard 12 in. square by cutting equal squares from the four corners and turning up the sides. Find the length of the side of the square to be cut out in order to obtain a box of the largest possible volume.

Solution Let x be the number of inches in the length of the side of the square to be cut out. Let $V(x)$ in.^{3} be the volume of the box. The number of inches in the dimensions of the box are then x, $(12 - 2x)$, and $(12 - 2x)$. Figure 4.3.1 represents a given piece of cardboard, and Fig. 4.3.2 represents the box.

The volume of the box is the product of the three dimensions; so

$$V(x) = x(12 - 2x)(12 - 2x) \tag{1}$$

$$V(0) = 0 \quad \text{and} \quad V(6) = 0$$

The value of x that we wish to find is in the closed interval $[0, 6]$. Because V is continuous on the closed interval $[0, 6]$, it follows from the extreme-value theorem that V has an absolute maximum value on this interval. We also know that this absolute maximum value of V must occur either at a critical number or at an endpoint of the interval. To find the critical numbers of V we find $V'(x)$ and then find the values of x for which either $V'(x) = 0$ or $V'(x)$ does not exist.

From (1),

$$V(x) = 144x - 48x^2 + 4x^3$$

Thus

$$V'(x) = 144 - 96x + 12x^2$$

$V'(x)$ exists for all values of x. If $V'(x) = 0$,

$$12(x^2 - 8x + 12) = 0$$

from which

$$x = 6 \quad \text{and} \quad x = 2$$

The critical numbers of V are 2 and 6, both of which are in the closed interval $[0, 6]$. The absolute maximum value of V on $[0, 6]$ must occur at either a critical number or at an endpoint of the interval. Because $V(0) = 0$ and $V(6) = 0$, while $V(2) = 128$, the absolute maximum value of V on $[0, 6]$ is 128, occurring at 2.

Therefore the largest possible volume is 128 in.^{3}, and this is obtained when the length of the side of the square cut out is 2 in.

Example 2 Points A and B are opposite each other on shores of a straight river 3 km wide. Point C is on the same shore as B but 2 km down the river from B. A telephone company wishes to lay a cable from A to C. If the cost per kilometer of the cable is 25 percent more under the water than it is on land, what line of cable would be least expensive for the company?

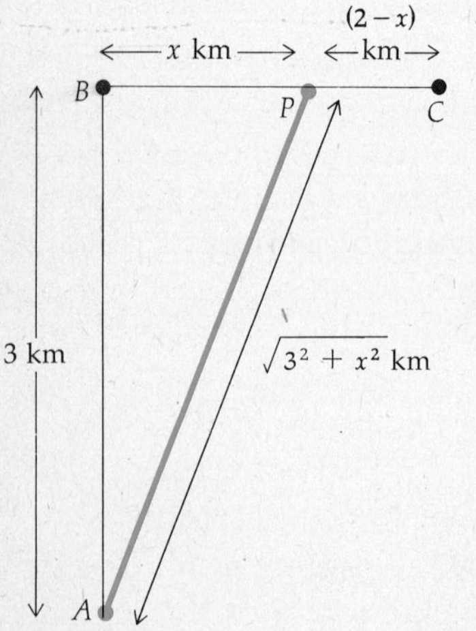

FIGURE 4.3.3

Solution Refer to Fig. 4.3.3. Let P be a point on the same shore as B and C and between B and C so that the cable will run from A to P to C. Let x km be the distance from B to P. Then $(2 - x)$ km is the distance from P to C, and $x \in [0, 2]$. Let k dollars be the cost per kilometer on land and $\frac{5}{4}k$ dollars be the cost per kilometer under the water (k is a constant). If $C(x)$ dollars is the total cost of running the cable from A to P and from P to C, then

$$C(x) = \tfrac{5}{4}k\sqrt{3^2 + x^2} + k(2 - x)$$

Because C is continuous on $[0, 2]$, the exteme-value theorem applies; thus C has both an absolute maximum value and an absolute minimum value on $[0, 2]$. We wish to find the absolute minimum value.

$$C'(x) = \frac{5kx}{4\sqrt{9 + x^2}} - k$$

$C'(x)$ exists for all values of x. Setting $C'(x) = 0$ and solving for x we have

$$\frac{5kx}{4\sqrt{9+x^2}} - k = 0$$ — k being a constant

$$5x = 4\sqrt{9+x^2} \tag{2}$$

$$25x^2 = 16(9+x^2)$$

$$9x^2 = 16 \cdot 9$$

$$x^2 = 16$$

$$x = \pm 4$$

The number -4 is an extraneous root of Eq. (2), and 4 is not in the interval [0, 2]. Therefore there are no critical numbers of C in [0, 2]. The absolute minimum value of C on [0, 2] must therefore occur at an endpoint of the interval. Computing $C(0)$ and $C(2)$ we get

$$C(0) = \tfrac{23}{4}k \quad \text{and} \quad C(2) = \tfrac{5}{4}k\sqrt{13}$$

Because $\frac{5}{4}k\sqrt{13} < \frac{23}{4}k$, the absolute minimum value of C on [0, 2] is $\frac{5}{4}k\sqrt{13}$, occurring when $x = 2$. Therefore, for the cost of the cable to be the least, the cable should go directly from A to C under the water.

Example 3 A rectangular field is to be fenced off along the bank of a river; no fence is required along the river. If the material for the fence costs \$8 per running foot for the two ends and \$12 per running foot for the side parallel to the river, find the dimensions of the field of largest possible area that can be enclosed with \$3600 worth of fence.

Solution Let x be the number of feet in the length of an end of the field, y be the number of feet in the length of the side parallel to the river, and A ft² be the area of the field. Hence Area = l × w

$$A = xy \tag{3}$$

Because the cost of the material for each end is \$8 per running foot and the length of an end is x ft, the total cost for the fence for each end is $8x$ dollars. Similarly, the total cost of the fence for the third side is $12y$ dollars. Then

$$8x + 8x + 12y = 3600 \tag{4}$$

To express A in terms of a single variable we solve (4) for y in terms of x and substitute this value into (3), yielding A as a function of x, and A=xy

$$A(x) = x(300 - \tfrac{4}{3}x) \tag{5}$$

From (4), if $y = 0$, $x = 225$, and if $x = 0$, $y = 300$. Because both x and y must be nonnegative, the value of x that will make A an absolute maximum is in the closed interval [0, 225]. Because A is continuous on the closed interval [0, 225], from the extreme-value theorem A has an absolute maximum value on this interval. From (5),

$$A(x) = 300x - \tfrac{4}{3}x^2$$

Hence

$$A'(x) = 300 - \tfrac{8}{3}x$$

Because $A'(x)$ exists for all x, the critical numbers of A are found by setting $A'(x) = 0$, which gives

$$x = 112\tfrac{1}{2}$$

The only critical number of A is $112\frac{1}{2}$, which is in the closed interval $[0, 225]$. Thus the absolute maximum value of A must occur at either 0, $112\frac{1}{2}$, or 225. Because $A(0) = 0$ and $A(225) = 0$, while $A(112\frac{1}{2}) = 16{,}875$, the absolute maximum value of A on $[0, 225]$ is 16,875, occurring when $x = 112\frac{1}{2}$ and $y = 150$ (obtained from (4) by substituting $112\frac{1}{2}$ for x).

Therefore the largest possible area that can be enclosed for \$3600 is 16,875 square feet, and this is obtained when the side parallel to the river is 150 ft long and the ends are each $112\frac{1}{2}$ ft long.

Example 4 In the planning of a coffee shop it is estimated that if there are places for 40 to 80 people, the daily gross profit will be \$8 per place. However, if the seating capacity is above 80 places, the daily gross profit on each place will be decreased by \$0.04 times the number of places above 80. What should be the seating capacity in order to yield the greatest daily gross profit?

Solution Let x be the number of places in the seating capacity and $P(x)$ dollars be the daily gross profit. $P(x)$ is obtained by multiplying x by the number of dollars in the profit per place. When $40 \le x \le 80$, \$8 is the profit per place; so $P(x) = 8x$. However, when $x > 80$, the number of dollars in the profit per place is $[8 - 0.04(x - 80)]$, thus giving $P(x) = x[8 - 0.04(x - 80)] = 11.20x - 0.04x^2$. So

$$P(x) = \begin{cases} 8x & \text{if } 40 \le x \le 80 \\ 11.20x - 0.04x^2 & \text{if } 80 < x \le 280 \end{cases}$$

The upper bound of 280 for x is obtained by noting that $11.20x - 0.04x^2 = 0$ when $x = 280$; and when $x > 280$, $11.20x - 0.04x^2$ is negative.

Even though x, by definition, is an integer, to have a continuous function we let x take on all real values in the interval $[40, 280]$. There is continuity at 80 because

$$\lim_{x \to 80^-} P(x) = \lim_{x \to 80^-} 8x = 640$$

and

$$\lim_{x \to 80^+} P(x) = \lim_{x \to 80^+} (11.20x - 0.04x^2) = 640$$

from which it follows that the two-sided limit $\lim_{x \to 80} P(x) = 640 = P(80)$. So P is continuous on the closed interval $[40, 280]$, and the extreme-value theorem guarantees an absolute maximum value of P on this interval.

When $40 < x < 80$, $P'(x) = 8$, and when $80 < x < 280$, $P'(x) = 11.20 - 0.08x$. $P'(80)$ does not exist since $P'_-(80) = 8$ and $P'_+(80) = 4.80$. Set $P'(x) = 0$.

$$11.20 - 0.08x = 0$$
$$x = 140$$

The critical numbers of P are then 80 and 140. We evaluate $P(x)$ at the endpoints of the interval [40, 280] and at the critical numbers.

$$P(40) = 320 \qquad P(80) = 640 \qquad P(140) = 784 \qquad P(280) = 0$$

The absolute maximum value of P, then, is 784, occurring when $x = 140$.

The seating capacity should be 140 places, which gives a daily gross profit of \$784.

Example 5 Find the dimensions of the right-circular cylinder of greatest volume that can be inscribed in a right-circular cone with a radius of 5 cm and a height of 12 cm.

Solution Let r be the number of centimeters in the radius of the cylinder, h be the number of centimeters in the height of the cylinder, and V be the number of cubic centimeters in the volume of the cylinder.

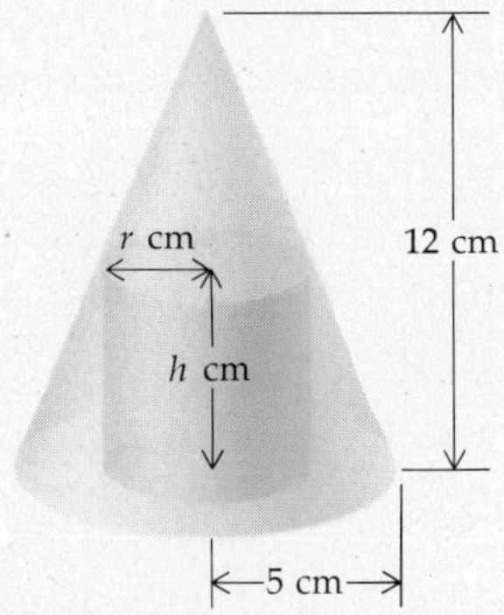

FIGURE 4.3.4

Figure 4.3.4 illustrates the cylinder inscribed in the cone, and Fig. 4.3.5 illustrates a plane section through the axis of the cone.

If $r = 0$ and $h = 12$ we have a degenerate cylinder, which is the axis of the cone. If $r = 5$ and $h = 0$, we also have a degenerate cylinder, which is a diameter of the base of the cone. r is in the closed interval [0, 5] and h is in the closed interval [0, 12].

The following formula expresses V in terms of r and h:

$$V = \pi r^2 h \tag{6}$$

To express V in terms of a single variable we need another equation involving r and h. From Fig. 4.3.5, and by similar triangles,

$$\frac{12 - h}{r} = \frac{12}{5}$$
$$h = \frac{60 - 12r}{5} \tag{7}$$

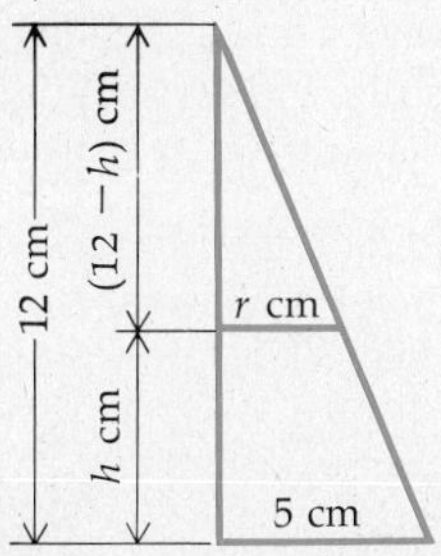

FIGURE 4.3.5

Substituting from (7) into formula (6), we obtain V as a function of r and write

$$V(r) = \tfrac{12}{5}\pi(5r^2 - r^3) \qquad \text{with } r \text{ in } [0, 5] \tag{8}$$

Because V is continuous on the closed interval [0, 5], it follows from the extreme-value theorem that V has an absolute maximum value on this interval. The values of r and h that give this absolute maximum value for V are the numbers to find.

$$V'(r) = \tfrac{12}{5}\pi(10r - 3r^2)$$

To find the critical numbers of V, set $V'(r) = 0$ and solve for r:

$$r(10 - 3r) = 0$$

from which

$$r = 0 \quad \text{and} \quad r = \tfrac{10}{3}$$

Because $V'(r)$ exists for all values of r, the only critical numbers of V are 0 and $\frac{10}{3}$, both of which are in the closed interval $[0, 5]$. The absolute maximum value of V on $[0, 5]$ must occur at either 0, $\frac{10}{3}$, or 5. From (8) we obtain

$$V(0) = 0 \qquad V(\tfrac{10}{3}) = \tfrac{400}{9}\pi \qquad V(5) = 0$$

Therefore the absolute maximum value of V is $\frac{400}{9}\pi$, and this occurs when $r = \frac{10}{3}$. When $r = \frac{10}{3}$, we find from (7) that $h = 4$.

Thus the greatest volume of an inscribed cylinder in the given cone is $\frac{400}{9}\pi$ cm^3, which occurs when the radius is $\frac{10}{3}$ cm and the height is 4 cm.

EXERCISES 4.3

1. Find the area of the largest rectangle having a perimeter of 200 cm.

2. Find the area of the largest isosceles triangle having a perimeter of 18 m. a=b·h

3. Find the number in the interval $[0, 1]$ such that the difference between the number and its square is a maximum.

4. Find the number in the interval $[\frac{1}{3}, 2]$ such that the sum of the number and its reciprocal is a maximum.

5. Find the area of the largest rectangle having two vertices on the x axis and two vertices on or above the x axis and on the parabola $y = 9 - x^2$.

6. Find the area of the largest rectangle that can be inscribed in a given circle of radius r.

7. A manufacturer of tin boxes wishes to make use of pieces of tin with dimensions 8 in. by 15 in. by cutting equal squares from the four corners and turning up the sides. Find the length of the side of the square to be cut out if an open box having the largest possible volume is to be obtained from each piece of tin.

8. A rectangular plot of ground is to be enclosed by a fence and then divided down the middle by another fence. If the fence down the middle costs \$2 per running foot and the other fence costs \$5 per running foot, find the dimensions of the plot of largest possible area that can be enclosed with \$960 worth of fence.

9. A rectangular plot of ground is to be enclosed by a fence on three sides with a river serving as a natural boundary for the fourth side. Under these conditions, find the dimensions of the largest plot that can be enclosed with 240 m of fence.

10. For a package to be mailed by parcel post, the sum of the length and girth (the perimeter of a cross section) must not be greater than 100 in. If a package is to be in the shape of a rectangular box with a square cross section, find the dimensions of the package having the greatest possible volume that can be sent by parcel post.

11. An island is at point A, 6 km offshore from the nearest point B on a straight beach. A woman on the island wishes to go to a point C, 9 km down the beach from B. The woman can rent a boat for \$2.50 per kilometer and travel by water to a point P between B and C, and then she can take a taxi at a cost of \$2 per kilometer and travel a straight road from P to C. Find the least expensive route from point A to point C.

12. Solve Exercise 11 if point C is only 7 km down the beach from B.

13. Solve Example 2 of this section if point C is 6 km down the river from B.

14. Example 2 and Exercises 11, 12, and 13 are special cases of the following more general problem. Let $f(x) = u\sqrt{a^2 + x^2} + v(b - x)$, where x is in $[0, b]$ and $u > v > 0$. Show that in order for the absolute minimum value of f to occur at a number in the open interval $(0, b)$ the following inequality must be satisfied: $av < b\sqrt{u^2 - v^2}$.

15. If R ft is the range of a projectile, then

$$R = \frac{v_0^2 \sin 2\theta}{g} \qquad 0 \le \theta \le \tfrac{1}{2}\pi$$

where v_0 ft/sec is the initial velocity, g ft/sec^2 is the acceleration due to gravity, and θ is the radian measure of the angle that the gun makes with the horizontal. Find the value of θ that makes the range a maximum.

16. If a body of weight W lb is dragged along a horizontal floor by means of a force of magnitude F lb and directed at an angle of θ radians with the plane of the floor, then F is given by the equation

$$F = \frac{kW}{k \sin \theta + \cos \theta}$$

where k is a constant called the coefficient of friction and $0 < k < 1$. If $0 \le \theta \le \frac{1}{2}\pi$, find $\cos \theta$ when F is least.

17. A school-sponsored trip will cost each student \$15 if not more than 150 students make the trip; however, the cost per student will be reduced \$0.05 for each student in excess of 150. How many students should make the trip in order for the school to receive the largest gross income?

18. Solve Exercise 17 if the reduction per student in excess of 150 is \$0.07.

19. A private club charges annual membership dues of \$100 per member less \$0.50 for each member over 600 and plus \$0.50 for each member less than 600. How many members will give the club the most revenue from annual dues?

20. A manufacturer can make a profit of \$20 on each item if not more than 800 items are produced each week. The profit decreases \$0.02 per item over 800. How many items should the manufacturer produce each week in order to have the greatest profit?

21. Orange trees grown in California produce 600 oranges per year if no more than 20 trees are planted per acre. For each additional tree planted per acre, the yield per tree decreases by 15 oranges. How many trees per acre should be planted in order to obtain the greatest number of oranges?

22. Show that the largest rectangle having a given perimeter of p units is a square.

23. Find the dimensions of the right-circular cylinder of greatest lateral surface area that can be inscribed in a sphere with a radius of 6 in.

24. Find the dimensions of the right-circular cylinder of greatest volume that can be inscribed in a sphere with a radius of 6 in.

25. Two products A and B are manufactured at a particular factory. If C dollars is the total cost of production for an 8-hr day, then $C = 3x^2 + 42y$, where x machines are used to produce product A and y machines are used to produce product B. If during an 8-hr day there are 15 machines working, determine how many of these machines should be used to produce A and how many should be used to produce B in order for the total cost to be least.

26. A manufacturer of boxes wishes to make open boxes from square pieces of sheet metal that measure k cm on a side by cutting equal squares from the four corners and turning up the sides. Find the length of the side of the square to be cut out in order to obtain a box of largest possible volume.

27. Given the circle having the equation $x^2 + y^2 = 9$, find (a) the shortest distance from the point (4, 5) to a point on the circle; (b) the longest distance from the point (4, 5) to a point on the circle.

28. Suppose a weight is to be held 10 ft below a horizontal line AB by a wire in the shape of a Y. If the points A and B are 8 ft apart, what is the shortest total length of wire that can be used?

29. A piece of wire 10 ft long is cut into two pieces. One piece is bent into the shape of a circle and the other into the shape of a square. How should the wire be cut so that (a) the combined area of the two figures is as small as possible; (b) the combined area of the two figures is as large as possible?

30. Solve Exercise 29 if one piece of wire is bent into the shape of an equilateral triangle and the other piece is bent into the shape of a square.

4.4 ROLLE'S THEOREM AND THE MEAN-VALUE THEOREM

Let f be a function that is continuous on the closed interval $[a, b]$, differentiable on the open interval (a, b), and such that $f(a) = 0$ and $f(b) = 0$. The French mathematician Michel Rolle (1652–1719) proved that if a function f satisfies these conditions, there is at least one number c between a and b for which $f'(c) = 0$.

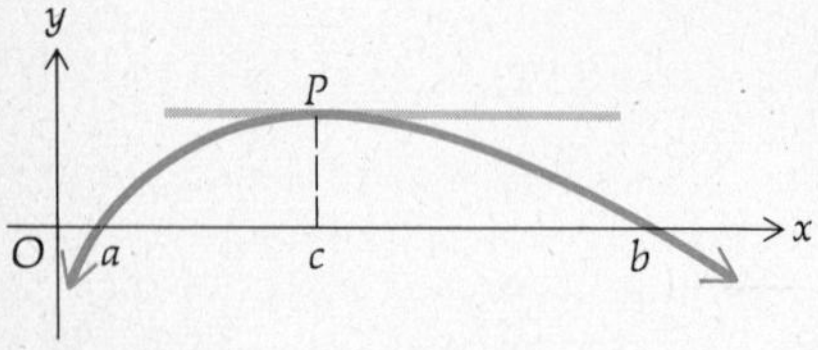

FIGURE 4.4.1

Let us see what this means geometrically. Figure 4.4.1 shows a sketch of the graph of a function f that satisfies the conditions in the preceding paragraph. We see intuitively that there is at least one point on the curve between the points $(a, 0)$ and $(b, 0)$ where the tangent line is parallel to the x axis; that is, the slope of the tangent line is zero. This situation is illustrated in Fig. 4.4.1 at the point P. So the abscissa of P is the c such that $f'(c) = 0$.

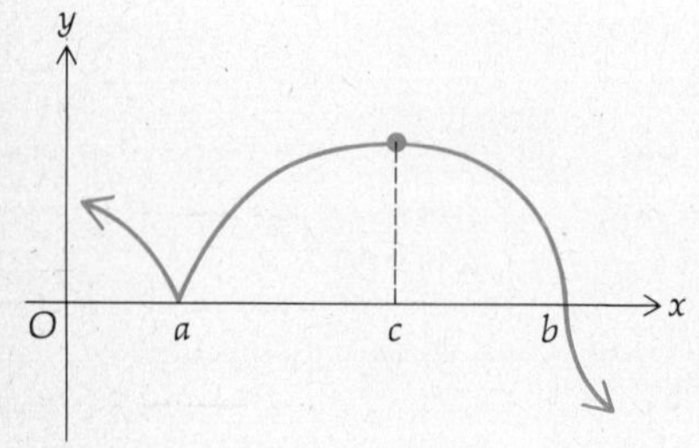

FIGURE 4.4.2

The function, whose graph is sketched in Fig. 4.4.1, not only is differentiable on the open interval (a, b) but also is differentiable at the endpoints of the interval. However, the condition that f be differentiable at the endpoints is not necessary for the graph to have a horizontal tangent line at some point in the interval; Fig. 4.4.2 illustrates this. We see in Fig. 4.4.2 that the function is not differentiable at a and b; there is, however, a horizontal tangent line at the point where $x = c$, and c is between a and b.

It is necessary, however, that the function be continuous at the endpoints of the interval to guarantee a horizontal tangent line at an interior point. Figure 4.4.3 shows a sketch of the graph of a function that is continuous on the interval $[a, b)$ but discontinuous at b; the function is differentiable on the open interval (a, b), and the function values are zero at both a and b. However, there is no point at which the graph has a horizontal tangent line.

We now state and prove Rolle's theorem.

4.4.1 Theorem **Rolle's Theorem**

Let f be a function such that

(i) it is continuous on the closed interval $[a, b]$;
(ii) it is differentiable on the open interval (a, b);
(iii) $f(a) = f(b) = 0$.

Then there is a number c in the open interval (a, b) such that

$$f'(c) = 0$$

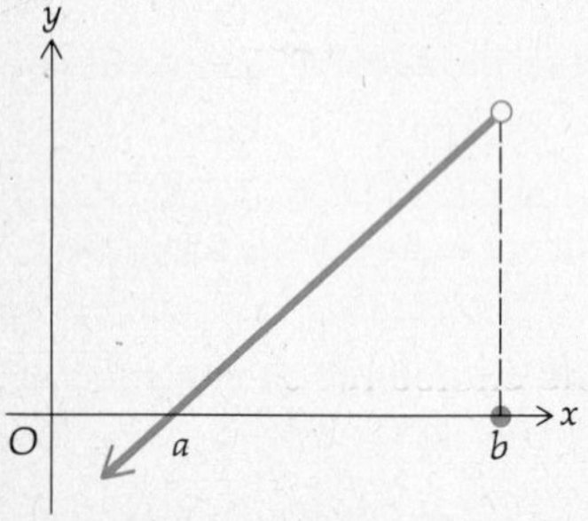

FIGURE 4.4.3

Proof Consider two cases.

Case 1: $f(x) = 0$ for all x in $[a, b]$.

Then $f'(x) = 0$ for all x in (a, b); therefore any number between a and b can be taken for c.

Case 2: $f(x)$ is not zero for some value of x in the open interval (a, b).

Because f is continuous on the closed interval $[a, b]$, from Theorem 4.2.9, f has an absolute maximum value on $[a, b]$ and an absolute minimum value on $[a, b]$. By hypothesis, $f(a) = f(b) = 0$. Furthermore, $f(x)$ is not zero for some x in (a, b). Hence f will have either a positive absolute maximum value at some c_1 in (a, b) or a negative absolute minimum value at some c_2 in (a, b), or both. Thus for $c = c_1$, or $c = c_2$ as the case may be, there is an absolute extremum at an interior point of the interval $[a, b]$. Therefore the absolute extremum $f(c)$ is also a relative extremum, and because $f'(c)$ exists by hypothesis, it follows from Theorem 4.2.3 that $f'(c) = 0$. This proves the theorem. ■

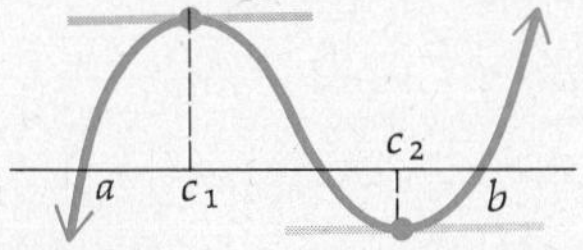

FIGURE 4.4.4

There may be more than one number in the open interval (a, b) for which the derivative of f is zero. This is illustrated geometrically in Fig. 4.4.4, where there is a horizontal tangent line at the point where $x = c_1$ and also at the point where $x = c_2$, so that both $f'(c_1) = 0$ and $f'(c_2) = 0$.

The converse of Rolle's theorem is not true. That is, we cannot conclude that if a function f is such that $f'(c) = 0$, with $a < c < b$, then the conditions (i), (ii), and (iii) must hold. (See Exercise 32.)

Example 1 Given $f(x) = 4x^3 - 9x$, verify that conditions (i), (ii), and (iii) of the hypothesis of Rolle's theorem are satisfied for each of the following intervals: $[-\frac{3}{2}, 0]$, $[0, \frac{3}{2}]$, and $[-\frac{3}{2}, \frac{3}{2}]$. Then find a suitable value for c in each of these intervals for which $f'(c) = 0$.

Solution $f(x) = 4x^3 - 9x$ and $f'(x) = 12x^2 - 9$; $f'(x)$ exists for all values of x, and so f is differentiable on $(-\infty, +\infty)$ and therefore continuous on $(-\infty, +\infty)$. Conditions (i) and (ii) of Rolle's theorem thus hold on any interval. To determine on which intervals condition (iii) holds, we find the values of x for which $f(x) = 0$. If $f(x) = 0$,

$$4x(x^2 - \tfrac{9}{4}) = 0$$

which gives

$$x = -\tfrac{3}{2} \qquad x = 0 \qquad x = \tfrac{3}{2}$$

With $a = -\frac{3}{2}$ and $b = 0$, Rolle's theorem holds on $[-\frac{3}{2}, 0]$. Similarly, Rolle's theorem holds on $[0, \frac{3}{2}]$ and $[-\frac{3}{2}, \frac{3}{2}]$.

To find the suitable values for c, set $f'(x) = 0$ and get

$$12x^2 - 9 = 0$$

which gives

$$x = -\tfrac{1}{2}\sqrt{3} \quad \text{and} \quad x = \tfrac{1}{2}\sqrt{3}$$

Therefore, in the interval $[-\frac{3}{2}, 0]$ a suitable choice for c is $-\frac{1}{2}\sqrt{3}$. In the interval $[0, \frac{3}{2}]$, take $c = \frac{1}{2}\sqrt{3}$. In the interval $[-\frac{3}{2}, \frac{3}{2}]$ there are two possibilities for c: either $-\frac{1}{2}\sqrt{3}$ or $\frac{1}{2}\sqrt{3}$.

We apply Rolle's theorem to prove one of the most important theorems in calculus—that known as the **mean-value theorem** (or law of the mean). The mean-value theorem is used to prove many theorems of both differential and integral calculus, as well as of other subjects, such as numerical analysis. You should become thoroughly familiar with the content of this theorem.

4.4.2 Theorem
Mean-Value Theorem

Let f be a function such that

(i) it is continuous on the closed interval $[a, b]$;
(ii) it is differentiable on the open interval (a, b).

Then there is a number c in the open interval (a, b) such that

$$f'(c) = \frac{f(b) - f(a)}{b - a}$$

Before proving this theorem, we interpret it geometrically. In a sketch of the graph of the function f, $[f(b) - f(a)]/(b - a)$ is the slope of the line segment joining the points $A(a, f(a))$ and $B(b, f(b))$. The mean-value theorem states that there is some point on the curve between A and B where the tangent line is parallel to the secant line through A and B; that is, there is some number c in (a, b) such that

$$f'(c) = \frac{f(b) - f(a)}{b - a}$$

Refer to Fig. 4.4.5.

Take the x axis along the line segment AB and observe that the mean-value theorem is a generalization of Rolle's theorem, which is used in its proof.

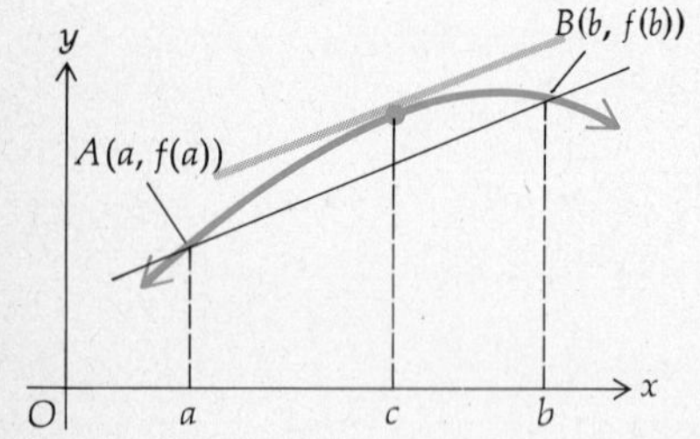

FIGURE 4.4.5

Proof An equation of the line through A and B in Fig. 4.4.5 is

$$y - f(a) = \frac{f(b) - f(a)}{b - a}(x - a)$$

or, equivalently,

$$y = \frac{f(b) - f(a)}{b - a}(x - a) + f(a)$$

Now if $F(x)$ measures the vertical distance between a point $(x, f(x))$ on the graph of the function f and the corresponding point on the secant line through A and B, then

$$F(x) = f(x) - \frac{f(b) - f(a)}{b - a}(x - a) - f(a) \tag{1}$$

We show that this function F satisfies the three conditions of the hypothesis of Rolle's theorem.

The function F is continuous on the closed interval $[a, b]$ because it is the sum of f and a linear polynomial function, both of which are continuous there. Therefore condition (i) is satisfied by F. Condition (ii) is satisfied by F because f is differentiable on (a, b). From (1) it follows that $F(a) = 0$ and $F(b) = 0$. Therefore condition (iii) of Rolle's theorem is satisfied by F.

The conclusion of Rolle's theorem states that there is a c in the open interval (a, b) such that $F'(c) = 0$. But

$$F'(x) = f'(x) - \frac{f(b) - f(a)}{b - a}$$

Thus

$$F'(c) = f'(c) - \frac{f(b) - f(a)}{b - a}$$

Therefore there is a number c in (a, b) such that

$$0 = f'(c) - \frac{f(b) - f(a)}{b - a}$$

or, equivalently,

$$f'(c) = \frac{f(b) - f(a)}{b - a}$$

which was to be proved. ■

Example 2 Given $f(x) = x^3 - 5x^2 - 3x$, verify that the hypothesis of the mean-value theorem is satisfied for $a = 1$ and $b = 3$. Then find all numbers c in the open interval $(1, 3)$ such that

$$f'(c) = \frac{f(3) - f(1)}{3 - 1}$$

Solution Because f is a polynomial function, f is continuous and differentiable for all values of x. Therefore the hypothesis of the mean-value theorem is satisfied for any a and b.

$$f(x) = x^3 - 5x^2 - 3x \qquad f'(x) = 3x^2 - 10x - 3$$

$$f(1) = -7 \quad \text{and} \quad f(3) = -27$$

Hence

$$\frac{f(3) - f(1)}{3 - 1} = \frac{-27 - (-7)}{2} = -10$$

Set $f'(c) = -10$ to obtain

$$3c^2 - 10c - 3 = -10$$

$$3c^2 - 10c + 7 = 0$$

$$(3c - 7)(c - 1) = 0$$

which gives

$$c = \tfrac{7}{3} \quad \text{and} \quad c = 1$$

Because 1 is not in the open interval $(1, 3)$, the only possible value for c is $\frac{7}{3}$.

Example 3 Given $f(x) = x^{2/3}$, draw a sketch of the graph of f. Show that there is no number c in the open interval $(-2, 2)$ such that

$$f'(c) = \frac{f(2) - f(-2)}{2 - (-2)}$$

Which condition of the hypothesis of the mean-value theorem fails to hold for f when $a = -2$ and $b = 2$?

Solution A sketch of the graph of f is shown in Fig. 4.4.6.

$$f(x) = x^{2/3} \qquad f'(x) = \tfrac{2}{3}x^{-1/3}$$

So

$$f'(c) = \frac{2}{3c^{1/3}}$$

$$\frac{f(2) - f(-2)}{2 - (-2)} = \frac{4^{1/3} - 4^{1/3}}{4}$$
$$= 0$$

There is no number c for which $\dfrac{2}{3c^{1/3}} = 0$.

The function f is continuous on the closed interval $[-2, 2]$; however, f is not differentiable on the open interval $(-2, 2)$ because $f'(0)$ does not exist. Therefore condition (ii) of the hypothesis of the mean-value theorem fails to hold for f when $a = -2$ and $b = 2$.

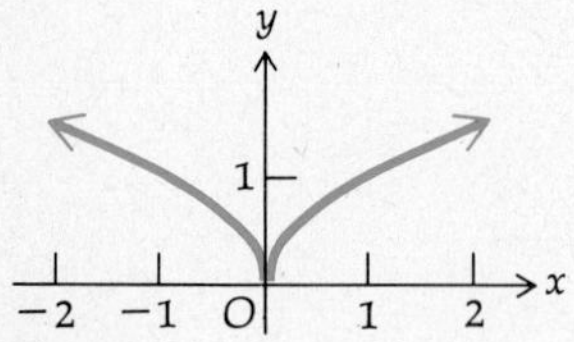

FIGURE 4.4.6

Prior to the statement of the mean-value theorem, we indicated that it is one of the most important theorems in the calculus because it is used to prove many other theorems. In such cases it is not necessary to find the value of the number c guaranteed by the theorem. The crucial fact of the theorem is that such a number c exists. To indicate the power of the mean-value theorem we show its use in the proof of the following theorem, which is needed in Chapter 5.

4.4.3 Theorem If f is a function such that $f'(x) = 0$ for all values of x in an interval I, then f is constant on I.

Proof Assume that f is not constant on the interval I. Then there exist two distinct numbers x_1 and x_2 in I, where $x_1 < x_2$, such that $f(x_1) \neq f(x_2)$. Because, by hypothesis, $f'(x) = 0$ for all x in I, then $f'(x) = 0$ for all x in the closed interval $[x_1, x_2]$. Hence f is differentiable at all x in $[x_1, x_2]$ and f is continuous on $[x_1, x_2]$. Therefore the hypothesis of the mean-value theorem is satisfied, and so there is a number c, with $x_1 < c < x_2$, such that

$$f'(c) = \frac{f(x_1) - f(x_2)}{x_1 - x_2} \tag{2}$$

But because $f'(x) = 0$ for all x in the interval $[x_1, x_2]$, then $f'(c) = 0$, and from (2) it follows that $f(x_1) = f(x_2)$. Yet the assumption was that $f(x_1) \neq f(x_2)$. Hence there is a contradiction, and so f is constant on I. ■

Another significant theorem that the mean-value theorem is used to prove is Theorem 4.5.3, given in the next section.

EXERCISES 4.4

In Exercises 1 through 4, verify that conditions (i), (ii), and (iii) of the hypothesis of Rolle's theorem are satisfied by the given function on the indicated interval. Then find a suitable value for c that satisfies the conclusion of Rolle's theorem.

1. $f(x) = x^2 - 4x + 3$; $[1, 3]$
2. $f(x) = x^3 - 2x^2 - x + 2$; $[1, 2]$
3. $f(x) = \sin 2x$; $[0, \frac{1}{2}\pi]$
4. $f(x) = 3\cos^2 x$; $[\frac{1}{2}\pi, \frac{3}{2}\pi]$

In Exercises 5 through 10, verify that the hypothesis of the mean-value theorem is satisfied for the given function on the indicated interval. Then find a suitable value for c that satisfies the conclusion of the mean-value theorem.

5. $f(x) = x^2 + 2x - 1$; $[0, 1]$
6. $f(x) = x^3 + x^2 - x$; $[-2, 1]$
7. $f(x) = x^{2/3}$; $[0, 1]$
8. $f(x) = \sqrt{1 - \sin x}$; $[0, \frac{1}{2}\pi]$
9. $f(x) = \sqrt{1 + \cos x}$; $[-\frac{1}{2}\pi, \frac{1}{2}\pi]$
10. $f(x) = \dfrac{x^2 + 4x}{x - 7}$; $[2, 6]$

In Exercises 11 through 16, (a) draw a sketch of the graph of the given function on the indicated interval; (b) test the three conditions (i), (ii), and (iii) of the hypothesis of Rolle's theorem and determine which conditions are satisfied and which, if any, are not satisfied; and (c) if the three conditions in part (b) are satisfied, determine a point at which there is a horizontal tangent line.

11. $f(x) = x^{4/3} - 3x^{1/3}$; $[0, 3]$
12. $f(x) = x^{3/4} - 2x^{1/4}$; $[0, 4]$
13. $f(x) = \dfrac{x^2 - x - 12}{x - 3}$; $[-3, 4]$
14. $f(x) = \begin{cases} 3x + 6 & \text{if } x < 1 \\ x - 4 & \text{if } 1 \le x \end{cases}$; $[-2, 4]$
15. $f(x) = \begin{cases} x^2 - 4 & \text{if } x < 1 \\ 5x - 8 & \text{if } 1 \le x \end{cases}$; $[-2, \frac{8}{5}]$
16. $f(x) = 1 - |x|$; $[-1, 1]$

The geometric interpretation of the mean-value theorem is that for a suitable c in the open interval (a, b), the tangent line to the curve y = f(x) at the point (c, f(c)) is parallel to the secant line through the points (a, f(a)) and (b, f(b)). In Exercises 17 through 20, find a value of c satisfying the conclusion of the mean-value theorem, draw a sketch of the graph on the closed interval [a, b], and show the tangent line and secant line.

17. $f(x) = x^2$; $a = 3$, $b = 5$
18. $f(x) = x^2$; $a = 2$, $b = 4$
19. $f(x) = \sin x$; $a = 0$, $b = \frac{1}{2}\pi$
20. $f(x) = 2\cos x$; $a = \frac{1}{3}\pi$, $b = \frac{2}{3}\pi$

For each of the functions in Exercises 21 through 24, there is no number c in the open interval (a, b) that satisfies the conclusion of the mean-value theorem. In each exercise, determine which part of the hypothesis of the mean-value theorem fails to hold. Draw a sketch of the graph of y = f(x) and the line through the points (a, f(a)) and (b, f(b)).

21. $f(x) = \dfrac{4}{(x - 3)^2}$; $a = 1$, $b = 6$
22. $f(x) = \dfrac{2x - 1}{3x - 4}$; $a = 1$, $b = 2$
23. $f(x) = 3(x - 4)^{2/3}$; $a = -4$, $b = 5$
24. $f(x) = \begin{cases} 2x + 3 & \text{if } x < 3 \\ 15 - 2x & \text{if } 3 \le x \end{cases}$; $a = -1$, $b = 5$

25. If $f(x) = x^4 - 2x^3 + 2x^2 - x$, then $f'(x) = 4x^3 - 6x^2 + 4x - 1$. Prove by Rolle's theorem that the equation $4x^3 - 6x^2 + 4x - 1 = 0$ has at least one real root in the open interval $(0, 1)$.

26. Prove by Rolle's theorem that the equation $x^3 + 2x + c = 0$, where c is any constant, cannot have more than one real root.

27. Use Rolle's theorem to prove that the equation $4x^5 + 3x^3 + 3x - 2 = 0$ has exactly one root that lies in the interval $(0, 1)$. (*Hint:* First show that there is at least one number in $(0, 1)$ that is a root of the equation. Then assume that there is more than one root of the equation in $(0, 1)$ and show that this leads to a contradiction.)

28. Suppose $s = f(t)$ is an equation of the motion of a particle moving in a straight line, where f satisfies the hypothesis of the mean-value theorem. Show that the conclusion of the mean-value theorem assures that there will be some instant during any time interval when the instantaneous velocity will equal the average velocity during that time interval.

29. If the equation of the motion in Exercise 28 is $s = t^2 - t + 4$ and $t \in [0, 3]$, find the value of t where the instantaneous velocity equals the average velocity over the given interval.

30. If $f(x) = \sin^2 x + \cos^2 x$, use Theorem 4.4.3 to show that $f(x) = 1$ for all x in $[-2\pi, 2\pi]$.

31. Suppose that the function f is continuous on $[a, b]$ and $f'(x) = 1$ for all x in (a, b). Prove that $f(x) = x - a + f(a)$ for all x in $[a, b]$.

32. The converse of Rolle's theorem is not true. Make up an example of a function for which the conclusion of Rolle's theorem is true and for which (a) condition (i) is not satisfied but conditions (ii) and (iii) are satisfied; (b) condition (ii) is not satisfied but conditions (i) and (iii) are satisfied; (c) condition (iii) is not satisfied but conditions (i) and (ii) are satisfied. Draw a sketch of the graph showing the horizontal tangent line for each case.

33. Use Rolle's theorem to prove that if every polynomial of the fourth degree has at most four real roots, then every polynomial of the fifth degree has at most five real roots. (*Hint:* Assume a polynomial of the fifth degree has six real roots and show that this leads to a contradiction.)

34. Use the method of Exercise 33 and mathematical induction to prove that a polynomial of the nth degree has at most n real roots.

4.5 INCREASING AND DECREASING FUNCTIONS AND THE FIRST-DERIVATIVE TEST

Suppose that Fig. 4.5.1 represents a sketch of the graph of a function f for all x in the closed interval $[x_1, x_7]$. In this sketch we have assumed that f is continuous on $[x_1, x_7]$.

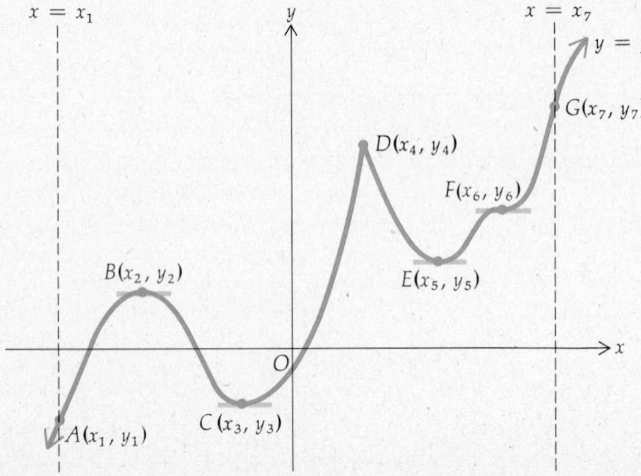

FIGURE 4.5.1

Figure 4.5.1 shows that as a point moves along the curve from A to B, the function values increase as the abscissa increases, and that as a point moves along the curve from B to C, the function values decrease as the abscissa increases. We say, then, that f is *increasing* on the closed interval $[x_1, x_2]$ and that f is *decreasing* on the closed interval $[x_2, x_3]$. Following are the precise definitions of a function increasing or decreasing on an interval.

4.5.1 Definition A function f defined on an interval is said to be **increasing** on that interval if and only if

$$f(x_1) < f(x_2) \quad \text{whenever} \quad x_1 < x_2$$

where x_1 and x_2 are any numbers in the interval.

The function of Fig. 4.5.1 is increasing on the following closed intervals: $[x_1, x_2]$; $[x_3, x_4]$; $[x_5, x_6]$; $[x_6, x_7]$; $[x_5, x_7]$.

4.5.2 Definition A function f defined on an interval is said to be **decreasing** on that interval if and only if

$$f(x_1) > f(x_2) \quad \text{whenever} \quad x_1 < x_2$$

where x_1 and x_2 are any numbers in the interval.

The function of Fig. 4.5.1 is decreasing on the following closed intervals: $[x_2, x_3]$; $[x_4, x_5]$.

If a function f is either increasing on an interval or decreasing on an interval, then f is said to be **monotonic** on the interval.

Before stating a theorem that gives a test for determining if a given function is monotonic on an interval, let us see what is happening geometrically. Refer to Fig. 4.5.1 and observe that when the slope of the tangent line is positive the function is increasing, and when the slope of the tangent line is negative the function is decreasing. Because $f'(x)$ is the slope of the tangent line to the curve $y = f(x)$, f is increasing when $f'(x) > 0$, and f is decreasing when $f'(x) < 0$. Also, because $f'(x)$ is the rate of change of the function values $f(x)$ with respect to x, when $f'(x) > 0$, the function values are increasing as x increases; and when $f'(x) < 0$, the function values are decreasing as x increases.

4.5.3 Theorem Let the function f be continuous on the closed interval $[a, b]$ and differentiable on the open interval (a, b):

(i) if $f'(x) > 0$ for all x in (a, b), then f is increasing on $[a, b]$;
(ii) if $f'(x) < 0$ for all x in (a, b), then f is decreasing on $[a, b]$.

Proof of (i) Let x_1 and x_2 be any two numbers in $[a, b]$ such that $x_1 < x_2$. Then f is continuous on $[x_1, x_2]$ and differentiable on (x_1, x_2). From the mean-value theorem it follows that there is some number c in (x_1, x_2) such that

$$f'(c) = \frac{f(x_2) - f(x_1)}{x_2 - x_1}$$

Because $x_1 < x_2$, then $x_2 - x_1 > 0$. Also, $f'(c) > 0$ by hypothesis. Therefore $f(x_2) - f(x_1) > 0$, and so $f(x_2) > f(x_1)$. We have shown, then, that $f(x_1) < f(x_2)$ whenever $x_1 < x_2$, where x_1 and x_2 are any numbers in the interval $[a, b]$. Therefore, by Definition 4.5.1 it follows that f is increasing on $[a, b]$.

The proof of part (ii) is similar and is left as an exercise (see Exercise 41). ■

An immediate application of Theorem 4.5.3 is in the proof of what is known as the **first-derivative test for relative extrema** of a function.

4.5.4 Theorem
First-Derivative Test for Relative Extrema

Let the function f be continuous at all points of the open interval (a, b) containing the number c, and suppose that f' exists at all points of (a, b) except possibly at c:

(i) if $f'(x) > 0$ for all values of x in some open interval having c as its right endpoint, and if $f'(x) < 0$ for all values of x in some open interval having c as its left endpoint, then f has a relative maximum value at c;

(ii) if $f'(x) < 0$ for all values of x in some open interval having c as its right endpoint, and if $f'(x) > 0$ for all values of x in some open interval having c as its left endpoint, then f has a relative minimum value at c.

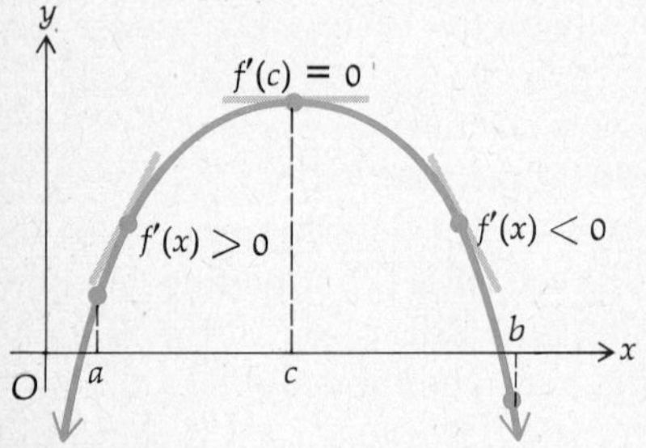

FIGURE 4.5.2

Proof of (i) Let (d, c) be the interval having c as its right endpoint for which $f'(x) > 0$ for all x in the interval. It follows from Theorem 4.5.3(i) that f is increasing on $[d, c]$. Let (c, e) be the interval having c as its left endpoint for which $f'(x) < 0$ for all x in the interval. By Theorem 4.5.3(ii), f is decreasing on $[c, e]$. Because f is increasing on $[d, c]$, it follows from Definition 4.5.1 that if x_1 is in $[d, c]$ and $x_1 \neq c$, then $f(x_1) < f(c)$. Also, because f is decreasing on $[c, e]$, it follows from Definition 4.5.2 that if x_2 is in $[c, e]$ and $x_2 \neq c$, then $f(c) > f(x_2)$. Therefore, from Definition 4.2.1, f has a relative maximum value at c.

The proof of part (ii) is similar to the proof of part (i), and it is left as an exercise (see Exercise 42). ■

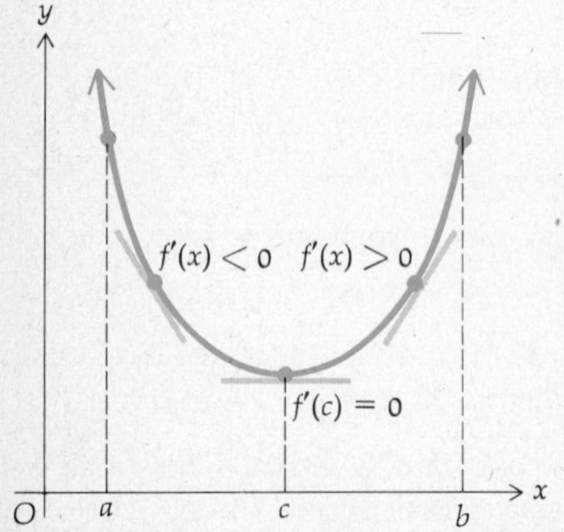

FIGURE 4.5.3

The first-derivative test for relative extrema essentially states that if f is continuous at c and $f'(x)$ changes algebraic sign from positive to negative as x increases through the number c, then f has a relative maximum value at c; and if $f'(x)$ changes algebraic sign from negative to positive as x increases through c, then f has a relative minimum value at c.

Figures 4.5.2 and 4.5.3 illustrate parts (i) and (ii), respectively, of Theorem 4.5.4 when $f'(c)$ exists. Figure 4.5.4 shows a sketch of the graph of a function f that has a relative maximum value at a number c, but $f'(c)$ does not exist; however, $f'(x) > 0$ when $x < c$, and $f'(x) < 0$ when $x > c$.

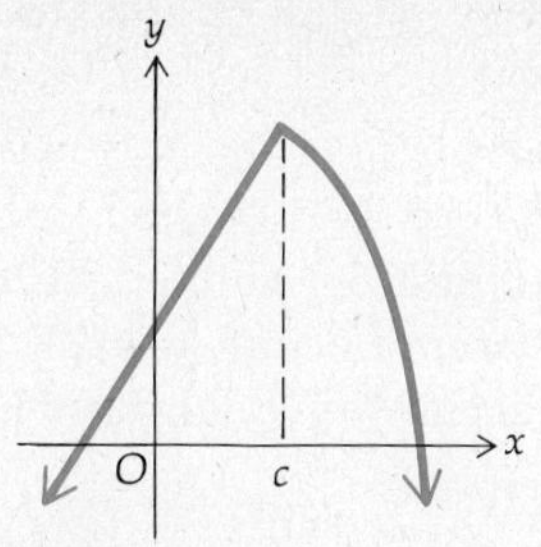

FIGURE 4.5.4

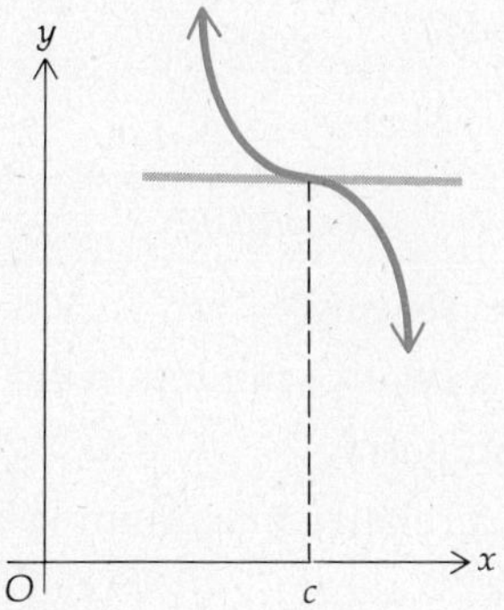

FIGURE 4.5.5

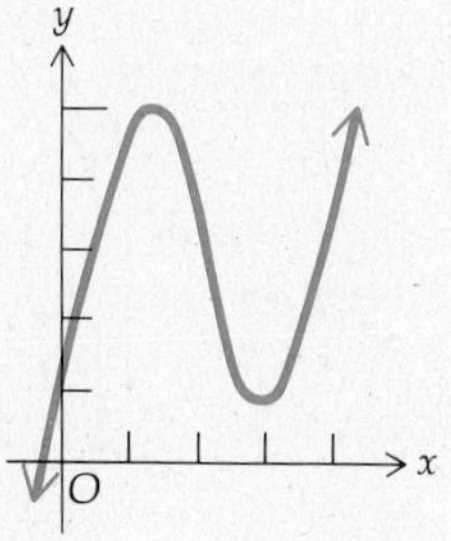

FIGURE 4.5.6

In Fig. 4.5.5 there is a sketch of the graph of a function f for which c is a critical number, and $f'(x) < 0$ when $x < c$, and $f'(x) < 0$ when $x > c$; f does not have a relative extremum at c.

Further illustrations of Theorem 4.5.4 occur in Fig. 4.5.1. At x_2 and x_4 the function has a relative maximum value, and at x_3 and x_5 the function has a relative minimum value; even though x_6 is a critical number for the function, there is no relative extremum at x_6.

In summary, to determine the relative extrema of a function f:

1. Find $f'(x)$.
2. Find the critical numbers of f, that is, the values of x for which $f'(x) = 0$ or for which $f'(x)$ does not exist.
3. Apply the first-derivative test (Theorem 4.5.4).

The following examples illustrate this procedure.

Example 1 Given $f(x) = x^3 - 6x^2 + 9x + 1$, find the relative extrema of f by applying the first-derivative test. Determine the values of x at which the relative extrema occur, as well as the intervals on which f is increasing and the intervals on which f is decreasing. Draw a sketch of the graph.

Solution $f(x) = x^3 - 6x^2 + 9x + 1$ and $f'(x) = 3x^2 - 12x + 9$. $f'(x)$ exists for all values of x. Set $f'(x) = 0$.

$$3x^2 - 12x + 9 = 0$$

$$3(x - 3)(x - 1) = 0$$

which gives

$$x = 3 \quad \text{and} \quad x = 1$$

Thus the critical numbers of f are 1 and 3. To determine whether f has a relative extremum at either of these numbers, apply the first-derivative test. The results are summarized in Table 4.5.1.

TABLE 4.5.1

	$f(x)$	$f'(x)$	*Conclusion*
$x < 1$		$+$	f is increasing
$x = 1$	5	0	f has a relative maximum value
$1 < x < 3$		$-$	f is decreasing
$x = 3$	1	0	f has a relative minimum value
$3 < x$		$+$	f is increasing

From the table, 5 is a relative maximum value of f occurring at 1, and 1 is a relative minimum value of f occurring at 3. A sketch of the graph is shown in Fig. 4.5.6.

Example 2 Given

$$f(x) = \begin{cases} x^2 - 4 & \text{if } x < 3 \\ 8 - x & \text{if } 3 \leq x \end{cases}$$

find the relative extrema of f by applying the first-derivative test. Determine the values of x at which the relative extrema occur, as well as the intervals on which f is increasing and the intervals on which f is decreasing. Draw a sketch of the graph.

Solution If $x < 3, f'(x) = 2x$. If $x > 3, f'(x) = -1$. Because $f'_-(3) = 6$ and $f'_+(3) = -1$, $f'(3)$ does not exist. Therefore 3 is a critical number of f.

Because $f'(x) = 0$ when $x = 0$, it follows that 0 is a critical number of f. Applying the first-derivative test, we summarize the results in Table 4.5.2. A sketch of the graph is shown in Fig. 4.5.7.

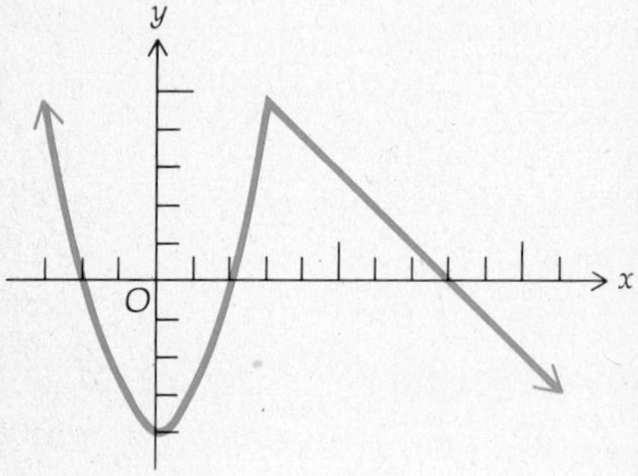

FIGURE 4.5.7

TABLE 4.5.2

	$f(x)$	$f'(x)$	*Conclusion*
$x < 0$		$-$	f is decreasing
$x = 0$	-4	0	f has a relative minimum value
$0 < x < 3$		$+$	f is increasing
$x = 3$	5	does not exist	f has a relative maximum value
$3 < x$		$-$	f is decreasing

Example 3 Given $f(x) = x^{4/3} + 4x^{1/3}$, find the relative extrema of f, determine the values of x at which the relative extrema occur, and determine the intervals on which f is increasing and the intervals on which f is decreasing. Draw a sketch of the graph.

Solution $f(x) = x^{4/3} + 4x^{1/3}$ and $f'(x) = \frac{4}{3}x^{1/3} + \frac{4}{3}x^{-2/3} = \frac{4}{3}x^{-2/3}(x + 1)$. Because $f'(x)$ does not exist when $x = 0$, and $f'(x) = 0$ when $x = -1$, the critical numbers of f are -1 and 0. We apply the first-derivative test and summarize the results in Table 4.5.3. A sketch of the graph is shown in Fig. 4.5.8.

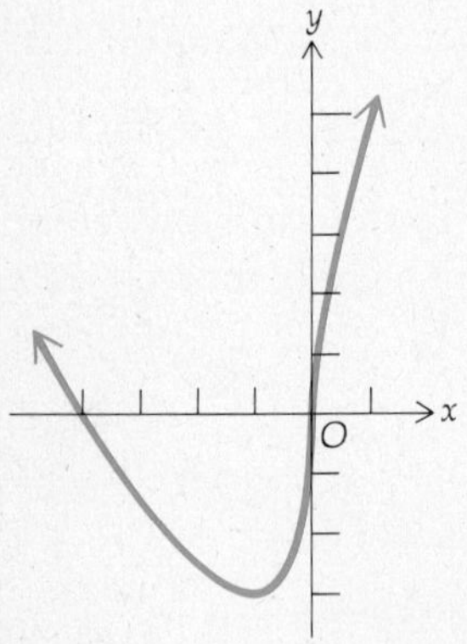

FIGURE 4.5.8

TABLE 4.5.3

	$f(x)$	$f'(x)$	*Conclusion*
$x < -1$		$-$	f is decreasing
$x = -1$	-3	0	f has a relative minimum value
$-1 < x < 0$		$+$	f is increasing
$x = 0$	0	does not exist	f does not have a relative extremum at $x = 0$
$0 < x$		$+$	f is increasing

EXERCISES 4.5

In Exercises 1 through 40, do each of the following: (a) find the relative extrema of f by applying the first-derivative test; (b) determine the values of x at which the relative extrema occur; (c) determine the intervals on which f is increasing; (d) determine the intervals on which f is decreasing; (e) draw a sketch of the graph.

1. $f(x) = x^2 - 4x - 1$

2. $f(x) = 3x^2 - 3x + 2$

3. $f(x) = x^3 - x^2 - x$

4. $f(x) = x^3 - 9x^2 + 15x - 5$

5. $f(x) = 2x^3 - 9x^2 + 2$

6. $f(x) = x^3 - 3x^2 - 9x$

7. $f(x) = 4 \sin \dfrac{x}{2}$

8. $f(x) = x^4 + 4x$

9. $f(x) = \frac{1}{4}x^4 - x^3 + x^2$

10. $f(x) = 2 \cos 3x$

11. $f(x) = \frac{1}{5}x^5 - \frac{5}{3}x^3 + 4x + 1$

12. $f(x) = x^5 - 5x^3 - 20x - 2$

13. $f(x) = \sqrt{x} - \dfrac{1}{\sqrt{x}}$

14. $f(x) = \dfrac{x-2}{x+2}$

15. $f(x) = x + \dfrac{1}{x^2}$

16. $f(x) = 2x + \dfrac{1}{2x}$

17. $f(x) = (1 - x)^2(1 + x)^3$

18. $f(x) = (x + 2)^2(x - 1)^2$

19. $f(x) = 2x\sqrt{3 - x}$

20. $f(x) = x\sqrt{5 - x^2}$

21. $f(x) = x - 3x^{1/3}$

22. $f(x) = 4x - 6x^{2/3}$

23. $f(x) = 2 - 3(x - 4)^{2/3}$

24. $f(x) = 2 - (x - 1)^{1/3}$

25. $f(x) = x^{2/3} - x^{1/3}$

26. $f(x) = x^{2/5} + 1$

27. $f(x) = (x + 1)^{2/3}(x - 2)^{1/3}$

28. $f(x) = x^{2/3}(x - 1)^2$

29. $f(x) = \begin{cases} 2x + 1 & \text{if } x \le 4 \\ 13 - x & \text{if } 4 < x \end{cases}$

30. $f(x) = \begin{cases} 5 - 2x & \text{if } x < 3 \\ 3x - 10 & \text{if } 3 \le x \end{cases}$

31. $f(x) = \begin{cases} 2x + 9 & \text{if } x \le -2 \\ x^2 + 1 & \text{if } -2 < x \end{cases}$

32. $f(x) = \begin{cases} \sqrt{25 - (x + 7)^2} & \text{if } -12 \le x \le -3 \\ 12 - x^2 & \text{if } -3 < x \end{cases}$

33. $f(x) = \begin{cases} 3x + 5 & \text{if } x < -1 \\ x^2 + 1 & \text{if } -1 \le x < 2 \\ 7 - x & \text{if } 2 \le x \end{cases}$

34. $f(x) = \begin{cases} 4 - (x + 5)^2 & \text{if } x < -4 \\ 12 - (x + 1)^2 & \text{if } -4 \le x \end{cases}$

35. $f(x) = \begin{cases} (x + 9)^2 - 8 & \text{if } x < -7 \\ -\sqrt{25 - (x + 4)^2} & \text{if } -7 \le x \le 0 \\ (x - 2)^2 - 7 & \text{if } 0 < x \end{cases}$

36. $f(x) = \begin{cases} 12 - (x + 5)^2 & \text{if } x \le -3 \\ 5 - x & \text{if } -3 < x \le -1 \\ \sqrt{100 - (x - 7)^2} & \text{if } -1 < x \le 17 \end{cases}$

37. $f(x) = x^{5/4} + 10x^{1/4}$

38. $f(x) = (x - a)^{2/5} + 1$

39. $f(x) = x^{1/3}(x + 4)^{-2/3}$

40. $f(x) = x^{5/3} - 10x^{2/3}$

41. Prove Theorem 4.5.3(ii).

42. Prove Theorem 4.5.4(ii).

43. Find a and b such that the function defined by $f(x) = x^3 + ax^2 + b$ will have a relative extremum at (2, 3).

44. Find a, b, and c such that the function defined by $f(x) = ax^2 + bx + c$ will have a relative maximum value of 7 at 1 and the graph of $y = f(x)$ will go through the point (2, −2).

45. Find a, b, c, and d such that the function defined by $f(x) = ax^3 + bx^2 + cx + d$ will have relative extrema at (1, 2) and (2, 3).

46. Given that the function f is continuous for all values of x, $f(3) = 2$, $f'(x) < 0$ if $x < 3$, and $f'(x) > 0$ if $x > 3$, draw a sketch of a possible graph of f in each of the following cases, where the additional condition is satisfied: (a) f' is continuous at 3; (b) $f'(x) = -1$ if $x < 3$, and $f'(x) = 1$ if $x > 3$; (c) $\lim_{x \to 3^-} f'(x) = -1$, $\lim_{x \to 3^+} f'(x) = 1$, and $f'(a) \neq f'(b)$ if $a \neq b$.

47. Given $f(x) = x^p(1 - x)^q$, where p and q are positive integers greater than 1, prove each of the following: (a) if p is even, f has a relative minimum value at 0; (b) if q is even, f has a relative minimum value at 1; (c) f has a relative maximum value at $p/(p + q)$ whether p and q are odd or even.

48. The function f is differentiable at each number in the closed interval $[a, b]$. Prove that if $f'(a) \cdot f'(b) < 0$, there is a number c in the open interval (a, b) such that $f'(c) = 0$.

49. If $f(x) = x^k$, where k is an odd positive integer, show that f has no relative extrema.

50. Prove that if f is increasing on $[a, b]$ and if g is increasing on $[f(a), f(b)]$, then if $g \circ f$ exists on $[a, b]$, $g \circ f$ is increasing on $[a, b]$.

51. The function f is increasing on the interval I. Prove that (a) if $g(x) = -f(x)$, then g is decreasing on I; (b) if $h(x) = 1/f(x)$ and $f(x) > 0$ on I, then h is decreasing on I.

4.6 THE SECOND-DERIVATIVE TEST FOR RELATIVE EXTREMA AND FURTHER TREATMENT OF ABSOLUTE EXTREMA

In Sec. 4.5 we learned how to determine whether a function f has a relative maximum value or a relative minimum value at a critical number c by checking the algebraic sign of f' at numbers in intervals to the left and right of c. Another test for relative extrema is one that involves only the critical number c. Before stating the test in the form of a theorem, we give an informal geometric discussion that should appeal to your intuition.

Suppose that f is a function such that f' and f'' exist on some open interval (a, b) containing c and that $f'(c) = 0$; furthermore suppose that f'' is negative on (a, b). From Theorem 4.5.3, because $f''(x) < 0$ on (a, b), then f' is decreasing on $[a, b]$. Since the value of f' at a point on the graph of f gives the slope of the tangent line at the point, it follows that the slope of the tangent line is decreasing on $[a, b]$. In Fig. 4.6.1 there is a sketch of the graph of a function f having these properties. Segments of a few tangent lines are shown in the figure. Observe that the slope of the tangent line is decreasing on $[a, b]$. Note that f has a relative maximum value at c where $f'(c) = 0$ and $f''(c) < 0$.

Now suppose that f is a function having the properties of the function in the previous paragraph except that f'' is positive on (a, b). Then from Theorem 4.5.3, because $f''(x) > 0$ on (a, b), it follows that f' is increasing on $[a, b]$. Thus the slope of the tangent line is increasing on $[a, b]$. Figure 4.6.2 shows a sketch of the graph of a function f having these properties. In the figure there are segments of a few tangent lines whose slopes are increasing on $[a, b]$. The function f has a relative minimum value at c where $f'(c) = 0$ and $f''(c) > 0$.

The facts in the preceding two paragraphs are given in the **second-derivative test for relative extrema,** which is now stated.

4.6.1 Theorem
Second-Derivative Test for Relative Extrema

Let c be a critical number of a function f at which $f'(c) = 0$, and let f' exist for all values of x in some open interval containing c. If $f''(c)$ exists and

(i) if $f''(c) < 0$, then f has a relative maximum value at c;
(ii) if $f''(c) > 0$, then f has a relative minimum value at c.

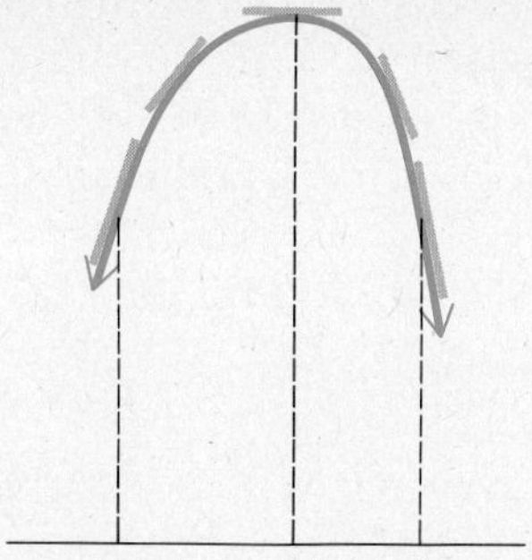

FIGURE 4.6.1

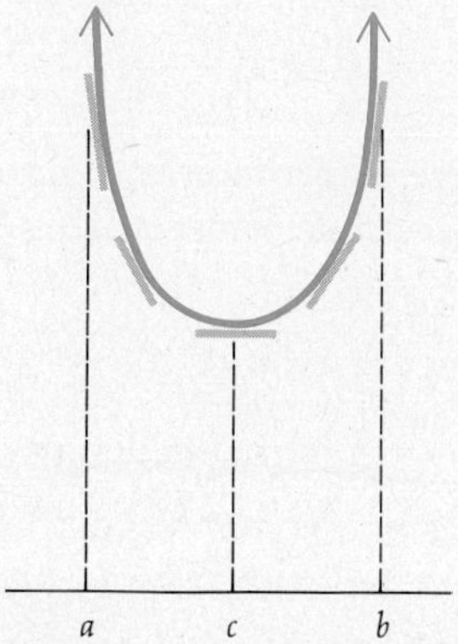

FIGURE 4.6.2

Proof of (i) By hypothesis, $f''(c)$ exists and is negative; so

$$f''(c) = \lim_{x \to c} \frac{f'(x) - f'(c)}{x - c} < 0$$

Therefore, by Theorem 2.10.2 there is an open interval I containing c such that

$$\frac{f'(x) - f'(c)}{x - c} < 0 \tag{1}$$

for every $x \neq c$ in the interval.

Let I_1 be the open interval containing all values of x in I for which $x < c$; therefore c is the right endpoint of the open interval I_1. Let I_2 be the open interval containing all values of x in I for which $x > c$; so c is the left endpoint of the open interval I_2.

Then if x is in I_1, $(x - c) < 0$, and it follows from inequality (1) that $[f'(x) - f'(c)] > 0$ or, equivalently, $f'(x) > f'(c)$. If x is in I_2, $(x - c) > 0$, and it follows from (1) that $[f'(x) - f'(c)] < 0$ or, equivalently, $f'(x) < f'(c)$.

But because $f'(c) = 0$, we conclude that if x is in I_1, $f'(x) > 0$, and if x is in I_2, $f'(x) < 0$. Therefore $f'(x)$ changes algebraic sign from positive to negative as x increases through c, and so, by Theorem 4.5.4, f has a relative maximum value at c.

The proof of part (ii) is similar and is left as an exercise (see Exercise 42). ■

Example 1 Given $f(x) = x^4 + \frac{4}{3}x^3 - 4x^2$, find the relative maxima and the relative minima of f by applying the second-derivative test.

Solution

$$f(x) = x^4 + \tfrac{4}{3}x^3 - 4x^2$$

$$f'(x) = 4x^3 + 4x^2 - 8x$$

$$f''(x) = 12x^2 + 8x - 8$$

Set $f'(x) = 0$.

$$4x(x + 2)(x - 1) = 0$$

$$x = 0 \qquad x = -2 \qquad x = 1$$

Thus the critical numbers of f are -2, 0, and 1. We determine whether or not there is a relative extremum at any of these critical numbers by finding the sign of the second derivative there. The results are summarized in Table 4.6.1.

TABLE 4.6.1

	$f(x)$	$f'(x)$	$f''(x)$	*Conclusion*
$x = -2$	$-\frac{32}{3}$	0	+	f has a relative minimum value
$x = 0$	0	0	−	f has a relative maximum value
$x = 1$	$-\frac{5}{3}$	0	+	f has a relative minimum value

Example 2 Find the relative extrema of the sine function by applying the second-derivative test.

Solution Let

$$f(x) = \sin x$$

Then

$$f'(x) = \cos x$$

$$f''(x) = -\sin x$$

$f'(x)$ exists for all x. The critical numbers are obtained by setting $f'(x) = 0$.

$$\cos x = 0$$

which gives

$$x = \tfrac{1}{2}\pi + n\pi \qquad n = 0, \pm 1, \pm 2, \ldots$$

We determine whether or not there is a relative extremum at any of these critical numbers by finding the sign of the second derivative there. If n is an even integer, then

$$\begin{aligned} f''(\tfrac{1}{2}\pi + n\pi) &= -\sin(\tfrac{1}{2}\pi + n\pi) \qquad n \text{ is an even integer} \\ &= -\sin \tfrac{1}{2}\pi \\ &= -1 \end{aligned}$$

If n is an odd integer, then

$$\begin{aligned} f''(\tfrac{1}{2}\pi + n\pi) &= -\sin(\tfrac{1}{2}\pi + n\pi) \qquad n \text{ is an odd integer} \\ &= -\sin(\tfrac{3}{2}\pi + (n-1)\pi) \qquad (n-1) \text{ is an even integer} \\ &= -\sin \tfrac{3}{2}\pi \\ &= 1 \end{aligned}$$

The results of applying the second-derivative test are summarized in Table 4.6.2.

TABLE 4.6.2

	$f(x)$	$f'(x)$	$f''(x)$	*Conclusion*
$x = \frac{1}{2}\pi + n\pi$ (n is an even integer)	1	0	$-$	f has a relative maximum value
$x = \frac{1}{2}\pi + n\pi$ (n is an odd integer)	-1	0	$+$	f has a relative minimum value

If $f''(c) = 0$, as well as $f'(c) = 0$, nothing can be concluded regarding a relative extremum of f at c. The following three illustrations justify this statement.

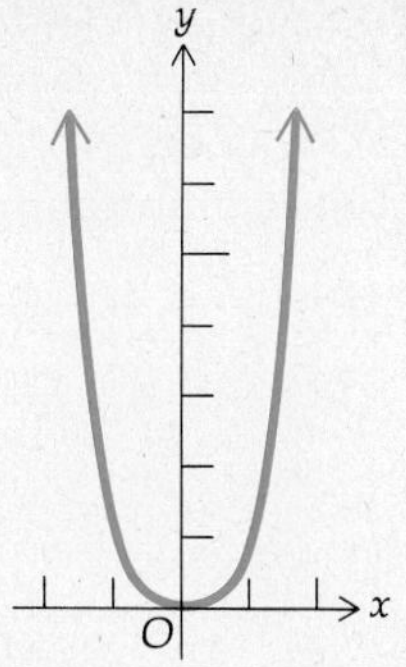

FIGURE 4.6.3

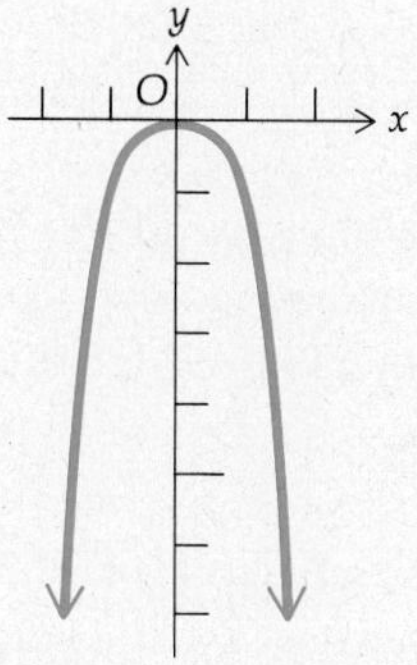

FIGURE 4.6.4

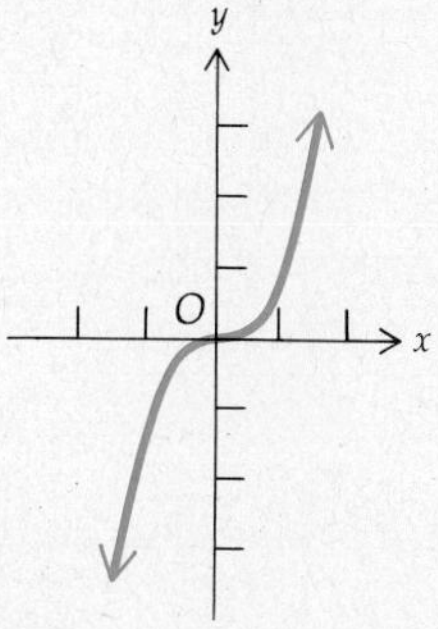

FIGURE 4.6.5

● ILLUSTRATION 1 If $f(x) = x^4$, then $f'(x) = 4x^3$ and $f''(x) = 12x^2$. Thus $f(0)$, $f'(0)$, and $f''(0)$ all have the value zero. By applying the first-derivative test, we see that f has a relative minimum value at 0. A sketch of the graph of f is shown in Fig. 4.6.3. ●

● ILLUSTRATION 2 If $g(x) = -x^4$, then $g'(x) = -4x^3$ and $g''(x) = -12x^2$. Hence $g(0) = g'(0) = g''(0) = 0$. In this case g has a relative maximum value at 0, as can be seen by applying the first-derivative test. A sketch of the graph of g is shown in Fig. 4.6.4. ●

● ILLUSTRATION 3 If $h(x) = x^3$, then $h'(x) = 3x^2$ and $h''(x) = 6x$; so $h(0) = h'(0) = h''(0) = 0$. The function h does not have a relative extremum at 0 because if $x < 0$, $h(x) < h(0)$; and if $x > 0$, $h(x) > h(0)$. A sketch of the graph of h is shown in Fig. 4.6.5. ●

In Illustrations 1, 2, and 3 we have examples of three functions, each of which has zero for its second derivative at a number for which its first derivative is zero; yet one function has a relative minimum value at that number, another function has a relative maximum value at that number, and the third function has neither a relative maximum value nor a relative minimum value at that number.

The extreme-value theorem (Theorem 4.2.9) guarantees an absolute maximum value and an absolute minimum value for a function that is continuous on a closed interval. We now consider some functions defined on intervals for which the extreme-value theorem does not apply and which may or may not have absolute extrema.

Example 3 Given

$$f(x) = \frac{x^2 - 27}{x - 6}$$

find the absolute extrema of f on the interval $[0, 6)$ if there are any.

Solution f is continuous on the interval $[0, 6)$ because the only discontinuity of f is at 6, which is not in the interval.

$$f'(x) = \frac{2x(x-6) - (x^2 - 27)}{(x-6)^2} = \frac{x^2 - 12x + 27}{(x-6)^2} = \frac{(x-3)(x-9)}{(x-6)^2}$$

$f'(x)$ exists for all values of x in $[0, 6)$, and $f'(x) = 0$ when $x = 3$ or 9; so the only critical number of f in the interval $[0, 6)$ is 3. The first-derivative test is applied to determine if f has a relative extremum at 3. The results are summarized in Table 4.6.3.

TABLE 4.6.3

	$f(x)$	$f'(x)$	*Conclusion*
$0 \le x < 3$		$+$	f is increasing
$x = 3$	6	0	f has a relative maximum value
$3 < x < 6$		$-$	f is decreasing

Because f has a relative maximum value at 3, and f is increasing on the interval [0,3) and decreasing on the interval (3, 6), then on [0, 6) f has an absolute maximum value at 3, and it is $f(3)$, which is 6. Note that $\lim_{x \to 6^-} f(x) = -\infty$, and conclude that there is no absolute minimum value of f on [0, 6).

Example 4 Given

$$f(x) = \frac{-x}{(x^2 + 6)^2}$$

find the absolute extrema of f on $(0, +\infty)$ if there are any.

Solution f is continuous for all values of x.

$$f'(x) = \frac{-1(x^2 + 6)^2 + 4x^2(x^2 + 6)}{(x^2 + 6)^4} = \frac{-(x^2 + 6) + 4x^2}{(x^2 + 6)^3} = \frac{3x^2 - 6}{(x^2 + 6)^3}$$

$f'(x)$ exists for all values of x. Setting $f'(x) = 0$ we obtain $x = \pm\sqrt{2}$; so $\sqrt{2}$ is the only critical number of f in $(0, +\infty)$. The first-derivative test is applied at $\sqrt{2}$, and the results are summarized in Table 4.6.4.

TABLE 4.6.4

	$f(x)$	$f'(x)$	*Conclusion*
$0 < x < \sqrt{2}$		$-$	f is decreasing
$x = \sqrt{2}$	$-\frac{1}{64}\sqrt{2}$	0	f has a relative minimum value
$\sqrt{2} < x$		$+$	f is increasing

Because f has a relative minimum value at $\sqrt{2}$ and because f is decreasing on $(0, \sqrt{2})$ and increasing on $(\sqrt{2}, +\infty)$, f has an absolute minimum value at $\sqrt{2}$ on $(0, +\infty)$. The absolute minimum value is $-\frac{1}{64}\sqrt{2}$. There is no absolute maximum value of f on $(0, +\infty)$.

For certain functions the following theorem is useful to determine if a relative extremum is an absolute extremum.

4.6.2 Theorem Let the function f be continuous on the interval I containing the number c. If $f(c)$ is a relative extremum of f on I and c is the only number in I for which f has a relative extremum, then $f(c)$ is an absolute extremum of f on I. Furthermore,

(i) if $f(c)$ is a relative maximum value of f on I, then $f(c)$ is an absolute maximum value of f on I;

(ii) if $f(c)$ is a relative minimum value of f on I, then $f(c)$ is an absolute minimum value of f on I.

Proof We prove part (i). The proof of part (ii) is similar.

Because $f(c)$ is a relative maximum value of f on I, then by Definition 4.2.1 there is an open interval J, where $J \subset I$, and where J contains c, such that

$$f(c) \geq f(x) \qquad \text{for all } x \in J$$

Because c is the only number in I for which f has a relative maximum value, it follows that

$$f(c) > f(k) \qquad \text{if } k \in J \text{ and } k \neq c \tag{2}$$

In order to show that $f(c)$ is an absolute maximum value of f on I, we show that if d is any number other than c in I, then $f(c) > f(d)$. We assume that

$$f(c) \leq f(d) \tag{3}$$

and show that this assumption leads to a contradiction. Because $d \neq c$, then either $c < d$ or $d < c$. We consider the case that $c < d$ (the proof is similar if $d < c$).

Because f is continuous on I, then f is continuous on the closed interval $[c, d]$. Therefore, by Theorem 4.2.9, f has an absolute minimum value on $[c, d]$. Assume this absolute minimum value occurs at e where $c \leq e \leq d$. From inequality (2) it follows that $e \neq c$, and from inequalities (2) and (3) it follows that $e \neq d$. Consequently, $c < e < d$, and hence f has a relative minimum value at e. But this statement contradicts the hypothesis that c is the only number in I for which f has a relative extremum. Thus our assumption that $f(c) \leq f(d)$ is false. Therefore $f(c) > f(d)$ if $d \in I$ and $d \neq c$, and consequently $f(c)$ is an absolute maximum value of f on I. ■

Example 5 Given $f(x) = 3x^4 - 8x^3 + 12x^2 - 12x + 3$, find the absolute extrema of f on $(-\infty, +\infty)$ if there are any.

Solution

$$f(x) = 3x^4 - 8x^3 + 12x^2 - 12x + 3$$
$$f'(x) = 12x^3 - 24x^2 + 24x - 12$$
$$f''(x) = 36x^2 - 48x + 24$$

$f'(x)$ exists for all values of x. Set $f'(x) = 0$ and obtain

$$12(x^3 - 2x^2 + 2x - 1) = 0 \tag{4}$$

We use synthetic division to determine that $x - 1$ is a factor of the left member of (4) and have

$$12(x - 1)(x^2 - x + 1) = 0 \tag{5}$$

Because the equation $x^2 - x + 1 = 0$ has only imaginary roots, the only real solution to (5) is 1. Therefore $f'(1) = 0$. To determine if $f(1)$ is a relative extremum we apply the second-derivative test; the results of this test are summarized in Table 4.6.5.

TABLE 4.6.5

	$f(x)$	$f'(x)$	$f''(x)$	*Conclusion*
$x = 1$	-2	0	+	f has a relative minimum value

The function f is continuous on $(-\infty, +\infty)$, and the one and only relative extremum of f on $(-\infty, +\infty)$ is at $x = 1$. Therefore it follows from Theorem 4.6.2(ii) that -2, the relative minimum value of f, is the absolute minimum value of f.

EXERCISES 4.6

In Exercises 1 through 24, find the relative extrema of the given function by using the second-derivative test, if it can be applied. If the second-derivative test cannot be applied, use the first-derivative test.

1. $f(x) = 3x^2 - 2x + 1$
2. $g(x) = 7 - 6x - 3x^2$
3. $f(x) = -4x^3 + 3x^2 + 18x$
4. $h(x) = 2x^3 - 9x^2 + 27$
5. $g(x) = \frac{1}{3}x^3 - x^2 + 3$
6. $f(y) = y^3 - 5y + 6$
7. $f(z) = (4 - z)^4$
8. $G(x) = (x + 2)^3$
9. $h(x) = x^4 - \frac{1}{3}x^3 - \frac{3}{2}x^2$
10. $f(x) = \frac{1}{5}x^5 - \frac{2}{3}x^3$
11. $F(t) = \cos 3t$
12. $f(x) = \sin x + \cos x$
13. $f(x) = x(x + 2)^3$
14. $g(t) = (t - 2)^{7/3}$
15. $f(x) = 4x^{1/2} + 4x^{-1/2}$
16. $f(x) = x(x - 1)^3$
17. $h(x) = x\sqrt{x + 3}$
18. $f(x) = x\sqrt{8 - x^2}$
19. $f(x) = x - \cos x$
20. $g(x) = \dfrac{9}{x} + \dfrac{x^2}{9}$
21. $F(x) = 6x^{1/3} - x^{2/3}$
22. $h(x) = (x + 2)(x - 1)^3$
23. $h(x) = 5x^3 - 3x^5$
24. $G(x) = x^{2/3}(x - 4)^2$

In Exercises 25 through 38, find the absolute extrema of the given function on the given interval if there are any.

25. $f(x) = x^2; (-3, 2]$
26. $g(x) = x^3 + 2x^2 - 4x + 1; (-3, 2)$
27. $F(x) = \dfrac{x + 2}{x - 2}; [-4, 4]$
28. $f(x) = \dfrac{x^2}{x + 3}; [-4, -1]$
29. $g(x) = 4x^2 - 2x + 1; (-\infty, +\infty)$
30. $f(x) = x^3 - 3x + 5; (-\infty, 0)$
31. $f(x) = (x - 1)^{1/3}; (-\infty, +\infty)$
32. $G(x) = (x - 5)^{2/3}; (-\infty, +\infty)$
33. $f(x) = x^4 + x^3 - \frac{9}{2}x^2 + 2x + 1; (-\infty, +\infty)$
34. $h(x) = 3x^4 - 2x^3 - 12x^2 - 6x - 1; (-\infty, +\infty)$
35. $g(x) = \dfrac{x^2 + 7}{x - 3}; [-3, 3)$
36. $f(x) = \dfrac{x^2 - 30}{x - 4}; (-\infty, 4)$
37. $f(x) = \dfrac{x}{(x^2 + 4)^{3/2}}; [0, +\infty)$
38. $g(x) = \sqrt{x^4 - 4x + 7}; (-\infty, +\infty)$

39. Suppose that $\frac{1}{2}\sqrt{2}$ and $-\frac{1}{2}\sqrt{3}$ are critical numbers of a function f and that $f''(x) = x[\![\frac{1}{2}x^2 + 1]\!]$. At each of these numbers, determine if f has a relative extremum, and if so, whether it is a relative maximum or a relative minimum.

40. Given $f(x) = x^r - rx + k$, where $r > 0$ and $r \neq 1$, prove that (a) if $0 < r < 1$, f has a relative maximum value at 1; (b) if $r > 1$, f has a relative minimum value at 1.

41. Given $f(x) = x^3 + 3rx + 5$, prove that (a) if $r > 0$, f has no relative extrema; (b) if $r < 0$, f has both a relative maximum value and a relative minimum value.

42. Prove Theorem 4.6.1(ii).

43. Given $f(x) = x^2 + rx^{-1}$, prove that regardless of the value of r, f has a relative minimum value and no relative maximum value.

44. If $f(x) = ax^2 + bx + c$, use the second-derivative test to show that f has a relative maximum value if $a < 0$. Find the number at which the relative maximum value occurs.

4.7 ADDITIONAL APPLICATIONS INVOLVING ABSOLUTE EXTREMA

In Sec. 4.3 the applications involved finding absolute extrema of functions continuous on a closed interval, and the extreme-value theorem was used in the solutions of the problems. We now deal with applications involving absolute extrema for which the extreme-value theorem cannot be applied. The examples show the procedure to follow in such situations.

Example 1 Find the shortest distance from the point $A(2, \frac{1}{2})$ to a point on the parabola $y = x^2$, and find the point on the parabola that is closest to A.

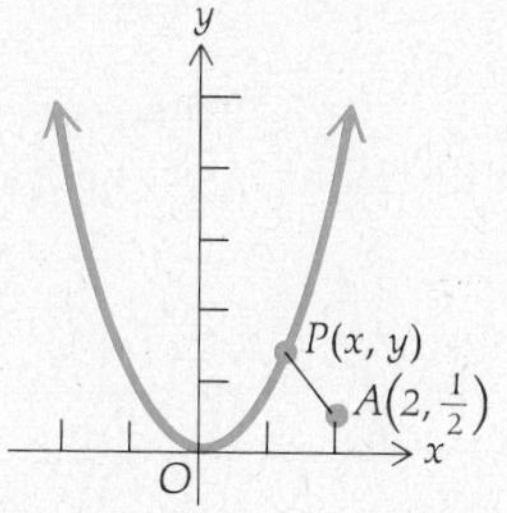

FIGURE 4.7.1

Solution Let z be the number of units in the distance from the point $A(2, \frac{1}{2})$ to a point $P(x, y)$ on the parabola. Refer to Fig. 4.7.1.

$$z = \sqrt{(x-2)^2 + (y - \tfrac{1}{2})^2} \tag{1}$$

Because $P(x, y)$ is on the parabola, its coordinates satisfy the equation $y = x^2$. Substituting this value of y into (1) we obtain z as a function of x and write

$$z(x) = \sqrt{(x-2)^2 + (x^2 - \tfrac{1}{2})^2}$$

$$z(x) = \sqrt{x^4 - 4x + \tfrac{17}{4}} \tag{2}$$

Because P can be any point on the parabola, x is any real number. We wish to find the absolute minimum value of z on $(-\infty, +\infty)$.

$$z'(x) = \tfrac{1}{2}(4x^3 - 4)(x^4 - 4x + \tfrac{17}{4})^{-1/2} = \frac{2(x-1)(x^2+x+1)}{\sqrt{x^4 - 4x + \frac{17}{4}}}$$

$z'(x)$ exists for all values of x because the denominator is $z(x)$, which is never zero. Consequently, the critical numbers are obtained by setting $z'(x) = 0$. The only real solution is 1; thus 1 is the only critical number of z. We apply the first-derivative test and summarize the results in Table 4.7.1.

TABLE 4.7.1

	$z(x)$	$z'(x)$	*Conclusion*
$x < 1$		$-$	z is decreasing
$x = 1$	$\frac{1}{2}\sqrt{5}$	0	z has a relative minimum value
$1 < x$		$+$	z is increasing

Because z has a relative minimum value at 1, and because z is decreasing on $(-\infty, 1)$ and increasing on $(1, +\infty)$, z has an absolute minimum value of $\frac{1}{2}\sqrt{5}$ at 1. So the point on the parabola closest to A is $(1, 1)$.

We can show that the point $A(2, \frac{1}{2})$ lies on the normal line at $(1, 1)$ of the graph of the parabola. Because the slope of the tangent line at any point (x, y) of the parabola is $2x$, the slope of the tangent line at $(1, 1)$ is 2. Therefore the slope of the normal line at $(1, 1)$ is $-\frac{1}{2}$, which is the same as the slope of the line through $A(2, \frac{1}{2})$ and $(1, 1)$.

Example 2 A closed box with a square base is to have a volume of 2000 cubic inches. The material for the top and bottom of the box is to cost \$3 per square inch, and the material for the sides is to cost \$1.50 per square inch. If the cost of the material is to be the least, find the dimensions of the box.

Solution Let x in. be the length of a side of the square base, y in. be the depth of the box, and C dollars be the cost of the material. See Fig. 4.7.2. The total number of square inches in the combined area of the top and bottom is $2x^2$, and for the sides it is $4xy$; so

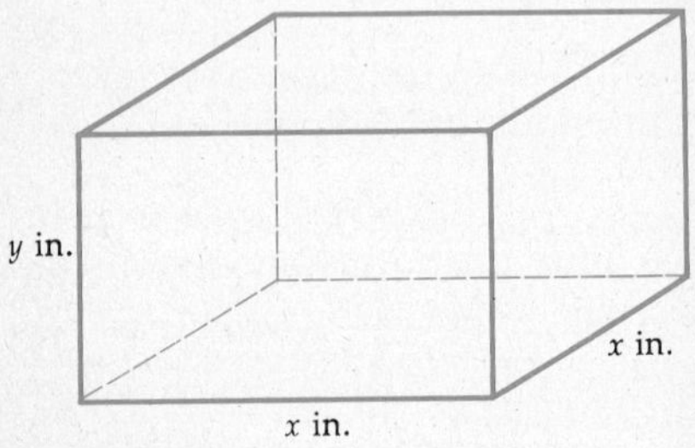

FIGURE 4.7.2

$$C = 3(2x^2) + \tfrac{3}{2}(4xy) \tag{3}$$

Because the volume of the box is the product of the area of the base and the depth,

$$x^2y = 2000 \tag{4}$$

Solving (4) for y in terms of x and substituting into (3), we obtain C as a function of x:

$$C(x) = 6x^2 + \frac{12{,}000}{x} \tag{5}$$

The domain of C is $(0, +\infty)$, and C is continuous on its domain. From (5),

$$C'(x) = 12x - \frac{12{,}000}{x^2} \tag{6}$$

$C'(x)$ does not exist when $x = 0$, but 0 is not in the domain of C. Therefore the only critical numbers will be those obtained by setting $C'(x) = 0$. The only real solution is 10; thus 10 is the only critical number. To determine if $x = 10$ makes C a relative minimum, we apply the second-derivative test. From (6) it follows that

$$C''(x) = 12 + \frac{24{,}000}{x^3}$$

The results of the second-derivative test are summarized in Table 4.7.2.

TABLE 4.7.2

	$C'(x)$	$C''(x)$	*Conclusion*
$x = 10$	0	$+$	C has a relative minimum value

Because C is continuous on its domain $(0, +\infty)$ and the one and only relative extremum of C on $(0, +\infty)$ is at $x = 10$, it follows from Theorem 4.6.2(ii) that this relative minimum value of C is the absolute minimum value of C. Thus the total cost of the material will be least when the side of the square base is 10 in. and the depth is 20 in.

In the preceding examples and in the exercises of Sec 4.3, the variable for which we wished to find an absolute extremum was expressed as a function of only one variable. Sometimes this procedure is either too difficult or too laborious, or occasionally even impossible. Often the given information enables us to obtain two equations involving three variables. Instead of eliminating one of the variables, it may be more advantageous to differentiate implicitly. The following example illustrates this method.

Example 3 If a closed tin can of specific volume is to be in the form of a right-circular cylinder, find the ratio of the height to the base radius if the least amount of material is to be used in its manufacture.

Solution We wish to find a relationship between the height and the base radius of the right-circular cylinder in order for the total surface area to be an absolute minimum for a fixed volume. Therefore we consider the volume of the cylinder a constant.

Let V be the number of cubic units in the volume of a cylinder (a constant).

We now define the variables.

Let r be the number of units in the base radius of the cylinder; $r > 0$. Let h be the number of units in the height of the cylinder; $h > 0$. Let S be the number of square units in the total surface area of the cylinder.

We have the following equations:

$$S = 2\pi r^2 + 2\pi rh \tag{7}$$

$$V = \pi r^2 h \tag{8}$$

Because V is a constant, we could solve (8) for either r or h in terms of the other and substitute into (7), which will give S as a function of one variable. The alternative method is to consider S as a function of two variables r and h; however, r and h are not independent of each other. That is, if we choose r as the independent variable, then S depends on r; also, h depends on r.

Differentiating S and V with respect to r and bearing in mind that h is a function of r, we have

$$D_rS = 4\pi r + 2\pi h + 2\pi r\, D_rh \tag{9}$$

and

$$D_rV = 2\pi rh + \pi r^2\, D_rh \tag{10}$$

Because V is a constant, $D_rV = 0$; therefore, from Eq. (10),

$$2\pi rh + \pi r^2\, D_rh = 0$$

with $r \neq 0$. Divide by r and solve for $D_r h$ to obtain

$$D_r h = -\frac{2h}{r} \tag{11}$$

Substituting from (11) into (9) we obtain

$$D_r S = 2\pi\left[2r + h + r\left(-\frac{2h}{r}\right)\right]$$

$$D_r S = 2\pi(2r - h) \tag{12}$$

To find when S has a relative minimum value, set $D_r S = 0$ and obtain $2r - h = 0$, which gives

$$r = \tfrac{1}{2}h$$

To determine if this relationship between r and h makes S a relative minimum, we apply the second-derivative test. Then from (12),

$$D_r^2 S = 2\pi(2 - D_r h) \tag{13}$$

Substituting from (11) into (13) we get

$$D_r^2 S = 2\pi\left[2 - \left(\frac{-2h}{r}\right)\right] = 2\pi\left(2 + \frac{2h}{r}\right)$$

The results of the second-derivative test are summarized in Table 4.7.3.

TABLE 4.7.3

	$D_r S$	$D_r^2 S$	*Conclusion*
$r = \frac{1}{2}h$	0	+	S has a relative minimum value

From (7) and (8), S is a continuous function of r on $(0, +\infty)$. Because the one and only relative extremum of S on $(0, +\infty)$ is at $r = \frac{1}{2}h$, we conclude from Theorem 4.6.2(ii) that S has an absolute minimum value at $r = \frac{1}{2}h$. Therefore the total surface area of the tin can will be least for a specific volume when the ratio of the height to the base radius is 2.

Geometric problems involving absolute extrema occasionally are more easily solved by using trigonometric functions. The following example illustrates this fact.

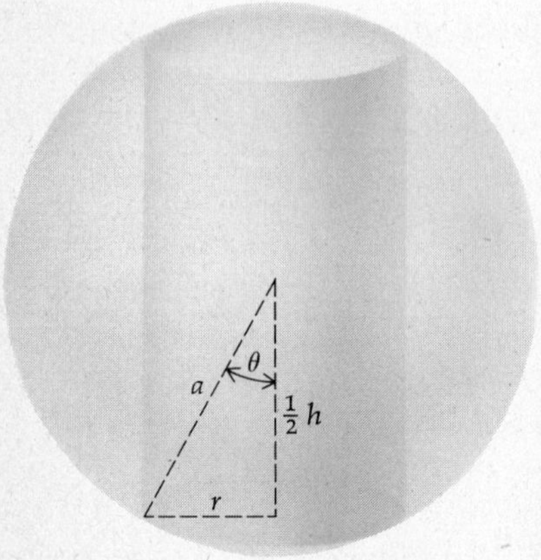

FIGURE 4.7.3

Example 4 A right-circular cylinder is to be inscribed in a sphere of given radius. Find the ratio of the altitude to the base radius of the cylinder having the largest lateral surface area.

Solution Refer to Fig. 4.7.3. The measure of the constant radius of the sphere is taken as a.

Let θ be the number of radians in the angle at the center of the sphere subtended by the radius of the cylinder, r be the number of units in the radius of the cylinder, h be the number of units in the altitude of the cylinder, and S be the number of square units in the lateral surface area of the cylinder.

From Fig. 4.7.3,

$$r = a \sin \theta \quad \text{and} \quad h = 2a \cos \theta$$

Because $S = 2\pi rh$,

$$S = 2\pi(a \sin \theta)(2a \cos \theta) = 2\pi a^2(2 \sin \theta \cos \theta) = 2\pi a^2 \sin 2\theta$$

So S is a function of θ and its domain is $(0, +\infty)$.

$$D_\theta S = 4\pi a^2 \cos 2\theta \quad \text{and} \quad D_\theta{}^2 S = -8\pi a^2 \sin 2\theta$$

Set $D_\theta S = 0$.

$$\cos 2\theta = 0$$

Because $0 < \theta < \frac{1}{2}\pi$,

$$\theta = \tfrac{1}{4}\pi$$

We apply the second derivative test for relative extrema and summarize the results in Table 4.7.4.

TABLE 4.7.4

	$D_\theta S$	$D_\theta{}^2 S$	*Conclusion*
$\theta = \frac{1}{4}\pi$	0	$-$	S has a relative maximum value

S is continuous on its domain, and because there is only one relative extremum, it follows from Theorem 4.6.2(i) that the relative maximum value of S is the absolute maximum value of S.

When $\theta = \frac{1}{4}\pi$,

$$r = a \sin \tfrac{1}{4}\pi \qquad h = 2a \cos \tfrac{1}{4}\pi$$
$$= \tfrac{1}{2}\sqrt{2}a \qquad = \sqrt{2}a$$

So for the cylinder having the largest lateral surface area, $h/r = 2$.

EXERCISES 4.7

1. A rectangular field having an area of 2700 m^2 is to be enclosed by a fence, and an additional fence is to be used to divide the field down the middle. If the cost of the fence down the middle is \$6 per meter, and the fence along the sides costs \$9 per meter, find the dimensions of the field such that the cost of the fencing will be the least.
2. A rectangular open tank is to have a square base, and its volume is to be 125 m^3. The cost per square meter for the bottom is \$20 and for the sides is \$10. Find the dimensions of the tank for the cost of the material to be the least.
3. A box manufacturer is to produce a closed box of specific volume whose base is a rectangle having a length that is three times its width. Find the most economical dimensions.
4. Solve Exercise 3 if the box is to have an open top.

5. Find an equation of the tangent line to the curve $y = x^3 - 3x^2 + 5x$ that has the least slope.

6. Find the point on the curve $y^2 - x^2 = 1$ that is closest to the point $(2, 0)$.

7. Find the point on the line $3x + y = 6$ that is closest to the origin.

8. A page of print is to contain 24 in.^{2} of printed material, a margin of $1\frac{1}{2}$ in. at the top and bottom, and a margin of 1 in. at the sides. What are the dimensions of the smallest page that will fill these requirements?

9. A one-story building having a rectangular floor space of 13,200 ft^2 is to be constructed where a walkway 22 ft wide is required in the front and back and a walkway 15 ft wide is required on each side. Find the dimensions of the lot having the least area on which this building can be located.

10. A Norman window consists of a rectangle surmounted by a semicircle. Find the shape of such a window that will admit the most light for a given perimeter.

11. Solve Exercise 10 if the window is such that the semicircle transmits only half as much light per square unit of area as the rectangle.

12. The strength of a rectangular beam is proportional to the breadth and the square of its depth. Find the dimensions of the strongest beam that can be cut from a log in the shape of a right-circular cylinder of radius a cm.

13. The stiffness of a rectangular beam is proportional to the breadth and the cube of its depth. Find the dimensions of the stiffest beam that can be cut from a log in the shape of a right-circular cylinder of radius a cm.

14. It is determined that if salaries are excluded, the number of dollars in the cost per kilometer for operating a truck is $8 + \frac{1}{300}x$, where x kilometers per hour is the speed of the truck. If the combined salary of the driver and the driver's assistant is \$27 per hour, what should be the average speed of the truck for the cost per kilometer to be least?

15. The number of dollars in the cost per hour of fuel for a cargo ship is $\frac{1}{50}v^3$, where v knots is the speed of the ship. If there are additional costs of \$400 per hour, what should be the average speed of the ship for the cost per nautical mile to be the least?

16. Two airplanes A and B are flying horizontally at the same altitude. The position of plane B is southwest of plane A and 20 km to the west and 20 km to the south of A. If plane A is flying due west at 16 km/min and plane B is flying due north at $\frac{64}{3}$ km/min, (a) in how many seconds will they be closest, and (b) what will be their closest distance?

17. An automobile traveling at the rate of 30 ft/sec is approaching an intersection. When the automobile is 120 ft from the intersection, a truck traveling at the rate of 40 ft/sec crosses the intersection. The automobile and the truck are on roads that are at right angles to each other. How long after the truck leaves the intersection are the two vehicles closest?

18. A direct current generator has an electromotive force of E volts and an internal resistance of r ohms, where E and r are constants. If R ohms is the external resistance, the total resistance is $(r + R)$ ohms, and if P watts is the power, then

$$P = \frac{E^2R}{(r + R)^2}$$

What external resistance will consume the most power?

19. A funnel of specific volume is to be in the shape of a right-circular cone. Find the ratio of the height to the base radius if the least amount of material is to be used in its manufacture.

20. A right-circular cone is to be inscribed in a sphere of given radius. Find the ratio of the altitude to the base radius of the cone of largest possible volume.

21. A right-circular cone is to be circumscribed about a sphere of given radius. Find the ratio of the altitude to the base radius of the cone of least possible volume.

22. Prove by the method of this section that the shortest distance from the point $P_1(x_1, y_1)$ to the line l having the equation $Ax + By + C = 0$ is $|Ax_1 + By_1 + C|/\sqrt{A^2 + B^2}$. (*Hint:* If s is the number of units from P_1 to a point $P(x, y)$ on l, then s will be an absolute minimum when s^2 is an absolute minimum.)

23. Find the altitude of the right-circular cone of largest possible volume that can be inscribed in a sphere of radius a units. Let 2θ be the radian measure of the vertical angle of the cone.

24. The cross section of a trough has the shape of an inverted isosceles triangle. If the lengths of the equal sides are 15 in., find the size of the vertex angle that will give maximum capacity for the trough.

4.8 CONCAVITY AND POINTS OF INFLECTION

Figure 4.8.1 shows a sketch of the graph of a function f whose first and second derivatives exist on the closed interval $[x_1, x_7]$. Because both f and f' are differentiable there, f and f' are continuous on $[x_1, x_7]$.

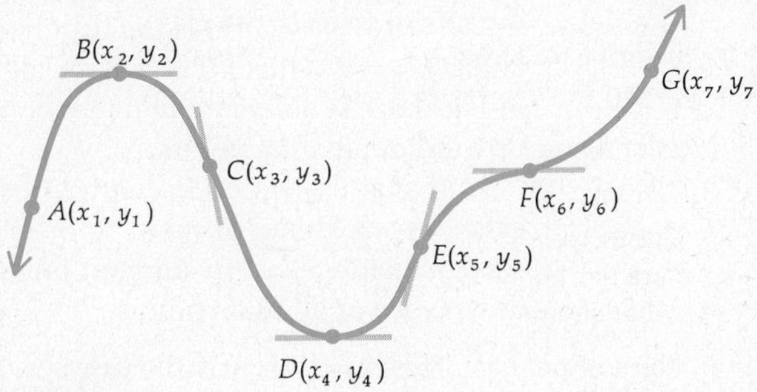

FIGURE 4.8.1

If we consider a point P moving along the graph of Fig. 4.8.1 from A to G, then the position of P varies as we increase x from x_1 to x_7. As P moves along the graph from A to B, the slope of the tangent line to the graph is positive and is decreasing; that is, the tangent line is turning clockwise, and the graph lies below the tangent line. When the point P is at B, the slope of the tangent line is zero and is still decreasing. As P moves along the graph from B to C, the slope of the tangent line is negative and is still decreasing; the tangent line is still turning clockwise, and the graph is below its tangent line. We say that the graph is *concave downward* from A to C. As P moves along the graph from C to D, the slope of the tangent line is negative and is increasing; that is, the tangent line is turning counterclockwise, and the graph is above its tangent line. At D the slope of the tangent line is zero and is still increasing. From D to E, the slope of the tangent line is positive and increasing; the tangent line is still turning counterclockwise, and the graph is above its tangent line. We say that the graph is *concave upward* from C to E. At the point C the graph changes from concave downward to concave upward. Point C is called a *point of inflection*. We have the following definitions.

4.8.1 Definition The graph of a function f is said to be **concave upward** at the point $(c, f(c))$ if $f'(c)$ exists and if there is an open interval I containing c such that for all values of $x \neq c$ in I the point $(x, f(x))$ on the graph is above the tangent line to the graph at $(c, f(c))$.

4.8.2 Definition The graph of a function f is said to be **concave downward** at the point $(c, f(c))$ if $f'(c)$ exists and if there is an open interval I containing c such that for all values of $x \neq c$ in I the point $(x, f(x))$ on the graph is below the tangent line to the graph at $(c, f(c))$.

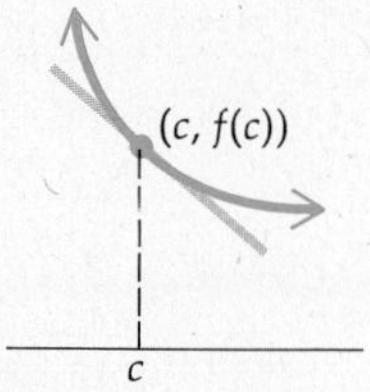

FIGURE 4.8.2

● ILLUSTRATION 1 Figure 4.8.2 shows a sketch of a portion of the graph of a function f that is concave upward at the point $(c, f(c))$, and Fig. 4.8.3 shows a sketch of a portion of the graph of a function f that is concave downward at the point $(c, f(c))$. ●

The graph of the function f of Fig. 4.8.1 is concave downward at all points $(x, f(x))$ for which x is in either of the following open intervals: (x_1, x_3) or (x_5, x_6). Similarly, the graph of the function f in Fig. 4.8.1 is concave upward at all points $(x, f(x))$ for which x is in either (x_3, x_5) or (x_6, x_7).

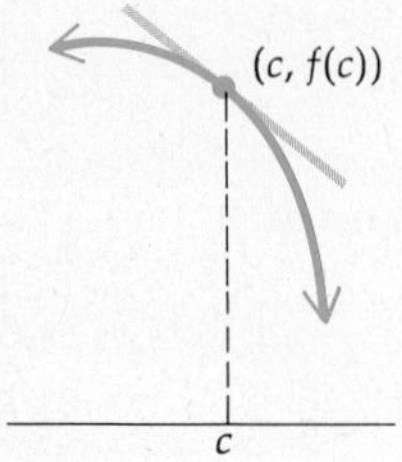

FIGURE 4.8.3

● ILLUSTRATION 2 If f is the function defined by $f(x) = x^2$, then $f'(x) = 2x$ and $f''(x) = 2$. Thus $f''(x) > 0$ for all x. Furthermore, the graph of f, shown in Fig. 4.8.4, is above all of its tangent lines. Thus the graph of f is concave upward at all of its points.

If g is the function defined by $g(x) = -x^2$, then $g'(x) = -2x$ and $g''(x) = -2$. Hence $g''(x) < 0$ for all x. Also, the graph of g, shown in Fig. 4.8.5, is below all of its tangent lines. Therefore the graph of g is concave downward at all of its points. ●

The function f of Illustration 2 is such that $f''(x) > 0$ for all x, and the graph of f is concave upward everywhere. For function g of Illustration 2, $g''(x) < 0$ for all x, and the graph of g is concave downward everywhere. These two situations are special cases of the following theorem.

4.8.3 Theorem Let f be a function which is differentiable on some open interval containing c. Then

(i) if $f''(c) > 0$, the graph of f is concave upward at $(c, f(c))$;
(ii) if $f''(c) < 0$, the graph of f is concave downward at $(c, f(c))$.

Proof of (i)

$$f''(c) = \lim_{x \to c} \frac{f'(x) - f'(c)}{x - c}$$

Because $f''(c) > 0$,

$$\lim_{x \to c} \frac{f'(x) - f'(c)}{x - c} > 0$$

Then, by Theorem 2.10.1 there is an open interval I containing c such that

$$\frac{f'(x) - f'(c)}{x - c} > 0 \tag{1}$$

for every $x \neq c$ in I.

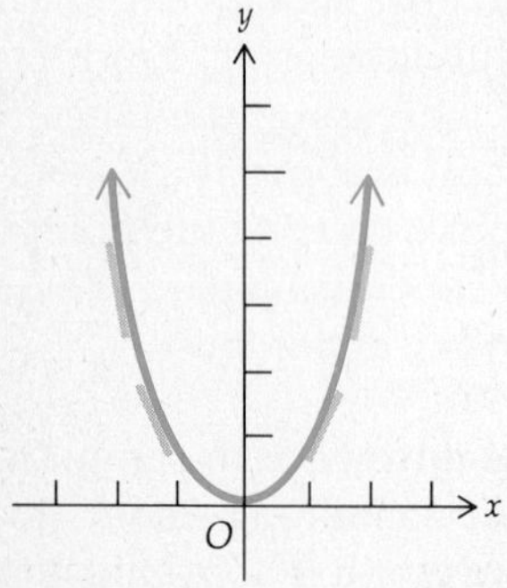

FIGURE 4.8.4

Now consider the tangent line to the graph of f at the point $(c, f(c))$.

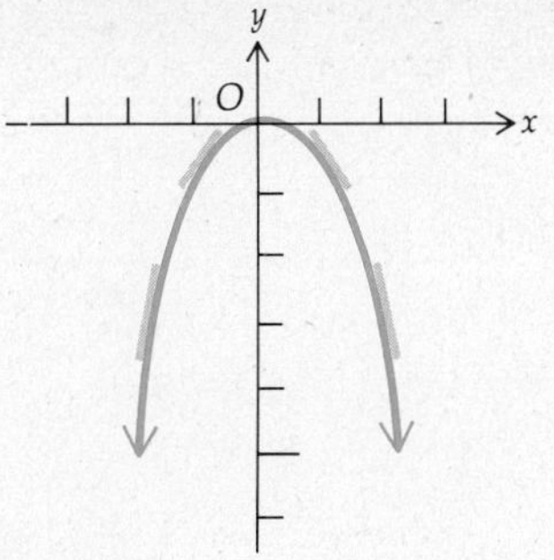

FIGURE 4.8.5

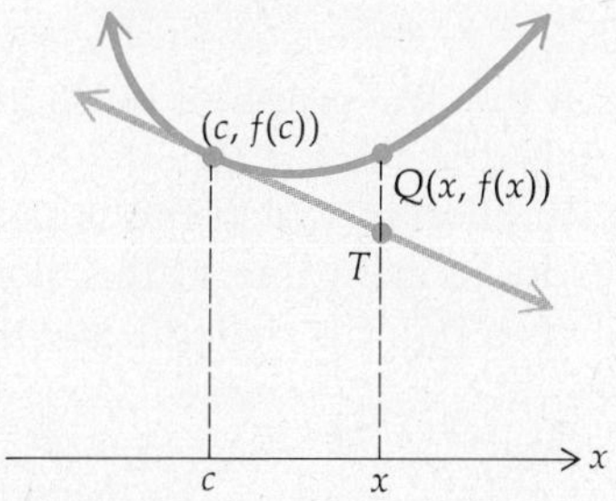

FIGURE 4.8.6

An equation of this tangent line is

$$y = f(c) + f'(c)(x - c) \tag{2}$$

Let x be a number in the interval I such that $x \neq c$, and let Q be the point on the graph of f whose abscissa is x. Through Q draw a line parallel to the y axis, and let T be the point of intersection of this line with the tangent line (see Fig. 4.8.6).

To prove that the graph of f is concave upward at $(c, f(c))$ we must show that the point Q is above the point T or, equivalently, that the directed distance $\overline{TQ} > 0$ for all values of $x \neq c$ in I. $\overline{TQ}$ equals the ordinate of Q minus the ordinate of T. The ordinate of Q is $f(x)$, and the ordinate of T is obtained from (2); so

$$\overline{TQ} = f(x) - [f(c) + f'(c)(x - c)]$$

$$\overline{TQ} = [f(x) - f(c)] - f'(c)(x - c) \tag{3}$$

From the mean-value theorem there exists some number d between x and c such that

$$f'(d) = \frac{f(x) - f(c)}{x - c}$$

That is,

$$f(x) - f(c) = f'(d)(x - c) \qquad \text{for some } d \text{ between } x \text{ and } c \tag{4}$$

Substituting from (4) into (3) we have

$$\overline{TQ} = f'(d)(x - c) - f'(c)(x - c)$$

$$\overline{TQ} = (x - c)[f'(d) - f'(c)] \tag{5}$$

Because d is between x and c, d is in the interval I, and so by taking $x = d$ in inequality (1) we obtain

$$\frac{f'(d) - f'(c)}{d - c} > 0 \tag{6}$$

To prove that $\overline{TQ} > 0$ we show that both of the factors on the right side of (5) have the same sign. If $(x - c) > 0$, then $x > c$. And because d is between x and c, then $d > c$; therefore, from inequality (6), $[f'(d) - f'(c)] > 0$. If $(x - c) < 0$, then $x < c$, and so $d < c$; therefore, from (6), $[f'(d) - f'(c)] < 0$. We conclude that $(x - c)$ and $[f'(d) - f'(c)]$ have the same sign; therefore $\overline{TQ}$ is a positive number. Thus the graph of f is concave upward at $(c, f(c))$.

The proof of part (ii) is similar and is omitted. ■

The converse of Theorem 4.8.3 is not true. For example, if f is the function defined by $f(x) = x^4$, the graph of f is concave upward at the point $(0, 0)$ but $f''(0) = 0$ (see Fig. 4.6.3). Accordingly, a sufficient condition for the graph of a function f to be concave upward at the point $(c, f(c))$ is that $f''(c) > 0$, but this is not a necessary condition. Similarly,

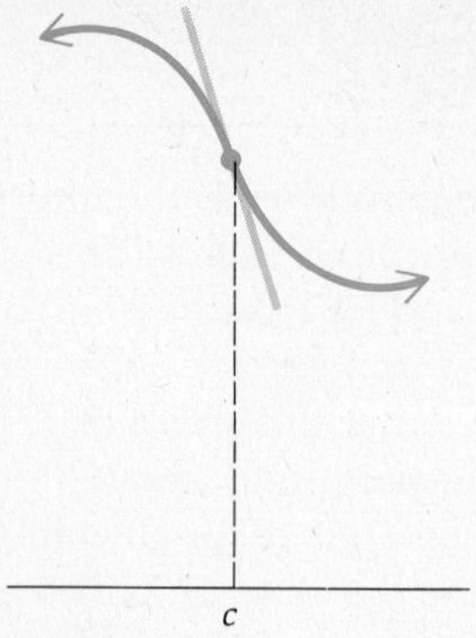

FIGURE 4.8.7

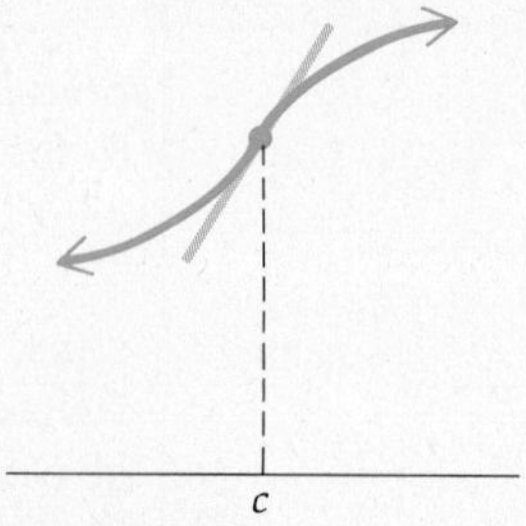

FIGURE 4.8.8

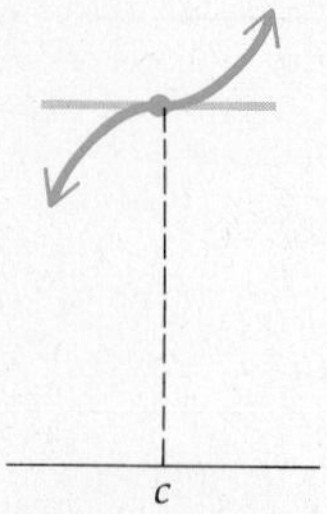

FIGURE 4.8.9

a sufficient—but not a necessary—condition that the graph of a function f be concave downward at the point $(c, f(c))$ is that $f''(c) < 0$.

If there is a point on the graph of a function at which the sense of concavity changes, then the graph crosses its tangent line at this point, as shown in Figs. 4.8.7, 4.8.8, and 4.8.9. Such a point is called a *point of inflection.*

4.8.4 Definition

The point $(c, f(c))$ is a **point of inflection** of the graph of the function f if the graph has a tangent line there, and if there exists an open interval I containing c such that if x is in I, then either

(i) $f''(x) < 0$ if $x < c$ and $f''(x) > 0$ if $x > c$, or
(ii) $f''(x) > 0$ if $x < c$ and $f''(x) < 0$ if $x > c$

● ILLUSTRATION 3 Figure 4.8.7 illustrates a point of inflection where condition (i) of Definition 4.8.4 holds; in this case the graph is concave downward at points immediately to the left of the point of inflection, and the graph is concave upward at points immediately to the right of the point of inflection. Condition (ii) is illustrated in Fig. 4.8.8, where the sense of concavity changes from upward to downward at the point of inflection. Figure 4.8.9 is another illustration of condition (i), where the sense of concavity changes from downward to upward at the point of inflection. Note that in Fig. 4.8.9 there is a horizontal tangent line at the point of inflection. ●

For the graph in Fig. 4.8.1 there are points of inflection at C, E, and F.

● ILLUSTRATION 4 Suppose it is estimated that t hours after starting work at 7 A.M. a factory worker on an assembly line has performed a particular task on $f(t)$ units and

$$f(t) = 21t + 9t^2 - t^3 \qquad 0 \le t \le 5 \tag{7}$$

In Table 4.8.1 there are function values for integer values of t from 1 through 5, and a sketch of the graph of f on $[0, 5]$ is shown in Fig. 4.8.10. From (7),

$$f'(t) = 21 + 18t - 3t^2$$

TABLE 4.8.1

t	1	2	3	4	5
$f(t)$	29	70	117	164	205

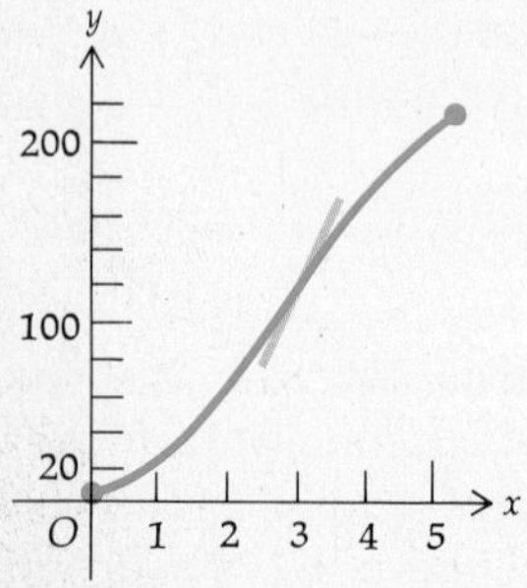

FIGURE 4.8.10

and

$$f''(t) = 18 - 6t$$
$$= 6(3 - t)$$

Observe that $f''(t) > 0$ if $0 < t < 3$ and $f''(t) < 0$ if $3 < t < 5$. From Definition 4.8.4(ii) it follows that the graph of f has a point of inflection at $t = 3$. From Theorem 4.5.3, because $f''(t) > 0$ when $0 < t < 3$, $f'(t)$ is increasing on $[0, 3]$, and because $f''(t) < 0$ when $3 < t < 5$, $f'(t)$ is decreasing on $[3, 5]$. Therefore, since $f'(t)$ is the rate of change of $f(t)$ with respect to t, we conclude that in the first three hours (from 7 A.M. until 10 A.M.) the worker is performing the task at an increasing rate, and during the remaining two hours (from 10 A.M. until noon) the worker is performing the task at a decreasing rate. At $t = 3$ (10 A.M.) the worker is producing most efficiently, and when $3 < t < 5$ (after 10 A.M.) there is a reduction in the worker's production rate. The point at which the worker is producing most efficiently is called the *point of diminishing returns;* this point is a point of inflection of the graph of f. ●

Definition 4.8.4 indicates nothing about the value of the second derivative of f at a point of inflection. The following theorem states that if the second derivative exists at a point of inflection, it must be zero there.

4.8.5 Theorem If the function f is differentiable on some open interval containing c, and if $(c, f(c))$ is a point of inflection of the graph of f, then if $f''(c)$ exists, $f''(c) = 0$.

Proof Let g be the function such that $g(x) = f'(x)$; then $g'(x) = f''(x)$. Because $(c, f(c))$ is a point of inflection of the graph of f, then $f''(x)$ changes sign at c and so $g'(x)$ changes sign at c. Therefore, by the first-derivative test (Theorem 4.5.4), g has a relative extremum at c, and c is a critical number of g. Because $g'(c) = f''(c)$, and since by hypothesis $f''(c)$ exists, it follows that $g'(c)$ exists. Therefore, by Theorem 4.2.3, $g'(c) = 0$ and $f''(c) = 0$, which is what we wanted to prove. ■

The converse of Theorem 4.8.5 is not true. That is, if the second derivative of a function is zero at a number c, it is not necessarily true that the graph of the function has a point of inflection where $x = c$. This fact is shown in the following illustration.

● ILLUSTRATION 5 Consider the function f defined by $f(x) = x^4$. $f'(x) = 4x^3$ and $f''(x) = 12x^2$. Further, $f''(0) = 0$; but because $f''(x) > 0$ if $x < 0$ and $f''(x) > 0$ if $x > 0$, the graph is concave upward at points on the graph immediately to the left of $(0, 0)$ and at points immediately to the right of $(0, 0)$. Consequently, $(0, 0)$ is not a point of inflection. In Illustration 1 of Sec. 4.6 we showed that this function f has a relative minimum value at zero. Furthermore, the graph is concave upward at the point $(0, 0)$ (see Fig. 4.6.3). ●

Example 1 For the function in Example 1 of Sec. 4.5, find the points of inflection of the graph of the function, and determine where the graph is concave upward and where it is concave downward.

Solution

$$f(x) = x^3 - 6x^2 + 9x + 1$$

$$f'(x) = 3x^2 - 12x + 9$$

$$f''(x) = 6x - 12$$

$f''(x)$ exists for all values of x; so the only possible point of inflection is where $f''(x) = 0$, which occurs at $x = 2$. To determine whether there is a point of inflection at $x = 2$, we must check to see if $f''(x)$ changes sign; at the same time we determine the concavity of the graph for the respective intervals. The results are summarized in Table 4.8.2.

TABLE 4.8.2

	$f(x)$	$f'(x)$	$f''(x)$	*Conclusion*
$x < 2$			$-$	graph is concave downward
$x = 2$	3	-3	0	graph has a point of inflection
$2 < x$			$+$	graph is concave upward

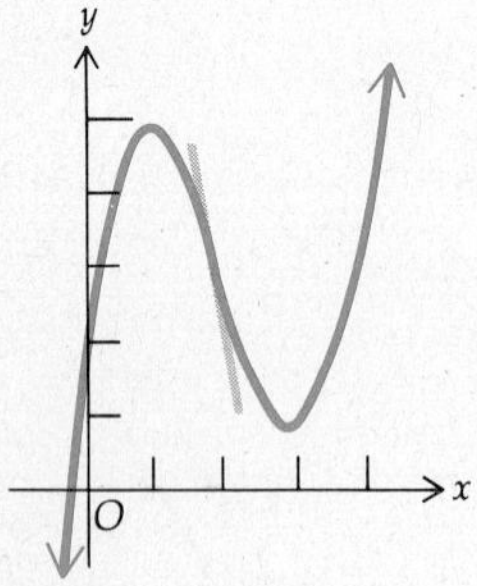

FIGURE 4.8.11

In Example 1 of Sec 4.5 we showed that f has a relative maximum value at 1 and a relative minimum value at 3. A sketch of the graph showing a segment of the inflectional tangent is shown in Fig. 4.8.11.

The graph of a function may have a point of inflection where the second derivative fails to exist. This situation is shown in the next example.

Example 2 Given $f(x) = x^{1/3}$, find the points of inflection of the graph of f, and determine where the graph is concave upward and where it is concave downward. Draw a sketch of the graph of f.

Solution

$$f(x) = x^{1/3}$$

$$f'(x) = \tfrac{1}{3}x^{-2/3}$$

$$f''(x) = -\tfrac{2}{9}x^{-5/3}$$

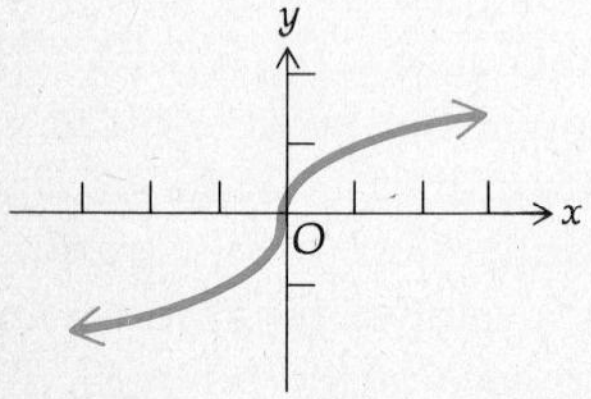

FIGURE 4.8.12

$f''(0)$ does not exist. However, $f''(x) > 0$ if $x < 0$, and $f''(x) < 0$ if $x > 0$. Hence, from Definition 4.8.4(ii), f has a point of inflection at (0, 0). The concavity of the graph is determined from the sign of $f''(x)$. The results are summarized in Table 4.8.3.

A sketch of the graph of f is shown in Fig. 4.8.12. Note that for this function $f'(0)$ also does not exist. The tangent line to the graph at (0, 0) is the y axis.

TABLE 4.8.3

	$f(x)$	$f'(x)$	$f''(x)$	*Conclusion*
$x < 0$		$+$	$+$	f is increasing; graph is concave upward
$x = 0$	0	does not exist	does not exist	graph has a point of inflection
$0 < x$		$+$	$-$	f is increasing; graph is concave downward

Example 3 If $f(x) = (1 - 2x)^3$, find the points of inflection of the graph of f, and determine where the graph is concave upward and where it is concave downward. Draw a sketch of the graph of f.

Solution

$$f(x) = (1 - 2x)^3$$

$$f'(x) = -6(1 - 2x)^2$$

$$f''(x) = 24(1 - 2x)$$

Because $f''(x)$ exists for all values of x, the only possible point of inflection is where $f''(x) = 0$, that is, at $x = \frac{1}{2}$. From the results summarized in Table 4.8.4, $f''(x)$ changes sign from $+$ to $-$ at $x = \frac{1}{2}$; so the graph has a point of inflection there. Note also that because $f'(\frac{1}{2}) = 0$, the graph has a horizontal tangent line at the point of inflection. A sketch of the graph is shown in Fig. 4.8.13.

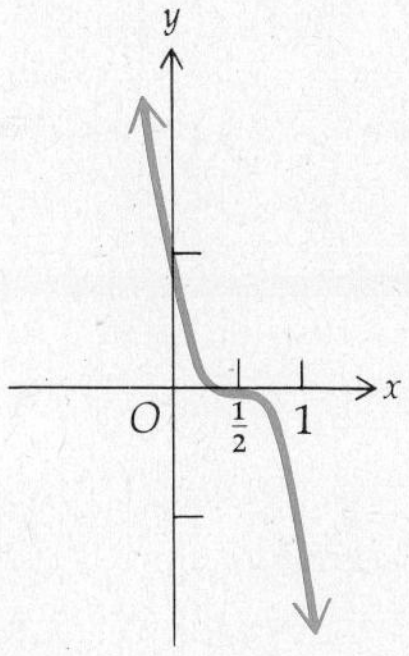

FIGURE 4.8.13

TABLE 4.8.4

	$f(x)$	$f'(x)$	$f''(x)$	*Conclusion*
$x < \frac{1}{2}$			$+$	graph is concave upward
$x = \frac{1}{2}$	0	0	0	graph has a point of inflection
$\frac{1}{2} < x$			$-$	graph is concave downward

Example 4 Find the points of inflection of the graph of the sine function. Also find the slopes of the inflectional tangents. Draw a portion of the graph of the sine function at the point of inflection having the smallest positive abscissa. Show a segment of the inflectional tangent at this point of inflection.

Solution Let

$$f(x) = \sin x$$

Then

$$f'(x) = \cos x$$

$$f''(x) = -\sin x$$

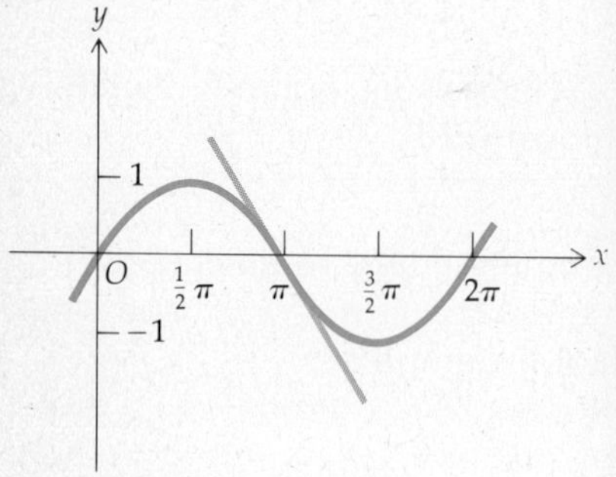

FIGURE 4.8.14

$f''(x)$ exists for all x. To determine the points of inflection set $f''(x) = 0$.

$$-\sin x = 0$$

which gives

$$x = n\pi \qquad n \text{ is any integer}$$

Because $f''(x)$ changes sign at each of these values of x, the graph has a point of inflection at every point having these abscissas. At each point of inflection,

$$f'(n\pi) = \cos n\pi \qquad n \text{ is any integer}$$

$$= \begin{cases} 1 & \text{if } n \text{ is an even integer} \\ -1 & \text{if } n \text{ is an odd integer} \end{cases}$$

Therefore the slopes of the inflectional tangents are either $+1$ or -1. Figure 4.8.14 shows the portion of the graph of the sine function in the interval $[0, 2\pi]$. At the point of inflection $(\pi, 0)$ a segment of the inflectional tangent is shown.

EXERCISES 4.8

In Exercises 1 through 20, determine where the graph of the given function is concave upward and where it is concave downward, and find the points of inflection, if there are any.

1. $f(x) = x^3 + 9x$

2. $g(x) = x^3 + 3x^2 - 3x - 3$

3. $g(x) = 2x^3 + 3x^2 - 7x + 1$

4. $f(x) = \frac{1}{12}x^4 + \frac{1}{6}x^3 - x^2$

5. $F(x) = x^4 - 8x^3 + 24x^2$

6. $f(x) = 16x^4 + 32x^3 + 24x^2 - 5x - 20$

7. $g(x) = \dfrac{x}{x^2 - 1}$

8. $G(x) = \dfrac{2x}{(x^2 + 4)^{3/2}}$

9. $f(x) = \cos 2x;\ x \in [-\pi, \pi]$

10. $g(x) = 2 \sin 3x;\ x \in [-\pi, \pi]$

11. $f(x) = (x - 2)^{1/5}$

12. $F(x) = (2x - 6)^{3/2} + 1$

13. $f(x) = \begin{cases} x^2 & \text{if } x < 0 \\ -x^2 & \text{if } 0 \le x \end{cases}$

14. $f(x) = \begin{cases} x^2 & \text{if } x < 1 \\ x^3 - 4x^2 + 7x - 3 & \text{if } 1 \le x \end{cases}$

15. $g(x) = \dfrac{x - 2}{x + 4}$

16. $f(x) = \dfrac{x + 5}{x - 3}$

17. $F(x) = (x - 3)^3(2 - x)^2$

18. $g(x) = \dfrac{(1 - x)^2}{(4 - x)^3}$

19. $f(x) = \dfrac{x}{\sqrt{x - 2}}$

20. $f(x) = x\sqrt{x^2 + 4}$

21. If $f(x) = ax^3 + bx^2$, determine a and b so that the graph of f will have a point of inflection at $(1, 2)$.

22. If $f(x) = ax^3 + bx^2 + cx$, determine a, b, and c so that the graph of f will have a point of inflection at $(1, 2)$ and so that the slope of the inflectional tangent there will be -2.

23. If $f(x) = ax^3 + bx^2 + cx + d$, determine a, b, c, and d so that f will have a relative extremum at $(0, 3)$ and so that the graph of f will have a point of inflection at $(1, -1)$.

24. If $f(x) = ax^4 + bx^3 + cx^2 + dx + e$, determine the values of a, b, c, d, and e so the graph of f will have a point of inflection at $(1, -1)$, have the origin on it, and be symmetric with respect to the y axis.

In Exercises 25 through 38, draw a portion of the graph of a function f through the point where $x = c$ if the given conditions are satisfied. It is assumed that f is continuous on some open interval containing c.

25. $f'(x) > 0$ if $x < c$; $f'(x) < 0$ if $x > c$; $f''(x) < 0$ if $x < c$; $f''(x) < 0$ if $x > c$.

26. $f'(x) > 0$ if $x < c$; $f'(x) > 0$ if $x > c$; $f''(x) > 0$ if $x < c$; $f''(x) < 0$ if $x > c$.

27. $f'(x) > 0$ if $x < c$; $f'(x) < 0$ if $x > c$; $f''(x) > 0$ if $x < c$; $f''(x) > 0$ if $x > c$.

28. $f'(x) < 0$ if $x < c$; $f'(x) > 0$ if $x > c$; $f''(x) > 0$ if $x < c$; $f''(x) < 0$ if $x > c$.

29. $f''(c) = 0$; $f'(c) = 0$; $f''(x) > 0$ if $x < c$; $f''(x) < 0$ if $x > c$.

30. $f'(c) = 0$; $f'(x) > 0$ if $x < c$; $f''(x) > 0$ if $x > c$.

31. $f''(c) = 0$; $f'(c) = 0$; $f''(x) > 0$ if $x < c$; $f''(x) > 0$ if $x > c$.

32. $f'(c) = 0$; $f'(x) < 0$ if $x < c$; $f''(x) > 0$ if $x > c$.

33. $f''(c) = 0$; $f'(c) = -1$; $f''(x) < 0$ if $x < c$; $f''(x) > 0$ if $x > c$.

34. $f''(c) = 0$; $f'(c) = \frac{1}{2}$; $f''(x) > 0$ if $x < c$; $f''(x) < 0$ if $x > c$.

35. $f'(c)$ does not exist; $f''(x) > 0$ if $x < c$; $f''(x) > 0$ if $x > c$.

36. $f'(c)$ does not exist; $f''(c)$ does not exist; $f''(x) < 0$ if $x < c$; $f''(x) > 0$ if $x > c$.

37. $\lim_{x \to c^-} f'(x) = +\infty$; $\lim_{x \to c^+} f'(x) = 0$; $f''(x) > 0$ if $x < c$; $f''(x) < 0$ if $x > c$.

38. $\lim_{x \to c^-} f'(x) = +\infty$; $\lim_{x \to c^+} f'(x) = -\infty$; $f''(x) > 0$ if $x < c$; $f''(x) > 0$ if $x > c$.

39. Draw a sketch of the graph of a function f for which $f(x)$, $f'(x)$, and $f''(x)$ exist and are positive for all x.

40. Draw a sketch of the graph of a function f for which $f(x)$, $f'(x)$, and $f''(x)$ exist and are negative for all x.

41. If $f(x) = x^5$, show that 0 is a critical number of f but that $f(0)$ is not a relative extremum. Is the origin a point of inflection of the graph of f? Prove your answer.

42. If $f(x) = 3x^2 + x|x|$, prove that $f''(0)$ does not exist but the graph of f is concave upward everywhere.

43. It is estimated that a worker in a shop that makes picture frames can paint y frames x hours after starting work at 8 A.M., and

$$y = 3x + 8x^2 - x^3 \qquad 0 \le x \le 4$$

Find at what time the worker is working most efficiently (that is, at what time does the worker reach the point of diminishing returns?). (*Hint*: See Illustration 4.)

44. Suppose that f is a function for which $f''(x)$ exists for all values of x in some open interval I and that at a number c in I, $f''(c) = 0$ and $f'''(c)$ exists and is not zero. Prove that the point $(c, f(c))$ is a point of inflection of the graph of f. (*Hint*: The proof is similar to the proof of the second-derivative test (Theorem 4.6.1).)

4.9 APPLICATIONS TO DRAWING A SKETCH OF THE GRAPH OF A FUNCTION

We now apply the discussions in Secs. 4.5, 4.6, and 4.8 to drawing a sketch of the graph of a function. If you are given $f(x)$ and wish to draw a sketch of the graph of f, proceed as follows. First find $f'(x)$ and $f''(x)$. Then the critical numbers of f are the values of x in the domain of f for which either $f'(x)$ does not exist or $f'(x) = 0$. Next apply the first-derivative test (Theorem 4.5.4) or the second-derivative test (Theorem 4.6.1) to determine whether at a critical number there is a relative maximum value, a relative minimum value, or neither. To determine the intervals on which f is increasing, find the values of x for which $f'(x)$ is positive; to determine the intervals on which f is decreasing, find the values of x for which $f'(x)$ is negative. In determining the intervals on which f is monotonic, also check the critical numbers at which f does not have a relative extremum. The values of x for which $f''(x) = 0$ or $f''(x)$ does not exist give the possible points of inflection; check to see if $f''(x)$ changes sign at each of these values of x to determine whether there actually is a point of inflection. The values of x for which $f''(x)$ is positive and those for which $f''(x)$ is negative will give points at which the graph is concave upward and points at which the graph is concave downward. It is also helpful to find the slope of each inflectional tangent. It is suggested that all the information so obtained be incorporated into a table, as in the following examples.

Example 1 Given $f(x) = x^3 - 3x^2 + 3$, find the relative extrema of f; the points of inflection of the graph of f; the intervals on which f is increasing; the intervals on which f is decreasing; where the graph is concave upward; where the graph is concave downward; and the slope of any inflectional tangent. Draw a sketch of the graph.

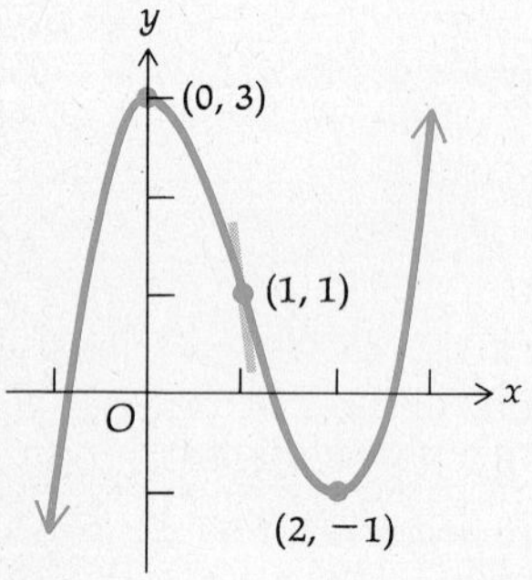

FIGURE 4.9.1

Solution $f(x) = x^3 - 3x^2 + 3$ and $f'(x) = 3x^2 - 6x$; $f''(x) = 6x - 6$. Set $f'(x) = 0$ to obtain $x = 0$ and $x = 2$. From $f''(x) = 0$ we obtain $x = 1$. In making the table, consider the points at which $x = 0$, $x = 1$, and $x = 2$, and the intervals excluding these values of x:

$$x < 0 \qquad 0 < x < 1 \qquad 1 < x < 2 \qquad 2 < x$$

TABLE 4.9.1

	$f(x)$	$f'(x)$	$f''(x)$	*Conclusion*
$x < 0$		$+$	$-$	f is increasing; graph is concave downward
$x = 0$	3	0	$-$	f has a relative maximum value; graph is concave downward
$0 < x < 1$		$-$	$-$	f is decreasing; graph is concave downward
$x = 1$	1	-3	0	f is decreasing; graph has a point of inflection
$1 < x < 2$		$-$	$+$	f is decreasing; graph is concave upward
$x = 2$	-1	0	$+$	f has a relative minimum value; graph is concave upward
$2 < x$		$+$	$+$	f is increasing; graph is concave upward

From the information in Table 4.9.1 and by plotting a few points, we obtain the sketch of the graph shown in Fig. 4.9.1.

Example 2 Given

$$f(x) = \frac{x^2}{x^2 - 4}$$

find the relative extrema of f; the points of inflection of the graph of f; the intervals on which f is increasing; the intervals on which f is decreasing; where the graph is concave upward; where the graph is concave downward; and the slope of any inflectional tangent. Also find the horizontal and vertical asymptotes. Draw a sketch of the graph.

Solution

$$f(x) = \frac{x^2}{x^2 - 4}$$

$$f'(x) = \frac{2x(x^2 - 4) - 2x(x^2)}{(x^2 - 4)^2} = \frac{-8x}{(x^2 - 4)^2}$$

$$f''(x) = \frac{-8(x^2 - 4)^2 + 8x[2(x^2 - 4)(2x)]}{(x^2 - 4)^4} = \frac{24x^2 + 32}{(x^2 - 4)^3}$$

Set $f'(x) = 0$ to obtain $x = 0$; $f''(x)$ is never zero. For Table 4.9.2 consider the points at which $x = 0$ and $x = \pm 2$ because 2 and -2 are not in the domain of f. Also for the table consider the intervals excluding these values of x:

$$x < -2 \qquad -2 < x < 0 \qquad 0 < x < 2 \qquad 2 < x$$

Because 2 and -2 are excluded from the domain of f, we compute the following limits:

$$\lim_{x \to 2^+} \frac{x^2}{x^2 - 4} = +\infty$$

$$\lim_{x \to 2^-} \frac{x^2}{x^2 - 4} = -\infty$$

$$\lim_{x \to -2^+} \frac{x^2}{x^2 - 4} = -\infty$$

$$\lim_{x \to -2^-} \frac{x^2}{x^2 - 4} = +\infty$$

Therefore $x = 2$ and $x = -2$ are vertical asymptotes of the graph.

$$\lim_{x \to +\infty} \frac{x^2}{x^2 - 4} = \lim_{x \to +\infty} \frac{1}{1 - \frac{4}{x^2}}$$

$= 1$

$$\lim_{x \to -\infty} \frac{x^2}{x^2 - 4} = \lim_{x \to -\infty} \frac{1}{1 - \dfrac{4}{x^2}}$$

$$= 1$$

Hence $y = 1$ is a horizontal asymptote of the graph.

TABLE 4.9.2

	$f(x)$	$f'(x)$	$f''(x)$	*Conclusion*
$x < -2$		$+$	$+$	f is increasing; graph is concave upward
$x = -2$	does not exist	does not exist	does not exist	
$-2 < x < 0$		$+$	$-$	f is increasing; graph is concave downward
$x = 0$	0	0	$-$	f has a relative maximum value
$0 < x < 2$		$-$	$-$	f is decreasing; graph is concave downward
$x = 2$	does not exist	does not exist	does not exist	
$2 < x$		$-$	$+$	f is decreasing; graph is concave upward

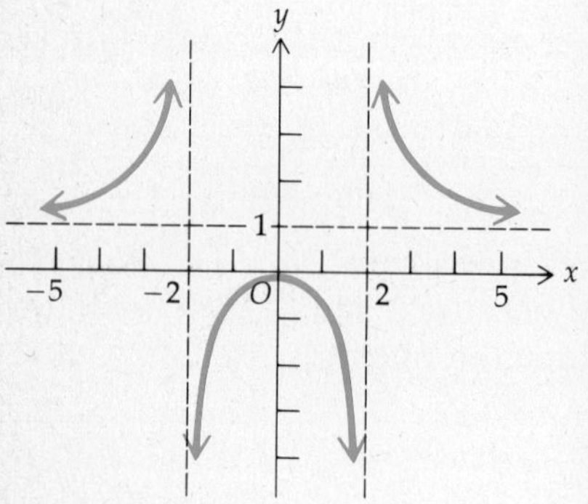

FIGURE 4.9.2

With the facts from Table 4.9.2, the asymptotes as guides, and the plotting of a few points, we obtain the sketch of the graph of f shown in Fig. 4.9.2.

Example 3 Given $f(x) = 5x^{2/3} - x^{5/3}$, find the relative extrema of f; the points of inflection of the graph of f; the intervals on which f is increasing; the intervals on which f is decreasing; where the graph is concave upward; where the graph is concave downward; and the slope of any inflectional tangent. Draw a sketch of the graph.

Solution

$$f(x) = 5x^{2/3} - x^{5/3}$$

$$f'(x) = \tfrac{10}{3}x^{-1/3} - \tfrac{5}{3}x^{2/3}$$

$$f''(x) = -\tfrac{10}{9}x^{-4/3} - \tfrac{10}{9}x^{-1/3}$$

$f'(x)$ does not exist when $x = 0$. Set $f'(x) = 0$ to obtain $x = 2$. Therefore the critical numbers of f are 0 and 2. $f''(x)$ does not exist when $x = 0$.

From $f''(x) = 0$, we obtain $x = -1$. In making the table, consider the points at which $x = -1$, $x = 0$, and $x = 2$, and the following intervals:

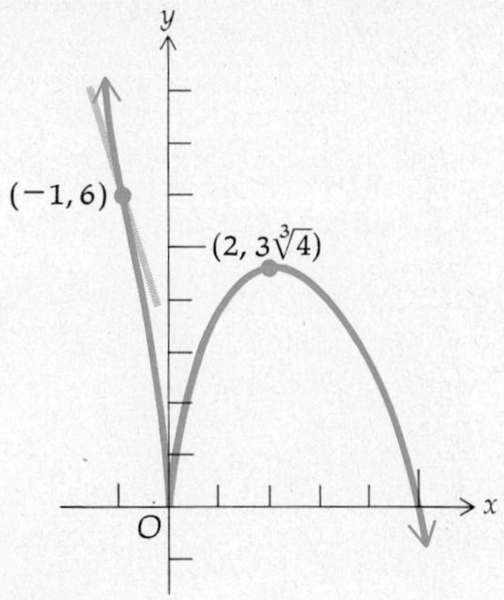

FIGURE 4.9.3

$x < -1 \qquad -1 < x < 0 \qquad 0 < x < 2 \qquad 2 < x$

A sketch of the graph, drawn from the information in Table 4.9.3 and by plotting a few points, is shown in Fig. 4.9.3.

TABLE 4.9.3

	$f(x)$	$f'(x)$	$f''(x)$	*Conclusion*
$x < -1$		$-$	$+$	f is decreasing; graph is concave upward
$x = -1$	6	-5	0	f is decreasing; graph has a point of inflection
$-1 < x < 0$		$-$	$-$	f is decreasing; graph is concave downward
$x = 0$	0	does not exist	does not exist	f has a relative minimum value
$0 < x < 2$		$+$	$-$	f is increasing; graph is concave downward
$x = 2$	$3\sqrt[3]{4} \approx 4.8$	0	$-$	f has a relative maximum value; graph is concave downward
$2 < x$		$-$	$-$	f is decreasing; graph is concave downward

EXERCISES 4.9

For each of the following functions find the relative extrema of f; the points of inflection of the graph of f; the intervals on which f is increasing; the intervals on which f is decreasing; where the graph is concave upward; where the graph is concave downward; the slope of any inflectional tangent. Find horizontal and vertical asymptotes, if there are any. Draw a sketch of the graph.

1. $f(x) = 2x^3 - 6x + 1$

2. $f(x) = x^3 + x^2 - 5x$

3. $f(x) = x^4 - 2x^3$

4. $f(x) = 3x^4 + 2x^3$

5. $f(x) = x^3 + 5x^2 + 3x - 4$

6. $f(x) = 2x^3 - \frac{1}{2}x^2 - 12x + 1$

7. $f(x) = x^4 - 3x^3 + 3x^2 + 1$

8. $f(x) = x^4 - 4x^3 + 16x$

9. $f(x) = \frac{1}{4}x^4 - \frac{1}{3}x^3 - x^2 + 1$

10. $f(x) = \frac{1}{4}x^4 - x^3$

11. $f(x) = \frac{1}{2}x^4 - 2x^3 + 3x^2 + 2$

12. $f(x) = 3x^4 + 4x^3 + 6x^2 - 4$

13. $f(x) = \begin{cases} x^2 & \text{if } x < 0 \\ 2x^2 & \text{if } 0 \le x \end{cases}$

14. $f(x) = \begin{cases} -x^3 & \text{if } x < 0 \\ x^3 & \text{if } 0 \le x \end{cases}$

15. $f(x) = \begin{cases} -x^4 & \text{if } x < 0 \\ x^4 & \text{if } 0 \le x \end{cases}$

16. $f(x) = \begin{cases} 2(x-1)^3 & \text{if } x < 1 \\ (x-1)^4 & \text{if } 1 \le x \end{cases}$

17. $f(x) = \begin{cases} 3(x-2)^2 & \text{if } x \le 2 \\ (2-x)^3 & \text{if } 2 < x \end{cases}$

18. $f(x) = x^2(x+4)^3$

19. $f(x) = (x+1)^3(x-2)^2$

20. $f(x) = 3x^5 + 5x^3$

21. $f(x) = 3x^5 + 5x^4$

22. $f(x) = |4 - x^2|$

23. $f(x) = 3\cos 2x;\ x \in [-\pi, \pi]$

24. $f(x) = 2\sin 3x;\ x \in [-\pi, \pi]$

25. $f(x) = \begin{cases} \sin x & \text{if } 0 \le x < \frac{1}{2}\pi \\ \sin(x - \frac{1}{2}\pi) & \text{if } \frac{1}{2}\pi \le x \le \pi \end{cases}$

26. $f(x) = \begin{cases} \cos x & \text{if } -\pi \le x \le 0 \\ \cos(\pi - x) & \text{if } 0 < x \le \pi \end{cases}$

27. $f(x) = \dfrac{x^2}{x-1}$

28. $f(x) = \dfrac{x}{x^2-4}$

29. $f(x) = \dfrac{x^2+1}{x^2-1}$

30. $f(x) = \dfrac{x^2+1}{x-3}$

31. $f(x) = \dfrac{2x}{x^2+1}$

32. $f(x) = \dfrac{x^2-4}{x^2-9}$

33. $f(x) = 3x^{2/3} - 2x$

34. $f(x) = x^{1/3} + 2x^{4/3}$

35. $f(x) = 3x^{4/3} - 4x$

36. $f(x) = 3x^{1/3} - x$

37. $f(x) = 2 + (x-3)^{1/3}$

38. $f(x) = 2 + (x-3)^{4/3}$

39. $f(x) = 2 + (x-3)^{5/3}$

40. $f(x) = 2 + (x-3)^{2/3}$

41. $f(x) = x^2\sqrt{4-x}$

42. $f(x) = x\sqrt{9-x^2}$

43. $f(x) = (x+2)\sqrt{-x}$

44. $f(x) = \dfrac{9x}{x^2+9}$

4.10 APPLICATIONS OF THE DERIVATIVE IN ECONOMICS

In economics the variation of one quantity with respect to another quantity may be described by either an *average* concept or a *marginal* concept. The average concept expresses the variation of one quantity over a specified range of values of a second quantity, whereas the marginal concept is the instantaneous change in the first quantity that results from a very small unit change in the second quantity. We begin our examples in economics with the definitions of average cost and marginal cost. To define a marginal concept precisely we use the notion of a limit, and this leads to the derivative.

Suppose $C(x)$ is the number of dollars in the total cost of producing x units of a commodity. The function C is called a **total cost function.** In normal circumstances x and $C(x)$ are positive. Since x represents the number of units of a commodity, x is usually a nonnegative integer. However, in order to apply the calculus we assume that x is a nonnegative real number to give the continuity requirements for the function C.

The **average cost** of producing each unit of a commodity is obtained by dividing the total cost by the number of units produced. If $Q(x)$ is the number of dollars in the average cost,

$$Q(x) = \frac{C(x)}{x}$$

and Q is called an **average cost function.**

Now suppose that the number of units in a particular output is x_1, and this is changed by Δx. Then the change in the total cost is given by $C(x_1 + \Delta x) - C(x_1)$, and the average change in the total cost with respect to the change in the number of units produced is given by

$$\frac{C(x_1 + \Delta x) - C(x_1)}{\Delta x} \qquad (1)$$

Economists use the term *marginal cost* for the limit of the quotient (1) as Δx approaches zero, provided the limit exists. This limit, being the derivative of C at x_1, gives the following definition.

4.10.1 Definition If $C(x)$ is the number of dollars in the total cost of producing x units of a commodity, then the **marginal cost,** when $x = x_1$, is given by $C'(x_1)$, if it exists. The function C' is called the **marginal cost function.**

In Definition 4.10.1, $C'(x_1)$ may be interpreted as the rate of change of the total cost when x_1 units are produced.

● ILLUSTRATION 1 Suppose that $C(x)$ is the number of dollars in the total cost of manufacturing x toys, and

$$C(x) = 110 + 4x + 0.02x^2$$

(a) The marginal cost function is C', and

$$C'(x) = 4 + 0.04x$$

(b) The marginal cost when $x = 50$ is given by $C'(50)$, and

$$\begin{aligned} C'(50) &= 4 + 0.04(50) \\ &= 6 \end{aligned}$$

Therefore the rate of change of the total cost, when 50 toys are manufactured, is \$6 per toy.

(c) The number of dollars in the actual cost of manufacturing the fifty-first toy is $C(51) - C(50)$, and

$$\begin{aligned} C(51) - C(50) &= [110 + 4(51) + 0.02(51)^2] - [110 + 4(50) + 0.02(50)^2] \\ &= 366.02 - 360 \\ &= 6.02 \end{aligned}$$

Note that the answers in (b) and (c) differ by 0.02. This discrepancy occurs because the marginal cost is the instantaneous rate of change of $C(x)$ with respect to a unit change in x. Hence $C'(50)$ is the approximate number of dollars in the cost of producing the fifty-first toy. ●

Observe that the computation of $C'(50)$ in Illustration 1 is simpler than computing $C(51) - C(50)$. Generally it is easier to compute one value of the marginal cost function than to compute two values of the total cost function. For this reason economists frequently approximate the cost of producing one additional unit by using the marginal cost function.

Specifically, $C'(k)$ dollars is the approximate cost of the $(k + 1)$st unit after the first k units have been produced.

Certain economic restrictions are imposed on the total cost function C in normal situations. They are as follows:

1. $C(0) \geq 0$. That is, when nothing is produced, the total cost must be positive or zero. $C(0)$ is called the **overhead cost** of production.
2. $C'(x) > 0$ for all x. That is, the total cost must increase as the number of units produced increases.
3. $C''(x) \geq 0$ for x greater than some positive number N. When the number of units of the commodity produced is large, the marginal cost will eventually be increasing or zero. So unless $C''(x) = 0$, the graph of the total cost function is concave upward for $x > N$. However, the marginal cost may decrease for some values of x; hence, for these values of x, $C''(x) < 0$, and therefore the graph of the total cost function will be concave downward for these values of x (see Example 2 for this situation).

• ILLUSTRATION 2 Consider a linear total cost function.

$$C(x) = mx + b$$

Observe that b represents the overhead cost. The marginal cost is given by $C'(x) = m$. If Q is the average cost function,

$$Q(x) = m + \frac{b}{x}$$

and

$$Q'(x) = -\frac{b}{x^2}$$

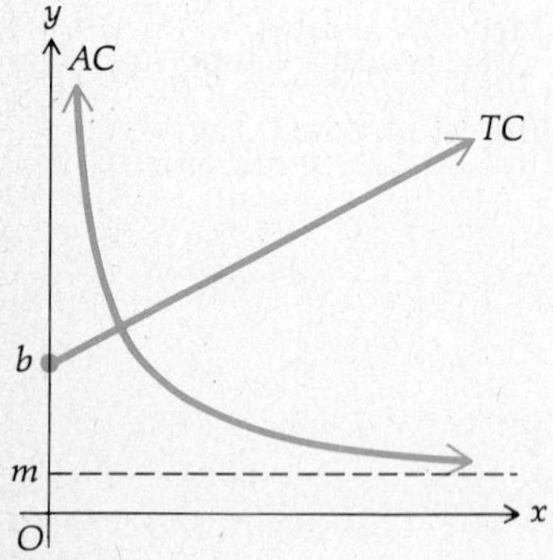

FIGURE 4.10.1

Refer to Fig. 4.10.1 for sketches of the total cost curve (labeled TC) and the average cost curve (labeled AC). The total cost curve is a segment of a straight line in the first quadrant having slope m and y intercept b. The average cost curve is a branch of an equilateral hyperbola in the first quadrant that has the line $y = m$ as horizontal asymptote. Because $Q'(x)$ is always negative, the average cost function is always decreasing, and as x increases, the value of $Q(x)$ gets closer and closer to m. The concept of the average cost approaching a constant frequently occurs in large manufacturing situations where the number of units produced is very great. •

• ILLUSTRATION 3 Suppose a quadratic total cost function

$$C(x) = ax^2 + bx + c$$

where a and c are positive. Here c is the number of dollars in the overhead cost. The total cost curve is a parabola opening upward. Because $C'(x) = 2ax + b$, a critical number of C is $-b/2a$, which gives the vertex of the parabola. We distinguish two cases: $b \geq 0$ and $b < 0$.

Case 1: $b \geq 0$. $-b/2a$ is then either negative or zero, and the vertex of the parabola is either to the left of the y axis or on the y axis. Hence the domain of C is the set of all nonnegative numbers. A sketch of TC for which $b > 0$ is shown in Fig. 4.10.2.

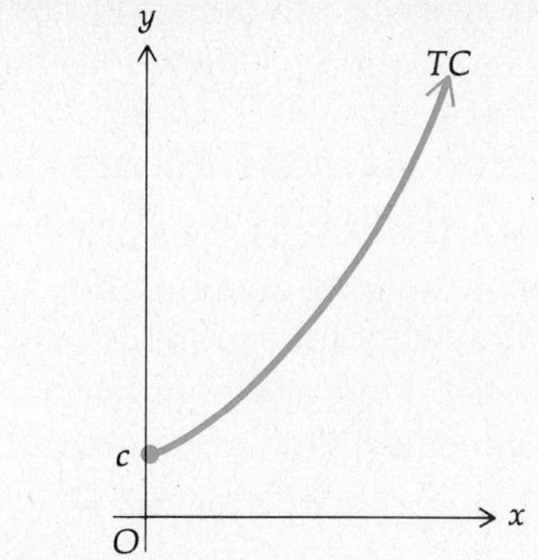

FIGURE 4.10.2

Case 2: $b < 0$. $-b/2a$ is positive; so the vertex of the parabola is to the right of the y axis, and the domain of C is restricted to numbers in the interval $[-b/2a, +\infty)$. A sketch of TC for which $b < 0$ is shown in Fig. 4.10.3. ●

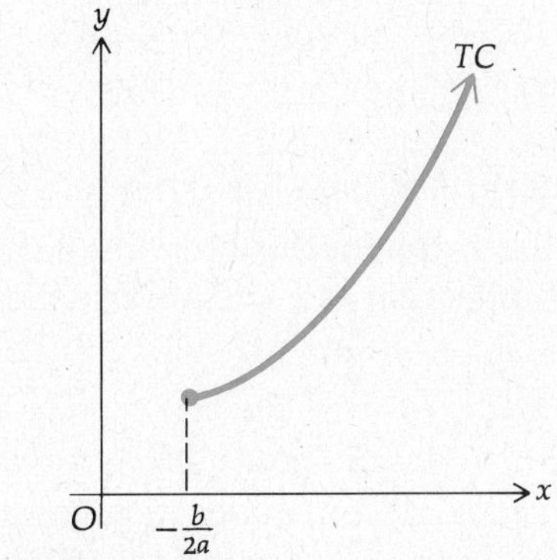

FIGURE 4.10.3

Example 1 Suppose that $100C(x)$ dollars is the total cost of producing $100x$ units of a commodity, and $C(x) = \frac{1}{2}x^2 - 2x + 5$. Find the function giving (a) the average cost and (b) the marginal cost. (c) Find the absolute minimum average unit cost. (d) Draw sketches of the total cost curve, average cost curve, and marginal cost curve on the same set of axes.

Solution $C(x) = \frac{1}{2}x^2 - 2x + 5$.

(a) If Q is the average cost function,

$$Q(x) = \frac{C(x)}{x} = \frac{1}{2}x - 2 + \frac{5}{x}$$

(b) The marginal cost function is C', and

$$C'(x) = x - 2$$

(c) We compute $Q'(x)$ and $Q''(x)$.

$$Q'(x) = \frac{1}{2} - \frac{5}{x^2} \qquad Q''(x) = \frac{10}{x^3}$$

Setting $Q'(x) = 0$, we obtain $\sqrt{10}$ as a critical number of Q, and $Q(\sqrt{10}) = \sqrt{10} - 2 \approx 1.16$. Because $Q''(\sqrt{10}) > 0$, Q has a relative minimum value of approximately 1.16 at $x = \sqrt{10}$. From the equation defining $Q(x)$ we see that Q is continuous on $(0, +\infty)$. Because the only relative extremum of Q on $(0, +\infty)$ is at $x = \sqrt{10}$, it follows from Theorem 4.6.2(ii) that Q has an absolute minimum value there. When $x = \sqrt{10} \approx 3.16$, $100x \approx 316$; so we conclude that the absolute minimum average unit cost is \$1.16 when 316 units are produced.

The sketches of the curves TC, AC, and MC (the marginal cost curve) are shown in Fig. 4.10.4.

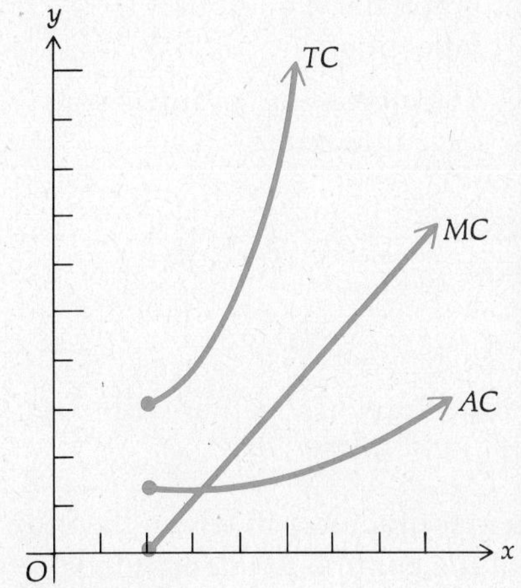

FIGURE 4.10.4

In Fig. 4.10.4, observe that the lowest point on curve AC occurs at the point of intersection of the curves AC and MC, which is the point where the average cost and the marginal cost are equal. This is true in general, and it follows from the fact that $Q'(x) = 0$ here. Because $Q(x) = C(x)/x$, then

$$Q'(x) = \frac{xC'(x) - C(x)}{x^2}$$

Thus $Q'(x) = 0$ when $xC'(x) - C(x) = 0$ or, equivalently, when

$$C'(x) = \frac{C(x)}{x}$$

You should note the economic significance that when the marginal cost and the average cost are equal, the commodity is being produced at the very lowest average unit cost.

The following example involving a cubic cost function illustrates the case in which the concavity of the graph of the total cost function changes.

Example 2 Draw a sketch of the graph of the total cost function C for which $C(x) = x^3 - 6x^2 + 13x + 1$. Determine where the graph is concave upward and where it is concave downward. Find any points of inflection and an equation of any inflectional tangent. Draw a segment of the inflectional tangent.

Solution

$$C(x) = x^3 - 6x^2 + 13x + 1$$
$$C'(x) = 3x^2 - 12x + 13$$
$$C''(x) = 6x - 12$$

$C'(x)$ can be written as $3(x-2)^2 + 1$. Hence $C'(x) > 0$ for all x. $C''(x) = 0$ when $x = 2$. To determine the concavity of the graph for the intervals $(0, 2)$ and $(2, +\infty)$ and if the graph has a point of inflection at $x = 2$, we use the results summarized in Table 4.10.1.

TABLE 4.10.1

	$C(x)$	$C'(x)$	$C''(x)$	*Conclusion*
$0 < x < 2$		+	−	C is increasing; graph is concave downward
$x = 2$	11	1	0	graph has a point of inflection
$2 < x$		+	+	C is increasing; graph is concave upward

An equation of the inflectional tangent is

$$y - 11 = 1(x - 2)$$
$$x - y + 9 = 0$$

A sketch of the graph of the total cost function together with a segment of the inflectional tangent is shown in Fig. 4.10.5.

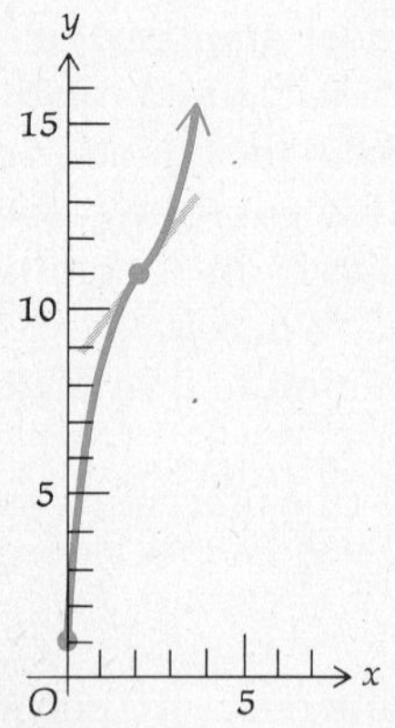

FIGURE 4.10.5

Consider circumstances affecting a manufacturer in which the only variables are the price and the quantity of the commodity demanded. Let p dollars be the price of one unit of the commodity, and let x be the number of units of the commodity demanded.

Upon reflection, it should seem reasonable that the amount of the commodity demanded in the marketplace by consumers will depend on the price of the commodity. As the price falls, consumers generally demand more of the commodity. Should the price rise, the opposite will occur: Consumers will demand less.

An equation giving the relationship between the amount, given by x, of a commodity demanded and the price, given by p, is called a **demand equation.** Such an equation is arrived at by the application of statistical methods to economic data. If the equation is solved for p, we have

$$p = f(x) \tag{2}$$

The function f in this equation is called the **price function,** and $f(x)$ dollars is the price of one unit of the commodity when x units are demanded. In normal economic situations, the domain of the price function consists, as you would expect, of nonnegative numbers.

The graph of the demand equation is called the **demand curve.** When drawing a sketch of the demand curve, it is customary in economics to use the vertical axis to represent the price and the horizontal axis to represent the demand. Because the given demand equation may apply for only certain values of x and p, it is often necessary to restrict them to closed intervals; that is, $x \in [0, a]$ and $p \in [0, b]$. Even though in actual practice quantities and prices usually take on rational values only, we allow x and p to be any real numbers within these closed intervals.

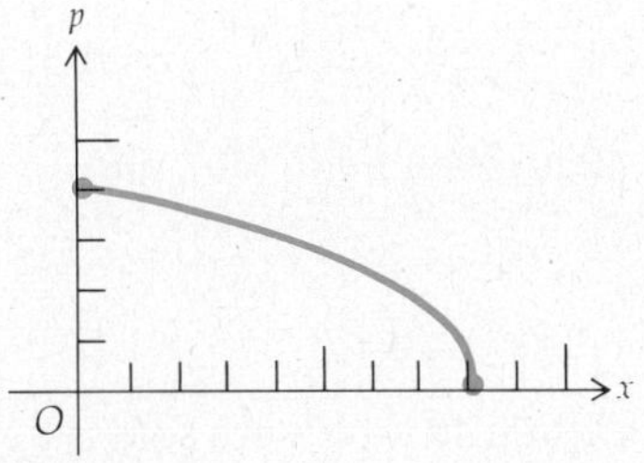

FIGURE 4.10.6

• ILLUSTRATION 4 Consider the following demand equation:

$$p^2 + 2x - 16 = 0 \tag{3}$$

Because in normal economic situations the variables x and p will be nonnegative, when (3) is solved for p we reject the negative values of p and get

$$p = \sqrt{16 - 2x} \tag{4}$$

which is of the form of (2). Hence the price function for the demand equation (3) is the function f for which $f(x) = \sqrt{16 - 2x}$. A sketch of the demand curve is shown in Fig. 4.10.6. The graph is restricted to the first quadrant because of the requirement that x and p be nonnegative. From (4) we see that $p \leq 4$ and $16 - 2x \geq 0$ or, equivalently, $x \leq 8$. Therefore $x \in [0, 8]$ and $p \in [0, 4]$. •

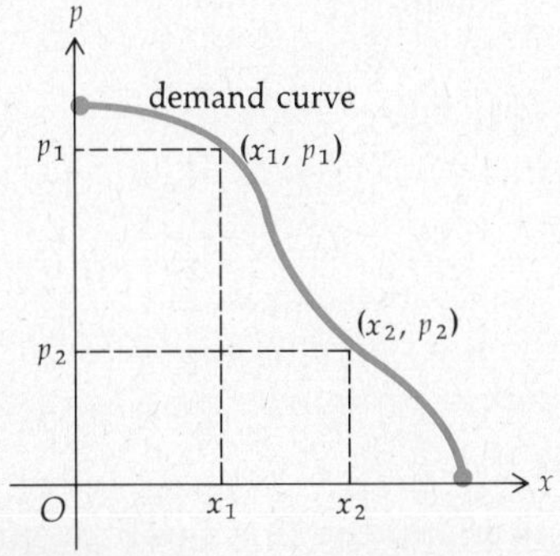

FIGURE 4.10.7

In addition to the restriction that x and p be nonnegative under normal circumstances, we impose the condition that as the price per unit decreases, the demand for a commodity increases, and as the price per unit increases, the demand for a commodity decreases; that is, if p_1 is the number of dollars in the price of x_1 units of a commodity and p_2 is the number of dollars in the price per unit of x_2 units, then $x_2 > x_1$ if and only if $p_2 < p_1$. This condition, in turn, reflects nothing more than "common economic sense" and is illustrated in Fig. 4.10.7.

Another function important in economics is the **total revenue function,** denoted by R, and

$$R(x) = px$$

Because p and x are nonnegative under normal circumstances, so is $R(x)$.

When $x \neq 0$, from the above equation,

$$\frac{R(x)}{x} = p$$

which shows that the revenue per unit (the average revenue) and the price per unit are equal.

4.10.2 Definition If $R(x)$ is the number of dollars in the total revenue obtained when x units of a commodity are demanded, then the **marginal revenue,** when $x = x_1$, is given by $R'(x_1)$, if it exists. The function R' is called the **marginal revenue function.**

$R'(x_1)$ may be positive, negative, or zero, and it may be interpreted as the rate of change of the total revenue when x_1 units are demanded. Just as $C'(k)$ dollars is the approximate cost of the $(k + 1)$st unit after the first k units have been produced, $R'(k)$ is the approximate revenue from the sale of the $(k + 1)$st unit after the first k units have been sold.

● ILLUSTRATION 5 Suppose that $R(x)$ dollars is the total revenue received from the sale of x tables, and

$$R(x) = 300x - \frac{x^2}{2}$$

(a) The marginal revenue function is R' and

$$R'(x) = 300 - x$$

(b) The marginal revenue when $x = 40$ is given by $R'(40)$, and

$$\begin{aligned} R'(40) &= 300 - 40 \\ &= 260 \end{aligned}$$

Thus the rate of change of the total revenue when 40 tables are sold is \$260 per table.

(c) The number of dollars in the actual revenue from the sale of the forty-first table is $R(41) - R(40)$, and

$$\begin{aligned} R(41) - R(40) &= \left[300(41) - \frac{(41)^2}{2}\right] - \left[300(40) - \frac{(40)^2}{2}\right] \\ &= [12{,}300 - 840.50] - [12{,}000 - 800] \\ &= 11{,}459.50 - 11{,}200 \\ &= 259.50 \end{aligned}$$

Hence the actual revenue from the sale of the forty-first table is \$259.50. In part (b) we obtained $R'(40) = 260$, and \$260 is an approximation of the revenue received from the sale of the forty-first table ●

The graphs of the total revenue function and the marginal revenue function are called the **total revenue curve** (labeled TR) and the **marginal revenue curve** (labeled MR), respectively.

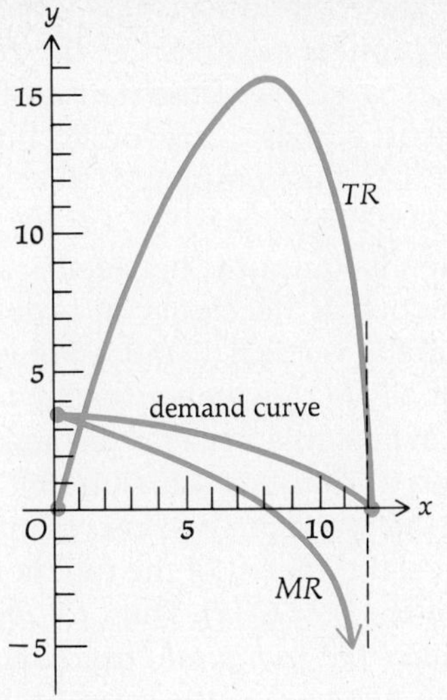

FIGURE 4.10.8

Example 3 The demand equation for a particular commodity is $p^2 + x - 12 = 0$. Find the total revenue function and the marginal revenue function. Draw sketches of the demand curve, total revenue curve, and marginal revenue curve on the same set of axes.

Solution If the demand equation is solved for p, we find $p = \pm\sqrt{12 - x}$. Because $R(x) = px$ and $p \geq 0$,

$$R(x) = x\sqrt{12 - x}$$

and

$$R'(x) = \sqrt{12 - x} - \frac{x}{2\sqrt{12 - x}}$$

$$R'(x) = \frac{24 - 3x}{2\sqrt{12 - x}}$$

Setting $R'(x) = 0$ we obtain $x = 8$. From the information in Table 4.10.2, the required sketches are drawn as shown in Fig. 4.10.8.

TABLE 4.10.2

x	p	$R(x)$	$R'(x)$
0	$\sqrt{12}$	0	$\sqrt{12}$
3	3	9	$\frac{5}{2}$
8	2	16	0
11	1	11	$-\frac{9}{2}$
12	0	0	does not exist

The total profit earned by a business is defined to be the difference between the total revenue and the total cost. That is, if $P(x)$ dollars is the total profit obtained by producing and selling x units of a commodity, then

$$P(x) = R(x) - C(x) \tag{5}$$

where $R(x)$ dollars is the total revenue and $C(x)$ dollars is the total cost. The function P is called the **total profit function.** From (5) we obtain

$$P'(x) = R'(x) - C'(x) \tag{6}$$

The function P' is called the **marginal profit function,** and $P'(k)$ is the approximate profit realized from the $(k + 1)$st unit after k units have been produced and sold. From (6), the marginal profit is the marginal revenue minus the marginal cost.

In Fig. 4.10.9(a) there are sketches of the graphs of a total cost function and a total revenue function of a particular company, and in Fig. 4.10.9(b) there is the sketch of the graph of the company's corresponding total profit function (denoted by TP). Observe in Fig. 4.10.9(a) that when curve TC is above curve TR (when the cost is greater than the revenue), then in Fig. 4.10.9(b) curve TP is below the x axis (the profit is negative; i.e., the company shows a loss); this occurs when $0 \leq x < x_2$ and when

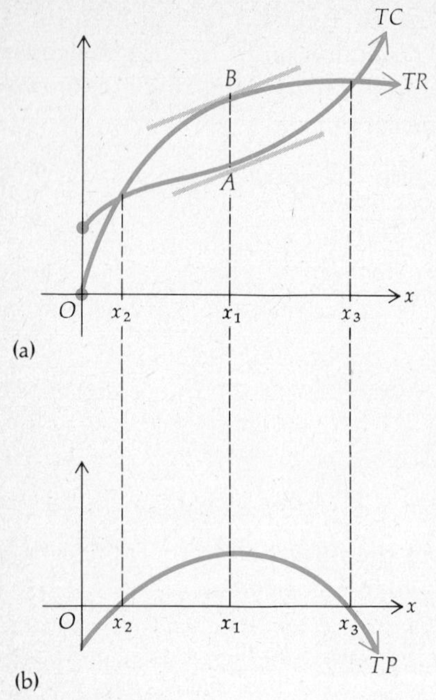

FIGURE 4.10.9

$x > x_3$. When curve *TC* is below curve *TR* (when the cost is less than the revenue), curve *TP* is above the x axis (the profit is positive; i.e., the company shows a profit); this occurs when $x_2 < x < x_3$. At the points of intersection of curves *TC* and *TR* (when the cost equals the revenue), curve *TP* intersects the x axis (the profit is zero; i.e., the company is breaking even); this occurs at x_2 and x_3.

The company is showing a profit when its output is between x_2 and x_3. Let us determine what level of production is necessary in order to obtain the maximum profit. In Fig. 4.10.9(a) the vertical distance between curves *TR* and *TC* for a particular value of x is $P(x)$, which gives the total profit corresponding to that value of x. When this vertical distance is largest, $P(x)$ has its maximum value. Observe from Fig. 4.10.9(a) that the distance AB is the largest vertical distance between the two curves in the interval $[x_2, x_3]$, and this occurs at the critical number x_1 of the function P. From (6) we see that $P'(x) = 0$ if and only if $R'(x) = C'(x)$. Thus, *for maximum profit the marginal revenue must equal the marginal cost.* Note in Fig. 4.10.9(a) that at points B and A on the two curves the tangent lines are parallel, and hence $R'(x_1) = C'(x_1)$.

From (6),

$$P''(x) = R''(x) - C''(x) \tag{7}$$

Because P will have a relative maximum function value at a number x for which $P'(x) = 0$ and $P''(x) < 0$, we can conclude from (6) and (7) that this will occur for a value of x for which $R'(x) = C'(x)$ and $R''(x) < C''(x)$. Therefore the profit function will have a relative maximum function value when the marginal revenue equals the marginal cost and the slope of the graph of the marginal revenue function is less than the slope of the graph of the marginal cost function. These concepts are illustrated in the following example, in which we consider *perfect competition.* When a company is operating under **perfect competition,** there is a great number of small firms and so any one firm cannot affect price by increasing production. Therefore, under perfect competition the price of a commodity is constant, and the company can sell as much as it wishes to sell at this constant price. The demand curve in such a case is horizontal; it is a line parallel to the x axis.

Example 4 Under perfect competition a firm can sell at a price of \$100 per unit all of a particular commodity it produces. If x units are produced each day, the number of dollars in the total cost of each day's production is $x^2 + 20x + 700$. Find the number of units that should be produced each day in order for the firm to have the greatest daily total profit.

Solution If $R(x)$ dollars is the total revenue from the sale of x units, then

$$R(x) = 100x \tag{8}$$

If C is the total cost function, then

$$C(x) = x^2 + 20x + 700 \tag{9}$$

If P is the total profit function, then

$$P(x) = R(x) - C(x)$$

$$P(x) = 100x - (x^2 + 20x + 700)$$

$$P(x) = -x^2 + 80x - 700 \qquad (10)$$

Observe that x is in $[0, +\infty)$.

From (8) and (9),

$$R'(x) = 100 \quad \text{and} \quad R''(x) = 0$$

$$C'(x) = 2x + 20 \quad \text{and} \quad C''(x) = 2$$

Equating $C'(x)$ and $R'(x)$ we get

$$2x + 20 = 100$$

$$x = 40$$

Because $R''(x) < C''(x)$ for all x, $P''(x) = R''(x) - C''(x) < 0$. Therefore $P(40)$ is a relative maximum value. It is also an absolute maximum value, because P is continuous on $[0, +\infty)$ and $P(40)$ is the only extremum of P for x in $[0, +\infty)$.

Therefore the daily total profit is greatest when 40 units are produced for the day.

An alternate method of solution is to find $P'(x)$ and $P''(x)$ from (10) and proceed as customary in extremum problems.

For a given demand equation, the amount demanded by the consumer depends only upon the price of the commodity. Under a **monopoly,** which means that there is only one producer of a certain commodity, price and hence demand can be controlled by regulating the quantity of the commodity produced. Generally the price decreases as the producer increases the output. The producer under a monopoly is called a **monopolist.** The monopolist wishes to control the quantity produced and hence the price per unit (determined from the demand equation) so that the profit will be as large as possible.

Example 5 Suppose that under a monopoly the demand equation for a certain commodity is $p = 4 - 0.0002x$, where x units are produced each day and p dollars is the price of each unit. The number of dollars in the total cost of producing x units is $600 + 3x$. If the daily profit is to be as large as possible, find the number of units that the monopolist should produce each day, the price of each unit, and the daily profit.

Solution Because the demand equation is $p = 4 - 0.0002x$ and x and p must be nonnegative, it follows that x is in the closed interval $[0, 20{,}000]$. Because $R(x) = xp$, we have

$$R(x) = 4x - 0.0002x^2 \qquad x \in [0, 20{,}000] \qquad (11)$$

The total cost function is given by

$$C(x) = 600 + 3x \tag{12}$$

If $P(x)$ dollars is the total profit, $P(x) = R(x) - C(x)$, and so

$$P(x) = x - 0.0002x^2 - 600 \qquad x \in [0, 20{,}000]$$

From (11) and (12),

$$R'(x) = 4 - 0.0004x \quad \text{and} \quad R''(x) = -0.0004$$

$$C'(x) = 3 \quad \text{and} \quad C''(x) = 0$$

Equating $R'(x)$ and $C'(x)$ we get

$$4 - 0.0004x = 3$$

$$x = 2500$$

Because $R''(x)$ is always less than $C''(x)$, $P(2500)$ is an absolute maximum value (observe that $P(x)$ is negative for $x = 0$ and $x = 20{,}000$). $P(2500) = 650$, and when $x = 2500$, $p = 3.50$.

Therefore, to have the greatest daily profit, 2500 units should be produced each day to be sold at \$3.50 each for a total profit of \$650.

Example 6 Solve Example 5 if a tax of \$0.20 is levied by the government on the monopolist for each unit produced.

Solution With the added tax, the total cost function is now given by

$$C(x) = (600 + 3x) + 0.20x$$

and so the marginal cost function is given by

$$C'(x) = 3.20$$

Equating $R'(x)$ and $C'(x)$ we get

$$4 - 0.0004x = 3.20$$

$$x = 2000$$

As in Example 5, it follows that when $x = 2000$, P has an absolute maximum value on $[0, 20{,}000]$. From $P(x) = 0.8x - 0.0002x^2 - 600$ we have $P(2000) = 200$. From the demand equation $p = 4 - 0.0002x$ we obtain $p = 3.60$ when $x = 2000$.

Therefore, if the tax of \$0.20 per unit is levied, the monopolist should produce only 2000 units each day, and they should be sold at \$3.60 per unit in order to attain a maximum daily total profit of \$200.

It is interesting to note that in comparing the results of Examples 5 and 6, the entire \$0.20 increase should not be passed on to the consumer to achieve the greatest daily profit. That is, it is most profitable to raise the unit price by only \$0.10. The economic significance of this result is that consumers are sensitive to price changes, which prohibits the monopolist from passing on the tax completely to the consumer.

A further note of interest in Examples 5 and 6 is how the overhead cost of a company does not affect the determination of the number of units to be produced or the unit price such that maximum profit is obtained. Regardless of whether anything is produced, the overhead cost must be met. In Examples 5 and 6, because $C(x) = 600 + 3x$ and $600 + 3.2x$, respectively, the overhead cost is \$600. If the 600 in the expression for $C(x)$ is replaced by any constant k, $C'(x)$ is not affected; hence the value of x for which the marginal cost equals the marginal revenue is not affected by any such change. Of course, a change in overhead cost affects the unit cost and hence the actual profit; however, if a company is to have the greatest profit possible, a change in its overhead cost will not affect the number of units to be produced nor the price per unit.

EXERCISES 4.10

1. The number of dollars in the total cost of manufacturing x watches in a certain plant is given by $C(x) = 1500 + 3x + x^2$. Find (a) the marginal cost function; (b) the marginal cost when $x = 40$; (c) the actual cost of manufacturing the forty-first watch.

2. If $C(x)$ dollars is the total cost of manufacturing x paperweights and

$$C(x) = 200 + \frac{50}{x} + \frac{x^2}{5}$$

find (a) the marginal cost function; (b) the marginal cost when $x = 10$; (c) the actual cost of manufacturing the eleventh paperweight.

3. Suppose a liquid is produced by a certain chemical process and the total cost function C is given by $C(x) = 6 + 4\sqrt{x}$, where $C(x)$ dollars is the total cost of producing x liters of the liquid. Find (a) the marginal cost when 16 liters are produced and (b) the number of liters produced when the marginal cost is \$0.40 per liter.

4. The number of dollars in the total cost of producing x units of a certain commodity is given by $C(x) = 40 + 3x + 9\sqrt{2x}$. Find (a) the marginal cost when 50 units are produced and (b) the number of units produced when the marginal cost is \$4.50.

5. The number of dollars in the total cost of producing x units of a commodity is given by $C(x) = x^2 + 4x + 8$. Find (a) the average cost function and (b) the marginal cost function. (c) Find the absolute minimum average unit cost. (d) Draw sketches of the total cost, average cost, and marginal cost curves on the same set of axes. Observe that the average cost and marginal cost are equal when the average cost has its least value.

6. Follow the instructions of Exercise 5 if $C(x) = 3x^2 + x + 3$.

7. If $C(x)$ dollars is the total cost of producing x units of a commodity and $C(x) = 3x^2 - 6x + 4$, find (a) the average cost function and (b) the marginal cost function. (c) What is the range of C? (d) Find the absolute minimum average unit cost. (e) Draw sketches of the total cost, average cost, and marginal cost curves on the same set of axes. Observe that the average cost and marginal cost are equal when the average cost has its least value.

8. If $C(x)$ dollars is the total cost of producing x units of a commodity and $C(x) = 2x^2 - 8x + 18$, find (a) the domain and range of C, (b) the average cost function, (c) the absolute minimum average unit cost, and (d) the marginal cost function. (e) Draw sketches of the total cost, average cost, and marginal cost curves on the same set of axes.

9. The total cost function C is given by $C(x) = \frac{1}{3}x^3 - 2x^2 + 5x + 2$. (a) Determine the range of C. (b) Find the marginal cost function. (c) Find the interval on which the marginal cost function is decreasing and the interval on which it is increasing. (d) Draw a sketch of the graph of the total cost function; determine where the graph is concave upward and where it is concave downward, and find the points of inflection and an equation of any inflectional tangent.

10. The number of hundreds of dollars in the total cost of producing $100x$ radios per day in a certain factory is given by $C(x) = 4x + 5$. Find (a) the average cost function and (b) the marginal cost function. (c) Show that there is no absolute minimum average unit cost. (d) What is the smallest number of radios that the factory must produce in a day so that the average cost per radio is less than \$7? (e) Draw sketches of the total cost, average cost, and marginal cost curves on the same set of axes.

11. The overhead cost of a manufacturer of children's toys is \$400 per week, and other costs amount to \$3 for each toy produced. Find (a) the total cost function, (b) the average cost function, and (c) the marginal cost function. (d) Show that there is no absolute minimum average unit cost. (e) What is the smallest number of toys that must be produced so that the average cost per toy is less than \$3.42? (f) Draw sketches of the graphs of the functions in (a), (b), and (c) on the same set of axes.

12. The total revenue received from the sale of x desks is $R(x)$ dollars and $R(x) = 200x - \frac{1}{3}x^2$. Find (a) the marginal revenue function; (b) the marginal revenue when $x = 30$; (c) the actual revenue from the sale of the thirty-first desk.

13. If $R(x)$ dollars is the total revenue received from the sale of x television sets and $R(x) = 600x - \frac{1}{20}x^3$, find (a) the marginal revenue function; (b) the marginal revenue when $x = 20$; (c) the actual revenue from the sale of the twenty-first television set.

14. If the demand equation for a particular commodity is $3x + 4p = 12$, find (a) the price function, (b) the total revenue function, and (c) the marginal revenue function. Draw sketches of the demand, total revenue, and marginal revenue curves on the same set of axes. Verify that the marginal revenue curve intersects the x axis at the point whose abscissa is the value of x for which the total revenue is greatest and that the demand curve intersects the x axis at the point whose abscissa is twice that.

15. The demand equation for a particular commodity is $x^2 + p^2 = 36$, where $100x$ units are demanded when p dollars is the price per unit. Find (a) the total revenue function and (b) the marginal revenue function. (c) Find the absolute maximum total revenue.

16. Follow the instructions of Exercise 15 if the demand equation is $px^2 + 9p - 18 = 0$.

17. A company that builds and sells desks is operating under perfect competition and can sell at a price of \$200 per desk all the desks it produces. If x desks are produced and sold each week and $C(x)$ dollars is the total cost of the week's production, then $C(x) = x^2 + 40x + 3000$. Determine how many desks should be built each week in order for the manufacturer to have the greatest weekly total profit. What is the greatest weekly total profit?

18. A firm operating under perfect competition manufactures and sells portable radios. The firm can sell at a price of \$75 per radio all the radios it produces. If x radios are manufactured each day and $C(x)$ dollars is the daily total cost of production, then $C(x) = x^2 + 25x + 100$. How many radios should be produced each day in order for the firm to have the greatest daily total profit? What is the greatest daily total profit?

19. Under a monopoly the demand equation for a certain commodity is $x + p = 140$, where x is the number of units produced daily and p is the number of dollars in the price of each unit. The number of dollars in the total cost of producing x units is given by $C(x) = x^2 + 20x + 300$, and x is in the closed interval $[0, 140]$. (a) Find the total profit function, and draw a sketch of its graph. (b) On a set of axes different from that in (a) draw sketches of the total revenue and total cost curves, and show the geometrical interpretation of the total profit function. (c) Find the marginal revenue and marginal cost functions. (d) Find the maximum daily profit. (e) Draw sketches of the graphs of the marginal revenue function and the marginal cost function on the same set of axes, and show that they intersect at the point for which the value of x makes the total profit a maximum.

20. Follow the instructions of Exercise 19 if the demand equation is $x^2 + p = 320$, $C(x) = 20x$, and x is in the closed interval $[0, 8\sqrt{5}]$.

21. Suppose that under a monopoly the demand equation of a particular article of merchandise is $p = 6 - \frac{1}{5}\sqrt{x - 100}$, where p dollars is the price per article when x articles are demanded and $x \in [100, 1000]$. If $C(x)$ dollars is the total cost of producing x articles, then $C(x) = 2x + 100$. (a) Find the marginal revenue and marginal cost functions. (b) Find the value of x that yields the maximum profit.

22. The demand equation for a certain commodity is $p = (8 - \frac{1}{100}x)^2$, where p dollars is the price per unit when x units are demanded and $x \in [0, 800]$. The total cost function is given by $C(x) = 18x - \frac{1}{100}x^2$, where $C(x)$ dollars is the total cost of producing x units. (a) Find the marginal revenue and marginal cost functions. (b) Find the value of x that yields the maximum total profit.

23. A monopolist determines that if $C(x)$ cents is the total cost of producing x units of a certain commodity, then $C(x) = 20{,}000 + 25x$. The demand equation is $p = 100 - 0.02x$, where p cents is the price per unit when x units are demanded. If the total profit is to be maximized, find (a) the number of units that should be produced, (b) the price of each unit, and (c) the total profit.

24. Solve Exercise 23 if the government levies a tax on the monopolist of 10 cents per unit produced.

25. Solve Exercise 23 if the government imposes a 10 percent tax based on the consumer's price.

26. For the monopolist of Exercise 23, determine the amount of tax that should be levied by the government on each unit produced in order for the tax revenue received by the government to be maximized.

27. Under a monopoly the demand equation for a certain commodity is $x + 2p = 24$, where x units are demanded when p dollars is the price per unit. If $C(x)$ dollars is the total cost of producing x units, then $C(x) = \frac{1}{100}x^3 + \frac{3}{10}x^2$. (a) Find the price per unit if the total profit is to be maximized. (b) Solve part (a) if the government levies on the monopolist a tax of \$2 for each unit produced.

28. Find the maximum tax revenue that can be received by the government if an additive tax for each unit produced is levied on a monopolist for which the demand equation is $x + 3p = 75$, where p dollars is the price of one unit when x units are demanded; and $C(x) = 3x + 100$, where $C(x)$ dollars is the total cost of producing x units.

29. The demand equation for a certain commodity produced by a monopolist is $p = a - bx$, where x units are demanded when p dollars is the price per unit, and the total cost, $C(x)$ dollars, of producing x units is determined by $C(x) = c + dx$. The numbers a, b, c, and d are positive constants. If the government levies a tax on the monopolist of t dollars per unit produced, show that in order to maximize profits the monopolist should pass on to the consumer only one-half of the tax; that is, the monopolist should increase the unit price by $\frac{1}{2}t$ dollars. *Note:* The demand equation and the total cost function of Example 6 and Exercise 24 are special cases of this more general situation.

30. The demand equation for a monopolist is $(100 - x)^2 = 100p$, where x units are demanded when p dollars is the price per unit. The total cost function is given by $C(x) = 55x - \frac{4}{5}x^2$, where $C(x)$ dollars is the total cost of producing x units. (a) Find the price the monopolist should charge in order to make the greatest profit. (b) If the government imposes on the monopolist a tax of \$9 for each unit produced, determine the price that should be charged for the greatest total profit. (c) Show that the increase in price from part (a) to part (b) is greater than the tax levied by the government.

REVIEW EXERCISES FOR CHAPTER 4

In Exercises 1 through 10, find the absolute extrema of the given function on the indicated interval, if there are any, and find the values of x at which the absolute extrema occur. Draw a sketch of the graph of the function on the interval.

1. $f(x) = \sqrt{5 + x}$; $[-5, +\infty)$

2. $f(x) = \sqrt{9 - x^2}$; $(-3, 3)$

3. $f(x) = \frac{5}{2}x^6 - 3x^5$; $[-1, 2]$

4. $f(x) = x^4 - 12x^2 + 36$; $[-2, 6]$

5. $f(x) = |9 - x^2|$; $[-2, 3]$

6. $f(x) = \dfrac{3}{x - 2}$; $[0, 4]$

7. $f(x) = x^4 - 12x^2 + 36$; $[0, 5]$

8. $f(x) = \begin{cases} 2x + 3 & \text{if } -2 \le x < 1 \\ x^2 + 4 & \text{if } 1 \le x \le 2 \end{cases}$; $[-2, 2]$

9. $f(x) = 2 \sin 3x$; $[-\frac{1}{3}\pi, \frac{1}{3}\pi]$

10. $f(x) = 4 \cos^2 2x$; $[0, \frac{3}{4}\pi]$

In Exercises 11 through 20, find the relative extrema of f; the points of inflection of the graph of f; the intervals on which f is increasing; the intervals on which f is decreasing; where the graph is concave upward; where the graph is concave downward; the slope of any inflectional tangent. Draw a sketch of the graph.

11. $f(x) = x^3 + 3x^2 - 4$

12. $f(x) = (x + 2)^{4/3}$

13. $f(x) = (x - 4)^2(x + 2)^3$

14. $f(x) = \begin{cases} x^3 - 3x & \text{if } x < 2 \\ 6 - x^2 & \text{if } 2 \le x \end{cases}$

15. $f(x) = (x - 3)^{5/3} + 1$

16. $f(x) = x\sqrt{25 - x^2}$

17. $f(x) = \dfrac{x^2}{x - 3}$

18. $f(x) = \dfrac{x - 3}{x + 1}$

19. $f(x) = \sin 2x - \cos 2x;\ x \in [-\frac{3}{8}\pi, \frac{5}{8}\pi]$

20. $f(x) = \dfrac{x^2}{x^2 - 9}$

21. Find two nonnegative numbers whose sum is 12 such that the sum of their squares is an absolute minimum.

22. Find the shortest distance from the point $A(0, 4)$ to a point on the hyperbola $x^2 - y^2 = 16$, and find the point on the hyperbola that is closest to A.

23. Find the shortest distance from the point $(\frac{9}{2}, 0)$ to the curve $y = \sqrt{x}$.

24. The demand in a certain market for a particular kind of breakfast cereal is given by the demand equation $px + 25p - 2000 = 0$, where p cents is the price of one box and x thousands of boxes is the quantity demanded per week. If the current price of the cereal is 80 cents per box and the price per box is increasing at the rate of 0.4 cent each week, find the rate of change in the demand.

25. Prove that among all the rectangles having a given perimeter, the square has the greatest area.

26. Prove that among all the rectangles of a given area, the square has the least perimeter.

27. Two particles start their motion at the same time. One particle is moving along a horizontal line and its equation of motion is $x = t^2 - 2t$, where x cm is the directed distance of the particle from the origin at t sec. The other particle is moving along a vertical line that intersects the horizontal line at the origin, and its equation of motion is $y = t^2 - 2$, where y cm is the directed distance of the particle from the origin at t sec. Find when the directed distance between the two particles is least, and their velocities at that time.

28. A man 6 ft tall is walking toward a building at the rate of 4 ft/sec. If there is a light on the ground 40 ft from the building, how fast is the man's shadow on the building growing shorter when he is 30 ft from the building?

29. A ship leaves a port at noon and travels due west at 20 knots. At noon the next day a second ship leaves the same port and travels northwest at 15 knots. How fast are the two ships separating when the second ship has traveled 90 nautical miles?

30. A reservoir is 80 m long and its cross section is an isosceles trapezoid having equal sides of 10 m, an upper base of 17 m, and a lower base of 5 m. At the instant when the water is 5 m deep, find the rate at which the water is leaking out if the water level is falling at the rate of 0.1 m/hr.

31. A funnel in the form of a cone is 10 in. across the top and 8 in. deep. Water is flowing into the funnel at the rate of 12 in.^{3}/sec and out at the rate of 4 in.^{3}/sec. How fast is the surface of the water rising when it is 5 in. deep?

32. As the last car of a train passes under a bridge, an automobile crosses the bridge on a roadway perpendicular to the track and 30 ft above it. The train is traveling at the rate of 80 ft/sec and the automobile is traveling at the rate of 40 ft/sec. How fast are the train and the automobile separating after 2 sec?

In Exercises 33 and 34, verify that the three conditions of the hypothesis of Rolle's theorem are satisfied by the given function on the indicated interval. Then find a suitable value for c that satisfies the conclusion of Rolle's theorem. Draw a sketch of the graph of f on the interval, and use the sketch to give the geometric interpretation of Rolle's theorem.

33. $f(x) = x^3 - x^2 - 4x + 4;\ [-2, 1]$

34. $f(x) = 2 \sin 3x;\ [0, \frac{1}{3}\pi]$

In Exercises 35 and 36, verify that the hypothesis of the mean-value theorem is satisfied for the given function on the indicated interval. Then find a suitable value for c that satisfies the conclusion of the mean-value theorem. Draw a sketch of the graph of f on the interval, and use the sketch to give the geometric interpretation of the mean-value theorem.

35. $f(x) = \sqrt{3 - x}$; $[-6, -1]$ **36.** $f(x) = x^3$; $[-2, 2]$

37. (a) If f is a polynomial function and $f(a)$, $f(b)$, $f'(a)$, and $f'(b)$ are zero, prove that there are at least two numbers in the open interval (a, b) that are roots of the equation $f''(x) = 0$. (b) Show that the function defined by the equation $f(x) = (x^2 - 4)^2$ satisfies part (a) for the open interval $(-2, 2)$.

38. If f is the function defined by $f(x) = |2x - 4| - 6$, then $f(-1) = 0$ and $f(5) = 0$. However, f' never has the value 0. Show why Rolle's theorem does not hold.

39. Suppose that f and g are two functions that satisfy the hypothesis of the mean-value theorem on $[a, b]$. Furthermore, suppose that $f'(x) = g'(x)$ for all x in the open interval (a, b). Prove that $f(x) - g(x) = f(a) - g(a)$ for all x in the closed interval $[a, b]$.

40. If f is a polynomial function, show that between any two consecutive roots of the equation $f'(x) = 0$ there is at most one root of the equation $f(x) = 0$.

41. Find the absolute maximum value attained by the function f if $f(x) = A \sin kx + B \cos kx$, where A, B, and k are positive constants.

42. Find the point (or points) of inflection of the graph of the function defined by $f(x) = x^{1/5}$, and determine where the graph is concave upward and where it is concave downward.

43. Find the point (or points) of inflection of the graph of the function defined by

$$f(x) = \begin{cases} 2 - x^2 & \text{if } x < 0 \\ x^2 + 2 & \text{if } 0 \leq x \end{cases}$$

and determine where the graph is concave upward and where it is concave downward.

44. If $f(x) = ax^3 + bx^2$, determine a and b so that the graph of f will have a point of inflection at $(2, 16)$.

45. If $f(x) = ax^3 + bx^2 + cx$, determine a, b, and c so that the graph of f will have a point of inflection at $(1, -1)$ and so that the slope of the inflectional tangent there will be -3.

46. If $f(x) = \dfrac{x + 1}{x^2 + 1}$, prove that the graph of f has three points of inflection that are collinear. Draw a sketch of the graph.

47. If $f(x) = x|x|$, show that the graph of f has a point of inflection at the origin.

48. Let $f(x) = x^n$ where n is a positive integer. Prove that the graph of f has a point of inflection at the origin if and only if n is an odd integer and $n > 1$. Furthermore show that if n is even, f has a relative minimum value at 0.

49. If $f(x) = (x^2 + a^2)^p$, where p is a rational number and $p \neq 0$, prove that if $p < \frac{1}{2}$ the graph of f has two points of inflection and if $p \geq \frac{1}{2}$ the graph of f has no points of inflection.

50. Suppose the graph of a function has a point of inflection at $x = c$. What can you conclude, if anything, about (a) the continuity of f at c; (b) the continuity of f' at c; (c) the continuity of f'' at c?

51. Two towns A and B are to get their water supply from the same pumping station to be located on the bank of a straight river that is 15 km from town A and 10 km from town B. If the points on the river nearest to A and B are 20 km apart and A and B are on the same side of the river, where should the pumping station be located so that the least amount of piping is required?

52. A manufacturer offers to deliver to a dealer 300 tables at \$90 per table and to reduce the price per table on the entire order by \$0.25 for each additional table over 300. Find the dollar total involved in the largest possible transaction between the manufacturer and the dealer under these circumstances.

53. A ladder is to reach over a fence h meters high to a wall w meters behind the fence. Find the length of the shortest ladder that may be used.

54. Find the volume of the largest right circular cylinder that can be inscribed in a right-circular cone having a radius of 4 in. and a height of 8 in.

55. Find the dimensions of the right-circular cone of least volume that can be circumscribed about a right-circular cylinder of radius r cm and altitude h cm.

56. A tent is to be in the shape of a cone. Find the ratio of the measure of the radius to the measure of the altitude for a tent of given volume to require the least material.

57. A large sign is to contain 50 m^2 of printed material. If margins of 4 m at the top and bottom and 2 m on each side are required, find the dimensions of the smallest sign that will meet these specifications.

58. An open box having a square base is to have a volume of k cm^3. Find the dimensions of the box that can be constructed with the least amount of material.

59. A piece of wire 20 cm long is cut into two pieces, and each piece is bent into the shape of a square. How should the wire be cut so that the total area of the two squares is as small as possible?

60. A piece of wire 80 cm long is bent to form a rectangle. Find the dimensions of the rectangle so that its area is as large as possible.

61. Suppose C millions of dollars is the capitalization of a certain corporation, P millions of dollars is the corporation's annual profit, and $P = 0.05C - 0.004C^2$. If the capitalization is increasing at the rate of \$400,000 per year, find the rate of change of the corporation's annual profit if its current capitalization is (a) \$3 million and (b) \$6 million.

62. A property development company rents each apartment at p dollars per month when x apartments are rented and $p = 30\sqrt{300 - 2x}$. How many apartments must be rented before the marginal revenue is zero?

63. The demand equation for a particular commodity is $(p + 4)(x + 3) = 48$, where $100x$ units are demanded when p dollars is the price per unit. Find (a) the price function, (b) the total revenue function, (c) the marginal revenue function. (d) Find the absolute maximum total revenue. (e) Draw sketches of the demand, total revenue, and marginal revenue curves on the same set of axes.

64. The demand equation for a certain commodity is $p + 2\sqrt{x - 1} = 6$ and the total cost function is given by $C(x) = 2x - 1$. (a) Determine the permissible values of x. (b) Find the marginal revenue and marginal cost functions. (c) Find the value of x that yields the maximum profit.

65. The demand equation for a certain commodity is $10^6 px = 10^9 - 2 \cdot 10^6 x + 18 \cdot 10^3 x^2 - 6x^3$, where x units are produced weekly and p dollars is the price of each unit; and $x \geq 100$. The number of dollars in the average cost of producing each unit is given by $Q(x) = \frac{1}{50}x - 24 + 11 \cdot 10^3 x^{-1}$. Find the number of units that should be produced each week and the price of each unit in order for the weekly profit to be maximized.

66. When $1000x$ boxes of a certain kind of material are produced, the number of dollars in the total cost of production is given by $C(x) = 135x^{1/3} + 450$. Find (a) the marginal cost when 8000 boxes are produced and (b) the number of boxes produced when the marginal cost (per thousand boxes) is \$20.

67. (a) If $f(x) = 3|x| + 4|x - 1|$, prove that f has an absolute minimum value of 3. (b) If $g(x) = 4|x| + 3|x - 1|$, prove that g has an absolute minimum value of 3. (c) If $h(x) = a|x| + b|x - 1|$, where $a > 0$ and $b > 0$, prove that h has an absolute minimum value that is the smaller of the two numbers a and b,

68. If $f(x) = |x|^a \cdot |x - 1|^b$, where a and b are positive rational numbers, prove that f has a relative maximum value of $a^a b^b/(a + b)^{a+b}$.

69. Let f and g be two functions that are differentiable at every number in the closed interval $[a, b]$. Suppose further that $f(a) = g(a)$ and $f(b) = g(b)$. Prove that there exists a number c in the open interval (a, b) such that $f'(c) = g'(c)$. (*Hint:* Consider the function $f - g$.)

70. Draw a sketch of the graph of a function f on the interval I in each case: (a) I is the open interval $(0, 2)$ and f is continuous on I. At 1, f has a relative maximum value but $f'(1)$ does not exist. (b) I is the closed interval $[0, 2]$. The function f has a relative minimum value at 1, but the absolute minimum value of f is at 0. (c) I is the open interval $(0, 2)$, and f' has a relative maximum value at 1.

CHAPTER 5 The Differential and Antidifferentiation

5.1 THE DIFFERENTIAL

Suppose that the function f is defined by

$$y = f(x)$$

Then, when $f'(x)$ exists,

$$f'(x) = \lim_{\Delta x \to 0} \frac{\Delta y}{\Delta x} \tag{1}$$

where $\Delta y = f(x + \Delta x) - f(x)$. From (1) it follows that for any $\epsilon > 0$ there exists a $\delta > 0$ such that

$$\left| \frac{\Delta y}{\Delta x} - f'(x) \right| < \epsilon \quad \text{whenever} \quad 0 < |\Delta x| < \delta$$

which is equivalent to

$$\frac{|\Delta y - f'(x)\Delta x|}{|\Delta x|} < \epsilon \quad \text{whenever} \quad 0 < |\Delta x| < \delta$$

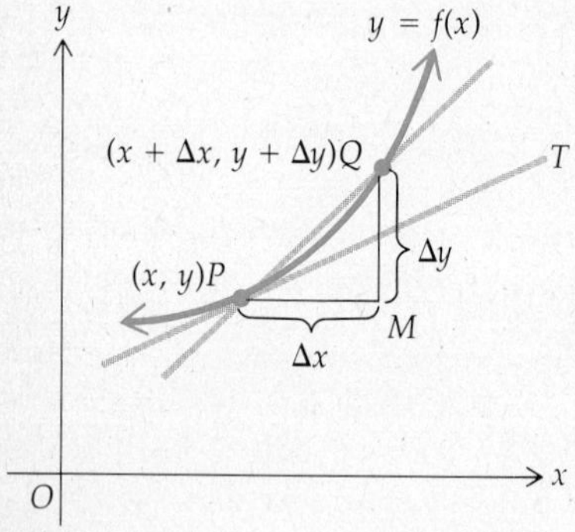

FIGURE 5.1.1

This means that $|\Delta y - f'(x)\Delta x|$ is small compared to $|\Delta x|$. That is, for a sufficiently small $|\Delta x|$, $f'(x)\Delta x$ is a good approximation to the value of Δy. Using the symbol $\approx$, meaning "is approximately equal to," we say then that

$$\Delta y \approx f'(x)\Delta x \tag{2}$$

if $|\Delta x|$ is sufficiently small.

To see what statement (2) means in terms of geometry, refer to Fig. 5.1.1. In the figure, an equation of the curve is $y = f(x)$. The line PT is tangent to the curve at $P(x, y)$, Q is the point $(x + \Delta x, y + \Delta y)$ and the directed distance $\overline{MQ}$ is Δy. In the figure, Δx and Δy are both positive; however, they could be negative. For a small value of $|\Delta x|$, the slope of the secant line PQ and the slope of the tangent line at P are approximately equal; that is,

$$\frac{\Delta y}{\Delta x} \approx f'(x)$$

or

$$\Delta y \approx f'(x)\Delta x$$

which is statement (2).

The right side of statement (2) is defined to be the *differential* of y.

5.1.1 Definition If the function f is defined by $y = f(x)$, then the **differential of** y, denoted by dy, is given by

$$dy = f'(x)\Delta x \tag{3}$$

where x is in the domain of f' and Δx is an arbitrary increment of x.

This concept of the differential involves a special type of function of two variables (a detailed study of such functions is given in Chapter 18). The notation df may be used to represent this function, where the symbol df is regarded as a single entity. The variable x can be any number in the domain of f', and Δx can be any number whatsoever. To state that df is a function of the two independent variables x and Δx means that to each ordered pair $(x, \Delta x)$ in the domain of df there corresponds one and only one number in the range of df, and this number can be represented by $df(x, \Delta x)$ so that

$$df(x, \Delta x) = f'(x)\Delta x \tag{4}$$

Comparing (3) and (4) we see that when $y = f(x)$, dy and $df(x, \Delta x)$ are two different notations for $f'(x)\Delta x$. The dy symbolism is used in subsequent discussions.

● ILLUSTRATION 1 If $y = 3x^2 - x$, then $f(x) = 3x^2 - x$; so $f'(x) = 6x - 1$. Therefore, from Definition 5.1.1,

$$dy = (6x - 1)\Delta x$$

In particular, if $x = 2$, then $dy = 11\Delta x$. ●

When $y = f(x)$, Definition 5.1.1 indicates what is meant by dy, the differential of the dependent variable. We also wish to define the differential of the independent variable, or dx. To arrive at a suitable definition for dx that is consistent with the definition of dy, we consider the identity function, which is the function f defined by $f(x) = x$. Then $f'(x) = 1$ and $y = x$; so $dy = 1 \cdot \Delta x = \Delta x$. Because $y = x$, we want dx to be equal to dy for this particular function; that is, for this function we want $dx = \Delta x$. It is this reasoning that leads to the following definition.

5.1.2 Definition If the function f is defined by $y = f(x)$, then the **differential of** x, denoted by dx, is given by

$$dx = \Delta x \tag{5}$$

where Δx is an arbitrary increment of x, and x is any number in the domain of f'.

From (3) and (5),

$$dy = f'(x)\, dx \tag{6}$$

Dividing on both sides of the above by dx, we have

$$\frac{dy}{dx} = f'(x) \qquad \text{if } dx \neq 0 \tag{7}$$

Equation (7) expresses the derivative as the quotient of two differentials. The notation dy/dx often denotes the derivative of y with respect to x; we have avoided this symbolism until now.

Example 1 Given $y = 4x^2 - 3x + 1$, find Δy, dy, and $\Delta y - dy$ for (a) any x and Δx; (b) $x = 2$, $\Delta x = 0.1$; (c) $x = 2$, $\Delta x = 0.01$; (d) $x = 2$, $\Delta x = 0.001$.

Solution

(a) Because $y = 4x^2 - 3x + 1$, then

$$y + \Delta y = 4(x + \Delta x)^2 - 3(x + \Delta x) + 1$$

$$(4x^2 - 3x + 1) + \Delta y = 4x^2 + 8x\,\Delta x + 4(\Delta x)^2 - 3x - 3\,\Delta x + 1$$

$$\Delta y = (8x - 3)\,\Delta x + 4(\Delta x)^2$$

Also, if $y = f(x)$,

$$dy = f'(x)dx$$

Thus

$$dy = (8x - 3)\,dx$$

$$= (8x - 3)\Delta x$$

$$\Delta y - dy = 4(\Delta x)^2$$

The results for parts (b), (c), and (d) are given in Table 5.1.1, where $\Delta y = (8x - 3)\Delta x + 4(\Delta x)^2$ and $dy = (8x - 3)\Delta x$.

TABLE 5.1.1

x	Δx	Δy	dy	$\Delta y - dy$
2	0.1	1.34	1.3	0.04
2	0.01	0.1304	0.13	0.0004
2	0.001	0.013004	0.013	0.000004

Note from Table 5.1.1 that the closer Δx is to zero, the smaller is the difference between Δy and dy. Furthermore, observe that for each value of Δx, the corresponding value of $\Delta y - dy$ is smaller than the value of Δx. More generally, dy is an approximation of Δy when Δx is small, and the approximation is of better accuracy than the size of Δx.

For a fixed value of x, say x_0,

$$dy = f'(x_0)dx \tag{8}$$

That is, dy is a linear function of dx; consequently, dy is usually easier to compute than Δy (this was seen in Example 1). Because $f(x_0 + \Delta x) - f(x_0) = \Delta y$,

$$f(x_0 + \Delta x) = f(x_0) + \Delta y$$

Thus

$$f(x_0 + \Delta x) \approx f(x_0) + dy \tag{9}$$

Our results are illustrated in Fig. 5.1.2. The equation of the curve in the figure is $y = f(x)$, and the graph is concave upward. The line PT is tangent to the curve at $P(x_0, f(x_0))$; Δx and dx are equal and are represented by the directed distance $\overline{PM}$, where M is the point $(x_0 + \Delta x, f(x_0))$. We let Q be the point $(x_0 + \Delta x, f(x_0 + \Delta x))$, and the directed distance $\overline{MQ}$ is Δy or, equivalently, $f(x_0 + \Delta x) - f(x_0)$. The slope of PT is $f'(x) = dy/dx$. Also, the slope of PT is $\overline{MR}/\overline{PM}$, and because $\overline{PM} = dx$, we have $dy = \overline{MR}$ and $\overline{RQ} = \Delta y - dy$. Note that the smaller the value of dx (i.e., the closer the point Q is to the point P), then the smaller will be the value of $\Delta y - dy$ (i.e., the smaller will be the length of the line segment RQ). An equation of the tangent line PT is

$$y = f(x_0) + f'(x_0)(x - x_0)$$

Thus, if $\bar{y}$ is the ordinate of R, then

$$\bar{y} = f(x_0) + dy \qquad (10)$$

By comparing (9) and (10), observe that when using $f(x_0) + dy$ to approximate the value of $f(x_0 + \Delta x)$, we are approximating the ordinate of the point $Q(x_0 + \Delta x, f(x_0 + \Delta x))$ on the curve by the ordinate of the point $R(x_0 + \Delta x, f(x_0) + dy)$ on the line that is tangent to the curve at $P(x_0, f(x_0))$.

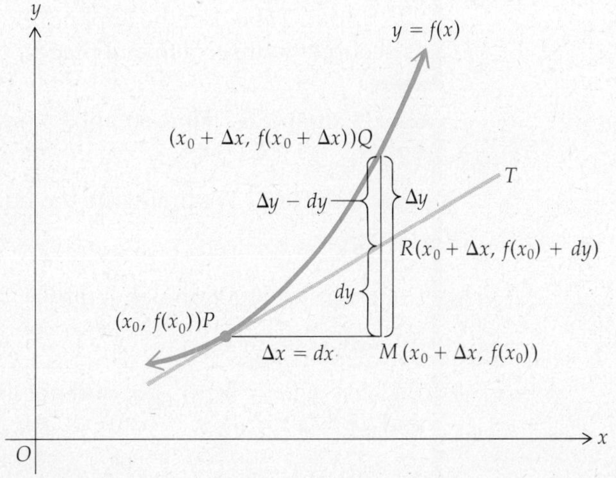

FIGURE 5.1.2

In Fig. 5.1.2 the graph is concave upward and the function is increasing. In Exercises 1 and 2 you are asked to draw similar figures. In Exercise 1 the graph is to be concave downward and the function increasing, while in Exercise 2 the graph is to be concave upward and the function decreasing.

The calculations in Examples 2 and 3 are intended to illustrate the concept of the differential.

Example 2 Use differentials to find an approximate value for $\sqrt[3]{28}$.

Solution Consider the function f defined by $f(x) = \sqrt[3]{x}$, and let $y = f(x)$. Hence

$$y = \sqrt[3]{x}$$

and

$$dy = f'(x)\,dx$$
$$= \frac{1}{3x^{2/3}}\,dx$$

The nearest perfect cube to 28 is 27. Thus we compute dy with $x = 27$ and $dx = \Delta x = 1$.

$$dy = \frac{1}{3(27)^{2/3}}(1) = \frac{1}{27}$$

By applying formula (9) with $x_0 = 27$, $\Delta x = 1$, and $dy = \frac{1}{27}$,

$$f(27 + 1) \approx f(27) + \tfrac{1}{27}$$
$$\sqrt[3]{27 + 1} \approx \sqrt[3]{27} + \tfrac{1}{27}$$
$$\sqrt[3]{28} \approx 3 + \tfrac{1}{27}$$

Therefore $\sqrt[3]{28} \approx 3.037$.

From Table 1 in the appendix, $\sqrt[3]{28} = 3.037$. Thus the approximation is accurate to three decimal places.

Example 3 Find an approximate value for cos 61° by using differentials.

Solution We first write the equivalent radian measure for 61°.

$$61° \approx \tfrac{61}{180}\pi \text{ rad}$$

Consider the function defined by $y = \cos x$. Then

$$dy = -\sin x\,dx$$

Because $\frac{61}{180}\pi = \frac{1}{3}\pi + \frac{1}{180}\pi$, and $\cos \frac{1}{3}\pi = 0.5$, we compute dy with $x = \frac{1}{3}\pi$ and $dx = \Delta x = \frac{1}{180}\pi$.

$$dy = (-\sin \tfrac{1}{3}\pi)(\tfrac{1}{180}\pi)$$
$$= -\frac{\sqrt{3}}{2}\left(\frac{\pi}{180}\right)$$
$$\approx -0.0151$$

We apply formula (9) with $x_0 = \frac{1}{3}\pi$, $\Delta x = \frac{1}{180}\pi$, and $dy = -0.0151$, and we have

$$\cos(\tfrac{1}{3}\pi + \tfrac{1}{180}\pi) \approx \cos(\tfrac{1}{3}\pi) + (-0.0151)$$

$$\cos \tfrac{61}{180}\pi \approx \cos \tfrac{1}{3}\pi - 0.0151$$

$$\cos \tfrac{61}{180}\pi \approx 0.5000 - 0.0151$$

Therefore $\cos 61° \approx 0.4849$.

From Table 5 in the Appendix, $\cos 61° = 0.4848$. Therefore the approximation is accurate to three decimal places.

Example 4 Find the approximate volume of a spherical shell whose inner radius is 4 in. and whose thickness is $\frac{1}{16}$ in.

Solution We consider the volume of the spherical shell as an increment of the volume of a sphere. Let r be the number of inches in the radius of a sphere; V be the number of cubic inches in the volume of a sphere; and ΔV be the number of cubic inches in the volume of a spherical shell.

$$V = \tfrac{4}{3}\pi r^3 \qquad dV = 4\pi r^2\, dr$$

Substituting $r = 4$ and $dr = \frac{1}{16}$ into the above we obtain

$$dV = 4\pi(4)^2 \tfrac{1}{16}$$
$$= 4\pi$$

Therefore $\Delta V \approx 4\pi$, and we conclude that the volume of the spherical shell is approximately 4π in.^{3}.

Suppose that y is a function of x and that x, in turn, is a function of a third variable t; that is,

$$y = f(x) \quad \text{and} \quad x = g(t) \tag{11}$$

The two equations in (11) together define y as a function of t. For example, suppose that $y = x^3$ and $x = 2t^2 - 1$. Combining these two equations we get $y = (2t^2 - 1)^3$. In general, if the two equations in (11) are combined, we obtain

$$y = f(g(t)) \tag{12}$$

The derivative of y with respect to t can be found by the chain rule, which yields

$$D_t y = D_x y\, D_t x \tag{13}$$

Equation (13) expresses $D_t y$ as a function of x and t because $D_x y$ is a function of x and $D_t x$ is a function of t.

● ILLUSTRATION 2 If $y = x^3$ and $x = 2t^2 - 1$, then

$$D_t y = D_x y\, D_t x$$
$$= 3x^2(4t)$$
$$= 12x^2 t$$

●

Because (12) defines y as a function of the independent variable t, the differential of y is obtained from (6):

$$dy = D_t y \, dt \tag{14}$$

Equation (14) expresses dy as a function of t and dt. Substituting from (13) into (14) we get

$$dy = D_x y \, D_t x \, dt \tag{15}$$

Now, because x is a function of the independent variable t, (6) can be applied to obtain the differential of x, and we have

$$dx = D_t x \, dt \tag{16}$$

Equation (16) expresses dx as a function of t and dt. From (15) and (16) we get

$$dy = D_x y \, dx \tag{17}$$

You should bear in mind that in (17) dy is a function of t and dt, and that dx is a function of t and dt. If $D_x y$ in (17) is replaced by $f'(x)$, we have

$$dy = f'(x) \, dx \tag{18}$$

Equation (18) resembles (6). However, in (6) x is the independent variable, and dy is expressed in terms of x and dx, whereas in (18) t is the independent variable, and both dy and dx are expressed in terms of t and dt. Thus we have the following theorem.

5.1.3 Theorem If $y = f(x)$, then when $f'(x)$ exists,

$$dy = f'(x) \, dx$$

whether or not x is an independent variable.

If on both sides of (18) we divide by dx (provided $dx \neq 0$), we obtain

$$f'(x) = \frac{dy}{dx} \qquad dx \neq 0 \tag{19}$$

Equation (19) states that if $y = f(x)$, then $f'(x)$ is the quotient of the two differentials dy and dx, even though x may not be an independent variable.

● ILLUSTRATION 3 Suppose $x = \cos t$, $y = \sin t$, and $0 < t < \pi$. Then

$$dx = -\sin t \, dt \quad \text{and} \quad dy = \cos t \, dt$$

Therefore

$$\frac{dy}{dx} = \frac{\cos t \, dt}{-\sin t \, dt} = -\frac{\cos t}{\sin t} = -\frac{x}{y} \tag{20}$$

Because $x^2 + y^2 = \cos^2 t + \sin^2 t$, then

$$x^2 + y^2 = 1$$

$$y^2 = 1 - x^2 \tag{21}$$

Furthermore, $y > 0$ because $y = \sin t$ and $0 < t < \pi$. Thus from (21),

$$y = \sqrt{1 - x^2} \tag{22}$$

Substituting the value of y from (22) into (20) we have

$$\frac{dy}{dx} = -\frac{x}{\sqrt{1 - x^2}} \tag{23}$$

This result can also be obtained by finding dy/dx from (22). ●

We now proceed to write the chain rule for differentiation by expressing the derivatives as quotients of differentials. If $y = f(u)$ and $D_u y$ exists, and if $u = g(x)$ and $D_x u$ exists, then the chain rule gives

$$D_x y = D_u y\, D_x u \tag{24}$$

But $D_x y = dy/dx$ if $dx \neq 0$; $D_u y = dy/du$ if $du \neq 0$; and $D_x u = du/dx$ if $dx \neq 0$. From (24), therefore, we have

$$\frac{dy}{dx} = \left(\frac{dy}{du}\right) \cdot \left(\frac{du}{dx}\right) \qquad \text{if } du \neq 0 \text{ and } dx \neq 0$$

Observe the convenient form the chain rule takes with this notation.

The German mathematician Gottfried Wilhelm Leibniz (1646–1716) was the first to use the notation dy/dx for the derivative of y with respect to x. The concept of a derivative was introduced in the seventeenth century, almost simultaneously by Leibniz and Sir Isaac Newton (1642–1727), who were working independently. Leibniz probably thought of dx and dy as small changes in the variables x and y and of the derivative of y with respect to x as the ratio of dy to dx as dy and dx become small. The concept of a limit as we know it today was not known to Leibniz.

Corresponding to the Leibniz notation dy/dx for the first derivative of y with respect to x we have the symbol d^2y/dx^2 for the second derivative of y with respect to x. However, d^2y/dx^2 must not be thought of as a quotient because in this text the differential of a differential is not considered. Similarly, d^ny/dx^n is a notation for the nth derivative of y with respect to x. The symbolism f', f'', f''', and so on, for the successive derivatives of a function f was introduced by the French mathematician Joseph Louis Lagrange (1736–1813) in the eighteenth century.

In Chapter 3 we derived formulas for finding derivatives. These formulas are now stated using the Leibniz notation. Along with the formula for the derivative there is a corresponding formula for the differential. In these formulas, u and v are functions of x, and it is understood that the formulas hold if $D_x u$ and $D_x v$ exist. When c appears, it is a constant.

I $\dfrac{d(c)}{dx} = 0$ I′ $d(c) = 0$

II $\dfrac{d(x^n)}{dx} = nx^{n-1}$ II′ $d(x^n) = nx^{n-1}\,dx$

III $\dfrac{d(cu)}{dx} = c\dfrac{du}{dx}$ III′ $d(cu) = c\,du$

IV $\dfrac{d(u+v)}{dx} = \dfrac{du}{dx} + \dfrac{dv}{dx}$ IV′ $d(u+v) = du + dv$

V $\dfrac{d(uv)}{dx} = u\dfrac{dv}{dx} + v\dfrac{du}{dx}$ V′ $d(uv) = u\,dv + v\,du$

VI $\dfrac{d\left(\dfrac{u}{v}\right)}{dx} = \dfrac{v\dfrac{du}{dx} - u\dfrac{dv}{dx}}{v^2}$ VI′ $d\left(\dfrac{u}{v}\right) = \dfrac{v\,du - u\,dv}{v^2}$

VII $\dfrac{d(u^n)}{dx} = nu^{n-1}\dfrac{du}{dx}$ VII′ $d(u^n) = nu^{n-1}\,du$

The operation of differentiation is extended to include the process of finding the differential as well as finding the derivative. If $y = f(x)$, dy can be found either by applying formulas I′–VII′ or by finding $f'(x)$ and multiplying it by dx.

Example 5 Given

$$y = \frac{\sqrt{x^2+1}}{2x+1}$$

find dy.

Solution From formula VI′,

$$dy = \frac{(2x+1)\,d(\sqrt{x^2+1}) - \sqrt{x^2+1}\,d(2x+1)}{(2x+1)^2} \tag{25}$$

From formula VII′,

$$d(\sqrt{x^2+1}) = \tfrac{1}{2}(x^2+1)^{-1/2}\,2x\,dx = x(x^2+1)^{-1/2}\,dx \tag{26}$$

and

$$d(2x+1) = 2\,dx \tag{27}$$

Substituting values from (26) and (27) into (25) we get

$$dy = \frac{x(2x+1)(x^2+1)^{-1/2}\,dx - 2(x^2+1)^{1/2}\,dx}{(2x+1)^2}$$

$$= \frac{(2x^2+x)\,dx - 2(x^2+1)\,dx}{(2x+1)^2(x^2+1)^{1/2}}$$

$$= \frac{x-2}{(2x+1)^2\sqrt{x^2+1}}\,dx$$

Example 6 Given $2x^2y^2 - 3x^3 + 5y^3 + 6xy^2 = 5$, where x and y are functions of a third variable, find dy/dx by finding the differential term by term.

Solution This is a problem in implicit differentiation. Taking the differential term by term we get

$$4xy^2\,dx + 4x^2y\,dy - 9x^2\,dx + 15y^2\,dy + 6y^2\,dx + 12xy\,dy = 0$$

Dividing by dx, if $dx \neq 0$, we have

$$(4x^2y + 15y^2 + 12xy)\frac{dy}{dx} = -4xy^2 + 9x^2 - 6y^2$$

$$\frac{dy}{dx} = \frac{9x^2 - 6y^2 - 4xy^2}{4x^2y + 15y^2 + 12xy}$$

EXERCISES 5.1

1. Draw a figure similar to Fig. 5.1.2 if the graph is concave downward and the function is increasing. Indicate the line segments whose lengths represent the following quantities: Δx, Δy, dx, and dy.

2. Draw a figure similar to Fig. 5.1.2 if the graph is concave upward, the function is decreasing, and $\Delta x < 0$. Indicate the line segments whose lengths represent the following quantities: $|\Delta x|$, Δy, $|dx|$, and dy.

In Exercises 3 through 10, find (a) Δy; (b) dy; (c) $\Delta y - dy$.

3. $y = x^2$
4. $y = 3x^2 - x$
5. $y = 6 - 3x - 2x^2$
6. $y = x^3 - x^2$
7. $y = \sqrt{x}$
8. $y = \dfrac{1}{x^2 + 1}$
9. $y = \dfrac{2}{x - 1}$
10. $y = \dfrac{1}{\sqrt{x}}$

In Exercises 11 through 16, find, for the given values, (a) Δy; (b) dy; (c) $\Delta y - dy$.

11. $y = x^2 - 3x$; $x = 2$; $\Delta x = 0.03$
12. $y = x^2 - 3x$; $x = -1$; $\Delta x = 0.02$
13. $y = \dfrac{1}{x}$; $x = -2$; $\Delta x = -0.1$
14. $y = \dfrac{1}{x}$; $x = 3$; $\Delta x = -0.2$
15. $y = x^3 + 1$; $x = 1$; $\Delta x = -0.5$
16. $y = x^3 + 1$; $x = -1$; $\Delta x = 0.1$

In Exercises 17 through 24, find dy.

17. $y = (3x^2 - 2x + 1)^3$
18. $y = \sqrt{4 - x^2}$
19. $y = x^2\sqrt[3]{2x + 3}$
20. $y = \dfrac{3x}{x^2 + 2}$
21. $y = \dfrac{2 + \cos x}{2 - \sin x}$
22. $y = x^2 \sin\dfrac{1}{x} - x\cos\dfrac{1}{x}$
23. $y = \sqrt{\dfrac{x - 1}{x + 1}}$
24. $y = \sqrt{3x + 4}\,\sqrt[3]{x^2 - 1}$

In Exercises 25 through 30, x and y are functions of a third variable. Find dy/dx by finding the differential term by term (see Example 6).

25. $3x^2 + 4y^2 = 48$
26. $8x^2 - y^2 = 32$
27. $\sqrt{x} + \sqrt{y} = 4$
28. $2x^2y - 3xy^3 + 6y^2 = 1$
29. $\sin x \cos y - \cos x \sin y = 1$
30. $3\sin^2 x + 4\cos^2 y = 1$

In Exercises 31 through 36, find dy/dt.

31. $y = 3x^3 - 5x^2 + 1$; $x = t^2 - 1$
32. $y = \sqrt[3]{5x - 1}$; $x = \sqrt{2t + 3}$

33. $y = x^2 - 5x + 1; x = s^3 - 2s + 1; s = \sqrt{t^2 + 1}$ **34.** $y = \dfrac{x^2 - 1}{x^2 + 1}; x = \dfrac{2s + 2}{\sqrt{s^2 + 2}}; s = t^2 - 4t + 5$

35. $x^3 - 3x^2y + y^3 = 5; x = 4t^2 + 1$ **36.** $3x^2y - 4xy^2 + 7y^3 = 0; 2x^3 - 3xt^2 + t^3 = 1$

In Exercises 37 through 40, do the following: (a) Find dx and dy in terms of t and dt; (b) use the results of part (a) to find dy/dx; (c) express y in terms of x by eliminating t between the pair of equations and then find dy/dx by using theorems of differentiation.

37. $x = 2t^2, y = 3t$ **38.** $x = 1 + t, y = 1 - t^2$

39. $x = 2 \cos t, y = 3 \sin t, 0 < t < \pi$ **40.** $x = 1 - \cos t, y = 2 + \sin t, 0 < t < \pi$

In Exercises 41 through 48, use differentials to find an approximate value for the given quantity. Express each answer to three significant digits.

41. $\sqrt{37.5}$ **42.** $\sqrt{82}$ **43.** $\sqrt[4]{82}$ **44.** $\sqrt[3]{7.5}$

45. $\sqrt[3]{0.00098}$ **46.** sin 46° **47.** cos 148° **48.** $\sqrt{0.042}$

49. The measurement of an edge of a cube is found to be 15 cm with a possible error of 0.01 cm. Using differentials find the approximate error in computing from this measurement (a) the volume; (b) the area of one of the faces.

50. A metal box in the form of a cube is to have an interior volume of 1000 cm³. The six sides are to be made of metal $\frac{1}{2}$ cm thick. If the cost of the metal to be used is $0.20 per cubic centimeter, use differentials to find the approximate cost of the metal to be used in the manufacture of the box.

51. An open cylindrical tank is to have an outside coating of thickness 2 cm. If the inner radius is 6 m and the altitude is 10 m, find by differentials the approximate amount of coating material to be used.

52. The altitude of a right-circular cone is twice the radius of the base. The altitude is measured as 12 in., with a possible error of 0.005 in. Find the approximate error in the calculated volume of the cone.

53. If the possible error in the measurement of the volume of a gas is 0.1 ft³ and the allowable error in the pressure is $0.001C$ lb/ft², find the size of the smallest container for which Boyle's law (Exercise 30 in Exercises 3.5) holds.

54. If t sec is the time for one complete swing of a simple pendulum of length l ft, then $4\pi^2 l = gt^2$, where $g = 32.2$. A clock having a pendulum of length 1 ft gains 5 min each day. Find the approximate amount by which the pendulum should be lengthened in order to correct the inaccuracy.

55. The measure of the electrical resistance of a wire is proportional to the measure of its length and inversely proportional to the square of the measure of its diameter. Suppose the resistance of a wire of given length is computed from a measurement of the diameter with a possible 2% error. Find the possible percent error in the computed value of the resistance.

56. A contractor agrees to paint both sides of 1000 circular signs each of radius 3 m. Upon receiving the signs it is discovered that the radius is 1 cm too large. Use differentials to find the approximate percent increase of paint that will be needed.

57. For the adiabatic law for the expansion of air of Exercise 12 in Sec. 4.1, prove that $dP/P = -1.4\, dV/V$.

5.2 ANTIDIFFERENTIATION

You are already familiar with *inverse operations.* Addition and subtraction are inverse operations; multiplication and division are also inverse operations, as well as raising to powers and extracting roots. In this section we develop the inverse operation of differentiation called **antidifferentiation.**

5.2.1 Definition A function F is called an **antiderivative** of a function f on an interval I if $F'(x) = f(x)$ for every value of x in I.

• ILLUSTRATION 1 If F is defined by $F(x) = 4x^3 + x^2 + 5$, then $F'(x) = 12x^2 + 2x$. Thus if f is the function defined by $f(x) = 12x^2 + 2x$, we state that f is the derivative of F and that F is an antiderivative of f. If G is the function defined by $G(x) = 4x^3 + x^2 - 17$, then G is also an antiderivative of f because $G'(x) = 12x^2 + 2x$. Actually, any function whose function value is given by $4x^3 + x^2 + C$, where C is any constant, is an antiderivative of f. •

In general, if a function F is an antiderivative of a function f on an interval I and if G is defined by

$$G(x) = F(x) + C$$

where C is an arbitrary constant, then

$$G'(x) = F'(x) = f(x)$$

and G is also an antiderivative of f on the interval I.

We now proceed to prove that if F is any particular antiderivative of f on an interval I, then all possible antiderivatives of f on I are defined by $F(x) + C$, where C is an arbitrary constant. First, a preliminary theorem is needed.

5.2.2 Theorem If f and g are two functions such that $f'(x) = g'(x)$ for all values of x in the interval I, then there is a constant K such that

$$f(x) = g(x) + K \qquad \text{for all } x \text{ in } I \tag{1}$$

Proof Let h be the function defined on I by

$$h(x) = f(x) - g(x)$$

so that for all values of x in I,

$$h'(x) = f'(x) - g'(x)$$

But, by hypothesis, $f'(x) = g'(x)$ for all values of x in I. Therefore

$$h'(x) = 0 \qquad \text{for all values of } x \text{ in } I$$

Thus Theorem 4.4.3 applies to the function h, and there is a constant K such that

$$h(x) = K \qquad \text{for all values of } x \text{ in } I$$

Replacing $h(x)$ by $f(x) - g(x)$ we have

$$f(x) = g(x) + K \qquad \text{for all values of } x \text{ in } I$$

and the theorem is proved. ■

The next theorem follows immediately from Theorem 5.2.2.

5.2.3 Theorem If F is a specific antiderivative of f on an interval I, then the most general antiderivative of f on I is given by

$$F(x) + C \tag{2}$$

where C is an arbitrary constant, and any antiderivative of f on I can be obtained from (2) by assigning a particular value to C.

Proof Let G be any antiderivative of f on I; then

$$G'(x) = f(x) \qquad \text{on } I \tag{3}$$

Because F is also an antiderivative of f on I,

$$F'(x) = f(x) \qquad \text{on } I \tag{4}$$

From (3) and (4) it follows that

$$G'(x) = F'(x) \qquad \text{on } I$$

Therefore, from Theorem 5.2.2, there is a constant K such that

$$G(x) = F(x) + K \qquad \text{for all } x \text{ in } I$$

Because G is any antiderivative of f on I, it follows that any antiderivative of f can be obtained from $F(x) + C$, where C is an arbitrary constant. Hence the theorem is proved. ■

If F is an antiderivative of f, then $F'(x) = f(x)$, and so

$$d(F(x)) = f(x)\,dx \tag{5}$$

The inverse operation of finding the differential of a function is that of finding the most general function having a given differential. This inverse operation is called **antidifferentiation** and is denoted by the symbol

$$\int$$

We write

$$\int f(x)\,dx = F(x) + C \tag{6}$$

where F is any function such that $d(F(x)) = f(x)\,dx$.

Substituting from (5) in the left side of (6) we have

$$\int d(F(x)) = F(x) + C \tag{7}$$

Taking the differential of each member of (6) gives

$$d\int f(x)\,dx = d(F(x) + C)$$

$$d\int f(x)\,dx = d(F(x)) \tag{8}$$

Substituting from (5) in the right member of (8) gives

$$d \int f(x)\,dx = f(x)\,dx \tag{9}$$

From (7) and (9) we observe that the operations defined by the symbols d and $\int$ are inverse operations.

If $F(x) + C$ is the most general function whose differential is $f(x)\,dx$, it is also the most general function whose derivative is $f(x)$. Therefore antidifferentiation is also considered as the operation of finding the most general function having a given derivative.

Because antidifferentiation is the inverse operation of differentiation, antidifferentiation theorems can be obtained from differentiation theorems. Thus the following theorems can be proved from the corresponding ones for differentiation.

5.2.4 Theorem

$$\int dx = x + C$$

5.2.5 Theorem

$$\int af(x)\,dx = a \int f(x)\,dx$$

where a is a constant.

Theorem 5.2.5 states that to find an antiderivative of a constant times a function, first find an antiderivative of the function and then multiply it by the constant.

5.2.6 Theorem If f_1 and f_2 are defined on the same interval, then

$$\int [f_1(x) + f_2(x)]\,dx = \int f_1(x)\,dx + \int f_2(x)\,dx$$

Theorem 5.2.6 states that to find an antiderivative of the sum of two functions, find an antiderivative of each of the functions separately and then add the results, it being understood that both functions are defined on the same interval. Theorem 5.2.6 can be extended to any finite number of functions. Combining Theorem 5.2.6 with Theorem 5.2.5, we have the following theorem.

5.2.7 Theorem If $f_1, f_2, \ldots, f_n$ are defined on the same interval.

$$\int [c_1 f_1(x) + c_2 f_2(x) + \cdots + c_n f_n(x)]\,dx$$
$$= c_1 \int f_1(x)\,dx + c_2 \int f_2(x)\,dx + \cdots + c_n \int f_n(x)\,dx$$

where $c_1, c_2, \ldots, c_n$ are constants.

5.2.8 Theorem

$$\int x^n\,dx = \frac{x^{n+1}}{n+1} + C \qquad \text{if } n \neq -1$$

As stated above, these theorems follow from the corresponding theorems for finding the derivative. Following is the proof of Theorem 5.2.8:

$$D_x\left(\frac{x^{n+1}}{n+1} + C\right) = \frac{(n+1)x^n}{n+1}$$
$$= x^n$$

Applications of the above theorems are illustrated in the following examples.

Example 1 Evaluate $\int (3x + 5)\,dx$.

Solution

$$\int (3x+5)\,dx = 3\int x\,dx + 5\int dx \quad \text{(by Theorem 5.2.7)}$$
$$= 3\left(\frac{x^2}{2} + C_1\right) + 5(x + C_2) \quad \text{(by Theorems 5.2.8 and 5.2.4)}$$
$$= \tfrac{3}{2}x^2 + 5x + 3C_1 + 5C_2$$

Because $3C_1 + 5C_2$ is an arbitrary constant, it may be denoted by C; so the answer is

$$\tfrac{3}{2}x^2 + 5x + C$$

The answer may be checked by finding its derivative.

$$D_x(\tfrac{3}{2}x^2 + 5x + C) = 3x + 5$$

Example 2 Evaluate $\int \sqrt[3]{x^2}\,dx$.

Solution

$$\int \sqrt[3]{x^2}\,dx = \int x^{2/3}\,dx$$
$$= \frac{x^{2/3+1}}{\frac{2}{3}+1} + C \quad \text{(from Theorem 5.2.8)}$$
$$= \tfrac{3}{5}x^{5/3} + C$$

Example 3 Evaluate

$$\int \left(\frac{1}{x^4} + \frac{1}{\sqrt[4]{x}}\right)dx$$

Solution

$$\int\left(\frac{1}{x^4}+\frac{1}{\sqrt[4]{x}}\right)dx = \int (x^{-4}+x^{-1/4})\,dx$$

$$= \frac{x^{-4+1}}{-4+1}+\frac{x^{-1/4+1}}{-\frac{1}{4}+1}+C$$

$$= \frac{x^{-3}}{-3}+\frac{x^{3/4}}{\frac{3}{4}}+C$$

$$= -\frac{1}{3x^3}+\frac{4}{3}x^{3/4}+C$$

Many antiderivatives cannot be found directly by applying theorems. However, sometimes it is possible to find an antiderivative by the theorems after changing the variable.

● ILLUSTRATION 2 Suppose that we wish to find

$$\int 2x\sqrt{1+x^2}\,dx \tag{10}$$

If we make the substitution $u = 1 + x^2$, then $du = 2x\,dx$, and (10) becomes

$$\int u^{1/2}\,du$$

which by Theorem 5.2.8 gives

$$\tfrac{2}{3}u^{3/2}+C$$

Then, replacing u by $(1 + x^2)$, we have as the result

$$\tfrac{2}{3}(1+x)^{3/2}+C$$

●

Justification of the procedure used in Illustration 2 is provided by the following theorem, which is analogous to the chain rule for differentiation and hence may be called the *chain rule for antidifferentiation.*

5.2.9 Theorem
Chain Rule for Antidifferentiation

Let g be a differentiable function of x, and let the range of g be an interval I. Suppose that f is a function defined on I and that F is an antiderivative of f on I. Then if $u = g(x)$,

$$\int f(g(x))g'(x)\,dx = \int f(u)\,du = F(u)+C = F(g(x))+C$$

Proof Because $u = g(x)$, then u is in I; and because F is an antiderivative of f on I, it follows that

$$\frac{dF(u)}{du} = f(u) \tag{11}$$

and

$$\int f(u)\,du = F(u) + C \tag{12}$$

Also,

$$\frac{dF(g(x))}{dx} = \frac{dF(u)}{dx} \tag{13}$$

Applying the chain rule for differentiation to the right side of (13) we obtain

$$\frac{dF(g(x))}{dx} = \frac{dF(u)}{du} \cdot \frac{du}{dx} \tag{14}$$

Substituting from (11) into (14) gives

$$\frac{dF(g(x))}{dx} = f(u) \cdot \frac{du}{dx} \tag{15}$$

Because $u = g(x)$, from (15),

$$\frac{dF(g(x))}{dx} = f(g(x)) \cdot g'(x)$$

from which

$$\int f(g(x)) \cdot g'(x)\,dx = F(g(x)) + C \tag{16}$$

Because $u = g(x)$,

$$F(g(x)) + C = F(u) + C \tag{17}$$

Therefore, from (12), (16), and (17),

$$\int f(g(x))g'(x)\,dx = \int f(u)\,du = F(u) + C = F(g(x)) + C$$

which is what we wished to prove. ■

As a particular case of Theorem 5.2.9, from Theorem 5.2.8 we have the generalized power formula for antiderivatives, which we now state.

5.2.10 Theorem If g is a differentiable function, then if $u = g(x)$,

$$\int [g(x)]^n g'(x)\,dx = \int u^n\,du = \frac{u^{n+1}}{n+1} + C = \frac{[g(x)]^{n+1}}{n+1} + C$$

where $n \neq -1$.

Examples 4, 5, and 6 illustrate the application of Theorem 5.2.10.

Example 4 Evaluate $\int \sqrt{3x+4}\, dx$.

Solution To apply Theorem 5.2.10 we make the substitution $u = 3x + 4$; then $du = 3\,dx$, or $\frac{1}{3}du = dx$. Hence

$$\begin{aligned}\int \sqrt{3x+4}\, dx &= \int u^{1/2} \cdot \frac{du}{3} \\ &= \frac{1}{3}\int u^{1/2}\, du \\ &= \frac{1}{3} \cdot \frac{u^{3/2}}{\frac{3}{2}} + C \\ &= \tfrac{2}{9}u^{3/2} + C \\ &= \tfrac{2}{9}(3x+4)^{3/2} + C\end{aligned}$$

Example 5 Evaluate $\int t(5 + 3t^2)^8\, dt$.

Solution Observe that $d(5 + 3t^2) = 6t\, dt$. Thus we make the substitution

$$u = 5 + 3t^2$$

and then

$$du = 6t\, dt$$

$$\tfrac{1}{6}du = t\, dt$$

Therefore

$$\begin{aligned}\int t(5+3t^2)^8\, dt &= \int (5+3t^2)^8(t\, dt) \\ &= \int u^8(\tfrac{1}{6}\, du) \\ &= \frac{1}{6}\int u^8\, du \\ &= \frac{1}{6} \cdot \frac{u^9}{9} + C \\ &= \tfrac{1}{54}(5+3t^2)^9 + C\end{aligned}$$

Example 6 Evaluate $\int x^2 \sqrt[5]{7 - 4x^3}\, dx$.

Solution Because $d(7 - 4x^3) = -12x^2\, dx$, we make the substitution

$$u = 7 - 4x^3$$

Then

$$du = -12x^2\,dx$$

$$-\tfrac{1}{12}\,du = x^2\,dx$$

We have, therefore,

$$\begin{aligned}\int x^2\sqrt[5]{7-4x^3}\,dx &= \int (7-4x^3)^{1/5}(x^2\,dx)\\ &= \int u^{1/5}(-\tfrac{1}{12}\,du)\\ &= -\frac{1}{12}\int u^{1/5}\,du\\ &= -\frac{1}{12}\frac{u^{6/5}}{\frac{6}{5}} + C\\ &= -\tfrac{5}{72}(7-4x^3)^{6/5} + C\end{aligned}$$

The details of the solutions of Examples 4, 5, and 6 can be shortened by not explicitly stating the substitution of u. The solution of Example 4 takes the following form:

$$\begin{aligned}\int \sqrt{3x+4}\,dx &= \frac{1}{3}\int (3x+4)^{1/2}(3\,dx)\\ &= \frac{1}{3}\cdot\frac{(3x+4)^{3/2}}{\frac{3}{2}} + C\\ &= \tfrac{2}{9}(3x+4)^{3/2} + C\end{aligned}$$

The solution of Example 5 can be written as

$$\begin{aligned}\int t(5+3t^2)^8\,dt &= \frac{1}{6}\int (5+3t^2)^8(6t\,dt)\\ &= \frac{1}{6}\cdot\frac{(5+3t^2)^9}{9} + C\\ &= \tfrac{1}{54}(5+3t^2)^9 + C\end{aligned}$$

and the solution of Example 6 can be shortened as follows:

$$\begin{aligned}\int x^2\sqrt[5]{7-4x^3}\,dx &= -\frac{1}{12}\int (7-4x^3)^{1/5}(-12x^2\,dx)\\ &= -\frac{1}{12}\cdot\frac{(7-4x^3)^{6/5}}{\frac{6}{5}} + C\\ &= -\tfrac{5}{72}(7-4x^3)^{6/5} + C\end{aligned}$$

Example 7 Evaluate

$$\int \frac{4x^2\,dx}{(1-8x^3)^4}$$

Solution Because $d(1-8x^3) = -24x^2\,dx$, we write

$$\begin{aligned}
\int \frac{4x^2\,dx}{(1-8x^3)^4} &= 4\int (1-8x^3)^{-4}(x^2\,dx) \\
&= 4\left(-\frac{1}{24}\right)\int (1-8x^3)^{-4}(-24x^2\,dx) \\
&= -\frac{1}{6}\cdot\frac{(1-8x^3)^{-3}}{-3} + C \\
&= \frac{1}{18(1-8x^3)^3} + C
\end{aligned}$$

Example 8 Evaluate $\int x^2\sqrt{1+x}\,dx$.

Solution Let $u = 1 + x$. Then $x = u - 1$ and $dx = du$. Making these substitutions we have

$$\begin{aligned}
\int x^2\sqrt{1+x}\,dx &= \int (u-1)^2u^{1/2}\,du \\
&= \int (u^2-2u+1)u^{1/2}\,du \\
&= \int u^{5/2}\,du - 2\int u^{3/2}\,du + \int u^{1/2}\,du \\
&= \frac{u^{7/2}}{\frac{7}{2}} - 2\cdot\frac{u^{5/2}}{\frac{5}{2}} + \frac{u^{3/2}}{\frac{3}{2}} + C \\
&= \tfrac{2}{7}(1+x)^{7/2} - \tfrac{4}{5}(1+x)^{5/2} + \tfrac{2}{3}(1+x)^{3/2} + C
\end{aligned}$$

● ILLUSTRATION 3 An alternate method for the solution of Example 8 is to make the substitution $v = \sqrt{1+x}$. Then $v^2 = 1 + x$, and therefore $x = v^2 - 1$ and $dx = 2v\,dv$. With these substitutions the computation takes the following form:

$$\begin{aligned}
\int x^2\sqrt{1+x}\,dx &= \int (v^2-1)^2\cdot v\cdot(2v\,dv) \\
&= 2\int v^6\,dv - 4\int v^4\,dv + 2\int v^2\,dv \\
&= \tfrac{2}{7}v^7 - \tfrac{4}{5}v^5 + \tfrac{2}{3}v^3 + C \\
&= \tfrac{2}{7}(1+x)^{7/2} - \tfrac{4}{5}(1+x)^{5/2} + \tfrac{2}{3}(1+x)^{3/2} + C
\end{aligned}$$

●

The results for each of the above examples can be checked by finding the derivative (or the differential) of the answer.

● ILLUSTRATION 4 In Example 5 we have

$$\int t(5+3t^2)^8\,dt = \tfrac{1}{54}(5+3t^2)^9 + C$$

Checking by differentiation gives

$$D_t[\tfrac{1}{54}(5+3t^2)^9] = \tfrac{1}{54}\cdot 9(5+3t^2)^8\cdot 6t$$
$$= t(5+3t^2)^8$$ ●

● ILLUSTRATION 5 In Example 8 we have

$$\int x^2\sqrt{1+x}\,dx = \tfrac{2}{7}(1+x)^{7/2} - \tfrac{4}{5}(1+x)^{5/2} + \tfrac{2}{3}(1+x)^{3/2} + C$$

Checking by differentiation gives

$$D_x[\tfrac{2}{7}(1+x)^{7/2} - \tfrac{4}{5}(1+x)^{5/2} + \tfrac{2}{3}(1+x)^{3/2}]$$
$$= (1+x)^{5/2} - 2(1+x)^{3/2} + (1+x)^{1/2}$$
$$= (1+x)^{1/2}[(1+x)^2 - 2(1+x) + 1]$$
$$= (1+x)^{1/2}[1 + 2x + x^2 - 2 - 2x + 1]$$
$$= x^2\sqrt{1+x}$$ ●

EXERCISES 5.2

In Exercises 1 through 46, find the most general antiderivative. In Exercises 1 through 4, 11 through 14, and 21 through 24, check by finding the derivative of your answer.

1. $\int 3x^4\,dx$
2. $\int (3x^5 - 2x^3)\,dx$
3. $\int (3 - 2t + t^2)\,dt$
4. $\int (4x^3 - 3x^2 + 6x - 1)\,dx$
5. $\int (8x^4 + 4x^3 - 6x^2 - 4x + 5)\,dx$
6. $\int (2 + 3x^2 - 8x^3)\,dx$
7. $\int \sqrt{x}(x+1)\,dx$
8. $\int (ax^2 + bx + c)\,dx$
9. $\int (x^{3/2} - x)\,dx$
10. $\int \left(\sqrt{x} - \frac{1}{\sqrt{x}}\right)dx$
11. $\int \left(\frac{2}{x^3} + \frac{3}{x^2} + 5\right)dx$
12. $\int \left(3 - \frac{1}{x^4} + \frac{1}{x^2}\right)dx$
13. $\int \frac{x^2 + 4x - 4}{\sqrt{x}}\,dx$
14. $\int \frac{y^4 + 2y^2 - 1}{\sqrt{y}}\,dy$
15. $\int \left(\sqrt{2x} - \frac{1}{\sqrt{2x}}\right)dx$
16. $\int \frac{27t^3 - 1}{\sqrt[3]{t}}\,dt$
17. $\int \sqrt{1 - 4y}\,dy$
18. $\int \sqrt[3]{3x - 4}\,dx$
19. $\int \sqrt[3]{6 - 2x}\,dx$
20. $\int \sqrt{5r + 1}\,dr$
21. $\int x\sqrt{x^2 - 9}\,dx$
22. $\int 3x\sqrt{4 - x^2}\,dx$
23. $\int x^2\sqrt{x^3 - 1}\,dx$
24. $\int x(2x^2 + 1)^6\,dx$

25. $\int 5x\sqrt[3]{(9-4x^2)^2}\,dx$

26. $\int \frac{x\,dx}{\sqrt[3]{x^2+1}}$

27. $\int \frac{y^3}{(1-2y^4)^5}\,dy$

28. $\int (x^2-4x+4)^{4/3}\,dx$

29. $\int \frac{s\,ds}{\sqrt{3s^2+1}}$

30. $\int \frac{2r\,dr}{(1-r)^{2/3}}$

31. $\int \frac{t\,dt}{\sqrt{t+3}}$

32. $\int \sqrt{\frac{1}{t}-1}\,\frac{dt}{t^2}$

33. $\int \sqrt{1+\frac{1}{3x}}\,\frac{dx}{x^2}$

34. $\int x^4\sqrt{3x^5-5}\,dx$

35. $\int \sqrt{3-x}\,x^2\,dx$

36. $\int (x^3+3)^{1/4}x^5\,dx$

37. $\int \frac{(x^2+2x)\,dx}{\sqrt{x^3+3x^2+1}}$

38. $\int x(x^2+1)\sqrt{4-2x^2-x^4}\,dx$

39. $\int \frac{x(3x^2+1)\,dx}{(3x^4+2x^2+1)^2}$

40. $\int \sqrt{3+s}\,(s+1)^2\,ds$

41. $\int \frac{y+3}{(3-y)^{2/3}}\,dy$

42. $\int (2t^2+1)^{1/3}t^3\,dt$

43. $\int \frac{(r^{1/3}+2)^4}{\sqrt[3]{r^2}}\,dr$

44. $\int \left(t+\frac{1}{t}\right)^{3/2}\left(\frac{t^2-1}{t^2}\right)dt$

45. $\int \frac{x^3}{(x^2+4)^{3/2}}\,dx$

46. $\int \frac{x^3}{\sqrt{1-2x^2}}\,dx$

47. Evaluate $\int (2x+1)^3\,dx$ by two methods: (a) expand $(2x+1)^3$ by the binomial theorem, and apply Theorems 5.2.4, 5.2.7, and 5.2.8; (b) make the substitution $u = 2x+1$. Explain the difference in appearance of the answers obtained in (a) and (b).

48. Evaluate $\int x(x^2+2)^2\,dx$ by two methods: (a) expand $(x^2+2)^2$ and multiply the result by x; then apply Theorems 5.2.4, 5.2.7, and 5.2.8; (b) make the substitution $u = x^2+2$. Explain the difference in appearance of the answers obtained in (a) and (b).

49. Evaluate $\int \frac{(\sqrt{x}-1)^2}{\sqrt{x}}\,dx$ by two methods: (a) expand $(\sqrt{x}-1)^2$ and multiply the result by $x^{-1/2}$; then apply Theorems 5.2.4, 5.2.7, and 5.2.8; (b) make the substitution $u = \sqrt{x}-1$. Explain the difference in appearance of the answers obtained in (a) and (b).

50. Evaluate $\int \sqrt{x-1}\,x^2\,dx$ by two methods: (a) make the substitution $u = x-1$; (b) make the substitution $v = \sqrt{x-1}$.

51. Let $f(x) = 1$ for all x in $(-1, 1)$ and let

$$g(x) = \begin{cases} -1 & \text{if } -1 < x \le 0 \\ 1 & \text{if } 0 < x < 1 \end{cases}$$

Then $f'(x) = 0$ for all x in $(-1, 1)$ and $g'(x) = 0$ whenever g' exists in $(-1, 1)$. However, $f(x) \neq g(x) + K$ for x in $(-1, 1)$. Why doesn't Theorem 5.2.2 hold?

52. Show that the unit step function U (Exercise 31 in Exercises 1.6) does not have an antiderivative on $(-\infty, +\infty)$. (*Hint:* Assume that U has an antiderivative F on $(-\infty, +\infty)$, and a contradiction is obtained by showing that it follows from the mean-value theorem that there exists a number k such that $F(x) = x + k$ if $x > 0$, and $F(x) = k$ if $x < 0$.)

53. Let $f(x) = |x|$ and let F be defined by

$$F(x) = \begin{cases} -\frac{1}{2}x^2 & \text{if } x < 0 \\ \frac{1}{2}x^2 & \text{if } x \ge 0 \end{cases}$$

Show that F is an antiderivative of f on $(-\infty, +\infty)$.

54. Let $f(x) = \operatorname{sgn} x$ and $F(x) = |x|$. Show that $F'(x) = f(x)$ if $x \neq 0$. Is F an antiderivative of f on $(-\infty, +\infty)$? Explain.

5.3 ANTIDIFFERENTIATION OF POWERS OF SINE AND COSINE

The theorems for the antiderivative of the sine and cosine functions follow immediately from the corresponding ones for differentiation.

5.3.1 Theorem

$$\int \sin u \, du = -\cos u + C$$

Proof $D_u(-\cos u) = \sin u.$ ■

5.3.2 Theorem

$$\int \cos u \, du = \sin u + C$$

Proof $D_u(\sin u) = \cos u.$ ■

Example 1 Find $\int 3 \sin 2x \, dx$.

Solution Let $u = 2x$. Then $du = 2\, dx$. Thus

$$\begin{aligned}\int 3 \sin 2x \, dx &= \frac{3}{2}\int \sin 2x \,(2\, dx)\\ &= \frac{3}{2}\int \sin u \, du\\ &= \tfrac{3}{2}(-\cos u) + C\\ &= -\tfrac{3}{2}\cos 2x + C\end{aligned}$$

Example 2 Find

$$\int \frac{\cos \sqrt{x}}{\sqrt{x}}\, dx$$

Solution Let $u = \sqrt{x}$. Then

$$du = \frac{1}{2\sqrt{x}}\, dx$$

Therefore

$$\begin{aligned}\int \frac{\cos \sqrt{x}}{\sqrt{x}}\, dx &= 2\int \cos \sqrt{x}\left(\frac{1}{2\sqrt{x}}\, dx\right)\\ &= 2\int \cos u \, du\\ &= 2 \sin u + C\\ &= 2 \sin \sqrt{x} + C\end{aligned}$$

Example 3 Find $\int \sin x \sqrt{1 - \cos x}\, dx$.

Solution Let $u = 1 - \cos x$. Then $du = \sin x\, dx$. So

$$\begin{aligned}\int \sin x\sqrt{1-\cos x}\, dx &= \int u^{1/2}\, du\\ &= \tfrac{2}{3}u^{3/2} + C\\ &= \tfrac{2}{3}(1-\cos x)^{3/2} + C\end{aligned}$$

We now consider four cases of antiderivatives involving powers of sine and cosine. The method used in each case is shown by an illustration.

Case 1: $\int \sin^n u\, du$ or $\int \cos^n u\, du$, where n is an odd integer.

● ILLUSTRATION 1

$$\begin{aligned}\int \cos^3 x\, dx &= \int \cos^2 x\,(\cos x\, dx)\\ &= \int (1-\sin^2 x)(\cos x\, dx)\end{aligned}$$

So

$$\int \cos^3 x\, dx = \int \cos x\, dx - \int \sin^2 x \cos x\, dx \qquad (1)$$

To evaluate the second antiderivative on the right side of (1) we let $u = \sin x$; then $du = \cos x\, dx$. This gives

$$\begin{aligned}\int \sin^2 x\,(\cos x\, dx) &= \int u^2\, du\\ &= \tfrac{1}{3}u^3 + C_1\\ &= \tfrac{1}{3}\sin^3 x + C_1\end{aligned}$$

Because the first antiderivative on the right side of (1) is $\sin x + C_2$,

$$\int \cos^3 x\, dx = \sin x - \tfrac{1}{3}\sin^3 x + C$$ ●

Example 4 Find $\int \sin^5 x\, dx$.

Solution

$$\begin{aligned}\int \sin^5 x\, dx &= \int (\sin^2 x)^2 \sin x\, dx\\ &= \int (1-\cos^2 x)^2 \sin x\, dx\\ &= \int (1 - 2\cos^2 x + \cos^4 x)\sin x\, dx\end{aligned}$$

Therefore

$$\int \sin^5 x\,dx = \int \sin x\,dx - 2\int \cos^2 x \sin x\,dx + \int \cos^4 x(\sin x\,dx) \quad (2)$$

To evaluate the second and third antiderivatives in (2), let $u = \cos x$ and $du = -\sin x\,dx$. Therefore

$$\int \sin^5 x\,dx = -\cos x + 2\int u^2\,du - \int u^4\,du$$

$$= -\cos x + \tfrac{2}{3}u^3 - \tfrac{1}{5}u^5 + C$$

$$= -\cos x + \tfrac{2}{3}\cos^3 x - \tfrac{1}{5}\cos^5 x + C$$

Case 2: $\int \sin^n u\,du$ and $\int \cos^n u\,du$, where n is an even integer.

The method used in Case 1 does not work in this case. We use the following formulas, which are obtained from formulas (19) and (18) of Sec. 1.7 by substituting $2x$ for t.

$$\sin^2 x = \frac{1 - \cos 2x}{2}$$

$$\cos^2 x = \frac{1 + \cos 2x}{2}$$

● ILLUSTRATION 2

$$\int \sin^2 x\,dx = \int \frac{1 - \cos 2x}{2}\,dx$$

$$= \tfrac{1}{2}x - \tfrac{1}{4}\sin 2x + C$$

●

Example 5 Find $\int \cos^4 x\,dx$.

Solution

$$\int \cos^4 x\,dx = \int (\cos^2 x)^2\,dx = \int \left(\frac{1 + \cos 2x}{2}\right)^2 dx$$

$$= \frac{1}{4}\int dx + \frac{1}{2}\int \cos 2x\,dx + \frac{1}{4}\int \cos^2 2x\,dx$$

$$= \frac{1}{4}x + \frac{1}{4}\sin 2x + \frac{1}{4}\int \frac{1 + \cos 4x}{2}\,dx$$

$$= \tfrac{1}{4}x + \tfrac{1}{4}\sin 2x + \tfrac{1}{8}x + \tfrac{1}{32}\sin 4x + C$$

$$= \tfrac{3}{8}x + \tfrac{1}{4}\sin 2x + \tfrac{1}{32}\sin 4x + C$$

Case 3: $\int \sin^n x \cos^m x\,dx$, where at least one of the exponents is odd.

The solution of this case is similar to the method used for Case 1.

● ILLUSTRATION 3

$$\begin{aligned}\int \sin^3 x \cos^4 x\, dx &= \int \sin^2 x \cos^4 x(\sin x\, dx)\\ &= \int (1 - \cos^2 x)\cos^4 x(\sin x\, dx)\\ &= \int \cos^4 x \sin x\, dx - \int \cos^6 x \sin x\, dx\\ &= -\tfrac{1}{5}\cos^5 x + \tfrac{1}{7}\cos^7 x + C\end{aligned}$$ ●

Case 4: $\int \sin^n x \cos^m x\, dx$, where both m and n are even.

The solution of this case is similar to the method used in Case 2.

● ILLUSTRATION 4

$$\begin{aligned}\int \sin^2 x \cos^4 x\, dx &= \int \left(\frac{1 - \cos 2x}{2}\right)\left(\frac{1 + \cos 2x}{2}\right)^2 dx\\ &= \frac{1}{8}\int dx + \frac{1}{8}\int \cos 2x\, dx\\ &\qquad -\frac{1}{8}\int \cos^2 2x\, dx - \frac{1}{8}\int \cos^3 2x\, dx\\ &= \frac{1}{8}x + \frac{1}{16}\sin 2x - \frac{1}{8}\int \frac{1 + \cos 4x}{2}\, dx\\ &\qquad -\frac{1}{8}\int (1 - \sin^2 2x)\cos 2x\, dx\\ &= \frac{x}{8} + \frac{\sin 2x}{16} - \frac{x}{16} - \frac{\sin 4x}{64}\\ &\qquad -\frac{1}{8}\int \cos 2x\, dx + \frac{1}{8}\int \sin^2 2x \cos 2x\, dx\\ &= \frac{x}{16} + \frac{\sin 2x}{16} - \frac{\sin 4x}{64} - \frac{\sin 2x}{16} + \frac{\sin^3 2x}{48} + C\\ &= \frac{x}{16} + \frac{\sin^3 2x}{48} - \frac{\sin 4x}{64} + C\end{aligned}$$ ●

Example 6 Find $\int \sin^4 x \cos^4 x\, dx$.

Solution If we make use of the formula $\sin x \cos x = \frac{1}{2}\sin 2x$, we have

$$\int \sin^4 x \cos^4 x\, dx = \frac{1}{16}\int \sin^4 2x\, dx$$

$$= \frac{1}{16}\int \left(\frac{1 - \cos 4x}{2}\right)^2 dx$$

$$= \frac{1}{64}\int dx - \frac{1}{32}\int \cos 4x\, dx + \frac{1}{64}\int \cos^2 4x\, dx$$

$$= \frac{x}{64} - \frac{\sin 4x}{128} + \frac{1}{64}\int \frac{1 + \cos 8x}{2}\, dx$$

$$= \frac{x}{64} - \frac{\sin 4x}{128} + \frac{x}{128} + \frac{\sin 8x}{1024} + C$$

$$= \frac{3x}{128} - \frac{\sin 4x}{128} + \frac{\sin 8x}{1024} + C$$

The following example illustrates another type of antiderivative involving a product of a sine and a cosine.

Example 7 Find $\int \sin 3x \cos 2x\, dx$.

Solution We use formula (22) of Sec. 1.7, which is

$$\sin a \cos b = \tfrac{1}{2}\sin(a - b) + \tfrac{1}{2}\sin(a + b)$$

So

$$\int \sin 3x \cos 2x\, dx = \int \left[\tfrac{1}{2}\sin x + \tfrac{1}{2}\sin 5x\right] dx$$

$$= \frac{1}{2}\int \sin x\, dx + \frac{1}{2}\int \sin 5x\, dx$$

$$= -\tfrac{1}{2}\cos x - \tfrac{1}{10}\cos 5x + C$$

EXERCISES 5.3

In Exercises 1 through 36, find the most general antiderivative.

1. $\int \sin 4x\, dx$ **2.** $\int \sin \frac{1}{3}x\, dx$ **3.** $\int 4 \cos \frac{1}{2}x\, dx$ **4.** $\int \frac{1}{2}\cos 4x\, dx$

5. $\int (3 \sin t + \cos 3t)\, dt$ **6.** $\int (3 \cos 2x - 2 \sin 3x)\, dx$ **7.** $\int \cos x(2 + \sin x)^3\, dx$ **8.** $\int \frac{\sin x}{(1 + \cos x)^2}\, dx$

9. $\int 2 \sin x \sqrt[3]{1 + \cos x}\, dx$ **10.** $\int \sin 2x \sqrt{2 - \cos 2x}\, dx$ **11.** $\int \frac{\cos 3x}{\sqrt{1 - 2 \sin 3x}}\, dx$ **12.** $\int \frac{\sin 3\sqrt{t}}{\sqrt{t}}\, dt$

13. $\int \sin x \cdot \sin(\cos x) dx$ **14.** $\int \frac{\frac{1}{2}\cos \frac{1}{4}x}{\sqrt{\sin \frac{1}{4}x}}\, dx$ **15.** $\int \cos^2 t \sin t\, dt$ **16.** $\int \sin^3 \theta \cos \theta\, d\theta$

17. $\int \sin^3 x\, dx$ 18. $\int \sin^2 3x\, dx$ 19. $\int \sin^4 z\, dz$ 20. $\int \cos^5 x\, dx$

21. $\int \cos^2 \frac{1}{2}x\, dx$ 22. $\int \sin^3 x \cos^3 x\, dx$ 23. $\int \sin^2 x \cos^3 x\, dx$ 24. $\int \cos^6 x\, dx$

25. $\int \sin^5 x \cos^2 x\, dx$ 26. $\int \sin^2 2t \cos^4 2t\, dt$ 27. $\int \sin^2 3t \cos^2 3t\, dt$ 28. $\int \sqrt{\cos z}\, \sin^3 z\, dz$

29. $\int \frac{\cos^3 3x}{\sqrt[3]{\sin 3x}}\, dx$ 30. $\int \sin^3 \frac{1}{2}y \cos^2 \frac{1}{2}y\, dy$ 31. $\int \cos 4x \cos 3x\, dx$ 32. $\int \sin 2x \cos 4x\, dx$

33. $\int \sin 3y \cos 5y\, dy$ 34. $\int \cos t \cos 3t\, dt$ 35. $\int (\sin 3t - \sin 2t)^2\, dt$ 36. $\int \sin x \sin 3x \sin 5x\, dx$

37. Evaluate $\int 2 \sin x \cos x\, dx$ by three methods: (a) make the substitution $u = \sin x$; (b) make the substitution $u = \cos x$; (c) use the identity $2 \sin x \cos x = \sin 2x$. Explain the difference in appearance of the answers obtained in (a), (b), and (c).

38. If q coulombs is the charge of electricity received by a condenser from an electric circuit of i amperes at t sec, then $i = D_t q$. Suppose $i = 5 \sin 60t$ and $q = 0$ when $t = \frac{1}{2}\pi$. Find the greatest positive charge on the condenser.

5.4 DIFFERENTIAL EQUATIONS WITH VARIABLES SEPARABLE

An equation containing derivatives is called a **differential equation.** Some simple differential equations are

$$\frac{dy}{dx} = 2x \tag{1}$$

$$\frac{dy}{dx} = \frac{2x^2}{3y^3} \tag{2}$$

$$\frac{d^2y}{dx^2} = 4x + 3 \tag{3}$$

The **order** of a differential equation is the order of the derivative of highest order that appears in the equation. Therefore (1) and (2) are first-order differential equations and (3) is of the second order. The simplest type of differential equation is a first-order equation of the form

$$\frac{dy}{dx} = f(x) \tag{4}$$

for which (1) is a particular example. Writing (4) with differentials we have

$$dy = f(x)\, dx \tag{5}$$

Another type of differential equation of the first order is one of the form

$$\frac{dy}{dx} = \frac{g(x)}{h(y)} \tag{6}$$

Equation (2) is a particular example of an equation of this type. If (6) is written with differentials, we have

$$h(y)\,dy = g(x)\,dx \tag{7}$$

In both (5) and (7), the left side involves only the variable y and the right side involves only the variable x. Thus the variables are separated, and we say that these are differential equations with variables separable.

Consider (5). To solve this equation we must find all functions G for which $y = g(x)$ such that the equation is satisfied. So if F is an antiderivative of f, all functions G are defined by $G(x) = F(x) + C$, where C is an arbitrary constant. That is, if $d(G(x)) = d(F(x) + C) = f(x)\,dx$, then what is called the **complete solution** of (5) is given by

$$y = F(x) + C \tag{8}$$

Equation (8) represents a family of functions depending on an arbitrary constant C. This is called a **one-parameter family.** The graphs of these functions form a one-parameter family of curves in the plane, and through any particular point (x_1, y_1) there passes just one curve of the family.

● ILLUSTRATION 1 Suppose we wish to find the complete solution of the differential equation

$$dy = 2x\,dx \tag{9}$$

The most general antiderivative for the left side of (9) is $(y + C_1)$ and the most general antiderivative of $2x$ is $(x^2 + C_2)$. Thus

$$y + C_1 = x^2 + C_2$$

Because $(C_2 - C_1)$ is an arbitrary constant if C_2 and C_1 are arbitrary, we can replace $(C_2 - C_1)$ by C, thereby obtaining

$$y = x^2 + C \tag{10}$$

which is the complete solution of the given differential equation.

Equation (10) represents a one-parameter family of functions. Figure 5.4.1 shows sketches of the graphs of the functions corresponding to $C = -4$, $C = -1$, $C = 0$, $C = 1$, and $C = 2$. ●

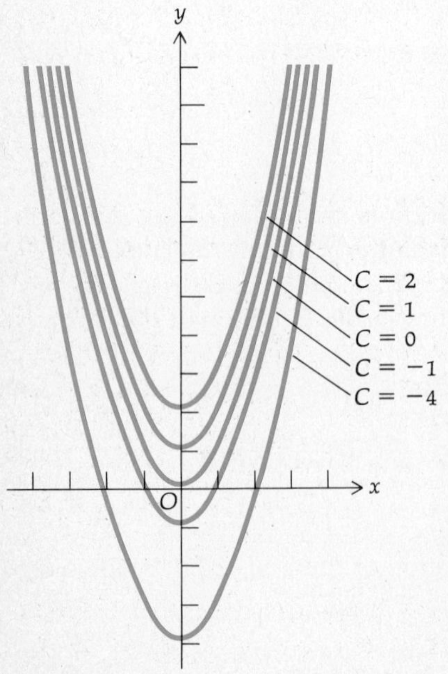

FIGURE 5.4.1

Now consider (7), which is

$$h(y)\,dy = g(x)\,dx$$

If we antidifferentiate on both sides of the equation, we have

$$\int h(y)\,dy = \int g(x)\,dx$$

If H is an antiderivative of h and G is an antiderivative of g, the complete solution of (7) is given by

$$H(y) = G(x) + C$$

Example 1 Find the complete solution of the differential equation

$$\frac{dy}{dx} = \frac{2x^2}{3y^3}$$

Solution If the given equation is written with differentials, we have

$$3y^3\,dy = 2x^2\,dx$$

and the variables are separated. We antidifferentiate on both sides of the equation and obtain

$$\int 3y^3\,dy = \int 2x^2\,dx$$

$$\frac{3y^4}{4} = \frac{2x^3}{3} + \frac{C}{12} \tag{11}$$

$$9y^3 = 8x^3 + C$$

which is the complete solution.

In (11) the arbitrary constant was written as $C/12$ so that when both sides of the equation are multiplied by 12, the arbitrary constant becomes C.

Often in problems involving differential equations it is desired to find particular solutions that satisfy certain conditions called **boundary conditions** or **initial conditions.** For example, if a first-order differential equation is given as well as the boundary condition that $y = y_1$ when $x = x_1$, then after the complete solution is found, if x and y in the complete solution are replaced by x_1 and y_1, a particular value of C is determined. When this value of C is substituted back into the complete solution, a particular solution is obtained.

● ILLUSTRATION 2 To find the particular solution of differential equation (9) satisfying the boundary condition that $y = 6$ when $x = 2$, we substitute these values in (10) and solve for C, giving $6 = 4 + C$, or $C = 2$. Substituting this value of C in (10) we obtain

$$y = x^2 + 2$$

which is the particular solution desired. ●

Equation (3) is an example of a particular type of differential equation of the second order

$$\frac{d^2y}{dx^2} = f(x) \tag{12}$$

Two successive antidifferentiations are necessary to solve (12), and two arbitrary constants occur in the complete solution. The complete solution of (12) therefore represents a **two-parameter family** of functions, and the graphs of these functions form a two-parameter family of curves in the plane. The following example shows the method of obtaining the complete solution of an equation of this kind.

Example 2 Find the complete solution of the differential equation

$$\frac{d^2y}{dx^2} = 4x + 3$$

Solution Because $d^2y/dx^2 = dy'/dx$, the given equation can be written as

$$\frac{dy'}{dx} = 4x + 3$$

Writing this in differential form we have

$$dy' = (4x + 3)\,dx$$

Antidifferentiating we have

$$\int dy' = \int (4x + 3)\,dx$$

from which we get

$$y' = 2x^2 + 3x + C_1$$

Because $y' = dy/dx$, we make this substitution in the above equation and get

$$\frac{dy}{dx} = 2x^2 + 3x + C_1$$

Using differentials we have

$$dy = (2x^2 + 3x + C_1)\,dx$$

Antidifferentiating we obtain

$$\int dy = \int (2x^2 + 3x + C_1)\,dx$$

from which we get

$$y = \tfrac{2}{3}x^3 + \tfrac{3}{2}x^2 + C_1x + C_2$$

which is the complete solution.

Example 3 Find the particular solution of the differential equation in Example 2 for which $y = 2$ and $y' = -3$ when $x = 1$.

Solution Because $y' = 2x^2 + 3x + C_1$, we substitute -3 for y' and 1 for x, giving $-3 = 2 + 3 + C_1$, or $C_1 = -8$. Substituting this value of C_1 into the complete solution gives

$$y = \tfrac{2}{3}x^3 + \tfrac{3}{2}x^2 - 8x + C_2$$

Because $y = 2$ when $x = 1$, we substitute these values in the above equation and get $2 = \frac{2}{3} + \frac{3}{2} - 8 + C_2$, from which we obtain $C_2 = \frac{47}{6}$. The particular solution desired, then, is

$$y = \tfrac{2}{3}x^3 + \tfrac{3}{2}x^2 - 8x + \tfrac{47}{6}$$

Example 4 A business firm has made an analysis of its production facilities and its personnel. With the present equipment and number of workers, the firm can produce 3000 units per day. It was estimated that without any change in investment, the rate of change of the number of units produced per day with respect to a change in the number of additional workers is $80 - 6x^{1/2}$, where x is the number of additional workers. Find the daily production if 25 workers are added to the labor force.

Solution Let $y =$ the number of units produced per day. Then

$$\frac{dy}{dx} = 80 - 6x^{1/2}$$

from which we get

$$dy = (80 - 6x^{1/2})\,dx$$

Antidifferentiating gives

$$y = 80x - 4x^{3/2} + C$$

Because $y = 3000$ when $x = 0$, we get $C = 3000$. Hence

$$y = 80x - 4x^{3/2} + 3000$$

Letting y_{25} be the value of y when $x = 25$ we have

$$\begin{aligned} y_{25} &= 2000 - 500 + 3000 \\ &= 4500 \end{aligned}$$

Therefore 4500 units are produced per day if the labor force is increased by 25 workers.

Example 5 The slope of the tangent line to a curve at any point (x, y) on the curve is equal to $3x^2y^2$. Find an equation of the curve if it contains the point $(2, 1)$.

Solution Because the slope of the tangent line to a curve at any point (x, y) is the value of the derivative at that point, we have

$$\frac{dy}{dx} = 3x^2y^2$$

This is a first-order differential equation, in which the variables can be separated, and we have

$$\frac{dy}{y^2} = 3x^2\,dx$$

Antidifferentiating we have

$$\int y^{-2}\,dy = \int 3x^2\,dx$$

from which we obtain

$$-\frac{1}{y} = x^3 + C$$

This is an equation of a one-parameter family of curves. To find the particular curve of this family that contains the point (2, 1), we substitute 2 for x and 1 for y, which gives $-1 = 8 + C$, from which we obtain $C = -9$. Therefore the required curve has the equation

$$-\frac{1}{y} = x^3 - 9$$

EXERCISES 5.4

In Exercises 1 through 20, find the complete solution of the differential equation.

1. $\frac{dy}{dx} = 4x - 5$ **2.** $\frac{dy}{dx} = 6 - 3x^2$ **3.** $\frac{dy}{dx} = 3x^2 + 2x - 7$ **4.** $\frac{dy}{dx} = x^4 + 2x^2 - \frac{1}{x^2}$

5. $\frac{du}{dv} = \frac{2v^2 - 3}{v^2}$ **6.** $\frac{dy}{dx} = (3x + 1)^3$ **7.** $\frac{dy}{dx} = 3\sin 2x$ **8.** $\frac{ds}{dt} = 5\sqrt{s}$

9. $\frac{dy}{dx} = 3xy^2$ **10.** $\frac{dy}{dx} = \frac{\sqrt{x} + x}{\sqrt{y} - y}$ **11.** $\frac{dy}{dx} = \frac{3x\sqrt{1 + y^2}}{y}$ **12.** $\frac{dy}{dx} = \frac{x^2\sqrt{x^3 - 3}}{y^2}$

13. $\frac{dy}{dx} = \frac{\cos x}{\sin^2 y}$ **14.** $\frac{du}{dv} = \frac{\cos^3 v}{\sin u}$ **15.** $\frac{d^2y}{dx^2} = 5x^2 + 1$ **16.** $\frac{d^2y}{dx^2} = \sqrt{2x - 3}$

17. $\frac{d^2u}{dv^2} = \sqrt{3v + 1}$ **18.** $\frac{d^2s}{dt^2} = \sin 3t + \cos 3t$ **19.** $\frac{d^2y}{dx^2} = 3\sin^3 x - \cos x$ **20.** $\frac{d^2y}{dx^2} = \cos^2 x$

In Exercises 21 through 30, find the particular solution of the given differential equation determined by the boundary conditions.

21. $\frac{dy}{dx} = x^2 - 2x - 4$; $y = -6$ when $x = 3$

22. $\frac{ds}{dt} = \cos \frac{1}{2}t$; $s = 3$ when $t = \frac{1}{3}\pi$

23. $\frac{dx}{y} = \frac{4\,dy}{x}$; $y = -2$ when $x = 4$

24. $\frac{dy}{dx} = (x + 1)(x + 2)$; $y = -\frac{3}{2}$ when $x = -3$

25. $\frac{dy}{dx} = \frac{\cos 3x}{\sin 2y}$; $y = \frac{1}{3}\pi$ when $x = \frac{1}{2}\pi$

26. $\frac{dy}{dx} = \frac{x}{4\sqrt{(1 + x^2)^3}}$; $y = 0$ when $x = 1$

27. $\frac{d^2y}{dx^2} = x^2 + 3x$; $y = 2$ and $y' = 1$ when $x = 1$

28. $\frac{d^2y}{dx^2} = -\frac{3}{x^4}$; $y = \frac{1}{2}$ and $y' = -1$ when $x = 1$

29. $\frac{d^2u}{dv^2} = 4(1 + 3v)^2$; $u = -1$ and $D_v u = -2$ when $v = -1$

30. $\frac{d^2y}{dx^2} = \sqrt[3]{3x - 1}$; $y = 2$ and $y' = 5$ when $x = 3$

31. The point (3, 2) is on a curve, and at any point (x, y) on the curve the tangent line has a slope equal to $2x - 3$. Find an equation of the curve.

32. The slope of the tangent line at any point (x, y) on a curve is $3\sqrt{x}$. If the point (9, 4) is on the curve, find an equation of the curve.

33. The points $(-1, 3)$ and $(0, 2)$ are on a curve, and at any point (x, y) on the curve $D_x^2 y = 2 - 4x$. Find an equation of the curve.

34. An equation of the tangent line to a curve at the point (1, 3) is $y = x + 2$. If at any point (x, y) on the curve, $D_x^2 y = 6x$, find an equation of the curve.

35. At any point (x, y) on a curve, $D_x^2 y = 1 - x^2$, and an equation of the tangent line to the curve at the point (1, 1) is $y = 2 - x$. Find an equation of the curve.

36. At any point (x, y) on a curve, $D_x^3 y = 2$, and (1, 3) is a point of inflection at which the slope of the inflectional tangent is -2. Find an equation of the curve.

37. The cost of a certain piece of machinery is \$700, and its value depreciates with time according to the formula $D_t V = -500(t + 1)^{-2}$, where V dollars is its value t years after its purchase. What is its value three years after its purchase?

38. The volume of a balloon is increasing according to the formula $D_t V = \sqrt{t + 1} + \frac{2}{3}t$, where V cm³ is the volume of the balloon at t sec. If $V = 33$ when $t = 3$, find (a) a formula for V in terms of t; (b) the volume of the balloon at 8 sec.

39. The volume of water in a tank is V cubic meters when the depth of the water is h meters. If the rate of change of V with respect to h is given by $D_h V = \pi(2h + 3)^2$, find the volume of water in the tank when the depth is 3 m.

40. The efficiency of a factory worker is expressed as a percent. For instance, if the worker's efficiency at a particular time is given as 70 percent, then the worker is performing at 70 percent of her full potential. Suppose that E percent is a factory worker's efficiency t hours after beginning work, and the rate at which E is changing is $(35 - 8t)$ percent per hour. If the worker's efficiency is 81 percent after working 3 hr, find her efficiency after working (a) 4 hr and (b) 8 hr.

41. The equation $x^2 = 4ay$ represents a one-parameter family of parabolas. Find an equation of another one-parameter family of curves such that at any point (x, y) there is a curve of each family through it and the tangent lines to the two curves at this point are perpendicular. (*Hint:* First show that the slope of the tangent line at any point (x, y), not on the y axis, of the parabola of the given family through that point is $2y/x$.)

42. Solve Exercise 41 if the given one-parameter family of curves has the equation $x^3 + y^3 = a^3$.

5.5 ANTIDIFFERENTIATION AND RECTILINEAR MOTION

We learned in Secs. 3.5 and 3.10 that in considering the motion of a particle along a straight line, when an equation of motion, $s = f(t)$, is given, then the instantaneous velocity and the instantaneous acceleration can be determined from the equations

$$v = D_t s = f'(t)$$

and

$$a = D_t^2 s = D_t v = f''(t)$$

Hence, if we are given v or a as a function of t, as well as some boundary conditions, it is possible to determine the equation of motion by antidifferentiation. This is illustrated in the following example.

Example 1 A particle is moving on a straight line; s is the number of feet in the directed distance of the particle from the origin at t sec of time, v is the number of feet per second in the velocity of the particle at t

sec, and a is the number of feet per second per second in the acceleration of the particle at t sec. If $a = 2t - 1$ and $v = 3$, and $s = 4$ when $t = 1$, express v and s as functions of t.

Solution Because $a = D_t v$,

$$\frac{dv}{dt} = 2t - 1$$

which, expressed in differential form, gives

$$dv = (2t - 1)\,dt$$

Antidifferentiating we have

$$\int dv = \int (2t - 1)\,dt$$

$$v = t^2 - t + C_1 \tag{1}$$

Substituting $v = 3$ and $t = 1$ we have

$$3 = 1 - 1 + C_1$$

and hence $C_1 = 3$. Therefore, substituting this value of C_1 into (1) we obtain

$$v = t^2 - t + 3 \tag{2}$$

which expresses v as a function of t.

Now, letting $v = D_t s$ in (2) we have

$$\frac{ds}{dt} = t^2 - t + 3$$

$$ds = (t^2 - t + 3)\,dt$$

$$\int ds = \int (t^2 - t + 3)\,dt$$

$$s = \tfrac{1}{3}t^3 - \tfrac{1}{2}t^2 + 3t + C_2 \tag{3}$$

Letting $s = 4$ and $t = 1$ in (3) and solving for C_2, we obtain $C_2 = \frac{7}{6}$. Therefore, by substituting $\frac{7}{6}$ for C_2 in (3) we have s expressed as a function of t, which is

$$s = \tfrac{1}{3}t^3 - \tfrac{1}{2}t^2 + 3t + \tfrac{7}{6}$$

If an object is moving freely in a vertical line and is being pulled toward the earth by a force of gravity, the acceleration of *gravity*, denoted by g ft/sec^2, varies with the distance of the object from the center of the earth. However, for small changes of distances the acceleration of gravity is almost constant, and an approximate value of g, if the object is near sea level, is 32.

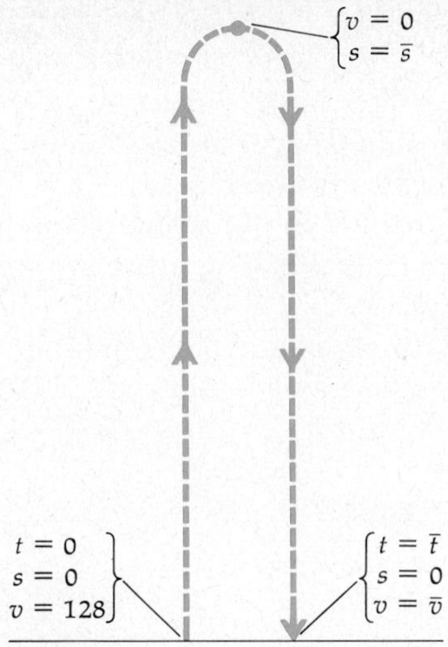

FIGURE 5.5.1

TABLE 5.5.1

t	s	v
0	0	128
	$\bar{s}$	0
$\bar{t}$	0	$\bar{v}$

Example 2 A stone is thrown vertically upward from the ground with an initial velocity of 128 ft/sec. If the only force considered is that attributed to the acceleration of gravity, find how high the stone will rise and the speed with which it will strike the ground. Also, find how long it will take for the stone to strike the ground.

Solution The motion of the stone is illustrated in Fig. 5.5.1. The positive direction is taken as upward.

Let t be the number of seconds in the time that has elapsed since the stone was thrown; s be the number of feet in the distance of the stone from the ground at t sec of time; v be the number of feet per second in the velocity of the stone at t sec of time; and $|v|$ be the number of feet per second in the speed of the stone at t sec of time.

The stone will be at its highest point when the velocity is zero. Let $\bar{s}$ be the particular value of s when $v = 0$. When the stone strikes the ground, $s = 0$. Let $\bar{t}$ and $\bar{v}$ be the particular values of t and v when $s = 0$, and $t \neq 0$. Table 5.5.1 is a table of boundary conditions for this problem.

The positive direction of the stone from the starting point is taken as upward. Because the only acceleration is due to gravity, which is in the downward direction, the acceleration has a constant value of -32 ft/sec². Because the acceleration is given by the derivative of v with respect to t, we have

$$\frac{dv}{dt} = -32$$

$$dv = -32\,dt$$

$$\int dv = \int -32\,dt$$

$$v = -32t + C_1$$

Because $v = 128$ when $t = 0$, we substitute these values in the above and get $C_1 = 128$. Therefore

$$v = -32t + 128 \tag{4}$$

Because v is the derivative of s with respect to t, then $v = ds/dt$; so

$$\frac{ds}{dt} = -32t + 128$$

$$ds = (-32t + 128)\,dt$$

$$\int ds = \int (-32t + 128)\,dt$$

$$s = -16t^2 + 128t + C_2$$

Because $s = 0$ when $t = 0$, then $C_2 = 0$, and substituting 0 for C_2 in the above equation gives

$$s = -16t^2 + 128t \tag{5}$$

Substituting $\bar{t}$ for t and 0 for s we get

$$0 = -16\bar{t}(\bar{t} - 8)$$

from which $\bar{t} = 0$ and $\bar{t} = 8$. However, the value 0 occurs when the stone is thrown; so it takes 8 sec for the stone to strike the ground.

To obtain $\bar{v}$, substitute 8 for t and $\bar{v}$ for v in (4), which gives $\bar{v} = (-32)(8) + 128$, from which $\bar{v} = -128$. So $|\bar{v}| = 128$. Therefore the stone strikes the ground with a speed of 128 ft/sec.

To find $\bar{s}$ we first find the value of t for which v is 0. From (4), $t = 4$ when $v = 0$. In (5) we substitute 4 for t and $\bar{s}$ for s and obtain $\bar{s} = 256$. Thus the stone will rise 256 ft.

EXERCISES 5.5

In Exercises 1 through 12, a particle is moving on a straight line, s ft is the directed distance of the particle from the origin at t sec, v ft/sec is the velocity of the particle at t sec, and a ft/sec² is the acceleration of the particle at t sec.

(Hint for Exercises 9 through 12: $a = \dfrac{dv}{dt} = \dfrac{dv}{ds}\dfrac{ds}{dt} = v\dfrac{dv}{ds}$.*)*

1. $v = \sqrt{2t + 4}$; $s = 0$ when $t = 0$. Express s in terms of t.

2. $v = 4 - t$; $s = 0$ when $t = 2$. Express s in terms of t.

3. $a = 5 - 2t$; $v = 2$ and $s = 0$ when $t = 0$. Express v and s in terms of t.

4. $a = 17$; $v = 0$ and $s = 0$ when $t = 0$. Express v and s in terms of t.

5. $a = t^2 + 2t$; $s = 1$ when $t = 0$ and $s = -3$ when $t = 2$. Express v and s in terms of t.

6. $a = 3t - t^2$; $v = \frac{7}{6}$ and $s = 1$ when $t = 1$. Express v and s in terms of t.

7. $a = -4\sqrt{2}\cos(2t - \frac{1}{4}\pi)$; $v = 2$ and $s = 1$ when $t = 0$. Express v and s in terms of t.

8. $a = 18 \sin 3t$; $v = -6$ and $s = 4$ when $t = 0$. Express v and s in terms of t.

9. $a = 800$; $v = 20$ when $s = 1$. Find an equation involving v and s.

10. $a = 500$; $v = 10$ when $s = 5$. Find an equation involving v and s.

11. $a = 5s + 2$; $v = 4$ when $s = 2$. Find an equation involving v and s.

12. $a = 2s + 1$; $v = 2$ when $s = 1$. Find an equation involving v and s.

In Exercises 13 through 20, the only force considered is that due to the acceleration of gravity, which we take as 32 ft/sec² in the downward direction.

13. A stone is thrown vertically upward from the ground with an initial velocity of 20 ft/sec. (a) How long will it take the stone to strike the ground? (b) With what speed will the stone strike the ground? (c) How long will the stone be going upward? (d) How high will the stone go?

14. A stone is thrown vertically upward from the ground with an initial velocity of 24 ft/sec. (a) How long will it take the stone to strike the ground, and (b) with what speed will the stone strike the ground?

15. A ball is dropped from the top of the Washington Monument, which is 555 ft high. (a) How long will it take the ball to reach the ground, and (b) with what speed will it strike the ground?

16. A stone is thrown vertically upward from the ground with an initial velocity of 16 ft/sec. (a) How long will the stone be going upward, and (b) how high will the stone go?

17. A man in a balloon drops his binoculars when the balloon is 150 ft above the ground and rising at the rate of 10 ft/sec. (a) How long will it take the binoculars to strike the ground, and (b) what is their speed on impact?

18. A ball is thrown downward from a window that is 80 ft above the ground with an initial velocity of -64 ft/sec. (a) When does the ball strike the ground, and (b) with what speed will it strike the ground?

19. A ball is thrown vertically upward with an initial velocity of 40 ft/sec from a point 20 ft above the ground. (a) If v ft/sec is the velocity of the ball when it is s ft from the starting point, express v in terms of s. (b) What is the velocity of the ball when it is 36 ft from the ground and rising? (*Hint:* See hint for Exercises 9 through 12.)

20. A stone is thrown vertically upward from the top of a house 60 ft above the ground with an initial velocity of 40 ft/sec. (a) At what time will the stone reach its greatest height, and (b) what is its greatest height? (c) How long will it take the stone to pass the top of the house on its way down, and (d) what is its velocity at that instant? (e) How long will it take the stone to strike the ground, and (f) with what velocity will it strike the ground?

21. If the driver of an automobile wishes to increase the speed from 40 km/hr to 100 km/hr while traveling a distance of 200 m, what constant acceleration should be maintained?

22. If a ball is rolled across level ground with an initial velocity of 20 ft/sec, and if the speed of the ball is decreasing at the rate of 6 ft/sec^2 due to friction, how far will the ball roll?

23. If the brakes are applied on a car traveling 100 km/hr and the brakes can give the car a constant negative acceleration of 8 m/sec^2, (a) how long will it take the car to come to a stop, and (b) how far will the car travel before stopping?

24. What constant negative acceleration will enable a driver to decrease the speed from 120 km/hr to 60 km/hr while traveling a distance of 100 m?

25. If the brakes on a car can give the car a constant negative acceleration of 8 m/sec^2, what is the greatest speed it may be going if it is necessary to be able to stop the car within 25 m after the brake is applied?

26. A ball started upward from the bottom of an inclined plane with an initial velocity of 6 ft/sec. If there was a downward acceleration of 4 ft/sec^2, how far up the plane did the ball go before rolling down?

27. A block of ice slides down a chute with a constant acceleration of 3 m/sec^2. The chute is 36 m long and it takes 4 sec for the ice to reach the bottom. (a) What is the initial velocity of the ice? (b) What is the speed of the ice after it has traveled 12 m? (c) How long does it take the ice to go the 12 m?

5.6 APPLICATIONS OF ANTIDIFFERENTIATION IN ECONOMICS

In Illustration 1 of Sec. 4.10, $C(x)$ is the number of dollars in the total cost of manufacturing x toys, and

$$C(x) = 110 + 4x + 0.02x^2$$

We showed in Sec. 4.10 that an approximation to $C(51) - C(50)$, which is the number of dollars in the cost of producing the fifty-first toy, is given by $C'(50)$, the value of the marginal cost function at 50. This fact can be proved by using the concept of the differential. Let

$$y = C(x)$$

Then

$$\Delta y = C(x + \Delta x) - C(x) \tag{1}$$

and

$$dy = C'(x)\,dx \tag{2}$$

If $x = 50$, and $dx = \Delta x = 1$, then from (1),

$$\Delta y = C(51) - C(50)$$

and from (2),

$$dy = [C'(50)](1) = C'(50)$$

Because dy is an approximation of Δy, $C'(50)$ is an approximation of $C(51) - C(50)$.

Because the marginal cost function C' and the marginal revenue function R' are the first derivatives of the total cost function C and the total revenue function R, respectively, C and R can be obtained from C' and R' by antidifferentiation. When finding the function C from C', the arbitrary constant can be evaluated if we know either the overhead cost (i.e., the cost when no units are produced) or the cost of production of a specific number of units of the commodity. Because it is generally true that the total revenue is zero when the number of units produced is zero, this fact may be used to evaluate the arbitrary constant when finding the function R from R'.

Example 1 The marginal cost function C' is given by $C'(x) = 4x - 8$, where $C(x)$ dollars is the total cost of producing x units. If the cost of producing 5 units is \$20, find the total cost function. Draw sketches of the total cost curve and the marginal cost curve on the same set of axes.

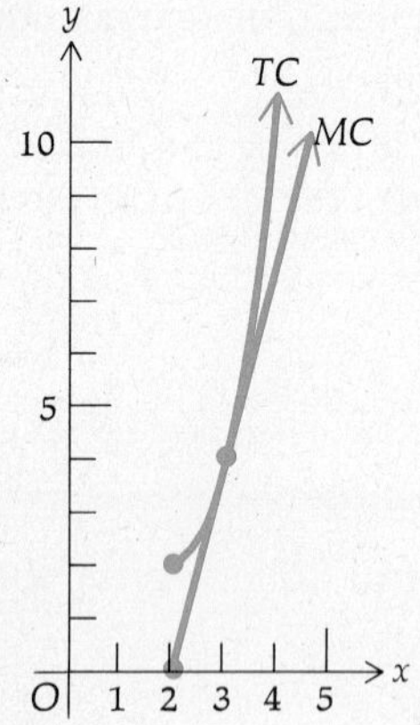

FIGURE 5.6.1

Solution The marginal cost must be nonnegative. Hence $4x - 8 \geq 0$, and therefore the permissible values of x are $x \geq 2$. Because

$$C'(x) = 4x - 8$$

$$\begin{aligned} C(x) &= \int (4x - 8)\,dx \\ &= 2x^2 - 8x + k \end{aligned}$$

Because $C(5) = 20$, we obtain $k = 10$. Hence

$$C(x) = 2x^2 - 8x + 10 \qquad x \geq 2$$

The required sketches are shown in Fig. 5.6.1.

Example 2 If the marginal revenue is given by $R'(x) = 27 - 12x + x^2$, find the total revenue function and the demand equation. Draw sketches of the demand curve, the total revenue curve, and the marginal revenue curve on the same set of axes.

Solution If R is the total revenue function, and

$$R'(x) = 27 - 12x + x^2$$

then

$$R(x) = \int (27 - 12x + x^2)\,dx$$

$$= 27x - 6x^2 + \tfrac{1}{3}x^3 + C$$

Because $R(0) = 0$, we get $C = 0$. Hence

$$R(x) = 27x - 6x^2 + \tfrac{1}{3}x^3$$

If f is the price function, $R(x) = xf(x)$; so

$$f(x) = 27 - 6x + \tfrac{1}{3}x^2$$

If p dollars is the price of one unit of the commodity when x units are demanded, then because $p = f(x)$, the demand equation is

$$3p = 81 - 18x + x^2$$

To determine the permissible values of x we use the facts that $x \geq 0, p \geq 0$, and f is a decreasing function. Because

$$f'(x) = -6 + \tfrac{2}{3}x$$

f is decreasing when $x < 9$ (i.e., when $f'(x) < 0$). Also, when $x = 9, p = 0$; so the permissible values of x are the numbers in the closed interval $[0, 9]$. The required sketches are shown in Fig. 5.6.2.

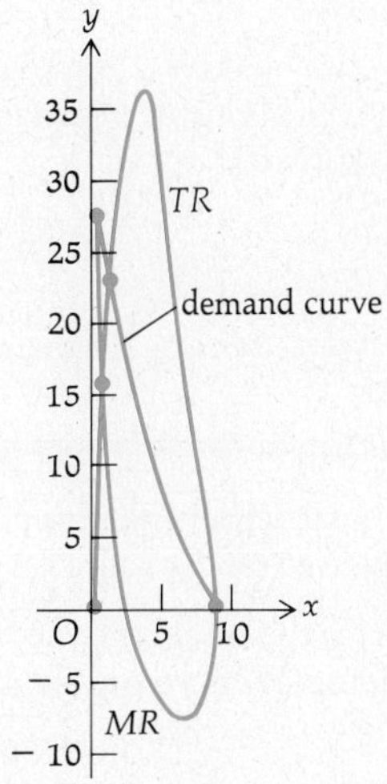

FIGURE 5.6.2

Example 3 After experimentation, a certain manufacturer determined that if x units of a certain article of merchandise are produced per week, the marginal cost is given by $C'(x) = 0.3x - 11$, where $C(x)$ dollars is the total cost of producing x units. If the selling price of the article is fixed at \$19 per unit and the overhead cost is \$100 per week, find the maximum weekly total profit that can be obtained.

Solution Let $R(x)$ dollars be the total revenue obtained by selling x units and $P(x)$ dollars be the total profit obtained by selling x units. Because x units are sold at \$19 per unit,

$$R(x) = 19x$$

Thus the marginal revenue is given by

$$R'(x) = 19$$

We are given that

$$C'(x) = 0.3x - 11$$

The total profit will be a maximum when the marginal revenue equals the marginal cost (see Sec. 4.10). Equating $R'(x)$ and $C'(x)$, we get $x = 100$. So 100 units should be produced each week for maximum total profit.

$$C(x) = \int (0.3x - 11)\,dx$$

$$= 0.15x^2 - 11x + k$$

Because the overhead cost is \$100, $C(0) = 100$, and hence $k = 100$.

Therefore

$$C(x) = 0.15x^2 - 11x + 100$$

Because $P(x) = R(x) - C(x)$,

$$P(x) = 19x - (0.15x^2 - 11x + 100)$$
$$= -0.15x^2 + 30x - 100$$

Hence $P(100) = 1400$. Therefore the maximum weekly total profit is \$1400, which is obtained if 100 units are produced weekly.

EXERCISES 5.6

1. The slope of the total revenue curve is given by $12 - 3x$ and $p = 6$ when $x = 4$. Find (a) the total revenue function and (b) the demand equation. (c) Draw sketches of the total revenue curve and the demand curve on the same set of axes.

2. The slope of the total revenue curve at any point is given by $15 - 4x$ and $R(0) = 0$. Find (a) the total revenue function and (b) the demand equation. (c) Draw sketches of the total revenue curve and the demand curve on the same set of axes.

3. The rate of change of the slope of the total cost curve of a particular company is the constant 2, and the total cost curve contains the points (2, 12) and (3, 18). Find the total cost function.

4. The marginal cost function is given by $C'(x) = 3\sqrt{2x + 4}$. If the overhead cost is zero, find the total cost function.

5. The marginal cost function is given by $C'(x) = 3(5x + 4)^{-1/2}$. If the overhead cost is \$10, find the total cost function.

6. Find the price function for a commodity for which the marginal revenue function is given by $R'(x) = 14 - 9x + x^2$.

7. Find the demand equation of a commodity for which the marginal revenue function is given by $R'(x) = 4 + 10(x + 5)^{-2}$.

8. For a particular commodity, the rate of change of the total cost function per unit change in x is $3x$ and the total cost curve contains the point (5, 45). Find the total cost function.

9. The marginal cost function for a particular commodity is given by $C'(x) = 6x - 17$. If the cost of producing 2 units is \$25, find the total cost function.

10. If the marginal revenue function is given by $R'(x) = 3x^2 - 12x + 10$, find (a) the total revenue function and (b) the price function.

11. The marginal revenue function is given by $R'(x) = 16 - 3x^2$. Find (a) the total revenue function and (b) the demand equation. (c) Draw sketches of the demand curve, the total revenue curve, and the marginal revenue curve on the same set of axes.

12. Find (a) the total revenue function and (b) the demand equation if the marginal revenue function is given by $R'(x) = \frac{3}{4}x^2 - 10x + 12$. (c) Draw sketches of the demand curve, the total revenue curve, and the marginal revenue curve on the same set of axes.

13. The marginal cost function is given by $C'(x) = 3x^2 + 8x + 4$, and the overhead cost is \$6. If $C(x)$ dollars is the total cost of x units, (a) find the total cost function, and (b) draw sketches of the total cost curve and the marginal cost curve on the same set of axes.

14. The marginal cost function is defined by $C'(x) = 6x$, where $C(x)$ is the number of hundreds of dollars in the total cost of x hundred units of a certain commodity. If the cost of 200 units is \$2000, find (a) the total cost function and (b) the overhead cost. (c) Draw sketches of the total cost curve and the marginal cost curve on the same set of axes.

15. A company has determined that the marginal cost function for the production of a particular commodity is given by $C'(x) = 125 + 10x + \frac{1}{9}x^2$, where $C(x)$ dollars is the total cost of producing x units of the commodity. If the overhead cost is \$250, what is the cost of producing 15 units?

16. The marginal revenue function is given by $R'(x) = ab(x + b)^{-2} - c$. Find (a) the total revenue function and (b) the demand equation.

17. The marginal cost function for a manufacturer of vases is $C'(x) = 4 - 9\sqrt{3x}/2x$, where $C(x)$ dollars is the total cost of producing x vases. The overhead cost is \$54. If 27 vases are produced, find (a) the marginal cost and (b) the average unit cost.

18. The marginal cost function for a particular article of merchandise is $C'(x) = 30x^{1/2}$, where $C(x)$ dollars is the total cost of producing x articles. The overhead cost is \$100. If 25 units are produced, find (a) the marginal cost and (b) the average unit cost.

19. A manufacturer of children's toys has a new toy coming on the market and wishes to determine a selling price for the toy such that the total profit will be a maximum. From analyzing the price and demand of another similar toy, it is anticipated that if x toys are demanded when p dollars is the price per toy, then $D_x p = -p^2/30{,}000$, and the demand should be 1800 when the price is \$10. If $C(x)$ dollars is the total cost of producing x toys, then $C(x) = x + 7500$. Find the price that should be charged for the manufacturer's total profit to be a maximum.

20. Suppose that a particular company estimates its growth in income from sales by the formula $D_t S = 2(t - 1)^{2/3}$, where S millions of dollars is the gross income from sales t years hence. If the gross income from the current year's sales is \$8 million, what should be the expected gross income from sales two years from now?

21. A firm estimates that its growth in income from sales for the next ten years is given by the formula $D_t S = \frac{1}{5}(t + 1)^{1/3}$, where S millions of dollars is the gross income from sales t years hence, and $0 \le t \le 10$. If this year's gross income from sales is \$1 million, what is the expected gross income from sales seven years hence?

22. A producer of a certain commodity determines that the marginal cost is given by $C'(x) = 8 + 2\cos(\frac{1}{100}\pi x)$, where $C(x)$ dollars is the total cost of producing x units. If the overhead is \$75, what is the total cost of producing 50 units?

REVIEW EXERCISES FOR CHAPTER 5

In Exercises 1 through 24, find the most general antiderivative.

1. $\int (2x^3 - x^2 + 3)\,dx$ **2.** $\int (5x^4 + 3x - 1)\,dx$ **3.** $\int (4y + 6\sqrt{y})\,dy$ **4.** $\int 3z^{-2}\,dz$

5. $\int \sin 3t\,dt$ **6.** $\int \frac{\cos\sqrt{w}}{\sqrt{w}}\,dw$ **7.** $\int \cos^2 x \sin x\,dx$ **8.** $\int \sqrt{x}(1 + x^2)\,dx$

9. $\int \left(\frac{2}{x^4} - \frac{5}{x^2}\right) dx$ **10.** $\int \left(\sqrt[3]{t} - \frac{1}{\sqrt[3]{t}}\right) dt$ **11.** $\int 5x\sqrt{2 + 3x^2}\,dx$ **12.** $\int x^4\sqrt{x^5 - 1}\,dx$

13. $\int \left(\sqrt{3x} + \frac{1}{\sqrt{5x}}\right) dx$ **14.** $\int \sqrt[3]{7w + 3}\,dw$ **15.** $\int \frac{x^3 + x}{\sqrt{x^4 + 2x^2}}\,dx$ **16.** $\int (x^3 + x)\sqrt{x^2 + 3}\,dx$

17. $\int \frac{s}{\sqrt{2s + 3}}\,ds$ **18.** $\int \sin^3 x \cos^2 x\,dx$ **19.** $\int \sin^4 \frac{1}{2}t\,dt$ **20.** $\int \sin^2 4x \cos^2 4x\,dx$

21. $\int \sin^3 2x \cos^5 2x\,dx$ **22.** $\int \sin 3y \cos 5y\,dy$ **23.** $\int \sqrt{4x + 3}\,(x^2 + 1)\,dx$ **24.** $\int \left(\frac{x^3 + 2}{x^3}\right)\sqrt{x - \frac{1}{x^2}}\,dx$

In Exercises 25 through 30, find the complete solution of the given differential equation.

25. $x^2y\dfrac{dy}{dx} = (y^2 - 1)^2$ **26.** $\dfrac{dy}{dx} = \dfrac{x+1}{\sqrt{x}}$ **27.** $y^2\,dx + y^2\,dy = dy$

28. $\dfrac{d^2y}{dx^2} = 12x^2 - 30x$ **29.** $\dfrac{d^2y}{d^2x} = \sqrt{2x-1}$ **30.** $y\sqrt{2x^2+1}\,dy = x\sqrt{1-y^2}\,dx$

31. If $y = 80x - 16x^2$, find the difference $\Delta y - dy$ if (a) $x = 2$ and $\Delta x = 0.1$; (b) $x = 4$ and $\Delta x = -0.2$.

32. Use differentials to find an approximate value to three decimal places of $\sqrt[3]{126}$.

33. Use differentials to find an approximate value to four decimal places of $\sqrt{0.17}$.

34. Use differentials to find an approximate value of sin 121°.

35. Use differentials to approximate the volume of material needed to make a rubber ball if the radius of the hollow inner core is 2 in. and the thickness of the rubber is $\frac{1}{8}$ in.

36. The slope of the tangent line at any point (x, y) on a curve is $10 - 4x$, and the point $(1, -1)$ is on the curve. Find an equation of the curve.

37. At any point (x, y) on a curve, $D_x^2y = 4 - x^2$ and an equation of the tangent line to the curve at the point $(1, -1)$ is $2x - 3y = 3$. Find an equation of the curve.

38. The points $(2, -3)$ and $(0, 1)$ are on a curve, and at any point (x, y) on the curve, $D_x^2y = -3$. Find an equation of the curve.

39. If $x^3 + y^3 - 3xy^2 + 1 = 0$, find dy at the point $(1, 1)$ if $dx = 0.1$.

40. Find dy/dt if $y = x^3 - 2x + 1$ and $x^3 + t^3 - 3t = 0$.

41. Find dy/dt if $8x^3 + 27y^3 - 4xy = 0$ and $t^3 + t^2x - x^3 = 4$.

42. Evaluate $\int (x^4)^6 4x^3\,dx$ as $\int u^6\,du$ and as $\int 4x^{27}\,dx$, and compare the results.

43. Evaluate $\int (x^3+1)^2 x^2\,dx$ by two methods: (a) make the substitution $u = x^3 + 1$; (b) first expand $(x^3+1)^2$. Compare the results and explain the difference in the appearance of the answers obtained in (a) and (b).

44. Find the particular solution of the differential equation $x^2\,dy = y^3\,dx$ for which $y = 1$ when $x = 4$.

45. Find the particular solution of the differential equation $D_x^2y = \sqrt{x+4}$ for which $y = 3$ and $y' = 2$ when $x = 4$.

In Exercises 46 and 47, a particle is moving on a straight line, s ft is the directed distance of the particle from the origin at t sec, v ft/sec is the velocity of the particle at t sec, and a ft/sec² is the acceleration of the particle at t sec.

46. $a = 3t + 4$; $v = 5$ and $s = 0$, when $t = 0$. Express v and s in terms of t.

47. $a = 6\cos 2t$; $v = 3$ and $s = 4$ when $t = \frac{1}{2}\pi$. Express v and s in terms of t.

48. Neglecting air resistance, if an object is dropped from an airplane at a height of 30,000 ft above the ocean, how long would it take the object to strike the water?

49. Suppose a bullet is fired directly downward from the airplane in Exercise 48 with a muzzle velocity of 2500 ft/sec. If air resistance is neglected, how long would it take the bullet to reach the ocean?

50. A ball is thrown vertically upward from the top of a house 64 ft above the ground, and the initial velocity is 48 ft/sec. (a) How long will it take the ball to reach its greatest height, and (b) what is its greatest height? (c) How long will it take the ball to strike the ground, and (d) with what velocity will it strike the ground?

51. Suppose the ball in Exercise 50 is thrown downward with an initial velocity of 48 ft/sec. (a) How long will it take the ball to strike the ground, and (b) with what velocity will it strike the ground?

52. Suppose the ball in Exercise 50 is dropped from the top of the house. (a) How long will it take for the ball to strike the ground, and (b) with what velocity will it strike the ground?

53. A company estimated that the rate of change of the number of units produced per day with respect to a change in the number of additional workers is $100 - 9x^{1/2}$, where x is the number of additional workers. Find the daily production when 9 workers are added to the labor force if 4000 units are produced when there are no additional workers.

54. The cost of a certain piece of machinery is \$900, and its value depreciates with time according to the formula $D_tV = -300(t + 1)^{-2}$, where V dollars is its value t years after its purchase. (a) What is its value two years after its purchase? (b) Find $\lim_{t \to +\infty} V$.

55. A container in the form of a cube having a volume of 1000 in.^{3} is to be made by using 6 equal squares of material costing 2 cents per square inch. How accurately must the side of each square be made so that the total cost of the material shall be correct to within 50 cents?

56. The measure of the radius of a right-circular cone is $\frac{4}{3}$ times the measure of the altitude. How accurately must the altitude be measured if the error in the computed volume is not to exceed 3%?

57. If t sec is the time for one complete swing of a simple pendulum of length l ft, then $4\pi^2 l = gt^2$, where $g = 32.2$. What is the effect upon the time if an error of 0.01 ft is made in measuring the length of the pendulum?

58. An automobile traveling at a constant speed of 60 mi/hr along a straight highway fails to stop at a stop sign. If 3 sec later a highway patrol car starts from rest from the stop sign and maintains a constant acceleration of 8 ft/sec^2, how long will it take the patrol car to overtake the automobile, and how far from the stop sign will this occur? Also determine the speed of the patrol car when it overtakes the automobile.

59. If $f(x) = x + |x - 1|$, and

$$F(x) = \begin{cases} x & \text{if } x < 1 \\ x^2 - x + 1 & \text{if } x \geq 1 \end{cases}$$

show that F is an antiderivative of f on $(-\infty, +\infty)$.

60. Let f and g be two functions such that for all x in $(-\infty, +\infty)$, $f'(x) = g(x)$ and $g'(x) = -f(x)$. Furthermore, suppose that $f(0) = 0$ and $g(0) = 1$. Prove that $[f(x)]^2 + [g(x)]^2 = 1$. (*Hint:* Consider the functions F and G where $F(x) = [f(x)]^2$ and $G(x) = -[g(x)]^2$, and show that $F'(x) = G'(x)$ for all x.)

61. The marginal cost for a particular commodity is given by $C'(x) = 4(3x + 1)^{-1/2}$, where $C(x)$ dollars is the total cost of producing x units. If the cost of producing 5 units is \$20, find the total cost function.

62. The marginal revenue for a particular commodity is given by $R'(x) = x^2 - 16x + 55$, where $R(x)$ dollars is the total revenue received from the sale of x units of the commodity. Find (a) the total revenue function and (b) the demand equation. (c) Draw sketches of the demand curve, the total revenue curve, and the marginal revenue curve on the same set of axes.

CHAPTER 6 The Definite Integral

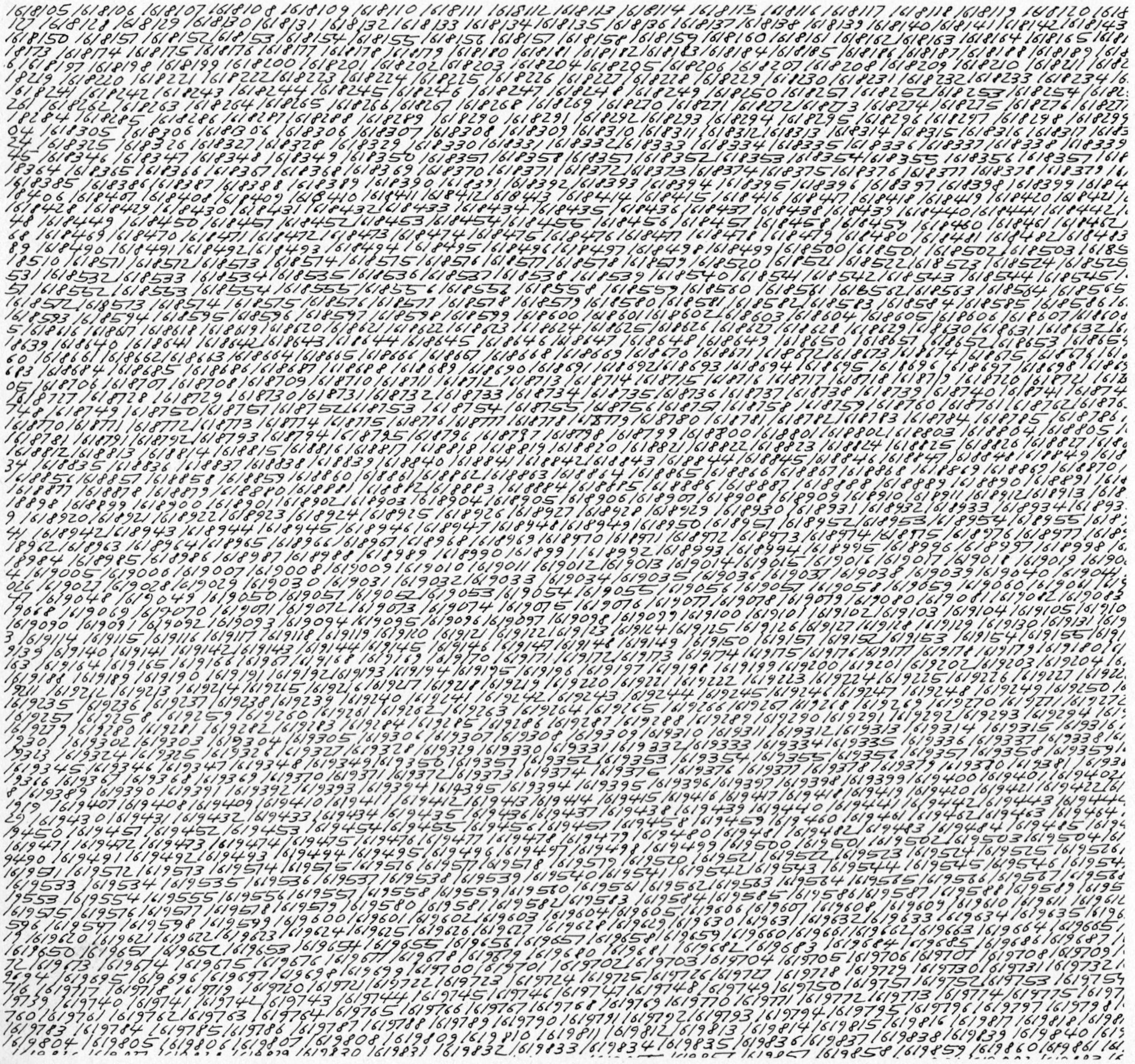

6.1 AREA

In this chapter we are concerned with the sums of many terms, and so a notation called the **sigma notation** is introduced to facilitate writing these sums. This notation involves the use of the symbol Σ, the capital sigma of the Greek alphabet, which corresponds to the letter S. Some examples of the sigma notation are given in the following illustration.

● ILLUSTRATION 1

$$\sum_{i=1}^{5} i^2 = 1^2 + 2^2 + 3^2 + 4^2 + 5^2$$

$$\sum_{i=-2}^{2} (3i + 2) = [3(-2) + 2] + [3(-1) + 2] + [3 \cdot 0 + 2] + [3 \cdot 1 + 2] + [3 \cdot 2 + 2]$$

$$= (-4) + (-1) + 2 + 5 + 8$$

$$\sum_{j=1}^{n} j^3 = 1^3 + 2^3 + 3^3 + \cdots + n^3$$

$$\sum_{k=3}^{8} \frac{1}{k} = \frac{1}{3} + \frac{1}{4} + \frac{1}{5} + \frac{1}{6} + \frac{1}{7} + \frac{1}{8}$$ ●

We have the formal definition of the sigma notation.

6.1.1 Definition

$$\sum_{i=m}^{n} F(i) = F(m) + F(m + 1) + F(m + 2) + \cdots + F(n - 1) + F(n) \quad (1)$$

where m and n are integers, and $m \leq n$.

The right side of formula (1) consists of the sum of $(n - m + 1)$ terms, the first of which is obtained by replacing i by m in $F(i)$, the second by replacing i by $(m + 1)$ in $F(i)$, and so on, until the last term is obtained by replacing i by n in $F(i)$.

The number m is called the **lower limit** of the sum, and n is called the **upper limit.** The symbol i is called the **index of summation.** It is a "dummy" symbol because any other letter can be used for this purpose. For example,

$$\sum_{k=3}^{5} k^2 = 3^2 + 4^2 + 5^2$$

is equivalent to

$$\sum_{i=3}^{5} i^2 = 3^2 + 4^2 + 5^2$$

● ILLUSTRATION 2 From Definition 6.1.1,

$$\sum_{i=3}^{6} \frac{i^2}{i+1} = \frac{3^2}{3+1} + \frac{4^2}{4+1} + \frac{5^2}{5+1} + \frac{6^2}{6+1}$$ ●

Sometimes the terms of a sum involve subscripts, as shown in the next illustration.

● ILLUSTRATION 3

$$\sum_{i=1}^{n} A_i = A_1 + A_2 + \cdots + A_n$$

$$\sum_{k=4}^{9} kb_k = 4b_4 + 5b_5 + 6b_6 + 7b_7 + 8b_8 + 9b_9$$

$$\sum_{i=1}^{5} f(x_i)\,\Delta x = f(x_1)\,\Delta x + f(x_2)\,\Delta x + f(x_3)\,\Delta x + f(x_4)\,\Delta x + f(x_5)\,\Delta x$$ ●

The following theorems involving the sigma notation are useful for computation and are easily proved.

6.1.2 Theorem $\sum_{i=1}^{n} c = cn$, where c is any constant

Proof

$$\sum_{i=1}^{n} c = c + c + \cdots + c \qquad (n \text{ terms})$$

$$= cn$$ ■

6.1.3 Theorem $\sum_{i=1}^{n} c \cdot F(i) = c \sum_{i=1}^{n} F(i)$, where c is any constant

Proof

$$\sum_{i=1}^{n} c \cdot F(i) = c \cdot F(1) + c \cdot F(2) + c \cdot F(3) + \cdots + c \cdot F(n)$$

$$= c[F(1) + F(2) + F(3) + \cdots + F(n)]$$

$$= c \sum_{1}^{n} F(i)$$ ■

6.1.4 Theorem $\sum_{i=1}^{n} [F(i) + G(i)] = \sum_{i=1}^{n} F(i) + \sum_{i=1}^{n} G(i)$

The proof is left as an exercise (see Exercise 45). Theorem 6.1.4 can be extended to the sum of any number of functions.

6.1.5 Theorem

$$\sum_{i=a}^{b} F(i) = \sum_{i=a+c}^{b+c} F(i-c) \tag{2}$$

and

$$\sum_{i=a}^{b} F(i) = \sum_{i=a-c}^{b-c} F(i+c) \tag{3}$$

The proof of Theorem 6.1.5 is left as an exercise (see Exercise 46). The following illustration shows the application of this theorem.

● ILLUSTRATION 4 From Eq. (2) of Theorem 6.1.5,

$$\sum_{i=3}^{10} F(i) = \sum_{i=5}^{12} F(i-2) \quad \text{and} \quad \sum_{i=6}^{11} i^2 = \sum_{i=7}^{12} (i-1)^2$$

From Eq. (3) of Theorem 6.1.5,

$$\sum_{i=3}^{10} F(i) = \sum_{i=1}^{8} F(i+2) \quad \text{and} \quad \sum_{i=6}^{11} i^2 = \sum_{i=1}^{6} (i+5)^2$$

●

6.1.6 Theorem

$$\sum_{i=1}^{n} [F(i) - F(i-1)] = F(n) - F(0)$$

Proof

$$\sum_{i=1}^{n} [F(i) - F(i-1)] = \sum_{i=1}^{n} F(i) - \sum_{i=1}^{n} F(i-1)$$

On the right side of this equation we write the first summation in another form and apply Eq. (3) of Theorem 6.1.5 with $c = 1$ to the second summation. Then

$$\begin{aligned}
\sum_{i=1}^{n} [F(i) - F(i-1)] &= \left(\sum_{i=1}^{n-1} F(i) + F(n)\right) - \sum_{i=1-1}^{n-1} F[(i+1)-1] \\
&= \sum_{i=1}^{n-1} F(i) + F(n) - \sum_{i=0}^{n-1} F(i) \\
&= \sum_{i=1}^{n-1} F(i) + F(n) - \left(F(0) + \sum_{i=1}^{n-1} F(i)\right) \\
&= F(n) - F(0)
\end{aligned}$$

■

Example 1 Evaluate $\sum_{i=1}^{n} (4^i - 4^{i-1})$.

Solution From Theorem 6.1.6, where $F(i) = 4^i$, it follows that

$$\sum_{i=1}^{n} (4^i - 4^{i-1}) = 4^n - 4^0$$
$$= 4^n - 1$$

In the following theorem there are four formulas that are useful for computation with sigma notation. They are numbered for future reference.

6.1.7 Theorem If n is a positive integer, then

$$\sum_{i=1}^{n} i = \frac{n(n+1)}{2} \qquad \text{(Formula 1)}$$

$$\sum_{i=1}^{n} i^2 = \frac{n(n+1)(2n+1)}{6} \qquad \text{(Formula 2)}$$

$$\sum_{i=1}^{n} i^3 = \frac{n^2(n+1)^2}{4} \qquad \text{(Formula 3)}$$

$$\sum_{i=1}^{n} i^4 = \frac{n(n+1)(6n^3+9n^2+n-1)}{30} \qquad \text{(Formula 4)}$$

Proof We prove Formula 1 and leave the proofs of Formulas 2, 3, and 4 as exercises (see Exercises 9, 10, and 11).

$$\sum_{i=1}^{n} i = 1 + 2 + 3 + \cdots + (n-1) + n$$

and

$$\sum_{i=1}^{n} i = n + (n-1) + (n-2) + \cdots + 2 + 1$$

If these two equations are added term by term, the left side is

$$2\sum_{i=1}^{n} i$$

and on the right side are n terms, each having the value $(n+1)$. Hence

$$2\sum_{i=1}^{n} i = (n+1) + (n+1) + (n+1) + \cdots + (n+1) \qquad n \text{ terms}$$
$$= n(n+1)$$

Therefore

$$\sum_{i=1}^{n} i = \frac{n(n+1)}{2}$$

The four formulas of Theorem 6.1.7 can also be proved by mathematical induction.

Example 2 Evaluate $\sum_{i=1}^{n} i(3i-2)$.

Solution

$$\begin{aligned}\sum_{i=1}^{n} i(3i-2) &= \sum_{i=1}^{n} (3i^2 - 2i)\\ &= \sum_{i=1}^{n} (3i^2) + \sum_{i=1}^{n} (-2i) \quad \text{(by Theorem 6.1.4)}\\ &= 3\sum_{i=1}^{n} i^2 - 2\sum_{i=1}^{n} i \quad \text{(by Theorem 6.1.3)}\\ &= 3\cdot\frac{n(n+1)(2n+1)}{6} - 2\cdot\frac{n(n+1)}{2} \quad \text{(by Formulas 2 and 1)}\\ &= \frac{2n^3 + 3n^2 + n - 2n^2 - 2n}{2}\\ &= \frac{2n^3 + n^2 - n}{2}\end{aligned}$$

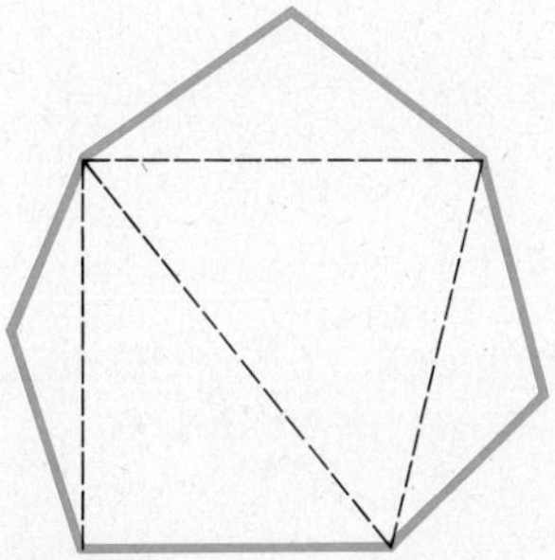

FIGURE 6.1.1

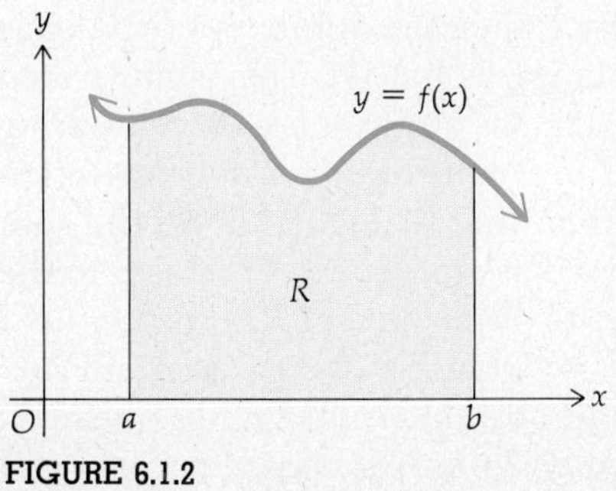

FIGURE 6.1.2

The word *measure* is used extensively throughout the book. A measure refers to a number (no units are included). For example, if the area of a triangle is 20 cm², we say that the square-centimeter measure of the area of the triangle is 20. You probably have an intuitive idea of what is meant by the measure of the area of certain geometrical figures; it is a number that in some way gives the size of the region enclosed by the figure. The area of a rectangle is the product of its length and width, and the area of a triangle is half the product of the lengths of the base and the altitude.

The area of a polygon can be defined as the sum of the areas of triangles into which it is decomposed, and it can be proved that the area thus obtained is independent of how the polygon is decomposed into triangles (see Fig. 6.1.1). However, how do we define the area of a region in a plane if the region is bounded by a curve? Are we even certain that such a region has an area?

Consider a region R in the plane as shown in Fig. 6.1.2. The region R is bounded by the x axis, the lines $x = a$ and $x = b$, and the curve having the equation $y = f(x)$, where f is a function continuous on the closed interval $[a, b]$. For simplicity, take $f(x) \geq 0$ for all x in $[a, b]$. We wish to assign a number A to be the measure of the area of R, and we use a limiting process similar to the one used in defining the area of a circle: The area of a circle is defined as the limit of the areas of inscribed regular polygons as the number of sides increases without bound. We realize intuitively that, whatever number is chosen to represent A, that number must be at least as great as the measure of the area of any polygonal region contained in R, and it must be no greater than the measure of the area of any polygonal region containing R.

We first define a polygonal region contained in R. Divide the closed interval $[a, b]$ into n subintervals. For simplicity, we now take each of these subintervals as being of equal length, for instance, Δx. Therefore $\Delta x = (b - a)/n$. Denote the endpoints of these subintervals by $x_0, x_1, x_2, \ldots, x_{n-1}, x_n$, where $x_0 = a$, $x_1 = a + \Delta x, \ldots, x_i = a + i\,\Delta x, \ldots, x_{n-1} = a + (n - 1)\,\Delta x$, $x_n = b$. Let the ith subinterval be denoted by $[x_{i-1}, x_i]$. Because f is continuous on the closed interval $[a, b]$, it is continuous on each closed subinterval. By the extreme-value theorem (4.2.9) there is a number in each subinterval for which f has an absolute minimum value. In the ith subinterval, let this number be c_i, so that $f(c_i)$ is the absolute minimum value of f on the subinterval $[x_{i-1}, x_i]$. Consider n rectangles, each having a width Δx units and an altitude $f(c_i)$ units (see Fig. 6.1.3). Let the sum of the areas of these n rectangles be given by S_n square units; then

$$S_n = f(c_1)\,\Delta x + f(c_2)\,\Delta x + \cdots + f(c_i)\,\Delta x + \cdots + f(c_n)\,\Delta x$$

or, with the sigma notation,

$$S_n = \sum_{i=1}^{n} f(c_i)\,\Delta x \tag{4}$$

The summation on the right side of (4) gives the sum of the measures of the areas of n inscribed rectangles. Thus, however we define A, it must be such that

$$A \geq S_n$$

In Fig. 6.1.3 the shaded region has an area of S_n square units. Now, let n increase. Specifically, multiply n by 2; then the number of rectangles is doubled, and the width of each rectangle is halved. This is illustrated in Fig. 6.1.4, showing twice as many rectangles as Fig. 6.1.3. By comparing the two figures, notice that the shaded region in Fig. 6.1.4 appears to approximate the region R more nearly than that of Fig. 6.1.3. So the sum of the measures of the areas of the rectangles in Fig. 6.1.4 is closer to the number we wish to represent the measure of the area of R.

As n increases, the values of S_n found from Eq. (4) increase, and successive values of S_n differ from each other by amounts that become arbitrarily small. This is proved in advanced calculus by a theorem that states that if f is continuous on $[a, b]$, then as n increases without bound, the

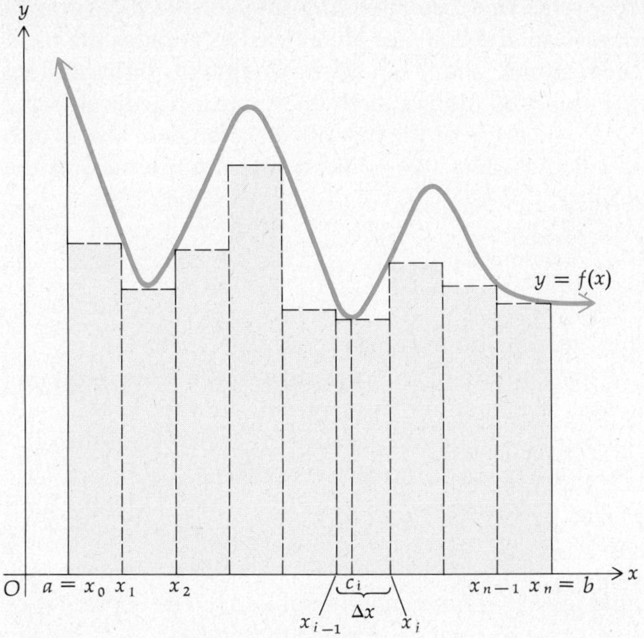

FIGURE 6.1.3

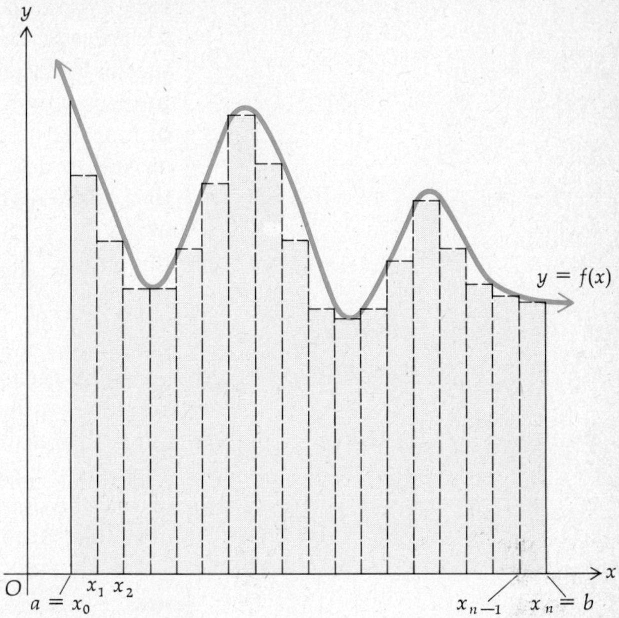

FIGURE 6.1.4

value of S_n given by (4) approaches a limit. It is this limit that we take as the definition of the measure of the area of region R.

6.1.8 Definition Suppose that the function f is continuous on the closed interval $[a, b]$, with $f(x) \geq 0$ for all x in $[a, b]$, and that R is the region bounded by the curve $y = f(x)$, the x axis, and the lines $x = a$ and $x = b$. Divide the interval $[a, b]$ into n subintervals, each of length $\Delta x = (b - a)/n$, and denote the ith subinterval by $[x_{i-1}, x_i]$. Then if $f(c_i)$ is the absolute minimum function value on the ith subinterval, the measure of the area of region R is given by

$$A = \lim_{n \to +\infty} \sum_{i=1}^{n} f(c_i)\, \Delta x \tag{5}$$

Equation (5) means that for any $\epsilon > 0$ there is a number $N > 0$ such that

$$\left| \sum_{i=1}^{n} f(c_i)\, \Delta x - A \right| < \epsilon \quad \text{whenever} \quad n > N$$

and n is a positive integer.

We could take circumscribed rectangles instead of inscribed rectangles. In this case we take as the measures of the altitudes of the rectangles the absolute maximum value of f on each subinterval. The existence of an absolute maximum value of f on each subinterval is guaranteed by the extreme-value theorem. The corresponding sums of the measures of the areas of the circumscribed rectangles are at least as great as the measure of the area of the region R, and it can be shown that the limit of these sums as n increases without bound is exactly the same as the limit of the sum of the measures of the areas of the inscribed rectangles. This is also proved in advanced calculus. Thus we could define the measure of the area of the region R by

$$A = \lim_{n \to +\infty} \sum_{i=1}^{n} f(d_i)\, \Delta x \tag{6}$$

where $f(d_i)$ is the absolute maximum value of f on $[x_{i-1}, x_i]$.

The measure of the altitude of the rectangle in the ith subinterval actually can be taken as the function value of any number in that subinterval, and the limit of the sum of the measures of the areas of the rectangles is the same no matter what numbers are selected. In Sec. 6.2 we extend the definition of the measure of the area of a region to be the limit of such a sum.

Example 3 Find the area of the region bounded by the curve $y = x^2$, the x axis, and the line $x = 3$ by taking inscribed rectangles.

Solution Figure 6.1.5 shows the region and the ith inscribed rectangle. We apply Definition 6.1.8. Divide the closed interval $[0, 3]$ into n subintervals, each of length Δx: $x_0 = 0$, $x_1 = \Delta x$, $x_2 = 2\,\Delta x, \ldots,$ $x_i = i\,\Delta x, \ldots, x_{n-1} = (n-1)\,\Delta x$, $x_n = 3$.

$$\Delta x = \frac{3-0}{n} = \frac{3}{n} \qquad f(x) = x^2$$

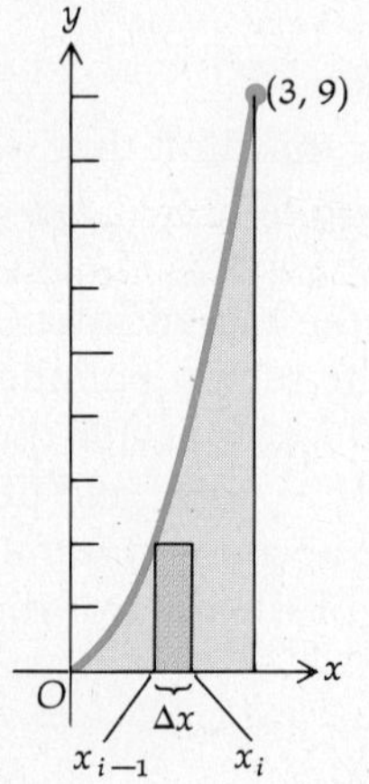

FIGURE 6.1.5

Because f is increasing on $[0, 3]$, the absolute minimum value of f on the ith subinterval $[x_{i-1}, x_i]$ is $f(x_{i-1})$. Therefore, from (5),

$$A = \lim_{n \to +\infty} \sum_{i=1}^{n} f(x_{i-1})\, \Delta x \tag{7}$$

Because $x_{i-1} = (i-1)\,\Delta x$ and $f(x) = x^2$,

$$f(x_{i-1}) = [(i-1)\,\Delta x]^2$$

Therefore

$$\sum_{i=1}^{n} f(x_{i-1})\, \Delta x = \sum_{i=1}^{n} (i-1)^2 (\Delta x)^3$$

But $\Delta x = 3/n$; so

$$\sum_{i=1}^{n} f(x_{i-1})\, \Delta x = \sum_{i=1}^{n} (i-1)^2 \frac{27}{n^3} = \frac{27}{n^3} \sum_{i=1}^{n} (i-1)^2$$

$$= \frac{27}{n^3}\left[\sum_{i=1}^{n} i^2 - 2\sum_{i=1}^{n} i + \sum_{i=1}^{n} 1\right]$$

and using Formulas 2 and 1 and Theorem 6.1.2 we get

$$\sum_{i=1}^{n} f(x_{i-1})\,\Delta x = \frac{27}{n^3}\left[\frac{n(n+1)(2n+1)}{6} - 2\cdot\frac{n(n+1)}{2} + n\right]$$

$$= \frac{27}{n^3}\cdot\frac{2n^3 + 3n^2 + n - 6n^2 - 6n + 6n}{6}$$

$$= \frac{9}{2}\cdot\frac{2n^2 - 3n + 1}{n^2}$$

Then, from (7),

$$A = \lim_{n\to+\infty}\left[\frac{9}{2}\cdot\frac{2n^2 - 3n + 1}{n^2}\right]$$

$$= \frac{9}{2}\cdot\lim_{n\to+\infty}\left(2 - \frac{3}{n} + \frac{1}{n^2}\right)$$

$$= \tfrac{9}{2}(2 - 0 + 0)$$

$$= 9$$

Therefore the area of the region is 9 square units.

Example 4 Find the area of the region in Example 3 by taking circumscribed rectangles.

Solution With circumscribed rectangles the measure of the altitude of the ith rectangle is the absolute maximum value of f on the ith subinterval $[x_{i-1}, x_i]$, which is $f(x_i)$. From (6),

$$A = \lim_{n\to+\infty}\sum_{i=1}^{n} f(x_i)\,\Delta x \tag{8}$$

Because $x_i = i\,\Delta x$, then $f(x_i) = (i\,\Delta x)^2$, and so

$$\sum_{i=1}^{n} f(x_i)\,\Delta x = \sum_{i=1}^{n} i^2\,(\Delta x)^3 = \frac{27}{n^3}\sum_{i=1}^{n} i^2$$

$$= \frac{27}{n^3}\left[\frac{n(n+1)(2n+1)}{6}\right]$$

$$= \frac{9}{2}\cdot\frac{2n^2 + 3n + 1}{n^2}$$

Therefore, from Eq. (8),

$$A = \lim_{n\to+\infty}\frac{9}{2}\cdot\left(2 + \frac{3}{n} + \frac{1}{n^2}\right)$$

$$= 9 \quad \text{(as in Example 3)}$$

Example 5 Find the area of the trapezoid which is the region bounded by the line $2x + y = 8$, the x axis, and the lines $x = 1$ and $x = 3$. Take inscribed rectangles.

Solution The region and the ith inscribed rectangle are shown in Fig. 6.1.6. The closed interval $[1, 3]$ is divided into n subintervals, each of length Δx; $x_0 = 1$, $x_1 = 1 + \Delta x$, $x_2 = 1 + 2\,\Delta x$, . . . , $x_i = 1 + i\,\Delta x$, . . . , $x_{n-1} = 1 + (n-1)\,\Delta x$, $x_n = 3$.

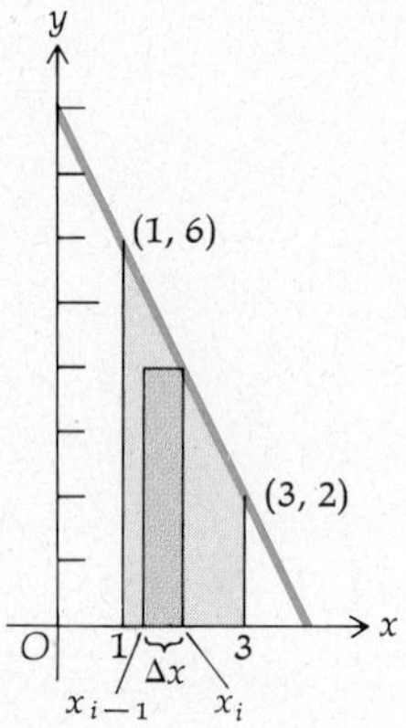

FIGURE 6.1.6

$$\Delta x = \frac{3-1}{n} = \frac{2}{n}$$

Solving the equation of the line for y we obtain $y = -2x + 8$. Therefore $f(x) = -2x + 8$, and because f is decreasing on $[1, 3]$, the absolute minimum value of f on the ith subinterval $[x_{i-1}, x_i]$ is $f(x_i)$. Because $x_i = 1 + i\,\Delta x$ and $f(x) = -2x + 8$, then $f(x_i) = -2(1 + i\,\Delta x) + 8 = 6 - 2i\,\Delta x$. From (5),

$$\begin{aligned} A &= \lim_{n\to+\infty} \sum_{i=1}^{n} f(x_i)\,\Delta x \\ &= \lim_{n\to+\infty} \sum_{i=1}^{n} (6 - 2i\,\Delta x)\,\Delta x \\ &= \lim_{n\to+\infty} \sum_{i=1}^{n} [6\,\Delta x - 2i(\Delta x)^2] \\ &= \lim_{n\to+\infty} \sum_{i=1}^{n} \left[6\left(\frac{2}{n}\right) - 2i\left(\frac{2}{n}\right)^2\right] \\ &= \lim_{n\to+\infty} \left[\frac{12}{n}\sum_{i=1}^{n} 1 - \frac{8}{n^2}\sum_{i=1}^{n} i\right] \end{aligned}$$

From Theorem 6.1.2 and Formula 1,

$$\begin{aligned} A &= \lim_{n\to+\infty} \left[\frac{12}{n}\cdot n - \frac{8}{n^2}\cdot\frac{n(n+1)}{2}\right] \\ &= \lim_{n\to+\infty} \left(8 - \frac{4}{n}\right) \\ &= 8 \end{aligned}$$

Therefore the area is 8 square units. Using the formula from plane geometry for the area of a trapezoid, $A = \frac{1}{2}h(b_1 + b_2)$, where h, b_1, and b_2 are, respectively, the number of units in the lengths of the altitude and the two bases, we get $A = \frac{1}{2}(2)(6 + 2) = 8$, which agrees with the result.

EXERCISES 6.1

In Exercises 1 through 8, find the given sum.

1. $\sum_{i=1}^{6} (3i - 2)$
2. $\sum_{i=1}^{7} (i + 1)^2$
3. $\sum_{i=2}^{5} \frac{i}{i - 1}$
4. $\sum_{j=3}^{6} \frac{2}{j(j - 2)}$
5. $\sum_{i=-2}^{3} 2^i$
6. $\sum_{i=0}^{3} \frac{1}{1 + i^2}$
7. $\sum_{k=1}^{4} \frac{(-1)^{k+1}}{k}$
8. $\sum_{k=-2}^{3} \frac{k}{k + 3}$

9. Prove Formula 2 of Theorem 6.1.7. (*Hint:* $i^3 - (i - 1)^3 = 3i^2 - 3i + 1$; so $\sum_{i=1}^{n} [i^3 - (i - 1)^3] = \sum_{i=1}^{n} (3i^2 - 3i + 1)$. On the left side of this equation use Theorem 6.1.6; on the right side use Theorems 6.1.2, 6.1.3, and 6.1.4 and Formula 1.)
10. Prove Formula 3 of Theorem 6.1.7. (*Hint:* $i^4 - (i - 1)^4 = 4i^3 - 6i^2 + 4i - 1$, and use a method similar to the one for Exercise 9.)
11. Prove Formula 4 of Theorem 6.1.7. (See hints for Exercises 9 and 10 above.)

In Exercises 12 through 20, evaluate the indicated sum by using Theorems 6.1.2 through 6.1.7.

12. $\sum_{i=1}^{20} 3i(i^2 + 2)$
13. $\sum_{i=1}^{25} 2i(i - 1)$
14. $\sum_{k=1}^{n} (2^{k-1} - 2^k)$
15. $\sum_{i=1}^{n} (10^{i+1} - 10^i)$
16. $\sum_{i=1}^{40} [\sqrt{2i + 1} - \sqrt{2i - 1}]$
17. $\sum_{k=1}^{100} \left[\frac{1}{k} - \frac{1}{k + 1}\right]$
18. $\sum_{i=1}^{n} 2i(1 + i^2)$
19. $\sum_{i=1}^{n} 4i^2(i - 2)$
20. $\sum_{k=1}^{n} [(3^{-k} - 3^k)^2 - (3^{k-1} + 3^{-k-1})^2]$

In Exercises 21 through 36, use the method of this section to find the area of the given region; use inscribed or circumscribed rectangles as indicated. For each exercise draw a figure showing the region and the ith rectangle.

21. The region bounded by $y = x^2$, the x axis, and the line $x = 2$; inscribed rectangles.
22. The region of Exercise 21; circumscribed rectangles.
23. The region bounded by $y = 2x$, the x axis, and the lines $x = 1$ and $x = 4$; circumscribed rectangles.
24. The region of Exercise 23; inscribed rectangles.
25. The region above the x axis and to the right of the line $x = 1$ bounded by the x axis, the line $x = 1$, and the curve $y = 4 - x^2$; inscribed rectangles.
26. The region of Exercise 25; circumscribed rectangles.
27. The region lying to the left of the line $x = 1$ bounded by the curve and lines of Exercise 25; circumscribed rectangles.
28. The region of Exercise 27; inscribed rectangles.
29. The region bounded by $y = 3x^4$, the x axis, and the line $x = 1$; inscribed rectangles.
30. The region of Exercise 29; circumscribed rectangles.
31. The region bounded by $y = x^3$, the x axis, and the lines $x = -1$ and $x = 2$; inscribed rectangles.
32. The region of Exercise 31; circumscribed rectangles.
33. The region bounded by $y = x^3 + x$, the x axis, and the lines $x = -2$ and $x = 1$; circumscribed rectangles.

34. The region of Exercise 33; inscribed rectangles.

35. The region bounded by $y = mx$, with $m > 0$, the x axis, and the lines $x = a$ and $x = b$, with $b > a > 0$; circumscribed rectangles.

36. The region of Exercise 35; inscribed rectangles.

37. Use the method of this section to find the area of an isosceles trapezoid whose bases have measures b_1 and b_2 and whose altitude has measure h.

38. The graph of $y = 4 - |x|$ and the x axis from $x = -4$ to $x = 4$ form a triangle. Use the method of this section to find the area of this triangle.

In Exercises 39 through 44, find the area of the region by taking as the measure of the altitude of the ith rectangle $f(m_i)$, where m_i is the midpoint of the ith subinterval. (Hint: $m_i = \frac{1}{2}(x_i + x_{i-1})$.)

39. The region of Example 3.

40. The region of Exercise 21.

41. The region of Exercise 23.

42. The region of Exercise 25.

43. The region of Exercise 27.

44. The region of Exercise 29.

45. Prove Theorem 6.1.4.

46. Prove Theorem 6.1.5.

6.2 THE DEFINITE INTEGRAL

In the preceding section the measure of the area of a region was defined as the following limit:

$$\lim_{n \to +\infty} \sum_{i=1}^{n} f(c_i)\,\Delta x \tag{1}$$

To lead up to this definition we divided the closed interval $[a, b]$ into subintervals of equal length and then took c_i as the point in the ith subinterval for which f has an absolute minimum value. We also restricted the function values $f(x)$ to be nonnegative on $[a, b]$ and further required f to be continuous on $[a, b]$.

To define the definite integral we need to consider a new kind of limiting process, of which the limit given in (1) is a special case. Let f be a function defined on the closed interval $[a, b]$. Divide this interval into n subintervals by choosing *any* $(n - 1)$ intermediate points between a and b. Let $x_0 = a$ and $x_n = b$, and let $x_1, x_2, \ldots, x_{n-1}$ be the intermediate points so that

$$x_0 < x_1 < x_2 < \cdots < x_{n-1} < x_n$$

The points $x_0, x_1, x_2, \ldots, x_{n-1}, x_n$ are not necessarily equidistant. Let $\Delta_1 x$ be the length of the first subinterval so that $\Delta_1 x = x_1 - x_0$; let $\Delta_2 x$ be the length of the second subinterval so that $\Delta_2 x = x_2 - x_1$; and so forth so that the length of the ith subinterval is $\Delta_i x$, and

$$\Delta_i x = x_i - x_{i-1}$$

A set of all such subintervals of the interval $[a, b]$ is called a **partition** of the interval $[a, b]$. Let Δ be such a partition. Figure 6.2.1 illustrates one such partition Δ of $[a, b]$.

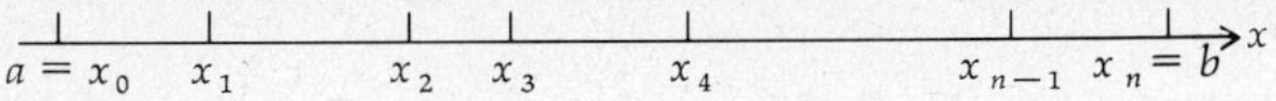

FIGURE 6.2.1

The partition Δ contains n subintervals. One of these subintervals is longest; however, there may be more than one such subinterval. The length of the longest subinterval of the partition Δ, called the **norm** of the partition, is denoted by $\|\Delta\|$.

Choose a point in each subinterval of the partition Δ: Let ξ_1 be the point chosen in $[x_0, x_1]$ so that $x_0 \le \xi_1 \le x_1$. Let ξ_2 be the point chosen in $[x_1, x_2]$ so that $x_1 \le \xi_2 \le x_2$, and so forth, so that ξ_i is the point chosen in $[x_{i-1}, x_i]$, and $x_{i-1} \le \xi_i \le x_i$. Form the sum

$$f(\xi_1)\,\Delta_1 x + f(\xi_2)\,\Delta_2 x + \cdots + f(\xi_i)\,\Delta_i x + \cdots + f(\xi_n)\,\Delta_n x$$

or

$$\sum_{i=1}^{n} f(\xi_i)\,\Delta_i x$$

Such a sum is called a **Riemann sum,** named for the mathematician Georg Friedrich Bernhard Riemann (1826–1866).

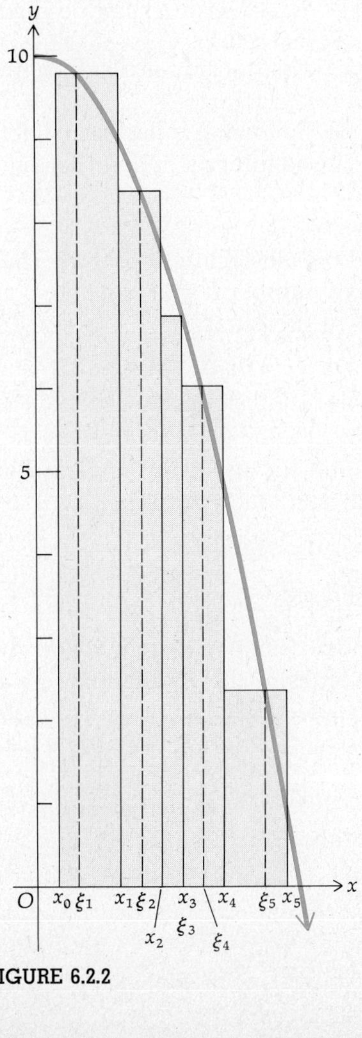

FIGURE 6.2.2

● ILLUSTRATION 1 Suppose $f(x) = 10 - x^2$, with $\frac{1}{4} \le x \le 3$. We will find the Riemann sum for the function f on $[\frac{1}{4}, 3]$ for the partition Δ: $x_0 = \frac{1}{4}$, $x_1 = 1$, $x_2 = 1\frac{1}{2}$, $x_3 = 1\frac{3}{4}$, $x_4 = 2\frac{1}{4}$, $x_5 = 3$, and $\xi_1 = \frac{1}{2}$, $\xi_2 = 1\frac{1}{4}$, $\xi_3 = 1\frac{3}{4}$, $\xi_4 = 2$, $\xi_5 = 2\frac{3}{4}$.

Figure 6.2.2 shows a sketch of the graph of f on $[\frac{1}{4}, 3]$ and the five rectangles, the measures of whose areas are the terms of the Riemann sum.

$$\begin{aligned}\sum_{i=1}^{5} f(\xi_i)\,\Delta_i x &= f(\xi_1)\,\Delta_1 x + f(\xi_2)\,\Delta_2 x + f(\xi_3)\,\Delta_3 x + f(\xi_4)\,\Delta_4 x + f(\xi_5)\,\Delta_5 x\\ &= f(\tfrac{1}{2})(1 - \tfrac{1}{4}) + f(\tfrac{5}{4})(1\tfrac{1}{2} - 1) + f(\tfrac{7}{4})(1\tfrac{3}{4} - 1\tfrac{1}{2})\\ &\qquad + f(2)(2\tfrac{1}{4} - 1\tfrac{3}{4}) + f(\tfrac{11}{4})(3 - 2\tfrac{1}{4})\\ &= (9\tfrac{3}{4})(\tfrac{3}{4}) + (8\tfrac{7}{16})(\tfrac{1}{2}) + (6\tfrac{15}{16})(\tfrac{1}{4}) + (6)(\tfrac{1}{2}) + (2\tfrac{7}{16})(\tfrac{3}{4})\\ &= 18\tfrac{3}{32}\end{aligned}$$

The norm of Δ is the length of the longest subinterval. Hence $\|\Delta\| = \frac{3}{4}$. ●

Because the function values $f(x)$ are not restricted to nonnegative values, some of the $f(\xi_i)$ could be negative. In such a case the geometric interpretation of the Riemann sum would be the sum of the measures of the areas of the rectangles lying above the x axis and the negatives of the measures of the areas of the rectangles lying below the x axis. This situation is illustrated in Fig. 6.2.3. Here

$$\sum_{i=1}^{10} f(\xi_i)\,\Delta_i x = A_1 + A_2 - A_3 - A_4 - A_5 + A_6 + A_7 - A_8 - A_9 - A_{10}$$

because $f(\xi_3)$, $f(\xi_4)$, $f(\xi_5)$, $f(\xi_8)$, $f(\xi_9)$, and $f(\xi_{10})$ are negative numbers.

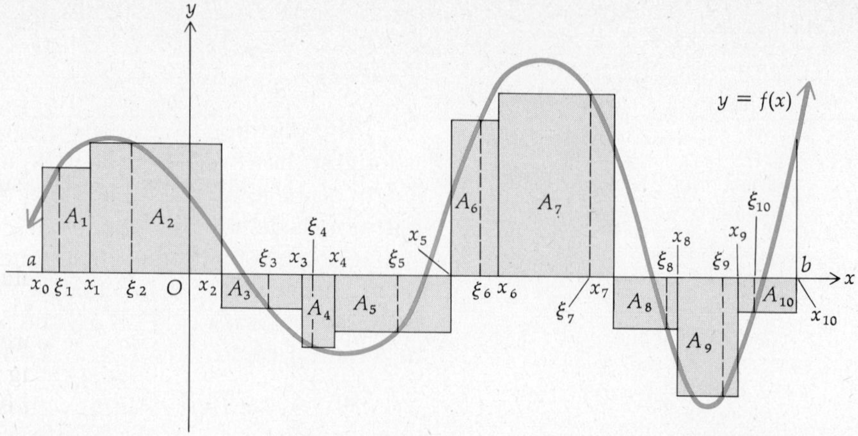

FIGURE 6.2.3

Let f be a function whose domain includes the closed interval $[a, b]$. Suppose there is a number L such that $\left|\sum_{i=1}^{n} f(\xi_i)\,\Delta_i x - L\right|$ can be made as small as we please for all partitions Δ whose norms are sufficiently small, and for any ξ_i in the closed interval $[x_{i-1}, x_i]$, $i = 1, 2, \ldots, n$. In such a case f is said to be *integrable* on $[a, b]$. Following is the formal definition of a function being integrable on a closed interval; it involves ϵ and δ notation.

6.2.1 Definition Let f be a function whose domain includes the closed interval $[a, b]$. Then f is said to be **integrable** on $[a, b]$ if there is a number L satisfying the condition that, for any $\epsilon > 0$, there exists a $\delta > 0$ such that

$$\left|\sum_{i=1}^{n} f(\xi_i)\,\Delta_i x - L\right| < \epsilon$$

for every partition Δ for which $\|\Delta\| < \delta$, and for any ξ_i in the closed interval $[x_{i-1}, x_i]$, $i = 1, 2, \ldots, n$. For such a situation we write

$$\lim_{\|\Delta\| \to 0} \sum_{i=1}^{n} f(\xi_i)\,\Delta_i x = L \tag{2}$$

Definition 6.2.1 states that, for a given function f defined on the closed interval $[a, b]$, we can make the values of the Riemann sums as close to L as we please by taking the norms $\|\Delta\|$ of all partitions Δ of $[a, b]$ sufficiently small for all possible choices of the numbers ξ_i for which $x_{i-1} \le \xi_i \le x_i$.

The limiting process given by (2) is different from that discussed in Chapter 2. From Definition 6.2.1 the number L in (2) exists if for $\epsilon > 0$ there exists a $\delta > 0$ such that

$$\left|\sum_{i=1}^{n} f(\xi_i)\,\Delta_i x - L\right| < \epsilon$$

for every partition Δ for which $\|\Delta\| < \delta$, and for any ξ_i in the closed interval $[x_{i-1}, x_i]$, $i = 1, 2, \ldots, n$.

In Definition 2.1.1 we had

$$\lim_{x \to a} f(x) = L \tag{3}$$

if for any $\epsilon > 0$ there exists a $\delta > 0$ such that

$$|f(x) - L| < \epsilon \quad \text{whenever} \quad 0 < |x - a| < \delta$$

In limiting process (2), for a particular $\delta > 0$ there are infinitely many partitions Δ having norm $\|\Delta\| < \delta$. This is analogous to the fact that in limiting process (3), for a given $\delta > 0$ there are infinitely many values of x for which $0 < |x - a| < \delta$. However, in limiting process (2), for each partition Δ there are infinitely many choices of ξ_i. It is in this respect that the two limiting processes differ.

In Chapter 2 (Theorem 2.1.2) we showed that if the number L in limiting process (3) exists, it is unique. In a similar manner we can show that if there is a number L satisfying Definition 6.2.1, then it is unique. Now we can define the *definite integral.*

6.2.2 Definition If f is a function defined on the closed interval $[a, b]$, then the **definite integral** of f from a to b, denoted by $\int_a^b f(x)\,dx$, is given by

$$\int_a^b f(x)\,dx = \lim_{\|\Delta\| \to 0} \sum_{i=1}^{n} f(\xi_i)\,\Delta_i x \tag{4}$$

if the limit exists.

Note that the statement "the function f is integrable on the closed interval $[a, b]$" is synonymous with the statement "the definite integral of f from a to b exists."

In the notation for the definite integral $\int_a^b f(x)\,dx$, $f(x)$ is called the **integrand,** a is called the **lower limit,** and b is called the **upper limit.** The symbol

$$\int$$

is called an **integral sign.** The integral sign resembles a capital S, which is appropriate because the definite integral is the limit of a sum. It is the same symbol we used in Chapter 5 to indicate the operation of antidifferentiation. The reason for the common symbol is that a theorem (6.5.2), called the fundamental theorem of the calculus, enables us to evaluate a definite integral by finding an antiderivative (also called an **indefinite integral**).

A question that now arises is as follows: Under what conditions does a number L satisfying Definition 6.2.1 exist; that is, under what conditions is a function integrable? An answer to this question is given by the following theorem.

6.2.3 Theorem If a function f is continuous on the closed interval $[a, b]$, then f is integrable on $[a, b]$.

The proof of this theorem is beyond the scope of this book and is given in advanced calculus texts. The condition that f is continuous on $[a, b]$, while being sufficient to guarantee that f is integrable on $[a, b]$, is not a necessary condition for the existence of $\int_a^b f(x)\,dx$. That is, if f is continuous on $[a, b]$, then Theorem 6.2.3 assures us that $\int_a^b f(x)\,dx$ exists; however, it is possible for the integral to exist even if the function is discontinuous at some numbers in $[a, b]$. The following example gives a function that is discontinuous and yet integrable on a closed interval.

Example 1 Let f be the function defined by

$$f(x) = \begin{cases} 0 & \text{if } x \neq 0 \\ 1 & \text{if } x = 0 \end{cases}$$

Let $[a, b]$ be any interval such that $a < 0 < b$. Show that f is discontinuous on $[a, b]$ and yet integrable on $[a, b]$.

Solution Because $\lim\limits_{x \to 0} f(x) = 0 \neq f(0)$, f is discontinuous at 0 and hence discontinuous on $[a, b]$.

To prove that f is integrable on $[a, b]$ we show that Definition 6.2.1 is satisfied. Consider the Riemann sum

$$\sum_{i=1}^{n} f(\xi_i)\,\Delta_i x$$

If none of the numbers $\xi_1, \xi_2, \ldots, \xi_n$ is zero, then the Riemann sum is zero. Suppose that $\xi_j = 0$. Then

$$\sum_{i=1}^{n} f(\xi_i)\,\Delta_i x = 1 \cdot \Delta_j x$$

In either case,

$$\left| \sum_{i=1}^{n} f(\xi_i)\,\Delta_i x \right| \leq \|\Delta\|$$

Hence

$$\left| \sum_{i=1}^{n} f(\xi_i)\,\Delta_i x - 0 \right| < \epsilon \quad \text{whenever} \quad \|\Delta\| < \epsilon$$

Comparing the above with Definition 6.2.1 where $\delta = \epsilon$ and $L = 0$, we see that f is integrable on $[a, b]$.

At the beginning of this section we stated that the limit used in Definition 6.1.9 to define the measure of the area of a region is a special case of the limit used in Definition 6.2.2 to define the definite integral. In the discussion of area, the interval $[a, b]$ was divided into n subintervals of equal length. Such a partition of the interval $[a, b]$ is called a **regular partition.** If Δx is the length of each subinterval in a regular partition, then each $\Delta_i x = \Delta x$, and the norm of the partition is Δx. Making these substitutions in (4) we have

$$\int_a^b f(x)\,dx = \lim_{\Delta x \to 0} \sum_{i=1}^{n} f(\xi_i)\,\Delta x \tag{5}$$

Furthermore,

$$\Delta x = \frac{b-a}{n} \tag{6}$$

and

$$n = \frac{b-a}{\Delta x} \tag{7}$$

So from (6),

$$\lim_{n \to +\infty} \Delta x = 0 \tag{8}$$

and from (7), because $b > a$ and Δx approaches zero through positive values (because $\Delta x > 0$),

$$\lim_{\Delta x \to 0} n = +\infty \tag{9}$$

From limits (8) and (9) we conclude that

$$\Delta x \to 0 \quad \text{is equivalent to} \quad n \to +\infty \tag{10}$$

Thus we have from Eq. (5) and statement (10),

$$\int_a^b f(x)\,dx = \lim_{n \to +\infty} \sum_{i=1}^{n} f(\xi_i)\,\Delta x \tag{11}$$

It should be remembered that ξ_i can be any point in the ith subinterval $[x_{i-1}, x_i]$.

In applications of the definite integral, regular partitions are often used; therefore formulas (5) and (11) are especially important.

Comparing the limit used in Definition 6.1.8, which gives the measure of the area of a region, with the limit on the right side of (11), we have in the first case

$$\lim_{n \to +\infty} \sum_{i=1}^{n} f(c_i)\,\Delta x \tag{12}$$

where $f(c_i)$ is the absolute minimum function value on $[x_{i-1}, x_i]$. In the second case we have

$$\lim_{n \to +\infty} \sum_{i=1}^{n} f(\xi_i)\,\Delta x \tag{13}$$

where ξ_i is any number in $[x_{i-1}, x_i]$.

Because the function f is continuous on $[a, b]$, by Theorem 6.2.3, $\int_a^b f(x)\,dx$ exists; therefore this definite integral is the limit of all Riemann sums of f on $[a, b]$ including those in (12) and (13). Because of this, we redefine the area of a region in a more general way.

6.2.4 Definition Let the function f be continuous on $[a, b]$ and $f(x) \geq 0$ for all x in $[a, b]$. Let R be the region bounded by the curve $y = f(x)$, the x axis, and the lines $x = a$ and $x = b$. Then the measure of the area of region R is given by

$$A = \lim_{\|\Delta\| \to 0} \sum_{i=1}^{n} f(\xi_i)\,\Delta_i x = \int_a^b f(x)\,dx$$

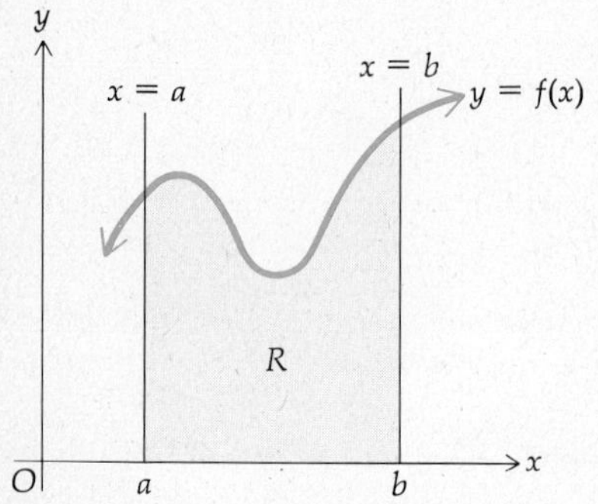

FIGURE 6.2.4

The above definition states that if $f(x) \geq 0$ for all x in $[a, b]$, the definite integral $\int_a^b f(x)\,dx$ can be interpreted geometrically as the measure of the area of the region R shown in Fig. 6.2.4.

Equation (11) can be used to find the exact value of a definite integral, as illustrated in the following example.

Example 2 Find the exact value of the definite integral $\int_1^3 x^2\,dx$. Interpret the result geometrically.

Solution Consider a regular partition of the closed interval $[1, 3]$ into n subintervals. Then $\Delta x = 2/n$.

If we choose ξ_i as the right endpoint of each subinterval, we have

$$\xi_1 = 1 + \frac{2}{n},\ \xi_2 = 1 + 2\left(\frac{2}{n}\right),\ \xi_3 = 1 + 3\left(\frac{2}{n}\right),\ \ldots,$$
$$\xi_i = 1 + i\left(\frac{2}{n}\right),\ \ldots,\ \xi_n = 1 + n\left(\frac{2}{n}\right)$$

Because $f(x) = x^2$,

$$f(\xi_i) = \left(1 + \frac{2i}{n}\right)^2 = \left(\frac{n + 2i}{n}\right)^2$$

Therefore, by using (11) and applying theorems from Sec. 6.1, we get

$$\int_1^3 x^2\,dx = \lim_{n \to +\infty} \sum_{i=1}^{n} \left(\frac{n + 2i}{n}\right)^2 \frac{2}{n}$$

$$= \lim_{n \to +\infty} \frac{2}{n^3} \sum_{i=1}^{n} (n^2 + 4ni + 4i^2)$$

$$= \lim_{n \to +\infty} \frac{2}{n^3} \left[n^2 \sum_{i=1}^{n} 1 + 4n \sum_{i=1}^{n} i + 4 \sum_{i=1}^{n} i^2 \right]$$

$$= \lim_{n \to +\infty} \frac{2}{n^3} \left[n^2 n + 4n \cdot \frac{n(n + 1)}{2} + \frac{4n(n + 1)(2n + 1)}{6} \right]$$

$$= \lim_{n \to +\infty} \frac{2}{n^3}\left[n^3 + 2n^3 + 2n^2 + \frac{2n(2n^2 + 3n + 1)}{3}\right]$$

$$= \lim_{n \to +\infty} \left[6 + \frac{4}{n} + \frac{8n^2 + 12n + 4}{3n^2}\right]$$

$$= \lim_{n \to +\infty} \left[6 + \frac{4}{n} + \frac{8}{3} + \frac{4}{n} + \frac{4}{3n^2}\right]$$

$$= 6 + 0 + \tfrac{8}{3} + 0 + 0$$

$$= 8\tfrac{2}{3}$$

We interpret the result geometrically. Because $x^2 \geq 0$ for all x in $[1, 3]$, the region bounded by the curve $y = x^2$, the x axis, and the lines $x = 1$ and $x = 3$ has an area of $8\frac{2}{3}$ square units. The region is shown in Fig. 6.2.5.

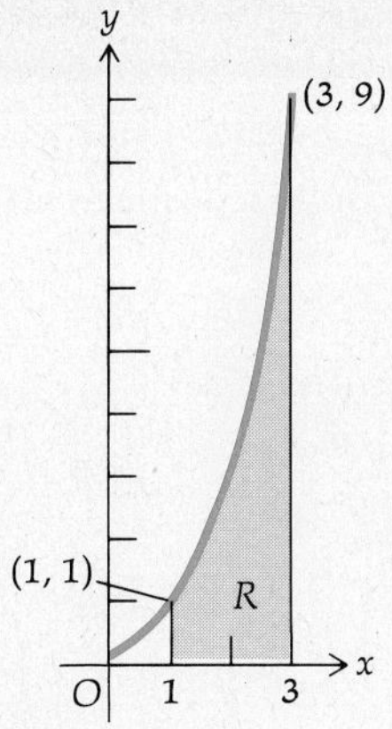

FIGURE 6.2.5

In Definition 6.2.2 the closed interval $[a, b]$ is given, and so we assume that $a < b$. To consider the definite integral of a function f from a to b when $a > b$, or when $a = b$, we have the following definitions.

6.2.5 Definition If $a > b$, then

$$\int_a^b f(x)\,dx = -\int_b^a f(x)\,dx$$

if $\int_b^a f(x)\,dx$ exists.

● ILLUSTRATION 2 In Example 2 we showed that $\int_1^3 x^2\,dx = 8\frac{2}{3}$. Therefore, from Definition 6.2.5,

$$\int_3^1 x^2\,dx = -\int_1^3 x^2\,dx = -8\tfrac{2}{3}$$ ●

6.2.6 Definition If $f(a)$ exists, then

$$\int_a^a f(x)\,dx = 0$$

● ILLUSTRATION 3 From Definition 6.2.6,

$$\int_1^1 x^2\,dx = 0$$ ●

EXERCISES 6.2

In Exercises 1 through 9, find the Riemann sum for the function on the interval, using the given partition Δ and the given values of ξ_i. Draw a sketch of the graph of the function on the given interval, and show the rectangles the measure of whose areas are the terms of the Riemann sum. (See Illustration 1 and Fig. 6.2.2.)

1. $f(x) = x^2, 0 \le x \le 3$; for Δ: $x_0 = 0, x_1 = \frac{1}{2}, x_2 = 1\frac{1}{4}, x_3 = 2\frac{1}{4}, x_4 = 3$; $\xi_1 = \frac{1}{4}, \xi_2 = 1, \xi_3 = 1\frac{1}{2}, \xi_4 = 2\frac{1}{2}$
2. $f(x) = x^2, 0 \le x \le 3$; for Δ: $x_0 = 0, x_1 = \frac{3}{4}, x_2 = 1\frac{1}{4}, x_3 = 2, x_4 = 2\frac{3}{4}, x_5 = 3$; $\xi_1 = \frac{1}{2}, \xi_2 = 1, \xi_3 = 1\frac{3}{4}, \xi_4 = 2\frac{1}{4}, \xi_5 = 2\frac{3}{4}$
3. $f(x) = 1/x, 1 \le x \le 3$; for Δ: $x_0 = 1, x_1 = 1\frac{2}{3}, x_2 = 2\frac{1}{4}, x_3 = 2\frac{2}{3}, x_4 = 3$; $\xi_1 = 1\frac{1}{4}, \xi_2 = 2, \xi_3 = 2\frac{1}{2}, \xi_4 = 2\frac{3}{4}$
4. $f(x) = x^3, -1 \le x \le 2$; for Δ: $x_0 = -1, x_1 = -\frac{1}{3}, x_2 = \frac{1}{2}, x_3 = 1, x_4 = 1\frac{1}{4}, x_5 = 2$; $\xi_1 = -\frac{1}{2}, \xi_2 = 0, \xi_3 = \frac{2}{3}, \xi_4 = 1, \xi_5 = 1\frac{1}{2}$
5. $f(x) = x^2 - x + 1, 0 \le x \le 1$; for Δ: $x_0 = 0, x_1 = 0.2, x_2 = 0.5, x_3 = 0.7, x_4 = 1$; $\xi_1 = 0.1, \xi_2 = 0.4, \xi_3 = 0.6, \xi_4 = 0.9$
6. $f(x) = 1/(x + 2), -1 \le x \le 3$; for Δ: $x_0 = -1, x_1 = -\frac{1}{4}, x_2 = 0, x_3 = \frac{1}{2}, x_4 = 1\frac{1}{4}, x_5 = 2, x_6 = 2\frac{1}{4}, x_7 = 2\frac{3}{4}, x_8 = 3$; $\xi_1 = -\frac{3}{4}, \xi_2 = 0, \xi_3 = \frac{1}{4}, \xi_4 = 1, \xi_5 = 1\frac{1}{2}, \xi_6 = 2, \xi_7 = 2\frac{1}{2}, \xi_8 = 3$
7. $f(x) = \sin x, 0 \le x \le \pi$; for Δ: $x_0 = 0, x_1 = \frac{1}{4}\pi, x_2 = \frac{1}{2}\pi, x_3 = \frac{2}{3}\pi, x_4 = \frac{3}{4}\pi, x_5 = \pi$; $\xi_1 = \frac{1}{6}\pi, \xi_2 = \frac{1}{3}\pi, \xi_3 = \frac{1}{2}\pi, \xi_4 = \frac{3}{4}\pi, \xi_5 = \frac{5}{6}\pi$.
8. $f(x) = 3 \cos \frac{1}{2}x, -\pi \le x \le \pi$; for Δ: $x_0 = -\pi, x_1 = -\frac{1}{2}\pi, x_2 = -\frac{1}{3}\pi, x_3 = \frac{1}{3}\pi, x_4 = \frac{7}{12}\pi, x_5 = \pi$; $\xi_1 = -\frac{2}{3}\pi, \xi_2 = -\frac{1}{3}\pi, \xi_3 = 0, \xi_4 = \frac{1}{2}\pi, \xi_5 = \frac{2}{3}\pi$
9. $f(x) = [\![x]\!] + 2, -3 \le x \le 3$; for Δ: $x_0 = -3, x_1 = -1, x_2 = 0, x_3 = 2, x_4 = 3$; $\xi_1 = -2.5, \xi_2 = -0.5, \xi_3 = 1, \xi_4 = 2.5$

In Exercises 10 through 18, find the exact value of the definite integral. Use the method of Example 2 of this section.

10. $\int_2^7 3x\,dx$
11. $\int_0^2 x^2\,dx$
12. $\int_2^4 x^2\,dx$
13. $\int_1^2 x^3\,dx$
14. $\int_0^5 (x^3 - 1)\,dx$
15. $\int_1^4 (x^2 + 4x + 5)\,dx$
16. $\int_0^4 (x^2 + x - 6)\,dx$
17. $\int_{-2}^2 (x^3 + 1)\,dx$
18. $\int_{-2}^1 x^4\,dx$

In Exercises 19 through 28, find the exact area of the region in the following way: (a) Express the measure of the area as the limit of a Riemann sum with regular partitions; (b) express this limit with definite integral notation; (c) evaluate the definite integral by the method of this section and a suitable choice of ξ_i. Draw a figure showing the region.

19. Bounded by the line $y = 2x - 1$, the x axis, and the lines $x = 1$ and $x = 5$.
20. Bounded by the line $y = 2x - 6$, the x axis, and the lines $x = 4$ and $x = 7$.
21. Bounded by the line $y = -3x + 2$, the x axis, and the lines $x = -5$ and $x = -1$.
22. Bounded by the line $y = 3x + 2$, the x axis, and the lines $x = -4$ and $x = -1$.
23. Bounded by the curve $y = 4 - x^2$, the x axis, and the lines $x = 1$ and $x = 2$.
24. Bounded by the curve $y = (x + 3)^2$, the x axis, and the lines $x = -3$ and $x = 0$.
25. Bounded by the curve $y = 12 - x - x^2$, the x axis, and the lines $x = -3$ and $x = 2$.
26. Bounded by the curve $y = 10 + x - x^2$, the x axis, and the lines $x = -2$ and $x = 3$.
27. Bounded by the curve $y = x^3 - 4$, the x axis, and the lines $x = -2$ and $x = -1$.
28. Bounded by the curve $y = 6x + x^2 - x^3$, the x axis, and the lines $x = -1$ and $x = 3$.

29. Express as a definite integral: $\lim_{n \to +\infty} \sum_{i=1}^{n} (8i^2/n^3)$. (*Hint:* Consider the function f for which $f(x) = x^2$.)

30. Express as a definite integral: $\lim_{n \to +\infty} \sum_{i=1}^{n} 1/(n+i)$. (*Hint:* Consider the function f for which $f(x) = 1/x$ on $[1, 2]$.)

31. Express as a definite integral: $\lim_{n \to +\infty} \sum_{i=1}^{n} (n/i^2)$. (*Hint:* Consider the function f for which $f(x) = 1/x^2$.)

32. Let $[a, b]$ be any interval such that $a < 0 < b$. Prove that even though the unit step function (Exercise 31 in Exercises 1.6) is discontinuous on $[a, b]$, it is integrable on $[a, b]$ and $\int_a^b U(x)\,dx = b$.

33. Prove that the signum function is discontinuous on $[-1, 1]$ and yet integrable on $[-1, 1]$. Furthermore show that $\int_{-1}^{1} \operatorname{sgn} x\,dx = 0$.

34. Prove that the greatest integer function is discontinuous on $[0, \frac{3}{2}]$ and yet integrable on $[0, \frac{3}{2}]$. Furthermore show that $\int_0^{3/2} [\![x]\!]\,dx = \frac{1}{2}$.

6.3 PROPERTIES OF THE DEFINITE INTEGRAL

Evaluating a definite integral from the definition, by actually finding the limit of a sum as we did in Sec. 6.2, is usually quite tedious and frequently almost impossible. To establish a much simpler method we first need to develop some properties of the definite integral. First the following two theorems about Riemann sums are needed.

6.3.1 Theorem If Δ is any partition of the closed interval $[a, b]$, then

$$\lim_{\|\Delta\| \to 0} \sum_{i=1}^{n} \Delta_i x = b - a$$

Proof

$$\sum_{i=1}^{n} \Delta_i x - (b - a) = (b - a) - (b - a) = 0$$

Hence, for any $\epsilon > 0$, any choice of $\delta > 0$ guarantees the inequality

$$\left| \sum_{i=1}^{n} \Delta_i x - (b - a) \right| < \epsilon \quad \text{whenever} \quad \|\Delta\| < \delta$$

and so by Definition 6.2.1,

$$\lim_{\|\Delta\| \to 0} \sum_{i=1}^{n} \Delta_i x = b - a$$

■

6.3.2 Theorem If f is defined on the closed interval $[a, b]$, and if

$$\lim_{\|\Delta\| \to 0} \sum_{i=1}^{n} f(\xi_i)\,\Delta_i x$$

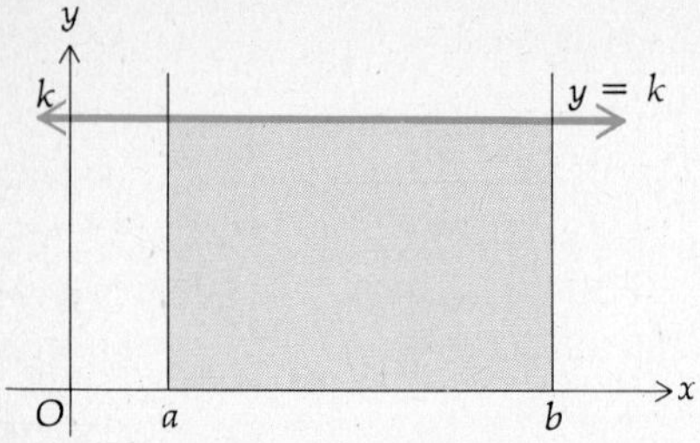

FIGURE 6.3.1

exists, where Δ is any partition of $[a, b]$, then if k is any constant,

$$\lim_{\|\Delta\|\to 0} \sum_{i=1}^{n} kf(\xi_i)\,\Delta_i x = k \lim_{\|\Delta\|\to 0} \sum_{i=1}^{n} f(\xi_i)\,\Delta_i x$$

The proof of this theorem is left as an exercise (see Exercise 43).

• ILLUSTRATION 1 Refer to Fig. 6.3.1. If $k > 0$, the definite integral $\int_a^b k\,dx$ gives the measure of the area of the shaded region, which is a rectangle whose dimensions are k units and $(b - a)$ units. This fact is a geometric interpretation of the following theorem when $k > 0$. •

6.3.3 Theorem If k is any constant, then

$$\int_a^b k\,dx = k(b - a)$$

Proof From Definition 6.2.2,

$$\int_a^b f(x)\,dx = \lim_{\|\Delta\|\to 0} \sum_{i=1}^{n} f(\xi_i)\,\Delta_i x$$

If $f(x) = k$ for all x in $[a, b]$, we have from this equation

$$\int_a^b k\,dx = \lim_{\|\Delta\|\to 0} \sum_{i=1}^{n} k\,\Delta_i x$$

$$= k \lim_{\|\Delta\|\to 0} \sum_{i=1}^{n} \Delta_i x \quad \text{(by Theorem 6.3.2)}$$

$$= k(b - a) \quad \text{(by Theorem 6.3.1)}$$

■

Example 1 Evaluate $\int_{-3}^{5} 4\,dx$.

Solution We apply Theorem 6.3.3.

$$\int_{-3}^{5} 4\,dx = 4[5 - (-3)]$$

$$= 4(8)$$

$$= 32$$

6.3.4 Theorem If the function f is integrable on the closed interval $[a, b]$, and if k is any constant, then

$$\int_a^b kf(x)\,dx = k \int_a^b f(x)\,dx$$

Proof Because f is integrable on $[a, b]$, $\lim_{\|\Delta\| \to 0} \sum_{i=1}^{n} f(\xi_i)\,\Delta_i x$ exists; so by Theorem 6.3.2,

$$\lim_{\|\Delta\| \to 0} \sum_{i=1}^{n} kf(\xi_i)\,\Delta_i x = k \lim_{\|\Delta\| \to 0} \sum_{i=1}^{n} f(\xi_i)\,\Delta_i x$$

Therefore

$$\int_a^b kf(x)\,dx = k \int_a^b f(x)\,dx$$

■

Observe the similarity of the following theorem to Limit Theorem 4 (2.2.4) about the limit of the sum of two functions. Also notice that the proofs of the two theorems are alike.

6.3.5 Theorem If the functions f and g are integrable on $[a, b]$, then $f + g$ is integrable on $[a, b]$ and

$$\int_a^b [f(x) + g(x)]\,dx = \int_a^b f(x)\,dx + \int_a^b g(x)\,dx$$

Proof The functions f and g are integrable on $[a, b]$; thus let

$$\int_a^b f(x)\,dx = M \quad \text{and} \quad \int_a^b g(x)\,dx = N$$

To prove that $f + g$ is integrable on $[a, b]$ and that $\int_a^b [f(x) + g(x)]\,dx = M + N$, we must show that for any $\epsilon > 0$ there exists a $\delta > 0$ such that

$$\left| \sum_{i=1}^{n} [f(\xi_i) + g(\xi_i)]\,\Delta_i x - (M + N) \right| < \epsilon$$

for all partitions Δ for which $\|\Delta\| < \delta$ and for any ξ_i in $[x_{i-1}, x_i]$. Because

$$M = \lim_{\|\Delta\| \to 0} \sum_{i=1}^{n} f(\xi_i)\,\Delta_i x \quad \text{and} \quad N = \lim_{\|\Delta\| \to 0} \sum_{i=1}^{n} g(\xi_i)\,\Delta_i x$$

it follows that for any $\epsilon > 0$ there exist a $\delta_1 > 0$ and a $\delta_2 > 0$ such that

$$\left| \sum_{i=1}^{n} f(\xi_i)\,\Delta_i x - M \right| < \frac{\epsilon}{2} \quad \text{and} \quad \left| \sum_{i=1}^{n} g(\xi_i)\,\Delta_i x - N \right| < \frac{\epsilon}{2}$$

for all partitions Δ for which $\|\Delta\| < \delta_1$ and $\|\Delta\| < \delta_2$, and for any ξ_i in $[x_{i-1}, x_i]$. Therefore, if $\delta = \min(\delta_1, \delta_2)$, then for any $\epsilon > 0$,

$$\left| \sum_{i=1}^{n} f(\xi_i)\,\Delta_i x - M \right| + \left| \sum_{i=1}^{n} g(\xi_i)\,\Delta_i x - N \right| < \frac{\epsilon}{2} + \frac{\epsilon}{2} = \epsilon \tag{1}$$

for all partitions Δ for which $\|\Delta\| < \delta$ and for any ξ_i in $[x_{i-1}, x_i]$.

By the triangle inequality we have

$$\left|\left(\sum_{i=1}^{n} f(\xi_i)\,\Delta_i x - M\right) + \left(\sum_{i=1}^{n} g(\xi_i)\,\Delta_i x - \mathrm{N}\right)\right| \\ \leq \left|\sum_{i=1}^{n} f(\xi_i)\,\Delta_i x - M\right| + \left|\sum_{i=1}^{n} g(\xi_i)\,\Delta_i x - N\right| \tag{2}$$

From inequalities (1) and (2) we have

$$\left|\left(\sum_{i=1}^{n} f(\xi_i)\,\Delta_i x + \sum_{i=1}^{n} g(\xi_i)\,\Delta_i x\right) - (M + N)\right| < \epsilon \tag{3}$$

From Theorem 6.1.4,

$$\sum_{i=1}^{n} f(\xi_i)\,\Delta_i x + \sum_{i=1}^{n} g(\xi_i)\,\Delta_i x = \sum_{i=1}^{n} [f(\xi_i) + g(\xi_i)]\,\Delta_i x \tag{4}$$

So by substituting from (4) into (3) we are able to conclude that for any $\epsilon > 0$,

$$\left|\sum_{i=1}^{n} [f(\xi_i) + g(\xi_i)]\,\Delta_i x - (M + N)\right| < \epsilon$$

for all partitions Δ for which $\|\Delta\| < \delta$, where $\delta = \min(\delta_1, \delta_2)$ and for any ξ_i in $[x_{i-1}, x_i]$. This proves that $f + g$ is integrable on $[a, b]$ and that

$$\int_a^b [f(x) + g(x)]\,dx = \int_a^b f(x)\,dx + \int_a^b g(x)\,dx$$

■

The plus sign in the statement of Theorem 6.3.5 can be replaced by a minus sign as a result of applying Theorem 6.3.4, where $k = -1$.

Theorem 6.3.5 can be extended to any number of functions. That is, if the functions $f_1, f_2, \ldots, f_n$ are all integrable on $[a, b]$, then $(f_1 \pm f_2 \pm \cdots \pm f_n)$ is integrable on $[a, b]$ and

$$\int_a^b [f_1(x) \pm f_2(x) \pm \cdots \pm f_n(x)]\,dx \\ = \int_a^b f_1(x)\,dx \pm \int_a^b f_2(x)\,dx \pm \cdots \pm \int_a^b f_n(x)\,dx$$

Example 2 Use the result of Example 2, Sec. 6.2, and the fact that $\int_1^3 x\,dx = 4$ to evaluate $\int_1^3 (3x^2 - 5x + 2)\,dx$.

Solution In Example 2, Sec. 6.2, we had the result

$$\int_1^3 x^2\,dx = \tfrac{26}{3}$$

From properties of the definite integral,

$$\int_1^3 (3x^2 - 5x + 2)\,dx = \int_1^3 3x^2\,dx - \int_1^3 5x\,dx + \int_1^3 2\,dx$$
$$= 3\int_1^3 x^2\,dx - 5\int_1^3 x\,dx + 2(3-1)$$
$$= 3(\tfrac{26}{3}) - 5(4) + .4$$
$$= 26 - 20 + 4$$
$$= 10$$

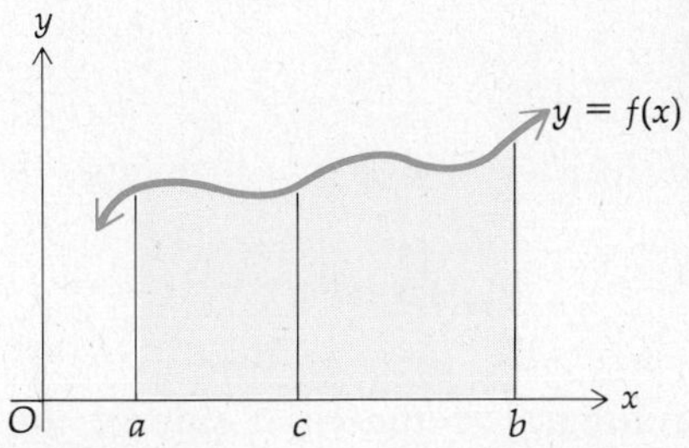

FIGURE 6.3.2

● ILLUSTRATION 2 A geometric interpretation of Theorem 6.3.6 that follows is shown in Fig. 6.3.2 where $f(x) \geq 0$. For all x in $[a, b]$, the measure of the area of the region bounded by the curve $y = f(x)$ and the x axis from a to b is equal to the sum of the measures of the areas of the regions from a to c and from c to b. ●

6.3.6 Theorem If the function f is integrable on the closed intervals $[a, b]$, $[a, c]$, and $[c, b]$,

$$\int_a^b f(x)\,dx = \int_a^c f(x)\,dx + \int_c^b f(x)\,dx$$

where $a < c < b$.

Proof Let Δ be a partition of $[a, b]$. Form the partition Δ' of $[a, b]$ in the following way. If c is one of the partitioning points of Δ (i.e., $c = x_i$ for some i), then Δ' is exactly the same as Δ. If c is not one of the partitioning points of Δ but is contained in the subinterval $[x_{i-1}, x_i]$, then the partition Δ' has as its partitioning points all the partitioning points of Δ and, in addition, the point c. Therefore the subintervals of the partition Δ' are the same as the subintervals of Δ, with the exception that the subinterval $[x_{i-1}, x_i]$ of Δ is divided into the two subintervals $[x_{i-1}, c]$ and $[c, x_i]$.

If $\|\Delta'\|$ is the norm of Δ' and if $\|\Delta\|$ is the norm of Δ, then

$$\|\Delta'\| \leq \|\Delta\|$$

If in the partition Δ' the interval $[a, c]$ is divided into r subintervals and the interval $[c, b]$ is divided into $(n - r)$ subintervals, then the part of the partition Δ' from a to c gives a Riemann sum of the form

$$\sum_{i=1}^{r} f(\xi_i)\,\Delta_i x$$

and the other part of the partition Δ', from c to b, gives a Riemann sum of the form

$$\sum_{i=r+1}^{n} f(\xi_i)\,\Delta_i x$$

Using the definition of the definite integral and properties of the sigma notation we have

$$\int_a^b f(x)\,dx = \lim_{\|\Delta\|\to 0} \sum_{i=1}^{n} f(\xi_i)\,\Delta_i x$$

$$= \lim_{\|\Delta\|\to 0} \left[\sum_{i=1}^{r} f(\xi_i)\,\Delta_i x + \sum_{i=r+1}^{n} f(\xi_i)\,\Delta_i x\right]$$

$$= \lim_{\|\Delta\|\to 0} \sum_{i=1}^{r} f(\xi_i)\,\Delta_i x + \lim_{\|\Delta\|\to 0} \sum_{i=r+1}^{n} f(\xi_i)\,\Delta_i x$$

Because $0 < \|\Delta'\| \leq \|\Delta\|$, we can replace $\|\Delta\| \to 0$ by $\|\Delta'\| \to 0$, giving

$$\int_a^b f(x)\,dx = \lim_{\|\Delta'\|\to 0} \sum_{i=1}^{r} f(\xi_i)\,\Delta_i x + \lim_{\|\Delta'\|\to 0} \sum_{i=r+1}^{n} f(\xi_i)\,\Delta_i x$$

Applying the definition of the definite integral to the right side of the above equation we have

$$\int_a^b f(x)\,dx = \int_a^c f(x)\,dx + \int_c^b f(x)\,dx$$ ■

The result of Theorem 6.3.6 is true for any ordering of the numbers a, b, and c. This is stated as another theorem.

6.3.7 Theorem If f is integrable on a closed interval containing the three numbers a, b, and c, then

$$\int_a^b f(x)\,dx = \int_a^c f(x)\,dx + \int_c^b f(x)\,dx \tag{5}$$

regardless of the order of a, b, and c.

Proof If a, b, and c are distinct, there are six possible orderings of these three numbers: $a < b < c$, $a < c < b$, $b < a < c$, $b < c < a$, $c < a < b$, and $c < b < a$. The second ordering, $a < c < b$, is Theorem 6.3.6. We make use of Theorem 6.3.6 in proving that Eq. (5) holds for the other orderings.

Suppose that $a < b < c$; then from Theorem 6.3.6 we have

$$\int_a^b f(x)\,dx + \int_b^c f(x)\,dx = \int_a^c f(x)\,dx \tag{6}$$

From Definition 6.2.5,

$$\int_b^c f(x)\,dx = -\int_c^b f(x)\,dx \tag{7}$$

Substituting from (7) into (6) we obtain

$$\int_a^b f(x)\,dx - \int_c^b f(x)\,dx = \int_a^c f(x)\,dx$$

Thus

$$\int_a^b f(x)\,dx = \int_a^c f(x)\,dx + \int_c^b f(x)\,dx$$

which is the desired result.

The proofs for the other four orderings are similar and left as exercises (see Exercises 44 through 47).

There is also the possibility that two of the three numbers are equal; for example, $a = c < b$. Then

$$\int_a^c f(x)\,dx = \int_a^a f(x)\,dx = 0 \quad \text{(by Definition 6.2.6)}$$

Also, because $a = c$,

$$\int_c^b f(x)\,dx = \int_a^b f(x)\,dx$$

Therefore

$$\int_a^c f(x)\,dx + \int_c^b f(x)\,dx = 0 + \int_a^b f(x)\,dx$$

which is the desired result. ■

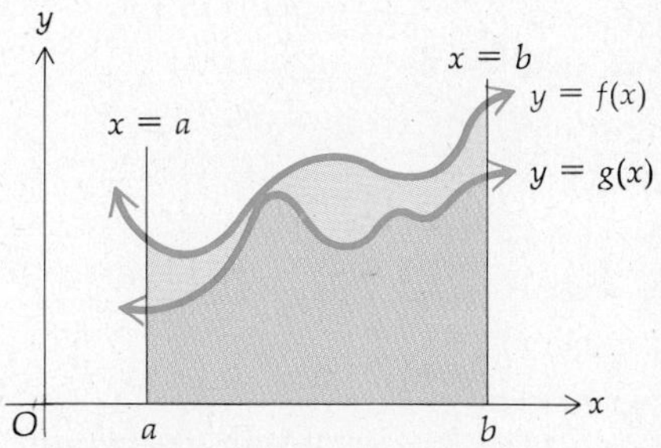

FIGURE 6.3.3

● ILLUSTRATION 3 In Fig. 6.3.3, $f(x) \geq g(x) \geq 0$ for all x in $[a, b]$. The definite integral $\int_a^b f(x)\,dx$ gives the measure of the area of the region bounded by the curve $y = f(x)$, the x axis, and the lines $x = a$ and $x = b$; $\int_a^b g(x)\,dx$ gives the measure of the area of the region bounded by the curve $y = g(x)$, the x axis, and the lines $x = a$ and $x = b$. In the figure we see that the first area is greater than the second area. We have given a geometric interpretation of the following theorem when $f(x)$ and $g(x)$ are nonnegative on $[a, b]$. ●

6.3.8 Theorem If the functions f and g are integrable on the closed interval $[a, b]$, and if $f(x) \geq g(x)$ for all x in $[a, b]$, then

$$\int_a^b f(x)\,dx \geq \int_a^b g(x)\,dx$$

Proof Because f and g are integrable on $[a, b]$, $\int_a^b f(x)\,dx$ and $\int_a^b g(x)\,dx$ both exist. Therefore, by Theorem 6.3.4,

$$\int_a^b f(x)\,dx - \int_a^b g(x)\,dx = \int_a^b f(x)\,dx + \int_a^b [-g(x)]\,dx$$

$$= \int_a^b [f(x) - g(x)]\,dx \quad \text{(by Theorem 6.3.5)}$$

Let h be the function defined by

$$h(x) = f(x) - g(x)$$

Then $h(x) \geq 0$ for all x in $[a, b]$ because $f(x) \geq g(x)$ for all x in $[a, b]$.

We wish to prove that $\int_a^b [f(x) - g(x)]\, dx \geq 0$.

$$\int_a^b [f(x) - g(x)]\, dx = \int_a^b h(x)\, dx = \lim_{\|\Delta\| \to 0} \sum_{i=1}^{n} h(\xi_i)\, \Delta_i x \tag{8}$$

Assume that

$$\lim_{\|\Delta\| \to 0} \sum_{i=1}^{n} h(\xi_i)\, \Delta_i x = L < 0 \tag{9}$$

Then by Definition 6.2.1, with $\epsilon = -L$, there exists a $\delta > 0$ such that

$$\left| \sum_{i=1}^{n} h(\xi_i)\, \Delta_i x - L \right| < -L \quad \text{whenever} \quad \|\Delta\| < \delta \tag{10}$$

But because

$$\sum_{i=1}^{n} h(\xi_i)\, \Delta_i x - L \leq \left| \sum_{i=1}^{n} h(\xi_i)\, \Delta_i x - L \right|$$

from inequality (10) we have

$$\sum_{i=1}^{n} h(\xi_i)\, \Delta_i x - L < -L \quad \text{whenever} \quad \|\Delta\| < \delta$$

or

$$\sum_{i=1}^{n} h(\xi_i)\, \Delta_i x < 0 \quad \text{whenever} \quad \|\Delta\| < \delta \tag{11}$$

But statement (11) is impossible, because every $h(\xi_i)$ is nonnegative and every $\Delta_i x > 0$; thus we have a contradiction to our assumption (9). Therefore (9) is false, and

$$\lim_{\|\Delta\| \to 0} \sum_{i=1}^{n} h(\xi_i)\, \Delta_i x \geq 0 \tag{12}$$

From (8) and (12) we have

$$\int_a^b [f(x) - g(x)]\, dx \geq 0$$

Hence

$$\int_a^b f(x)\, dx - \int_a^b g(x)\, dx \geq 0$$

and so

$$\int_a^b f(x)\, dx \geq \int_a^b g(x)\, dx$$

■

● ILLUSTRATION 4 In Fig. 6.3.4, $f(x) \geq 0$ for all x in $[a, b]$ and m and M are, respectively, the absolute minimum and absolute maximum function values of f on $[a, b]$. The integral $\int_a^b f(x)\,dx$ gives the measure of the area of the region bounded by the curve $y = f(x)$, the x axis, and the lines $x = a$ and $x = b$. This area is greater than that of the rectangle whose dimensions are m and $(b - a)$ and less than that of the rectangle whose dimensions are M and $(b - a)$. Thus we have a geometric interpretation of the next theorem if $f(x) \geq 0$ for all x in $[a, b]$. ●

6.3.9 Theorem Suppose that the function f is continuous on the closed interval $[a, b]$. If m and M are, respectively, the absolute minimum and absolute maximum function values of f on $[a, b]$ so that

$$m \leq f(x) \leq M \qquad \text{for } a \leq x \leq b$$

then

$$m(b - a) \leq \int_a^b f(x)\,dx \leq M(b - a)$$

Proof Because f is continuous on $[a, b]$, the extreme-value theorem (4.2.9) guarantees the existence of m and M.

By Theorem 6.3.3,

$$\int_a^b m\,dx = m(b - a) \tag{13}$$

and

$$\int_a^b M\,dx = M(b - a) \tag{14}$$

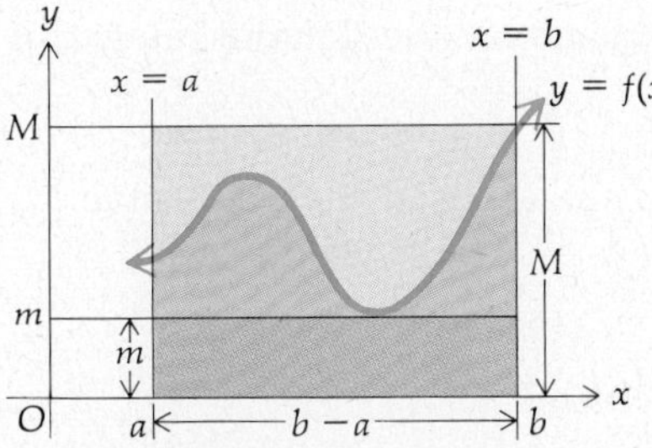

FIGURE 6.3.4

Because f is continuous on $[a, b]$, it follows from Theorem 6.2.3 that f is integrable on $[a, b]$. Then because $f(x) \geq m$ for all x in $[a, b]$, we have from Theorem 6.3.8

$$\int_a^b f(x)\,dx \geq \int_a^b m\,dx$$

which from (13) gives

$$\int_a^b f(x)\,dx \geq m(b - a) \tag{15}$$

Similarly, because $M \geq f(x)$ for all x in $[a, b]$, it follows from Theorem 6.3.8 that

$$\int_a^b M\,dx \geq \int_a^b f(x)\,dx$$

which from (14) gives

$$M(b - a) \geq \int_a^b f(x)\,dx \tag{16}$$

Combining inequalities (15) and (16) we have

$$m(b-a) \le \int_a^b f(x)\,dx \le M(b-a) \qquad \blacksquare$$

Example 3 Apply Theorem 6.3.9 to find a closed interval containing the value of $\int_{1/2}^{4} (x^3 - 6x^2 + 9x + 1)\,dx$. Use the results of Example 1, Sec. 4.5.

Solution If $f(x) = x^3 - 6x^2 + 9x + 1$, from Example 1 in Sec. 4.5, f has a relative minimum value of 1 at $x = 3$ and a relative maximum value of 5 at $x = 1$. $f(\frac{1}{2}) = \frac{33}{8}$ and $f(4) = 5$. Hence the absolute minimum value of f on $[\frac{1}{2}, 4]$ is 1, and the absolute maximum value is 5. Taking $m = 1$ and $M = 5$ in Theorem 6.3.9 we have

$$1(4 - \tfrac{1}{2}) \le \int_{1/2}^{4} (x^3 - 6x^2 + 9x + 1)\,dx \le 5(4 - \tfrac{1}{2})$$

$$\tfrac{7}{2} \le \int_{1/2}^{4} (x^3 - 6x^2 + 9x + 1)\,dx \le \tfrac{35}{2}$$

Therefore the closed interval $[\frac{7}{2}, \frac{35}{2}]$ contains the value of the definite integral. In Example 2, Sec. 6.5, the exact value of the definite integral is shown to be $10\frac{39}{64}$.

Example 4 Apply Theorem 6.3.9 to find a closed interval containing the value of $\int_{\pi/4}^{3\pi/4} \sqrt{\sin x}\,dx$.

Solution If $f(x) = \sqrt{\sin x}$, then

$$f'(x) = \frac{\cos x}{2\sqrt{\sin x}}$$

For x in $[\frac{1}{4}\pi, \frac{3}{4}\pi]$, $f'(x) = 0$ when $x = \frac{1}{2}\pi$. Since $f'(x) > 0$ when $\frac{1}{4}\pi < x < \frac{1}{2}\pi$, and $f'(x) < 0$ when $\frac{1}{2}\pi < x < \frac{3}{4}\pi$, it follows that f has a relative maximum value at $\frac{1}{2}\pi$; and $f(\frac{1}{2}\pi) = 1$. Furthermore, $f(\frac{1}{4}\pi) = \sqrt[4]{2}/\sqrt{2} \approx 0.266$, and $f(\frac{3}{4}\pi) \approx 0.266$. Thus, on $[\frac{1}{4}\pi, \frac{3}{4}\pi]$ the absolute minimum value of f is 0.266 and the absolute maximum value is 1. So, with $m = 0.266$ and $M = 1$ in Theorem 6.3.9,

$$0.266[\tfrac{3}{4}\pi - \tfrac{1}{4}\pi] \le \int_{\pi/4}^{3\pi/4} \sqrt{\sin x}\,dx \le 1[\tfrac{3}{4}\pi - \tfrac{1}{4}\pi]$$

$$0.133\pi \le \int_{\pi/4}^{3\pi/4} \sqrt{\sin x}\,dx \le 0.5\pi$$

$$0.418 \le \int_{\pi/4}^{3\pi/4} \sqrt{\sin x}\,dx \le 1.571$$

The value of the definite integral is therefore in the closed interval [0.418, 1.571].

EXERCISES 6.3

In Exercises 1 through 6, evaluate the definite integral.

1. $\int_2^5 4\,dx$ **2.** $\int_{-3}^4 7\,dx$ **3.** $\int_{-2}^2 \sqrt{5}\,dx$ **4.** $\int_5^{-1} 6\,dx$ **5.** $\int_{-5}^{-10} dx$ **6.** $\int_3^3 dx$

In Exercises 7 through 18, evaluate the given definite integral by using the following results:

$$\int_{-1}^2 x^2\,dx = 3 \qquad \int_{-1}^2 x\,dx = \tfrac{3}{2} \qquad \int_0^\pi \sin x\,dx = 2 \qquad \int_0^\pi \cos x\,dx = 0 \qquad \int_0^\pi \sin^2 x\,dx = \tfrac{1}{2}\pi$$

7. $\int_{-1}^2 (2x^2 - 4x + 5)\,dx$ **8.** $\int_{-1}^2 (8 - x^2)\,dx$ **9.** $\int_{-1}^2 (2 - 5x + \tfrac{1}{2}x^2)\,dx$

10. $\int_{-1}^2 (3x^2 - 4x - 1)\,dx$ **11.** $\int_2^{-1} (2x + 1)^2\,dx$ **12.** $\int_{-1}^2 (5x^2 + \tfrac{1}{3}x - \tfrac{1}{2})\,dx$

13. $\int_{-1}^2 (x - 1)(2x + 3)\,dx$ **14.** $\int_2^{-1} 3x(x - 4)\,dx$ **15.** $\int_0^\pi (2\sin x + 3\cos x + 1)\,dx$

16. $\int_0^\pi 3\cos^2 x\,dx$ **17.** $\int_0^\pi (\cos x + 4)^2\,dx$ **18.** $\int_\pi^0 (\sin x - 2)^2\,dx$

In Exercises 19 through 34, apply Theorem 6.3.9 to find a closed interval containing the value of the given definite integral.

19. $\int_3^7 2x\,dx$ **20.** $\int_2^5 3x\,dx$ **21.** $\int_0^4 x^2\,dx$ **22.** $\int_{-2}^1 x^3\,dx$

23. $\int_{-3}^6 \sqrt{3 + x}\,dx$ **24.** $\int_{-2}^1 (x + 1)^{2/3}\,dx$ **25.** $\int_{-4}^0 (x^4 - 8x^2 + 16)\,dx$ **26.** $\int_{-1}^4 (x^4 - 8x^2 + 16)\,dx$

27. $\int_{\pi/6}^{\pi/3} \sin x\,dx$ **28.** $\int_{-\pi/3}^{2\pi/3} \cos x\,dx$ **29.** $\int_1^4 |x - 2|\,dx$ **30.** $\int_{-1}^2 \sqrt{x^2 + 5}\,dx$

31. $\int_{-1}^2 \frac{x}{x + 2}\,dx$ **32.** $\int_{-5}^2 \frac{x + 5}{x - 3}\,dx$ **33.** $\int_{-\pi/2}^{\pi/2} (4\cos^3 x - 9\cos x)\,dx$ **34.** $\int_{-\pi/2}^{\pi/2} 3\sin^3 x\,dx$

In Exercises 35 through 38, use Theorem 6.3.8 to determine which of the symbols $\geq$ or $\leq$ should be inserted in the blank to make the inequality correct.

35. $\int_{-1}^3 (2x^2 - 4)\,dx$ _____ $\int_{-1}^3 (x^2 - 6)\,dx$ **36.** $\int_4^5 \sqrt{6 - x}\,dx$ _____ $\int_4^5 \sqrt{x - 2}\,dx$

37. $\int_{3\pi/4}^{5\pi/4} \sin^2 x\,dx$ _____ $\int_{3\pi/4}^{5\pi/4} \cos^2 x\,dx$ **38.** $\int_0^{\pi/4} \cos x\,dx$ _____ $\int_0^{\pi/4} \sin x\,dx$

39. Show that if f is continuous on $[-3, 4]$, then

$$\int_3^{-1} f(x)\,dx + \int_4^3 f(x)\,dx + \int_{-3}^4 f(x)\,dx + \int_{-1}^{-3} f(x)\,dx = 0$$

40. Show that if f is continuous on $[-1, 2]$, then

$$\int_{-1}^2 f(x)\,dx + \int_2^0 f(x)\,dx + \int_0^1 f(x)\,dx + \int_1^{-1} f(x)\,dx = 0$$

41. Show that

$$\int_0^1 x\,dx \geq \int_0^1 x^2\,dx$$

but

$$\int_1^2 x\,dx \leq \int_1^2 x^2\,dx$$

Do not evaluate the definite integrals.

42. If f is continuous on $[a, b]$, prove that

$$\left|\int_a^b f(x)\,dx\right| \leq \int_a^b |f(x)|\,dx$$

(*Hint:* $-|f(x)| \leq f(x) \leq |f(x)|$.)

43. Prove Theorem 6.3.2.

In Exercises 44 through 49, prove that Theorem 6.3.7 is valid for the given ordering of a, b, and c; in each case use the result of Theorem 6.3.6.

44. $b < a < c$ **45.** $b < c < a$ **46.** $c < a < b$ **47.** $c < b < a$ **48.** $a < c = b$ **49.** $a = b < c$

6.4 THE MEAN-VALUE THEOREM FOR INTEGRALS

We are now ready to state and prove the mean-value theorem for integrals. This theorem is used in the proof of many important theorems, the most prominent of which is the *fundamental theorem of the calculus,* given in Sec. 6.5. Before the statement of the mean-value theorem for integrals, we give an illustration that interprets the theorem geometrically.

• ILLUSTRATION 1 Consider $f(x) \geq 0$ for all values of x in $[a, b]$. Then $\int_a^b f(x)\,dx$ gives the measure of the area of the region bounded by the curve whose equation is $y = f(x)$, the x axis, and the lines $x = a$ and $x = b$. See Fig. 6.4.1. The mean-value theorem for integrals states that there is a number X in $[a, b]$ such that the area of the rectangle $AEFB$ of height $f(\text{X})$ units and width $(b - a)$ units is equal to the area of the region $ADCB$. •

6.4.1 Theorem
Mean-Value Theorem for Integrals

If the function f is continuous on the closed interval $[a, b]$, then there exists a number X in $[a, b]$ such that

$$\int_a^b f(x)\,dx = f(\text{X})(b - a)$$

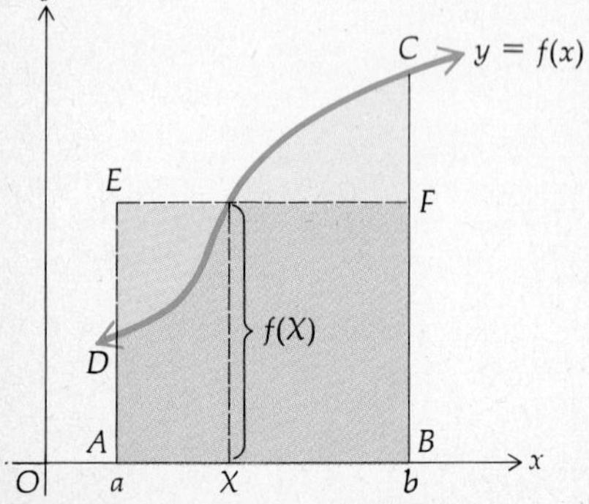

FIGURE 6.4.1

Proof Because f is continuous on $[a, b]$, from the extreme-value theorem (4.2.9) f has an absolute maximum value and an absolute minimum value on $[a, b]$.

Let m be the absolute minimum value occurring at $x = x_m$. Thus

$$f(x_m) = m \qquad a \leq x_m \leq b \tag{1}$$

Let M be the absolute maximum value, occurring at $x = x_M$. Thus

$$f(x_M) = M \qquad a \leq x_M \leq b \tag{2}$$

We have, then,

$$m \leq f(x) \leq M \qquad \text{for all } x \text{ in } [a, b]$$

From Theorem 6.3.9 it follows that

$$m(b - a) \leq \int_a^b f(x)\,dx \leq M(b - a)$$

Dividing by $(b - a)$ and noting that $(b - a)$ is positive because $b > a$, we get

$$m \leq \frac{\int_a^b f(x)\,dx}{b - a} \leq M$$

But from (1) and (2), $m = f(x_m)$ and $M = f(x_M)$; so we have

$$f(x_m) \leq \frac{\int_a^b f(x)\,dx}{b - a} \leq f(x_M) \tag{3}$$

From inequalities (3) and the intermediate-value theorem there is some number X in a closed interval containing x_m and x_M such that

$$f(\mathrm{X}) = \frac{\int_a^b f(x)\,dx}{b - a}$$

or

$$\int_a^b f(x)\,dx = f(\mathrm{X})(b - a) \qquad a \leq \mathrm{X} \leq b$$ ■

The value of X in Theorem 6.4.1 is not necessarily unique. The theorem does not provide a method for finding X, but it states that a value of X exists, and this fact is used to prove other theorems. In some particular cases we can find the value of X guaranteed by the theorem, as we illustrate in the following example.

Example 1 Find the value of X such that $\int_1^3 f(x)\,dx = f(\mathrm{X})(3 - 1)$ if $f(x) = x^2$. Use the result of Example 2 in Sec. 6.2.

Solution In Example 2 of Sec. 6.2 we obtained

$$\int_1^3 x^2\,dx = 8\tfrac{2}{3}$$

Therefore we wish to find X such that

$$f(\mathrm{X}) \cdot (2) = \tfrac{26}{3}$$

that is,

$$\mathrm{X}^2 = \tfrac{13}{3}$$

Therefore

$$\mathrm{X} = \pm\tfrac{1}{3}\sqrt{39}$$

We reject $-\frac{1}{3}\sqrt{39}$ since it is not in the interval [1, 3], and we have

$$\int_1^3 f(x)\,dx = f(\tfrac{1}{3}\sqrt{39})(3-1)$$

The value $f(X)$ given by Theorem 6.4.1 is called the *mean value* (or *average value*) of f on the interval $[a, b]$. It is a generalization of the arithmetic mean of a finite set of numbers. That is, if $\{f(x_1), f(x_2), \ldots, f(x_n)\}$ is a set of n numbers, then the arithmetic mean is given by

$$\frac{\sum_{i=1}^{n} f(x_i)}{n}$$

To generalize this definition, consider a regular partition of the closed interval $[a, b]$, which is divided into n subintervals of equal length $\Delta x = (b-a)/n$. Let ξ_i be any point in the ith subinterval. Form the sum:

$$\frac{\sum_{i=1}^{n} f(\xi_i)}{n} \tag{4}$$

The quotient (4) corresponds to the arithmetic mean of n numbers. Because $\Delta x = (b-a)/n$, we have

$$n = \frac{b-a}{\Delta x} \tag{5}$$

Substituting from (5) into (4) we obtain

$$\frac{\sum_{i=1}^{n} f(\xi_i)}{\dfrac{(b-a)}{\Delta x}} \quad \text{or, equivalently,} \quad \frac{\sum_{i=1}^{n} f(\xi_i)\,\Delta x}{b-a}$$

Taking the limit as $n \to +\infty$ (or $\Delta x \to 0$) we have, if the limit exists,

$$\lim_{n\to+\infty} \frac{\sum_{i=1}^{n} f(\xi_i)\,\Delta x}{b-a} = \frac{\int_a^b f(x)\,dx}{b-a}$$

This leads to the following definition.

6.4.2 Definition If the function f is integrable on the closed interval $[a, b]$, the **average value** of f on $[a, b]$ is

$$\frac{\int_a^b f(x)\,dx}{b-a}$$

Example 2 (a) If $f(x) = x^2$, find the average value of f on the interval $[1, 3]$. (b) Interpret the result of part (a) geometrically.

Solution

(a) In Example 2, Sec. 6.2, we obtained

$$\int_1^3 x^2\,dx = \tfrac{26}{3}$$

So if A.V. is the average value of f on $[1, 3]$, we have

$$\text{A.V.} = \frac{\frac{26}{3}}{3-1}$$

$$= \tfrac{13}{3}$$

(b) In Example 1 we found for this function

$$f(\tfrac{1}{3}\sqrt{39}) = \tfrac{13}{3}$$

Therefore the average value of f occurs at $\frac{1}{3}\sqrt{39}$. In Fig. 6.4.2 there is a sketch of the graph of f on $[1, 3]$ and the line segment from the point $E(\frac{1}{3}\sqrt{39}, 0)$ on the x axis to the point $F(\frac{1}{3}\sqrt{39}, \frac{13}{3})$ on the graph of f. The area of rectangle $AGHB$ having height $\frac{13}{3}$ and width 2 is equal to the area of the region $ACDB$. Consequently, the area of the shaded region CGF is equal to the area of the shaded region FDH.

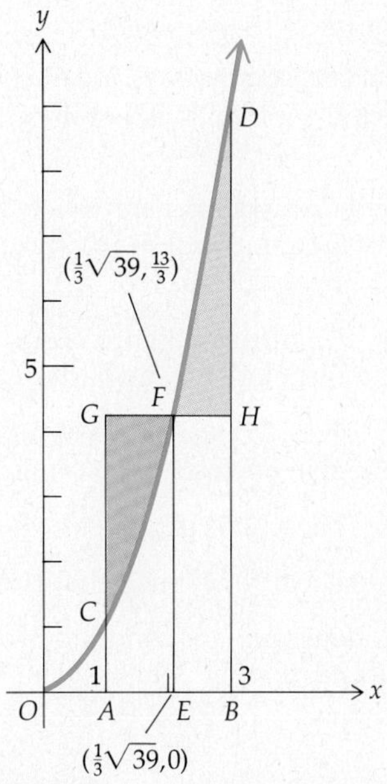

FIGURE 6.4.2

An important application of the average value of a function occurs in physics and engineering in connection with the concept of center of mass. This is discussed in Chap. 11.

In economics, Definition 6.4.2 can be used to find an average total cost or an average total revenue.

● ILLUSTRATION 2 If the total cost function C is given by $C(x) = x^2$, where $C(x)$ thousands of dollars is the total cost of producing $100x$ units of a certain commodity, then the number of thousands of dollars in the average total cost, when the number of units produced takes on all values from 100 to 300, is given by the average value of C on $[1, 3]$. From Example 2 this is $\frac{13}{3} = 4.333$. Hence the average total cost is \$4333. ●

EXERCISES 6.4

In Exercises 1 through 8, find the value of X satisfying the mean-value theorem for integrals. For the value of the definite integral use the result of the indicated exercise in Exercises 6.2. Draw a figure illustrating the application of the theorem.

1. $\int_0^2 x^2\,dx$; Exercise 11

2. $\int_2^4 x^2\,dx$; Exercise 12

3. $\int_1^2 x^3\,dx$; Exercise 13

4. $\int_0^5 (x^3 - 1)\,dx$; Exercise 14

5. $\int_1^4 (x^2 + 4x + 5)\,dx$; Exercise 15

6. $\int_0^4 (x^2 + x - 6)\,dx$; Exercise 16

7. $\int_{-2}^2 (x^3 + 1)\,dx$; Exercise 17

8. $\int_{-2}^1 x^4\,dx$; Exercise 18

In Exercises 9 through 12, use the mean-value theorem for integrals to prove the inequality.

9. $\int_0^2 \frac{1}{x^2+4}\,dx \le \frac{1}{2}$ **10.** $\int_{-3}^3 \frac{1}{x^2+6}\,dx \le 1$ **11.** $\int_{-\pi/6}^{\pi/6} \cos x^2\,dx \le \frac{\pi}{3}$ **12.** $\int_0^{\pi} \sin \sqrt{x}\,dx \le \pi$

In Exercises 13 through 16, use the mean-value theorem for integrals and Theorem 6.3.8 to prove the inequality.

13. $0 \le \int_2^5 \frac{1}{x^3+1}\,dx \le \frac{1}{3}$ **14.** $0 \le \int_5^9 \frac{1}{\sqrt{x-1}}\,dx \le 2$ **15.** $0 \le \int_0^2 \sin \frac{1}{2}\pi x\,dx \le 2$ **16.** $0 \le \int_{-1/2}^{1/2} \cos \pi x \le 1$

17. Given that $\int_{-1}^2 x\,dx = \frac{3}{2}$, find the average value of the identity function on the interval $[-1, 2]$. Also find the value of x at which the average value occurs. Give a geometric interpretation of the results.

18. Find the average value of the function f defined by $f(x) = x^2$ on the interval $[-1, 2]$ given that $\int_{-1}^2 x^2\,dx = 3$. Also find the value of x at which the average value occurs. Give a geometric interpretation of the results.

19. Given that $\int_0^{\pi} \sin x\,dx = 2$, find the average value of the sine function on the interval $[0, \pi]$.

20. Suppose that $C(x)$ hundreds of dollars is the total cost of producing $10x$ units of a certain commodity, and $C(x) = x^2$. Find the average total cost when the number of units produced takes on all values from 0 to 20. Use the result of Exercise 11 in Exercises 6.2 for the value of the definite integral.

21. Suppose a ball is dropped from rest and after t sec its velocity is v ft/sec. Neglecting air resistance, express v in terms of t as $v = f(t)$, and find the average value of f on $[0, 2]$. (*Hint:* Find the value of the definite integral by interpreting it as the measure of the area of a region enclosed by a triangle.)

22. Find the average value of the function f defined by $f(x) = \sqrt{49 - x^2}$ on the interval $[0, 7]$. Draw a figure. (*Hint:* Find the value of the definite integral by interpreting it as the measure of the area of a region enclosed by a quarter-circle.)

23. Find the average value of the function f defined by $f(x) = \sqrt{16 - x^2}$ on the interval $[-4, 4]$. Draw a figure. (*Hint:* Find the value of the definite integral by interpreting it as the measure of the area of a region enclosed by a semicircle.)

24. Suppose that f is integrable on $[-4, 7]$. If the average value of f on the interval $[-4, 7]$ is $\frac{17}{4}$, find $\int_{-4}^7 f(x)\,dx$.

25. If f is continuous on $[a, b]$ and $\int_a^b f(x)\,dx = 0$, prove that there is at least one number c in $[a, b]$ such that $f(c) = 0$.

26. The following theorem is a generalization of Theorem 6.4.1: If f and g are two functions that are continuous on the closed interval $[a, b]$ and $g(x) > 0$ for all x in the open interval (a, b), then there exists a number X in $[a, b]$such that

$$\int_a^b f(x)g(x)\,dx = f(X)\int_a^b g(x)\,dx$$

Prove this theorem by a method similar to that for Theorem 6.4.1: Obtain the inequality $m \le f(x) \le M$ and then conclude that $mg(x) \le f(x)g(x) \le Mg(x)$; apply Theorem 6.3.8 and proceed as in the proof of Theorem 6.4.1.

27. Show that when $g(x) = 1$, the theorem of Exercise 26 becomes Theorem 6.4.1.

In Exercises 28 through 32, use the theorem of Exercise 26 to prove the inequality.

28. $\int_0^4 \frac{x\,dx}{x^3+2} < \int_0^4 x\,dx$ **29.** $\int_{-1}^1 \frac{x^2\,dx}{\sqrt{x^2+4}} < \int_{-1}^1 x^2\,dx$

30. $\int_0^{\pi} e^x \sin x\,dx \le \int_0^{\pi} e^x\,dx$ **31.** $\int_{-1/2}^{1/2} \sin^2 \pi x \cos \pi x\,dx \le \int_{-1/2}^{1/2} \cos \pi x\,dx$

32. $\int_0^1 \frac{x \cos x}{x^2+1}\,dx \le \int_0^1 x\,dx$

6.5 THE FUNDAMENTAL THEOREM OF THE CALCULUS

Historically the basic concepts of the definite integral were used by the ancient Greeks, principally Archimedes (287–212 B.C.), more than 2000 years ago, which was many years before the differential calculus was discovered.

In the seventeenth century, almost simultaneously but working independently, Newton and Leibniz showed how the calculus could be used to find the area of a region bounded by a curve or a set of curves, by evaluating a definite integral by antidifferentiation. The procedure involves what is known as the *fundamental theorem of the calculus*. Before we state and prove this important theorem, we discuss definite integrals having a variable upper limit and we give a preliminary theorem.

Let f be a function continuous on the closed interval $[a, b]$. Then the value of the definite integral $\int_a^b f(x)\,dx$ depends only on the function f and the numbers a and b, and not on the symbol x, which is used here as the independent variable. In Example 2, Sec. 6.2, we found the value of $\int_1^3 x^2\,dx$ to be $8\frac{2}{3}$. Any other symbol instead of x could have been used; for example,

$$\int_1^3 t^2\,dt = \int_1^3 u^2\,du = \int_1^3 r^2\,dr = 8\tfrac{2}{3}$$

If f is continuous on the closed interval $[a, b]$, then by Theorem 6.2.3 the definite integral $\int_a^b f(t)\,dt$ exists. We previously stated that if the definite integral exists, it is a unique number. If x is a number in $[a, b]$, then f is continuous on $[a, x]$ because it is continuous on $[a, b]$. Consequently, $\int_a^x f(t)\,dt$ exists and is a unique number whose value depends on x. Therefore $\int_a^x f(t)\,dt$ defines a function F having as its domain all numbers in the closed interval $[a, b]$ and whose function value at any number x in $[a, b]$ is given by

$$F(x) = \int_a^x f(t)\,dt \tag{1}$$

As a notational observation, if the limits of the definite integral are variables, different symbols are used for these limits and for the independent variable in the integrand. Hence, in (1), because x is the upper limit, we use the letter t as the independent variable in the integrand.

If, in Eq. (1), $f(t) \geq 0$ for all values of t in $[a, b]$, then the function value $F(x)$ can be interpreted geometrically as the measure of the area of the region bounded by the curve whose equation is $y = f(t)$, the t axis, and the lines $t = a$ and $t = x$. (See Fig. 6.5.1.) Note that $F(a) = \int_a^a f(t)\,dt$, which by Definition 6.2.6 equals 0.

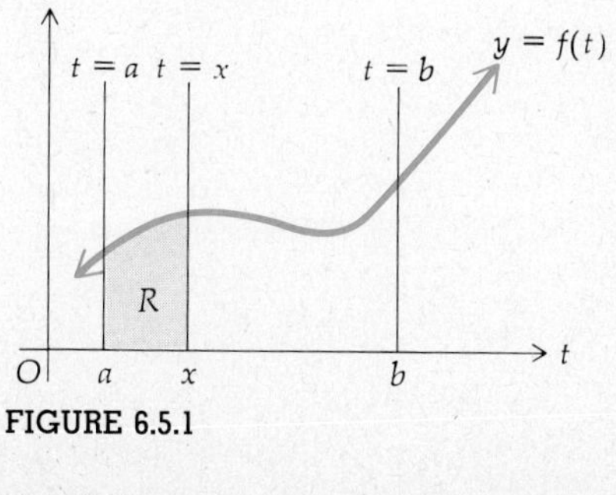

FIGURE 6.5.1

We now state and prove an important theorem giving the derivative of a function F defined as a definite integral having a variable upper limit. This theorem is sometimes called "the first fundamental theorem of the calculus," and in that case our Theorem 6.5.2 is called "the second fundamental theorem of the calculus."

6.5.1 Theorem Let the function f be continuous on the closed interval $[a, b]$ and let x be any number in $[a, b]$. If F is the function defined by

$$F(x) = \int_a^x f(t)\,dt$$

then

$$F'(x) = f(x) \tag{2}$$

(If $x = a$, the derivative in (2) may be a derivative from the right, and if $x = b$, the derivative in (2) may be a derivative from the left.)

Proof Consider two numbers x_1 and $(x_1 + \Delta x)$ in $[a, b]$. Then

$$F(x_1) = \int_a^{x_1} f(t)\,dt$$

and

$$F(x_1 + \Delta x) = \int_a^{x_1+\Delta x} f(t)\,dt$$

so that

$$F(x_1 + \Delta x) - F(x_1) = \int_a^{x_1+\Delta x} f(t)\,dt - \int_a^{x_1} f(t)\,dt \tag{3}$$

By Theorem 6.3.7,

$$\int_a^{x_1} f(t)\,dt + \int_{x_1}^{x_1+\Delta x} f(t)\,dt = \int_a^{x_1+\Delta x} f(t)\,dt$$

or, equivalently,

$$\int_a^{x_1+\Delta x} f(t)\,dt - \int_a^{x_1} f(t)\,dt = \int_{x_1}^{x_1+\Delta x} f(t)\,dt \tag{4}$$

Substituting from (4) into (3) we get

$$F(x_1 + \Delta x) - F(x_1) = \int_{x_1}^{x_1+\Delta x} f(t)\,dt \tag{5}$$

By the mean-value theorem for integrals (6.4.1) there is some number X in the closed interval bounded by x_1 and $(x_1 + \Delta x)$ such that

$$\int_{x_1}^{x_1+\Delta x} f(t)\,dt = f(\mathrm{X})\,\Delta x \tag{6}$$

From (5) and (6) we obtain

$$F(x_1 + \Delta x) - F(x_1) = f(\mathrm{X})\,\Delta x$$

or, if we divide by Δx,

$$\frac{F(x_1 + \Delta x) - F(x_1)}{\Delta x} = f(\mathrm{X})$$

Taking the limit as Δx approaches zero we have

$$\lim_{\Delta x \to 0} \frac{F(x_1 + \Delta x) - F(x_1)}{\Delta x} = \lim_{\Delta x \to 0} f(X) \tag{7}$$

The left side of (7) is $F'(x_1)$. To determine $\lim_{\Delta x \to 0} f(X)$, recall that X is in the closed interval bounded by x_1 and $x_1 + \Delta x$, and because

$$\lim_{\Delta x \to 0} x_1 = x_1 \quad \text{and} \quad \lim_{\Delta x \to 0} (x_1 + \Delta x) = x_1$$

it follows from the squeeze theorem (2.9.1) that $\lim_{\Delta x \to 0} X = x_1$. Because f is continuous at x_1, we have $\lim_{\Delta x \to 0} f(X) = \lim_{X \to x_1} f(X) = f(x_1)$; thus, from (7) we get

$$F'(x_1) = f(x_1) \tag{8}$$

If the function f is not defined for values of x less than a but is continuous from the right at a, then in the above argument, if $x_1 = a$ in (7), Δx must approach 0 from the right. Hence the left side of (8) will be $F'_+(x_1)$. Similarly, if f is not defined for values of x greater than b but is continuous from the left at b, then if $x_1 = b$ in (7), Δx must approach 0 from the left. Hence we have $F'_-(x_1)$ on the left side of (8).

Because x_1 is any number in $[a, b]$, Eq. (8) states what we wished to prove. ■

Theorem 6.5.1 states that the definite integral $\int_a^x f(t)\,dt$, with variable upper limit x, is an antiderivative of f.

Example 1 Apply Theorem 6.5.1 to find $D_x \int_1^x \sqrt{2 + t}\,dt$.

Solution With the notation of the statement of Theorem 6.5.1,

$$F(x) = \int_1^x \sqrt{2 + t}\,dt \qquad f(t) = \sqrt{2 + t}$$

and $a = 1$. Then from the theorem,

$$F'(x) = f(x)$$

Thus

$$D_x \int_1^x \sqrt{2 + t}\,dt = \sqrt{2 + x}$$

We now use Theorem 6.5.1 to prove the fundamental theorem of the calculus.

6.5.2 Theorem
Fundamental Theorem of the Calculus

Let the function f be continuous on the closed interval $[a, b]$ and let g be a function such that

$$g'(x) = f(x) \tag{9}$$

for all x in $[a, b]$. Then

$$\int_a^b f(t)\, dt = g(b) - g(a)$$

(If $x = a$, the derivative in (9) may be a derivative from the right, and if $x = b$, the derivative in (9) may be a derivative from the left.)

Proof If f is continuous at all numbers in $[a, b]$, we know from Theorem 6.5.1 that the definite integral $\int_a^x f(t)\, dt$, with variable upper limit x, defines a function F whose derivative on $[a, b]$, is f. Because by hypothesis $g'(x) = f(x)$, it follows from Theorem 5.2.2 that

$$g(x) = \int_a^x f(t)\, dt + k \tag{10}$$

where k is some constant.

Letting $x = b$ and $x = a$, successively, in (10) we get

$$g(b) = \int_a^b f(t)\, dt + k \tag{11}$$

and

$$g(a) = \int_a^a f(t)\, dt + k \tag{12}$$

From (11) and (12),

$$g(b) - g(a) = \int_a^b f(t)\, dt - \int_a^a f(t)\, dt$$

But, by Definition 6.2.6, $\int_a^a f(t)\, dt = 0$; so

$$g(b) - g(a) = \int_a^b f(t)\, dt$$

which is what we wished to prove.

If f is not defined for values of x greater than b but is continuous from the left at b, the derivative in (9) is a derivative from the left, and we have $g'_-(b) = F'_-(b)$, from which (11) follows. Similarly, if f is not defined for values of x less than a but is continuous from the right at a, then the derivative in (9) is a derivative from the right, and we have $g'_+(a) = F'_+(a)$, from which (12) follows. ■

We are now in a position to find the exact value of a definite integral by applying Theorem 6.5.2. In applying the theorem we denote

$$[g(b) - g(a)] \quad \text{by} \quad g(x)\Big]_a^b$$

● ILLUSTRATION 1 We apply the fundamental theorem of the calculus to evaluate

$$\int_1^3 x^2\, dx$$

Here $f(x) = x^2$. An antiderivative of x^2 is $\frac{1}{3}x^3$. From this we choose

$$g(x) = \frac{x^3}{3}$$

Therefore, from Theorem 6.5.2,

$$\int_1^3 x^2\, dx = \frac{x^3}{3}\Big]_1^3 = 9 - \tfrac{1}{3}$$
$$= 8\tfrac{2}{3}$$

Compare this result with that of Example 2, Sec. 6.2.

Because of the connection between definite integrals and antiderivatives, we used the integral sign $\int$ for the notation $\int f(x)\, dx$ for an antiderivative. We now dispense with the terminology of antiderivatives and antidifferentiation and begin to call $\int f(x)\, dx$ the **indefinite integral** of "f of x, dx." The process of evaluating an indefinite integral or a definite integral is called **integration.**

The difference between an indefinite integral and a definite integral should be emphasized. The indefinite integral $\int f(x)\, dx$ is defined as a function g such that its derivative $D_x[g(x)] = f(x)$. However, the definite integral $\int_a^b f(x)\, dx$ is a number whose value depends on the function f and the numbers a and b, and it is defined as the limit of a Riemann sum. The definition of the definite integral makes no reference to differentiation.

The general indefinite integral involves an arbitrary constant; for instance,

$$\int x^2\, dx = \frac{x^3}{3} + C$$

This arbitrary constant C is called a **constant of integration**. In applying the fundamental theorem to evaluate a definite integral, we do not need to include the arbitrary constant C in the expression for $g(x)$ because the fundamental theorem permits us to select *any* antiderivative, including the one for which $C = 0$.

Example 2 Evaluate $\displaystyle\int_{1/2}^4 (x^3 - 6x^2 + 9x + 1)\, dx$.

Solution

$$\int_{1/2}^4 (x^3 - 6x^2 + 9x + 1)\, dx = \int_{1/2}^4 x^3\, dx - 6\int_{1/2}^4 x^2\, dx + 9\int_{1/2}^4 x\, dx + \int_{1/2}^4 dx$$
$$= \frac{x^4}{4} - 6\cdot\frac{x^3}{3} + 9\cdot\frac{x^2}{2} + x\Big]_{1/2}^4$$
$$= (64 - 128 + 72 + 4) - (\tfrac{1}{64} - \tfrac{1}{4} + \tfrac{9}{8} + \tfrac{1}{2})$$
$$= \tfrac{679}{64}$$

Example 3 Evaluate $\int_{-1}^{1} (x^{4/3} + 4x^{1/3})\,dx$.

Solution

$$\int_{-1}^{1} (x^{4/3} + 4x^{1/3})\,dx = \tfrac{3}{7}x^{7/3} + 4 \cdot \tfrac{3}{4}x^{4/3}\Big]_{-1}^{1}$$

$$= \tfrac{3}{7} + 3 - (-\tfrac{3}{7} + 3)$$

$$= \tfrac{6}{7}$$

Example 4 Evaluate $\int_{0}^{2} 2x^2\sqrt{x^3 + 1}\,dx$.

Solution

$$\int_{0}^{2} 2x^2\sqrt{x^3 + 1}\,dx = \frac{2}{3}\int_{0}^{2} \sqrt{x^3 + 1}\,(3x^2\,dx)$$

$$= \frac{2}{3} \cdot \frac{(x^3 + 1)^{3/2}}{\frac{3}{2}}\Bigg]_{0}^{2}$$

$$= \tfrac{4}{9}(8 + 1)^{3/2} - \tfrac{4}{9}(0 + 1)^{3/2}$$

$$= \tfrac{4}{9}(27 - 1)$$

$$= \tfrac{104}{9}$$

Example 5 Evaluate $\int_{0}^{3} x\sqrt{1 + x}\,dx$.

Solution To evaluate the indefinite integral $\int x\sqrt{1 + x}\,dx$ we let

$$u = \sqrt{1 + x} \qquad u^2 = 1 + x \qquad x = u^2 - 1 \qquad dx = 2u\,du$$

Substituting we have

$$\int x\sqrt{1 + x}\,dx = \int (u^2 - 1)u(2u\,du)$$

$$= 2\int (u^4 - u^2)\,du$$

$$= \tfrac{2}{5}u^5 - \tfrac{2}{3}u^3 + C$$

$$= \tfrac{2}{5}(1 + x)^{5/2} - \tfrac{2}{3}(1 + x)^{3/2} + C$$

Therefore the definite integral

$$\int_{0}^{3} x\sqrt{1 + x}\,dx = \tfrac{2}{5}(1 + x)^{5/2} - \tfrac{2}{3}(1 + x)^{3/2}\Big]_{0}^{3}$$

$$= \tfrac{2}{5}(4)^{5/2} - \tfrac{2}{3}(4)^{3/2} - \tfrac{2}{5}(1)^{5/2} + \tfrac{2}{3}(1)^{3/2}$$

$$= \tfrac{64}{5} - \tfrac{16}{3} - \tfrac{2}{5} + \tfrac{2}{3}$$

$$= \tfrac{116}{15}$$

Another method for evaluating the definite integral in Example 5 involves changing the limits of the definite integral to values of u. The procedure is shown in the following illustration. Often this second method is shorter and its justification follows immediately from Theorems 5.2.9 and 6.5.2.

● ILLUSTRATION 2 Because $u = \sqrt{1+x}$, we see that when $x = 0$, $u = 1$; and when $x = 3$, $u = 2$. Thus we have

$$\int_0^3 x\sqrt{1+x}\,dx = 2\int_1^2 (u^4 - u^2)\,du$$

$$= \tfrac{2}{5}u^5 - \tfrac{2}{3}u^3\Big]_1^2$$

$$= \tfrac{64}{5} - \tfrac{16}{3} - \tfrac{2}{5} + \tfrac{2}{3}$$

$$= \tfrac{116}{15}$$

●

Example 6 Evaluate $\displaystyle\int_0^{\pi/2} \cos^3 x\,dx.$

Solution

$$\int_0^{\pi/2} \cos^3 x\,dx = \int_0^{\pi/2} \cos^2 x\,(\cos x\,dx)$$

$$= \int_0^{\pi/2} (1 - \sin^2 x)(\cos x\,dx)$$

Let $u = \sin x$; then $du = \cos x\,dx$. When $x = 0$, $u = 0$; when $x = \frac{1}{2}\pi$, $u = 1$. Therefore

$$\int_0^{\pi/2} \cos^3 x\,dx = \int_0^1 (1 - u^2)\,du$$

$$= u - \tfrac{1}{3}u^3\Big]_0^1$$

$$= 1 - \tfrac{1}{3}$$

$$= \tfrac{2}{3}$$

Example 7 Evaluate $\displaystyle\int_{-3}^{4} |x+2|\,dx.$

Solution If $f(x) = |x+2|$, instead of finding an antiderivative of f directly, write $f(x)$ as

$$f(x) = \begin{cases} x + 2 & \text{if } x \geq -2 \\ -x - 2 & \text{if } x \leq -2 \end{cases}$$

From Theorem 6.3.7,

$$\int_{-3}^{4} |x + 2|\, dx = \int_{-3}^{-2} (-x - 2)\, dx + \int_{-2}^{4} (x + 2)\, dx$$

$$= \left[-\frac{x^2}{2} - 2x\right]_{-3}^{-2} + \left[\frac{x^2}{2} + 2x\right]_{-2}^{4}$$

$$= \tfrac{1}{2} + 18$$

$$= \tfrac{37}{2}$$

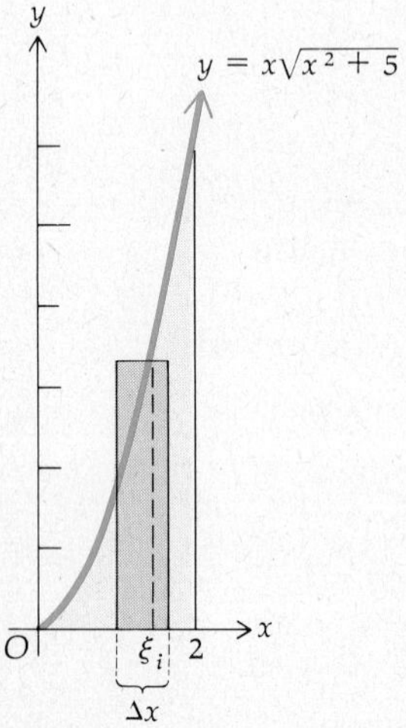

FIGURE 6.5.2

Example 8 Find the area of the region in the first quadrant bounded by the curve whose equation is $y = x\sqrt{x^2 + 5}$, the x axis, and the line $x = 2$. Make a sketch.

Solution See Fig. 6.5.2. The region is shown together with one of the rectangular elements of area.

We take a regular partition of the interval [0, 2]. The width of each rectangle is Δx units, and the altitude of the ith rectangle is $\xi_i\sqrt{\xi_i^2 + 5}$ units, where ξ_i is any number in the ith subinterval. Therefore the measure of the area of the rectangular element is $\xi_i\sqrt{\xi_i^2 + 5}\,\Delta x$. The sum of the measures of the areas of n such rectangles is

$$\sum_{i=1}^{n} \xi_i\sqrt{\xi_i^2 + 5}\,\Delta x$$

which is a Riemann sum. The limit of this sum as Δx approaches zero (or $n \to +\infty$) gives the measure of the desired area. The limit of the Riemann sum is a definite integral, which we evaluate by the fundamental theorem of the calculus.

Let A be the number of square units in the area of the region. Then

$$A = \lim_{\Delta x \to 0} \sum_{i=1}^{n} \xi_i\sqrt{\xi_i^2 + 5}\,\Delta x$$

$$= \int_0^2 x\sqrt{x^2 + 5}\, dx$$

$$= \frac{1}{2}\int_0^2 \sqrt{x^2 + 5}\,(2x\, dx)$$

$$= \tfrac{1}{2} \cdot \tfrac{2}{3}(x^2 + 5)^{3/2}\Big]_0^2$$

$$= \tfrac{1}{3}[(9)^{3/2} - (5)^{3/2}]$$

$$= \tfrac{1}{3}(27 - 5\sqrt{5})$$

Hence the area is $\frac{1}{3}(27 - 5\sqrt{5})$ square units.

Example 9 Given $f(x) = \sin^2 x$, find the average value of f on the interval $[-\frac{1}{2}\pi, \frac{1}{2}\pi]$.

Solution Let A.V. be the average value of f on $[-\frac{1}{2}\pi, \frac{1}{2}\pi]$. From Definition 6.4.2,

$$\begin{aligned}
\text{A.V.} &= \frac{1}{\frac{1}{2}\pi - (-\frac{1}{2}\pi)} \int_{-\pi/2}^{\pi/2} \sin^2 x\, dx \\
&= \frac{1}{\pi} \int_{-\pi/2}^{\pi/2} \frac{1 - \cos 2x}{2}\, dx \\
&= \frac{1}{\pi}\left[\frac{1}{2}\int_{-\pi/2}^{\pi/2} dx - \frac{1}{4}\int_{-\pi/2}^{\pi/2} \cos 2x(2\, dx)\right] \\
&= \frac{1}{\pi}\left[\tfrac{1}{2}x - \tfrac{1}{4}\sin 2x\right]_{-\pi/2}^{\pi/2} \\
&= \frac{1}{\pi}\left[\tfrac{1}{4}\pi - \tfrac{1}{4}\sin\pi - \tfrac{1}{2}(-\tfrac{1}{2}\pi) + \tfrac{1}{4}\sin(-\pi)\right] \\
&= \frac{1}{\pi}\left[\tfrac{1}{4}\pi - 0 + \tfrac{1}{4}\pi + 0\right] \\
&= \tfrac{1}{2}
\end{aligned}$$

EXERCISES 6.5

In Exercises 1 through 36, evaluate the definite integral by using the fundamental theorem of the calculus.

1. $\int_0^3 (3x^2 - 4x + 1)\, dx$

2. $\int_0^4 (x^3 - x^2 + 1)\, dx$

3. $\int_3^6 (x^2 - 2x)\, dx$

4. $\int_{-1}^3 (3x^2 + 5x - 1)\, dx$

5. $\int_1^2 \frac{x^2 + 1}{x^2}\, dx$

6. $\int_{-3}^5 (y^3 - 4y)\, dy$

7. $\int_0^1 \frac{z}{(z^2 + 1)^3}\, dz$

8. $\int_1^4 \sqrt{x}(2 + x)\, dx$

9. $\int_1^{10} \sqrt{5x - 1}\, dx$

10. $\int_0^{\sqrt{5}} t\sqrt{t^2 + 1}\, dt$

11. $\int_{-2}^0 3w\sqrt{4 - w^2}\, dw$

12. $\int_{-1}^3 \frac{dy}{(y + 2)^3}$

13. $\int_0^{\pi/2} \sin 2x\, dx$

14. $\int_0^{\pi} \cos^2 x\, dx$

15. $\int_1^2 t^2\sqrt{t^3 + 1}\, dt$

16. $\int_1^3 \frac{x\, dx}{(3x^2 - 1)^3}$

17. $\int_0^1 \frac{(y^2 + 2y)\, dy}{\sqrt[3]{y^3 + 3y^2 + 4}}$

18. $\int_2^4 \frac{w^4 - w}{w^3}\, dw$

19. $\int_0^{15} \frac{w\, dw}{(1 + w)^{3/4}}$

20. $\int_4^5 x^2\sqrt{x - 4}\, dx$

21. $\int_{-2}^5 |x - 3|\, dx$

22. $\int_{-4}^4 |x - 2|\, dx$

23. $\int_{-1}^1 \sqrt{|x| - x}\, dx$

24. $\int_{-3}^3 \sqrt{3 + |x|}\, dx$

25. $\int_0^3 (x + 2)\sqrt{x + 1}\, dx$

26. $\int_{-2}^1 (x + 1)\sqrt{x + 3}\, dx$

27. $\int_0^1 \frac{x^3 + 1}{x + 1}\, dx$ (*Hint:* Divide the numerator by the denominator.)

28. $\int_1^4 \frac{x^5 - x}{3x^3}\, dx$

29. $\displaystyle\int_1^{64}\left(\sqrt{t}-\frac{1}{\sqrt{t}}+\sqrt[3]{t}\right)dt$ **30.** $\displaystyle\int_0^1 \sqrt{x}\sqrt{1+x\sqrt{x}}\,dx$

31. $\displaystyle\int_1^2 \frac{x^3+2x^2+x+2}{(x+1)^2}dx$ **32.** $\displaystyle\int_{-3}^2 \frac{3x^3-24x^2+48x+5}{x^2-8x+16}dx$

(*Hint* for Exercises 31 and 32: Divide the numerator by the denominator.)

33. $\displaystyle\int_0^1 \sin^2 \pi x \cos^2 \pi x\,dx$ **34.** $\displaystyle\int_0^1 \sin^3 \tfrac{1}{2}\pi t\,dt$ **35.** $\displaystyle\int_0^{\pi/3} \sin^3 t\cos^2 t\,dt$ **36.** $\displaystyle\int_0^{\pi/6} \sin 2x\cos 4x\,dx$

In Exercises 37 through 42, use Theorem 6.5.1 to find the indicated derivative.

37. $\displaystyle D_x\int_0^x \sqrt{4+t^6}\,dt$ **38.** $\displaystyle D_x\int_2^x \frac{dt}{t^4+4}$ **39.** $\displaystyle D_x\int_x^3 \sqrt{\sin t}\,dt$

40. $\displaystyle D_x\int_x^5 \sqrt{1+t^4}\,dt$ **41.** $\displaystyle D_x\int_{-x}^x \frac{dt}{3+t^2}$ **42.** $\displaystyle D_x\int_1^{x^2} \sqrt{1+t^4}\,dt$

In Exercises 43 through 48, find the area of the region bounded by the given curve and lines. Draw a figure showing the region and a rectangular element of area. Express the measure of the area as the limit of a Riemann sum, and then with definite integral notation. Evaluate the definite integral by the fundamental theorem of the calculus.

43. $y=4-x^2$; x axis **44.** $y=x^2-2x+3$; x axis; $x=-2$; $x=1$ **45.** $y=4x-x^2$; x axis; $x=1$; $x=3$

46. $y=6-x-x^2$; x axis **47.** $y=\sqrt{x+1}$; x axis; y axis; $x=8$ **48.** $y=\dfrac{1}{x^2}-x$; x axis; $x=2$; $x=3$

49. Find the area of the region bounded by one arch of the sine curve.

In Exercises 50 through 53, find the average value of the function f on the given interval [a, b]. In Exercises 50 and 51, find the value of x at which the average value of f occurs and make a sketch.

50. $f(x)=9-x^2$; $[a,b]=[0,3]$ **51.** $f(x)=8x-x^2$; $[a,b]=[0,4]$

52. $f(x)=3x\sqrt{x^2-16}$; $[a,b]=[4,5]$ **53.** $f(x)=x^2\sqrt{x-3}$; $[a,b]=[7,12]$

54. The demand equation for a certain commodity is $x^2+p^2=100$, where p dollars is the price of one unit when $100x$ units are demanded. Find the average total revenue when the number of units demanded takes on all values from 600 to 800.

55. In an electric circuit, suppose that E volts is the electromotive force at t sec and $E=2\sin 3t$. Find the average value of E from $t=0$ to $t=\frac{1}{3}\pi$.

56. For the electric circuit of Exercise 55, find the square root of the average value of E^2 from $t=0$ to $t=\frac{1}{3}\pi$.

57. A ball is dropped from rest and after t sec its velocity is v ft/sec. Neglecting air resistance, show that the average velocity during the first $\frac{1}{2}T$ sec is one-third of the average velocity during the next $\frac{1}{2}T$ sec.

58. A stone is thrown downward with an initial velocity of v_0 ft/sec. Neglect air resistance. (a) Show that if v ft/sec is the velocity of the stone after falling s ft, then $v=\sqrt{v_0^2+2gs}$. (b) Find the average velocity during the first 100 ft of fall if the initial velocity is 60 ft/sec. (Take $g=32$ and downward as the positive direction.)

59. Find $\displaystyle\int_4^{16}\left[D_x\int_5^x (2\sqrt{t}-1)\,dt\right]dx$.

60. Let f be a function whose derivative f' is continuous on $[a,b]$. Find the average value of the slope of the tangent line of the graph of f on $[a,b]$ and give a geometric interpretation of the result.

In Exercises 61 through 63, m and n are any integers except zero; show that the given formula is true.

61. $\int_{-1}^{1} \cos n\pi x \cos m\pi x \, dx = \begin{cases} 0 & \text{if } m \neq n \\ 1 & \text{if } m = n \end{cases}$

62. $\int_{-1}^{1} \cos n\pi x \sin m\pi x \, dx = 0$

63. $\int_{-1}^{1} \sin n\pi x \sin m\pi x \, dx = \begin{cases} 0 & \text{if } m \neq n \\ 1 & \text{if } m = n \end{cases}$

64. Evaluate $\int_{0}^{\pi/2} |\cos x - \sin x| \, dx$.

6.6 APPROXIMATE INTEGRATION

In evaluating a definite integral by the fundamental theorem of the calculus, it is necessary to obtain an indefinite integral. In Sec. 6.5 we found the indefinite integral by techniques presented in Secs. 5.2 and 5.3. Even though we shall learn in Chapter 10 additional methods for obtaining indefinite integrals, there will still remain many functions for which a definite integral cannot be evaluated exactly by using elementary functions. Examples of such definite integrals are $\int_0^1 \sqrt{1 + x^4}\, dx$ and $\int_0^{\pi/2} \sin x^2 \, dx$. Other examples are given in Exercises 29 through 34. In this section we learn two methods for computing an approximate value of a definite integral of a function continuous on a closed interval. These devices can often give fairly good accuracy and can be used for evaluating a definite integral by numerical calculations. The first method is known as the *trapezoidal rule.*

Let f be a function continuous on a closed interval $[a, b]$. The definite integral $\int_a^b f(x)\, dx$ is the limit of a Riemann sum; that is,

$$\int_a^b f(x)\, dx = \lim_{\|\Delta\| \to 0} \sum_{i=1}^{n} f(\xi_i)\, \Delta_i x$$

The geometric interpretation of the Riemann sum

$$\sum_{i=1}^{n} f(\xi_i)\, \Delta_i x$$

is that it is equal to the sum of the measures of the areas of the rectangles lying above the x axis plus the negative of the measures of the areas of the rectangles lying below the x axis (see Fig. 6.2.3).

To approximate the measure of the area of a region we shall use trapezoids instead of rectangles. Let us also use regular partitions and function values at equally spaced points.

Thus for the definite integral $\int_a^b f(x)\, dx$ we divide the interval $[a, b]$ into n subintervals, each of length $\Delta x = (b - a)/n$. This gives the following $(n + 1)$ points: $x_0 = a, x_1 = a + \Delta x, x_2 = a + 2\,\Delta x, \ldots, x_i = a + i\,\Delta x, \ldots, x_{n-1} = a + (n - 1)\,\Delta x, x_n = b$. Then the definite integral $\int_a^b f(x)\, dx$ may be expressed as the sum of n definite integrals as follows:

$$\int_a^b f(x)\,dx = \int_a^{x_1} f(x)\,dx + \int_{x_1}^{x_2} f(x)\,dx + \cdots + \int_{x_{i-1}}^{x_i} f(x)\,dx + \cdots + \int_{x_{n-1}}^{b} f(x)\,dx \quad (1)$$

To interpret (1) geometrically, refer to Fig. 6.6.1, in which $f(x) \geq 0$ for all x in $[a, b]$; however, (1) holds for any function that is continuous on $[a, b]$.

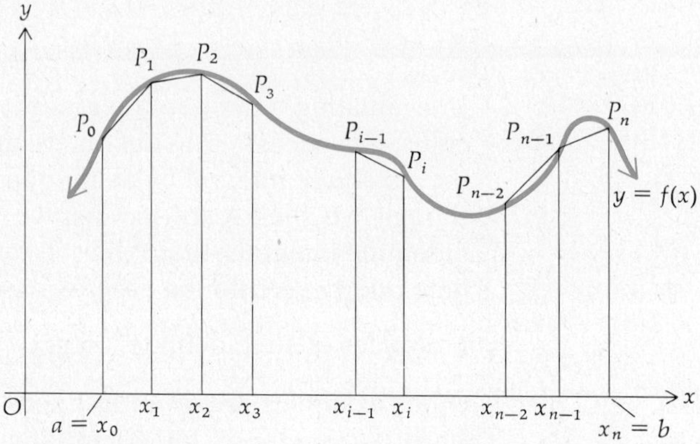

FIGURE 6.6.1

Then the integral $\int_a^{x_1} f(x)\,dx$ is the measure of the area of the region bounded by the x axis, the lines $x = a$ and $x = x_1$, and the portion of the curve from P_0 to P_1. This integral may be approximated by the measure of the area of the trapezoid formed by the lines $x = a$, $x = x_1$, P_0P_1, and the x axis. By a formula from geometry, the measure of the area of this trapezoid is

$$\tfrac{1}{2}[f(x_0) + f(x_1)]\,\Delta x$$

Similarly, the other integrals on the right side of (1) may be approximated by the measure of the area of a trapezoid. For the ith integral,

$$\int_{x_{i-1}}^{x_i} f(x)\,dx \approx \tfrac{1}{2}[f(x_{i-1}) + f(x_i)]\,\Delta x \quad (2)$$

So, using (2) for each of the integrals on the right side of (1), we have

$$\int_a^b f(x)\,dx \approx \tfrac{1}{2}[f(x_0) + f(x_1)]\,\Delta x + \tfrac{1}{2}[f(x_1) + f(x_2)]\,\Delta x + \cdots + \tfrac{1}{2}[f(x_{n-2}) + f(x_{n-1})]\,\Delta x + \tfrac{1}{2}[f(x_{n-1}) + f(x_n)]\,\Delta x$$

Thus

$$\int_a^b f(x)\,dx \approx \tfrac{1}{2}\Delta x[f(x_0) + 2f(x_1) + 2f(x_2) + \cdots + 2f(x_{n-1}) + f(x_n)] \quad (3)$$

Formula (3) is known as the *trapezoidal rule* and is stated formally in the following theorem.

6.6.1 Theorem
The Trapezoidal Rule

If the function f is continuous on the closed interval $[a, b]$ and the numbers $a = x_0, x_1, x_2, \ldots, x_n = b$ form a regular partition of $[a, b]$, then

$$\int_a^b f(x)\,dx \approx \frac{b-a}{2n}[f(x_0) + 2f(x_1) + 2f(x_2) + \cdots + 2f(x_{n-1}) + f(x_n)]$$

Example 1 Compute

$$\int_0^3 \frac{dx}{16 + x^2}$$

by using the trapezoidal rule with $n = 6$. Express the result to three decimal places.

Solution Because $[a, b] = [0, 3]$ and $n = 6$,

$$\Delta x = \frac{b-a}{n} = \frac{3}{6} = 0.5 \quad \text{and} \quad \frac{b-a}{2n} = 0.25$$

Therefore

$$\int_0^3 \frac{dx}{16 + x^2}$$

$$\approx 0.25\,[f(x_0) + 2f(x_1) + 2f(x_2) + 2f(x_3) + 2f(x_4) + 2f(x_5) + f(x_6)]$$

where $f(x) = 1/(16 + x^2)$. The computation of the sum in brackets in the above is shown in Table 6.6.1. So

$$\int_0^3 \frac{dx}{16 + x^2} \approx 0.25(0.6427)$$

$$\int_0^3 \frac{dx}{16 + x^2} \approx 0.1607$$

$$\int_0^3 \frac{dx}{16 + x^2} \approx 0.161$$

The exact value of this definite integral can be found by the use of the inverse tangent function from trigonometry. You will learn this technique in Sec. 9.5. The exact value to four decimal places is 0.1609.

TABLE 6.6.1

i	x_i	$f(x_i)$	k_i	$k_i \cdot f(x_i)$
0	0	0.0625	1	0.0625
1	0.5	0.0615	2	0.1230
2	1	0.0588	2	0.1176
3	1.5	0.0548	2	0.1096
4	2	0.0500	2	0.1000
5	2.5	0.0450	2	0.0900
6	3	0.0400	1	0.0400
		$\sum_{i=0}^{6} k_i f(x_i) = 0.6427$		

To consider the accuracy of the approximation of a definite integral by the trapezoidal rule, we prove first that as Δx approaches zero and n increases without bound, the limit of the approximation by the trapezoidal rule is the exact value of the definite integral. Let

$$T = \tfrac{1}{2}\Delta x[f(x_0) + 2f(x_1) + \cdots + 2f(x_{n-1}) + f(x_n)]$$

Then

$$T = [f(x_1) + f(x_2) + \cdots + f(x_n)]\,\Delta x + \tfrac{1}{2}[f(x_0) - f(x_n)]\,\Delta x$$

or, equivalently,

$$T = \sum_{i=1}^{n} f(x_i)\,\Delta x + \tfrac{1}{2}[f(a) - f(b)]\,\Delta x$$

Therefore, if $n \to +\infty$ and $\Delta x \to 0$,

$$\lim_{\Delta x \to 0} T = \lim_{\Delta x \to 0} \sum_{i=1}^{n} f(x_i)\,\Delta x + \lim_{\Delta x \to 0} \tfrac{1}{2}[f(a) - f(b)]\,\Delta x$$

$$= \int_a^b f(x)\,dx + 0$$

Thus we can make the difference between T and the value of the definite integral as small as we please by taking n sufficiently large (and consequently Δx sufficiently small).

The following theorem, which is proved in numerical analysis, gives a method for estimating the error obtained when using the trapezoidal rule. The error is denoted by ϵ_T.

6.6.2 Theorem Let the function f be continuous on the closed interval $[a, b]$, and f' and f'' both exist on $[a, b]$. If

$$\epsilon_T = \int_a^b f(x)\,dx - T$$

where T is the approximate value of $\int_a^b f(x)\,dx$ found by the trapezoidal rule, then there is some number η in $[a, b]$ such that

$$\epsilon_T = -\tfrac{1}{12}(b - a)f''(\eta)(\Delta x)^2 \qquad (4)$$

Example 2 Find the bounds for the error in the result of Example 1.

Solution We first find the absolute minimum and absolute maximum values of $f''(x)$ on $[0, 3]$.

$$f(x) = (16 + x^2)^{-1}$$

$$f'(x) = -2x(16 + x^2)^{-2}$$

$$f''(x) = 8x^2(16 + x^2)^{-3} - 2(16 + x^2)^{-2} = (6x^2 - 32)(16 + x^2)^{-3}$$

$$f'''(x) = -6x(6x^2 - 32)(16 + x^2)^{-4} + 12x(16 + x^2)^{-3}$$

$$= 24x(16 - x^2)(16 + x^2)^{-4}$$

Because $f'''(x) > 0$ for all x in the open interval $(0, 3)$, then f'' is increasing on the open interval $(0, 3)$. Therefore the absolute minimum value of f'' on $[0, 3]$ is $f''(0)$, and the absolute maximum value of f'' on $[0, 3]$ is $f''(3)$.

$$f''(0) = -\frac{1}{128} \quad \text{and} \quad f''(3) = \frac{22}{15{,}625}$$

Taking $\eta = 0$ on the right side of (4) we get

$$-\frac{3}{12}\left(-\frac{1}{128}\right)\frac{1}{4} = \frac{1}{2048}$$

Taking $\eta = 3$ on the right side of (4) we have

$$-\frac{3}{12}\left(\frac{22}{15{,}625}\right)\frac{1}{4} = -\frac{11}{125{,}000}$$

Therefore if ϵ_T is the error in the result of Example 1,

$$-\frac{11}{125{,}000} \leq \epsilon_T \leq \frac{1}{2048}$$

$$-0.0001 \leq \epsilon_T \leq 0.0005$$

If in Theorem 6.6.2, $f(x) = mx + b$, then $f''(x) = 0$ for all x. Therefore $\epsilon_T = 0$; so the trapezoidal rule gives the exact value of the definite integral of a linear function.

Another method for approximating the value of a definite integral is provided by *Simpson's rule* (sometimes referred to as the *parabolic rule*). For a given partition of the closed interval $[a, b]$, Simpson's rule usually gives a better approximation than the trapezoidal rule. In the trapezoidal rule, successive points on the graph of $y = f(x)$ are connected by segments of straight lines, whereas in Simpson's rule the points are connected by segments of parabolas. Before Simpson's rule is developed, we state and prove a theorem that will be needed.

6.6.3 Theorem If $P_0(x_0, y_0)$, $P_1(x_1, y_1)$, and $P_2(x_2, y_2)$ are three noncollinear points on the parabola having the equation $y = Ax^2 + Bx + C$, where $y_0 \geq 0$, $y_1 \geq 0$, $y_2 \geq 0$, $x_1 = x_0 + h$, and $x_2 = x_0 + 2h$, then the measure of the area of the region bounded by the parabola, the x axis, and the lines $x = x_0$ and $x = x_2$ is given by

$$\tfrac{1}{3}h(y_0 + 4y_1 + y_2)$$

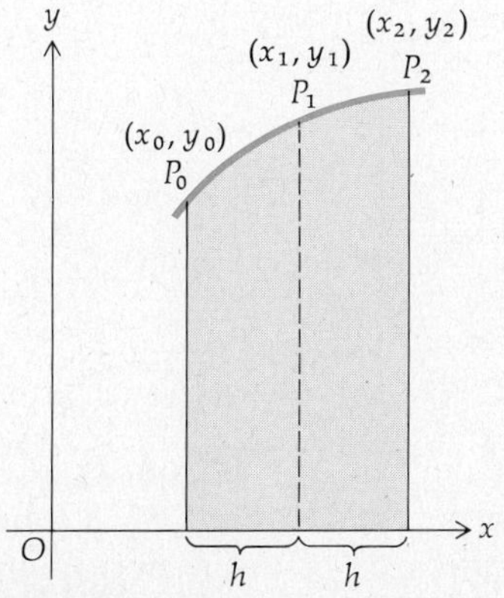

FIGURE 6.6.2

Proof The parabola whose equation is $y = Ax^2 + Bx + C$ has a vertical axis. Refer to Fig. 6.6.2, which shows the region bounded by the parabola, the x axis, and the lines $x = x_0$ and $x = x_2$.

Because P_0, P_1, and P_2 are points on the parabola, their coordinates satisfy the equation of the parabola. So when we replace x_1 by $x_0 + h$, and x_2 by $x_0 + 2h$, we have

$$y_0 = Ax_0^2 + Bx_0 + C$$

$$y_1 = A(x_0 + h)^2 + B(x_0 + h) + C$$

$$= A(x_0^2 + 2hx_0 + h^2) + B(x_0 + h) + C$$

$$y_2 = A(x_0 + 2h)^2 + B(x_0 + 2h) + C$$

$$= A(x_0^2 + 4hx_0 + 4h^2) + B(x_0 + 2h) + C$$

Therefore

$$y_0 + 4y_1 + y_2 = A(6x_0^2 + 12hx_0 + 8h^2) + B(6x_0 + 6h) + 6C \qquad (5)$$

Now if K square units is the area of the region, then K can be computed by the limit of a Riemann sum, and we have

$$\begin{aligned} K &= \lim_{\|\Delta\| \to 0} \sum_{i=1}^{n} (A\xi_i^2 + B\xi_i + C)\,\Delta x \\ &= \int_{x_0}^{x_0+2h} (Ax^2 + Bx + C)\,dx \\ &= \tfrac{1}{3}Ax^3 + \tfrac{1}{2}Bx^2 + Cx\Big]_{x_0}^{x_0+2h} \\ &= \tfrac{1}{3}A(x_0 + 2h)^3 + \tfrac{1}{2}B(x_0 + 2h)^2 + C(x_0 + 2h) \\ &\qquad - (\tfrac{1}{3}Ax_0^3 + \tfrac{1}{2}Bx_0^2 + Cx_0) \\ &= \tfrac{1}{3}h[A(6x_0^2 + 12hx_0 + 8h^2) + B(6x_0 + 6h) + 6C] \qquad (6) \end{aligned}$$

Substituting from (5) in (6) we get

$$K = \tfrac{1}{3}h[y_0 + 4y_1 + y_2]$$

■

Let the function f be continuous on the closed interval $[a, b]$. Consider a regular partition of the interval $[a, b]$ of $2n$ subintervals ($2n$ is used instead of n because we want an even number of subintervals). The length of each subinterval is given by $\Delta x = (b - a)/2n$. Let the points on the curve $y = f(x)$ having these partitioning points as abscissas be denoted by $P_0(x_0, y_0)$, $P_1(x_1, y_1)$, . . . , $P_{2n}(x_{2n}, y_{2n})$; see Fig. 6.6.3, where $f(x) \geq 0$ for all x in $[a, b]$.

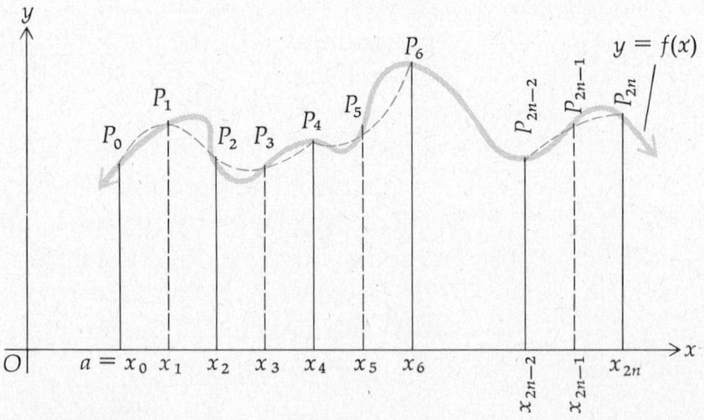

FIGURE 6.6.3

We approximate the segment of the curve $y = f(x)$ from P_0 to P_2 by the segment of the parabola with a vertical axis and through P_0, P_1, and P_2. Then by Theorem 6.6.3 the measure of the area of the region bounded by this parabola, the x axis, and the lines $x = x_0$ and $x = x_2$, with $h = \Delta x$, is given by

$$\tfrac{1}{3}\Delta x(y_0 + 4y_1 + y_2) \quad \text{or} \quad \tfrac{1}{3}\Delta x[f(x_0) + 4f(x_1) + f(x_2)]$$

In a similar manner we approximate the segment of the curve $y = f(x)$ from P_2 to P_4 by the segment of the parabola with a vertical axis and through P_2, P_3, and P_4. The measure of the area of the region bounded by this parabola, the x axis, and the lines $x = x_2$ and $x = x_4$ is given by

$$\tfrac{1}{3}\Delta x(y_2 + 4y_3 + y_4) \quad \text{or} \quad \tfrac{1}{3}\Delta x[f(x_2) + 4f(x_3) + f(x_4)]$$

This process is continued until there are n such regions, and the measure of the area of the last region is given by

$$\tfrac{1}{3}\Delta x(y_{2n-2} + 4y_{2n-1} + y_{2n}) \quad \text{or} \quad \tfrac{1}{3}\Delta x[f(x_{2n-2}) + 4f(x_{2n-1}) + f(x_{2n})]$$

The sum of the measures of the areas of these regions approximates the measure of the area of the region bounded by the curve whose equation is $y = f(x)$, the x axis, and the lines $x = a$ and $x = b$. The measure of the area of this region is given by the definite integral $\int_a^b f(x)\,dx$. So we have as an approximation to the definite integral

$$\tfrac{1}{3}\Delta x[f(x_0) + 4f(x_1) + f(x_2)] + \tfrac{1}{3}\Delta x[f(x_2) + 4f(x_3) + f(x_4)] + \cdots +$$
$$\tfrac{1}{3}\Delta x[f(x_{2n-4}) + 4f(x_{2n-3}) + f(x_{2n-2})] + \tfrac{1}{3}\Delta x[f(x_{2n-2}) + 4f(x_{2n-1}) + f(x_{2n})]$$

Thus

$$\int_a^b f(x)\,dx \approx \tfrac{1}{3}\Delta x[f(x_0) + 4f(x_1) + 2f(x_2) + 4f(x_3) + 2f(x_4) + \cdots$$
$$+ 2f(x_{2n-2}) + 4f(x_{2n-1}) + f(x_{2n})] \qquad (7)$$

where $\Delta x = (b - a)/2n$.

Formula (7) is known as *Simpson's rule*. It is stated formally in the next theorem.

6.6.4 Theorem **Simpson's Rule**

If the function f is continuous on the closed interval $[a, b]$, $2n$ is an even integer, and the numbers $a = x_0, x_1, x_2, \ldots, x_{2n-1}, x_{2n} = b$ form a regular partition of $[a, b]$, then

$$\int_a^b f(x)\,dx \approx \frac{b-a}{6n}[f(x_0) + 4f(x_1) + 2f(x_2) + 4f(x_3) + 2f(x_4) + \cdots$$
$$+ 2f(x_{2n-2}) + 4f(x_{2n-1}) + f(x_{2n})]$$

Example 3 Use Simpson's rule to approximate the value of

$$\int_0^1 \frac{dx}{x+1}$$

with $2n = 4$. Give the result to four decimal places.

Solution Applying Simpson's rule with $2n = 4$ we have

$$\Delta x = \frac{b-a}{2n} = \frac{1}{4} \quad \text{and} \quad \frac{b-a}{6n} = \frac{1}{12}$$

Therefore, if $f(x) = 1/(x+1)$,

$$\int_0^1 \frac{dx}{x+1} \approx \tfrac{1}{12}[f(x_0) + 4f(x_1) + 2f(x_2) + 4f(x_3) + f(x_4)] \tag{8}$$

The computation of the expression in brackets on the right side of (8) is shown in Table 6.6.2.

TABLE 6.6.2

i	x_i	$f(x_i)$	k_i	$k_i \cdot f(x_i)$
0	0	1.00000	1	1.00000
1	0.25	0.80000	4	3.20000
2	0.5	0.66667	2	1.33334
3	0.75	0.57143	4	2.28572
4	1	0.50000	1	0.50000
				$\sum_{i=0}^{4} k_i f(x_i) = 8.31906$

Substituting the sum from Table 6.6.2 in (8) we get

$$\int_0^1 \frac{dx}{x+1} \approx \frac{1}{12}(8.31906) \approx 0.69325^+$$

Rounding off the result to four decimal places gives

$$\int_0^1 \frac{dx}{x+1} \approx 0.6933$$

The exact value of this definite integral to four decimal places is 0.6931, which agrees with our approximation in the first three places; and the error in the approximation is -0.0002. The procedure for computing the exact value of $\int_0^1 dx/(x+1)$ will be learned in Sec. 8.1.

In applying Simpson's rule, the larger we take the value of $2n$, the smaller will be the value of Δx; so geometrically it seems evident that the greater will be the accuracy of the approximation, because a parabola passing through three points of a curve that are close to each other will be close to the curve throughout the subinterval of width $2\,\Delta x$.

The following theorem, which is proved in numerical analysis, gives a method for determining the error in applying Simpson's rule. The error is denoted by ϵ_S.

6.6.5 Theorem Let the function f be continuous on the closed interval $[a, b]$, and f', f'', f''', and $f^{(iv)}$ all exist on $[a, b]$. If

$$\epsilon_S = \int_a^b f(x)\,dx - S$$

where S is the approximate value of $\int_a^b f(x)\,dx$ found by Simpson's rule, then there is some number η in $[a, b]$ such that

$$\epsilon_S = -\tfrac{1}{180}(b-a)f^{(iv)}(\eta)(\Delta x)^4 \tag{9}$$

Example 4 Find the bounds for the error in Example 3.

Solution

$$f(x) = (x+1)^{-1}$$
$$f'(x) = -1(x+1)^{-2}$$
$$f''(x) = 2(x+1)^{-3}$$
$$f'''(x) = -6(x+1)^{-4}$$
$$f^{(iv)}(x) = 24(x+1)^{-5}$$
$$f^{(v)}(x) = -120(x+1)^{-6}$$

Because $f^{(v)}(x) < 0$ for all x in $[0, 1]$, $f^{(iv)}$ is decreasing on $[0, 1]$. Thus the absolute minimum value of $f^{(iv)}$ is at the right endpoint 1, and the absolute maximum value of $f^{(iv)}$ on $[0, 1]$ is at the left endpoint 0.

$$f^{(iv)}(0) = 24 \quad \text{and} \quad f^{(iv)}(1) = \tfrac{3}{4}$$

Substituting 0 for η in the right side of (9) we get

$$-\tfrac{1}{180}(b-a)f^{(iv)}(0)(\Delta x)^4 = -\tfrac{1}{180}(24)(\tfrac{1}{4})^4 = -\tfrac{1}{1920} = -0.00052$$

Substituting 1 for η in the right side of (9) we have

$$-\frac{1}{180}(b-a)f^{(iv)}(1)(\Delta x)^4 = -\frac{1}{180}\cdot\frac{3}{4}\left(\frac{1}{4}\right)^4 = -\frac{1}{61{,}440} = -0.00002$$

So

$$-0.00052 \le \epsilon_S \le -0.00002 \tag{10}$$

The inequality (10) agrees with the discussion in Example 3 regarding the error in the approximation of $\int_0^1 dx/(x+1)$ by Simpson's rule because $-0.00052 < -0.0002 < -0.00002$.

If $f(x)$ is a polynomial of degree three or less, then $f^{(iv)}(x) \equiv 0$ and therefore $\epsilon_S = 0$. In other words, Simpson's rule gives an exact result for a polynomial of the third degree or lower. This statement is geometrically obvious if $f(x)$ is of the second or first degree because in the first case the graph of $y = f(x)$ is a parabola, and in the second case the graph is a straight line (a degenerate parabola).

We apply Simpson's rule to the definite integral $\int_a^b f(x)\,dx$, where $f(x)$ is a third-degree polynomial, and take $2n = 2$. Here Simpson's rule gives an exact value for the definite integral. Because $2n = 2$, $x_0 = a$, $x_1 = \frac{1}{2}(a+b)$, and $x_2 = b$; and $\Delta x = \frac{1}{2}(b-a)$. So

$$\int_a^b f(x)\,dx = \frac{b-a}{6}\left[f(a) + 4f\left(\frac{a+b}{2}\right) + f(b)\right] \tag{11}$$

Formula (11) is known as the **prismoidal formula,** and in Sec. 7.4 we show how it can be applied to find the volume of certain solids.

Example 5 Evaluate $\int_{-2}^{1} (x^3 + 2x - 1)\,dx$ by two methods: (a) use the prismoidal formula; (b) use the fundamental theorem of the calculus.

Solution

(a) If $f(x) = x^3 + 2x - 1$, f is of the third degree and so the prismoidal formula gives the exact value of the definite integral. Using this formula with $b = 1$, $a = -2$, and $\frac{1}{2}(a + b) = -\frac{1}{2}$, we have

$$\int_{-2}^{1} (x^3 + 2x - 1)\,dx = \tfrac{3}{6}[f(-2) + 4f(-\tfrac{1}{2}) + f(1)]$$
$$= \tfrac{1}{2}[-13 - \tfrac{17}{2} + 2]$$
$$= -\tfrac{39}{4}$$

(b) $$\int_{-2}^{1} (x^3 + 2x - 1)\,dx = \frac{x^4}{4} + x^2 - x\Big]_{-2}^{1}$$
$$= \tfrac{1}{4} + 1 - 1 - (4 + 4 + 2)$$
$$= -\tfrac{39}{4}$$

EXERCISES 6.6

In Exercises 1 through 12, compute the approximate value of the given definite integral by the trapezoidal rule for the indicated value of n. Express the result to three decimal places. In Exercises 1, 2, 5, and 6, find the exact value of the definite integral, and compare the result with the approximation.

1. $\int_0^2 x^3\,dx;\ n = 4$ **2.** $\int_0^2 x\sqrt{4 - x^2}\,dx;\ n = 8$ **3.** $\int_1^2 \frac{dx}{x};\ n = 5$ **4.** $\int_2^{10} \frac{dx}{1 + x};\ n = 8$

5. $\int_0^{\pi} \cos x\,dx;\ n = 4$ **6.** $\int_0^{\pi} \sin x\,dx;\ n = 6$ **7.** $\int_0^1 \frac{dx}{\sqrt{1 + x^2}};\ n = 5$ **8.** $\int_2^3 \sqrt{1 + x^2}\,dx;\ n = 6$

9. $\int_{\pi/2}^{3\pi/2} \frac{\sin x}{x}\,dx;\ n = 6$ **10.** $\int_0^1 \sqrt{1 + x^3}\,dx;\ n = 4$ **11.** $\int_0^2 \sqrt{1 + x^4}\,dx;\ n = 6$ **12.** $\int_0^{\pi} \frac{\sin x}{1 + x}\,dx;\ n = 6$

In Exercises 13 through 18, find the bounds for the error in the approximation of the indicated exercise.

13. Exercise 1 **14.** Exercise 5 **15.** Exercise 3 **16.** Exercise 4 **17.** Exercise 6 **18.** Exercise 8

19. (a) Approximate $\int_0^2 x^3\,dx$ by Simpson's rule with $2n = 4$, and compare the result with that obtained in Exercise 1. (b) Find the exact value of the definite integral in part (a).

20. (a) Approximate $\int_0^{\pi} \sin x\,dx$ by Simpson's rule with $2n = 6$, and compare the result with that obtained in Exercise 6. (b) Find the exact value of the definite integral in part (a).

In Exercises 21 through 24, approximate the definite integral by Simpson's rule with the indicated value of 2n. Express the answer to three decimal places.

21. $\int_{-1}^{0} \frac{dx}{1 - x};\ 2n = 4$ **22.** $\int_{-1/2}^{0} \frac{dx}{\sqrt{1 - x^2}};\ 2n = 4$ **23.** $\int_0^1 \frac{dx}{x^2 + x + 1};\ 2n = 4$ **24.** $\int_1^2 \frac{dx}{x + 1};\ 2n = 8$

In Exercises 25 through 28, find bounds for the error in the indicated exercise.

25. Exercise 19(a) **26.** Exercise 20(a) **27.** Exercise 21 **28.** Exercise 24

Each of the definite integrals in Exercises 29 through 34 cannot be evaluated exactly in terms of elementary functions. Use Simpson's rule, with the indicated value of 2n, to find an approximate value of the given definite integral. Express the result to three decimal places.

29. $\int_{\pi/2}^{3\pi/2} \frac{\sin x}{x}\, dx;\ 2n = 6$ **30.** $\int_0^2 \sqrt{1 + x^4}\, dx;\ 2n = 6$ **31.** $\int_1^{1.8} \sqrt{1 + x^3}\, dx;\ 2n = 4$

32. $\int_0^1 \sqrt[3]{1 - x^2}\, dx;\ 2n = 4$ **33.** $\int_0^2 \frac{dx}{\sqrt{1 + x^3}};\ 2n = 8$ **34.** $\int_0^{\pi/2} \sqrt{\sin x}\, dx;\ 2n = 6$

35. Show that the exact value of the integral $\int_0^2 \sqrt{4 - x^2}\, dx$ is π by interpreting it as the measure of the area of a region. Approximate the definite integral by the trapezoidal rule with $n = 8$. Give the result to three decimal places, and compare the value so obtained with the exact value.

36. Show that the exact value of the integral $\int_0^1 4\sqrt{1 - x^2}\, dx$ is π by interpreting it as the measure of the area of a region. Use Simpson's rule with $2n = 6$ to get an approximate value of the definite integral to three decimal places. Compare the results.

In Exercises 37 through 40, evaluate the definite integral by two methods: (a) use the prismoidal formula; (b) use the fundamental theorem of the calculus.

37. $\int_1^3 (4x^3 - 3x^2 + 1)\, dx$ **38.** $\int_2^6 (2x^3 - 2x - 3)\, dx$

39. $\int_{-1}^5 (x^3 + 3x^2 - 2x - 6)\, dx$ **40.** $\int_{-2}^2 (x^3 + x^2 - 4x - 2)\, dx$

41. Suppose that f is a function continuous on the closed interval $[a, b]$. Let T be the approximate value of $\int_a^b f(x)\, dx$ by using the trapezoidal rule, and let S be the approximate value of $\int_a^b f(x)\, dx$ by using Simpson's rule, where the same partition of the interval $[a, b]$ is used for both approximations. Show that

$$S = \tfrac{2}{3}[T + \Delta x(f(x_1) + f(x_3) + f(x_5) + \cdots + f(x_{n-1}))]$$

where n is even.

REVIEW EXERCISES FOR CHAPTER 6

In Exercises 1 through 4, find the exact value of the definite integral by using the definition of the definite integral; do not use the fundamental theorem of the calculus.

1. $\int_0^2 (3x^2 - 1)\, dx$ **2.** $\int_0^1 (2x^2 + 4x + 1)\, dx$ **3.** $\int_{-2}^1 (x^3 + 2x)\, dx$ **4.** $\int_{-1}^3 (x^2 - 1)^2\, dx$

In Exercises 5 through 18, evaluate the definite integral by using the fundamental theorem of the calculus.

5. $\int_{-2}^2 (t^3 - 3t)\, dt$ **6.** $\int_2^4 \frac{4\, dx}{5x^4}$ **7.** $\int_1^5 \frac{dx}{\sqrt{3x - 1}}$ **8.** $\int_{-5}^5 2x\sqrt[3]{x^2 + 2}\, dx$

9. $\int_0^{\pi/6} \frac{\sin t\, dt}{\cos^2 t}$ **10.** $\int_0^{\pi} \sin^2 \tfrac{1}{2}x \cos^3 \tfrac{1}{2}x\, dx$ **11.** $\int_0^2 x\sqrt{3x + 4}\, dx$ **12.** $\int_1^2 \frac{y\, dy}{\sqrt{5 - y}}$

13. $\displaystyle\int_{-1}^{7} \frac{x^2\,dx}{\sqrt{x+2}}$ **14.** $\displaystyle\int_{0}^{1} (x^2+4x)\sqrt{x^3+6x^2+1}\,dx$ **15.** $\displaystyle\int_{-3}^{3} |x-2|^3\,dx$ **16.** $\displaystyle\int_{-2}^{2} x|x-3|\,dx$

17. $\displaystyle\int_{0}^{\pi/2} \sin 5x \cos 3x\,dx$ **18.** $\displaystyle\int_{\pi/4}^{\pi/2} (\sin^4 2x - 2\sin 4x)\,dx$

In Exercises 19 and 20, use the fundamental theorem of the calculus to evaluate the definite integral. Then find the value of X satisfying the mean-value theorem for integrals.

19. $\displaystyle\int_{0}^{3} (x^2+1)\,dx$ **20.** $\displaystyle\int_{1}^{4} \sqrt{x}\,dx$

In Exercises 21 and 22, find the sum.

21. $\displaystyle\sum_{i=1}^{100} 2i(i^3-1)$ **22.** $\displaystyle\sum_{i=1}^{41} (\sqrt[3]{3i-1} - \sqrt[3]{3i+2})$

23. Prove that $\displaystyle\sum_{i=1}^{n} i^3 = \left(\sum_{i=1}^{n} i\right)^2$ and verify the formula for $n = 1, 2$, and 3.

24. Express as a definite integral and evaluate the definite integral: $\displaystyle\lim_{n\to+\infty} \sum_{i=1}^{n} (8\sqrt{i}\,/n^{3/2})$ (*Hint*: Consider the function f for which $f(x) = \sqrt{x}$.)

25. Show that each of the following inequalities holds: (a) $\displaystyle\int_{-2}^{-1} \frac{dx}{x-3} \geq \int_{-2}^{-1} \frac{dx}{x}$; (b) $\displaystyle\int_{1}^{2} \frac{dx}{x} \geq \int_{1}^{2} \frac{dx}{x-3}$; (c) $\displaystyle\int_{4}^{5} \frac{dx}{x-3} \geq \int_{4}^{5} \frac{dx}{x}$

26. If

$$f(x) = \begin{cases} 0 & \text{if } a \leq x < c \\ k & \text{if } x = c \\ 1 & \text{if } c < x \leq b \end{cases}$$

prove that f is integrable on $[a, b]$ and $\displaystyle\int_{a}^{b} f(x)\,dx = b - c$ regardless of the value of k.

In Exercises 27 through 30, apply Theorem 6.3.9 to find a closed interval containing the value of the given definite integral.

27. $\displaystyle\int_{1}^{4} \sqrt{5+4x-x^2}\,dx$ **28.** $\displaystyle\int_{-2}^{1} \sqrt[3]{2x^3-3x^2+1}\,dx$ **29.** $\displaystyle\int_{-\pi/2}^{\pi/2} \sqrt{\cos t}\,dt$ **30.** $\displaystyle\int_{0}^{\pi/6} \frac{dt}{\cos^2 t}$

31. Find $D_x \displaystyle\int_{x}^{4} (3t^2-4)^{3/2}\,dt$. **32.** Find $D_x \displaystyle\int_{-x}^{x} \frac{4}{1+t^2}\,dt$.

In Exercises 33 through 38, find the area of the region bounded by the given curve and lines. Draw a figure showing the region and a rectangular element of area. Express the measure of the area as the limit of a Riemann sum, and then with definite integral notation. Evaluate the definite integral by the fundamental theorem of the calculus.

33. $y = 9 - x^2$; x axis; y axis; $x = 3$ **34.** $y = 3\cos\frac{1}{2}x$; x axis; $x = -\frac{1}{2}\pi$; $x = \frac{1}{2}\pi$

35. $y = 2\sqrt{x-1}$; x axis; $x = 5$; $x = 17$ **36.** $y = 16 - x^2$; x axis

37. $y = x\sqrt{x+5}$; x axis; $x = -1$; $x = 4$ **38.** $y = \dfrac{4}{x^2} - x$; x axis; $x = -2$; $x = -1$

39. Find the average value of the cosine function on the closed interval $[a, a + 2\pi]$.

40. Interpret the mean-value theorem for integrals (6.4.1) in terms of an average function value.

41. If $f(x) = x^2\sqrt{x-3}$, find the average value of f on $[7, 12]$.

42. Suppose $C(x)$ hundreds of dollars is the total cost of producing $100x$ units of a certain commodity, and $C(x) = 2x^2 + 5$. Find the average cost when the number of units produced takes on all values from 0 to 400.

43. A body falls from rest and travels a distance of s ft before striking the ground. If the only force acting is that of gravity, which gives the body an acceleration of g ft/sec² toward the ground, show that the average value of the velocity, expressed as a function of distance, while traveling this distance is $\frac{2}{3}\sqrt{2gs}$ ft/sec, and that this average velocity is two-thirds of the final velocity.

44. Suppose a ball is dropped from rest and after t sec its directed distance from the starting point is s ft and its velocity is v ft/sec. Neglect air resistance. When $t = t_1$, $s = s_1$, and $v = v_1$. (a) Express v as a function of t as $v = f(t)$, and find the average value of f on $[0, t_1]$. (b) Express v as a function of s as $v = h(s)$, and find the average velocity of h on $[0, s_1]$. (c) Write the results of parts (a) and (b) in terms of t_1, and determine which average velocity is larger.

45. The demand equation of a certain commodity is $p^2 + x - 10 = 0$, where p dollars is the price of one unit when $100x$ units of the commodity are demanded. Find the average total revenue when the number of units demanded takes on all values from 600 to 800.

In Exercises 46 and 47, find an approximate value for the given integral by using the trapezoidal rule with $n = 4$. Express the result to three decimal places.

46. $\displaystyle\int_1^{9/5} \sqrt{1 + x^3}\, dx$ **47.** $\displaystyle\int_{1/10}^{1/2} \frac{\cos x}{x}\, dx$

48. Find an approximate value for the integral of Exercise 46 by using Simpson's rule with $2n = 4$. Express the result to three decimal places.

49. Find an approximate value for the integral of Exercise 47 by using Simpson's rule with $2n = 4$. Express the result to three decimal places.

50. Given the integral $\displaystyle\int_d^{d+6} (x^2 + bx + c)\, dx$, where b, c, and d are constants. Suppose this integral is approximated by using the trapezoidal rule with $n = k$. (a) Show that the error in the approximation is exactly $-36/k^2$. (b) What is the smallest value of k such that the approximation is accurate to 1 decimal place?

51. Evaluate $\displaystyle\int_0^{\pi} |\cos x + \tfrac{1}{2}|\, dx$.

52. If n is any positive integer, prove that $\displaystyle\int_0^{\pi} \sin^2 nx\, dx = \tfrac{1}{2}\pi$.

53. (a) Find the average value of the function f defined by $f(x) = 1/x^2$ on the interval $[1, r]$. (b) If A is the average value found in part (a), find $\displaystyle\lim_{r \to +\infty} A$.

54. Given $F(x) = \displaystyle\int_x^{2x} \frac{1}{t}\, dt$ and $x > 0$. Prove that F is a constant function by showing that $F'(x) = 0$. (*Hint*: Use Theorem 6.5.1 after writing the given integral as the difference of two integrals.)

55. Let f be continuous on $[a, b]$ and $\displaystyle\int_a^b f(t)\, dt \neq 0$. Show that for any number k in $(0, 1)$ there is a number c in (a, b) such that $\displaystyle\int_a^c f(t)\, dt = k\int_a^b f(t)\, dt$. (*Hint*: Consider the function F for which $F(x) = \displaystyle\int_a^x f(t)\, dt \Big/ \int_a^b f(t)\, dt$ and apply the intermediate-value theorem.)

56. Make up an example of a discontinuous function for which the mean-value theorem for integrals (a) does not hold, and (b) does hold.

CHAPTER 7 Applications of the Definite Integral

7.1 AREA OF A REGION IN A PLANE

From Definition 6.2.4, if f is a function continuous on the closed interval $[a, b]$ and if $f(x) \geq 0$ for all x in $[a, b]$, then the number of square units in the area of the region bounded by the curve $y = f(x)$, the x axis, and the lines $x = a$ and $x = b$ is

$$\lim_{\|\Delta\| \to 0} \sum_{i=1}^{n} f(\xi_i)\, \Delta_i x$$

which is the definite integral $\int_a^b f(x)\, dx$.

In Chapter 6 we considered the area of a region for which $f(x) \geq 0$ on $[a, b]$. For instance, refer to Example 8, Sec. 6.5. Suppose now that $f(x) < 0$ for all x in $[a, b]$. Then each $f(\xi_i)$ is a negative number; so we define the number of square units in the area of the region bounded by $y = f(x)$, the x axis, and the lines $x = a$ and $x = b$ to be

$$\lim_{\|\Delta\| \to 0} \sum_{i=1}^{n} [-f(\xi_i)]\, \Delta_i x$$

which equals

$$-\int_a^b f(x)\, dx$$

Example 1 Find the area of the region bounded by the curve $y = x^2 - 4x$, the x axis, and the lines $x = 1$ and $x = 3$.

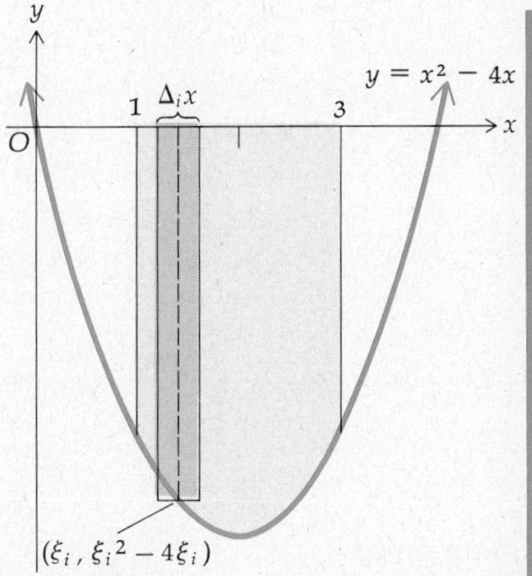

FIGURE 7.1.1

Solution The region, together with a rectangular element of area, is shown in Fig. 7.1.1.

We take a partition of the interval $[1, 3]$; the width of the ith rectangle is $\Delta_i x$. Because $x^2 - 4x < 0$ on $[1, 3]$, the altitude of the ith rectangle is $-(\xi_i^2 - 4\xi_i) = 4\xi_i - \xi_i^2$. Hence the sum of the measures of the areas of n rectangles is given by

$$\sum_{i=1}^{n} (4\xi_i - \xi_i^2)\, \Delta_i x$$

The measure of the desired area is given by the limit of this sum as $\|\Delta\|$ approaches 0; so if A square units is the area of the region,

$$\begin{aligned} A &= \lim_{\|\Delta\| \to 0} \sum_{i=1}^{n} (4\xi_i - \xi_i^2)\, \Delta_i x \\ &= \int_1^3 (4x - x^2)\, dx \\ &= 2x^2 - \tfrac{1}{3}x^3 \Big]_1^3 \\ &= \tfrac{22}{3} \end{aligned}$$

Thus the area of the region is $\frac{22}{3}$ square units.

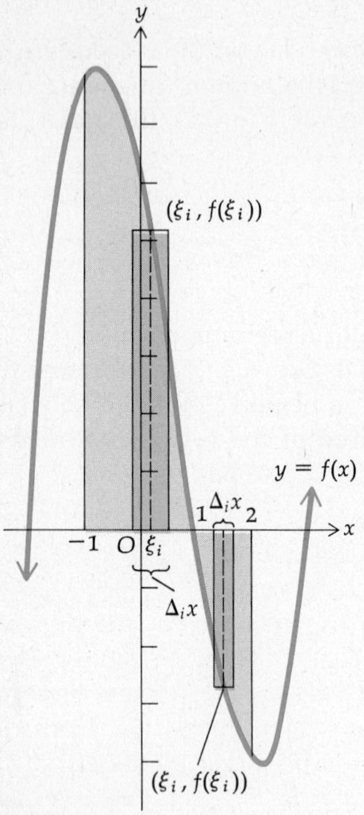

FIGURE 7.1.2

Example 2 Find the area of the region bounded by the curve $y = x^3 - 2x^2 - 5x + 6$, the x axis, and the lines $x = -1$ and $x = 2$.

Solution The region is shown in Fig. 7.1.2. Let $f(x) = x^3 - 2x^2 - 5x + 6$. Because $f(x) \geq 0$ when x is in the closed interval $[-1, 1]$ and $f(x) \leq 0$ when x is in the closed interval $[1, 2]$, we separate the region into two parts. Let A_1 be the number of square units in the area of the region when x is in $[-1, 1]$, and let A_2 be the number of square units in the area of the region when x is in $[1, 2]$. Then

$$\begin{aligned} A_1 &= \lim_{\|\Delta\| \to 0} \sum_{i=1}^{n} f(\xi_i)\, \Delta_i x \\ &= \int_{-1}^{1} f(x)\, dx \\ &= \int_{-1}^{1} (x^3 - 2x^2 - 5x + 6)\, dx \end{aligned}$$

and

$$A_2 = \lim_{\|\Delta\| \to 0} \sum_{i=1}^{n} [-f(\xi_i)]\, \Delta_i x = \int_{1}^{2} -(x^3 - 2x^2 - 5x + 6)\, dx$$

If A square units is the area of the entire region, then

$$\begin{aligned} A &= A_1 + A_2 \\ &= \int_{-1}^{1} (x^3 - 2x^2 - 5x + 6)\, dx - \int_{1}^{2} (x^3 - 2x^2 - 5x + 6)\, dx \\ &= \left[\tfrac{1}{4}x^4 - \tfrac{2}{3}x^3 - \tfrac{5}{2}x^2 + 6x\right]_{-1}^{1} - \left[\tfrac{1}{4}x^4 - \tfrac{2}{3}x^3 - \tfrac{5}{2}x^2 + 6x\right]_{1}^{2} \\ &= [(\tfrac{1}{4} - \tfrac{2}{3} - \tfrac{5}{2} + 6) - (\tfrac{1}{4} + \tfrac{2}{3} - \tfrac{5}{2} - 6)] \\ &\qquad - [(4 - \tfrac{16}{3} - 10 + 12) - (\tfrac{1}{4} - \tfrac{2}{3} - \tfrac{5}{2} + 6] \\ &= \tfrac{32}{3} - (-\tfrac{29}{12}) \\ &= \tfrac{157}{12} \end{aligned}$$

The area of the region is therefore $\frac{157}{12}$ square units.

Now consider two functions f and g continuous on the closed interval $[a, b]$ and such that $f(x) \geq g(x)$ for all x in $[a, b]$. We wish to find the area of the region bounded by the two curves $y = f(x)$ and $y = g(x)$ and the two lines $x = a$ and $x = b$. Such a situation is shown in Fig. 7.1.3.

Take a partition of the interval $[a, b]$, with the ith subinterval having a length of $\Delta_i x$. In each subinterval choose a point ξ_i. Consider the rectangle having altitude $[f(\xi_i) - g(\xi_i)]$ units and width $\Delta_i x$ units. A rectangle is shown in Fig. 7.1.3. There are n such rectangles, one associated with each subinterval. The sum of the measures of the areas of these n rectangles is given by the following Riemann sum:

$$\sum_{i=1}^{n} [f(\xi_i) - g(\xi_i)]\, \Delta_i x$$

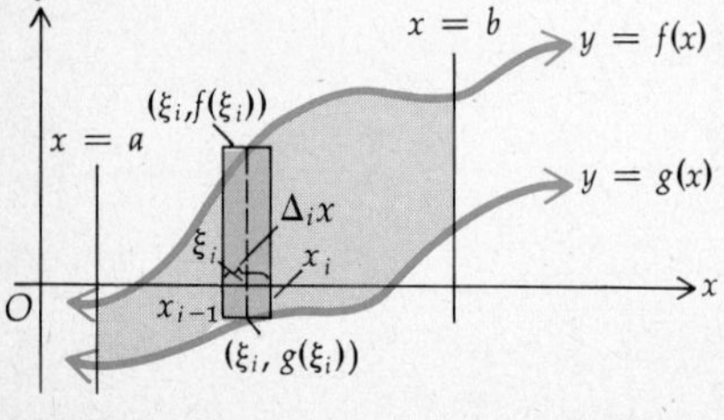

FIGURE 7.1.3

This Riemann sum is an approximation to what we intuitively think of as the number representing the "measure of the area" of the region. The smaller the value of $\|\Delta\|$, the better is this approximation. If A square units is the area of the region, we define

$$A = \lim_{\|\Delta\| \to 0} \sum_{i=1}^{n} [f(\xi_i) - g(\xi_i)]\, \Delta_i x \tag{1}$$

Because f and g are continuous on $[a, b]$, so too is $(f - g)$; therefore the limit in (1) exists and is equal to the definite integral

$$\int_a^b [f(x) - g(x)]\, dx$$

Example 3 Find the area of the region bounded by the curves $y = x^2$ and $y = -x^2 + 4x$.

Solution To find the points of intersection of the two curves we solve the equations simultaneously and obtain the points (0, 0) and (2, 4). The region is shown in Fig. 7.1.4.

Let $f(x) = -x^2 + 4x$, and $g(x) = x^2$. Therefore, in the interval [0, 2] the curve $y = f(x)$ is above the curve $y = g(x)$. We draw a vertical rectangular element of area, having altitude $[f(\xi_i) - g(\xi_i)]$ units and width $\Delta_i x$ units. The measure of the area of this rectangle then is given by $[f(\xi_i) - g(\xi_i)]\, \Delta_i x$. The sum of the measures of the areas of n such rectangles is given by the Riemann sum

$$\sum_{i=1}^{n} [f(\xi_i) - g(\xi_i)]\, \Delta_i x$$

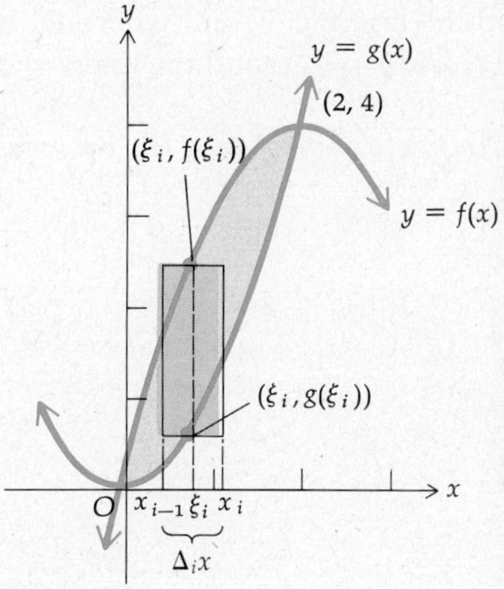

FIGURE 7.1.4

If A square units is the area of the region, then

$$A = \lim_{\|\Delta\| \to 0} \sum_{i=1}^{n} [f(\xi_i) - g(\xi_i)]\, \Delta_i x$$

and the limit of the Riemann sum is a definite integral. Hence

$$\begin{aligned} A &= \int_0^2 [f(x) - g(x)]\, dx \\ &= \int_0^2 [(-x^2 + 4x) - x^2]\, dx \\ &= \int_0^2 (-2x^2 + 4x)\, dx \\ &= \left[-\tfrac{2}{3}x^3 + 2x^2\right]_0^2 \\ &= -\tfrac{16}{3} + 8 - 0 \\ &= \tfrac{8}{3} \end{aligned}$$

The area of the region is $\frac{8}{3}$ square units.

Example 4 Find the area of the region bounded by the parabola $y^2 = 2x - 2$ and the line $y = x - 5$.

Solution The two curves intersect at the points $(3, -2)$ and $(9, 4)$. The region is shown in Fig. 7.1.5.

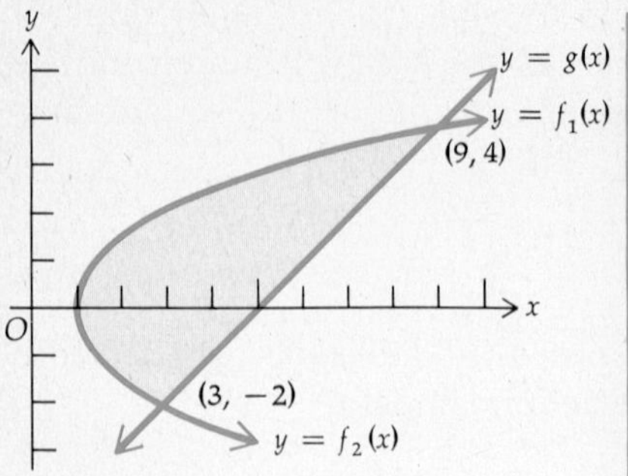

FIGURE 7.1.5

The equation $y^2 = 2x - 2$ is equivalent to the two equations

$$y = \sqrt{2x - 2} \quad \text{and} \quad y = -\sqrt{2x - 2}$$

with the first equation giving the upper half of the parabola and the second equation giving the bottom half. If we let $f_1(x) = \sqrt{2x - 2}$ and $f_2(x) = -\sqrt{2x - 2}$, the equation of the top half of the parabola is $y = f_1(x)$, and the equation of the bottom half of the parabola is $y = f_2(x)$. If we let $g(x) = x - 5$, the equation of the line is $y = g(x)$.

In Fig. 7.1.6 we see two vertical rectangular elements of area. Each rectangle has the upper base on the curve $y = f_1(x)$. Because the base of the first rectangle is on the curve $y = f_2(x)$, the altitude is $[f_1(\xi_i) - f_2(\xi_i)]$ units. Because the base of the second rectangle is on the curve $y = g(x)$, its altitude is $[f_1(\xi_i) - g(\xi_i)]$ units. If we wish to solve this problem by using vertical rectangular elements of area, we must divide the region into two separate regions, for instance R_1 and R_2, where R_1 is the region bounded by the curves $y = f_1(x)$ and $y = f_2(x)$ and the line $x = 3$, and where R_2 is the region bounded by the curves $y = f_1(x)$ and $y = g(x)$ and the line $x = 3$ (see Fig. 7.1.7).

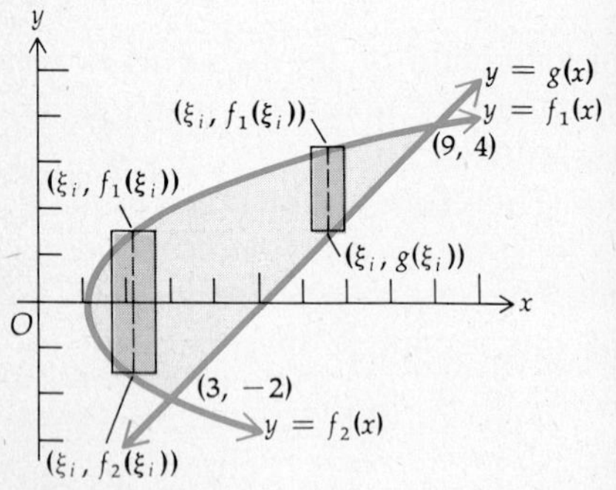

FIGURE 7.1.6

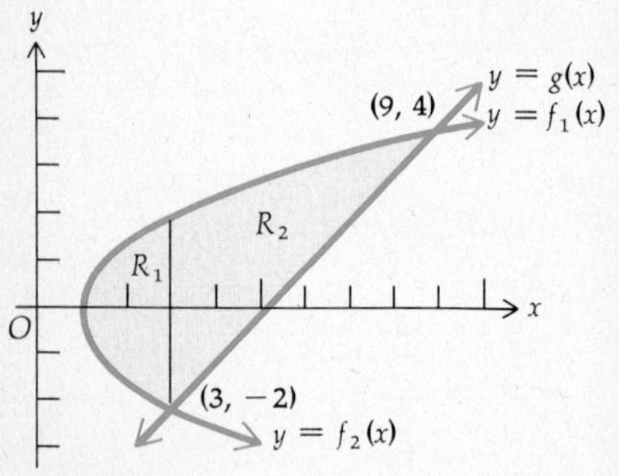

FIGURE 7.1.7

If A_1 is the number of square units in the area of region R_1, we have

$$\begin{aligned} A_1 &= \lim_{\|\Delta\| \to 0} \sum_{i=1}^{n} [f_1(\xi_i) - f_2(\xi_i)]\, \Delta_i x \\ &= \int_1^3 [f_1(x) - f_2(x)]\, dx \\ &= \int_1^3 [\sqrt{2x - 2} + \sqrt{2x - 2}]\, dx \\ &= 2 \int_1^3 \sqrt{2x - 2}\, dx \\ &= \tfrac{2}{3}(2x - 2)^{3/2} \Big]_1^3 \\ &= \tfrac{16}{3} \end{aligned}$$

If A_2 is the number of square units in the area of region R_2, we have

$$\begin{aligned} A_2 &= \lim_{\|\Delta\| \to 0} \sum_{i=1}^{n} [f_1(\xi_i) - g(\xi_i)]\, \Delta_i x \\ &= \int_3^9 [f_1(x) - g(x)]\, dx \\ &= \int_3^9 [\sqrt{2x - 2} - (x - 5)]\, dx \end{aligned}$$

$$= \left[\tfrac{1}{3}(2x - 2)^{3/2} - \tfrac{1}{2}x^2 + 5x\right]_3^9$$

$$= [\tfrac{64}{3} - \tfrac{81}{2} + 45] - [\tfrac{8}{3} - \tfrac{9}{2} + 15]$$

$$= \tfrac{38}{3}$$

Hence $A_1 + A_2 = \frac{16}{3} + \frac{38}{3} = 18$. Therefore the area of the entire region is 18 square units.

Example 5 Find the area of the region in Example 4 by taking horizontal rectangular elements of area.

Solution Figure 7.1.8 illustrates the region with a horizontal rectangular element of area.

If in the equations of the parabola and the line we solve for x,

$$x = \tfrac{1}{2}(y^2 + 2) \quad \text{and} \quad x = y + 5$$

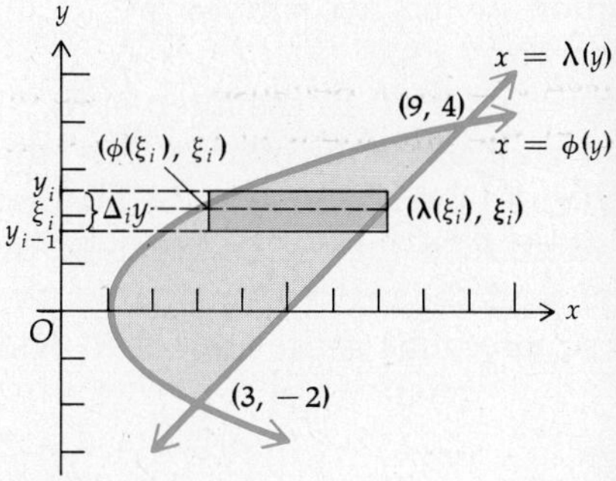

FIGURE 7.1.8

Letting $\phi(y) = \frac{1}{2}(y^2 + 2)$ and $\lambda(y) = y + 5$, the equation of the parabola may be written as $x = \phi(y)$ and the equation of the line as $x = \lambda(y)$. Consider the closed interval $[-2, 4]$ on the y axis, and take a partition of this interval. The ith subinterval will have a length of $\Delta_i y$. In the ith subinterval $[y_{i-1}, y_i]$, choose a point ξ_i. Then the length of the ith rectangular element is $[\lambda(\xi_i) - \phi(\xi_i)]$ units and the width is $\Delta_i y$ units. The measure of the area of the region can be approximated by the Riemann sum

$$\sum_{i=1}^{n} [\lambda(\xi_i) - \phi(\xi_i)]\, \Delta_i y$$

If A square units is the area of the region, then

$$A = \lim_{\|\Delta\| \to 0} \sum_{i=1}^{n} [\lambda(\xi_i) - \phi(\xi_i)]\, \Delta_i y$$

Because λ and ϕ are continuous on $[-2, 4]$, so too is $(\lambda - \phi)$, and the limit of the Riemann sum is a definite integral:

$$A = \int_{-2}^{4} [\lambda(y) - \phi(y)]\, dy$$

$$= \int_{-2}^{4} [(y + 5) - \tfrac{1}{2}(y^2 + 2)]\, dy$$

$$= \frac{1}{2}\int_{-2}^{4} (-y^2 + 2y + 8)\, dy$$

$$= \tfrac{1}{2}\left[-\tfrac{1}{3}y^3 + y^2 + 8y\right]_{-2}^{4}$$

$$= \tfrac{1}{2}[(-\tfrac{64}{3} + 16 + 32) - (\tfrac{8}{3} + 4 - 16)]$$

$$= 18$$

Comparing the solutions in Examples 4 and 5, we see that in the first case there are two definite integrals to evaluate, whereas in the second case there is only one. In general, if possible, the rectangular elements of area should be constructed so that a single definite integral is obtained. The following example illustrates a situation where two definite integrals are necessary.

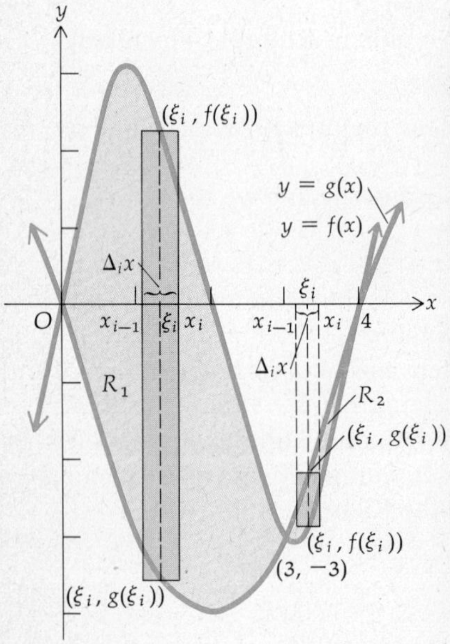

FIGURE 7.1.9

Example 6 Find the area of the region bounded by the two curves $y = x^3 - 6x^2 + 8x$ and $y = x^2 - 4x$.

Solution The points of intersection of the two curves are (0, 0), (3, −3), and (4, 0). The region is shown in Fig. 7.1.9.

Let $f(x) = x^3 - 6x^2 + 8x$ and $g(x) = x^2 - 4x$. In the interval [0, 3] the curve $y = f(x)$ is above the curve $y = g(x)$, and in the interval [3, 4] the curve $y = g(x)$ is above the curve $y = f(x)$. So the region must be divided into two separate regions R_1 and R_2, where R_1 is the region bounded by the two curves in the interval [0, 3] and R_2 is the region bounded by the two curves in the interval [3, 4]. If A_1 square units is the area of R_1 and A_2 square units is the area of R_2,

$$A_1 = \lim_{\|\Delta\| \to 0} \sum_{i=1}^{n} [f(\xi_i) - g(\xi_i)] \Delta_i x$$

and

$$A_2 = \lim_{\|\Delta\| \to 0} \sum_{i=1}^{n} [g(\xi_i) - f(\xi_i)] \Delta_i x$$

so that

$$A_1 + A_2 = \int_0^3 [(x^3 - 6x^2 + 8x) - (x^2 - 4x)]\, dx + \int_3^4 [(x^2 - 4x) - (x^3 - 6x^2 + 8x)]\, dx$$

$$= \int_0^3 (x^3 - 7x^2 + 12x)\, dx + \int_3^4 (-x^3 + 7x^2 - 12x)\, dx$$

$$= \left[\tfrac{1}{4}x^4 - \tfrac{7}{3}x^3 + 6x^2\right]_0^3 + \left[-\tfrac{1}{4}x^4 + \tfrac{7}{3}x^3 - 6x^2\right]_3^4$$

$$= \tfrac{45}{4} + \tfrac{7}{12}$$

$$= \tfrac{71}{6}$$

Therefore the required area is $\frac{71}{6}$ square units.

EXERCISES 7.1

In Exercises 1 through 26, find the area of the region bounded by the given curves. In each problem do the following: (a) Draw a figure showing the region and a rectangular element of area; (b) express the area of the region as the limit of a Riemann sum; (c) find the limit in part (b) by evaluating a definite integral by the fundamental theorem of the calculus.

1. $x^2 = -y$; $y = -4$ **2.** $y^2 = -x$; $x = -2$; $x = -4$

3. $x^2 + y + 4 = 0$; $y = -8$. Take the elements of area perpendicular to the y axis.

4. The same region as in Exercise 3. Take the elements of area parallel to the y axis.

5. $x^2 - y + 1 = 0$; $x - y + 1 = 0$. Take the elements of area perpendicular to the x axis.

6. The same region as in Exercise 5. Take the elements of area parallel to the x axis.

7. $x^3 = 2y^2$; $x = 0$, $y = -2$ **8.** $y^3 = 4x$; $x = 0$; $y = -2$ **9.** $y = 2 - x^2$; $y = -x$

10. $y = x^2$; $y = x^4$ **11.** $y^2 = x - 1$; $x = 3$ **12.** $y = x^2$; $x^2 = 18 - y$

13. $y = \sqrt{x}$; $y = x^3$ **14.** $x = 4 - y^2$; $x = 4 - 4y$ **15.** $y^3 = x^2$; $x - 3y + 4 = 0$

16. $xy^2 = y^2 - 1$; $x = 1$; $y = 1$; $y = 4$ **17.** $x = y^2 - 2$; $x = 6 - y^2$ **18.** $x = y^2 - y$; $x = y - y^2$

19. $y = 2x^3 - 3x^2 - 9x$; $y = x^3 - 2x^2 - 3x$ **20.** $3y = x^3 - 2x^2 - 15x$; $y = x^3 - 4x^2 - 11x + 30$

21. $y = x^3 + 3x^2 + 2x$; $y = 2x^2 + 4x$ **22.** $y = |x - 1| + 3$; $y = 0$; $x = -2$; $x = 4$

23. $y = \cos x - \sin x$; $x = 0$; $y = 0$ **24.** $y = \sin x$; $y = -\sin x$; $x = -\frac{1}{2}\pi$; $x = \frac{1}{2}\pi$

25. $y = |x|$, $y = x^2 - 1$, $x = -1$, $x = 1$ **26.** $y = |x + 1| + |x|$, $y = 0$, $x = -2$, $x = 3$

27. Find by integration the area of the triangle having vertices at (5, 1), (1, 3), and (−1, −2).

28. Find by integration the area of the triangle having vertices at (3, 4), (2, 0), and (0, 1).

29. Find the area of the region bounded by the curve $x^3 - x^2 + 2xy - y^2 = 0$ and the line $x = 4$. (*Hint:* Solve the cubic equation for y in terms of x, and express y as two functions of x.)

30. Find the area of the region bounded by the three curves $y = x^2$, $x = y^3$, and $x + y = 2$.

31. Find the area of the region bounded by the three curves $y = x^2$, $y = 8 - x^2$, and $4x - y + 12 = 0$.

32. Find by integration the area of the trapezoid having vertices at (−1, −1), (2, 2), (6, 2), and (7, −1).

33. Find the area of the region bounded by the curve $y = \sin x$, the line $y = 1$, and the y axis.

34. Find the area of the region bounded by the two curves $y = \sin x$ and $y = \cos x$ between two consecutive points of intersection.

35. Find the area of the region bounded by the two parabolas $y^2 = 4px$ and $x^2 = 4py$.

36. Find the area of the region above the parabola $x^2 = 4py$ and inside the triangle formed by the x axis and the lines $y = x + 8p$ and $y = -x + 8p$.

37. Find the rate of change of the measure of the area of Exercise 35 with respect to p when $p = 3$.

38. Find the rate of change of the measure of the area of Exercise 36 with respect to p when $p = \frac{3}{8}$.

39. If A square units is the area of the region bounded by the parabola $y^2 = 4x$ and the line $y = mx$ ($m > 0$), find the rate of change of A with respect to m.

40. Determine m so that the region above the line $y = mx$ and below the parabola $y = 2x - x^2$ has an area of 36 square units.

41. Determine m so that the region above the curve $y = mx^2$ ($m > 0$), to the right of the y axis, and below the line $y = m$ has an area of K square units, where $K > 0$.

7.2 VOLUME OF A SOLID OF REVOLUTION: CIRCULAR-DISK AND CIRCULAR-RING METHODS

We have developed a method for finding the area of a plane region. Now this process is extended to find the volume of a **solid of revolution,** which is a solid obtained by revolving a region in a plane about a line in the plane, called the **axis of revolution,** which either touches the boundary of the region or does not intersect the region at all. For example, if the region bounded by a semicircle and its diameter is revolved about the diameter, a sphere is generated (see Fig. 7.2.1). A right-circular cone is generated if the region bounded by a right triangle is revolved about one of its legs (see Fig. 7.2.2).

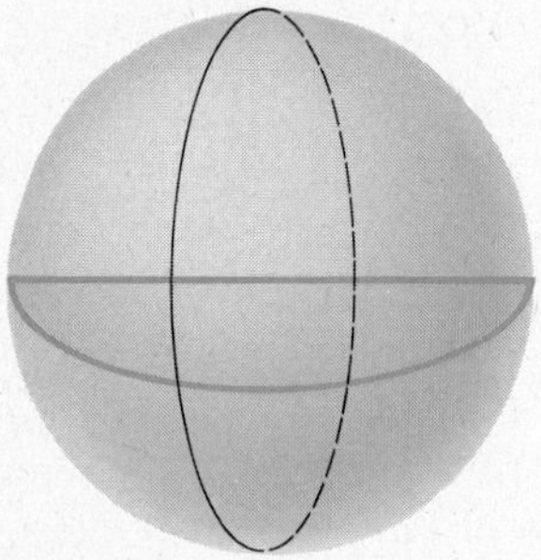
FIGURE 7.2.1

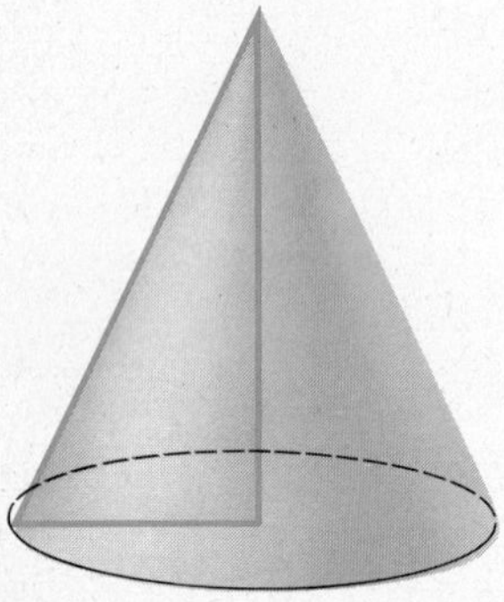
FIGURE 7.2.2

We must first define what is meant by the *volume* of a solid of revolution; that is, we wish to assign a number, for example V, to what we intuitively think of as the measure of the volume of such a solid. We define the measure of the volume of a right-circular cylinder by $\pi r^2 h$, where r and h are, respectively, the number of units in the base radius and altitude.

Consider first the case where the axis of revolution is a boundary of the region that is revolved. Let the function f be continuous on the closed interval $[a, b]$ and assume that $f(x) \geq 0$ for all x in $[a, b]$. Let R be the region bounded by the curve $y = f(x)$, the x axis, and the lines $x = a$ and $x = b$. Let S be the solid of revolution obtained by revolving the region R about the x axis. We proceed to find a suitable definition for the number V that gives the measure of the volume of S.

Let Δ be a partition of the closed interval $[a, b]$ given by

$$a = x_0 < x_1 < x_2 < \cdots < x_{n-1} < x_n = b$$

Then there are n subintervals of the form $[x_{i-1}, x_i]$, where $i = 1, 2, \ldots, n$, with the length of the ith subinterval being $\Delta_i x = x_i - x_{i-1}$. Choose any point ξ_i, with $x_{i-1} \leq \xi_i \leq x_i$, in each subinterval, and draw the rectangles having widths $\Delta_i x$ units and altitudes $f(\xi_i)$ units. Figure 7.2.3 shows the region R together with the ith rectangle.

When the ith rectangle is revolved about the x axis, we obtain a circular disk in the form of a right-circular cylinder whose base radius is given by $f(\xi_i)$ units and whose altitude is given by $\Delta_i x$ units, as shown in Fig. 7.2.4. The measure of the volume of this circular disk, which we denote by $\Delta_i V$, is

$$\Delta_i V = \pi[f(\xi_i)]^2 \Delta_i x \tag{1}$$

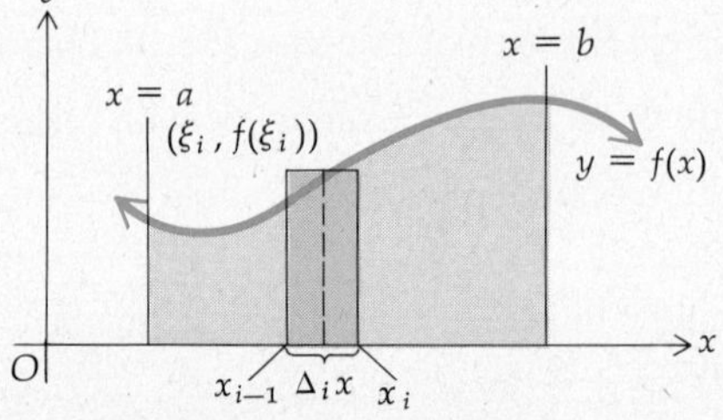

FIGURE 7.2.3

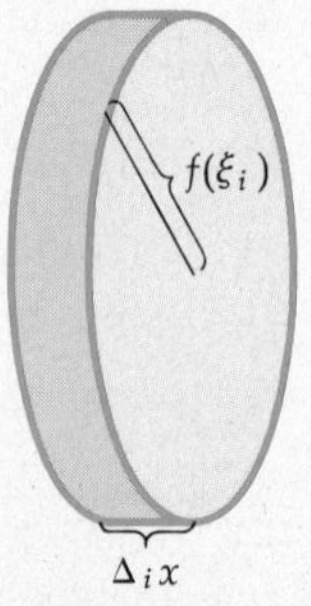

FIGURE 7.2.4

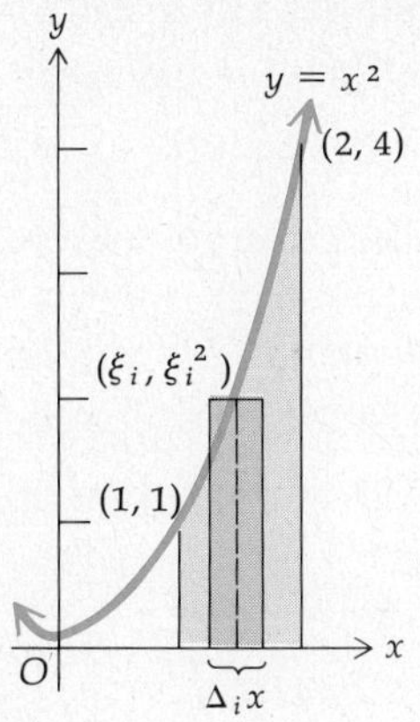

FIGURE 7.2.5

Because there are n rectangles, n circular disks are obtained in this way, and the sum of the measures of the volumes of these n circular disks is

$$\sum_{i=1}^{n} \Delta_i V = \sum_{i=1}^{n} \pi[f(\xi_i)]^2 \, \Delta_i x \tag{2}$$

which is a Riemann sum.

The Riemann sum in (2) is an approximation of what we intuitively think of as the number of cubic units in the volume of the solid of revolution. The smaller we take the norm $\|\Delta\|$ of the partition, the larger will be n, and the closer this approximation will be to the number V we wish to assign to the measure of the volume. We therefore define V to be the limit of the Riemann sum in (2) as $\|\Delta\|$ approaches zero. This limit exists because f^2 is continuous on $[a, b]$, which is true because f is continuous there. We have, then, the following definition.

7.2.1 Definition Let the function f be continuous on the closed interval $[a, b]$, and assume that $f(x) \geq 0$ for all x in $[a, b]$. If S is the solid of revolution obtained by revolving about the x axis the region bounded by the curve $y = f(x)$, the x axis, and the lines $x = a$ and $x = b$, and if V is the number of cubic units in the volume of S, then

$$V = \lim_{\|\Delta\| \to 0} \sum_{i=1}^{n} \pi[f(\xi_i)]^2 \, \Delta_i x = \pi \int_a^b [f(x)]^2 \, dx \tag{3}$$

● ILLUSTRATION 1 We find the volume of the solid of revolution generated when the region bounded by the curve $y = x^2$, the x axis, and the lines $x = 1$ and $x = 2$ is revolved about the x axis. Refer to Fig. 7.2.5, showing the region and a rectangular element of area. Figure 7.2.6 shows an element of volume and the solid of revolution. The measure of the volume of the circular disk is given by

$$\Delta_i V = \pi(\xi_i^{\,2})^2 \, \Delta_i x = \pi \xi_i^{\,4} \, \Delta_i x$$

Then

$$\begin{aligned} V &= \lim_{\|\Delta\| \to 0} \sum_{i=1}^{n} \pi \xi_i^{\,4} \, \Delta_i x \\ &= \pi \int_1^2 x^4 \, dx \\ &= \pi(\tfrac{1}{5}x^5) \Big]_1^2 \\ &= \tfrac{31}{5}\pi \end{aligned}$$

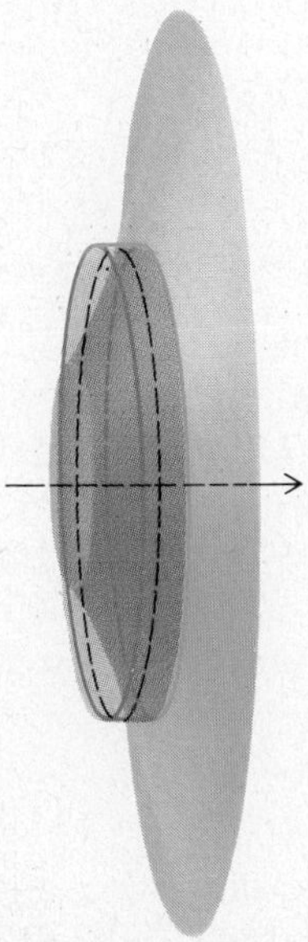
FIGURE 7.2.6

Therefore the volume of the solid of revolution is $\frac{31}{5}\pi$ cubic units. ●

A definition similar to Definition 7.2.1 applies when both the axis of revolution and a boundary of the revolved region is the y axis or any line parallel to either the x axis or the y axis.

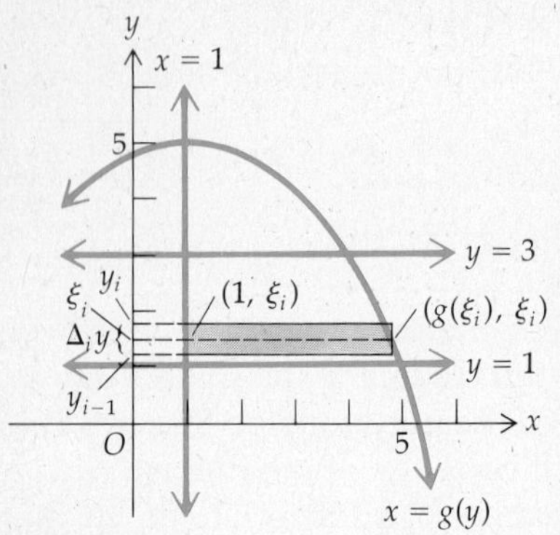

FIGURE 7.2.7

Example 1 Find the volume of the solid generated by revolving about the line $x = 1$ the region bounded by the curve $(x - 1)^2 = 20 - 4y$ and the lines $x = 1$, $y = 1$, and $y = 3$ and to the right of $x = 1$.

Solution The region and a rectangular element of area are shown in Fig. 7.2.7. An element of volume and the solid of revolution appear in Fig. 7.2.8.

In the equation of the curve we solve for x and have

$$x = \sqrt{20 - 4y} + 1$$

Let $g(y) = \sqrt{20 - 4y} + 1$. We take a partition of the interval $[1, 3]$ on the y axis. Then if $\Delta_i V$ cubic units is the volume of the ith circular disk,

$$\begin{aligned}\Delta_i V &= \pi[g(\xi_i) - 1]^2\, \Delta_i y\\ &= \pi[(\sqrt{20 - 4\xi_1} + 1) - 1]^2\, \Delta_i y\\ &= \pi(20 - 4\xi_i)\, \Delta_i y\end{aligned}$$

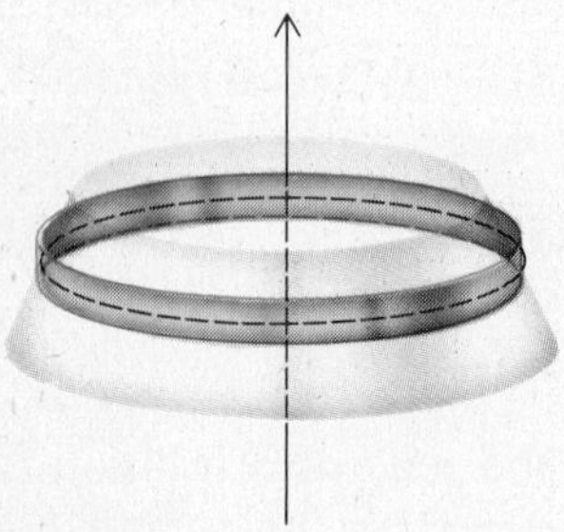

FIGURE 7.2.8

If V cubic units is the volume of the solid of revolution,

$$\begin{aligned}V &= \lim_{\|\Delta\| \to 0} \sum_{i=1}^{n} \pi(20 - 4\xi_i)\, \Delta_i y\\ &= \pi \int_1^3 (20 - 4y)\, dy\\ &= \pi\Big[20y - 2y^2\Big]_1^3\\ &= \pi[(60 - 18) - (20 - 2)]\\ &= 24\pi\end{aligned}$$

The volume of the solid of revolution is therefore 24π cubic units.

Now suppose that the axis of revolution is not a boundary of the region being revolved. Let f and g be two continuous functions on the closed interval $[a, b]$, and assume that $f(x) \geq g(x) \geq 0$ for all x in $[a, b]$. Let R be the region bounded by the curves $y = f(x)$ and $y = g(x)$ and the lines $x = a$ and $x = b$. Let S be the solid of revolution generated by revolving the region R about the x axis. If V cubic units is the volume of S, we wish to find a value for V.

Let Δ be a partition of the interval $[a, b]$, given by

$$a = x_0 < x_1 < x_2 < \cdots < x_{n-1} < x_n = b$$

and the ith subinterval $[x_{i-1}, x_i]$ has length $\Delta_i x = x_i - x_{i-1}$. In the ith subinterval choose any ξ_i, with $x_{i-1} \leq \xi_i \leq x_i$. Consider the n rectangular

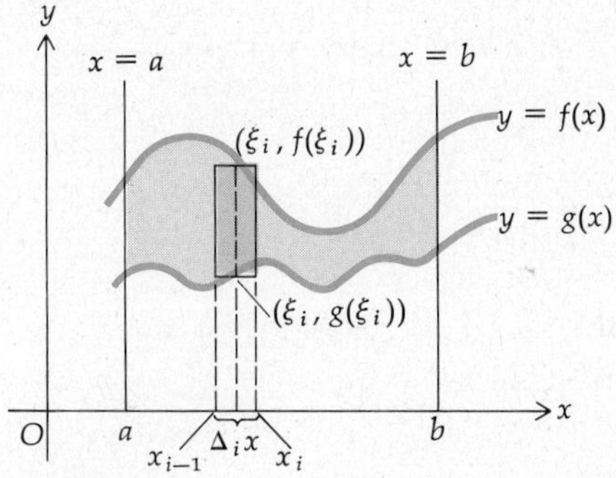

FIGURE 7.2.9

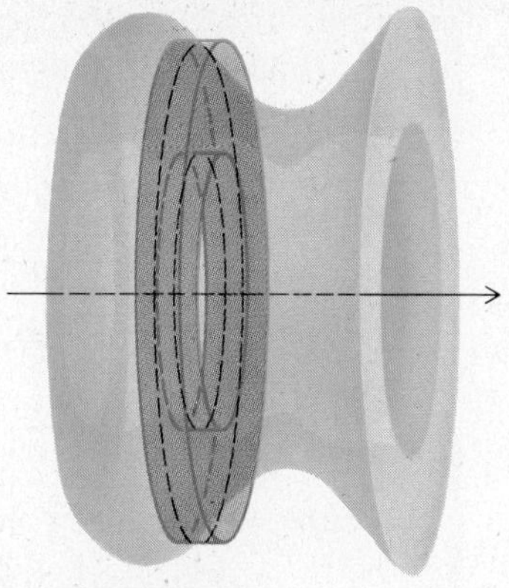
FIGURE 7.2.10

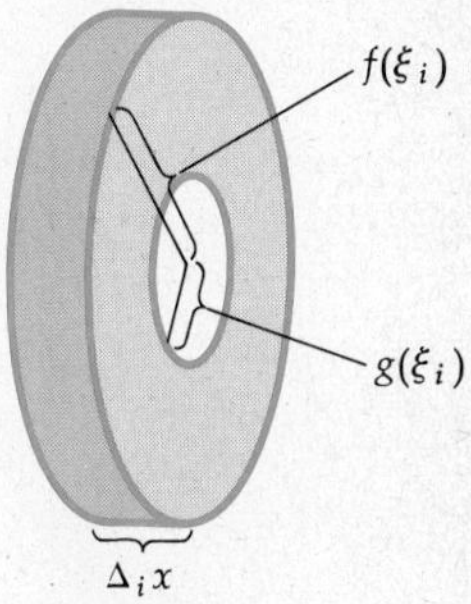

FIGURE 7.2.11

elements of area for the region R. See Fig. 7.2.9 illustrating the region and the ith rectangle, and Fig. 7.2.10 showing the solid of revolution.

When the ith rectangle is revolved about the x axis, a circular ring (or "washer"), as shown in Fig. 7.2.11, is obtained. The number giving the difference of the measures of the areas of the two circular regions is $(\pi[f(\xi_i)]^2 - \pi[g(\xi_i)]^2)$ and the thickness is $\Delta_i x$ units. Therefore the measure of the volume of the circular ring is given by

$$\Delta_i V = \pi([f(\xi_i)]^2 - [g(\xi_i)]^2)\,\Delta_i x$$

The sum of the measures of the volumes of the n circular rings formed by revolving the n rectangular elements of area about the x axis is

$$\sum_{i=1}^{n} \Delta_i V = \sum_{i=1}^{n} \pi([f(\xi_i)]^2 - [g(\xi_i)]^2)\,\Delta_i x \tag{4}$$

The number of cubic units in the volume, then, is defined to be the limit of the Riemann sum in (4) as $\|\Delta\|$ approaches zero. The limit exists since f^2 and g^2 are continuous on $[a, b]$ because f and g are continuous there.

7.2.2 Definition Let the functions f and g be continuous on the closed interval $[a, b]$, and assume that $f(x) \geq g(x) \geq 0$ for all x in $[a, b]$. Then if V cubic units is the volume of the solid of revolution generated by revolving about the x axis the region bounded by the curves $y = f(x)$ and $y = g(x)$ and the lines $x = a$ and $x = b$,

$$V = \lim_{\|\Delta\| \to 0} \sum_{i=1}^{n} \pi([f(\xi_i)]^2 - [g(\xi_i)]^2)\,\Delta_i x = \pi \int_a^b ([f(x)]^2 - [g(x)]^2)\,dx \tag{5}$$

As before, a similar definition applies when the axis of revolution is the y axis or any line parallel to either the x axis or the y axis.

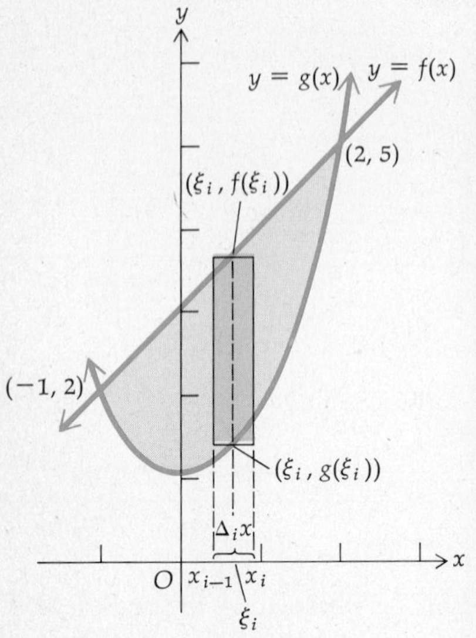

FIGURE 7.2.12

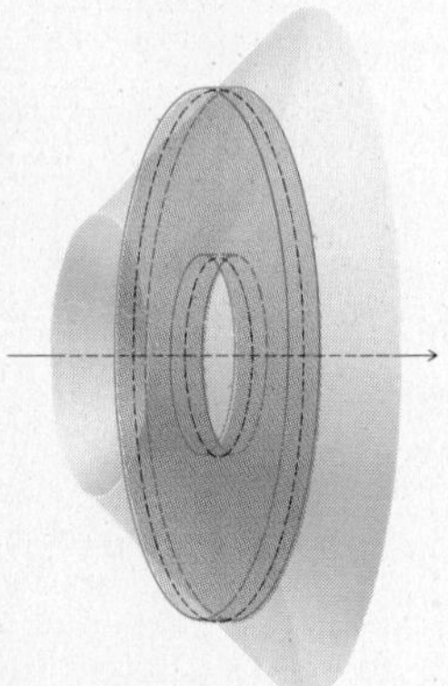
FIGURE 7.2.13

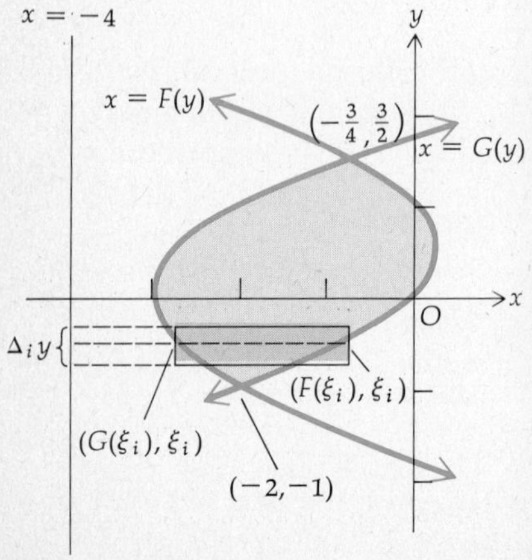

FIGURE 7.2.14

Example 2 Find the volume of the solid generated by revolving about the x axis the region bounded by the parabola $y = x^2 + 1$ and the line $y = x + 3$.

Solution The points of intersection are $(-1, 2)$ and $(2, 5)$. Figure 7.2.12 shows the region and a rectangular element of area. An element of volume and the solid of revolution are shown in Fig. 7.2.13.

If $f(x) = x + 3$ and $g(x) = x^2 + 1$, the measure of the volume of the circular ring is

$$\Delta_i V = \pi([f(\xi_i)]^2 - [g(\xi_i)]^2)\, \Delta_i x$$

If V cubic units is the volume of the solid, then

$$\begin{aligned}
V &= \lim_{\|\Delta\| \to 0} \sum_{i=1}^{n} \pi([f(\xi_i)]^2 - [g(\xi_i)]^2)\, \Delta_i x \\
&= \pi \int_{-1}^{2} ([f(x)]^2 - [g(x)]^2)\, dx \\
&= \pi \int_{-1}^{2} [(x + 3)^2 - (x^2 + 1)^2]\, dx \\
&= \pi \int_{-1}^{2} [-x^4 - x^2 + 6x + 8]\, dx \\
&= \pi \Big[-\tfrac{1}{5}x^5 - \tfrac{1}{3}x^3 + 3x^2 + 8x \Big]_{-1}^{2} \\
&= \pi[(-\tfrac{32}{5} - \tfrac{8}{3} + 12 + 16) - (\tfrac{1}{5} + \tfrac{1}{3} + 3 - 8)] \\
&= \tfrac{117}{5}\pi
\end{aligned}$$

Therefore the volume of the solid of revolution is $\frac{117}{5}\pi$ cubic units.

Example 3 Find the volume of the solid generated by revolving about the line $x = -4$ the region bounded by the two parabolas $x = y - y^2$ and $x = y^2 - 3$.

Solution The curves intersect at the points $(-2, -1)$ and $(-\frac{3}{4}, \frac{3}{2})$. The region and a rectangular element of area are shown in Fig. 7.2.14. Figure 7.2.15 shows the solid of revolution as well as an element of volume, which is a circular ring.

Let $F(y) = y - y^2$ and $G(y) = y^2 - 3$. The number of cubic units in the volume of the circular ring is

$$\Delta_i V = \pi([4 + F(\xi_i)]^2 - [4 + G(\xi_i)]^2)\, \Delta_i y$$

Thus

$$\begin{aligned}
V &= \lim_{\|\Delta\| \to 0} \sum_{i=1}^{n} \pi([4 + F(\xi_i)]^2 - [4 + G(\xi_i)]^2)\, \Delta_i y \\
&= \pi \int_{-1}^{3/2} [(4 + y - y^2)^2 - (4 + y^2 - 3)^2]\, dy
\end{aligned}$$

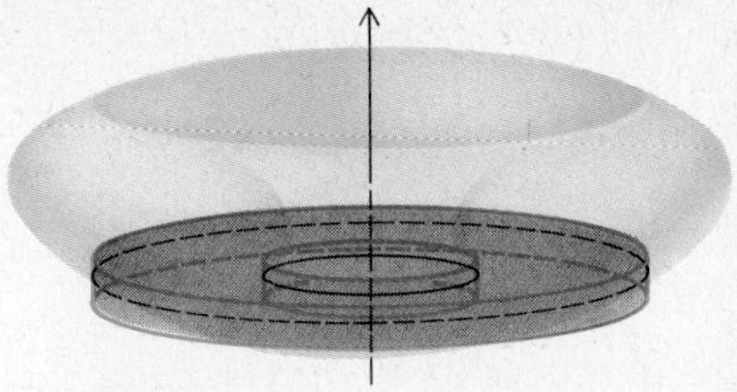

FIGURE 7.2.15

$$= \pi \int_{-1}^{3/2} (-2y^3 - 9y^2 + 8y + 15)\, dy$$

$$= \pi \left[-\tfrac{1}{2}y^4 - 3y^3 + 4y^2 + 15y \right]_{-1}^{3/2}$$

$$= \tfrac{875}{32}\pi$$

The volume of the solid of revolution is then $\frac{875}{32}\pi$ cubic units.

EXERCISES 7.2

In Exercises 1 through 8, find the volume of the solid of revolution when the given region of Fig. 7.2.16 is revolved about the indicated line. An equation of the curve in the figure is $y^2 = x^3$.

1. OAC about the x axis
2. OAC about the line AC
3. OAC about the line BC
4. OAC about the y axis
5. OBC about the y axis
6. OBC about the line BC
7. OBC about the line AC
8. OBC about the x axis

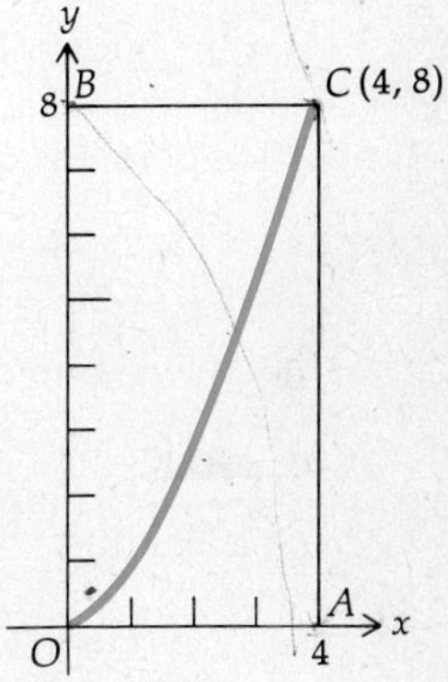

FIGURE 7.2.16

In Exercises 9 through 12, find the volume of the solid of revolution generated by revolving about the indicated line the region bounded by the curve $x = y^2$, the x axis, and the line $x = 4$.

9. the line $x = 4$ **10.** the x axis **11.** the y axis **12.** the line $y = 2$

13. Find the volume of the solid generated by revolving about the x axis the region bounded by the curve $y = x^3$, the x axis, and the line $x = 2$.
14. Find the volume of the solid generated by revolving the region in Exercise 13 about the y axis.
15. Find the volume of the sphere generated by revolving about a diameter the region enclosed by the circle $x^2 + y^2 = r^2$.
16. Find by integration the volume of a right-circular cone of altitude h units and base radius a units.
17. Find the volume of the solid of revolution generated if the region bounded by one arch of the sine curve is revolved about the x axis.
18. Find the volume of the solid generated if the region bounded by the curve $y = \sin^2 x$ and the x axis from $x = 0$ to $x = \pi$ is revolved about the x axis.
19. Find the volume of the solid generated if the region of Exercise 18 is revolved about the line $y = 1$.
20. The region bounded by the y axis and the curves $y = \sin x$ and $y = \cos x$ for $0 \le x \le \frac{1}{4}\pi$ is revolved about the x axis. Find the volume of the solid of revolution generated.
21. Find the volume of the solid generated by revolving about the line $x = -4$ the region bounded by that line and the parabola $x = 4 + 6y - 2y^2$.

22. Find the volume of the solid generated by revolving about the x axis the region bounded by the parabola $y^2 = 4x$ and the line $y = x$.

23. Find the volume of the solid generated by revolving the region of Exercise 22 about the line $x = 4$.

24. Find the volume of the solid generated by revolving about the y axis the region bounded by the line through (1, 3) and (3, 7), and the lines $y = 3$, $y = 7$, and $x = 0$.

25. Find the volume of the solid generated by revolving about the line $y = -3$ the region bounded by the two parabolas $y = x^2$ and $y = 1 + x - x^2$.

26. Find the volume of the solid generated by revolving about the x axis the region bounded by the graph of $y = |x - 3|$, and the lines $x = 1$, $x = 4$, and $y = 0$.

27. Find the volume of the solid generated by revolving about the x axis the region bounded by the lines $y = [\![\sqrt{x+1}]\!]$, $x = 3$, $x = 7$, and $y = 0$.

28. An oil tank in the shape of a sphere has a diameter of 60 ft. How much oil does the tank contain if the depth of the oil is 25 ft?

29. A paraboloid of revolution is obtained by revolving the parabola $y^2 = 4px$ about the x axis. Find the volume bounded by a paraboloid of revolution and a plane perpendicular to its axis if the plane is 10 cm from the vertex, and if the plane section of intersection is a circle having a radius of 6 cm.

30. Find the volume of the solid generated by revolving about the x axis the region bounded by the loop of the curve whose equation is $2y^2 = x(x^2 - 4)$.

31. Find the volume of the solid generated when the region bounded by one loop of the curve whose equation is $x^2y^2 = (x^2 - 9)(1 - x^2)$ is revolved about the x axis.

32. The region bounded by a pentagon having vertices at $(-4, 4)$, $(-2, 0)$, $(0, 8)$, $(2, 0)$, and $(4, 4)$ is revolved about the x axis. Find the volume of the solid generated.

33. A solid of revolution is formed by revolving about the x axis the region bounded by the curve $y = \sqrt{2x + 4}$, the x axis, the y axis, and the line $x = c$ $(c > 0)$. For what value of c will the volume be 12π cubic units?

34. The region in the first quadrant bounded by the curve $y = \cos x$ and the lines $y = 1$ and $x = \frac{1}{2}\pi$ is revolved about the x axis. Find the volume of the solid generated.

7.3 VOLUME OF A SOLID OF REVOLUTION: CYLINDRICAL-SHELL METHOD

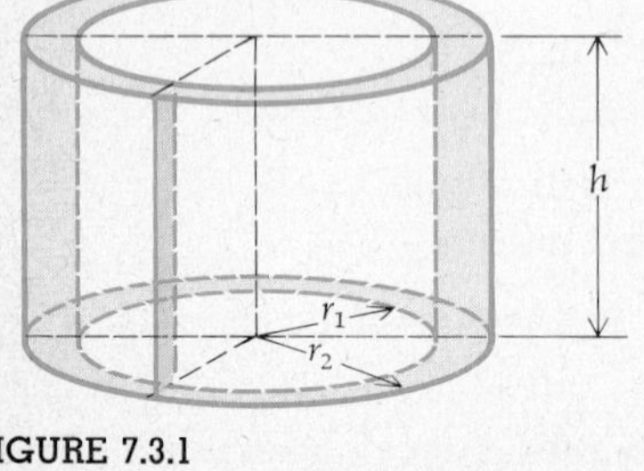

FIGURE 7.3.1

In the preceding section we found the volume of a solid of revolution by taking the rectangular elements of area perpendicular to the axis of revolution, and the element of volume was either a circular disk or a circular ring. If a rectangular element of area is parallel to the axis of revolution, then when this element of area is revolved about the axis of revolution, a *cylindrical shell* is obtained. A **cylindrical shell** is a solid contained between two cylinders having the same center and axis. Such a cylindrical shell is shown in Fig. 7.3.1.

If the cylindrical shell has an inner radius r_1 units, outer radius r_2 units, and altitude h units, then its volume V cubic units is given by

$$V = \pi r_2^2 h - \pi r_1^2 h \tag{1}$$

Let R be the region bounded by the curve $y = f(x)$, the x axis, and the lines $x = a$ and $x = b$, where f is continuous on the closed interval $[a, b]$ and $f(x) \geq 0$ for all x in $[a, b]$; furthermore assume that $a \geq 0$. Such

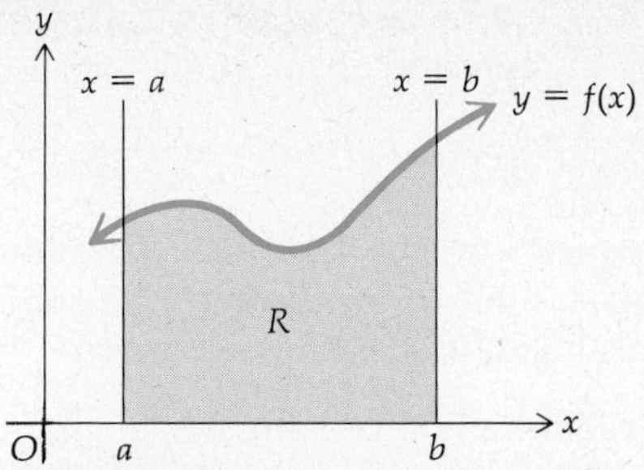

FIGURE 7.3.2

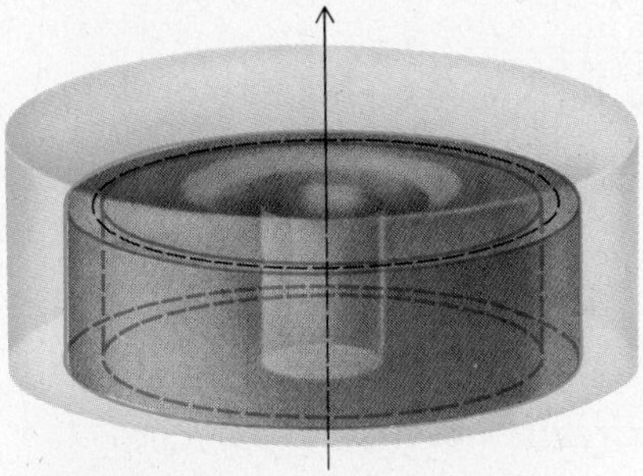
FIGURE 7.3.3

a region is shown in Fig. 7.3.2. If R is revolved about the y axis, a solid of revolution S is generated. Such a solid is shown in Fig. 7.3.3. To find the volume of S when the rectangular elements of area are taken parallel to the y axis, we proceed in the following manner.

Let Δ be a partition of the closed interval $[a, b]$ given by

$$a = x_0 < x_1 < x_2 < \cdots < x_{n-1} < x_n = b$$

Let m_i be the midpoint of the ith subinterval $[x_{i-1}, x_i]$. Then $m_i = \frac{1}{2}(x_{i-1} + x_i)$. Consider the rectangle having altitude $f(m_i)$ units and width $\Delta_i x$ units. If this rectangle is revolved about the y axis, a cylindrical shell is obtained. Figure 7.3.3 shows the cylindrical shell generated by the rectangular element of area.

If $\Delta_i V$ gives the measure of the volume of this cylindrical shell, we have, from formula (1), where $r_1 = x_{i-1}$, $r_2 = x_i$, and $h = f(m_i)$,

$$\Delta_i V = \pi x_i^2 f(m_i) - \pi x_{i-1}^2 f(m_i)$$

$$= \pi(x_i^2 - x_{i-1}^2) f(m_i)$$

$$\Delta_i V = \pi(x_i - x_{i-1})(x_i + x_{i-1}) f(m_i) \tag{2}$$

Because $x_i - x_{i-1} = \Delta_i x$, and because $x_i + x_{i-1} = 2m_i$, then from (2),

$$\Delta_i V = 2\pi m_i f(m_i)\, \Delta_i x \tag{3}$$

If n rectangular elements of area are revolved about the y axis, n cylindrical shells are obtained. The sum of the measures of their volumes is

$$\sum_{i=1}^{n} \Delta_i V = \sum_{i=1}^{n} 2\pi m_i f(m_i)\, \Delta_i x \tag{4}$$

which is a Riemann sum.

Then we define the measure of the volume of the solid of revolution to be the limit of the Riemann sum in (4) as $\|\Delta\|$ approaches zero. The limit exists because if f is continuous on $[a, b]$, so is the function having function values $2\pi x f(x)$.

7.3.1 Definition Let the function f be continuous on the closed interval $[a, b]$, where $a \geq 0$. Assume that $f(x) \geq 0$ for all x in $[a, b]$. If R is the region bounded by the curve $y = f(x)$, the x axis, and the lines $x = a$ and $x = b$, if S is the solid of revolution obtained by revolving R about the y axis, and if V cubic units is the volume of S, then

$$V = \lim_{\|\Delta\| \to 0} \sum_{i=1}^{n} 2\pi m_i f(m_i)\, \Delta_i x = 2\pi \int_a^b x f(x)\, dx \tag{5}$$

Definition 7.3.1 is consistent with Definitions 7.2.1 and 7.2.2 for the measure of the volume of a solid of revolution for which the definitions apply.

The formula for the measure of the volume of the shell is easily remembered by noticing that $2\pi m_i$, $f(m_i)$, and $\Delta_i x$ are, respectively, the numbers giving the circumference of the circle having as radius the mean of the inner and outer radii of the shell, the altitude of the shell, and the thickness.

Example 1 The region bounded by the curve $y = x^2$, the x axis, and the line $x = 2$ is revolved about the y axis. Find the volume of the solid generated. Take the elements of area parallel to the axis of revolution.

Solution Figure 7.3.4 shows the region and a rectangular element of area. Figure 7.3.5 shows the solid of revolution and the cylindrical shell obtained by revolving the rectangular element of area about the y axis.

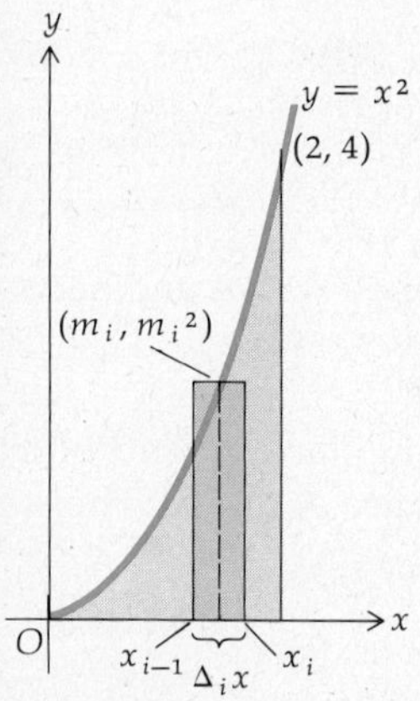

FIGURE 7.3.4

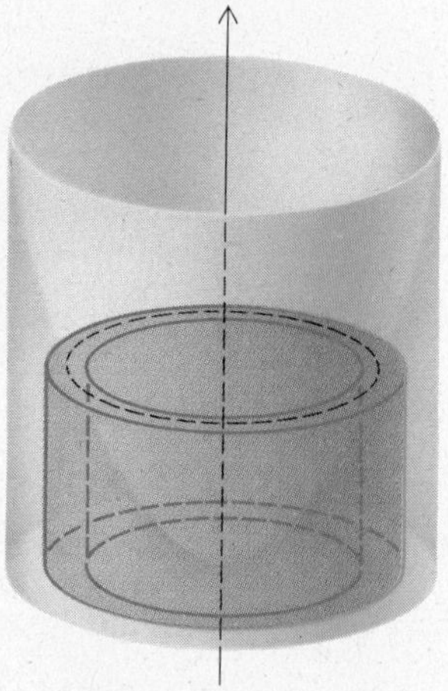

FIGURE 7.3.5

The element of volume is a cylindrical shell the measure of whose volume is

$$\Delta_i V = 2\pi m_i(m_i^2)\,\Delta_i x = 2\pi m_i^3\,\Delta_i x$$

Thus

$$\begin{aligned} V &= \lim_{\|\Delta\| \to 0} \sum_{i=1}^{n} 2\pi m_i^3\,\Delta_i x \\ &= 2\pi \int_0^2 x^3\,dx \\ &= 2\pi\left(\tfrac{1}{4}x^4\right)\Big]_0^2 \\ &= 8\pi \end{aligned}$$

Therefore the volume of the solid of revolution is 8π cubic units.

Example 2 The region bounded by the curve $y = x^2$ and the lines $y = 1$ and $x = 2$ is revolved about the line $y = -3$. Find the volume of the solid generated by taking the rectangular elements of area parallel to the axis of revolution.

Solution The region and a rectangular element of area are shown in Fig. 7.3.6.

The equation of the curve is $y = x^2$. Solving for x we obtain $x = \pm\sqrt{y}$. Because $x > 0$ for the given region, $x = \sqrt{y}$.

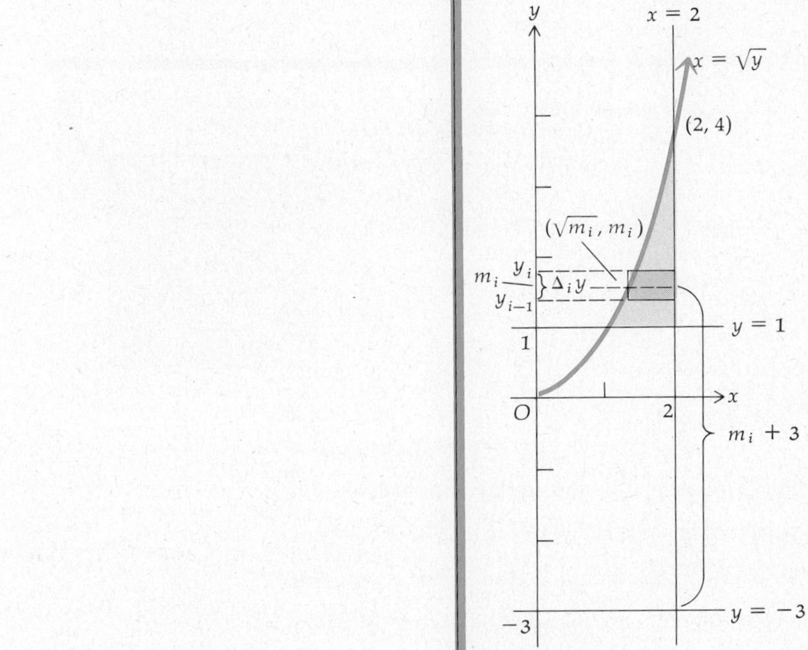

FIGURE 7.3.6

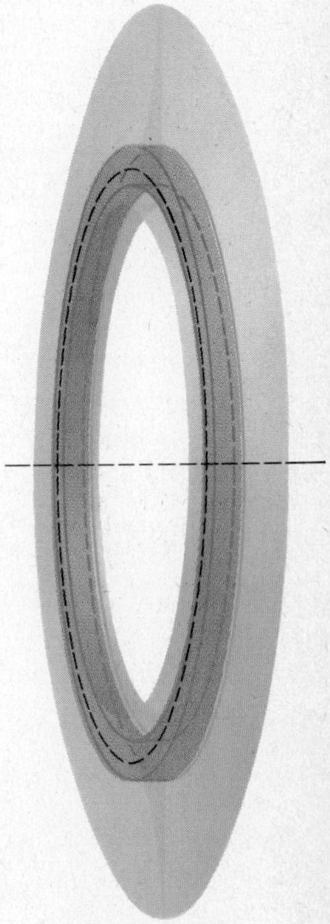

FIGURE 7.3.7

The solid of revolution as well as a cylindrical shell element of volume is shown in Fig. 7.3.7. The outer radius of the cylindrical shell is $(y_i + 3)$ units and the inner radius is $(y_{i-1} + 3)$ units. Hence the mean of the inner and outer radii is $(m_i + 3)$ units. Because the altitude and thickness of the cylindrical shell are, respectively, $(2 - \sqrt{m_i})$ units and $\Delta_i y$ units,

$$\Delta_i V = 2\pi(m_i + 3)(2 - \sqrt{m_i})\,\Delta_i y$$

Hence, if V cubic units is the volume of the solid of revolution,

$$V = \lim_{\|\Delta\| \to 0} \sum_{i=1}^{n} 2\pi(m_i + 3)(2 - \sqrt{m_i})\,\Delta_i y$$

$$= \int_1^4 2\pi(y+3)(2-\sqrt{y})\,dy$$

$$= 2\pi \int_1^4 (-y^{3/2} + 2y - 3y^{1/2} + 6)\,dy$$

$$= 2\pi \left[-\tfrac{2}{5}y^{5/2} + y^2 - 2y^{3/2} + 6y \right]_1^4$$

$$= \tfrac{66}{5}\pi$$

Therefore the volume is $\frac{66}{5}\pi$ cubic units.

EXERCISES 7.3

1–12. Solve Exercises 1 through 12 in Sec. 7.2 by the cylindrical-shell method.

In Fig. 7.3.8 the region bounded by the x axis, the line $x = 1$, and the curve $y = x^2$ is denoted by R_1; the region bounded by the two curves $y = x^2$ and $y^2 = x$ is denoted by R_2; the region bounded by the y axis, the line $y = 1$, and the curve $y^2 = x$ is denoted by R_3. In Exercises 13 through 20, find the volume of the solid generated when the indicated region is revolved about the given line.

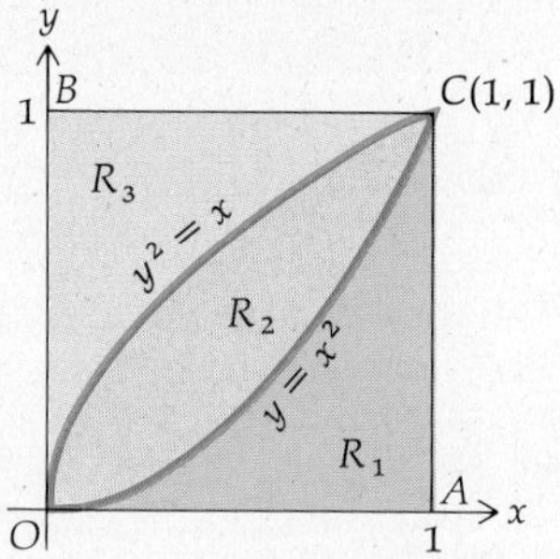

FIGURE 7.3.8

13. R_1 is revolved about the y axis; the rectangular elements are parallel to the axis of revolution.

14. Same as Exercise 13, but the rectangular elements are perpendicular to the axis of revolution.

15. R_2 is revolved about the x axis; the rectangular elements are parallel to the axis of revolution.

16. Same as Exercise 15, but the rectangular elements are perpendicular to the axis of revolution.

17. R_3 is revolved about the line $y = 2$; the rectangular elements are parallel to the axis of revolution.

18. Same as Exercise 17, but the rectangular elements are perpendicular to the axis of revolution.

19. R_2 is revolved about the line $x = -2$; the rectangular elements are parallel to the axis of revolution.

20. Same as Exercise 19, but the rectangular elements are perpendicular to the axis of revolution.

In Exercises 21 through 24, the region of Exercise 17 in Exercises 7.1 is revolved about the indicated axis. Find the volume of the solid generated.

21. the x axis **22.** the y axis **23.** the line $x = 2$ **24.** the line $y = 2$

25. Find the volume of the solid generated if the region bounded by the parabola $y^2 = 4ax$ $(a > 0)$ and the line $x = a$ is revolved about $x = a$.

26. Find the volume of the solid generated by revolving about the x axis the region bounded by the curves $y = x^3$ and $x = y^3$. Take the rectangular elements of area parallel to the axis of revolution.

27. Find the volume of the solid generated by revolving about the line $y = 1$ the region bounded by that line and the parabola $x^2 = 4y$. Take the rectangular elements of area parallel to the axis of revolution.

28. Find the volume of the solid generated by revolving about the y axis the region bounded by the curve $x^{2/3} + y^{2/3} = a^{2/3}$.

29. Find the volume of the solid generated by revolving about the y axis the region bounded by the curve $y = \sin x^2$, the x axis, and the lines $x = \frac{1}{2}\sqrt{\pi}$ and $x = \sqrt{\pi}$.

30. The region bounded by the y axis and the curve $x = \cos y^2$ on the interval $[0, 1]$ on the x axis is revolved about the x axis. Find the volume of the solid of revolution generated.

31. Find the volume of the solid generated by revolving about the y axis the region outside the curve $y = x^2$ and between the lines $y = 2x - 1$ and $y = x + 2$.

32. A hole of radius 2 cm is bored through a spherical shaped solid of radius 6 cm, and the axis of the hole is a diameter of the sphere. Find the volume of the part of the solid that remains.

33. A hole of radius $2\sqrt{3}$ in. is bored through the center of a spherical shaped solid of radius 4 in. Find the volume of the portion of the solid cut out.

34. Find the volume of the solid generated if the region bounded by the parabola $y^2 = 4px$ and the line $x = p$ is revolved about the y axis.

35. Find the volume of the solid generated by revolving about the y axis the region bounded by the graph of $y = |x - 3|$, and the lines $x = 1$, $x = 5$, and $y = 0$. Take the rectangular elements of area parallel to the axis of revolution.

36. The region in the first quadrant bounded by the curve $y = \cos x$ and the lines $y = 1$ and $x = \frac{1}{2}\pi$ is revolved about the line $x = \frac{1}{2}\pi$. Find the volume of the solid generated.

37. A solid of revolution is formed by revolving about the y axis the region bounded by the curve $y = \sqrt[3]{x}$, the x axis, and the line $x = c$ $(c > 0)$. Take the rectangular elements of area parallel to the axis of revolution to determine the value of c that will give a volume of 12π cubic units.

38. Find the volume of the solid generated by revolving about the x axis the region bounded by the x axis and the lines $y = [\![\sqrt{2x + 1}]\!]$, $x = 4$, and $x = 7$. Take the rectangular elements of area parallel to the axis of revolution.

7.4 VOLUME OF A SOLID HAVING KNOWN PARALLEL PLANE SECTIONS

Let S be a solid. By a **plane section** of S is meant a plane region formed by the intersection of a plane with S. In Sec. 7.2 we learned how to find the volume of a solid of revolution for which all plane sections perpendicular to the axis of revolution are circular. We now generalize this method to find the volume of a solid for which it is possible to express the area of any plane section perpendicular to a fixed line in terms of the perpendicular distance of the plane section from a fixed point. We first define a *cylinder*.

7.4.1 Definition A solid is a **right cylinder** if it is bounded by two congruent plane regions R_1 and R_2 lying in parallel planes and by a lateral surface generated by a line segment, having its endpoints on the boundaries of R_1 and R_2, which moves so that it is always perpendicular to the planes of R_1 and R_2.

Figure 7.4.1 illustrates a right cylinder. The **height** of the cylinder is the perpendicular distance between the planes of R_1 and R_2, and the base is either R_1 or R_2. If the base of the right cylinder is a region enclosed by a circle, we have a **right-circular cylinder** (see Fig. 7.4.2); if the base is a region enclosed by a rectangle, we have a **rectangular parallelepiped** (see Fig. 7.4.3).

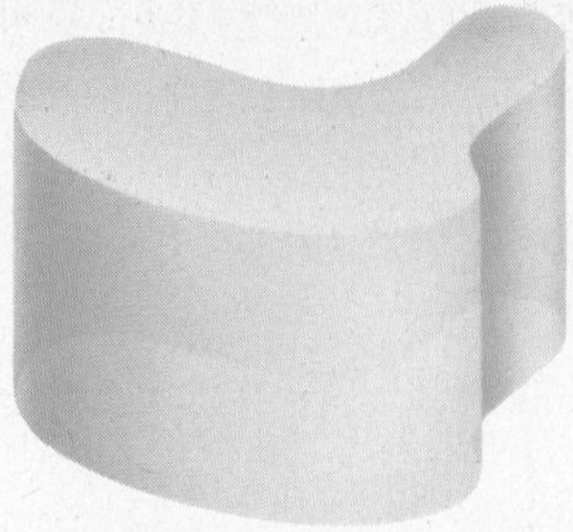

FIGURE 7.4.1

FIGURE 7.4.2

FIGURE 7.4.3

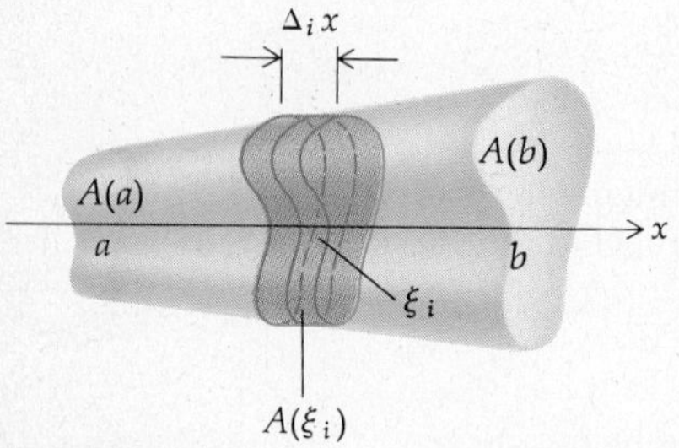

FIGURE 7.4.4

If the area of the base of a right cylinder is A square units and the height is h units, then by definition the volume of the right cylinder is measured by the product of A and h. As stated above, we are considering solids for which the area of any plane section that is perpendicular to a fixed line is a function of the perpendicular distance of the plane section from a fixed point. The solid S of Fig. 7.4.4 is such a solid, and it lies between the planes perpendicular to the x axis at a and b. We represent the number of square units in the area of the plane section of S in the plane perpendicular to the x axis at x by $A(x)$, where A is continuous on $[a, b]$.

Let Δ be a partition of the closed interval $[a, b]$ given by

$$a = x_0 < x_1 < x_2 < \cdots < x_n = b$$

There are, then, n subintervals of the form $[x_{i-1}, x_i]$, where $i = 1, 2, \ldots, n$, with the length of the ith subinterval being $\Delta_i x = x_i - x_{i-1}$. Choose any number ξ_i, with $x_{i-1} \le \xi_i \le x_i$, in each subinterval, and construct the right cylinders of heights $\Delta_i x$ units and plane section areas $A(\xi_i)$ square units. The volume of the ith cylinder is $A(\xi_i)\,\Delta_i x$ cubic units. We obtain n right cylinders, and the sum of the measures of the volumes of these n cylinders is given by the Riemann sum

$$\sum_{i=1}^{n} A(\xi_i)\,\Delta_i x \tag{1}$$

This Riemann sum is an approximation of what we intuitively think of as the measure of the volume of S, and the smaller we take the norm $\|\Delta\|$ of the partition Δ, the larger n will be, and the closer this approximation will be to the number we wish to assign to the measure of the volume. We have, then, the following definition.

7.4.2 Definition Let S be a solid such that S lies between planes drawn perpendicular to the x axis at a and b. If the measure of the area of the plane section of S drawn perpendicular to the x axis at x is given by $A(x)$, where A is continuous on $[a, b]$, then the measure of the volume of S is given by

$$V = \lim_{\|\Delta\| \to 0} \sum_{i=1}^{n} A(\xi_i)\,\Delta_i x = \int_a^b A(x)\,dx \tag{2}$$

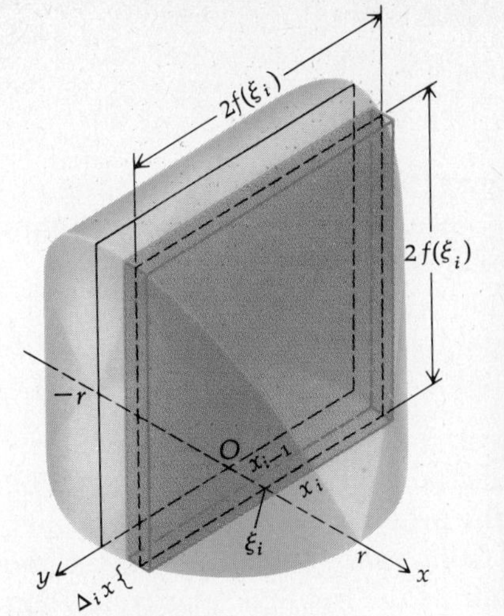

FIGURE 7.4.5

Definitions 7.2.1 and 7.2.2 are special cases of Definition 7.4.2. If in (2) we take $A(x) = \pi[f(x)]^2$, we have Eq. (3) of Sec. 7.2. If $A(x) = \pi([f(x)]^2 - [g(x)]^2)$, we have Eq. (5) of Sec. 7.2.

Example 1 If the base of a solid is the region enclosed by a circle with a radius of r units, and if all plane sections perpendicular to a fixed diameter of the base are squares, find the volume of the solid.

Solution Take the circle in the xy plane, the center at the origin, and the fixed diameter along the x axis. Therefore an equation of the circle is $x^2 + y^2 = r^2$. Figure 7.4.5 shows the solid and an element of volume, which is a right cylinder of altitude $\Delta_i x$ units and with an area of the base given by $[2f(\xi_i)]^2$ square units, where $f(x)$ is obtained by solving the equation of the circle for y and setting $y = f(x)$. This computation gives $f(x) = \sqrt{r^2 - x^2}$. Therefore, if V cubic units is the volume of the solid,

$$V = \lim_{\|\Delta\| \to 0} \sum_{i=1}^{n} [2f(\xi_i)]^2 \Delta_i x$$

$$= 4 \int_{-r}^{r} (r^2 - x^2)\, dx$$

$$= 4 \left[r^2 x - \tfrac{1}{3} x^3 \right]_{-r}^{r}$$

$$= \tfrac{16}{3} r^3$$

Example 2 A wedge is cut from a right-circular cylinder with a radius of r cm by two planes, one perpendicular to the axis of the cylinder and the other intersecting the first along a diameter of the circular plane section at an angle of measurement 60°. Find the volume of the wedge.

Solution The wedge is shown in Fig. 7.4.6. The xy plane is taken as the plane perpendicular to the axis of the cylinder, and the origin is at the point of perpendicularity. An equation of the circular plane section is then $x^2 + y^2 = r^2$. Every plane section of the wedge perpendicular to the x axis is a right triangle. An element of volume is a right cylinder having altitude $\Delta_i x$ cm, and area of the base given by $\frac{1}{2}\sqrt{3}[f(\xi_i)]^2$ cm^2, where $f(x)$ is obtained by solving the equation of the circle for y and setting $y = f(x)$, giving $f(x) = \sqrt{r^2 - x^2}$. Therefore, if V cm^3 is the volume of the wedge,

$$V = \lim_{\|\Delta\| \to 0} \sum_{i=1}^{n} \tfrac{1}{2}\sqrt{3}\,(r^2 - \xi_i^2)\, \Delta_i x$$

$$= \tfrac{1}{2}\sqrt{3} \int_{-r}^{r} (r^2 - x^2)\, dx$$

$$= \tfrac{1}{2}\sqrt{3} \left[r^2 x - \tfrac{1}{3} x^3 \right]_{-r}^{r}$$

$$= \tfrac{2}{3}\sqrt{3} r^3$$

Hence the volume of the wedge is $\frac{2}{3}\sqrt{3} r^3$ cm^3.

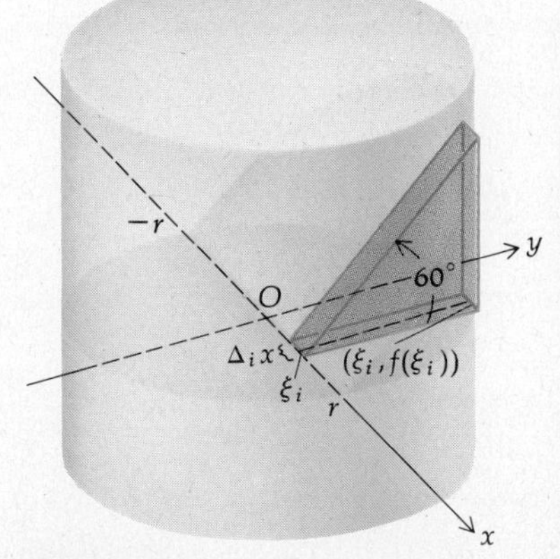

FIGURE 7.4.6

In Sec. 6.6 we obtained the prismoidal formula, which is

$$\int_a^b f(x)\,dx = \frac{b-a}{6}\left[f(a) + 4f\left(\frac{a+b}{2}\right) + f(b)\right]$$

where f is a polynomial of the third degree or less. In the following example the prismoidal formula is used to evaluate a definite integral that gives the measure of the volume of a solid.

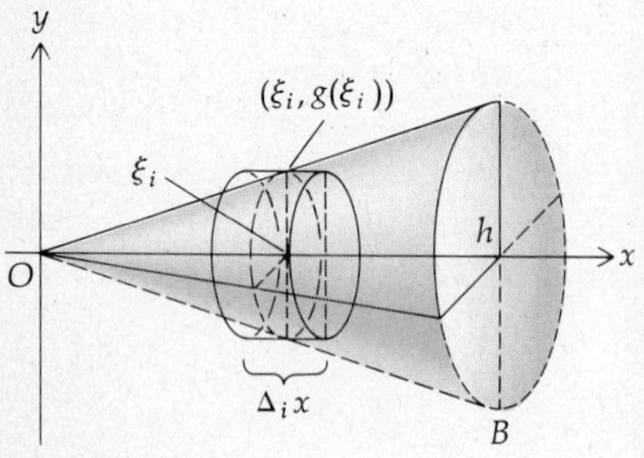

FIGURE 7.4.7

Example 3 Find by the prismoidal formula the volume of a right-circular cone of height h and base radius r.

Solution Figure 7.4.7 illustrates a right-circular cone having its vertex at the origin and its axis along the positive side of the x axis. The area of the plane section ξ_i units from the origin is $\pi[g(\xi_i)]^2$ square units, where $g(x) = (r/h)x$, which is obtained from an equation of line OA in the xy plane. So if V cubic units is the volume of the right-circular cone, then

$$V = \lim_{\|\Delta\| \to 0} \sum_{i=1}^{n} \pi[g(\xi_i)]^2 \Delta_i x$$

$$= \int_0^h \pi[g(x)]^2\,dx$$

$$= \int_0^h \pi \frac{r^2}{h^2} x^2\,dx$$

$$= \frac{\pi r^2}{h^2} \int_0^h x^2\,dx$$

Evaluating $\int_0^h x^2\,dx$ by the prismoidal formula with $f(x) = x^2$, $b = h$, $a = 0$, and $\frac{1}{2}(a + b) = \frac{1}{2}h$, we have

$$V = \frac{\pi r^2}{h^2} \cdot \frac{h}{6}\left(0 + 4 \cdot \frac{h^2}{4} + h^2\right) = \frac{\pi r^2}{6h}(2h^2) = \frac{1}{3}\pi r^2 h$$

EXERCISES 7.4

1. The base of a solid is the region enclosed by an ellipse having the equation $3x^2 + y^2 = 6$. Find the volume of the solid if all plane sections perpendicular to the x axis are squares.
2. The base of a solid is the region enclosed by the hyperbola $25x^2 - 4y^2 = 100$ and the line $x = 4$. Find the volume of the solid if all plane sections perpendicular to the x axis are squares.
3. The base of a solid is the region enclosed by a circle having a radius of 7 cm. Find the volume of the solid if all plane sections perpendicular to a fixed diameter of the base are equilateral triangles.
4. The base of a solid is the region of Exercise 2. Find the volume of the solid if all plane sections perpendicular to the x axis are equilateral triangles.
5. The base of a solid is the region of Exercise 3. Find the volume of the solid if all plane sections perpendicular to a fixed diameter of the base are isosceles triangles of height equal to the distance of the plane section from the center of the circle.

6. The base of a solid is the region enclosed by a circle with a radius of r units, and all plane sections perpendicular to a fixed diameter of the base are isosceles right triangles having the hypotenuse in the plane of the base. Find the volume of the solid.

7. Solve Exercise 6 if the isosceles right triangles have one leg in the plane of the base.

8. Find the volume of a right pyramid having a height of h units and a square base of side a units.

9. Find the volume of a tetrahedron having 3 mutually perpendicular faces and 3 mutually perpendicular edges whose lengths are 3 in., 4 in., and 7 in.

10. Solve Exercise 9 if the edges of the tetrahedron have lengths of measures a, b, and c.

11. The base of a solid is the region enclosed by an ellipse having the equation $2x^2 + y^2 = 8$. Find the volume of the solid if all plane sections perpendicular to the x axis are isosceles triangles whose height and length of base are equal.

12. The base of a solid is the region enclosed by a circle with a radius of 4 in., and each plane section perpendicular to a fixed diameter of the base is an isosceles triangle having an altitude of 10 in. and a chord of the circle as a base. Find the volume of the solid.

13. The base of a solid is the region enclosed by a circle with a radius of 9 cm, and each plane section perpendicular to a fixed diameter of the base is a square having a chord of the circle as a diagonal. Find the volume of the solid.

14. The base of a solid is the region enclosed by a circle having a radius of r units. Find the volume of the solid if all plane sections perpendicular to a fixed diameter of the base are rectangles having a height equal to one-half the length of the chord of the circle forming the base of the rectangle.

15. The base of a solid is the region enclosed by the curve $x = 2\sqrt{y}$ and the lines $x + y = 0$ and $y = 9$. Find the volume of the solid if all plane sections perpendicular to the y axis are squares having a diagonal with one endpoint on the line $x + y = 0$ and the other endpoint on the curve $x = 2\sqrt{y}$.

16. Two right-circular cylinders, each having a radius of r units, have axes that intersect at right angles. Find the volume of the solid common to the two cylinders.

17. A wedge is cut from a solid in the shape of a right-circular cylinder with a radius of r cm by a plane through a diameter of the base and inclined to the plane of the base at an angle of measurement 45°. Find the volume of the wedge.

18. A wedge is cut from a solid in the shape of a right-circular cone having a base radius of 5 ft and an altitude of 20 ft by two half planes through the axis of the cone. The angle between the two planes has a measurement of 30°. Find the volume of the wedge cut out.

In Exercises 19 through 22, use the prismoidal formula to find the volume of the given solid.

19. A right-circular cylinder of height h and base radius r.

20. A sphere of radius r.

21. A frustum of a right-circular cone of height h and base radii r_1 and r_2.

22. A right pyramid of height h and base a square having a side of length s.

7.5 WORK

The *work* done by a force acting on an object is defined in physics as "force times displacement." For example, suppose that an object is moving to the right along the x axis from a point a to a point b, and a constant force of F lb is acting on the object in the direction of motion. Then if the displacement is measured in feet, $(b - a)$ is the number of feet in the displacement. And if W is the number of foot-pounds of work done by the force, W is defined by

$$W = F(b - a) \tag{1}$$

● ILLUSTRATION 1 If W is the number of foot-pounds in the work necessary to lift a 70-lb weight to a height of 3 ft, then

$$W = 70 \cdot 3 = 210$$ ●

The unit of measurement for work depends on the units of force and distance. In the English system, where the force is measured in pounds and the distance is measured in feet, work is measured in foot-pounds. In the metric system, if the unit of mass is the kilogram and distance is measured in meters, then the unit of force is the *newton*, and the unit of work is a newton-meter, which is called a *joule*. If in the metric system the unit of mass is the gram and distance is measured in centimeters, then the unit of force is the *dyne*, and the unit of work is a dyne-centimeter, which is called an *erg*. For conversion purposes, 1 newton is 10^5 dynes, and 1 joule is 10^7 ergs.

The following illustration shows a computation for work using units in the metric system.

● ILLUSTRATION 2 We wish to find the work done in lifting a rock of mass 8 kg a distance of 4 m. We use the formula $F = Ma$, where F newtons (nt) is the force required to give a mass of M kilograms an acceleration of a meters per second per second. The force in this case is the force of gravity and the acceleration is that due to gravity, which is 9.81 m/sec². The mass is 8 kg. Therefore $M = 8$, $a = 9.81$, and

$$\begin{aligned} F &= 8(9.81) \\ &= 78.48 \end{aligned}$$

Thus we wish to find the work done by a force of 78.48 nt and a displacement of 4 m. If W joules is the work,

$$\begin{aligned} W &= (78.48)(4) \\ &= 313.92 \end{aligned}$$

Hence the work done is 313.92 joules. ●

We now consider the work done by a variable force, which is a function of the position of the object on which the force is acting. We wish to define what is meant by the term *work* in such a case.

Suppose that $f(x)$, where f is continuous on $[a, b]$, is the number of units in the force acting in the direction of motion on an object as it moves to the right along the x axis from point a to point b. Let Δ be a partition of the closed interval $[a, b]$:

$$a = x_0 < x_1 < x_2 < \cdots < x_{n-1} < x_n = b$$

The ith subinterval is $[x_{i-1}, x_i]$; and if x_{i-1} is close to x_i, the force is almost constant in this subinterval. If we assume that the force is constant in the ith subinterval and if ξ_i is any point such that $x_{i-1} \le \xi_i \le x_i$, then if $\Delta_i W$

is the number of units of work done on the object as it moves from the point x_{i-1} to the point x_i, from formula (1) we have

$$\Delta_i W = f(\xi_i)(x_i - x_{i-1})$$

Replacing $x_i - x_{i-1}$ by $\Delta_i x$ we have

$$\Delta_i W = f(\xi_i)\,\Delta_i x$$

and

$$\sum_{i=1}^{n} \Delta_i W = \sum_{i=1}^{n} f(\xi_i)\,\Delta_i x \tag{2}$$

The smaller we take the norm of the partition Δ, the larger n will be and the closer the Riemann sum in (2) will be to what we intuitively think of as the measure of the total work done. We therefore define the measure of the total work as the limit of the Riemann sum in (2).

7.5.1 Definition Let the function f be continuous on the closed interval $[a, b]$ and $f(x)$ be the number of units in the force acting on an object at the point x on the x axis. Then if W units is the **work** done by the force as the object moves from a to b, W is given by

$$W = \lim_{\|\Delta\| \to 0} \sum_{i=1}^{n} f(\xi_i)\,\Delta_i x = \int_a^b f(x)\,dx \tag{3}$$

Example 1 A particle is moving along the x axis under the action of a force of $f(x)$ lb when the particle is x ft from the origin. If $f(x) = x^2 + 4$, find the work done as the particle moves from the point where $x = 2$ to the point where $x = 4$.

Solution We take a partition of the closed interval $[2, 4]$. If W ft-lb is the work done as the particle moves from the point where $x = 2$ to the point where $x = 4$, then from Definition 7.5.1,

$$\begin{aligned}
W &= \lim_{\|\Delta\| \to 0} \sum_{i=1}^{n} f(\xi_i)\,\Delta_i x \\
&= \int_2^4 f(x)\,dx \\
&= \int_2^4 (x^2 + 4)\,dx \\
&= \frac{x^3}{3} + 4x \Big]_2^4 \\
&= \tfrac{64}{3} + 16 - (\tfrac{8}{3} + 8) \\
&= 26\tfrac{2}{3}
\end{aligned}$$

Therefore the work done is $26\frac{2}{3}$ ft-lb.

In the following example we use **Hooke's law,** which states that if a spring is stretched x units beyond its natural length, it is pulled back with a force equal to kx units, where k is a constant depending on the material and size of the spring.

Example 2 A spring has a natural length of 14 cm. If a force of 50 dynes is required to keep the spring stretched 2 cm, how much work is done in stretching the spring from its natural length to a length of 18 cm?

Solution Place the spring along the x axis with the origin at the point where the stretching starts. See Fig. 7.5.1.

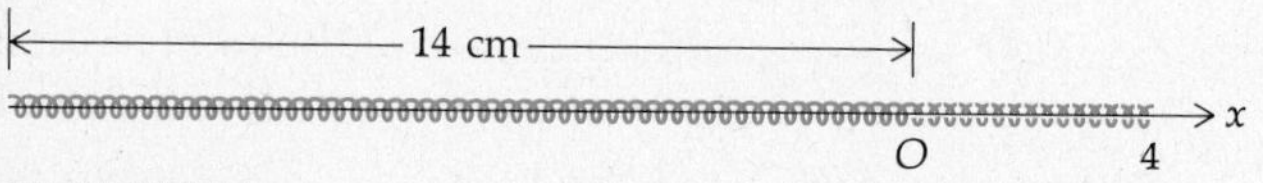

FIGURE 7.5.1

Let x be the number of centimeters the spring is stretched and $f(x)$ be the number of dynes in the force acting on the spring x cm beyond its natural length. Then, by Hooke's law,

$$f(x) = kx$$

Because $f(2) = 50$, we have

$$50 = k \cdot 2$$
$$k = 25$$

Therefore

$$f(x) = 25x$$

If W is the number of ergs (dyne-centimeters) of work done in stretching the spring from its natural length of 14 cm to a length of 18 cm,

$$\begin{aligned} W &= \lim_{\|\Delta\| \to 0} \sum_{i=1}^{n} f(\xi_i)\, \Delta_i x \\ &= \int_0^4 f(x)\, dx \\ &= \int_0^4 25x\, dx \\ &= \tfrac{25}{2}x^2 \Big]_0^4 \\ &= 200 \end{aligned}$$

Therefore the work done in stretching the spring is 200 ergs.

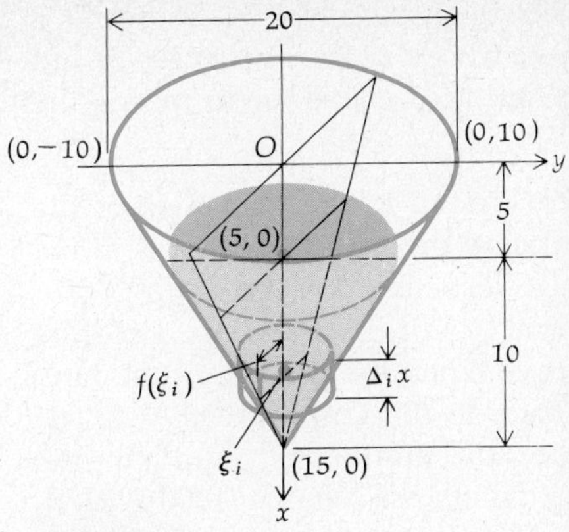

FIGURE 7.5.2

Example 3 A water tank in the form of an inverted right-circular cone is 20 ft across the top and 15 ft deep. If the surface of the water is 5 ft below the top of the tank, find the work done in pumping the water to the top of the tank.

Solution Refer to Fig. 7.5.2. The positive x axis is chosen in the downward direction because the motion is vertical. Take the origin at the top of the tank. We consider a partition of the closed interval $[5, 15]$ on the x axis and let ξ_i be any point in the ith subinterval $[x_{i-1}, x_i]$. An element of volume is a circular disk having thickness $\Delta_i x$ ft and radius $f(\xi_i)$ ft, where the function f is determined by an equation of the line through the points $(0, 10)$ and $(15, 0)$ in the form $y = f(x)$. The number of cubic feet in the volume of this element is given by $\Delta_i V = \pi[f(\xi_i)]^2 \Delta_i x$. If w is the number of pounds in the weight of 1 ft^3 of water, then the number of pounds in the weight of this element is $w\pi[f(\xi_i)]^2 \Delta_i x$, which is the force acting on the element. If x_{i-1} is close to x_i, then the distance through which this element moves is approximately ξ_i ft. Thus, if $\Delta_i W$ ft-lb is the work done in pumping the element to the top of the tank, $\Delta_i W$ is approximately $(w\pi[f(\xi_i)]^2 \Delta_i x) \cdot \xi_i$; so if W is the number of foot-pounds of work done,

$$W = \lim_{\|\Delta\| \to 0} \sum_{i=1}^{n} w\pi[f(\xi_i)]^2 \cdot \xi_i \, \Delta_i x$$

$$= w\pi \int_5^{15} [f(x)]^2 x \, dx$$

To determine $f(x)$, we find an equation of the line through the points $(15, 0)$ and $(0, 10)$ by using the intercept form:

$$\frac{x}{15} + \frac{y}{10} = 1 \quad \text{or} \quad y = -\tfrac{2}{3}x + 10$$

Therefore $f(x) = -\frac{2}{3}x + 10$, and

$$W = w\pi \int_5^{15} (-\tfrac{2}{3}x + 10)^2 x \, dx$$

$$= w\pi \int_5^{15} (\tfrac{4}{9}x^3 - \tfrac{40}{3}x^2 + 100x) \, dx$$

$$= w\pi \left[\tfrac{1}{9}x^4 - \tfrac{40}{9}x^3 + 50x^2 \right]_5^{15}$$

$$= \frac{10{,}000}{9} \pi w$$

Therefore the work done is $(10{,}000/9)\pi w$ ft-lb.

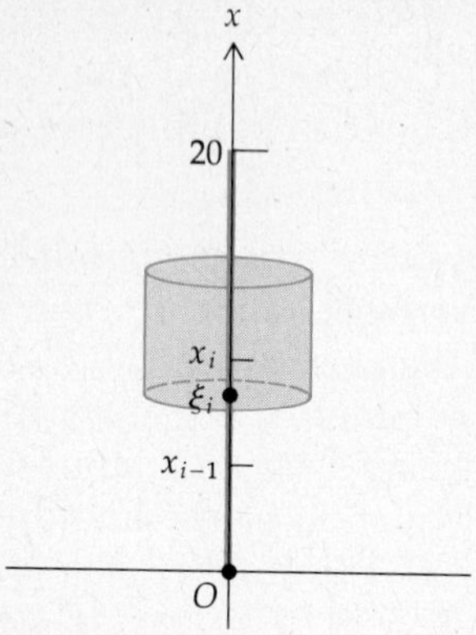

FIGURE 7.5.3

Example 4 As a water tank is being raised, water spills out at a constant rate of 2 ft³ per foot of rise. If the weight of the tank is 200 lb and it originally contains 1000 ft³ of water, find the work done in raising the tank 20 ft.

Solution Refer to Fig. 7.5.3. Here we are taking the origin at the starting point of the bottom of the tank and the positive x axis is in the upward direction because the motion is vertically upward from O. We consider a partition of the closed interval $[0, 20]$ on the x axis. Let ξ_i be any point in the ith subinterval $[x_{i-1}, x_i]$. When the bottom of the tank is at ξ_i, the number of cubic feet of water in the tank is $1000 - 2\xi_i$. If w lb is the weight of 1 ft³ of water, then the number of pounds in the weight of the tank and its contents when it is at ξ_i is $[200 + w(1000 - 2\xi_i)]$, which is the force acting on the tank. If $\Delta_i W$ ft-lb is the work done in raising the tank through the ith subinterval, then $\Delta_i W$ is approximately $[200 + w(1000 - 2\xi_i)]\ \Delta_i x$. We use the terminology "approximately" because we are assuming that the amount of water in the tank is constant throughout this subinterval. If W ft-lb is the work done in raising the tank 20 ft, we have

$$W = \lim_{\|\Delta\| \to 0} \sum_{i=1}^{n} [200 + w(1000 - 2\xi_i)]\ \Delta_i x$$

$$= \int_0^{20} [200 + w(1000 - 2x)]\ dx$$

$$= \Big[200x + w(1000x - x^2)\Big]_0^{20}$$

$$= 4000 + 19{,}600w$$

Therefore the work done is $(4000 + 19{,}600w)$ ft-lb.

EXERCISES 7.5

In Exercises 1 and 2, a particle is moving along the x axis under the action of a force of f(x) lb when the particle is x ft from the origin. Find the work done as the particle moves from the point where x = a to the point where x = b.

1. $f(x) = (2x + 1)^2$; $a = 1$; $b = 3$ **2.** $f(x) = x^2\sqrt{x^3 + 1}$; $a = 0$; $b = 2$

In Exercises 3 and 4, a particle is moving along the x axis under the action of a force of f(x) newtons when the particle is x meters from the origin. Find the work done as the particle moves from the point where x = a to the point where x = b.

3. $f(x) = x\sqrt{x + 1}$; $a = 3$; $b = 8$ **4.** $f(x) = (4x - 1)^2$; $a = 1$; $b = 4$

5. An object is moving along the x axis under the action of a force of $f(x)$ dynes when the object is x cm from the origin. If 96 ergs is the work done in moving the object from the origin to the point where $x = K$, and $f(x) = 2x - 3$, find K if $K > 0$.

6. Solve Exercise 5 if 90 ergs is the work done, and $f(x) = 4x - 3$.

7. A spring has a natural length of 8 in. If a force of 20 lb stretches the spring $\frac{1}{2}$ in., find the work done in stretching the spring from 8 in. to 11 in.

8. A spring has a natural length of 10 in., and a 30-lb force stretches it to $11\frac{1}{2}$ in. (a) Find the work done in stretching the spring from 10 in. to 12 in. (b) Find the work done in stretching the spring from 12 in. to 14 in.

9. A force of 8 nt stretches a spring of natural length 4 m to an additional 50 cm. Find the work done in stretching the spring from its natural length to 5 m.

10. A force of 500 dynes stretches a spring from its natural length of 20 cm to a length of 24 cm. Find the work done in stretching the spring from its natural length to a length of 28 cm.

11. A spring has a natural length of 12 cm. A force of 600 dynes compresses the spring to 10 cm. Find the work done in compressing the spring from 12 cm to 9 cm. Hooke's law holds for compression as well as for extension.

12. A spring has a natural length of 6 in. A 1200-lb force compresses it to $5\frac{1}{2}$ in. Find the work done in compressing it from 6 in. to $4\frac{1}{2}$ in.

13. A tank full of water is in the form of a rectangular parallelepiped 5 ft deep, 15 ft wide, and 25 ft long. Find the work required to pump the water in the tank up to a level 1 ft above the surface of the tank.

14. A trough full of water is 10 ft long, and its cross section is in the shape of an isosceles triangle 2 ft wide across the top and 2 ft high. How much work is done in pumping all the water out of the trough over the top?

15. A hemispherical tank with a radius of 6 ft is filled with water to a depth of 4 ft. Find the work done in pumping the water to the top of the tank.

16. A right-circular cylindrical tank with a depth of 12 ft and a radius of 4 ft is half full of oil weighing 60 lb/ft^3. Find the work done in pumping the oil to a height 6 ft above the tank.

17. A cable 200 ft long and weighing 4 lb/ft is hanging vertically down a well. If a weight of 100 lb is suspended from the lower end of the cable, find the work done in pulling the cable and weight to the top of the well.

18. A bucket weighing 20 lb containing 60 lb of sand is attached to the lower end of a chain 100 ft long and weighing 10 lb that is hanging in a deep well. Find the work done in raising the bucket to the top of the well.

19. Solve Exercise 18 if the sand is leaking out of the bucket at a constant rate and has all leaked out just as soon as the bucket is at the top of the well.

20. As a flour sack is being raised a distance of 9 ft, flour leaks out at such a rate that the number of pounds lost is directly proportional to the square root of the distance traveled. If the sack originally contained 60 lb of flour and it loses a total of 12 lb while being raised the 9 ft, find the work done in raising the sack.

21. A tank in the form of an inverted right-circular cone is 8 m across the top and 10 m deep. The tank is filled to a height of 8 m with oil weighing 950 kg/m^3. Find the work done in pumping the oil to the top of the tank. (*Hint:* The number of newtons of force necessary to lift an element is the product of the number of kilograms of mass (the same as the number of kilograms of weight) and 9.81, the number of m/sec^2 in the acceleration due to gravity.)

22. If the tank of Exercise 21 is filled to a height of 9 m with water, find the work done in pumping the water to the top of the tank. The weight of 1 m^3 of water at 0°C is 1000 kg. (See hint for Exercise 21.)

23. A right-circular cylindrical tank with a depth of 10 m and a radius of 5 m is half filled with water. Find the work necessary to pump the water to the top of the tank. The weight of 1 m^3 of water at 0°C is 1000 kg. (See hint for Exercise 21.)

24. If in Exercise 22 only half of the water is to be pumped to the top of the tank, find the work.

25. A 1-horsepower motor can do 550 ft-lb of work per second. If a 0.1 hp motor is used to pump water from a full tank in the shape of a rectangular parallelepiped 2 ft deep, 2 ft wide, and 6 ft long to a point 5 ft above the top of the tank, how long will it take?

26. A meteorite is a miles from the center of the earth and falls to the surface of the earth. The force of gravity is inversely proportional to the square of the distance of a body from the center of the earth. Find the work done by gravity if the weight of the meteorite is w lb at the surface of the earth. Let R miles be the radius of the earth.

27. A tank in the form of a rectangular parallelepiped 6 ft deep, 4 ft wide, and 12 ft long is full of oil weighing 50 lb/ft³. When one-third of the work necessary to pump the oil to the top of the tank has been done, find by how much the surface of the oil is lowered.

28. A cylindrical tank 10 ft high and 5 ft in radius is standing on a platform 50 ft high. Find the depth of the water when one-half of the work required to fill the tank from the ground level through a pipe in the bottom has been done.

7.6 LENGTH OF ARC OF A PLANE CURVE

Let the function f be continuous on the closed interval $[a, b]$. Consider the graph of the equation $y = f(x)$, of which a sketch is shown in Fig. 7.6.1.

The portion of the curve from the point $A(a, f(a))$ to the point $B(b, f(b))$ is called an arc. We wish to assign a number to what we intuitively think of as the length of such an arc. If the arc is a line segment from the point (x_1, y_1) to the point (x_2, y_2), we know from the formula for the distance between two points that its length is given by $\sqrt{(x_1 - x_2)^2 + (y_1 - y_2)^2}$. We use this formula for defining the length of an arc in general. Recall from geometry that the circumference of a circle is defined as the limit of the perimeters of regular polygons inscribed in the circle. For other curves we proceed in a similar way.

Let Δ be a partition of the closed interval $[a, b]$ formed by dividing the interval into n subintervals by choosing any $(n - 1)$ intermediate numbers between a and b. Let $x_0 = a$ and $x_n = b$, and let $x_1, x_2, x_3, \ldots,$ x_{n-1} be the intermediate numbers so that $x_0 < x_1 < x_2 < \cdots < x_{n-1} < x_n$. Then the ith subinterval is $[x_{i-1}, x_i]$; its length, denoted by $\Delta_i x$, is $x_i - x_{i-1}$, where $i = 1, 2, 3, \ldots, n$. Then if $\|\Delta\|$ is the norm of the partition Δ, each $\Delta_i x \leq \|\Delta\|$.

Associated with each point $(x_i, 0)$ on the x axis is a point $P_i(x_i, f(x_i))$ on the curve. Draw a line segment from each point P_{i-1} to the next point P_i, as shown in Fig. 7.6.2. The length of the line segment from P_{i-1} to P_i is denoted by $|\overline{P_{i-1}P_i}|$ and is given by the distance formula

$$|\overline{P_{i-1}P_i}| = \sqrt{(x_i - x_{i-1})^2 + (y_i - y_{i-1})^2} \tag{1}$$

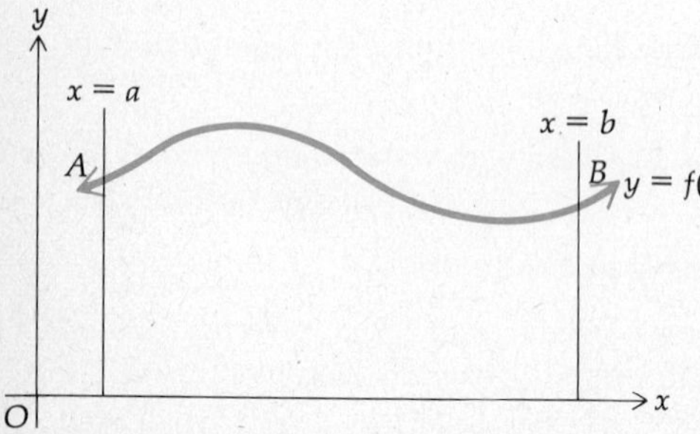

FIGURE 7.6.1

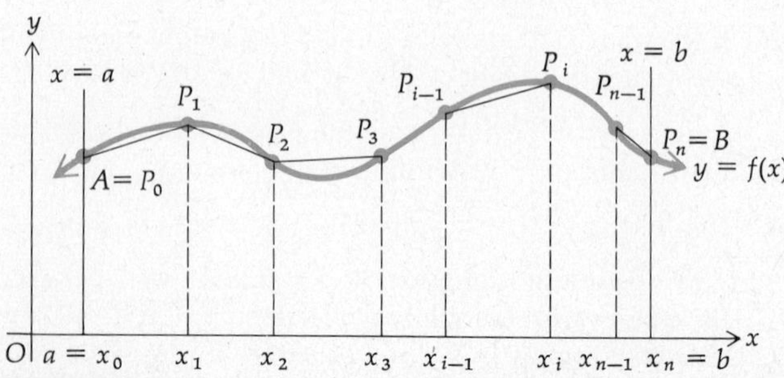

FIGURE 7.6.2

The sum of the lengths of the line segments is

$$|\overline{P_0P_1}| + |\overline{P_1P_2}| + |\overline{P_2P_3}| + \cdots + |\overline{P_{i-1}P_i}| + \cdots + |\overline{P_{n-1}P_n}|$$

which can be written with sigma notation as

$$\sum_{i=1}^{n} |\overline{P_{i-1}P_i}| \tag{2}$$

It seems plausible that if n is sufficiently large, the sum in (2) will be "close to" what we would intuitively think of as the length of the arc AB. So we define the length of the arc as the limit of the sum in (2) as the norm of Δ approaches zero, in which case n increases without bound. We have, then, the following definition.

7.6.1 Definition If the function f is continuous on the closed interval $[a, b]$ and if there exists a number L having the property that for any $\epsilon > 0$ there is a $\delta > 0$ such that

$$\left| \sum_{i=1}^{n} |\overline{P_{i-1}P_i}| - L \right| < \epsilon$$

for every partition Δ of the interval $[a, b]$ for which $\|\Delta\| < \delta$, then we write

$$L = \lim_{\|\Delta\| \to 0} \sum_{i=1}^{n} |\overline{P_{i-1}P_i}| \tag{3}$$

and L is called the **length of the arc** of the curve $y = f(x)$ from the point $A(a, f(a))$ to the point $B(b, f(b))$.

If the limit in (3) exists, the arc is said to be **rectifiable.**

We derive a formula for finding the length L of an arc that is rectifiable. The derivation requires that the derivative of f be continuous on $[a, b]$; such a function is said to be *smooth* on $[a, b]$.

7.6.2 Definition A function f is said to be **smooth** on an interval I if f' is continuous on I.

Refer now to Fig. 7.6.3. If P_{i-1} has coordinates (x_{i-1}, y_{i-1}) and P_i has coordinates (x_i, y_i), then the length of the chord $P_{i-1}P_i$ is given by formula (1).

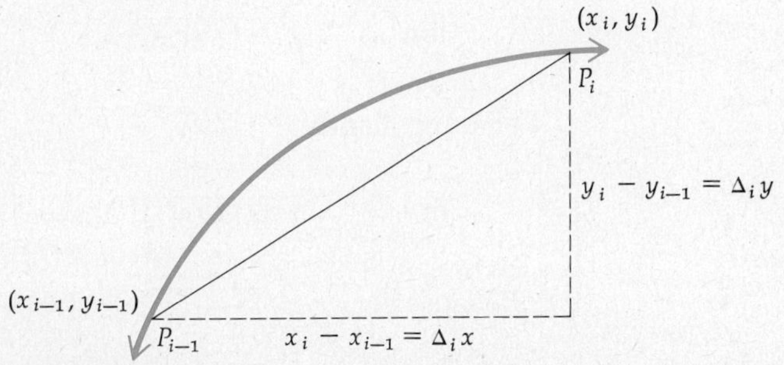

FIGURE 7.6.3

Letting $x_i - x_{i-1} = \Delta_i x$ and $y_i - y_{i-1} = \Delta_i y$ we have

$$|\overline{P_{i-1}P_i}| = \sqrt{(\Delta_i x)^2 + (\Delta_i y)^2} \tag{4}$$

or, equivalently, because $\Delta_i x \neq 0$,

$$|\overline{P_{i-1}P_i}| = \sqrt{1 + \left(\frac{\Delta_i y}{\Delta_i x}\right)^2}\,(\Delta_i x) \tag{5}$$

Because we required that f' be continuous on $[a, b]$, the hypothesis of the mean-value theorem (4.4.2) is satisfied by f; so there is a number z_i in the open interval (x_{i-1}, x_i) such that

$$f(x_i) - f(x_{i-1}) = f'(z_i)(x_i - x_{i-1}) \tag{6}$$

Because $\Delta_i y = f(x_i) - f(x_{i-1})$ and $\Delta_i x = x_i - x_{i-1}$, from (6) we have

$$\frac{\Delta_i y}{\Delta_i x} = f'(z_i) \tag{7}$$

Substituting from (7) into (5) we get

$$|\overline{P_{i-1}P_i}| = \sqrt{1 + [f'(z_i)]^2}\,\Delta_i x \tag{8}$$

where $x_{i-1} < z_i < x_i$.

For each i from 1 to n there is an equation of the form (8).

$$\sum_{i=1}^{n} |\overline{P_{i-1}P_i}| = \sum_{i=1}^{n} \sqrt{1 + [f'(z_i)]^2}\,\Delta_i x \tag{9}$$

Taking the limit on both sides of (9) as $||\Delta||$ approaches zero we obtain

$$\lim_{||\Delta|| \to 0} \sum_{i=1}^{n} |\overline{P_{i-1}P_i}| = \lim_{||\Delta|| \to 0} \sum_{i=1}^{n} \sqrt{1 + [f'(z_i)]^2}\,\Delta_i x \tag{10}$$

if this limit exists.

To show that the limit on the right side of (10) exists, let g be the function defined by

$$g(x) = \sqrt{1 + [f'(x)]^2}$$

Because f' is continuous on $[a, b]$, g is continuous on $[a, b]$. Therefore, because $x_{i-1} < z_i < x_i$, for $i = 1, 2, \ldots, n$,

$$\lim_{||\Delta|| \to 0} \sum_{i=1}^{n} \sqrt{1 + [f'(z_i)]^2}\,\Delta_i x = \lim_{||\Delta|| \to 0} \sum_{i=1}^{n} g(z_i)\,\Delta_i x$$

or, equivalently

$$\lim_{||\Delta|| \to 0} \sum_{i=1}^{n} \sqrt{1 + [f'(z_i)]^2}\,\Delta_i x = \int_a^b g(x)\,dx \tag{11}$$

Because $g(x) = \sqrt{1 + [f'(x)]^2}$, from (11) we have

$$\lim_{||\Delta|| \to 0} \sum_{i=1}^{n} \sqrt{1 + [f'(z_i)]^2}\,\Delta_i x = \int_a^b \sqrt{1 + [f'(x)]^2}\,dx \tag{12}$$

Substituting from (12) into (10) we get

$$\lim_{\|\Delta\|\to 0} \sum_{i=1}^{n} |\overline{P_{i-1}P_i}| = \int_a^b \sqrt{1 + [f'(x)]^2}\, dx \tag{13}$$

Then from (3) and (13) we obtain

$$L = \int_a^b \sqrt{1 + [f'(x)]^2}\, dx \tag{14}$$

In this way we have proved the following theorem.

7.6.3 Theorem If the function f and its derivative f' are continuous on the closed interval $[a, b]$, then the length of arc of the curve $y = f(x)$ from the point $(a, f(a))$ to the point $(b, f(b))$ is given by

$$L = \int_a^b \sqrt{1 + [f'(x)]^2}\, dx$$

We also have the following theorem, which gives the length of the arc of a curve when x is expressed as a function of y.

7.6.4 Theorem If the function F and its derivative F' are continuous on the closed interval $[c, d]$, then the length of arc of the curve $x = F(y)$ from the point $(F(c), c)$ to the point $(F(d), d)$ is given by

$$L = \int_c^d \sqrt{1 + [F'(y)]^2}\, dy$$

The proof of Theorem 7.6.4 is identical with that of Theorem 7.6.3; here we interchange x and y as well as the functions f and F.

The definite integral obtained when applying Theorem 7.6.3 or Theorem 7.6.4 is often difficult to evaluate. Because our techniques of integration have so far been limited to integration of powers, we are further restricted in finding equations of curves for which we can evaluate the resulting definite integrals to find the length of an arc.

Example 1 Find the length of the arc of the curve $y = x^{2/3}$ from the point (1, 1) to (8, 4) by using Theorem 7.6.3.

Solution See Fig. 7.6.4. Because $f(x) = x^{2/3}$, $f'(x) = \frac{2}{3}x^{-1/3}$. From Theorem 7.6.3,

$$L = \int_1^8 \sqrt{1 + \frac{4}{9x^{2/3}}}\, dx$$

$$= \frac{1}{3}\int_1^8 \frac{\sqrt{9x^{2/3} + 4}}{x^{1/3}}\, dx$$

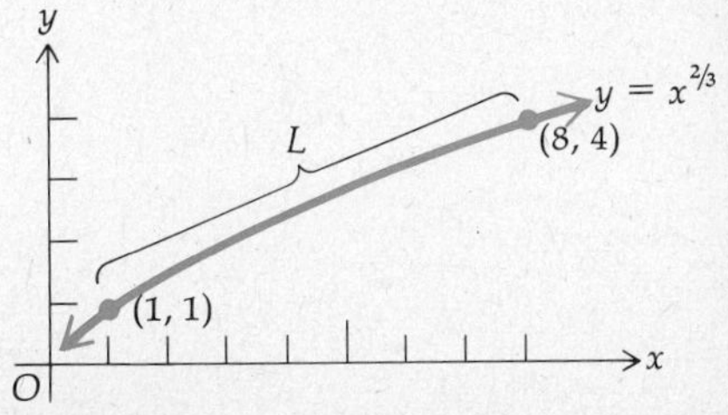

FIGURE 7.6.4

To evaluate this definite integral, let $u = 9x^{2/3} + 4$; then $du = 6x^{-1/3}\, dx$. When $x = 1$, $u = 13$; when $x = 8$, $u = 40$. Therefore

$$L = \frac{1}{18}\int_{13}^{40} u^{1/2}\, du$$

$$= \frac{1}{18}\left[\frac{2}{3}u^{3/2}\right]_{13}^{40}$$

$$= \tfrac{1}{27}(40^{3/2} - 13^{3/2})$$

$$\approx 7.6$$

Example 2 Find the length of the arc in Example 1 by using Theorem 7.6.4.

Solution Because $y = x^{2/3}$ and $x > 0$, we solve for x and obtain $x = y^{3/2}$. Letting $F(y) = y^{3/2}$ we have

$$F'(y) = \tfrac{3}{2}y^{1/2}$$

Then, from Theorem 7.6.4,

$$L = \int_{1}^{4} \sqrt{1 + \tfrac{9}{4}y}\, dy$$

$$= \frac{1}{2}\int_{1}^{4} \sqrt{4 + 9y}\, dy$$

$$= \frac{1}{18}\left[\frac{2}{3}(4 + 9y)^{3/2}\right]_{1}^{4}$$

$$= \tfrac{1}{27}(40^{3/2} - 13^{3/2})$$

$$\approx 7.6$$

If f' is continuous on $[a, b]$, the definite integral $\int_a^x \sqrt{1 + [f'(t)]^2}\, dt$ is a function of x; and it gives the length of the arc of the curve $y = f(x)$ from the point $(a, f(a))$ to the point $(x, f(x))$, where x is any number in the closed interval $[a, b]$. Let s denote the length of this arc; thus s is a function of x, and

$$s(x) = \int_a^x \sqrt{1 + [f'(t)]^2}\, dt$$

From Theorem 6.5.1,

$$s'(x) = \sqrt{1 + [f'(x)]^2}$$

or, because $s'(x) = ds/dx$ and $f'(x) = dy/dx$,

$$\frac{ds}{dx} = \sqrt{1 + \left(\frac{dy}{dx}\right)^2}$$

Multiplying by dx we obtain

$$ds = \sqrt{1 + \left(\frac{dy}{dx}\right)^2}\, dx \qquad (15)$$

Similarly, for the length of arc of the curve $x = g(y)$ from $(g(c), c)$ to $(g(y), y)$,

$$ds = \sqrt{\left(\frac{dx}{dy}\right)^2 + 1}\, dy \tag{16}$$

Squaring on both sides of (15) gives

$$(ds)^2 = (dx)^2 + (dy)^2 \tag{17}$$

From (17) we get the geometric interpretation of ds, which is shown in Fig. 7.6.5.

In Fig. 7.6.5, line T is tangent to the curve $y = f(x)$ at the point P. $|\overline{PM}| = \Delta x = dx$; $|\overline{MQ}| = \Delta y$; $|\overline{MR}| = dy$; $|\overline{PR}| = ds$; the length of arc $PQ = \Delta s$.

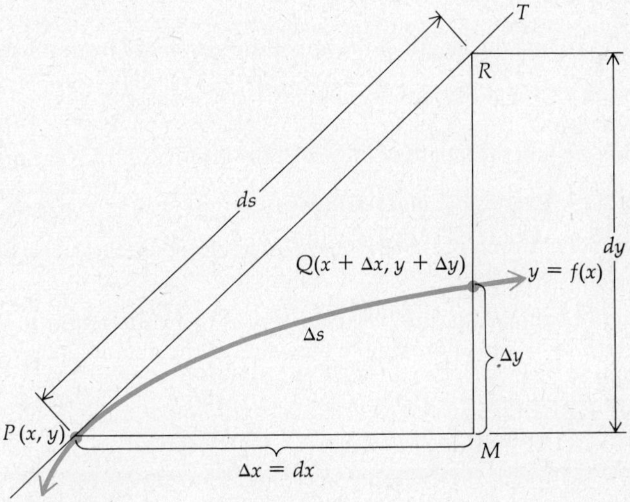

FIGURE 7.6.5

EXERCISES 7.6

1. Find the length of the arc of the curve $9y^2 = 4x^3$ from the origin to the point $(3, 2\sqrt{3})$.
2. Find the length of the arc of the curve $x^2 = (2y + 3)^3$ from $(1, -1)$ to $(7\sqrt{7}, 2)$.
3. Find the length of the arc of the curve $8y = x^4 + 2x^{-2}$ from the point where $x = 1$ to the point where $x = 2$.
4. Use Theorem 7.6.3 to find the length of the arc of the curve $y^3 = 8x^2$ from the point $(1, 2)$ to the point $(27, 18)$.
5. Solve Exercise 4 by using Theorem 7.6.4.
6. Find the length of the arc of the curve $y = \frac{2}{3}(x - 5)^{3/2}$ from the point where $x = 6$ to the point where $x = 8$.
7. Find the length of the arc of the curve $y = \frac{1}{3}(x^2 + 2)^{3/2}$ from the point where $x = 0$ to the point where $x = 3$.

8. Find the length of the arc of the curve $6xy = y^4 + 3$ from the point where $y = 1$ to the point where $y = 2$.

9. Find the length of the arc of the curve $y = \frac{1}{3}\sqrt{x}(3x - 1)$ from the point where $x = 1$ to the point where $x = 4$.

10. Find the length of the arc of the curve $y = \frac{1}{6}x^3 + \frac{1}{2}x^{-1}$ from the point $(2, \frac{19}{12})$ to the point $(5, \frac{314}{15})$.

11. Find the length of the arc of the curve $x^{2/3} + y^{2/3} = 1$ from the point where $x = \frac{1}{8}$ to the point where $x = 1$.

12. Find the length of the arc of the curve $x^{2/3} + y^{2/3} = a^{2/3}$ (a is a constant, $a > 1$) from the point where $x = 1$ to the point where $x = a$.

13. Find the length of the curve $(x/a)^{2/3} + (y/b)^{2/3} = 1$ in the first quadrant from the point where $x = \frac{1}{8}a$ to the point where $x = a$.

14. Find the length of the curve $9y^2 = x^2(2x + 3)$ in the second quadrant from the point where $x = -1$ to the point where $x = 0$.

15. Find the length of the curve $9y^2 = x(x - 3)^2$ in the first quadrant from the point where $x = 1$ to the point where $x = 3$.

16. Find the length of the curve $9y^2 = 4(1 + x^2)^3$ in the first quadrant from the point where $x = 0$ to the point where $x = 2\sqrt{2}$.

17. If $f(x) = \int_0^x \sqrt{\cos t}\, dt$, find the length of the arc of the graph of f from the point where $x = 0$ to the point where $x = \pi$. (*Hint*: Use Theorem 6.5.1 and the identity $\cos^2 \frac{1}{2}x = \frac{1}{2}(1 + \cos x)$.)

18. If $f(x) = \int_0^x \sqrt{\sin t}\, dt$, find the length of the arc of the graph of f from the point where $x = 0$ to the point where $x = \frac{1}{2}\pi$. (*Hint*: Use the hint for Exercise 17 and the identity $\sin x = \cos(\frac{1}{2}\pi - x)$.)

19. Express the length of one arch of the sine curve as a definite integral. Use Simpson's rule with $2n = 4$ to approximate the value of the definite integral.

20. Express the length of the arc of the parabola $y = x^2$ from the origin to the point $(1, 1)$ as a definite integral. Use Simpson's rule with $2n = 4$ to approximate the value of the definite integral.

REVIEW EXERCISES FOR CHAPTER 7

1. Find the area of the region bounded by the curves $x = y^2$ and $x = y^3$.

2. Find the area of the region bounded by the curve $y = x^2 - 7x$, the x axis, and the lines $x = 2$ and $x = 4$.

3. Find the area of the region bounded by the curve $x^2 + y - 5 = 0$ and the line $y = -4$. Take the elements of area perpendicular to the y axis.

4. Find the area of the region bounded by the loop of the curve $y^2 = x^2(4 - x)$.

5. Find the area of the region bounded by the curves $y = x^3 + 3x^2 + 2x$ and $y = 2x^2 + 4x$.

6. The region in the first quadrant bounded by the curves $x = y^2$ and $x = y^4$ is revolved about the y axis. Find the volume of the solid generated.

7. Find the volume of the solid generated by revolving about the y axis the region bounded by the parabola $x = y^2 + 2$ and the line $x = y + 8$.

8. Find the volume of the solid of revolution generated when the region bounded by the parabola $y^2 = x$, the x axis, and the line $x = 4$ is revolved about the line $x = 4$. Take the elements of area parallel to the axis of revolution.

9. The base of a solid is the region bounded by the parabola $y^2 = 8x$ and the line $x = 8$. Find the volume of the solid if every plane section perpendicular to the axis of the base is a square.

10. The base of a solid is the region enclosed by a circle having a radius of r units, and every plane section perpendicular to a fixed diameter of the base is a square for which a chord of the circle is a diagonal. Find the volume of the solid.

11. Find the area of the region bounded by the curves $y = |x - 1|$, $y = x^2 - 2x$, the y axis, and the line $x = 2$.

12. Find the area of the region bounded by the curves $y = |x| + |x - 1|$ and $y = x + 1$.

13. Find the volume of the solid generated by revolving the region bounded by the curve $y = |x - 2|$, the x axis, and the lines $x = 1$ and $x = 4$ about the x axis.

14. Find the length of arc of the curve $ay^2 = x^3$ from the origin to $(4a, 8a)$.

15. Find the length of the curve $6y^2 = x(x - 2)^2$ from $(2, 0)$ to $(8, 4\sqrt{3})$.

16. Use integration to find the volume of a segment of a sphere if the sphere has a radius of r units and the altitude of the segment is h units.

17. Find the volume of the solid generated by revolving about the line $y = -1$ the region above the x axis bounded by the line $2y = x + 3$ and the curves $y^2 + x = 0$ and $y^2 - 4x = 0$ from $x = -1$ to $x = 1$.

18. Find the area of the region bounded by the curves $y = \sin 2x$ and $y = \sin x$ from $x = 0$ to $x = \frac{1}{3}\pi$.

19. Find the area of the region bounded by the curves $y = \cos x$ and $y = \sin x$ from $x = \frac{1}{4}\pi$ to $x = \frac{5}{4}\pi$.

20. A force of 500 lb is required to compress a spring whose natural length is 10 in. to a length of 9 in. Find the work done to compress the spring to a length of 8 in.

21. A force of 600 dynes stretches a spring from its natural length of 30 cm to a length of 35 cm. Find the work done in stretching the spring from its natural length to a length of 40 cm.

22. A trough full of water is 6 ft long, and its cross section is in the shape of a semicircle with a diameter of 2 ft at the top. How much work is required to pump the water out over the top?

23. A cable 20 ft long and weighing 2 lb/ft is hanging vertically from the top of a pole. Find the work done in raising the entire cable to the top of the pole.

24. The work necessary to stretch a spring from 9 in. to 10 in. is $\frac{3}{2}$ times the work necessary to stretch it from 8 in. to 9 in. What is the natural length of the spring?

25. A tank full of water is in the form of a rectangular parallelepiped 4 m deep, 15 m wide, and 30 m long. Find the work required to pump the water in the tank up to a level 50 cm above the top of the tank. The weight of 1 m^3 of water at 0°C is 1000 kg. (*Hint:* See the hint for Exercise 21 in Exercises 7.5.)

26. A hemispherical tank having a diameter of 10 m is filled with water to a depth of 3 m. Find the work done in pumping the water to the top of the tank. The weight of 1 m^3 of water at 0°C is 1000 kg. (*Hint:* See the hint for Exercise 21 in Exercises 7.5.)

27. A water tank is in the shape of a hemisphere surmounted by a right-circular cylinder. The radius of both the hemisphere and the cylinder is 4 ft, and the altitude of the cylinder is 8 ft. If the tank is full of water, find the work necessary to empty the tank by pumping it through an outlet at the top of the tank.

28. A container has the same shape and dimensions as a solid of revolution formed by revolving about the y axis the region in the first quadrant bounded by the parabola $x^2 = 4py$, the y axis, and the line $y = p$. If the container is full of water, find the work done in pumping all the water up to a point $3p$ ft above the top of the container.

29. The surface of a tank is the same as that of a paraboloid of revolution obtained by revolving the parabola $y = x^2$ about the y axis. The vertex of the parabola is at the bottom of the tank, and the tank is 36 ft high. If the tank is filled with water to a depth of 20 ft, find the work done in pumping all of the water out over the top.

30. A wedge is cut from a right-circular cylinder with a radius of r units by two planes, one perpendicular to the axis of the cylinder and the other intersecting the first along a diameter of the circular plane section at an angle of measurement 30°. Find the volume of the wedge.

31. A church steeple is 30 ft high, and every horizontal plane section is a square having sides of length one-tenth of the distance of the plane section from the top of the steeple. Find the volume of the steeple.

32. A sphere of radius 10 cm is intersected by two parallel planes on the same side of the center of the sphere. The distance from the center of the sphere to one of the planes is 1 cm, and the distance between the two planes is 6 cm. Find the volume of the solid portion of the sphere between the two planes.

33. Solve Exercise 32 if the two planes lie on opposite sides of the center of the sphere but the other facts are the same.

34. The region bounded by the curve $y = \sin^2 x$ and the x axis from $x = 0$ to $x = \pi$ is revolved about the x axis. Find the volume of the solid of revolution generated.

35. A solid is formed by revolving about the y axis the region bounded by the curve $y^3 = x$, the x axis, and the line $x = c$, where $c > 0$. For what value of c will the volume of the solid be 12π cubic units?

36. Find the length of the arc of the curve $3y = (x^2 - 2)^{3/2}$ from the point where $x = 3$ to the point where $x = 6$.

37. Find the length of the curve $9x^{2/3} + 4y^{2/3} = 36$ in the second quadrant from the point where $x = -1$ to the point where $x = -\frac{1}{8}$.

38. Find by the prismoidal formula the volume of the right-circular cone of height 5 units and base radius 7 units.

39. The region from $x = 0$ to $x = \frac{1}{2}\pi$ bounded by the curve $y = \sin x$, the line $y = 1$, and the y axis is revolved about the x axis. Find the volume of the solid generated.

40. Express as a definite integral the length of the arc of the curve $y = \frac{1}{3}x^3$ from the point where $x = 0$ to the point where $x = 1$. Use Simpson's rule with $2n = 4$ to approximate the value of the definite integral.

CHAPTER 8 Inverse Functions, Logarithmic Functions, and Exponential Functions

8.1 THE INVERSE OF A FUNCTION

Consider the function f defined by

$$f(x) = x^2$$

A sketch of the graph of f is shown in Fig. 8.1.1. The domain of f is the set R of real numbers, and the range of f is the interval $[0, +\infty)$. For each value of x in the domain there corresponds one and only one number in the range. For instance, because $f(2) = 4$, the number in the range that corresponds to the number 2 in the domain is 4. However, because $f(-2) = 4$, the number corresponding to the number -2 in the domain is also the number 4 in the range. So 4 is the function value of two distinct numbers in the domain. Furthermore, every number except 0 in the range of this function is the function value of two distinct numbers in the domain. In particular, $\frac{25}{4}$ is the function value of both $\frac{5}{2}$ and $-\frac{5}{2}$, 1 is the function value of both 1 and -1, and 9 is the function value of both 3 and -3.

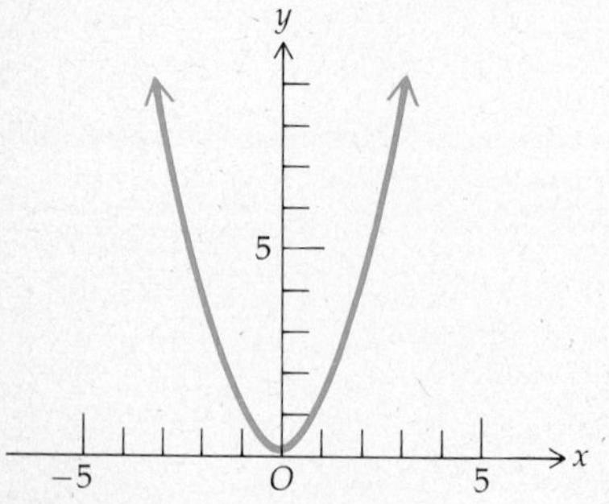

FIGURE 8.1.1

A different situation occurs with the function g defined by

$$g(x) = x^3 \qquad -2 \leq x \leq 2$$

The domain of g is the closed interval [–2, 2], and the range is $[-8, 8]$. A sketch of the graph of g is shown in Fig. 8.1.2. This function is one for which a number in its range is the function value of one and only one number in the domain. Such a function is called *one-to-one*.

8.1.1 Definition A function f is said to be **one-to-one** if and only if whenever x_1 and x_2 are any two distinct numbers in the domain of f, then $f(x_1) \neq f(x_2)$.

● ILLUSTRATION 1 As discussed in the first paragraph of this section, for the function f defined by

$$f(x) = x^2$$

every number except 0 in the range of f is the function value of two distinct numbers in the domain. Therefore, by Definition 8.1.1, this function is not one-to-one. ●

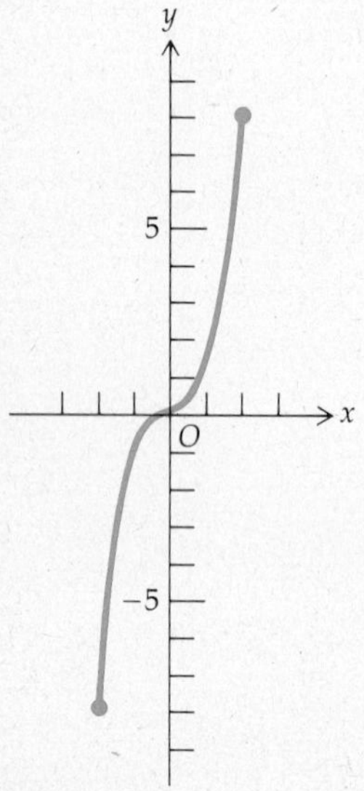

FIGURE 8.1.2

We have seen that a vertical line can intersect the graph of a function in only one point. Observe that this is the situation in Fig. 8.1.1 for any vertical line and the function defined by $f(x) = x^2$ and in Fig. 8.1.2 for a line $x = a$ with $-2 \leq a \leq 2$, and the function defined by $g(x) = x^3$ where $-2 \leq x \leq 2$. For a one-to-one function it is also true that a horizontal line can intersect the graph in only one point. You can verify this fact for the one-to-one function whose graph appears in Fig. 8.1.2. Furthermore, note that for the function of Fig. 8.1.1 a line $y = b$, with $b > 0$, intersects the graph in two points, and this function is not one-to-one.

Example 1 (a) If $f(x) = 4x - 3$, prove that f is one-to-one. Draw a sketch of the graph of f. (b) If $h(x) = 4 - x^2$, prove that h is not one-to-one. Draw a sketch of the graph of h.

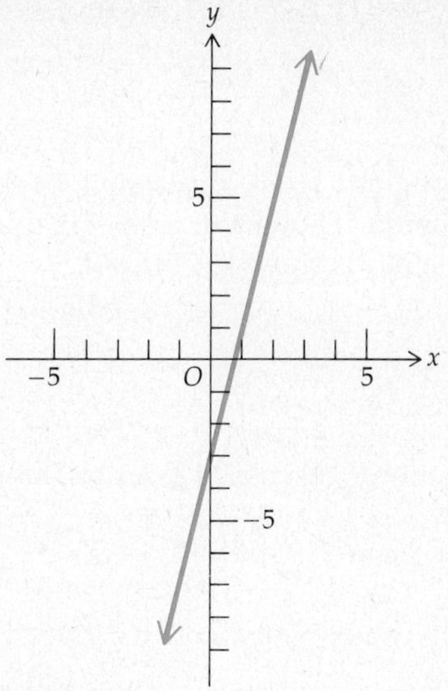

FIGURE 8.1.3

Solution

(a) The domain of f is the set R of real numbers. If x_1 and x_2 are two real numbers and $x_1 \neq x_2$, then

$$4x_1 \neq 4x_2$$
$$4x_1 - 3 \neq 4x_2 - 3$$
$$f(x_1) \neq f(x_2)$$

Therefore, by Definition 8.1.1, f is one-to-one. A sketch of the graph of f is shown in Fig. 8.1.3. Observe that a horizontal line intersects the graph in only one point.

(b) The domain of h is the set R of real numbers. The numbers 1 and -1 are in the domain of h, and

$$h(1) = 3 \quad \text{and} \quad h(-1) = 3$$

Because $1 \neq -1$ and $h(1) = h(-1)$, it follows from Definition 8.1.1 that h is not one-to-one. A sketch of the graph of h appears in Fig. 8.1.4. Note that a horizontal line $y = b$, with $b < 4$, intersects the graph in two points.

It is not always a simple matter to use Definition 8.1.1 to prove that a function is one-to-one. The following theorem gives a test that can often be used.

8.1.2 Theorem A function that is either increasing on an interval or decreasing on an interval is one-to-one on the interval.

Proof Assume that the function f is increasing on an interval. If x_1 and x_2 are two numbers in the interval and $x_1 \neq x_2$, then either $x_1 < x_2$ or $x_2 < x_1$. If $x_1 < x_2$, it follows from the definition of an increasing function (Definition 4.5.1) that $f(x_1) < f(x_2)$; so $f(x_1) \neq f(x_2)$. If $x_2 < x_1$, then $f(x_2) < f(x_1)$, and so again $f(x_1) \neq f(x_2)$. Therefore it follows from Definition 8.1.1 that f is one-to-one on the interval. The proof is similar if f is decreasing on an interval. ■

To apply Theorem 8.1.2 it is necessary first to determine if the function is monotonic (increasing or decreasing) on an interval. Theorem 4.5.3 can usually be used for this purpose.

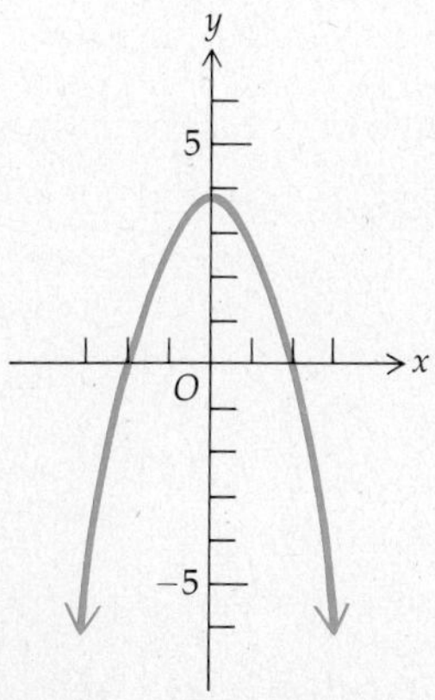

FIGURE 8.1.4

Example 2 If

$$f(x) = \frac{2x + 3}{x - 1}$$

prove that f is one-to-one. Draw a sketch of the graph of f.

Solution The domain of f is the set of all real numbers except 1. We are given

$$f(x) = \frac{2x + 3}{x - 1}$$

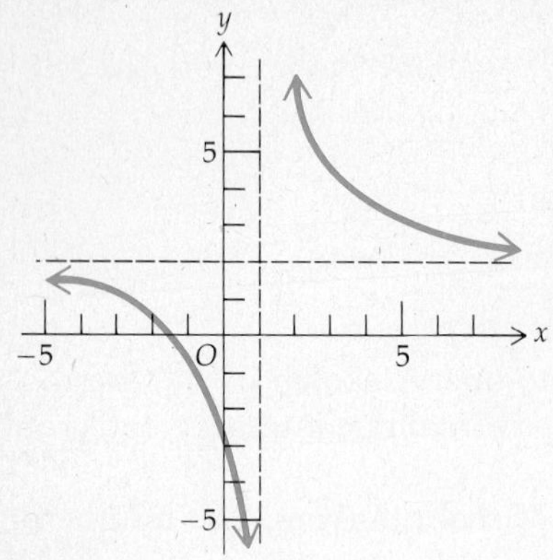

FIGURE 8.1.5

Thus

$$f'(x) = -\frac{5}{(x-1)^2}$$

Because $f'(x) < 0$ for all $x \neq 1$, we can conclude from Theorem 4.5.3 that f is decreasing on each interval in its domain. Therefore, from Theorem 8.1.2, f is one-to-one. A sketch of the graph of f is shown in Fig. 8.1.5.

● ILLUSTRATION 2 Consider the equation

$$y = x^3 \qquad -2 \leq x \leq 2 \tag{1}$$

This equation defines the one-to-one function g discussed prior to Definition 8.1.1, where

$$g(x) = x^3 \qquad -2 \leq x \leq 2 \tag{2}$$

Function g is the set of ordered pairs (x, y) satisfying (1). If we solve Eq. (1) for x, we obtain

$$x = \sqrt[3]{y} \qquad -8 \leq y \leq 8 \tag{3}$$

which defines a function G where

$$G(y) = \sqrt[3]{y} \qquad -8 \leq y \leq 8 \tag{4}$$

The function G is the set of ordered pairs (y, x) satisfying (3). ●

The function G of Illustration 2 is called the *inverse* of the function g. The notation g^{-1} denotes the inverse of g; it is read "g inverse." Note that -1 to denote the inverse of a function should not be confused with the exponent -1. We have the following formal definition of the inverse of a function.

8.1.3 Definition If f is a one-to-one function that is the set of ordered pairs (x, y), then there is a function f^{-1}, called the **inverse** of f, where f^{-1} is the set of ordered pairs (y, x) defined by

$$x = f^{-1}(y) \quad \text{if and only if} \quad y = f(x) \tag{5}$$

The domain of f^{-1} is the range of f and the range of f^{-1} is the domain of f.

We eliminate y between the equations in (5) by writing the equation

$$f^{-1}(y) = x$$

and replacing y by $f(x)$ to obtain

$$f^{-1}(f(x)) = x \tag{6}$$

where x is in the domain of f.

We eliminate x between the same pair of equations by writing the equation

$$f(x) = y$$

and replacing x by $f^{-1}(y)$. We get

$$f(f^{-1}(y)) = y \tag{7}$$

where y is in the domain of f^{-1}. Because the symbol used for the independent variable is arbitrary, we can replace y by x in (7) to obtain

$$f(f^{-1}(x)) = x \tag{8}$$

where x is in the domain of f^{-1}.

From (6) and (8) we see that if the inverse of the function f is the function f^{-1}, then the inverse of f^{-1} is f. We state this fact as well as (6) and (8) formally as the following theorem.

8.1.4 Theorem If f is a one-to-one function having f^{-1} as its inverse, then f^{-1} is a one-to-one function having f as its inverse. Furthermore,

$$f^{-1}(f(x)) = x \qquad \text{for } x \text{ in the domain of } f$$

and

$$f(f^{-1}(x)) = x \qquad \text{for } x \text{ in the domain of } f^{-1}$$

We use the terminology *inverse functions* when referring to a function and its inverse.

● ILLUSTRATION 3 Because the functions g and G defined by (2) and (4) in Illustration 2 are inverse functions, g^{-1} can be written in place of G. We have from (2),

$$g(x) = x^3 \qquad -2 \leq x \leq 2$$

and from (4),

$$g^{-1}(y) = \sqrt[3]{y} \qquad -8 \leq y \leq 8$$

or, equivalently, if we replace y by x,

$$g^{-1}(x) = \sqrt[3]{x} \qquad -8 \leq x \leq 8$$

Observe that the domain of g is $[-2, 2]$, which is the range of g^{-1}; also the range of g is $[-8, 8]$, which is the domain of g^{-1}. ●

If a function f is defined by the equation

$$y = f(x)$$

and if f has an inverse, then f^{-1} can be found by solving the equation for x in terms of y and obtaining

$$x = f^{-1}(y)$$

Then if we interchange x and y in this equation, we get

$$y = f^{-1}(x)$$

This procedure is used in the following example.

Example 3 Find $f^{-1}(x)$ for the function f of Example 1(a). Verify the equations of Theorem 8.1.4 for f and f^{-1}. Draw a sketch of the graph of f^{-1}.

Solution In Example 1(a) we showed that the function f defined by

$$f(x) = 4x - 3$$

is one-to-one. Therefore f^{-1} exists. To find $f^{-1}(x)$, write the equation

$$y = 4x - 3$$

and solve for x. We obtain

$$x = \frac{y+3}{4}$$

Therefore

$$f^{-1}(y) = \frac{y+3}{4} \quad \text{or, equivalently,} \quad f^{-1}(x) = \frac{x+3}{4}$$

We verify the equations of Theorem 8.1.4.

$$\begin{aligned} f^{-1}(f(x)) &= f^{-1}(4x-3) \\ &= \frac{(4x-3)+3}{4} \\ &= \frac{4x}{4} \\ &= x \end{aligned} \qquad \begin{aligned} f(f^{-1}(x)) &= f\left(\frac{x+3}{4}\right) \\ &= 4\left(\frac{x+3}{4}\right) - 3 \\ &= (x+3) - 3 \\ &= x \end{aligned}$$

A sketch of the graph of f^{-1} is shown in Fig. 8.1.6.

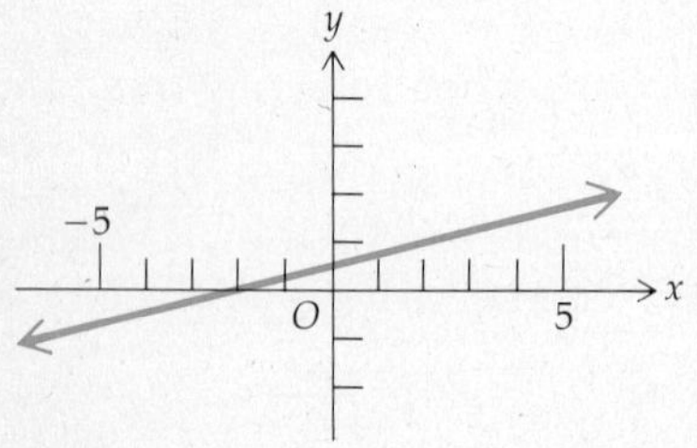

FIGURE 8.1.6

In Fig. 8.1.3 there is a sketch of the graph of the function f of Example 3, and in Fig. 8.1.6 there is a sketch of the graph of f^{-1}. In Fig. 8.1.7 the graphs of f and f^{-1} are shown on the same set of axes. In this figure it appears intuitively true that if the point $Q(u, v)$ is on the graph of f, then the point $R(v, u)$ is on the graph of f^{-1}. That this situation actually occurs is demonstrated in the following paragraph.

In general, if Q is the point (u, v) and R is the point (v, u), the line segment QR is perpendicular to the line $y = x$ and is bisected by it. We state that the point Q is a *reflection of the point R* with respect to the line $y = x$, and the point R is a *reflection of the point Q* with respect to the line $y = x$. If x and y are interchanged in the equation $y = f(x)$, we obtain the equation $x = f(y)$, and the graph of the equation $x = f(y)$ is said to be a *reflection of the graph* of the equation $y = f(x)$ with respect to the line $y = x$. Because the equation $x = f(y)$ is equivalent to the equation $y = f^{-1}(x)$, we conclude that the graph of the equation $y = f^{-1}(x)$ is a reflection of the graph of the equation $y = f(x)$ with respect to the line $y = x$. Therefore, if a function has an inverse, the graphs of the functions are reflections of each other with respect to the line $y = x$.

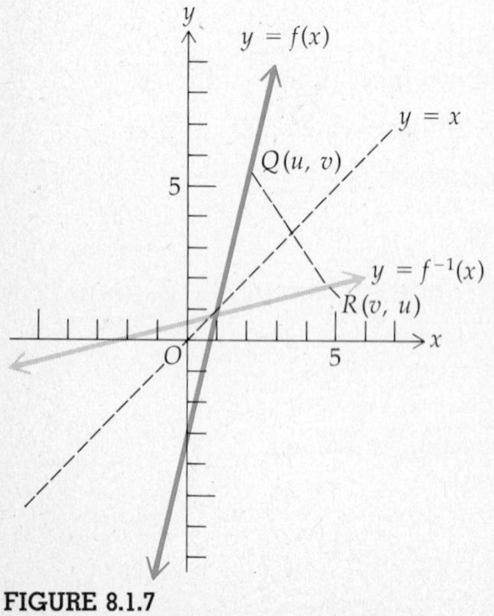

FIGURE 8.1.7

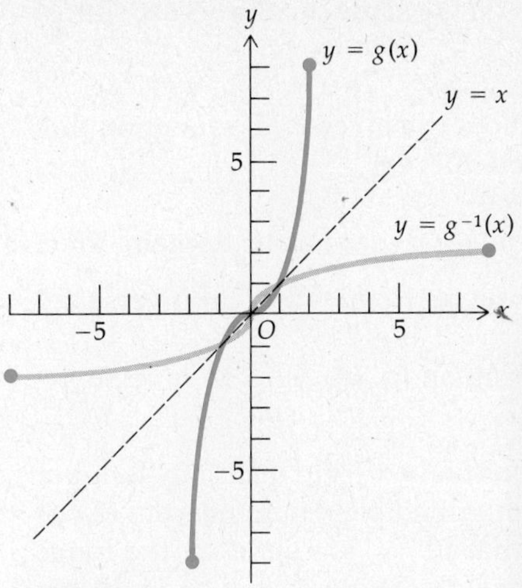

FIGURE 8.1.8

● ILLUSTRATION 4 Functions g and g^{-1} of Illustration 3 are defined by

$$g(x) = x^3 \qquad -2 \le x \le 2$$

and

$$g^{-1}(x) = \sqrt[3]{x} \qquad -8 \le x \le 8$$

Sketches of the graphs of g and g^{-1} are shown on the same set of axes in Fig. 8.1.8. Observe that these graphs are reflections of each other with respect to the line $y = x$. ●

Example 4 Find $f^{-1}(x)$ for the function f of Example 2. Verify the equations of Theorem 8.1.4 for f and f^{-1}. Draw sketches of the graphs of f and f^{-1} on the same set of axes.

Solution The function f of Example 2 is defined by

$$f(x) = \frac{2x+3}{x-1}$$

In Example 2 we showed that f is one-to-one. Therefore f has an inverse f^{-1}. To find $f^{-1}(x)$, let $y = f(x)$ and solve for x, thus giving $x = f^{-1}(y)$. So we have

$$y = \frac{2x+3}{x-1}$$

$$xy - y = 2x + 3$$

$$x(y-2) = y + 3$$

$$x = \frac{y+3}{y-2}$$

Therefore

$$f^{-1}(y) = \frac{y+3}{y-2} \quad \text{or, equivalently,} \quad f^{-1}(x) = \frac{x+3}{x-2}$$

The domain of f^{-1} is the set of all real numbers except 2. We verify the equations of Theorem 8.1.4.

$$f^{-1}(f(x)) = f^{-1}\left(\frac{2x+3}{x-1}\right) \qquad\qquad f(f^{-1}(x)) = f\left(\frac{x+3}{x-2}\right)$$

$$= \frac{\left(\frac{2x+3}{x-1}\right) + 3}{\left(\frac{2x+3}{x-1}\right) - 2} \qquad\qquad = \frac{2\left(\frac{x+3}{x-2}\right) + 3}{\left(\frac{x+3}{x-2}\right) - 1}$$

$$= \frac{(2x+3) + 3(x-1)}{(2x+3) - 2(x-1)} \qquad\qquad = \frac{2(x+3) + 3(x-2)}{(x+3) - (x-2)}$$

$$= \frac{5x}{5} \qquad\qquad = \frac{5x}{5}$$

$$= x \qquad\qquad = x$$

Sketches of the graphs of f and f^{-1} on the same set of axes are shown in Fig. 8.1.9.

We have the following theorem about the inverse of a function that is continuous and monotonic on a closed interval.

8.1.5 Theorem Suppose the function f has the closed interval $[a, b]$ as its domain. Then

(i) if f is continuous and increasing on $[a, b]$, f has an inverse f^{-1} that is defined on $[f(a), f(b)]$.
(ii) if f is continuous and decreasing on $[a, b]$, f has an inverse f^{-1} that is defined on $[f(b), f(a)]$.

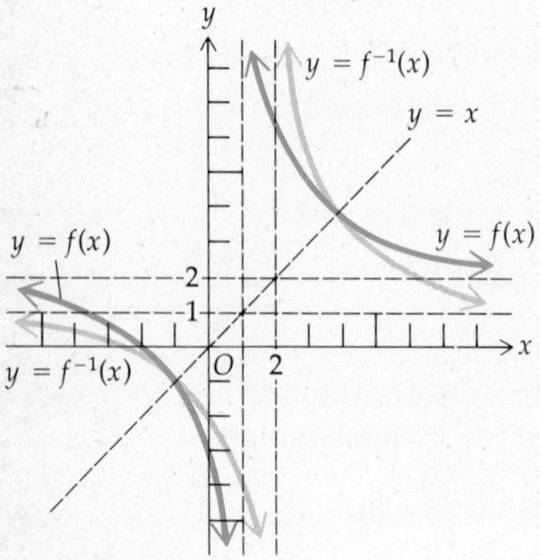

FIGURE 8.1.9

Proof For part (i), if f is continuous on $[a, b]$ and if k is any number such that $f(a) < k < f(b)$, then by the intermediate-value theorem (2.8.6) there exists a number c in (a, b) such that $f(c) = k$. Therefore the range of f is the closed interval $[f(a), f(b)]$. Because f is increasing on $[a, b]$, f is one-to-one and so f has an inverse f^{-1}. Because the domain of f^{-1} is the range of f, f^{-1} is defined on $[f(a), f(b)]$.

The proof of part (ii) is similar. However, because f is decreasing on $[a, b]$, $f(a) > f(b)$; so the range of f is $[f(b), f(a)]$. Thus f^{-1} is defined on $[f(b), f(a)]$. ■

• ILLUSTRATION 5 Let f be the function defined by

$$f(x) = x^2 - 1$$

A sketch of the graph of f is shown in Fig. 8.1.10. Because f is not one-to-one, f does not have an inverse.

However, let us restrict the domain to the closed interval $[0, c]$ and consider the function f_1 defined by

$$f_1(x) = x^2 - 1 \qquad \text{and } x \text{ in } [0, c] \tag{9}$$

Also consider the function f_2 for which the domain is restricted to the closed interval $[-c, 0]$; that is,

$$f_2(x) = x^2 - 1 \qquad \text{and } x \text{ in } [-c, 0] \tag{10}$$

The functions f_1 and f_2 defined by (9) and (10) are distinct functions because their domains are different. We find the derivatives of f_1 and f_2 and obtain

$$f_1'(x) = 2x \qquad \text{and } x \text{ in } [0, c]$$

and

$$f_2'(x) = 2x \quad \text{and } x \text{ in } [-c, 0]$$

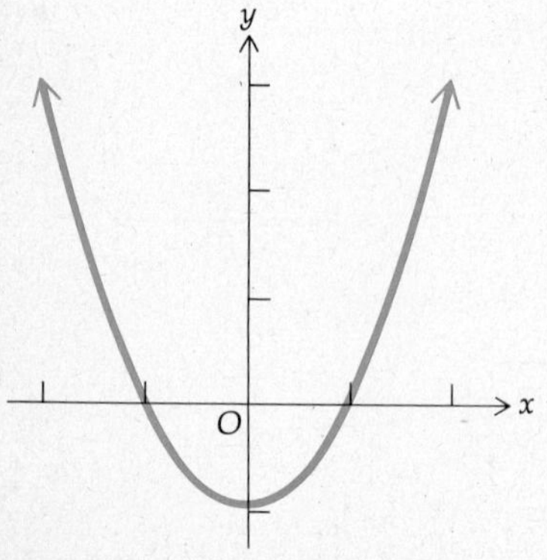

FIGURE 8.1.10

Because f_1 is continuous on $[0, c]$ and $f_1'(x) > 0$ for all x in $(0, c)$, by Theorem 4.5.3(i) f_1 is increasing on $[0, c]$. Therefore, by Theorem 8.1.5(i), f_1 has an inverse f_1^{-1} that is defined on $[f_1(0), f_1(c)]$ which is $[-1, c^2 - 1]$. We obtain $f_1^{-1}(x)$ by replacing $f_1(x)$ by y in (9) and solving for x, where $x \geq 0$. We get

$$y = x^2 - 1$$
$$x^2 = y + 1$$
$$x = \sqrt{y+1}$$

Therefore $f_1^{-1}(y) = \sqrt{y+1}$, and if we replace y by x,

$$f_1^{-1}(x) = \sqrt{x+1} \quad \text{and } x \text{ in } [-1, c^2 - 1] \tag{11}$$

Because f_2 is continuous on $[-c, 0]$ and because $f_2'(x) < 0$ for all x in $(-c, 0)$, by Theorem 4.5.3(ii) f_2 is decreasing on $[-c, 0]$. Thus, by Theorem 8.1.5(ii), f_2 has an inverse f_2^{-1} that is defined on $[f_2(0), f_2(-c)]$ which is $[-1, c^2 - 1]$. If in (10) we replace $f_2(x)$ by y and solve for x, where $x \leq 0$, we obtain

$$x = -\sqrt{y+1}$$

Therefore $f_2^{-1}(y) = -\sqrt{y+1}$, and

$$f_2^{-1}(x) = -\sqrt{x+1} \quad \text{and } x \text{ in } [-1, c^2 - 1] \tag{12}$$

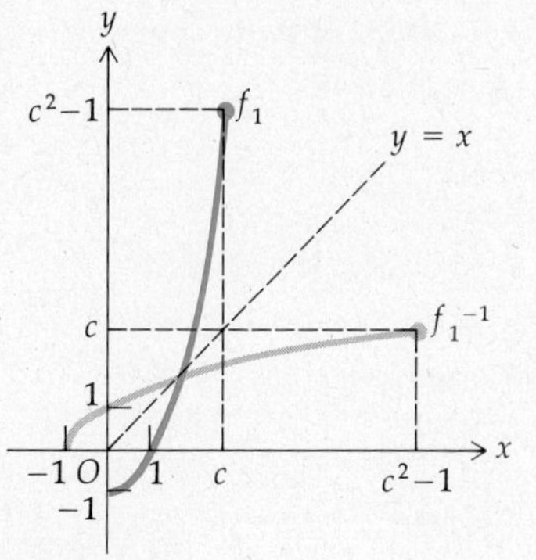

FIGURE 8.1.11

In Fig. 8.1.11 there are sketches of the graphs of f_1 and its inverse f_1^{-1}, as defined in (9) and (11), plotted on the same set of axes. Sketches of the graphs of f_2 and its inverse f_2^{-1}, as defined in (10) and (12), are plotted on the same set of axes in Fig. 8.1.12. In both figures observe that the graphs of the function and its inverse are reflections of each other with respect to the line $y = x$. ●

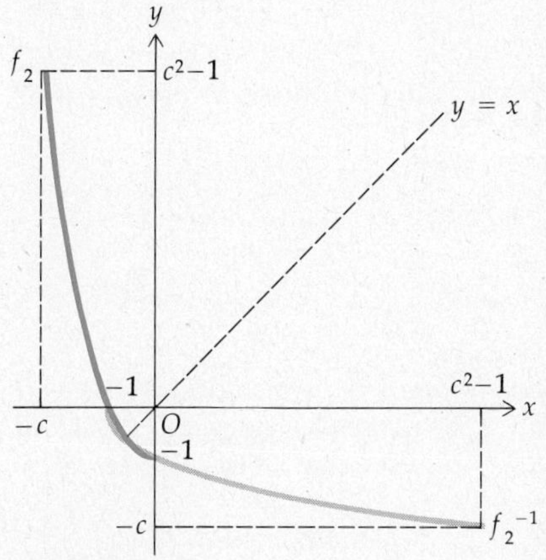

FIGURE 8.1.12

There are functions that have an inverse but for which we cannot obtain an equation defining the inverse function explicitly. For example, if

$$f(x) = x^5 + 5x^3 + 2x - 4 \tag{13}$$

then

$$f'(x) = 5x^4 + 15x^2 + 2$$

Because $f'(x) > 0$ for all x, f is an increasing function, and therefore f has an inverse. However, if we replace $f(x)$ by y in (13), we obtain a fifth-degree equation in x. Equations of the fifth degree and higher in general do not have solutions that can be expressed by a formula. Nevertheless we can determine some properties of the inverse function. We discuss these properties in Sec. 8.2.

EXERCISES 8.1

In Exercises 1 through 14, determine if the given function is one-to-one. Draw a sketch of the graph of the function.

1. $f(x) = 2x + 3$
2. $g(x) = 8 - 4x$
3. $f(x) = \frac{1}{2}x^2 - 2$
4. $f(x) = 3 - x^2$
5. $g(x) = 4 - x^3$
6. $h(x) = \frac{1}{2}x^3 + 1$

7. $f(x) = \sqrt{x+3}$ **8.** $g(x) = \sqrt{1-x^2}$ **9.** $h(x) = 2\sin x, -\frac{1}{2}\pi \le x \le \frac{1}{2}\pi$

10. $f(x) = 1 - \cos x, 0 \le x \le \pi$ **11.** $f(x) = \dfrac{2}{x+3}$ **12.** $g(x) = 5$

13. $g(x) = |x-2|$ **14.** $f(x) = \dfrac{1}{2x-4}$

In Exercises 15 through 36, determine if the given function has an inverse. If the inverse exists, do the following: (a) Find it and state its domain and range; (b) draw sketches of the graphs of the function and its inverse on the same set of axes. If the function does not have an inverse, show that a horizontal line intersects the graph of the function in more than one point.

15. $f(x) = 5x - 7$ **16.** $f(x) = 3x + 6$ **17.** $g(x) = 1 - x^2$

18. $g(x) = x^5$ **19.** $f(x) = (4-x)^3$ **20.** $F(x) = 3(x^2+1)$

21. $h(x) = \sqrt{2x-6}$ **22.** $g(x) = \sqrt{1-x^2}$ **23.** $F(x) = \sqrt[3]{x+1}$

24. $f(x) = |x| + x$ **25.** $f(x) = (x+2)^4$ **26.** $g(x) = 3\sqrt[3]{x} + 1$

27. $f(x) = 2\sqrt[5]{x}$ **28.** $f(x) = \dfrac{2x-1}{x}$ **29.** $f(x) = \dfrac{x-3}{x+1}$

30. $g(x) = \dfrac{8}{x^3+1}$ **31.** $g(x) = x^2 + 5, x \ge 0$ **32.** $f(x) = (2x-1)^2, x \le \frac{1}{2}$

33. $f(x) = (2x+1)^3, -\frac{1}{2} \le x \le \frac{1}{2}$ **34.** $f(x) = \frac{1}{8}x^3, -1 \le x \le 1$ **35.** $F(x) = \sqrt{9-x^2}, 0 \le x \le 3$

36. $G(x) = \sqrt{4x^2-9}, x \ge \frac{3}{2}$

In Exercises 37 through 42, (a) prove that the given function f has an inverse, (b) find $f^{-1}(x)$, and (c) verify the equations of Theorem 8.1.4 for f and f^{-1}.

37. $f(x) = 4x - 3$ **38.** $f(x) = 5x + 2$ **39.** $f(x) = x^3 + 2$

40. $f(x) = (x+2)^3$ **41.** $f(x) = \dfrac{3x+1}{2x+4}$ **42.** $f(x) = \dfrac{x-3}{3x-6}$

In Exercises 43 through 46, do the following: (a) Show that the given function f is not one-to-one and hence does not have an inverse; (b) restrict the domain and obtain two one-to-one functions f_1 and f_2 having the same range as f; (c) find $f_1^{-1}(x)$ and $f_2^{-1}(x)$ and state the domains of f_1^{-1} and f_2^{-1}; (d) draw sketches of the graphs of f_1 and f_1^{-1} on the same set of axes; (e) draw sketches of the graphs of f_2 and f_2^{-1} on the same set of axes.

43. $f(x) = x^2 + 4$ **44.** $f(x) = 2x^2 - 6$

45. $f(x) = \sqrt{9-x^2}$ **46.** $f(x) = \frac{1}{2}\sqrt{x^2-16}$

47. Given

$$f(x) = \begin{cases} x & \text{if } x < 1 \\ x^2 & \text{if } 1 \le x \le 9 \\ 27\sqrt{x} & \text{if } 9 < x \end{cases}$$

Prove that f has an inverse function and find $f^{-1}(x)$.

48. Determine the value of the constant k so that the function defined by

$$f(x) = \frac{x+5}{x+k}$$

will be its own inverse.

8.2 INVERSE FUNCTION THEOREM AND DERIVATIVE OF THE INVERSE OF A FUNCTION

At the conclusion of Sec. 8.1 we stated that a function f can have an inverse f^{-1} which is not defined explicitly by an equation. However, information about the continuity and differentiability of f^{-1} can be acquired from properties of f.

• ILLUSTRATION 1 In Example 3 of Sec. 8.1 we had the function f defined by

$$f(x) = 4x - 3$$

and its inverse is f^{-1} defined by

$$f^{-1}(x) = \frac{x+3}{4}$$

In Fig. 8.1.7 the graphs of f and f^{-1} are shown on the same set of axes. We see that both f and f^{-1} are continuous and increasing on their domains. •

• ILLUSTRATION 2 In Illustration 4 of Sec. 8.1 the function g and its inverse g^{-1} are defined by

$$g(x) = x^3 \qquad -2 \le x \le 2$$

and

$$g^{-1}(x) = \sqrt[3]{x} \qquad -8 \le x \le 8$$

and the graphs of g and g^{-1} are shown in Fig. 8.1.8. Each of these functions is continuous and increasing on its domain. •

In Illustrations 1 and 2 there are examples of a function and its inverse that satisfy the conditions in the following theorem, known as the *inverse function theorem* for increasing functions. The existence of the inverse function in the statement of the theorem is guaranteed by Theorem 8.1.5.

8.2.1 Theorem (Inverse Function Theorem)

Suppose that the function f is continuous and increasing on the closed interval $[a, b]$. Then if f^{-1} is its inverse, which is defined on $[f(a), f(b)]$,

(i) f^{-1} is increasing on $[f(a), f(b)]$.
(ii) f^{-1} is continuous on $[f(a), f(b)]$.

Proof of (i) To prove that f^{-1} is increasing on $[f(a), f(b)]$ we must show that if y_1 and y_2 are two numbers in $[f(a), f(b)]$ such that $y_1 < y_2$, then $f^{-1}(y_1) < f^{-1}(y_2)$. Because f^{-1} is defined on $[f(a), f(b)]$, there exist numbers x_1 and x_2 in $[a, b]$ such that $y_1 = f(x_1)$ and $y_2 = f(x_2)$. Therefore

$$f^{-1}(y_1) = f^{-1}(f(x_1)) = x_1 \tag{1}$$

and

$$f^{-1}(y_2) = f^{-1}(f(x_2)) = x_2 \tag{2}$$

If $x_2 < x_1$, then because f is increasing on $[a, b]$, $f(x_2) < f(x_1)$ or, equivalently, $y_2 < y_1$. But $y_1 < y_2$; therefore x_2 cannot be less than x_1.

If $x_2 = x_1$, then because f is a function, $f(x_1) = f(x_2)$ or, equivalently, $y_1 = y_2$, but this also contradicts the fact that $y_1 < y_2$. Therefore $x_2 \neq x_1$.

So if x_2 is not less than x_1 and $x_2 \neq x_1$, it follows that $x_1 < x_2$; hence, from (1) and (2), $f^{-1}(y_1) < f^{-1}(y_2)$. Thus we have proved that f^{-1} is increasing on $[f(a), f(b)]$.

Proof of (ii) To prove that f^{-1} is continuous on the closed interval $[f(a), f(b)]$ we must show that if r is any number in the open interval $(f(a), f(b))$, then f^{-1} is continuous at r, and f^{-1} is continuous from the right at $f(a)$, and f^{-1} is continuous from the left at $f(b)$.

We prove that f^{-1} is continuous at any r in the open interval $(f(a), f(b))$ by showing that Theorem 2.7.5 holds at r. We wish to show that, for any $\epsilon > 0$ small enough so that $f^{-1}(r) - \epsilon$ and $f^{-1}(r) + \epsilon$ are both in $[a, b]$, there exists a $\delta > 0$ such that

$$|f^{-1}(y) - f^{-1}(r)| < \epsilon \quad \text{whenever} \quad |y - r| < \delta$$

Let $f^{-1}(r) = s$. Then $f(s) = r$. Because from (ii) f^{-1} is increasing on $[f(a), f(b)]$, we conclude that $a < s < b$. Therefore

$$a \leq s - \epsilon < s < s + \epsilon \leq b$$

Because f is increasing on $[a, b]$,

$$f(a) \leq f(s - \epsilon) < r < f(s + \epsilon) \leq f(b) \tag{3}$$

Let δ be the smaller of the two numbers $r - f(s - \epsilon)$ and $f(s + \epsilon) - r$; so $\delta \leq r - f(s - \epsilon)$ and $\delta \leq f(s + \epsilon) - r$ or, equivalently,

$$f(s - \epsilon) \leq r - \delta \tag{4}$$

and

$$r + \delta \leq f(s + \epsilon) \tag{5}$$

Whenever $|y - r| < \delta$, we have $-\delta < y - r < \delta$ or, equivalently,

$$r - \delta < y < r + \delta \tag{6}$$

From (3), (4), (5), and (6) we have, whenever $|y - r| < \delta$,

$$f(a) \leq f(s - \epsilon) < y < f(s + \epsilon) \leq f(b)$$

Because f^{-1} is increasing on $[f(a), f(b)]$, it follows from the above that

$$f^{-1}(f(s - \epsilon)) < f^{-1}(y) < f^{-1}(f(s + \epsilon)) \quad \text{whenever} \quad |y - r| < \delta$$

or, equivalently,

$$s - \epsilon < f^{-1}(y) < s + \epsilon \quad \text{whenever} \quad |y - r| < \delta$$

or, equivalently,

$$-\epsilon < f^{-1}(y) - s < \epsilon \quad \text{whenever} \quad |y - r| < \delta$$

or, equivalently,

$$|f^{-1}(y) - f^{-1}(r)| < \epsilon \quad \text{whenever} \quad |y - r| < \delta$$

So f^{-1} is continuous on the open interval $(f(a), f(b))$.

The proofs that f^{-1} is continuous from the right at $f(a)$ and continuous from the left at $f(b)$ are left as an exercise (see Exercise 37). ■

The inverse function theorem for decreasing functions is now stated. The proof is similar to that for Theorem 8.2.1 and is left as an exercise (See Exercises 38 and 39).

8.2.2 Theorem (Inverse Function Theorem) Suppose that the function f is continuous and decreasing on the closed interval $[a, b]$. Then if f^{-1} is its inverse, which is defined on $[f(b), f(a)]$.

(i) f^{-1} is decreasing on $[f(b), f(a)]$.

(ii) f^{-1} is continuous on $[f(b), f(a)]$.

The inverse function theorem is used to prove the following theorem that expresses a relationship between the derivative of a function and the derivative of its inverse if the function has an inverse.

8.2.3 Theorem Suppose that the function f is continuous and monotonic on the closed interval $[a, b]$, and let $y = f(x)$. If f is differentiable on $[a, b]$ and $f'(x) \neq 0$ for any x in $[a, b]$, then the derivative of the inverse function f^{-1}, defined by $x = f^{-1}(y)$, is given by

$$D_y x = \frac{1}{D_x y}$$

Proof Because f is continuous and monotonic on $[a, b]$, then by Theorems 8.2.1 and 8.2.2, f has an inverse that is continuous and monotonic on $[f(a), f(b)]$ (or $[f(b), f(a)]$ if $f(b) < f(a)$).

If x is a number in $[a, b]$, let Δx be an increment of x, $\Delta x \neq 0$, such that $x + \Delta x$ is also in $[a, b]$. Then the corresponding increment of y is given by

$$\Delta y = f(x + \Delta x) - f(x) \tag{7}$$

$\Delta y \neq 0$ since $\Delta x \neq 0$ and f is monotonic on $[a, b]$; that is, either

$$f(x + \Delta x) < f(x) \quad \text{or} \quad f(x + \Delta x) > f(x) \qquad \text{on } [a, b]$$

If x is in $[a, b]$ and $y = f(x)$, then y is in $[f(a), f(b)]$ (or $[f(b), f(a)]$). Also, if $x + \Delta x$ is in $[a, b]$, then $y + \Delta y$ is in $[f(a), f(b)]$ (or $[f(b), f(a)]$) because $y + \Delta y = f(x + \Delta x)$ by (7). So

$$x = f^{-1}(y) \tag{8}$$

and

$$x + \Delta x = f^{-1}(y + \Delta y) \tag{9}$$

It follows from (8) and (9) that

$$\Delta x = f^{-1}(y + \Delta y) - f^{-1}(y) \tag{10}$$

From the definition of a derivative,

$$D_y x = \lim_{\Delta y \to 0} \frac{f^{-1}(y + \Delta y) - f^{-1}(y)}{\Delta y}$$

Substituting from (7) and (10) into the above equation we get

$$D_y x = \lim_{\Delta y \to 0} \frac{\Delta x}{f(x + \Delta x) - f(x)}$$

and because $\Delta x \neq 0$,

$$D_y x = \lim_{\Delta y \to 0} \frac{1}{\dfrac{f(x + \Delta x) - f(x)}{\Delta x}} \tag{11}$$

Before we find the limit in (11) we show that under the hypothesis of this theorem $\Delta x \to 0$ is equivalent to $\Delta y \to 0$. First we show that $\lim_{\Delta y \to 0} \Delta x = 0$. From (10),

$$\lim_{\Delta y \to 0} \Delta x = \lim_{\Delta y \to 0} [f^{-1}(y + \Delta y) - f^{-1}(y)]$$

Because f^{-1} is continuous on $[f(a), f(b)]$ (or $[f(b), f(a)]$), the limit on the right side of the above equation is zero. So

$$\lim_{\Delta y \to 0} \Delta x = 0 \tag{12}$$

Now we demonstrate that $\lim_{\Delta x \to 0} \Delta y = 0$. From (7),

$$\lim_{\Delta x \to 0} \Delta y = \lim_{\Delta x \to 0} [f(x + \Delta x) - f(x)]$$

Because f is continuous on $[a, b]$, the limit on the right side of the above equation is zero, and therefore

$$\lim_{\Delta x \to 0} \Delta y = 0 \tag{13}$$

From (12) and (13) it follows that

$$\Delta x \to 0 \quad \text{if and only if} \quad \Delta y \to 0 \tag{14}$$

Thus, applying the limit theorem (regarding the limit of a quotient) to (11) and using (14), we have

$$D_y x = \frac{1}{\lim_{\Delta x \to 0} \dfrac{f(x + \Delta x) - f(x)}{\Delta x}}$$

Because f is differentiable on $[a, b]$, the limit in the denominator of the above is $f'(x)$ or, equivalently, $D_x y$, and so

$$D_y x = \frac{1}{D_x y}$$ ■

● ILLUSTRATION 3 We verify Theorem 8.2.3 for the function f defined by $f(x) = \sqrt{x}$. If we let $y = f(x)$, we have the equation

$$y = \sqrt{x} \qquad x \geq 0, y \geq 0 \tag{15}$$

Because f is one-to-one, f^{-1} exists and is defined by $f^{-1}(y) = y^2$. If we let $x = f^{-1}(y)$, we have the equation

$$x = y^2 \qquad y \geq 0, x \geq 0 \tag{16}$$

From (15),

$$D_x y = \frac{1}{2\sqrt{x}}$$

and from (16),

$$D_y x = 2y \tag{17}$$

Substituting from (15) into (17) we have

$$\begin{aligned} D_y x &= 2\sqrt{x} \\ &= \frac{1}{\frac{1}{2\sqrt{x}}} \\ &= \frac{1}{D_x y} \end{aligned} \tag{18}$$

When $x = 0$, $D_x y$ does not exist; thus (18) is not satisfied for this value of x. Because the domain of f is the closed interval $[0, +\infty)$, the theorem is valid for any x in the open interval $(0, +\infty)$. ●

Example 1 Show that Theorem 8.2.3 holds for the function f of Examples 2 and 3 in Sec. 8.1.

Solution The function f is defined by $(2x + 3)/(x - 1)$. If we let $y = f(x)$, we have

$$y = \frac{2x + 3}{x - 1} \tag{19}$$

$$D_x y = -\frac{5}{(x - 1)^2}$$

In the solution of Example 3 in Sec. 8.1 we showed that

$$x = f^{-1}(y) = \frac{y + 3}{y - 2}$$

We compute $D_y x$ from this equation and get

$$D_y x = (f^{-1})'(y) = -\frac{5}{(y - 2)^2}$$

In the above equation we substitute the value of y from (19) and obtain

$$D_y x = -\frac{5}{\left(\frac{2x+3}{x-1} - 2\right)^2}$$

$$= -\frac{5(x-1)^2}{(2x+3-2x+2)^2}$$

$$= -\tfrac{5}{25}(x-1)^2$$

$$= -\tfrac{1}{5}(x-1)^2$$

$$= \frac{1}{D_x y}$$

Example 2 Given that f is the function defined by $f(x) = x^3 + x$, determine if f has an inverse. If it does, find the derivative of the inverse function.

Solution Because $f(x) = x^3 + x$,

$$f'(x) = 3x^2 + 1$$

Therefore $f'(x) > 0$ for all real numbers; so f is increasing on its domain. Thus f is a one-to-one function, and hence it has an inverse f^{-1}. Let $y = f(x)$ and then $x = f^{-1}(y)$. By Theorem 8.2.3,

$$D_y x = \frac{1}{D_x y}$$

$$= \frac{1}{3x^2 + 1}$$

When using Theorem 8.2.3 to compute the value of the derivative of the inverse of a function at a particular number, it is more convenient to have the statement of the theorem with f' and $(f^{-1})'$ notations for the derivatives. With this notation we restate the theorem as Theorem 8.2.4.

8.2.4 Theorem Suppose the function f is continuous and monotonic on a closed interval $[a, b]$ containing the number c, and let $f(c) = d$. If $f'(c)$ exists and $f'(c) \neq 0$, then $(f^{-1})'(d)$ exists and

$$(f^{-1})'(d) = \frac{1}{f'(c)}$$

● ILLUSTRATION 4 We show that Theorem 8.2.4 holds for a particular function and particular values of c and d. If f is the function of Illustration 3,

$$f(x) = \sqrt{x} \quad \text{and} \quad f'(x) = \frac{1}{2\sqrt{x}}$$

$$f^{-1}(x) = x^2 \quad \text{and} \quad (f^{-1})'(x) = 2x \qquad x \geq 0$$

The function f is continuous and monotonic on any closed interval $[a, b]$ for which $0 \leq a < b$. Let $c = 9$, and then $d = f(9) = 3$. Theorem 8.2.4 states that

$$(f^{-1})'(3) = \frac{1}{f'(9)} \tag{20}$$

Because $(f^{-1})'(3) = 6$ and $f'(9) = \frac{1}{6}$, Eq. (20) is verified. ●

● ILLUSTRATION 5 The function f mentioned at the conclusion of Sec. 8.1 is defined by

$$f(x) = x^5 + 5x^3 + 2x - 4 \tag{21}$$

and as stated there, f has an inverse f^{-1}. But we do not have an equation defining explicitly the function value of f^{-1}. Nevertheless it is possible to compute the derivative of f^{-1} at a particular point on the graph of $y = f^{-1}(x)$. For instance, because $(1, 4)$ is on the graph of $y = f(x)$, the point $(4, 1)$ is on the graph of $y = f^{-1}(x)$. We compute the value of $(f^{-1})'(4)$ by using Theorem 8.2.4, which states that

$$(f^{-1})'(4) = \frac{1}{f'(1)}$$

We first find $f'(x)$ from (21).

$$f'(x) = 5x^4 + 15x^2 + 2$$

Therefore

$$f'(1) = 22$$

Thus

$$(f^{-1})'(4) = \tfrac{1}{22}$$ ●

Example 3 (a) On the same set of axes draw sketches of the graphs of the functions f and f^{-1} of Example 2. (b) Find the slope of the tangent line to the graph of f at the point $(1, 2)$. (c) Find the slope of the tangent line to the graph of f^{-1} at the point $(2, 1)$.

Solution

(a) For the function f,

$$f(x) = x^3 + x$$

$$f'(x) = 3x^2 + 1$$

$$f''(x) = 6x$$

The function is increasing on its domain. Because $f''(0) = 0$ and $f''(x)$ changes sign at $x = 0$, the graph of f has a point of inflection at the origin. The slope of the inflectional tangent at the origin is $f'(0) = 1$. We use this information to draw the sketch of the graph of f shown in Fig. 8.2.1. This figure also shows a sketch of the graph of f^{-1}, which is obtained from the fact that it is the reflection of the graph of f with respect to the line $y = x$.

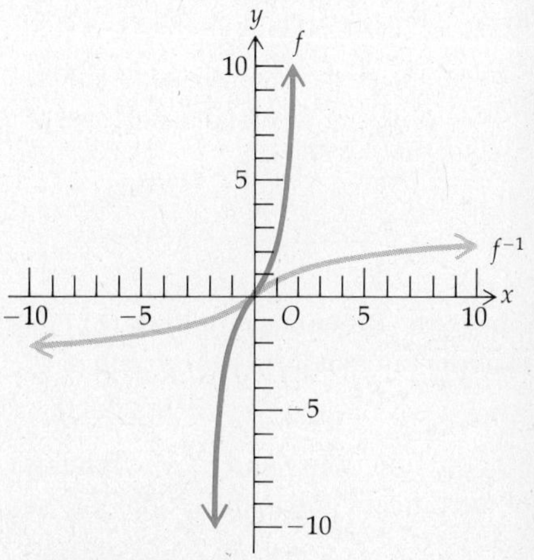

FIGURE 8.2.1

Observe that to find an equation defining $f^{-1}(x)$ explicitly it is necessary to solve the third degree equation $y = x^3 + x$.

(b) The slope of the tangent line to the graph of f at the point $(1, 2)$ is $f'(1) = 4$.

(c) The slope of the tangent line to the graph of f^{-1} at the point $(2, 1)$ is $(f^{-1})'(2)$. From Theorem 8.2.4,

$$(f^{-1})'(2) = \frac{1}{f'(1)} = \tfrac{1}{4}$$

EXERCISES 8.2

In Exercises 1 through 10, let $y = f(x)$ and $x = f^{-1}(y)$, and verify that $D_y x = \dfrac{1}{D_x y}$.

1. $f(x) = 4x - 3$
2. $f(x) = 7 - 2x$
3. $f(x) = \sqrt{x+1}$
4. $f(x) = 8x^3$
5. $f(x) = \frac{1}{5}x^5$
6. $f(x) = \sqrt{4-3x}$
7. $f(x) = \sqrt[3]{x-8}$
8. $f(x) = \sqrt[5]{x}$
9. $f(x) = \dfrac{2x-3}{x+2}$
10. $f(x) = \dfrac{3x+4}{2x+6}$

In Exercises 11 through 24, find $(f^{-1})'(d)$ for the given f and d.

11. $f(x) = \sqrt{3x+1}$; $d = 1$
12. $f(x) = x^5 + 2$; $d = 1$
13. $f(x) = x^2 - 16, x \geq 0$; $d = 9$
14. $f(x) = \sqrt{4-x}$; $d = 3$
15. $f(x) = x^3 + 5$; $d = -3$
16. $f(x) = 4x^3 + 2x$; $d = 6$
17. $f(x) = 3x^5 + 2x^3$; $d = 5$
18. $f(x) = \sin x$; $-\frac{1}{2}\pi \leq x \leq \frac{1}{2}\pi$; $d = \frac{1}{2}$
19. $f(x) = \frac{1}{2}\cos^2 x$; $0 \leq x \leq \frac{1}{2}\pi$; $d = \frac{1}{4}$
20. $f(x) = x^2 - 6x + 7, x \leq 3$; $d = 0$
21. $f(x) = 2x^2 + 8x + 7, x \leq -2$; $d = 1$
22. $f(x) = 2x^3 + x + 20$; $d = 2$
23. $f(x) = \int_{-3}^{x} \sqrt{t+3}\, dt, x > -3$; $d = 18$
24. $f(x) = \int_{x}^{2} t\, dt, x < 0$; $d = -6$
25. Given $f(x) = \int_{1}^{x} \sqrt{16 - t^4}\, dt$, $-2 \leq x \leq 2$. Prove that f has an inverse f^{-1}, and compute $(f^{-1})'(0)$.
26. Given $f(x) = \int_{2}^{x} \sqrt{9 + t^4}\, dt$. Prove that f has an inverse f^{-1}, and compute $(f^{-1})'(0)$.
27. Given $f(x) = x^3 + 3x - 1$. (a) On the same set of axes draw sketches of the graphs of the functions f and f^{-1}. (b) Find the slope of the tangent line to the graph of f at the point $(1, 3)$. (c) Find the slope of the tangent line to the graph of f^{-1} at the point $(3, 1)$.
28. Given $f(x) = 6 - x - x^3$. (a) On the same set of axes draw sketches of the graphs of the functions f and f^{-1}. (b) Find the slope of the tangent line to the graph of f at the point $(2, -4)$. (c) Find the slope of the tangent line to the graph of f^{-1} at the point $(-4, 2)$.

In Exercises 29 through 34, do the following: (a) Solve the equation for y in terms of x and express y as one or more functions of x; (b) for each of the functions obtained in (a) determine if the function has an inverse, and, if it does, determine the domain of the inverse function; (c) use implicit differentiation to find $D_x y$ and $D_y x$, and determine the values of x and y for which $D_x y$ and $D_y x$ are reciprocals.

29. $x^2 + y^2 = 9$ **30.** $x^2 - 4y^2 = 16$ **31.** $xy = 4$

32. $9y^2 - 8x^3 = 0$ **33.** $2x^2 - 3xy + 1 = 0$ **34.** $2x^2 + 2y + 1 = 0$

35. Show that the formula of Theorem 8.2.3 can be written as

$$(f^{-1})'(x) = \frac{1}{f'(f^{-1}(x))}$$

36. Use the formula of Exercise 35 to show that

$$(f^{-1})''(x) = -\frac{f''(f^{-1}(x))}{[f'(f^{-1}(x))]^3}$$

37. Given that the function f is continuous and increasing on the closed interval $[a, b]$, by assuming Theorem 8.2.1(i) prove that f^{-1} is continuous from the right at $f(a)$ and continuous from the left at $f(b)$.

38. Prove Theorem 8.2.2(i). **39.** Prove Theorem 8.2.2(ii).

8.3 THE NATURAL LOGARITHMIC FUNCTION

The definition of the logarithmic function that you encountered in algebra is based on exponents. The properties of logarithms are then proved from corresponding properties of exponents. One such property of exponents is

$$a^x \cdot a^y = a^{x+y} \tag{1}$$

If the exponents, x and y, are positive integers and a is any real number, (1) follows from the definition of a positive integer exponent and mathematical induction. If the exponents are allowed to be any integers, either positive, negative, or zero, and $a \neq 0$, then (1) will hold if a zero exponent and a negative integer exponent are defined by

$$a^0 = 1$$

and

$$a^{-n} = \frac{1}{a^n} \qquad n > 0$$

If the exponents are rational numbers and $a \geq 0$, then (1) holds when $a^{m/n}$ is defined by

$$a^{m/n} = (\sqrt[n]{a})^m$$

It is not quite so simple to define a^x when x is an irrational number. For example, what is meant by $2^{\sqrt{3}}$? The definition of the logarithmic function, as given in elementary algebra, is based on the assumption that a^x exists if a is any positive number and x is any real number.

This definition states that the equation

$$a^x = N$$

where a is any positive number except 1 and N is any positive number, can be solved for x, and x is uniquely determined by

$$x = \log_a N$$

When logarithms are used mainly as an aid to computation, the number a (called the *base*) is taken as 10. The following properties of logarithms are proved from those of exponents:

$$\log_a 1 = 0 \tag{2}$$

$$\log_a MN = \log_a \mathrm{M} + \log_a N \tag{3}$$

$$\log_a \frac{M}{N} = \log_a M - \log_a N \tag{4}$$

$$\log_a M^n = n \log_a M \tag{5}$$

$$\log_a a = 1 \tag{6}$$

In this chapter we define the logarithmic function by using calculus and prove the properties of logarithms by means of this definition. Then the exponential function is defined in terms of the logarithmic function. This definition enables us to define a^x when x is any real number and $a > 0$. The properties of exponents will then be proved if the exponent is any real number.

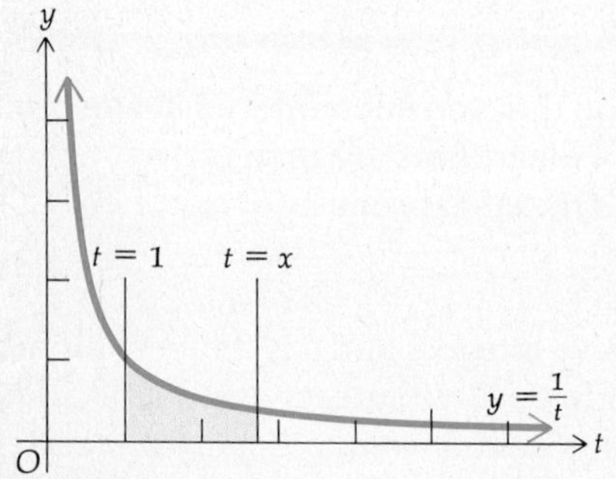

FIGURE 8.3.1

Recall the formula

$$\int t^n\, dt = \frac{t^{n+1}}{n+1} + C \qquad n \neq -1$$

This formula does not hold when $n = -1$. Consider the function defined by the equation $y = t^{-1}$, where t is positive. A sketch of the graph of this equation is shown in Fig. 8.3.1.

Let R be the region bounded by the curve $y = 1/t$, by the t axis, on the left by the line $t = 1$, and on the right by the line $t = x$, where x is greater than 1. This region R is shown in Fig. 8.3.1. The measure of the area of R is a function of x; call it $A(x)$ and define it as a definite integral by

$$A(x) = \int_1^x \frac{1}{t}\, dt \tag{7}$$

Now consider the integral in (7) if $0 < x < 1$. From Definition 6.2.5,

$$\int_1^x \frac{1}{t}\, dt = -\int_x^1 \frac{1}{t}\, dt$$

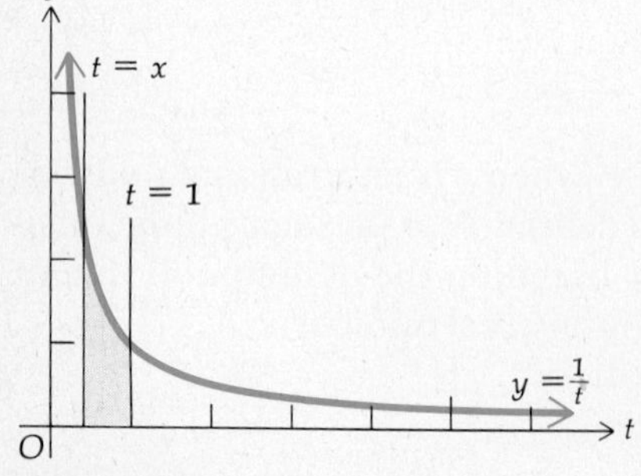

FIGURE 8.3.2

Then the integral $\int_x^1 (1/t)\, dt$ represents the measure of the area of the region bounded by the curve $y = 1/t$, by the t axis, on the left by the line $t = x$, and on the right by the line $t = 1$. So the integral $\int_1^x (1/t)\, dt$ is then the negative of the measure of the area of the region shown in Fig. 8.3.2.

If $x = 1$, the integral $\int_1^x (1/t)\, dt$ becomes $\int_1^1 (1/t)\, dt$, which equals 0 by Definition 6.2.6. In this case the left and right boundaries of the region are the same and the measure of the area is 0.

Thus the integral $\int_1^x (1/t)\,dt$ for $x > 0$ can be interpreted in terms of the measure of the area of a region. Its value depends on x and is therefore a function of x. This integral is used to define the *natural logarithmic function.*

8.3.1 Definition The **natural logarithmic function** is the function defined by

$$\ln x = \int_1^x \frac{1}{t}\,dt \qquad x > 0$$

The domain of the natural logarithmic function is the set of all positive numbers. We read $\ln x$ as "the natural logarithm of x."

The natural logarithmic function is differentiable because by applying Theorem 6.5.1,

$$D_x(\ln x) = D_x\left(\int_1^x \frac{1}{t}\,dt\right) = \frac{1}{x}$$

From this result and the chain rule we have the following theorem.

8.3.2 Theorem If u is a differentiable function of x and $u(x) > 0$, then

$$D_x(\ln u) = \frac{1}{u} \cdot D_x u$$

Example 1 Given $y = \ln(3x^2 - 6x + 8)$, find dy/dx.

Solution From Theorem 8.3.2,

$$\frac{dy}{dx} = \frac{1}{3x^2 - 6x + 8} \cdot (6x - 6) = \frac{6x - 6}{3x^2 - 6x + 8}$$

Example 2 Given $y = \ln[(4x^2 + 3)(2x - 1)]$, find $D_x y$.

Solution Applying Theorem 8.3.2 we get

$$D_x y = \frac{1}{(4x^2 + 3)(2x - 1)} \cdot [8x(2x - 1) + 2(4x^2 + 3)]$$

$$= \frac{24x^2 - 8x + 6}{(4x^2 + 3)(2x - 1)} \qquad (8)$$

Example 3 Given

$$y = \ln\left(\frac{x}{x + 1}\right)$$

find dy/dx.

Solution From Theorem 8.3.2,

$$\frac{dy}{dx} = \frac{1}{\frac{x}{x+1}} \cdot \frac{(x+1) - x}{(x+1)^2}$$

$$= \frac{x+1}{x} \cdot \frac{1}{(x+1)^2}$$

$$= \frac{1}{x(x+1)}$$

It should be emphasized that when using Theorem 8.3.2, $u(x)$ must be positive; that is, a number in the domain of the derivative must be in the domain of the given function, $\ln u$.

● ILLUSTRATION 1 In Example 1 the domain of the given function is the set of all real numbers because $3x^2 - 6x + 8 > 0$ for all x. This can be seen from the fact that the parabola having equation $y = 3x^2 - 6x + 8$ has its vertex at (1, 5) and opens upward. Hence $(6x - 6)/(3x^2 - 6x + 8)$ is the derivative for all values of x.

In Example 2, because $(4x^2 + 3)(2x - 1) > 0$ only when $x > \frac{1}{2}$, the domain of the given function is the interval $(\frac{1}{2}, +\infty)$. Therefore it is understood that fraction (8) is the derivative only if $x > \frac{1}{2}$.

Because $x/(x + 1) > 0$ when either $x < -1$ or $x > 0$, the domain of the function in Example 3 is $(-\infty, -1) \cup (0, +\infty)$; so $1/x(x + 1)$ is the derivative if either $x < -1$ or $x > 0$. ●

We show that the natural logarithmic function obeys the properties of logarithms given earlier.

8.3.3 Theorem

$$\ln 1 = 0$$

Proof If $x = 1$ in Definition 8.3.1,

$$\ln 1 = \int_1^1 \frac{1}{t}\, dt$$

The right side of the above is zero by Definition 6.2.6. Thus

$$\ln 1 = 0$$

■

Theorem 8.3.3 corresponds to the property of logarithms given by (2). The next three theorems correspond to the properties of logarithms given by (3), (4), and (5).

8.3.4 Theorem If a and b are any positive numbers, then

$$\ln(ab) = \ln a + \ln b$$

Proof If $a > 0$, then from Theorem 8.3.2,

$$D_x(\ln ax) = \frac{1}{ax} D_x(ax) = \frac{1}{ax}(a) = \frac{1}{x}$$

Therefore the functions $\ln x$ and $\ln ax$ have the same derivative. Thus from Theorem 5.2.2 there is a constant K such that

$$\ln ax = \ln x + K \qquad \text{for all } x > 0 \tag{9}$$

To determine K, let $x = 1$ in (9) and obtain

$$\ln a = \ln 1 + K$$

Because $\ln 1 = 0$, it follows that $K = \ln a$. Replacing K by $\ln a$ in (9) we get

$$\ln ax = \ln x + \ln a$$

Now let $x = b$ in this equation, and we have

$$\ln ab = \ln a + \ln b$$

■

8.3.5 Theorem If a and b are any positive numbers, then

$$\ln \frac{a}{b} = \ln a - \ln b$$

Proof Because $a = (a/b) \cdot b$,

$$\ln a = \ln\left(\frac{a}{b} \cdot b\right)$$

Applying theorem 8.3.4 to the right side of the above equation we get

$$\ln a = \ln \frac{a}{b} + \ln b$$

$$\ln \frac{a}{b} = \ln a - \ln b$$

■

8.3.6 Theorem If a is any positive number and r is any rational number, then

$$\ln a^r = r \ln a$$

Proof From Theorem 8.3.2, if r is any rational number and $x > 0$,

$$D_x(\ln x^r) = \frac{1}{x^r} \cdot rx^{r-1} = \frac{r}{x}$$

and

$$D_x(r \ln x) = \frac{r}{x}$$

Therefore

$$D_x(\ln x^r) = D_x(r \ln x)$$

From the above equation the derivatives of $\ln x^r$ and $r \ln x$ are equal; so it follows from Theorem 5.2.2 that there is a constant K such that

$$\ln x^r = r \ln x + K \qquad \text{for all } x > 0 \tag{10}$$

To determine K, substitute 1 for x in (10) and get

$$\ln 1^r = r \ln 1 + K$$

But $\ln 1 = 0$; hence $K = 0$. Replacing K by 0 in (10) gives

$$\ln x^r = r \ln x \qquad \text{for all } x > 0$$

from which it follows that if $x = a$, where a is any positive number, then

$$\ln a^r = r \ln a$$ ■

● ILLUSTRATION 2 In Example 2, if Theorem 8.3.4 is applied before finding the derivative, we have

$$y = \ln(4x^2 + 3) + \ln(2x - 1) \tag{11}$$

The domain of the function defined by the above equation is the interval $(\frac{1}{2}, +\infty)$, which is the same as the domain of the given function. From (11),

$$D_x y = \frac{8x}{4x^2 + 3} + \frac{2}{2x - 1}$$

and combining the fractions gives

$$D_x y = \frac{8x(2x - 1) + 2(4x^2 + 3)}{(4x^2 + 3)(2x - 1)}$$

which is the same as the first line of the solution of Example 2. ●

● ILLUSTRATION 3 If we apply Theorem 8.3.5 before finding the derivative in Example 3, we have

$$\ln y = \ln x - \ln(x + 1) \tag{12}$$

$\ln x$ is defined only when $x > 0$, and $\ln(x + 1)$ is defined only when $x > -1$. Therefore the domain of the function defined by (12) is the interval $(0, +\infty)$. But the domain of the function given in Example 3 consists of the two intervals $(-\infty, -1)$ and $(0, +\infty)$. Finding $D_x y$ from (12) we have

$$D_x y = \frac{1}{x} - \frac{1}{x + 1} = \frac{1}{x(x + 1)}$$

but remember here that x must be greater than 0, whereas in the solution of Example 3 values of x less than -1 are also included. ●

Illustration 3 shows the care that must be taken when applying Theorems 8.3.4, 8.3.5, and 8.3.6 to functions involving the natural logarithm.

Example 4 Given $y = \ln(2x - 1)^3$, find $D_x y$.

Solution From Theorem 8.3.6,

$$y = \ln(2x - 1)^3 = 3 \ln(2x - 1)$$

Note that $\ln(2x - 1)^3$ and $3 \ln(2x - 1)$ both have the same domain, $x > \frac{1}{2}$. Applying Theorem 8.3.2 gives

$$D_x y = 3 \cdot \frac{1}{2x - 1} \cdot 2 = \frac{6}{2x - 1}$$

To draw a sketch of the graph of the natural logarithmic function we must first consider some properties of this function.

Let f be the function defined by

$$f(x) = \ln x = \int_1^x \frac{1}{t}\, dt \qquad x > 0$$

The domain of f is the set of all positive numbers. The function f is differentiable for all x in its domain, and

$$f'(x) = \frac{1}{x} \tag{13}$$

Because f is differentiable for all $x > 0$, f is continuous for all $x > 0$. From (13), $f'(x) > 0$ for all $x > 0$, and therefore f is an increasing function.

$$f''(x) = -\frac{1}{x^2} \tag{14}$$

From (14), $f''(x) < 0$ when $x > 0$. Therefore the graph of $y = f(x)$ is concave downward at every point. Because $f(1) = \ln 1 = 0$, the x intercept of the graph is 1.

$$f(2) = \ln 2 = \int_1^2 \frac{1}{t}\, dt$$

To determine an approximate numerical value for ln 2, the definite integral $\int_1^2 (1/t)\, dt$ is interpreted as the number of square units in the area of the region shown in Fig. 8.3.3.

From Fig. 8.3.3 we see that ln 2 is between the measures of the areas of the rectangles, each having a base of length 1 unit and altitudes of lengths $\frac{1}{2}$ unit and 1 unit; that is,

$$0.5 < \ln 2 < 1 \tag{15}$$

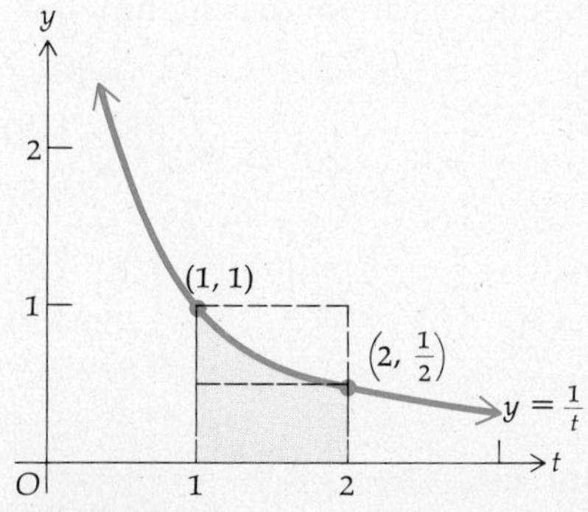

FIGURE 8.3.3

An inequality can be obtained analytically, using Theorem 6.3.8, by proceeding as follows. Let $f(t) = 1/t$ and $g(t) = \frac{1}{2}$. Then $f(t) \geq g(t)$ for all t in $[1, 2]$. Because f and g are continuous on $[1, 2]$, they are integrable on $[1, 2]$, and from Theorem 6.3.8,

$$\int_1^2 \frac{1}{t}\, dt \geq \int_1^2 \frac{1}{2}\, dt$$

or, equivalently,

$$\ln 2 \geq \tfrac{1}{2} \tag{16}$$

Similarly, if $f(t) = 1/t$ and $h(t) = 1$, then $h(t) \geq f(t)$ for all t in $[1, 2]$. Because h and f are continuous on $[1, 2]$, they are integrable there; and again using Theorem 6.3.8, we obtain

$$\int_1^2 1\, dt \geq \int_1^2 \frac{1}{t}\, dt$$

or, equivalently,

$$1 \geq \ln 2 \tag{17}$$

Combining (16) and (17) we get

$$0.5 \leq \ln 2 \leq 1 \tag{18}$$

The number 0.5 is a lower bound of ln 2 and 1 is an upper bound of ln 2. In a similar manner we obtain a lower and upper bound for the natural logarithm of any positive real number. Later we learn, by applying infinite series, how to compute the natural logarithm of any positive real number to any desired number of decimal places.

The value of ln 2 to five decimal places is given by

$$\ln 2 \approx 0.69315$$

Using this value of ln 2 and applying Theorem 8.3.6 we can find an approximate value for the natural logarithm of any power of 2. In particular,

$$\begin{aligned}
\ln 4 = \ln 2^2 &= 2 \ln 2 \approx 1.38630 \\
\ln 8 = \ln 2^3 &= 3 \ln 2 \approx 2.07945 \\
\ln \tfrac{1}{2} = \ln 2^{-1} &= -1 \ln 2 \approx -0.69315 \\
\ln \tfrac{1}{4} = \ln 2^{-2} &= -2 \ln 2 \approx -1.38630
\end{aligned}$$

In the appendix there is a table of natural logarithms giving the values to four decimal places.

We now determine the behavior of the natural logarithmic function for large values of x by considering $\lim\limits_{x \to +\infty} \ln x$.

Because the natural logarithmic function is an increasing function, if we take p as any positive number,

$$\ln x > \ln 2^p \quad \text{whenever} \quad x > 2^p \tag{19}$$

From Theorem 8.3.5 we have

$$\ln 2^p = p \ln 2 \tag{20}$$

Substituting from (20) into (19) we get

$$\ln x > p \ln 2 \quad \text{whenever} \quad x > 2^p$$

Because from (18) $\ln 2 \geq \frac{1}{2}$, we have from the above

$$\ln x > \tfrac{1}{2}p \quad \text{whenever} \quad x > 2^p$$

Letting $p = 2n$, where $n > 0$, we have

$$\ln x > n \quad \text{whenever} \quad x > 2^{2n} \tag{21}$$

It follows from (21), by taking $N = 2^{2n}$, that for any $n > 0$,

$$\ln x > n \quad \text{whenever} \quad x > N$$

So we may conclude that

$$\lim_{n \to +\infty} \ln x = +\infty \tag{22}$$

To determine the behavior of the natural logarithmic function for positive values of x near zero we investigate $\lim_{x \to 0^+} \ln x$. Because $\ln x = \ln(x^{-1})^{-1}$,

$$\ln x = -\ln \frac{1}{x} \tag{23}$$

The expression "$x \to 0^+$" is equivalent to "$1/x \to +\infty$"; so from (23) we write

$$\lim_{x \to 0^+} \ln x = -\lim_{1/x \to +\infty} \ln \frac{1}{x} \tag{24}$$

From (22) we have

$$\lim_{1/x \to +\infty} \ln \frac{1}{x} = +\infty \tag{25}$$

Therefore, from (24) and (25) we get

$$\lim_{x \to 0^+} \ln x = -\infty \tag{26}$$

From (26) and (22) and the intermediate-value theorem (2.8.6) we conclude that the range of the natural logarithmic function is the set of all real numbers. From (26) we conclude that the graph of the natural logarithmic function is asymptotic to the negative part of the y axis through the fourth quadrant. The properties of the natural logarithmic function are summarized as follows:

(i) The domain is the set of all positive numbers.
(ii) The range is the set of all real numbers.
(iii) The function is increasing on its entire domain.
(iv) The function is continuous at all numbers in its domain.
(v) The graph of the function is concave downward at all points.
(vi) The graph of the function is asymptotic to the negative part of the y axis through the fourth quadrant.

By using these properties and plotting a few points with a piece of the tangent line at the points, we can draw a sketch of the graph of the natural logarithmic function. In Fig. 8.3.4 we have plotted the points having abscissas of $\frac{1}{4}$, $\frac{1}{2}$, 1, 2, 4, and 8. The slope of the tangent line is found from $D_x(\ln x) = 1/x$.

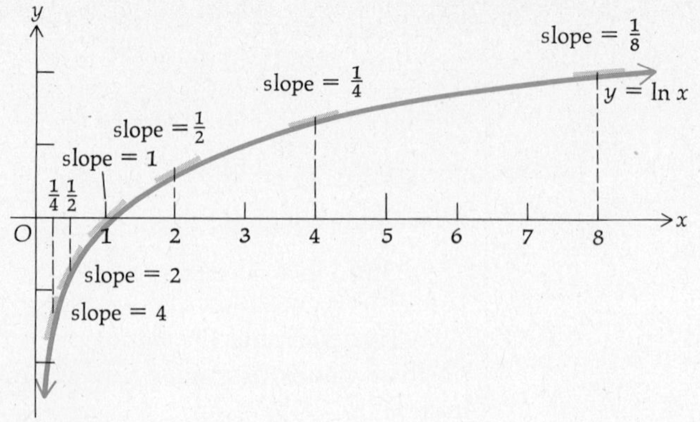

FIGURE 8.3.4

EXERCISES 8.3

In Exercises 1 through 22, differentiate the given function and simplify the result.

1. $f(x) = \ln(4 + 5x)$
2. $g(x) = \ln(1 + 4x^2)$
3. $h(x) = \ln\sqrt{4 + 5x}$
4. $f(x) = \ln(8 - 2x)$
5. $f(t) = \ln(3t + 1)^2$
6. $h(x) = \ln(8 - 2x)^5$
7. $g(t) = \ln^2(3t + 1)$
8. $G(x) = \ln\sqrt{1 + 4x^2}$
9. $f(x) = \ln\sqrt[3]{4 - x^2}$
10. $g(y) = \ln(\ln y)$
11. $F(y) = \ln(\sin 5y)$
12. $f(x) = x \ln x$
13. $f(x) = \cos(\ln x)$
14. $f(t) = \ln\sqrt[4]{\dfrac{t^2 - 1}{t^2 + 1}}$
15. $f(w) = \ln\sqrt{\dfrac{3w + 1}{2w - 5}}$
16. $f(x) = \ln[(5x - 3)^4(2x^2 + 7)^3]$
17. $h(x) = \dfrac{x}{\ln x}$
18. $g(x) = \ln(\cos 2x + \sin 2x)$
19. $g(x) = \ln\sqrt[3]{\dfrac{x + 1}{x^2 + 1}}$
20. $f(x) = \sqrt[3]{\ln x^3}$
21. $F(x) = \sqrt{x + 1} - \ln(1 + \sqrt{x + 1})$
22. $G(x) = x \ln(x + \sqrt{1 + x^2}) - \sqrt{1 + x^2}$

In Exercises 23 through 28, find $D_x y$ by implicit differentiation.

23. $\ln xy + x + y = 2$
24. $\ln \dfrac{y}{x} + xy = 1$
25. $x = \ln(x + y + 1)$
26. $\ln(x + y) - \ln(x - y) = 4$
27. $x + \ln x^2y + 3y^2 = 2x^2 - 1$
28. $x \ln y + y \ln x = xy$
29. Draw a sketch of the graph of $y = \ln x$ by plotting the points having the abscissas $\frac{1}{9}$, $\frac{1}{3}$, 1, 3, and 9, and use $\ln 3 \approx 1.09861$. At each of the five points find the slope of the tangent line and draw a piece of the tangent line.

In Exercises 30 through 37, draw a sketch of the graph of the curve having the given equation.

30. $x = \ln y$
31. $y = \ln(-x)$
32. $y = \ln|x|$
33. $y = \ln(x + 1)$
34. $y = \ln \dfrac{1}{x - 1}$
35. $y = x - \ln x$
36. $y = x \ln x$
37. $y = \ln \sin x$
38. Find an equation of the tangent line to the curve $y = \ln x$ at the point whose abscissa is 2.
39. Find an equation of the normal line to the curve $y = \ln x$ that is parallel to the line $x + 2y - 1 = 0$.

40. Find an equation of the normal line to the graph of $y = x \ln x$ that is perpendicular to the line with equation $x - y + 7 = 0$.

41. Find an equation of the tangent line to the graph of $y = \ln(4x^2 - 3)^5$ at the point whose abscissa is 1.

42. A particle is moving on a straight line according to the equation of motion $s = (t + 1)^2 \ln(t + 1)$, where s is the directed distance of the particle from the starting point at t sec. Find the velocity and acceleration when $t = 3$.

43. In a telegraph cable, the measure of the speed of the signal is proportional to $x^2 \ln(1/x)$, where x is the ratio of the measure of the radius of the core of the cable to the measure of the thickness of the cable's winding. Find the value of $\ln x$ for which the speed of the signal is greatest.

44. A manufacturer of electric generators began operations on Jan. 1, 1971. During the first year there were no sales because the company concentrated on product development and research. After the first year the sales increased steadily according to the equation $y = x \ln x$, where x is the number of years during which the company has been operating and y is the number of millions of dollars in the sales volume. (a) Draw a sketch of the graph of the equation. Determine the rate at which the sales were increasing on (b) Jan. 1, 1975, and (c) Jan. 1, 1981.

45. A particular company has determined that when its weekly advertising expense is x dollars, then if S dollars is its total weekly income from sales, $S = 4000 \ln x$. (a) Determine the rate of change of sales income with respect to advertising expense when \$800 is the weekly advertising budget. (b) If the weekly advertising budget is increased to \$950, what is the approximate increase in the total weekly income from sales?

46. Prove that $x - 1 - \ln x > 0$ and $1 - \ln x - 1/x < 0$ for all $x > 0$ and $x \neq 1$, thus establishing the inequality

$$1 - \frac{1}{x} < \ln x < x - 1$$

for all $x > 0$ and $x \neq 1$. (*Hint:* Let $f(x) = x - 1 - \ln x$ and $g(x) = 1 - \ln x - 1/x$, and determine the signs of $f'(x)$ and $g'(x)$ on the intervals $(0, 1)$ and $(1, +\infty)$.)

47. Show the geometric interpretation of the inequality in Exercise 46 by drawing on the same pair of axes sketches of the graphs of the functions f, g, and h defined by the equations

$$f(x) = 1 - \frac{1}{x} \qquad g(x) = \ln x \qquad h(x) = x - 1$$

48. Use the result of Exercise 46 to prove

$$\lim_{x \to 0} \frac{\ln(1 + x)}{x} = 1$$

49. Establish the limit of Exercise 48 by using the definition of the derivative to find $F'(0)$ for the function F for which $F(x) = \ln(1 + x)$.

8.4 LOGARITHMIC DIFFERENTIATION AND INTEGRALS YIELDING THE NATURAL LOGARITHMIC FUNCTION

In the discussion that follows we need to make use of $D_x(\ln |x|)$. To find this by using Theorem 8.3.2, $\sqrt{x^2}$ is substituted for $|x|$, and so

$$\begin{aligned} D_x(\ln |x|) &= D_x(\ln \sqrt{x^2}) \\ &= \frac{1}{\sqrt{x^2}} \cdot D_x(\sqrt{x^2}) \\ &= \frac{1}{\sqrt{x^2}} \cdot \frac{x}{\sqrt{x^2}} \\ &= \frac{x}{x^2} \end{aligned}$$

Hence

$$D_x(\ln|x|) = \frac{1}{x} \tag{1}$$

From formula (1) and the chain rule we obtain the following theorem.

8.4.1 Theorem If u is a differentiable function of x,

$$D_x(\ln|u|) = \frac{1}{u} \cdot D_x u$$

The following example illustrates how the properties of the natural logarithmic function, given in Theorems 8.3.4, 8.3.5, and 8.3.6, can simplify the work involved in differentiating complicated expressions involving products, quotients, and powers.

Example 1 Given

$$y = \frac{\sqrt[3]{x+1}}{(x+2)\sqrt{x+3}}$$

find $D_x y$.

Solution From the given equation,

$$|y| = \left|\frac{\sqrt[3]{x+1}}{(x+2)\sqrt{x+3}}\right| = \frac{|\sqrt[3]{x+1}|}{|x+2||\sqrt{x+3}|}$$

Taking the natural logarithm and applying the properties of logarithms, we obtain

$$\ln|y| = \tfrac{1}{3}\ln|x+1| - \ln|x+2| - \tfrac{1}{2}\ln|x+3|$$

Differentiating on both sides implicitly with respect to x and applying Theorem 8.4.1, we get

$$\frac{1}{y}D_x y = \frac{1}{3(x+1)} - \frac{1}{x+2} - \frac{1}{2(x+3)}$$

Multiplying on both sides by y we obtain

$$D_x y = y \cdot \frac{2(x+2)(x+3) - 6(x+1)(x+3) - 3(x+1)(x+2)}{6(x+1)(x+2)(x+3)}$$

Replacing y by its given value we obtain $D_x y$ equal to

$$\frac{(x+1)^{1/3}}{(x+2)(x+3)^{1/2}} \cdot \frac{2x^2 + 10x + 12 - 6x^2 - 24x - 18 - 3x^2 - 9x - 6}{6(x+1)(x+2)(x+3)}$$

and so

$$D_x y = \frac{-7x^2 - 23x - 12}{6(x+1)^{2/3}(x+2)^2(x+3)^{3/2}}$$

The process illustrated in Example 1 is called **logarithmic differentiation,** which was developed in 1697 by Johann Bernoulli (1667–1748).

From Theorem 8.4.1 we obtain the following one for indefinite integration.

8.4.2 Theorem

$$\int \frac{1}{u}\,du = \ln|u| + C$$

From Theorems 8.4.2 and 5.2.8, for n any real number,

$$\int u^n\,du = \begin{cases} \dfrac{u^{n+1}}{n+1} + C & \text{if } n \neq -1 \\ \ln|u| + C & \text{if } n = -1 \end{cases} \tag{2}$$

Example 2 Evaluate

$$\int \frac{x^2\,dx}{x^3+1}$$

Solution

$$\int \frac{x^2\,dx}{x^3+1} = \frac{1}{3}\int \frac{3x^2\,dx}{x^3+1} = \frac{1}{3}\ln|x^3+1| + C$$

Example 3 Evaluate

$$\int_0^2 \frac{x^2+2}{x+1}\,dx$$

Solution Because $(x^2+2)/(x+1)$ is an improper fraction, we divide the numerator by the denominator and obtain

$$\frac{x^2+2}{x+1} = x - 1 + \frac{3}{x+1}$$

Therefore

$$\begin{aligned}\int_0^2 \frac{x^2+2}{x+1}\,dx &= \int_0^2 \left(x - 1 + \frac{3}{x+1}\right) dx \\ &= \tfrac{1}{2}x^2 - x + 3\ln|x+1|\Big]_0^2 \\ &= 2 - 2 + 3\ln 3 - 3\ln 1 \\ &= 3\ln 3 - 3\cdot 0 \\ &= 3\ln 3\end{aligned}$$

The answer in Example 3 also can be written as ln 27 because, by Theorem 8.3.6, $3\ln 3 = \ln 3^3$.

Example 4 Evaluate

$$\int \frac{\ln x}{x}\,dx$$

Solution Let $u = \ln x$; then $du = dx/x$; therefore

$$\int \frac{\ln x}{x}\,dx = \int u\,du = \frac{1}{2}u^2 + C = \frac{1}{2}(\ln x)^2 + C$$

EXERCISES 8.4

In Exercises 1 through 4, use Theorem 8.4.1 to find $D_x y$.

1. $y = \ln|x^3 + 1|$ **2.** $y = \ln\left|\frac{3x}{x^2 + 4}\right|$ **3.** $y = \ln|\cos 3x|$ **4.** $y = \sin(\ln|2x + 1|)$

In Exercises 5 through 14, find $D_x y$ by logarithmic differentiation.

5. $y = x^2(x^2 - 1)^3(x + 1)^4$ **6.** $y = \frac{x^5(x + 2)}{x - 3}$ **7.** $y = \frac{x(x - 1)(x + 2)}{(x - 4)^3}$ **8.** $y = (5x - 4)(x^2 + 3)(3x^3 - 5)$

9. $y = \frac{x^3 + 2x}{\sqrt[5]{x^7 + 1}}$ **10.** $y = \frac{x\sqrt{x + 1}}{\sqrt[3]{x - 1}}$ **11.** $y = \frac{3x}{\sqrt{(x + 1)(x + 2)}}$ **12.** $y = \frac{\sqrt{1 - x^2}}{(x + 1)^{2/3}}$

13. $y = \sqrt{x^2 + 1}\ln(x^2 - 1)$ **14.** $y = \frac{x\sqrt[3]{4 + 3x}}{\sqrt{3 + 2x}}$

In Exercises 15 through 30, evaluate the indefinite integral.

15. $\int \frac{dx}{3 - 2x}$ **16.** $\int \frac{dx}{7x + 10}$ **17.** $\int \frac{3x}{x^2 + 4}\,dx$ **18.** $\int \frac{x}{2 - x^2}\,dx$ **19.** $\int \frac{3x^2}{5x^3 - 1}\,dx$

20. $\int \frac{2x - 1}{x(x - 1)}\,dx$ **21.** $\int \frac{\cos t}{1 + 2\sin t}\,dt$ **22.** $\int \frac{\sin 3t}{\cos 3t - 1}\,dt$ **23.** $\int \frac{2x^3}{x^2 - 4}\,dx$ **24.** $\int \frac{5 - 4y^2}{3 + 2y}\,dy$

25. $\int \frac{dx}{x\ln x}$ **26.** $\int \frac{dx}{x(1 + x)}$ **27.** $\int \frac{\ln^2 3x}{x}\,dx$ **28.** $\int \frac{(2 + \ln^2 x)}{x(1 - \ln x)}\,dx$ **29.** $\int \frac{2\ln x + 1}{x[(\ln x)^2 + \ln x]}\,dx$

30. $\int \frac{3x^5 - 2x^3 + 5x^2 - 2}{x^3 + 1}\,dx$

In Exercises 31 through 36, evaluate the definite integral.

31. $\int_3^5 \frac{2x}{x^2 - 5}\,dx$ **32.** $\int_4^5 \frac{x}{4 - x^2}\,dx$ **33.** $\int_1^3 \frac{2t + 3}{t + 1}\,dt$ **34.** $\int_1^5 \frac{4z^3 - 1}{2z - 1}\,dz$ **35.** $\int_2^4 \frac{dx}{x\ln^2 x}$ **36.** $\int_2^4 \frac{\ln x}{x}\,dx$

In Exercises 37 through 43, give the exact value of the number to be found and then give an approximation of this number to three decimal places by using the table of natural logarithms in the appendix.

37. Find the area of the region bounded by the curve $y = 2/(x - 3)$, the x axis, and the lines $x = 4$ and $x = 5$.

38. Find the area of the region bounded by the curve $y = x/(2x^2 + 4)$, the x axis, the y axis, and the line $x = 4$.

39. If $f(x) = 1/x$, find the average value of f on the interval $[1, 5]$.

40. If $f(x) = (x + 2)/(x - 3)$, find the average value of f on the interval $[4, 6]$.

41. Use Boyle's law for the expansion of a gas (see Exercise 30 in Exercises 3.5) to find the average pressure with respect to the volume as the volume increases from 4 ft^3 to 8 ft^3 and the pressure is 2000 lb/ft^2 when the volume is 4 ft^3.

42. Find the volume of the solid of revolution generated when the region bounded by the curve $y = 1 - 3/x$, the x axis, and the line $x = 1$ is revolved about the x axis.

43. Find the volume of the solid of revolution generated when the region bounded by the curve $y = 1 + 2/\sqrt{x}$, the x axis, and the lines $x = 1$ and $x = 4$ is revolved about the x axis.

44. Prove: $\lim_{x \to +\infty} \dfrac{\ln x}{x} = 0$. (*Hint:* First prove that $\int_1^x \dfrac{1}{\sqrt{t}}\,dt \geq \int_1^x \dfrac{1}{t}\,dt$ by using Theorem 6.3.8.)

8.5 THE EXPONENTIAL FUNCTION

Because the natural logarithmic function is increasing on its entire domain, then by Theorems 8.1.5 and 8.2.1 it has an inverse that is also an increasing function. The inverse of the natural logarithmic function is called the *exponential function,* which we now define.

8.5.1 Definition The **exponential function** is the inverse of the natural logarithmic function, and it is defined by

$$\exp(x) = y \quad \text{if and only if} \quad x = \ln y \tag{1}$$

The exponential function $\exp(x)$ is read as "the value of the exponential function at x."

Because the range of the natural logarithmic function is the set of all real numbers, the domain of the exponential function is the set of all real numbers. The range of the exponential function is the set of positive numbers because this is the domain of the natural logarithmic function.

Because the natural logarithmic function and the exponential function are inverses of each other, it follows from Theorem 8.1.4 that

$$\ln(\exp(x)) = x \tag{2}$$

and

$$\exp(\ln x) = x \tag{3}$$

We now wish to define what is meant by a^x, where a is a positive number and x is an irrational number. To arrive at a reasonable definition, consider the case a^r, where $a > 0$ and r is a rational number. We have from (3), where x is a^r,

$$a^r = \exp[\ln(a^r)] \tag{4}$$

But by Theorem 8.3.5, $\ln a^r = r \ln a$; so from (4),

$$a^r = \exp(r \ln a) \tag{5}$$

Because the right side of (5) has a meaning if r is any real number, we use it for our definition.

8.5.2 Definition If a is any positive number and x is any real number, we define

$$a^x = \exp(x \ln a) \tag{6}$$

Theorem 8.3.5 states that if a is any positive number and r is any rational number, then $\ln a^r = r \ln a$. Because of Definition 8.5.2, this equation is also valid if r is any real number, and we now state this fact as a theorem.

8.5.3 Theorem If a is any positive number and x is any real number,

$$\ln a^x = x \ln a$$

Proof From Definition 8.5.2,

$$a^x = \exp(x \ln a)$$

Hence, from Definition 8.5.1,

$$\ln a^x = x \ln a$$ ■

Following is the definition of one of the most important numbers in mathematics.

8.5.4 Definition The number e is defined by the formula

$$e = \exp 1$$

The letter e was chosen because of the Swiss mathematician and physicist Leonhard Euler (1707–1783).

The number e is a *transcendental* number; that is, it cannot be expressed as the root of any polynomial with integer coefficients. The number π is another example of a transcendental number. The proof that e is transcendental was first given in 1873 by Charles Hermite, and its value can be expressed to any required degree of accuracy. In Chapter 15 we learn a method for doing this. The value of e to seven decimal places is 2.7182818. Thus

$$e \approx 2.7182818$$

In the appendix there is a table giving powers of e. The importance of the number e will become apparent as we proceed through this chapter.

8.5.5 Theorem

$$\ln e = 1$$

Proof By Definition 8.5.4,

$$e = \exp 1$$

Therefore

$$\ln e = \ln(\exp 1)$$

Because the natural logarithmic function and the exponential function are inverses, it follows that the right side of this equation is 1. Thus

$$\ln e = 1$$ ■

8.5.6 Theorem For all values of x,

$$\exp(x) = e^x$$

Proof By Definition 8.5.2, with $a = e$,

$$e^x = \exp(x \ln e) \tag{7}$$

But by Theorem 8.5.5, $\ln e = 1$, and substituting in (17) we obtain

$$e^x = \exp(x)$$ ■

From now on we write e^x in place of $\exp(x)$, and so from Definition 8.5.1,

$$e^x = y \quad \text{if and only if} \quad x = \ln y \tag{8}$$

With e^x in place of $\exp(x)$, (2) and (3) become

$$\ln e^x = x \tag{9}$$

and

$$e^{\ln x} = x \tag{10}$$

If we replace $\exp(x \ln a)$ by $e^{x \ln a}$ in (6) of Definition 8.5.2, we have

$$a^x = e^{x \ln a} \qquad \text{for every } a > 0 \tag{11}$$

Equation (11) and a table of powers of e can be used to compute a^x if x is any irrational number. The following example shows the procedure.

Example 1 Compute the value of $2^{\sqrt{3}}$ to two decimal places.

Solution From (11),

$$\begin{aligned} 2^{\sqrt{3}} &= e^{\sqrt{3} \ln 2} \\ &= e^{1.732(0.6931)} \\ &= e^{1.200} \\ &\approx 3.32 \end{aligned}$$

Because $0 = \ln 1$, we have from statement (8)

$$e^0 = 1$$

We now state some properties of the exponential function as theorems.

8.5.7 Theorem If a and b are any real numbers, then

$$e^a \cdot e^b = e^{a+b}$$

Proof Let $A = e^a$ and $B = e^b$. Then from statement (8),

$$\ln A = a \tag{12}$$

and

$$\ln B = b \tag{13}$$

From Theorem 8.3.4,

$$\ln AB = \ln A + \ln B \tag{14}$$

Substituting from (12) and (13) into (14) we obtain

$$\ln AB = a + b$$

Thus

$$e^{\ln AB} = e^{a+b} \tag{15}$$

From (10) it follows that the left side of (15) is AB; so

$$AB = e^{a+b}$$

Replacing A and B by their values we get

$$e^a \cdot e^b = e^{a+b}$$ ■

8.5.8 Theorem If a and b are any real numbers, then

$$e^a \div e^b = e^{a-b}$$

The proof is analogous to the proof of Theorem 8.5.7, where Theorem 8.3.4 is replaced by Theorem 8.3.5.

8.5.9 Theorem If a and b are any real numbers, then

$$(e^a)^b = e^{ab}$$

Proof If in (10) $x = (e^a)^b$, we have

$$(e^a)^b = e^{\ln(e^a)^b}$$

Applying Theorem 8.5.3 to the exponent in the right side of this equation we obtain

$$(e^a)^b = e^{b \ln e^a}$$

But from (9), $\ln e^a = a$, and therefore

$$(e^a)^b = e^{ab}$$ ■

We now obtain the theorem for the derivative of the exponential function. Let

$$y = e^x$$

Then from statement (8),

$$x = \ln y$$

On both sides of this equation we differentiate implicitly with respect to x to get

$$1 = \frac{1}{y} \cdot D_x y$$

Thus

$$D_x y = y$$

Replacing y by e^x we obtain

$$D_x(e^x) = e^x \tag{16}$$

The next theorem follows from (16) and the chain rule.

8.5.10 Theorem If u is a differentiable function of x,

$$D_x(e^u) = e^u D_x u$$

Observe that the derivative of the function defined by $f(x) = ke^x$, where k is a constant, is itself. The only other function we have previously encountered that has this property is the constant function zero; actually, this is the special case of $f(x) = ke^x$ when $k = 0$. It can be proved that the most general function which is its own derivative is given by $f(x) = ke^x$ (see Exercise 48).

Example 2 Given $y = e^{1/x^2}$, find dy/dx.

Solution From Theorem 8.5.10,

$$\frac{dy}{dx} = e^{1/x^2}\left(-\frac{2}{x^3}\right) = -\frac{2e^{1/x^2}}{x^3}$$

The indefinite integration formula given in the following theorem is a consequence of Theorem 8.5.10.

8.5.11 Theorem

$$\int e^u \, du = e^u + C$$

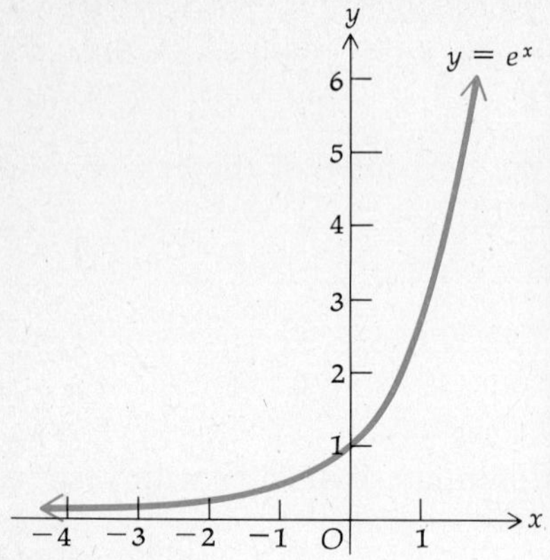

FIGURE 8.5.1

Example 3 Find

$$\int \frac{e^{\sqrt{x}}}{\sqrt{x}}\,dx$$

Solution Let $u = \sqrt{x}$; then $du = \frac{1}{2}x^{-1/2}\,dx$; so

$$\int \frac{e^{\sqrt{x}}}{\sqrt{x}}\,dx = 2\int e^u\,du$$

$$= 2e^u + C$$

$$= 2e^{\sqrt{x}} + C$$

Because from (8) $e^x = y$ if and only if $x = \ln y$, the graph of $y = e^x$ is identical to the graph of $x = \ln y$. So we can obtain the graph of $y = e^x$ by interchanging the x and y axes in Fig. 8.3.4 (see Fig. 8.5.1).

The graph of $y = e^x$ can be obtained without referring to the graph of the natural logarithmic function. Because the range of the exponential function is the set of all positive numbers, it follows that $e^x > 0$ for all values of x. So the graph lies entirely above the x axis. $D_x y = e^x > 0$ for x; so the function is increasing for all x. $D_x{}^2 y = e^x > 0$ for all x; so the graph is concave upward at all points.

We have the following two limits:

$$\lim_{x \to +\infty} e^x = +\infty \tag{17}$$

and

$$\lim_{x \to -\infty} e^x = 0 \tag{18}$$

The proofs of (17) and (18) are left as exercises (see Exercises 56 and 57). To plot some specific points use the table in the appendix giving powers of e.

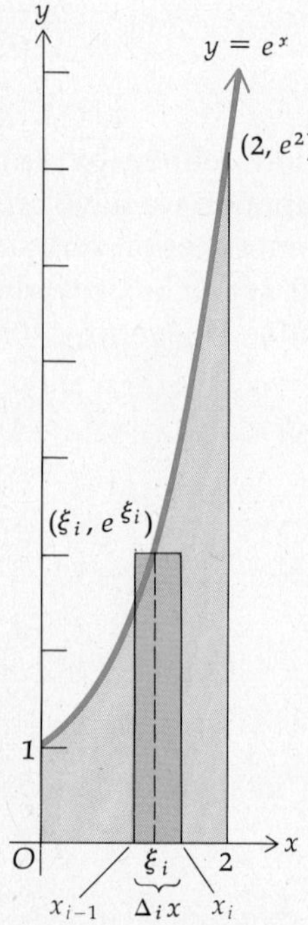

FIGURE 8.5.2

Example 4 Find the area of the region bounded by the curve $y = e^x$, the coordinate axes, and the line $x = 2$.

Solution The region is shown in Fig. 8.5.2.

$$A = \lim_{\|\Delta\| \to 0} \sum_{i=1}^{n} e^{\xi_i}\,\Delta_i x$$

$$= \int_0^2 e^x\,dx$$

$$= e^x\Big]_0^2$$

$$= e^2 - e^0$$

$$= e^2 - 1$$

From a table of powers of e we see that the value of e^2 to two decimal places is 7.39. Therefore the area of the region is approximately 6.39 square units.

Example 5 Given $y = e^{2x+\ln x}$, find $D_x y$.

Solution Because $e^{2x+\ln x} = e^{2x}e^{\ln x} = e^{2x}(x)$,

$$y = xe^{2x}$$

Therefore

$$D_x y = e^{2x} + 2xe^{2x}$$

In Definition 8.5.4 the number e was defined as the value of the exponential function at 1; that is, $e = \exp 1$. To arrive at another way of defining e we consider the natural logarithmic function

$$f(x) = \ln x$$

We know that the derivative of f is given by $f'(x) = 1/x$; hence $f'(1) = 1$. However, let us apply the definition of the derivative to find $f'(1)$. We have

$$\begin{aligned} f'(1) &= \lim_{\Delta x \to 0} \frac{f(1 + \Delta x) - f(1)}{\Delta x} \\ &= \lim_{\Delta x \to 0} \frac{\ln(1 + \Delta x) - \ln 1}{\Delta x} \\ &= \lim_{\Delta x \to 0} \frac{1}{\Delta x} \ln(1 + \Delta x) \end{aligned}$$

Therefore

$$\lim_{\Delta x \to 0} \frac{1}{\Delta x} \ln(1 + \Delta x) = 1$$

Replacing Δx by h we have from the above equation and Theorem 8.5.3

$$\lim_{h \to 0} \ln(1 + h)^{1/h} = 1 \tag{19}$$

Now, because the exponential function and the natural logarithmic function are inverse functions, we have

$$\lim_{h \to 0} (1 + h)^{1/h} = \lim_{h \to 0} \exp[\ln(1 + h)^{1/h}] \tag{20}$$

Because the exponential function is continuous and $\lim_{h \to 0} \ln(1 + h)^{1/h}$ exists and equals 1 as shown in Eq. (19), we can apply Theorem 2.7.6 to the right side of (20) and get

$$\lim_{h \to 0} (1 + h)^{1/h} = \exp\left[\lim_{h \to 0} \ln(1 + h)^{1/h}\right] = \exp 1$$

Hence

$$\lim_{h \to 0} (1 + h)^{1/h} = e \tag{21}$$

Equation (21) is sometimes used to define e; however, to use this as a definition it is necessary to prove that the limit exists.

Let us consider the function F defined by

$$F(h) = (1 + h)^{1/h} \tag{22}$$

and determine the function values for some values of h close to zero. These values are given in Table 8.5.1.

TABLE 8.5.1

h	1	0.5	0.05	0.01	0.001	−0.001	−0.01	−0.05	−0.5
$F(h) = (1 + h)^{1/h}$	2	2.25	2.65	2.70	2.7169	2.7196	2.73	2.79	4

Table 8.5.1 leads us to suspect that $\lim_{h \to 0} (1 + h)^{1/h}$ is probably a number that lies between 2.7169 and 2.7196. As previously mentioned, in Chapter 15 we learn a method for finding the value of e to any desired number of decimal places.

EXERCISES 8.5

In Exercises 1 through 16, find dy/dx.

1. $y = e^{5x}$
2. $y = e^{-7x}$
3. $y = e^{-3x^2}$
4. $y = e^{x^2-3}$
5. $y = e^{\cos x}$
6. $y = e^{2 \sin 3x}$
7. $y = e^x \sin e^x$
8. $y = \dfrac{e^x}{x}$
9. $y = \dfrac{e^{2x}}{x^2}$
10. $y = e^{e^x}$
11. $y = \dfrac{e^x - e^{-x}}{e^x + e^{-x}}$
12. $y = \ln \dfrac{e^{4x} - 1}{e^{4x} + 1}$
13. $y = x^5 e^{-3 \ln x}$
14. $y = \ln(e^x + e^{-x})$
15. $y = e^{x \ln x}$
16. $y = e^{x/\sqrt{4+x^2}}$

In Exercises 17 through 20, find $D_x y$ by implicit differentiation.

17. $e^x + e^y = e^{x+y}$
18. $e^y = \ln(x^3 + 3y)$
19. $y^2 e^{2x} + xy^3 = 1$
20. $ye^{2x} + xe^{2y} = 1$

In Exercises 21 through 28, evaluate the indefinite integral.

21. $\int e^{2-5x}\, dx$
22. $\int e^{2x+1}\, dx$
23. $\int \dfrac{1 + e^{2x}}{e^x}\, dx$
24. $\int e^{3x} e^{2x}\, dx$
25. $\int \dfrac{e^{3x}}{(1 - 2e^{3x})^2}\, dx$
26. $\int x^2 e^{2x^3}\, dx$
27. $\int \dfrac{e^{2x}}{e^x + 3}\, dx$
28. $\int \dfrac{dx}{1 + e^x}$

In Exercises 29 through 36, evaluate the definite integral.

29. $\int_0^1 e^2\, dx$
30. $\int_1^{e^2} \dfrac{dx}{x}$
31. $\int_e^{e^3} \dfrac{dx}{x}$
32. $\int_1^e \dfrac{\ln x}{x}\, dx$

33. $\displaystyle\int_e^{e^2} \frac{dx}{x(\ln x)^2}$ **34.** $\displaystyle\int_0^3 \frac{e^x + e^{-x}}{2}\, dx$ **35.** $\displaystyle\int_0^2 xe^{4-x^2}\, dx$ **36.** $\displaystyle\int_1^2 \frac{e^x}{e^x + e}\, dx$

37. Draw a sketch of the graph of $y = e^{-x}$. **38.** Draw a sketch of the graph of $y = e^{|x|}$.

In Exercises 39 and 40, find the relative extrema of f, the points of inflection of the graph of f, the intervals on which f is increasing, the intervals on which f is decreasing, where the graph is concave upward, where the graph is concave downward, and the slope of any inflectional tangent. Draw a sketch of the graph of f.

39. $f(x) = xe^{-x}$ **40.** $f(x) = e^{-x^2}$

In Exercises 41 through 44, compute the value to two decimal places.

41. $2^{\sqrt{2}}$ **42.** $(\sqrt{2})^{\sqrt{2}}$ **43.** $(\sqrt{2})^e$ **44.** e^e

45. Find an equation of the tangent line to the curve $y = e^{-x}$ that is perpendicular to the line $2x - y = 5$.

46. Find the area of the region bounded by the curve $y = e^x$ and the line through the points $(0, 1)$ and $(1, e)$.

47. A solid has as its base the region bounded by the curves $y = e^x$ and $y = e^{-x}$ and the line $x = 1$. If every plane section perpendicular to the x axis is a square, find the volume of the solid.

48. Prove that the most general function that is equal to its derivative is given by $f(x) = ke^x$. (*Hint*: Let $y = f(x)$ and solve the differential equation $dy/dx = y$.)

49. If p lb/ft² is the atmospheric pressure at a height of h ft above sea level, then $p = 2116e^{-0.0000318h}$. Find the time rate of change of the atmospheric pressure outside an airplane that is 5000 ft high and rising at the rate of 160 ft/sec.

50. At a certain height the gauge on an airplane indicates that the atmospheric pressure is 1500 lb/ft². Applying the formula of Exercise 49, approximate by differentials how much higher the airplane must rise so that the pressure will be 1480 lb/ft².

51. If l ft is the length of an iron rod when t degrees is its temperature, then $l = 60e^{0.00001t}$. Use differentials to find the approximate increase in l when t increases from 0 to 10.

52. A simple electric circuit containing no condensers, a resistance of R ohms, and an inductance of L henrys has the electromotive force cut off when the current is I_0 amperes. The current dies down so that at t sec the current is i amperes, and $i = I_0e^{-(R/L)t}$. Show that the rate of change of the current is proportional to the current.

53. A body is moving along a straight line and at t sec the velocity is v ft/sec, where $v = e^3 - e^{2t}$. Find the distance traveled by the particle while $v > 0$ after $t = 0$.

54. An advertising agency determined statistically that if a breakfast food manufacturer increases its budget for television commercials by x thousand dollars, there will be an increase in the total profit of $25x^2e^{-0.2x}$ hundred dollars. What should be the advertising budget increase in order for the manufacturer to have the greatest profit? What will be the corresponding increase in the company's profit?

55. A tank is in the shape of the solid of revolution formed by rotating about the x axis the region bounded by the curve $y^2x = e^{-2x}$ and the lines $x = 1$ and $x = 4$. If the tank is full of water, find the work done in pumping all the water to a point 1 ft above the top of the tank. Distance is measured in feet.

56. Prove that $\lim_{x \to +\infty} e^x = +\infty$ by showing that for any $N > 0$ there exists an $M > 0$ such that $e^x > N$ whenever $x > M$.

57. Prove that $\lim_{x \to -\infty} e^x = 0$ by showing that for any $\epsilon > 0$ there exists an $N < 0$ such that $e^x < \epsilon$ whenever $x < N$.

58. Draw a sketch of the graph of F if $F(h) = (1 + h)^{1/h}$.

59. Draw a sketch of the graph of the function f for which $f(x) = e^{-x} \sin x$. Show that $\lim_{x \to +\infty} fx) = 0$ but that the line $y = 0$ is not a horizontal asymptote of the graph of f. Why does Definition 2.5.4(i) not hold for this function f?

60. An important function in statistics is the probability density function defined by

$$f(x) = \frac{1}{\sigma\sqrt{2\pi}} \exp\left[-\frac{(x-\mu)^2}{2\sigma^2}\right]$$

where σ and μ are constants such that $\sigma > 0$ and μ is any real number. Find (a) the relative extrema of f; (b) the points of inflection of the graph of f; (c) $\lim_{x\to+\infty} f(x)$; (d) $\lim_{x\to-\infty} f(x)$. (e) Draw a sketch of the graph of f.

8.6 OTHER EXPONENTIAL AND LOGARITHMIC FUNCTIONS

From Definition 8.5.2, if $a > 0$,

$$a^x = e^{x \ln a} \tag{1}$$

The expression on the left side of (1) is called the *exponential function to the base a.*

8.6.1 Definition If a is any positive number and x is any real number, then the function f defined by

$$f(x) = a^x$$

is called the **exponential function to the base *a*.**

The exponential function to the base a satisfies the same properties as the exponential function to the base e.

● ILLUSTRATION 1 If x and y are any real numbers and a is positive, then from (1),

$$\begin{aligned} a^x a^y &= e^{x\ln a} e^{y \ln a} \\ &= e^{x \ln a + y \ln a} \\ &= e^{(x+y)\ln a} \\ &= a^{x+y} \end{aligned}$$ ●

From Illustration 1 we have the property

$$a^x a^y = a^{x+y} \tag{2}$$

We also have the following properties:

$$a^x \div a^y = a^{x-y} \tag{3}$$

$$(a^x)^y = a^{xy} \tag{4}$$

$$(ab)^x = a^x b^x \tag{5}$$

$$a^0 = 1 \tag{6}$$

The proofs of (3) through (6) are left as exercises (see Exercises 37 to 40).

To find the derivative of the exponential function to the base a we set $a^x = e^{x \ln a}$ and apply the chain rule.

$$a^x = e^{x \ln a}$$

$$D_x(a^x) = e^{x \ln a} D_x(x \ln a)$$

$$= e^{x \ln a}(\ln a)$$

$$= a^x \ln a$$

Therefore we have the following theorem.

8.6.2 Theorem If a is any positive number and u is a differentiable function of x,

$$D_x(a^u) = a^u \ln a \, D_x u$$

● ILLUSTRATION 2 If $y = 3^{x^2}$, then from Theorem 8.6.2,

$$D_x y = 3^{x^2} \ln 3(2x) = 2(\ln 3)x3^{x^2}$$ ●

The next theorem giving the indefinite integration formula for the exponential function to the base a follows from Theorem 8.6.2.

8.6.3 Theorem If a is any positive number other than 1,

$$\int a^u \, du = \frac{a^u}{\ln a} + C$$

Example 1 Find $\int \sqrt{10^{3x}} \, dx$.

Solution $\int \sqrt{10^{3x}} \, dx = \int 10^{3x/2} \, dx$. Let $u = \frac{3}{2}x$; then $du = \frac{3}{2}dx$; thus $\frac{2}{3} du = dx$. Therefore

$$\int 10^{3x/2} \, dx = \int 10^u \left(\tfrac{2}{3} \, du\right)$$

$$= \frac{2}{3} \cdot \frac{10^u}{\ln 10} + C$$

$$= \frac{2 \cdot 10^{3x/2}}{3 \ln 10} + C$$

Example 2 Draw sketches of the graphs of $y = 2^x$ and $y = 2^{-x}$ on the same set of axes. Find the area of the region bounded by these two graphs and the line $x = 2$.

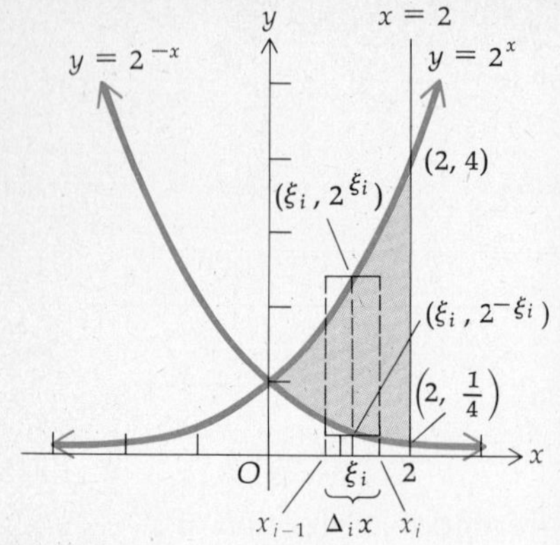

FIGURE 8.6.1

Solution The required sketches are shown in Fig. 8.6.1. The region is shaded in the figure. If A square units is the desired area,

$$A = \lim_{\|\Delta\| \to 0} \sum_{i=1}^{n} [2^{\xi_i} - 2^{-\xi_i}] \Delta_i x$$

$$= \int_0^2 (2^x - 2^{-x})\, dx$$

$$= \frac{2^x}{\ln 2} + \frac{2^{-x}}{\ln 2}\Big]_0^2$$

$$= \frac{4}{\ln 2} + \frac{\frac{1}{4}}{\ln 2} - \frac{1}{\ln 2} - \frac{1}{\ln 2}$$

$$= \frac{9}{4 \ln 2}$$

$$\approx 3.25$$

We can now define the *logarithmic function to the base a* if a is any positive number other than 1.

8.6.4 Definition If a is any positive number except 1, the **logarithmic function to the base a** is the inverse of the exponential function to the base a; we write

$$y = \log_a x \quad \text{if and only if} \quad a^y = x \tag{7}$$

The above is the definition of the logarithmic function to the base a usually given in elementary algebra; however, (7) has meaning for y any real number because a^y has been precisely defined. It should be noted that if $a = e$, we have the logarithmic function to the base e, which is the natural logarithmic function.

$\log_a x$ is read as "the logarithm of x to the base a."

The logarithmic function to the base a obeys the same laws as the natural logarithmic function. We list them.

$$\log_a(xy) = \log_a x + \log_a y \tag{8}$$

$$\log_a(x \div y) = \log_a x - \log_a y \tag{9}$$

$$\log_a 1 = 0 \tag{10}$$

$$\log_a x^y = y \log_a x \tag{11}$$

The proofs of (8) through (11) are left as exercises (see Exercises 41 to 44).

A relationship between logarithms to the base a and natural logarithms follows easily. Let

$$y = \log_a x$$

Then

$$a^y = x$$
$$\ln a^y = \ln x$$
$$y \ln a = \ln x$$
$$y = \frac{\ln x}{\ln a}$$

Replacing y by $\log_a x$ we obtain

$$\log_a x = \frac{\ln x}{\ln a} \tag{12}$$

Equation (12) sometimes is used as the definition of the logarithmic function to the base a. Because the natural logarithmic function is continuous at all $x > 0$, it follows from (12) that the logarithmic function to the base a is continuous at all $x > 0$.

If in (12) $x = e$, we have

$$\log_a e = \frac{\ln e}{\ln a}$$

or, equivalently,

$$\log_a e = \frac{1}{\ln a} \tag{13}$$

We now find the derivative of the logarithmic function to the base a. We differentiate on both sides of (12) with respect to x, and we obtain

$$D_x(\log_a x) = \frac{1}{\ln a} D_x(\ln x)$$

$$D_x(\log_a x) = \frac{1}{\ln a} \cdot \frac{1}{x} \tag{14}$$

Substituting from (13) into (14) we get

$$D_x(\log_a x) = \frac{\log_a e}{x} \tag{15}$$

From (15) and the chain rule we have the following theorem.

8.6.5 Theorem If u is a differentiable function of x,

$$D_x(\log_a u) = \frac{\log_a e}{u} D_x u$$

If in Theorem 8.6.5 $a = e$, then we have

$$D_x(\log_e u) = \frac{\log_e e}{u} D_x u$$

or, equivalently,

$$D_x(\ln u) = \frac{1}{u} D_x u$$

which is Theorem 8.3.2 for the derivative of the natural logarithmic function.

Example 3 Given

$$y = \log_{10} \frac{x+1}{x^2+1}$$

find dy/dx.

Solution Using (9) we write

$$y = \log_{10}(x+1) - \log_{10}(x^2+1)$$

From Theorem 8.6.5,

$$\frac{dy}{dx} = \frac{\log_{10} e}{x+1} - \frac{\log_{10} e}{x^2+1} \cdot 2x$$

$$= \log_{10} e \left(\frac{1}{x+1} - \frac{2x}{x^2+1} \right)$$

$$= \frac{\log_{10} e(1 - 2x - x^2)}{(x+1)(x^2+1)}$$

Because x^n has been defined for any real number n, we can now prove the formula for finding the derivative of the power function if the exponent is any real number.

8.6.6 Theorem If n is any real number and the function f is defined by

$$f(x) = x^n \qquad \text{where } x > 0$$

then

$$f'(x) = nx^{n-1}$$

Proof Let $y = x^n$. Then from (1),

$$y = e^{n \ln x}$$

Thus

$$D_xy = e^{n \ln x} D_x(n \ln x)$$
$$= e^{n \ln x}\left(\frac{n}{x}\right)$$
$$= x^n \cdot \frac{n}{x}$$
$$= nx^{n-1}$$

■

Theorem 8.6.6 enables us to find the derivative of a variable to a constant power. Previously in this section we learned how to differentiate a constant to a variable power. We now consider the derivative of a function whose function value is a variable to a variable power.

Example 4 If $y = x^x$, where $x > 0$, find dy/dx.

Solution From Definition 8.5.2, if $x > 0$, $x^x = e^{x \ln x}$. Therefore

$$y = e^{x \ln x}$$
$$\frac{dy}{dx} = e^{x \ln x} D_x(x \ln x)$$
$$= e^{x \ln x}\left(x \cdot \frac{1}{x} + \ln x\right)$$
$$= x^x(1 + \ln x)$$

The method of logarithmic differentiation can also be used to find the derivative of a function whose function value is a variable to a variable power, as shown in the following example.

Example 5 Find dy/dx in Example 4 by using logarithmic differentiation.

Solution We are given $y = x^x$ with $x > 0$. We take the natural logarithm on both sides of the equation and obtain

$$\ln y = \ln x^x$$
$$\ln y = x \ln x$$

Differentiating on both sides of the above equation with respect to x gives

$$\frac{1}{y} \cdot \frac{dy}{dx} = x \cdot \frac{1}{x} + \ln x$$
$$\frac{dy}{dx} = y(1 + \ln x)$$
$$= x^x(1 + \ln x)$$

EXERCISES 8.6

In Exercises 1 through 24, find the derivative of the given function.

1. $f(x) = 3^{5x}$ **2.** $f(x) = 6^{-3x}$ **3.** $f(t) = 4^{3t^2}$ **4.** $g(x) = 10^{x^2-2x}$

5. $f(x) = 4^{\sin 2x}$ **6.** $f(z) = 2^z \cos z$ **7.** $g(x) = 2^{5x}3^{4x^2}$ **8.** $f(x) = (x^3 + 3)2^{-7x}$

9. $h(x) = \dfrac{\log_{10} x}{x}$ **10.** $f(t) = \log_{10} \dfrac{t}{t+1}$ **11.** $f(x) = \sqrt{\log_a x}$ **12.** $f(x) = \log_a[\log_a(\log_a x)]$

13. $f(x) = \log_{10}[\log_{10}(x + 1)]$ **14.** $g(w) = \sin 2^{3w}$ **15.** $f(t) = \cos 3^{t^2}$ **16.** $f(x) = x^{x^2}$; $x > 0$

17. $f(x) = x^{\sqrt{x}}$; $x > 0$ **18.** $f(x) = x^{\ln x}$; $x > 0$ **19.** $g(z) = z^{\sin z}$; $z > 0$ **20.** $g(t) = (\cos t)^t$; $\cos t > 0$

21. $h(x) = (\cos x)^{\sin x}$; $\cos x > 0$ **22.** $f(x) = (x)^{x^x}$; $x > 0$ **23.** $f(x) = x^{e^x}$; $x > 0$ **24.** $f(x) = (\ln x)^{\ln x}$; $x > 1$

In Exercises 25 through 32, evaluate the indefinite integral.

25. $\int 3^{2x}\, dx$ **26.** $\int a^{nx}\, dx$ **27.** $\int a^t e^t\, dt$ **28.** $\int 5^{x^4+2x}(2x^3 + 1)\, dx$ **29.** $\int x^2 10^{x^3}\, dx$

30. $\int a^{z \ln z}(\ln z + 1)\, dz$ **31.** $\int e^y 2^{e^y} 3^{e^y}\, dy$ **32.** $\int \dfrac{4^{\ln(1/x)}}{x}\, dx$

In Exercises 33 and 34, compute the value of the given logarithm.

33. $\log_5 e$ **34.** $\log_2 10$

In Exercises 35 and 36, use differentials to find an approximate value of the given logarithm, and express the answer to three decimal places.

35. $\log_{10} 997$ **36.** $\log_{10} 1.015$

In Exercises 37 through 40, prove the given property if a and b are any positive numbers and x and y are any real numbers.

37. $a^x \div a^y = a^{x-y}$ **38.** $a^0 = 1$ **39.** $(ab)^x = a^x b^x$ **40.** $(a^x)^y = a^{xy}$

In Exercises 41 through 44, prove the given property if a is any positive number except 1 and x and y are any positive numbers.

41. $\log_a(xy) = \log_a x + \log_a y$ **42.** $\log_a 1 = 0$ **43.** $\log_a x^y = y \log_a x$ **44.** $\log_a(x \div y) = \log_a x - \log_a y$

45. A particle is moving along a straight line according to the equation of motion $s = A2^{kt} + B2^{-kt}$, where s ft is the directed distance of the particle from the starting point at t sec. Prove that if a ft/sec² is the acceleration at t sec, then a is proportional to s.

46. Find an equation of the tangent line to the curve $y^2 4^y = x2^x$ at (4, 2).

47. A company has learned that when it initiates a new sales campaign, the number of sales per day increases. However, the number of extra daily sales per day decreases as the impact of the campaign wears off. For a specific campaign the company has determined that if S is the number of extra daily sales as a result of the campaign and x is the number of days that have elapsed since the campaign ended, then

$$S = 1000 \cdot 3^{-x/2}$$

Find the rate at which the extra daily sales are decreasing when (a) $x = 4$ and (b) $x = 10$.

48. A particle moves along a straight line according to the equation of motion $s = t^{1/t}$, where s ft is the directed distance of the particle from the starting point at t sec. Find the velocity and acceleration at 2 sec.

49. An historically important abstract painting was purchased in 1920 for \$200, and its value has doubled every 10 years since its purchase. If y dollars is the value of the painting t years after its purchase, (a) define y in terms of t. (b) What was the value of the painting in 1980? (c) Find the rate at which the value of the painting was increasing in 1980.

50. Draw sketches of the graphs of $y = \log_{10} x$ and $y = \ln x$ on the same set of axes.

51. Draw sketches of the graphs of $y = e^x$ and $y = 2^x$ on the same set of axes. Find the area of the region bounded by these two graphs and the line $x = 1$.

52. Find the area of the region bounded by the graph of $y = 3^x$ and the lines $x = 1$ and $y = 1$.

53. Find the volume of the solid generated if the region of Exercise 52 is revolved about the x axis.

54. Given $f(x) = \frac{1}{2}(a^x + a^{-x})$. Prove that $f(b + c) + f(b - c) = 2f(b)f(c)$.

8.7 LAWS OF GROWTH AND DECAY

The laws of growth and decay provide applications of the exponential function in fields such as chemistry, physics, biology, business, and economics. Such a situation arises when the rate of change of the amount of a quantity with respect to time is proportional to the amount of the quantity present at a given instant. For example, it may be that the rate of growth of the population of a community is proportional to the actual population at any given instant. In biology, under certain circumstances, the rate of growth of a culture of bacteria is proportional to the amount of bacteria present at any specific time. In a chemical reaction it is often the case that the rate of the reaction is proportional to the quantity of the substance present; for instance, it is known from experiments that the rate of decay of radium is proportional to the amount of radium present at a given moment. An interesting application in business occurs when interest is compounded continuously.

In such cases, if the time is represented by t units, and A units represents the amount of the quantity present at any time, then

$$\frac{dA}{dt} = kA$$

where k is a constant. If A increases as t increases, then $k > 0$, and we have the **law of natural growth.** If A decreases as t increases, then $k < 0$, and we have the **law of natural decay.**

If by definition A is a nonnegative integer (for instance, if A is the population of a certain community), we assume that A can be any nonnegative number in order for A to be a continuous function of t.

Example 1 The rate of increase of the population of a certain city is proportional to the population. If the population in 1930 was 50,000 and in 1960 it was 75,000, what is the expected population in 1990?

Solution Let t be the number of years in the time since 1930. Let A be the population in t years. We have the initial conditions given in Table 8.7.1.

TABLE 8.7.1

t	0	30	60
A	50,000	75,000	A_{60}

The differential equation is

$$\frac{dA}{dt} = kA$$

Separating the variables we obtain

$$\frac{dA}{A} = k\,dt$$

Integrating we have

$$\int \frac{dA}{A} = k \int dt$$

$$\ln |A| = kt + \bar{c}$$

$$|A| = e^{kt+\bar{c}} = e^{\bar{c}} \cdot e^{kt}$$

Letting $e^{\bar{c}} = C$ we have $|A| = Ce^{kt}$, and because A is nonnegative, we can omit the absolute-value bars, thereby giving

$$A = Ce^{kt}$$

Because $A = 50{,}000$ when $t = 0$, we obtain $C = 50{,}000$. Thus

$$A = 50{,}000e^{kt} \tag{1}$$

Because $A = 75{,}000$ when $t = 30$, we get

$$75{,}000 = 50{,}000e^{30k}$$

$$e^{30k} = \tfrac{3}{2} \tag{2}$$

When $t = 60$, $A = A_{60}$. Therefore

$$A_{60} = 50{,}000e^{60k}$$

$$A_{60} = 50{,}000(e^{30k})^2 \tag{3}$$

Substituting from (2) into (3) we obtain

$$A_{60} = 50{,}000(\tfrac{3}{2})^2$$

$$= 112{,}500$$

Therefore the expected population in 1990 is 112,500.

Example 2 In a certain culture the rate of growth of bacteria is proportional to the amount present. If there are 1000 bacteria present initially and the amount doubles in 12 min, how long will it take before there will be 1,000,000 bacteria present?

Solution Let t be the number of minutes in the time from now, and let A be the number of bacteria present at t min. Table 8.7.2 gives the initial conditions. The differential equation is

TABLE 8.7.2

t	0	12	T
A	1,000	2,000	1,000,000

$$\frac{dA}{dt} = kA$$

The differential equation is the same as we had in Example 1; hence, as above, the general solution is

$$A = Ce^{kt}$$

When $t = 0, A = 1000$; hence $C = 1000$, which gives

$$A = 1000e^{kt}$$

From the condition that $A = 2000$ when $t = 12$ we obtain

$$e^{12k} = 2$$

and so

$$k = \tfrac{1}{12}\ln 2 = 0.05776$$

Hence we have

$$A = 1000e^{0.05776t}$$

Replacing t by T and A by 1,000,000 we get

$$1{,}000{,}000 = 1000e^{0.05776T}$$

$$e^{0.05776T} = 1000$$

$$0.05776T = \ln 1000$$

$$T = \frac{\ln 1000}{0.05776} = 119.6$$

Therefore there will be 1,000,000 bacteria present in 1 hr, 59 min, 36 sec.

In problems involving the law of natural decay, the **half-life** of a substance is the time required for half of it to decay.

Example 3 The rate of decay of radium is proportional to the amount present at any time. If 60 mg of radium are present now and its half-life is 1690 years, how much radium will be present 100 years from now?

Solution Let t be the number of years in the time from now. Let A be the number of milligrams of radium present at t years. We have the initial conditions given in Table 8.7.3. The differential equation is

TABLE 8.7.3

t	0	1690	100
A	60	30	A_{100}

$$\frac{dA}{dt} = kA$$

As in Example 1 the general solution is

$$A = Ce^{kt}$$

Because $A = 60$ when $t = 0$, we obtain $60 = C$. Therefore

$$A = 60e^{kt} \tag{4}$$

Because $A = 30$ when $t = 1690$, we get $30 = 60e^{1690k}$ or

$$0.5 = e^{1690k}$$

So

$$\ln 0.5 = 1690k$$

and

$$k = \frac{\ln 0.5}{1690} = \frac{-0.6931}{1690} = -0.000410$$

Substituting this value of k into (4) we obtain

$$A = 60e^{-0.000410t}$$

When $t = 100$, $A = A_{100}$, and we have

$$A_{100} = 60e^{-0.0410} = 57.6$$

Therefore there will be 57.6 mg of radium present 100 years from now.

Example 4 There are 100 million liters of fluoridated water in the reservoir containing a city's water supply, and the water contains 700 kg of fluoride. In order to decrease the fluoride content, fresh water runs into the reservoir at the rate of 3 million liters per day, and the mixture of water and fluoride, kept uniform, runs out of the reservoir at the same rate. How many kilograms of fluoride are in the reservoir 60 days after the pure water started to flow into the reservoir?

Solution Let t be the number of days that have elapsed since the pure water started to flow into the reservoir. Let x be the number of kilograms of fluoride in the reservoir at t days.

Because 100 million liters of fluoridated water are in the tank at all times, at t days the number of kilograms of fluoride per million liters is $x/100$. Three million liters of the fluoridated water run out of the reservoir each day; so the reservoir loses $3(x/100)$ kg of fluoride per day. Because $D_t x$ is the rate of change of x with respect to t, and x is decreasing as t increases, we have the differential equation

$$\frac{dx}{dt} = -\frac{3x}{100}$$

TABLE 8.7.4

t	0	60
x	700	x_{60}

We also have the initial conditions given in Table 8.7.4. Separating the variables and integrating we have

$$\int \frac{dx}{x} = -0.03 \int dt$$

$$\ln |x| = -0.03t + \bar{c}$$

$$x = Ce^{-0.03t}$$

When $t = 0$, $x = 700$; so $C = 700$. Letting $t = 60$ and $x = x_{60}$ we have

$$\begin{aligned} x_{60} &= 700e^{-1.8} \\ &= 700(0.1653) \\ &= 115.71 \end{aligned}$$

Hence there are 115.71 kg of fluoride in the reservoir 60 days after the pure water started to flow into the reservoir.

Example 5 Newton's law of cooling states that the rate at which a body changes temperature is proportional to the difference between its temperature and that of the surrounding medium. If a body is in air of temperature 35° and the body cools from 120° to 60° in 40 min, find the temperature of the body after 100 min.

Solution Let t be the number of minutes in the time that has elapsed since the body started to cool. Let x be the number of degrees in the temperature of the body at t minutes.

TABLE 8.7.5

t	0	40	100
x	120	60	x_{100}

Table 8.7.5 gives the initial conditions. From Newton's law of cooling we have

$$\frac{dx}{dt} = k(x - 35)$$

Separating the variables we obtain

$$\frac{dx}{x - 35} = k\,dt$$

Thus

$$\int \frac{dx}{x - 35} = k \int dt$$

$$\ln|x - 35| = kt + \bar{c}$$

$$x - 35 = e^{kt+\bar{c}}$$

$$x = e^{\bar{c}}e^{kt} + 35$$

$$= Ce^{kt} + 35$$

When $t = 0$, $x = 120$; so $C = 85$. Therefore

$$x = 85e^{kt} + 35$$

When $t = 40$, $x = 60$, and we obtain

$$60 = 85e^{40k} + 35$$

$$40k = \ln \tfrac{5}{17}$$

$$k = \tfrac{1}{40}(\ln 5 - \ln 17)$$

$$= \tfrac{1}{40}(1.6094 - 2.8332)$$

$$= -0.0306$$

So

$$x = 85e^{-0.0306t} + 35$$

Then

$$x_{100} = 85e^{-3.06} + 35 = 39$$

Therefore the temperature of the body is 39° after 100 min.

The calculus is often very useful to the economist for evaluating certain business decisions. However, to use the calculus we must be concerned with continuous functions. Consider, for example, the following formula, which gives A, the number of dollars in the amount after t years, if P dollars is invested at an annual rate of $100i$ percent, compounded m times per year:

$$A = P\left(1 + \frac{i}{m}\right)^{mt} \tag{5}$$

Let us conceive of a situation in which the interest is continuously compounding; that is, consider formula (5), where we let the number of interest periods per year increase without bound. Then going to the limit in formula (5) we have

$$A = P \lim_{m \to +\infty} \left(1 + \frac{i}{m}\right)^{mt}$$

which can be written as

$$A = P \lim_{m \to +\infty} \left[\left(1 + \frac{i}{m}\right)^{m/i}\right]^{it} \tag{6}$$

Now consider

$$\lim_{m \to +\infty} \left(1 + \frac{i}{m}\right)^{m/i}$$

Letting $h = i/m$ we have $m/i = 1/h$; and because $m \to +\infty$ is equivalent to $h \to 0^+$, we have

$$\lim_{m \to +\infty} \left(1 + \frac{i}{m}\right)^{m/i} = \lim_{h \to 0^+} (1 + h)^{1/h} = e$$

Hence, using Theorem 2.7.6 we have

$$\lim_{m \to +\infty} \left[\left(1 + \frac{i}{m}\right)^{m/i}\right]^{it} = \left[\lim_{m \to +\infty} \left(1 + \frac{i}{m}\right)^{m/i}\right]^{it} = e^{it}$$

and so (6) becomes

$$A = Pe^{it} \tag{7}$$

By letting t vary through the set of nonnegative real numbers, we see that (7) expresses A as a continuous function of t.

Another way of looking at the same situation is to consider an investment of P dollars, which increases at a rate proportional to its size. This is the law of natural growth. Then if A dollars is the amount at t years, we have

$$\frac{dA}{dt} = kA$$

$$\int \frac{dA}{A} = k \int dt$$

$$\ln |A| = kt + \bar{c}$$

$$A = Ce^{kt}$$

When $t = 0$, $A = P$; so $C = P$. Therefore we have

$$A = Pe^{kt} \tag{8}$$

Comparing (8) with (7), we see that they are the same if $k = i$. So if an investment increases at a rate proportional to its size, we say that the interest is **compounded continuously,** and the annual interest rate is the constant of proportionality.

● ILLUSTRATION 1 If P dollars is invested at a rate of 8% per year compounded continuously, and A dollars is the amount of the investment at t years,

$$\frac{dA}{dt} = 0.08A$$

and

$$A = Pe^{0.08t}$$

If in (7) we take $P = 1$, $i = 1$, and $t = 1$, we get $A = e$, which gives a justification for the economist's interpretation of the number e as the yield on an investment of $1 for a year at an interest rate of 100% compounded continuously.

In the following example we use the terminology *effective rate* of interest, which is the rate that gives the same amount of interest compounded once a year.

Example 6 If $5000 is borrowed at an interest rate of 12% per year, compounded continuously, and the loan is to be repaid in one payment at the end of a year, how much must the borrower repay? Also, find the effective annual rate of interest.

Solution Let A dollars be the amount to be repaid, and because $P = 5000$, $i = 0.12$, and $t = 1$, we have from (7)

$$\begin{aligned} A &= 5000e^{0.12} \\ &= 5000(1.1275) \\ &= 5637.50 \end{aligned}$$

Hence the borrower must repay $5637.50. Letting j be the effective annual rate of interest we have

$$\begin{aligned} 5000(1 + j) &= 5000e^{0.12} \\ 1 + j &= e^{0.12} \\ j &= 1.1275 - 1 \\ &= 0.1275 \\ &= 12.75\% \end{aligned}$$

EXERCISES 8.7

1. The rate of natural increase of the population of a certain city is proportional to the population. If the population increases from 40,000 to 60,000 in 40 years, when will the population be 80,000?

2. The population of a certain city doubled in the 60 years from 1890 to 1950. If the rate of natural increase of the population at any time is proportional to the population at the time, and the population in 1950 was 60,000, estimate the population in the year 2000.

3. After an automobile is one year old, its rate of depreciation at any time is proportional to its value at that time. If an automobile was purchased on June 1, 1979, and its values on June 1, 1980, and June 1, 1981, were respectively \$7000 and \$5800, what is its expected value on June 1, 1985?

4. Sugar decomposes in water at a rate proportional to the amount still unchanged. If there were 50 lb of sugar present initially and at the end of 5 hr this is reduced to 20 lb, how long will it take until 90% of the sugar is decomposed?

5. Bacteria grown in a certain culture increase at a rate proportional to the amount present. If there are 1000 bacteria present initially and the amount doubles in 1 hr, how many bacteria will there be in $3\frac{1}{2}$ hr?

6. In a certain culture where the rate of growth of bacteria is proportional to the amount present, the number triples in 3 hr. If at the end of 12 hr there were 10 million bacteria, how many bacteria were present initially?

7. In a certain chemical reaction the rate of conversion of a substance is proportional to the amount of the substance still untransformed at that time. After 10 min one-third of the original amount of the substance has been converted, and 20 g has been converted after 15 min. What was the original amount of the substance?

8. There are 100 gal of brine in a tank and the brine contains 70 lb of dissolved salt. Fresh water runs into the tank at the rate of 3 gal/min, and the mixture, kept uniform by stirring, runs out at the same rate. How many pounds of salt are there in the tank at the end of 1 hr?

9. Suppose that the value of a certain antique collection increases with age and its rate of appreciation at any time is proportional to its value at that time. If the value of the collection was \$25,000 ten years ago and its present value is \$35,000, in how many years is its value expected to be \$50,000?

10. If a body in air at a temperature of 0° cools from 200° to 100° in 40 min, use Newton's law of cooling (see Example 5) to determine how many more minutes it will take for the body to cool to 50°.

11. Under the conditions of Example 5, after how many minutes will the temperature of the body be 45°?

12. When a simple electric circuit, containing no capacitors but having inductance and resistance, has the electromotive force removed, the rate of decrease of the current is proportional to the current. The current is i amperes t sec after the cutoff, and $i = 40$ when $t = 0$. If the current dies down to 15 amperes in 0.01 sec, find i in terms of t.

13. If a thermometer is taken from a room in which the temperature is 75° into the open, where the temperature is 35°, and the reading of the thermometer is 65° after 30 sec, (a) how long after the removal will the reading be 50°? (b) What is the thermometer reading 3 min after the removal? Use Newton's law of cooling (see Example 5).

14. Thirty percent of a radioactive substance disappears in 15 years. Find the half-life of the substance.

15. If the half-life of radium is 1690 years, what percent of the amount present now will be remaining after (a) 100 years and (b) 1000 years?

16. A tank contains 200 liters of brine in which there are 3 kg of salt per liter. It is desired to dilute this solution by adding brine containing 1 kg of salt per liter, which flows into the tank at the rate of 4 liters/min and runs out at the same rate. When will the tank contain $1\frac{1}{2}$ kg of salt per liter?

17. A tank contains 100 liters of fresh water and brine containing 2 kg of salt per gallon flows into the tank at the rate of 3 liters/min. If the mixture, kept uniform by stirring, flows out at the same rate, how many kilograms of salt are there in the tank at the end of 30 min?

18. A loan of $100 is repaid in one payment at the end of a year. If the interest rate is 8% compounded continuously, determine (a) the total amount repaid and (b) the effective rate of interest.

19. (a) Find the interest earned on an investment of $500 at the end of one year if the annual interest rate is 10% compounded continuously. (b) What is the effective annual rate of interest?

20. If the purchasing power of a dollar is decreasing at the rate of 10% annually, compounded continuously, how long will it take for the purchasing power to be $0.50?

21. If an amount of money invested doubles itself in 10 years at interest compounded continuously, how long will it take for the original amount to triple itself?

22. How long will it take for $500 to accumulate to $1000 if money is invested at 8 percent per year compounded continuously?

23. Professor Willard Libby of University of California at Los Angeles was awarded the Nobel prize in chemistry for discovering a method of determining the date of death of a once-living object. Professor Libby made use of the fact that the tissue of a living organism is composed of two kinds of carbons, a radioactive carbon A and a stable carbon B, in which the ratio of the amount of A to the amount of B is approximately constant. When the organism dies, the law of natural decay applies to A. If it is determined that the amount of A in a piece of charcoal is only 15% of its original amount and the half-life of A is 5500 years, when did the tree from which the charcoal came die?

REVIEW EXERCISES FOR CHAPTER 8

In Exercises 1 through 6, determine if the given function has an inverse. If the inverse exists, do the following: (a) Define it and state its domain and range; (b) draw sketches of the graphs of the function and its inverse on the same set of axes. If the function does not have an inverse, show that a horizontal line intersects the graph of the function in more than one point.

1. $f(x) = x^3 - 4$ **2.** $f(x) = 2\sqrt[3]{x} - 1$ **3.** $f(x) = 9 - x^2$

4. $f(x) = \sqrt{4 - x^2}$ **5.** $f(x) = \dfrac{3x - 4}{x}$ **6.** $f(x) = |2x - 3|$

In Exercises 7 and 8, (a) prove that the given function f has an inverse, (b) find $f^{-1}(x)$, and (c) verify the equations of Theorem 8.1.4 for f and f^{-1}.

7. $f(x) = \sqrt[3]{x + 1}$ **8.** $f(x) = \dfrac{2x - 1}{2x + 1}$

In Exercises 9 through 12, find $(f^{-1})'(d)$ for the given f and d.

9. $f(x) = x^2 - 6x + 8, x \geq 3; d = 3$ **10.** $f(x) = \sqrt{3x + 4}; d = 5$

11. $f(x) = 8x^3 + 6x; d = 4$ **12.** $f(x) = x^5 + x - 22; d = 12$

In Exercises 13 through 24, differentiate the given function.

13. $f(x) = (\ln x^2)^2$ **14.** $f(t) = 3 \sin (e^{4t})$ **15.** $g(x) = 2^{\cos 4x}$ **16.** $f(x) = \dfrac{e^x}{(e^x + e^{-x})^2}$

17. $f(x) = e^{x/(4+x^2)}$ **18.** $f(x) = 10^{-5x^2}$ **19.** $f(x) = \sqrt{\log_{10} \dfrac{1 + x}{1 - x}}$ **20.** $g(x) = \ln \sqrt{\dfrac{2x + 1}{x - 3}}$

21. $f(t) = t^{3/\ln t}$ **22.** $F(x) = \log_{10} (x^2 2^{x^2+1})$ **23.** $f(x) = (x)^{xe^x}; x > 0$ **24.** $f(x) = 3^{x e^{x^2}}$

In Exercises 25 through 32, evaluate the indefinite integral.

25. $\displaystyle\int \frac{3e^{2x}}{1 + e^{2x}}\, dx$ **26.** $\displaystyle\int e^{2x^2-4x}(x - 1)\, dx$ **27.** $\displaystyle\int (e^{3x} + 2^{3x})\, dx$ **28.** $\displaystyle\int \frac{10^{\ln x^2}}{x}\, dx$

29. $\int \frac{xe^{6x^2}}{\sqrt{1+e^{6x^2}}}\,dx$ **30.** $\int (x+1)e^x 7^{xe^x}\,dx$ **31.** $\int \frac{2^x\,dx}{\sqrt{3\cdot 2^x+4}}$ **32.** $\int \frac{10^x+1}{10^x-1}\,dx$

In Exercises 33 through 38, evaluate the definite integral.

33. $\int_0^2 x^2 e^{x^3}\,dx$ **34.** $\int_0^1 (e^{2x}+1)^2\,dx$ **35.** $\int_1^8 \frac{x^{1/3}}{x^{4/3}+4}\,dx$

36. $\int_{1/3}^{1/2} \frac{4x^{-3}+2}{x^{-2}-x}\,dx$ **37.** $\int_0^{\ln 2} \frac{e^{2x}}{e^x-5}\,dx$ **38.** $\int_e^{e^2} \frac{dx}{x(\ln x)}$

39. Find $D_x y$ if $ye^x + xe^y + x + y = 0$. **40.** If $f(x) = \log_{(e^x)}(x+1)$, find $f'(x)$.

41. Use differentials to find an approximate value to three decimal places of $\log_{10} 100937$. Use the fact that $\log_{10} e = 0.434$, with accuracy to three decimal places.

42. Find an equation of the tangent line to the curve $y = x^{x-1}$ at $(2, 2)$.

43. A particle is moving on a straight line where s ft is the directed distance of the particle from the origin, v ft/sec is the velocity of the particle, and a ft/sec² is the acceleration of the particle at t sec. If $a = e^t + e^{-t}$ and $v = 1$ and $s = 2$ when $t = 0$, find v and s in terms of t.

44. The area of the region, bounded by the curve $y = e^{-x}$, the coordinate axes, and the line $x = b\,(b > 0)$, is a function of b. If f is this function, find $f(b)$. Also find $\lim_{b\to+\infty} f(b)$.

45. The volume of the solid of revolution obtained by revolving the region in Exercise 44 about the x axis is a function of b. If g is this function, find $g(b)$. Also find $\lim_{b\to+\infty} g(b)$.

46. The rate of natural increase of the population of a certain city is proportional to the population. If the population doubles in 60 years and if the population in 1950 was 60,000, estimate the population in the year 2000.

47. The rate of decay of a radioactive substance is proportional to the amount present. If half of a given deposit of the substance disappears in 1900 years, how long will it take for 95% of the deposit to disappear?

48. Prove that if a rectangle is to have its base on the x axis and two of its vertices on the curve $y = e^{-x^2}$, then the rectangle will have the largest possible area if the two vertices are at the points of inflection of the graph.

49. Given $f(x) = \ln|x|$ and $x < 0$. Show that f has an inverse function. If g is the inverse function, find $g(x)$ and the domain of g.

50. Prove that if $x < 1$, $\ln x < x$. (*Hint:* Let $f(x) = x - \ln x$ and show that f is decreasing on $(0, 1)$ and find $f(1)$.)

51. When a gas undergoes an adiabatic (no gain or loss of heat) expansion or compression, then the rate of change of the pressure with respect to the volume varies directly as the pressure and inversely as the volume. If the pressure is p lb/in.² when the volume is v in.³, and the initial pressure and volume are p_0 lb/in.² and v_0 in.³, show that $pv^k = p_0 v_0{}^k$.

52. If W in.-lb is the work done by a gas expanding against a piston in a cylinder and P lb/in.² is the pressure of the gas when the volume of the gas is V in.³, show that if V_1 in.³ and V_2 in.³ are the initial and final volumes, respectively, then

$$W = \int_{V_1}^{V_2} P\,dV$$

53. Suppose a piston compresses a gas in a cylinder from an initial volume of 60 in.³ to a volume of 40 in.³. If Boyle's law (Exercise 30 in Exercises 3.5) holds, and the initial pressure is 50 lb/in.², find the work done by the piston. (Use the result of Exercise 52.)

54. The charge of electricity on a spherical surface leaks off at a rate proportional to the charge. Initially, the charge of electricity was 8 coulombs and one-fourth leaks off in 15 min. When will there be only 2 coulombs remaining?

55. How long will it take for an investment to double itself if interest is paid at the rate of 8% per year compounded continuously?

56. A tank contains 60 gal of salt water with 120 lb of dissolved salt. Salt water with 3 lb of salt per gallon flows into the tank at the rate of 2 gal/min and the mixture, kept uniform by stirring, flows out at the same rate. How long will it be before there are 200 lb of salt in the tank?

57. Find the point on the graph of $f(x) = e^x$ for which the tangent line to the graph there passes through the origin.

58. Find the volume of the solid of revolution generated if the region bounded by the curve $y = 2^{-x}$ and the lines $x = 1$ and $x = 4$ is revolved about the x axis.

59. Find $\int_0^t e^{-|x|}\,dx$ if t is any real number.

60. Prove that if $x > 0$, and $\int_1^x t^{h-1}\,dt = 1$, then $\lim_{h\to 0} x = \lim_{h\to 0} (1 + h)^{1/h}$.

61. Prove that

$$\lim_{x\to 0} \frac{\log_a(1 + x)}{x} = \log_a e$$

(*Note*: Compare with Exercise 48 in Exercises 8.3.)

62. Prove that

$$\lim_{x\to 0} \frac{a^x - 1}{x} = \ln a$$

(*Hint*: Let $y = a^x - 1$ and express $(a^x - 1)/x$ as a function of y, say $g(y)$. Then show that $y \to 0$ as $x \to 0$, and find $\lim_{y\to 0} g(y)$.)

63. Use the results of Exercises 61 and 62 to prove that

$$\lim_{x\to 0} \frac{x^b - 1}{x - 1} = b$$

(*Hint*: Write

$$\frac{x^b - 1}{x - 1} = \frac{e^{b\ln x} - 1}{b \ln x} \cdot \frac{b \ln x}{x - 1}$$

Then let $s = b \ln x$ and $t = x - 1$.)

64. Prove that

$$\lim_{x\to 0} \frac{e^{ax} - 1}{x} = a$$

(*Hint*: Let $f(x) = e^{ax}$ and find $f'(0)$ by two methods.)

65. If the domain of f is the set of all real numbers and $f'(x) = cf(x)$ for all x where c is a constant, prove that there is a constant k for which $f(x) = ke^{cx}$ for all x. (*Hint*: Consider the function g for which $g(x) = f(x)e^{-cx}$ and find $g'(x)$.)

66. Prove that

$$D_x{}^n(\ln x) = (-1)^{n-1}\frac{(n - 1)!}{x^n}$$

(*Hint*: Use mathematical induction.)

67. Do Exercise 25 in the Review Exercises for Chapter 6 by evaluating each integral.

CHAPTER 9 Trigonometric and Hyperbolic Functions

9.1 THE TANGENT, COTANGENT, SECANT, AND COSECANT FUNCTIONS

The sine and cosine functions were introduced in Sec. 1.7, and in subsequent sections their derivatives and antiderivatives were discussed. We now deal with the other trigonometric functions, tangent, cotangent, secant, and cosecant. They are defined in terms of the sine and cosine.

9.1.1 Definition The **tangent** and **secant** functions are defined by

$$\tan x = \frac{\sin x}{\cos x} \qquad \sec x = \frac{1}{\cos x}$$

for all real numbers x for which $\cos x \neq 0$.

The **cotangent** and **cosecant** functions are defined by

$$\cot x = \frac{\cos x}{\sin x} \qquad \csc x = \frac{1}{\sin x}$$

for all real numbers x for which $\sin x \neq 0$.

The tangent and secant functions are not defined when $\cos x = 0$, which occurs when x is $\frac{1}{2}\pi$, $\frac{3}{2}\pi$, or $\frac{1}{2}\pi + n\pi$, where n is any positive or negative integer or zero. Therefore the domain of the tangent and secant functions is the set of all real numbers except numbers of the form $\frac{1}{2}\pi + n\pi$, where n is any integer. Similarly, because $\cot x$ and $\csc x$ are not defined when $\sin x = 0$, the domain of the cotangent and cosecant functions is the set of all real numbers except numbers of the form $n\pi$, where n is any integer.

By Theorem 2.7.1(iv), if the functions f and g are continuous at the number a, then f/g is continuous at a, provided that $g(a) \neq 0$. Because the sine and cosine functions are continuous at all real numbers, it follows that the tangent, cotangent, secant, and cosecant functions are continuous at all numbers in their domain.

By using the identity

$$\cos^2 x + \sin^2 x = 1 \tag{1}$$

and Definition 9.1.1, we obtain two other important identities. If on both sides of (1) we divide by $\cos^2 x$ when $\cos x \neq 0$, we get

$$\frac{\cos^2 x}{\cos^2 x} + \frac{\sin^2 x}{\cos^2 x} = \frac{1}{\cos^2 x}$$

and because $\sin x/\cos x = \tan x$ and $1/\cos x = \sec x$, we have the identity

$$1 + \tan^2 x = \sec^2 x \tag{2}$$

By dividing on both sides of (1) by $\sin^2 x$ when $\sin x \neq 0$, we obtain in a similar way the identity

$$\cot^2 x + 1 = \csc^2 x \tag{3}$$

Three other important formulas that follow immediately from Definition 9.1.1 are

$$\sin x \csc x = 1 \tag{4}$$

$$\cos x \sec x = 1 \tag{5}$$

and

$$\tan x \cot x = 1 \tag{6}$$

Formulas (4), (5), and (6) are valid for all values of x for which the functions are defined, and therefore they are identities.

Another formula that we need is one that expresses the tangent of the difference of two numbers in terms of the tangents of the two numbers. By using formulas (13) and (10) from Sec. 1.7 we have

$$\tan(a - b) = \frac{\sin(a - b)}{\cos(a - b)} = \frac{\sin a \cos b - \cos a \sin b}{\cos a \cos b + \sin a \sin b}$$

Dividing both the numerator and denominator by $\cos a \cos b$ we get

$$\tan(a - b) = \frac{\dfrac{\sin a \cos b}{\cos a \cos b} - \dfrac{\cos a \sin b}{\cos a \cos b}}{\dfrac{\cos a \cos b}{\cos a \cos b} + \dfrac{\sin a \sin b}{\cos a \cos b}} = \frac{\dfrac{\sin a}{\cos a} - \dfrac{\sin b}{\cos b}}{1 + \dfrac{\sin a}{\cos a} \cdot \dfrac{\sin b}{\cos b}}$$

Therefore we have the identity

$$\tan(a - b) = \frac{\tan a - \tan b}{1 + \tan a \tan b} \tag{7}$$

Taking $a = 0$ in formula (7) gives

$$\tan(0 - b) = \frac{\tan 0 - \tan b}{1 + \tan 0 \tan b}$$

Because $\tan 0 = 0$, we have from the above equation

$$\tan(-b) = -\tan b \tag{8}$$

Therefore the tangent is an odd function.

Taking $b = -b$ in formula (7) gives

$$\tan(a - (-b)) = \frac{\tan a - \tan(-b)}{1 + \tan a \tan(-b)}$$

and replacing $\tan(-b)$ by $-\tan b$ we get the identity

$$\tan(a + b) = \frac{\tan a + \tan b}{1 - \tan a \tan b} \tag{9}$$

The derivatives of the tangent, cotangent, secant, and cosecant functions are obtained from those of the sine and cosine functions and theorems on differentiation.

$$D_x(\tan x) = D_x\left(\frac{\sin x}{\cos x}\right) = \frac{\cos x \cdot D_x(\sin x) - \sin x \cdot D_x(\cos x)}{\cos^2 x}$$

$$= \frac{(\cos x)(\cos x) - (\sin x)(-\sin x)}{\cos^2 x}$$

$$= \frac{\cos^2 x + \sin^2 x}{\cos^2 x}$$

$$= \frac{1}{\cos^2 x}$$

$$= \sec^2 x \tag{10}$$

From (10) and the chain rule we have the following theorem.

9.1.2 Theorem If u is a differentiable function of x,

$$D_x(\tan u) = \sec^2 u\, D_x u$$

Example 1 Find $f'(x)$ if $f(x) = 2 \tan \frac{1}{2}x - x$.

Solution $f'(x) = 2 \sec^2 \frac{1}{2}x(\frac{1}{2}) - 1 = \sec^2 \frac{1}{2}x - 1$. If we use identity (2), this simplifies to

$$f'(x) = \tan^2 \tfrac{1}{2}x$$

The derivative of the cotangent function is obtained by a method analogous to that for the tangent function. The result is

$$D_x(\cot x) = -\csc^2 x \tag{11}$$

The derivation of (11) is left as an exercise (see Exercise 1).

The next theorem follows from (11) and the chain rule.

9.1.3 Theorem If u is a differentiable function of x,

$$D_x(\cot u) = -\csc^2 u\, D_x u$$

We now proceed to find the derivative of the secant function.

$$D_x(\sec x) = D_x[(\cos x)^{-1}]$$

$$= -1(\cos x)^{-2}(-\sin x)$$

$$= \frac{1}{\cos^2 x} \cdot \sin x$$

$$= \frac{1}{\cos x} \cdot \frac{\sin x}{\cos x}$$

$$= \sec x \tan x \tag{12}$$

By applying (12) and the chain rule, the following theorem is obtained.

9.1.4 Theorem If u is a differentiable function of x,

$$D_x(\sec u) = \sec u \tan u \, D_x u$$

Example 2 Find $f'(x)$ if $f(x) = \sec^4 3x$.

Solution

$$\begin{aligned} f'(x) &= 4 \sec^3 3x \cdot D_x(\sec 3x) \\ &= 4 \sec^3 3x(\sec 3x \tan 3x)(3) \\ &= 12 \sec^4 3x \tan 3x \end{aligned}$$

In a manner similar to that for the secant, the formula for the derivative of the cosecant function may be derived, and we obtain

$$D_x(\csc x) = -\csc x \cot x \tag{13}$$

The derivation of (13) is left as an exercise (see Exercise 2). We obtain the following theorem from (13) and the chain rule.

9.1.5 Theorem If u is a differentiable function of x,

$$D_x(\csc u) = -\csc u \cot u \, D_x u$$

Example 3 Find $f'(x)$ if $f(x) = \cot x \csc x$.

Solution

$$\begin{aligned} f'(x) &= \cot x \cdot D_x(\csc x) + \csc x \cdot D_x(\cot x) \\ &= \cot x(-\csc x \cot x) + \csc x(-\csc^2 x) \\ &= -\csc x \cot^2 x - \csc^3 x \end{aligned}$$

Example 4 An airplane is flying west at 500 ft/sec at an altitude of 4000 ft. The airplane is in a vertical plane with a searchlight on the ground. If the light is to be kept on the plane, how fast is the searchlight revolving when the airplane is due east of the searchlight at an airline distance of 2000 ft?

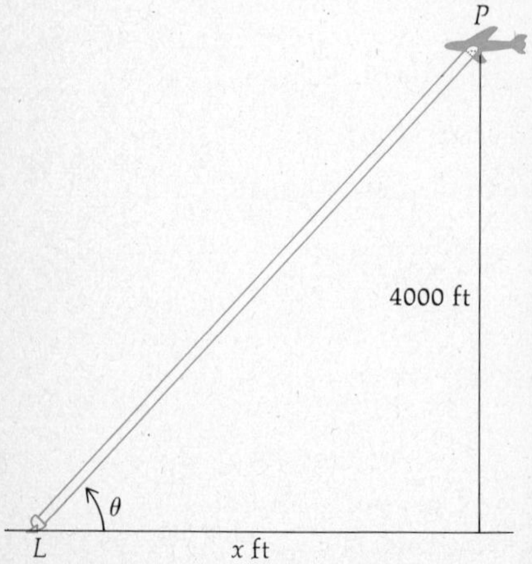

FIGURE 9.1.1

Solution Refer to Fig. 9.1.1. The searchlight is at point L, and at a particular instant of time the airplane is at point P.

Let t be the number of seconds in the time; x be the number of feet due east in the airline distance of the airplane from the searchlight at time t sec; and θ be the number of radians in the angle of elevation of the airplane at the searchlight at time t sec.

We are given $D_t x = -500$, and we wish to find $D_t \theta$ when $x = 2000$.

$$\tan \theta = \frac{4000}{x} \tag{14}$$

Differentiating on both sides of (14) with respect to t we obtain

$$\sec^2 \theta \, D_t\theta = -\frac{4000}{x^2} D_t x$$

Substituting $D_t x = -500$ in the above and dividing by $\sec^2 \theta$ gives

$$D_t\theta = \frac{2{,}000{,}000}{x^2 \sec^2 \theta} \tag{15}$$

When $x = 2000$, $\tan \theta = 2$. Therefore $\sec^2 \theta = 1 + \tan^2 \theta = 5$. Substituting these values into (15) we have, when $x = 2000$,

$$D_t\theta = \frac{2{,}000{,}000}{4{,}000{,}000(5)} = \frac{1}{10}$$

We conclude that at the given instant the measurement of the angle is increasing at the rate of $\frac{1}{10}$ rad/sec, and this is how fast the searchlight is revolving.

We now consider the graph of the tangent function. Because from (8), $\tan(-x) = -\tan x$, the graph is symmetric with respect to the origin. Furthermore,

$$\begin{aligned}
\tan(x + \pi) &= \frac{\sin(x + \pi)}{\cos(x + \pi)} \\
&= \frac{\sin x \cos \pi + \cos x \sin \pi}{\cos x \cos \pi - \sin x \sin \pi} \\
&= \frac{\sin x(-1) + \cos x(0)}{\cos x(-1) - \sin x(0)} \\
&= \frac{-\sin x}{-\cos x} \\
&= \tan x
\end{aligned}$$

and so by Definition 1.7.3 the tangent function is periodic with period π.

The tangent function is continuous at all numbers in its domain, which is the set of all real numbers except those of the form $\frac{1}{2}\pi + n\pi$, where n is any integer. However,

$$\lim_{x \to (\pi/2 + n\pi)^-} \tan x = +\infty \quad \text{and} \quad \lim_{x \to (\pi/2 + n\pi)^+} \tan x = -\infty$$

so the lines having equations $x = \frac{1}{2}\pi + n\pi$ are vertical asymptotes of the graph.

If n is any integer, $\sin n\pi = 0$ and $\cos n\pi$ is either $+1$ or -1; so $\tan n\pi = 0$. Therefore the graph intersects the x axis at the points $(n\pi, 0)$.

To find the relative extrema of the tangent function and the points of inflection of its graph, we consider the first and second derivatives. So if $f(x) = \tan x$, then $f'(x) = \sec^2 x$ and $f''(x) = 2 \sec^2 x \tan x$.

Setting $f'(x) = 0$ gives $\sec^2 x = 0$. Because $\sec^2 x \geq 1$ for all x, we conclude that there are no relative extrema. Setting $f''(x) = 0$, we obtain

$2\sec^2 x \tan x = 0$, from which we get $\tan x = 0$ because $\sec^2 x \neq 0$. Therefore $f''(x) = 0$ if $x = n\pi$, where n is any integer. At these values of x, $f''(x)$ changes sign; so the points of inflection are the points $(n\pi, 0)$, which are the points where the graph intersects the x axis. Because $f'(n\pi) = \sec^2 n\pi = 1$, the slopes of the inflectional tangents are 1.

Consider now the open interval $(-\frac{1}{2}\pi, \frac{1}{2}\pi)$ on which the tangent function is defined everywhere. Because $f'(x) = \sec^2 x > 0$ for all values of x, it follows that the tangent is an increasing function on this interval. When $-\frac{1}{2}\pi < x < 0$, $f''(x) = 2\sec^2 x \tan x < 0$; hence the graph is concave downward on the open interval $(-\frac{1}{2}\pi, 0)$. When $0 < x < \frac{1}{2}\pi$, $f''(x) > 0$; hence it follows that the graph is concave upward on the open interval $(0, \frac{1}{2}\pi)$.

TABLE 9.1.1

x	0	$\frac{1}{6}\pi$	$\frac{1}{4}\pi$	$\frac{1}{3}\pi$
$y = \tan x$	0	$\frac{1}{3}\sqrt{3} \approx 0.58$	1	$\sqrt{3} \approx 1.73$

In Table 9.1.1 there are some corresponding values of x and y satisfying the equation $y = \tan x$. By plotting the points having as coordinates the number pairs (x, y) and using the above information, a sketch of the graph of the tangent function may be drawn; it is shown in Fig. 9.1.2.

The graph of the cotangent function can be obtained from the graph of the tangent function by using an identity that will now be proved.

$$\begin{aligned}\tan(\tfrac{1}{2}\pi + x) &= \frac{\sin(\frac{1}{2}\pi + x)}{\cos(\frac{1}{2}\pi + x)}\\ &= \frac{\sin\frac{1}{2}\pi \cos x + \cos\frac{1}{2}\pi \sin x}{\cos\frac{1}{2}\pi \cos x - \sin\frac{1}{2}\pi \sin x}\\ &= \frac{(1)\cos x + (0)\sin x}{(0)\cos x - (1)\sin x}\\ &= \frac{\cos x}{-\sin x}\\ &= -\cot x\end{aligned}$$

Therefore

$$\cot x = -\tan(\tfrac{1}{2}\pi + x)$$

From the above identity it follows that the graph of the cotangent function is obtained from the graph of the tangent function by translating the y axis $\frac{1}{2}\pi$ units to the right and then taking a reflection of the graph with respect to the x axis. A sketch of the graph of the cotangent function is in Fig. 9.1.3.

The secant function is periodic with period 2π because

$$\sec(x + 2\pi) = \frac{1}{\cos(x + 2\pi)} = \frac{1}{\cos x} = \sec x$$

The domain of the secant function is the set of all real numbers except

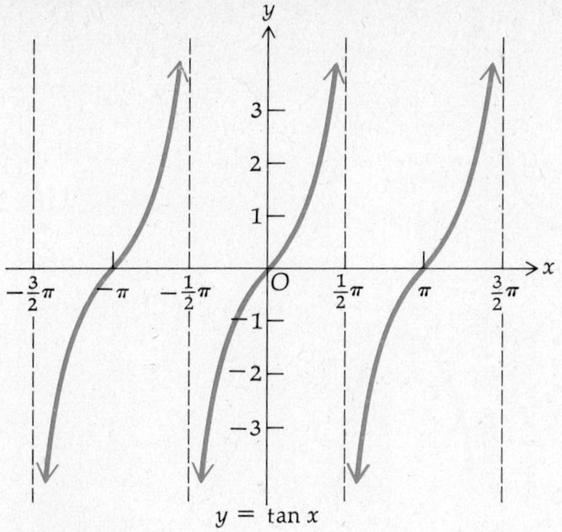

FIGURE 9.1.2

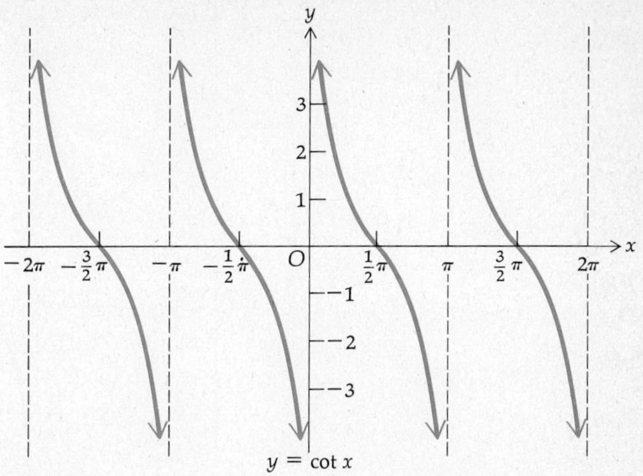

FIGURE 9.1.3

those of the form $\frac{1}{2}\pi + n\pi$, and the function is continuous on its domain. Because $\lim_{x \to \pi/2 + n\pi} \sec x = \pm\infty$, the graph of the secant function has the lines $x = \frac{1}{2}\pi + n\pi$ as vertical asymptotes. There is no intersection of the graph with the x axis because $\sec x$ is never zero. The secant function is even because

$$\sec(-x) = \frac{1}{\cos(-x)} = \frac{1}{\cos x} = \sec x$$

so the graph of the secant function is symmetric with respect to the y axis.

If $f(x) = \sec x$, it follows that $f'(x) = \sec x \tan x$ and $f''(x) = \sec x(2 \tan^2 x + 1)$. Setting $f'(x) = 0$ gives $\sec x \tan x = 0$. Because $\sec x \neq 0$, $f'(x) = 0$ when $\tan x = 0$, which is when $x = n\pi$, where n is any integer. Also, $f''(n\pi) = \sec n\pi[2 \tan^2 n\pi + 1]$. If n is an even integer, $f''(n\pi) = 1 \cdot (0 + 1) = 1$; and if n is an odd integer, $f''(n\pi) = (-1)(0 + 1) = -1$. Therefore, when $x = n\pi$ and n is an even integer, f has a relative minimum value; and when $x = n\pi$ and n is an odd integer, f has a relative maximum value. There are no points of inflection because for all x, $f''(x) \neq 0$.

Using the above information and plotting a few points, we get a sketch of the graph of the secant function shown in Fig. 9.1.4.

Because

$$\sec(x + \tfrac{1}{2}\pi) = \frac{1}{\cos(x + \frac{1}{2}\pi)} = \frac{1}{-\sin x} = -\csc x$$

then

$$\csc x = -\sec(x + \tfrac{1}{2}\pi)$$

so the graph of the cosecant function is obtained from that of the secant function by translating the y axis $\frac{1}{2}\pi$ units to the right and taking a reflection of the graph with respect to the x axis. A sketch of the graph of the cosecant function is in Fig. 9.1.5.

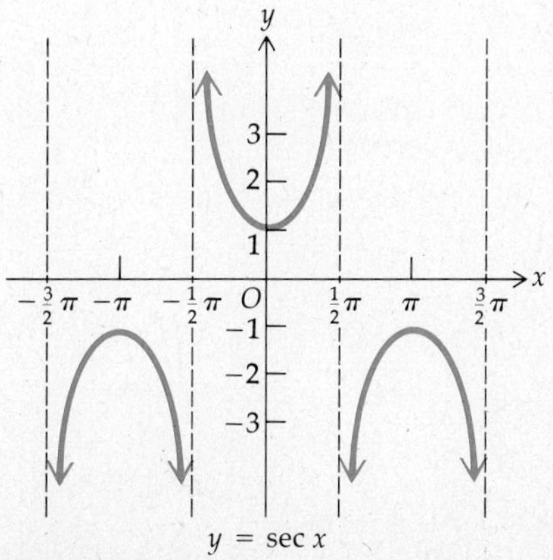

FIGURE 9.1.4

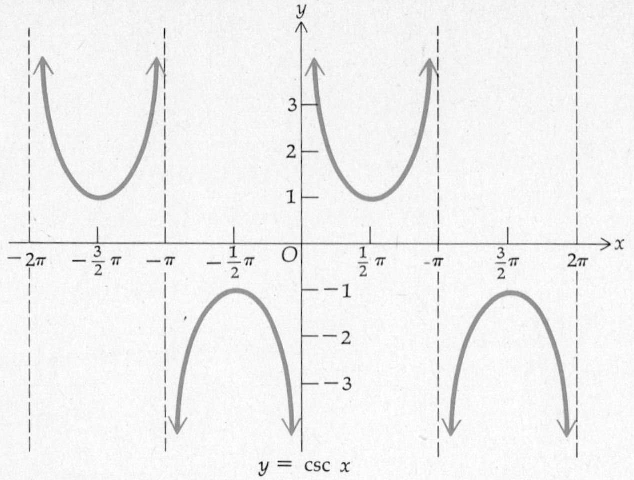

FIGURE 9.1.5

EXERCISES 9.1

1. Derive: $D_x(\cot x) = -\csc^2 x$.
2. Derive: $D_x(\csc x) = -\csc x \cot x$.

In Exercises 3 through 30, find the derivative of the given function.

3. $f(x) = \csc 4x$
4. $g(x) = 3 \cot 4x$
5. $h(x) = \tan^2 x$
6. $f(x) = \sqrt{\sec x}$
7. $f(x) = \sec x^2$
8. $g(x) = \ln \csc^2 x$
9. $g(x) = 3 \cot e^x$
10. $f(w) = 2 \tan(w^2 - 1)$
11. $F(x) = \ln |\sec 2x|$
12. $h(x) = \ln |\cot \frac{1}{2}x|$
13. $G(r) = \sqrt{\cot 3r}$
14. $f(x) = \sin x \tan x$
15. $g(x) = \sec^2 x \tan^2 x$
16. $h(t) = \sec^2 2t - \tan^2 2t$
17. $f(t) = \csc(t^3 + 1)$
18. $f(x) = e^{\sec 5x}$
19. $f(x) = 2^{\csc 3x}$
20. $g(t) = 2 \sec \sqrt{t}$
21. $H(t) = \cot^4 t - \csc^4 t$
22. $F(x) = \dfrac{\cot^2 2x}{1 + x^2}$
23. $f(x) = \ln |\sec 5x + \tan 5x|$
24. $f(x) = \frac{1}{3} \sec^3 2x - \sec 2x$
25. $g(z) = \cot^3 z \cdot \csc z$
26. $g(x) = 3^x \sec x$
27. $f(x) = \dfrac{\tan x}{\cos x - 4}$
28. $G(x) = (\tan x)^x$; $\tan x > 0$
29. $F(x) = (\sin x)^{\tan x}$; $\sin x > 0$
30. $G(x) = \log_{10} |\csc x - \cot x|$

In Exercises 31 through 33, use logarithmic differentiation to find $D_x y$.

31. $y = \dfrac{\sec x \sqrt{1 - \cot^2 x}}{\csc^3 x}$
32. $y = \dfrac{\tan^2 x}{\sqrt{1 + \sec^2 x}}$
33. $y = \dfrac{\sec^5 x \sqrt[3]{\tan x + 1}}{\sqrt{\csc^3 x - 4}}$

In Exercises 34 through 38, find $D_x y$ by implicit differentiation.

34. $y = \tan(x + y)$
35. $\sec^2 x + \csc^2 y = 4$
36. $\cot xy + xy = 0$
37. $\sec^2 y + \cot(x - y) = \tan^2 x$
38. $\csc(x - y) + \sec(x + y) = x$

In Exercises 39 through 42, draw a sketch of the graph of the given function.

39. $f(x) = 3 \tan 2x$ **40.** $f(x) = \frac{1}{2} \sec 2x$ **41.** $f(x) = 2\,|\csc \frac{1}{2}x|$ **42.** $f(x) = \frac{1}{2}\,|\cot 3x|$

43. Find an equation of the tangent line to the curve $y = \sec x$ at the point $(\frac{1}{4}\pi, \sqrt{2})$.

44. Find equations of the tangent line and normal line to the graph of $f(x) = \sec 4x \cdot \csc 2x$ at the point $(\frac{1}{4}\pi, -1)$.

45. An airplane is flying at a constant speed at an altitude of 10,000 ft on a line that will take it directly over an observer on the ground. At a given instant the observer notes that the angle of elevation of the airplane is $\frac{1}{3}\pi$ radians and is increasing at the rate of $\frac{1}{60}$ rad/sec. Find the speed of the airplane.

46. A radar antenna is located on a ship 4 mi from a straight shore, and it is rotating at 32 rpm. How fast is the radar beam moving along the shoreline when the beam makes an angle of 45° with the shore?

47. A steel girder 27 ft long is moved horizontally along a passageway 8 ft wide and into a corridor at right angles to the passageway. How wide must the corridor be for the girder to go around the corner? Neglect the horizontal width of the girder.

48. If two corridors at right angles to each other are 10 ft and 15 ft wide, respectively, what is the length of the longest steel girder that can be moved horizontally around the corner? Neglect the horizontal width of the girder.

9.2 AN APPLICATION OF THE TANGENT FUNCTION TO THE SLOPE OF A LINE

The tangent function can be used in connection with the slope of a straight line. We first define the *angle of inclination* of a line.

9.2.1 Definition The **angle of inclination** of a line not parallel to the x axis is the smallest angle measured counterclockwise from the positive direction of the x axis to the line. The angle of inclination of a line parallel to the x axis is defined to have measure zero.

If α is the degree measure of the angle of inclination of a line, α may be any number in the interval $0 \leq \alpha < 180$. Figure 9.2.1 shows a line L for which $0 < \alpha < 90$, and Fig. 9.2.2 shows one for which $90 < \alpha < 180$.

9.2.2 Theorem If α is the degree measure of the angle of inclination of line L, not parallel to the y axis, then the slope m of L is given by

$$m = \tan \alpha°$$

Proof Refer to Figs. 9.2.1 and 9.2.2, which show the given line L whose angle of inclination has degree measure α and whose slope is m. The line L' that passes through the origin and is parallel to L also has slope m and an angle of inclination whose degree measure is α. The point $P(\cos \alpha°, \sin \alpha°)$, at the intersection of L and the unit circle $x^2 + y^2 = 1$, lies on L'. And because the point (0, 0) also lies on L', it follows from Definition 1.4.1 that

$$m = \frac{\sin \alpha° - 0}{\cos \alpha° - 0} = \frac{\sin \alpha°}{\cos \alpha°} = \tan \alpha°$$

■

If the line L is parallel to the y axis, the degree measure of the angle of inclination of L is 90 and $\tan 90°$ does not exist. This is consistent with the fact that a vertical line has no slope.

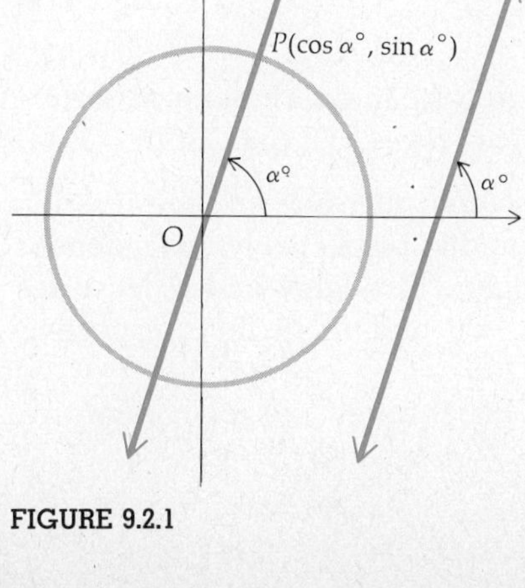

FIGURE 9.2.1

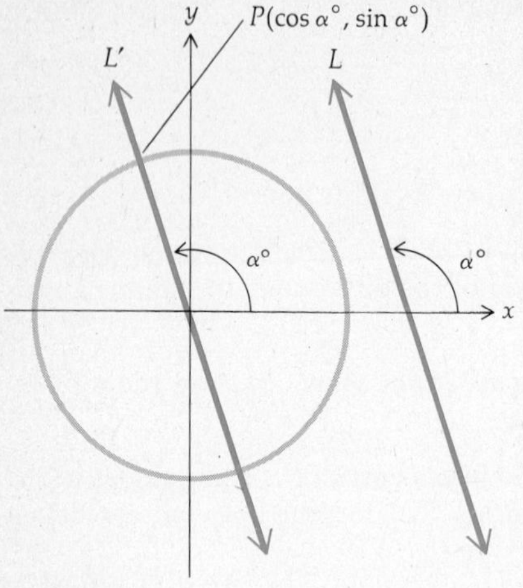

FIGURE 9.2.2

Theorem 9.2.2 is used to obtain a formula for finding the angle between two nonvertical intersecting lines. If two lines intersect, two supplementary angles are formed at their point of intersection. To distinguish these two angles, let L_2 be the line with the greater angle of inclination of degree measure α_2 and let L_1 be the other line for which the degree measure of its angle of inclination is α_1. If θ is the degree measure of the angle between the two lines, then we define

$$\theta = \alpha_2 - \alpha_1 \tag{1}$$

If L_1 and L_2 are parallel, then $\alpha_1 = \alpha_2$ and the angle between the two lines has degree measure 0. Thus if L_1 and $L_2$are two distinct lines, $0 \le \theta < 180$. Refer to Figs. 9.2.3 and 9.2.4.

The following theorem enables us to find θ when the slopes of L_1 and L_2 are known.

9.2.3 Theorem Let L_1 and L_2 be two nonvertical lines, which intersect and are not perpendicular, and let L_2 be the line having the greater angle of inclination. Then if m_1 is the slope of L_1, m_2 is the slope of L_2, and θ is the degree measure of the angle between L_1 and L_2,

$$\tan \theta° = \frac{m_2 - m_1}{1 + m_1 m_2}$$

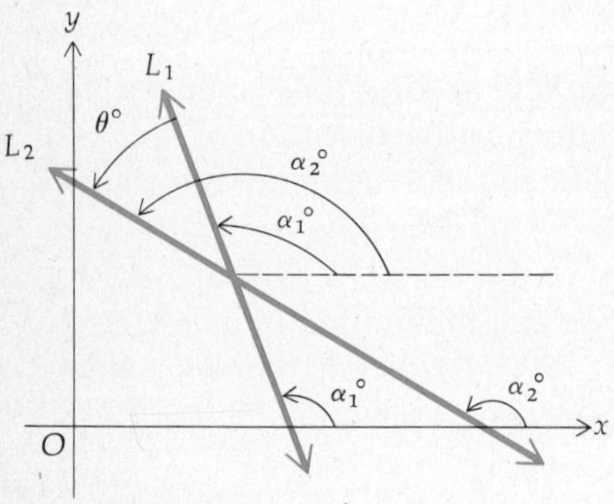

FIGURE 9.2.3

Proof If α_1 and α_2 are degree measures of the angles of inclination of L_1 and L_2, respectively, from Eq. (1),

$$\theta = \alpha_2 - \alpha_1$$

and so

$$\tan \theta° = \tan(\alpha_2° - \alpha_1°)$$

Applying formula (7) of Sec. 9.1 to the right side of the above we get

$$\tan \theta° = \frac{\tan \alpha_2° - \tan \alpha_1°}{1 + \tan \alpha_2° \tan \alpha_1°}$$

$$= \frac{m_2 - m_1}{1 + m_1 m_2}$$

■

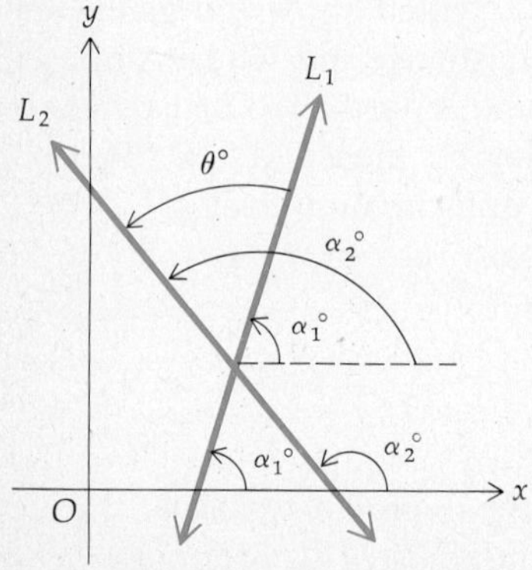

FIGURE 9.2.4

Example 1 Find to the nearest degree the measurements of the interior angles of the triangle having vertices $B(-2, 1)$, $C(2, 2)$, $D(-3, 4)$.

Solution Let β, γ, and δ be the degree measures of the interior angles of the triangle at the vertices B, C, and D, respectively. Let u denote the line through B and C, let v denote the line through C and D, let w denote the line through B and D, and let m_u, m_v, and m_w be their respective slopes. Refer to Fig. 9.2.5.

Using the formula for the slope of a line through two given points we get

$$m_u = \tfrac{1}{4} \qquad m_v = -\tfrac{2}{5} \qquad m_w = -3$$

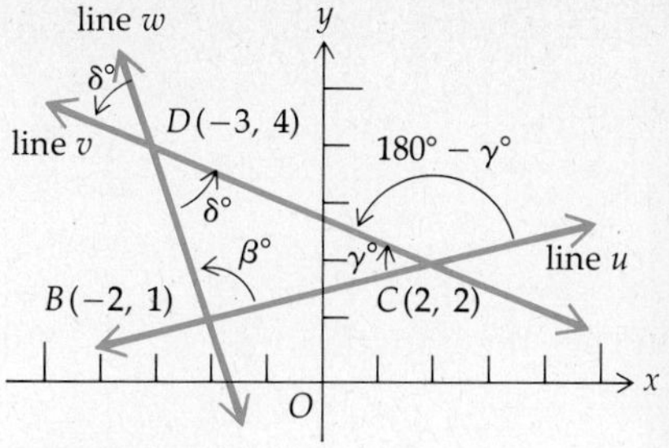

FIGURE 9.2.5

To determine β, we observe that line w has a greater angle of inclination than line u; so in formula (2), $m_2 = m_w = -3$ and $m_1 = m_u = \frac{1}{4}$. The degree measure of the angle between lines u and v is β. Thus, from Theorem 9.2.3,

$$\tan \beta° = \frac{-3 - \frac{1}{4}}{1 + \frac{1}{4}(-3)} = -13$$

Therefore

$$\beta = 94$$

To determine γ, because line v has a greater angle of inclination than line u, $m_2 = m_v = -\frac{2}{5}$ and $m_1 = m_u = \frac{1}{4}$. The degree measure of the angle between lines u and v is, by definition, $180 - \gamma$. From Theorem 9.2.3,

$$\tan(180° - \gamma°) = \frac{-\frac{2}{5} - \frac{1}{4}}{1 + \frac{1}{4}(-\frac{2}{5})} = -\frac{13}{18}$$

Hence

$$180 - \gamma = 144$$
$$\gamma = 36$$

To determine δ, because line v has a greater angle of inclination than line w, $m_2 = m_v = -\frac{2}{5}$ and $m_1 = m_w = -3$. The degree measure of the angle between lines v and w is then δ. From Theorem 9.2.3,

$$\tan \delta° = \frac{-\frac{2}{5} - (-3)}{1 + (-\frac{2}{5})(-3)} = \frac{13}{11}$$

Therefore

$$\delta = 50$$

The sum of the degree measures of the angles of a triangle must be 180. We use this to check the results, and we have

$$\beta + \gamma + \delta = 94 + 36 + 50 = 180$$

9.2.4 Definition The **angle between two curves** at a point of intersection is the angle between the tangent lines to the curves at the point.

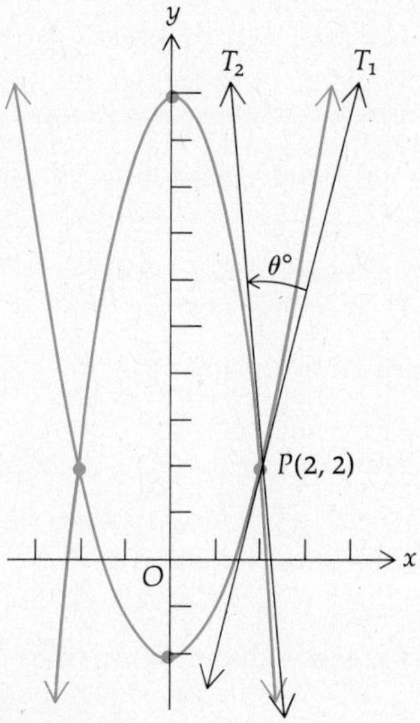

FIGURE 9.2.6

Example 2 Find to the nearest degree the measurement of the angle between the curves $y = x^2 - 2$ and $y = -2x^2 + 10$ at their points of intersection.

Solution Solving the two equations simultaneously we obtain (2, 2) and (−2, 2) as the points of intersection. In Fig. 9.2.6 sketches of the two curves are shown together with the tangent lines at one point of intersection. Because of the symmetry of the two curves with respect to the y axis, the angle between the curves at (2, 2) has the same measure as the angle between them at (−2, 2). Let P be the point (2, 2), T_1 be the tangent line to $y = x^2 - 2$ at P, and T_2 be the tangent line to $y = -2x^2 + 10$ at P.

The line T_2 has a greater angle of inclination than T_1. So if θ is the degree measure of the angle between T_1 and T_2,

$$\tan \theta° = \frac{m_2 - m_1}{1 + m_1 m_2}$$

where m_2 is the slope of T_2 and m_1 is the slope of T_1. From the equation $y = x^2 - 2$ we get $D_x y = 2x$; so $m_1 = 4$. From the equation $y = -2x^2 + 10$ we get $D_x y = -4x$; thus $m_2 = -8$. Hence

$$\tan \theta° = \frac{-8 - 4}{1 + (4)(-8)} = \frac{12}{31} \approx 0.387$$

Therefore, to the nearest degree, $\theta° = 21°$.

EXERCISES 9.2

1. Find $\tan \theta°$ if $\theta°$ is the measurement of the angle between the lines whose slopes are (a) 1 and $\frac{1}{3}$; (b) 5 and $-\frac{5}{3}$; (c) $-\frac{2}{3}$ and $\frac{1}{2}$; (d) $-\frac{1}{3}$ and $-\frac{1}{10}$; (e) $-\frac{4}{3}$ and $-\frac{5}{2}$.

2. Find, to the nearest degree, the measurement of the angle between the lines whose slopes are (a) 5 and $-\frac{7}{9}$; (b) -3 and 2; (c) $\frac{3}{2}$ and $\frac{1}{4}$; (d) $-\frac{3}{2}$ and $-\frac{1}{4}$.

3. Find, to the nearest degree, the measurements of the interior angles of the triangle formed by the lines that have equations $2x + y - 6 = 0$, $3x - y - 4 = 0$, and $3x + 4y + 8 = 0$.

4. Find, to the nearest degree, the measurements of the interior angles of the triangle having vertices at $(1, 0)$, $(-3, 2)$ and $(2, 3)$.

5. Find an equation of a line through the point $(-1, 4)$ making an angle of radian measure $\frac{1}{4}\pi$ with the line having equation $2x + y - 5 = 0$ (two solutions).

6. Find an equation of a line through the point $(-3, -2)$ making an angle of radian measure $\frac{1}{3}\pi$ with the line having equation $3x - 2y - 7 = 0$.

7. Use Theorem 9.2.3 to prove that the triangle having vertices at $(2, 3)$, $(6, 2)$, and $(3, -1)$ is isosceles.

8. Use Theorem 9.2.3 to prove that the triangle having vertices at $(-3, 0)$, $(-1, 0)$, and $(-2, \sqrt{3})$ is equilateral.

9. Find the slope of the bisector of the angle at A in the triangle having the vertices $A(4, 1)$; $B(6, 5)$; and $C(-1, 8)$.

10. Find the slope of the bisector of the angle at A in the triangle having the vertices $A(-3, 5)$; $B(2, -4)$; and $C(-1, 7)$.

11. Let A, B, and C be the vertices of an oblique triangle (no right angle), and let α, β, and γ be the radian measures of the interior angles at vertices A, B, and C, respectively. Prove that $\tan \alpha + \tan \beta + \tan \gamma = \tan \alpha \tan \beta \tan \gamma$. (*Hint*: First show that $\tan(\alpha + \beta) = -\tan \gamma$.)

12. Find the tangents of the measurements of the interior angles of the triangle having vertices at $(-3, -2)$, $(-6, 3)$, and $(5, 1)$, and check by applying the result of Exercise 11.

13. Find the points of intersection of the graphs of the sine and cosine functions, and find the measurement of the angle between their tangent lines at their points of intersection.

14. Find, to the nearest 10 minutes, the measurement of the angle between the curves $y = 2x^2 - 11$ and $y = x^2 - 2$ at their points of intersection.

15. Find $\tan \theta°$ if $\theta°$ is the measurement of the angle between the curves $y = a^x$ and $y = a^{-x}$, where $a > 1$, at their point of intersection.

16. For what value of a are the curves of Exercise 15 orthogonal?

9.3 INTEGRALS INVOLVING THE TANGENT, COTANGENT, SECANT, AND COSECANT

A formula for the indefinite integral of the tangent function is derived as follows. Because

$$\int \tan u\, du = \int \frac{\sin u}{\cos u}\, du$$

we let $v = \cos u$, $dv = -\sin u\, du$, and we obtain

$$\int \tan u\, du = -\int \frac{dv}{v} = -\ln |v| + C$$

Hence

$$\int \tan u\, du = -\ln |\cos u| + C \tag{1}$$

Because $-\ln |\cos u| = \ln |(\cos u)^{-1}| = \ln |\sec u|$, we have the following theorem.

9.3.1 Theorem

$$\int \tan u\, du = \ln |\sec u| + C$$

● ILLUSTRATION 1

$$\begin{aligned}\int \tan 3x\, dx &= \frac{1}{3}\int \tan 3x\,(3\, dx)\\ &= \tfrac{1}{3}\ln |\sec 3x| + C\end{aligned}$$

●

The theorem giving the indefinite integral of the cotangent function is proved in a way similar to the proof of Theorem 9.3.1. See Exercise 41.

9.3.2 Theorem

$$\int \cot u\, du = \ln |\sin u| + C$$

To obtain the formula for $\int \sec u\, du$ we multiply the numerator and denominator of the integrand by $(\sec u + \tan u)$, and we have

$$\int \sec u\, du = \int \frac{\sec u(\sec u + \tan u)}{\sec u + \tan u}\, du = \int \frac{(\sec^2 u + \sec u \tan u)}{\sec u + \tan u}\, du$$

Let $v = \sec u + \tan u$. Then $dv = (\sec u \tan u + \sec^2 u)\, du$; so we have

$$\int \sec u\, du = \int \frac{dv}{v} = \ln |v| + C$$

We have proved the following theorem.

9.3.3 Theorem

$$\int \sec u\, du = \ln |\sec u + \tan u| + C$$

The formula for $\int \csc u\, du$ is derived by multiplying the numerator and denominator of the integrand by $(\csc u - \cot u)$ and proceeding as above. This derivation is left as an exercise (see Exercise 42). The formula is given in the next theorem.

9.3.4 Theorem

$$\int \csc u\, du = \ln|\csc u - \cot u| + C$$

● ILLUSTRATION 2

$$\int \frac{dx}{\sin 2x} = \int \csc 2x\, dx = \frac{1}{2}\int \csc 2x\,(2dx)$$

$$= \tfrac{1}{2}\ln|\csc 2x - \cot 2x| + C$$ ●

The indefinite integral formulas in the next two theorems follow immediately from the corresponding differentiation formulas.

9.3.5 Theorem

$$\int \sec^2 u\, du = \tan u + C$$

9.3.6 Theorem

$$\int \sec u \tan u\, du = \sec u + C$$

Many integrals involving powers of tangent and secant can be evaluated by applying these two theorems and the identity given by formula (2) in Sec. 9.1, which is

$$1 + \tan^2 u = \sec^2 u \tag{2}$$

There are similar theorems involving the cotangent and cosecant.

9.3.7 Theorem

$$\int \csc^2 u\, du = -\cot u + C$$

Proof $D_u(-\cot u) = -(-\csc^2 u) = \csc^2 u$ ■

9.3.8 Theorem

$$\int \csc u \cot u\, du = -\csc u + C$$

Proof $D_u(-\csc u) = -(-\csc u \cot u) = \csc u \cot u$ ■

● ILLUSTRATION 3

$$\int \frac{2 + 3\cos u}{\sin^2 u}\, du = 2\int \frac{1}{\sin^2 u}\, du + 3\int \frac{1}{\sin u}\cdot\frac{\cos u}{\sin u}\, du$$

$$= 2\int \csc^2 u\, du + 3\int \csc u \cot u\, du$$

$$= -2\cot u - 3\csc u + C$$ ●

Formula (3) in Sec. 9.1 is the identity

$$1 + \cot^2 u = \csc^2 u \tag{3}$$

Theorems 9.3.5 and 9.3.6 and formula (2) are used to evaluate integrals of the form

$$\int \tan^m u \sec^n u \, du$$

and Theorems 9.3.7 and 9.3.8 and formula (3) are used to evaluate integrals of the form

$$\int \cot^m u \csc^n u \, du$$

where m and n are positive integers.

We distinguish various cases.

Case 1: $\int \tan^n u \, du$ or $\int \cot^n u \, du$, where n is a positive integer.

We write

$$\begin{aligned} \tan^n u &= \tan^{n-2} u \tan^2 u \\ &= \tan^{n-2} u(\sec^2 u - 1) \end{aligned}$$

$$\begin{aligned} \cot^n u &= \cot^{n-2} u \cot^2 u \\ &= \cot^{n-2} u(\csc^2 u - 1) \end{aligned}$$

Example 1 Find $\int \tan^3 x \, dx$.

Solution

$$\int \tan^3 x \, dx = \int \tan x(\sec^2 x - 1) \, dx = \int \tan x \sec^2 x \, dx - \int \tan x \, dx$$

$$= \tfrac{1}{2} \tan^2 x + \ln|\cos x| + C$$

Example 2 Find $\int \cot^4 3x \, dx$.

Solution

$$\begin{aligned} \int \cot^4 3x \, dx &= \int \cot^2 3x(\csc^2 3x - 1) \, dx \\ &= \int \cot^2 3x \csc^2 3x \, dx - \int \cot^2 3x \, dx \\ &= \tfrac{1}{9}(-\cot^3 3x) - \int (\csc^2 3x - 1) \, dx \\ &= -\tfrac{1}{9} \cot^3 3x + \tfrac{1}{3} \cot 3x + x + C \end{aligned}$$

Case 2: $\int \sec^n u\, du$ or $\int \csc^n u\, du$, where n is a positive even integer.

We write

$$\sec^n u = \sec^{n-2} u \sec^2 u = (\tan^2 u + 1)^{(n-2)/2} \sec^2 u$$

$$\csc^n u = \csc^{n-2} u \csc^2 u = (\cot^2 u + 1)^{(n-2)/2} \csc^2 u$$

Example 3 Find $\int \csc^6 x\, dx$.

Solution

$$\int \csc^6 x\, dx = \int (\cot^2 x + 1)^2 \csc^2 x\, dx$$

$$= \int \cot^4 x \csc^2 x\, dx + 2\int \cot^2 x \csc^2 x\, dx + \int \csc^2 x\, dx$$

$$= -\tfrac{1}{5}\cot^5 x - \tfrac{2}{3}\cot^3 x - \cot x + C$$

To integrate $\int \sec^n u\, du$ or $\int \csc^n u\, du$ when n is a positive odd integer, we must use integration by parts, which is discussed in Sec. 10.2.

Case 3: $\int \tan^m u \sec^n u\, du$ or $\int \cot^m u \csc^n u\, du$, where n is a positive even integer.

This case is illustrated by the following example.

Example 4 Find $\int \tan^5 x \sec^4 x\, dx$.

Solution

$$\int \tan^5 x \sec^4 x\, dx = \int \tan^5 x(\tan^2 x + 1)\sec^2 x\, dx$$

$$= \int \tan^7 x \sec^2 x\, dx + \int \tan^5 x \sec^2 x\, dx$$

$$= \tfrac{1}{8}\tan^8 x + \tfrac{1}{6}\tan^6 x + C$$

Case 4: $\int \tan^m u \sec^n u\, du$ or $\int \cot^m u \csc^n u\, du$, where m is a positive odd integer.

The following example illustrates this case.

Example 5 Find $\int \tan^5 x \sec^7 x\, dx$.

Solution

$$\int \tan^5 x \sec^7 x\, dx = \int \tan^4 x \sec^6 x \sec x \tan x\, dx$$

$$= \int (\sec^2 x - 1)^2 \sec^6 x(\sec x \tan x\, dx)$$

$$= \int \sec^{10} x(\sec x \tan x\, dx)$$

$$- 2\int \sec^8 x(\sec x \tan x\, dx) + \int \sec^6 x(\sec x \tan x\, dx)$$

$$= \tfrac{1}{11}\sec^{11} x - \tfrac{2}{9}\sec^9 x + \tfrac{1}{7}\sec^7 x + C$$

Integration by parts (Sec. 10.2) is used to integrate $\int \tan^m u \sec^n u\, du$ and $\int \cot^m u \csc^n u\, du$ when m is a positive even integer and n is a positive odd integer.

EXERCISES 9.3

In Exercises 1 through 34, evaluate the indefinite integral.

1. $\int \sec^2 5x\, dx$ **2.** $\int \cot(3x + 1)\, dx$ **3.** $\int \tan 2w\, dw$ **4.** $\int \csc^2 4t\, dt$ **5.** $\int \csc 3x \cot 3x\, dx$

6. $\int \sec 3y\, dy$ **7.** $\int \csc 10t\, dt$ **8.** $\int e^x \csc^2 e^x\, dx$ **9.** $\int x \csc^2 5x^2\, dx$ **10.** $\int \frac{\tan \sqrt{x}}{\sqrt{x}}\, dx$

11. $\int e^t \cot e^t\, dt$ **12.** $\int \sec x \tan x \tan(\sec x)\, dx$ **13.** $\int y \sec 3y^2\, dy$ **14.** $\int t^2 \sec^2 t^3\, dt$ **15.** $\int \cot^3 x\, dx$

16. $\int \csc^4 x\, dx$ **17.** $\int \cot^2 6r\, dr$ **18.** $\int \frac{\tan^2(\ln x)}{x}\, dx$

19. $\int \tan^6 x \sec^4 x\, dx$ **20.** $\int \tan^5 x \sec^3 x\, dx$ **21.** $\int \cot^2 3x \csc^4 3x\, dx$ **22.** $\int (\sec 5x + \csc 5x)^2\, dx$

23. $\int (\tan 2x + \cot 2x)^2\, dx$ **24.** $\int \frac{dx}{1 + \cos x}$ **25.** $\int \frac{2 \sin w - 1}{\cos^2 w}\, dw$ **26.** $\int \tan^4 \frac{1}{2}t\, dt$

27. $\int \tan^5 3x\, dx$ **28.** $\int \frac{\tan^4 y}{\sec^5 y}\, dy$ **29.** $\int \frac{du}{1 + \sec \frac{1}{2}u}$ **30.** $\int \frac{\csc^4 x}{\cot^2 x}\, dx$

31. $\int \frac{\sec^3 x}{\tan^4 x}\, dx$ **32.** $\int \frac{\sin^2 \pi x}{\cos^6 \pi x}\, dx$ **33.** $\int \frac{\tan^3(\ln x) \sec^6(\ln x)}{x}\, dx$ **34.** $\int \frac{\sec^4 w}{\sqrt{\tan w}}\, dw$

In Exercises 35 through 40, evaluate the definite integral.

35. $\int_{\pi/16}^{\pi/12} \tan 4x\, dx$ **36.** $\int_{\pi/8}^{\pi/4} 3 \csc 2x\, dx$ **37.** $\int_{-\pi/4}^{\pi/4} \sec^6 x\, dx$

38. $\int_0^{\pi/3} \frac{\tan^3 x}{\sec x}\, dx$ **39.** $\int_{\pi/4}^{\pi/2} \frac{\cos^4 t}{\sin^6 t}\, dt$ **40.** $\int_{\pi/6}^{\pi/4} \cot^3 w\, dw$

41. Derive the formula $\int \cot u\, du = \ln |\sin u| + C$. **42.** Derive the formula $\int \csc u\, du = \ln |\csc u - \cot u| + C$.

43. Prove that $\int \csc u\, du = -\ln|\csc u + \cot u| + C$.

44. Prove: $\int \cot x \csc^n x\, dx = -\dfrac{\csc^n x}{n} + C$ if $n \neq 0$.

45. Prove: $\int \tan^n x\, dx = \dfrac{\tan^{n-1} x}{n-1} - \int \tan^{n-2} x\, dx$ if n is a positive integer greater than 1.

46. Derive a formula similar to that in Exercise 44 for $\int \tan x \sec^n x\, dx$ if $n \neq 0$.

47. Derive a formula similar to that in Exercise 45 for $\int \cot^n x\, dx$, if n is a positive integer greater than 1.

48. We can integrate $\int \sec^2 x \tan x\, dx$ in two ways as follows: $\int \sec^2 x \tan x\, dx = \int \tan x(\sec^2 x\, dx) = \frac{1}{2}\tan^2 x + C$ and $\int \sec^2 x \tan x\, dx = \int \sec x(\sec x \tan x\, dx) = \frac{1}{2}\sec^2 x + C$. Explain the difference in the appearance of the two answers.

49. Find the length of the arc of the curve $y = \ln \csc x$ from $x = \frac{1}{6}\pi$ to $x = \frac{1}{2}\pi$.

50. Find the area of the region bounded by the curve $y = \tan^2 x$, the x axis, and the line $x = \frac{1}{4}\pi$.

51. Find the volume of the solid of revolution generated if the region bounded by the curve $y = \sec^2 x$, the axes, and the line $x = \frac{1}{4}\pi$ is revolved about the x axis.

52. Find the volume of the solid of revolution generated if the region bounded by the curve $y = 3\csc^3 x$, the x axis, and the lines $x = \frac{1}{6}\pi$ and $x = \frac{1}{2}\pi$ is revolved about the x axis.

9.4 INVERSE TRIGONOMETRIC FUNCTIONS

Figure 9.4.1 shows a sketch of the graph of the sine function. This function is not one-to-one because every number in its range is the function value of more than one number in its domain. Therefore the sine function does not have an inverse. However, observe from Fig. 9.4.1 that the sine is increasing on the interval $[-\frac{1}{2}\pi, \frac{1}{2}\pi]$. This fact follows from Theorem 4.5.3, because if $f(x) = \sin x$, then $f'(x) = \cos x > 0$ for all x in $(-\frac{1}{2}\pi, \frac{1}{2}\pi)$. Therefore, even though the sine function does not have an inverse, it follows from Theorem 8.1.5 that the function F for which

$$F(x) = \sin x \quad \text{and} \quad -\tfrac{1}{2}\pi \leq x \leq \tfrac{1}{2}\pi \qquad (1)$$

does have an inverse function. The domain of F is $[-\frac{1}{2}\pi, \frac{1}{2}\pi]$, and its range is $[-1, 1]$. A sketch of the graph of F is shown in Fig. 9.4.2. The inverse of the function defined by (1) is called the *inverse sine function* and is denoted by the symbol $\sin^{-1}$. Following is the formal definition.

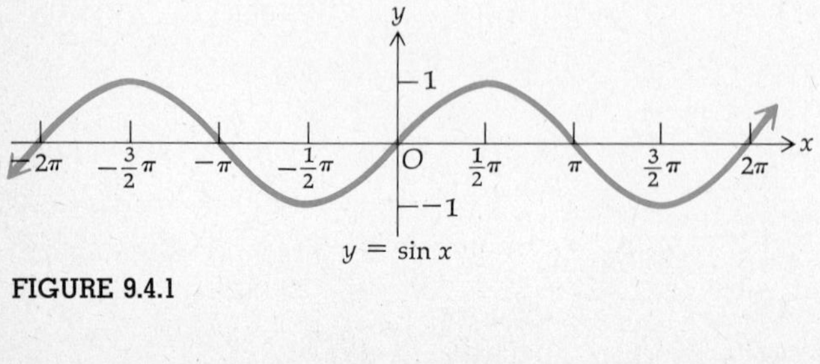

FIGURE 9.4.1

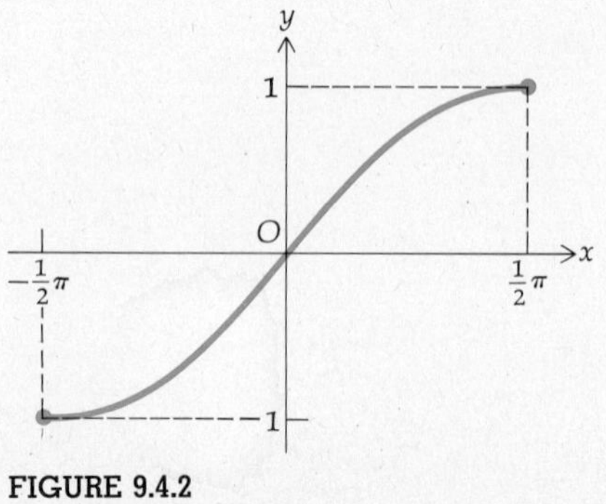

FIGURE 9.4.2

9.4.1 Definition The **inverse sine function,** denoted by $\sin^{-1}$, is defined as follows:

$$y = \sin^{-1} x \quad \text{if and only if} \quad x = \sin y \text{ and } -\tfrac{1}{2}\pi \le y \le \tfrac{1}{2}\pi$$

The domain of $\sin^{-1}$ is the closed interval $[-1, 1]$, and the range is the closed interval $[-\frac{1}{2}\pi, \frac{1}{2}\pi]$. A sketch of the graph is shown in Fig. 9.4.3.

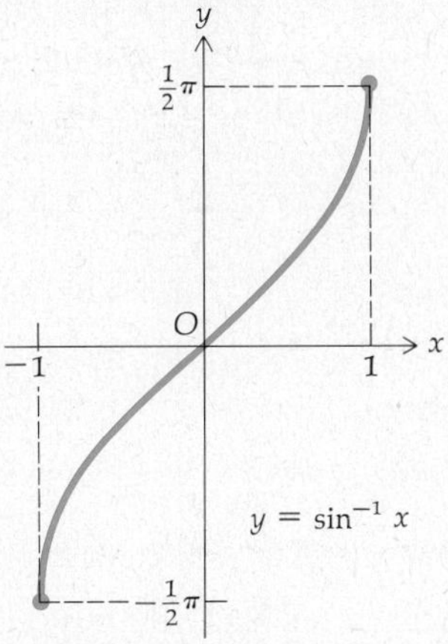

FIGURE 9.4.3

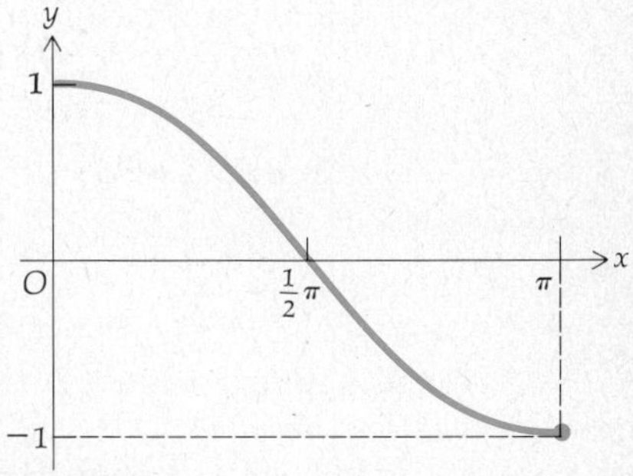

FIGURE 9.4.4

It follows from Definition 9.4.1 that

$$\sin(\sin^{-1} x) = x \quad \text{for } x \text{ in } [-1, 1] \tag{2}$$

and

$$\sin^{-1}(\sin y) = y \quad \text{for } y \text{ in } [-\tfrac{1}{2}\pi, \tfrac{1}{2}\pi] \tag{3}$$

In (1) the domain of F is restricted to the closed interval $[-\frac{1}{2}\pi, \frac{1}{2}\pi]$ so that the function is monotonic on its domain and therefore has an inverse function. However, the sine function has period 2π and is increasing on other intervals, for instance, $[-\frac{5}{2}\pi, -\frac{3}{2}\pi]$ and $[\frac{3}{2}\pi, \frac{5}{2}\pi]$. Also, the function is decreasing on certain closed intervals, in particular the intervals $[-\frac{3}{2}\pi, -\frac{1}{2}\pi]$ and $[\frac{1}{2}\pi, \frac{3}{2}\pi]$. Any one of these intervals could just as well be chosen for the domain of the function F of Eq. (1). The choice of the interval $[-\frac{1}{2}\pi, \frac{1}{2}\pi]$, however, is customary because it is the largest interval containing the number 0 on which the function is monotonic.

The cosine function does not have an inverse function because it also is not one-to-one. To define the inverse cosine function we restrict the cosine to an interval on which the function is monotonic. We choose the interval $[0, \pi]$ on which the cosine is decreasing. So consider the function G defined by

$$G(x) = \cos x \quad \text{and} \quad 0 \le x \le \pi \tag{4}$$

The range of this function G is the closed interval $[-1, 1]$, and its domain is the closed interval $[0, \pi]$. A sketch of the graph of G is shown in Fig. 9.4.4. Because this function is continuous and decreasing on its domain, it has an inverse function which is called the *inverse cosine function* and is denoted by $\cos^{-1}$.

9.4.2 Definition The **inverse cosine function,** denoted by $\cos^{-1}$, is defined as follows:

$$y = \cos^{-1} x \quad \text{if and only if} \quad x = \cos y \text{ and } 0 \le y \le \pi$$

The domain of $\cos^{-1}$ is the closed interval $[-1, 1]$, and the range is the closed interval $[0, \pi]$. A sketch of its graph is shown in Fig. 9.4.5.

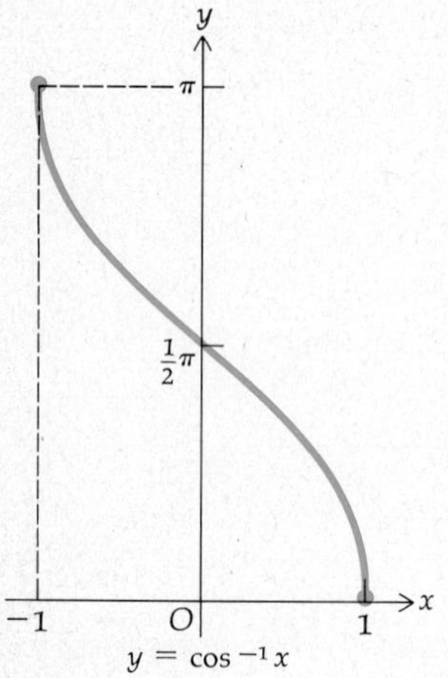

FIGURE 9.4.5

From Definition 9.4.2 it follows that

$$\cos(\cos^{-1} x) = x \qquad \text{for } x \text{ in } [-1, 1] \tag{5}$$

and

$$\cos^{-1}(\cos y) = y \qquad \text{for } y \text{ in } [0, \pi] \tag{6}$$

An interesting equation relating $\sin^{-1}$ and $\cos^{-1}$ is given in the following example.

Example 1 Prove: $\cos^{-1} x = \frac{1}{2}\pi - \sin^{-1} x$ for $|x| \le 1$.

Solution Let x be in $[-1, 1]$ and $w = \frac{1}{2}\pi - \sin^{-1} x$. Then

$$\sin^{-1} x = \tfrac{1}{2}\pi - w$$

and so

$$\sin(\sin^{-1} x) = \sin(\tfrac{1}{2}\pi - w)$$

and using (2) we get

$$x = \sin(\tfrac{1}{2}\pi - w)$$

Applying (11) in Sec. 1.7 to the right side of the above equation we get

$$x = \cos w \tag{7}$$

Because $-\frac{1}{2}\pi \le \sin^{-1} x \le \frac{1}{2}\pi$, by adding $-\frac{1}{2}\pi$ to each member we have

$$-\pi \le -\tfrac{1}{2}\pi + \sin^{-1} x \le 0$$

Multiplying each member of the above inequality by -1 and reversing the direction of the inequality signs gives

$$0 \le \tfrac{1}{2}\pi - \sin^{-1} x \le \pi$$

Replacing $\frac{1}{2}\pi - \sin^{-1} x$ by w we obtain

$$0 \le w \le \pi \tag{8}$$

From (7), (8), and Definition 9.4.2 it follows that

$$w = \cos^{-1} x \qquad \text{for } |x| \le 1$$

and replacing w by $\frac{1}{2}\pi - \sin^{-1} x$ we get

$$\tfrac{1}{2}\pi - \sin^{-1} x = \cos^{-1} x \qquad \text{for } |x| \le 1 \tag{9}$$

which is what we wished to prove.

Observe in the above solution that the relation given by (9) depends on choosing the range of the inverse cosine function to be $[0, \pi]$.

To obtain the inverse tangent function we first restrict the tangent function to the open interval $(-\frac{1}{2}\pi, \frac{1}{2}\pi)$ on which the function is continuous and increasing. We let H be the function defined by

$$H(x) = \tan x \quad \text{and} \quad -\tfrac{1}{2}\pi < x < \tfrac{1}{2}\pi \tag{10}$$

The domain of H is the open interval $(-\frac{1}{2}\pi, \frac{1}{2}\pi)$, and the range is the set of all real numbers. A sketch of its graph is shown in Fig. 9.4.6. Because the function H is continuous and increasing on its domain, it has an inverse function, which is called the *inverse tangent function* and is denoted by $\tan^{-1}$.

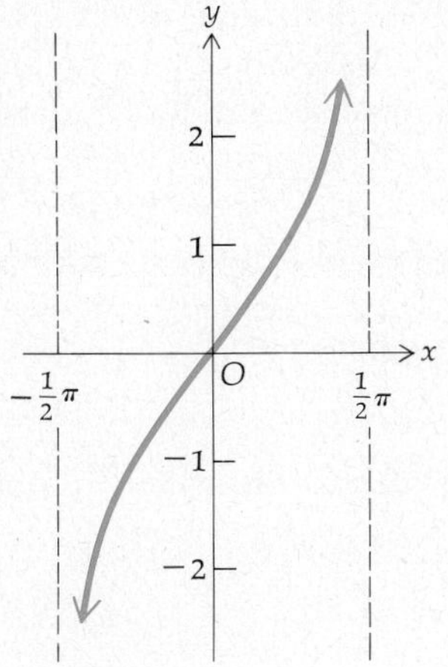

FIGURE 9.4.6

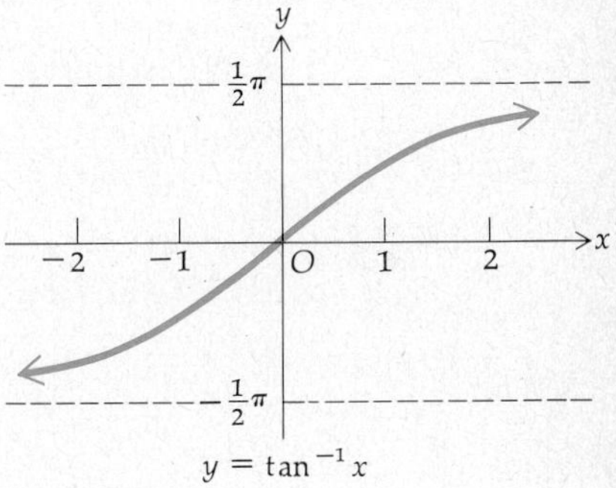

FIGURE 9.4.7

9.4.3 Definition The **inverse tangent function,** denoted by $\tan^{-1}$, is defined as follows:

$$y = \tan^{-1} x \quad \text{if and only if} \quad x = \tan y \text{ and } -\tfrac{1}{2}\pi < y < \tfrac{1}{2}\pi$$

The domain of $\tan^{-1}$ is the set of all real numbers, and the range is the open interval $(-\frac{1}{2}\pi, \frac{1}{2}\pi)$. A sketch of its graph is shown in Fig. 9.4.7.

Before defining the inverse cotangent function, we refer back to Example 1, in which we proved the relationship between $\cos^{-1}$ and $\sin^{-1}$ given by (9). This equation can be used to define the inverse cosine function, and then it can be proved that the range of $\cos^{-1}$ is $[0, \pi]$. We use this kind of procedure in discussing the inverse cotangent function.

9.4.4 Definition The **inverse cotangent function,** denoted by $\cot^{-1}$, is defined by

$$\cot^{-1} x = \tfrac{1}{2}\pi - \tan^{-1} x \qquad \text{where } x \text{ is any real number} \tag{11}$$

It follows from the definition that the domain of $\cot^{-1}$ is the set of all real numbers. To obtain the range we write (11) as

$$\tan^{-1} x = \tfrac{1}{2}\pi - \cot^{-1} x \tag{12}$$

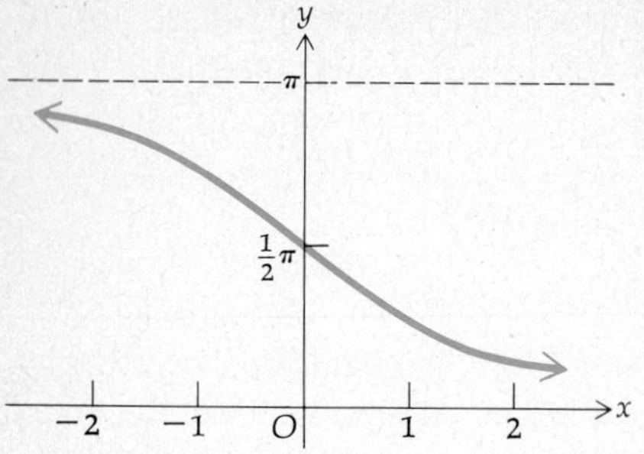

$y = \cot^{-1} x$

FIGURE 9.4.8

Because

$$-\tfrac{1}{2}\pi < \tan^{-1} x < \tfrac{1}{2}\pi$$

by substituting from (12) into this inequality we get

$$-\tfrac{1}{2}\pi < \tfrac{1}{2}\pi - \cot^{-1} x < \tfrac{1}{2}\pi$$

Subtracting $\frac{1}{2}\pi$ from each member we get

$$-\pi < -\cot^{-1} x < 0$$

and now multiplying each member by -1 and reversing the direction of the inequality signs we obtain

$$0 < \cot^{-1} x < \pi$$

Therefore the range of the inverse cotangent function is the open interval $(0, \pi)$. A sketch of its graph is in Fig. 9.4.8.

The inverse secant and the inverse cosecant functions are defined in terms of $\cos^{-1}$ and $\sin^{-1}$, respectively.

9.4.5 Definition The **inverse secant function,** denoted by $\sec^{-1}$, is defined by

$$\sec^{-1} x = \cos^{-1}\left(\frac{1}{x}\right) \qquad \text{for } |x| \geq 1$$

9.4.6 Definition The **inverse cosecant function,** denoted by $\csc^{-1}$, is defined by

$$\csc^{-1} x = \sin^{-1}\left(\frac{1}{x}\right) \qquad \text{for } |x| \geq 1$$

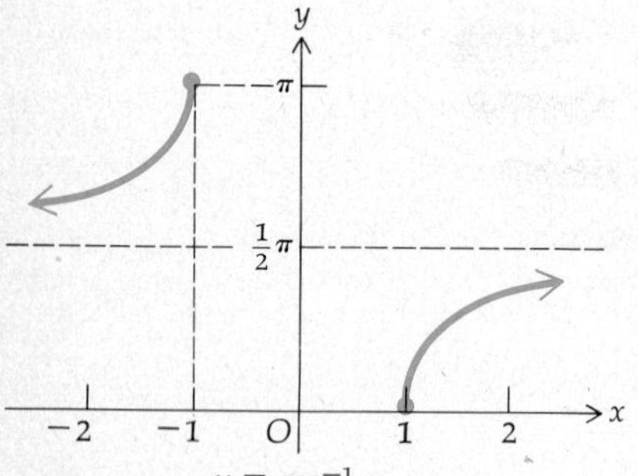

$y = \sec^{-1} x$

FIGURE 9.4.9

From Definition 9.4.5 it follows that the domain of $\sec^{-1}$ is $(-\infty, -1] \cup [1, +\infty)$, and the range is $[0, \frac{1}{2}\pi) \cup (\frac{1}{2}\pi, \pi]$. Note that the number $\frac{1}{2}\pi$ is not in the range because $\sec \frac{1}{2}\pi$ is not defined. A sketch of the graph of $\sec^{-1}$ is shown in Fig. 9.4.9.

It follows from Definition 9.4.6 that the domain of $\csc^{-1}$ is $(-\infty, -1] \cup [1, +\infty)$, and the range is $[-\frac{1}{2}\pi, 0) \cup (0, \frac{1}{2}\pi]$. The number 0 is not in the range because csc 0 is not defined. A sketch of the graph of $\csc^{-1}$ is shown in Fig. 9.4.10.

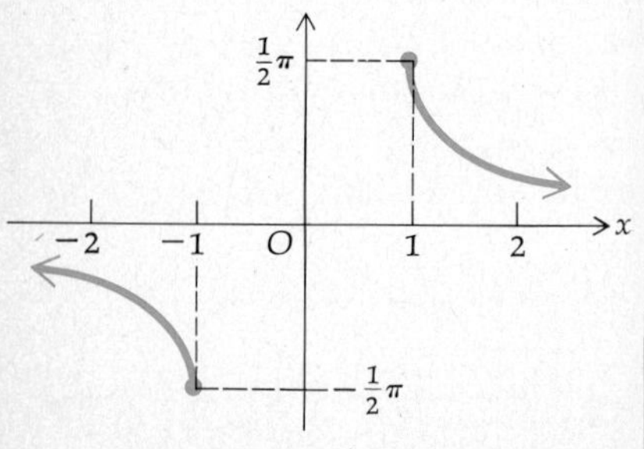

$y = \csc^{-1} x$

FIGURE 9.4.10

Example 2 Find the exact value of $\sec[\sin^{-1}(-\frac{3}{4})]$.

Solution Let $t = \sin^{-1}(-\frac{3}{4})$. Then $\sin t = -\frac{3}{4}$. Because the range of $\sin^{-1}$ is $[-\frac{1}{2}\pi, \frac{1}{2}\pi]$, and because $\sin t$ is negative, it follows that $-\frac{1}{2}\pi < t < 0$. Figure 9.4.11 shows an angle having radian measure t that satisfies these requirements. From the figure we see that $\sec t = \frac{4}{7}\sqrt{7}$; so we conclude that $\sec[\sin^{-1}(-\frac{3}{4})] = \frac{4}{7}\sqrt{7}$.

Example 3 Find the exact value of $\sin[2 \cos^{-1}(-\frac{3}{5})]$.

Solution Let $t = \cos^{-1}(-\frac{3}{5})$. So we wish to find the exact value of $\sin 2t$. From formula (14) of Sec. 1.7,

$$\sin 2t = 2 \sin t \cos t \tag{13}$$

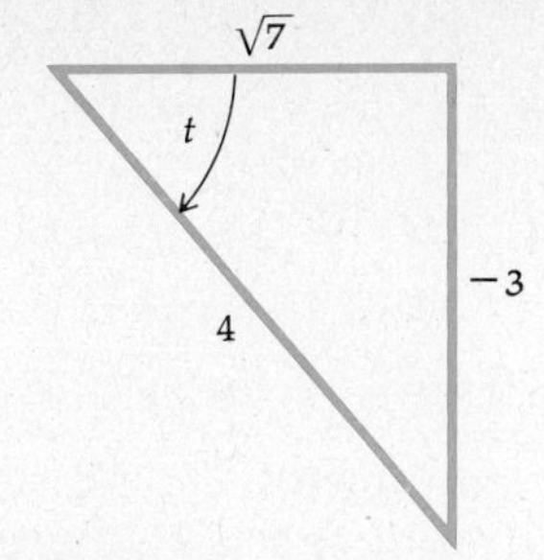

FIGURE 9.4.11

Because $t = \cos^{-1}(-\frac{3}{5})$, it follows that $\cos t = -\frac{3}{5}$ and $\frac{1}{2}\pi < t < \pi$. From the identity, $\sin^2 t + \cos^2 t = 1$, and because $\sin t > 0$, since t is in $(\frac{1}{2}\pi, \pi)$,

$$\sin t = \sqrt{1 - \cos^2 t} = \sqrt{1 - (-\tfrac{3}{5})^2} = \tfrac{4}{5}$$

Therefore, from formula (13),

$$\sin 2t = 2(\tfrac{4}{5})(-\tfrac{3}{5}) = -\tfrac{24}{25}$$

from which we conclude that

$$\sin[2\cos^{-1}(-\tfrac{3}{5})] = -\tfrac{24}{25}$$

EXERCISES 9.4

In Exercises 1 through 4, find the function values.

1. (a) $\sin^{-1}\dfrac{\sqrt{2}}{2}$; (b) $\cos^{-1}\dfrac{\sqrt{2}}{2}$; (c) $\sin^{-1}\left(-\dfrac{\sqrt{2}}{2}\right)$; (d) $\cos^{-1}\left(-\dfrac{\sqrt{2}}{2}\right)$; (e) $\sec^{-1}\sqrt{2}$; (f) $\csc^{-1}(-\sqrt{2})$.

2. (a) $\sin^{-1}\left(-\dfrac{\sqrt{3}}{2}\right)$; (b) $\cos^{-1}\left(-\dfrac{\sqrt{3}}{2}\right)$; (c) $\tan^{-1}(-\sqrt{3})$; (d) $\cot^{-1}(-\sqrt{3})$; (e) $\sec^{-1}\left(-\dfrac{2}{\sqrt{3}}\right)$; (f) $\csc^{-1}\left(-\dfrac{2}{\sqrt{3}}\right)$.

3. (a) $\sin^{-1}(-1)$; (b) $\cos^{-1}(-1)$; (c) $\tan^{-1}(-1)$; (d) $\cot^{-1}(-1)$; (e) $\sec^{-1}(-1)$; (f) $\csc^{-1}(-1)$.

4. (a) $\sin^{-1}\frac{1}{2}$; (b) $\sin^{-1}(-\frac{1}{2})$; (c) $\cos^{-1}\frac{1}{2}$; (d) $\cos^{-1}(-\frac{1}{2})$; (e) $\sec^{-1} 2$; (f) $\csc^{-1}(-2)$.

5. Given $y = \sin^{-1}\frac{1}{3}$. Find the exact value of each of the following: (a) $\cos y$; (b) $\tan y$; (c) $\cot y$; (d) $\sec y$; (e) $\csc y$.

6. Given $y = \cos^{-1}(-\frac{2}{3})$. Find the exact value of each of the following: (a) $\sin y$; (b) $\tan y$; (c) $\cot y$; (d) $\sec y$; (e) $\csc y$.

7. Given $y = \tan^{-1}(-2)$. Find the exact value of each of the following: (a) $\sin y$; (b) $\cos y$; (c) $\cot y$; (d) $\sec y$; (e) $\csc y$.

8. Given $y = \cot^{-1}(-\frac{1}{2})$. Find the exact value of each of the following: (a) $\sin y$; (b) $\cos y$; (c) $\tan y$; (d) $\sec y$; (e) $\csc y$.

9. Given $y = \csc^{-1}(-\frac{3}{2})$. Find the exact value of each of the following: (a) $\sin y$; (b) $\cos y$; (c) $\tan y$; (d) $\cot y$; (e) $\sec y$.

10. Given $y = \sec^{-1} 3$. Find the exact value of each of the following: (a) $\sin y$; (b) $\cos y$; (c) $\tan y$; (d) $\cot y$; (e) $\csc y$.

In Exercises 11 through 14, find the exact value of the given quantities.

11. (a) $\tan[\sin^{-1}\frac{1}{2}\sqrt{3}]$; (b) $\sin[\tan^{-1}\frac{1}{2}\sqrt{3}]$

12. (a) $\cos[\tan^{-1}(-3)]$; (b) $\tan[\sec^{-1}(-3)]$

13. (a) $\cos[\sin^{-1}(-\frac{1}{2})]$; (b) $\sin[\cos^{-1}(-\frac{1}{2})]$

14. (a) $\tan[\cot^{-1}(-1)]$; (c) $\cot[\tan^{-1}(-1)]$

15. Find the exact value of each of the following: (a) $\sin^{-1}[\tan\frac{1}{4}\pi]$; (b) $\tan^{-1}[\sin(-\frac{1}{2}\pi)]$; (c) $\sec^{-1}[\tan(-\frac{1}{4}\pi)]$; (d) $\cot^{-1}[\csc(-\frac{1}{2}\pi)]$.

16. Find the exact value of each of the following: (a) $\cos^{-1}[\sin(-\frac{1}{6}\pi)]$; (b) $\sin^{-1}[\cos(-\frac{1}{6}\pi)]$; (c) $\cos^{-1}[\sin\frac{2}{3}\pi]$; (d) $\sin^{-1}[\cos\frac{2}{3}\pi]$.

In Exercises 17 through 26, find the exact value of the given quantity.

17. $\cos[2\sin^{-1}(-\frac{5}{13})]$

18. $\tan[2\sec^{-1}(-\frac{5}{4})]$

19. $\sin[\sin^{-1}\frac{2}{3} + \cos^{-1}\frac{1}{3}]$

20. $\cos[\sin^{-1}(-\frac{1}{2}) + \sin^{-1}\frac{1}{4}]$

21. $\cos[\sin^{-1}\frac{2}{3} + 2\sin^{-1}(-\frac{1}{3})]$

22. $\tan[\tan^{-1}(-\frac{2}{5}) + \cos^{-1}(-\frac{1}{2}\sqrt{2})]$

23. $\tan(\tan^{-1}\frac{3}{4} - \sin^{-1}\frac{1}{2})$

24. $\tan[\sec^{-1}\frac{5}{3} + \csc^{-1}(-\frac{13}{12})]$

25. $\cos(\sin^{-1} \frac{1}{3} - \tan^{-1} \frac{1}{2})$

26. $\sin[\cos^{-1}(-\frac{2}{3}) + 2 \sin^{-1}(-\frac{1}{3})]$

27. Prove: $\cos^{-1} (3/\sqrt{10}) + \cos^{-1}(2/\sqrt{5}) = \frac{1}{4}\pi$.

28. Prove: $2 \tan^{-1} \frac{1}{3} - \tan^{-1}(-\frac{1}{7}) = \frac{1}{4}\pi$.

In Exercises 29 through 36, draw a sketch of the graph of the given equation.

29. $y = \frac{1}{2} \sin^{-1} x$ **30.** $y = \sin^{-1} \frac{1}{2}x$ **31.** $y = \tan^{-1} 2x$ **32.** $y = 2 \tan^{-1} x$

33. $y = \cos^{-1} 3x$ **34.** $y = \frac{1}{2} \sec^{-1} 2x$ **35.** $y = 2 \cot^{-1} \frac{1}{2}x$ **36.** $y = \frac{1}{2} \csc^{-1} \frac{1}{2}x$

37. Given $\cot A = \tan(\frac{1}{2}\pi - A)$, show that $\tan^{-1} x + \tan^{-1}\left(\frac{1}{x}\right) = \frac{1}{2}\pi$, if $x > 0$.

9.5 DERIVATIVES OF THE INVERSE TRIGONOMETRIC FUNCTIONS

Because the inverse trigonometric functions are continuous and monotonic on their domains, it follows from Theorem 8.2.3 that they have derivatives. To derive the formula for the derivative of the inverse sine function, let

$$y = \sin^{-1} x$$

which is equivalent to

$$x = \sin y \quad \text{and} \quad y \text{ in } [-\tfrac{1}{2}\pi, \tfrac{1}{2}\pi] \tag{1}$$

Differentiating on both sides of (1) with respect to y we obtain

$$D_y x = \cos y \quad \text{and} \quad y \text{ in } [-\tfrac{1}{2}\pi, \tfrac{1}{2}\pi] \tag{2}$$

From the identity $\sin^2 y + \cos^2 y = 1$, and replacing $\sin y$ by x, we obtain

$$\cos^2 y = 1 - x^2 \tag{3}$$

If y is in $[-\frac{1}{2}\pi, \frac{1}{2}\pi]$, $\cos y$ is nonnegative; so from (3) it follows that

$$\cos y = \sqrt{1 - x^2} \qquad \text{if } y \text{ is in } [-\tfrac{1}{2}\pi, \tfrac{1}{2}\pi] \tag{4}$$

Substituting from (4) into (2) we get

$$D_y x = \sqrt{1 - x^2}$$

From Theorem 8.2.3, $D_x y = 1/D_y x$; hence

$$D_x(\sin^{-1} x) = \frac{1}{\sqrt{1 - x^2}} \tag{5}$$

The domain of the derivative of the inverse sine function is the open interval $(-1, 1)$.

From (5) and the chain rule we have the following theorem.

9.5.1 Theorem If u is a differentiable function of x,

$$D_x(\sin^{-1} u) = \frac{1}{\sqrt{1 - u^2}} D_x u$$

● ILLUSTRATION 1 If $y = \sin^{-1} x^2$, then

$$\frac{dy}{dx} = \frac{1}{\sqrt{1 - (x^2)^2}}(2x) = \frac{2x}{\sqrt{1 - x^4}}$$

●

To derive the formula for the derivative of the inverse cosine function we use (9) of Sec. 9.4, which is

$$\cos^{-1} x = \tfrac{1}{2}\pi - \sin^{-1} x$$

Differentiating with respect to x we have

$$D_x(\cos^{-1} x) = D_x(\tfrac{1}{2}\pi - \sin^{-1} x)$$

and so

$$D_x(\cos^{-1} x) = -\frac{1}{\sqrt{1 - x^2}} \tag{6}$$

where x is in $(-1, 1)$.

The next theorem follows from (6) and the chain rule.

9.5.2 Theorem If u is a differentiable function of x,

$$D_x(\cos^{-1} u) = -\frac{1}{\sqrt{1 - u^2}} D_x u$$

We now derive the formula for the derivative of the inverse tangent function. If

$$y = \tan^{-1} x$$

then

$$x = \tan y \quad \text{and} \quad y \text{ is in } (-\tfrac{1}{2}\pi, \tfrac{1}{2}\pi)$$

Differentiating on both sides of the above equation with respect to y we obtain

$$D_y x = \sec^2 y \quad \text{and} \quad y \text{ is in } (-\tfrac{1}{2}\pi, \tfrac{1}{2}\pi) \tag{7}$$

From the identity $\sec^2 y = 1 + \tan^2 y$, and replacing $\tan y$ by x, we have

$$\sec^2 y = 1 + x^2 \tag{8}$$

Substituting from (8) into (7) we get

$$D_y x = 1 + x^2$$

So from Theorem 8.2.3,

$$D_x(\tan^{-1} x) = \frac{1}{1 + x^2} \tag{9}$$

The domain of the derivative of the inverse tangent function is the set of all real numbers.

From (9) and the chain rule the next theorem is obtained.

9.5.3 Theorem If u is a differentiable function of x,

$$D_x(\tan^{-1} u) = \frac{1}{1 + u^2} D_x u$$

● ILLUSTRATION 2 If $f(x) = \tan^{-1} \dfrac{1}{x+1}$, then

$$f'(x) = \frac{1}{1 + \dfrac{1}{(x+1)^2}} \cdot \frac{-1}{(x+1)^2}$$

$$= \frac{-1}{(x+1)^2 + 1}$$

$$= \frac{-1}{x^2 + 2x + 2}$$

From Definition 9.4.4,

$$\cot^{-1} x = \tfrac{1}{2}\pi - \tan^{-1} x$$

Differentiating with respect to x we obtain the formula

$$D_x(\cot^{-1} x) = -\frac{1}{1 + x^2}$$

From this formula and the chain rule we have the following theorem.

9.5.4 Theorem If u is a differentiable function of x,

$$D_x(\cot^{-1} u) = -\frac{1}{1 + u^2} D_x u$$

Example 1 Given $y = x^3 \cot^{-1} \frac{1}{3}x$, find dy/dx.

Solution

$$\frac{dy}{dx} = 3x^2 \cot^{-1} \frac{1}{3}x + x^3 \cdot \frac{-1}{1 + \frac{1}{9}x^2} \cdot \frac{1}{3}$$

$$= 3x^2 \cot^{-1} \frac{1}{3}x - \frac{3x^3}{9 + x^2}$$

Example 2 Given $\ln(x + y) = \tan^{-1} \left(\dfrac{x}{y}\right)$, find $D_x y$.

Solution Differentiating implicitly on both sides of the given equation with respect to x we get

$$\frac{1}{x + y}(1 + D_x y) = \frac{1}{1 + \dfrac{x^2}{y^2}} \cdot \frac{y - x\, D_x y}{y^2}$$

$$\frac{1 + D_x y}{x + y} = \frac{y - x\, D_x y}{y^2 + x^2}$$

Setting $D_x\theta = 0$ we obtain

$$9(16^2 + x^2) - 16(9^2 + x^2) = 0$$
$$-7x^2 + 9 \cdot 16(16 - 9) = 0$$
$$x^2 = 9 \cdot 16$$
$$x = 12$$

The -12 is rejected because it is not in the interval $(0, +\infty)$. The results of the first-derivative test are shown in Table 9.5.1. Because the relative maximum value of θ is an absolute maximum value, we conclude that the observer should stand 12 ft from the wall.

TABLE 9.5.1

	$D_x\theta$	Conclusion
$0 < x < 12$	$+$	
$x = 12$	0	θ has a relative maximum value
$12 < x < +\infty$	$-$	

EXERCISES 9.5

In Exercises 1 through 26, find the derivative of the given function.

1. $f(x) = \sin^{-1} \frac{1}{2}x$ **2.** $f(x) = \cos^{-1} 3x$ **3.** $g(x) = \tan^{-1} 2x$

4. $f(x) = \csc^{-1} 2x$ **5.** $F(x) = 2\cos^{-1} \sqrt{x}$ **6.** $g(x) = \frac{1}{2}\sin^{-1} x^2$

7. $g(t) = \sec^{-1} 5t$ **8.** $f(y) = \cot^{-1} e^y$ **9.** $f(x) = \sin^{-1}\sqrt{1 - x^2}$

10. $f(w) = 2\tan^{-1}\dfrac{1}{w}$ **11.** $F(x) = \cot^{-1}\dfrac{2}{x} + \tan^{-1}\dfrac{x}{2}$ **12.** $f(x) = \ln(\tan^{-1} 3x)$

13. $h(y) = y \sin^{-1} 2y$ **14.** $f(t) = t^2 \cos^{-1} t$ **15.** $g(x) = x \csc^{-1}\dfrac{1}{x}$

16. $g(s) = \cos^{-1} s + \dfrac{s}{1 - s^2}$ **17.** $f(x) = \cos^{-1}(\sin x)$ **18.** $h(x) = \tan^{-1}\dfrac{2x}{1 - x^2}$

19. $F(x) = \ln(\tan^{-1} x^2)$ **20.** $F(x) = \sec^{-1}\sqrt{x^2 + 4}$ **21.** $f(x) = 4\sin^{-1}\frac{1}{2}x + x\sqrt{4 - x^2}$

22. $f(t) = a\sin^{-1}\dfrac{t}{a} + \sqrt{a^2 - t^2}$ **23.** $h(x) = \sec^{-1} x + \csc^{-1} x$ **24.** $f(x) = \sin^{-1} x + \cos^{-1} x$

25. $G(x) = x\cot^{-1} x + \ln\sqrt{1 + x^2}$ **26.** $f(t) = \sin^{-1}\dfrac{t - 1}{t + 1}$

In Exercises 27 through 30, find $D_x y$ by implicit differentiation.

27. $e^x + y = \cos^{-1} x$ **28.** $\ln(\sin^2 3x) = e^x + \cot^{-1} y$ **29.** $x \sin y + x^3 = \tan^{-1} y$ **30.** $\sin^{-1}(xy) = \cos^{-1}(x + y)$

31. A light is 3 km from a straight beach. If the light revolves and makes 2 rpm, find the speed of the spot of light along the beach when the spot is 2 km from the point on the beach nearest the light.

32. A ladder 25 ft long is leaning against a vertical wall. If the bottom of the ladder is pulled horizontally away from the wall so that the top is sliding down at 3 ft/sec, how fast is the measure of the angle between the ladder and the ground changing when the bottom of the ladder is 15 ft from the wall?

33. A picture 40 cm high is placed on a wall with its base 30 cm above the level of the eye of an observer. If the observer is approaching the wall at the rate of 40 cm/sec, how fast is the measure of the angle subtended at the observer's eye by the picture changing when the observer is 1 m from the wall?

34. A man on a dock is pulling in at the rate of 2 ft/sec a rowboat by means of a rope. The man's hands are 20 ft above the level of the point where the rope is attached to the boat. How fast is the measure of the angle of depression of the rope changing when there are 52 ft of rope out?

35. A woman is walking at the rate of 5 ft/sec along the diameter of a circular courtyard. A light at one end of a diameter perpendicular to her path casts a shadow on the circular wall. How fast is the shadow moving along the wall when the distance from the woman to the center of the courtyard is $\frac{1}{2}r$, where r ft is the radius of the courtyard?

36. In Exercise 35, how far is the woman from the center of the courtyard when the speed of her shadow along the wall is 9 ft/sec?

37. Use differentials to find an approximate value of $\sin^{-1} 0.52$ to three decimal places.

38. Find equations of the tangent line and normal line to the graph of the equation $y = \sec^{-1}(2x + 1)$ at the point $(\frac{1}{2}, \frac{1}{3}\pi)$.

39. A rope is attached to a weight and it passes over a hook that is 8 ft above the ground. The rope is pulled over the hook at the rate of $\frac{3}{4}$ ft/sec and drags the weight along level ground. If the length of the rope between the weight and the hook is x ft when the radian measure of the angle between the rope and the floor is θ, find the time rate of change of θ in terms of x.

40. Given: $f(x) = \tan^{-1}(1/x) - \cot^{-1} x$. (a) Show that $f'(x) = 0$ for all x in the domain of f. (b) Prove that there is no constant C for which $f(x) = C$ for all x in its domain. (c) Why doesn't the statement in part (b) contradict Theorem 5.2.2?

41. Do Example 1 of Sec. 9.4 by showing that $\sin^{-1} x$ and $-\cos^{-1} x$ differ by a constant and then by evaluating the constant.

42. Do Exercise 37 of Exercises 9.4 by showing that $\tan^{-1} x$ and $-\tan^{-1}\left(\frac{1}{x}\right)$ differ by a constant and then by evaluating the constant.

9.6 INTEGRALS YIELDING INVERSE TRIGONOMETRIC FUNCTIONS

From the formulas for the derivatives of the inverse trigonometric functions we obtain the following theorem giving some indefinite integral formulas.

9.6.1 Theorem

$$\int \frac{du}{\sqrt{1 - u^2}} = \sin^{-1} u + C \tag{1}$$

$$\int \frac{du}{1 + u^2} = \tan^{-1} u + C \tag{2}$$

$$\int \frac{du}{u\sqrt{u^2 - 1}} = \sec^{-1} |u| + C \tag{3}$$

Proof Formulas (1) and (2) follow directly from those for the derivatives of $\sin^{-1} u$ and $\tan^{-1} u$. Formula (3) needs an explanation.

If $u > 0$, then $u = |u|$, and

$$\int \frac{du}{u\sqrt{u^2 - 1}} = \int \frac{du}{|u|\sqrt{u^2 - 1}} = \sec^{-1} u + C = \sec^{-1}|u| + C$$

If $u < 0$, then $u = -|u|$, and

$$\begin{aligned}\int \frac{du}{u\sqrt{u^2 - 1}} &= \int \frac{du}{-|u|\sqrt{u^2 - 1}}\\ &= \int \frac{-du}{|u|\sqrt{u^2 - 1}}\\ &= \int \frac{d(-u)}{|-u|\sqrt{(-u)^2 - 1}}\\ &= \sec^{-1}(-u) + C\\ &= \sec^{-1}|u| + C\end{aligned}$$

Therefore, if $u > 0$ or $u < 0$, we obtain (3). ■

We also have the formulas given in the next theorem.

9.6.2 Theorem

$$\int \frac{du}{\sqrt{a^2 - u^2}} = \sin^{-1}\frac{u}{a} + C \qquad \text{where } a > 0 \tag{4}$$

$$\int \frac{du}{a^2 + u^2} = \frac{1}{a}\tan^{-1}\frac{u}{a} + C \qquad \text{where } a \neq 0 \tag{5}$$

$$\int \frac{du}{u\sqrt{u^2 - a^2}} = \frac{1}{a}\sec^{-1}\left|\frac{u}{a}\right| + C \qquad \text{where } a > 0 \tag{6}$$

Proof These formulas can be proved by finding the derivative of the right side and obtaining the integrand. We prove formula (4)

$$\begin{aligned}D_u\left(\sin^{-1}\frac{u}{a}\right) &= \frac{1}{\sqrt{1 - \left(\frac{u}{a}\right)^2}} D_u\left(\frac{u}{a}\right)\\ &= \frac{\sqrt{a^2}}{\sqrt{a^2 - u^2}} \cdot \frac{1}{a}\\ &= \frac{a}{\sqrt{a^2 - u^2}} \cdot \frac{1}{a} \qquad \text{if } a > 0\\ &= \frac{1}{\sqrt{a^2 - u^2}} \qquad \text{if } a > 0\end{aligned}$$

The proofs of (5) and (6) are left as exercises (see Exercises 38 and 39). ■

Example 1 Evaluate

$$\int \frac{dx}{\sqrt{4 - 9x^2}}$$

Solution

$$\int \frac{dx}{\sqrt{4 - 9x^2}} = \frac{1}{3}\int \frac{d(3x)}{\sqrt{4 - (3x)^2}}$$

$$= \frac{1}{3}\sin^{-1}\frac{3x}{2} + C$$

Example 2 Evaluate

$$\int \frac{dx}{3x^2 - 2x + 5}$$

Solution

$$\int \frac{dx}{3x^2 - 2x + 5} = \int \frac{dx}{3(x^2 - \frac{2}{3}x) + 5}$$

To complete the square of $x^2 - \frac{2}{3}x$ we add $\frac{1}{9}$, and because $\frac{1}{9}$ is multiplied by 3, we actually add $\frac{1}{3}$ to the denominator, and so we also subtract $\frac{1}{3}$ from the denominator. Therefore we have

$$\int \frac{dx}{3x^2 - 2x + 5} = \int \frac{dx}{3(x^2 - \frac{2}{3}x + \frac{1}{9}) + 5 - \frac{1}{3}}$$

$$= \int \frac{dx}{3(x - \frac{1}{3})^2 + \frac{14}{3}} = \frac{1}{3}\int \frac{dx}{(x - \frac{1}{3})^2 + \frac{14}{9}}$$

$$= \frac{1}{3}\cdot\frac{3}{\sqrt{14}}\tan^{-1}\left(\frac{x - \frac{1}{3}}{\frac{1}{3}\sqrt{14}}\right) + C$$

$$= \frac{1}{\sqrt{14}}\tan^{-1}\left(\frac{3x - 1}{\sqrt{14}}\right) + C$$

Example 3 Evaluate

$$\int \frac{(2x + 7)\,dx}{x^2 + 2x + 5}$$

Solution Because $d(x^2 + 2x + 5) = (2x + 2)\,dx$, we write the numerator as $(2x + 2)\,dx + 5\,dx$ and express the original integral as the sum of two integrals.

$$\int \frac{(2x + 7)\,dx}{x^2 + 2x + 5} = \int \frac{(2x + 2)\,dx}{x^2 + 2x + 5} + 5\int \frac{dx}{x^2 + 2x + 5}$$

$$= \ln |x^2 + 2x + 5| + 5 \int \frac{dx}{(x+1)^2 + 4}$$

$$= \ln(x^2 + 2x + 5) + \frac{5}{2} \tan^{-1} \frac{x+1}{2} + C$$

Note: Because $x^2 + 2x + 5 > 0$, for all x, $|x^2 + 2x + 5| = x^2 + 2x + 5$.

Example 4 Evaluate

$$\int \frac{3\,dx}{(x+2)\sqrt{x^2 + 4x + 3}}$$

Solution

$$\int \frac{3\,dx}{(x+2)\sqrt{x^2 + 4x + 3}} = \int \frac{3\,dx}{(x+2)\sqrt{(x^2 + 4x + 4) - 1}}$$

$$= 3 \int \frac{d(x+2)}{(x+2)\sqrt{(x+2)^2 - 1}}$$

$$= 3 \sec^{-1} |x + 2| + C$$

Example 5 Find the area of the region in the first quadrant bounded by the curve

$$y = \frac{1}{1 + x^2}$$

the x axis, the y axis, and the line $x = 1$.

Solution Figure 9.6.1 shows the region and a rectangular element of area. If A square units is the area of the region,

$$A = \lim_{\|\Delta\| \to 0} \sum_{i=1}^{n} \frac{1}{1 + \xi_i^{\,2}} \Delta_i x$$

$$= \int_0^1 \frac{dx}{1 + x^2}$$

$$= \tan^{-1} x \Big]_0^1$$

$$= \tan^{-1} 1 - \tan^{-1} 0$$

$$= \tfrac{1}{4}\pi - 0$$

$$= \tfrac{1}{4}\pi$$

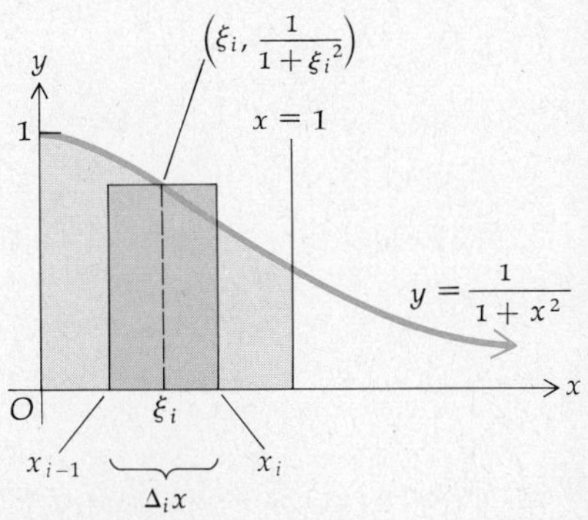

FIGURE 9.6.1

EXERCISES 9.6

In Exercises 1 through 25, evaluate the indefinite integral.

1. $\int \frac{dx}{\sqrt{1-4x^2}}$
2. $\int \frac{dx}{x^2+25}$
3. $\int \frac{dx}{9x^2+16}$
4. $\int \frac{dt}{\sqrt{1-16t^2}}$
5. $\int \frac{dx}{4+(x-1)^2}$
6. $\int \frac{dx}{9+(3-x)^2}$
7. $\int \frac{dx}{4x\sqrt{x^2-16}}$
8. $\int \frac{x\,dx}{x^4+16}$
9. $\int \frac{dx}{\sqrt{2-5x^2}}$
10. $\int \frac{3\,dx}{x\sqrt{x^2-9}}$
11. $\int \frac{r\,dr}{\sqrt{16-9r^4}}$
12. $\int \frac{du}{u\sqrt{16u^2-9}}$
13. $\int \frac{e^x\,dx}{7+e^{2x}}$
14. $\int \frac{\sin x\,dx}{\sqrt{2-\cos^2 x}}$
15. $\int \frac{dx}{(1+x)\sqrt{x}}$
16. $\int \frac{ds}{\sqrt{2s-s^2}}$
17. $\int \frac{dx}{x^2-x+2}$
18. $\int \frac{dx}{\sqrt{3x-x^2-2}}$
19. $\int \frac{dx}{\sqrt{15+2x-x^2}}$
20. $\int \frac{dx}{2x^2+2x+3}$
21. $\int \frac{x\,dx}{\sqrt{3-2x-x^2}}$
22. $\int \frac{x\,dx}{x^2+x+1}$
23. $\int \frac{(2+x)\,dx}{\sqrt{4-2x-x^2}}$
24. $\int \frac{2\,dt}{(t-3)\sqrt{t^2-6t+5}}$
25. $\int \frac{2x^3\,dx}{2x^2-4x+3}$

In Exercises 26 through 33, evaluate the definite integral.

26. $\int_0^{1/2} \frac{dx}{\sqrt{1-x^2}}$
27. $\int_0^1 \frac{1+x}{1+x^2}\,dx$
28. $\int_0^{\sqrt{3}} \frac{x\,dx}{\sqrt{1+x^2}}$
29. $\int_{-4}^{-2} \frac{dt}{\sqrt{-t^2-6t-5}}$
30. $\int_2^5 \frac{dx}{x^2-4x+13}$
31. $\int_0^1 \frac{dx}{e^x+e^{-x}}$
32. $\int_{1/\sqrt{2}}^1 \frac{dx}{x\sqrt{4x^2-1}}$
33. $\int_1^e \frac{dx}{x[1+(\ln x)^2]}$

34. Find the area of the region bounded by the curve $y = 8/(x^2+4)$, the x axis, the y axis, and the line $x = 2$.
35. Find the area of the region bounded by the curve $y = 1/\sqrt{5-4x-x^2}$, the x axis, and the lines $x = -\frac{7}{2}$ and $x = -\frac{1}{2}$.
36. Find the area of the region bounded by the curves $x^2 = 4ay$ and $y = 8a^3/(x^2+4a^2)$.
37. Find the circumference of the circle $x^2 + y^2 = r^2$ by integration.

In Exercises 38 through 42, prove the given formula by showing that the derivative of the right side is equal to the integrand.

38. $\int \frac{du}{a^2+u^2} = \frac{1}{a}\tan^{-1}\frac{u}{a} + C$
39. $\int \frac{du}{u\sqrt{u^2-a^2}} = \frac{1}{a}\sec^{-1}\left|\frac{u}{a}\right| + C \quad \text{if } a > 0$

    Take two cases: $u > 0$ and $u < 0$.
40. $\int \frac{du}{\sqrt{1-u^2}} = -\cos^{-1} u + C$

    Is this formula equivalent to formula (1)? Why?
41. $\int \frac{du}{1+u^2} = -\cot^{-1} u + C$

    Is this formula equivalent to formula (2)? Why?
42. $\int \frac{du}{u\sqrt{u^2-1}} = -\csc^{-1}|u| + C$

    Take two cases: $u > 0$ and $u < 0$. Is this formula equivalent to formula (3)? Why?
43. A particle moving in a straight line is said to have *simple harmonic motion* if the measure of its acceleration is always proportional to the measure of its displacement from a fixed point on the line and its acceleration and displacement

are oppositely directed. So if at t sec, s cm is the directed distance of the particle from the origin and v cm/sec is the velocity of the particle, then a differential equation for simple harmonic motion is

$$\frac{dv}{dt} = -k^2 s \tag{7}$$

where k^2 is the constant of proportionality and the minus sign indicates that the acceleration is opposite in direction from the displacement. Because $dv/dt = (dv/ds)(ds/dt) = v(dv/ds)$, (7) may be written as

$$v\frac{dv}{ds} = -k^2 s \tag{8}$$

(a) Solve (8) for v to get $v = \pm k\sqrt{a^2 - s^2}$. *Note:* Take a^2k^2 as the arbitrary constant of integration and justify this choice. (b) Letting $v = ds/dt$ in the solution of part (a) we obtain the differential equation

$$\frac{ds}{dt} = \pm k\sqrt{a^2 - s^2} \tag{9}$$

Taking $t = 0$ at the instant when $v = 0$ (and hence $s = a$), solve (9) to obtain

$$s = a \cos kt \tag{10}$$

(c) Show that the largest value for $|s|$ is a. The number a is called the *amplitude* of the motion. (d) The particle will oscillate between the points where $s = a$ and $s = -a$. If T sec is the time for the particle to go from a to $-a$ and return, show that $T = 2\pi/k$. The number T is called the *period* of the motion.

44. A particle is moving in a straight line according to the equation of motion $s = 5 - 10 \sin^2 2t$, where s cm is the directed distance of the particle from the origin at t sec. Use the result of part (b) of Exercise 43 to show that the motion is simple harmonic. Find the amplitude and period of this motion.

45. Show that the motion of Exercise 44 is simple harmonic by showing that differential equation (7) is satisfied.

9.7 THE HYPERBOLIC FUNCTIONS

Certain combinations of e^x and e^{-x} appear so frequently in applications of mathematics that they are given special names. Two of these functions are the *hyperbolic sine function* and the *hyperbolic cosine function*. The function values are related to the coordinates of the points of an equilateral hyperbola in a manner similar to that in which the values of the corresponding trigonometric functions are related to the coordinates of points of a circle (see Sec. 16.4). Following are the definitions of the hyperbolic sine function and the hyperbolic cosine function.

9.7.1 Definition The **hyperbolic sine function** is defined by

$$\sinh x = \frac{e^x - e^{-x}}{2}$$

The domain and range are the set of all real numbers.

9.7.2 Definition The **hyperbolic cosine function** is defined by

$$\cosh x = \frac{e^x + e^{-x}}{2}$$

The domain is the set of all real numbers and the range is the set of all numbers in the interval $[1, +\infty)$.

Because

$$\sinh(-x) = \frac{e^{-x} - e^{x}}{2} = -\frac{e^{x} - e^{-x}}{2} = -\sinh x$$

it follows that the hyperbolic sine function is an odd function. Because

$$\cosh(-x) = \frac{e^{-x} + e^{x}}{2} = \cosh x$$

the hyperbolic cosine function is even.

The formulas for the derivatives of the hyperbolic sine and hyperbolic cosine functions may be obtained by applying Definitions 9.7.1 and 9.7.2 and differentiating the resulting expressions involving exponential functions. Thus

$$D_x(\sinh x) = D_x\left(\frac{e^{x} - e^{-x}}{2}\right) = \frac{e^{x} + e^{-x}}{2} = \cosh x$$

$$D_x(\cosh x) = D_x\left(\frac{e^{x} + e^{-x}}{2}\right) = \frac{e^{x} - e^{-x}}{2} = \sinh x$$

From these formulas and the chain rule we have the following theorem.

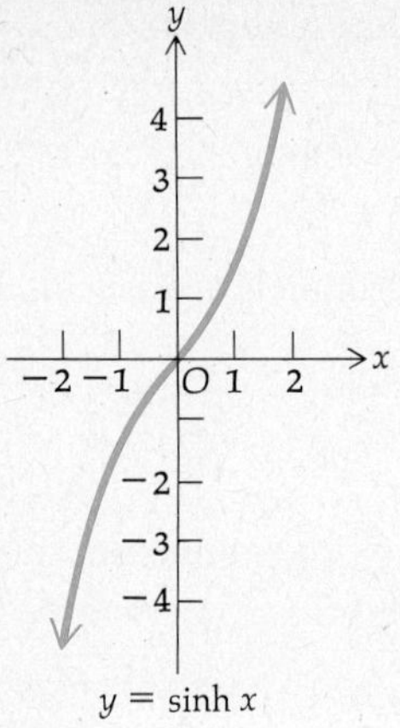

FIGURE 9.7.1

9.7.3 Theorem If u is a differentiable function of x,

$$D_x(\sinh u) = \cosh u \, D_x u$$

$$D_x(\cosh u) = \sinh u \, D_x u$$

Because $D_x(\sinh x) > 0$ for all x, the hyperbolic sine function is increasing on its entire domain. With this information, the knowledge that it is an odd function, and some values obtained from the table of hyperbolic functions in the appendix, we have the graph of the hyperbolic sine function shown in Fig. 9.7.1.

The hyperbolic cosine function is decreasing on the interval $(-\infty, 0]$ because $D_x(\cosh x) < 0$ if $x < 0$, and it is increasing on the interval $[0, +\infty)$ because $D_x(\cosh x) > 0$ if $x > 0$. Furthermore, the hyperbolic cosine is an even function. With these facts and some values of $\cosh x$ from the table in the appendix, we obtain the graph of the hyperbolic cosine function shown in Fig. 9.7.2.

The remaining four hyperbolic functions may be defined in terms of the hyperbolic sine and hyperbolic cosine functions. You should note that each satisfies a relation analogous to one satisfied by corresponding trigonometric functions.

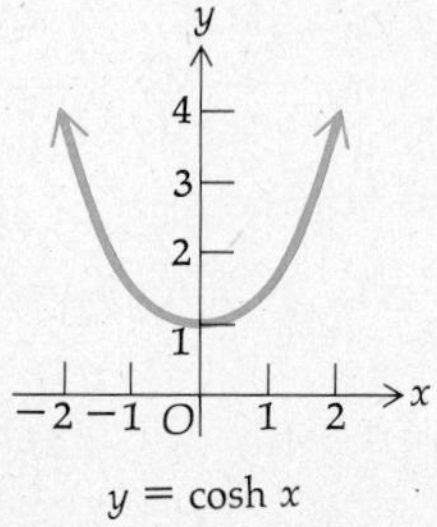

FIGURE 9.7.2

9.7.4 Definition The **hyperbolic tangent function, hyperbolic cotangent function, hyperbolic secant function,** and **hyperbolic cosecant function** are defined, respectively, as follows:

$$\tanh x = \frac{\sinh x}{\cosh x} = \frac{e^x - e^{-x}}{e^x + e^{-x}} \tag{1}$$

$$\coth x = \frac{\cosh x}{\sinh x} = \frac{e^x + e^{-x}}{e^x - e^{-x}} \tag{2}$$

$$\operatorname{sech} x = \frac{1}{\cosh x} = \frac{2}{e^x + e^{-x}} \tag{3}$$

$$\operatorname{csch} x = \frac{1}{\sinh x} = \frac{2}{e^x - e^{-x}} \tag{4}$$

A sketch of the graph of the hyperbolic tangent function is shown in Fig. 9.7.3. Notice from Figs. 9.7.1, 9.7.2, and 9.7.3 that neither of these functions is periodic, while the corresponding trigonometric functions are periodic.

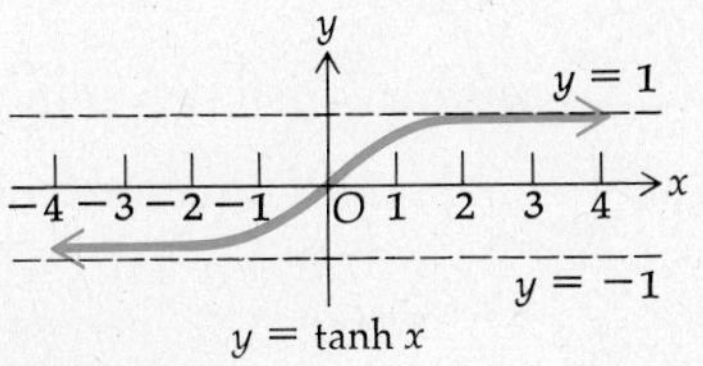

FIGURE 9.7.3

There are identities that are satisfied by the hyperbolic functions which are similar to those satisfied by the trigonometric functions. Four of the fundamental identities are given in Definition 9.7.4. The other four fundamental identities are as follows:

$$\tanh x = \frac{1}{\coth x} \tag{5}$$

$$\cosh^2 x - \sinh^2 x = 1 \tag{6}$$

$$1 - \tanh^2 x = \operatorname{sech}^2 x \tag{7}$$

$$1 - \coth^2 x = -\operatorname{csch}^2 x \tag{8}$$

Equation (5) follows immediately from (1) and (2). Following is the proof of (6).

$$\begin{aligned}\cosh^2 x - \sinh^2 x &= \left(\frac{e^x + e^{-x}}{2}\right)^2 - \left(\frac{e^x - e^{-x}}{2}\right)^2 \\ &= \tfrac{1}{4}(e^{2x} + 2e^0 + e^{-2x} - e^{2x} + 2e^0 - e^{-2x}) \\ &= 1\end{aligned}$$

Equation (7) can be proved by using the definitions of tanh x and sech x in terms of e^x and e^{-x} as in the proof above, or an alternate proof is obtained by using other identities as follows:

$$1 - \tanh^2 x = 1 - \frac{\sinh^2 x}{\cosh^2 x} = \frac{\cosh^2 x - \sinh^2 x}{\cosh^2 x} = \frac{1}{\cosh^2 x} = \operatorname{sech}^2 x$$

The proof of (8) is left as an exercise (see Exercise 1). From the eight fundamental identities it is possible to prove others. Some of these are given in the exercises that follow. Other identities, such as the hyperbolic functions of the sum or difference of two numbers and the hyperbolic functions of twice a number or one-half a number, are similar to the corresponding trigonometric identities. Sometimes it is helpful to make use of the following two relations that follow from Definitions 9.7.1 and 9.7.2.

$$\cosh x + \sinh x = e^{x} \tag{9}$$

$$\cosh x - \sinh x = e^{-x} \tag{10}$$

We use them in proving the following identity:

$$\sinh(x + y) = \sinh x \cosh y + \cosh x \sinh y \tag{11}$$

From Definition 9.7.1,

$$\sinh(x + y) = \frac{e^{x+y} - e^{-(x+y)}}{2} = \frac{e^{x}e^{y} - e^{-x}e^{-y}}{2}$$

Applying (9) and (10) to the right side of the above equation we obtain

$$\sinh(x + y) = \tfrac{1}{2}[(\cosh x + \sinh x)(\cosh y + \sinh y) - (\cosh x - \sinh x)(\cosh y - \sinh y)]$$

Expanding the right side of the above equation and combining terms we have formula (11).

In a similar manner we can prove

$$\cosh(x + y) = \cosh x \cosh y + \sinh x \sinh y \tag{12}$$

If in (11) and (12) y is replaced by x, the following two formulas are obtained:

$$\sinh 2x = 2 \sinh x \cosh x \tag{13}$$

$$\cosh 2x = \cosh^2 x + \sinh^2 x \tag{14}$$

Formula (14) combined with the identity (6) gives two alternate formulas for $\cosh 2x$, which are

$$\cosh 2x = 2 \sinh^2 x + 1 \tag{15}$$

$$\cosh 2x = 2 \cosh^2 x - 1 \tag{16}$$

Solving (15) and (16) for $\sinh x$ and $\cosh x$, respectively, and replacing x by $\frac{1}{2}x$, we get

$$\sinh \frac{x}{2} = \pm \sqrt{\frac{\cosh x - 1}{2}} \tag{17}$$

$$\cosh \frac{x}{2} = \sqrt{\frac{\cosh x + 1}{2}} \tag{18}$$

We do not have a $\pm$ sign on the right side of (18) since the range of the hyperbolic cosine function is $[1, +\infty)$. The details of the proofs of (12) through (18) are left as exercises (see Exercises 2, 4, and 5).

To find the derivative of the hyperbolic tangent function we use some of the identities.

$$D_x(\tanh x) = D_x\left(\frac{\sinh x}{\cosh x}\right) = \frac{\cosh^2 x - \sinh^2 x}{\cosh^2 x} = \frac{1}{\cosh^2 x} = \operatorname{sech}^2 x$$

The formulas for the derivatives of the remaining hyperbolic functions are as follows: $D_x(\coth x) = -\operatorname{csch}^2 x$; $D_x(\operatorname{sech} x) = -\operatorname{sech} x \tanh x$; $D_x(\operatorname{csch} x) = -\operatorname{csch} x \coth x$. The proofs of these formulas are left as exercises (see Exercises 15 and 16).

From these formulas and the chain rule we have the following theorem.

9.7.5 Theorem If u is a differentiable function of x,

$$D_x(\tanh u) = \operatorname{sech}^2 u\, D_x u$$

$$D_x(\coth u) = -\operatorname{csch}^2 u\, D_x u$$

$$D_x(\operatorname{sech} u) = -\operatorname{sech} u \tanh u\, D_x u$$

$$D_x(\operatorname{csch} u) = -\operatorname{csch} u \coth u\, D_x u$$

You should notice that the formulas for the derivatives of the hyperbolic sine, cosine, and tangent all have a plus sign, whereas those for the derivatives of the hyperbolic cotangent, secant, and cosecant all have a minus sign. Otherwise the formulas are similar to the corresponding ones for the derivatives of the trigonometric functions.

Example 1 Find $D_x y$ if $y = \tanh(1 - x^2)$.

Solution $D_x y = -2x \operatorname{sech}^2(1 - x^2)$.

Example 2 Find dy/dx if $y = \ln \sinh x$.

Solution $$\frac{dy}{dx} = \frac{1}{\sinh x} \cdot \cosh x = \frac{\cosh x}{\sinh x} = \coth x.$$

The differentiation formulas in Theorems 9.7.3 and 9.7.5 give rise to the indefinite integration formulas in the next theorem.

9.7.6 Theorem

$$\int \sinh u\, du = \cosh u + C$$

$$\int \cosh u\, du = \sinh u + C$$

$$\int \operatorname{sech}^2 u\, du = \tanh u + C$$

$$\int \operatorname{csch}^2 u\, du = -\coth u + C$$

$$\int \operatorname{sech} u \tanh u\, du = -\operatorname{sech} u + C$$

$$\int \operatorname{csch} u \coth u\, du = -\operatorname{csch} u + C$$

The methods used to integrate powers of the hyperbolic functions are similar to those used for powers of the trigonometric functions. The following two examples illustrate the method.

Example 3 Evaluate $\int \sinh^3 x \cosh^2 x\, dx$.

Solution

$$\begin{aligned}\int \sinh^3 x \cosh^2 x\, dx &= \int \sinh^2 x \cosh^2 x(\sinh x\, dx)\\ &= \int (\cosh^2 x - 1)\cosh^2 x(\sinh x\, dx)\\ &= \int \cosh^4 x(\sinh x\, dx) - \int \cosh^2 x(\sinh x\, dx)\\ &= \tfrac{1}{5}\cosh^5 x - \tfrac{1}{3}\cosh^3 x + C\end{aligned}$$

Example 4 Evaluate $\int \operatorname{sech}^4 x\, dx$.

Solution

$$\begin{aligned}\int \operatorname{sech}^4 x\, dx &= \int \operatorname{sech}^2 x(\operatorname{sech}^2 x\, dx)\\ &= \int (1 - \tanh^2 x)(\operatorname{sech}^2 x\, dx)\\ &= \int \operatorname{sech}^2 x\, dx - \int \tanh^2 x(\operatorname{sech}^2 x\, dx)\\ &= \tanh x - \tfrac{1}{3}\tanh^3 x + C\end{aligned}$$

A **catenary** is the curve formed by a homogeneous flexible cable hanging from two points under its own weight. If the lowest point of the catenary is the point $(0, a)$, it can be shown that an equation of it is

$$y = a \cosh\left(\frac{x}{a}\right) \qquad a > 0 \tag{19}$$

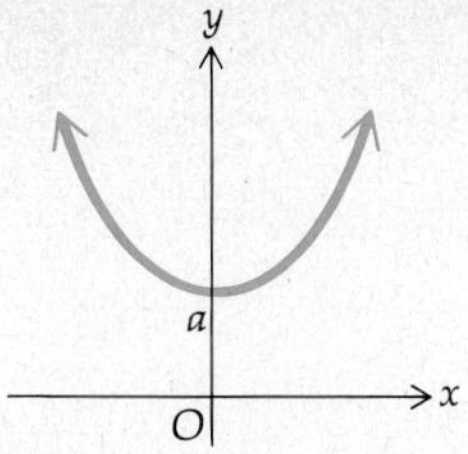

FIGURE 9.7.4

A sketch of the graph of Eq. (19) is shown in Fig. 9.7.4.

Example 5 Find the length of arc of the catenary having Eq. (19) from the point $(0, a)$ to the point (x_1, y_1) where $x_1 > 0$.

Solution If

$$f(x) = a \cosh\left(\frac{x}{a}\right)$$

then

$$f'(x) = a \sinh\left(\frac{x}{a}\right) \cdot \left(\frac{1}{a}\right) = \sinh\left(\frac{x}{a}\right)$$

and using Theorem 7.6.3, if L units is the length of the given arc, we have

$$\begin{aligned} L &= \int_0^{x_1} \sqrt{1 + [f'(x)]^2}\, dx \\ &= \int_0^{x_1} \sqrt{1 + \sinh^2\left(\frac{x}{a}\right)}\, dx \qquad (20) \end{aligned}$$

From formula (6), $1 + \sinh^2 x = \cosh^2 x$, and because $\cosh(x/a) \geq 1$, it follows that

$$\sqrt{1 + \sinh^2\left(\frac{x}{a}\right)} = \sqrt{\cosh^2\left(\frac{x}{a}\right)} = \cosh\left(\frac{x}{a}\right)$$

and substituting from the above into (20) we get

$$\begin{aligned} L &= \int_0^{x_1} \cosh\left(\frac{x}{a}\right) dx \\ &= a \sinh\left(\frac{x}{a}\right)\Big]_0^{x_1} \\ &= a \sinh\left(\frac{x_1}{a}\right) - a \sinh 0 \\ &= a \sinh\left(\frac{x_1}{a}\right) \end{aligned}$$

EXERCISES 9.7

In Exercises 1 through 10, prove the identities.

1. $1 - \coth^2 x = -\operatorname{csch}^2 x$ **2.** $\cosh(x + y) = \cosh x \cosh y + \sinh x \sinh y$ **3.** $\tanh(x + y) = \dfrac{\tanh x + \tanh y}{1 + \tanh x \tanh y}$

4. (a) $\sinh 2x = 2 \sinh x \cosh x$; (b) $\cosh 2x = \cosh^2 x + \sinh^2 x = 2 \sinh^2 x + 1 = 2 \cosh^2 x - 1$

5. (a) $\cosh \frac{1}{2}x = \sqrt{\dfrac{\cosh x + 1}{2}}$; (b) $\sinh \frac{1}{2}x = \pm \sqrt{\dfrac{\cosh x - 1}{2}}$ **6.** $\operatorname{csch} 2x = \frac{1}{2} \operatorname{sech} x \operatorname{csch} x$

7. (a) $\tanh(-x) = -\tanh x$; (b) $\operatorname{sech}(-x) = \operatorname{sech} x$

8. $\sinh 3x = 3 \sinh x + 4 \sinh^3 x$

9. $\cosh 3x = 4 \cosh^3 x - 3 \cosh x$

10. $\sinh^2 x - \sinh^2 y = \sinh(x + y) \sinh(x - y)$

11. Prove: $(\sinh x + \cosh x)^n = \cosh nx + \sinh nx$, if n is any positive integer. (*Hint*: Use formula (9).)

12. Prove that the hyperbolic tangent function is an odd function and the hyperbolic secant function is an even function.

13. Prove: $\tanh(\ln x) = \dfrac{x^2 - 1}{x^2 + 1}$

14. Prove: $\dfrac{1 + \tanh x}{1 - \tanh x} = e^{2x}$

15. Prove: (a) $D_x(\sinh x) = \cosh x$; (b) $D_x(\coth x) = -\operatorname{csch}^2 x$.

16. Prove: (a) $D_x(\operatorname{sech} x) = -\operatorname{sech} x \tanh x$; (b) $D_x(\operatorname{csch} x) = -\operatorname{csch} x \coth x$.

In Exercises 17 through 30, find the derivative of the given function.

17. $f(x) = \tanh \dfrac{4x + 1}{5}$

18. $g(t) = \cosh t^3$

19. $f(y) = \sinh e^{2y}$

20. $f(x) = \coth(\ln x)$

21. $f(w) = \operatorname{sech}^2 4w$

22. $g(x) = \ln(\sinh x^3)$

23. $f(x) = e^x \cosh x$

24. $h(x) = \coth \dfrac{1}{x}$

25. $h(t) = \ln(\tanh t)$

26. $F(x) = \tan^{-1}(\sinh x^2)$

27. $G(x) = \sin^{-1}(\tanh x^2)$

28. $f(x) = \ln(\coth 3x - \operatorname{csch} 3x)$

29. $f(x) = x^{\sinh x}$; $x > 0$

30. $g(x) = (\cosh x)^x$

31. Prove: (a) $\int \tanh u \, du = \ln|\cosh u| + C$; (b) $\int \coth u \, du = \ln|\sinh u| + C$.

32. Prove: $\int \operatorname{sech} u \, du = 2 \tan^{-1} e^u + C$.

33. Prove: $\int \operatorname{csch} u \, du = \ln|\tanh \frac{1}{2}u| + C$.

In Exercises 34 through 43, evaluate the indefinite integral.

34. $\int e^x \operatorname{csch}^2 e^x \, dx$

35. $\int \dfrac{\sinh \sqrt{x}}{\sqrt{x}} \, dx$

36. $\int \tanh^2 3x \, dx$

37. $\int \cosh^4 7t \, dt$

38. $\int \sinh^4 x \cosh^3 x \, dx$

39. $\int \tanh x \ln(\cosh x) \, dx$

40. $\int x \sinh^2 x^2 \, dx$

41. $\int \dfrac{e^{\cosh t}}{\operatorname{csch} t} \, dt$

42. $\int \operatorname{sech} x \tanh^3 x \, dx$

43. $\int \operatorname{sech}^4 3x \, dx$

In Exercises 44 through 47, evaluate the definite integral.

44. $\int_0^{\ln 2} \tanh z \, dz$

45. $\int_0^{\ln 3} \operatorname{sech}^2 t \, dt$

46. $\int_{-1}^{1} \cosh^2 x \, dx$

47. $\int_0^2 \sinh^3 x \, dx$

48. Draw a sketch of the graph of the hyperbolic cotangent function. Find equations of the asymptotes.

49. Draw a sketch of the graph of the hyperbolic secant function. Find the relative extrema of the function, the points of inflection of the graph, the intervals on which the graph is concave upward, and the intervals on which the graph is concave downward.

50. Prove that a catenary is concave upward at each point.

51. Find the area of the region bounded by the catenary of Example 5, the y axis, the x axis, and the line $x = x_1$, where $x_1 > 0$.

52. Find the volume of the solid of revolution if the region of Exercise 51 is revolved about the x axis.

53. A particle is moving along a straight line according to the equation of motion $s = e^{-ct/2}(A \sinh t + B \cosh t)$ where s cm is the directed distance of the particle from the origin at t sec. If v cm/sec and a cm/sec^2 are the velocity and acceleration, respectively, of the particle at t sec, find v and a. Also show that a is the sum of two numbers, one of which is proportional to s and the other proportional to v.

54. Prove that the hyperbolic sine function is continuous and increasing on its entire domain.

55. Prove that the hyperbolic tangent function is continuous and increasing on its entire domain.

56. Prove that the hyperbolic cosine function is continuous on its entire domain but is not monotonic on its entire domain. Find the intervals on which the function is increasing and the intervals on which the function is decreasing.

9.8 THE INVERSE HYPERBOLIC FUNCTIONS

From the graph of the hyperbolic sine function (Fig. 9.7.1) we see that a horizontal line intersects the graph in one and only one point; therefore the hyperbolic sine function is one-to-one. Furthermore, the hyperbolic sine function is continuous and increasing on its domain (see Exercise 54 in Exercises 9.7). Therefore, by Theorem 8.1.5, the function has an inverse.

9.8.1 Definition The **inverse hyperbolic sine function,** denoted by $\sinh^{-1}$, is defined as follows:

$$y = \sinh^{-1} x \quad \text{if and only if} \quad x = \sinh y$$

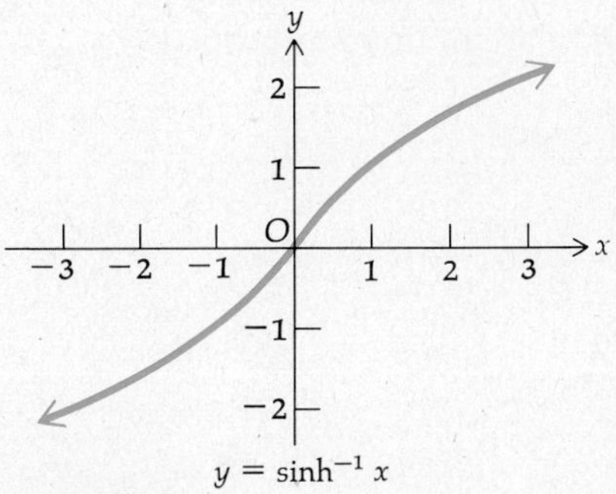

FIGURE 9.8.1

A sketch of the graph of the inverse hyperbolic sine function is shown in Fig. 9.8.1. Its domain is the set R of all real numbers, and its range is also the set R.

It follows from Definition 9.8.1 that

$$\sinh(\sinh^{-1} x) = x \tag{1}$$

and

$$\sinh^{-1}(\sinh y) = y \tag{2}$$

From Fig. 9.7.2 we notice that a horizontal line, $y = k$, where $k > 1$, intersects the graph of the hyperbolic cosine function in two points. Therefore, for each number greater than 1 in the range of this function there correspond two numbers in the domain. So the hyperbolic cosine function is not one-to-one and does not have an inverse. However, we define a function F as follows:

$$F(x) = \cosh x \qquad \text{for } x \geq 0 \tag{3}$$

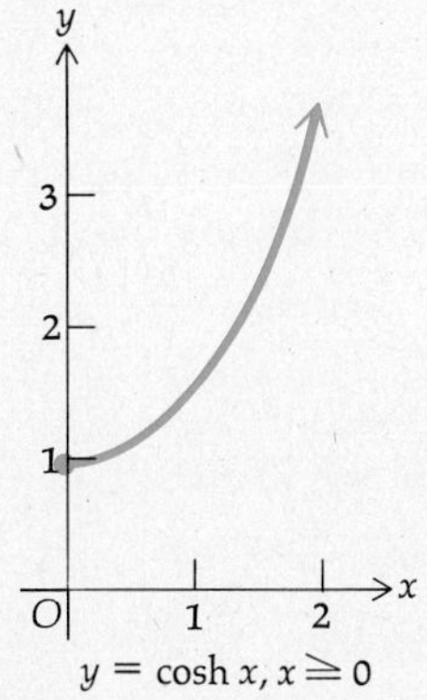

FIGURE 9.8.2

The domain of F is the interval $[0, +\infty)$, and the range is the interval $[1, +\infty)$. A sketch of the graph of F is shown in Fig. 9.8.2. The function F is continuous and increasing on its domain (see Exercise 1), and so by Theorem 8.1.5 F has an inverse, which is called the *inverse hyperbolic cosine function.*

9.8.2 Definition The **inverse hyperbolic cosine function,** denoted by $\cosh^{-1}$, is defined as follows:

$$y = \cosh^{-1} x \quad \text{if and only if} \quad x = \cosh y \text{ and } y \geq 0$$

A sketch of the graph of the inverse hyperbolic cosine function is

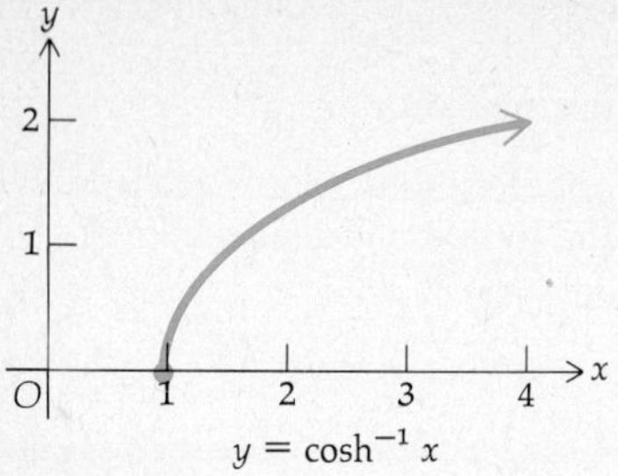

FIGURE 9.8.3

shown in Fig. 9.8.3. The domain of this function is the interval $[1, +\infty)$, and the range is the interval $[0, +\infty)$. From Definition 9.8.2,

$$\cosh(\cosh^{-1} x) = x \qquad \text{if } x \geq 1 \tag{4}$$

and

$$\cosh^{-1}(\cosh y) = y \qquad \text{if } y \geq 0 \tag{5}$$

As with the hyperbolic sine function, a horizontal line intersects each of the graphs of the hyperbolic tangent function, the hyperbolic cotangent function, and the hyperbolic cosecant function at one and only one point. For the hyperbolic tangent function this may be seen in Fig. 9.7.3. Each of these three functions then is one-to-one; furthermore, each is continuous and monotonic on its domain. Thus each has an inverse function.

9.8.3 Definition The **inverse hyperbolic tangent function,** the **inverse hyperbolic cotangent function,** and the **inverse hyperbolic cosecant function,** denoted respectively by $\tanh^{-1}$, $\coth^{-1}$, and csch^{-1}, are defined as follows:

$$y = \tanh^{-1} x \quad \text{if and only if} \quad x = \tanh y$$

$$y = \coth^{-1} x \quad \text{if and only if} \quad x = \coth y$$

$$y = \operatorname{csch}^{-1} x \quad \text{if and only if} \quad x = \operatorname{csch} y$$

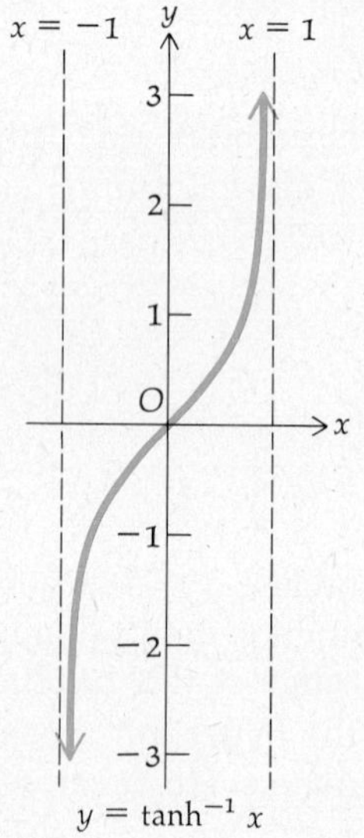

FIGURE 9.8.4

A sketch of the graph of the inverse hyperbolic tangent function is in Fig. 9.8.4.

Because $\operatorname{sech} x = 1/\cosh x$ and the hyperbolic cosine function is not one-to-one, it follows that the hyperbolic secant function is not one-to-one. Therefore it does not have an inverse. However, we proceed as we did with the hyperbolic cosine function and define a new function that has an inverse function. Let the function G be defined by

$$G(x) = \operatorname{sech} x \qquad \text{for } x \geq 0 \tag{6}$$

It can be shown that G is continuous and monotonic on its domain (see Exercise 2); therefore G has an inverse, which is called the *inverse hyperbolic secant function.*

9.8.4 Definition The **inverse hyperbolic secant function,** denoted by sech^{-1}, is defined as follows:

$$y = \operatorname{sech}^{-1} x \quad \text{if and only if} \quad x = \operatorname{sech} y \text{ and } y \geq 0 \tag{7}$$

The inverse hyperbolic functions can be expressed in terms of natural logarithms. This should not be surprising because the hyperbolic functions were defined in terms of the exponential function, and the natural logarithmic function is the inverse of the exponential function.

Following are these expressions for $\sinh^{-1}$, $\cosh^{-1}$, $\tanh^{-1}$, and $\coth^{-1}$.

$$\sinh^{-1} x = \ln(x + \sqrt{x^2 + 1}) \qquad x \text{ any real number} \tag{8}$$

$$\cosh^{-1} x = \ln(x + \sqrt{x^2 - 1}) \qquad x \geq 1 \tag{9}$$

$$\tanh^{-1} x = \frac{1}{2} \ln \frac{1 + x}{1 - x} \qquad |x| < 1 \tag{10}$$

$$\coth^{-1} x = \frac{1}{2} \ln \frac{x + 1}{x - 1} \qquad |x| > 1 \tag{11}$$

We prove (9) and leave the proofs of the other three formulas as exercises (see Exercises 3 through 5).

To prove (9), let $y = \cosh^{-1} x$, $x \geq 1$. Then, by Definition 9.8.2, $x = \cosh y$, $y \geq 0$. Applying Definition 9.7.2 to $\cosh y$ we get

$$x = \frac{e^y + e^{-y}}{2} \qquad y \geq 0$$

from which we obtain

$$e^{2y} - 2xe^y + 1 = 0 \qquad y \geq 0$$

Solving this equation for e^y by the quadratic formula we obtain

$$e^y = \frac{2x \pm \sqrt{4x^2 - 4}}{2} = x \pm \sqrt{x^2 - 1} \qquad y \geq 0 \tag{12}$$

We know that $y \geq 0$ and $x \geq 1$. Therefore $e^y \geq 1$. When $x = 1$, $x + \sqrt{x^2 - 1} = x - \sqrt{x^2 - 1} = 1$. Furthermore, when $x > 1$, $0 < x - 1 < x + 1$; so $\sqrt{x - 1} < \sqrt{x + 1}$. Therefore

$$\sqrt{x - 1}\,\sqrt{x - 1} < \sqrt{x - 1}\,\sqrt{x + 1}$$

giving $x - 1 < \sqrt{x^2 - 1}$. Hence when $x > 1$, $x - \sqrt{x^2 - 1} < 1$. Consequently, we can safely reject the minus sign in (12). And since $y = \cosh^{-1} x$, we have

$$\cosh^{-1} x = \ln(x + \sqrt{x^2 - 1}) \qquad x \geq 1$$

which is (9).

Example 1 Express each of the following in terms of a natural logarithm: (a) $\tanh^{-1}(-\frac{4}{5})$; (b) $\sinh^{-1} 2$.

Solution

(a) From (10),

$$\tanh^{-1}(-\tfrac{4}{5}) = \tfrac{1}{2} \ln \frac{\frac{1}{5}}{\frac{9}{5}} = \tfrac{1}{2} \ln \tfrac{1}{9} = \ln \tfrac{1}{3} = -\ln 3$$

(b) From (8),

$$\sinh^{-1} 2 = \ln(2 + \sqrt{5})$$

To obtain a formula for the derivative of the inverse hyperbolic sine function, let $y = \sinh^{-1} x$, and then $x = \sinh y$. Thus, because $D_y x = \cosh y$ and $D_x y = 1/D_y x$,

$$D_x y = \frac{1}{\cosh y} \tag{13}$$

From the identity $\cosh^2 y - \sinh^2 y = 1$, it follows that $\cosh y = \sqrt{\sinh^2 y + 1}$. (*Note:* When taking the square root, the minus sign is rejected because $\cosh y \geq 1$.)

Hence, because $x = \sinh y$, $\cosh y = \sqrt{x^2 + 1}$. Substituting this in (13) we obtain

$$D_x(\sinh^{-1} x) = \frac{1}{\sqrt{x^2 + 1}} \tag{14}$$

Formula (14) can also be derived by using (8), as follows:

$$\begin{aligned} D_x(\sinh^{-1} x) &= D_x \ln(x + \sqrt{x^2 + 1}) \\ &= \frac{1 + \dfrac{x}{\sqrt{x^2 + 1}}}{x + \sqrt{x^2 + 1}} \\ &= \frac{\sqrt{x^2 + 1} + x}{\sqrt{x^2 + 1}\,(x + \sqrt{x^2 + 1})} \\ &= \frac{1}{\sqrt{x^2 + 1}} \end{aligned}$$

The formulas for the derivatives of the other five inverse hyperbolic functions are obtained in a way analogous to that for the derivative of the inverse hyperbolic sine function. Their derivations are left as exercises (see Exercises 6 through 10). From these formulas and the chain rule the next theorem follows.

9.8.5 Theorem If u is a differentiable function of x,

$$D_x(\sinh^{-1} u) = \frac{1}{\sqrt{u^2 + 1}} D_x u \tag{15}$$

$$D_x(\cosh^{-1} u) = \frac{1}{\sqrt{u^2 - 1}} D_x u \qquad u > 1 \tag{16}$$

$$D_x(\tanh^{-1} u) = \frac{1}{1 - u^2} D_x u \qquad |u| < 1 \tag{17}$$

$$D_x(\coth^{-1} u) = \frac{1}{1 - u^2} D_x u \qquad |u| > 1 \tag{18}$$

$$D_x(\operatorname{sech}^{-1} u) = -\frac{1}{u\sqrt{1-u^2}} D_x u \qquad 0 < u < 1 \tag{19}$$

$$D_x(\operatorname{csch}^{-1} u) = -\frac{1}{|u|\sqrt{1+u^2}} D_x u \qquad u \neq 0 \tag{20}$$

Example 2 Find $f'(x)$ if $f(x) = \tanh^{-1}(\cos 2x)$.

Solution From formula (17),

$$f'(x) = \frac{-2\sin 2x}{1-\cos^2 2x} = \frac{-2\sin 2x}{\sin^2 2x} = -2\csc 2x$$

Example 3 Verify that $D_x(x\cosh^{-1} x - \sqrt{x^2-1}) = \cosh^{-1} x$.

Solution

$$D_x(x\cosh^{-1} x - \sqrt{x^2-1}) = \cosh^{-1} x + x \cdot \frac{1}{\sqrt{x^2-1}} - \frac{x}{\sqrt{x^2-1}}$$
$$= \cosh^{-1} x$$

The main use of the inverse hyperbolic functions is in connection with integration. The discussion of this topic is postponed until Sec. 10.6.

EXERCISES 9.8

1. If $F(x) = \cosh x$ and $x \geq 0$, prove that F is continuous and increasing on its domain.
2. If $G(x) = \operatorname{sech} x$ and $x \geq 0$, prove that G is continuous and monotonic on its domain.

In Exercises 3 through 10, prove the indicated formula of this section.

3. Formula (8)
4. Formula (10)
5. Formula (11)
6. Formula (16)
7. Formula (17)
8. Formula (18)
9. Formula (19)
10. Formula (20)

In Exercises 11 through 14, express the given quantity in terms of a natural logarithm.

11. $\sinh^{-1} \frac{1}{4}$
12. $\cosh^{-1} 3$
13. $\tanh^{-1} \frac{1}{2}$
14. $\coth^{-1}(-2)$

In Exercises 15 through 32, find the derivative of the given function.

15. $f(x) = \sinh^{-1} x^2$
16. $G(x) = \cosh^{-1} \frac{1}{3}x$
17. $F(x) = \tanh^{-1} 4x$
18. $h(w) = \tanh^{-1} w^3$
19. $g(x) = \coth^{-1}(3x+1)$
20. $f(r) = \operatorname{csch}^{-1} \frac{1}{2}r^2$
21. $f(x) = x^2 \cosh^{-1} x^2$
22. $h(x) = (\operatorname{sech}^{-1} x)^2$
23. $f(x) = \sinh^{-1}(\tan x)$
24. $g(x) = \tanh^{-1}(\sin 3x)$
25. $h(x) = \cosh^{-1}(\csc x)$
26. $F(x) = \coth^{-1}(\cosh x)$
27. $f(z) = (\coth^{-1} z^2)^3$
28. $f(t) = \sinh^{-1} e^{2t}$
29. $g(x) = \tanh^{-1}(\sec e^x)$
30. $h(x) = \cosh^{-1}(\ln x)$
31. $G(x) = x\sinh^{-1} x - \sqrt{1+x^2}$
32. $H(x) = \ln\sqrt{x^2+1} - x\tanh^{-1} x$

In Exercises 33 through 36, draw a sketch of the graph of the indicated function.

33. The inverse hyperbolic tangent function.
34. The inverse hyperbolic cotangent function.
35. The inverse hyperbolic secant function.
36. The inverse hyperbolic cosecant function.

REVIEW EXERCISES FOR CHAPTER 9

In Exercises 1 through 27, find the derivative of the given function.

1. $f(x) = \tan^2 3x$
2. $g(x) = \sqrt{\csc 3x}$
3. $h(x) = \ln |\cot x|$
4. $f(w) = w \tan \dfrac{1}{w}$
5. $g(x) = e^{\tan x} + \sec^2 x$
6. $G(x) = 4^{\csc 2x}$
7. $F(x) = \tan^{-1} 2^x$
8. $g(x) = \cot 3x \sqrt{\sin 2x}$
9. $f(x) = \cosh^3 2x$
10. $g(t) = \ln \operatorname{sech} t^2$
11. $f(x) = \log_{10} |\sec 5x + \tan 5x|$
12. $h(x) = e^x(\cosh x + \sinh x)$
13. $h(w) = w^2 \sinh^{-1} 2w$
14. $g(x) = x^2 \sin^{-1} 2x$
15. $f(x) = \dfrac{1 + \sec x}{1 + \tan x}$
16. $f(t) = \sec^3 t + 3^{\sec t}$
17. $g(x) = \dfrac{\cot^2 x}{1 + x^2}$
18. $f(x) = \dfrac{\cot 2x}{x^4}$
19. $G(w) = \tan^{-1}\left(\dfrac{w - 3}{3w + 1}\right)$
20. $F(x) = \cot^3 x \cdot \csc x$
21. $g(x) = \tan^{-1}(\tanh \sqrt{x})$
22. $h(x) = \tan(\cosh 3x)$
23. $f(x) = \tanh^{-1}(\sec e^x)$
24. $f(x) = \tan^{-1}(\tan^{-1} x)$
25. $f(t) = \dfrac{\ln t}{\csc t}$
26. $h(x) = (\sec x)^{e^x}$; $\sec x > 0$
27. $f(x) = (\tan x)^{1/x^2}$; $\tan x > 0$

28. The gudermannian is the function defined by $\operatorname{gd} x = \tan^{-1}(\sinh x)$. Show that $D_x(\operatorname{gd} x) = \operatorname{sech} x$.

29. Find to the nearest degree the measurements of the four interior angles of the quadrilateral having vertices at (5, 6), (−2, 4), (−2, 1), and (3, 1), and verify that the sum is 360°.

In Exercises 30 through 32, evaluate the given limit if it exists.

30. $\lim\limits_{x \to 0} \dfrac{\tan x}{2x}$
31. $\lim\limits_{x \to 0} \dfrac{\tan^4 2x}{4x^4}$
32. $\lim\limits_{x \to 0} \dfrac{\csc 3x}{\cot x}$

In Exercises 33 through 36, find $D_x y$ by implicit differentiation.

33. $\tan(x + y) = xy$
34. $\sec(x + y) - \sec(x - y) = 1$
35. $\cot^{-1} \dfrac{x^2}{3y} + xy^2 = 0$
36. $\tan^{-1}(x + 3y) = \ln y$

In Exercises 37 through 52, evaluate the indefinite integral.

37. $\int x \tan 3x^2 \, dx$
38. $\int \dfrac{\sec(1/x)}{x^2} \, dx$
39. $\int \dfrac{3x}{\sqrt{1 - x^4}} \, dx$
40. $\int (\csc 2t - \cot 2t) \, dt$
41. $\int e^x \cot^3 e^x \csc^4 e^x \, dx$
42. $\int \tan^3 x \sec^3 x \, dx$
43. $\int \dfrac{dx}{2x^2 + 3x + 5}$
44. $\int \dfrac{dx}{\sqrt{7 + 5x - 2x^2}}$
45. $\int \tanh^3 x \, dx$
46. $\int x \coth \tfrac{1}{2}x^2 \, dx$
47. $\int \dfrac{2 \sin w - 1}{\cos^2 w} \, dw$
48. $\int 2^x \sec^6 2^x \, dx$
49. $\int \dfrac{dx}{\sqrt{e^{2x} - 8}}$
50. $\int \dfrac{e^y}{\sqrt{e^{2y} + 9}} \, dy$
51. $\int \dfrac{\sec^4 \sqrt{t} \, dt}{\sqrt{t}}$
52. $\int \dfrac{\cosh^3 x}{\sqrt{\sinh x}} \, dx$

In Exercises 53 through 58, evaluate the definite integral.

53. $\int_0^{\pi/3} \sec^3 x \tan^5 x\, dx$ **54.** $\int_0^{\pi/6} \tan^4 2x\, dx$ **55.** $\int_1^2 \frac{(t+2)\, dt}{\sqrt{4t - t^2}}$

56. $\int_{-1}^1 \frac{2x+6}{x^2+2x+5}\, dx$ **57.** $\int_0^1 \operatorname{sech}^2 x\, dx$ **58.** $\int_0^1 \sqrt{1 + \cosh y}\, dy$

59. Evaluate: $\int_{-\pi/4}^{\pi/4} |\tan x + 1|\, dx.$ **60.** Evaluate: $\int_{-\pi/3}^{\pi/3} |\sec x - 2|\, dx.$

61. Find the length of the arc of the curve $y = \ln \sec x$ from the origin to the point $(\frac{1}{3}\pi, \ln 2)$.

62. Find the average value of the tangent function on the closed interval $[0, \frac{1}{4}\pi]$.

63. Find the area of the region in the first quadrant bounded by the y axis and the curves $y = \sec^2 x$ and $y = 2\tan^2 x$.

64. Find the area of the region bounded by the curve $y = 9/\sqrt{9 - x^2}$, the two coordinate axes, and the line $x = 2\sqrt{2}$.

65. Find an equation of the normal line to the curve $y = \sec x$ at the point $(\frac{2}{3}\pi, -2)$.

66. Find the length of the catenary $y = \cosh x$ from $(\ln 2, \frac{5}{4})$ to $(\ln 3, \frac{5}{3})$.

67. Show that $\cosh(\ln x) = \frac{x^2+1}{2x}$. **68.** Prove: (a) $\lim_{x \to +\infty} \coth x = 1$; (b) $\lim_{x \to +\infty} \operatorname{csch} x = 0$.

69. Find the volume of the solid of revolution generated when the region bounded by the curve $y = \sinh x$, the x axis, and the lines $x = 0$ and $x = \ln 2$ is revolved about the x axis.

70. Draw a sketch of the graph of the function defined by $f(x) = \frac{1}{2}\sin^{-1} 3x$.

71. A searchlight is $\frac{1}{2}$ km from a straight road and it keeps a light trained on an automobile that is traveling at a constant speed of 60 km/hr. Find the rate at which the light beam is changing direction (a) when the car is at the point on the road nearest the searchlight and (b) when the car is $\frac{1}{2}$ km down the road from this point.

72. A helicopter leaves the ground at a point 800 ft from an observer and rises vertically at 25 ft/sec. Find the time rate of change of the measure of the observer's angle of elevation of the helicopter when the helicopter is 600 ft above the ground.

73. An airplane is flying at a speed of 300 mi/hr at an altitude of 4 mi. If an observer is on the ground, find the time rate of change of the measure of the observer's angle of elevation of the airplane when the airplane is directly over a point on the ground 2 mi from the observer.

74. A picture 5 ft high is placed on a wall with its base 7 ft above the level of the eye of an observer. If the observer is approaching the wall at the rate of 3 ft/sec, how fast is the measure of the angle subtended at her eye by the picture changing when the observer is 10 ft from the wall?

75. A particle is moving along a straight line and $s = \sin(4t + \frac{1}{3}\pi) + \sin(4t + \frac{1}{6}\pi)$ where s cm is the directed distance of the particle from the origin at t sec. Show that the motion is simple harmonic, and find the amplitude of the motion. (See Exercise 43 in Exercises 9.6.)

76. If an equation of motion is $s = \cos 2t + 2\sin 2t$, prove that the motion is simple harmonic. (See Exercise 43 in Exercises 9.6.)

77. Two points A and B are diametrically opposite each other on the shores of a circular lake 1 km in diameter. A man desires to go from point A to point B. He can row at the rate of $1\frac{1}{2}$ km/hr and walk at the rate of 5 km/hr. Find the least amount of time it can take him to get from point A to point B.

78. Solve Exercise 77 if the rates of rowing and walking are, respectively, 2 km/hr and 4 km/hr.

79. The graph of the equation $x = a \sinh^{-1}\sqrt{\frac{a^2}{y^2} - 1} - \sqrt{a^2 - y^2}$ is called a tractrix. Prove that the slope of the curve at any point (x, y) is $-y/\sqrt{a^2 - y^2}$.

CHAPTER 10 Techniques of Integration

10.1 INTRODUCTION

In Chapter 6 the definite integral of a function f from a to b was defined as the limit of a Riemann sum as follows:

$$\int_a^b f(x)\,dx = \lim_{\|\Delta\| \to 0} \sum_{i=1}^{n} f(\xi_i)\,\Delta_i x \tag{1}$$

if the limit exists. We stated a theorem, proved in advanced calculus, that the limit on the right side of (1) exists if f is continuous on the closed interval $[a, b]$.

The exact value of a definite integral may be calculated by the fundamental theorem of the calculus, provided that an antiderivative of the integrand can be found. The process of finding the most general antiderivative of a given integrand is called **indefinite integration.** We use the term **indefinite integral** to mean the most general antiderivative of a given integrand. In practice, it is not always possible to find the indefinite integral. That is, we may have a definite integral that exists, but the integrand has no antiderivative that can be expressed in terms of elementary functions. An example of this is $\int_0^{1/2} e^{-t^2}\,dt$ (a method for computing the value of this definite integral to any required degree of accuracy by using infinite series is given in Sec. 15.9). However, many of the definite integrals that occur in practice can be evaluated by finding an antiderivative of the integrand. Some methods for doing this were given previously, and additional ones are presented in this chapter. Often you may find it desirable to resort to a table of integrals instead of performing a complicated integration. (A short table of integrals can be found on the front and back endpapers.) However, it may be necessary to employ some of the techniques of integration in order to express the integrand in a form that is found in a table. Therefore you should acquire proficiency in recognizing which technique to apply to a given integral. Furthermore, development of computational skills is important in all branches of mathematics, and the exercises in this chapter provide a good training ground. For these reasons you are advised to use a table of integrals only after you have mastered integration.

The standard indefinite integration formulas that you learned in previous chapters and that are used frequently are listed below.

$$\int du = u + C$$

$$\int a\,du = au + C \qquad \text{where } a \text{ is any constant}$$

$$\int [f(u) + g(u)]\,du = \int f(u)\,du + \int g(u)\,du$$

$$\int u^n\,du = \frac{u^{n+1}}{n+1} + C \qquad n \neq -1$$

$$\int \frac{du}{u} = \ln|u| + C$$

$$\int a^u \, du = \frac{a^u}{\ln a} + C \qquad \text{where } a > 0 \text{ and } a \neq 1$$

$$\int e^u \, du = e^u + C$$

$$\int \sin u \, du = -\cos u + C$$

$$\int \cos u \, du = \sin u + C$$

$$\int \sec^2 u \, du = \tan u + C$$

$$\int \csc^2 u \, du = -\cot u + C$$

$$\int \sec u \tan u \, du = \sec u + C$$

$$\int \csc u \cot u \, du = -\csc u + C$$

$$\int \tan u \, du = \ln |\sec u| + C$$

$$\int \cot u \, du = \ln |\sin u| + C$$

$$\int \sec u \, du = \ln |\sec u + \tan u| + C$$

$$\int \csc u \, du = \ln |\csc u - \cot u| + C$$

$$\int \frac{du}{\sqrt{a^2 - u^2}} = \sin^{-1} \frac{u}{a} + C \qquad \text{where } a > 0$$

$$\int \frac{du}{a^2 + u^2} = \frac{1}{a} \tan^{-1} \frac{u}{a} + C \qquad \text{where } a \neq 0$$

$$\int \frac{du}{u\sqrt{u^2 - a^2}} = \frac{1}{a} \sec^{-1} \left|\frac{u}{a}\right| + C \qquad \text{where } a > 0$$

$$\int \sinh u \, du = \cosh u + C$$

$$\int \cosh u \, du = \sinh u + C$$

$$\int \operatorname{sech}^2 u \, du = \tanh u + C$$

$$\int \operatorname{csch}^2 u\, du = -\coth u + C$$

$$\int \operatorname{sech} u \tanh u\, du = -\operatorname{sech} u + C$$

$$\int \operatorname{csch} u \coth u\, du = -\operatorname{csch} u + C$$

10.2 INTEGRATION BY PARTS

A method of integration that is quite useful is *integration by parts.* It depends on the formula for the differential of a product.

$$d(uv) = u\, dv + v\, du$$

or, equivalently,

$$u\, dv = d(uv) - v\, du \tag{1}$$

Integrating on both sides of (1) we have

$$\int u\, dv = uv - \int v\, du \tag{2}$$

Formula (2) is called the **formula for integration by parts.** This formula expresses the integral $\int u\, dv$ in terms of another integral, $\int v\, du$. By a suitable choice of u and dv, it may be easier to integrate the second integral than the first. When choosing the substitutions for u and dv, we usually want dv to be the most complicated factor of the integrand that can be integrated directly and u to be a function whose derivative is a simpler function. The method is shown by the following illustrations and examples.

● ILLUSTRATION 1 We wish to evaluate

$$\int x^3 e^{x^2}\, dx$$

To determine the substitutions for u and dv, bear in mind that to find v we must be able to integrate dv. This suggests letting $dv = xe^{x^2}\, dx$, and then $u = x^2$. We integrate to find

$$v = \tfrac{1}{2}e^{x^2} + C_1$$

Also,

$$du = 2x\, dx$$

From formula (2),

$$\int x^3 e^{x^2}\, dx = x^2(\tfrac{1}{2}e^{x^2} + C_1) - \int (\tfrac{1}{2}e^{x^2} + C_1)2x\, dx$$

$$= \tfrac{1}{2}x^2e^{x^2} + C_1x^2 - \int xe^{x^2}\,dx - 2C_1\int x\,dx$$

$$= \tfrac{1}{2}x^2e^{x^2} + C_1x^2 - \tfrac{1}{2}e^{x^2} - C_1x^2 + C_2$$

$$= \tfrac{1}{2}e^{x^2}(x^2 - 1) + C_2$$

In Illustration 1 observe that the first constant of integration C_1 does not appear in the final result. This is true in general, and we prove it as follows: By writing $v + C_1$ in formula 2, we have

$$\int u\,dv = u(v + C_1) - \int (v + C_1)\,du$$

$$= uv + C_1u - \int v\,du - C_1\int du$$

$$= uv + C_1u - \int v\,du - C_1u$$

$$= uv - \int v\,du$$

Therefore it is not necessary to write C_1 when finding v from dv.

Example 1 Find $\int x \ln x\,dx$.

Solution Let $u = \ln x$ and $dv = x\,dx$. Then

$$du = \frac{dx}{x} \qquad v = \frac{x^2}{2}$$

So

$$\int x \ln x\,dx = \frac{x^2}{2}\ln x - \int \frac{x^2}{2}\cdot\frac{dx}{x}$$

$$= \frac{x^2}{2}\ln x - \frac{1}{2}\int x\,dx$$

$$= \tfrac{1}{2}x^2 \ln x - \tfrac{1}{4}x^2 + C$$

Example 2 Find $\int x \sin x\,dx$.

Solution Let $u = x$ and $dv = \sin x\,dx$. Then

$$du = dx \quad \text{and} \quad v = -\cos x$$

Therefore

$$\int x \sin x\,dx = -x\cos x + \int \cos x\,dx$$

$$= -x\cos x + \sin x + C$$

● ILLUSTRATION 2 In Example 2, if instead of our choices of u and dv as above, we let

$$u = \sin x \quad \text{and} \quad dv = x\,dx$$

we get

$$du = \cos x\,dx \quad \text{and} \quad v = \tfrac{1}{2}x^2$$

So

$$\int x \sin x\,dx = \frac{x^2}{2}\sin x - \frac{1}{2}\int x^2 \cos x\,dx$$

The integral on the right is more complicated than the integral with which we started, thereby indicating that these are not desirable choices for u and dv. ●

It may happen that a particular integral requires repeated applications of integration by parts. This is illustrated in the following example.

Example 3 Find $\int x^2 e^x\,dx$.

Solution Let $u = x^2$ and $dv = e^x\,dx$. Then

$$du = 2x\,dx \quad \text{and} \quad v = e^x$$

We have, then,

$$\int x^2 e^x\,dx = x^2 e^x - 2\int x e^x\,dx$$

We now apply integration by parts to the integral on the right. Let

$$\bar{u} = x \quad \text{and} \quad d\bar{v} = e^x\,dx$$

Then

$$d\bar{u} = dx \quad \text{and} \quad \bar{v} = e^x$$

So we obtain

$$\begin{aligned}\int x e^x\,dx &= x e^x - \int e^x\,dx \\ &= x e^x - e^x + \bar{C}\end{aligned}$$

Therefore

$$\begin{aligned}\int x^2 e^x\,dx &= x^2 e^x - 2(x e^x - e^x + \bar{C}) \\ &= x^2 e^x - 2x e^x + 2e^x + C\end{aligned}$$

Integration by parts often is used when the integrand involves logarithms, inverse trigonometric functions, and products of functions.

Example 4 Find $\int \tan^{-1} x\,dx$.

Solution Let $u = \tan^{-1} x$ and $dv = dx$. Then

$$du = \frac{dx}{1 + x^2} \qquad v = x$$

Thus

$$\int \tan^{-1} dx = x \tan^{-1} x - \int \frac{x\,dx}{1 + x^2}$$
$$= x \tan^{-1} x - \tfrac{1}{2} \ln (1 + x^2) + C$$

A situation that sometimes occurs when using integration by parts is shown in Example 5.

Example 5 Find $\int e^x \sin x\,dx$.

Solution Let $u = e^x$ and $dv = \sin x\,dx$. Then

$$du = e^x\,dx \qquad v = -\cos x$$

Therefore

$$\int e^x \sin x\,dx = -e^x \cos x + \int e^x \cos x\,dx$$

The integral on the right is similar to the first integral except it has $\cos x$ in place of $\sin x$. We apply integration by parts again by letting

$$\bar{u} = e^x \quad \text{and} \quad d\bar{v} = \cos x\,dx$$

So

$$d\bar{u} = e^x\,dx \quad \text{and} \quad \bar{v} = \sin x$$

Thus

$$\int e^x \sin x\,dx = -e^x \cos x + \left(e^x \sin x - \int e^x \sin x\,dx\right)$$

Now on the right there is the same integral as on the left. So we add $\int e^x \sin x\,dx$ to both sides of the equation and obtain

$$2 \int e^x \sin x\,dx = -e^x \cos x + e^x \sin x + \bar{C}$$

Dividing on both sides by 2 we have

$$\int e^x \sin x\,dx = \tfrac{1}{2} e^x (\sin x - \cos x) + C$$

In applying integration by parts to a specific integral, one pair of choices for u and dv may work while another pair may not. We saw this in Illustration 2, and another case occurs in Illustration 3.

• ILLUSTRATION 3 In Example 5, in the step where we have

$$\int e^x \sin x\, dx = -e^x \cos x + \int e^x \cos x\, dx$$

if we evaluate the integral on the right by letting

$$\bar{u} = \cos x \quad \text{and} \quad d\bar{v} = e^x\, dx$$

$$d\bar{u} = -\sin x\, dx \quad \text{and} \quad \bar{v} = e^x$$

we get

$$\int e^x \sin x\, dx = -e^x \cos x + \left(e^x \cos x + \int e^x \sin x\, dx\right)$$

$$\int e^x \sin x\, dx = \int e^x \sin x\, dx$$

•

In Sec. 9.2 we noted that in order to integrate odd powers of the secant and cosecant, integration by parts is used. This process is illustrated in Example 6.

Example 6 Find $\int \sec^3 x\, dx$.

Solution Let $u = \sec x$ and $dv = \sec^2 x\, dx$. Then

$$du = \sec x \tan x\, dx \quad \text{and} \quad v = \tan x$$

Therefore

$$\int \sec^3 x\, dx = \sec x \tan x - \int \sec x \tan^2 x\, dx$$

$$\int \sec^3 x\, dx = \sec x \tan x - \int \sec x(\sec^2 x - 1)\, dx$$

$$\int \sec^3 x\, dx = \sec x \tan x - \int \sec^3 x\, dx + \int \sec x\, dx$$

Adding $\int \sec^3 x\, dx$ to both sides we get

$$2\int \sec^3 x\, dx = \sec x \tan x + \ln|\sec x + \tan x| + 2C$$

So

$$\int \sec^3 x\, dx = \tfrac{1}{2}\sec x \tan x + \tfrac{1}{2}\ln|\sec x + \tan x| + C$$

EXERCISES 10.2

In Exercises 1 through 24, evaluate the indefinite integral.

1. $\int xe^{3x}\,dx$
2. $\int x\cos 2x\,dx$
3. $\int x\sec x\tan x\,dx$
4. $\int x\,3^x\,dx$
5. $\int \ln x\,dx$
6. $\int \sin^{-1} w\,dw$
7. $\int (\ln x)^2\,dx$
8. $\int x\sec^2 x\,dx$
9. $\int x\tan^{-1} x\,dx$
10. $\int x^2\ln x\,dx$
11. $\int \dfrac{xe^x}{(x+1)^2}\,dx$
12. $\int x^2\sin 3x\,dx$
13. $\int \sin x\ln(\cos x)\,dx$
14. $\int \sin(\ln x)\,dx$
15. $\int e^x\cos x\,dx$
16. $\int x^3 e^{x^2}\,dx$
17. $\int \dfrac{x^3\,dx}{\sqrt{1-x^2}}$
18. $\int \dfrac{\sin 2x}{e^x}\,dx$
19. $\int x^2\sinh x\,dx$
20. $\int \dfrac{e^{2x}}{\sqrt{1-e^x}}\,dx$
21. $\int \sec^5 x\,dx$
22. $\int \csc^3 t\,dt$
23. $\int \dfrac{\cot^{-1}\sqrt{z}}{\sqrt{z}}\,dz$
24. $\int \tanh^{-1} x\,dx$

In Exercises 25 through 34, evaluate the definite integral.

25. $\int_0^2 x^2\,3^x\,dx$
26. $\int_{-1}^2 \ln(x+2)\,dx$
27. $\int_0^{\pi/3} \sin 3x\cos x\,dx$
28. $\int_0^{\pi^2/2} \cos\sqrt{2x}\,dx$
29. $\int_0^2 xe^{2x}\,dx$
30. $\int_{-\pi}^{\pi} z^2\cos 2z\,dz$
31. $\int_0^{\pi/4} e^{3x}\sin 4x\,dx$
32. $\int_0^1 x\sin^{-1} x\,dx$
33. $\int_1^4 \sec^{-1}\sqrt{t}\,dt$
34. $\int_{\pi/4}^{3\pi/4} x\cot x\csc x\,dx$

35. Find the area of the region bounded by the curve $y = \ln x$, the x axis, and the line $x = e^2$.
36. Find the volume of the solid generated by revolving the region in Exercise 35 about the x axis.
37. Find the volume of the solid generated by revolving the region in Exercise 35 about the y axis.
38. Find the area of the region bounded by the curve $y = x\csc^2 x$, the x axis, and the lines $x = \frac{1}{6}\pi$ and $x = \frac{1}{4}\pi$.
39. Find the area of the region bounded by the curve $y = 2xe^{-x/2}$, the x axis, and the line $x = 4$.
40. Find the volume of the solid of revolution generated by revolving about the x axis the region of Exercise 39.
41. A water tank full of water is in the shape of the solid of revolution formed by rotating about the x axis the region bounded by the curve $y = e^{-x}$, the coordinate axes, and the line $x = 4$. Find the work done in pumping all the water to the top of the tank. Distance is measured in feet. Take the positive x axis vertically downward.
42. The marginal cost function is C' and $C'(x) = \ln x$, where $x \geq 1$. Find the total cost function if $C(x)$ dollars is the total cost of producing x units and $C(1) = 5$.
43. A manufacturer has discovered that if x hundreds of units of a particular commodity are produced per week, the marginal cost is determined by $x2^{x/2}$ and the marginal revenue is determined by $8 \cdot 2^{-x/2}$, where both the production cost and the revenue are in thousands of dollars. If the weekly fixed costs amount to \$2000, find the maximum weekly profit that can be obtained.
44. A particle is moving along a straight line and s ft is the directed distance of the particle from the origin at t sec. If v ft/sec is the velocity at t sec, $s = 0$ when $t = 0$, and $v \cdot s = t\sin t$, find s in terms of t and also s when $t = \frac{1}{2}\pi$.

45. (a) Show that

$$\int \sec^m x\,dx = \frac{\sec^{m-2} x \tan x}{m-1} + \frac{m-2}{m-1}\int \sec^{m-2} x\,dx$$

(b) Use the formula in part (a) to find $\int \sec^6 x\,dx$.

46. (a) Derive the following formula, where r is any real number:

$$\int x^r \ln x\,dx = \begin{cases} \dfrac{x^{r+1}}{r+1}\ln x - \dfrac{x^{r+1}}{(r+1)^2} + C & \text{if } r \neq -1 \\ \frac{1}{2}(\ln x)^2 + C & \text{if } r = -1 \end{cases}$$

(b) Use the formula in part (a) to find $\int x^3 \ln x\,dx$.

47. (a) Derive the following formula, where r and q are any real numbers:

$$\int x^r(\ln x)^q\,dx = \begin{cases} \dfrac{x^{r+1}(\ln x)^q}{r+1} - \dfrac{q}{r+1}\displaystyle\int x^r(\ln x)^{q-1}\,dx & \text{if } r \neq -1 \\ \dfrac{(\ln x)^{q+1}}{q+1} + C & \text{if } r = -1 \text{ and } q \neq -1 \end{cases}$$

(b) Use the formula derived in (a) to find $\int x^4 (\ln x)^2\,dx$.

48. (a) Derive the following formula, where r is any real number:

$$\int x^r e^x\,dx = x^r e^x - r\int x^{r-1} e^x\,dx$$

(b) Use the formula derived in (a) to find $\int x^4 e^x\,dx$.

10.3 INTEGRATION BY TRIGONOMETRIC SUBSTITUTION

If the integrand contains an expression of the form $\sqrt{a^2 - u^2}$, $\sqrt{a^2 + u^2}$, or $\sqrt{u^2 - a^2}$, where $a > 0$, it is often possible to perform the integration by making a trigonometric substitution that results in an integral involving trigonometric functions. We consider each form as a separate case.

Case 1: The integrand contains an expression of the form $\sqrt{a^2 - u^2}$, where $a > 0$.

We introduce a new variable θ by letting $u = a \sin\theta$, where

$$0 \le \theta \le \tfrac{1}{2}\pi \quad \text{if } u \ge 0 \quad \text{and} \quad -\tfrac{1}{2}\pi \le \theta < 0 \quad \text{if } u < 0$$

Then $du = a\cos\theta\,d\theta$, and

$$\sqrt{a^2 - u^2} = \sqrt{a^2 - a^2\sin^2\theta} = \sqrt{a^2(1 - \sin^2\theta)} = a\sqrt{\cos^2\theta}$$

Because $-\frac{1}{2}\pi \le \theta \le \frac{1}{2}\pi$, $\cos\theta \ge 0$. Hence $\sqrt{\cos^2\theta} = \cos\theta$, and

$$\sqrt{a^2 - u^2} = a\cos\theta$$

Because $\sin\theta = u/a$ and $-\frac{1}{2}\pi \le \theta \le \frac{1}{2}\pi$, it follows that

$$\theta = \sin^{-1}\frac{u}{a}$$

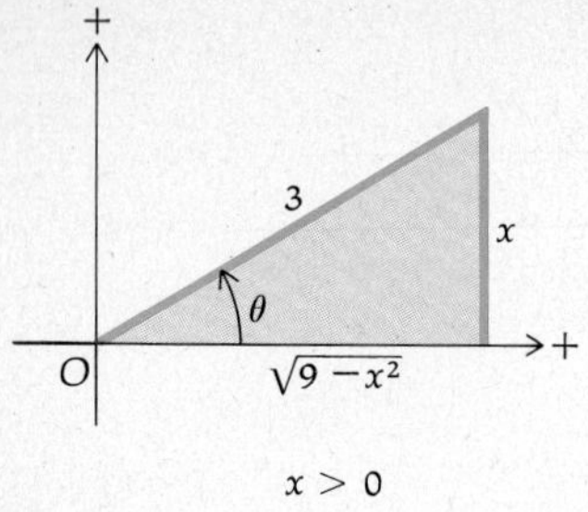

FIGURE 10.3.1

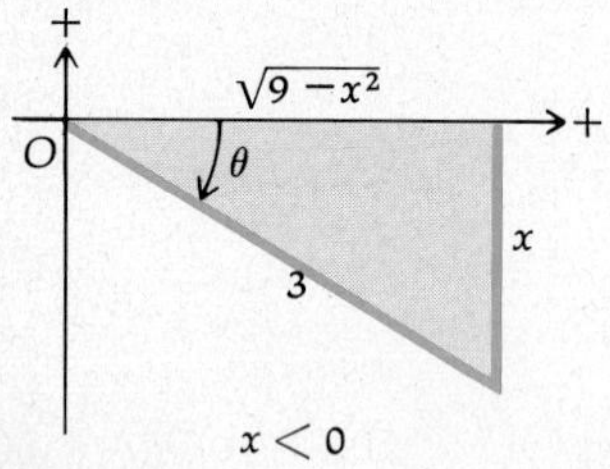

FIGURE 10.3.2

Example 1 Evaluate

$$\int \frac{\sqrt{9-x^2}}{x^2}\,dx$$

Solution Let $x = 3\sin\theta$, where $0 < \theta \le \frac{1}{2}\pi$ if $x > 0$ and $-\frac{1}{2}\pi \le \theta < 0$ if $x < 0$. Then $dx = 3\cos\theta\,d\theta$ and

$$\sqrt{9-x^2} = \sqrt{9-9\sin^2\theta} = 3\sqrt{\cos^2\theta} = 3\cos\theta$$

Therefore

$$\begin{aligned}\int \frac{\sqrt{9-x^2}}{x^2}\,dx &= \int \frac{3\cos\theta}{9\sin^2\theta}(3\cos\theta\,d\theta)\\ &= \int \cot^2\theta\,d\theta\\ &= \int (\csc^2\theta - 1)\,d\theta\\ &= -\cot\theta - \theta + C\end{aligned}$$

Because $\sin\theta = \frac{1}{3}x$ and $-\frac{1}{2}\pi \le \theta \le \frac{1}{2}\pi$, $\theta = \sin^{-1}\frac{1}{3}x$. To find $\cot\theta$, refer to Figs. 10.3.1 (for $x > 0$) and 10.3.2 (for $x < 0$). We see that $\cot\theta = \sqrt{9-x^2}/x$. Therefore

$$\int \frac{\sqrt{9-x^2}}{x^2}\,dx = -\frac{\sqrt{9-x^2}}{x} - \sin^{-1}\frac{x}{3} + C$$

Case 2: The integrand contains an expression of the form $\sqrt{a^2+u^2}$, where $a > 0$.

We introduce a new variable θ by letting $u = a\tan\theta$, where

$$0 \le \theta < \tfrac{1}{2}\pi \quad \text{if } u \ge 0 \quad \text{and} \quad -\tfrac{1}{2}\pi < \theta < 0 \quad \text{if } u < 0$$

Then $du = a\sec^2\theta\,d\theta$, and

$$\sqrt{a^2+u^2} = \sqrt{a^2+a^2\tan^2\theta} = a\sqrt{1+\tan^2\theta} = a\sqrt{\sec^2\theta}$$

Because $-\frac{1}{2}\pi < \theta < \frac{1}{2}\pi$, $\sec\theta \ge 1$. Thus $\sqrt{\sec^2\theta} = \sec\theta$, and

$$\sqrt{a^2+u^2} = a\sec\theta$$

Because $\tan\theta = u/a$ and $-\frac{1}{2}\pi < \theta < \frac{1}{2}\pi$, it follows that

$$\theta = \tan^{-1}\frac{u}{a}$$

Example 2 Evaluate $\int \sqrt{x^2+5}\,dx$.

Solution Substitute $x = \sqrt{5}\tan\theta$, where $0 \le \theta < \frac{1}{2}\pi$ if $x \ge 0$ and $-\frac{1}{2}\pi < \theta < 0$ if $x < 0$. Then $dx = \sqrt{5}\sec^2\theta\,d\theta$ and

$$\sqrt{x^2+5} = \sqrt{5\tan^2\theta + 5} = \sqrt{5}\sqrt{\sec^2\theta} = \sqrt{5}\sec\theta$$

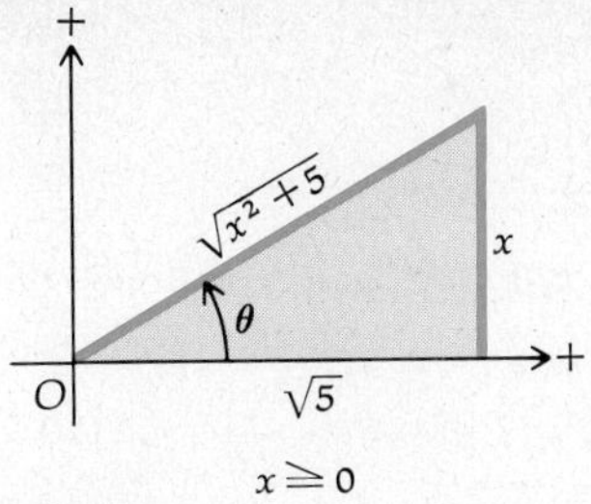

FIGURE 10.3.3

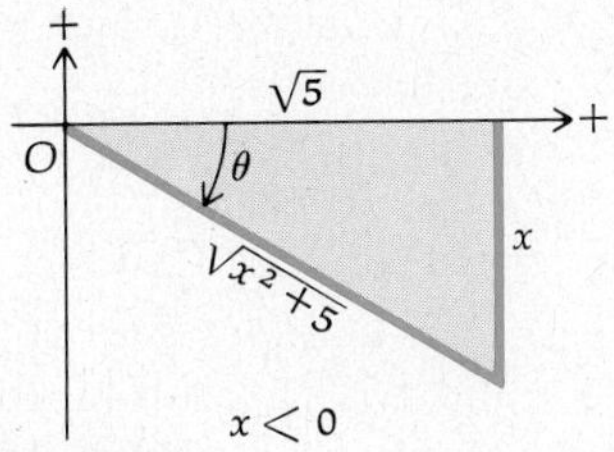

FIGURE 10.3.4

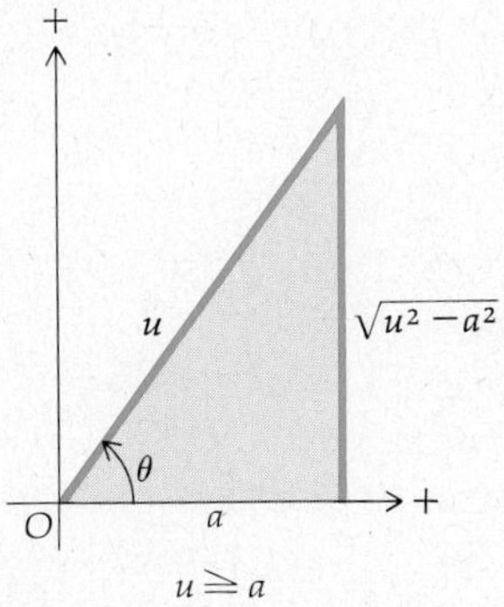

FIGURE 10.3.5

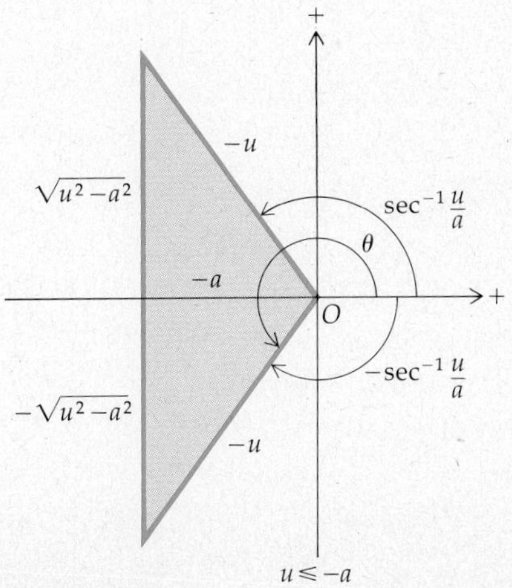

FIGURE 10.3.6

Therefore

$$\int \sqrt{x^2+5}\, dx = \int \sqrt{5}\sec\theta(\sqrt{5}\sec^2\theta\, d\theta) = 5\int \sec^3\theta\, d\theta$$

Using the result of Example 6 of Sec. 10.2 we have

$$\int \sqrt{x^2+5}\, dx = \tfrac{5}{2}\sec\theta\tan\theta + \tfrac{5}{2}\ln|\sec\theta + \tan\theta| + C$$

We find $\sec\theta$ from Figs. 10.3.3 (for $x \geq 0$) and 10.3.4 (for $x < 0$), where $\tan\theta = x/\sqrt{5}$. We see that $\sec\theta = \sqrt{x^2+5}/\sqrt{5}$. Hence

$$\begin{aligned}\int \sqrt{x^2+5}\, dx &= \frac{5}{2}\cdot\frac{\sqrt{x^2+5}}{\sqrt{5}}\cdot\frac{x}{\sqrt{5}} + \frac{5}{2}\ln\left|\frac{\sqrt{x^2+5}}{\sqrt{5}} + \frac{x}{\sqrt{5}}\right| + C\\ &= \tfrac{1}{2}x\sqrt{x^2+5} + \tfrac{5}{2}\ln|\sqrt{x^2+5}+x| - \tfrac{5}{2}\ln\sqrt{5} + C\\ &= \tfrac{1}{2}x\sqrt{x^2+5} + \tfrac{5}{2}\ln(\sqrt{x^2+5}+x) + C_1\end{aligned}$$

Observe that we replaced $-\frac{5}{2}\ln\sqrt{5} + C$ by the arbitrary constant C_1. Furthermore, because $\sqrt{x^2+5} + x > 0$, we drop the absolute-value bars.

Case 3: The integrand contains an expression of the form $\sqrt{u^2-a^2}$, where $a > 0$.

We introduce a new variable by letting $u = a\sec\theta$, where

$$0 \leq \theta < \tfrac{1}{2}\pi \quad \text{if } u \geq a \quad \text{and} \quad \pi \leq \theta < \tfrac{3}{2}\pi \quad \text{if } u \leq -a$$

Then $du = a\sec\theta\tan\theta\, d\theta$ and

$$\sqrt{u^2-a^2} = \sqrt{a^2\sec^2\theta - a^2} = \sqrt{a^2(\sec^2\theta - 1)} = a\sqrt{\tan^2\theta}$$

Because either $0 \leq \theta < \frac{1}{2}\pi$ or $\pi \leq \theta < \frac{3}{2}\pi$, then $\tan\theta \geq 0$. Thus $\sqrt{\tan^2\theta} = \tan\theta$, and we have

$$\sqrt{u^2-a^2} = a\tan\theta$$

Observe in Fig. 10.3.5 that

$$u \geq a \qquad \sec\theta = \frac{u}{a} \geq 1 \qquad 0 \leq \theta < \tfrac{1}{2}\pi$$

Therefore

$$\theta = \sec^{-1}\frac{u}{a} \quad \text{when} \quad u \geq a$$

Because the inverse secant function has a range of $[0, \frac{1}{2}\pi) \cup (\frac{1}{2}\pi, \pi]$, then $\theta \neq \sec^{-1}(u/a)$ when $u \leq -a$. To determine the relationship between θ and $\sec^{-1}(u/a)$ when $u \leq -a$ we note in Fig. 10.3.6 that

$$u \leq -a \qquad \sec\theta = \frac{u}{a} \leq -1 \qquad \pi \leq \theta < \tfrac{3}{2}\pi$$

Because when $u \leq -a$, $\frac{1}{2}\pi < \sec^{-1}(u/a) \leq \pi$, it follows that

$$\theta = 2\pi - \sec^{-1}\frac{u}{a} \quad \text{when} \quad u \leq -a$$

Example 3 Evaluate

$$\int \frac{dx}{x^3\sqrt{x^2-9}}$$

Solution Let $x = 3\sec\theta$, where $0 < \theta < \frac{1}{2}\pi$ if $x > 3$ and $\pi < \theta < \frac{3}{2}\pi$ if $x < -3$. Then $dx = 3\sec\theta\tan\theta\, d\theta$ and

$$\sqrt{x^2-9} = \sqrt{9\sec^2\theta - 9} = 3\sqrt{\tan^2\theta} = 3\tan\theta$$

Hence

$$\int \frac{dx}{x^3\sqrt{x^2-9}} = \int \frac{3\sec\theta\tan\theta\, d\theta}{27\sec^3\theta \cdot 3\tan\theta} = \frac{1}{27}\int \cos^2\theta\, d\theta$$

$$= \frac{1}{54}\int (1+\cos 2\theta)\, d\theta = \frac{1}{54}(\theta + \tfrac{1}{2}\sin 2\theta) + C$$

$$= \tfrac{1}{54}(\theta + \sin\theta\cos\theta) + C$$

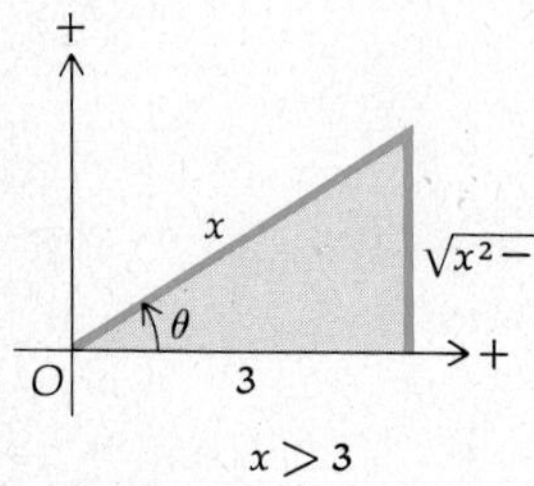

FIGURE 10.3.7

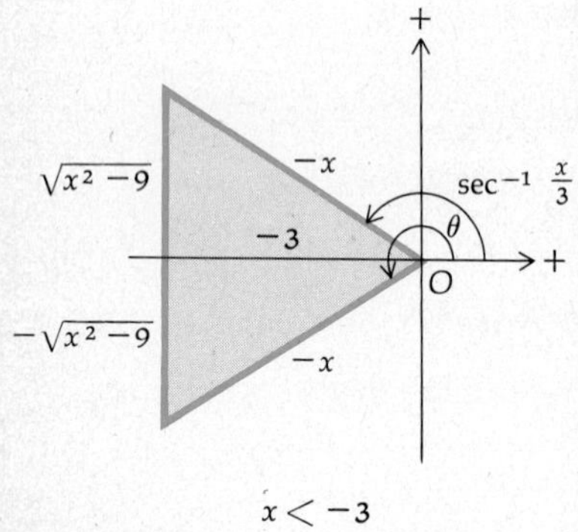

FIGURE 10.3.8

When $x > 3$, $0 < \theta < \frac{1}{2}\pi$, and we obtain $\sin\theta$ and $\cos\theta$ from Fig. 10.3.7. When $x < -3$, $\pi < \theta < \frac{3}{2}\pi$, and we obtain $\sin\theta$ and $\cos\theta$ from Fig. 10.3.8. In either case, $\sin\theta = \sqrt{x^2-9}/x$ and $\cos\theta = 3/x$. To express θ in terms of x we must distinguish between the two cases: $x > 3$ and $x < -3$. When $x > 3$, $\theta = \sec^{-1}\frac{1}{3}x$. However, when $x < -3$, $\theta = 2\pi - \sec^{-1}\frac{1}{3}x$ (refer to Fig. 10.3.8). We have, then,

$$\int \frac{dx}{x^3\sqrt{x^2-9}} = \begin{cases} \dfrac{1}{54}\left(\sec^{-1}\dfrac{x}{3} + \dfrac{\sqrt{x^2-9}}{x}\cdot\dfrac{3}{x}\right) + C & \text{if } x > 3 \\ \dfrac{1}{54}\left(2\pi - \sec^{-1}\dfrac{x}{3} + \dfrac{\sqrt{x^2-9}}{x}\cdot\dfrac{3}{x}\right) + C_1 & \text{if } x < -3 \end{cases}$$

and letting $C = C_1 + \frac{1}{27}\pi$ we get

$$\int \frac{dx}{x^3\sqrt{x^2-9}} = \begin{cases} \dfrac{1}{54}\sec^{-1}\dfrac{x}{3} + \dfrac{\sqrt{x^2-9}}{18x^2} + C & \text{if } x > 3 \\ -\dfrac{1}{54}\sec^{-1}\dfrac{x}{3} + \dfrac{\sqrt{x^2-9}}{18x^2} + C & \text{if } x < -3 \end{cases}$$

Example 4 Evaluate

$$\int \frac{dx}{\sqrt{x^2-25}}$$

Solution Let $x = 5\sec\theta$, where $0 < \theta < \frac{1}{2}\pi$ if $x > 5$ and $\pi < \theta < \frac{3}{2}\pi$ if $x < -5$. Then $dx = 5\sec\theta\tan\theta\, d\theta$ and

$$\sqrt{x^2-25} = \sqrt{25\sec^2\theta - 25} = 5\sqrt{\tan^2\theta} = 5\tan\theta$$

Therefore

$$\int \frac{dx}{\sqrt{x^2-25}} = \int \frac{5\sec\theta\tan\theta\, d\theta}{5\tan\theta} = \int \sec\theta\, d\theta$$

$$= \ln|\sec\theta + \tan\theta| + C$$

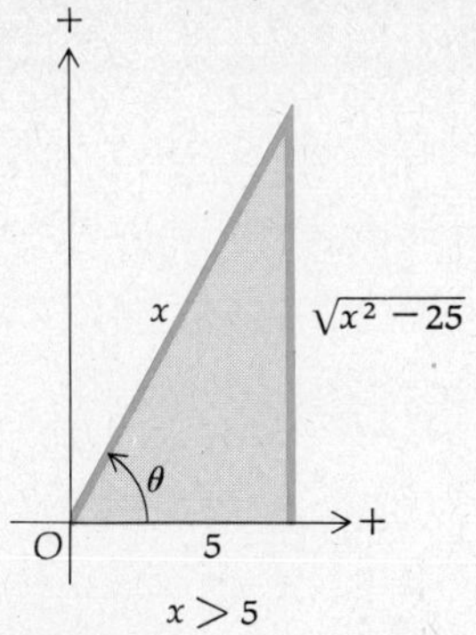

FIGURE 10.3.9

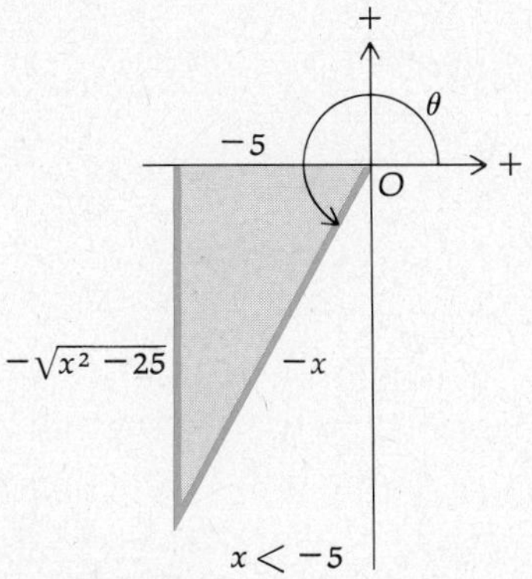

FIGURE 10.3.10

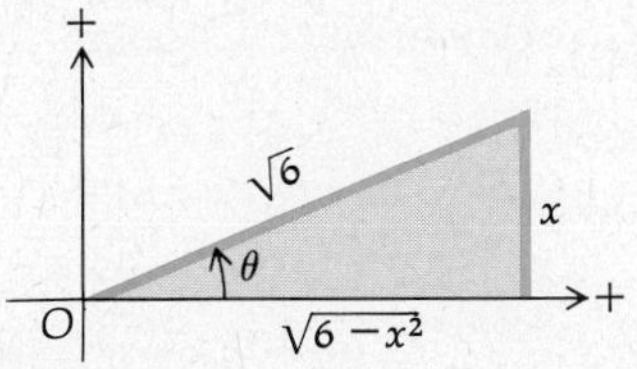

FIGURE 10.3.11

To find $\sec\theta$ and $\tan\theta$ refer to Fig. 10.3.9 (for $x > 5$) and Fig. 10.3.10 (for $x < -5$). In either case, $\sec\theta = \frac{1}{5}x$ and $\tan\theta = \frac{1}{5}\sqrt{x^2 - 25}$. We have, then,

$$\int \frac{dx}{\sqrt{x^2 - 25}} = \ln\left|\frac{x}{5} + \frac{\sqrt{x^2 - 25}}{5}\right| + C$$
$$= \ln|x + \sqrt{x^2 - 25}| - \ln 5 + C$$
$$= \ln|x + \sqrt{x^2 - 25}| + C_1$$

Example 5 Evaluate

$$\int_1^2 \frac{dx}{(6 - x^2)^{3/2}}$$

Solution To evaluate the indefinite integral $\int dx/(6 - x^2)^{3/2}$ we make the substitution $x = \sqrt{6}\sin\theta$. In this case we can restrict θ to the interval $0 < \theta < \frac{1}{2}\pi$ because we are evaluating a definite integral for which $x > 0$ because x is in $[1, 2]$. So we have $x = \sqrt{6}\sin\theta, 0 < \theta < \frac{1}{2}\pi$, and $dx = \sqrt{6}\cos\theta\, d\theta$. Furthermore,

$$(6 - x^2)^{3/2} = (6 - 6\sin^2\theta)^{3/2}$$
$$= 6\sqrt{6}(1 - \sin^2\theta)^{3/2}$$
$$= 6\sqrt{6}(\cos^2\theta)^{3/2}$$
$$= 6\sqrt{6}\cos^3\theta$$

Hence

$$\int \frac{dx}{(6 - x^2)^{3/2}} = \int \frac{\sqrt{6}\cos\theta\, d\theta}{6\sqrt{6}\cos^3\theta}$$
$$= \frac{1}{6}\int \frac{d\theta}{\cos^2\theta}$$
$$= \frac{1}{6}\int \sec^2\theta\, d\theta$$
$$= \tfrac{1}{6}\tan\theta + C$$

We find $\tan\theta$ from Fig. 10.3.11, in which $\sin\theta = x/\sqrt{6}$ and $0 < \theta < \frac{1}{2}\pi$. Thus $\tan\theta = x/\sqrt{6 - x^2}$, and so

$$\int \frac{dx}{(6 - x^2)^{3/2}} = \frac{x}{6\sqrt{6 - x^2}} + C$$

Therefore

$$\int_1^2 \frac{dx}{(6 - x^2)^{3/2}} = \frac{x}{6\sqrt{6 - x^2}}\bigg]_1^2$$
$$= \frac{1}{3\sqrt{2}} - \frac{1}{6\sqrt{5}}$$
$$= \frac{2\sqrt{5} - \sqrt{2}}{6\sqrt{10}}$$

EXERCISES 10.3

In Exercises 1 through 24, evaluate the indefinite integral.

1. $\int \frac{dx}{x^2\sqrt{4-x^2}}$
2. $\int \frac{\sqrt{4-x^2}}{x^2}\,dx$
3. $\int \frac{dx}{x\sqrt{x^2+4}}$
4. $\int \frac{x^2\,dx}{\sqrt{x^2+6}}$
5. $\int \frac{dx}{x\sqrt{25-x^2}}$
6. $\int \sqrt{1-u^2}\,du$
7. $\int \frac{dx}{\sqrt{x^2-a^2}}$
8. $\int \frac{dw}{w^2\sqrt{w^2-7}}$
9. $\int \frac{x^2\,dx}{(x^2+4)^2}$
10. $\int \frac{dx}{(4+x^2)^{3/2}}$
11. $\int \frac{dx}{(4x^2-9)^{3/2}}$
12. $\int \frac{dx}{x^4\sqrt{16+x^2}}$
13. $\int \frac{2\,dt}{t\sqrt{t^4+25}}$
14. $\int \frac{x^3\,dx}{(25-x^2)^2}$
15. $\int \frac{dx}{\sqrt{4x+x^2}}$
16. $\int \frac{dx}{\sqrt{4x-x^2}}$
17. $\int \frac{dx}{(5-4x-x^2)^{3/2}}$
18. $\int \frac{dx}{x\sqrt{x^4-4}}$
19. $\int \frac{\sec^2 x\,dx}{(4-\tan^2 x)^{3/2}}$
20. $\int \frac{e^{-x}\,dx}{(9e^{-2x}+1)^{3/2}}$
21. $\int \frac{\ln^3 w\,dw}{w\sqrt{\ln^2 w-4}}$
22. $\int \frac{dz}{(z^2-6z+18)^{3/2}}$
23. $\int \frac{e^t\,dt}{(e^{2t}+8e^t+7)^{3/2}}$
24. $\int \frac{\sqrt{16-e^{2x}}}{e^x}\,dx$

In Exercises 25 through 32, evaluate the definite integral.

25. $\int_0^2 \frac{x^3\,dx}{\sqrt{16-x^2}}$
26. $\int_0^4 \frac{dx}{(16+x^2)^{3/2}}$
27. $\int_{\sqrt{3}}^{3\sqrt{3}} \frac{dx}{x^2\sqrt{x^2+9}}$
28. $\int_0^1 \frac{x^2\,dx}{\sqrt{4-x^2}}$
29. $\int_4^6 \frac{dx}{x\sqrt{x^2-4}}$
30. $\int_1^3 \frac{dx}{x^4\sqrt{x^2+3}}$
31. $\int_0^5 x^2\sqrt{25-x^2}\,dx$
32. $\int_4^8 \frac{dw}{(w^2-4)^{3/2}}$

33. Use methods previous to this section (that is, without a trigonometric substitution) to evaluate the integrals:

    (a) $\int \frac{3\,dx}{x\sqrt{4x^2-9}}$

    (b) $\int \frac{5x\,dx}{\sqrt{3-2x^2}}$

34. Find the area of the region bounded by the curve $y = \sqrt{x^2-9}/x^2$, the x axis, and the line $x = 5$.
35. Find the length of the arc of the curve $y = \ln x$ from $x = 1$ to $x = 3$.
36. Find the volume of the solid of revolution generated by revolving the region of Exercise 34 about the y axis.
37. Find the volume of the solid of revolution generated when the region to the right of the y axis bounded by the curve $y = x\sqrt[4]{9-x^2}$ and the x axis is revolved about the x axis.
38. Use integration to prove that the area of the region enclosed by a circle of radius r is πr^2.
39. A manufacturer who began operations four years ago has determined that income from sales has increased steadily at the rate of $\frac{t^3+3t^2+6t+7}{t^2+3t+2}$ millions of dollars per year, where t is the number of years that the company has been operating. It is estimated that the total income from sales will increase at the same rate for the next 2 years. If the total income from sales for the year just ended was \$6 million, what is the total income from sales expected for the period ending 1 year from now? Give the answer to the nearest \$100.

10.4 INTEGRATION OF RATIONAL FUNCTIONS BY PARTIAL FRACTIONS WHEN THE DENOMINATOR HAS ONLY LINEAR FACTORS

In Sec. 1.6 a rational function was defined as one that can be expressed as the quotient of two polynomial functions. That is, the function H is a rational function if $H(x) = P(x)/Q(x)$, where $P(x)$ and $Q(x)$ are polynomials. We saw previously that if the degree of the numerator is not less than the degree of the denominator, we have an improper fraction, and in that case we divide the numerator by the denominator until we obtain a proper fraction, one in which the degree of the numerator is less than the degree of the denominator. For example,

$$\frac{x^4 - 10x^2 + 3x + 1}{x^2 - 4} = x^2 - 6 + \frac{3x - 23}{x^2 - 4}$$

So if we wish to integrate

$$\int \frac{x^4 - 10x^2 + 3x + 1}{x^2 - 4}\, dx$$

the problem is reduced to integrating

$$\int (x^2 - 6)\, dx + \int \frac{3x - 23}{x^2 - 4}\, dx$$

In general, then, we are concerned with the integration of expressions of the form

$$\int \frac{P(x)}{Q(x)}\, dx$$

where the degree of $P(x)$ is less than the degree of $Q(x)$.

To do this it is often necessary to write $P(x)/Q(x)$ as the sum of *partial fractions*. The denominators of the partial fractions are obtained by factoring $Q(x)$ into a product of linear and quadratic factors. Sometimes it may be difficult to find these factors of $Q(x)$; however, a theorem from advanced algebra states that theoretically this can always be done. We state this theorem without proof.

10.4.1 Theorem Any polynomial with real coefficients can be expressed as a product of linear and quadratic factors in such a way that each of the factors has real coefficients.

After $Q(x)$ has been factored into products of linear and quadratic factors, the method of determining the partial fractions depends on the nature of these factors. We consider various cases separately. The results of advanced algebra, which are not proved here, provide us with the form of the partial fractions in each case.

Case 1: The factors of $Q(x)$ are all linear, and none is repeated. That is,

$$Q(x) = (a_1x + b_1)(a_2x + b_2) \cdots (a_nx + b_n)$$

where no two of the factors are identical. In this case we write

$$\frac{P(x)}{Q(x)} \equiv \frac{A_1}{a_1x + b_1} + \frac{A_2}{a_2x + b_2} + \cdots + \frac{A_n}{a_nx + b_n} \tag{1}$$

where $A_1, A_2, \ldots, A_n$ are constants to be determined.

Note that we used $\equiv$ (read as "identically equal") instead of $=$ in (1). This is because (1) is an identity for the value of each A_i.

The following illustration shows how the values of A_i are found.

● ILLUSTRATION 1 To evaluate

$$\int \frac{(x-1)\,dx}{x^3 - x^2 - 2x}$$

we factor the denominator and have

$$\frac{x-1}{x^3 - x^2 - 2x} \equiv \frac{x-1}{x(x-2)(x+1)}$$

So

$$\frac{x-1}{x(x-2)(x+1)} \equiv \frac{A}{x} + \frac{B}{x-2} + \frac{C}{x+1} \tag{2}$$

Equation (2) is an identity for all x (except $x = 0, 2, -1$). From (2),

$$x - 1 \equiv A(x-2)(x+1) + Bx(x+1) + Cx(x-2) \tag{3}$$

Equation (3) is an identity that is true for all values of x including 0, 2, and -1. We wish to find the constants A, B, and C. Substituting 0 for x in (3) we obtain

$$-1 = -2A \quad \text{or} \quad A = \tfrac{1}{2}$$

Substituting 2 for x in (3) we get

$$1 = 6B \quad \text{or} \quad B = \tfrac{1}{6}$$

Substituting -1 for x in (3) we obtain

$$-2 = 3C \quad \text{or} \quad C = -\tfrac{2}{3}$$

There is another method for finding the values of A, B, and C. If on the right side of (3) we combine terms,

$$x - 1 \equiv (A + B + C)x^2 + (-A + B - 2C)x - 2A \tag{4}$$

For (4) to be an identity, the coefficients on the left must equal the corresponding coefficients on the right. Hence

$$A + B + C = 0$$
$$-A + B - 2C = 1$$
$$-2A = -1$$

Solving these equations simultaneously we get $A = \frac{1}{2}$, $B = \frac{1}{6}$, and $C = -\frac{2}{3}$. Substituting these values in (2) we get

$$\frac{x-1}{x(x-2)(x+1)} \equiv \frac{\frac{1}{2}}{x} + \frac{\frac{1}{6}}{x-2} + \frac{-\frac{2}{3}}{x+1}$$

So the given integral can be expressed as follows:

$$\int \frac{x-1}{x^3-x^2-2x}\,dx = \frac{1}{2}\int\frac{dx}{x} + \frac{1}{6}\int\frac{dx}{x-2} - \frac{2}{3}\int\frac{dx}{x+1}$$

$$= \tfrac{1}{2}\ln|x| + \tfrac{1}{6}\ln|x-2| - \tfrac{2}{3}\ln|x+1| + \tfrac{1}{6}\ln C$$

$$= \tfrac{1}{6}(3\ln|x| + \ln|x-2| - 4\ln|x+1| + \ln C)$$

$$= \frac{1}{6}\ln\left|\frac{Cx^3(x-2)}{(x+1)^4}\right|$$

Case 2: The factors of $Q(x)$ are all linear, and some are repeated.

Suppose that (a_ix+b_i) is a p-fold factor. Then, corresponding to this factor there will be the sum of p partial fractions

$$\frac{A_1}{(a_ix+b_i)^p} + \frac{A_2}{(a_ix+b_i)^{p-1}} + \cdots + \frac{A_{p-1}}{(a_ix+b_i)^2} + \frac{A_p}{a_ix+b_i}$$

where $A_1, A_2, \ldots, A_p$ are constants to be determined.

Example 1 following illustrates this case and the method of determining each A_i.

Example 1 Find

$$\int\frac{(x^3-1)\,dx}{x^2(x-2)^3}$$

Solution The fraction in the integrand is written as a sum of partial fractions as follows:

$$\frac{x^3-1}{x^2(x-2)^3} \equiv \frac{A}{x^2} + \frac{B}{x} + \frac{C}{(x-2)^3} + \frac{D}{(x-2)^2} + \frac{E}{x-2} \tag{5}$$

The above is an identity for all x (except $x = 0, 2$). Multiplying on both sides of (5) by the lowest common denominator we get

$$x^3 - 1 \equiv A(x-2)^3 + Bx(x-2)^3 + Cx^2 + Dx^2(x-2) + Ex^2(x-2)^2 \tag{6}$$

We substitute 2 for x in (6) and obtain

$$7 = 4C \quad\text{or}\quad C = \tfrac{7}{4}$$

Substituting 0 for x in (6) we get

$$-1 = -8A \quad\text{or}\quad A = \tfrac{1}{8}$$

We substitute these values for A and C in (6) and expand the powers of the binomials, and we have

$$x^3 - 1 \equiv \tfrac{1}{8}(x^3 - 6x^2 + 12x - 8) + Bx(x^3 - 6x^2 + 12x - 8) + \tfrac{7}{4}x^2 + Dx^3 - 2Dx^2 + Ex^2(x^2 - 4x + 4)$$

$$x^3 - 1 \equiv (B+E)x^4 + (\tfrac{1}{8} - 6B + D - 4E)x^3 + (-\tfrac{3}{4} + 12B + \tfrac{7}{4} - 2D + 4E)x^2 + (\tfrac{3}{2} - 8B)x - 1$$

Equating the coefficients of like powers of x we obtain

$$B + E = 0$$

$$\tfrac{1}{8} - 6B + D - 4E = 1$$

$$-\tfrac{3}{4} + 12B + \tfrac{7}{4} - 2D + 4E = 0$$

$$\tfrac{3}{2} - 8B = 0$$

Solving, we get

$$B = \tfrac{3}{16} \qquad D = \tfrac{5}{4} \qquad E = -\tfrac{3}{16}$$

Therefore, from (5),

$$\frac{x^3 - 1}{x^2(x-2)^3} \equiv \frac{\frac{1}{8}}{x^2} + \frac{\frac{3}{16}}{x} + \frac{\frac{7}{4}}{(x-2)^3} + \frac{\frac{5}{4}}{(x-2)^2} + \frac{-\frac{3}{16}}{x-2}$$

Thus

$$\int \frac{x^3 - 1}{x^2(x-2)^3}\,dx$$

$$= \frac{1}{8}\int \frac{dx}{x^2} + \frac{3}{16}\int \frac{dx}{x} + \frac{7}{4}\int \frac{dx}{(x-2)^3} + \frac{5}{4}\int \frac{dx}{(x-2)^2} - \frac{3}{16}\int \frac{dx}{x-2}$$

$$= -\frac{1}{8x} + \frac{3}{16}\ln|x| - \frac{7}{8(x-2)^2} - \frac{5}{4(x-2)} - \frac{3}{16}\ln|x-2| + C$$

$$= \frac{-11x^2 + 17x - 4}{8x(x-2)^2} + \frac{3}{16}\ln\left|\frac{x}{x-2}\right| + C$$

Example 2 Find

$$\int \frac{du}{u^2 - a^2}$$

Solution

$$\frac{1}{u^2 - a^2} \equiv \frac{A}{u-a} + \frac{B}{u+a}$$

Multiplying by $(u - a)(u + a)$ we get

$$1 \equiv A(u + a) + B(u - a)$$

$$1 \equiv (A + B)u + Aa - Ba$$

Equating coefficients we have

$$A + B = 0$$

$$Aa - Ba = 1$$

Solving simultaneously we get

$$A = \frac{1}{2a} \quad \text{and} \quad B = -\frac{1}{2a}$$

Therefore

$$\int \frac{du}{u^2 - a^2} = \frac{1}{2a}\int \frac{du}{u - a} - \frac{1}{2a}\int \frac{du}{u + a}$$

$$= \frac{1}{2a}\ln |u - a| - \frac{1}{2a}\ln |u + a| + C$$

or, equivalently,

$$\int \frac{du}{u^2 - a^2} = \frac{1}{2a}\ln \left|\frac{u - a}{u + a}\right| + C$$

The type of integral of the above example occurs frequently enough for it to be listed as a formula. It is not necessary to memorize it because an integration by partial fractions is fairly simple.

$$\int \frac{du}{u^2 - a^2} = \frac{1}{2a}\ln \left|\frac{u - a}{u + a}\right| + C \qquad (7)$$

If we have $\int du/(a^2 - u^2)$, we write

$$\int \frac{du}{a^2 - u^2} = -\int \frac{du}{u^2 - a^2}$$

$$= -\frac{1}{2a}\ln \left|\frac{u - a}{u + a}\right| + C$$

$$= \frac{1}{2a}\ln \left|\frac{u + a}{u - a}\right| + C$$

This is also listed as a formula.

$$\int \frac{du}{a^2 - u^2} = \frac{1}{2a}\ln \left|\frac{u + a}{u - a}\right| + C \qquad (8)$$

In chemistry, the *law of mass action* affords an application of integration that leads to the use of partial fractions. Under certain conditions it is found that a substance A reacts with a substance B to form a third substance C in such a way that the rate of change of the amount of C is proportional to the product of the amounts of A and B remaining at any given time.

Suppose that initially there are α grams of A and β grams of B and that r grams of A combine with s grams of B to form $(r + s)$ grams of C. If x is the number of grams of substance C present at t units of time, then C contains $rx/(r + s)$ grams of A and $sx/(r + s)$ grams of B. The number of grams of substance A remaining is then $\alpha - rx/(r + s)$, and the number of grams of substance B remaining is $\beta - sx/(r + s)$. Therefore the law of mass action gives

$$\frac{dx}{dt} = K\left(\alpha - \frac{rx}{r + s}\right)\left(\beta - \frac{sx}{r + s}\right)$$

where K is the constant of proportionality. This equation can be written as

$$\frac{dx}{dt} = \frac{Krs}{(r+s)^2}\left(\frac{r+s}{r}\alpha - x\right)\left(\frac{r+s}{s}\beta - x\right) \tag{9}$$

Letting

$$k = \frac{Krs}{(r+s)^2} \qquad a = \frac{r+s}{r}\alpha \qquad b = \frac{r+s}{s}\beta$$

Eq. (9) becomes

$$\frac{dx}{dt} = k(a-x)(b-x) \tag{10}$$

We can separate the variables in (10) and get

$$\frac{dx}{(a-x)(b-x)} = k\,dt$$

If $a = b$, then the left side of the above equation can be integrated by the power formula. If $a \neq b$, partial fractions can be used for the integration.

Example 3 A chemical reaction causes a substance A to combine with a substance B to form a substance C so that the law of mass action is obeyed. If in Eq. (10) $a = 8$ and $b = 6$, and 2 g of substance C are formed in 10 min, how many grams of C are formed in 15 min?

TABLE 10.4.1

t	0	10	15
x	0	2	x_{15}

Solution Letting x grams be the amount of substance C present at t minutes, we have the initial conditions shown in Table 10.4.1. Equation (10) becomes

$$\frac{dx}{dt} = k(8-x)(6-x)$$

Separating the variables we have

$$\int \frac{dx}{(8-x)(6-x)} = k\int dt \tag{11}$$

Writing the integrand as the sum of partial fractions gives

$$\frac{1}{(8-x)(6-x)} \equiv \frac{A}{8-x} + \frac{B}{6-x}$$

$$1 \equiv A(6-x) + B(8-x)$$

Substituting 6 for x gives $B = \frac{1}{2}$, and substituting 8 for x gives $A = -\frac{1}{2}$. Hence (11) is written as

$$-\frac{1}{2}\int \frac{dx}{8-x} + \frac{1}{2}\int \frac{dx}{6-x} = k\int dt$$

Integrating, we have

$$\tfrac{1}{2}\ln|8-x| - \tfrac{1}{2}\ln|6-x| + \tfrac{1}{2}\ln|C| = kt$$

$$\ln\left|\frac{6-x}{C(8-x)}\right| = -2kt$$

$$\frac{6-x}{8-x} = Ce^{-2kt} \tag{12}$$

Substituting $x = 0$, $t = 0$ in (12) gives $C = \frac{3}{4}$. Hence

$$\frac{6-x}{8-x} = \frac{3}{4}e^{-2kt} \tag{13}$$

Substituting $x = 2$, $t = 10$ in (13) we have

$$\tfrac{4}{6} = \tfrac{3}{4}e^{-20k}$$

$$e^{-20k} = \tfrac{8}{9}$$

Substituting $x = x_{15}$, $t = 15$ into (13) we get

$$\frac{6-x_{15}}{8-x_{15}} = \frac{3}{4}e^{-30k}$$

$$4(6-x_{15}) = 3(e^{-20k})^{3/2}(8-x_{15})$$

$$24 - 4x_{15} = 3(\tfrac{8}{9})^{3/2}(8-x_{15})$$

$$24 - 4x_{15} = \frac{16\sqrt{2}}{9}(8-x_{15})$$

$$x_{15} = \frac{54-32\sqrt{2}}{9-4\sqrt{2}}$$

$$\approx 2.6$$

Therefore 2.6 g of substance C will be formed in 15 min.

EXERCISES 10.4

In Exercises 1 through 20, evaluate the indefinite integral.

1. $\displaystyle\int \frac{dx}{x^2-4}$

2. $\displaystyle\int \frac{x^2\,dx}{x^2+x-6}$

3. $\displaystyle\int \frac{5x-2}{x^2-4}\,dx$

4. $\displaystyle\int \frac{(4x-2)\,dx}{x^3-x^2-2x}$

5. $\displaystyle\int \frac{4w-11}{2w^2+7w-4}\,dw$

6. $\displaystyle\int \frac{9t^2-26t-5}{3t^2-5t-2}\,dt$

7. $\displaystyle\int \frac{6x^2-2x-1}{4x^3-x}\,dx$

8. $\displaystyle\int \frac{x^2+x+2}{x^2-1}\,dx$

9. $\displaystyle\int \frac{dx}{x^3+3x^2}$

10. $\displaystyle\int \frac{x^2+4x-1}{x^3-x}\,dx$

11. $\displaystyle\int \frac{dx}{x^2(x+1)^2}$

12. $\displaystyle\int \frac{3x^2-x+1}{x^3-x^2}\,dx$

13. $\displaystyle\int \frac{x^2-3x-7}{(2x+3)(x+1)^2}\,dx$

14. $\displaystyle\int \frac{dt}{(t+2)^2(t+1)}$

15. $\displaystyle\int \frac{3z+1}{(z^2-4)^2}\,dz$

16. $\displaystyle\int \frac{(5x^2-11x+5)\,dx}{x^3-4x^2+5x-2}$

17. $\int \frac{x^4 + 3x^3 - 5x^2 - 4x + 17}{x^3 + x^2 - 5x + 3}\,dx$ **18.** $\int \frac{2x^4 - 2x + 1}{2x^5 - x^4}\,dx$

19. $\int \frac{-24x^3 + 30x^2 + 52x + 17}{9x^4 - 6x^3 - 11x^2 + 4x + 4}\,dx$ **20.** $\int \frac{dx}{16x^4 - 8x^2 + 1}$

In Exercises 21 through 28, evaluate the definite integral.

21. $\int_1^2 \frac{x-3}{x^3 + x^2}\,dx$ **22.** $\int_0^4 \frac{(x-2)\,dx}{2x^2 + 7x + 3}$ **23.** $\int_1^3 \frac{x^2 - 4x + 3}{x(x+1)^2}\,dx$ **24.** $\int_1^4 \frac{(2x^2 + 13x + 18)\,dx}{x^3 + 6x^2 + 9x}$

25. $\int_1^2 \frac{5x^2 - 3x + 18}{9x - x^3}\,dx$ **26.** $\int_0^1 \frac{(3x^2 + 7x)\,dx}{x^3 + 6x^2 + 11x + 6}$ **27.** $\int_0^5 \frac{(x^2 - 3)\,dx}{x^3 + 4x^2 + 5x + 2}$ **28.** $\int_0^4 \frac{x^2\,dx}{2x^3 + 9x^2 + 12x + 4}$

29. Find the area of the region bounded by the curve $y = (x - 1)/(x^2 - 5x + 6)$, the x axis, and the lines $x = 4$ and $x = 6$.

30. Find the area of the region in the first quadrant bounded by the curve $(x + 2)^2y = 4 - x$.

31. Find the volume of the solid of revolution generated by revolving the region in Exercise 29 about the y axis.

32. Find the volume of the solid of revolution generated if the region in Exercise 30 is revolved about the x axis.

33. Suppose in Example 3 that $a = 5$ and $b = 4$ and 1 g of substance C is formed in 5 min. How many grams of C are formed in 10 min?

34. Suppose in Example 3 that $a = 6$ and $b = 3$ and 1 g of substance C is formed in 4 min. How long will it take for 2 g of substance C to be formed?

35. At any instant the rate at which a substance dissolves is proportional to the product of the amount of the substance present at that instant and the difference between the concentration of the substance in solution at that instant and the concentration of the substance in a saturated solution. A quantity of insoluble material is mixed with 10 lb of salt initially, and the salt is dissolving in a tank containing 20 gal of water. If 5 lb of salt dissolves in 10 min and the concentration of salt in a saturated solution is 3 lb/gal, how much salt will dissolve in 20 min?

36. A particle is moving along a straight line so that if v ft/sec is the velocity of the particle at t sec, then

$$v = \frac{t + 3}{t^2 + 3t + 2}$$

Find the distance traveled by the particle from the time when $t = 0$ to the time when $t = 2$.

37. For the *law of logistic growth,* it is assumed that at t units of time the rate of growth $f'(t)$ of a quantity $f(t)$ is given by $f'(t) = Af(t)[B - f(t)]$, where A and B are constants. If $f(0) = k$, show that

$$f(t) = \frac{Bk}{k + (B - k)e^{-ABt}}$$

10.5 INTEGRATION OF RATIONAL FUNCTIONS BY PARTIAL FRACTIONS WHEN THE DENOMINATOR CONTAINS QUADRATIC FACTORS

The discussion of integration of rational functions by partial fractions continues with the two cases where the denominator contains quadratic factors.

Case 3: The factors of $Q(x)$ are linear and quadratic, and none of the quadratic factors is repeated.

Corresponding to the quadratic factor $ax^2 + bx + c$ in the denominator is the partial fraction of the form

$$\frac{Ax + B}{ax^2 + bx + C}$$

Example 1 Find

$$\int \frac{(x^2 - 2x - 3)\,dx}{(x-1)(x^2 + 2x + 2)}$$

Solution The fraction in the integrand is written as a sum of partial fractions as follows:

$$\frac{x^2 - 2x - 3}{(x-1)(x^2+2x+2)} \equiv \frac{Ax + B}{x^2 + 2x + 2} + \frac{C}{x-1} \tag{1}$$

Multiplying on both sides of (1) by the lowest common denominator we have

$$x^2 - 2x - 3 \equiv (Ax + B)(x - 1) + C(x^2 + 2x + 2) \tag{2}$$

We can compute C by substituting 1 for x in (2), and we have

$$-4 = 5C \quad \text{or} \quad C = -\tfrac{4}{5}$$

We replace C by $-\frac{4}{5}$ in (2) and multiply on the right side to obtain

$$x^2 - 2x - 3 \equiv (A - \tfrac{4}{5})x^2 + (B - A - \tfrac{8}{5})x + (-\tfrac{8}{5} - B)$$

Equating coefficients of like powers of x gives

$$A - \tfrac{4}{5} = 1$$

$$B - A - \tfrac{8}{5} = -2$$

$$-\tfrac{8}{5} - B = -3$$

Therefore

$$A = \tfrac{9}{5} \qquad B = \tfrac{7}{5}$$

Substituting the values of A, B, and C in (1) we obtain

$$\frac{x^2 - 2x - 3}{(x-1)(x^2+2x+2)} \equiv \frac{\frac{9}{5}x + \frac{7}{5}}{x^2 + 2x + 2} + \frac{-\frac{4}{5}}{x-1}$$

So

$$\int \frac{x^2 - 2x - 3}{(x-1)(x^2+2x+2)}\,dx$$

$$= \frac{9}{5}\int \frac{x\,dx}{x^2+2x+2} + \frac{7}{5}\int \frac{dx}{x^2+2x+2} - \frac{4}{5}\int \frac{dx}{x-1} \tag{3}$$

To integrate $\int (x\,dx)/(x^2 + 2x + 2)$, we see that the differential of the denominator is $2(x+1)\,dx$; so we add and subtract 1 in the numerator, thereby giving

$$\frac{9}{5}\int \frac{x\,dx}{x^2+2x+2} = \frac{9}{5}\int \frac{(x+1)\,dx}{x^2+2x+2} - \frac{9}{5}\int \frac{dx}{x^2+2x+2} \tag{4}$$

Substituting from (4) into (3) and combining terms we get

$$\int \frac{x^2 - 2x - 3}{(x-1)(x^2+2x+2)}\,dx$$

$$= \frac{9}{5} \cdot \frac{1}{2} \int \frac{2(x+1)\,dx}{x^2+2x+2} - \frac{2}{5} \int \frac{dx}{x^2+2x+2} - \frac{4}{5} \int \frac{dx}{x-1} \tag{5}$$

$$= \frac{9}{10} \ln |x^2+2x+2| - \frac{2}{5} \int \frac{dx}{(x+1)^2+1} - \frac{4}{5} \ln |x-1|$$

$$= \tfrac{9}{10} \ln |x^2+2x+2| - \tfrac{2}{5} \tan^{-1}(x+1) - \tfrac{8}{10} \ln |x-1| + \tfrac{1}{10} \ln C$$

$$= \frac{1}{10} \ln \left| \frac{C(x^2+2x+2)^9}{(x-1)^8} \right| - \frac{2}{5} \tan^{-1}(x+1)$$

● ILLUSTRATION 1 In Example 1 some steps are saved if instead of (1) the original fraction is expressed as

$$\frac{x^2-2x-3}{(x-1)(x^2+2x+2)} \equiv \frac{D(2x+2)+E}{x^2+2x+2} + \frac{F}{x-1}$$

Note: We write $D(2x+2)+E$ instead of $Ax+B$ because

$$2x+2 = D_x(x^2+2x+2)$$

Then, solving for D, E, and F we obtain

$$D = \tfrac{9}{10} \qquad E = -\tfrac{2}{5} \qquad F = -\tfrac{4}{5}$$

giving (5) directly. ●

Case 4: The factors of $Q(x)$ are linear and quadratic, and some of the quadratic factors are repeated.

If ax^2+bx+c is a p-fold quadratic factor of $Q(x)$, then, corresponding to this factor $(ax^2+bx+c)^p$, we have the sum of the following p partial fractions:

$$\frac{A_1x+B_1}{(ax^2+bx+c)^p} + \frac{A_2x+B_2}{(ax^2+bx+c)^{p-1}} + \cdots + \frac{A_px+B_p}{(ax^2+bx+c)}$$

● ILLUSTRATION 2 If the denominator contains the factor $(x^2-5x+2)^3$, we have, corresponding to this factor,

$$\frac{Ax+B}{(x^2-5x+2)^3} + \frac{Cx+D}{(x^2-5x+2)^2} + \frac{Ex+F}{x^2-5x+2}$$

or, more conveniently,

$$\frac{A(2x-5)+B}{(x^2-5x+2)^3} + \frac{C(2x-5)+D}{(x^2-5x+2)^2} + \frac{E(2x-5)+F}{x^2-5x+2}$$ ●

Example 2 Find

$$\int \frac{(x-2)\,dx}{x(x^2-4x+5)^2}$$

Solution

$$\frac{x-2}{x(x^2-4x+5)^2} \equiv \frac{A}{x} + \frac{B(2x-4)+C}{(x^2-4x+5)^2} + \frac{D(2x-4)+E}{x^2-4x+5} \tag{6}$$

Multiplying on both sides of (6) by the lowest common denominator we have

$$x - 2 \equiv A(x^2 - 4x + 5)^2 + x(2Bx - 4B + C) + x(x^2 - 4x + 5)(2Dx - 4D + E) \quad (7)$$

$$x - 2 \equiv Ax^4 + 16Ax^2 + 25A - 8Ax^3 + 10Ax^2 - 40Ax + 2Bx^2 - 4Bx + Cx + 2Dx^4 - 12Dx^3 + Ex^3 + 26Dx^2 - 4Ex^2 - 20Dx + 5Ex$$

$$x - 2 \equiv (A + 2D)x^4 + (-8A - 12D + E)x^3 + (26A + 2B + 26D - 4E)x^2 + (-40A - 4B + C - 20D + 5E)x + 25A \quad (8)$$

The value of A can be computed from (7) by substituting 0 for x. If we equate coefficients in (8) and solve the resulting equations simultaneously, we obtain

$$A = -\tfrac{2}{25} \qquad B = \tfrac{1}{5} \qquad C = \tfrac{1}{5} \qquad D = \tfrac{1}{25} \qquad E = -\tfrac{4}{25}$$

Therefore

$$\int \frac{(x-2)\,dx}{x(x^2 - 4x + 5)^2}$$

$$= -\frac{2}{25}\int \frac{dx}{x} + \frac{1}{5}\int \frac{(2x-4)\,dx}{(x^2 - 4x + 5)^2} + \frac{1}{5}\int \frac{dx}{(x^2 - 4x + 5)^2} + \frac{1}{25}\int \frac{(2x-4)\,dx}{x^2 - 4x + 5} - \frac{4}{25}\int \frac{dx}{x^2 - 4x + 5}$$

$$= -\frac{2}{25}\ln|x| - \frac{1}{5(x^2 - 4x + 5)} + \frac{1}{5}\int \frac{dx}{[(x^2 - 4x + 4) + 1]^2} + \frac{1}{25}\ln|x^2 - 4x + 5| - \frac{4}{25}\int \frac{dx}{(x^2 - 4x + 4) + 1} \quad (9)$$

We evaluate separately the integrals in the third and fifth terms on the right side of (9).

$$\int \frac{dx}{[(x^2 - 4x + 4) + 1]^2} = \int \frac{dx}{[(x-2)^2 + 1]^2}$$

Let $x - 2 = \tan\theta$, where $0 \le \theta < \frac{1}{2}\pi$, if $x \ge 2$ and $-\frac{1}{2}\pi < \theta < 0$ if $x < 2$. Then $dx = \sec^2\theta\, d\theta$ and $(x-2)^2 + 1 = \tan^2\theta + 1 = \sec^2\theta$. Hence

$$\int \frac{dx}{[(x-2)^2 + 1]^2} = \int \frac{\sec^2\theta\, d\theta}{(\tan^2\theta + 1)^2}$$

$$= \int \frac{\sec^2\theta\, d\theta}{\sec^4\theta}$$

$$= \int \frac{d\theta}{\sec^2\theta}$$

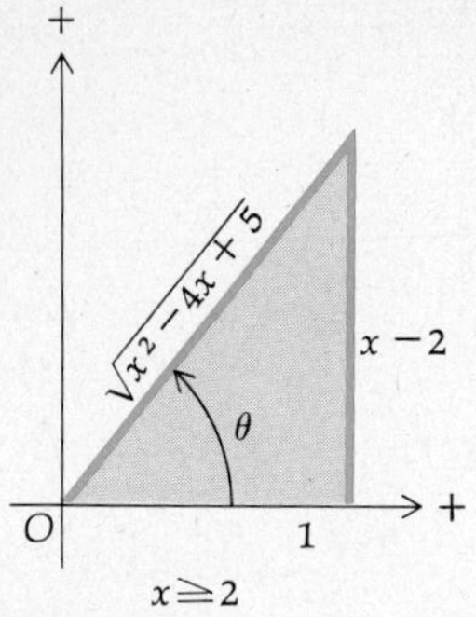

FIGURE 10.5.1

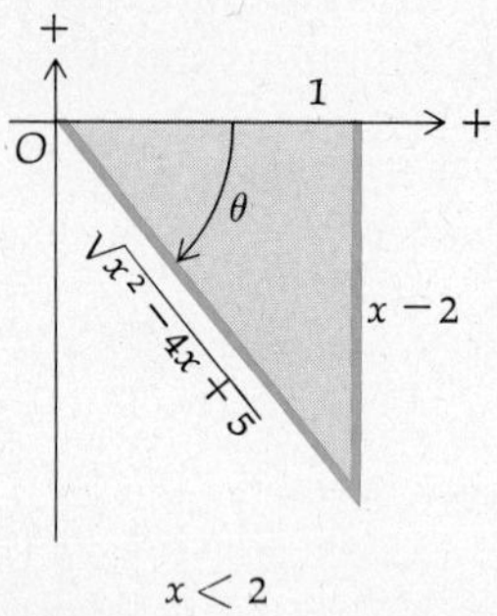

FIGURE 10.5.2

$$= \int \cos^2 \theta \, d\theta$$

$$= \int \frac{1 + \cos 2\theta}{2} \, d\theta$$

$$= \frac{\theta}{2} + \frac{1}{4} \sin 2\theta + C_1$$

$$= \frac{\theta}{2} + \frac{1}{2} \sin \theta \cos \theta + C_1$$

Because $\tan \theta = x - 2$ and $-\frac{1}{2}\pi < \theta < \frac{1}{2}\pi$, $\theta = \tan^{-1}(x - 2)$. We find $\sin \theta$ and $\cos \theta$ from Figs. 10.5.1 (if $x \geq 2$) and 10.5.2 (if $x < 2$). In either case

$$\sin \theta = \frac{x-2}{\sqrt{x^2 - 4x + 5}} \quad \text{and} \quad \cos \theta = \frac{1}{\sqrt{x^2 - 4x + 5}}$$

So

$$\int \frac{dx}{[(x-2)^2 + 1]^2} = \frac{1}{2} \tan^{-1}(x-2) + \frac{1}{2} \cdot \frac{x-2}{\sqrt{x^2-4x+5}} \cdot \frac{1}{\sqrt{x^2-4x+5}} + C_1$$

Thus

$$\int \frac{dx}{[(x-2)^2 + 1]^2} = \frac{1}{2} \tan^{-1}(x-2) + \frac{x-2}{2(x^2-4x+5)} + C_1 \qquad (10)$$

Now, considering the other integral on the right side of (9), we have

$$\int \frac{dx}{(x^2 - 4x + 4) + 1} = \int \frac{dx}{(x-2)^2 + 1} = \tan^{-1}(x-2) + C_2 \qquad (11)$$

Substituting from (10) and (11) into (9) we get

$$\int \frac{(x-2)\,dx}{x(x^2-4x+5)^2}$$

$$= -\frac{2}{25} \ln |x| - \frac{1}{5(x^2-4x+5)} + \frac{1}{10} \tan^{-1}(x-2) + \frac{x-2}{10(x^2-4x+5)}$$

$$+ \frac{1}{25} \ln |x^2 - 4x + 5| - \frac{4}{25} \tan^{-1}(x-2) + C$$

$$= \frac{1}{25} \ln \left| \frac{x^2-4x+5}{x^2} \right| - \frac{3}{50} \tan^{-1}(x-2) + \frac{x-4}{10(x^2-4x+5)} + C$$

EXERCISES 10.5

In Exercises 1 through 20, evaluate the indefinite integral.

1. $\int \frac{dx}{2x^3 + x}$

2. $\int \frac{(x+4)\,dx}{x(x^2+4)}$

3. $\int \frac{dx}{16x^4 - 1}$

4. $\int \frac{(x^2 - 4x - 4)\,dx}{x^3 - 2x^2 + 4x - 8}$

5. $\int \frac{(t^2 + t + 1)\,dt}{(2t+1)(t^2+1)}$

6. $\int \frac{3w^3 + 13w + 4}{w^3 + 4w} \, dw$

7. $\int \frac{(x^2 + x)\,dx}{x^3 - x^2 + x - 1}$
8. $\int \frac{dx}{9x^4 + x^2}$
9. $\int \frac{dx}{x^3 + x^2 + x}$
10. $\int \frac{(x + 3)\,dx}{4x^4 + 4x^3 + x^2}$
11. $\int \frac{(2x^2 - x + 2)\,dx}{x^5 + 2x^3 + x}$
12. $\int \frac{(2x^3 + 9x)\,dx}{(x^2 + 3)(x^2 - 2x + 3)}$
13. $\int \frac{(5z^3 - z^2 + 15z - 10)\,dz}{(z^2 - 2z + 5)^2}$
14. $\int \frac{dt}{(t^2 + 1)^3}$
15. $\int \frac{(x^2 + 2x - 1)\,dx}{27x^3 - 1}$
16. $\int \frac{e^{5x}\,dx}{(e^{2x} + 1)^2}$
17. $\int \frac{18\,dx}{(4x^2 + 9)^2}$
18. $\int \frac{(2x^2 + 3x + 2)\,dx}{x^3 + 4x^2 + 6x + 4}$
19. $\int \frac{(\sec^2 x + 1)\sec^2 x\,dx}{1 + \tan^3 x}$
20. $\int \frac{(6w^4 + 4w^3 + 9w^2 + 24w + 32)\,dw}{(w^3 + 8)(w^2 + 3)}$

In Exercises 21 through 29, evaluate the definite integral.

21. $\int_1^4 \frac{(4 + 5x^2)\,dx}{x^3 + 4x}$
22. $\int_0^1 \frac{x\,dx}{x^3 + 2x^2 + x + 2}$
23. $\int_3^4 \frac{(5x^3 - 4x)\,dx}{x^4 - 16}$
24. $\int_0^1 \frac{9\,dx}{8x^3 + 1}$
25. $\int_{-1}^0 \frac{x^2\,dx}{(2x^2 + 2x + 1)^2}$
26. $\int_0^{1/2} \frac{(x + 1)\,dx}{x^3 - 1}$
27. $\int_0^1 \frac{(x^2 + 3x + 3)\,dx}{x^3 + x^2 + x + 1}$
28. $\int_{\pi/6}^{\pi/2} \frac{\cos x\,dx}{\sin x + \sin^3 x}$
29. $\int_{\ln 2}^{\ln 5} \frac{12\,dt}{e^{2t} + 16}$

30. Find the area of the region bounded by the curve $y(x^2 + 1)^3 = x^3$, the x axis, the y axis, and the line $x = 1$.

31. Find the area of the region bounded by the curve $y(x^3 + 8) = 4$, the x axis, the y axis, and the line $x = 1$.

32. Find the volume of the solid of revolution generated by revolving the region of Exercise 31 about the y axis.

33. A particle is moving along a straight line so that if v cm/sec is the velocity of the particle at t sec, then

$$v = \frac{t^2 - t + 1}{(t + 2)^2(t^2 + 1)}$$

Find a formula for the distance traveled by the particle from the time when $t = 0$ to the time when $t = t_1$.

34. Use methods previous to this section (that is, without partial fractions) to evaluate the integrals:

(a) $\int \frac{(x^2 - 4x + 6)\,dx}{x^3 - 6x^2 + 18x}$ (b) $\int \frac{3x + 1}{(x + 2)^4}\,dx$

10.6 INTEGRALS YIELDING INVERSE HYPERBOLIC FUNCTIONS

The inverse hyperbolic functions can be applied in integration, and sometimes their use shortens the computation considerably. However, no new types of integrals are evaluated by this procedure. Only new forms of the results are obtained.

From formula (15) in Sec. 9.8,

$$D_x(\sinh^{-1} u) = \frac{1}{\sqrt{u^2 + 1}} D_x u$$

from which we obtain the integration formula

$$\int \frac{du}{\sqrt{u^2 + 1}} = \sinh^{-1} u + C$$

If formula (8) of Sec. 9.8 is used to express $\sinh^{-1} u$ as a natural logarithm, the following theorem is obtained.

10.6.1 Theorem

$$\int \frac{du}{\sqrt{u^2+1}} = \sinh^{-1} u + C = \ln(u + \sqrt{u^2+1}) + C$$

Formula (16) in Sec. 9.8 is

$$D_x(\cosh^{-1} u) = \frac{1}{\sqrt{u^2-1}} D_x u \qquad \text{where } u > 1$$

from which it follows that

$$\int \frac{du}{\sqrt{u^2-1}} = \cosh^{-1} u + C \qquad \text{if } u > 1$$

By combining this with formula (9) in Sec. 9.8, we obtain the following theorem.

10.6.2 Theorem

$$\int \frac{du}{\sqrt{u^2-1}} = \cosh^{-1} u + C = \ln(u + \sqrt{u^2-1}) + C \qquad \text{if } u > 1$$

Formulas (17) and (18) of Sec. 9.8 are, respectively,

$$D_x(\tanh^{-1} u) = \frac{1}{1-u^2} D_x u \qquad \text{where } |u| < 1$$

$$D_x(\coth^{-1} u) = \frac{1}{1-u^2} D_x u \qquad \text{where } |u| > 1$$

From the above two formulas we get

$$\int \frac{du}{1-u^2} = \begin{cases} \tanh^{-1} u + C & \text{if } |u| < 1 \\ \coth^{-1} u + C & \text{if } |u| > 1 \end{cases}$$

With this formula and (10) and (11) of Sec. 9.8 we have the next theorem.

10.6.3 Theorem

$$\int \frac{du}{1-u^2} = \begin{cases} \tanh^{-1} u + C & \text{if } |u| < 1 \\ \coth^{-1} u + C & \text{if } |u| > 1 \end{cases} = \frac{1}{2} \ln \left| \frac{1+u}{1-u} \right| + C \quad \text{if } u \neq 1$$

We also have the three formulas given in the following theorem.

10.6.4 Theorem

$$\int \frac{du}{\sqrt{u^2+a^2}} = \sinh^{-1} \frac{u}{a} + C = \ln(u + \sqrt{u^2+a^2}) + C \quad \text{if } a > 0 \qquad (1)$$

$$\int \frac{du}{\sqrt{u^2-a^2}} = \cosh^{-1} \frac{u}{a} + C = \ln(u + \sqrt{u^2-a^2}) + C \quad \text{if } u > a > 0 \qquad (2)$$

$$\int \frac{du}{a^2 - u^2} = \left\{ \begin{matrix} \frac{1}{a}\tanh^{-1}\frac{u}{a} + C & \text{if } |u| < a \\ \frac{1}{a}\coth^{-1}\frac{u}{a} + C & \text{if } |u| > a \end{matrix} \right\} = \frac{1}{2a}\ln\left|\frac{a+u}{a-u}\right| + C \qquad (3)$$

if $u \neq a$ and $a \neq 0$

Proof The formulas can be proved by finding the derivative of the right side and obtaining the integrand, or else more directly by using a hyperbolic function substitution. The proof of (1) by differentiating the right side is as follows:

$$D_u\left(\sinh^{-1}\frac{u}{a}\right) = \frac{1}{\sqrt{\left(\frac{u}{a}\right)^2 + 1}} \cdot \frac{1}{a} = \frac{\sqrt{a^2}}{\sqrt{u^2 + a^2}} \cdot \frac{1}{a}$$

and because $a > 0$, $\sqrt{a^2} = a$; thus

$$D_u\left(\sinh^{-1}\frac{u}{a}\right) = \frac{1}{\sqrt{u^2 + a^2}}$$

To obtain the natural logarithm representation we use formula (8) of Sec. 9.8 and we have

$$\begin{aligned}\sinh^{-1}\frac{u}{a} &= \ln\left(\frac{u}{a} + \sqrt{\left(\frac{u}{a}\right)^2 + 1}\right) \\ &= \ln\left(\frac{u}{a} + \frac{\sqrt{u^2 + a^2}}{a}\right) \\ &= \ln(u + \sqrt{u^2 + a^2}) - \ln a\end{aligned}$$

Therefore

$$\begin{aligned}\sinh^{-1}\frac{u}{a} + C &= \ln(u + \sqrt{u^2 + a^2}) - \ln a + C \\ &= \ln(u + \sqrt{u^2 + a^2}) + C_1\end{aligned}$$

where $C_1 = C - \ln a$.

The proof of (2) by a hyperbolic function substitution makes use of the identity $\cosh^2 x - \sinh^2 x = 1$. Let $u = a \cosh x$, where $x > 0$; then $du = a \sinh x\, dx$, and $x = \cosh^{-1}(u/a)$. Substituting in the integrand we get

$$\begin{aligned}\int \frac{du}{\sqrt{u^2 - a^2}} &= \int \frac{a \sinh x\, dx}{\sqrt{a^2\cosh^2 x - a^2}} \\ &= \int \frac{a \sinh x\, dx}{\sqrt{a^2}\sqrt{\cosh^2 x - 1}} \\ &= \int \frac{a \sinh x\, dx}{\sqrt{a^2}\sqrt{\sinh^2 x}}\end{aligned}$$

Because $a > 0$, $\sqrt{a^2} = a$. Furthermore, because $x > 0$, $\sinh x > 0$; so $\sqrt{\sinh^2 x} = \sinh x$. Then

$$\int \frac{du}{\sqrt{u^2 - a^2}} = \int \frac{a \sinh x \, dx}{a \sinh x}$$

$$= \int dx$$

$$= x + C$$

$$= \cosh^{-1} \frac{u}{a} + C$$

The natural logarithm representation may be obtained in a way similar to the way we obtained the one for formula (1). The proof of formula (3) is left as an exercise (see Exercise 23). ■

The formulas of Theorems 10.6.1 through 10.6.4 give alternate representations of the integral in question. In evaluating a definite integral in which one of these forms occurs and a table of hyperbolic functions is available, the inverse hyperbolic function representation may be easier to use. Also the inverse hyperbolic function representation is a less cumbersome one to write for the indefinite integral. Observe that the natural logarithm form of formula (3) was obtained in Sec. 10.4 by using partial fractions.

Example 1 Evaluate

$$\int \frac{dx}{\sqrt{x^2 - 6x + 13}}$$

Solution We apply formula (1) after rewriting the denominator.

$$\int \frac{dx}{\sqrt{x^2 - 6x + 13}} = \int \frac{dx}{\sqrt{(x^2 - 6x + 9) + 4}}$$

$$= \int \frac{dx}{\sqrt{(x-3)^2 + 4}}$$

$$= \sinh^{-1}\left(\frac{x-3}{2}\right) + C$$

$$= \ln(x - 3 + \sqrt{x^2 - 6x + 13}) + C$$

Example 2 Evaluate

$$\int_6^{10} \frac{dx}{\sqrt{x^2 - 25}}$$

Solution From formula (2),

$$\int_6^{10} \frac{dx}{\sqrt{x^2 - 25}} = \cosh^{-1} \frac{x}{5}\bigg]_6^{10} = \cosh^{-1} 2 - \cosh^{-1} 1.2$$

Using Table 4 in the appendix we find

$$\cosh^{-1} 2 - \cosh^{-1}(1.2) = 1.32 - 0.62 = 0.70$$

Instead of applying the formulas, integrals of the forms of those in Theorems 10.6.1 through 10.6.4 can be obtained by using a hyperbolic function substitution and proceeding in a manner similar to that using a trigonometric substitution.

Example 3 Evaluate the integral of Example 1 without using a formula but by using a hyperbolic function substitution.

Solution In the solution of Example 1 the given integral was rewritten as

$$\int \frac{dx}{\sqrt{(x-3)^2+4}}$$

If we let $x - 3 = 2 \sinh u$, then $dx = 2 \cosh u\, du$ and $u = \sinh^{-1} \frac{1}{2}(x - 3)$. Therefore

$$\begin{aligned}
\int \frac{dx}{\sqrt{(x-3)^2+4}} &= \int \frac{2 \cosh u\, du}{\sqrt{4 \sinh^2 u + 4}} \\
&= \int \frac{2 \cosh u\, du}{2\sqrt{\sinh^2 u + 1}} \\
&= \int \frac{\cosh u\, du}{\sqrt{\cosh^2 u}} \\
&= \int \frac{\cosh u\, du}{\cosh u} \\
&= \int du \\
&= u + C \\
&= \sinh^{-1}\left(\frac{x-3}{2}\right) + C
\end{aligned}$$

which agrees with the result of Example 1.

EXERCISES 10.6

In Exercises 1 through 16, express the indefinite integral in terms of an inverse hyperbolic function and as a natural logarithm.

1. $\int \frac{dx}{\sqrt{4+x^2}}$ **2.** $\int \frac{dx}{\sqrt{4x^2-9}}$ **3.** $\int \frac{x\, dx}{\sqrt{x^4-1}}$ **4.** $\int \frac{dx}{25-x^2}$

5. $\int \frac{dx}{9x^2-16}$ **6.** $\int \frac{dx}{4e^x-e^{-x}}$ **7.** $\int \frac{\cos x\, dx}{\sqrt{4-\cos^2 x}}$ **8.** $\int \frac{dx}{\sqrt{x^2-4x+1}}$

9. $\int \frac{dt}{\sqrt{5-e^{-2t}}}$ **10.** $\int \frac{dw}{4w-w^2-3}$ **11.** $\int \frac{dx}{2-4x-x^2}$ **12.** $\int \frac{dx}{\sqrt{25x^2+9}}$

13. $\int \frac{dx}{x^2+10x+24}$ **14.** $\int \frac{dz}{\sqrt{9z^2-6z-8}}$ **15.** $\int \frac{3\, dw}{w\sqrt{4\ln^2 w+9}}$ **16.** $\int \frac{3x\, dx}{\sqrt{x^4+6x^2+5}}$

In Exercises 17 through 22, evaluate the definite integral and express the answer in terms of a natural logarithm.

17. $\int_3^5 \frac{dx}{\sqrt{x^2-4}}$ **18.** $\int_{-4}^{-3} \frac{dx}{1-x^2}$ **19.** $\int_{-1/2}^{1/2} \frac{dx}{1-x^2}$

20. $\int_1^2 \frac{dx}{\sqrt{x^2+2x}}$ **21.** $\int_2^3 \frac{dx}{\sqrt{9x^2-12x-5}}$ **22.** $\int_{-2}^2 \frac{dx}{\sqrt{16+x^2}}$

23. Prove formula (3) by using a hyperbolic function substitution.

24. A curve goes through the point $(0, a)$, $a > 0$, and the slope at any point is $\sqrt{y^2/a^2 - 1}$. Prove that the curve is a catenary.

25. A man wearing a parachute falls out of an airplane and when the parachute opens his velocity is 200 ft/sec. If v ft/sec is his velocity t sec after the parachute opens.

$$\frac{324}{g} \cdot \frac{dv}{dt} = 324 - v^2$$

Solve this differential equation to obtain

$$t = \frac{18}{g}\left(\coth^{-1}\frac{v}{18} - \coth^{-1}\frac{100}{9}\right)$$

26. Show that $\sinh^{-1} u = \ln(u + \sqrt{u^2+1})$ by using a trigonometric substitution to evaluate $\int \frac{du}{\sqrt{u^2+1}}$.

27. The region bounded by the curve $y = (16 - x^2)^{-1/2}$, the x axis, and the lines $x = -2$ and $x = 3$ is revolved about the x axis. Show that the volume of the solid generated is $\frac{1}{4}\pi[\tanh^{-1}\frac{3}{4} - \tanh^{-1}(-\frac{1}{2})]$.

10.7 INTEGRATION OF RATIONAL FUNCTIONS OF SINE AND COSINE

If an integrand is a rational function of $\sin x$ and $\cos x$, it can be reduced to a rational function of z by the substitution

$$z = \tan \tfrac{1}{2}x \tag{1}$$

as will be shown by an example. From the identity $\cos 2y \equiv 2\cos^2 y - 1$ with $y = \frac{1}{2}x$,

$$\cos x = 2\cos^2 \tfrac{1}{2}x - 1 = \frac{2}{\sec^2 \frac{1}{2}x} - 1 = \frac{2}{1+\tan^2 \frac{1}{2}x} - 1 = \frac{2}{1+z^2} - 1$$

So

$$\cos x = \frac{1-z^2}{1+z^2} \tag{2}$$

In a similar manner, from the identity $\sin 2y \equiv 2 \sin y \cos y$ with $y = \frac{1}{2}x$

$$\sin x = 2 \sin \tfrac{1}{2}x \cos \tfrac{1}{2}x = 2 \cdot \frac{\sin \frac{1}{2}x \cos^2 \frac{1}{2}x}{\cos \frac{1}{2}x}$$

$$= 2 \tan \tfrac{1}{2}x \cdot \frac{1}{\sec^2 \frac{1}{2}x} = \frac{2 \tan \frac{1}{2}x}{1 + \tan^2 \frac{1}{2}x}$$

Thus

$$\sin x = \frac{2z}{1+z^2} \tag{3}$$

Because $z = \tan \frac{1}{2}x$,

$$dz = \tfrac{1}{2} \sec^2 \tfrac{1}{2}x \, dx = \tfrac{1}{2}(1 + \tan^2 \tfrac{1}{2}x) \, dx$$

Hence

$$dx = \frac{2\,dz}{1 + z^2} \tag{4}$$

Example 1 Find

$$\int \frac{dx}{1 - \sin x + \cos x}$$

Solution Let $z = \tan \frac{1}{2}x$ and use (2), (3), and (4).

$$\begin{aligned}
\int \frac{dx}{1 - \sin x + \cos x} &= \int \frac{\dfrac{2\,dz}{1 + z^2}}{1 - \dfrac{2z}{1 + z^2} + \dfrac{1 - z^2}{1 + z^2}} \\
&= 2 \int \frac{dz}{(1 + z^2) - 2z + (1 - z^2)} \\
&= 2 \int \frac{dz}{2 - 2z} \\
&= \int \frac{dz}{1 - z} \\
&= -\ln |1 - z| + C \\
&= -\ln |1 - \tan \tfrac{1}{2}x| + C
\end{aligned}$$

Example 2 Find $\int \sec x \, dx$ by using the substitution of this section.

Solution

$$\int \sec x \, dx = \int \frac{dx}{\cos x}$$

Letting $z = \tan \frac{1}{2}x$ and using (2) and (4) we get

$$\int \frac{dx}{\cos x} = \int \frac{2\,dz}{1 + z^2} \cdot \frac{1 + z^2}{1 - z^2} = 2 \int \frac{dz}{1 - z^2}$$

Using formula (8) from Sec. 10.4 we have

$$2 \int \frac{dz}{1 - z^2} = \ln \left| \frac{1 + z}{1 - z} \right| + C$$

Therefore

$$\int \sec x\,dx = \ln\left|\frac{1+\tan\frac{1}{2}x}{1-\tan\frac{1}{2}x}\right| + C \tag{5}$$

Equation (5) can be written in another form by noting that $1 = \tan\frac{1}{4}\pi$ and using the identity $\tan(\alpha+\beta) = (\tan\alpha + \tan\beta)/(1 - \tan\alpha\tan\beta)$. So we have

$$\int \sec x\,dx = \ln\left|\frac{\tan\frac{1}{4}\pi + \tan\frac{1}{2}x}{1-\tan\frac{1}{4}\pi\cdot\tan\frac{1}{2}x}\right| + C$$

or, equivalently,

$$\int \sec x\,dx = \ln\left|\tan\left(\tfrac{1}{4}\pi + \tfrac{1}{2}x\right)\right| + C \tag{6}$$

Formula (6) is an alternative form of the formula

$$\int \sec x\,dx = \ln|\sec x + \tan x| + C$$

which is obtained by the trick of multiplying the numerator and denominator of the integrand by $\sec x + \tan x$.

It is worth noting that still another form for $\int \sec x\,dx$ can be obtained as follows:

$$\int \sec x\,dx = \int \frac{dx}{\cos x} = \int \frac{\cos x\,dx}{\cos^2 x} = \int \frac{\cos x\,dx}{1-\sin^2 x}$$

Substituting $u = \sin x$, $du = \cos x\,dx$, we have

$$\int \sec x\,dx = \int \frac{du}{1-u^2} = \frac{1}{2}\ln\left|\frac{1+u}{1-u}\right| + C = \ln\left|\frac{1+\sin x}{1-\sin x}\right|^{1/2} + C$$

Because $-1 \le \sin x \le 1$ for all x, $1 + \sin x$ and $1 - \sin x$ are nonnegative; hence the absolute-value bars can be removed and we have

$$\int \sec x\,dx = \ln\left(\frac{1+\sin x}{1-\sin x}\right)^{1/2} + C = \ln\sqrt{\frac{1+\sin x}{1-\sin x}} + C \tag{7}$$

EXERCISES 10.7

In Exercises 1 through 18, evaluate the indefinite integral.

1. $\displaystyle\int \frac{3\,dx}{8+7\cos x}$

2. $\displaystyle\int \frac{dx}{1+\sin x}$

3. $\displaystyle\int \frac{3\,dx}{7+8\cos x}$

4. $\displaystyle\int \frac{dx}{\sin x - \cos x + 2}$

5. $\displaystyle\int \frac{dx}{\sin x + \tan x}$

6. $\displaystyle\int \frac{dx}{5+4\cos x}$

7. $\displaystyle\int \frac{dx}{3-5\sin x}$

8. $\displaystyle\int \frac{dx}{\tan x - 1}$

9. $\displaystyle\int \frac{dx}{4\sin x - 3\cos x}$

10. $\displaystyle\int \frac{dx}{3\cos x - 2\sin x + 3}$

11. $\displaystyle\int \frac{8\,dx}{3\cos 2x + 1}$

12. $\displaystyle\int \frac{\cos x\,dx}{3\cos x - 5}$

13. $\displaystyle\int \frac{\cos x\, dx}{1 + 2\cos x}$ **14.** $\displaystyle\int \frac{dx}{\sin x - \tan x}$ **15.** $\displaystyle\int \frac{dx}{\cot x(6 + 7\cos 2x)}$ **16.** $\displaystyle\int \frac{dx}{\cot 2x(1 - \cos 2x)}$

17. $\displaystyle\int \frac{dx}{2\sin x + 2\cos x + 3}$ **18.** $\displaystyle\int \frac{5\, dx}{6 + 4\sec x}$

In Exercises 19 through 24, evaluate the definite integral.

19. $\displaystyle\int_0^{\pi/2} \frac{dx}{5\sin x + 3}$ **20.** $\displaystyle\int_0^{\pi/4} \frac{8\, dx}{\tan x + 1}$ **21.** $\displaystyle\int_{-\pi/3}^{\pi/2} \frac{3\, dx}{2\cos x + 1}$

22. $\displaystyle\int_0^{\pi/2} \frac{dx}{3 + \cos 2x}$ **23.** $\displaystyle\int_{\pi/6}^{\pi/3} \frac{3\, dx}{2\sin 2x + 1}$ **24.** $\displaystyle\int_0^{\pi/2} \frac{\sin 2x\, dx}{2 + \cos x}$

25. Show that formula (7) of this section is equivalent to the formula $\int \sec x\, dx = \ln|\sec x + \tan x| + C$. (*Hint:* Multiply the numerator and denominator under the radical sign by $(1 + \sin x)$.)

26. By using the substitution $z = \tan \frac{1}{2}x$, prove that

$$\int \csc x\, dx = \ln \sqrt{\frac{1 - \cos x}{1 + \cos x}} + C$$

27. Show that the result in Exercise 26 is equivalent to the formula $\int \csc x\, dx = \ln|\csc x - \cot x| + C$. (*Hint:* Use a method similar to that suggested in the hint for Exercise 25.)

28. Evaluate the integral

$$\int \frac{\tan \frac{1}{2}x}{\sin x}\, dx$$

by two methods: (a) let $z = \tan \frac{1}{2}x$; (b) let $u = \frac{1}{2}x$ and obtain an integral involving trigonometric functions of u.

29. Use the substitution of this section, $z = \tan \frac{1}{2}x$, to show that $\int \sin x\, dx = -\cos x + C$.

10.8 MISCELLANEOUS SUBSTITUTIONS

If an integrand involves fractional powers of a variable x, the integrand can be simplified by the substitution

$$x = z^n$$

where n is the lowest common denominator of the denominators of the exponents. This is illustrated by the following example.

Example 1 Find

$$\int \frac{\sqrt{x}\, dx}{1 + \sqrt[3]{x}}$$

Solution We let $x = z^6$; then $dx = 6z^5\, dz$. So

$$\int \frac{x^{1/2}\, dx}{1 + x^{1/3}} = \int \frac{z^3(6z^5\, dz)}{1 + z^2} = 6\int \frac{z^8}{z^2 + 1}\, dz$$

Dividing the numerator by the denominator we have

$$\int \frac{x^{1/2}\, dx}{1 + x^{1/3}} = 6\int \left(z^6 - z^4 + z^2 - 1 + \frac{1}{z^2 + 1}\right) dz$$

$$= 6\left(\tfrac{1}{7}z^7 - \tfrac{1}{5}z^5 + \tfrac{1}{3}z^3 - z + \tan^{-1} z\right) + C$$

$$= \tfrac{6}{7}x^{7/6} - \tfrac{6}{5}x^{5/6} + 2x^{1/2} - 6x^{1/6} + 6\tan^{-1} x^{1/6} + C$$

No general rule can be given to determine a substitution that will result in a simpler integrand. Sometimes a substitution that does not rationalize the given integrand may still result in a simpler integrand.

Example 2 Evaluate $\int x^5\sqrt{x^2+4}\,dx$.

Solution Let $z = \sqrt{x^2+4}$. Then $z^2 = x^2 + 4$, and $2z\,dz = 2x\,dx$. So

$$\int x^5\sqrt{x^2+4}\,dx = \int (x^2)^2\sqrt{x^2+4}\,x\,dx = \int (z^2-4)^2 z(z\,dz)$$

$$= \int (z^6 - 8z^4 + 16z^2)\,dz = \tfrac{1}{7}z^7 - \tfrac{8}{5}z^5 + \tfrac{16}{3}z^3 + C$$

$$= \tfrac{1}{105}z^3[15z^4 - 168z^2 + 560] + C$$

$$= \tfrac{1}{105}(x^2+4)^{3/2}[15(x^2+4)^2 - 168(x^2+4) + 560] + C$$

$$= \tfrac{1}{105}(x^2+4)^{3/2}[15x^4 - 48x^2 + 128] + C$$

Example 3 Evaluate

$$\int \frac{dx}{x^2\sqrt{27x^2+6x-1}}$$

by using the reciprocal substitution $x = 1/z$.

Solution If $x = 1/z$, then $dx = -dz/z^2$. Therefore

$$\int \frac{dx}{x^2\sqrt{27x^2+6x-1}} = \int \frac{\dfrac{-dz}{z^2}}{\left(\dfrac{1}{z^2}\right)\sqrt{\dfrac{27}{z^2} + \dfrac{6}{z} - 1}}$$

$$= -\int \frac{\sqrt{z^2}\,dz}{\sqrt{27+6z-z^2}}$$

$$= -\int \frac{|z|\,dz}{\sqrt{27+6z-z^2}}$$

$$= \begin{cases} -\displaystyle\int \frac{z\,dz}{\sqrt{27+6z-z^2}} & \text{if } z > 0 \\[2ex] \displaystyle\int \frac{z\,dz}{\sqrt{27+6z-z^2}} & \text{if } z < 0 \end{cases} \qquad (1)$$

We evaluate $\int z\,dz/\sqrt{27+6z-z^2}$.

$$\int \frac{z\,dz}{\sqrt{27+6z-z^2}} = -\frac{1}{2}\int \frac{(-2z+6)\,dz}{\sqrt{27+6z-z^2}} + 3\int \frac{dz}{\sqrt{27+6z-z^2}}$$

$$= -\frac{1}{2}\cdot 2\sqrt{27+6z-z^2} + 3\int \frac{dz}{\sqrt{27+9-(z^2-6z+9)}}$$

$$= -\sqrt{27 + 6z - z^2} + 3\int \frac{dz}{\sqrt{36 - (z-3)^2}}$$

$$= -\sqrt{27 + 6z - z^2} + 3\sin^{-1}\frac{z-3}{6} + C$$

$$= -\sqrt{27 + \frac{6}{x} - \frac{1}{x^2}} + 3\sin^{-1}\frac{\frac{1}{x} - 3}{6} + C$$

$$= -\frac{\sqrt{27x^2 + 6x - 1}}{\sqrt{x^2}} + 3\sin^{-1}\frac{1-3x}{6x} + C$$

$$= -\frac{\sqrt{27x^2 + 6x - 1}}{|x|} + 3\sin^{-1}\frac{1-3x}{6x} + C \qquad (2)$$

Substituting from (2) into (1) we get

$$\int \frac{dx}{x^2\sqrt{27x^2 + 6x - 1}}$$

$$= \begin{cases} \dfrac{\sqrt{27x^2+6x-1}}{|x|} - 3\sin^{-1}\dfrac{1-3x}{6x} + C & \text{if } x > 0 \\[2ex] -\dfrac{\sqrt{27x^2+6x-1}}{|x|} + 3\sin^{-1}\dfrac{1-3x}{6x} + C & \text{if } x < 0 \end{cases}$$

$$= \begin{cases} \dfrac{\sqrt{27x^2+6x-1}}{x} - 3\sin^{-1}\dfrac{1-3x}{6x} + C & \text{if } x > 0 \\[2ex] \dfrac{\sqrt{27x^2+6x-1}}{x} + 3\sin^{-1}\dfrac{1-3x}{6x} + C & \text{if } x < 0 \end{cases}$$

EXERCISES 10.8

In Exercises 1 through 19, evaluate the indefinite integral. In Exercises 14 through 19, evaluate the indefinite integral by using the reciprocal substitution $x = 1/z$.

1. $\displaystyle\int \frac{x\,dx}{3 + \sqrt{x}}$ **2.** $\displaystyle\int \frac{dx}{\sqrt[3]{x} - x}$ **3.** $\displaystyle\int \frac{dx}{x\sqrt{1 + 4x}}$ **4.** $\displaystyle\int x(1 + x)^{2/3}\,dx$

5. $\displaystyle\int \frac{\sqrt{1 + x}}{1 - x}\,dx$ **6.** $\displaystyle\int \frac{dx}{3 + \sqrt{x + 2}}$ **7.** $\displaystyle\int \frac{dx}{1 + \sqrt[3]{x - 2}}$ **8.** $\displaystyle\int \frac{dx}{2\sqrt[4]{x} + \sqrt{x}}$

9. $\displaystyle\int \frac{dx}{\sqrt{2x} - \sqrt{x + 4}}$ **10.** $\displaystyle\int \frac{x^3\,dx}{\sqrt{5x^2 + 4}}$ **11.** $\displaystyle\int \frac{(2x^5 + 3x^2)\,dx}{\sqrt{1 + 2x^3}}$ **12.** $\displaystyle\int \frac{dx}{\sqrt{\sqrt{x} + 1}}$

13. $\displaystyle\int \frac{dx}{\sqrt{x}\sqrt[3]{x}(1 + \sqrt[3]{x})^2}$ **14.** $\displaystyle\int \frac{dx}{x\sqrt{x^2 + 4x - 4}}$ **15.** $\displaystyle\int \frac{dx}{x\sqrt{1 + x + x^2}}$ **16.** $\displaystyle\int \frac{dx}{x^2\sqrt{1 + 2x + 3x^2}}$

17. $\displaystyle\int \frac{dx}{x\sqrt{9x^2 + 4x - 1}}$ **18.** $\displaystyle\int \frac{dx}{x\sqrt{x^2 + 2x - 1}}$ **19.** $\displaystyle\int \frac{dx}{x^2\sqrt{5x^2 + 8x - 1}}$

20. Evaluate the integral in Exercise 18 by using the substitution $\sqrt{x^2 + 2x - 1} = z - x$.

In Exercises 21 through 28, evaluate the definite integral.

21. $\displaystyle\int_0^4 \frac{dx}{1+\sqrt{x}}$ **22.** $\displaystyle\int_0^1 \frac{x^{3/2}}{x+1}\,dx$ **23.** $\displaystyle\int_{1/2}^2 \frac{dx}{\sqrt{2x}(\sqrt{2x}+9)}$ **24.** $\displaystyle\int_{16}^{18} \frac{dx}{\sqrt{x}-\sqrt[4]{x^3}}$

25. $\displaystyle\int_{1/3}^3 \frac{\sqrt[3]{x-x^3}}{x^4}\,dx$ **26.** $\displaystyle\int_1^2 \frac{dx}{x\sqrt{x^2+4x-4}}$ **27.** $\displaystyle\int_0^1 \frac{\sqrt{x}}{1+\sqrt[3]{x}}\,dx$ **28.** $\displaystyle\int_2^{11} \frac{x^3\,dx}{\sqrt[3]{x^2+4}}$

29. Evaluate $\displaystyle\int \frac{dx}{x-\sqrt{x}}$ by two methods: (a) let $x = z^2$; (b) write $x - \sqrt{x} = \sqrt{x}(\sqrt{x}-1)$ and let $u = \sqrt{x} - 1$.

REVIEW EXERCISES FOR CHAPTER 10 AND REVIEW OF INTEGRATION

In Exercises 1 through 62, evaluate the indefinite integral.

1. $\displaystyle\int \tan^2 4x \cos^4 4x\,dx$ **2.** $\displaystyle\int \frac{5x^2-3}{x^3-x}\,dx$ **3.** $\displaystyle\int \frac{e^x\,dx}{\sqrt{4-e^x}}$ **4.** $\displaystyle\int \frac{dx}{x^2\sqrt{a^2+x^2}}$

5. $\displaystyle\int \tan^{-1}\sqrt{x}\,dx$ **6.** $\displaystyle\int \frac{dt}{t^4+1}$ **7.** $\displaystyle\int \cos^2 \tfrac{1}{3}x\,dx$ **8.** $\displaystyle\int \frac{\sqrt{x+1}+1}{\sqrt{x+1}-1}\,dx$

9. $\displaystyle\int \frac{x^2+1}{(x-1)^3}\,dx$ **10.** $\displaystyle\int \frac{dy}{\sqrt{y}+1}$ **11.** $\displaystyle\int \sin x \sin 3x\,dx$ **12.** $\displaystyle\int \cos\theta \cos 2\theta\,d\theta$

13. $\displaystyle\int \frac{dx}{x+x^{4/3}}$ **14.** $\displaystyle\int t\sqrt{2t-t^2}\,dt$ **15.** $\displaystyle\int (\sec 3x + \csc 3x)^2\,dx$ **16.** $\displaystyle\int \frac{dx}{\sqrt{e^x-1}}$

17. $\displaystyle\int \frac{2t^3+11t+8}{t^3+4t^2+4t}\,dt$ **18.** $\displaystyle\int x^3 e^{3x}\,dx$ **19.** $\displaystyle\int \frac{x^4+1}{x^4-1}\,dx$ **20.** $\displaystyle\int \frac{\sqrt{x^2-4}}{x^2}\,dx$

21. $\displaystyle\int \sin^4 3x \cos^2 3x\,dx$ **22.** $\displaystyle\int t \sin^2 2t\,dt$ **23.** $\displaystyle\int \frac{dr}{\sqrt{3-4r-r^2}}$ **24.** $\displaystyle\int \frac{4x^2+x-2}{x^3-5x^2+8x-4}\,dx$

25. $\displaystyle\int x^3 \cos x^2\,dx$ **26.** $\displaystyle\int \frac{y\,dy}{9+16y^4}$ **27.** $\displaystyle\int e^{t/2}\cos 2t\,dt$ **28.** $\displaystyle\int \frac{du}{u^{5/8}-u^{1/8}}$

29. $\displaystyle\int \frac{\sin x \cos x}{4+\sin^4 x}\,dx$ **30.** $\displaystyle\int \frac{\sqrt{w-a}}{w}\,dw,\ a>0$ **31.** $\displaystyle\int \sin^5 nx\,dx$ **32.** $\displaystyle\int \frac{dx}{x\ln x(\ln x - 1)}$

33. $\displaystyle\int \csc^5 x\,dx$ **34.** $\displaystyle\int \frac{dx}{5+4\sec x}$ **35.** $\displaystyle\int \frac{2y^2+1}{y^3-6y^2+12y-8}\,dy$ **36.** $\displaystyle\int \frac{x^5\,dx}{(x^2-a^2)^3}$

37. $\displaystyle\int \frac{\sin x\,dx}{1+\cos^2 x}$ **38.** $\displaystyle\int \frac{dx}{x\sqrt{x^2+x+1}}$ **39.** $\displaystyle\int \sqrt{4t-t^2}\,dt$ **40.** $\displaystyle\int \frac{dx}{\sqrt{1-x+3x^2}}$

41. $\displaystyle\int \frac{dx}{x^4-x}$ **42.** $\displaystyle\int \frac{\sqrt{t}-1}{\sqrt{t}+1}\,dt$ **43.** $\displaystyle\int \frac{e^x\,dx}{\sqrt{4-9e^{2x}}}$ **44.** $\displaystyle\int \frac{dx}{5+4\cos 2x}$

45. $\displaystyle\int \cot^2 3x \csc^4 3x\,dx$ **46.** $\displaystyle\int \frac{\cot x\,dx}{3+2\sin x}$ **47.** $\displaystyle\int x^2 \sin^{-1} x\,dx$ **48.** $\displaystyle\int \frac{dx}{x\sqrt{5x-6-x^2}}$

49. $\displaystyle\int \frac{dx}{\sin x - 2\csc x}$ **50.** $\displaystyle\int \cos x \ln(\sin x)\,dx$ **51.** $\displaystyle\int \frac{\cos 3t\,dt}{\sin 3t\sqrt{\sin^2 3t - \frac{1}{4}}}$ **52.** $\displaystyle\int \frac{dx}{(x^2+6x+34)^2}$

53. $\displaystyle\int \frac{\sqrt{x^2+a^2}}{x^4}\,dx$ **54.** $\displaystyle\int \tan x \sin x\,dx$ **55.** $\displaystyle\int \frac{\sin^{-1}\sqrt{2t}}{\sqrt{1-2t}}\,dt$ **56.** $\displaystyle\int \ln(x^2+1)\,dx$

57. $\displaystyle\int \frac{dx}{\sqrt{2+\sqrt{x-1}}}$ **58.** $\displaystyle\int \frac{dx}{2+2\sin x+\cos x}$ **59.** $\displaystyle\int \sqrt{\tan x}\,dx$ **60.** $\displaystyle\int \frac{dx}{\sqrt{1+\sqrt[3]{x}}}$

61. $\displaystyle\int x^n \ln x\,dx$ **62.** $\displaystyle\int \tan^n x \sec^4 x\,dx,\ n>0$

In Exercises 63 through 94, evaluate the definite integral.

63. $\displaystyle\int_0^{\pi} \sqrt{2+2\cos x}\,dx$ **64.** $\displaystyle\int_{1/2}^{1} \sqrt{\frac{1-x}{x}}\,dx$ **65.** $\displaystyle\int_1^2 \frac{2x^2+x+4}{x^3+4x^2}\,dx$ **66.** $\displaystyle\int_0^1 \frac{dx}{e^x+e^{-x}}$

67. $\displaystyle\int_0^2 \frac{t^3\,dt}{\sqrt{4+t^2}}$ **68.** $\displaystyle\int_0^{\pi/2} \sin^3 t\cos^3 t\,dt$ **69.** $\displaystyle\int_{-2}^{2\sqrt{3}} \frac{x^2\,dx}{(16-x^2)^{3/2}}$ **70.** $\displaystyle\int_0^1 \frac{xe^x\,dx}{(1+x)^2}$

71. $\displaystyle\int_0^{\pi/4} \sec^4 x\,dx$ **72.** $\displaystyle\int_0^2 \frac{(1-x)\,dx}{x^2+3x+2}$ **73.** $\displaystyle\int_{\pi/12}^{\pi/8} \cot^3 2y\,dy$ **74.** $\displaystyle\int_0^2 (2^x+x^2)\,dx$

75. $\displaystyle\int_a^{a/2} \frac{\sqrt{(a^2-x^2)^3}}{x^2}\,dx$ **76.** $\displaystyle\int_1^2 (\ln x)^2\,dx$ **77.** $\displaystyle\int_{\sqrt{3}/3}^{1} \frac{(2x^2-2x+1)\,dx}{x^3+x}$ **78.** $\displaystyle\int_{\sqrt{2}/2}^{1} \frac{x^3\,dx}{\sqrt{2-x^2}}$

79. $\displaystyle\int_1^{10} \log_{10}\sqrt{ex}\,dx$ **80.** $\displaystyle\int_0^{2\pi} |\sin x-\cos x|\,dx$ **81.** $\displaystyle\int_1^2 \frac{x+2}{(x+1)^2}\,dx$ **82.** $\displaystyle\int_0^{\sqrt{\pi/2}} xe^{x^2}\cos x^2\,dx$

83. $\displaystyle\int_0^{\pi} |\cos^3 x|\,dx$ **84.** $\displaystyle\int_{-\pi/4}^{\pi/4} |\tan^5 x|\,dx$ **85.** $\displaystyle\int_0^{1/2} \frac{2x\,dx}{x^3-x^2-x+1}$ **86.** $\displaystyle\int_0^1 x^3\sqrt{1+x^2}\,dx$

87. $\displaystyle\int_0^{1/2} \frac{x\,dx}{\sqrt{1-4x^4}}$ **88.** $\displaystyle\int_0^{\pi/12} \frac{dx}{\cos^4 3x}$ **89.** $\displaystyle\int_0^1 \sqrt{2y+y^2}\,dy$ **90.** $\displaystyle\int_0^4 \frac{x^2dx}{x^3+4x^2+5x+2}$

91. $\displaystyle\int_0^{\pi/2} \frac{dt}{12+13\cos t}$ **92.** $\displaystyle\int_0^3 \frac{dr}{(r+2)\sqrt{r+1}}$ **93.** $\displaystyle\int_0^{16} \sqrt{4-\sqrt{x}}\,dx$ **94.** $\displaystyle\int_{2\pi/3}^{\pi} \frac{\sin\frac{1}{2}t}{1+\cos\frac{1}{2}t}\,dt$

In Exercises 95 and 96, obtain the given result by a hyperbolic function substitution.

95. $\displaystyle\int \frac{dx}{x^2\sqrt{a^2+x^2}} = -\frac{1}{a^2}\coth\left(\sinh^{-1}\frac{x}{a}\right)+C,\ a>0$ **96.** $\displaystyle\int \frac{dx}{(a^2-x^2)^{3/2}} = \frac{1}{a^2}\sinh\left(\tanh^{-1}\frac{x}{a}\right)+C,\ a>0$

97. Find the length of the arc of the parabola $y^2 = 6x$ from $x = 6$ to $x = 12$.

98. Find the area of the region bounded by the curve $y = \sin^{-1} 2x$, the line $x = \frac{1}{4}\sqrt{3}$, and the x axis.

99. Find the area of the region enclosed by one loop of the curve $x^2 = y^4(1-y^2)$.

100. Find the length of the arc of the curve $y = \ln x$ from $x = 1$ to $x = e$.

101. Find the volume of the solid of revolution generated by revolving about the y axis the region bounded by the curve $y = \ln 2x$, the x axis, and the line $x = e$.

102. The region in the first quadrant bounded by the curve $y = \dfrac{5-x}{(x+1)^2}$, the x axis, and the y axis is revolved about the x axis. Find the volume of the solid generated.

103. Two chemicals A and B react to form a chemical C, and the rate of change of the amount of C is proportional to the product of the amounts of A and B remaining at any given time. Initially there are 60 lb of chemical A and 60 lb of chemical B, and to form 5 lb of C, 3 lb of A and 2 lb of B are required. After 1 hr, 15 lb of C are formed. (a) If x lb of C are formed at t hr, find an expression for x in terms of t. (b) Find the amount of C after 3 hr.

104. A tank is in the shape of the solid of revolution formed by rotating about the x axis the region bounded by the curve $y = \ln x$, the x axis, and the lines $x = e$ and $x = e^2$. If the tank is full of water, find the work done in pumping all the water to the top of the tank. Distance is measured in feet. Take the positive x axis vertically downward.

CHAPTER 11 Additional Applications of Integration

11.1 CENTER OF MASS OF A ROD

In Sec. 6.4 we learned that if a function f is continuous on the closed interval $[a, b]$, the average value of f on $[a, b]$ is given by

$$\frac{\int_a^b f(x)\,dx}{b-a}$$

An important application of the average value of a function occurs in physics in connection with the concept of *center of mass.*

To arrive at a definition of *mass,* consider a particle that is set into motion along an axis by a force exerted on the particle. Assume that all velocities are small compared with the velocity of light. So long as the force is acting on the particle, the velocity of the particle is increasing; that is, the particle is accelerating. The ratio of the force to the acceleration is constant regardless of the magnitude of the force, and this constant ratio is called the **mass** of the particle. Therefore, if the force is F units, the acceleration is a units, and the mass is M units, then

$$M = \frac{F}{a}$$

We shall be measuring force, mass, and acceleration in units of both the British engineering system and the metric system. We now discuss these units.

In the British engineering system we first have a unit of force that is 1 lb and a unit of acceleration that is 1 ft/sec². The unit of mass is then defined as the mass of a particle whose acceleration is 1 ft/sec² when the particle is subjected to a force of 1 lb. This unit of mass is called 1 *slug.* Hence

$$1 \text{ slug} = \frac{1 \text{ lb}}{1 \text{ ft/sec}^2}$$

• ILLUSTRATION 1 The acceleration of a certain particle moving on a horizontal line is 10 ft/sec² when the force is 30 lb. Therefore the mass of the particle is

$$\frac{30 \text{ lb}}{10 \text{ ft/sec}^2} = \frac{3 \text{ lb}}{1 \text{ ft/sec}^2}$$

Thus for every 1 ft/sec² of acceleration, a force of 3 lb must be exerted on the particle. The mass of the particle is then 3 slugs. •

In the meter-kilogram-second (mks) system we have the standard unit of mass, the kilogram. The unit of acceleration in the mks system is 1 m/sec². Therefore the unit of force is the force that provides a mass of 1 kg an acceleration of 1 m/sec². This force is called 1 *newton,* written as 1 nt.

● ILLUSTRATION 2 A particle of mass 6 kg is subjected to a constant horizontal force of 3 nt. The accleration of the particle is obtained by dividing the force by the mass, and so it is

$$\frac{3 \text{ nt}}{6 \text{ kg}} = 0.5 \text{ m/sec}^2$$

With symbols, if M kg is the mass, F nt is the force, and a m/sec² is the acceleration, then

$$a = \frac{F}{M} = \frac{3}{6} = 0.5$$ ●

In the mks system, the acceleration due to gravity near the surface of the earth is 9.81 m/sec². If M kg is the mass of an object, and if F nt is the force on the object due to gravity near the surface of the earth, then

$$F = 9.81M$$

In the centimeter-gram-second (cgs) system, the standard unit of mass is the gram, and 1 g = 0.001 kg. The unit of acceleration in the cgs system is 1 cm/sec². Therefore the unit of force is the force that gives a particle of mass 1 g an acceleration of 1 cm/sec². This force is called 1 *dyne*. Because

$$1 \text{ kg} = 10^3 \text{ g} \quad \text{and} \quad 1 \text{ m/sec}^2 = 10^2 \text{ cm/sec}^2$$

then

$$1 \text{ nt} = 10^5 \text{ dynes}$$

We summarize the discussion of units in the British engineering system and the metric system in Table 11.1.1. This table gives the units of force, mass, and acceleration in each system.

TABLE 11.1.1

System of Units	*Force*	*Mass*	*Acceleration*
British engineering	pound	slug	ft/sec²
mks	newton	kilogram	m/sec²
cgs	dyne	gram	cm/sec²

Consider now a horizontal rod, of negligible weight and thickness, placed on the x axis. On the rod is a system of n particles located at points $x_1, x_2, \ldots, x_n$. The ith particle ($i = 1, 2, \ldots, n$) is at a directed distance x_i m from the origin and its mass is m_i kg. See Fig. 11.1.1. The number of kilograms in the total mass of the system is $\sum_{i=1}^{n} m_i$. We define the number of kilogram-meters in the *moment of mass* of the ith particle with respect to the origin as $m_i x_i$. The **moment of mass** for the system is defined as

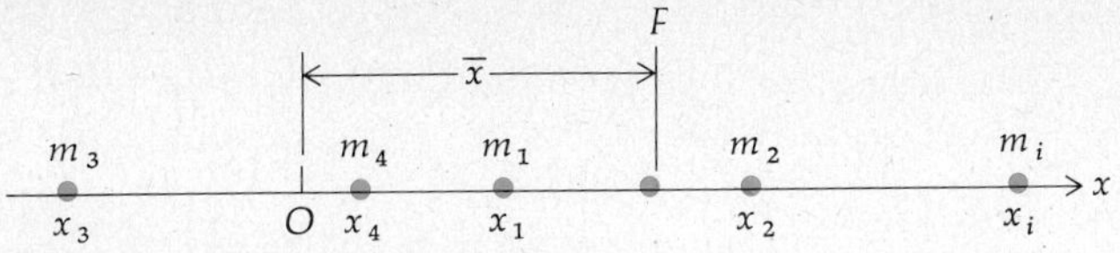

FIGURE 11.1.1

the sum of the moments of mass of all the particles. Hence, if M_0 kg-m is the moment of mass of the system with respect to the origin, then

$$M_0 = \sum_{i=1}^{n} m_i x_i$$

If the measurement of the distance is in feet and of the mass is in slugs, then the moment of mass is measured in slug-ft.

We wish to find a point $\bar{x}$ such that if the total mass of the system were concentrated there, its moment of mass with respect to the origin would be equal to the moment of mass of the system with respect to the origin. Then $\bar{x}$ must satisfy the equation

$$\bar{x} \sum_{i=1}^{n} m_i = \sum_{i=1}^{n} m_i x_i$$

and so

$$\bar{x} = \frac{\sum_{i=1}^{n} m_i x_i}{\sum_{i=1}^{n} m_i} \tag{1}$$

The point $\bar{x}$ is called the **center of mass** of the system, and it is the point where the system will balance. The position of the center of mass is independent of the position of the origin; that is, the location of the center of mass relative to the positions of the particles does not change when the origin is changed. The center of mass is important because the behavior of an entire system of particles can be described by the behavior of the center of mass of the system.

Example 1 Given four particles of masses 2, 3, 1, and 5 kg located on the x axis at the points having coordinates 5, 2, -3, and -4, respectively, where distance measurement is in meters, find the center of mass of this system.

Solution If $\bar{x}$ is the coordinate of the center of mass, we have from formula (1)

$$\bar{x} = \frac{2(5) + 3(2) + (1)(-3) + 5(-4)}{2 + 3 + 1 + 5} = -\frac{7}{11}$$

Thus the center of mass is $\frac{7}{11}$ m to the left of the origin.

The preceding discussion is now extended to a rigid horizontal rod having a continuously distributed mass. The rod is said to be **homogeneous** if it has constant linear density, that is, if its mass is directly proportional to its length. In other words, if the segment of the rod whose length is $\Delta_i x$ meters has a mass of $\Delta_i m$ kg, and $\Delta_i m = k\Delta_i x$, then the rod is homogeneous. The number k is a constant, and k kg/m is called the **linear density** of the rod.

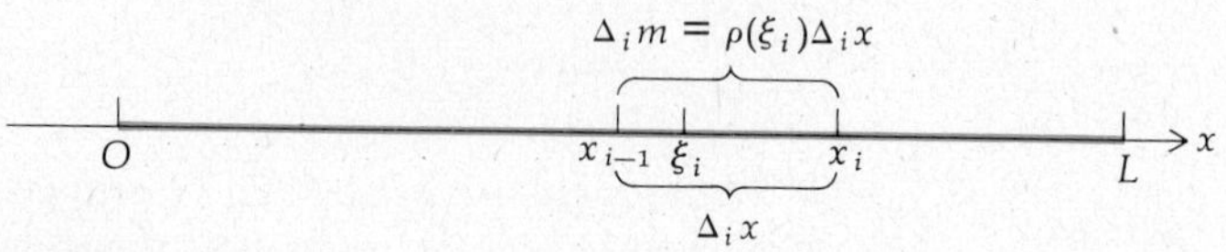

FIGURE 11.1.2

Suppose that we have a nonhomogeneous rod, in which case the linear density varies along the rod. Let L m be the length of the rod, and place the rod on the x axis so the left endpoint of the rod is at the origin and the right endpoint is at L. See Fig. 11.1.2. The linear density at any point x on the rod is $\rho(x)$ kg/m, where ρ is continuous on $[0, L]$. To find the total mass of the rod we consider a partition Δ of the closed interval $[0, L]$ into n subintervals. The ith subinterval is $[x_{i-1}, x_i]$, and its length is $\Delta_i x$ m. If ξ_i is any point in $[x_{i-1}, x_i]$, an approximation to the mass of the part of the rod contained in the ith subinterval is $\Delta_i m$ kg, where

$$\Delta_i m = \rho(\xi_i)\, \Delta_i x$$

The number of kilograms in the total mass of the rod is approximated by

$$\sum_{i=1}^{n} \Delta_i m = \sum_{i=1}^{n} \rho(\xi_i)\, \Delta_i x \tag{2}$$

The smaller we take the norm of the partition Δ, the closer the Riemann sum in (2) will be to what we intuitively think of as the measure of the mass of the rod, and so we define the measure of the mass as the limit of the Riemann sum in (2).

11.1.1 Definition A rod of length L m has its left endpoint at the origin. If the number of kilograms per meter in the linear density at a point x m from the origin is $\rho(x)$, where ρ is continuous on $[0, L]$, then the total **mass** of the rod is M kg, where

$$M = \lim_{\|\Delta\| \to 0} \sum_{i=1}^{n} \rho(\xi_i)\, \Delta_i x = \int_0^L \rho(x)\, dx \tag{3}$$

In Definition 11.1.1, if the distance is measured in feet and the mass is measured in slugs, then the density is measured in slugs per foot.

Example 2 The density at any point of a rod 4 m long varies directly as the distance from the point to an external point in the line of the rod and 2 m from an end, where the density is 5 kg/m. Find the total mass of the rod.

Solution Figure 11.1.3 shows the rod placed on the x axis. If $\rho(x)$ is the number of kilograms per meter in the density of the rod at the point x m from the end having the greater density, then

$$\rho(x) = c(6 - x)$$

where c is the constant of proportionality. Because $\rho(4) = 5$, then $5 = 2c$ or $c = \frac{5}{2}$. Hence $\rho(x) = \frac{5}{2}(6 - x)$. Therefore, if M kg is the total mass of the rod, we have from Definition 11.1.1

$$\begin{aligned} M &= \lim_{\|\Delta\| \to 0} \sum_{i=1}^{n} \tfrac{5}{2}(6 - \xi_i)\, \Delta_i x \\ &= \int_0^4 \tfrac{5}{2}(6 - x)\, dx \\ &= \frac{5}{2}\Big[6x - \tfrac{1}{2}x^2\Big]_0^4 \\ &= 40 \end{aligned}$$

The total mass of the rod is therefore 40 kg.

We now proceed to define the center of mass of the rod of Definition 11.1.1. However, first we must define the moment of mass of the rod with respect to the origin.

As before, place the rod on the x axis with the left endpoint at the origin and the right endpoint at L. See Fig. 11.1.4. Let Δ be a partition of $[0, L]$ into n subintervals, with the ith subinterval $[x_{i-1}, x_i]$ having length $\Delta_i x$ m. If ξ_i is any point in $[x_{i-1}, x_i]$, an approximation to the moment of mass with respect to the origin of the part of the rod contained in the ith subinterval is $\xi_i\, \Delta_i m$ kg-m, where $\Delta_i m = \rho(\xi_i)\, \Delta_i x$. The number of kilogram-meters in the moment of mass of the entire rod is approximated by

$$\sum_{i=1}^{n} \xi_i\, \Delta_i m = \sum_{i=1}^{n} \xi_i\, \rho(\xi_i)\, \Delta_i x \tag{4}$$

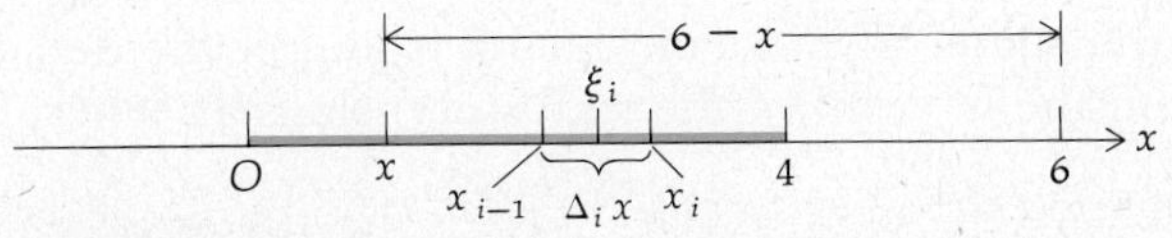

FIGURE 11.1.3

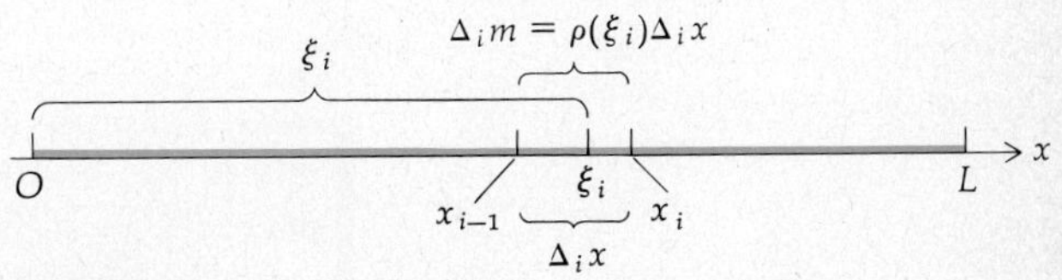

FIGURE 11.1.4

The smaller we take the norm of the partition Δ, the closer the Riemann sum in (4) will be to what we intuitively think of as the measure of the moment of mass of the rod with respect to the origin. We have, then, the following definition.

11.1.2 Definition A rod of length L m has its left endpoint at the origin and the number of kilograms per meter in the linear density at a point x m from the origin is $\rho(x)$, where ρ is continuous on $[0, L]$. The **moment of mass** of the rod with respect to the origin is M_0 kg-m, where

$$M_0 = \lim_{\|\Delta\| \to 0} \sum_{i=1}^{n} \xi_i \rho(\xi_i)\, \Delta_i x = \int_0^L x\rho(x)\, dx \tag{5}$$

The center of mass of the rod is at the point $\bar{x}$ such that if M kg is the total mass of the rod, $\bar{x}M = M_0$. Thus, from (3) and (5),

$$\bar{x} = \frac{\int_0^L x\rho(x)\, dx}{\int_0^L \rho(x)\, dx} \tag{6}$$

Example 3 Find the center of mass for the rod in Example 2.

Solution In Example 2, $M = 40$. From (6) with $\rho(x) = \frac{5}{2}(6 - x)$,

$$\bar{x} = \frac{\int_0^4 \frac{5}{2}x(6 - x)\, dx}{40} = \frac{1}{16}\left[3x^2 - \frac{1}{3}x^3\right]_0^4 = \frac{5}{3}$$

Therefore the center of mass is at $\frac{5}{3}$ m from the end having the greater density.

● ILLUSTRATION 3 If a rod is of uniform density k kg/m, where k is a constant, then from formula (6),

$$\bar{x} = \frac{\int_0^L xk\, dx}{\int_0^L k\, dx} = \frac{\left.\frac{kx^2}{2}\right]_0^L}{\left.kx\right]_0^L} = \frac{\frac{kL^2}{2}}{kL} = \frac{L}{2}$$

Thus the center of mass is at the center of the rod, as is to be expected. ●

Example 4 Find the center of mass of a rod 2 ft long if the density at the point x ft from the left end of the rod is $\rho(x)$ slugs/ft, where $\rho(x) = \sqrt{x^2 + 5}$.

Solution The rod is placed on the x axis with its left endpoint at the origin. Then if $\bar{x}$ is the center of mass, we have from formula (6)

$$\bar{x} = \frac{\displaystyle\int_0^2 x\sqrt{x^2+5}\,dx}{\displaystyle\int_0^2 \sqrt{x^2+5}\,dx} \tag{7}$$

In Example 2 of Sec. 10.3 we obtained by using a trigonometric substitution

$$\int \sqrt{x^2+5}\,dx = \tfrac{1}{2}x\sqrt{x^2+5} + \tfrac{5}{2}\ln(\sqrt{x^2+5}+x) + C$$

We use this result to evaluate the integral in the denominator of (7).

$$\begin{aligned}\int_0^2 \sqrt{x^2+5}\,dx &= \left[\tfrac{1}{2}x\sqrt{x^2+5} + \tfrac{5}{2}\ln(\sqrt{x^2+5}) + x)\right]_0^2\\ &= \sqrt{9} + \tfrac{5}{2}\ln(\sqrt{9}+2) - (\tfrac{1}{2}\cdot 0 + \tfrac{5}{2}\ln\sqrt{5})\\ &= 3 + \tfrac{5}{2}\ln 5 - \tfrac{5}{4}\ln 5\\ &= 3 + \tfrac{5}{4}\ln 5 \end{aligned} \tag{8}$$

We compute the integral in the numerator of (7).

$$\begin{aligned}\int_0^2 x\sqrt{x^2+5}\,dx &= \tfrac{1}{3}(x^2+5)^{3/2}\Big]_0^2\\ &= \tfrac{1}{3}(27 - 5\sqrt{5})\end{aligned} \tag{9}$$

Substituting from (8) and (9) into (7) we have

$$\bar{x} = \frac{1}{3}\cdot\frac{27-5\sqrt{5}}{3+\frac{5}{4}\ln 5} \approx 1.05$$

Thus the center of mass is 1.05 ft from the left end of the rod. This is just a little beyond the center of the rod.

EXERCISES 11.1

In Exercises 1 through 4, a particle is moving on a horizontal line. Find the force exerted on the particle if it has the given mass and acceleration.

1. Mass is 50 slugs; acceleration is 5 ft/sec^2.

2. Mass is 10 kg; acceleration is 6 m/sec^2.

3. Mass is 80 g; acceleration is 50 cm/sec^2.

4. Mass is 22 slugs; acceleration is 4 ft/sec^2.

In Exercises 5 through 8, a particle is subjected to the given horizontal force, and either the mass or acceleration of the particle is given. Find the other quantity.

5. Force is 6 nt; mass is 4 kg.

6. Force is 32 lb; mass is 8 slugs.

7. Force is 24 lb; acceleration is 9 ft/sec^2.

8. Force is 700 dynes; acceleration is 80 cm/sec^2.

In Exercises 9 through 12, a system of particles is located on the x axis. The number of kilograms in the mass of each particle and the coordinate of its position are given. Distance is measured in meters. Find the center of mass of each system.

9. $m_1 = 5$ at 2; $m_2 = 6$ at 3; $m_3 = 4$ at 5; $m_4 = 3$ at 8.

10. $m_1 = 2$ at -4; $m_2 = 8$ at -1; $m_3 = 4$ at 2; $m_4 = 2$ at 3.

11. $m_1 = 2$ at -3; $m_2 = 4$ at -2; $m_3 = 20$ at 4; $m_4 = 10$ at 6; $m_5 = 30$ at 9.

12. $m_1 = 5$ at -7; $m_2 = 3$ at -2; $m_3 = 5$ at 0; $m_4 = 1$ at 2; $m_5 = 8$ at 10.

In Exercises 13 through 21, find the total mass of the given rod and the center of mass.

13. The length of a rod is 6 m and the linear density of the rod at a point x m from one end is $(2x + 3)$ kg/m.

14. The length of a rod is 20 cm and the linear density of the rod at a point x cm from one end is $(3x + 2)$ g/cm.

15. The length of a rod is 9 in. and the linear density of the rod at a point x in. from one end is $(4x + 1)$ slugs/in.

16. The length of a rod is 3 ft, and the linear density of the rod at a point x ft from one end is $(5 + 2x)$ slugs/ft.

17. The length of a rod is 12 cm, and the measure of the linear density at a point is a linear function of the measure of the distance from the left end of the rod. The linear density at the left end is 3 g/cm and at the right end is 4 g/cm.

18. The length of a rod is 10 m and the measure of the linear density at a point is a linear function of the measure of the distance of the point from the left end of the rod. The linear density at the left end is 2 kg/m and at the right end is 3 kg/m.

19. The measure of the linear density at any point of a rod 6 m long varies directly as the distance from the point to an external point in the line of the rod and 4 m from an end, where the density is 3 kg/m.

20. A rod is 10 ft long, and the measure of the linear density at a point is a linear function of the measure of the distance from the center of the rod. The linear density at each end of the rod is 5 slugs/ft and at the center the linear density is $3\frac{1}{2}$ slugs/ft.

21. The measure of the linear density at a point of a rod varies directly as the third power of the measure of the distance of the point from one end. The length of the rod is 4 ft and the linear density is 2 slugs/ft at the center.

22. The linear density at any point of a rod 5 m long varies directly as the distance from the point to an external point in the line of the rod and 2 m from an end, where the density is K kg/m. Find K if the total mass of the rod is 135 kg.

23. The linear density at any point of a rod 3 m long varies directly as the distance from a point to an external point in the line of the rod and 1 m from an end where the density is 2 kg/m. If the total mass of the rod is 15 kg, find the center of mass of the rod.

24. The measure of the linear density at a point on a rod varies directly as the fourth power of the measure of the distance of the point from one end. The length of the rod is 2 m and the linear density is 2 kg/m at the center. If the total mass of the rod is $\frac{64}{5}$ kg, find the center of mass of the rod.

25. Find the center of mass of a rod 8 cm long if the linear density at a point x cm from the left end of the rod is $\rho(x)$ g/cm, where $\rho(x) = \sqrt{x^2 + 36}$.

26. The linear density of a rod at a point x m from one end is $\sqrt{9 + x^2}$ kg/m. If the rod is 3 m long, find the mass and center of mass of the rod.

27. The linear density of a rod at a point x m from one end is $2e^{-x}$ kg/m. If the rod is 6 m long, find the mass and center of mass of the rod.

28. The total mass of a rod of length L ft is M slugs, and the measure of the linear density at a point x ft from the left end is proportional to the measure of the distance of the point from the right end. Show that the linear density at a point on the rod x ft from the left end is $2M(L - x)/L^2$ slugs/ft.

29. The linear density of a rod at a point x cm from one end is $2/(1 + x)$ g/cm. If the rod is 15 cm long, find the mass and center of mass of the rod.

30. A rod is 6 m long and its mass is 24 kg. If the measure of the linear density at any point of the rod varies directly as the square of the distance of the point from one end, find the largest value of the linear density.

31. A rod is L ft long and the center of mass of the rod is at the point $\frac{3}{4}L$ ft from the left end. If the measure of the linear density at a point is proportional to a power of the measure of the distance of the point from the left end and the linear density at the right end is 20 slugs/ft, find the linear density at a point x ft from the left end. The mass is in slugs.

11.2 CENTROID OF A PLANE REGION

Let the masses of n particles located at the points (x_1, y_1), (x_2, y_2), . . . , (x_n, y_n) in the xy plane be measured by $m_1, m_2, \ldots, m_n$, and consider the problem of finding the center of mass of this system. We may imagine the particles being supported by a sheet of negligible weight and negligible thickness and may assume that each particle has its position at exactly one point. The center of mass is the point where the sheet will balance. Refer to Fig. 11.2.1, which shows eight particles placed on the sheet. The label of the ith particle in the figure is m_i, which is the measure of its mass. The sheet will balance on a fulcrum located at the center of mass denoted by $(\bar{x}, \bar{y})$. To determine the center of mass we first define the moment of mass of a system of particles with respect to an axis.

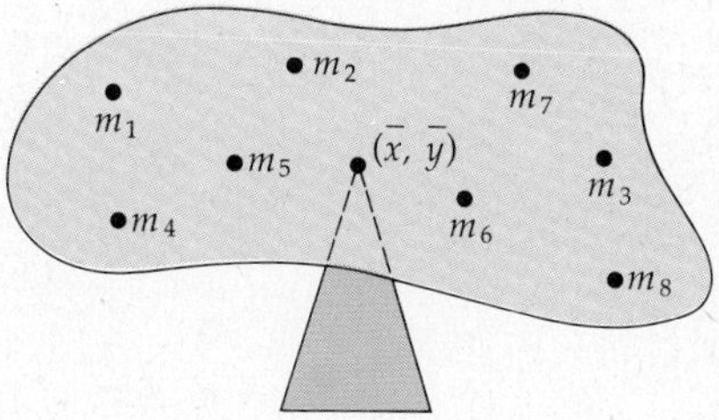

FIGURE 11.2.1

Suppose a particle at a distance of d meters from an axis has a mass of m kg. If M_1 kg-m is the moment of mass of the particle with respect to the axis, then

$$M_1 = md \tag{1}$$

If the ith particle, having mass m_i kg, is located at the point (x_i, y_i), its distance from the y axis is x_i meters; thus, from formula (1), the moment of mass of this particle with respect to the y axis is $m_i x_i$ kg-m. Similarly, the moment of mass of the particle with respect to the x axis is $m_i y_i$ kg-m. The moment of the system of n particles with respect to the y axis is M_y kg-m, where

$$M_y = \sum_{i=1}^{n} m_i x_i \tag{2}$$

and the moment of the system with respect to the x axis is M_x kg-m, where

$$M_x = \sum_{i=1}^{n} m_i y_i \tag{3}$$

The total mass of the system is M kg, where

$$M = \sum_{i=1}^{n} m_i \tag{4}$$

The center of mass of the system is at the point $(\bar{x}, \bar{y})$, where

$$\bar{x} = \frac{M_y}{M} \tag{5}$$

and

$$\bar{y} = \frac{M_x}{M} \tag{6}$$

The point $(\bar{x}, \bar{y})$ can be interpreted as the point such that, if the total mass M kg of the system were concentrated there, its moment of mass with respect to the y axis, M_y kg-m, would be determined by $M_y = M\bar{x}$, and its moment of mass with respect to the x axis, M_x kg-m, would be determined by $M_x = M\bar{y}$.

Example 1 Find the center of mass of the four particles having masses 2, 6, 4, and 1 kg located at the points $(5, -2)$, $(-2, 1)$, $(0, 3)$, and $(4, -1)$, respectively.

Solution

$$M_y = \sum_{i=1}^{4} m_i x_i = 2(5) + 6(-2) + 4(0) + 1(4) = 2$$

$$M_x = \sum_{i=1}^{4} m_i y_i = 2(-2) + 6(1) + 4(3) + 1(-1) = 13$$

$$M = \sum_{i=1}^{4} m_i = 2 + 6 + 4 + 1 = 13$$

Therefore

$$\bar{x} = \frac{M_y}{M} = \frac{2}{13} \quad \text{and} \quad \bar{y} = \frac{M_x}{M} = \frac{13}{13} = 1$$

The center of mass is at $(\frac{2}{13}, 1)$.

Consider now a thin sheet of continuously distributed mass, for example, a piece of paper or a flat strip of tin. We regard such sheets as being two dimensional and call such a plane region a *lamina*. In this section we confine our discussion to homogeneous laminae, that is, laminae having constant area density. Laminae of variable area density are considered in Chapter 20 in connection with applications of multiple integrals.

Let a homogeneous lamina of area A m² have a mass of M kg. Then if the constant area density is k kg/m², $M = kA$. If the homogeneous lamina is a rectangle, its center of mass is defined to be at the center of the rectangle. We use this definition to define the center of mass of a more general homogeneous lamina.

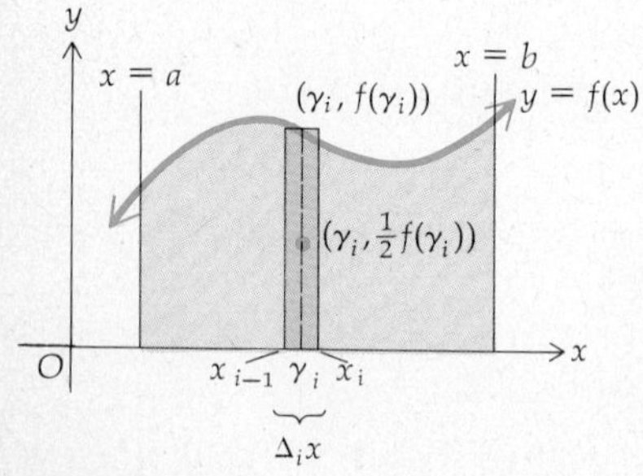

FIGURE 11.2.2

Let L be the homogeneous lamina whose constant area density is k kg/m², and which is bounded by the curve $y = f(x)$, the x axis, and the lines $x = a$ and $x = b$. The function f is continuous on the closed interval $[a, b]$, and $f(x) \geq 0$ for all x in $[a, b]$. See Fig. 11.2.2. Let Δ be a partition of the interval $[a, b]$ into n subintervals. The ith subinterval is $[x_{i-1}, x_i]$, and $\Delta_i x = x_i - x_{i-1}$. The midpoint of $[x_{i-1}, x_i]$ is γ_i. Associated with each subinterval is a rectangular lamina whose width, altitude, and area density are given by $\Delta_i x$ m, $f(\gamma_i)$ m, and k kg/m², respectively, and whose center of mass is at the point $(\gamma_i, \frac{1}{2}f(\gamma_i))$. The area of the rectangular lamina is

$f(\gamma_i)\,\Delta_i x$ m²; hence $kf(\gamma_i)\,\Delta_i x$ kg is its mass. Consequently, if $\Delta_i M_y$ kg-m is the moment of mass of this rectangular element with respect to the y axis.

$$\Delta_i M_y = \gamma_i k f(\gamma_i)\,\Delta_i x$$

The sum of the measures of the moments of mass of n such rectangular laminae with respect to the y axis is given by the Riemann sum

$$\sum_{i=1}^{n} k\gamma_i f(\gamma_i)\,\Delta_i x$$

We define the *moment of mass of L with respect to the y axis* as the limit of this Riemann sum as $\|\Delta\| \to 0$. We state this formally in Definition 11.2.1.

Similarly, if $\Delta_i M_x$ kg-m is the moment of mass of the ith rectangular lamina with respect to the x axis,

$$\Delta_i M_x = \tfrac{1}{2} f(\gamma_i) k f(\gamma_i)\,\Delta_i x$$

and the sum of the measures of the moments of mass of n such rectangular laminae with respect to the x axis is given by the Riemann sum

$$\sum_{i=1}^{n} \tfrac{1}{2} k [f(\gamma_i)]^2\,\Delta_i x$$

The limit of this Riemann sum as $\|\Delta\| \to 0$ is defined in Definition 11.2.1 as the *moment of mass of L with respect to the x axis.*

The mass of the ith rectangular lamina is $kf(\gamma_i)\,\Delta_i x$ kg; so the sum of the measures of the masses of n rectangular laminae is given by

$$\sum_{i=1}^{n} k f(\gamma_i)\,\Delta_i x$$

In Definition 11.2.1, the limit of this Riemann sum is defined as the *total mass of L.*

11.2.1 Definition Let L be the homogeneous lamina whose constant area density is k kg/m² and which is bounded by the curve $y = f(x)$, the x axis, and the lines $x = a$ and $x = b$. The function f is continuous on $[a, b]$ and $f(x) \geq 0$ for all x in $[a, b]$. If M_y kg-m is the **moment of mass of the lamina L with respect to the y axis,** then

$$M_y = \lim_{\|\Delta\| \to 0} \sum_{i=1}^{n} k\gamma_i f(\gamma_i)\,\Delta_i x = k \int_a^b x f(x)\,dx \qquad (7)$$

If M_x kg-m is the **moment of mass of the lamina L with respect to the x axis,** then

$$M_x = \lim_{\|\Delta\| \to 0} \sum_{i=1}^{n} \tfrac{1}{2} k [f(\gamma_i)]^2\,\Delta_i x = \tfrac{1}{2} k \int_a^b [f(x)]^2\,dx \qquad (8)$$

If M kg is the **total mass of the lamina L,** then

$$M = \lim_{\|\Delta\| \to 0} \sum_{i=1}^{n} kf(\gamma_i)\, \Delta_i x = k \int_a^b f(x)\, dx \tag{9}$$

If $(\bar{x}, \bar{y})$ is the **center of mass of the lamina L,** then

$$\bar{x} = \frac{M_y}{M} \quad \text{and} \quad \bar{y} = \frac{M_x}{M} \tag{10}$$

Substituting from (7), (8), and (9) into (10) we have

$$\bar{x} = \frac{k \displaystyle\int_a^b xf(x)\, dx}{k \displaystyle\int_a^b f(x)\, dx} \quad \text{and} \quad \bar{y} = \frac{\frac{1}{2}k \displaystyle\int_a^b [f(x)]^2\, dx}{k \displaystyle\int_a^b f(x)\, dx}$$

Dividing both the numerator and denominator by k we get

$$\bar{x} = \frac{\displaystyle\int_a^b xf(x)\, dx}{\displaystyle\int_a^b f(x)\, dx} \tag{11}$$

and

$$\bar{y} = \frac{\frac{1}{2} \displaystyle\int_a^b [f(x)]^2\, dx}{\displaystyle\int_a^b f(x)\, dx} \tag{12}$$

In formulas (11) and (12) the denominator is the number of square units in the area of the region; so we have expressed a physical problem in terms of a geometric one. That is, $\bar{x}$ and $\bar{y}$ can be considered as the average abscissa and the average ordinate, respectively, of a geometric region. In such a case $\bar{x}$ and $\bar{y}$ depend only on the region, not on the mass of the lamina. So we refer to the center of mass of a plane region instead of to the center of mass of a homogeneous lamina. In such a case the center of mass is called the *centroid* of the region. Instead of moments of mass, we consider moments of the region.

11.2.2 Definition Let R be the region bounded by the curve $y = f(x)$, the x axis, and the lines $x = a$ and $x = b$. The function f is continuous on $[a, b]$ and $f(x) \geq 0$ for all x in $[a, b]$. If M_x denotes the **moment of R with respect to the x axis** and M_y denotes the **moment of R with respect to the y axis,** then

$$M_x = \lim_{\|\Delta\| \to 0} \sum_{i=1}^{n} \tfrac{1}{2}[f(\gamma_i)]^2\, \Delta_i x = \frac{1}{2} \int_a^b [f(x)]^2\, dx$$

$$M_y = \lim_{\|\Delta\| \to 0} \sum_{i=1}^{n} \gamma_i f(\gamma_i)\, \Delta_i x = \int_a^b xf(x)\, dx$$

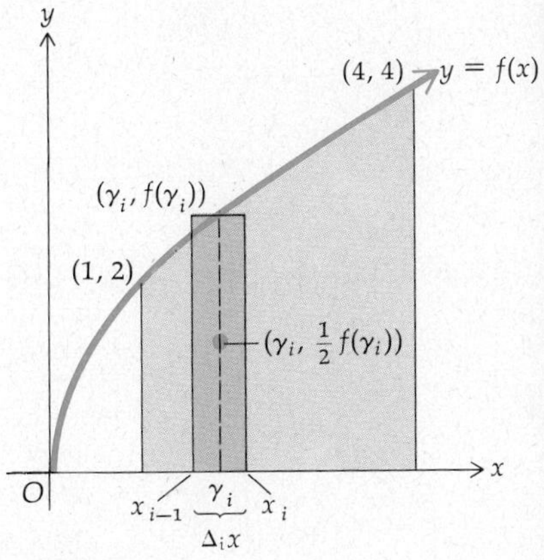

FIGURE 11.2.3

If $(\bar{x}, \bar{y})$ is the **centroid** of the plane region R whose area is A square units, and M_x and M_y are defined as above

$$\bar{x} = \frac{M_y}{A} \quad \text{and} \quad \bar{y} = \frac{M_x}{A}$$

Example 2 Find the centroid of the first quadrant region bounded by the curve $y^2 = 4x$, the x axis, and the lines $x = 1$ and $x = 4$.

Solution Let $f(x) = 2x^{1/2}$. The equation of the curve is then $y = f(x)$. In Fig. 11.2.3 the region is shown together with the ith rectangular element. The centroid of the rectangle is at $(\gamma_i, \frac{1}{2}f(\gamma_i))$. The area A square units of the region is given by

$$\begin{aligned} A &= \lim_{\|\Delta\| \to 0} \sum_{i=1}^{n} f(\gamma_i)\, \Delta_i x \\ &= \int_1^4 f(x)\, dx \\ &= \int_1^4 2x^{1/2}\, dx \\ &= \tfrac{4}{3} x^{3/2} \Big]_1^4 \\ &= \tfrac{28}{3} \end{aligned}$$

We now compute M_y and M_x.

$$\begin{aligned} M_y &= \lim_{\|\Delta\| \to 0} \sum_{i=1}^{n} \gamma_i f(\gamma_i)\, \Delta_i x \\ &= \int_1^4 x f(x)\, dx \\ &= \int_1^4 x(2x^{1/2})\, dx \\ &= 2\int_1^4 x^{3/2}\, dx \\ &= \tfrac{4}{5} x^{5/2} \Big]_1^4 \\ &= \tfrac{124}{5} \end{aligned}$$

$$\begin{aligned} M_x &= \lim_{\|\Delta\| \to 0} \sum_{i=1}^{n} \tfrac{1}{2} f(\gamma_i) \cdot f(\gamma_i)\, \Delta_i x \\ &= \frac{1}{2}\int_1^4 [f(x)]^2\, dx \\ &= \frac{1}{2}\int_1^4 4x\, dx \\ &= x^2 \Big]_1^4 \\ &= 15 \end{aligned}$$

Hence

$$\bar{x} = \frac{M_y}{A} = \frac{\frac{124}{5}}{\frac{28}{3}} = \frac{93}{35} \quad \text{and} \quad \bar{y} = \frac{M_x}{A} = \frac{15}{\frac{28}{3}} = \frac{45}{28}$$

Therefore the centroid is at the point $(\frac{93}{35}, \frac{45}{28})$.

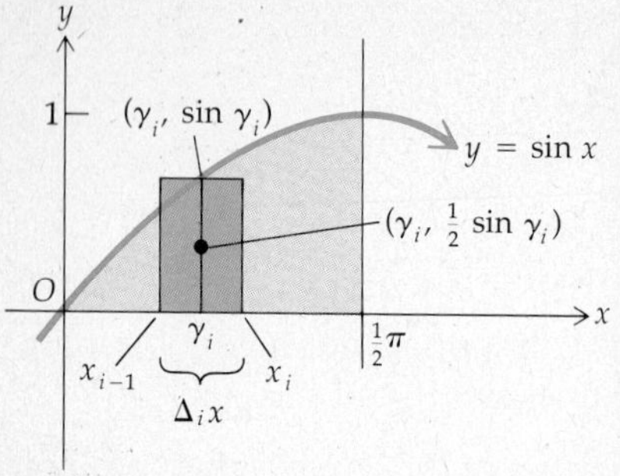

FIGURE 11.2.4

Example 3 Find the centroid of the region in the first quadrant to the left of the line $x = \frac{1}{2}\pi$ and bounded by the curve $y = \sin x$, the x axis, and the line $x = \frac{1}{2}\pi$.

Solution The region and the ith rectangular element are shown in Fig. 11.2.4. We first compute the area of the region.

$$\begin{aligned} A &= \lim_{\|\Delta\| \to 0} \sum_{i=1}^{n} \sin \gamma_i \, \Delta_i x \\ &= \int_0^{\pi/2} \sin x \, dx \\ &= -\cos x \Big]_0^{\pi/2} \\ &= 1 \end{aligned}$$

We compute M_x and M_y. To evaluate the integral for M_y we use integration by parts with $u = x$ and $dv = \sin x \, dx$.

$$\begin{aligned} M_x &= \lim_{\|\Delta\| \to 0} \sum_{i=1}^{n} \tfrac{1}{2}[\sin \gamma_i]^2 \, \Delta_i x \\ &= \frac{1}{2} \int_0^{\pi/2} \sin^2 x \, dx \\ &= \frac{1}{2} \int_0^{\pi/2} \frac{1 - \cos 2x}{2} \, dx \\ &= \tfrac{1}{4}\left[x - \tfrac{1}{2} \sin 2x\right]_0^{\pi/2} \\ &= \tfrac{1}{4}\left[\tfrac{1}{2}\pi\right] \\ &= \tfrac{1}{8}\pi \end{aligned}$$

$$\begin{aligned} M_y &= \lim_{\|\Delta\| \to 0} \sum_{i=1}^{n} \gamma_i \sin \gamma_i \, \Delta_i x \\ &= \int_0^{\pi/2} x \sin x \, dx \\ &= -x \cos x \Big]_0^{\pi/2} + \int_0^{\pi/2} \cos x \, dx \\ &= -x \cos x + \sin x \Big]_0^{\pi/2} \\ &= -\tfrac{1}{2}\pi \cos \tfrac{1}{2}\pi + \sin \tfrac{1}{2}\pi \\ &= 1 \end{aligned}$$

Therefore

$$\bar{x} = \frac{M_y}{A} = \frac{1}{1} = 1 \quad \text{and} \quad \bar{y} = \frac{M_x}{A} = \frac{\frac{1}{8}\pi}{1} = \tfrac{1}{8}\pi$$

Thus the centroid is at the point $(1, \frac{1}{8}\pi)$.

In the following example the region is bounded by two curves instead of one curve and a coordinate axis. The method for finding the centroid is the same as before, but the equations for M_x and M_y are now dependent upon the equations defining the two curves. The procedure to be followed is shown in the example.

Example 4 Find the centroid of the region bounded by the curves $y = x^2$ and $y = 2x + 3$.

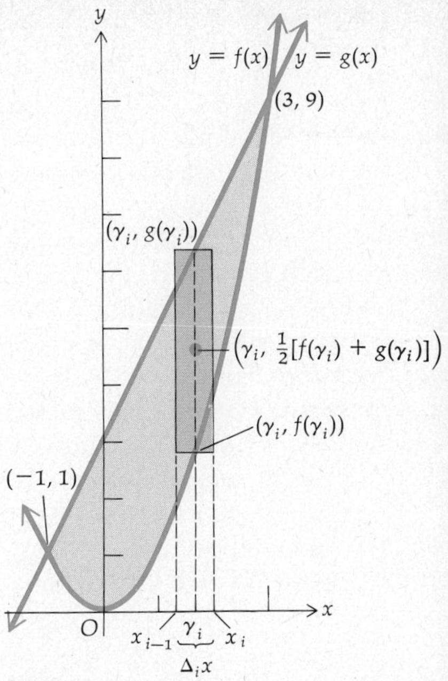

FIGURE 11.2.5

Solution The points of intersection of the two curves are $(-1, 1)$ and $(3, 9)$. The region is shown in Fig. 11.2.5, together with the ith rectangular element.

Let $f(x) = x^2$ and $g(x) = 2x + 3$. The centroid of the ith rectangular element is at the point $(\gamma_i, \frac{1}{2}[f(\gamma_i) + g(\gamma_i)])$, where γ_i is the midpoint of the ith subinterval $[x_{i-1}, x_i]$. The measure of the area of the region is given by

$$A = \lim_{\|\Delta\| \to 0} \sum_{i=1}^{n} [g(\gamma_i) - f(\gamma_i)]\, \Delta_i x$$

$$= \int_{-1}^{3} [g(x) - f(x)]\, dx$$

$$= \int_{-1}^{3} [2x + 3 - x^2]\, dx$$

$$= \tfrac{32}{3}$$

We now compute M_y and M_x.

$$M_y = \lim_{\|\Delta\| \to 0} \sum_{i=1}^{n} \gamma_i [g(\gamma_i) - f(\gamma_i)]\, \Delta_i x$$

$$= \int_{-1}^{3} x[g(x) - f(x)]\, dx$$

$$= \int_{-1}^{3} x[2x + 3 - x^2]\, dx$$

$$= \tfrac{32}{3}$$

$$M_x = \lim_{\|\Delta\| \to 0} \sum_{i=1}^{n} \tfrac{1}{2}[g(\gamma_i) + f(\gamma_i)][g(\gamma_i) - f(\gamma_i)]\, \Delta_i x$$

$$= \frac{1}{2}\int_{-1}^{3} [g(x) + f(x)][g(x) - f(x)]\, dx$$

$$= \frac{1}{2}\int_{-1}^{3} [2x + 3 + x^2][2x + 3 - x^2]\, dx$$

$$= \frac{1}{2}\int_{-1}^{3} [4x^2 + 12x + 9 - x^4]\, dx$$

$$= \tfrac{544}{15}$$

Therefore

$$\bar{x} = \frac{M_y}{A} = \frac{\frac{32}{3}}{\frac{32}{3}} = 1 \quad \text{and} \quad \bar{y} = \frac{M_x}{A} = \frac{\frac{544}{15}}{\frac{32}{3}} = \frac{17}{5}$$

Hence the centroid is at the point $(1, \frac{17}{5})$.

If a plane region has an axis of symmetry, the centroid of the region lies on the axis of symmetry. This is now stated and proved as a theorem.

11.2.3 Theorem If the plane region R has the line L as an axis of symmetry, the centroid of R lies on L.

Proof Choose the coordinate axes so that L is on the y axis and the origin is in the region R. Figure 11.2.6 illustrates an example of this situation. In the figure, R is the region CDE, C is the point $(-a, 0)$, E is the point $(a, 0)$, and an equation of the curve CDE is $y = f(x)$.

Consider a partition of the interval $[0, a]$. Let γ_i be the midpoint of the ith subinterval $[x_{i-1}, x_i]$. The moment with respect to the y axis of the rectangular element having an altitude $f(\gamma_i)$ and a width $\Delta_i x$ is $\gamma_i[f(\gamma_i)\,\Delta_i x]$. Because of symmetry, for a similar partition of the interval $[-a, 0]$ there is a corresponding element having as its moment with respect to the y axis $-\gamma_i f(\gamma_i)\,\Delta_i x$. The sum of these two moments is 0; therefore $M_y = 0$. Because $\bar{x} = M_y/A$, we conclude that $\bar{x} = 0$. Thus the centroid of the region R lies on the y axis, which is what was to be proved. ■

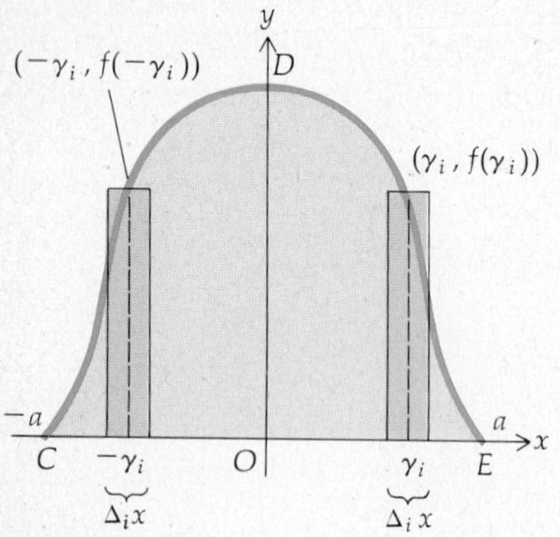

FIGURE 11.2.6

By applying the preceding theorem, we can simplify the problem of finding the centroid of a plane region that can be divided into regions having axes of symmetry.

Example 5 Find the centroid of the region bounded by the semicircle $y = \sqrt{4 - x^2}$ and the x axis.

Solution The region is shown in Fig. 11.2.7.

Because the y axis is an axis of symmetry, the centroid lies on the y axis; so $\bar{x} = 0$.

The moment of the region with respect to the x axis is given by

$$M_x = \lim_{\|\Delta\| \to 0} \sum_{i=1}^{n} \tfrac{1}{2}[\sqrt{4 - \gamma_i^2}]^2\,\Delta_i x$$

$$= 2 \cdot \frac{1}{2} \int_0^2 (4 - x^2)\,dx$$

$$= 4x - \tfrac{1}{3}x^3 \Big]_0^2$$

$$= \tfrac{16}{3}$$

The area of the region is 2π square units; so

$$\bar{y} = \frac{\frac{16}{3}}{2\pi} = \frac{8}{3\pi}$$

The centroid is at the point $(0, 8/3\pi)$.

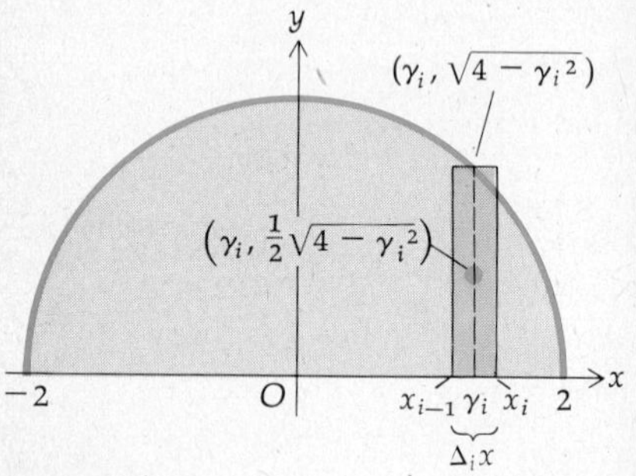

FIGURE 11.2.7

EXERCISES 11.2

1. Find the center of mass of the three particles having masses of 1, 2, and 3 kg and located at the points $(-1, 3)$, $(2, 1)$, and $(3, -1)$, respectively.

2. Find the center of mass of the four particles having masses of 2, 3, 3, and 4 kg and located at the points $(-1, -2)$, $(1, 3)$, $(0, 5)$, and $(2, 1)$, respectively.

3. The y coordinate of the center of mass of four particles is 5. The particles have masses 2, 5, 4, and m slugs and are located at the points (3, 2), (−1, 0), (0, 20), and (2, −2), respectively. Find m.

4. Find the center of mass of the three particles having masses 3, 7, and 2 slugs located at the points (2, 3), (−1, 4), and (0, 2), respectively.

5. Find the center of mass of three particles, of equal mass, located at the points (4, −2), (−3, 0), and (1, 5).

6. Prove that the center of mass of three particles, of equal mass, in a plane lies at the point of intersection of the medians of the triangle having as vertices the points at which the particles are located.

In Exercises 7 through 19, find the centroid of the region with the indicated boundaries.

7. The parabola $y = 4 - x^2$ and the x axis.
8. The parabola $x = 2y - y^2$ and the y axis.
9. The parabola $y = x^2$ and the line $y = 4$.
10. The parabola $y^2 = 4x$, the y axis, and the line $y = 4$.
11. The curves $y = x^3$ and $y = 4x$ in the first quadrant.
12. The lines $y = 2x + 1$, $x + y = 7$, and $x = 8$.
13. The curves $y = x^2 - 4$ and $y = 2x - x^2$.
14. The curves $y = x^2$ and $y = x^3$.
15. The curve $y = e^x$, the coordinate axes, and the line $x = 3$.
16. The curves $y = \sin x$ and $y = \cos x$, and the y axis, in the first quadrant.
17. The curve $yx^2 = \sqrt{x^2 - 9}$, the x axis, and the line $x = 5$.
18. The curve $y = (x - 1)/(x^2 - 5x + 6)$, the x axis, and the lines $x = 4$ and $x = 6$.
19. The curve $(x + 2)^2 y = 4 - x$, in the first quadrant.

20. Find the abscissa of the centroid of the region bounded by the curve $y = 4/(x^3 + 8)$, the x axis, the y axis, and the line $x = 2$.

21. If the centroid of the region bounded by the parabola $y^2 = 4px$ and the line $x = a$ is to be at the point $(p, 0)$, find the value of a.

22. Prove that the distance from the centroid of a triangle to any side of the triangle is equal to one-third of the length of the altitude to that side.

23. Find the center of mass of the lamina bounded by the parabola $2y^2 = 18 - 3x$ and the y axis if the area density at any point (x, y) is $\sqrt{6 - x}$ kg/m².

24. Solve Exercise 23 if the area density at any point (x, y) is x kg/m².

25. Find the mass of the lamina in the shape of the region of Exercise 20 if the area density at any point (x, y) is kx^2 kg/m².

26. Let R be the region bounded by the curves $y = f_1(x)$ and $y = f_2(x)$ (see Fig. 11.2.8). If A is the measure of the area of R and if $\bar{y}$ is the ordinate of the centroid of R, prove that the measure of the volume, V, of the solid of revolution obtained by revolving R about the x axis is given by $V = 2\pi\bar{y}A$. Stating this equation in words we have:

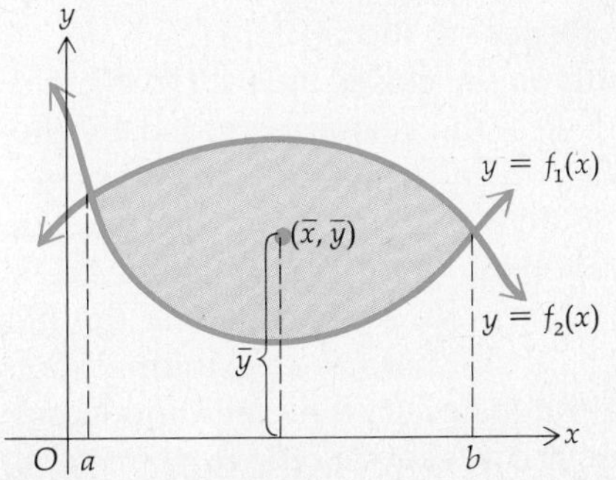

FIGURE 11.2.8

If a plane region is revolved about a line in its plane that does not cut the region, then the measure of the volume of the solid of revolution generated is equal to the product of the measure of the area of the region and the measure of the distance traveled by the centroid of the region.

The above statement is known as the *theorem of Pappus* for volumes of solids of revolution.

27. Use the theorem of Pappus to find the centroid of the region bounded by a semicircle and its diameter.

28. Use the theorem of Pappus to find the volume of the torus (doughnut shape) generated by revolving a circle with a radius of r units about a line in its plane at a distance of b units from its center, where $b > r$.

29. Use the theorem of Pappus to find the volume of a right-circular cone with base radius r units and height h units.

30. Use the theorem of Pappus to find the volume of a sphere with a radius of r units.

31. Let R be the region bounded by the semicircle $y = \sqrt{r^2 - x^2}$ and the x axis. Use the theorem of Pappus to find the moment of R with respect to the line $y = -r$.

32. If R is the region of Exercise 31, use the theorem of Pappus to find the volume of the solid of revolution generated by revolving R about the line $x - y = r$. (*Hint*: Use the result of Exercise 22 in Exercises 4.7.)

11.3 CENTROID OF A SOLID OF REVOLUTION

To find the center of mass of a solid, in general we must make use of multiple integration. This procedure is taken up in Chapter 20 as an application of multiple integrals. However, if the shape of the solid is that of a solid of revolution and its volume density is constant, we find the center of mass by a method similar to the one used to obtain the center of mass of a homogeneous lamina. Following is the procedure for finding the center of mass of a homogeneous solid of revolution, with the assumption that the center of mass is on the axis of revolution.

We first set up a three-dimensional coordinate system. The x and y axes are taken as in two dimensions, and the third axis, the z axis, is taken perpendicular to them at the origin. A point in three dimensions is then given by (x, y, z). The plane containing the x and y axes is called the xy plane, and the xz plane and the yz plane are defined similarly.

Suppose that the x axis is the axis of revolution. Then under the assumption that the center of mass lies on the axis of revolution, the y and z coordinates of the center of mass are each zero; so it is necessary only to find the x coordinate, which we call $\bar{x}$. To find $\bar{x}$ we make use of the moment of the solid of revolution with respect to the yz plane.

Let f be a function that is continuous on the closed interval $[a, b]$, and assume that $f(x) \geq 0$ for all x in $[a, b]$. R is the region bounded by the curve $y = f(x)$, the x axis, and the lines $x = a$ and $x = b$; S is the homogeneous solid of revolution whose volume density is k kg/m³, where k is a constant, and which is generated by revolving the region R about the x axis. Take a partition Δ of the closed interval $[a, b]$, and denote the ith subinterval by $[x_{i-1}, x_i]$ (with $i = 1, 2, \ldots, n$). Let γ_i be the midpoint of $[x_{i-1}, x_i]$. Form n rectangles having altitudes of $f(\gamma_i)$ meters and bases whose width is $\Delta_i x$ meters. Refer to Fig. 11.3.1, showing the region R and

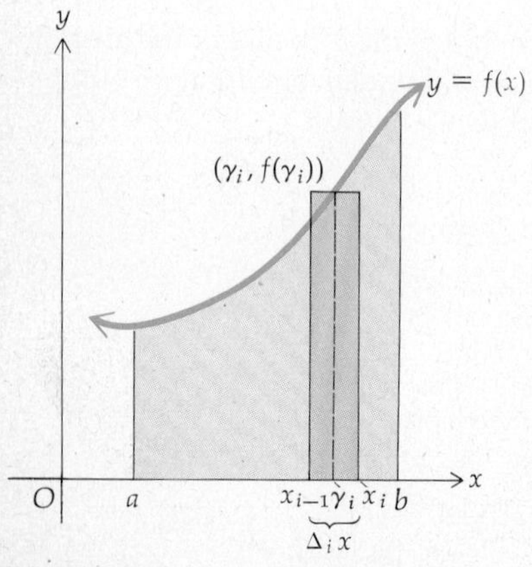

FIGURE 11.3.1

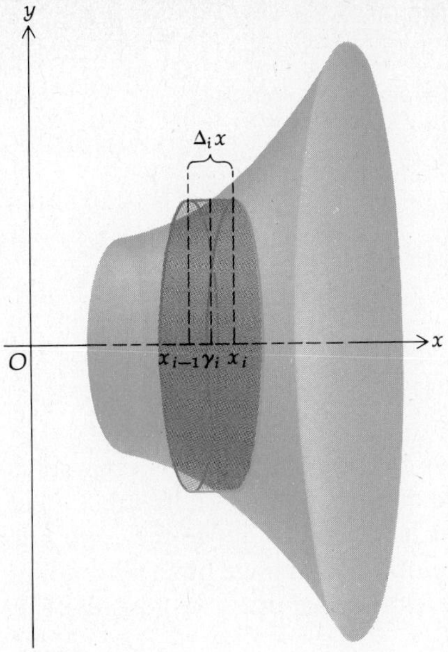

FIGURE 11.3.2

the ith rectangle. If each of the n rectangles is revolved about the x axis, n circular disks are generated. The ith rectangle generates a circular disk having a radius of $f(\gamma_i)$ m and a thickness of $\Delta_i x$ m; its volume is $\pi[f(\gamma_i)]^2 \Delta_i x$ m^3, and its mass is $k\pi[f(\gamma_i)]^2 \Delta_i x$ kg. Figure 11.3.2 shows the solid of revolution S and the ith circular disk.

The center of mass of the circular disk lies on the axis of revolution at the center of the disk: $(\gamma_i, 0, 0)$. The moment of mass of the disk with respect to the yz plane is then $\Delta_i M_{yz}$ kg-m, where

$$\Delta_i M_{yz} = \gamma_i (k\pi[f(\gamma_i)]^2 \Delta_i x)$$

The sum of the measures of the moments of mass of the n circular disks with respect to the yz plane is given by the Riemann sum

$$\sum_{i=1}^{n} \gamma_i k\pi[f(\gamma_i)]^2 \Delta_i x$$

The limit of this sum as $\|\Delta\|$ approaches zero is defined to be the *moment of mass of S with respect to the yz plane.* We state this formally in Definition 11.3.1.

The volume, V m^3, of S was defined in Sec. 7.2 by

$$V = \lim_{\|\Delta\| \to 0} \sum_{i=1}^{n} \pi[f(\gamma_i)]^2 \Delta_i x = \pi \int_a^b [f(x)]^2 \, dx \qquad (1)$$

The *mass* of S is defined as kV.

11.3.1 Definition The region R is bounded by the curve $y = f(x)$, the x axis, and the lines $x = a$ and $x = b$, where the function f is continuous on the closed interval $[a, b]$, and $f(x) \geq 0$ for all x in $[a, b]$. The homogeneous solid of revolution S, whose volume density is k kg/m^3, where k is a constant, is generated by revolving the region R about the x axis. The number of kilogram-meters in the **moment of mass of S with respect to the yz plane** is denoted by M_{yz}, and

$$M_{yz} = \lim_{\|\Delta\| \to 0} \sum_{i=1}^{n} \gamma_i k\pi[f(\gamma_i)]^2 \Delta_i x = k\pi \int_a^b x[f(x)]^2 \, dx \qquad (2)$$

The **mass,** M kgm, of solid S is defined by

$$M = \lim_{\|\Delta\| \to 0} \sum_{i=1}^{n} k\pi[f(\gamma_i)]^2 \Delta_i x = k\pi \int_a^b [f(x)]^2 \, dx \qquad (3)$$

The **center of mass** of S is the point $(\bar{x}, 0, 0)$ such that

$$\bar{x} = \frac{M_{yz}}{M} \qquad (4)$$

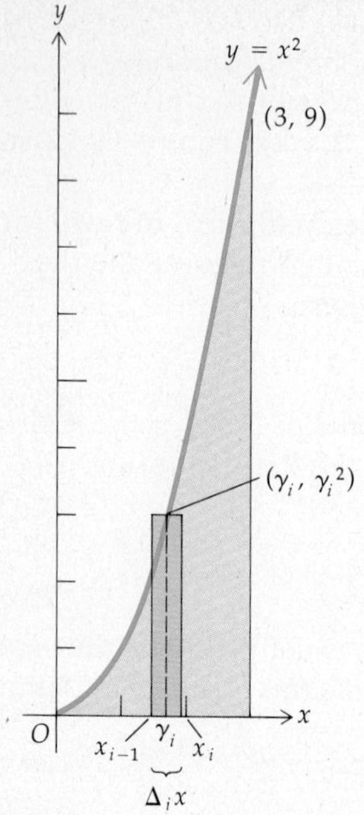

FIGURE 11.3.3

If we substitute from (2) and (3) into (4), we get

$$\bar{x} = \frac{k\pi \int_a^b x[f(x)]^2\,dx}{k\pi \int_a^b [f(x)]^2\,dx}$$

or, equivalently,

$$\bar{x} = \frac{\int_a^b x[f(x)]^2\,dx}{\int_a^b [f(x)]^2\,dx} \tag{5}$$

From (5) we observe that the center of mass of a homogeneous solid of revolution depends only on the shape of the solid, not on its substance. Therefore, as we did for a homogeneous lamina, we refer to the center of mass as the *centroid* of the solid of revolution. When we have a homogeneous solid of revolution, instead of the moment of mass we consider the *moment of the solid*.

11.3.2 Definition If S is the solid of revolution of Definition 11.3.1, then the **moment of the solid S with respect to the yz plane** is given by

$$M_{yz} = \lim_{\|\Delta\|\to 0} \sum_{i=1}^{n} \gamma_i \pi [f(\gamma_i)]^2\,\Delta_i x = \pi \int_a^b x[f(x)]^2\,dx \tag{6}$$

The **centroid** of S, whose volume is V m³, is $(\bar{x}, 0, 0)$, where

$$\bar{x} = \frac{M_{yz}}{V} \tag{7}$$

Equation (7) is obtained from (1), (5), and (6).

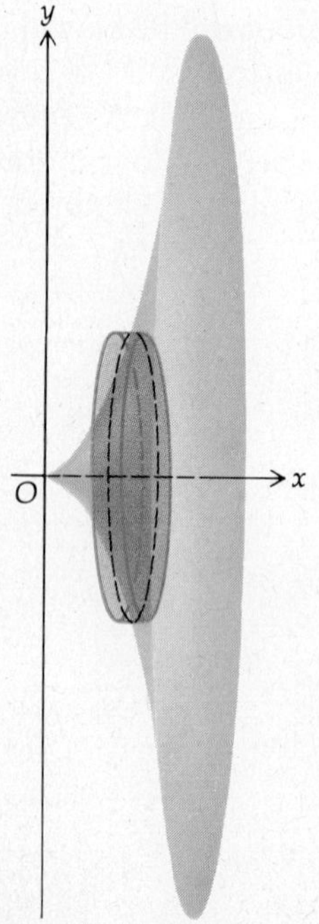

FIGURE 11.3.4

Example 1 Find the centroid of the solid of revolution generated by revolving about the x axis the region bounded by the curve $y = x^2$, the x axis, and the line $x = 3$.

Solution The region and a rectangular element are shown in Fig. 11.3.3. The solid of revolution and an element of volume are shown in Fig. 11.3.4.

$$\begin{aligned} M_{yz} &= \lim_{\|\Delta\|\to 0} \sum_{i=1}^{n} \gamma_i \pi (\gamma_i^{\,2})^2\,\Delta_i x \\ &= \pi \int_0^3 x^5\,dx \\ &= \tfrac{243}{2}\pi \end{aligned} \qquad \begin{aligned} V &= \lim_{\|\Delta\|\to 0} \sum_{i=1}^{n} \pi (\gamma_i^{\,2})^2\,\Delta_i x \\ &= \pi \int_0^3 x^4\,dx \\ &= \tfrac{243}{5}\pi \end{aligned}$$

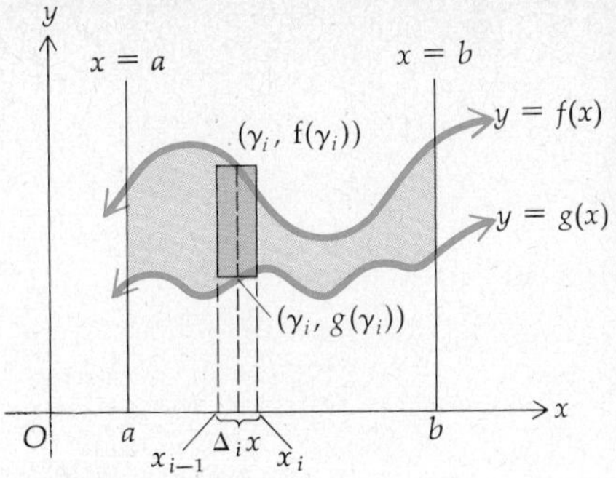

FIGURE 11.3.5

Therefore

$$\bar{x} = \frac{M_{yz}}{V} = \frac{\frac{243}{2}\pi}{\frac{243}{5}\pi} = \frac{5}{2}$$

Thus the centroid is at the point $(\frac{5}{2}, 0, 0)$.

In the preceding discussion and example the element of volume is a circular disk. When the element of volume is a circular ring, the concepts involved are the same, but the equations used to define the moment of the solid and the volume of the solid are obtained by considering the Riemann sum of volumes of circular rings. We proceed to derive these equations.

The functions f and g are continuous on the closed interval $[a, b]$, where $f(x) \geq g(x) \geq 0$ for all x in $[a, b]$. Let R be the region bounded above by the curve $y = f(x)$, below by the curve $y = g(x)$, and on the sides by the lines $x = a$ and $x = b$. The rectangular elements of area are taken perpendicular to the x axis; so the element of volume is a circular ring. Figure 11.3.5 shows the region R and a rectangular element of area. The ith rectangle has a width of $\Delta_i x = x_i - x_{i-1}$, and γ_i is the midpoint of the interval $[x_{i-1}, x_i]$. The centroid of the circular ring obtained by revolving the rectangular element of area about the x axis is at the center of the ring, which is the point $(\gamma_i, 0, 0)$. Let S be the solid of revolution obtained by revolving R about the x axis. Figure 11.3.6 shows S and the ith circular ring. The moment M_{yz} of S with respect to the yz plane is given by

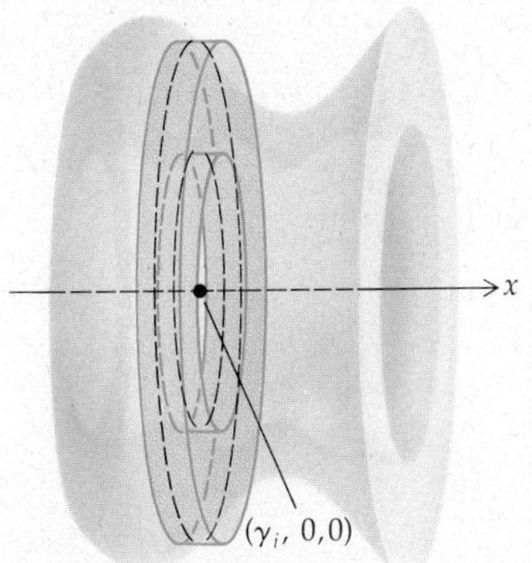

FIGURE 11.3.6

$$M_{yz} = \lim_{\|\Delta\| \to 0} \sum_{i=1}^{n} \gamma_i \pi([f(\gamma_i)]^2 - [g(\gamma_i)]^2)\,\Delta_i x$$

$$= \pi \int_a^b x([f(x)]^2 - [g(x)]^2)\,dx \qquad (8)$$

If V cubic units is the volume of S, then

$$V = \lim_{\|\Delta\| \to 0} \sum_{i=1}^{n} \pi([f(\gamma_i)]^2 - [g(\gamma_i)]^2)\,\Delta_i x$$

$$= \pi \int_a^b ([f(x)]^2 - [g(x)])^2\,dx \qquad (9)$$

If the centroid of S is at the point $(\bar{x}, 0, 0)$, then

$$\bar{x} = \frac{M_{yz}}{V}$$

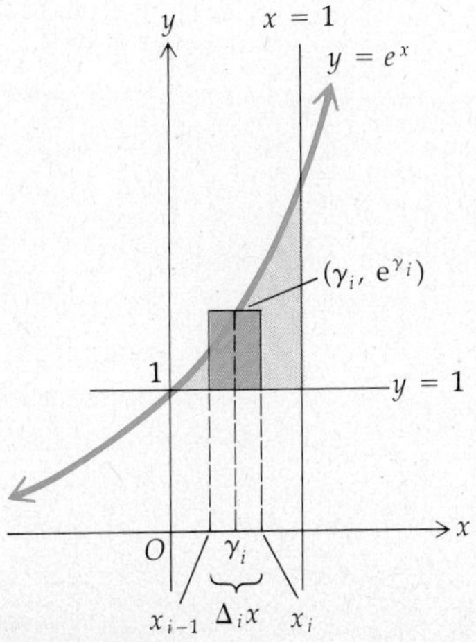

FIGURE 11.3.7

Example 2 Find the centroid of the solid of revolution generated by revolving about the x axis the region bounded by the curve $y = e^x$ and the lines $x = 1$ and $y = 1$.

Solution Figure 11.3.7 shows the region and a rectangular element of area, and Fig. 11.3.8 shows the solid of revolution and an element of volume, which is a circular ring. From (8),

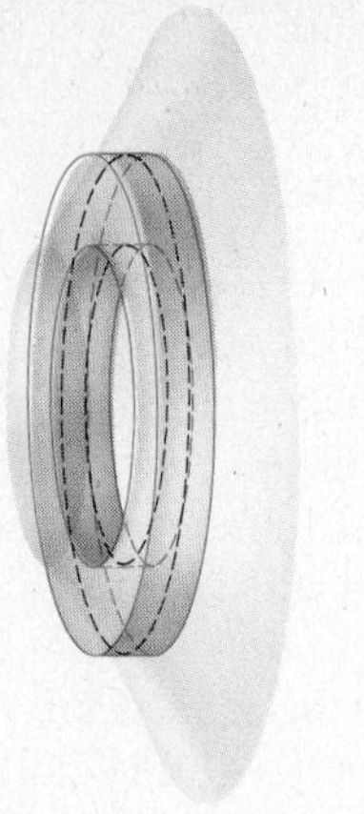

FIGURE 11.3.8

$$M_{yz} = \lim_{\|\Delta\|\to 0} \sum_{i=1}^{n} \gamma_i \pi[(e^{\gamma_i})^2 - 1^2]\,\Delta_i x$$
$$= \pi\int_0^1 xe^{2x}\,dx - \pi\int_0^1 x\,dx$$

To evaluate the first integral we use integration by parts with $u = x$ and $dv = e^{2x}\,dx$.

$$M_{yz} = \pi\left[\frac{1}{2}xe^{2x} - \frac{1}{2}\int e^{2x}\,dx\right]_0^1 - \pi\left[\frac{x^2}{2}\right]_0^1$$
$$= \pi\left[\frac{1}{2}xe^{2x} - \frac{1}{4}e^{2x}\right]_0^1 - \frac{\pi}{2}$$
$$= \pi\left[\frac{1}{2}e^2 - \frac{1}{4}e^2 + \frac{1}{4}\right] - \frac{\pi}{2}$$
$$= \frac{\pi}{4}(e^2 - 1)$$

$$V = \lim_{\|\Delta\|\to 0} \sum_{i=1}^{n} \pi[(e^{\gamma_i})^2 - 1^2]\,\Delta_i x$$
$$= \pi\int_0^1 (e^{2x} - 1)\,dx$$
$$= \pi\left[\tfrac{1}{2}e^{2x} - x\right]_0^1$$
$$= \pi[(\tfrac{1}{2}e^2 - 1) - \tfrac{1}{2}]$$
$$= \pi(\tfrac{1}{2}e^2 - \tfrac{3}{2})$$
$$= \frac{\pi}{2}(e^2 - 3)$$

Therefore

$$\bar{x} = \frac{M_{yz}}{V} = \frac{\frac{1}{4}\pi(e^2 - 1)}{\frac{1}{2}\pi(e^2 - 3)} \approx 0.694$$

The centroid is at the point (0.69, 0, 0).

The centroid of a solid of revolution also can be found by the cylindrical-shell method. Let R be the region bounded by the curve $y = f(x)$, where f is continuous and $f(x) \geq 0$ on $[a, b]$, the x axis, and the lines $x = a$ and $x = b$. Let S be the solid of revolution generated by revolving R about the y axis. If the rectangular elements are taken parallel to the y axis, then the element of volume is a cylindrical shell. Let the ith rectangle have a width of $\Delta_i x = x_i - x_{i-1}$, and let γ_i be the midpoint of the interval $[x_{i-1}, x_i]$. The centroid of the cylindrical shell obtained by revolving this rectangle about the y axis is at the center of the cylindrical shell, which is the point $(0, \frac{1}{2}f(\gamma_i), 0)$. Figure 11.3.9 shows the region R and

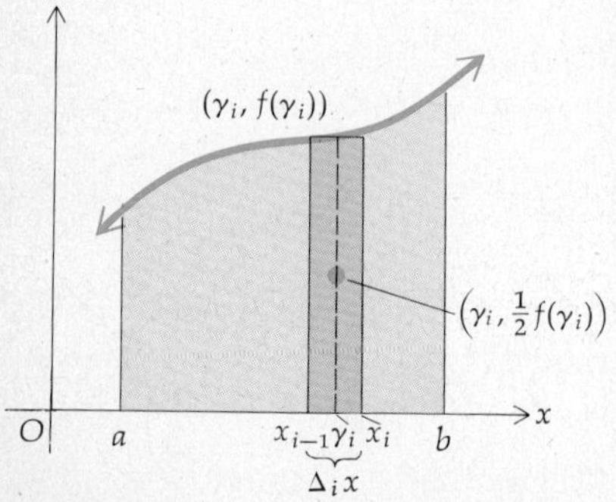

FIGURE 11.3.9

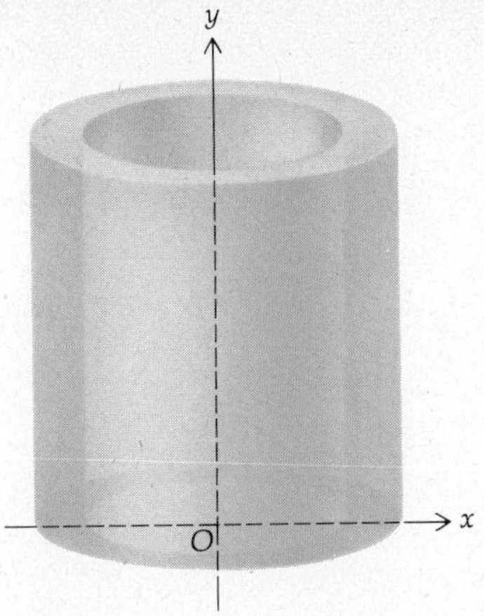

FIGURE 11.3.10

a rectangular element of area, and Fig. 11.3.10 shows the cylindrical shell. The moment, M_{xz}, of S with respect to the xz plane is given by

$$M_{xz} = \lim_{\|\Delta\| \to 0} \sum_{i=1}^{n} \tfrac{1}{2} f(\gamma_i) 2\pi \gamma_i f(\gamma_i)\, \Delta_i x = \pi \int_a^b x[f(x)]^2\, dx$$

If V cubic units is the volume of S,

$$V = \lim_{\|\Delta\| \to 0} \sum_{i=1}^{n} 2\pi \gamma_i f(\gamma_i)\, \Delta_i x$$

$$= 2\pi \int_a^b x f(x)\, dx$$

Then the centroid of S is at the point $(0, \bar{y}, 0)$, where

$$\bar{y} = \frac{M_{xz}}{V}$$

Example 3 Use the cylindrical-shell method to find the centroid of the solid of revolution generated by revolving the region in Example 1 about the y axis.

Solution Figure 11.3.11 shows the region and a rectangular element of area. The solid of revolution and a cylindrical-shell element of volume are shown in Fig. 11.3.12.

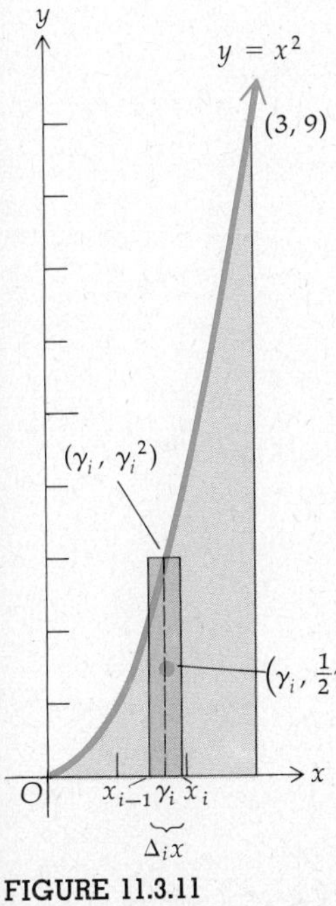

FIGURE 11.3.11

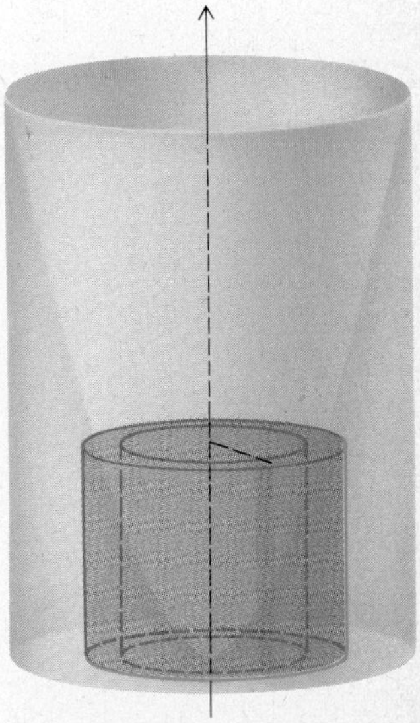

FIGURE 11.3.12

$$M_{xz} = \lim_{\|\Delta\| \to 0} \sum_{i=1}^{n} \tfrac{1}{2}\gamma_i^2\, 2\pi\gamma_i\gamma_i^2\, \Delta_i x \qquad V = \lim_{\|\Delta\| \to 0} \sum_{i=1}^{n} 2\pi\gamma_i\gamma_i^2\, \Delta_i x$$

$$= \pi \int_0^3 x^5\, dx \qquad\qquad = 2\pi \int_0^3 x^3\, dx$$

$$= \tfrac{243}{2}\pi \qquad\qquad = \tfrac{81}{2}\pi$$

Therefore

$$\bar{y} = \frac{M_{xz}}{V} = \frac{\frac{243}{2}\pi}{\frac{81}{2}\pi} = 3$$

Hence the centroid is at the point (0, 3, 0).

Example 4 Solve Example 1 by the cylindrical-shell method.

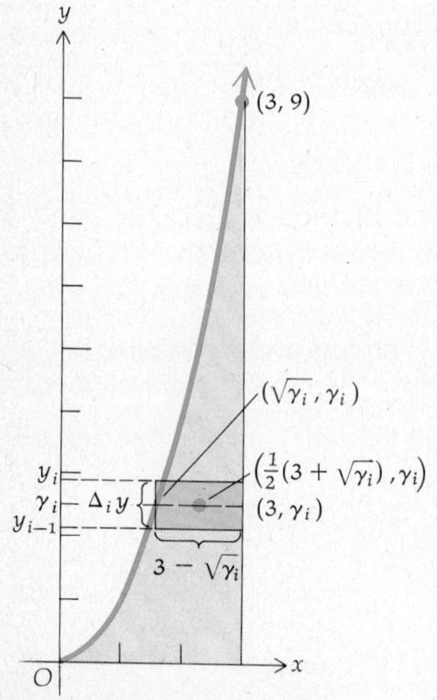

FIGURE 11.3.13

Solution Figure 11.3.13 shows the region and a rectangular element of area, and Fig. 11.3.14 shows the cylindrical-shell element of volume obtained by revolving the rectangle about the x axis. The centroid of the cylindrical shell is at its center, which is the point $(\frac{1}{2}(3 + \sqrt{\gamma_i}), 0, 0)$. The centroid of the solid of revolution is at $(\bar{x}, 0, 0)$.

$$M_{yz} = \lim_{\|\Delta\| \to 0} \sum_{i=1}^{n} \tfrac{1}{2}(3 + \sqrt{\gamma_i})\, 2\pi\gamma_i(3 - \sqrt{\gamma_i})\, \Delta_i y$$

$$= \pi \int_0^9 y(3 + \sqrt{y})(3 - \sqrt{y})\, dy$$

$$= \pi \int_0^9 (9y - y^2)\, dy$$

$$= \tfrac{243}{2}\pi$$

Finding V by the cylindrical-shell method we have

$$V = \lim_{\|\Delta\| \to 0} \sum_{i=1}^{n} 2\pi\gamma_i(3 - \sqrt{\gamma_i})\, \Delta_i y$$

$$= 2\pi \int_0^9 y(3 - y^{1/2})\, dy$$

$$= \tfrac{243}{5}\pi$$

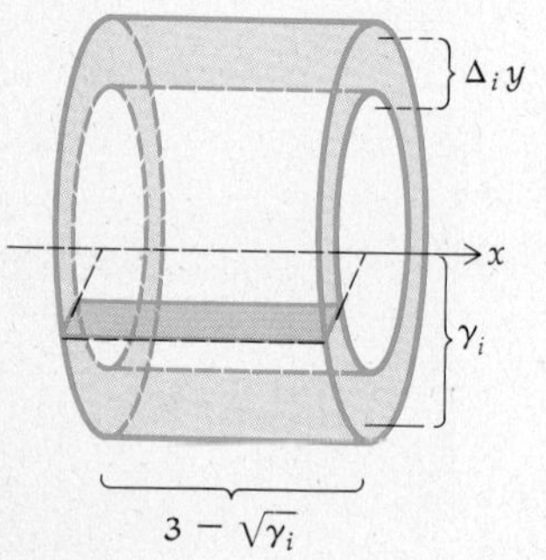

FIGURE 11.3.14

Thus

$$\bar{x} = \frac{M_{yz}}{V} = \frac{5}{2}$$

and the centroid is at $(\frac{5}{2}, 0, 0)$. The result agrees with that of Example 1.

EXERCISES 11.3

In Exercises 1 through 20, find the centroid of the solid of revolution generated by revolving the given region about the indicated line.

1. The region bounded by $y = 4x - x^2$ and the x axis, about the y axis. Take the rectangular elements parallel to the axis of revolution.
2. Same as Exercise 1, but take the rectangular elements perpendicular to the axis of revolution.
3. The region bounded by $x + 2y = 2$, the x axis, and the y axis, about the x axis. Take the rectangular elements perpendicular to the axis of revolution.
4. Same as Exercise 3, but take the rectangular elements parallel to the axis of revolution.
5. The region bounded by $y^2 = x^3$ and $x = 4$, about the x axis. Take the rectangular elements parallel to the axis of revolution.
6. Same as Exercise 5, but take the rectangular elements perpendicular to the axis of revolution.
7. The region bounded by $y = x^3$, $x = 2$, and the x axis, about the line $x = 2$. Take the rectangular elements perpendicular to the axis of revolution.
8. Same as Exercise 7, but take the rectangular elements parallel to the axis of revolution.
9. The region bounded by $x^4y = 1$, $y = 1$, and $y = 4$, about the y axis.
10. The region in Exercise 9, about the x axis.
11. The region bounded by the lines $y = x$, $y = 2x$, and $x + y = 6$, about the y axis.
12. The region bounded by the portion of the circle $x^2 + y^2 = 4$ in the first quadrant, the portion of the line $2x - y = 4$ in the fourth quadrant, and the y axis, about the y axis.
13. The region bounded by $y = x^2$ and $y = x + 2$, about the line $y = 4$.
14. The region bounded by $y^2 = 4x$ and $y^2 = 16 - 4x$, about the x axis.
15. The region bounded by $y = \sqrt{4px}$, the x axis, and the line $x = p$, about the line $x = p$.
16. The region of Exercise 15, about the line $y = 2p$.
17. The region bounded by $y = \sin x$, the x axis, and the line $x = \frac{1}{2}\pi$, about the x axis.
18. The region bounded by $yx^2 = \sqrt{x^2 - 9}$, the x axis, and the line $x = 5$, about the y axis.
19. The region bounded by $y = e^{-x}$, the coordinate axes, and the line $x = 1$, about the x axis.
20. The region of Exercise 19, about the y axis.
21. Find the centroid of the right-circular cone of altitude h units and base radius r units.
22. The region in the first quadrant bounded by a line through the origin with positive slope, the x axis, and the line $x = c$ $(c > 0)$ is revolved about the x axis. Show that the centroid of the solid of revolution is at the point $(\frac{3}{4}c, 0, 0)$.
23. The region bounded by the curve $y = x^2$, the x axis, and the line $x = b$ $(b > 0)$ is revolved about the x axis. Its moment of mass with respect to the yz plane is $\frac{32}{3}k\pi$ kg-m, where the volume density of the solid is k kg/m^3. Find the value of b.
24. Suppose that a cylindrical hole with a radius of r units is bored through a solid wooden hemisphere of radius $2r$ units such that the axis of the cylinder is the same as the axis of the hemisphere. Find the centroid of the solid remaining.
25. Find the center of mass of the solid of revolution of Exercise 3 if the measure of the volume density of the solid at any point is equal to a constant k times the measure of the distance of the point from the yz plane.

26. Find the center of mass of the solid of revolution of Exercise 5 if the measure of the volume density of the solid at any point is equal to a constant k times the measure of the distance of the point from the yz plane.

27. The region bounded by $y = x^{5/2}$, the x axis, and the line $x = 1$ is revolved about the x axis. Find the center of mass of the solid of revolution if the measure of the volume density of the solid at any point is equal to a constant k times the square of the measure of the distance of the point from the yz plane.

11.4 LIQUID PRESSURE

Another application of the definite integral in physics is to find the force caused by liquid pressure on a plate submerged in the liquid or on a side of a container holding the liquid. First suppose that a flat plate is inserted horizontally into a liquid in a container. The weight of the liquid exerts a force on the plate. The force per square unit of area exerted by the liquid on the plate is called the **pressure** of the liquid.

Let ρ lb/ft³ be the weight density of the liquid and h be the number of feet in the depth of a point below the surface of the liquid. If P is the number of pounds per square foot of pressure exerted by the liquid at the point, then

$$P = \rho h \tag{1}$$

If A is the number of square feet in the area of a flat plate that is submerged horizontally in the liquid, and F is the number of pounds in the force caused by liquid pressure acting on the upper face of the plate, then

$$F = PA \tag{2}$$

Substituting from (1) into (2) gives

$$F = \rho h A \tag{3}$$

From (1) it follows that the size of the container is immaterial so far as liquid pressure is concerned. For example, at a depth of 5 ft in a swimming pool filled with salt water the pressure is the same as at a depth of 5 ft in the Pacific Ocean, assuming the density of the water is the same.

● ILLUSTRATION 1 A rectangular sheet of tin 8 ft × 12 ft is submerged in a tank of water at a depth of 10 ft. If P lb/ft² is the pressure exerted by the water at a point on the upper face of the sheet of tin,

$$P = 10\rho \tag{4}$$

The area of the piece of tin is 96 ft². So if F lb is the force caused by liquid pressure acting on the upper face of the sheet of tin,

$$F = 96P \tag{5}$$

Substituting from (4) into (5) we get

$$F = 960\rho$$

Letting $\rho = 62.5$ we have $F = 60{,}000$. Therefore the force due to water pressure on the upper face of the sheet of tin is 60,000 lb. ●

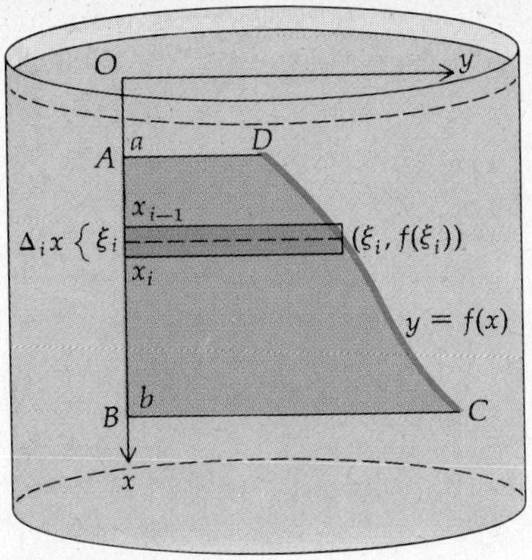

FIGURE 11.4.1

Now suppose that the plate is submerged vertically in the liquid. Then at points on the plate at different depths the pressure, computed from (1), will be different and will be greater at the bottom of the plate than at the top. We now proceed to define the force caused by liquid pressure when the plate is submerged vertically in the liquid. We use Pascal's principle: At any point in a liquid, the pressure is the same in all directions.

In Fig. 11.4.1 let $ABCD$ be the region bounded by the x axis, the lines $x = a$ and $x = b$, and the curve $y = f(x)$, where the function f is continuous and $f(x) \geq 0$ on the closed interval $[a, b]$. Choose the coordinate axes so the y axis lies along the line of the surface of the liquid. Take the x axis vertical with the positive direction downward. The length of the plate at a depth x ft is given by $f(x)$ ft.

Let Δ be a partition of the closed interval $[a, b]$ that divides the interval into n subintervals. Choose a point ξ_i in the ith subinterval, with $x_{i-1} \leq \xi_i \leq x_i$. Draw n horizontal rectangles. The ith rectangle has a length of $f(\xi_i)$ ft and a width of $\Delta_i x$ ft (see Fig. 11.4.1).

If we rotate each rectangular element through an angle of 90°, each element becomes a plate submerged in the liquid at a depth of ξ_i ft below the surface of the liquid and perpendicular to the region $ABCD$. Then the force on the ith rectangular element is given by $\rho\xi_i f(\xi_i)\,\Delta_i x$ lb. An approximation to F, the number of pounds in the total force on the vertical plate, is given by

$$\sum_{i=1}^{n} \rho\xi_i f(\xi_i)\,\Delta_i x \qquad (6)$$

which is a Riemann sum.

The smaller we take $\|\Delta\|$, the larger n will be and the closer the approximation given by (6) will be to what we wish to be the measure of the total force. We have, then, the following definition.

11.4.1 Definition Suppose that a flat plate is submerged vertically in a liquid of weight ρ pounds per cubic unit. The length of the plate at a depth of x units below the surface of the liquid is $f(x)$ units, where f is continuous on the closed interval $[a, b]$ and $f(x) \geq 0$ on $[a, b]$. Then F, the number of pounds of **force caused by liquid pressure** on the plate, is given by

$$F = \lim_{\|\Delta\| \to 0} \sum_{i=1}^{n} \rho\xi_i f(\xi_i)\,\Delta_i x = \int_a^b \rho x f(x)\,dx \qquad (7)$$

Example 1 A trough having a trapezoidal cross section is full of water. If the trapezoid is 3 ft wide at the top, 2 ft wide at the bottom, and 2 ft deep, find the total force owing to water pressure on one end of the trough.

Solution Figure 11.4.2 illustrates one end of the trough together with a rectangular element of area. An equation of line AB is $y = \frac{3}{2} - \frac{1}{4}x$. Let $f(x) = \frac{3}{2} - \frac{1}{4}x$. If we rotate the rectangular element through 90°, the

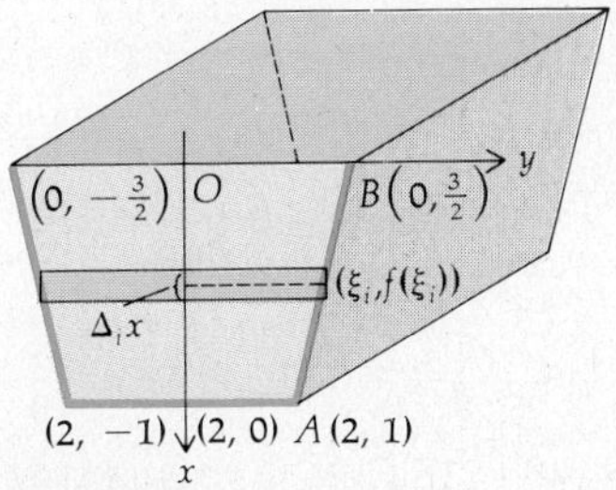

FIGURE 11.4.2

force on the element is given by $2\rho\xi_i f(\xi_i)\ \Delta_i x$ lb. If F is the number of pounds in the total force on the side of the trough,

$$\begin{aligned}F &= \lim_{\|\Delta\|\to 0} \sum_{i=1}^{n} 2\rho\xi_i f(\xi_i)\ \Delta_i x \\ &= 2\rho \int_0^2 x f(x)\, dx \\ &= 2\rho \int_0^2 x(\tfrac{3}{2} - \tfrac{1}{4}x)\, dx \\ &= 2\rho\left[\tfrac{3}{4}x^2 - \tfrac{1}{12}x^3\right]_0^2 \\ &= \tfrac{14}{3}\rho\end{aligned}$$

Taking $\rho = 62.5$, we find that the total force is 291.2 lb.

Example 2 The ends of a trough are semicircular regions, each with a radius of 2 ft. Find the force caused by liquid pressure on one end if the trough is full of water.

Solution Figure 11.4.3 shows one end of the trough together with a rectangular element of area. An equation of the semicircle is $x^2 + y^2 = 4$. Solving for y gives $y = \sqrt{4 - x^2}$. The force on the rectangular element is given by $2\rho\xi_i\sqrt{4 - \xi_i^2}\ \Delta_i x$. So if F pounds is the total force on the side of the trough,

$$\begin{aligned}F &= \lim_{\|\Delta\|\to 0} \sum_{i=1}^{n} 2\rho\xi_i\sqrt{4 - \xi_i^2}\ \Delta_i x \\ &= 2\rho \int_0^2 x\sqrt{4 - x^2}\, dx \\ &= -\tfrac{2}{3}\rho(4 - x^2)^{3/2}\Big]_0^2 \\ &= \tfrac{16}{3}\rho\end{aligned}$$

Therefore the total force is $\frac{16}{3}\rho$ lb.

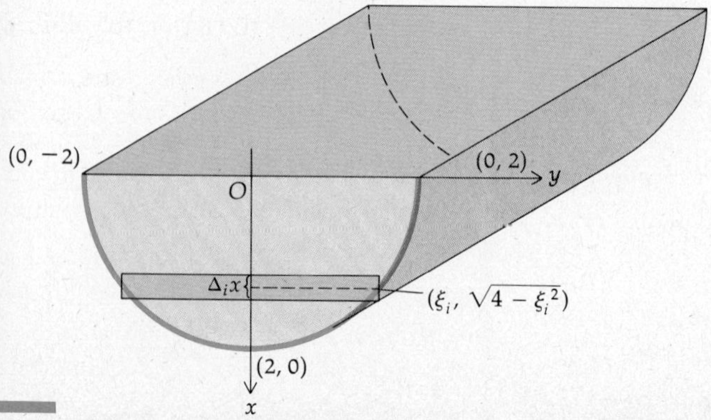

FIGURE 11.4.3

There is a useful relation between the force caused by liquid pressure on a plane region and the location of the centroid of the region. Refer to Fig. 11.4.1, where $ABCD$ is the region bounded by the x axis, the lines $x = a$ and $x = b$, and the curve $y = f(x)$, where f is continuous and $f(x) \geq 0$ on the closed interval $[a, b]$. We consider the region $ABCD$ as a vertical plate immersed in a liquid having weight density ρ pounds per cubic unit of the liquid. If F lb is the force owing to liquid pressure on the vertical plate, from (7),

$$F = \rho \int_a^b x f(x)\, dx \tag{8}$$

If $\bar{x}$ is the abscissa of the centroid of the region $ABCD$, then $\bar{x} = M_y/A$. Because $M_y = \int_a^b x f(x)\, dx$,

$$\bar{x} = \frac{\displaystyle\int_a^b x f(x)\, dx}{A}$$

so

$$\int_a^b x f(x)\, dx = \bar{x}A \tag{9}$$

Substituting from (9) into (8) we obtain

$$F = \rho \bar{x} A \tag{10}$$

From (10) it follows that the total force owing to liquid pressure against a vertical plane region is the same as it would be if the region were horizontal at a depth $\bar{x}$ units below the surface of a liquid.

● ILLUSTRATION 2 Consider a trough full of water having as ends semicircular regions each with a radius of 2 ft. Using the result of Example 5 in Sec 11.2 we find that the centroid of the region is at a depth of $8/(3\pi)$ ft. Therefore, using (10) we see that if F lb is the force on one end of the trough,

$$F = \rho \cdot \frac{8}{3\pi} \cdot 2\pi = \frac{16}{3}\rho$$

This agrees with the result found in Example 2. ●

For various simple plane regions, the centroid may be found in a table. When both the area of the region and the centroid of the region may be obtained directly, (10) is easy to apply and is used in such cases by engineers to find the force caused by liquid pressure.

If mass is measured in kilograms, then pressure is measured in newtons per square meter and force is measured in newtons. If P nt/m^2 is the pressure, h is the number of meters in the depth of a point below the surface of the liquid, and ρ kg/m^3 is the density of the liquid, then instead of Eq. (1) we have

$$P = \rho h(9.81) \tag{11}$$

where 9.81 m/sec^2 is the acceleration due to gravity. From (2), if A square meters is the area of a flat plate that is submerged horizontally in the liquid and F nt is the force caused by liquid pressure acting on the upper face of the plate, then

$$F = PA \tag{12}$$

Substituting from (11) into (12) gives

$$F = 9.81\rho hA \tag{13}$$

Thus in the mks system, Eq. (7) of Definition 11.4.1 becomes

$$F = \lim_{\|\Delta\| \to 0} \sum_{i=1}^{n} 9.81\rho\xi_i f(\xi_i)\,\Delta_i x = \int_a^b 9.81\rho x f(x)\,dx$$

where the length of the plate x m below the surface of the liquid is $f(x)$ m. The density of water is approximately 1000 kg/m^3. So if the liquid causing the pressure is water, we take $\rho = 1000$.

Example 3 A container in the shape of a right-circular cylinder having a base of radius 3 m is on its side at the bottom of a tank full of water. The depth of the tank is 13 m. Find the total force due to water pressure on one end of the container.

Solution Figure 11.4.4 shows one end of the container in the tank and a rectangular element of area. The coordinate system has been chosen so that the origin is at the center of the circle. An equation of the circle is $x^2 + y^2 = 9$. Solving for x gives $x = \sqrt{9 - y^2}$. The force on the rectangular element is given by

$$9.81\rho(10 - \xi_i)[2\sqrt{9 - \xi_i^2}]\,\Delta_i y$$

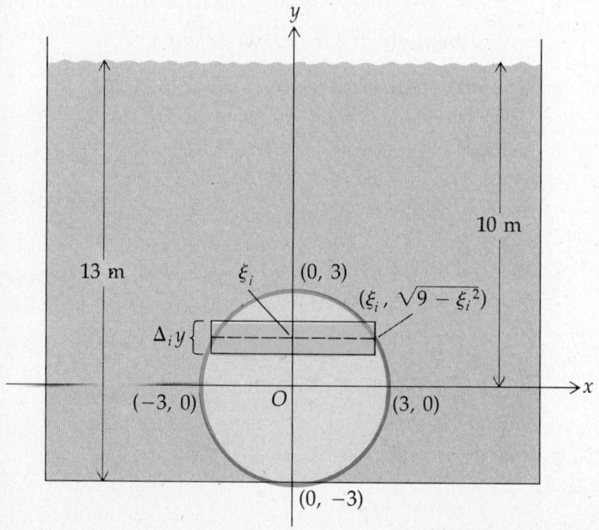

FIGURE 11.4.4

So if F nt is the total force on the end of the container,

$$F = \lim_{\|\Delta\| \to 0} \sum_{i=1}^{n} 9.81\rho(10 - \xi_i)[2\sqrt{9 - \xi_i^2}]\,\Delta_i y$$

$$= 19.62\rho \int_{-3}^{3} (10 - y)\sqrt{9 - y^2}\,dy$$

Because $\rho = 1000$, we have

$$F = 196{,}200 \int_{-3}^{3} \sqrt{9 - y^2}\,dy - 19{,}620 \int_{-3}^{3} y\sqrt{9 - y^2}\,dy \qquad (14)$$

To evaluate $\int_{-3}^{3} \sqrt{9 - y^2}\,dy$ we can use the trigonometric substitution $y = 3\sin\theta$, or we can consider $\int_{-3}^{3} \sqrt{9 - y^2}\,dy$ as the measure of the area of the region enclosed by a semicircle of radius 3. Therefore $\int_{-3}^{3} \sqrt{9 - y^2}\,dy = \frac{1}{2}\pi(3)^2 = \frac{9}{2}\pi$. Substituting this value into (14) and evaluating the second integral we have

$$F = 196{,}200(\tfrac{9}{2}\pi) + 19{,}620\left[\tfrac{1}{3}(9 - y^2)^{3/2}\right]_{-3}^{3}$$

$$= 882{,}900\pi$$

Thus the total force is $882{,}900\pi$ nt.

EXERCISES 11.4

1. A plate in the shape of a rectangle is submerged vertically in a tank of water, with the upper edge lying in the surface. If the width of the plate is 10 ft and the depth is 8 ft, find the force due to liquid pressure on one side of the plate.

2. A square plate of side 4 ft is submerged vertically in a tank of water and its center is 2 ft below the surface. Find the force due to liquid pressure on one side of the plate.

3. Solve Exercise 2 if the center of the plate is 4 ft below the surface.

4. A plate in the shape of an isosceles right triangle is submerged vertically in a tank of water, with one leg lying in the surface. The legs are each 6 ft long. Find the force due to liquid pressure on one side of the plate.

5. A rectangular tank full of water is 2 ft wide and 18 in. deep. Find the force due to liquid pressure on one end of the tank.

6. The ends of a trough are equilateral triangles having sides with lengths of 2 ft. If the water in the trough is 1 ft deep, find the force due to liquid pressure on one end.

7. The face of the gate of a dam is in the shape of an isosceles triangle 4 m wide at the top and 3 m high. If the upper edge of the face of the gate is 15 m below the surface of the water, find the total force due to water pressure on the gate.

8. The face of a gate of a dam is vertical and in the shape of an isosceles trapezoid 3 m wide at the top, 4 m wide at the bottom, and 3 m high. If the upper base is 20 m below the surface of the water, find the total force due to water pressure on the gate.

9. The face of a dam adjacent to the water is vertical, and its shape is in the form of an isosceles triangle 250 m wide across the top and 100 m high in the center. If the water is 10 m deep in the center, find the total force on the dam due to water pressure.

10. An oil tank is in the shape of a right-circular cylinder 4 m in diameter, and its axis is horizontal. If the tank is half full of oil having density 750 kg/m^3, find the total force on one end due to liquid pressure.

11. A board is in the shape of a region bounded by a straight line and one arch of the sine curve. If the board is submerged vertically in water so that the straight line is the lower boundary 2 ft below the surface of the water, find the force on the board due to water pressure.

12. The face of a dam is in the shape of one arch of the curve $y = 100 \cos \frac{1}{200}\pi x$, and the surface of the water is at the top of the dam. Find the force due to water pressure on the face of the dam if distance is measured in feet.

13. A horizontal cylindrical pipe has a 4-ft inner diameter and is closed at one end by a circular gate that just fits over the pipe. If the pipe contains water at a depth of 3 ft, find the force on the gate due to water pressure.

14. An automobile's gasoline tank is in the shape of a right-circular cylinder of radius 8 in. with a horizontal axis. Find the total force on one end when the gasoline is 12 in. deep and ρ $oz/in.^3$ is the weight density of gasoline.

15. An oil tank is in the shape of a right-circular cylinder with a radius of r m, and its axis is horizontal. If the tank is half full of oil having density 750 kg/m^3, find r if the total force on one end of the tank due to liquid pressure is 80,000 nt.

16. A gate in an irrigation ditch is in the shape of a segment of a circle of radius 4 ft. The top of the gate is horizontal and 3 ft above the lowest point on the gate. If the water level is 2 ft above the top of the gate, find the force on the gate due to the water pressure.

17. The face of a dam adjacent to the water is inclined at an angle of 30° from the vertical. The shape of the face is a rectangle of width 50 ft and slant height 30 ft. If the dam is full of water, find the total force due to water pressure on the face.

18. Solve Exercise 17 if the face of the dam is an isosceles trapezoid 120 ft wide at the top, 80 ft wide at the bottom, and with a slant height of 40 ft.

19. The bottom of a swimming pool is an inclined plane. The pool is 2 ft deep at one end and 8 ft deep at the other. If the width of the pool is 25 ft and the length is 40 ft, find the total force due to water pressure on the bottom.

20. If the end of a water tank is in the shape of a rectangle and the tank is full, show that the measure of the force due to water pressure on the end is the product of the measure of the area of the end and the measure of the force at the geometrical center.

21. Solve Exercise 5 by using Eq. (10).

22. Solve Exercise 4 by using Eq. (10). 23. Solve Exercise 6 by using Eq. (10).

24. The face of a dam adjacent to the water is vertical and is in the shape of an isosceles trapezoid 90 ft wide at the top, 60 ft wide at the bottom, and 20 ft high. Use Eq. (10) to find the total force due to water pressure on the face of the dam.

25. Find the moment about the lower base of the trapezoid of the force in Exercise 24.

REVIEW EXERCISES FOR CHAPTER 11

1. Three particles of masses 4, 2, and 7 kg are located on the x axis at the points having coordinates -5, 4, and 2, respectively, where the distance is measured in meters. Find the center of mass of the system.

2. Three particles having masses 5, 2, and 8 slugs are located, respectively, at the points $(-1, 3)$, $(2, -1)$, and $(5, 2)$. Find the center of mass, where distance is measured in feet.

3. Find the coordinates of the center of mass of the four particles having equal masses located at the points $(3, 0)$, $(2, 2)$, $(2, 4)$, and $(-1, 2)$.

4. Three particles, each having the same mass, are located on the x axis at the points having coordinates -4, 1, and 5, where the distance is measured in meters. Find the coordinates of the center of mass of the system.

5. The length of a rod is 8 in. and the linear density of the rod at a point x in. from the left end is $2\sqrt{x+1}$ slugs/in. Find the total mass of the rod and the center of mass.

6. The length of a rod is 4 m and the linear density of the rod at a point x m from the left end is $(3x + 1)$ kg/m. Find the total mass of the rod and the center of mass.

7. The linear density of a rod 3 m long at a point x m from one end is ke^{-3x} kg/m. Find the mass and center of mass of the rod.

8. Find the center of mass of a rod 4 m long if the linear density at the point x m from the left end is $\sqrt{9 + x^2}$ kg/m.

9. Find the centroid of the region in the first quadrant bounded by the coordinate axes and the parabola $y = 9 - x^2$.

10. Find the centroid of the region bounded by the parabola $y^2 = x$ and the line $y = x - 2$.

11. Find the centroid of the region bounded by the curve $y = x \ln x$, the x axis, and the line $x = e$.

12. Find the centroid of the region in the first quadrant bounded by the coordinate axes and the curve $y = \cos x$.

13. Find the centroid of the region bounded by the y axis and the curves $y = \sin x - \cos x$ and $y = \sin x + \cos x$ from $x = 0$ to $x = \frac{1}{2}\pi$.

14. Find the centroid of the region bounded by the loop of the curve $y^2 = x^2 - x^3$.

15. Find the centroid of the region enclosed by one loop of the curve $x^2 = y^4(1 - y^2)$.

16. Give an example to show that the centroid of a plane region is not necessarily a point within the region.

17. Find the centroid of the region bounded above by the parabola $4x^2 = 36 - 9y$ and below by the x axis.

18. Find the center of mass of the region of Exercise 17 if the measure of the area density at any point (x, y) is $\sqrt{4 - y}$.

19. Use the theorem of Pappus to find the volume of a sphere of radius 4 m.

20. Use the theorem of Pappus to find the volume of a right-circular cone with base radius 2 m and height 3 m.

21. Find the centroid of the solid of revolution generated by revolving about the y axis the region bounded by the curve $y = e^x$, the coordinate axes, and the line $x = 1$.

22. The region bounded by the parabola $x^2 = 4y$, the x axis, and the line $x = 4$ is revolved about the y axis. Find the centroid of the solid of revolution formed.

23. Find the center of mass of the solid of revolution of Exercise 22 if the measure of the volume density of the solid at any point is equal to the measure of the distance of the point from the xz plane.

24. The region bounded by the curve $x = \sqrt{y}$, the y axis, and the line $y = 4$ is revolved about the x axis. Find the centroid of the solid of revolution formed.

25. Find the centroid of the solid of revolution generated by revolving about the x axis the region in the first quadrant bounded by the curve $y = \cos x$ and the coordinate axes.

26. If the region in Exercise 25 is revolved about the y axis, find the centroid of the resulting solid of revolution.

27. A plate in the shape of a region bounded by the parabola $x^2 = 6y$ and the line $2y = 3$ is placed in a water tank with its vertex downward and the line in the surface of the water. Find the force due to water pressure on one side of the plate if distance is measured in feet.

28. A semicircular plate with a radius of 3 ft is submerged vertically in a tank of water, with its diameter lying in the surface. Use Eq. (10) of Sec. 11.4 to find the force due to water pressure on one side of the plate.

29. A cylindrical tank is half full of gasoline having weight density 42 lb/ft^3. If the axis is horizontal and the diameter is 6 ft, find the force on an end due to liquid pressure.

30. The face of a dam adjacent to the water is inclined at an angle of 45° from the vertical. The face is a rectangle of width 80 ft and slant height 40 ft. If the dam is full of water, find the total force due to water pressure on the face.

31. The vertical end of a water trough is 3 ft wide at the top, 2 ft deep, and has the form of the region bounded by the x axis and one arch of the curve $y = 2 \sin \frac{1}{3}\pi x$. If the trough is full of water, find the force due to water pressure on the end.

CHAPTER 12 | Polar Coordinates

12.1 THE POLAR COORDINATE SYSTEM

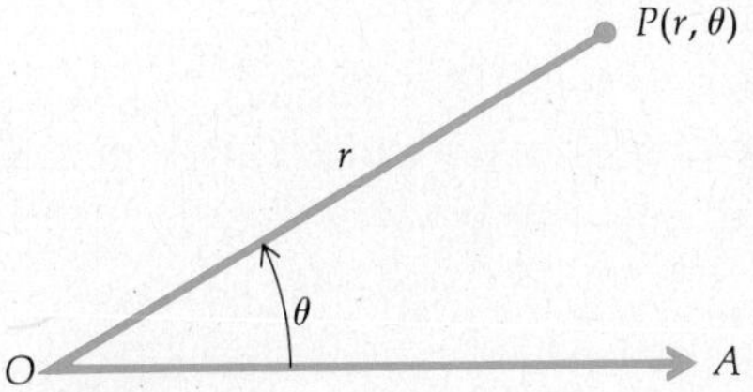

FIGURE 12.1.1

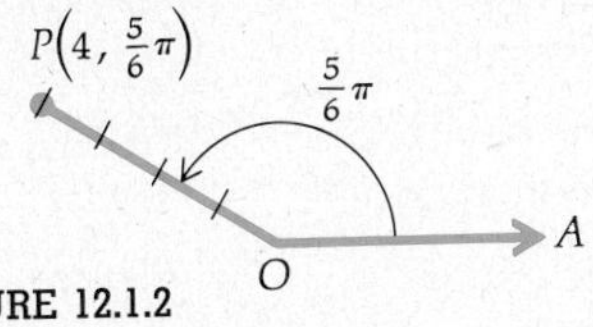

FIGURE 12.1.2

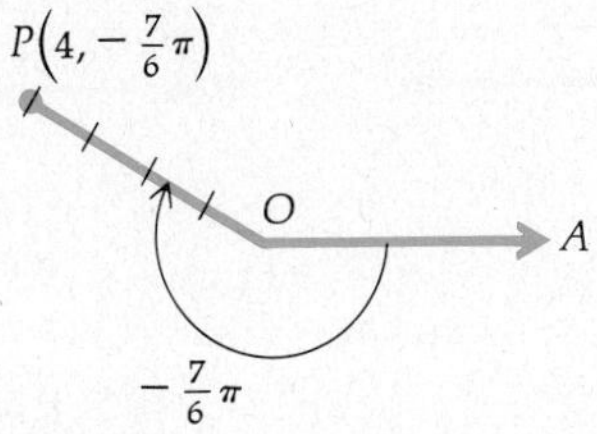

FIGURE 12.1.3

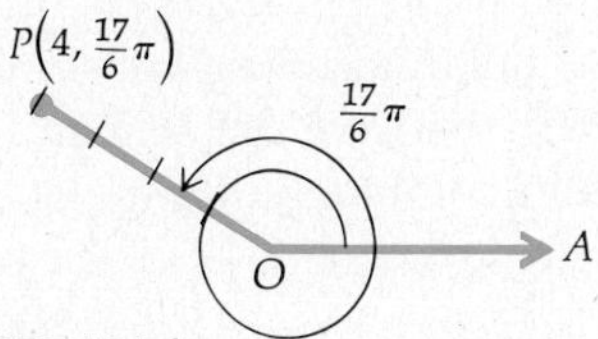

FIGURE 12.1.4

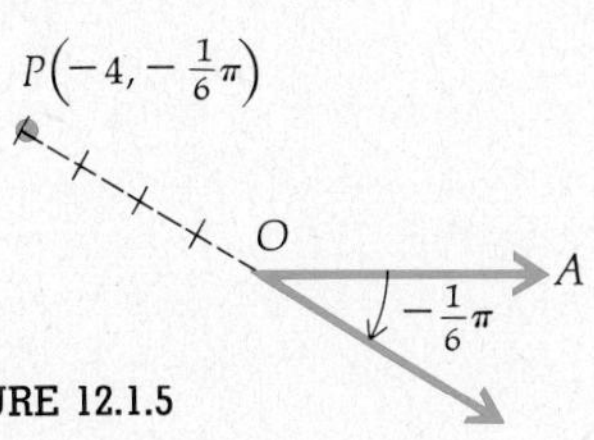

FIGURE 12.1.5

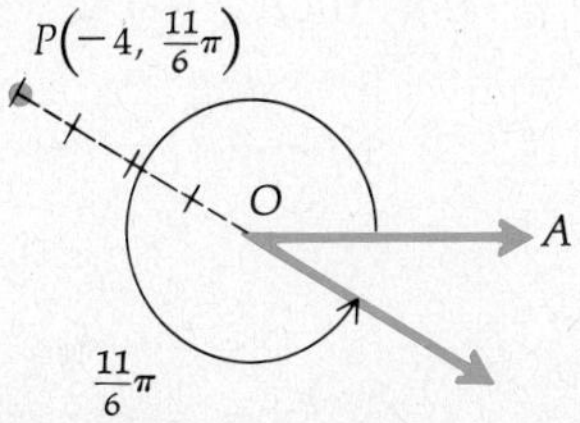

FIGURE 12.1.6

Until now we have located a point in a plane by its rectangular cartesian coordinates. There are other coordinate systems that can be used. Probably the next in importance to the cartesian coordinate system is the **polar coordinate system.** In the cartesian coordinate system the coordinates are numbers called the abscissa and the ordinate, and these numbers are directed distances from two fixed lines. In the polar coordinate system the coordinates consist of a directed distance and the measure of an angle relative to a fixed point and a fixed ray (or half line).

The fixed point is called the **pole** (or origin), and it is designated by the letter O. The fixed ray is called the **polar axis** (or polar line), which we label OA. The ray OA is usually drawn horizontally and to the right, and it extends indefinitely (see Fig. 12.1.1).

Let P be any point in the plane distinct from O. Let θ be the radian measure of the directed angle AOP, positive when measured counterclockwise and negative when measured clockwise, having as its initial side the ray OA and as its terminal side the ray OP. Then if r is the undirected distance from O to P (i.e., $r = |\overline{OP}|$), one set of polar coordinates of P is given by r and θ, and we write these coordinates as (r, θ).

• ILLUSTRATION 1 The point $P(4, \frac{5}{6}\pi)$ is determined by first drawing the angle with radian measure $\frac{5}{6}\pi$, having its vertex at the pole and its initial side along the polar axis. Then the point on the terminal side that is four units from the pole is the point P (see Fig. 12.1.2). Another set of polar coordinates for this point is $(4, -\frac{7}{6}\pi)$; see Fig. 12.1.3. Furthermore, the polar coordinates $(4, \frac{17}{6}\pi)$ also yield the same point, as shown in Fig. 12.1.4. •

Actually the coordinates $(4, \frac{5}{6}\pi + 2n\pi)$, where n is any integer, give the same point as $(4, \frac{5}{6}\pi)$. So a given point has an unlimited number of sets of polar coordinates. This is unlike the rectangular cartesian coordinate system because there is a one-to-one correspondence between the rectangular cartesian coordinates and the position of points in the plane, whereas there is no such one-to-one correspondence between the polar coordinates and the position of points in the plane. A further example is obtained by considering sets of polar coordinates for the pole. If $r = 0$ and θ is any real number, we have the pole, which is designated by $(0, \theta)$.

We consider polar coordinates for which r is negative. In this case, instead of the point being on the terminal side of the angle, it is on the extension of the terminal side, which is the ray from the pole extending in the direction opposite to the terminal side. So if P is on the extension of the terminal side of the angle of radian measure θ, a set of polar coordinates of P is (r, θ), where $r = -|\overline{OP}|$.

• ILLUSTRATION 2 The point $(-4, -\frac{1}{6}\pi)$ shown in Fig. 12.1.5 is the same point as $(4, \frac{5}{6}\pi)$, $(4, -\frac{7}{6}\pi)$, and $(4, \frac{17}{6}\pi)$ in Illustration 1. Still another set of polar coordinates for this point is $(-4, \frac{11}{6}\pi)$; see Fig. 12.1.6. •

The angle is usually measured in radians; thus a set of polar coordinates of a point is an ordered pair of real numbers. For each ordered pair of real numbers there is a unique point having this set of polar coordinates. However, we have seen that a particular point can be given by an unlimited number of ordered pairs of real numbers. If the point P is not the pole, and r and θ are restricted so that $r > 0$ and $0 \leq \theta < 2\pi$, then there is a unique set of polar coordinates for P.

Example 1 (a) Plot the point having polar coordinates $(3, -\frac{2}{3}\pi)$. Find another set of polar coordinates of this point for which (b) $r < 0$ and $0 < \theta < 2\pi$; (c) $r > 0$ and $0 < \theta < 2\pi$; (d) $r < 0$ and $-2\pi < \theta < 0$.

Solution

(a) The point is plotted by drawing the angle of radian measure $-\frac{2}{3}\pi$ in a clockwise direction from the polar axis. Because $r > 0$, P is on the terminal side of the angle, three units from the pole; see Fig. 12.1.7(a).

The answers to (b), (c), and (d) are, respectively, $(-3, \frac{1}{3}\pi)$, $(3, \frac{4}{3}\pi)$, and $(-3, -\frac{5}{3}\pi)$. They are illustrated in Fig. 12.1.7(b), (c), and (d).

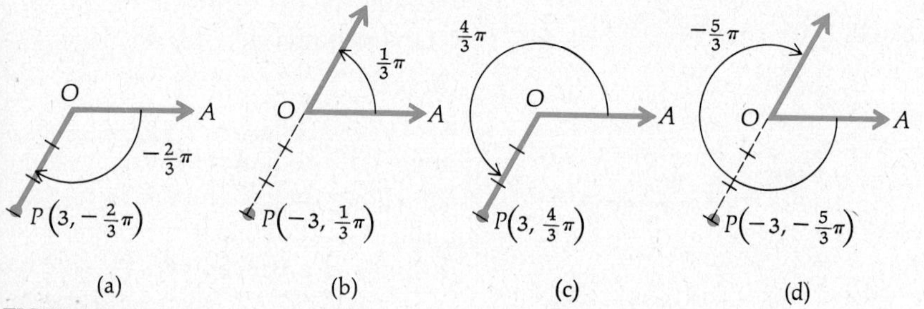

FIGURE 12.1.7

Often we wish to refer to both the rectangular cartesian coordinates and the polar coordinates of a point. To do this we take the origin of the first system and the pole of the second system coincident, the polar axis as the positive side of the x axis, and the ray for which $\theta = \frac{1}{2}\pi$ as the positive side of the y axis.

Suppose that P is a point whose representation in the rectangular cartesian coordinate system is (x, y) and (r, θ) is a polar coordinate representation of P. We distinguish two cases: $r > 0$ and $r < 0$. In the first case, if $r > 0$, then the point P is on the terminal side of the angle of θ radians, and $r = |\overline{OP}|$. Such a case is shown in Fig. 12.1.8. Then $\cos\theta = x/|\overline{OP}| = x/r$ and $\sin\theta = y/|\overline{OP}| = y/r$; so

$$x = r\cos\theta \quad \text{and} \quad y = r\sin\theta \tag{1}$$

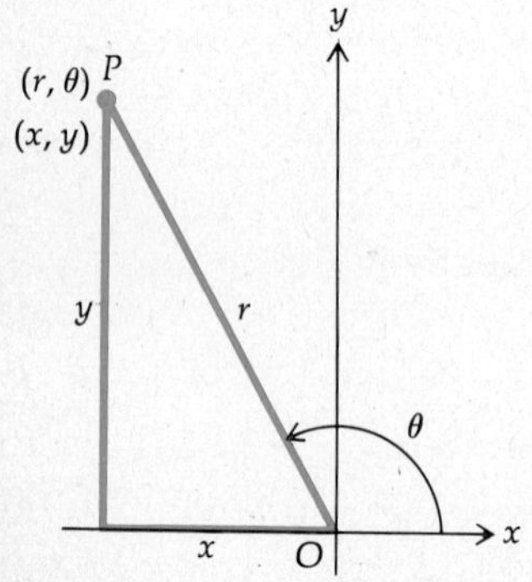

FIGURE 12.1.8

In the second case, if $r < 0$, then the point P is on the extension of the terminal side and $r = -|\overline{OP}|$ (see Fig. 12.1.9). Then if Q is the point $(-x, -y)$,

$$\cos\theta = \frac{-x}{|\overline{OQ}|} = \frac{-x}{|\overline{OP}|} = \frac{-x}{-r} = \frac{x}{r}$$

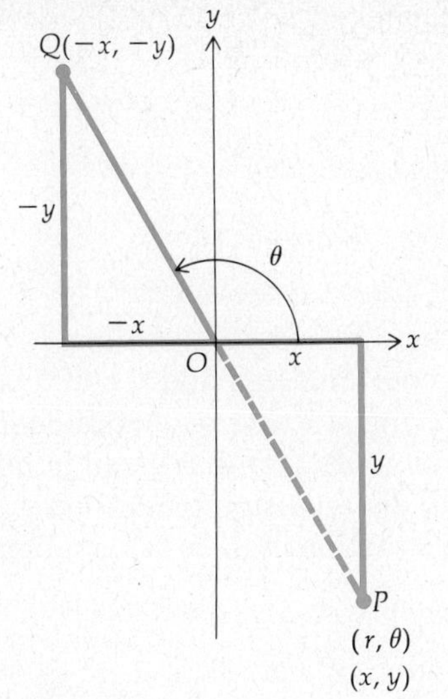

FIGURE 12.1.9

So

$$x = r\cos\theta \tag{2}$$

Also,

$$\sin\theta = \frac{-y}{|\overline{OQ}|} = \frac{-y}{|\overline{OP}|} = \frac{-y}{-r} = \frac{y}{r}$$

Hence

$$y = r\sin\theta \tag{3}$$

Equations (2) and (3) are the same as the equations in (1); thus the equations (1) hold in all cases.

From Eqs. (1) we can obtain the rectangular cartesian coordinates of a point when its polar coordinates are known. Also, from the equations we can obtain a polar equation of a curve if a rectangular cartesian equation is known.

To obtain equations that give a set of polar coordinates of a point when its rectangular cartesian coordinates are known, we square on both sides of each equation in (1) and obtain

$$x^2 = r^2\cos^2\theta \quad \text{and} \quad y^2 = r^2\sin^2\theta$$

Equating the sum of the left members of the above to the sum of the right members we have

$$x^2 + y^2 = r^2\cos^2\theta + r^2\sin^2\theta$$

$$x^2 + y^2 = r^2(\sin^2\theta + \cos^2\theta)$$

$$x^2 + y^2 = r^2$$

and so

$$r = \pm\sqrt{x^2 + y^2} \tag{4}$$

From the equations in (1) and dividing, we have

$$\frac{r\sin\theta}{r\cos\theta} = \frac{y}{x}$$

or, equivalently,

$$\tan\theta = \frac{y}{x} \tag{5}$$

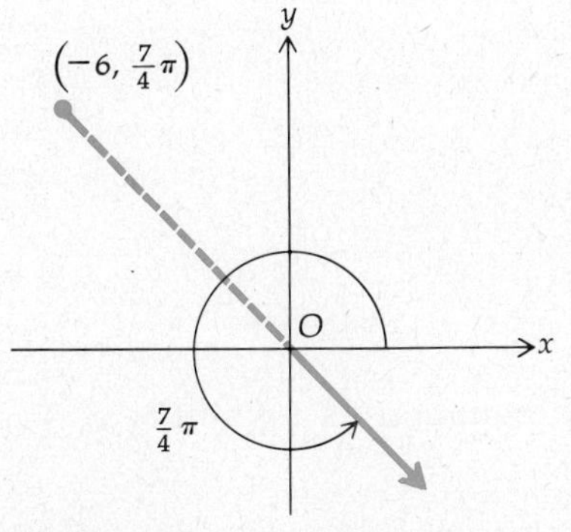

FIGURE 12.1.10

● ILLUSTRATION 3 The point whose polar coordinates are $(-6, \frac{7}{4}\pi)$ is plotted in Fig. 12.1.10. We find its rectangular cartesian coordinates. From (1),

$$x = r\cos\theta \qquad\qquad y = r\sin\theta$$
$$= -6\cos\tfrac{7}{4}\pi \qquad\qquad = -6\sin\tfrac{7}{4}\pi$$
$$= -6\cdot\frac{\sqrt{2}}{2} \qquad\qquad = -6\left(-\frac{\sqrt{2}}{2}\right)$$
$$= -3\sqrt{2} \qquad\qquad = 3\sqrt{2}$$

So the point is $(-3\sqrt{2}, 3\sqrt{2})$.

The graph of an equation in polar coordinates r and θ consists of all those points and only those points P having at least one pair of coordinates that satisfy the equation. If an equation of a graph is given in polar coordinates, it is called a *polar equation* to distinguish it from a *cartesian equation,* which is the term used when an equation is given in rectangular cartesian coordinates.

Example 2 Given that a polar equation of a graph is $r^2 = 4\sin 2\theta$, find a cartesian equation.

Solution Because $r^2 = x^2 + y^2$ and $\sin 2\theta = 2\sin\theta\cos\theta = 2(y/r)(x/r)$, we have, upon substituting in the given polar equation,

$$x^2 + y^2 = 4(2)\frac{y}{r}\cdot\frac{x}{r}$$

$$x^2 + y^2 = \frac{8xy}{r^2}$$

$$x^2 + y^2 = \frac{8xy}{x^2 + y^2}$$

$$(x^2 + y^2)^2 = 8xy$$

Example 3 Find (r, θ) if $r > 0$ and $0 \le \theta < 2\pi$ for the point whose rectangular cartesian coordinate representation is $(-\sqrt{3}, -1)$.

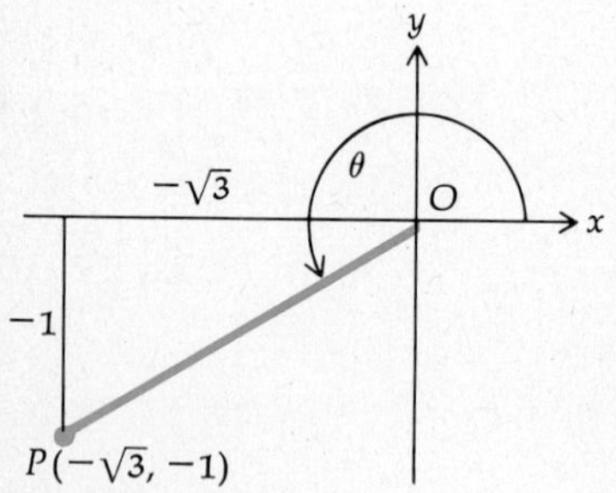

FIGURE 12.1.11

Solution The point $(-\sqrt{3}, -1)$ is plotted in Fig. 12.1.11. From (4), because $r > 0$,

$$r = \sqrt{1 + 3} = 2$$

From (5), $\tan\theta = -1/(-\sqrt{3})$, and since $\pi < \theta < \frac{3}{2}\pi$,

$$\theta = \tfrac{7}{6}\pi$$

So the point is $(2, \frac{7}{6}\pi)$.

Example 4 Given that a cartesian equation of a graph is $x^2 + y^2 - 4x = 0$, find a polar equation.

Solution Substituting $x = r\cos\theta$ and $y = r\sin\theta$ in

$$x^2 + y^2 - 4x = 0$$

we have

$$r^2 \cos^2 \theta + r^2 \sin^2 \theta - 4r \cos \theta = 0$$
$$r^2 - 4r \cos \theta = 0$$
$$r(r - 4 \cos \theta) = 0$$

Therefore

$$r = 0 \quad \text{or} \quad r - 4 \cos \theta = 0$$

The graph of $r = 0$ is the pole. However, the pole is a point on the graph of $r - 4 \cos \theta = 0$ because $r = 0$ when $\theta = \frac{1}{2}\pi$. Therefore a polar equation of the graph is

$$r = 4 \cos \theta$$

The graph of $x^2 + y^2 - 4x = 0$ is a circle. The equation may be written in the form

$$(x - 2)^2 + y^2 = 4$$

which is an equation of the circle with center at (2, 0) and radius 2.

EXERCISES 12.1

In Exercises 1 through 6, plot the point having the given set of polar coordinates; then find another set of polar coordinates for the same point for which (a) $r < 0$ and $0 \le \theta < 2\pi$; (b) $r > 0$ and $-2\pi < \theta \le 0$; (c) $r < 0$ and $-2\pi < \theta \le 0$.

1. $(4, \frac{1}{4}\pi)$ **2.** $(3, \frac{5}{6}\pi)$ **3.** $(2, \frac{1}{2}\pi)$ **4.** $(3, \frac{3}{2}\pi)$ **5.** $(\sqrt{2}, \frac{7}{4}\pi)$ **6.** $(2, \frac{4}{3}\pi)$

7. Plot the point having the polar coordinates $(2, -\frac{1}{4}\pi)$. Find another set of polar coordinates for this point for which (a) $r < 0$ and $0 \le \theta < 2\pi$; (b) $r < 0$ and $-2\pi < \theta \le 0$; (c) $r > 0$ and $2\pi \le \theta < 4\pi$.

8. Plot the point having the polar coordinates $(-3, -\frac{2}{3}\pi)$. Find another set of polar coordinates for this point for which (a) $r > 0$ and $0 \le \theta < 2\pi$; (b) $r > 0$ and $-2\pi < \theta \le 0$; (c) $r < 0$ and $2\pi \le \theta < 4\pi$.

In Exercises 9 through 16, plot the point having the given set of polar coordinates; then give two other sets of polar coordinates of the same point, one with the same value of r and one with an r having opposite sign.

9. $(3, -\frac{2}{3}\pi)$ **10.** $(\sqrt{2}, -\frac{1}{4}\pi)$ **11.** $(-4, \frac{5}{6}\pi)$ **12.** $(-2, \frac{4}{3}\pi)$

13. $(-2, -\frac{5}{4}\pi)$ **14.** $(-3, -\pi)$ **15.** $(2, 6)$ **16.** $(5, \frac{1}{6}\pi)$

In Exercises 17 and 18, find the rectangular cartesian coordinates of the points whose polar coordinates are given.

17. (a) $(3, \pi)$; (b) $(\sqrt{2}, -\frac{3}{4}\pi)$; (c) $(-4, \frac{2}{3}\pi)$; (d) $(-1, -\frac{7}{6}\pi)$ **18.** (a) $(-2, -\frac{1}{2}\pi)$; (b) $(-1, \frac{1}{4}\pi)$; (c) $(2, -\frac{7}{6}\pi)$; (d) $(2, \frac{7}{4}\pi)$

In Exercises 19 and 20, find a set of polar coordinates of the points whose rectangular cartesian coordinates are given. Take $r > 0$ and $0 \le \theta < 2\pi$.

19. (a) $(1, -1)$; (b) $(-\sqrt{3}, 1)$; (c) $(2, 2)$; (d) $(-5, 0)$ **20.** (a) $(3, -3)$; (b) $(-1, \sqrt{3})$; (c) $(0, -2)$; (d) $(-2, -2\sqrt{3})$

In Exercises 21 through 30, find a polar equation of the graph having the given cartesian equation.

21. $x^2 + y^2 = a^2$ **22.** $x + y = 1$ **23.** $y^2 = 4(x + 1)$ **24.** $x^3 = 4y^2$ **25.** $x^2 = 6y - y^2$

26. $x^2 - y^2 = 16$ **27.** $(x^2 + y^2)^2 = 4(x^2 - y^2)$ **28.** $2xy = a^2$ **29.** $x^3 + y^3 - 3axy = 0$ **30.** $y = \dfrac{2x}{x^2 + 1}$

In Exercises 31 through 40, find a cartesian equation of the graph having the given polar equation.

31. $r^2 = 2 \sin 2\theta$ **32.** $r^2 \cos 2\theta = 10$ **33.** $r^2 = \cos \theta$ **34.** $r^2 = 4 \cos 2\theta$ **35.** $r^2 = \theta$

36. $r = 2 \sin 3\theta$ **37.** $r \cos \theta = -1$ **38.** $r^6 = r^2 \cos^2 \theta$ **39.** $r = \dfrac{6}{2 - 3 \sin \theta}$ **40.** $r = \dfrac{4}{3 - 2 \cos \theta}$

12.2 GRAPHS OF EQUATIONS IN POLAR COORDINATES

Let $r = f(\theta)$ be a polar equation of a curve. We first derive a formula for finding the slope of a tangent line to a polar curve at a point (r, θ) on the curve. Consider a rectangular cartesian coordinate system and a polar coordinate system in the same plane and having the positive side of the x axis coincident with the polar axis. In Sec. 12.1 we saw that the two sets of coordinates are related by the equations

$$x = r \cos \theta \quad \text{and} \quad y = r \sin \theta$$

x and y can be considered as functions of θ because $r = f(\theta)$. If we differentiate with respect to θ on both sides of these equations we get, by applying the chain rule,

$$\frac{dx}{d\theta} = \cos \theta \frac{dr}{d\theta} - r \sin \theta \tag{1}$$

and

$$\frac{dy}{d\theta} = \sin \theta \frac{dr}{d\theta} + r \cos \theta \tag{2}$$

If α is the radian measure of the inclination of the tangent line, then

$$\tan \alpha = \frac{dy}{dx}$$

and if $dx/d\theta \neq 0$,

$$\frac{dy}{dx} = \frac{\dfrac{dy}{d\theta}}{\dfrac{dx}{d\theta}} \tag{3}$$

Substituting from (1) and (2) into (3) we obtain

$$\frac{dy}{dx} = \frac{\sin \theta \dfrac{dr}{d\theta} + r \cos \theta}{\cos \theta \dfrac{dr}{d\theta} - r \sin \theta} \tag{4}$$

If $\cos \theta \neq 0$, we divide the numerator and the denominator of the right side of (4) by $\cos\theta$ and replace dy/dx by $\tan\alpha$, thereby giving

$$\tan\alpha = \frac{\tan\theta \dfrac{dr}{d\theta} + r}{\dfrac{dr}{d\theta} - r\tan\theta} \tag{5}$$

Example 1 Find the slope of the tangent line to the curve whose equation is $r = 1 - \cos\theta$ at the point $(1 - \frac{1}{2}\sqrt{2}, \frac{1}{4}\pi)$.

Solution Since $r = 1 - \cos\theta$, $dr/d\theta = \sin\theta$. So at the point $(1 - \frac{1}{2}\sqrt{2}, \frac{1}{4}\pi)$, $dr/d\theta = \frac{1}{2}\sqrt{2}$ and $\tan\theta = 1$; so from (5),

$$\tan\alpha = \frac{(1)\frac{1}{2}\sqrt{2} + (1 - \frac{1}{2}\sqrt{2})}{\frac{1}{2}\sqrt{2} - (1 - \frac{1}{2}\sqrt{2})(1)} = \frac{1}{\sqrt{2} - 1} = \sqrt{2} + 1$$

If we are to draw a sketch of the graph of a polar equation, it will be helpful to consider properties of symmetry of the graph. In Sec 1.2 (Definition 1.2.4) we learned that two points P and Q are symmetric with respect to a line if and only if the line is the perpendicular bisector of the line segment PQ, and that two points P and Q are symmetric with respect to a third point if and only if the third point is the midpoint of the line segment PQ. Therefore the points $(2, \frac{1}{3}\pi)$ and $(2, \frac{2}{3}\pi)$ are symmetric with respect to the $\frac{1}{2}\pi$ axis and the points $(2, \frac{1}{3}\pi)$ and $(2, -\frac{2}{3}\pi)$ are symmetric with respect to the pole. We also learned (Definition 1.2.5) that the graph of an equation is symmetric with respect to a line l if and only if for every point P on the graph there is a point Q, also on the graph, such that P and Q are symmetric with respect to l. Similarly, the graph of an equation is symmetric with respect to a point R if and only if for every point P on the graph there is a point S, also on the graph, such that P and S are symmetric with respect to R. We have three theorems giving tests for symmetry of graphs of polar equations.

12.2.1 Theorem If for an equation in polar coordinates an equivalent equation is obtained when (r, θ) is replaced by either $(r, -\theta + 2n\pi)$ or $(-r, \pi - \theta + 2n\pi)$, where n is any integer, the graph of the equation is symmetric with respect to the polar axis.

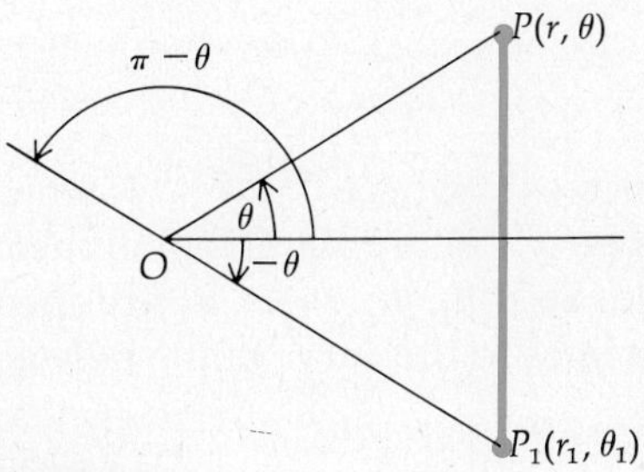

FIGURE 12.2.1

Proof If the point $P(r, \theta)$ is a point on the graph of an equation, then the graph is symmetric with respect to the polar axis if there is a point $P_1(r_1, \theta_1)$ on the graph such that the polar axis is the perpendicular bisector of the line segment P_1P (see Fig. 12.2.1). So if $r_1 = r$, then θ_1 must equal $-\theta + 2n\pi$, where n is an integer. And if $r_1 = -r$, then θ_1 must be $\pi - \theta + 2n\pi$. ■

12.2.2 Theorem If for an equation in polar coordinates an equivalent equation is obtained when (r, θ) is replaced by either $(r, \pi - \theta + 2n\pi)$ or $(-r, -\theta + 2n\pi)$, where

n is any integer, the graph of the equation is symmetric with respect to the $\frac{1}{2}\pi$ axis.

12.2.3 Theorem If for an equation in polar coordinates an equivalent equation is obtained when (r, θ) is replaced by either $(-r, \theta + 2n\pi)$ or $(r, \pi + \theta + 2n\pi)$, where n is any integer, the graph of the equation is symmetric with respect to the pole.

The proofs of Theorems 12.2.2 and 12.2.3 are similar to the proof of Theorem 12.2.1 and are left as exercises (see Exercises 45 and 46).

● ILLUSTRATION 1 We test for symmetry the graph of the equation $r = 4 \cos 2\theta$ with respect to the polar axis, the $\frac{1}{2}\pi$ axis, and the pole.

Using Theorem 12.2.1 to test for symmetry with respect to the polar axis, we replace (r, θ) by $(r, -\theta)$ and obtain $r = 4 \cos(-2\theta)$, which is equivalent to $r = 4 \cos 2\theta$. So the graph is symmetric with respect to the polar axis.

Using Theorem 12.2.2 to test for symmetry with respect to the $\frac{1}{2}\pi$ axis, we replace (r, θ) by $(r, \pi - \theta)$ and get $r = 4 \cos(2(\pi - \theta))$ or, equivalently, $r = 4 \cos(2\pi - 2\theta)$, which is equivalent to the equation $r = 4 \cos 2\theta$. Therefore the graph is symmetric with respect to the $\frac{1}{2}\pi$ axis.

To test for symmetry with respect to the pole, we replace (r, θ) by $(-r, \theta)$ and obtain the equation $-r = 4 \cos 2\theta$, which is not equivalent to the given equation. But we must also determine if the other set of coordinates works. We replace (r, θ) by $(r, \pi + \theta)$ and obtain $r = 4 \cos 2(\pi + \theta)$ or, equivalently, $r = 4 \cos(2\pi + 2\theta)$, which is equivalent to the equation $r = 4 \cos 2\theta$. Therefore the graph is symmetric with respect to the pole. ●

When drawing a sketch of a graph, it is desirable to determine if the pole is on the graph. This is done by substituting 0 for r and solving for θ. Also, it is advantageous to plot the points for which r has a relative maximum or relative minimum value. As a further aid in plotting, if a curve contains the pole, it is sometimes helpful to find equations of the tangent lines to the graph at the pole.

To find the slope of the tangent line at the pole we use (5) with $r = 0$. If $dr/d\theta \neq 0$,

$$\tan \alpha = \tan \theta \tag{6}$$

From (6) we conclude that the values of θ, where $0 \leq \theta < \pi$, which satisfy a polar equation of the curve when $r = 0$ are the inclinations of the tangent lines to the curve at the pole. So if $\theta_1, \theta_2, \ldots, \theta_k$ are these values of θ, then equations of the tangent lines to the curve at the pole are

$$\theta = \theta_1, \theta = \theta_2, \ldots, \theta = \theta_k$$

Example 2 Draw a sketch of the graph of $r = 1 - 2 \cos \theta$.

Solution Replacing (r, θ) by $(r, -\theta)$ we obtain an equivalent equation. Thus the graph is symmetric with respect to the polar axis.

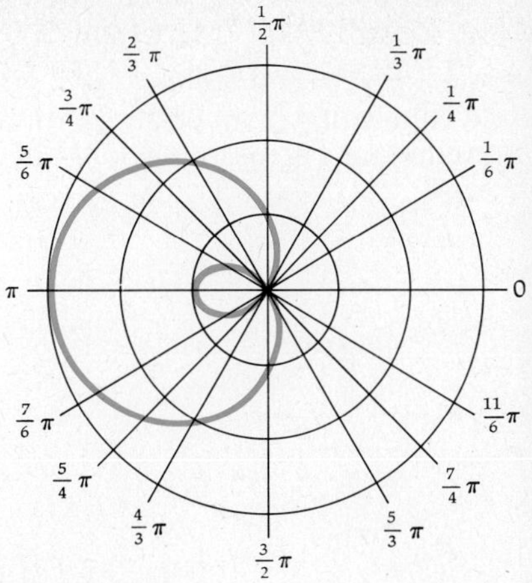

FIGURE 12.2.2

Table 12.2.1 gives the coordinates of some points on the graph. From these points we draw half of the graph; the remainder is drawn from its symmetry with respect to the polar axis.

TABLE 12.2.1

θ	0	$\frac{1}{6}\pi$	$\frac{1}{3}\pi$	$\frac{1}{2}\pi$	$\frac{2}{3}\pi$	$\frac{5}{6}\pi$	π
r	-1	$1-\sqrt{3}$	0	1	2	$1+\sqrt{3}$	3

If $r = 0$, we obtain $\cos\theta = \frac{1}{2}$, and if $0 \le \theta < \pi$, then $\theta = \frac{1}{3}\pi$. Therefore the point $(0, \frac{1}{3}\pi)$ is on the graph, and an equation of the tangent line there is $\theta = \frac{1}{3}\pi$. A sketch of the graph is shown in Fig. 12.2.2. It is called a *limaçon.*

The graph of an equation of the form

$$r = a \pm b\cos\theta$$

or

$$r = a \pm b\sin\theta$$

is a **limaçon.** If $b > a$, as is the case in Example 2, the limaçon has a loop. If $a = b$, the limaçon is a **cardioid,** which is heart-shaped. If $a > b$, the limaçon has a shape similar to the one in Example 3.

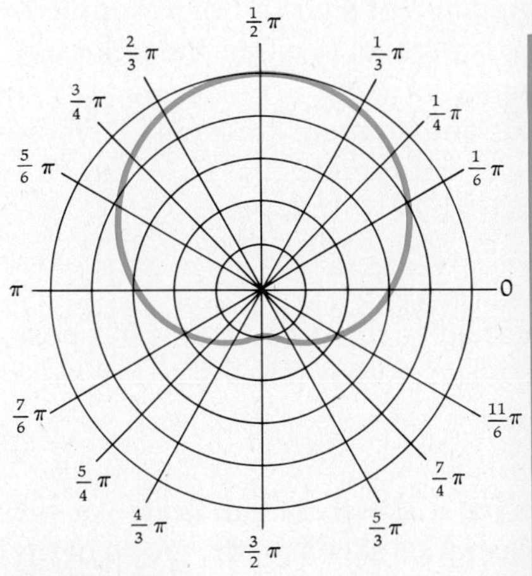

FIGURE 12.2.3

Example 3 Draw a sketch of the graph of $r = 3 + 2\sin\theta$.

Solution The graph is symmetric with respect to the $\frac{1}{2}\pi$ axis because if (r, θ) is replaced by $(r, \pi - \theta)$, an equivalent equation is obtained.

Table 12.2.2 gives the coordinates of some of the points on the graph. A sketch of the graph is shown in Fig. 12.2.3. It is drawn by plotting the points whose coordinates are given in Table 12.2.2 and using the symmetry property.

TABLE 12.2.2

θ	0	$\frac{1}{6}\pi$	$\frac{1}{3}\pi$	$\frac{1}{2}\pi$	π	$\frac{7}{6}\pi$	$\frac{4}{3}\pi$	$\frac{3}{2}\pi$
r	3	4	$3+\sqrt{3}$	5	3	2	$3-\sqrt{3}$	1

The graph of an equation of the form

$$r = a\cos n\theta \quad \text{or} \quad r = a\sin n\theta$$

is a **rose,** having n leaves if n is odd and $2n$ leaves if n is even.

Example 4 Draw a sketch of the four-leafed rose $r = 4\cos 2\theta$.

Solution In Illustration 1 we proved that the graph is symmetric with respect to the polar axis, the $\frac{1}{2}\pi$ axis, and the pole. Substituting 0 for r in the given equation we get

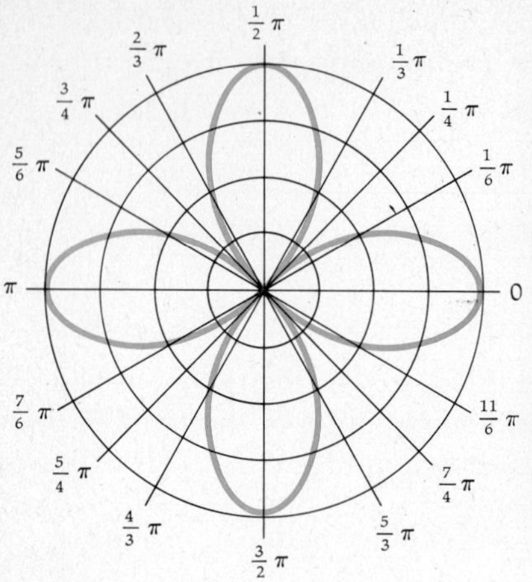

FIGURE 12.2.4

$$\cos 2\theta = 0$$

from which we obtain, for $0 \leq \theta < 2\pi$, $\theta = \frac{1}{4}\pi, \frac{3}{4}\pi, \frac{7}{4}\pi$, and $\frac{11}{4}\pi$.

Table 12.2.3 gives values of r for some values of θ from 0 to $\frac{1}{2}\pi$. From these values and the symmetry properties we draw a sketch of the graph, as shown in Fig. 12.2.4.

TABLE 12.2.3

θ	0	$\frac{1}{12}\pi$	$\frac{1}{6}\pi$	$\frac{1}{4}\pi$	$\frac{1}{3}\pi$	$\frac{5}{12}\pi$	$\frac{1}{2}\pi$
r	4	$2\sqrt{3}$	2	0	-2	$-2\sqrt{3}$	-4

The graph of the equation

$$\theta = C$$

where C is any constant, is a straight line through the pole and makes an angle of C radians with the polar axis. The same line is given by the equation

$$\theta = C \pm n\pi$$

where n is any integer.

In general, the polar form of an equation of a line is not as simple as the cartesian form. However, if the line is parallel to either the polar axis or the $\frac{1}{2}\pi$ axis, the equation is fairly simple.

If a line is parallel to the polar axis and contains the point B whose cartesian coordinates are $(0, b)$ and polar coordinates are $(b, \frac{1}{2}\pi)$, then a cartesian equation is $y = b$. If we replace y by $r \sin \theta$, we have

$$r \sin \theta = b$$

which is a polar equation of any line parallel to the polar axis. If b is positive, the line is above the polar axis. If b is negative, it is below the polar axis.

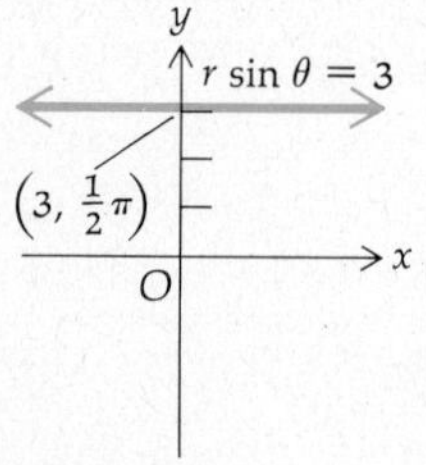

FIGURE 12.2.5

● ILLUSTRATION 2 In Fig. 12.2.5 we have a sketch of the graph of the equation $r \sin \theta = 3$, and in Fig. 12.2.6 we have a sketch of the graph of the equation $r \sin \theta = -3$. ●

If a line is parallel to the $\frac{1}{2}\pi$ axis or, equivalently, perpendicular to the polar axis, and goes through the point A whose cartesian coordinates are $(a, 0)$ and polar coordinates are $(a, 0)$, a cartesian equation is $x = a$. Replacing x by $r \cos \theta$ we obtain

$$r \cos \theta = a$$

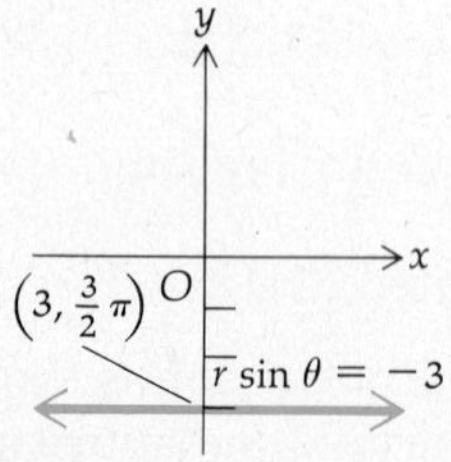

FIGURE 12.2.6

which is an equation of any line perpendicular to the polar axis. If a is positive, the line is to the right of the $\frac{1}{2}\pi$ axis. If a is negative, the line is to the left of the $\frac{1}{2}\pi$ axis.

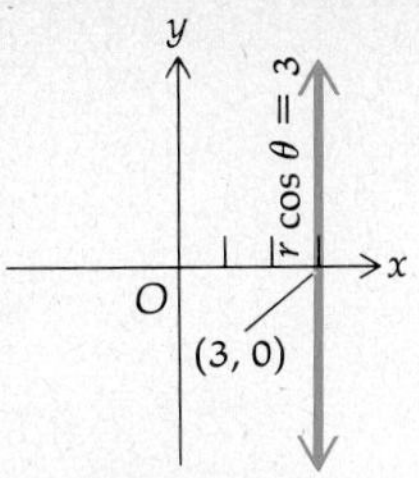

FIGURE 12.2.7

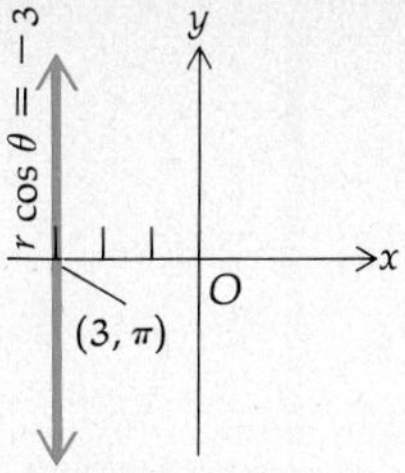

FIGURE 12.2.8

• ILLUSTRATION 3 Figure 12.2.7 shows a sketch of the graph of the equation $r \cos \theta = 3$, and Fig. 12.2.8 shows a sketch of the graph of the equation $r \cos \theta = -3$. •

The graph of the equation

$$r = C$$

where C is any constant, is a circle whose center is at the pole and radius is $|C|$. The same circle is given by the equation

$$r = -C$$

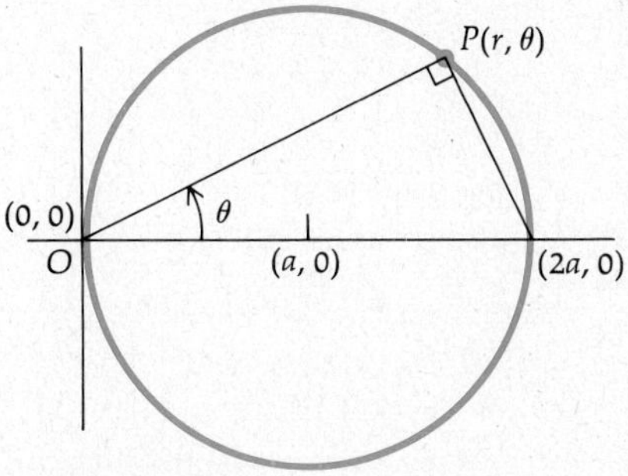

FIGURE 12.2.9

As was the case with the straight line, the general polar equation of a circle is not as simple as the cartesian form. However, there are special cases of an equation of a circle that are worth considering in polar form.

If a circle has its center on the polar axis at the point having polar coordinates $(a, 0)$ and is tangent to the $\frac{1}{2}\pi$ axis, then its radius is $|a|$. The points of intersection of such a circle with the polar axis are the pole $(0, 0)$ and the point $(2a, 0)$ (see Fig. 12.2.9). Then if $P(r, \theta)$ is any point on the circle, the angle at P inscribed in the circle is a right angle. Therefore $\cos \theta = r/2a$ or, equivalently,

$$r = 2a \cos \theta$$

which is a polar equation of the circle having its center on the polar axis or its extension, and tangent to the $\frac{1}{2}\pi$ axis. If a is a positive number, the circle is to the right of the pole. If a is a negative number, the circle is to the left of the pole.

In a similar manner it may be shown that

$$r = 2b \sin \theta$$

is a polar equation of the circle having its center on the $\frac{1}{2}\pi$ axis or its extension, and tangent to the polar axis. If b is positive, the circle is above the pole. If b is negative, the circle is below the pole.

Example 5 Draw a sketch of the graph of $r = \theta, \theta \geq 0$.

Solution When $\theta = n\pi$, where n is any integer, the graph intersects the polar axis or its extension, and when $\theta = \frac{1}{2}n\pi$, where n is any odd integer, the graph intersects the $\frac{1}{2}\pi$ axis or its extension. When $r = 0$,

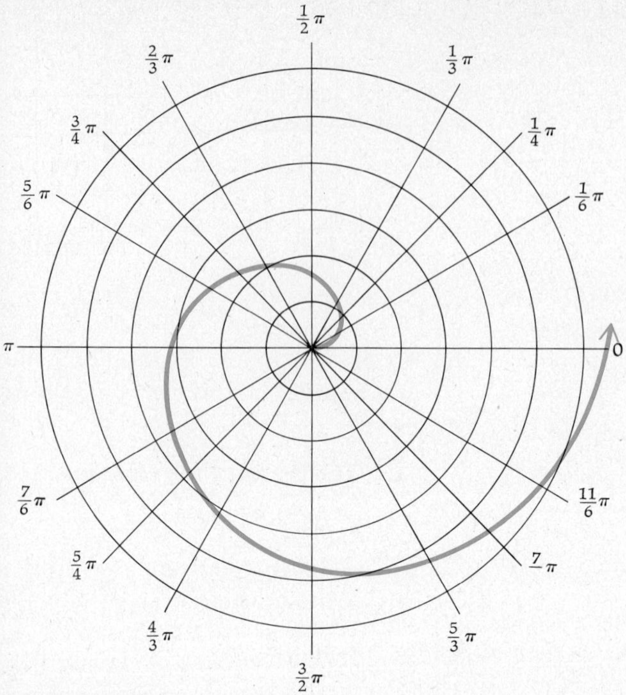

FIGURE 12.2.10

$\theta = 0$; thus the tangent line to the curve at the pole is the polar axis. A sketch of the graph is shown in Fig. 12.2.10. The curve is called a *spiral of Archimedes.*

EXERCISES 12.2

In Exercises 1 through 40, draw a sketch of the graph of the given equation.

1. $r \cos \theta = 4$

2. $r \sin \theta = 2$

3. $r = 4 \cos \theta$

4. $r = 2 \sin \theta$

5. $\theta = 5$

6. $\theta = -4$

7. $r = 5$

8. $r = -4$

9. $r \sin \theta = -4$

10. $r \cos \theta = -5$

11. $r = -4 \sin \theta$

12. $r = -5 \cos \theta$

13. $r = e^{\theta}$ (logarithmic spiral)

14. $r = e^{\theta/3}$ (logarithmic spiral)

15. $r = 1/\theta$ (reciprocal spiral)

16. $r = 2\theta$ (spiral of Archimedes)

17. $r = 2 \sin 3\theta$ (three-leafed rose)

18. $r = 3 \cos 2\theta$ (four-leafed rose)

19. $r = 2 \cos 4\theta$ (eight-leafed rose)

20. $r = 4 \sin 5\theta$ (five-leafed rose)

21. $r = 4 \sin 2\theta$ (four-leafed rose)

22. $r = 3 \cos 3\theta$ (three-leafed rose)

23. $r = 2 + 2 \sin \theta$ (cardioid)

24. $r = 3 + 3 \cos \theta$ (cardioid)

25. $r = 4 - 4\cos\theta$ (cardioid)

26. $r = 3 - 3\sin\theta$ (cardioid)

27. $r = 4 - 3\cos\theta$ (limaçon)

28. $r = 3 - 4\cos\theta$ (limaçon)

29. $r = 4 + 2\sin\theta$ (limaçon)

30. $r = 2 - 3\sin\theta$ (limaçon)

31. $r^2 = 9\sin 2\theta$ (lemniscate)

32. $r^2 = -4\sin 2\theta$ (lemniscate)

33. $r^2 = -25\cos 2\theta$ (lemniscate)

34. $r^2 = 16\cos 2\theta$ (lemniscate)

35. $r = 2\sin\theta\tan\theta$ (cissoid)

36. $(r - 2)^2 = 8\theta$ (parabolic spiral)

37. $r = 2\sec\theta - 1$ (conchoid of Nicomedes)

38. $r = 2\csc\theta + 3$ (conchoid of Nicomedes)

39. $r = |\sin 2\theta|$

40. $r = 2|\cos\theta|$

In Exercises 41 through 44, find a polar equation of each of the tangent lines to the given curve at the pole.

41. $r = 4\cos\theta + 2$ **42.** $r = 2\sin 3\theta$ **43.** $r^2 = 4\cos 2\theta$ **44.** $r^2 = 9\sin 2\theta$

45. Prove Theorem 12.2.2. **46.** Prove Theorem 12.2.3.

47. Find the slope of the tangent line to the curve $r = 4$ at $(4, \frac{1}{4}\pi)$.

48. Find a polar equation of the tangent line to the curve $r = -6\sin\theta$ at the point $(6, \frac{3}{2}\pi)$.

12.3 INTERSECTION OF GRAPHS IN POLAR COORDINATES

To find the points of intersection of two curves whose equations are in cartesian coordinates, we solve the two equations simultaneously. The common solutions give all the points of intersection. However, because a point has an unlimited number of sets of polar coordinates, it is possible to have as the intersection of two curves a point for which no single pair of polar coordinates satisfies both equations. This is illustrated in the following example.

Example 1 Draw sketches of the graphs of $r = 2\sin 2\theta$ and $r = 1$, and find the points of intersection.

Solution The graph of $r = 2\sin 2\theta$ is a four-leafed rose. The graph of $r = 1$ is the circle with its center at the pole and radius 1. Sketches of the graphs are shown in Fig. 12.3.1. Solving the two equations simultaneously we have $1 = 2\sin 2\theta$ or $\sin 2\theta = \frac{1}{2}$. Therefore we get

$$2\theta = \tfrac{1}{6}\pi, \tfrac{5}{6}\pi, \tfrac{13}{6}\pi, \text{ and } \tfrac{17}{6}\pi$$

$$\theta = \tfrac{1}{12}\pi, \tfrac{5}{12}\pi, \tfrac{13}{12}\pi, \text{ and } \tfrac{17}{12}\pi$$

Hence we obtain the points of intersection $(1, \frac{1}{12}\pi)$, $(1, \frac{5}{12}\pi)$, $(1, \frac{13}{12}\pi)$, and $(1, \frac{17}{12}\pi)$. We notice in Fig. 12.3.1 that eight points of intersection are shown. The other four points are obtained if we take another form of the equation of the circle $r = 1$; that is, consider the equation $r = -1$, which is the same circle. Solving this equation simultaneously with the equation of the four-leafed rose we have $\sin 2\theta = -\frac{1}{2}$. Then we get

$$2\theta = \tfrac{7}{6}\pi, \tfrac{11}{6}\pi, \tfrac{19}{6}\pi, \text{ and } \tfrac{23}{6}\pi$$

$$\theta = \tfrac{7}{12}\pi, \tfrac{11}{12}\pi, \tfrac{19}{12}\pi, \text{ and } \tfrac{23}{12}\pi$$

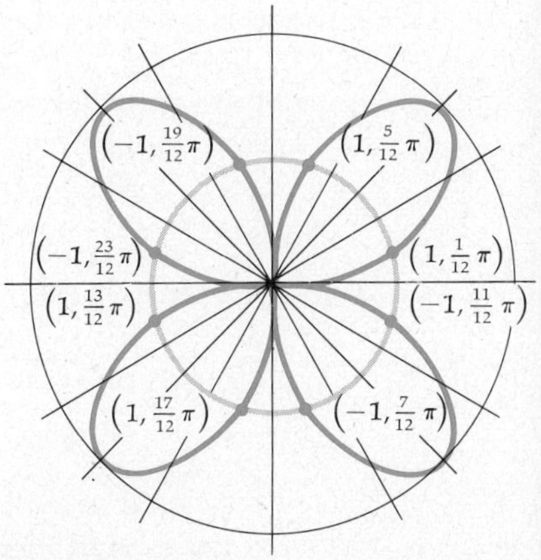

FIGURE 12.3.1

Thus we have the four points $(-1, \frac{7}{12}\pi)$, $(-1, \frac{11}{12}\pi)$, $(-1, \frac{19}{12}\pi)$, and $(-1, \frac{23}{12}\pi)$. Incidentally, $(-1, \frac{7}{12}\pi)$ also can be written as $(1, \frac{19}{12}\pi)$, $(-1, \frac{11}{12}\pi)$ can be written as $(1, \frac{23}{12}\pi)$, $(-1, \frac{19}{12}\pi)$ can be written as $(1, \frac{7}{12}\pi)$, and $(-1, \frac{23}{12}\pi)$ can be written as $(1, \frac{11}{12}\pi)$.

Because $(0, \theta)$ represents the pole for any θ, we determine if the pole is a point of intersection by setting $r = 0$ in each equation and solving for θ.

Often the coordinates of the points of intersection of two curves can be found directly from their graphs. However, the following is a general method.

If an equation of a curve in polar coordinates is given by

$$r = f(\theta) \tag{1}$$

then the same curve is given by

$$(-1)^n r = f(\theta + n\pi) \tag{2}$$

where n is any integer.

● ILLUSTRATION 1 Consider the curves of Example 1. The graph of $r = 2 \sin 2\theta$ also has the equation (by taking $n = 1$ in (2))

$$(-1)r = 2 \sin 2(\theta + \pi) \quad \text{or, equivalently,} \quad -r = 2 \sin 2\theta$$

If we take $n = 2$ in (2), the graph of $r = 2 \sin 2\theta$ also has the equation

$$(-1)^2 r = 2 \sin 2(\theta + 2\pi) \quad \text{or, equivalently,} \quad r = 2 \sin 2\theta$$

which is the same as the original equation. Taking n any other integer, we get either $r = 2 \sin 2\theta$ or $r = -2 \sin 2\theta$. The graph of the equation $r = 1$ also has the equation (by taking $n = 1$ in (2))

$$(-1)r = 1 \quad \text{or, equivalently,} \quad r = -1$$

Other integer values of n in (2) applied to the equation $r = 1$ give either $r = 1$ or $r = -1$. ●

If we are given the two equations $r = f(\theta)$ and $r = g(\theta)$, we obtain all the points of intersection of the graphs of the equations by doing the following.

(a) Use (2) to determine all the distinct equations of the two curves:

$$r = f_1(\theta), r = f_2(\theta), r = f_3(\theta), \ldots \tag{3}$$

$$r = g_1(\theta), r = g_2(\theta), r = g_3(\theta), \ldots \tag{4}$$

(b) Solve each equation in (3) simultaneously with each equation in (4).

(c) Check to see if the pole is a point of intersection by setting $r = 0$ in each equation, thereby giving

$$f(\theta) = 0 \quad \text{and} \quad g(\theta) = 0 \tag{5}$$

If Eqs. (5) each have a solution for θ, not necessarily the same, then the pole lies on both curves.

Example 2 Find the points of intersection of the two curves $r = 2 - 2\cos\theta$ and $r = 2\cos\theta$. Draw sketches of their graphs.

Solution To find other equations of the curve represented by $r = 2 - 2\cos\theta$, we have

$$(-1)r = 2 - 2\cos(\theta + \pi)$$
$$-r = 2 + 2\cos\theta$$

and

$$(-1)^2 r = 2 - 2\cos(\theta + 2\pi)$$
$$r = 2 - 2\cos\theta$$

which is the same as the original equation.

In a similar manner we find other equations of the curve given by $r = 2\cos\theta$:

$$(-1)r = 2\cos(\theta + \pi)$$
$$-r = -2\cos\theta$$
$$r = 2\cos\theta$$

which is the same as the original equation.

So there are two possible equations for the first curve, $r = 2 - 2\cos\theta$ and $-r = 2 + 2\cos\theta$, and one equation for the second curve, $r = 2\cos\theta$. Solving simultaneously $r = 2 - 2\cos\theta$ and $r = 2\cos\theta$ gives

$$2\cos\theta = 2 - 2\cos\theta$$
$$4\cos\theta = 2$$
$$\cos\theta = \tfrac{1}{2}$$

Thus $\theta = \frac{1}{3}\pi$ and $\frac{5}{3}\pi$, giving the points $(1, \frac{1}{3}\pi)$ and $(1, \frac{5}{3}\pi)$. Solving simultaneously $-r = 2 + 2\cos\theta$ and $r = 2\cos\theta$ yields

$$2 + 2\cos\theta = -2\cos\theta$$
$$4\cos\theta = -2$$
$$\cos\theta = -\tfrac{1}{2}$$

Hence $\theta = \frac{2}{3}\pi$ and $\theta = \frac{4}{3}\pi$, giving the points $(-1, \frac{2}{3}\pi)$ and $(-1, \frac{4}{3}\pi)$. However, $(-1, \frac{2}{3}\pi)$ is the same point as $(1, \frac{5}{3}\pi)$, and $(-1, \frac{4}{3}\pi)$ is the same point as $(1, \frac{1}{3}\pi)$.

Checking to see if the pole is on the first curve by substituting $r = 0$ in the equation $r = 2 - 2\cos\theta$, we have

$$0 = 2 - 2\cos\theta$$
$$\cos\theta = 1$$
$$\theta = 0$$

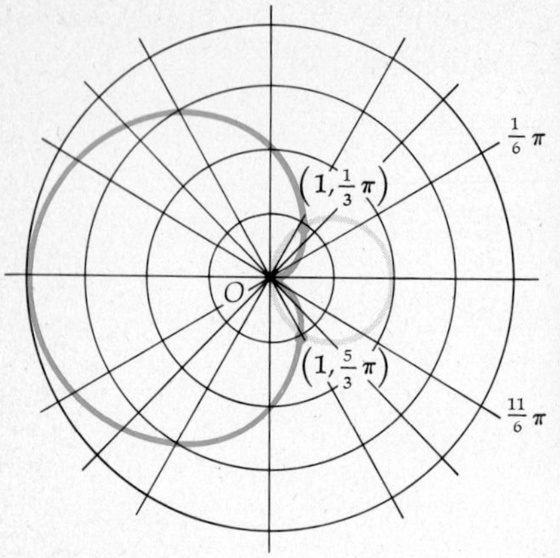

FIGURE 12.3.2

Therefore the pole lies on the first curve. In a similar fashion, by substituting $r = 0$ in $r = 2 \cos \theta$ we get

$$0 = 2 \cos \theta$$
$$\cos \theta = 0$$
$$\theta = \tfrac{1}{2}\pi \quad \text{or} \quad \tfrac{3}{2}\pi$$

So the pole lies on the second curve.

Therefore the points of intersection of the two curves are $(1, \frac{1}{3}\pi)$, $(1, \frac{5}{3}\pi)$, and the pole. Sketches of the two curves are shown in Fig. 12.3.2.

EXERCISES 12.3

In Exercises 1 through 21, find the points of intersection of the graphs of the given pair of equations. Draw a sketch of each pair of graphs with the same pole and polar axis.

1. $\begin{cases} 2r = 3 \\ r = 3 \sin \theta \end{cases}$ **2.** $\begin{cases} 2r = 3 \\ r = 1 + \cos \theta \end{cases}$ **3.** $\begin{cases} r = 2 \cos \theta \\ r = 2 \sin \theta \end{cases}$ **4.** $\begin{cases} r = 2 \cos 2\theta \\ r = 2 \sin \theta \end{cases}$ **5.** $\begin{cases} r = 4\theta \\ r = \frac{1}{2}\pi \end{cases}$

6. $\begin{cases} \theta = \frac{1}{6}\pi \\ r = 2 \end{cases}$ **7.** $\begin{cases} r = \cos \theta - 1 \\ r = \cos 2\theta \end{cases}$ **8.** $\begin{cases} r = 1 - \sin \theta \\ r = \cos 2\theta \end{cases}$ **9.** $\begin{cases} r = \sin 2\theta \\ r = \cos 2\theta \end{cases}$ **10.** $\begin{cases} r^2 = 2 \cos \theta \\ r = 1 \end{cases}$

11. $\begin{cases} r = \tan \theta \\ r = 4 \sin \theta \end{cases}$ **12.** $\begin{cases} r = 4 \tan \theta \sin \theta \\ r = 4 \cos \theta \end{cases}$ **13.** $\begin{cases} r \sin \theta = 4 \\ r \cos \theta = 4 \end{cases}$ **14.** $\begin{cases} r = 4 \cos \theta \\ \theta = \frac{1}{3}\pi \end{cases}$ **15.** $\begin{cases} r = 4 \\ r\theta = 4 \end{cases}$

16. $\begin{cases} r = 2 \cos \theta \\ r = 2\sqrt{3} \sin \theta \end{cases}$ **17.** $\begin{cases} r = 3 \\ r = 2(1 + \cos \theta) \end{cases}$ **18.** $\begin{cases} r = 4 \sin 2\theta \\ r = -2 \end{cases}$ **19.** $\begin{cases} r = \sin \theta \\ r = \sin 2\theta \end{cases}$ **20.** $\begin{cases} r = 4(1 + \sin \theta) \\ r(1 - \sin \theta) = 3 \end{cases}$

21. $\begin{cases} r^2 \sin 2\theta = 8 \\ r \cos \theta = 2 \end{cases}$

22. For each pair of equations state the number of points of intersection: (a) $r \sin \theta = 3$, $r \cos \theta = -2$; (b) $r = 4$, $r = -3$; (c) $r = 5$, $r = -4 \cos \theta$; (d) $r = 2$, $r = 4 \sin 3\theta$; (e) $r = 4(1 - \cos \theta)$, $r = 4 \cos \theta$.

In Exercises 23 and 24, the graph of the given equation intersects itself. Find the points at which this occurs.

23. $r = \sin \frac{3}{2}\theta$ **24.** $r = 1 + 2 \cos 2\theta$

12.4 TANGENT LINES OF POLAR CURVES

In Sec. 12.2 we derived a formula for finding the slope of a tangent line to a polar curve whose equation is $r = f(\theta)$. If α radians is the measure of the inclination of the tangent line to the curve at (r, θ),

$$\tan \alpha = \frac{\tan \theta \dfrac{dr}{d\theta} + r}{\dfrac{dr}{d\theta} - r \tan \theta} \tag{1}$$

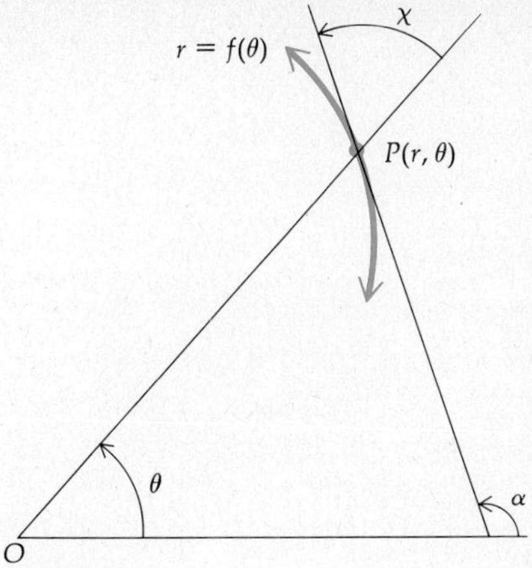

FIGURE 12.4.1

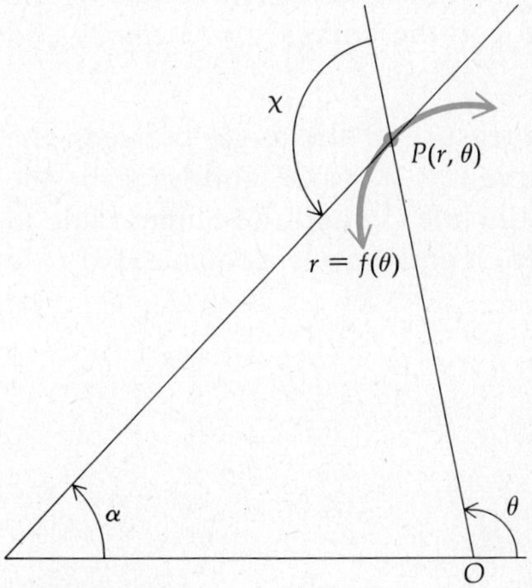

FIGURE 12.4.2

Formula (1) is generally complicated to apply. A simpler formula is obtained by considering the angle between the line OP and the tangent line. This angle will have radian measure χ and is measured from the line OP counterclockwise to the tangent line, $0 \le \chi < \pi$.

There are two possible cases: $\alpha \ge \theta$ and $\alpha < \theta$. These two cases are illustrated in Figs. 12.4.1 and 12.4.2. In Fig. 12.4.1, $\alpha > \theta$ and $\chi = \alpha - \theta$. In Fig. 12.4.2, $\alpha < \theta$ and $\chi = \pi - (\theta - \alpha)$. In each case

$$\tan \chi = \tan(\alpha - \theta)$$

$$\tan \chi = \frac{\tan \alpha - \tan \theta}{1 + \tan \alpha \tan \theta} \tag{2}$$

Substituting the value of $\tan \alpha$ from (1) in (2) we get

$$\tan \chi = \frac{\left(\tan \theta \dfrac{dr}{d\theta} + r\right)\Big/\left(\dfrac{dr}{d\theta} - r \tan \theta\right) - \tan \theta}{1 + \left[\left(\tan \theta \dfrac{dr}{d\theta} + r\right)\Big/\left(\dfrac{dr}{d\theta} - r \tan \theta\right)\right](\tan \theta)}$$

$$= \frac{\tan \theta \dfrac{dr}{d\theta} + r - \tan \theta \dfrac{dr}{d\theta} + r \tan^2 \theta}{\dfrac{dr}{d\theta} - r \tan \theta + \tan^2 \theta \dfrac{dr}{d\theta} + r \tan \theta}$$

$$= \frac{r(1 + \tan^2 \theta)}{(1 + \tan^2 \theta)\dfrac{dr}{d\theta}}$$

$$= \frac{r}{\dfrac{dr}{d\theta}}$$

We have proved the following theorem.

12.4.1 Theorem Let χ be the radian measure of the angle between the line OP and the tangent line to the graph of $r = f(\theta)$ at the point $P(r, \theta)$, where the angle is measured from OP counterclockwise to the tangent line and $0 \le \chi < \pi$. Then

$$\tan \chi = \frac{r}{\dfrac{dr}{d\theta}} \tag{3}$$

Comparing (3) with (1) you can see why it is more desirable to consider χ instead of α when working with polar coordinates.

Example 1 Find χ for the cardioid $r = a + a \sin \theta$ at the point $(\frac{3}{2}a, \frac{1}{6}\pi)$.

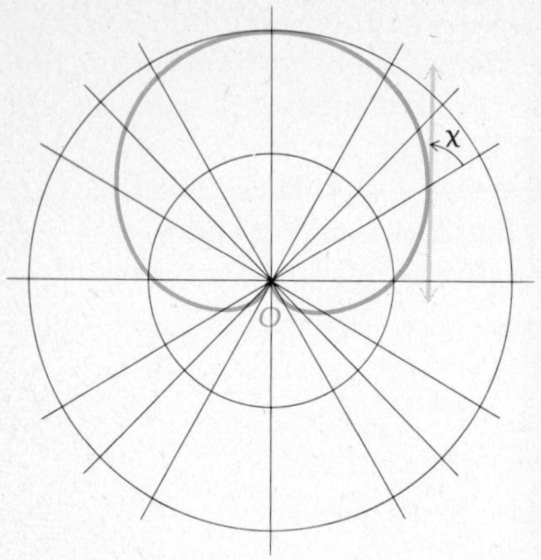

FIGURE 12.4.3

Solution See Fig. 12.4.3. Because $r = a + a \sin\theta$,

$$\frac{dr}{d\theta} = a\cos\theta$$

From Theorem 12.4.1,

$$\tan\chi = \frac{a + a\sin\theta}{a\cos\theta} = \frac{1+\sin\theta}{\cos\theta}$$

Therefore, at the point $(\frac{3}{2}a, \frac{1}{6}\pi)$,

$$\tan\chi = \frac{1+\frac{1}{2}}{\frac{1}{2}\sqrt{3}} = \frac{3}{\sqrt{3}} = \sqrt{3}$$

So

$$\chi = \tfrac{1}{3}\pi$$

Example 2 Find to the nearest 10 min the measurement of the smaller angle between the tangent lines to the curves $r = 3\cos 2\theta$ and $r = 3\sin 2\theta$ at the point $P(\frac{3}{2}\sqrt{2}, \frac{1}{8}\pi)$.

Solution Let χ_1 be the radian measure of the angle between the line OP and the tangent line to the curve $r = 3\cos 2\theta$, and let χ_2 be the radian measure of the angle between the line OP and the tangent line to the curve $r = 3\sin 2\theta$. The graph of each equation is a four-leafed rose (see Fig. 12.4.4).

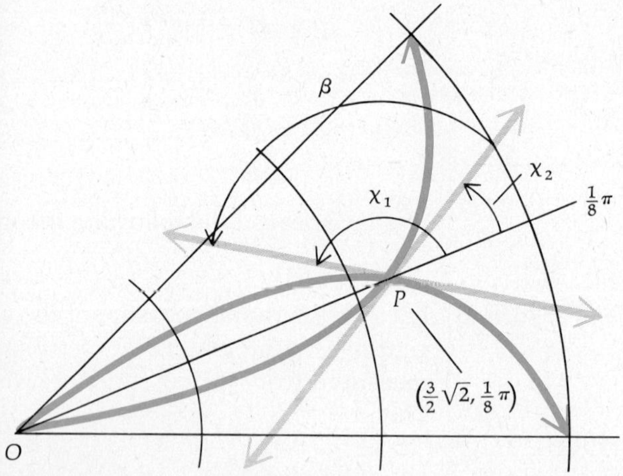

FIGURE 12.4.4

If β is the radian measure of the angle between the tangent lines at $P(\frac{3}{2}\sqrt{2}, \frac{1}{8}\pi)$, then

$$\beta = \chi_1 - \chi_2$$

So

$$\tan\beta = \tan(\chi_1 - \chi_2)$$

or, equivalently,

$$\tan \beta = \frac{\tan \chi_1 - \tan \chi_2}{1 + \tan \chi_1 \tan \chi_2} \tag{4}$$

$\tan \chi_1$ and $\tan \chi_2$ are found from Theorem 12.4.1. For $\tan \chi_1$, $r = 3 \cos 2\theta$ and $dr/d\theta = -6 \sin 2\theta$. So

$$\tan \chi_1 = \frac{3 \cos 2\theta}{-6 \sin 2\theta} = -\frac{1}{2} \cot 2\theta$$

When $\theta = \frac{1}{8}\pi$, $\tan \chi_1 = -\frac{1}{2} \cot \frac{1}{4}\pi = -\frac{1}{2}$.

For $\tan \chi_2$, $r = 3 \sin 2\theta$ and $dr/d\theta = 6 \cos 2\theta$. So

$$\tan \chi_2 = \frac{3 \sin 2\theta}{6 \cos 2\theta} = \frac{1}{2} \tan 2\theta$$

When $\theta = \frac{1}{8}\pi$, $\tan \chi_2 = \frac{1}{2} \tan \frac{1}{4}\pi = \frac{1}{2}$.

Substituting $\tan \chi_1 = -\frac{1}{2}$ and $\tan \chi_2 = \frac{1}{2}$ in (4) we obtain

$$\tan \beta = \frac{-\frac{1}{2} - \frac{1}{2}}{1 + (-\frac{1}{2})(\frac{1}{2})} = \frac{-1}{1 - \frac{1}{4}} = -\frac{4}{3}$$

The angle of radian measure β has a measurement of 126°50′ to the nearest 10 min. So the measurement of the smaller angle between the two tangent lines is 180° − 126°50′ = 53°10′.

EXERCISES 12.4

In Exercises 1 through 9, find χ at the point indicated.

1. $r\theta = a$; $(a, 1)$ **2.** $r = a\theta$; $(\frac{5}{2}\pi a, \frac{5}{2}\pi)$ **3.** $r = a \sec 2\theta$; $(\sqrt{2}a, -\frac{1}{8}\pi)$ **4.** $r = a \sin \frac{1}{2}\theta$; $(\frac{1}{2}a, \frac{1}{3}\pi)$

5. $r = \theta^2$; $(\frac{1}{4}\pi^2, \frac{1}{2}\pi)$ **6.** $r = \cos 2\theta$; $(\frac{1}{2}\sqrt{3}, \frac{1}{12}\pi)$ **7.** $r^2 = a^2 \cos 2\theta$; $\left(\frac{a}{\sqrt{2}}, \frac{\pi}{6}\right)$ **8.** $r = a(1 - \sin \theta)$; (a, π)

9. $r = a(\sqrt{3} - \cos \theta)$; $(\sqrt{3}a, \frac{1}{2}\pi)$

In Exercises 10 through 14, find the measurement of the smaller angle between the tangent lines of the given pair of curves at the indicated point of intersection.

10. $\begin{cases} r = a \\ r = 2a \sin \theta \end{cases}$; $(a, \frac{1}{6}\pi)$ **11.** $\begin{cases} r = a \cos \theta \\ r = a \sin \theta \end{cases}$; $(\frac{1}{2}\sqrt{2}a, \frac{1}{4}\pi)$ **12.** $\begin{cases} r = -a \sin \theta \\ r = a \cos 2\theta \end{cases}$; the pole

13. $\begin{cases} r = 4 \cos \theta \\ r = 4 \cos^2 \theta - 3 \end{cases}$; $(-2, \frac{2}{3}\pi)$ **14.** $\begin{cases} r = 2(1 - \cos \theta) \\ r = 2 \cos \theta \end{cases}$; $(1, \frac{1}{3}\pi)$

In Exercises 15 through 18, find a measurement of the angle between the tangent lines of the given pair of curves at all points of intersection.

15. $\begin{cases} r = 1 - \sin \theta \\ r = 1 + \sin \theta \end{cases}$ **16.** $\begin{cases} r = 3 \cos \theta \\ r = 1 + \cos \theta \end{cases}$ **17.** $\begin{cases} r = \cos \theta \\ r = \sin 2\theta \end{cases}$ **18.** $\begin{cases} r = 2 \sec \theta \\ r = \csc^2 \frac{1}{2}\theta \end{cases}$

19. Given the curve $r = k$, where k is a nonzero constant and α radians is the measure of the inclination of the tangent line to the curve at the point (r, θ). Show that $\tan \alpha \tan \theta = -1$ for all θ for which $\tan \theta \neq 0$.

20. For the curve $r \cos \theta = 4$ find $\lim_{\theta \to \pi/2} \chi$.

21. Prove that $\tan \chi = \tan \frac{1}{2}\theta$ at all points of the cardioid $r = 2(1 - \cos \theta)$.

22. Prove that at each point of the logarithmic spiral $r = be^{a\theta}$, χ is the same.

23. Prove that at the points of intersection of the cardioids $r = a(1 + \sin \theta)$ and $r = b(1 - \sin \theta)$ their tangent lines are perpendicular for all values of a and b.

24. Prove that at the points of intersection of the two curves $r = a \sec^2 \frac{1}{2}\theta$ and $r = b \csc^2 \frac{1}{2}\theta$ their tangent lines are perpendicular.

12.5 AREA OF A REGION IN POLAR COORDINATES

We now develop a method for finding the area of a region bounded by two lines through the pole and a curve whose equation is given in polar coordinates.

Let the function f be continuous and nonnegative on the closed interval $[\alpha, \beta]$. Let R be the region bounded by the curve whose equation is $r = f(\theta)$ and by the lines $\theta = \alpha$ and $\theta = \beta$. Then the region R is the region AOB shown in Fig. 12.5.1.

Consider a partition Δ of $[\alpha, \beta]$ defined by $\alpha = \theta_0 < \theta_1 < \theta_2 < \cdots < \theta_{i-1} < \theta_i < \cdots < \theta_{n-1} < \theta_n = \beta$. Therefore we have n subintervals of the form $[\theta_{i-1}, \theta_i]$, where $i = 1, 2, \ldots, n$. Let ξ_i be a value of θ in the ith subinterval $[\theta_{i-1}, \theta_i]$. See Fig. 12.5.2, where the ith subinterval is shown together with $\theta = \xi_i$. The radian measure of the angle between the lines $\theta = \theta_{i-1}$ and $\theta = \theta_i$ is denoted by $\Delta_i\theta$. The number of square units in the area of the circular sector of radius $f(\xi_i)$ units and central angle of radian measure $\Delta_i\theta$ is given by

$$\tfrac{1}{2}[f(\xi_i)]^2 \, \Delta_i\theta$$

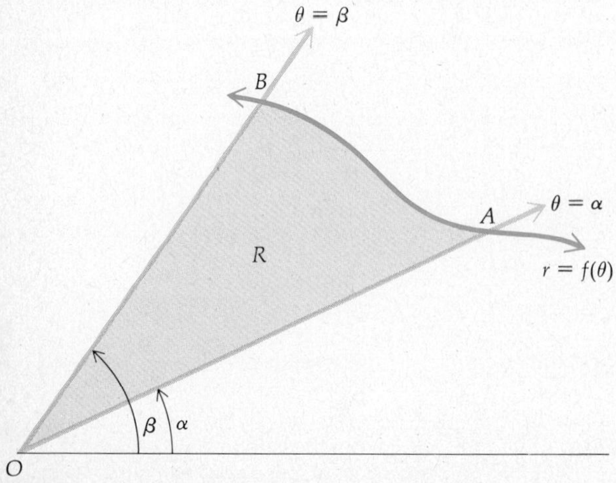

FIGURE 12.5.1

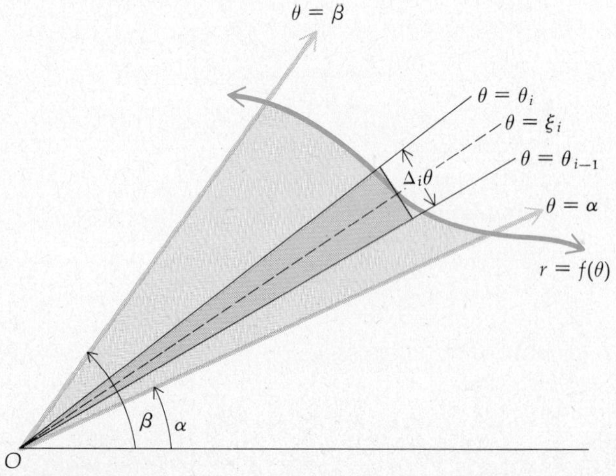

FIGURE 12.5.2

There is such a circular sector for each of the n subintervals. The sum of the measures of the areas of these n circular sectors is

$$\tfrac{1}{2}[f(\xi_1)]^2\,\Delta_1\theta + \tfrac{1}{2}[f(\xi_2)]^2\,\Delta_2\theta + \cdots + \tfrac{1}{2}[f(\xi_i)]^2\,\Delta_i\theta + \cdots + \tfrac{1}{2}[f(\xi_n)]^2\,\Delta_n\theta$$

which can be written, using sigma notation, as

$$\sum_{i=1}^{n} \tfrac{1}{2}[f(\xi_i)]^2\,\Delta_i\theta \tag{1}$$

Let $\|\Delta\|$ be the norm of the partition Δ; that is, $\|\Delta\|$ is the measure of the largest $\Delta_i\theta$. Then if we let A square units be the area of the region R, we define A as the limit of Riemann sum (1) as $\|\Delta\|$ approaches 0, which is a definite integral. This definition is stated formally.

12.5.1 Definition Let R be the region bounded by the lines $\theta = \alpha$ and $\theta = \beta$ and the curve whose equation is $r = f(\theta)$, where f is continuous and nonnegative on the closed interval $[\alpha, \beta]$. Then if A square units is the area of region R,

$$A = \lim_{\|\Delta\|\to 0} \sum_{i=1}^{n} \tfrac{1}{2}[f(\xi_i)]^2\,\Delta_i\theta = \frac{1}{2}\int_{\alpha}^{\beta} [f(\theta)]^2\,d\theta$$

Example 1 Find the area of the region bounded by the graph of $r = 2 + 2\cos\theta$.

Solution The region together with an element of area is shown in Fig. 12.5.3. Because the curve is symmetric with respect to the polar axis, we take the θ limits from 0 to π that determine the area of the region

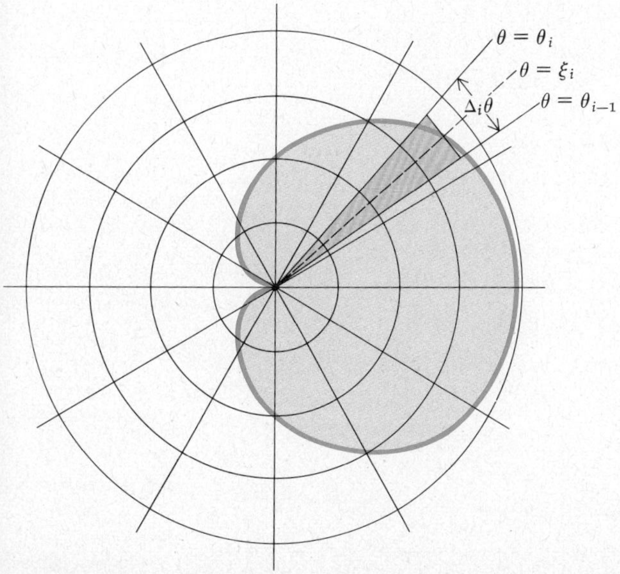

FIGURE 12.5.3

bounded by the curve above the polar axis. Then the area of the entire region is determined by multiplying that area by 2. Thus, if A square units is the required area,

$$\begin{aligned} A &= 2 \lim_{\|\Delta\| \to 0} \sum_{i=1}^{n} \tfrac{1}{2}(2 + 2\cos \xi_i)^2 \Delta_i\theta \\ &= 2\int_0^{\pi} \tfrac{1}{2}(2 + 2\cos\theta)^2 \, d\theta \\ &= 4\int_0^{\pi} (1 + 2\cos\theta + \cos^2\theta)\, d\theta \\ &= 4\Big[\theta + 2\sin\theta + \tfrac{1}{2}\theta + \tfrac{1}{4}\sin 2\theta\Big]_0^{\pi} \\ &= 4(\pi + 0 + \tfrac{1}{2}\pi + 0 - 0) \\ &= 6\pi \end{aligned}$$

Therefore the area is 6π square units.

Consider now the region bounded by the lines $\theta = \alpha$ and $\theta = \beta$ and the two curves whose equations are $r = f(\theta)$ and $r = g(\theta)$, where f and g are continuous on the closed interval $[\alpha, \beta]$ and $f(\theta) \geq g(\theta)$ on $[\alpha, \beta]$. See Fig. 12.5.4. We wish to find the area of this region. We take a partition of the interval $[\alpha, \beta]$ with ξ_i a value of θ in the ith subinterval $[\theta_{i-1}, \theta_i]$. The measure of the area of an element of area is the difference of the measures of the areas of two circular sectors.

$$\tfrac{1}{2}[f(\xi_i)]^2 \Delta_i\theta - \tfrac{1}{2}[g(\xi_i)]^2 \Delta_i\theta = \tfrac{1}{2}([f(\xi_i)]^2 - [g(\xi_i)]^2)\Delta_i\theta$$

The sum of the measures of the areas of n such elements is given by

$$\sum_{i=1}^{n} \tfrac{1}{2}([f(\xi_i)]^2 - [g(\xi_i)]^2)\Delta_i\theta$$

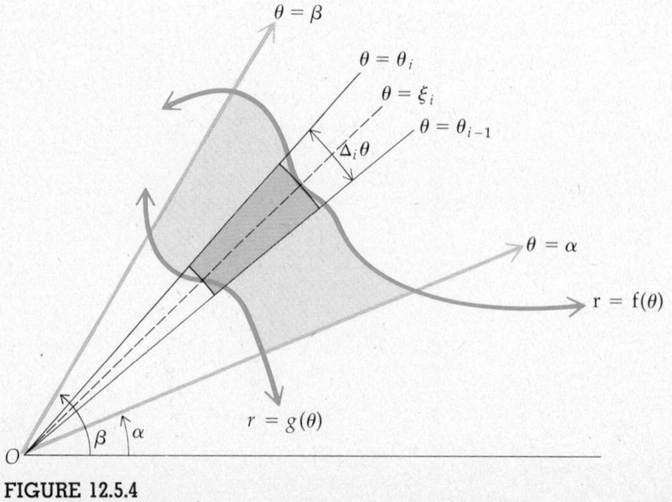

FIGURE 12.5.4

Hence, if A square units is the area of the region desired, we have

$$A = \lim_{\|\Delta\| \to 0} \sum_{i=1}^{n} \tfrac{1}{2}([f(\xi_i)]^2 - [g(\xi_i)]^2)\, \Delta_i \theta \tag{2}$$

Because f and g are continuous on $[\alpha, \beta]$, so also is $(f - g)$; therefore the limit in (2) exists and is equal to a definite integral. Thus

$$A = \frac{1}{2} \int_{\alpha}^{\beta} ([f(\theta)]^2 - [g(\theta)]^2)\, d\theta$$

Example 2 Find the area of the region inside the circle $r = 3 \sin \theta$ and outside the limaçon $r = 2 - \sin \theta$.

Solution To find the points of intersection we set

$$3 \sin \theta = 2 - \sin \theta$$

$$\sin \theta = \tfrac{1}{2}$$

So

$$\theta = \tfrac{1}{6}\pi \quad \text{and} \quad \theta = \tfrac{5}{6}\pi$$

The curves are sketched and the region is shown together with an element of area in Fig. 12.5.5.

If we let $f(\theta) = 3 \sin \theta$ and $g(\theta) = 2 - \sin \theta$, then the equation of the circle is $r = f(\theta)$, and the equation of the limaçon is $r = g(\theta)$.

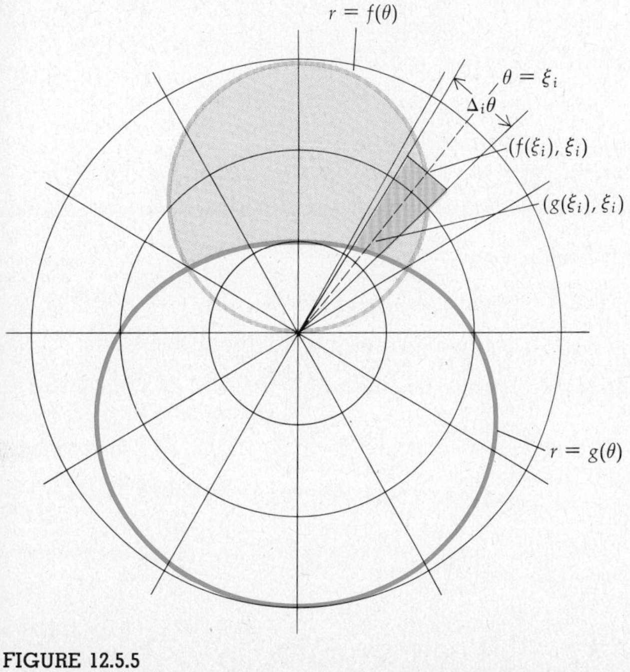

FIGURE 12.5.5

Instead of taking the limits $\frac{1}{6}\pi$ to $\frac{5}{6}\pi$ we use the property of symmetry with respect to the $\frac{1}{2}\pi$ axis and take the limits from $\frac{1}{6}\pi$ to $\frac{1}{2}\pi$ and multiply by 2. Then, if A square units is the area of the given region,

$$
\begin{aligned}
A &= 2 \lim_{\|\Delta\| \to 0} \sum_{i=1}^{n} \tfrac{1}{2}([f(\xi_i)]^2 - [g(\xi_i)]^2)\, \Delta_i \theta \\
&= 2 \cdot \frac{1}{2} \int_{\pi/6}^{\pi/2} ([f(\theta)]^2 - [g(\theta)]^2)\, d\theta \\
&= \int_{\pi/6}^{\pi/2} [9 \sin^2 \theta - (2 - \sin \theta)^2]\, d\theta \\
&= 8 \int_{\pi/6}^{\pi/2} \sin^2 \theta\, d\theta + 4 \int_{\pi/6}^{\pi/2} \sin \theta\, d\theta - 4 \int_{\pi/6}^{\pi/2} d\theta \\
&= 8 \int_{\pi/6}^{\pi/2} (1 - \cos 2\theta)\, d\theta + \Big[-4 \cos \theta - 4\theta \Big]_{\pi/6}^{\pi/2} \\
&= 4\theta - 2 \sin 2\theta - 4 \cos \theta - 4\theta \Big]_{\pi/6}^{\pi/2} \\
&= -2 \sin 2\theta - 4 \cos \theta \Big]_{\pi/6}^{\pi/2} \\
&= (-2 \sin \pi - 4 \cos \tfrac{1}{2}\pi) - (-2 \sin \tfrac{1}{3}\pi - 4 \cos \tfrac{1}{6}\pi) \\
&= 2 \cdot \tfrac{1}{2}\sqrt{3} + 4 \cdot \tfrac{1}{2}\sqrt{3} \\
&= 3\sqrt{3}
\end{aligned}
$$

Therefore the area is $3\sqrt{3}$ square units.

EXERCISES 12.5

In Exercises 1 through 6, find the area of the region enclosed by the graph of the given equation.

1. $r = 3 \cos \theta$ **2.** $r = 2 - \sin \theta$ **3.** $r = 4 \cos 3\theta$ **4.** $r = 4 \sin^2 \frac{1}{2}\theta$ **5.** $r^2 = 4 \sin 2\theta$ **6.** $r = 4 \sin^2 \theta \cos \theta$

7. Find the area of the region enclosed by the graph of the equation $r = \theta$ from $\theta = 0$ to $\theta = \frac{3}{2}\pi$.

8. Find the area of the region enclosed by the graph of $r = e^\theta$ and the lines $\theta = 0$ and $\theta = 1$.

In Exercises 9 through 12, find the area of the region enclosed by one loop of the graph of the given equation.

9. $r = 3 \cos 2\theta$ **10.** $r = a(1 - 2 \cos \theta)$ **11.** $r = 1 + 3 \sin \theta$ **12.** $r = a \sin 3\theta$

In Exercises 13 through 16, find the area of the intersection of the regions enclosed by the graphs of the two given equations.

13. $\begin{cases} r = 2 \\ r = 3 - 2\cos\theta \end{cases}$ **14.** $\begin{cases} r = 4\sin\theta \\ r = 4\cos\theta \end{cases}$ **15.** $\begin{cases} r = 3\sin 2\theta \\ r = 3\cos 2\theta \end{cases}$ **16.** $\begin{cases} r^2 = 2\cos 2\theta \\ r = 1 \end{cases}$

In Exercises 15 through 21, find the area of the region that is inside the graph of the first equation and outside the graph of the second equation.

17. $\begin{cases} r = a \\ r = a(1 - \cos\theta) \end{cases}$ **18.** $\begin{cases} r^2 = 4\sin 2\theta \\ r = \sqrt{2} \end{cases}$ **19.** $\begin{cases} r = 2\sin\theta \\ r = \sin\theta + \cos\theta \end{cases}$ **20.** $\begin{cases} r = 2a\sin\theta \\ r = a \end{cases}$ **21.** $\begin{cases} r = a(1 + \cos\theta) \\ r = 2a\cos\theta \end{cases}$

22. Determine the value of a for which the area of the region enclosed by the cardioid $r = a(1 - \cos\theta)$ is 9π square units.

23. The face of a bow tie is the region enclosed by the graph of the equation $r^2 = 4\cos 2\theta$. How much material is necessary to cover the face of the tie?

24. Find the area of the region swept out by the radius vector of the spiral $r = a\theta$ during its second revolution that was not swept out during its first revolution.

25. Find the area of the region swept out by the radius vector of the curve of Exercise 24 during its third revolution that was not swept out during its second revolution.

REVIEW EXERCISES FOR CHAPTER 12

In Exercises 1 through 4, find a polar equation of the graph having the given cartesian equation.

1. $4x^2 - 9y^2 = 36$ **2.** $2xy = 1$ **3.** $x^2 + y^2 - 9x + 8y = 0$ **4.** $y^4 = x^2(a^2 - y^2)$

In Exercises 5 through 8, find a cartesian equation of the graph having the given polar equation.

5. $r = 9\sin^2 \frac{1}{2}\theta$ **6.** $r(1 - \cos\theta) = 2$ **7.** $r = 4\cos 3\theta$ **8.** $r = a\tan^2\theta$

9. Draw a sketch of the graph of (a) $r\theta = 3$ (reciprocal spiral) and (b) $3r = \theta$ (spiral of Archimedes).

10. Show that the equations $r = 1 + \sin\theta$ and $r = \sin\theta - 1$ have the same graph.

11. Draw a sketch of the graph of $r = \sqrt{|\cos\theta|}$. **12.** Draw a sketch of the graph of $r = \sqrt{|\cos 2\theta|}$.

13. Draw a sketch of the graph of $r = 2(1 + \cos\theta)$. **14.** Draw a sketch of the graph of $r^2 = 16\cos\theta$.

15. Find the slope of the tangent line to the graph of $r = \frac{1}{2} + \sin\theta$ at the point $(1, \frac{1}{6}\pi)$.

16. Find an equation of each of the tangent lines at the pole to the four-leafed rose $r = 4\cos 2\theta$.

17. Find an equation of each of the tangent lines at the pole to the cardioid $r = 3 - 3\cos\theta$.

18. Find the area of one leaf of the rose $r = 2\sin 3\theta$.

19. Find the area of the region inside the graph of $r = 2a\sin\theta$ and outside the graph of $r = a$.

20. Find the area of the region inside the graph of the lemniscate $r^2 = 2\sin 2\theta$ and outside the graph of the circle $r = 1$.

21. Find the area of the region swept out by the radius vector of the logarithmic spiral $r = e^{k\theta}$ $(k > 0)$ as θ varies from 0 to 2π.

22. Find the area of the intersection of the regions enclosed by the graphs of the two equations $r = a\cos\theta$ and $r = a(1 - \cos\theta)$, $a > 0$.

23. Find a polar equation of the tangent line to the curve $r = \theta$ at the point (π, π).

In Exercises 24 and 25, find all the points of intersection of the graphs of the two given equations.

24. $\begin{cases} r\cos\theta = 1 \\ r = 1 + 2\cos\theta \end{cases}$ **25.** $\begin{cases} r = 2(1 + \cos\theta) \\ r^2 = 2\cos\theta \end{cases}$

26. Prove that the graphs of $r = a\theta$ and $r\theta = a$ have an unlimited number of points of intersection. Also prove that the tangent lines are perpendicular at only two of these points of intersection, and find these points.

27. Find the radian measure of the angle between the tangent lines at each point of intersection of the graphs of the equations $r = -6 \cos \theta$ and $r = 2(1 - \cos \theta)$.

28. Find the slope of the tangent line to the curve $r = 6 \cos \theta - 2$ at the point $(1, \frac{5}{3}\pi)$.

29. Find the area of the region enclosed by the loop of the limaçon $r = 4(1 + 2 \cos \theta)$ and also the area of the region enclosed by the outer part of the limaçon.

30. Find the area enclosed by one loop of the curve $r = a \sin n\theta$, where n is a positive integer.

31. Prove that the distance between the two points $P_1(r_1, \theta_1)$ and $P_2(r_2, \theta_2)$ is $\sqrt{r_1^2 + r_2^2 - 2r_1r_2 \cos(\theta_2 - \theta_1)}$.

32. Find the points of intersection of the graphs of the equations $r = \tan \theta$ and $r = \cot \theta$.

33. Find a polar equation of the circle having its center at (r_0, θ_0) and a radius of a units. (*Hint:* Apply the law of cosines to the triangle having vertices at the pole, (r_0, θ_0), and (r, θ).)

CHAPTER 13 The Conic Sections

13.1 THE PARABOLA

A **conic section** (or **conic**) is a curve of intersection of a plane with a right-circular cone. There are three types of curves that occur in this way: the *parabola*, the *ellipse* (including the circle as a special case), and the *hyperbola*. The resulting curve depends on the inclination of the axis of the cone to the cutting plane. We discuss this geometry of conic sections in Sec. 13.3. In this section we study the parabola.

13.1.1 Definition

A **parabola** is the set of all points in a plane equidistant from a fixed point and a fixed line. The fixed point is called the **focus**, and the fixed line is called the **directrix**.

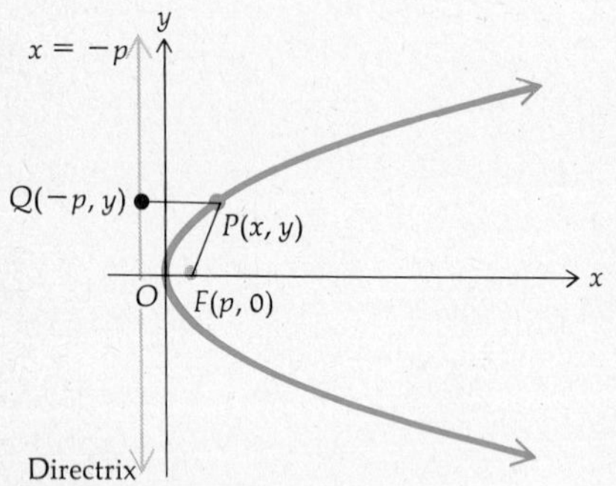

FIGURE 13.1.1

We now derive an equation of a parabola from the definition. For this equation to be as simple as possible, we choose the x axis as perpendicular to the directrix and containing the focus. The origin is taken as the point on the x axis midway between the focus and the directrix. It should be stressed that we are choosing the axes (*not* the parabola) in a special way. See Fig. 13.1.1.

Let p be the directed distance $\overline{OF}$. The focus is the point $F(p, 0)$, and the directrix is the line having the equation $x = -p$. A point $P(x, y)$ is on the parabola if and only if P is equidistant from F and the directrix. That is, if $Q(-p, y)$ is the foot of the perpendicular line from P to the directrix, then P is on the parabola if and only if

$$|\overline{FP}| = |\overline{QP}|$$

Because

$$|\overline{FP}| = \sqrt{(x-p)^2 + y^2}$$

and

$$|\overline{QP}| = \sqrt{(x+p)^2 + (y-y)^2}$$

P is on the parabola if and only if

$$\sqrt{(x-p)^2 + y^2} = \sqrt{(x+p)^2}$$

By squaring on both sides of the equation we obtain

$$x^2 - 2px + p^2 + y^2 = x^2 + 2px + p^2$$
$$y^2 = 4px$$

This result is stated as a theorem.

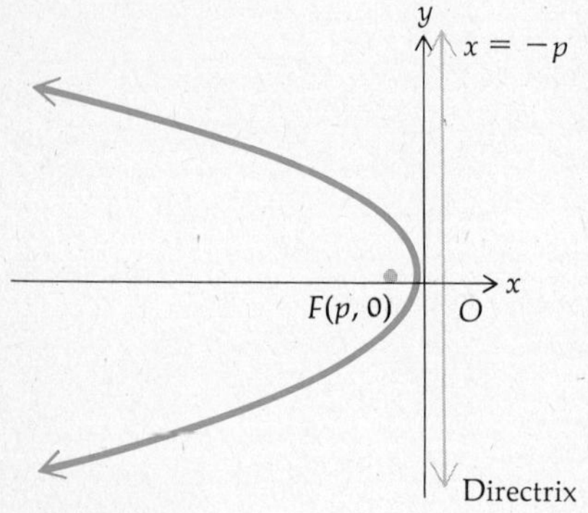

FIGURE 13.1.2

13.1.2 Theorem

An equation of the parabola having its focus at $(p, 0)$ and as its directrix the line $x = -p$ is

$$y^2 = 4px \qquad (1)$$

In Fig. 13.1.1, p is positive; p may be negative, however, because it is the directed distance $\overline{OF}$. Figure 13.1.2 shows a parabola for $p < 0$.

From Figs. 13.1.1 and 13.1.2 we see that for the equation $y^2 = 4px$ the parabola opens to the right if $p > 0$ and to the left if $p < 0$. The point midway between the focus and the directrix on the parabola is called the **vertex.** The vertex of the parabolas in Figs. 13.1.1 and 13.1.2 is the origin. The line through the vertex and the focus is called the **axis** of the parabola. The axis of the parabolas in Figs. 13.1.1 and 13.1.2 is the x axis.

In the above derivation, if the x axis and the y axis are interchanged, then the focus is at the point $F(0, p)$, and the directrix is the line having the equation $y = -p$. An equation of this parabola is $x^2 = 4py$, and the result is stated as a theorem.

13.1.3 Theorem An equation of the parabola having its focus at $(0, p)$ and as its directrix the line $y = -p$ is

$$x^2 = 4py \tag{2}$$

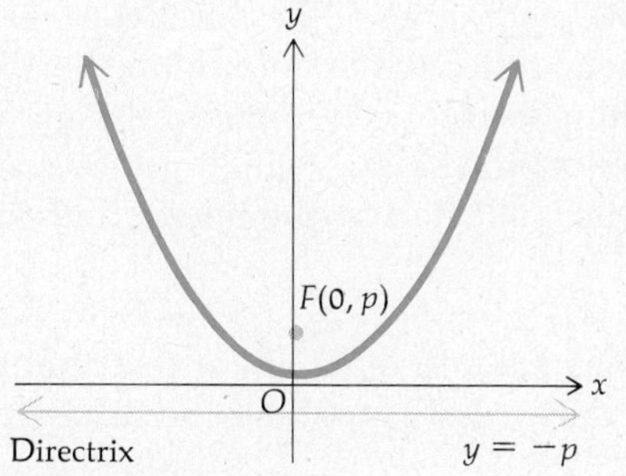

FIGURE 13.1.3

If $p > 0$, the parabola opens upward, as shown in Fig. 13.1.3; if $p < 0$, the parabola opens downward, as shown in Fig. 13.1.4. In each case the vertex is at the origin, and the y axis is the axis of the parabola.

When drawing a sketch of the graph of a parabola, it is helpful to consider the chord through the focus, perpendicular to the axis of the parabola, because the endpoints of this chord give two points on the parabola. Such a chord is called the **latus rectum** of the parabola. The length of the latus rectum is $|4p|$. (See Exercise 20.)

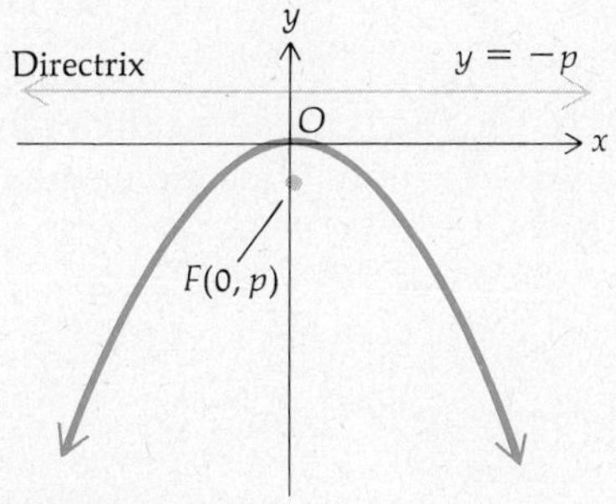

FIGURE 13.1.4

Example 1 Find an equation of the parabola having its focus at $(0, -3)$ and as its directrix the line $y = 3$. Draw a sketch of the graph.

Solution Because the focus is on the y axis and is also below the directrix, the parabola opens downward, and $p = -3$. Hence an equation of the parabola is

$$x^2 = -12y$$

The length of the latus rectum is

$$|4(-3)| = 12$$

A sketch of the graph is shown in Fig. 13.1.5.

Any point on the parabola is equidistant from the focus and the directrix. In Fig. 13.1.5 three such points (P_1, P_2, and P_3) are shown, and we have

$$|\overline{FP_1}| = |\overline{P_1Q_1}| \qquad |\overline{FP_2}| = |\overline{P_2Q_2}| \qquad |\overline{FP_3}| = |\overline{P_3Q_3}|$$

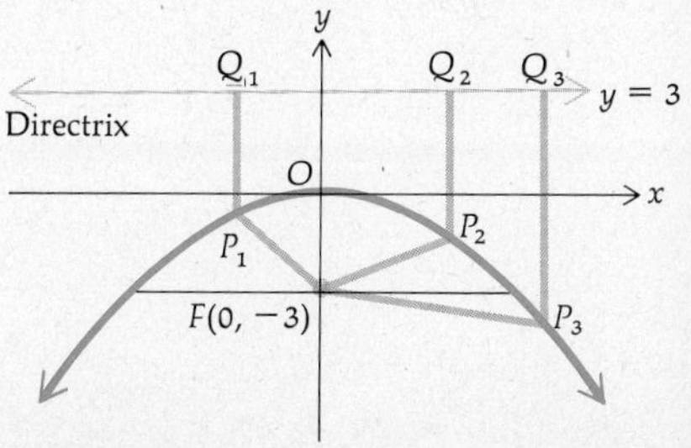

FIGURE 13.1.5

Example 2 Given the parabola having the equation $y^2 = 7x$, find the coordinates of the focus, an equation of the directrix, and the length of the latus rectum. Draw a sketch of the graph.

Solution The given equation is of the form of (1); so

$$4p = 7$$

$$p = \tfrac{7}{4}$$

Because $p > 0$, the parabola opens to the right. The focus is at the point $F(\frac{7}{4}, 0)$. An equation of the directrix is $x = -\frac{7}{4}$: The length of the latus rectum is 7. A sketch of the graph is shown in Fig. 13.1.6.

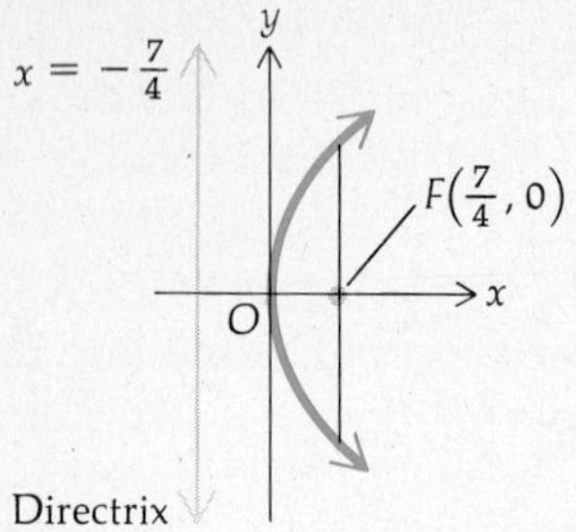

FIGURE 13.1.6

There is an interesting property of parabolas that has applications to the construction of searchlights, automobile headlights, and telescopes. In Fig. 13.1.7 the line PT is the tangent line at point P to the graph of a parabola. The point F is the focus of the parabola, and α is the measure of the angle between the line segment FP and the tangent line PT. Line PR is parallel to the axis of the parabola, and β is the measure of the angle between PR and PT. In Exercise 32 you are asked to prove that $\alpha = \beta$. Because of this equality, for a parabolic mirror in a searchlight, rays of light from a source at the focus are reflected along a line parallel to the axis. A similar principle involves parabolic reflectors in automobile headlights. For a parabolic mirror in a reflecting telescope, a converse situation occurs, where rays of light from an object in the sky, which go toward the mirror and are parallel to the axis, are reflected by the mirror and go through the focus.

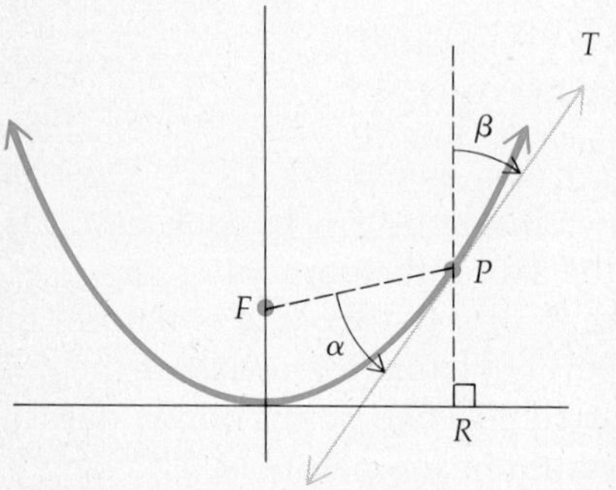

FIGURE 13.1.7

Example 3 A parabolic mirror has a depth of 12 cm at the center and the distance across the top of the mirror is 32 cm. Find the distance from the vertex to the focus.

Solution See Fig. 13.1.8. The coordinate axes are chosen so that the parabola has its vertex at the origin and its axis along the y axis, and it opens upward. Therefore an equation of the parabola is of the form

$$x^2 = 4py$$

where p cm is the distance from the vertex to the focus. Because the point $(16, 12)$ is on the parabola, its coordinates satisfy the equation, and we have

$$16^2 = 4p(12)$$

$$p = \frac{16}{3}$$

Therefore the distance from the vertex to the focus is $5\frac{1}{3}$ cm.

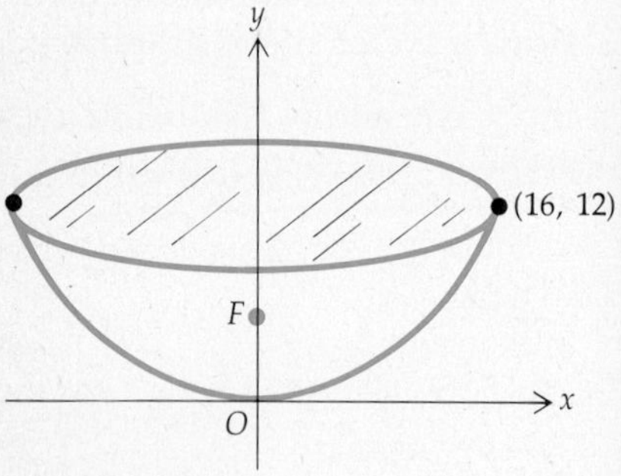

FIGURE 13.1.8

There are other practical applications of parabolas. The path of a projectile is a parabola if motion is considered to be in a plane and air resistance is neglected. Arches are sometimes parabolic in shape, and the cable of a suspension bridge hangs in the shape of a parabola.

EXERCISES 13.1

For each of the parabolas in Exercises 1 through 8, find the coordinates of the focus, an equation of the directrix, and the length of the latus rectum. Draw a sketch of the curve.

1. $x^2 = 4y$ **2.** $y^2 = 6x$ **3.** $y^2 = -8x$ **4.** $x^2 = -16y$

5. $x^2 + y = 0$ **6.** $y^2 + 5x = 0$ **7.** $2y^2 - 9x = 0$ **8.** $3x^2 + 4y = 0$

In Exercises 9 through 17, find an equation of the parabola having the given properties.

9. Focus, (5, 0); directrix, $x = -5$.

10. Focus, (0, 4); directrix, $y = -4$.

11. Focus, (0, -2); directrix, $y - 2 = 0$.

12. Focus, ($-\frac{5}{3}$, 0); directrix, $5 - 3x = 0$.

13. Focus, ($\frac{1}{2}$, 0); directrix, $2x + 1 = 0$.

14. Focus, (0, $\frac{2}{3}$); directrix, $3y + 2 = 0$.

15. Vertex, (0, 0); opens to the left; length of latus rectum is 6.

16. Vertex, (0, 0); opens upward; length of latus rectum is 3.

17. Vertex, (0, 0); directrix, $2x = -5$.

18. Find an equation of the parabola having its vertex at the origin, the x axis as its axis, and passing through the point (2, -4).

19. Find an equation of the parabola having its vertex at the origin, the y axis as its axis, and passing through the point (-2, -4).

20. Prove that the length of the latus rectum of a parabola is $|4p|$.

21. The endpoints of the latus rectum of a parabola are (5, k) and (-5, k). If the vertex of the parabola is at the origin and the parabola opens downward, find (a) the value of k; (b) an equation of the parabola.

22. Assume that water issuing from the end of a horizontal pipe 25 ft above the ground describes a parabolic curve, the vertex of the parabola being at the end of the pipe. If at a point 8 ft below the line of the pipe the flow of water has curved outward 10 ft beyond a vertical line through the end of the pipe, how far beyond this vertical line will the water strike the ground?

23. The cable of a suspension bridge hangs in the form of a parabola when the load is uniformly distributed horizontally. The distance between two towers is 150 m, the points of support of the cable on the towers are 22 m above the roadway, and the lowest point on the cable is 7 m above the roadway. Find the vertical distance to the cable from a point in the roadway 15 m from the foot of a tower.

24. A parabolic arch has a height of 20 m and a width of 36 m at the base. If the vertex of the parabola is at the top of the arch, at which height above the base is it 18 m wide?

25. Prove that on a parabola, the point closest to the focus is the vertex.

26. The orbit of a comet is a parabola having the sun at the focus. When the comet is 100 million km from the sun, the angle between the axis of the parabola and the line from the sun to the comet is 45°. Use the result of Exercise 25 to determine the shortest distance from the comet to the sun.

27. A reflecting telescope has a parabolic mirror for which the distance from the vertex to the focus is 30 ft. If the distance across the top of the mirror is 64 in., how deep is the mirror at the center?

28. Use Definition 13.1.1 to find an equation of the parabola having as its directrix the line $y = 4$ and as its focus the point (-3, 8).

29. Use Definition 13.1.1 to find an equation of the parabola having as its directrix the line $x = -3$ and as its focus the point (2, 5).

30. Find all points on the parabola $y^2 = 8x$ such that the foot of the perpendicular drawn from the point to the directrix, the focus, and the point itself are vertices of an equilateral triangle.

31. Find an equation of the circle passing through the vertex and the endpoints of the latus rectum of the parabola $x^2 = -8y$.

32. In Fig. 13.1.7, prove $\alpha = \beta$. (*Hint:* Choose the coordinate axes so that the parabola has its vertex at the origin and its axis along the y axis, and opens upward. Let Q be the point of intersection of the tangent line PT with the y axis. Prove $\alpha = \beta$ by showing that ΔQPF is isosceles.)

33. An equation of the directrix of a parabola is $x + y = 0$ and its focus is at the point (1, 1). Find (a) an equation of the axis of the parabola, (b) the coordinates of the vertex, and (c) the length of the latus rectum.

34. Prove analytically that the circle having as its diameter the latus rectum of a parabola is tangent to the directrix of the parabola.

35. A focal chord of a parabola is a line segment through the focus and with its endpoints on the parabola. If A and B are the endpoints of a focal chord of a parabola, and if C is the point of intersection of the directrix with a line through the vertex and point A, prove that the line through C and B is parallel to the axis of the parabola.

36. Prove that the distance from the midpoint of a focal chord (see Exercise 35) of a parabola to the directrix is half the length of the focal chord.

13.2 TRANSLATION OF AXES

The shape of a curve is not affected by the position of the coordinate axes; however, an equation of the curve is affected.

● ILLUSTRATION 1 If a circle with a radius of 3 has its center at the point $(4, -1)$, then an equation of this circle is

$$(x-4)^2 + (y+1)^2 = 9$$

$$x^2 + y^2 - 8x + 2y + 8 = 0$$

However, if the origin is at the center, the same circle has a simpler equation, namely,

$$x^2 + y^2 = 9$$ ●

If the coordinate axes may be taken as we please, they are generally chosen in such a way that the equations will be as simple as possible. If the axes are given, however, we often wish to find a simpler equation of a given curve referred to another set of axes.

In general, if in the plane with given x and y axes new coordinate axes are chosen parallel to the given ones, we say that there has been a *translation of axes* in the plane.

In particular, let the given x and y axes be translated to the x' and y' axes, having origin (h, k) with respect to the given axes. Also, assume that the positive numbers are on the same side of the origin on the x' and y' axes as they are on the x and y axes (see Fig. 13.2.1).

A point P in the plane, having coordinates (x, y) with respect to the given coordinate axes, will have coordinates (x', y') with respect to the new axes. To obtain relationships between these two sets of coordinates, draw a line through P parallel to the y axis and the y' axis, and also a line through P parallel to the x axis and the x' axis. Let the first line intersect the x axis at the point A and the x' axis at the point A', and the second line intersect the y axis at the point B and the y' axis at the point B'.

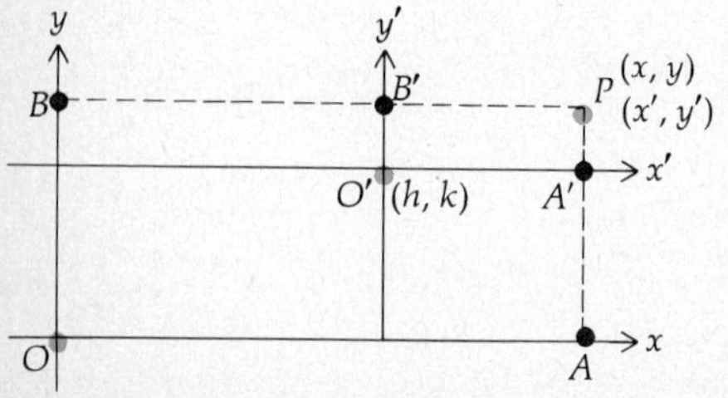

FIGURE 13.2.1

With respect to the x and y axes, the coordinates of P are (x, y), the coordinates of A are $(x, 0)$, and the coordinates of A' are (x, k). Because $\overline{A'P} = \overline{AP} - \overline{AA'}$,

$$y' = y - k$$

$$y = y' + k$$

With respect to the x and y axes, the coordinates of B are $(0, y)$, and the coordinates of B' are (h, y). Because $\overline{B'P} = \overline{BP} - \overline{BB'}$,

$$x' = x - h$$

$$x = x' + h$$

These results are stated as a theorem.

13.2.1 Theorem If (x, y) represents a point P with respect to a given set of axes, and (x', y') is a representation of P after the axes are translated to a new origin having coordinates (h, k) with respect to the given axes, then

$$x = x' + h \quad \text{and} \quad y = y' + k \tag{1}$$

or, equivalently,

$$x' = x - h \quad \text{and} \quad y' = y - k \tag{2}$$

Equations (1) or (2) are called the *equations of translating the axes.*

If an equation of a curve is given in x and y, then an equation in x' and y' is obtained by replacing x by $(x' + h)$ and y by $(y' + k)$. The graph of the equation in x and y, with respect to the x and y axes, is exactly the same set of points as the graph of the corresponding equation in x' and y' with respect to the x' and y' axes.

Example 1 Given the equation $x^2 + 10x + 6y + 19 = 0$, find an equation of the graph with respect to the x' and y' axes after a translation of axes to the new origin $(-5, 1)$.

Solution A point P, represented by (x, y) with respect to the old axes, has the representation (x', y') with respect to the new axes. Then by (1), with $h = -5$ and $k = 1$,

$$x = x' - 5 \quad \text{and} \quad y = y' + 1$$

Substituting these values of x and y into the given equation we obtain

$$(x' - 5)^2 + 10(x' - 5) + 6(y' + 1) + 19 = 0$$

$$x'^2 - 10x' + 25 + 10x' - 50 + 6y' + 6 + 19 = 0$$

$$x'^2 = -6y'$$

The graph of this equation with respect to the x' and y' axes is a parabola with its vertex at the origin, opening downward, and with $4p = -6$. The graph with respect to the x and y axes is, then, a parabola having its vertex at $(-5, 1)$, its focus at $(-5, -\frac{1}{2})$, and as its directrix the line $y = \frac{5}{2}$ (see Fig. 13.2.2).

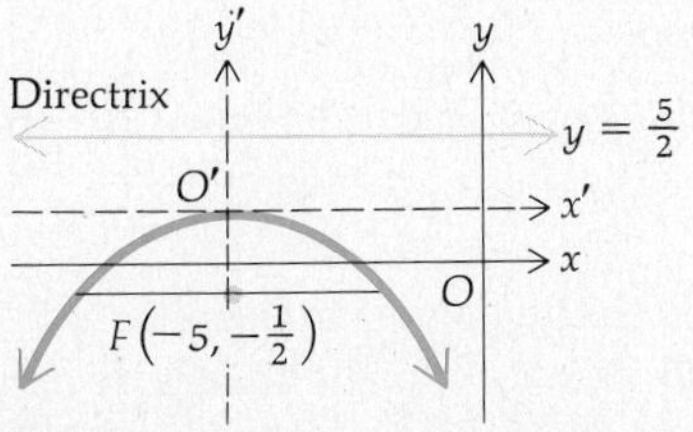

FIGURE 13.2.2

The above example illustrates how an equation can be reduced to a simpler form by a suitable translation of axes. In general, equations of the second degree that contain no term involving xy can be simplified by a translation of axes. This is illustrated in the following example.

Example 2 Given the equation $9x^2 + 4y^2 - 18x + 32y + 37 = 0$, translate the axes so the equation of the graph with respect to the x' and y' axes contains no first-degree terms.

Solution We rewrite the given equation:

$$9(x^2 - 2x) + 4(y^2 + 8y) = -37$$

Completing the squares of the terms in parentheses by adding $9 \cdot 1$ and $4 \cdot 16$ on both sides of the equation, we have

$$9(x^2 - 2x + 1) + 4(y^2 + 8y + 16) = -37 + 9 + 64$$
$$9(x - 1)^2 + 4(y + 4)^2 = 36$$

Then, if we let $x' = x - 1$ and $y' = y + 4$, we obtain

$$9x'^2 + 4y'^2 = 36$$

From (2) we see that the substitutions of $x' = x - 1$ and $y' = y + 4$ result in a translation of axes to a new origin of $(1, -4)$. In Fig. 13.2.3 there is a sketch of the graph of the equation in x' and y' with respect to the x' and y' axes.

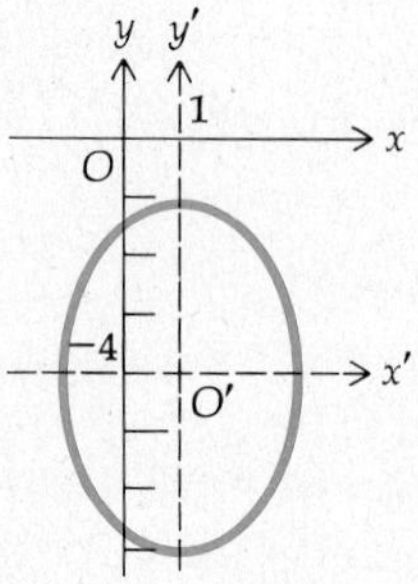

FIGURE 13.2.3

We now apply the translation of axes to finding the general equation of a parabola having its directrix parallel to a coordinate axis and its vertex at the point (h, k). In particular, let the directrix be parallel to the y axis. If the vertex is at the point $V(h, k)$, then the directrix has the equation $x = h - p$, and the focus is at the point $F(h + p, k)$. Let the x' and y' axes be such that the origin O' is at $V(h, k)$ (see Fig. 13.2.4).

An equation of the parabola in Fig. 13.2.4 with respect to the x' and y' axes is

$$y'^2 = 4px'$$

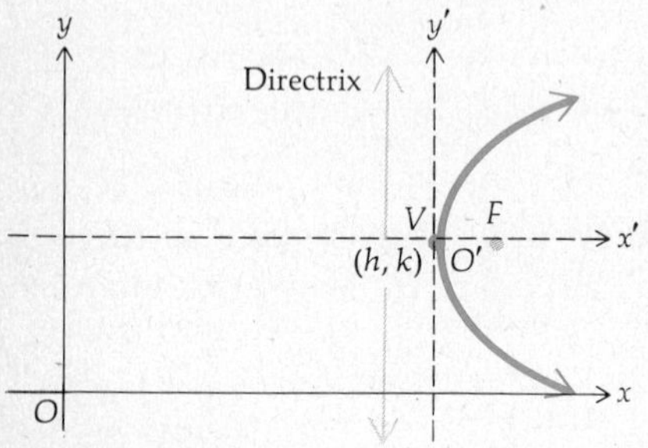

FIGURE 13.2.4

To obtain an equation of this parabola with respect to the x and y axes, we replace x' by $(x - h)$ and y' by $(y - k)$ from (2), which gives

$$(y - k)^2 = 4p(x - h)$$

The axis of this parabola is parallel to the x axis.

Similarly, if the directrix of a parabola is parallel to the x axis and the vertex is at $V(h, k)$, then its focus is at $F(h, k + p)$ and the directrix has the equation $y = k - p$, and an equation of the parabola with respect to the x and y axes is

$$(x - h)^2 = 4p(y - k)$$

The axis of this parabola is parallel to the y axis. We have proved, then, the following theorem.

13.2.2 Theorem If p is the directed distance from the vertex to the focus, an equation of the parabola with its vertex at (h, k) and with its axis parallel to the x axis is

$$(y - k)^2 = 4p(x - h) \qquad (3)$$

A parabola with the same vertex and with its axis parallel to the y axis has for an equation

$$(x - h)^2 = 4p(y - k) \tag{4}$$

Example 3 Find an equation of the parabola having as its directrix the line $y = 1$ and as its focus the point $F(-3, 7)$.

Solution Because the directrix is parallel to the x axis, the axis will be parallel to the y axis, and the equation will have the form (4).

Because the vertex V is halfway between the directrix and the focus, V has coordinates $(-3, 4)$. The directed distance from the vertex to the focus is p; so

$$p = 7 - 4 = 3$$

Therefore an equation is

$$(x + 3)^2 = 12(y - 4)$$

Squaring and simplifying we have

$$x^2 + 6x - 12y + 57 = 0$$

A sketch of the graph of this parabola is shown in Fig. 13.2.5.

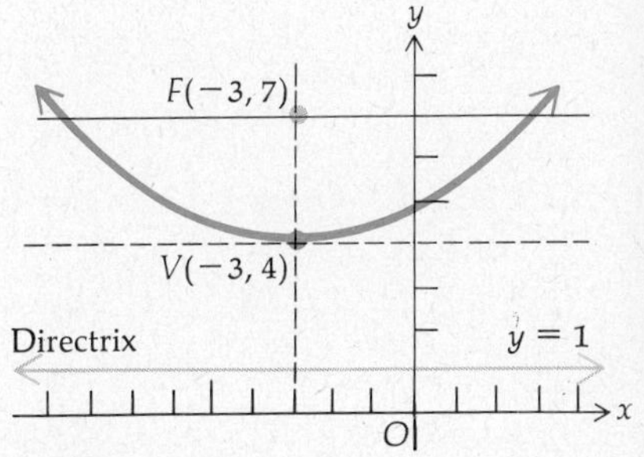

FIGURE 13.2.5

Example 4 Given the parabola having the equation $y^2 + 6x + 8y + 1 = 0$, find the vertex, the focus, an equation of the directrix, an equation of the axis, and the length of the latus rectum, and draw a sketch of the graph.

Solution Rewrite the given equation as

$$y^2 + 8y = -6x - 1$$

Completing the square of the terms involving y on the left side of this equation by adding 16 on both sides we obtain

$$y^2 + 8y + 16 = -6x + 15$$

$$(y + 4)^2 = -6(x - \tfrac{5}{2})$$

Comparing this equation with (3) we let

$$k = -4 \qquad h = \tfrac{5}{2}$$

and

$$4p = -6 \quad \text{or} \quad p = -\tfrac{3}{2}$$

Therefore the vertex is at $(\frac{5}{2}, -4)$, an equation of the axis is $y = -4$, the focus is at $(1, -4)$, an equation of the directrix is $x = 4$, and the length of the latus rectum is 6. A sketch of the graph is shown in Fig. 13.2.6.

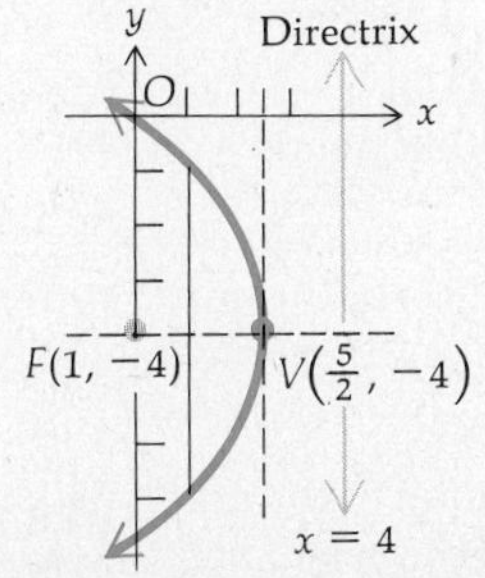

FIGURE 13.2.6

EXERCISES 13.2

In Exercises 1 through 12, find a new equation of the graph of the given equation after a translation of axes to the new origin as indicated. Draw the original and the new axes and a sketch of the graph.

1. $x^2 + y^2 + 6x + 4y = 0; (-3, -2)$

2. $x^2 + y^2 - 10x + 4y + 13 = 0; (5, -2)$

3. $y^2 - 6x + 9 = 0; (\frac{3}{2}, 0)$

4. $y^2 + 3x - 2y + 7 = 0; (-2, 1)$

5. $x^2 + 4y^2 + 4x + 8y + 4 = 0; (-2, -1)$

6. $25x^2 + y^2 - 50x + 20y - 500 = 0; (1, -10)$

7. $y - 4 = 2(x - 1)^3; (1, 4)$

8. $(y + 1)^2 = 4(x - 2)^3; (2, -1)$

9. $y = \ln(x - 5) - 2; (5, -2)$

10. $y = \sin(x + \frac{1}{2}\pi) + 3; (-\frac{1}{2}\pi, 3)$

11. $\cosh(x - 2) = y + 1; (2, -1)$

12. $2e^{2x-2} = y + 2; (1, -2)$

In Exercises 13 through 16, translate the axes so that an equation of the graph with respect to the new axes will contain no first-degree terms. Draw the original and the new axes and a sketch of the graph.

13. $x^2 + 4y^2 - 16x + 24y + 84 = 0$

14. $16x^2 + 25y^2 - 32x - 100y - 284 = 0$

15. $3x^2 - 2y^2 + 6x - 8y - 11 = 0$

16. $x^2 - y^2 + 14x - 8y - 35 = 0$

In Exercises 17 through 22, find the vertex, the focus, an equation of the axis, and an equation of the directrix of the given parabola. Draw a sketch of the graph.

17. $x^2 + 6x + 4y + 8 = 0$

18. $4x^2 - 8x + 3y - 2 = 0$

19. $y^2 + 6x + 10y + 19 = 0$

20. $3y^2 - 8x - 12y - 4 = 0$

21. $2y^2 = 4y - 3x$

22. $y = 3x^2 - 3x + 3$

In Exercises 23 through 32, find an equation of the parabola having the given properties. Draw a sketch of the graph.

23. Vertex at $(2, 4)$; focus at $(-3, 4)$.

24. Vertex at $(1, -3)$; directrix, $y = 1$.

25. Focus at $(-1, 7)$; directrix, $y = 3$.

26. Focus at $(-\frac{3}{4}, 4)$; directrix, $x = -\frac{5}{4}$.

27. Vertex at $(3, -2)$; axis, $x = 3$; length of the latus rectum is 6.

28. Axis parallel to the x axis; through the points $(1, 2)$, $(5, 3)$, and $(11, 4)$.

29. Vertex at $(-4, 2)$; axis, $y = 2$; through the point $(0, 6)$.

30. Directrix, $x = -2$; axis, $y = 4$; length of the latus rectum is 8.

31. Directrix, $x = 4$; axis, $y = 4$; through the point $(9, 7)$.

32. Endpoints of the latus rectum are $(1, 3)$ and $(7, 3)$.

33. Given the parabola having the equation $y = ax^2 + bx + c$, with $a \neq 0$, find the coordinates of the vertex.

34. Find the coordinates of the focus of the parabola in Exercise 25.

35. Find an equation of every parabola containing the points $A(-3, -4)$ and $B(5, -4)$, such that points A and B are each 5 units from the focus.

36. Given the equation $4x^3 - 12x^2 + 12x - 3y - 10 = 0$, translate the axes so that the equation of the graph with respect to the new axes will contain no second-degree term and no constant term. Draw a sketch of the graph and the two sets of axes. (*Hint:* Let $x = x' + h$ and $y = y' + k$ in the given equation.)

37. Given the equation $x^3 + 3x^2 - y^2 + 3x + 4y - 3 = 0$, translate the axes so that the equation of the graph with respect to the new axes will contain no first-degree term and no constant term. Draw a sketch of the graph and the two sets of axes (see hint for Exercise 36).

38. If a parabola has its focus at the origin and the x axis is its axis, prove that it must have an equation of the form $y^2 = 4kx + 4k^2$, $k \neq 0$.

39. If the axes are translated to a new origin having coordinates (h, k), show that after the translation the equation $y = \sin x$ becomes $y' = A \sin x' + B \cos x' + C$, where A, B, and C are constants. Find A, B, and C in terms of h and k. Show that $A^2 + B^2 = 1$.

40. Show that after a translation of axes to the new origin $(-\frac{1}{4}\pi, 1)$ the equation $y = \dfrac{1}{\sqrt{2}}(\sin x + \cos x) + 1$ becomes $y' = \sin x'$.

41. Refer to the results of Exercises 39 and 40 and determine a translation of axes so that the equation

$$y = \frac{1}{\sqrt{2}}(\sin x - \cos x) - 3$$

becomes $y' = \sin x'$.

42. Refer to the results of Exercises 39 and 40 and determine a translation of axes so that the equation

$$y = \tfrac{1}{2}(\sqrt{3} \sin x + \cos x) + 2$$

becomes $y' = \sin x'$.

13.3 SOME PROPERTIES OF CONICS

In Sec. 13.1 we stated that a *conic section* (or *conic*) is a curve of intersection of a plane with a right-circular cone, and three types of curves of intersection that occur are the *parabola*, the *ellipse*, and the *hyperbola*. The Greek mathematician Apollonius studied conic sections, in terms of geometry, by using this concept. We studied the parabola in Sec. 13.1, where an analytic definition (13.1.1) of a parabola was given. In this section an analytic definition of a conic section is given and the three types of curves are obtained as special cases of this definition.

In a consideration of the geometry of conic sections, a cone is regarded as having two nappes, extending indefinitely far in both directions. A portion of a right-circular cone of two nappes is shown in Fig. 13.3.1. A **generator** (or element) of the cone is a line lying in the cone, and all the generators of a cone contain the point V, called the **vertex** of the cone.

In Fig. 13.3.2 we have a cone and a cutting plane that is parallel to

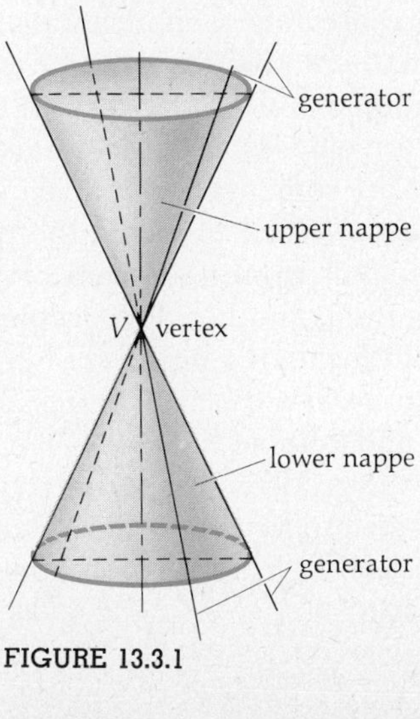

FIGURE 13.3.1

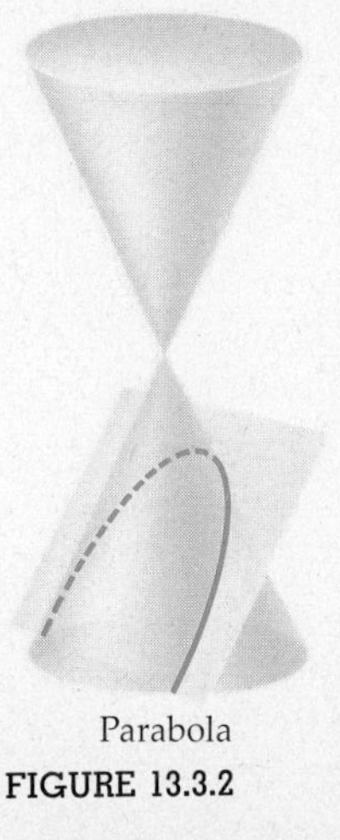

FIGURE 13.3.2

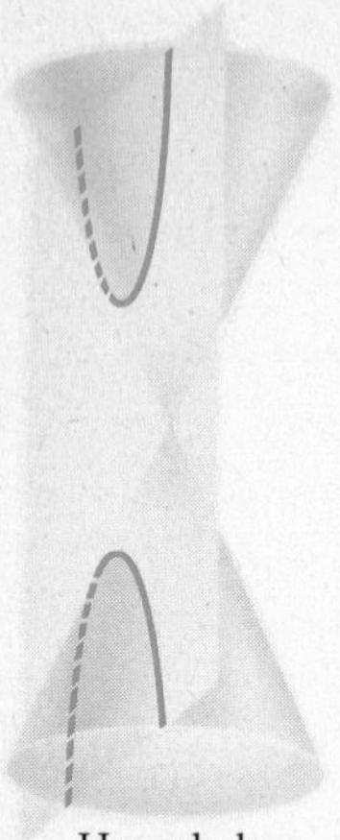

FIGURE 13.3.3

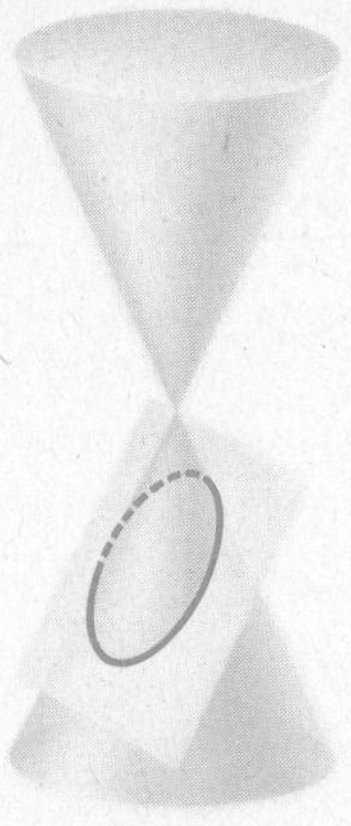

FIGURE 13.3.4

FIGURE 13.3.5

FIGURE 13.3.6

FIGURE 13.3.7

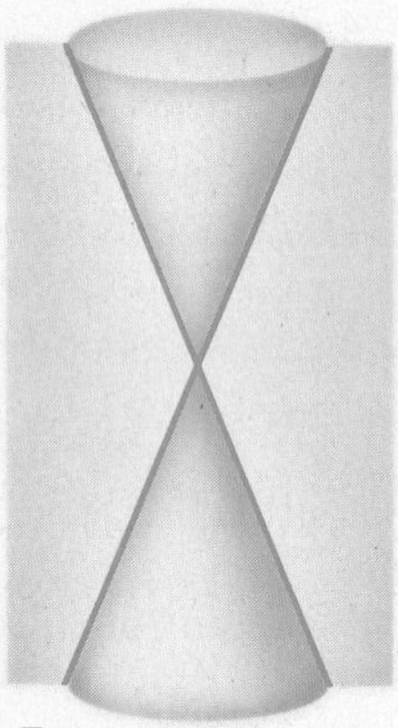

FIGURE 13.3.8

one and only one generator of the cone. This conic is a *parabola.* If the cutting plane is parallel to two generators, it intersects both nappes of the cone and we have a *hyperbola,* shown in Fig. 13.3.3. An *ellipse* is obtained if the cutting plane is parallel to no generator, in which case the cutting plane intersects each generator, as in Fig. 13.3.4.

A special case of the ellipse is a *circle,* shown in Fig. 13.3.5, and it is formed if the cutting plane that intersects each generator is also perpendicular to the axis of the cone. Degenerate cases of the conic sections include a point, a straight line, and two intersecting straight lines. In Fig. 13.3.6 a point is obtained if the cutting plane contains the vertex of the cone but does not contain a generator. This is a degenerate ellipse. If the cutting plane contains the vertex of the cone and only one generator, as in Fig. 13.3.7, then a straight line is obtained, and this is a degenerate parabola. A degenerate hyperbola is formed when the cutting plane contains the vertex of the cone and two generators, thereby giving two intersecting straight lines, as shown in Fig. 13.3.8.

There are many applications of conic sections to both pure and applied mathematics. In Sec. 13.1 we stated some for parabolas, and we now mention a few for ellipses and hyperbolas. The orbits of planets and satellites are ellipses. Ellipses are used in making machine gears. Arches of bridges are sometimes elliptical in shape. In some navigational systems hyperbolas are used to locate the position of oceanic traffic by the use of radio signals from a network of pairs of transmitters at a known distance from one another. Hyperbolas are also used in combat in "sound ranging" to locate the position of enemy guns by the sound of the firing of those guns. If a quantity varies inversely as another quantity, such as pressure and volume in Boyle's law for a perfect gas at a constant temperature, the graph is a hyperbola.

The geometric properties of conics as plane curves form the basis of our discussion. We start with a definition that gives a property common to all conics.

13.3.1 Definition A **conic** is the set of all points P in a plane such that the undirected distance of P from a fixed point is in a constant ratio to the undirected distance of P from a fixed line that does not contain the fixed point.

The constant ratio in the above definition is called the **eccentricity** of the conic and is denoted by e. The eccentricity e is a nonnegative number because it is the ratio of two undirected distances. Actually for nondegenerate conics, $e > 0$. (Later we learn that when $e = 0$, we have a point.) If $e = 1$, we see by comparing Definitions 13.3.1 and 13.1.1 that the conic is a parabola; if $e < 1$, the conic is an ellipse; and if $e > 1$, the conic is a hyperbola.

The fixed point, referred to in Definition 13.3.1, is called a **focus** of the conic, and the fixed line is called the corresponding **directrix.** We learned in Sec. 13.1 that a parabola has one focus and one directrix. Later we see that an ellipse and a hyperbola each have two foci and two directrices, with each focus corresponding to a particular directrix.

The line through a focus of a conic perpendicular to its directrix is called the **principal axis** of the conic. The points of intersection of the conic and its principal axis are called the **vertices** of the conic. From our study of the parabola, we know that a parabola has one vertex. However, both the ellipse and the hyperbola have two vertices. This is proved in the following theorem.

13.3.2 Theorem If e is the eccentricity of a nondegenerate conic, then if $e \neq 1$, the conic has two vertices; if $e = 1$, the conic has only one vertex.

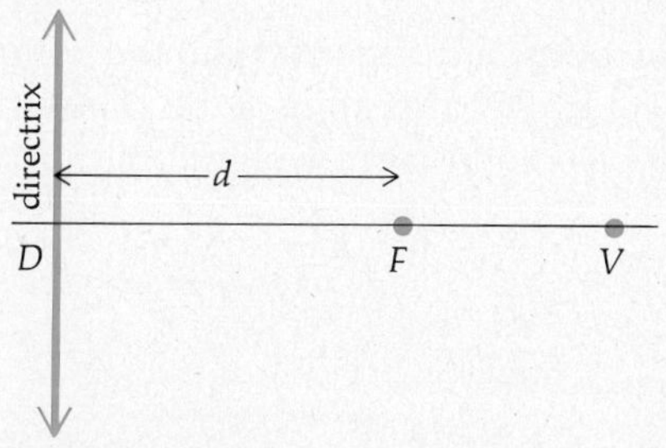

FIGURE 13.3.9

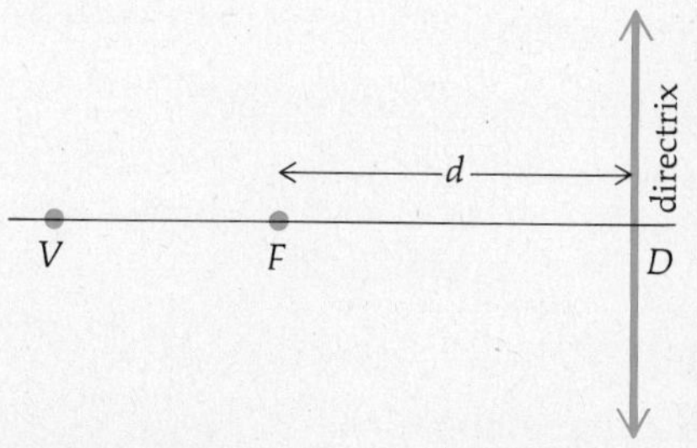

FIGURE 13.3.10

Proof Let F denote the focus of the conic and D denote the point of intersection of the directrix and the principal axis. Let d denote the undirected distance between the focus and its directrix. In Fig. 13.3.9 F is to the right of the directrix, and in Fig. 13.3.10 F is to the left of the directrix. Let V denote a vertex of the conic. We wish to show that if $e \neq 1$, there are two possible vertices V, and if $e = 1$, there is only one vertex V.

From Definition 13.3.1,

$$|\overline{FV}| = e|\overline{DV}|$$

Removing the absolute-value bars we obtain

$$\overline{FV} = \pm e(\overline{DV}) \tag{1}$$

Because D, F, and V are all on the principal axis,

$$\overline{DV} = \overline{DF} + \overline{FV}$$

If F is to the right of D, $\overline{DF} = d$. Thus, from the above,

$$\overline{DV} = d + \overline{FV} \tag{2}$$

Substituting from (1) into (2) gives

$$\overline{DV} = d \pm e(\overline{DV})$$

from which we get

$$\overline{DV} = \frac{d}{1 \pm e} \tag{3}$$

If $e = 1$, the minus sign in the above equations must be rejected because it involves dividing by zero. So if $e \neq 1$, two points V are obtained; if $e = 1$, we obtain only one point V.

If F is to the left of D, $\overline{DF} = -d$. Instead of (3) we get

$$\overline{DV} = \frac{-d}{1 \mp e}$$

from which the same conclusion follows. So the theorem is proved. ■

The point on the principal axis of an ellipse or a hyperbola that lies halfway between the two vertices is called the **center** of the conic. Hence the ellipse and the hyperbola are called **central conics** in contrast to the parabola, which has no center because it has only one vertex.

13.3.3 Theorem A conic is symmetric with respect to its principal axis.

The proof of this theorem is left as an exercise (see Exercise 18).

Example 1 Use Definition 13.3.1 to find an equation of the conic whose eccentricity is $\frac{4}{5}$ and having a focus at the point $F(1, -2)$ with the line $4y + 17 = 0$ as the corresponding directrix.

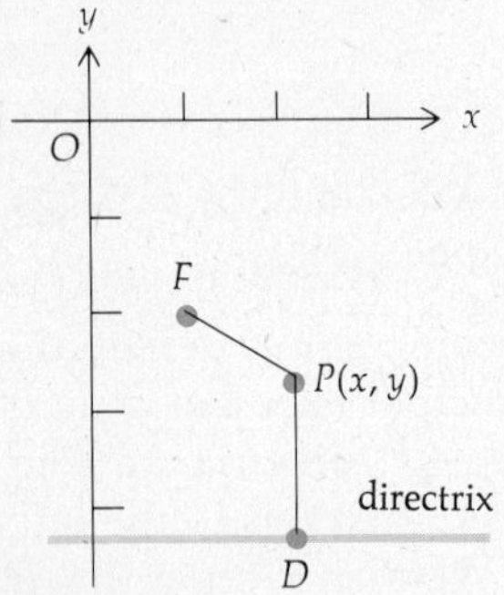

FIGURE 13.3.11

Solution Figure 13.3.11 shows the focus, the directrix, and the point $P(x, y)$ representing any point on the conic. Let the point D be the foot of the perpendicular from P to the directrix, and from Definition 13.3.1,

$$\frac{|\overline{PF}|}{|\overline{PD}|} = \frac{4}{5}$$

$$|\overline{PF}| = \tfrac{4}{5}|\overline{PD}|$$

Using the distance formula to find $|\overline{PF}|$ and $|\overline{PD}|$, and substituting into the above equation, we get

$$\sqrt{(x-1)^2 + (y+2)^2} = \tfrac{4}{5}|y + \tfrac{17}{4}|$$

Squaring on both sides of this equation and simplifying we get

$$x^2 - 2x + 1 + y^2 + 4y + 4 = \tfrac{16}{25}(y^2 + \tfrac{17}{2}y + \tfrac{289}{16})$$

$$25x^2 - 50x + 25y^2 + 100y + 125 = 16y^2 + 136y + 289$$

$$25x^2 + 9y^2 - 50x - 36y = 164$$

Because $e = \frac{4}{5} < 1$, the equation is that of an ellipse.

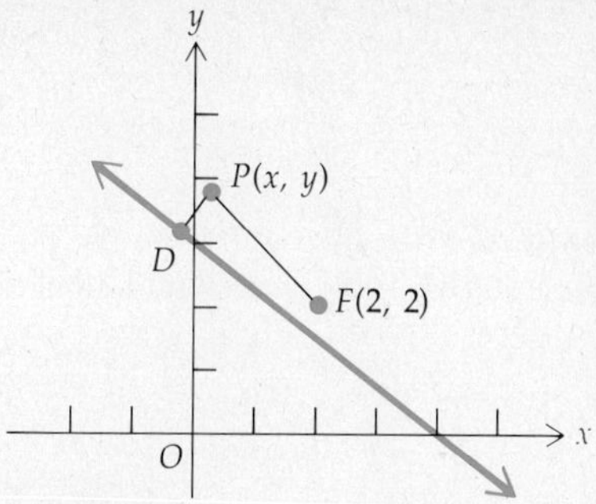

FIGURE 13.3.12

Example 2 Use Definition 13.3.1 and the formula of Exercise 22 of Exercises 4.7 to find an equation of the conic whose eccentricity is 3 and having a focus at (2, 2) with the line $3x + 4y = 12$ as the corresponding directrix.

Solution The focus, the directrix, and the point $P(x, y)$ representing any point on the conic are shown in Fig. 13.3.12. The point D in the figure is the foot of the perpendicular from P to the directrix.

From Exercise 22 of Exercises 4.7, the distance from the point $P_1(x_1, y_1)$ to the line having the equation $Ax + By + C = 0$ is

$$\frac{|Ax_1 + By_1 + C|}{\sqrt{A^2 + B^2}}$$

Therefore the distance from $P(x, y)$ to the directrix, denoted by $|\overline{PD}|$, is given by

$$|\overline{PD}| = \frac{|3x + 4y - 12|}{\sqrt{3^2 + 4^2}} = \tfrac{3}{5}x + \tfrac{4}{5}y - \tfrac{12}{5}$$

From Definition 13.3.1,

$$|\overline{PF}| = 3|\overline{PD}|$$

Thus

$$\sqrt{(x-2)^2 + (y-2)^2} = 3|\tfrac{3}{5}x + \tfrac{4}{5}y - \tfrac{12}{5}|$$

$$x^2 - 4x + 4 + y^2 - 4y + 4 = \tfrac{9}{25}(9x^2 + 24xy + 16y^2 - 72x - 96y + 144)$$

$$25x^2 + 25y^2 - 100x - 100y + 400 = 81x^2 + 216xy + 144y^2 - 648x - 864y + 1296$$

$$56x^2 + 216xy + 119y^2 - 548x - 764y + 896 = 0$$

This is an equation of a hyperbola because $e = 3 > 1$.

EXERCISES 13.3

In Exercises 1 through 12, use Definition 13.3.1 and the formula of Exercise 22 of Exercises 4.7 to find an equation of the conic having the given focus, corresponding directrix, and eccentricity. Identify the conic as a parabola, an ellipse, or a hyperbola.

1. Focus at (2, 0); directrix, $x = -4$; $e = \frac{1}{2}$.

2. Focus at (0, −4); directrix, $y = -2$; $e = \frac{3}{4}$.

3. Focus at (0, −3); directrix, $y = 3$; $e = 2$.

4. Focus at $(\frac{3}{2}, 0)$; directrix, $x = 0$; $e = \frac{4}{3}$.

5. Focus at (−3, 2); directrix, $x = 1$; $e = 3$.

6. Focus at (1, 3); directrix, $y = 8$; $e = \frac{3}{2}$.

7. Focus at $(-4, -\frac{1}{2})$; directrix, $y = -2$; $e = 1$.

8. Focus at (2, −5); directrix, $x = -\frac{5}{3}$; $e = 1$.

9. Focus at (1, 1); directrix, $3x + y = 0$; $e = 5$.

10. Focus at (2, −3); directrix, $x + y = 0$; $e = \frac{1}{4}$.

11. Focus at (4, −3); directrix, $2x - y - 2 = 0$; $e = \frac{1}{2}$.

12. Focus at (−1, 4); directrix, $2x - y + 3 = 0$; $e = 2$.

In Exercises 13 through 17, find the eccentricity of the conic having the given focus and corresponding directrix if the given point P is on the conic.

13. Focus at (3, 0); directrix, $3x = 25$; $P = (0, 4)$.

14. Focus at (2, 0); directrix, $x = -4$; $P = (8, 0)$.

15. Focus at $(4, -2)$; directrix, $3x - 4y = 4$; $P = (2, -2)$.

16. Focus at $(2, -3)$; directrix, $2x - 2y - 1 = 0$; $P = (1, -2)$.

17. Focus at $(-1, 0)$; directrix, $x + y + 1 = 0$; $P = (2, 3)$.

18. Prove Theorem 13.3.3.

19. Find an equation whose graph consists of all points P in a plane such that the undirected distance of P from the point $(ae, 0)$ is in a constant ratio e to the undirected distance of P from the line $x = a/e$. Let $a^2(1 - e^2) = \pm b^2$ and consider the three cases: $e = 1$, $e < 1$, and $e > 1$.

20. Solve Exercise 19 if the point is $(-ae, 0)$ and the line is $x = -a/e$.

13.4 POLAR EQUATIONS OF THE CONICS

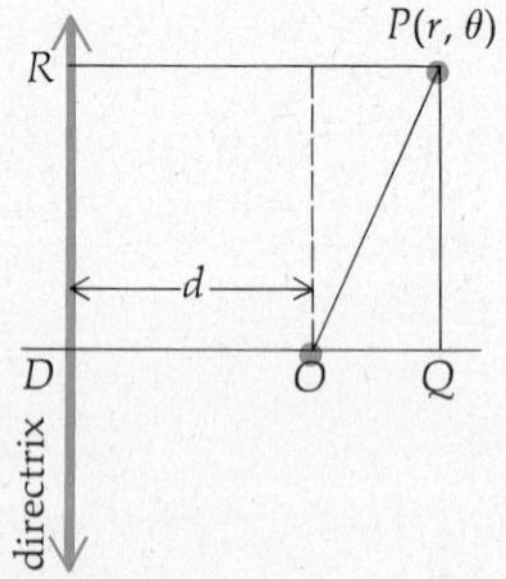

FIGURE 13.4.1

Fairly simple polar equations of conics are obtained by taking the pole at a focus and the polar axis and its extension along the principal axis. We first consider this situation when the directrix corresponding to the focus at the pole is to the left of the focus. Let D be the point of intersection of this directrix with the principal axis, and let d denote the undirected distance between a focus and its directrix (refer to Fig. 13.4.1). Let $P(r, \theta)$ be any point on the conic to the right of the directrix and on the terminal side of the angle of measure θ. Draw perpendiculars PQ and PR to the principal axis and the directrix, respectively. By Definition 13.3.1, the point P is on the conic if and only if

$$|\overline{OP}| = e|\overline{RP}| \tag{1}$$

Because P is to the right of the directrix, $\overline{RP} > 0$; thus $|\overline{RP}| = \overline{RP}$. $|\overline{OP}| = r$ because $r > 0$. So from (1),

$$r = e(\overline{RP}) \tag{2}$$

However, $\overline{RP} = \overline{DQ} = \overline{DO} + \overline{OQ} = d + r \cos \theta$. Substituting this expression for $\overline{RP}$ in (2) we get

$$r = e(d + r \cos \theta)$$

from which we obtain

$$r = \frac{ed}{1 - e \cos \theta} \tag{3}$$

In a similar manner we can derive an equation of a conic if the directrix corresponding to the focus at the pole is to the right of the focus, and we obtain

$$r = \frac{ed}{1 + e \cos \theta} \tag{4}$$

The derivation of (4) is left as an exercise (see Exercise 25).

If a focus of a conic is at the pole and the polar axis is parallel to the corresponding directrix, then the $\frac{1}{2}\pi$ axis and its extension are along the principal axis. We obtain the equation

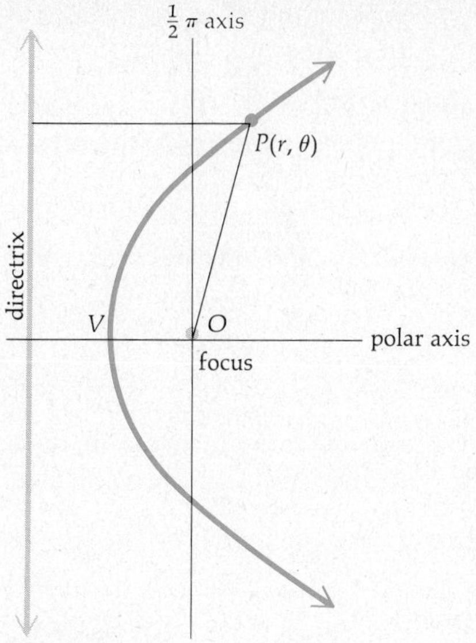

FIGURE 13.4.2

$$r = \frac{ed}{1 \pm e \sin \theta} \tag{5}$$

where e and d are, respectively, the eccentricity and the undirected distance between the focus and the corresponding directrix. The plus sign is taken when the directrix corresponding to the focus at the pole is above the focus, and the minus sign is taken when it is below the focus. The derivations of (5) are left as exercises (see Exercises 26 and 27).

We now discuss the graph of Eq. (3) for each of the three cases: $e = 1$, $e < 1$, and $e > 1$. Similar discussions apply to the graphs of Eqs. (4) and (5).

Case 1: $e = 1$. The conic is a parabola.

Equation (3) becomes

$$r = \frac{d}{1 - \cos \theta} \tag{6}$$

r is not defined when $\theta = 0$; however, r is defined for all other values of θ for which $0 < \theta < 2\pi$. Differentiating in (6) we obtain

$$D_\theta r = \frac{-d \sin \theta}{(1 - \cos \theta)^2}$$

Setting $D_\theta r = 0$, for values of θ in the interval $(0, 2\pi)$ we obtain $\theta = \pi$. Because when $0 < \theta < \pi$, $D_\theta r < 0$, and when $\pi < \theta < 2\pi$, $D_\theta r > 0$, r has an absolute minimum value at $\theta = \pi$. When $\theta = \pi$, $r = \frac{1}{2}d$, and the point $(\frac{1}{2}d, \pi)$ is the vertex. Therefore the undirected distance from the focus to a point on the parabola is shortest when the point on the parabola is the vertex.

By Theorem 13.3.3, the parabola is symmetric with respect to its principal axis; by Theorem 13.3.2, there is one vertex. Note that the curve does not contain the pole because there is no value of θ that gives a value of 0 for r. A sketch of the parabola is shown in Fig. 13.4.2.

Example 1 A parabola has its focus at the pole and its vertex at $(4, \pi)$. Find an equation of the parabola and an equation of the directrix. Draw a sketch of the parabola and the directrix.

Solution Because the focus is at the pole and the vertex is at $(4, \pi)$, the polar axis and its extension are along the principal axis of the parabola. Furthermore, the vertex is to the left of the focus; so the directrix is also to the left of the focus. Hence an equation of the parabola is of the form of (3). Because the vertex is at $(4, \pi)$, $\frac{1}{2}d = 4$; thus $d = 8$. The eccentricity $e = 1$, and therefore we obtain the equation

$$r = \frac{8}{1 - \cos \theta}$$

An equation of the directrix is given by $r \cos \theta = -d$, and because $d = 8$, then $r \cos \theta = -8$. Figure 13.4.3 shows a sketch of the parabola and the directrix.

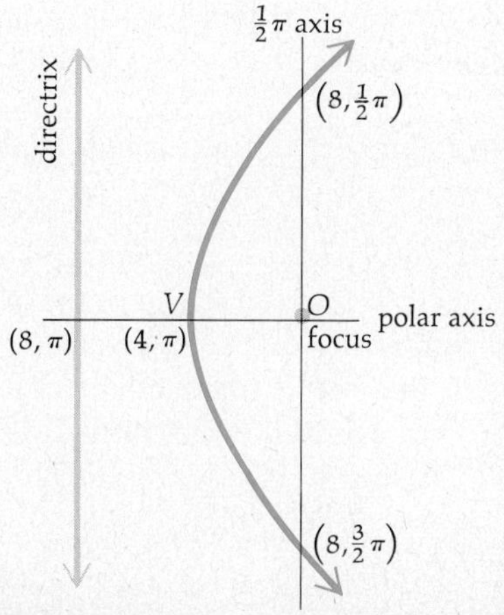

FIGURE 13.4.3

Example 2 Follow the instructions of Example 1 if the parabola has its focus at the pole and its vertex at $(3, \frac{3}{2}\pi)$.

Solution The vertex of the parabola is below the focus, and the $\frac{1}{2}\pi$ axis and its extension are along the principal axis. Therefore the directrix is also below the focus, and an equation of the parabola is of the form of (5) with the minus sign. The undirected distance from the focus to the vertex is $\frac{1}{2}d = 3$; thus $d = 6$. Therefore, from (5), an equation of the parabola is

$$r = \frac{6}{1 - \sin\theta}$$

An equation of the directrix is given by $r \sin\theta = -d$. Because $d = 6$, then $r \sin\theta = -6$. Sketches of the parabola and the directrix are shown in Fig. 13.4.4.

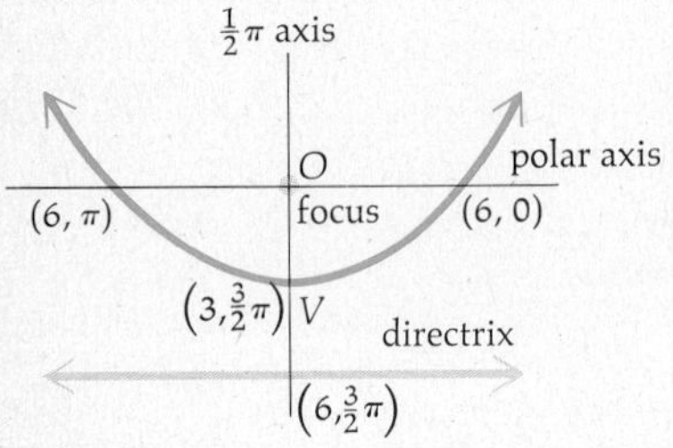

FIGURE 13.4.4

Case 2: $e < 1$. The conic is an ellipse.

When $e < 1$, the denominator of the fraction in (3) is never zero; so r exists for all values of θ. To find the absolute extrema of r on the interval $[0, 2\pi)$, we find $D_\theta r$ and obtain

$$D_\theta r = \frac{-e^2 d \sin\theta}{(1 - e\cos\theta)^2} \tag{7}$$

For θ in the interval $[0, 2\pi)$, $D_\theta r = 0$ when $\theta = 0$ and π. When $-\frac{1}{2}\pi < \theta < 0$, $D_\theta r > 0$; and when $0 < \theta < \frac{1}{2}\pi$, $D_\theta r < 0$. Hence, when $\theta = 0$, r has an absolute maximum value of $ed/(1 - e)$. When $\frac{1}{2}\pi < \theta < \pi$, $D_\theta r < 0$, and when $\pi < \theta < \frac{3}{2}\pi$, $D_\theta r > 0$. Thus r has an absolute minimum value of $ed/(1 + e)$ when $\theta = \pi$. The points $(ed/(1 - e), 0)$ and $(ed/(1 + e), \pi)$ are the vertices of the ellipse. We denote them by V_1 and V_2, respectively. It follows, then, that the undirected distance from the focus $(0, 0)$ to a point on the ellipse is greatest when the point on the ellipse is V_1 and smallest when it is V_2.

By Theorem 13.3.3, the ellipse is symmetric with respect to its principal axis. There is no value of θ that gives a value of 0 for r; so the curve does not contain the pole. A sketch of the ellipse is shown in Fig. 13.4.5.

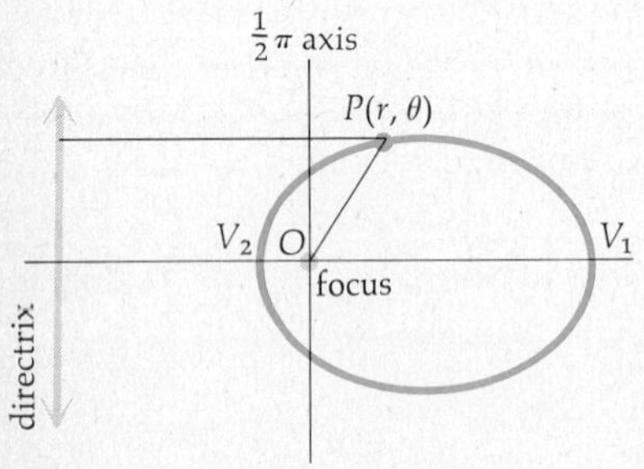

FIGURE 13.4.5

Example 3 Find an equation of the ellipse having a focus at the pole and vertices at $(5, 0)$ and $(2, \pi)$. Write an equation of the directrix corresponding to the focus at the pole. Draw a sketch of the ellipse.

Solution Because the vertices are at $(5, 0)$ and $(2, \pi)$, the polar axis and its extension are along the principal axis of the ellipse. The directrix corresponding to the focus at the pole is to the left of the focus because the vertex closest to this focus is at $(2, \pi)$. The required equation is therefore of the form of (3). The vertex V_1 is at $(5, 0)$ and V_2 is at $(2, \pi)$. It follows from the discussion of Case 2 that

$$\frac{ed}{1 - e} = 5 \quad \text{and} \quad \frac{ed}{1 + e} = 2$$

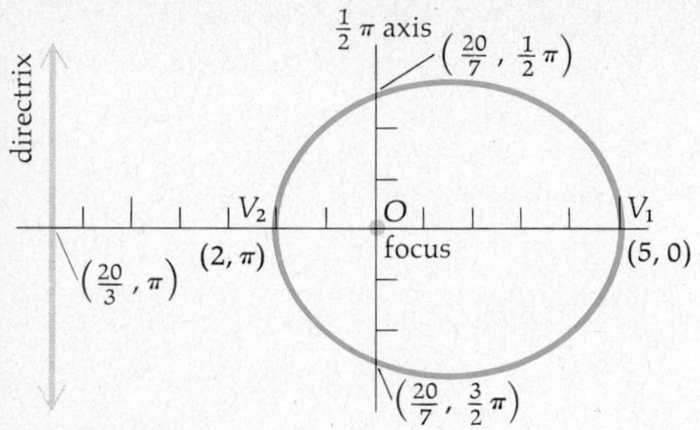

FIGURE 13.4.6

Solving these two equations simultaneously we obtain $e = \frac{3}{7}$ and $d = \frac{20}{3}$. Hence, from (3) an equation of the ellipse is

$$r = \frac{\frac{3}{7} \cdot \frac{20}{3}}{1 - \frac{3}{7}\cos\theta}$$

$$r = \frac{20}{7 - 3\cos\theta}$$

Because $d = \frac{20}{3}$, an equation of the directrix corresponding to the focus at the pole is $r\cos\theta = -\frac{20}{3}$ or, equivalently, $3r\cos\theta = -20$. A sketch of the ellipse is shown in Fig. 13.4.6.

Example 4 An equation of a conic is

$$r = \frac{5}{3 + 2\sin\theta}$$

Find the eccentricity, identify the conic, write an equation of the directrix corresponding to the focus at the pole, find the vertices, and draw a sketch of the curve.

Solution Dividing the numerator and denominator of the fraction in the given equation by 3 we obtain

$$r = \frac{\frac{5}{3}}{1 + \frac{2}{3}\sin\theta}$$

which is of the form of (5) with the plus sign. The eccentricity $e = \frac{2}{3} < 1$; so the conic is an ellipse. Because $ed = \frac{5}{3}$, $d = \frac{5}{3} \div \frac{2}{3} = \frac{5}{2}$. The $\frac{1}{2}\pi$ axis and its extension are along the principal axis. The directrix corresponding to the focus at the pole is above the focus, and an equation of it is $r\sin\theta = \frac{5}{2}$. When $\theta = \frac{1}{2}\pi$, $r = 1$. And when $\theta = \frac{3}{2}\pi$, $r = 5$. The vertices are therefore at $(1, \frac{1}{2}\pi)$ and $(5, \frac{3}{2}\pi)$. A sketch of the ellipse is shown in Fig. 13.4.7.

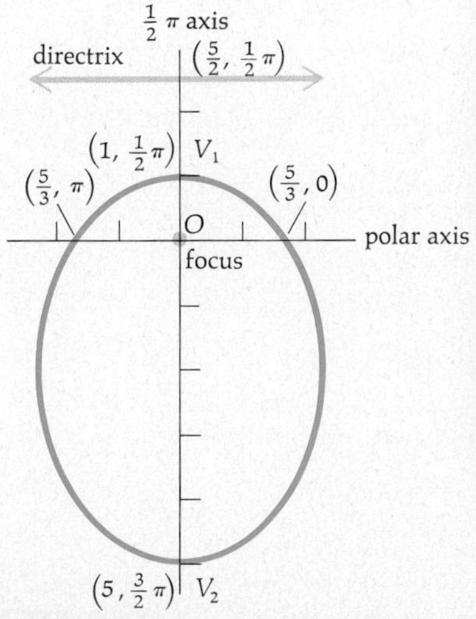

FIGURE 13.4.7

Case 3: $e > 1$. The conic is a hyperbola.

In the derivation of (3) we assumed that the point P on the conic is to the right of the given directrix and that P is on the terminal side of the angle of θ radians, thus making r positive. In the previous two cases, when the conic is an ellipse or a parabola, we obtain the entire curve with these assumptions. However, when $e > 1$, we shall see that we obtain only one of two branches of the hyperbola when $r > 0$, and this branch is to the right of the directrix. There is also a branch to the left of the given directrix; this is obtained by letting r assume negative as well as positive values as θ takes on values in the interval $[0, 2\pi)$.

If $e > 1$ in (3), r is undefined when $\cos\theta = 1/e$. Two values of θ in $[0, 2\pi)$ satisfy this equation. They are

$$\theta_1 = \cos^{-1}\frac{1}{e} \quad \text{and} \quad \theta_2 = 2\pi - \cos^{-1}\frac{1}{e}$$

We investigate the values of r as θ increases from 0 to θ_1. When $\theta = 0$, $r = ed/(1 - e)$, and because $e > 1, r < 0$. So the point $(ed/(1 - e), 0)$ is the left vertex of the hyperbola. The cosine function is decreasing in the interval $(0, \theta_1)$; therefore, when $0 < \theta < \theta_1$, $\cos\theta > \cos\theta_1 = 1/e$ or, equivalently, $1 - e\cos\theta < 0$. Hence, for values of θ in the interval $(0, \theta_1)$, $r = ed/(1 - e\cos\theta) < 0$. When $\theta = \theta_1$, r is undefined; however,

$$\lim_{\theta \to \theta_1^-} r = \lim_{\theta \to \theta_1^-} \frac{ed}{1 - e\cos\theta} = -\infty$$

We conclude, then, that as θ increases from 0 to θ_1, $r < 0$ and $|r|$ increases without bound. From these values of θ we obtain the lower half of the left branch of the curve. See Fig. 13.4.8. We show later that a hyperbola has two asymptotes. For this hyperbola one of the asymptotes is parallel to the line $\theta = \theta_1$.

Now we consider the points on the hyperbola as θ increases through values between θ_1 and θ_2. When $\theta_1 < \theta < \theta_2$, $\cos\theta < \cos\theta_1 = 1/e$ or, equivalently, $1 - e\cos\theta > 0$. Therefore, for values of θ in the interval (θ_1, θ_2), $r > 0$.

$$\lim_{\theta \to \theta_1^+} r = \lim_{\theta \to \theta_1^+} \frac{ed}{1 - e\cos\theta} = +\infty$$

When $\theta = \pi$, $r = ed/(1 + e)$; so the point $(ed/(1 + e), \pi)$ is the right vertex of the hyperbola. Finding $D_\theta r$ from (3) we get

$$D_\theta r = -\frac{e^2 d \sin\theta}{(1 - e\cos\theta)^2} \tag{8}$$

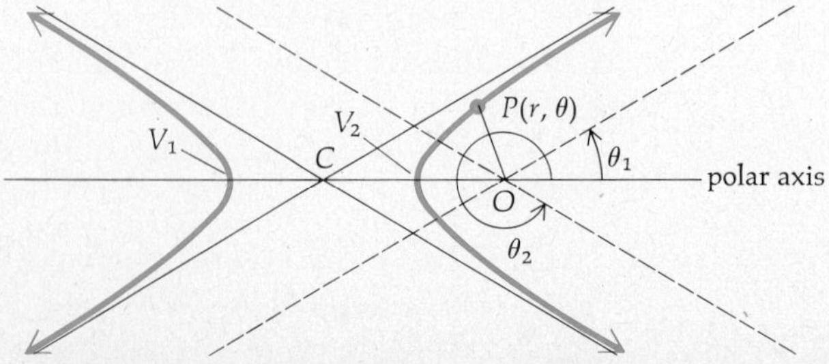

FIGURE 13.4.8

When $\theta_1 < \theta < \pi$, $\sin\theta > 0$; hence $D_\theta r < 0$, from which we conclude that r is decreasing when θ is in the interval (θ_1, π). Therefore, for values of θ in the interval (θ_1, π) we obtain the top half of the right branch of the hyperbola shown in Fig. 13.4.8.

When $\pi < \theta < \theta_2$, $\sin\theta < 0$; so from (8), $D_\theta r > 0$. Hence r is increasing when θ is in the interval (π, θ_2). Furthermore,

$$\lim_{\theta\to\theta_2^-} r = \lim_{\theta\to\theta_2^-} \frac{ed}{1 - e\cos\theta} = +\infty$$

Consequently, for values of θ in the interval (π, θ_2) we have the lower half of the right branch of the hyperbola. The other asymptote of the hyperbola is parallel to the line $\theta = \theta_2$.

The cosine function is increasing in the interval $(\theta_2, 2\pi)$. So when $\theta_2 < \theta < 2\pi$, $\cos\theta > \cos\theta_2 = 1/e$ or, equivalently, $1 - e\cos\theta < 0$. Thus, from (3), for values of θ in the interval $(\theta_2, 2\pi)$, $r < 0$. Also,

$$\lim_{\theta\to\theta_2^+} r = \lim_{\theta\to\theta_2^+} \frac{ed}{1 - e\cos\theta} = -\infty$$

It follows from (8) that $D_\theta r > 0$; hence r is increasing when $\theta_2 < \theta < 2\pi$. When $r < 0$ and r is increasing, $|r|$ is decreasing. Thus, for values of θ in the interval $(\theta_2, 2\pi)$ we obtain the upper half of the left branch of the hyperbola; so the curve of Fig. 13.4.8 is complete.

Example 5 The polar axis and its extension are along the principal axis of a hyperbola having a focus at the pole. The corresponding directrix is to the left of the focus. If the hyperbola contains the point $(1, \frac{2}{3}\pi)$ and $e = 2$, find (a) an equation of the hyperbola; (b) the vertices; (c) the center; (d) an equation of the directrix corresponding to the focus at the pole. Draw a sketch of the hyperbola.

Solution An equation of the hyperbola is of the form of (3) with $e = 2$. We have, then,

$$r = \frac{2d}{1 - 2\cos\theta}$$

Because the point $(1, \frac{2}{3}\pi)$ lies on the hyperbola, its coordinates satisfy the equation. Therefore

$$1 = \frac{2d}{1 - 2(-\frac{1}{2})}$$

from which we obtain $d = 1$. Hence an equation of the hyperbola is

$$r = \frac{2}{1 - 2\cos\theta} \tag{9}$$

The vertices are the points on the hyperbola for which $\theta = 0$ and $\theta = \pi$. From (9), when $\theta = 0$, $r = -2$; and when $\theta = \pi$, $r = \frac{2}{3}$. Consequently, the left vertex V_1 is at the point $(-2, 0)$, and the right vertex V_2 is at the point $(\frac{2}{3}, \pi)$.

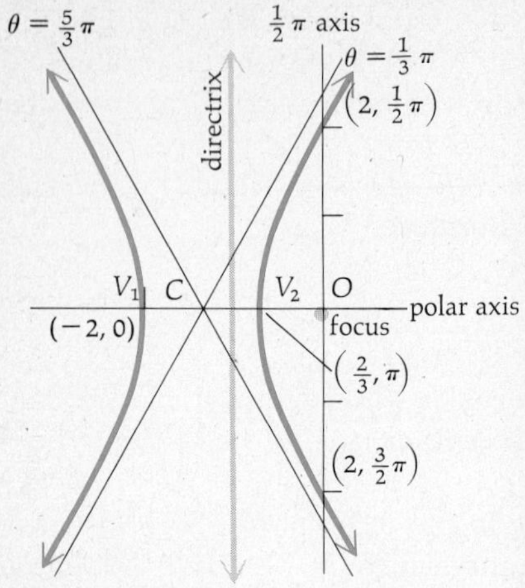

FIGURE 13.4.9

The center of the hyperbola is the point on the principal axis halfway between the two vertices. This is the point $(\frac{4}{3}, \pi)$.

An equation of the directrix corresponding to the focus at the pole is given by $r \cos \theta = -d$. Because $d = 1$, this equation is $r \cos \theta = -1$.

As an aid in drawing a sketch of the hyperbola, we first draw the two asymptotes. In the discussion of Case 3 we stated that these are lines through the center of the hyperbola that are parallel to the lines $\theta = \theta_1$ and $\theta = \theta_2$, where θ_1 and θ_2 are the values of θ in the interval $[0, 2\pi)$ for which r is not defined. From (9), r is not defined when $1 - 2 \cos \theta = 0$. Therefore $\theta_1 = \frac{1}{3}\pi$ and $\theta_2 = \frac{5}{3}\pi$. Figure 13.4.9 shows a sketch of the hyperbola as well as the two asymptotes, and the directrix corresponding to the focus at the pole.

Example 6 Find a polar equation of the hyperbola having the line $r \cos \theta = 4$ as the directrix corresponding to a focus at the pole, and $e = \frac{3}{2}$.

Solution The given directrix is perpendicular to the polar axis and is four units to the right of the focus at the pole. Therefore an equation of the hyperbola is of the form of (4), which is

$$r = \frac{ed}{1 + e \cos \theta}$$

Because $d = 4$ and $e = \frac{3}{2}$, the equation becomes

$$r = \frac{\frac{3}{2} \cdot 4}{1 + \frac{3}{2} \cos \theta}$$

$$r = \frac{12}{2 + 3 \cos \theta}$$

EXERCISES 13.4

In Exercises 1 and 2, each polar equation represents a conic having a focus at the pole. Identify the conic.

1. (a) $r = \dfrac{3}{1 - \cos \theta}$; (b) $r = \dfrac{6}{4 + 5 \sin \theta}$; (c) $r = \dfrac{5}{4 - \cos \theta}$; (d) $r = \dfrac{4}{1 + \sin \pi}$

2. (a) $r = \dfrac{1}{1 - \sin \theta}$; (b) $r = \dfrac{2}{3 + \sin \theta}$; (c) $r = \dfrac{3}{2 + 4 \cos \theta}$; (d) $r = \dfrac{5}{1 - \cos \pi}$

In Exercises 3 through 14, the equation is that of a conic having a focus at the pole. In each exercise, (a) find the eccentricity; (b) identify the conic; (c) write an equation of the directrix that corresponds to the focus at the pole; (d) draw a sketch of the curve.

3. $r = \dfrac{2}{1 - \cos \theta}$

4. $r = \dfrac{4}{1 + \cos \theta}$

5. $r = \dfrac{5}{2 + \sin \theta}$

6. $r = \dfrac{4}{1 - 3 \cos \theta}$

7. $r = \dfrac{6}{3 - 2 \cos \theta}$

8. $r = \dfrac{1}{2 + \sin \theta}$

9. $r = \dfrac{9}{5 - 6 \sin \theta}$

10. $r = \dfrac{1}{1 - 2 \sin \theta}$

11. $r = \dfrac{10}{7 - 2 \sin \theta}$

12. $r = \dfrac{7}{3 + 4 \cos \theta}$

13. $r = \dfrac{10}{4 + 5 \cos \theta}$

14. $r = \dfrac{1}{5 - 3 \sin \theta}$

In Exercises 15 through 20, find a polar equation of the conic having a focus at the pole and satisfying the given conditions.

15. Parabola; vertex at $(4, \frac{3}{2}\pi)$. **16.** Ellipse; $e = \frac{1}{2}$; a vertex at $(4, \pi)$.

17. Hyperbola; $e = \frac{4}{3}$; $r \cos \theta = 9$ is the directrix corresponding to the focus at the pole.

18. Hyperbola; vertices at $(1, \frac{1}{2}\pi)$ and $(3, \frac{1}{2}\pi)$. **19.** Ellipse; vertices at $(3, 0)$ and $(1, \pi)$.

20. Parabola; vertex at $(6, \frac{1}{2}\pi)$.

21. (a) Find a polar equation of the hyperbola having a focus at the pole and the corresponding directrix to the left of the focus if the point $(2, \frac{4}{3}\pi)$ is on the hyperbola and $e = 3$. (b) Write an equation of the directrix that corresponds to the focus at the pole.

22. (a) Find a polar equation of the hyperbola for which $e = 3$ and which has the line $r \sin \theta = 3$ as the directrix corresponding to a focus at the pole. (b) Find the polar equations of the two lines through the pole that are parallel to the asymptotes of the hyperbola.

23. Find the area of the region inside the ellipse $r = 6/(2 - \sin \theta)$ and above the parabola $r = 3/(1 + \sin \theta)$.

24. For the ellipse and parabola of Exercise 23, find the area of the region inside the ellipse and below the parabola.

25. Show that an equation of a conic having its principal axis along the polar axis and its extension, a focus at the pole, and the corresponding directrix to the right of the focus is $r = ed/(1 + e \cos \theta)$.

26. Show that an equation of a conic having its principal axis along the $\frac{1}{2}\pi$ axis and its extension, a focus at the pole, and the corresponding directrix above the focus is $r = ed/(1 + e \sin \theta)$.

27. Show that an equation of a conic having its principal axis along the $\frac{1}{2}\pi$ axis and its extension, a focus at the pole, and the corresponding directrix below the focus is $r = ed/(1 - e \sin \theta)$.

28. Show that the equation $r = k \csc^2 \frac{1}{2}\theta$, where k is a constant, is a polar equation of a parabola.

29. A comet is moving in a parabolic orbit around the sun at the focus of the parabola. When the comet is 80 million miles from the sun the line segment from the sun to the comet makes an angle of $\frac{1}{3}\pi$ radians with the axis of the orbit. (a) Find an equation of the comet's orbit. (b) How close does the comet come to the sun?

30. The orbit of a planet is in the form of an ellipse having the equation $r = p/(1 + e \cos \theta)$ where the pole is at the sun. Find the average measure of the distance of the planet from the sun with respect to θ.

31. Use Definition 13.3.1 to find a polar equation of a central conic for which the center is at the pole, the principal axis is along the polar axis and its extension, and the distance from the pole to a directrix is a/e.

32. Show that the tangent lines at the points of intersection of the parabolas $r = a/(1 + \cos \theta)$ and $r = b/(1 - \cos \theta)$ are perpendicular.

13.5 CARTESIAN EQUATIONS OF THE CONICS

In Sec. 13.4 we learned that a polar equation of a conic is

$$r = \frac{ed}{1 - e \cos \theta}$$

or, equivalently,

$$r = e(r \cos \theta + d) \tag{1}$$

A cartesian representation of (1) can be obtained by replacing r by $\pm\sqrt{x^2 + y^2}$ and $r \cos \theta$ by x. Making these substitutions we get

$$\pm\sqrt{x^2 + y^2} = e(x + d)$$

Squaring on both sides of the above equation gives

$$x^2 + y^2 = e^2x^2 + 2e^2dx + e^2d^2$$

$$y^2 + x^2(1 - e^2) = 2e^2dx + e^2d^2 \tag{2}$$

If $e = 1$, (2) becomes

$$y^2 = 2d(x + \tfrac{1}{2}d) \tag{3}$$

If we translate the origin to the point $(-\frac{1}{2}d, 0)$ by replacing x by $\bar{x} - \frac{1}{2}d$ and y by $\bar{y}$, (3) becomes

$$\bar{y}^2 = 2d\bar{x} \tag{4}$$

Equation (4) resembles (1) of Sec. 13.1,

$$y^2 = 4px$$

which is an equation of the parabola having its focus at $(p, 0)$ and its directrix the line $x = -p$. Therefore, relative to the $\bar{x}$ and $\bar{y}$ axes, (4) is an equation of a parabola having its focus at $(\frac{1}{2}d, 0)$, which is the origin relative to the x and y axes. The parabola having its axis parallel to one of the coordinate axes was discussed in detail earlier in Secs. 13.1 and 13.2.

If $e \neq 1$, we divide on both sides of (2) by $1 - e^2$ and obtain

$$x^2 - \frac{2e^2d}{1 - e^2}x + \frac{1}{1 - e^2}y^2 = \frac{e^2d^2}{1 - e^2}$$

Completing the square for the terms involving x by adding $e^4d^2/(1 - e^2)^2$ on both sides of the above equation, we get

$$\left(x - \frac{e^2d}{1 - e^2}\right)^2 + \frac{1}{1 - e^2}y^2 = \frac{e^2d^2}{(1 - e^2)^2} \tag{5}$$

If the origin is translated to the point $(e^2d/(1 - e^2), 0)$, (5) becomes

$$\bar{x}^2 + \frac{1}{1 - e^2}\bar{y}^2 = \frac{e^2d^2}{(1 - e^2)^2}$$

$$\frac{\bar{x}^2}{\dfrac{e^2d^2}{(1 - e^2)^2}} + \frac{\bar{y}^2}{\dfrac{e^2d^2}{1 - e^2}} = 1 \tag{6}$$

Now let $e^2d^2/(1 - e^2)^2 = a^2$, where $a > 0$. Then

$$a = \begin{cases} \dfrac{ed}{1 - e^2} & \text{if } 0 < e < 1 \\[2ex] \dfrac{ed}{e^2 - 1} & \text{if } e > 1 \end{cases} \tag{7}$$

Then (6) can be written as

$$\frac{\bar{x}^2}{a^2} + \frac{\bar{y}^2}{a^2(1 - e^2)} = 1$$

Replacing $\bar{x}$ and $\bar{y}$ by x and y we get

$$\frac{x^2}{a^2} + \frac{y^2}{a^2(1 - e^2)} = 1 \tag{8}$$

Equation (8) is a standard form of a cartesian equation of a central conic having its principal axis on the x axis. Because (5) is an equation of a conic having a focus at the origin (pole), and (8) is obtained from (5) by translating the origin to the point $(e^2d/(1 - e^2), 0)$, it follows that the central conic having Eq. (8) has a focus at the point $(-e^2d/(1 - e^2), 0)$. However, from (7) it follows that

$$\frac{-e^2d}{1 - e^2} = \begin{cases} -ae & \text{if } 0 < e < 1 \\ ae & \text{if } e > 1 \end{cases}$$

Therefore we conclude that if (8) is an equation of an ellipse $(0 < e < 1)$, there is a focus at the point $(-ae, 0)$. If (8) is an equation of a hyperbola $(e > 1)$, there is a focus at $(ae, 0)$. In each case the corresponding directrix is d units to the left of the focus. So if the graph of (8) is an ellipse, the directrix corresponding to the focus at $(-ae, 0)$ has as an equation

$$x = -ae - d$$

Because when $0 < e < 1$, $d = a(1 - e^2)/e$, this equation becomes

$$x = -ae - \frac{a(1 - e^2)}{e}$$

$$x = -\frac{a}{e}$$

Similarly, if the graph of (8) is a hyperbola, the directrix corresponding to the focus at $(ae, 0)$ has as an equation

$$x = ae - d$$

When $e > 1$, $d = a(e^2 - 1)/e$; thus the above equation of the directrix can be written as

$$x = \frac{a}{e}$$

Hence we have shown that if (8) is an equation of an ellipse, a focus and its corresponding directrix are $(-ae, 0)$ and $x = -a/e$; and if (8) is an equation of a hyperbola, a focus and its corresponding directrix are $(ae, 0)$ and $x = a/e$.

Because (8) contains only even powers of x and y, its graph is symmetric with respect to both the x and y axes. Therefore, if there is a focus at $(-ae, 0)$ having a corresponding directrix of $x = -a/e$, by symmetry there is also a focus at $(ae, 0)$ having a corresponding directrix of $x = a/e$. Similarly, for a focus at $(ae, 0)$ and a corresponding directrix of $x = a/e$, there is also a focus at $(-ae, 0)$ and a corresponding directrix of $x = -a/e$. These results are summarized in the following theorem.

13.5.1 Theorem The central conic having as an equation

$$\frac{x^2}{a^2} + \frac{y^2}{a^2(1 - e^2)} = 1 \tag{9}$$

where $a > 0$, has a focus at $(-ae, 0)$, whose corresponding directrix is $x = -a/e$, and a focus at $(ae, 0)$, whose corresponding directrix is $x = a/e$.

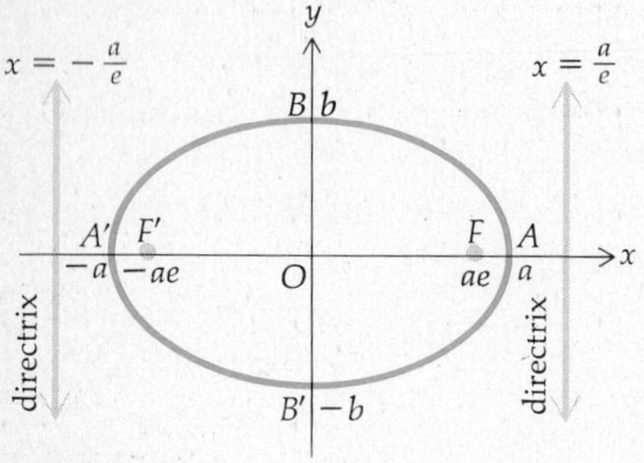

FIGURE 13.5.1

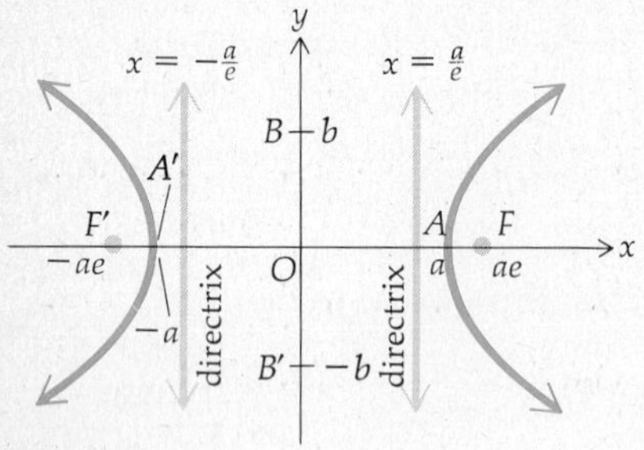

FIGURE 13.5.2

Figures 13.5.1 and 13.5.2 show sketches of the graph of (9) together with the foci and directrices in the respective cases of an ellipse and a hyperbola.

Suppose that in Eq. (9) $0 < e < 1$. Then the conic is an ellipse. Because in this case $1 - e^2 > 0$, we conclude that $\sqrt{1 - e^2}$ is a real number. Therefore we let

$$b = a\sqrt{1 - e^2} \tag{10}$$

Because $a > 0$, it follows that $b > 0$. Substituting from (10) into (9) we obtain

$$\frac{x^2}{a^2} + \frac{y^2}{b^2} = 1 \tag{11}$$

Equation (11) is an equation of an ellipse having its principal axis on the x axis. Because the vertices are the points of intersection of the ellipse with its principal axis, they are at $(-a, 0)$ and $(a, 0)$. The center of the ellipse is at the origin because it is the point midway between the vertices. Figure 13.5.1 shows a sketch of this ellipse. Refer to this figure as you read the next two paragraphs.

If we denote the vertices $(-a, 0)$ and $(a, 0)$ by A' and A, respectively, the segment $A'A$ of the principal axis is called the **major axis** of the ellipse, and its length is $2a$ units. Then we state that a is the number of units in the length of the semimajor axis of the ellipse.

The ellipse having Eq. (11) intersects the y axis at the points $(-b, 0)$ and $(b, 0)$, which we denote by B' and B, respectively. The segment $B'B$ is called the **minor axis** of the ellipse. Its length is $2b$ units. Hence b is the number of units in the length of the semiminor axis of the ellipse. Because a and b are positive numbers and $0 < e < 1$, it follows from (10) that $b < a$.

Example 1 Given the ellipse having the equation

$$\frac{x^2}{25} + \frac{y^2}{16} = 1$$

find the vertices, foci, directrices, eccentricity, and extremities of the minor axis. Draw a sketch of the ellipse, and show the foci and directrices.

Solution From the equation of the ellipse, $a^2 = 25$ and $b^2 = 16$; thus $a = 5$ and $b = 4$. Therefore the vertices are at the points $A'(-5, 0)$

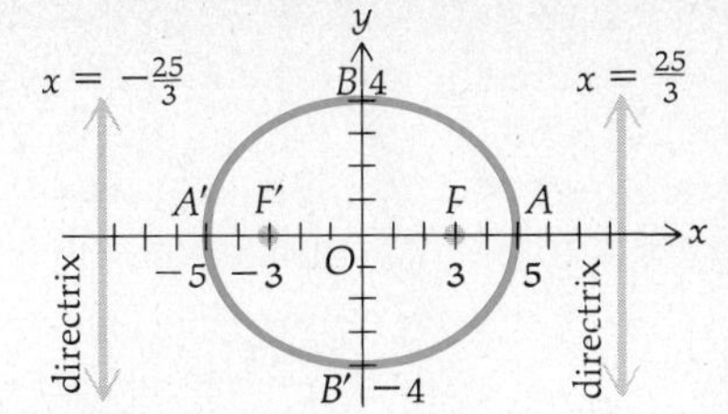

FIGURE 13.5.3

and $A(5, 0)$, and the extremities of the minor axis are at the points $B'(0, -4)$ and $B(0, 4)$. From (10), $4 = 5\sqrt{1 - e^2}$, from which it follows that $e = \frac{3}{5}$. The foci and the directrices are obtained by applying Theorem 13.5.1. Because $ae = 3$ and $a/e = \frac{25}{3}$, we conclude that one focus F is at $(3, 0)$ and its corresponding directrix has the equation $x = \frac{25}{3}$; the other focus F' is at $(-3, 0)$ and its corresponding directrix has the equation $x = -\frac{25}{3}$. A sketch of the ellipse, the foci F and F', and the directrices are in Fig. 13.5.3. From the definition of a conic (13.3.1) it follows that if P is any point on this ellipse, the ratio of the undirected distance of P from a focus to the undirected distance of P from the corresponding directrix is equal to the eccentricity $\frac{3}{5}$.

Example 2 An arch is in the form of a semiellipse. It is 48 ft wide at the base and has a height of 20 ft. How wide is the arch at a height of 10 ft above the base?

Solution Figure 13.5.4 shows a sketch of the arch and the coordinate axes chosen so that the x axis is along the base and the origin is at the midpoint of the base. Then the ellipse has its principal axis on the x axis, its center at the origin, $a = 24$, and $b = 20$. An equation of the ellipse is of the form of (11):

$$\frac{x^2}{576} + \frac{y^2}{400} = 1$$

Let $2\bar{x}$ be the number of feet in the width of the arch at a height of 10 ft above the base. Therefore the point $(\bar{x}, 10)$ lies on the ellipse. Thus

$$\frac{\bar{x}^2}{576} + \frac{100}{400} = 1$$

and so

$$\bar{x}^2 = 432$$

$$\bar{x} = 12\sqrt{3}$$

Hence, at a height of 10 ft above the base the width of the arch is $24\sqrt{3}$ ft.

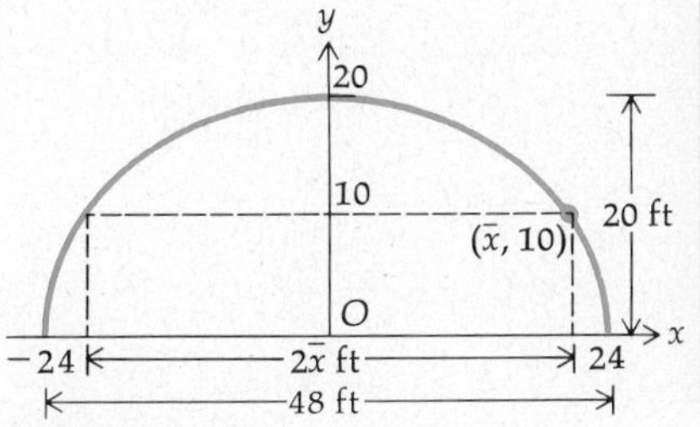

FIGURE 13.5.4

Consider now the central conic of Theorem 13.5.1 when $e > 1$, that is, when the conic is a hyperbola. Because $e > 1$, $1 - e^2 < 0$; hence $e^2 - 1 > 0$. Therefore, for a hyperbola let

$$b = a\sqrt{e^2 - 1} \tag{12}$$

It follows from (12) that b is a positive number. Also, from (12), if $1 < e < \sqrt{2}$, $b < a$; if $e = \sqrt{2}$, $b = a$; and if $e > \sqrt{2}$, $b > a$. Substituting from (12) into (9) we get

$$\frac{x^2}{a^2} - \frac{y^2}{b^2} = 1 \tag{13}$$

Equation (13) is an equation of a hyperbola having its principal axis on the x axis, its vertices at the points $(-a, 0)$ and $(a, 0)$, and its center at the origin. A sketch of this hyperbola is in Fig. 13.5.2. If we denote the vertices $(-a, 0)$ and $(a, 0)$ by A' and A, the segment $A'A$ is called the **transverse axis** of the hyperbola. The length of the transverse axis is $2a$ units; so a is the number of units in the length of the semitransverse axis.

Substituting 0 for x in (13) we obtain $y^2 = -b^2$, which has no real roots. Consequently, the hyperbola does not intersect the y axis. However, the line segment having extremities at the points $(0, -b)$ and $(0, b)$ is called the **conjugate axis** of the hyperbola, and its length is $2b$ units. Hence b is the number of units in the length of the semiconjugate axis.

Solving (13) for y in terms of x we obtain

$$y = \pm\frac{b}{a}\sqrt{x^2 - a^2} \tag{14}$$

We conclude from (14) that if $|x| < a$, there is no real value of y. Therefore there are no points (x, y) on the hyperbola for which $-a < x < a$. We also see from (14) that if $|x| \geq a$, then y has two real values. As previously learned, the hyperbola has two branches. One branch contains the vertex $A(a, 0)$ and extends indefinitely far to the right of A. The other branch contains the vertex $A'(-a, 0)$ and extends indefinitely far to the left of A'.

Example 3 Given the hyperbola

$$\frac{x^2}{9} - \frac{y^2}{16} = 1$$

find the vertices, foci, directrices, eccentricity, and lengths of the transverse and conjugate axes. Draw a sketch of the hyperbola, and show the foci and the directrices.

Solution The given equation is of the form of (13); thus $a = 3$ and $b = 4$. The vertices are therefore at the points $A'(-3, 0)$ and $A(3, 0)$. The number of units in the length of the transverse axis is $2a = 6$, and the number of units in the length of the conjugate axis is $2b = 8$. Because from (12) $b = a\sqrt{e^2 - 1}$, we have $4 = 3\sqrt{e^2 - 1}$; so $e = \frac{5}{3}$. Therefore $ae = 5$ and $a/e = \frac{9}{5}$. Hence the foci are at $F'(-5, 0)$ and $F(5, 0)$. The corresponding directrices have, respectively, the equations $x = -\frac{9}{5}$ and $x = \frac{9}{5}$. A sketch of the hyperbola and the foci and directrices are in Fig. 13.5.5. From Definition 13.3.1 it follows that if P is any point on this hyperbola, the ratio of the undirected distance of P from a focus to the undirected distance of P from the corresponding directrix is equal to the eccentricity $\frac{5}{3}$.

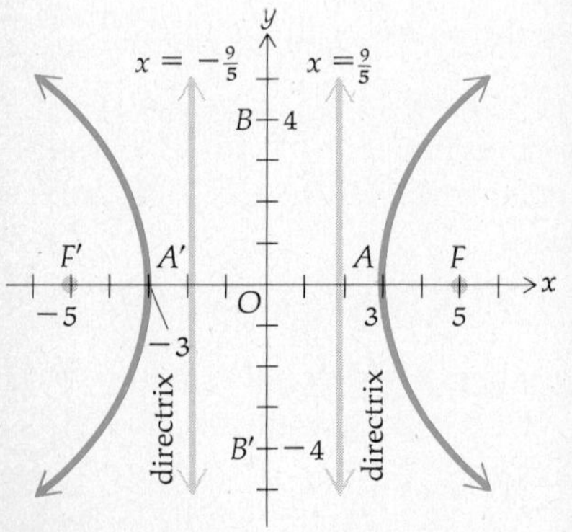

FIGURE 13.5.5

Example 4 Find an equation of the hyperbola having the ends of its conjugate axis at $(0, -2)$ and $(0, 2)$ and an eccentricity equal to $\frac{3}{5}\sqrt{5}$.

Solution Because the ends of the conjugate axis are at $(0, -2)$ and $(0, 2)$, then $b = 2$, the transverse axis is along the x axis, and the center is at the origin. Hence an equation is of the form

$$\frac{x^2}{a^2} - \frac{y^2}{b^2} = 1$$

To find a we use the equation $b = a\sqrt{e^2 - 1}$ and let $b = 2$ and $e = \frac{3}{5}\sqrt{5}$, thereby yielding $2 = a\sqrt{\frac{9}{5} - 1}$. Therefore $a = \sqrt{5}$, and an equation of the hyperbola is

$$\frac{x^2}{5} - \frac{y^2}{4} = 1$$

EXERCISES 13.5

In Exercises 1 through 4, find the vertices, foci, directrices, eccentricity, and ends of the minor axis of the given ellipse. Draw a sketch of the curve, and show the foci and the directrices.

1. $4x^2 + 9y^2 = 36$ **2.** $4x^2 + 9y^2 = 4$ **3.** $2x^2 + 3y^2 = 18$ **4.** $3x^2 + 4y^2 = 9$

In Exercises 5 through 9, find the vertices, foci, directrices, eccentricity, and lengths of the transverse and conjugate axes of the given hyperbola. Draw a sketch of the curve, and show the foci and the directrices.

5. $9x^2 - 4y^2 = 36$ **6.** $x^2 - 9y^2 = 9$ **7.** $9x^2 - 16y^2 = 1$ **8.** $25x^2 - 25y^2 = 1$

In Exercises 9 through 14, find an equation of the given conic satisfying the given conditions, and draw a sketch of the graph.

9. Hyperbola having vertices at $(-2, 0)$ and $(2, 0)$ and a conjugate axis of length 6.

10. Ellipse having foci at $(-5, 0)$ and $(5, 0)$ and one directrix the line $x = -20$.

11. Ellipse having vertices at $(-\frac{5}{2}, 0)$ and $(\frac{5}{2}, 0)$ and one focus at $(\frac{3}{2}, 0)$.

12. Hyperbola having the ends of its conjugate axis at $(0, -3)$ and $(0, 3)$ and $e = 2$.

13. Hyperbola having its center at the origin, its foci on the x axis, and passing through the points $(4, -2)$ and $(7, -6)$.

14. Ellipse having its center at the origin, its foci on the x axis, the length of the major axis equal to three times the length of the minor axis, and passing through the point $(3, 3)$.

15. Find an equation of the tangent line to the ellipse $4x^2 + 9y^2 = 72$ at the point $(3, 2)$.

16. Find an equation of the normal line to the hyperbola $4x^2 - 3y^2 = 24$ at the point $(3, 2)$.

17. Find an equation of the tangent line to the hyperbola $k^2x^2 - y^2 = -1$ at the point $(1/k, \sqrt{2})$, where $k \neq 0$.

18. (a) Find a cartesian equation of the parabola having the polar equation $r = 4/(1 + \sin\theta)$. (b) Find the cartesian coordinates of the vertex and the focus.

19. Find an equation of the hyperbola whose foci are the vertices of the ellipse $7x^2 + 11y^2 = 77$ and whose vertices are the foci of this ellipse.

20. Find an equation of the ellipse whose foci are the vertices of the hyperbola $11x^2 - 7y^2 = 77$ and whose vertices are the foci of this hyperbola.

21. The ceiling in a hallway 10 m wide is in the shape of a semiellipse and is 9 m high in the center and 6 m high at the side walls. Find the height of the ceiling 2 m from either wall.

22. A football is 12 in. long, and a plane section containing a seam is an ellipse, of which the length of the minor axis is 7 in. Find the volume of the football if the leather is so stiff that every cross section is a square.

23. Solve Exercise 22 if every cross section is a circle.

24. The orbit of the earth around the sun is elliptical in shape with the sun at one focus, a semimajor axis of length 92.9 million miles, and an eccentricity of 0.017. Find (a) how close the earth gets to the sun and (b) the greatest possible distance between the earth and the sun.

25. The cost of production of a commodity is \$12 less per unit at a point A than it is at a point B, and the distance between A and B is 100 km. Assuming that the route of delivery of the commodity is along a straight line, and that the delivery cost is 20 cents per unit per kilometer, find the curve at any point of which the commodity can be supplied from either A or B at the same total cost. (*Hint:* Take points A and B at $(-50, 0)$ and $(50, 0)$, respectively.)

26. Prove that there is no tangent line to the hyperbola $x^2 - y^2 = 1$ that passes through the origin.

27. Find the volume of the solid of revolution generated by revolving the region bounded by the hyperbola $x^2/a^2 - y^2/b^2 = 1$ and the line $x = 2a$ about the x axis.

28. Find the centroid of the solid of revolution of Exercise 27.

29. A tank has a horizontal axis of length 20 ft and its ends are semiellipses. The width across the top of the tank is 10 ft and the depth is 6 ft. If the tank is full of water, how much work is necessary to pump all the water to the top of the tank?

30. Suppose that the orbit of a planet is in the shape of an ellipse with an eccentricity of 0.8 and a major axis whose length is 500 million km. Find an equation of the orbit.

13.6 THE ELLIPSE

In Sec. 13.5 we considered ellipses for which the center is at the origin and the principal axis is on the x axis. Now more general equations of ellipses will be discussed.

If an ellipse has its center at the origin and its principal axis on the y axis, then an equation of the ellipse is of the form

$$\frac{y^2}{a^2} + \frac{x^2}{b^2} = 1 \tag{1}$$

which is obtained from (11) of Sec. 13.5 by interchanging x and y.

● ILLUSTRATION 1 Because for an ellipse $a > b$, it follows that the ellipse having the equation

$$\frac{x^2}{16} + \frac{y^2}{25} = 1$$

has its foci on the y axis. This ellipse has the same shape as the ellipse of Example 1 in Sec. 13.5. The vertices are at $(0, -5)$ and $(0, 5)$, and the foci are at $(0, -3)$ and $(0, 3)$, and their corresponding directrices have the equations $y = -\frac{25}{3}$ and $y = \frac{25}{3}$, respectively. ●

If the center of an ellipse is at the point (h, k) rather than at the origin, and if the principal axis is parallel to one of the coordinate axes, then by a translation of axes so that the point (h, k) is the new origin, an equation of the ellipse is $\bar{x}^2/a^2 + \bar{y}^2/b^2 = 1$ if the principal axis is horizontal, and $\bar{y}^2/a^2 + \bar{x}^2/b^2 = 1$ if the principal axis is vertical. Because $\bar{x} = x - h$ and $\bar{y} = y - k$, these equations become the following in x and y:

$$\frac{(x-h)^2}{a^2} + \frac{(y-k)^2}{b^2} = 1 \tag{2}$$

if the principal axis is horizontal, and

$$\frac{(y-k)^2}{a^2} + \frac{(x-h)^2}{b^2} = 1 \tag{3}$$

if the principal axis is vertical.

In Sec. 1.3 we discussed the general equation of the second degree in two variables:

$$Ax^2 + Bxy + Cy^2 + Dx + Ey + F = 0 \tag{4}$$

where $B = 0$ and $A = C$. In such a case, the graph of (4) is either a circle or a degenerate case of a circle, which is either a point-circle or the empty set. If we eliminate the fractions and combine terms in (2) and (3), we obtain an equation of the form

$$Ax^2 + Cy^2 + Dx + Ey + F = 0 \tag{5}$$

where $A \neq C$ if $a \neq b$ and $AC > 0$. It can be shown by completing the squares in x and y that an equation of the form (5) can be put in the form

$$\frac{(x-h)^2}{\frac{1}{A}} + \frac{(y-k)^2}{\frac{1}{C}} = G \tag{6}$$

If $AC > 0$, then A and C have the same sign. If G has the same sign as A and C, then (6) can be written in the form of (2) or (3). Thus the graph of (5) is an ellipse.

● ILLUSTRATION 2 Suppose we have the equation

$$6x^2 + 9y^2 - 24x - 54y + 51 = 0$$

which can be written as

$$6(x^2 - 4x) + 9(y^2 - 6y) = -51$$

Completing the squares in x and y we get

$$6(x^2 - 4x + 4) + 9(y^2 - 6y + 9) = -51 + 24 + 81$$

$$6(x-2)^2 + 9(y-3)^2 = 54$$

$$\frac{(x-2)^2}{\frac{1}{6}} + \frac{(y-3)^2}{\frac{1}{9}} = 54$$

This is an equation of the form of (6). By dividing on both sides by 54 we have

$$\frac{(x-2)^2}{9} + \frac{(y-3)^2}{6} = 1$$

which has the form of (2). ●

If in (6) G has a sign opposite to that of A and C, then (6) is not satisfied by any real values of x and y. Hence the graph of (5) is the empty set.

● ILLUSTRATION 3 Suppose that (5) is

$$6x^2 + 9y^2 - 24x - 54y + 115 = 0$$

Then, upon completing the squares in x and y, we get

$$6(x - 2)^2 + 9(y - 3)^2 = -115 + 24 + 81$$

$$\frac{(x-2)^2}{\frac{1}{6}} + \frac{(y-3)^2}{\frac{1}{9}} = -10 \qquad (7)$$

This is of the form of (6), where $G = -10$, $A = 6$, and $C = 9$. For all values of x and y the left side of (7) is nonnegative; hence the graph of (7) is the empty set. ●

If $G = 0$ in (6), then the equation is satisfied by only the point (h, k). Therefore the graph of (5) is a single point, which is called a **point-ellipse.**

● ILLUSTRATION 4 Because the equation

$$6x^2 + 9y^2 - 24x - 54y + 105 = 0$$

can be written as

$$\frac{(x-2)^2}{\frac{1}{6}} + \frac{(y-3)^2}{\frac{1}{9}} = 0$$

its graph is the point (2, 3). ●

If the graph of (5) is a point-ellipse or the empty set, the graph is said to be degenerate.

If $A = C$ in (5), we have either a circle or a degenerate circle, as mentioned above. A circle is a limiting form of an ellipse. This can be shown by considering the formula relating a, b, and e for an ellipse:

$$b^2 = a^2(1 - e^2)$$

From this formula it is seen that as e approaches zero, b^2 approaches a^2. If $b^2 = a^2$, (2) and (3) become

$$(x - h)^2 + (y - k)^2 = a^2$$

which is an equation of a circle having its center at (h, k) and radius a. The results of Sec. 1.3 for a circle are the same as those obtained for (5) applied to an ellipse.

The results of the above discussion are summarized in the following theorem.

13.6.1 Theorem If in the general second-degree equation (4) $B = 0$ and $AC > 0$, then the graph is either an ellipse, a point-ellipse, or the empty set. In addition, if $A = C$, then the graph is either a circle, a point-circle, or the empty set.

Example 1 Determine the graph of the equation $25x^2 + 16y^2 + 150x - 128y - 1119 = 0$.

Solution From Theorem 13.6.1, because $B = 0$ and $AC = (25)(16) = 400 > 0$, the graph is either an ellipse or is degenerate. Completing the squares in x and y we have

$$25(x^2 + 6x + 9) + 16(y^2 - 8y + 16) = 1119 + 225 + 256$$

$$25(x + 3)^2 + 16(y - 4)^2 = 1600$$

$$\frac{(x + 3)^2}{64} + \frac{(y - 4)^2}{100} = 1 \tag{8}$$

Equation (8) is of the form of (3); so the graph is an ellipse having its principal axis parallel to the y axis and its center at $(-3, 4)$.

Example 2 For the ellipse of Example 1, find the vertices, foci, directrices, eccentricity, and extremities of the minor axis. Draw a sketch of the ellipse, and show the foci and the directrices.

Solution From (8) it follows that $a = 10$ and $b = 8$. Because the center of the ellipse is at $(-3, 4)$ and the principal axis is vertical, the vertices are at the points $V'(-3, -6)$ and $V(-3, 14)$. The extremities of the minor axis are at the points $B'(-11, 4)$ and $B(5, 4)$. Because $b = a\sqrt{1 - e^2}$,

$$8 = 10\sqrt{1 - e^2}$$

and, solving for e, we get $e = \frac{3}{5}$. Consequently, $ae = 6$ and $a/e = \frac{50}{3}$. Therefore the foci are at the points $F'(-3, -2)$ and $F(-3, 10)$. The corresponding directrices have, respectively, the equations $y = -\frac{38}{3}$ and $y = \frac{62}{3}$. A sketch of the ellipse, the foci, and the directrices are in Fig. 13.6.1.

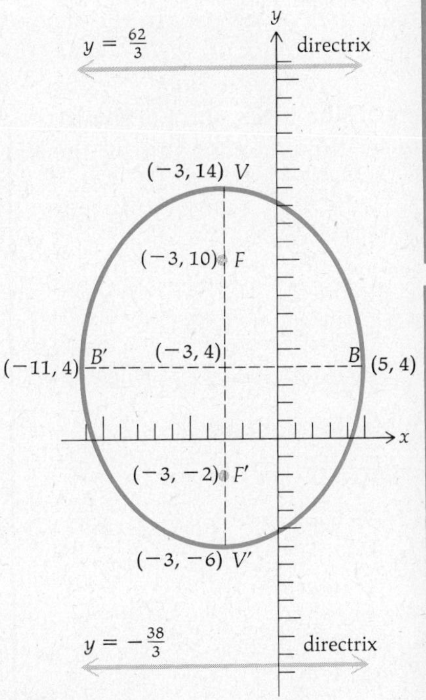

FIGURE 13.6.1

Example 3 Find an equation of the ellipse for which the foci are at $(-8, 2)$ and $(4, 2)$ and the eccentricity is $\frac{2}{3}$. Draw a sketch of the ellipse.

Solution The center of the ellipse is halfway between the foci and is the point $(-2, 2)$. The distance between the foci of any ellipse is $2ae$, and the distance between $(-8, 2)$ and $(4, 2)$ is 12. Therefore we have the equation $2ae = 12$. Replacing e by $\frac{2}{3}$ we get $2a(\frac{2}{3}) = 12$; so $a = 9$. Because $b = a\sqrt{1 - e^2}$,

$$b = 9\sqrt{1 - (\tfrac{2}{3})^2} = 9\sqrt{1 - \tfrac{4}{9}} = 3\sqrt{5}$$

The principal axis is parallel to the x axis; hence an equation of the ellipse is of the form of (2). Because $(h, k) = (-2, 2)$, $a = 9$, and $b = 3\sqrt{5}$, the required equation is

$$\frac{(x + 2)^2}{81} + \frac{(y - 2)^2}{45} = 1$$

A sketch of this ellipse is shown in Fig. 13.6.2.

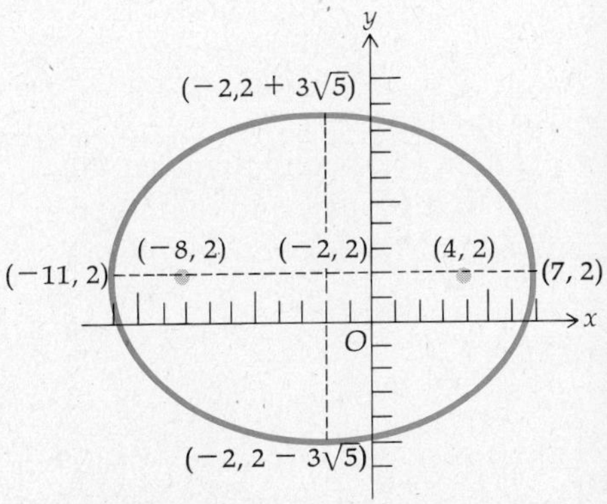

FIGURE 13.6.2

The following theorem gives an alternative definition of an ellipse and is based on a characteristic property of the ellipse.

13.6.2 Theorem An ellipse can be defined as the set of points such that the sum of the distances from any point of the set to two given points (the foci) is a constant.

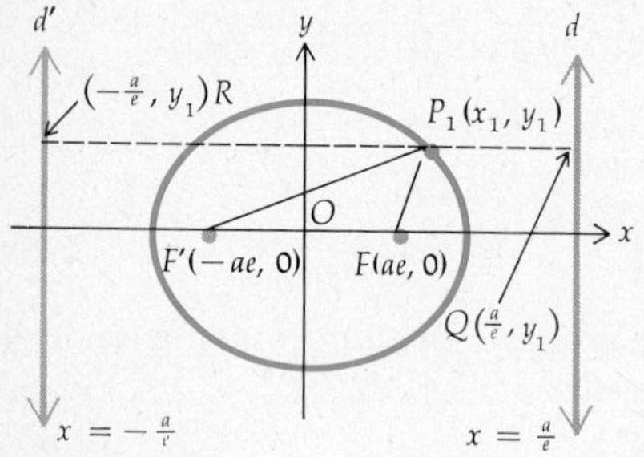

FIGURE 13.6.3

Proof The proof consists of two parts. In the first part we show that the set of points defined in the theorem is an ellipse. In the second part we prove that any ellipse is such a set of points. Refer to Fig. 13.6.3 in both parts of the proof.

Let the point $P_1(x_1, y_1)$ be any point in the given set. Let the foci be the points $F'(-ae, 0)$ and $F(ae, 0)$, and let the constant sum of the distances be $2a$. Then

$$|\overline{FP_1}| + |\overline{F'P_1}| = 2a$$

From the distance formula,

$$\sqrt{(x_1 - ae)^2 + y_1^2} + \sqrt{(x_1 + ae)^2 + y_1^2} = 2a$$

$$\sqrt{(x_1 - ae)^2 + y_1^2} = 2a - \sqrt{(x_1 + ae)^2 + y_1^2}$$

Squaring on both sides of the above equation we obtain

$$x_1^2 - 2aex_1 + a^2e^2 + y_1^2 = 4a^2 - 4a\sqrt{(x_1 + ae)^2 + y_1^2} + x_1^2 + 2aex_1 + a^2e^2 + y_1^2$$

$$\sqrt{(x_1 + ae)^2 + y_1^2} = a + ex_1$$

Squaring on both sides again we get

$$x_1^2 + 2aex_1 + a^2e^2 + y_1^2 = a^2 + 2aex_1 + e^2x_1^2$$

$$(1 - e^2)x_1^2 + y_1^2 = a^2(1 - e^2)$$

$$\frac{x_1^2}{a^2} + \frac{y_1^2}{a^2(1 - e^2)} = 1$$

Because $b^2 = a^2(1 - e^2)$, and replacing x_1 and y_1 by x and y, respectively, we obtain

$$\frac{x^2}{a^2} + \frac{y^2}{b^2} = 1$$

which is the required form of an equation of an ellipse.

Now consider the ellipse in Fig. 13.6.3; $P_1(x_1, y_1)$ is any point on the ellipse. Through P_1 draw a line that is parallel to the x axis and that intersects the directrices d and d' at the points Q and R, respectively. From Definition 13.3.1,

$$|\overline{F'P_1}| = e|\overline{RP_1}| \quad \text{and} \quad |\overline{FP_1}| = e|\overline{P_1Q}|$$

Therefore

$$|\overline{F'P_1}| + |\overline{FP_1}| = e(|\overline{RP_1}| + |\overline{P_1Q}|) \tag{9}$$

$\overline{RP_1}$ and $\overline{P_1Q}$ are both positive because an ellipse lies between its directrices. Hence

$$|\overline{RP_1}| + |\overline{P_1Q}| = \overline{RP_1} + \overline{P_1Q} = \overline{RQ} = \frac{a}{e} - \left(-\frac{a}{e}\right) = \frac{2a}{e} \tag{10}$$

Substituting from (10) into (9) we get

$$|\overline{F'P_1}| + |\overline{FP_1}| = 2a$$

which proves that an ellipse is a set of points as described in the theorem. ■

● ILLUSTRATION 5 The ellipse of Example 1 has foci at $(-3, -2)$ and $(-3, 10)$, and $2a = 20$. Therefore, by Theorem 13.6.2 this ellipse can be defined as the set of points such that the sum of the distances from any point of the set to the points $(-3, -2)$ and $(-3, 10)$ is equal to 20. Similarly, the ellipse of Example 3 can be defined as the set of points such that the sum of the distances from any point of the set to the points $(-8, 2)$ and $(4, 2)$ is equal to 18. ●

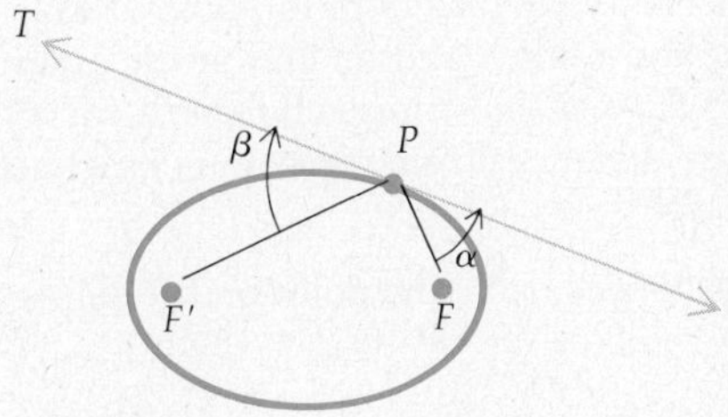

FIGURE 13.6.4

There is a reflective property of the ellipse that is analogous to the one shown for the parabola in Fig. 13.1.7. For the ellipse refer to Fig. 13.6.4, where the line PT is the tangent line at P to the graph of the ellipse having foci at F and F'. The measure of the angle between the line segment FP and the tangent line PT is α, and the measure of the angle between the line segment $F'P$ and the tangent line PT is β. In Exercise 26 you are asked to prove that $\alpha = \beta$. Therefore a light ray from a source at one focus of an elliptical mirror that hits the mirror is reflected along a line through the other focus. This property of ellipses is used in so-called whispering galleries, where the ceilings have cross sections that are arcs of ellipses with common foci. A person located at one focus F can hear another person whispering at the other focus F' because the sound waves originating from the whisperer at F' hit the ceiling and are reflected by the ceiling to the listener at F.

EXERCISES 13.6

In Exercises 1 through 8, find the eccentricity, center, foci, and directrices of the given ellipse, and draw a sketch of the graph.

1. $6x^2 + 9y^2 - 24x - 54y + 51 = 0$ **2.** $9x^2 + 4y^2 - 18x + 16y - 11 = 0$ **3.** $5x^2 + 3y^2 - 3y - 12 = 0$

4. $2x^2 + 2y^2 - 2x + 18y + 33 = 0$ **5.** $4x^2 + 4y^2 + 20x - 32y + 89 = 0$ **6.** $3x^2 + 4y^2 - 30x + 16y + 100 = 0$

7. $3x^2 + 5y^2 - 6x - 12 = 0$ **8.** $2x^2 + 3y^2 - 4x + 12y + 2 = 0$

In Exercises 9 and 10, determine whether the graph of the given equation is an ellipse, a point-ellipse, or the empty set.

9. $4x^2 + y^2 - 8x + 2y + 5 = 0$ **10.** $2x^2 + 3y^2 + 8x - 6y + 20 = 0$

In Exercises 11 through 18, find an equation of the ellipse satisfying the given conditions, and draw a sketch of the graph.

11. Foci at (5, 0) and (−5, 0); one directrix is the line $x = -20$.

12. Vertices at (0, 5) and (0, −5) and through the point $(2, -\frac{5}{3}\sqrt{5})$.

13. Vertices at (2, 0) and (−2, 0) and through the point $(-1, \frac{1}{2}\sqrt{3})$.

14. Foci at (0, 3) and (0, −3); one directrix is the line $y = 9$.

15. Center at (4, −2), a vertex at (9, −2), and one focus at (0, −2).

16. Foci at (2, 3) and (2, −7) and eccentricity of $\frac{2}{3}$.

17. Foci at (−1, −1) and (−1, 7) and the semimajor axis of length 8 units.

18. Directrices the lines $y = 3 \pm \frac{169}{12}$ and a focus at (0, −2).

19. Theorem 13.6.2 gives a procedure for drawing the graph of an ellipse. To apply the method to the ellipse $4x^2 + 9y^2 = 36$, first determine the points of intersection with the coordinate axes. Obtain the foci on the x axis by using a compass with its center at one of the points of intersection with the y axis and with radius of 3. Then fasten thumbtacks at each focus. Take a piece of string of length 6, which is $2a$, and attach an end at one thumbtack and an end at the other thumbtack. Place a pencil against the string and make it tight. Move the pencil against the string and trace a curve. This curve is an ellipse because the pencil traces a set of points the sum of whose distances from the two tacks is the constant 6.

20. Use Theorem 13.6.2 to find an equation of the ellipse for which the sum of the distances from any point on the ellipse to (4, −1) and (4, 7) is equal to 12.

21. Solve Exercise 20 if the sum of the distances from (−4, −5) and (6, −5) is equal to 16.

22. If one focus of the ellipse $x^2/a^2 + y^2/b^2 = 1$ is at the point $(c, 0)$, show that $a^2 + b^2 = c^2$.

23. For the ellipse whose equation is

$$\frac{(x-h)^2}{a^2} + \frac{(y-k)^2}{b^2} = 1$$

where $a > b > 0$, find the coordinates of the foci in terms of h, k, a, and b.

24. A plate is in the shape of the region bounded by the ellipse having a semimajor axis of length 3 ft and a semiminor axis of length 2 ft. If the plate is lowered vertically in a tank of water until the minor axis lies in the surface of the water, find the force due to water pressure on one side of the submerged portion of the plate.

25. If the plate of Exercise 24 is lowered until the center is 3 ft below the surface of the water, find the force due to water pressure on one side of the plate. The minor axis is still horizontal.

26. In Fig. 13.6.4 prove that $\alpha = \beta$. (*Hint:* Choose the coordinate axes so that the center of the ellipse is at the origin and the axes of the ellipse are along the coordinate axes. Then use Theorem 9.2.3.)

13.7 THE HYPERBOLA

In Sec. 13.4 we learned that an equation of a hyperbola having its center at the origin and its principal axis on the x axis is of the form $x^2/a^2 - y^2/b^2 = 1$. Interchanging x and y in this equation we obtain

$$\frac{y^2}{a^2} - \frac{x^2}{b^2} = 1 \qquad (1)$$

which is an equation of a hyperbola having its center at the origin and its principal axis on the y axis.

● ILLUSTRATION 1 The hyperbola with an equation

$$\frac{y^2}{9} - \frac{x^2}{16} = 1$$

has its foci on the y axis because the equation is of the form (1). ●

There is no general inequality involving a and b corresponding to the inequality $a > b$ for an ellipse. That is, for a hyperbola it is possible to have $a < b$, as in Illustration 1, where $a = 3$ and $b = 4$; or it is possible to have $a > b$, as for the hyperbola having the equation $y^2/25 - x^2/16 = 1$, where $a = 5$ and $b = 4$. If, for a hyperbola, $a = b$, then the hyperbola is said to be *equilateral.*

We stated in Sec. 13.3 that a hyperbola has asymptotes, and we now show how to obtain equations of these asymptotes. In Secs. 2.4 and 2.5 horizontal and vertical asymptotes of the graph of a function were defined. What follows is a more general definition, of which the definitions in Secs. 2.4 and 2.5 are special cases.

13.7.1 Definition The graph of the equation $y = f(x)$ has the line $y = mx + b$ as an **asymptote** if either of the following statements is true:

(i) $\lim_{x \to +\infty} [f(x) - (mx + b)] = 0$, and for some number $M > 0$, $f(x) \neq mx + b$ whenever $x > M$.
(ii) $\lim_{x \to -\infty} [f(x) - (mx + b)] = 0$, and for some number $M < 0$, $f(x) \neq mx + b$ whenever $x < M$.

Statement (i) indicates that for any $\epsilon > 0$ there exists a number $N > 0$ such that

$$|f(x) - (mx + b)| < \epsilon \quad \text{whenever} \quad x > N$$

that is, we can make the function value $f(x)$ as close to the value of $mx + b$ as we please by taking x large enough. This is consistent with our intuitive notion of an asymptote of a graph. A similar statement may be made for part (ii) of Definition 13.7.1.

For the hyperbola $x^2/a^2 - y^2/b^2 = 1$, upon solving for y we get

$$y = \pm\frac{b}{a}\sqrt{x^2 - a^2}$$

So if

$$f(x) = \frac{b}{a}\sqrt{x^2 - a^2}$$

then

$$\lim_{x \to +\infty}\left[f(x) - \frac{b}{a}x\right] = \lim_{x \to +\infty}\left[\frac{b}{a}\sqrt{x^2 - a^2} - \frac{b}{a}x\right]$$

$$= \frac{b}{a}\lim_{x \to +\infty}\frac{(\sqrt{x^2 - a^2} - x)(\sqrt{x^2 - a^2} + x)}{\sqrt{x^2 - a^2} + x}$$

$$= \frac{b}{a} \lim_{x \to +\infty} \frac{-a^2}{\sqrt{x^2 - a^2} + x}$$

$$= 0$$

Therefore, by Definition 13.7.1, the line $y = bx/a$ is an asymptote of the graph of $y = b\sqrt{x^2 - a^2}/a$. Similarly, it can be shown that the line $y = bx/a$ is an asymptote of the graph of $y = -b\sqrt{x^2 - a^2}/a$. Consequently, the line $y = bx/a$ is an asymptote of the hyperbola $x^2/a^2 - y^2/b^2 = 1$. In an analogous manner we can demonstrate that the line $y = -bx/a$ is an asymptote of this same hyperbola. We have, then, the following theorem.

13.7.2 Theorem The lines

$$y = \frac{b}{a}x \quad \text{and} \quad y = -\frac{b}{a}x$$

are asymptotes of the hyperbola

$$\frac{x^2}{a^2} - \frac{y^2}{b^2} = 1$$

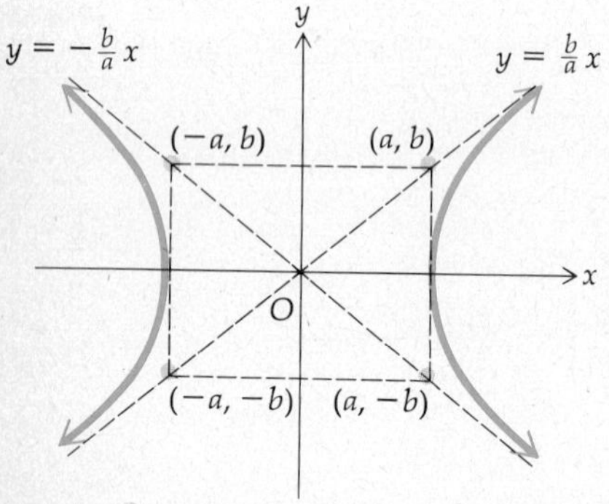

FIGURE 13.7.1

Figure 13.7.1 shows a sketch of the hyperbola of Theorem 13.7.2 together with its asymptotes. In the figure note that the diagonals of the rectangle having vertices at (a, b), $(a, -b)$, $(-a, b)$, and $(-a, -b)$ are on the asymptotes of the hyperbola. This rectangle is called the *auxiliary rectangle* of the hyperbola. The vertices of the hyperbola are the points of intersection of the principal axis and the auxiliary rectangle. A fairly good sketch of a hyperbola can be made by first drawing the auxiliary rectangle and then drawing the branch of the hyperbola through each vertex tangent to the side of the auxiliary rectangle there and approaching asymptotically the lines on which the diagonals of the rectangle lie.

There is a mnemonic device for obtaining equations of the asymptotes of a hyperbola. For example, for the hyperbola having the equation $x^2/a^2 - y^2/b^2 = 1$, if the right side is replaced by zero, we obtain $x^2/a^2 - y^2/b^2 = 0$. Upon factoring, this equation becomes

$$\left(\frac{x}{a} - \frac{y}{b}\right)\left(\frac{x}{a} + \frac{y}{b}\right) = 0$$

which is equivalent to the two equations $x/a - y/b = 0$ and $x/a + y/b = 0$, which, by Theorem 13.7.2, are equations of the asymptotes of the given hyperbola. Using this device for the hyperbola having Eq. (1), we see that the asymptotes are the lines having equations $y/a - x/b = 0$ and $y/a + x/b = 0$, which are the same lines as the asymptotes of the hyperbola with the equation $x^2/b^2 - y^2/a^2 = 1$. These two hyperbolas are called **conjugate hyperbolas.**

The asymptotes of an equilateral hyperbola ($a = b$) are perpendicular to each other. The auxiliary rectangle for such a hyperbola is a square, and the transverse and conjugate axes have equal lengths.

If the center of a hyperbola is at (h, k) and its principal axis is parallel to the x axis, then if the axes are translated so that the point (h, k) is the new origin, an equation of the hyperbola relative to this new coordinate system is $\bar{x}^2/a^2 - \bar{y}^2/b^2 = 1$. If we replace $\bar{x}$ by $x - h$ and $\bar{y}$ by $y - k$, this equation becomes

$$\frac{(x-h)^2}{a^2} - \frac{(y-k)^2}{b^2} = 1 \tag{2}$$

Similarly, an equation of a hyperbola having its center at (h, k) and its principal axis parallel to the y axis is

$$\frac{(y-k)^2}{a^2} - \frac{(x-h)^2}{b^2} = 1 \tag{3}$$

Example 1 The vertices of a hyperbola are at $(-5, -3)$ and $(-5, -1)$, and the eccentricity is $\sqrt{5}$. Find an equation of the hyperbola and equations of the asymptotes. Draw a sketch of the hyperbola and the asymptotes.

Solution The distance between the vertices is $2a$; so $a = 1$. For a hyperbola, $b = a\sqrt{e^2 - 1}$, and therefore $b = 1\sqrt{5 - 1} = 2$. Because the principal axis is parallel to the y axis, an equation of the hyperbola is of the form (3). The center (h, k) is halfway between the vertices and is therefore at the point $(-5, -2)$. Hence an equation of the hyperbola is

$$\frac{(y+2)^2}{1} - \frac{(x+5)^2}{4} = 1$$

Using the mnemonic device to obtain equations of the asymptotes we have

$$\left(\frac{y+2}{1} - \frac{x+5}{2}\right)\left(\frac{y+2}{1} + \frac{x+5}{2}\right) = 0$$

which gives

$$y + 2 = \tfrac{1}{2}(x + 5) \quad \text{and} \quad y + 2 = -\tfrac{1}{2}(x + 5)$$

A sketch of the hyperbola and the asymptotes are in Fig. 13.7.2.

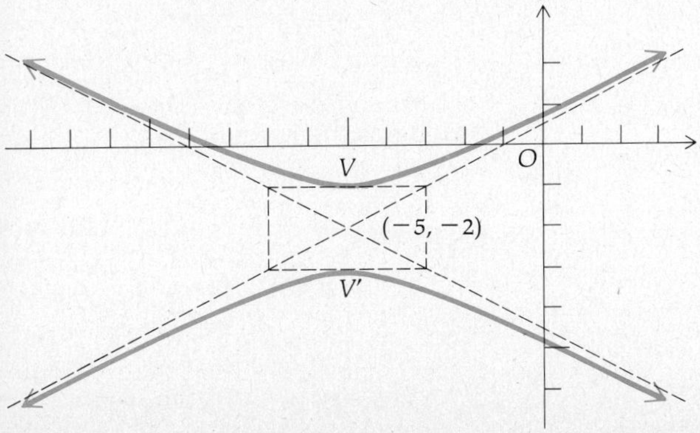

FIGURE 13.7.2

If in (2) and (3) we eliminate fractions and combine terms, the resulting equations are of the form

$$Ax^2 + Cy^2 + Dx + Ey + F = 0 \tag{4}$$

where A and C have different signs; that is, $AC < 0$. We now wish to show that the graph of an equation of the form (4), where $AC < 0$, is either a hyperbola or it degenerates. Completing the squares in x and y in (4), where $AC < 0$, the resulting equation has the form

$$\alpha^2(x - h)^2 - \beta^2(y - k)^2 = H \tag{5}$$

If $H > 0$, (5) can be written as

$$\frac{(x - h)^2}{\frac{H}{\alpha^2}} - \frac{(y - k)^2}{\frac{H}{\beta^2}} = 1$$

which has the form of (2).

● ILLUSTRATION 2 The equation

$$4x^2 - 12y^2 + 24x + 96y - 181 = 0$$

can be written as

$$4(x^2 + 6x) - 12(y^2 - 8y) = 181$$

and upon completing the squares in x and y we have

$$4(x^2 + 6x + 9) - 12(y^2 - 8y + 16) = 181 + 36 - 192$$

$$4(x + 3)^2 - 12(y - 4)^2 = 25$$

This has the form of (5), where $H = 25 > 0$. It may be written as

$$\frac{(x + 3)^2}{\frac{25}{4}} - \frac{(y - 4)^2}{\frac{25}{12}} = 1$$

which has the form of (2). ●

If in (5), $H < 0$, then (5) may be written as

$$\frac{(y - k)^2}{\frac{|H|}{\alpha^2}} - \frac{(x - h)^2}{\frac{|H|}{\beta^2}} = 1$$

which has the form of (3).

● ILLUSTRATION 3 Suppose that (4) is

$$4x^2 - 12y^2 + 24x + 96y - 131 = 0$$

Upon completing the squares in x and y we get

$$4(x + 3)^2 - 12(y - 4)^2 = -25$$

This has the form of (5), where $H = -25 < 0$, and it may be written as

$$\frac{(y-4)^2}{\frac{25}{12}} - \frac{(x+3)^2}{\frac{25}{4}} = 1$$

which has the form of (3). •

If $H = 0$ in (5), then (5) is equivalent to the two equations

$$\alpha(x-h) - \beta(y-k) = 0 \quad \text{and} \quad \alpha(x-h) + \beta(y-k) = 0$$

which are equations of two straight lines through the point (h, k). This is the degenerate case of the hyperbola.

The following theorem summarizes the results of the above discussion.

13.7.3 Theorem If in the general second-degree equation

$$Ax^2 + Bxy + Cy^2 + Dx + Ey + F = 0$$

$B = 0$ and $AC < 0$, then the graph is either a hyperbola or two intersecting straight lines.

Example 2 Determine the graph of the equation $9x^2 - 4y^2 - 18x - 16y + 29 = 0$.

Solution From Theorem 13.7.3, because $B = 0$ and $AC = (9)(-4) = -36 < 0$, the graph is either a hyperbola or two intersecting straight lines. Completing the squares in x and y we get

$$9(x^2 - 2x + 1) - 4(y^2 + 4y + 4) = -29 + 9 - 16$$

$$9(x-1)^2 - 4(y+2)^2 = -36$$

$$\frac{(y+2)^2}{9} - \frac{(x-1)^2}{4} = 1 \qquad (6)$$

Equation (6) has the form of (3); so the graph is a hyperbola whose principal axis is parallel to the y axis and whose center is at $(1, -2)$.

Example 3 For the hyperbola of Example 2, find the eccentricity, the vertices, the foci, and the directrices. Draw a sketch of the hyperbola, and show the foci and the directrices.

Solution From (6) we see that $a = 3$ and $b = 2$. For a hyperbola, $b = a\sqrt{e^2 - 1}$; thus $2 = 3\sqrt{e^2 - 1}$ and solving for e we obtain $e = \frac{1}{3}\sqrt{13}$. Because the center is at $(1, -2)$, the principal axis is vertical, and $a = 3$, it follows that the vertices are at the points $V'(1, -5)$ and $V(1, 1)$. Because $ae = \sqrt{13}$, the foci are at the points $F'(1, -2 - \sqrt{13})$ and $F(1, -2 + \sqrt{13})$. Furthermore, $a/e = \frac{9}{13}\sqrt{13}$; hence the directrix corresponding to the focus at F' has as an equation $y = -2 - \frac{9}{13}\sqrt{13}$, and the directrix corresponding to the focus at F has as an equation $y = -2 + \frac{9}{13}\sqrt{13}$. Figure 13.7.3 shows a sketch of the hyperbola and the foci and directrices.

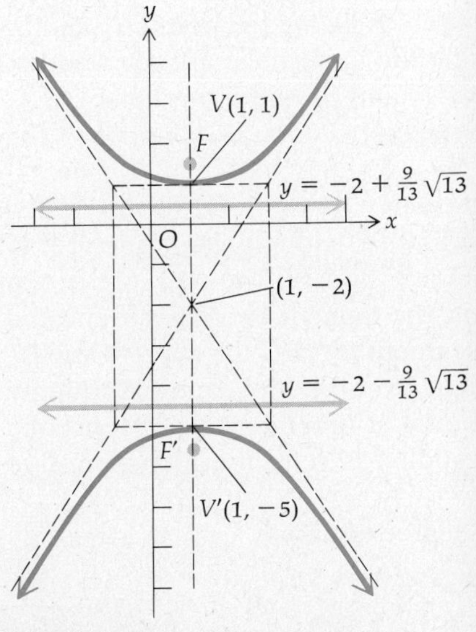

FIGURE 13.7.3

Just as Theorem 13.6.2 gives an alternate definition of an ellipse, the following theorem gives an alternate definition of a hyperbola.

13.7.4 Theorem A hyperbola can be defined as the set of points such that the absolute value of the difference of the distances from any point of the set to two given points (the foci) is a constant.

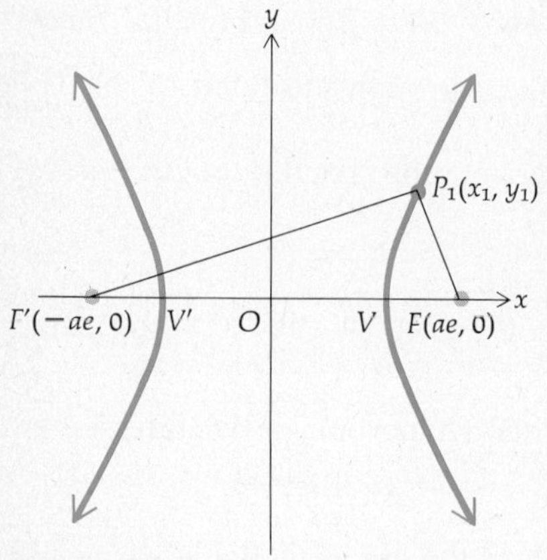

FIGURE 13.7.4

The proof of this theorem is similar to the proof of Theorem 13.6.2 and is left as an exercise (see Exercises 19 and 20). In the proof, take the foci at the points $F'(-ae, 0)$ and $F(ae, 0)$, and let the constant absolute value of the difference of the distances be $2a$. See Fig. 13.7.4.

● ILLUSTRATION 4 Because the hyperbola of Example 2 has foci at $(1, -2 - \sqrt{13})$ and $(1, -2 + \sqrt{13})$ and $2a = 6$, it follows from Theorem 13.7.4 that this hyperbola can be defined as the set of points such that the absolute value of the difference of the distances from any point of the set to the points $(1, -2 - \sqrt{13})$ and $(1, -2 + \sqrt{13})$ is 6. ●

The property of the hyperbola given in Theorem 13.7.4 forms the basis of several important navigational systems. These systems involve a network of pairs of radio transmitters at fixed positions at a known distance from one another. The transmitters send out radio signals that are received by a navigator. The difference in arrival time of the two signals determines the difference $2a$ of the distances from the navigator. Thus the navigator's position is known to be somewhere along one arc of a hyperbola having foci at the locations of the two transmitters. One arc, rather than both, is determined because of the signal delay between the two transmitters that is built into the system. The procedure is then repeated for a different pair of radio transmitters, and another arc of a hyperbola that contains the navigator's position is determined. The point of intersection of the two hyperbolic arcs is the actual position. For example, in Fig. 13.7.5 suppose a pair of transmitters is located at points T_1 and S_1 and the signals from this pair determine the hyperbolic arc A_1. Another pair of transmitters is located at points T_2 and S_2 and hyperbolic arc A_2 is determined from their signals. Then the intersection of A_1 and A_2 is the position of the navigator.

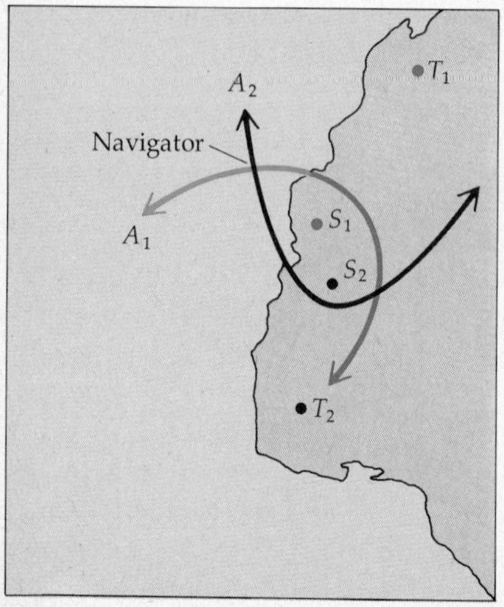

FIGURE 13.7.5

The hyperbola has a reflective property that is used in the design of certain telescopes. In Fig. 13.7.6 the line PT is the tangent line at P to the graph of the hyperbola having foci at F and F'. The measure of the angle between the line segment FP and the tangent line PT is α and the measure of the angle between the line segment $F'P$ and the tangent line PT is β. You are asked to prove that $\alpha = \beta$ in Exercise 25. From this equality it follows that a ray of light from a source at one focus of a hyperbolic mirror (one with hyperbolic cross sections) is reflected along the line through the other focus.

The graph of the general second-degree equation

$$Ax^2 + Bxy + Cy^2 + Dx + Ey + F = 0 \tag{7}$$

where $B = 0$ and $AC > 0$, was discussed in Sec. 13.6. In this section we

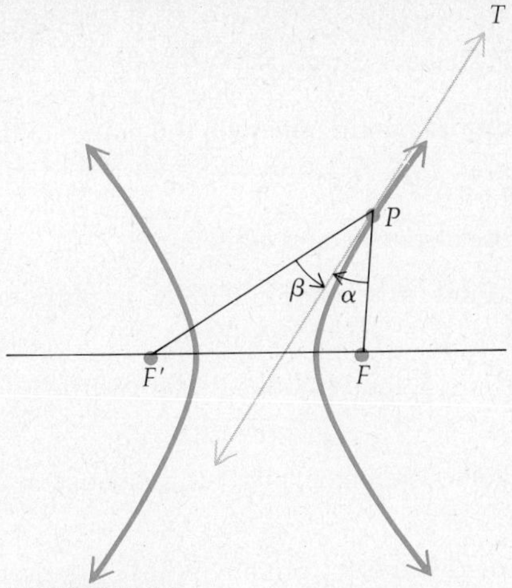

FIGURE 13.7.6

have so far considered the equation when $B = 0$ and $AC < 0$. Now if for Eq. (7) $B = 0$ and $AC = 0$, then either $A = 0$ or $C = 0$ (we do not consider $A = B = C = 0$ because then (7) would not be a quadratic equation). Suppose that in (7), $B = 0, A = 0$, and $C \neq 0$. The equation becomes

$$Cy^2 + Dx + Ey + F = 0 \tag{8}$$

If $D \neq 0$, (8) is an equation of a parabola. If $D = 0$, then the graph of (8) may be two parallel lines, one line, or the empty set. These are the degenerate cases of the parabola.

● ILLUSTRATION 5 The graph of the equation $4y^2 - 9 = 0$ is two parallel lines; $9y^2 + 6y + 1 = 0$ is an equation of one line; and $2y^2 + y + 1 = 0$ is satisfied by no real values of y. ●

A similar discussion holds if $B = 0$, $C = 0$, and $A \neq 0$. The results are summarized in the following theorem.

13.7.5 Theorem If in the general second-degree equation (7) $B = 0$ and either $A = 0$ and $C \neq 0$ or $C = 0$ and $A \neq 0$, then the graph is one of the following: a parabola, two parallel lines, one line, or the empty set.

From Theorems 13.6.1, 13.7.3, and 13.7.5 it may be concluded that the graph of the general quadratic equation in two unknowns when $B = 0$ is either a conic or a degenerate conic. The type of conic can be determined from the product of A and C. We have the following theorem.

13.7.6 Theorem The graph of the equation $Ax^2 + Cy^2 + Dx + Ey + F = 0$, where A and C are not both zero, is either a conic or a degenerate conic; if it is a conic then the graph is

(i) a *parabola* if either $A = 0$ or $C = 0$, that is, if $AC = 0$;
(ii) an *ellipse* if A and C have the same sign, that is, if $AC > 0$;
(iii) a *hyperbola* if A and C have opposite signs, that is, if $AC < 0$.

A discussion of the graph of the general quadratic equation, where $B \neq 0$, is given in Sec. 13.8.

EXERCISES 13.7

In Exercises 1 through 8, find the eccentricity, center, foci, directrices, and equations of the asymptotes of the given hyperbola, and draw a sketch of the graph.

1. $x^2 - y^2 + 8x - 2y - 21 = 0$
2. $4x^2 - y^2 - 8x - 12 = 0$
3. $9x^2 - 18y^2 + 54x - 36y + 79 = 0$
4. $x^2 - y^2 + 6x + 10y - 4 = 0$
5. $3y^2 - 4x^2 - 8x - 24y - 40 = 0$
6. $4x^2 - y^2 + 56x + 2y + 195 = 0$
7. $4y^2 - 9x^2 + 16y + 18x = 29$
8. $y^2 - x^2 + 2y - 2x - 1 = 0$

In Exercises 7 through 15, find an equation of the hyperbola satisfying the given conditions, and draw a sketch of the graph.

9. One focus at (26, 0) and asymptotes the lines $12y = \pm 5x$.

10. Center at $(3, -5)$, a vertex at $(7, -5)$, and a focus at $(8, -5)$.

11. Center at $(-2, -1)$, a focus at $(-2, 14)$, and a directrix the line $5y = -53$.

12. Foci at $(3, 6)$ and $(3, 0)$ and passing through the point $(5, 3 + 6/\sqrt{5})$.

13. One focus at $(-3 - 3\sqrt{13}, 1)$, asymptotes intersecting at $(-3, 1)$, and one asymptote passing through the point $(1, 7)$.

14. Foci at $(-1, 4)$ and $(7, 4)$ and eccentricity of 3.

15. One focus at $(1, -2 + \sqrt{13})$, with the line $y = -2 + \frac{9}{13}\sqrt{13}$ as the corresponding directrix, and center at $(1, -2)$.

16. The vertices of a hyperbola are at $(2, 7)$ and $(2, -1)$, and the eccentricity is $\sqrt{3}$. Find (a) an equation of the hyperbola and (b) equations of the asymptotes.

17. The vertices of a hyperbola are at $(-3, -1)$ and $(-1, -1)$, and the eccentricity is $\sqrt{5}$. Find (a) an equation of the hyperbola and (b) equations of the asymptotes.

18. For an equilateral hyperbola having its center at (h, k) and principal axis parallel to the x axis, prove that the asymptotes are perpendicular.

19. Prove that the set of points, such that the absolute value of the difference of the distances from any point of the set to two given points (the foci) is a constant, is a hyperbola.

20. Prove that any hyperbola is a set of points such that the absolute value of the difference of the distances from any point of the set to two given points (the foci) is a constant.

21. Use Theorem 13.7.4 to find an equation of the hyperbola for which the difference of the distances from any point on the hyperbola to $(2, 1)$ and $(2, 9)$ is equal to 4.

22. Solve Exercise 21 if the difference of the distances from any point on the hyperbola to $(-8, -4)$ and $(2, -4)$ is equal to 6.

23. Three listening posts are located at the points $A(0, 0)$, $B(0, \frac{21}{4})$, and $C(\frac{25}{3}, 0)$, the unit being 1 km. Microphones located at these points show that a gun is $\frac{5}{3}$ km closer to A than to C and $\frac{7}{4}$ km closer to B than to A. Locate the position of the gun by use of Theorem 13.7.4.

24. Prove that the eccentricity of an equilateral hyperbola is equal to $\sqrt{2}$.

25. In Fig. 13.7.6 prove that $\alpha = \beta$. (*Hint:* Choose the coordinate axes so that the center of the hyperbola is at the origin and the axes of the hyperbola are along the coordinate axes. Then prove that the tangent line PT bisects the angle formed by the line segments FP and $F'P$.)

13.8 ROTATION OF AXES

We have shown how a translation of coordinate axes can simplify the form of certain equations. A translation of axes gives a new coordinate system whose axes are parallel to the original axes. We now consider a rotation of coordinate axes that enables us to transform a second-degree equation having an xy term into one having no such term.

Suppose that we have two rectangular cartesian coordinate systems with the same origin. Let one system be the xy system and the other the $\bar{x}\bar{y}$ system. Suppose further that the $\bar{x}$ axis makes an angle of radian measure α with the x axis. Then of course the $\bar{y}$ axis makes an angle of radian measure α with the y axis. In such a case we state that the xy system of coordinates is *rotated* through an angle of radian measure α to form the $\bar{x}\bar{y}$ system of coordinates. A point P having coordinates (x, y) with respect to the original coordinate system will have coordinates $(\bar{x}, \bar{y})$ with respect to the new one. We now obtain relationships between these two sets of coordinates. To do this we introduce two polar coordinate systems, each system having the pole at the origin. In the first polar coordinate system the positive side of the x axis is taken as the polar axis; in the second polar

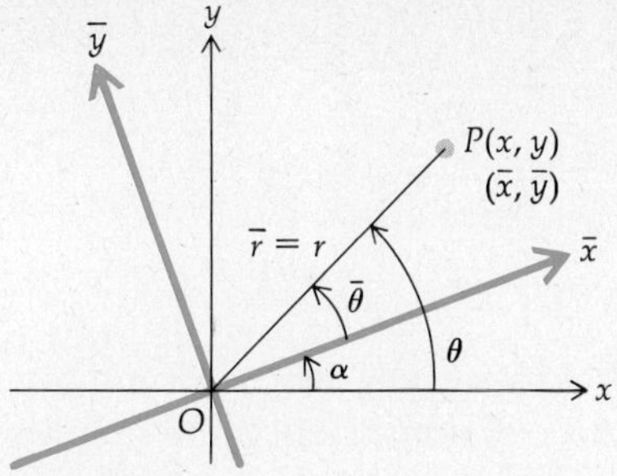

FIGURE 13.8.1

coordinate system the positive side of the $\bar{x}$ axis is taken as the polar axis (see Fig. 13.8.1). Point P has two sets of polar coordinates, (r, θ) and $(\bar{r}, \bar{\theta})$, where

$$\bar{r} = r \quad \text{and} \quad \bar{\theta} = \theta - \alpha \tag{1}$$

The following equations hold:

$$\bar{x} = \bar{r} \cos \bar{\theta} \quad \text{and} \quad \bar{y} = \bar{r} \sin \bar{\theta} \tag{2}$$

Substituting from (1) into (2) we get

$$\bar{x} = r \cos(\theta - \alpha) \quad \text{and} \quad \bar{y} = r \sin(\theta - \alpha) \tag{3}$$

Using the trigonometric identities for the sine and cosine of the difference of two numbers, Eqs. (3) become

$$\bar{x} = r \cos \theta \cos \alpha + r \sin \theta \sin \alpha$$

and

$$\bar{y} = r \sin \theta \cos \alpha - r \cos \theta \sin \alpha$$

Because $r \cos \theta = x$ and $r \sin \theta = y$, we obtain from the above two equations

$$\bar{x} = x \cos \alpha + y \sin \alpha \tag{4}$$

and

$$\bar{y} = -x \sin \alpha + y \cos \alpha \tag{5}$$

Solving (4) and (5) simultaneously for x and y in terms of $\bar{x}$ and $\bar{y}$ we obtain

$$x = \bar{x} \cos \alpha - \bar{y} \sin \alpha \tag{6}$$

and

$$y = \bar{x} \sin \alpha + \bar{y} \cos \alpha \tag{7}$$

It is left as an exercise to fill in the steps going from (4) and (5) to (6) and (7). (See Exercise 3.) The results are stated as a theorem.

13.8.1 Theorem If (x, y) represents a point P with respect to a given set of axes, and $(\bar{x}, \bar{y})$ is a representation of P after the axes have been rotated through an angle of radian measure α, then

$$x = \bar{x} \cos \alpha - \bar{y} \sin \alpha \quad \text{and} \quad y = \bar{x} \sin \alpha + \bar{y} \cos \alpha$$

Example 1 Given the equation $xy = 1$, find an equation of the graph with respect to the $\bar{x}$ and $\bar{y}$ axes after a rotation of axes through an angle of radian measure $\frac{1}{4}\pi$.

Solution Taking $\alpha = \frac{1}{4}\pi$ in Theorem 13.8.1 we obtain

$$x = \frac{1}{\sqrt{2}}\bar{x} - \frac{1}{\sqrt{2}}\bar{y} \quad \text{and} \quad y = \frac{1}{\sqrt{2}}\bar{x} + \frac{1}{\sqrt{2}}\bar{y}$$

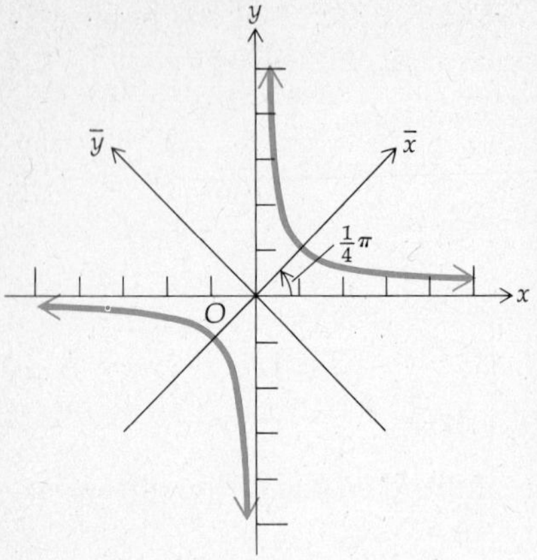

FIGURE 13.8.2

Substituting these expressions for x and y in the equation $xy = 1$ we get

$$\left(\frac{1}{\sqrt{2}}\bar{x} - \frac{1}{\sqrt{2}}\bar{y}\right)\left(\frac{1}{\sqrt{2}}\bar{x} + \frac{1}{\sqrt{2}}\bar{y}\right) = 1$$

$$\frac{\bar{x}^2}{2} - \frac{\bar{y}^2}{2} = 1$$

This is an equation of an equilateral hyperbola whose asymptotes are the bisectors of the quadrants in the $\bar{x}\bar{y}$ system. Hence the graph of the equation $xy = 1$ is an equilateral hyperbola lying in the first and third quadrants whose asymptotes are the x and y axes (see Fig. 13.8.2).

In Sec. 13.7 we showed that when $B = 0$ and A and C are not both zero, the graph of the general second-degree equation in two unknowns,

$$Ax^2 + Bxy + Cy^2 + Dx + Ey + F = 0 \tag{8}$$

is either a conic or a degenerate conic. We now show that if $B \neq 0$, then any equation of the form (8) can be transformed by a suitable rotation of axes into an equation of the form

$$\bar{A}\bar{x}^2 + \bar{C}\bar{y}^2 + \bar{D}\bar{x} + \bar{E}\bar{y} + \bar{F} = 0 \tag{9}$$

where $\bar{A}$ and $\bar{C}$ are not both zero.

If the xy system is rotated through an angle of radian measure α, then to obtain an equation of the graph of (8) with respect to the $\bar{x}\bar{y}$ system, we replace x by $\bar{x}\cos\alpha - \bar{y}\sin\alpha$ and y by $\bar{x}\sin\alpha + \bar{y}\cos\alpha$. We get

$$\bar{A}\bar{x}^2 + \bar{B}\bar{x}\bar{y} + \bar{C}\bar{y}^2 + \bar{D}\bar{x} + \bar{E}\bar{y} + \bar{F} = 0 \tag{10}$$

where

$$\begin{aligned} \bar{A} &= A\cos^2\alpha + B\sin\alpha\cos\alpha + C\sin^2\alpha \\ \bar{B} &= -2A\sin\alpha\cos\alpha + B(\cos^2\alpha - \sin^2\alpha) + 2C\sin\alpha\cos\alpha \\ \bar{C} &= A\sin^2\alpha - B\sin\alpha\cos\alpha + C\cos^2\alpha \end{aligned} \tag{11}$$

We wish to find an α so that the rotation transforms (8) into an equation of the form (9). Setting $\bar{B}$ from (11) equal to zero we have

$$B(\cos^2\alpha - \sin^2\alpha) + (C - A)(2\sin\alpha\cos\alpha) = 0$$

or, equivalently, with trigonometric identities,

$$B\cos 2\alpha + (C - A)\sin 2\alpha = 0$$

Because $B \neq 0$, this gives

$$\cot 2\alpha = \frac{A - C}{B} \tag{12}$$

We have shown, then, that an equation of the form (8), where $B \neq 0$, can be transformed to an equation of the form (9) by a rotation of axes through an angle of radian measure α satisfying (12). We wish to show that $\bar{A}$ and $\bar{C}$ in (9) are not both zero. To prove this, notice that (10) is obtained from (8) by rotating the axes through the angle of radian measure α. Also, (8) can be obtained from (10) by rotating the axes back through the angle of radian measure $(-\alpha)$. If $\bar{A}$ and $\bar{C}$ in (10) are both zero, then the substitutions

$$\bar{x} = x \cos \alpha + y \sin \alpha \quad \text{and} \quad \bar{y} = -x \sin \alpha + y \cos \alpha$$

in (10) would result in the equation

$$\bar{D}(x \cos \alpha + y \sin \alpha) + \bar{E}(-x \sin \alpha + y \cos \alpha) + \bar{F} = 0$$

which is an equation of the first degree and hence different from (8) because we have assumed that at least $B \neq 0$. The following theorem has, therefore, been proved.

13.8.2 Theorem If $B \neq 0$, the equation $Ax^2 + Bxy + Cy^2 + Dx + Ey + F = 0$ can be transformed into the equation $\bar{A}\bar{x}^2 + \bar{C}\bar{y}^2 + \bar{D}\bar{x} + \bar{E}\bar{y} + \bar{F} = 0$, where $\bar{A}$ and $\bar{C}$ are not both zero, by a rotation of axes through an angle of radian measure α for which $\cot 2\alpha = (A - C)/B$.

By Theorems 13.8.2 and 13.7.6 it follows that the graph of an equation of the form (8) is either a conic or a degenerate conic. To determine which type of conic is the graph of a particular equation, we use the fact that A, B, and C of (8) and $\bar{A}$, $\bar{B}$, and $\bar{C}$ of (10) satisfy the relation

$$B^2 - 4AC = \bar{B}^2 - 4\bar{A}\bar{C} \tag{13}$$

which can be proved by substituting the expressions for $\bar{A}$, $\bar{B}$, and $\bar{C}$ given in Eqs. (11) in the right side of (13). This is left as an exercise (see Exercise 19).

The expression $B^2 - 4AC$ is called the **discriminant** of (8). Equation (13) states that the discriminant of the general quadratic equation in two variables is *invariant* under a rotation of axes. If the angle of rotation is chosen so that $\bar{B} = 0$, then (13) becomes

$$B^2 - 4AC = -4\bar{A}\bar{C} \tag{14}$$

From Theorem 13.7.6 it follows that if the graph of (9) is not degenerate, then it is a parabola if $\bar{A}\bar{C} = 0$, an ellipse if $\bar{A}\bar{C} > 0$, and a hyperbola if $\bar{A}\bar{C} < 0$. So we conclude that the graph of (9) is a parabola, an ellipse, or a hyperbola depending on whether $-4\bar{A}\bar{C}$ is zero, negative, or positive. Because the graph of (8) is the same as the graph of (9), it follows from (14) that if the graph of (8) is not degenerate, then it is a parabola, an ellipse, or a hyperbola depending on whether the discriminant $B^2 - 4AC$ is zero, negative, or positive. We have proved the following theorem.

13.8.3 Theorem The graph of the equation

$$Ax^2 + Bxy + Cy^2 + Dx + Ey + F = 0$$

is either a conic or a degenerate conic. If it is a conic, then it is

(i) a *parabola* if $B^2 - 4AC = 0$;
(ii) an *ellipse* if $B^2 - 4AC < 0$;
(iii) a *hyperbola* if $B^2 - 4AC > 0$.

Example 2 Given the equation $17x^2 - 12xy + 8y^2 - 80 = 0$, simplify the equation by a rotation of axes. Draw a sketch of the graph of the equation showing both sets of axes.

Solution $B^2 - 4AC = (-12)^2 - 4(17)(8) = -400 < 0$. Therefore, by Theorem 13.8.3 the graph is an ellipse or else it is degenerate. To eliminate the xy term by a rotation of axes we must choose an α such that

$$\cot 2\alpha = \frac{A - C}{B} = \frac{17 - 8}{-12} = -\frac{3}{4}$$

There is a 2α in the interval $(0, \pi)$ for which $\cot 2\alpha = -\frac{3}{4}$. Therefore α is in the interval $(0, \frac{1}{2}\pi)$. To apply Theorem 13.8.1 it is not necessary to find α so long as we find $\cos \alpha$ and $\sin \alpha$. These functions can be found from the value of $\cot 2\alpha$ by the trigonometric identities

$$\cos \alpha = \sqrt{\frac{1 + \cos 2\alpha}{2}} \quad \text{and} \quad \sin \alpha = \sqrt{\frac{1 - \cos 2\alpha}{2}}$$

Because $\cot 2\alpha = -\frac{3}{4}$ and $0 < \alpha < \frac{1}{2}\pi$, it follows that $\cos 2\alpha = -\frac{3}{5}$. So

$$\cos \alpha = \sqrt{\frac{1 - \frac{3}{5}}{2}} = \frac{1}{\sqrt{5}} \quad \text{and} \quad \sin \alpha = \sqrt{\frac{1 + \frac{3}{5}}{2}} = \frac{2}{\sqrt{5}}$$

Substituting $x = \bar{x}/\sqrt{5} - 2\bar{y}/\sqrt{5}$ and $y = 2\bar{x}/\sqrt{5} + \bar{y}/\sqrt{5}$ in the given equation we obtain

$$17\left(\frac{\bar{x}^2 - 4\bar{x}\bar{y} + 4\bar{y}^2}{5}\right) - 12\left(\frac{2\bar{x}^2 - 3\bar{x}\bar{y} - 2\bar{y}^2}{5}\right) + 8\left(\frac{4\bar{x}^2 + 4\bar{x}\bar{y} + \bar{y}^2}{5}\right) - 80 = 0$$

Upon simplification this equation becomes

$$\bar{x}^2 + 4\bar{y}^2 = 16$$

$$\frac{\bar{x}^2}{16} + \frac{\bar{y}^2}{4} = 1$$

So the graph is an ellipse whose major axis is 8 units long and whose minor axis is 4 units long. A sketch of the ellipse with both sets of axes is shown in Fig. 13.8.3.

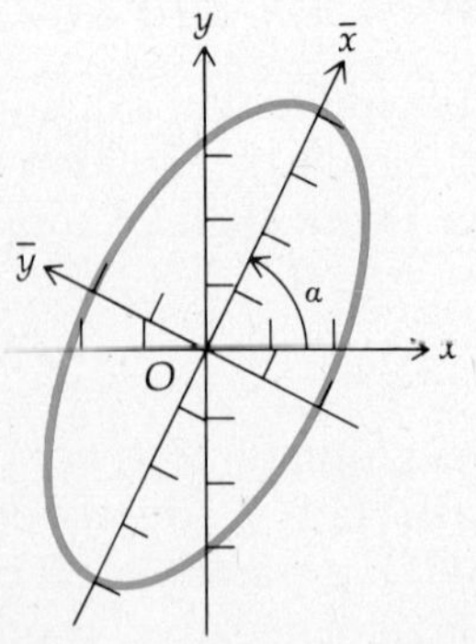

FIGURE 13.8.3

EXERCISES 13.8

In Exercises 1 and 2, an equation is given. Find an equation of the graph with respect to the $\bar{x}$ and $\bar{y}$ axes after a rotation of the axes through an angle of radian measure $\frac{1}{4}\pi$.

1. $xy = -4$ **2.** $x^2 - y^2 = 8$

3. Derive (6) and (7) of this section by solving (4) and (5) for x and y in terms of $\bar{x}$ and $\bar{y}$.

In Exercises 4 through 10, remove the xy term from the given equation by a rotation of axes. Draw a sketch of the graph and show both sets of axes.

4. $x^2 + xy + y^2 = 3$ **5.** $24xy - 7y^2 + 36 = 0$ **6.** $4xy + 3x^2 = 4$ **7.** $xy = 8$

8. $6x^2 + 20\sqrt{3}xy + 26y^2 = 324$ **9.** $31x^2 + 10\sqrt{3}xy + 21y^2 = 144$ **10.** $5x^2 + 6xy + 5y^2 = 9$

In Exercises 11 through 18, simplify the given equation by a rotation and translation of axes. Draw a sketch of the graph and show the three sets of axes.

11. $x^2 + xy + y^2 - 3y - 6 = 0$ **12.** $x^2 - 10xy + y^2 + x + y + 1 = 0$

13. $17x^2 - 12xy + 8y^2 - 68x + 24y - 12 = 0$ **14.** $19x^2 + 6xy + 11y^2 - 26x + 38y + 31 = 0$

15. $x^2 + 2xy + y^2 + x - y - 4 = 0$ **16.** $3x^2 - 4xy + 8x - 1 = 0$

17. $11x^2 - 24xy + 4y^2 + 30x + 40y - 45 = 0$ **18.** $16x^2 - 24xy + 9y^2 - 60x - 80y + 400 = 0$

19. Show that for the general second-degree equation in two variables, the discriminant $B^2 - 4AC$ is invariant under a rotation of axes.

20. Given the equation $(a^2 + b^2)xy = 1$, where $a > 0$ and $b > 0$, find an equation of the graph with respect to the $\bar{x}$ and $\bar{y}$ axes after a rotation of the axes through an angle of radian measure $\tan^{-1}(b/a)$.

REVIEW EXERCISES FOR CHAPTER 13

In Exercises 1 and 2, find an equation of the parabola having the given focus and directrix. Also find the length of the latus rectum, and draw a sketch of the curve.

1. Focus at $(0, -3)$; directrix, $y = 3$. **2.** Focus at $(6, 0)$; directrix, $x = -6$.

3. Given the parabola having the equation $y^2 + 8x - 6y - 7 = 0$, find (a) the vertex, (b) the focus, (c) an equation of the axis, and (d) an equation of the directrix.

4. Find an equation of the parabola having its vertex at $(-3, 5)$ and its focus at $(-3, -1)$.

In Exercises 5 through 7, use the definition of a conic (Definition 13.3.1) to find an equation of the conic having the given eccentricity, focus, and corresponding directrix. Also identify the conic.

5. $e = 2$; focus at $(1, -4)$; directrix, $y = -3$. **6.** $e = 1$; focus at $(3, -2)$; directrix, $x = 0$.

7. $e = \frac{2}{3}$; focus at $(5, 3)$; directrix, $2x - 3 = 0$. **8.** Find the focus of the parabola having the equation $y = ax^2 + bx$.

In Exercises 9 through 14, find a cartesian equation of the conic satisfying the given conditions, and draw a sketch of the graph.

9. Vertices at $(1, 8)$ and $(1, -4)$; $e = \frac{2}{3}$. **10.** Vertex at $(4, 2)$; focus at $(4, -3)$; $e = 1$.

11. Center at the origin; foci on the x axis; $e = 2$; containing the point $(2, 3)$.

12. Foci at $(-5, 1)$ and $(1, 1)$; one vertex at $(-4, 1)$. **13.** A focus at $(-5, 3)$; directrix, $x = -3$; $e = \frac{3}{4}$.

14. A focus at $(4, -2)$; directrix, $y = 2$; $e = \frac{3}{2}$.

In Exercises 15 through 18, find a polar equation of the conic satisfying the given conditions, and draw a sketch of the graph.

15. A focus at the pole; vertices at $(2, \pi)$ and $(4, \pi)$. **16.** A focus at the pole; a vertex at $(6, \frac{1}{2}\pi)$; $e = \frac{3}{4}$.

17. A focus at the pole; a vertex at $(3, \frac{3}{2}\pi)$; $e = 1$.

18. The line $r \sin \theta = 6$ is the directrix corresponding to the focus at the pole and $e = \frac{5}{3}$.

In Exercises 19 through 22, the equation is that of a conic having a focus at the pole. In each exercise, (a) find the eccentricity; (b) identify the conic; (c) write an equation of the directrix that corresponds to the focus at the pole; (d) draw a sketch of the curve.

19. $r = \dfrac{2}{2 - \sin \theta}$ **20.** $r = \dfrac{5}{3 + 3 \sin \theta}$ **21.** $r = \dfrac{4}{2 + 3 \cos \theta}$ **22.** $r = \dfrac{4}{3 - 2 \cos \theta}$

In Exercises 23 through 26, the equation is that of either an ellipse or a hyperbola. Find the eccentricity, center, foci, and directrices, and draw a sketch of the graph. If it is a hyperbola, also find equations of the asymptotes.

23. $4x^2 + y^2 + 24x - 16y + 84 = 0$ **24.** $3x^2 - 2y^2 + 6x - 8y + 11 = 0$

25. $25x^2 - y^2 + 50x + 6y - 9 = 0$ **26.** $4x^2 + 9y^2 + 32x - 18y + 37 = 0$

In Exercises 27 and 28, simplify the given equation by a rotation and translation of axes. Draw a sketch of the graph and show the three sets of axes.

27. $3x^2 - 3xy - y^2 - 6y = 0$ **28.** $4x^2 + 3xy + y^2 - 6x + 12y = 0$

29. Find a polar equation of the parabola containing the point $(2, \frac{1}{3}\pi)$, whose focus is at the pole and whose vertex is on the extension of the polar axis.

30. If the distance between the two directrices of an ellipse is three times the distance between the foci, find the eccentricity.

31. Find the volume of the solid of revolution generated if the region bounded by the hyperbola $x^2/a^2 - y^2/b^2 = 1$ and the line $x = 2a$ is revolved about the y axis.

32. Show that the hyperbola $x^2 - y^2 = 4$ has the same foci as the ellipse $x^2 + 9y^2 = 9$.

33. Find an equation of the parabola having vertex at (5, 1), axis parallel to the y axis, and through the point (9, 3).

34. Show that any equation of the form $xy + ax + by + c = 0$ can always be written in the form $x'y' = k$ by a translation of the axes, and determine the value of k.

35. Any section of a parabolic mirror made by passing a plane through the axis of the mirror is a segment of a parabola. The altitude of the segment is 12 cm and the length of the base is 18 cm. A section of the mirror made by a plane perpendicular to its axis is a circle. Find the circumference of the circular plane section if the plane perpendicular to the axis is 3 cm from the vertex.

36. The directrix of the parabola $y^2 = 4px$ is tangent to a circle having the focus of the parabola as its center. Find an equation of the circle and the points of intersection of the two curves.

37. A satellite is traveling around the earth in an elliptical orbit having the earth at one focus and an eccentricity of $\frac{1}{3}$. The closest distance that the satellite gets to the earth is 300 mi. Find the farthest distance that the satellite gets from the earth.

38. The orbit of the planet Mercury around the sun is elliptical in shape with the sun at one focus, a semimajor axis of length 36 million miles, and an eccentricity of 0.206. Find (a) how close Mercury gets to the sun and (b) the greatest possible distance between Mercury and the sun.

39. A comet is moving in a parabolic orbit around the sun at the focus F of the parabola. An observation of the comet is made when it is at point P_1, 15 million miles from the sun, and a second observation is made when it is at point P_2, 5 million miles from the sun. The line segments FP_1 and FP_2 are perpendicular. With this information there are two possible orbits for the comet. Find how close the comet comes to the sun for each orbit.

40. The arch of a bridge is in the shape of a semiellipse having a horizontal span of 40 m and a height of 16 m at its center. How high is the arch 9 m to the right or left of the center?

41. Points A and B are 1000 m apart, and it is determined from the sound of an explosion heard at these points at different times that the location of the explosion is 600 m closer to A than to B. Show that the location of the explosion is restricted to a particular curve, and find an equation of it.

42. Find the area of the region bounded by the two parabolas $r = 2/(1 - \cos\theta)$ and $r = 2/(1 + \cos\theta)$.

43. Prove that the line $y = 2x$ is an asymptote of the graph of the function f for which $f(x) = 2\sqrt{x^2 - 1}$.

44. A focal chord of a conic is divided into two segments by the focus. Prove that the sum of the reciprocals of the measures of the lengths of the two segments is the same, regardless of what chord is taken. (*Hint:* Use polar coordinates.)

45. A focal chord of a conic is a line segment passing through a focus and having its endpoints on the conic. Prove that if two focal chords of a parabola are perpendicular, the sum of the reciprocals of the measures of their lengths is a constant. (*Hint:* Use polar coordinates.)

46. Prove that the midpoints of all chords parallel to a fixed chord of a parabola lie on a line that is parallel to the axis of the parabola.

47. Show that $\sqrt{x} \pm \sqrt{y} = \pm\sqrt{a}$ represents a one-parameter family of conics, and determine the type of conic. Draw sketches of the conics for $a = 1$, $a = 2$, and $a = 4$.

48. The graph of the equation $(1 - e^2)x^2 + y^2 - 2px + p^2 = 0$ is a central conic having eccentricity e and a focus at $(p, 0)$. Simplify the equation by a translation of axes. Find the new origin with respect to the x and y axes, and determine the type of conic.

CHAPTER 14 Indeterminate Forms, Improper Integrals, and Taylor's Formula

14.1 THE INDETERMINATE FORM 0/0

Limit Theorem 9 (2.2.9) states that if $\lim_{x \to a} f(x)$ and $\lim_{x \to a} g(x)$ both exist, then

$$\lim_{x \to a} \frac{f(x)}{g(x)} = \frac{\lim_{x \to a} f(x)}{\lim_{x \to a} g(x)}$$

provided that $\lim_{x \to a} g(x) \neq 0$.

There are various situations for which this theorem cannot be used. In particular, if $\lim_{x \to a} g(x) = 0$ and $\lim_{x \to a} f(x) = k$, where k is a constant not equal to 0, then Limit Theorem 12 (2.4.4) can be applied. Consider the case when both $\lim_{x \to a} f(x) = 0$ and $\lim_{x \to a} g(x) = 0$. Some limits of this type have previously been discussed.

● ILLUSTRATION 1 We wish to find

$$\lim_{x \to 4} \frac{x^2 - x - 12}{x^2 - 3x - 4}$$

Here, $\lim_{x \to 4} (x^2 - x - 12) = 0$ and $\lim_{x \to 4} (x^2 - 3x - 4) = 0$. However, the numerator and denominator can be factored, which gives

$$\lim_{x \to 4} \frac{x^2 - x - 12}{x^2 - 3x - 4} = \lim_{x \to 4} \frac{(x - 4)(x + 3)}{(x - 4)(x + 1)}$$

Because $x \neq 4$, the numerator and denominator can be divided by $(x - 4)$, and we obtain

$$\lim_{x \to 4} \frac{x^2 - x - 12}{x^2 - 3x - 4} = \lim_{x \to 4} \frac{x + 3}{x + 1} = \frac{7}{5}$$ ●

● ILLUSTRATION 2 $\lim_{x \to 0} \sin x = 0$ and $\lim_{x \to 0} x = 0$; and by Theorem 2.9.2,

$$\lim_{x \to 0} \frac{\sin x}{x} = 1$$ ●

14.1.1 Definition If f and g are two functions such that $\lim_{x \to a} f(x) = 0$ and $\lim_{x \to a} g(x) = 0$, then the function f/g has the **indeterminate form 0/0 at *a*.**

● ILLUSTRATION 3 From Definition 14.1.1, $(x^2 - x - 12)/(x^2 - 3x - 4)$ has the indeterminate form 0/0 at 4; however, we saw in Illustration 1 that

$$\lim_{x \to 4} \frac{x^2 - x - 12}{x^2 - 3x - 4} = \frac{7}{5}$$

Also, $(\sin x/x)$ has the indeterminate form 0/0 at 0, while $\lim_{x \to 0} (\sin x/x) = 1$, as shown in Illustration 2. ●

We now consider a general method for finding the limit, if it exists, of a function at a number where it has the indeterminate form 0/0. This method is attributed to the French mathematician Guillaume François de L'Hôpital (1661–1707), who wrote the first calculus textbook, published in 1696. It is known as *L'Hôpital's rule.*

14.1.2 Theorem
L'Hôpital's Rule

Let f and g be functions that are differentiable on an open interval I, except possibly at the number a in I. Suppose that for all $x \neq a$ in I, $g'(x) \neq 0$. Then if $\lim_{x \to a} f(x) = 0$ and $\lim_{x \to a} g(x) = 0$, and if

$$\lim_{x \to a} \frac{f'(x)}{g'(x)} = L$$

it follows that

$$\lim_{x \to a} \frac{f(x)}{g(x)} = L$$

The theorem is valid if all the limits are right-hand limits or all the limits are left-hand limits.

Before giving the proof of Theorem 14.1.2, we show the use of the theorem by an illustration and examples.

● ILLUSTRATION 4 We use L'Hôpital's rule to evaluate the limits in Illustrations 1 and 2. In Illustration 1, because $\lim_{x \to 4} (x^2 - x - 12) = 0$ and $\lim_{x \to 4} (x^2 - 3x - 4) = 0$, we can apply L'Hôpital's rule and obtain

$$\lim_{x \to 4} \frac{x^2 - x - 12}{x^2 - 3x - 4} = \lim_{x \to 4} \frac{2x - 1}{2x - 3} = \frac{7}{5}$$

L'Hôpital's rule can be applied in Illustration 2 because $\lim_{x \to 0} \sin x = 0$ and $\lim_{x \to 0} x = 0$. Then

$$\lim_{x \to 0} \frac{\sin x}{x} = \lim_{x \to 0} \frac{\cos x}{1} = 1$$

●

Example 1 Find

$$\lim_{x \to 0} \frac{x}{1 - e^x}$$

if it exists.

Solution Because $\lim_{x \to 0} x = 0$ and $\lim_{x \to 0} (1 - e^x) = 0$, L'Hôpital's rule can be applied. Thus

$$\lim_{x \to 0} \frac{x}{1 - e^x} = \lim_{x \to 0} \frac{1}{-e^x} = \frac{1}{-1} = -1$$

Example 2 Find

$$\lim_{x \to 1} \frac{1 - x + \ln x}{x^3 - 3x + 2}$$

if it exists.

Solution

$$\lim_{x \to 1} (1 - x + \ln x) = 1 - 1 + 0 = 0$$

and

$$\lim_{x \to 1} (x^3 - 3x + 2) = 1 - 3 + 2 = 0$$

Therefore, applying L'Hôpital's rule we have

$$\lim_{x \to 1} \frac{1 - x + \ln x}{x^3 - 3x + 2} = \lim_{x \to 1} \frac{-1 + \frac{1}{x}}{3x^2 - 3}$$

Now, because $\lim_{x \to 1} (-1 + 1/x) = 0$ and $\lim_{x \to 1} (3x^2 - 3) = 0$, we apply L'Hôpital's rule again, giving

$$\lim_{x \to 1} \frac{-1 + \frac{1}{x}}{3x^2 - 3} = \lim_{x \to 1} \frac{-\frac{1}{x^2}}{6x} = -\frac{1}{6}$$

Therefore

$$\lim_{x \to 1} \frac{1 - x + \ln x}{x^3 - 3x + 2} = -\frac{1}{6}$$

To prove Theorem 14.1.2 we need to use the theorem known as *Cauchy's mean-value theorem,* which extends to two functions the mean-value theorem (4.4.2) for a single function. This theorem is attributed to the French mathematician Augustin L. Cauchy (1789–1857).

14.1.3 Theorem
Cauchy's Mean-Value Theorem

If f and g are two functions such that

(i) f and g are continuous on the closed interval $[a, b]$;
(ii) f and g are differentiable on the open interval (a, b);
(iii) for all x in the open interval (a, b), $g'(x) \neq 0$

then there exists a number z in the open interval (a, b) such that

$$\frac{f(b) - f(a)}{g(b) - g(a)} = \frac{f'(z)}{g'(z)}$$

Proof We first show that $g(b) \neq g(a)$. Assume $g(b) = g(a)$. Because g satisfies the two conditions in the hypothesis of the mean-value theorem (4.4.2), there is some number c in (a, b) such that $g'(c) = [g(b) - g(a)]/(b - a)$. But if $g(b) = g(a)$, then there is some number c in (a, b) such that $g'(c) = 0$. But condition (iii) of the hypothesis of this theorem states that for all x in (a, b), $g'(x) \neq 0$. Therefore there is a contradiction. Hence the assumption that $g(b) = g(a)$ is false. So $g(b) \neq g(a)$, and consequently $g(b) - g(a) \neq 0$.

Now consider the function h defined by

$$h(x) = f(x) - f(a) - \left[\frac{f(b) - f(a)}{g(b) - g(a)}\right][g(x) - g(a)]$$

Then

$$h'(x) = f'(x) - \left[\frac{f(b) - f(a)}{g(b) - g(a)}\right]g'(x) \qquad (1)$$

Therefore h is differentiable on (a, b) because f and g are differentiable there, and h is continuous on $[a, b]$ because f and g are continuous there.

$$h(a) = f(a) - f(a) - \left[\frac{f(b) - f(a)}{g(b) - g(a)}\right][g(a) - g(a)] = 0$$

$$h(b) = f(b) - f(a) - \left[\frac{f(b) - f(a)}{g(b) - g(a)}\right][g(b) - g(a)] = 0$$

Hence the three conditions of the hypothesis of Rolle's theorem are satisfied by the function h. So there exists a number z in the open interval (a, b) such that $h'(z) = 0$. Thus, from (1),

$$f'(z) - \frac{f(b) - f(a)}{g(b) - g(a)}g'(z) = 0 \qquad (2)$$

Because $g'(z) \neq 0$ on (a, b), we have from (2)

$$\frac{f(b) - f(a)}{g(b) - g(a)} = \frac{f'(z)}{g'(z)}$$

where z is some number in (a, b). This proves the theorem. ■

If g is the function such that $g(x) = x$, then the conclusion of Cauchy's mean-value theorem becomes the conclusion of the former mean-value theorem because then $g'(z) = 1$. So the former mean-value theorem is a special case of Cauchy's mean-value theorem.

A geometric interpretation of Cauchy's mean-value theorem is given in Sec. 16.4. It is postponed until then because parametric equations are needed.

● ILLUSTRATION 5 Suppose $f(x) = 3x^2 + 3x - 1$ and $g(x) = x^3 - 4x + 2$. We find a number z in $(0, 1)$, predicted by Theorem 14.1.3.

$$f'(x) = 6x + 3 \qquad g'(x) = 3x^2 - 4$$

Thus f and g are differentiable and continuous everywhere, and for all x in $(0, 1)$, $g'(x) \neq 0$. Hence, by Theorem 14.1.3, there exists a z in $(0, 1)$ such that

$$\frac{f(1) - f(0)}{g(1) - g(0)} = \frac{6z + 3}{3z^2 - 4}$$

Substituting $f(1) = 5$, $g(1) = -1$, $f(0) = -1$, and $g(0) = 2$, and solving for z, we have

$$\frac{5-(-1)}{-1-2} = \frac{6z+3}{3z^2-4}$$

$$6z^2 + 6z - 5 = 0$$

$$z = \frac{-6 \pm \sqrt{36+120}}{12}$$

$$= \frac{-6 \pm 2\sqrt{39}}{12}$$

$$= \frac{-3 \pm \sqrt{39}}{6}$$

Only one of these numbers is in (0, 1), namely $z = \frac{1}{6}(-3 + \sqrt{39})$. ●

We are now in a position to prove Theorem 14.1.2. We distinguish three cases: (i) $x \to a^+$; (ii) $x \to a^-$; (iii) $x \to a$.

Proof of Theorem 14.1.2(i) Because in the hypothesis it is not assumed that f and g are defined at a, we consider two new functions F and G for which

$$F(x) = f(x) \quad \text{if } x \neq a \text{ and } F(a) = 0$$
$$G(x) = g(x) \quad \text{if } x \neq a \text{ and } G(a) = 0 \tag{3}$$

Let b be the right endpoint of the open interval I given in the hypothesis. Because f and g are both differentiable on I, except possibly at a, we conclude that F and G are both differentiable on the interval $(a, x]$, where $a < x < b$. Therefore F and G are both continuous on $(a, x]$. The functions F and G are also both continuous from the right at a because $\lim_{x\to a^+} F(x) = \lim_{x\to a^+} f(x) = 0 = F(a)$, and $\lim_{x\to a^+} G(x) = \lim_{x\to a^+} g(x) = 0 = G(a)$. Therefore F and G are continuous on the closed interval $[a, x]$. So F and G satisfy the three conditions of the hypothesis of Cauchy's mean-value theorem (Theorem 14.1.3) on the interval $[a, x]$. Hence

$$\frac{F(x) - F(a)}{G(x) - G(a)} = \frac{F'(z)}{G'(z)} \tag{4}$$

where z is some number such that $a < z < x$. From (3) and (4) we have

$$\frac{f(x)}{g(x)} = \frac{f'(z)}{g'(z)} \tag{5}$$

Because $a < z < x$, it follows that as $x \to a^+$, $z \to a^+$; therefore

$$\lim_{x\to a^+} \frac{f(x)}{g(x)} = \lim_{x\to a^+} \frac{f'(z)}{g'(z)} = \lim_{z\to a^+} \frac{f'(z)}{g'(z)} \tag{6}$$

But by hypothesis, the limit on the right side of (6) is L. Therefore

$$\lim_{x\to a^+} \frac{f(x)}{g(x)} = L$$ ■

The proof of case (ii) is similar to the proof of case (i) and is left as an exercise (see Exercise 34). The proof of case (iii) is based on the results of cases (i) and (ii) and is also left as an exercise (see Exercise 35).

L'Hôpital's rule also holds if either x increases without bound or x decreases without bound, as given in the next theorem.

14.1.4 Theorem L'Hôpital's Rule

Let f and g be functions that are differentiable for all $x > N$, where N is a positive constant, and suppose that for all $x > N$, $g'(x) \neq 0$. Then if $\lim_{x \to +\infty} f(x) = 0$ and $\lim_{x \to +\infty} g(x) = 0$ and if

$$\lim_{x \to +\infty} \frac{f'(x)}{g'(x)} = L$$

it follows that

$$\lim_{x \to +\infty} \frac{f(x)}{g(x)} = L$$

The theorem is also valid if $x \to +\infty$ is replaced by $x \to -\infty$.

Proof We prove the theorem for $x \to +\infty$. The proof for $x \to -\infty$ is left as an exercise (see Exercise 37).

For all $x > N$, let $x = 1/t$; then $t = 1/x$. Let F and G be the functions defined by $F(t) = f(1/t)$ and $G(t) = g(1/t)$, if $t \neq 0$. Then $f(x) = F(t)$ and $g(x) = G(t)$, where $x > N$ and $0 < t < 1/N$. From Definitions 2.5.1 and 2.3.1 it may be shown that the statements

$$\lim_{x \to +\infty} f(x) = M \quad \text{and} \quad \lim_{t \to 0^+} F(t) = M$$

have the same meaning. It is left as an exercise to prove this (see Exercise 36). Because, by hypothesis, $\lim_{x \to +\infty} f(x) = 0$ and $\lim_{x \to +\infty} g(x) = 0$, we can conclude that

$$\lim_{t \to 0^+} F(t) = 0 \quad \text{and} \quad \lim_{t \to 0^+} G(t) = 0 \tag{7}$$

Considering the quotient $F'(t)/G'(t)$, we have, using the chain rule,

$$\frac{F'(t)}{G'(t)} = \frac{-\dfrac{1}{t^2} f'\left(\dfrac{1}{t}\right)}{-\dfrac{1}{t^2} g'\left(\dfrac{1}{t}\right)} = \frac{f'\left(\dfrac{1}{t}\right)}{g'\left(\dfrac{1}{t}\right)} = \frac{f'(x)}{g'(x)}$$

Because by hypothesis $\lim_{x \to +\infty} f'(x)/g'(x) = L$, it follows from the above that

$$\lim_{t \to 0^+} \frac{F'(t)}{G'(t)} = L \tag{8}$$

Because for all $x > N$, $g'(x) \neq 0$,

$$G'(t) \neq 0 \qquad \text{for all } 0 < t < \frac{1}{N} \tag{9}$$

From (7), (8), and (9) it follows from Theorem 14.1.2 that

$$\lim_{t \to 0^+} \frac{F(t)}{G(t)} = L$$

But because $F(t)/G(t) = f(x)/g(x)$ for all $x > N$ and $t \neq 0$, then

$$\lim_{x \to +\infty} \frac{f(x)}{g(x)} = L$$

and so the theorem is proved. ■

Theorems 14.1.2 and 14.1.4 also hold if L is replaced by $+\infty$ or $-\infty$. The proofs for these cases are omitted.

Example 3 Find

$$\lim_{x \to +\infty} \frac{\sin \frac{1}{x}}{\tan^{-1}\left(\frac{1}{x}\right)}$$

if it exists.

Solution $\lim_{x \to +\infty} \sin(1/x) = 0$ and $\lim_{x \to +\infty} \tan^{-1}(1/x) = 0$. So from L'Hôpital's rule,

$$\lim_{x \to +\infty} \frac{\sin \frac{1}{x}}{\tan^{-1}\left(\frac{1}{x}\right)} = \lim_{x \to +\infty} \frac{\cos \frac{1}{x} \cdot \left(-\frac{1}{x^2}\right)}{\frac{1}{1 + \frac{1}{x^2}} \cdot \left(-\frac{1}{x^2}\right)} = \lim_{x \to +\infty} \frac{\cos \frac{1}{x}}{\frac{x^2}{x^2 + 1}}$$

Because

$$\lim_{x \to +\infty} \cos \frac{1}{x} = 1 \quad \text{and} \quad \lim_{x \to +\infty} \frac{x^2}{x^2 + 1} = \lim_{x \to +\infty} \frac{1}{1 + \frac{1}{x^2}} = 1$$

it follows that

$$\lim_{x \to +\infty} \frac{\cos \frac{1}{x}}{\frac{x^2}{x^2 + 1}} = 1$$

Therefore the given limit is 1.

EXERCISES 14.1

In Exercises 1 through 5, find all values of z in the given interval (a, b) satisfying the conclusion of Theorem 14.1.3 for the given pair of functions.

1. $f(x) = x^3, g(x) = x^2; (a, b) = (0, 2)$

2. $f(x) = \dfrac{2x}{1 + x^2}, g(x) = \dfrac{1 - x^2}{1 + x^2}; (a, b) = (0, 2)$

3. $f(x) = \sin x, g(x) = \cos x; (a, b) = (0, \pi)$

4. $f(x) = \cos 2x, g(x) = \sin x; (a, b) = (0, \frac{1}{2}\pi)$

5. $f(x) = \sqrt{x + 5}, g(x) = x + 3; (a, b) = (-4, -1)$

In Exercises 6 through 30, evaluate the limit, if it exists.

6. $\displaystyle\lim_{x \to \pi/2} \frac{3 \cos x}{2x - \pi}$

7. $\displaystyle\lim_{x \to 0} \frac{x}{\tan x}$

8. $\displaystyle\lim_{x \to 0} \frac{\tan x - x}{x - \sin x}$

9. $\displaystyle\lim_{x \to +\infty} \frac{\sin \frac{2}{x}}{\frac{1}{x}}$

10. $\displaystyle\lim_{x \to 2} \frac{\sin \pi x}{2 - x}$

11. $\displaystyle\lim_{x \to -\infty} \frac{2^x}{e^x}$

12. $\displaystyle\lim_{x \to 0} \frac{e^x - \cos x}{x \sin x}$

13. $\displaystyle\lim_{x \to \pi/2} \frac{\ln(\sin x)}{(\pi - 2x)^2}$

14. $\displaystyle\lim_{x \to 0} \frac{\sin^{-1} x}{x}$

15. $\displaystyle\lim_{x \to 0} \frac{2^x - 3^x}{x}$

16. $\displaystyle\lim_{t \to 2} \frac{t^n - 2^n}{t - 2}$

17. $\displaystyle\lim_{\theta \to 0} \frac{\theta - \sin \theta}{\tan^3 \theta}$

18. $\displaystyle\lim_{x \to 0} \frac{e^{2x^2} - 1}{\sin^2 x}$

19. $\displaystyle\lim_{z \to +\infty} \frac{1 - e^{1/z}}{-\frac{3}{z}}$

20. $\displaystyle\lim_{y \to 0} \frac{y^2}{1 - \cosh y}$

21. $\displaystyle\lim_{t \to 0} \frac{\sin t}{\ln(2e^t - 1)}$

22. $\displaystyle\lim_{z \to 0} \frac{5z}{5^z - e^z}$

23. $\displaystyle\lim_{t \to 0} \frac{\sin^2 t}{\sin t^2}$

24. $\displaystyle\lim_{x \to 0} \frac{\tanh 2x}{\tanh x}$

25. $\displaystyle\lim_{x \to 0} \frac{(1 + x)^{1/5} - (1 - x)^{1/5}}{(1 + x)^{1/3} - (1 - x)^{1/3}}$

26. $\displaystyle\lim_{x \to 0} \frac{e^x - 10^x}{x}$

27. $\displaystyle\lim_{x \to \pi} \frac{1 + \cos 2x}{1 - \sin x}$

28. $\displaystyle\lim_{x \to +\infty} \frac{\frac{1}{x^2} - 2 \tan^{-1} \frac{1}{x}}{\frac{1}{x}}$

29. $\displaystyle\lim_{x \to 0} \frac{\cos x - \cosh x}{x^2}$

30. $\displaystyle\lim_{x \to 0} \frac{\sinh x - \sin x}{\sin^3 x}$

31. An electrical circuit has a resistance of R ohms, an inductance of L henries, and an electromotive force of E volts, where R, L, and E are positive. If I amperes is the current flowing in the circuit t sec after a switch is turned on, then

$$I = \frac{E}{R}(1 - e^{-Rt/L})$$

If t, E, and L are constants find $\lim_{R \to 0^+} I$.

32. In a geometric progression, if a is the first term, r is the common ratio of two successive terms, and S is the sum of the first n terms, then if $r \neq 1$,

$$S = \frac{a(r^n - 1)}{r - 1}$$

Find $\lim_{r \to 1} S$. Is the result consistent with the sum of the first n terms if $r = 1$?

33. Find values for a and b such that

$$\lim_{x \to 0} \frac{\sin 3x + ax + bx^3}{x^3} = 0$$

34. Prove Theorem 14.1.2(ii). **35.** Prove Theorem 14.1.2(iii).

36. Suppose that f is a function defined for all $x > N$, where N is a positive constant. If $t = 1/x$ and $F(t) = f(1/t)$, where $t \neq 0$, prove that the statements $\lim_{x \to +\infty} f(x) = M$ and $\lim_{t \to 0^+} F(t) = M$ have the same meaning.

37. Prove Theorem 14.1.4 for $x \to -\infty$.

14.2 OTHER INDETERMINATE FORMS

Suppose that we wish to determine if $\lim_{x \to \pi/2} (\sec^2 x/\sec^2 3x)$ exists. We cannot apply the theorem involving the limit of a quotient because $\lim_{x \to \pi/2} \sec^2 x = +\infty$ and $\lim_{x \to \pi/2} \sec^2 3x = +\infty$. In this case we say that the function defined by $\sec^2 x/\sec^2 3x$ has the indeterminate form $(+\infty/+\infty)$ at $x = \frac{1}{2}\pi$. L'Hôpital's rule also applies to an indeterminate form of this type as well as to $(-\infty/-\infty)$, $(-\infty/+\infty)$ and $(+\infty/-\infty)$. This is given by the following theorems, for which the proofs are omitted because they are beyond the scope of this book.

14.2.1 Theorem
L'Hôpital's Rule

Let f and g be functions that are differentiable on an open interval I, except possibly at the number a in I, and suppose that for all $x \neq a$ in I, $g'(x) \neq 0$. Then if $\lim_{x \to a} f(x) = +\infty$ or $-\infty$, and $\lim_{x \to a} g(x) = +\infty$ or $-\infty$, and if

$$\lim_{x \to a} \frac{f'(x)}{g'(x)} = L$$

it follows that

$$\lim_{x \to a} \frac{f(x)}{g(x)} = L$$

The theorem is valid if all the limits are right-hand limits or if all the limits are left-hand limits.

Example 1 Find

$$\lim_{x \to 0^+} \frac{\ln x}{\dfrac{1}{x}}$$

if it exists.

Solution Because $\lim_{x \to 0^+} \ln x = -\infty$ and $\lim_{x \to 0^+} (1/x) = +\infty$, we apply L'Hôpital's rule and get

$$\lim_{x \to 0^+} \frac{\ln x}{\dfrac{1}{x}} = \lim_{x \to 0^+} \frac{\dfrac{1}{x}}{-\dfrac{1}{x^2}} = \lim_{x \to 0^+} (-x) = 0$$

14.2.2 Theorem L'Hôpital's Rule

Let f and g be functions that are differentiable for all $x > N$, where N is a positive constant, and suppose that for all $x > N$, $g'(x) \neq 0$. Then if $\lim_{x \to +\infty} f(x) = +\infty$ or $-\infty$, and $\lim_{x \to +\infty} g(x) = +\infty$ or $-\infty$, and if

$$\lim_{x \to +\infty} \frac{f'(x)}{g'(x)} = L$$

it follows that

$$\lim_{x \to +\infty} \frac{f(x)}{g(x)} = L$$

The theorem is also valid if $x \to +\infty$ is replaced by $x \to -\infty$.

Theorems 14.2.1 and 14.2.2 also hold if L is replaced by $+\infty$ or $-\infty$, and the proofs for these cases are also omitted.

Example 2 Find

$$\lim_{x \to +\infty} \frac{\ln(2 + e^x)}{3x}$$

if it exists.

Solution Because $\lim_{x \to +\infty} \ln(2 + e^x) = +\infty$ and $\lim_{x \to +\infty} 3x = +\infty$, by applying L'Hôpital's rule we obtain

$$\lim_{x \to +\infty} \frac{\ln(2 + e^x)}{3x} = \lim_{x \to +\infty} \frac{\dfrac{1}{2 + e^x} \cdot e^x}{3}$$

$$= \lim_{x \to +\infty} \frac{e^x}{6 + 3e^x}$$

Now because $\lim_{x \to +\infty} e^x = +\infty$ and $\lim_{x \to +\infty} (6 + 3e^x) = +\infty$, we apply L'Hôpital's rule again and get

$$\lim_{x \to +\infty} \frac{e^x}{6 + 3e^x} = \lim_{x \to +\infty} \frac{e^x}{3e^x} = \lim_{x \to +\infty} \frac{1}{3} = \frac{1}{3}$$

Therefore

$$\lim_{x \to +\infty} \frac{\ln(2 + e^x)}{3x} = \frac{1}{3}$$

Example 3 Find

$$\lim_{x \to \pi/2^-} \frac{\sec^2 x}{\sec^2 3x}$$

if it exists.

Solution $\lim\limits_{x \to \pi/2^-} \sec^2 x = +\infty$ and $\lim\limits_{x \to \pi/2^-} \sec^2 3x = +\infty$. So from L'Hôpital's rule,

$$\lim_{x \to \pi/2^-} \frac{\sec^2 x}{\sec^2 3x} = \lim_{x \to \pi/2^-} \frac{2 \sec^2 x \tan x}{6 \sec^2 3x \tan 3x}$$

$$\lim_{x \to \pi/2^-} 2 \sec^2 x \tan x = +\infty \quad \text{and} \quad \lim_{x \to \pi/2^-} 6 \sec^2 3x \tan 3x = +\infty$$

It may be seen that further applications of L'Hôpital's rule will not help us. However, the original quotient may be rewritten and we have

$$\lim_{x \to \pi/2^-} \frac{\sec^2 x}{\sec^2 3x} = \lim_{x \to \pi/2^-} \frac{\cos^2 3x}{\cos^2 x}$$

Now, because $\lim\limits_{x \to \pi/2^-} \cos^2 3x = 0$ and $\lim\limits_{x \to \pi/2^-} \cos^2 x = 0$, we may apply L'Hôpital's rule, giving

$$\begin{aligned}
\lim_{x \to \pi/2^-} \frac{\cos^2 3x}{\cos^2 x} &= \lim_{x \to \pi/2^-} \frac{-6 \cos 3x \sin 3x}{-2 \cos x \sin x} \\
&= \lim_{x \to \pi/2^-} \frac{3(2 \cos 3x \sin 3x)}{(2 \cos x \sin x)} \\
&= \lim_{x \to \pi/2^-} \frac{3 \sin 6x}{\sin 2x}
\end{aligned}$$

Because $\lim\limits_{x \to \pi/2^-} 3 \sin 6x = 0$ and $\lim\limits_{x \to \pi/2^-} \sin 2x = 0$, we use L'Hôpital's rule again and have

$$\lim_{x \to \pi/2^-} \frac{3 \sin 6x}{\sin 2x} = \lim_{x \to \pi/2^-} \frac{18 \cos 6x}{2 \cos 2x} = \frac{18(-1)}{2(-1)} = 9$$

Therefore

$$\lim_{x \to \pi/2^-} \frac{\sec^2 x}{\sec^2 3x} = 9$$

The limit in Example 3 can be evaluated without L'Hôpital's rule by using Eqs. (11) in Sec. 1.7 and Theorem 2.9.2. This is left as an exercise (see Exercise 38).

In addition to $0/0$ and $\pm\infty/\pm\infty$, other indeterminate forms are $0 \cdot (+\infty)$, $+\infty - (+\infty)$, 0^0, $(\pm\infty)^0$, and $1^{\pm\infty}$. These indeterminate forms are defined analogously to the other two. For instance, if $\lim\limits_{x \to a} f(x) = +\infty$ and $\lim\limits_{x \to a} g(x) = 0$, then the function defined by $f(x)^{g(x)}$ has the indeterminate form $(+\infty)^0$ at a. To find the limit of a function having one of these indeterminate forms, it must be changed to either the form $0/0$ or $\pm\infty/\pm\infty$ before L'Hôpital's rule can be applied. The following examples illustrate the method.

Example 4 Find $\lim_{x \to 0^+} \sin^{-1} x \csc x$ if it exists.

Solution Because $\lim_{x \to 0^+} \sin^{-1} x = 0$ and $\lim_{x \to 0^+} \csc x = +\infty$, the function defined by $\sin^{-1} x \csc x$ has the indeterminate form $0 \cdot (+\infty)$ at 0. Before we can apply L'Hôpital's rule we rewrite $\sin^{-1} x \csc x$ as $\sin^{-1} x/\sin x$, and consider $\lim_{x \to 0^+} (\sin^{-1} x/\sin x)$. Now $\lim_{x \to 0^+} \sin^{-1} x = 0$ and $\lim_{x \to 0^+} \sin x = 0$; so we have the indeterminate form $0/0$. Therefore, from L'Hôpital's rule we obtain

$$\lim_{x \to 0^+} \frac{\sin^{-1} x}{\sin x} = \lim_{x \to 0^+} \frac{\dfrac{1}{\sqrt{1 - x^2}}}{\cos x} = \frac{1}{1} = 1$$

Example 5 Find

$$\lim_{x \to 0} \left(\frac{1}{x^2} - \frac{1}{x^2 \sec x} \right)$$

if it exists.

Solution Because

$$\lim_{x \to 0} \frac{1}{x^2} = +\infty \quad \text{and} \quad \lim_{x \to 0} \frac{1}{x^2 \sec x} = +\infty$$

we have the indeterminate form $+\infty - (+\infty)$. Rewriting the expression we have

$$\lim_{x \to 0} \left(\frac{1}{x^2} - \frac{1}{x^2 \sec x} \right) = \lim_{x \to 0} \frac{\sec x - 1}{x^2 \sec x}$$

$\lim_{x \to 0} (\sec x - 1) = 0$ and $\lim_{x \to 0} (x^2 \sec x) = 0$; so we apply L'Hôpital's rule and obtain

$$\lim_{x \to 0} \frac{\sec x - 1}{x^2 \sec x} = \lim_{x \to 0} \frac{\sec x \tan x}{2x \sec x + x^2 \sec x \tan x}$$

$$= \lim_{x \to 0} \frac{\tan x}{2x + x^2 \tan x}$$

$$\lim_{x \to 0} \tan x = 0 \quad \text{and} \quad \lim_{x \to 0} (2x + x^2 \tan x) = 0$$

and so we apply the rule again and obtain

$$\lim_{x \to 0} \frac{\tan x}{2x + x^2 \tan x} = \lim_{x \to 0} \frac{\sec^2 x}{2 + 2x \tan x + x^2 \sec^2 x} = \frac{1}{2}$$

Therefore

$$\lim_{x \to 0} \left(\frac{1}{x^2} - \frac{1}{x^2 \sec x} \right) = \frac{1}{2}$$

For any one of the indeterminate forms 0^0, $(\pm\infty)^0$, $1^{\pm\infty}$, the procedure for evaluating the limit is illustrated in Example 6.

Example 6 Find $\lim_{x\to 0^+} (x+1)^{\cot x}$ if it exists.

Solution Because $\lim_{x\to 0^+} (x+1) = 1$ and $\lim_{x\to 0^+} \cot x = +\infty$, we have the indeterminate form $1^{+\infty}$. Let

$$y = (x+1)^{\cot x} \tag{1}$$

Then

$$\ln y = \cot x \ln(x+1)$$

$$\ln y = \frac{\ln(x+1)}{\tan x}$$

So

$$\lim_{x\to 0^+} \ln y = \lim_{x\to 0^+} \frac{\ln(x+1)}{\tan x} \tag{2}$$

Because $\lim_{x\to 0^+} \ln(x+1) = 0$ and $\lim_{x\to 0^+} \tan x = 0$, we may apply L'Hôpital's rule to the right side of (2) and obtain

$$\lim_{x\to 0^+} \frac{\ln(x+1)}{\tan x} = \lim_{x\to 0^+} \frac{\frac{1}{x+1}}{\sec^2 x} = 1$$

Therefore, substituting 1 on the right side of (2) we have

$$\lim_{x\to 0^+} \ln y = 1 \tag{3}$$

Because the exponential function is continuous on its entire domain, which is the set of all real numbers, we may apply Theorem 2.7.6, and so

$$\lim_{x\to 0^+} \exp(\ln y) = \exp\left(\lim_{x\to 0^+} \ln y\right) \tag{4}$$

Therefore it follows from (3) and (4) that

$$\lim_{x\to 0^+} y = e^1$$

But from (1), $y = (x+1)^{\cot x}$, and therefore

$$\lim_{x\to 0^+} (x+1)^{\cot x} = e$$

EXERCISES 14.2

In Exercises 1 through 34, evaluate the limit, if it exists.

1. $\lim_{x\to +\infty} \frac{x^2}{e^x}$ **2.** $\lim_{x\to \pi/2^-} \frac{\ln(\cos x)}{\ln(\tan x)}$ **3.** $\lim_{x\to 1/2^-} \frac{\ln(1-2x)}{\tan \pi x}$

4. $\lim\limits_{x\to+\infty} \dfrac{(\ln x)^3}{x}$

5. $\lim\limits_{x\to0^+} x \csc x$

6. $\lim\limits_{x\to1/2^+} (2x-1)\tan \pi x$

7. $\lim\limits_{x\to+\infty} \dfrac{\ln x}{\sqrt{x}}$

8. $\lim\limits_{x\to0^+} \tan^{-1} x \cot x$

9. $\lim\limits_{x\to0^+} \tan x(\ln x)$

10. $\lim\limits_{x\to0^+} \left(\dfrac{1}{\sin x} - \dfrac{1}{x}\right)$

11. $\lim\limits_{x\to1} \left(\dfrac{1}{\ln x} - \dfrac{1}{x-1}\right)$

12. $\lim\limits_{x\to0^+} (\sinh x)^{\tan x}$

13. $\lim\limits_{x\to0^+} x^{\sin x}$

14. $\lim\limits_{x\to0} (x + e^{2x})^{1/x}$

15. $\lim\limits_{x\to+\infty} (x^2 - \sqrt{x^4 - x^2 + 2})$

16. $\lim\limits_{x\to2} \left(\dfrac{5}{x^2+x-6} - \dfrac{1}{x-2}\right)$

17. $\lim\limits_{x\to+\infty} x^{1/x}$

18. $\lim\limits_{x\to0^+} (1+x)^{\ln x}$

19. $\lim\limits_{x\to+\infty} (e^x + x)^{2/x}$

20. $\lim\limits_{x\to0^+} x^{1/\ln x}$

21. $\lim\limits_{x\to0} (1+ax)^{1/x};\ a \neq 0$

22. $\lim\limits_{x\to+\infty} \dfrac{x^2+2x}{e^{3x}-1}$

23. $\lim\limits_{x\to0^+} (\sin x)^{x^2}$

24. $\lim\limits_{x\to+\infty} \left(1 + \dfrac{1}{2x}\right)^{x^2}$

25. $\lim\limits_{x\to0} (1 + \sinh x)^{2/x}$

26. $\lim\limits_{x\to2} (x-2) \tan \frac{1}{4}\pi x$

27. $\lim\limits_{x\to0} [(\cos x)e^{x^2/2}]^{4/x^4}$

28. $\lim\limits_{x\to0} (\cos x)^{1/x^2}$

29. $\lim\limits_{x\to+\infty} [(x^6 + 3x^5 + 4)^{1/6} - x]$

30. $\lim\limits_{x\to+\infty} \dfrac{\ln(x+e^x)}{3x}$

31. $\lim\limits_{x\to0^+} \dfrac{e^{-1/x}}{x}$

32. $\lim\limits_{x\to0^+} x^{x^x}$

33. $\lim\limits_{x\to+\infty} \dfrac{x}{\sqrt{1+x^2}}$

34. $\lim\limits_{x\to+\infty} (x - \sqrt{x^2 + x})$

35. If

$$f(x) = \begin{cases} (1-e^{4x})^x & \text{if } x < 0 \\ k & \text{if } 0 \le x \end{cases}$$

find k so that f is continuous at $x = 0$.

36. If $\lim\limits_{x\to+\infty} \left(\dfrac{nx+1}{nx-1}\right)^x = 9$, find n.

37. (a) Prove that $\lim\limits_{x\to0} (e^{-1/x^2}/x^n) = 0$ for any positive integer n. (b) If $f(x) = e^{-1/x^2}$, use the result of (a) to prove that the limit of f and all of its derivatives, as x approaches 0, is 0.

38. Evaluate the limit in Example 3 without using L'Hôpital's rule but by using Eqs. (11) in Sec. 1.7 and Theorem 2.9.2.

39. Suppose $f(x) = \displaystyle\int_1^x e^{3t}\sqrt{9t^4+1}\, dt$ and $g(x) = x^n e^{3x}$. If $\lim\limits_{x\to+\infty} \left[\dfrac{f'(x)}{g'(x)}\right] = 1$, find n.

14.3 IMPROPER INTEGRALS WITH INFINITE LIMITS OF INTEGRATION

In defining the definite integral $\int_a^b f(x)\, dx$, the function f was assumed to be defined on the closed interval $[a, b]$. We now extend the definition of the definite integral to consider an infinite interval of integration and call such an integral an **improper integral.** In Sec. 14.4 we discuss another kind of improper integral.

● ILLUSTRATION 1 Consider the problem of finding the area of the region bounded by the curve $y = e^{-x}$, the x axis, the y axis, and the line $x = b$, where $b > 0$. This region is shown in Fig. 14.3.1. If A square units is the area of the region,

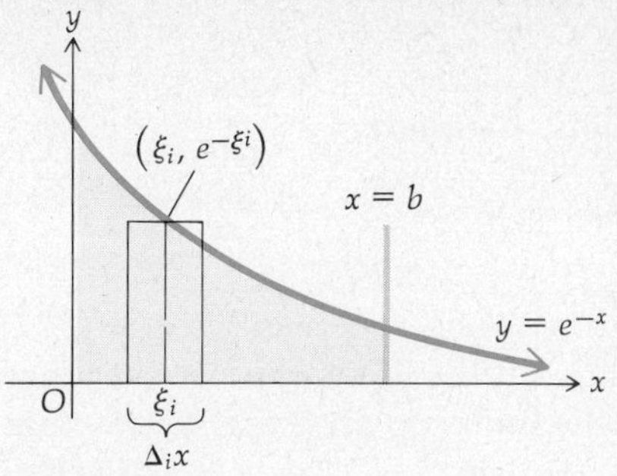

FIGURE 14.3.1

$$A = \lim_{\|\Delta\| \to 0} \sum_{i=1}^{n} e^{-\xi_i} \Delta_i x$$

$$= \int_0^b e^{-x}\, dx$$

$$= -e^{-x}\Big]_0^b$$

$$= 1 - e^{-b}$$

If we let b increase without bound, then

$$\lim_{b \to +\infty} \int_0^b e^{-x}\, dx = \lim_{b \to +\infty} (1 - e^{-b})$$

$$\lim_{b \to +\infty} \int_0^b e^{-x}\, dx = 1 \tag{1}$$

From (1) it follows that no matter how large a value we take for b, the area of the region shown in Fig. 14.3.1 will always be less than 1 square unit. ●

Equation (1) states that if $b > 0$, for any $\epsilon > 0$ there exists an $N > 0$ such that

$$\left| \int_0^b e^{-x}\, dx - 1 \right| < \epsilon \quad \text{whenever} \quad b > N$$

In place of (1) we write

$$\int_0^{+\infty} e^{-x}\, dx = 1$$

In general, we have the following definition.

14.3.1 Definition If f is continuous for all $x \geq a$, then

$$\int_a^{+\infty} f(x)\, dx = \lim_{b \to +\infty} \int_a^b f(x)\, dx$$

if this limit exists.

If the lower limit of integration is infinite, we have the following definition.

14.3.2 Definition If f is continuous for all $x \leq b$, then

$$\int_{-\infty}^b f(x)\, dx = \lim_{a \to -\infty} \int_a^b f(x)\, dx$$

if this limit exists.

Finally, we have the case when both limits of integration are infinite.

14.3.3 Definition If f is continuous for all values of x, and c is any real number, then

$$\int_{-\infty}^{+\infty} f(x)\,dx = \lim_{a\to-\infty} \int_a^c f(x)\,dx + \lim_{b\to+\infty} \int_c^b f(x)\,dx$$

if these limits exist.

When Definition 14.3.3 is applied, c is usually taken as 0.

In Definitions 14.3.1, 14.3.2, and 14.3.3, if the limits exist, we say that the improper integral is **convergent.** If the limits do not exist, we say that the improper integral is **divergent.**

Example 1 Evaluate

$$\int_{-\infty}^{2} \frac{dx}{(4-x)^2}$$

if it converges.

Solution

$$\int_{-\infty}^{2} \frac{dx}{(4-x)^2} = \lim_{a\to-\infty} \int_a^2 \frac{dx}{(4-x)^2}$$

$$= \lim_{a\to-\infty} \left[\frac{1}{4-x}\right]_a^2$$

$$= \lim_{a\to-\infty} \left(\frac{1}{2} - \frac{1}{4-a}\right)$$

$$= \tfrac{1}{2} - 0$$

$$= \tfrac{1}{2}$$

Example 2 Evaluate if they exist:

(a) $\displaystyle\int_{-\infty}^{+\infty} x\,dx$; (b) $\displaystyle\lim_{r\to+\infty} \int_{-r}^{r} x\,dx$.

Solution

(a) From Definition 14.3.3 with $c = 0$ we have

$$\int_{-\infty}^{+\infty} x\,dx = \lim_{a\to-\infty} \int_a^0 x\,dx + \lim_{b\to+\infty} \int_0^b x\,dx$$

$$= \lim_{a\to-\infty} \left[\tfrac{1}{2}x^2\right]_a^0 + \lim_{b\to+\infty} \left[\tfrac{1}{2}x^2\right]_0^b$$

$$= \lim_{a\to-\infty} (-\tfrac{1}{2}a^2) + \lim_{b\to+\infty} \tfrac{1}{2}b^2$$

Because neither of these two limits exists, the improper integral diverges.

$$\text{(b)} \lim_{r \to +\infty} \int_{-r}^{r} x\,dx = \lim_{r \to +\infty} \left[\tfrac{1}{2}x^2\right]_{-r}^{r}$$

$$= \lim_{r \to +\infty} \left(\tfrac{1}{2}r^2 - \tfrac{1}{2}r^2\right)$$

$$= \lim_{r \to +\infty} 0$$

$$= 0$$

Example 2 illustrates why we do not use the limit in (b) to determine the convergence of an improper integral where both limits of integration are infinite. That is, the improper integral in (a) is divergent, but the limit in (b) exists and is zero.

Example 3 Evaluate

$$\int_{-\infty}^{+\infty} \frac{dx}{x^2 + 6x + 12}$$

if it converges.

Solution

$$\int_{-\infty}^{+\infty} \frac{dx}{x^2 + 6x + 12} = \lim_{a \to -\infty} \int_{a}^{0} \frac{dx}{(x+3)^2 + 3} + \lim_{b \to +\infty} \int_{0}^{b} \frac{dx}{(x+3)^2 + 3}$$

$$= \lim_{a \to -\infty} \left[\frac{1}{\sqrt{3}} \tan^{-1} \frac{x+3}{\sqrt{3}}\right]_{a}^{0} + \lim_{b \to +\infty} \left[\frac{1}{\sqrt{3}} \tan^{-1} \frac{x+3}{\sqrt{3}}\right]_{0}^{b}$$

$$= \lim_{a \to -\infty} \left(\frac{1}{\sqrt{3}} \tan^{-1} \sqrt{3} - \frac{1}{\sqrt{3}} \tan^{-1} \frac{a+3}{\sqrt{3}}\right)$$

$$+ \lim_{b \to +\infty} \left(\frac{1}{\sqrt{3}} \tan^{-1} \frac{b+3}{\sqrt{3}} - \frac{1}{\sqrt{3}} \tan^{-1} \sqrt{3}\right)$$

$$= \frac{1}{\sqrt{3}} \left[\lim_{a \to -\infty} \left(-\tan^{-1} \frac{a+3}{\sqrt{3}}\right) + \lim_{b \to +\infty} \tan^{-1} \frac{b+3}{\sqrt{3}}\right]$$

$$= \frac{1}{\sqrt{3}} \left[-\left(-\frac{\pi}{2}\right) + \frac{\pi}{2}\right]$$

$$= \frac{\pi}{\sqrt{3}}$$

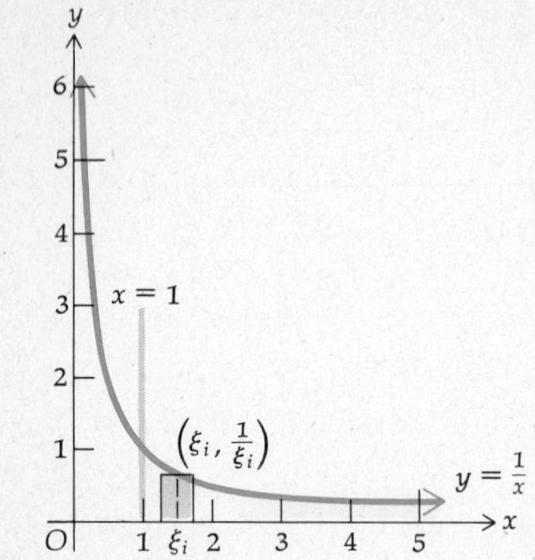

FIGURE 14.3.2

Example 4 Is it possible to assign a finite number to represent the measure of the area of the region bounded by the graphs of the equations $y = 1/x$, $y = 0$, and $x = 1$?

Solution The region is shown in Fig. 14.3.2. Let L be the number we wish to assign to the measure of the area, if possible. Let A be the measure of the area of the region bounded by the graphs of the equations $y = 1/x$, $y = 0$, $x = 1$, and $x = b$ where $b > 1$. Then

$$A = \lim_{\|\Delta\| \to 0} \sum_{i=1}^{n} \frac{1}{\xi_i} \Delta_i x = \int_1^b \frac{1}{x}\, dx$$

So we shall let $L = \lim_{b \to +\infty} A$ if this limit exists. But

$$\begin{aligned}\lim_{b \to +\infty} A &= \lim_{b \to +\infty} \int_1^b \frac{1}{x}\, dx \\ &= \lim_{b \to +\infty} [\ln b - \ln 1] \\ &= +\infty\end{aligned}$$

Therefore it is not possible to assign a finite number to represent the measure of the area of the region.

Example 5 Is it possible to assign a finite number to represent the measure of the volume of the solid formed by revolving the region in Example 4 about the x axis?

Solution The element of volume is a circular disk having a thickness of $\Delta_i x$ and a base radius of $1/\xi_i$. Let L be the number we wish to assign to the measure of the volume, and let V be the measure of the volume of the solid formed by revolving about the x axis the region bounded by the graphs of the equations $y = 1/x$, $y = 0$, $x = 1$, and $x = b$ where $b > 1$. Then

$$V = \lim_{\|\Delta\| \to 0} \sum_{i=1}^{n} \pi \left(\frac{1}{\xi_i}\right)^2 \Delta_i x = \pi \int_1^b \frac{1}{x^2}\, dx$$

We shall let $L = \lim_{b \to +\infty} V$, if the limit exists.

$$\begin{aligned}\lim_{b \to +\infty} V &= \lim_{b \to +\infty} \pi \int_1^b \frac{dx}{x^2} \\ &= \pi \lim_{b \to +\infty} \left[-\frac{1}{x}\right]_1^b \\ &= \pi \lim_{b \to +\infty} \left(-\frac{1}{b} + 1\right) \\ &= \pi\end{aligned}$$

Therefore we assign π to represent the measure of the volume of the solid.

Example 6 Determine if $\int_0^{+\infty} \sin x\, dx$ is convergent or divergent.

Solution

$$\int_0^{+\infty} \sin x\, dx = \lim_{b \to +\infty} \int_0^b \sin x\, dx$$
$$= \lim_{b \to +\infty} \Big[-\cos x\Big]_0^b$$
$$= \lim_{b \to +\infty} (-\cos b + 1)$$

For any integer n, as b takes on all values from $n\pi$ to $2n\pi$, $\cos b$ takes on all values from -1 to 1. Hence $\lim_{b \to +\infty} \cos b$ does not exist. Therefore the improper integral is divergent.

Example 6 illustrates the case for which an improper integral is divergent where the limit is not infinite.

EXERCISES 14.3

In Exercises 1 through 18, determine whether the improper integral is convergent or divergent. If it is convergent, evaluate it.

1. $\int_0^{+\infty} e^{-x}\, dx$
2. $\int_{-\infty}^{1} e^{x}\, dx$
3. $\int_{-\infty}^{0} x\, 5^{-x^2}\, dx$
4. $\int_1^{+\infty} 2^{-x}\, dx$
5. $\int_0^{+\infty} xe^{-x}\, dx$
6. $\int_5^{+\infty} \dfrac{dx}{\sqrt{x-1}}$
7. $\int_{-\infty}^{+\infty} x \cosh x\, dx$
8. $\int_{-\infty}^{0} x^2 e^{x}\, dx$
9. $\int_5^{+\infty} \dfrac{x\, dx}{\sqrt[3]{9-x^2}}$
10. $\int_{-\infty}^{+\infty} \dfrac{3x\, dx}{(3x^2+2)^3}$
11. $\int_{\sqrt{3}}^{+\infty} \dfrac{3\, dx}{x^2+9}$
12. $\int_e^{+\infty} \dfrac{dx}{x \ln x}$
13. $\int_{-\infty}^{+\infty} e^{-|x|}\, dx$
14. $\int_{-\infty}^{+\infty} xe^{-x^2}\, dx$
15. $\int_e^{+\infty} \dfrac{dx}{x(\ln x)^2}$
16. $\int_{-\infty}^{+\infty} \dfrac{dx}{16+x^2}$
17. $\int_1^{+\infty} \ln x\, dx$
18. $\int_0^{+\infty} e^{-x} \cos x\, dx$
19. Evaluate if they exist: (a) $\int_{-\infty}^{+\infty} \sin x\, dx$; (b) $\lim_{r \to +\infty} \int_{-r}^{r} \sin x\, dx$.
20. Prove that if $\int_{-\infty}^{b} f(x)\, dx$ is convergent, then $\int_{-b}^{+\infty} f(-x)\, dx$ is also convergent and has the same value.
21. Show that the improper integral $\int_{-\infty}^{+\infty} x(1+x^2)^{-2}\, dx$ is convergent and the improper integral $\int_{-\infty}^{+\infty} x(1+x^2)^{-1}\, dx$ is divergent.
22. Prove that the improper integral $\int_1^{+\infty} \dfrac{dx}{x^n}$ is convergent if and only if $n > 1$.
23. Determine if it is possible to assign a finite number to represent the measure of the area of the region bounded by the curve whose equation is $y = 1/(e^x + e^{-x})$ and the x axis. If a finite number can be assigned, find it.
24. Determine if it is possible to assign a finite number to represent the measure of the area of the region bounded by the x axis, the line $x = 2$, and the curve whose equation is $y = 1/(x^2 - 1)$. If a finite number can be assigned, find it.
25. Determine if it is possible to assign a finite number to represent the measure of the volume of the solid formed by revolving about the x axis the region to the right of the line $x = 1$ and bounded by the curve whose equation is $y = 1/x^{3/2}$ and the x axis. If a finite number can be assigned, find it.

26. Determine if it is possible to assign a finite number to represent the measure of the volume of the solid formed by revolving about the x axis the region bounded by the x axis, the y axis, and the curve whose equation is $y = e^{-2x}$. If a finite number can be assigned, find it.

Exercises 27, 28, and 29 show an interesting application of improper integrals in the field of economics. Suppose there is a continuous flow of income for which interest is compounded continuously at the annual rate of 100i percent and f(t) dollars is the income per year at any time t years. If the income continues indefinitely, the present value, V dollars, of all future income is defined by

$$V = \int_0^{+\infty} f(t)e^{-it}\, dt$$

27. A continuous flow of income is decreasing with time, and at t years the number of dollars in the annual income is $1000 \cdot 2^{-t}$. Find the present value of this income if it continues indefinitely using an interest rate of 8 percent compounded continuously.

28. The continuous flow of profit for a company is increasing with time, and at t years the number of dollars in the profit per year is proportional to t. Show that the present value of the company is inversely proportional to i^2, where $100i$ percent is the interest rate compounded continuously.

29. The British Consol is a bond with no maturity (i.e., it never comes due), and it affords the holder an annual lump-sum payment. By finding the present value of a flow of payments of R dollars annually and using the current interest rate $100i$ percent, compounded continuously, show that the fair selling price of a British Consol is R/i dollars.

30. Determine the values of n for which the following improper integral is convergent: $\displaystyle\int_e^{+\infty} \frac{dx}{x(\ln x)^n}$.

31. Determine a value of n for which the improper integral $\displaystyle\int_1^{+\infty} \left(\frac{n}{x+1} - \frac{3x}{2x^2+n}\right) dx$ is convergent, and evaluate the integral for this value of n.

32. Determine a value of n for which the improper integral $\displaystyle\int_1^{+\infty} \left(\frac{nx^2}{x^3+1} - \frac{1}{3x+1}\right) dx$ is convergent, and evaluate the integral for this value of n.

14.4 OTHER IMPROPER INTEGRALS

Figure 14.4.1 shows the region bounded by the curve whose equation is $y = 1/\sqrt{x}$, the x axis, the y axis, and the line $x = 4$. If it is possible to assign a finite number to represent the measure of the area of this region, it would be given by

$$\lim_{\|\Delta\| \to 0} \sum_{i=1}^{n} \frac{1}{\sqrt{\xi_i}}\, \Delta_i x$$

If this limit exists, it is the definite integral denoted by

$$\int_0^4 \frac{dx}{\sqrt{x}} \tag{1}$$

However, the integrand is discontinuous at the lower limit zero. Furthermore, $\lim_{x \to 0^+} 1/\sqrt{x} = +\infty$, and so we state that the integrand has an infinite discontinuity at the lower limit. Such an integral is improper, and its existence can be determined from the following definition.

14.4.1 Definition If f is continuous at all x in the interval half open on the left $(a, b]$, and if $\lim_{x \to a^+} f(x) = \pm\infty$, then

$$\int_a^b f(x)\,dx = \lim_{\epsilon \to 0^+} \int_{a+\epsilon}^b f(x)\,dx$$

if this limit exists.

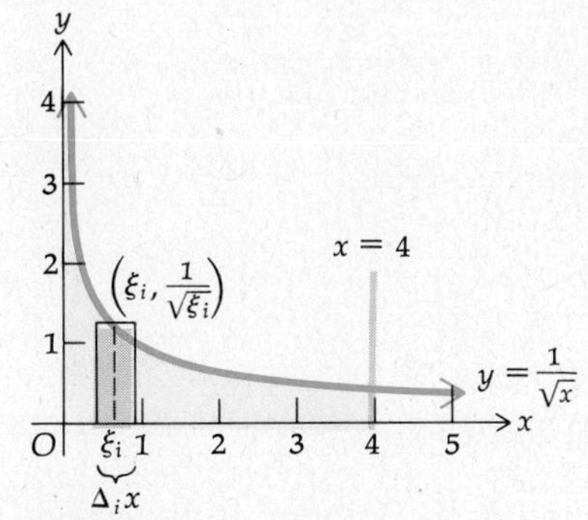

FIGURE 14.4.1

● ILLUSTRATION 1 We determine if it is possible to assign a finite number to represent the measure of the area of the region shown in Fig. 14.4.1. From the discussion preceding Definition 14.4.1, the measure of the area of the given region will be the improper integral (1) if it exists. By Definition 14.4.1,

$$\begin{aligned}\int_0^4 \frac{dx}{\sqrt{x}} &= \lim_{\epsilon \to 0^+} \int_\epsilon^4 \frac{dx}{\sqrt{x}} \\ &= \lim_{\epsilon \to 0^+} 2x^{1/2}\Big]_\epsilon^4 \\ &= \lim_{\epsilon \to 0^+} [4 - 2\sqrt{\epsilon}] \\ &= 4 - 0 \\ &= 4\end{aligned}$$

Therefore we assign 4 to the measure of the area of the given region. ●

If the integrand has an infinite discontinuity at the upper limit of integration, we use the following definition to determine the existence of the improper integral.

14.4.2 Definition If f is continuous at all x in the interval half open on the right $[a, b)$ and if $\lim_{x \to b^-} f(x) = \pm\infty$, then

$$\int_a^b f(x)\,dx = \lim_{\epsilon \to 0^+} \int_a^{b-\epsilon} f(x)\,dx$$

if this limit exists.

If there is an infinite discontinuity at an interior point of the interval of integration, the existence of the improper integral is determined from the following definition.

14.4.3 Definition If f is continuous at all x in the interval $[a, b]$ except c where $a < c < b$ and if $\lim_{x \to c} |f(x)| = +\infty$, then

$$\int_a^b f(x)\,dx = \lim_{\epsilon \to 0^+} \int_a^{c-\epsilon} f(x)\,dx + \lim_{\delta \to 0^+} \int_{c+\delta}^b f(x)\,dx$$

if these limits exist.

If $\int_a^b f(x)\,dx$ is an improper integral, it is convergent if the corresponding limit exists; otherwise it is divergent.

Example 1 Evaluate

$$\int_0^2 \frac{dx}{(x-1)^2}$$

if it is convergent.

Solution The integrand has an infinite discontinuity at 1. Applying Definition 14.4.3 we have

$$\begin{aligned}\int_0^2 \frac{dx}{(x-1)^2} &= \lim_{\epsilon \to 0^+} \int_0^{1-\epsilon} \frac{dx}{(x-1)^2} + \lim_{\delta \to 0^+} \int_{1+\delta}^2 \frac{dx}{(x-1)^2} \\ &= \lim_{\epsilon \to 0^+} \left[-\frac{1}{x-1} \right]_0^{1-\epsilon} + \lim_{\delta \to 0^+} \left[-\frac{1}{x-1} \right]_{1+\delta}^2 \\ &= \lim_{\epsilon \to 0^+} \left[\frac{1}{\epsilon} - 1 \right] + \lim_{\delta \to 0^+} \left[-1 + \frac{1}{\delta} \right]\end{aligned}$$

Because neither of these limits exist, the improper integral is divergent.

● ILLUSTRATION 2 Suppose that in evaluating the integral in Example 1 we had failed to note the infinite discontinuity of the integrand at 1. We would have obtained

$$-\frac{1}{x-1}\bigg]_0^2 = -\frac{1}{1} + \frac{1}{-1} = -2$$

This is obviously an incorrect result. Because the integrand $1/(x-1)^2$ is never negative, the integral from 0 to 2 could not possibly be a negative number. ●

Example 2 Evaluate $\int_0^1 x \ln x \, dx$ if it is convergent.

Solution The integrand has a discontinuity at the lower limit. From Definition 14.4.1,

$$\begin{aligned}\int_0^1 x \ln x \, dx &= \lim_{\epsilon \to 0^+} \int_\epsilon^1 x \ln x \, dx \\ &= \lim_{\epsilon \to 0^+} \left[\tfrac{1}{2}x^2 \ln x - \tfrac{1}{4}x^2 \right]_\epsilon^1 \\ &= \lim_{\epsilon \to 0^+} \left[\tfrac{1}{2} \ln 1 - \tfrac{1}{4} - \tfrac{1}{2}\epsilon^2 \ln \epsilon - \tfrac{1}{4}\epsilon^2 \right]\end{aligned}$$

Hence

$$\int_0^1 x \ln x \, dx = 0 - \tfrac{1}{4} - \tfrac{1}{2} \lim_{\epsilon \to 0^+} \epsilon^2 \ln \epsilon - 0 \qquad (2)$$

To evaluate

$$\lim_{\epsilon \to 0^+} \epsilon^2 \ln \epsilon = \lim_{\epsilon \to 0^+} \frac{\ln \epsilon}{\dfrac{1}{\epsilon^2}}$$

we apply L'Hôpital's rule, because $\lim_{\epsilon \to 0^+} \ln \epsilon = -\infty$ and $\lim_{\epsilon \to 0^+} 1/\epsilon^2 = +\infty$. We have

$$\lim_{\epsilon \to 0^+} \frac{\ln \epsilon}{\frac{1}{\epsilon^2}} = \lim_{\epsilon \to 0^+} \frac{\frac{1}{\epsilon}}{-\frac{2}{\epsilon^3}} = \lim_{\epsilon \to 0^+} \left[-\frac{\epsilon^2}{2}\right] = 0$$

Therefore, from (2),

$$\int_0^1 x \ln x \, dx = -\tfrac{1}{4}$$

Example 3 Evaluate

$$\int_1^{+\infty} \frac{dx}{x\sqrt{x^2 - 1}}$$

if it is convergent.

Solution For this integral there is both an infinite upper limit and an infinite discontinuity of the integrand at the lower limit. We proceed as follows.

$$\begin{aligned}
\int_1^{+\infty} \frac{dx}{x\sqrt{x^2-1}} &= \lim_{a \to 1^+} \int_a^2 \frac{dx}{x\sqrt{x^2-1}} + \lim_{b \to +\infty} \int_2^b \frac{dx}{x\sqrt{x^2-1}} \\
&= \lim_{a \to 1^+} \left[\sec^{-1} x\right]_a^2 + \lim_{b \to +\infty} \left[\sec^{-1} x\right]_2^b \\
&= \lim_{a \to 1^+} (\sec^{-1} 2 - \sec^{-1} a) + \lim_{b \to +\infty} (\sec^{-1} b - \sec^{-1} 2) \\
&= -\lim_{a \to 1^+} \sec^{-1} a + \lim_{b \to +\infty} \sec^{-1} b \\
&= 0 + \tfrac{1}{2}\pi \\
&= \tfrac{1}{2}\pi
\end{aligned}$$

EXERCISES 14.4

In Exercises 1 through 25, determine whether the improper integral is convergent or divergent. If it is convergent, evaluate it.

1. $\int_0^1 \frac{dx}{\sqrt{1-x}}$ **2.** $\int_0^{16} \frac{dx}{x^{3/4}}$ **3.** $\int_{-5}^{-3} \frac{x\,dx}{\sqrt{x^2-9}}$ **4.** $\int_0^4 \frac{x\,dx}{\sqrt{16-x^2}}$ **5.** $\int_2^4 \frac{dt}{\sqrt{16-t^2}}$

6. $\int_{-4}^1 \frac{dz}{(z+3)^3}$ **7.** $\int_{\pi/4}^{\pi/2} \sec\theta\,d\theta$ **8.** $\int_{-2}^0 \frac{dx}{\sqrt{4-x^2}}$ **9.** $\int_0^{+\infty} \frac{dx}{x^3}$ **10.** $\int_0^{\pi/2} \tan\theta\,d\theta$

11. $\int_0^{\pi/2} \frac{dt}{1-\sin t}$ **12.** $\int_0^2 \frac{dx}{(x-1)^{2/3}}$ **13.** $\int_0^4 \frac{dx}{x^2-2x-3}$ **14.** $\int_2^{+\infty} \frac{dx}{x\sqrt{x^2-4}}$ **15.** $\int_0^{+\infty} \ln x\,dx$

16. $\int_0^2 \frac{dx}{\sqrt{2x-x^2}}$ **17.** $\int_{-2}^0 \frac{dt}{(t+1)^{1/3}}$ **18.** $\int_{-1}^1 \frac{dx}{x^2}$ **19.** $\int_{-2}^2 \frac{dx}{x^3}$ **20.** $\int_0^{+\infty} \frac{e^{-\sqrt{x}}}{\sqrt{x}}\,dx$

21. $\displaystyle\int_{1/2}^{2} \frac{dt}{t(\ln t)^{1/5}}$ **22.** $\displaystyle\int_{0}^{2} \frac{x\,dx}{1-x}$ **23.** $\displaystyle\int_{1}^{2} \frac{dx}{x\sqrt{x^2-1}}$ **24.** $\displaystyle\int_{0}^{1} \frac{dx}{x\sqrt{4-x^2}}$ **25.** $\displaystyle\int_{1}^{3} \frac{dy}{\sqrt[3]{y-2}}$

26. Evaluate, if they exist: (a) $\displaystyle\int_{-1}^{1} \frac{dx}{x}$; (b) $\displaystyle\lim_{r\to 0^+}\left[\int_{-1}^{-r} \frac{dx}{x} + \int_{r}^{1} \frac{dx}{x}\right]$.

In Exercises 27 through 29, find the values of n for which the improper integral converges, and evaluate the integral for these values of n.

27. $\displaystyle\int_0^1 x^n\,dx$ **28.** $\displaystyle\int_0^1 x^n \ln x\,dx$ **29.** $\displaystyle\int_0^1 x^n \ln^2 x\,dx$

30. Show that it is possible to assign a finite number to represent the measure of the area of the region bounded by the curve whose equation is $y = 1/\sqrt{x}$, the line $x = 1$, and the x and y axes, but that it is not possible to assign a finite number to represent the measure of the volume of the solid of revolution generated if this region is revolved about the x axis.

31. Determine if it is possible to assign a finite number to represent the measure of the volume of the solid formed by revolving about the x axis the region bounded by the curve whose equation is $y = x^{-1/3}$, the line $x = 8$, and the x and y axes.

32. Given the improper integral $\displaystyle\int_a^b \frac{dx}{(x-a)^n}$ where $b > a$. Determine if the integral is convergent or divergent in each case: (a) $0 < n < 1$; (b) $n = 1$; (c) $n > 1$. If the integral is convergent, evaluate it.

33. Use integration to prove that the circumference of the circle $x^2 + y^2 = a^2$ is $2\pi a$.

14.5 TAYLOR'S FORMULA

Values of polynomial functions can be found by performing a finite number of additions and multiplications. However, there are other functions, such as the logarithmic, exponential, and trigonometric functions, that cannot be evaluated as easily. We show in this section that many functions can be approximated by polynomials and that the polynomial, instead of the original function, can be used for computations when the difference between the actual function value and the polynomial approximation is sufficiently small.

There are various methods of approximating a given function by polynomials. One of the most widely used is that involving **Taylor's formula,** named in honor of the English mathematician Brook Taylor (1685–1731). The following theorem, which can be considered as a generalization of the mean-value theorem (4.4.2), gives Taylor's formula.

14.5.1 Theorem Let f be a function such that f and its first n derivatives are continuous on the closed interval $[a, b]$. Furthermore, let $f^{(n+1)}(x)$ exist for all x in the open interval (a, b). Then there is a number ξ in the open interval (a, b) such that

$$f(b) = f(a) + \frac{f'(a)}{1!}(b-a) + \frac{f''(a)}{2!}(b-a)^2 + \cdots + \frac{f^{(n)}(a)}{n!}(b-a)^n + \frac{f^{(n+1)}(\xi)}{(n+1)!}(b-a)^{n+1} \quad (1)$$

Equation (1) also holds if $b < a$; in such a case $[a, b]$ is replaced by $[b, a]$, and (a, b) is replaced by (b, a).

Before proving this theorem, note that when $n = 0$, (1) becomes

$$f(b) = f(a) + f'(\xi)(b - a)$$

where ξ is between a and b. This is the mean-value theorem (4.4.2).

There are several known proofs of Theorem 14.5.1, although none is very well motivated. The one following makes use of Cauchy's mean-value theorem (14.1.3).

Proof of Theorem 14.5.1 Let F and G be two functions defined by

$$F(x) = f(b) - f(x) - f'(x)(b - x) - \frac{f''(x)}{2!}(b - x)^2 - \cdots - \frac{f^{(n-1)}(x)}{(n-1)!}(b - x)^{n-1} - \frac{f^{(n)}(x)}{n!}(b - x)^n \quad (2)$$

and

$$G(x) = \frac{(b - x)^{n+1}}{(n + 1)!} \quad (3)$$

It follows that $F(b) = 0$ and $G(b) = 0$. Differentiating in (2) we get

$$F'(x) = -f'(x) + f'(x) - f''(x)(b - x) + \frac{2f''(x)(b - x)}{2!} - \frac{f'''(x)(b - x)^2}{2!} + \frac{3f'''(x)(b - x)^2}{3!} - \frac{f^{(\text{iv})}(x)(b - x)^3}{3!} + \cdots + \frac{(n-1)f^{(n-1)}(x)(b - x)^{n-2}}{(n-1)!} - \frac{f^{(n)}(x)(b - x)^{n-1}}{(n-1)!} + \frac{nf^{(n)}(x)(b - x)^{n-1}}{n!} - \frac{f^{(n+1)}(x)(b - x)^n}{n!}$$

Combining terms we see that the sum of every odd-numbered term with the following even-numbered term is zero; so only the last term remains. Therefore

$$F'(x) = -\frac{f^{(n+1)}(x)}{n!}(b - x)^n \quad (4)$$

Differentiating in (3) we obtain

$$G'(x) = -\frac{1}{n!}(b - x)^n \quad (5)$$

Checking the hypothesis of Theorem 14.1.3 we see that

(i) F and G are continuous on $[a, b]$;
(ii) F and G are differentiable on (a, b);
(iii) for all x in (a, b), $G'(x) \neq 0$.

So by the conclusion of Theorem 14.1.3,

$$\frac{F(b) - F(a)}{G(b) - G(a)} = \frac{F'(\xi)}{G'(\xi)}$$

where ξ is in (a, b). But $F(b) = 0$ and $G(b) = 0$. So

$$F(a) = \frac{F'(\xi)}{G'(\xi)} G(a) \tag{6}$$

for some ξ in (a, b).

Letting $x = a$ in (3), $x = \xi$ in (4), and $x = \xi$ in (5), and substituting into (6), we obtain

$$F(a) = -\frac{f^{(n+1)}(\xi)}{n!}(b - \xi)^n \left[-\frac{n!}{(b - \xi)^n} \right] \frac{(b - a)^{n+1}}{(n + 1)!}$$

$$F(a) = \frac{f^{(n+1)}(\xi)}{(n + 1)!}(b - a)^{n+1} \tag{7}$$

If $x = a$ in (2), we obtain

$$F(a) = f(b) - f(a) - f'(a)(b - a) - \frac{f''(a)}{2!}(b - a)^2 - \cdots - \frac{f^{(n-1)}(a)}{(n - 1)!}(b - a)^{n-1} - \frac{f^{(n)}(a)}{n!}(b - a)^n \tag{8}$$

Substituting from (7) into (8) we get

$$f(b) = f(a) + f'(a)(b - a) + \frac{f''(a)}{2!}(b - a)^2 + \cdots + \frac{f^{(n)}(a)}{n!}(b - a)^n + \frac{f^{(n+1)}(\xi)}{(n + 1)!}(b - a)^{n+1}$$

which is the desired result. The theorem holds if $b < a$ because the conclusion of Theorem 14.1.3 is unaffected if a and b are interchanged. ■

If in (1) b is replaced by x, Taylor's formula is obtained. It is

$$f(x) = f(a) + \frac{f'(a)}{1!}(x - a) + \frac{f''(a)}{2!}(x - a)^2 + \cdots + \frac{f^{(n)}(a)}{n!}(x - a)^n + \frac{f^{(n+1)}(\xi)}{(n + 1)!}(x - a)^{n+1} \tag{9}$$

where ξ is between a and x.

The condition under which (9) holds is that f and its first n derivatives must be continuous on a closed interval containing a and x, and the

$(n + 1)$st derivative of f must exist at all points of the corresponding open interval. Formula (9) may be written as

$$f(x) = P_n(x) + R_n(x) \tag{10}$$

where

$$P_n(x) = f(a) + \frac{f'(a)}{1!}(x - a) + \frac{f''(a)}{2!}(x - a)^2 + \cdots + \frac{f^{(n)}(a)}{n!}(x - a)^n \tag{11}$$

and

$$R_n(x) = \frac{f^{(n+1)}(\xi)}{(n + 1)!}(x - a)^{n+1} \qquad \text{where } \xi \text{ is between } a \text{ and } x \tag{12}$$

$P_n(x)$ is called the nth-degree **Taylor polynomial** of the function f at the number a, and $R_n(x)$ is called the *remainder*. The term $R_n(x)$ as given in (12) is called the *Lagrange form* of the remainder, named in honor of the French mathematician Joseph L. Lagrange (1736–1813).

Example 1 Find the third-degree Taylor polynomial of the cosine function at $\frac{1}{4}\pi$ and the Lagrange form of the remainder.

Solution Let $f(x) = \cos x$. Then from (11),

$$P_3(x) = f\left(\frac{\pi}{4}\right) + f'\left(\frac{\pi}{4}\right)\left(x - \frac{\pi}{4}\right) + \frac{f''(\frac{1}{4}\pi)}{2!}\left(x - \frac{\pi}{4}\right)^2 + \frac{f'''(\frac{1}{4}\pi)}{3!}\left(x - \frac{\pi}{4}\right)^3$$

Because $f(x) = \cos x$, $f(\frac{1}{4}\pi) = \frac{1}{2}\sqrt{2}$; $f'(x) = -\sin x$, $f'(\frac{1}{4}\pi) = -\frac{1}{2}\sqrt{2}$; $f''(x) = -\cos x$, $f''(\frac{1}{4}\pi) = -\frac{1}{2}\sqrt{2}$; $f'''(x) = \sin x$, $f'''(\frac{1}{4}\pi) = \frac{1}{2}\sqrt{2}$. Therefore

$$P_3(x) = \tfrac{1}{2}\sqrt{2} - \tfrac{1}{2}\sqrt{2}(x - \tfrac{1}{4}\pi) - \tfrac{1}{4}\sqrt{2}(x - \tfrac{1}{4}\pi)^2 + \tfrac{1}{12}\sqrt{2}(x - \tfrac{1}{4}\pi)^3$$

Because $f^{(\text{iv})}(x) = \cos x$, we obtain from (12)

$$R_3(x) = \tfrac{1}{24}(\cos \xi)(x - \tfrac{1}{4}\pi)^4 \qquad \text{where } \xi \text{ is between } \tfrac{1}{4}\pi \text{ and } x$$

Because $|\cos \xi| \leq 1$, we may conclude that for all x, $|R_3(x)| \leq \frac{1}{24}(x - \frac{1}{4}\pi)^4$.

Taylor's formula may be used to approximate a function by means of a Taylor polynomial. From (10),

$$|R_n(x)| = |f(x) - P_n(x)| \tag{13}$$

If $P_n(x)$ is used to approximate $f(x)$, we can obtain an upper bound for the error of this approximation if we can find a number $E > 0$ such that $|R_n(x)| \leq E$ or, because of (13), such that $|f(x) - P_n(x)| \leq E$ or, equivalently,

$$P_n(x) - E \leq f(x) \leq P_n(x) + E$$

Example 2 Use the result of Example 1 to compute an approximate value of cos 47° and determine the accuracy of the result.

Solution $47° \sim \frac{47}{180}\pi$ radians. Thus in the solution of Example 1, take $x = \frac{47}{180}\pi$ and $x - \frac{1}{4}\pi = \frac{1}{90}\pi$, and

$$\cos 47° = \tfrac{1}{2}\sqrt{2}[1 - \tfrac{1}{90}\pi - \tfrac{1}{2}(\tfrac{1}{90}\pi)^2 + \tfrac{1}{6}(\tfrac{1}{90}\pi)^3] + R_3(\tfrac{47}{180}\pi)$$

where

$$R_3(\tfrac{47}{180}\pi) = \tfrac{1}{24}\cos\xi\,(\tfrac{1}{90}\pi)^4 \qquad \text{with } \tfrac{1}{4}\pi < \xi < \tfrac{47}{180}\pi$$

Because $0 < \cos\xi < 1$,

$$0 < R_3(\tfrac{47}{180}\pi) < \tfrac{1}{24}(\tfrac{1}{90}\pi)^4 < 0.00000007$$

Taking $\frac{1}{90}\pi \approx 0.0349066$ we obtain

$$\cos 47° \approx 0.681998$$

which is accurate to six decimal places.

Example 3 Use a Taylor polynomial at zero to find the value of $\sqrt{e}$ accurate to four decimal places.

Solution If $f(x) = e^x$, all the derivatives of f at x are e^x and all the derivatives evaluated at zero are 1. Therefore, from (11),

$$P_n(x) = 1 + x + \frac{x^2}{2!} + \frac{x^3}{3!} + \cdots + \frac{x^n}{n!}$$

and from (12),

$$R_n(x) = \frac{e^\xi}{(n+1)!}x^{n+1} \qquad \text{where } \xi \text{ is between 0 and } x$$

We want $|R_n(\frac{1}{2})|$ to be less than 0.00005. If $x = \frac{1}{2}$ in the above, and because $e^{1/2} < 2$,

$$|R_n(\tfrac{1}{2})| < \frac{e^{1/2}}{2^{n+1}(n+1)!} < \frac{2}{2^{n+1}(n+1)!} = \frac{1}{2^n(n+1)!}$$

$|R_n(\frac{1}{2})|$ will be less than 0.00005 if $1/2^n(n+1)! < 0.00005$. When $n = 5$,

$$\frac{1}{2^n(n+1)!} = \frac{1}{(32)(720)} = 0.00004 < 0.00005$$

So

$$\sqrt{e} \approx P_5(\tfrac{1}{2}) = 1 + \tfrac{1}{2} + \tfrac{1}{8} + \tfrac{1}{48} + \tfrac{1}{384} + \tfrac{1}{3840}$$

from which we obtain $\sqrt{e} \approx 1.64870$.

A special case of Taylor's formula is obtained by taking $a = 0$ in (9), and we get

$$f(x) = f(0) + \frac{f'(0)}{1!}x + \frac{f''(0)}{2!}x^2 + \cdots + \frac{f^{(n)}(0)}{n!}x^n + \frac{f^{(n+1)}(\xi)}{(n+1)!}x^{n+1} \tag{14}$$

where ξ is between 0 and x. Formula (14) is called **Maclaurin's formula,** named in honor of the Scottish mathematician Colin Maclaurin (1698–1746). However, the formula was obtained earlier by Taylor and by another British mathematician, James Stirling (1692–1770).

There are other forms of the remainder in Taylor's formula. Depending on the function, one form of the remainder may be more desirable to use than another. The following theorem expresses the remainder as an integral, and it is known as *Taylor's formula with integral form of the remainder.*

14.5.2 Theorem If f is a function whose first $n + 1$ derivatives are continuous on a closed interval containing a and x, then $f(x) = P_n(x) + R_n(x)$, where $P_n(x)$ is the nth-degree Taylor polynomial of f at a and $R_n(x)$ is the remainder given by

$$R_n(x) = \frac{1}{n!}\int_a^x (x - t)^n f^{(n+1)}(t)\, dt \tag{15}$$

The proof of this theorem is left as an exercise (see Exercise 30).

EXERCISES 14.5

In Exercises 1 through 14, find the Taylor polynomial of degree n with the Lagrange form of the remainder at the number a for the function defined by the given equation.

1. $f(x) = \dfrac{1}{x-2}$; $a = 1$; $n = 3$
2. $f(x) = e^{-x}$; $a = 0$; $n = 4$
3. $f(x) = x^{3/2}$; $a = 4$; $n = 3$
4. $f(x) = \tan x$; $a = 0$; $n = 3$
5. $f(x) = \sin x$; $a = \frac{1}{6}\pi$; $n = 3$
6. $f(x) = \cosh x$; $a = 0$; $n = 4$
7. $f(x) = \sinh x$; $a = 0$; $n = 4$
8. $f(x) = \sqrt{x}$; $a = 4$; $n = 4$
9. $f(x) = \ln x$; $a = 1$; $n = 3$
10. $f(x) = \ln(x + 2)$; $a = -1$; $n = 3$
11. $f(x) = \ln \cos x$; $a = \frac{1}{3}\pi$; $n = 3$
12. $f(x) = e^{-x^2}$; $a = 0$; $n = 3$
13. $f(x) = (1 + x)^{3/2}$; $a = 0$; $n = 3$
14. $f(x) = (1 - x)^{-1/2}$; $a = 0$; $n = 3$
15. Compute the value of e correct to five decimal places, and prove that your answer has the required accuracy.
16. Use the Taylor polynomial in Exercise 8 to compute $\sqrt{5}$ accurate to as many places as is justified when R_4 is neglected.
17. Estimate the error that results when $\cos x$ is replaced by $1 - \frac{1}{2}x^2$ if $|x| < 0.1$.
18. Estimate the error that results when $\sqrt{1 + x}$ is replaced by $1 + \frac{1}{2}x$ if $0 < x < 0.01$.
19. Compute sin 31° accurate to three decimal places by using the Taylor polynomial in Exercise 5 at $\frac{1}{6}\pi$. (Use the approximation $\frac{1}{180}\pi \approx 0.0175$.)

20. Use a Taylor polynomial at 0 for the function defined by $f(x) = \ln(1 + x)$ to compute the value of ln 1.2 accurate to four decimal places.

21. Use a Taylor polynomial at 0 for the function defined by

$$f(x) = \ln \frac{1 + x}{1 - x}$$

to compute the value of ln 1.2 accurate to four decimal places. Compare the computation with that of Exercise 20.

22. Show that if $0 \leq x \leq \frac{1}{2}$,

$$\sin x = x - \frac{x^3}{3!} + R(x)$$

where $|R(x)| < \frac{1}{3840}$.

23. Use the result of Exercise 22 to find an approximate value of $\int_0^{1/\sqrt{2}} \sin x^2 \, dx$, and estimate the error.

24. Show that the formula $(1 + x)^{3/2} \approx 1 + \frac{3}{2}x$ is accurate to three decimal places if $-0.03 \leq x \leq 0$.

25. Show that the formula $(1 + x)^{-1/2} \approx 1 - \frac{1}{2}x$ is accurate to two decimal places if $-0.1 \leq x \leq 0$.

26. Draw sketches of the graphs of $y = \sin x$ and $y = mx$ on the same set of axes. Note that if m is positive and close to zero, then the graphs intersect at a point whose abscissa is close to π. By finding the second-degree Taylor polynomial at π for the function f defined by $f(x) = \sin x - mx$, show that an approximate solution of the equation $\sin x = mx$, when m is positive and close to zero, is given by $x \approx \pi/(1 + m)$.

27. Use the method described in Exercise 26 to find an approximate solution of the equation $\cot x = mx$ when m is positive and close to zero.

28. (a) Use the first-degree Taylor polynomial at 0 to approximate e^k if $0 < k < 0.01$. (b) Estimate the error in terms of k.

29. Apply Taylor's formula to express the polynomial $P(x) = x^4 - x^3 + 2x^2 - 3x + 1$ as a polynomial in powers of $(x - 1)$.

30. Prove Theorem 14.5.2. (*Hint:* Let $\int_a^x f'(t)\, dt = f(x) - f(a)$. Solve for $f(x)$ and integrate $\int_a^x f'(t)\, dt$ by parts by letting $u = f'(t)$ and $dv = dt$. Repeat this process, and the desired result follows by mathematical induction.)

REVIEW EXERCISES FOR CHAPTER 14

In Exercises 1 through 18, evaluate the limit, if it exists.

1. $\displaystyle\lim_{x \to 0^+} \frac{\tanh 2x}{\sinh^2 x}$

2. $\displaystyle\lim_{x \to \pi/2} \left(\frac{1}{1 - \sin x} - \frac{2}{\cos^2 x}\right)$

3. $\displaystyle\lim_{x \to 0} (\csc^2 x - x^{-2})$

4. $\displaystyle\lim_{x \to 0} \frac{e - (1 + x)^{1/x}}{x}$

5. $\displaystyle\lim_{t \to +\infty} \frac{\ln(1 + e^{2t}/t)}{t^{1/2}}$

6. $\displaystyle\lim_{x \to +\infty} x \ln \frac{x + 1}{x - 1}$

7. $\displaystyle\lim_{x \to \pi/2} (\sin^2 x)^{\tan x}$

8. $\displaystyle\lim_{t \to +\infty} \frac{t^{100}}{e^t}$

9. $\displaystyle\lim_{y \to 0} \frac{\sin 2y}{y - \sin 5y}$

10. $\displaystyle\lim_{z \to \pi/2^-} \frac{z \cos 3z}{\cos^2 z}$

11. $\displaystyle\lim_{t \to 0^-} (1 + 4t)^{3/t}$

12. $\displaystyle\lim_{y \to +\infty} (1 + e^{2y})^{-2/y}$

13. $\displaystyle\lim_{x \to +\infty} \frac{\ln(\ln x)}{\ln(x - \ln x)}$

14. $\displaystyle\lim_{x \to 0} \frac{x - \tan^{-1} x}{4x^3}$

15. $\displaystyle\lim_{\theta \to \pi/2} \frac{\tan \theta + 3}{\sec \theta - 1}$

16. $\displaystyle\lim_{x \to 0} \left(\frac{\sin x}{x}\right)^{1/x}$

17. $\displaystyle\lim_{x \to +\infty} (e^x - x)^{1/x}$

18. $\displaystyle\lim_{x \to 0^+} \left(\frac{1}{x}\right)^{\tan x}$

In Exercises 19 through 32, determine whether the improper integral is convergent or divergent. If it is convergent, evaluate it.

19. $\int_{-2}^{0} \frac{dx}{2x+3}$ **20.** $\int_{0}^{+\infty} \frac{dx}{\sqrt{e^x}}$ **21.** $\int_{-\infty}^{0} \frac{dx}{(x-2)^2}$ **22.** $\int_{2}^{4} \frac{x\,dx}{\sqrt{x-2}}$ **23.** $\int_{0}^{\pi/4} \cot^2\theta\,d\theta$

24. $\int_{1}^{+\infty} \frac{dt}{t^4+t^2}$ **25.** $\int_{-\infty}^{3} 4^x\,dx$ **26.** $\int_{-\infty}^{0} x\,e^x\,dx$ **27.** $\int_{0}^{1} \frac{(\ln x)^2}{x}\,dx$ **28.** $\int_{0}^{+\infty} \frac{3^{-\sqrt{x}}}{\sqrt{x}}\,dx$

29. $\int_{-\infty}^{+\infty} \frac{dx}{4x^2+4x+5}$ **30.** $\int_{0}^{1} \frac{\ln x}{x}\,dx$ **31.** $\int_{0}^{1} \frac{dx}{x+x^3}$ **32.** $\int_{-3}^{0} \frac{dx}{\sqrt{3-2x-x^2}}$

33. Given:

$$f(x) = \begin{cases} \dfrac{e^x - 1}{x} & \text{if } x \neq 0 \\ 1 & \text{if } x = 0 \end{cases}$$

(a) Prove that f is continuous at 0. (b) Prove that f is differentiable at 0 and find $f'(0)$.

34. For the function of Exercise 33, find, if they exist: (a) $\lim_{x\to+\infty} f(x)$; (b) $\lim_{x\to-\infty} f(x)$.

35. Given:

$$f(x) = \begin{cases} \dfrac{e^x - e^{-x}}{e^{2x} - e^{-2x}} & \text{if } x \neq 0 \\ 1 & \text{if } x = 0 \end{cases}$$

(a) Is f continuous at 0? (b) Find $\lim_{x\to+\infty} f(x)$, if it exists.

36. If the normal line to the curve $y = \ln x$ at the point $(x_1, \ln x_1)$ intersects the x axis at the point having an abscissa of a, prove that $\lim_{x_1\to+\infty} (a - x_1) = 0$.

37. Evaluate if they exist: (a) $\int_{-\infty}^{+\infty} \sinh x\,dx$; (b) $\lim_{r\to+\infty} \int_{-r}^{r} \sinh x\,dx$.

38. (a) Prove that $\lim_{x\to+\infty} (x^n/e^x) = 0$ for n any positive integer. (b) Find $\lim_{x\to 0} (e^{-1/x}/x^n)$, where $x > 0$ and n is any positive integer, by letting $x = 1/t$ and using the result of part (a).

In Exercises 39 through 41, find the Taylor polynomial of degree n with the Lagrange form of the remainder at the number a for the function defined by the given equation.

39. $f(x) = \cos x$; $a = 0$; $n = 6$ **40.** $f(x) = (1 + x^2)^{-1}$; $a = 1$; $n = 3$ **41.** $f(x) = x^{-1/2}$; $a = 9$; $n = 4$

42. Express both e^{x^2} and $\cos x$ as a Taylor polynomial of degree 4 at 0, and use them to evaluate

$$\lim_{x\to 0} \frac{e^{x^2} - \cos x}{x^4}$$

43. Evaluate the limit in Exercise 42 by L'Hôpital's rule.

44. Apply Taylor's formula to express the polynomial $P(x) = 4x^3 + 5x^2 - 2x + 1$ as a polynomial in powers of $(x + 2)$.

45. Determine if it is possible to assign a finite number to represent the measure of the area of the region to the right of the y axis bounded by the curve $4y^2 - xy^2 - x^2 = 0$ and its asymptote. If a finite number can be assigned, find it.

46. Determine if it is possible to assign a finite number to represent the measure of the area of the region bounded by the curve whose equation is $2xy - y = 1$, the x axis, and the line $x = 1$. If a finite number can be assigned, find it.

47. Determine if it is possible to assign a finite number to represent the measure of the area of the region in the first quadrant and below the curve having the equation $y = e^{-x}$. If a finite number can be assigned, find it.

48. Find the values of n for which the improper integral $\int_1^{+\infty} \frac{\ln x}{x^n}\, dx$ converges, and evaluate the integral for those values of n.

49. Assuming a continuous flow of income for a particular business and that at t years from now the number of dollars in the income per year is given by $1000t - 300$, what is the present value of all future income if 8% is the interest rate compounded continuously? (*Hint:* See the paragraph preceding Exercise 27 in Sec. 14.3.)

50. If the first four terms of the Taylor polynomial at 0 are used to approximate the value of $\sqrt{e}$, how accurate is the result?

CHAPTER 15 Infinite Series

15.1 SEQUENCES

You have undoubtedly encountered sequences of numbers in your previous study of mathematics. For example, the numbers 5, 7, 9, 11, 13, 15 define a sequence. This sequence is said to be **finite** because there is a first and last number. If the set of numbers that defines a sequence does not have both a first and last number, the sequence is said to be **infinite.** For example, the sequence defined by

$$\tfrac{1}{3}, \tfrac{2}{5}, \tfrac{3}{7}, \tfrac{4}{9}, \ldots \qquad (1)$$

is infinite because the three dots with no number following indicate that there is no last number. We are concerned here with infinite sequences, and when the word "sequence" is used, it is understood that we are referring to an infinite sequence. A sequence is defined as a particular kind of function.

15.1.1 Definition A **sequence** is a function whose domain is the set of positive integers.

The numbers in the range of the sequence, which are called the **elements** of the sequence, are restricted to real numbers in this book.

If the nth element is given by $f(n)$, then the sequence is the set of ordered pairs of the form $(n, f(n))$, where n is a positive integer.

● ILLUSTRATION 1 If $f(n) = n/(2n+1)$, then

$$f(1) = \tfrac{1}{3} \qquad f(2) = \tfrac{2}{5} \qquad f(3) = \tfrac{3}{7} \qquad f(4) = \tfrac{4}{9}$$

and so on. The range of f consists of the elements of sequence (1). Some of the ordered pairs in the sequence f are $(1, \tfrac{1}{3})$, $(2, \tfrac{2}{5})$, $(3, \tfrac{3}{7})$, $(4, \tfrac{4}{9})$, and $(5, \tfrac{5}{11})$. A sketch of the graph of this sequence is shown in Fig. 15.1.1. ●

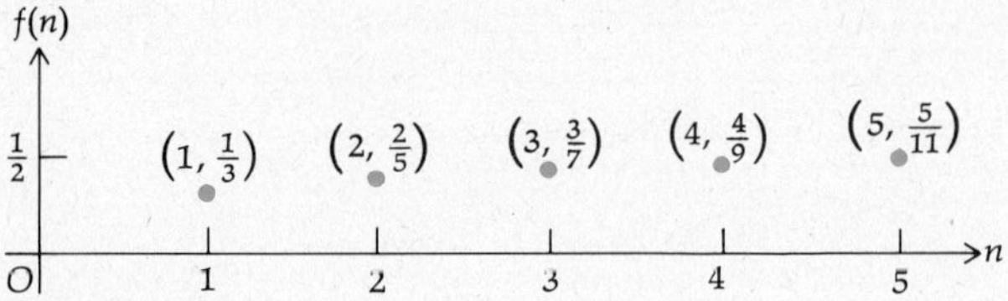

FIGURE 15.1.1

Usually the nth element $f(n)$ of the sequence is stated when the elements are listed in order. Thus the elements of sequence (1) can be written as

$$\frac{1}{3}, \frac{2}{5}, \frac{3}{7}, \frac{4}{9}, \ldots, \frac{n}{2n+1}, \ldots$$

Because the domain of every sequence is the same, the notation $\{f(n)\}$ may be used to denote a sequence. So the sequence (1) can be denoted by $\{n/(2n+1)\}$. The subscript notation $\{a_n\}$ is also used to denote the sequence for which $f(n) = a_n$.

You should distinguish between the elements of a sequence and the sequence itself, as shown in the following illustration.

● ILLUSTRATION 2 The sequence $\{1/n\}$ has as its elements the reciprocals of the positive integers

$$1, \frac{1}{2}, \frac{1}{3}, \frac{1}{4}, \ldots, \frac{1}{n}, \ldots \tag{2}$$

The sequence for which

$$f(n) = \begin{cases} 1 & \text{if } n \text{ is odd} \\ \dfrac{2}{n+2} & \text{if } n \text{ is even} \end{cases}$$

has as its elements

$$1, \tfrac{1}{2}, 1, \tfrac{1}{3}, 1, \tfrac{1}{4}, \ldots \tag{3}$$

The elements of sequences (2) and (3) are the same; however, the sequences are different. Sketches of the graphs of sequences (2) and (3) are shown in Figs. 15.1.2 and 15.1.3, respectively. ●

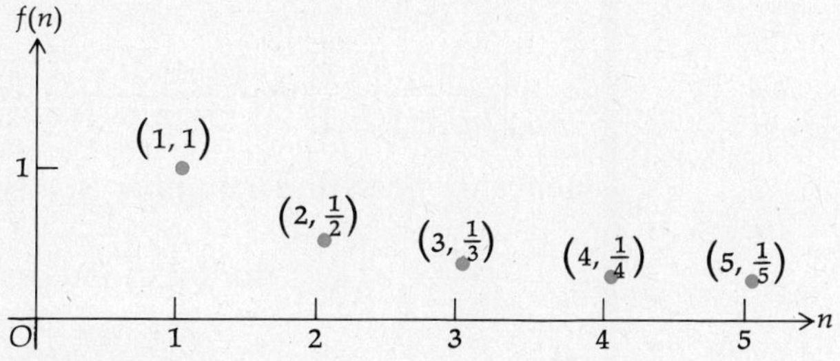

FIGURE 15.1.3

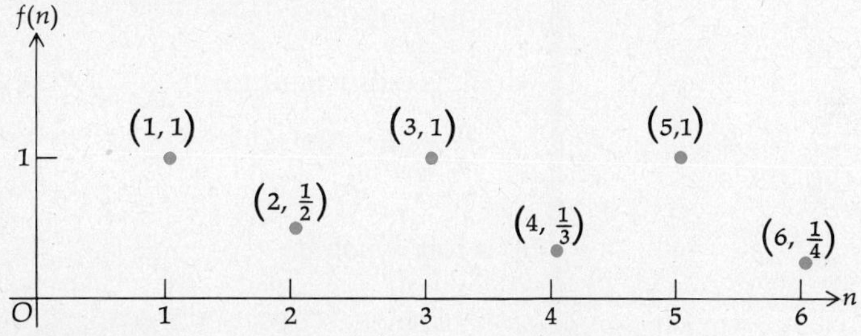

FIGURE 15.1.2

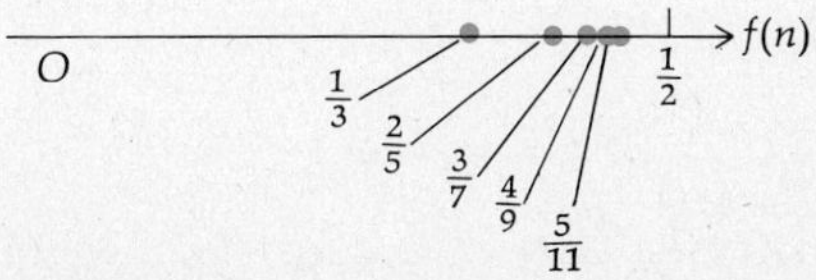

FIGURE 15.1.4

We now plot on a horizontal axis the points corresponding to successive elements of a sequence. This is done in Fig. 15.1.4 for sequence (1), which is $\{n/(2n+1)\}$. Observe that the successive elements of the sequence get closer and closer to $\frac{1}{2}$, even though no element in the sequence has the value $\frac{1}{2}$. Intuitively we see that the element will be as close to $\frac{1}{2}$ as we please by taking the number of the element sufficiently large. Or stating this another way, $|n/(2n+1) - \frac{1}{2}|$ can be made less than any given positive ϵ by taking n large enough. Because of this we state that the limit of the sequence $\{n/(2n+1)\}$ is $\frac{1}{2}$.

In general, if there is a number L such that $|a_n - L|$ is arbitrarily small for n sufficiently large, the sequence $\{a_n\}$ is said to have the limit L. Following is the precise definition of the limit of a sequence.

15.1.2 Definition A sequence $\{a_n\}$ has the limit L if for any $\epsilon > 0$ there exists a number $N > 0$ such that $|a_n - L| < \epsilon$ for every integer $n > N$; we write

$$\lim_{n \to +\infty} a_n = L$$

Example 1 Use Definition 15.1.2 to prove that the sequence

$$\left\{\frac{n}{2n+1}\right\}$$

has the limit $\frac{1}{2}$.

Solution We must show that for any $\epsilon > 0$ there exists a number $N > 0$ such that

$$\left|\frac{n}{2n+1} - \frac{1}{2}\right| < \epsilon \qquad \text{for every integer } n > N$$

$$\left|\frac{n}{2n+1} - \frac{1}{2}\right| = \left|\frac{2n - 2n - 1}{2(2n+1)}\right| = \left|\frac{-1}{4n+2}\right| = \frac{1}{4n+2}$$

Hence we must find a number $N > 0$ such that

$$\frac{1}{4n+2} < \epsilon \qquad \text{for every integer } n > N$$

But

$$\frac{1}{4n+2} < \epsilon \quad \text{is equivalent to} \quad 2n + 1 > \frac{1}{2\epsilon}$$

which is equivalent to

$$n > \frac{1 - 2\epsilon}{4\epsilon}$$

So it follows that

$$\left|\frac{n}{2n+1} - \frac{1}{2}\right| < \epsilon \qquad \text{for every integer } n > \frac{1 - 2\epsilon}{4\epsilon}$$

Therefore, if $N = (1 - 2\epsilon)/4\epsilon$, Definition 15.1.2 holds.

In particular, if $\epsilon = \frac{1}{8}$, $N = (1 - \frac{1}{4})/\frac{1}{2} = \frac{3}{2}$. So

$$\left|\frac{n}{2n+1} - \frac{1}{2}\right| < \frac{1}{8} \qquad \text{for every integer } n > \frac{3}{2}$$

For instance, if $n = 4$,

$$\left|\frac{n}{2n+1} - \frac{1}{2}\right| = \left|\frac{4}{9} - \frac{1}{2}\right| = \left|\frac{-1}{18}\right| = \frac{1}{18} < \frac{1}{8}$$

● ILLUSTRATION 3 Consider the sequence $\{(-1)^{n+1}/n\}$. Note that the nth element of this sequence is $(-1)^{n+1}/n$, and $(-1)^{n+1}$ is equal to $+1$ when n is odd and to -1 when n is even. Hence the elements of the sequence can be written

$$1, -\frac{1}{2}, \frac{1}{3}, -\frac{1}{4}, \frac{1}{5}, \ldots, \frac{(-1)^{n+1}}{n}, \ldots$$

In Fig. 15.1.5 are plotted points corresponding to successive elements of this sequence. In the figure, $a_1 = 1$, $a_2 = -\frac{1}{2}$, $a_3 = \frac{1}{3}$, $a_4 = -\frac{1}{4}$, $a_5 = \frac{1}{5}$, $a_6 = -\frac{1}{6}$, $a_7 = \frac{1}{7}$, $a_8 = -\frac{1}{8}$, $a_9 = \frac{1}{9}$, $a_{10} = -\frac{1}{10}$. The limit of the sequence is 0, and the elements oscillate about 0. ●

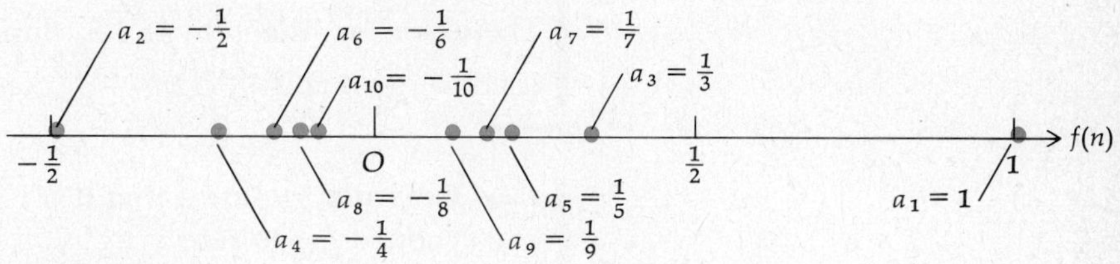

FIGURE 15.1.5

Compare Definition 15.1.2 with Definition 2.5.1 of the limit of $f(x)$ as x increases without bound. The two definitions are almost identical; however, when we state that $\lim_{x \to +\infty} f(x) = L$, the function f is defined for all real numbers greater than some real number R, while when we consider $\lim_{n \to +\infty} a_n$, n is restricted to positive integers. There is, however, the following theorem that follows immediately from Definition 2.5.1.

15.1.3 Theorem If $\lim_{x \to +\infty} f(x) = L$, and f is defined for every positive integer, then also $\lim_{n \to +\infty} f(n) = L$ when n is any positive integer.

The proof is left as an exercise (see Exercise 26).

● ILLUSTRATION 4 We verify Theorem 15.1.3 for the sequence of Example 1. For that sequence $f(n) = n/(2n + 1)$. Hence $f(x) = x/(2x + 1)$ and

$$\lim_{x \to +\infty} \frac{x}{2x + 1} = \lim_{x \to +\infty} \frac{1}{2 + \frac{1}{x}} = \frac{1}{2}$$

It follows, then, from Theorem 15.1.3 that $\lim_{n \to +\infty} f(n) = \frac{1}{2}$ when n is any positive integer. This agrees with the solution of Example 1. ●

15.1.4 Definition If a sequence $\{a_n\}$ has a limit, the sequence is said to be **convergent,** and a_n **converges** to that limit. If the sequence is not convergent, it is **divergent.**

Example 2 Determine if the sequence

$$\left\{\frac{4n^2}{2n^2+1}\right\}$$

is convergent or divergent.

Solution We wish to determine if $\lim_{n\to+\infty} 4n^2/(2n^2+1)$ exists. Let $f(x) = 4x^2/(2x^2+1)$, and investigate $\lim_{x\to+\infty} f(x)$.

$$\lim_{x\to+\infty}\frac{4x^2}{2x^2+1} = \lim_{x\to+\infty}\frac{4}{2+\dfrac{1}{x^2}} = 2$$

Therefore, by Theorem 15.1.3, $\lim_{n\to+\infty} f(n) = 2$. Thus the given sequence is convergent and $4n^2/(2n^2+1)$ converges to 2.

Example 3 Prove that if $|r| < 1$, the sequence $\{r^n\}$ is convergent and r^n converges to zero.

Solution First of all, if $r = 0$, the sequence is $\{0\}$ and $\lim_{x\to+\infty} 0 = 0$. Hence the sequence is convergent and the nth element converges to zero.

If $0 < |r| < 1$, we consider the function f defined by $f(x) = r^x$, where x is any positive number, and show that $\lim_{x\to+\infty} r^x = 0$. Then from Theorem 15.1.3 it will follow that $\lim_{n\to+\infty} r^n = 0$ when n is any positive integer.

To prove that $\lim_{x\to+\infty} r^x = 0$ $(0 < |r| < 1)$, we shall show that for any $\epsilon > 0$ there exists a number $N > 0$ such that

$$|r^x - 0| < \epsilon \quad \text{whenever} \quad x > N \tag{4}$$

Statement (4) is equivalent to

$$|r|^x < \epsilon \quad \text{whenever} \quad x > N$$

which is true if and only if

$$\ln |r|^x < \ln \epsilon \quad \text{whenever} \quad x > N$$

or, equivalently,

$$x \ln |r| < \ln \epsilon \quad \text{whenever} \quad x > N \tag{5}$$

Because $0 < |r| < 1$, $\ln |r| < 0$. Thus (5) is equivalent to

$$x > \frac{\ln \epsilon}{\ln |r|} \quad \text{whenever} \quad x > N$$

Therefore, if $N = \ln \epsilon / \ln |r|$, we may conclude (4). Consequently, $\lim_{x\to+\infty} r^x = 0$, and so $\lim_{n\to+\infty} r^n = 0$ if n is any positive integer. Hence, by Definitions 15.1.2 and 15.1.4, $\{r^n\}$ is convergent and r^n converges to zero.

Example 4 Determine if the sequence $\{(-1)^n + 1\}$ is convergent or divergent.

Solution The elements of this sequence are 0, 2, 0, 2, 0, 2, . . . , $(-1)^n + 1$, ... Because $a_n = 0$ if n is odd and $a_n = 2$ if n is even, it appears that the sequence is divergent. To prove this, let us assume that the sequence is convergent and show that this assumption leads to a contradiction. If the sequence has the limit L, then by Definition 15.1.2, for any $\epsilon > 0$ there exists a number $N > 0$ such that $|a_n - L| < \epsilon$ for every integer $n > N$. In particular, when $\epsilon = \frac{1}{2}$, there exists a number $N > 0$ such that

$$|a_n - L| < \tfrac{1}{2} \qquad \text{for every integer } n > N$$

or, equivalently,

$$-\tfrac{1}{2} < a_n - L < \tfrac{1}{2} \qquad \text{for every integer } n > N \tag{6}$$

Because $a_n = 0$ if n is odd and $a_n = 2$ if n is even, it follows from (6) that

$$-\tfrac{1}{2} < -L < \tfrac{1}{2} \quad \text{and} \quad -\tfrac{1}{2} < 2 - L < \tfrac{1}{2}$$

But if $-L > -\frac{1}{2}$, then $2 - L > \frac{3}{2}$; hence $2 - L$ cannot be less than $\frac{1}{2}$. So there is a contradiction, and therefore the given sequence is divergent.

Example 5 Determine if the sequence

$$\left\{ n \sin \frac{\pi}{n} \right\}$$

is convergent or divergent.

Solution We wish to determine if $\lim_{n \to +\infty} n \sin(\pi/n)$ exists. Let $f(x) = x \sin(\pi/x)$ and investigate $\lim_{x \to +\infty} f(x)$. Because $f(x)$ can be written as $[\sin(\pi/x)]/(1/x)$ and $\lim_{x \to +\infty} \sin(\pi/x) = 0$ and $\lim_{x \to +\infty} (1/x) = 0$, L'Hôpital's rule can be applied to obtain

$$\lim_{x \to +\infty} f(x) = \lim_{x \to +\infty} \frac{-\dfrac{\pi}{x^2} \cos \dfrac{\pi}{x}}{-\dfrac{1}{x^2}} = \lim_{x \to +\infty} \pi \cos \frac{\pi}{x} = \pi$$

Therefore $\lim_{n \to +\infty} f(n) = \pi$ when n is a positive integer. So the given sequence is convergent and $n \sin(\pi/n)$ converges to π.

There are limit theorems for sequences that are analogous to limit theorems for functions given in Chapter 2. The statement of these theorems uses the terminology of sequences. The proofs are omitted because they are similar to the proofs of the corresponding theorems given in Chapter 2.

15.1.5 Theorem If $\{a_n\}$ and $\{b_n\}$ are convergent sequences and c is a constant, then

(i) the constant sequence $\{c\}$ has c as its limit;

(ii) $\lim\limits_{n \to +\infty} ca_n = c \lim\limits_{n \to +\infty} a_n$;

(iii) $\lim\limits_{n \to +\infty} (a_n \pm b_n) = \lim\limits_{n \to +\infty} a_n \pm \lim\limits_{n \to +\infty} b_n$;

(iv) $\lim\limits_{n \to +\infty} a_n b_n = \left(\lim\limits_{n \to +\infty} a_n\right)\left(\lim\limits_{n \to +\infty} b_n\right)$;

(v) $\lim\limits_{n \to +\infty} \dfrac{a_n}{b_n} = \dfrac{\lim\limits_{n \to +\infty} a_n}{\lim\limits_{n \to +\infty} b_n}$ if $\lim\limits_{n \to +\infty} b_n \neq 0$.

Example 6 Use Theorem 15.1.5 to prove that the sequence

$$\left\{\frac{n^2}{2n+1} \sin \frac{\pi}{n}\right\}$$

is convergent, and find the limit of the sequence.

Solution

$$\frac{n^2}{2n+1} \sin \frac{\pi}{n} = \frac{n}{2n+1} \cdot n \sin \frac{\pi}{n}$$

In Example 1 the sequence $\{n/(2n+1)\}$ was shown to be convergent and $\lim\limits_{n \to +\infty} [n/(2n+1)] = \frac{1}{2}$. In Example 5 we showed that the sequence $\{n \sin(\pi/n)\}$ is convergent and $\lim\limits_{n \to +\infty} [n \sin(\pi/n)] = \pi$. Hence, by Theorem 15.1.5(iv),

$$\lim_{n \to +\infty} \left[\frac{n}{2n+1} \cdot n \sin \frac{\pi}{n}\right] = \lim_{n \to +\infty} \frac{n}{2n+1} \cdot \lim_{n \to +\infty} n \sin \frac{\pi}{n} = \frac{1}{2} \cdot \pi$$

Thus the given sequence is convergent, and its limit is $\frac{1}{2}\pi$.

EXERCISES 15.1

In Exercises 1 through 19, write the first four elements of the sequence and determine if the sequence is convergent or divergent. If the sequence converges, find its limit.

1. $\left\{\dfrac{n+1}{2n-1}\right\}$ **2.** $\left\{\dfrac{2n^2+1}{3n^2-n}\right\}$ **3.** $\left\{\dfrac{n^2+1}{n}\right\}$ **4.** $\left\{\dfrac{3n^3+1}{2n^2+n}\right\}$ **5.** $\left\{\dfrac{3-2n^2}{n^2-1}\right\}$ **6.** $\left\{\dfrac{e^n}{n}\right\}$

7. $\left\{\dfrac{\ln n}{n^2}\right\}$ **8.** $\left\{\dfrac{\log_b n}{n}\right\}, b > 1$ **9.** $\{\tanh n\}$ **10.** $\{\sinh n\}$ **11.** $\left\{\dfrac{n}{n+1} \sin \dfrac{n\pi}{2}\right\}$ **12.** $\left\{\dfrac{\sinh n}{\sin n}\right\}$

13. $\left\{\dfrac{1}{\sqrt{n^2+1}-n}\right\}$ **14.** $\{\sqrt{n+1}-\sqrt{n}\}$ **15.** $\left\{\left(1+\dfrac{1}{3n}\right)^n\right\}$ (*Hint:* Use $\lim\limits_{x \to 0} (1+x)^{1/x} = e$.) **16.** $\left\{\left(1+\dfrac{2}{n}\right)^n\right\}$ See Hint for Exercise 15.

17. $\{r^{1/n}\}$ and $r > 0$. (*Hint:* Consider two cases: $r \leq 1$ and $r > 1$.) **18.** $\{\cos n\pi\}$ **19.** $\left\{\dfrac{n}{c^n}\right\}, c > 1$

In Exercises 20 through 25, use Definition 15.1.2 to prove that the given sequence has the limit L.

20. $\left\{\dfrac{4}{2n-1}\right\}; L = 0$ **21.** $\left\{\dfrac{3}{n-1}\right\}; L = 0$ **22.** $\left\{\dfrac{1}{\sqrt{n}}\right\}; L = 0$

Theorem 15.2.6 states that if $\{a_n\}$ is a bounded monotonic sequence, there exists a number L such that $\lim_{n \to +\infty} a_n = L$, but it does not state how to find L. For this reason Theorem 15.2.6 is called an *existence theorem.* Many important concepts in mathematics are based on existence theorems. In particular, there are many sequences for which the limit cannot be found by direct use of the definition or by use of limit theorems, but the knowledge that such a limit exists can be of value to a mathematician.

In the proof of Theorem 15.2.6 the limit of the bounded increasing sequence is the least upper bound B of the sequence. Hence, if D is an upper bound of the sequence, $\lim_{n \to +\infty} a_n = B \le D$. We have, then, the following theorem.

15.2.7 Theorem Let $\{a_n\}$ be an increasing sequence, and suppose that D is an upper bound of this sequence. Then $\{a_n\}$ is convergent, and

$$\lim_{n \to +\infty} a_n \le D$$

In proving Theorem 15.2.6 for the case when the bounded monotonic sequence is decreasing, the limit of the sequence is the greatest lower bound. The following theorem follows in a way similar to that of Theorem 15.2.7.

15.2.8 Theorem Let $\{a_n\}$ be a decreasing sequence, and suppose that C is a lower bound of this sequence. Then $\{a_n\}$ is convergent, and

$$\lim_{n \to +\infty} a_n \ge C$$

Example 2 Use Theorem 15.2.6 to prove that the sequence

$$\left\{\frac{2^n}{n!}\right\}$$

is convergent.

Solution The elements of the given sequence are

$$\frac{2^1}{1!}, \frac{2^2}{2!}, \frac{2^3}{3!}, \frac{2^4}{4!}, \ldots, \frac{2^n}{n!}, \frac{2^{n+1}}{(n+1)!}, \ldots$$

$1! = 1$, $2! = 1 \cdot 2 = 2$, $3! = 1 \cdot 2 \cdot 3 = 6$, $4! = 1 \cdot 2 \cdot 3 \cdot 4 = 24$. Hence the elements of the sequence can be written as

$$2, 2, \frac{4}{3}, \frac{2}{3}, \ldots, \frac{2^n}{n!}, \frac{2^{n+1}}{(n+1)!}, \ldots$$

Then $a_1 = a_2 > a_3 > a_4$; so the given sequence may be decreasing. We must check to see if $a_n \ge a_{n+1}$; that is, it must be determined if

$$\frac{2^n}{n!} \ge \frac{2^{n+1}}{(n+1)!} \tag{8}$$

which is equivalent to

$$2^n(n+1)! \ge 2^{n+1}n!$$

which is equivalent to

$$2^n n!(n + 1) \geq 2 \cdot 2^n n!$$

which is equivalent to

$$n + 1 \geq 2 \tag{9}$$

When $n = 1$, inequality (9) becomes $2 = 2$, and (9) obviously holds when $n > 2$. Because inequality (8) is equivalent to (9), it follows that the given sequence is decreasing and hence monotonic. An upper bound for the given sequence is 2, and a lower bound is 0. Therefore the sequence is bounded.

The sequence $\{2^n/n!\}$ is therefore a bounded monotonic sequence, and by Theorem 15.2.6 it is convergent.

Theorem 15.2.6 states that a sufficient condition for a monotonic sequence to be convergent is that it be bounded. This is also a necessary condition and is given in the following theorem.

15.2.9 Theorem A convergent monotonic sequence is bounded.

Proof We prove the theorem for the case when the monotonic sequence is increasing. Let the sequence be $\{a_n\}$.

To prove that $\{a_n\}$ is bounded, it must be shown that it has a lower bound and an upper bound. Because $\{a_n\}$ is an increasing sequence, its first element serves as a lower bound. We must now find an upper bound.

Because $\{a_n\}$ is convergent, the sequence has a limit; call this limit L. Therefore $\lim_{n \to +\infty} a_n = L$, and so by Definition 15.1.2, for any $\epsilon > 0$ there exists a number $N > 0$ such that

$$|a_n - L| < \epsilon \qquad \text{for every positive integer } n > N$$

or, equivalently,

$$-\epsilon < a_n - L < \epsilon \quad \text{whenever} \quad n > N$$

$$L - \epsilon < a_n < L + \epsilon \quad \text{whenever} \quad n > N$$

Because $\{a_n\}$ is increasing,

$$a_n < L + \epsilon \qquad \text{for all positive integers } n$$

Therefore $L + \epsilon$ will serve as an upper bound of the sequence $\{a_n\}$.

To prove the theorem when $\{a_n\}$ is a decreasing sequence, do as suggested in the proof of Theorem 15.2.6: Consider the sequence $\{-a_n\}$, which will be increasing, and apply the above results. You are asked to do this proof in Exercise 18. ■

EXERCISES 15.2

In Exercises 1 through 16, determine if the given sequence is increasing, decreasing, or not monotonic.

1. $\left\{\dfrac{3n - 1}{4n + 5}\right\}$ **2.** $\left\{\dfrac{2n - 1}{4n - 1}\right\}$ **3.** $\left\{\dfrac{1 - 2n^2}{n^2}\right\}$ **4.** $\{\sin n\pi\}$

5. $\{\cos \frac{1}{3}n\pi\}$ **6.** $\left\{\frac{n^3 - 1}{n}\right\}$ **7.** $\left\{\frac{1}{n + \sin n^2}\right\}$ **8.** $\left\{\frac{2^n}{1 + 2^n}\right\}$

9. $\left\{\frac{5^n}{1 + 5^{2n}}\right\}$ **10.** $\left\{\frac{(2n)!}{5^n}\right\}$ **11.** $\left\{\frac{n!}{3^n}\right\}$ **12.** $\left\{\frac{n}{2^n}\right\}$

13. $\left\{\frac{n^n}{n!}\right\}$ **14.** $\{n^2 + (-1)^n n\}$ **15.** $\left\{\frac{n!}{1 \cdot 3 \cdot 5 \cdot \ldots \cdot (2n-1)}\right\}$ **16.** $\left\{\frac{1 \cdot 3 \cdot 5 \cdot \ldots \cdot (2n-1)}{2^n \cdot n!}\right\}$

17. Use the fact that Theorem 15.2.6 holds for an increasing sequence to prove that the theorem holds when $\{a_n\}$ is a decreasing sequence. (*Hint:* Consider the sequence $\{-a_n\}$.)

18. Prove Theorem 15.2.9 when $\{a_n\}$ is a decreasing sequence by a method similar to that used in Exercise 17.

In Exercises 19 and 20, determine if the given sequence is bounded.

19. $\left\{\frac{n^2 + 3}{n + 1}\right\}$ **20.** $\{3 - (-1)^{n-1}\}$

In Exercises 21 through 30, prove that the given sequence is convergent by using Theorem 15.2.6.

21. The sequence of Exercise 1. **22.** $\left\{\frac{n}{3^{n+1}}\right\}$ **23.** $\left\{\frac{1 \cdot 3 \cdot 5 \cdot \ldots \cdot (2n-1)}{2 \cdot 4 \cdot 6 \cdot \ldots \cdot (2n)}\right\}$

24. The sequence of Exercise 8. **25.** The sequence of Exercise 9. **26.** The sequence of Exercise 12.

27. The sequence of Exercise 15. **28.** The sequence of Exercise 16. **29.** $\left\{\frac{n^2}{2^n}\right\}$ **30.** $\{k^{1/n}\}, k > 1$

31. Give an example of a sequence that is bounded and convergent but not monotonic.

32. Given the sequence $\{a_n\}$ where $a_n > 0$ for all n, and $a_{n+1} < ka_n$ with $0 < k < 1$. Prove that $\{a_n\}$ is convergent.

15.3 INFINITE SERIES OF CONSTANT TERMS

The familiar operation of addition applies only to a finite set of numbers. To extend addition to infinitely many numbers it is necessary to define what is meant by such a sum. To carry this out we deal with a limiting process by considering sequences.

15.3.1 Definition If $\{u_n\}$ is a sequence and

$$s_n = \sum_{i=1}^{n} u_i = u_1 + u_2 + u_3 + \cdots + u_n$$

then the sequence $\{s_n\}$ is called an **infinite series.**

The numbers $u_1, u_2, u_3, \ldots$ are called the **terms** of the infinite series. The following symbolism is used to denote an infinite series:

$$\sum_{n=1}^{+\infty} u_n = u_1 + u_2 + u_3 + \cdots + u_n + \cdots \tag{1}$$

Given the infinite series denoted by (1), $s_1 = u_1$, $s_2 = u_1 + u_2$, $s_3 = u_1 + u_2 + u_3$, and in general

$$s_k = \sum_{i=1}^{k} u_i = u_1 + u_2 + u_3 + \cdots + u_k \tag{2}$$

where s_k, defined by (2), is called the kth **partial sum** of the given series, and the sequence $\{s_n\}$ is a sequence of partial sums. Thus Definition 15.3.1 states that an infinite series is a sequence of partial sums.

Because $s_{n-1} = u_1 + u_2 + \cdots + u_{n-1}$ and $s_n = u_1 + u_2 + \cdots + u_{n-1} + u_n$, then

$$s_n = s_{n-1} + u_n \tag{3}$$

Example 1 Given the infinite series

$$\sum_{n=1}^{+\infty} u_n = \sum_{n=1}^{+\infty} \frac{1}{n(n+1)}$$

find the first four elements of the sequence of partial sums $\{s_n\}$, and find a formula for s_n in terms of n.

Solution From formula (3),

$$s_1 = u_1 = \frac{1}{1 \cdot 2} = \frac{1}{2}$$

$$s_2 = s_1 + u_2 = \frac{1}{2} + \frac{1}{2 \cdot 3} = \frac{2}{3}$$

$$s_3 = s_2 + u_3 = \frac{2}{3} + \frac{1}{3 \cdot 4} = \frac{3}{4}$$

$$s_4 = s_3 + u_4 = \frac{3}{4} + \frac{1}{4 \cdot 5} = \frac{4}{5}$$

By partial fractions,

$$u_k = \frac{1}{k(k+1)} = \frac{1}{k} - \frac{1}{k+1}$$

Therefore

$$u_1 = 1 - \frac{1}{2}, u_2 = \frac{1}{2} - \frac{1}{3}, u_3 = \frac{1}{3} - \frac{1}{4}, \ldots, u_{n-1} = \frac{1}{n-1} - \frac{1}{n},$$

$$u_n = \frac{1}{n} - \frac{1}{n+1}$$

Thus, because $s_n = u_1 + u_2 + \cdots + u_{n-1} + u_n$,

$$s_n = \left(1 - \frac{1}{2}\right) + \left(\frac{1}{2} - \frac{1}{3}\right) + \left(\frac{1}{3} - \frac{1}{4}\right) + \cdots + \left(\frac{1}{n-1} - \frac{1}{n}\right) + \left(\frac{1}{n} - \frac{1}{n+1}\right)$$

Upon removing parentheses and combining terms we obtain

$$s_n = 1 - \frac{1}{n+1} = \frac{n}{n+1}$$

By taking $n = 1, 2, 3$, and 4, we see that the previous results agree.

The method of solution of the above example applies only to a special case. In general, it is not possible to obtain such an expression for s_n.

We now define the *sum* of an infinite series.

15.3.2 Definition Let $\sum_{n=1}^{+\infty} u_n$ be a given infinite series, and let $\{s_n\}$ be the sequence of partial sums defining this infinite series. Then if $\lim_{n \to +\infty} s_n$ exists and is equal to S, the given series is said to be **convergent** and S is the **sum** of the given infinite series. If $\lim_{n \to +\infty} s_n$ does not exist, the series is **divergent** and the series does not have a sum.

Essentially Definition 15.3.2 states that an infinite series is convergent if and only if the corresponding sequence of partial sums is convergent.

If an infinite series has a sum S, we also say that the series converges to S.

Observe that the sum of a convergent series is the limit of a sequence of partial sums and is not obtained by ordinary addition. For a convergent series the symbolism

$$\sum_{n=1}^{+\infty} u_n$$

is used to denote both the series and the sum of the series. The use of the same symbol should not be confusing because the correct interpretation is apparent from the context in which it is employed.

Example 2 Determine if the infinite series of Example 1 has a sum.

Solution In the solution of Example 1 it was shown that the sequence of partial sums for the given series is $\{s_n\} = \{n/(n + 1)\}$. Therefore

$$\lim_{n \to +\infty} s_n = \lim_{n \to +\infty} \frac{n}{n + 1} = \lim_{n \to +\infty} \frac{1}{1 + \frac{1}{n}} = 1$$

So the infinite series has a sum equal to 1, and we write

$$\sum_{n=1}^{+\infty} \frac{1}{n(n + 1)} = \frac{1}{2} + \frac{1}{6} + \frac{1}{12} + \frac{1}{20} + \cdots + \frac{1}{n(n + 1)} + \cdots = 1$$

As mentioned above, in most cases it is not possible to obtain an expression for s_n in terms of n; so we must have other methods for determining whether or not a given infinite series has a sum or, equivalently, whether a given infinite series is convergent or divergent.

15.3.3 Theorem If the infinite series $\sum_{n=1}^{+\infty} u_n$ is convergent, then $\lim_{n \to +\infty} u_n = 0$.

Proof Let $\{s_n\}$ be the sequence of partial sums for the given series, and denote the sum of the series by S. From Definition 15.3.2, $\lim_{n \to +\infty} s_n = S$. Thus, for any $\epsilon > 0$ there exists a number $N > 0$ such that $|S - s_n| < \frac{1}{2}\epsilon$ for every integer $n > N$. Also, $|S - s_{n+1}| < \frac{1}{2}\epsilon$ for these integers $n > N$. Then

$$|u_{n+1}| = |s_{n+1} - s_n| = |S - s_n + s_{n+1} - S| \leq |S - s_n| + |s_{n+1} - S|$$

So

$$|u_{n+1}| < \tfrac{1}{2}\epsilon + \tfrac{1}{2}\epsilon = \epsilon \qquad \text{for every integer } n > N$$

Therefore, $\lim_{n \to +\infty} u_n = 0$. ■

From Theorem 15.3.3 it follows that if $\lim_{n \to +\infty} u_n \neq 0$, then the series $\sum_{n=1}^{+\infty} u_n$ is divergent.

Example 3 Prove that the following two series are divergent.

(a) $\displaystyle\sum_{n=1}^{+\infty} \frac{n^2 + 1}{n^2} = 2 + \frac{5}{4} + \frac{10}{9} + \frac{17}{16} + \cdots$

(b) $\displaystyle\sum_{n=1}^{+\infty} (-1)^{n+1}3 = 3 - 3 + 3 - 3 + \cdots$

Solution

(a)
$$\begin{aligned}
\lim_{n \to +\infty} u_n &= \lim_{n \to +\infty} \frac{n^2 + 1}{n^2} \\
&= \lim_{n \to +\infty} \frac{1 + \dfrac{1}{n^2}}{1} \\
&= 1 \\
&\neq 0
\end{aligned}$$

Therefore, by Theorem 15.3.3, the series is divergent.

(b) $\lim_{n \to +\infty} u_n = \lim_{n \to +\infty} (-1)^{n+1}3$, which does not exist. Therefore, by Theorem 15.3.3, the series is divergent.

The converse of Theorem 15.3.3 is false. That is, if $\lim_{n \to +\infty} u_n = 0$, it does not follow that the series is necessarily convergent. In other words, it is possible to have a divergent series for which $\lim_{n \to +\infty} u_n = 0$. An example of such a series is the one known as the **harmonic series,** which is

$$\sum_{n=1}^{+\infty} \frac{1}{n} = 1 + \frac{1}{2} + \frac{1}{3} + \frac{1}{4} + \cdots + \frac{1}{n} + \cdots \tag{4}$$

Clearly, $\lim_{n \to +\infty} 1/n = 0$. In Illustration 1 the harmonic series is proved to diverge and we use the following theorem, which states that the difference between two partial sums s_R and s_T of a convergent series can be made as small as we please by taking R and T sufficiently large.

15.3.4 Theorem Let $\{s_n\}$ be the sequence of partial sums for a given convergent series $\sum_{n=1}^{+\infty} u_n$. Then for any $\epsilon > 0$ there exists a number N such that

$$|s_R - s_T| < \epsilon \quad \text{whenever} \quad R > N \text{ and } T > N$$

Proof Because the series $\sum_{n=1}^{+\infty} u_n$ is convergent, call its sum S. Then for any $\epsilon > 0$ there exists an $N > 0$ such that $|S - s_n| < \frac{1}{2}\epsilon$ whenever $n > N$. Therefore, if $R > N$ and $T > N$,

$$|s_R - s_T| = |s_R - S + S - s_T| \le |s_R - S| + |S - s_T| < \tfrac{1}{2}\epsilon + \tfrac{1}{2}\epsilon$$

So

$$|s_R - s_T| < \epsilon \quad \text{whenever} \quad R > N \text{ and } T > N$$

■

• ILLUSTRATION 1 We prove that the harmonic series (4) is divergent. For this series,

$$s_n = 1 + \frac{1}{2} + \cdots + \frac{1}{n}$$

and

$$s_{2n} = 1 + \frac{1}{2} + \cdots + \frac{1}{n} + \frac{1}{n+1} + \cdots + \frac{1}{2n}$$

So

$$s_{2n} - s_n = \frac{1}{n+1} + \frac{1}{n+2} + \frac{1}{n+3} + \cdots + \frac{1}{2n} \tag{5}$$

If $n > 1$,

$$\frac{1}{n+1} + \frac{1}{n+2} + \frac{1}{n+3} + \cdots + \frac{1}{2n} > \frac{1}{2n} + \frac{1}{2n} + \frac{1}{2n} + \cdots + \frac{1}{2n} \tag{6}$$

There are n terms on each side of the inequality sign in (6); so the right side is $n(1/2n) = \frac{1}{2}$. Therefore, from (5) and (6),

$$s_{2n} - s_n > \tfrac{1}{2} \quad \text{whenever} \quad n > 1 \tag{7}$$

But Theorem 15.3.4 states that if the given series is convergent, then

$s_{2n} - s_n$ may be made as small as we please by taking n large enough; that is, if $\epsilon = \frac{1}{2}$, there exists an N such that $s_{2n} - s_n < \frac{1}{2}$ whenever $2n > N$ and $n > N$. But this contradicts (7). Therefore the harmonic series is divergent even though $\lim_{n \to +\infty} 1/n = 0$. ●

A **geometric series** is a series of the form

$$\sum_{n=1}^{+\infty} ar^{n-1} = a + ar + ar^2 + \cdots + ar^{n-1} + \cdots \tag{8}$$

The nth partial sum of this series is given by

$$s_n = a(1 + r + r^2 + \cdots + r^{n-1}) \tag{9}$$

From the identity

$$1 - r^n = (1 - r)(1 + r + r^2 + \cdots + r^{n-1})$$

(9) can be written as

$$s_n = \frac{a(1 - r^n)}{1 - r} \qquad \text{if } r \neq 1 \tag{10}$$

15.3.5 Theorem The geometric series converges to the sum $a/(1 - r)$ if $|r| < 1$, and the geometric series diverges if $|r| \geq 1$.

Proof In Example 3, Sec. 15.1, it was shown that $\lim_{n \to +\infty} r^n = 0$ if $|r| < 1$. Therefore, from (10) we can conclude that if $|r| < 1$,

$$\lim_{n \to +\infty} s_n = \frac{a}{1 - r}$$

So if $|r| < 1$, the geometric series converges, and its sum is $a/(1 - r)$.

If $r = 1$, $s_n = na$. Then $\lim_{n \to +\infty} s_n = +\infty$ if $a > 0$, and $\lim_{n \to +\infty} s_n = -\infty$ if $a < 0$.

If $r = -1$, then the geometric series becomes $a - a + a - \cdots + (-1)^{n-1}a + \cdots$; so $s_n = 0$ if n is even, and $s_n = a$ if n is odd. Therefore $\lim_{n \to +\infty} s_n$ does not exist. Hence the geometric series diverges when $|r| = 1$.

If $|r| > 1$, $\lim_{n \to +\infty} ar^{n-1} = a \lim_{n \to +\infty} r^{n-1}$. Clearly, $\lim_{n \to +\infty} r^{n-1} \neq 0$ because $|r^{n-1}|$ can be made as large as we please by taking n large enough. Therefore, by Theorem 15.3.3, the series is divergent. This completes the proof. ■

● ILLUSTRATION 2

$$\sum_{n=1}^{+\infty} \frac{1}{2^{n-1}} = 1 + \frac{1}{2} + \frac{1}{4} + \frac{1}{8} + \cdots + \frac{1}{2^{n-1}} + \cdots$$

which is the geometric series, with $a = 1$ and $r = \frac{1}{2}$. Therefore, by Theorem 15.3.5, the series is convergent. Because $a/(1 - r) = 1/(1 - \frac{1}{2}) = 2$, the sum of the series is 2. ●

The following example illustrates how Theorem 15.3.5 can be used to express a nonterminating repeating decimal as a common fraction.

Example 4 Express the decimal 0.3333... as a common fraction.

Solution

$$0.3333\ldots = \frac{3}{10} + \frac{3}{100} + \frac{3}{1{,}000} + \frac{3}{10{,}000} + \cdots + \frac{3}{10^n} + \cdots$$

This is a geometric series in which $a = \frac{3}{10}$ and $r = \frac{1}{10}$. Because $|r| < 1$, it follows from Theorem 15.3.5 that the series converges and its sum is $a/(1 - r)$. Therefore

$$0.3333\ldots = \frac{\frac{3}{10}}{1 - \frac{1}{10}} = \frac{1}{3}$$

The next theorem states that the convergence or divergence of an infinite series is not affected by changing a finite number of terms.

15.3.6 Theorem If $\sum\limits_{n=1}^{+\infty} a_n$ and $\sum\limits_{n=1}^{+\infty} b_n$ are two infinite series, differing only in their first m terms (i.e., $a_k = b_k$ if $k > m$), then either both series converge or both series diverge.

Proof Let $\{s_n\}$ and $\{t_n\}$ be the sequences of partial sums of the series $\sum\limits_{n=1}^{+\infty} a_n$ and $\sum\limits_{n=1}^{+\infty} b_n$, respectively. Then

$$s_n = a_1 + a_2 + \cdots + a_m + a_{m+1} + a_{m+2} + \cdots + a_n$$

and

$$t_n = b_1 + b_2 + \cdots + b_m + b_{m+1} + b_{m+2} + \cdots + b_n$$

Because $a_k = b_k$ if $k > m$, then if $n \geq m$,

$$s_n - t_n = (a_1 + a_2 + \cdots + a_m) - (b_1 + b_2 + \cdots + b_m)$$

So

$$s_n - t_n = s_m - t_m \quad \text{whenever} \quad n \geq m \tag{11}$$

We wish to show that either both $\lim\limits_{n \to +\infty} s_n$ and $\lim\limits_{n \to +\infty} t_n$ exist or do not exist.

Suppose that $\lim\limits_{n \to +\infty} t_n$ exists. Then from (11),

$$s_n = t_n + (s_m - t_m) \quad \text{whenever} \quad n \geq m$$

and so

$$\lim_{n \to +\infty} s_n = \lim_{n \to +\infty} t_n + (s_m - t_m)$$

Hence, when $\lim\limits_{n \to +\infty} t_n$ exists, $\lim\limits_{n \to +\infty} s_n$ also exists and both series converge.

Now suppose that $\lim\limits_{n \to +\infty} t_n$ does not exist and $\lim\limits_{n \to +\infty} s_n$ exists. From (11),

$$t_n = s_n + (t_m - s_m) \quad \text{whenever} \quad n \geq m$$

Because $\lim_{n \to +\infty} s_n$ exists, it follows that

$$\lim_{n \to +\infty} t_n = \lim_{n \to +\infty} s_n + (t_m - s_m)$$

and so $\lim_{n \to +\infty} t_n$ has to exist, which is a contradiction. Hence, if $\lim_{n \to +\infty} t_n$ does not exist, then $\lim_{n \to +\infty} s_n$ does not exist, and both series diverge. ■

Example 5 Determine whether the infinite series

$$\sum_{n=1}^{+\infty} \frac{1}{n+4}$$

is convergent or divergent.

Solution The given series is

$$\frac{1}{5} + \frac{1}{6} + \frac{1}{7} + \cdots + \frac{1}{n+4} + \cdots$$

which can be written as

$$0 + 0 + 0 + 0 + \frac{1}{5} + \frac{1}{6} + \frac{1}{7} + \cdots + \frac{1}{n} + \cdots \tag{12}$$

Now the harmonic series, which is known to be divergent, is

$$1 + \frac{1}{2} + \frac{1}{3} + \frac{1}{4} + \frac{1}{5} + \frac{1}{6} + \frac{1}{7} + \cdots + \frac{1}{n} + \cdots \tag{13}$$

Series (12) differs from series (13) only in the first four terms. Hence, by Theorem 15.3.6, series (12) is also divergent.

Example 6 Determine whether the following infinite series is convergent or divergent:

$$\sum_{n=1}^{+\infty} \frac{\left[\!\left[\cos \frac{3}{n}\pi + 2\right]\!\right]}{3^n}$$

Solution The given series can be written as

$$\frac{[\![\cos 3\pi + 2]\!]}{3} + \frac{[\![\cos \frac{3}{2}\pi + 2]\!]}{3^2} + \frac{[\![\cos \pi + 2]\!]}{3^3} + \frac{[\![\cos \frac{3}{4}\pi + 2]\!]}{3^4} + \frac{[\![\cos \frac{3}{5}\pi + 2]\!]}{3^5} + \frac{[\![\cos \frac{1}{2}\pi + 2]\!]}{3^6} + \frac{[\![\cos \frac{3}{7}\pi + 2]\!]}{3^7} + \cdots$$

$$= \frac{1}{3} + \frac{2}{3^2} + \frac{1}{3^3} + \frac{1}{3^4} + \frac{1}{3^5} + \frac{2}{3^6} + \frac{2}{3^7} + \frac{2}{3^8} + \cdots \tag{14}$$

Consider the geometric series with $a = \frac{2}{3}$ and $r = \frac{1}{3}$:

$$\frac{2}{3} + \frac{2}{3^2} + \frac{2}{3^3} + \frac{2}{3^4} + \frac{2}{3^5} + \frac{2}{3^6} + \frac{2}{3^7} + \frac{2}{3^8} + \cdots \tag{15}$$

This series is convergent by Theorem 15.3.5. Because series (14) differs from series (15) only in the first five terms, it follows from Theorem 15.3.6 that series (14) is also convergent.

As a consequence of Theorem 15.3.6, for a given infinite series a finite number of terms can be added or subtracted without affecting its convergence or divergence. For instance, in Example 5 the given series may be thought of as being obtained from the harmonic series by subtracting the first four terms. And because the harmonic series is divergent, the given series is divergent. In Example 6 we could consider the convergent geometric series

$$\frac{2}{3^6} + \frac{2}{3^7} + \frac{2}{3^8} + \cdots \tag{16}$$

and obtain the given series (14) by adding five terms. Because series (16) is convergent, it follows that series (14) is convergent.

The following theorem states that if an infinite series is multiplied term by term by a nonzero constant, its convergence or divergence is not affected.

15.3.7 Theorem Let c be any nonzero constant.

(i) If the series $\sum_{n=1}^{+\infty} u_n$ is convergent and its sum is S, then the series $\sum_{n=1}^{+\infty} cu_n$ is also convergent and its sum is $c \cdot S$.

(ii) If the series $\sum_{n=1}^{+\infty} u_n$ is divergent, then the series $\sum_{n=1}^{+\infty} cu_n$ is also divergent.

Proof Let the nth partial sum of the series $\sum_{n=1}^{+\infty} u_n$ be s_n. Therefore $s_n = u_1 + u_2 + \cdots + u_n$. The nth partial sum of the series $\sum_{n=1}^{+\infty} cu_n$ is $c(u_1 + u_2 + \cdots + u_n) = cs_n$.

Proof of (i) If the series $\sum_{n=1}^{+\infty} u_n$ is convergent, then $\lim_{n \to +\infty} s_n$ exists and is S. Therefore

$$\lim_{n \to +\infty} cs_n = c \lim_{n \to +\infty} s_n = c \cdot S$$

Hence the series $\sum_{n=1}^{+\infty} cu_n$ is convergent and its sum is $c \cdot S$.

Proof of (ii) If the series $\sum_{n=1}^{+\infty} u_n$ is divergent, then $\lim_{n \to +\infty} s_n$ does not exist. Now suppose that the series $\sum_{n=1}^{+\infty} cu_n$ is convergent. Then $\lim_{n\to+\infty} cs_n$ exists. But $s_n = cs_n/c$; so

$$\lim_{n\to+\infty} s_n = \lim_{n\to+\infty} \frac{1}{c}(cs_n) = \frac{1}{c} \lim_{n\to+\infty} cs_n$$

Thus $\lim_{n\to+\infty} s_n$ must exist, which is a contradiction. Therefore the series $\sum_{n=1}^{+\infty} cu_n$ is divergent. ■

Example 7 Determine whether the infinite series

$$\sum_{n=1}^{+\infty} \frac{1}{4n}$$

is convergent or divergent.

Solution

$$\sum_{n=1}^{+\infty} \frac{1}{4n} = \frac{1}{4} + \frac{1}{8} + \frac{1}{12} + \frac{1}{16} + \cdots + \frac{1}{4n} + \cdots$$

Because $\sum_{n=1}^{+\infty} 1/n$ is the harmonic series that is divergent, then by Theorem 15.3.7(ii) with $c = \frac{1}{4}$, the given series is divergent.

Theorem 15.3.7(i) is an extension to convergent infinite series of the following property of finite sums:

$$\sum_{k=1}^{n} ca_k = c \sum_{k=1}^{n} a_k$$

Another property of finite sums is

$$\sum_{k=1}^{n} (a_k \pm b_k) = \sum_{k=1}^{n} a_k \pm \sum_{k=1}^{n} b_k$$

and its extension to convergent infinite series is given by the following theorem.

15.3.8 Theorem If $\sum_{n=1}^{+\infty} a_n$ and $\sum_{n=1}^{+\infty} b_n$ are convergent infinite series whose sums are S and R, respectively, then

(i) $\sum_{n=1}^{+\infty} (a_n + b_n)$ is a convergent series and its sum is $S + R$;

(ii) $\sum_{n=1}^{+\infty} (a_n - b_n)$ is a convergent series and its sum is $S - R$.

The proof of this theorem is left as an exercise (see Exercise 30).

The next theorem is a corollary of the above theorem and is sometimes used to prove that a series is divergent.

15.3.9 Theorem If the series $\sum_{n=1}^{+\infty} a_n$ is convergent and the series $\sum_{n=1}^{+\infty} b_n$ is divergent, then the series $\sum_{n=1}^{+\infty} (a_n + b_n)$ is divergent.

Proof Assume that $\sum_{n=1}^{+\infty} (a_n + b_n)$ is convergent and its sum is S. Let the sum of the series $\sum_{n=1}^{+\infty} a_n$ be R. Then because

$$\sum_{n=1}^{+\infty} b_n = \sum_{n=1}^{+\infty} [(a_n + b_n) - a_n]$$

it follows from Theorem 15.3.8(ii) that $\sum_{n=1}^{+\infty} b_n$ is convergent and its sum is $S - R$. But this is a contradiction to the hypothesis that $\sum_{n=1}^{+\infty} b_n$ is divergent. Hence $\sum_{n=1}^{+\infty} (a_n + b_n)$ is divergent. ■

Example 8 Determine whether the infinite series

$$\sum_{n=1}^{+\infty} \left(\frac{1}{4n} + \frac{1}{4^n}\right)$$

is convergent or divergent.

Solution In Example 7 the series $\sum_{n=1}^{+\infty} 1/4n$ was proved to be divergent. Because the series $\sum_{n=1}^{+\infty} 1/4^n$ is a geometric series with $|r| = \frac{1}{4} < 1$, it is convergent. Hence, by Theorem 15.3.9, the given series is divergent.

If both series $\sum_{n=1}^{+\infty} a_n$ and $\sum_{n=1}^{+\infty} b_n$ are divergent, the series $\sum_{n=1}^{+\infty} (a_n + b_n)$ may or may not be convergent. For example, if $a_n = 1/n$ and $b_n = 1/n$, then $a_n + b_n = 2/n$ and $\sum_{n=1}^{+\infty} 2/n$ is divergent. But if $a_n = 1/n$ and $b_n = -1/n$, then $a_n + b_n = 0$ and $\sum_{n=1}^{+\infty} 0$ is convergent.

EXERCISES 15.3

In Exercises 1 through 8, find the first four elements of the sequence of partial sums $\{s_n\}$, and find a formula for s_n in terms of n. Also determine if the infinite series is convergent or divergent; if it is convergent, find its sum.

1. $\sum_{n=1}^{+\infty} \frac{1}{(2n-1)(2n+1)}$ **2.** $\sum_{n=1}^{+\infty} n$ **3.** $\sum_{n=1}^{+\infty} \frac{5}{(3n+1)(3n-2)}$ **4.** $\sum_{n=1}^{+\infty} \frac{2}{(4n-3)(4n+1)}$

5. $\sum_{n=1}^{+\infty} \ln \frac{n}{n+1}$ **6.** $\sum_{n=1}^{+\infty} \frac{2n+1}{n^2(n+1)^2}$ **7.** $\sum_{n=1}^{+\infty} \frac{2}{5^{n-1}}$ **8.** $\sum_{n=1}^{+\infty} \frac{2^{n-1}}{3^n}$

In Exercises 9 through 13, find the infinite series that is the given sequence of partial sums. Also determine if the infinite series is convergent or divergent; if it is convergent, find its sum.

9. $\{s_n\} = \left\{\frac{2n}{3n+1}\right\}$ **10.** $\{s_n\} = \left\{\frac{n^2}{n+1}\right\}$ **11.** $\{s_n\} = \left\{\frac{1}{2^n}\right\}$ **12.** $\{s_n\} = \{3^n\}$ **13.** $\{s_n\} = \{\ln(2n+1)\}$

In Exercises 14 through 29, write the first four terms of the given infinite series and determine if the series is convergent or divergent. If the series is convergent, find its sum.

14. $\sum_{n=1}^{+\infty} \frac{n}{n+1}$ **15.** $\sum_{n=1}^{+\infty} \frac{2n+1}{3n+2}$ **16.** $\sum_{n=1}^{+\infty} \frac{2}{3n}$ **17.** $\sum_{n=1}^{+\infty} \left(\frac{2}{3}\right)^n$ **18.** $\sum_{n=1}^{+\infty} \frac{3n^2}{n^2+1}$

19. $\sum_{n=1}^{+\infty} \ln \frac{1}{n}$ **20.** $\sum_{n=1}^{+\infty} \frac{2}{3^{n-1}}$ **21.** $\sum_{n=1}^{+\infty} (-1)^{n+1} \frac{3}{2^n}$ **22.** $\sum_{n=1}^{+\infty} \tan^n \frac{\pi}{6}$ **23.** $\sum_{n=1}^{+\infty} e^{-n}$

24. $\sum_{n=1}^{+\infty} \frac{\sinh n}{n}$ **25.** $\sum_{n=1}^{+\infty} \cos \pi n$ **26.** $\sum_{n=1}^{+\infty} \sin \pi n$ **27.** $\sum_{n=1}^{+\infty} (2^{-n} + 3^{-n})$ **28.** $\sum_{n=1}^{+\infty} [1 + (-1)^n]$

29. $\sum_{n=1}^{+\infty} \left(\frac{1}{2n} - \frac{1}{3n}\right)$ **30.** Prove Theorem 15.3.8.

In Exercises 31 through 34, express the given nonterminating repeating decimal as a common fraction.

31. 0.27 27 27 . . . **32.** 2.045 45 45 . . . **33.** 1.234 234 234 . . . **34.** 0.4653 4653 4653 . . .

35. The path of each swing, after the first, of a pendulum bob is 0.93 as long as the path of the previous swing (from one side to the other side). If the path of the first swing is 56 cm long, and air resistance eventually brings the pendulum to rest, how far does the bob travel before it comes to rest?

36. A ball is dropped from a height of 12 m. Each time it strikes the ground, it bounces back to a height of three-fourths the distance from which it fell. Find the total distance traveled by the ball before it comes to rest.

37. Give an example to show that even if each of the series $\sum_{n=1}^{+\infty} a_n$ and $\sum_{n=1}^{+\infty} b_n$ is divergent, it is possible for the series $\sum_{n=1}^{+\infty} a_n b_n$ to be convergent.

15.4 INFINITE SERIES OF POSITIVE TERMS

If all the terms of an infinite series are positive, the sequence of partial sums is increasing. Thus the following theorem follows immediately from Theorems 15.2.6 and 15.2.9.

15.4.1 Theorem An infinite series of positive terms is convergent if and only if its sequence of partial sums has an upper bound.

Proof For an infinite series of positive terms, the sequence of partial sums has a lower bound of 0. If the sequence of partial sums also has an upper bound, it is bounded. Furthermore, the sequence of partial sums of an infinite series of positive terms is increasing. It follows, then, from Theorem 15.2.6 that the sequence of partial sums is convergent, and therefore the infinite series is convergent.

Suppose now that an infinite series of positive terms is convergent. Then the sequence of partial sums is also convergent. It follows from Theorem 15.2.9 that the sequence of partial sums is bounded, and so it has an upper bound. ■

Example 1 Prove that the series

$$\sum_{n=1}^{+\infty} \frac{1}{n!}$$

is convergent by using Theorem 15.4.1.

Solution We must find an upper bound for the sequence of partial sums of the series $\sum_{n=1}^{+\infty} 1/n!$.

$$s_1 = 1, s_2 = 1 + \frac{1}{1 \cdot 2}, s_3 = 1 + \frac{1}{1 \cdot 2} + \frac{1}{1 \cdot 2 \cdot 3},$$

$$s_4 = 1 + \frac{1}{1 \cdot 2} + \frac{1}{1 \cdot 2 \cdot 3} + \frac{1}{1 \cdot 2 \cdot 3 \cdot 4}, \ldots,$$

$$s_n = 1 + \frac{1}{1 \cdot 2} + \frac{1}{1 \cdot 2 \cdot 3} + \frac{1}{1 \cdot 2 \cdot 3 \cdot 4} + \cdots + \frac{1}{1 \cdot 2 \cdot 3 \cdot \ldots \cdot n} \tag{1}$$

Now consider the first n terms of the geometric series with $a = 1$ and $r = \frac{1}{2}$:

$$\sum_{k=1}^{n} \frac{1}{2^{k-1}} = 1 + \frac{1}{2} + \frac{1}{2^2} + \frac{1}{2^3} + \cdots + \frac{1}{2^{n-1}} \tag{2}$$

By Theorem 15.3.5 the geometric series with $a = 1$ and $r = \frac{1}{2}$ has the sum $a/(1 - r) = 1/(1 - \frac{1}{2}) = 2$. Hence summation (2) is less than 2. Observe that each term of summation (1) is less than or equal to the corresponding term of summation (2); that is,

$$\frac{1}{k!} \leq \frac{1}{2^{k-1}}$$

This is true because $k! = 1 \cdot 2 \cdot 3 \cdot \ldots \cdot k$, which in addition to the factor 1 contains $k - 1$ factors each greater than or equal to 2. Hence

$$s_n = \sum_{k=1}^{n} \frac{1}{k!} \le \sum_{k=1}^{n} \frac{1}{2^{k-1}} < 2$$

From the above, s_n has an upper bound of 2. Therefore, by Theorem 15.4.1, the given series is convergent.

In the above example the terms of the given series were compared with those of a known convergent series. This is a particular case of the following theorem known as the *comparison test.*

15.4.2 Theorem
Comparison Test

Let the series $\sum_{n=1}^{+\infty} u_n$ be a series of positive terms.

(i) If $\sum_{n=1}^{+\infty} v_n$ is a series of positive terms that is known to be convergent, and $u_n \le v_n$ for all positive integers n, then $\sum_{n=1}^{+\infty} u_n$ is convergent.

(ii) If $\sum_{n=1}^{+\infty} w_n$ is a series of positive terms that is known to be divergent, and $u_n \ge w_n$ for all positive integers n, then $\sum_{n=1}^{+\infty} u_n$ is divergent.

Proof of (i) Let $\{s_n\}$ be the sequence of partial sums for the series $\sum_{n=1}^{+\infty} u_n$ and $\{t_n\}$ be the sequence of partial sums for the series $\sum_{n=1}^{+\infty} v_n$. Because $\sum_{n=1}^{+\infty} v_n$ is a series of positive terms that is convergent, it follows from Theorem 15.4.1 that the sequence $\{t_n\}$ has an upper bound; call it B. Because $u_n \le v_n$ for all positive integers n, we can conclude that $s_n \le t_n \le B$ for all positive integers n. Therefore B is an upper bound of the sequence $\{s_n\}$. And because the terms of the series $\sum_{n=1}^{+\infty} u_n$ are all positive, it follows from Theorem 15.4.1 that $\sum_{n=1}^{+\infty} u_n$ is convergent.

Proof of (ii) Assume that $\sum_{n=1}^{+\infty} u_n$ is convergent. Then because both $\sum_{n=1}^{+\infty} u_n$ and $\sum_{n=1}^{+\infty} w_n$ are infinite series of positive terms and $w_n \le u_n$ for all positive integers n, it follows from part (i) that $\sum_{n=1}^{+\infty} w_n$ is convergent. However, this contradicts the hypothesis; so our assumption is false. Therefore $\sum_{n=1}^{+\infty} u_n$ is divergent. ■

As stated in Sec. 15.3, as a result of Theorem 15.3.6, the convergence or divergence of an infinite series is not affected by discarding a finite number of terms. Therefore, when applying the comparison test, if $u_i \leq v_i$ or $u_i \geq w_i$ when $i > m$, the test is valid regardless of how the first m terms of the two series compare.

Example 2 Determine whether the infinite series

$$\sum_{n=1}^{+\infty} \frac{4}{3^n + 1}$$

is convergent or divergent.

Solution The given series is

$$\frac{4}{4} + \frac{4}{10} + \frac{4}{28} + \frac{4}{82} + \cdots + \frac{4}{3^n + 1} + \cdots$$

Comparing the nth term of this series with the nth term of the convergent geometric series

$$\frac{4}{3} + \frac{4}{9} + \frac{4}{27} + \frac{4}{81} + \cdots + \frac{4}{3^n} + \cdots \qquad r = \tfrac{1}{3} < 1$$

we have

$$\frac{4}{3^n + 1} < \frac{4}{3^n}$$

for every positive integer n. Therefore, by the comparison test, Theorem 15.4.2(i), the given series is convergent.

Example 3 Determine whether the infinite series $\sum_{n=1}^{+\infty} 1/\sqrt{n}$ is convergent or divergent.

Solution The given series is

$$\sum_{n=1}^{+\infty} \frac{1}{\sqrt{n}} = \frac{1}{\sqrt{1}} + \frac{1}{\sqrt{2}} + \frac{1}{\sqrt{3}} + \cdots + \frac{1}{\sqrt{n}} + \cdots$$

Comparing the nth term of this series with the nth term of the divergent harmonic series we have

$$\frac{1}{\sqrt{n}} \geq \frac{1}{n} \qquad \text{for every positive integer } n$$

So by Theorem 15.4.2(ii) the given series $\sum_{n=1}^{+\infty} 1/\sqrt{n}$ is divergent.

The following theorem, known as the *limit comparison test,* is a consequence of Theorem 15.4.2 and is often easier to apply.

15.4.3 Theorem Limit Comparison Test Let $\sum_{n=1}^{+\infty} u_n$ and $\sum_{n=1}^{+\infty} v_n$ be two series of positive terms.

(i) If $\lim_{n \to +\infty} (u_n/v_n) = c > 0$, then the two series either both converge or both diverge.

(ii) If $\lim_{n \to +\infty} (u_n/v_n) = 0$, and if $\sum_{n=1}^{+\infty} v_n$ converges, then $\sum_{n=1}^{+\infty} u_n$ converges.

(iii) If $\lim_{n \to +\infty} (u_n/v_n) = +\infty$, and if $\sum_{n=1}^{+\infty} v_n$ diverges, then $\sum_{n=1}^{+\infty} u_n$ diverges.

Proof of (i) Because $\lim_{n \to +\infty} (u_n/v_n) = c$, it follows that there exists an $N > 0$ such that

$$\left|\frac{u_n}{v_n} - c\right| < \frac{c}{2} \qquad \text{for all } n > N$$

or, equivalently,

$$-\frac{c}{2} < \frac{u_n}{v_n} - c < \frac{c}{2} \qquad \text{for all } n > N$$

$$\frac{c}{2} < \frac{u_n}{v_n} < \frac{3c}{2} \qquad \text{for all } n > N \tag{3}$$

From the right-hand inequality (3),

$$u_n < \tfrac{3}{2}cv_n \tag{4}$$

If $\sum_{n=1}^{+\infty} v_n$ is convergent, so is $\sum_{n=1}^{+\infty} \frac{3}{2}cv_n$. It follows from inequality (4) and the comparison test that $\sum_{n=1}^{+\infty} u_n$ is convergent.

From the left-hand inequality (3),

$$v_n < \frac{2}{c}u_n \tag{5}$$

If $\sum_{n=1}^{+\infty} u_n$ is convergent, so is $\sum_{n=1}^{+\infty} \frac{2}{c}u_n$. From inequality (5) and the comparison test it follows that $\sum_{n=1}^{+\infty} v_n$ is convergent.

If $\sum_{n=1}^{+\infty} v_n$ is divergent, $\sum_{n=1}^{+\infty} u_n$ can be shown to be divergent by assuming that $\sum_{n=1}^{+\infty} u_n$ is convergent and getting a contradiction by applying inequality (5) and the comparison test.

In a similar manner, if $\sum_{n=1}^{+\infty} u_n$ is divergent, it follows that $\sum_{n=1}^{+\infty} v_n$ is divergent because a contradiction is obtained from inequality (4) and the comparison test if $\sum_{n=1}^{+\infty} v_n$ is assumed to be convergent.

We have therefore proved part (i). The proofs of parts (ii) and (iii) are left as exercises (see Exercises 30 and 31). ■

A word of caution is in order regarding part (ii) of Theorem 15.4.3. Note that when $\lim_{n \to +\infty} (u_n/v_n) = 0$, the divergence of the series $\sum_{n=1}^{+\infty} v_n$ does *not* imply that the series $\sum_{n=1}^{+\infty} u_n$ diverges.

Example 4 Solve Example 2 by using the limit comparison test.

Solution Let u_n be the nth term of the given series $\sum_{n=1}^{+\infty} 4/(3^n + 1)$ and v_n be the nth term of the convergent geometric series $\sum_{n=1}^{+\infty} 4/3^n$. Therefore

$$\lim_{n \to +\infty} \frac{u_n}{v_n} = \lim_{n \to +\infty} \frac{\dfrac{4}{3^n + 1}}{\dfrac{4}{3^n}}$$

$$= \lim_{n \to +\infty} \frac{3^n}{3^n + 1}$$

$$= \lim_{n \to +\infty} \frac{1}{1 + 3^{-n}}$$

$$= 1$$

Hence, by part (i) of the limit comparison test it follows that the given series is convergent.

Example 5 Solve Example 3 by using the limit comparison test.

Solution Let u_n be the nth term of the given series $\sum_{n=1}^{+\infty} 1/\sqrt{n}$ and v_n be the nth term of the divergent harmonic series. Then

$$\lim_{n\to+\infty} \frac{u_n}{v_n} = \lim_{n\to+\infty} \frac{\dfrac{1}{\sqrt{n}}}{\dfrac{1}{n}} = \lim_{n\to+\infty} \sqrt{n} = +\infty$$

Therefore, by part (iii) of the limit comparison test the given series is divergent.

Example 6 Determine whether the series $\sum_{n=1}^{+\infty} n^3/n!$ is convergent or divergent.

Solution In Example 1 we proved that the series $\sum_{n=1}^{+\infty} 1/n!$ is convergent. By the limit comparison test with $u_n = n^3/n!$ and $v_n = 1/n!$,

$$\lim_{n\to+\infty} \frac{u_n}{v_n} = \lim_{n\to+\infty} \frac{\dfrac{n^3}{n!}}{\dfrac{1}{n!}} = \lim_{n\to+\infty} n^3 = +\infty$$

Part (iii) of the limit comparison test is not applicable because $\sum_{n=1}^{+\infty} v_n$ converges. However, there is a way that the limit comparison test can be used. The given series can be written as

$$\frac{1^3}{1!} + \frac{2^3}{2!} + \frac{3^3}{3!} + \frac{4^3}{4!} + \frac{5^3}{5!} + \cdots + \frac{n^3}{n!} + \cdots$$

Because Theorem 15.3.6 allows us to subtract a finite number of terms without affecting the behavior (convergence or divergence) of a series, we discard the first three terms and obtain

$$\frac{4^3}{4!} + \frac{5^3}{5!} + \frac{6^3}{6!} + \cdots + \frac{(n+3)^3}{(n+3)!} + \cdots$$

Now let $u_n = (n+3)^3/(n+3)!$ and, as before, let $v_n = 1/n!$. Then

$$\begin{aligned}\lim_{n\to+\infty} \frac{u_n}{v_n} &= \lim_{n\to+\infty} \frac{\dfrac{(n+3)^3}{(n+3)!}}{\dfrac{1}{n!}}\\ &= \lim_{n\to+\infty} \frac{(n+3)^3 n!}{(n+3)!}\\ &= \lim_{n\to+\infty} \frac{(n+3)^3 n!}{n!(n+1)(n+2)(n+3)}\\ &= \lim_{n\to+\infty} \frac{(n+3)^2}{(n+1)(n+2)}\end{aligned}$$

$$= \lim_{n \to +\infty} \frac{n^2 + 6n + 9}{n^2 + 3n + 2}$$

$$= \lim_{n \to +\infty} \frac{1 + \frac{6}{n} + \frac{9}{n^2}}{1 + \frac{3}{n} + \frac{2}{n^2}}$$

$$= 1$$

It follows from part (i) of the Limit Comparison Test that the given series is convergent.

● ILLUSTRATION 1 Consider the geometric series

$$1 + \frac{1}{2} + \frac{1}{4} + \frac{1}{8} + \frac{1}{16} + \frac{1}{32} + \cdots + \frac{1}{2^{n-1}} + \cdots \tag{6}$$

which converges to 2 as shown in Illustration 2 of Sec. 15.3. Regroup the terms of this series to obtain

$$\left(1 + \frac{1}{2}\right) + \left(\frac{1}{4} + \frac{1}{8}\right) + \left(\frac{1}{16} + \frac{1}{32}\right) + \cdots + \left(\frac{1}{4^{n-1}} + \frac{1}{2 \cdot 4^{n-1}}\right) + \cdots$$

which is the series

$$\frac{3}{2} + \frac{3}{8} + \frac{3}{32} + \cdots + \frac{3}{2 \cdot 4^{n-1}} + \cdots \tag{7}$$

Series (7) is the geometric series with $a = \frac{3}{2}$ and $r = \frac{1}{4}$. Thus, by Theorem 15.3.5 it is convergent, and its sum is

$$\frac{a}{1 - r} = \frac{\frac{3}{2}}{1 - \frac{1}{4}} = 2$$

Therefore series (7), which is obtained from the convergent series (6) by regrouping the terms, is also convergent. Its sum is the same as that of series (6). ●

Illustration 1 gives a particular case of the following theorem.

15.4.4 Theorem If $\sum_{n=1}^{+\infty} u_n$ is a given convergent series of positive terms, its terms can be grouped in any manner, and the resulting series also will be convergent and will have the same sum as the given series.

Proof Let $\{s_n\}$ be the sequence of partial sums for the given convergent series of positive terms. Then $\lim_{n \to +\infty} s_n$ exists; let this limit be S. Consider a series $\sum_{n=1}^{+\infty} v_n$ whose terms are obtained by grouping the terms of $\sum_{n=1}^{+\infty} u_n$ in some manner. For example, $\sum_{n=1}^{+\infty} v_n$ may be the series

$$u_1 + (u_2 + u_3) + (u_4 + u_5 + u_6) + (u_7 + u_8 + u_9 + u_{10}) + \cdots$$

or it may be the series

$$(u_1 + u_2) + (u_3 + u_4) + (u_5 + u_6) + (u_7 + u_8) + \cdots$$

and so forth. Let $\{t_m\}$ be the sequence of partial sums for the series $\sum_{n=1}^{+\infty} v_n$. Each partial sum of the sequence $\{t_m\}$ is also a partial sum of the sequence $\{s_n\}$. Therefore, as m increases without bound, so does n. Because $\lim_{n \to +\infty} s_n = S$, we conclude that $\lim_{m \to +\infty} t_m = S$. This proves the theorem. ■

Theorem 15.4.4 and the next theorem state properties of the sum of a convergent series of positive terms that are similar to properties that hold for the sum of a finite number of terms.

15.4.5 Theorem If $\sum_{n=1}^{+\infty} u_n$ is a given convergent series of positive terms, the order of the terms can be rearranged, and the resulting series also will be convergent and will have the same sum as the given series.

Proof Let $\{s_n\}$ be the sequence of partial sums for the given convergent series of positive terms, and let $\lim_{n \to +\infty} s_n = S$. Let $\sum_{n=1}^{+\infty} v_n$ be a series formed by rearranging the order of the terms of $\sum_{n=1}^{+\infty} u_n$. For example, $\sum_{n=1}^{+\infty} v_n$ may be the series

$$u_4 + u_3 + u_7 + u_1 + u_9 + u_5 + \cdots$$

Let $\{t_n\}$ be the sequence of partial sums for the series $\sum_{n=1}^{+\infty} v_n$. Each partial sum of the sequence $\{t_n\}$ will be less than S because it is the sum of n terms of the infinite series $\sum_{n=1}^{+\infty} u_n$. Therefore S is an upper bound of the sequence $\{t_n\}$. Furthermore, because all the terms of the series $\sum_{n=1}^{+\infty} v_n$ are positive, $\{t_n\}$ is a monotonic increasing sequence. Hence, by Theorem 15.2.7 the sequence $\{t_n\}$ is convergent, and $\lim_{n \to +\infty} t_n = T \leq S$. Now because the given series $\sum_{n=1}^{+\infty} u_n$ can be obtained from the series $\sum_{n=1}^{+\infty} v_n$ by rearranging the order of the terms, we can use the same argument and conclude that $S \le T$. If both inequalities, $T \le S$ and $S \le T$, must hold, it follows that $S = T$. This proves the theorem. ■

A series that is often used in the comparison test is the one known as the *p* **series,** or the **hyperharmonic series.** It is

$$\frac{1}{1^p} + \frac{1}{2^p} + \frac{1}{3^p} + \cdots + \frac{1}{n^p} + \cdots \qquad \text{where } p \text{ is a constant} \tag{8}$$

In the following illustration we prove that the *p* series diverges if $p \le 1$ and converges if $p > 1$.

• ILLUSTRATION 2 If $p = 1$, the *p* series is the harmonic series, which diverges. If $p < 1$, then $n^p \le n$; so

$$\frac{1}{n^p} \ge \frac{1}{n} \qquad \text{for every positive integer } n$$

Hence, by Theorem 15.4.2(ii) the *p* series is divergent if $p < 1$.

If $p > 1$, group the terms as follows:

$$\frac{1}{1^p} + \left(\frac{1}{2^p} + \frac{1}{3^p}\right) + \left(\frac{1}{4^p} + \frac{1}{5^p} + \frac{1}{6^p} + \frac{1}{7^p}\right) + \left(\frac{1}{8^p} + \frac{1}{9^p} + \cdots + \frac{1}{15^p}\right) + \cdots \tag{9}$$

Consider the series

$$\frac{1}{1^p} + \frac{2}{2^p} + \frac{4}{4^p} + \frac{8}{8^p} + \cdots + \frac{2^{n-1}}{(2^{n-1})^p} + \cdots \tag{10}$$

This is a geometric series whose ratio is $2/2^p = 1/2^{p-1}$, which is a positive number less than 1. Hence series (10) is convergent. Rewrite the terms of series (10) to get

$$\frac{1}{1^p} + \left(\frac{1}{2^p} + \frac{1}{2^p}\right) + \left(\frac{1}{4^p} + \frac{1}{4^p} + \frac{1}{4^p} + \frac{1}{4^p}\right) + \left(\frac{1}{8^p} + \frac{1}{8^p} + \cdots + \frac{1}{8^p}\right) + \cdots \tag{11}$$

By comparing series (9) and series (11) it is seen that the group of terms in each set of parentheses after the first group is less in sum for (9) than it is for (11). Therefore, by the comparison test, series (9) is convergent. Because (9) is merely a regrouping of the terms of the *p* series when $p > 1$, it follows from Theorem 15.4.4 that the *p* series is convergent if $p > 1$. •

Note that the series in Example 3 is the *p* series where $p = \frac{1}{2} < 1$; therefore it is divergent.

Example 7 Determine whether the infinite series

$$\sum_{n=1}^{+\infty} \frac{1}{(n^2 + 2)^{1/3}}$$

is convergent or divergent.

Solution Because for large values of n the number $n^2 + 2$ is close to the number n^2, so is the number $1/(n^2 + 2)^{1/3}$ close to the number

$1/n^{2/3}$. The series $\sum_{n=1}^{+\infty} 1/n^{2/3}$ is divergent because it is the p series with $p = \frac{2}{3} < 1$. From the limit comparison test with $u_n = 1/(n^2 + 2)^{1/3}$ and $v_n = 1/n^{2/3}$,

$$\lim_{n\to+\infty} \frac{u_n}{v_n} = \lim_{n\to+\infty} \frac{\dfrac{1}{(n^2+2)^{1/3}}}{\dfrac{1}{n^{2/3}}}$$

$$= \lim_{n\to+\infty} \frac{n^{2/3}}{(n^2+2)^{1/3}}$$

$$= \lim_{n\to+\infty} \left(\frac{n^2}{n^2+2}\right)^{1/3}$$

$$= \lim_{n\to+\infty} \left(\frac{1}{1+\dfrac{2}{n^2}}\right)^{1/3}$$

$$= 1$$

Therefore the given series is divergent.

EXERCISES 15.4

In Exercises 1 through 26, determine if the given series is convergent or divergent.

1. $\sum_{n=1}^{+\infty} \frac{1}{n2^n}$
2. $\sum_{n=1}^{+\infty} \frac{1}{\sqrt{2n+1}}$
3. $\sum_{n=1}^{+\infty} \frac{1}{n^n}$
4. $\sum_{n=1}^{+\infty} \frac{n^2}{4n^3+1}$
5. $\sum_{n=1}^{+\infty} \frac{3n+1}{2n^2+5}$
6. $\sum_{n=1}^{+\infty} \frac{3}{\sqrt{n^3+n}}$
7. $\sum_{n=1}^{+\infty} \frac{\cos^2 n}{3^n}$
8. $\sum_{n=1}^{+\infty} \frac{1}{\ln(n+1)}$
9. $\sum_{n=1}^{+\infty} \frac{1}{\sqrt{n^2+4n}}$
10. $\sum_{n=1}^{+\infty} \frac{|\sin n|}{n^2}$
11. $\sum_{n=1}^{+\infty} \frac{n!}{(n+2)!}$
12. $\sum_{n=1}^{+\infty} \frac{1}{\sqrt{n^3+1}}$
13. $\sum_{n=1}^{+\infty} \frac{n}{5n^2+3}$
14. $\sum_{n=1}^{+\infty} \frac{(n-1)!}{(n+1)!}$
15. $\sum_{n=1}^{+\infty} \frac{n!}{(2n)!}$
16. $\sum_{n=1}^{+\infty} \sin\frac{1}{n}$
17. $\sum_{n=1}^{+\infty} \frac{|\csc n|}{n}$
18. $\sum_{n=1}^{+\infty} \frac{1}{n+\sqrt{n}}$
19. $\sum_{n=2}^{+\infty} \frac{1}{n\sqrt{n^2-1}}$
20. $\sum_{n=1}^{+\infty} \frac{2^n}{n!}$
21. $\sum_{n=1}^{+\infty} \frac{3}{2n-\sqrt{n}}$
22. $\sum_{n=1}^{+\infty} \frac{\sqrt{n}}{n^2+1}$
23. $\sum_{n=1}^{+\infty} \frac{\ln n}{n^2+2}$
24. $\sum_{n=1}^{+\infty} \frac{1}{3^n - \cos n}$
25. $\sum_{n=1}^{+\infty} \frac{(n+1)^2}{(n+2)!}$
26. $\sum_{n=1}^{+\infty} \frac{1}{(n+2)(n+4)}$

27. Use the series $\sum_{n=1}^{+\infty} (-1)^{n+1}$ to show that Theorem 15.4.1 does not apply to an infinite series of both positive and negative terms; that is, show that the sequence of partial sums has an upper bound but the series is not convergent.

28. Suppose f is a function such that $f(n) > 0$ for n any positive integer. Furthermore, suppose that if p is any positive number $\lim_{n \to +\infty} n^p f(n)$ exists and is positive. Prove that the series $\sum_{n=1}^{+\infty} f(n)$ is convergent if $p > 1$ and divergent if $0 < p \leq 1$.

29. If $\sum_{n=1}^{+\infty} a_n$ and $\sum_{n=1}^{+\infty} b_n$ are two convergent series of positive terms, use the Limit Comparison Test to prove that the series $\sum_{n=1}^{+\infty} a_n b_n$ is also convergent.

30. Prove Theorem 15.4.3(ii). **31.** Prove Theorem 15.4.3(iii).

15.5 THE INTEGRAL TEST

The theorem known as the *integral test* makes use of the theory of improper integrals to test an infinite series of positive terms for convergence.

15.5.1 Theorem Integral Test Let f be a function that is continuous, decreasing, and positive valued for all $x \geq 1$. Then the infinite series

$$\sum_{n=1}^{+\infty} f(n) = f(1) + f(2) + f(3) + \cdots + f(n) + \cdots$$

is convergent if the improper integral

$$\int_1^{+\infty} f(x)\,dx$$

exists, and it is divergent if $\lim_{b \to +\infty} \int_1^b f(x)\,dx = +\infty$.

Proof If i is a positive integer and $i \geq 2$, then by the mean-value theorem for integrals (6.4.1) there exists a number X such that $i - 1 \leq X \leq i$ and

$$\int_{i-1}^{i} f(x)\,dx = f(X) \cdot 1 \tag{1}$$

Because f is a decreasing function,

$$f(i-1) \geq f(X) \geq f(i)$$

and so from (1),

$$f(i-1) \geq \int_{i-1}^{i} f(x)\,dx \geq f(i)$$

Therefore, if n is a positive integer and $n \geq 2$,

$$\sum_{i=2}^{n} f(i-1) \geq \sum_{i=2}^{n} \int_{i-1}^{i} f(x)\,dx \geq \sum_{i=2}^{n} f(i)$$

or, equivalently,

$$\sum_{i=1}^{n-1} f(i) \geq \int_1^n f(x)\,dx \geq \sum_{i=1}^{n} f(i) - f(1) \tag{2}$$

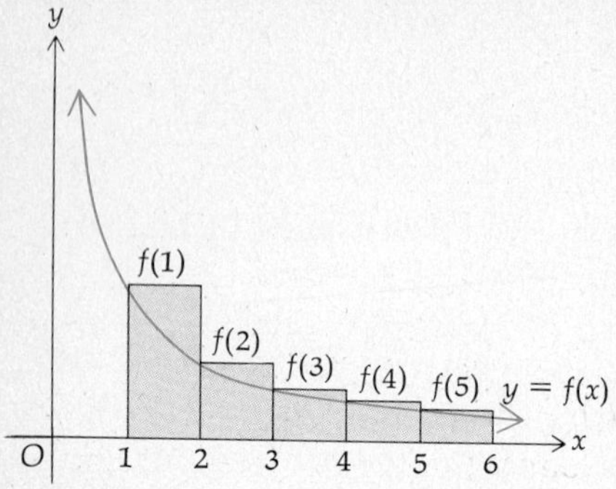

FIGURE 15.5.1

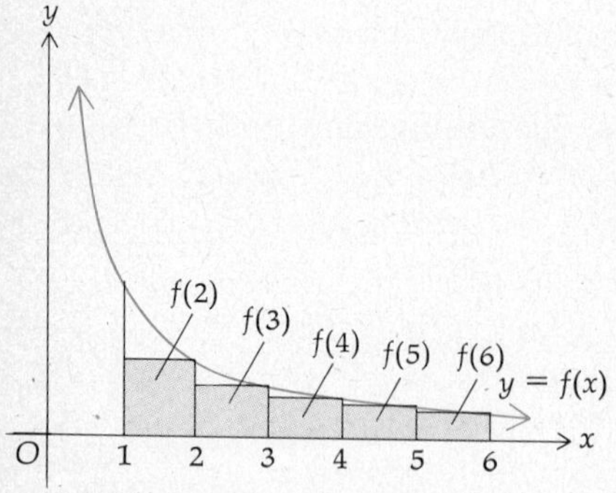

FIGURE 15.5.2

Figures 15.5.1 and 15.5.2 show the geometric interpretation of the above discussion for $n = 6$. In Fig. 15.5.1 there is a sketch of the graph of a function f satisfying the hypothesis. The sum of the measures of the areas of the shaded rectangles is $f(1) + f(2) + f(3) + f(4) + f(5)$, which is the left member of inequality (2) when $n = 6$. Clearly, the sum of the measures of the areas of these rectangles is greater than the measure of the area given by the definite integral when $n = 6$. In Fig. 15.5.2 the sum of the measures of the areas of the shaded rectangles is $f(2) + f(3) + f(4) + f(5) + f(6)$, which is the right member of the inequality (2) when $n = 6$. This sum is less than the value of the definite integral when $n = 6$.

If the given improper integral exists, let L be its value. Then

$$\int_1^n f(x)\,dx \le L \tag{3}$$

From the second and third members of the inequality (2) and from (3),

$$\sum_{i=1}^{n} f(i) \le f(1) + \int_1^n f(x)\,dx \le f(1) + L \tag{4}$$

Consider now the infinite series $\sum_{n=1}^{+\infty} f(n)$. Let the sequence of partial sums of this series be $\{s_n\}$, where $s_n = \sum_{i=1}^{n} f(i)$. From (4), $\{s_n\}$ has an upper bound of $f(1) + L$. Hence, by Theorem 15.4.1, $\sum_{n=1}^{+\infty} f(n)$ is convergent.

Suppose that $\lim_{b\to+\infty} \int_1^b f(x)\,dx = +\infty$. From (2)

$$\sum_{i=1}^{n-1} f(i) \ge \int_1^n f(x)\,dx$$

for all positive integers n. Therefore $\lim_{n\to+\infty} s_n = \lim_{n\to+\infty} \sum_{i=1}^{n} f(i) = +\infty$. Hence $\sum_{n=1}^{+\infty} f(n)$ is divergent. ■

Example 1 Use the integral test to show that the p series diverges if $p \le 1$ and converges if $p > 1$.

Solution The p series is $\sum_{n=1}^{+\infty} 1/n^p$. If $f(x) = 1/x^p$, then f is continuous and positive valued for all $x \ge 1$. Furthermore, if $1 \le x_1 < x_2$, then $1/x_1^p > 1/x_2^p$, and so f is decreasing for all $x \ge 1$. Therefore the hypothesis of Theorem 15.5.1 is satisfied by the function f. We consider the improper integral and have

$$\int_1^{+\infty} \frac{dx}{x^p} = \lim_{b\to+\infty} \int_1^b \frac{dx}{x^p}$$

If $p = 1$, the above integral gives

$$\lim_{b \to +\infty} \ln x \Big]_1^b = \lim_{b \to +\infty} \ln b = +\infty$$

If $p \neq 1$, the integral gives

$$\lim_{b \to +\infty} \frac{x^{1-p}}{1-p} \Big]_1^b = \lim_{b \to +\infty} \frac{b^{1-p} - 1}{1-p}$$

This limit is $+\infty$ when $p < 1$; it is $-1/(1-p)$ if $p > 1$. Therefore, by the integral test it follows that the p series converges for $p > 1$ and diverges for $p \leq 1$.

Example 2 Determine if the infinite series $\sum_{n=1}^{+\infty} ne^{-n}$ is convergent or divergent.

Solution Let $f(x) = xe^{-x}$. Then

$$f'(x) = e^{-x} - xe^{-x} = e^{-x}(1 - x)$$

Because $f'(x) < 0$ if $x > 1$, it follows from Theorem 4.5.3 that f is decreasing if $x \geq 1$. Furthermore, f is continuous and positive valued for all $x \geq 1$. Thus the hypothesis of the integral test is satisfied. By applying integration by parts,

$$\int xe^{-x}\, dx = -e^{-x}(x + 1) + C$$

Hence

$$\int_1^{+\infty} xe^{-x}\, dx = \lim_{b \to +\infty} \Big[-e^{-x}(x + 1)\Big]_1^b = \lim_{b \to +\infty} \left[-\frac{b + 1}{e^b} + \frac{2}{e}\right]$$

Because $\lim_{b \to +\infty} (b + 1) = +\infty$ and $\lim_{b \to +\infty} e^b = +\infty$, L'Hôpital's rule can be used to obtain

$$\lim_{b \to +\infty} \frac{b + 1}{e^b} = \lim_{b \to +\infty} \frac{1}{e^b} = 0$$

Therefore

$$\int_1^{+\infty} xe^{-x}\, dx = \frac{2}{e}$$

so the given series is convergent.

Example 3 Determine if the series

$$\sum_{n=1}^{+\infty} \frac{1}{(n + 1)\sqrt{\ln(n + 1)}}$$

is convergent or divergent.

Solution The function f defined by

$$f(x) = \frac{1}{(x+1)\sqrt{\ln(x+1)}}$$

is continuous and positive valued for all $x \geq 1$. Also, if $1 \leq x_1 < x_2$, then $f(x_1) > f(x_2)$; so f is decreasing for all $x \geq 1$. Therefore the integral test can be applied.

$$\begin{aligned}\int_1^{+\infty} \frac{dx}{(x+1)\sqrt{\ln(x+1)}} &= \lim_{b\to+\infty} \int_1^b [\ln(x+1)]^{-1/2} \frac{dx}{x+1} \\ &= \lim_{b\to+\infty} \left[2\sqrt{\ln(x+1)}\right]_1^b \\ &= \lim_{b\to+\infty} [2\sqrt{\ln(b+1)} - 2\sqrt{\ln 2}] \\ &= +\infty\end{aligned}$$

Thus the given series is divergent.

EXERCISES 15.5

In Exercises 1 through 8, use the integral test to determine if the given series is convergent or divergent.

1. $\sum_{n=1}^{+\infty} \frac{1}{2n+1}$ **2.** $\sum_{n=1}^{+\infty} \frac{2}{(3n+5)^2}$ **3.** $\sum_{n=1}^{+\infty} \frac{1}{(n+2)^{3/2}}$ **4.** $\sum_{n=2}^{+\infty} \frac{n}{n^2-2}$

5. $\sum_{n=3}^{+\infty} \frac{4}{n^2-4}$ **6.** $\sum_{n=1}^{+\infty} \frac{2n+3}{(n^2+3n)^2}$ **7.** $\sum_{n=1}^{+\infty} e^{-5n}$ **8.** $\sum_{n=1}^{+\infty} \frac{2n}{n^4+1}$

In Exercises 9 through 22, determine if the given series is convergent or divergent.

9. $\sum_{n=1}^{+\infty} \frac{\ln n}{n}$ **10.** $\sum_{n=2}^{+\infty} \frac{1}{n \ln n}$ **11.** $\sum_{n=1}^{+\infty} \frac{\tan^{-1} n}{n^2+1}$ **12.** $\sum_{n=1}^{+\infty} ne^{-n^2}$ **13.** $\sum_{n=1}^{+\infty} n^2e^{-n}$ **14.** $\sum_{n=1}^{+\infty} ne^{-n}$

15. $\sum_{n=2}^{+\infty} \frac{\ln n}{n^3}$ **16.** $\sum_{n=1}^{+\infty} \cot^{-1} n$ **17.** $\sum_{n=1}^{+\infty} \operatorname{csch} n$ **18.** $\sum_{n=1}^{+\infty} \frac{e^{\tan^{-1} n}}{n^2+1}$ **19.** $\sum_{n=1}^{+\infty} \frac{e^{1/n}}{n^2}$ **20.** $\sum_{n=1}^{+\infty} \operatorname{sech}^2 n$

21. $\sum_{n=1}^{+\infty} \ln\left(\frac{n+3}{n}\right)$ **22.** $\sum_{n=2}^{+\infty} \frac{1}{n(\ln n)^3}$

23. Prove that the series $\sum_{n=2}^{+\infty} \frac{1}{n(\ln n)^p}$ is convergent if and only if $p > 1$.

24. Prove that the series $\sum_{n=3}^{+\infty} \frac{1}{n(\ln n)[\ln(\ln n)]^p}$ is convergent if and only if $p > 1$.

25. Prove that the series $\sum_{n=1}^{+\infty} \frac{\ln n}{n^p}$ is convergent if and only if $p > 1$.

26. If s_k is the kth partial sum of the harmonic series, prove that $\ln(k + 1) < s_k < 1 + \ln k$. (*Hint:* $\dfrac{1}{m+1} \le \dfrac{1}{x} \le \dfrac{1}{m}$ if $0 < m \le x \le m + 1$. Integrate each member of the inequality from m to $m + 1$; let m take on successively the values $1, 2, \ldots, n - 1$, and add the results.)

27. Use the result of Exercise 26 to estimate the sum $\displaystyle\sum_{m=50}^{100} \frac{1}{m} = \frac{1}{50} + \frac{1}{51} + \cdots + \frac{1}{100}$.

15.6 INFINITE SERIES OF POSITIVE AND NEGATIVE TERMS

In this section infinite series having both positive and negative terms are considered. The first type of such a series that we discuss is one whose terms are alternately positive and negative—an *alternating series.*

15.6.1 Definition

If $a_n > 0$ for all positive integers n, then the series

$$\sum_{n=1}^{+\infty} (-1)^{n+1} a_n = a_1 - a_2 + a_3 - a_4 + \cdots + (-1)^{n+1} a_n + \cdots$$

and the series

$$\sum_{n=1}^{+\infty} (-1)^{n} a_n = -a_1 + a_2 - a_3 + a_4 - \cdots + (-1)^{n} a_n + \cdots$$

are called **alternating series.**

The following theorem gives a test for the convergence of an alternating series.

15.6.2 Theorem
Alternating-Series Test

If the numbers $u_1, u_2, u_3, \ldots, u_n, \ldots$ are alternately positive and negative, $|u_{n+1}| < |u_n|$ for all positive integers n, and $\lim\limits_{n \to +\infty} u_n = 0$, then the alternating series $\displaystyle\sum_{n=1}^{+\infty} u_n$ is convergent.

Proof Assume that the odd-numbered terms of the given series are positive and the even-numbered terms are negative. This assumption is not a loss of generality because if this is not the case, then we consider the series whose first term is u_2 because discarding a finite number of terms does not affect the convergence of the series. So $u_{2n-1} > 0$ and $u_{2n} < 0$ for every positive integer n. Consider the partial sum

$$s_{2n} = (u_1 + u_2) + (u_3 + u_4) + \cdots + (u_{2n-1} + u_{2n}) \tag{1}$$

The first term of each quantity in parentheses in (1) is positive and the second term is negative. Because by hypothesis $|u_{n+1}| < |u_n|$, then each quantity in parentheses is positive. Therefore

$$0 < s_2 < s_4 < s_6 < \cdots < s_{2n} < \cdots \tag{2}$$

We also can write s_{2n} as

$$s_{2n} = u_1 + (u_2 + u_3) + (u_4 + u_5) + \cdots + (u_{2n-2} + u_{2n-1}) + u_{2n} \tag{3}$$

Because $|u_{n+1}| < |u_n|$, each quantity in parentheses in (3) is negative and so also is u_{2n}. Therefore

$$s_{2n} < u_1 \qquad \text{for every positive integer } n \tag{4}$$

From (2) and (4),

$$0 < s_{2n} < u_1 \qquad \text{for every positive integer } n$$

So the sequence $\{s_{2n}\}$ is bounded. Furthermore, from (2) the sequence $\{s_{2n}\}$ is monotonic. Therefore, by Theorem 15.2.6 the sequence $\{s_{2n}\}$ is convergent. Let $\lim_{n \to +\infty} s_{2n} = S$, and from Theorem 15.2.7, $S \leq u_1$.

Because $s_{2n+1} = s_{2n} + u_{2n+1}$,

$$\lim_{n \to +\infty} s_{2n+1} = \lim_{n \to +\infty} s_{2n} + \lim_{n \to +\infty} u_{2n+1}$$

but, by hypothesis, $\lim_{n \to +\infty} u_{2n+1} = 0$; so $\lim_{n \to +\infty} s_{2n+1} = \lim_{n \to +\infty} s_{2n}$. Therefore the sequence of partial sums of the even-numbered terms and the sequence of partial sums of the odd-numbered terms have the same limit S.

We now show that $\lim_{n \to +\infty} s_n = S$. Because $\lim_{n \to +\infty} s_{2n} = S$, then for any $\epsilon > 0$ there exists an integer $N_1 > 0$ such that

$$|s_{2n} - S| < \epsilon \quad \text{whenever} \quad 2n \geq N_1$$

And because $\lim_{n \to +\infty} s_{2n+1} = S$, there exists an integer $N_2 > 0$ such that

$$|s_{2n+1} - S| < \epsilon \quad \text{whenever} \quad 2n + 1 \geq N_2$$

If N is the larger of the two integers N_1 and N_2, it follows that if n is any integer, either odd or even, then

$$|s_n - S| < \epsilon \quad \text{whenever} \quad n \geq N$$

Therefore $\lim_{n \to +\infty} s_n = S$; so the series $\sum_{n=1}^{+\infty} u_n$ is convergent. ■

Example 1 Prove that the alternating series

$$\sum_{n=1}^{+\infty} \frac{(-1)^{n+1}}{n}$$

is convergent.

Solution The given series is

$$1 - \frac{1}{2} + \frac{1}{3} - \frac{1}{4} + \cdots + (-1)^{n+1}\frac{1}{n} + (-1)^{n+2}\frac{1}{n+1} + \cdots$$

Because $1/(n+1) < 1/n$ for all positive integers n, and $\lim_{n \to +\infty} (1/n) = 0$, it follows from Theorem 15.6.2 that the given alternating series is convergent.

Example 2 Determine if the series

$$\sum_{n=1}^{+\infty} (-1)^n \frac{n+2}{n(n+1)}$$

is convergent or divergent.

Solution The given series is an alternating series, with

$$u_n = (-1)^n \frac{n+2}{n(n+1)} \quad \text{and} \quad u_{n+1} = (-1)^{n+1} \frac{n+3}{(n+1)(n+2)}$$

$$\lim_{n \to +\infty} \frac{n+2}{n(n+1)} = \lim_{n \to +\infty} \frac{\frac{1}{n} + \frac{2}{n^2}}{1 + \frac{1}{n}} = 0$$

Before the alternating-series test can be applied, we must also show that $|u_{n+1}| < |u_n|$ or, equivalently, $|u_{n+1}|/|u_n| < 1$.

$$\frac{|u_{n+1}|}{|u_n|} = \frac{n+3}{(n+1)(n+2)} \cdot \frac{n(n+1)}{n+2} = \frac{n(n+3)}{(n+2)^2} = \frac{n^2+3n}{n^2+4n+4} < 1$$

Then it follows from Theorem 15.6.2 that the given series is convergent.

15.6.3 Definition If an infinite series $\sum_{n=1}^{+\infty} u_n$ is convergent and its sum is S, then the **remainder** obtained by approximating the sum of the series by the kth partial sum s_k is denoted by R_k and

$$R_k = S - s_k$$

15.6.4 Theorem Suppose $\sum_{n=1}^{+\infty} u_n$ is an alternating series, $|u_{n+1}| < |u_n|$, and $\lim_{n \to +\infty} u_n = 0$. Then if R_k is the remainder obtained by approximating the sum of the series by the sum of the first k terms, $|R_k| < |u_{k+1}|$.

Proof Assume that the odd-numbered terms of the given series are positive and the even-numbered terms are negative. Then, from (2) in the proof of Theorem 15.6.2, the sequence $\{s_{2n}\}$ is increasing. So if S is the sum of the given series,

$$s_{2k} < s_{2k+2} < S \quad \text{for all } k \geq 1 \tag{5}$$

To show that the sequence $\{s_{2n-1}\}$ is decreasing we write

$$s_{2n-1} = u_1 + (u_2 + u_3) + (u_4 + u_5) + \cdots + (u_{2n-2} + u_{2n-1}) \tag{6}$$

The first term of each quantity in parentheses in (6) is negative and the second term is positive, and because $|u_{n+1}| < |u_n|$, it follows that each quantity in parentheses is negative. Therefore, because $u_1 > 0$, it follows that $s_1 > s_3 > s_5 > \cdots > s_{2n-1} > \cdots$; and so the sequence $\{s_{2n-1}\}$ is decreasing. Thus

$$S < s_{2k+1} < s_{2k-1} \quad \text{for all } k \geq 1 \tag{7}$$

From (7), $S - s_{2k} < s_{2k+1} - s_{2k} = u_{2k+1}$, and from (5), $0 < S - s_{2k}$. Therefore

$$0 < S - s_{2k} < u_{2k+1} \quad \text{for all } k \geq 1 \tag{8}$$

From (5), $-S < -s_{2k}$; hence $s_{2k-1} - S < s_{2k-1} - s_{2k} = -u_{2k}$. Because from (7) it follows that $0 < s_{2k-1} - S$, then

$$0 < s_{2k-1} - S < -u_{2k} \quad \text{for all } k \geq 1 \tag{9}$$

Because from Definition 15.6.3 $R_k = S - s_k$, then (8) can be written as $0 < |R_{2k}| < |u_{2k+1}|$ and (9) can be written as $0 < |R_{2k-1}| < |u_{2k}|$. Hence $|R_k| < |u_{k+1}|$ for all $k \geq 1$, and the theorem is proved. ■

Example 3 A series for computing $\ln(1 - x)$ if x is in the open interval $(-1, 1)$ is

$$\ln(1 - x) = \sum_{n=1}^{+\infty} \left(-\frac{x^n}{n}\right)$$

Find an estimate of the error when the first three terms of this series are used to approximate the value of ln 1.1.

Solution Use the given series with $x = -0.1$ to obtain

$$\ln 1.1 = 0.1 - \frac{(0.1)^2}{2} + \frac{(0.1)^3}{3} - \frac{(0.1)^4}{4} + \cdots$$

This series satisfies the conditions of Theorem 15.6.4; so if R_3 is the difference between the actual value of ln 1.1 and the sum of the first three terms, then

$$|R_3| < |u_4| = 0.000025$$

Thus the sum of the first three terms will yield a value of ln 1.1 accurate at least to four decimal places. Using the first three terms we get

$$\ln 1.1 \approx 0.0953$$

Associated with each infinite series is the series whose terms are the absolute values of the terms of that series.

15.6.5 Definition The infinite series $\sum_{n=1}^{+\infty} u_n$ is said to be **absolutely convergent** if the series $\sum_{n=1}^{+\infty} |u_n|$ is convergent.

● ILLUSTRATION 1 Consider the series

$$\sum_{n=1}^{+\infty} (-1)^{n+1} \frac{2}{3^n} = \frac{2}{3} - \frac{2}{3^2} + \frac{2}{3^3} - \frac{2}{3^4} + \cdots + (-1)^{n+1} \frac{2}{3^n} + \cdots \quad (10)$$

This series will be absolutely convergent if the series

$$\sum_{n=1}^{+\infty} \frac{2}{3^n} = \frac{2}{3} + \frac{2}{3^2} + \frac{2}{3^3} + \frac{2}{3^4} + \cdots + \frac{2}{3^n} + \cdots$$

is convergent. Because this is the geometric series with $r = \frac{1}{3} < 1$, it is convergent. Therefore series (10) is absolutely convergent. ●

● ILLUSTRATION 2 A convergent series that is not absolutely convergent is the series

$$\sum_{n=1}^{+\infty} \frac{(-1)^{n+1}}{n}$$

In Example 1 this series was proved to be convergent. The series is not absolutely convergent because the series of absolute values is the harmonic series, which is divergent. ●

The series of Illustration 2 is an example of a *conditionally convergent* series.

15.6.6 Definition A series that is convergent, but not absolutely convergent, is said to be **conditionally convergent.**

It is possible, then, for a series to be convergent but not absolutely convergent. If a series is absolutely convergent, it must be convergent, however, and this is given by the next theorem.

15.6.7 Theorem If the infinite series $\sum_{n=1}^{+\infty} u_n$ is absolutely convergent, it is convergent and

$$\left| \sum_{n=1}^{+\infty} u_n \right| \leq \sum_{n=1}^{+\infty} |u_n|$$

Proof Consider the three infinite series $\sum_{n=1}^{+\infty} u_n$, $\sum_{n=1}^{+\infty} |u_n|$, and $\sum_{n=1}^{+\infty} (u_n + |u_n|)$, and let their sequences of partial sums be $\{s_n\}$, $\{t_n\}$, and $\{r_n\}$, respectively. For every positive integer n, $u_n + |u_n|$ is either 0 or $2|u_n|$; so we have the inequality

$$0 \leq u_n + |u_n| \leq 2|u_n| \quad (11)$$

Because $\sum_{n=1}^{+\infty} |u_n|$ is convergent, it has a sum, which we denote by T. $\{t_n\}$ is an increasing sequence of positive numbers; so $t_n \leq T$ for all positive integers n. From (11) it follows that

$$0 \leq r_n \leq 2t_n \leq 2T$$

Therefore the sequence $\{r_n\}$ has an upper bound of $2T$. Thus, by Theorem 15.4.1 the series $\sum_{n=1}^{+\infty} (u_n + |u_n|)$ is convergent. Let its sum be R. Because from (11), $\{r_n\}$ is an increasing sequence, it may be concluded from Theorem 15.2.7 that $R \leq 2T$.

Each of the series $\sum_{n=1}^{+\infty} (u_n + |u_n|)$ and $\sum_{n=1}^{+\infty} |u_n|$ is convergent; hence, from Theorem 15.3.8 the series

$$\sum_{n=1}^{+\infty} [(u_n + |u_n|) - |u_n|] = \sum_{n=1}^{+\infty} u_n$$

is also convergent.

Let the sum of the series $\sum_{n=1}^{+\infty} u_n$ be S. Then, also from Theorem 15.3.8, $S = R - T$. And because $R \leq 2T$, $S \leq 2T - T = T$.

Because $\sum_{n=1}^{+\infty} u_n$ is convergent and has the sum S, it follows from Theorem 15.3.7 that $\sum_{n=1}^{+\infty} (-u_n)$ is convergent and has the sum $-S$. Because $\sum_{n=1}^{+\infty} |-u_n| = \sum_{n=1}^{+\infty} |u_n| = T$, we can replace $\sum_{n=1}^{+\infty} u_n$ by $\sum_{n=1}^{+\infty} (-u_n)$ in the above discussion and show that $-S \leq T$. Because $S \leq T$ and $-S \leq T$, we have $|S| \leq T$; therefore $\left|\sum_{n=1}^{+\infty} u_n\right| \leq \sum_{n=1}^{+\infty} |u_n|$, and the theorem is proved. ■

Example 4 Determine if the series

$$\sum_{n=1}^{+\infty} \frac{\cos \frac{1}{3}n\pi}{n^2}$$

is convergent or divergent.

Solution Denote the given series by $\sum_{n=1}^{+\infty} u_n$. Therefore,

$$\sum_{n=1}^{+\infty} u_n = \frac{\frac{1}{2}}{1^2} - \frac{\frac{1}{2}}{2^2} - \frac{1}{3^2} - \frac{\frac{1}{2}}{4^2} + \frac{\frac{1}{2}}{5^2} + \frac{1}{6^2} + \frac{\frac{1}{2}}{7^2} - \cdots + \frac{\cos \frac{1}{3}n\pi}{n^2} + \cdots$$

$$= \tfrac{1}{2} - \tfrac{1}{8} - \tfrac{1}{9} - \tfrac{1}{32} + \tfrac{1}{50} + \tfrac{1}{36} + \tfrac{1}{98} - \cdots$$

This is a series of positive and negative terms. We can prove this series is convergent if we can show that it is absolutely convergent.

$$\sum_{n=1}^{+\infty} |u_n| = \sum_{n=1}^{+\infty} \frac{|\cos \frac{1}{3}n\pi|}{n^2}$$

Because

$$|\cos \tfrac{1}{3}n\pi| \leq 1 \qquad \text{for all } n$$

$$\frac{|\cos \frac{1}{3}n\pi|}{n^2} \leq \frac{1}{n^2} \qquad \text{for all positive integers } n$$

The series $\sum_{n=1}^{+\infty} 1/n^2$ is the p series, with $p = 2$, and is therefore convergent.

So by the comparison test $\sum_{n=1}^{+\infty} |u_n|$ is convergent. The given series is therefore absolutely convergent; hence, by Theorem 15.6.7 it is convergent.

Observe that the terms of the series $\sum_{n=1}^{+\infty} |u_n|$ neither increase monotonically nor decrease monotonically. For example, $|u_4| = \frac{1}{32}$, $|u_5| = \frac{1}{50}$, $|u_6| = \frac{1}{36}$; and so $|u_5| < |u_4|$, but $|u_6| > |u_5|$.

The *ratio test*, given in the next theorem, is used frequently to determine whether a given series is absolutely convergent.

15.6.8 Theorem
Ratio Test

Let $\sum_{n=1}^{+\infty} u_n$ be a given infinite series for which every u_n is nonzero. Then

(i) if $\lim_{n \to +\infty} |u_{n+1}/u_n| = L < 1$, the given series is absolutely convergent;

(ii) if $\lim_{n \to +\infty} |u_{n+1}/u_n| = L > 1$ or if $\lim_{n \to +\infty} |u_{n+1}/u_n| = +\infty$, the series is divergent;

(iii) if $\lim_{n \to +\infty} |u_{n+1}/u_n| = 1$, no conclusion regarding convergence may be made from this test.

Proof of (i) It is given that $L < 1$. Let R be a number such that $L < R < 1$. Let $R - L = \epsilon < 1$. Because $\lim_{n \to +\infty} |u_{n+1}/u_n| = L$, there exists an integer $N > 0$ such that

$$\left| \left| \frac{u_{n+1}}{u_n} \right| - L \right| < \epsilon \quad \text{whenever} \quad n \geq N$$

Therefore

$$0 < \left| \frac{u_{n+1}}{u_n} \right| < L + \epsilon = R \quad \text{whenever} \quad n \geq N \tag{12}$$

Let n take on the successive values $N, N+1, N+2, \ldots$, and so forth. We obtain from (12)

$$|u_{N+1}| < R|u_N|$$

$$|u_{N+2}| < R|u_{N+1}| < R^2|u_N|$$

$$|u_{N+3}| < R|u_{N+2}| < R^3|u_N|$$

$$\cdots$$

In general,

$$|u_{N+k}| < R^k|u_N| \qquad \text{for every positive integer } k \tag{13}$$

The series

$$\sum_{k=1}^{+\infty} |u_N|R^k = |u_N|R + |u_N|R^2 + \cdots + |u_N|R^n + \cdots$$

is convergent because it is a geometric series whose ratio is less than 1. So from (13) and the comparison test it follows that the series $\sum_{k=1}^{+\infty} |u_{N+k}|$ is convergent. The series $\sum_{k=1}^{+\infty} |u_{N+k}|$ differs from the series $\sum_{n=1}^{+\infty} |u_n|$ in only the first N terms. Therefore $\sum_{n=1}^{+\infty} |u_n|$ is convergent; so the given series is absolutely convergent.

Proof of (ii) If $\lim_{n\to+\infty} |u_{n+1}/u_n| = L > 1$ or $\lim_{n\to+\infty} |u_{n+1}/u_n| = +\infty$, then in either case there is an integer $N > 0$ such that $|u_{n+1}/u_n| > 1$ for all $n \geq N$. Let n take on the successive values $N, N+1, N+2, \ldots$, and so on. We obtain

$$|u_{N+1}| > |u_N|$$

$$|u_{N+2}| > |u_{N+1}| > |u_N|$$

$$|u_{N+3}| > |u_{N+2}| > |u_N|$$

$$\cdots$$

Thus $|u_n| > |u_N|$ for all $n > N$. Hence $\lim_{n\to+\infty} u_n \neq 0$; so the given series is divergent.

Proof of (iii) If the ratio test is applied to the p series, we have

$$\lim_{n\to+\infty}\left|\frac{u_{n+1}}{u_n}\right| = \lim_{n\to+\infty}\left|\frac{\dfrac{1}{(n+1)^p}}{\dfrac{1}{n^p}}\right| = \lim_{n\to+\infty}\left|\left(\frac{n}{n+1}\right)^p\right| = 1$$

Because the p series diverges if $p \leq 1$ and converges if $p > 1$, we have shown that it is possible to have both convergent and divergent series for which $\lim_{n\to+\infty} |u_{n+1}/u_n| = 1$. This proves part (iii). ■

Example 5 Determine if the series

$$\sum_{n=1}^{+\infty} \frac{(-1)^{n+1}n}{2^n}$$

is convergent or divergent.

Solution $u_n = (-1)^{n+1}n/2^n$ and $u_{n+1} = (-1)^{n+2}(n+1)/2^{n+1}$. Therefore

$$\left|\frac{u_{n+1}}{u_n}\right| = \frac{n+1}{2^{n+1}} \cdot \frac{2^n}{n} = \frac{n+1}{2n}$$

So

$$\lim_{n \to +\infty} \left|\frac{u_{n+1}}{u_n}\right| = \lim_{n \to +\infty} \frac{1 + \frac{1}{n}}{2} = \frac{1}{2} < 1$$

Therefore, by the ratio test, the given series is absolutely convergent and hence, by Theorem 15.6.7, it is convergent.

Example 6 In Example 2 the series

$$\sum_{n=1}^{+\infty} (-1)^n \frac{n+2}{n(n+1)}$$

was shown to be convergent. Is this series absolutely convergent or conditionally convergent?

Solution To test for absolute convergence we apply the ratio test. In the solution of Example 2 it was shown that the ratio $|u_{n+1}|/|u_n| = (n^2 + 3n)/(n^2 + 4n + 4)$. Hence

$$\lim_{n \to +\infty} \left|\frac{u_{n+1}}{u_n}\right| = \lim_{n \to +\infty} \frac{1 + \frac{3}{n}}{1 + \frac{4}{n} + \frac{4}{n^2}} = 1$$

So the ratio test fails. Because

$$|u_n| = \frac{n+2}{n(n+1)} = \frac{n+2}{n+1} \cdot \frac{1}{n} > \frac{1}{n}$$

the comparison test can be applied. And because the series $\sum_{n=1}^{+\infty} 1/n$ is the harmonic series, which diverges, we conclude that the series $\sum_{n=1}^{+\infty} |u_n|$ is divergent and hence $\sum_{n=1}^{+\infty} u_n$ is not absolutely convergent. Therefore the series is conditionally convergent.

It should be noted that the ratio test does not include all possibilities for $\lim_{n \to +\infty} |u_{n+1}/u_n|$ because it is possible that the limit does not exist and is not $+\infty$. The discussion of such cases is beyond the scope of this book.

To conclude the discussion of infinite series of constant terms we suggest a possible procedure to follow for determining the convergence or divergence of a given series. First of all, if $\lim_{n \to +\infty} u_n \neq 0$, the series is divergent. If $\lim_{n \to +\infty} u_n = 0$ and the series is an alternating series, then try the alternating-series test. If this test is not applicable, try the ratio test. If in applying the ratio test, $L = 1$, then another test must be used. The integral test may work when the ratio test does not; this was shown for the p series. Also, the comparison test can be tried.

EXERCISES 15.6

In Exercises 1 through 10, determine if the given alternating series is convergent or divergent.

1. $\sum_{n=2}^{+\infty} (-1)^n \frac{1}{\ln n}$ **2.** $\sum_{n=1}^{+\infty} (-1)^{n+1} \sin \frac{\pi}{n}$ **3.** $\sum_{n=1}^{+\infty} (-1)^{n+1} \frac{n^2}{n^3 + 2}$ **4.** $\sum_{n=1}^{+\infty} (-1)^{n+1} \frac{\ln n}{n}$ **5.** $\sum_{n=1}^{+\infty} (-1)^{n+1} \frac{\ln n}{n^2}$

6. $\sum_{n=1}^{+\infty} (-1)^n \frac{e^n}{n}$ **7.** $\sum_{n=1}^{+\infty} (-1)^n \frac{3^n}{n^2}$ **8.** $\sum_{n=1}^{+\infty} (-1)^n \frac{\sqrt{n}}{3n - 1}$ **9.** $\sum_{n=1}^{+\infty} (-1)^n \frac{n}{2^n}$ **10.** $\sum_{n=1}^{+\infty} (-1)^{n+1} \frac{3^n}{1 + 3^{2n}}$

In Exercises 11 through 14, find the error if the sum of the first four terms is used as an approximation to the sum of the given infinite series.

11. $\sum_{n=1}^{+\infty} (-1)^{n+1} \frac{1}{n}$ **12.** $\sum_{n=1}^{+\infty} (-1)^{n+1} \frac{1}{n^n}$ **13.** $\sum_{n=1}^{+\infty} (-1)^{n+1} \frac{1}{(2n - 1)^2}$ **14.** $\sum_{n=1}^{+\infty} (-1)^n \frac{1}{n!}$

In Exercises 15 through 18, find the sum of the given infinite series, accurate to three decimal places.

15. $\sum_{n=1}^{+\infty} (-1)^{n+1} \frac{1}{(2n)^3}$ **16.** $\sum_{n=1}^{+\infty} (-1)^n \frac{1}{(2n + 1)^3}$ **17.** $\sum_{n=1}^{+\infty} (-1)^{n+1} \frac{1}{n2^n}$ **18.** $\sum_{n=1}^{+\infty} (-1)^{n+1} \frac{1}{(2n)!}$

In Exercises 19 through 32, determine if the given series is absolutely convergent, conditionally convergent, or divergent. Prove your answer.

19. $\sum_{n=1}^{+\infty} (-1)^{n+1} \frac{2^n}{n!}$ **20.** $\sum_{n=1}^{+\infty} n \left(\frac{2}{3}\right)^n$ **21.** $\sum_{n=1}^{+\infty} \frac{n^2}{n!}$ **22.** $\sum_{n=1}^{+\infty} (-1)^{n+1} \frac{1}{(2n - 1)!}$

23. $\sum_{n=1}^{+\infty} (-1)^n \frac{n!}{2^{n+1}}$ **24.** $\sum_{n=1}^{+\infty} (-1)^{n+1} \frac{1}{n(n + 2)}$ **25.** $\sum_{n=1}^{+\infty} \frac{1 - 2 \sin n}{n^3}$ **26.** $\sum_{n=1}^{+\infty} (-1)^n \frac{1}{(n + 1)^3}$

27. $\sum_{n=1}^{+\infty} (-1)^{n+1} \frac{3^n}{n!}$ **28.** $\sum_{n=1}^{+\infty} (-1)^n \frac{n^2 + 1}{n^3}$ **29.** $\sum_{n=2}^{+\infty} (-1)^{n+1} \frac{1}{n(\ln n)^2}$ **30.** $\sum_{n=1}^{+\infty} (-1)^n \frac{\cos n}{n^2}$

31. $\sum_{n=1}^{+\infty} \frac{n^n}{n!}$ **32.** $\sum_{n=1}^{+\infty} \frac{1 \cdot 3 \cdot 5 \cdot \ldots \cdot (2n - 1)}{1 \cdot 4 \cdot 7 \cdot \ldots \cdot (3n - 2)}$

33. If $|r| < 1$, prove that the series $\sum_{n=1}^{+\infty} r^n \sin nt$ is absolutely convergent for all values of t.

34. Prove that if $\sum_{n=1}^{+\infty} u_n$ is absolutely convergent and $u_n \neq 0$ for all n, then $\sum_{n=1}^{+\infty} 1/|u_n|$ is divergent.

35. Prove that if $\sum_{n=1}^{+\infty} u_n$ is absolutely convergent, then $\sum_{n=1}^{+\infty} u_n{}^2$ is convergent.

36. Show by means of an example that the converse of Exercise 35 is not true.

37. Prove by mathematical induction that $1/n! \leq 1/2^{n-1}$.

15.7 POWER SERIES

An important type of series of variable terms called *power series* is now introduced.

15.7.1 Definition A **power series** in $(x - a)$ is a series of the form

$$c_0 + c_1(x - a) + c_2(x - a)^2 + \cdots + c_n(x - a)^n + \cdots \tag{1}$$

We use the notation $\sum_{n=0}^{+\infty} c_n(x - a)^n$ to represent series (1). (Note that we take $(x - a)^0 = 1$, even when $x = a$, for convenience in writing the general term.) If x is a particular number, the power series (1) becomes an infinite series of constant terms, as was discussed in previous sections. A special case of (1) is obtained when $a = 0$ and the series becomes a power series in x, which is

$$\sum_{n=0}^{+\infty} c_n x^n = c_0 + c_1 x + c_2 x^2 + \cdots + c_n x^n + \cdots \tag{2}$$

In addition to power series in $(x - a)$ and x, there are power series of the form

$$\sum_{n=0}^{+\infty} c_n[\phi(x)]^n = c_0 + c_1\phi(x) + c_2[\phi(x)]^2 + \cdots + c_n[\phi(x)]^n + \cdots$$

where ϕ is a function of x. Such a series is called a power series in $\phi(x)$. In this book we are concerned exclusively with power series of the forms (1) and (2), and when the term "power series" is used, we mean either of these forms. The discussion of the theory of power series is confined to series (2). The more general power series (1) can be obtained from (2) by the translation $x = \bar{x} - a$; therefore our results can be applied to series (1) as well.

In dealing with an infinite series of constant terms we were concerned with the question of convergence or divergence of the series. In considering a power series we ask, For what values of x does the power series converge? For each value of x for which the power series converges, the series represents the number that is the sum of the series. Therefore, a power series defines a function. The function f, with function values

$$f(x) = \sum_{n=0}^{+\infty} c_n x^n \tag{3}$$

has as its domain all values of x for which the power series in (3) converges. It is apparent that every power series (2) is convergent for $x = 0$. There are some series (see Example 3) that are convergent for no other value of x, and there are also series that converge for every value of x (see Example 2).

The following three examples illustrate how the ratio test can be used to determine the values of x for which a power series is convergent. When $n!$ is used in representing the nth term of a power series (as in Example 2), we take $0! = 1$ so that the expression for the nth term will hold when $n = 0$.

Example 1 Find the values of x for which the power series

$$\sum_{n=1}^{+\infty} (-1)^{n+1} \frac{2^n x^n}{n 3^n}$$

is convergent.

Solution For the given series,

$$u_n = (-1)^{n+1} \frac{2^n x^n}{n 3^n} \quad \text{and} \quad u_{n+1} = (-1)^{n+2} \frac{2^{n+1} x^{n+1}}{(n+1)3^{n+1}}$$

So

$$\lim_{n \to +\infty} \left| \frac{u_{n+1}}{u_n} \right| = \lim_{n \to +\infty} \left| \frac{2^{n+1} x^{n+1}}{(n+1)3^{n+1}} \cdot \frac{n 3^n}{2^n x^n} \right| = \lim_{n \to +\infty} \frac{2}{3} |x| \frac{n}{n+1} = \frac{2}{3} |x|$$

Therefore the power series is absolutely convergent when $\frac{2}{3}|x| < 1$ or, equivalently, when $|x| < \frac{3}{2}$. The series is divergent when $\frac{2}{3}|x| > 1$ or, equivalently, when $|x| > \frac{3}{2}$. When $\frac{2}{3}|x| = 1$ (i.e., when $x = \pm\frac{3}{2}$), the ratio test fails. When $x = \frac{3}{2}$, the given power series becomes

$$\frac{1}{1} - \frac{1}{2} + \frac{1}{3} - \frac{1}{4} + \cdots + (-1)^{n+1} \frac{1}{n} + \cdots$$

which is convergent, as was shown in Example 1 of Sec. 15.6. When $x = -\frac{3}{2}$, we have

$$-\frac{1}{1} - \frac{1}{2} - \frac{1}{3} - \frac{1}{4} - \cdots - \frac{1}{n} - \cdots$$

which by Theorem 15.3.7 is divergent. We conclude, then, that the given power series is convergent when $-\frac{3}{2} < x \leq \frac{3}{2}$. The series is absolutely convergent when $-\frac{3}{2} < x < \frac{3}{2}$ and is conditionally convergent when $x = \frac{3}{2}$. If $x \leq -\frac{3}{2}$ or $x > \frac{3}{2}$, the series is divergent.

Example 2 Find the values of x for which the power series $\sum_{n=0}^{+\infty} \frac{x^n}{n!}$ is convergent.

Solution For the given series, $u_n = x^n/n!$ and $u_{n+1} = x^{n+1}/(n+1)!$. So by applying the ratio test,

$$\lim_{n \to +\infty} \left| \frac{u_{n+1}}{u_n} \right| = \lim_{n \to +\infty} \left| \frac{x^{n+1}}{(n+1)!} \cdot \frac{n!}{x^n} \right| = |x| \lim_{n \to +\infty} \frac{1}{n+1} = 0 < 1$$

Therefore the given power series is absolutely convergent for all values of x.

Example 3 Find the values of x for which the power series $\sum_{n=0}^{+\infty} n!x^n$ is convergent.

Solution For the given series, $u_n = n!x^n$ and $u_{n+1} = (n+1)!x^{n+1}$. Applying the ratio test we have

$$\begin{aligned} \lim_{n \to +\infty} \left| \frac{u_{n+1}}{u_n} \right| &= \lim_{n \to +\infty} \left| \frac{(n+1)!x^{n+1}}{n!x^n} \right| \\ &= \lim_{n \to +\infty} |(n+1)x| \\ &= \begin{cases} 0 & \text{if } x = 0 \\ +\infty & \text{if } x \neq 0 \end{cases} \end{aligned}$$

It follows that the series is divergent for all values of x except 0.

15.7.2 Theorem If the power series $\sum_{n=0}^{+\infty} c_n x^n$ is convergent for $x = x_1$ ($x_1 \neq 0$), then it is absolutely convergent for all values of x for which $|x| < |x_1|$.

Proof If $\sum_{n=0}^{+\infty} c_n {x_1}^n$ is convergent, then $\lim_{n \to +\infty} c_n {x_1}^n = 0$. Therefore, if we take $\epsilon = 1$ in Definition 2.5.1, there exists an integer $N > 0$ such that

$$|c_n {x_1}^n| < 1 \quad \text{whenever} \quad n \geq N$$

Now if x is any number such that $|x| < |x_1|$,

$$|c_n x^n| = \left| c_n {x_1}^n \frac{x^n}{{x_1}^n} \right| = |c_n {x_1}^n| \left| \frac{x}{x_1} \right|^n < \left| \frac{x}{x_1} \right|^n \quad \text{whenever} \quad n \geq N \tag{4}$$

The series

$$\sum_{n=N}^{+\infty} \left| \frac{x}{x_1} \right|^n \tag{5}$$

is convergent because it is a geometric series with $r = |x/x_1| < 1$ (because

$|x| < |x_1|$). Compare the series $\sum_{n=N}^{+\infty} |c_n x^n|$, where $|x| < |x_1|$, with series (5). From (4) and the comparison test, $\sum_{n=N}^{+\infty} |c_n x^n|$ is convergent for $|x| < |x_1|$. So the given power series is absolutely convergent for all values of x for which $|x| < |x_1|$. ■

● ILLUSTRATION 1 An illustration of Theorem 15.7.2 is given in Example 1. The power series is convergent for $x = \frac{3}{2}$ and is absolutely convergent for all values of x for which $|x| < \frac{3}{2}$. ●

The following theorem is a corollary of Theorem 15.7.2.

15.7.3 Theorem If the power series $\sum_{n=0}^{+\infty} c_n x^n$ is divergent for $x = x_2$, it is divergent for all values of x for which $|x| > |x_2|$.

Proof Suppose that the given power series is convergent for some number x for which $|x| > |x_2|$. Then by Theorem 15.7.2 the series must converge when $x = x_2$. However, this contradicts the hypothesis. Therefore the given power series is divergent for all values of x for which $|x| > |x_2|$. ■

● ILLUSTRATION 2 To illustrate Theorem 15.7.3, consider again the power series of Example 1. It is divergent for $x = -\frac{3}{2}$ and is also divergent for all values of x for which $|x| > |-\frac{3}{2}|$. ●

From Theorems 15.7.2 and 15.7.3 the following important theorem can be proved.

15.7.4 Theorem Let $\sum_{n=0}^{+\infty} c_n x^n$ be a given power series. Then exactly one of the following conditions holds:

(i) the series converges only when $x = 0$;
(ii) the series is absolutely convergent for all values of x;
(iii) there exists a number $R > 0$ such that the series is absolutely convergent for all values of x for which $|x| < R$ and is divergent for all values of x for which $|x| > R$.

Proof If x is replaced by zero in the given power series, we have $c_0 + 0 + 0 + \cdots$, which is obviously convergent. Therefore every power series of the form $\sum_{n=0}^{+\infty} c_n x^n$ is convergent when $x = 0$. If this is the only value of x for which the series converges, then condition (i) holds.

Suppose that the given series is convergent for $x = x_1$ where $x_1 \neq 0$. Then it follows from Theorem 15.7.2 that the series is absolutely convergent for all values of x for which $|x| < |x_1|$. Now if in addition there is no

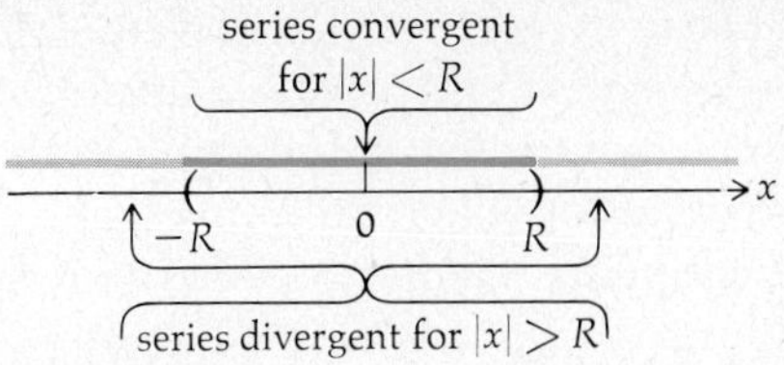

FIGURE 15.7.1

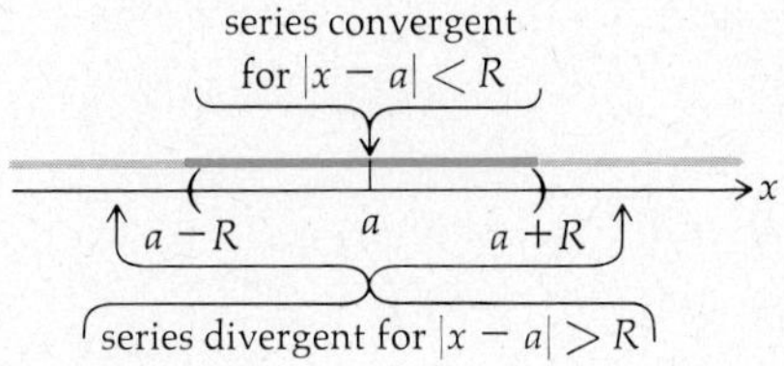

FIGURE 15.7.2

value of x for which the given series is divergent, then the series is absolutely convergent for all values of x. This is condition (ii).

If the given series is convergent for $x = x_1$ where $x_1 \neq 0$ and is divergent for $x = x_2$ where $|x_2| > |x_1|$, it follows from Theorem 15.7.3 that the series is divergent for all values of x for which $|x| > |x_2|$. Hence $|x_2|$ is an upper bound of the set of values of $|x|$ for which the series is absolutely convergent. Therefore, by the axiom of completeness (15.2.5) this set of numbers has a least upper bound, which is the number R of condition (iii). This proves that exactly one of the three conditions holds. ■

Theorem 15.7.4(iii) can be illustrated on the number line. See Fig. 15.7.1.

If instead of the power series $\sum_{n=0}^{+\infty} c_n x^n$ we have the series $\sum_{n=0}^{+\infty} c_n(x-a)^n$, in conditions (i) and (iii) of Theorem 15.7.4 x is replaced by $x - a$. The conditions become

(i) the series converges only when $x = a$;

(iii) there exists a number $R > 0$ such that the series is absolutely convergent for all values of x for which $|x - a| < R$ and is divergent for all values of x for which $|x - a| > R$. (See Fig. 15.7.2 for an illustration of this on the number line.)

The set of all values of x for which a given power series is convergent is called the **interval of convergence** of the power series. The number R of condition (iii) of Theorem 15.7.4 is called the **radius of convergence** of the power series. If condition (i) holds, $R = 0$; if condition (ii) holds, we write $R = +\infty$.

• ILLUSTRATION 3 For the power series of Example 1, $R = \frac{3}{2}$ and the interval of convergence is $(-\frac{3}{2}, \frac{3}{2}]$. In Example 2, $R = +\infty$, and the interval of convergence is written as $(-\infty, +\infty)$. •

If R is the radius of convergence of the power series $\sum_{n=0}^{+\infty} c_n x^n$, the interval of convergence is one of the following intervals: $(-R, R)$, $[-R, R]$, $(-R, R]$, or $[-R, R)$. For the more general power series $\sum_{n=0}^{+\infty} c_n(x-a)^n$, the interval of convergence is one of the following: $(a - R, a + R)$, $[a - R, a + R]$, $(a - R, a + R]$, or $[a - R, a + R)$.

A given power series defines a function having the interval of convergence as its domain. The most useful method at our disposal for determining the interval of convergence of a power series is the ratio test. However, the ratio test will not reveal anything about the convergence or divergence of the power series at the endpoints of the interval of convergence. At an endpoint, a power series may be either absolutely convergent, conditionally convergent, or divergent. If a power series converges absolutely at one endpoint, it follows from the definition of absolute convergence that the series is absolutely convergent at each endpoint (see Exercise 33). If a power series converges at one endpoint and diverges at the other, the series is conditionally convergent at the endpoint at which it converges (see Exercise 34). There are cases for which the convergence or divergence of a power series at the endpoints cannot be determined by the methods of elementary calculus.

Example 4 Determine the interval of convergence of the power series $\sum_{n=1}^{+\infty} n(x-2)^n$.

Solution The given series is

$$(x-2) + 2(x-2)^2 + \cdots + n(x-2)^n + (n+1)(x-2)^{n+1} + \cdots$$

Applying the ratio test we have

$$\begin{aligned}\lim_{n\to+\infty}\left|\frac{u_{n+1}}{u_n}\right| &= \lim_{n\to+\infty}\left|\frac{(n+1)(x-2)^{n+1}}{n(x-2)^n}\right| \\ &= |x-2| \lim_{n\to+\infty}\frac{n+1}{n} \\ &= |x-2|\end{aligned}$$

The given series then will be absolutely convergent if $|x-2| < 1$ or, equivalently, $-1 < x-2 < 1$ or, equivalently, $1 < x < 3$.

When $x = 1$, the series is $\sum_{n=1}^{+\infty} (-1)^n n$, which is divergent because $\lim_{n\to+\infty} u_n \neq 0$. When $x = 3$, the series is $\sum_{n=1}^{+\infty} n$, which is also divergent because $\lim_{n\to+\infty} u_n \neq 0$. Therefore the interval of convergence is $(1, 3)$. So the given power series defines a function having the interval $(1, 3)$ as its domain.

Example 5 Determine the interval of convergence of the power series

$$\sum_{n=1}^{+\infty} \frac{x^n}{2+n^2}$$

Solution The given series is

$$\frac{x}{2+1^2} + \frac{x^2}{2+2^2} + \frac{x^3}{2+3^2} + \cdots + \frac{x^n}{2+n^2} + \frac{x^{n+1}}{2+(n+1)^2} + \cdots$$

Applying the ratio test we have

$$\lim_{n\to+\infty}\left|\frac{u_{n+1}}{u_n}\right| = \lim_{n\to+\infty}\left|\frac{x^{n+1}}{2+(n+1)^2}\cdot\frac{2+n^2}{x^n}\right|$$

$$= |x| \lim_{n \to +\infty} \frac{2 + n^2}{2 + n^2 + 2n + 1}$$

$$= |x|$$

So the given series will be absolutely convergent if $|x| < 1$ or, equivalently, $-1 < x < 1$. When $x = 1$, the series is

$$\frac{1}{2 + 1^2} + \frac{1}{2 + 2^2} + \frac{1}{2 + 3^2} + \cdots + \frac{1}{2 + n^2} + \cdots$$

Because $1/(2 + n^2) < 1/n^2$ for all positive integers n, and because $\sum_{n=1}^{+\infty} 1/n^2$ is a convergent p series, it follows from the comparison test that the given power series is convergent when $x = 1$. When $x = -1$, the series is $\sum_{n=1}^{+\infty} (-1)^n/(2 + n^2)$, which is convergent because we have just seen that it is absolutely convergent. Hence the interval of convergence of the given power series is $[-1, 1]$.

EXERCISES 15.7

In Exercises 1 through 28, find the interval of convergence of the given power series.

1. $\sum_{n=0}^{+\infty} \frac{x^n}{n+1}$
2. $\sum_{n=0}^{+\infty} \frac{x^n}{n^2+1}$
3. $\sum_{n=0}^{+\infty} \frac{x^n}{n^2-3}$
4. $\sum_{n=0}^{+\infty} \frac{n^2 x^n}{2^n}$
5. $\sum_{n=1}^{+\infty} \frac{2^n x^n}{n^2}$
6. $\sum_{n=1}^{+\infty} \frac{x^n}{2^n \sqrt{n}}$
7. $\sum_{n=1}^{+\infty} \frac{n x^n}{3^n}$
8. $\sum_{n=1}^{+\infty} (-1)^n \frac{x^{2n}}{(2n)!}$
9. $\sum_{n=1}^{+\infty} (-1)^{n+1} \frac{x^{2n-1}}{(2n-1)!}$
10. $\sum_{n=1}^{+\infty} \frac{n+1}{n^{2n}} x^n$
11. $\sum_{n=1}^{+\infty} n! x^n$
12. $\sum_{n=0}^{+\infty} \frac{x^n}{(n+1)5^n}$
13. $\sum_{n=1}^{+\infty} (-1)^n \frac{x^n}{(2n-1)3^{2n-1}}$
14. $\sum_{n=1}^{+\infty} (-1)^{n+1} \frac{(n+1)x}{n!}$
15. $\sum_{n=1}^{+\infty} (-1)^{n+1} \frac{(x-1)^n}{n}$
16. $\sum_{n=1}^{+\infty} \frac{(x+2)^n}{(n+1)2^n}$
17. $\sum_{n=0}^{+\infty} (\sinh 2n) x^n$
18. $\sum_{n=1}^{+\infty} \frac{x^n}{\ln(n+1)}$
19. $\sum_{n=2}^{+\infty} (-1)^{n+1} \frac{x^n}{n(\ln n)^2}$
20. $\sum_{n=1}^{+\infty} \frac{(x+5)^{n-1}}{n^2}$
21. $\sum_{n=1}^{+\infty} \frac{n^2}{5^n} (x-1)^n$
22. $\sum_{n=0}^{+\infty} \frac{4^{n+1} x^{2n}}{n+3}$
23. $\sum_{n=1}^{+\infty} \frac{\ln n (x-5)^n}{n+1}$
24. $\sum_{n=1}^{+\infty} \frac{x^n}{n^n}$
25. $\sum_{n=1}^{+\infty} (-1)^n \frac{1 \cdot 3 \cdot 5 \cdot \ldots \cdot (2n-1)}{2 \cdot 4 \cdot 6 \cdot \ldots \cdot 2n} x^{2n+1}$
26. $\sum_{n=1}^{+\infty} n^n (x-3)^n$
27. $\sum_{n=1}^{+\infty} \frac{n! x^n}{n^n}$
28. $\sum_{n=1}^{+\infty} \frac{(-1)^{n+1} 1 \cdot 3 \cdot 5 \cdot \ldots \cdot (2n-1)}{2 \cdot 4 \cdot 6 \cdot \ldots \cdot 2n} x^n$
29. If a and b are positive integers, find the radius of convergence of the power series $\sum_{n=1}^{+\infty} \frac{(n+a)!}{n!(n+b)!} x^n$.

30. If $\sum_{n=1}^{+\infty} a_n$ is an absolutely convergent series, prove that the power series $\sum_{n=1}^{+\infty} a_n x^n$ is absolutely convergent when $|x| \le 1$.

31. Prove that if the radius of convergence of the power series $\sum_{n=1}^{+\infty} u_n x^n$ is r, then the radius of convergence of the series $\sum_{n=1}^{+\infty} u_n x^{2n}$ is $\sqrt{r}$.

32. Prove that if $\lim_{n \to +\infty} \sqrt[n]{|u_n|} = L$ ($L \neq 0$), then the radius of convergence of the power series $\sum_{n=1}^{+\infty} u_n x^n$ is $1/L$.

33. Prove that if a power series converges absolutely at one endpoint of its interval of convergence, then the power series is absolutely convergent at each endpoint.

34. Prove that if a power series converges at one endpoint of its interval of convergence and diverges at the other endpoint, then the power series is conditionally convergent at the endpoint at which it converges.

15.8 DIFFERENTIATION OF POWER SERIES

A power series $\sum_{n=0}^{+\infty} c_n x^n$ defines a function whose domain is the interval of convergence of the series.

• ILLUSTRATION 1 Consider the geometric series with $a = 1$ and $r = x$, which is $\sum_{n=0}^{+\infty} x^n$. By Theorem 15.3.5 this series converges to the sum $1/(1 - x)$ if $|x| < 1$. Therefore the power series $\sum_{n=0}^{+\infty} x^n$ defines the function f for which $f(x) = 1/(1 - x)$ and $|x| < 1$. Hence

$$1 + x + x^2 + x^3 + \cdots + x^n + \cdots = \frac{1}{1 - x} \quad \text{if } |x| < 1 \tag{1}$$

•

The series in (1) can be used to form other power series whose sums can be determined.

• ILLUSTRATION 2 If in (1) x is replaced by $-x$, we have

$$1 - x + x^2 - x^3 + \cdots + (-1)^n x^n + \cdots = \frac{1}{1 + x} \quad \text{if } |x| < 1 \tag{2}$$

Let $x = x^2$ in (1), and get

$$1 + x^2 + x^4 + x^6 + \cdots + x^{2n} + \cdots = \frac{1}{1 - x^2} \quad \text{if } |x| < 1 \tag{3}$$

If x is replaced by $-x^2$ in (1) we obtain

$$1 - x^2 + x^4 - x^6 + \cdots + (-1)^n x^{2n} + \cdots = \frac{1}{1 + x^2} \quad \text{if } |x| < 1 \tag{4}$$

•

In this section and the next, other interesting series are obtained from those like the above by differentiation and integration. We prove that if R (where $R \neq 0$) is the radius of convergence of a power series that defines a function f, then f is differentiable on the open interval $(-R, R)$

The power series obtained from the given series by differentiating term by term is

$$\sum_{n=0}^{+\infty} \frac{(n+1)x^n}{(n+1)^2} = \sum_{n=0}^{+\infty} \frac{x^n}{n+1}$$

$$= 1 + \frac{x}{2} + \frac{x^2}{3} + \frac{x^3}{4} + \cdots + \frac{x^n}{n+1} + \frac{x^{n+1}}{n+2} + \cdots$$

Applying the ratio test for this power series we have

$$\lim_{n \to +\infty} \left| \frac{u_{n+1}}{u_n} \right| = \lim_{n \to +\infty} \left| \frac{(n+1)x^{n+1}}{(n+2)x^n} \right| = |x| \lim_{n \to +\infty} \left| \frac{n+1}{n+2} \right| = |x|$$

This power series is convergent when $|x| < 1$; thus its radius of convergence $R' = 1$. Because $R = R'$, Theorem 15.8.1 is verified. ●

15.8.2 Theorem If the radius of convergence of the power series $\sum_{n=0}^{+\infty} c_n x^n$ is $R > 0$, then R is also the radius of convergence of the series $\sum_{n=2}^{+\infty} n(n-1)c_n x^{n-2}$.

Proof If Theorem 15.8.1 is applied to the series $\sum_{n=1}^{+\infty} nc_n x^{n-1}$, the desired result follows. ■

We are now in a position to prove the theorem regarding term-by-term differentiation of a power series.

15.8.3 Theorem Let $\sum_{n=0}^{+\infty} c_n x^n$ be a power series whose radius of convergence is $R > 0$. Then if f is the function defined by

$$f(x) = \sum_{n=0}^{+\infty} c_n x^n \tag{11}$$

$f'(x)$ exists for every x in the open interval $(-R, R)$, and it is given by

$$f'(x) = \sum_{n=1}^{+\infty} nc_n x^{n-1}$$

Proof Let x and a be two distinct numbers in the open interval $(-R, R)$. Taylor's formula (formula (9) in Sec. 14.5), with $n = 1$, is

$$f(x) = f(a) + \frac{f'(a)}{1!}(x-a) + \frac{f''(\xi)}{2!}(x-a)^2$$

From this formula with $f(x) = x^n$ it follows that for every positive integer n,

$$x^n = a^n + na^{n-1}(x-a) + \tfrac{1}{2}n(n-1)(\xi_n)^{n-2}(x-a)^2 \tag{12}$$

where ξ_n is between a and x for every positive integer n. From (11),

$$\begin{aligned} f(x) - f(a) &= \sum_{n=0}^{+\infty} c_n x^n - \sum_{n=0}^{+\infty} c_n a^n \\ &= c_0 + \sum_{n=1}^{+\infty} c_n x^n - c_0 - \sum_{n=1}^{+\infty} c_n a^n \\ &= \sum_{n=1}^{+\infty} c_n (x^n - a^n) \end{aligned}$$

Dividing by $x - a$ (because $x \neq a$) and using (12) we have from the above equation

$$\frac{f(x) - f(a)}{x - a} = \frac{1}{x - a} \sum_{n=1}^{+\infty} c_n [na^{n-1}(x - a) + \tfrac{1}{2}n(n - 1)(\xi_n)^{n-2}(x - a)^2]$$

So

$$\frac{f(x) - f(a)}{x - a} = \sum_{n=1}^{+\infty} nc_n a^{n-1} + \tfrac{1}{2}(x - a) \sum_{n=2}^{+\infty} n(n - 1)c_n(\xi_n)^{n-2} \tag{13}$$

Because a is in $(-R, R)$, it follows from Theorem 15.8.1 that $\sum_{n=1}^{+\infty} nc_n a^{n-1}$ is absolutely convergent.

Because both a and x are in $(-R, R)$, there is some number $K > 0$ such that $|a| < K < R$ and $|x| < K < R$. It follows from Theorem 15.8.2 that

$$\sum_{n=2}^{+\infty} n(n - 1)c_n K^{n-2}$$

is absolutely convergent. Then because

$$|n(n - 1)c_n(\xi_n)^{n-2}| < |n(n - 1)c_n K^{n-2}| \tag{14}$$

for each ξ_n, we can conclude from the comparison test that

$$\sum_{n=2}^{+\infty} n(n - 1)c_n(\xi_n)^{n-2}$$

is absolutely convergent.

It follows from (13) that

$$\left|\frac{f(x) - f(a)}{x - a} - \sum_{n=1}^{+\infty} nc_n a^{n-1}\right| = \left|\tfrac{1}{2}(x - a) \sum_{n=2}^{+\infty} n(n - 1)c_n(\xi_n)^{n-2}\right| \tag{15}$$

However, from Theorem 15.6.7, if $\sum_{n=1}^{+\infty} u_n$ is absolutely convergent, then

$$\left|\sum_{n=1}^{+\infty} u_n\right| \leq \sum_{n=1}^{+\infty} |u_n|$$

Applying this to the right side of (15) we obtain

$$\left|\frac{f(x) - f(a)}{x - a} - \sum_{n=1}^{+\infty} nc_n a^{n-1}\right| \leq \tfrac{1}{2}|x - a| \sum_{n=2}^{+\infty} n(n - 1)|c_n||\xi_n|^{n-2} \tag{16}$$

From (14) and 16),

$$\left|\frac{f(x) - f(a)}{x - a} - \sum_{n=1}^{+\infty} nc_n a^{n-1}\right| \leq \tfrac{1}{2}|x - a| \sum_{n=2}^{+\infty} n(n - 1)|c_n|K^{n-2} \tag{17}$$

where $0 < K < R$. Because the series on the right side of (17) is absolutely convergent, the limit of the right side, as x approaches a, is zero. Therefore, from (17) and the squeeze theorem (2.9.1) it follows that

$$\lim_{x \to a} \frac{f(x) - f(a)}{x - a} = \sum_{n=1}^{+\infty} nc_n a^{n-1}$$

or, equivalently,

$$f'(a) = \sum_{n=1}^{+\infty} nc_n a^{n-1}$$

and because a may be any number in the open interval $(-R, R)$, the theorem is proved. ■

Example 1 Let f be the function defined by the power series of Illustration 3. (a) Find the domain of f; (b) write the power series that defines the function f' and find the domain of f'.

Solution

(a) $f(x) = \displaystyle\sum_{n=0}^{+\infty} \frac{x^{n+1}}{(n + 1)^2}$

The domain of f is the interval of convergence of the power series. In Illustration 3 we showed that the radius of convergence of the power series is 1; that is, the series converges when $|x| < 1$. Consider now the power series when $|x| = 1$. When $x = 1$, the series is

$$1 + \frac{1}{4} + \frac{1}{9} + \cdots + \frac{1}{(n + 1)^2} + \cdots$$

which is convergent because it is the p series with $p = 2$. When $x = -1$, we have the series $\displaystyle\sum_{n=0}^{+\infty} (-1)^{n+1}/(n + 1)^2$, which is convergent because it is absolutely convergent. Hence the domain of f is the interval $[-1, 1]$.

(b) From Theorem 15.8.3 it follows that f' is defined by

$$f'(x) = \sum_{n=0}^{+\infty} \frac{x^n}{n+1} \tag{18}$$

and that $f'(x)$ exists for every x in the open interval $(-1, 1)$. In Illustration 3 the radius of convergence of the power series in (18) was shown to be 1. We now consider the power series in (18) when $x = \pm 1$. When $x = 1$, the series is

$$1 + \frac{1}{2} + \frac{1}{3} + \frac{1}{4} + \cdots + \frac{1}{n+1} + \cdots$$

which is the harmonic series and hence is divergent. When $x = -1$, the series is

$$1 - \frac{1}{2} + \frac{1}{3} - \frac{1}{4} + \cdots + (-1)^n \frac{1}{n+1} + \cdots$$

which is a convergent alternating series. Therefore the domain of f' is the interval $[-1, 1)$.

Example 1 illustrates the fact that if a function f is defined by a power series and this power series is differentiated term by term, the resulting power series, which defines f', has the same radius of convergence but not necessarily the same interval of convergence.

Example 2 Obtain a power-series representation of

$$\frac{1}{(1-x)^2}$$

Solution From (1),

$$\frac{1}{1-x} = 1 + x + x^2 + x^3 + \cdots + x^n + \cdots \qquad \text{if } |x| < 1$$

Using Theorem 15.8.3 and differentiating on both sides of the above we get

$$\frac{1}{(1-x)^2} = 1 + 2x + 3x^2 + \cdots + nx^{n-1} + \cdots \qquad \text{if } |x| < 1$$

Example 3 Show that

$$e^x = \sum_{n=0}^{+\infty} \frac{x^n}{n!} = 1 + x + \frac{x^2}{2!} + \frac{x^3}{3!} + \cdots$$

for all real values of x.

Solution In Example 2 of Sec. 15.7 we showed that the power series $\sum_{n=0}^{+\infty} x^n/n!$ is absolutely convergent for all real values of x. Therefore, if f is the function defined by

$$f(x) = \sum_{n=0}^{+\infty} \frac{x^n}{n!} \tag{19}$$

the domain of f is the set of all real numbers; that is, the interval of convergence is $(-\infty, +\infty)$. It follows from Theorem 15.8.3 that for all real values of x,

$$f'(x) = \sum_{n=1}^{+\infty} \frac{nx^{n-1}}{n!} \tag{20}$$

Because $n/n! = 1/(n-1)!$, (20) can be written as

$$f'(x) = \sum_{n=1}^{+\infty} \frac{x^{n-1}}{(n-1)!}$$

or, equivalently,

$$f'(x) = \sum_{n=0}^{+\infty} \frac{x^n}{n!} \tag{21}$$

From (19) and (21), $f'(x) = f(x)$ for all real values of x. Therefore the function f satisfies the differential equation

$$\frac{dy}{dx} = y$$

for which the general solution is $y = Ce^x$. Hence, for some constant C, $f(x) = Ce^x$. From (19), $f(0) = 1$. (Remember that we take $x^0 = 1$ even when $x = 0$ for convenience in writing the general term.) Therefore $C = 1$; so $f(x) = e^x$, and we have the desired result.

Example 4 Use the result of Example 3 to find a power-series representation of e^{-x}.

Solution If x is replaced by $-x$ in the series for e^x, it follows that

$$e^{-x} = 1 - x + \frac{x^2}{2!} - \frac{x^3}{3!} + \cdots + (-1)^n \frac{x^n}{n!} + \cdots$$

for all real values of x.

Example 5 Use the series of Example 4 to find the value of e^{-1} correct to five decimal places.

Solution If $x = 1$ in the series for e^{-x},

$$e^{-1} = 1 - 1 + \frac{1}{2!} - \frac{1}{3!} + \frac{1}{4!} - \frac{1}{5!} + \frac{1}{6!} - \frac{1}{7!} + \frac{1}{8!} - \frac{1}{9!} + \frac{1}{10!} - \cdots$$

$$= 1 - 1 + \frac{1}{2} - \frac{1}{6} + \frac{1}{24} - \frac{1}{120} + \frac{1}{720} - \frac{1}{5{,}040} + \frac{1}{40{,}320} - \frac{1}{362{,}880}$$

$$+\frac{1}{3{,}628{,}800} - \cdots$$

$$\approx 1 - 1 + 0.5 - 0.166667 + 0.041667 - 0.008333 + 0.001389$$
$$- 0.000198 + 0.000025 - 0.000003 + 0.0000003 - \cdots$$

This is a convergent alternating series for which $|u_{n+1}| < |u_n|$. So if the first ten terms are used to approximate the sum, by Theorem 15.6.4 the error is less than the absolute value of the eleventh term. Adding the first ten terms we obtain 0.367880. Rounding off to five decimal places gives

$$e^{-1} \approx 0.36788$$

In computation with infinite series two kinds of errors occur. One is the error given by the remainder after the first n terms. The other is the round-off error that occurs when each term of the series is approximated by a decimal with a finite number of places. In particular, in Example 5 we wanted the result accurate to five decimal places; so each term was rounded off to six decimal places. After computing the sum, we rounded off this result to five decimal places. Of course, the error given by the remainder can be reduced by considering additional terms of the series, whereas the round-off error can be reduced by using more decimal places.

EXERCISES 15.8

In Exercises 1 through 10, a function f is defined by a power series. In each exercise do the following: (a) Find the radius of convergence of the given power series and the domain of f; (b) write the power series that defines the function f′ and find its radius of convergence by using methods of Sec. 15.7 (thus verifying Theorem 15.8.1); (c) find the domain of f′.

1. $f(x) = \sum_{n=1}^{+\infty} \frac{x^n}{n^2}$ **2.** $f(x) = \sum_{n=1}^{+\infty} (-1)^{n-1}\frac{x^n}{n}$ **3.** $f(x) = \sum_{n=1}^{+\infty} \frac{x^n}{\sqrt{n}}$

4. $f(x) = \sum_{n=2}^{+\infty} \frac{(x-2)^n}{\sqrt{n-1}}$ **5.** $f(x) = \sum_{n=1}^{+\infty} (-1)^{n-1}\frac{x^{2n-1}}{(2n-1)!}$ **6.** $f(x) = \sum_{n=0}^{+\infty} \frac{x^{2n}}{(n!)^2}$

7. $f(x) = \sum_{n=1}^{+\infty} (n+1)(3x-1)^n$ **8.** $f(x) = \sum_{n=1}^{+\infty} \frac{x^{2n-2}}{(2n-2)!}$ **9.** $f(x) = \sum_{n=1}^{+\infty} \frac{(x-1)^n}{n3^n}$

10. $f(x) = \sum_{n=2}^{+\infty} (-1)^n \frac{(x-3)^n}{n(n-1)}$

11. Use the result of Example 2 to find a power-series representation of $1/(1-x)^3$.

12. Use the result of Example 3 to find a power-series representation of $e^{\sqrt{x}}$.

13. Obtain a power-series representation of $1/(1+x)^2$ if $|x| < 1$ by differentiating series (2) term by term.

14. Obtain a power-series representation of $x/(1+x^2)^2$ if $|x| <$ by differentiating series (4) term by term.

15. (a) Use series (1) to find a power-series representation for $1/(1-2x)$. (b) Differentiate term by term the series found in part (a) to find a power-series representation for $2/(1-2x)^2$.

16. (a) Use series (2) to find a power-series representation for $1/(1+x^3)$. (b) Differentiate term by term the series found in part (a) to find a power-series representation for $-3x^2/(1+x^3)^2$.

17. (a) Use the result of Example 3 to find a power-series representation for e^{x^2}. (b) Differentiate term by term the series found in part (a) to find a power-series representation for xe^{x^2}.

18. Let f be the function defined by $f(x) = \sum_{n=0}^{+\infty} (-1)^n \dfrac{x^n}{3^n(n+2)}$. (a) Find the domain of f. (b) Find $f'(x)$ and find the domain of f'.

19. Use the result of Example 4 to find the value of $1/\sqrt{e}$ correct to five decimal places.

20. If $f(x) = \sum_{n=0}^{+\infty} (-1)^n \dfrac{x^{2n}}{3^n}$, find $f'(\frac{1}{2})$ correct to four decimal places.

21. Use the results of Examples 3 and 4 to find a power-series representation of (a) $\sinh x$ and (b) $\cosh x$.

22. Show that each of the power series in parts (a) and (b) of Exercise 21 can be obtained from the other by term-by-term differentiation.

23. Use the result of Example 2 to find the sum of the series $\sum_{n=1}^{+\infty} \dfrac{n}{2^n}$.

24. (a) Find a power-series representation for $(e^x - 1)/x$. (b) By differentiating term by term the power series in part (a), show that $\sum_{n=1}^{+\infty} \dfrac{n}{(n+1)!} = 1$.

25. (a) Find a power-series representation for x^2e^{-x}. (b) By differentiating term by term the power series in part (a), show that $\sum_{n=1}^{+\infty} (-2)^{n+1} \dfrac{n+2}{n!} = 4$.

26. (a) Find a power-series representation for e^{-x^2}. (b) By differentiating the power series in part (a) twice term by term, show that $\sum_{n=1}^{+\infty} (-1)^{n+1} \dfrac{2n+1}{2^n n!} = 1$.

27. Suppose a function f has the power-series representation $\sum_{n=0}^{+\infty} c_n x^n$, where the radius of convergence $R > 0$. If $f'(x) = f(x)$ and $f(0) = 1$, find the power series by using only properties of power series and nothing about the exponential function.

28. (a) Use only properties of power series to find a power-series representation of the function f for which $f(x) > 0$ and $f'(x) = 2xf(x)$ for all x, and $f(0) = 1$. (b) Verify your result in part (a) by solving the differential equation $D_x y = 2xy$ with the boundary condition $y = 1$ when $x = 0$.

29. Suppose a function f has the power-series representation $\sum_{n=0}^{+\infty} c_n x^n$. If f is an even function, show that $c_n = 0$ when n is odd.

30. Assume that the constant 0 has a power-series representation $\sum_{n=0}^{+\infty} c_n x^n$, where the radius of convergence $R > 0$. Prove that $c_n = 0$ for all n.

15.9 INTEGRATION OF POWER SERIES

The theorem regarding the term-by-term integration of a power series is a consequence of Theorem 15.8.3.

15.9.1 Theorem Let $\sum_{n=0}^{+\infty} c_n x^n$ be a power series whose radius of convergence is $R > 0$. Then if f is the function defined by

$$f(x) = \sum_{n=0}^{+\infty} c_n x^n$$

f is integrable on every closed subinterval of $(-R, R)$, and the integral of f is evaluated by integrating the given power series term by term; that is, if x is in $(-R, R)$, then

$$\int_0^x f(t)\,dt = \sum_{n=0}^{+\infty} \frac{c_n}{n+1} x^{n+1}$$

Furthermore, R is the radius of convergence of the resulting series.

Proof Let g be the function defined by

$$g(x) = \sum_{n=0}^{+\infty} \frac{c_n}{n+1} x^{n+1}$$

Because the terms of the power-series representation of $f(x)$ are the derivatives of the terms of the power-series representation of $g(x)$, the two series have, by Theorem 15.8.1, the same radius of convergence. By Theorem 15.8.3 it follows that

$$g'(x) = f(x) \qquad \text{for every } x \text{ in } (-R, R)$$

By Theorem 15.8.2 it follows that $f'(x) = g''(x)$ for every x in $(-R, R)$. Because f is differentiable on $(-R, R)$, f is continuous there; consequently, f is continuous on every closed subinterval of $(-R, R)$. From Theorem 6.5.2 it follows that if x is in $(-R, R)$, then

$$\int_0^x f(t)\,dt = g(x) - g(0) = g(x)$$

or, equivalently,

$$\int_0^x f(t)\,dt = \sum_{n=0}^{+\infty} \frac{c_n}{n+1} x^{n+1} \qquad \blacksquare$$

Theorem 15.9.1 often is used to compute a definite integral that cannot be evaluated directly by finding an antiderivative of the integrand. Example 1 and 2 illustrate the technique. The definite integral $\int_0^x e^{-t^2}\,dt$ appearing in these two examples is similar to the one that represents the measure of the area of a region under the "normal probability curve."

Example 1 Find a power-series representation of $\int_0^x e^{-t^2}\,dt$.

Solution From Example 4 of Sec. 15.8,

$$e^{-x} = \sum_{n=0}^{+\infty} \frac{(-1)^n x^n}{n!}$$

for all values of x. If x is replaced by t^2,

$$e^{-t^2} = 1 - t^2 + \frac{t^4}{2!} - \frac{t^6}{3!} + \cdots + (-1)^n \frac{t^{2n}}{n!} + \cdots \qquad \text{for all values of } t$$

Applying Theorem 15.9.1 we integrate term by term and obtain

$$\int_0^x e^{-t^2}\,dt = \sum_{n=0}^{+\infty} \int_0^x (-1)^n \frac{t^{2n}}{n!}\,dt$$

$$= x - \frac{x^3}{3} + \frac{x^5}{2!\cdot 5} - \frac{x^7}{3!\cdot 7} + \cdots + (-1)^n \frac{x^{2n+1}}{n!(2n+1)} + \cdots$$

The power series represents the integral for all values of x.

Example 2 Use the result of Example 1 to compute accurate to three decimal places the value of $\int_0^{1/2} e^{-t^2}\,dt$.

Solution Replace x by $\frac{1}{2}$ in the power series obtained in Example 1 to obtain

$$\int_0^{1/2} e^{-t^2}\,dt = \tfrac{1}{2} - \tfrac{1}{24} + \tfrac{1}{320} - \tfrac{1}{5376} + \cdots$$

$$\approx 0.5 - 0.0417 + 0.0031 - 0.0002 + \cdots$$

This is a convergent alternating series with $|u_{n+1}| < |u_n|$. Thus, if we use the first three terms to approximate the sum, by Theorem 15.6.4 the error is less than the absolute value of the fourth term. From the first three terms,

$$\int_0^{1/2} e^{-t^2}\,dt \approx 0.461$$

Example 3 Obtain a power-series representation of $\ln(1 + x)$.

Solution Consider the function f defined by $f(t) = 1/(1 + t)$. A power-series representation of this function is given by series (2) in Sec. 15.8, which is

$$\frac{1}{1+t} = 1 - t + t^2 - t^3 + \cdots + (-1)^n t^n + \cdots \qquad \text{if } |t| < 1$$

Apply Theorem 15.9.1 and integrate term by term to obtain

$$\int_0^x \frac{dt}{1+t} = \sum_{n=0}^{+\infty} \int_0^x (-1)^n t^n \, dt \qquad \text{if } |x| < 1$$

Therefore

$$\ln(1+x) = x - \frac{x^2}{2} + \frac{x^3}{3} - \frac{x^4}{4} + \cdots + (-1)^n \frac{x^{n+1}}{n+1} + \cdots \qquad \text{if } |x| < 1$$

or, equivalently,

$$\ln(1+x) = \sum_{n=1}^{+\infty} (-1)^{n-1} \frac{x^n}{n} \qquad \text{if } |x| < 1 \tag{1}$$

Because $|x| < 1$, $|1 + x| = (1 + x)$. Thus the absolute-value bars are not needed when writing $\ln(1 + x)$.

In Example 3, Theorem 15.9.1 allows us to conclude that the power series in (1) represents the function only for values of x in the open interval $(-1, 1)$. However, the power series is convergent at the right endpoint 1, as was shown in Example 1 of Sec. 15.6. When $x = -1$, the power series becomes the negative of the harmonic series and is divergent. Hence the interval of convergence of the power series in (1) is $(-1, 1]$.

In the following illustration we show that the power series in (1) represents $\ln(1 + x)$ at $x = 1$ by proving that the sum of the series $\sum_{n=1}^{+\infty} (-1)^{n-1}/n$ is ln 2.

● ILLUSTRATION 1 For the infinite series $\sum_{n=1}^{+\infty} (-1)^{n-1}/n$, the nth partial sum is

$$s_n = 1 - \frac{1}{2} + \frac{1}{3} - \frac{1}{4} + \cdots + (-1)^{n-1} \frac{1}{n} \tag{2}$$

It follows from Definition 15.3.2 that if we show $\lim_{n \to +\infty} s_n = \ln 2$, we will have proved that the sum of the series is ln 2.

From algebra comes the following formula for the sum of a geometric progression:

$$a + ar + ar^2 + ar^3 + \cdots + ar^{n-1} = \frac{a - ar^n}{1 - r}$$

From this formula with $a = 1$ and $r = -t$,

$$1 - t + t^2 - t^3 + \cdots + (-t)^{n-1} = \frac{1 - (-t)^n}{1 + t}$$

which may be written as

$$1 - t + t^2 - t^3 + \cdots + (-1)^{n-1}t^{n-1} = \frac{1}{1+t} + (-1)^{n+1}\frac{t^n}{1+t}$$

Integrating from 0 to 1 we get

$$\int_0^1 [1 - t + t^2 - t^3 + \cdots + (-1)^{n-1}t^{n-1}]\,dt$$

$$= \int_0^1 \frac{dt}{1+t} + (-1)^{n+1}\int_0^1 \frac{t^n}{1+t}\,dt$$

which gives

$$1 - \frac{1}{2} + \frac{1}{3} - \frac{1}{4} + \cdots + (-1)^{n-1}\frac{1}{n} = \ln 2 + (-1)^{n+1}\int_0^1 \frac{t^n}{1+t}\,dt \quad (3)$$

Referring to (2), we see that the left side of (3) is s_n. Letting

$$R_n = (-1)^{n+1}\int_0^1 \frac{t^n}{1+t}\,dt$$

(3) may be written as

$$s_n = \ln 2 + R_n \quad (4)$$

Because $t^n/(1+t) \le t^n$ for all t in $[0, 1]$ it follows from Theorem 6.3.8 that

$$\int_0^1 \frac{t^n}{1+t}\,dt \le \int_0^1 t^n\,dt$$

Hence

$$0 \le |R_n| = \int_0^1 \frac{t^n}{1+\text{t}}\,dt \le \int_0^1 t^n\,dt = \frac{1}{n+1}$$

Because $\lim_{n\to+\infty} 1/(n+1) = 0$, it follows from the above inequality and the squeeze theorem (2.9.1) that $\lim_{n\to+\infty} R_n = 0$. Therefore, from (4),

$$\lim_{n\to+\infty} s_n = \ln 2 + \lim_{n\to+\infty} R_n$$

$$= \ln 2$$

Thus

$$\sum_{n=1}^{+\infty} (-1)^{n-1}\frac{1}{n} = 1 - \frac{1}{2} + \frac{1}{3} - \frac{1}{4} + \cdots = \ln 2 \quad (5)$$

The solution of Example 3 shows that the power series in (1) represents $\ln(x+1)$ if $|x| < 1$. Hence, with the result of Illustration 1 we can conclude that the power series in (1) represents $\ln(x+1)$ for all x in its interval of convergence $(-1, 1]$.

Although it is interesting that the sum of the series in (5) is ln 2, this series converges too slowly to use it to calculate ln 2. We now proceed to obtain a power series for computation of natural logarithms.

From (1),

$$\ln(1+x) = x - \frac{x^2}{2} + \frac{x^3}{3} - \cdots + (-1)^{n-1}\frac{x^n}{n} + \cdots \qquad \text{for } x \text{ in } (-1, 1] \qquad (6)$$

If x is replaced by $-x$ in this series,

$$\ln(1-x) = -x - \frac{x^2}{2} - \frac{x^3}{3} - \frac{x^4}{4} - \cdots - \frac{x^n}{n} - \cdots \qquad \text{for } x \text{ in } [-1, 1) \qquad (7)$$

Subtracting term by term (7) from (6) we obtain

$$\ln\frac{1+x}{1-x} = 2\left(x + \frac{x^3}{3} + \frac{x^5}{5} + \cdots + \frac{x^{2n-1}}{2n-1} + \cdots\right) \qquad \text{if } |x| < 1 \qquad (8)$$

The series in (8) can be used to compute the value of the natural logarithm of any positive number.

● ILLUSTRATION 2 If y is any positive number, let

$$y = \frac{1+x}{1-x} \quad \text{and then} \quad x = \frac{y-1}{y+1} \quad \text{and } |x| < 1$$

For instance, if $y = 2$, then $x = \frac{1}{3}$. From (8),

$$\ln 2 = 2\left(\frac{1}{3} + \frac{1}{3^4} + \frac{1}{5 \cdot 3^5} + \frac{1}{7 \cdot 3^7} + \frac{1}{9 \cdot 3^9} + \frac{1}{11 \cdot 3^{11}} + \cdots\right)$$

$$= 2\left(\frac{1}{3} + \frac{1}{81} + \frac{1}{1{,}215} + \frac{1}{15{,}309} + \frac{1}{177{,}147} + \frac{1}{1{,}948{,}617} + \cdots\right)$$

$$\approx 2(0.333333 + 0.012346 + 0.000823 + 0.000065 + 0.000006 + 0.000001 + \cdots)$$

Using the first six terms in parentheses, multiplying by 2, and rounding off to five decimal places, we get

$$\ln 2 \approx 0.69315$$

●

Example 4 Obtain a power-series representation of $\tan^{-1} x$.

Solution From series (4) in Sec. 15.8,

$$\frac{1}{1+x^2} = 1 - x^2 + x^4 - x^6 + \cdots + (-1)^n x^{2n} + \cdots \qquad \text{if } |x| < 1$$

Apply Theorem 15.9.1 and integrate term by term to obtain

$$\int_0^x \frac{1}{1+t^2}\,dt = x - \frac{x^3}{3} + \frac{x^5}{5} - \cdots + (-1)^n \frac{x^{2n+1}}{2n+1} + \cdots$$

Therefore

$$\tan^{-1} x = \sum_{n=0}^{+\infty} (-1)^n \frac{x^{2n+1}}{2n+1} \qquad \text{if } |x| < 1 \tag{9}$$

Although Theorem 15.9.1 allows us to conclude that the power series in (9) represents $\tan^{-1} x$ only for values of x such that $|x| < 1$, it can be shown that the interval of convergence of the power series is $[-1, 1]$ and that the power series is a representation of $\tan^{-1} x$ for all x in its interval of convergence. (You are asked to do this in Exercise 28.) Therefore,

$$\tan^{-1} x = \sum_{n=0}^{+\infty} (-1)^n \frac{x^{2n+1}}{2n+1} = x - \frac{x^3}{3} + \frac{x^5}{5} - \cdots \qquad \text{if } |x| \le 1 \tag{10}$$

● ILLUSTRATION 3 If $x = 1$ in (10),

$$\frac{\pi}{4} = 1 - \frac{1}{3} + \frac{1}{5} - \frac{1}{7} + \cdots + (-1)^n \frac{1}{2n+1} + \cdots$$ ●

The series in Illustration 3 is not suitable for computing π because it converges too slowly. The following example gives a better method.

Example 5 Prove that $\frac{1}{4}\pi = \tan^{-1} \frac{1}{2} + \tan^{-1} \frac{1}{3}$. Use this formula and the power series for $\tan^{-1} x$ of Example 4 to compute the value of π accurate to four decimal places.

Solution Let $\alpha = \tan^{-1} \frac{1}{2}$ and $\beta = \tan^{-1} \frac{1}{3}$. Then

$$\begin{aligned}\tan(\alpha + \beta) &= \frac{\tan \alpha + \tan \beta}{1 - \tan \alpha \tan \beta} \\ &= \frac{\frac{1}{2} + \frac{1}{3}}{1 - \frac{1}{2} \cdot \frac{1}{3}} \\ &= \frac{3 + 2}{6 - 1} \\ &= 1 \\ &= \tan \tfrac{1}{4}\pi\end{aligned}$$

Therefore

$$\tfrac{1}{4}\pi = \alpha + \beta = \tan^{-1} \tfrac{1}{2} + \tan^{-1} \tfrac{1}{3} \tag{11}$$

From formula (10) with $x = \frac{1}{2}$,

$$\tan^{-1} \frac{1}{2} = \frac{1}{2} - \frac{1}{3}\left(\frac{1}{2}\right)^3 + \frac{1}{5}\left(\frac{1}{2}\right)^5 - \frac{1}{7}\left(\frac{1}{2}\right)^7 + \frac{1}{9}\left(\frac{1}{2}\right)^9 - \frac{1}{11}\left(\frac{1}{2}\right)^{11} + \frac{1}{13}\left(\frac{1}{2}\right)^{13} - \frac{1}{15}\left(\frac{1}{2}\right)^{15} + \cdots$$

$$= \frac{1}{2} - \frac{1}{24} + \frac{1}{160} - \frac{1}{896} + \frac{1}{4{,}608} - \frac{1}{22{,}528} + \frac{1}{106{,}492} - \frac{1}{491{,}520} + \cdots$$

$$\approx 0.5 - 0.04167 + 0.00625 - 0.00112 + 0.00022 - 0.00004 + 0.00001 - 0.000002 + \cdots$$

Because the series is alternating and $|u_{n+1}| < |u_n|$, it follows from Theorem 15.6.4 that if the first seven terms are used to approximate the sum of the series, the error is less than the absolute value of the eighth term. Therefore

$$\tan^{-1} \tfrac{1}{2} \approx 0.46365 \qquad (12)$$

From formula (10) with $x = \frac{1}{3}$,

$$\tan^{-1} \frac{1}{3} = \frac{1}{3} - \frac{1}{3}\left(\frac{1}{3}\right)^3 + \frac{1}{5}\left(\frac{1}{3}\right)^5 - \frac{1}{7}\left(\frac{1}{3}\right)^7 + \frac{1}{9}\left(\frac{1}{3}\right)^9 - \frac{1}{11}\left(\frac{1}{3}\right)^{11} + \cdots$$

$$= \frac{1}{3} - \frac{1}{81} + \frac{1}{1{,}215} - \frac{1}{15{,}309} + \frac{1}{177{,}147} - \frac{1}{1{,}948{,}617} + \cdots$$

$$\approx 0.33333 - 0.01235 + 0.00082 - 0.00007 + 0.00001 - 0.000001 + \cdots$$

If the first five terms are used to approximate the sum,

$$\tan^{-1} \tfrac{1}{3} \approx 0.32174 \qquad (13)$$

By substituting from (12) and (13) into (11),

$$\tfrac{1}{4}\pi \approx 0.46365 + 0.32174 = 0.78539$$

Multiplying by 4 and rounding off to four decimal places gives $\pi \approx 3.1416$.

EXERCISES 15.9

In Exercises 1 through 4, compute the value of the definite integral to three decimal places by two methods: (a) use the fundamental theorem of the calculus; (b) use a power series. Compare the results.

1. $\int_{1/3}^{1/2} \frac{dx}{1 - x}$ **2.** $\int_0^1 \frac{dx}{x^2 + 4}$ **3.** $\int_0^1 e^x \, dx$ **4.** $\int_0^{1/3} \ln(1 + x) \, dx$

In Exercises 5 through 16, compute the value of the given integral accurate to three decimal places, by using series.

5. $\int_0^{1/2} \frac{dx}{1 + x^3}$ **6.** $\int_0^{1/2} e^{-x^3} \, dx$ **7.** $\int_0^1 e^{-x^2} \, dx$ **8.** $\int_0^{1/3} \frac{dx}{1 + x^4}$

9. $\int_0^{1/2} \tan^{-1} x^2 \, dx$ **10.** $\int_0^{1/2} \cosh x^2 \, dx$ **11.** $\int_0^1 x \sinh \sqrt{x} \, dx$

12. $\int_0^1 h(x)\,dx$, where $h(x) = \begin{cases} \dfrac{\sinh x}{x} & \text{if } x \neq 0 \\ 1 & \text{if } x = 0 \end{cases}$

13. $\int_0^{1/4} g(x)\,dx$, where $g(x) = \begin{cases} \dfrac{\tan^{-1} x}{x} & \text{if } x \neq 0 \\ 1 & \text{if } x = 0 \end{cases}$

14. $\int_0^{1/2} f(x)\,dx$, where $f(x) = \begin{cases} \dfrac{\ln(1+x)}{x} & \text{if } x \neq 0 \\ 1 & \text{if } x = 0 \end{cases}$

15. $\int_0^{1} f(x)\,dx$, where $f(x) = \begin{cases} \dfrac{e^x - 1}{x} & \text{if } x \neq 0 \\ 1 & \text{if } x = 0 \end{cases}$

16. $\int_0^{1} g(x)\,dx$, where $g(x) = \begin{cases} \dfrac{\cosh x - 1}{x} & \text{if } x \neq 0 \\ 0 & \text{if } x = 0 \end{cases}$

17. Given $\cosh x = \sum_{n=0}^{+\infty} \dfrac{x^{2n}}{(2n)!}$ for all x. Obtain a power-series representation for $\sinh x$ by integrating term by term from 0 to x the given series.

18. Find a power-series representation for $\ln(1 + ax)$ by integrating term by term from 0 to x a power-series representation for $1/(1 + at)$.

19. Use the power series in Eq. (9) to compute $\tan^{-1} \frac{1}{4}$ accurate to four decimal places.

20. Use the power series in Eq. (8) to compute ln 3 accurate to four decimal places

21. If $f'(x) = \sum_{n=0}^{+\infty} (-1)^n \dfrac{(x-1)^n}{n!}$, find $f(\frac{5}{4})$ accurate to three decimal places.

22. If $g'(x) = \sum_{n=0}^{+\infty} (-1)^n \dfrac{x^n}{n^2 + 3}$, find $g(1)$ accurate to two decimal places.

23. Find a power-series representation for $\tanh^{-1} x$ by integrating term by term from 0 to x a power-series representation for $(1 - t^2)^{-1}$.

24. Find a power series for xe^x by multiplying the series for e^x by x, and then integrate the resulting series term by term from 0 to 1 and show that $\sum_{n=1}^{+\infty} \dfrac{1}{n!(n+2)} = \frac{1}{2}$.

25. By integrating term by term from 0 to x a power-series representation for $\ln(1 - t)$, show that

$$\sum_{n=2}^{+\infty} \frac{x^n}{(n-1)n} = x + (1 - x)\ln(1 - x)$$

26. By integrating term by term from 0 to x a power-series representation for $\ln(1 + t)$, show that

$$\sum_{n=0}^{+\infty} \frac{(-1)^n}{(n+1)(n+2)} = 2 \ln 2 - 1$$

27. By integrating term by term from 0 to x a power-series representation for $t \tan^{-1} t$, show that

$$\sum_{n=1}^{+\infty} (-1)^{n+1} \frac{x^{2n+1}}{(2n-1)(2n+1)} = \tfrac{1}{2}[(x^2 + 1)\tan^{-1} x - x]$$

28. Show that the interval of convergence of the power series in Eq. (9) is $[-1, 1]$ and that the power series is a representation of $\tan^{-1} x$ for all x in its interval of convergence.

29. Find the power series in x of $f(x)$ if $f''(x) = -f(x)$, $f(0) = 0$, and $f'(0) = 1$. Also, find the radius of convergence of the resulting series.

30. Integrate term by term from 0 to x a power-series representation for $(1 - t^2)^{-1}$ to obtain the power series in Eq. (8) for $\ln[(1 + x)/(1 - x)]$.

15.10 TAYLOR SERIES

If f is the function defined by

$$f(x) = \sum_{n=0}^{+\infty} c_n x^n = c_0 + c_1 x + c_2 x^2 + c_3 x^3 + \cdots + c_n x^n + \cdots \tag{1}$$

whose radius of convergence is $R > 0$, it follows from successive applications of Theorem 15.8.3 that f has derivatives of all orders on $(-R, R)$. Such a function is said to be *infinitely differentiable* on $(-R, R)$. Successive differentiations of the function in (1) give

$$f'(x) = c_1 + 2c_2 x + 3c_3 x^2 + 4c_4 x^3 + \cdots + nc_n x^{n-1} + \cdots \tag{2}$$

$$f''(x) = 2c_2 + 2 \cdot 3c_3 x + 3 \cdot 4c_4 x^2 + \cdots + (n-1)nc_n x^{n-2} + \cdots \tag{3}$$

$$f'''(x) = 2 \cdot 3c_3 + 2 \cdot 3 \cdot 4c_4 x + \cdots + (n-2)(n-1)nc_n x^{n-3} + \cdots \tag{4}$$

$$f^{(iv)}(x) = 2 \cdot 3 \cdot 4c_4 + \cdots + (n-3)(n-2)(n-1)nc_n x^{n-4} + \cdots \tag{5}$$

etc. If $x = 0$ in (1),

$$f(0) = c_0$$

If $x = 0$ in (2),

$$f'(0) = c_1$$

If in (3) $x = 0$,

$$f''(0) = 2c_2 \quad \text{and so} \quad c_2 = \frac{f''(0)}{2!}$$

From (4), if $x = 0$,

$$f'''(0) = 2 \cdot 3c_3 \quad \text{and so} \quad c_3 = \frac{f'''(0)}{3!}$$

In a similar manner from (5), if $x = 0$,

$$f^{(iv)}(0) = 2 \cdot 3 \cdot 4c_4 \quad \text{and so} \quad c_4 = \frac{f^{(iv)}(0)}{4!}$$

In general,

$$c_n = \frac{f^{(n)}(0)}{n!} \qquad \text{for every positive integer } n \tag{6}$$

Formula (6) also holds when $n = 0$ if we take $f^{(0)}(0)$ to be $f(0)$ and $0! = 1$. So from (1) and (6) the power series of f in x can be written as

$$\sum_{n=0}^{+\infty} \frac{f^{(n)}(0)}{n!} x^n = f(0) + f'(0)x + \frac{f''(0)}{2!} x^2 + \cdots + \frac{f^{(n)}(0)}{n!} x^n + \cdots \tag{7}$$

In a more general sense, consider the function f as a power series in $(x - a)$; that is,

$$f(x) = \sum_{n=0}^{+\infty} c_n(x-a)^n$$
$$= c_0 + c_1(x-a) + c_2(x-a)^2 + \cdots + c_n(x-a)^n + \cdots \tag{8}$$

If the radius of convergence of this series is R, then f is infinitely differentiable on $(a - R, a + R)$. Successive differentiations of the function in (8) give

$$f'(x) = c_1 + 2c_2(x-a) + 3c_3(x-a)^2 + 4c_4(x-a)^3 + \cdots + nc_n(x-a)^{n-1} + \cdots$$
$$f''(x) = 2c_2 + 2\cdot 3c_3(x-a) + 3\cdot 4c_4(x-a)^2 + \cdots + (n-1)nc_n(x-a)^{n-2} + \cdots$$
$$f'''(x) = 2\cdot 3c_3 + 2\cdot 3\cdot 4c_4(x-a) + \cdots + (n-2)(n-1)nc_n(x-a)^{n-3} + \cdots$$

etc. Letting $x = a$ in the power-series representations of f and its derivatives we get

$$c_0 = f(a) \qquad c_1 = f'(a) \qquad c_2 = \frac{f''(a)}{2!} \qquad c_3 = \frac{f'''(a)}{3!}$$

and in general

$$c_n = \frac{f^{(n)}(a)}{n!} \tag{9}$$

From (8) and (9) the power series of f in $(x - a)$ can be written as

$$\sum_{n=0}^{+\infty} \frac{f^{(n)}(a)}{n!}(x-a)^n = f(a) + f'(a)(x-a) + \frac{f''(a)}{2!}(x-a)^2 + \cdots + \frac{f^{(n)}(a)}{n!}(x-a)^n + \cdots \tag{10}$$

The series in (10) is called the **Taylor series** of f at a. The special case of (10), when $a = 0$, is (7), which is called the **Maclaurin series.**

Example 1 Find the Maclaurin series for e^x.

Solution If $f(x) = e^x$, $f^{(n)}(x) = e^x$ for all x; therefore $f^{(n)}(0) = 1$ for all n. So from (7) we have the Maclaurin series for e^x:

$$1 + x + \frac{x^2}{2!} + \frac{x^3}{3!} + \cdots + \frac{x^n}{n!} + \cdots \tag{11}$$

Series (11) is the same as the series for e^x obtained in Example 3 of Sec. 15.8.

Example 2 Find the Taylor series for $\sin x$ at a.

Solution If $f(x) = \sin x$, then $f'(x) = \cos x$, $f''(x) = -\sin x$, $f'''(x) = -\cos x$, $f^{(\text{iv})}(x) = \sin x$, and so forth. Thus, from formula (9), $c_0 = \sin a$, $c_1 = \cos a$, $c_2 = (-\sin a)/2!$, $c_3 = (-\cos a)/3!$, $c_4 = (\sin a)/4!$, and so on. The required Taylor series is obtained from (10), and it is

$$\sin a + \cos a(x-a) - \sin a\frac{(x-a)^2}{2!} - \cos a\frac{(x-a)^3}{3!} + \sin a\frac{(x-a)^4}{4!} + \cdots$$

We can deduce that a power-series representation of a function is unique. That is, if two functions have the same function values in some interval containing the number a, and if both functions have a power-series representation in $(x - a)$, then these series must be the same because the coefficients in the series are obtained from the values of the functions and their derivatives at a. Therefore, if a function has a power-series representation in $(x - a)$, this series must be its Taylor series at a. Hence the Taylor series for a given function does not have to be obtained by using formula (10). Any method that gives a power series in $(x - a)$ representing the function will be the Taylor series of the function at a.

● ILLUSTRATION 1 To find the Taylor series for e^x at a, write $e^x = e^a e^{x-a}$ and then use series (11), where x is replaced by $(x - a)$. Then

$$e^x = e^a\left[1 + (x-a) + \frac{(x-a)^2}{2!} + \frac{(x-a)^3}{3!} + \cdots + \frac{(x-a)^n}{n!} + \cdots\right]$$ ●

● ILLUSTRATION 2 The series for $\ln(1 + x)$ found in Example 3 of Sec. 15.9 can be used to find the Taylor series for $\ln x$ at a ($a > 0$) by writing

$$\ln x = \ln[a + (x-a)] = \ln a + \ln\left(1 + \frac{x-a}{a}\right) \tag{12}$$

Because

$$\ln(1+t) = \sum_{n=1}^{+\infty} (-1)^{n-1}\frac{t^n}{n} \qquad \text{if } -1 < t \le 1$$

then

$$\ln\left(1 + \frac{x-a}{a}\right) = \frac{x-a}{a} - \frac{(x-a)^2}{2a^2} + \frac{(x-a)^3}{3a^3} - \cdots$$

Therefore, from (12),

$$\ln x = \ln a + \frac{x-a}{a} - \frac{(x-a)^2}{2a^2} + \frac{(x-a)^3}{3a^3} - \cdots$$

and the series represents $\ln x$ if $-1 < (x - a)/a \le 1$ or, equivalently, $0 < x \le 2a$. ●

A natural question that arises is: If a function has a Taylor series in $(x - a)$ having radius of convergence $R > 0$, does this series represent the function for all values of x in the interval $(a - R, a + R)$? For most elementary functions the answer is yes. However, there are functions for which the answer is no. The following example shows this.

Example 3 Let f be the function defined by

$$f(x) = \begin{cases} e^{-1/x^2} & \text{if } x \neq 0 \\ 0 & \text{if } x = 0 \end{cases}$$

Find the Maclaurin series for f, and show that it converges for all values of x but that it represents $f(x)$ only when $x = 0$.

Solution To find $f'(0)$, use the definition of a derivative.

$$f'(0) = \lim_{x \to 0} \frac{e^{-1/x^2} - 0}{x - 0} = \lim_{x \to 0} \frac{\frac{1}{x}}{e^{1/x^2}}$$

Because $\lim\limits_{x \to 0} (1/x) = +\infty$ and $\lim\limits_{x \to 0} e^{1/x^2} = +\infty$, L'Hôpital's rule can be used. Therefore

$$f'(0) = \lim_{x \to 0} \frac{-\frac{1}{x^2}}{e^{1/x^2}\left(-\frac{2}{x^3}\right)} = \lim_{x \to 0} \frac{x}{2e^{1/x^2}} = 0$$

By a similar method, using the definition of a derivative and L'Hôpital's rule, we get 0 for every derivative. So $f^{(n)}(0) = 0$ for all n. Therefore the Maclaurin series for the given function is $0 + 0 + 0 + \cdots + 0 + \cdots$. This series converges to 0 for all x; however, if $x \neq 0$, $f(x) = e^{-1/x^2} \neq 0$.

A theorem that gives a test for determining whether a function is represented by its Taylor series is the following.

15.10.1 Theorem Let f be a function such that f and all of its derivatives exist in some interval $(a - r, a + r)$. Then the function is represented by its Taylor series

$$\sum_{n=0}^{+\infty} \frac{f^{(n)}(a)}{n!}(x - a)^n$$

for all x such that $|x - a| < r$ if and only if

$$\lim_{n \to +\infty} R_n(x) = \lim_{n \to +\infty} \frac{f^{(n+1)}(\xi_n)}{(n + 1)!}(x - a)^{n+1} = 0$$

where each ξ_n is between x and a.

Proof In the interval $(a - r, a + r)$, the function f satisfies the hypothesis of Theorem 14.5.1 for which

$$f(x) = P_n(x) + R_n(x) \tag{13}$$

where $P_n(x)$ is the nth-degree Taylor polynomial of f at a and $R_n(x)$ is the remainder, given by

$$R_n(x) = \frac{f^{(n+1)}(\xi_n)}{(n+1)!}(x-a)^{n+1} \tag{14}$$

where each ξ_n is between x and a.

Now $P_n(x)$ is the nth partial sum of the Taylor series of f at a. So if we show that $\lim\limits_{n \to +\infty} P_n(x)$ exists and equals $f(x)$ if and only if $\lim\limits_{n \to +\infty} R_n(x) = 0$, the theorem will be proved. From (13),

$$P_n(x) = f(x) - R_n(x) \tag{15}$$

If $\lim\limits_{n \to +\infty} R_n(x) = 0$, it follows from (15) that

$$\begin{aligned}\lim_{n \to +\infty} P_n(x) &= f(x) - \lim_{n \to +\infty} R_n(x)\\ &= f(x) - 0\\ &= f(x)\end{aligned}$$

Now under the hypothesis that $\lim\limits_{n \to +\infty} P_n(x) = f(x)$ we wish to show that $\lim\limits_{n \to +\infty} R_n(x) = 0$. From (13),

$$R_n(x) = f(x) - P_n(x)$$

and so

$$\begin{aligned}\lim_{n \to +\infty} R_n(x) &= f(x) - \lim_{n \to +\infty} P_n(x)\\ &= f(x) - f(x)\\ &= 0\end{aligned}$$

This proves the theorem. ∎

Theorem 15.10.1 also holds for other forms of the remainder $R_n(x)$ besides the Lagrange form.

It is often difficult to apply Theorem 15.10.1 in practice because the values of ξ_n are arbitrary. However, sometimes an upper bound for $R_n(x)$ can be found, and it may be possible to prove that the limit of the upper bound is zero as $n \to +\infty$. The following limit is helpful in some cases:

$$\lim_{n \to +\infty} \frac{x^n}{n!} = 0 \qquad \text{for all } x \tag{16}$$

This follows from Example 2 of Sec. 15.7, where it was shown that the power series $\sum\limits_{n=0}^{+\infty} x^n/n!$ is convergent for all values of x and hence the limit of its nth term must be zero. In a similar manner, because $\sum\limits_{n=0}^{+\infty} (x-a)^n/n!$ is convergent for all values of x,

$$\lim_{n \to +\infty} \frac{(x-a)^n}{n!} = 0 \qquad \text{for all } x \tag{17}$$

Example 4 Use Theorem 15.10.1 to show that the Maclaurin series for e^x, found in Example 1, represents the function for all values of x.

Solution The Maclaurin series for e^x is series (11) and

$$R_n(x) = \frac{e^{\xi_n}}{(n+1)!} x^{n+1}$$

where each ξ_n is between 0 and x.

We must show that $\lim_{n \to +\infty} R_n = 0$ for all x. There are three cases: $x > 0$, $x < 0$, and $x = 0$.

If $x > 0$, then $0 < \xi_n < x$; hence $e^{\xi_n} < e^x$. So

$$0 < \frac{e^{\xi_n}}{(n+1)!} x^{n+1} < e^x \frac{x^{n+1}}{(n+1)!} \tag{18}$$

From (16) it follows that $\lim_{n \to +\infty} x^{n+1}/(n+1)! = 0$, and so

$$\lim_{n \to +\infty} e^x \frac{x^{n+1}}{(n+1)!} = 0$$

Therefore, from (18) and the squeeze theorem (2.9.1) it follows that $\lim_{n \to +\infty} R_n(x) = 0$.

If $x < 0$, then $x < \xi_n < 0$ and $0 < e^{\xi_n} < 1$. Therefore, if $x^{n+1} > 0$,

$$0 < \frac{e^{\xi_n}}{(n+1)!} x^{n+1} < \frac{x^{n+1}}{(n+1)!}$$

and if $x^{n+1} < 0$,

$$\frac{x^{n+1}}{(n+1)!} < \frac{e^{\xi_n}}{(n+1)!} x^{n+1} < 0$$

In either case, because $\lim_{n \to +\infty} x^{n+1}/(n+1)! = 0$, it follows that $\lim_{n \to +\infty} R_n = 0$.

Finally, if $x = 0$, the series has the sum of 1, which is e^0. Hence series (11) represents e^x for all values of x.

From the results of the above example we can write

$$e^x = \sum_{n=0}^{+\infty} \frac{x^n}{n!} = 1 + x + \frac{x^2}{2!} + \frac{x^3}{3!} + \cdots \qquad \text{for all } x \tag{19}$$

and this agrees with Example 3 of Sec. 15.8.

Example 5 Show that the Taylor series for $\sin x$ at a, found in Example 2, represents the function for all values of x.

Solution We use Theorem 15.10.1; so we must show that

$$\lim_{n \to +\infty} R_n(x) = \lim_{n \to +\infty} \frac{f^{(n+1)}(\xi_n)}{(n+1)!} (x-a)^{n+1} = 0$$

Because $f(x) = \sin x$, $f^{(n+1)}(\xi_n)$ will be one of the following numbers: $\cos \xi_n$, $\sin \xi_n$, $-\cos \xi_n$, or $-\sin \xi_n$. In any case, $|f^{(n+1)}(\xi_n)| \leq 1$. Hence

$$0 < |R_n(x)| \leq \frac{|x-a|^{n+1}}{(n+1)!} \tag{20}$$

From (17), $\lim\limits_{n \to +\infty} |x-a|^{n+1}/(n+1)! = 0$. Thus, by the squeeze theorem (2.9.1) and (20) it follows that $\lim\limits_{n \to +\infty} R_n(x) = 0$.

Example 6 Compute the value of sin 47° accurate to four decimal places.

Solution In Examples 2 and 5 the Taylor series for $\sin x$ at a was obtained. It represents the function for all values of x. The series is

$$\sin x = \sin a + \cos a(x-a) - \sin a\,\frac{(x-a)^2}{2!} - \cos a\,\frac{(x-a)^3}{3!} + \cdots$$

To make $(x - a)$ small we must choose a value of a near the value of x for which the function value is being computed. The sine and cosine of a must also be known. We therefore choose $a = \frac{1}{4}\pi$ and have

$$\sin x = \sin \tfrac{1}{4}\pi + \cos \tfrac{1}{4}\pi(x - \tfrac{1}{4}\pi) - \sin \tfrac{1}{4}\pi\,\frac{(x-\frac{1}{4}\pi)^2}{2!} - \cos \tfrac{1}{4}\pi\,\frac{(x-\frac{1}{4}\pi)^3}{3!} + \cdots \tag{21}$$

Because 47° is equivalent to $\frac{47}{180}\pi$ radians $= (\frac{1}{4}\pi + \frac{1}{90}\pi)$ radians, from Eq. (21),

$$\sin \tfrac{47}{180}\pi = \tfrac{1}{2}\sqrt{2} + \tfrac{1}{2}\sqrt{2}\cdot\tfrac{1}{90}\pi - \tfrac{1}{2}\sqrt{2}\cdot\tfrac{1}{2}(\tfrac{1}{90}\pi)^2 - \tfrac{1}{2}\sqrt{2}\cdot\tfrac{1}{6}(\tfrac{1}{90}\pi)^3 + \cdots$$

$$\approx \tfrac{1}{2}\sqrt{2}(1 + 0.03490 - 0.00061 - 0.000002 + \cdots)$$

Taking $\sqrt{2} = 1.41421$ and using the first three terms of the series we get

$$\sin \tfrac{47}{180}\pi \approx (0.70711)(1.03429) = 0.73136$$

Rounding off to four decimal places gives $\sin 47° \approx 0.7314$. The error introduced by using the first three terms is $R_2(\frac{47}{180}\pi)$, and from (20),

$$\left|R_2\left(\frac{47}{180}\pi\right)\right| \leq \frac{(\frac{1}{90}\pi)^3}{3!} \approx 0.00001$$

The result, then, is accurate to four decimal places.

The following Maclaurin series represent the given function for all values of x:

$$\sin x = \sum_{n=0}^{+\infty} \frac{(-1)^n x^{2n+1}}{(2n+1)!} = x - \frac{x^3}{3!} + \frac{x^5}{5!} - \frac{x^7}{7!} + \cdots \tag{22}$$

$$\cos x = \sum_{n=0}^{+\infty} \frac{(-1)^n x^{2n}}{(2n)!} = 1 - \frac{x^2}{2!} + \frac{x^4}{4!} - \frac{x^6}{6!} + \cdots \tag{23}$$

$$\sinh x = \sum_{n=0}^{+\infty} \frac{x^{2n+1}}{(2n+1)!} = x + \frac{x^3}{3!} + \frac{x^5}{5!} + \frac{x^7}{7!} + \cdots \tag{24}$$

$$\cosh x = \sum_{n=0}^{+\infty} \frac{x^{2n}}{(2n)!} = 1 + \frac{x^2}{2!} + \frac{x^4}{4!} + \frac{x^6}{6!} + \cdots \tag{25}$$

Series (22) is a direct result of Examples 2 and 5 with $a = 0$. You are asked to verify series (23), (24), and (25) in Exercises 1, 2, and 3.

Example 7 Evaluate

$$\int_{1/2}^{1} \frac{\sin x}{x}\, dx$$

accurate to six decimal places.

Solution An antiderivative of the integrand in terms of elementary functions cannot be found. However, from series (22),

$$\begin{aligned}\frac{\sin x}{x} &= \frac{1}{x} \cdot \sin x \\ &= \frac{1}{x}\left(x - \frac{x^3}{3!} + \frac{x^5}{5!} - \frac{x^7}{7!} + \frac{x^9}{9!} - \cdots\right) \\ &= 1 - \frac{x^2}{3!} + \frac{x^4}{5!} - \frac{x^6}{7!} + \frac{x^8}{9!} - \cdots\end{aligned}$$

which is true for all $x \neq 0$. Using term-by-term integration we get

$$\begin{aligned}\int_{1/2}^{1} \frac{\sin x}{x}\, dx &= x - \frac{x^3}{3 \cdot 3!} + \frac{x^5}{5 \cdot 5!} - \frac{x^7}{7 \cdot 7!} + \frac{x^9}{9 \cdot 9!} - \cdots \Big]_{1/2}^{1} \\ &\approx (1 - 0.0555555 + 0.0016667 - 0.0000283 + 0.0000003 - \cdots) \\ &\quad -(0.5 - 0.0069444 + 0.0000521 - 0.0000002 + \cdots)\end{aligned}$$

In each set of parentheses there is a convergent alternating series with $|u_{n+1}| < |u_n|$. In the first set of parentheses we use the first four terms because the error obtained is less than 0.0000003. In the second set of parentheses we use the first three terms where the error obtained is less than 0.0000002. Doing the arithmetic and rounding off to six decimal places we get

$$\int_{1/2}^{1} \frac{\sin x}{x}\, dx \approx 0.452975$$

EXERCISES 15.10

1. Prove that the series
$$\sum_{n=0}^{+\infty} \frac{(-1)^n x^{2n}}{(2n)!}$$
represents $\cos x$ for all values of x.

2. Prove that the series
$$\sum_{n=0}^{+\infty} \frac{x^{2n+1}}{(2n+1)!}$$
represents $\sinh x$ for all values of x.

3. Prove that the series
$$\sum_{n=0}^{+\infty} \frac{x^{2n}}{(2n)!}$$
represents $\cosh x$ for all values of x.

4. Obtain the Maclaurin series for the cosine function by differentiating the Maclaurin series for the sine function. Also obtain the Maclaurin series for the sine function by differentiating the one for the cosine function.

5. Obtain the Maclaurin series for the hyperbolic sine function by differentiating the Maclaurin series for the hyperbolic cosine function. Also differentiate the Maclaurin series for the hyperbolic sine function to obtain the one for the hyperbolic cosine function.

6. Find the Taylor series for e^x at 3 by using the Maclaurin series for e^x.

7. Use the Maclaurin series for $\ln(1 + x)$ to find the Taylor series for $\ln x$ at 2.

8. Given $\ln 2 = 0.6931$, use the series obtained in Exercise 7 to find $\ln 3$ accurate to four decimal places.

In Exercises 9 through 14, find a power-series representation for the given function at the number a, and determine its radius of convergence.

9. $f(x) = \ln(x + 1)$; $a = 1$
10. $f(x) = \sqrt[3]{x}$; $a = 1$
11. $f(x) = \sqrt{x}$; $a = 4$
12. $f(x) = \dfrac{1}{x}$; $a = 1$
13. $f(x) = \cos x$; $a = \frac{1}{3}\pi$
14. $f(x) = 2^x$; $a = 0$

15. Find the Maclaurin series for $\sin^2 x$. (*Hint:* Use $\sin^2 x = \frac{1}{2}(1 - \cos 2x)$.)

16. Find the Maclaurin series for $\cos^2 x$. (*Hint:* Use $\cos^2 x = \frac{1}{2}(1 + \cos 2x)$.)

17. (a) Find the first three nonzero terms of the Maclaurin series for $\tan x$. (b) Use the result of part (a) and term-by-term differentiation to find the first three nonzero terms of the Maclaurin series for $\sec^2 x$. (c) Use the result of part (a) and term-by-term integration to find the first three nonzero terms of the Maclaurin series for $\ln|\sec x|$.

18. (a) Find the first three nonzero terms of the Taylor series for $\cot x$ at $\frac{1}{2}\pi$. (b) Use the result of part (a) and term-by-term integration to find the first three nonzero terms of the Taylor series for $\ln \sin x$ at $\frac{1}{2}\pi$.

In Exercises 19 through 24, use a power series to compute the value of the given quantity to the indicated accuracy.

19. $\cos 58°$; four decimal places.
20. $\sqrt[5]{e}$; four decimal places.
21. $\sqrt[5]{30}$; five decimal places.
22. $\sinh \frac{1}{2}$; five decimal places.
23. $\ln(0.8)$; four decimal places.
24. $\sqrt[3]{29}$; three decimal places.

25. Compute the value of e correct to seven decimal places, and prove that your answer has the required accuracy.

In Exercises 26 through 31, compute the value of the definite integral accurate to three decimal places.

26. $\displaystyle\int_0^1 \sqrt{x}e^{-x^2}\,dx$
27. $\displaystyle\int_0^{1/2} \sin x^2\,dx$
28. $\displaystyle\int_0^1 \cos\sqrt{x}\,dx$
29. $\displaystyle\int_0^{0.1} \ln(1 + \sin x)\,dx$

30. $\displaystyle\int_0^{1/3} f(x)\,dx$, where $f(x) = \begin{cases} \dfrac{\sin x}{x} & \text{if } x \neq 0 \\ 1 & \text{if } x = 0 \end{cases}$

31. $\displaystyle\int_0^1 g(x)\,dx$, where $g(x) = \begin{cases} \dfrac{1 - \cos x}{x} & \text{if } x \neq 0 \\ 0 & \text{if } x = 0 \end{cases}$

32. The function E defined by

$$E(x) = \frac{2}{\sqrt{\pi}} \int_0^x e^{-t^2}\, dt$$

is called the *error function,* and it is important in mathematical statistics. Find the Maclaurin series for the error function.

33. Determine a_n $(n = 0, 1, 2, 3, 4)$ so that the polynomial $f(x) = 3x^4 - 17x^3 + 35x^2 - 32x + 17$ is written in the form

$$f(x) = a_4(x-1)^4 + a_3(x-1)^3 + a_2(x-1)^2 + a_1(x-1) + a_0$$

34. Determine $a_n (n = 0, 1, 2, 3)$ so that the polynomial $f(x) = 4x^3 - 5x^2 + 2x - 3$ is written in the form

$$f(x) = a_3(x+2)^3 + a_2(x+2)^2 + a_1(x+2) + a_0$$

15.11 THE BINOMIAL SERIES

In elementary algebra you learned that the binomial theorem expresses $(a + b)^m$ as a sum of powers of a and b, where m is a positive integer, as follows:

$$(a+b)^m = a^m + ma^{m-1}b + \frac{m(m-1)}{2!}a^{m-2}b^2 + \cdots + \frac{m(m-1)\cdot \ldots \cdot (m-k+1)}{k!}a^{m-k}b^k + \cdots + b^m$$

We now take $a = 1$ and $b = x$ and apply the binomial theorem to the expression $(1 + x)^m$, where m is not a positive integer. We obtain the power series

$$1 + mx + \frac{m(m-1)}{2!}x^2 + \frac{m(m-1)(m-2)}{3!}x^3 + \cdots + \frac{m(m-1)(m-2)\cdot \ldots \cdot (m-n+1)}{n!}x^n + \cdots \tag{1}$$

Series (1) is the Maclaurin series for $(1 + x)^m$. It is called a **binomial series.** To find the radius of convergence of series (1), apply the ratio test and get

$$\lim_{n\to+\infty}\left|\frac{u_{n+1}}{u_n}\right| = \lim_{n\to+\infty}\left|\frac{\dfrac{m(m-1)\cdot \ldots \cdot (m-n+1)(m-n)}{(n+1)!}x^{n+1}}{\dfrac{m(m-1)\cdot \ldots \cdot (m-n+1)}{n!}x^n}\right|$$

$$= \lim_{n\to+\infty}\left|\frac{m-n}{n+1}\right||x|$$

$$= \lim_{n\to+\infty}\left|\frac{\dfrac{m}{n} - 1}{1 + \dfrac{1}{n}}\right||x|$$

$$= |x|$$

So the series is convergent if $|x| < 1$. We now prove that series (1) represents $(1 + x)^m$ for all real numbers m if x is in the open interval $(-1, 1)$. This is not done by calculating $R_n(x)$ and showing that its limit is zero because this is quite difficult, as you will soon see if you attempt to do so. Instead the following method is used. Let

$$f(x) = 1 + \sum_{n=1}^{+\infty} \frac{m(m-1) \cdot \ldots \cdot (m-n+1)}{n!} x^n \qquad |x| < 1 \tag{2}$$

We wish to show that $f(x) = (1 + x)^m$, where $|x| < 1$. By Theorem 15.8.3,

$$f'(x) = \sum_{n=1}^{+\infty} \frac{m(m-1) \cdot \ldots \cdot (m-n+1)}{(n-1)!} x^{n-1} \qquad |x| < 1 \tag{3}$$

Multiplying on both sides of (3) by x we get from Theorem 15.3.8

$$xf'(x) = \sum_{n=1}^{+\infty} \frac{m(m-1) \cdot \ldots \cdot (m-n+1)}{(n-1)!} x^n \tag{4}$$

By rewriting the right side of (3),

$$f'(x) = m + \sum_{n=2}^{+\infty} \frac{m(m-1) \cdot \ldots \cdot (m-n+1)}{(n-1)!} x^{n-1} \tag{5}$$

The summation in (5) is rewritten with the lower limit decreased by 1 and n replaced by $n + 1$. Then

$$f'(x) = m + \sum_{n=1}^{+\infty} (m-n) \frac{m(m-1) \cdot \ldots \cdot (m-n+1)}{n!} x^n \tag{6}$$

In (4) the numerator and the denominator are multiplied by n to obtain

$$xf'(x) = \sum_{n=1}^{+\infty} n \frac{m(m-1) \cdot \ldots \cdot (m-n+1)}{n!} x^n \tag{7}$$

Because the series in (6) and (7) are absolutely convergent for $|x| < 1$, then by Theorem 15.3.8 they can be added term by term, and the resulting series will be absolutely convergent for $|x| < 1$. Therefore, from the addition,

$$(1 + x)f'(x) = m\left[1 + \sum_{n=1}^{+\infty} \frac{m(m-1) \cdot \ldots \cdot (m-n+1)}{n!} x^n\right]$$

Because by (2) the expression in brackets is $f(x)$, then

$$(1 + x)f'(x) = mf(x)$$

or, equivalently,

$$\frac{f'(x)}{f(x)} = \frac{m}{1 + x}$$

The left side of the above equation is $D_x[\ln f(x)]$; thus

$$D_x[\ln f(x)] = \frac{m}{1+x}$$

However, we also know that

$$D_x[\ln(1+x)^m] = \frac{m}{1+x}$$

Because $\ln f(x)$ and $\ln(1+x)^m$ have the same derivative, they differ by a constant. Hence

$$\ln f(x) = \ln(1+x)^m + C$$

From (2), $f(0) = 1$. Therefore $C = 0$; so

$$f(x) = (1+x)^m$$

We have proved the general binomial theorem, which is now stated.

15.11.1 Theorem
Binomial Theorem

If m is any real number, then

$$(1+x)^m = 1 + \sum_{n=1}^{+\infty} \frac{m(m-1)(m-2)\cdot \ldots \cdot(m-n+1)}{n!} x^n \qquad (8)$$

for all values of x such that $|x| < 1$.

If m is a positive integer, the binomial series will terminate after a finite number of terms.

Example 1 Express

$$\frac{1}{\sqrt{1+x}}$$

as a power series in x.

Solution From Theorem 15.11.1, when $|x| < 1$,

$$\begin{aligned}(1+x)^{-1/2} &= 1 - \frac{1}{2}x + \frac{(-\frac{1}{2})(-\frac{1}{2}-1)}{2!}x^2 + \frac{(-\frac{1}{2})(-\frac{1}{2}-1)(-\frac{1}{2}-2)}{3!}x^3 \\ &\quad + \cdots + \frac{(-\frac{1}{2})(-\frac{3}{2})(-\frac{5}{2})\cdot \ldots \cdot(-\frac{1}{2}-n+1)}{n!}x^n + \cdots \\ &= 1 - \frac{1}{2}x + \frac{1\cdot 3}{2^2\cdot 2!}x^2 - \frac{1\cdot 3\cdot 5}{2^3\cdot 3!}x^3 + \cdots \\ &\quad + (-1)^n\frac{1\cdot 3\cdot 5\cdot \ldots \cdot(2n-1)}{2^n n!}x^n + \cdots\end{aligned}$$

Example 2 From the result of Example 1 obtain a binomial series for $(1 - x^2)^{-1/2}$, and use it to find a power series for $\sin^{-1} x$.

Solution Replace x by $-x^2$ in the series for $(1 + x)^{-1/2}$ and get for $|x| < 1$

$$(1 - x^2)^{-1/2} = 1 + \frac{1}{2}x^2 + \frac{1 \cdot 3}{2^2 \cdot 2!}x^4 + \frac{1 \cdot 3 \cdot 5}{2^3 \cdot 3!}x^6 + \cdots + \frac{1 \cdot 3 \cdot 5 \cdot \ldots \cdot (2n-1)}{2^n n!}x^{2n} + \cdots$$

Apply Theorem 15.9.1 and integrate term by term to obtain

$$\int_0^x \frac{dt}{\sqrt{1-t^2}} = x + \frac{1}{2} \cdot \frac{x^3}{3} + \frac{1 \cdot 3}{2^2 \cdot 2!} \cdot \frac{x^5}{5} + \frac{1 \cdot 3 \cdot 5}{2^3 \cdot 3!} \cdot \frac{x^7}{7} + \cdots + \frac{1 \cdot 3 \cdot 5 \cdot \ldots \cdot (2n-1)}{2^n n!} \cdot \frac{x^{2n+1}}{2n+1} + \cdots$$

Therefore

$$\sin^{-1} x = x + \sum_{n=1}^{+\infty} \frac{1 \cdot 3 \cdot 5 \cdot \ldots \cdot (2n-1)}{2^n n!} \cdot \frac{x^{2n+1}}{2n+1} \qquad \text{for } |x| < 1$$

Example 3 Compute the value of $\sqrt[3]{25}$ accurate to three decimal places by using the binomial series for $(1 + x)^{1/3}$.

Solution From Eq. (8),

$$(1 + x)^{1/3} = 1 + \frac{1}{3}x + \left(\frac{1}{3}\right)\left(-\frac{2}{3}\right)\frac{x^2}{2!} + \left(\frac{1}{3}\right)\left(-\frac{2}{3}\right)\left(-\frac{5}{3}\right)\frac{x^3}{3!} + \cdots \qquad (9)$$

$$\text{if } |x| < 1$$

We can write

$$\sqrt[3]{25} = \sqrt[3]{27}\,\sqrt[3]{\tfrac{25}{27}} = 3\sqrt[3]{1 - \tfrac{2}{27}} = 3(1 - \tfrac{2}{27})^{1/3} \qquad (10)$$

From Eq. (9) with $x = -\frac{2}{27}$,

$$\left(1 - \frac{2}{27}\right)^{1/3} = 1 + \frac{1}{3}\left(-\frac{2}{27}\right) - \frac{2}{3^2 \cdot 2!}\left(-\frac{2}{27}\right)^2 + \frac{2 \cdot 5}{3^3 \cdot 3!}\left(-\frac{2}{27}\right)^3 + \cdots$$

$$\approx 1 - 0.0247 - 0.0006 - 0.00003 - \cdots \qquad (11)$$

If the first three terms of the above series are used, it follows from (14) of Sec. 15.10 that the remainder is

$$R_2\left(-\frac{2}{27}\right) = \frac{f'''(\xi_2)}{3!}\left(-\frac{2}{27}\right)^3$$

$$= \left(\frac{1}{3}\right)\left(-\frac{2}{3}\right)\left(-\frac{5}{3}\right)\left(\frac{1}{3!}\right)(1 + \xi_2)^{(1/3)-3}\left(-\frac{2}{27}\right)^3$$

where $-\frac{2}{27} < \xi_2 < 0$. Therefore

$$\left|R_2\left(-\frac{2}{27}\right)\right| = \left(\frac{2 \cdot 5}{3^3 \cdot 3!}\right)\frac{(1+\xi_2)^{1/3}}{(1+\xi_2)^3}\left(\frac{2}{27}\right)^3 \tag{12}$$

Because $-\frac{2}{27} < \xi_2 < 0$, it follows that

$$(1+\xi_2)^{1/3} < 1^{1/3} = 1 \quad \text{and} \quad \frac{1}{(1+\xi_2)^3} < \frac{1}{(\frac{25}{27})^3} \tag{13}$$

Furthermore,

$$\frac{2 \cdot 5}{3^3 \cdot 3!} = \frac{2 \cdot 5}{3 \cdot 3 \cdot 3 \cdot 1 \cdot 2 \cdot 3} = \frac{2}{3} \cdot \frac{5}{6} \cdot \frac{1}{9} < \frac{1}{9} \tag{14}$$

and so from inequalities (13) and (14) in (12),

$$\left|R_2\left(-\frac{2}{27}\right)\right| < \frac{1}{9} \cdot \frac{1}{(\frac{25}{27})^3} \cdot \left(\frac{2}{27}\right)^3 = \frac{8}{140{,}625} < 0.00006$$

Using, then, the first three terms of series (11) gives

$$\left(1 - \frac{2}{27}\right)^{1/3} \approx 0.9747$$

with an error less than 0.00006. From (10) we obtain

$$\sqrt[3]{25} \approx 3(0.9747) = 2.9241$$

with an error less than $3(0.00006) = 0.00018$. Rounding off to three decimal places gives $\sqrt[3]{25} \approx 2.924$.

EXERCISES 15.11

In Exercises 1 through 10, use a binomial series to find the Maclaurin series for the given function. Determine the radius of convergence of the resulting series.

1. $f(x) = \sqrt{1+x}$
2. $f(x) = (3-x)^{-2}$
3. $f(x) = (4+x)^{-1/2}$
4. $f(x) = \sqrt[3]{8+x}$
5. $f(x) = \sqrt[3]{1-x^3}$
6. $f(x) = (4+x^2)^{-1}$
7. $f(x) = (9+x^4)^{-1/2}$
8. $f(x) = \dfrac{x}{\sqrt{1-x}}$
9. $f(x) = \dfrac{x^2}{\sqrt{1+x}}$
10. $f(x) = \dfrac{x}{\sqrt[3]{1+x^2}}$

In Exercises 11 through 16, compute the value of the given quantity accurate to three decimal places by using a binomial series.

11. $\sqrt{24}$
12. $\sqrt{51}$
13. $\sqrt[4]{630}$
14. $\sqrt[3]{66}$
15. $\dfrac{1}{\sqrt[3]{128}}$
16. $\dfrac{1}{\sqrt[5]{31}}$

17. Integrate term by term from 0 to x the binomial series for $(1+t^2)^{-1/2}$ to obtain the Maclaurin series for $\sinh^{-1} x$. Determine the radius of convergence.
18. Use a method similar to that of Exercise 17 to find the Maclaurin series for $\tanh^{-1} x$. Determine the radius of convergence.

In Exercises 19 through 24, compute the value of the definite integral accurate to four decimal places.

19. $\int_0^1 \sqrt[3]{8 + x^2}\, dx$ **20.** $\int_0^{1/2} \sqrt{1 - x^3}\, dx$ **21.** $\int_0^{1/2} \frac{dx}{\sqrt{1 + x^4}}$ **22.** $\int_0^{1/3} \frac{dx}{\sqrt[3]{x^2 + 1}}$ **23.** $\int_0^{1/2} \frac{dx}{\sqrt{1 - x^3}}$

24. $\int_0^{1/2} f(x)\, dx$, where $f(x) = \begin{cases} \dfrac{\sin^{-1} x}{x} & \text{if } x \neq 0 \\ 1 & \text{if } x = 0 \end{cases}$

25. Find the Maclaurin series for $\int_0^x \frac{t^p}{\sqrt{1 - t^2}}\, dt$ if $p \geq 0$. Determine the radius of convergence of the series.

REVIEW EXERCISES FOR CHAPTER 15

In Exercises 1 through 8, write the first four numbers of the sequence and find the limit of the sequence, if it exists.

1. $\left\{\frac{3n}{n + 2}\right\}$ **2.** $\left\{\frac{(-1)^{n-1}}{(n + 1)^2}\right\}$ **3.** $\left\{\frac{n^2 - 1}{n^2 + 1}\right\}$ **4.** $\left\{\frac{n + 3n^2}{4 + 2n^3}\right\}$

5. $\{2 + (-1)^n\}$ **6.** $\left\{\frac{n^2}{\ln(n + 1)}\right\}$ **7.** $\left\{\left(1 + \frac{1}{n}\right)^{2n}\right\}$ **8.** $\left\{\frac{(n + 2)^2}{n + 4} - \frac{(n + 2)^2}{n}\right\}$

In Exercises 9 and 10, find the first four elements of the sequence of partial sums $\{s_n\}$, and find a formula for s_n in terms of n. Also determine if the infinite series is convergent or divergent; if it is convergent, find its sum.

9. $\sum_{n=1}^{+\infty} \frac{3}{4^{n+1}}$ **10.** $\sum_{n=1}^{+\infty} \ln\left(\frac{2n - 1}{2n + 1}\right)$

In Exercises 11 through 17, determine if the series is convergent or divergent. If the series is convergent, find its sum.

11. $\sum_{n=1}^{+\infty} \left(\frac{3}{4}\right)^n$ **12.** $\sum_{n=1}^{+\infty} e^{-2n}$ **13.** $\sum_{n=1}^{+\infty} \frac{n - 1}{n + 1}$ **14.** $\sum_{n=0}^{+\infty} [(-1)^n + (-1)^{n+1}]$ **15.** $\sum_{n=0}^{+\infty} \sin^n \frac{1}{3}\pi$ **16.** $\sum_{n=0}^{+\infty} \cos^n \frac{1}{3}\pi$

17. $\sum_{n=1}^{+\infty} \frac{1}{(3n - 1)(3n + 2)}$ (*Hint:* To find the sum, first find the sequence of partial sums.)

In Exercises 18 through 32, determine if the series is convergent or divergent.

18. $\sum_{n=1}^{+\infty} \frac{1}{(2n + 1)^3}$ **19.** $\sum_{n=1}^{+\infty} \frac{2}{n^2 + 6n}$ **20.** $\sum_{n=1}^{+\infty} \frac{3 + \sin n}{n^2}$ **21.** $\sum_{n=1}^{+\infty} \cos\left(\frac{\pi}{2n^2 - 1}\right)$ **22.** $\sum_{n=1}^{+\infty} \frac{n}{\sqrt{3n + 2}}$

23. $\sum_{n=1}^{+\infty} \frac{(n!)^2}{(2n)!}$ **24.** $\sum_{n=1}^{+\infty} \frac{(-1)^{n+1}}{1 + \sqrt{n}}$ **25.** $\sum_{n=1}^{+\infty} (-1)^n \ln \frac{1}{n}$ **26.** $\sum_{n=1}^{+\infty} \frac{\ln n}{n^2}$ **27.** $\sum_{n=2}^{+\infty} \frac{1}{n(\ln n)^2}$

28. $\sum_{n=0}^{+\infty} \frac{n!}{10^n}$ **29.** $\sum_{n=1}^{+\infty} \left(\frac{2}{5n} - \frac{3}{2n}\right)$ **30.** $\sum_{n=1}^{+\infty} \frac{|\sec n|}{n^{3/4}}$ **31.** $\sum_{n=1}^{+\infty} \frac{1}{1 + 2 \ln n}$ **32.** $\sum_{n=1}^{+\infty} n3^{-n^2}$

33. Express the nonterminating repeating decimal 1.324 24 24 . . . as a common fraction.

34. A ball is dropped from a height of 18 m. Each time it strikes the ground, it bounces back to a height of two-thirds the distance from which it fell. Find the total distance traveled by the ball before it comes to rest.

In Exercises 35 through 42, determine if the given series is absolutely convergent, conditionally convergent, or divergent. Prove your answer.

35. $\sum_{n=0}^{+\infty} (-1)^n \frac{n^2}{3^n}$ **36.** $\sum_{n=0}^{+\infty} (-1)^n \frac{5^{2n+1}}{(2n+1)!}$ **37.** $\sum_{n=1}^{+\infty} (-1)^{n-1} \frac{1}{(n+1)^{3/4}}$

38. $\sum_{n=1}^{+\infty} (-1)^{n-1} \frac{6^n}{5^{n+1}}$ **39.** $\sum_{n=1}^{+\infty} (-1)^n \frac{n!}{10n}$ **40.** $\sum_{n=1}^{+\infty} (-1)^n \frac{\sqrt{2n-1}}{n}$

41. $\sum_{n=1}^{+\infty} c_n$, where $c_n = \begin{cases} -\frac{1}{n} & \text{if } n \text{ is a perfect square} \\ \frac{1}{n^2} & \text{if } n \text{ is not a perfect square} \end{cases}$

42. $\sum_{n=1}^{+\infty} c_n$, where $c_n = \begin{cases} -\frac{1}{n} & \text{if } \frac{1}{4}n \text{ is an integer} \\ \frac{1}{n^2} & \text{if } \frac{1}{4}n \text{ is not an integer} \end{cases}$

In Exercises 43 through 52, find the interval of convergence of the given power series.

43. $\sum_{n=1}^{+\infty} \frac{x^n}{\sqrt{n}}$ **44.** $\sum_{n=1}^{+\infty} \frac{(x-2)^n}{n}$ **45.** $\sum_{n=1}^{+\infty} \frac{x^n}{3^n(n^2+n)}$ **46.** $\sum_{n=0}^{+\infty} \frac{x^n}{2^n}$ **47.** $\sum_{n=0}^{+\infty} \frac{n!}{2^n}(x-3)^n$

48. $\sum_{n=1}^{+\infty} \frac{(-1)^{n-1}x^{2n-1}}{(2n-1)!}$ **49.** $\sum_{n=1}^{+\infty} \frac{n^2}{6^n}(x+1)^n$ **50.** $\sum_{n=1}^{+\infty} n(2x-1)^n$ **51.** $\sum_{n=1}^{+\infty} \frac{(-1)^{n-1}(x-1)^n}{n2^n}$ **52.** $\sum_{n=1}^{+\infty} n^n x^n$

In Exercises 53 through 62, use a power series to compute the value of the given quantity accurate to four decimal places.

53. $\tan^{-1} \frac{1}{5}$ **54.** $\sin 0.3$ **55.** $\sqrt[3]{130}$ **56.** $\sin^{-1} 1$ **57.** $\cos 3°$

58. $\sqrt[4]{e}$ **59.** $\ln 5$ **60.** $\int_0^1 \cos x^3\, dx$ **61.** $\int_0^{1/4} \sqrt{x} \sin x\, dx$ **62.** $\int_0^{1/2} \frac{dx}{1+x^5}$

In Exercises 63 through 65, find the Maclaurin series for the given function, and find its interval of convergence.

63. $f(x) = a^x (a > 0)$ **64.** $f(x) = \frac{1}{2-x}$ **65.** $f(x) = \sin^3 x$

In Exercises 66 through 69, find the Taylor series for the given function at the given number.

66. $f(x) = e^{x-2}$; at 2 **67.** $f(x) = \sin 3x$; at $-\frac{1}{3}\pi$ **68.** $f(x) = \frac{1}{x}$; at 2 **69.** $f(x) = \ln|x|$; at -1

Exercises 70 through 74 pertain to the functions J_0 and J_1 defined by power series as follows:

$$J_0(x) = \sum_{n=0}^{+\infty} (-1)^n \frac{x^{2n}}{n!n!2^{2n}} \qquad J_1(x) = \sum_{n=0}^{+\infty} (-1)^n \frac{x^{2n+1}}{n!(n+1)!2^{2n+1}}$$

The functions J_0 and J_1 are called Bessel functions of the first kind *of orders zero and one, respectively.*

70. Show that both J_0 and J_1 converge for all real values of x.

71. Show that $J_0'(x) = -J_1(x)$.

72. Show that $y = J_0(x)$ is a solution of the differential equation

$$x\frac{d^2y}{dx^2} + \frac{dy}{dx} + xy = 0$$

73. Show that $y = J_1(x)$ is a solution of the differential equation

$$x^2\frac{d^2y}{dx^2} + x\frac{dy}{dx} + (x^2-1)y = 0$$

74. Show that $D_x(xJ_1(x)) = xJ_0(x)$.

CHAPTER 16 Vectors in the Plane and Parametric Equations

16.1 VECTORS IN THE PLANE

The applications of mathematics to physics and engineering are often concerned with quantities that possess both *magnitude* and *direction;* examples of these are *force, velocity, acceleration,* and *displacement.* Such quantities may be represented geometrically by a *directed line segment.* Physicists and engineers refer to a directed line segment as a *vector,* and the quantities that have both magnitude and direction are called **vector quantities.** The study of vectors is called **vector analysis.**

The approach to vector analysis can be on either a geometric or an analytic basis. If the geometric approach is taken, we first define a directed line segment as a line segment from a point P to a point Q and denote this directed line segment by $\overrightarrow{PQ}$. The point P is called the **initial point,** and the point Q is called the **terminal point.** Then two directed line segments $\overrightarrow{PQ}$ and $\overrightarrow{RS}$ are said to be equal if they have the same *length* and *direction,* and we write $\overrightarrow{PQ} = \overrightarrow{RS}$ (see Fig. 16.1.1). The directed line segment PQ is called the **vector** from P to Q. A vector is denoted by a single letter, set in boldface type, such as **A**. In some books, a letter in lightface type, with an arrow above it, is used to indicate a vector, for example, $\vec{A}$.

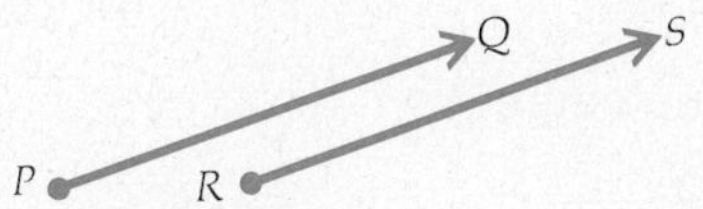

$\overrightarrow{PQ} = \overrightarrow{RS}$

FIGURE 16.1.1

Continuing with the geometric approach to vector analysis, note that if the directed line segment $\overrightarrow{PQ}$ is the vector **A**, and $\overrightarrow{PQ} = \overrightarrow{RS}$, the directed line segment $\overrightarrow{RS}$ is also the vector **A**. Then a vector is considered to remain unchanged if it is moved parallel to itself. With this interpretation of a vector, we can assume for convenience that every vector has its initial point at some fixed reference point. By taking this point as the origin of a rectangular cartesian coordinate system, a vector can be defined analytically in terms of real numbers. Such a definition permits the study of vector analysis from a purely mathematical viewpoint.

In this book we use the analytic approach; however, the geometric interpretation is used for illustrative purposes. A vector in the plane is denoted by an ordered pair of real numbers and the notation $\langle x, y\rangle$ is used instead of (x, y) to avoid confusing the notation for a vector with the notation for a point. V_2 is the set of all such ordered pairs. Following is the formal definition.

16.1.1 Definition A **vector in the plane** is an ordered pair of real numbers $\langle x, y\rangle$. The numbers x and y are called the **components** of the vector $\langle x, y\rangle$.

There is a one-to-one correspondence between the vectors $\langle x, y\rangle$ in the plane and the points (x, y) in the plane. Let the vector **A** be the ordered pair of real numbers $\langle a_1, a_2\rangle$. If A is the point (a_1, a_2), then the vector **A** may be represented geometrically by the directed line segment $\overrightarrow{OA}$. Such a directed line segment is called a **representation** of vector **A**. Any directed line segment that is equal to $\overrightarrow{OA}$ is also a representation of vector **A**. The particular representation of a vector that has its initial point at the origin is called the **position representation** of the vector.

● ILLUSTRATION 1 The vector $\langle 2, 3\rangle$ has as its position representation the directed line segment from the origin to the point (2, 3). The representation of the vector $\langle 2, 3\rangle$ whose initial point is (h, k) has as its terminal point $(h + 2, k + 3)$; refer to Fig. 16.1.2. ●

The vector $\langle 0, 0\rangle$ is called the **zero vector,** and it is denoted by **0**; that is,

$$\mathbf{0} = \langle 0, 0\rangle$$

Any point is a representation of the zero vector.

16.1.2 Definition The **magnitude** of a vector is the length of any of its representations, and the **direction** of a nonzero vector is the direction of any of its representations.

The magnitude of the vector **A** is denoted by $|\mathbf{A}|$.

16.1.3 Theorem If **A** is the vector $\langle a_1, a_2\rangle$, then $|\mathbf{A}| = \sqrt{a_1{}^2 + a_2{}^2}$

Proof Because by Definition 16.1.2 $|\mathbf{A}|$ is the length of any of the representations of **A**, then $|\mathbf{A}|$ will be the length of the position representation of **A**, which is the distance from the origin to the point (a_1, a_2). So from the formula for the distance between two points,

$$|\mathbf{A}| = \sqrt{(a_1 - 0)^2 + (a_2 - 0)^2} = \sqrt{a_1{}^2 + a_2{}^2}$$ ■

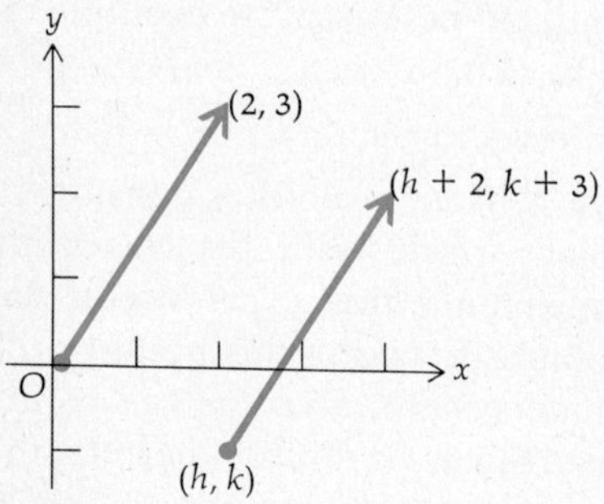

FIGURE 16.1.2

Observe that $|\mathbf{A}|$ is a nonnegative number and not a vector. From Theorem 16.1.3 it follows that $|\mathbf{0}| = 0$.

● ILLUSTRATION 2 If $\mathbf{A} = \langle -3, 5\rangle$, then

$$|\mathbf{A}| = \sqrt{(-3)^2 + 5^2} = \sqrt{34}$$ ●

The direction of any nonzero vector is given by the radian measure θ of the angle from the positive x axis counterclockwise to the position representation of the vector: $0 \le \theta < 2\pi$. So if $\mathbf{A} = \langle a_1, a_2\rangle$, then $\tan\theta = a_2/a_1$ if $a_1 \neq 0$. If $a_1 = 0$ and $a_2 > 0$, then $\theta = \frac{1}{2}\pi$. If $a_1 = 0$ and $a_2 < 0$, then $\theta = \frac{3}{2}\pi$. Figures 16.1.3, 16.1.4, and 16.1.5 show the angle of radian measure θ for specific vectors $\langle a_1, a_2\rangle$ whose position representations are drawn.

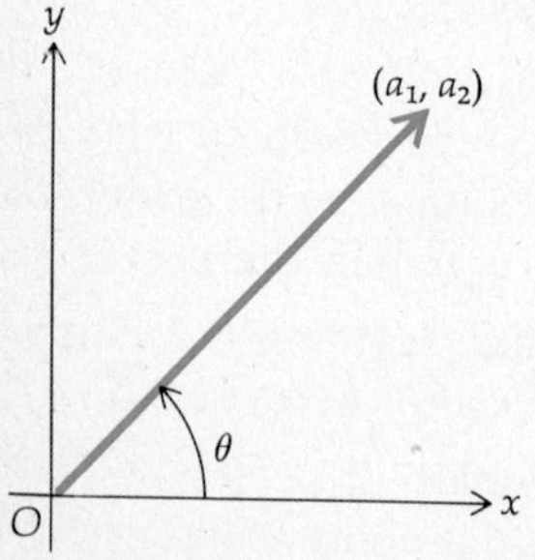

FIGURE 16.1.3

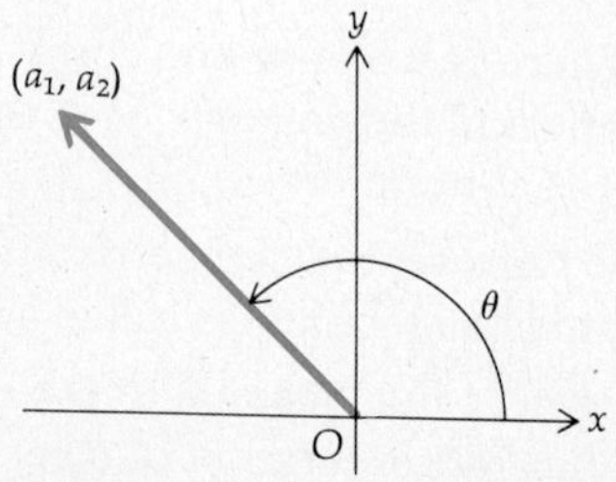

FIGURE 16.1.4

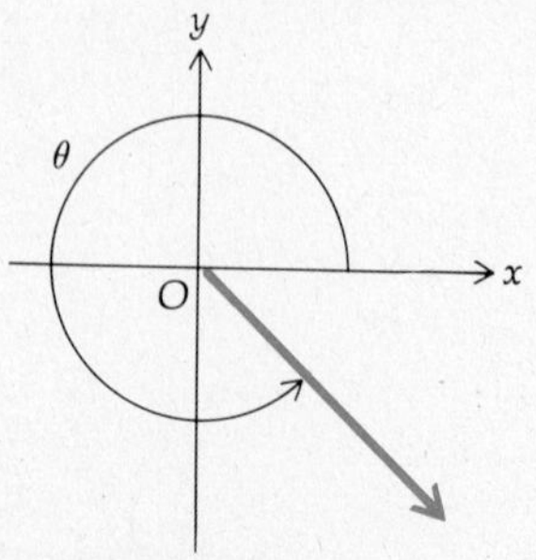

FIGURE 16.1.5

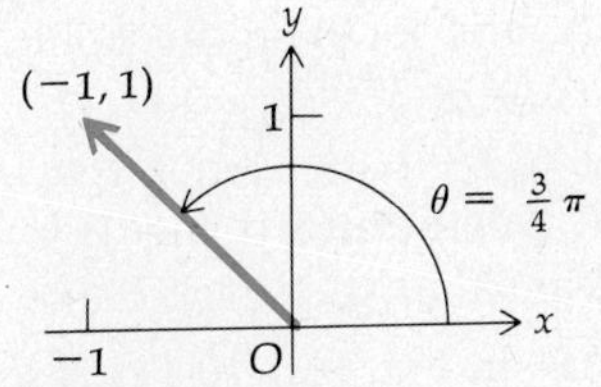

FIGURE 16.1.6

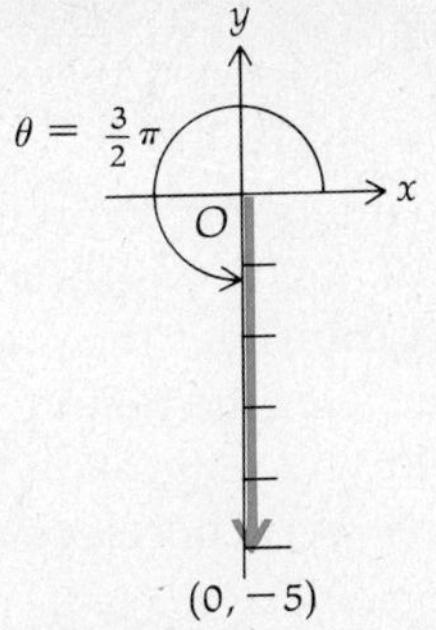

FIGURE 16.1.7

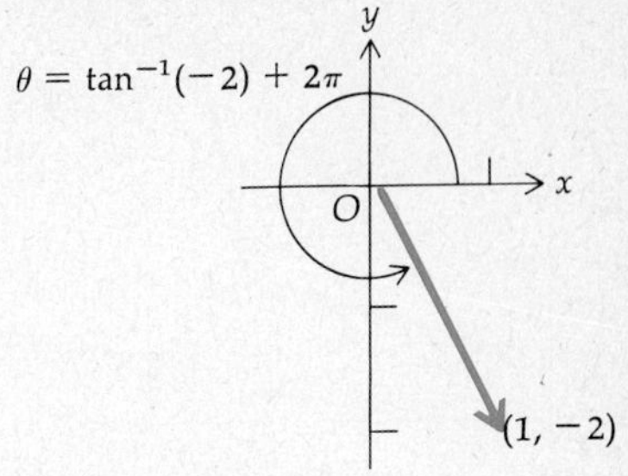

FIGURE 16.1.8

Example 1 Find the radian measure of the angle giving the direction of each of the following vectors: (a) $\langle -1, 1\rangle$; (b) $\langle 0, -5\rangle$; (c) $\langle -1, 2\rangle$.

Solution The position representation of each of the vectors in (a), (b), and (c) is shown in Figs. 16.1.6, 16.1.7, and 16.1.8, respectively. (a) $\tan\theta = -1$; so $\theta = \frac{3}{4}\pi$; (b) $\tan\theta$ does not exist and $a_2 < 0$; $\theta = \frac{3}{2}\pi$; (c) $\tan\theta = -2$; $\theta = \tan^{-1}(-2) + 2\pi$.

If the vector $\mathbf{A} = \langle a_1, a_2\rangle$, then the representation of **A** whose initial point is (x, y) has as its endpoint $(x + a_1, y + a_2)$. In this way a vector may be thought of as a translation of the plane into itself. Figure 16.1.9 illustrates five representations of the vector $\mathbf{A} = \langle a_1, a_2\rangle$. In each case **A** translates the point (x_i, y_i) into the point $(x_i + a_1, y_i + a_2)$.

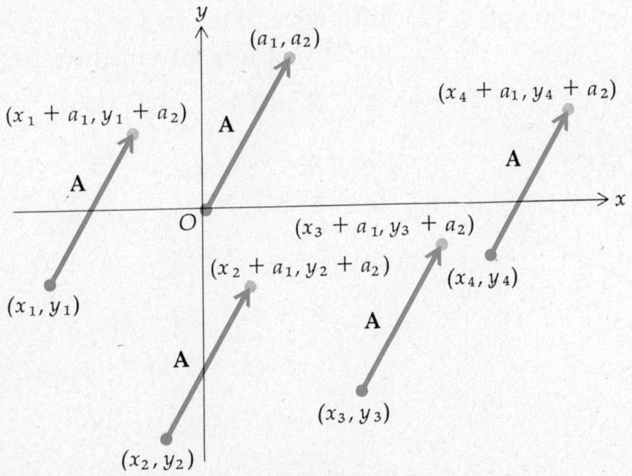

FIGURE 16.1.9

The following definition gives the method for adding two vectors.

16.1.4 Definition The **sum** of two vectors $\mathbf{A} = \langle a_1, a_2\rangle$ and $\mathbf{B} = \langle b_1, b_2\rangle$ is the vector $\mathbf{A} + \mathbf{B}$ defined by

$$\mathbf{A} + \mathbf{B} = \langle a_1 + b_1, a_2 + b_2\rangle$$

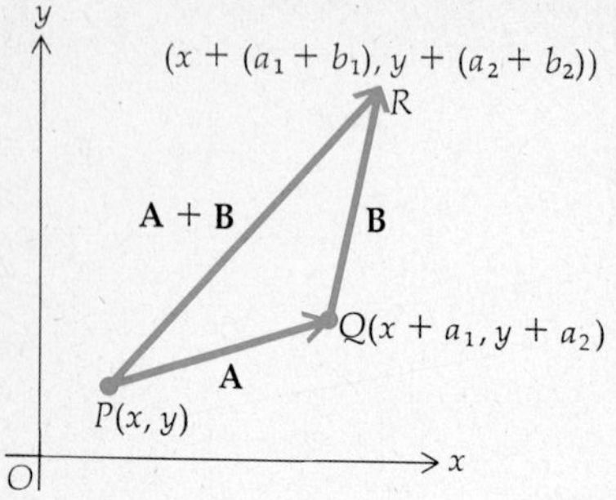

FIGURE 16.1.10

● ILLUSTRATION 3 If $\mathbf{A} = \langle 3, -1 \rangle$ and $\mathbf{B} = \langle -4, 5 \rangle$, then

$$\mathbf{A} + \mathbf{B} = \langle 3 + (-4), -1 + 5 \rangle = \langle -1, 4 \rangle$$

●

The geometric interpretation of the sum of two vectors is shown in Fig. 16.1.10. Let $\mathbf{A} = \langle a_1, a_2 \rangle$ and $\mathbf{B} = \langle b_1, b_2 \rangle$, and let P be the point (x, y). Then **A** translates the point P into the point $(x + a_1, y + a_2) = Q$. The vector **B** translates the point Q into the point $((x + a_1) + b_1, (y + a_2) + b_2)$ or, equivalently, $(x + (a_1 + b_1), y + (a_2 + b_2)) = R$. Furthermore, $\mathbf{A} + \mathbf{B} = \langle a_1 + b_1, a_2 + b_2 \rangle$. Therefore the vector $\mathbf{A} + \mathbf{B}$ translates the point P into the point $(x + (a_1 + b_1), y + (a_2 + b_2)) = R$. Thus, in Fig. 16.1.10 $\overrightarrow{PQ}$ is a representation of the vector **A**, $\overrightarrow{QR}$ is a representation of the vector **B**, and $\overrightarrow{PR}$ is a representation of the vector $\mathbf{A} + \mathbf{B}$. The representations of the vectors **A** and **B** are adjacent sides of a parallelogram, and the representation of the vector $\mathbf{A} + \mathbf{B}$ is a diagonal of the parallelogram. Thus the rule for the addition of vectors is sometimes referred to as the **parallelogram law.**

16.1.5 Definition If $\mathbf{A} = \langle a_1, a_2 \rangle$, then the vector $\langle -a_1, -a_2 \rangle$ is defined to be the **negative** of **A**, denoted by $-\mathbf{A}$.

If the directed line segment $\overrightarrow{PQ}$ is a representation of the vector **A**, then the directed line segment $\overrightarrow{QP}$ is a representation of $-\mathbf{A}$. Any directed line segment that is parallel to $\overrightarrow{PQ}$, has the same length as $\overrightarrow{PQ}$, and has a direction opposite to that of $\overrightarrow{PQ}$ is also a representation of $-\mathbf{A}$ (see Fig. 16.1.11). We now define subtraction of two vectors.

16.1.6 Definition The **difference** of the two vectors **A** and **B**, denoted by $\mathbf{A} - \mathbf{B}$, is the vector obtained by adding **A** to the negative of **B**; that is,

$$\mathbf{A} - \mathbf{B} = \mathbf{A} + (-\mathbf{B})$$

So if $\mathbf{A} = \langle a_1, a_2 \rangle$ and $\mathbf{B} = \langle b_1, b_2 \rangle$, then $-\mathbf{B} = \langle -b_1, -b_2 \rangle$, and so

$$\mathbf{A} - \mathbf{B} = \langle a_1 - b_1, a_2 - b_2 \rangle$$

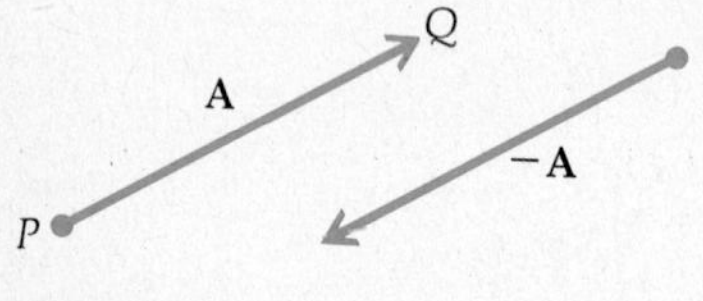

FIGURE 16.1.11

● ILLUSTRATION 4 If $\mathbf{A} = \langle 4, -2 \rangle$ and $\mathbf{B} = \langle 6, -3 \rangle$, then

$$\begin{aligned} \mathbf{A} - \mathbf{B} &= \langle 4, -2 \rangle - \langle 6, -3 \rangle \\ &= \langle 4, -2 \rangle + \langle -6, 3 \rangle \\ &= \langle -2, 1 \rangle \end{aligned}$$

●

To interpret the difference of two vectors geometrically, let the representations of the vectors **A** and **B** have the same initial point. Then the directed line segment from the endpoint of the representation of **B** to the endpoint of the representation of **A** is a representation of the vector $\mathbf{A} - \mathbf{B}$. This obeys the parallelogram law $\mathbf{B} + (\mathbf{A} - \mathbf{B}) = \mathbf{A}$ (see Fig. 16.1.12).

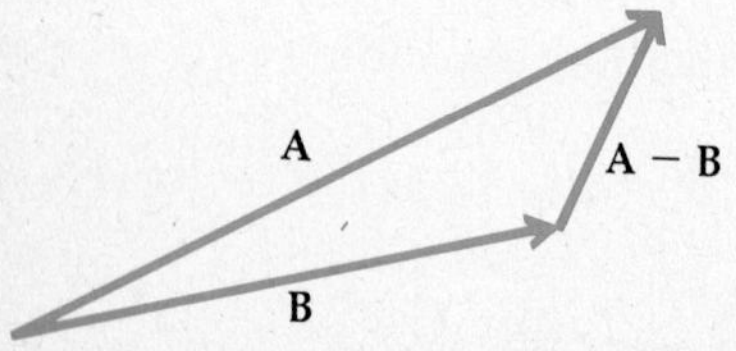

FIGURE 16.1.12

Another operation with vectors is *scalar multiplication.* A **scalar** is a real number. Following is the definition of the multiplication of a vector by a scalar.

16.1.7 Definition If c is a scalar and $\mathbf{A}$ is the vector $\langle a_1, a_2\rangle$, then the **product** of c and $\mathbf{A}$, denoted by $c\mathbf{A}$, is a vector and is given by

$$c\mathbf{A} = c\langle a_1, a_2\rangle = \langle ca_1, ca_2\rangle$$

● ILLUSTRATION 5 If $\mathbf{A} = \langle 4, -5\rangle$, then

$$3\mathbf{A} = 3\langle 4, -5\rangle = \langle 12, -15\rangle$$ ●

Example 2 If $\mathbf{A}$ is any vector and c is any scalar, show that $0(\mathbf{A}) = \mathbf{0}$ and $c(\mathbf{0}) = \mathbf{0}$.

Solution From Definition 16.1.7,

$$0(\mathbf{A}) = 0\langle a_1, a_2\rangle = \langle 0, 0\rangle = \mathbf{0}$$

and

$$c(\mathbf{0}) = c\langle 0, 0\rangle = \langle 0, 0\rangle = \mathbf{0}$$

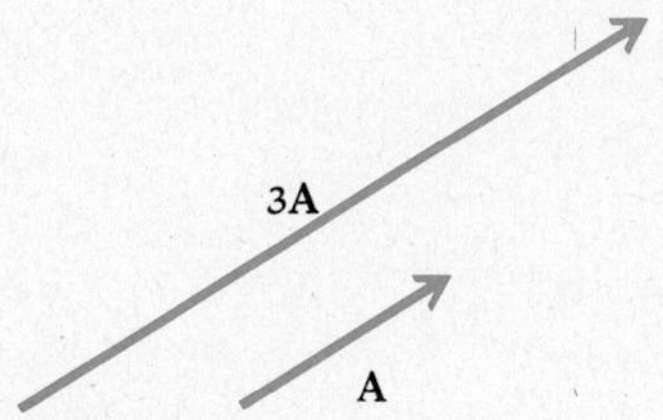

FIGURE 16.1.13

The magnitude of the vector $c\mathbf{A}$ is computed as follows.

$$\begin{aligned} c\mathbf{A} &= \sqrt{(ca_1)^2 + (ca_2)^2} \\ &= \sqrt{c^2(a_1{}^2 + a_2{}^2)} \\ &= \sqrt{c^2}\,\sqrt{a_1{}^2 + a_2{}^2} \\ &= |c||\mathbf{A}| \end{aligned}$$

Therefore the magnitude of $c\mathbf{A}$ is the absolute value of c times the magnitude of $\mathbf{A}$.

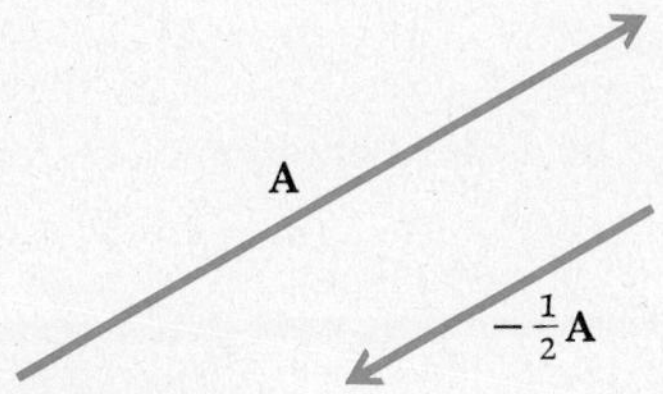

FIGURE 16.1.14

The geometric interpretation of the vector $c\mathbf{A}$ is given in Figs. 16.1.13 and 16.1.14. If $c > 0$, then $c\mathbf{A}$ is a vector whose representation has a length c times the magnitude of $\mathbf{A}$ and the same direction as $\mathbf{A}$; an example of this is shown in Fig. 16.1.13, where $c = 3$. If $c < 0$, then $c\mathbf{A}$ is a vector whose representation has a length that is $|c|$ times the magnitude of $\mathbf{A}$ and a direction opposite to that of $\mathbf{A}$. This is shown in Fig. 16.1.14, where $c = -\frac{1}{2}$.

EXERCISES 16.1

In Exercises 1 through 6, draw the position representation of the given vector **A** *and also the particular representation through the given point P; find the magnitude of* **A**.

1. $\mathbf{A} = \langle 3, 4\rangle$; $P = (2, 1)$
2. $\mathbf{A} = \langle -2, 5\rangle$; $P = (3, -4)$
3. $\mathbf{A} = \langle 0, -2\rangle$; $P = (-3, 4)$
4. $\mathbf{A} = \langle 4, 0\rangle$; $P = (2, 6)$
5. $\mathbf{A} = \langle 3, \sqrt{2}\rangle$; $P = (4, -\sqrt{2})$
6. $\mathbf{A} = \langle e, -\frac{1}{2}\rangle$; $P = (-2, -e)$

In Exercises 7 through 12, find the vector **A** *having* $\overrightarrow{PQ}$ *as a representation. Draw* $\overrightarrow{PQ}$ *and the position representation of* **A**.

7. $P = (3, 7)$; $Q = (5, 4)$
8. $P = (5, 4)$; $Q = (3, 7)$
9. $P = (-3, 5)$; $Q = (-5, -2)$
10. $P = (0, \sqrt{3})$; $Q = (2, 3\sqrt{3})$
11. $P = (-5, -3)$; $Q = (0, 3)$
12. $P = (-\sqrt{2}, 0)$; $Q = (0, 0)$

In Exercises 13 through 16, find the point S so that $\overrightarrow{PQ}$ and $\overrightarrow{RS}$ are each representations of the same vector.

13. $P = (2, 5)$; $Q = (1, 6)$; $R = (-3, 2)$ **14.** $P = (-1, 4)$; $Q = (2, -3)$; $R = (-5, -2)$

15. $P = (0, 3)$; $Q = (5, -2)$; $R = (7, 0)$ **16.** $P = (-2, 0)$; $Q = (-3, -4)$; $R = (4, 2)$

In Exercises 17 through 22, find the sum of the given pairs of vectors and illustrate geometrically.

17. $\langle 2, 4\rangle$; $\langle -3, 5\rangle$ **18.** $\langle 0, 3\rangle$; $\langle -2, 3\rangle$ **19.** $\langle -3, 0\rangle$; $\langle 4, -5\rangle$

20. $\langle 2, 3\rangle$; $\langle -\sqrt{2}, -1\rangle$ **21.** $\langle 0, 0\rangle$; $\langle -2, 2\rangle$ **22.** $\langle 2, 5\rangle$; $\langle 2, 5\rangle$

In Exercises 23 through 28, subtract the second vector from the first and illustrate geometrically.

23. $\langle 4, 5\rangle$; $\langle -3, 2\rangle$ **24.** $\langle 0, 5\rangle$; $\langle 2, 8\rangle$ **25.** $\langle -3, -4\rangle$; $\langle 6, 0\rangle$

26. $\langle 1, e\rangle$; $\langle -3, 2e\rangle$ **27.** $\langle 0, \sqrt{3}\rangle$; $\langle -\sqrt{2}, 0\rangle$ **28.** $\langle 3, 7\rangle$; $\langle 3, 7\rangle$

29. Given $\mathbf{A} = \langle 2, -5\rangle$; $\mathbf{B} = \langle 3, 1\rangle$; $\mathbf{C} = \langle -4, 2\rangle$. (a) Find $\mathbf{A} + (\mathbf{B} + \mathbf{C})$ and illustrate geometrically. (b) Find $(\mathbf{A} + \mathbf{B}) + \mathbf{C}$ and illustrate geometrically.

In Exercises 30 through 35, let $\mathbf{A} = \langle 2, 4\rangle$, $\mathbf{B} = \langle 4, -3\rangle$ and $\mathbf{C} = \langle -3, 2\rangle$.

30. Find $\mathbf{A} + \mathbf{B}$ **31.** Find $\mathbf{A} - \mathbf{B}$ **32.** Find $|\mathbf{C}|$

33. Find $|\mathbf{C} - \mathbf{B}|$ **34.** Find $2\mathbf{A} + 3\mathbf{B}$ **35.** Find $|7\mathbf{A} - \mathbf{B}|$

36. Given $\mathbf{A} = \langle 3, 2\rangle$; $\mathbf{C} = \langle 8, 8\rangle$; $\mathbf{A} + \mathbf{B} = \mathbf{C}$; find $|\mathbf{B}|$.

37. Given $\mathbf{A} = \langle 6, -9\rangle$; $\mathbf{C} = \langle 10, -8\rangle$; $2\mathbf{A} - \mathbf{B} = \mathbf{C}$; find $|\mathbf{B}|$.

38. Given $\mathbf{A} = \langle -3, 2\rangle$; $\mathbf{B} = \langle 4, -1\rangle$; $\mathbf{C} = \langle 8, 3\rangle$; find scalars h and k such that $h\mathbf{A} + k\mathbf{B} = \mathbf{C}$.

39. Given $\mathbf{A} = \langle 5, -2\rangle$; $\mathbf{B} = \langle -4, 3\rangle$; $\mathbf{C} = \langle -6, 8\rangle$; find scalars h and k such that $\mathbf{B} = h\mathbf{C} - k\mathbf{A}$.

40. Prove analytically the triangle inequality for vectors $|\mathbf{A} + \mathbf{B}| \le |\mathbf{A}| + |\mathbf{B}|$.

41. Let $\overrightarrow{PQ}$ be a representation of vector $\mathbf{A}$, $\overrightarrow{QR}$ be a representation of vector $\mathbf{B}$, and $\overrightarrow{RS}$ be a representation of vector $\mathbf{C}$. Prove that if $\overrightarrow{PQ}$, $\overrightarrow{QR}$, and $\overrightarrow{RS}$ are sides of a triangle, then $\mathbf{A} + \mathbf{B} + \mathbf{C} = \mathbf{0}$.

16.2 PROPERTIES OF VECTOR ADDITION AND SCALAR MULTIPLICATION

The following theorem gives laws satisfied by the operations of vector addition and scalar multiplication of any vectors in V_2.

16.2.1 Theorem If **A**, **B**, and **C** are any vectors in V_2, and c and d are any scalars, then vector addition and scalar multiplication satisfy the following properties:

(i) $\mathbf{A} + \mathbf{B} = \mathbf{B} + \mathbf{A}$ (commutative law)
(ii) $\mathbf{A} + (\mathbf{B} + \mathbf{C}) = (\mathbf{A} + \mathbf{B}) + \mathbf{C}$ (associative law)
(iii) There is a vector $\mathbf{0}$ in V_2 for which $\mathbf{A} + \mathbf{0} = \mathbf{A}$ (existence of additive identity)
(iv) There is a vector $-\mathbf{A}$ in V_2 such that $\mathbf{A} + (-\mathbf{A}) = \mathbf{0}$ (existence of negative)
(v) $(cd)\mathbf{A} = c(d\mathbf{A})$ (associative law)
(vi) $c(\mathbf{A} + \mathbf{B}) = c\mathbf{A} + c\mathbf{B}$ (distributive law)
(vii) $(c + d)\mathbf{A} = c\mathbf{A} + d\mathbf{A}$ (distributive law)
(viii) $1(\mathbf{A}) = \mathbf{A}$ (existence of scalar multiplicative identity)

Proof We give the proofs of (i), (iv), and (vi) and leave the others as exercises (see Exercises 32 through 35). Let $\mathbf{A} = \langle a_1, a_2 \rangle$ and $\mathbf{B} = \langle b_1, b_2 \rangle$.

Proof of (i) By the commutative law for real numbers, $a_1 + b_1 = b_1 + a_1$ and $a_2 + b_2 = b_2 + a_2$. Therefore

$$\begin{aligned} \mathbf{A} + \mathbf{B} &= \langle a_1, a_2 \rangle + \langle b_1, b_2 \rangle \\ &= \langle a_1 + b_1, a_2 + b_2 \rangle \\ &= \langle b_1 + a_1, b_2 + a_2 \rangle \\ &= \langle b_1, b_2 \rangle + \langle a_1, a_2 \rangle \\ &= \mathbf{B} + \mathbf{A} \end{aligned}$$

Proof of (iv) The vector $-\mathbf{A}$ is given by Definition 16.1.5 and

$$\begin{aligned} \mathbf{A} + (-\mathbf{A}) &= \langle a_1, a_2 \rangle + \langle -a_1, -a_2 \rangle \\ &= \langle a_1 + (-a_1), a_2 + (-a_2) \rangle \\ &= \langle 0, 0 \rangle \\ &= \mathbf{0} \end{aligned}$$

Proof of (vi)

$$\begin{aligned} c(\mathbf{A} + \mathbf{B}) &= c(\langle a_1, a_2 \rangle + \langle b_1, b_2 \rangle) \\ &= c(\langle a_1 + b_1, a_2 + b_2 \rangle) \\ &= \langle c(a_1 + b_1), c(a_2 + b_2) \rangle \\ &= \langle ca_1 + cb_1, ca_2 + cb_2 \rangle \\ &= \langle ca_1, ca_2 \rangle + \langle cb_1, cb_2 \rangle \\ &= c\langle a_1, a_2 \rangle + c\langle b_1, b_2 \rangle \\ &= c\mathbf{A} + c\mathbf{B} \end{aligned}$$

■

Example 1 Verify (ii), (iii), (v), (vii), and (viii) of Theorem 17.2.1 if $\mathbf{A} = \langle 3, 4 \rangle$, $\mathbf{B} = \langle -2, 1 \rangle$, $\mathbf{C} = \langle 5, -3 \rangle$, $c = 2$, and $d = -6$.

Solution

$$\begin{aligned} \mathbf{A} + (\mathbf{B} + \mathbf{C}) &= \langle 3, 4 \rangle + (\langle -2, 1 \rangle + \langle 5, -3 \rangle) \\ &= \langle 3, 4 \rangle + \langle 3, -2 \rangle \\ &= \langle 6, 2 \rangle \\ (\mathbf{A} + \mathbf{B}) + \mathbf{C} &= (\langle 3, 4 \rangle + \langle -2, 1 \rangle) + \langle 5, -3 \rangle \\ &= \langle 1, 5 \rangle + \langle 5, -3 \rangle \\ &= \langle 6, 2 \rangle \end{aligned}$$

Therefore $\mathbf{A} + (\mathbf{B} + \mathbf{C}) = (\mathbf{A} + \mathbf{B}) + \mathbf{C}$, and so (ii) holds.

$$\mathbf{A} + \mathbf{0} = \langle 3, 4 \rangle + \langle 0, 0 \rangle = \langle 3, 4 \rangle = \mathbf{A}$$

Hence (iii) holds.

$$\begin{aligned}(cd)\mathbf{A} &= [(2)(-6)]\langle 3, 4\rangle \\ &= (-12)\langle 3, 4\rangle \\ &= \langle -36, -48\rangle \\ c(d\mathbf{A}) &= 2(-6\langle 3, 4\rangle) \\ &= 2\langle -18, -24\rangle \\ &= \langle -36, -48\rangle\end{aligned}$$

Thus $(cd)\mathbf{A} = c(d\mathbf{A})$, and so (v) holds.

$$(c + d)\mathbf{A} = [2 + (-6)]\langle 3, 4\rangle = (-4)\langle 3, 4\rangle = \langle -12, -16\rangle$$

$$c\mathbf{A} + d\mathbf{A} = 2\langle 3, 4\rangle + (-6)\langle 3, 4\rangle = \langle 6, 8\rangle + \langle -18, -24\rangle = \langle -12, -16\rangle$$

Therefore $(c + d)\mathbf{A} = c\mathbf{A} + d\mathbf{A}$, and so (vii) holds.

$$1\mathbf{A} = 1\langle 3, 4\rangle = \langle (1)(3), (1)(4)\rangle = \langle 3, 4\rangle = \mathbf{A}$$

and so (viii) holds.

Theorem 16.2.1 is important because every algebraic law for the operations of vector addition and scalar multiplication of vectors in V_2 can be derived from the eight properties stated in the theorem. These laws are similar to the laws of arithmetic of real numbers. Furthermore, in linear algebra, a *real vector space* is defined as a set of vectors together with a set of real numbers (*scalars*) and the two operations of vector addition and scalar multiplication that satisfy the eight properties given in Theorem 16.2.1. Following is the formal definition.

16.2.2 Definition

A **real vector space V** is a set of elements, called *vectors*, together with a set of real numbers, called *scalars*, with two operations called *vector addition* and *scalar multiplication* such that for every pair of vectors **A** and **B** in V and for every scalar c, a vector $\mathbf{A} + \mathbf{B}$ and a vector $c\mathbf{A}$ are defined so that properties (i)–(viii) of Theorem 16.2.1 are satisfied.

From Definition 16.2.2 and Theorem 16.2.1 it follows that V_2 is a real vector space.

We now take an arbitrary vector in V_2 and write it in a special form.

$$\mathbf{A} = \langle a_1, a_2\rangle = \langle a_1, 0\rangle + \langle 0, a_2\rangle = a_1\langle 1, 0\rangle + a_2\langle 0, 1\rangle$$

Because the magnitude of each of the two vectors $\langle 1, 0\rangle$ and $\langle 0, 1\rangle$ is one unit, they are called **unit vectors.** We introduce the following notations for these two unit vectors:

$$\mathbf{i} = \langle 1, 0\rangle \quad \text{and} \quad \mathbf{j} = \langle 0, 1\rangle$$

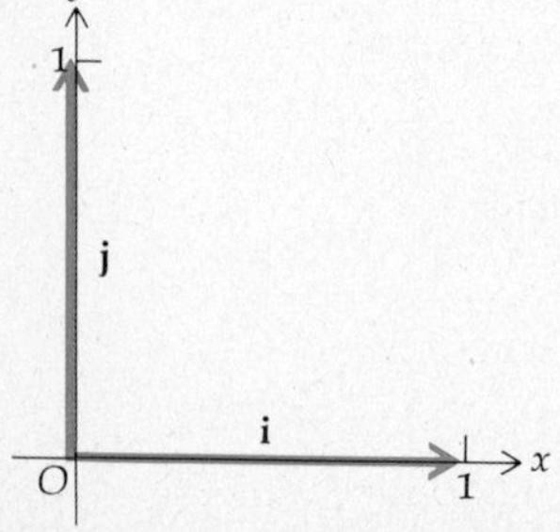

FIGURE 16.2.1

The position representation of each of these unit vectors is shown in Fig. 16.2.1. Because

$$\langle a_1, a_2 \rangle = a_1\langle 1, 0 \rangle + a_2\langle 0, 1 \rangle = a_1\mathbf{i} + a_2\mathbf{j}$$

it follows that any vector in V_2 can be written as a linear combination of the two vectors **i** and **j**. Because of this the vectors **i** and **j** are said to form a **basis** for the vector space V_2. The number of elements in a basis of a vector space is called the **dimension** of the vector space. Hence V_2 is a two-dimensional vector space.

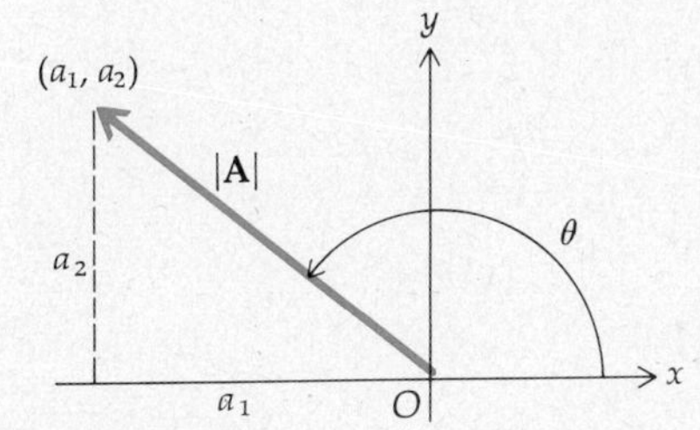

FIGURE 16.2.2

● ILLUSTRATION 1 We express the vector $\langle 3, -4 \rangle$ in terms of **i** and **j**.

$$\langle 3, -4 \rangle = 3\langle 1, 0 \rangle + (-4)\langle 0, 1 \rangle = 3\mathbf{i} - 4\mathbf{j}$$ ●

Let **A** be the vector $\langle a_1, a_2 \rangle$ and θ be the radian measure of the angle giving the direction of **A** (see Fig. 16.2.2, where (a_1, a_2) is in the second quadrant). $a_1 = |\mathbf{A}| \cos \theta$ and $a_2 = |\mathbf{A}| \sin \theta$. Because $\mathbf{A} = a_1\mathbf{i} + a_2\mathbf{j}$, we can write

$$\mathbf{A} = |\mathbf{A}| \cos \theta\mathbf{i} + |\mathbf{A}| \sin \theta\mathbf{j}$$

or, equivalently,

$$\mathbf{A} = |\mathbf{A}|(\cos \theta\mathbf{i} + \sin \theta\mathbf{j}) \tag{1}$$

Equation (1) expresses the vector **A** in terms of its magnitude, the cosine and sine of the radian measure of the angle giving the direction of **A**, and the unit vectors **i** and **j**.

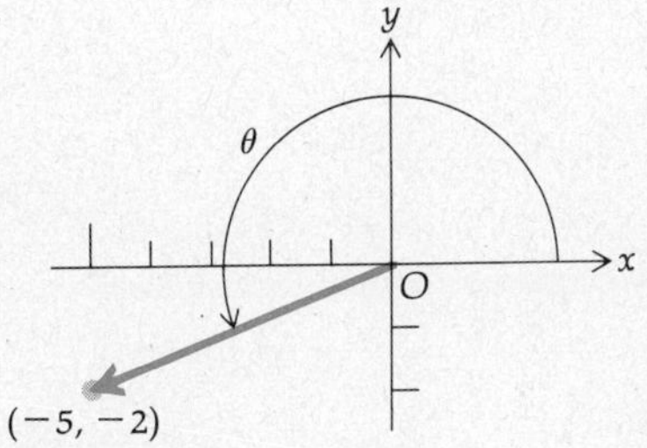

FIGURE 16.2.3

Example 2 Express the vector $\langle -5, -2 \rangle$ in the form of Eq. (1).

Solution Refer to Fig. 16.2.3, which shows the position representation of the vector $\langle -5, -2 \rangle$.

$$|\langle -5, -2 \rangle| = \sqrt{(-5)^2 + (-2)^2} = \sqrt{25 + 4} = \sqrt{29}$$

$$\sin \theta = -\frac{2}{\sqrt{29}} \quad \text{and} \quad \cos \theta = -\frac{5}{\sqrt{29}}$$

So from (1) we have

$$\langle -5, -2 \rangle = \sqrt{29}\left(-\frac{5}{\sqrt{29}}\mathbf{i} - \frac{2}{\sqrt{29}}\mathbf{j}\right)$$

16.2.3 Theorem If the nonzero vector $\mathbf{A} = a_1\mathbf{i} + a_2\mathbf{j}$, then the unit vector **U** having the same direction as **A** is given by

$$\mathbf{U} = \frac{a_1}{|\mathbf{A}|}\mathbf{i} + \frac{a_2}{|\mathbf{A}|}\mathbf{j} \tag{2}$$

Proof From (2),

$$\mathbf{U} = \frac{1}{|\mathbf{A}|}(a_1\mathbf{i} + a_2\mathbf{j}) = \frac{1}{|\mathbf{A}|}(\mathbf{A})$$

Therefore **U** is a positive scalar times the vector **A**; so the direction of **U** is the same as the direction of **A**. Furthermore,

$$|\mathbf{U}| = \sqrt{\left(\frac{a_1}{|\mathbf{A}|}\right)^2 + \left(\frac{a_2}{|\mathbf{A}|}\right)^2}$$
$$= \frac{\sqrt{a_1^2 + a_2^2}}{|\mathbf{A}|}$$
$$= \frac{|\mathbf{A}|}{|\mathbf{A}|}$$
$$= 1$$

Therefore **U** is the unit vector having the same direction as **A**, and the theorem is proved. ■

Example 3 Given $\mathbf{A} = \langle 3, 1 \rangle$ and $\mathbf{B} = \langle -2, 4 \rangle$, find the unit vector having the same direction as $\mathbf{A} - \mathbf{B}$.

Solution $\mathbf{A} - \mathbf{B} = \langle 3, 1 \rangle - \langle -2, 4 \rangle = \langle 5, -3 \rangle$. Therefore

$$\mathbf{A} - \mathbf{B} = 5\mathbf{i} - 3\mathbf{j}$$

Then

$$|\mathbf{A} - \mathbf{B}| = \sqrt{5^2 + (-3)^2} = \sqrt{34}$$

By Theorem 16.2.3, the desired unit vector is given by

$$\mathbf{U} = \frac{5}{\sqrt{34}}\mathbf{i} - \frac{3}{\sqrt{34}}\mathbf{j}$$

EXERCISES 16.2

In Exercises 1 through 16, find the given vector or scalar if $\mathbf{A} = 2\mathbf{i} + 3\mathbf{j}$ *and* $\mathbf{B} = 4\mathbf{i} - \mathbf{j}$.

1. $5\mathbf{A}$ **2.** $-2\mathbf{A}$ **3.** $-6\mathbf{B}$ **4.** $3\mathbf{B}$ **5.** $\mathbf{A} + \mathbf{B}$ **6.** $\mathbf{A} - \mathbf{B}$

7. $|\mathbf{A} + \mathbf{B}|$ **8.** $|\mathbf{A} - \mathbf{B}|$ **9.** $|\mathbf{A}| + |\mathbf{B}|$ **10.** $|\mathbf{A}| - |\mathbf{B}|$ **11.** $5\mathbf{A} - 6\mathbf{B}$ **12.** $3\mathbf{B} - 2\mathbf{A}$

13. $|5\mathbf{A} - 6\mathbf{B}|$ **14.** $|3\mathbf{B} - 2\mathbf{A}|$ **15.** $|5\mathbf{A}| - |6\mathbf{B}|$ **16.** $|3\mathbf{B}| - |2\mathbf{A}|$

In Exercises 17 and 18, $\mathbf{A} = -4\mathbf{i} + 2\mathbf{j}$, $\mathbf{B} = -\mathbf{i} + 3\mathbf{j}$, *and* $\mathbf{C} = 5\mathbf{i} - \mathbf{j}$.

17. Find: (a) $5\mathbf{A} - 2\mathbf{B} - 2\mathbf{C}$ and (b) $|5\mathbf{A} - 2\mathbf{B} - \mathbf{C}|$. **18.** Find: (a) $3\mathbf{B} - 2\mathbf{A} - \mathbf{C}$ and (b) $|3\mathbf{B} - 2\mathbf{A} - \mathbf{C}|$.

In Exercises 19 and 20, $\mathbf{A} = 8\mathbf{i} + 5\mathbf{j}$ *and* $\mathbf{B} = 3\mathbf{i} - \mathbf{j}$.

19. Find a unit vector having the same direction as $\mathbf{A} + \mathbf{B}$. **20.** Find a unit vector having the same direction as $\mathbf{A} - \mathbf{B}$.

In Exercises 21 through 28, (a) write the given vector in the form $r(\cos\theta\,\mathbf{i} + \sin\theta\,\mathbf{j})$, *where r is the magnitude of the vector and θ is the radian measure of the angle giving the direction of the vector, and (b) find a unit vector having the same direction.*

21. $3\mathbf{i} - 4\mathbf{j}$ **22.** $8\mathbf{i} + 6\mathbf{j}$ **23.** $2\mathbf{i} + 2\mathbf{j}$ **24.** $3\mathbf{i} - 3\mathbf{j}$

25. $-4\mathbf{i} + 4\sqrt{3}\mathbf{j}$ **26.** $2\sqrt{5}\mathbf{i} + 4\mathbf{j}$ **27.** $-16\mathbf{i}$ **28.** $2\mathbf{j}$

29. Given $\mathbf{A} = -2\mathbf{i} + \mathbf{j}$, $\mathbf{B} = 3\mathbf{i} - 2\mathbf{j}$, and $\mathbf{C} = 5\mathbf{i} - 4\mathbf{j}$; find scalars h and k such that $\mathbf{C} = h\mathbf{A} + k\mathbf{B}$.

30. Given $\mathbf{A} = 2\mathbf{i} - 5\mathbf{j}$, $\mathbf{B} = \mathbf{i} + 3\mathbf{j}$, and $\mathbf{C} = 4\mathbf{i} - \mathbf{j}$; find scalars h and k such that $\mathbf{C} = h\mathbf{A} + k\mathbf{B}$.

31. Given $\mathbf{A} = \mathbf{i} - 2\mathbf{j}$, $\mathbf{B} = -2\mathbf{i} + 4\mathbf{j}$, and $\mathbf{C} = 7\mathbf{i} - 5\mathbf{j}$; show that $\mathbf{C}$ cannot be written in the form $h\mathbf{A} + k\mathbf{B}$, where h and k are scalars.

32. Prove Theorem 16.2.1(ii).

33. Prove Theorem 16.2.1(v).

34. Prove Theorem 16.2.1(vii).

35. Prove Theorem 16.2.1(iii) and (viii).

36. Two vectors are said to be *independent* if and only if their position representations are not collinear. Furthermore, two vectors $\mathbf{A}$ and $\mathbf{B}$ are said to form a *basis* for the vector space V_2 if and only if any vector in V_2 can be written as a linear combination of $\mathbf{A}$ and $\mathbf{B}$. A theorem can be proved which states that two vectors form a basis for the vector space V_2 if they are independent. Show that this theorem holds for the two vectors $\langle 2, 5\rangle$ and $\langle 3, -1\rangle$ by doing the following: (a) Verify that the vectors are independent by showing that their position representations are not collinear; (b) verify that the vectors form a basis by showing that any vector $a_1\mathbf{i} + a_2\mathbf{j}$ can be written as $c(2\mathbf{i} + 5\mathbf{j}) + d(3\mathbf{i} - \mathbf{j})$, where c and d are scalars. (*Hint:* Find c and d in terms of a_1 and a_2.)

37. Refer to the first two sentences of Exercise 36. A theorem can be proved which states that two vectors form a basis for the vector space V_2 only if they are independent. Show that this theorem holds for the two vectors $\langle 3, -2\rangle$ and $\langle -6, 4\rangle$ by doing the following: (a) Verify that the vectors are dependent (not independent) by showing that their position representations are collinear; (b) verify that the vectors do not form a basis by taking a particular vector and showing that it cannot be written as $c(3\mathbf{i} - 2\mathbf{j}) + d(-6\mathbf{i} + 4\mathbf{j})$, where c and d are scalars.

38. A set of vectors $\mathbf{V}_1, \mathbf{V}_2, \mathbf{V}_3, \ldots, \mathbf{V}_n$ is said to be linearly dependent if and only if there are scalars $k_1, k_2, k_3, \ldots, k_n$, not all zero, such that

$$k_1\mathbf{V}_1 + k_2\mathbf{V}_2 + k_3\mathbf{V}_3 + \cdots + k_n\mathbf{V}_n = \mathbf{0}$$

Show that if $\mathbf{V}_1 = 3\mathbf{i} - 2\mathbf{j}$, $\mathbf{V}_2 = \mathbf{i} + 4\mathbf{j}$, and $\mathbf{V}_3 = 2\mathbf{i} + 5\mathbf{j}$, then $\mathbf{V}_1$, $\mathbf{V}_2$, and $\mathbf{V}_3$ are linearly dependent.

16.3 DOT PRODUCT

In Sec. 16.1 addition and subtraction of vectors and multiplication of a vector by a scalar were defined. However, the multiplication of two vectors was not considered. We now define a multiplication operation on two vectors that gives what is called the *dot product.*

16.3.1 Definition If $\mathbf{A} = \langle a_1, a_2\rangle$ and $\mathbf{B} = \langle b_1, b_2\rangle$ are two vectors in V_2, then the **dot product** of $\mathbf{A}$ and $\mathbf{B}$, denoted by $\mathbf{A}\cdot\mathbf{B}$, is given by

$$\mathbf{A}\cdot\mathbf{B} = \langle a_1, a_2\rangle\cdot\langle b_1, b_2\rangle = a_1b_1 + a_2b_2$$

The dot product of two vectors is a real number (or scalar) and not a vector. It is sometimes called the **scalar product** or **inner product.**

● ILLUSTRATION 1 If $\mathbf{A} = \langle 2, -3\rangle$ and $\mathbf{B} = \langle -\frac{1}{2}, 4\rangle$, then

$$\mathbf{A}\cdot\mathbf{B} = \langle 2, -3\rangle\cdot\langle -\tfrac{1}{2}, 4\rangle = (2)(-\tfrac{1}{2}) + (-3)(4) = -13$$ ●

The following dot products are useful and are easily verified (see Exercise 5).

$$\mathbf{i}\cdot\mathbf{i} = 1 \qquad (1)$$

$$\mathbf{j}\cdot\mathbf{j} = 1 \qquad (2)$$

$$\mathbf{i}\cdot\mathbf{j} = 0 \qquad (3)$$

The next theorem states that dot multiplication is commutative and distributive with respect to vector addition.

16.3.2 Theorem If $\mathbf{A}$, $\mathbf{B}$, and $\mathbf{C}$ are any vectors in V_2, then

(i) $\mathbf{A} \cdot \mathbf{B} = \mathbf{B} \cdot \mathbf{A}$ (commutative law)
(ii) $\mathbf{A} \cdot (\mathbf{B} + \mathbf{C}) = \mathbf{A} \cdot \mathbf{B} + \mathbf{A} \cdot \mathbf{C}$ (distributive law)

The proofs of (i) and (ii) are left as exercises (see Exercises 6 and 7).

Because $\mathbf{A} \cdot \mathbf{B}$ is a scalar, the expression $(\mathbf{A} \cdot \mathbf{B}) \cdot \mathbf{C}$ is meaningless. Hence we do not consider associativity of dot multiplication.

Some other laws of dot multiplication are given in the following theorem.

16.3.3 Theorem If $\mathbf{A}$ and $\mathbf{B}$ are any vectors in V_2 and c is any scalar, then

(i) $c(\mathbf{A} \cdot \mathbf{B}) = (c\mathbf{A}) \cdot \mathbf{B}$;
(ii) $\mathbf{0} \cdot \mathbf{A} = 0$;
(iii) $\mathbf{A} \cdot \mathbf{A} = |\mathbf{A}|^2$.

The proofs are left as exercises (see Exercises 8 through 10).

We now consider what is meant by the angle between two vectors, and this leads to another expression for the dot product of two vectors.

16.3.4 Definition Let $\mathbf{A}$ and $\mathbf{B}$ be two nonzero vectors such that $\mathbf{A}$ is not a scalar multiple of $\mathbf{B}$. If $\overrightarrow{OP}$ is the position representation of $\mathbf{A}$ and $\overrightarrow{OQ}$ is the position representation of $\mathbf{B}$, then the **angle between the vectors A and B** is defined to be the angle of positive measure between $\overrightarrow{OP}$ and $\overrightarrow{OQ}$ interior to the triangle determined by the points O, P, and Q. If $\mathbf{A} = c\mathbf{B}$, where c is a scalar, then if $c > 0$ the angle between the vectors has radian measure 0; if $c < 0$, the angle between the vectors has radian measure π.

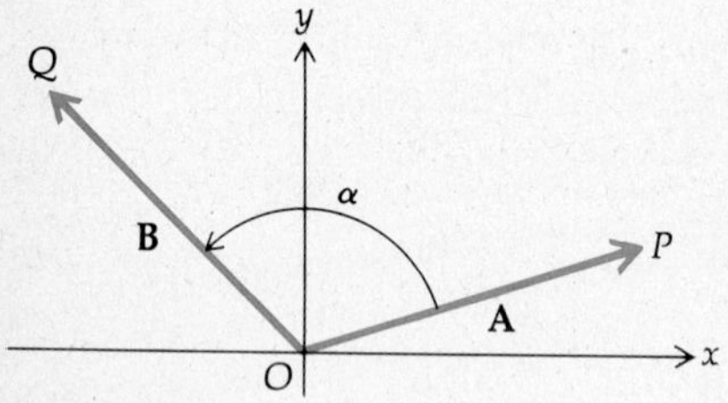

FIGURE 16.3.1

It follows from Definition 16.3.4 that if α is the radian measure of the angle between two vectors, then $0 \leq \alpha \leq \pi$. Figure 16.3.1 shows the angle between two vectors if $\mathbf{A}$ is not a scalar multiple of $\mathbf{B}$.

The following theorem is perhaps the most important fact about the dot product of two vectors.

16.3.5 Theorem If α is the radian measure of the angle between the two nonzero vectors $\mathbf{A}$ and $\mathbf{B}$, then

$$\mathbf{A} \cdot \mathbf{B} = |\mathbf{A}||\mathbf{B}| \cos \alpha \tag{4}$$

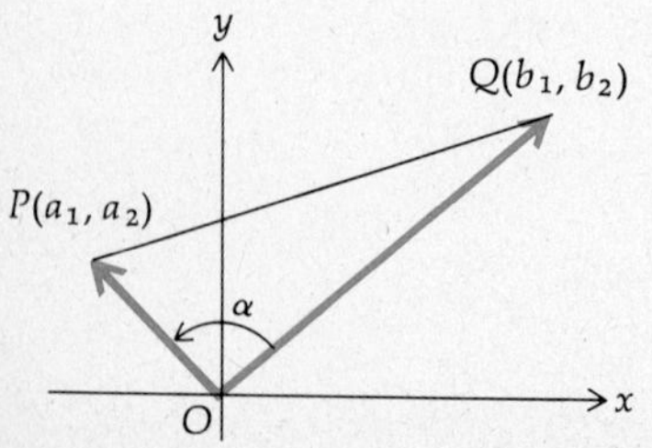

FIGURE 16.3.2

Proof Let $\mathbf{A} = a_1\mathbf{i} + a_2\mathbf{j}$ and $\mathbf{B} = b_1\mathbf{i} + b_2\mathbf{j}$. Let $\overrightarrow{OP}$ be the position representation of $\mathbf{A}$ and OQ be the position representation of $\mathbf{B}$. Then the angle between the vectors $\mathbf{A}$ and $\mathbf{B}$ is the angle at the origin in triangle POQ (see Fig. 16.3.2); P is the point (a_1, a_2) and Q is the point (b_1, b_2). In triangle POQ, $|\mathbf{A}|$ is the length of the side OP and $|\mathbf{B}|$ is the length of the side OQ. So from the law of cosines,

$$\cos \alpha = \frac{|\mathbf{A}|^2 + |\mathbf{B}|^2 - |\overline{PQ}|^2}{2|\mathbf{A}||\mathbf{B}|}$$

$$= \frac{(a_1{}^2 + a_2{}^2) + (b_1{}^2 + b_2{}^2) - [(a_1 - b_1)^2 + (a_2 - b_2)^2]}{2|\mathbf{A}||\mathbf{B}|}$$

$$= \frac{2a_1b_1 + 2a_2b_2}{2|\mathbf{A}||\mathbf{B}|}$$

$$= \frac{a_1b_1 + a_2b_2}{|\mathbf{A}||\mathbf{B}|}$$

Hence

$$\cos \alpha = \frac{\mathbf{A} \cdot \mathbf{B}}{|\mathbf{A}||\mathbf{B}|}$$

from which we obtain

$$\mathbf{A} \cdot \mathbf{B} = |\mathbf{A}||\mathbf{B}| \cos \alpha$$ ■

Theorem 16.3.5 states that the dot product of two vectors is the product of the magnitudes of the vectors and the cosine of the radian measure of the angle between them.

● ILLUSTRATION 2 If $\mathbf{A} = 3\mathbf{i} - 2\mathbf{j}$, $\mathbf{B} = 2\mathbf{i} + \mathbf{j}$, and α is the radian measure of the angle between **A** and **B**, then from Theorem 16.3.5,

$$\cos \alpha = \frac{\mathbf{A} \cdot \mathbf{B}}{|\mathbf{A}||\mathbf{B}|}$$

$$= \frac{(3)(2) + (-2)(1)}{\sqrt{9 + 4}\sqrt{4 + 1}}$$

$$= \frac{6 - 2}{\sqrt{13}\sqrt{5}}$$

$$= \frac{4}{\sqrt{65}}$$ ●

We learned in Sec. 16.2 that if two nonzero vectors are scalar multiples of each other, they have either the same or opposite directions. We have, then, the following definition.

16.3.6 Definition Two vectors are said to be **parallel** if and only if one of the vectors is a scalar multiple of the other.

● ILLUSTRATION 3 The vectors $\langle 3, -4 \rangle$ and $\langle \frac{3}{4}, -1 \rangle$ are parallel because $\langle 3, -4 \rangle = 4\langle \frac{3}{4}, -1 \rangle$. ●

If **A** is any vector, $\mathbf{0} = 0\mathbf{A}$; thus, from Definition 16.3.6 it follows that the zero vector is parallel to any vector.

It is left as an exercise for you to show that two nonzero vectors are parallel if and only if the radian measure of the angle between them is 0 or π (see Exercise 37).

If **A** and **B** are nonzero vectors, then from (4) it follows that

$$\cos \alpha = 0 \quad \text{if and only if} \quad \mathbf{A} \cdot \mathbf{B} = 0$$

Because $0 \le \alpha \le \pi$, it follows from this statement that

$$\alpha = \tfrac{1}{2}\pi \quad \text{if and only if} \quad \mathbf{A} \cdot \mathbf{B} = 0$$

We have, then, the following definition.

16.3.7 Definition Two vectors **A** and **B** are said to be **orthogonal (perpendicular)** if and only if $\mathbf{A} \cdot \mathbf{B} = 0$.

● ILLUSTRATION 4 The vectors $\langle -4, 5\rangle$ and $\langle 10, 8\rangle$ are orthogonal because

$$\begin{aligned}\langle -4, 5\rangle \cdot \langle 10, 8\rangle &= (-4)(10) + (5)(8) \\ &= 0\end{aligned}$$

●

If **A** is any vector, $\mathbf{0} \cdot \mathbf{A} = 0$, and therefore it follows from Definition 16.3.7 that the zero vector is orthogonal to any vector.

Example 1 Given $\mathbf{A} = 3\mathbf{i} + 2\mathbf{j}$ and $\mathbf{B} = 2\mathbf{i} + k\mathbf{j}$, where k is a scalar, find (a) k such that **A** and **B** are orthogonal; (b) k such that **A** and **B** are parallel.

Solution

(a) By Definition 16.3.7, **A** and **B** are orthogonal if and only if $\mathbf{A} \cdot \mathbf{B} = 0$; that is,

$$(3)(2) + 2(k) = 0$$

Hence

$$k = -3$$

(b) From Definition 16.3.6, **A** and **B** are parallel if and only if there is some scalar c such that $\langle 3, 2\rangle = c\langle 2, k\rangle$; that is,

$$3 = 2c \quad \text{and} \quad 2 = ck$$

Solving these two equations simultaneously we obtain $k = \frac{4}{3}$.

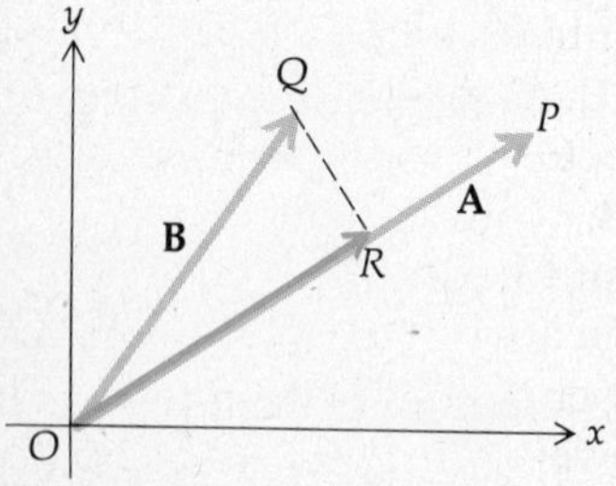

FIGURE 16.3.3

A geometric interpretation of the dot product is obtained by considering the *projection* of a vector onto another vector. Let $\overrightarrow{OP}$ and $\overrightarrow{OQ}$ be representations of the vectors **A** and **B**, respectively. See Fig. 16.3.3. The projection of $\overrightarrow{OQ}$ in the direction of $\overrightarrow{OP}$ is the directed line segment $\overrightarrow{OR}$, where R is the foot of the perpendicular from Q to the line containing $\overrightarrow{OP}$. Then the vector for which $\overrightarrow{OR}$ is a representation is called the **vector projection** of the vector **B** onto the vector **A**. The **scalar projection** of **B** onto **A** is defined to be $|\mathbf{B}| \cos \alpha$, where α is the radian measure of the angle

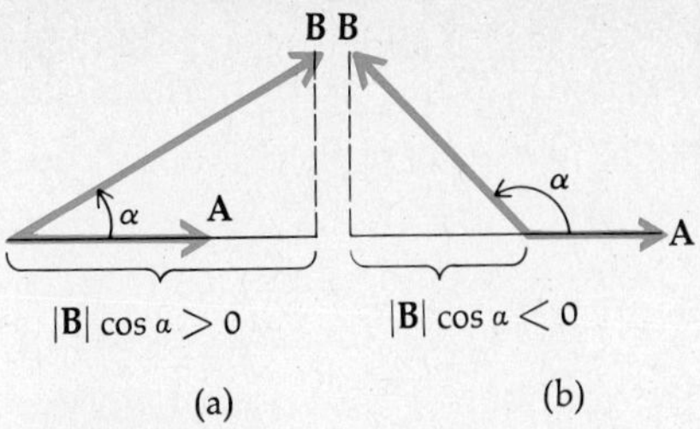

FIGURE 16.3.4

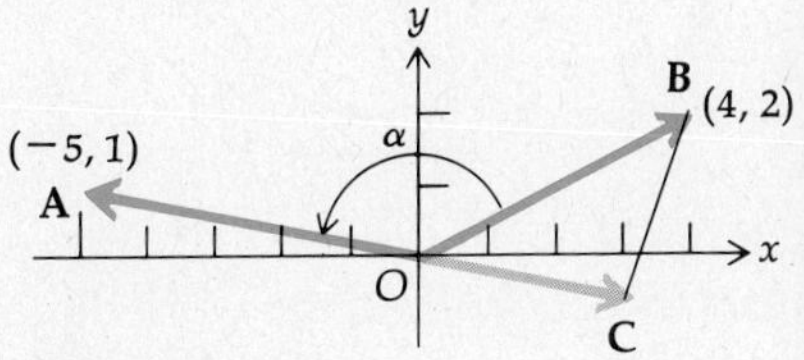

FIGURE 16.3.5

between **A** and **B**. Note that $|\mathbf{B}| \cos \alpha$ may be either positive or negative depending on α. Because $\mathbf{A} \cdot \mathbf{B} = |\mathbf{A}||\mathbf{B}| \cos \alpha$, it follows that

$$\mathbf{A} \cdot \mathbf{B} = |\mathbf{A}|(|\mathbf{B}| \cos \alpha) \tag{5}$$

Hence the dot product of **A** and **B** is the magnitude of **A** multiplied by the scalar projection of **B** onto **A**. See Fig. 16.3.4(a) and (b). Because dot multiplication is commutative, $\mathbf{A} \cdot \mathbf{B}$ is also the magnitude of **B** multiplied by the scalar projection of **A** onto **B**.

Example 2 Given $\mathbf{A} = -5\mathbf{i} + \mathbf{j}$ and $\mathbf{B} = 4\mathbf{i} + 2\mathbf{j}$, find the vector projection of **B** onto **A**.

Solution Figure 16.3.5 shows the position representations of vectors **A** and **B** as well as that of **C**, which is the vector projection of **B** onto **A**. From (5),

$$|\mathbf{B}| \cos \alpha = \frac{\mathbf{A} \cdot \mathbf{B}}{|\mathbf{A}|} = \frac{(-5)(4) + (1)(2)}{\sqrt{26}} = \frac{-18}{\sqrt{26}}$$

Therefore $|\mathbf{C}| = 18/\sqrt{26}$. Because $\cos \alpha < 0$, $\frac{1}{2}\pi < \alpha < \pi$; so the direction of **C** is opposite that of **A**. Hence $\mathbf{C} = c\mathbf{A}$, and $c < 0$. Then

$$\mathbf{C} = -5c\mathbf{i} + c\mathbf{j}$$

Because $|\mathbf{C}| = 18/\sqrt{26}$,

$$\frac{18}{\sqrt{26}} = \sqrt{25c^2 + c^2}$$

so $c = -\frac{9}{13}$, from which it follows that

$$\mathbf{C} = \frac{45}{13}\mathbf{i} - \frac{9}{13}\mathbf{j}$$

If $\mathbf{A} = a_1\mathbf{i} + a_2\mathbf{j}$, then

$$\mathbf{A} \cdot \mathbf{i} = a_1 \quad \text{and} \quad \mathbf{A} \cdot \mathbf{j} = a_2$$

Hence the dot product of **A** and **i** gives the component of **A** in the direction of **i** and the dot product of **A** and **j** gives the component of **A** in the direction of **j**. To generalize this result let **U** be any unit vector. Then from (5),

$$\mathbf{A} \cdot \mathbf{U} = |\mathbf{A}|(|\mathbf{U}| \cos \alpha) = |\mathbf{A}| \cos \alpha$$

and so $\mathbf{A} \cdot \mathbf{U}$ is the scalar projection of **A** onto **U**, which is called the *component* of the vector **A** in the direction of **U**. More generally, the **component** of a vector **A** in the direction of a vector **B** is the scalar projection of **A** onto a unit vector in the direction of **B**.

In Sec. 7.5 we stated that if a constant force of F pounds moves an object a distance d feet along a straight line and the force is acting in the direction of motion, then if W is the number of foot-pounds in the work done by the force, $W = Fd$. Suppose, however, that the constant force is not directed along the line of motion. In this case the physicist defines the **work** done as the *product of the component of the force along the line of motion times the displacement.* If the object moves from the point A to the point B, we call the vector, having $\overrightarrow{AB}$ as a representation, the **displacement vector** and denote it by $\mathbf{V}(\overrightarrow{AB})$. So if the magnitude of a constant force vector $\mathbf{F}$ is expressed in pounds and the distance from A to B is expressed in feet, and α is the radian measure of the angle between the vectors $\mathbf{F}$ and $\mathbf{V}(\overrightarrow{AB})$, then if W is the number of foot-pounds in the work done by the force $\mathbf{F}$ in moving an object from A to B,

$$\begin{aligned} W &= (|\mathbf{F}| \cos \alpha)|\mathbf{V}(\overrightarrow{AB})| \\ &= |\mathbf{F}||\mathbf{V}(\overrightarrow{AB})| \cos \alpha \\ &= \mathbf{F} \cdot \mathbf{V}(\overrightarrow{AB}) \end{aligned}$$

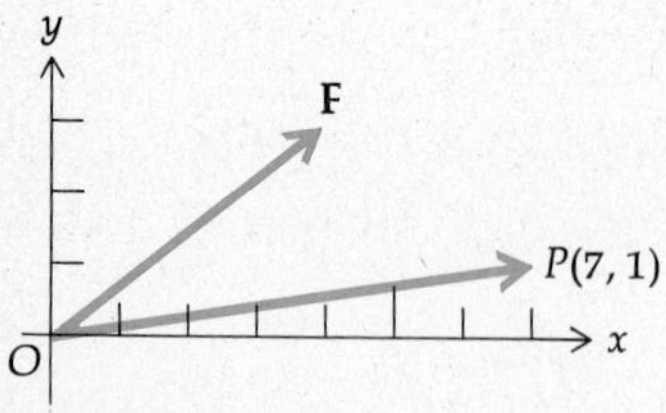

FIGURE 16.3.6

Example 3 Suppose that a force $\mathbf{F}$ has a magnitude of 6 lb and $\frac{1}{6}\pi$ is the radian measure of the angle giving its direction. Find the work done by $\mathbf{F}$ in moving an object along a straight line from the origin to the point $P(7, 1)$, where distance is measured in feet.

Solution Figure 16.3.6 shows the position representations of $\mathbf{F}$ and $\mathbf{V}(\overrightarrow{OP})$. If W ft-lb is the work done, then

$$W = \mathbf{F} \cdot \mathbf{V}(\overrightarrow{OP})$$

$\mathbf{F} = \langle 6 \cos \frac{1}{6}\pi, 6 \sin \frac{1}{6}\pi \rangle = \langle 3\sqrt{3}, 3 \rangle$, and $\mathbf{V}(\overrightarrow{OP}) = \langle 7, 1 \rangle$. So

$$W = \langle 3\sqrt{3}, 3 \rangle \cdot \langle 7, 1 \rangle = 21\sqrt{3} + 3 \approx 39.37$$

Therefore the work done is 39.37 ft-lb.

Vectors have geometric representations that are independent of the coordinate system used. Because of this, vector analysis can be used to prove certain theorems of plane geometry. This is illustrated in the following example.

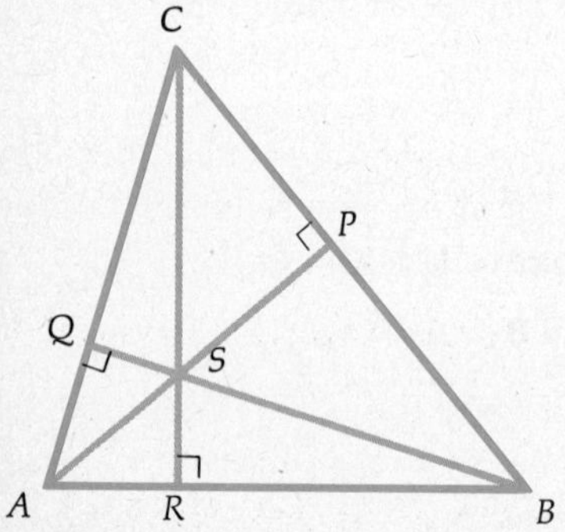

FIGURE 16.3.7

Example 4 Prove by vector analysis that the altitudes of a triangle meet in a point.

Solution Let ABC be a triangle having altitudes AP and BQ intersecting at point S. Draw a line through C and S intersecting AB at point R. We wish to prove that RC is perpendicular to AB (see Fig. 16.3.7).

Let $\overrightarrow{AB}, \overrightarrow{BC}, \overrightarrow{AC}, \overrightarrow{AS}, \overrightarrow{BS}, \overrightarrow{CS}$ be representations of vectors. Let $\mathbf{V}(\overrightarrow{AB})$ be the vector having directed line segment $\overrightarrow{AB}$ as a representation. In a similar manner let $\mathbf{V}(\overrightarrow{BC})$, $\mathbf{V}(\overrightarrow{AC})$, $\mathbf{V}(\overrightarrow{AS})$, $\mathbf{V}(\overrightarrow{BS})$, and $\mathbf{V}(\overrightarrow{CS})$ be the vectors having the directed line segment in parentheses as a representation.

Because AP is an altitude of the triangle,

$$\mathbf{V}(\overrightarrow{AS}) \cdot \mathbf{V}(\overrightarrow{BC}) = 0 \tag{6}$$

Also, because BQ is an altitude of the triangle,

$$\mathbf{V}(\overrightarrow{BS}) \cdot \mathbf{V}(\overrightarrow{AC}) = 0 \tag{7}$$

To prove that RC is perpendicular to AB we shall show that

$$\mathbf{V}(\overrightarrow{CS}) \cdot \mathbf{V}(\overrightarrow{AB}) = 0$$

$$\begin{aligned}\mathbf{V}(\overrightarrow{CS}) \cdot \mathbf{V}(\overrightarrow{AB}) &= \mathbf{V}(\overrightarrow{CS}) \cdot [\mathbf{V}(\overrightarrow{AC}) + \mathbf{V}(\overrightarrow{CB})] \\ &= \mathbf{V}(\overrightarrow{CS}) \cdot \mathbf{V}(\overrightarrow{AC}) + \mathbf{V}(\overrightarrow{CS}) \cdot \mathbf{V}(\overrightarrow{CB}) \\ &= [\mathbf{V}(\overrightarrow{CB}) + \mathbf{V}(\overrightarrow{BS})] \cdot \mathbf{V}(\overrightarrow{AC}) \\ &\qquad + [\mathbf{V}(\overrightarrow{CA}) + \mathbf{V}(\overrightarrow{AS})] \cdot \mathbf{V}(\overrightarrow{CB}) \\ &= \mathbf{V}(\overrightarrow{CB}) \cdot \mathbf{V}(\overrightarrow{AC}) + \mathbf{V}(\overrightarrow{BS}) \cdot \mathbf{V}(\overrightarrow{AC}) \\ &\qquad + \mathbf{V}(\overrightarrow{CA}) \cdot \mathbf{V}(\overrightarrow{CB}) + \mathbf{V}(\overrightarrow{AS}) \cdot \mathbf{V}(\overrightarrow{CB})\end{aligned}$$

Replacing $\mathbf{V}(\overrightarrow{CA})$ by $-\mathbf{V}(\overrightarrow{AC})$ and using (6) and (7), we obtain

$$\begin{aligned}\mathbf{V}(\overrightarrow{CS}) \cdot \mathbf{V}(\overrightarrow{AB}) &= \mathbf{V}(\overrightarrow{CB}) \cdot \mathbf{V}(\overrightarrow{AC}) + 0 + [-\mathbf{V}(\overrightarrow{AC})] \cdot \mathbf{V}(\overrightarrow{CB}) + 0 \\ &= 0\end{aligned}$$

Therefore altitudes AP, BQ, and RC meet in a point.

EXERCISES 16.3

In Exercises 1 through 4, find $\mathbf{A} \cdot \mathbf{B}$.

1. $\mathbf{A} = \langle -1, 2 \rangle$; $\mathbf{B} = \langle -4, 3 \rangle$
2. $\mathbf{A} = \langle \frac{1}{3}, -\frac{1}{2} \rangle$; $\mathbf{B} = \langle \frac{5}{2}, \frac{4}{3} \rangle$
3. $\mathbf{A} = 2\mathbf{i} - \mathbf{j}$; $\mathbf{B} = \mathbf{i} + 3\mathbf{j}$
4. $\mathbf{A} = -2\mathbf{i}$; $\mathbf{B} = -\mathbf{i} + \mathbf{j}$
5. Show that $\mathbf{i} \cdot \mathbf{i} = 1$; $\mathbf{j} \cdot \mathbf{j} = 1$; $\mathbf{i} \cdot \mathbf{j} = 0$.
6. Prove Theorem 16.3.2(i).
7. Prove Theorem 16.3.2(ii).
8. Prove Theorem 16.3.3(i).
9. Prove Theorem 16.3.3(ii).
10. Prove Theorem 16.3.3(iii).

In Exercises 11 through 14, if α is the radian measure of the angle between **A** *and* **B**, *find* $\cos \alpha$.

11. $\mathbf{A} = \langle 4, 3 \rangle$; $\mathbf{B} = \langle 1, -1 \rangle$
12. $\mathbf{A} = \langle -2, -3 \rangle$; $\mathbf{B} = \langle 3, 2 \rangle$
13. $\mathbf{A} = 5\mathbf{i} - 12\mathbf{j}$; $\mathbf{B} = 4\mathbf{i} + 3\mathbf{j}$
14. $\mathbf{A} = 2\mathbf{i} + 4\mathbf{j}$; $\mathbf{B} = -5\mathbf{j}$
15. Find k such that the radian measure of the angle between the vectors in Example 1 of this section is $\frac{1}{4}\pi$.
16. Given $\mathbf{A} = k\mathbf{i} - 2\mathbf{j}$ and $\mathbf{B} = k\mathbf{i} + 6\mathbf{j}$, where k is a scalar. Find k such that **A** and **B** are orthogonal.
17. Given $\mathbf{A} = 5\mathbf{i} - k\mathbf{j}$ and $\mathbf{B} = k\mathbf{i} + 6\mathbf{j}$, where k is a scalar. Find (a) k such that **A** and **B** are orthogonal; (b) k such that **A** and **B** are parallel.
18. Find k such that the vectors given in Exercise 16 have opposite directions.
19. Given $\mathbf{A} = 5\mathbf{i} + 12\mathbf{j}$; $\mathbf{B} = \mathbf{i} + k\mathbf{j}$, where k is a scalar. Find k such that the radian measure of the angle between **A** and **B** is $\frac{1}{3}\pi$.

20. Find two unit vectors each having a representation whose initial point is (2, 4) and which is tangent to the parabola $y = x^2$ there.

21. Find two unit vectors each having a representation whose initial point is (2, 4) and which is normal to the parabola $y = x^2$ there.

22. If $\mathbf{A} = 2\mathbf{i} - 7\mathbf{j}$, find the unit vectors that are orthogonal to **A**.

23. If **A** is the vector $a_1\mathbf{i} + a_2\mathbf{j}$, find the unit vectors that are orthogonal to **A**.

24. If $\mathbf{A} = 5\mathbf{i} - 9c\mathbf{j}$ and $\mathbf{B} = 7\mathbf{i} - 4c\mathbf{j}$, show that there is no real value for c such that **A** and **B** are orthogonal.

25. If $\mathbf{A} = -8\mathbf{i} + 4\mathbf{j}$ and $\mathbf{B} = 7\mathbf{i} - 6\mathbf{j}$, find the vector projection of **A** onto **B**.

26. Find the vector projection of **B** onto **A** for the vectors of Exercise 25.

27. Find the component of the vector $\mathbf{A} = 5\mathbf{i} - 6\mathbf{j}$ in the direction of the vector $\mathbf{B} = 7\mathbf{i} + \mathbf{j}$.

28. For the vectors **A** and **B** of Exercise 27, find the component of the vector **B** in the direction of vector **A**.

29. A vector **F** represents a force that has a magnitude of 8 lb and $\frac{1}{3}\pi$ is the radian measure of the angle giving its direction. Find the work done by the force in moving an object (a) along the x axis from the origin to the point (6, 0) and (b) along the y axis from the origin to the point (0, 6). Distance is measured in feet.

30. A vector **F** represents a force that has a magnitude of 10 lb and $\frac{1}{4}\pi$ is the radian measure of the angle giving its direction. Find the work done by the force in moving an object along the y axis from the point (0, −2) to the point (0, 5). Distance is measured in feet.

31. A vector **F** represents a force that has a magnitude of 9 lb and $\frac{2}{3}\pi$ is the radian measure of the angle giving its direction. Find the work done by the force in moving an object from the origin to the point (−4, −2). Distance is measured in feet.

32. Two forces represented by the vectors $\mathbf{F}_1$ and $\mathbf{F}_2$ act on a particle and cause it to move along a straight line from the point (2, 5) to the point (7, 3). If $\mathbf{F}_1 = 3\mathbf{i} - \mathbf{j}$ and $\mathbf{F}_2 = -4\mathbf{i} + 5\mathbf{j}$, the magnitudes of the forces are measured in pounds, and distance is measured in feet, find the work done by the two forces acting together.

33. If **A** and **B** are vectors, prove that

$$(\mathbf{A} + \mathbf{B}) \cdot (\mathbf{A} + \mathbf{B}) = \mathbf{A} \cdot \mathbf{A} + 2\mathbf{A} \cdot \mathbf{B} + \mathbf{B} \cdot \mathbf{B}$$

34. Prove by vector analysis that the medians of a triangle meet in a point.

35. Prove by vector analysis that the line segment joining the midpoints of two sides of a triangle is parallel to the third side and its length is one-half the length of the third side.

36. Prove by vector analysis that the line segment joining the midpoints of the nonparallel sides of a trapezoid is parallel to the parallel sides and its length is one-half the sum of the lengths of the parallel sides.

37. Prove that two nonzero vectors are parallel if and only if the radian measure of the angle between them is 0 or π.

16.4.1 VECTOR-VALUED FUNCTIONS AND PARAMETRIC EQUATIONS

We now consider a function whose domain is a set of real numbers and whose range is a set of vectors. Such a function is called a *vector-valued function.*

16.4.1 Definition

Let f and g be two real-valued functions of a real variable t. Then for every number t in the domain common to f and g there is a vector **R** defined by

$$\mathbf{R}(t) = f(t)\mathbf{i} + g(t)\mathbf{j} \tag{1}$$

and **R** is called a **vector-valued function.**

● ILLUSTRATION 1 Suppose

$$\mathbf{R}(t) = \sqrt{t-2}\,\mathbf{i} + (t-3)^{-1}\mathbf{j}$$

Let

$$f(t) = \sqrt{t-2} \text{ and } g(t) = (t-3)^{-1}$$

The domain of **R** is the set of values of t for which both $f(t)$ and $g(t)$ are defined. The function value $f(t)$ is defined for $t \geq 2$ and $g(t)$ is defined for all real numbers except 3. Therefore the domain of **R** is $\{t \mid t \geq 2, t \neq 3\}$. ●

If **R** is the vector-valued function defined by (1), as t assumes all values in the domain of **R**, the endpoint of the position representation of the vector $\mathbf{R}(t)$ traces a curve C. For each such value of t there is a point (x, y) on C for which

$$x = f(t) \quad \text{and} \quad y = g(t) \tag{2}$$

The curve C may be defined by (1) or (2). Equation (1) is called a **vector equation** of C, and Eqs. (2) are called **parametric equations** of C. The variable t is a **parameter.** The curve C is also called a **graph;** that is, the set of all points (x, y) satisfying (2) is the graph of the vector-valued function **R**.

A vector equation of a curve, as well as parametric equations of a curve, gives the curve a direction at each point. That is, if we think of the curve as being traced by a particle, we can consider the positive direction along a curve as the direction in which the particle moves as the parameter t increases. In such a case as this, t may be taken to be the measure of the time, and the vector $\mathbf{R}(t)$ is called the **position vector.** Sometimes $\mathbf{R}(t)$ is referred to as the **radius vector.**

If the parameter t is eliminated from the pair of Eqs. (2), we obtain one equation in x and y, which is called a **cartesian equation** of C. It may happen that elimination of the parameter leads to a cartesian equation whose graph contains more points than the graph defined by either the vector equation or the parametric equations. This situation occurs in Example 4.

Example 1 Given the vector equation $\mathbf{R}(t) = 2\cos t\mathbf{i} + 2\sin t\mathbf{j}$. (a) Draw a sketch of the graph of this equation, and (b) find a cartesian equation of the graph.

Solution The domain of **R** is the set of all real numbers. The values of x and y for particular values of t can be tabulated. We find the magnitude of the position vector. For every t,

$$|\mathbf{R}(t)| = \sqrt{4\cos^2 t + 4\sin^2 t} = 2\sqrt{\cos^2 t + \sin^2 t} = 2$$

Therefore the endpoint of the position representation of each vector $\mathbf{R}(t)$ is two units from the origin. By letting t take on all numbers in the closed interval $[0, 2\pi]$, we obtain a circle having its center at the origin and radius 2. This is the entire graph because any value of t will give a point on this circle. A sketch of the circle is shown in Fig. 16.4.1. Parametric equations of the graph are

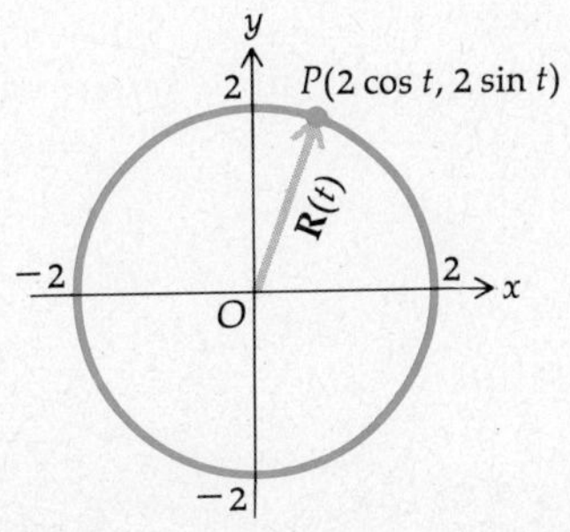

FIGURE 16.4.1

$$x = 2\cos t \quad \text{and} \quad y = 2\sin t$$

A cartesian equation of the graph can be found by eliminating t from the two parametric equations, which when squaring on both sides of each equation and adding gives

$$x^2 + y^2 = 4$$

As previously stated, upon eliminating t from parametric equations (2) we obtain a cartesian equation. The cartesian equation either implicitly or explicitly defines y as one or more functions of x. That is, if $x = f(t)$ and $y = g(t)$, then $y = h(x)$. If h is a differentiable function of x and f is a differentiable function of t, it follows from the chain rule that

$$D_t y = (D_x y)(D_t x)$$

or

$$g'(t) = (h'(x))(f'(t))$$

or, by using differential notation,

$$\frac{dy}{dt} = \frac{dy}{dx}\frac{dx}{dt}$$

If $dx/dt \neq 0$, we can divide on both sides of the above equation by dx/dt and obtain

$$\frac{dy}{dx} = \frac{\frac{dy}{dt}}{\frac{dx}{dt}} \tag{3}$$

Equation (3) enables us to find the derivative of y with respect to x directly from the parametric equations.

Because $d^2y/dx^2 = d(y')/dx$, then

$$\frac{d^2y}{dx^2} = \frac{\frac{d(y')}{dt}}{\frac{dx}{dt}} \tag{4}$$

Example 2 Given $x = 3t^2$ and $y = 4t^3$, find dy/dx and d^2y/dx^2 without eliminating t.

Solution From (3),

$$\frac{dy}{dx} = \frac{\frac{dy}{dt}}{\frac{dx}{dt}} = \frac{12t^2}{6t} = 2t$$

TABLE 16.4.1

t	x	y
0	0	0
$\frac{1}{2}$	$\frac{3}{4}$	$\frac{1}{2}$
1	3	4
2	12	32
$-\frac{1}{2}$	$\frac{3}{4}$	$-\frac{1}{2}$
-1	3	-4
-2	12	-32

Because $y' = 2t$, $d(y')/dt = 2$. Thus, from (4),

$$\frac{d^2y}{dx^2} = \frac{d(y')}{dx} = \frac{\frac{d(y')}{dt}}{\frac{dx}{dt}} = \frac{2}{6t} = \frac{1}{3t}$$

Example 3 (a) Draw a sketch of the graph of the curve defined by the parametric equations of Example 2, and (b) find a cartesian equation of the graph in (a).

Solution Because $x = 3t^2$, x is never negative. Table 16.4.1 gives values of x and y for particular values of t. Because $D_xy = 2t$, then when $t = 0$, $D_xy = 0$; hence the tangent line is horizontal at the point (0, 0). A sketch of the graph is shown in Fig. 16.4.2. From the two parametric equations $x = 3t^2$ and $y = 4t^3$ we get $x^3 = 27t^6$ and $y^2 = 16t^6$. Therefore

$$\frac{x^3}{27} = \frac{y^2}{16}$$

or, equivalently,

$$16x^3 = 27y^2 \tag{5}$$

which is the cartesian equation desired.

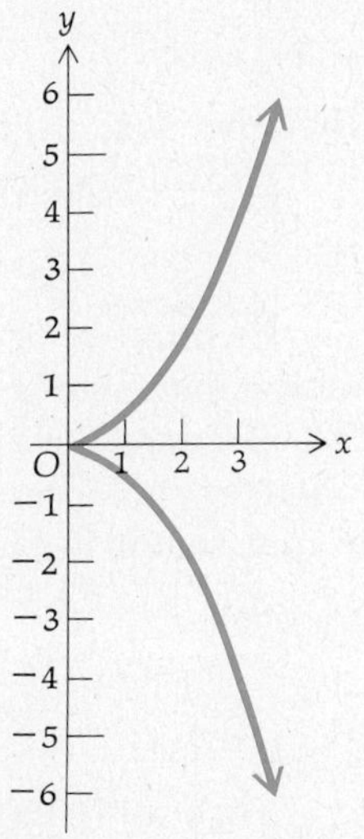

FIGURE 16.4.2

● ILLUSTRATION 2 If in Eq. (5) we differentiate implicitly,

$$48x^2 = 54y\frac{dy}{dx}$$

$$\frac{dy}{dx} = \frac{8x^2}{9y}$$

Substituting for x and y in terms of t from the given parametric equations, we obtain

$$\frac{dy}{dx} = \frac{8(3t^2)^2}{9(4t^3)} = 2t$$

which agrees with the value of dy/dx found in Example 2. ●

Example 4 Draw a sketch of the graph of the curve defined by the parametric equations

$$x = \cosh t \quad \text{and} \quad y = \sinh t \tag{6}$$

Also find a cartesian equation of the graph.

Solution Squaring on both sides of the given equations and subtracting we have

$$x^2 - y^2 = \cosh^2 t - \sinh^2 t$$

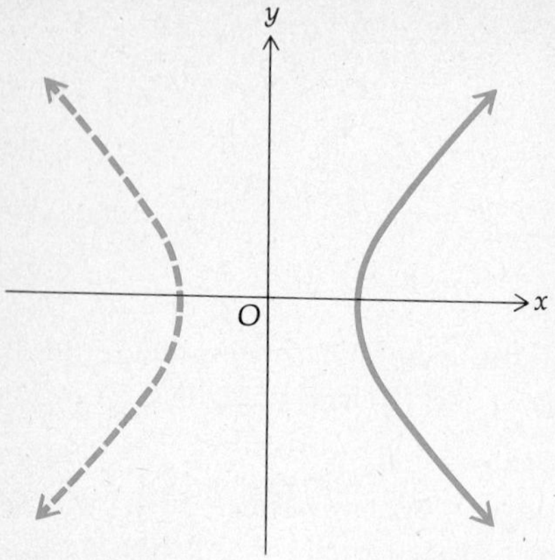

FIGURE 16.4.3

From the identity $\cosh^2 t - \sinh^2 t = 1$ this equation becomes

$$x^2 - y^2 = 1$$

This is an equation of an equilateral hyperbola. Note that for t any real number, $\cosh t$ is never less than 1. Thus the curve defined by parametric equations (6) consists of only the points on the right branch of the hyperbola. A sketch of this curve is shown in Fig. 16.4.3. A cartesian equation is $x^2 - y^2 = 1$, where $x \geq 1$.

The results of Example 4 can be used to show how the function values of the hyperbolic sine and hyperbolic cosine functions have the same relationship to the equilateral hyperbola as the trigonometric sine and cosine functions have to the circle. The equations

$$x = \cos t \quad \text{and} \quad y = \sin t \tag{7}$$

are a set of parametric equations of the unit circle because if t is eliminated from them by squaring on both sides of each and adding, we obtain

$$x^2 + y^2 = \cos^2 t + \sin^2 t = 1$$

The parameter t in Eqs. (7) can be interpreted as the number of radians in the measure of the angle between the x axis and a line from the origin to $P(\cos t, \sin t)$ on the unit circle. See Fig. 16.4.4. Because the area of a circular sector of radius r units and a central angle of radian measure t is given by $\frac{1}{2}r^2t$ square units, the area of the circular sector in Fig. 16.4.4 is $\frac{1}{2}t$ square units because $r = 1$.

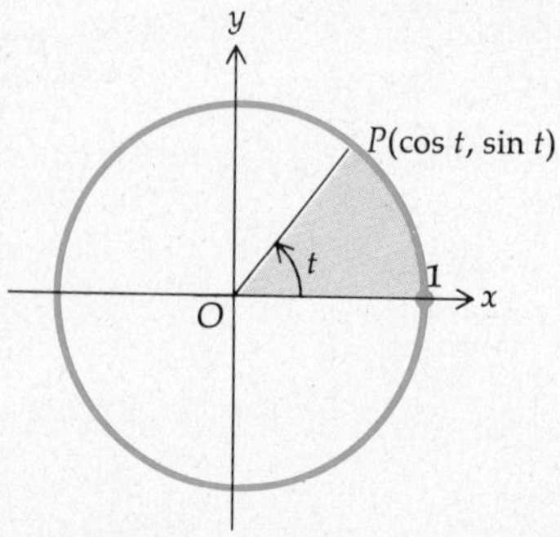

FIGURE 16.4.4

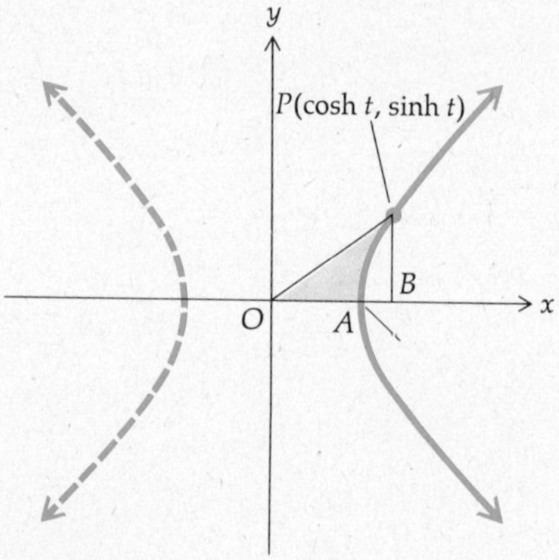

FIGURE 16.4.5

In Example 4 we showed that parametric equations (6) are a set of parametric equations of the right branch of the equilateral hyperbola $x^2 - y^2 = 1$. This hyperbola is called the **unit hyperbola.** Let $P(\cosh t, \sinh t)$ be a point on this curve, and calculate the area of the sector AOP shown in Fig. 16.4.5. The sector AOP is the region bounded by the x axis, the line OP, and the arc AP of the unit hyperbola. If A_1 square units is the area of sector AOP, A_2 square units is the area of triangle OBP, and A_3 square units is the area of region ABP,

$$A_1 = A_2 - A_3 \tag{8}$$

From the formula for determining the area of a triangle,

$$A_2 = \tfrac{1}{2}\cosh t \sinh t \tag{9}$$

We find A_3 by integration:

$$\begin{aligned} A_3 &= \int_0^t \sinh u \, d(\cosh u) \\ &= \int_0^t \sinh^2 u \, du \\ &= \tfrac{1}{2}\int_0^t (\cosh 2u - 1)\, du \\ &= \tfrac{1}{4}\sinh 2u - \tfrac{1}{2}u \Big]_0^t \end{aligned}$$

and so

$$A_3 = \tfrac{1}{2}\cosh t \sinh t - \tfrac{1}{2}t \tag{10}$$

Substituting from (9) and (10) into (8) we have

$$\begin{aligned} A_1 &= \tfrac{1}{2}\cosh t \sinh t - (\tfrac{1}{2}\cosh t \sinh t - \tfrac{1}{2}t) \\ &= \tfrac{1}{2}t \end{aligned}$$

Thus the number of square units in the area of circular sector AOP of Fig. 16.4.4 and the number of square units in the area of sector AOP of Fig. 16.4.5 is, in each case, one-half the value of the parameter associated with the point P. For the unit circle, the parameter t is the radian measure of the angle AOP. The parameter t for the unit hyperbola is not interpreted as the measure of an angle; however, sometimes the term *hyperbolic radian* is used in connection with t.

In Sec. 14.1, where Cauchy's mean-value theorem (14.1.3) was stated and proved, we indicated that a geometric interpretation would be given in this section because parametric equations are needed. Recall that the theorem states that if f and g are two functions such that (i) f and g are continuous on $[a, b]$, (ii) f and g are differentiable on (a, b), and (iii) for all x in (a, b) $g'(x) \neq 0$, then there exists a number z in the open interval (a, b) such that

$$\frac{f(b) - f(a)}{g(b) - g(a)} = \frac{f'(z)}{g'(z)}$$

Figure 16.4.6 shows a curve having the parametric equations $x = g(t)$ and $y = f(t)$, where $a \leq t \leq b$. The slope of the curve in the figure at a particular point is given by

$$\frac{dy}{dx} = \frac{f'(t)}{g'(t)}$$

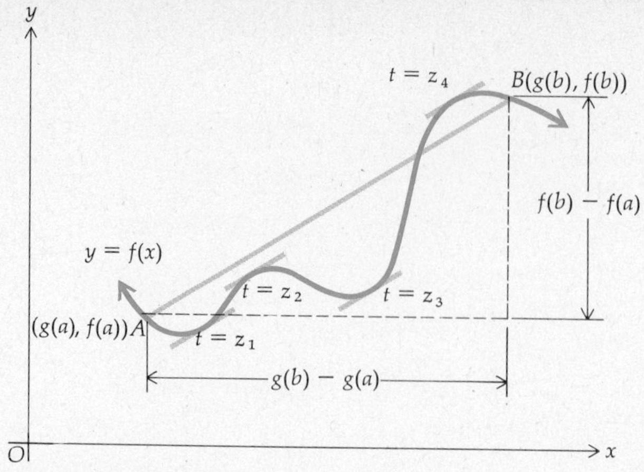

FIGURE 16.4.6

and the slope of the line segment through the points $A(g(a), f(a))$ and $B(g(b), f(b))$ is given by

$$\frac{f(b) - f(a)}{g(b) - g(a)}$$

Cauchy's mean-value theorem states that the slopes are equal for at least one value of t between a and b. For the curve shown in Fig. 16.4.6 there are four values of t satisfying the conclusion of the theorem: $t = z_1$, $t = z_2$, $t = z_3$, and $t = z_4$.

We now show how parametric equations can be used to define a curve that is described by a physical motion. The curve we consider is a **cycloid,** which is the curve traced by a point on the circumference of a circle as the circle rolls along a straight line. Suppose the circle has radius a. Let the fixed straight line on which the circle rolls be the x axis, and let the origin be one of the points at which the given point P comes in contact with the x axis. See Fig. 16.4.7, which shows the circle after it has rolled through an angle of t radians.

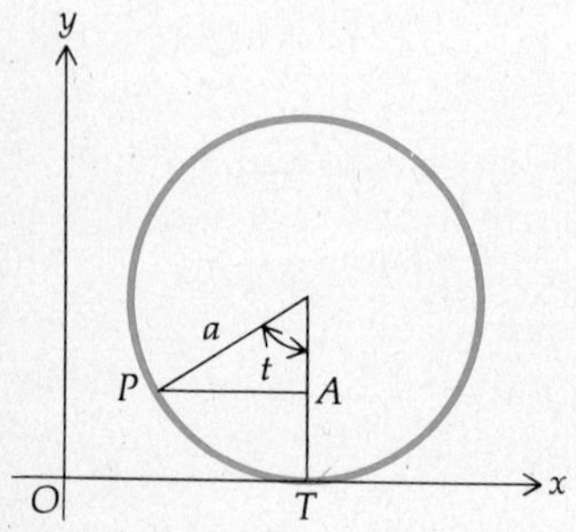

FIGURE 16.4.7

From Fig. 16.4.7,

$$\mathbf{V}(\overrightarrow{OT}) + \mathbf{V}(\overrightarrow{TA}) + \mathbf{V}(\overrightarrow{AP}) = \mathbf{V}(\overrightarrow{OP}) \tag{11}$$

$|\mathbf{V}(\overrightarrow{OT})|$ = length of the arc $PT = at$. Because the direction of $\mathbf{V}(\overrightarrow{OT})$ is along the positive x axis,

$$\mathbf{V}(\overrightarrow{OT}) = at\mathbf{i} \tag{12}$$

Also, $|\mathbf{V}(\overrightarrow{TA})| = a - a \cos t$. And because the direction of $\mathbf{V}(\overrightarrow{TA})$ is the same as the direction of $\mathbf{j}$,

$$\mathbf{V}(\overrightarrow{TA}) = a(1 - \cos t)\mathbf{j} \tag{13}$$

$|\mathbf{V}(\overrightarrow{AP})| = a \sin t$, and the direction of $\mathbf{V}(\overrightarrow{AP})$ is the same as the direction of $-\mathbf{i}$; thus

$$\mathbf{V}(\overrightarrow{AP}) = -a \sin t\mathbf{i} \tag{14}$$

Substituting from (12), (13), and (14) into (11) we obtain

$$at\mathbf{i} + a(1 - \cos t)\mathbf{j} - a \sin t\mathbf{i} = \mathbf{V}(\overrightarrow{OP})$$

or, equivalently,

$$\mathbf{V}(\overrightarrow{OP}) = a(t - \sin t)\mathbf{i} + a(1 - \cos t)\mathbf{j} \tag{15}$$

Equation (15) is a vector equation of the cycloid. So parametric equations of the cycloid are

$$x = a(t - \sin t) \quad \text{and} \quad y = a(1 - \cos t) \tag{16}$$

where t is any real number. A sketch of a portion of the cycloid is shown in Fig. 16.4.8.

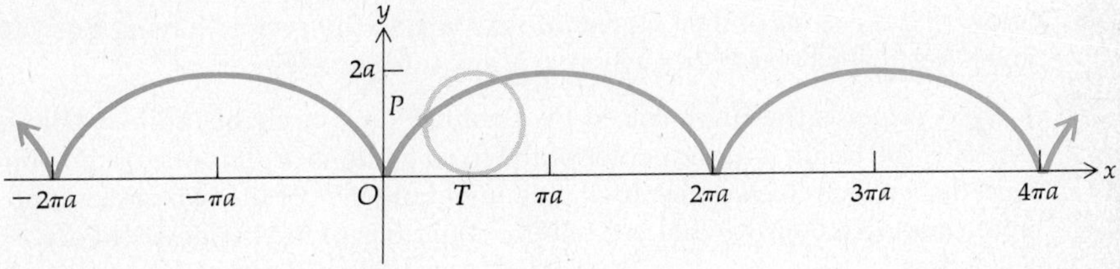

FIGURE 16.4.8

EXERCISES 16.4

In Exercises 1 through 6, find the domain of the vector-valued function **R**.

1. $\mathbf{R}(t) = (1/t)\mathbf{i} + \sqrt{4 - t}\mathbf{j}$
2. $\mathbf{R}(t) = (t^2 + 3)\mathbf{i} + (t - 1)\mathbf{j}$
3. $\mathbf{R}(t) = (\sin^{-1} t)\mathbf{i} + (\cos^{-1} t)\mathbf{j}$
4. $\mathbf{R}(t) = \ln(t + 1)\mathbf{i} + (\tan^{-1} t)\mathbf{j}$
5. $\mathbf{R}(t) = \sqrt{t^2 - 9}\mathbf{i} + \sqrt{t^2 + 2t - 8}\mathbf{j}$
6. $\mathbf{R}(t) = \sqrt{t - 4}\mathbf{i} + \sqrt{4 - t}\mathbf{j}$

In Exercises 7 through 12, find dy/dx and d^2y/dx^2 without eliminating the parameter.

7. $x = 3t, y = 2t^2$
8. $x = 1 - t^2, y = 1 + t$
9. $x = t^2e^t, y = t \ln t$
10. $x = e^{2t}, y = 1 + \cos t$
11. $x = a \cos t, y = b \sin t$
12. $x = a \cosh t, y = b \sinh t$

In Exercises 13 through 19, draw a sketch of the graph of the given vector equation and find a cartesian equation of the graph.

13. $\mathbf{R}(t) = t^2\mathbf{i} + (t + 1)\mathbf{j}$
14. $\mathbf{R}(t) = (t - 2)\mathbf{i} + (t^2 + 4)\mathbf{j}$
15. $\mathbf{R}(t) = 3 \cosh t\mathbf{i} + 5 \sinh t\mathbf{j}$
16. $\mathbf{R}(t) = \dfrac{4}{t^2}\mathbf{i} + \dfrac{4}{t}\mathbf{j}$
17. $\mathbf{R}(t) = \sec t\mathbf{i} + \tan t\mathbf{j}$; t in $(-\frac{1}{2}\pi, \frac{1}{2}\pi)$
18. $\mathbf{R}(t) = \cos t\mathbf{i} + \cos t\mathbf{j}$; t in $[0, \frac{1}{2}\pi]$
19. $\mathbf{R}(t) = 4 \cos t\mathbf{i} + 3 \sin t\mathbf{j}$; t in $[0, 2\pi]$

20. Find an equation of the tangent line to the curve $x = 1 + 3 \sin t$, $y = 2 - 5 \cos t$, at the point where $t = \frac{1}{6}\pi$.

21. Find an equation of the tangent line to the curve $x = 2 \sin t$, $y = 5 \cos t$, at the point where $t = \frac{1}{3}\pi$.

In Exercises 22 through 24, find equations of the horizontal tangent lines by finding the values of t for which $dy/dt = 0$, and find equations of the vertical tangent lines by finding the values of t for which $dx/dt = 0$. Then draw a sketch of the graph of the given pair of parametric equations.

22. $x = t^2 + t, y = t^2 - t$ **23.** $x = 4t^2 - 4t, y = 1 - 4t^2$ **24.** $x = \dfrac{3at}{1 + t^3}, y = \dfrac{3at^2}{1 + t^3}$

25. Parametric equations for the *trochoid* are

$$x = at - b \sin t \quad \text{and} \quad y = a - b \cos t$$

Show that the trochoid has no vertical tangent line if $a > b > 0$.

26. A projectile moves so that the coordinates of its position at any time t are given by the equations $x = 60t$ and $y = 80t - 16t^2$. Draw a sketch of the path of the projectile.

27. Find dy/dx, d^2y/dx^2, and d^3y/dx^3 at the point on the cycloid having Eqs. (16) for which y has its largest value when x is in the closed interval $[0, 2\pi a]$.

28. Show that the slope of the tangent line at $t = t_1$ to the cycloid having Eqs. (16) is $\cot \frac{1}{2}t_1$. Deduce then that the tangent line is vertical when $t = 2n\pi$, where n is any integer.

29. A *hypocycloid* is the curve traced by a point P on a circle of radius b that is rolling inside a fixed circle of radius a, $a > b$. If the origin is at the center of the fixed circle, $A(a, 0)$ is one of the points at which the point P comes in contact with the fixed circle, B is the moving point of tangency of the two circles, and the parameter t is the number of radians in the angle AOB, prove that parametric equations of the hypocycloid are

$$x = (a - b) \cos t + b \cos \frac{a - b}{b} t$$

and

$$y = (a - b) \sin t - b \sin \frac{a - b}{b} t$$

30. If $a = 4b$ in Exercise 29, we have a *hypocycloid of four cusps*. Show that parametric equations of this curve are $x = a \cos^3 t$ and $y = a \sin^3 t$.

31. Use the parametric equations of Exercise 30 to find a cartesian equation of the hypocycloid of four cusps, and draw a sketch of the graph of the resulting equation.

32. Parametric equations for the *tractrix* are

$$x = t - a \tanh \frac{t}{a} \qquad y = a \operatorname{sech} \frac{t}{a}$$

Draw a sketch of the curve for $a = 4$.

33. Prove that the parameter t in the parametric equations of a tractrix (see Exercise 32) is the x intercept of the tangent line.

34. Show that the tractrix of Exercise 32 is a curve such that the length of the segment of every tangent line from the point of tangency to the point of intersection with the x axis is constant and equal to a.

35. Find the area of the region bounded by the x axis and one arch of the cycloid, having Eqs. (16).

36. Find the centroid of the region of Exercise 35.

16.5 CALCULUS OF VECTOR-VALUED FUNCTIONS

We now discuss limits, continuity, and derivatives of vector-valued functions.

16.5.1 Definition Let **R** be a vector-valued function whose function values are given by

$$\mathbf{R}(t) = f(t)\mathbf{i} + g(t)\mathbf{j}$$

Then the **limit of R (*t*) as *t* approaches t_1** is defined by

$$\lim_{t \to t_1} \mathbf{R}(t) = \left[\lim_{t \to t_1} f(t)\right]\mathbf{i} + \left[\lim_{t \to t_1} g(t)\right]\mathbf{j}$$

if $\lim\limits_{t \to t_1} f(t)$ and $\lim\limits_{t \to t_1} g(t)$ both exist.

● ILLUSTRATION 1 If $\mathbf{R}(t) = \cos t\mathbf{i} + 2e^t\mathbf{j}$, then

$$\lim_{t \to 0} \mathbf{R}(t) = (\lim_{t \to 0} \cos t)\mathbf{i} + (\lim_{t \to 0} 2e^t)\mathbf{j} = \mathbf{i} + 2\mathbf{j}$$ ●

16.5.2 Definition The vector-valued function **R** is **continuous** at t_1 if and only if the following three conditions are satisfied:

(i) $\mathbf{R}(t_1)$ exists;

(ii) $\lim\limits_{t \to t_1} \mathbf{R}(t)$ exists;

(iii) $\lim\limits_{t \to t_1} \mathbf{R}(t) = \mathbf{R}(t_1)$.

From Definitions 16.5.1 and 16.5.2 it follows that the vector-valued function **R**, defined by $\mathbf{R}(t) = f(t)\mathbf{i} + g(t)\mathbf{j}$, is continuous at t_1 if and only if f and g are continuous there.

In the following definition the expression

$$\frac{\mathbf{R}(t + \Delta t) - \mathbf{R}(t)}{\Delta t}$$

is used to indicate the division of a vector by a scalar. This expression means

$$\frac{1}{\Delta t}[\mathbf{R}(t + \Delta t) - \mathbf{R}(t)]$$

16.5.3 Definition If **R** is a vector-valued function, then the **derivative** of **R** is a vector-valued function, denoted by $\mathbf{R}'$ and defined by

$$\mathbf{R}'(t) = \lim_{\Delta t \to 0} \frac{\mathbf{R}(t + \Delta t) - \mathbf{R}(t)}{\Delta t}$$

if this limit exists.

The notation $D_t\mathbf{R}(t)$ is sometimes used in place of $\mathbf{R}'(t)$.

The following theorem follows from Definition 16.5.3 and the definition of the derivative of a real-valued function.

16.5.4 Theorem If **R** is a vector-valued function defined by

$$\mathbf{R}(t) = f(t)\mathbf{i} + g(t)\mathbf{j} \qquad (1)$$

then

$$\mathbf{R}'(t) = f'(t)\mathbf{i} + g'(t)\mathbf{j}$$

if $f'(t)$ and $g'(t)$ exist.

Proof From Definition 16.5.3,

$$\begin{aligned}\mathbf{R}'(t) &= \lim_{\Delta t \to 0} \frac{\mathbf{R}(t + \Delta t) - \mathbf{R}(t)}{\Delta t} \\ &= \lim_{\Delta t \to 0} \frac{[f(t + \Delta t)\mathbf{i} + g(t + \Delta t)\mathbf{j}] - [f(t)\mathbf{i} + g(t)\mathbf{j}]}{\Delta t} \\ &= \lim_{\Delta t \to 0} \frac{[f(t + \Delta t) - f(t)]}{\Delta t}\mathbf{i} + \lim_{\Delta t \to 0} \frac{[g(t + \Delta t) - g(t)]}{\Delta t}\mathbf{j} \\ &= f'(t)\mathbf{i} + g'(t)\mathbf{j}\end{aligned}$$

■

The direction of $\mathbf{R}'(t)$ is along the tangent line at the point $(f(t), g(t))$ to the graph of $\mathbf{R}(t)$ having vector equation (1). That is, the direction of $\mathbf{R}'(t)$ is given by θ $(0 \leq \theta < 2\pi)$, where

$$\tan\theta = \frac{g'(t)}{f'(t)} = \frac{\frac{dy}{dt}}{\frac{dx}{dt}} = \frac{dy}{dx}$$

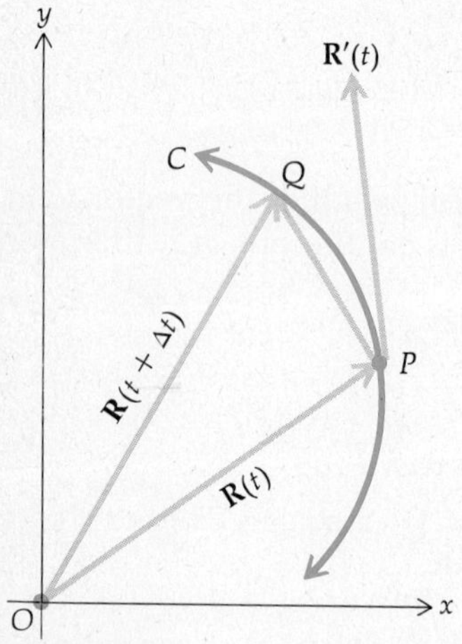

FIGURE 16.5.1

A geometric interpretation of Definition 16.5.3 is obtained by considering representations of the vectors $\mathbf{R}(t)$, $\mathbf{R}(t + \Delta t)$, and $\mathbf{R}'(t)$. Refer to Fig. 16.5.1. The curve C is traced by the endpoint of the position representation of $\mathbf{R}(t)$ as t assumes all values in the domain of $\mathbf{R}$. Let $\overrightarrow{OP}$ be the position representation of $\mathbf{R}(t)$ and OQ be the position representation of $\mathbf{R}(t + \Delta T)$. Then $\mathbf{R}(t + \Delta t) - \mathbf{R}(t)$ is a vector for which $\overrightarrow{PQ}$ is a representation. If the vector $\mathbf{R}(t + \Delta t) - \mathbf{R}(t)$ is multiplied by the scalar $1/\Delta t$, we obtain a vector having the same direction and whose magnitude is $1/|\Delta t|$ times the magnitude of $\mathbf{R}(t + \Delta t) - \mathbf{R}(t)$. As Δt approaches zero, the vector $[\mathbf{R}(t + \Delta t) - \mathbf{R}(t)]/\Delta t$ approaches a vector having one of its representations tangent to the curve C at the point P.

● ILLUSTRATION 2 If $\mathbf{R}(t) = (2 + \sin t)\mathbf{i} + \cos t\mathbf{j}$, then

$$\mathbf{R}'(t) = \cos t\mathbf{i} - \sin t\mathbf{j}$$

●

Higher order derivatives of vector-valued functions are defined as for higher order derivatives of real-valued functions. So if $\mathbf{R}$ is a vector-valued function defined by $\mathbf{R}(t) = f(t)\mathbf{i} + g(t)\mathbf{j}$, the second derivative of $\mathbf{R}$, denoted by $\mathbf{R}''(t)$, is given by

$$\mathbf{R}''(t) = D_t[\mathbf{R}'(t)]$$

The notation $D_t^2\mathbf{R}(t)$ can be used in place of $\mathbf{R}''(t)$. By applying Theorem 16.5.4 to $\mathbf{R}'(t)$,

$$\mathbf{R}''(t) = f''(t)\mathbf{i} + g''(t)\mathbf{j}$$

if $f''(t)$ and $g''(t)$ exist.

● ILLUSTRATION 3 If $\mathbf{R}(t) = (\ln t)\mathbf{i} + \left(\frac{1}{t}\right)\mathbf{j}$, then

$$\mathbf{R}'(t) = \frac{1}{t}\mathbf{i} - \frac{1}{t^2}\mathbf{j}$$

and

$$\mathbf{R}''(t) = -\frac{1}{t^2}\mathbf{i} + \frac{2}{t^3}\mathbf{j}$$ ●

16.5.5 Definition A vector-valued function $\mathbf{R}$ is said to be **differentiable** on an interval if $\mathbf{R}'(t)$ exists for all values of t in the interval.

The following theorems give differentiation formulas for vector-valued functions. The proofs are based on Theorem 16.5.4 and theorems on differentiation of real-valued functions.

16.5.6 Theorem If $\mathbf{R}$ and $\mathbf{Q}$ are differentiable vector-valued functions on an interval, then $\mathbf{R} + \mathbf{Q}$ is differentiable on the interval, and

$$D_t[\mathbf{R}(t) + \mathbf{Q}(t)] = D_t\mathbf{R}(t) + D_t\mathbf{Q}(t)$$

The proof of this theorem is left as an exercise (see Exercise 21).

Example 1 If $\mathbf{R}(t) = t^2\mathbf{i} + (t - 1)\mathbf{j}$ and $\mathbf{Q}(t) = \sin t\mathbf{i} + \cos t\mathbf{j}$, verify Theorem 16.5.6.

Solution

$$\begin{aligned} D_t[\mathbf{R}(t) + \mathbf{Q}(t)] &= D_t([t^2\mathbf{i} + (t-1)\mathbf{j}] + [\sin t\mathbf{i} + \cos t\mathbf{j}]) \\ &= D_t[(t^2 + \sin t)\mathbf{i} + (t - 1 + \cos t)\mathbf{j}] \\ &= (2t + \cos t)\mathbf{i} + (1 - \sin t)\mathbf{j} \\ D_t\mathbf{R}(t) + D_t\mathbf{Q}(t) &= D_t[t^2\mathbf{i} + (t-1)\mathbf{j}] + D_t(\sin t\mathbf{i} + \cos t\mathbf{j}) \\ &= (2t\mathbf{i} + \mathbf{j}) + (\cos t\mathbf{i} - \sin t\mathbf{j}) \\ &= (2t + \cos t)\mathbf{i} + (1 - \sin t)\mathbf{j} \end{aligned}$$

Hence $D_t[\mathbf{R}(t) + \mathbf{Q}(t)] = D_t\mathbf{R}(t) + D_t\mathbf{Q}(t)$.

16.5.7 Theorem If $\mathbf{R}$ and $\mathbf{Q}$ are differentiable vector-valued functions on an interval, then $\mathbf{R} \cdot \mathbf{Q}$ is differentiable on the interval, and

$$D_t[\mathbf{R}(t) \cdot \mathbf{Q}(t)] = [D_t\mathbf{R}(t)] \cdot \mathbf{Q}(t) + \mathbf{R}(t) \cdot [D_t\mathbf{Q}(t)]$$

Proof Let $\mathbf{R}(t) = f_1(t)\mathbf{i} + g_1(t)\mathbf{j}$ and $\mathbf{Q}(t) = f_2(t)\mathbf{i} + g_2(t)\mathbf{j}$. Then by Theorem 16.5.4,

$$D_t\mathbf{R}(t) = f_1'(t)\mathbf{i} + g_1'(t)\mathbf{j} \quad \text{and} \quad D_t\mathbf{Q}(t) = f_2'(t)\mathbf{i} + g_2'(t)\mathbf{j}$$

$$\mathbf{R}(t) \cdot \mathbf{Q}(t) = [f_1(t)][f_2(t)] + [g_1(t)][g_2(t)]$$

So

$$\begin{aligned} &D_t[\mathbf{R}(t) \cdot \mathbf{Q}(t)] \\ &= [f_1'(t)][f_2(t)] + [f_1(t)][f_2'(t)] + [g_1'(t)][g_2(t)] + [g_1(t)][g_2'(t)] \\ &= \{[f_1'(t)][f_2(t)] + [g_1'(t)][g_2(t)]\} + \{[f_1(t)][f_2'(t)] + [g_1(t)][g_2'(t)]\} \\ &= [D_t\mathbf{R}(t)] \cdot \mathbf{Q}(t) + \mathbf{R}(t) \cdot [D_t\mathbf{Q}(t)] \end{aligned}$$

■

Example 2 Verify Theorem 16.5.7 for the vectors given in Example 1.

Solution $\mathbf{R}(t) \cdot \mathbf{Q}(t) = t^2 \sin t + (t - 1) \cos t$. Therefore

$$D_t[\mathbf{R}(t) \cdot \mathbf{Q}(t)] = 2t \sin t + t^2 \cos t + \cos t + (t - 1)(-\sin t)$$

$$D_t[\mathbf{R}(t) \cdot \mathbf{Q}(t)] = (t + 1) \sin t + (t^2 + 1) \cos t \qquad (2)$$

Because

$$D_t\mathbf{R}(t) = D_t[t^2\mathbf{i} + (t - 1)\mathbf{j}] = 2t\mathbf{i} + \mathbf{j}$$

$$\begin{aligned} [D_t\mathbf{R}(t)] \cdot \mathbf{Q}(t) &= (2t\mathbf{i} + \mathbf{j}) \cdot (\sin t\mathbf{i} + \cos t\mathbf{j}) \\ &= 2t \sin t + \cos t \end{aligned}$$

Because

$$D_t\mathbf{Q}(t) = D_t[\sin t\mathbf{i} + \cos t\mathbf{j}] = \cos t\mathbf{i} - \sin t\mathbf{j}$$

$$\begin{aligned} \mathbf{R}(t) \cdot [D_t\mathbf{Q}(t)] &= [t^2\mathbf{i} + (t - 1)\mathbf{j}] \cdot (\cos t\mathbf{i} - \sin t\mathbf{j}) \\ &= t^2 \cos t - (t - 1) \sin t \end{aligned}$$

Therefore

$$[D_t\mathbf{R}(t)] \cdot \mathbf{Q}(t) + \mathbf{R}(t) \cdot [D_t\mathbf{Q}(t)] = (2t \sin t + \cos t) + [t^2 \cos t - (t - 1) \sin t]$$

Thus

$$[D_t\mathbf{R}(t)] \cdot \mathbf{Q}(t) + \mathbf{R}(t) \cdot [D_t\mathbf{Q}(t)] = (t + 1) \sin t + (t^2 + 1) \cos t \qquad (3)$$

Comparing (2) and (3) we see that Theorem 16.5.7 holds.

16.5.8 Theorem If $\mathbf{R}$ is a differentiable vector-valued function on an interval and f is a differentiable real-valued function on the interval, then

$$D_t\{[f(t)][\mathbf{R}(t)]\} = [D_t f(t)]\mathbf{R}(t) + f(t)\, D_t\mathbf{R}(t)$$

The proof is left as an exercise (see Exercise 22).

The following theorem is the chain rule for vector-valued functions. The proof, which is left as an exercise (see Exercise 23), is based on Theorems 2.7.7 and 3.7.1, which involve the analogous conclusions for real-valued functions.

16.5.9 Theorem Suppose that **F** is a vector-valued function, h is a real-valued function such that $\phi = h(t)$, and $\mathbf{G}(t) = \mathbf{F}(h(t))$. If h is continuous at t and **F** is continuous at $h(t)$, then **G** is continuous at t. Furthermore, if $D_t\phi$ and $D_\phi\mathbf{G}(t)$ exist, then $D_t\mathbf{G}(t)$ exists and is given by

$$D_t\mathbf{G}(t) = [D_\phi\mathbf{G}(t)]D_t\phi$$

We now define an indefinite integral (or antiderivative) of a vector-valued function.

16.5.10 Definition If **Q** is the vector-valued function given by

$$\mathbf{Q}(t) = f(t)\mathbf{i} + g(t)\mathbf{j}$$

then the **indefinite integral of** $\mathbf{Q}(t)$ is defined by

$$\int \mathbf{Q}(t)\,dt = \mathbf{i}\int f(t)\,dt + \mathbf{j}\int g(t)\,dt \tag{4}$$

This definition is consistent with the definition of an indefinite integral of a real-valued function because if we take the derivative on both sides of (4) with respect to t,

$$D_t \int \mathbf{Q}(t)\,dt = \mathbf{i}D_t \int f(t)\,dt + \mathbf{j}D_t \int g(t)\,dt$$

which gives

$$D_t \int \mathbf{Q}(t)\,dt = \mathbf{i}f(t) + \mathbf{j}g(t)$$

For each of the indefinite integrals on the right side of (4) there occurs an arbitrary scalar constant. When each of these scalars is multiplied by either **i** or **j**, there occurs an arbitrary constant vector in the sum. So

$$\int \mathbf{Q}(t)\,dt = \mathbf{R}(t) + \mathbf{C}$$

where $D_t\mathbf{R}(t) = \mathbf{Q}(t)$ and **C** is an arbitrary constant vector.

Example 3 Find the most general vector-valued function whose derivative is $\mathbf{Q}(t) = \sin t\mathbf{i} - 3\cos t\mathbf{j}$.

Solution If $D_t\mathbf{R}(t) = \mathbf{Q}(t)$, then $\mathbf{R}(t) = \int \mathbf{Q}(t)\,dt$, or

$$\mathbf{R}(t) = \mathbf{i}\int \sin t\,dt - 3\mathbf{j}\int \cos t\,dt$$

$$= \mathbf{i}(-\cos t + C_1) - 3\mathbf{j}(\sin t + C_2)$$

$$= -\cos t\mathbf{i} - 3\sin t\mathbf{j} + (C_1\mathbf{i} - 3C_2\mathbf{j})$$

$$= -\cos t\mathbf{i} - 3\sin t\mathbf{j} + \mathbf{C}$$

Example 4 Find the vector $\mathbf{R}(t)$ for which $D_t\mathbf{R}(t) = e^{-t}\mathbf{i} + e^t\mathbf{j}$ and for which $\mathbf{R}(0) = \mathbf{i} + \mathbf{j}$.

Solution

$$\mathbf{R}(t) = \mathbf{i}\int e^{-t}\,dt + \mathbf{j}\int e^t\,dt$$

So

$$\mathbf{R}(t) = \mathbf{i}(-e^{-t} + C_1) + \mathbf{j}(e^t + C_2)$$

Because $\mathbf{R}(0) = \mathbf{i} + \mathbf{j}$,

$$\mathbf{i} + \mathbf{j} = \mathbf{i}(-1 + C_1) + \mathbf{j}(1 + C_2)$$

Therefore

$$C_1 - 1 = 1 \quad \text{and} \quad C_2 + 1 = 1$$

$$C_1 = 2 \qquad\qquad C_2 = 0$$

Hence

$$\mathbf{R}(t) = (-e^{-t} + 2)\mathbf{i} + e^t\mathbf{j}$$

The following theorem will be useful later.

16.5.11 Theorem If $\mathbf{R}$ is a differentiable vector-valued function on an interval and $|\mathbf{R}(t)|$ is constant for all t in the interval, then the vectors $\mathbf{R}(t)$ and $D_t\mathbf{R}(t)$ are orthogonal.

Proof Let $|\mathbf{R}(t)| = k$. Then by Theorem 16.3.3(iii),

$$\mathbf{R}(t) \cdot \mathbf{R}(t) = k^2$$

Differentiating on both sides with respect to t and using Theorem 16.5.7 we obtain

$$[D_t\mathbf{R}(t)] \cdot \mathbf{R}(t) + \mathbf{R}(t) \cdot [D_t\mathbf{R}(t)] = 0$$

Therefore

$$2\mathbf{R}(t) \cdot D_t\mathbf{R}(t) = 0$$

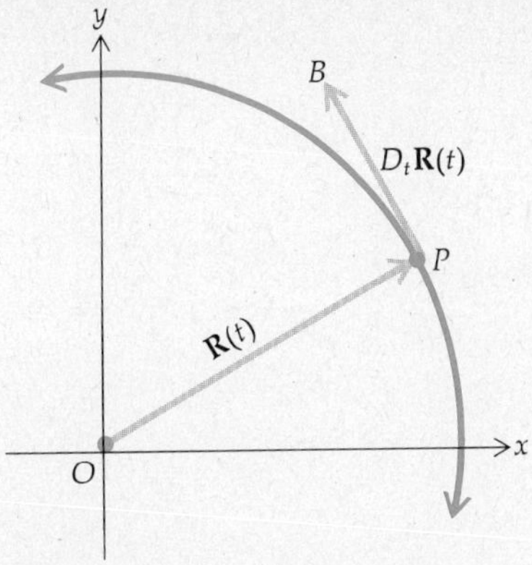

FIGURE 16.5.2

Because the dot product of $\mathbf{R}(t)$ and $D_t\mathbf{R}(t)$ is zero, it follows from Definition 16.3.7 that $\mathbf{R}(t)$ and $D_t\mathbf{R}(t)$ are orthogonal. ■

The geometric interpretation of Theorem 16.5.11 is evident. If the vector $\mathbf{R}(t)$ has constant magnitude, then the position representation $\overrightarrow{OP}$ of $\mathbf{R}(t)$ has its terminal point P on the circle with its center at the origin and radius k. So the graph of $\mathbf{R}$ is this circle. Because $D_t\mathbf{R}(t)$ and $\mathbf{R}(t)$ are orthogonal, $\overrightarrow{OP}$ is perpendicular to a representation of $D_t\mathbf{R}(t)$. Figure 16.5.2 shows a sketch of a quarter circle, the position representation $\overrightarrow{OP}$ of $\mathbf{R}(t)$, and the representation $\overrightarrow{PB}$ of $D_t\mathbf{R}(t)$.

EXERCISES 16.5

In Exercises 1 through 5, find the indicated limit, if it exists.

1. $\mathbf{R}(t) = (3t - 2)\mathbf{i} + t^2\mathbf{j};\ \lim\limits_{t\to 2}\mathbf{R}(t)$

2. $\mathbf{R}(t) = (t - 2)\mathbf{i} + \dfrac{t^2 - 4}{t - 2}\mathbf{j};\ \lim\limits_{t\to 2}\mathbf{R}(t)$

3. $\mathbf{R}(t) = 2\sin t\mathbf{i} + \cos t\mathbf{j};\ \lim\limits_{t\to \pi/2}\mathbf{R}(t)$

4. $\mathbf{R}(t) = \dfrac{t^2 - 2t - 3}{t - 3}\mathbf{i} + \dfrac{t^2 - 5t + 6}{t - 3}\mathbf{j};\ \lim\limits_{t\to 3}\mathbf{R}(t)$

5. $\mathbf{R}(t) = e^{t+1}\mathbf{i} + |t + 1|\mathbf{j};\ \lim\limits_{t\to -1}\mathbf{R}(t)$

In Exercises 6 through 14, find $\mathbf{R}'(t)$ and $\mathbf{R}''(t)$.

6. $\mathbf{R}(t) = (t^2 - 3)\mathbf{i} + (2t + 1)\mathbf{j}$

7. $\mathbf{R}(t) = e^{2t}\mathbf{i} + \ln t\mathbf{j}$

8. $\mathbf{R}(t) = \cos 2t\mathbf{i} + \tan t\mathbf{j}$

9. $\mathbf{R}(t) = \tan^{-1} t\mathbf{i} + 2^t\mathbf{j}$

10. $\mathbf{R}(t) = \dfrac{t - 1}{t + 1}\mathbf{i} + \dfrac{t - 2}{t}\mathbf{j}$

11. $\mathbf{R}(t) = (t^2 + 4)^{-1}\mathbf{i} + \sqrt{1 - 5t}\mathbf{j}$

12. $\mathbf{R}(t) = \sqrt{2t + 1}\mathbf{i} + (t - 1)^2\mathbf{j}$

13. $\mathbf{R}(t) = 5\sin 2t\mathbf{i} - \sec 4t\mathbf{j}$

14. $\mathbf{R}(t) = (e^{3t} + 2)\mathbf{i} + 2e^{3t}\mathbf{j}$

In Exercises 15 and 16, find $D_t|\mathbf{R}(t)|$.

15. $\mathbf{R}(t) = (t - 1)\mathbf{i} + (2 - t)\mathbf{j}$

16. $\mathbf{R}(t) = (e^t + 1)\mathbf{i} + (e^t - 1)\mathbf{j}$

In Exercises 17 through 20, find $\mathbf{R}'(t) \cdot \mathbf{R}''(t)$.

17. $\mathbf{R}(t) = (2t^2 - 1)\mathbf{i} + (t^2 + 3)\mathbf{j}$

18. $\mathbf{R}(t) = \ln(t - 1)\mathbf{i} - 3t^{-1}\mathbf{j}$

19. $\mathbf{R}(t) = e^{2t}\mathbf{i} + e^{-2t}\mathbf{j}$

20. $\mathbf{R}(t) = -\cos 2t\mathbf{i} + \sin 2t\mathbf{j}$

21. Prove Theorem 16.5.6.

22. Prove Theorem 16.5.8.

23. Prove Theorem 16.5.9.

In Exercises 24 through 29, find the most general vector whose derivative has the given function value.

24. $(t^2 - 9)\mathbf{i} + (2t - 5)\mathbf{j}$

25. $\tan t\mathbf{i} - \dfrac{1}{t}\mathbf{j}$

26. $3^t\mathbf{i} - 2^t\mathbf{j}$

27. $e^{3t}\mathbf{i} + \dfrac{1}{t - 1}\mathbf{j}$

28. $\dfrac{1}{4 + t^2}\mathbf{i} - \dfrac{4}{1 - t^2}\mathbf{j}$

29. $\ln t\mathbf{i} + t^2\mathbf{j}$

30. If $\mathbf{R}'(t) = t^2\mathbf{i} + \dfrac{1}{t-2}\mathbf{j}$, and $\mathbf{R}(3) = 2\mathbf{i} - 5\mathbf{j}$, find $\mathbf{R}(t)$.

31. If $\mathbf{R}'(t) = \sin^2 t\mathbf{i} + 2\cos^2 t\mathbf{j}$, and $\mathbf{R}(\pi) = \mathbf{0}$, find $\mathbf{R}(t)$.

32. If $\mathbf{R}'(t) = e^t \sin t\mathbf{i} + e^t \cos t\mathbf{j}$, and $\mathbf{R}(0) = \mathbf{i} - \mathbf{j}$, find $\mathbf{R}(t)$.

In Exercises 33 and 34, for the given vector equation find a cartesian equation of the curve that is traced by the endpoint of the position representation of $\mathbf{R}'(t)$. Find $\mathbf{R}(t) \cdot \mathbf{R}'(t)$. Interpret the result geometrically.

33. $\mathbf{R}(t) = \cos t\mathbf{i} + \sin t\mathbf{j}$ **34.** $\mathbf{R}(t) = \cosh t\mathbf{i} - \sinh t\mathbf{j}$

In Exercises 35 and 36, if $\alpha(t)$ is the radian measure of the angle between $\mathbf{R}(t)$ and $\mathbf{Q}(t)$, find $D_t\alpha(t)$.

35. $\mathbf{R}(t) = 3e^{2t}\mathbf{i} - 4e^{2t}\mathbf{j}$ and $\mathbf{Q}(t) = 6e^{3t}\mathbf{j}$ **36.** $\mathbf{R}(t) = 2t\mathbf{i} + (t^2 - 1)\mathbf{j}$ and $\mathbf{Q}(t) = 3t\mathbf{i}$

37. Suppose $\mathbf{R}$ and $\mathbf{R}'$ are vector-valued functions defined on an interval and $\mathbf{R}'$ is differentiable on the interval. Prove

$$D_t[\mathbf{R}'(t) \cdot \mathbf{R}(t)] = |\mathbf{R}'(t)|^2 + \mathbf{R}(t) \cdot \mathbf{R}''(t)$$

38. If $|\mathbf{R}(t)| = h(t)$, prove that $\mathbf{R}(t) \cdot \mathbf{R}'(t) = [h(t)][h'(t)]$.

39. If the vector-valued function $\mathbf{R}$ and the real-valued function f are both differentiable on an interval and $f(t) \neq 0$ on the interval, prove that $\mathbf{R}/f$ is also differentiable on the interval and

$$D_t\left[\frac{\mathbf{R}(t)}{f(t)}\right] = \frac{f(t)\mathbf{R}'(t) - f'(t)\mathbf{R}(t)}{[f(t)]^2}$$

40. Prove that if $\mathbf{A}$ and $\mathbf{B}$ are constant vectors and f and g are integrable functions, then

$$\int [\mathbf{A}f(t) + \mathbf{B}g(t)]\, dt = \mathbf{A} \int f(t)\, dt + \mathbf{B} \int g(t)\, dt$$

(*Hint:* Express $\mathbf{A}$ and $\mathbf{B}$ in terms of $\mathbf{i}$ and $\mathbf{j}$.)

16.6 LENGTH OF ARC

In Sec. 7.6 we obtained a formula for finding the length of arc of a curve having an equation of the form $y = f(x)$. This is a special kind of curve because the graph of a function f cannot be intersected by a vertical line in more than one point.

We now develop a method for finding the length of arc of some other kinds of curves. Let C be the curve having parametric equations

$$x = f(t) \quad \text{and} \quad y = g(t) \tag{1}$$

and suppose that f and g are continuous on the closed interval $[a, b]$. We wish to assign a number L to represent the number of units in the length of arc of C from $t = a$ to $t = b$. We proceed as in Sec. 7.6.

Let Δ be a partition of the closed interval $[a, b]$ formed by dividing the interval into n subintervals by choosing $n - 1$ numbers between a and b. Let $t_0 = a$ and $t_n = b$, and let $t_1, t_2, \ldots, t_{n-1}$ be intermediate numbers:

$$t_0 < t_1 < \cdots < t_{n-1} < t_n$$

The ith subinterval is $[t_{i-1}, t_i]$ and the number of units in its length, denoted by $\Delta_i t$, is $t_i - t_{i-1}$, where $i = 1, 2, \ldots, n$. Let $\|\Delta\|$ be the norm of the partition; so each $\Delta_i t \leq \|\Delta\|$.

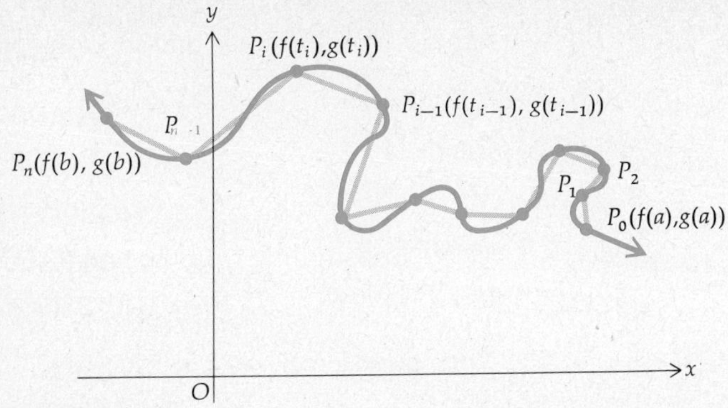

FIGURE 16.6.1

Associated with each number t_i is a point $P_i(f(t_i), g(t_i))$ on C. From each point P_{i-1} draw a line segment to the next point P_i (see Fig. 16.6.1). The number of units in the length of the line segment from P_{i-1} to P_i is denoted by $|\overline{P_{i-1}P_i}|$. From the distance formula we have

$$|\overline{P_{i-1}P_i}| = \sqrt{[f(t_i) - f(t_{i-1})]^2 + [g(t_i) - g(t_{i-1})]^2} \tag{2}$$

The sum of the numbers of units of lengths of the n line segments is

$$\sum_{i=1}^{n} |\overline{P_{i-1}P_i}| \tag{3}$$

Our intuitive notion of the length of the arc from $t = a$ to $t = b$ leads us to define the number of units of the length of arc as the limit of the sum in (3) as $\|\Delta\|$ approaches zero.

16.6.1 Definition Let the curve C have parametric equations $x = f(t)$ and $y = g(t)$. Then if there exists a number L having the property that for any $\epsilon > 0$ there is a $\delta > 0$ such that

$$\left| \sum_{i=1}^{n} |\overline{P_{i-1}P_i}| - L \right| < \epsilon$$

for every partition Δ of the interval $[a, b]$ for which $\|\Delta\| < \delta$, we write

$$L = \lim_{\|\Delta\| \to 0} \sum_{i=1}^{n} |\overline{P_{i-1}P_i}| \tag{4}$$

and L units is called the **length of arc** of the curve C from the point $(f(a), g(a))$ to the point $(f(b), g(b))$.

The arc of the curve is rectifiable if the limit in (4) exists.

If f' and g' are continuous on $[a, b]$, we proceed as follows to find a formula for evaluating the limit in (4).

Because f' and g' are continuous on $[a, b]$, they are continuous on each subinterval of the partition Δ. So the hypothesis of the mean-value theorem (Theorem 4.4.2) is satisfied by f and g on each $[t_{i-1}, t_i]$; therefore there are numbers z_i and w_i in the open interval (t_{i-1}, t_i) such that

$$f(t_i) - f(t_{i-1}) = f'(z_i)\,\Delta_i t \tag{5}$$

and

$$g(t_i) - g(t_{i-1}) = g'(w_i)\,\Delta_i t \tag{6}$$

Substituting from (5) and (6) into (2) we obtain

$$|\overline{P_{i-1}P_i}| = \sqrt{[f'(z_i)\,\Delta_i t]^2 + [g'(w_i)\,\Delta_i t]^2}$$

or, equivalently,

$$|\overline{P_{i-1}P_i}| = \sqrt{[f'(z_i)]^2 + [g'(w_i)]^2}\,\Delta_i t \tag{7}$$

where z_i and w_i are in the open interval (t_{i-1}, t_i). Then from (4) and (7), if the limit exists,

$$L = \lim_{\|\Delta\| \to 0} \sum_{i=1}^{n} \sqrt{[f'(z_i)]^2 + [g'(w_i)]^2}\,\Delta_i t \tag{8}$$

The sum in (8) is not a Riemann sum because z_i and w_i are not necessarily the same numbers. So we cannot apply the definition of a definite integral to evaluate the limit in (8). However, there is a theorem that can be applied to evaluate this limit. We state the theorem, but a proof is not given because it is beyond the scope of this book. You can find a proof in an advanced calculus text.

16.6.2 Theorem If the functions F and G are continuous on the closed interval $[a, b]$, then the function $\sqrt{F^2 + G^2}$ is also continuous on $[a, b]$, and if Δ is a partition of the interval $[a, b]$ (Δ: $a = t_0 < t_1 < \cdots < t_{i-1} < t_i < \cdots < t_n = b$), and z_i and w_i are any numbers in (t_{i-1}, t_i), then

$$\lim_{\|\Delta\| \to 0} \sum_{i=1}^{n} \sqrt{[F(z_i)]^2 + [G(w_i)]^2}\,\Delta_i t = \int_a^b \sqrt{[F(t)]^2 + [G(t)]^2}\,dt \tag{9}$$

Applying (9) to (8), where F is f' and G is g', we have

$$L = \int_a^b \sqrt{[f'(t)]^2 + [g'(t)]^2}\,dt$$

This result is stated as a theorem.

16.6.3 Theorem Let the curve C have parametric equations $x = f(t)$ and $y = g(t)$, and suppose that f' and g' are continuous on the closed interval $[a, b]$. Then the length of arc L units of the curve C from the point $(f(a), g(a))$ to the point $(f(b), g(b))$ is determined by

$$L = \int_a^b \sqrt{[f'(t)]^2 + [g'(t)]^2}\,dt \tag{10}$$

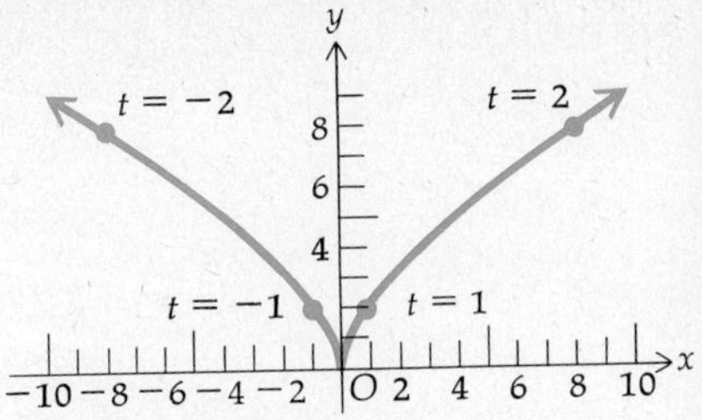

FIGURE 16.6.2

Example 1 Find the length of the arc of the curve having parametric equations $x = t^3$ and $y = 2t^2$ in each of the following cases: (a) from $t = 0$ to $t = 1$; (b) from $t = -2$ to $t = 0$.

Solution A sketch of the curve is shown in Fig. 16.6.2.

(a) Let $x = f(t)$; then $f'(t) = D_t x = 3t^2$. Let $y = g(t)$; then $g'(t) = D_t y = 4t$. So from Theorem 16.6.3, if L units is the length of the arc of the curve from $t = 0$ to $t = 1$.

$$\begin{aligned} L &= \int_0^1 \sqrt{9t^4 + 16t^2}\, dt \\ &= \int_0^1 t\sqrt{9t^2 + 16}\, dt \\ &= \tfrac{1}{18} \cdot \tfrac{2}{3}(9t^2 + 16)^{3/2}\Big]_0^1 \\ &= \tfrac{1}{27}[(25)^{3/2} - (16)^{3/2}] \\ &= \tfrac{1}{27}(125 - 64) \\ &= \tfrac{61}{27} \end{aligned}$$

(b) If L units is the length of the arc of the curve from $t = -2$ to $t = 0$, then from Theorem 16.6.3,

$$L = \int_{-2}^{0} \sqrt{9t^4 + 16t^2}\, dt = \int_{-2}^{0} \sqrt{t^2}\,\sqrt{9t^2 + 16}\, dt$$

Because $-2 \le t \le 0$, $\sqrt{t^2} = -t$. Therefore

$$\begin{aligned} L &= \int_{-2}^{0} -t\sqrt{9t^2 + 16}\, dt \\ &= -\tfrac{1}{27}(9t^2 + 16)^{3/2}\Big]_{-2}^{0} \\ &= -\tfrac{1}{27}[(16)^{3/2} - (52)^{3/2}] \\ &= \tfrac{1}{27}(104\sqrt{13} - 64) \\ &\approx 11.5 \end{aligned}$$

The curve C has parametric equations (1). Let s units be the length of arc of C from the point $(f(t_0), g(t_0))$ to the point $(f(t), g(t))$, and let s increase as t increases. Then s is a function of t and is given by

$$s = \int_{t_0}^{t} \sqrt{[f'(u)]^2 + [g'(u)]^2}\, du \tag{11}$$

From Theorem 7.6.1,

$$\frac{ds}{dt} = \sqrt{[f'(t)]^2 + [g'(t)]^2} \tag{12}$$

A vector equation of C is

$$\mathbf{R}(t) = f(t)\mathbf{i} + g(t)\mathbf{j} \tag{13}$$

Because

$$\mathbf{R}'(t) = f'(t)\mathbf{i} + g'(t)\mathbf{j}$$

then

$$|\mathbf{R}'(t)| = \sqrt{[f'(t)]^2 + [g'(t)]^2} \tag{14}$$

Substituting from (14) into (12) we obtain

$$|\mathbf{R}'(t)| = \frac{ds}{dt} \tag{15}$$

From (15) it follows that if s units is the length of arc of curve C having vector equation (13) measured from some fixed point to the point $(f(t), g(t))$ where s increases as t increases, then the derivative of s with respect to t is the magnitude of the derivative of the position vector at the point $(f(t), g(t))$.

We substitute from (14) into (10) and obtain $L = \int_a^b |\mathbf{R}'(t)|\, dt$. So Theorem 16.6.3 can be stated in terms of vectors in the following way.

16.6.4 Theorem Let the curve C have the vector equation $\mathbf{R}(t) = f(t)\mathbf{i} + g(t)\mathbf{j}$, and suppose that f' and g' are continuous on the closed interval $[a, b]$. Then the length of arc of C, traced by the terminal point of the position representation of $\mathbf{R}(t)$ as t increases from a to b, is determined by

$$L = \int_a^b |\mathbf{R}'(t)|\, dt \tag{16}$$

Example 2 Find the length of the arc traced by the terminal point of the position representation of $\mathbf{R}(t)$ as t increases from 1 to 4 if $\mathbf{R}(t) = e^t \sin t\mathbf{i} + e^t \cos t\mathbf{j}$.

Solution $\mathbf{R}'(t) = (e^t \sin t + e^t \cos t)\mathbf{i} + (e^t \cos t - e^t \sin t)\mathbf{j}$. Therefore

$$\begin{aligned}|\mathbf{R}'(t)| &= e^t\sqrt{\sin^2 t + 2\sin t\cos t + \cos^2 t + \cos^2 t - 2\sin t\cos t + \sin^2 t}\\ &= e^t\sqrt{2}\end{aligned}$$

From (16),

$$L = \int_1^4 \sqrt{2}e^t\, dt = \sqrt{2}e^t\Big]_1^4 = \sqrt{2}(e^4 - e)$$

An alternate form of formula (10) for the length of an arc of a curve C, having parametric equations $x = f(t)$ and $y = g(t)$, is obtained by replacing $f'(t)$ by dx/dt and $g'(t)$ by dy/dt. This form is

$$L = \int_a^b \sqrt{\left(\frac{dx}{dt}\right)^2 + \left(\frac{dy}{dt}\right)^2}\, dt \tag{17}$$

Now suppose that we wish to find the length of arc of a curve C whose polar equation is $r = F(\theta)$. If (x, y) is the cartesian representation of a point P on C and (r, θ) is a polar representation of P, then

$$x = r\cos\theta \quad \text{and} \quad y = r\sin\theta \tag{18}$$

Replacing r by $F(\theta)$ in Eqs. (18) we have

$$x = F(\theta)\cos\theta \quad \text{and} \quad y = F(\theta)\sin\theta \tag{19}$$

Equations (19) can be considered as parametric equations of C where θ is the parameter instead of t. Therefore, if F' is continuous on the closed interval $[\alpha, \beta]$, the formula for the length of arc of the curve C whose polar equation is $r = F(\theta)$ is obtained from (17) by taking $t = \theta$. So

$$L = \int_\alpha^\beta \sqrt{\left(\frac{dx}{d\theta}\right)^2 + \left(\frac{dy}{d\theta}\right)^2}\, d\theta \tag{20}$$

From (18),

$$\frac{dx}{d\theta} = \cos\theta\frac{dr}{d\theta} - r\sin\theta \quad \text{and} \quad \frac{dy}{d\theta} = \sin\theta\frac{dr}{d\theta} + r\cos\theta$$

Therefore

$$\sqrt{\left(\frac{dx}{d\theta}\right)^2 + \left(\frac{dy}{d\theta}\right)^2} = \sqrt{\left(\cos\theta\frac{dr}{d\theta} - r\sin\theta\right)^2 + \left(\sin\theta\frac{dr}{d\theta} + r\cos\theta\right)^2}$$

$$= \sqrt{\cos^2\theta\left(\frac{dr}{d\theta}\right)^2 - 2r\sin\theta\cos\theta\frac{dr}{d\theta} + r^2\sin^2\theta + \sin^2\theta\left(\frac{dr}{d\theta}\right)^2 + 2r\sin\theta\cos\theta\frac{dr}{d\theta} + r^2\cos^2\theta}$$

$$= \sqrt{(\cos^2\theta + \sin^2\theta)\left(\frac{dr}{d\theta}\right)^2 + (\sin^2\theta + \cos^2\theta)r^2}$$

$$= \sqrt{\left(\frac{dr}{d\theta}\right)^2 + r^2}$$

Substituting this into (20) we obtain

$$L = \int_\alpha^\beta \sqrt{\left(\frac{dr}{d\theta}\right)^2 + r^2}\, d\theta \tag{21}$$

Example 3 Find the length of the cardioid $r = 2(1 + \cos\theta)$.

Solution A sketch of the curve is shown in Fig. 16.6.3. To obtain the length of the entire curve we can let θ take on values from 0 to 2π or we can make use of the symmetry of the curve and find half the length by letting θ take on values from 0 to π.

Because $r = 2(1 + \cos\theta)$, $dr/d\theta = -2\sin\theta$. Substituting into (21), integrating from 0 to π, and multiplying by 2, we have

$$L = 2\int_0^\pi \sqrt{(-2\sin\theta)^2 + 4(1 + \cos\theta)^2}\, d\theta$$

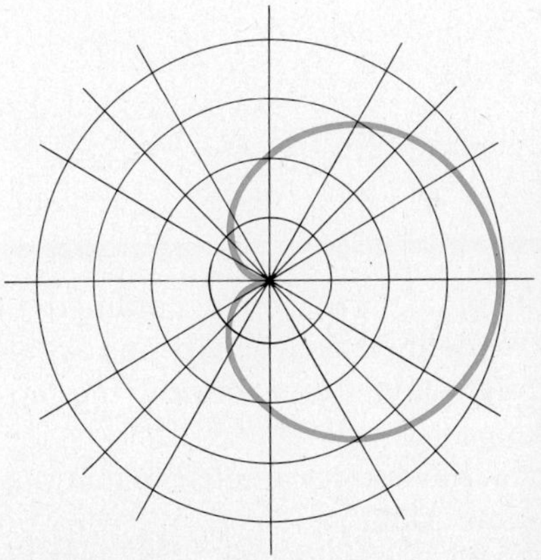

FIGURE 16.6.3

$$= 4\int_0^{\pi} \sqrt{\sin^2\theta + 1 + 2\cos\theta + \cos^2\theta}\, d\theta$$

$$= 4\sqrt{2}\int_0^{\pi} \sqrt{1 + \cos\theta}\, d\theta$$

To evaluate this integral, use the identity $\cos^2 \frac{1}{2}\theta = \frac{1}{2}(1 + \cos\theta)$, which gives $\sqrt{1+\cos\theta} = \sqrt{2}\,|\cos\frac{1}{2}\theta|$. Because $0 \le \theta \le \pi$, $0 \le \frac{1}{2}\theta \le \frac{1}{2}\pi$; thus $\cos\frac{1}{2}\theta \ge 0$. Therefore $\sqrt{1+\cos\theta} = \sqrt{2}\cos\frac{1}{2}\theta$. So

$$L = 4\sqrt{2}\int_0^{\pi} \sqrt{2}\cos\tfrac{1}{2}\theta\, d\theta = 16\sin\tfrac{1}{2}\theta\Big]_0^{\pi} = 16$$

EXERCISES 16.6

Find the length of arc in each of the following exercises. When a appears, a > 0.

1. $x = \frac{1}{2}t^2 + t$, $y = \frac{1}{2}t^2 - t$; from $t = 0$ to $t = 1$.
2. $x = t^3$, $y = 3t^2$; from $t = -2$ to $t = 0$.
3. $x = t^2 + 2t$, $y = t^2 - 2t$; from $t = 0$ to $t = 2$.
4. $\mathbf{R}(t) = (t^2 + 3)\mathbf{i} + 3t^2\mathbf{j}$; from $t = 1$ to $t = 4$.
5. $\mathbf{R}(t) = 2t^2\mathbf{i} + 2t^3\mathbf{j}$; from $t = 1$ to $t = 2$.
6. $\mathbf{R}(t) = t\mathbf{i} + \cosh t\mathbf{j}$; from $t = 0$ to $t = 3$.
7. $\mathbf{R}(t) = 3e^{2t}\mathbf{i} - 4e^{2t}\mathbf{j}$; from $t = 0$ to $t = \ln 5$.
8. $\mathbf{R}(t) = a(\cos t + t\sin t)\mathbf{i} + a(\sin t - t\cos t)\mathbf{j}$; from $t = 0$ to $t = \frac{1}{3}\pi$.
9. The entire hypocycloid of four cusps: $x = a\cos^3 t$, $y = a\sin^3 t$.
10. $x = e^{-t}\cos t$, $y = e^{-t}\sin t$; from $t = 0$ to $t = \pi$.
11. $x = 4\sin 2t$, $y = 4\cos 2t$; from $t = 0$ to $t = \pi$.
12. One arch of the cycloid: $x = a(t - \sin t)$, $y = a(1 - \cos t)$.
13. The tractrix $x = t - a\tanh\dfrac{t}{a}$, $y = a\,\text{sech}\,\dfrac{t}{a}$ from $t = -a$ to $t = 2a$.
14. The circumference of the circle: $\mathbf{R}(t) = a\cos t\mathbf{i} + a\sin t\mathbf{j}$.
15. The circumference of the circle: $r = 5\cos\theta$.
16. The circumference of the circle: $r = a\sin\theta$.
17. The circumference of the circle: $r = a$.
18. The entire curve: $r = 1 - \sin\theta$.
19. The entire curve: $r = 3\cos^2\frac{1}{2}\theta$.
20. $r = a\theta$; from $\theta = 0$ to $\theta = 2\pi$.
21. $r = e^{2\theta}$; from $\theta = 0$ to $\theta = 4$.
22. $r = a\theta^2$; from $\theta = 0$ to $\theta = \pi$.
23. $r = a\sin^3\frac{1}{3}\theta$; from $\theta = 0$ to $\theta = \theta_1$.
24. $r = \sin^2\frac{1}{2}\theta$; from $\theta = 0$ to $\theta = \frac{1}{2}\pi$.
25. Find the distance traveled by a thumbtack in the tread of a bicycle tire if the radius of the tire is 40 cm and the bicycle goes a distance of 50π m. (*Hint:* The path of the thumbtack is a cycloid.)

16.7 PLANE MOTION

The previous discussion of the motion of a particle was confined to straight-line motion. In this connection we defined the velocity and acceleration of a particle moving along a straight line. Consider now the motion of a particle along a curve in the plane.

Suppose that C is the plane curve having parametric equations $x = f(t)$, $y = g(t)$, where t units denotes time. Then

$$\mathbf{R}(t) = f(t)\mathbf{i} + g(t)\mathbf{j}$$

is a vector equation of C. As t varies, the endpoint $P(f(t), g(t))$ of $\overrightarrow{OP}$ moves along the curve C. The position at time t units of a particle moving along C is the point $P(f(t), g(t))$. The **velocity vector** of the particle at time t units is defined to be $\mathbf{R}'(t)$ and is denoted by the symbol $\mathbf{V}(t)$.

16.7.1 Definition Let C be the curve having parametric equations $x = f(t)$ and $y = g(t)$. If a particle is moving along C so that its position at any time t units is the point (x, y), then the **instantaneous velocity** of the particle at time t units is determined by the velocity vector

$$\mathbf{V}(t) = f'(t)\mathbf{i} + g'(t)\mathbf{j}$$

if $f'(t)$ and $g'(t)$ exist.

In Sec. 16.5 we learned that the direction of $\mathbf{R}'(t)$ at the point $P(f(t), g(t))$ is along the tangent line to the curve C at P. Therefore the velocity vector $\mathbf{V}(t)$ has this direction at P.

The magnitude of the velocity vector is a measure of the **speed** of the particle at time t and is given by

$$|\mathbf{V}(t)| = \sqrt{[f'(t)]^2 + [g'(t)]^2} \tag{1}$$

Note that the velocity is a vector and the speed is a scalar. As shown in Sec. 16.6, the expression on the right side of (1) is ds/dt. So the speed is the rate of change of s with respect to t, and we write

$$|\mathbf{V}(t)| = \frac{ds}{dt} \tag{2}$$

The **acceleration vector** of the particle at time t units is denoted by $\mathbf{A}(t)$ and is defined to be the derivative of the velocity vector or, equivalently, the second derivative of the position vector.

16.7.2 Definition The **instantaneous acceleration** at time t units of a particle moving along a curve C, having parametric equations $x = f(t)$ and $y = g(t)$, is determined by the acceleration vector

$$\mathbf{A}(t) = \mathbf{V}'(t) = \mathbf{R}''(t)$$

where $\mathbf{R}(t) = f(t)\mathbf{i} + g(t)\mathbf{j}$ and $\mathbf{R}''(t)$ exists.

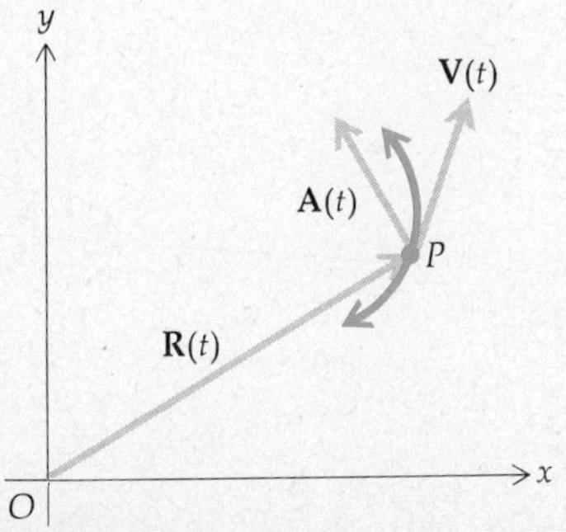

FIGURE 16.7.1

Figure 16.7.1 shows the representations of the velocity vector and the acceleration vector whose initial point is the point P on C.

Example 1 A particle is moving along the curve having parametric equations $x = 4 \cos \frac{1}{2}t$ and $y = 4 \sin \frac{1}{2}t$. If t is the number of seconds in the time and x and y represent number of centimeters, find the speed and magnitude of the particle's acceleration vector at time t sec. Draw a sketch of the particle's path, and also draw the representations of the velocity and acceleration vectors having initial point where $t = \frac{1}{3}\pi$.

Solution A vector equation of C is

$$\mathbf{R}(t) = 4\cos\tfrac{1}{2}t\mathbf{i} + 4\sin\tfrac{1}{2}t\mathbf{j}$$

$$\mathbf{V}(t) = \mathbf{R}'(t) = -2\sin\tfrac{1}{2}t\mathbf{i} + 2\cos\tfrac{1}{2}t\mathbf{j}$$

$$\mathbf{A}(t) = \mathbf{V}'(t) = -\cos\tfrac{1}{2}t\mathbf{i} - \sin\tfrac{1}{2}t\mathbf{j}$$

$$\begin{aligned}|\mathbf{V}(t)| &= \sqrt{(-2\sin\tfrac{1}{2}t)^2 + (2\cos\tfrac{1}{2}t)^2}\\ &= \sqrt{4\sin^2\tfrac{1}{2}t + 4\cos^2\tfrac{1}{2}t}\\ &= 2\end{aligned}$$

$$|\mathbf{A}(t)| = \sqrt{(-\cos\tfrac{1}{2}t)^2 + (-\sin\tfrac{1}{2}t)^2} = 1$$

Therefore the speed of the particle is constant and is 2 cm/sec. The magnitude of the acceleration vector is also constant and is 1 cm/sec^2.

Eliminating t between the parametric equations of C we obtain the cartesian equation $x^2 + y^2 = 16$, which is a circle with its center at 0 and radius 4.

When $t = \frac{1}{3}\pi$, the direction of $\mathbf{V}(t)$ is given by $\frac{1}{2}\pi < \theta < \pi$ and

$$\tan\theta = -\frac{\cos\frac{1}{6}\pi}{\sin\frac{1}{6}\pi} = -\cot\tfrac{1}{6}\pi = -\sqrt{3}$$

When $t = \frac{1}{3}\pi$, the direction of $\mathbf{A}(t)$ is given by $\pi < \theta < \frac{3}{2}\pi$ and

$$\tan\theta = \frac{\sin\frac{1}{6}\pi}{\cos\frac{1}{6}\pi} = \tan\tfrac{1}{6}\pi = \frac{1}{\sqrt{3}}$$

Figure 16.7.2 shows the representations of the velocity and acceleration vectors having initial point where $t = \frac{1}{3}\pi$.

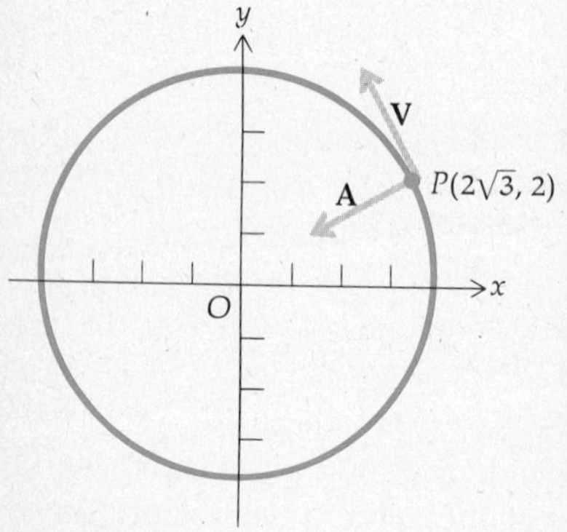

FIGURE 16.7.2

Example 2 The position of a moving particle at time t units is given by the vector equation $\mathbf{R}(t) = e^{-2t}\mathbf{i} + 3e^t\mathbf{j}$. Find $\mathbf{V}(t)$, $\mathbf{A}(t)$, $|\mathbf{V}(t)|$, $|\mathbf{A}(t)|$. Draw a sketch of the path of the particle and the representations of the velocity and acceleration vectors having initial point where $t = \frac{1}{2}$.

Solution

$$\mathbf{V}(t) = \mathbf{R}'(t) = -2e^{-2t}\mathbf{i} + 3e^t\mathbf{j}$$

$$\mathbf{A}(t) = \mathbf{V}'(t) = 4e^{-2t}\mathbf{i} + 3e^t\mathbf{j}$$

$$|\mathbf{V}(t)| = \sqrt{4e^{-4t} + 9e^{2t}}$$

$$|\mathbf{A}(t)| = \sqrt{16e^{-4t} + 9e^{2t}}$$

$$|\mathbf{V}(\tfrac{1}{2})| = \sqrt{4e^{-2} + 9e} \approx 5.00$$

$$|\mathbf{A}(\tfrac{1}{2})| = \sqrt{16e^{-2} + 9e} \approx 5.16$$

Parametric equations of the path of the particle are $x = e^{-2t}$ and $y = 3e^t$. Eliminating t between these two equations we obtain

$$xy^2 = 9$$

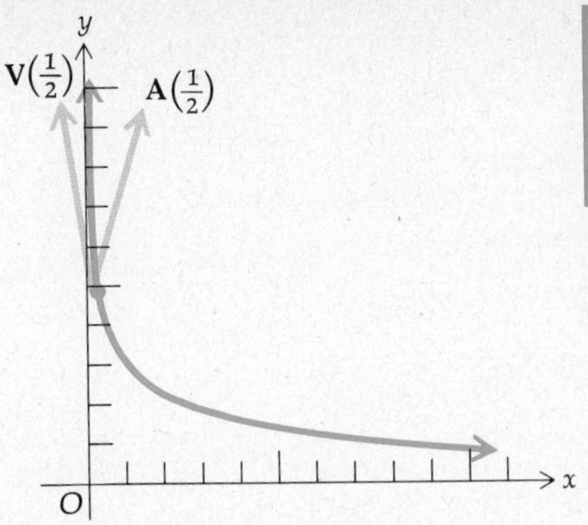

FIGURE 16.7.3

Because $x > 0$ and $y > 0$, the path of the particle is the portion of the curve $xy^2 = 9$ in the first quadrant. Figure 16.7.3 shows the path of the particle and the velocity and acceleration vectors when $t = \frac{1}{2}$. The slope of $\mathbf{V}(\frac{1}{2})$ is $-\frac{3}{2}e^{3/2} \approx -6.7$, and the slope of $\mathbf{A}(\frac{1}{2})$ is $\frac{3}{4}e^{3/2} \approx 3.4$.

We now derive the equations of motion of a projectile by assuming that the projectile is moving in a vertical plane. We also assume that the only force acting on the projectile is its weight, which has a downward direction and a magnitude of mg lb, where m slugs is its mass and g ft/sec² is the constant of acceleration caused by gravity. We are neglecting the force attributed to air resistance (which for heavy bodies traveling at small speeds has no noticeable effect). The positive direction is taken as vertically upward and horizontally to the right.

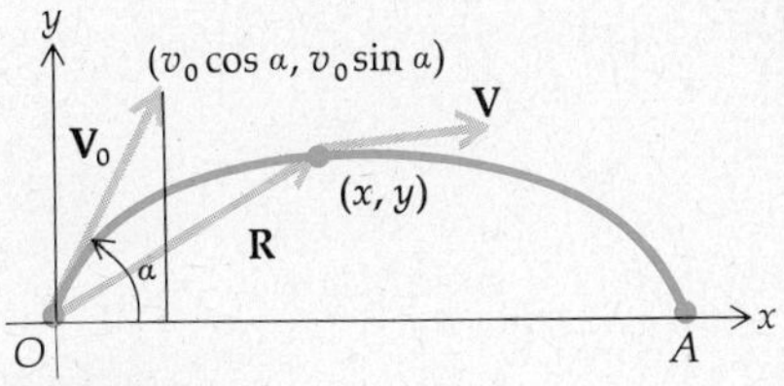

FIGURE 16.7.4

Suppose, then, that a projectile is shot from a gun having an angle of elevation of radian measure α. Let the number of feet per second in the initial speed, or *muzzle speed,* be denoted by v_0. The coordinate axes are set up so that the gun is located at the origin. Refer to Fig. 16.7.4. The initial velocity vector $\mathbf{V}_0$, of the projectile is given by

$$\mathbf{V}_0 = v_0 \cos \alpha \mathbf{i} + v_0 \sin \alpha \mathbf{j} \tag{3}$$

Let t sec be the time that has elapsed since the gun was fired, x ft be the horizontal distance of the projectile from the starting point at t sec, and y ft be the vertical distance of the projectile from the starting point at t sec. $\mathbf{R}(t)$ is the position vector of the projectile at t sec, $\mathbf{V}(t)$ is the velocity vector of the projectile at t sec, and $\mathbf{A}(t)$ is the acceleration vector of the projectile at t sec.

Because x is a function of t, we write $x(t)$. Similarly, y is a function of t and we write $y(t)$. Then

$$\mathbf{R}(t) = x(t)\mathbf{i} + y(t)\mathbf{j} \tag{4}$$

$$\mathbf{V}(t) = \mathbf{R}'(t) \tag{5}$$

$$\mathbf{A}(t) = \mathbf{V}'(t) \tag{6}$$

Because the only force acting on the projectile has a magnitude of mg lb and is in the downward direction, then if $\mathbf{F}$ denotes this force,

$$\mathbf{F} = -mg\mathbf{j} \tag{7}$$

Newton's second law of motion states that the net force acting on a body is its "mass times acceleration." So

$$\mathbf{F} = m\mathbf{A} \tag{8}$$

From (7) and (8),

$$m\mathbf{A} = -mg\mathbf{j}$$

$$\mathbf{A} = -g\mathbf{j} \tag{9}$$

Because $\mathbf{A}(t) = \mathbf{V}'(t)$, we have from (9)

$$\mathbf{V}'(t) = -g\mathbf{j} \tag{10}$$

Integrating on both sides of (10) with respect to t we obtain

$$\mathbf{V}(t) = -gt\mathbf{j} + \mathbf{C}_1 \tag{11}$$

where $\mathbf{C}_1$ is a vector constant of integration.

When $t = 0$, $\mathbf{V} = \mathbf{V}_0$. So $\mathbf{C}_1 = \mathbf{V}_0$. Therefore, from (11),

$$\mathbf{V}(t) = -gt\mathbf{j} + \mathbf{V}_0$$

or, because $\mathbf{V}(t) = \mathbf{R}'(t)$,

$$\mathbf{R}'(t) = -gt\mathbf{j} + \mathbf{V}_0 \tag{12}$$

Integrating on both sides of the vector equation (12) with respect to t we obtain

$$\mathbf{R}(t) = -\tfrac{1}{2}gt^2\mathbf{j} + \mathbf{V}_0 t + \mathbf{C}_2$$

where $\mathbf{C}_2$ is a vector constant of integration.

When $t = 0$, $\mathbf{R} = \mathbf{0}$ because the projectile is at the origin at the start. So $\mathbf{C}_2 = \mathbf{0}$. Therefore

$$\mathbf{R}(t) = -\tfrac{1}{2}gt^2\mathbf{j} + \mathbf{V}_0 t \tag{13}$$

Substituting the value of $\mathbf{V}_0$ from (3) into (13) we obtain

$$\mathbf{R}(t) = -\tfrac{1}{2}gt^2\mathbf{j} + (v_0 \cos\alpha\mathbf{i} + v_0 \sin\alpha\mathbf{j})t$$

$$\mathbf{R}(t) = tv_0 \cos\alpha\mathbf{i} + (tv_0 \sin\alpha - \tfrac{1}{2}gt^2)\mathbf{j} \tag{14}$$

Equation (14) gives the position vector of the projectile at time t sec. From this equation we can discuss the motion of the projectile. We are usually concerned with the following questions:

1. What is the range of the projectile? The range is the distance $|\overline{OA}|$ along the x axis (see Fig. 16.7.4).
2. What is the total time of flight, that is, the time it takes the projectile to go from O to A?
3. What is the maximum height of the projectile?
4. What is a cartesian equation of the curve traveled by the projectile?
5. What is the velocity vector of the projectile at impact?

These questions are answered in the following example.

Example 3 A projectile is shot from a gun at an angle of elevation of radian measure $\frac{1}{6}\pi$. Its muzzle speed is 480 ft/sec. Find (a) the position vector of the projectile at any time; (b) the time of flight; (c) the range; (d) the maximum height; (e) the velocity vector of the projectile at impact; (f) the position vector and the velocity vector at 2 sec; (g) the speed at 2 sec; (h) a cartesian equation of the curve traveled by the projectile.

Solution $v_0 = 480$ and $\alpha = \frac{1}{6}\pi$. So

$$\mathbf{V}_0 = 480 \cos \tfrac{1}{6}\pi\mathbf{i} + 480 \sin \tfrac{1}{6}\pi\mathbf{j}$$

$$= 240\sqrt{3}\mathbf{i} + 240\mathbf{j}$$

(a) The position vector at t sec is given by Eq. (14), which in this case is

$$\mathbf{R}(t) = 240\sqrt{3}t\mathbf{i} + (240t - \tfrac{1}{2}gt^2)\mathbf{j} \tag{15}$$

So if (x, y) is the position of the projectile at t sec, $x = 240\sqrt{3}t$ and $y = 240t - \frac{1}{2}gt^2$.

(b) To find the time of flight, determine t when $y = 0$. In part (a), if $y = 0$,

$$240t - \tfrac{1}{2}gt^2 = 0$$

$$t(240 - \tfrac{1}{2}gt) = 0$$

$$t = 0 \quad \text{and} \quad t = \frac{480}{g}$$

The value $t = 0$ is when the projectile is fired because then $y = 0$. If $g = 32$, the time of flight is determined by $t = 480/g = 480/32 = 15$. So the time of flight is 15 sec.

(c) To find the range, determine x when $t = 15$. From part (a), $x = 240\sqrt{3}t$. So when $t = 15$, $x = 3600\sqrt{3}$. Therefore the range is 6235 ft (approximately).

(d) The maximum height is attained when $D_t y = 0$, that is, when the vertical component of the velocity vector is 0. Because

$$y = 240t - \tfrac{1}{2}gt^2$$

$$D_t y = 240 - gt$$

So $D_t y = 0$ when $t = 240/g$. If $g = 32$, the maximum height is attained when $t = 7\frac{1}{2}$, which is half the total time of flight. When $t = 7\frac{1}{2}$, $y = 900$. So the maximum height attained is 900 ft.

(e) Because the time of flight is 15 sec, the velocity vector at impact is $\mathbf{V}(15)$. Finding $\mathbf{V}(t)$ by using (15) we have

$$\mathbf{V}(t) = D_t\mathbf{R}(t) = 240\sqrt{3}\mathbf{i} + (240 - gt)\mathbf{j}$$

If $g = 32$,

$$\mathbf{V}(15) = 240\sqrt{3}\mathbf{i} - 240\mathbf{j}$$

(f) If $t = 2$ in Eq. (15),

$$\mathbf{R}(2) = 480\sqrt{3}\mathbf{i} + 416\mathbf{j}$$

Because $\mathbf{V}(t) = \mathbf{R}'(t)$,

$$\mathbf{V}(t) = 240\sqrt{3}\mathbf{i} + (240 - gt)\mathbf{j}$$

So

$$\mathbf{V}(2) = 240\sqrt{3}\mathbf{i} + 176\mathbf{j}$$

(g) The speed when $t = 2$ is determined by

$$\begin{aligned} |\mathbf{V}(2)| &= \sqrt{(240\sqrt{3})^2 + (176)^2} \\ &= 32\sqrt{199} \end{aligned}$$

So at 2 sec the speed is approximately 451.4 ft/sec.

(h) To find a cartesian equation, eliminate t between the parametric equations

$$x = 240\sqrt{3}t$$

$$y = 240t - \tfrac{1}{2}gt^2$$

Solving the first equation for t and substituting into the second equation we obtain

$$y = \frac{1}{\sqrt{3}}x - \frac{1}{10{,}800}x^2$$

which is an equation of a parabola.

EXERCISES 16.7

In Exercises 1 through 8, a particle is moving along the curve having the given parametric equations, where t sec is the time. Find: (a) the velocity vector $\mathbf{V}(t)$*; (b) the acceleration vector* $\mathbf{A}(t)$*; (c) the speed at* $t = t_1$*; (d) the magnitude of the acceleration vector at* $t = t_1$*. Draw a sketch of the path of the particle and the representations of the velocity vector and the acceleration vector at* $t = t_1$*.*

1. $x = t^2 + 4, y = t - 2; t_1 = 3$

2. $x = \ln(t - 2), y = t^3 - 1; t_1 = 3$

3. $x = 5\cos 2t, y = 3\sin 2t; t_1 = \frac{1}{4}\pi$

4. $x = 2/t, y = -\frac{1}{4}t; t_1 = 4$

5. $x = t, y = \ln \sec t; t_1 = \frac{1}{4}\pi$

6. $x = 2\cos t, y = 3\sin t; t_1 = \frac{1}{3}\pi$

7. $x = \sin t, y = \tan t; t_1 = \frac{1}{6}\pi$

8. $x = e^{2t}, y = e^{3t}; t_1 = 0$

In Exercises 9 through 16, the position of a moving particle at t sec is determined from a vector equation. Find: (a) $\mathbf{V}(t_1)$*; (b)* $\mathbf{A}(t_1)$*; (c)* $|\mathbf{V}(t_1)|$*; (d)* $|\mathbf{A}(t_1)|$*. Draw a sketch of a portion of the path of the particle containing the position of the particle at* $t = t_1$*, and draw the representations of* $\mathbf{V}(t_1)$ *and* $\mathbf{A}(t_1)$ *having initial point where* $t = t_1$*.*

9. $\mathbf{R}(t) = (2t - 1)\mathbf{i} + (t^2 + 1)\mathbf{j}; t_1 = 3$

10. $\mathbf{R}(t) = (1 - t)\mathbf{i} + (t^2 - 1)\mathbf{j}; t_1 = -1$

11. $\mathbf{R}(t) = e^t\mathbf{i} + e^{2t}\mathbf{j}; t_1 = \ln 2$

12. $\mathbf{R}(t) = (t^2 + 3t)\mathbf{i} + (1 - 3t^2)\mathbf{j}; t_1 = \frac{1}{2}$

13. $\mathbf{R}(t) = \cos 2t\mathbf{i} - 3\sin t\mathbf{j}; t_1 = \pi$

14. $\mathbf{R}(t) = e^{-t}\mathbf{i} + e^{2t}\mathbf{j}; t_1 = \ln 2$

15. $\mathbf{R}(t) = 2(1 - \cos t)\mathbf{i} + 2(1 - \sin t)\mathbf{j}; t_1 = \frac{5}{6}\pi$

16. $\mathbf{R}(t) = \ln(t + 2)\mathbf{i} + \frac{1}{3}t^2\mathbf{j}; t_1 = 1$

In Exercises 17 through 20, find the position vector $\mathbf{R}(t)$*.*

17. $\mathbf{V}(t) = \dfrac{1}{(t-1)^2}\mathbf{i} - (t + 1)\mathbf{j}$, and $\mathbf{R}(0) = 3\mathbf{i} + 2\mathbf{j}$

18. $\mathbf{V}(t) = (2t - 1)\mathbf{i} + 3t^{-2}\mathbf{j}$, and $\mathbf{R}(1) = 4\mathbf{i} - 3\mathbf{j}$

19. $\mathbf{A}(t) = e^{-t}\mathbf{i} + 2e^{2t}\mathbf{j}$, $\mathbf{V}(0) = 2\mathbf{i} + \mathbf{j}$, and $\mathbf{R}(0) = 3\mathbf{j}$

20. $\mathbf{A}(t) = 2\cos 2t\mathbf{i} + 2\sin 2t\mathbf{j}$, $\mathbf{V}(0) = \mathbf{i} + \mathbf{j}$, and $\mathbf{R}(0) = \frac{1}{2}\mathbf{i} - \frac{1}{2}\mathbf{j}$

21. A projectile is shot from a gun at an angle of elevation of 45° with a muzzle speed of 2500 ft/sec. Find (a) the range of the projectile; (b) the maximum height reached; (c) the velocity at impact.

22. A projectile is shot from a gun at an angle of elevation of 60°. The muzzle speed is 160 ft/sec. Find (a) the position vector of the projectile at t sec; (b) the time of flight; (c) the range; (d) the maximum height reached; (e) the velocity at impact; (f) the speed at 4 sec.

23. A projectile is shot from the top of a building 96 ft high from a gun at an angle of 30° with the horizontal. If the muzzle speed is 1600 ft/sec, find the time of flight and the distance from the base of the building to the point where the projectile lands.

24. The muzzle speed of a gun is 160 ft/sec. At what angle of elevation should the gun be fired so that a projectile will hit an object on the same level as the gun and a distance of 400 ft from it?

25. What is the muzzle speed of a gun if a projectile fired from it has a range of 2000 ft and reaches a maximum height of 1000 ft?

26. A child throws a ball horizontally from the top of a cliff 256 ft high with an initial speed of 50 ft/sec. Find the time of flight of the ball and the distance from the base of the cliff to the point where the ball lands.

27. A child throws a ball with an initial speed of 60 ft/sec at an angle of elevation of 60° toward a tall building that is 25 ft from the child. If the child's hand is 5 ft from the ground, show that the ball hits the building, and find the direction of the ball when it hits the building.

16.8 THE UNIT TANGENT AND UNIT NORMAL VECTORS AND ARC LENGTH AS A PARAMETER

A unit vector is a vector having a magnitude of 1, examples of which are the two unit vectors **i** and **j**. With each point on a curve in the plane we now associate two other unit vectors, the *unit tangent vector* and the *unit normal vector*.

16.8.1 Definition If $\mathbf{R}(t)$ is the position vector of curve C at a point P on C, then the **unit tangent vector** of C at P, denoted by $\mathbf{T}(t)$, is the unit vector in the direction of $D_t\mathbf{R}(t)$ if $D_t\mathbf{R}(t) \neq 0$.

The unit vector in the direction of $D_t\mathbf{R}(t)$ is given by $D_t\mathbf{R}(t)/|D_t\mathbf{R}(t)|$; thus

$$\mathbf{T}(t) = \frac{D_t\mathbf{R}(t)}{|D_t\mathbf{R}(t)|} \tag{1}$$

Because $\mathbf{T}(t)$ is a unit vector, it follows from Theorem 16.5.11 that $D_t\mathbf{T}(t)$ must be orthogonal to $\mathbf{T}(t)$. $D_t\mathbf{T}(t)$ is not necessarily a unit vector. However, the vector $D_t\mathbf{T}(t)/|D_t\mathbf{T}(t)|$ is of unit magnitude and has the same direction as $D_t\mathbf{T}(t)$. Therefore $D_t\mathbf{T}(t)/|D_t\mathbf{T}(t)|$ is a unit vector that is orthogonal to $\mathbf{T}(t)$, and it is called the *unit normal vector.*

16.8.2 Definition If $\mathbf{T}(t)$ is the unit tangent vector of curve C at a point P on C, then the **unit normal vector,** denoted by $\mathbf{N}(t)$, is the unit vector in the direction of $D_t\mathbf{T}(t)$.

From Definition 16.8.2 and the previous discussion,

$$\mathbf{N}(t) = \frac{D_t\mathbf{T}(t)}{|D_t\mathbf{T}(t)|} \tag{2}$$

Example 1 Given the curve having parametric equations $x = t^3 - 3t$ and $y = 3t^2$, find $\mathbf{T}(t)$ and $\mathbf{N}(t)$. Draw a sketch of a portion of the curve at $t = 2$ and draw the representations of $\mathbf{T}(2)$ and $\mathbf{N}(2)$ having their initial point at $t = 2$.

Solution A vector equation of the curve is

$$\mathbf{R}(t) = (t^3 - 3t)\mathbf{i} + 3t^2\mathbf{j}$$

So

$$D_t\mathbf{R}(t) = (3t^2 - 3)\mathbf{i} + 6t\mathbf{j}$$

Then

$$|D_t\mathbf{R}(t)| = \sqrt{(3t^2 - 3)^2 + 36t^2} = \sqrt{9(t^4 + 2t^2 + 1)} = 3(t^2 + 1)$$

From (1),

$$\mathbf{T}(t) = \frac{t^2 - 1}{t^2 + 1}\mathbf{i} + \frac{2t}{t^2 + 1}\mathbf{j}$$

Differentiating $\mathbf{T}(t)$ with respect to t we obtain

$$D_t\mathbf{T}(t) = \frac{4t}{(t^2 + 1)^2}\mathbf{i} + \frac{2 - 2t^2}{(t^2 + 1)^2}\mathbf{j}$$

Therefore

$$\begin{aligned}|D_t\mathbf{T}(t)| &= \sqrt{\frac{16t^2}{(t^2 + 1)^4} + \frac{4 - 8t^2 + 4t^4}{(t^2 + 1)^4}} \\ &= \sqrt{\frac{4 + 8t^2 + 4t^4}{(t^2 + 1)^4}} \\ &= \sqrt{\frac{4(t^2 + 1)^2}{(t^2 + 1)^4}} \\ &= \frac{2}{t^2 + 1}\end{aligned}$$

From (2),

$$\mathbf{N}(t) = \frac{2t}{t^2 + 1}\mathbf{i} + \frac{1 - t^2}{t^2 + 1}\mathbf{j}$$

We find $\mathbf{R}(t)$, $\mathbf{T}(t)$, and $\mathbf{N}(t)$, when $t = 2$.

$$\mathbf{R}(2) = 2\mathbf{i} + 12\mathbf{j} \qquad \mathbf{T}(2) = \tfrac{3}{5}\mathbf{i} + \tfrac{4}{5}\mathbf{j} \qquad \mathbf{N}(2) = \tfrac{4}{5}\mathbf{i} - \tfrac{3}{5}\mathbf{j}$$

The required sketch is shown in Fig. 16.8.1.

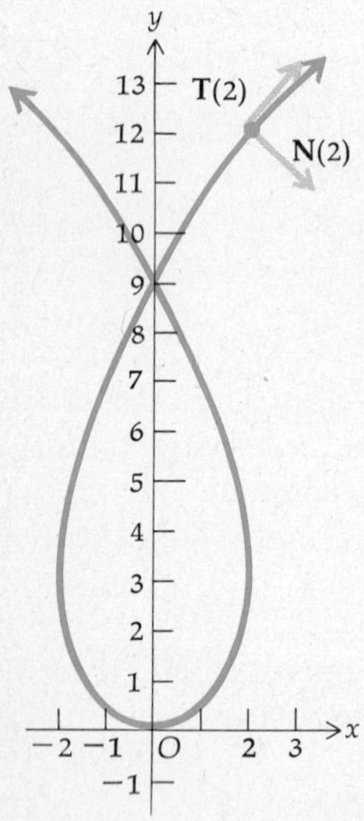

FIGURE 16.8.1

From (1),

$$D_t\mathbf{R}(t) = |D_t\mathbf{R}(t)|\mathbf{T}(t) \tag{3}$$

The right side of (3) is the product of a scalar and a vector. To differentiate this product, apply Theorem 16.5.8.

$$D_t^2\mathbf{R}(t) = [D_t|D_t\mathbf{R}(t)|]\mathbf{T}(t) + |D_t\mathbf{R}(t)|[D_t\mathbf{T}(t)] \tag{4}$$

From (2),

$$D_t\mathbf{T}(t) = |D_t\mathbf{T}(t)|\mathbf{N}(t) \tag{5}$$

Substituting from (5) into (4) we get

$$D_t^2\mathbf{R}(t) = [D_t|D_t\mathbf{R}(t)|]\mathbf{T}(t) + |D_t\mathbf{R}(t)||D_t\mathbf{T}(t)|\mathbf{N}(t) \tag{6}$$

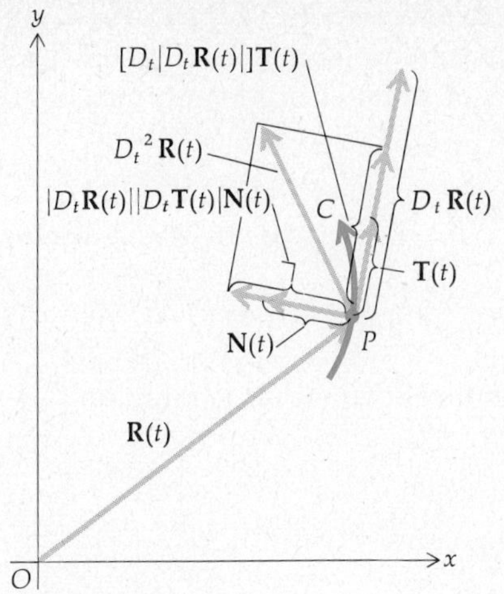

FIGURE 16.8.2

Equation (3) expresses the vector $D_t\mathbf{R}(t)$ as a scalar times the unit tangent vector, and (6) expresses the vector $D_t^2\mathbf{R}(t)$ as a scalar times the unit tangent vector plus a scalar times the unit normal vector. The coefficient of $\mathbf{T}(t)$ on the right side of (6) is the component of the vector $D_t^2\mathbf{R}(t)$ in the direction of the unit tangent vector. The coefficient of $\mathbf{N}(t)$ on the right side of (6) is the component of $D_t^2\mathbf{R}(t)$ in the direction of the unit normal vector. Figure 16.8.2 shows a portion of a curve C with the position representation of $\mathbf{R}(t)$ and the representations of $\mathbf{T}(t)$, $\mathbf{N}(t)$, $D_t\mathbf{R}(t)$, $D_t^2\mathbf{R}(t)$, $[D_t|D_t\mathbf{R}(t)|]\mathbf{T}(t)$, and $|D_t\mathbf{R}(t)||D_t\mathbf{T}(t)|\mathbf{N}(t)$, all of whose initial points are at the point P on C. Note that the representation of the unit normal vector $\mathbf{N}$ is on the concave side of the curve. This is proved in general in Sec. 16.9.

Sometimes instead of a parameter t, we wish to use as a parameter the number of units of arc length s from an arbitrarily chosen point $P_0(x_0, y_0)$ on curve C to the point $P(x, y)$ on C. Let s increase as t increases so that s is positive if the length of arc is measured in the direction of increasing t and s is negative if the length of arc is measured in the opposite direction. Therefore s units is a directed distance. Also, $D_ts > 0$. To each value of s there corresponds a unique point P on the curve C. Consequently, the coordinates of P are functions of s, and s is a function of t. In Sec. 16.6 we showed that

$$|D_t\mathbf{R}(t)| = D_ts \tag{7}$$

Substituting from (7) into (1) we get

$$\mathbf{T}(t) = \frac{D_t\mathbf{R}(t)}{D_ts}$$

$$D_t\mathbf{R}(t) = D_ts\mathbf{T}(t)$$

If the parameter is s instead of t, we have from the above equation, by taking $t = s$ and noting that $D_ss = 1$,

$$D_s\mathbf{R}(s) = \mathbf{T}(s)$$

This result is stated as a theorem.

16.8.3 Theorem If the vector equation of a curve C is $\mathbf{R}(s) = f(s)\mathbf{i} + g(s)\mathbf{j}$, where s units is the length of arc measured from a particular point P_0 on C to the point P, then the unit tangent vector of C at P is given by

$$\mathbf{T}(s) = D_s\mathbf{R}(s)$$

if it exists.

Now suppose that the parametric equations of a curve C involve a parameter t, and we wish to find parametric equations of C, with s, the number of units of arc length measured from some fixed point, as the parameter. Often the operations involved are quite complicated. However, the method used is illustrated in the following example.

Example 2 Suppose that parametric equations of the curve C are $x = t^3$ and $y = t^2$, where $t \geq 0$. Find parametric equations of C having s as a parameter, where s is the number of units of arc length measured from the point where $t = 0$.

Solution If P_0 is the point where $t = 0$, P_0 is the origin. A vector equation of C is

$$\mathbf{R}(t) = t^3\mathbf{i} + t^2\mathbf{j}$$

Because $D_t s = |D_t\mathbf{R}(t)|$, we differentiate the above vector and obtain

$$D_t\mathbf{R}(t) = 3t^2\mathbf{i} + 2t\mathbf{j}$$

So

$$|D_t\mathbf{R}(t)| = \sqrt{9t^4 + 4t^2} = \sqrt{t^2}\sqrt{9t^2 + 4}$$

Because $t \geq 0$, $\sqrt{t^2} = t$. Thus

$$|D_t\mathbf{R}(t)| = t\sqrt{9t^2 + 4}$$

Therefore

$$D_t s = t\sqrt{9t^2 + 4}$$

and so

$$\begin{aligned} s &= \int_0^t u\sqrt{9u^2 + 4}\, du \\ &= \frac{1}{18}\int_0^t 18u\sqrt{9u^2 + 4}\, du \\ &= \tfrac{1}{27}(9u^2 + 4)^{3/2}\Big]_0^t \end{aligned}$$

Hence

$$s = \tfrac{1}{27}(9t^2 + 4)^{3/2} - \tfrac{8}{27} \tag{8}$$

Solving (8) for t in terms of s we have

$$(9t^2 + 4)^{3/2} = 27s + 8$$

$$9t^2 + 4 = (27s + 8)^{2/3}$$

Because $t \geq 0$,

$$t = \tfrac{1}{3}\sqrt{(27s + 8)^{2/3} - 4}$$

Substituting this value of t into the given parametric equations for C we obtain

$$x = \tfrac{1}{27}[(27s + 8)^{2/3} - 4]^{3/2} \quad \text{and} \quad y = \tfrac{1}{9}[(27s + 8)^{2/3} - 4] \tag{9}$$

Now because $D_s\mathbf{R}(s) = \mathbf{T}(s)$, it follows that if $\mathbf{R}(s) = x(s)\mathbf{i} + y(s)\mathbf{j}$, then $\mathbf{T}(s) = (D_s x)\mathbf{i} + (D_s y)\mathbf{j}$. And because $\mathbf{T}(s)$ is a unit vector,

$$(D_s x)^2 + (D_s y)^2 = 1 \tag{10}$$

Equation (10) can be used to check Eqs. (9). This check is left as an exercise (see Exercise 17).

EXERCISES 16.8

In Exercises 1 through 10, for the given curve find $\mathbf{T}(t)$ *and* $\mathbf{N}(t)$, *and at* $t = t_1$ *draw a sketch of a portion of the curve and draw the representations of* $\mathbf{T}(t_1)$ *and* $\mathbf{N}(t_1)$ *having initial point at* $t = t_1$.

1. $x = \frac{1}{3}t^3 - t, y = t^2; t_1 = 2$

2. $x = \frac{1}{2}t^2, y = \frac{1}{3}t^3; t_1 = 1$

3. $\mathbf{R}(t) = e^t\mathbf{i} + e^{-t}\mathbf{j}; t_1 = 0$

4. $\mathbf{R}(t) = e^{-2t}\mathbf{i} + e^{2t}\mathbf{j}; t_1 = 0$

5. $\mathbf{R}(t) = 3\cos t\mathbf{i} + 3\sin t\mathbf{j}; t_1 = \frac{1}{2}\pi$

6. $x = e^t \sin t, y = e^t \cos t; t_1 = 0$

7. $x = \cos kt, y = \sin kt, k > 0; t_1 = \pi/k$

8. $x = t - \sin t, y = 1 - \cos t; t_1 = \pi$

9. $\mathbf{R}(t) = \ln \cos t\mathbf{i} + \ln \sin t\mathbf{j}, 0 < t < \frac{1}{2}\pi; t_1 = \frac{1}{4}\pi$

10. $\mathbf{R}(t) = t\cos t\mathbf{i} + t\sin t\mathbf{j}; t_1 = 0$

In Exercises 11 through 13, find $\mathbf{T}(t)$ *and* $\mathbf{N}(t)$ *for the curve having the given vector equation.*

11. $\mathbf{R}(t) = t^2\mathbf{i} + e^t\mathbf{j}$

12. $\mathbf{R}(t) = t\mathbf{i} + \cosh t\mathbf{j}; t \geq 0$

13. $\mathbf{R}(t) = \sin^3 t\mathbf{i} + \cos^3 t\mathbf{j}; 0 \leq t \leq \frac{1}{2}\pi$

14. If the vector equation of curve C is $\mathbf{R}(t) = 2\sin t\mathbf{i} + \sin 2t\mathbf{j}$, find the cosine of the measure of the angle between the vectors $\mathbf{R}(\frac{1}{6}\pi)$ and $\mathbf{T}(\frac{1}{6}\pi)$.

15. If the vector equation of curve C is $\mathbf{R}(t) = 3t^2\mathbf{i} + (t^3 - 3t)\mathbf{j}$, find the cosine of the measure of the angle between the vectors $\mathbf{R}(2)$ and $\mathbf{T}(2)$.

16. If the vector equation of curve C is $\mathbf{R}(t) = (4 - 3t^2)\mathbf{i} + (t^3 - 3t)\mathbf{j}$, find the radian measure of the angle between the vectors $\mathbf{N}(1)$ and $D_t^2\mathbf{R}(1)$.

17. Check Eqs. (9) of the solution of Example 2 by using Eq. (10).

In Exercises 18 through 23, find parametric equations of the curve having arc length s as a parameter, where s is measured from the point where $t = 0$. *Check your result by using Eq. (10).*

18. $x = a\cos t, y = a\sin t$

19. $x = 2 + \cos t, y = 3 + \sin t$

20. $x = 2(\cos t + t\sin t), y = 2(\sin t - t\cos t)$

21. $y = x^{3/2}$

22. $x = t^2 - 1, y = \frac{1}{3}t^3$

23. One cusp of the hypocycloid of four cusps: $\mathbf{R}(t) = a\cos^3 t\mathbf{i} + a\sin^3 t\mathbf{j}, 0 \leq t \leq \frac{1}{2}\pi$.

24. Given the cycloid $x = 2(t - \sin t), y = 2(1 - \cos t)$, express the arc length s as a function of t, where s is measured from the point where $t = 0$.

25. For the curve having parametric equations $x = e^t \cos t$ and $y = e^t \sin t$, express the arc length s as a function of t, where s is measured from the point where $t = 0$.

26. Prove that parametric equations of the catenary $y = a\cosh(x/a)$, where the parameter s is the number of units in the length of the arc from the point $(0, a)$ to the point (x, y) and $s \geq 0$ when $x \geq 0$ and $s < 0$ when $x < 0$, are

$$x = a\sinh^{-1}\frac{s}{a} \quad \text{and} \quad y = \sqrt{a^2 + s^2}$$

16.9 CURVATURE

Let ϕ be the radian measure of the angle giving the direction of the unit tangent vector associated with a curve C. Therefore ϕ is the radian measure of the angle from the direction of the positive x axis counterclockwise to the direction of the unit tangent vector $\mathbf{T}(t)$. See Fig. 16.9.1.

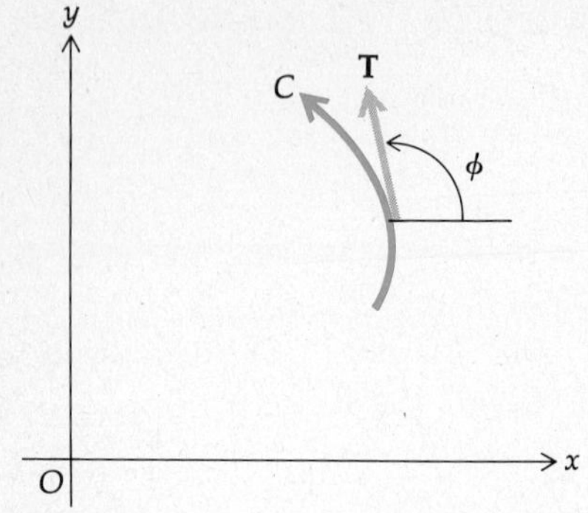

FIGURE 16.9.1

Because $|\mathbf{T}(t)| = 1$, it follows from Eq. (1) in Sec. 16.2 that

$$\mathbf{T}(t) = \cos\phi\mathbf{i} + \sin\phi\mathbf{j} \tag{1}$$

Differentiating with respect to ϕ we obtain

$$D_\phi\mathbf{T}(t) = -\sin\phi\mathbf{i} + \cos\phi\mathbf{j} \tag{2}$$

Because $|D_\phi\mathbf{T}(t)| = \sqrt{(-\sin\phi)^2 + (\cos\phi)^2} = 1$, $D_\phi\mathbf{T}(t)$ is a unit vector. Because $\mathbf{T}(t)$ has constant magnitude, it follows from Theorem 16.5.11 that $D_\phi\mathbf{T}(t)$ is orthogonal to $\mathbf{T}(t)$.

Replacing $-\sin\phi$ by $\cos(\frac{1}{2}\pi + \phi)$ and $\cos\phi$ by $\sin(\frac{1}{2}\pi + \phi)$, we write (2) as

$$D_\phi\mathbf{T}(t) = \cos(\tfrac{1}{2}\pi + \phi)\mathbf{i} + \sin(\tfrac{1}{2}\pi + \phi)\mathbf{j} \tag{3}$$

From (3) and the previous discussion, the vector $D_\phi\mathbf{T}(t)$ is a unit vector orthogonal to $\mathbf{T}(t)$ in the direction $\frac{1}{2}\pi$ counterclockwise from the direction of $\mathbf{T}(t)$. The unit normal vector $\mathbf{N}(t)$ is also orthogonal to $\mathbf{T}(t)$. By the chain rule (Theorem 16.5.9),

$$D_t\mathbf{T}(t) = [D_\phi\mathbf{T}(t)]D_t\phi \tag{4}$$

Because the direction of $\mathbf{N}(t)$ is the same as the direction of $D_t\mathbf{T}(t)$, it follows from (4) that the direction of $\mathbf{N}(t)$ is the same as the direction of $D_\phi\mathbf{T}(t)$ if $D_t\phi > 0$ (i.e., if $\mathbf{T}(t)$ turns counterclockwise as t increases), and the direction of $\mathbf{N}(t)$ is opposite that of $D_\phi\mathbf{T}(t)$ if $D_t\phi < 0$ (i.e., if $\mathbf{T}(t)$ turns clockwise as t increases). Because both $D_\phi\mathbf{T}(t)$ and $\mathbf{N}(t)$ are unit vectors, we conclude that

$$D_\phi\mathbf{T}(t) = \begin{cases} \mathbf{N}(t) & \text{if } D_t\phi > 0 \\ -\mathbf{N}(t) & \text{if } D_t\phi < 0 \end{cases} \tag{5}$$

• ILLUSTRATION 1 In Fig. 16.9.2(a), (b), (c), and (d) various cases are shown; in (a) and (b), $D_t\phi > 0$, and in (c) and (d), $D_t\phi < 0$. The positive direction along the curve C is indicated by the tip of the arrow on C. In each figure are shown the angle of radian meaure ϕ and representations of the vectors $\mathbf{T}(t)$, $D_\phi\mathbf{T}(t)$, and $\mathbf{N}(t)$. *The representation of the unit normal vector* $\mathbf{N}(t)$ *is always on the concave side of the curve.* •

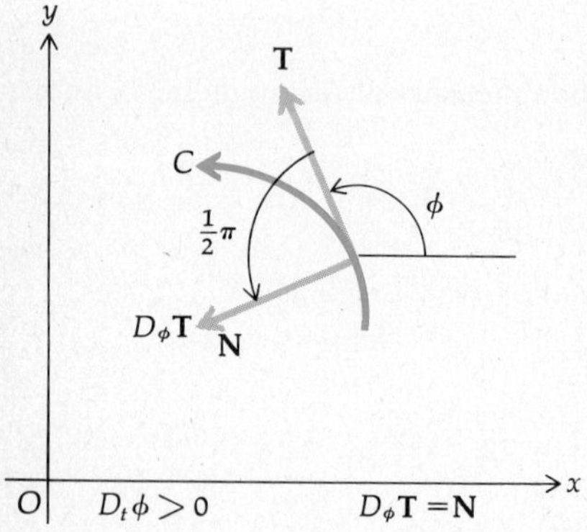

(a)

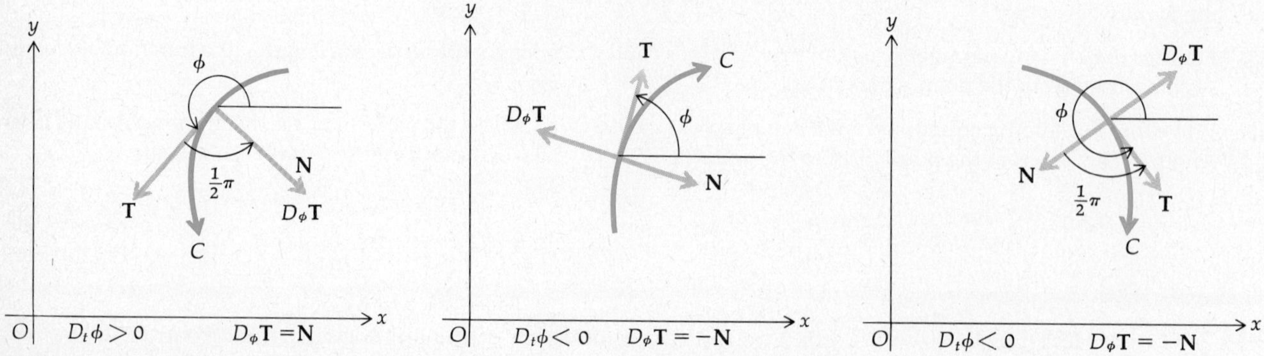

(b) (c) (d)

FIGURE 16.9.2

Consider now $D_s\mathbf{T}(t)$, where s units is the arc length measured from an abitrarily chosen point on C to point P and s increases as t increases. By the chain rule,

$$D_s\mathbf{T}(t) = D_\phi\mathbf{T}(t)D_s\phi$$

Hence

$$|D_s\mathbf{T}(t)| = |D_\phi\mathbf{T}(t)D_s\phi| = |D_\phi\mathbf{T}(t)||D_s\phi|$$

But because $D_\phi\mathbf{T}(t)$ is a unit vector, $|D_\phi\mathbf{T}(t)| = 1$; thus

$$|D_s\mathbf{T}(t)| = |D_s\phi| \tag{6}$$

The number $|D_s\phi|$ is the absolute value of the rate of change of the measure of the angle giving the direction of the unit tangent vector $\mathbf{T}(t)$ at a point on a curve with respect to the measure of arc length along the curve. This number is called the *curvature* of the curve at the point. Before giving the formal definition of curvature, we show that taking it as this number is consistent with what we intuitively think of as the curvature. For example, at point P on C, ϕ is the radian measure of the angle giving the direction of the vector $\mathbf{T}(t)$, and s units is the arc length from a point P_0 on C to P. Let Q be a point on C for which the radian measure of the angle giving the direction of $\mathbf{T}(t + \Delta t)$ at Q is $\phi + \Delta\phi$ and $s + \Delta s$ units is the arc length from P_0 to Q. Then the arc length from P to Q is Δs units, and the ratio $\Delta\phi/\Delta s$ seems like a good measure of what we would intuitively think of as the *average curvature* along arc PQ.

● ILLUSTRATION 2 See Fig. 16.9.3(a), (b), (c), and (d): In (a), $\Delta\phi > 0$ and $\Delta s > 0$; in (b), $\Delta\phi > 0$ and $\Delta s < 0$; in (c), $\Delta\phi < 0$ and $\Delta s > 0$; and in (d), $\Delta\phi < 0$ and $\Delta s < 0$. ●

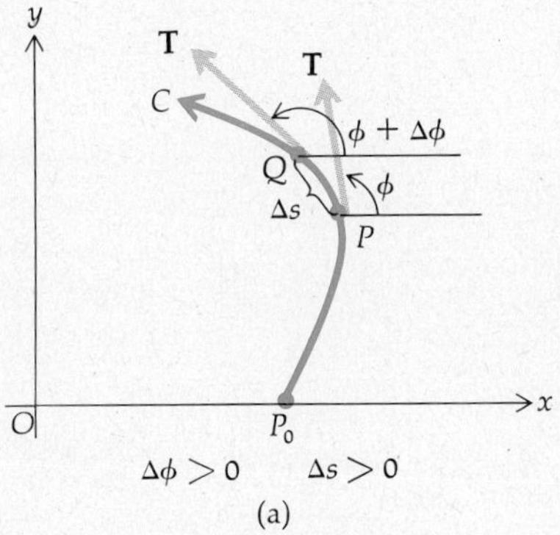

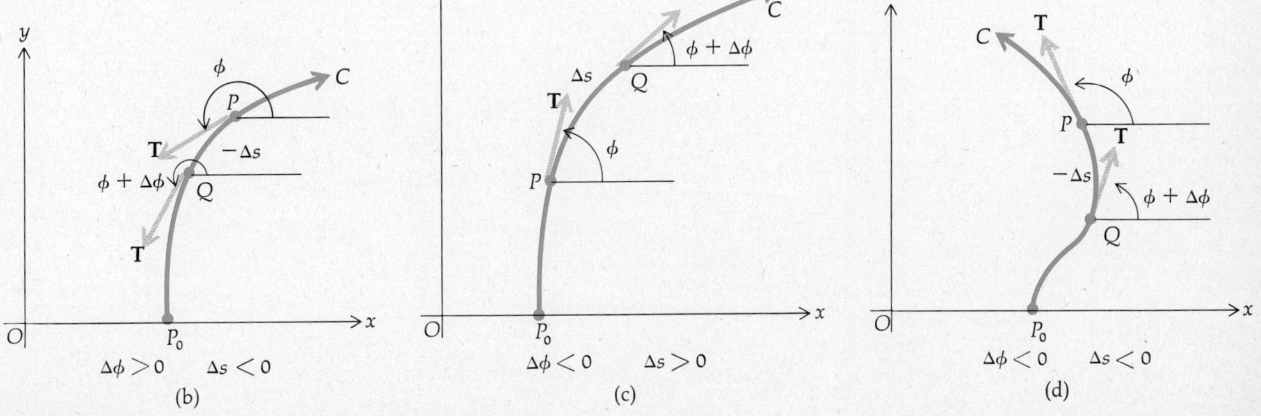

FIGURE 16.9.3

16.9.1 Definition If $\mathbf{T}(t)$ is the unit tangent vector to a curve C at a point P, s is the arc length measured from an arbitrarily chosen point on C to P, and s increases as t increases, then the **curvature vector** of C at P, denoted by $\mathbf{K}(t)$, is given by

$$\mathbf{K}(t) = D_s\mathbf{T}(t) \tag{7}$$

The **curvature** of C at P, denoted by $K(t)$, is the magnitude of the curvature vector.

To find the curvature vector and the curvature for a particular curve, it is convenient to have a formula expressing the curvature vector in terms of derivatives with repect to t. By the chain rule,

$$D_t\mathbf{T}(t) = [D_s\mathbf{T}(t)]D_ts$$

Replacing D_ts by $|D_t\mathbf{R}(t)|$ and dividing by this we obtain $D_s\mathbf{T}(t) = D_t\mathbf{T}(t)/|D_t\mathbf{R}(t)|$. So from (7),

$$\mathbf{K}(t) = \frac{D_t\mathbf{T}(t)}{|D_t\mathbf{R}(t)|} \tag{8}$$

Because $K(t) = |\mathbf{K}(t)|$, the curvature is given by

$$K(t) = \left|\frac{D_t\mathbf{T}(t)}{|D_t\mathbf{R}(t)|}\right| \tag{9}$$

Example 1 Given the circle $x = a\cos t$, $y = a\sin t$, where $a > 0$, find the curvature vector and the curvature at any t.

Solution The vector equation of the circle is

$$\mathbf{R}(t) = a\cos t\mathbf{i} + a\sin t\mathbf{j}$$

So,

$$D_t\mathbf{R}(t) = -a\sin t\mathbf{i} + a\cos t\mathbf{j}$$

$$|D_t\mathbf{R}(t)| = \sqrt{(-a\sin t)^2 + (a\cos t)^2} = a$$

Therefore

$$\mathbf{T}(t) = \frac{D_t\mathbf{R}(t)}{|D_t\mathbf{R}(t)|} = -\sin t\mathbf{i} + \cos t\mathbf{j}$$

$$D_t\mathbf{T}(t) = -\cos t\mathbf{i} - \sin t\mathbf{j}$$

$$\frac{D_t\mathbf{T}(t)}{|D_t\mathbf{R}(t)|} = -\frac{\cos t}{a}\mathbf{i} - \frac{\sin t}{a}\mathbf{j}$$

So the curvature vector

$$\mathbf{K}(t) = -\frac{1}{a}\cos t\mathbf{i} - \frac{1}{a}\sin t\mathbf{j}$$

and the curvature

$$K(t) = |\mathbf{K}(t)| = \frac{1}{a}$$

The result of Example 1 states that the curvature of a circle is constant and is the reciprocal of the radius.

Suppose we are given a curve C and at a particular point P the curvature exists and is $K(t)$, where $K(t) \neq 0$. Consider the circle that is tangent to curve C at P and has a curvature $K(t)$ at P. From Example 1, the radius of this circle is $1/K(t)$ and its center is on a line perpendicular to the tangent line in the direction of $\mathbf{N}(t)$. This circle is called the *circle of curvature*, and its radius is the *radius of curvature* of C at P. The circle of curvature is sometimes referred to as the *osculating circle*.

16.9.2 Definition If $K(t)$ is the curvature of a curve C at point P and $K(t) \neq 0$, then the **radius of curvature** of C at P, denoted by $\rho(t)$, is defined by

$$\rho(t) = \frac{1}{K(t)}$$

Example 2 Given that a vector equation of a curve C is $\mathbf{R}(t) = 2t\mathbf{i} + (t^2 - 1)\mathbf{j}$, find the curvature and the radius of curvature at $t = 1$. Draw a sketch of a portion of the curve, the unit tangent vector, and the circle of curvature at $t = 1$.

Solution

$$\mathbf{R}(t) = 2t\mathbf{i} + (t^2 - 1)\mathbf{j}$$

$$D_t\mathbf{R}(t) = 2\mathbf{i} + 2t\mathbf{j}$$

$$|D_t\mathbf{R}(t)| = 2\sqrt{1 + t^2}$$

$$\mathbf{T}(t) = \frac{D_t\mathbf{R}(t)}{|D_t\mathbf{R}(t)|} = \frac{1}{\sqrt{1 + t^2}}\mathbf{i} + \frac{t}{\sqrt{1 + t^2}}\mathbf{j}$$

$$D_t\mathbf{T}(t) = -\frac{t}{(1 + t^2)^{3/2}}\mathbf{i} + \frac{1}{(1 + t^2)^{3/2}}\mathbf{j}$$

$$\frac{D_t\mathbf{T}(t)}{|D_t\mathbf{R}(t)|} = -\frac{t}{2(1 + t^2)^2}\mathbf{i} + \frac{t}{2(1 + t^2)^2}\mathbf{j}$$

$$K(t) = \left|\frac{D_t\mathbf{T}(t)}{|D_t\mathbf{R}(t)|}\right| = \sqrt{\frac{t^2}{4(1 + t^2)^4} + \frac{1}{4(1 + t^2)^4}}$$

$$K(t) = \frac{1}{2(1 + t^2)^{3/2}}$$

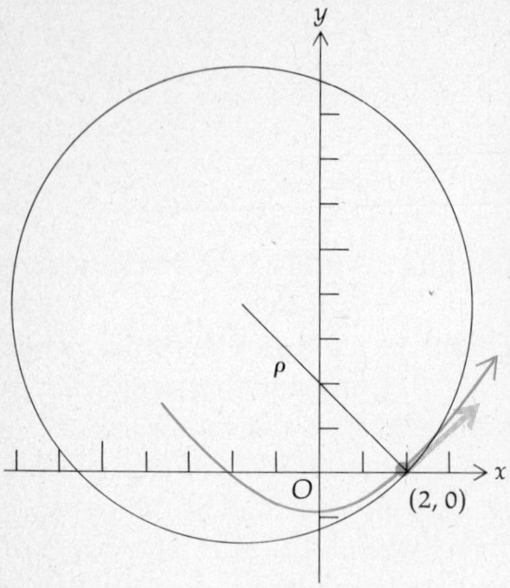

FIGURE 16.9.4

TABLE 16.9.1

t	x	y
-2	-4	3
-1	-2	0
0	0	-1
1	2	0
2	4	3

Thus

$$K(1) = \frac{1}{4\sqrt{2}} \qquad \rho(1) = 4\sqrt{2}$$

and

$$\mathbf{T}(1) = \frac{1}{\sqrt{2}}\mathbf{i} + \frac{1}{\sqrt{2}}\mathbf{j}$$

Figure 16.9.4 shows the sketch that is required. The accompanying Table 16.9.1 gives the corresponding values of x and y for $t = -2, -1, 0, 1,$ and 2.

We now find a formula for computing the curvature directly from parametric equations of the curve, $x = f(t)$ and $y = g(t)$. From Eq. (6), $|D_s\mathbf{T}(t)| = |D_s\phi|$; so

$$K(t) = |D_s\phi|$$

Assuming that s and t increase together we have

$$D_s\phi = \frac{d\phi}{ds} = \frac{\dfrac{d\phi}{dt}}{\dfrac{ds}{dt}} = \frac{\dfrac{d\phi}{dt}}{\sqrt{[f'(t)]^2 + [g'(t)]^2}}$$

Thus

$$D_s\phi = \frac{\dfrac{d\phi}{dt}}{\sqrt{\left(\dfrac{dx}{dt}\right)^2 + \left(\dfrac{dy}{dt}\right)^2}} \tag{10}$$

To find $d\phi/dt$ we observe that because ϕ is the radian measure of the angle giving the direction of the unit tangent vector,

$$\tan\phi = \frac{dy}{dx} = \frac{\dfrac{dy}{dt}}{\dfrac{dx}{dt}} \tag{11}$$

Differentiating implicitly with respect to t the left and right members of (11), we obtain

$$\sec^2\phi\,\frac{d\phi}{dt} = \frac{\left(\dfrac{dx}{dt}\right)\left(\dfrac{d^2y}{dt^2}\right) - \left(\dfrac{dy}{dt}\right)\left(\dfrac{d^2x}{dt^2}\right)}{\left(\dfrac{dx}{dt}\right)^2}$$

So

$$\frac{d\phi}{dt} = \frac{\left(\frac{dx}{dt}\right)\left(\frac{d^2y}{dt^2}\right) - \left(\frac{dy}{dt}\right)\left(\frac{d^2x}{dt^2}\right)}{\sec^2 \phi \left(\frac{dx}{dt}\right)^2} \tag{12}$$

But

$$\sec^2 \phi = 1 + \tan^2 \phi = 1 + \frac{\left(\frac{dy}{dt}\right)^2}{\left(\frac{dx}{dt}\right)^2}$$

Substituting this expression for $\sec^2 \phi$ in (12) we get

$$\frac{d\phi}{dt} = \frac{\left(\frac{dx}{dt}\right)\left(\frac{d^2y}{dt^2}\right) - \left(\frac{dy}{dt}\right)\left(\frac{d^2x}{dt^2}\right)}{\left(\frac{dx}{dt}\right)^2 + \left(\frac{dy}{dt}\right)^2} \tag{13}$$

Substituting from (13) into (10), and because $K(t) = |D_s \phi|$, we have

$$K(t) = \frac{\left|\left(\frac{dx}{dt}\right)\left(\frac{d^2y}{dt^2}\right) - \left(\frac{dy}{dt}\right)\left(\frac{d^2x}{dt^2}\right)\right|}{\left[\left(\frac{dx}{dt}\right)^2 + \left(\frac{dy}{dt}\right)^2\right]^{3/2}} \tag{14}$$

Example 3 Find the curvature of the curve in Example 2 by using formula (14).

Solution Parametric equations of C are $x = 2t$ and $y = t^2 - 1$. Hence

$$\frac{dx}{dt} = 2 \qquad \frac{d^2x}{dt^2} = 0 \qquad \frac{dy}{dt} = 2t \qquad \frac{d^2y}{dt^2} = 2$$

Therefore, from (14),

$$K(t) = \frac{|2(2) - 2t(0)|}{[(2)^2 + (2t)^2]^{3/2}} = \frac{4}{(4 + 4t^2)^{3/2}} = \frac{1}{2(1 + t^2)^{3/2}}$$

Suppose a cartesian equation of a curve is given either in the form $y = F(x)$ or $x = G(y)$. Special cases of formula (14) can be used to find the curvature of a curve in such situations.

If $y = F(x)$ is an equation of a curve C, a set of parametric equations of C is $x = t$ and $y = F(t)$. Then $dx/dt = 1$, $d^2x/dt^2 = 0$, $dy/dt = dy/dx$, and $d^2y/dt^2 = d^2y/dx^2$. Substituting into (14) we obtain

$$K = \frac{\left|\dfrac{d^2y}{dx^2}\right|}{\left[1 + \left(\dfrac{dy}{dx}\right)^2\right]^{3/2}} \tag{15}$$

Similarly, if an equation of a curve C is $x = G(y)$,

$$K = \frac{\left|\dfrac{d^2x}{dy^2}\right|}{\left[1 + \left(\dfrac{dx}{dy}\right)^2\right]^{3/2}} \tag{16}$$

Example 4 If the curve C has an equation $xy = 1$, find the radius of curvature of C at the point $(1, 1)$ and draw a sketch of the curve and the circle of curvature at $(1, 1)$.

Solution Solving for y we obtain $y = 1/x$. So $dy/dx = -1/x^2$ and $d^2y/dx^2 = 2/x^3$. From formula (15),

$$K = \frac{\left|\dfrac{2}{x^3}\right|}{\left[1 + \left(\dfrac{1}{x^4}\right)\right]^{3/2}} = \frac{2x^6}{|x^3|(x^4 + 1)^{3/2}} = \frac{2x^4}{|x|(x^4 + 1)^{3/2}}$$

Because $\rho = 1/K$,

$$\rho = \frac{|x|(x^4 + 1)^{3/2}}{2x^4}$$

So at $(1, 1)$, $\rho = \sqrt{2}$. The required sketch is shown in Fig. 16.9.5.

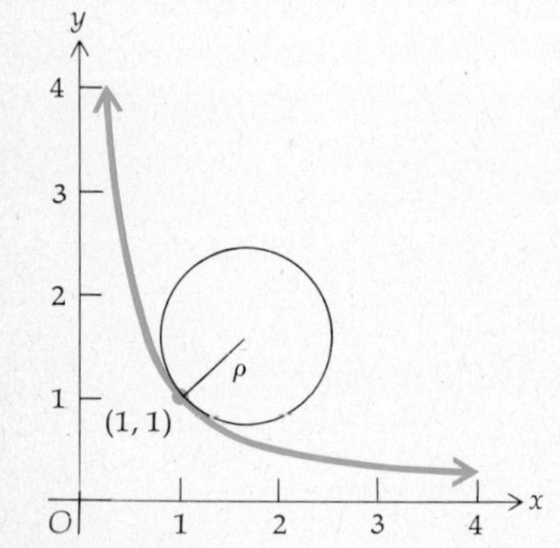

FIGURE 16.9.5

EXERCISES 16.9

In Exercises 1 through 4, find the curvature K and the radius of curvature ρ at the point where $t = t_1$. Use formula (9) to find K. Draw a sketch showing a portion of the curve, the unit tangent vector, and the circle of curvature at $t = t_1$.

1. $\mathbf{R}(t) = t^2\mathbf{i} + (2t + 1)\mathbf{j}$; $t_1 = 1$

2. $\mathbf{R}(t) = (t^2 - 2t)\mathbf{i} + (t^3 - t)\mathbf{j}$; $t_1 = 1$

3. $\mathbf{R}(t) = 2e^t\mathbf{i} + 2e^{-t}\mathbf{j}$; $t_1 = 0$

4. $\mathbf{R}(t) = \sin t\mathbf{i} + \sin 2t\mathbf{j}$; $t_1 = \frac{1}{2}\pi$

In Exercises 5 and 6, find the curvature K by using formula (14). Then find K and ρ at the point where $t = t_1$ and draw a sketch showing a portion of the curve, the unit tangent vector, and the circle of curvature at $t = t_1$.

5. $x = \dfrac{1}{1 + t}, y = \dfrac{1}{1 - t}$; $t_1 = 0$

6. $x = e^t + e^{-t}, y = e^t - e^{-t}$; $t_1 = 0$

In Exercises 7 through 14, find the curvature K and the radius of curvature ρ at the given point. Draw a sketch showing a portion of the curve, a piece of the tangent line, and the circle of curvature at the given point.

7. $y = 2\sqrt{x}$; $(0, 0)$ **8.** $y^2 = x^3$; $(\frac{1}{4}, \frac{1}{8})$ **9.** $y = e^x$; $(0, 1)$ **10.** $y = \ln x$; $(e, 1)$

11. $x = \sin y$; $(\frac{1}{2}, \frac{1}{6}\pi)$ **12.** $4x^2 + 9y^2 = 36$; $(0, 2)$ **13.** $x = \sqrt{y - 1}$; $(2, 5)$ **14.** $x = \tan y$; $(1, \frac{1}{4}\pi)$

In Exercises 15 through 22, find the radius of curvature at any point on the given curve.

15. $y = \sin^{-1} x$ **16.** $y = \ln \sec x$ **17.** $4x^2 - 9y^2 = 16$

18. $x = \tan^{-1} y$ **19.** $x^{1/2} + y^{1/2} = a^{1/2}$ **20.** $\mathbf{R}(t) = e^t \sin t\mathbf{i} + e^t \cos t\mathbf{j}$

21. The cycloid $x = a(t - \sin t)$, $y = a(1 - \cos t)$ **22.** The tractrix $x = t - a \tanh \dfrac{t}{a}$, $y = a \operatorname{sech} \dfrac{t}{a}$

23. Show that the curvature of the catenary $y = a \cosh (x/a)$ at any point (x, y) on the curve is a/y^2. Draw the circle of curvature at $(0, a)$. Show that the curvature K is an absolute maximum at the point $(0, a)$ without referring to $K'(x)$.

In Exercises 24 through 28, find a point on the given curve at which the curvature is an absolute maximum.

24. $y = e^x$ **25.** $y = 6x - x^2$ **26.** $y = \sin x$ **27.** $\mathbf{R}(t) = (2t - 3)\mathbf{i} + (t^2 - 1)\mathbf{j}$ **28.** $y = x^2 - 2x + 3$

29. Find an equation of the circle of curvature for $y = e^x$ at the point $(0, 1)$.

30. If a polar equation of a curve is $r = F(\theta)$, prove that the curvature K is given by the formula

$$K = \frac{|r^2 + 2(dr/d\theta)^2 - r(d^2r/d\theta^2)|}{[r^2 + (dr/d\theta)^2]^{3/2}}$$

In Exercises 31 through 34, find the curvature K and the radius of curvature ρ at the indicated point. Use the formula of Exercise 30 to find K.

31. $r = 4 \cos 2\theta$; $\theta = \frac{1}{12}\pi$ **32.** $r = 1 - \sin \theta$; $\theta = 0$ **33.** $r = a \sec^2 \frac{1}{2}\theta$; $\theta = \frac{2}{3}\pi$ **34.** $r = a\theta$; $\theta = 1$

35. The center of the circle of curvature of a curve C at a point P is called the *center of curvature* at P. Prove that the coordinates of the center of curvature of a curve at $P(x, y)$ are given by

$$x_c = x - \frac{(dy/dx)[1 + (dy/dx)^2]}{d^2y/dx^2} \qquad y_c = y + \frac{(dy/dx)^2 + 1}{d^2y/dx^2}$$

In Exercises 36 through 38, find the curvature K, the radius of curvature ρ, and the center of curvature at the given point. Draw a sketch of the curve and the circle of curvature.

36. $y = \ln x$; $(1, 0)$ **37.** $y = x^4 - x^2$; $(0, 0)$ **38.** $y = \cos x$; $(\frac{1}{3}\pi, \frac{1}{2})$

In Exercises 39 through 42, find the coordinates of the center of curvature at any point.

39. $y^2 = 4px$ **40.** $y^3 = a^2x$ **41.** $\mathbf{R}(t) = a \cos t\mathbf{i} + b \sin t\mathbf{j}$ **42.** $\mathbf{R}(t) = a \cos^3 t\mathbf{i} + a \sin^3 t\mathbf{j}$

43. Show that the curvature of a straight line is zero at every point.

16.10 TANGENTIAL AND NORMAL COMPONENTS OF ACCELERATION

If a particle is moving along a curve C having the vector equation

$$\mathbf{R}(t) = f(t)\mathbf{i} + g(t)\mathbf{j} \tag{1}$$

from Definition 16.7.1 the velocity vector at P is given by

$$\mathbf{V}(t) = D_t\mathbf{R}(t) \tag{2}$$

From Sec. 16.8, if $\mathbf{T}(t)$ is the unit tangent vector at P, s is the length of arc of C from a fixed point P_0 to P, and s increases as t increases,

$$D_t\mathbf{R}(t) = D_ts[\mathbf{T}(t)] \tag{3}$$

Substituting from (3) into (2) we have

$$\mathbf{V}(t) = D_ts[\mathbf{T}(t)] \tag{4}$$

Equation (4) expresses the velocity vector at a point as a scalar times the unit tangent vector at the point. We now proceed to express the acceleration vector at a point in terms of the unit tangent and unit normal vectors at the point. From Definition 16.7.2 the acceleration vector at P is given by

$$\mathbf{A}(t) = D_t^2\mathbf{R}(t) \tag{5}$$

From (6) in Sec. 16.8,

$$D_t^2\mathbf{R}(t) = [D_t|D_t\mathbf{R}(t)|]\mathbf{T}(t) + |D_t\mathbf{R}(t)||D_t\mathbf{T}(t)|\mathbf{N}(t) \tag{6}$$

Because

$$D_ts = |D_t\mathbf{R}(t)| \tag{7}$$

we differentiate with respect to t to obtain

$$D_t^2s = D_t|D_t\mathbf{R}(t)| \tag{8}$$

Furthermore,

$$|D_t\mathbf{R}(t)||D_t\mathbf{T}(t)| = |D_t\mathbf{R}(t)|^2\left|\frac{D_t\mathbf{T}(t)}{|D_t\mathbf{R}(t)|}\right| \tag{9}$$

Applying (7) above and Eq. (9) of Sec. 16.9 to the right side of (9) wc have

$$|D_t\mathbf{R}(t)||D_t\mathbf{T}(t)| = (D_ts)^2K(t) \tag{10}$$

Substituting from (5), (8), and (10) into (6) we obtain

$$\mathbf{A}(t) = (D_t^2s)\mathbf{T}(t) + (D_ts)^2K(t)\mathbf{N}(t) \tag{11}$$

Equation (11) expresses the acceleration vector as the sum of a scalar times the unit tangent vector and a scalar times the unit normal vector. The coefficient of $\mathbf{T}(t)$ is called the **tangential component** of the acceleration vector and is denoted by $A_T(t)$, whereas the coefficient of $\mathbf{N}(t)$ is called the **normal component** of the acceleration vector and is denoted by $A_N(t)$. So

$$A_T(t) = D_t^2s \tag{12}$$

and

$$A_N(t) = (D_ts)^2K(t) \tag{13}$$

or, equivalently,

$$A_N(t) = \frac{(D_ts)^2}{\rho(t)} \tag{14}$$

Because the number of units in the speed of the particle at time t units is $|\mathbf{V}(t)| = D_t s$, $A_T(t)$ is the derivative of the measure of the speed of the particle and $A_N(t)$ is the square of the measure of the speed divided by the radius of curvature.

From Newton's second law of motion,

$$\mathbf{F} = m\mathbf{A}$$

where $\mathbf{F}$ is the force vector applied to the moving object, m is the measure of the mass of the object, and $\mathbf{A}$ is the acceleration vector of the object. In curvilinear motion the normal component of $\mathbf{F}$ is the force normal to the curve necessary to keep the object on the curve. For example, if an automobile is going around a curve at a high speed, then the normal force must have a large magnitude to keep the car on the road. Also, if the curve is sharp, the radius of curvature is a small number; so the magnitude of the normal force must be a large number.

Equation (4) indicates that the tangential component of the velocity vector is $D_t s$ and that the normal component of the velocity vector is zero.

Substituting from (12) and (13) into (11) we have

$$\mathbf{A}(t) = A_T(t)\mathbf{T}(t) + A_N(t)\mathbf{N}(t) \tag{15}$$

from which it follows that

$$|\mathbf{A}(t)| = \sqrt{[A_T(t)]^2 + [A_N(t)]^2}$$

Solving for $A_N(t)$, noting from (13) that $A_N(t)$ is nonnegative, we have

$$A_N(t) = \sqrt{|\mathbf{A}(t)|^2 - [A_T(t)]^2} \tag{16}$$

Example 1 A particle is moving along the curve having the vector equation $\mathbf{R}(t) = (t^2 - 1)\mathbf{i} + (\frac{1}{3}t^3 - t)\mathbf{j}$. Find each of the following vectors: $\mathbf{V}(t)$, $\mathbf{A}(t)$, $\mathbf{T}(t)$, and $\mathbf{N}(t)$. Also, find the following scalars: $|\mathbf{V}(t)|$, $A_T(t)$, $A_N(t)$, and $K(t)$. Find the particular values when $t = 2$. Draw a sketch showing a portion of the curve at the point where $t = 2$, and representations of $\mathbf{V}(2)$, $\mathbf{A}(2)$, $A_T(2)\mathbf{T}(2)$, and $A_N(2)\mathbf{N}(2)$, having their initial point at $t = 2$.

Solution

$$\mathbf{V}(t) = D_t\mathbf{R}(t) = 2t\mathbf{i} + (t^2 - 1)\mathbf{j}$$

$$\mathbf{A}(t) = D_t\mathbf{V}(t) = 2\mathbf{i} + 2t\mathbf{j}$$

$$|\mathbf{V}(t)| = \sqrt{4t^2 + t^4 - 2t^2 + 1} = \sqrt{t^4 + 2t^2 + 1} = t^2 + 1$$

$$|\mathbf{A}(t)| = \sqrt{4 + 4t^2} = 2\sqrt{1 + t^2}$$

$$D_t s = |\mathbf{V}(t)| = t^2 + 1$$

$$A_T(t) = D_t^2 s = 2t$$

From (16),

$$A_N(t) = \sqrt{|\mathbf{A}(t)|^2 - [A_T(t)]^2} = \sqrt{4 + 4t^2 - 4t^2} = 2$$

$$\mathbf{T}(t) = \frac{\mathbf{V}(t)}{|\mathbf{V}(t)|} = \frac{2t}{t^2 + 1}\mathbf{i} + \frac{t^2 - 1}{t^2 + 1}\mathbf{j}$$

To find $\mathbf{N}(t)$ we use the following formula that comes from (11):

$$\mathbf{N}(t) = \frac{1}{(D_t s)^2 K(t)}[\mathbf{A}(t) - (D_t^2 s)\mathbf{T}(t)] \tag{17}$$

$$\mathbf{A}(t) - (D_t^2 s)\mathbf{T}(t) = 2\mathbf{i} + 2t\mathbf{j} - 2t\left(\frac{2t}{t^2+1}\mathbf{i} + \frac{t^2-1}{t^2+1}\mathbf{j}\right)$$

$$\mathbf{A}(t) - (D_t^2 s)\mathbf{T}(t) = \frac{2}{t^2+1}[(1-t^2)\mathbf{i} + 2t\mathbf{j}] \tag{18}$$

From (17), $\mathbf{N}(t)$ is a scalar times the vector in (18). Because $\mathbf{N}(t)$ is a unit vector, $\mathbf{N}(t)$ can be obtained by dividing the vector in (18) by its magnitude. Thus

$$\mathbf{N}(t) = \frac{(1-t^2)\mathbf{i} + 2t\mathbf{j}}{\sqrt{(1-t^2)^2 + (2t)^2}} = \frac{1-t^2}{1+t^2}\mathbf{i} + \frac{2t}{1+t^2}\mathbf{j}$$

$K(t)$ is found from (13), and we have

$$K(t) = \frac{A_N(t)}{(D_t s)^2} = \frac{2}{(t^2+1)^2}$$

When $t = 2$, we obtain $\mathbf{V}(2) = 4\mathbf{i} + 3\mathbf{j}$, $\mathbf{A}(2) = 2\mathbf{i} + 4\mathbf{j}$, $|\mathbf{V}(2)| = 5$, $A_T(2) = 4$, $A_N(2) = 2$, $\mathbf{T}(2) = \frac{4}{5}\mathbf{i} + \frac{3}{5}\mathbf{j}$, $\mathbf{N}(2) = -\frac{3}{5}\mathbf{i} + \frac{4}{5}\mathbf{j}$, $K(2) = \frac{2}{25}$. The required sketch is shown in Fig. 16.10.1.

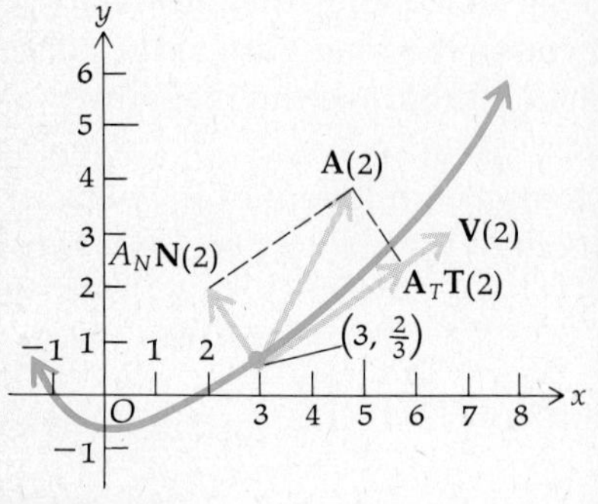

FIGURE 16.10.1

EXERCISES 16.10

In Exercises 1 through 3, a particle is moving along the curve having the given vector equation. In each exercise, find the vectors $\mathbf{V}(t)$ and $\mathbf{A}(t)$ and the scalars $A_T(t)$ and $A_N(t)$.

1. $\mathbf{R}(t) = (t^3 - 3t)\mathbf{i} + 3t^2\mathbf{j}$ **2.** $\mathbf{R}(t) = 2\sin 4t\mathbf{i} + 2\cos 4t\mathbf{j}$ **3.** $\mathbf{R}(t) = (\cos t + t\sin t)\mathbf{i} + (\sin t - t\cos t)\mathbf{j};\ t \ge 0$

In Exercises 4 and 5, a particle is moving along the curve having the given vector equation. In each exercise, find $\mathbf{V}(t_1)$, $\mathbf{A}(t_1)$, $A_T(t_1)$, and $A_N(t_1)$ for the given value of t_1.

4. $\mathbf{R}(t) = e^{-2t}\mathbf{i} + e^{2t}\mathbf{j};\ t_1 = \ln 2$ **5.** $\mathbf{R}(t) = \sin^3 t\mathbf{i} + \cos^3 t\mathbf{j};\ t_1 = \frac{1}{4}\pi$

6. If $\mathbf{R}(t) = (t^2 + 4)\mathbf{i} + (2t - 5)\mathbf{j}$, find the following scalars: $|\mathbf{V}(t)|$, $A_T(t)$, $A_N(t)$, and $K(t)$.

In Exercises 7 through 12, a particle is moving along the curve having the given vector equation. In each exercise, find the vectors $\mathbf{V}(t)$, $\mathbf{A}(t)$, $\mathbf{T}(t)$, and $\mathbf{N}(t)$, and the following scalars for an arbitrary value of t: $|\mathbf{V}(t)|$, $A_T(t)$, $A_N(t)$, and $K(t)$. Also find the particular values when $t = t_1$. At $t = t_1$, draw a sketch of a portion of the curve and representations of the vectors $\mathbf{V}(t_1)$, $\mathbf{A}(t_1)$, $A_T(t_1)\mathbf{T}(t_1)$, and $A_N(t_1)\mathbf{N}(t_1)$.

7. $\mathbf{R}(t) = (2t + 3)\mathbf{i} + (t^2 - 1)\mathbf{j};\ t_1 = 2$ **8.** $\mathbf{R}(t) = (t - 1)\mathbf{i} + t^2\mathbf{j};\ t_1 = 1$

9. $\mathbf{R}(t) = 5\cos 3t\mathbf{i} + 5\sin 3t\mathbf{j};\ t_1 = \frac{1}{3}\pi$ **10.** $\mathbf{R}(t) = 3t^2\mathbf{i} + 2t^3\mathbf{j};\ t_1 = 1$

11. $\mathbf{R}(t) = e^t\mathbf{i} + e^{-t}\mathbf{j};\ t_1 = 0$ **12.** $\mathbf{R}(t) = \cos t^2\mathbf{i} + \sin t^2\mathbf{j};\ t_1 = \frac{1}{2}\sqrt{\pi}$

In Exercises 13 and 14, a particle is moving along the curve having the given cartesian equation. At the given point find (a) the position vector, (b) the velocity vector, (c) the acceleration vector, (d) A_T, and (e) A_N.

13. $y = 4x^2;\ (1, 4)$ **14.** $y^2 = x^3;\ (4, 8)$

15. A particle is moving along the parabola $y^2 = 8x$ and its speed is constant. Find each of the following when the particle is at (2, 4): the position vector, the velocity vector, the acceleration vector, the unit tangent vector, the unit normal vector, A_T, and A_N.

16. A particle is moving along the top branch of the hyperbola $y^2 - x^2 = 9$, such that $D_t x$ is a positive constant. Find each of the following when the particle is at (4, 5): the position vector, the velocity vector, the acceleration vector, the unit tangent vector, the unit normal vector, A_T, and A_N.

REVIEW EXERCISES FOR CHAPTER 16

In Exercises 1 through 14, $\mathbf{A} = 4\mathbf{i} - 6\mathbf{j}$, $\mathbf{B} = \mathbf{i} + 7\mathbf{j}$, *and* $\mathbf{C} = 9\mathbf{i} - 5\mathbf{j}$.

1. Find $3\mathbf{B} - 7\mathbf{A}$. **2.** Find $5\mathbf{B} - 3\mathbf{C}$. **3.** Find $|3\mathbf{B} - 7\mathbf{A}|$. **4.** Find $|5\mathbf{B} - 3\mathbf{C}|$.

5. Find $|3\mathbf{B}| - |7\mathbf{A}|$. **6.** Find $|5\mathbf{B}| - |3\mathbf{C}|$. **7.** Find $(\mathbf{A} - \mathbf{B}) \cdot \mathbf{C}$. **8.** Find $(\mathbf{A} \cdot \mathbf{B})\mathbf{C}$.

9. Find a unit vector having the same direction as $2\mathbf{A} + \mathbf{B}$. **10.** Find the unit vectors that are orthogonal to $\mathbf{B}$.

11. Find scalars h and k such that $\mathbf{A} = h\mathbf{B} + k\mathbf{C}$. **12.** Find the vector projection of $\mathbf{C}$ onto $\mathbf{A}$.

13. Find the component of $\mathbf{A}$ in the direction of $\mathbf{B}$.

14. Find $\cos \alpha$ if α is the radian measure of the angle between $\mathbf{A}$ and $\mathbf{C}$.

In Exercises 15 and 16, for the vector-valued function, find (a) the domain of $\mathbf{R}$; *(b)* $\lim_{t \to 1} \mathbf{R}(t)$; *(c)* $D_t\mathbf{R}(t)$.

15. $\mathbf{R}(t) = \dfrac{1}{t+1}\mathbf{i} + \dfrac{\sqrt{t} - 1}{t - 1}\mathbf{j}$ **16.** $\mathbf{R}(t) = |t - 1|\mathbf{i} + \ln t\mathbf{j}$

In Exercises 17 and 18, find dy/dx *and* d^2y/dx^2 *without eliminating the parameter.*

17. $x = 9t^2 - 1, y = 3t + 1$ **18.** $x = e^{2t}, y = e^{-3t}$

In Exercises 19 and 20, find equations of the horizontal and vertical tangent lines, and then draw a sketch of the graph of the given pair of parametric equations.

19. $x = 12 - t^2, y = 12t - t^3$ **20.** $x = \dfrac{2at^2}{1 + t^2}, y = \dfrac{2at^3}{1 + t^2}, a > 0$ (the cissoid of Diocles)

21. If $\mathbf{R}(t) = \ln(t^2 - 1)\mathbf{i} - 2t^{-3}\mathbf{j}$, find $\mathbf{R}'(t) \cdot \mathbf{R}''(t)$.

22. Find the length of the arc of the curve having parametric equations $x = t^2, y = t^3$, from $t = 1$ to $t = 2$.

23. Find the length of the arc of the curve $\mathbf{R}(t) = (2 - t)\mathbf{i} + t^2\mathbf{j}$ from $t = 0$ to $t = 3$.

24. Find the length of the arc of the curve $r = 3 \sec \theta$ from $\theta = 0$ to $\theta = \frac{1}{4}\pi$.

25. (a) Show that the curve defined by the parametric equations $x = a \sin t$ and $y = b \cos t$ is an ellipse. (b) If s is the measure of the length of arc of the ellipse of part (a), show that

$$s = 4 \int_0^{\pi/2} a\sqrt{1 - k^2 \sin^2 t}\, dt$$

where $k^2 = (a^2 - b^2)/a^2 < 1$. This integral is called an *elliptic integral* and cannot be evaluated exactly in terms of elementary functions. Tables are available that give the value of the integral in terms of k.

26. Draw a sketch of the graph of the vector equation $\mathbf{R}(t) = e^t\mathbf{i} + e^{-t}\mathbf{j}$, and find a cartesian equation of the graph.

27. Show that the curvature of the curve $y = \ln x$ at any point (x, y) is $x/(x^2 + 1)^{3/2}$. Also show that the absolute maximum curvature is $2/3\sqrt{3}$, which occurs at the point $(\frac{1}{2}\sqrt{2}, -\frac{1}{2}\ln 2)$.

28. Find the curvature at any point of the branch of the hyperbola defined by $x = a \cosh t$, $y = b \sinh t$. Also show that the curvature is an absolute maximum at the vertex.

29. Find the radius of curvature at any point on the curve $x = a(\cos t + t \sin t)$, $y = a(\sin t - t \cos t)$.

30. Find the curvature, the radius of curvature, and the center of curvature of the curve $y = e^{-x}$ at the point $(0, 1)$.

31. Find the curvature and radius of curvature of the curve $\mathbf{R}(t) = 3t^2\mathbf{i} + (t^3 - 3t)\mathbf{j}$ at the point where $t = 2$.

32. If $\mathbf{R}(t) = e^{\lambda t}\mathbf{i} + e^{-\lambda t}\mathbf{j}$, where λ is a constant, show that $\mathbf{R}(t)$ satisfies the equation $\mathbf{R}''(t) - \lambda^2\mathbf{R}(t) = \mathbf{0}$.

33. For the hypocycloid of four cusps $x = a \cos^3 t$ and $y = a \sin^3 t$, find dy/dx and d^2y/dx^2 without eliminating the parameter.

34. A particle is moving along a curve having the vector equation $\mathbf{R}(t) = 3t\mathbf{i} + (4t - t^2)\mathbf{j}$. Find a cartesian equation of the path of the particle. Also find the velocity vector and the acceleration vector at $t = 1$.

35. Find the tangential and normal components of the acceleration vector for the particle of Exercise 34.

36. If a particle is moving along a curve, under what conditions will the acceleration vector and the unit tangent vector have the same or opposite directions?

In Exercises 37 and 38, for the given curve find $\mathbf{T}(t)$ and $\mathbf{N}(t)$, and at $t = t_1$ draw a sketch of a portion of the curve and draw the representations of $\mathbf{T}(t_1)$ and $\mathbf{N}(t_1)$ having initial point at $t = t_1$.

37. $\mathbf{R}(t) = (e^t + e^{-t})\mathbf{i} + 2t\mathbf{j}$; $t_1 = 2$

38. $\mathbf{R}(t) = 3(\cos t + t \sin t)\mathbf{i} + 3(\sin t - t \cos t)\mathbf{j}$, $t > 0$; $t_1 = \frac{1}{2}\pi$

39. Given the curve having parametric equations $x = 4t$, $y = \frac{1}{3}(2t + 1)^{3/2}$, $t \geq 0$, find parametric equations having the measure of arc length s as a parameter, where arc length is measured from the point where $t = 0$. Check your result by using Eq. (10) of Sec. 16.8.

40. Find the radian measure of the angle of elevation at which a gun should be fired in order to obtain the maximum range for a given muzzle speed.

41. Find a formula for obtaining the maximum height reached by a projectile fired from a gun having a given muzzle speed of v_0 ft/sec and an angle of elevation of radian measure α.

42. A girl throws a ball horizontally from the top of a cliff 288 ft high at an initial speed of 32 ft/sec. Find (a) the time of flight, and (b) the distance from the base of the cliff to the point where the ball hits the ground.

43. Prove by vector analysis that the diagonals of a parallelogram bisect each other.

44. Find the position vector $\mathbf{R}(t)$ if the acceleration vector $\mathbf{A}(t) = t^2\mathbf{i} - \dfrac{1}{t^2}\mathbf{j}$ and $\mathbf{V}(1) = \mathbf{j}$, and $\mathbf{R}(1) = -\frac{1}{4}\mathbf{i} + \frac{1}{2}\mathbf{j}$.

45. An *epicycloid* is the curve traced by a point P on the circumference of a circle of radius b which is rolling externally on a fixed circle of radius a. If the origin is at the center of the fixed circle, $A(a, 0)$ is one of the points at which the given point P comes in contact with the fixed circle, B is the moving point of tangency of the two circles, and the parameter t is the radian measure of the angle AOB, prove that parametric equations of the epicycloid are

$$x = (a + b) \cos t - b \cos \frac{a + b}{b} t$$

and

$$y = (a + b) \sin t - b \sin \frac{a + b}{b} t$$

46. Given triangle ABC, points D, E, and F are on sides AB, BC, and AC, respectively, and $\mathbf{V}(\overrightarrow{AD}) = \frac{1}{3}\mathbf{V}(\overrightarrow{AB})$; $\mathbf{V}(\overrightarrow{BE}) = \frac{1}{3}\mathbf{V}(\overrightarrow{BC})$; $\mathbf{V}(\overrightarrow{CF}) = \frac{1}{3}\mathbf{V}(\overrightarrow{CA})$. Prove $\mathbf{V}(\overrightarrow{AE}) + \mathbf{V}(\overrightarrow{BF}) + \mathbf{V}(\overrightarrow{CD}) = \mathbf{0}$.

In Exercises 47 and 48, find the velocity and acceleration vectors, the speed, and the tangential and normal components of acceleration.

47. $\mathbf{R}(t) = \cosh 2t\mathbf{i} + \sinh 2t\mathbf{j}$

48. $\mathbf{R}(t) = (2 \tan^{-1} t - t)\mathbf{i} + \ln(1 + t^2)\mathbf{j}$

CHAPTER 17 Vectors in Three-Dimensional Space and Solid Analytic Geometry

17.1 R^3, THE THREE-DIMENSIONAL NUMBER SPACE

In Chapter 1 the number line R^1 (the one-dimensional number space) and the number plane R^2 (the two-dimensional number space) were discussed. We identified the real numbers in R^1 with points on a horizontal axis and the real number pairs in R^2 with points in a geometric plane. In an analogous fashion we now introduce the set of all ordered triples of real numbers.

17.1.1 Definition The set of all ordered triples of real numbers is called the **three-dimensional number space** and is denoted by R^3. Each ordered triple (x, y, z) is called a **point** in the three-dimensional number space.

To represent R^3 in a geometric three-dimensional space, consider the directed distances of a point from three mutually perpendicular planes. The planes are formed by first considering three mutually perpendicular lines that intersect at a point that we call the *origin* and denote by the letter O. These lines, called the coordinate axes, are designated as the x axis, the y axis, and the z axis. Usually the x axis and the y axis are taken in a horizontal plane, and the z axis is vertical. A positive direction is selected on each axis. If the positive directions are chosen as in Fig. 17.1.1, the coordinate system is called a *right-handed system.* This terminology follows from the fact that if the right hand is placed so the thumb is pointed in the positive direction of the x axis and the index finger is pointed in the positive direction of the y axis, then the middle finger is pointed in the positive direction of the z axis. If the middle finger is pointed in the negative direction of the z axis, then the coordinate system is called *left handed.* A left-handed system is shown in Fig. 17.1.2. In general, we use a right-handed system. The three axes determine three coordinate planes: the xy plane containing the x and y axes, the xz plane containing the x and z axes, and the yz plane containing the y and z axes.

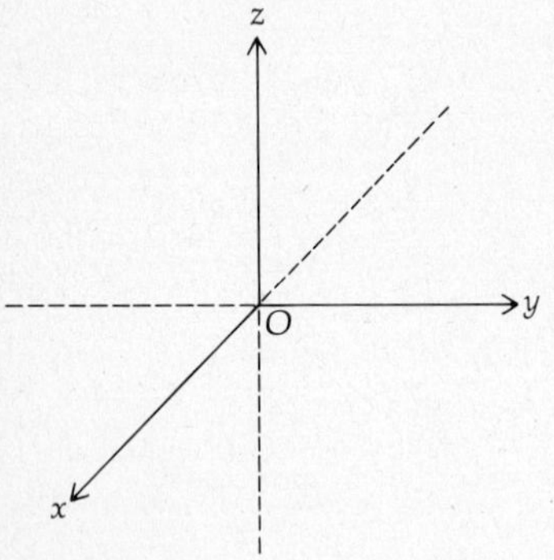

FIGURE 17.1.1

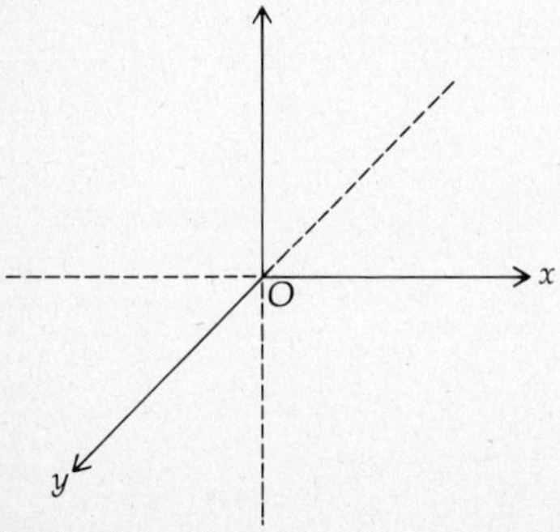

FIGURE 17.1.2

An ordered triple of real numbers (x, y, z) is associated with each point P in a geometric three-dimensional space. The directed distance of P from the yz plane is called the *x coordinate,* the directed distance of P from the xz plane is called the *y coordinate,* and the *z coordinate* is the directed distance of P from the xy plane. These three coordinates are called the *rectangular cartesian coordinates* of the point, and there is a one-to-one correspondence (called a *rectangular cartesian coordinate system*) between all such ordered triples of real numbers and the points in a geometric three-dimensional space. Hence we identify R^3 with the geometric three-dimensional space, and we call an ordered triple (x, y, z) a point. The point $(3, 2, 4)$ is shown in Fig. 17.1.3, and the point $(4, -2, -5)$ is shown in Fig. 17.1.4. The three coordinate planes divide the space into eight parts, called *octants.* The first octant is the one in which all three coordinates are positive.

A line is parallel to a plane if and only if the distance from any point on the line to the plane is the same.

• ILLUSTRATION 1 A line parallel to the yz plane, one parallel to the xz plane, and one parallel to the xy plane are shown in Fig. 17.1.5(a), (b), and (c), respectively. •

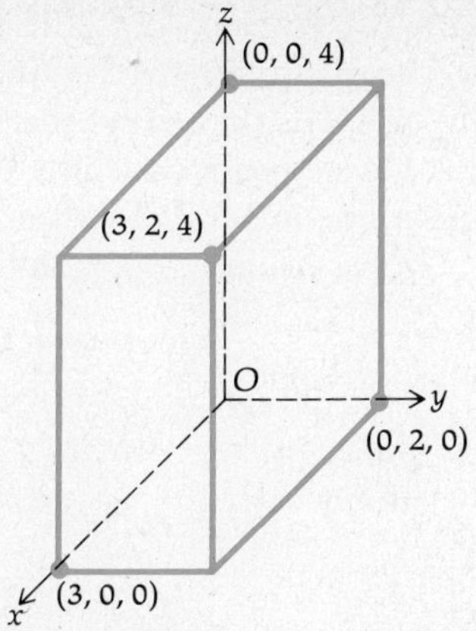

FIGURE 17.1.3

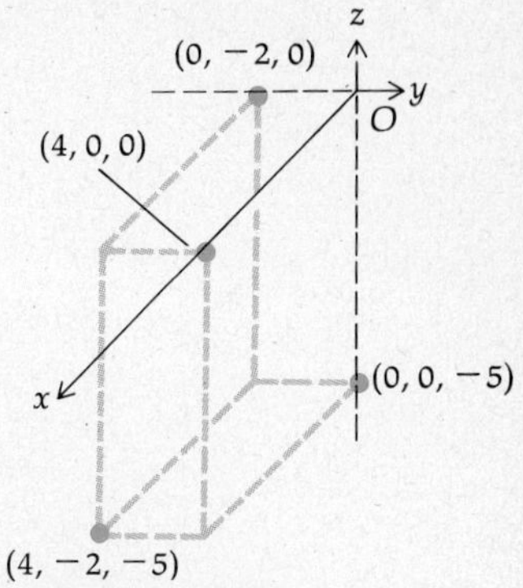

FIGURE 17.1.4

We consider all lines lying in a given plane as being parallel to the plane, in which case the distance from any point on the line to the plane is zero. The following theorem follows immediately.

17.1.2 Theorem

(i) A line is parallel to the yz plane if and only if all points on the line have equal x coordinates.
(ii) A line is parallel to the xz plane if and only if all points on the line have equal y coordinates.
(iii) A line is parallel to the xy plane if and only if all points on the line have equal z coordinates.

In three-dimensional space, if a line is parallel to each of two intersecting planes, it is parallel to the line of intersection of the two planes. Also, if a given line is parallel to a second line, then the given line is parallel to any plane containing the second line. Theorem 17.1.3 follows from these two facts from solid geometry and from Theorem 17.1.2.

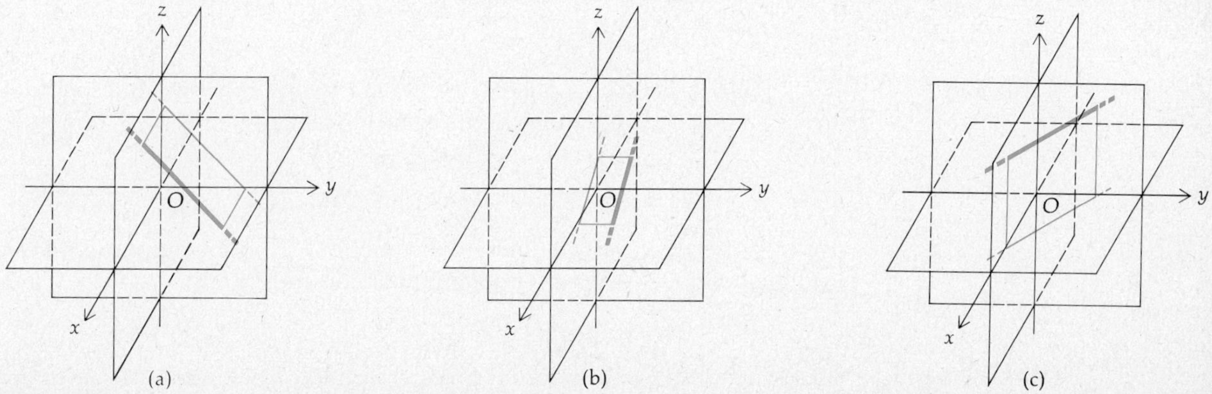

FIGURE 17.1.5

17.1.3 Theorem

(i) A line is parallel to the x axis if and only if all points on the line have equal y coördinates and equal z coordinates.
(ii) A line is parallel to the y axis if and only if all points on the line have equal x coordinates and equal z coordinates.
(iii) A line is parallel to the z axis if and only if all points on the line have equal x coordinates and equal y coordinates.

● ILLUSTRATION 2 A line parallel to the x axis, a line parallel to the y axis, and a line parallel to the z axis are shown in Fig. 17.1.6(a), (b), and (c), respectively. ●

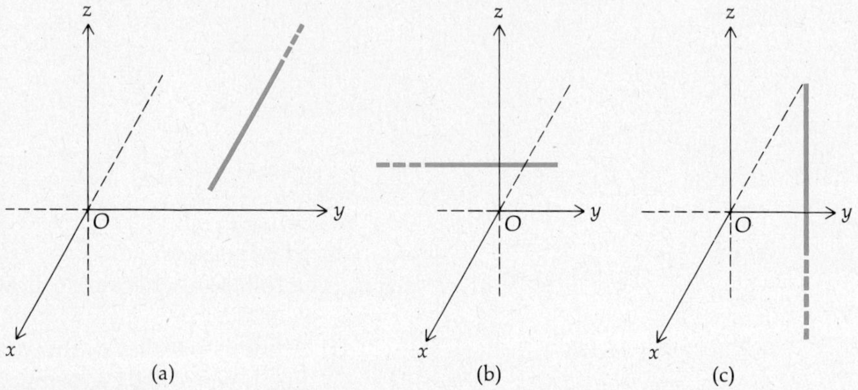

FIGURE 17.1.6

The formulas for finding the directed distance from one point to another on a line parallel to a coordinate axis follow from the definition of directed distance given in Sec. 1.3 and are stated in the following theorem.

17.1.4 Theorem

(i) If $A(x_1, y, z)$ and $B(x_2, y, z)$ are two points on a line parallel to the x axis, then the directed distance from A to B, denoted by $\overline{AB}$, is given by

$$\overline{AB} = x_2 - x_1$$

(ii) If $C(x, y_1, z)$ and $D(x, y_2, z)$ are two points on a line parallel to the y axis, then the directed distance from C to D, denoted by $\overline{CD}$, is given by

$$\overline{CD} = y_2 - y_1$$

(iii) If $E(x, y, z_1)$ and $F(x, y, z_2)$ are two points on a line parallel to the z axis, then the directed distance from E to F, denoted by $\overline{EF}$, is given by

$$\overline{EF} = z_2 - z_1$$

● ILLUSTRATION 3 The directed distance $\overline{PQ}$ from the point $P(2, -5, -4)$ to the point $Q(2, -3, -4)$ is given by Theorem 17.1.4(ii).

$$\overline{PQ} = (-3) - (-5) = 2$$ ●

The following theorem gives a formula for finding the undirected distance between any two points in three-dimensional space.

17.1.5 Theorem The undirected distance between the two points $P_1(x_1, y_1, z_1)$ and $P_2(x_2, y_2, z_2)$ is given by

$$|\overline{P_1P_2}| = \sqrt{(x_2 - x_1)^2 + (y_2 - y_1)^2 + (z_2 - z_1)^2}$$

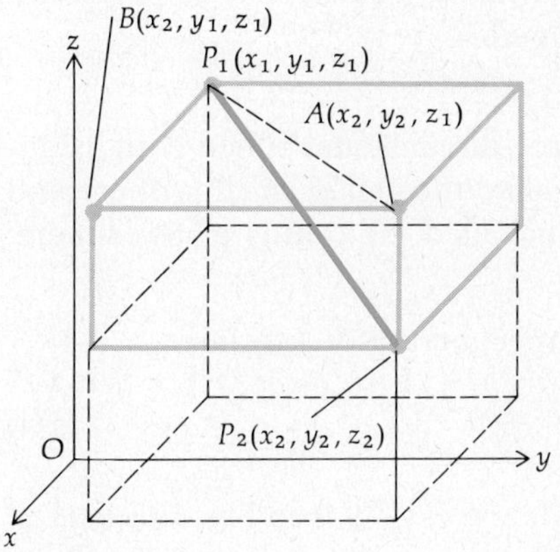

FIGURE 17.1.7

Proof Construct a rectangular parallelepiped having P_1 and P_2 as opposite vertices and faces parallel to the coordinate planes (see Fig. 17.1.7).

By the Pythagorean theorem,

$$|\overline{P_1P_2}|^2 = |\overline{P_1A}|^2 + |\overline{AP_2}|^2 \tag{1}$$

Because

$$|\overline{P_1A}|^2 = |\overline{P_1B}|^2 + |\overline{BA}|^2 \tag{2}$$

we obtain, by substituting from (2) into (1),

$$|\overline{P_1P_2}|^2 = |\overline{P_1B}|^2 + |\overline{BA}|^2 + |\overline{AP_2}|^2 \tag{3}$$

Applying Theorem 17.1.4(i), (ii), and (iii) to the right side of (3) we obtain

$$|\overline{P_1P_2}|^2 = (x_2 - x_1)^2 + (y_2 - y_1)^2 + (z_2 - z_1)^2$$

So

$$|\overline{P_1P_2}| = \sqrt{(x_2 - x_1)^2 + (y_2 - y_1)^2 + (z_2 - z_1)^2}$$

and the theorem is proved. ■

Example 1 Find the undirected distance between the points $P(-3, 4, -1)$ and $Q(2, 5, -4)$.

Solution From Theorem 17.1.5,

$$|\overline{PQ}| = \sqrt{(2 + 3)^2 + (5 - 4)^2 + (-4 + 1)^2} = \sqrt{25 + 1 + 9} = \sqrt{35}$$

The formula for the distance between two points in R^3 is merely an extension of the corresponding formula for the distance between two points in R^2 given in Theorem 1.3.1. It is noteworthy that the undirected distance between two points x_2 and x_1 in R^1 is given by

$$|x_2 - x_1| = \sqrt{(x_2 - x_1)^2}$$

The formulas for the coordinates of the midpoint of a line segment are derived by forming congruent triangles and proceeding in a manner analogous to the two-dimensional case. These formulas are given in Theorem 17.1.6, and the proof is left as an exercise (see Exercise 18).

17.1.6 Theorem The coordinates of the midpoint of the line segment having endpoints $P_1(x_1, y_1, z_1)$ and $P_2(x_2, y_2, z_2)$ are given by

$$\bar{x} = \frac{x_1 + x_2}{2} \qquad \bar{y} = \frac{y_1 + y_2}{2} \qquad \bar{z} = \frac{z_1 + z_2}{2}$$

17.1.7 Definition The **graph of an equation** in R^3 is the set of all points (x, y, z) whose coordinates are numbers satisfying the equation.

The graph of an equation in R^3 is called a **surface.** One particular surface is the sphere, which is now defined.

17.1.8 Definition A **sphere** is the set of all points in three-dimensional space equidistant from a fixed point. The fixed point is called the **center** of the sphere and the measure of the constant distance is called the **radius** of the sphere.

17.1.9 Theorem An equation of the sphere of radius r and center at (h, k, l) is

$$(x - h)^2 + (y - k)^2 + (z - l)^2 = r^2 \tag{4}$$

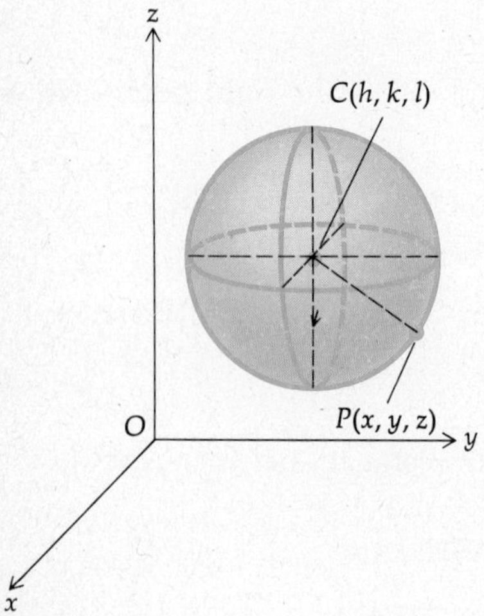

FIGURE 17.1.8

Proof Let the point (h, k, l) be denoted by C (see Fig. 17.1.8). The point $P(x, y, z)$ is a point on the sphere if and only if

$$|\overline{CP}| = r$$

or, equivalently,

$$\sqrt{(x - h)^2 + (y - k)^2 + (z - l)^2} = r$$

Squaring on both sides of the above equation we obtain the desired result. ■

If the center of the sphere is at the origin, then $h = k = l = 0$; so an equation of this sphere is

$$x^2 + y^2 + z^2 = r^2$$

If we expand the terms of (4) and regroup the terms, we have

$$x^2 + y^2 + z^2 - 2hx - 2ky - 2lz + (h^2 + k^2 + l^2 - r^2) = 0$$

This equation is of the form

$$x^2 + y^2 + z^2 + Gx + Hy + Iz + J = 0 \tag{5}$$

where G, H, I, and J are constants. Equation (5) is called the *general form* of an equation of a sphere, whereas (4) is called the *center-radius* form. Because every sphere has a center and a radius, its equation can be put in the center-radius form and hence the general form.

It can be shown that any equation of the form (5) can be put in the form

$$(x - h)^2 + (y - k)^2 + (z - l)^2 = K \tag{6}$$

where

$$h = -\tfrac{1}{2}G \qquad k = -\tfrac{1}{2}H \qquad l = -\tfrac{1}{2}I \qquad K = \tfrac{1}{4}(G^2 + H^2 + I^2 - 4J)$$

It is left as an exercise to show this (see Exercise 19).

If $K > 0$, then (6) is of the form of Eq. (4); so the graph of the equation is a sphere having its center at (h, k, l) and radius $\sqrt{K}$. If $K = 0$, the graph of the equation is the point (h, k, l). This is called a *point-sphere.* If $K < 0$, the graph is the empty set because the sum of the squares of three real numbers is nonnegative. We state this result as a theorem.

17.1.10 Theorem The graph of any second-degree equation in x, y, and z, of the form

$$x^2 + y^2 + z^2 + Gx + Hy + Iz + J = 0$$

is either a sphere, a point-sphere, or the empty set.

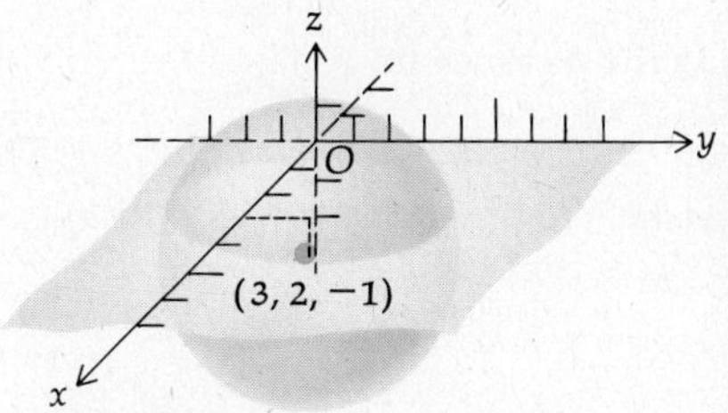

FIGURE 17.1.9

Example 2 Draw a sketch of the graph of the equation $x^2 + y^2 + z^2 - 6x - 4y + 2z = 2$.

Solution Regrouping terms and completing the squares we have

$$x^2 - 6x + 9 + y^2 - 4y + 4 + z^2 + 2z + 1 = 2 + 9 + 4 + 1$$

$$(x - 3)^2 + (y - 2)^2 + (z + 1)^2 = 16$$

So the graph is a sphere having its center at $(3, 2, -1)$ and radius 4. A sketch of the graph is shown in Fig. 17.1.9.

Example 3 Find an equation of the sphere having the points $A(-5, 6, -2)$ and $B(9, -4, 0)$ as endpoints of a diameter.

Solution The center of the sphere will be the midpoint of the line segment AB. Let this point be $C(\bar{x}, \bar{y}, \bar{z})$. By Theorem 17.1.6 we get

$$\bar{x} = \frac{9 - 5}{2} = 2 \qquad \bar{y} = \frac{-4 + 6}{2} = 1 \qquad \bar{z} = \frac{0 - 2}{2} = -1$$

So C is the point $(2, 1, -1)$. The radius of the sphere is given by

$$r = |\overline{CB}| = \sqrt{(9 - 2)^2 + (-4 - 1)^2 + (0 + 1)^2} = \sqrt{75}$$

Therefore, from Theorem 17.1.9 an equation of the sphere is

$$(x - 2)^2 + (y - 1)^2 + (z + 1)^2 = 75$$

or, equivalently,

$$x^2 + y^2 + z^2 - 4x - 2y + 2z - 69 = 0$$

EXERCISES 17.1

In Exercises 1 through 5, the given points A and B are opposite vertices of a rectangular parallelepiped having its faces parallel to the coordinate planes. In each exercise, (a) draw a sketch of the figure, (b) find the coordinates of the other six vertices, (c) find the length of the diagonal AB.

1. $A(0, 0, 0)$; $B(7, 2, 3)$ **2.** $A(1, 1, 1)$; $B(3, 4, 2)$ **3.** $A(-1, 1, 2)$; $B(2, 3, 5)$

4. $A(2, -1, -3)$; $B(4, 0, -1)$ **5.** $A(1, -1, 0)$; $B(3, 3, 5)$

6. The vertex opposite one corner of a room is 18 ft east, 15 ft south, and 12 ft up from the first corner. (a) Draw a sketch of the figure; (b) determine the length of the diagonal joining two opposite vertices; (c) find the coordinates of all eight vertices of the room.

In Exercises 6 through 9, find (a) the undirected distance between the points A and B and (b) the midpoint of the line segment joining A and B.

7. $A(3, 4, 2)$; $B(1, 6, 3)$ **8.** $A(4, -3, 2)$; $B(-2, 3, -5)$ **9.** $A(2, -4, 1)$; $B(\frac{1}{2}, 2, 3)$

10. $A(-2, -\frac{1}{2}, 5)$; $B(5, 1, -4)$ **11.** $A(-5, 2, 1)$; $B(3, 7, -2)$

12. Prove that the three points $(1, -1, 3)$, $(2, 1, 7)$, and $(4, 2, 6)$ are the vertices of a right triangle, and find its area.

13. A line is drawn through the point $(6, 4, 2)$ perpendicular to the yz plane. Find the coordinates of the points on this line at a distance of 10 units from the point $(0, 4, 0)$.

14. Solve Exercise 13 if the line is drawn perpendicular to the xy plane.

15. Prove that the three points $(-3, 2, 4)$, $(6, 1, 2)$, and $(-12, 3, 6)$ are collinear by using the distance formula.

16. Find the vertices of the triangle whose sides have midpoints at $(3, 2, 3)$, $(-1, 1, 5)$, and $(0, 3, 4)$.

17. For the triangle having vertices at $A(2, -5, 3)$, $B(-1, 7, 0)$, and $C(-4, 9, 7)$ find (a) the length of each side and (b) the midpoint of each side.

18. Prove Theorem 17.1.6.

19. Show that any equation of the form $x^2 + y^2 + z^2 + Gx + Hy + Iz + J = 0$ can be put in the form $(x - h)^2 + (y - k)^2 + (z - l)^2 = K$.

In Exercises 20 through 25, determine the graph of the given equation.

20. $x^2 + y^2 + z^2 - 8y + 6z - 25 = 0$ **21.** $x^2 + y^2 + z^2 - 8x + 4y + 2z - 4 = 0$

22. $x^2 + y^2 + z^2 - x - y - 3z + 2 = 0$ **23.** $x^2 + y^2 + z^2 - 6z + 9 = 0$

24. $x^2 + y^2 + z^2 - 8x + 10y - 4z + 13 = 0$ **25.** $x^2 + y^2 + z^2 - 6x + 2y - 4z + 19 = 0$

In Exercises 26 through 28, find an equation of the sphere satisfying the given conditions.

26. A diameter is the line segment having endpoints at $(6, 2, -5)$ and $(-4, 0, 7)$.

27. It is concentric with the sphere having equation $x^2 + y^2 + z^2 - 2y + 8z - 9 = 0$ and has radius 3.

28. It contains the points $(0, 0, 4)$, $(2, 1, 3)$, and $(0, 2, 6)$ and has its center in the yz plane.

29. Prove analytically that the four diagonals joining opposite vertices of a rectangular parallelepiped bisect each other.

30. If P, Q, R, and S are four points in three-dimensional space and A, B, C, and D are the midpoints of PQ, QR, RS, and SP, respectively, prove analytically that $ABCD$ is a parallelogram.

31. Prove analytically that the four diagonals of a cube have the same length.

17.2 VECTORS IN THREE-DIMENSIONAL SPACE

The presentation of topics in solid analytic geometry is simplified by the use of vectors in three-dimensional space. The definitions and theorems given in Secs. 16.1 and 16.2 for vectors in the plane are easily extended.

17.2.1 Definition A **vector in three-dimensional space** is an ordered triple of real numbers $\langle x, y, z\rangle$. The numbers x, y, and z are called the **components** of the vector $\langle x, y, z\rangle$.

We let V_3 be the set of all ordered triples $\langle x, y, z\rangle$ for which x, y, and z are real numbers. In this chapter a vector is always in V_3 unless otherwise stated.

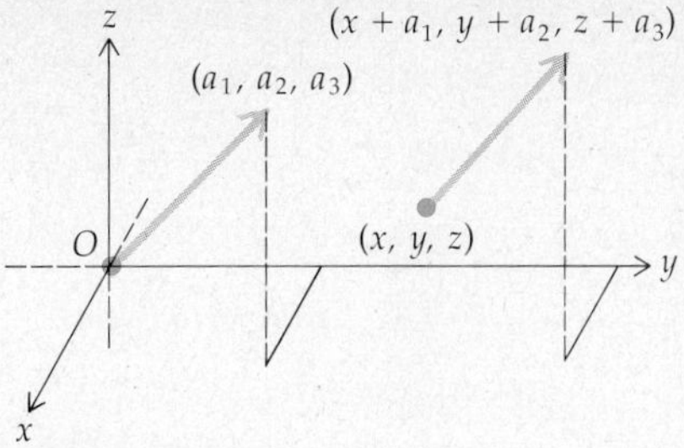

FIGURE 17.2.1

Just as for vectors in V_2, a vector in V_3 can be represented by a directed line segment. If $\mathbf{A} = \langle a_1, a_2, a_3 \rangle$, then the directed line segment having its initial point at the origin and its terminal point at the point (a_1, a_2, a_3) is called the **position representation of** A. A directed line segment having its initial point at (x, y, z) and its terminal point at $(x + a_1, y + a_2, z + a_3)$ is also a representation of the vector **A**. See Fig. 17.2.1.

The **zero vector** is the vector $\langle 0, 0, 0 \rangle$ and is denoted by **0**. Any point is a representation of the zero vector.

The **magnitude** of a vector is the length of any of its representations. If the vector $\mathbf{A} = \langle a_1, a_2, a_3 \rangle$, the magnitude of **A** is denoted by $|\mathbf{A}|$, and it follows that

$$|\mathbf{A}| = \sqrt{a_1^2 + a_2^2 + a_3^2}$$

The **direction** of a nonzero vector in V_3 is given by three angles, called the *direction angles* of the vector.

17.2.2 Definition The **direction angles** of a nonzero vector are the three angles that have the smallest nonnegative radian measures α, β, and γ measured from the positive x, y, and z axes, respectively, to the position representation of the vector.

The radian measure of each direction angle of a vector is greater than or equal to 0 and less than or equal to π. The direction angles having radian measures α, β, and γ of the vector $\mathbf{A} = \langle a_1, a_2, a_3 \rangle$ are shown in Fig. 17.2.2. In this figure the components of **A** are all positive numbers, and the direction angles of this vector all have positive radian measure less than $\frac{1}{2}\pi$. From the figure we see that triangle POR is a right triangle and

$$\cos \alpha = \frac{|\overrightarrow{OR}|}{|\overrightarrow{OP}|} = \frac{a_1}{|\mathbf{A}|}$$

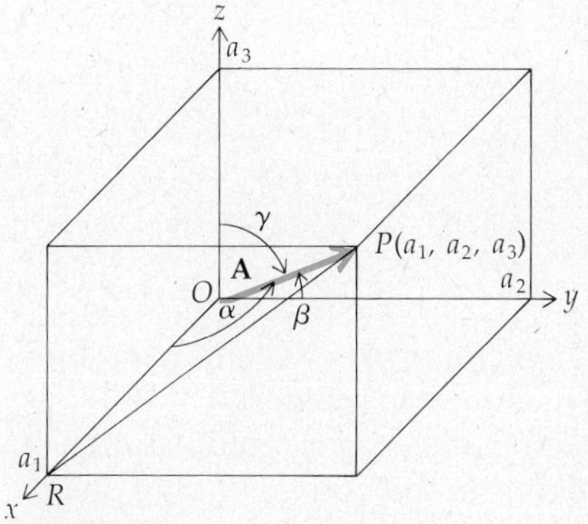

FIGURE 17.2.2

It can be shown that the same formula holds if $\frac{1}{2}\pi \le \alpha \le \pi$. Similar formulas can be found for $\cos \beta$ and $\cos \gamma$, and we have

$$\cos \alpha = \frac{a_1}{|\mathbf{A}|} \qquad \cos \beta = \frac{a_2}{|\mathbf{A}|} \qquad \cos \gamma = \frac{a_3}{|\mathbf{A}|} \tag{1}$$

The three numbers $\cos \alpha$, $\cos \beta$, and $\cos \gamma$ are called the **direction cosines** of vector **A**. The zero vector has no direction angles and hence no direction cosines.

• ILLUSTRATION 1 We find the magnitude and direction cosines of the vector $\mathbf{A} = \langle 3, 2, -6 \rangle$.

$$|\mathbf{A}| = \sqrt{(3)^2 + (2)^2 + (-6)^2} = \sqrt{9 + 4 + 36} = \sqrt{49} = 7$$

From Eqs. (1),

$$\cos \alpha = \tfrac{3}{7} \qquad \cos \beta = \tfrac{2}{7} \qquad \cos \gamma = -\tfrac{6}{7}$$ •

If the magnitude of a vector and its direction cosines are known, the vector is uniquely determined because from (1) it follows that

$$a_1 = |\mathbf{A}| \cos \alpha \qquad a_2 = |\mathbf{A}| \cos \beta \qquad a_3 = |\mathbf{A}| \cos \gamma \tag{2}$$

The three direction cosines of a vector are not independent of each other, as we see by the following theorem.

17.2.3 Theorem If $\cos \alpha$, $\cos \beta$, and $\cos \gamma$ are the direction cosines of a vector, then

$$\cos^2 \alpha + \cos^2 \beta + \cos^2 \gamma = 1$$

Proof If $\mathbf{A} = \langle a_1, a_2, a_3 \rangle$, then the direction cosines of $\mathbf{A}$ are given by (1) and

$$\begin{aligned}\cos^2 \alpha + \cos^2 \beta + \cos^2 \gamma &= \frac{a_1{}^2}{|\mathbf{A}|^2} + \frac{a_2{}^2}{|\mathbf{A}|^2} + \frac{a_3{}^2}{|\mathbf{A}|^2} \\ &= \frac{a_1{}^2 + a_2{}^2 + a_3{}^2}{|\mathbf{A}|^2} \\ &= \frac{|\mathbf{A}|^2}{|\mathbf{A}|^2} \\ &= 1\end{aligned}$$

■

● ILLUSTRATION 2 We verify Theorem 17.2.3 for the vector of Illustration 1.

$$\begin{aligned}\cos^2 \alpha + \cos^2 \beta + \cos^2 \gamma &= (\tfrac{3}{7})^2 + (\tfrac{2}{7})^2 + (-\tfrac{6}{7})^2 \\ &= \tfrac{9}{49} + \tfrac{4}{49} + \tfrac{36}{49} \\ &= \tfrac{49}{49} \\ &= 1\end{aligned}$$

●

The vector $\mathbf{A} = \langle a_1, a_2, a_3 \rangle$ is a unit vector if $|\mathbf{A}| = 1$, and from Eqs. (1) the components of a unit vector are its direction cosines.

The operations of addition, subtraction, and scalar multiplication of vectors in V_3 are given definitions analogous to the corresponding definitions for vectors in V_2.

17.2.4 Definition If $\mathbf{A} = \langle a_1, a_2, a_3 \rangle$ and $\mathbf{B} = \langle b_1, b_2, b_3 \rangle$, then the sum of these vectors is given by

$$\mathbf{A} + \mathbf{B} = \langle a_1 + b_1, a_2 + b_2, a_3 + b_3 \rangle$$

Example 1 Given $\mathbf{A} = \langle 5, -2, 6 \rangle$ and $\mathbf{B} = \langle 8, -5, -4 \rangle$, find $\mathbf{A} + \mathbf{B}$.

Solution $\mathbf{A} + \mathbf{B} = \langle 5 + 8, (-2) + (-5), 6 + (-4) \rangle = \langle 13, -7, 2 \rangle$.

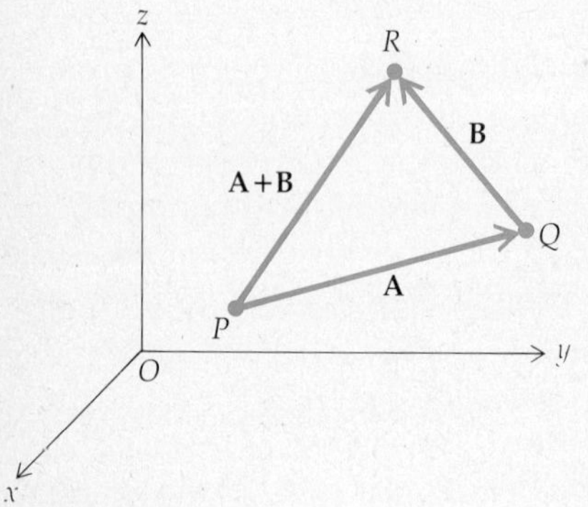

FIGURE 17.2.3

The geometric interpretation of the sum of two vectors in V_3 is similar to that for vectors in V_2. See Fig. 17.2.3. If P is the point (x, y, z), $\mathbf{A} =$

$\langle a_1, a_2, a_3\rangle$ and $\overrightarrow{PQ}$ is a representation of $\mathbf{A}$, then Q is the point $(x + a_1, y + a_2, z + a_3)$. Let $\mathbf{B} = \langle b_1, b_2, b_3\rangle$ and let $\overrightarrow{QR}$ be a representation of $\mathbf{B}$. Then R is the point $(x + (a_1 + b_1), y + (a_2 + b_2), z + (a_3 + b_3))$. Therefore $\overrightarrow{PR}$ is a representation of the vector $\mathbf{A} + \mathbf{B}$, and the parallelogram law holds

17.2.5 Definition If $\mathbf{A} = \langle a_1, a_2, a_3\rangle$, then the vector $\langle -a_1, -a_2, -a_3\rangle$ is defined to be the **negative** of $\mathbf{A}$, denoted by $-\mathbf{A}$.

17.2.6 Definition The **difference** of the two vectors $\mathbf{A}$ and $\mathbf{B}$, denoted by $\mathbf{A} - \mathbf{B}$, is defined by

$$\mathbf{A} - \mathbf{B} = \mathbf{A} + (-\mathbf{B})$$

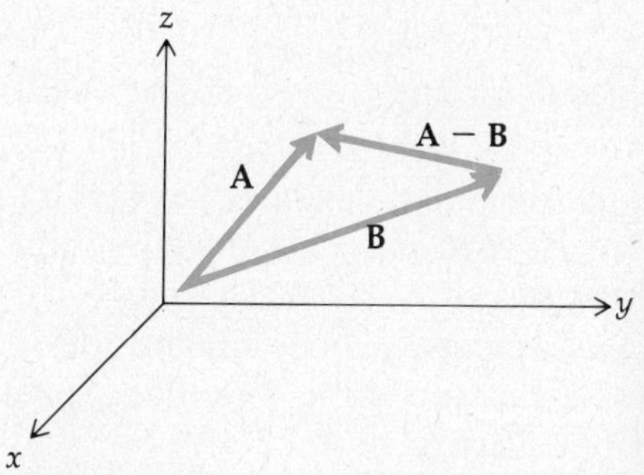

FIGURE 17.2.4

From Definitions 17.2.5 and 17.2.6 it follows that if $\mathbf{A} = \langle a_1, a_2, a_3\rangle$ and $\mathbf{B} = \langle b_1, b_2, b_3\rangle$, then $-\mathbf{B} = \langle -b_1, -b_2, -b_3\rangle$ and

$$\mathbf{A} - \mathbf{B} = \langle a_1 - b_1, a_2 - b_2, a_3 - b_3\rangle$$

Example 2 For the vectors $\mathbf{A}$ and $\mathbf{B}$ of Example 1 find $\mathbf{A} - \mathbf{B}$.

Solution

$$\begin{aligned}\mathbf{A} - \mathbf{B} &= \langle 5, -2, 6\rangle - \langle 8, -5, -4\rangle \\ &= \langle 5, -2, 6\rangle + \langle -8, 5, 4\rangle \\ &= \langle -3, 3, 10\rangle\end{aligned}$$

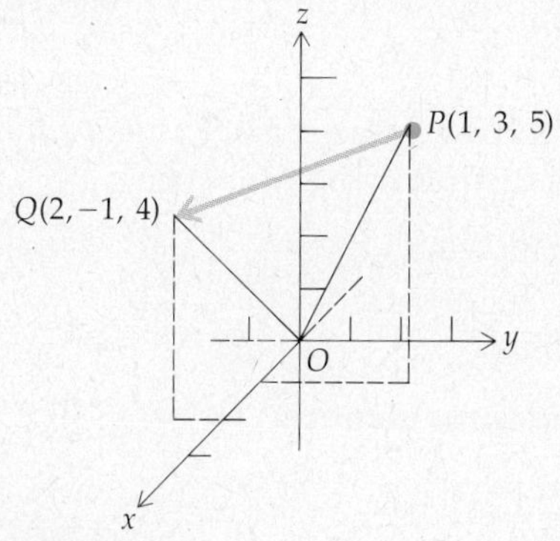

FIGURE 17.2.5

The difference of two vectors in V_3 is also interpreted geometrically as it is in V_2. See Fig. 17.2.4. A representation of the vector $\mathbf{A} - \mathbf{B}$ is obtained by choosing representations of $\mathbf{A}$ and $\mathbf{B}$ having the same initial point. Then a representation of the vector $\mathbf{A} - \mathbf{B}$ is the directed line segment from the terminal point of the representation of $\mathbf{B}$ to the terminal point of the representation of $\mathbf{A}$.

● ILLUSTRATION 3 Given the points $P(1, 3, 5)$ and $Q(2, -1, 4)$, Fig. 17.2.5 shows $\overrightarrow{PQ}$ as well as $\overrightarrow{OP}$ and $\overrightarrow{OQ}$. Observe from the figure that $\mathbf{V}(\overrightarrow{PQ}) = \mathbf{V}(\overrightarrow{OQ}) - \mathbf{V}(\overrightarrow{OP})$. Hence

$$\mathbf{V}(\overrightarrow{PQ}) = \langle 2, -1, 4\rangle - \langle 1, 3, 5\rangle = \langle 1, -4, -1\rangle$$ ●

17.2.7 Definition If c is a scalar and $\mathbf{A}$ is the vector $\langle a_1, a_2, a_3\rangle$, then the product of c and $\mathbf{A}$, denoted by $c\mathbf{A}$, is a vector and is given by

$$c\mathbf{A} = c\langle a_1, a_2, a_3\rangle = \langle ca_1, ca_2, ca_3\rangle$$

Example 3 Given $\mathbf{A} = \langle -4, 7, -2\rangle$, find $3\mathbf{A}$ and $-5\mathbf{A}$.

Solution

$$\begin{aligned}3\mathbf{A} &= 3\langle -4, 7, -2\rangle \\ &= \langle -12, 21, -6\rangle\end{aligned} \qquad \begin{aligned}-5\mathbf{A} &= (-5)\langle -4, 7, -2\rangle \\ &= \langle 20, -35, 10\rangle\end{aligned}$$

Suppose that $\mathbf{A} = \langle a_1, a_2, a_3 \rangle$ is a nonzero vector having direction cosines $\cos\alpha$, $\cos\beta$, and $\cos\gamma$, and c is any nonzero scalar. Then $c\mathbf{A} = \langle ca_1, ca_2, ca_3 \rangle$; and if $\cos\alpha_1$, $\cos\beta_1$, and $\cos\gamma_1$ are the direction cosines of $c\mathbf{A}$, we have from Eqs. (1)

$$\cos\alpha_1 = \frac{ca_1}{|c\mathbf{A}|} \qquad \cos\beta_1 = \frac{ca_2}{|c\mathbf{A}|} \qquad \cos\gamma_1 = \frac{ca_3}{|c\mathbf{A}|}$$

or, equivalently,

$$\cos\alpha_1 = \frac{c}{|c|}\frac{a_1}{|\mathbf{A}|} \qquad \cos\beta_1 = \frac{c}{|c|}\frac{a_2}{|\mathbf{A}|} \qquad \cos\gamma_1 = \frac{c}{|c|}\frac{a_3}{|\mathbf{A}|}$$

from which

$$\cos\alpha_1 = \frac{c}{|c|}\cos\alpha \qquad \cos\beta_1 = \frac{c}{|c|}\cos\beta \qquad \cos\gamma_1 = \frac{c}{|c|}\cos\gamma \tag{3}$$

So if $c > 0$, it follows from Eqs. (3) that the direction cosines of vector $c\mathbf{A}$ are the same as the direction cosines of $\mathbf{A}$. And if $c < 0$, the direction cosines of $c\mathbf{A}$ are the negatives of the direction cosines of $\mathbf{A}$. Therefore, if c is a nonzero scalar, then the vector $c\mathbf{A}$ is a vector whose magnitude is $|c|$ times the magnitude of $\mathbf{A}$. If $c > 0$, $c\mathbf{A}$ has the same direction as $\mathbf{A}$, whereas if $c < 0$, the direction of $c\mathbf{A}$ is opposite that of $\mathbf{A}$.

The operations of vector addition and scalar multiplication of any vectors in V_3 satisfy properties identical with those given in Theorem 16.2.1. These are given in the following theorem, and the proofs are left as exercises (see Exercises 15 through 20).

17.2.8 Theorem If $\mathbf{A}$, $\mathbf{B}$, and $\mathbf{C}$ are any vectors in V_3 and c and d are any scalars, then vector addition and scalar multiplication satisfy the following properties:

(i) $\mathbf{A} + \mathbf{B} = \mathbf{B} + \mathbf{A}$ (commutative law)
(ii) $\mathbf{A} + (\mathbf{B} + \mathbf{C}) = (\mathbf{A} + \mathbf{B}) + \mathbf{C}$ (associative law)
(iii) There is a vector $\mathbf{0}$ in V_3 for which

$$\mathbf{A} + \mathbf{0} = \mathbf{A} \qquad \text{(existence of additive identity)}$$

(iv) There is a vector $-\mathbf{A}$ in V_3 such that

$$\mathbf{A} + (-\mathbf{A}) = \mathbf{0} \qquad \text{(existence of negative)}$$

(v) $(cd)\mathbf{A} = c(d\mathbf{A})$ (associative law)
(vi) $c(\mathbf{A} + \mathbf{B}) = c\mathbf{A} + c\mathbf{B}$ (distributive law)
(vii) $(c + d)\mathbf{A} = c\mathbf{A} + d\mathbf{A}$ (distributive law)
(viii) $1(\mathbf{A}) = \mathbf{A}$ (existence of scalar multiplicative identity)

From Definition 16.2.2 and Theorem 17.2.8 it follows that V_3 is a real vector space. The three unit vectors

$$\mathbf{i} = \langle 1, 0, 0 \rangle \qquad \mathbf{j} = \langle 0, 1, 0 \rangle \qquad \mathbf{k} = \langle 0, 0, 1 \rangle$$

form a basis for the vector space V_3 because any vector $\langle a_1, a_2, a_3 \rangle$ can be written in terms of them as follows:

$$\langle a_1, a_2, a_3 \rangle = a_1\langle 1, 0, 0 \rangle + a_2\langle 0, 1, 0 \rangle + a_3\langle 0, 0, 1 \rangle$$

Hence, if $\mathbf{A} = \langle a_1, a_2, a_3 \rangle$, we also can write

$$\mathbf{A} = a_1\mathbf{i} + a_2\mathbf{j} + a_3\mathbf{k} \tag{4}$$

Because there are three elements in a basis, V_3 is a three-dimensional vector space.

Substituting from (2) into (4) we have

$$\mathbf{A} = |\mathbf{A}| \cos \alpha \mathbf{i} + |\mathbf{A}| \cos \beta \mathbf{j} + |\mathbf{A}| \cos \gamma \mathbf{k}$$

or, equivalently,

$$\mathbf{A} = |\mathbf{A}|(\cos \alpha \mathbf{i} + \cos \beta \mathbf{j} + \cos \gamma \mathbf{k}) \tag{5}$$

Equation (5) enables us to express any nonzero vector in terms of its magnitude and direction cosines.

Example 4 Express the vector of Illustration 1 in terms of its magnitude and direction cosines.

Solution In Illustration 1, $\mathbf{A} = \langle 3, 2, -6 \rangle$, $|\mathbf{A}| = 7$, $\cos \alpha = \frac{3}{7}$, $\cos \beta = \frac{2}{7}$, and $\cos \gamma = -\frac{6}{7}$. Hence, from (5),

$$\mathbf{A} = 7(\tfrac{3}{7}\mathbf{i} + \tfrac{2}{7}\mathbf{j} - \tfrac{6}{7}\mathbf{k})$$

17.2.9 Theorem If the nonzero vector $\mathbf{A} = a_1\mathbf{i} + a_2\mathbf{j} + a_3\mathbf{k}$, then the unit vector $\mathbf{U}$ having the same direction as $\mathbf{A}$ is given by

$$\mathbf{U} = \frac{a_1}{|\mathbf{A}|}\mathbf{i} + \frac{a_2}{|\mathbf{A}|}\mathbf{j} + \frac{a_3}{|\mathbf{A}|}\mathbf{k}$$

The proof of Theorem 17.2.9 is analogous to the proof of Theorem 16.2.3 for a vector in V_2 and is left as an exercise (see Exercise 38).

Example 5 Given the points $R(2, -1, 3)$ and $S(3, 4, 6)$, find the unit vector having the same direction as $\mathbf{V}(\overrightarrow{RS})$.

Solution

$$\begin{aligned}\mathbf{V}(\overrightarrow{RS}) &= \langle 3, 4, 6 \rangle - \langle 2, -1, 3 \rangle = \langle 1, 5, 3 \rangle \\ &= \mathbf{i} + 5\mathbf{j} + 3\mathbf{k}\end{aligned}$$

So

$$|\mathbf{V}(\overrightarrow{RS})| = \sqrt{1^2 + 5^2 + 3^2} = \sqrt{35}$$

Therefore, by Theorem 17.2.9 the desired unit vector is

$$\mathbf{U} = \frac{1}{\sqrt{35}}\mathbf{i} + \frac{5}{\sqrt{35}}\mathbf{j} + \frac{3}{\sqrt{35}}\mathbf{k}$$

EXERCISES 17.2

In Exercises 1 through 14, let $\mathbf{A} = \langle 1, 2, 3\rangle$, $\mathbf{B} = \langle 4, -3, -1\rangle$, $\mathbf{C} = \langle -5, -3, 5\rangle$, *and* $\mathbf{D} = \langle -2, 1, 6\rangle$.

1. Find $\mathbf{A} + 5\mathbf{B}$.
2. Find $2\mathbf{A} - \mathbf{C}$.
3. Find $7\mathbf{C} - 5\mathbf{D}$.
4. Find $|2\mathbf{A}| - |\mathbf{C}|$.
5. Find $|7\mathbf{C}| - |5\mathbf{D}|$.
6. Find $4\mathbf{B} + 6\mathbf{C} - 2\mathbf{D}$.
7. Find $|7\mathbf{C} - 5\mathbf{D}|$.
8. Find $|4\mathbf{B}| + |6\mathbf{C}| - |2\mathbf{D}|$.
9. Find $\mathbf{C} + 3\mathbf{D} - 8\mathbf{A}$.
10. Find $3\mathbf{A} - 2\mathbf{B} + \mathbf{C} - 12\mathbf{D}$.
11. Find $|\mathbf{A}||\mathbf{B}|(\mathbf{C} - \mathbf{D})$.
12. Find $|\mathbf{A}|\mathbf{C} - |\mathbf{B}|\mathbf{D}$.
13. Find scalars a and b such that $a(\mathbf{A} + \mathbf{B}) + b(\mathbf{C} + \mathbf{D}) = \mathbf{0}$.
14. Find scalars a, b, and c such that $a\mathbf{A} + b\mathbf{B} + c\mathbf{C} = \mathbf{D}$.
15. Prove Theorem 17.2.8(i).
16. Prove Theorem 17.2.8(ii).
17. Prove Theorem 17.2.8(iii), (iv), and (viii).
18. Prove Theorem 17.2.8(v).
19. Prove Theorem 17.2.8(vi).
20. Prove Theorem 17.2.8(vii).

In Exercises 21 through 24, find the direction cosines of the vector $\mathbf{V}(\overrightarrow{P_1P_2})$ *and check the answers by verifying that the sum of their squares is 1.*

21. $P_1(3, -1, -4)$; $P_2(7, 2, 4)$
22. $P_1(1, 3, 5)$; $P_2(2, -1, 4)$
23. $P_1(4, -3, -1)$; $P_2(-2, -4, -8)$
24. $P_1(-2, 6, 5)$; $P_2(2, 4, 1)$
25. Use the points P_1 and P_2 of Exercise 21 and find the point Q such that $\mathbf{V}(\overrightarrow{P_1P_2}) = 3\mathbf{V}(\overrightarrow{P_1Q})$.
26. Use the points P_1 and P_2 of Exercise 22 and find the point R such that $\mathbf{V}(\overrightarrow{P_1R}) = -2\mathbf{V}(\overrightarrow{P_2R})$.
27. Given $P_1(3, 2, -4)$ and $P_2(-5, 4, 2)$, find the point P_3 such that $4\mathbf{V}(\overrightarrow{P_1P_2}) = -3\mathbf{V}(\overrightarrow{P_2P_3})$.
28. Given $P_1(7, 0, -2)$ and $P_2(2, -3, 5)$, find the point P_3 such that $\mathbf{V}(\overrightarrow{P_1P_3}) = 5\mathbf{V}(\overrightarrow{P_2P_3})$.

In Exercises 29 through 32, express the given vector in terms of its magnitude and direction cosines.

29. $-6\mathbf{i} + 2\mathbf{j} + 3\mathbf{k}$
30. $2\mathbf{i} - 2\mathbf{j} + \mathbf{k}$
31. $-2\mathbf{i} + \mathbf{j} - 3\mathbf{k}$
32. $3\mathbf{i} + 4\mathbf{j} - 5\mathbf{k}$

In Exercises 33 through 36, find the unit vector having the same direction as $\mathbf{V}(\overrightarrow{P_1P_2})$.

33. $P_1(4, -1, -6)$;$P_2(5, 7, -2)$
34. $P_1(3, 0, -1)$;$P_2(-3, 8, -1)$
35. $P_1(-2, 5, 3)$;$P_2(-4, 7, 5)$
36. $P_1(-8, -5, 2)$;$P_2(-3, -9, 4)$
37. If the radian measure of each direction angle of a vector is the same, what is it?
38. Prove Theorem 17.2.9.
39. Three vectors in V_3 are said to be *independent* if and only if their position representations do not lie in a plane, and three vectors $\mathbf{E}_1$, $\mathbf{E}_2$, and $\mathbf{E}_3$ are said to form a *basis* for the vector space V_3 if and only if any vector in V_3 can be written as a linear combination of $\mathbf{E}_1$, $\mathbf{E}_2$, and $\mathbf{E}_3$. A theorem can be proved which states that three vectors form a basis for the vector space V_3 if they are independent. Show that this theorem holds for the three vectors $\langle 1, 0, 0\rangle$, $\langle 1, 1, 0\rangle$, and $\langle 1, 1, 1\rangle$ by doing the following: (a) Verify that the vectors are independent by showing that their position representations are not coplanar; (b) verify that the vectors form a basis by showing that any vector $\mathbf{A}$ can be written

$$\mathbf{A} = r\langle 1, 0, 0\rangle + s\langle 1, 1, 0\rangle + t\langle 1, 1, 1\rangle \tag{6}$$

where r, s, and t are scalars. (c) If $\mathbf{A} = \langle 6, -2, 5\rangle$, find the particular values of r, s, and t so that (6) holds.

40. (a) Verify that the vectors $\langle 2, 0, 1\rangle$, $\langle 0, -1, 0\rangle$, and $\langle 1, -1, 0\rangle$ form a basis for V_3 by showing that any vector $\mathbf{A}$ can be written

$$\mathbf{A} = r\langle 2, 0, 1\rangle + s\langle 0, -1, 0\rangle + t\langle 1, -1, 0\rangle \tag{7}$$

where r, s, and t are scalars. (b) If $\mathbf{A} = \langle -2, 3, 5\rangle$, find the particular values of r, s, and t such that (7) holds.

41. Refer to the first sentence of Exercise 39. A theorem can be proved which states that three vectors form a basis for the vector space V_3 only if they are independent. Show that this theorem holds for the three vectors $\mathbf{F}_1 = \langle 1, 0, 1\rangle$, $\mathbf{F}_2 = \langle 1, 1, 1\rangle$, and $\mathbf{F}_3 = \langle 2, 1, 2\rangle$ by doing the following: (a) Verify that $\mathbf{F}_1$, $\mathbf{F}_2$, and $\mathbf{F}_3$ are not independent by showing that their position representations are coplanar; (b) verify that the vectors do not form a basis by showing that every vector in V_3 cannot be written as a linear combination of $\mathbf{F}_1$, $\mathbf{F}_2$, and $\mathbf{F}_3$.

17.3 THE DOT PRODUCT IN V_3

The definition of the dot product of two vectors in V_3 is an extension of the definition for vectors in V_2.

17.3.1 Definition If $\mathbf{A} = \langle a_1, a_2, a_3\rangle$ and $\mathbf{B} = \langle b_1, b_2, b_3\rangle$, then the **dot product** of $\mathbf{A}$ and $\mathbf{B}$, denoted by $\mathbf{A}\cdot\mathbf{B}$, is given by

$$\mathbf{A}\cdot\mathbf{B} = \langle a_1, a_2, a_3\rangle\cdot\langle b_1, b_2, b_3\rangle = a_1b_1 + a_2b_2 + a_3b_3$$

● ILLUSTRATION 1 If $\mathbf{A} = \langle 4, 2, -6\rangle$ and $\mathbf{B} = \langle -5, 3, -2\rangle$, then

$$\begin{aligned}\mathbf{A}\cdot\mathbf{B} &= \langle 4, 2, -6\rangle\cdot\langle -5, 3, -2\rangle\\ &= 4(-5) + 2(3) + (-6)(-2)\\ &= -20 + 6 + 12\\ &= -2\end{aligned}$$

●

For the unit vectors $\mathbf{i}$, $\mathbf{j}$, and $\mathbf{k}$,

$$\mathbf{i}\cdot\mathbf{i} = \mathbf{j}\cdot\mathbf{j} = \mathbf{k}\cdot\mathbf{k} = 1$$

and

$$\mathbf{i}\cdot\mathbf{j} = \mathbf{i}\cdot\mathbf{k} = \mathbf{j}\cdot\mathbf{k} = 0$$

Laws of dot multiplication that are given in Theorem 17.3.2 are the same as those in Theorems 16.3.2 and 16.3.3 for vectors in V_2. The proofs are left as exercises (see Exercises 15 through 18).

17.3.2 Theorem If $\mathbf{A}$, $\mathbf{B}$, and $\mathbf{C}$ are any vectors in V_3 and c is a scalar, then

(i) $\mathbf{A}\cdot\mathbf{B} = \mathbf{B}\cdot\mathbf{A}$ (commutative law)
(ii) $\mathbf{A}\cdot(\mathbf{B} + \mathbf{C}) = \mathbf{A}\cdot\mathbf{B} + \mathbf{A}\cdot\mathbf{C}$ (distributive law)
(iii) $c(\mathbf{A}\cdot\mathbf{B}) = (c\mathbf{A})\cdot\mathbf{B}$
(iv) $\mathbf{0}\cdot\mathbf{A} = 0$
(v) $\mathbf{A}\cdot\mathbf{A} = |\mathbf{A}|^2$

Before giving a geometric representation of the dot product for vectors in V_3, we do as was done with vectors in V_2. We define the angle between two vectors and then express the dot product in terms of the cosine of the radian measure of this angle.

17.3.3 Definition Let **A** and **B** be two nonzero vectors in V_3 such that **A** is not a scalar multiple of **B**. If $\overrightarrow{OP}$ is the position representation of **A** and $\overrightarrow{OQ}$ is the position representation of **B**, then the **angle between the vectors A and B** is defined to be the angle of positive measure between $\overrightarrow{OP}$ and $\overrightarrow{OQ}$ interior to the triangle POQ. If $\mathbf{A} = c\mathbf{B}$, where c is a scalar, then if $c > 0$, the angle between the vectors has radian measure 0, and if $c < 0$, the angle between the vectors has radian measure π.

Figure 17.3.1 shows the angle of radian measure θ between the two vectors if **A** is not a scalar multiple of **B**.

17.3.4 Theorem If θ is the radian measure of the angle between the two nonzero vectors **A** and **B** in V_3, then

$$\mathbf{A}\cdot\mathbf{B} = |\mathbf{A}||\mathbf{B}|\cos\theta \qquad (1)$$

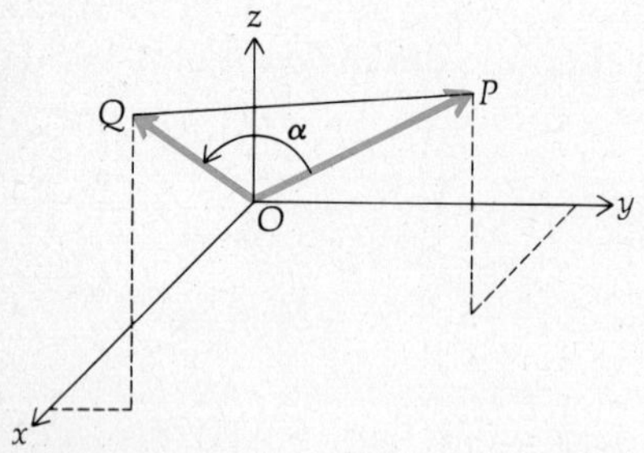

FIGURE 17.3.1

The proof of Theorem 17.3.4 is analogous to the proof of Theorem 16.3.5 for vectors in V_2 and is left as an exercise (see Exercise 19).

If **U** is a unit vector in the direction of **A**, then from (1),

$$\mathbf{U}\cdot\mathbf{B} = |\mathbf{U}||\mathbf{B}|\cos\theta = |\mathbf{B}|\cos\theta$$

As with vectors in V_2, $|\mathbf{B}|\cos\theta$ is the *scalar projection* of **B** onto **A** and the *component* of **B** in the direction of **A**. It follows from Theorem 17.3.4 that the dot product of two vectors **A** and **B** is the product of the magnitude, $|\mathbf{A}|$, of one vector by the scalar projection, $|\mathbf{B}|\cos\theta$, of the second vector onto the first.

Example 1 Given the vectors $\mathbf{A} = 6\mathbf{i} - 3\mathbf{j} + 2\mathbf{k}$ and $\mathbf{B} = 2\mathbf{i} + \mathbf{j} - 3\mathbf{k}$, find (a) the component of **B** in the direction of **A**; (b) the vector projection of **B** onto **A**; and (c) $\cos\theta$ if θ is the radian measure of the angle between **A** and **B**.

Solution

(a) $|\mathbf{A}| = \sqrt{6^2 + (-3)^2 + 2^2} = \sqrt{36 + 9 + 4} = 7$. Hence a unit vector in the direction of **A** is

$$\mathbf{U} = \tfrac{6}{7}\mathbf{i} - \tfrac{3}{7}\mathbf{j} + \tfrac{2}{7}\mathbf{k}$$

Because $\mathbf{U}\cdot\mathbf{B} = |\mathbf{B}|\cos\theta$, the component of **B** in the direction of **A** is

$$\begin{aligned}\mathbf{U}\cdot\mathbf{B} &= (\tfrac{6}{7}\mathbf{i} - \tfrac{3}{7}\mathbf{j} + \tfrac{2}{7}\mathbf{k})\cdot(2\mathbf{i} + \mathbf{j} - 3\mathbf{k})\\ &= \tfrac{12}{7} - \tfrac{3}{7} - \tfrac{6}{7}\\ &= \tfrac{3}{7}\end{aligned}$$

(b) The vector projection of **B** onto **A** is therefore

$$\tfrac{3}{7}\mathbf{U} = \tfrac{18}{49}\mathbf{i} - \tfrac{9}{49}\mathbf{j} + \tfrac{6}{49}\mathbf{k}$$

(c) From (1),

$$\cos\theta = \frac{\mathbf{A}\cdot\mathbf{B}}{|\mathbf{A}||\mathbf{B}|} = \frac{\mathbf{U}\cdot\mathbf{B}}{|\mathbf{B}|}$$

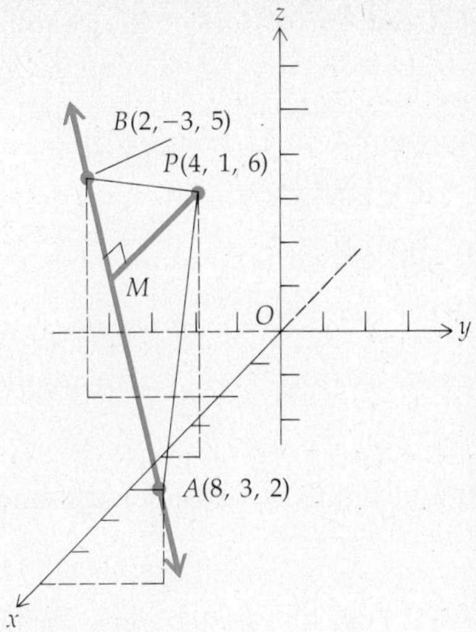

FIGURE 17.3.2

Because $\mathbf{U}\cdot\mathbf{B} = \frac{3}{7}$ and $|\mathbf{B}| = \sqrt{2^2 + 1^2 + (-3)^2} = \sqrt{14}$,

$$\cos\theta = \frac{3}{7\sqrt{14}} = \frac{3\sqrt{14}}{98}$$

Example 2 Find the distance from the point $P(4, 1, 6)$ to the line through the points $A(8, 3, 2)$ and $B(2, -3, 5)$.

Solution Figure 17.3.2 shows the point P and a sketch of the line through A and B. The point M is the foot of the perpendicular line from P to the line through A and B. Let d units be the distance $|\overline{PM}|$. To find d we find $|\overline{AM}|$ and $|\overline{AP}|$ and use the Pythagorean theorem. $|\overline{AP}|$ is the magnitude of the vector $\mathbf{V}(\overrightarrow{AP})$.

$$\begin{aligned}\mathbf{V}(\overrightarrow{AP}) &= \mathbf{V}(\overrightarrow{OP}) - \mathbf{V}(\overrightarrow{OA})\\ &= \langle 4, 1, 6\rangle - \langle 8, 3, 2\rangle\\ &= \langle -4, -2, 4\rangle\end{aligned}$$

Hence

$$|\overline{AP}| = \sqrt{(-4)^2 + (-2)^2 + 4^2} = \sqrt{36} = 6$$

To find $|\overline{AM}|$ we find the scalar projection of $\mathbf{V}(\overrightarrow{AP})$ onto $\mathbf{V}(\overrightarrow{AB})$.

$$\begin{aligned}\mathbf{V}(\overrightarrow{AB}) &= \mathbf{V}(\overrightarrow{OB}) - \mathbf{V}(\overrightarrow{OA})\\ &= \langle 2, -3, 5\rangle - \langle 8, 3, 2\rangle\\ &= \langle -6, -6, 3\rangle\end{aligned}$$

The scalar projection of $\mathbf{V}(\overrightarrow{AP})$ onto $\mathbf{V}(\overrightarrow{AB})$ is then

$$\begin{aligned}\frac{\mathbf{V}(\overrightarrow{AP})\cdot\mathbf{V}(\overrightarrow{AB})}{|\mathbf{V}(\overrightarrow{AB})|} &= \frac{\langle -4, -2, 4\rangle\cdot\langle -6, -6, 3\rangle}{\sqrt{36 + 36 + 9}}\\ &= \frac{24 + 12 + 12}{\sqrt{81}}\\ &= \tfrac{48}{9}\end{aligned}$$

Thus $|\overline{AM}| = \frac{48}{9}$; so

$$\begin{aligned}d &= \sqrt{|\overline{AP}|^2 - |\overline{AM}|^2}\\ &= \sqrt{36 - (\tfrac{48}{9})^2}\\ &= 6\sqrt{1 - \tfrac{64}{81}}\\ &= \tfrac{2}{3}\sqrt{17}\end{aligned}$$

The definition of parallel vectors in V_3 is analogous to Definition 16.3.6 for vectors in V_2.

17.3.5 Definition Two vectors in V_3 are said to be **parallel** if and only if one of the vectors is a scalar multiple of the other.

The following theorem follows from Definition 17.3.5 and Theorem 17.3.4. Its proof is left as an exercise (see Exercise 20).

17.3.6 Theorem Two nonzero vectors in V_3 are parallel if and only if the radian measure of the angle between them is 0 or π.

The following definition of orthogonal vectors in V_3 corresponds to Definition 16.3.7 for vectors in V_2.

17.3.7 Definition If **A** and **B** are two vectors in V_3, **A** and **B** are said to be **orthogonal** if and only if $\mathbf{A}\cdot\mathbf{B} = 0$.

Example 3 Prove by using vectors that the points $A(4, 9, 1)$, $B(-2, 6, 3)$, and $C(6, 3, -2)$ are the vertices of a right triangle.

Solution Triangle CAB is shown in Fig. 17.3.3. From the figure it looks as if the angle at A is the one that may be a right angle. We shall find $\mathbf{V}(\overrightarrow{AB})$ and $\mathbf{V}(\overrightarrow{AC})$, and if the dot product of these two vectors is zero, the angle is a right angle.

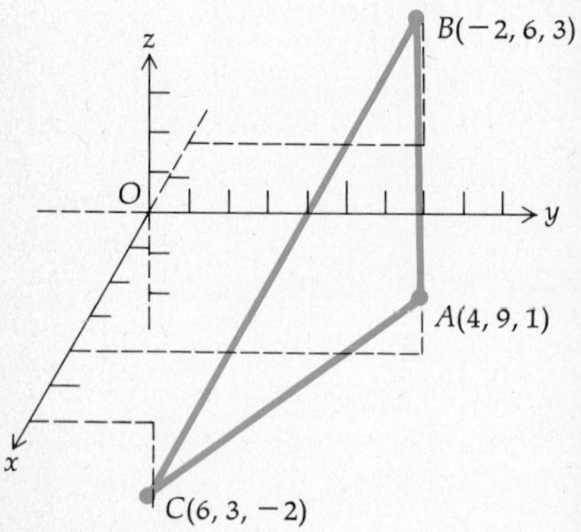

FIGURE 17.3.3

$$\begin{aligned}\mathbf{V}(\overrightarrow{AB}) &= \mathbf{V}(\overrightarrow{OB}) - \mathbf{V}(\overrightarrow{OA})\\ &= \langle -2, 6, 3\rangle - \langle 4, 9, 1\rangle\\ &= \langle -6, -3, 2\rangle\end{aligned}$$

$$\begin{aligned}\mathbf{V}(\overrightarrow{AC}) &= \mathbf{V}(\overrightarrow{OC}) - \mathbf{V}(\overrightarrow{OA})\\ &= \langle 6, 3, -2\rangle - \langle 4, 9, 1\rangle\\ &= \langle 2, -6, -3\rangle\end{aligned}$$

$$\mathbf{V}(\overrightarrow{AB})\cdot\mathbf{V}(\overrightarrow{AC}) = \langle -6, -3, 2\rangle\cdot\langle 2, -6, -3\rangle = -12 + 18 - 6 = 0$$

Therefore $\mathbf{V}(\overrightarrow{AB})$ and $\mathbf{V}(\overrightarrow{AC})$ are orthogonal; so the angle at A in triangle CAB is a right angle.

EXERCISES 17.3

In Exercises 1 through 14, let $\mathbf{A} = \langle -4, -2, 4\rangle$, $\mathbf{B} = \langle 2, 7, -1\rangle$, $\mathbf{C} = \langle 6, -3, 0\rangle$, *and* $\mathbf{D} = \langle 5, 4, -3\rangle$.

1. Find $\mathbf{A}\cdot(\mathbf{B}+\mathbf{C})$. **2.** Find $\mathbf{A}\cdot\mathbf{B} + \mathbf{A}\cdot\mathbf{C}$. **3.** Find $(\mathbf{A}\cdot\mathbf{B})(\mathbf{C}\cdot\mathbf{D})$. **4.** Find $(\mathbf{A}\cdot\mathbf{D})(\mathbf{B}\cdot\mathbf{C})$.

5. Find $\mathbf{A}\cdot\mathbf{D} - \mathbf{B}\cdot\mathbf{C}$. **6.** Find $(\mathbf{A}\cdot\mathbf{B})\mathbf{C} + (\mathbf{B}\cdot\mathbf{C})\mathbf{D}$. **7.** Find $(\mathbf{B}\cdot\mathbf{D})\mathbf{A} - (\mathbf{D}\cdot\mathbf{A})\mathbf{B}$. **8.** Find $(2\mathbf{A}+3\mathbf{B})\cdot(4\mathbf{C}-\mathbf{D})$.

9. Find the cosine of the measure of the angle between **A** and **B**.

10. Find the cosine of the measure of the angle between **C** and **D**.

11. Find (a) the component of **C** in the direction of **A** and (b) the vector projection of **C** onto **A**.

12. Find (a) the component of **B** in the direction of **D** and (b) the vector projection of **B** onto **D**.

13. Find (a) the component of **A** in the direction of **B** and (b) the vector projection of **A** onto **B**.

14. Find (a) the component of **D** in the direction of **C** and (b) the vector projection of **D** onto **C**.

15. Prove Theorem 17.3.2(i). **16.** Prove Theorem 17.3.2(ii). **17.** Prove Theorem 17.3.2(iii).

18. Prove Theorem 17.3.2(iv) and (v). **19.** Prove Theorem 17.3.4. **20.** Prove Theorem 17.3.6.

21. Find the distance from the point $(2, -1, -4)$ to the line through the points $(3, -2, 2)$ and $(-9, -6, 6)$.

22. Find the distance from the point $(3, 2, 1)$ to the line through the points $(1, 2, 9)$ and $(-3, -6, -3)$.

23. Prove by using vectors that the points $(2, 2, 2)$, $(2, 0, 1)$, $(4, 1, -1)$, and $(4, 3, 0)$ are the vertices of a rectangle.

24. Prove by using vectors that the points $(2, 2, 2)$, $(0, 1, 2)$, $(-1, 3, 3)$, and $(3, 0, 1)$ are the vertices of a parallelogram.

25. Find the area of the triangle having vertices at $(-2, 3, 1)$, $(1, 2, 3)$, and $(3, -1, 2)$.

26. Prove by using vectors that the points $(-2, 1, 6)$, $(2, 4, 5)$, and $(-1, -2, 1)$ are the vertices of a right triangle, and find the area of the triangle.

27. If $\mathbf{A} = 3\mathbf{i} + 5\mathbf{j} - 2\mathbf{k}$, $\mathbf{B} = -\mathbf{i} - 2\mathbf{j} + 3\mathbf{k}$, and $\mathbf{C} = 2\mathbf{i} - \mathbf{j} + 4\mathbf{k}$, find the component of **B** in the direction of $\mathbf{A} - 2\mathbf{C}$.

28. Find the cosines of the angles of the triangle having vertices at $A(0, 0, 0)$, $B(4, -1, 3)$ and $C(1, 2, 3)$.

29. If a force has the vector representation $\mathbf{F} = 3\mathbf{i} - 2\mathbf{j} + \mathbf{k}$, find the work done by the force in moving an object from the point $P_1(-2, 4, 3)$ along a straight line to the point $P_2(1, -3, 5)$. The magnitude of the force is measured in pounds and distance is measured in feet. (*Hint:* Review Sec. 16.3.)

30. If a force has the vector representation $\mathbf{F} = 5\mathbf{i} - 3\mathbf{k}$, find the work done by the force in moving an object from the point $P_1(4, 1, 3)$ along a straight line to the point $P_2(-5, 6, 2)$. The magnitude of the force is measured in pounds and distance is measured in feet. (See hint for Exercise 29.)

31. A force is represented by the vector **F**, it has a magnitude of 10 lb, and direction cosines of **F** are $\cos \alpha = \frac{1}{6}\sqrt{6}$ and $\cos \beta = \frac{1}{3}\sqrt{6}$. If the force moves an object from the origin along a straight line to the point $(7, -4, 2)$, find the work done. Distance is measured in feet. (See hint for Exercise 29.)

32. If **A** and **B** are nonzero vectors, prove that the vector $\mathbf{A} - c\mathbf{B}$ is orthogonal to **B** if $c = \mathbf{A} \cdot \mathbf{B}/|\mathbf{B}|^2$.

33. If $\mathbf{A} = 12\mathbf{i} + 9\mathbf{j} - 5\mathbf{k}$ and $\mathbf{B} = 4\mathbf{i} + 3\mathbf{j} - 5\mathbf{k}$, use the result of Exercise 32 to find the value of the scalar c so that the vector $\mathbf{B} - c\mathbf{A}$ is orthogonal to **A**.

34. For the vectors of Exercise 33, use the result of Exercise 32 to find the value of the scalar d so that the vector $\mathbf{A} - d\mathbf{B}$ is orthogonal to **B**.

35. Prove that if **A** and **B** are any vectors, then the vectors $|\mathbf{B}|\mathbf{A} + |\mathbf{A}|\mathbf{B}$ and $|\mathbf{B}|\mathbf{A} - |\mathbf{A}|\mathbf{B}$ are orthogonal.

36. Prove that if **A** and **B** are any nonzero vectors and $\mathbf{C} = |\mathbf{B}|\mathbf{A} + |\mathbf{A}|\mathbf{B}$, then the angle between **A** and **C** has the same measure as the angle between **B** and **C**.

17.4 PLANES

The graph of an equation in two variables, x and y, is a curve in the xy plane. The simplest kind of curve in two-dimensional space is a straight line, and the general equation of a straight line is of the form $Ax + By + C = 0$, which is an equation of the first degree. In three-dimensional space the graph of an equation in three variables, x, y, and z, is a surface. The simplest kind of surface is a *plane,* and we shall see that an equation of a plane is an equation of the first degree in three variables.

17.4.1 Definition If $\mathbf{N}$ is a given nonzero vector and P_0 is a given point, then the set of all points P for which $\mathbf{V}(\overrightarrow{P_0P})$ and $\mathbf{N}$ are orthogonal is defined to be a **plane** through P_0 having $\mathbf{N}$ as a **normal vector.**

Figure 17.4.1 shows a portion of a plane through the point $P_0(x_0, y_0, z_0)$ and the representation of the normal vector $\mathbf{N}$ having its initial point at P_0.

In plane analytic geometry we can obtain an equation of a line if a point on the line and its direction (slope) are given. In an analogous manner, in solid analytic geometry an equation of a plane can be determined by knowing a point in the plane and the direction of a normal vector.

17.4.2 Theorem If $P_0(x_0, y_0, z_0)$ is a point in a plane and a normal vector to the plane is $\mathbf{N} = \langle a, b, c\rangle$, then an equation of the plane is

$$a(x - x_0) + b(y - y_0) + c(z - z_0) = 0 \quad (1)$$

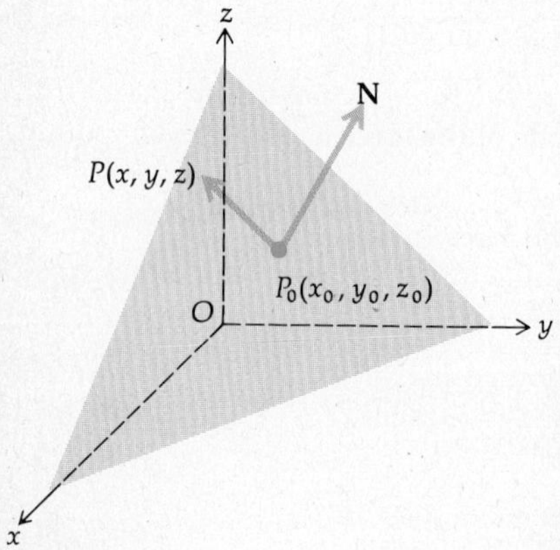

FIGURE 17.4.1

Proof Refer to Fig. 17.4.1. Let $P(x, y, z)$ be any point in the plane. $\mathbf{V}(\overrightarrow{P_0P})$ is the vector having $\overrightarrow{P_0P}$ as a representation; so

$$\mathbf{V}(\overrightarrow{P_0P}) = \langle x - x_0, y - y_0, z - z_0\rangle \quad (2)$$

From Definitions 17.4.1 and 17.3.7 it follows that

$$\mathbf{V}(\overrightarrow{P_0P})\cdot\mathbf{N} = 0$$

Because $\mathbf{N} = \langle a, b, c\rangle$, from (2) and the above equation,

$$a(x - x_0) + b(y - y_0) + c(z - z_0) = 0 \quad (3)$$

which is the desired equation. ■

Example 1 Find an equation of the plane containing the point (2, 1, 3) and having $3\mathbf{i} - 4\mathbf{j} + \mathbf{k}$ as a normal vector.

Solution Using (1) with the point $(x_0, y_0, z_0) = (2, 1, 3)$ and the vector $\langle a, b, c\rangle = \langle 3, -4, 1\rangle$, we have as an equation of the required plane

$$3(x - 2) - 4(y - 1) + (z - 3) = 0$$

or, equivalently,

$$3x - 4y + z - 5 = 0$$

17.4.3 Theorem If a, b, and c are not all zero, the graph of an equation of the form

$$ax + by + cz + d = 0 \quad (4)$$

is a plane and $\langle a, b. c\rangle$ is a normal vector to the plane.

Proof Suppose that $b \neq 0$. Then the point $(0, -d/b, 0)$ is on the graph of the equation because its coordinates satisfy the equation. The given equation can be written as

$$a(x - 0) + b\left(y + \frac{d}{b}\right) + c(z - 0) = 0$$

which from Theorem 17.4.2 is an equation of a plane through the point $(0, -d/b, 0)$ and for which $\langle a, b, c\rangle$ is a normal vector. This proves the theorem if $b \neq 0$. A similar argument holds if $b = 0$ and either $a \neq 0$ or $c \neq 0$. ■

Equations (1) and (4) are called *cartesian equations* of a plane. Equation (1) is analogous to the point-slope form of an equation of a line in two dimensions. Equation (4) is the general first-degree equation in three variables and it is called a *linear equation.*

A plane is determined by three noncollinear points, by a line and a point not on the line, by two intersecting lines, or by two parallel lines. To draw a sketch of a plane from its equation it is convenient to find the points at which the plane intersects each of the coordinate axes. The x coordinate of the point at which the plane intersects the x axis is called the *x intercept* of the plane; the y coordinate of the point at which the plane intersects the y axis is called the *y intercept* of the plane; and the *z intercept* of the plane is the z coordinate of the point at which the plane intersects the z axis.

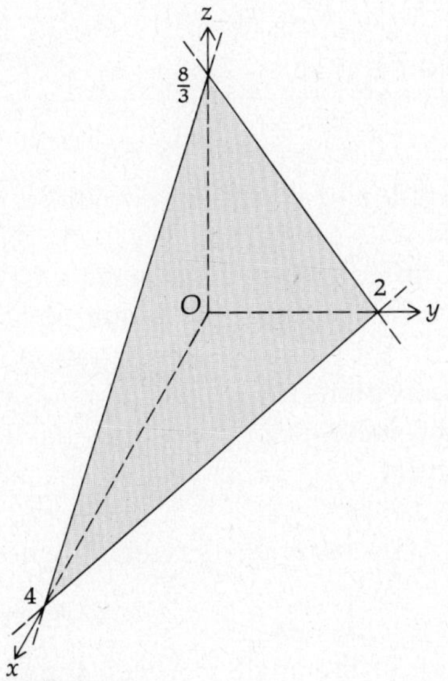

FIGURE 17.4.2

● ILLUSTRATION 1 We wish to draw a sketch of the plane having the equation

$$2x + 4y + 3z = 8$$

By substituting zero for y and z we obtain $x = 4$; so the x intercept of the plane is 4. The y intercept and the z intercept are obtained in a similar manner; they are 2 and $\frac{8}{3}$, respectively. Plotting the points corresponding to these intercepts and connecting them with lines we have the sketch of the plane shown in Fig. 17.4.2. Note that only a portion of the plane is shown in the figure. ●

● ILLUSTRATION 2 To draw a sketch of the plane having the equation

$$3x + 2y - 6z = 0$$

first notice that because the equation is satisfied when x, y, and z are all zero, the plane intersects each of the axes at the origin. If $x = 0$ in the given equation, we obtain $y - 3z = 0$, which is a line in the yz plane; this is the line of intersection of the yz plane with the given plane. Similarly, the line of intersection of the xz plane with the given plane is obtained by setting $y = 0$, and we get $x - 2z = 0$. Drawing a sketch of each of these two lines and drawing a line segment from a point on one of the lines to a point on the other line, we obtain Fig. 17.4.3. ●

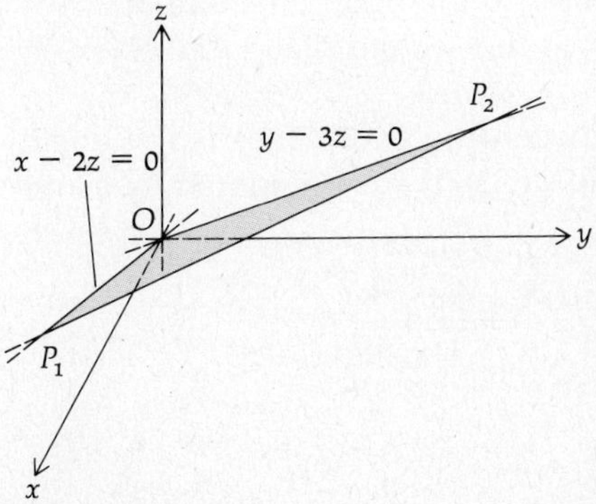

FIGURE 17.4.3

In Illustration 2 the line in the yz plane and the line in the xz plane used to draw the sketch of the plane are called the **traces** of the given plane in the yz plane and the xz plane, respectively. The equation $x = 0$ is an equation of the yz plane because the point (x, y, z) is in the yz plane if and

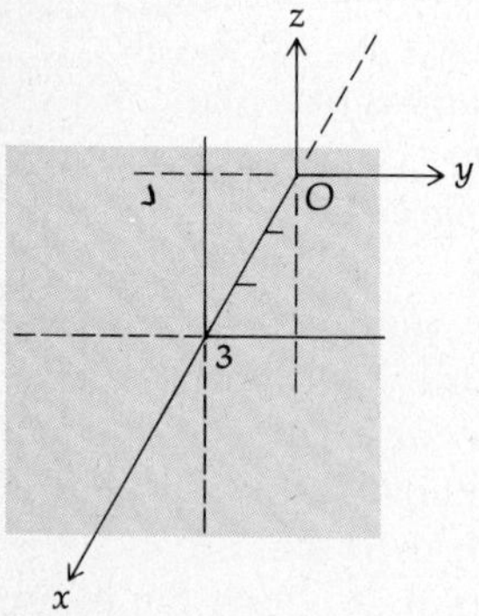

FIGURE 17.4.4

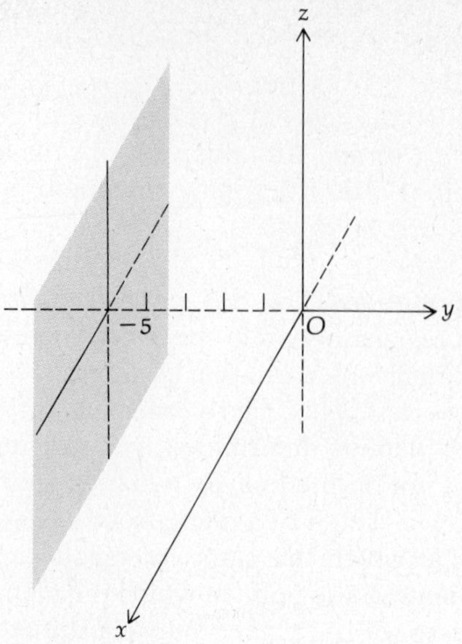

FIGURE 17.4.5

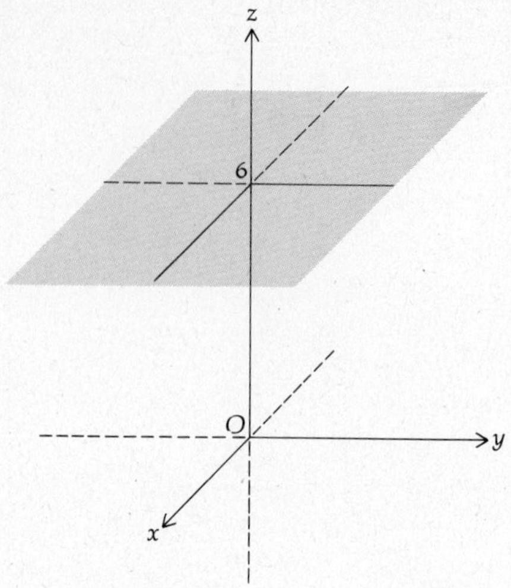

FIGURE 17.4.6

only if $x = 0$. Similarly, the equations $y = 0$ and $z = 0$ are equations of the xz plane and the xy plane, respectively.

A plane parallel to the yz plane has an equation of the form $x = k$, where k is a constant. Figure 17.4.4 shows a sketch of the plane having the equation $x = 3$. A plane parallel to the xz plane has an equation of the form $y = k$, and a plane parallel to the xy plane has an equation of the form $z = k$. Figures 17.4.5 and 17.4.6 show sketches of the planes having the equations $y = -5$ and $z = 6$, respectively.

17.4.4 Definition The **angle between two planes** is defined to be the angle between the normal vectors of the two planes.

Example 2 Find the radian measure of the angle between the two planes $5x - 2y + 5z - 12 = 0$ and $2x + y - 7z + 11 = 0$.

Solution Let $\mathbf{N}_1$ be a normal vector to the first plane and $\mathbf{N}_1 = 5\mathbf{i} - 2\mathbf{j} + 5\mathbf{k}$. Let $\mathbf{N}_2$ be a normal vector to the second plane and $\mathbf{N}_2 = 2\mathbf{i} + \mathbf{j} - 7\mathbf{k}$.

By Definition 17.4.4 the angle between the two planes is the angle between $\mathbf{N}_1$ and $\mathbf{N}_2$, and so by Theorem 17.3.4, if θ is the radian measure of this angle,

$$\cos\theta = \frac{\mathbf{N}_1 \cdot \mathbf{N}_2}{|\mathbf{N}_1||\mathbf{N}_2|} = \frac{(5\mathbf{i} - 2\mathbf{j} + 5\mathbf{k}) \cdot (2\mathbf{i} + \mathbf{j} - 7\mathbf{k})}{\sqrt{25 + 4 + 25}\,\sqrt{4 + 1 + 49}} = \frac{-27}{54} = -\frac{1}{2}$$

Therefore

$$\theta = \tfrac{2}{3}\pi$$

17.4.5 Definition Two planes are **parallel** if and only if their normal vectors are parallel.

From Definitions 17.4.5 and 17.3.5 it follows that if we have two planes with equations

$$a_1x + b_1y + c_1z + d_1 = 0 \tag{5}$$

and

$$a_2x + b_2y + c_2z + d_2 = 0 \tag{6}$$

and normal vectors $\mathbf{N}_1 = \langle a_1, b_1, c_1 \rangle$ and $\mathbf{N}_2 = \langle a_2, b_2, c_2 \rangle$, respectively, then the two planes are parallel if and only if

$$\mathbf{N}_1 = k\mathbf{N}_2 \qquad \text{where } k \text{ is a constant}$$

Figure 17.4.7 shows sketches of two parallel planes and representations of some of their normal vectors.

17.4.6 Definition Two planes are **perpendicular** if and only if their normal vectors are orthogonal.

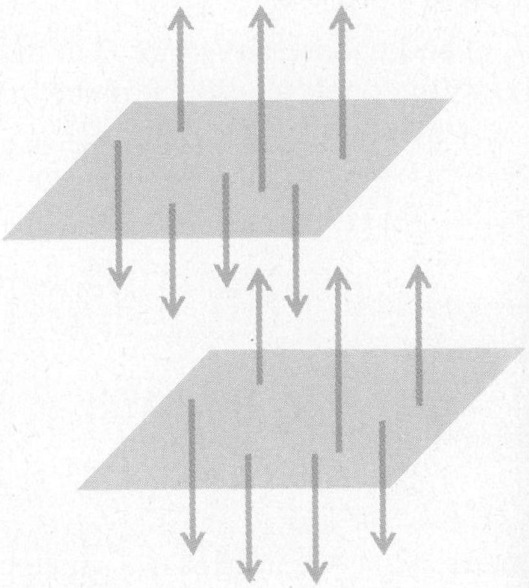

FIGURE 17.4.7

From Definitions 17.4.6 and 17.3.7 it follows that two planes having normal vectors $\mathbf{N}_1$ and $\mathbf{N}_2$ are perpendicular if and only if

$$\mathbf{N}_1 \cdot \mathbf{N}_2 = 0 \tag{7}$$

Example 3 Find an equation of the plane perpendicular to each of the planes $x - y + z = 0$ and $2x + y - 4z - 5 = 0$ and containing the point $(4, 0, -2)$.

Solution Let M be the required plane and $\langle a, b, c \rangle$ be a normal vector of M. Let M_1 be the plane having equation $x - y + z = 0$. By Theorem 17.4.3, a normal vector of M_1 is $\langle 1, -1, 1 \rangle$. Because M and M_1 are perpendicular, it follows from (7) that

$$\langle a, b, c \rangle \cdot \langle 1, -1, 1 \rangle = 0$$

or, equivalently,

$$a - b + c = 0 \tag{8}$$

Let M_2 be the plane having the equation $2x + y - 4z - 5 = 0$. A normal vector of M_2 is $\langle 2, 1, -4 \rangle$. Because M and M_2 are perpendicular,

$$\langle a, b, c \rangle \cdot \langle 2, 1, -4 \rangle = 0$$

or, equivalently,

$$2a + b - 4c = 0 \tag{9}$$

Solving (8) and (9) simultaneously for b and c in terms of a we get $b = 2a$ and $c = a$. Therefore a normal vector of M is $\langle a, 2a, a \rangle$. Because $(4, 0, -2)$ is a point in M, it follows from Theorem 17.4.2 that an equation of M is

$$a(x - 4) + 2a(y - 0) + a(z + 2) = 0$$

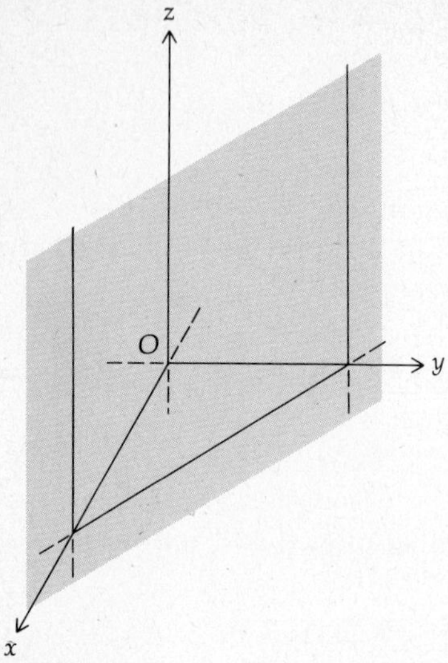

FIGURE 17.4.8

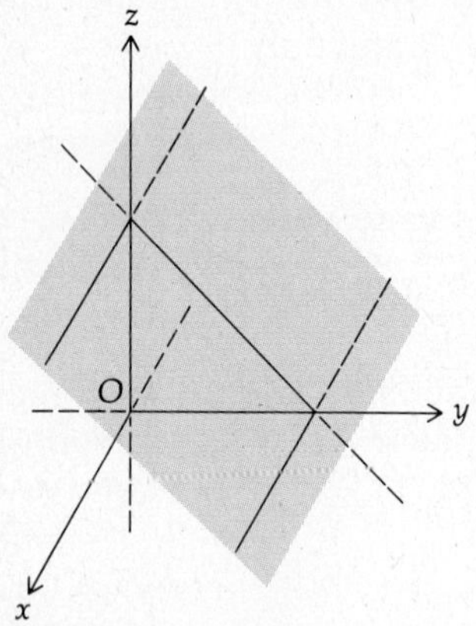

FIGURE 17.4.9

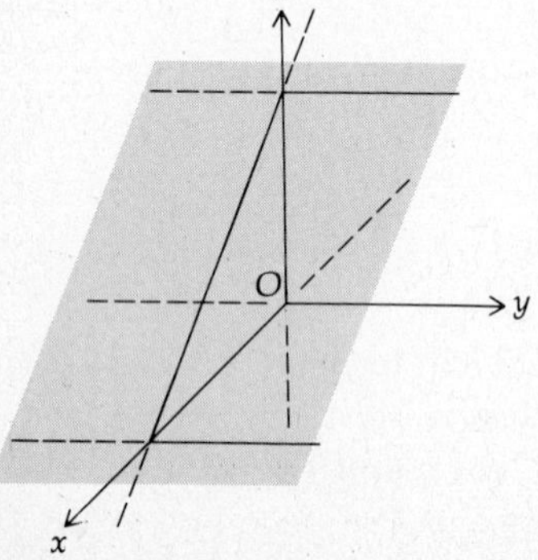

FIGURE 17.4.10

or, equivalently,

$$x + 2y + z - 2 = 0$$

Consider now the plane having the equation $ax + by + d = 0$ and the xy plane whose equation is $z = 0$. Normal vectors to these planes are $\langle a, b, 0\rangle$ and $\langle 0, 0, 1\rangle$, respectively. Because $\langle a, b, 0\rangle \cdot \langle 0, 0, 1\rangle = 0$, the two planes are perpendicular. This means that a plane having an equation with no z term is perpendicular to the xy plane. Figure 17.4.8 illustrates this. In a similar manner we can conclude that a plane having an equation with no x term is perpendicular to the yz plane (see Fig. 17.4.9), and a plane having an equation with no y term is perpendicular to the xz plane (see Fig. 17.4.10).

An important application of the use of vectors is in finding the undirected distance from a plane to a point. The following example illustrates this.

Example 4 Find the distance from the plane $2x - y + 2z + 10 = 0$ to the point $(1, 4, 6)$.

Solution Let P be the point $(1, 4, 6)$ and choose any point Q in the plane. For simplicity choose the point Q as the point where the plane intersects the x axis, that is, the point $(-5, 0, 0)$. The vector having $\overrightarrow{QP}$ as a representation is given by

$$\mathbf{V}(\overrightarrow{QP}) = 6\mathbf{i} + 4\mathbf{j} + 6\mathbf{k}$$

A normal vector to the given plane is

$$\mathbf{N} = 2\mathbf{i} - \mathbf{j} + 2\mathbf{k}$$

The negative of $\mathbf{N}$ is also a normal vector to the given plane and

$$-\mathbf{N} = -2\mathbf{i} + \mathbf{j} - 2\mathbf{k}$$

We are not certain which of the two vectors, $\mathbf{N}$ or $-\mathbf{N}$, makes the smaller angle with vector $\mathbf{V}(\overrightarrow{QP})$. Let $\mathbf{N}'$ be the one of the two vectors $\mathbf{N}$ or $-\mathbf{N}$ that makes an angle of radian measure $\theta < \frac{1}{2}\pi$ with$\mathbf{V}(\overrightarrow{QP})$. In Fig. 17.4.11 there is a portion of the given plane containing the point $Q(-5, 0, 0,)$, the representation of the vector $\mathbf{N}'$ having its initial point at Q, the point $P(1, 4, 6)$, the directed line segment $\overrightarrow{QP}$, and the point R, which is the foot of the perpendicular from P to the plane. For simplicity the coordinate axes are not included in this figure. The distance $|\overrightarrow{RP}|$ is the required distance, which we call d. We see from Fig. 17.4.11 that

$$d = |\mathbf{V}(\overrightarrow{QP})| \cos\theta \tag{10}$$

Because θ is the radian measure of the angle between $\mathbf{N}'$ and $\mathbf{V}(\overrightarrow{QP})$,

$$\cos\theta = \frac{\mathbf{N}' \cdot \mathbf{V}(\overrightarrow{QP})}{|\mathbf{N}||\mathbf{V}(\overrightarrow{QP})|} \tag{11}$$

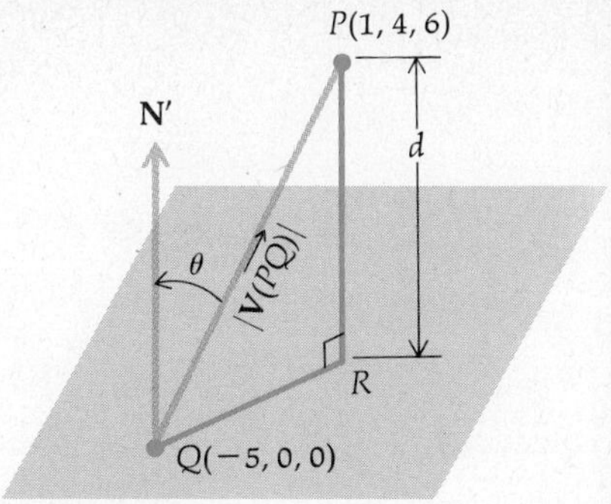

FIGURE 17.4.11

Substituting from (11) into (10) and replacing $|\mathbf{N}'|$ by $|\mathbf{N}|$ we obtain

$$d = \frac{|\mathbf{V}(\overrightarrow{QP})|(\mathbf{N}' \cdot \mathbf{V}(\overrightarrow{QP}))}{|\mathbf{N}||\mathbf{V}(\overrightarrow{QP})|} = \frac{\mathbf{N}' \cdot \mathbf{V}(\overrightarrow{QP})}{|\mathbf{N}|}$$

Because d is an undirected distance, it is nonnegative; hence we can replace the numerator in the above expression by the absolute value of the dot product of $\mathbf{N}$ and $\mathbf{V}(\overrightarrow{QP})$. Therefore

$$d = \frac{|\mathbf{N} \cdot \mathbf{V}(\overrightarrow{QP})|}{|\mathbf{N}|} = \frac{|(2\mathbf{i} - \mathbf{j} + 2\mathbf{k}) \cdot (6\mathbf{i} + 4\mathbf{j} + 6\mathbf{k})|}{\sqrt{4 + 1 + 4}} = \frac{20}{3}$$

EXERCISES 17.4

In Exercises 1 through 6, find an equation of the plane containing the given point P and having the given vector $\mathbf{N}$ as a normal vector.

1. $P(3, 1, 2); \mathbf{N} = \langle 1, 2, -3 \rangle$ **2.** $P(-3, 2, 5); \mathbf{N} = \langle 6, -3, -2 \rangle$ **3.** $P(0, -1, 2); \mathbf{N} = \langle 0, 1, -1 \rangle$

4. $P(-1, 8, 3); \mathbf{N} = \langle -7, -1, 1 \rangle$ **5.** $P(2, 1, -1); \mathbf{N} = -\mathbf{i} + 3\mathbf{j} + 4\mathbf{k}$ **6.** $P(1, 0, 0); \mathbf{N} = \mathbf{i} + \mathbf{k}$

In Exercises 7 and 8, find an equation of the plane containing the given three points.

7. $(3, 4, 1), (1, 7, 1), (-1, -2, 5)$ **8.** $(0, 0, 2), (2, 4, 1), (-2, 3, 3)$

In Exercises 9 through 14, draw a sketch of the given plane and find two unit vectors that are normal to the plane.

9. $2x - y + 2z - 6 = 0$ **10.** $4x - 4y + 2z - 9 = 0$ **11.** $4x + 3y - 12z = 0$

12 $y + 2z - 4 = 0$ **13.** $3x + 2z - 6 = 0$ **14.** $z = 5$

In Exercises 15 through 20, find an equation of the plane satisfying the given conditions.

15. Perpendicular to the line through the points $(2, 2, -4)$ and $(7, -1, 3)$ and containing the point $(-5, 1, 2)$.

16. Parallel to the plane $4x - 2y + z - 1 = 0$ and containing the point $(2, 6, -1)$.

17. Perpendicular to the plane $x + 3y - z - 7 = 0$ and containing the points $(2, 0, 5)$ and $(0, 2, -1)$.

18. Perpendicular to each of the planes $x - y + z = 0$ and $2x + y - 4z - 5 = 0$ and containing the point $(4, 0, -2)$.

19. Perpendicular to the yz plane, containing the point $(2, 1, 1)$, and making an angle of radian measure $\cos^{-1}(\frac{2}{3})$ with the plane $2x - y + 2z - 3 = 0$.

20. Containing the point $P(-3, 5, -2)$ and perpendicular to the representations of the vector $\mathbf{V}(\overrightarrow{OP})$.

In Exercises 21 through 23, find the cosine of the measure of the angle between the two given planes.

21. $2x - y - 2z - 5 = 0$ and $6x - 2y + 3z + 8 = 0$ **22.** $2x - 5y + 3z - 1 = 0$ and $y - 5z + 3 = 0$

23. $3x + 4y = 0$ and $4x - 7y + 4z - 6 = 0$

24. Find the distance from the plane $2x + 2y - z - 6 = 0$ to the point $(2, 2, -4)$.

25. Find the distance from the plane $5x + 11y + 2z - 30 = 0$ to the point $(-2, 6, 3)$.

26. Find the perpendicular distance between the parallel planes $4x - 8y - z + 9 = 0$ and $4x - 8y - z - 6 = 0$.

27. Find the perpendicular distance between the parallel planes $4y - 3z - 6 = 0$ and $8y - 6z - 27 = 0$.

28. Prove that the undirected distance from the plane $ax + by + cz + d = 0$ to the point (x_0, y_0, z_0) is given by

$$\frac{|ax_0 + by_0 + cz_0 + d|}{\sqrt{a^2 + b^2 + c^2}}$$

29. Prove that the perpendicular distance between the two parallel planes $ax + by + cz + d_1 = 0$ and $ax + by + cz + d_2 = 0$ is given by

$$\frac{|d_1 - d_2|}{\sqrt{a^2 + b^2 + c^2}}$$

30. If a, b, and c are nonzero and are the x intercept, y intercept, and z intercept, respectively, of a plane, prove that an equation of the plane is

$$\frac{x}{a} + \frac{y}{b} + \frac{z}{c} = 1$$

This is called the *intercept form* of an equation of a plane.

17.5 LINES IN R^3

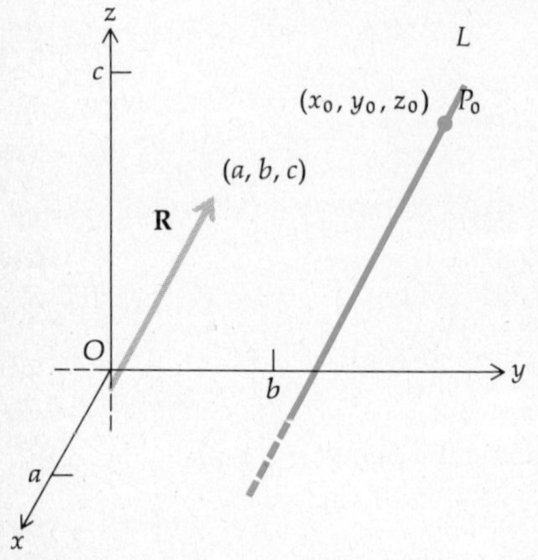

FIGURE 17.5.1

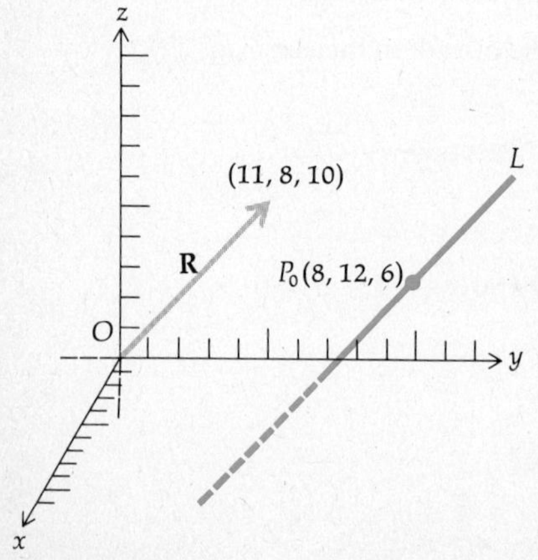

FIGURE 17.5.2

Let L be a line in R^3 such that it contains a given point $P_0(x_0, y_0, z_0)$ and is parallel to the representations of a given vector $\mathbf{R} = \langle a, b, c\rangle$. Figure 17.5.1 shows a sketch of line L and the position representation of vector **R**. Line L is the set of points $P(x, y, z)$ such that $\mathbf{V}(\overrightarrow{P_0P})$ is parallel to the vector **R**. So P is on the line L if and only if there is a nonzero scalar t such that

$$\mathbf{V}(\overrightarrow{P_0P}) = t\mathbf{R} \tag{1}$$

Because $\mathbf{V}(\overrightarrow{P_0P}) = \langle x - x_0, y - y_0, z - z_0\rangle$, we obtain from (1)

$$\langle x - x_0, y - y_0, z - z_0\rangle = t\langle a, b, c\rangle$$

from which it follows that

$$x - x_0 = ta \qquad y - y_0 = tb \qquad z - z_0 = tc$$

or, equivalently,

$$x = x_0 + ta \qquad y = y_0 + tb \qquad z = z_0 + tc \tag{2}$$

Letting the parameter t be any real number (i.e., t takes on all values in the interval $(-\infty, +\infty)$), the point P may be any point on the line L. Therefore Eqs. (2) represent the line L; these equations are called **parametric equations** of the line.

• ILLUSTRATION 1 From Eqs. (2), parametric equations of the line that is parallel to the representations of the vector $\mathbf{R} = \langle 11, 8, 10\rangle$ and that contains the point (8, 12, 6) are

$$x = 8 + 11t \qquad y = 12 + 8t \qquad z = 6 + 10t$$

Figure 17.5.2 shows a sketch of the line and the position representation of **R**. •

If none of the numbers a, b, or c is zero, we can eliminate t from Eqs. (2) and obtain

$$\frac{x - x_0}{a} = \frac{y - y_0}{b} = \frac{z - z_0}{c} \qquad (3)$$

These equations are called **symmetric equations** of the line.

The vector $\mathbf{R} = \langle a, b, c\rangle$ determines the direction of the line, and the numbers a, b, and c are called **direction numbers** of the line. Any vector parallel to **R** has either the same or the opposite direction as **R**; hence such a vector can be used in place of **R** in the above discussion. Because the components of any vector parallel to **R** are proportional to the components of **R**, any set of three numbers proportional to a, b, and c also can serve as a set of direction numbers of the line. So a line has an unlimited number of sets of direction numbers. A set of direction numbers of a line is written in brackets as $[a, b, c]$.

● ILLUSTRATION 2 If $[2, 3, -4]$ represents a set of direction numbers of a line, other sets of direction numbers of the same line can be represented as $[4, 6, -8]$, $[1, \frac{3}{2}, -2]$, and $[2/\sqrt{29}, 3/\sqrt{29}, -4/\sqrt{29}]$. ●

● ILLUSTRATION 3 A set of direction numbers of the line of Illustration 1 is $[11, 8, 10]$, and the line contains the point $(8, 12, 6)$. Thus, from (3), symmetric equations of this line are

$$\frac{x - 8}{11} = \frac{y - 12}{8} = \frac{z - 6}{10}$$ ●

Example 1 Find two sets of symmetric equations of the line through the two points $(-3, 2, 4)$ and $(6, 1, 2)$.

Solution Let P_1 be the point $(-3, 2, 4)$ and P_2 be the point $(6, 1, 2)$. Then the required line is parallel to the representations of the vector $\mathbf{V}(\overrightarrow{P_1P_2})$, and so the components of this vector constitute a set of direction numbers of the line. $\mathbf{V}(\overrightarrow{P_1P_2}) = \langle 9, -1, -2\rangle$. Taking P_0 as the point $(-3, 2, 4)$ we have from (3) the equations

$$\frac{x + 3}{9} = \frac{y - 2}{-1} = \frac{z - 4}{-2}$$

Another set of symmetric equations of this line, obtained by taking P_0 as the point $(6, 1, 2)$, is

$$\frac{x - 6}{9} = \frac{y - 1}{-1} = \frac{z - 2}{-2}$$

Equations (3) are equivalent to the system of three equations

$$\begin{aligned} b(x - x_0) &= a(y - y_0) \\ c(x - x_0) &= a(z - z_0) \\ c(y - y_0) &= b(z - z_0) \end{aligned} \qquad (4)$$

Actually, the three equations in (4) are not independent because any one of them can be derived from the other two. Each of the equations in (4) is

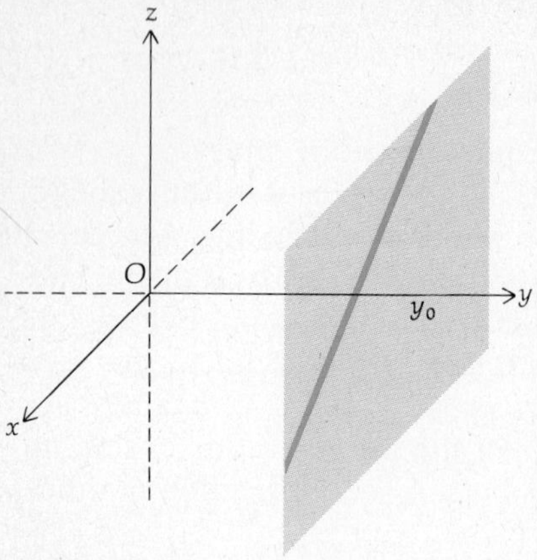

FIGURE 17.5.3

an equation of a plane containing the line L represented by Eqs. (3). Any two of these planes have as their intersection the line L; hence any two of the Eqs. (4) define the line. However, there is an unlimited number of planes containing a given line and because any two of them will determine the line, there is an unlimited number of pairs of equations that represent a line.

If one of the numbers a, b, or c is zero, we do not use symmetric equations (3). However, suppose for example that $b = 0$ and neither a nor c is zero. Then equations of the line are

$$\frac{x - x_0}{a} = \frac{z - z_0}{c} \quad \text{and} \quad y = y_0 \tag{5}$$

A line having symmetric equations (5) lies in the plane $y = y_0$ and hence is parallel to the xz plane. Figure 17.5.3 shows such a line.

Example 2 Given the two planes $x + 3y - z - 9 = 0$ and $2x - 3y + 4z + 3 = 0$. For the line of intersection of these two planes, find (a) a set of symmetric equations and (b) a set of parametric equations.

Solution If the two given equations are solved for x and y in terms of z, we obtain

$$x = -z + 2 \qquad y = \tfrac{2}{3}z + \tfrac{7}{3}$$

from which we get

$$\frac{x - 2}{-1} = \frac{y - \frac{7}{3}}{\frac{2}{3}} = \frac{z - 0}{1}$$

or, equivalently,

$$\frac{x - 2}{-3} = \frac{y - \frac{7}{3}}{2} = \frac{z - 0}{3}$$

which is a set of symmetric equations of the line. A set of parametric equations can be obtained by setting each of the above ratios equal to t, and we have

$$x = 2 - 3t \qquad y = \tfrac{7}{3} + 2t \qquad z = 3t$$

Example 3 Find the direction cosines of a vector whose representations are parallel to the line of Example 2.

Solution From the symmetric equations of the line in Example 2, a set of direction numbers of the line is $[-3, 2, 3]$. Therefore the vector $\langle -3, 2, 3 \rangle$ is a vector whose representations are parallel to the line. The direction cosines of this vector are as follows: $\cos\alpha = -3/\sqrt{22}$, $\cos\beta = 2/\sqrt{22}$, $\cos\gamma = 3/\sqrt{22}$.

Example 4 Find equations of the line through the point $(1, -1, 1)$, perpendicular to the line $3x = 2y = z$ and parallel to the plane $x + y - z = 0$.

Solution Let $[a, b, c]$ be a set of direction numbers of the required line. The equations $3x = 2y = z$ can be written as

$$\frac{x-0}{\frac{1}{3}} = \frac{y-0}{\frac{1}{2}} = \frac{z-0}{1}$$

which are symmetric equations of a line. A set of direction numbers of this line is $[\frac{1}{3}, \frac{1}{2}, 1]$. Because the required line is perpendicular to this line, it follows that the vectors $\langle a, b, c\rangle$ and $\langle \frac{1}{3}, \frac{1}{2}, 1\rangle$ are orthogonal. So

$$\langle a, b, c\rangle \cdot \langle \tfrac{1}{3}, \tfrac{1}{2}, 1\rangle = 0$$

or, equivalently,

$$\tfrac{1}{3}a + \tfrac{1}{2}b + c = 0 \tag{6}$$

A normal vector to the plane $x + y - z = 0$ is $\langle 1, 1, -1\rangle$. Because the required line is parallel to this plane, it is perpendicular to representations of the normal vector. Hence the vectors $\langle a, b, c\rangle$ and $\langle 1, 1, -1\rangle$ are orthogonal, and so

$$\langle a, b, c\rangle \cdot \langle 1, 1, -1\rangle = 0$$

or, equivalently,

$$a + b - c = 0 \tag{7}$$

Solving (6) and (7) simultaneously for a and b in terms of c we get $a = 9c$ and $b = -8c$. The required line then has the set of direction numbers $[9c, -8c, c]$ and contains the point $(1, -1, 1)$. Therefore symmetric equations of the line are

$$\frac{x-1}{9c} = \frac{y+1}{-8c} = \frac{z-1}{c}$$

or, equivalently,

$$\frac{x-1}{9} = \frac{y+1}{-8} = \frac{z-1}{1}$$

Example 5 If l_1 is the line through $A(1, 2, 7)$ and $B(-2, 3, -4)$ and l_2 is the line through $C(2, -1, 4)$ and $D(5, 7, -3)$, prove that l_1 and l_2 are skew lines (i.e., they do not lie in one plane).

Solution To show that two lines do not lie in one plane we demonstrate that they do not intersect and are not parallel. Parametric equations of a line are

$$x = x_0 + ta \qquad y = y_0 + tb \qquad z = z_0 + tc$$

where $[a, b, c]$ is a set of direction numbers of the line and (x_0, y_0, z_0) is any point on the line. Because $\mathbf{V}(\overrightarrow{AB}) = \langle -3, 1, -11\rangle$, a set of direction numbers of l_1 is $[-3, 1, -11]$. Taking A as the point P_0 we have as parametric equations of l_1

$$x = 1 - 3t \qquad y = 2 + t \qquad z = 7 - 11t \tag{8}$$

Because $\mathbf{V}(\overrightarrow{CD}) = \langle 3, 8, -7\rangle$ and l_2 contains the point C, parametric equations of l_2 are

$$x = 2 + 3s \qquad y = -1 + 8s \qquad z = 4 - 7s \tag{9}$$

Because the sets of direction numbers are not proportional, l_1 and l_2 are not parallel. For the lines to intersect, there have to be a value of t and a value of s that give the same point (x_1, y_1, z_1) in both sets of Eqs. (8) and (9). Therefore we equate the right sides of the respective equations and obtain

$$1 - 3t = 2 + 3s$$

$$2 + t = -1 + 8s$$

$$7 - 11t = 4 - 7s$$

Solving the first two equations simultaneously we obtain $s = \frac{8}{27}$ and $t = -\frac{17}{27}$. This set of values does not satisfy the third equation; hence the two lines do not intersect. Thus l_1 and l_2 are skew lines.

EXERCISES 17.5

In Exercises 1 through 8, find parametric and symmetric equations for the line satisfying the given conditions.

1. Through the two points $(1, 2, 1)$ and $(5, -1, 1)$.

2. Through the point $(5, 3, 2)$ with direction numbers $[4, 1, -1]$.

3. Through the origin and perpendicular to the line $\frac{1}{4}(x - 10) = \frac{1}{3}y = \frac{1}{2}z$ at their intersection.

4. Through the origin and perpendicular to the lines having direction numbers $[4, 2, 1]$ and $[-3, -2, 1]$.

5. Perpendicular to the lines having direction numbers $[-5, 1, 2]$ and $[2, -3, -4]$ at the point $(-2, 0, 3)$.

6. Through the point $(-3, 1, -5)$ and perpendicular to the plane $4x - 2y + z - 7 - 0$.

7. Through the point $(4, -5, 20)$ and perpendicular to the plane $x + 3y - 6z - 8 = 0$.

8. Through the point $(2, 0, -4)$ and parallel to each of the planes $2x + y - z = 0$ and $x + 3y + 5z = 0$.

9. Find a set of symmetric equations for the line

$$\begin{cases} 4x - 3y + z - 2 = 0 \\ 2x + 5y - 3z + 4 = 0 \end{cases}$$

10. Show that the lines

$$\frac{x+1}{2} = \frac{y+4}{-5} = \frac{z-2}{3} \quad \text{and} \quad \frac{x-3}{-2} = \frac{y+14}{5} = \frac{z-8}{-3}$$

are coincident.

11. Prove that the line $\frac{1}{2}(x - 3) = \frac{1}{3}(y + 2) = \frac{1}{4}(z + 1)$ lies in the plane $x - 2y + z = 6$.

12. Prove that the line $x + 1 = -\frac{1}{2}(y - 6) = z$ lies in the plane $3x + y - z = 3$.

The planes through a line which are perpendicular to the coordinate planes are called the **projecting planes** *of the line. In Exercises 13 through 16, find equations of the projecting planes of the given line and draw a sketch of the line.*

13. $\begin{cases} 3x - 2y + 5z - 30 = 0 \\ 2x + 3y - 10z - 6 = 0 \end{cases}$ **14.** $\begin{cases} x + y - 3z + 1 = 0 \\ 2x - y - 3z + 14 = 0 \end{cases}$

15. $\begin{cases} x - 2y - 3z + 6 = 0 \\ x + y + z - 1 = 0 \end{cases}$ **16.** $\begin{cases} 2x - y + z - 7 = 0 \\ 4x - y + 3z - 13 = 0 \end{cases}$

17. Find the cosine of the measure of the smallest angle between the two lines $x = 2y + 4$, $z = -y + 4$, and $x = y + 7$, $2z = y + 2$.

18. Find an equation of the plane containing the point $(6, 2, 4)$ and the line $\frac{1}{5}(x - 1) = \frac{1}{6}(y + 2) = \frac{1}{7}(z - 3)$.

In Exercises 19 and 20, find an equation of the plane containing the given intersecting lines.

19. $\dfrac{x-2}{4} = \dfrac{y+3}{-1} = \dfrac{z+2}{3}$ and $\begin{cases} 3x + 2y + z + 2 = 0 \\ x - y + 2z - 1 = 0 \end{cases}$

20. $\dfrac{x}{2} = \dfrac{y-2}{3} = \dfrac{z-1}{1}$ and $\dfrac{x}{1} = \dfrac{y-2}{-1} = \dfrac{z-1}{1}$

21. Show that the lines

$$\begin{cases} 3x - y - z = 0 \\ 8x - 2y - 3z + 1 = 0 \end{cases} \quad \text{and} \quad \begin{cases} x - 3y + z + 3 = 0 \\ 3x - y - z + 5 = 0 \end{cases}$$

are parallel, and find an equation of the plane determined by these lines.

22. Show that the lines

$$\frac{x+2}{5} = \frac{y-1}{-2} = z + 4$$

and

$$\frac{x-3}{-5} = \frac{y+4}{2} = \frac{z-3}{-1}$$

are parallel, and find an equation of the plane determined by these lines.

23. Find the coordinates of the point of intersection of the plane $5x - y + 2z - 12 = 0$ and the line $\frac{1}{4}(x - 2) = -\frac{1}{2}(y + 3) = \frac{1}{7}(z - 1)$.

24. Find equations of the line through the point $(1, -1, 1)$, perpendicular to the line $3x = 2y = z$, and parallel to the plane $x + y - z = 0$.

25. Find equations of the line through the point $(3, 6, 4)$, intersecting the z axis, and parallel to the plane $x - 3y + 5z - 6 = 0$.

26. Find the perpendicular distance from the origin to the line

$$x = -2 + \tfrac{6}{7}t \qquad y = 7 - \tfrac{2}{7}t \qquad z = 4 + \tfrac{3}{7}t$$

27. Find the perpendicular distance from the point $(-1, 3, -1)$ to the line $x - 2z = 7$, $y = 1$.

28. Find equations of the line through the origin, perpendicular to the line $x = y - 5$, $z = 2y - 3$, and intersecting the line $y = 2x + 1$, $z = x + 2$.

29. Prove that the lines

$$\frac{x-1}{5} = \frac{y-2}{-2} = \frac{z+1}{-3} \quad \text{and} \quad \frac{x-2}{1} = \frac{y+1}{-3} = \frac{z+3}{2}$$

are skew lines.

30. Find equations of the line through the point $(3, -4, -5)$ that intersects each of the skew lines of Exercise 29.

17.6 CROSS PRODUCT

Let **A** and **B** be two nonparallel vectors. Representations of these two vectors having the same initial point determine a plane as shown in Fig. 17.6.1. We show that a vector whose representations are perpendicular to this plane is given by the vector operation called the *cross product* of the two vectors **A** and **B**. The cross product is a vector operation for vectors in V_3 that we did not have for vectors in V_2. We first define this operation and then consider its algebraic and geometric properties.

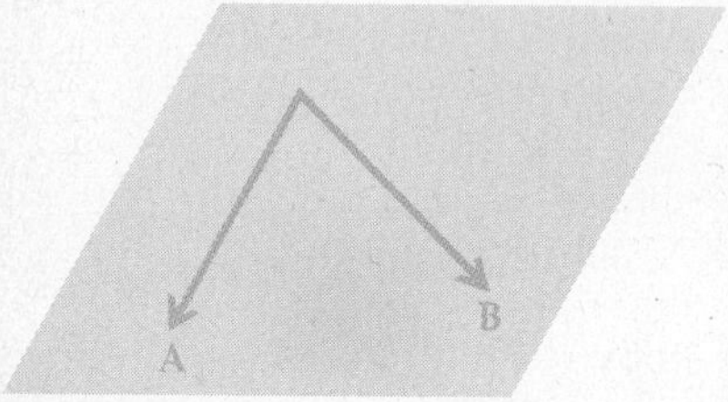

FIGURE 17.6.1

17.6.1 Definition If $\mathbf{A} = \langle a_1, a_2, a_3\rangle$ and $\mathbf{B} = \langle b_1, b_2, b_3\rangle$, then the **cross product** of **A** and **B**, denoted by $\mathbf{A} \times \mathbf{B}$, is given by

$$\mathbf{A} \times \mathbf{B} = \langle a_2b_3 - a_3b_2, a_3b_1 - a_1b_3, a_1b_2 - a_2b_1\rangle \qquad (1)$$

Because the cross product of two vectors is a vector, the cross product also is called the **vector product.** The operation of obtaining the cross product is called **vector multiplication.**

● ILLUSTRATION 1 If $\mathbf{A} = \langle 2, 1, -3\rangle$ and $\mathbf{B} = \langle 3, -1, 4\rangle$, then from Definition 17.6.1,

$$\begin{aligned}\mathbf{A} \times \mathbf{B} &= \langle 2, 1, -3\rangle \times \langle 3, -1, 4\rangle \\ &= \langle (1)(4) - (-3)(-1), (-3)(3) - (2)(4), (2)(-1) - (1)(3)\rangle \\ &= \langle 4 - 3, -9 - 8, -2 - 3\rangle \\ &= \langle 1, -17, -5\rangle \\ &= \mathbf{i} - 17\mathbf{j} - 5\mathbf{k}\end{aligned}$$ ●

There is a mnemonic device for remembering formula (1) that makes use of determinant notation. A second-order determinant is defined by the equation

$$\begin{vmatrix} a & b \\ c & d \end{vmatrix} = ad - bc$$

where a, b, and c are real numbers. For example,

$$\begin{vmatrix} 3 & 6 \\ -2 & 5 \end{vmatrix} = 3(5) - (6)(-2) = 27$$

Therefore formula (1) can be written as

$$\mathbf{A} \times \mathbf{B} = \begin{vmatrix} a_2 & a_3 \\ b_2 & b_3 \end{vmatrix}\mathbf{i} - \begin{vmatrix} a_1 & a_3 \\ b_2 & b_3 \end{vmatrix}\mathbf{j} + \begin{vmatrix} a_1 & a_2 \\ b_1 & b_2 \end{vmatrix}\mathbf{k}$$

The right side of the above expression can be written symbolically as

$$\begin{vmatrix} \mathbf{i} & \mathbf{j} & \mathbf{k} \\ a_1 & a_2 & a_3 \\ b_1 & b_2 & b_3 \end{vmatrix}$$

which is the notation for a third-order determinant. However, observe that the first row contains vectors and not real numbers as is customary with determinant notation.

● ILLUSTRATION 2 We use the mnemonic device employing determinant notation to find the cross product of the vectors of Illustration 1.

$$\mathbf{A} \times \mathbf{B} = \begin{vmatrix} \mathbf{i} & \mathbf{j} & \mathbf{k} \\ 2 & 1 & -3 \\ 3 & -1 & 4 \end{vmatrix}$$

$$= \begin{vmatrix} 1 & -3 \\ -1 & 4 \end{vmatrix} \mathbf{i} - \begin{vmatrix} 2 & -3 \\ 3 & 4 \end{vmatrix} \mathbf{j} + \begin{vmatrix} 2 & 1 \\ 3 & -1 \end{vmatrix} \mathbf{k}$$

$$= [(1)(4) - (-3)(-1)]\mathbf{i} - [(2)(4) - (-3)(3)]\mathbf{j} + [(2)(-1) - (1)(3)]\mathbf{k}$$

$$= \mathbf{i} - 17\mathbf{j} - 5\mathbf{k}$$ ●

17.6.2 Theorem If $\mathbf{A}$ is any vector in V_3, then

(i) $\mathbf{A} \times \mathbf{A} = \mathbf{0}$
(ii) $\mathbf{0} \times \mathbf{A} = \mathbf{0}$
(iii) $\mathbf{A} \times \mathbf{0} = \mathbf{0}$

Proof of (i) If $\mathbf{A} = \langle a_1, a_2, a_3 \rangle$, then by Definition 17.6.1,

$$\mathbf{A} \times \mathbf{A} = \langle a_2a_3 - a_3a_2, a_3a_1 - a_1a_3, a_1a_2 - a_2a_1 \rangle$$
$$= \langle 0, 0, 0 \rangle$$
$$= \mathbf{0}$$

The proofs of (ii) and (iii) are left as exercises (see Exercise 13). ■

By applying Definition 17.6.1 to pairs of unit vectors **i**, **j**, and **k** we obtain the following:

$$\mathbf{i} \times \mathbf{i} = \mathbf{j} \times \mathbf{j} = \mathbf{k} \times \mathbf{k} = \mathbf{0}$$

$$\mathbf{i} \times \mathbf{j} = \mathbf{k} \qquad \mathbf{j} \times \mathbf{k} = \mathbf{i} \qquad \mathbf{k} \times \mathbf{i} = \mathbf{j}$$

$$\mathbf{j} \times \mathbf{i} = -\mathbf{k} \qquad \mathbf{k} \times \mathbf{j} = -\mathbf{i} \qquad \mathbf{i} \times \mathbf{k} = -\mathbf{j}$$

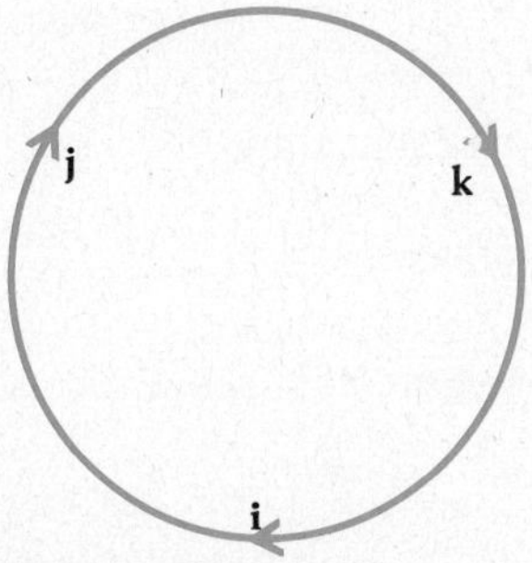

FIGURE 17.6.2

As an aid in remembering the above cross products, first notice that the cross product of any one of the unit vectors **i**, **j**, or **k** with itself is the zero vector. The other six cross products can be obtained from Fig. 17.6.2 by applying the following rule: The cross product of two consecutive vectors in the clockwise direction is the next vector; and the cross product of two consecutive vectors in the counterclockwise direction is the negative of the next vector.

It can be seen easily that cross multiplication of two vectors is not commutative because in particular $\mathbf{i} \times \mathbf{j} \neq \mathbf{j} \times \mathbf{i}$. However, $\mathbf{i} \times \mathbf{j} = \mathbf{k}$ and $\mathbf{j} \times \mathbf{i} = -\mathbf{k}$; so $\mathbf{i} \times \mathbf{j} = -(\mathbf{j} \times \mathbf{i})$. It is true in general that $\mathbf{A} \times \mathbf{B} = -(\mathbf{B} \times \mathbf{A})$, which we state and prove as a theorem.

17.6.3 Theorem If $\mathbf{A}$ and $\mathbf{B}$ are any vectors in V_3,

$$\mathbf{A} \times \mathbf{B} = -(\mathbf{B} \times \mathbf{A})$$

Proof If $\mathbf{A} = \langle a_1, a_2, a_3 \rangle$ and $\mathbf{B} = \langle b_1, b_2, b_3 \rangle$, then by Definition 17.6.1,

$$\begin{aligned}\mathbf{A} \times \mathbf{B} &= \langle a_2b_3 - a_3b_2, a_3b_1 - a_1b_3, a_1b_2 - a_2b_1 \rangle \\ &= -1\langle a_3b_2 - a_2b_3, a_1b_3 - a_3b_1, a_2b_1 - a_1b_2 \rangle \\ &= -(\mathbf{B} \times \mathbf{A})\end{aligned}$$

■

Cross multiplication of vectors is not associative. This is shown by the following example:

$$\mathbf{i} \times (\mathbf{i} \times \mathbf{j}) = \mathbf{i} \times \mathbf{k} = -\mathbf{j}$$

$$(\mathbf{i} \times \mathbf{i}) \times \mathbf{j} = \mathbf{0} \times \mathbf{j} = \mathbf{0}$$

So

$$\mathbf{i} \times (\mathbf{i} \times \mathbf{j}) \neq (\mathbf{i} \times \mathbf{i}) \times \mathbf{j}$$

Cross multiplication of vectors is distributive with respect to vector addition, as given by the following theorem.

17.6.4 Theorem If **A**, **B**, and **C** are any vectors in V_3, then

$$\mathbf{A} \times (\mathbf{B} + \mathbf{C}) = \mathbf{A} \times \mathbf{B} + \mathbf{A} \times \mathbf{C} \tag{2}$$

Theorem 17.6.4 can be proved by letting $\mathbf{A} = \langle a_1, a_2, a_3 \rangle$, $\mathbf{B} = \langle b_1, b_2, b_3 \rangle$, and $\mathbf{C} = \langle c_1, c_2, c_3 \rangle$, and then showing that the components of the vector on the left side of (2) are the same as the components of the vector on the right side of (2). The details are left as an exercise (see Exercise 14).

17.6.5 Theorem If **A** and **B** are any two vectors in V_3 and c is a scalar, then

(i) $(c\mathbf{A}) \times \mathbf{B} = \mathbf{A} \times (c\mathbf{B})$;
(ii) $(c\mathbf{A}) \times \mathbf{B} = c(\mathbf{A} \times \mathbf{B})$.

The proof of Theorem 17.6.5 is left as exercises (see Exercises 15 and 16).

Repeated applications of Theorems 17.6.4 and 17.6.5 can be used to compute the cross product of two vectors by using laws of algebra, provided the order of the vectors in cross multiplication is not changed, because it is prohibited by Theorem 17.6.3. The following illustration demonstrates this procedure.

● ILLUSTRATION 3 We find the cross product of the vectors in Illustration 1 by applying Theorems 17.6.4 and 17.6.5.

$$\begin{aligned}\mathbf{A} \times \mathbf{B} &= (2\mathbf{i} + \mathbf{j} - 3\mathbf{k}) \times (3\mathbf{i} - \mathbf{j} + 4\mathbf{k}) \\ &= 6(\mathbf{i} \times \mathbf{i}) - 2(\mathbf{i} \times \mathbf{j}) + 8(\mathbf{i} \times \mathbf{k}) + 3(\mathbf{j} \times \mathbf{i}) - 1(\mathbf{j} \times \mathbf{j}) \\ &\quad + 4(\mathbf{j} \times \mathbf{k}) - 9(\mathbf{k} \times \mathbf{i}) + 3(\mathbf{k} \times \mathbf{j}) - 12(\mathbf{k} \times \mathbf{k})\end{aligned}$$

$$= 6(\mathbf{0}) - 2(\mathbf{k}) + 8(-\mathbf{j}) + 3(-\mathbf{k}) - 1(\mathbf{0}) + 4(\mathbf{i}) - 9(\mathbf{j}) + 3(-\mathbf{i}) - 12(\mathbf{0})$$

$$= -2\mathbf{k} - 8\mathbf{j} - 3\mathbf{k} + 4\mathbf{i} - 9\mathbf{j} - 3\mathbf{i}$$

$$= \mathbf{i} - 17\mathbf{j} - 5\mathbf{k}$$

The method used in Illustration 3 gives a way of finding the cross product without having to remember formula (1) or to use determinant notation. Actually all the steps shown in the solution need not be included because the various cross products of the unit vectors can be obtained immediately by using Fig. 17.6.2 and the corresponding rule.

The following theorem gives a formula that is useful in proving Theorem 17.6.7, from which a geometric interpretation of the cross product is obtained.

17.6.6 Theorem If $\mathbf{A}$ and $\mathbf{B}$ are any two vectors in V_3, then

$$|\mathbf{A} \times \mathbf{B}|^2 = |\mathbf{A}|^2|\mathbf{B}|^2 - (\mathbf{A} \cdot \mathbf{B})^2$$

Proof Let $\mathbf{A} = \langle a_1, a_2, a_3 \rangle$ and $\mathbf{B} = \langle b_1, b_2, b_3 \rangle$. Then

$$|\mathbf{A} \times \mathbf{B}|^2 = (a_2b_3 - a_3b_2)^2 + (a_3b_1 - a_1b_3)^2 + (a_1b_2 - a_2b_1)^2$$
$$= a_2^2b_3^2 - 2a_2a_3b_2b_3 + a_3^2b_2^2 + a_3^2b_1^2 - 2a_1a_3b_1b_3 + a_1^2b_3^2 + a_1^2b_2^2 - 2a_1a_2b_1b_2 + a_2^2b_1^2$$

$$|\mathbf{A}|^2|\mathbf{B}|^2 - (\mathbf{A} \cdot \mathbf{B})^2 = (a_1^2 + a_2^2 + a_3^2)(b_1^2 + b_2^2 + b_3^2) - (a_1b_1 + a_2b_2 + a_3b_3)^2$$
$$= a_1^2b_2^2 + a_1^2b_3^2 + a_2^2b_1^2 + a_2^2b_3^2 + a_3^2b_1^2 + a_3^2b_2^2 - 2a_1a_3b_1b_3 - 2a_2a_3b_2b_3 - 2a_1a_2b_1b_2$$

From the two expressions,

$$|\mathbf{A} \times \mathbf{B}|^2 = |\mathbf{A}|^2|\mathbf{B}|^2 - (\mathbf{A} \cdot \mathbf{B})^2$$

17.6.7 Theorem If $\mathbf{A}$ and $\mathbf{B}$ are two vectors in V_3 and θ is the radian measure of the angle between $\mathbf{A}$ and $\mathbf{B}$, then

$$|\mathbf{A} \times \mathbf{B}| = |\mathbf{A}||\mathbf{B}| \sin \theta \tag{3}$$

Proof From Theorem 17.6.6,

$$|\mathbf{A} \times \mathbf{B}|^2 = |\mathbf{A}|^2|\mathbf{B}|^2 - (\mathbf{A} \cdot \mathbf{B})^2 \tag{4}$$

From Theorem 17.3.4, if θ is the radian measure of the angle between $\mathbf{A}$ and $\mathbf{B}$,

$$\mathbf{A} \cdot \mathbf{B} = |\mathbf{A}||\mathbf{B}| \cos \theta \tag{5}$$

Substituting from (5) into (4) we get

$$|\mathbf{A} \times \mathbf{B}|^2 = |\mathbf{A}|^2|\mathbf{B}|^2 - |\mathbf{A}|^2|\mathbf{B}|^2 \cos^2 \theta$$
$$= |\mathbf{A}|^2|\mathbf{B}|^2(1 - \cos^2 \theta)$$

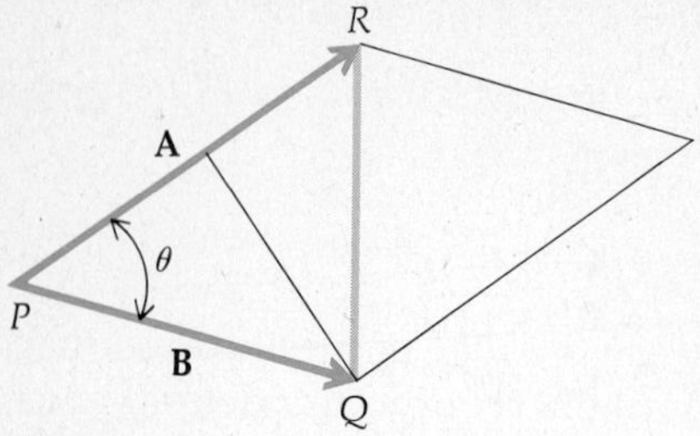

FIGURE 17.6.3

So

$$|\mathbf{A} \times \mathbf{B}|^2 = |\mathbf{A}|^2|\mathbf{B}|^2 \sin^2 \theta \tag{6}$$

Because $0 \le \theta \le \pi$, $\sin \theta \ge 0$. Therefore, from (6),

$$|\mathbf{A} \times \mathbf{B}| = |\mathbf{A}||\mathbf{B}| \sin \theta$$

■

We consider now a geometric interpretation of $|\mathbf{A} \times \mathbf{B}|$. Let $\overrightarrow{PR}$ be a representation of **A** and let $\overrightarrow{PQ}$ be a representation of **B**. Then the angle between the vectors **A** and **B** is the angle at P in triangle RPQ (see Fig. 17.6.3). Let the radian measure of this angle be θ. Therefore the number of square units in the area of the parallelogram having $\overrightarrow{PR}$ and $\overrightarrow{PQ}$ as adjacent sides is $|\mathbf{A}||\mathbf{B}| \sin \theta$ because the altitude of the parallelogram has length $|\mathbf{B}| \sin \theta$ units and the length of the base is $|\mathbf{A}|$ units. So from (3) it follows that $|\mathbf{A} \times \mathbf{B}|$ square units is the area of this parallelogram.

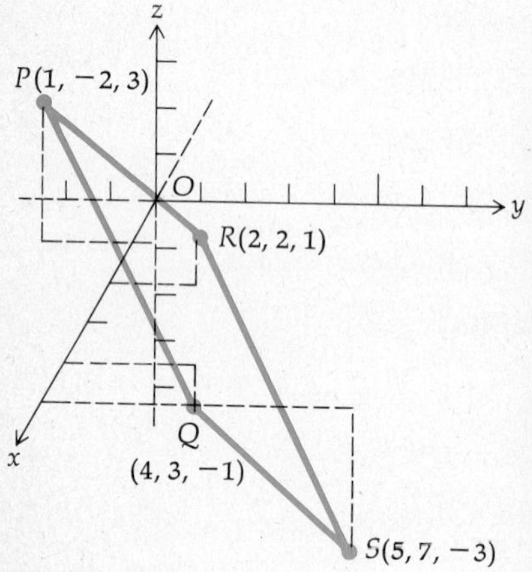

FIGURE 17.6.4

Example 1 Show that the quadrilateral having vertices at $P(1, -2, 3)$, $Q(4, 3, -1)$, $R(2, 2, 1)$, and $S(5, 7, -3)$ is a parallelogram, and find its area.

Solution Figure 17.6.4 shows the quadrilateral $PQSR$.

$$\mathbf{V}(\overrightarrow{PQ}) = \langle 4 - 1, 3 - (-2), (-1) - 3 \rangle = \langle 3, 5, -4 \rangle$$

$$\mathbf{V}(\overrightarrow{PR}) = \langle 2 - 1, 2 - (-2), 1 - 3 \rangle = \langle 1, 4, -2 \rangle$$

$$\mathbf{V}(\overrightarrow{RS}) = \langle 5 - 2, 7 - 2, -3 - 1 \rangle = \langle 3, 5, -4 \rangle$$

$$\mathbf{V}(\overrightarrow{QS}) = \langle 5 - 4, 7 - 3, -3 - (-1) \rangle = \langle 1, 4, -2 \rangle$$

Because $\mathbf{V}(\overrightarrow{PQ}) = \mathbf{V}(\overrightarrow{RS})$ and $\mathbf{V}(\overrightarrow{PR}) = \mathbf{V}(\overrightarrow{QS})$, it follows that $\overrightarrow{PQ}$ is parallel to $\overrightarrow{RS}$ and $\overrightarrow{PR}$ is parallel to $\overrightarrow{QS}$. Therefore $PQSR$ is a parallelogram.

Let $\mathbf{A} = \mathbf{V}(\overrightarrow{PR})$ and $\mathbf{B} = \mathbf{V}(\overrightarrow{PQ})$; then

$$\begin{aligned}
\mathbf{A} \times \mathbf{B} &= (\mathbf{i} + 4\mathbf{j} - 2\mathbf{k}) \times (3\mathbf{i} + 5\mathbf{j} - 4\mathbf{k}) \\
&= 3(\mathbf{i} \times \mathbf{i}) + 5(\mathbf{i} \times \mathbf{j}) - 4(\mathbf{i} \times \mathbf{k}) + 12(\mathbf{j} \times \mathbf{i}) + 20(\mathbf{j} \times \mathbf{j}) \\
&\qquad - 16(\mathbf{j} \times \mathbf{k}) - 6(\mathbf{k} \times \mathbf{i}) - 10(\mathbf{k} \times \mathbf{j}) + 8(\mathbf{k} \times \mathbf{k}) \\
&= 3(\mathbf{0}) + 5(\mathbf{k}) - 4(-\mathbf{j}) + 12(-\mathbf{k}) + 20(\mathbf{0}) - 16(\mathbf{i}) \\
&\qquad - 6(\mathbf{j}) - 10(-\mathbf{i}) + 8(\mathbf{0}) \\
&= -6\mathbf{i} - 2\mathbf{j} - 7\mathbf{k}
\end{aligned}$$

Hence

$$|\mathbf{A} \times \mathbf{B}| = \sqrt{36 + 4 + 49} = \sqrt{89}$$

The area of the parallelogram is therefore $\sqrt{89}$ square units.

The following theorem, which gives a method for determining if two vectors in V_3 are parallel, follows from Theorem 17.6.7.

17.6.8 Theorem If **A** and **B** are two vectors in V_3, **A** and **B** are parallel if and only if $\mathbf{A} \times \mathbf{B} = \mathbf{0}$.

Proof If either **A** or **B** is the zero vector, then from Theorem 17.6.2, $\mathbf{A} \times \mathbf{B} = \mathbf{0}$. Because the zero vector is parallel to any vector, the thereom holds.

If neither **A** nor **B** is the zero vector, $|\mathbf{A}| \neq 0$ and $|\mathbf{B}| \neq 0$. Therefore, from (3), $|\mathbf{A} \times \mathbf{B}| = 0$ if and only if $\sin\theta = 0$. Because $|\mathbf{A} \times \mathbf{B}| = 0$ if and only if $\mathbf{A} \times \mathbf{B} = \mathbf{0}$ and $\sin\theta = 0$ $(0 \leq \theta \leq \pi)$ if and only if $\theta = 0$ or π, we can conclude that

$$\mathbf{A} \times \mathbf{B} = \mathbf{0} \quad \text{if and only if} \quad \theta = 0 \text{ or } \pi$$

However, from Theorem 17.3.6, two nonzero vectors are parallel if and only if the radian measure of the angle between the two vectors is 0 or π. Thus the theorem follows. ■

The product $\mathbf{A} \cdot (\mathbf{B} \times \mathbf{C})$ is called the **triple scalar product** of the vectors **A**, **B**, and **C**. Actually, the parentheses are not needed because $\mathbf{A} \cdot \mathbf{B}$ is a scalar, and therefore $\mathbf{A} \cdot \mathbf{B} \times \mathbf{C}$ can be interpreted only in one way.

17.6.9 Theorem If **A**, **B**, and **C** are vectors in V_3, then

$$\mathbf{A} \cdot \mathbf{B} \times \mathbf{C} = \mathbf{A} \times \mathbf{B} \cdot \mathbf{C} \tag{7}$$

Theorem 17.6.9 can be proved by letting $\mathbf{A} = \langle a_1, a_2, a_3 \rangle$, $\mathbf{B} = \langle b_1, b_2, b_3 \rangle$, and $\mathbf{C} = \langle c_1, c_2, c_3 \rangle$ and then by showing that the number on the left side of (7) is the same as the number on the right. The details are left as an exercise (see Exercise 17).

● ILLUSTRATION 4 We verify Theorem 17.6.9 if $\mathbf{A} = \langle 1, -1, 2 \rangle$, $\mathbf{B} = \langle 3, 4, -2 \rangle$, and $\mathbf{C} = \langle -5, 1, -4 \rangle$.

$$\begin{aligned}
\mathbf{B} \times \mathbf{C} &= (3\mathbf{i} + 4\mathbf{j} - 2\mathbf{k}) \times (-5\mathbf{i} + \mathbf{j} - 4\mathbf{k}) \\
&= 3\mathbf{k} - 12(-\mathbf{j}) - 20(-\mathbf{k}) - 16\mathbf{i} + 10\mathbf{j} - 2(-\mathbf{i}) \\
&= -14\mathbf{i} + 22\mathbf{j} + 23\mathbf{k}
\end{aligned}$$

$$\begin{aligned}
\mathbf{A} \cdot (\mathbf{B} \times \mathbf{C}) &= \langle 1, -1, 2 \rangle \cdot \langle -14, 22, 23 \rangle \\
&= -14 - 22 + 46 \\
&= 10
\end{aligned}$$

$$\begin{aligned}
\mathbf{A} \times \mathbf{B} &= (\mathbf{i} - \mathbf{j} + 2\mathbf{k}) \times (3\mathbf{i} + 4\mathbf{j} - 2\mathbf{k}) \\
&= 4\mathbf{k} - 2(-\mathbf{j}) - 3(-\mathbf{k}) + 2\mathbf{i} + 6\mathbf{j} + 8(-\mathbf{i}) \\
&= -6\mathbf{i} + 8\mathbf{j} + 7\mathbf{k}
\end{aligned}$$

$$\begin{aligned}
(\mathbf{A} \times \mathbf{B}) \cdot \mathbf{C} &= \langle -6, 8, 7 \rangle \cdot \langle -5, 1, -4 \rangle \\
&= 30 + 8 - 28 \\
&= 10
\end{aligned}$$

This verifies the theorem for these three vectors. ●

17.6.10 Theorem If **A** and **B** are two vectors in V_3, then the vector $\mathbf{A} \times \mathbf{B}$ is orthogonal to both **A** and **B**.

Proof From Theorem 17.6.9,

$$\mathbf{A} \cdot \mathbf{A} \times \mathbf{B} = \mathbf{A} \times \mathbf{A} \cdot \mathbf{B}$$

From Theorem 17.6.2(i), $\mathbf{A} \times \mathbf{A} = \mathbf{0}$. Therefore, from the above equation,

$$\mathbf{A} \cdot \mathbf{A} \times \mathbf{B} = \mathbf{0} \cdot \mathbf{B} = 0$$

Because the dot product of **A** and $\mathbf{A} \times \mathbf{B}$ is zero, it follows from Definition 17.3.7 that **A** and $\mathbf{A} \times \mathbf{B}$ are orthogonal.

Also from Theorem 17.6.9,

$$\mathbf{A} \times \mathbf{B} \cdot \mathbf{B} = \mathbf{A} \cdot \mathbf{B} \times \mathbf{B}$$

Again applying Theorem 17.6.2(i) we get $\mathbf{B} \times \mathbf{B} = \mathbf{0}$, and so from the above equation,

$$\mathbf{A} \times \mathbf{B} \cdot \mathbf{B} = \mathbf{A} \cdot \mathbf{0} = 0$$

Therefore, because the dot product of $\mathbf{A} \times \mathbf{B}$ and **B** is zero, $\mathbf{A} \times \mathbf{B}$ and **B** are orthogonal and the theorem is proved. ■

From Theorem 17.6.10 we can conclude that if representations of the vectors **A**, **B**, and $\mathbf{A} \times \mathbf{B}$ have the same initial point, then the representation of $\mathbf{A} \times \mathbf{B}$ is perpendicular to the plane formed by the representations of **A** and **B**.

Example 2 Given the points $P(-1, -2, -3)$, $Q(-2, 1, 0)$, and $R(0, 5, 1)$, find a unit vector whose representations are perpendicular to the plane through the points P, Q, and R.

Solution Let $\mathbf{A} = \mathbf{V}(\overrightarrow{PQ})$ and $\mathbf{B} = \mathbf{V}(\overrightarrow{PR})$. Then

$$\mathbf{A} = \langle -2 - (-1), 1 - (-2), 0 - (-3) \rangle = \langle -1, 3, 3 \rangle$$

$$\mathbf{B} = \langle 0 - (-1), 5 - (-2), 1 - (-3) \rangle = \langle 1, 7, 4 \rangle$$

The plane through P, Q, and R is the plane formed by $\overrightarrow{PQ}$ and $\overrightarrow{PR}$, which are, respectively, representations of vectors **A** and **B**. Therefore any representation of the vector $\mathbf{A} \times \mathbf{B}$ is perpendicular to this plane.

$$\mathbf{A} \times \mathbf{B} = (-\mathbf{i} + 3\mathbf{j} + 3\mathbf{k}) \times (\mathbf{i} + 7\mathbf{j} + 4\mathbf{k}) = -9\mathbf{i} + 7\mathbf{j} - 10\mathbf{k}$$

The desired vector is a unit vector parallel to $\mathbf{A} \times \mathbf{B}$. To find this unit vector we apply Theorem 17.2.9 and divide $\mathbf{A} \times \mathbf{B}$ by $|\mathbf{A} \times \mathbf{B}|$ and obtain

$$\frac{\mathbf{A} \times \mathbf{B}}{|\mathbf{A} \times \mathbf{B}|} = -\frac{9}{\sqrt{230}}\mathbf{i} + \frac{7}{\sqrt{230}}\mathbf{j} - \frac{10}{\sqrt{230}}\mathbf{k}$$

Example 3 Find an equation of the plane through the points $P(1, 3, 2)$, $Q(3, -2, 2)$, and $R(2, 1, 3)$.

Solution $\mathbf{V}(\overrightarrow{QR}) = -\mathbf{i} + 3\mathbf{j} + \mathbf{k}$ and $\mathbf{V}(\overrightarrow{PR}) = \mathbf{i} - 2\mathbf{j} + \mathbf{k}$. A normal vector to the required plane is the cross product $\mathbf{V}(\overrightarrow{QR}) \times \mathbf{V}(\overrightarrow{PR})$, which is

$$(-\mathbf{i} + 3\mathbf{j} + \mathbf{k}) \times (\mathbf{i} - 2\mathbf{j} + \mathbf{k}) = 5\mathbf{i} + 2\mathbf{j} - \mathbf{k}$$

So if $P_0 = (1, 3, 2)$ and $\mathbf{N} = \langle 5, 2, -1 \rangle$, from Theorem 17.4.2 an equation of the required plane is

$$5(x - 1) + 2(y - 3) - (z - 2) = 0$$

or, equivalently,

$$5x + 2y - z - 9 = 0$$

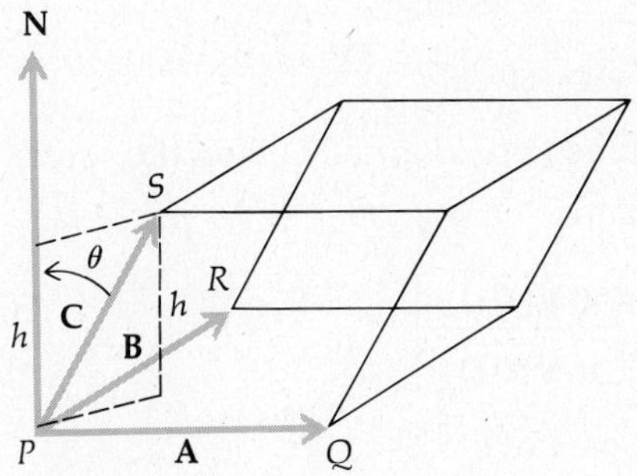

FIGURE 17.6.5

A geometric interpretation of the triple scalar product is obtained by considering a parallelepiped having edges $\overrightarrow{PQ}$, $\overrightarrow{PR}$, and $\overrightarrow{PS}$, and letting $\mathbf{A} = \mathbf{V}(\overrightarrow{PQ})$, $\mathbf{B} = \mathbf{V}(\overrightarrow{PR})$, and $\mathbf{C} = \mathbf{V}(\overrightarrow{PS})$. See Fig. 17.6.5. The vector $\mathbf{A} \times \mathbf{B}$ is a normal vector to the plane of $\overrightarrow{PQ}$ and $\overrightarrow{PR}$. The vector $-(\mathbf{A} \times \mathbf{B})$ is also a normal vector to this plane. We are not certain which of the two vectors, $(\mathbf{A} \times \mathbf{B})$ or $-(\mathbf{A} \times \mathbf{B})$, makes the smaller angle with $\mathbf{C}$. Let $\mathbf{N}$ be the one of the two vectors $(\mathbf{A} \times \mathbf{B})$ or $-(\mathbf{A} \times \mathbf{B})$ that makes an angle of radian measure $\theta < \frac{1}{2}\pi$ with $\mathbf{C}$. Then the representations of $\mathbf{N}$ and $\mathbf{C}$ having their initial points at P are on the same side of the plane of $\overrightarrow{PQ}$ and $\overrightarrow{PR}$ as shown in Fig. 17.6.5. The area of the base of the parallelepiped is $|\mathbf{A} \times \mathbf{B}|$ square units. If h units is the length of the altitude of the parallelepiped, and if V cubic units is the volume of the parallelepiped,

$$V = |\mathbf{A} \times \mathbf{B}|h \tag{8}$$

Consider now the dot product $\mathbf{N} \cdot \mathbf{C}$. By Theorem 17.3.4, $\mathbf{N} \cdot \mathbf{C} = |\mathbf{N}||\mathbf{C}| \cos \theta$. But $h = |\mathbf{C}| \cos \theta$, and so $\mathbf{N} \cdot \mathbf{C} = |\mathbf{N}|h$. Because $\mathbf{N}$ is either $(\mathbf{A} \times \mathbf{B})$ or $-(\mathbf{A} \times \mathbf{B})$, it follows that $|\mathbf{N}| = |\mathbf{A} \times \mathbf{B}|$. Hence

$$\mathbf{N} \cdot \mathbf{C} = |\mathbf{A} \times \mathbf{B}|h \tag{9}$$

Comparing (8) and (9) we have

$$\mathbf{N} \cdot \mathbf{C} = V$$

It follows that the measure of the volume of the parallelepiped is either $(\mathbf{A} \times \mathbf{B}) \cdot \mathbf{C}$ or $-(\mathbf{A} \times \mathbf{B}) \cdot \mathbf{C}$; that is, the measure of the volume of the parallelepiped is the absolute value of the triple scalar product $\mathbf{A} \times \mathbf{B} \cdot \mathbf{C}$.

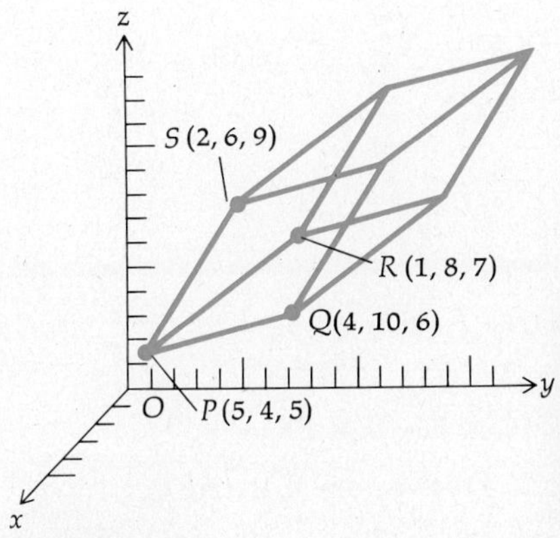

FIGURE 17.6.6

Example 4 Find the volume of the parallelepiped having vertices $P(5, 4, 5)$, $Q(4, 10, 6)$, $R(1, 8, 7)$, and $S(2, 6, 9)$ and edges $\overrightarrow{PQ}$, $\overrightarrow{PR}$, and $\overrightarrow{PS}$.

Solution Figure 17.6.6 shows the parallelepiped. Let $\mathbf{A} = \mathbf{V}(\overrightarrow{PQ}) = \langle -1, 6, 1 \rangle$, $\mathbf{B} = \mathbf{V}(\overrightarrow{PR}) = \langle -4, 4, 2 \rangle$, and $\mathbf{C} = \mathbf{V}(\overrightarrow{PS}) = \langle -3, 2, 4 \rangle$. Then

$$\mathbf{A} \times \mathbf{B} = (-\mathbf{i} + 6\mathbf{j} + \mathbf{k}) \times (-4\mathbf{i} + 4\mathbf{j} + 2\mathbf{k}) = 8\mathbf{i} - 2\mathbf{j} + 20\mathbf{k}$$

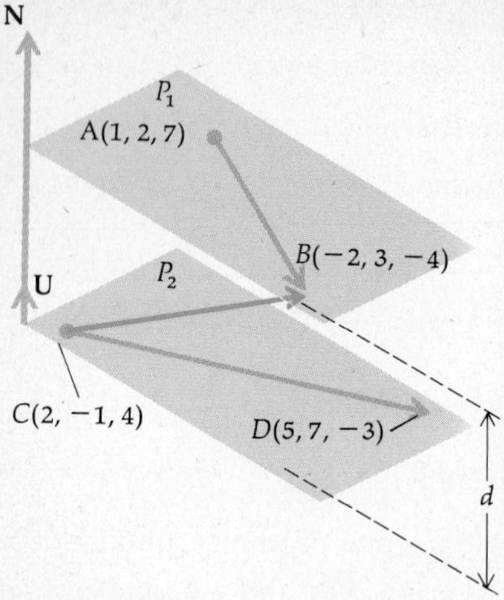

FIGURE 17.6.7

Therefore

$$(\mathbf{A} \times \mathbf{B}) \cdot \mathbf{C} = \langle 8, -2, 20 \rangle \cdot \langle -3, 2, 4 \rangle = -24 - 4 + 80 = 52$$

Thus the volume is 52 cubic units.

Example 5 Find the distance between the two skew lines, l_1 and l_2, of Example 5 in Sec. 17.5.

Solution Because l_1 and l_2 are skew lines, there are parallel planes P_1 and P_2 containing the lines l_1 and l_2, respectively. See Fig. 17.6.7. Let d units be the distance between planes P_1 and P_2. The distance between l_1 and l_2 is also d units. A normal vector to the two planes is $\mathbf{N} = \mathbf{V}(\overrightarrow{AB}) \times \mathbf{V}(\overrightarrow{CD})$. Let $\mathbf{U}$ be a unit normal vector in the direction of $\mathbf{N}$. Then

$$\mathbf{U} = \frac{\mathbf{V}(\overrightarrow{AB}) \times \mathbf{V}(\overrightarrow{CB})}{|\mathbf{V}(\overrightarrow{AB}) \times \mathbf{V}(\overrightarrow{CD})|}$$

Now we take two points, one in each plane (e.g., B and C). Then the scalar projection of $\mathbf{V}(\overrightarrow{CB})$ on $\mathbf{N}$ is $\mathbf{V}(\overrightarrow{CB}) \cdot \mathbf{U}$, and

$$d = |\mathbf{V}(\overrightarrow{CB}) \cdot \mathbf{U}| = \left| \mathbf{V}(\overrightarrow{CB}) \cdot \frac{\mathbf{V}(\overrightarrow{AB}) \times \mathbf{V}(\overrightarrow{CD})}{|\mathbf{V}(\overrightarrow{AB}) \times \mathbf{V}(\overrightarrow{CD})|} \right|$$

Performing the computations required we have

$$\mathbf{V}(\overrightarrow{AB}) = -3\mathbf{i} + \mathbf{j} - 11\mathbf{k} \qquad \mathbf{V}(\overrightarrow{CD}) = 3\mathbf{i} + 8\mathbf{j} - 7\mathbf{k}$$

$$\mathbf{N} = \mathbf{V}(\overrightarrow{AB}) \times \mathbf{V}(\overrightarrow{CD}) = \begin{vmatrix} \mathbf{i} & \mathbf{j} & \mathbf{k} \\ -3 & 1 & -11 \\ 3 & 8 & -7 \end{vmatrix} = 27(3\mathbf{i} - 2\mathbf{j} - \mathbf{k})$$

$$\mathbf{U} = \frac{27(3\mathbf{i} - 2\mathbf{j} - \mathbf{k})}{\sqrt{27^2(3^2 + 2^2 + 1^2)}} = \frac{3\mathbf{i} - 2\mathbf{j} - \mathbf{k}}{\sqrt{14}}$$

Finally, $\mathbf{V}(\overrightarrow{CB}) = -4\mathbf{i} + 4\mathbf{j} - 8\mathbf{k}$, and so

$$d = |\mathbf{V}(\overrightarrow{CB}) \cdot \mathbf{U}| = \frac{1}{\sqrt{14}}|-12 - 8 + 8| = \frac{12}{\sqrt{14}} = \frac{6}{7}\sqrt{14}$$

EXERCISES 17.6

In Exercises 1 through 12, let $\mathbf{A} = \langle 1, 2, 3 \rangle$, $\mathbf{B} = \langle 4, -3, -1 \rangle$, $\mathbf{C} = \langle -5, -3, 5 \rangle$, $\mathbf{D} = \langle -2, 1, 6 \rangle$, $\mathbf{E} = \langle 4, 0, -7 \rangle$, *and* $\mathbf{F} = \langle 0, 2, 1 \rangle$.

1. Find $\mathbf{A} \times \mathbf{B}$.

2. Find $\mathbf{D} \times \mathbf{E}$.

3. Find $(\mathbf{C} \times \mathbf{D}) \cdot (\mathbf{E} \times \mathbf{F})$.

4. Find $(\mathbf{C} \times \mathbf{E}) \cdot (\mathbf{D} \times \mathbf{F})$.

5. Verify Theorem 17.6.3 for vectors **A** and **B**.

6. Verify Theorem 17.6.4 for vectors **A**, **B**, and **C**.

7. Verify Theorem 17.6.5(i) for vectors **A** and **B** and $c = 3$.

8. Verify Theorem 17.6.5(ii) for vectors **A** and **B** and $c = 3$.

9. Verify Theorem 17.6.9 for vectors **A**, **B**, and **C**.

10. Find $(\mathbf{A} \times \mathbf{B}) \times \mathbf{C}$ and $\mathbf{A} \times (\mathbf{B} \times \mathbf{C})$.

11. Find $(\mathbf{A} + \mathbf{B}) \times (\mathbf{C} - \mathbf{D})$ and $(\mathbf{D} - \mathbf{C}) \times (\mathbf{A} + \mathbf{B})$ and verify that they are equal.

12. Find $|\mathbf{A} \times \mathbf{B}||\mathbf{C} \times \mathbf{D}|$. **13.** Prove Theorem 17.6.2(ii) and (iii). **14.** Prove Theorem 17.6.4.

15. Prove Theorem 17.6.5(i). **16.** Prove Theorem 17.6.5(ii). **17.** Prove Theorem 17.6.9.

18. Given the two unit vectors $\mathbf{A} = \frac{4}{9}\mathbf{i} + \frac{7}{9}\mathbf{j} - \frac{4}{9}\mathbf{k}$ and $\mathbf{B} = -\frac{2}{3}\mathbf{i} + \frac{2}{3}\mathbf{j} + \frac{1}{3}\mathbf{k}$. If θ is the radian measure of the angles between $\mathbf{A}$ and $\mathbf{B}$, find $\sin\theta$ in two ways: (a) by using the cross product (formula (3) of this section); (b) by using the dot product and a trigonometric identity.

19. Follow the instructions of Exercise 18 for the two unit vectors

$$\mathbf{A} = \frac{1}{\sqrt{3}}\mathbf{i} - \frac{1}{\sqrt{3}}\mathbf{j} + \frac{1}{\sqrt{3}}\mathbf{k} \quad \text{and} \quad \mathbf{B} = \frac{1}{3\sqrt{3}}\mathbf{i} + \frac{5}{3\sqrt{3}}\mathbf{j} + \frac{1}{3\sqrt{3}}\mathbf{k}$$

20. Show that the quadrilateral having vertices at $(1, 1, 3)$, $(-2, 1, -1)$, $(-5, 4, 0)$ and $(-8, 4, -4)$ is a parallelogram, and find its area.

21. Show that the quadrilateral having vertices at $(1, -2, 3)$, $(4, 3, -1)$, $(2, 2, 1)$, and $(5, 7, -3)$ is a parallelogram, and find its area.

22. Find the area of the parallelogram $PQRS$ if $\mathbf{V}(\overrightarrow{PQ}) = 3\mathbf{i} - 2\mathbf{j}$ and $\mathbf{V}(\overrightarrow{PS}) = 3\mathbf{j} + 4\mathbf{k}$.

23. Find the area of the triangle having vertices at $(0, 2, 2)$, $(8, 8, -2)$, and $(9, 12, 6)$.

24. Find the area of the triangle having vertices at $(4, 5, 6)$, $(4, 4, 5)$, and $(3, 5, 5)$.

In Exercises 25 and 26, use the cross product to find an equation of the plane containing the given three points.

25. $(-2, 2, 2)$, $(-8, 1, 6)$, $(3, 4, -1)$ **26.** $(a, b, 0)$, $(a, 0, c)$ $(0, b, c)$

27. Let $\overrightarrow{OP}$ be the position representation of vector $\mathbf{A}$, $\overrightarrow{OQ}$ be the position representation of vector $\mathbf{B}$, and $\overrightarrow{OR}$ be the position representation of vector $\mathbf{C}$. Prove that the area of triangle PQR is $\frac{1}{2}|(\mathbf{B} - \mathbf{A}) \times (\mathbf{C} - \mathbf{A})|$.

28. Find a unit vector whose representations are perpendicular to the plane containing $\overrightarrow{PQ}$ and $\overrightarrow{PR}$ if $\overrightarrow{PQ}$ is a representation of the vector $\mathbf{i} + 3\mathbf{j} - 2\mathbf{k}$ and $\overrightarrow{PR}$ is a representation of the vector $2\mathbf{i} - \mathbf{j} - \mathbf{k}$.

In Exercises 29 through 31, find a unit vector whose representations are perpendicular to the plane through the given points P, Q, and R.

29. $P(5, 2, -1)$, $Q(2, 4, -2)$, $R(11, 1, 4)$ **30.** $P(-2, 1, 0)$, $Q(2, -2, -1)$, $R(-5, 0, 2)$ **31.** $P(1, 4, 2)$, $Q(3, 2, 4)$, $R(4, 3, 1)$

32. Find the volume of the parallelepiped having edges $\overrightarrow{PQ}$, $\overrightarrow{PR}$, and $\overrightarrow{PS}$ if the points P, Q, R, and S are, respectively, $(1, 3, 4)$, $(3, 5, 3)$, $(2, 1, 6)$, and $(2, 2, 5)$.

33. Find the volume of the parallelepiped $PQRS$ if the vectors $\mathbf{V}(\overrightarrow{PQ})$, $\mathbf{V}(\overrightarrow{PR})$, and $\mathbf{V}(\overrightarrow{PS})$ are, respectively, $\mathbf{i} + 3\mathbf{j} + 2\mathbf{k}$, $2\mathbf{i} + \mathbf{j} - \mathbf{k}$, and $\mathbf{i} - 2\mathbf{j} + \mathbf{k}$.

34. If $\mathbf{A}$ and $\mathbf{B}$ are any two vectors in V_3, prove that $(\mathbf{A} - \mathbf{B}) \times (\mathbf{A} + \mathbf{B}) = 2(\mathbf{A} \times \mathbf{B})$.

In Exercises 35 and 36, find the perpendicular distance between the two given skew lines.

35. $\dfrac{x-1}{5} = \dfrac{y-2}{3} = \dfrac{z+1}{2}$ and $\dfrac{x+2}{4} = \dfrac{y+1}{2} = \dfrac{z-3}{-3}$ **36.** $\dfrac{x+1}{2} = \dfrac{y+2}{-4} = \dfrac{z-1}{-3}$ and $\dfrac{x-1}{5} = \dfrac{y-1}{3} = \dfrac{z+1}{2}$

37. Let P, Q, and R be three noncollinear points in R^3 and let $\overrightarrow{OP}$, $\overrightarrow{OQ}$, and $\overrightarrow{OR}$ be the position representations of vectors $\mathbf{A}$, $\mathbf{B}$, and $\mathbf{C}$, respectively. Prove that the representations of the vector $\mathbf{A} \times \mathbf{B} + \mathbf{B} \times \mathbf{C} + \mathbf{C} \times \mathbf{A}$ are perpendicular to the plane containing the points P, Q, and R.

38. Find an equation of the plane containing the endpoints of the position representations of the vectors $2\mathbf{i} - \mathbf{j} + 3\mathbf{k}$, $-\mathbf{i} + \mathbf{j} + 2\mathbf{k}$, and $5\mathbf{i} + \mathbf{j} - \mathbf{k}$.

17.7 CYLINDERS AND SURFACES OF REVOLUTION

As mentioned previously, the graph of an equation in three variables is a *surface*. A surface is represented by an equation if the coordinates of every point on the surface satisfy the equation and if every point whose coordinates satisfy the equation lies on the surface. We have already discussed two kinds of surfaces, a plane and a sphere. Another kind of surface that is fairly simple is a cylinder. You are probably familiar with right-circular cylinders from previous experience. We now consider a more general cylindrical surface.

17.7.1 Definition A **cylinder** is a surface that is generated by a line moving along a given plane curve in such a way that it always remains parallel to a fixed line not lying in the plane of the given curve. The moving line is called a **generator** of the cylinder and the given plane curve is called a **directrix** of the cylinder. Any position of a generator is called a **ruling** of the cylinder.

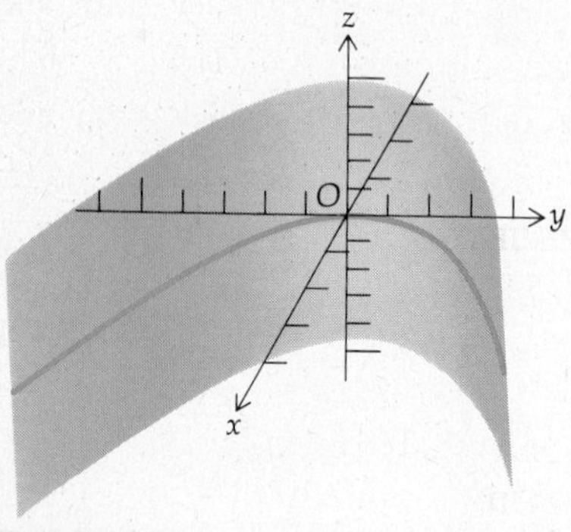

FIGURE 17.7.1

This discussion is confined to cylinders having a directrix in a coordinate plane and rulings perpendicular to that plane. If the rulings of a cylinder are perpendicular to the plane of a directrix, the cylinder is said to be perpendicular to the plane.

The familiar right-circular cylinder is one for which a directrix is a circle in a plane perpendicular to the cylinder.

• ILLUSTRATION 1 In Fig. 17.7.1 there is a cylinder whose directrix is the parabola $y^2 = 8x$ in the xy plane and whose rulings are parallel to the z axis. The cylinder is called a *parabolic cylinder*. An *elliptic cylinder* is shown in Fig. 17.7.2; its directrix is the ellipse $9x^2 + 16y^2 = 144$ in the xy plane and its rulings are parallel to the z axis. Figure 17.7.3 shows a *hyperbolic cylinder* having as a directrix the hyperbola $25x^2 - 4y^2 = 100$ in the xy plane and rulings parallel to the z axis. •

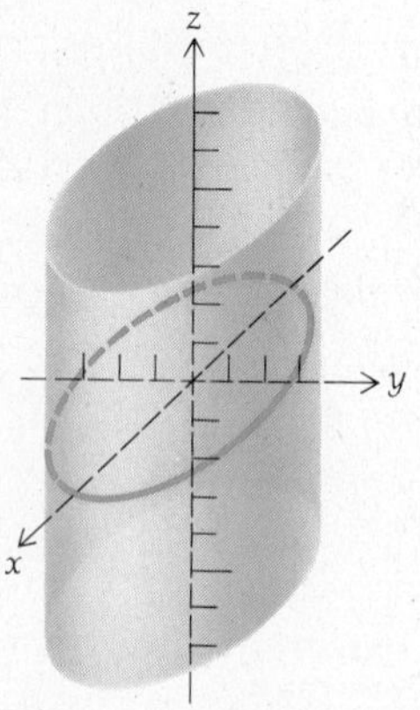

FIGURE 17.7.2

Consider the problem of finding an equation of a cylinder having a directrix in a coordinate plane and rulings parallel to the coordinate axis not in that plane. To be specific, take the directrix in the xy plane and the rulings parallel to the z axis. Refer to Fig. 17.7.4. Suppose that an equation of the directrix in the xy plane is $y = f(x)$. If the point $(x_0, y_0, 0)$ in the xy plane satisfies this equation, any point (x_0, y_0, z) in three-dimensional space, where z is any real number, will satisfy the same equation because z does not appear in the equation. The points having representations (x_0, y_0, z) all lie on the line parallel to the z axis through the point $(x_0, y_0, 0)$. This line is a ruling of the cylinder. Hence any point whose x and y coordinates satisfy the equation $y = f(x)$ lies on the cylinder. Conversely, if the point $P(x, y, z)$ lies on the cylinder (see Fig. 17.7.5), then the point $(x, y, 0)$ lies on the directrix of the cylinder in the xy plane, and hence the x and y coordinates of P satisfy the equation $y = f(x)$. Therefore, if $y = f(x)$ is considered as an equation of a graph in three-dimensional space, the graph is a cylinder whose rulings are parallel to the z axis and which has as a directrix the curve $y = f(x)$ in the plane $z = 0$. A similar discussion pertains when the directrix is in either of the other coordinate planes. The results are summarized in the following theorem.

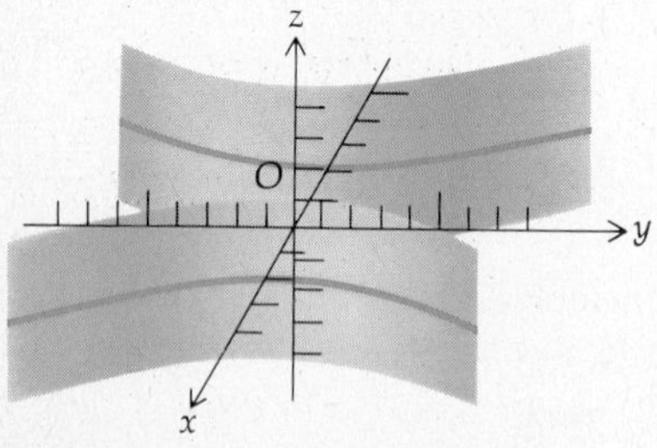

FIGURE 17.7.3

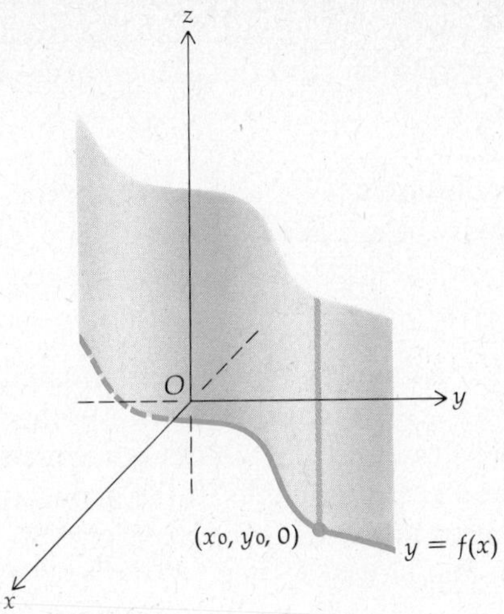

FIGURE 17.7.4

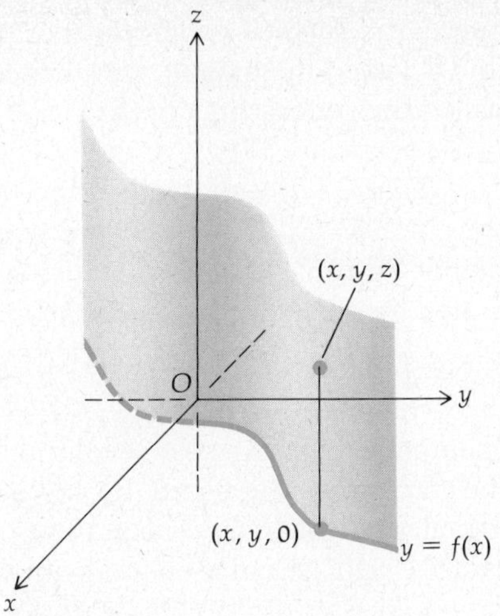

FIGURE 17.7.5

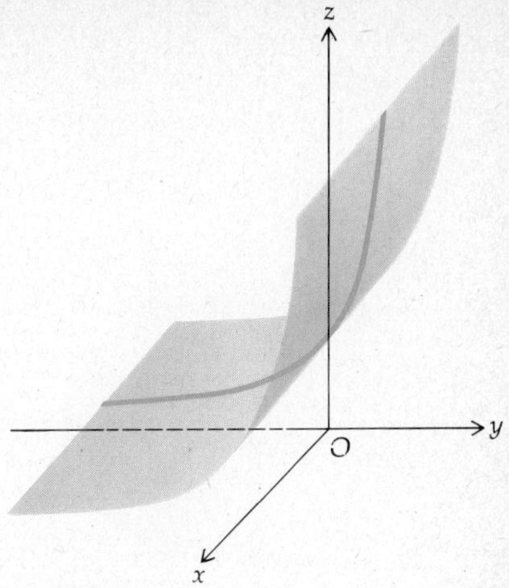

FIGURE 17.7.6

17.7.2 Theorem In three-dimensional space, the graph of an equation in two of the three variables x, y, and z is a cylinder whose rulings are parallel to the axis associated with the missing variable and whose directrix is a curve in the plane associated with the two variables appearing in the equation.

• ILLUSTRATION 2 It follows from Theorem 17.7.2 that an equation of the parabolic cylinder of Fig. 17.7.1 is $y^2 = 8x$, considered as an equation in R^3. Similarly, equations of the elliptic cylinder of Fig. 17.7.2 and the hyperbolic cylinder of Fig. 17.7.3 are, respectively, $9x^2 + 16y^2 = 144$ and $25x^2 - 4y^2 = 100$, both considered as equations in R^3. •

A **cross section** of a surface in a plane is the set of all points of the surface that lie in the given plane. If a plane is parallel to the plane of the directrix of a cylinder, the cross section of the cylinder is the same as the directrix. For example, the cross section of the elliptic cylinder of Fig. 17.7.2 in any plane parallel to the xy plane is an ellipse.

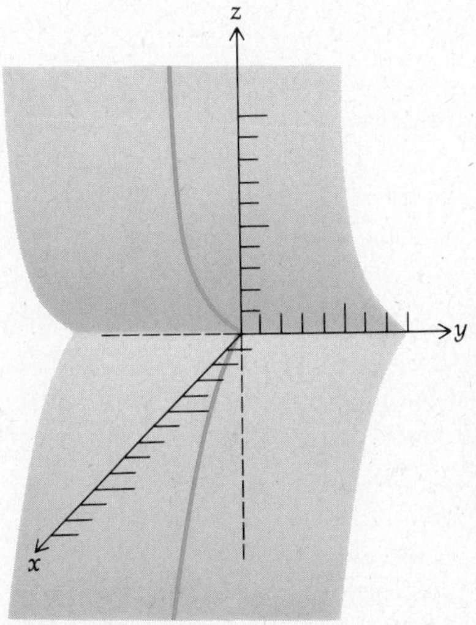

FIGURE 17.7.7

Example 1 Draw a sketch of the graph of each of the following equations: (a) $y = \ln z$; (b) $z^2 = x^3$.

Solution

(a) The graph is a cylinder whose directrix in the yz plane is the curve $y = \ln z$ and whose rulings are parallel to the x axis. A sketch of the graph is shown in Fig. 17.7.6.

(b) The graph is a cylinder whose directrix is in the xz plane and whose rulings are parallel to the y axis. An equation of the directrix is the curve $z^2 = x^3$ in the xz plane. A sketch of the graph is shown in Fig. 17.7.7.

17.7.3 Definition If a plane curve is revolved about a fixed line lying in the plane of the curve, the surface generated is called a **surface of revolution**. The fixed line is called the **axis** of the surface of revolution, and the plane curve is called the **generating curve**.

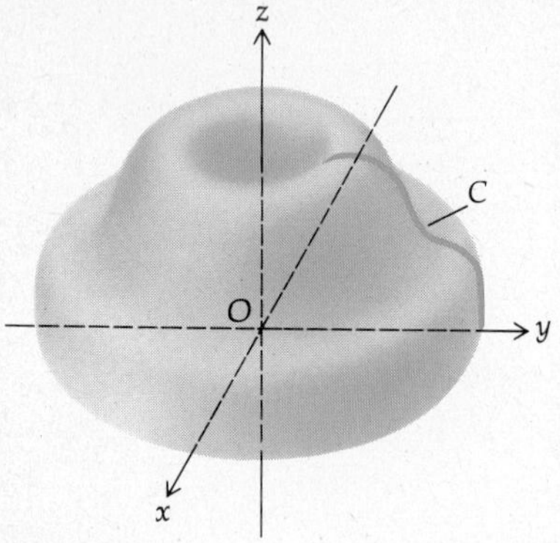

FIGURE 17.7.8

Figure 17.7.8 shows a surface of revolution whose generating curve is the curve C in the yz plane and whose axis is the z axis. A sphere is a particular example of a surface of revolution because a sphere can be generated by revolving a semicircle about a diameter.

● ILLUSTRATION 3 Figure 17.7.9 shows a sphere that can be generated by revolving the semicircle $y^2 + z^2 = r^2$, $z \geq 0$, about the y axis. Another example of a surface of revolution is a right-circular cylinder for which the generating curve and the axis are parallel straight lines. If the generating curve is the line $z = k$ in the xz plane and the axis is the x axis, we obtain the right-circular cylinder shown in Fig. 17.7.10. ●

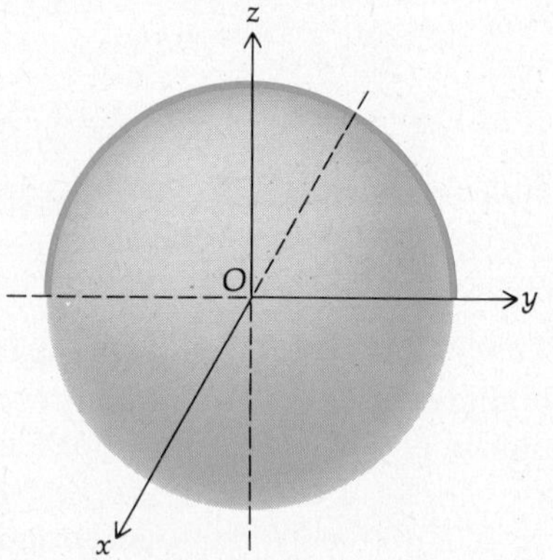

FIGURE 17.7.9

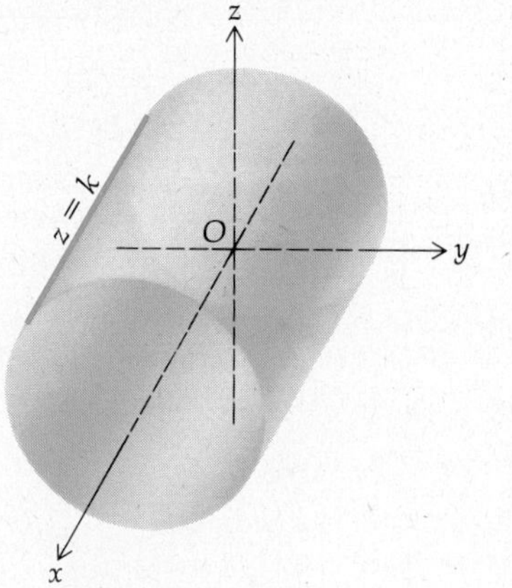

FIGURE 17.7.10

We now find an equation of the surface generated by revolving about the y axis the curve in the yz plane having the two-dimensional equation

$$z = f(y) \tag{1}$$

Refer to Fig. 17.7.11. Let $P(x, y, z)$ be any point on the surface of revolution. Through P, pass a plane perpendicular to the y axis. Denote the point of intersection of this plane with the y axis by $Q(0, y, 0)$, and let $P_0(0, y, z_0)$ be the point of intersection of the plane with the generating curve. Because the cross section of the surface with the plane through P is a circle, P is on the surface if and only if

$$|\overline{QP}|^2 = |\overline{QP_0}|^2 \tag{2}$$

Because $|\overline{QP}| = \sqrt{x^2 + z^2}$ and $|\overline{QP_0}| = z_0$, we obtain from (2)

$$x^2 + z^2 = z_0{}^2 \tag{3}$$

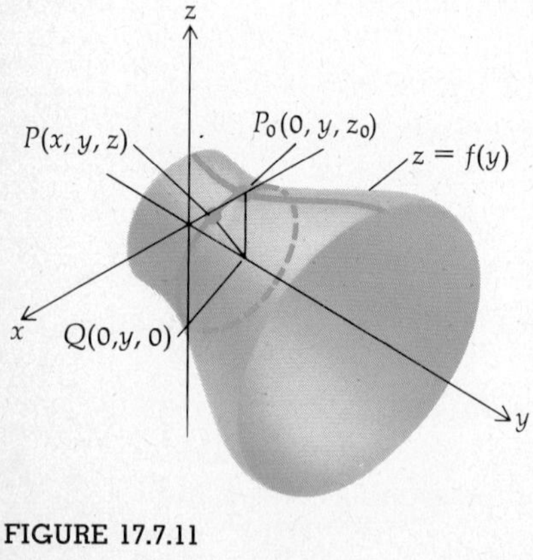

FIGURE 17.7.11

The point P_0 is on the generating curve; so its coordinates must satisfy (1). Therefore

$$z_0 = f(y) \tag{4}$$

From (3) and (4), the point P is on the surface of revolution if and only if

$$x^2 + z^2 = [f(y)]^2 \tag{5}$$

Equation (5) is the desired equation of the surface of revolution. Because (5) is equivalent to

$$\pm \sqrt{x^2 + z^2} = f(y)$$

we can obtain (5) by replacing z in (1) by $\pm \sqrt{x^2 + z^2}$.

In a similar manner we can show that if the curve in the yz plane having the two-dimensional equation

$$y = g(z) \tag{6}$$

is revolved about the z axis, an equation of the surface of revolution generated is obtained by replacing y in (6) by $\pm \sqrt{x^2 + y^2}$. Analogous remarks hold when a curve in any coordinate plane is revolved about either one of the coordinate axes in that plane. In summary, the graphs of any of the following equations are surfaces of revolution having the indicated axis: $x^2 + y^2 = [F(z)]^2$–z axis; $x^2 + z^2 = [F(y)]^2$–y axis; $y^2 + z^2 = [F(x)]^2$–x axis. In each case, cross sections of the surface in planes perpendicular to the axis are circles having centers on the axis.

Example 2 Find an equation of the surface of revolution generated by revolving the parabola $y^2 = 4x$ in the xy plane about the x axis. Draw a sketch of the graph of the surface.

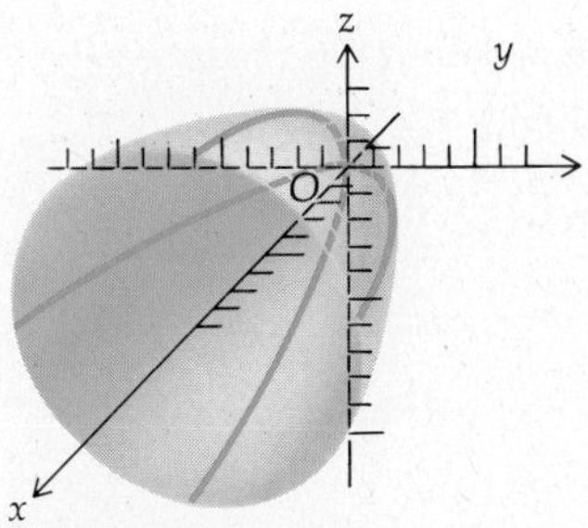

FIGURE 17.7.12

Solution In the equation of the parabola we replace y by $\pm \sqrt{y^2 + z^2}$ and obtain

$$y^2 + z^2 = 4x$$

A sketch of the graph is shown in Fig. 17.7.12. The same surface is generated if the parabola $z^2 = 4x$ in the xz plane is revolved about the x axis.

The surface obtained in Example 2 is called a *paraboloid of revolution*. If an ellipse is revolved about one of its axes, the surface obtained is called an *ellipsoid of revolution*. A *hyperboloid of revolution* is obtained when a hyperbola is revolved about an axis.

Example 3 Draw a sketch of the surface $x^2 + z^2 - 4y^2 = 0$, if $y \geq 0$.

Solution The given equation is of the form $x^2 + z^2 = [F(y)]^2$; so its graph is a surface of revolution having the y axis as axis. We solve the

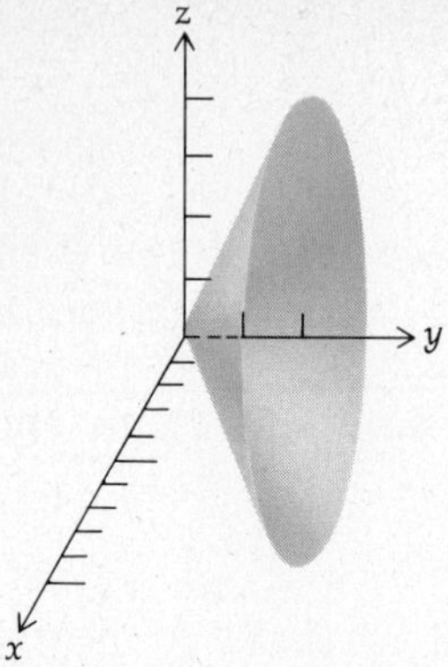

FIGURE 17.7.13

given equation for y and obtain

$$2y = \pm\sqrt{x^2 + z^2}$$

Hence the generating curve can be either the straight line $2y = x$ in the xy plane or the straight line $2y = z$ in the yz plane. By drawing sketches of the two possible generating curves and using the fact that cross sections of the surface in planes perpendicular to the y axis are circles having centers on the y axis, we obtain the surface shown in Fig. 17.7.13 (note that because $y \geq 0$ there is only one nappe of the cone).

The surface obtained in Example 3 is called a *right-circular cone.*

EXERCISES 17.7

In Exercises 1 through 3, draw a sketch of the cross section of the given cylinder in the given plane.

1. $4z^2 - y^2 = 4$; yz plane **2.** $x = |y|$; xy plane **3.** $z = e^x$; xz plane

In Exercises 4 through 11, draw a sketch of the cylinder having the given equation.

4. $x^2 - z^2 = 4$ **5.** $4x^2 + 9y^2 = 36$ **6.** $z = \sin y$ **7.** $y = |z|$

8. $x^2 = y^3$ **9.** $z = 2x^2$ **10.** $z^2 = 4y^2$ **11.** $y = \cosh x$

In Exercises 12 through 18, find an equation of the surface of revolution generated by revolving the given plane curve about the indicated axis. Draw a sketch of the surface.

12. $x^2 = 4y$ in the xy plane, about the x axis.

13. $x^2 = 4y$ in the xy plane, about the y axis.

14. $x^2 + 4z^2 = 16$ in the xz plane, about the z axis.

15. $x^2 + 4z^2 = 16$ in the xz plane, about the x axis.

16. $y^2 = z^3$ in the yz plane, about the z axis.

17. $y = \sin x$ in the xy plane, about the x axis.

18. $9y^2 - 4z^2 = 144$ in the yz plane, about the z axis.

In Exercises 19 through 24, find a generating curve and the axis for the given surface of revolution. Draw a sketch of the surface.

19. $x^2 + y^2 - z^2 = 4$ **20.** $y^2 + z^2 = e^{2x}$ **21.** $x^2 + z^2 = |y|$

22. $4x^2 + 9y^2 + 4z^2 = 36$ **23.** $9x^2 - y^2 + 9z^2 = 0$ **24.** $4x^2 + 4y^2 - z = 9$

25. The tractrix

$$x = t - a\tanh\frac{t}{a} \qquad y = a\,\mathrm{sech}\,\frac{t}{a}$$

from $x = -a$ to $x = 2a$ is revolved about the x axis. Draw a sketch of the surface of revolution.

17.8 QUADRIC SURFACES

The graph of a second-degree equation in three variables x, y, and z is called a **quadric surface**. These surfaces correspond to the conics in the plane.

The simplest types of quadric surfaces are the parabolic, elliptic, and hyperbolic cylinders, which were discussed in the preceding section. There are six other types of quadric surfaces, which are now considered. We choose the coordinate axes so the equations are in their simplest form. In the discussion of each of these surfaces we refer to the cross sections of the surfaces in planes parallel to the coordinate planes. These cross sections help to visualize the surface.

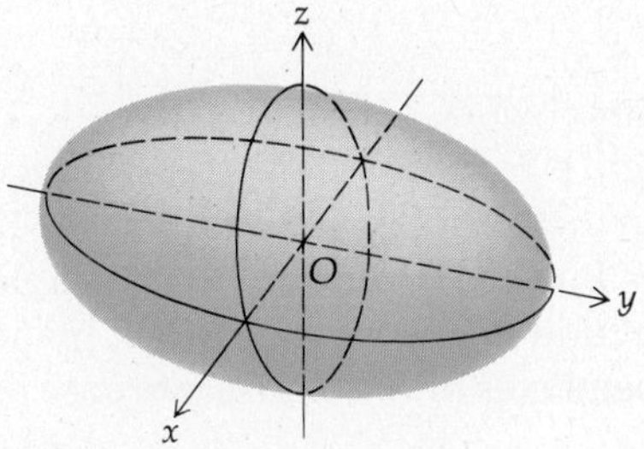

FIGURE 17.8.1

The ellipsoid

$$\frac{x^2}{a^2} + \frac{y^2}{b^2} + \frac{z^2}{c^2} = 1 \tag{1}$$

where a, b, and c are positive (see Fig. 17.8.1).

If in (1) z is replaced by zero, we obtain the cross section of the ellipsoid in the xy plane, which is the ellipse $x^2/a^2 + y^2/b^2 = 1$. To obtain the cross sections of the surface with the planes $z = k$, replace z by k in the equation of the ellipsoid and get

$$\frac{x^2}{a^2} + \frac{y^2}{b^2} = 1 - \frac{k^2}{c^2}$$

If $|k| < c$, the cross section is an ellipse and the lengths of the semiaxes decrease to zero as $|k|$ increases to the value c. If $|k| = c$, the intersection of a plane $z = k$ with the ellipsoid is the single point $(0, 0, k)$. If $|k| > c$, there is no intersection. The discussion is similar if we consider cross sections formed by planes parallel to either of the other coordinate planes.

The numbers a, b, and c are the lengths of the semiaxes of the ellipsoid. If any two of these three numbers are equal, we have an ellipsoid of revolution, which is also called a *spheroid*. A spheroid for which the third number is greater than the two equal numbers is said to be *prolate*. A prolate spheroid is shaped like a football. An *oblate* spheroid is obtained if the third number is less than the two equal numbers. If all three numbers a, b, and c in the equation of an ellipsoid are equal, the ellipsoid is a *sphere*.

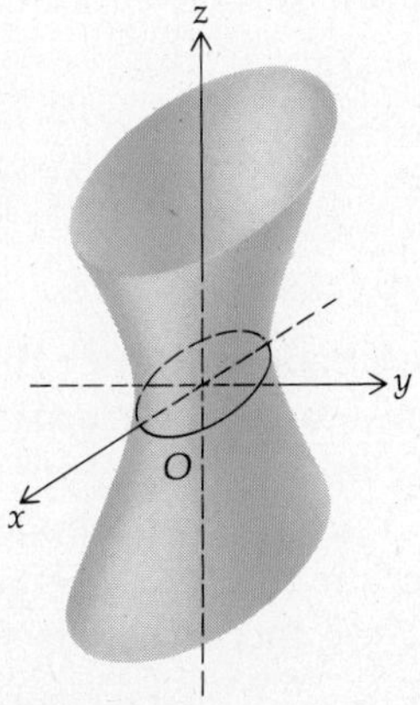

FIGURE 17.8.2

The elliptic hyperboloid of one sheet

$$\frac{x^2}{a^2} + \frac{y^2}{b^2} - \frac{z^2}{c^2} = 1 \tag{2}$$

where a, b, and c are positive (see Fig. 17.8.2).

The cross sections in the planes $z = k$ are ellipses $x^2/a^2 + y^2/b^2 = 1 + k^2/c^2$. When $k = 0$, the lengths of the semiaxes of the ellipse are smallest, and these lengths increase as $|k|$ increases. The cross sections in the planes $x = k$ are hyperbolas $y^2/b^2 - z^2/c^2 = 1 - k^2/a^2$. If $|k| < a$, the transverse axis of the hyperbola is parallel to the y axis, and if $|k| > a$, the transverse axis is parallel to the z axis. If $k = a$, the hyperbola degenerates into two straight lines: $y/b - z/c = 0$ and $y/b + z/c = 0$. In an analogous manner, the cross sections in the planes $y = k$ are also hyperbolas. The axis of this hyperboloid is the z axis.

If $a = b$, the surface is a hyperboloid of revolution for which the axis is the line containing the conjugate axis.

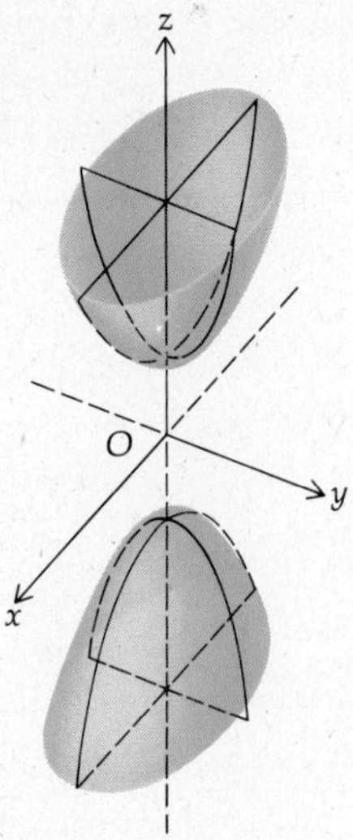

FIGURE 17.8.3

The elliptic hyperboloid of two sheets

$$-\frac{x^2}{a^2} - \frac{y^2}{b^2} + \frac{z^2}{c^2} = 1 \qquad (3)$$

where a, b, and c are positive (see Fig. 17.8.3).

Replacing z by k in (3) we obtain $x^2/a^2 + y^2/b^2 = k^2/c^2 - 1$. If $|k| < c$, there is no intersection of the plane $z = k$ with the surface; hence there are no points of the surface between the planes $z = -c$ and $z = c$. If $|k| = c$, the intersection of the plane $z = k$ with the surface is the single point $(0, 0, k)$. When $|k| > c$, the cross section of the surface in the plane $z = k$ is an ellipse and the lengths of the semiaxes of the ellipse increase as $|k|$ increases.

The cross sections of the surface in the planes $x = k$ are the hyperbolas $z^2/c^2 - y^2/b^2 = 1 + k^2/a^2$ whose transverse axes are parallel to the z axis. In a similar fashion, the cross sections in the planes $y = k$ are the hyperbolas given by $z^2/c^2 - x^2/a^2 = 1 + k^2/b^2$ for which the transverse axes are also parallel to the z axis.

If $a = b$, the surface is a hyperboloid of revolution in which the axis is the line containing the transverse axis of the hyperbola.

Each of the above three quadric surfaces is symmetric with respect to each of the coordinate planes and symmetric with respect to the origin. Their graphs are called *central quadrics* and their center is at the origin. The graph of any equation of the form

$$\pm\frac{x^2}{a^2} \pm \frac{y^2}{b^2} \pm \frac{z^2}{c^2} = 1$$

where a, b, and c are positive, is a central quadric.

Example 1 Draw a sketch of the graph of the equation $4x^2 - y^2 + 25z^2 = 100$, and name the surface.

Solution The given equation can be written as

$$\frac{x^2}{25} - \frac{y^2}{100} + \frac{z^2}{4} = 1$$

which is of the form of Eq. (2) with y and z interchanged. Hence the surface is an elliptic hyperboloid of one sheet whose axis is the y axis. The cross sections in the planes $y = k$ are the ellipses $x^2/25 + z^2/4 = 1 + k^2/100$. The cross sections in the planes $x = k$ are the hyperbolas $z^2/4 - y^2/100 = 1 - k^2/25$, and the cross sections in the planes $z = k$ are the hyperbolas $x^2/25 - y^2/100 = 1 - k^2/4$. A sketch of the surface is shown in Fig. 17.8.4.

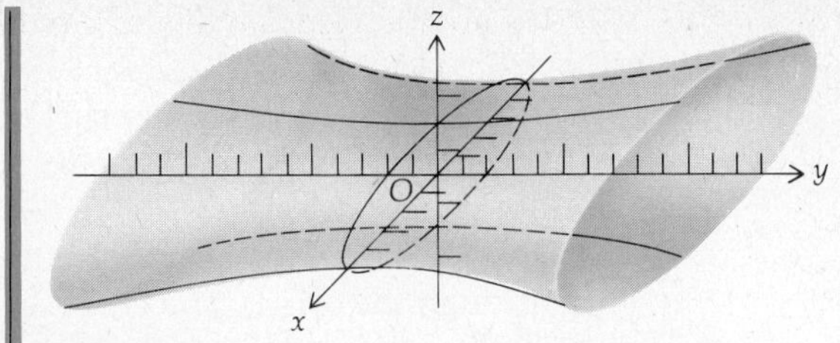

FIGURE 17.8.4

Example 2 Draw a sketch of the graph of the equation $4x^2 - 25y^2 - z^2 = 100$, and name the surface.

Solution The given equation can be written as

$$\frac{x^2}{25} - \frac{y^2}{4} - \frac{z^2}{100} = 1$$

which is of the form of Eq. (3) with x and z interchanged; thus the surface is an elliptic hyperboloid of two sheets whose axis is the x axis. The cross sections in the planes $x = k$, where $|k| > 5$, are the ellipses $y^2/4 + z^2/100 = k^2/25 - 1$. The planes $x = k$, where $|k| < 5$, do not intersect the surface. The cross sections in the planes $y = k$ are the hyperbolas $x^2/25 - z^2/100 = 1 + k^2/4$, and the cross sections in the planes $z = k$ are the hyperbolas $x^2/25 - y^2/4 = 1 + k^2/100$. The required sketch is shown in Fig. 17.8.5.

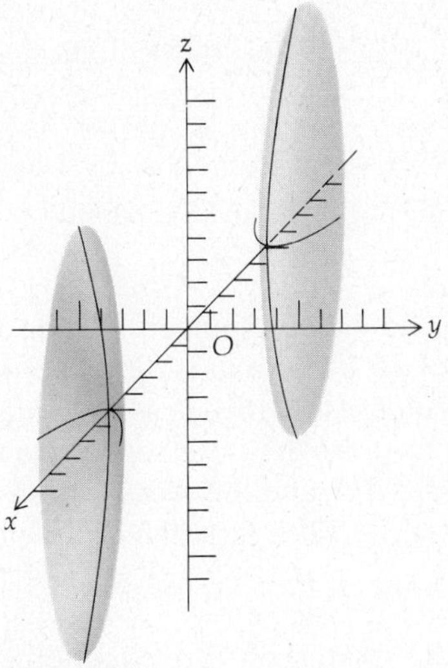

FIGURE 17.8.5

The following two quadrics are called noncentral quadrics.

The elliptic paraboloid

$$\frac{x^2}{a^2} + \frac{y^2}{b^2} = \frac{z}{c} \tag{4}$$

where a and b are positive and $c \neq 0$. Figure 17.8.6 shows the surface if $c > 0$.

Substituting k for z in (4) we obtain $x^2/a^2 + y^2/b^2 = k/c$. When $k = 0$, this equation becomes $x^2/a^2 + y^2/b^2 = 0$, which represents a single point, the origin. If $k \neq 0$ and k and c have the same sign, the equation is that of an ellipse. So we conclude that cross sections of the surface in the planes $z = k$, where k and c have the same sign, are ellipses and the lengths of the semiaxes increase as $|k|$ increases. If k and c have opposite signs, the planes $z = k$ do not intersect the surface. The cross sections of the surface with the planes $x = k$ and $y = k$ are parabolas. When $c > 0$, the parabolas open upward, as shown in Fig. 17.8.6; when $c < 0$, the parabolas open downward.

If $a = b$, the surface is a paraboloid of revolution.

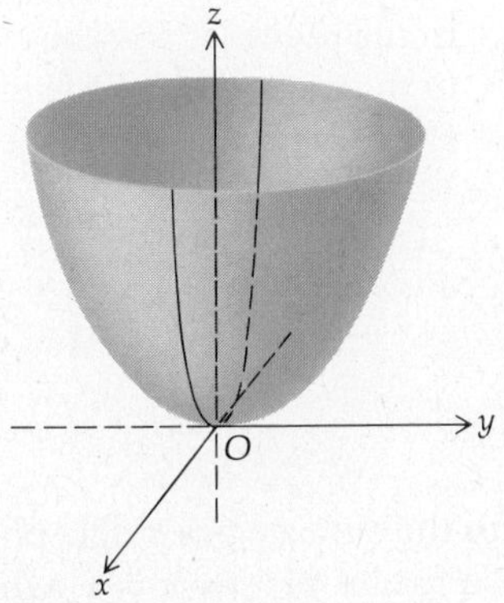

FIGURE 17.8.6

The hyperbolic paraboloid

$$\frac{y^2}{b^2} - \frac{x^2}{a^2} = \frac{z}{c} \tag{5}$$

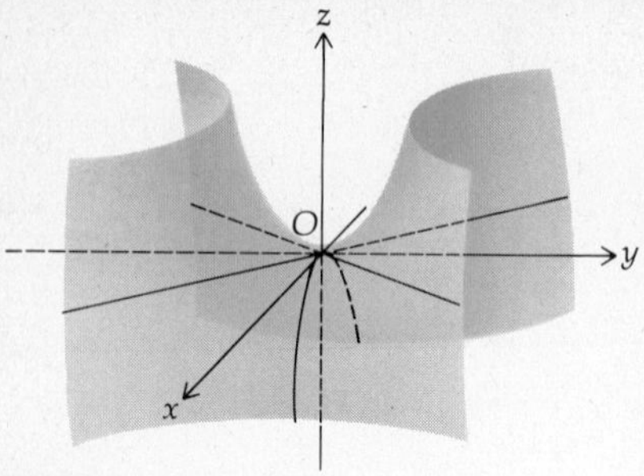

FIGURE 17.8.7

where a and b are positive and $c \neq 0$. The surface is shown in Fig. 17.8.7 for $c > 0$.

The cross sections of the surface in the planes $z = k$, where $k \neq 0$, are hyperbolas having their transverse axes parallel to the y axis if k and c have the same sign and parallel to the x axis if k and c have opposite signs. The cross section of the surface in the plane $z = 0$ consists of two straight lines through the origin. The cross sections in the planes $x = k$ are parabolas opening upward if $c > 0$ and opening downward if $c < 0$. The cross sections in the planes $y = k$ are parabolas opening downward if $c > 0$ and opening upward if $c < 0$.

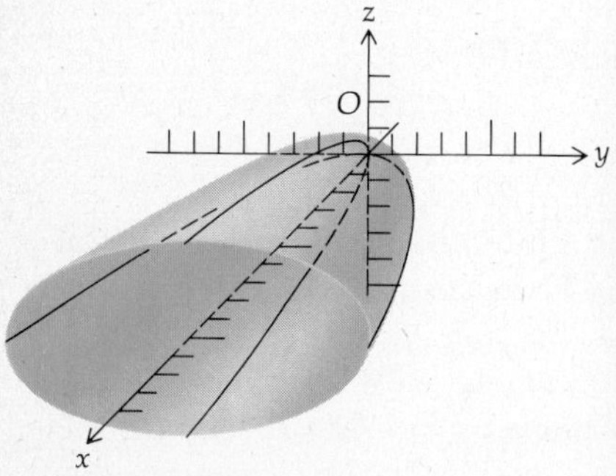

FIGURE 17.8.8

Example 3 Draw a sketch of the graph of the equation $3y^2 + 12z^2 = 16x$, and name the surface.

Solution The given equation can be written as

$$\frac{y^2}{16} + \frac{z^2}{4} = \frac{x}{3}$$

which is of the form of Eq. (4) with x and z interchanged. Hence the graph of the equation is an elliptic paraboloid whose axis is the x axis. The cross sections in the planes $x = k > 0$ are ellipses $y^2/16 + z^2/4 = k/3$, and the planes $x = k < 0$ do not intersect the surface. The cross sections in the planes $y = k$ are the parabolas $12z^2 = 16x - 3k^2$, and the cross sections in the planes $z = k$ are the parabolas $3y^2 = 16x - 12k^2$. A sketch of the elliptic paraboloid is shown in Fig. 17.8.8.

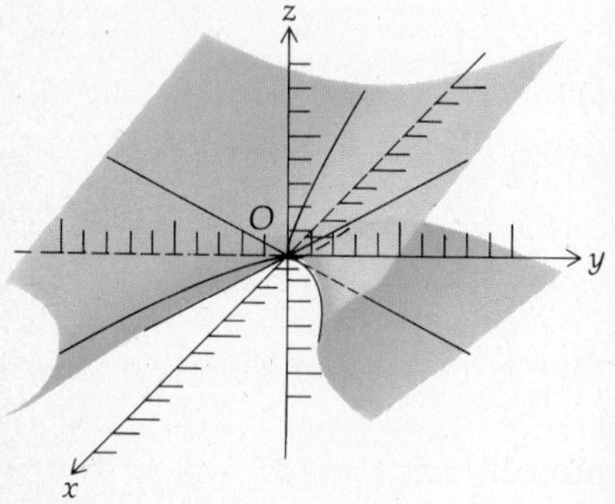

FIGURE 17.8.9

Example 4 Draw a sketch of the graph of the equation $3y^2 - 12z^2 = 16x$, and name the surface.

Solution If the given equation is written as

$$\frac{y^2}{16} - \frac{z^2}{4} = \frac{x}{3}$$

it is of the form of (5) with x and z interchanged. The surface is therefore a hyperbolic paraboloid. The cross sections in the planes $x = k \neq 0$ are the hyperbolas $y^2/16 - z^2/4 = k/3$. The cross section in the yz plane ($x = 0$) consists of the two lines $y = 2z$ and $y = -2z$. In the planes $z = k$ the cross sections are the parabolas $3y^2 = 16x + 12k^2$; in the planes $y = k$ the cross sections are the parabolas $12z^2 = 3k^2 - 16x$. Figure 17.8.9 shows a sketch of the hyperbolic paraboloid.

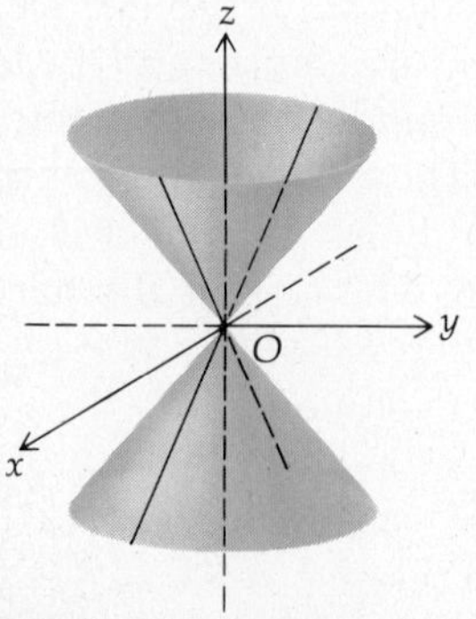

FIGURE 17.8.10

The elliptic cone

$$\frac{x^2}{a^2} + \frac{y^2}{b^2} - \frac{z^2}{c^2} = 0 \tag{6}$$

where a, b, and c are positive (see Fig. 17.8.10).

The intersection of the plane $z = 0$ with the surface is a single point, the origin. The cross sections of the surface in the planes $z = k$, where $k \neq 0$, are ellipses, and the lengths of the semiaxes increase as k increases.

Cross sections in the planes $x = 0$ and $y = 0$ are pairs of intersecting lines. In the planes $x = k$ and $y = k$, where $k \neq 0$, the cross sections are hyperbolas.

Example 5 Draw a sketch of the graph of the equation $4x^2 - y^2 + 25z^2 = 0$, and name the surface.

Solution The given equation can be written as

$$\frac{x^2}{25} - \frac{y^2}{100} + \frac{z^2}{4} = 0$$

which is of the form of (6) with y and z interchanged. Therefore the surface is an elliptic cone having the y axis as its axis. The surface intersects the xz plane ($y = 0$) at the origin only. The intersection of the surface with the yz plane ($x = 0$) is the pair of intersecting lines $y = \pm 5z$, and the intersection with the xy plane ($z = 0$) is the pair of intersecting lines $y = \pm 2x$. The cross sections in the planes $y = k \neq 0$ are the ellipses $x^2/25 + z^2/4 = k^2/100$. In the planes $x = k \neq 0$ and $z = k \neq 0$ the cross sections are, respectively, the hyperbolas $y^2/100 - z^2/4 = k^2/25$ and $y^2/100 - x^2/25 = k^2/4$. A sketch of the surface is shown in Fig. 17.8.11.

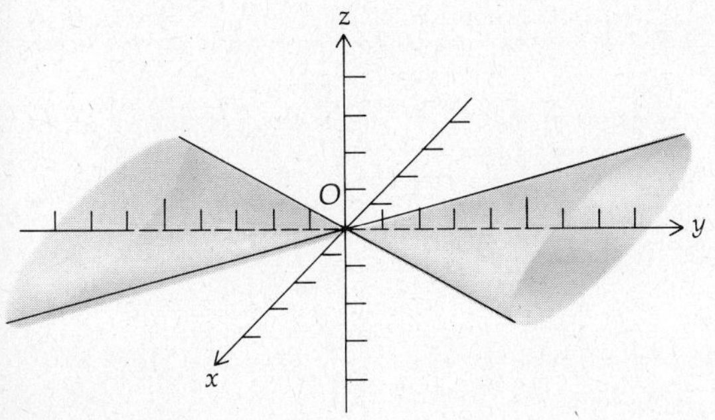

FIGURE 17.8.11

The general equation of the second degree in x, y, and z is of the form

$$ax^2 + by^2 + cz^2 + dxy + exz + fyz + gx + hy + iz + j = 0$$

where $a, b, \ldots, j$ are constants. It can be shown that by translation and rotation of the three-dimensional coordinate axes (the study of which is beyond the scope of this book) this equation can be reduced to one of the following two forms:

$$Ax^2 + By^2 + Cz^2 + J = 0 \qquad (7)$$

or

$$Ax^2 + By^2 + Iz = 0 \qquad (8)$$

Graphs of the equations of the second degree will either be one of the above six types of quadrics or else will degenerate into a cylinder, plane, line, point, or the empty set.

The nondegenerate curves associated with equations of the form (7) are the central quadrics and the elliptic cone, whereas those associated with equations of the form (8) are the noncentral quadrics. Following are examples of some degenerate cases:

$x^2 - y^2 = 0$; two planes, $x - y = 0$ and $x + y = 0$

$z^2 = 0$; one plane, the xy plane

$x^2 + y^2 = 0$; one line, the z axis

$x^2 + y^2 + z^2 = 0$; a single point, the origin

$x^2 + y^2 + z^2 + 1 = 0$; the empty set

EXERCISES 17.8

In Exercises 1 through 6, name the surface having the given equation.

1. $9x^2 - 4y^2 + 36z^2 = 36$ **2.** $4x^2 - 16y^2 + 9z^2 = 0$ **3.** $5x^2 - 2z^2 - 3y = 0$

4. $25x^2 = 4y^2 + z^2 + 100$ **5.** $4y^2 - 25x^2 = 100$ **6.** $3y^2 + 7z^2 = 6x$

In Exercises 7 through 18, draw a sketch of the graph of the given equation and name the surface.

7. $4x^2 + 9y^2 + z^2 = 36$ **8.** $4x^2 - 9y^2 - z^2 = 36$ **9.** $4x^2 + 9y^2 - z^2 = 36$ **10.** $4x^2 - 9y^2 + z^2 = 36$

11. $x^2 = y^2 - z^2$ **12.** $x^2 = y^2 + z^2$ **13.** $\dfrac{x^2}{36} + \dfrac{z^2}{25} = 4y$ **14.** $\dfrac{y^2}{25} + \dfrac{x^2}{36} = 4$

15. $\dfrac{x^2}{36} - \dfrac{z^2}{25} = 9y$ **16.** $x^2 = 2y + 4z$ **17.** $x^2 + 16z^2 = 4y^2 - 16$ **18.** $9y^2 - 4z^2 + 18x = 0$

19. Find the values of k for which the intersection of the plane $x + ky = 1$ and the elliptic hyperboloid of two sheets $y^2 - x^2 - z^2 = 1$ is (a) an ellipse and (b) a hyperbola.

20. Find the vertex and focus of the parabola that is the intersection of the plane $y = 2$ with the hyperbolic paraboloid $y^2/16 - x^2/4 = z/9$.

21. Find the vertex and focus of the parabola that is the intersection of the plane $x = 1$ with the hyperbolic paraboloid $z^2/4 - x^2/9 = y/3$.

22. Find the area of the plane section formed by the intersection of the plane $y = 3$ with the solid bounded by the ellipsoid $x^2/9 + y^2/25 + z^2/4 = 1$.

23. Show that the intersection of the surface $x^2 - 4y^2 - 9z^2 = 36$ and the plane $x + z = 9$ is an ellipse.

24. Show that the intersection of the hyperbolic paraboloid $y^2/b^2 - x^2/a^2 = z/c$ and the plane $z = bx + ay$ consists of two intersecting straight lines.

In Exercises 25 through 27, use the method of parallel plane sections (Sec. 7.4) to find the volume of the given solid. The measure of the area of the region enclosed by the ellipse having semiaxes a and b is πab.

25. The solid bounded by the ellipsoid $36x^2 + 9y^2 + 4z^2 = 36$.

26. The solid bounded by the ellipsoid $x^2/a^2 + y^2/b^2 + z^2/c^2 = 1$.

27. The solid bounded by the plane $z = h$, where $h > 0$, and the elliptic paraboloid $x^2/a^2 + y^2/b^2 = z/c$, where $c > 0$.

17.9 CURVES IN R^3

We consider vector-valued functions in three-dimensional space.

17.9.1 Definition Let f_1, f_2, and f_3 be three real-valued functions of a real variable t. Then for every number t in the domain common to f_1, f_2, and f_3 there is a vector **R** defined by

$$\mathbf{R}(t) = f_1(t)\mathbf{i} + f_2(t)\mathbf{j} + f_3(t)\mathbf{k} \qquad (1)$$

and **R** is called a **vector-valued function.**

The graph of a vector-valued function in three-dimensional space is obtained analogously to the way the graph of a vector-valued function in two dimensions was obtained in Sec. 16.4. As t assumes all values in the domain of **R**, the terminal point of the position representation of the vector $\mathbf{R}(t)$ traces a curve C, and this curve is called the graph of (1). A point on the curve C has the cartesian representation (x, y, z), where

$$x = f_1(t) \qquad y = f_2(t) \qquad z = f_3(t) \qquad (2)$$

Equations (2) are called **parametric equations** of C, whereas Eq. (1) is called a **vector equation** of C. By eliminating t from Eqs. (2) we obtain two equations in x, y, and z. These equations are called **cartesian equations** of C. Each cartesian equation is an equation of a surface, and curve C is the intersection of the two surfaces. The equations of any two surfaces containing C may be taken as the cartesian equations defining C.

TABLE 17.9.1

t	x	y	z
0	a	0	0
$\frac{\pi}{4}$	$\frac{a}{\sqrt{2}}$	$\frac{b}{\sqrt{2}}$	$\frac{\pi}{4}$
$\frac{\pi}{2}$	0	b	$\frac{\pi}{2}$
$\frac{3\pi}{4}$	$-\frac{a}{\sqrt{2}}$	$\frac{b}{\sqrt{2}}$	$\frac{3\pi}{4}$
π	$-a$	0	π
$\frac{3\pi}{2}$	0	$-b$	$\frac{3\pi}{2}$

● ILLUSTRATION 1 We draw a sketch of the curve having the vector equation

$$\mathbf{R}(t) = a\cos t\mathbf{i} + b\sin t\mathbf{j} + t\mathbf{k}$$

Parametric equations of the given curve are

$$x = a\cos t \qquad y = b\sin t \qquad z = t$$

To eliminate t from the first two equations we write them as

$$x^2 = a^2\cos^2 t \quad \text{and} \quad y^2 = b^2\sin^2 t$$

from which we get

$$\frac{x^2}{a^2} = \cos^2 t \quad \text{and} \quad \frac{y^2}{b^2} = \sin^2 t$$

Adding corresponding members of these two equations we obtain

$$\frac{x^2}{a^2} + \frac{y^2}{b^2} = 1$$

Therefore the curve lies entirely on the elliptical cylinder whose directrix is an ellipse in the xy plane and whose rulings are parallel to the z axis. Table 17.9.1 gives sets of values of x, y, and z for specific values of t. A sketch of the curve is shown in Fig. 17.9.1. ●

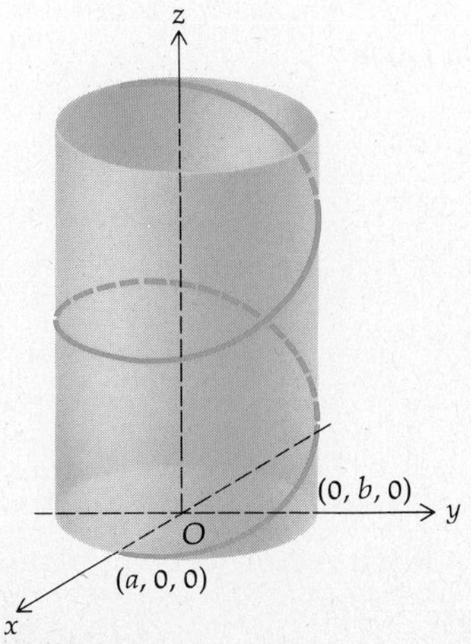

FIGURE 17.9.1

The curve of Illustration 1 is called a *helix*. If $a = b$, the helix is a *circular helix* and it lies on the right-circular cylinder $x^2 + y^2 = a^2$.

● ILLUSTRATION 2 The curve having the vector equation

$$\mathbf{R}(t) = t\mathbf{i} + t^2\mathbf{j} + t^3\mathbf{k}$$

is called a *twisted cubic*. Parametric equations of the twisted cubic are

$$x = t \qquad y = t^2 \qquad z = t^3$$

Eliminating t from the first two of these equations yields $y = x^2$, which is a cylinder whose directrix in the xy plane is a parabola. The twisted cubic lies on this cylinder. Figure 17.9.2 shows a sketch of the cylinder and the portion of the twisted cubic from $t = 0$ to $t = 2$. ●

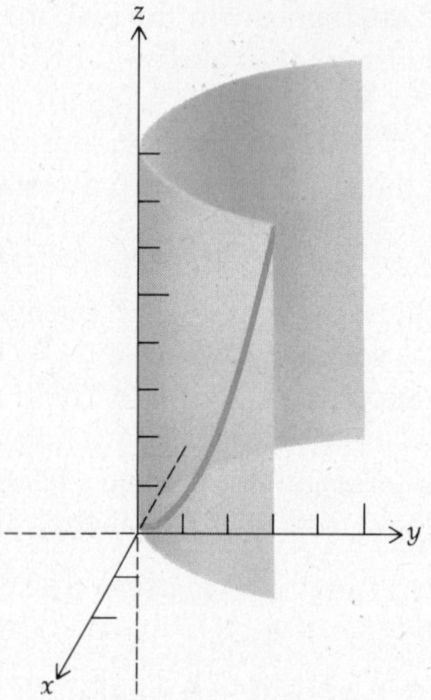

FIGURE 17.9.2

Many of the definitions and theorems pertaining to vector-valued functions in two dimensions can be extended to vector-valued functions in three dimensions.

17.9.2 Definition If $\mathbf{R}(t) = f_1(t)\mathbf{i} + f_2(t)\mathbf{j} + f_3(t)\mathbf{k}$, then

$$\lim_{t \to t_1} \mathbf{R}(t) = \lim_{t \to t_1} f_1(t)\mathbf{i} + \lim_{t \to t_1} f_2(t)\mathbf{j} + \lim_{t \to t_1} f_3(t)\mathbf{k}$$

if $\lim_{t \to t_1} f_1(t)$, $\lim_{t \to t_1} f_2(t)$, and $\lim_{t \to t_1} f_3(t)$ all exist.

17.9.3 Definition The vector-valued function $\mathbf{R}$ is continuous at t_1 if and only if

(i) $\mathbf{R}(t_1)$ exists;
(ii) $\lim_{t \to t_1} \mathbf{R}(t)$ exists;
(iii) $\lim_{t \to t_1} \mathbf{R}(t) = \mathbf{R}(t_1)$.

17.9.4 Definition The derivative of the vector-valued function $\mathbf{R}$ is a vector-valued function, denoted by $\mathbf{R}'$ and defined by

$$\mathbf{R}'(t) = \lim_{\Delta t \to 0} \frac{\mathbf{R}(t + \Delta t) - \mathbf{R}(t)}{\Delta t}$$

if this limit exists.

17.9.5 Theorem If $\mathbf{R}$ is the vector-valued function defined by

$$\mathbf{R}(t) = f_1(t)\mathbf{i} + f_2(t)\mathbf{j} + f_3(t)\mathbf{k}$$

and $\mathbf{R}'(t)$ exists, then

$$\mathbf{R}'(t) = f_1'(t)\mathbf{i} + f_2'(t)\mathbf{j} + f_3'(t)\mathbf{k}$$

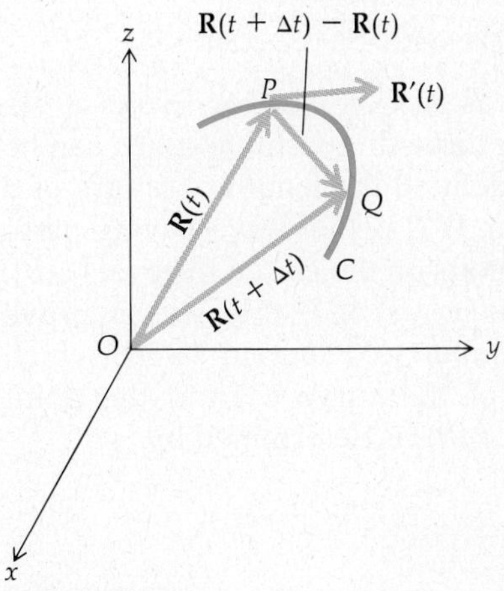

FIGURE 17.9.3

The proof of Theorem 17.9.5 is left as an exercise (see Exercise 13).

The geometric interpretation for the derivative of $\mathbf{R}$ is the same as that for the derivative of a vector-valued function in R^2. Figure 17.9.3 shows a portion of the curve C, which is the graph of $\mathbf{R}$. In the figure $\overrightarrow{OP}$ is the position representation of $\mathbf{R}(t)$, $\overrightarrow{OQ}$ is the position representation of $\mathbf{R}(t + \Delta t)$, and so $\overrightarrow{PQ}$ is a representation of the vector $\mathbf{R}(t + \Delta t) - \mathbf{R}(t)$. As Δt approaches zero, the vector $[\mathbf{R}(t + \Delta t) - \mathbf{R}(t)]/\Delta t$ has a representation approaching a directed line segment tangent to the curve C at P.

The definition of the unit tangent vector is analogous to Definition 16.8.1 for vectors in the plane. So if $\mathbf{T}(t)$ denotes the unit tangent vector to curve C having vector equation (1), then

$$\mathbf{T}(t) = \frac{D_t\mathbf{R}(t)}{|D_t\mathbf{R}(t)|} \qquad (3)$$

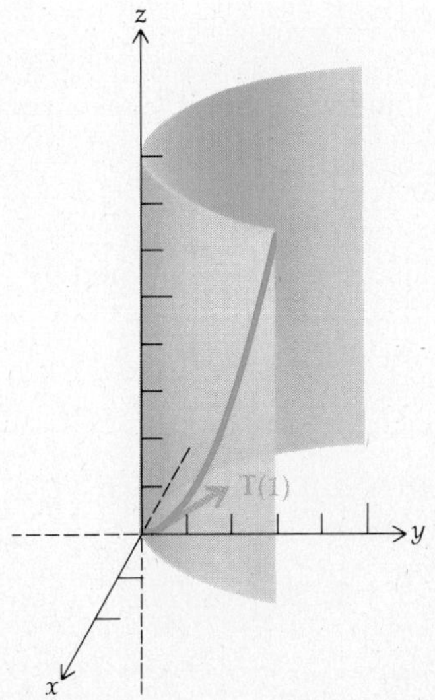

FIGURE 17.9.4

● ILLUSTRATION 3 We find the unit tangent vector for the twisted cubic of Illustration 2.

Because $\mathbf{R}(t) = t\mathbf{i} + t^2\mathbf{j} + t^3\mathbf{k}$,

$$D_t\mathbf{R}(t) = \mathbf{i} + 2t\mathbf{j} + 3t^2\mathbf{k}$$

$$|D_t\mathbf{R}(t)| = \sqrt{1 + 4t^2 + 9t^4}$$

From (3), then,

$$\mathbf{T}(t) = \frac{1}{\sqrt{1 + 4t^2 + 9t^4}}(\mathbf{i} + 2t\mathbf{j} + 3t^2\mathbf{k})$$

Therefore, in particular,

$$\mathbf{T}(1) = \frac{1}{\sqrt{14}}\mathbf{i} + \frac{2}{\sqrt{14}}\mathbf{j} + \frac{3}{\sqrt{14}}\mathbf{k}$$

Figure 17.9.4 shows the representation of $\mathbf{T}(1)$ at the point (1, 1, 1). ●

Theorems 16.5.6, 16.5.7, and 16.5.8 regarding derivatives of sums and products of two-dimensional vector-valued functions also hold for vectors in three dimensions. The following theorem regarding the derivative of the cross product of two vector-valued functions is similar to the corresponding formula for the derivative of the product of real-valued functions; however, it is important to maintain the correct order of the vector-valued functions because the cross product is not commutative.

17.9.6 Theorem If **R** and **Q** are vector-valued functions, then

$$D_t[\mathbf{R}(t) \times \mathbf{Q}(t)] = \mathbf{R}(t) \times \mathbf{Q}'(t) + \mathbf{R}'(t) \times \mathbf{Q}(t)$$

for all values of t for which $\mathbf{R}'(t)$ and $\mathbf{Q}'(t)$ exist.

The proof of Theorem 17.9.6 is left as an exercise (see Exercise 14).

The length of an arc of a curve C in three-dimensional space can be defined in exactly the same way as we defined the length of an arc of a curve in the plane (see Definition 16.6.1). If C is the curve having parametric equations (2), f_1', f_2', f_3' are continuous on the closed interval $[a, b]$, and no two values of t give the same point (x, y, z) on C, then we can prove (as we did for the plane) a theorem similar to Theorem 16.6.3, which states that the length of arc, L units, of the curve C from the point $(f_1(a), f_2(a), f_3(a))$ to the point $(f_1(b), f_2(b), f_3(b))$ is determined by

$$L = \int_a^b \sqrt{[f_1'(t)]^2 + [f_2'(t)]^2 + [f_3'(t)]^2}\, dt \tag{4}$$

If s is the measure of the length of arc of C from the fixed point $(f_1(t_0), f_2(t_0), f_3(t_0))$ to the variable point $(f_1(t), f_2(t), f_3(t))$ and s increases as t increases, then s is a function of t and is given by

$$s = \int_{t_0}^t \sqrt{[f_1'(u)]^2 + [f_2'(u)]^2 + [f_3'(u)]^2}\, du \tag{5}$$

As shown in Sec. 16.6 for plane curves, we can show that if (1) is a vector equation of C, then

$$D_t s = |D_t\mathbf{R}(t)| \tag{6}$$

and the length of arc, L units, given by (4), also can be determined by

$$L = \int_a^b |D_t\mathbf{R}(t)|\, dt \tag{7}$$

Example 1 Given the circular helix $\mathbf{R}(t) = a \cos t\mathbf{i} + a \sin t\mathbf{j} + t\mathbf{k}$, where $a > 0$, find the length of arc from $t = 0$ to $t = 2\pi$.

Solution $D_t\mathbf{R}(t) = -a \sin t\mathbf{i} + a \cos t\mathbf{j} + \mathbf{k}$. So from (7),

$$L = \int_0^{2\pi} \sqrt{(-a \sin t)^2 + (a \cos t)^2 + 1}\, dt$$

$$= \int_0^{2\pi} \sqrt{a^2 + 1}\, dt = 2\pi\sqrt{a^2 + 1}$$

Thus the length of arc is $2\pi\sqrt{a^2 + 1}$ units.

The definitions of the *curvature vector* $\mathbf{K}(t)$ and the *curvature* $K(t)$ at a point P on a curve C in R^3 are the same as for plane curves given in Definition 16.9.1. Hence, if $\mathbf{T}(t)$ is the unit tangent vector to C at P and s is the measure of the arc length from an arbitrarily chosen point on C to P, where s increases as t increases, then

$$\mathbf{K}(t) = D_s\mathbf{T}(t)$$

or, equivalently,

$$\mathbf{K}(t) = \frac{D_t\mathbf{T}(t)}{|D_t\mathbf{R}(t)|} \tag{8}$$

and

$$K(t) = |D_s\mathbf{T}(t)|$$

or, equivalently,

$$K(t) = \left|\frac{D_t\mathbf{T}(t)}{|D_t\mathbf{R}(t)|}\right| \tag{9}$$

Taking the dot product of $\mathbf{K}(t)$ and $\mathbf{T}(t)$ and using (8) we get

$$\mathbf{K}(t)\cdot\mathbf{T}(t) = \frac{D_t\mathbf{T}(t)}{|D_t\mathbf{R}(t)|}\cdot\mathbf{T}(t) = \frac{1}{|D_t\mathbf{R}(t)|}D_t\mathbf{T}(t)\cdot\mathbf{T}(t) \tag{10}$$

Theorem 16.5.11 states that if a vector-valued function in a plane has a constant magnitude, it is orthogonal to its derivative. This theorem and its proof also hold for vectors in three dimensions. Therefore, because $|\mathbf{T}(t)| = 1$, we can conclude from (10) that $\mathbf{K}(t)\cdot\mathbf{T}(t) = 0$. And so the curvature vector and the unit tangent vector of a curve at a point are orthogonal.

The *unit normal vector* is defined as the unit vector having the same direction as the curvature vector, provided that the curvature vector is not the zero vector. So if $\mathbf{N}(t)$ denotes the unit normal vector to a curve C at a point P, then if $\mathbf{K}(t) \neq \mathbf{0}$,

$$\mathbf{N}(t) = \frac{\mathbf{K}(t)}{|\mathbf{K}(t)|} \tag{11}$$

From (11) and the previous discussion it follows that the unit normal vector and the unit tangent vector are orthogonal. Thus the angle between these two vectors has a radian measure of $\frac{1}{2}\pi$, and from Theorem 17.6.6,

$$|\mathbf{T}(t) \times \mathbf{N}(t)| = |\mathbf{T}(t)||\mathbf{N}(t)| \sin \tfrac{1}{2}\pi = (1)(1)(1) = 1$$

Therefore the cross product of $\mathbf{T}(t)$ and $\mathbf{N}(t)$ is a unit vector. By Theorem 17.6.9, $\mathbf{T}(t) \times \mathbf{N}(t)$ is orthogonal to both $\mathbf{T}(t)$ and $\mathbf{N}(t)$; hence the vector $\mathbf{B}(t)$, defined by

$$\mathbf{B}(t) = \mathbf{T}(t) \times \mathbf{N}(t) \tag{12}$$

is a unit vector orthogonal to $\mathbf{T}(t)$ **and** $\mathbf{N}(t)$ and is called the *unit binormal vector* to the curve C at P.

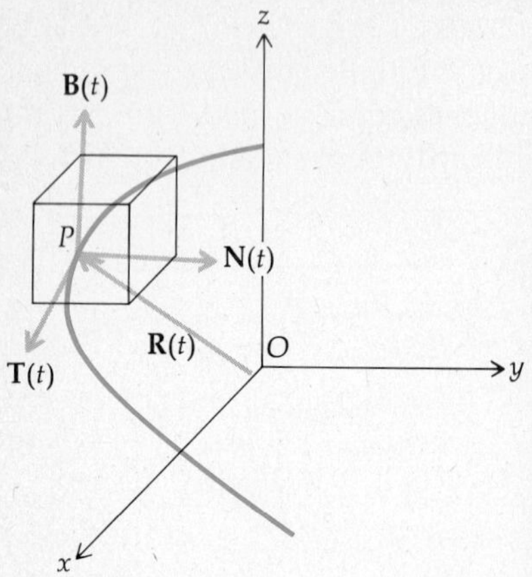

FIGURE 17.9.5

The three mutually orthogonal unit vectors, $\mathbf{T}(t)$, $\mathbf{N}(t)$, and $\mathbf{B}(t)$, of a curve C are called the **moving trihedral** of C (see Fig. 17.9.5).

Example 2 Find the moving trihedral and the curvature at any point of the circular helix of Example 1.

Solution A vector equation of the circular helix is

$$\mathbf{R}(t) = a \cos t\mathbf{i} + a \sin t\mathbf{j} + t\mathbf{k}$$

So $D_t\mathbf{R}(t) = -a \sin t\mathbf{i} + a \cos t\mathbf{j} + \mathbf{k}$ and $|D_t\mathbf{R}(t)| = \sqrt{a^2 + 1}$. From (3),

$$\mathbf{T}(t) = \frac{1}{\sqrt{a^2 + 1}}(-a \sin t\mathbf{i} + a \cos t\mathbf{j} + \mathbf{k})$$

So

$$D_t\mathbf{T}(t) = \frac{1}{\sqrt{a^2 + 1}}(-a \cos t\mathbf{i} - a \sin t\mathbf{j})$$

Applying (8) we obtain

$$\mathbf{K}(t) = \frac{1}{a^2 + 1}(-a \cos t\mathbf{i} - a \sin t\mathbf{j})$$

The curvature, then, is given by

$$K(t) = |\mathbf{K}(t)| = \frac{a}{a^2 + 1}$$

so the curvature of the circular helix is constant. From (11),

$$\mathbf{N}(t) = -\cos t\mathbf{i} - \sin t\mathbf{j}$$

Applying (12) we have

$$\begin{aligned}\mathbf{B}(t) &= \frac{1}{\sqrt{a^2 + 1}}(-a \sin t\mathbf{i} + a \cos t\mathbf{j} + \mathbf{k}) \times (-\cos t\mathbf{i} - \sin t\mathbf{j}) \\ &= \frac{1}{\sqrt{a^2 + 1}}(\sin t\mathbf{i} - \cos t\mathbf{j} + a\mathbf{k})\end{aligned}$$

A thorough study of curves and surfaces by means of calculus forms the subject of *differential geometry.* The use of the calculus of vectors further enhances this subject. The previous discussion has been but a short introduction.

We now consider briefly the motion of a particle along a curve in three-dimensional space. If the parameter t in the vector equation (1) measures time, then the position at t of a particle moving along the curve C, having vector equation (1), is the point $P(f_1(t), f_2(t), f_3(t))$. The *velocity vector,* $\mathbf{V}(t)$, and the *acceleration vector,* $\mathbf{A}(t)$, are defined as in the plane. The vector $\mathbf{R}(t)$ is called the *position vector,* and

$$\mathbf{V}(t) = D_t\mathbf{R}(t) \tag{13}$$

and

$$\mathbf{A}(t) = D_t\mathbf{V}(t) = D_t^{\,2}\mathbf{R}(t) \tag{14}$$

The *speed* of the particle at t is the magnitude of the velocity vector. By applying (6) we can write

$$|\mathbf{V}(t)| = D_t s$$

Example 3 A particle is moving along the curve having parametric equations $x = 3t$, $y = t^2$, and $z = \frac{2}{3}t^3$. Find the velocity and acceleration vectors and the speed of the particle at $t = 1$. Draw a sketch of a portion of the curve at $t = 1$, and draw representations of the velocity and acceleration vectors there.

Solution A vector equation of the curve is

$$\mathbf{R}(t) = 3t\mathbf{i} + t^2\mathbf{j} + \tfrac{2}{3}t^3\mathbf{k}$$

Therefore

$$\mathbf{V}(t) = D_t\mathbf{R}(t) = 3\mathbf{i} + 2t\mathbf{j} + 2t^2\mathbf{k}$$

and

$$\mathbf{A}(t) = D_t\mathbf{V}(t) = 2\mathbf{j} + 4t\mathbf{k}$$

Also,

$$|\mathbf{V}(t)| = \sqrt{9 + 4t^2 + 4t^4}$$

So when $t = 1$, $\mathbf{V} = 3\mathbf{i} + 2\mathbf{j} + 2\mathbf{k}$, $\mathbf{A} = 2\mathbf{j} + 4\mathbf{k}$, and $|\mathbf{V}(1)| = \sqrt{17}$. The required sketch is shown in Fig. 17.9.6.

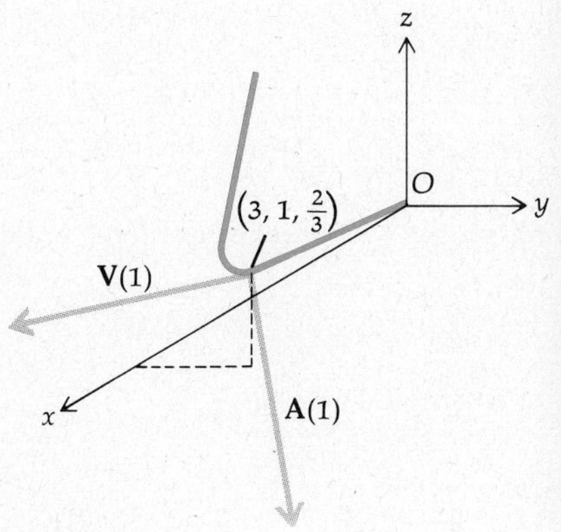

FIGURE 17.9.6

EXERCISES 17.9

In Exercises 1 through 5, find the unit tangent vector for the curve having the given vector equation.

1. $\mathbf{R}(t) = (t + 1)\mathbf{i} - t^2\mathbf{j} + (1 - 2t)\mathbf{k}$ **2.** $\mathbf{R}(t) = \sin 2t\mathbf{i} + \cos 2t\mathbf{j} + 2t^{3/2}\mathbf{k}$ **3.** $\mathbf{R}(t) = e^t \cos t\mathbf{i} + e^t \sin t\mathbf{j} + e^t\mathbf{k}$

4. $\mathbf{R}(t) = t^2\mathbf{i} + (t + \frac{1}{3}t^3)\mathbf{j} + (t - \frac{1}{3}t^3)\mathbf{k}$ **5.** $\mathbf{R}(t) = 2t \cos t\mathbf{i} + 5t\mathbf{j} + 2t \sin t\mathbf{k}$

6. Find the unit tangent vector for the curve having the vector equation $\mathbf{R}(t) = 4 \cosh 2t\mathbf{i} + 4 \sinh 2t\mathbf{j} + 6t\mathbf{k}$ at the point where $t = 0$.

In Exercises 7 through 11, find the length of arc of the curve from t_1 to t_2.

7. The curve of Exercise 1; $t_1 = -1$; $t_2 = 2$. **8.** The curve of Exercise 2; $t_1 = 0$; $t_2 = 1$.

9. The curve of Exercise 3; $t_1 = 0$; $t_2 = 3$. **10.** The curve of Exercise 4; $t_1 = 0$; $t_2 = 1$.

11. $\mathbf{R}(t) = 4t^{3/2}\mathbf{i} - 3 \sin t\mathbf{j} + 3 \cos t\mathbf{k}$; $t_1 = 0$; $t_2 = 2$.

12. Prove that the unit tangent vector of the circular helix of Example 1 makes an angle of constant radian measure with the unit vector $\mathbf{k}$.

13. Prove Theorem 17.9.5. **14.** Prove Theorem 17.9.6.

15. Write a vector equation of the curve of intersection of the surfaces $y = e^x$ and $z = xy$.

16. Write a vector equation for the curve of intersection of the surfaces $x = \ln(z^2 + 2)$ and $y = xz^3$.

17. Find the cosine of the angle between the vector $\mathbf{j}$ and the unit tangent vector to the curve $\mathbf{R}(t) = \cos 2t\mathbf{i} - 3t\mathbf{j} + 2\sin 2t\mathbf{k}$ at the point where $t = \pi$.

18. Find the curvature at the point where $t = 1$ of the curve $\mathbf{R}(t) = t^2\mathbf{i} + (4 + t)\mathbf{j} + (3 - 2t)\mathbf{k}$.

19. Find the moving trihedral and the curvature at the point where $t = 1$ of the twisted cubic of Illustration 2.

20. Find the moving trihedral and the curvature at any point of the curve $\mathbf{R}(t) = \cosh t\mathbf{i} + \sinh t\mathbf{j} + t\mathbf{k}$.

In Exercises 21 through 24, find the moving trihedral and the curvature of the given curve at $t = t_1$, if they exist.

21. The curve of Exercise 1; $t_1 = -1$.

22. The curve of Exercise 2; $t_1 = 0$.

23. The curve of Exercise 3; $t_1 = 0$.

24. The curve of Exercise 4; $t_1 = 1$.

In Exercises 25 through 28, a particle is moving along the given curve. Find the velocity vector, the acceleration vector, and the speed at $t = t_1$. Draw a sketch of a portion of the curve at $t = t_1$, and draw the velocity and acceleration vectors there.

25. The circular helix of Example 1; $t_1 = \frac{1}{2}\pi$.

26. $x = t, y = \frac{1}{2}t^2, z = \frac{1}{3}t^3$; $t_1 = 2$.

27. $x = e^{2t}, y = e^{-2t}, z = te^{2t}$; $t_1 = 1$.

28. $x = 1/2(t^2 + 1), y = \ln(1 + t^2), z = \tan^{-1} t$; $t_1 = 1$.

29. Prove that if $\mathbf{R}(t) = f_1(t)\mathbf{i} + f_2(t)\mathbf{j} + f_3(t)\mathbf{k}$ is a vector equation of curve C, and $K(t)$ is the curvature of C, then

$$K(t) = \frac{|D_t\mathbf{R}(t) \times D_t^2\mathbf{R}(t)|}{|D_t\mathbf{R}(t)|^3}$$

30. Use the formula of Exercise 29 to show that the curvature of the circular helix of Example 1 is $a/(a^2 + 1)$.

In Exercises 31 and 32, find the curvature of the given curve at the indicated point.

31. $x = t, y = t^2, z = t^3$; the origin.

32. $x = e^t, y = e^{-t}, z = t$; $t = 0$.

33. Prove that if $\mathbf{R}(t) = f_1(t)\mathbf{i} + f_2(t)\mathbf{j} + f_3(t)\mathbf{k}$ is a vector equation of curve C, $K(t)$ is the curvature of C at a point P, and s units is the arc length measured from an arbitrarily chosen point on C to P, then

$$D_s\mathbf{R}(t) \cdot D_s^3\mathbf{R}(t) = -[K(t)]^2$$

34. A particle is moving along a curve with vector equation $\mathbf{R}(t) = \tan t\mathbf{i} + \sinh 2t\mathbf{j} + \text{sech}\, t\mathbf{k}$. Prove that the velocity vector and the acceleration vector are orthogonal at $t = 0$.

35. Prove that if the speed of a moving particle is constant, its acceleration vector is always orthogonal to its velocity vector.

36. Prove that for the twisted cubic of Illustration 2, if $t \neq 0$, no two of the vectors $\mathbf{R}(t)$, $\mathbf{V}(t)$, and $\mathbf{A}(t)$ are orthogonal.

17.10 CYLINDRICAL AND SPHERICAL COORDINATES

Cylindrical and *spherical coordinates* are generalizations of polar coordinates to three-dimensional space. The cylindrical coordinate representation of a point P is (r, θ, z), where r and θ are the polar coordinates of the projection of P on a polar plane and z is the directed distance from this polar plane to P. See Fig. 17.10.1.

Example 1 Draw a sketch of the graph of each of the following equations where c is a constant: (a) $r = c$; (b) $\theta = c$; (c) $z = c$.

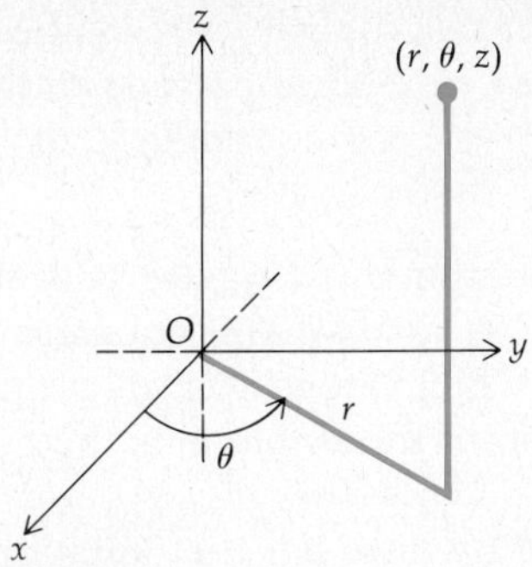

FIGURE 17.10.1

Solution

(a) For a point $P(r, \theta, z)$ on the graph of $r = c$, θ and z can have any values and r is a constant. The graph is a right-circular cylinder having radius $|c|$ and the z axis as its axis. A sketch of the graph is shown in Fig. 17.10.2.

(b) For all points $P(r, \theta, z)$ on the graph of $\theta = c$, r and z can assume any value while θ remains constant. The graph is a plane through the z axis. See Fig. 17.10.3 for a sketch of the graph.

(c) The graph of $z = c$ is a plane parallel to the polar plane at a directed distance of c units from it. Figure 17.10.4 shows a sketch of the graph.

The name "cylindrical coordinates" comes from the fact that the graph of $r = c$ is a right-circular cylinder as in Example 1(a). Cylindrical coordinates are often used in a physical problem when there is an axis of symmetry.

Suppose that a cartesian coordinate system and a cylindrical coordinate system are placed so the xy plane is the polar plane of the cylindrical coordinate system and the positive side of the x axis is the polar axis as shown in Fig. 17.10.5. Then the point P has (x, y, z) and (r, θ, z) as two sets of coordinates that are related by the equations

$$x = r\cos\theta \qquad y = r\sin\theta \qquad z = z \tag{1}$$

and

$$r^2 = x^2 + y^2 \qquad \tan\theta = \frac{y}{x} \quad \text{if } x \neq 0 \qquad z = z \tag{2}$$

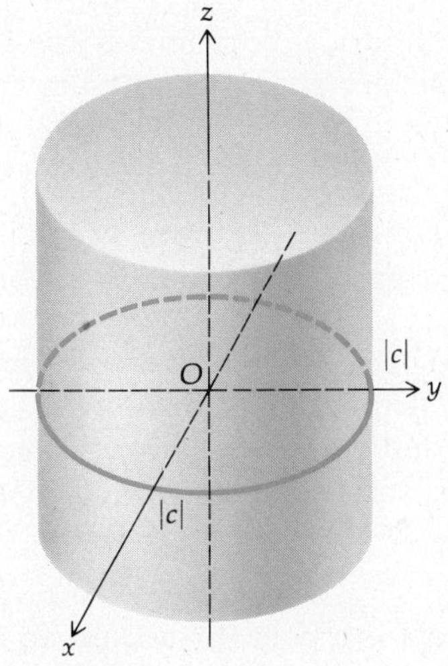

FIGURE 17.10.2

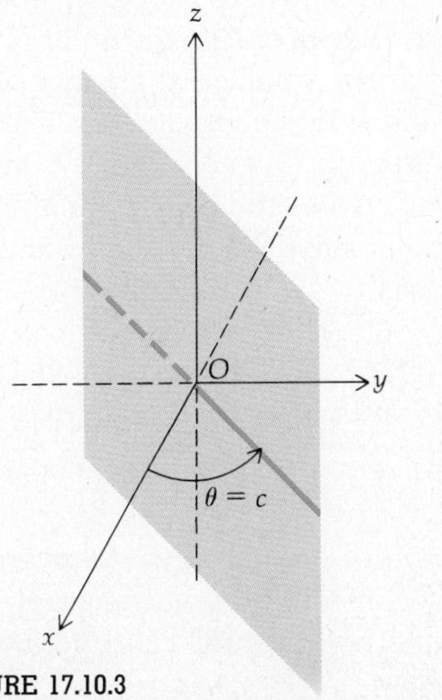

FIGURE 17.10.3

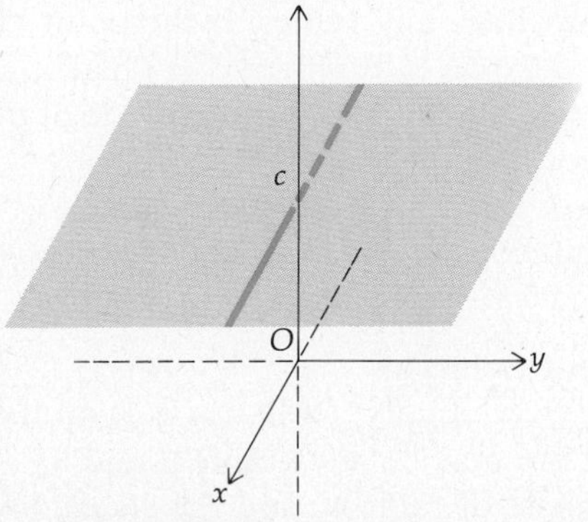

FIGURE 17.10.4

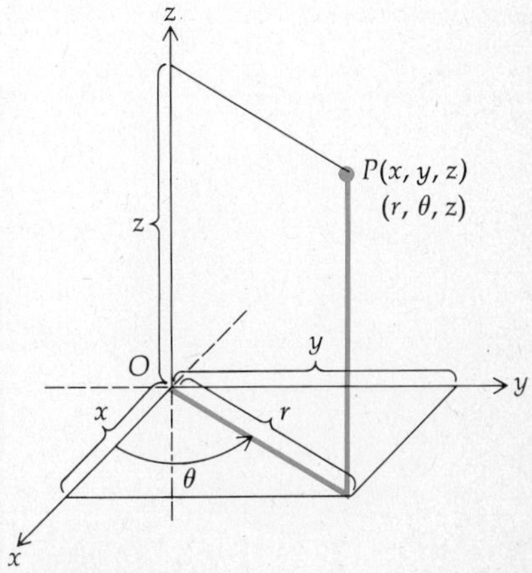

FIGURE 17.10.5

Example 2 Find an equation in cartesian coordinates of the following surfaces whose equations are given in cylindrical coordinates, and identify the surface: (a) $r = 6 \sin \theta$; (b) $r(3 \cos \theta + 2 \sin \theta) + 6z = 0$.

Solution

(a) Multiplying on both sides of the equation by r we get $r^2 = 6r \sin \theta$. Because $r^2 = x^2 + y^2$ and $r \sin \theta = y$, then $x^2 + y^2 = 6y$. This equation can be written in the form $x^2 + (y - 3)^2 = 9$, which shows that its graph is a right-circular cylinder whose cross section in the xy plane is the circle with its center at (0, 3) and radius 3.

(b) Replacing $r \cos \theta$ by x and $r \sin \theta$ by y we get the equation $3x + 2y + 6z = 0$. Hence the graph is a plane through the origin and has $\langle 3, 2, 6 \rangle$ as a normal vector.

Example 3 Find an equation in cylindrical coordinates for each of the following surfaces whose equations are given in cartesian coordinates, and identify the surface: (a) $x^2 + y^2 = z$; (b) $x^2 - y^2 = z$.

Solution

(a) The equation is similar to Eq. (4) of Sec. 17.8; so the graph is an elliptic paraboloid. If $x^2 + y^2$ is replaced by r^2, the equation becomes $r^2 = z$.

(b) The equation is similar to Eq. (5) of Sec. 17.8 with x and y interchanged. The graph is therefore a hyperbolic paraboloid having the z axis as its axis. When x is replaced by $r \cos \theta$ and y is replaced by $r \sin \theta$, the equation becomes $r^2 \cos^2 \theta - r^2 \sin^2 \theta = z$; and because $\cos^2 \theta - \sin^2 \theta = \cos 2\theta$, we can write this as $z = r^2 \cos 2\theta$.

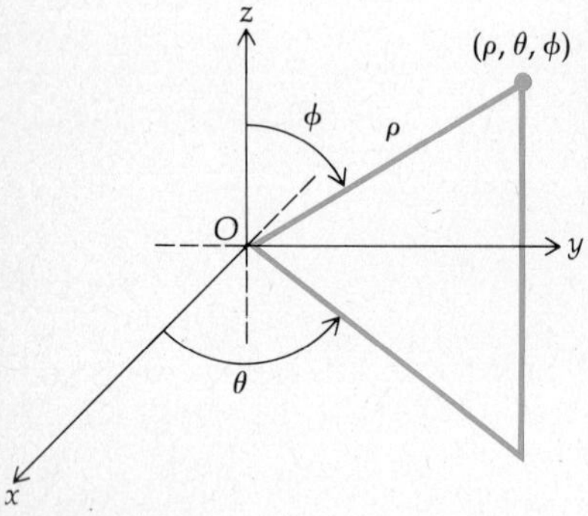

FIGURE 17.10.6

In a spherical coordinate system there is a polar plane and an axis perpendicular to the polar plane, with the origin of the z axis at the pole of the polar plane. A point is located by three numbers, and the spherical coordinate representation of a point P is (ρ, θ, ϕ), where $\rho = |\overline{OP}|$, θ is the radian measure of the polar angle of the projection of P on the polar plane, and ϕ is the nonnegative radian measure of the smallest angle measured from the positive side of the z axis to the line OP. See Fig. 17.10.6. The origin has the spherical coordinate representation $(0, \theta, \phi)$, where θ and ϕ may have any values. If the point $P(\rho, \theta, \phi)$ is not the origin, then $\rho > 0$ and $0 \le \phi \le \pi$, where $\phi = 0$ if P is on the positive side of the z axis and $\phi = \pi$ if P is on the negative side of the z axis.

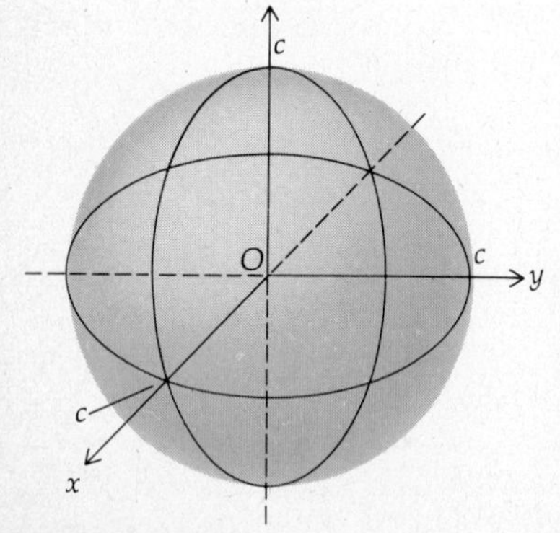

FIGURE 17.10.7

Example 4 Draw a sketch of the graph of each of the following equations where c is a constant: (a) $\rho = c$, and $c > 0$; (b) $\theta = c$; (c) $\phi = c$, and $0 < c < \pi$.

Solution

(a) Every point $P(\rho, \theta, \phi)$ on the graph of $\rho = c$ has the same value of ρ, θ may be any number, and $0 \le \phi \le \pi$. It follows that the graph is a sphere of radius c and has its center at the pole. Figure 17.10.7 shows a sketch of the sphere.

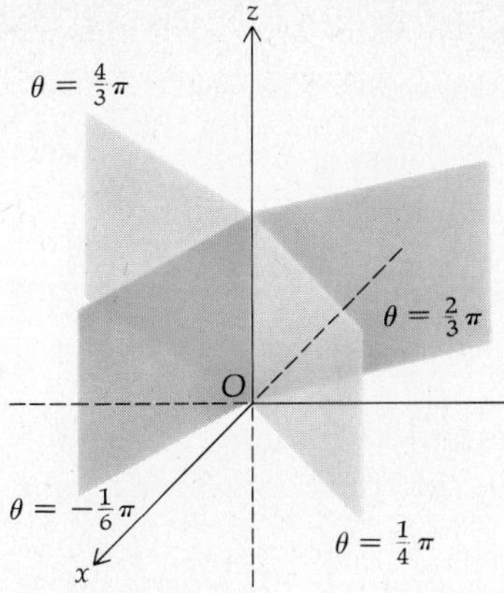

FIGURE 17.10.8

(b) For any point $P(\rho, \theta, \phi)$ on the graph of $\theta = c$, ρ may be any nonnegative number, ϕ may be any number in the closed interval $[0, \pi]$, and θ is constant. The graph is a half plane containing the z axis and is obtained by rotating about the z axis through an angle of c radians that half of the xz plane for which $x \geq 0$. Figure 17.10.8 shows sketches of the half planes for $\theta = \frac{1}{4}\pi$, $\theta = \frac{2}{3}\pi$, $\theta = \frac{4}{3}\pi$, and $\theta = -\frac{1}{6}\pi$.

(c) The graph of $\phi = c$ contains all the points $P(\rho, \theta, \phi)$ for which ρ is any nonnegative number, θ is any number, and ϕ is the constant c. The graph is half of a cone having its vertex at the origin and the z axis as its axis. Figure 17.10.9(a) and (b) each show a sketch of the half cone for $0 < c < \frac{1}{2}\pi$ and $\frac{1}{2}\pi < c < \pi$, respectively.

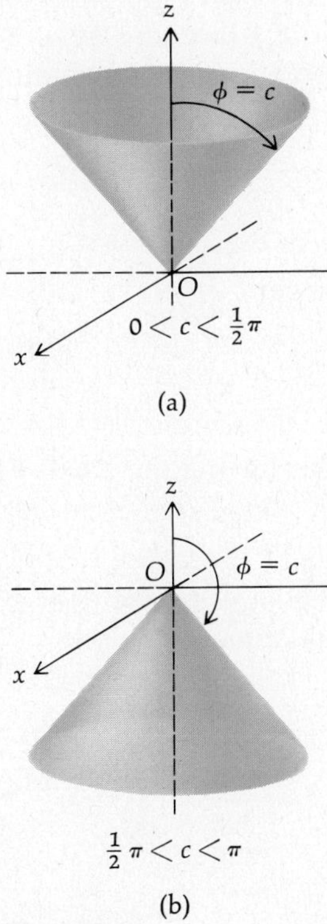

FIGURE 17.10.9

Because the graph of $\rho = c$ is a sphere as seen in Example 4(a), we have the name "spherical coordinates." In a physical problem when there is a point that is a center of symmetry, spherical coordinates are often used.

By placing a spherical coordinate system and a cartesian coordinate system together as shown in Fig. 17.10.10, we obtain relationships between the spherical coordinates and the cartesian coordinates of a point P from

$$x = |\overline{OQ}| \cos\theta \qquad y = |\overline{OQ}| \sin\theta \qquad z = |\overline{QP}|$$

Because $|\overline{OQ}| = \rho \sin\phi$ and $|\overline{QP}| = \rho \cos\phi$, these equations become

$$x = \rho \sin\phi \cos\theta \qquad y = \rho \sin\phi \sin\theta \qquad z = \rho \cos\phi \tag{3}$$

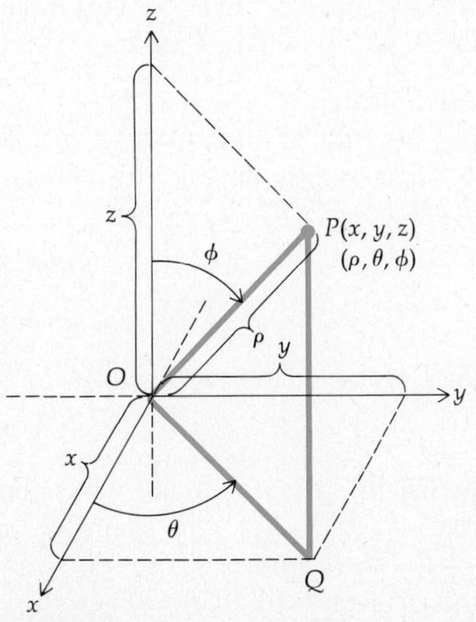

FIGURE 17.10.10

By squaring each of the equations in (3) and adding,

$$x^2 + y^2 + z^2 = \rho^2 \sin^2 \phi \cos^2 \theta + \rho^2 \sin^2 \phi \sin^2 \theta + \rho^2 \cos^2 \phi$$

$$x^2 + y^2 + z^2 = \rho^2 \sin^2 \phi(\cos^2 \theta + \sin^2 \theta) + \rho^2 \cos^2 \phi$$

$$x^2 + y^2 + z^2 = \rho^2(\sin^2 \phi + \cos^2 \phi)$$

$$x^2 + y^2 + z^2 = \rho^2$$

Example 5 Find an equation in cartesian coordinates of the following surfaces whose equations are given in spherical coordinates, and identify the surface: (a) $\rho \cos \phi = 4$; (b) $\rho \sin \phi = 4$.

Solution

(a) Because $z = \rho \cos \phi$, the equation becomes $z = 4$. Hence the graph is a plane parallel to the xy plane and 4 units above it.

(b) For spherical coordinates $\rho \geq 0$ and $\sin \phi \geq 0$ (because $0 \leq \phi \leq \pi$); therefore, by squaring on both sides of the given equation we obtain the equivalent equation $\rho^2 \sin^2 \phi = 16$, which in turn is equivalent to

$$\rho^2(1 - \cos^2 \phi) = 16$$

$$\rho^2 - \rho^2 \cos^2 \phi = 16$$

Replacing ρ^2 by $x^2 + y^2 + z^2$ and $\rho \cos \phi$ by z we get

$$x^2 + y^2 + z^2 - z^2 = 16$$

$$x^2 + y^2 = 16$$

Therefore the graph is the right-circular cylinder having the z axis as its axis and radius 4.

Example 6 Find an equation in spherical coordinates for (a) the elliptic paraboloid of Example 3(a); (b) the plane of Example 2(b).

Solution

(a) A cartesian equation of the elliptic paraboloid of Example 3(a) is $x^2 + y^2 = z$. Replacing x by $\rho \sin \phi \cos \theta$, y by $\rho \sin \phi \sin \theta$, and z by $\rho \cos \phi$, we get

$$\rho^2 \sin^2 \phi \cos^2 \theta + \rho^2 \sin^2 \phi \sin^2 \theta = \rho \cos \phi$$

$$\rho^2 \sin^2 \phi(\cos^2 \theta + \sin^2 \theta) = \rho \cos \phi$$

which is equivalent to the two equations

$$\rho = 0 \quad \text{and} \quad \rho \sin^2 \phi = \cos \phi$$

The origin is the only point whose coordinates satisfy $\rho = 0$. Because the origin $(0, \theta, \frac{1}{2}\pi)$ lies on $\rho \sin^2 \phi = \cos \phi$, we can disregard the equation $\rho = 0$. Furthermore, $\sin \phi \neq 0$ because there is no value of ϕ for which both $\sin \phi$ and $\cos \phi$ are 0. Therefore the equation $\rho \sin^2 \phi = \cos \phi$ can be written as $\rho = \csc^2 \phi \cos \phi$, or, equivalently, $\rho = \csc \phi \cot \phi$.

(b) A cartesian equation of the plane of Example 2(b) is $3x + 2y + 6z = 0$. By using Eqs. (3) this equation becomes

$$3\rho \sin \phi \cos \theta + 2\rho \sin \phi \sin \theta + 6\rho \cos \phi = 0$$

EXERCISES 17.10

1. Find the cartesian coordinates of the point having the given cylindrical coordinates: (a) $(3, \frac{1}{2}\pi, 5)$; (b) $(7, \frac{2}{3}\pi, -4)$; (c) $(1, 1, 1)$.

2. Find a set of cylindrical coordinates of the point having the given cartesian coordinates: (a) $(4, 4, -2)$; (b) $(-3\sqrt{3}, 3, 6)$; (c) $(1, 1, 1)$.

3. Find the cartesian coordinates of the point having the given spherical coordinates: (a) $(4, \frac{1}{6}\pi, \frac{1}{4}\pi)$; (b) $(4, \frac{1}{2}\pi, \frac{1}{3}\pi)$; (c) $(\sqrt{6}, \frac{1}{3}\pi, \frac{3}{4}\pi)$.

4. Find a set of spherical coordinates of the point having the given cartesian coordinates: (a) $(1, -1, -\sqrt{2})$; (b) $(-1, \sqrt{3}, 2)$; (c) $(2, 2, 2)$.

5. Find a set of cylindrical coordinates of the point having the given spherical coordinates: (a) $(4, \frac{2}{3}\pi, \frac{5}{6}\pi)$; (b) $(\sqrt{2}, \frac{3}{4}\pi, \pi)$; (c) $(2\sqrt{3}, \frac{1}{3}\pi, \frac{1}{4}\pi)$.

6. Find a set of spherical coordinates of the point having the given cylindrical coordinates: (a) $(3, \frac{1}{6}\pi, 3)$; (b) $(3, \frac{1}{2}\pi, 2)$; (c) $(2, \frac{5}{6}\pi, -4)$.

In Exercises 7 through 12, find an equation in cylindrical coordinates of the given surface, and identify the surface.

7. $x^2 + y^2 + 4z^2 = 16$ 8. $x^2 - y^2 = 9$ 9. $x^2 + y^2 = 3z$

10. $9x^2 + 4y^2 = 36$ 11. $x^2 - y^2 = 3z^2$ 12. $x^2 + y^2 = z^2$

In Exercises 13 through 17, find an equation in spherical coordinates of the given surface, and identify the surface.

13. $x^2 + y^2 + z^2 - 9z = 0$ 14. $x^2 + y^2 = z^2$ 15. $x^2 + y^2 = 9$ 16. $x^2 + y^2 = 2z$ 17. $x^2 + y^2 + z^2 - 8x = 0$

In Exercises 18 through 22, find an equation in cartesian coordinates for the surface whose equation is given in cylindrical coordinates. In Exercises 18 and 19, identify the surface.

18. $r = 3 \cos \theta$ 19. (a) $r = 4$; (b) $\theta = \frac{1}{4}\pi$ 20. $r = 3 + 2 \cos \theta$ 21. $r^2 \cos 2\theta = z^3$ 22. $z^2 \sin^3 \theta = r^3$

In Exercises 23 through 27, find an equation in cartesian coordinates for the surface whose equation is given in spherical coordinates. In Exercise 23 through 25, identify the surface.

23. (a) $\rho = 9$; (b) $\theta = \frac{1}{4}\pi$; (c) $\phi = \frac{1}{4}\pi$ 24. $\rho = 9 \sec \phi$ 25. $\rho = 6 \csc \phi$

26. $\rho = 3 \cos \phi$ 27. $\rho = 2 \tan \theta$ 28. $\rho = 6 \sin \phi \sin \theta + 3 \cos \phi$

29. A curve C in R^3 has the following parametric equations in cylindrical coordinates:

$$r = F_1(t) \qquad \theta = F_2(t) \qquad z = F_3(t)$$

Use formula (4) of Sec. 17.9 and formulas (1) of this section to prove that if L units is the length of arc of C from the point where $t = a$ to the point where $t = b$, then

$$L = \int_a^b \sqrt{(D_t r)^2 + r^2(D_t \theta)^2 + (D_t z)^2}\, dt$$

30. A curve C in R^3 has the following parametric equations in spherical coordinates:

$$\rho = G_1(t) \qquad \theta = G_2(t) \qquad \phi = G_3(t)$$

Use formula (4) of Sec. 17.9 and formulas (3) of this section to prove that if L units is the length of arc of C from the point where $t = a$ to the point where $t = b$, then

$$L = \int_a^b \sqrt{(D_t\rho)^2 + \rho^2 \sin^2 \phi (D_t\theta)^2 + \rho^2(D_t\phi)^2}\, dt$$

31. (a) Show that parametric equations for the circular helix of Example 1, Sec. 17.9, are

$$r = a \qquad \theta = t \qquad z = t$$

(b) Use the formula of Exercise 29 to find the length of arc of the circular helix of part (a) from $t = 0$ to $t = 2\pi$. Check your result with that of Example 1, Sec. 17.9.

32. A *conical helix* winds around a cone in a way similar to that in which a circular helix winds around a cylinder. Use the formula of Exercise 30 to find the length of arc from $t = 0$ to $t = 2\pi$ of the conical helix having parametric equations

$$\rho = t \qquad \theta = t \qquad \phi = \tfrac{1}{4}\pi$$

REVIEW EXERCISES FOR CHAPTER 17

1. Draw a sketch of the graph of $x = 3$ in R^1, R^2, and R^3.

2. Draw a sketch of the set of points satisfying the simultaneous equations $x = 6$ and $y = 3$ in R^2 and R^3.

In Exercises 3 through 11, describe in words the set of points in R^3 satisfying the given equation or the given pair of equations. Draw a sketch of the graph.

3. $\begin{cases} y = 0 \\ z = 0 \end{cases}$ **4.** $\begin{cases} x = z \\ y = z \end{cases}$ **5.** $\begin{cases} x^2 + z^2 = 4 \\ y = 0 \end{cases}$ **6.** $y^2 - z^2 = 0$ **7.** $x = y$

8. $x^2 + z^2 = 4$ **9.** $x^2 + y^2 = 9z$ **10.** $x^2 + y^2 = z^2$ **11.** $x^2 - y^2 = z^2$

In Exercises 12 through 17, let $\mathbf{A} = -\mathbf{i} + 3\mathbf{j} + 2\mathbf{k}$, $\mathbf{B} = 2\mathbf{i} + \mathbf{j} - 4\mathbf{k}$, $\mathbf{C} = \mathbf{i} + 2\mathbf{j} - 2\mathbf{k}$, $\mathbf{D} = 3\mathbf{j} - \mathbf{k}$, $\mathbf{E} = 5\mathbf{i} - 2\mathbf{j}$, *and find the indicated vector or scalar.*

12. $3\mathbf{A} - 2\mathbf{B} + \mathbf{C}$ **13.** $6\mathbf{C} + 4\mathbf{D} - \mathbf{E}$ **14.** $2\mathbf{B}\cdot\mathbf{C} + 3\mathbf{D}\cdot\mathbf{E}$

15. $\mathbf{D}\cdot\mathbf{B} \times \mathbf{C}$ **16.** $(\mathbf{A} \times \mathbf{C}) - (\mathbf{D} \times \mathbf{E})$ **17.** $|\mathbf{A} \times \mathbf{B}||\mathbf{D} \times \mathbf{E}|$

In Exercises 18 through 23, there is only one way that a meaningful expression can be obtained by inserting parentheses. Insert the parentheses and find the indicated vector or scalar if $\mathbf{A} = \langle 3, -2, 4\rangle$, $\mathbf{B} = \langle -5, 7, 2\rangle$, *and* $\mathbf{C} = \langle 4, 6, -1\rangle$.

18. $\mathbf{A}\cdot\mathbf{B}\,\mathbf{C}$ **19.** $\mathbf{A}\,\mathbf{B}\cdot\mathbf{C}$ **20.** $\mathbf{B}\cdot\mathbf{A} - \mathbf{C}$ **21.** $\mathbf{A} + \mathbf{B}\cdot\mathbf{C}$ **22.** $\mathbf{A} \times \mathbf{B}\cdot\mathbf{C} \times \mathbf{A}$ **23.** $\mathbf{A} \times \mathbf{B}\cdot\mathbf{A} + \mathbf{B} - \mathbf{C}$

24. If $\mathbf{A}$ is any vector, prove that $\mathbf{A} = (\mathbf{A}\cdot\mathbf{i})\mathbf{i} + (\mathbf{A}\cdot\mathbf{j})\mathbf{j} + (\mathbf{A}\cdot\mathbf{k})\mathbf{k}$.

25. A line is drawn through the point $(-3, 5, 1)$ perpendicular to the xz plane. Find the coordinates of the points on this line at a distance of 13 units from $(-2, 0, 0)$.

26. Find an equation of the sphere having as a diameter the line segment with endpoints $(3, 5, -4)$ and $(-1, 7, 4)$.

27. Find an equation of the sphere concentric with the sphere $x^2 + y^2 + z^2 + 4x + 2y - 6z + 10 = 0$ and containing the point $(-4, 2, 5)$.

28. Prove that the points $(4, 1, -1)$, $(2, 0, 1)$, and $(4, 3, 0)$ are the vertices of a right triangle, and find the area of the triangle.

29. Find a generating curve and the axis for the surface of revolution having the equation $x^2 + z^2 = e^{4y}$.

30. Find an equation of the surface of revolution generated by revolving the ellipse $9x^2 + 4z^2 = 36$ in the xz plane about the x axis. Draw a sketch of the surface.

31. Determine the value of c such that the vectors $3\mathbf{i} + c\mathbf{j} - 3\mathbf{k}$ and $5\mathbf{i} - 4\mathbf{j} + \mathbf{k}$ are orthogonal.

32. Show that there are representations of the three vectors $\mathbf{A} = 5\mathbf{i} + \mathbf{j} - 3\mathbf{k}$, $\mathbf{B} = \mathbf{i} + 3\mathbf{j} - 2\mathbf{k}$, and $\mathbf{C} = -4\mathbf{i} + 2\mathbf{j} + \mathbf{k}$ which form a triangle.

33. Given the points $A(5, 9, -3)$ and $B(-2, 4, -5)$, find (a) the direction cosines of $\mathbf{V}(\overrightarrow{AB})$ and (b) the unit vector having the same direction as $\mathbf{V}(\overrightarrow{AB})$.

34. If $\mathbf{A} = \mathbf{i} + \mathbf{j} - \mathbf{k}$, $\mathbf{B} = 2\mathbf{i} - \mathbf{j} + \mathbf{k}$, $\mathbf{C} = 3\mathbf{i} - 2\mathbf{j} + 4\mathbf{k}$, and $\mathbf{D} = 5\mathbf{i} + 6\mathbf{j} - 8\mathbf{k}$, find scalars a, b, and c such that $a\mathbf{A} + b\mathbf{B} + c\mathbf{C} = \mathbf{D}$.

35. If $\mathbf{A} = \langle 7, -1, 5\rangle$ and $\mathbf{B} = \langle -2, 3, 1\rangle$, find (a) the component of $\mathbf{B}$ in the direction of $\mathbf{A}$ and (b) the vector projection of $\mathbf{B}$ onto $\mathbf{A}$.

36. Find an equation of the plane containing the points $(1, 7, -3)$ and $(3, 1, 2)$ and which does not intersect the x axis.

37. Find an equation of the plane through the three points $(-1, 2, 1)$, $(1, 4, 0)$, and $(1, -1, 3)$ by two methods: (a) using the cross product; (b) without using the cross product.

38. Find an equation of the plane that contains the point $(7, -2, -5)$ and is perpendicular to the line through the points $(-3, 0, 4)$ and $(3, 2, 1)$.

39. Find the distance from the origin to the plane through the point $(-6, 3, -2)$ and having $5\mathbf{i} - 3\mathbf{j} + 4\mathbf{k}$ as a normal vector.

40. Find two unit vectors orthogonal to $\mathbf{i} - 3\mathbf{j} + 4\mathbf{k}$ and whose representations are parallel to the yz plane.

41. Find the distance from the point $P(4, 6, -4)$ to the line through the two points $A(2, 2, 1)$ and $B(4, 3, -1)$.

42. Find the distance from the plane $9x - 2y + 6z + 44 = 0$ to the point $(-3, 2, 0)$.

43. If θ is the radian measure of the angle between the vectors $\mathbf{A} = 2\mathbf{i} + \mathbf{j} + \mathbf{k}$ and $\mathbf{B} = 4\mathbf{i} - 3\mathbf{j} + 5\mathbf{k}$, find $\cos\theta$ in two ways: (a) by using the dot product; (b) by using the cross product and a trigonometric identity.

44. Prove that the lines

$$\frac{x-1}{1} = \frac{y+2}{2} = \frac{z-2}{2} \quad \text{and} \quad \frac{x-2}{2} = \frac{y-5}{3} = \frac{z-5}{1}$$

are skew lines, and find the distance between them.

45. Find symmetric and parametric equations of the line through the origin and perpendicular to each of the lines of Exercise 44.

46. Find symmetric and parametric equations of the line through the two points $(-3, 5, 2)$ and $(-1, -3, 4)$.

47. Show that the lines

$$\frac{x-2}{3} = \frac{y+3}{-1} = \frac{z-5}{4} \quad \text{and} \quad \frac{x+1}{-6} = \frac{y+2}{2} = \frac{z-1}{-8}$$

are coincident.

48. Find an equation of the plane containing the line $\frac{1}{2}(x - 3) = -(y + 5) = -\frac{1}{3}(z + 2)$ and the point $(5, 0, -4)$.

49. Find the area of the cross section of the ellipsoid $x^2/4 + y^2/9 + z^2/25 = 1$ in the plane $z = 4$.

50. Find the area of the parallelogram two of whose sides are the position representations of the vectors $2\mathbf{j} - 3\mathbf{k}$ and $5\mathbf{i} + 4\mathbf{k}$.

51. Find the volume of the parallelepiped having vertices at $(1, 3, 0)$, $(2, -1, 3)$, $(-2, 2, -1)$, and $(-1, 1, 2)$.

52. Find the length of the arc of the curve

$$x = t \cos t \qquad y = t \sin t \qquad z = t$$

from $t = 0$ to $t = \frac{1}{2}\pi$.

53. Find the length of the arc of the curve $\mathbf{R}(t) = (2 - 3t)\mathbf{i} + (4t - 1)\mathbf{j} + t^2\mathbf{k}$ from $t = 0$ to $t = \frac{5}{2}$.

54. A particle is moving along the curve with parametric equations

$$x = \ln(t^2 + 1) \qquad y = t^2 + 1 \qquad z = \tan^{-1} t$$

Find the velocity vector, the acceleration vector, and the speed at $t = 0$.

55. A particle is moving along the curve of Exercise 52. Find the velocity vector, the acceleration vector, and the speed at $t = \frac{1}{2}\pi$. Draw a sketch of a portion of the curve at $t = \frac{1}{2}\pi$, and draw the representations of the velocity and acceleration vectors there.

56. Find the unit tangent vector and the curvature at any point on the curve having the vector equation $\mathbf{R}(t) = e^t\mathbf{i} + e^{-t}\mathbf{j} + 2t\mathbf{k}$.

57. Find the moving trihedral of the curve $\mathbf{R}(t) = 3 \sin 2t\mathbf{i} - 4t\mathbf{j} - 3 \cos 2t\mathbf{k}$ at the point where $t = \frac{1}{2}\pi$.

58. Find a set of spherical coordinates for the point having cartesian coordinates $(-3, \sqrt{3}, 2)$.

59. Find a set of cylindrical coordinates for the point having spherical coordinates $(3, \pi, \frac{1}{3}\pi)$.

60. Find an equation in spherical coordinates of the graph of each of the equations: (a) $x^2 + y^2 + 4z^2 = 4$; (b) $4x^2 - 4y^2 + 9z^2 = 36$.

61. Find an equation in cylindical coordinates of the graph of each of the equations: (a) $(x + y)^2 + 1 = z$; (b) $25x^2 + 4y^2 = 100$.

62. If **R**, **Q**, and **W** are three vector-valued functions whose derivatives with respect to t exist, prove that

$$D_t[\mathbf{R}(t) \cdot \mathbf{Q}(t) \times \mathbf{W}(t)] = D_t\mathbf{R}(t) \cdot \mathbf{Q}(t) \times \mathbf{W}(t) + \mathbf{R}(t) \cdot D_t\mathbf{Q}(t) \times \mathbf{W}(t) + \mathbf{R}(t) \cdot \mathbf{Q}(t) \times D_t\mathbf{W}(t)$$

CHAPTER 18 Differential Calculus of Functions of Several Variables

18.1 FUNCTIONS OF MORE THAN ONE VARIABLE

We now extend the concept of a function to functions of n variables, and in succeeding sections we extend to functions of n variables the concepts of the *limit* of a function, *continuity* of a function, and *derivative* of a function. A thorough treament of these topics belongs to a course in advanced calculus. In this book most of our discussion of functions of more than one variable is confined to those of two and three variables; however, we make the definitions for functions of n variables and then show the applications of these definitions to functions of two and three variables. We also show that when each of these definitions is applied to a function of one variable, we have the definition previously given.

To extend the concept of a function to functions of any number of variables, we must first consider points in n-dimensional number space. Just as we denoted a point in R^1 by a real number x, a point in R^2 by an ordered pair of real numbers (x, y), and a point in R^3 by an ordered triple of real numbers (x, y, z), a point in n-dimensional number space, R^n, is represented by an ordered n-tuple of real numbers customarily denoted by $P = (x_1, x_2, \ldots, x_n)$. In particular, if $n = 1$, let $P = x$; if $n = 2$, $P = (x, y)$; if $n = 3$, $P = (x, y, z)$; if $n = 6$, $P = (x_1, x_2, x_3, x_4, x_5, x_6)$.

18.1.1 Definition The set of all ordered n-tuples of real numbers is called the ***n*-dimensional number space** and is denoted by R^n. Each ordered n-tuple $(x_1, x_2, \ldots, x_n)$ is called a **point** in the n-dimensional number space.

18.1.2 Definition A **function of *n* variables** is a set of ordered pairs of the form (P, w) in which no two distinct ordered pairs have the same first element. P is a point in n-dimensional number space and w is a real number. The set of all possible values of P is called the **domain** of the function, and the set of all possible values of w is called the **range** of the function.

From this definition, the domain of a function of n variables is a set of points in R^n and the range is a set of real numbers or, equivalently, a set of points in R^1. When $n = 1$, we have a function of one variable; thus the domain is a set of points in R^1 or, equivalently, a set of real numbers, and the range is a set of real numbers. Hence Definition 1.5.1 is a special case of Definition 18.1.2. If $n = 2$, we have a function of two variables, and the domain is a set of points in R^2 or, equivalently, a set of ordered pairs of real numbers (x, y). The range is a set of real numbers.

● ILLUSTRATION 1 Let the function f of two variables x and y be the set of all ordered pairs of the form (P, z) such that

$$z = \sqrt{25 - x^2 - y^2}$$

The domain of f is the set of all ordered pairs (x, y) for which $25 - x^2 - y^2 \geq 0$. This is the set of all points in the xy plane on the circle $x^2 + y^2 = 25$ and in the interior region bounded by the circle.

Because $z = \sqrt{25 - (x^2 + y^2)}$, then $0 \leq z \leq 5$; therefore the range of f is the set of all real numbers in the closed interval $[0, 5]$. In Fig. 18.1.1 there is a sketch showing as a shaded region in R^2 the set of points in the domain of f. ●

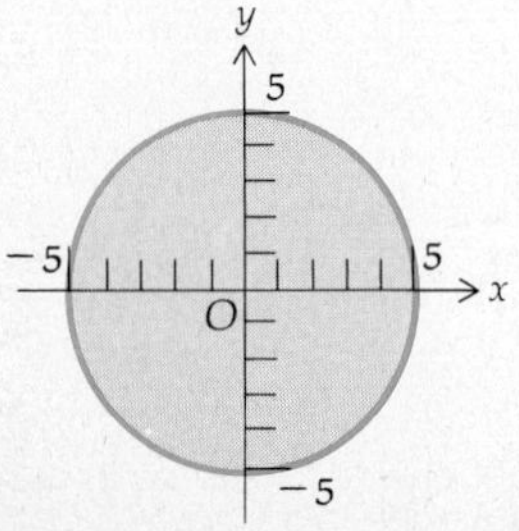

FIGURE 18.1.1

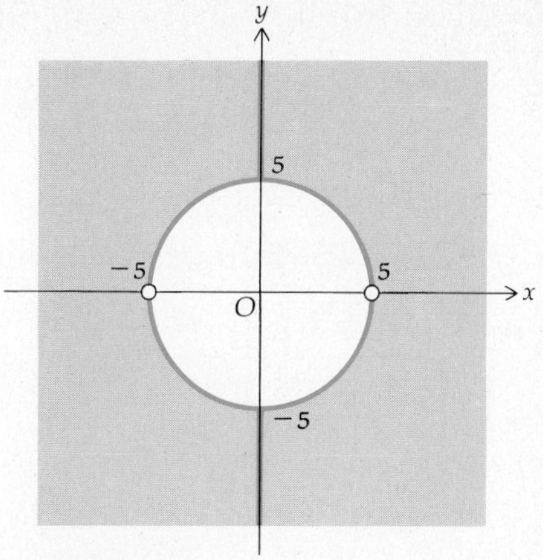

FIGURE 18.1.2

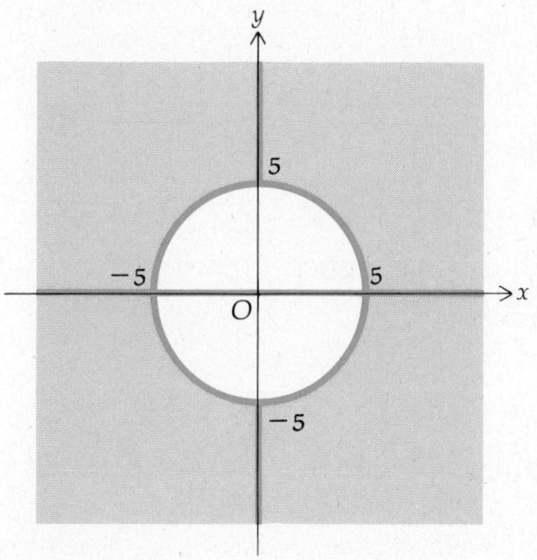

FIGURE 18.1.3

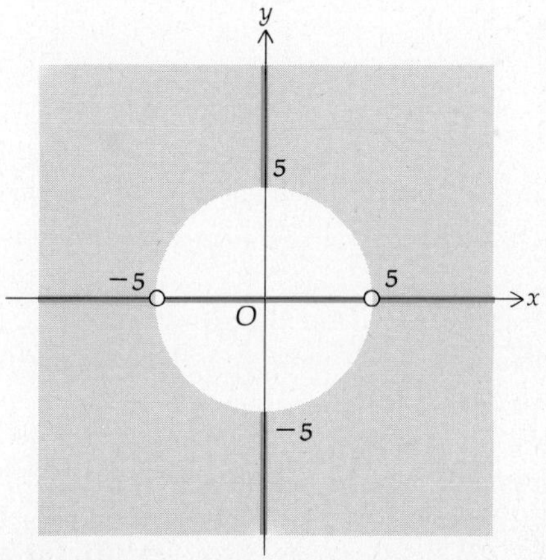

FIGURE 18.1.4

● ILLUSTRATION 2 The function g of two variables x and y is the set of all ordered pairs of the form (P, z) such that

$$z = \frac{\sqrt{x^2 + y^2 - 25}}{y}$$

The domain of g is the set $\{(x, y) | x^2 + y^2 \geq 25 \text{ and } y \neq 0\}$. This is the set of points, not on the x axis, that are either on the circle $x^2 + y^2 = 25$ or in the exterior region bounded by the circle. Figure 18.1.2 is a sketch showing as a shaded region in R^2 the set of points in the domain of g. ●

● ILLUSTRATION 3 The function f of two variables x and y is the set of all ordered pairs of the form (P, z) such that

$$z = y\sqrt{x^2 + y^2 - 25}$$

If $y = 0$, then $z = 0$ regardless of the value of x. However, if $y \neq 0$, then $x^2 + y^2 - 25$ must be nonnegative in order for z to be defined. Therefore the domain of F consists of all ordered pairs (x, y) for which either $y = 0$ or $x^2 + y^2 - 25 \geq 0$. This is the set of all points on the circle $x^2 + y^2 = 25$, all points in the exterior region bounded by the circle, and all points on the x axis for which $-5 < x < 5$. In Fig. 18.1.3 we have a sketch showing as a shaded region in R^2 the set of points in the domain of F. ●

● ILLUSTRATION 4 The function G of two variables x and y is the set of all ordered pairs of the form (P, z) such that

$$z = \frac{y}{\sqrt{x^2 + y^2 - 25}}$$

If $y = 0$, then $z = 0$ provided that $x^2 + y^2 - 25 \neq 0$. If $y \neq 0$, then $x^2 + y^2 - 25$ must be positive in order for z to be defined. Hence the domain of G consists of all ordered pairs (x, y) for which $x^2 + y^2 - 25 > 0$ and those for which $y = 0$ and $x \neq \pm 5$. These are all the points in the exterior region bounded by the circle $x^2 + y^2 = 25$ and the points on the x axis for which $-5 < x < 5$. Figure 18.1.4 is a sketch showing as a shaded region in R^2 the set of points in the domain of G. ●

If f is a function of n variables, then according to Definition 18.1.2, f is a set of ordered pairs of the form (P, w), where $P = (x_1, x_2, \ldots, x_n)$ is a point in R^n and w is a real number. The particular value of w that corresponds to a point P is denoted by the symbol $f(P)$ or $f(x_1, x_2, \ldots, x_n)$. In particular, if $n = 2$ and $P = (x, y)$, we can represent the function value by either $f(P)$ or $f(x, y)$. Similarly, if $n = 3$ and $P = (x, y, z)$, we denote the function value by either $f(P)$ or $f(x, y, z)$. Note that if $n = 1$, $P = x$; hence if f is a function of one variable, $f(P) = f(x)$. Therefore this notation is consistent with the notation for function values of one variable.

A function f of n variables can be defined by the equation

$$w = f(x_1, x_2, \ldots, x_n)$$

The variables $x_1, x_2, \ldots, x_n$ are called the *independent variables,* and w is called the *dependent variable.*

● ILLUSTRATION 5 Let f be the function of Illustration 1; that is,

$$f(x, y) = \sqrt{25 - x^2 - y^2}$$

Then

$$f(3, -4) = \sqrt{25 - (3)^2 - (-4)^2} = \sqrt{25 - 9 - 16} = 0$$
$$f(-2, 1) = \sqrt{25 - (-2)^2 - (1)^2} = \sqrt{25 - 4 - 1} = 2\sqrt{5}$$
$$f(u, 3v) = \sqrt{25 - u^2 - (3v)^2} = \sqrt{25 - u^2 - 9v^2}$$ ●

Example 1 The function g is defined by $g(x, y, z) = x^2 - 5xz + yz^2$. Find (a) $g(1, 4, -2)$; (b) $g(2a, -b, 3c)$; (c) $g(x^2, y^2, z^2)$; (d) $g(y, z, -x)$.

Solution

(a) $g(1, 4, -2) = 1^2 - 5(1)(-2) + 4(-2)^2 = 1 + 10 + 16 = 27$

(b) $g(2a, -b, 3c) = (2a)^2 - 5(2a)(3c) + (-b)(3c)^2$
$= 4a^2 - 30ac - 9bc^2$

(c) $g(x^2, y^2, z^2) = (x^2)^2 - 5(x^2)(z^2) + (y^2)(z^2)^2 = x^4 - 5x^2z^2 + y^2z^4$

(d) $g(y, z, -x) = y^2 - 5y(-x) + z(-x)^2 = y^2 + 5xy + x^2z$

18.1.3 Definition If f is a function of a single variable and g is a function of two variables, then the **composite function** $f \circ g$ is the function of two variables defined by

$$(f \circ g)(x, y) = f(g(x, y))$$

and the domain of $f \circ g$ is the set of all points (x, y) in the domain of g such that $g(x, y)$ is in the domain of f.

Example 2 Given $f(t) = \ln t$ and $g(x, y) = x^2 + y$, find $h(x, y)$ if $h = f \circ g$, and find the domain of h.

Solution

$$\begin{aligned} h(x, y) &= (f \circ g)(x, y) \\ &= f(g(x, y)) \\ &= f(x^2 + y) \\ &= \ln(x^2 + y) \end{aligned}$$

The domain of g is the set of all points in R^2, and the domain of f is $(0, +\infty)$. Therefore the domain of h is the set $\{(x, y) | x^2 + y \geq 0\}$.

Definition 18.1.3 can be extended to a composite function of n variables as follows.

18.1.4 Definition If f is a function of a single variable and g is a function of n variables, then the **composite function** $f \circ g$ is the function of n variables defined by

$$(f \circ g)(x_1, x_2, \ldots, x_n) = f(g(x_1, x_2, \ldots, x_n))$$

and the domain of $f \circ g$ is the set of all points $(x_1, x_2, \ldots, x_n)$ in the domain of g such that $g(x_1, x_2, \ldots, x_n)$ is in the domain of f.

Example 3 Given $F(x) = \sin^{-1} x$ and

$$G(x, y, z) = \sqrt{x^2 + y^2 + z^2 - 4}$$

find the function $F \circ G$ and its domain.

Solution

$$\begin{aligned}(F \circ G)(x, y, z) &= F(G(x, y, z)) \\ &= F(\sqrt{x^2 + y^2 + z^2 - 4}) \\ &= \sin^{-1} \sqrt{x^2 + y^2 + z^2 - 4}\end{aligned}$$

The domain of G is the set $\{(x, y, z) | x^2 + y^2 + z^2 - 4 \geq 0\}$, and the domain of F is $[-1, 1]$. So the domain of $F \circ G$ is the set of all points (x, y, z) in R^3 such that $0 \leq x^2 + y^2 + z^2 - 4 \leq 1$ or, equivalently, $4 \leq x^2 + y^2 + z^2 \leq 5$.

A **polynomial function** of two variables x and y is a function f such that $f(x, y)$ is the sum of terms of the form $cx^n y^m$, where c is a real number and n and m are nonnegative integers. The **degree** of the polynomial function is determined by the largest sum of the exponents of x and y appearing in any one term. Hence the function f defined by

$$f(x, y) = 6x^3y^2 - 5xy^3 + 7x^2y - 2x^2 + y$$

is a polynomial function of degree 5.

The graph of a function f of a single variable consists of the set of points (x, y) in R^2 for which $y = f(x)$. Similarly, the graph of a function of two variables is a set of points in R^3.

18.1.5 Definition If f is a function of two variables, then the **graph** of f is the set of all points (x, y, z) in R^3 for which (x, y) is a point in the domain of f and $z = f(x, y)$.

Hence the graph of a function f of two variables is a surface that is the set of all points in three-dimensional space whose cartesian coordinates are given by the ordered triples of real numbers (x, y, z). Because the domain of f is a set of points in the xy plane, and because for each ordered pair (x, y) in the domain of f there corresponds a unique value of z, no line perpendicular to the xy plane can intersect the graph of f in more than one point.

● ILLUSTRATION 6 The function of Illustration 1 is the function f that is the set of all ordered pairs of the form (P, z) such that

$$z = \sqrt{25 - x^2 - y^2}$$

So the graph of f is the hemisphere on and above the xy plane having a radius of 5 and its center at the origin. A sketch of the graph of this hemisphere is shown in Fig. 18.1.5. ●

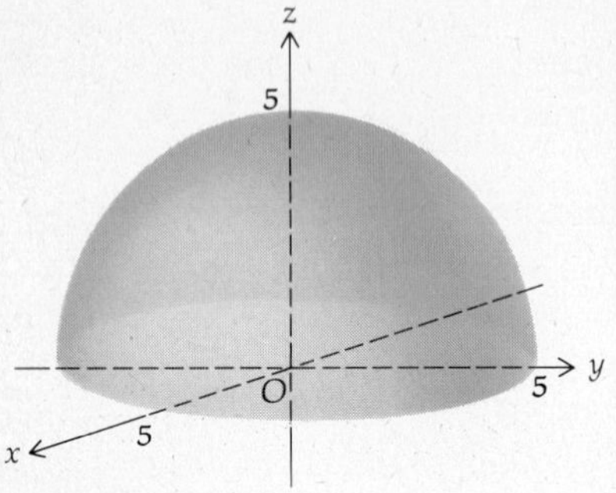

FIGURE 18.1.5

Example 4 Draw a sketch of the graph of the function f having function values $f(x, y) = x^2 + y^2$.

Solution The graph of f is the surface having the equation $z = x^2 + y^2$. The trace of the surface in the xy plane is found by using the equation $z = 0$ simultaneously with the equation of the surface. We obtain $x^2 + y^2 = 0$, which is the origin. The traces in the xz and yz planes are found by using the equations $y = 0$ and $x = 0$, respectively, with the equation $z = x^2 + y^2$. These traces are the parabolas $z = x^2$ and $z = y^2$. The cross section of the surface in a plane $z = k$, parallel to the xy plane, is a circle with its center on the z axis and radius $\sqrt{k}$. With this information we have the required sketch shown in Fig. 18.1.6.

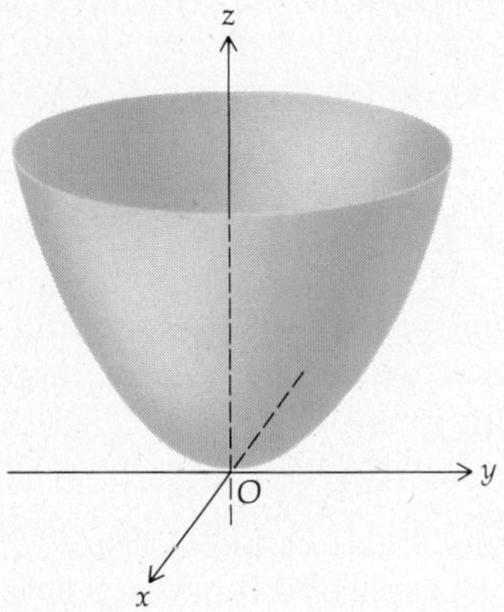

FIGURE 18.1.6

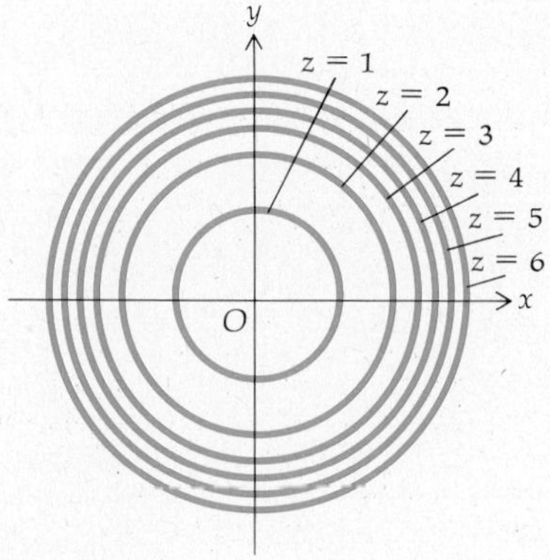

FIGURE 18.1.7

Another useful method of representing a function of two variables geometrically is similar to that of representing a three-dimensional landscape by a two-dimensional topographical map. Suppose that the surface $z = f(x, y)$ is intersected by the plane $z = k$, and the curve of intersection is projected onto the xy plane. This projected curve has $f(x, y) = k$ as an equation, and the curve is called the **level curve** (or **contour curve**) of the function f at k. Each point on the level curve corresponds to the unique point on the surface that is k units above it if k is positive, or k units below it if k is negative. By considering different values for the constant k we obtain a set of level curves called a **contour map**. The set of all possible values of k is the range of the function f, and each level curve, $f(x, y) = k$, in the contour map consists of the points (x, y) in the domain of f having equal function values of k. For example, for the function f of Example 4 the level curves are circles with the center at the origin. The particular level curves for $z = 1, 2, 3, 4, 5$, and 6 are shown in Fig. 18.1.7.

A contour map shows the variation of z with x and y. The level curves are usually shown for values of z at constant intervals, and the values of z are changing more rapidly when the level curves are close together than when they are far apart; that is, when the level curves are close together the surface is steep, and when the level curves are far apart the elevation of the surface is changing slowly. On a two-dimensional topographical map of a landscape, a general notion of its steepness is obtained by considering the spacing of its level curves. Also on a topographical map if the path of a level curve is followed, the elevation remains constant.

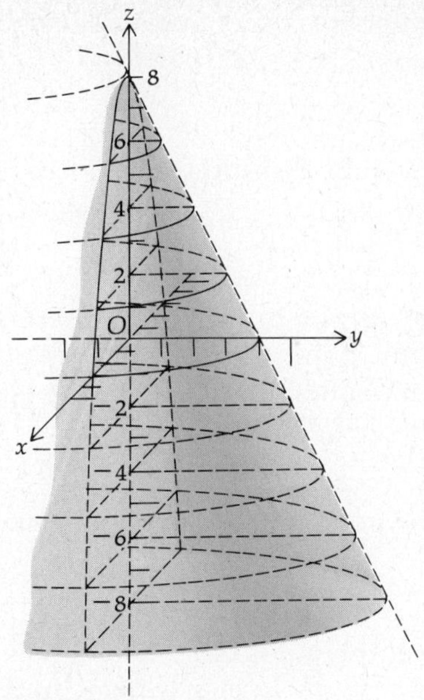

FIGURE 18.1.8

Example 5 Let f be the function for which $f(x, y) = 8 - x^2 - 2y$. Draw a sketch of the graph of f and a contour map of f showing the level curves of f at 10, 8, 6, 4, 2, 0, −2, −4, −6, and −8.

Solution A sketch of the graph of f is shown in Fig. 18.1.8. This is the surface $z = 8 - x^2 - 2y$. The trace in the xy plane is obtained by setting $z = 0$, which gives the parabola $x^2 = -2(y - 4)$. Setting $y = 0$ and $x = 0$, we obtain the traces in the xz and yz planes, which are, respectively, the parabola $x^2 = -(z - 8)$ and the line $2y + z = 8$. The cross section of the surface made by the plane $z = k$ is a parabola having its vertex on the line $2y + z = 8$ in the yz plane and opening to the left. The cross sections for $z = 8, 6, 4, 2, -2, -4, -6$, and -8 are shown in the figure.

The level curves of f are the parabolas $x^2 = -2(y - 4 + \frac{1}{2}k)$. The contour map of f with sketches of the required level curves is shown in Fig. 18.1.9.

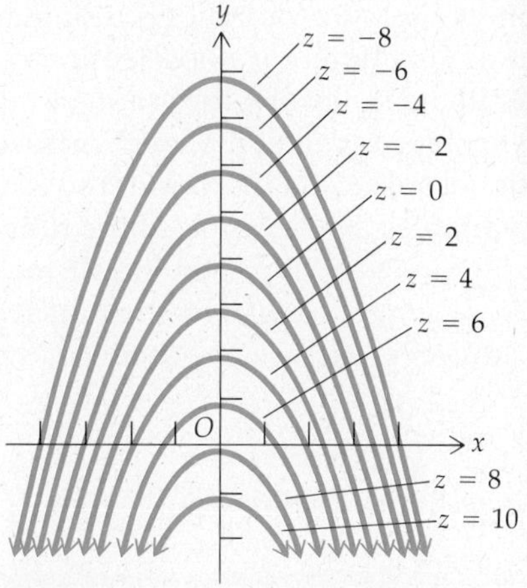

FIGURE 18.1.9

To illustrate a use of level curves, suppose that the temperature at any point of a flat metal plate is given by the function f; that is, if t degrees is the temperature, then at the point (x, y), $t = f(x, y)$. Then the curves having equations of the form $f(x, y) = k$, where k is a constant, are curves on which the temperature is constant. These are the level curves of f and are called **isothermals.** Furthermore, if V volts gives the amount of electric potential at any point (x, y) of the xy plane, and $V = f(x, y)$, then the level curves of f are called **equipotential curves** because the electric potential at each point of such a curve is the same.

For an application of level curves in economics, consider the productivity (or output) of an industry that is dependent on several inputs. Among the inputs may be the number of machines used in production, the number of person-hours available, the amount of working capital to be had, the quantity of material used, and the amount of land available. Suppose the amounts of the inputs are given by x and y, the amount of the output is given by z, and $z = f(x, y)$. Such a function is called a **production function,** and the level curves of f, having equations of the form $f(x, y) = k$, where k is a constant, are called **constant product curves.**

Example 6 Let f be the production function for which $f(x, y) = 2x^{1/2}y^{1/2}$. Draw a contour map of f showing the constant product curves at 8, 6, 4, and 2.

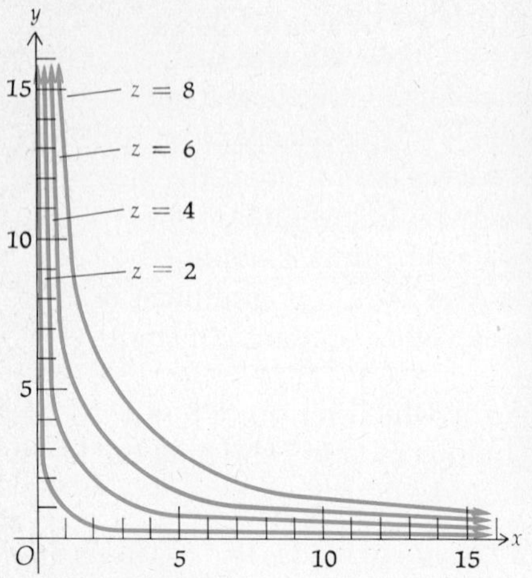

FIGURE 18.1.10

Solution The contour map consists of the curves that are the intersection of the surface

$$z = 2x^{1/2}y^{1/2} \tag{1}$$

with the planes $z = k$, where $k = 8, 6, 4, 2$, and 1. Substituting $z = 8$ in Eq. (1) we obtain $4 = x^{1/2}y^{1/2}$ or, equivalently,

$$xy = 16 \qquad x > 0 \quad \text{and} \quad y > 0 \tag{2}$$

The curve in the xy plane represented by (2) is a branch of a hyperbola lying in the first quadrant. With each of the numbers 6, 4, and 2 we also obtain a branch of a hyperbola in the first quadrant. These are the constant product curves, and they are shown in Fig. 18.1.10.

The following definition extends the notion of the graph of a function to a function of n variables.

18.1.6 Definition If f is a function of n variables, then the **graph** of f is the set of all points $(x_1, x_2, \ldots, x_n, w)$ in R^{n+1} for which $(x_1, x_2, \ldots, x_n)$ is a point in the domain of f and $w = f(x_1, x_2, \ldots, x_n)$.

Analogous to level curves for a function of two variables is a similar situation for functions of three variables. If f is a function whose domain is a set of points in R^3, then if k is a number in the range of f, the graph of the equation $f(x, y, z) = k$ is a surface. This surface is called the **level surface** of f at k. Every surface in three-dimensional space can be considered as a level surface of some function of three variables. For example, if the function g is defined by the equation $g(x, y, z) = x^2 + y^2 - z$, then the surface shown in Fig. 18.1.6 is the level surface of g at 0. Similarly, the surface having equation $z - x^2 - y^2 + 5 = 0$ is the level surface of g at 5.

EXERCISES 18.1

1. Let the function f of two variables x and y be the set of all ordered pairs of the form (P, z) such that $z = (x + y)/(x - y)$. Find: (a) $f(-3, 4)$; (b) $f(x^2, y^2)$; (c) $[f(x, y)]^2$; (d) $f(-x, y) - f(x, -y)$; (e) the domain of f; (f) the range of f. Draw a sketch showing as a shaded region in R^2 the set of points not in the domain of f.

2. Let the function g of three variables, x, y, and z, be the set of all ordered pairs of the form (P, w) such that $w = \sqrt{4 - x^2 - y^2 - z^2}$. Find: (a) $g(1, -1, -1)$; (b) $g(-a, 2b, \frac{1}{2}c)$; (c) $g(y, -x, -y)$; (d) the domain of g; (e) the range of g; (f) $[g(x, y, z)]^2 - [g(x + 2, y + 2, z)]^2$. Draw a sketch showing as a shaded solid in R^3 the set of points in the domain of g.

In Exercises 3 through 16, find the domain and range of the function f, and draw a sketch showing as a shaded region in R^2 the set of points in the domain of f.

3. $f(x, y) = \sqrt{25 - x^2 - y^2}$

4. $f(x, y) = \sqrt{16 - x^2 - 4y^2}$

5. $f(x, y) = \dfrac{\sqrt{25 - x^2 - y^2}}{x}$

6. $f(x, y) = \dfrac{\sqrt{16 - x^2 - 4y^2}}{y}$ **7.** $f(x, y) = x\sqrt{25 - x^2 - y^2}$ **8.** $f(x, y) = y\sqrt{16 - x^2 - 4y^2}$

9. $f(x, y) = \dfrac{x}{\sqrt{25 - x^2 - y^2}}$ **10.** $f(x, y) = \dfrac{y}{\sqrt{16 - x^2 - 4y^2}}$ **11.** $f(x, y) = \dfrac{x^2 - y^2}{x - y}$

12. $f(x, y) = \sqrt{\dfrac{x - y}{x + y}}$ **13.** $f(x, y) = \dfrac{x}{|y|}$ **14.** $f(x, y) = \dfrac{x + y}{xy}$

15. $f(x, y) = \ln(xy - 1)$ **16.** $f(x, y) = \sin^{-1}(x + y)$

In Exercises 17 through 19, find the domain and range of the function f.

17. $f(x, y, z) = (x + y)\sqrt{z - 2}$ **18.** $f(x, y, z) = |x|e^{y/z}$ **19.** $f(x, y, z) = \sin^{-1} x + \cos^{-1} y + \tan^{-1} z$

In Exercises 20 through 26, find the domain and range of the function f and draw a sketch of the graph.

20. $f(x, y) = \sqrt{x + y}$ **21.** $f(x, y) = 4x^2 + 9y^2$ **22.** $f(x, y) = 144 - 9x^2 - 16y^2$

23. $f(x, y) = 16 - x^2 - y^2$ **24.** $f(x, y) = \sqrt{100 - 25x^2 - 4y^2}$ **25.** $f(x, y) = \sqrt{10 - x - y^2}$

26. $f(x, y) = \begin{cases} 2 & \text{if } x \neq y \\ 0 & \text{if } x = y \end{cases}$

In Exercises 27 through 32, draw a sketch of a contour map of the function f showing the level curves of f at the given numbers.

27. The function of Exercise 21 at 36, 25, 16, 9, 4, 1, and 0. **28.** The function of Exercise 20 at 5, 4, 3, 2, 1, and 0.

29. The function of Exercise 23 at 8, 4, 0, −4, and −8. **30.** The function of Exercise 24 at 10, 8, 6, 5, and 0.

31. The function f for which $f(x, y) = \frac{1}{2}(x^2 + y^2)$ at 8, 6, 4, 2, and 0.

32. The function f for which $f(x, y) = (x - 3)/(y + 2)$ at 4, 2, 1, $\frac{1}{2}$, $\frac{1}{4}$, 0, $-\frac{1}{4}$, $-\frac{1}{2}$, −1, −2, and −4.

In Exercises 33 and 34, find $h(x, y)$ if $h = f \circ g$; also find the domain of h.

33. $f(t) = \sin^{-1} t$; $g(x, y) = \sqrt{1 - x^2 - y^2}$ **34.** $f(t) = e^t$; $g(x, y) = y \ln x$

35. Given $f(x, y) = x - y$, $g(t) = \sqrt{t}$, $h(s) = s^2$. Find (a) $(g \circ f)(5, 1)$; (b) $f(h(3), g(9))$; (c) $f(g(x), h(y))$; (d) $g((h \circ f)(x, y))$; (e) $(g \circ h)(f(x, y))$.

36. Given $f(x, y) = x/y^2$, $g(x) = x^2$, $h(x) = \sqrt{x}$. Find (a) $(h \circ f)(2, 1)$; (b) $f(g(2), h(4))$; (c) $f(g(\sqrt{x}), h(x^2))$; (d) $h((g \circ f)(x, y))$; (e) $(h \circ g)(f(x, y))$.

37. The electric potential at a point (x, y) of the xy plane is V volts and $V = 4/\sqrt{9 - x^2 - y^2}$. Draw the equipotential curves for $V = 16$, 12, 8, 4, and 1.

38. The production function f for a certain commodity has function values $f(x, y) = 4x^{1/3}y^{2/3}$, where x and y give the amounts of two inputs. Draw a contour map of f showing the constant product curves at 16, 12, 8, 4, and 2.

39. Suppose the number of units of a commodity produced is z, and $z = 6xy$, where x is the number of machines used in production and y is the number of person-hours available. Then the function f defined by $f(x, y) = 6xy$ is a production function. Draw a contour map of f showing the constant product curves for $z = 30$, 24, 18, 12, and 6.

40. The temperature at a point (x, y) of a flat metal plate is t degrees and $t = 4x^2 + 2y^2$. Draw the isothermals for $t = 12$, 8, 4, 1, and 0.

In Exercises 41 and 42, draw sketches of the level surfaces of the function f at the given numbers.

41. $f(x, y, z) = x^2 + y^2 - 4z$ at 8, 4, 0, −4, and −8. **42.** $f(x, y, z) = x^2 + y^2 + z^2$ at 9, 4, 1, and 0.

18.2 LIMITS OF FUNCTIONS OF MORE THAN ONE VARIABLE

In R^1 the distance between two points is the absolute value of the difference of two real numbers. That is, $|x - a|$ is the distance between the points x and a. In R^2 the distance between the two points $P(x, y)$ and $P_0(x_0, y_0)$ is given by $\sqrt{(x - x_0)^2 + (y - y_0)^2}$. In R^3 the distance between the two points $P(x, y, z)$ and $P_0(x_0, y_0, z_0)$ is given by

$$\sqrt{(x - x_0)^2 + (y - y_0)^2 + (z - z_0)^2}$$

In R^n the distance between two points is defined analogously.

18.2.1 Definition If $P(x_1, x_2, \ldots, x_n)$ and $A(a_1, a_2, \ldots, a_n)$ are two points in R^n, then the distance between P and A, denoted by $\|P - A\|$, is given by

$$\|P - A\| = \sqrt{(x_1 - a_1)^2 + (x_2 - a_2)^2 + \cdots + (x_n - a_n)^2} \tag{1}$$

If in R^1, $P = x$ and $A = a$, (1) becomes

$$\|x - a\| = \sqrt{(x - a)^2} = |x - a| \tag{2}$$

If in R^2, $P = (x, y)$ and $A = (x_0, y_0)$, (1) becomes

$$\|(x, y) - (x_0, y_0)\| = \sqrt{(x - x_0)^2 + (y - y_0)^2} \tag{3}$$

And if in R^3, $P = (x, y, z)$ and $A = (x_0, y_0, z_0)$, (1) becomes

$$\|(x, y, z) - (x_0, y_0, z_0)\| = \sqrt{(x - x_0)^2 + (y - y_0)^2 + (z - z_0)^2} \tag{4}$$

$\|P - A\|$ is read as "the distance between P and A." It is a nonnegative number.

18.2.2 Definition If A is a point in R^n and r is a positive number, then the **open ball** $B(A; r)$ is defined to be the set of all points P in R^n such that $\|P - A\| < r$.

18.2.3 Definition If A is a point in R^n and r is a positive number, then the **closed ball** $B[A; r]$ is defined to be the set of all points P in R^n such that $\|P - A\| \le r$.

To illustrate these definitions, we show what they mean in R^1, R^2, and R^3. First of all, if a is a point in R^1, then the open ball $B(a; r)$ is the set of all points x in R^1 such that

$$|x - a| < r \tag{5}$$

The set of all points x satisfying (5) is the set of all points in the open interval $(a - r, a + r)$; so the open ball $B(a; r)$ in R^1 (see Fig. 18.2.1) is simply an open interval having its midpoint at a and its endpoints at $a - r$ and $a + r$. The closed ball $B[a; r]$ in R^1 (Fig. 18.2.2) is the closed interval $[a - r, a + r]$.

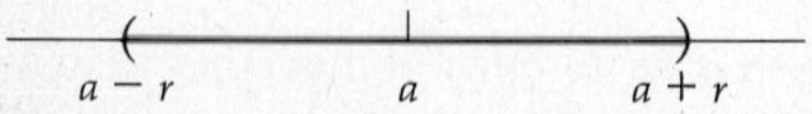

open ball $B(a; r)$ in R^1

FIGURE 18.2.1

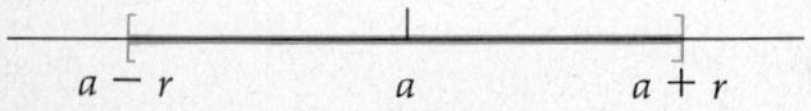

closed ball $B[a; r]$ in R^1

FIGURE 18.2.2

If (x_0, y_0) is a point in R^2, then the open ball $B((x_0, y_0); r)$ is the set of all points (x, y) in R^2 such that

$$\|(x, y) - (x_0, y_0)\| < r \tag{6}$$

From (3) it follows that (6) is equivalent to

$$\sqrt{(x - x_0)^2 + (y - y_0)^2} < r$$

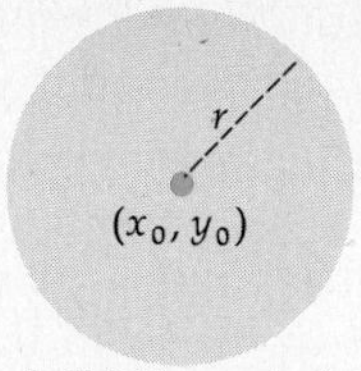

open ball $B((x_0, y_0); r)$ in R^2
FIGURE 18.2.3

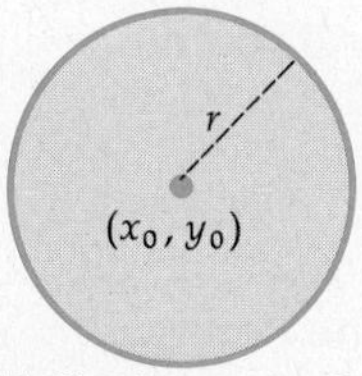

closed ball $B[(x_0, y_0); r]$ in R^2
FIGURE 18.2.4

So the open ball $B((x_0, y_0); r)$ in R^2 (Fig. 18.2.3) consists of all points in the interior region bounded by the circle having its center at (x_0, y_0) and radius r. An open ball in R^2 is sometimes called an *open disk*. The closed ball, or closed disk, $B[(x_0, y_0); r]$ in R^2 (Fig. 18.2.4) is the set of all points in the open ball $B((x_0, y_0); r)$ and on the circle having its center at (x_0, y_0) and radius r.

If (x_0, y_0, z_0) is a point in R^3, then the open ball $B((x_0, y_0, z_0); r)$ is the set of all points (x, y, z) in R^3 such that

$$\|(x, y, z) - (x_0, y_0, z_0)\| < r \tag{7}$$

From (4) and (7) it follows that the open ball $B((x_0, y_0, z_0); r)$ in R^3 (Fig. 18.2.5) consists of all points in the interior region bounded by the sphere having its center at P_0 and radius r. Similarly, the closed ball $B[(x_0, y_0, z_0); r]$ in R^3 (Fig. 18.2.6) consists of all points in the open ball $B((x_0, y_0, z_0); r)$ and on the sphere having its center at (x_0, y_0, z_0) and radius r. We are now in a position to define what is meant by the limit of a function of n variables.

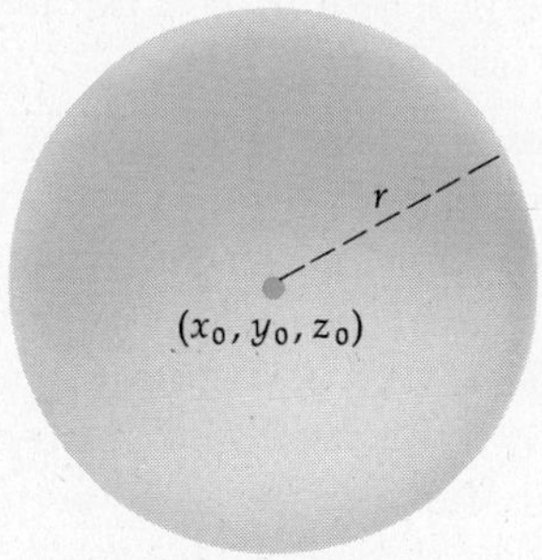

open ball $B((x_0, y_0, z_0); r)$ in R^3
FIGURE 18.2.5

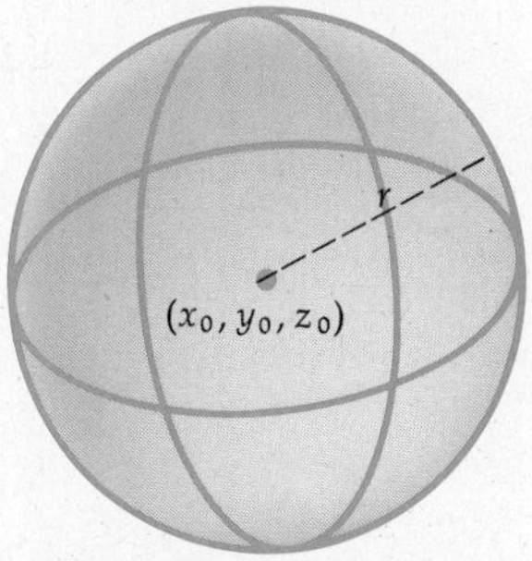

closed ball $B[(x_0, y_0, z_0); r]$ in R^3
FIGURE 18.2.6

18.2.4 Definition Let f be a function of n variables that is defined on some open ball $B(A; r)$, except possibly at the point A itself. Then the **limit of $f(P)$ as P approaches A is L,** written as

$$\lim_{P \to A} f(P) = L \tag{8}$$

if for any $\epsilon > 0$, however small, there exists a $\delta > 0$ such that

$$|f(P) - L| < \epsilon \quad \text{whenever} \quad 0 < \|P - A\| < \delta \tag{9}$$

If f is a function of one variable and if in the above definition $A = a$ in R^1 and $P = x$, then (8) becomes

$$\lim_{x \to a} f(x) = L$$

and (9) becomes

$$|f(x) - L| < \epsilon \quad \text{whenever} \quad 0 < |x - a| < \delta$$

So the definition (2.1.1) of the limit of a function of one variable is a special case of Definition 18.2.4.

We now state the definition of the limit of a function of two variables. It is the special case of Definition 18.2.4, where A is the point (x_0, y_0) and P is the point (x, y).

18.2.5 Definition Let f be a function of two variables that is defined on some open disk $B((x_0, y_0); r)$, except possibly at the point (x_0, y_0) itself. Then the **limit of $f(x, y)$ as (x, y) approaches (x_0, y_0) is L**, written as

$$\lim_{(x,y)\to(x_0,y_0)} f(x, y) = L$$

if for any $\epsilon > 0$, however small, there exists a $\delta > 0$ such that

$$|f(x, y) - L| < \epsilon \quad \text{whenever} \quad 0 < \sqrt{(x - x_0)^2 + (y - y_0)^2} < \delta \tag{10}$$

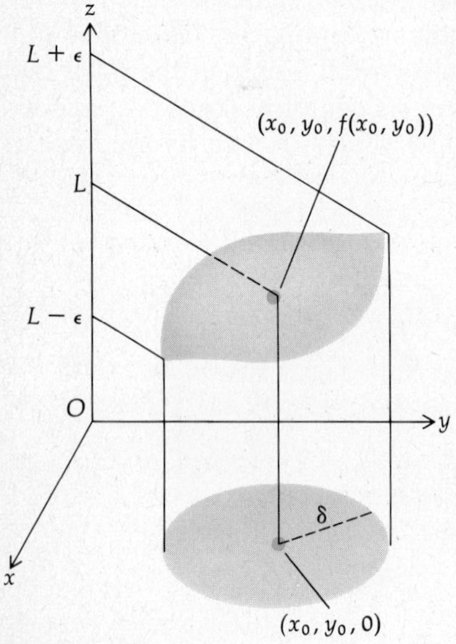

FIGURE 18.2.7

In words, Definition 18.2.5 states that the function values $f(x, y)$ approach a limit L as the point (x, y) approaches the point (x_0, y_0) if the absolute value of the difference between $f(x, y)$ and L can be made arbitrarily small by taking the point (x, y) sufficiently close to (x_0, y_0) but not equal to (x_0, y_0). In Definition 18.2.5 nothing is said about the function value at the point (x_0, y_0); that is, it is not necessary that the function be defined at (x_0, y_0) in order for $\lim_{(x,y)\to(x_0,y_0)} f(x, y)$ to exist.

A geometric interpretation of Definition 18.2.5 is illustrated in Fig. 18.2.7. The portion above the open disk $B((x_0, y_0); \delta)$ of the surface having equation $z = f(x, y)$ is shown. We see that $f(x, y)$ on the z axis will lie between $L - \epsilon$ and $L + \epsilon$ whenever the point (x, y) in the xy plane is in the open disk $B((x_0, y_0); \delta)$. Another way of stating this is that $f(x, y)$ on the z axis can be restricted to lie between $L - \epsilon$ and $L + \epsilon$ by restricting the point (x, y) in the xy plane to be in the open disk $B((x_0, y_0); \delta)$.

● ILLUSTRATION 1 We apply Definition 18.2.5 to prove that

$$\lim_{(x,y)\to(1,3)} (2x + 3y) = 11$$

The first requirement of the definition is that $2x + 3y$ must be defined on some open disk having its center at the point $(1, 3)$, except possibly at $(1, 3)$. Because $2x + 3y$ is defined at every point (x, y), any open disk having its center at $(1, 3)$ will satisfy this requirement. Now, we must show that for any $\epsilon > 0$ there exists a $\delta > 0$ such that

$$|(2x + 3y) - 11| < \epsilon \quad \text{whenever} \quad 0 < \sqrt{(x - 1)^2 + (y - 3)^2} < \delta$$

From the triangle inequality,

$$|2x + 3y - 11| = |2x - 2 + 3y - 9| \leq 2|x - 1| + 3|y - 3|$$

Because

$$|x - 1| \leq \sqrt{(x - 1)^2 + (y - 3)^2}$$

and

$$|y - 3| \leq \sqrt{(x - 1)^2 + (y - 3)^2}$$

then

$$2|x - 1| + 3|y - 3| < 2\delta + 3\delta$$

whenever

$$0 < \sqrt{(x-1)^2 + (y-3)^2} < \delta$$

So if we take $\delta = \frac{1}{5}\epsilon$,

$$|2x + 3y - 11| \leq 2|x - 1| + 3|y - 3| < 5\delta = \epsilon$$

whenever

$$0 < \sqrt{(x-1)^2 + (y-3)^2} < \delta$$

This proves that $\lim\limits_{(x,y)\to(1,3)} (2x + 3y) = 11$. ●

Example 1 Prove that $\lim\limits_{(x,y)\to(1,2)} (3x^2 + y) = 5$ by applying Definition 18.2.5.

Solution Because $3x^2 + y$ is defined at every point (x, y), any open disk having its center at $(1, 2)$ will satisfy the first requirement of Definition 18.2.5.

We must show that for any $\epsilon > 0$ there exists a $\delta > 0$ such that

$$|(3x^2 + y) - 5| < \epsilon \quad \text{whenever} \quad 0 < \sqrt{(x-1)^2 + (y-2)^2} < \delta$$

From the triangle inequality,

$$|3x^2 + y - 5| = |3x^2 - 3 + y - 2| \leq 3|x - 1||x + 1| + |y - 2| \tag{11}$$

If we require the δ, for which we are looking, to be less than or equal to 1, then $|x - 1| < \delta \leq 1$ and $|y - 2| < \delta \leq 1$ whenever

$$0 < \sqrt{(x-1)^2 + (y-2)^2} < \delta \leq 1$$

Furthermore, whenever $|x - 1| < 1$, then $-1 < x - 1 < 1$, and so $1 < x + 1 < 3$. Hence

$$3|x - 1||x + 1| + |y - 2| < 3 \cdot \delta \cdot 3 + \delta = 10\delta \tag{12}$$

whenever

$$0 < \sqrt{(x-1)^2 + (y-2)^2} < \delta \leq 1$$

So if for any $\epsilon > 0$ we take $\delta = \min(1, \frac{1}{10}\epsilon)$, then from (11) and (12),

$$|3x^2 + y - 5| < 10\delta \leq \epsilon \quad \text{whenever} \quad 0 < \sqrt{(x-1)^2 + (y-2)^2} < \delta$$

This proves that $\lim\limits_{(x,y)\to(1,2)} (3x^2 + y) = 5$.

The limit theorems of Chapter 2 and their proofs, with minor modifications, apply to functions of several variables. We use these theorems without restating them and their proofs.

● ILLUSTRATION 2 By applying the limit theorems on sums and products,

$$\begin{aligned}\lim_{(x,y)\to(-2,1)} (x^3 + 2x^2y - y^2 + 2) &= (-2)^3 + 2(-2)^2(1) - (1)^2 + 2 \\ &= 1\end{aligned}$$

●

Example 2 Find

$$\lim_{(x,y)\to(1,3)} \frac{3x - y}{27x^3 - y^3}$$

Solution

$$\begin{aligned}
\lim_{(x,y)\to(1,3)} \frac{3x - y}{27x^3 - y^3} &= \lim_{(x,y)\to(1,3)} \frac{3x - y}{(3x - y)(9x^2 + 3xy + y^2)} \\
&= \lim_{(x,y)\to(1,3)} \frac{1}{9x^2 + 3xy + y^2} \\
&= \frac{1}{\lim_{(x,y)\to(1,3)} (9x^2 + 3xy + y^2)} \\
&= \frac{1}{9 + 9 + 9} \\
&= \tfrac{1}{27}
\end{aligned}$$

Analogous to Theorem 2.7.6 for functions of a single variable is the following theorem regarding the limit of a composite function of two variables.

18.2.6 Theorem If g is a function of two variables and $\lim_{(x,y)\to(x_0,y_0)} g(x, y) = b$, and f is a function of a single variable continuous at b, then

$$\lim_{(x,y)\to(x_0,y_0)} (f \circ g)(x, y) = f(b)$$

or, equivalently,

$$\lim_{(x,y)\to(x_0,y_0)} f(g(x, y)) = f\left(\lim_{(x,y)\to(x_0,y_0)} g(x, y)\right)$$

The proof of this theorem is similar to the proof of Theorem 2.7.6 and is left as an exercise (see Exercise 49).

Example 3 Use Theorem 18.2.6 to find $\lim_{(x,y)\to(2,1)} \ln(xy - 1)$.

Solution Let g be the function such that $g(x, y) = xy - 1$, and let f be the function such that $f(t) = \ln t$.

$$\lim_{(x,y)\to(2,1)} (xy - 1) = 1$$

and because f is continuous at 1, from Theorem 18.2.6,

$$\begin{aligned}
\lim_{(x,y)\to(2,1)} \ln(xy - 1) &= \ln\left(\lim_{(x,y)\to(2,1)} (xy - 1)\right) \\
&= \ln 1 \\
&= 0
\end{aligned}$$

We now introduce the concept of an *accumulation point*, which is needed in order to continue the discussion of limits of functions of two variables.

18.2.7 Definition A point P_0 is said to be an **accumulation point** of a set S of points in R^n if every open ball $B(P_0; r)$ contains infinitely many points of S.

● ILLUSTRATION 3 If S is the set of all points in R^2 on the positive side of the x axis, the origin will be an accumulation point of S because no matter how small we take the value of r, every open disk having its center at the origin and radius r will contain infinitely many points of S. This is an example of a set having an accumulation point for which the accumulation point is not a point of the set. Any point of this set S also will be an accumulation point of S. ●

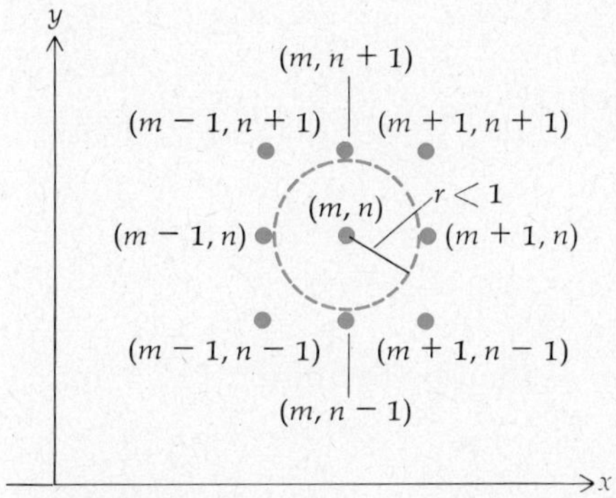

FIGURE 18.2.8

● ILLUSTRATION 4 If S is the set of all points in R^2 for which the cartesian coordinates are positive integers, then this set has no accumulation point. This can be seen by considering the point (m, n), where m and n are positive integers. Then an open disk having its center at (m, n) and radius less than 1 will contain no points of S other than (m, n); therefore Definition 18.2.7 will not be satisfied (see Fig. 18.2.8). ●

We now consider the limit of a function of two variables as a point (x, y) approaches a point (x_0, y_0), where (x, y) is restricted to a specific set of points.

18.2.8 Definition Let f be a function defined on a set of points S in R^2, and let (x_0, y_0) be an accumulation point of S. Then the **limit of $f(x, y)$ as (x, y) approaches (x_0, y_0) in S is L**, written as

$$\lim_{\substack{(x,y)\to(x_0,y_0)\\(P\text{ in }S)}} f(x, y) = L \qquad (13)$$

if for any $\epsilon > 0$, however small, there exists a $\delta > 0$ such that

$$|f(x, y) - L| < \epsilon \quad \text{whenever} \quad 0 < \|(x, y) - (x_0, y_0)\| < \delta$$

and (x, y) is in S.

A special case of (13) occurs when S is a set of points on a curve containing (x_0, y_0). In such cases the limit in (13) becomes the limit of a function of one variable. For example, consider $\lim_{(x,y)\to(0,0)} f(x, y)$. Then if S_1 is the set of all points on the positive side of the x axis,

$$\lim_{\substack{(x,y)\to(0,0)\\(P\text{ in }S_1)}} f(x, y) = \lim_{x\to 0^+} f(x, 0)$$

If S_2 is the set of all points on the negative side of the y axis,

$$\lim_{\substack{(x,y)\to(0,0)\\(P\text{ in }S_2)}} f(x, y) = \lim_{y\to 0^-} f(0, y)$$

If S_3 is the set of all points on the x axis,

$$\lim_{\substack{(x,y)\to(0,0)\\(P \text{ in } S_3)}} f(x, y) = \lim_{x\to 0} f(x, 0)$$

If S_4 is the set of all points on the parabola $y = x^2$,

$$\lim_{\substack{(x,y)\to(0,0)\\(P \text{ in } S_4)}} f(x, y) = \lim_{x\to 0} f(x, x^2)$$

18.2.9 Theorem Suppose that the function f is defined for all points on an open disk having its center at (x_0, y_0), except possibly at (x_0, y_0) itself, and

$$\lim_{(x,y)\to(x_0,y_0)} f(x, y) = L$$

Then if S is any set of points in R^2 having (x_0, y_0) as an accumulation point,

$$\lim_{\substack{(x,y)\to(x_0,y_0)\\(P \text{ in } S)}} f(x, y)$$

exists and always has the value L.

Proof Because $\lim_{(x,y)\to(x_0,y_0)} f(x, y) = L$, then by Definition 18.2.5, for any $\epsilon > 0$ there exists a $\delta > 0$ such that

$$|f(x, y) - L| < \epsilon \quad \text{whenever} \quad 0 < \|(x, y) - (x_0, y_0)\| < \delta$$

The above will be true if we further restrict (x, y) by the requirement that (x, y) be in a set S, where S is any set of points having (x_0, y_0) as an accumulation point. Therefore, by Definition 18.2.8,

$$\lim_{\substack{(x,y)\to(x_0,y_0)\\(P \text{ in } S)}} f(x, y) = L$$

and L does not depend on the set S through which (x, y) is approaching (x_0, y_0). This proves the theorem. ■

The following is an immediate consequence of Theorem 18.2.9.

18.2.10 Theorem If the function f has different limits as (x, y) approaches (x_0, y_0) through two distinct sets of points having (x_0, y_0) as an accumulation point, then $\lim_{(x,y)\to(x_0,y_0)} f(x, y)$ does not exist.

Proof Let S_1 and S_2 be two distinct sets of points in R^2 having (x_0, y_0) as an accumulation point, and let

$$\lim_{\substack{(x,y)\to(x_0,y_0)\\(P \text{ in } S_1)}} f(x, y) = L_1 \quad \text{and} \quad \lim_{\substack{(x,y)\to(x_0,y_0)\\(P \text{ in } S_2)}} f(x, y) = L_2$$

Now assume that $\lim_{(x,y)\to(x_0,y_0)} f(x, y)$ exists. Then by Theorem 18.2.9 L_1 must equal L_2, but by hypothesis $L_1 \neq L_2$, and so we have a contradiction. Therefore $\lim_{(x,y)\to(x_0,y_0)} f(x, y)$ does not exist. ■

Example 4 Given

$$f(x, y) = \frac{xy}{x^2 + y^2}$$

find $\lim_{(x,y)\to(0,0)} f(x, y)$ if it exists.

Solution The function f is defined at all points in R^2 except at (0, 0). Let S_1 be the set of all points on the x axis. Then

$$\lim_{\substack{(x,y)\to(0,0)\\(P \text{ in } S_1)}} f(x, y) = \lim_{x\to 0} f(x, 0) = \lim_{x\to 0} \frac{0}{x^2 + 0} = 0$$

Let S_2 be the set of all points on the line $y = x$. Then

$$\lim_{\substack{(x,y)\to(0,0)\\(P \text{ in } S_2)}} f(x, y) = \lim_{x\to 0} \frac{x^2}{x^2 + x^2} = \lim_{x\to 0} \frac{1}{2} = \frac{1}{2}$$

Because

$$\lim_{\substack{(x,y)\to(0,0)\\(P \text{ in } S_1)}} f(x, y) \neq \lim_{\substack{(x,y)\to(0,0)\\(P \text{ in } S_2)}} f(x, y)$$

it follows from Theorem 18.2.10 that $\lim_{(x,y)\to(0,0)} f(x, y)$ does not exist.

Example 5 Given

$$f(x, y) = \frac{3x^2y}{x^2 + y^2}$$

find $\lim_{(x,y)\to(0,0)} f(x, y)$ if it exists.

Solution The function f is defined everywhere in R^2 except at the origin. Let S_1 be the set of all points on the x axis. Then

$$\lim_{\substack{(x,y)\to(0,0)\\(P \text{ in } S_1)}} f(x, y) = \lim_{x\to 0} \frac{0}{x^2 + 0} = 0$$

Let S_2 be the set of all points on any line through the origin; that is, for any point (x, y) in S_2, $y = mx$.

$$\lim_{\substack{(x,y)\to(0,0)\\(P \text{ in } S_2)}} f(x, y) = \lim_{x\to 0} \frac{3x^2(mx)}{x^2 + m^2x^2} = \lim_{x\to 0} \frac{3mx}{1 + m^2} = 0$$

Even though the same limit of 0 is obtained if (x, y) approaches (0, 0) through a set of points on any line through the origin, we cannot conclude that $\lim_{(x,y)\to(0,0)} f(x, y)$ exists and is zero (see Example 6). However, let us attempt to prove that $\lim_{(x,y)\to(0,0)} f(x, y) = 0$. Any open disk having its center at the origin will satisfy the first requirement of Definition 18.2.5. If we can show that for any $\epsilon > 0$ there exists a $\delta > 0$ such that

$$\left|\frac{3x^2y}{x^2 + y^2}\right| < \epsilon \quad \text{whenever} \quad 0 < \sqrt{x^2 + y^2} < \delta \tag{14}$$

then we have proved that $\lim_{(x,y)\to(0,0)} f(x, y) = 0$.

Because $x^2 \le x^2 + y^2$ and $|y| \le \sqrt{x^2 + y^2}$,

$$\left|\frac{3x^2y}{x^2+y^2}\right| = \frac{3x^2|y|}{x^2+y^2} \le \frac{3(x^2+y^2)\sqrt{x^2+y^2}}{x^2+y^2} = 3\sqrt{x^2+y^2}$$

So if $\delta = \frac{1}{3}\epsilon$,

$$\left|\frac{3x^2y}{x^2+y^2}\right| < \epsilon \quad \text{whenever} \quad 0 < \sqrt{x^2+y^2} < \delta$$

which is (14). Hence we have proved that $\lim_{(x,y)\to(0,0)} f(x, y) = 0$.

Example 6 Given

$$f(x, y) = \frac{x^2y}{x^4 + y^2}$$

find $\lim_{(x,y)\to(0,0)} f(x, y)$ if it exists.

Solution The function f is defined everywhere in R^2 except at $(0, 0)$. Let S_1 be the set of all points on either the x axis or the y axis. So if (x, y) is in S_1, $xy = 0$. Therefore

$$\lim_{\substack{(x,y)\to(0,0)\\(P\text{ in }S_1)}} f(x, y) = 0$$

Let S_2 be the set of all points on any line through the origin; so if (x, y) is a point in S_2, $y = mx$. Then

$$\lim_{\substack{(x,y)\to(0,0)\\(P\text{ in }S_2)}} f(x, y) = \lim_{x\to0} \frac{mx^3}{x^4 + m^2x^2} = \lim_{x\to0} \frac{mx}{x^2 + m^2} = 0$$

Let S_3 be the set of all points on the parabola $y = x^2$. Then

$$\lim_{\substack{(x,y)\to(0,0)\\(P\text{ in }S_3)}} f(x, y) = \lim_{x\to0} \frac{x^4}{x^4 + x^4} = \lim_{x\to0} \frac{1}{2} = \frac{1}{2}$$

Because

$$\lim_{\substack{(x,y)\to(0,0)\\(P\text{ in }S_3)}} f(x, y) \ne \lim_{\substack{(x,y)\to(0,0)\\(P\text{ in }S_1)}} f(x, y)$$

it follows that $\lim_{(x,y)\to(0,0)} f(x, y)$ does not exist.

Example 7 Given

$$f(x, y) = \begin{cases} (x + y)\sin\dfrac{1}{x} & \text{if } x \ne 0 \\ 0 & \text{if } x = 0 \end{cases}$$

find $\lim_{(x,y)\to(0,0)} f(x, y)$ if it exists.

Solution The function f is defined at all points in R^2. Let S_1 be the set of all points on the y axis. Then

$$\lim_{\substack{(x,y)\to(0,0)\\(P\text{ in }S_1)}} f(x, y) = \lim_{x\to 0} 0 = 0$$

Let S_2 be the set of all points on any line through the origin except points on the y axis; that is, if (x, y) is a point in S_2, $y = kx$, where $x \neq 0$. Then

$$\lim_{\substack{(x,y)\to(0,0)\\(P\text{ in }S_2)}} f(x, y) = \lim_{x\to 0} (x + kx) \sin \frac{1}{x}$$

To find the above limit we make use of the fact that $\lim_{x\to 0} (x + kx) = 0$. Because $\lim_{x\to 0} \sin(1/x)$ does not exist, the theorem about the limit of a product cannot be applied. However, because $\lim_{x\to 0} (x + kx) = 0$, it follows that for any $\epsilon > 0$ there exists a $\delta > 0$ such that

$$|x + kx| < \epsilon \quad \text{whenever} \quad 0 < |x| < \delta$$

The δ is, in fact, $\epsilon/|1 + k|$. But

$$\left|(x + kx) \sin \frac{1}{x}\right| = |x + kx| \left|\sin \frac{1}{x}\right| \leq |x + kx| \cdot 1$$

Hence, for any $\epsilon > 0$ there exists a $\delta > 0$ such that

$$\left|(x + kx) \sin \frac{1}{x}\right| < \epsilon \quad \text{whenever} \quad 0 < |x| < \delta$$

Thus

$$\lim_{\substack{(x,y)\to(0,0)\\(P\text{ in }S_2)}} f(x, y) = 0 \tag{15}$$

Let S_3 be the set of all points (x, y) for which $y = kx^n$, where n is any positive integer and $x \neq 0$. Then by an argument similar to that used for proving (15) it follows that

$$\lim_{\substack{(x,y)\to(0,0)\\(P\text{ in }S_3)}} f(x, y) = \lim_{x\to 0} (x + kx^n) \sin \frac{1}{x} = 0$$

We now attempt to find a $\delta > 0$ for any $\epsilon > 0$ such that

$$|f(x, y) - 0| < \epsilon \quad \text{whenever} \quad 0 < \|(x, y) - (0, 0)\| < \delta \tag{16}$$

which will prove that $\lim_{(x,y)\to(0,0)} f(x, y) = 0$. We distinguish two cases: $x = 0$ and $x \neq 0$.

Case 1: If $x = 0$, $|f(x, y) - 0| = |0 - 0| = 0$, which is less than ϵ for any $\delta > 0$.

Case 2: If $x \neq 0$, $|f(x, y) - 0| = |(x + y) \sin(1/x)|$.

$$\begin{aligned}\left|(x + y) \sin \frac{1}{x}\right| &= |x + y| \left|\sin \frac{1}{x}\right| \\ &\leq |x + y|(1) \\ &\leq |x| + |y| \\ &\leq \sqrt{x^2 + y^2} + \sqrt{x^2 + y^2}\end{aligned}$$

That is,

$$\left|(x + y) \sin \frac{1}{x}\right| \le 2\sqrt{x^2 + y^2}$$

Then

$$\left|(x + y) \sin \frac{1}{x}\right| < 2 \cdot \tfrac{1}{2}\epsilon \quad \text{whenever} \quad 0 < \sqrt{x^2 + y^2} < \tfrac{1}{2}\epsilon$$

So take $\delta = \frac{1}{2}\epsilon$.

Therefore, in both cases we have found a $\delta > 0$ for any $\epsilon > 0$ such that (16) holds, and this proves that $\lim_{(x,y)\to(0,0)} f(x, y) = 0$.

EXERCISES 18.2

In Exercises 1 through 8, evaluate the given limit by the use of limit theorems.

1. $\lim_{(x,y)\to(2,3)} (3x^2 + xy - 2y^2)$ **2.** $\lim_{(x,y)\to(1,1)} \dfrac{x^2 - y^2}{x - y}$ **3.** $\lim_{(x,y)\to(2,-1)} \dfrac{x^3 + 8y^3}{x + 2y}$ **4.** $\lim_{(x,y)\to(-2,4)} y\sqrt[3]{x^3 + 2y}$

5. $\lim_{(x,y)\to(0,0)} \dfrac{e^x + e^y}{\cos x + \sin y}$ **6.** $\lim_{(x,y)\to(0,0)} \dfrac{e^x - e^y}{e^{-x} - e^{-y}}$ **7.** $\lim_{(x,y)\to(0,1)} \dfrac{\sqrt{x + y} - \sqrt{y}}{x}$ **8.** $\lim_{(x,y)\to(1,1)} \dfrac{\sqrt[3]{x} - \sqrt[3]{y}}{x - y}$

In Exercises 9 through 16, establish the limit by finding a $\delta > 0$ for any $\epsilon > 0$ such that Definition 18.2.5 holds.

9. $\lim_{(x,y)\to(3,2)} (3x - 4y) = 1$ **10.** $\lim_{(x,y)\to(2,4)} (5x - 3y) = -2$ **11.** $\lim_{(x,y)\to(-1,3)} (3x - 2y) = -9$

12. $\lim_{(x,y)\to(-2,1)} (5x + 4y) = -6$ **13.** $\lim_{(x,y)\to(1,1)} (x^2 + y^2) = 2$ **14.** $\lim_{(x,y)\to(2,3)} (2x^2 - y^2) = -1$

15. $\lim_{(x,y)\to(2,4)} (x^2 + 2x - y) = 4$ **16.** $\lim_{(x,y)\to(3,-1)} (x^2 + y^2 - 4x + 2y) = -4$

In Exercises 17 through 22, prove that for the given function f, $\lim_{(x,y)\to(0,0)} f(x, y)$ does not exist.

17. $f(x, y) = \dfrac{x^2 - y^2}{x^2 + y^2}$ **18.** $f(x, y) = \dfrac{x^2}{x^2 + y^2}$ **19.** $f(x, y) = \dfrac{x^4y^4}{(x^2 + y^4)^3}$

20. $f(x, y) = \dfrac{x^4 + 3x^2y^2 + 2xy^3}{(x^2 + y^2)^2}$ **21.** $f(x, y) = \dfrac{x^9y}{(x^6 + y^2)^2}$ **22.** $f(x, y) = \dfrac{x^2y^2}{x^3 + y^3}$

In Exercises 23 through 26, prove that $\lim_{(x,y)\to(0,0)} f(x, y)$ exists.

23. $f(x, y) = \dfrac{xy}{\sqrt{x^2 + y^2}}$ **24.** $f(x, y) = \dfrac{x^3 + y^3}{x^2 + y^2}$

25. $f(x, y) = \begin{cases} (x + y) \sin \dfrac{1}{x} \sin \dfrac{1}{y} & \text{if } x \ne 0 \text{ and } y \ne 0 \\ 0 & \text{if either } x = 0 \text{ or } y = 0 \end{cases}$ **26.** $f(x, y) = \begin{cases} \dfrac{1}{x} \sin (xy) & \text{if } x \ne 0 \\ y & \text{if } x = 0 \end{cases}$

In Exercises 27 through 29, show the application of Theorem 18.2.6 to find the indicated limit.

27. $\lim_{(x,y)\to(2,2)} \tan^{-1} \dfrac{y}{x}$ **28.** $\lim_{(x,y)\to(-2,3)} [\![5x + \tfrac{1}{4}y^2]\!]$ **29.** $\lim_{(x,y)\to(4,2)} \sqrt{\dfrac{1}{3x - 4y}}$

In Exercises 30 through 35, determine if the indicated limit exists.

30. $\lim\limits_{(x,y)\to(0,0)} \dfrac{x^2y^2}{x^4+y^4}$ **31.** $\lim\limits_{(x,y)\to(0,0)} \dfrac{x^2y^2}{x^2+y^2}$ **32.** $\lim\limits_{(x,y)\to(2,-2)} \dfrac{\sin(x+y)}{x+y}$ **33.** $\lim\limits_{(x,y)\to(0,0)} \dfrac{x^2+y}{x^2+y^2}$

34. $f(x, y) = \begin{cases} \dfrac{xy}{x^2+y^2} + y\sin\dfrac{1}{x} & \text{if } x \neq 0 \\ 0 & \text{if } x = 0 \end{cases}$; $\lim\limits_{(x,y)\to(0,0)} f(x, y)$

35. $f(x, y) = \begin{cases} x\sin\dfrac{1}{y} + y\sin\dfrac{1}{x} & \text{if } x \neq 0 \text{ and } y \neq 0 \\ 0 & \text{if either } x = 0 \text{ or } y = 0 \end{cases}$; $\lim\limits_{(x,y)\to(0,0)} f(x, y)$

36. (a) Give a definition, similar to Definition 18.2.5, of the limit of a function of three variables as a point (x, y, z) approaches a point (x_0, y_0, z_0). (b) Give a definition, similar to Definition 18.2.8, of the limit of a function of three variables as a point (x, y, z) approaches a point (x_0, y_0, z_0) in a specific set of points S in R^3.

37. (a) State and prove a theorem similar to Theorem 18.2.9 for a function f of three variables. (b) State and prove a theorem similar to Theorem 18.2.10 for a function f of three variables.

In Exercises 38 through 41, evaluate the given limit by the use of limit theorems.

38. $\lim\limits_{(x,y,z)\to(-2,1,4)} (4x^2y - 3xyz^2 + 7y^2z^3)$ **39.** $\lim\limits_{(x,y,z)\to(\pi/3,1,\pi)} \dfrac{\sec xy + \sec yz}{y - \sec z}$

40. $\lim\limits_{(x,y,z)\to(1,-1,1)} \dfrac{x^3 + y^3z^3}{x + yz}$ **41.** $\lim\limits_{(x,y,z)\to(0,0,0)} \dfrac{e^xe^z - e^ye^z}{e^{2x} - e^{2y}}$

In Exercises 42 through 45, use the definitions and theorems of Exercises 36 and 37 to prove that $\lim\limits_{(x,y,z)\to(0,0,0)} f(x, y, z)$ *does not exist.*

42. $f(x, y, z) = \dfrac{x^3 + yz^2}{x^4 + y^2 + z^4}$ **43.** $f(x, y, z) = \dfrac{x^2 + y^2 - z^2}{x^2 + y^2 + z^2}$

44. $f(x, y, z) = \dfrac{x^4 + yx^3 + z^2x^2}{x^4 + y^4 + z^4}$ **45.** $f(x, y, z) = \dfrac{x^2y^2z^2}{x^6 + y^6 + z^6}$

In Exercises 46 and 47, use the definition in Exercise 36(a) to prove that $\lim\limits_{(x,y,z)\to(0,0,0)} f(x, y, z)$ exists.

46. $f(x, y, z) = \dfrac{y^3 + xz^2}{x^2 + y^2 + z^2}$

47. $f(x, y, z) = \begin{cases} (x + y + z)\sin\dfrac{1}{x}\sin\dfrac{1}{y} & \text{if } x \neq 0 \text{ and } y \neq 0 \\ 0 & \text{if either } x = 0 \text{ or } y = 0 \end{cases}$

48. Suppose that f and g are functions of two variables satisfying the following conditions:
(i) $f(tx, ty) = t^nf(x, y)$; $g(tx, ty) = t^ng(x, y)$ for some n and for all t;
(ii) $g(1, 1) \neq 0$ and $g(1, 0) \neq 0$;
(iii) $g(1, 1) \cdot f(1, 0) \neq g(1, 0) \cdot f(1, 1)$.

Show that $\lim\limits_{(x,y)\to(0,0)} \dfrac{f(x, y)}{g(x, y)}$ does not exist.

49. Prove Theorem 18.2.6.

18.3 CONTINUITY OF FUNCTIONS OF MORE THAN ONE VARIABLE

Following is the definition of continuity of a function of n variables at a point in R^n.

18.3.1 Definition Suppose that f is a function of n variables and A is a point in R^n. Then f is said to be **continuous** at the point A if and only if the following three conditions are satisfied:

(i) $f(A)$ exists;
(ii) $\lim\limits_{P \to A} f(P)$ exists;
(iii) $\lim\limits_{P \to A} f(P) = f(A)$.

If one or more of these three conditions fails to hold at the point A, then f is said to be **discontinuous** at A.

Definition 2.6.1 of continuity of a function of one variable at a number a is a special case of Definition 18.3.1.

If f is a function of two variables, A is the point (x_0, y_0), and P is a point (x, y), then Definition 18.3.1 becomes the following.

18.3.2 Definition The function f of two variables x and y is said to be **continuous** at the point (x_0, y_0) if and only if the following three conditions are satisfied:

(i) $f(x_0, y_0)$ exists;
(ii) $\lim\limits_{(x,y) \to (x_0,y_0)} f(x, y)$ exists;
(iii) $\lim\limits_{(x,y) \to (x_0,y_0)} f(x, y) = f(x_0, y_0)$.

Example 1 Given

$$f(x, y) = \begin{cases} \dfrac{3x^2y}{x^2 + y^2} & \text{if } f(x, y) \neq (0, 0) \\ 0 & \text{if } f(x, y) = (0, 0) \end{cases}$$

determine if f is continuous at $(0, 0)$.

Solution We check the three conditions of Definition 18.3.2 at the point $(0, 0)$

(i) $f(0, 0) = 0$. Therefore condition (i) holds.

(ii) $\lim\limits_{(x,y) \to (0,0)} f(x, y) = \lim\limits_{(x,y) \to (0,0)} \dfrac{3x^2y}{x^2 + y^2} = 0$, which was proved in Example 5, Sec. 18.2.

(iii) $\lim\limits_{(x,y) \to (0,0)} f(x, y) = f(0, 0)$.

Therefore f is continuous at $(0, 0)$.

Example 2 Let the function f be defined by

$$f(x, y) = \begin{cases} \dfrac{xy}{x^2 + y^2} & \text{if } f(x, y) \neq (0, 0) \\ 0 & \text{if } f(x, y) = (0, 0) \end{cases}$$

Determine if f is continuous at $(0, 0)$.

Solution Checking the conditions of Definition 18.3.2, we have the following.

(i) $f(0, 0) = 0$; so condition (i) holds.
(ii) When $(x, y) \neq (0, 0)$, $f(x, y) = xy/(x^2 + y^2)$. In Example 4, Sec. 18.2, we showed that $\lim\limits_{(x,y)\to(0,0)} xy/(x^2 + y^2)$ does not exist and so $\lim\limits_{(x,y)\to(0,0)} f(x, y)$ does not exist. Therefore condition (ii) fails to hold. Thus f is discontinuous at $(0, 0)$.

If a function f of two variables is discontinuous at the point (x_0, y_0) but $\lim\limits_{(x,y)\to(x_0,y_0)} f(x, y)$ exists, then f is said to have a **removable discontinuity** at (x_0, y_0) because if f is redefined at (x_0, y_0) so that $f(x_0, y_0) = \lim\limits_{(x,y)\to(x_0,y_0)} f(x, y)$, then f becomes continuous at (x_0, y_0). If the discontinuity is not removable, it is called an **essential discontinuity.**

● ILLUSTRATION 1

(a) If $g(x, y) = 3x^2y/(x^2 + y^2)$, then g is discontinuous at the origin because $g(0, 0)$ is not defined. However, in Example 5, Sec. 18.2, we showed that $\lim\limits_{(x,y)\to(0,0)} 3x^2y/(x^2 + y^2) = 0$. Therefore the discontinuity is removable if $g(0, 0)$ is defined to be 0. (Refer to Example 1.)

(b) Let $h(x, y) = xy/(x^2 + y^2)$. Then h is discontinuous at the origin because $h(0, 0)$ is not defined. In Example 4, Sec. 18.2, we showed that $\lim\limits_{(x,y)\to(0,0)} xy/(x^2 + y^2)$ does not exist. Therefore the discontinuity is essential. (Refer to Example 2.) ●

The theorems about continuity for functions of a single variable can be extended to functions of two variables.

18.3.3 Theorem If f and g are two functions that are continuous at the point (x_0, y_0), then

(i) $f + g$ is continuous at (x_0, y_0);
(ii) $f - g$ is continuous at (x_0, y_0);
(iii) fg is continuous at (x_0, y_0);
(iv) f/g is continuous at (x_0, y_0), provided that $g(x_0, y_0) \neq 0$.

The proof of this theorem is analogous to the proof of the corresponding theorem (2.7.1) for functions of one variable, and hence it is omitted.

18.3.4 Theorem A polynomial function of two variables is continuous at every point in R^2.

Proof Every polynomial function is the sum of products of the functions defined by $f(x, y) = x$, $g(x, y) = y$, and $h(x, y) = c$, where c is a real number. Because f, g, and h are continuous at every point in R^2, the theorem follows by repeated applications of Theorem 18.3.3, parts (i) and (iii). ■

18.3.5 Theorem A rational function of two variables is continuous at every point in its domain.

Proof A rational function is the quotient of two polynomial functions f and g that are continuous at every point in R^2, by Theorem 18.3.4. If (x_0, y_0) is any point in the domain of f/g, then $g(x_0, y_0) \neq 0$; so by Theorem 18.3.3(iv) f/g is continuous there. ■

Example 3 Let the function f be defined by

$$f(x, y) = \begin{cases} x^2 + y^2 & \text{if } x^2 + y^2 \leq 1 \\ 0 & \text{if } x^2 + y^2 > 1 \end{cases}$$

Discuss the continuity of f. What is the region of continuity of f?

Solution The function f is defined at all points in R^2. Therefore condition (i) of Definition 18.3.2 holds for every point (x_0, y_0).

Consider the points (x_0, y_0) if $x_0^2 + y_0^2 \neq 1$. If $x_0^2 + y_0^2 < 1$, then

$$\lim_{(x,y)\to(x_0,y_0)} f(x, y) = \lim_{(x,y)\to(x_0,y_0)} (x^2 + y^2) = x_0^2 + y_0^2 = f(x_0, y_0)$$

If $x_0^2 + y_0^2 > 1$, then

$$\lim_{(x,y)\to(x_0,y_0)} f(x, y) = \lim_{(x,y)\to(x_0,y_0)} 0 = 0 = f(x_0, y_0)$$

Thus f is continuous at all points (x_0, y_0) for which $x_0^2 + y_0^2 \neq 1$.

To determine the continuity of f at points (x_0, y_0) for which $x_0^2 + y_0^2 = 1$, we determine if $\lim_{(x,y)\to(x_0,y_0)} f(x, y)$ exists and equals 1.

Let S_1 be the set of all points (x, y) such that $x^2 + y^2 \leq 1$. Then

$$\lim_{\substack{(x,y)\to(x_0,y_0) \\ (P \text{ in } S_1)}} f(x, y) = \lim_{\substack{(x,y)\to(x_0,y_0) \\ (P \text{ in } S_1)}} (x^2 + y^2) = x_0^2 + y_0^2 = 1$$

Let S_2 be the set of all points (x, y) such that $x^2 + y^2 > 1$. Then

$$\lim_{\substack{(x,y)\to(x_0,y_0) \\ (P \text{ in } S_2)}} f(x, y) = \lim_{\substack{(x,y)\to(x_0,y_0) \\ (P \text{ in } S_2)}} 0 = 0$$

Because

$$\lim_{\substack{(x,y)\to(x_0,y_0) \\ (P \text{ in } S_1)}} f(x, y) \neq \lim_{\substack{(x,y)\to(x_0,y_0) \\ (P \text{ in } S_2)}} f(x, y)$$

we conclude that $\lim_{(x,y)\to(x_0,y_0)} f(x, y)$ does not exist. Hence f is discontinuous at all points (x_0, y_0) for which $x_0^2 + y_0^2 = 1$. The region of continuity of f consists of all points in the xy plane except those on the circle $x^2 + y^2 = 1$.

18.3.6 Definition The function f of n variables is said to be **continuous on an open ball** if it is continuous at every point of the open ball.

As an illustration of the above definition, the function of Example 3 is continuous on every open disk that does not contain a point of the circle $x^2 + y^2 = 1$.

The following theorem states that a continuous function of a continuous function is continuous. It is analogous to Theorem 2.7.7.

18.3.7 Theorem Suppose that f is a function of a single variable and g is a function of two variables. Suppose further that g is continuous at (x_0, y_0) and f is continuous at $g(x_0, y_0)$. Then the composite function $f \circ g$ is continuous at (x_0, y_0).

The proof of this theorem, which makes use of Theorem 18.2.6, is similar to the proof of Theorem 2.7.7 and is left as an exercise (see Exercise 15).

● ILLUSTRATION 2 Let

$$h(x, y) = \ln (xy - 1)$$

We discuss the continuity of h. If g is the function defined by $g(x, y) = xy - 1$, g is continuous at all points in R^2. The natural logarithmic function is continuous on its entire domain, which is the set of all positive numbers. So if f is the function defined by $f(t) = \ln t$, f is continuous for all $t > 0$. Then the function h is the composite function $f \circ g$ and, by Theorem 18.3.7, is continuous at all points (x, y) in R^2 for which $xy - 1 > 0$ or, equivalently, $xy > 1$. The shaded region of Fig. 18.3.1 is the region of continuity of h. ●

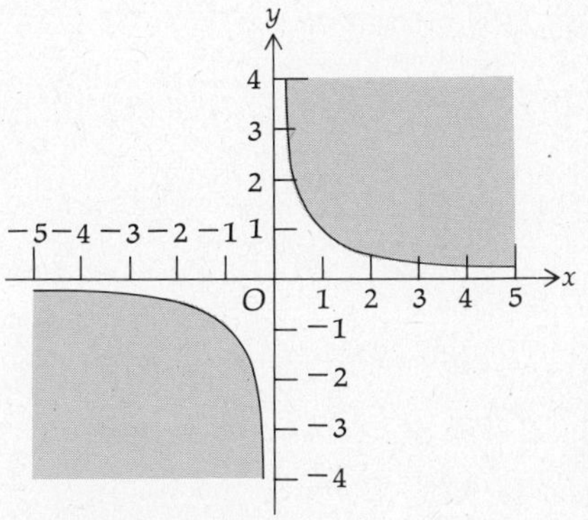

FIGURE 18.3.1

Example 4 Given that

$$f(x, y) = \frac{y}{\sqrt{x^2 + y^2 - 25}}$$

discuss the continuity of f and draw a sketch showing as a shaded region in R the region of continuity of f.

Solution This is the same function as the function G of Illustration 4 in Sec. 18.1. The domain of this function is the set of all points in the exterior region bounded by the circle $x^2 + y^2 = 25$ and the points on the x axis for which $-5 < x < 5$.

The function f is the quotient of the functions g and h for which $g(x, y) = y$ and $h(x, y) = \sqrt{x^2 + y^2 - 25}$. The function g is a polynomial function and therefore is continuous everywhere. It follows from Theorem 18.3.7 that h is continuous at all points in R^2 for which $x^2 + y^2 > 25$.

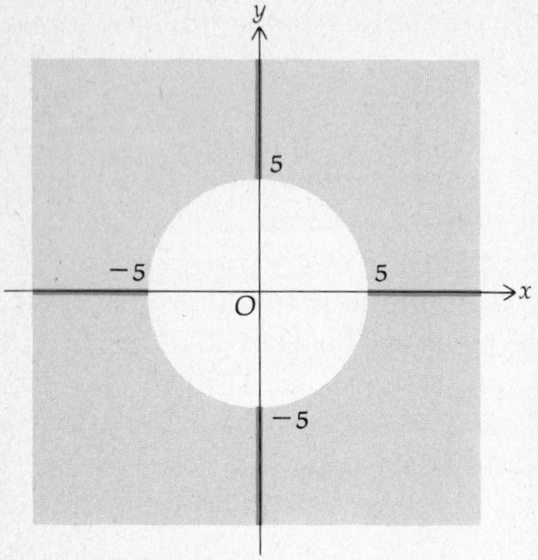

FIGURE 18.3.2

Therefore, by Theorem 18.3.3(iv), f is continuous at all points in the exterior region bounded by the circle $x^2 + y^2 = 25$.

Now consider the points on the x axis for which $-5 < x < 5$, that is, the points $(a, 0)$ where $-5 < a < 5$. If S_1 is the set of points on the line $x = a$, then $(a, 0)$ is an accumulation point of S_1, but

$$\lim_{\substack{(x,y)\to(a,0)\\(P \text{ in } S_1)}} f(x, y) = \lim_{y\to 0} \frac{y}{\sqrt{a^2 + y^2 - 25}}$$

which does not exist because $y/\sqrt{a^2 + y^2 - 25}$ is not defined if $|y| \leq \sqrt{25 - a^2}$. Therefore f is discontinuous at the points on the x axis for which $-5 < x < 5$. The shaded region of Fig. 18.3.2 is the region of continuity of f.

EXERCISES 18.3

In Exercises 1 through 14, discuss the continuity of the given function.

1. $f(x, y) = \dfrac{x^2}{y - 1}$ **2.** $F(x, y) = \dfrac{1}{x - y}$ **3.** $h(x, y) = \sin \dfrac{y}{x}$

4. $f(x, y) = \ln xy^2$ **5.** $f(x, y) = \dfrac{4x^2y + 3y^2}{2x - y}$ **6.** $g(x, y) = \dfrac{5xy^2 + 2y}{16 - x^2 - 4y^2}$

7. $g(x, y) = \ln(25 - x^2 - y^2)$ **8.** $f(x, y) = \cos^{-1}(x + y)$

9. $f(x, y) = \begin{cases} \dfrac{xy}{\sqrt{x^2 + y^2}} & \text{if } (x, y) \neq (0, 0) \\ 0 & \text{if } (x, y) = (0, 0) \end{cases}$ (*Hint:* See Exercise 23, Exercises 18.2.)

10. $h(x, y) = \begin{cases} \dfrac{x^2y}{x^4 + y^2} & \text{if } (x, y) \neq (0, 0) \\ 0 & \text{if } (x, y) = (0, 0) \end{cases}$ (*Hint:* See Example 6, Sec. 18.2.)

11. $f(x, y) = \begin{cases} \dfrac{x + y}{x^2 + y^2} & \text{if } (x, y) \neq (0, 0) \\ 0 & \text{if } (x, y) = (0, 0) \end{cases}$ **12.** $f(x, y) = \begin{cases} \dfrac{x^3 + y^3}{x^2 + y^2} & \text{if } (x, y) \neq (0, 0) \\ 0 & \text{if } (x, y) = (0, 0) \end{cases}$

13. $G(x, y) = \begin{cases} \dfrac{xy}{|x| + |y|} & \text{if } (x, y) \neq (0, 0) \\ 0 & \text{if } (x, y) = (0, 0) \end{cases}$ **14.** $F(x, y) = \begin{cases} \dfrac{x^2y^2}{|x^3| + |y^3|} & \text{if } (x, y) \neq (0, 0) \\ 0 & \text{if } (x, y) = (0, 0) \end{cases}$

15. Prove Theorem 18.3.7.

In Exercises 16 through 25, determine the region of continuity of f, and draw a sketch showing as a shaded region in R^2 the region of continuity of f.

16. $f(x, y) = \dfrac{y}{\sqrt{x^2 - y^2 - 4}}$ **17.** $f(x, y) = \dfrac{xy}{\sqrt{16 - x^2 - y^2}}$

18. $f(x, y) = \dfrac{x^2 + y^2}{\sqrt{9 - x^2 - y^2}}$ **19.** $f(x, y) = \dfrac{x}{\sqrt{4x^2 + 9y^2 - 36}}$

20. $f(x, y) = \ln(x^2 + y^2 - 9) - \ln(1 - x^2 - y^2)$

21. $f(x, y) = \sec^{-1}(xy)$

22. $f(x, y) = \sin^{-1}(xy)$

23. $f(x, y) = \sin^{-1}(x + y) + \ln(xy)$

24. $f(x, y) = \begin{cases} \dfrac{x^2 - y^2}{x - y} & \text{if } x \neq y \\ x - y & \text{if } x = y \end{cases}$

25. $f(x, y) = \begin{cases} \dfrac{\sin(x + y)}{x + y} & \text{if } x + y \neq 0 \\ 1 & \text{if } x + y = 0 \end{cases}$

In Exercises 26 through 31, the function is discontinuous at the origin because $f(0, 0)$ does not exist. Determine if the discontinuity is removable or essential. If the discontinuity is removable, define $f(0, 0)$ so that the discontinuity is removed.

26. $f(x, y) = \dfrac{x}{x^2 + y^2}$

27. $f(x, y) = (x + y)\sin\dfrac{x}{y}$

28. $f(x, y) = \dfrac{x^2y^2}{x^2 + y^2}$

29. $f(x, y) = \dfrac{x^3y^2}{x^6 + y^4}$

30. $f(x, y) = \dfrac{2y^2 - 3xy}{\sqrt{x^2 + y^2}}$

31. $f(x, y) = \dfrac{x^3 - 4xy^2}{x^2 + y^2}$

32. The function F is defined by

$$F(x, y) = \begin{cases} x^2 - 3y^2 & \text{if } x^2 - 3y^2 \leq 1 \\ 2 & \text{if } x^2 - 3y^2 > 1 \end{cases}$$

Show that the region of continuity of F consists of all points in R^2 except those on the hyperbola $x^2 - 3y^2 = 1$.

33. The function G is defined by

$$G(x, y) = \begin{cases} x^2 + 4y^2 & \text{if } x^2 + 4y^2 \leq 5 \\ 3 & \text{if } x^2 + 4y^2 > 5 \end{cases}$$

Show that the region of continuity of G consists of all points in R^2 except those on the ellipse $x^2 + 4y^2 = 5$.

34. (a) Give a definition of continuity at a point for a function of three variables, similar to Definition 18.3.2. (b) State theorems for functions of three variables similar to Theorems 18.3.3 and 18.3.7. (c) Define a polynomial function of three variables and a rational function of three variables.

In Exercises 35 through 38, use the definitions and theorems of Exercise 34 to discuss the continuity of the given function.

35. $f(x, y, z) = \dfrac{xz}{\sqrt{x^2 + y^2 + z^2 - 1}}$

36. $f(x, y, z) = \ln(36 - 4x^2 - y^2 - 9z^2)$

37. $f(x, y, z) = \begin{cases} \dfrac{3xyz}{x^2 + y^2 + z^2} & \text{if } (x, y, z) \neq (0, 0, 0) \\ 0 & \text{if } (x, y, z) = (0, 0, 0) \end{cases}$

38. $f(x, y, z) = \begin{cases} \dfrac{xz - y^2}{x^2 + y^2 + z^2} & \text{if } (x, y, z) \neq (0, 0, 0) \\ 0 & \text{if } (x, y, z) = (0, 0, 0) \end{cases}$

18.4 PARTIAL DERIVATIVES

The discussion of the differentiation of real-valued functions of n variables, is reduced to the one-dimensional case by treating a function of n variables as a function of one variable at a time and holding the others fixed. This leads to the concept of a *partial derivative*. We first define the partial derivative of a function of two variables.

18.4.1 Definition Let f be a function of two variables, x and y. The **partial derivative of f with respect to x** is that function, denoted by D_1f, such that its function value at any point (x, y) in the domain of f is given by

$$D_1 f(x, y) = \lim_{\Delta x \to 0} \frac{f(x + \Delta x, y) - f(x, y)}{\Delta x} \tag{1}$$

if this limit exists. Similarly, the **partial derivative of f with respect to y** is that function, denoted by $D_2 f$, such that its function value at any point (x, y) in the domain of f is given by

$$D_2 f(x, y) = \lim_{\Delta y \to 0} \frac{f(x, y + \Delta y) - f(x, y)}{\Delta y} \tag{2}$$

if this limit exists.

The process of finding a partial derivative is called **partial differentiation.**

$D_1 f$ is read as "D sub 1 of f" and this denotes the function that is the partial derivative of f with respect to the first variable. $D_1 f(x, y)$ is read as "D sub 1 of f of x and y," and this denotes the function value of $D_1 f$ at the point (x, y). Other notations for $D_1 f$ are f_1, f_x, and $\partial f/\partial x$. Other notations for $D_1 f(x, y)$ are $f_1(x, y)$, $f_x(x, y)$, and $\partial f(x, y)/\partial x$. Similarly, other notations for $D_2 f$ are f_2, f_y, and $\partial f/\partial y$; other notations for $D_2 f(x, y)$ are $f_2(x, y)$, $f_y(x, y)$, and $\partial f(x, y)/\partial y$. If $z = f(x, y)$, we can write $\partial z/\partial x$ for $D_1 f(x, y)$. A partial derivative cannot be thought of as a ratio of ∂z and ∂x because neither of these symbols has a separate meaning. The notation dy/dx can be regarded as the quotient of two differentials when y is a function of the single variable x, but there is not a similar interpretation for $\partial z/\partial x$.

Example 1 Given $f(x, y) = 3x^2 - 2xy + y^2$, find $D_1 f(x, y)$ and $D_2 f(x, y)$ by applying Definition 18.4.1.

Solution

$$\begin{aligned}
D_1 f(x, y) &= \lim_{\Delta x \to 0} \frac{f(x + \Delta x, y) - f(x, y)}{\Delta x} \\
&= \lim_{\Delta x \to 0} \frac{3(x + \Delta x)^2 - 2(x + \Delta x)y + y^2 - (3x^2 - 2xy + y^2)}{\Delta x} \\
&= \lim_{\Delta x \to 0} \frac{3x^2 + 6x\,\Delta x + 3(\Delta x)^2 - 2xy - 2y\,\Delta x + y^2 - 3x^2 + 2xy - y^2}{\Delta x} \\
&= \lim_{\Delta x \to 0} \frac{6x\,\Delta x + 3(\Delta x)^2 - 2y\,\Delta x}{\Delta x} \\
&= \lim_{\Delta x \to 0} (6x + 3\,\Delta x - 2y) \\
&= 6x - 2y
\end{aligned}$$

$$\begin{aligned}
D_2 f(x, y) &= \lim_{\Delta y \to 0} \frac{f(x, y + \Delta y) - f(x, y)}{\Delta y} \\
&= \lim_{\Delta y \to 0} \frac{3x^2 - 2x(y + \Delta y) + (y + \Delta y)^2 - (3x^2 - 2xy + y^2)}{\Delta y}
\end{aligned}$$

$$= \lim_{\Delta y \to 0} \frac{3x^2 - 2xy - 2x\,\Delta y + y^2 + 2y\,\Delta y + (\Delta y)^2 - 3x^2 + 2xy - y^2}{\Delta y}$$

$$= \lim_{\Delta y \to 0} \frac{-2x\,\Delta y + 2y\,\Delta y + (\Delta y)^2}{\Delta y}$$

$$= \lim_{\Delta y \to 0} (-2x + 2y + \Delta y)$$

$$= -2x + 2y$$

If (x_0, y_0) is a particular point in the domain of f, then

$$D_1 f(x_0, y_0) = \lim_{\Delta x \to 0} \frac{f(x_0 + \Delta x, y_0) - f(x_0, y_0)}{\Delta x} \tag{3}$$

if this limit exists, and

$$D_2 f(x_0, y_0) = \lim_{\Delta y \to 0} \frac{f(x_0, y_0 + \Delta y) - f(x_0, y_0)}{\Delta y} \tag{4}$$

if this limit exists.

● ILLUSTRATION 1 We apply formula (3) to find $D_1 f(3, -2)$ for the function f of Example 1.

$$D_1 f(3, -2) = \lim_{\Delta x \to 0} \frac{f(3 + \Delta x, -2) - f(3, -2)}{\Delta x}$$

$$= \lim_{\Delta x \to 0} \frac{3(3 + \Delta x)^2 - 2(3 + \Delta x)(-2) + (-2)^2 - (27 + 12 + 4)}{\Delta x}$$

$$= \lim_{\Delta x \to 0} \frac{27 + 18\,\Delta x + 3(\Delta x)^2 + 12 + 4\,\Delta x + 4 - 43}{\Delta x}$$

$$= \lim_{\Delta x \to 0} (18 + 3\,\Delta x + 4)$$

$$= 22$$ ●

Alternate formulas to (3) and (4) for $D_1 f(x_0, y_0)$ and $D_2 f(x_0, y_0)$ are given by

$$D_1 f(x_0, y_0) = \lim_{x \to x_0} \frac{f(x, y_0) - f(x_0, y_0)}{x - x_0} \tag{5}$$

if this limit exists, and

$$D_2 f(x_0, y_0) = \lim_{y \to y_0} \frac{f(x_0, y) - f(x_0, y_0)}{y - y_0} \tag{6}$$

if this limit exists.

● ILLUSTRATION 2 We apply formula (5) to find $D_1 f(3, -2)$ for the function f of Example 1.

$$\begin{aligned} D_1 f(3, -2) &= \lim_{x \to 3} \frac{f(x, -2) - f(3, -2)}{x - 3} \\ &= \lim_{x \to 3} \frac{3x^2 + 4x + 4 - 43}{x - 3} \\ &= \lim_{x \to 3} \frac{3x^2 + 4x - 39}{x - 3} \\ &= \lim_{x \to 3} \frac{(3x + 13)(x - 3)}{x - 3} \\ &= \lim_{x \to 3} (3x + 13) \\ &= 22 \end{aligned}$$

●

● ILLUSTRATION 3 In Example 1 we showed that

$$D_1 f(x, y) = 6x - 2y$$

Therefore

$$\begin{aligned} D_1 f(3, -2) &= 18 + 4 \\ &= 22 \end{aligned}$$

This result agrees with those of Illustrations 1 and 2. ●

Comparing Definition 18.4.1 with the definition of an ordinary derivative (3.2.1), we see that $D_1 f(x, y)$ is the ordinary derivative of f if f is considered as a function of one variable x (i.e., y is held constant), and $D_2 f(x, y)$ is the ordinary derivative of f if f is considered as a function of one variable y (and x is held constant). So the results in Example 1 can be obtained more easily by applying the theorems for ordinary differentiation if y is considered constant when finding $D_1 f(x, y)$ and if x is considered constant when finding $D_2 f(x, y)$. The following example illustrates this.

Example 2 Given $f(x, y) = 3x^3 - 4x^2y + 3xy^2 + 7x - 8y$, find $D_1 f(x, y)$ and $D_2 f(x, y)$.

Solution Treating f as a function of x and holding y constant we have

$$D_1 f(x, y) = 9x^2 - 8xy + 3y^2 + 7$$

Considering f as a function of y and holding x constant we have

$$D_2 f(x, y) = -4x^2 + 6xy - 8$$

Example 3 Given

$$f(x, y) = \begin{cases} \dfrac{xy(x^2 - y^2)}{x^2 + y^2} & \text{if } (x, y) \neq (0, 0) \\ 0 & \text{if } (x, y) = (0, 0) \end{cases}$$

find: (a) $f_1(0, y)$; (b) $f_2(x, 0)$.

Solution

(a) If $y \neq 0$, from (5),

$$\begin{aligned} f_1(0, y) &= \lim_{x \to 0} \frac{f(x, y) - f(0, y)}{x - 0} \\ &= \lim_{x \to 0} \frac{\dfrac{xy(x^2 - y^2)}{x^2 + y^2} - 0}{x} \\ &= \lim_{x \to 0} \frac{y(x^2 - y^2)}{x^2 + y^2} \\ &= -\frac{y^3}{y^2} \\ &= -y \end{aligned}$$

If $y = 0$, we have

$$f_1(0, 0) = \lim_{x \to 0} \frac{f(x, 0) - f(0, 0)}{x - 0} = \lim_{x \to 0} \frac{0 - 0}{x} = 0$$

Because $f_1(0, y) = -y$ if $y \neq 0$ and $f_1(0, 0) = 0$, we can conclude that $f_1(0, y) = -y$ for all y.

(b) If $x \neq 0$, from (6),

$$\begin{aligned} f_2(x, 0) &= \lim_{y \to 0} \frac{f(x, y) - f(x, 0)}{y - 0} \\ &= \lim_{y \to 0} \frac{\dfrac{xy(x^2 - y^2)}{x^2 + y^2} - 0}{y} \\ &= \lim_{y \to 0} \frac{x(x^2 - y^2)}{x^2 + y^2} \\ &= \frac{x^3}{x^2} \\ &= x \end{aligned}$$

If $x = 0$, we have

$$f_2(0, 0) = \lim_{y \to 0} \frac{f(0, y) - f(0, 0)}{y - 0} = \lim_{y \to 0} \frac{0 - 0}{y} = 0$$

Because $f_2(x, 0) = x$, if $x \neq 0$ and $f_2(0, 0) = 0$, then $f_2(x, 0) = x$ for all x.

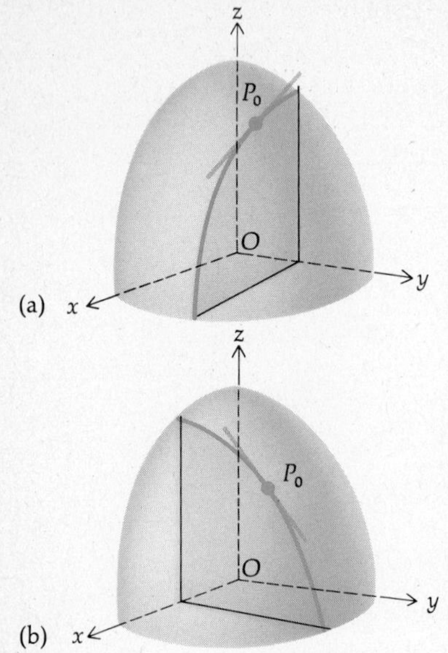

FIGURE 18.4.1

Geometric interpretations of the partial derivatives of a function of two variables are similar to those of a function of one variable. The graph of a function f of two variables is a surface having equation $z = f(x, y)$. If y is held constant (say, $y = y_0$), then $z = f(x, y_0)$ is an equation of the trace of this surface in the plane $y = y_0$. The curve can be represented by the two equations

$$y = y_0 \quad \text{and} \quad z = f(x, y) \tag{7}$$

because the curve is the intersection of these two surfaces.

Then $D_1 f(x_0, y_0)$ is the slope of the tangent line to the curve given by Eqs. (7) at the point $P_0(x_0, y_0, f(x_0, y_0))$ in the plane $y = y_0$. In an analogous fashion, $D_2 f(x_0, y_0)$ represents the slope of the tangent line to the curve having equations

$$x = x_0 \quad \text{and} \quad z = f(x, y)$$

at the point P_0 in the plane $x = x_0$. Figure 18.4.1(a) and (b) shows the portions of the curves and the tangent lines.

Example 4 Find the slope of the tangent line to the curve of intersection of the surface $z = \frac{1}{2}\sqrt{24 - x^2 - 2y^2}$ with the plane $y = 2$ at the point $(2, 2, \sqrt{3})$.

Solution The required slope is the value of $\partial z/\partial x$ at the point $(2, 2, \sqrt{3})$.

$$\frac{\partial z}{\partial x} = \frac{-x}{2\sqrt{24 - x^2 - 2y^2}}$$

So at $(2, 2, \sqrt{3})$,

$$\frac{\partial z}{\partial x} = \frac{-2}{2\sqrt{12}} = -\frac{1}{2\sqrt{3}}$$

Because every derivative is a measure of a rate of change, a partial derivative can be so interpreted. If f is a function of the two variables x and y, the partial derivative of f with respect to x at the point $P_0(x_0, y_0)$ gives the instantaneous rate of change, at P_0, of $f(x, y)$ per unit change in x (x alone varies and y is held fixed at y_0). Similarly, the partial derivative of f with respect to y at P_0 gives the instantaneous rate of change, at P_0, of $f(x, y)$ per unit change in y.

Example 5 According to the *ideal gas law* for a confined gas if P newtons per square centimeter is the pressure, V cubic centimeters is the volume, and T degrees is the temperature, we have the formula

$$PV = kT \tag{8}$$

where k is a constant of proportionality. Suppose that the volume of a gas in a certain container is 100 cm³ and the temperature is 90° and $k = 8$.

(a) Find the instantaneous rate of change of P per unit change in T if V remains fixed at 100. (b) Use the result of part (a) to approximate the change in the pressure if the temperature is increased to 92°. (c) Find the instantaneous rate of change of V per unit change in P if T remains fixed at 90. (d) Suppose that the temperature is held constant. Use the result of part (c) to find the approximate change in the volume necessary to produce the same change in the pressure as obtained in part (b).

Solution Substituting $V = 100$, $T = 90$, and $k = 8$ in Eq. (8) we obtain $P = 7.2$.

(a) Solving (8) for P when $k = 8$ we get

$$P = \frac{8T}{V}$$

The instantaneous rate of change of P per unit change in T if V remains fixed is $\partial P/\partial t$, and

$$\frac{\partial P}{\partial T} = \frac{8}{V}$$

When $T = 90$ and $V = 100$, $\partial P/\partial T = 0.08$, which is the answer required.

(b) From the result of part (a), when T is increased by 2 (and V remains fixed), an approximate increase in P is $2(0.08) = 0.16$. We conclude, then, that if the temperature is increased from 90° to 92°, the increase in the pressure is approximately 0.16 nt/cm².

(c) Solving (8) for V when $k = 8$ we obtain

$$V = \frac{8T}{P}$$

The instantaneous rate of change of V per unit change in P if T remains fixed is $\partial V/\partial P$, and

$$\frac{\partial V}{\partial P} = -\frac{8T}{P^2}$$

When $T = 90$ and $P = 7.2$,

$$\frac{\partial V}{\partial P} = -\frac{8(90)}{(7.2)^2} = -\frac{125}{9}$$

which is the instantaneous rate of change of V per unit change in P when $T = 90$ and $P = 7.2$ if T remains fixed at 90.

(d) If P is to be increased by 0.16 and T is held fixed, then from the result of part (c) the change in V should be approximately $(0.16)(-\frac{125}{9}) = -\frac{20}{9}$. Hence the volume should be decreased by approximately $\frac{20}{9}$ cm³ if the pressure is to be increased from 7.2 nt/cm² to 7.36 nt/cm².

We now extend the concept of partial derivative to functions of n variables.

18.4.2 Definition Let $P(x_1, x_2, \ldots, x_n)$ be a point in R^n, and let f be a function of the n variables $x_1, x_2, \ldots, x_n$. Then the partial derivative of f with respect to x_k is that function, denoted by $D_k f$, such that its function value at any point

P in the domain of f is given by

$$D_k f(x_1, x_2, \ldots, x_n) = \lim_{\Delta x_k \to 0} \frac{f(x_1, x_2, \ldots, x_{k-1}, x_k + \Delta x_k, x_{k+1}, \ldots, x_n) - f(x_1, x_2, \ldots, x_n)}{\Delta x_k}$$

if this limit exists.

In particular, if f is a function of the three variables x, y, and z, then the partial derivatives of f are given by

$$D_1 f(x, y, z) = \lim_{\Delta x \to 0} \frac{f(x + \Delta x, y, z) - f(x, y, z)}{\Delta x}$$

$$D_2 f(x, y, z) = \lim_{\Delta y \to 0} \frac{f(x, y + \Delta y, z) - f(x, y, z)}{\Delta y}$$

and

$$D_3 f(x, y, z) = \lim_{\Delta z \to 0} \frac{f(x, y, z + \Delta z) - f(x, y, z)}{\Delta z}$$

if these limits exist.

Example 6 Given $f(x, y, z) = x^2y + yz^2 + z^3$, verify that $xf_1(x, y, z) + yf_2(x, y, z) + zf_3(x, y, z) = 3f(x, y, z)$.

Solution Holding y and z constant we get

$$f_1(x, y, z) = 2xy$$

Holding x and z constant we obtain

$$f_2(x, y, z) = x^2 + z^2$$

Holding x and y constant we get

$$f_3(x, y, z) = 2yz + 3z^2$$

Therefore

$$\begin{aligned} xf_1(x, y, z) + yf_2(x, y, z) + zf_3(x, y, z) &= x(2xy) + y(x^2 + z^2) + z(2yz + 3z^2) \\ &= 2x^2y + x^2y + yz^2 + 2yz^2 + 3z^3 \\ &= 3(x^2y + yz^2 + z^3) \\ &= 3f(x, y, z) \end{aligned}$$

EXERCISES 18.4

In Exercises 1 through 6, apply Definition 18.4.1 to find each of the partial derivatives.

1. $f(x, y) = 6x + 3y - 7$; $D_1 f(x, y)$

2. $f(x, y) = 4x^2 - 3xy$; $D_1 f(x, y)$

3. $f(x, y) = 3xy + 6x - y^2$; $D_2 f(x, y)$

4. $f(x, y) = xy^2 - 5y + 6$; $D_2 f(x, y)$

5. $f(x, y) = \sqrt{x^2 + y^2}$; $D_1 f(x, y)$

6. $f(x, y) = \dfrac{x + 2y}{x^2 - y}$; $D_2 f(x, y)$

In Exercises 7 through 10, apply Definition 18.4.2 to find each of the partial derivatives.

7. $f(x, y, z) = x^2y - 3xy^2 + 2yz$; $D_2f(x, y, z)$

8. $f(x, y, z) = x^2 + 4y^2 + 9z^2$; $D_1f(x, y, z)$

9. $f(x, y, z, r, t) = xyr + yzt + yrt + zrt$; $D_4f(x, y, z, r, t)$

10. $f(r, s, t, u, v, w) = 3r^2st + st^2v - 2tuv^2 - tvw + 3uw^2$; $D_5f(r, s, t, u, v, w)$

11. Given $f(x, y) = x^2 - 9y^2$. Find $D_1f(2, 1)$ by (a) applying formula (3); (b) applying formula (5); (c) applying formula (1) and then replacing x and y by 2 and 1, respectively.

12. For the function in Exercise 11, find $D_2f(2, 1)$ by (a) applying formula (4); (b) applying formula (6); (c) applying formula (2) and then replacing x and y by 2 and 1, respectively.

In Exercises 13 through 24, find the indicated partial derivative by holding all but one of the variables constant and applying theorems for ordinary differentiation.

13. $f(x, y) = 4y^3 + \sqrt{x^2 + y^2}$; $D_1f(x, y)$

14. $f(x, y) = \dfrac{x + y}{\sqrt{y^2 - x^2}}$; $D_2f(x, y)$

15. $f(\theta, \phi) = \sin 3\theta \cos 2\phi$; $D_2f(\theta, \phi)$

16. $f(r, \theta) = r^2 \cos \theta - 2r \tan \theta$; $D_2f(r, \theta)$

17. $z = e^{y/x} \ln \dfrac{x^2}{y}$; $\dfrac{\partial z}{\partial y}$

18. $r = e^{-\theta} \cos(\theta + \phi)$; $\dfrac{\partial r}{\partial \theta}$

19. $u = (x^2 + y^2 + z^2)^{-1/2}$; $\dfrac{\partial u}{\partial z}$

20. $u = \tan^{-1}(xyzw)$; $\dfrac{\partial u}{\partial w}$

21. $f(x, y, z) = 4xyz + \ln(2xyz)$; $f_3(x, y, z)$

22. $f(x, y, z) = e^{xy} \sinh 2z - e^{xy} \cosh 2z$; $f_3(x, y, z)$

23. $f(x, y, z) = e^{xyz} + \tan^{-1} \dfrac{3xy}{z^2}$; $f_2(x, y, z)$

24. $f(r, \theta, \phi) = 4r^2 \sin \theta + 5e^r \cos \theta \sin \phi - 2 \cos \phi$; $f_2(r, \theta, \phi)$

25. If $f(r, \theta) = r \tan \theta - r^2 \sin \theta$, find (a) $f_1(\sqrt{2}, \frac{1}{4}\pi)$; (b) $f_2(3, \pi)$.

26. If $f(x, y, z) = e^{xy^2} + \ln(y + z)$, find (a) $f_1(3, 0, 17)$; (b) $f_z(1, 0, 2)$; (c) $f_3(0, 0, 1)$.

In Exercises 27 and 28, find $f_x(x, y)$ and $f_y(x, y)$.

27. $f(x, y) = \displaystyle\int_x^y \ln \sin t \, dt$

28. $f(x, y) = \displaystyle\int_x^y e^{\cos t} \, dt$

29. Given $u = \sin \dfrac{r}{t} + \ln \dfrac{t}{r}$. Verify $t\dfrac{\partial u}{\partial t} + r\dfrac{\partial u}{\partial r} = 0$.

30. Given $w = x^2y + y^2z + z^2x$. Verify $\dfrac{\partial w}{\partial x} + \dfrac{\partial w}{\partial y} + \dfrac{\partial w}{\partial z} = (x + y + z)^2$.

31. Given $f(x, y) = \begin{cases} \dfrac{x^3 + y^3}{x^2 + y^2} & \text{if } (x, y) \neq (0, 0) \\ 0 & \text{if } (x, y) = (0, 0) \end{cases}$

Find (a) $f_1(0, 0)$; (b) $f_2(0, 0)$.

32. Given $f(x, y) = \begin{cases} \dfrac{x^2 - xy}{x + y} & \text{if } (x, y) \neq (0, 0) \\ 0 & \text{if } (x, y) = (0, 0) \end{cases}$

Find (a) $f_1(0, y)$ if $y \neq 0$; (b) $f_1(0, 0)$.

33. For the function of Exercise 32 find (a) $f_2(x, 0)$ if $x \neq 0$; (b) $f_2(0, 0)$.

34. Find the slope of the tangent line to the curve of intersection of the surface $36x^2 - 9y^2 + 4z^2 + 36 = 0$ with the plane $x = 1$ at the point $(1, \sqrt{12}, -3)$. Interpret this slope as a partial derivative.

35. Find the slope of the tangent line to the curve of intersection of the surface $z = x^2 + y^2$ with the plane $y = 1$ at the point $(2, 1, 5)$. Draw a sketch. Interpret this slope as a partial derivative.

36. Find equations of the tangent line to the curve of intersection of the surface $x^2 + y^2 + z^2 = 9$ with the plane $y = 2$ at the point $(1, 2, 2)$.

37. The temperature at any point (x, y) of a flat plate is T degrees and $T = 54 - \frac{2}{3}x^2 - 4y^2$. If distance is measured in centimeters, find the rate of change of the temperature with respect to the distance moved along the plate in the directions of the positive x and y axes, respectively, at the point $(3, 1)$.

38. Use the ideal gas law for a confined gas (see Example 5) to show that

$$\frac{\partial V}{\partial T} \cdot \frac{\partial T}{\partial P} \cdot \frac{\partial P}{\partial V} = -1$$

39. If V dollars is the present value of an ordinary annuity of equal payments of \$100 per year for t years at an interest rate of $100i$ percent per year, then

$$V = 100 \left[\frac{1 - (1 + i)^{-t}}{i}\right]$$

(a) Find the instantaneous rate of change of V per unit change in i if t remains fixed at 8. (b) Use the result of part (a) to find the approximate change in the present value if the interest rate changes from 6% to 7% and the time remains fixed at 8 years. (c) Find the instantaneous rate of change of V per unit change in t if i remains fixed at 0.06. (d) Use the result of part (c) to find the approximate change in the present value if the time is decreased from 8 to 7 years and the interest rate remains fixed at 6%.

40. Suppose that x is the number of tens of thousands of dollars in the inventory carried in a store, y is the number of clerks in the store, P is the number of dollars in the weekly profit of the store, and

$$P = 3000 + 240y + 20y(x - 2y) - 10(x - 12)^2$$

where $15 \le x \le 25$ and $5 \le y \le 12$. At present the inventory is \$180,000 and there are 8 clerks. (a) Find the instantaneous rate of change of P per unit change in x if y remains fixed at 8. (b) Use the result of part (a) to find the approximate change in the weekly profit if the inventory changes from \$180,000 to \$200,000 and the number of clerks remains fixed at 8. (c) Find the instantaneous rate of change of P per unit change in y if x remains fixed at 18. (d) Use the result of part (c) to find the approximate change in the weekly profit if the number of clerks is increased from 8 to 10 and the inventory remains fixed at \$180,000.

18.5 DIFFERENTIABILITY AND THE TOTAL DIFFERENTIAL

In Sec. 3.7 in the proof of the chain rule we showed that if f is a differentiable function of the single variable x and $y = f(x)$, then the increment Δy of the dependent variable can be expressed as

$$\Delta y = f'(x)\,\Delta x + \eta\,\Delta x$$

where η is a function of Δx and $\eta \to 0$ as $\Delta x \to 0$.

From the above it follows that if the function f is differentiable at x_0, the increment of f at x_0, denoted by $\Delta f(x_0)$, is given by

$$\Delta f(x_0) = f'(x_0)\,\Delta x + \eta\,\Delta x \tag{1}$$

where $\lim_{\Delta x \to 0} \eta = 0$.

For functions of two or more variables an equation corresponding to Eq. (1) is used to define *differentiability* of a function. And from the definition we determine criteria for a function to be differentiable at a point. We give the details for a function of two variables and begin by defining the increment of such a function.

18.5.1 Definition If f is a function of two variables x and y, then the **increment of** f at the point (x_0, y_0), denoted by $\Delta f(x_0, y_0)$, is given by

$$\Delta f(x_0, y_0) = f(x_0 + \Delta x, y_0 + \Delta y) - f(x_0, y_0) \tag{2}$$

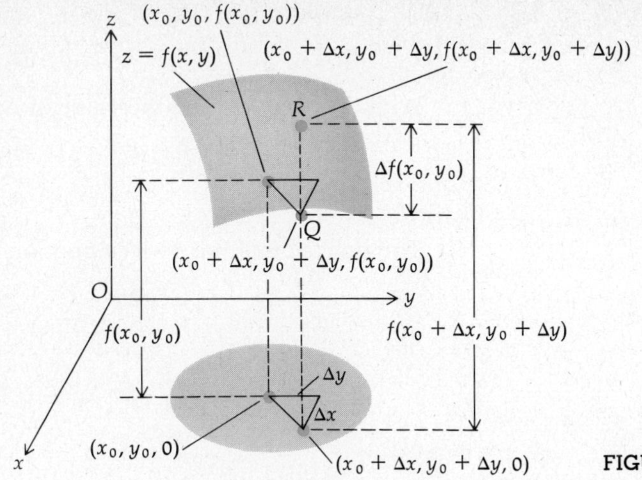

FIGURE 18.5.1

Figure 18.5.1 illustrates Eq. (2) for a function that is continuous on an open disk containing the points (x_0, y_0) and $(x_0 + \Delta x, y_0 + \Delta y)$. The figure shows a portion of the surface $z = f(x, y)$. $\Delta f(x_0, y_0) = \overline{QR}$, where Q is the point $(x_0 + \Delta x, y_0 + \Delta y, f(x_0, y_0))$ and R is the point $(x_0 + \Delta x, y_0 + \Delta y, f(x_0 + \Delta x, y_0 + \Delta y))$.

• ILLUSTRATION 1 For the function f defined by

$$f(x, y) = 3x - xy^2$$

we find the increment of f at any point (x_0, y_0).

$$\begin{aligned}\Delta f(x_0, y_0) &= f(x_0 + \Delta x, y_0 + \Delta y) - f(x_0, y_0)\\ &= 3(x_0 + \Delta x) - (x_0 + \Delta x)(y_0 + \Delta y)^2 - (3x_0 - x_0 y_0{}^2)\\ &= 3x_0 + 3\,\Delta x - x_0 y_0{}^2 - y_0{}^2\,\Delta x - 2x_0 y_0\,\Delta y - 2y_0\,\Delta x\,\Delta y\\ &\qquad - x_0(\Delta y)^2 - \Delta x(\Delta y)^2 - 3x_0 + x_0 y_0{}^2\\ &= 3\,\Delta x - y_0{}^2\,\Delta x - 2x_0 y_0\,\Delta y - 2y_0\,\Delta x\,\Delta y - x_0(\Delta y)^2 - \Delta x(\Delta y)^2\end{aligned}$$ •

18.5.2 Definition If f is a function of two variables x and y and the increment of f at (x_0, y_0) can be written as

$$\Delta f(x_0, y_0) = D_1 f(x_0, y_0)\,\Delta x + D_2 f(x_0, y_0)\,\Delta y + \epsilon_1\,\Delta x + \epsilon_2\,\Delta y \tag{3}$$

where ϵ_1 and ϵ_2 are functions of Δx and Δy such that $\epsilon_1 \to 0$ and $\epsilon_2 \to 0$ as $(\Delta x, \Delta y) \to (0, 0)$, then f is said to be **differentiable** at (x_0, y_0).

• ILLUSTRATION 2 We use Definition 18.5.2 to prove that the function of Illustration 1 is differentiable at all points in R^2. We must show that for all points (x_0, y_0) in R^2 we can find an ϵ_1 and an ϵ_2 such that

$$\Delta f(x_0, y_0) - D_1 f(x_0, y_0)\,\Delta x - D_2 f(x_0, y_0)\,\Delta y = \epsilon_1\,\Delta x + \epsilon_2\,\Delta y$$

and $\epsilon_1 \to 0$ and $\epsilon_2 \to 0$ as $(\Delta x, \Delta y) \to (0, 0)$

Because $f(x, y) = 3x - xy^2$,

$$D_1 f(x_0, y_0) = 3 - y_0^2 \quad \text{and} \quad D_2 f(x_0, y_0) = -2x_0 y_0$$

With these values and the value of $\Delta f(x_0, y_0)$ from Illustration 1,

$$\Delta f(x_0, y_0) - D_1 f(x_0, y_0)\,\Delta x - D_2 f(x_0, y_0)\,\Delta y = -x_0(\Delta y)^2 - 2y_0\,\Delta x\,\Delta y - \Delta x(\Delta y)^2$$

The right side of the above equation can be written in the following ways:

$$[-2y_0\,\Delta y - (\Delta y)^2]\,\Delta x + (-x_0\,\Delta y)\,\Delta y$$

or

$$(-2y_0\,\Delta y)\,\Delta x + (-\Delta x\,\Delta y - x_0\,\Delta y)\,\Delta y$$

or

$$[-(\Delta y)^2]\,\Delta x + (-2y_0\,\Delta x - x_0\,\Delta y)\,\Delta y$$

or

$$0 \cdot \Delta x + [-2y_0\,\Delta x - \Delta x\,\Delta y - x_0\,\Delta y]\,\Delta y$$

So there are four possible pairs of values for ϵ_1 and ϵ_2:

$$\epsilon_1 = -2y_0\,\Delta y - (\Delta y)^2 \quad \text{and} \quad \epsilon_2 = -x_0\,\Delta y$$

or

$$\epsilon_1 = -2y_0\,\Delta y \quad \text{and} \quad \epsilon_2 = -\Delta x\,\Delta y - x_0\,\Delta y$$

or

$$\epsilon_1 = -(\Delta y)^2 \quad \text{and} \quad \epsilon_2 = -2y_0\,\Delta x - x_0\,\Delta y$$

or

$$\epsilon_1 = 0 \quad \text{and} \quad \epsilon_2 = -2y_0\,\Delta x - \Delta x\,\Delta y - x_0\,\Delta y$$

For each pair,

$$\lim_{(\Delta x, \Delta y) \to (0,0)} \epsilon_1 = 0 \quad \text{and} \quad \lim_{(\Delta x, \Delta y) \to (0,0)} \epsilon_2 = 0$$

It should be noted that it is only necessary to find one pair of values for ϵ_1 and ϵ_2. ●

18.5.3 Theorem If a function f of two variables is differentiable at a point, it is continuous at that point.

Proof If f is differentiable at the point (x_0, y_0), it follows from Definition 18.5.2 that

$$f(x_0 + \Delta x, y_0 + \Delta y) - f(x_0, y_0) = D_1 f(x_0, y_0)\,\Delta x + D_2 f(x_0, y_0)\,\Delta y + \epsilon_1\,\Delta x + \epsilon_2\,\Delta y$$

where $\epsilon_1 \to 0$ and $\epsilon_2 \to 0$ as $(\Delta x, \Delta y) \to (0, 0)$. Therefore

$$f(x_0 + \Delta x, y_0 + \Delta y) = f(x_0, y_0) + D_1 f(x_0, y_0)\,\Delta x + D_2 f(x_0, y_0)\,\Delta y + \epsilon_1\,\Delta x + \epsilon_2\,\Delta y$$

Taking the limit on both sides of the above as $(\Delta x, \Delta y) \to (0, 0)$ we obtain

$$\lim_{(\Delta x, \Delta y) \to (0,0)} f(x_0 + \Delta x, y_0 + \Delta y) = f(x_0, y_0) \tag{4}$$

If we let $x_0 + \Delta x = x$ and $y_0 + \Delta y = y$, then "$(\Delta x, \Delta y) \to (0, 0)$" is equivalent to "$(x, y) \to (x_0, y_0)$." Thus, from (4),

$$\lim_{(x,y) \to (x_0,y_0)} f(x, y) = f(x_0, y_0)$$

which proves that f is continuous at (x_0, y_0), ■

Theorem 18.5.3 states that for a function of two variables *differentiability implies continuity.* However, the mere existence of the partial derivatives $D_1 f$ and $D_2 f$ at a point does not imply differentiability at that point. The following example illustrates this.

Example 1 Given

$$f(x, y) = \begin{cases} \dfrac{xy}{x^2 + y^2} & \text{if } (x, y) \neq (0, 0) \\ 0 & \text{if } (x, y) = (0, 0) \end{cases}$$

prove that $D_1 f(0, 0)$ and $D_2 f(0, 0)$ exist but that f is not differentiable at $(0, 0)$.

Solution

$$D_1 f(0, 0) = \lim_{x \to 0} \frac{f(x, 0) - f(0, 0)}{x - 0} = \lim_{x \to 0} \frac{0 - 0}{x} = 0$$

$$D_2 f(0, 0) = \lim_{y \to 0} \frac{f(0, y) - f(0, 0)}{y - 0} = \lim_{y \to 0} \frac{0 - 0}{y} = 0$$

Therefore both $D_1 f(0, 0)$ and $D_2 f(0, 0)$ exist.

In Example 4 of Sec. 18.2 we demonstrated that for this function $\lim_{(x,y) \to (0,0)} f(x, y)$ does not exist; hence f is not continuous at $(0, 0)$. Because f is not continuous at $(0, 0)$, it follows from Theorem 18.5.3 that f is not differentiable there.

Before stating a theorem that gives conditions for which a function is differentiable at a point, we consider a theorem needed in its proof. It is the mean-value theorem for a function of a single variable applied to a function of two variables.

18.5.4 Theorem Let f be a function of two variables defined for all x in the closed interval $[a, b]$ and all y in the closed interval $[c, d]$.

(i) If $D_1 f(x, y_0)$ exists for some y_0 in $[c, d]$ and for all x in $[a, b]$, then there is a number ξ_1 in the open interval (a, b) such that

$$f(b, y_0) - f(a, y_0) = (b - a) D_1 f(\xi_1, y_0) \tag{5}$$

(ii) If $D_2 f(x_0, y)$ exists for some x_0 in [a, b] and for all y in $[c, d]$, then there is a number ξ_2 in the open interval (c, d) such that

$$f(x_0, d) - f(x_0, c) = (d - c) D_2 f(x_0, \xi_2) \tag{6}$$

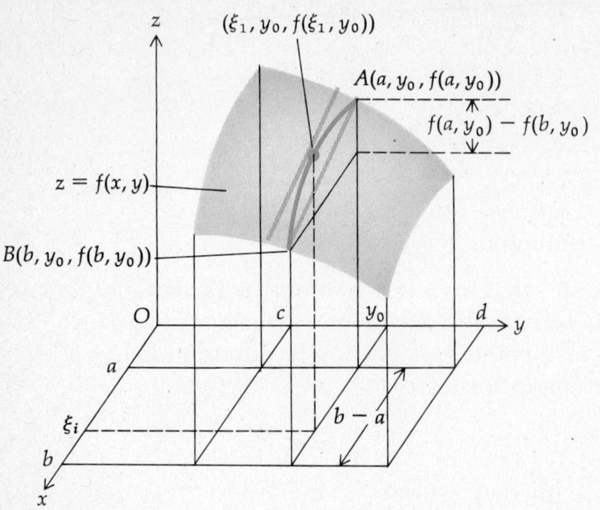

FIGURE 18.5.2

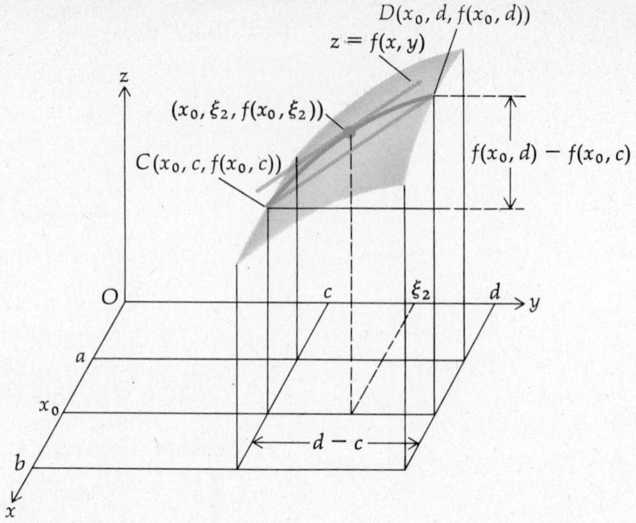

FIGURE 18.5.3

Before proving this theorem, we interpret it geometrically. For part (i) refer to Fig. 18.5.2, which shows the portion of the surface $z = f(x, y)$ above the rectangular region in the xy plane bounded by the lines $x = a$, $x = b$, $y = c$, and $y = d$. The plane $y = y_0$ intersects the surface in the curve represented by the two equations $y = y_0$ and $z = f(x, y)$. The slope of the line through the points $A(a, y_0, f(a, y_0))$ and $B(b, y_0, f(b, y_0))$ is

$$\frac{f(b, y_0) - f(a, y_0)}{b - a}$$

Theorem 18.5.4(i) states that there is some point $(\xi_1, y_0, f(\xi_1, y_0))$ on the curve between the points A and B where the tangent line is parallel to the secant line through A and B; that is, there is some number ξ_1 in (a, b) such that $D_1 f(\xi_1, y_0) = [f(b, y_0) - f(a, y_0)]/(b - a)$, and this is illustrated in the figure, for which $D_1 f(\xi_1, y_0) < 0$.

Figure 18.5.3 illustrates part (ii) of Theorem 18.5.4. The plane $x = x_0$ intersects the surface $z = f(x, y)$ in the curve represented by the two equations $x = x_0$ and $z = f(x, y)$. The slope of the line through the points $C(x_0, c, f(x_0, c))$ and $D(x_0, d, f(x_0, d))$ is $[f(x_0, d) - f(x_0, c)]/(d - c)$, and Theorem 18.5.4(ii) states that there is some point $(x_0, \xi_2, f(x_0, \xi_2))$ on the curve between the points C and D where the tangent line is parallel to the secant line through C and D; that is, there is some number ξ_2 in (c, d) such that $D_2 f(x_0, \xi_2) = [f(x_0, d) - f(x_0, c)]/(d - c)$.

Proof of Theorem 18.5.4(i) Let g be the function of one variable x defined by

$$g(x) = f(x, y_0)$$

Then

$$g'(x) = D_1 f(x, y_0)$$

Because $D_1 f(x, y_0)$ exists for all x in $[a, b]$, it follows that $g'(x)$ exists for all x in $[a, b]$, and therefore g is continuous on $[a, b]$. So by the mean-value theorem (4.4.2) for ordinary derivatives there exists a number ξ_1 in (a, b) such that

$$g'(\xi_1) = \frac{g(b) - g(a)}{b - a}$$

or, equivalently,

$$D_1 f(\xi_1, y_0) = \frac{f(b, y_0) - f(a, y_0)}{b - a}$$

from which we obtain

$$f(b, y_0) - f(a, y_0) = (b - a)D_1 f(\xi_1, y_0)$$

The proof of part (ii) is similar to the proof of part (i) and is left as an exercise (see Exercise 21). ■

Equation (5) can be written in the form

$$f(x_0 + h, y_0) - f(x_0, y_0) = hD_1 f(\xi_1, y_0) \tag{7}$$

where ξ_1 is between x_0 and $x_0 + h$ and h is either positive or negative (see Exercise 22).

Equation (6) can be written in the form

$$f(x_0, y_0 + k) - f(x_0, y_0) = kD_2 f(x_0, \xi_2) \tag{8}$$

where ξ_2 is between y_0 and $y_0 + k$ and k is either positive or negative (see Exercise 23).

Example 2 Given

$$f(x, y) = \frac{2xy}{3 + x}$$

find a ξ_1 required by Theorem 18.5.4 if x is in $[2, 5]$ and $y = 4$.

Solution By Theorem 18.5.4 there is a number ξ_1 in the open interval $(2, 5)$ such that

$$f(5, 4) - f(2, 4) = (5 - 2)D_1 f(\xi_1, 4)$$

So

$$5 - \frac{16}{5} = 3 \cdot \frac{24}{(3 + \xi_1)^2}$$

$$\frac{9}{5} = \frac{72}{(3 + \xi_1)^2}$$

$$(3 + \xi_1)^2 = 40$$

Therefore

$$3 + \xi_1 = \pm 2\sqrt{10}$$

But because $2 < \xi_1 < 5$, we take only the + sign and obtain

$$\xi_1 = 2\sqrt{10} - 3$$

The following theorem states that a function having continuous partial derivatives at a point is necessarily differentiable at the point.

18.5.5 Theorem Let f be a function of two variables x and y. Suppose that $D_1 f$ and $D_2 f$ exist on an open disk $B(P_0; r)$, where P_0 is the point (x_0, y_0). Then if $D_1 f$ and $D_2 f$ are continuous at P_0, f is differentiable at P_0.

Proof Choose the point $(x_0 + \Delta x, y_0 + \Delta y)$ so that it is in $B(P_0; r)$. Then

$$\Delta f(x_0, y_0) = f(x_0 + \Delta x, y_0 + \Delta y) - f(x_0, y_0)$$

Subtracting and adding $f(x_0 + \Delta x, y_0)$ to the right side of the above equation we get

$$\Delta f(x_0, y_0) = [f(x_0 + \Delta x, y_0 + \Delta y) - f(x_0 + \Delta x, y_0)] + [f(x_0 + \Delta x, y_0) - f(x_0, y_0)] \quad (9)$$

Because $D_1 f$ and $D_2 f$ exist on $B(P_0; r)$ and $(x_0 + \Delta x, y_0 + \Delta y)$ is in $B(P_0; r)$, it follows from (8) that

$$f(x_0 + \Delta x, y_0 + \Delta y) - f(x_0 + \Delta x, y_0) = (\Delta y)D_2 f(x_0 + \Delta x, \xi_2) \quad (10)$$

where ξ_2 is between y_0 and $y_0 + \Delta y$.

From (7) it follows that

$$f(x_0 + \Delta x, y_0) - f(x_0, y_0) = (\Delta x)D_1 f(\xi_1, y_0) \quad (11)$$

where ξ_1 is between x_0 and $x_0 + \Delta x$.

Substituting from (10) and (11) in (9) we obtain

$$\Delta f(x_0, y_0) = (\Delta y)D_2 f(x_0 + \Delta x, \xi_2) + (\Delta x)D_1 f(\xi_1, y_0) \quad (12)$$

Because $(x_0 + \Delta x, y_0 + \Delta y)$ is in $B(P_0; r)$, ξ_2 is between y_0 and $y_0 + \Delta y$, and $D_2 f$ is continuous at P_0, it follows that

$$\lim_{(\Delta x, \Delta y) \to (0,0)} D_2 f(x_0 + \Delta x, \xi_2) = D_2 f(x_0, y_0) \quad (13)$$

and because ξ_1 is between x_0 and $x_0 + \Delta x$ and $D_1 f$ is continuous at P_0,

$$\lim_{(\Delta x, \Delta y) \to (0,0)} D_1 f(\xi_1, y_0) = D_1 f(x_0, y_0) \quad (14)$$

If

$$\epsilon_1 = D_1 f(\xi_1, y_0) - D_1 f(x_0, y_0) \quad (15)$$

it follows from (14) that

$$\lim_{(\Delta x, \Delta y) \to (0,0)} \epsilon_1 = 0 \quad (16)$$

and if

$$\epsilon_2 = D_2 f(x_0 + \Delta x, \xi_2) - D_2 f(x_0, y_0) \quad (17)$$

it follows from (13) that

$$\lim_{(\Delta x, \Delta y) \to (0,0)} \epsilon_2 = 0 \tag{18}$$

Substituting from (15) and (17) into (12) we get

$$\Delta f(x_0, y_0) = \Delta y[D_2 f(x_0, y_0) + \epsilon_2] + \Delta x[D_1 f(x_0, y_0) + \epsilon_1]$$

from which we obtain

$$\Delta f(x_0, y_0) = D_1 f(x_0, y_0)\, \Delta x + D_2 f(x_0, y_0)\, \Delta y + \epsilon_1\, \Delta x + \epsilon_2\, \Delta y \tag{19}$$

From (16), (18), and (19), Definition 18.5.2 holds; so f is differentiable at (x_0, y_0). ■

A function satisfying the hypothesis of Theorem 18.5.5 is said to be **continuously differentiable** at the point P_0.

Example 3 Given

$$f(x, y) = \begin{cases} \dfrac{x^2y^2}{x^2 + y^2} & \text{if } (x, y) \neq (0, 0) \\ 0 & \text{if } (x, y) = (0, 0) \end{cases}$$

use Theorem 18.5.5 to prove that f is differentiable at (0, 0).

Solution To find $D_1 f$ consider two cases: $(x, y) = (0, 0)$ and $(x, y) \neq (0, 0)$. If $(x, y) = (0, 0)$, we have

$$D_1 f(0, 0) = \lim_{x \to 0} \frac{f(x, 0) - f(0, 0)}{x - 0} = \lim_{x \to 0} \frac{0 - 0}{x} = 0$$

If $(x, y) \neq (0, 0)$, $f(x, y) = x^2y^2/(x^2 + y^2)$. To find $D_1 f(x, y)$ we use the theorem for the ordinary derivative of a quotient and consider y as a constant.

$$D_1 f(x, y) = \frac{2xy^2(x^2 + y^2) - 2x(x^2y^2)}{(x^2 + y^2)^2}$$

$$= \frac{2xy^4}{(x^2 + y^2)^2}$$

The function $D_1 f$ is therefore defined by

$$D_1 f(x, y) = \begin{cases} \dfrac{2xy^4}{(x^2 + y^2)^2} & \text{if } (x, y) \neq (0, 0) \\ 0 & \text{if } (x, y) = (0, 0) \end{cases}$$

In the same manner we obtain the function $D_2 f$ defined by

$$D_2 f(x, y) = \begin{cases} \dfrac{2x^4y}{(x^2 + y^2)^2} & \text{if } (x, y) \neq (0, 0) \\ 0 & \text{if } (x, y) = (0, 0) \end{cases}$$

Both $D_1 f$ and $D_2 f$ exist on every open disk having its center at the origin. It remains to show that $D_1 f$ and $D_2 f$ are continuous at (0, 0).

Because $D_1 f(0, 0) = 0$, $D_1 f$ will be continuous at $(0, 0)$ if

$$\lim_{(x,y)\to(0,0)} D_1 f(x, y) = 0$$

Therefore we must show that for any $\epsilon > 0$ there exists a $\delta > 0$ such that

$$\left|\frac{2xy^4}{(x^2 + y^2)^2}\right| < \epsilon \quad \text{whenever} \quad 0 < \sqrt{x^2 + y^2} < \delta \tag{20}$$

$$\left|\frac{2xy^4}{(x^2 + y^2)^2}\right| = \frac{2|x|y^4}{(x^2 + y^2)^2} \leq \frac{2\sqrt{x^2 + y^2}(\sqrt{x^2 + y^2})^4}{(x^2 + y^2)^2} = 2\sqrt{x^2 + y^2}$$

Therefore

$$\left|\frac{2xy^4}{(x^2 + y^2)^2}\right| < 2\delta \quad \text{whenever} \quad 0 < \sqrt{x^2 + y^2} < \delta$$

So if $\delta = \frac{1}{2}\epsilon$, we have (20). Hence $D_1 f$ is continuous at $(0, 0)$. In the same way it can be shown that $D_2 f$ is continuous at $(0, 0)$. It follows from Theorem 18.5.5 that f is differentiable at $(0, 0)$.

Refer back to Eq. (3). The expression involving the first two terms on the right side, $D_1 f(x_0, y_0)\,\Delta x + D_2 f(x_0, y_0)\,\Delta y$, is called the *principal part* of $\Delta f(x_0, y_0)$ or the *total differential* of the function f at (x_0, y_0). We make this as a formal definition.

18.5.6 Definition If f is a function of two variables x and y, and f is differentiable at (x, y), then the **total differential** of f is the function df having function values given by

$$df(x, y, \Delta x, \Delta y) = D_1 f(x, y)\,\Delta x + D_2 f(x, y)\,\Delta y \tag{21}$$

Note that df is a function of the four variables x, y, Δx, and Δy. If $z = f(x, y)$, we sometimes write dz in place of $df(x, y, \Delta x, \Delta y)$, and then Eq. (21) is written as

$$dz = D_1 f(x, y)\,\Delta x + D_2 f(x, y)\,\Delta y \tag{22}$$

If in particular $f(x, y) = x$, then $z = x$, $D_1 f(x, y) = 1$, and $D_2 f(x, y) = 0$, and so Eq. (22) gives $dz = \Delta x$. Because $z = x$, for this function $dx = \Delta x$. In a similar fashion, if we take $f(x, y) = y$, then $z = y$, $D_1 f(x, y) = 0$, and $D_2 f(x, y) = 1$, and so (22) gives $dz = \Delta y$. Because $z = y$, then for this function $dy = \Delta y$. Hence we define the differentials of the independent variables as $dx = \Delta x$ and $dy = \Delta y$. Then (22) can be written as

$$dz = D_1 f(x, y)\,dx + D_2 f(x, y)\,dy \tag{23}$$

and at the point (x_0, y_0),

$$dz = D_1 f(x_0, y_0)\,dx + D_2 f(x_0, y_0)\,dy \tag{24}$$

In Eq. (3) let $\Delta z = \Delta f(x_0, y_0)$, $dx = \Delta x$, and $dy = \Delta y$. Then

$$\Delta z = D_1 f(x_0, y_0)\, dx + D_2 f(x_0, y_0)\, dy + \epsilon_1\, dx + \epsilon_2\, dy \tag{25}$$

By comparing (24) and (25), observe that when dx (i.e., Δx) and dy (i.e., Δy) are close to zero, and because then ϵ_1 and ϵ_2 also will be close to zero, dz is an approximation to Δz. Because dz is often easier to calculate than Δz, we make use of the fact that $dz \approx \Delta z$ in certain situations. Before showing this in an example, we write (23) with the notation $\partial z/\partial x$ and $\partial z/\partial y$ instead of $D_1 f(x, y)$ and $D_2 f(x, y)$, respectively:

$$dz = \frac{\partial z}{\partial x}\, dx + \frac{\partial z}{\partial y}\, dy \tag{26}$$

Example 4 A closed metal container in the shape of a right-circular cylinder is to have an inside height of 6 in., an inside radius of 2 in., and a thickness of 0.1 in. If the cost of the metal to be used is 10 cents per in.^{3}, find by differentials the approximate cost of the metal to be used in the manufacture of the container.

Solution The formula for the volume of a right-circular cylinder, where the volume is V in.^{3}, the radius is r in., and the height is h in., is

$$V = \pi r^2 h \tag{27}$$

The exact volume of metal in the container is the difference between the volumes of two right-circular cylinders for which $r = 2.1$, $h = 6.2$, and $r = 2$, $h = 2$, respectively.

ΔV would give the exact volume of metal, but because only an approximate value is wanted, we find dV instead. From (26),

$$dV = \frac{\partial V}{\partial r}\, dr + \frac{\partial V}{\partial h}\, dh \tag{28}$$

From (27) it follows that

$$\frac{\partial V}{\partial r} = 2\pi r h \quad \text{and} \quad \frac{\partial V}{\partial h} = \pi r^2$$

Substituting these values into (28) gives

$$dV = 2\pi r h\, dr + \pi r^2\, dh$$

Because $r = 2$, $h = 6$, $dr = 0.1$, and $dh = 0.2$,

$$\begin{aligned} dV &= 2\pi(2)(6)(0.1) + \pi(2)^2(0.2) \\ &= 3.2\pi \end{aligned}$$

Hence $\Delta V \approx 3.2\pi$, and so there is approximately 3.2π in.^{3} of metal in the container. Because the cost of the metal is 10 cents per cubic inch and $10 \cdot 3.2\pi = 32\pi \approx 100.53$, the approximate cost of the metal to be used in the manufacture of the container is \$1.

The concepts of differentiability and the total differential are now extended to a function of n variables.

18.5.7 Definition If f is a function of the n variables $x_1, x_2, \ldots, x_n$, and $\overline{P}$ is the point $(\overline{x}, \overline{x}_2, \ldots, \overline{x}_n)$, then the **increment of** f at $\overline{P}$ is given by

$$\Delta f(\overline{P}) = f(\overline{x}_1 + \Delta x_1, \overline{x}_2 + \Delta x_2, \ldots, \overline{x}_n + \Delta x_n) - f(\overline{P}) \tag{29}$$

18.5.8 Definition If f is a function of the n variables $x_1, x_2, \ldots, x_n$, and the increment of f at the point $\overline{P}$ can be written as

$$\Delta f(\overline{P}) = D_1 f(\overline{P})\,\Delta x_1 + D_2 f(\overline{P})\,\Delta x_2 + \cdots + D_n f(\overline{P})\,\Delta x_n + \epsilon_1\,\Delta x_1 + \epsilon_2\,\Delta x_2 + \cdots + \epsilon_n\,\Delta x_n \tag{30}$$

where $\epsilon_1 \to 0, \epsilon_2 \to 0, \ldots, \epsilon_n \to 0$, as

$$(\Delta x_1, \Delta x_2, \ldots, \Delta x_n) \to (0, 0, \ldots, 0)$$

then f is said to be **differentiable** at $\overline{P}$.

Analogously to Theorem 18.5.5 it can be proved that sufficient conditions for a function f of n variables to be differentiable at a point $\overline{P}$ are that $D_1 f, D_2 f, \ldots, D_n f$ all exist on an open ball $B(\overline{P}; r)$ and that $D_1 f, D_2 f, \ldots, D_n f$ are all continuous at $\overline{P}$. As was the case for functions of two variables, it follows that for functions of n variables differentiability implies continuity. However, the existence of the partial derivatives $D_1 f, D_2 f, \ldots, D_n f$ at a point does not imply differentiability of the function at the point.

18.5.9 Definition If f is a function of the n variables $x_1, x_2, \ldots, x_n$ and f is differentiable at P, then the **total differential** of f is the function df having function values given by

$$df(P, \Delta x_1, \Delta x_2, \ldots, \Delta x_n) = D_1 f(P)\,\Delta x_1 + D_2 f(P)\,\Delta x_2 + \cdots + D_n f(P)\,\Delta x_n \tag{31}$$

Letting $w = f(x_1, x_2, \ldots, x_n)$, defining $dx_1 = \Delta x_1$, $dx_2 = \Delta x_2, \ldots,$ $dx_n = \Delta x_n$, and using the notation $\partial w/\partial x_i$ instead of $D_i f(P)$, we can write (31) as

$$dw = \frac{\partial w}{\partial x_1}\,dx_1 + \frac{\partial w}{\partial x_2}\,dx_2 + \cdots + \frac{\partial w}{\partial x_n}\,dx_n \tag{32}$$

Example 5 The dimensions of a box are measured to be 10 cm, 12 cm, and 15 cm, and the measurements are correct to 0.02 cm. Find approximately the greatest error if the volume of the box is calculated from the given measurements. Also find the approximate percent error.

Solution Let V cm³ be the volume of a box whose dimensions are x cm, y cm, and z cm. Then

$$V = xyz$$

The exact value of the error is found from ΔV; however, we use dV as an approximation to ΔV. From Eq. (32) for three independent variables,

$$dV = \frac{\partial V}{\partial x}dx + \frac{\partial V}{\partial y}dy + \frac{\partial V}{\partial z}dz$$

and so

$$dV = yz\,dx + xz\,dy + xy\,dz \tag{33}$$

From the given information $|\Delta x| \le 0.02$, $|\Delta y| \le 0.02$, and $|\Delta z| \le 0.02$. To find the greatest error in the volume we take the greatest error in the measurements of the three dimensions. So taking $dx = 0.02$, $dy = 0.02$, $dz = 0.02$, and $x = 10$, $y = 12$, $z = 15$, we have from Eq. (33)

$$\begin{aligned} dV &= (12)(15)(0.02) + (10)(15)(0.02) + (10)(12)(0.02) \\ &= 9 \end{aligned}$$

Thus $\Delta V \approx 9$, and therefore the greatest possible error in the calculation of the volume from the given measurements is approximately 9 cm^3.

The relative error is found by dividing the error by the actual value. Hence the relative error in computing the volume from the given measurements is $\Delta V/V \approx dV/V = \frac{9}{1800} = \frac{1}{200} = 0.005$. So the approximate percent error is 0.5%.

EXERCISES 18.5

1. If $f(x, y) = 3x^2 + 2xy - y^2$, find: (a) $\Delta f(1, 4)$, the increment of f at $(1, 4)$; (b) $\Delta f(1, 4)$ when $\Delta x = 0.03$ and $\Delta y = -0.02$; (c) $df(1, 4, \Delta x, \Delta y)$, the total differential of f at $(1, 4)$; (d) $df(1, 4, 0.03, -0.02)$.

2. If $f(x, y) = 2x^2 + 5xy + 4y^2$, find: (a) $\Delta f(2, -1)$, the increment of f at $(2, -1)$; (b) $\Delta f(2, -1)$ when $\Delta x = -0.01$ and $\Delta y = 0.02$; (c) $df(2, -1, \Delta x, \Delta y)$, the total differential of f at $(2, -1)$; (d) $df(2, -1, -0.01, 0.02)$.

3. If $g(x, y) = xye^{xy}$, find: (a) $\Delta g(2, -4)$, the increment of g at $(2, -4)$; (b) $\Delta g(2, -4)$ when $\Delta x = -0.1$ and $\Delta y = 0.2$; (c) $dg(2, -4, \Delta x, \Delta y)$, the total differential of g at $(2, -4)$; (d) $dg(2, -4, -0.1, 0.2)$.

4. If $h(x, y) = (x + y)/(x - y)$, find: (a) $\Delta h(3, 0)$, the increment of h at $(3, 0)$; (b) $\Delta h(3, 0)$ when $\Delta x = 0.04$ and $\Delta y = 0.03$; (c) $dh(3, 0, \Delta x, \Delta y)$, the total differential of h at $(3, 0)$; (d) $dh(3, 0, 0.04, 0.03)$.

5. If $F(x, y, z) = xy + \ln(yz)$, find: (a) $\Delta F(4, 1, 5)$, the increment of F at $(4, 1, 5)$; (b) $\Delta F(4, 1, 5)$ when $\Delta x = 0.02$, $\Delta y = 0.04$, and $\Delta z = -0.03$; (c) $dF(4, 1, 5, \Delta x, \Delta y, \Delta z)$, the total differential of F at $(4, 1, 5)$; (d) $dF(4, 1, 5, 0.02, 0.04, -0.03)$.

6. If $G(x, y, z) = x^2y + 2xyz - z^3$, find: (a) $\Delta G(-3, 0, 2)$, the increment of G at $(-3, 0, 2)$; (b) $\Delta G(-3, 0, 2)$ when $\Delta x = 0.01$, $\Delta y = 0.03$, and $\Delta z = -0.01$; (c) $dG(-3, 0, 2, \Delta x, \Delta y, \Delta z)$, the total differential of G at $(-3, 0, 2)$; (d) $dG(-3, 0, 2, 0.01, 0.03, -0.01)$.

In Exercises 7 through 14, find the total differential dw.

7. $w = 4x^3 - xy^2 + 3y - 7$
8. $w = y \tan x^2 - 2xy$
9. $w = x \cos y - y \sin x$
10. $w = xe^{2y} + e^{-y}$
11. $w = \ln(x^2 + y^2 + z^2)$
12. $w = \dfrac{xyz}{x + y + z}$
13. $w = x \tan^{-1} z - \dfrac{y^2}{z}$
14. $w = e^{yz} - \cos xz$

In Exercises 15 through 18, prove that f is differentiable at all points in its domain by doing each of the following: (a) Find $\Delta f(x_0, y_0)$ for the given function; (b) find an ϵ_1 and an ϵ_2 so that Eq. (3) holds; (c) show that the ϵ_1 and the ϵ_2 found in part (b) both approach zero as $(\Delta x, \Delta y) \to (0, 0)$.

15. $f(x, y) = x^2y - 2xy$ **16.** $f(x, y) = 2x^2 + 3y^2$ **17.** $f(x, y) = \dfrac{x^2}{y}$ **18.** $f(x, y) = \dfrac{y}{x}$

19. Given $f(x, y) = \begin{cases} x + y - 2 & \text{if } x = 1 \text{ or } y = 1 \\ 2 & \text{if } x \neq 1 \text{ and } y \neq 1 \end{cases}$

Prove that $D_1f(1, 1)$ and $D_2f(1, 1)$ exist but f is not differentiable at $(1, 1)$.

20. Given $f(x, y) = \begin{cases} \dfrac{3x^2y^2}{x^4 + y^4} & \text{if } (x, y) \neq (0, 0) \\ 0 & \text{if } (x, y) = (0, 0) \end{cases}$

Prove that $D_1f(0, 0)$ and $D_2f(0, 0)$ exist but f is not differentiable at $(0, 0)$.

21. Prove Theorem 18.5.4.(ii).

22. Show that Eq. (5) may be written in the form (7) where ξ_1 is between x_0 and $x_0 + h$.

23. Show that Eq. (6) may be written in the form (8) where ξ_2 is between y_0 and $y_0 + k$.

In Exercises 24 through 27, use Theorem 18.5.4 to find either a ξ_1 or a ξ_2, whichever applies.

24. $f(x, y) = x^2 + 3xy - y^2$; x is in $[1, 3]$; $y = 4$ **25.** $f(x, y) = x^3 - y^2$; x is in $[2, 6]$; $y = 3$

26. $f(x, y) = \dfrac{4x}{x + y}$; y is in $[-2, 2]$; $x = 4$ **27.** $f(x, y) = \dfrac{2x - y}{2y + x}$; y is in $[0, 4]$; $x = 2$

28. Given $f(x, y) = \begin{cases} \dfrac{3x^2y}{x^2 + y^2} & \text{if } (x, y) \neq (0, 0) \\ 0 & \text{if } (x, y) = (0, 0) \end{cases}$

This function is continuous at $(0, 0)$ (see Example 5, Sec. 18.2, and Illustration 1, Sec. 18.3). Prove that $D_1f(0, 0)$ and $D_2f(0, 0)$ exist but D_1f and D_2f are not continuous at $(0, 0)$.

29. Given $f(x, y) = \begin{cases} \dfrac{xy(x^2 - y^2)}{x^2 + y^2} & \text{if } (x, y) \neq (0, 0) \\ 0 & \text{if } (x, y) = (0, 0) \end{cases}$

Prove that f is differentiable at $(0, 0)$ by using Theorem 18.5.5.

In Exercises 30 through 32, prove that f is differentiable at all points in R^3 by doing each of the following: (a) Find $\Delta f(x_0, y_0, z_0)$; (b) find an ϵ_1, ϵ_2, and ϵ_3 such that Eq. (30) holds; (c) show that the ϵ_1, ϵ_2, and ϵ_3 found in (b) all approach zero as $(\Delta x, \Delta y, \Delta z)$ approaches $(0, 0, 0)$.

30. $f(x, y, z) = 3x + 2y - 4z$ **31.** $f(x, y, z) = xy - xz + z^2$ **32.** $f(x, y, z) = 2x^2z - 3yz^2$

33. Given $f(x, y, z) = \begin{cases} \dfrac{xy^2z}{x^4 + y^4 + z^4} & \text{if } (x, y, z) \neq (0, 0, 0) \\ 0 & \text{if } (x, y, z) = (0, 0, 0) \end{cases}$

(a) Show that $D_1f(0, 0, 0)$, $D_2f(0, 0, 0)$, and $D_3f(0, 0, 0)$ exist; (b) make use of the fact that differentiability implies continuity to prove that f is not differentiable at $(0, 0, 0)$.

34. Given $f(x, y, z) = \begin{cases} \dfrac{xyz^2}{x^2 + y^2 + z^2} & \text{if } (x, y, z) \neq (0, 0, 0) \\ 0 & \text{if } (x, y, z) = (0, 0, 0) \end{cases}$

Prove that f is differentiable at $(0, 0, 0)$.

35. A closed container in the shape of a rectangular solid is to have an inside length of 8 m, an inside width of 5 m, an inside height of 4 m, and a thickness of 4 cm. Use differentials to approximate the amount of material needed to construct the container.

36. Use the total differential to find approximately the greatest error in calculating the area of a right triangle from the lengths of the legs if they are measured to be 6 cm and 8 cm, respectively, with a possible error of 0.1 cm for each measurement. Also find the approximate percent error.

37. Find approximately, by using the total differential, the greatest error in calculating the length of the hypotenuse of the right triangle from the measurements of Exercise 36. Also find the approximate percent error.

38. If the ideal gas law (see Example 5, Sec. 18.4) is used to find P when T and V are given, but there is an error of 0.3% in measuring T and an error of 0.8% in measuring V, find approximately the greatest percent error in P.

39. The specific gravity s of an object is given by the formula

$$s = \frac{A}{A - W}$$

where A is the number of pounds in the weight of the object in air and W is the number of pounds in the weight of the object in water. If the weight of an object in air is read as 20 lb with a possible error of 0.01 lb and its weight in water is read as 12 lb with a possible error of 0.02 lb, find approximately the largest possible error in calculating s from these measurements. Also find the largest possible relative error.

40. A wooden box is to be made of lumber that is $\frac{2}{3}$ in. thick. The inside length is to be 6 ft, the inside width is to be 3 ft, the inside depth is to be 4 ft, and the box is to have no top. Use the total differential to find the approximate amount of lumber to be used in the box.

41. A company has contracted to manufacture 10,000 closed wooden crates having dimensions 3 m, 4 m, and 5 m. The cost of the wood to be used is $1 per square meter. If the machines that are used to cut the pieces of wood have a possible error of 0.5 cm in each dimension, find approximately, by using the total differential, the greatest possible error in the estimate of the cost of the wood.

In Exercises 42 through 45, we show that a function may be differentiable at a point even though it is not continuously differentiable there. Hence the conditions of Theorem 18.5.5 are sufficient but not necessary for differentiability. The function f in these exercises is defined by

$$f(x, y) = \begin{cases} (x^2 + y^2) \sin \dfrac{1}{\sqrt{x^2 + y^2}} & \text{if } (x, y) \neq (0, 0) \\ 0 & \text{if } (x, y) = (0, 0) \end{cases}$$

42. Find $\Delta f(0, 0)$. **43.** Find $D_1 f(x, y)$ and $D_2 f(x, y)$.

44. Prove that f is differentiable at (0, 0) by using Definition 18.5.2 and the results of Exercises 42 and 43.

45. Prove that $D_1 f$ and $D_2 f$ are not continuous at (0, 0).

18.6 THE CHAIN RULE

In Sec. 3.7 we had the following chain rule (Theorem 3.7.1) for functions of a single variable: If y is a function of u, defined by $y = f(u)$, and $D_u y$ exists; and u is a function of x, defined by $u = g(x)$, and $D_x u$ exists; then y is a function of x, and $D_x y$ exists and is given by

$$D_x y = D_u y \, D_x u$$

or, equivalently, with Leibniz notation,

$$\frac{dy}{dx} = \frac{dy}{du}\frac{du}{dx} \tag{1}$$

We now consider the chain rule for a function of two variables, where each of these variables is also a function of two variables.

18.6.1 Theorem The Chain Rule If u is a differentiable function of x and y, defined by $u = f(x, y)$, and $x = F(r, s)$, $y = G(r, s)$, and $\partial x/\partial r$, $\partial x/\partial s$, $\partial y/\partial r$, and $\partial y/\partial s$ all exist, then u is a function of r and s and

$$\frac{\partial u}{\partial r} = \left(\frac{\partial u}{\partial x}\right)\left(\frac{\partial x}{\partial r}\right) + \left(\frac{\partial u}{\partial y}\right)\left(\frac{\partial y}{\partial r}\right) \tag{2}$$

$$\frac{\partial u}{\partial s} = \left(\frac{\partial u}{\partial x}\right)\left(\frac{\partial x}{\partial s}\right) + \left(\frac{\partial u}{\partial y}\right)\left(\frac{\partial y}{\partial s}\right) \tag{3}$$

Proof We prove (2). The proof of (3) is similar.

If s is held fixed and r is changed by an amount Δr, then x is changed by an amount Δx and y is changed by an amount Δy. Thus

$$\Delta x = F(r + \Delta r, s) - F(r, s) \tag{4}$$

and

$$\Delta y = G(r + \Delta r, s) - G(r, s) \tag{5}$$

Because f is differentiable,

$$\Delta f(x, y) = D_1 f(x, y)\, \Delta x + D_2 f(x, y)\, \Delta y + \epsilon_1\, \Delta x + \epsilon_2\, \Delta y \tag{6}$$

where ϵ_1 and ϵ_2 both approach zero as $(\Delta x, \Delta y)$ approaches $(0, 0)$. Furthermore, we require that $\epsilon_1 = 0$ and $\epsilon_2 = 0$ when $\Delta x = \Delta y = 0$. We make this requirement so that ϵ_1 and ϵ_2, which are functions of Δx and Δy, will be continuous at $(\Delta x, \Delta y) = (0, 0)$.

If in (6) we replace $\Delta f(x, y)$ by Δu, $D_1 f(x, y)$ by $\partial u/\partial x$, and $D_2 f(x, y)$ by $\partial u/\partial y$ and divide on both sides by Δr ($\Delta r \neq 0$), we obtain

$$\frac{\Delta u}{\Delta r} = \frac{\partial u}{\partial x}\frac{\Delta x}{\Delta r} + \frac{\partial u}{\partial y}\frac{\Delta y}{\Delta r} + \epsilon_1 \frac{\Delta x}{\Delta r} + \epsilon_2 \frac{\Delta y}{\Delta r}$$

Taking the limit on both sides of the above as Δr approaches zero we get

$$\lim_{\Delta r \to 0} \frac{\Delta u}{\Delta r} = \frac{\partial u}{\partial x} \lim_{\Delta r \to 0} \frac{\Delta x}{\Delta r} + \frac{\partial u}{\partial y} \lim_{\Delta r \to 0} \frac{\Delta y}{\Delta r} + \left(\lim_{\Delta r \to 0} \epsilon_1\right) \lim_{\Delta r \to 0} \frac{\Delta x}{\Delta r} + \left(\lim_{\Delta r \to 0} \epsilon_2\right) \lim_{\Delta r \to 0} \frac{\Delta y}{\Delta r} \tag{7}$$

Because u is a function of x and y and both x and y are functions of r and s, u is a function of r and s. Because s is held fixed and r is changed by an amount Δr,

$$\lim_{\Delta r \to 0} \frac{\Delta u}{\Delta r} = \lim_{\Delta r \to 0} \frac{u(r + \Delta r, s) - u(r, s)}{\Delta r} = \frac{\partial u}{\partial r} \tag{8}$$

Also,

$$\lim_{\Delta r \to 0} \frac{\Delta x}{\Delta r} = \lim_{\Delta r \to 0} \frac{F(r + \Delta r, s) - F(r, s)}{\Delta r} = \frac{\partial x}{\partial r} \tag{9}$$

and

$$\lim_{\Delta r \to 0} \frac{\Delta y}{\Delta r} = \lim_{\Delta r \to 0} \frac{G(r + \Delta r, s) - G(r, s)}{\Delta r} = \frac{\partial y}{\partial r} \tag{10}$$

Because $\partial x/\partial r$ and $\partial y/\partial r$ exist, F and G are each continuous with respect to the variable r. (*Note:* The existence of the partial derivatives of a function does not imply continuity with respect to all of the variables simultaneously, as we saw in the preceding section, but as with functions of a single variable it does imply continuity of the function with respect to each variable separately.) Hence, from (4),

$$\begin{aligned}\lim_{\Delta r \to 0} \Delta x &= \lim_{\Delta r \to 0} [F(r + \Delta r, s) - F(r, s)] \\ &= F(r, s) - F(r, s) \\ &= 0\end{aligned}$$

and from (5),

$$\begin{aligned}\lim_{\Delta r \to 0} \Delta y &= \lim_{\Delta r \to 0} [G(r + \Delta r, s) - G(r, s)] \\ &= G(r, s) - G(r, s) \\ &= 0\end{aligned}$$

Therefore, as Δr approaches zero, both Δx and Δy approach zero. And because both ϵ_1 and ϵ_2 approach zero as $(\Delta x, \Delta y)$ approaches $(0, 0)$, we can conclude that

$$\lim_{\Delta r \to 0} \epsilon_1 = 0 \quad \text{and} \quad \lim_{\Delta r \to 0} \epsilon_2 = 0 \tag{11}$$

Now, it is possible that for certain values of Δr, $\Delta x = \Delta y = 0$. Because we required in such a case that $\epsilon_1 = \epsilon_2 = 0$, the limits in (11) are still zero. Substituting from (8), (9), (10), and (11) into (7) we obtain

$$\frac{\partial u}{\partial r} = \left(\frac{\partial u}{\partial x}\right)\left(\frac{\partial x}{\partial r}\right) + \left(\frac{\partial u}{\partial y}\right)\left(\frac{\partial y}{\partial r}\right)$$

which proves (2). ■

Example 1 Given

$$u = \ln \sqrt{x^2 + y^2}$$

$x = re^s$, and $y = re^{-s}$, find $\partial u/\partial r$ and $\partial u/\partial s$.

Solution

$$\frac{\partial u}{\partial x} = \frac{x}{x^2 + y^2} \qquad \frac{\partial u}{\partial y} = \frac{y}{x^2 + y^2} \qquad \frac{\partial x}{\partial r} = e^s$$

$$\frac{\partial x}{\partial s} = re^s \qquad \frac{\partial y}{\partial r} = e^{-s} \qquad \frac{\partial y}{\partial s} = -re^{-s}$$

From (2) we get

$$\frac{\partial u}{\partial r} = \frac{x}{x^2 + y^2}(e^s) + \frac{y}{x^2 + y^2}(e^{-s}) = \frac{xe^s + ye^{-s}}{x^2 + y^2}$$

From (3) we get

$$\frac{\partial u}{\partial s} = \frac{x}{x^2 + y^2}(re^s) + \frac{y}{x^2 + y^2}(-re^{-s}) = \frac{r(xe^s - ye^{-s})}{x^2 + y^2}$$

As mentioned earlier, the symbols $\partial u/\partial r$, $\partial u/\partial s$, $\partial u/\partial x$, $\partial u/\partial y$, and so forth, must not be considered as fractions. The symbols ∂u, ∂x, and so on, have no meaning by themselves. For functions of one variable, the chain rule, given by Eq. (1), is easily remembered by thinking of an ordinary derivative as the quotient of two differentials, but there is no similar interpretation for partial derivatives.

Another troublesome notational problem arises when considering u as a function of x and y and then as a function of r and s. If $u = f(x, y)$, $x = F(r, s)$, and $y = G(r, s)$, then $u = f(F(r, s), G(r, s))$. [It is incorrect to write $u = f(r, s)$.]

● ILLUSTRATION 1 In Example 1,

$$u = f(x, y) = \ln \sqrt{x^2 + y^2}$$

$$x = F(r, s) = re^s$$

$$y = G(r, s) = re^{-s}$$

and so

$$u = f(F(r, s), G(r, s)) = \ln \sqrt{r^2e^{2s} + r^2e^{-2s}}$$

$[f(r, s) - \ln \sqrt{r^2 + s^2} \neq u.]$ ●

If we let $f(F(r, s), G(r, s)) = h(r, s)$, then (2) and (3) can be written respectively as

$$h_1(r, s) = f_1(x, y)F_1(r, s) + f_2(x, y)G_1(r, s)$$

and

$$h_2(r, s) = f_1(x, y)F_2(r, s) + f_2(x, y)G_2(r, s)$$

In the statement of Theorem 18.6.1 the independent variables are r and s, and u is the dependent variable. The variables x and y can be called the intermediate variables. We now extend the chain rule to n intermediate variables and m independent variables.

18.6.2 Theorem
The General Chain Rule

Suppose that u is a differentiable function of the n variables $x_1, x_2, \ldots, x_n$, and each of these variables is in turn a function of the m variables $y_1, y_2, \ldots, y_m$. Suppose further that each of the partial derivatives $\partial x_i/\partial y_j$ ($i = 1, 2, \ldots, n$; $j = 1, 2, \ldots, m$) exists. Then u is a function of $y_1, y_2, \ldots, y_m$, and

$$\frac{\partial u}{\partial y_1} = \left(\frac{\partial u}{\partial x_1}\right)\left(\frac{\partial x_1}{\partial y_1}\right) + \left(\frac{\partial u}{\partial x_2}\right)\left(\frac{\partial x_2}{\partial y_1}\right) + \cdots + \left(\frac{\partial u}{\partial x_n}\right)\left(\frac{\partial x_n}{\partial y_1}\right)$$

$$\frac{\partial u}{\partial y_2} = \left(\frac{\partial u}{\partial x_1}\right)\left(\frac{\partial x_1}{\partial y_2}\right) + \left(\frac{\partial u}{\partial x_2}\right)\left(\frac{\partial x_2}{\partial y_2}\right) + \cdots + \left(\frac{\partial u}{\partial x_n}\right)\left(\frac{\partial x_n}{\partial y_2}\right)$$

$$\vdots$$

$$\frac{\partial u}{\partial y_m} = \left(\frac{\partial u}{\partial x_1}\right)\left(\frac{\partial x_1}{\partial y_m}\right) + \left(\frac{\partial u}{\partial x_2}\right)\left(\frac{\partial x_2}{\partial y_m}\right) + \cdots + \left(\frac{\partial u}{\partial x_n}\right)\left(\frac{\partial x_n}{\partial y_m}\right)$$

The proof is an extension of the proof of Theorem 18.6.1.

Observe that in the general chain rule there are as many terms on the right side of each equation as there are intermediate variables.

Example 2 Given $u = xy + xz + yz$, $x = r$, $y = r\cos t$, and $z = r\sin t$, find $\partial u/\partial r$ and $\partial u/\partial t$.

Solution From the chain rule,

$$\begin{aligned}
\frac{\partial u}{\partial r} &= \left(\frac{\partial u}{\partial x}\right)\left(\frac{\partial x}{\partial r}\right) + \left(\frac{\partial u}{\partial y}\right)\left(\frac{\partial y}{\partial r}\right) + \left(\frac{\partial u}{\partial z}\right)\left(\frac{\partial z}{\partial r}\right) \\
&= (y+z)(1) + (x+z)(\cos t) + (x+y)(\sin t) \\
&= y + z + x\cos t + z\cos t + x\sin t + y\sin t \\
&= r\cos t + r\sin t + r\cos t + (r\sin t)(\cos t) + r\sin t + (r\cos t)(\sin t) \\
&= 2r(\cos t + \sin t) + r(2\sin t\cos t) \\
&= 2r(\cos t + \sin t) + r\sin 2t
\end{aligned}$$

$$\begin{aligned}
\frac{\partial u}{\partial t} &= \left(\frac{\partial u}{\partial x}\right)\left(\frac{\partial x}{\partial t}\right) + \left(\frac{\partial u}{\partial y}\right)\left(\frac{\partial y}{\partial t}\right) + \left(\frac{\partial u}{\partial z}\right)\left(\frac{\partial z}{\partial t}\right) \\
&= (y+z)(0) + (x+z)(-r\sin t) + (s+y)(r\cos t) \\
&= (r + r\sin t)(-r\sin t) + (r + r\cos t)(r\cos t) \\
&= -r^2\sin t - r^2\sin^2 t + r^2\cos t + r^2\cos^2 t \\
&= r^2(\cos t - \sin t) + r^2(\cos^2 t - \sin^2 t) \\
&= r^2(\cos t - \sin t) + r^2\cos 2t
\end{aligned}$$

Now suppose that u is a differentiable function of the two variables x and y, and both x and y are differentiable functions of the single variable t. Then u is a function of the single variable t, and so instead of the partial derivative of u with respect to t, we have the ordinary derivative of u with respect to t, which is given by

$$\frac{du}{dt} = \left(\frac{\partial u}{\partial x}\right)\left(\frac{dx}{dt}\right) + \left(\frac{\partial u}{\partial y}\right)\left(\frac{dy}{dt}\right) \tag{12}$$

We call du/dt given by Eq. (12) the **total derivative** of u with respect to t. If u is a differentiable function of the n variables $x_1, x_2, \ldots, x_n$ and each x_i is a differentiable function of the single variable t, then u is a function of t and the total derivative of u with respect to t is given by

$$\frac{du}{dt} = \left(\frac{\partial u}{\partial x_1}\right)\left(\frac{dx_1}{dt}\right) + \left(\frac{\partial u}{\partial x_2}\right)\left(\frac{dx_2}{dt}\right) + \cdots + \left(\frac{\partial u}{\partial x_n}\right)\left(\frac{dx_n}{dt}\right)$$

Example 3 Given $u = x^2 + 2xy + y^2$, $x = t\cos t$, and $y = t\sin t$, find du/dt by two methods: (a) using the chain rule; (b) expressing u in terms of t before differentiating.

Solution

(a) $\partial u/\partial x = 2x + 2y$; $\partial u/\partial y = 2x + 2y$; $dx/dt = \cos t - t\sin t$; $dy/dt = \sin t + t\cos t$. So from (12),

$$\begin{aligned}\frac{du}{dt} &= (2x + 2y)(\cos t - t\sin t) + (2x + 2y)(\sin t + t\cos t)\\ &= 2(x + y)(\cos t - t\sin t + \sin t + t\cos t)\\ &= 2(t\cos t + t\sin t)(\cos t - t\sin t + \sin t + t\cos t)\\ &= 2t(\cos^2 t - t\sin t\cos t + \sin t\cos t + t\cos^2 t + \sin t\cos t\\ &\qquad - t\sin^2 t + \sin^2 t + t\sin t\cos t)\\ &= 2t[1 + 2\sin t\cos t + t(\cos^2 t - \sin^2 t)]\\ &= 2t(1 + \sin 2t + t\cos 2t)\end{aligned}$$

(b)
$$\begin{aligned}u &= (t\cos t)^2 + 2(t\cos t)(t\sin t) + (t\sin t)^2\\ &= t^2\cos^2 t + t^2(2\sin t\cos t) + t^2\sin^2 t\\ &= t^2 + t^2\sin 2t\end{aligned}$$

Thus

$$\frac{du}{dt} = 2t + 2t\sin 2t + 2t^2\cos 2t$$

Example 4 If f is a differentiable function and a and b are constants, prove that $z = f(\frac{1}{2}bx^2 - \frac{1}{3}ay^3)$ satisfies the partial differential equation

$$ay^2\frac{\partial z}{\partial x} + bx\frac{\partial z}{\partial y} = 0$$

Solution Let $u = \frac{1}{2}bx^2 - \frac{1}{3}ay^3$. We wish to show that $z = f(u)$ satisfies the given equation. By the chain rule,

$$\frac{\partial z}{\partial x} = \frac{dz}{du}\frac{\partial u}{\partial x} = f'(u)(bx) \quad\text{and}\quad \frac{\partial z}{\partial y} = \frac{dz}{du}\frac{\partial u}{\partial y} = f'(u)(-ay^2)$$

Therefore

$$ay^2 \frac{\partial z}{\partial x} + bx \frac{\partial z}{\partial y} = ay^2[f'(u)(bx)] + bx[f'(u)(-ay^2)] = 0$$

which is what we wished to prove.

Example 5 Use the ideal gas law (see Example 5, Sec. 18.4) with $k = 10$ to find the rate at which the temperature is changing at the instant when the volume of the gas is 120 cm^3 and the gas is under a pressure of 8 nt/cm^2 if the volume is increasing at the rate of 2 cm^3/min and the pressure is decreasing at the rate of 0.1 nt/cm^2 per min.

Solution Let t min be the time that has elapsed since the volume of the gas started to increase, T degrees be the temperature at t min, P newtons per square centimeter be the pressure at t min, and V cubic centimeters be the volume of the gas at t min.

From the ideal gas law,

$$T = \frac{PV}{10}$$

At the given instant, $P = 8$, $V = 120$, $dP/dt = -0.1$, and $dV/dt = 2$. From the chain rule,

$$\begin{aligned}\frac{dT}{dt} &= \frac{\partial T}{\partial P}\frac{dP}{dt} + \frac{\partial T}{\partial V}\frac{dV}{dt} \\ &= \frac{V}{10}\frac{dP}{dt} + \frac{P}{10}\frac{dV}{dt} \\ &= \tfrac{120}{10}(-0.1) + \tfrac{8}{10}(2) \\ &= -1.2 + 1.6 \\ &= 0.4\end{aligned}$$

Therefore the temperature is increasing at the rate of 0.4 degree per minute at the given instant.

EXERCISES 18.6

In Exercises 1 through 6, find the indicated partial derivative by two methods: (a) use the chain rule; (b) make the substitutions for x and y before differentiating.

1. $u = x^2 - y^2$; $x = 3r - s$; $y = r + 2s$; $\frac{\partial u}{\partial r}$; $\frac{\partial u}{\partial s}$

2. $u = 3x - 4y^2$; $x = 5pq$; $y = 3p^2 - 2q$; $\frac{\partial u}{\partial p}$; $\frac{\partial u}{\partial q}$

3. $u = 3x^2 + xy - 2y^2 + 3x - y$; $x = 2r - 3s$; $y = r + s$; $\frac{\partial u}{\partial r}$; $\frac{\partial u}{\partial s}$

4. $u = x^2 + y^2$; $x = \cosh r \cos t$; $y = \sinh r \sin t$; $\frac{\partial u}{\partial r}$; $\frac{\partial u}{\partial t}$

5. $u = e^{y/x}$; $x = 2r \cos t$; $y = 4r \sin t$; $\frac{\partial u}{\partial r}$; $\frac{\partial u}{\partial t}$

6. $V = \pi x^2 y$; $x = \cos z \sin t$; $y = z^2 e^t$; $\frac{\partial V}{\partial z}$; $\frac{\partial V}{\partial t}$

In Exercises 7 through 14, find the indicated partial derivative by using the chain rule.

7. $u = x^2 + xy; x = r^2 + s^2; y = 3r - 2s; \dfrac{\partial u}{\partial r}; \dfrac{\partial u}{\partial s}$

8. $u = xy + xz + yz; x = rs; y = r^2 - s^2; z = (r - s)^2; \dfrac{\partial u}{\partial r}; \dfrac{\partial u}{\partial s}$

9. $u = \sin^{-1}(3x + y); x = r^2 e^s; y = \sin rs; \dfrac{\partial u}{\partial r}; \dfrac{\partial u}{\partial s}$

10. $u = \sin(xy); x = 2ze^t; y = t^2 e^{-z}; \dfrac{\partial u}{\partial t}; \dfrac{\partial u}{\partial z}$

11. $u = \cosh \dfrac{y}{x}; x = 3r^2 s; y = 6se^r; \dfrac{\partial u}{\partial r}; \dfrac{\partial u}{\partial s}$

12. $u = xe^{-y}; x = \tan^{-1}(rst); y = \ln(3rs + 5st); \dfrac{\partial u}{\partial r}; \dfrac{\partial u}{\partial s}; \dfrac{\partial u}{\partial t}$

13. $u = x^2 + y^2 + z^2; x = r \sin \phi \cos \theta; y = r \sin \phi \sin \theta; z = r \cos \phi; \dfrac{\partial u}{\partial r}; \dfrac{\partial u}{\partial \phi}; \dfrac{\partial u}{\partial \theta}$

14. $u = x^2 yz; x = \dfrac{r}{s}; y = re^s; z = re^{-s}; \dfrac{\partial u}{\partial r}; \dfrac{\partial u}{\partial s}$

In Exercises 15 through 18, find the total derivative du/dt by two methods: (a) use the chain rule; (b) make the substitutions for x and y or for x, y, and z before differentiating.

15. $u = ye^x + xe^y; x = \cos t; y = \sin t$

16. $u = \ln xy + y^2; x = e^t; y = e^{-t}$

17. $u = \sqrt{x^2 + y^2 + z^2}; x = \tan t; y = \cos t; z = \sin t; 0 < t < \frac{1}{2}\pi$

18. $u = \dfrac{t + e^x}{y - e^t}; x = 3 \sin t; y = \ln t$

In Exercises 19 through 22, find the total derivative du/dt by using the chain rule; do not express u as a function of t before differentiating.

19. $u = \tan^{-1}\left(\dfrac{y}{x}\right); x = \ln t; y = e^t$

20. $u = xy + xz + yz; x = t \cos t; y = t \sin t; z = t$

21. $u = \dfrac{x + t}{y + t}; x = \ln t; y = \ln \dfrac{1}{t}$

22. $u = \ln(x^2 + y^2 + t^2); x = t \sin t; y = \cos t$

In Exercises 23 through 26, assume that the given equation defines z as a function of x and y. Differentiate implicitly to find ∂z/∂x and ∂z/∂y.

23. $3x^2 + y^2 + z^2 - 3xy + 4xz - 15 = 0$

24. $z = (x^2 + y^2) \sin xz$

25. $ye^{xyz} \cos 3xz = 5$

26. $ze^{yz} + 2xe^{xz} - 4e^{xy} = 3$

27. If f is a differentiable function of the variable u, let $u = bx - ay$ and prove that $z = f(bx - ay)$ satisfies the equation $a(\partial z/\partial x) + b(\partial z/\partial y) = 0$, where a and b are constants.

28. If f is differentiable function of two variables u and v, let $u = x - y$ and $v = y - x$; prove that $z = f(x - y, y - x)$ satisfies the equation $\partial z/\partial x + \partial z/\partial y = 0$.

29. If f is a differentiable function of x and y and $u = f(x, y)$, $x = r \cos \theta$, and $y = r \sin \theta$, show that

$$\frac{\partial u}{\partial x} = \frac{\partial u}{\partial r} \cos \theta - \frac{\partial u}{\partial \theta} \frac{\sin \theta}{r}$$

$$\frac{\partial u}{\partial y} = \frac{\partial u}{\partial r} \sin \theta + \frac{\partial u}{\partial \theta} \frac{\cos \theta}{r}$$

30. If f and g are differentiable functions of x and y and $u = f(x, y)$ and $v = g(x, y)$, such that $\partial u/\partial x = \partial v/\partial y$ and $\partial u/\partial y = -\partial v/\partial x$, then if $x = r \cos \theta$ and $y = r \sin \theta$, show that

$$\frac{\partial u}{\partial r} = \frac{1}{r} \frac{\partial v}{\partial \theta} \quad \text{and} \quad \frac{\partial v}{\partial r} = -\frac{1}{r} \frac{\partial u}{\partial \theta}$$

31. Suppose f is a differentiable function of x and y and $u = f(x, y)$. Then if $x = \cosh v \cos w$ and $y = \sinh v \sin w$, express $\partial u/\partial v$ and $\partial u/\partial w$ in terms of $\partial u/\partial x$ and $\partial u/\partial y$.

32. Suppose f is a differentiable function of x, y, and z and $u = f(x, y, z)$. Then if $x = r \sin \phi \cos \theta$, $y = r \sin \phi \sin \theta$, and $z = r \cos \phi$, express $\partial u/\partial r$, $\partial u/\partial \phi$, and $\partial u/\partial \theta$ in terms of $\partial u/\partial x$, $\partial u/\partial y$, *and* $\partial u/\partial z$.

33. At a given instant, the length of one leg of a right triangle is 10 cm and it is increasing at the rate of 1 cm/min and the length of the other leg of the right triangle is 12 cm and it is decreasing at the rate of 2 cm/min. Find the rate of change of the measure of the acute angle opposite the leg of length 12 cm at the given instant.

34. The height of a right-circular cone is increasing at the rate of 40 cm/min and the radius is decreasing at the rate of 15 cm/min. Find the rate of change of the volume at the instant when the height is 200 cm and the radius is 60 cm.

35. The height of a right-circular cylinder is decreasing at the rate of 10 cm/min and the radius is increasing at the rate of 4 cm/min. Find the rate of change of the volume at the instant when the height is 50 cm and the radius is 16 cm.

36. Water is flowing into a tank in the form of a right-circular cylinder at the rate of $\frac{4}{5}\pi$ m^3/min. The tank is stretching in such a way that even though it remains cylindrical, its radius is increasing at the rate of 0.2 cm/min. How fast is the surface of the water rising when the radius is 2 m and the volume of water in the tank is 20π m^3?

37. A quantity of gas obeys the ideal gas law (see Example 5, Sec. 18.4) with $k = 12$, and the gas is in a container that is being heated at a rate of 3° per minute. If at the instant when the temperature is 300°, the pressure is 6 nt/cm^2 and is decreasing at the rate of 0.1 nt/cm^2 per minute, find the rate of change of the volume at that instant.

38. A vertical wall makes an angle of radian measure $\frac{2}{3}\pi$ with the ground. A ladder of length 20 ft is leaning against the wall and its top is sliding down the wall at the rate of 3 ft/sec. How fast is the area of the triangle formed by the ladder, the wall, and the ground changing when the ladder makes an angle of $\frac{1}{6}\pi$ radians with the ground?

18.7 HIGHER ORDER PARTIAL DERIVATIVES

If f is a function of two variables, then in general $D_1 f$ and $D_2 f$ are also functions of two variables. And if the partial derivatives of these functions exist, they are called second partial derivatives of f. In contrast, $D_1 f$ and $D_2 f$ are called first partial derivatives of f. There are four second partial derivatives of a function of two variables. If f is a function of the two variables x and y, the notations

$$D_2(D_1 f) \qquad D_{12} f \qquad f_{12} \qquad f_{xy} \qquad \frac{\partial^2 f}{\partial y\, \partial x}$$

all denote the second partial derivative of f, which is obtained by first partial-differentiating f with respect to x and then partial-differentiating the result with respect to y. This second partial derivative is defined by

$$f_{12}(x, y) = \lim_{\Delta y \to 0} \frac{f_1(x, y + \Delta y) - f_1(x, y)}{\Delta y} \tag{1}$$

if this limit exists. The notations

$$D_1(D_1 f) \qquad D_{11} f \qquad f_{11} \qquad f_{xx} \qquad \frac{\partial^2 f}{\partial x^2}$$

all denote the second partial derivative of f, which is obtained by partial-differentiating twice with respect to x. We have the definition

$$f_{11}(x, y) = \lim_{\Delta x \to 0} \frac{f_1(x + \Delta x, y) - f_1(x, y)}{\Delta x} \tag{2}$$

if this limit exists. The other two second partial derivatives are defined in an analogous way.

$$f_{21}(x, y) = \lim_{\Delta x \to 0} \frac{f_2(x + \Delta x, y) - f_2(x, y)}{\Delta x} \tag{3}$$

and

$$f_{22}(x, y) = \lim_{\Delta y \to 0} \frac{f_2(x, y + \Delta y) - f_2(x, y)}{\Delta y} \tag{4}$$

if these limits exist.

The definitions of higher order partial derivatives are similar. Again there are various notations for a specific derivative. For example,

$$D_{112}f \qquad f_{112} \qquad f_{xxy} \qquad \frac{\partial^3 f}{\partial y\, \partial x\, \partial x} \qquad \frac{\partial^3 f}{\partial y\, \partial x^2}$$

all stand for the third partial derivative of f, which is obtained by partial-differentiating twice with respect to x and then once with respect to y. In the subscript notation, the order of partial differentiation is from left to right; in the notation $\partial^3 f/\partial y\, \partial x\, \partial x$, the order is from right to left.

Example 1 Given $f(x, y) = e^x \sin y + \ln xy$, find: (a) $D_{11}f(x, y)$; (b) $D_{12}f(x, y)$; (c) $\partial^3 f/\partial x\, \partial y^2$.

Solution

$$D_1 f(x, y) = e^x \sin y + \frac{1}{xy} y = e^x \sin y + \frac{1}{x}$$

So (a) $D_{11}f(x, y) = e^x \sin y - 1/x^2$; and (b) $D_{12}f(x, y) = e^x \cos y$. (c) To find $\partial^3 f/\partial x\, \partial y^2$ we partial-differentiate twice with respect to y and then once with respect to x. This gives

$$\frac{\partial f}{\partial y} = e^x \cos y + \frac{1}{y} \qquad \frac{\partial^2 f}{\partial y^2} = -e^x \sin y - \frac{1}{y^2} \qquad \frac{\partial^3 f}{\partial x\, \partial y^2} = -e^x \sin y$$

Higher order partial derivatives of a function of n variables have definitions that are analogous to the definitions of higher order partial derivatives of a function of two variables. If f is a function of n variables, there may be n^2 second partial derivatives of f at a particular point. That is, for a function of three variables, if all the second-order partial derivatives exist, there are nine of them: f_{11}, f_{12}, f_{13}, f_{21}, f_{22}, f_{23}, f_{31}, f_{32}, and f_{33}.

Example 2 Given $f(x, y, z) = \sin(xy + 2z)$, find $D_{132}f(x, y, z)$.

Solution

$$D_1 f(x, y, z) = y \cos(xy + 2z)$$

$$D_{13} f(x, y, z) = -2y \sin(xy + 2z)$$

$$D_{132} f(x, y, z) = -2 \sin(xy + 2z) - 2xy \cos(xy + 2z)$$

Example 3 Given $f(x, y) = x^3y - y \cosh xy$, find: (a) $D_{12}f(x, y)$; (b) $D_{21}f(x, y)$.

Solution

(a) $$D_1f(x, y) = 3x^2y - y^2 \sinh xy$$

$$D_{12}f(x, y) = 3x^2 - 2y \sinh xy - xy^2 \cosh xy$$

(b) $$D_2f(x, y) = x^3 - \cosh xy - xy \sinh xy$$

$$D_{21}f(x, y) = 3x^2 - y \sinh xy - y \sinh xy - xy^2 \cosh xy$$

$$= 3x^2 - 2y \sinh xy - xy^2 \cosh xy$$

Observe from the above results that for the function of Example 3 the "mixed" partial derivatives $D_{12}f(x, y)$ and $D_{21}f(x, y)$ are equal. So for this particular function, when finding the second partial derivative with respect to x and then y, the order of differentiation is immaterial. This condition holds for many functions. However, the following example shows that it is not always true.

Example 4 Let f be the function defined by

$$f(x, y) = \begin{cases} (xy)\dfrac{x^2 - y^2}{x^2 + y^2} & \text{if } (x, y) \neq (0, 0) \\ 0 & \text{if } (x, y) = (0, 0) \end{cases}$$

Find $f_{12}(0, 0)$ and $f_{21}(0, 0)$.

Solution In Example 3, Sec. 18.4, we showed that for this function

$$f_1(0, y) = -y \qquad \text{for all } y \tag{5}$$

and

$$f_2(x, 0) = x \qquad \text{for all } x \tag{6}$$

From formula (1),

$$f_{12}(0, 0) = \lim_{\Delta y \to 0} \frac{f_1(0, 0 + \Delta y) - f_1(0, 0)}{\Delta y}$$

But from (5), $f_1(0, \Delta y) = -\Delta y$ and $f_1(0, 0) = 0$, and so

$$f_{12}(0, 0) = \lim_{\Delta y \to 0} \frac{-\Delta y - 0}{\Delta y} = \lim_{\Delta y \to 0} (-1) = -1$$

From formula (3),

$$f_{21}(0, 0) = \lim_{\Delta x \to 0} \frac{f_2(0 + \Delta x, 0) - f_2(0, 0)}{\Delta x}$$

From (6), $f_2(\Delta x, 0) = \Delta x$ and $f_2(0, 0) = 0$. Therefore

$$f_{21}(0, 0) = \lim_{\Delta x \to 0} \frac{\Delta x - 0}{\Delta x} = \lim_{\Delta x \to 0} 1 = 1$$

For the function in Example 4 the mixed partial derivatives $f_{12}(x, y)$ and $f_{21}(x, y)$ are not equal at (0, 0). A set of conditions for which $f_{12}(x_0, y_0) = f_{21}(x_0, y_0)$ is given by Theorem 18.7.1, which follows. The function of Example 4 does not satisfy the hypothesis of this theorem because both f_{12} and f_{21} are discontinuous at (0, 0). It is left as an exercise to show this (see Exercise 24).

18.7.1 Theorem Suppose that f is a function of two variables x and y defined on an open disk $B((x_0, y_0); r)$ and f_x, f_y, f_{xy}, and f_{yx} also are defined on B. Furthermore, suppose that f_{xy} and f_{yx} are continuous on B. Then

$$f_{xy}(x_0, y_0) = f_{yx}(x_0, y_0)$$

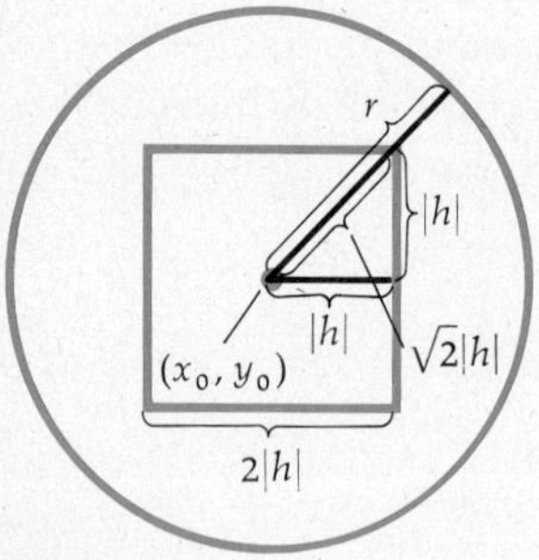

FIGURE 18.7.1

Proof Consider a square having its center at (x_0, y_0) and the length of its side $2|h|$ such that $0 < \sqrt{2}|h| < r$. Then all the points in the interior of the square and on the sides of the square are in the open disk B (see Fig. 18.7.1). So the points $(x_0 + h, y_0 + h)$, $(x_0 + h, y_0)$, and $(x_0, y_0 + h)$ are in B. Let Δ be defined by

$$\Delta = f(x_0 + h, y_0 + h) - f(x_0 + h, y_0) - f(x_0, y_0 + h) + f(x_0, y_0) \qquad (7)$$

Consider the function G defined by

$$G(x) = f(x, y_0 + h) - f(x, y_0) \qquad (8)$$

Then

$$G(x + h) = f(x + h, y_0 + h) - f(x + h, y_0)$$

So (7) can be written as

$$\Delta = G(x_0 + h) - G(x_0) \qquad (9)$$

From (8),

$$G'(x) = f_x(x, y_0 + h) - f_x(x, y_0) \qquad (10)$$

Now, because $f_x(x, y_0 + h)$ and $f_x(x, y_0)$ are defined on B, $G'(x)$ exists if x is in the closed interval having endpoints at x_0 and $x_0 + h$. Hence G is continuous if x is in this closed interval. By the mean-value theorem (4.4.2) there is a number c_1 between x_0 and $x_0 + h$ such that

$$G(x_0 + h) - G(x_0) = hG'(c_1) \qquad (11)$$

Substituting from (11) into (9) we get

$$\Delta = hG'(c_1) \qquad (12)$$

From (12) and (10),

$$\Delta = h[f_x(c_1, y_0 + h) - f_x(c_1, y_0)] \qquad (13)$$

Now if g is the function defined by

$$g(y) = f_x(c_1, y) \qquad (14)$$

we can write (13) as

$$\Delta = h[g(y_0 + h) - g(y_0)] \tag{15}$$

From (14),

$$g'(y) = f_{xy}(c_1, y) \tag{16}$$

Because $f_{xy}(c_1, y)$ is defined on B, $g'(y)$ exists if y is in the closed interval having endpoints at y_0 and $y_0 + h$; hence g is continuous if y is in this closed interval. Therefore, by the mean-value theorem there is a number d_1 between y_0 and $y_0 + h$ such that

$$g(y_0 + h) - g(y_0) = hg'(d_1) \tag{17}$$

Substituting from (17) into (15) we get $\Delta = h^2g'(d_1)$; so from (16) it follows that

$$\Delta = h^2 f_{xy}(c_1, d_1) \tag{18}$$

for some point (c_1, d_1) in the open disk B. We define a function ϕ by

$$\phi(y) = f(x_0 + h, y) - f(x_0, y) \tag{19}$$

and so $\phi(y + h) = f(x_0 + h, y + h) - f(x_0, y + h)$. Therefore (7) can be written as

$$\Delta = \phi(y_0 + h) - \phi(y_0) \tag{20}$$

From (19),

$$\phi'(y) = f_y(x_0 + h, y) - f_y(x_0, y) \tag{21}$$

ϕ' exists if y is in the closed interval having y_0 and $y_0 + h$ as endpoints because by hypothesis each term on the right side of (21) exists on B. Therefore ϕ is continuous on this closed interval. So by the mean-value theorem there is a number d_2 between y_0 and $y_0 + h$ such that

$$\phi(y_0 + h) - \phi(y_0) = h\phi'(d_2) \tag{22}$$

From (20), (21), and (22) it follows that

$$\Delta = h[f_y(x_0 + h, d_2) - f_y(x_0, d_2)] \tag{23}$$

Define the function χ by

$$\chi(x) = f_y(x, d_2) \tag{24}$$

and write (23) as

$$\Delta = h[\chi(x_0 + h) - \chi(x_0)] \tag{25}$$

From (24),

$$\chi'(x) = f_{yx}(x, d_2) \tag{26}$$

and by the mean-value theorem there is a number c_2 between x_0 and $x_0 + h$ such that

$$\chi(x_0 + h) - \chi(x_0) = h\chi'(c_2) \tag{27}$$

From (25), (26), and (27),

$$A = h^2 f_{yx}(c_2, d_2) \tag{28}$$

Equating the right sides of (18) and (28) we get

$$h^2 f_{xy}(c_1, d_1) = h^2 f_{yx}(c_2, d_2)$$

and because $h \neq 0$, we can divide by h^2, which gives

$$f_{xy}(c_1, d_1) = f_{yx}(c_2, d_2) \tag{29}$$

where (c_1, d_1) and (c_2, d_2) are in B.

Because c_1 and c_2 are each between x_0 and $x_0 + h$, we can write $c_1 = x_0 + \epsilon_1 h$, where $0 < \epsilon_1 < 1$, and $c_2 = x_0 + \epsilon_2 h$, where $0 < \epsilon_2 < 1$. Similarly, because both d_1 and d_2 are between y_0 and $y_0 + h$, we can write $d_1 = y_0 + \epsilon_3 h$, where $0 < \epsilon_3 < 1$, and $d_2 = y_0 + \epsilon_4 h$, where $0 < \epsilon_4 < 1$. Making these substitutions in (29) gives

$$f_{xy}(x_0 + \epsilon_1 h, y_0 + \epsilon_3 h) = f_{yx}(x_0 + \epsilon_2 h, y_0 + \epsilon_4 h) \tag{30}$$

Because f_{xy} and f_{yx} are continuous on B, upon taking the limit on both sides of (30) as h approaches zero we obtain

$$f_{xy}(x_0, y_0) = f_{yx}(x_0, y_0)$$ ■

As a result of the above theorem, if the function f of two variables has continuous partial derivatives on some open disk, then the order of partial differentiation can be changed without affecting the result; that is,

$$D_{112}f = D_{121}f = D_{211}f$$

$$D_{1122}f = D_{1212}f = D_{1221}f = D_{2112}f = D_{2121}f = D_{2211}f$$

and so forth. In particular, assuming that all of the partial derivatives are continuous on some open disk, we can prove that $D_{211}f = D_{112}f$ by applying Theorem 18.7.1 repeatedly. Doing this, we have

$$\begin{aligned} D_{211}f &= D_1(D_{21}f) = D_1(D_{12}f) = D_1[D_2(D_1f)] = D_2[D_1(D_1f)] \\ &= D_2(D_{11}f) = D_{112}f \end{aligned}$$

Example 5 Given that $u = f(x, y)$, $x = F(r, s)$, and $y = G(r, s)$, and assuming that $f_{xy} = f_{yx}$, prove by using the chain rule that

$$\begin{aligned} \frac{\partial^2 u}{\partial r^2} &= f_{xx}(x, y)[F_r(r, s)]^2 + 2f_{xy}(x, y)F_r(r, s)G_r(r, s) \\ &\quad + f_{yy}(x, y)[G_r(r, s)]^2 + f_x(x, y)F_{rr}(r, s) + f_y(x, y)G_{rr}(r, s) \end{aligned}$$

Solution From the chain rule (Theorem 18.6.1),

$$\frac{\partial u}{\partial r} = f_x(x, y)F_r(r, s) + f_y(x, y)G_r(r, s)$$

Taking the partial derivative again with respect to r, and using the formula for the derivative of a product and the chain rule, we obtain

$$\frac{\partial^2 u}{\partial r^2} = [f_{xx}(x, y)F_r(r, s) + f_{xy}(x, y)G_r(r, s)]F_r(r, s) + f_x(x, y)F_{rr}(r, s)$$

$$+ [f_{yx}(x, y)F_r(r, s) + f_{yy}(x, y)G_r(r, s)]G_r(r, s) + f_y(x, y)G_{rr}(r, s)$$

Multiplying and combining terms, and using the fact that $f_{xy}(x, y) = f_{yx}(x, y)$, we get

$$\frac{\partial^2 u}{\partial r^2} = f_{xx}(x, y)[F_r(r, s)]^2 + 2f_{xy}(x, y)F_r(r, s)G_r(r, s)$$

$$+ f_{yy}(x, y)[G_r(r, s)]^2 + f_x(x, y)F_{rr}(r, s) + f_y(x, y)G_{rr}(r, s)$$

which is what we wished to prove.

EXERCISES 18.7

In Exercises 1 through 10, do each of the following: (a) Find $D_{11}f(x, y)$; (b) find $D_{22}f(x, y)$; (c) show that $D_{12}f(x, y) = D_{21}f(x, y)$.

1. $f(x, y) = \dfrac{x^2}{y} - \dfrac{y}{x^2}$

2. $f(x, y) = 2x^3 - 3x^2y + xy^2$

3. $f(x, y) = e^{2x} \sin y$

4. $f(x, y) = e^{-x/y} + \ln \dfrac{y}{x}$

5. $f(x, y) = (x^2 + y^2) \tan^{-1} \dfrac{y}{x}$

6. $f(x, y) = \sin^{-1} \dfrac{3y}{x^2}$

7. $f(x, y) = 4x \sinh y + 3y \cosh x$

8. $f(x, y) = x \cos y - ye^x$

9. $f(x, y) = e^x \cos y + \tan^{-1} x \cdot \ln y$

10. $f(x, y) = 3x \cosh y - y \sin^{-1} e^x$

In Exercises 11 through 18, find the indicated partial derivatives.

11. $f(x, y) = 2x^3y + 5x^2y^2 - 3xy^2$; (a) $f_{121}(x, y)$; (b) $f_{211}(x, y)$

12. $G(x, y) = 3x^3y^2 + 5x^2y^3 + 2x$; (a) $G_{yyx}(x, y)$; (b) $G_{yxy}(x, y)$

13. $f(x, y, z) = ye^x + ze^y + e^z$; (a) $f_{xz}(x, y, z)$; (b) $f_{yz}(x, y, z)$

14. $g(x, y, z) = \sin(xyz)$; (a) $g_{23}(x, y, z)$; (b) $g_{12}(x, y, z)$

15. $f(w, z) = w^2 \cos e^z$; (a) $f_{121}(w, z)$; (b) $f_{212}(w, z)$

16. $f(u, v) = \ln \cos(u - v)$; (a) $f_{uuv}(u, v)$; (b) $f_{vuv}(u, v)$

17. $g(r, s, t) = \ln(r^2 + 4s^2 - 5t^2)$; (a) $g_{132}(r, s, t)$; (b) $g_{122}(r, s, t)$

18. $f(x, y, z) = \tan^{-1}(3xyz)$; (a) $f_{113}(x, y, z)$; (b) $f_{123}(x, y, z)$

In Exercises 19 through 22, show that $u(x, y)$ satisfies the equation

$$\frac{\partial^2 u}{\partial x^2} + \frac{\partial^2 u}{\partial y^2} = 0$$

which is known as Laplace's equation *in R^2.*

19. $u(x, y) = \ln(x^2 + y^2)$

20. $u(x, y) = e^x \sin y + e^y \cos x$

21. $u(x, y) = \tan^{-1} \dfrac{y}{x} + \dfrac{x}{x^2 + y^2}$

22. $u(x, y) = \tan^{-1} \dfrac{2xy}{x^2 - y^2}$

23. *Laplace's equation* in R^3 is

$$\frac{\partial^2 u}{\partial x^2} + \frac{\partial^2 u}{\partial y^2} + \frac{\partial^2 u}{\partial z^2} = 0$$

Show that $u(x, y, z) = (x^2 + y^2 + z^2)^{-1/2}$ satisfies this equation.

24. For the function of Example 4, show that f_{12} is discontinuous at (0, 0) and hence that the hypothesis of Theorem 18.7.1 is not satisfied if $(x_0, y_0) = (0, 0)$.

In Exercises 25 through 27, find $f_{12}(0, 0)$ and $f_{21}(0, 0)$, if they exist.

25. $f(x, y) = \begin{cases} \dfrac{2xy}{x^2 + y^2} & \text{if } (x, y) \neq (0, 0) \\ 0 & \text{if } (x, y) = (0, 0) \end{cases}$

26. $f(x, y) = \begin{cases} \dfrac{x^2y^2}{x^4 + y^4} & \text{if } (x, y) \neq (0, 0) \\ 0 & \text{if } (x, y) = (0, 0) \end{cases}$

27. $f(x, y) = \begin{cases} x^2 \tan^{-1} \dfrac{y}{x} - y^2 \tan^{-1} \dfrac{x}{y} & \text{if } x \neq 0 \text{ and } y \neq 0 \\ 0 & \text{if either } x = 0 \text{ or } y = 0 \end{cases}$

28. Given that $u = f(x, y)$, $x = F(t)$, and $y = G(t)$, and assuming that $f_{xy} = f_{yx}$, prove by using the chain rule that

$$\frac{d^2u}{dt^2} = f_{xx}(x, y)[F'(t)]^2 + 2f_{xy}(x, y)F'(t)G'(t) + f_{yy}(x, y)[G'(t)]^2 + f_x(x, y)F''(t) + f_y(x, y)G''(t)$$

29. Given that $u = f(x, y)$, $x = F(r, s)$, and $y = G(r, s)$, and assuming that $f_{xy} = f_{yx}$, prove by using the chain rule that

$$\frac{\partial^2 u}{\partial r\, \partial s} = f_{xx}(x, y)F_r(r, s)F_s(r, s) + f_{xy}(x, y)[F_s(r, s)G_r(r, s) + F_r(r, s)G_s(r, s)]$$
$$+ f_{yy}(x, y)G_r(r, s)G_s(r, s) + f_x(x, y)F_{sr}(r, s) + f_y(x, y)G_{sr}(r, s)$$

30. Given $u = e^y \cos x$, $x = 2t$, $y = t^2$. Find d^2u/dt^2 in three ways: (a) by first expressing u in terms of t; (b) by using the formula of Exercise 28; (c) by using the chain rule.

31. Given $u = 3xy - 4y^2$, $x = 2se^r$, $y = re^{-s}$. Find $\partial^2 u/\partial r^2$ in three ways: (a) by first expressing u in terms of r and s; (b) by using the formula of Example 5; (c) by using the chain rule.

32. For u, x, and y as given in Exercise 31, find $\partial^2 u/\partial s\, \partial r$ in three ways: (a) by first expressing u in terms of r and s; (b) by using the formula of Exercise 29; (c) by using the chain rule.

33. Given $u = 9x^2 + 4y^2$, $x = r \cos \theta$, $y = r \sin \theta$. Find $\partial^2 u/\partial r^2$ in three ways: (a) by first expressing u in terms of r and θ; (b) by using the formula of Example 5; (c) by using the chain rule.

34. For u, x, and y as given in Exercise 33, find $\partial^2 u/\partial \theta^2$ in three ways: (a) by first expressing u in terms of r and θ; (b) by using the formula of Example 5; (c) by using the chain rule.

35. For u, x, and y as given in Exercise 33, find $\partial^2 u/\partial r\, \partial \theta$ in three ways: (a) by first expressing u in terms of r and θ; (b) by using the formula of Exercise 29; (c) by using the chain rule.

36. If $u = f(x, y)$ and $v = g(x, y)$, then the equations

$$\frac{\partial u}{\partial x} = \frac{\partial v}{\partial y} \quad \text{and} \quad \frac{\partial v}{\partial x} = -\frac{\partial u}{\partial y}$$

are called the *Cauchy-Riemann equations.* If f and g and their first and second partial derivatives are continuous, prove that if u and v satisfy the Cauchy-Riemann equations, they also satisfy Laplace's equation (see Exercises 19 through 22).

37. The one-dimensional heat-conduction partial differential equation is

$$\frac{\partial u}{\partial t} = k^2 \frac{\partial^2 u}{\partial x^2}$$

Show that if f is a function of x satisfying the equation

$$\frac{d^2f}{dx^2} + \lambda^2 f(x) = 0$$

and g is a function of t satisfying the equation $dg/dt + k^2\lambda^2 g(t) = 0$, then if $u = f(x)g(t)$, the partial differential equation is satisfied. k and λ are constants.

38. The partial differential equation for a vibrating string is

$$\frac{\partial^2 u}{\partial t^2} = a^2 \frac{\partial^2 u}{\partial x^2}$$

Show that if f is a function of x satisfying the equation $d^2f/dx^2 + \lambda^2 f(x) = 0$ and g is a function of t satisfying the equation $d^2g/dt^2 + a^2\lambda^2 g(t) = 0$, then if $u = f(x)g(t)$, the partial differential equation is satisfied. a and λ are constants.

39. Prove that if f and g are two arbitrary functions of a real variable having continuous second derivatives and $u = f(x + at) + g(x - at)$, then u satisfies the partial differential equation of the vibrating string given in Exercise 38. (*Hint:* Let $v = x + at$ and $w = x - at$; then u is a function of v and w, and v and w are in turn functions of x and t.)

40. Prove that if f is a function of two variables and all the partial derivatives of f up to the fourth order are continuous on some open disk, then

$$D_{1122}f = D_{2121}f$$

REVIEW EXERCISES FOR CHAPTER 18

In Exercises 1 through 6, find the indicated partial derivatives.

1. $f(x, y) = \dfrac{x^2 - y}{3y^2}$; $D_1 f(x, y)$, $D_2 f(x, y)$, $D_{12} f(x, y)$

2. $F(x, y, z) = 2xy^2 + 3yz^2 - 5xz^3$; $D_1 f(x, y, z)$, $D_3 f(x, y, z)$, $D_{13} f(x, y, z)$

3. $g(s, t) = \sin(st^2) + te^s$; $D_1 g(s, t)$, $D_2 g(s, t)$, $D_{21} g(s, t)$

4. $h(x, y) = \tan^{-1}\dfrac{x^3}{y^2}$; $D_1 h(x, y)$, $D_2 h(x, y)$, $D_{11} h(x, y)$

5. $f(u, v, w) = \dfrac{\ln 4uv}{w^2}$; $D_1 f(u, v, w)$, $D_{13} f(u, v, w)$, $D_{131} f(u, v, w)$

6. $f(u, v, w) = w \cos 2v + 3v \sin u - 2uv \tan w$; $D_2 f(u, v, w)$, $D_1 f(u, v, w)$, $D_{131} f(u, v, w)$

In Exercises 7 and 8, find $\partial u/\partial t$ and $\partial u/\partial s$ by two methods.

7. $u = y \ln(x^2 + y^2)$, $x = 2s + 3t$, $y = 3t - 2s$

8. $u = e^{2x+y} \cos(2y - x)$, $x = 2s^2 - t^2$, $y = s^2 + 2t^2$

9. If $u = 3x^2y + 2xy - 3yz - 2z^2$, $x = e^{3rs}$, $y = r^3s^2$, and $z = \ln 4$, find $\partial u/\partial r$ by two methods: (a) use the chain rule; (b) make the substitutions for x, y, and z before differentiating.

10. If $u = e^{x^2+y^2} - \dfrac{3x}{y} + 3z$, $x = \sin\theta$, $y = \cos\theta$, and $z = \tan\theta$, find the total derivative $du/d\theta$ by two methods: (a) do not express u in terms of θ before differentiating; (b) express u in terms of θ before differentiating.

11. If $u = xy + x^2$, $x = 4\cos t$, and $y = 3\sin t$, find the value of the total derivative du/dt at $t = \frac{1}{4}\pi$ by two methods: (a) do not express u in terms of t before differentiating; (b) express u in terms of t before differentiating.

12. If $f(x, y) = x^2 + ye^x$, find: (a) $\Delta f(0, 2)$, the increment of f at $(0, 2)$; (b) $\Delta f(0, 2)$ when $\Delta x = -0.1$ and $\Delta y = 0.2$; (c) $df(0, 2, \Delta x, \Delta y)$, the total differential of f at $(0, 2)$; (d) $df(0, 2, -0.1, 0.2)$.

13. If $f(x, y, z) = 3xy^2 - 5xz^2 - 2xyz$, find: (a) $\Delta f(-1, 3, 2)$, the increment of f at $(-1, 3, 2)$; (b) $\Delta f(-1, 3, 2)$ when $\Delta x = 0.02$, $\Delta y = -0.01$, and $\Delta z = -0.02$; (c) $df(-1, 3, 2, \Delta x, \Delta y, \Delta z)$, the total differential of f at $(-1, 3, 2)$; (d) $df(-1, 3, 2, 0.02, -0.01, -0.02)$.

In Exercises 14 through 18, find the domain and range of the function f and draw a sketch showing as a shaded region in R^2 the set of points in the domain of f.

14. $f(x, y) = \sqrt{4x^2 - y}$

15. $f(x, y) = \tan^{-1}\sqrt{x^2 - y^2}$

16. $f(x, y) = \sin^{-1}\sqrt{1 - x^2 - y^2}$

17. $f(x, y) = \dfrac{x}{\sqrt{36 - 4x^2 - y^2}}$

18. $f(x, y) = [\![x]\!] + [\![\sqrt{1 - y^2}]\!]$

In Exercises 19 through 21, find the domain and range of the function f.

19. $f(x, y, z) = \dfrac{xy}{\sqrt{z}}$

20. $f(x, y, z) = \dfrac{xy + xz + yz}{xyz}$

21. $f(x, y, z) = \dfrac{x}{|y| - |z|}$

22. The temperature at a point (x, y) of a flat metal plate is t degrees and $t = x^2 + 2y$. Draw the isothermals for $t = 0, 2, 4, 6$, and 8.

23. Given $f(x) = x^2 + 1$, $g(x, y) = 2x/3y$, and $h(x) = 1/x$ find: (a) $(h \circ g)(-3, 4)$; (b) $g(f(3), h(\frac{1}{4}))$; (c) $g(f(x), h(y))$; (d) $f((h \circ g)(x, y))$.

In Exercises 24 and 25, evaluate the given limit by the use of limit theorems.

24. $\displaystyle\lim_{(x,y)\to(0,\pi/2)} \frac{xy^2 + e^x}{\cos x + \sin y}$

25. $\displaystyle\lim_{(x,y)\to(1,3)} \sin^{-1}\left(\frac{3x}{2y}\right)$

In Exercises 26 through 28, establish the limit by finding a $\delta > 0$ for any $\epsilon > 0$ such that Definition 18.2.5 holds.

26. $\displaystyle\lim_{(x,y)\to(4,-1)} (4x - 5y) = 21$

27. $\displaystyle\lim_{(x,y)\to(2,-2)} (3x^2 - 4y^2) = -4$

28. $\displaystyle\lim_{(x,y)\to(3,1)} (x^2 - y^2 + 2x - 4y) = 10$

In Exercises 29 through 32, determine if the indicated limit exists.

29. $\displaystyle\lim_{(x,y)\to(0,0)} \frac{x^3y^3}{x^2 + y^2}$

30. $\displaystyle\lim_{(x,y)\to(0,0)} \frac{x^4 - y^4}{x^4 + y^4}$

31. $\displaystyle\lim_{(x,y)\to(0,0)} \frac{x^9y}{(x^6 + y^2)^2}$

32. $\displaystyle\lim_{(x,y)\to(0,0)} \frac{2x^3 + 4x^2y}{x^2 + y^2}$

In Exercises 33 through 37, discuss the continuity of f.

33. $f(x, y) = \dfrac{x^2 + 4y^2}{x^2 - 4y^2}$

34. $f(x, y) = \dfrac{1}{\cos \frac{1}{2}\pi x} + \dfrac{1}{\cos \frac{1}{2}\pi y}$

35. $f(x, y) = \dfrac{1}{\cos^2 \frac{1}{2}\pi x + \cos^2 \frac{1}{2}\pi y}$

36. $f(x, y) = \begin{cases} \dfrac{x^4 - y^4}{x^4 + y^4} & \text{if } (x, y) \neq (0, 0) \\ 0 & \text{if } (x, y) = (0, 0) \end{cases}$ (*Hint:* See Exercise 30.)

37. $f(x, y) = \begin{cases} \dfrac{x^3y^3}{x^2 + y^2} & \text{if } (x, y) \neq (0, 0) \\ 0 & \text{if } (x, y) = (0, 0) \end{cases}$ (*Hint:* See Exercise 29.)

38. If $w = x^2y - y^2x + y^2z - z^2y + z^2x - x^2z$, show that

$$\frac{\partial w}{\partial x} + \frac{\partial w}{\partial y} + \frac{\partial w}{\partial z} = 0$$

39. Find (a) $f_2(x, 0)$ if $x \neq 0$ and (b) $f_2(0, 0)$ if

$$f(x, y) = \begin{cases} \dfrac{12x^2y - 3y^2}{x^2 + y} & \text{if } (x, y) \neq (0, 0) \\ 0 & \text{if } (x, y) = (0, 0) \end{cases}$$

In Exercises 40 and 41, prove that the function f is differentiable at all points in its domain by showing that Definition 18.5.2 holds.

40. $f(x, y) = 3xy^2 - 4x^2 + y^2$ **41.** $f(x, y) = \dfrac{2x + y}{y^2}$

42. Suppose α is the radian measure of an acute angle of a right triangle and $\sin \alpha$ is determined by a/c, where a cm is the length of the side opposite the angle and c cm is the length of the hypotenuse. If by measurement a is found to be 3.52 and c is found to be 7.14, and there is a possible error of 0.01 in each, find the possible error in the computation of $\sin \alpha$ from these measurements.

43. A painting contractor charges \$2 per square meter for painting the four walls and ceiling of a room. If the dimensions of the ceiling are measured to be 4 m and 5 m, the height of the room is measured to be 3 m, and these measurements are correct to 0.5 cm, find approximately, by using the total differential, the greatest error in estimating the cost of the job from these measurements.

44. At a given instant, the length of one side of a rectangle is 6 cm and it is increasing at the rate of 1 cm/sec and the length of another side of the rectangle is 10 cm and it is decreasing at the rate of 2 cm/sec. Find the rate of change of the area of the rectangle at the given instant.

45. The radius of a right-circular cylinder is decreasing at the rate of 5 cm/min and the height is increasing at the rate of 12 cm/min. Find the rate of change of the volume at the instant when the radius is 20 cm and the height is 40 cm.

46. Find the slope of the tangent line to the curve of intersection of the surface $25x^2 - 16y^2 + 9z^2 - 4 = 0$ with the plane $x = 4$ at the point (4, 9, 10).

47. Use the ideal gas law (see Example 5, Sec. 18.4) with $k = 5$ to find the rate of change of the pressure at the instant when the volume of the gas is 80 cm^3 and the temperature is 75° if the volume is increasing at the rate of 3 cm^3/min and the temperature is increasing at the rate of 0.3° per minute.

48. Verify that $u(x, y) = (\sinh x)(\sin y)$ satisfies Laplace's equation in R^2:

$$\frac{\partial^2 u}{\partial x^2} + \frac{\partial^2 u}{\partial y^2} = 0$$

49. If f is a differentiable function of the variable u, let $u = x^2 + y^2$ and prove that $z = xy + f(x^2 + y^2)$ satisfies the equation

$$y\frac{\partial z}{\partial x} - x\frac{\partial z}{\partial y} = y^2 - x^2$$

50. Laplace's equation in polar coordinates is

$$r^2\frac{\partial^2 u}{\partial r^2} + r\frac{\partial u}{\partial r} + \frac{\partial^2 u}{\partial \theta^2} = 0$$

Verify that $u(r, \theta) = r^n \sin n\theta$, where n is a constant, satisfies this equation.

51. Verify that $u(x, y, z) = e^{3x+4y} \sin 5z$ satisfies Laplace's equation in R^3:

$$\frac{\partial^2 u}{\partial x^2} + \frac{\partial^2 u}{\partial y^2} + \frac{\partial^2 u}{\partial z^2} = 0$$

52. Verify that $u(x, t) = A \cos(kat) \sin(kx)$, where A and k are arbitrary constants, satisfies the partial differential equation for a vibrating string:

$$\frac{\partial^2 u}{\partial t^2} = a^2\frac{\partial^2 u}{\partial x^2}$$

53. Verify that

$$u(x, t) = \sin \frac{n\pi x}{L} e^{(-n^2\pi^2 k^2/L^2)t}$$

satisfies the one-dimensional heat-conduction partial differential equation:

$$\frac{\partial u}{\partial t} = k^2 \frac{\partial^2 u}{\partial x^2}$$

54. Given

$$f(x, y) = \begin{cases} \dfrac{x^2 y}{x^4 + y^2} & \text{if } (x, y) \neq (0, 0) \\ 0 & \text{if } (x, y) = (0, 0) \end{cases}$$

Prove that $D_1 f(0, 0)$ and $D_2 f(0, 0)$ exist but that f is not differentiable at $(0, 0)$. (*Hint:* See Example 6, Sec. 18.2, and Exercise 10 in Exercises 18.3).

55. Given

$$f(x, y, z) = \begin{cases} \dfrac{x^2 y^2 z^2}{(x^2 + y^2 + z^2)^2} & \text{if } (x, y, z) \neq (0, 0, 0) \\ 0 & \text{if } (x, y, z) = (0, 0, 0) \end{cases}$$

Prove that f is differentiable at $(0, 0, 0)$.

56. Let f be the function defined by

$$f(x, y) = \begin{cases} \dfrac{e^{-1/x^2} y}{e^{-2/x^2} + y^2} & \text{if } x \neq 0 \\ 0 & \text{if } x = 0 \end{cases}$$

Prove that f is discontinuous at the origin.

57. For the function of Exercise 56, prove that $D_1 f(0, 0)$ and $D_2 f(0, 0)$ both exist.

58. If f is a differentiable function of x and y and $u = f(x, y)$, $x = r \cos \theta$, and $y = r \sin \theta$, show that

$$\left(\frac{\partial u}{\partial r}\right)^2 + \frac{1}{r^2}\left(\frac{\partial u}{\partial \theta}\right)^2 = \left(\frac{\partial u}{\partial x}\right)^2 + \left(\frac{\partial u}{\partial y}\right)^2$$

CHAPTER 19 Directional Derivatives, Gradients, Applications of Partial Derivatives, and Line Integrals

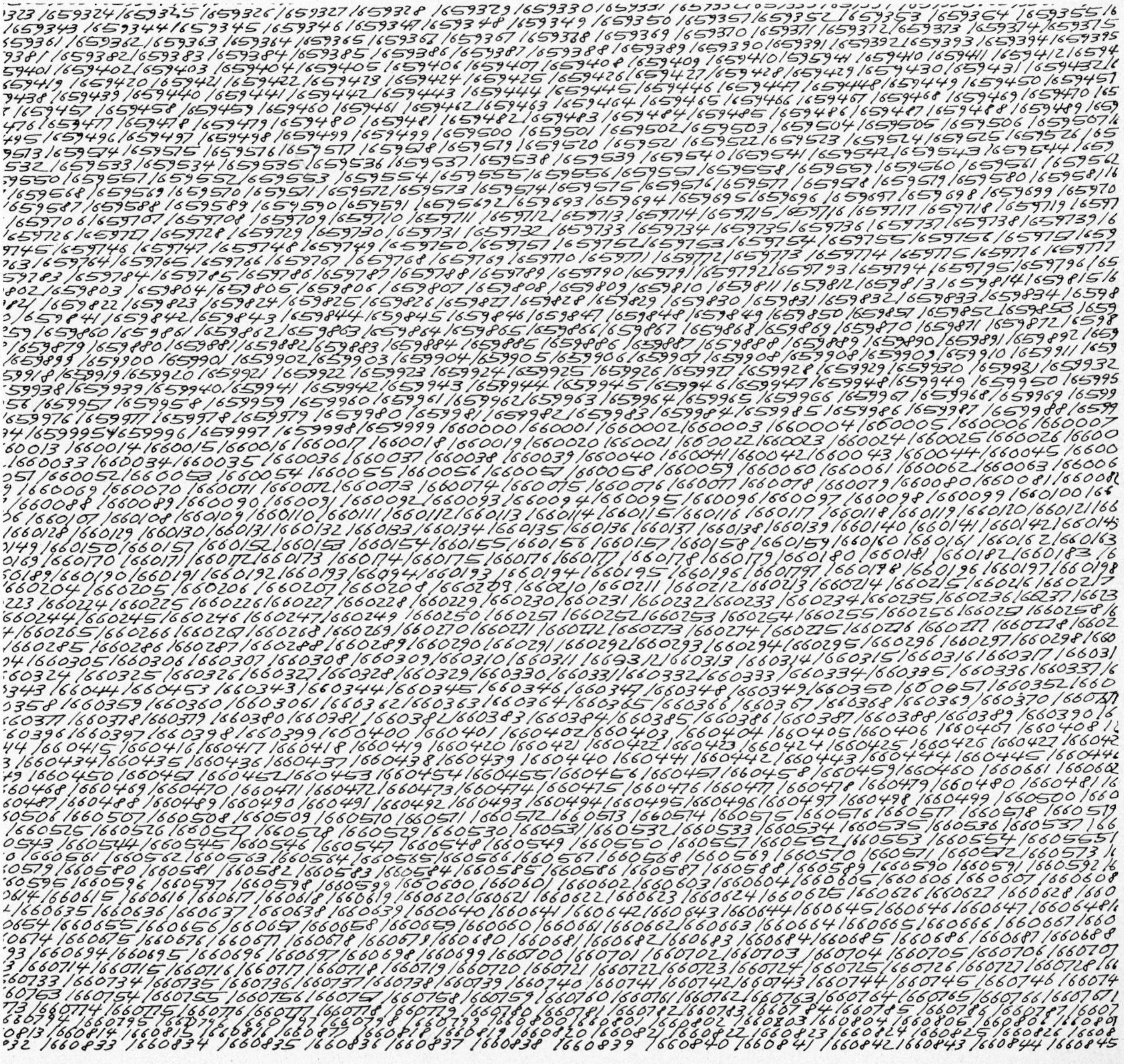

19.1 DIRECTIONAL DERIVATIVES AND GRADIENTS

We now generalize the definition of a partial derivative to obtain the rate of change of a function with respect to any direction. This leads to the concept of a *directional derivative.*

Let f be a function of the two variables x and y and let $P(x, y)$ be a point in the xy plane. Suppose that $\mathbf{U}$ is the unit vector making an angle of radian measure θ with the positive side of the x axis. Then

$$\mathbf{U} = \cos\theta\mathbf{i} + \sin\theta\mathbf{j}$$

Figure 19.1.1 shows the representation of $\mathbf{U}$ having its initial point at the point $P(x, y)$.

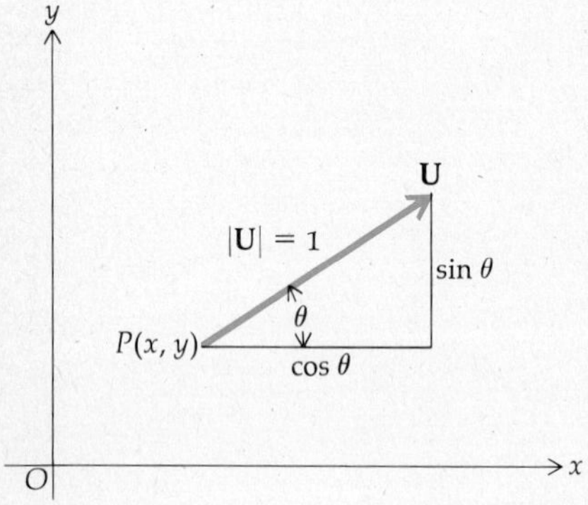

FIGURE 19.1.1

19.1.1 Definition Let f be a function of two variables x and y. If $\mathbf{U}$ is the unit vector $\cos\theta\mathbf{i} + \sin\theta\mathbf{j}$, then the **directional derivative** of f in the direction of $\mathbf{U}$, denoted by $D_{\mathbf{U}}f$, is given by

$$D_{\mathbf{U}}f(x, y) = \lim_{h\to 0} \frac{f(x + h\cos\theta, y + h\sin\theta) - f(x, y)}{h}$$

if this limit exists.

The directional derivative gives the rate of change of the function values $f(x, y)$ with respect to the direction of the unit vector $\mathbf{U}$. This is illustrated in Fig. 19.1.2. An equation of the surface S in the figure is $z = f(x, y)$. $P_0(x_0, y_0, z_0)$ is a point on the surface, and the points $R(x_0, y_0, 0)$ and $Q(x_0 + h\cos\theta, y_0 + h\sin\theta, 0)$ are points in the xy plane. The plane through R and Q, parallel to the z axis, makes an angle of θ radians with the positive direction on the x axis. This plane intersects the surface S in the curve C. The directional derivative $D_{\mathbf{U}}f$, evaluated at P_0, is the slope of the tangent line to the curve C at P_0 in the plane of R, Q, and P_0.

If $\mathbf{U} = \mathbf{i}$, then $\cos\theta = 1$ and $\sin\theta = 0$, and from Definition 19.1.1,

$$D_{\mathbf{i}}f(x, y) = \lim_{h\to 0} \frac{f(x + h, y) - f(x, y)}{h}$$

which is the partial derivative of f with respect to x.

If $\mathbf{U} = \mathbf{j}$, then $\cos\theta = 0$ and $\sin\theta = 1$, and

$$D_{\mathbf{j}}f(x, y) = \lim_{h\to 0} \frac{f(x, y + h) - f(x, y)}{h}$$

which is the partial derivative of f with respect to y.

So f_x and f_y are special cases of the directional derivative in the directions of the unit vectors $\mathbf{i}$ and $\mathbf{j}$, respectively.

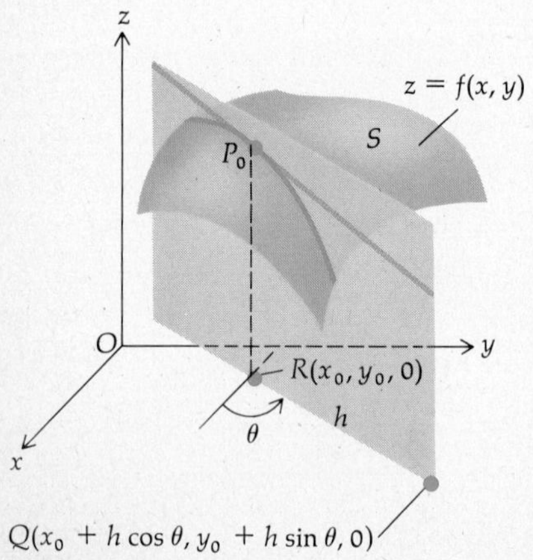

FIGURE 19.1.2

● ILLUSTRATION 1 We apply Definition 19.1.1 to find $D_{\mathbf{U}}f$ if

$$f(x, y) = 3x^2 - y^2 + 4x$$

and $\mathbf{U}$ is the unit vector in the direction $\frac{1}{6}\pi$. Then $\mathbf{U} = \cos\frac{1}{6}\pi\mathbf{i} + \sin\frac{1}{6}\pi\mathbf{j} = \frac{1}{2}\sqrt{3}\mathbf{i} + \frac{1}{2}\mathbf{j}$. Thus from Definition 19.1.1,

$$D_{\mathbf{U}}f(x, y) = \lim_{h \to 0} \frac{f(x + \frac{1}{2}\sqrt{3}h, y + \frac{1}{2}h) - f(x, y)}{h}$$

$$= \lim_{h \to 0} \frac{3(x + \frac{1}{2}\sqrt{3}h)^2 - (y + \frac{1}{2}h)^2 + 4(x + \frac{1}{2}\sqrt{3}h) - (3x^2 - y^2 + 4x)}{h}$$

$$= \lim_{h \to 0} \frac{3x^2 + 3\sqrt{3}hx + \frac{9}{4}h^2 - y^2 - hy - \frac{1}{4}h^2 + 4x + 2\sqrt{3}h - 3x^2 + y^2 - 4x}{h}$$

$$= \lim_{h \to 0} \frac{3\sqrt{3}hx + \frac{9}{4}h^2 - hy - \frac{1}{4}h^2 + 2\sqrt{3}h}{h}$$

$$= \lim_{h \to 0} (3\sqrt{3}x + \tfrac{9}{4}h - y - \tfrac{1}{4}h + 2\sqrt{3})$$

$$= 3\sqrt{3}x - y + 2\sqrt{3}$$

We now proceed to obtain a formula that will enable us to calculate a directional derivative in a way that is shorter than if we used the definition. Let g be the function of the single variable t, with x, y, and θ fixed, such that

$$g(t) = f(x + t\cos\theta, y + t\sin\theta) \tag{1}$$

and let $\mathbf{U} = \cos\theta\mathbf{i} + \sin\theta\mathbf{j}$. Then by the definition of an ordinary derivative,

$$g'(0) = \lim_{h \to 0} \frac{f(x + (0 + h)\cos\theta, y + (0 + h)\sin\theta) - f(x + 0\cos\theta, y + 0\sin\theta)}{h}$$

$$g'(0) = \lim_{h \to 0} \frac{f(x + h\cos\theta, y + h\sin\theta) - f(x, y)}{h}$$

Because the right side of the above is $D_{\mathbf{U}}f(x, y)$,

$$g'(0) = D_{\mathbf{U}}f(x, y) \tag{2}$$

We now find $g'(t)$ by applying the chain rule to the right side of (1), which gives

$$g'(t) = f_1(x + t\cos\theta, y + t\sin\theta)\frac{\partial(x + t\cos\theta)}{\partial t} + f_2(x + t\cos\theta, y + t\sin\theta)\frac{\partial(y + t\sin\theta)}{\partial t}$$

$$= f_1(x + t\cos\theta, y + t\sin\theta)\cos\theta + f_2(x + t\cos\theta, y + t\sin\theta)\sin\theta$$

Therefore

$$g'(0) = f_x(x, y)\cos\theta + f_y(x, y)\sin\theta \tag{3}$$

From (2) and (3) the following theorem is obtained.

19.1.2 Theorem If f is a differentiable function of x and y, and $\mathbf{U} = \cos\theta\mathbf{i} + \sin\theta\mathbf{j}$, then

$$D_{\mathbf{U}}f(x, y) = f_x(x, y)\cos\theta + f_y(x, y)\sin\theta$$

● ILLUSTRATION 2 For the function f and the unit vector $\mathbf{U}$ of Illustration 1, we find $D_{\mathbf{U}}f$ by Theorem 19.1.2.

Because $f(x, y) = 3x^2 - y^2 + 4x$, $f_x(x, y) = 6x + 4$ and $f_y(x, y) = -2y$. Because $\mathbf{U} = \cos\frac{1}{6}\pi\mathbf{i} + \sin\frac{1}{6}\pi\mathbf{j}$, then from Theorem 19.1.2,

$$D_{\mathbf{U}}f(x, y) = (6x + 4)\tfrac{1}{2}\sqrt{3} + (-2y)\tfrac{1}{2} = 3\sqrt{3}x + 2\sqrt{3} - y$$

which agrees with the result in Illustration 1. ●

The directional derivative can be written as the dot product of two vectors. Because

$$f_x(x, y)\cos\theta + f_y(x, y)\sin\theta = (\cos\theta\mathbf{i} + \sin\theta\mathbf{j}) \cdot [f_x(x, y)\mathbf{i} + f_y(x, y)\mathbf{j}]$$

it follows from Theorem 19.1.2 that

$$D_{\mathbf{U}}f(x, y) = (\cos\theta\mathbf{i} + \sin\theta\mathbf{j}) \cdot [f_x(x, y)\mathbf{i} + f_y(x, y)\mathbf{j}] \tag{4}$$

The second vector on the right side of (4) is a very important one, and it is called the *gradient* of the function f. The symbol used for the gradient of f is ∇f, where ∇ is an inverted capital delta and is read "del." Sometimes the abbreviation *grad f* is used.

19.1.3 Definition If f is a function of two variables x and y, and f_x and f_y exist, then the **gradient** of f, denoted by ∇f (read "del f"), is defined by

$$\nabla f(x, y) = f_x(x, y)\mathbf{i} + f_y(x, y)\mathbf{j}$$

From Definition 19.1.3, Eq. (4) can be written as

$$D_{\mathbf{U}}f(x, y) = \mathbf{U} \cdot \nabla f(x, y) \tag{5}$$

Therefore any directional derivative of a differentiable function can be obtained by dot-multiplying the gradient by a unit vector in the desired direction.

Example 1 If $f(x, y) = \frac{1}{16}x^2 + \frac{1}{9}y^2$, find the gradient of f at the point (4, 3). Also find the rate of change of $f(x, y)$ in the direction $\frac{1}{4}\pi$ at (4, 3).

Solution Because $f_x(x, y) = \frac{1}{8}x$ and $f_y(x, y) = \frac{2}{9}y$,

$$\nabla f(x, y) = \tfrac{1}{8}x\mathbf{i} + \tfrac{2}{9}y\mathbf{j}$$

Therefore

$$\nabla f(4, 3) = \tfrac{1}{2}\mathbf{i} + \tfrac{2}{3}\mathbf{j}$$

The rate of change of $f(x, y)$ in the direction $\frac{1}{4}\pi$ at (4, 3) is $D_{\mathbf{U}}f(4, 3)$, where $\mathbf{U}$ is the unit vector

$$\frac{1}{\sqrt{2}}\mathbf{i} + \frac{1}{\sqrt{2}}\mathbf{j}$$

This is found by dot-multiplying $\nabla f(4, 3)$ by $\mathbf{U}$. We have, then,

$$D_{\mathbf{U}}f(4, 3) = \left(\frac{1}{\sqrt{2}}\mathbf{i} + \frac{1}{\sqrt{2}}\mathbf{j}\right)\cdot\left(\frac{1}{2}\mathbf{i} + \frac{2}{3}\mathbf{j}\right) = \frac{7}{12}\sqrt{2}$$

If α is the radian measure of the angle between the two vectors $\mathbf{U}$ and ∇f, then

$$\mathbf{U} \cdot \nabla f(x, y) = |\mathbf{U}||\nabla f(x, y)| \cos \alpha \qquad (6)$$

From (5) and (6) it follows that

$$D_{\mathbf{U}}f(x, y) = |\mathbf{U}||\nabla f(x, y)| \cos \alpha \qquad (7)$$

We see from (7) that $D_{\mathbf{U}}f$ will be a maximum when $\cos \alpha = 1$, that is, when $\mathbf{U}$ is in the direction of ∇f; and in this case, $D_{\mathbf{U}}f = |\nabla f|$. Hence the gradient of a function is in the direction in which the function has its maximum rate of change. In particular, on a two-dimensional topographical map of a landscape where z units is the elevation at a point (x, y) and $z = f(x, y)$, the direction in which the rate of change of z is the greatest is given by $\nabla f(x, y)$; that is, $\nabla f(x, y)$ points in the direction of steepest ascent. This accounts for the name *gradient* (the grade is steepest in the direction of the gradient).

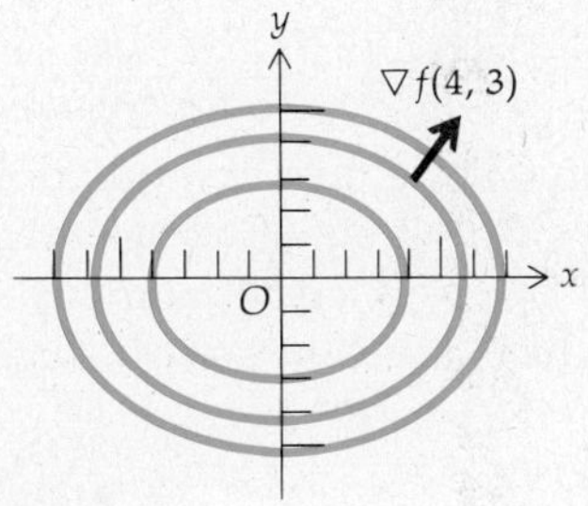

FIGURE 19.1.3

• ILLUSTRATION 3 In Fig. 19.1.3 there is a contour map showing the level curves of the function of Example 1 at 1, 2, and 3. The level curves are ellipses. The figure also shows the representation of $\nabla f(4, 3)$ having its initial point at (4, 3). •

Example 2 Given $f(x, y) = 2x^2 - y^2 + 3x - y$, find the maximum value of $D_{\mathbf{U}}f$ at the point where $x = 1$ and $y = -2$.

Solution $f_x(x, y) = 4x + 3$ and $f_y(x, y) = -2y - 1$. So

$$\nabla f(x, y) = (4x + 3)\mathbf{i} + (-2y - 1)\mathbf{j}$$

Therefore

$$\nabla f(1, -2) = 7\mathbf{i} + 3\mathbf{j}$$

Thus the maximum value of $D_{\mathbf{U}}f$ at the point $(1, -2)$ is

$$|\nabla f(1, -2)| = \sqrt{49 + 9} = \sqrt{58}$$

Example 3 The temperature at any point (x, y) of a rectangular plate lying in the xy plane is determined by $T(x, y) = x^2 + y^2$. (a) Find the rate of change of the temperature at the point (3, 4) in the direction making an angle of radian measure $\frac{1}{3}\pi$ with the positive x direction; (b) find the direction for which the rate of change of the temperature at the point $(-3, 1)$ is a maximum.

Solution

(a) We wish to find $D_{\mathbf{U}}T(x, y)$, where

$$\mathbf{U} = \cos \tfrac{1}{3}\pi\mathbf{i} + \sin \tfrac{1}{3}\pi\mathbf{j} = \tfrac{1}{2}\mathbf{i} + \tfrac{1}{2}\sqrt{3}\mathbf{j}$$

Because $T(x, y) = x^2 + y^2$, $T_x(x, y) = 2x$, and $T_y(x, y) = 2y$. Hence

$$\nabla T(x, y) = T_x(x, y)\mathbf{i} + T_y(x, y)\mathbf{j} = 2x\mathbf{i} + 2y\mathbf{j}$$

Therefore

$$\begin{aligned} D_{\mathbf{U}}T(x, y) &= \mathbf{U} \cdot \nabla T(x, y) \\ &= (\tfrac{1}{2}\mathbf{i} + \tfrac{1}{2}\sqrt{3}\mathbf{j}) \cdot (2x\mathbf{i} + 2y\mathbf{j}) \\ &= x + \sqrt{3}y \end{aligned}$$

Thus

$$D_{\mathbf{U}}T(3, 4) = 3 + 4\sqrt{3} \approx 3 + 4(1.732) = 9.93$$

So at (3, 4) the temperature is increasing at the rate of 9.93 units per unit change in the distance measured in the direction of **U**.

(b) $D_{\mathbf{U}}T(-3, 1)$ is a maximum when **U** is in the direction of $\nabla T(-3, 1)$. Because $\nabla T(-3, 1) = -6\mathbf{i} + 2\mathbf{j}$, the radian measure of the angle giving the direction of $\nabla T(-3, 1)$ is θ, where $\tan \theta = -\frac{1}{3}$. So $\theta = \pi - \tan^{-1} \frac{1}{3}$. Therefore the rate of change of the temperature at the point $(-3, 1)$ is a maximum in the direction making an angle of radian measure $\pi - \tan^{-1} \frac{1}{3}$ with the the positive side of the x axis.

We extend the definition of a directional derivative to a function of three variables. In three-dimensional space the direction of a vector is determined by its direction cosines. So we let $\cos \alpha$, $\cos \beta$, and $\cos \gamma$ be the direction cosines of the unit vector **U**; therefore $\mathbf{U} = \cos \alpha\mathbf{i} + \cos \beta\mathbf{j} + \cos \gamma\mathbf{k}$.

19.1.4 Definition Suppose that f is a function of three variables x, y, and z. If **U** is the unit vector $\cos \alpha\mathbf{i} + \cos \beta\mathbf{j} + \cos \gamma\mathbf{k}$, then the directional derivative of f in the dirction of **U**, denoted by $D_{\mathbf{U}}f$, is given by

$$D_{\mathbf{U}}f(x, y, z) = \lim_{h \to 0} \frac{f(x + h\cos \alpha, y + h\cos \beta, z + h\cos \gamma) - f(x, y, z)}{h}$$

if this limit exists.

The directional derivative of a function of three variables gives the rate of change of the function values $f(x, y, z)$ with respect to distance in three-dimensional space measured in the direction of the unit vector **U**.

The following theorem, which provides a method for calculating a directional derivative for a function of three variables, is proved in a manner similar to the proof of Theorem 19.1.2.

19.1.5 Theorem If f is a differentiable function of x, y, and z and

$$\mathbf{U} = \cos \alpha\mathbf{i} + \cos \beta\mathbf{j} + \cos \gamma\mathbf{k}$$

then

$$D_{\mathbf{U}}f(x, y, z) = f_x(x, y, z)\cos\alpha + f_y(x, y, z)\cos\beta + f_z(x, y, z)\cos\gamma \quad (8)$$

Example 4 Given $f(x, y, z) = 3x^2 + xy - 2y^2 - yz + z^2$, find the rate of change of $f(x, y, z)$ at $(1, -2, -1)$ in the direction of the vector $2\mathbf{i} - 2\mathbf{j} - \mathbf{k}$.

Solution The unit vector in the direction of $2\mathbf{i} - 2\mathbf{j} - \mathbf{k}$ is

$$\mathbf{U} = \tfrac{2}{3}\mathbf{i} - \tfrac{2}{3}\mathbf{j} - \tfrac{1}{3}\mathbf{k}$$

Also,

$$f(x, y, z) = 3x^2 + xy - 2y^2 - yz + z^2$$

So from (8),

$$D_{\mathbf{U}}f(x, y, z) = \tfrac{2}{3}(6x + y) - \tfrac{2}{3}(x - 4y - z) - \tfrac{1}{3}(-y + 2z)$$

Therefore the rate of change of $f(x, y, z)$ at $(1, -2, -1)$ in the direction of $\mathbf{U}$ is given by

$$D_{\mathbf{U}}f(1, -2, -1) = \tfrac{2}{3}(4) - \tfrac{2}{3}(10) - \tfrac{1}{3}(0) = -4$$

19.1.6 Definition If f is a function of three variables x, y, and z and the first partial derivatives f_x, f_y, and f_z exist, then the **gradient** of f, denoted by ∇f, is defined by

$$\nabla f(x, y, z) = f_x(x, y, z)\mathbf{i} + f_y(x, y, z)\mathbf{j} + f_z(x, y, z)\mathbf{k}$$

Just as for functions of two variables, it follows from Theorem 19.1.5 and Definition 19.1.6 that if $\mathbf{U} = \cos\alpha\mathbf{i} + \cos\beta\mathbf{j} + \cos\gamma\mathbf{k}$, then

$$D_{\mathbf{U}}f(x, y, z) = \mathbf{U} \cdot \nabla f(x, y, z) \quad (9)$$

Also, the directional derivative is a maximum when $\mathbf{U}$ is in the direction of the gradient, and the maximum directional derivative is the magnitude of the gradient.

Applications of the gradient occur in physics in problems in heat conduction and electricity. Suppose that $w = f(x, y, z)$. A level surface of this function f at the constant k is given by

$$f(x, y, z) = k \quad (10)$$

If w is the number of degrees in the temperature at point (x, y, z), then all points on the surface of Eq. (10) have the same temperature of k degrees, and the surface is called an **isothermal surface.** If w is the number of volts in the electric potential at point (x, y, z), then all points on the surface are at the same potential, and the surface is called an **equipotential surface.** The gradient vector at a point gives the direction of greatest rate of change of w. So if the level surface of Eq. (10) is an isothermal surface, $\nabla f(x, y, z)$ gives the direction of the greatest rate of change of temperature at (x, y, z). If (10) is an equation of an equipotential surface, then $\nabla f(x, y, z)$ gives the direction of the greatest rate of change of potential at (x, y, z).

Example 5 If V volts is the electric potential at any point (x, y, z) in three-dimensional space and $V = 1/\sqrt{x^2 + y^2 + z^2}$, find: (a) the rate of change of V at the point $(2, 2, -1)$ in the direction of the vector $2\mathbf{i} - 3\mathbf{j} + 6\mathbf{k}$; (b) the direction of the greatest rate of change of V at $(2, 2, -1)$.

Solution Let $f(x, y, z) = 1/\sqrt{x^2 + y^2 + z^2}$.

(a) A unit vector in the direction of $2\mathbf{i} - 3\mathbf{j} + 6\mathbf{k}$ is

$$\mathbf{U} = \tfrac{2}{7}\mathbf{i} - \tfrac{3}{7}\mathbf{j} + \tfrac{6}{7}\mathbf{k}$$

We wish to find $D_{\mathbf{U}}f(2, 2, -1)$.

$$\nabla f(x, y, z) = f_x(x, y, z)\mathbf{i} + f_y(x, y, z)\mathbf{j} + f_z(x, y, z)\mathbf{k}$$

$$= \frac{-x}{(x^2 + y^2 + z^2)^{3/2}}\mathbf{i} + \frac{-y}{(x^2 + y^2 + z^2)^{3/2}}\mathbf{j} + \frac{-z}{(x^2 + y^2 + z^2)^{3/2}}\mathbf{k}$$

Then

$$\begin{aligned} D_{\mathbf{U}}f(2, 2, -1) &= \mathbf{U} \cdot \nabla f(2, 2, -1) \\ &= (\tfrac{2}{7}\mathbf{i} - \tfrac{3}{7}\mathbf{j} + \tfrac{6}{7}\mathbf{k}) \cdot (-\tfrac{2}{27}\mathbf{i} - \tfrac{2}{27}\mathbf{j} + \tfrac{1}{27}\mathbf{k}) \\ &= -\tfrac{4}{189} + \tfrac{6}{189} + \tfrac{6}{189} \\ &= \tfrac{8}{189} \\ &\approx 0.042 \end{aligned}$$

Therefore, at $(2, 2, -1)$ the potential is increasing at the rate of 0.042 volt per unit change in the distance measured in the direction of $\mathbf{U}$.

(b) $\nabla f(2, 2, -1) = -\tfrac{2}{27}\mathbf{i} - \tfrac{2}{27}\mathbf{j} + \tfrac{1}{27}\mathbf{k}$. A unit vector in the direction of $\nabla f(2, 2, -1)$ is

$$\frac{\nabla f(2, 2, -1)}{|\nabla f(2, 2, -1)|} = \frac{-\tfrac{2}{27}\mathbf{i} - \tfrac{2}{27}\mathbf{j} + \tfrac{1}{27}\mathbf{k}}{\tfrac{3}{27}} = -\tfrac{2}{3}\mathbf{i} - \tfrac{2}{3}\mathbf{j} + \tfrac{1}{3}\mathbf{k}$$

The direction cosines of this vector are $-\tfrac{2}{3}$, $-\tfrac{2}{3}$, and $\tfrac{1}{3}$, which give the direction of the greatest rate of change of V at $(2, 2, -1)$.

EXERCISES 19.1

In Exercises 1 through 6, find the directional derivative of the given function in the direction of the given unit vector $\mathbf{U}$ *by using either Definition 19.1.1 or Definition 19.1.4, and then verify your result by applying either Theorem 19.1.2 or Theorem 19.1.5, whichever one applies.*

1. $f(x, y) = 2x^2 + 5y^2$; $\mathbf{U} = \cos \tfrac{1}{4}\pi\mathbf{i} + \sin \tfrac{1}{4}\pi\mathbf{j}$

2. $g(x, y) = 3x^2 - 4y^2$; $\mathbf{U} = \cos \tfrac{1}{3}\pi\mathbf{i} + \sin \tfrac{1}{3}\pi\mathbf{j}$

3. $h(x, y, z) = 3x^2 + y^2 - 4z^2$; $\mathbf{U} = \cos \tfrac{1}{3}\pi\mathbf{i} + \cos \tfrac{1}{4}\pi\mathbf{j} + \cos \tfrac{2}{3}\pi\mathbf{k}$

4. $f(x, y, z) = 6x^2 - 2xy + yz$; $\mathbf{U} = \tfrac{3}{7}\mathbf{i} + \tfrac{2}{7}\mathbf{j} + \tfrac{6}{7}\mathbf{k}$

5. $g(x, y) = \dfrac{1}{x - y}$; $\mathbf{U} = -\tfrac{12}{13}\mathbf{i} + \tfrac{5}{13}\mathbf{j}$

6. $f(x, y) = \dfrac{1}{x^2 + y^2}$; $\mathbf{U} = \tfrac{3}{5}\mathbf{i} - \tfrac{4}{5}\mathbf{j}$

In Exercises 7 through 14, find the gradient of the given function.

7. $f(x, y) = 4x^2 - 3xy + y^2$ **8.** $g(x, y) = \dfrac{xy}{x^2 + y^2}$ **9.** $g(x, y) = \ln \sqrt{x^2 + y^2}$

10. $f(x, y) = e^y \tan 2x$ **11.** $f(x, y, z) = \dfrac{x - y}{x + z}$ **12.** $f(x, y, z) = 3z \ln(x + y)$

13. $g(x, y, z) = xe^{-2y} \sec z$ **14.** $g(x, y, z) = e^{2z}(\sin x - \cos y)$

In Exercises 15 through 22, find the value of the directional derivative at the particular point P_0 for the given function in the direction of **U**.

15. $f(x, y) = x^2 - 2xy^2$; $\mathbf{U} = \cos \pi \mathbf{i} + \sin \pi \mathbf{j}$; $P_0 = (1, -2)$

16. $g(x, y) = 3x^3y + 4y^2 - xy$; $\mathbf{U} = \cos \frac{1}{4}\pi \mathbf{i} + \sin \frac{1}{4}\pi \mathbf{j}$; $P_0 = (0, 3)$

17. $g(x, y) = y^2 \tan^2 x$; $\mathbf{U} = -\frac{1}{2}\sqrt{3}\mathbf{i} + \frac{1}{2}\mathbf{j}$; $P_0 = (\frac{1}{3}\pi, 2)$ **18.** $f(x, y) = xe^{2y}$; $\mathbf{U} = \frac{1}{2}\mathbf{i} + \frac{1}{2}\sqrt{3}\mathbf{j}$; $P_0 = (2, 0)$

19. $h(x, y, z) = \cos(xy) + \sin(yz)$; $\mathbf{U} = -\frac{1}{3}\mathbf{i} + \frac{2}{3}\mathbf{j} + \frac{2}{3}\mathbf{k}$; $P_0 = (2, 0, -3)$

20. $f(x, y, z) = \ln(x^2 + y^2 + z^2)$; $\mathbf{U} = \dfrac{1}{\sqrt{3}}\mathbf{i} - \dfrac{1}{\sqrt{3}}\mathbf{j} - \dfrac{1}{\sqrt{3}}\mathbf{k}$; $P_0 = (1, 3, 2)$

21. $f(x, y) = e^{-3x} \cos 3y$; $\mathbf{U} = \cos(-\frac{1}{12}\pi)\mathbf{i} + \sin(-\frac{1}{12}\pi)\mathbf{j}$; $P_0 = (-\frac{1}{12}\pi, 0)$

22. $g(x, y, z) = \cos 2x \cos 3y \sinh 4z$; $\mathbf{U} = \dfrac{1}{\sqrt{3}}\mathbf{i} - \dfrac{1}{\sqrt{3}}\mathbf{j} + \dfrac{1}{\sqrt{3}}\mathbf{k}$; $P_0 = (\frac{1}{2}\pi, 0, 0)$

In Exercises 23 through 26, a function f, a point P, and a unit vector **U** *are given. Find (a) the gradient of f at P and (b) the rate of change of the function value in the direction of* **U** *at P.*

23. $f(x, y) = x^2 - 4y$; $P = (-2, 2)$; $\mathbf{U} = \cos \frac{1}{3}\pi \mathbf{i} + \sin \frac{1}{3}\pi \mathbf{j}$ **24.** $f(x, y) = e^{2xy}$; $P = (2, 1)$; $\mathbf{U} = \frac{4}{5}\mathbf{i} - \frac{3}{5}\mathbf{j}$

25. $f(x, y, z) = y^2 + z^2 - 4xz$; $P = (-2, 1, 3)$; $\mathbf{U} = \frac{2}{7}\mathbf{i} - \frac{6}{7}\mathbf{j} + \frac{3}{7}\mathbf{k}$

26. $f(x, y, z) = 2x^3 + xy^2 + xz^2$; $P = (1, 1, 1)$; $\mathbf{U} = \frac{1}{7}\sqrt{21}\mathbf{j} - \frac{2}{7}\sqrt{7}\mathbf{k}$

27. Draw a contour map showing the level curves of the function of Exercise 23 at 8, 4, 0, -4, and -8. Also show the representation of $\nabla f(-2, 2)$ having its initial point at $(-2, 2)$.

28. Draw a contour map showing the level curves of the function of Exercise 24, at e^8, e^4, 1, e^{-4}, and e^{-8}. Also show the representation of $\nabla f(2, 1)$ having its initial point at $(2, 1)$.

In Exercises 29 through 32, find $D_{\mathbf{U}}f$ at the given point P for which **U** *is a unit vector in the direction of $\overrightarrow{PQ}$. Also at P find $D_{\mathbf{U}}f$, if* **U** *is a unit vector for which $D_{\mathbf{U}}f$ is a maximum.*

29. $f(x, y) = e^x \tan^{-1} y$; $P(0, 1)$, $Q(3, 5)$ **30.** $f(x, y) = e^x \cos y + e^y \sin x$; $P(1, 0)$, $Q(-3, 3)$

31. $f(x, y, z) = x - 2y + z^2$; $P(3, 1, -2)$, $Q(10, 7, 4)$ **32.** $f(x, y, z) = x^2 + y^2 - 4xz$; $P(3, 1, -2)$, $Q(-6, 3, 4)$

33. Find the direction from the point $(1, 3)$ for which the value of f does not change if $f(x, y) = e^{2y} \tan^{-1}(y/3x)$.

34. The density is ρ slugs/ft^2 at any point (x, y) of a rectangular plate in the xy plane and $\rho = 1/\sqrt{x^2 + y^2 + 3}$. (a) Find the rate of change of the density at the point $(3, 2)$ in the direction of the unit vector $\cos \frac{2}{3}\pi \mathbf{i} + \sin \frac{2}{3}\pi \mathbf{j}$. (b) Find the direction and magnitude of the greatest rate of change of ρ at $(3, 2)$.

35. The temperature is T degrees at any point (x, y) of a rectangular plate lying in the xy plane, and $T = 3x^2 + 2xy$. Distance is measured in meters. (a) Find the maximum rate of change of the temperature at the point $(3, -6)$ on the plate. (b) Find the direction for which this maximum rate of change at $(3, -6)$ occurs.

36. The temperature is T degrees at any point (x, y, z) of a solid in three-dimensional space, and $T = 60/(x^2 + y^2 + z^2 + 3)$. Distance is measured in inches. (a) Find the rate of change of the temperature at the point $(3, -2, 2)$ in the direction of the vector $-2\mathbf{i} + 3\mathbf{j} - 6\mathbf{k}$. (b) Find the direction and magnitude of the greatest rate of change of T at $(3, -2, 2)$.

37. The electric potential is V volts at any point (x, y) in the xy plane, and $V = e^{-2x} \cos 2y$. Distance is measured in feet. (a) Find the rate of change of the potential at the point $(0, \frac{1}{4}\pi)$ in the direction of the unit vector $\cos \frac{1}{6}\pi\mathbf{i} + \sin \frac{1}{6}\pi\mathbf{j}$. (b) Find the direction and magnitude of the greatest rate of change of V at $(0, \frac{1}{4}\pi)$.

38. An equation of the surface of a mountain is $z = 1200 - 3x^2 - 2y^2$, where distance is measured in meters, the x axis points to the east, and the y axis points to the north. A mountain climber is at the point corresponding to $(-10, 5, 850)$. (a) What is the direction of steepest ascent? (b) If the climber moves in the east direction, is she ascending or descending, and what is her rate? (c) If the climber moves in the southwest direction, is she ascending or descending, and what is her rate? (d) In what direction is she traveling a level path?

19.2 TANGENT PLANES AND NORMALS TO SURFACES

Let S be the surface having the equation

$$F(x, y, z) = 0 \tag{1}$$

and suppose that $P_0(x_0, y_0, z_0)$ is a point on S. Then $F(x_0, y_0, z_0) = 0$. Suppose further that C is a curve on S through P_0 and a set of parametric equations of C is

$$x = f(t) \qquad y = g(t) \qquad z = h(t) \tag{2}$$

Let the value of the parameter t at the point P_0 be t_0. A vector equation of C is

$$\mathbf{R}(t) = f(t)\mathbf{i} + g(t)\mathbf{j} + h(t)\mathbf{k} \tag{3}$$

Because curve C is on surface S, we have, upon substituting from (2) in (1),

$$F(f(t), g(t), h(t)) = 0 \tag{4}$$

Let $G(t) = F(f(t), g(t), h(t))$. If F_x, F_y, and F_z are continuous and not all zero at P_0, and if $f'(t_0)$, $g'(t_0)$, and $h'(t_0)$ exist, then the total derivative of F with respect to t at P_0 is given by

$$G'(t_0) = F_x(x_0, y_0, z_0)f'(t_0) + F_y(x_0, y_0, z_0)g'(t_0) + F_z(x_0, y_0, z_0)h'(t_0)$$

which also can be written as

$$G'(t_0) = [F_x(x_0, y_0, z_0)\mathbf{i} + F_y(x_0, y_0, z_0)\mathbf{j} + F_z(x_0, y_0, z_0)\mathbf{k}] \cdot [f'(t_0)\mathbf{i} + g'(t_0)\mathbf{j} + h'(t_0)\mathbf{k}]$$

$$G'(t_0) = \nabla F(x_0, y_0, z_0) \cdot D_t\mathbf{R}(t_0)$$

Because $G'(t) = 0$ for all t under consideration (because of (4)), $G'(t_0) = 0$; so it follows from the above that

$$\nabla F(x_0, y_0, z_0) \cdot D_t\mathbf{R}(t_0) = 0 \tag{5}$$

From Sec. 17.9 we know that $D_t\mathbf{R}(t_0)$ has the same direction as the unit tangent vector to curve C at P_0. Therefore, from (5) we can conclude that the gradient vector of F at P_0 is orthogonal to the unit tangent vector of every curve C on S through the point P_0. We are led, then, to the following definition.

19.2.1 Definition A vector that is orthogonal to the unit tangent vector of every curve C through a point P_0 on a surface S is called a **normal vector** to S at P_0.

From this definition and the preceding discussion we have the following theorem.

19.2.2 Theorem If an equation of a surface S is $F(x, y, z) = 0$, and F_x, F_y, and F_z are continuous and not all zero at the point $P_0(x_0, y_0, z_0)$ on S, then a normal vector to S at P_0 is $\nabla F(x_0, y_0, z_0)$.

The concept of a normal vector is used to define the *tangent plane* to a surface at a point.

19.2.3 Definition If an equation of a surface S is $F(x, y, z) = 0$, then the **tangent plane** of S at a point $P_0(x_0, y_0, z_0)$ is the plane through P_0 having as a normal vector $\nabla F(x_0, y_0, z_0)$.

An equation of the tangent plane of the above definition is

$$F_x(x_0, y_0, z_0)(x - x_0) + F_y(x_0, y_0, z_0)(y - y_0) + F_z(x_0, y_0, z_0)(z - z_0) = 0 \quad (6)$$

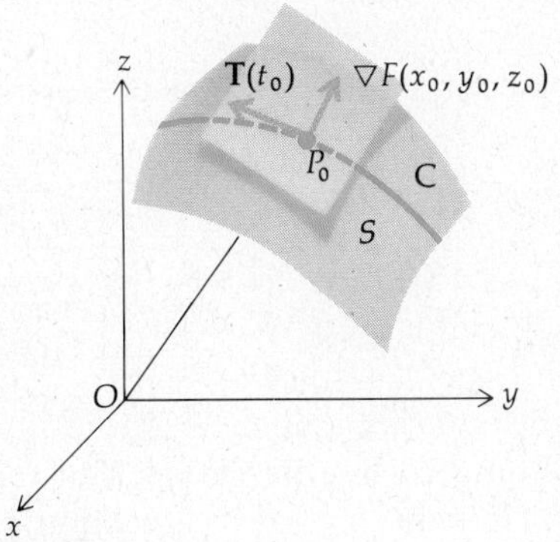

FIGURE 19.2.1

Refer to Fig. 19.2.1, which shows the tangent plane to the surface S at P_0 and the representation of the gradient vector having its initial point at P_0.

A vector equation of the tangent plane given by (6) is

$$\nabla F(x_0, y_0, z_0) \cdot [(x - x_0)\mathbf{i} + (y - y_0)\mathbf{j} + (z - z_0)\mathbf{k}] = 0 \quad (7)$$

Example 1 Find an equation of the tangent plane to the elliptic paraboloid $4x^2 + y^2 - 16z = 0$ at the point $(2, 4, 2)$.

Solution Let $F(x, y, z) = 4x^2 + y^2 - 16z$. Then $\nabla F(x, y, z) = 8x\mathbf{i} + 2y\mathbf{j} - 16\mathbf{k}$, and so $\nabla F(2, 4, 2) = 16\mathbf{i} + 8\mathbf{j} - 16\mathbf{k}$. From (7) it follows that an equation of the tangent plane is

$$16(x - 2) + 8(y - 4) - 16(z - 2) = 0$$

$$2x + y - 2z - 4 = 0$$

19.2.4 Definition The **normal line** to a surface S at a point P_0 on S is the line through P_0 having as a set of direction numbers the components of any normal vector to S at P_0.

If an equation of a surface S is $F(x, y, z) = 0$, it follows from Definition 19.2.4 that symmetric equations of the normal line to S at $P_0(x_0, y_0, z_0)$ are

$$\frac{x - x_0}{F_x(x_0, y_0, z_0)} = \frac{y - y_0}{F_y(x_0, y_0, z_0)} = \frac{z - z_0}{F_z(x_0, y_0, z_0)} \quad (8)$$

The denominators in (8) are components of $\nabla F(x_0, y_0, z_0)$, which is a normal vector to S at P_0; thus (8) follows from Definition 19.2.4. The normal line at a point on a surface is perpendicular to the tangent plane there.

Example 2 Find symmetric equations of the normal line to the surface of Example 1 at (2, 4, 2).

Solution Because $\nabla F(2, 4, 2) = 16\mathbf{i} + 8\mathbf{j} - 16\mathbf{k}$, it follows that symmetric equations of the required normal line are

$$\frac{x-2}{2} = \frac{y-4}{1} = \frac{z-2}{-2}$$

19.2.5 Definition The **tangent line** to a curve C at a point P_0 is the line through P_0 having as a set of direction numbers the components of the unit tangent vector to C at P_0.

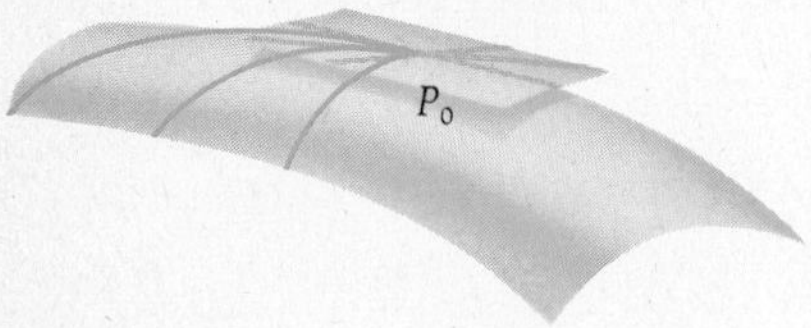

FIGURE 19.2.2

From Definitions 19.2.3 and 19.2.5, all the tangent lines at the point P_0 to the curves lying on a given surface lie in the tangent plane to the surface at P_0. Refer to Fig. 19.2.2, showing sketches of a surface and the tangent plane at P_0. Some of the curves through P_0 and their tangent lines are also sketched in the figure.

Consider a curve C that is the intersection of two surfaces having equations

$$F(x, y, z) = 0 \tag{9}$$

and

$$G(x, y, z) = 0 \tag{10}$$

respectively. We shall show how to obtain equations of the tangent line to C at a point $P_0(x_0, y_0, z_0)$. Because this tangent line lies in each of the tangent planes to the given surfaces at P_0, it is the line of intersection of the two tangent planes. A normal vector at P_0 to the surface having Eq. (9) is given by

$$\mathbf{N}_1 = \nabla F(x_0, y_0, z_0) = F_x(x_0, y_0, z_0)\mathbf{i} + F_y(x_0, y_0, z_0)\mathbf{j} + F_z(x_0, y_0, z_0)\mathbf{k}$$

and a normal vector at P_0 to the surface having Eq. (10) is given by

$$\mathbf{N}_2 = \nabla G(x_0, y_0, z_0) = G_x(x_0, y_0, z_0)\mathbf{i} + G_y(x_0, y_0, z_0)\mathbf{j} + G_z(x_0, y_0, z_0)\mathbf{k}$$

Both $\mathbf{N}_1$ and $\mathbf{N}_2$ are orthogonal to the unit tangent vector to C at P_0; so if $\mathbf{N}_1$ and $\mathbf{N}_2$ are not parallel, it follows from Theorem 17.6.9 that the unit tangent vector has the direction that is the same as, or opposite to, the direction of $\mathbf{N}_1 \times \mathbf{N}_2$. Therefore the components of $\mathbf{N}_1 \times \mathbf{N}_2$ serve as a set of direction numbers of the tangent line. From this set of direction numbers and the coordinates of P_0 we can obtain symmetric equations of the required tangent line. This is illustrated in the following example.

Example 3 Find symmetric equations of the tangent line to the curve of intersection of the surfaces $3x^2 + 2y^2 + z^2 = 49$ and $x^2 + y^2 - 2z^2 = 10$ at the point $(3, -3, 2)$.

Solution Let $F(x, y, z) = 3x^2 + 2y^2 + z^2 - 49$

and

$$G(x, y, z) = x^2 + y^2 - 2z^2 - 10$$

Then $\nabla F(x, y, z) = 6x\mathbf{i} + 4y\mathbf{j} + 2z\mathbf{k}$ and $\nabla G(x, y, z) = 2x\mathbf{i} + 2y\mathbf{j} - 4z\mathbf{k}$. So

$$\mathbf{N}_1 = \nabla F(3, -3, 2) = 18\mathbf{i} - 12\mathbf{j} + 4\mathbf{k} = 2(9\mathbf{i} - 6\mathbf{j} + 2\mathbf{k})$$

and

$$\mathbf{N}_2 = \nabla G(3, -3, 2) = 6\mathbf{i} - 6\mathbf{j} - 8\mathbf{k} = 2(3\mathbf{i} - 3\mathbf{j} - 4\mathbf{k})$$

$$\begin{aligned} \mathbf{N}_1 \times \mathbf{N}_2 &= 4(9\mathbf{i} - 6\mathbf{j} + 2\mathbf{k}) \times (3\mathbf{i} - 3\mathbf{j} - 4\mathbf{k}) \\ &= 4(30\mathbf{i} + 42\mathbf{j} - 9\mathbf{k}) \\ &= 12(10\mathbf{i} + 14\mathbf{j} - 3\mathbf{k}) \end{aligned}$$

Therefore a set of direction numbers of the required tangent line is $[10, 14, -3]$. Symmetric equations of the line are, then,

$$\frac{x - 3}{10} = \frac{y + 3}{14} = \frac{z - 2}{-3}$$

If two surfaces have a common tangent plane at a point, the two surfaces are said to be *tangent* at that point.

EXERCISES 19.2

In Exercises 1 through 12, find an equation of the tangent plane and equations of the normal line to the given surface at the indicated point.

1. $x^2 + y^2 + z^2 = 17$; $(2, -2, 3)$
2. $4x^2 + y^2 + 2z^2 = 26$; $(1, -2, 3)$
3. $x^2 + y^2 - 3z = 2$; $(-2, -4, 6)$
4. $x^2 + y^2 - z^2 = 6$; $(3, -1, 2)$
5. $y = e^x \cos z$; $(1, e, 0)$
6. $z = e^{3x} \sin 3y$; $(0, \frac{1}{6}\pi, 1)$
7. $x^2 = 12y$; $(6, 3, 3)$
8. $z = x^{1/2} + y^{1/2}$; $(1, 1, 2)$
9. $x^{1/2} + y^{1/2} + z^{1/2} = 4$; $(4, 1, 1)$
10. $zx^2 - xy^2 - yz^2 = 18$; $(0, -2, 3)$
11. $x^{2/3} + y^{2/3} + z^{2/3} = 14$; $(-8, 27, 1)$
12. $x^{1/2} + z^{1/2} = 8$; $(25, 2, 9)$

In Exercises 13 through 20, if the two given surfaces intersect in a curve, find equations of the tangent line to the curve of intersection at the given point; if the two given surfaces are tangent at the given point, prove it.

13. $x^2 + y^2 - z = 8$, $x - y^2 + z^2 = -2$; $(2, -2, 0)$
14. $x^2 + y^2 - 2z + 1 = 0$, $x^2 + y^2 - z^2 = 0$; $(0, 1, 1)$
15. $y = x^2$, $y = 16 - z^2$; $(4, 16, 0)$
16. $x = 2 + \cos \pi yz$, $y = 1 + \sin \pi xz$; $(3, 1, 2)$
17. $y = e^x \sin 2\pi z + 2$, $z = y^2 - \ln(x + 1) - 3$; $(0, 2, 1)$
18. $x^2 - 3xy + y^2 = z$, $2x^2 + y^2 - 3z + 27 = 0$; $(1, -2, 11)$
19. $x^2 + z^2 + 4y = 0$, $x^2 + y^2 + z^2 - 6z + 7 = 0$; $(0, -1, 2)$
20. $x^2 + y^2 + z^2 = 8$, $yz = 4$; $(0, 2, 2)$
21. Show that the spheres $x^2 + y^2 + z^2 = a^2$ and $(x - b)^2 + y^2 + z^2 = (b - a)^2$ are tangent at the point $(a, 0, 0)$.
22. Prove that every normal line to the sphere $x^2 + y^2 + z^2 = a^2$ passes through the center of the sphere.
23. Two surfaces are said to be *perpendicular* at a point P_0 of intersection if the normal vectors to the surfaces at P_0 are orthogonal. Show that the surface $x^2 - 2yz + y^3 = 4$ is perpendicular to every member of the family of surfaces $x^2 + (4c - 2)y^2 - cz^2 + 1 = 0$ at the point $(1, -1, 2)$.

19.3 EXTREMA OF FUNCTIONS OF TWO VARIABLES

An important application of the derivative of a function of a single variable is in the study of extreme values of a function, which leads to a variety of problems involving maximum and minimum. This was discussed in Chapter 4, where we proved theorems involving the first and second derivatives, from which relative maximum and minimum values of a function of one variable were determined. In extending the theory to functions of two variables, you will see that it is similar to the one-variable case; however, more complications arise.

19.3.1 Definition The function f of two variables is said to have a **relative maximum value** at the point (x_0, y_0) if there exists an open disk $B((x_0, y_0); r)$ such that $f(x_0, y_0) \geq f(x, y)$ for all (x, y) in B.

19.3.2 Definition The function f of two variables is said to have a **relative minimum value** at the point (x_0, y_0) if there exists an open disk $B((x_0, y_0); r)$ such that $f(x_0, y_0) \leq f(x, y)$ for all (x, y) in B.

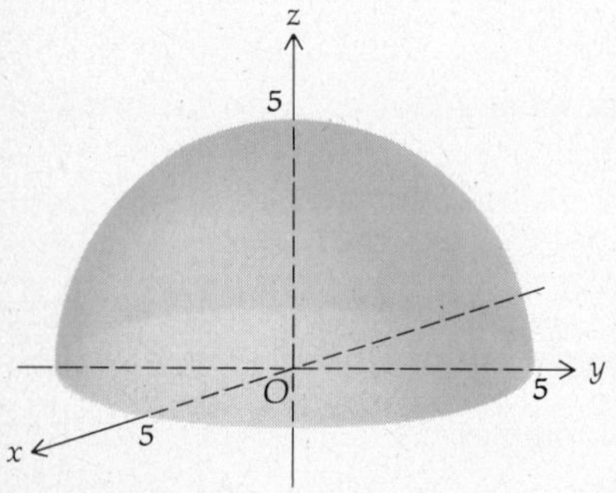

FIGURE 19.3.1

● ILLUSTRATION 1 In Fig. 19.3.1 there is the graph of the function f defined by

$$f(x, y) = \sqrt{25 - x^2 - y^2}$$

Let B be any open disk $((0, 0); r)$ for which $r \leq 5$. From Definition 19.3.1 it follows that f has a relative maximum value of 5 at the point where $x = 0$ and $y = 0$.

In Fig. 19.3.2 there appears a sketch of the graph of the function g for which

$$g(x, y) = x^2 + y^2$$

Let B be any open disk $((0, 0); r)$. Then from Definition 19.3.2, g has a relative minimum value of 0 at the origin. ●

Analogous to Theorem 4.2.3 for functions of a single variable, is the following one for functions of two variables.

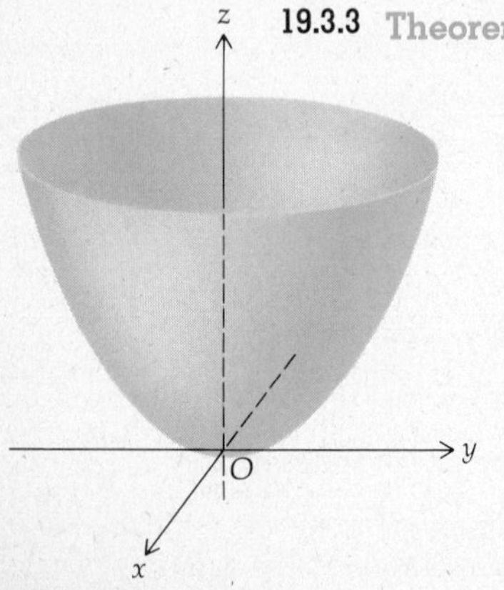

FIGURE 19.3.2

19.3.3 Theorem If $f(x, y)$ exists at all points in some open disk $B((x_0, y_0); r)$ and if f has a relative extremum at (x_0, y_0), then if $f_x(x_0, y_0)$ and $f_y(x_0, y_0)$ exist,

$$f_x(x_0, y_0) = f_y(x_0, y_0) = 0$$

Before proving Theorem 19.3.3, we give an informal geometric argument. Let f be a function satisfying the hypothesis of Theorem 19.3.3, and let f have a relative maximum value at (x_0, y_0). Consider the curve of intersection of the plane $y = y_0$ with the surface $z = f(x, y)$ (refer to Fig. 19.3.3). This curve is represented by the equations

$$y = y_0 \quad \text{and} \quad z = f(x, y) \tag{1}$$

Because f has a relative maximum value at the point $(x_0, y_0, f(x_0, y_0))$, it follows that the curve having Eqs. (1) has a horizontal tangent line in the plane $y = y_0$ at $(x_0, y_0, f(x_0, y_0))$. The slope of this tangent line is $f_x(x_0, y_0)$;

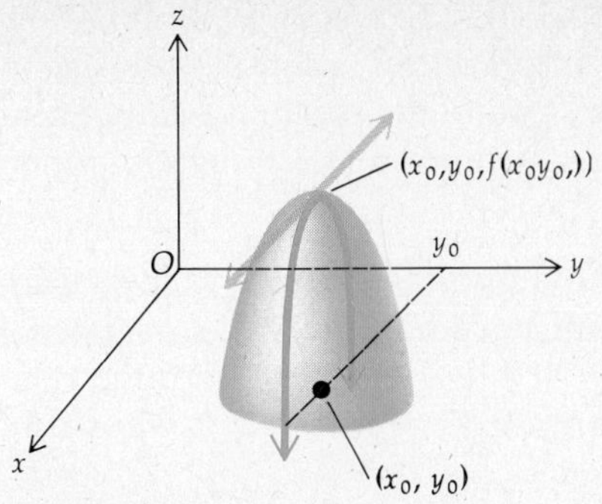

FIGURE 19.3.3

so $f_x(x_0, y_0) = 0$. In a similar way we can consider the curve of intersection of the plane $x = x_0$ with the surface $z = f(x, y)$ and obtain $f_y(x_0, y_0) = 0$. A similar discussion can be given if f has a relative minimum value at (x_0, y_0). Following is the formal proof.

Proof of Theorem 19.3.3 We prove that if f has a relative maximum value at (x_0, y_0) and if $f_x(x_0, y_0)$ exists, then $f_x(x_0, y_0) = 0$. By the definition of a partial derivative,

$$f_x(x_0, y_0) = \lim_{\Delta x \to 0} \frac{f(x_0 + \Delta x, y_0) - f(x_0, y_0)}{\Delta x}$$

Because f has a relative maximum value at (x_0, y_0), by Definition 19.3.1 it follows that

$$f(x_0 + \Delta x, y_0) - f(x_0, y_0) \le 0$$

whenever Δx is sufficiently small so that $(x_0 + \Delta x, y_0)$ is in B. If Δx approaches zero from the right, $\Delta x > 0$; therefore

$$\frac{f(x_0 + \Delta x, y_0) - f(x_0, y_0)}{\Delta x} \le 0$$

Hence, by Theorem 2.10.3, if $f_x(x_0, y_0)$ exists, $f_x(x_0, y_0) \le 0$.

Similarly, if Δx approaches zero from the left, $\Delta x < 0$; so

$$\frac{f(x_0 + \Delta x, y_0) - f(x_0, y_0)}{\Delta x} \ge 0$$

Therefore, by Theorem 2.10.4, if $f_x(x_0, y_0)$ exists, $f_x(x_0, y_0) \ge 0$. We conclude, then, that because $f_x(x_0, y_0)$ exists, both inequalities, $f_x(x_0, y_0) \le 0$ and $f_x(x_0, y_0) \ge 0$, must hold. Consequently, $f_x(x_0, y_0) = 0$.

The proof that $f_y(x_0, y_0) = 0$, if $f_y(x_0, y_0)$ exists and f has a relative maximum value at (x_0, y_0), is analogous and is left as an exercise (see Exercise 29). The proof of the theorem when $f(x_0, y_0)$ is a relative minimum value is also left as an exercise (see Exercise 30). ■

19.3.4 Definition A point (x_0, y_0) for which both $f_x(x_0, y_0) = 0$ and $f_y(x_0, y_0) = 0$ is called a **critical point.**

Theorem 19.3.3 states that a necessary condition for a function of two variables to have a relative extremum at a point, where its first partial derivatives exist, is that this point be a critical point. It is possible for a function of two variables to have a relative extremum at a point at which the partial derivatives do not exist, but we do not consider this situation in this book. Furthermore, the vanishing of the first partial derivatives of a function of two variables is not a sufficient condition for the function to have a relative extremum at the point. Such a situation occurs at a point called a *saddle point.*

● ILLUSTRATION 2 A simple example of a function that has a saddle point is the one defined by

$$f(x, y) = y^2 - x^2$$

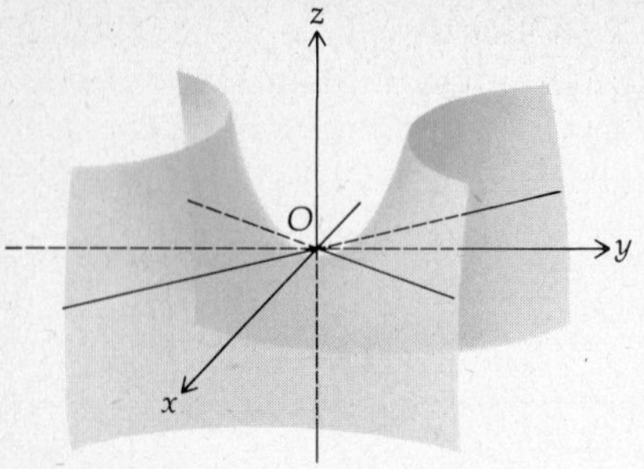

FIGURE 19.3.4

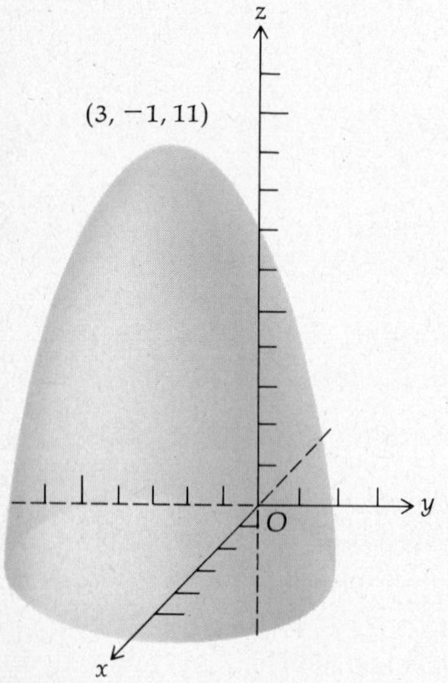

FIGURE 19.3.5

For this function we see that $f_x(x, y) = -2x$ and $f_y(x, y) = 2y$. Both $f_x(0, 0)$ and $f_y(0, 0)$ equal zero. A sketch of the graph of the function is shown in Fig. 19.3.4; it is saddle shaped at points close to the origin. It is apparent that this function f does not satisfy either Definition 19.3.1 or 19.3.2 when $(x_0, y_0) = (0, 0)$. ●

There is a second-derivative test that gives conditions that guarantee a function to have a relative extremum at a point where the first partial derivatives vanish. However, sometimes it is possible to determine relative extrema of a function by Definitions 19.3.1 and 19.3.2, as shown in the following illustration.

● ILLUSTRATION 3 Let f be the function defined by

$$f(x, y) = 6x - 4y - x^2 - 2y^2$$

We determine if f has any relative extrema.

Because f and its first partial derivatives exist at all (x, y) in R^2, Theorem 19.3.3 is applicable. Differentiating we obtain

$$f_x(x, y) = 6 - 2x \quad \text{and} \quad f_y(x, y) = -4 - 4y$$

Setting $f_x(x, y)$ and $f_y(x, y)$ equal to zero we get $x = 3$ and $y = -1$. See Fig. 19.3.5 for a sketch of the graph of the equation

$$z = 6x - 4y - x^2 - 2y^2$$

It is a paraboloid having a vertical axis, with vertex at $(3, -1, 11)$ and opening downward. We can conclude that $f(x, y) \le f(3, -1)$ for all (x, y); hence, by Definition 19.3.1, $f(3, -1) = 11$ is a relative maximum function value. ●

The basic test for determining relative maxima and minima for functions of two variables is the second-derivative test, which is given in the next theorem.

19.3.5 Theorem
Second-Derivative Test

Let f be a function of two variables such that f and its first- and second-order partial derivatives are continuous on some open disk $B((a, b); r)$. Suppose further that $f_x(a, b) = f_y(a, b) = 0$. Then

(i) f has a relative minimum value at (a, b) if

$$f_{xx}(a, b)f_{yy}(a, b) - f_{xy}^2(a, b) > 0 \quad \text{and} \quad f_{xx}(a, b) > 0$$

(ii) f has a relative maximum value at (a, b) if

$$f_{xx}(a, b)f_{yy}(a, b) - f_{xy}^2(a, b) > 0 \quad \text{and} \quad f_{xx}(a, b) < 0$$

(iii) $f(a, b)$ is not a relative extremum if

$$f_{xx}(a, b)f_{yy}(a, b) - f_{xy}^2(a, b) < 0$$

(iv) We can make no conclusion if

$$f_{xx}(a, b)f_{yy}(a, b) - f_{xy}^2(a, b) = 0$$

Proof of (i) For simplicity of notation let us define

$$\phi(x, y) = f_{xx}(x, y)f_{yy}(x, y) - f_{xy}^2(x, y)$$

We are given $\phi(a, b) > 0$ and $f_{xx}(a, b) > 0$, and we wish to prove that $f(a, b)$ is a relative minimum function value. Because f_{xx}, f_{xy}, and f_{yy} are continuous on $B((a, b); r)$, it follows that ϕ is also continuous on B. Hence there exists an open disk $B'((a, b); r')$, where $r' \leq r$, such that $\phi(x, y) > 0$ and $f_{xx}(x, y) > 0$ for every point (x, y) in B'. Let h and k be constants, not both zero, such that the point $(a + h, b + k)$ is in B'. Then the two equations

$$x = a + ht \quad \text{and} \quad y = b + kt \qquad 0 \leq t \leq 1$$

define all the points on the line segment from (a, b) to $(a + h, b + k)$, and all these points are in B'. Let F be the function of one variable defined by

$$F(t) = f(a + ht, b + kt) \tag{2}$$

By Taylor's formula (formula (9), Sec. 14.5),

$$F(t) = F(0) + F'(0)t + \frac{F''(\xi)}{2!}t^2 \tag{3}$$

where ξ is between 0 and t. If $t = 1$ in Eq. (3), we get

$$F(1) = F(0) + F'(0) + \tfrac{1}{2}F''(\xi) \tag{4}$$

where $0 < \xi < 1$. Because $F(0) = f(a, b)$ and $F(1) = f(a + h, b + k)$, it follows from (4) that

$$f(a + h, b + k) = f(a, b) + F'(0) + \tfrac{1}{2}F''(\xi) \tag{5}$$

where $0 < \xi < 1$.

To find $F'(t)$ and $F''(t)$ from (2) we use the chain rule and obtain

$$F'(t) = hf_x(a + ht, b + kt) + kf_y(a + ht, b + kt) \tag{6}$$

and

$$F''(t) = h^2 f_{xx} + hkf_{yx} + hkf_{xy} + k^2 f_{yy}$$

where each second partial derivative is evaluated at $(a + ht, b + kt)$. From Theorem 18.7.1 it follows that $f_{xy}(x, y) = f_{yx}(x, y)$ for all (x, y) in B'. So

$$F''(t) = h^2 f_{xx} + 2hkf_{xy} + k^2 f_{yy} \tag{7}$$

where each second partial derivative is evaluated at $(a + ht, b + kt)$. Substituting 0 for t in Eq. (6) and ξ for t in Eq. (7) we get

$$F'(0) = hf_x(a, b) + kf_y(a, b) = 0 \tag{8}$$

and

$$F''(\xi) = h^2 f_{xx} + 2hkf_{xy} + k^2 f_{yy} \tag{9}$$

where each second partial derivative is evaluated at $(a + h\xi, b + k\xi)$, where $0 < \xi < 1$. Substituting from (8) and (9) into (5) we obtain

$$f(a + h, b + k) - f(a, b) = \tfrac{1}{2}(h^2 f_{xx} + 2hkf_{xy} + k^2 f_{yy}) \tag{10}$$

The terms in parentheses on the right side of (10) can be written as

$$h^2 f_{xx} + 2hkf_{xy} + k^2 f_{yy} = f_{xx}\left[h^2 + 2hk\frac{f_{xy}}{f_{xx}} + \left(k\frac{f_{xy}}{f_{xx}}\right)^2 - \left(k\frac{f_{xy}}{f_{xx}}\right)^2 + k^2\frac{f_{yy}}{f_{xx}}\right]$$

So from (10),

$$f(a+h, b+k) - f(a, b) = \frac{f_{xx}}{2}\left[\left(h + \frac{f_{xy}}{f_{xx}} k\right)^2 + \frac{f_{xx}f_{yy} - f_{xy}^2}{f_{xx}^2} k^2\right] \tag{11}$$

Because $f_{xx}f_{yy} - f_{xy}^2$ evaluated at $(a + h\xi, b + k\xi)$ equals

$$\phi(a + h\xi, b + k\xi) > 0$$

it follows that the expression in brackets on the right side of Eq. (11) is positive. Furthermore, because $f_{xx}(a + h\xi, b + k\xi) > 0$, it follows from (11) that $f(a + h, b + k) - f(a, b) > 0$. Hence we have proved that

$$f(a + h, b + k) > f(a, b)$$

for every point $(a + h, b + k) \neq (a, b)$ in B'. Therefore, by Definition 19.3.2, $f(a, b)$ is a relative minimum value of f.

The proof of part (ii) is similar and is left as an exercise (see Exercise 31). The proof of part (iii) is also left as an exercise (see Exercise 32). Part (iv) is included to cover all possible cases. ■

Example 1 If $f(x, y) = 2x^4 + y^2 - x^2 - 2y$, determine the relative extrema of f if there are any.

Solution To apply the second-derivative test, we find the first and second partial derivatives of f.

$$f(x, y) = 2x^4 + y^2 - x^2 - 2y$$

$$f_x(x, y) = 8x^3 - 2x \qquad f_y(x, y) = 2y - 2$$

$$f_{xx}(x, y) = 24x^2 - 2 \qquad f_{yy}(x, y) = 2 \qquad f_{xy}(x, y) = 0$$

Setting $f_x(x, y) = 0$ we get $x = -\frac{1}{2}$, $x = 0$, and $x = \frac{1}{2}$. Setting $f_y(x, y) = 0$ we obtain $y = 1$. Therefore f_x and f_y are both 0 at the points $(-\frac{1}{2}, 1)$, $(0, 1)$, and $(\frac{1}{2}, 1)$, and these are the critical points of f. The results of applying the second-derivative test at these points are summarized in Table 19.3.1.

TABLE 19.3.1

Critical point	f_{xx}	f_{yy}	f_{xy}	$f_{xx}f_{yy} - f_{xy}^2$	*Conclusion*
$(-\frac{1}{2}, 1)$	4	2	0	8	f has a relative minimum value
$(0, 1)$	-2	2	0	-4	f does not have a relative extremum
$(\frac{1}{2}, 1)$	4	2	0	8	f has a relative minimum value

At the point $(-\frac{1}{2}, 1)$, $f_{xx} > 0$ and $f_{xx}f_{yy} - f_{xy}^2 > 0$; thus from Theorem 19.3.5(i), f has a relative minimum value at $(-\frac{1}{2}, 1)$. At $(0, 1)$, $f_{xx}f_{yy} - f_{xy}^2 < 0$; so from Theorem 19.3.5(iii), f does not have a relative extremum at $(0, 1)$. Because $f_{xx} > 0$ and $f_{xx}f_{yy} - f_{xy}^2 > 0$ at $(\frac{1}{2}, 1)$, f has a relative minimum value there by Theorem 19.3.5(i).

Because $f(-\frac{1}{2}, 1) = -\frac{9}{8}$ and $f(\frac{1}{2}, 1) = -\frac{9}{8}$, we conclude that f has a relative minimum value of $-\frac{9}{8}$ at each of the points $(-\frac{1}{2}, 1)$ and $(\frac{1}{2}, 1)$.

We now discuss absolute extrema of functions of two variables.

19.3.6 Definition The function f of two variables is said to have an **absolute maximum value** on its domain D in the xy plane if there is some point (x_0, y_0) in D such that $f(x_0, y_0) \geq f(x, y)$ for all points (x, y) in D. In such a case, $f(x_0, y_0)$ is the absolute maximum value of f on D.

19.3.7 Definition The function f of two variables is said to have an **absolute minimum value** on its domain D in the xy plane if there is some point (x_0, y_0) in D such that $f(x_0, y_0) \leq f(x, y)$ for all (x, y) in D. In such a case, $f(x_0, y_0)$ is the absolute minimum value of f on D.

For functions of a single variable we had the following theorem (4.2.9), called the extreme-value theorem: If the function f is continuous on the closed interval $[a, b]$, then f has an absolute maximum value and an absolute minimum value on $[a, b]$. We learned that an absolute extremum of a function continuous on a closed interval must be either a relative extremum or a function value at an endpoint of the interval. We have a corresponding situation for functions of two variables. In the statement of the extreme-value theorem for functions of two variables we refer to a *closed region* in the xy plane. By a **closed region** we mean that the region includes its *boundary*. In the following illustration we give some closed regions and identify the boundary of each region.

● ILLUSTRATION 4

(a) A closed disk is a closed region. The boundary is the circumference of the disk. See Fig. 19.3.6.

(b) The sides of a triangle together with the region enclosed by the triangle is a closed region. The boundary consists of the sides of the triangle. See Fig. 19.3.7.

(c) The edges of a rectangle together with the region enclosed by the rectangle is a closed region. The boundary consists of the edges of the rectangle. See Fig. 19.3.8. ●

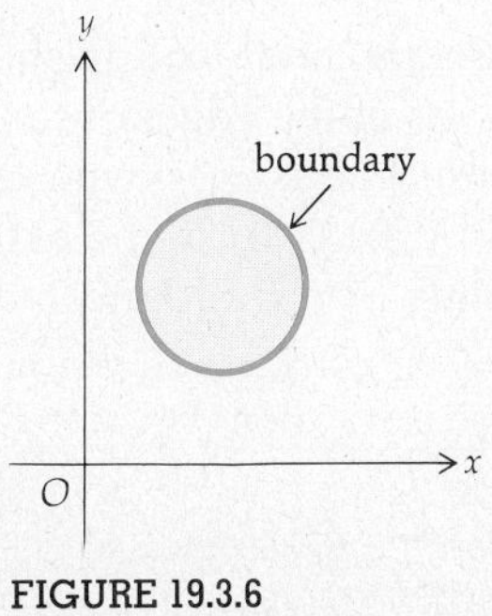

FIGURE 19.3.6

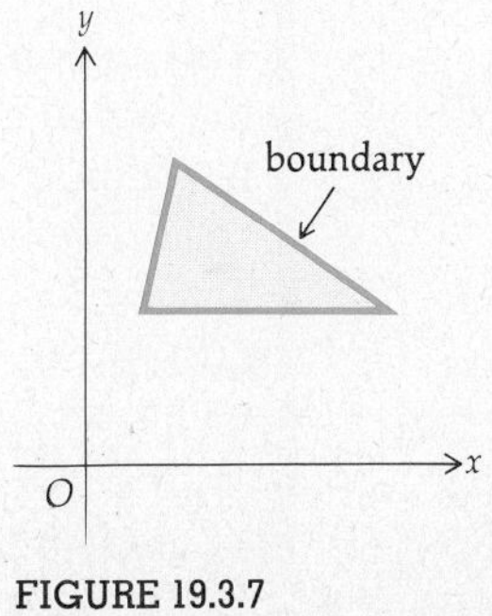

FIGURE 19.3.7

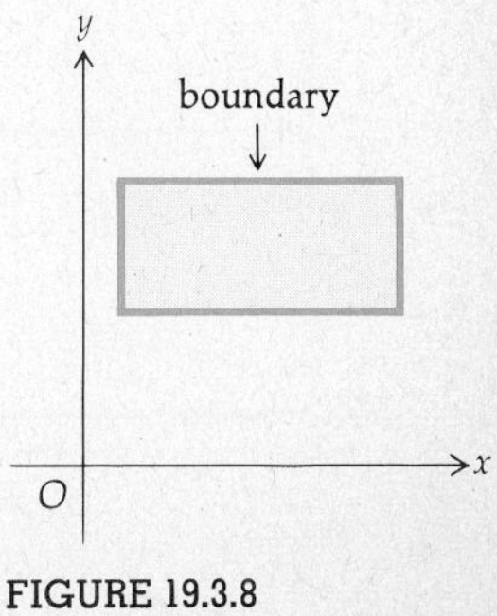

FIGURE 19.3.8

19.3.8 Theorem
The Extreme-Value Theorem for Functions of Two Variables

Let R be a closed region in the xy plane, and let f be a function of two variables that is continuous on R. Then there is at least one point in R where f has an absolute maximum value and at least one point in R where f has an absolute minimum value.

The proof of Theorem 19.3.8 is omitted because it is beyond the scope of this book.

If f is a function satisfying Theorem 19.3.8 and if both $f_x(x, y)$ and $f_y(x, y)$ exist at all points of R, then the absolute extrema of f occur at either a point (x_0, y_0), where $f_x(x_0, y_0) = f_y(x_0, y_0) = 0$, or at a point on the boundary of R.

Example 2 A manufacturer who is a monopolist makes two types of lamps. From experience the manufacturer has determined that if x lamps of the first type and y lamps of the second type are made, they can be sold for $(100 - 2x)$ dollars each and $(125 - 3y)$ dollars each, respectively. The cost of manufacturing x lamps of the first type and y lamps of the second type is $(12x + 11y + 4xy)$ dollars. How many lamps of each type should be produced in order to realize the greatest profit, and what is the greatest profit?

Solution The number of dollars in the revenue received from the lamps of the first type is $x(100 - 2x)$, and the number of dollars in the revenue received from the lamps of the second type is $y(125 - 3y)$. Hence, if $f(x, y)$ dollars is the manufacturer's profit,

$$f(x, y) = x(100 - 2x) + y(125 - 3y) - (12x + 11y + 4xy)$$

$$f(x, y) = 88x + 114y - 2x^2 - 3y^2 - 4xy \qquad (12)$$

Because x and y both represent the number of lamps, we require that $x \geq 0$ and $y \geq 0$ and allow x and y to be any nonnegative real numbers. Furthermore, $(100 - 2x)$ dollars is the selling price of lamps of the first type. Thus we require that $100 - 2x \geq 0$ or, equivalently, $x \leq 50$. Similarly, because $(125 - 3y)$ dollars is the selling price of lamps of the second type, we require that $y \leq \frac{125}{3}$. Therefore the domain of f is the closed region defined by the set

$$\{(x, y) | 0 \leq x \leq 50 \text{ and } 0 \leq y \leq \tfrac{125}{3}\} \qquad (13)$$

This region is rectangular and is shown in Fig. 19.3.9. The boundary of the region consists of the edges of the rectangle. Because f is a polynomial function, it is continuous everywhere. Hence f is continuous on the closed region defined by set (13); so the extreme-value theorem can be applied. The critical points of f are found by determining where $f_x(x, y) = 0$ and $f_y(x, y) = 0$.

$$f_x(x, y) = 88 - 4x - 4y$$

$$f_y(x, y) = 114 - 6y - 4x$$

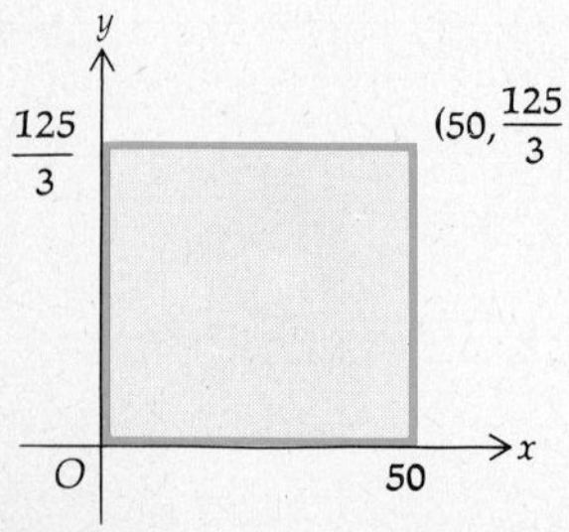

FIGURE 19.3.9

Setting $f_x(x, y) = 0$ and $f_y(x, y) = 0$ we have

$$x + y = 22$$
$$2x + 3y = 57$$

Solving these two equations simultaneously we obtain $x = 9$ and $y = 13$. To apply the second-derivative test we find the second partial derivatives.

$$f_{xx}(x, y) = -4 \qquad f_{yy}(x, y) = -6 \qquad f_{xy}(x, y) = -4$$

At the point (9, 13),

$$f_{xx}(9, 13) = -4 < 0$$
$$f_{xx}(9, 13)f_{yy}(9, 13) - f_{xy}{}^2(9, 13) = (-4)(-6) - (-4)^2$$
$$= 8 > 0$$

It follows, then, by Theorem 19.3.5(ii) that f will have a relative maximum value at (9, 13).

From Eq. (12),

$$f(x, y) = x(88 - 2x) + y(114 - 3y) - 4xy \tag{14}$$

Thus

$$f(9, 13) = 9(70) + 13(75) - 468$$
$$= 1137$$

The absolute maximum value of f must occur at either (9, 13) or on the boundary of the domain of f. Let us compare $f(9, 13)$ with the function values on the boundary.

For that part of the boundary on the x axis with $x \in [0, 50]$ we have the function values computed from (14) as follows:

$$f(x, 0) = 88x - 2x^2$$

Let

$$g(x) = 88x - 2x^2 \qquad x \in [0, 50]$$

Then

$$g'(x) = 88 - 4x \quad \text{and} \quad g''(x) = -4$$

Because $g'(22) = 0$ and $g''(22) < 0$, g has a relative maximum value of 968 at $x = 22$. Furthermore, $g(0) = 0$ and $g(50) < 0$. Because $f(9, 13) = 1137 > 968$, the absolute maximum value of f does not occur on the x axis.

For that part of the boundary on the y axis with $y \in [0, \frac{125}{3}]$, from Eq. (14),

$$f(0, y) = 114y - 3y^2$$

Let

$$h(y) = 114y - 3y^2 \qquad y \in [0, \tfrac{125}{3}]$$

Then

$$h'(y) = 114 - 6y \quad \text{and} \quad h''(y) = -6$$

Because $h'(19) = 0$ and $h''(19) < 0$, h has a relative maximum value of 1083 at $y = 19$. Moreover, $h(0) = 0$ and $h(\frac{125}{3}) < 0$. Because $f(9, 13) = 1137 > 1083$, the absolute maximum value of f does not occur on the y axis.

We now consider the part of the boundary on the line $x = 50$ with $y \in [0, \frac{125}{3}]$. From (14),

$$f(50, y) = y(114 - 3y) - 600 - 200y \tag{15}$$

$$f(0, y) = y(114 - 3y) \tag{16}$$

By comparing (15) and (16),

$$f(50, y) < f(0, y) \tag{17}$$

Because $f(9, 13) > f(0, y)$ for all y in $[0, \frac{125}{3}]$, then from inequality (17)

$$f(9, 13) > f(50, y) \qquad \text{for } y \in [0, \tfrac{125}{3}]$$

Hence the absolute maximum value of f does not occur on the line $x = 50$.

Finally, we have that part of the boundary on the line $y = \frac{125}{3}$ with $x \in [0, 50]$. From (14)

$$f(x, \tfrac{125}{3}) = x(88 - 2x) - \tfrac{1375}{3} - \tfrac{500}{3}x \tag{18}$$

$$f(x, 0) = x(88 - 2x) \tag{19}$$

From (18) and (19) it follows that $f(x, \frac{125}{3}) < f(x, 0)$. Therefore, because $f(9, 13) > f(x, 0)$ for all x in $[0, 50]$, we can conclude that it is also greater than $f(x, \frac{125}{3})$ for all x in $[0, 50]$. Thus the absolute maximum value cannot occur on the line $y = \frac{125}{3}$.

Therefore the absolute maximum value of f is not on the boundary, and thus it is at the point (9, 13). We conclude then that 9 lamps of the first type and 13 lamps of the second type should be produced for the greatest profit of \$1137.

Example 3 Determine the relative dimensions of a rectangular box, without a top and having a specific volume, if the least amount of material is to be used in its manufacture.

Solution Let x units be the length of the base of the box, y units be the width of the base of the box, z units be the depth of the box, and S square units be the surface area of the box. The box has a specific volume; therefore, if V cubic units is the volume of the box, V is a constant. Figure 19.3.10 shows the box.

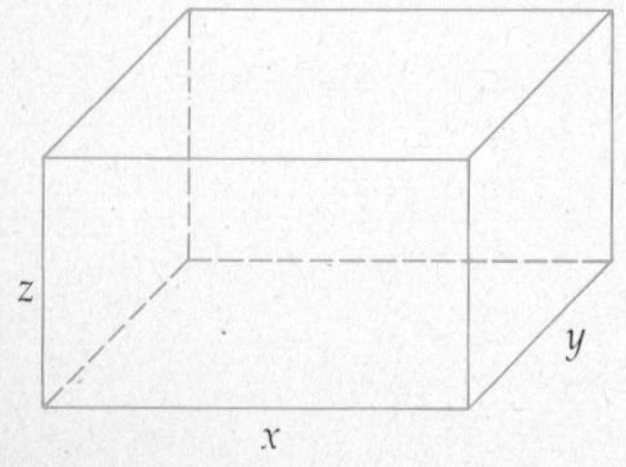

FIGURE 19.3.10

Each of the variables x, y, and z is in the interval $(0, +\infty)$. We have the equations

$$S = xy + 2xz + 2yz \quad \text{and} \quad V = xyz$$

Solving the second equation for z in terms of x, y, and the constant V, we get $z = V/xy$, and substituting this into the first equation gives

$$S = xy + \frac{2V}{y} + \frac{2V}{x} \tag{20}$$

Differentiating we get

$$\frac{\partial S}{\partial x} = y - \frac{2V}{x^2} \qquad \frac{\partial S}{\partial y} = x - \frac{2V}{y^2}$$

$$\frac{\partial^2 S}{\partial x^2} = \frac{4V}{x^3} \qquad \frac{\partial^2 S}{\partial y\, \partial x} = 1 \qquad \frac{\partial^2 S}{\partial y^2} = \frac{4V}{y^3}$$

Setting $\partial S/\partial x = 0$ and $\partial S/\partial y = 0$ and solving simultaneously we get

$$x^2y - 2V = 0$$

$$xy^2 - 2V = 0$$

from which it follows that $x = \sqrt[3]{2V}$ and $y = \sqrt[3]{2V}$. For these values of x and y,

$$\frac{\partial^2 S}{\partial x^2} = \frac{4V}{(\sqrt[3]{2V})^3} = \frac{4V}{2V} = 2 > 0$$

and

$$\frac{\partial^2 S}{\partial x^2} \cdot \frac{\partial^2 S}{\partial y^2} - \left(\frac{\partial^2 S}{\partial y\, \partial x}\right)^2 = \frac{4V}{(\sqrt[3]{2V})^3} \cdot \frac{4V}{(\sqrt[3]{2V})^3} - 1 = 3 > 0$$

From Theorem 19.3.5(i) it follows that S has a relative minimum value when $x = \sqrt[3]{2V}$ and $y = \sqrt[3]{2V}$. Recall that x and y are both in the interval $(0, +\infty)$, and notice from Eq. (20) that S is very large when x and y are either close to zero or very large. We therefore conclude that the relative minimum value of S is an absolute minimum value of S.

When $x = \sqrt[3]{2V}$ and $y = \sqrt[3]{2V}$,

$$z = \frac{V}{xy} = \frac{V}{\sqrt[3]{4V^2}} = \frac{\sqrt[3]{2V}}{2}$$

Hence the box should have a square base and a depth that is one-half of the length of a side of the base.

In Sec. 18.1 we referred to a production function where the independent variables can be any of several inputs, such as the number of machines used in production, the number of person-hours available, the amount of working capital to be had, the quantity of materials used, and the amount of land available. Consider now a production function f of two variables, where the amounts of the inputs are given by x and y, and z gives the amount of the output; then $z = f(x, y)$. Suppose that the prices of the two inputs are a dollars and b dollars per unit, respectively, and that the price of the output is c dollars (a, b, and c are constants). This situation could occur if there were so many producers in the market that a change in the output of any particular producer would not affect the price of the commodity. Such a market is called a *perfectly competitive market.* If P dollars is the total profit, and because the total profit is obtained by subtracting the total cost from the total revenue, then

$$P = cz - (ax + by)$$

and because $z = f(x, y)$,

$$P = cf(x, y) - ax - by$$

It is, of course, desired to maximize P. This is illustrated by an example.

Example 4 Suppose that the production of a certain commodity depends upon two inputs. The amounts of these are given by $100x$ and $100y$, whose prices are, respectively, \$7 and \$4. The amount of the output is given by $100z$, the price per unit of which is \$9. Furthermore, the production function f has the function values

$$f(x, y) = \frac{x}{3} + \frac{y}{3} + 5 - \frac{1}{x} - \frac{1}{y}$$

Determine the greatest profit.

Solution If P dollars is the profit,

$$P = 9(100z) - 7(100x) - 4(100y) \tag{21}$$

Letting $z = f(x, y)$ we get

$$P = 900\left(\frac{x}{3} + \frac{y}{3} + 5 - \frac{1}{x} - \frac{1}{y}\right) - 700x - 400y$$

$$= 4500 - \frac{900}{x} - \frac{900}{y} - 400x - 100y$$

x and y are both in the interval $(0, +\infty)$. Hence

$$\frac{\partial P}{\partial x} = \frac{900}{x^2} - 400 \quad \text{and} \quad \frac{\partial P}{\partial y} = \frac{900}{y^2} - 100$$

Also,

$$\frac{\partial^2 P}{\partial x^2} = -\frac{1800}{x^3} \qquad \frac{\partial^2 P}{\partial y^2} = -\frac{1800}{y^3} \qquad \frac{\partial^2 P}{\partial y\,\partial x} = 0$$

Setting $\partial P/\partial x = 0$ and $\partial P/\partial y = 0$ we have

$$\frac{900}{x^2} - 400 = 0 \quad \text{and} \quad \frac{900}{y^2} - 100 = 0$$

from which we obtain $x = \frac{3}{2}$ and $y = 3$ (the negative result is rejected because x and y must be positive). At $(\frac{3}{2}, 3)$,

$$\frac{\partial^2 P}{\partial x^2} \cdot \frac{\partial^2 P}{\partial y^2} - \left(\frac{\partial^2 P}{\partial y\,\partial x}\right)^2 = \left(-\frac{1800}{\frac{27}{8}}\right)\left(-\frac{1800}{27}\right) - (0)^2 > 0$$

From the above and the fact that at $(\frac{3}{2}, 3)$, $\partial^2 P/\partial x^2 < 0$, it follows from Theorem 19.3.5(ii) that P has a relative maximum value at $(\frac{3}{2}, 3)$. Because x and y are both in the interval $(0, +\infty)$, and P is a negative number when x and y are either close to zero or very large, we conclude that the relative maximum value of P is an absolute maximum value. Because $z = f(x, y)$, the value of z at $(\frac{3}{2}, 3)$ is $f(\frac{3}{2}, 3) = \frac{1}{2} + 1 + 5 - \frac{2}{3} - \frac{1}{3} = \frac{11}{2}$. Hence, from Eq. (21),

$$P_{\text{max}} = 900 \cdot \tfrac{11}{2} - 700 \cdot \tfrac{3}{2} - 400 \cdot 3 = 2700$$

The greatest profit then is \$2700.

The discussion of the extrema of functions of two variables can be extended to functions of three or more variables. The definitions of relative extrema and critical point are easily made. For example, if f is a function of the three variables x, y, and z, and $f_x(x_0, y_0, z_0) = f_y(x_0, y_0, z_0) = f_z(x_0, y_0, z_0) = 0$, then (x_0, y_0, z_0) is a critical point of f. Such a point is obtained by solving simultaneously three equations in three unknowns. For a function of n variables, the critical points are found by setting all the n first partial derivatives equal to zero and solving simultaneously the n equations in n unknowns. The extension of Theorem 19.3.5 to functions of three or more variables is given in advanced calculus texts.

EXERCISES 19.3

In Exercises 1 through 12, determine the relative extrema of f, if there are any.

1. $f(x, y) = x^3 + y^2 - 6x^2 + y - 1$

2. $f(x, y) = x^2 - 4xy + y^3 + 4y$

3. $f(x, y) = \dfrac{1}{x} - \dfrac{64}{y} + xy$

4. $f(x, y) = 18x^2 - 32y^2 - 36x - 128y - 110$

5. $f(x, y) = e^{xy}$

6. $f(x, y) = x^3 + y^3 - 18xy$

7. $f(x, y) = 4xy^2 - 2x^2y - x$

8. $f(x, y) = \dfrac{2x + 2y + 1}{x^2 + y^2 + 1}$

9. $f(x, y) = x^3 + y^3 + 3y^2 - 3x - 9y + 2$

10. $f(x, y) = \sin x + \sin y;\ 0 \le x \le \pi;\ 0 \le y \le \pi$

11. $f(x, y) = \sin(x + y) + \sin x + \sin y;\ 0 \le x \le 2\pi;\ 0 \le y \le 2\pi$

12. $f(x, y) = e^x \sin y$

13. Find the three positive numbers whose sum is 24 such that their product is as great as possible.

14. Find the three positive numbers whose product is 24 such that their sum is as small as possible.

15. Determine the relative dimensions of a rectangular box, without a top, to be made from a given amount of material in order for the box to have the greatest possible volume.

16. Prove that the box having the largest volume that can be placed inside a sphere is in the shape of a cube.

17. Find the volume of the largest rectangular parallelepiped that can be inscribed in the ellipsoid $36x^2 + 9y^2 + 4z^2 = 36$ if the edges are parallel to the coordinate axes.

18. A manufacturing plant has two classifications for its workers, A and B. Class A workers earn \$14 per run, and class B workers earn \$13 per run. For a certain production run it is determined that in addition to the salaries of the workers, if x class A workers and y class B workers are used, the number of dollars in the cost of the run is $y^3 + x^2 - 8xy + 600$. How many workers of each class should be used so that the cost of the run is a minimum if at least three workers of each class are required for a run?

19. Find the point in the plane $3x + 2y - z = 5$ that is closest to the point $(1, -2, 3)$, and find the minimum distance.

20. Find the points on the surface $y^2 - xz = 4$ that are closest to the origin, and find the minimum distance.

21. Find the points on the curve of intersection of the ellipsoid $x^2 + 4y^2 + 4z^2 = 4$ and the plane $x - 4y - z = 0$ that are closest to the origin, and find the minimum distance.

22. A rectangular box without a top is to be made at a cost of \$10 for the material. If the material for the bottom of the box costs \$0.15 per square foot and the material for the sides costs \$0.30 per square foot, find the dimensions of the box of greatest volume that can be made.

23. A closed rectangular box to contain 16 ft^3 is to be made of three kinds of material. The cost of the material for the top and the bottom is \$0.18 per ft^2, the cost of the material for the front and the back is \$0.16 per ft^2, and the cost of the material for the other two sides is \$0.12 per ft^2. Find the dimensions of the box such that the cost of the materials is a minimum.

24. Suppose that T degrees is the temperature at any point (x, y, z) on the sphere $x^2 + y^2 + z^2 = 4$, and $T = 100xy^2z$. Find the points on the sphere where the temperature is the greatest and also the points where the temperature is the least. Also find the temperature at these points.

25. Suppose that when the production of a particular commodity requires x machine-hours and y person-hours, the cost of production is given by $f(x, y)$, where $f(x, y) = 2x^3 - 6xy + y^2 + 500$. Determine the number of machine-hours and the number of person-hours needed to produce the commodity at the least cost.

26. A clothing store sells two kinds of sweaters that are similar but are made by different manufacturers. The cost to the store of the first kind is \$40 and the cost of the second kind is \$50. It has been determined by experience that if the selling price of the first kind is x dollars and the selling price of the second kind is y dollars, then the number sold monthly of the first kind is $3200 - 50x + 25y$, and the number sold monthly of the second kind is $25x - 25y$. What should be the selling price of each kind of sweater for the greatest gross profit?

27. The production function f for a certain commodity has function values $f(x, y) = 4 - \left(\frac{8}{xy}\right)$. The amounts of the two inputs are given by $100x$ and $100y$, whose prices per unit are, respectively, \$10 and \$5, and the amount of the output is given by $100z$, whose price per unit is \$20. Determine the greatest profit.

28. Solve Exercise 27 if $f(x, y) = x + \frac{5}{2}y - \frac{1}{8}x^2 - \frac{1}{4}y^2 - \frac{9}{8}$; the prices per unit of the two inputs are \$4 and \$8 (instead of \$10 and \$5, respectively), and the price per unit of the output is \$16.

29. Prove that $f_y(x_0, y_0) = 0$ if $f_y(x_0, y_0)$ exists and f has a relative maximum value at (x_0, y_0).

30. Prove Theorem 19.3.3 when $f(x_0, y_0)$ is a relative minimum value.

31. Prove Theorem 19.3.5(ii). **32.** Prove Theorem 19.3.5(iii).

19.4 LAGRANGE MULTIPLIERS

In the solution of Example 3 in Sec. 19.3 we minimized the function having function values $xy + 2xz + 2yz$, subject to the condition that x, y, and z satisfy the equation $xyz = V$. Compare this with Example 1 of Sec. 19.3, in which we found the relative extrema of the function f for which $f(x, y) = 2x^4 + y^2 - x^2 - 2y$. These are essentially two different kinds of problems because in the first case we had an additional condition, called a *constraint* (or *side condition*). Such a problem is called one in *constrained extrema*, whereas that of the second type is called a problem in *free extrema*.

The solution of Example 3 of Sec. 19.3 involved obtaining a function of the two variables x and y by replacing z in the first equation by its value from the second equation. Becaue it is not always feasible to solve the constraint for one of the variables in terms of the others, there is another procedure that can be used to solve a problem in constrained extrema. It is due to Joseph L. Lagrange (1736–1813) and is known as the method of

Lagrange multipliers. Before discussing the theory of this method, we outline the procedure and illustrate it by an example.

Suppose that we wish to find the relative extrema of a function f of the three variables x, y, and z, subject to the constraint $g(x, y, z) = 0$. We introduce a new variable λ, called a **Lagrange multiplier,** and form the auxiliary function F for which

$$F(x, y, z, \lambda) = f(x, y, z) + \lambda g(x, y, z)$$

The problem, then, becomes one of finding the critical points of the function F of the four variables x, y z, and λ. The values of x, y, and z that give the relative extrema of f are among these critical points. The critical points of F are the values of x, y, z, and λ for which the four first partial derivatives of F are zero:

$$F_x = 0 \qquad F_y = 0 \qquad F_z = 0 \qquad F_\lambda = 0$$

Example 1 Solve Example 3 of Sec. 19.3 by the method of Lagrange multipliers.

Solution The variables x, y, and z and the constant V are as defined in the solution of Example 3 of Sec. 19.3. Let

$$S = f(x, y, z) = xy + 2xz + 2yz$$

and

$$g(x, y, z) = xyz - V$$

We wish to minimize the function f subject to the constraint that

$$g(x, y, z) = 0$$

Form the function F for which

$$\begin{aligned} F(x, y, z, \lambda) &= f(x, y, z) + \lambda g(x, y, z) \\ &= xy + 2xz + 2yz + \lambda(xyz - V) \end{aligned}$$

To find the critical points of F we compute the four partial derivatives F_x, F_y, F_z, and F_λ and set the function values equal to zero.

$$F_x(x, y, z, \lambda) = y + 2z + \lambda yz = 0 \tag{1}$$

$$F_y(x, y, z, \lambda) = x + 2z + \lambda xz = 0 \tag{2}$$

$$F_z(x, y, z, \lambda) = 2x + 2y + \lambda xy = 0 \tag{3}$$

$$F_\lambda(x, y, z, \lambda) = xyz - V = 0 \tag{4}$$

Subtracting corresponding members of Eq. (2) from those of Eq. (1) we obtain

$$y - x + \lambda z(y - x) = 0$$

$$(y - x)(1 + \lambda z) = 0$$

giving the two equations

$$y = x \tag{5}$$

and

$$\lambda = -\frac{1}{z} \tag{6}$$

Substituting from (6) into (2) we get $x + 2z - x = 0$, giving $z = 0$, which is impossible because z is in the interval $(0, +\infty)$. Substituting from (5) into (3) gives

$$2x + 2x + \lambda x^2 = 0$$

$$x(4 + \lambda x) = 0$$

$$\lambda = -\frac{4}{x} \qquad \text{(because } x \neq 0\text{)}$$

If in Eq. (2), $\lambda = -4/x$, then

$$x + 2z - \frac{4}{x}(xz) = 0$$

$$x + 2z - 4z = 0$$

$$z = \frac{x}{2} \tag{7}$$

Substituting from (5) and (7) into (4) we get $\frac{1}{2}x^3 - V = 0$, from which $x = \sqrt[3]{2V}$. From (5) and (7) it follows that $y = \sqrt[3]{2V}$ and $z = \frac{1}{2}\sqrt[3]{2V}$. Therefore $(\sqrt[3]{2V}, \sqrt[3]{2V}, \frac{1}{2}\sqrt[3]{2V})$ is a critical point of the function F, and, as shown in Example 3 of Sec. 19.3, f has an absolute minimum value at this point.

Note in the solution that the equation $F_\lambda(x, y, z, \lambda) = 0$ is equivalent to the constraint given by the equation $V = xyz$.

The validity of the method of Lagrange multipliers can be shown by considering the general problem of constrained extrema. Suppose we wish to find the relative extrema of a function f of three variables x, y, and z, subject to the constraint

$$g(x, y, z) = 0 \tag{8}$$

Assume that (8) can be solved for z to obtain

$$z = h(x, y) \tag{9}$$

where h is defined on an open disk $B((x_0, y_0); r)$ and $f(x, y, h(x, y))$ has a relative extremum at $(x_0, y_0, h(x_0, y_0))$. Also assume that the first partial derivatives of f, g, and h exist on B and $g_3(x, y, h(x, y)) \neq 0$ on B. Because f has a relative extremum at $(x_0, y_0, h(x_0, y_0))$, the first partial derivatives of f are zero there. We compute these first partial derivatives by the chain rule.

$$\text{at } (x_0, y_0, h(x_0, y_0)) \qquad f_1 + f_3\frac{\partial h}{\partial x} = 0 \quad \text{and} \quad f_2 + f_3\frac{\partial h}{\partial y} = 0 \tag{10}$$

If in (8) we differentiate implicitly with respect to x and then y, and consider z as the differentiable function h of x and y, then at a point (x, y) in the open disk B

$$g_1 + g_3 \frac{\partial h}{\partial x} = 0 \quad \text{and} \quad g_2 + g_3 \frac{\partial h}{\partial y} = 0$$

or, equivalently, because $g_3 \neq 0$ on the open disk B,

$$\text{at } (x, y) \text{ in } B \qquad \frac{\partial h}{\partial x} = -\frac{g_1}{g_3} \quad \textit{and} \quad \frac{\partial h}{\partial y} = -\frac{g_2}{g_3} \tag{11}$$

where the function values of g_1, g_2, and g_3 are at $(x, y, h(x, y))$. If the values of $\partial h/\partial x$ and $\partial h/\partial y$ from Eqs. (11) are substituted in Eqs. (10), then at $(x_0, y_0, h(x_0, y_0))$,

$$f_1 + f_3\left(-\frac{g_1}{g_3}\right) = 0 \quad \text{and} \quad f_2 + f_3\left(-\frac{g_2}{g_3}\right) = 0$$

Furthermore,

$$f_3 - g_3\left(\frac{f_3}{g_3}\right) = 0$$

everywhere where $g_3 \neq 0$. Thus at $(x_0, y_0, h(x_0, y_0))$,

$$f_1 + g_1\left(-\frac{f_3}{g_3}\right) = 0 \qquad f_2 + g_2\left(-\frac{f_3}{g_3}\right) = 0 \qquad f_3 + g_3\left(-\frac{f_3}{g_3}\right) = 0 \tag{12}$$

If $\lambda = -f_3/g_3$ then Eqs. (12) can be written as

$$f_1 + \lambda g_1 = 0 \qquad f_2 + \lambda g_2 = 0 \qquad f_3 + \lambda g_3 = 0 \tag{13}$$

Furthermore, because f has a relative extremum at $(x_0, y_0, h(x_0, y_0))$ and this extremum is subject to the constraint $g(x, y, z) = 0$, then

$$g(x_0, y_0, h(x_0, y_0)) = 0 \tag{14}$$

If

$$F(x, y, z, \lambda) = f(x, y, z) + \lambda g(x, y, z) \tag{15}$$

and if $z_0 = h(x_0, y_0)$, then (13) and (14) are equivalent to the equations

$$F_x = 0 \qquad F_y = 0 \qquad F_z = 0 \qquad F_\lambda = 0 \qquad \text{at } (x_0, y_0, z_0) \tag{16}$$

Therefore we can conclude that a point (x_0, y_0, z_0) at which the function f has a relative extremum is among the critical points of the function F defined by (15).

Observe that Eqs. (13) can be written as the vector equation

$$\nabla f + \lambda \nabla g = \mathbf{0} \qquad \text{at } (x_0, y_0, z_0)$$

This vector equation together with the equation $g(x_0, y_0, z_0) = 0$ give another form of Eqs. (16).

Example 2 Use the method of Lagrange multipliers to find the shortest distance from the origin to the plane $Ax + By + Cz = D$.

Solution Let w units be the distance from the origin to a point (x, y, z) in the plane. Then

$$w = \sqrt{x^2 + y^2 + z^2}$$

Because w will be a minimum when w^2 is a minimum, we form the function f for which

$$f(x, y, z) = x^2 + y^2 + z^2$$

We wish to find the minimum value of f subject to the constraint

$$Ax + By + Cz - D = 0$$

With the assumption that there is such a minimum value it will occur at a critical point of the function F such that

$$F(x, y, z, \lambda) = x^2 + y^2 + z^2 + \lambda(Ax + By + Cz - D)$$

To find the critical points of F we compute the partial derivatives of F and set them equal to zero.

$$F_x(x, y, z, \lambda) = 2x + \lambda A = 0 \tag{17}$$

$$F_y(x, y, z, \lambda) = 2y + \lambda B = 0 \tag{18}$$

$$F_z(x, y, z, \lambda) = 2z + \lambda C = 0 \tag{19}$$

$$F_\lambda(x, y, z, \lambda) = Ax + By + Cz - D = 0 \tag{20}$$

From (17), (18), and (19),

$$x = -\tfrac{1}{2}\lambda A \qquad y = -\tfrac{1}{2}\lambda B \qquad z = -\tfrac{1}{2}\lambda C \tag{21}$$

Substituting these values of x, y, and z into (20) we get

$$-\tfrac{1}{2}\lambda(A^2 + B^2 + C^2) = D$$

$$-\tfrac{1}{2}\lambda = \frac{D}{A^2 + B^2 + C^2} \tag{22}$$

In Eqs. (21) we replace $-\frac{1}{2}\lambda$ by its value from (22) and obtain

$$x = \frac{AD}{A^2 + B^2 + C^2} \qquad y = \frac{BD}{A^2 + B^2 + C^2} \qquad z = \frac{CD}{A^2 + B^2 + C^2} \tag{23}$$

The point having these coordinates is the one and only critical point of F. Therefore the minimum distance from the origin to the plane is the distance from the origin to the point (x_0, y_0, z_0), where x_0, y_0, and z_0 are the values of x, y, and z in Eqs. (23). The minimum distance then is

$$\sqrt{x_0^2 + y_0^2 + z_0^2} = \sqrt{\frac{A^2D^2}{A^2 + B^2 + C^2} + \frac{B^2D^2}{A^2 + B^2 + C^2} + \frac{C^2D^2}{A^2 + B^2 + C^2}}$$

$$= \frac{|D|}{\sqrt{A^2 + B^2 + C^2}}$$

Example 3 Solve Example 4 of Sec. 19.3 by the method of Lagrange multipliers.

Solution We wish to maximize the function P defined by

$$P(x, y, z) = 900z - 700x - 400y$$

subject to the constraint given by the equation $z = x/3 + y/3 + 5 - 1/x - 1/y$, which we can write as

$$g(x, y, z) = \frac{1}{x} + \frac{1}{y} + z - \frac{x}{3} - \frac{y}{3} - 5 = 0$$

Let

$$F(x, y, z, \lambda) = P(x, y, z) + \lambda g(x, y, z)$$

$$= 900z - 700x - 400y + \lambda\left(\frac{1}{x} + \frac{1}{y} + z - \frac{x}{3} - \frac{y}{3} - 5\right)$$

We find the four partial derivatives F_x, F_y, F_z, and F_λ and set them equal to zero.

$$F_x(x, y, z, \lambda) = -700 - \frac{\lambda}{x^2} - \frac{\lambda}{3} = 0 \qquad F_y(x, y, z, \lambda) = -400 - \frac{\lambda}{y^2} - \frac{\lambda}{3} = 0$$

$$F_z(x, y, z, \lambda) = 900 + \lambda = 0 \qquad F_\lambda(x, y, z, \lambda) = \frac{1}{x} + \frac{1}{y} + z - \frac{x}{3} - \frac{y}{3} - 5 = 0$$

Solving these equations simultaneously we obtain

$$\lambda = -900 \qquad x = \tfrac{3}{2} \qquad y = 3 \qquad z = \tfrac{11}{2}$$

The values of x, y, and z agree with those found previously, and $P(\frac{3}{2}, 3, \frac{11}{2}) = 2700$. P is shown to have an absolute maximum value of 2700 in the same way as before.

If several constraints are imposed, the method of Lagrange multipliers can be extended by using several multipliers. In particular, if we wish to find critical points of the function having function values $f(x, y, z)$ subject to the two side conditions $g(x, y, z) = 0$ and $h(x, y, z) = 0$, we find the critical points of the function F of the five variables x, y, z, λ, and μ for which

$$F(x, y, z, \lambda, \mu) = f(x, y, z) + \lambda g(x, y, z) + \mu h(x, y, z)$$

The following example illustrates the method.

Example 4 Find the relative extrema of the function f if $f(x, y, z) = xz + yz$ and the point (x, y, z) lies on the intersection of the surfaces $x^2 + z^2 = 2$ and $yz = 2$.

Solution We form the function F for which

$$F(x, y, z, \lambda, \mu) = xz + yz + \lambda(x^2 + z^2 - 2) + \mu(yz - 2)$$

Finding the five partial derivatives and setting them equal to zero we have

$$F_x(x, y, z, \lambda, \mu) = z + 2\lambda x = 0 \tag{24}$$

$$F_y(x, y, z, \lambda, \mu) = z + \mu z = 0 \tag{25}$$

$$F_z(x, y, z, \lambda, \mu) = x + y + 2\lambda z + \mu y = 0 \tag{26}$$

$$F_\lambda(x, y, z, \lambda, \mu) = x^2 + z^2 - 2 = 0 \tag{27}$$

$$F_\mu(x, y, z, \lambda, \mu) = yz - 2 = 0 \tag{28}$$

From (25) we obtain $\mu = -1$ and $z = 0$. We reject $z = 0$ because this contradicts (28). From (24) we obtain

$$\lambda = -\frac{z}{2x} \tag{29}$$

Substituting from (29) and $\mu = -1$ into (26) we get

$$x + y - \frac{z^2}{x} - y = 0$$

and so

$$x^2 = z^2 \tag{30}$$

Substituting from (30) into (27) we have $2x^2 - 2 = 0$, or $x^2 = 1$. This gives two values for x, namely 1 and -1; and for each of these values of x we get, from (30), the two values 1 and -1 for z. Obtaining the corresponding values for y from (28) we have four sets of solutions for the five Eqs. (24) through (28). These solutions are

$$x = 1 \quad y = 2 \quad z = 1 \quad \lambda = -\tfrac{1}{2} \quad \mu = -1$$

$$x = 1 \quad y = -2 \quad z = -1 \quad \lambda = \tfrac{1}{2} \quad \mu = -1$$

$$x = -1 \quad y = 2 \quad z = 1 \quad \lambda = \tfrac{1}{2} \quad \mu = -1$$

$$x = -1 \quad y = -2 \quad z = -1 \quad \lambda = -\tfrac{1}{2} \quad \mu = -1$$

The first and fourth sets of solutions give $f(x, y, z) = 3$, and the second and third sets of solutions give $f(x, y, z) = 1$. Hence f has a relative maximum function value of 3 and a relative minimum function value of 1.

EXERCISES 19.4

In Exercises 1 through 4, use the method of Lagrange multipliers to find the critical points of the given function subject to the given constraint.

1. $f(x, y) = 25 - x^2 - y^2$ with constraint $x^2 + y^2 - 4y = 0$

2. $f(x, y) = 4x^2 + 2y^2 + 5$ with constraint $x^2 + y^2 - 2y = 0$

3. $f(x, y, z) = x^2 + y^2 + z^2$ with constraint $3x - 2y + z - 4 = 0$

4. $f(x, y, z) = x^2 + y^2 + z^2$ with constraint $y^2 - x^2 = 1$

In Exercises 5 through 8, use the method of Lagrange multipliers to find the relative extrema of f subject to the indicated constraint. Also find the points at which the extrema occur. Assume that the relative extrema exist.

5. $f(x, y) = x^2 + y$ with constraint $x^2 + y^2 = 9$

6. $f(x, y) = x^2y$ with constraint $x^2 + 8y^2 = 24$

7. $f(x, y, z) = xyz$ with constraint $x^2 + 2y^2 + 4z^2 = 4$

8. $f(x, y, z) = y^3 + xz^2$ with constraint $x^2 + y^2 + z^2 = 1$

In Exercises 9 and 10, find the minimum value of f subject to the indicated constraint. Assume that the minimum value exists.

9. $f(x, y, z) = x^2 + y^2 + z^2$ with constraint $xyz = 1$

10. $f(x, y, z) = xyz$ with constraint $x^2 + y^2 + z^2 = 1$

In Exercises 11 and 12, find the maximum value of f subject to the indicated constraint. Assume that the maximum value exists.

11. $f(x, y, z) = x + y + z$ with constraint $x^2 + y^2 + z^2 = 9$

12. $f(x, y, z) = xyz$ with constraint $2xy + 3xz + yz = 72$

13. Find the minimum value of the function f for which $f(x, y, z) = x^2 + 4y^2 + 16z^2$ with the constraint (a) $xyz = 1$; (b) $xy = 1$; (c) $x = 1$.

14. Use the method of Lagrange multipliers to find the shortest distance from the point (1, 3, 0) to the plane $4x + 2y - z = 5$.

15. Use the method of Lagrange multipliers to find the shortest distance from the point (1, −1, −1) to the plane $x + 4y + 3z = 2$.

16. Find the least and greatest distances from the origin to a point on the ellipse $x^2 + 4y^2 = 16$.

17. Find the least and greatest distances from the origin to a point on the ellipsoid $9x^2 + 4y^2 + z^2 = 36$.

18. If $f(x, y, z) = 2x^2 + 3y^2 + z^2$, use the method of Lagrange multipliers to find the point in the plane $x + y + z = 5$ at which $f(x, y, z)$ is a minimum.

19. Use the method of Lagrange multipliers to find a relative minimum function value of f if $f(x, y, z) = x^2 + y^2 + z^2$ with the two constraints $x + 2y + 3z = 6$ and $x - y - z = -1$.

20. Use the method of Lagrange multipliers to find a relative minimum function value of f if $f(x, y, z) = x^2 + y^2 + z^2$ with the two constraints $x + y + 2z = 1$ and $3x - 2y + z = -4$.

21. Use the method of Lagrange multipliers to find a relative maximum function value of f if $f(x, y, z) = xyz$ with the two constraints $x + y + z = 4$ and $x - y - z = 3$.

22. Use the method of Lagrange multipliers to find a relative maximum function value of f if $f(x, y, z) = x^3 + y^3 + z^3$ with the two constraints $x + y + z = 1$ and $x + y - z = 0$.

In Exercises 23 through 32, use the method of Lagrange multipliers to solve the indicated exercise of Exercises 19.3.

23. Exercise 13 **24.** Exercise 14 **25.** Exercise 15 **26.** Exercise 16 **27.** Exercise 19

28. Exercise 20 **29.** Exercise 21 **30.** Exercise 22 **31.** Exercise 23 **32.** Exercise 24

33. A circular disk is in the shape of the region bounded by the circle $x^2 + y^2 = 1$. If T degrees is the temperature at any point (x, y) of the disk and $T = 2x^2 + y^2 - y$, find the hottest and coldest points on the disk.

34. A company has three factories, each manufacturing the same product. If factory A produces x units, factory B produces y units, and factory C produces z units, their respective manufacturing costs are $(3x^2 + 200)$ dollars, $(y^2 + 400)$ dollars, and $(2z^2 + 300)$ dollars. If an order for 1100 units is to be filled, use the method of Lagrange multipliers to determine how the production should be distributed among the three factories in order to minimize the total manufacturing cost.

19.5 OBTAINING A FUNCTION FROM ITS GRADIENT

We consider the problem of how to obtain a function if its gradient is given; that is, we are given

$$\nabla f(x, y) = f_x(x, y)\mathbf{i} + f_y(x, y)\mathbf{j} \tag{1}$$

and we wish to find $f(x, y)$.

● ILLUSTRATION 1 Suppose

$$\nabla f(x, y) = (y^2 + 2x + 4)\mathbf{i} + (2xy + 4y - 5)\mathbf{j} \tag{2}$$

Then because Eq. (1) must be satisfied, it follows that

$$f_x(x, y) = y^2 + 2x + 4 \tag{3}$$

and

$$f_y(x, y) = 2xy + 4y - 5 \tag{4}$$

By integrating both members of (3) with respect to x,

$$f(x, y) = y^2x + x^2 + 4x + g(y) \tag{5}$$

Observe that the "constant" of integration is a function of y and independent of x because we are integrating with respect to x. If we now differentiate both members of (5) partially with respect to y, we obtain

$$f_y(x, y) = 2xy + g'(y) \tag{6}$$

Equations (4) and (6) give two expressions for $f_y(x, y)$. Hence

$$2xy + 4y - 5 = 2xy + g'(y)$$

Therefore

$$g'(y) = 4y - 5$$

$$g(y) = 2y^2 - 5y + C$$

Substituting this value of $g(y)$ into (5) we have

$$f(x, y) = y^2x + x^2 + 4x + 2y^2 - 5y + C$$

●

Example 1 Find $f(x, y)$ if $\nabla f(x, y) = e^{y^2} \cos x\mathbf{i} + 2ye^{y^2} \sin x\mathbf{j}$.

Solution Because Eq. (1) must hold,

$$f_x(x, y) = e^{y^2} \cos x \tag{7}$$

and

$$f_y(x, y) = 2ye^{y^2} \sin x \tag{8}$$

Integrating both members of (8) with respect to y we obtain

$$f(x, y) = e^{y^2} \sin x + g(x) \tag{9}$$

where $g(x)$ is independent of y. We now partially differentiate both members of (9) with respect to x and get

$$f_x(x, y) = e^{y^2} \cos x + g'(x) \tag{10}$$

By equating the right members of (7) and (10),

$$e^{y^2} \cos x = e^{y^2} \cos x + g'(x)$$

$$g'(x) = 0$$

$$g(x) = C$$

We substitute this value of $g(x)$ into (9) and obtain

$$f(x, y) = e^{y^2} \sin x + C$$

All vectors of the form $M(x, y)\mathbf{i} + N(x, y)\mathbf{j}$ are not necessarily gradients, as shown in the next illustration.

● ILLUSTRATION 2 We show that there is no function f such that

$$\nabla f(x, y) = 3y\mathbf{i} - 2x\mathbf{j} \tag{11}$$

Assume that there is such a function. Then it follows that

$$f_x(x, y) = 3y \tag{12}$$

and

$$f_y(x, y) = -2x \tag{13}$$

We integrate both members of (12) with respect to x and obtain

$$f(x, y) = 3xy + g(y)$$

We partially differentiate both members of this equation with respect to y and have

$$f_y(x, y) = 3x + g'(y) \tag{14}$$

Equating the right members of (13) and (14) we obtain

$$-2x = 3x + g'(y)$$

$$-5x = g'(y)$$

If both members of this equation are differentiated with respect to x, it must follow that

$$-5 = 0$$

which, of course, is not true. Thus the assumption that $3y\mathbf{i} - 2x\mathbf{j}$ is a gradient leads to a contradiction. ●

We now investigate a condition that must be satisfied in order for a vector to be a gradient.

Suppose that M_y and N_x are continuous on an open disk B in R^2. If

$$M(x, y)\mathbf{i} + N(x, y)\mathbf{j} \tag{15}$$

is a gradient on B, then there is a function f such that

$$f_x(x, y) = M(x, y) \tag{16}$$

and

$$f_y(x, y) = N(x, y) \tag{17}$$

for all (x, y) in B. Because $M_y(x, y)$ exists on B, then from (16) it follows that

$$M_y(x, y) = f_{xy}(x, y) \tag{18}$$

Furthermore, because $N_x(x, y)$ exists on B, it follows from (17) that

$$N_x(x, y) = f_{xy}(x, y) \tag{19}$$

Because M_y and N_x are continuous on B, their equivalents f_{xy} and f_{yx} are also continuous on B. Thus, from Theorem 18.7.1 it follows that $f_{xy}(x, y) = f_{yx}(x, y)$ at all points in B. Therefore the left members of (18) and (19) are equal at all points in B. We have proved that if M_y and N_x are continuous on an open disk B in R^2, a necessary condition for vector (15) to be a gradient on B is that

$$M_y(x, y) = N_x(x, y) \tag{20}$$

Equation (20) is also a sufficient condition for vector (15) to be a gradient on B. A rigorous proof of the sufficiency of the condition involves concepts beyond the scope of this book. If (20) holds, we can show how to find a function f such that vector (15) is a gradient. However, the proof that whenever (20) holds such a function exists belongs to a course in advanced calculus. The method for finding f is a generalization of that used in Illustration 1 and Example 1 and in Example 2 below. We have the following theorem.

19.5.1 Theorem Suppose that M and N are functions of two variables x and y defined on an open disk $B((x_0, y_0); r)$ in R^2, and M_y and N_x are continuous on B. Then the vector

$$M(x, y)\mathbf{i} + N(x, y)\mathbf{j}$$

is a gradient on B if and only if

$$M_y(x, y) = N_x(x, y)$$

at all points in B.

● ILLUSTRATION 3

(a) We apply Theorem 19.5.1 to the vector in the right member of Eq. (2) in Illustration 1. Let $M(x, y) = y^2 + 2x + 4$ and $N(x, y) = 2xy + 4y - 5$. Then

$$M_y(x, y) = 2y \quad \text{and} \quad N_x(x, y) = 2y$$

Thus $M_y(x, y) = N_x(x, y)$, and therefore the vector is a gradient.

(b) If we apply Theorem 19.5.1 to the vector in the right member of Eq. (11) in Illustration 2, with $M(x, y) = 3y$ and $N(x, y) = -2x$, we obtain

$$M_y(x, y) = 3 \quad \text{and} \quad N_x(x, y) = -2$$

Thus $M_y(x, y) \neq N_x(x, y)$; so the vector is not a gradient. ●

Example 2 Determine if the vector $(e^{-y} - 2x)\mathbf{i} - (xe^{-y} + \sin y)\mathbf{j}$ is a gradient $\nabla f(x, y)$, and if it is, then find $f(x, y)$.

Solution We apply Theorem 19.5.1. Let $M(x, y) = e^{-y} - 2x$ and $N(x, y) = -xe^{-y} - \sin y$. Then

$$M_y(x, y) = -e^{-y} \quad \text{and} \quad N_x(x, y) = -e^{-y}$$

Therefore $M_y(x, y) = N_x(x, y)$; so the given vector is a gradient $\nabla f(x, y)$. Furthermore,

$$f_x(x, y) = e^{-y} - 2x \tag{21}$$

and

$$f_y(x, y) = -xe^{-y} - \sin y \tag{22}$$

Integrating both members of (21) with respect to x we obtain

$$f(x, y) = xe^{-y} - x^2 + g(y) \tag{23}$$

where $g(y)$ is independent of x. We now partially differentiate both members of (23) with respect to y and have

$$f_y(x, y) = -xe^{-y} + g'(y) \tag{24}$$

We equate the right members of (24) and (22) and get

$$-xe^{-y} + g'(y) = -xe^{-y} - \sin y$$

$$g'(y) = -\sin y$$

$$g(y) = \cos y + C$$

We substitute this expression for $g(y)$ into (23) and have

$$f(x, y) = xe^{-y} - x^2 + \cos y + C$$

Theorem 19.5.1 can be extended to functions of three variables.

19.5.2 Theorem Let M, N, and R be functions of three variables x, y, and z defined on an open ball $B((x_0, y_0, z_0); r)$ in R^3, and M_y, M_z, N_x, N_z, R_x, and R_y are continuous on B. Then the vector $M(x, y, z)\mathbf{i} + N(x, y, z)\mathbf{j} + R(x, y, z)\mathbf{k}$ is a gradient on B if and only if $M_y(x, y, z) = N_x(x, y, z)$, $M_z(x, y, z) = R_x(x, y, z)$, and $N_z(x, y, z) = R_y(x, y, z)$.

The proof of the "only if" part of Theorem 19.5.2 is similar to the proof of the "only if" part of Theorem 19.5.1 and is left as an exercise (see Exercise 27). The proof of the "if" part is beyond the scope of this book.

Example 3 Determine if the following vector is a gradient $\nabla f(x, y, z)$, and if it is, then find $f(x, y, z)$:

$$(e^x \sin z + 2yz)\mathbf{i} + (2xz + 2y)\mathbf{j} + (e^x \cos z + 2xy + 3z^2)\mathbf{k}$$

Solution We apply Theorem 19.5.2. Let $M(x, y, z) = e^x \sin z + 2yz$, $N(x, y, z) = 2xz + 2y$, and $R(x, y, z) = e^x \cos z + 2xy + 3z^2$. Then

$$M_y(x, y, z) = 2z \qquad M_z(x, y, z) = e^x \cos z + 2y$$

$$N_x(x, y, z) = 2z \qquad N_z(x, y, z) = 2x$$

$$R_x(x, y, z) = e^x \cos z + 2y \qquad R_y(x, y, z) = 2x$$

Therefore $M_y(x, y, z) = N_x(x, y, z)$, $M_z(x, y, z) = R_x(x, y, z)$, and $N_z(x, y, z) = R_y(x, y, z)$. Thus the given vector is a gradient $\nabla f(x, y, z)$. Furthermore,

$$f_x(x, y, z) = e^x \sin z + 2yz \tag{25}$$

$$f_y(x, y, z) = 2xz + 2y \tag{26}$$

$$f_z(x, y, z) = e^x \cos z + 2xy + 3z^2 \tag{27}$$

Integrating both members of (25) with respect to x we have

$$f(x, y, z) = e^x \sin z + 2xyz + g(y, z) \tag{28}$$

where $g(y, z)$ is independent of x. We partial differentiate both members of (28) with respect to y and obtain

$$f_y(x, y, z) = 2xz + g_y(y, z) \tag{29}$$

Equating the right members of (29) and (26) we have

$$2xz + g_y(y, z) = 2xz + 2y$$

$$g_y(y, z) = 2y$$

We now integrate both members of this equation with respect to y and get

$$g(y, z) = y^2 + h(z) \tag{30}$$

where h is independent of x and y. Substituting from (30) into (28) we obtain

$$f(x, y, z) = e^x \sin z + 2xyz + y^2 + h(z) \tag{31}$$

We now partial differentiate with respect to z both members of (31).

$$f_z(x, y, z) = e^x \cos z + 2xy + h'(z) \tag{32}$$

Equating the right members of (32) and (27) we have

$$e^x \cos z + 2xy + h'(z) = e^x \cos z + 2xy + 3z^2$$

$$h'(z) = 3z^2$$

$$h(z) = z^3 + C \tag{33}$$

We substitute from (33) into (31) and obtain

$$f(x, y, z) = e^x \sin z + 2xyz + y^2 + z^3 + C$$

EXERCISES 19.5

In Exercises 1 through 26, determine if the vector is a gradient. If it is, find a function having the given gradient.

1. $4x\mathbf{i} - 3y\mathbf{j}$

2. $y^2\mathbf{i} + 3x^2\mathbf{j}$

3. $(6x - 5y)\mathbf{i} - (5x - 6y^2)\mathbf{j}$

4. $(4y^2 + 6xy - 2)\mathbf{i} + (3x^2 + 8xy + 1)\mathbf{j}$

5. $(6x^2y^2 - 14xy + 3)\mathbf{i} + (4x^3y - 7x^2 - 8)\mathbf{j}$

6. $3(2x^2 + 6xy)\mathbf{i} + 3(3x^2 + 8y)\mathbf{j}$

7. $(2xy + y^2 + 1)\mathbf{i} + (x^2 + 2xy + x)\mathbf{j}$

8. $\dfrac{2x - 1}{y}\mathbf{i} + \dfrac{x - x^2}{y^2}\mathbf{j}$

9. $\left(\frac{1}{x^2} + \frac{1}{y^2}\right)\mathbf{i} + \left(\frac{1 - 2x}{y^3}\right)\mathbf{j}$

10. $(2x + \ln y)\mathbf{i} + \left(y^2 + \frac{x}{y}\right)\mathbf{j}$

11. $(2x \cos y - 1)\mathbf{i} - x^2 \sin y\mathbf{j}$

12. $(ye^x + x)\mathbf{i} + (xe^y - y)\mathbf{j}$

13. $2x \sec 2y\mathbf{i} + 2x^2 \sec 2y \tan 2y\mathbf{j}$

14. $(2xy - y \sin x)\mathbf{i} + (x^2 + \cos x)\mathbf{j}$

15. $(e^y - 2xy)\mathbf{i} + (xe^y - x^2)\mathbf{j}$

16. $(\sin 2x - \tan y)\mathbf{i} - x \sec^2 y\mathbf{j}$

17. $(2y - 5z)\mathbf{i} + (2x + 8z)\mathbf{j} - (5x - 8y)\mathbf{k}$

18. $4x^3\mathbf{i} + 9y^2\mathbf{j} - 2z\mathbf{k}$

19. $e^y \sin z\mathbf{i} + xe^y \sin z\mathbf{j} + xe^y \cos z\mathbf{k}$

20. $(2xy + 7z^3)\mathbf{i} + (x^2 + 2y^2 - 3z)\mathbf{j} + (21xz^2 - 4y)\mathbf{k}$

21. $(4xy + 3yz - 2)\mathbf{i} + (2x^2 + 3xz - 5z^2)\mathbf{j} + (3xy - 10yz + 1)\mathbf{k}$

22. $(2y + z)\mathbf{i} + (2x - 3z + 4yz)\mathbf{j} + (x - 3y + 2y^2)\mathbf{k}$

23. $z \tan y\mathbf{i} + xz \sec^2 y\mathbf{j} + x \tan y\mathbf{k}$

24. $e^z \cos x\mathbf{i} + z \sin y\mathbf{j} + (e^z \sin x - \cos y)\mathbf{k}$

25. $e^x(e^z - \ln y)\mathbf{i} + (e^y \ln z - e^x y^{-1})\mathbf{j} + (e^{x+z} + e^y z^{-1})\mathbf{k}$

26. $\frac{1}{y+z}\mathbf{i} - \frac{x - z}{(y+z)^2}\mathbf{j} - \frac{x + y}{(y+z)^2}\mathbf{k}$

27. Prove the "only if" part of Theorem 19.5.2.

28. Prove that $\nabla f(x, y) = a\mathbf{i} + b\mathbf{j}$ if and only if $f(x, y) = ax + by + c$, where a, b, and c are constants.

19.6 LINE INTEGRALS

In Chapter 6 the concept of area was used to motivate the definition of the definite integral. To motivate the definition of an integral of a vector-valued function we use the physical concept of work.

We have seen that if a constant force of vector measure $\mathbf{F}$ moves an object along a straight line, then the measure of the work done is the product of the component of $\mathbf{F}$ along the line of motion times the vector measure of displacement. We showed in Sec. 16.3 that if the constant force of vector measure $\mathbf{F}$ moves an object from a point A to a point B, then if W is the measure of the work done,

$$W = \mathbf{F} \cdot \mathbf{V}(\overrightarrow{AB}) \tag{1}$$

Suppose now that the force vector is not constant and instead of the motion being along a straight line, it is along a curve. Let the force that is exerted on the object at the point (x, y) in some open disk B in R^2 be given by the force vector

$$\mathbf{F}(x, y) = M(x, y)\mathbf{i} + N(x, y)\mathbf{j}$$

where M and N are continuous on B. The vector-valued function $\mathbf{F}$ is called a **force field** on B or a **vector field** on B. Let C be a curve lying in B and having the vector equation

$$\mathbf{R}(t) = f(t)\mathbf{i} + g(t)\mathbf{j} \qquad a \leq t \leq b$$

We require the functions f and g be such that f' and g' are continuous on $[a, b]$. We wish to define the work done by the variable force of vector measure $\mathbf{F}$ in moving the object along C from the point $(f(a), g(a))$ to $(f(b), g(b))$. At a point $(f(t), g(t))$ on C the force vector is

$$\mathbf{F}(f(t), g(t)) = M(f(t), g(t))\mathbf{i} + N(f(t), g(t))\mathbf{j} \tag{2}$$

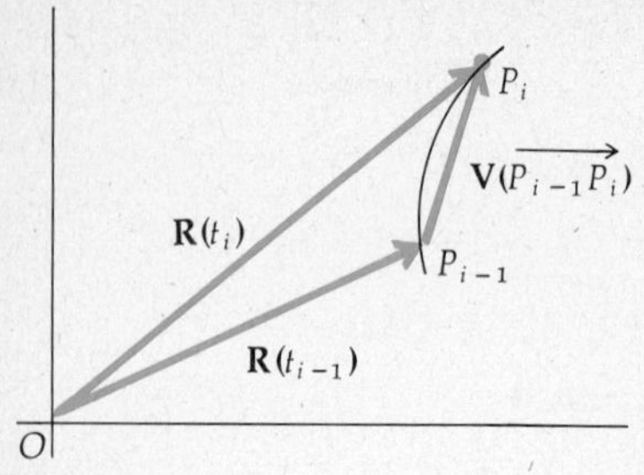

FIGURE 19.6.1

Let Δ be a partition of the interval $[a, b]$:

$$a = t_0 < t_1 < t_2 < \cdots < t_{n-1} < t_n = b$$

On C let P_i be the point $(x_i, y_i) = (f(t_i), g(t_i))$. Refer to Fig. 19.6.1. The vector $\mathbf{V}(\overrightarrow{P_{i-1}P_i}) = \mathbf{R}(t_i) - \mathbf{R}(t_{i-1})$; therefore

$$\mathbf{V}(\overrightarrow{P_{i-1}P_i}) = f(t_i)\mathbf{i} + g(t_i)\mathbf{j} - [f(t_{i-1})\mathbf{i} + g(t_{i-1})\mathbf{j}]$$

or, equivalently,

$$\mathbf{V}(\overrightarrow{P_{i-1}P_i}) = [f(t_i) - f(t_{i-1})]\mathbf{i} + [g(t_i) - g(t_{i-1})]\mathbf{j} \tag{3}$$

Because f' and g' are continuous on $[a, b]$, it follows from the mean-value theorem (4.4.2) that there are numbers c_i and d_i in the open interval (t_{i-1}, t_i) such that

$$f(t_i) - f(t_{i-1}) = f'(c_i)(t_i - t_{i-1}) \tag{4}$$

and

$$g(t_i) - g(t_{i-1}) = g'(d_i)(t_i - t_{i-1}) \tag{5}$$

Letting $\Delta_i t = t_i - t_{i-1}$, and substituting from (4) and (5) into (3), we obtain

$$\mathbf{V}(\overrightarrow{P_{i-1}P_i}) = [f'(c_i)\mathbf{i} + g'(d_i)\mathbf{j}]\,\Delta_i t \tag{6}$$

For each i consider the vector

$$\mathbf{F}_i = M(f(c_i), g(c_i))\mathbf{i} + N(f(d_i), g(d_i))\mathbf{j} \tag{7}$$

Each of the vectors $\mathbf{F}_i$ ($i = 1, 2, \ldots, n$) is an approximation to the force vector $\mathbf{F}(f(t), g(t))$, given by Eq. (2), along the arc of C from P_{i-1} to P_i. Observe that even though c_i and d_i are in general different numbers in the open interval (t_{i-1}, t_i), the values of the vectors $\mathbf{F}(f(t), g(t))$ are close to the vector $\mathbf{F}_i$. Furthermore, we approximate the arc of C from P_{i-1} to P_i by the line segment $\overrightarrow{P_{i-1}P_i}$. Thus we apply formula (1) and obtain an approximation for the work done by the vector $\mathbf{F}(f(t), g(t))$ in moving an object along the arc of C from P_{i-1} to P_i. Denoting this approximation by $\Delta_i W$, we have from formula (1) and Eqs. (7) and (6)

$$\Delta_i W = [M(f(c_i), g(c_i))\mathbf{i} + N(f(d_i), g(d_i))\mathbf{j}] \cdot [f'(c_i)\mathbf{i} + g'(d_i)\mathbf{j}]\,\Delta_i t$$

or, equivalently,

$$\Delta_i W = [M(f(c_i), g(c_i))f'(c_i)]\,\Delta_i t + [N(f(d_i), g(d_i))g'(d_i)]\,\Delta_i t$$

An approximation of the measure of the work done by $F(f(t), g(t))$ along C is $\displaystyle\sum_{i=1}^{n} \Delta_i W$ or, equivalently,

$$\sum_{i=1}^{n} [M(f(c_i), g(c_i))f'(c_i)]\,\Delta_i t + \sum_{i=1}^{n} [N(f(d_i), g(d_i))g'(d_i)]\,\Delta_i t$$

Each of these sums is a Riemann sum. The first summation is a Riemann sum for the function having function values $M(f(t), g(t))f'(t)$ and the second summation is a Riemann sum for the function having function values $N(f(t), g(t))g'(t)$. If $n \to +\infty$, these two sums approach the definite integral:

$$\int_a^b [M(f(t), g(t))f'(t) + N(f(t), g(t))g'(t)]\, dt$$

We therefore have the following definition.

19.6.1 Definition Let C be a curve lying in an open disk B in R^2 for which a vector equation of C is $\mathbf{R}(t) = f(t)\mathbf{i} + g(t)\mathbf{j}$, where f' and g' are continuous on $[a, b]$. Furthermore, let a force field on B be defined by $\mathbf{F}(x, y) = M(x, y)\mathbf{i} + N(x, y)\mathbf{j}$, where M and N are continuous on B. Then if W is the measure of the work done by a force of vector measure $\mathbf{F}$ in moving an object along C from $(f(a), g(a))$ to $(f(b), g(b))$,

$$W = \int_a^b [M(f(t), g(t))f'(t) + N(f(t), g(t))g'(t)]\, dt \tag{8}$$

or, equivalently, by using vector notation,

$$W = \int_a^b \langle M(f(t), g(t)), N(f(t), g(t))\rangle \cdot \langle f'(t), g'(t)\rangle\, dt \tag{9}$$

or, equivalently,

$$W = \int_a^b \mathbf{F}(f(t), g(t)) \cdot \mathbf{R}'(t)\, dt \tag{10}$$

Example 1 Suppose an object moves along the parabola $y = x^2$ from the point $(-1, 1)$ to the point $(2, 4)$. Find the total work done if the motion is caused by the force field $\mathbf{F}(x, y) = (x^2 + y^2)\mathbf{i} + 3x^2y\mathbf{j}$. Assume the arc is measured in meters and the force is measured in newtons.

Solution Parametric equations of the parabola are

$$x = t \quad \text{and} \quad y = t^2 \qquad -1 \le t \le 2$$

Thus a vector equation of the parabola is

$$\mathbf{R}(t) = t\mathbf{i} + t^2\mathbf{j}$$

Because $f(t) = t$, $g(t) = t^2$, and $\mathbf{F}(x, y) = \langle x^2 + y^2, 3x^2y\rangle$, then

$$\mathbf{F}(f(t), g(t)) = \mathbf{F}(t, t^2) = \langle t^2 + t^4, 3t^4\rangle$$

If W joules is the work done, then from formula (10),

$$\begin{aligned} W &= \int_{-1}^{2} \mathbf{F}(t, t^2) \cdot \mathbf{R}'(t)\, dt \\ &= \int_{-1}^{2} \langle t^2 + t^4, 3t^4\rangle \cdot \langle 1, 2t\rangle\, dt \end{aligned}$$

$$= \int_{-1}^{2} (t^2 + t^4 + 6t^5)\, dt$$

$$= \frac{t^3}{3} + \frac{t^5}{5} + t^6 \Big]_{-1}^{2}$$

$$= \tfrac{8}{3} + \tfrac{32}{5} + 64 - (-\tfrac{1}{3} - \tfrac{1}{5} + 1)$$

$$= \tfrac{363}{5}$$

Therefore the work done is $\frac{363}{5}$ joules.

The integral in Eq. (8) is called a *line integral.* A common notation for the line integral of Eq. (8) is

$$\int_C M(x, y)\, dx + N(x, y)\, dy$$

This notation is suggested by the fact that because parametric equations of C are $x = f(t)$ and $y = g(t)$, then $dx = f'(t)\, dt$ and $dy = g'(t)\, dt$. We have the following formal definition.

19.6.2 Definition Let M and N be functions of two variables x and y such that they are continuous on an open disk B in R^2. Let C be a curve lying in B and having parametric equations

$$x = f(t) \qquad y = g(t) \qquad a \le t \le b$$

such that f' and g' are continuous on $[a, b]$. Then the **line integral** of $M(x, y)\, dx + N(x, y)\, dy$ over C is given by

$$\int_C M(x, y)\, dx + N(x, y)\, dy = \int_a^b [M(f(t), g(t))f'(t) + N(f(t), g(t))g'(t)]\, dt$$

or, equivalently, by using vector notation,

$$\int_C M(x, y)\, dx + N(x, y)\, dy = \int_a^b \langle M(f(t), g(t)), N(f(t), g(t))\rangle \cdot \langle f'(t), g'(t)\rangle\, dt$$

● ILLUSTRATION 1 In Example 1, W is given by the line integral

$$\int_C (x^2 + y^2)\, dx + 3x^2 y\, dy$$

where C is the arc of the parabola $y = x^2$ from $(-1, 1)$ to $(2, 4)$ ●

If an equation of C is of the form $y = F(x)$, then x may be used as a parameter in place of t. Similarly, if an equation of C is of the form $x = G(y)$, then y may be used as a parameter in place of t.

● ILLUSTRATION 2 In Example 1 and Illustration 1, the equation of C is $y = x^2$, which is of the form $y = F(x)$. Therefore we can use x as a parameter instead of t. Thus, in the integral of Illustration 1 we can replace y by x^2 and dy by $2x\,dx$, and we have

$$W = \int_{-1}^{2} (x^2 + x^4)\,dx + 3x^2x^2(2x\,dx)$$

$$= \int_{-1}^{2} (x^2 + x^4 + 6x^5)\,dx$$

This integral is the same as the third one appearing in the solution of Example 1, except that the variable is x instead of t. ●

If the curve C in the definition of the line integral is the closed interval $[a, b]$ on the x axis, then $y = 0$ and $dy = 0$. Thus

$$\int_C M(x, y)\,dx + N(x, y)\,dy = \int_a^b M(x, 0)\,dx$$

Therefore, in such a case, the line integral reduces to a definite integral.

In the definition of a line integral, if C has the parametric equations

$$x = f(t) \qquad y = g(t) \qquad a \le t \le b$$

we required that f' and g' be continuous on $[a, b]$. If, in addition to this continuity requirement, $f'(t)$ and $g'(t)$ are not both zero at every point in the open interval (a, b), then C is said to be **smooth** on $[a, b]$. If an interval I can be partitioned into a finite number of subintervals on which C is smooth, then C is said to be **sectionally smooth** on I.

● ILLUSTRATION 3

(a) The curve defined by the parametric equations

$$x = 2\cos t \qquad y = 2\sin t \qquad 0 \le t \le 2\pi$$

is a circle with center at the origin and radius 2. See Fig. 19.6.2. For this curve,

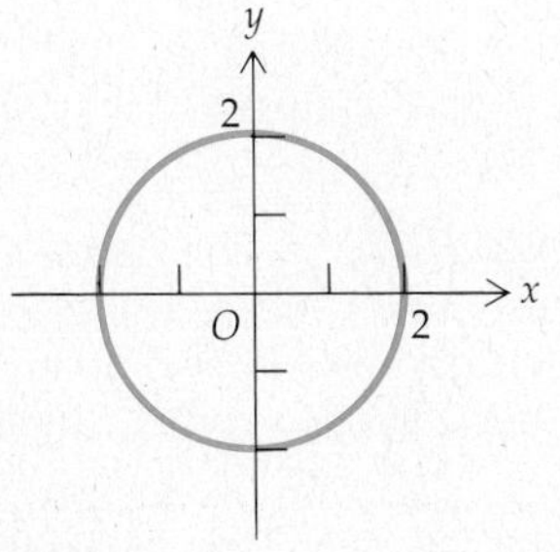

FIGURE 19.6.2

$$f(t) = 2\cos t \qquad g(t) = 2\sin t$$

$$f'(t) = -2\sin t \quad g'(t) = 2\cos t$$

Because f' and g' are continuous for all t and $f'(t)$ and $g'(t)$ are not both zero anywhere, the circle is a smooth curve.

(b) The curve defined by the parametric equations

$$x = a(t - \sin t) \qquad y = a(1 - \cos t) \qquad t \in R$$

is a cycloid. This curve was discussed in Sec. 16.4 and a sketch of its graph on the interval $[-2\pi a, 4\pi a]$ is shown in Fig. 19.6.3. For the cycloid,

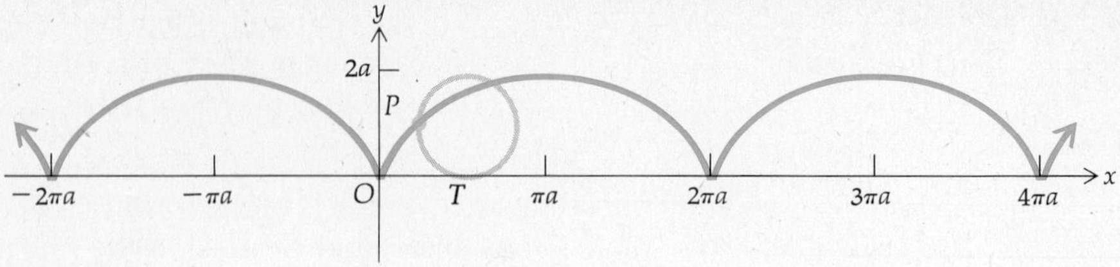

FIGURE 19.6.3

$$f(t) = a(t - \sin t) \qquad g(t) = a(1 - \cos t)$$

$$f'(t) = a(1 - \cos t) \qquad g'(t) = a \sin t$$

The functions f' and g' are continuous for all t but $f'(t)$ and $g'(t)$ are both zero if $t = 2\pi n$, where n is any integer. Therefore the cycloid is not smooth. However, the cycloid is sectionally smooth because it is smooth in each subinterval $[2\pi n, 2\pi(n + 1)]$, where n is any integer. ●

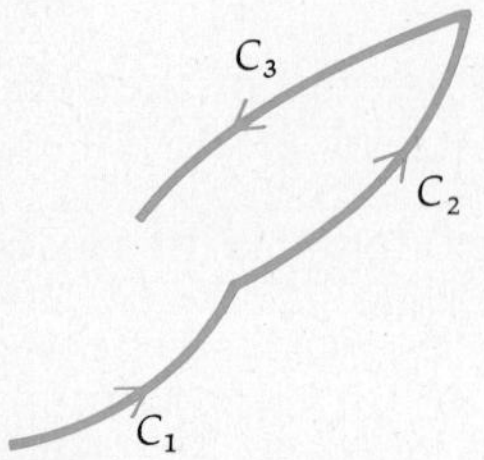

FIGURE 19.6.4

Figures 19.6.4 and 19.6.5 show two other sectionally smooth curves. We can extend the concept of a line integral to include curves that are sectionally smooth.

19.6.3 Definition Let the curve C consist of the smooth arcs $C_1, C_2, \ldots, C_n$. Then the **line integral** of $M(x, y)\,dx + N(x, y)\,dy$ over C is defined by

$$\int_C M(x, y)\,dx + N(x, y)\,dy = \sum_{i=1}^{n} \left(\int_{C_i} M(x, y)\,dx + N(x, y)\,dy \right)$$

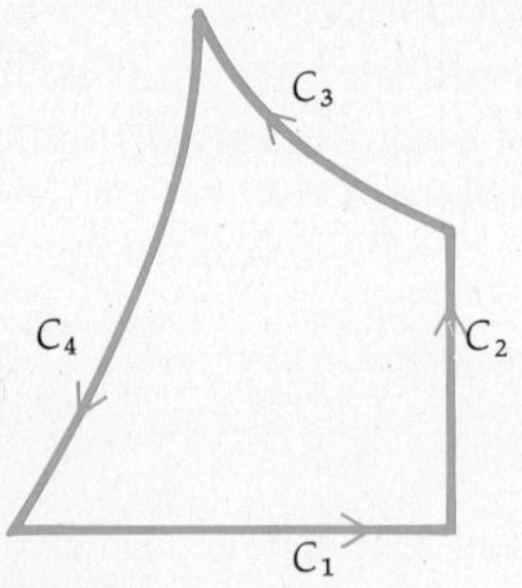

FIGURE 19.6.5

Example 2 Evaluate the line integral $\int_C 4xy\,dx + (2x^2 - 3xy)\,dy$ over the curve C consisting of the line segment from $(-3, -2)$ to $(1, 0)$ and the first quadrant arc of the circle $x^2 + y^2 = 1$ from $(1, 0)$ to $(0, 1)$, traversed in the counterclockwise direction.

Solution Figure 19.6.6 shows the curve C composed of arcs C_1 and C_2. The arc C_1 is the line segment. An equation of the line through $(-3, -2)$ and $(1, 0)$ is $x - 2y = 1$. Therefore C_1 can be represented parametrically by

$$x = 1 + 2t \qquad y = t \qquad -2 \le t \le 0$$

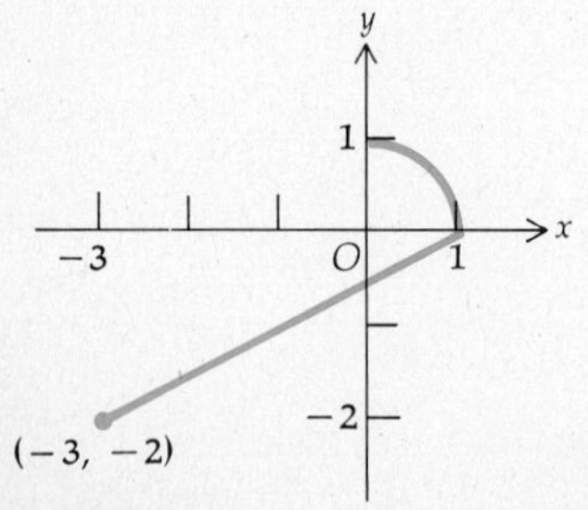

FIGURE 19.6.6

The arc C_2, which is the first quadrant arc of the circle $x^2 + y^2 = 1$, can be represented parametrically by

$$x = \cos t \qquad y = \sin t \qquad 0 \le t \le \tfrac{1}{2}\pi$$

Applying Definition 19.6.2 for each of the arcs C_1 and C_2, we have

$$\int_{C_1} 4xy\,dx + (2x^2 - 3xy)\,dy$$

$$= \int_{-2}^{0} 4(1 + 2t)t(2\,dt) + [2(1 + 2t)^2 - 3(1 + 2t)t]\,dt$$

$$= \int_{-2}^{0} (8t + 16t^2 + 2 + 8t + 8t^2 - 3t - 6t^2)\,dt$$

$$= \int_{-2}^{0} (18t^2 + 13t + 2)\,dt$$

$$= 6t^3 + \tfrac{13}{2}t^2 + 2t\Big]_{-2}^{0}$$

$$= -(-48 + 26 - 4)$$

$$= 26$$

and

$$\int_{C_2} 4xy\,dx + (2x^2 - 3xy)\,dy$$

$$= \int_{0}^{\pi/2} 4\cos t \sin t\,(-\sin t\,dt) + [2\cos^2 t - 3\cos t \sin t](\cos t\,dt)$$

$$= \int_{0}^{\pi/2} (-4\cos t \sin^2 t + 2\cos^3 t - 3\cos^2 t \sin t)\,dt$$

$$= \int_{0}^{\pi/2} [-4\cos t \sin^2 t + 2\cos t\,(1 - \sin^2 t) - 3\cos^2 t \sin t]\,dt$$

$$= \int_{0}^{\pi/2} (2\cos t - 6\cos t \sin^2 t - 3\cos^2 t \sin t)\,dt$$

$$= 2\sin t - 2\sin^3 t + \cos^3 t\Big]_{0}^{\pi/2}$$

$$= 2 - 2 - 1$$

$$= -1$$

Therefore, from Definition 19.6.3,

$$\int_{C} 4xy\,dx + (2x^2 - 3xy)\,dy = 26 + (-1)$$

$$= 25$$

The concept of a line integral can be extended to three dimensions.

19.6.4 Definition Let M, N, and R be functions of three variables x, y, and z such that they are continuous on an open disk B in R^3. Let C be a curve lying in B and having parametric equations

$$x = f(t) \qquad y = g(t) \qquad z = h(t) \qquad a \le t \le b$$

such that f', g', and h' are continuous on $[a, b]$. Then the **line integral** of $M(x, y, z)\,dx + N(x, y, z)\,dy + R(x, y, z)\,dz$ over C is given by

$$\int_C M(x, y, z)\,dx + N(x, y, z)\,dy + R(x, y, z)\,dz$$

$$= \int_a^b [M(f(t), g(t), h(t))f'(t) + N(f(t), g(t), h(t))g'(t) + R(f(t), g(t), h(t))h'(t)]\,dt$$

or, equivalently, by using vector notation,

$$\int_C M(x, y, z)\,dx + N(x, y, z)\,dy + R(x, y, z)\,dz$$

$$= \int_a^b \langle M(f(t), g(t), h(t)), N(f(t), g(t), h(t)), R(f(t), g(t), h(t))\rangle \cdot \langle f'(t), g'(t), h'(t)\rangle\,dt$$

We can define the work done by a force field in moving an object along a curve in R^3 as a line integral of the form of that in Definition 19.6.4.

Example 3 An object traverses the twisted cubic

$$R(t) = t\mathbf{i} + t^2\mathbf{j} + t^3\mathbf{k} \qquad 0 \le t \le 1$$

Find the total work done if the motion is caused by the force field $\mathbf{F}(x, y, z) = e^x\mathbf{i} + xe^z\mathbf{j} + x \sin \pi y^2\mathbf{k}$. Assume that the arc is measured in meters and the force is measured in newtons.

Solution Because $f(t) = t$, $g(t) = t^2$, $h(t) = t^3$, $M(x, y, z) = e^x$, $N(x, y, z) = xe^z$, and $R(x, y, z) = x \sin \pi y^2$, then $M(f(t), g(t), h(t)) = e^t$, $N(f(t), g(t), h(t)) = te^{t^3}$, and $R(f(t), g(t), h(t)) = t \sin \pi t^4$. Thus, if W joules is the work done, we have from the line integral of Definition 19.6.4

$$W = \int_0^1 [(e^t)(dt) + (te^{t^3})(2t\,dt) + (t \sin \pi t^4)(3t^2\,dt)]$$

$$= \int_0^1 (e^t + 2t^2e^{t^3} + 3t^3 \sin \pi t^4)\,dt$$

$$= e^t + \frac{2}{3}e^{t^3} - \frac{3}{4\pi}\cos \pi t^4 \Big]_0^1$$

$$= e + \frac{2}{3}e - \frac{3}{4\pi}\cos \pi - 1 - \frac{2}{3} + \frac{3}{4\pi}\cos 0$$

$$= \frac{5}{3}e + \frac{3}{2\pi} - \frac{5}{3}$$

Therefore the work done is $\left[\frac{5}{3}(e - 1) + \frac{3}{2\pi}\right]$ joules.

EXERCISES 19.6

In Exercises 1 through 22, evaluate the line integral over the given curve.

1. $\int_C (x^2 + xy)\,dx + (y^2 - xy)\,dy$; C: the line $y = x$ from the origin to the point $(2, 2)$.

2. The line integral of Exercise 1; C: the parabola $x^2 = 2y$ from the origin to the point $(2, 2)$.

3. The line integral of Exercise 1; C: the x axis from the origin to $(2, 0)$ and then the line $x = 2$ from $(2, 0)$ to $(2, 2)$.

4. $\int_C yx^2\,dx + (x + y)\,dy$; C: the line $y = -x$ from the origin to the point $(1, -1)$.

5. The line integral of Exercise 4; C: the curve $y = -x^3$ from the origin to the point $(1, -1)$.

6. The line integral of Exercise 4; C: the y axis from the origin to $(0, -1)$ and then the line $y = -1$ from $(0, -1)$ to $(1, -1)$.

7. $\int_C 3xy\,dx + (4x^2 - 3y)\,dy$; C: the line $y = 2x + 3$ from $(0, 3)$ to $(3, 9)$ and then the parabola $y = x^2$ from $(3, 9)$ to $(5, 25)$.

8. $\int_C 9x^2y\,dx + (5x^2 - y)\,dy$; C: the curve $y = x^3 + 1$ from $(1, 2)$ to $(3, 28)$.

9. $\int_C y\,dx + x\,dy$; C: $\mathbf{R}(t) = t\mathbf{i} + t^2\mathbf{j}$, $0 \le t \le 1$.

10. $\int_C (x - 2y)\,dx + xy\,dy$; C: $\mathbf{R}(t) = 3\cos t\mathbf{i} + 2\sin t\mathbf{j}$, $0 \le t \le \frac{1}{2}\pi$.

11. $\int_C 2xy\,dx + (x - 2y)\,dy$; C: $\mathbf{R}(t) = \sin t\mathbf{i} - 2\cos t\mathbf{j}$; $0 \le t \le \pi$.

12. $\int_C xy\,dx - y^2\,dy$; C: $\mathbf{R}(t) = t^2\mathbf{i} + t^3\mathbf{j}$, from the point $(1, 1)$ to the point $(4, -8)$.

13. $\int_C (x - y)\,dx + (y + x)\,dy$; C: the entire circle $x^2 + y^2 = 4$.

14. $\int_C y\sin x\,dx - \cos x\,dy$; C: the line segment from $(\frac{1}{2}\pi, 0)$ to $(\pi, 1)$.

15. $\int_C (xy + x^2)\,dx + x^2\,dy$; C: the parabola $y = 2x^2$ from the origin to the point $(1, 2)$.

16. $\int_C (xy - z)\,dx + e^x\,dy + y\,dz$; C: the line segment from $(1, 0, 0)$ to $(3, 4, 8)$.

17. $\int_C (x + y)\,dx + (y + z)\,dy + (x + z)\,dz$; C: the line segment from the origin to the point $(1, 2, 4)$.

18. The line integral of Exercise 16; C: $\mathbf{R}(t) = (t + 1)\mathbf{i} + t^2\mathbf{j} + t^3\mathbf{k}$, $0 \le t \le 2$.

19. $\int_C z\,dx + x\,dy + y\,dz$; C: the circular helix $\mathbf{R}(t) = a\cos t\mathbf{i} + a\sin t\mathbf{j} + t\mathbf{k}$; $0 \le t \le 2\pi$.

20. $\int_C 2xy\,dx + (6y^2 - xz)\,dy + 10z\,dz$; C: the twisted cubic $\mathbf{R}(t) = t\mathbf{i} + t^2\mathbf{j} + t^3\mathbf{k}$, $0 \le t \le 1$.

21. The line integral of Exercise 20; C: the line segment from the origin to the point (0, 0, 1); then the line segment from (0, 0, 1) to (0, 1, 1); then the line segment from (0, 1, 1) to (1, 1, 1).

22. The line integral of Exercise 20; C: the line segment from the origin to the point (1, 1, 1).

In Exercises 23 through 36, find the total work done in moving an object along the given arc C if the motion is caused by the given force field. Assume the arc is measured in meters and the force is meaured in newtons.

23. $\mathbf{F}(x, y) = 2xy\mathbf{i} + (x^2 + y^2)\mathbf{j}$; C: the line segment from the origin to the point (1, 1).

24. The force field of Exercise 23; C: the arc of the parabola $y^2 = x$ from the origin to the point (1, 1).

25. $\mathbf{F}(x, y) = (y - x)\mathbf{i} + x^2y\mathbf{j}$; C: the line segment from the point (1, 1) to (2, 4).

26. The force field of Exercise 25; C: the arc of the parabola $y = x^2$ from the point (1, 1) to (2, 4).

27. The force field of Exercise 25; C: the line segment from (1, 1) to (2, 2) and then the line segment from (2, 2) to (2, 4).

28. $\mathbf{F}(x, y) = -x^2y\mathbf{i} + 2y\mathbf{j}$; C: the line segment from $(a, 0)$ to $(0, a)$.

29. The force field of Exercise 28; C: $\mathbf{R}(t) = a \cos t\mathbf{i} + a \sin t\mathbf{j}$, $0 \le t \le \frac{1}{2}\pi$.

30. The force field of Exercise 28; C: the line segment from $(a, 0)$ to (a, a) and then the line segment from (a, a) to $(0, a)$.

31. $\mathbf{F}(x, y, z) = (y + z)\mathbf{i} + (x + z)\mathbf{j} + (x + y)\mathbf{k}$; C: the line segment from the origin to the point (1, 1, 1).

32. $\mathbf{F}(x, y, z) = z^2\mathbf{i} + y^2\mathbf{j} + xz\mathbf{k}$; C: the line segment from the origin to the point (4, 0, 3).

33. $\mathbf{F}(x, y, z) = e^x\mathbf{i} + e^y\mathbf{j} + e^z\mathbf{k}$; C: $\mathbf{R}(t) = t\mathbf{i} + t^2\mathbf{j} + t^3\mathbf{k}$, $0 \le t \le 2$.

34. $\mathbf{F}(x, y, z) = (xyz + x)\mathbf{i} + (x^2z + y)\mathbf{j} + (x^2y + z)\mathbf{k}$; C: the arc of Exercise 33.

35. The force field of Exercise 34; C: the line segment from the origin to the point (1, 0, 0); then the line segment from (1, 0, 0) to (1, 1, 0); then the line segment from (1, 1, 0) to (1, 1, 1).

36. $\mathbf{F}(x, y, z) = x\,\mathbf{i} + y\mathbf{j} + (yz - x)\mathbf{k}$; C: $\mathbf{R}(t) = 2t\mathbf{i} + t^2\mathbf{j} + 4t^3\mathbf{k}$, $0 \le t \le 1$.

19.7 LINE INTEGRALS INDEPENDENT OF THE PATH

We learned in Sec. 19.6 that the value of a line integral is determined by the integrand and a curve C between two points P_1 and P_2. However, under certain conditions the value of a line integral depends only on the integrand and the points P_1 and P_2 and not on the path from P_1 to P_2. Such a line integral is said to be *independent of the path.*

● ILLUSTRATION 1 Suppose a force field

$$\mathbf{F}(x, y) = (y^2 + 2x + 4)\mathbf{i} + (2xy + 4y - 5)\mathbf{j}$$

moves an object from the origin to the point (1, 1). We show that the total work done is the same if the path is along (a) the line segment from the origin to (1, 1); (b) the segment of the parabola $y = x^2$ from the origin to (1, 1); and (c) the segment of the curve $x = y^3$ from the origin to (1, 1).

If W is the measure of the work done, then

$$W = \int_C (y^2 + 2x + 4)\,dx + (2xy + 4y - 5)\,dy \tag{1}$$

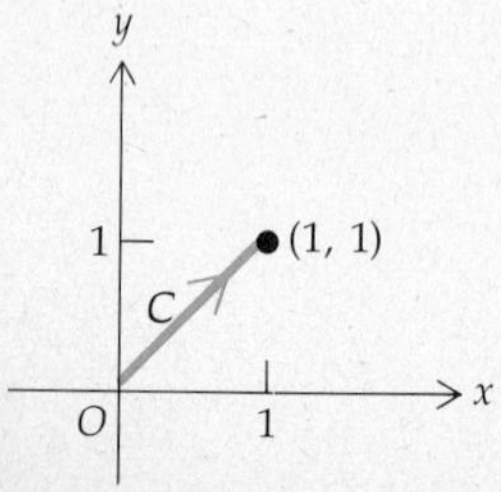

FIGURE 19.7.1

(a) See Fig. 19.7.1. An equation of C is $y = x$. We use x as the parameter and let $y = x$ and $dy = dx$ in Eq. (1). Then

$$W = \int_0^1 (x^2 + 2x + 4)\,dx + (2x^2 + 4x - 5)\,dx$$

$$= \int_0^1 (3x^2 + 6x - 1)\,dx$$

$$= x^3 + 3x^2 - x\Big]_0^1$$

$$= 3$$

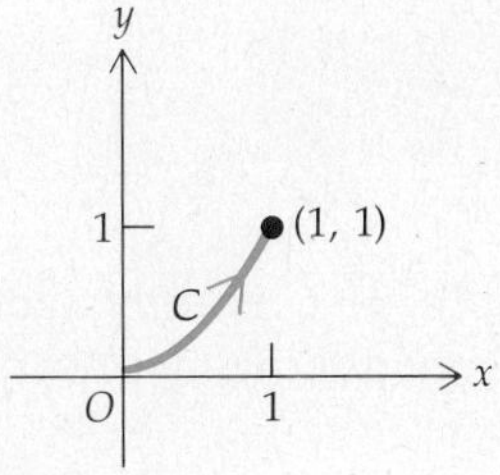

FIGURE 19.7.2

(b) See Fig. 19.7.2. An equation of C is $y = x^2$. Again taking x as the parameter and in Eq. (1) letting $y = x^2$ and $dy = 2x\,dx$, we have

$$W = \int_0^1 (x^4 + 2x + 4)\,dx + (2x^3 + 4x^2 - 5)2x\,dx$$

$$= \int_0^1 (5x^4 + 8x^3 - 8x + 4)\,dx$$

$$= x^5 + 2x^4 - 4x^2 + 4x\Big]_0^1$$

$$= 3$$

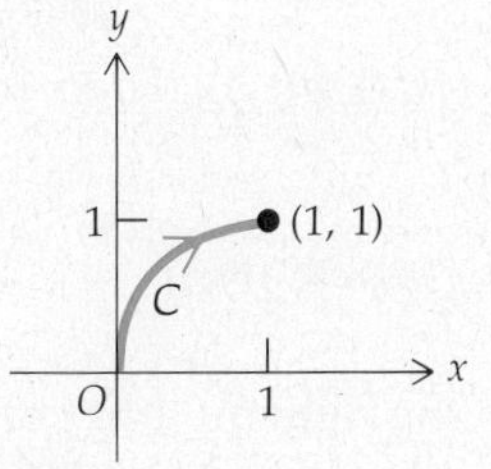

FIGURE 19.7.3

(c) See Fig. 19.7.3. An equation of C is $x = y^3$. We take y as the parameter and in Eq. (1) let $x = y^3$ and $dx = 3y^2\,dy$. Then

$$W = \int_0^1 (y^2 + 2y^3 + 4)3y^2\,dy + (2y^4 + 4y - 5)\,dy$$

$$= \int_0^1 (6y^5 + 5y^4 + 12y^2 + 4y - 5)\,dy$$

$$= y^6 + y^5 + 4y^3 + 2y^2 - 5y\Big]_0^1$$

$$= 3$$

In Illustration 1 we see that the value of the line integral is the same over three different paths from (0, 0) to (1, 1). Actually the value of the line integral is the same over any sectionally smooth curve from the origin to (1, 1); so this line integral is independent of the path. (This fact is proved in Illustration 2, which follows Theorem 19.7.1.)

We now state and prove Theorem 19.7.1, which not only gives conditions for which the value of a line integral is independent of the path but also gives a formula for finding the value of such a line integral.

19.7.1 Theorem Suppose that M and N are functions of two variables x and y defined on an open disk $B((x_0, y_0); r)$ in R^2, M_y and N_x are continuous on B, and ϕ is a function such that

$$\nabla\phi(x, y) = M(x, y)\mathbf{i} + N(x, y)\mathbf{j}$$

Assume that C is any sectionally smooth curve in B from the point (x_1, y_1) to the point (x_2, y_2). Then the line integral

$$\int_C M(x, y)\, dx + N(x, y)\, dy$$

is independent of the path C, and

$$\int_C M(x, y)\, dx + N(x, y)\, dy = \phi(x_2, y_2) - \phi(x_1, y_1)$$

Proof We give the proof if C is smooth. If C is only sectionally smooth, then consider each piece separately; the following proof applies to each smooth piece.

Let a vector equation of C be

$$\mathbf{R}(t) = f(t)\mathbf{i} + g(t)\mathbf{j} \quad t_1 \le t \le t_2$$

where $(x_1, y_1) = (f(t_1), g(t_1))$ and $(x_2, y_2) = (f(t_2), g(t_2))$. Then from the definition of a line integral,

$$\int_C M(x, y)\, dx + N(x, y)\, dy = \int_{t_1}^{t_2} \nabla\phi(f(t), g(t)) \cdot \mathbf{R}'(t)\, dt \tag{2}$$

We proceed to evaluate the definite integral in the right member of Eq. (2).

Because $\nabla\phi(x, y) = M(x, y)\mathbf{i} + N(x, y)\mathbf{j}$, then $\phi_x(x, y) = M(x, y)$ and $\phi_y(x, y) = N(x, y)$. Therefore

$$d\phi(x, y) = M(x, y)\, dx + N(x, y)\, dy$$

for all points (x, y) in B. Then if $(f(t), g(t))$ is a point on C,

$$d\phi(f(t), g(t)) = [M(f(t), g(t))f'(t) + N(f(t), g(t))g'(t)]\, dt$$

or, equivalently,

$$d\phi(f(t), g(t)) = [\nabla\phi(f(t), g(t)) \cdot \mathbf{R}'(t)]\, dt \tag{3}$$

for $t_1 \le t \le t_2$. Consider the function G of the single variable t for which

$$\phi(f(t), g(t)) = G(t) \qquad t_1 \le t \le t_2 \tag{4}$$

Then

$$d\phi(f(t), g(t)) = G'(t)\, dt \tag{5}$$

From (3) and (5) it follows that

$$\nabla\phi(f(t), g(t)) \cdot \mathbf{R}'(t)\, dt = G'(t)\, dt$$

Hence

$$\int_{t_1}^{t_2} \nabla\phi(f(t), g(t)) \cdot \mathbf{R}'(t)\, dt = \int_{t_1}^{t_2} G'(t)\, dt$$

But from the fundamental theorem of the calculus (Theorem 6.5.2) the right member of this equation is $G(t_2) - G(t_1)$. Furthermore, from (4), $G(t_2) - G(t_1) = \phi(f(t_2), g(t_2)) - \phi(f(t_1), g(t_1))$. Therefore

$$\int_{t_1}^{t_2} \nabla\phi(f(t), g(t)) \cdot \mathbf{R}'(t)\, dt = \phi(f(t_2), g(t_2)) - \phi(f(t_1), g(t_1)) \tag{6}$$

From (2) and (6) and because $(f(t_2), g(t_2)) = (x_2, y_2)$ and $(f(t_1), g(t_1)) = (x_1, y_1)$,

$$\int_C M(x, y)\, dx + N(x, y)\, dy = \phi(x_2, y_2) - \phi(x_1, y_1)$$

which is what we wished to prove. ■

Because of the resemblance of Theorem 19.7.1 to the fundamental theorem of the calculus, it is sometimes called the fundamental theorem for line integrals.

● ILLUSTRATION 2 We use Theorem 19.7.1 to evaluate the line integral in Illustration 1. The line integral is

$$\int_C (y^2 + 2x + 4)\, dx + (2xy + 4y - 5)\, dy$$

From Illustration 1, Sec. 19.5, it follows that if $\phi(x, y) = y^2x + x^2 + 4x + 2y^2 - 5y$, then

$$\nabla\phi(x, y) = (y^2 + 2x + 4)\mathbf{i} + (2xy + 4y - 5)\mathbf{j}$$

Thus the value of the line integral is independent of the path, and C can be any sectionally smooth curve from (0, 0) to (1, 1). From Theorem 19.7.1, then,

$$\begin{aligned}\int_C (y^2 + 2x + 4)\, dx + (2xy + 4y - 5)\, dy &= \phi(1, 1) - \phi(0, 0)\\ &= 3 - 0\\ &= 3\end{aligned}$$

This result agrees with that of Illustration 1. ●

If $\mathbf{F}$ is the vector-valued function defined by

$$\mathbf{F}(x, y) = M(x, y)\mathbf{i} + N(x, y)\mathbf{j} \tag{7}$$

and

$$\mathbf{F}(x, y) = \nabla\phi(x, y) \tag{8}$$

then $\mathbf{F}$ is called a **gradient field,** ϕ is called a **potential function,** and $\phi(x, y)$ is the potential of $\mathbf{F}$ at (x, y). Furthermore, if $\mathbf{F}$ is a vector field satisfying Eq. (8), then $\mathbf{F}$ is said to be a **conservative** vector field. Also, when a function $\mathbf{F}$, defined by Eq. (7) is a gradient field, then the expression $M(x, y)\, dx + N(x, y)\, dy$ is called an **exact differential;** this terminology is reasonable because $M(x, y)\, dx + N(x, y)\, dy = d\phi$.

● ILLUSTRATION 3 In Illustration 1 we have the force field

$$\mathbf{F}(x, y) = (y^2 + 2x + 4)\mathbf{i} + (2xy + 4y - 5)\mathbf{j}$$

In Illustration 2 the function ϕ is defined by

$$\phi(x, y) = y^2x + x^2 + 4x + 2y^2 - 5y$$

Because

$$\nabla\phi(x, y) = \mathbf{F}(x, y)$$

it follows that $\mathbf{F}$ is a conservative force field, and ϕ is a potential function of $\mathbf{F}$. In particular, because $\phi(2, 1) = 11$, the potential of $\mathbf{F}$ at $(2, 1)$ is 11.

Furthermore,

$$(y^2 + 2x + 4)\,dx + (2xy + 4y - 5)\,dy$$

is an exact differential and is equal to $d\phi$. ●

Example 1 Evaluate the line integral

$$\int_C (e^{-y} - 2x)\,dx - (xe^{-y} + \sin y)\,dy$$

if C is the first quadrant arc of the circle $\mathbf{R}(t) = \pi\cos t\mathbf{i} + \pi\sin t\mathbf{j}$, and $0 \le t \le \frac{1}{2}\pi$.

Solution In Example 2, Sec. 19.5, we showed that

$$\nabla(xe^{-y} - x^2 + \cos y) = (e^{-y} - 2x)\mathbf{i} - (xe^{-y} + \sin y)\mathbf{j}$$

Therefore we apply Theorem 19.7.1 with $\phi(x, y) = xe^{-y} - x^2 + \cos y$.

$$\begin{aligned}\int_C (e^{-y} - 2x)\,dx - (xe^{-y} + \sin y)\,dy &= \phi(0, \pi) - \phi(\pi, 0)\\ &= \cos\pi - (\pi - \pi^2 + 1)\\ &= \pi^2 - \pi - 2\end{aligned}$$

If the value of a line integral is independent of the path, it is not necessary to find a potential function ϕ. We show the procedure in the next example.

Example 2 Show that the value of the following line integral is independent of the path and evaluate it:

$$\int_C \frac{1}{y}\,dx - \frac{x}{y^2}\,dy$$

C is any sectionally smooth curve from the point $(5, -1)$ to the point $(9, -3)$.

Solution Let $M(x, y) = 1/y$ and $N(x, y) = -x/y^2$. Then

$$M_y(x, y) = -\frac{1}{y^2} \quad \text{and} \quad N_x(x, y) = -\frac{1}{y^2}$$

Because $M_y(x, y) = N_x(x, y)$, it follows from Theorem 19.5.1 that $(1/y)\mathbf{i} - (x/y^2)\mathbf{j}$ is a gradient, and therefore the value of the integral is independent of the path. We take for the path the line segment from $(5, -1)$ to $(9, -3)$. An equation of the line is $x + 2y = 3$. Thus parametric equations of the line are

$$x = 3 + 2t \qquad y = -t \qquad 1 \le t \le 3$$

We compute the value of the line integral by applying Definition 19.6.2.

$$\begin{aligned}\int_C \frac{1}{y}\,dx - \frac{x}{y^2}\,dy &= \int_1^3 \langle M((3 + 2t), -t), N((3 + 2t), -t)\rangle \cdot \langle 2, -1\rangle\,dt \\ &= \int_1^3 \left\langle -\frac{1}{t}, -\frac{3 + 2t}{t^2}\right\rangle \cdot \langle 2, -1\rangle\,dt \\ &= \int_1^3 \left(-\frac{2}{t} + \frac{3 + 2t}{t^2}\right) dt \\ &= \int_1^3 \frac{3}{t^2}\,dt \\ &= -\frac{3}{t}\Big]_1^3 \\ &= 2\end{aligned}$$

The results of this section can be extended to functions of three variables. The statement of the following theorem and its proof are analogous to Theorem 19.7.1. The proof is left as an exercise (see Exercise 31).

19.7.2 Theorem Suppose that M, N, and R are functions of three variables x, y, and z defined on an open ball $B((x_0, y_0, z_0); r)$ in R^3; M_y, M_z, N_x, N_z, R_x, and R_y are continuous on B; and

$$\nabla\phi(x, y, z) = M(x, y, z)\mathbf{i} + N(x, y, z)\mathbf{j} + R(x, y, z)\mathbf{k}$$

Suppose that C is any sectionally smooth curve in B from the point (x_1, y_1, z_1) to the point (x_2, y_2, z_2). Then the line integral

$$\int_C M(x, y, z)\,dx + N(x, y, z)\,dy + R(x, y, z)\,dz$$

is independent of the path C, and

$$\int_C M(x, y, z)\,dx + N(x, y, z)\,dy + R(x, y, z)\,dz = \phi(x_2, y_2, z_2) - \phi(x_1, y_1, z_1)$$

● ILLUSTRATION 4 In Example 3, Sec. 19.5, we showed that the vector

$$(e^x \sin z + 2yz)\mathbf{i} + (2xz + 2y)\mathbf{j} + (e^x \cos z + 2xy + 3z^2)\mathbf{k}$$

is a gradient $\nabla f(x, y, z)$ and

$$f(x, y, z) = e^x \sin z + 2xyz + y^2 + z^3 + C$$

Therefore, if C is any sectionally smooth curve from $(0, 0, 0)$ to $(1, -2, \pi)$, then it follows from Theorem 19.7.2 that the line integral

$$\int_C (e^x \sin z + 2yz)\,dx + (2xz + 2y)\,dy + (e^x \cos z + 2xy + 3z^2)\,dz$$

is independent of the path and its value is

$$\begin{aligned} f(1, -2, \pi) - f(0, 0, 0) &= (e \sin \pi - 4\pi + 4 + \pi^3) - 0 \\ &= \pi^3 - 4\pi + 4 \end{aligned}$$

●

As with functions of two variables, if $\mathbf{F}(x, y, z) = \nabla\phi(x, y, z)$, then $\mathbf{F}$ is a gradient field, ϕ is a potential function, and $\phi(x, y, z)$ is the potential of $\mathbf{F}$ at (x, y, z). If $\mathbf{F}$ is a gradient field, then $\mathbf{F}$ is a conservative vector field. Also, if the function $\mathbf{F}$, defined by $\mathbf{F}(x, y, z) = M(x, y, z)\mathbf{i} + N(x, y, z)\mathbf{j} + R(x, y, z)\mathbf{k}$, is a gradient field, then the expression $M(x, y, z)\,dx + N(x, y, z)\,dy + R(x, y, z)\,dz$ is an exact differential.

Example 3 If $\mathbf{F}$ is the vector field defined by $\mathbf{F}(x, y, z) = (z^2 + 1)\mathbf{i} + 2yz\mathbf{j} + (2xz + y^2)\mathbf{k}$, prove that $\mathbf{F}$ is conservative and find a potential function.

Solution To determine if $\mathbf{F}$ is conservative we apply Theorem 19.5.2 to find out if $\mathbf{F}$ is a gradient. Let $M(x, y, z) = z^2 + 1$, $N(x, y, z) = 2yz$, and $R(x, y, z) = 2xz + y^2$. Then

$$M_y(x, y, z) = 0 \qquad M_z(x, y, z) = 2z \qquad N_x(x, y, z) = 0$$

$$N_z(x, y, z) = 2y \qquad R_x(x, y, z) = 2z \qquad R_y(x, y, z) = 2y$$

Therefore $M_y(x, y, z) = N_x(x, y, z)$, $M_z(x, y, z) = R_x(x, y, z)$, and $N_z(x, y, z) = R_y(x, y, z)$. Thus, by Theorem 19.5.2, $\mathbf{F}$ is a gradient and hence a conservative vector field. We now find a potential function ϕ such that $\mathbf{F}(x, y, z) = \nabla\phi(x, y, z)$; hence

$$\phi_x(x, y, z) = z^2 + 1 \qquad \phi_y(x, y, z) = 2yz \qquad \phi_z(x, y, z) = 2xz + y^2 \tag{9}$$

Integrating with respect to x both members of the first of Eqs. (9), we have

$$\phi(x, y, z) = xz^2 + x + g(y, z) \tag{10}$$

We now partial differentiate with respect to y both members of Eq. (10) and obtain

$$\phi_y(x, y, z) = g_y(y, z) \tag{11}$$

Equating the right members of (11) and the second of Eqs. (9), we have

$$g_y(y, z) = 2yz$$

We now integrate with respect to y both members of this equation and get

$$g(y, z) = y^2 z + h(z) \tag{12}$$

Substituting from (12) into (10) we obtain

$$\phi(x, y, z) = xz^2 + x + y^2 z + h(z) \tag{13}$$

We partial differentiate with respect to z both members of (13) and get

$$\phi_z(x, y, z) = 2xz + y^2 + h'(z) \tag{14}$$

We equate the right members of (14) and the third of Eqs. (9) and have

$$2xz + y^2 + h'(z) = 2xz + y^2$$

$$h'(z) = 0$$

$$h(z) = C \tag{15}$$

Substituting from (15) into (13) we obtain the required potential function ϕ such that

$$\phi(x, y, z) = xz^2 + x + y^2 z + C$$

EXERCISES 19.7

In Exercises 1 through 10, prove that the given vector field is conservative and find a potential function.

1. $\mathbf{F}(x, y) = y\mathbf{i} + x\mathbf{j}$ **2.** $\mathbf{F}(x, y) = x\mathbf{i} + y\mathbf{j}$ **3.** $\mathbf{F}(x, y) = e^x \sin y\mathbf{i} + e^x \cos y\mathbf{j}$

4. $\mathbf{F}(x, y) = (\sin y \sinh x + \cos y \cosh x)\mathbf{i} + (\cos y \cosh x - \sin y \sinh x)\mathbf{j}$

5. $\mathbf{F}(x, y) = (2xy^2 - y^3)\mathbf{i} + (2x^2y - 3xy^2 + 2)\mathbf{j}$

6. $\mathbf{F}(x, y) = (3x^2 + 2y - y^2e^x)\mathbf{i} + (2x - 2ye^x)\mathbf{j}$

7. $\mathbf{F}(x, y, z) = (x^2 - y)\mathbf{i} - (x - 3z)\mathbf{j} + (z + 3y)\mathbf{k}$

8. $\mathbf{F}(x, y, z) = (2y^3 - 8xz^2)\mathbf{i} + (6xy^2 + 1)\mathbf{j} - (8x^2z + 3z^2)\mathbf{k}$

9. $\mathbf{F}(x, y, z) = (2x \cos y - 3)\mathbf{i} - (x^2 \sin y + z^2)\mathbf{j} - (2yz - 2)\mathbf{k}$

10. $\mathbf{F}(x, y, z) = (\tan y + 2xy \sec z)\mathbf{i} + (x \sec^2 y + x^2 \sec z)\mathbf{j} + \sec z(x^2y \tan z - \sec z)\mathbf{k}$

In Exercises 11 through 20, use the results of the indicated exercise to prove that the value of the given line integral is independent of the path. Then evaluate the line integral by applying either Theorem 19.7.1 or Theorem 19.7.2 and using the potential function found in the indicated exercise. In each exercise C is any sectionally smooth curve from the point A to the point B.

11. $\int_C y\,dx + x\,dy$; A is $(1, 4)$ and B is $(3, 2)$; Exercise 1

12. $\int_C x\,dx + y\,dy$; A is $(-5, 2)$ and B is $(1, 3)$; Exercise 2

13. $\int_C e^x \sin y\,dx + e^x \cos y\,dy$; A is $(0, 0)$ and B is $(2, \frac{1}{2}\pi)$; Exercise 3

14. $\int_C (\sin y \sinh x + \cos y \cosh x)\,dx + (\cos y \cosh x - \sin y \sinh x)\,dy$; A is $(1, 0)$ and B is $(2, \pi)$; Exercise 4

15. $\int_C (2xy^2 - y^3)\,dx + (2x^2y - 3xy^2 + 2)\,dy$; A is $(-3, -1)$ and B is $(1, 2)$; Exercise 5

16. $\int_C (3x^2 + 2y - y^2e^x)\,dx + (2x - 2ye^x)\,dy$; A is $(0, 2)$ and B is $(1, -3)$; Exercise 6

17. $\int_C (x^2 - y)\,dx - (x - 3z)\,dy + (z + 3y)\,dz$; A is $(-3, 1, 2)$ and B is $(3, 0, 4)$; Exercise 7

18. $\int_C (2y^3 - 8xz^2)\,dx + (6xy^2 + 1)\,dy - (8x^2z + 3z^2)\,dz$; A is $(2, 0, 0)$ and B is $(3, 2, 1)$; Exercise 8

19. $\int_C (2x \cos y - 3)\,dx - (x^2 \sin y + z^2)\,dy - (2yz - 2)\,dz$; A is $(-1, 0, 3)$ and B is $(1, \pi, 0)$; Exercise 9

20. $\int_C (\tan y + 2xy \sec z)\,dx + (x \sec^2 y + x^2 \sec z)\,dy + \sec z(x^2y \tan z - \sec z)\,dz$; A is $(2, \frac{1}{4}\pi, 0)$ and B is $(3, \pi, \pi)$; Exercise 10

In Exercises 21 through 30, show that the value of the line integral is independent of the path and compute the value in any convenient manner. In each exercise C is any sectionally smooth curve from the point A to the point B.

21. $\int_C (2y - x)\,dx + (y^2 + 2x)\,dy$; A is $(0, -1)$ and B is $(1, 2)$

22. $\int_C (\ln x + 2y)\,dx + (e^y + 2x)\,dy$; A is $(3, 1)$ and B is $(1, 3)$

23. $\int_C \tan y\,dx + x \sec^2 y\,dy$; A is $(-2, 0)$ and B is $(4, \frac{1}{4}\pi)$

24. $\int_C \sin y\,dx + (\sin y + x \cos y)\,dy$; A is $(-2, 0)$ and B is $(2, \frac{1}{6}\pi)$

25. $\int_C \frac{2y}{(xy + 1)^2}\,dx + \frac{2x}{(xy + 1)^2}\,dy$; A is $(0, 2)$ and B is $(1, 0)$

26. $\int_C \frac{x}{x^2 + y^2 + z^2}\,dx + \frac{y}{x^2 + y^2 + z^2}\,dy + \frac{z}{x^2 + y^2 + z^2}\,dz$; A is $(1, 0, 0)$ and B is $(1, 2, 3)$

27. $\int_C (y + z)\,dx + (x + z)\,dy + (x + y)\,dz$; A is $(0, 0, 0)$ and B is $(1, 1, 1)$

28. $\int_C (yz + x)\,dx + (xz + y)\,dy + (xy + z)\,dz$; A is $(0, 0, 0)$ and B is $(1, 1, 1)$

29. $\int_C (e^x \sin y + yz)\,dx + (e^x \cos y + z \sin y + xz)\,dy + (xy - \cos y)\,dz$; A is $(2, 0, 1)$ and B is $(0, \pi, 3)$

30. $\int_C (2x \ln yz - 5ye^x)\,dx - (5e^x - x^2y^{-1})\,dy + (x^2z^{-1} + 2z)\,dz$; A is $(2, 1, 1)$ and B is $(3, 1, e)$

31. Prove Theorem 19.7.2.

REVIEW EXERCISES FOR CHAPTER 19

In Exercises 1 through 5, find the value of the directional derivative at the particular point P_0 for the given function in the direction of **U**.

1. $f(x, y) = 3x^2 - 2xy + 1$; $\mathbf{U} = \frac{3}{5}\mathbf{i} - \frac{4}{5}\mathbf{j}$; $P_0 = (5, 10)$
2. $g(x, y) = \tan^{-1}\dfrac{y}{x}$; $\mathbf{U} = \dfrac{2}{\sqrt{13}}\mathbf{i} - \dfrac{3}{\sqrt{13}}\mathbf{j}$; $P_0 = (4, -4)$
3. $h(x, y) = e^x + y^2 \cos x$; $\mathbf{U} = \frac{1}{2}\sqrt{2}\mathbf{i} - \frac{1}{2}\sqrt{2}\mathbf{j}$; $P_0 = (0, 3)$
4. $f(x, y) = x^2 - 2x^2y + \ln x$; $\mathbf{U} = \cos \pi\mathbf{i} + \sin \pi\mathbf{j}$; $P_0 = (1, -2)$
5. $f(x, y, z) = xy^2z - 3xyz + 2xz^2$; $\mathbf{U} = -\frac{2}{3}\mathbf{i} + \frac{2}{3}\mathbf{j} - \frac{1}{3}\mathbf{k}$; $P_0 = (2, 1, 1)$

In Exercises 6 through 9, find (a) the gradient of f at P_0; (b) the rate of change of the function value in the direction of **U** *at P_0.*

6. $f(x, y) = 3x^2 - 2xy^3$; $\mathbf{U} = \cos \frac{1}{6}\pi\mathbf{i} + \sin \frac{1}{6}\pi\mathbf{j}$; $P_0 = (-3, 1)$
7. $f(x, y) = \frac{1}{2}\ln(x^2 + y^2)$; $\mathbf{U} = \frac{1}{2}\mathbf{i} + \frac{1}{2}\sqrt{3}\mathbf{j}$; $P_0 = (1, 1)$
8. $f(x, y, z) = yz - y^2 - xz$; $\mathbf{U} = \frac{6}{7}\mathbf{i} + \frac{3}{7}\mathbf{j} + \frac{2}{7}\mathbf{k}$; $P_0 = (1, 2, 3)$
9. $f(x, y, z) = x^3 + y^3 + 2xyz$; $\mathbf{U} = \dfrac{3}{\sqrt{14}}\mathbf{i} - \dfrac{2}{\sqrt{14}}\mathbf{j} + \dfrac{1}{\sqrt{14}}\mathbf{k}$; $P_0 = (2, -1, 0)$

In Exercises 10 through 13, determine if the vector is a gradient. If it is, then find a function having the given gradient.

10. $(e^x \tan y - \sec y)\mathbf{i} - \sec y(x \tan y - e^x \sec y)\mathbf{j}$
11. $2xe^{x^2} \ln y\mathbf{i} + \dfrac{e^{x^2}}{y}\mathbf{j}$
12. $y(\cos x - z \sin x)\mathbf{i} + z(\cos x + \sin y)\mathbf{j} - (\cos y - y \cos x)\mathbf{k}$
13. $\left(\dfrac{-y}{(x+z)^2} + \dfrac{1}{x^2}\right)\mathbf{i} + \dfrac{1}{x+z}\mathbf{j} + \left(\dfrac{-y}{(x+z)^2} + \dfrac{2}{z^2}\right)\mathbf{k}$

In Exercises 14 through 16, find an equation of the tangent plane and equations of the normal line to the given surface at the indicated point.

14. $3x^2 + 2xy - y^2 = 15$; $(2, 3, 4)$
15. $x^2 + 2y + z = 8$; $(2, 1, 2)$
16. $z = x^2 + 2xy$; $(1, 3, 7)$
17. Find symmetric equations of the tangent line to the curve of intersection of the surfaces $x^2 - 3xy + y^2 = z$ and $2x^2 + y^2 - 3z + 27 = 0$ at the point $(1, -2, 11)$.
18. Find equations of the tangent line to the curve of intersection of the surface $z = 3x^2 + y^2 + 1$ with the plane $x = 2$ at the point $(2, -1, 14)$.
19. An equation of the surface of a mountain is $z = 900 - 3xy$, where distance is measured in meters, the x axis points to the west, and the y axis points to the south. A mountain climber is at the point corresponding to (50, 4, 300). (a) What is the direction of steepest ascent? (b) Is the climber ascending or descending when moving in the north direction? (c) In what direction is the climber traveling a level path?
20. If $f(x, y, z) = \sinh(x + z) \cosh y$, find the rate of change of $f(x, y, z)$ with respect to distance in R^3 at the point $P(1, 1, 0)$ in the direction of $\overrightarrow{PQ}$ if Q is the point $(-1, 0, 2)$.

In Exercises 21 and 22, determine the relative extrema of f, if there are any.

21. $f(x, y) = x^3 + y^3 + 3xy$ **22.** $f(x, y) = 2x^2 - 3xy + 2y^2 + 10x - 11y$

In Exercises 23 through 26, use the method of Lagrange multipliers to find the critical point(s) of the given function subject to the indicated constraint. Determine if the function has a relative maximum or a relative minimum value at any critical point.

23. $f(x, y) = 5 + x^2 - y^2$ with constraint $x^2 - 2y^2 = 5$. **24.** $f(x, y, z) = x^2 + y^2 + z^2$ with constraint $x^2 - y^2 = 1$.

25. $f(x, y, z) = y + xz - 2x^2 - y^2 - z^2$ with constraint $z = 35 - x - y$.

26. $f(x, y, z) = xz^2 + y^3$ with constraint $x^2 + y^2 + z^2 = 1$.

27. Use the method of Lagrange multipliers to find the shortest distance from the point (4, 1, 2) to the plane $x - y + 2z = 0$.

28. Use the method of Lagrange multipliers to find the point on the surface $z = x^2 + y^2 + 2$ that is closest to the origin.

In Exercises 29 through 32, evaluate the line integral over the given curve.

29. $\int_C (2x + 3y)\,dx + xy\,dy$; C: $\mathbf{R}(t) = 4\sin t\mathbf{i} - \cos t\mathbf{j}$; $0 \le t \le \frac{1}{2}\pi$

30. $\int_C (2x + y)\,dx + (x - 2y)\,dy$; C: $x^2 + y^2 = 9$

31. $\int_C y^2\,dx + z^2\,dy + x^2\,dz$; C: $\mathbf{R}(t) = (t - 1)\mathbf{i} + (t + 1)\mathbf{j} + t^2\mathbf{k}$; $0 \le t \le 1$

32. $\int_C xe^y\,dx - xe^z\,dy + e^z\,dz$; C: $\mathbf{R}(t) = t\mathbf{i} + t^2\mathbf{j} + t^3\mathbf{k}$; $0 \le t \le 1$

In Exercises 33 through 36, find the total work done in moving an object along the given arc C if the motion is caused by the given force field. Assume the arc is measured in meters and the force is measured in newtons.

33. $F(x, y) = 2x^2y\mathbf{i} + (x^2 + 3y)\mathbf{j}$; C: the arc of the parabola $y = 3x^2 + 2x + 4$ from (0, 4) to (1, 9).

34. $F(x, y) = xy^2\mathbf{i} - x^2y\mathbf{j}$; C: the arc of the circle $x^2 + y^2 = 4$ from (2, 0) to (0, 2).

35. $F(x, y, z) = (xy - z)\mathbf{i} + y\mathbf{j} + z\mathbf{k}$; C: the line segment from the origin to the point (4, 1, 2).

36. $F(x, y, z) = xyz\mathbf{i} + e^y\mathbf{j} + (x + z)\mathbf{k}$; C: $\mathbf{R}(t) = 3t\mathbf{i} + t^2\mathbf{j} + 2t\mathbf{k}$; $0 \le t \le 3$.

In Exercises 37 through 40, prove that the value of the given line integral is independent of the path, and compute the value in any convenient manner. In each exercise C is any sectionally smooth curve from the point A to the point B.

37. $\int_C 2xe^y\,dx + x^2e^y\,dy$; A is (1, 0) and B is (3, 2).

38. $\int_C \left(\frac{1}{y} - y\right)dx + \left(-\frac{x}{y^2} - x\right)dy$; A is (0, 1) and B is (6, 3).

39. $\int_C 3y\,dx + (3x + 4y)\,dy - 2z\,dz$; A is (0, 1, −1) and B is (1, 2, 0).

40. $\int_C z\sin y\,dx + xz\cos y\,dy + x\sin y\,dz$; A is (0, 0, 0) and B is $(2, 3, \frac{1}{2}\pi)$.

41. Find the dimensions of the rectangular parallelepiped of greatest volume that can be inscribed in the ellipsoid $x^2 + 9y^2 + z^2 = 9$. Assume that the edges are parallel to the coordinate axes.

42. Find three numbers whose sum is 100 and the sum of whose squares is least.

43. The temperature is T degrees at any point on a heated circular plate and

$$T = \frac{44}{x^2 + y^2 + 9}$$

where distance is measured in centimeters from the origin at the center of the plate. (a) Find the rate of change of the temperature at the point (3, 2) in the direction of the vector $\cos \frac{1}{6}\pi\mathbf{i} + \sin \frac{1}{6}\pi\mathbf{j}$. (b) Find the direction and magnitude of the greatest rate of change of T at the point (3, 2).

44. The temperature is T degrees at any point (x, y) of the curve $4x^2 + 12y^2 = 1$, and $T = 4x^2 + 24y^2 - 2x$. Find the points on the curve where the temperature is the greatest and where it is the least. Also find the temperature at these points.

45. If $f(x, y, z) = xyz$, find the maximum and minimum values of f subject to the constraints $y = z$ and $x^2 + y^2 = 1$.

In Exercises 46 and 47, prove that the given vector field is conservative and find a potential function.

46. $\mathbf{F}(x, y) = (y \sin x - \sin y)\mathbf{i} - (x \cos y + \cos x)\mathbf{j}$

47. $\mathbf{F}(x, y, z) = z^2 \sec^2 x\mathbf{i} + 2ye^{3z}\mathbf{j} + (3y^2e^{3z} + 2z \tan x)\mathbf{k}$

48. A piece of wire L ft long is cut into three pieces. One piece is bent into the shape of a circle, a second piece is bent into the shape of a square, and the third piece is bent into the shape of an equilateral triangle. How should the wire be cut so that (a) the combined area of the three figures is as small as possible and (b) the combined area of the three figures is as large as possible?

49. Determine the relative dimensions of a rectangular box, without a top and having a specific surface area, if the volume is to be a maximum.

50. Find the greatest and least distances from the origin to the curve of intersection of the surfaces $x^2 = 2yz$ and $x^2 + 3y^2 + 2z^2 = 30$.

51. Find the value of the line integral

$$\int_C \frac{-y}{x^2 + y^2}\,dx + \frac{x}{x^2 + y^2}\,dy$$

if C is the arc of the circle $x^2 + y^2 = 4$ from $(\sqrt{2}, \sqrt{2})$ to $(-\sqrt{2}, \sqrt{2})$.

CHAPTER 20 | Multiple Integration

20.1 THE DOUBLE INTEGRAL

The definite integral of a function of a single variable can be extended to a function of several variables. We call an integral of a function of a single variable a **single integral** to distinguish it from a *multiple integral*, which involves a function of several variables. There are physical and geometric applications of multiple integrals similar to those given in Chapters 7 and 11 for single integrals.

In the discussion of a single integral we required that the function be defined on a closed interval in R^1. For the *double integral* of a function of two variables we require that the function be defined on a closed region in R^2. A closed region is one that includes its boundary. In this chapter, when we refer to a region, it is assumed to be closed. The simplest kind of closed region in R^2 is a closed rectangle, which we proceed to define. Consider two distinct points $A(a_1, a_2)$ and $B(b_1, b_2)$ such that $a_1 \leq b_1$ and $a_2 \leq b_2$. These two points determine a rectangle having sides parallel to the coordinate axes. Refer to Fig. 20.1.1. The two points, together with the points (b_1, a_2) and (a_1, b_2), are called the *vertices* of the rectangle. The line segments joining consecutive vertices are called the *edges* of the rectangle. The set of all points interior to the rectangle is called an *open rectangle,* and the set of all points in the open rectangle, together with the points on the edges of the rectangle, is called a *closed rectangle.*

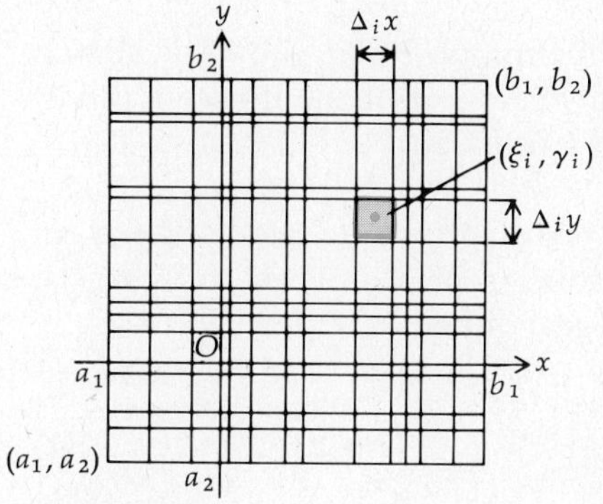

FIGURE 20.1.1

Let the closed rectangular region of Fig. 20.1.1 be denoted by R, and let f be a function defined on R. The region R can be considered as a **region of integration.** The first step is to define a **partition,** Δ, of R. We draw lines parallel to the coordinate axes and obtain a network of rectangular subregions that cover R. The **norm** of this partition, denoted by $\|\Delta\|$, is determined by the length of the longest diagonal of a rectangular subregion of the partition. Number the subregions in some arbitrary way and let the total be n. Denote the width of the ith subregion by $\Delta_i x$ units and its height by $\Delta_i y$ units. Then if $\Delta_i A$ square units is the area of the ith rectangular subregion,

$$\Delta_i A = \Delta_i x \, \Delta_i y$$

Let (ξ_i, γ_i) be an arbitrary point in the ith subregion and $f(\xi_i, \gamma_i)$ be the function value there. Consider the product $f(\xi_i, \gamma_i)\, \Delta_i A$. Associated with each of the n subregions is such a product, and their sum is

$$\sum_{i=1}^{n} f(\xi_i, \gamma_i)\, \Delta_i A \tag{1}$$

There are many sums of the form (1) because the norm of the partition can be any positive number and each point (ξ_i, γ_i) can be any point in the ith subregion. If all such sums can be made arbitrarily close to one number L by taking partitions with sufficiently small norms, then L is defined to be the limit of these sums as the norm of the partition of R approaches zero. We have the following definition.

20.1.1 Definition Let f be a function defined on a closed rectangular region R. The number

L is said to be the **limit** of sums of the form $\sum_{i=1}^{n} f(\xi_i, \gamma_i)\,\Delta_i A$ if L satisfies the property that for any $\epsilon > 0$ there exists a $\delta > 0$ such that

$$\left| \sum_{i=1}^{n} f(\xi_i, \gamma_i)\,\Delta_i A - L \right| < \epsilon$$

for every partition Δ for which $\|\Delta\| < \delta$ and for all possible selections of the point (ξ_i, γ_i) in the ith rectangle, $i = 1, 2, \ldots, n$. If such a number L exists, we write

$$\lim_{\|\Delta\| \to 0} \sum_{i=1}^{n} f(\xi_i, \gamma_i)\,\Delta_i A = L$$

If there is a number L satisfying Definition 20.1.1, it can be shown that it is unique. The proof is similar to the proof of the theorem (2.1.2) regarding the uniqueness of the limit of a function.

20.1.2 Definition A function f of two variables is said to be **integrable** on a rectangular region R if f is defined on R and the number L of Definition 20.1.1 exists. This number L is called the **double integral** of f on R, and we write

$$\lim_{\|\Delta\| \to 0} \sum_{i=1}^{n} f(\xi_i, \gamma_i)\,\Delta_i A = \iint_R f(x, y)\, dA \tag{2}$$

Other symbols for the double integral in (2) are

$$\iint_R f(x, y)\, dx\, dy \quad \text{and} \quad \iint_R f(x, y)\, dy\, dx$$

The following theorem, stated without proof, gives a sufficient condition for which a function of two variables is integrable.

20.1.3 Theorem If a function of two variables is continuous on a closed rectangular region R, then it is integrable on R.

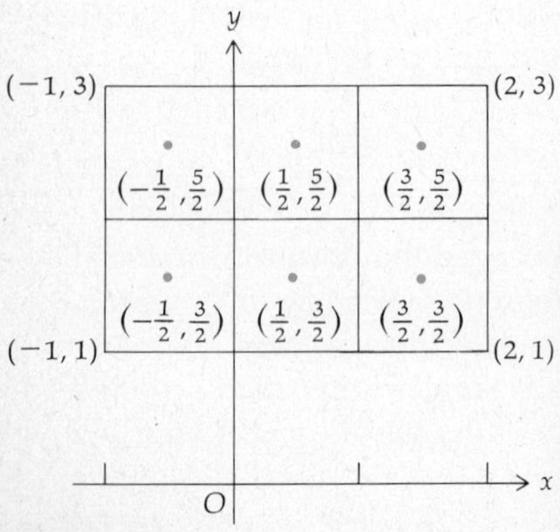

FIGURE 20.1.2

The approximation of the value of a double integral is shown in the following example.

Example 1 Find an approximate value of the double integral $\iint_R (2x^2 - 3y)\, dA$, where R is the rectangular region having vertices $(-1, 1)$ and $(2, 3)$. Take a partition of R formed by the lines $x = 0$, $x = 1$, and $y = 2$, and take (ξ_i, γ_i) at the center of the ith subregion.

Solution Refer to Fig. 20.1.2, which shows the region R partitioned into six subregions that are squares having sides one unit in length. So for each i, $\Delta_i A = 1$. In each of the subregions the point (ξ_i, γ_i) is at the center of the square. With $f(x, y) = 2x^2 - 3y$, an approximation to the given double integral is given by

$$\iint_R (2x^2 - 3y)\, dA \approx f(-\tfrac{1}{2}, \tfrac{3}{2}) \cdot 1 + f(\tfrac{1}{2}, \tfrac{3}{2}) \cdot 1 + f(\tfrac{3}{2}, \tfrac{3}{2}) \cdot 1$$

$$+ f(\tfrac{3}{2}, \tfrac{5}{2}) \cdot 1 + f(\tfrac{1}{2}, \tfrac{5}{2}) \cdot 1 + f(-\tfrac{1}{2}, \tfrac{5}{2}) \cdot 1$$

$$= -4 \cdot 1 - 4 \cdot 1 + 0 \cdot 1 - 3 \cdot 1 - 7 \cdot 1 - 7 \cdot 1$$

$$= -25$$

The exact value of the double integral in Example 1 is -24, as will be shown in Example 1 of Sec. 20.2.

We now consider the double integral of a function over a more general region. In Definition 7.6.2 a smooth function was defined as one that has a continuous derivative, and a smooth curve is the graph of a smooth function. Let R be a closed region whose boundary consists of a finite number of arcs of smooth curves that are joined together to form a closed curve. As we did with a rectangular region, draw lines parallel to the coordinate axes, which gives a rectangular partition of the region R. Discard the subregions that contain points not in R and consider only those that lie entirely in R (these are shaded in Fig. 20.1.3). Let the number of these shaded subregions be n and proceed in a manner analogous to the procedure used for a rectangular region. Definitions 20.1.1 and 20.1.2 apply when the region R is the more general one described above. You should intuitively realize that as the norm of the partition approaches zero, n increases without bound, and the area of the region omitted (i.e., the discarded rectangles) approaches zero. Actually, it can be proved that if a function is integrable on a region R, the limit of the approximating sums of the form (1) is the same no matter how we subdivide R, so long as each subregion has a shape to which an area can be assigned.

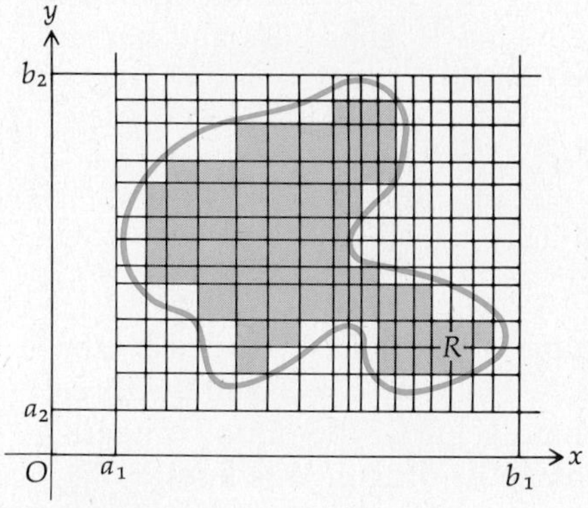

FIGURE 20.1.3

Just as the integral of a function of a single variable is interpreted geometrically as the measure of the area of a plane region, the double integral can be interpreted geometrically as the measure of the volume of a three-dimensional solid. Suppose that the function f is continuous on a closed region R in R^2. Furthermore, for simplicity in this discussion, assume that $f(x, y)$ is nonnegative on R. The graph of the equation $z = f(x, y)$ is a surface lying above the xy plane, as shown in Fig. 20.1.4. The figure shows a particular rectangular subregion of R, having dimensions of measures $\Delta_i x$ and $\Delta_i y$. The figure also shows a rectangular solid having this subregion as a base and $f(\xi_i, \gamma_i)$ as the measure of the altitude, where (ξ_i, γ_i) is a point in the ith subregion. The volume of the rectangular solid is determined by

$$\Delta_i V = f(\xi_i, \gamma_i)\, \Delta_i A = f(\xi_i, \gamma_i)\, \Delta_i x\, \Delta_i y \tag{3}$$

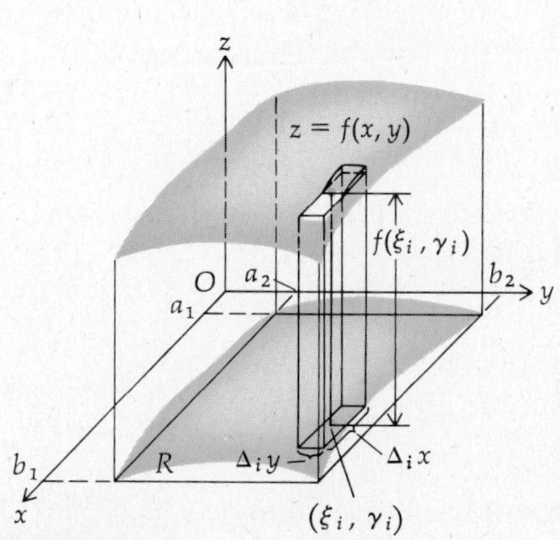

FIGURE 20.1.4

The number given in (3) is the measure of the volume of the thin rectangular solid shown in Fig. 20.1.4; thus the sum given in (1) is the sum of the measures of the volumes of n such solids. This sum approximates the measure of the volume of the three-dimensional solid shown in Fig. 20.1.4. The solid is bounded above by the graph of f and below by the region R in the xy plane. The sum in (1) also approximates the number given by the double integral

$$\iint_R f(x, y)\, dA$$

It can be proved that the volume of the three-dimensional solid of Fig. 20.1.4 is the value of the double integral. This fact is stated in the following theorem, for which a formal proof is not given.

20.1.4 Theorem Let f be a function of two variables that is continuous on a closed region R in the xy plane and $f(x, y) \geq 0$ for all (x, y) in R. If $V(S)$ is the measure of the volume of the solid S having the region R as its base and having an altitude of measure $f(x, y)$ at the point (x, y) in R, then

$$V(S) = \lim_{\|\Delta\| \to 0} \sum_{i=1}^{n} f(\xi_i, \gamma_i)\, \Delta_i A = \iint_R f(x, y)\, dA$$

Example 2 Approximate the volume of the solid bounded by the surface $f(x, y) = 4 - \frac{1}{9}x^2 - \frac{1}{16}y^2$, the planes $x = 3$ and $y = 2$, and the three coordinate planes. To find an approximate value of the double integral take a partition of the region in the xy plane by drawing the lines $x = 1$, $x = 2$, and $y = 1$, and take (ξ_i, γ_i) at the center of the ith subregion.

Solution The solid is shown in Fig. 20.1.5. The rectangular region R is the rectangle in the xy plane bounded by the coordinate axes and the lines $x = 3$ and $y = 2$. From Theorem 20.1.4, if V cubic units is the volume of the solid,

$$V = \iint_R \left(4 - \tfrac{1}{9}x^2 - \tfrac{1}{16}y^2\right) dA$$

Figure 20.1.5 shows R partitioned into six subregions that are squares having sides of length one unit. Therefore, for each i, $\Delta_i A = 1$. The point (ξ_i, γ_i) in each subregion is at the center of the square. Then an approximation of V is given by an approximation of the double integral. Therefore

$$\begin{aligned} V &\approx f(\tfrac{1}{2}, \tfrac{1}{2}) \cdot 1 + f(\tfrac{3}{2}, \tfrac{1}{2}) \cdot 1 + f(\tfrac{5}{2}, \tfrac{1}{2}) \cdot 1 + f(\tfrac{1}{2}, \tfrac{3}{2}) \cdot 1 + f(\tfrac{3}{2}, \tfrac{3}{2}) \cdot 1 + f(\tfrac{5}{2}, \tfrac{3}{2}) \cdot 1 \\ &= (4 - \tfrac{25}{576}) + (4 - \tfrac{17}{64}) + (4 - \tfrac{409}{576}) + (4 - \tfrac{97}{576}) + (4 - \tfrac{25}{64}) + (4 - \tfrac{481}{576}) \\ &= 24 - \tfrac{695}{288} \\ &\approx 21.59 \end{aligned}$$

Thus the volume is approximately 21.59 cubic units.

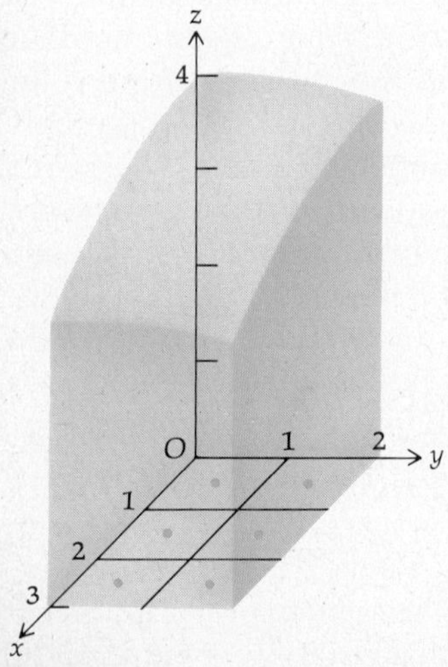

FIGURE 20.1.5

The exact value of the volume in Example 2 is shown to be 21.5 cubic units in Example 2 of Sec. 20.2.

Analogous to properties of the definite integral of a function of a single variable are several properties of the double integral, and the most important ones are given in the following theorems.

20.1.5 Theorem If c is a constant and the function f is integrable on a closed region R, then cf is integrable on R and

$$\iint_R cf(x, y)\, dA = c \iint_R f(x, y)\, dA$$

20.1.6 Theorem If the functions f and g are integrable on a closed region R, then the function $f + g$ is integrable on R and

$$\iint_R [f(x, y) + g(x, y)]\, dA = \iint_R f(x, y)\, dA + \iint_R g(x, y)\, dA$$

The result of Theorem 20.1.6 can be extended to any finite number of functions that are integrable. The proofs of Theorems 20.1.5 and 20.1.6 follow directly from the definition of a double integral. These proofs are left as exercises (see Exercises 23 and 24).

20.1.7 Theorem If the functions f and g are integrable on the closed region R and $f(x, y) \geq g(x, y)$ for all (x, y) in R, then

$$\iint_R f(x, y)\, dA \geq \iint_R g(x, y)\, dA$$

Theorem 20.1.7 is analogous to Theorem 6.3.8 for the definite integral of a function of a single variable. The proof is similar and is left as an exercise (see Exercise 25).

20.1.8 Theorem Let the function f be integrable on a closed region R, and suppose that m and M are two numbers such that $m \leq f(x, y) \leq M$ for all (x, y) in R. Then if A is the measure of the area of region R,

$$mA \leq \iint_R f(x, y)\, dA \leq MA$$

The proof of Theorem 20.1.8 is left as an exercise (see Exercise 26). The proof is similar to that of Theorem 6.3.9 and is based on Theorem 20.1.7.

20.1.9 Theorem Suppose that the function f is continuous on the closed region R and that region R is composed of the two subregions R_1 and R_2 that have no points in common except for points on parts of their boundaries. Then

$$\iint_R f(x, y)\, dA = \iint_{R_1} f(x, y)\, dA + \iint_{R_2} f(x, y)\, dA$$

The proof of Theorem 20.1.9 is also left as an exercise and depends on the definition of a double integral and limit theorems (see Exercise 27).

EXERCISES 20.1

1. Find an approximate value of the double integral $\iint_R (3x - 2y + 1)\, dA$ where R is the rectangular region having vertices $(0, -2)$ and $(3, 0)$. Take a partition of R formed by the lines $x = 1$, $x = 2$, and $y = -1$, and take (ξ_i, γ_i) at the center of the ith subregion.

2. Find an approximate value of the double integral $\iint_R (y^2 - 4x)\, dA$ where R is the rectangular region having vertices $(-1, 0)$ and $(1, 3)$. Take a partition of R formed by the lines $x = 0$, $y = 1$, and $y = 2$, and take (ξ_i, γ_i) at the center of the ith subregion.

In Exercises 3 through 8, find an approximate value of the given double integral where R is the rectangular region having the vertices P and Q, Δ is a partition of R, and (ξ_i, γ_i) is the midpoint of each subregion.

3. $\iint_R (x^2 + y)\, dA$; $P(0, 0)$; $Q(4, 2)$; Δ: $x_1 = 0, x_2 = 1, x_3 = 2, x_4 = 3, y_1 = 0, y_2 = 1$.

4. $\iint_R (2 - x - y)\, dA$; $P(0, 0)$; $Q(6, 4)$; Δ: $x_1 = 0, x_2 = 2, x_3 = 4, y_1 = 0, y_2 = 2$.

5. $\iint_R (xy + 3y^2)\, dA$; $P(-2, 0)$; $Q(4, 6)$; Δ: $x_1 = -2, x_2 = 0, x_3 = 2, y_1 = 0, y_2 = 2, y_3 = 4$.

6. $\iint_R (xy + 3y^2)\, dA$; $P(0, -2)$; $Q(6, 4)$; Δ: $x_1 = 0, x_2 = 2, x_3 = 4, y_1 = -2, y_2 = 0, y_3 = 2$.

7. $\iint_R (x^2y - 2xy^2)\, dA$; $P(-3, -2)$; $Q(1, 6)$; Δ: $x_1 = -3, x_2 = -1, y_1 = -2, y_2 = 0, y_3 = 2, y_4 = 4$.

8. $\iint_R (x^2y - 2xy^2)\, dA$; $P(-3, -2)$; $Q(1, 6)$; Δ: $x_1 = -3, x_2 = -2, x_3 = -1, x_4 = 0, y_1 = -2, y_2 = -1, y_3 = 0, y_4 = 1,$ $y_5 = 2, y_6 = 3, y_7 = 4, y_8 = 5$.

In Exercises 9 through 12, find an approximate value of the given double integral where R is the rectangular region having the vertices P and Q, Δ is a partition of R, and (ξ_i, γ_i) is an arbitrary point in each subregion.

9. The double integral, P, Q, and Δ are the same as in Exercise 3; $(\xi_1, \gamma_1) = (\frac{1}{4}, \frac{1}{2})$; $(\xi_2, \gamma_2) = (\frac{7}{4}, 0)$; $(\xi_3, \gamma_3) = (\frac{5}{2}, \frac{1}{4})$; $(\xi_4, \gamma_4) = (4, 1)$; $(\xi_5, \gamma_5) = (\frac{3}{4}, \frac{7}{4})$; $(\xi_6, \gamma_6) = (\frac{5}{4}, \frac{3}{2})$; $(\xi_7, \gamma_7) = (\frac{5}{2}, 2)$; $(\xi_8, \gamma_8) = (3, 1)$.

10. The double integral, P, Q, and Δ are the same as in Exercise 4; $(\xi_1, \gamma_1) = (\frac{1}{2}, \frac{3}{2})$; $(\xi_2, \gamma_2) = (3, 1)$; $(\xi_3, \gamma_3) = (\frac{11}{2}, \frac{1}{2})$; $(\xi_4, \gamma_4) = (2, 2)$; $(\xi_5, \gamma_5) = (2, 2)$; $(\xi_6, \gamma_6) = (5, 3)$.

11. The double integral, P, Q, and Δ are the same as in Exercise 5; $(\xi_1, \gamma_1) = (-\frac{1}{2}, \frac{1}{2})$; $(\xi_2, \gamma_2) = (1, \frac{3}{2})$; $(\xi_3, \gamma_3) = (\frac{5}{2}, 2)$; $(\xi_4, \gamma_4) = (-\frac{3}{2}, \frac{7}{2})$; $(\xi_5, \gamma_5) = (0, 3)$; $(\xi_6, \gamma_6) = (4, 4)$; $(\xi_7, \gamma_7) = (-1, \frac{9}{2})$; $(\xi_8, \gamma_8) = (1, \frac{9}{2})$; $(\xi_9, \gamma_9) = (3, \frac{9}{2})$.

12. The double integral, P, Q, and Δ are the same as in Exercise 5; $(\xi_1, \gamma_1) = (-2, 0)$; $(\xi_2, \gamma_2) = (0, 0)$; $(\xi_3, \gamma_3) = (2, 0)$; $(\xi_4, \gamma_4) = (-2, 2)$; $(\xi_5, \gamma_5) = (0, 2)$; $(\xi_6, \gamma_6) = (2, 2)$; $(\xi_7, \gamma_7) = (-2, 4)$; $(\xi_8, \gamma_8) = (0, 4)$; $(\xi_9, \gamma_9) = (2, 4)$.

13. Approximate the volume of the solid in the first octant bounded by the sphere $x^2 + y^2 + z^2 = 64$, the planes $x = 3$, $y = 3$, and the three coordinate planes. To find an approximate value of the double integral take a partition of the region in the xy plane formed by the lines $x = 1$, $x = 2$, $y = 1$, and $y = 2$, and take (ξ_i, γ_i) at the center of the ith subregion.

14. Approximate the volume of the solid bounded by the planes $z = 2x + y + 4$, $x = 2$, $y = 3$, and the three coordinate planes. To find an approximate value of the double integral take a partition of the region in the xy plane formed by the lines $x = 1$, $y = 1$, and $y = 2$, and take (ξ_i, γ_i) at the center of the ith subregion.

15. Approximate the volume of the solid bounded by the surface $z = 10 - \frac{1}{4}x^2 - \frac{1}{9}y^2$, the planes $x = 2$, $y = 2$, and the three coordinate planes. To find an approximate value of the double integral take a partition of the region in the xy plane formed by the lines $x = 1$ and $y = 1$, and take (ξ_i, γ_i) at the center of the ith subregion.

16. Approximate the volume of the solid bounded by the surface $100z = 300 - 25x^2 - 4y^2$, the planes $x = -1$, $x = 3$, $y = -3$, $y = 5$ and the xy plane. To find an approximate value of the double integral take a partition of the region in the xy plane formed by the lines $x = 1$, $y = -1$, $y = 1$, and $y = 3$, and take (ξ_i, γ_i) at the center of the ith subregion.

In Exercises 17 through 22, apply Theorem 20.1.8 to find a closed interval containing the value of the given double integral.

17. $\iint_R (2x + 5y)\, dA$ where R is the rectangular region having vertices (0, 0), (1, 0), (1, 2), and (0, 2).

18. $\iint_R (x^2 + y^2)\, dA$ where R is the rectangular region having vertices (0, 0), (1, 0), (1, 1), and (0, 1).

19. $\iint_R e^{xy}\, dA$ where R is the rectangular region having vertices (0, 0), (1, 0), (1, 1), and (0, 1).

20. $\iint_R (\sin x + \sin y)\, dA$ where R is the rectangular region having vertices $(0, 0)$, $(\pi, 0)$, (π, π), and $(0, \pi)$. (*Hint:* Use the result of Exercise 10 in Exercises 19.3.)

21. $\iint_R [\sin(x + y) + \sin x + \sin y]\, dA$ where R is the rectangular region having vertices $(0, 0)$, $(\pi, 0)$, (π, π), and $(0, \pi)$. (*Hint:* Use the result of Exercise 11 in Exercises 19.3.)

22. $\iint_R \dfrac{2x + 2y + 1}{x^2 + y^2 + 1}\, dA$ where R is the rectangular region having vertices $(-1, -1)$, $(1, -1)$, $(1, 1)$, and $(-1, 1)$. (*Hint:* Use the result of Exercise 8 in Exercises 19.3.)

23. Prove Theorem 20.1.5. **24.** Prove Theorem 20.1.6. **25.** Prove Theorem 20.1.7.

26. Prove Theorem 20.1.8. **27.** Prove Theorem 20.1.9.

20.2 EVALUATION OF DOUBLE INTEGRALS AND ITERATED INTEGRALS

For functions of a single variable, the fundamental theorem of the calculus provides a method for evaluating a definite integral by finding an antiderivative (or indefinite integral) of the integrand. There is a corresponding method for evaluating a double integral that involves performing successive single integrations. A rigorous development of this method belongs to a course in advanced calculus. Our discussion is an intuitive one, and we use the geometric interpretation of the double integral as the measure of a volume. We first develop the method for the double integral on a rectangular region.

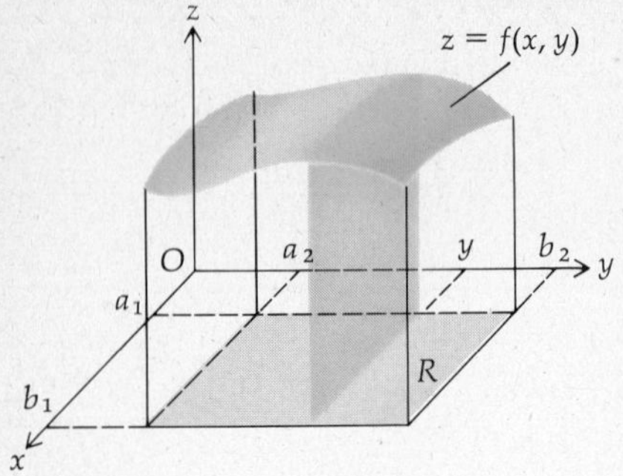

FIGURE 20.2.1

Let f be a given function that is integrable on a closed rectangular region R in the xy plane bounded by the lines $x = a_1$, $x = b_1$, $y = a_2$, and $y = b_2$. Assume that $f(x, y) \geq 0$ for all (x, y) in R. Refer to Fig. 20.2.1, which shows a sketch of the graph of the equation $z = f(x, y)$ when (x, y) is in R. The number that represents the value of the double integral

$$\iint_R f(x, y)\, dA$$

is the measure of the volume of the solid between the surface and the region R. This number can be found by the method of parallel plane sections, discussed in Sec. 7.4, as we now do.

Let y be a number in $[a_2, b_2]$. Consider the plane parallel to the xz plane through the point $(0, y, 0)$. Let $A(y)$ be the measure of the area of the plane region of intersection of this plane with the solid. By the method of parallel plane sections, the measure of the volume of the solid is expressed by

$$\int_{a_2}^{b_2} A(y)\, dy$$

Because the volume of the solid also is determined by the double integral,

$$\iint_R f(x, y)\, dA = \int_{a_2}^{b_2} A(y)\, dy \tag{1}$$

By using (1) we can find the value of the double integral of the function f on R by evaluating a single integral of $A(y)$. We now must find $A(y)$ when y is given. Because $A(y)$ is the measure of the area of a plane region, we can find it by integration. In Fig. 20.2.1, notice that the upper boundary of the plane region is the graph of the equation $z = f(x, y)$ when x is in $[a_1, b_1]$. Therefore $A(y) = \int_{a_1}^{b_1} f(x, y)\, dx$. Substituting from this equation into (1) we obtain

$$\iint_R f(x, y)\, dA = \int_{a_2}^{b_2} \left[\int_{a_1}^{b_1} f(x, y)\, dx \right] dy \tag{2}$$

The integral on the right side of (2) is called an **iterated integral.** Usually the brackets are omitted when writing an iterated integral. So (2) can be written as

$$\iint_R f(x, y)\, dA = \int_{a_2}^{b_2} \int_{a_1}^{b_1} f(x, y)\, dx\, dy \tag{3}$$

When evaluating the "inner integral" in (3), remember that x is the variable of integration and y is considered a constant. This is comparable to considering y as a constant when finding the partial derivative of $f(x, y)$ with respect to x.

By considering plane sections parallel to the yz plane, we obtain an iterated integral that interchanges the order of integration; we have

$$\iint_R f(x, y)\, dA = \int_{a_1}^{b_1} \int_{a_2}^{b_2} f(x, y)\, dy\, dx \tag{4}$$

A sufficient condition for (3) and (4) to be valid is that the function be continuous on the rectangular region R.

Example 1 Evaluate the double integral $\iint_R (2x^2 - 3y)\, dA$ if R is the region consisting of all points (x, y) for which $-1 \le x \le 2$ and $1 \le y \le 3$.

Solution $a_1 = -1$, $b_1 = 2$, $a_2 = 1$, and $b_2 = 3$. So from (3),

$$\begin{aligned}\iint_R (2x^2 - 3y)\, dA &= \int_1^3 \int_{-1}^2 (2x^2 - 3y)\, dx\, dy \\ &= \int_1^3 \left[\int_{-1}^2 (2x^2 - 3y)\, dx \right] dy \\ &= \int_1^3 \left[\tfrac{2}{3}x^3 - 3xy \right]_{-1}^2 dy \\ &= \int_1^3 (6 - 9y)\, dy \\ &= -24\end{aligned}$$

In Example 1 of Sec. 20.1 we found an approximate value of the double integral in the above example to be -25.

Example 2 Find the volume of the solid bounded by the surface $f(x, y) = 4 - \frac{1}{9}x^2 - \frac{1}{16}y^2$, the planes $x = 3$ and $y = 2$, and the three coordinate planes.

Solution Figure 20.2.2 shows the graph of the equation $z = f(x, y)$ in the first octant and the given solid. If V cubic units is the volume of the solid, then from Theorem 20.1.4,

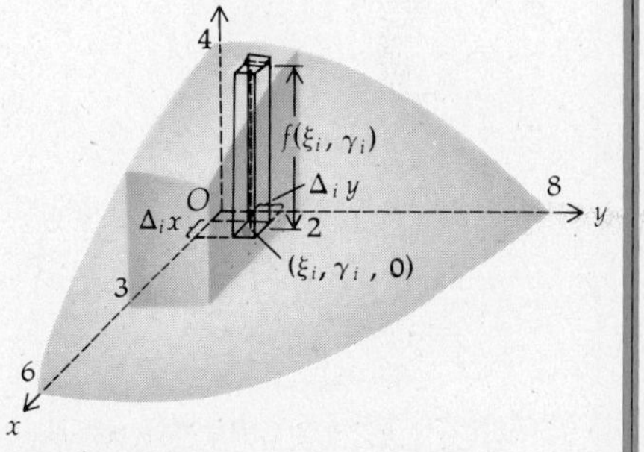

FIGURE 20.2.2

$$\begin{aligned}V &= \lim_{\|\Delta\| \to 0} \sum_{i=1}^{n} f(\xi_i, \gamma_i)\, \Delta_i A \\ &= \iint_R f(x, y)\, dA \\ &= \int_0^3 \int_0^2 (4 - \tfrac{1}{9}x^2 - \tfrac{1}{16}y^2)\, dy\, dx \\ &= \int_0^3 \left[4y - \tfrac{1}{9}x^2 y - \tfrac{1}{48}y^3 \right]_0^2 dx\end{aligned}$$

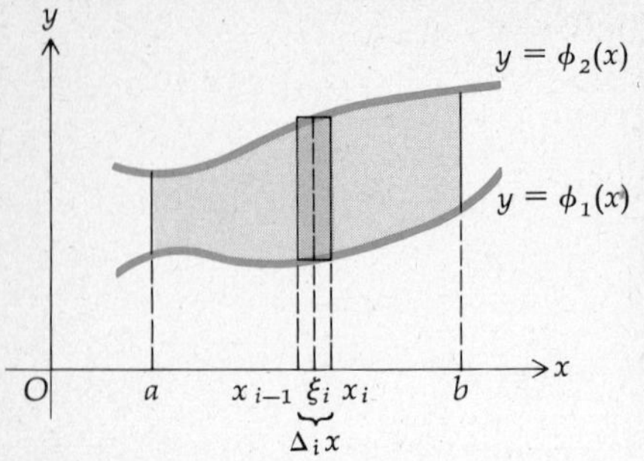

FIGURE 20.2.3

$$= \int_0^3 \left(\tfrac{47}{6} - \tfrac{2}{9}x^2\right) dx$$

$$= \tfrac{47}{6}x - \tfrac{2}{27}x^3 \Big]_0^3$$

$$= 21.5$$

The volume is therefore 21.5 cubic units.

In Example 2 of Sec. 20.1 an approximate value of the volume in Example 2 was found to be 21.59 cubic units.

Suppose now that R is a region in the xy plane that is bounded by the lines $x = a$ and $x = b$, where $a < b$, and by the curves $y = \phi_1(x)$ and $y = \phi_2(x)$, where ϕ_1 and ϕ_2 are two functions continuous on the closed interval $[a, b]$; furthermore, $\phi_1(x) \le \phi_2(x)$ whenever $a \le x \le b$ (see Fig. 20.2.3). Let Δ be a partition of the interval $[a, b]$ defined by Δ: $a = x_0 < x_1 < \cdots < x_n = b$. Consider the region R of Fig. 20.2.3 to be divided into vertical strips with widths of measure $\Delta_i x$. A particular strip is shown in the figure. The intersection of the surface $z = f(x, y)$ and a plane $x = \xi_i$, where $x_{i-1} \le \xi_i \le x_i$, is a curve. A segment of this curve is over the ith vertical strip. The region under this curve segment and above the xy plane is shown in Fig. 20.2.4, and the measure of the area of this region is given by

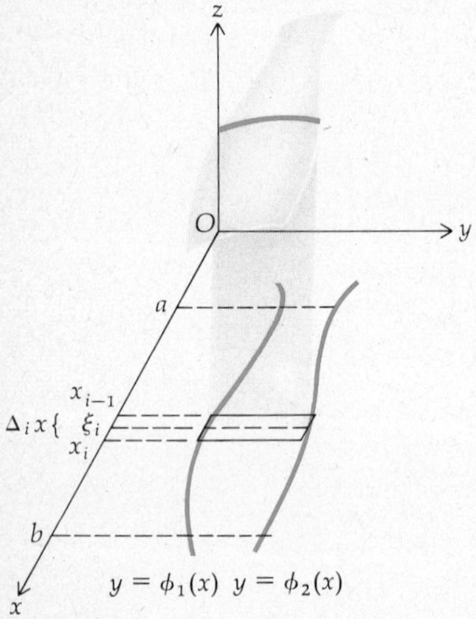

FIGURE 20.2.4

$$\int_{\phi_1(\xi_i)}^{\phi_2(\xi_i)} f(\xi_i, y)\, dy$$

The measure of the volume of the solid bounded above by the surface $z = f(x, y)$ and below by the ith vertical strip is approximately equal to

$$\left[\int_{\phi_1(\xi_i)}^{\phi_2(\xi_i)} f(\xi_i, y)\, dy\right] \Delta_i x$$

If we take the limit, as the norm of Δ approaches zero, of the sum of these measures of volume for n vertical strips of R from $x = a$ to $x = b$, we obtain the measure of the volume of the solid bounded above by the surface $z = f(x, y)$ and below by the region R in the xy plane. (See Fig. 20.2.5.) This is the double integral of f on R; that is,

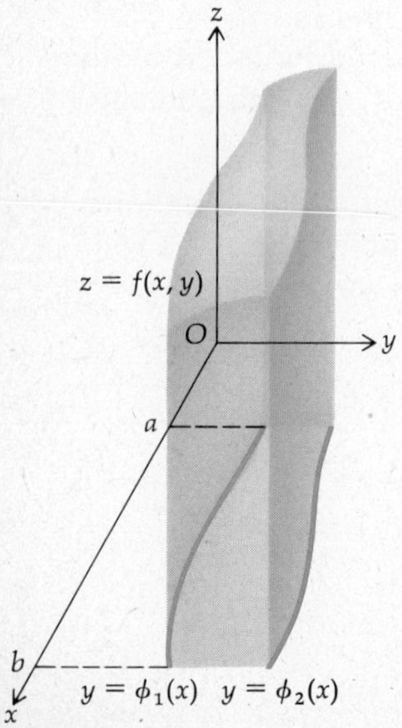

FIGURE 20.2.5

$$\lim_{\|\Delta\| \to 0} \sum_{i=1}^{n} \left[\int_{\phi_1(\xi_i)}^{\phi_2(\xi_i)} f(\xi_i, y)\, dy\right] \Delta_i x = \int_a^b \int_{\phi_1(x)}^{\phi_2(x)} f(x, y)\, dy\, dx$$

$$= \iint_R f(x, y)\, dy\, dx \qquad (5)$$

Sufficient conditions for (5) to be valid are that f be continuous on the closed region R and that ϕ_1 and ϕ_2 be smooth functions.

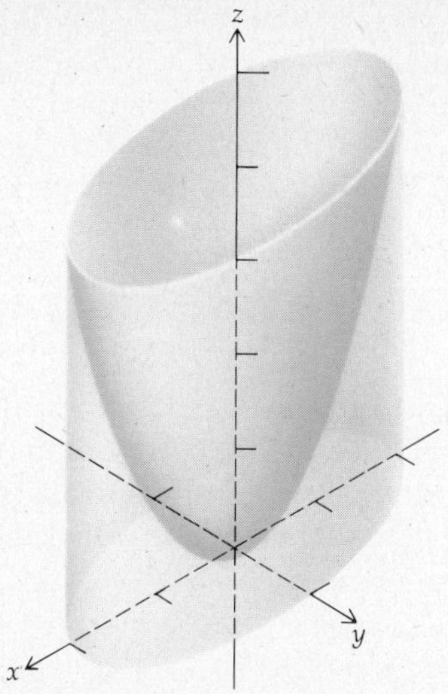

FIGURE 20.2.6

Example 3 Express as both a double integral and an iterated integral the measure of the volume of the solid above the xy plane bounded by the elliptic paraboloid $z = x^2 + 4y^2$ and the cylinder $x^2 + 4y^2 = 4$. Evaluate the iterated integral to find the volume of the solid.

Solution The solid is shown in Fig. 20.2.6. We find the volume of the portion of the solid in the first octant, which from properties of symmetry is one-fourth of the required volume. The region R in the xy plane is that bounded by the x and y axes and the ellipse $x^2 + 4y^2 = 4$. This region is shown in Fig. 20.2.7, which also shows the ith subregion of a rectangular partition of R, where (ξ_i, γ_i) is any point in this ith subregion. If V cubic units is the volume of the given solid, then by Theorem 20.1.4,

$$V = 4 \lim_{\|\Delta\| \to 0} \sum_{i=1}^{n} (\xi_i^2 + 4\gamma_i^2)\, \Delta_i A$$

$$= 4 \iint_R (x^2 + 4y^2)\, dA$$

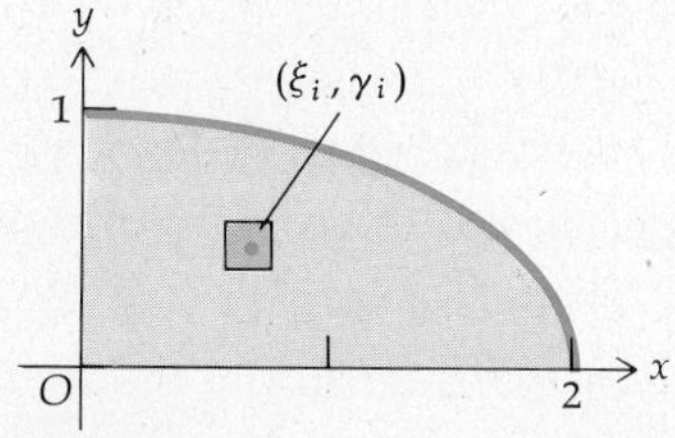

FIGURE 20.2.7

To express the measure of the volume as an iterated integral we divide the region R into n vertical strips. Figure 20.2.8 shows the region R and the ith vertical strip having width of $\Delta_i x$ units and length of $\frac{1}{2}\sqrt{4 - \xi_i^2}$ units, where $x_{i-1} \le \xi_i \le x_i$. From (5),

$$V = 4 \lim_{\|\Delta\| \to 0} \sum_{i=1}^{n} \left[\int_0^{\sqrt{4-\xi_i^2}/2} (\xi_i^2 + 4y^2)\, dy \right] \Delta_i x$$

$$= 4 \int_0^2 \int_0^{\sqrt{4-x^2}/2} (x^2 + 4y^2)\, dy\, dx$$

$$= 4 \int_0^2 \left[x^2 y + \tfrac{4}{3} y^3 \right]_0^{\sqrt{4-x^2}/2} dx$$

$$= 4 \int_0^2 \left[\tfrac{1}{2} x^2 \sqrt{4 - x^2} + \tfrac{1}{6}(4 - x^2)^{3/2}\right] dx$$

$$= \frac{4}{3} \int_0^2 (x^2 + 2)\sqrt{4 - x^2}\, dx$$

$$= -\tfrac{1}{3} x (4 - x^2)^{3/2} + 2x\sqrt{4 - x^2} + 8 \sin^{-1} \tfrac{1}{2} x \Big]_0^2$$

$$= 4\pi$$

Therefore the volume is 4π cubic units.

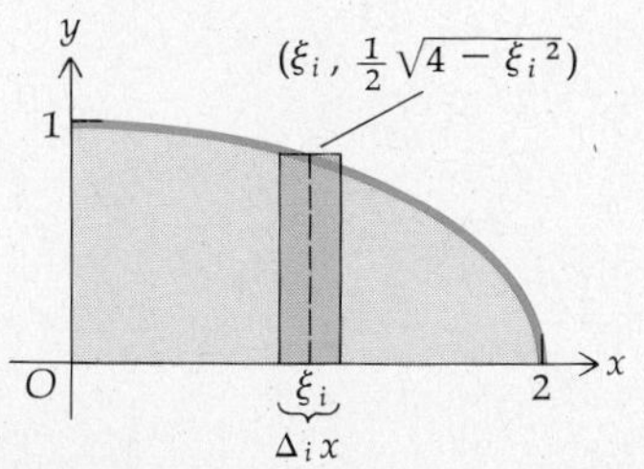

FIGURE 20.2.8

Suppose the region R is bounded by the curves $x = \lambda_1(y)$ and $x = \lambda_2(y)$ and the lines $y = c$ and $y = d$, where $c < d$, and λ_1 and λ_2 are two functions continuous on the closed interval $[c, d]$ for which $\lambda_1(y) \le \lambda_2(y)$ whenever $c \le y \le d$. Consider a partition Δ of the interval $[c, d]$ and divide the region into horizontal strips, the measures of whose widths

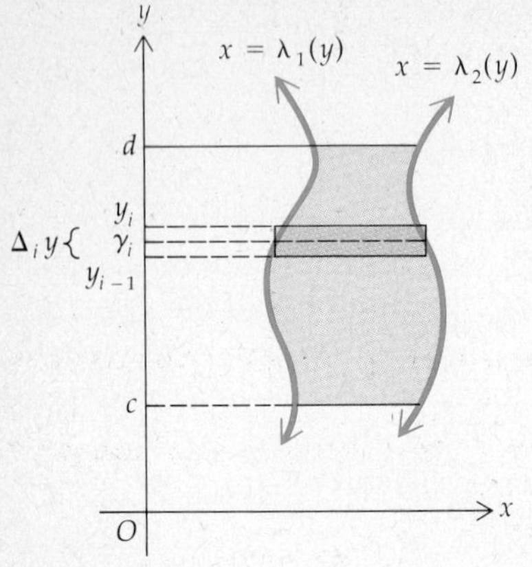

FIGURE 20.2.9

are $\Delta_i y$. See Fig. 20.2.9, which shows the ith horizontal strip. The intersection of the surface $z = f(x, y)$ and a plane $y = \gamma_i$, where $y_{i-1} \leq \gamma_i \leq y_i$, is a curve, and a segment of this curve is over the ith horizontal strip. Then, as in the derivation of (5), the measure of the volume of the solid bounded above by the surface $z = f(x, y)$ and below by the ith vertical strip is approximately equal to

$$\left[\int_{\lambda_1(\gamma_i)}^{\lambda_2(\gamma_i)} f(x, \gamma_i)\, dx\right] \Delta_i y$$

Taking the limit, as $||\Delta||$ approaches zero, of the sum of these measures of volume for n horizontal strips of R from $y = c$ to $y = d$, we obtain the measure of the volume of the solid bounded above by the surface $z = f(x, y)$ and below by the region R in the xy plane. This measure of volume is the double integral of f on R. Hence

$$\lim_{||\Delta|| \to 0} \sum_{i=1}^{n} \left[\int_{\lambda_1(\gamma_i)}^{\lambda_2(\gamma_i)} f(x, \gamma_i)\, dx\right] \Delta_i y = \int_c^d \int_{\lambda_1(y)}^{\lambda_2(y)} f(x, y)\, dx\, dy = \iint_R f(x, y)\, dx\, dy \qquad (6)$$

Sufficient conditions for (6) to be valid are that λ_1 and λ_2 be smooth functions and f be continuous on R. In applying both (5) and (6), sometimes it may be necessary to subdivide a region R into subregions on which these sufficient conditions hold.

Example 4 Express the volume of the solid of Example 3 by an iterated integral in which the order of integration is the reverse of that of Example 3. Compute the volume.

Solution Again we find the volume of the solid in the first octant and multiply the result by 4. Figure 20.2.10 shows the region R in the xy plane and the ith horizontal strip whose width has a measure of $\Delta_i y$ and whose length has a measure of $2\sqrt{1 - \gamma_i^2}$. Then by (6),

$$\begin{aligned} V &= 4 \lim_{||\Delta|| \to 0} \sum_{i=1}^{n} \left[\int_0^{2\sqrt{1-\gamma_i^2}} (x^2 + 4\gamma_i^2)\, dx\right] \Delta_i y \\ &= 4 \int_0^1 \int_0^{2\sqrt{1-y^2}} (x^2 + 4y^2)\, dx\, dy \\ &= 4 \int_0^1 \left[\tfrac{1}{3}x^3 + 4y^2 x\right]_0^{2\sqrt{1-y^2}} dy \\ &= 4 \int_0^1 \left[\tfrac{8}{3}(1 - y^2)^{3/2} + 8y^2\sqrt{1 - y^2}\right] dy \\ &= \frac{32}{3} \int_0^1 (2y^2 + 1)\sqrt{1 - y^2}\, dy \end{aligned}$$

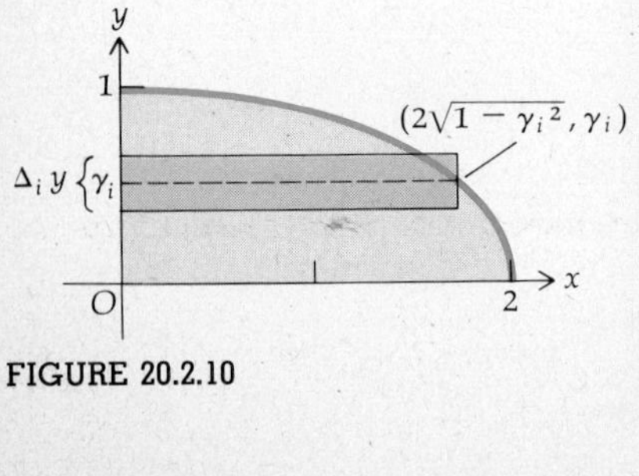

FIGURE 20.2.10

$$= -\tfrac{16}{3}y(1-y^2)^{3/2} + 8y\sqrt{1-y^2} + 8\sin^{-1} y\Big]_0^1$$

$$= 4\pi$$

Hence the volume is 4π cubic units, which agrees with the answer of Example 3.

From the solutions of Examples 3 and 4 we see that the double integral $\iint_R (x^2 + 4y^2)\, dA$ can be evaluated by either of the iterated integrals

$$\int_0^2 \int_0^{\sqrt{4-x^2}/2} (x^2 + 4y^2)\, dy\, dx \quad \text{or} \quad \int_0^1 \int_0^{2\sqrt{1-y^2}} (x^2 + 4y^2)\, dx\, dy$$

If in either (5) or (6), $f(x, y) = 1$ for all x and y, then the measure A of the area of a region R is expressed as a double integral. We have

$$A = \iint_R dy\, dx = \iint_R dx\, dy \tag{7}$$

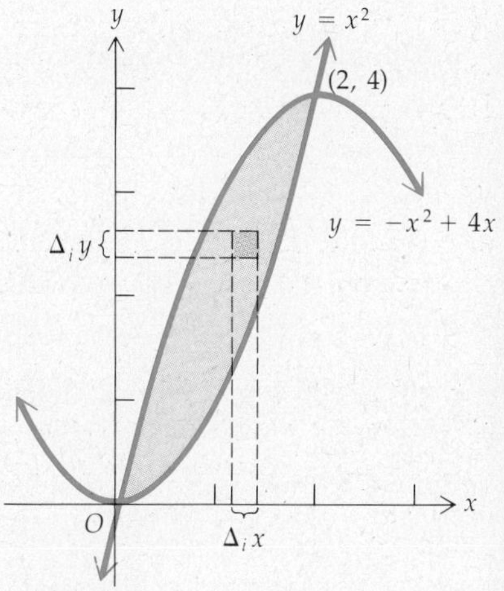

FIGURE 20.2.11

Example 5 Find by double integration the area of the region in the xy plane bounded by the curves $y = x^2$ and $y = 4x - x^2$.

Solution The region is shown in Fig. 20.2.11. From (7),

$$A = \iint_R dy\, dx$$

$$= \int_0^2 \int_{x^2}^{4x-x^2} dy\, dx$$

$$= \int_0^2 (4x - x^2 - x^2)\, dx$$

$$= 2x^2 - \tfrac{2}{3}x^3\Big]_0^2$$

$$= \tfrac{8}{3}$$

Hence the area of the region is $\frac{8}{3}$ square units.

EXERCISES 20.2

In Exercises 1 through 10, evaluate the iterated integral.

1. $\int_1^2 \int_0^{2x} xy^3\, dy\, dx$ **2.** $\int_0^4 \int_0^y dx\, dy$ **3.** $\int_0^4 \int_0^y \sqrt{9 + y^2}\, dx\, dy$ **4.** $\int_{-1}^1 \int_1^{e^x} \frac{x}{y}\, dy\, dx$

5. $\displaystyle\int_1^4\int_{y^2}^{y}\sqrt{\frac{y}{x}}\,dx\,dy$ **6.** $\displaystyle\int_1^4\int_{x^2}^{x}\sqrt{\frac{y}{x}}\,dy\,dx$ **7.** $\displaystyle\int_0^1\int_0^1|x-y|\,dy\,dx$ **8.** $\displaystyle\int_0^3\int_0^x x^2e^{xy}\,dy\,dx$

9. $\displaystyle\int_{\pi/2}^{\pi}\int_0^x \sin(4x-y)\,dy\,dx$ **10.** $\displaystyle\int_{\pi/2}^{\pi}\int_0^{y^2}\sin\frac{x}{y}\,dx\,dy$

In Exercises 11 through 18, find the exact value of the double integral.

11. The double integral is the same as in Exercise 1 in Exercises 20.1.

12. The double integral is the same as in Exercise 2 in Exercises 20.1.

13. The double integral is the same as in Exercise 3 in Exercises 20.1.

14. The double integral is the same as in Exercise 6 in Exercises 20.1.

15. $\displaystyle\iint_R \sin x\,dA$; R is the region bounded by the lines $y = 2x$, $y = \frac{1}{2}x$, and $x = \pi$.

16. $\displaystyle\iint_R \cos(x+y)\,dA$; R is the region bounded by the lines $y = x$ and $x = \pi$, and the x axis.

17. $\displaystyle\iint_R x^2\sqrt{9-y^2}\,dA$; R is the region bounded by the circle $x^2 + y^2 = 9$.

18. $\displaystyle\iint_R \frac{y^2}{x^2}\,dA$; R is the region bounded by the lines $y = x$ and $y = 2$, and the hyperbola $xy = 1$.

19. Find the volume of the solid under the plane $z = 4x$ and above the circle $x^2 + y^2 = 16$ in the xy plane. Draw a sketch of the solid.

20. Find the volume of the solid bounded by the planes $x = y + 2z + 1$, $x = 0$, $y = 0$, $z = 0$, and $3y + z - 3 = 0$. Draw a sketch of the solid.

21. Find the volume of the solid in the first octant bounded by the two cylinders $x^2 + y^2 = 4$ and $x^2 + z^2 = 4$. Draw a sketch of the solid.

22. Find the volume of the solid in the first octant bounded by the paraboloid $z = 9 - x^2 - 3y^2$. Draw a sketch of the solid.

23. Find the volume of the solid in the first octant bounded by the surfaces $x + z^2 = 1$, $x = y$, and $x = y^2$. Draw a sketch of the solid.

24. Find by double integration the volume of the portion of the solid bounded by the sphere $x^2 + y^2 + z^2 = 16$ that lies in the first octant. Draw a sketch of the solid.

In Exercises 25 through 28, use double integrals to find the area of the region bounded by the given curves in the xy plane. Draw a sketch of the region.

25. $y = x^3$ and $y = x^2$ **26.** $y^2 = 4x$ and $x^2 = 4y$ **27.** $y = x^2 - 9$ and $y = 9 - x^2$ **28.** $x^2 + y^2 = 16$ and $y^2 = 6x$

29. Express as an iterated integral the measure of the volume of the solid bounded by the ellipsoid

$$\frac{x^2}{a^2}+\frac{y^2}{b^2}+\frac{z^2}{c^2}=1$$

30. Use double integration to find the area of the region in the first quadrant bounded by the parabola $y^2 = 4x$, the circle $x^2 + y^2 = 5$, and the x axis by two methods: (1) integrate first with respect to x; (b) integrate first with respect to y. Compare the two methods of solution.

31. Find, by two methods, the volume of the solid below the plane $3x + 8y + 6z = 24$ and above the region in the xy plane bounded by the parabola $y^2 = 2x$, the line $2x + 3y = 10$, and the x axis: (a) integrate first with respect to x; (b) integrate first with respect to y. Compare the two methods of solution.

32. Given the iterated integral $\int_0^a \int_0^x \sqrt{a^2 - x^2}\, dy\, dx$. (a) Draw a sketch of the solid the measure of whose volume is represented by the given iterated integral; (b) evaluate the iterated integral; (c) write the iterated integral that gives the measure of the volume of the same solid with the order of integration reversed.

33. Given the iterated integral $\frac{2}{3}\int_0^a \int_0^{\sqrt{a^2-x^2}} (2x + y)\, dy\, dx$. The instructions are the same as for Exercise 32.

34. Use double integration to find the volume of the solid common to two right-circular cylinders of radius r units, whose axes intersect at right angles. (See Exercise 16 in Exercises 7.4.)

In Exercises 35 and 36, the iterated integral cannot be evaluated exactly in terms of elementary functions by the given order of integration. Reverse the order of integration and perform the computation.

35. $\int_0^4 \int_{\sqrt{x}}^2 \sin \pi y^3\, dy\, dx$ **36.** $\int_0^1 \int_y^1 e^{x^2}\, dx\, dy$

20.3 CENTER OF MASS AND MOMENTS OF INERTIA

In Chapter 11 we used single integrals to find the center of mass of a homogeneous lamina. In using single integrals we can consider only laminae of constant area density (except in special cases); however, with double integrals we can find the center of mass of either a homogeneous or a nonhomogeneous lamina.

Suppose we are given a lamina having the shape of a closed region R in the xy plane. Let $\rho(x, y)$ be the measure of the area density of the lamina at any point (x, y) of R where ρ is continuous on R. To find the total mass of the lamina we proceed as follows. Let Δ be a partition of R into n rectangles. If (ξ_i, γ_i) is any point in the ith rectangle having an area of measure $\Delta_i A$, then an approximation to the measure of the mass of the ith rectangle is given by $\rho(\xi_i, \gamma_i)\, \Delta_i A$, and the measure of the total mass of the lamina is approximated by

$$\sum_{i=1}^{n} \rho(\xi_i, \gamma_i)\, \Delta_i A$$

Taking the limit of the above sum as the norm of Δ approaches zero, we express the measure M of the mass of the lamina by

$$M = \lim_{\|\Delta\| \to 0} \sum_{i=1}^{n} \rho(\xi_i, \gamma_i)\, \Delta_i A = \iint_R \rho(x, y)\, dA \tag{1}$$

The measure of the moment of mass of the ith rectangle with respect to the x axis is approximated by $\gamma_i \rho(\xi_i, \gamma_i)\, \Delta_i A$. The sum of the measures of the moments of mass of the n rectangles with respect to the x axis is then approximated by the sum of n such terms. The measure M_x of the moment of mass with respect to the x axis of the entire lamina is given by

$$M_x = \lim_{\|\Delta\| \to 0} \sum_{i=1}^{n} \gamma_i \rho(\xi_i, \gamma_i)\, \Delta_i A = \iint_R y\rho(x, y)\, dA \tag{2}$$

Analogously, the measure M_y of its moment of mass with respect to the y axis is given by

$$M_y = \lim_{\|\Delta\| \to 0} \sum_{i=1}^{n} \xi_i \rho(\xi_i, \gamma_i)\, \Delta_i A = \iint_R x\rho(x, y)\, dA \tag{3}$$

The center of mass of the lamina is denoted by the point $(\bar{x}, \bar{y})$ and

$$\bar{x} = \frac{M_y}{M} \quad \text{and} \quad \bar{y} = \frac{M_x}{M}$$

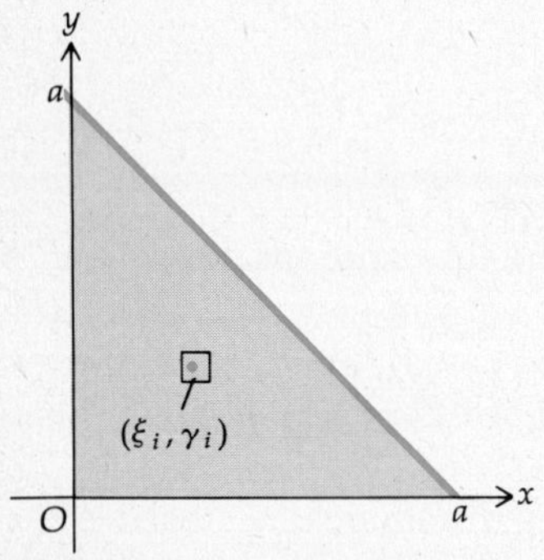

FIGURE 20.3.1

Example 1 A lamina in the shape of an isosceles right triangle has an area density that varies as the square of the distance from the vertex of the right angle. If mass is measured in kilograms and distance is measured in meters, find the mass and the center of mass of the lamina.

Solution Choose the coordinate axes so that the vertex of the right triangle is at the origin and the sides of length a meters of the triangle are along the coordinate axes (see Fig. 20.3.1). Let $\rho(x, y)$ be the number of kilograms per square meter in the area density of the lamina at the point (x, y). Then $\rho(x, y) = k(x^2 + y^2)$ where k is a constant. Therefore, if M kg is the mass of the lamina, we have from (1)

$$\begin{aligned}
M &= \lim_{\|\Delta\| \to 0} \sum_{i=1}^{n} k(\xi_i^2 + \gamma_i^2)\, \Delta_i A \\
&= k \iint_R (x^2 + y^2)\, dA \\
&= \int_0^a \int_0^{a-x} (x^2 + y^2)\, dy\, dx \\
&= k \int_0^a \left[yx^2 + \tfrac{1}{3}y^3 \right]_0^{a-x} dx \\
&= k \int_0^a (\tfrac{1}{3}a^3 - a^2x + 2ax^2 - \tfrac{4}{3}x^3)\, dx \\
&= k(\tfrac{1}{3}a^4 - \tfrac{1}{2}a^4 + \tfrac{2}{3}a^4 - \tfrac{1}{3}a^4) \\
&= \tfrac{1}{6}ka^4
\end{aligned}$$

To find the center of mass, observe that because of symmetry it must lie on the line $y = x$. Therefore, if we find $\bar{x}$, we also have $\bar{y}$. From (3),

$$\begin{aligned}
M_y &= \lim_{\|\Delta\| \to 0} \sum_{i=1}^{n} k\xi_i(\xi_i^2 + \gamma_i^2)\, \Delta_i A \\
&= k \iint_R x(x^2 + y^2)\, dA
\end{aligned}$$

$$= k \int_0^a \int_0^{a-x} x(x^2 + y^2)\, dy\, dx$$

$$= k \int_0^a \left[x^3 y + \tfrac{1}{3} x y^3 \right]_0^{a-x} dx$$

$$= k \int_0^a (\tfrac{1}{3}a^3 x - a^2 x^2 + 2ax^3 - \tfrac{4}{3}x^4)\, dx$$

$$= k(\tfrac{1}{6}a^5 - \tfrac{1}{3}a^5 + \tfrac{1}{2}a^5 - \tfrac{4}{15}a^5)$$

$$= \tfrac{1}{15}ka^5$$

Because $M\bar{x} = M_y$, then $M\bar{x} = \frac{1}{15}ka^5$; and because $M = \frac{1}{6}ka^4$, we get $\bar{x} = \frac{2}{5}a$. Therefore the center of mass is at the point $(\frac{2}{5}a, \frac{2}{5}a)$.

20.3.1 Definition The **moment of inertia** of a particle, whose mass is m kg, about an axis is defined to be mr^2 kg-m^2, where r meters is the perpendicular distance from the particle to the axis.

If we have a system of n particles, the moment of inertia of the system is defined as the sum of the moments of inertia of all the particles. That is, if the ith particle has a mass of m_i kg and is at a distance of r_i meters from the axis, then I kg-m^2 is the moment of inertia of the system where

$$I = \sum_{i=1}^{n} m_i r_i^2 \tag{4}$$

Extending this concept of moment of inertia to a continuous distribution of mass in a plane such as rods or laminae by processes similar to those previously used, we have the following definition.

20.3.2 Definition Suppose that we are given a continuous distribution of mass occupying a region R in the xy plane, and suppose that the measure of area density of this distribution at the point (x, y) is $\rho(x, y)$ kg/m^2, where ρ is continuous on R. Then the **moment of inertia I_x** kg-m^2 **about the x axis** of this distribution of mass is determined by

$$I_x = \lim_{\|\Delta\| \to 0} \sum_{i=1}^{n} \gamma_i^2 \rho(\xi_i, \gamma_i)\, \Delta_i A = \iint_R y^2 \rho(x, y)\, dA \tag{5}$$

Similarly, the measure I_y of the **moment of inertia about the y axis** is given by

$$I_y = \lim_{\|\Delta\| \to 0} \sum_{i=1}^{n} \xi_i^2 \rho(\xi_i, \gamma_i)\, \Delta_i A = \iint_R x^2 \rho(x, y)\, dA \tag{6}$$

and the measure I_0 of the **moment of inertia about the origin,** or the z axis, is given by

$$I_0 = \lim_{\|\Delta\| \to 0} \sum_{i=1}^{n} (\xi_i^2 + \gamma_i^2)\rho(\xi_i, \gamma_i)\, \Delta_i A = \iint_R (x^2 + y^2)\rho(x, y)\, dA \qquad (7)$$

The number I_0 is the measure of what is called the **polar moment of inertia.**

Example 2 A homogeneous straight wire has a constant linear density of ρ kg/m. Find the moment of inertia of the wire about an axis perpendicular to the wire and passing through one end.

Solution Let the wire be of length a meters, and suppose that it extends along the x axis from the origin. We find its moment of inertia about the y axis. Divide the wire into n segments; the length of the ith segment is $\Delta_i x$ meters. The mass of the ith segment is then $\rho\, \Delta_i x$ kg. Assume that the mass of the ith segment is concentrated at a single point ξ_i, where $x_{i-1} \le \xi_i \le x_i$. The moment of inertia of the ith segment about the y axis lies between $\rho x_{i-1}^2\, \Delta_i x$ kg-m^2 and $\rho x_i^2\, \Delta_i x$ kg-m^2 and is approximated by $\rho \xi_i^2\, \Delta_i x$ kg-m^2, where $x_{i-1} \le \xi_i \le x_i$. If the moment of inertia of the wire about the y axis is I_y kg-m^2, then

$$I_y = \lim_{\|\Delta\| \to 0} \sum_{i=1}^{n} \rho \xi_i^2\, \Delta_i x = \int_0^a \rho x^2\, dx = \tfrac{1}{3}\rho a^3$$

Therefore the moment of inertia is $\frac{1}{3}\rho a^3$ kg-m^2.

Example 3 A homogeneous rectangular lamina has constant area density of ρ slugs/ft^2. Find the moment of inertia of the lamina about one corner.

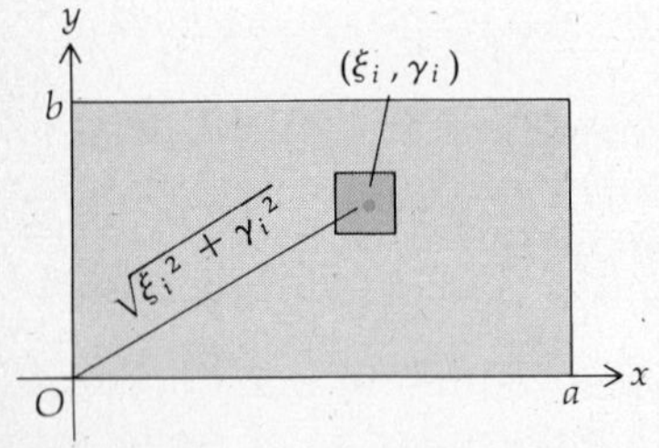

FIGURE 20.3.2

Solution Suppose that the lamina is bounded by the lines $x = a$, $y = b$, the x axis, and the y axis. See Fig. 20.3.2. If I_0 slug-ft^2 is the moment of inertia about the origin, then

$$\begin{aligned}
I_0 &= \lim_{\|\Delta\| \to 0} \sum_{i=1}^{n} \rho(\xi_i^2 + \gamma_i^2)\, \Delta_i A \\
&= \iint_R \rho(x^2 + y^2)\, dA \\
&= \rho \int_0^b \int_0^a (x^2 + y^2)\, dx\, dy \\
&= \rho \int_0^b \Big[\tfrac{1}{3}x^3 + xy^2\Big]_0^a dy \\
&= \rho \int_0^b (\tfrac{1}{3}a^3 + ay^2)\, dy \\
&= \tfrac{1}{3}\rho ab(a^2 + b^2)
\end{aligned}$$

The moment of inertia is therefore $\frac{1}{3}\rho ab(a^2 + b^2)$ slug-ft^2.

It is possible to find the distance from any axis L at which the mass of a lamina can be concentrated without affecting the moment of inertia of the lamina about L. The measure of this distance, denoted by r, is called the *radius of gyration* of the lamina about L. That is, if the mass M kg of a lamina is concentrated at a point r meters from L, the moment of inertia of the lamina about L is the same as that of a particle of mass M kg at a distance of r meters from L; this moment of inertia is Mr^2 kg-m^2. Thus we have the following definition.

20.3.3 Definition If I is the measure of the moment of inertia about an axis L of a distribution of mass in a plane and M is the measure of the total mass of the distribution, then the **radius of gyration** of the distribution about L has measure r, where

$$r^2 = \frac{I}{M}$$

Example 4 Suppose that a lamina is in the shape of a semicircle and the measure of the area density of the lamina at any point is proportional to the measure of the distance of the point from the diameter. If mass is measured in kilograms and distance is measured in meters, find the radius of gyration of the lamina about the x axis.

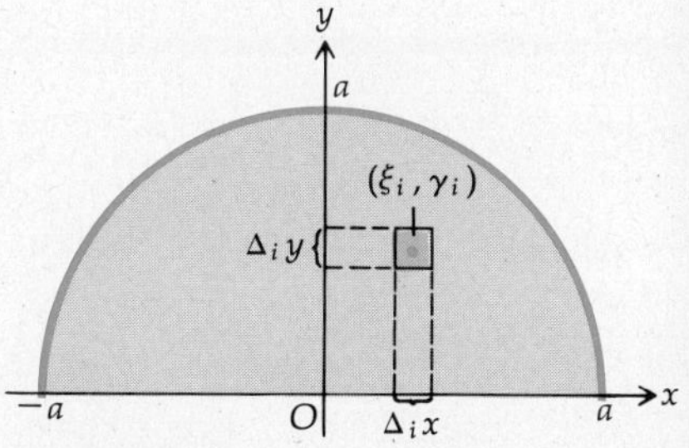

FIGURE 20.3.3

Solution Choose the x and y axes such that the semicircle is the top half of the circle $x^2 + y^2 = a^2$. See Fig. 20.3.3. The area density of the lamina at the point (x, y) is then ky kg/m^2. So if M kg is the mass of the lamina, we have

$$\begin{aligned}
M &= \lim_{\|\Delta\| \to 0} \sum_{i=1}^{n} k\gamma_i \, \Delta_i A \\
&= \iint_R ky \, dA \\
&= \int_0^a \int_{-\sqrt{a^2-y^2}}^{\sqrt{a^2-y^2}} ky \, dx \, dy \\
&= k \int_0^a \Big[yx \Big]_{-\sqrt{a^2-y^2}}^{\sqrt{a^2-y^2}} dy \\
&= 2k \int_0^a y\sqrt{a^2 - y^2} \, dy \\
&= -\tfrac{2}{3}k(a^2 - y^2)^{3/2} \Big]_0^a \\
&= \tfrac{2}{3}ka^3
\end{aligned}$$

If I_x kg-m^2 is the moment of inertia of the lamina about the x axis, then

$$I_x = \lim_{\|\Delta\| \to 0} \sum_{i=1}^{n} \gamma_i^2 (k\gamma_i) \, \Delta_i A$$

$$= \iint_R ky^3 \, dy \, dx$$

$$= \int_{-a}^{a} \int_0^{\sqrt{a^2 - x^2}} ky^3 \, dy \, dx$$

$$= k \int_{-a}^{a} \left[\tfrac{1}{4} y^4 \right]_0^{\sqrt{a^2 - x^2}} dx$$

$$= \tfrac{1}{4} k \int_{-a}^{a} (a^4 - 2a^2x^2 + x^4) \, dx$$

$$= \tfrac{1}{4} k (2a^5 - \tfrac{4}{3} a^5 + \tfrac{2}{5} a^5)$$

$$= \tfrac{4}{15} ka^5$$

Therefore, if r meters is the radius of gyration,

$$r^2 = \frac{\frac{4}{15} ka^5}{\frac{2}{3} ka^3} = \frac{2}{5} a^2$$

and so $r = \frac{1}{5}\sqrt{10}\,a$. The radius of gyration is therefore $\frac{1}{5}\sqrt{10}\,a$ meters.

EXERCISES 20.3

In Exercises 1 through 12, find the mass and center of mass of the given lamina if the area density is as indicated. Mass is measured in kilograms and distance is measured in meters.

1. A lamina in the shape of the rectangular region bounded by the lines $x = 3$ and $y = 2$ and the coordinate axes. The area density at any point is xy^2 kg/m^2.

2. A lamina in the shape of the rectangular region bounded by the lines $x = 4$ and $y = 5$ and the coordinate axes. The area density at any point is $(x^2 + y)$ kg/m^2.

3. A lamina in the shape of the triangular region whose sides are segments of the coordinate axes and the line $x + 2y = 6$. The area density at any point is y^2 kg/m^2.

4. A lamina in the shape of the region in the first quadrant bounded by the parabola $y = x^2$, the line $y = 1$, and the y axis. The area density at any point is $(x + y)$ kg/m^2.

5. A lamina in the shape of the region bounded by the parabola $x^2 = 8y$, the line $y = 2$, and the y axis. The area density varies as the distance from the line $y = -1$.

6. A lamina in the shape of the region bounded by the curve $y = e^x$, the line $x = 1$, and the coordinate axes. The area density varies as the distance from the x axis.

7. A lamina in the shape of the region in the first quadrant bounded by the circle $x^2 + y^2 = a^2$ and the coordinate axes. The area density varies as the sum of the distances from the two straight edges.

8. A lamina in the shape of the region bounded by the triangle whose sides are segments of the coordinate axes and the line $3x + 2y = 18$. The area density varies as the product of the distances from the coordinate axes.

9. A lamina in the shape of the region bounded by the curve $y = \sin x$ and the x axis from $x = 0$ to $x = \pi$. The area density varies as the distance from the x axis.

10. A lamina in the shape of the region bounded by the curve $y = \sqrt{x}$ and the line $y = x$. The area density varies as the distance from the y axis.

11. A lamina in the shape of the region in the first quadrant bounded by the circle $x^2 + y^2 = 4$ and the line $x + y = 2$. The area density at any point is xy kg/m².

12. A lamina in the shape of the region bounded by the circle $x^2 + y^2 = 1$ and the lines $x = 1$ and $y = 1$. The area density at any point is xy kg/m².

In Exercises 13 through 18, find the moment of inertia of the given homogeneous lamina about the indicated axis if the area density is ρ kg/m² and distance is measured in meters.

13. A lamina in the shape of the region bounded by $4y = 3x$, $x = 4$, and the x axis; about the x axis.

14. The lamina of Exercise 13; about the line $x = 4$.

15. A lamina in the shape of the region bounded by a circle of radius a units; about its center.

16. A lamina in the shape of the region bounded by the parabola $x^2 = 4 - 4y$ and the x axis; about the x axis.

17. The lamina of Exercise 16; about the origin.

18. A lamina in the shape of the region bounded by a triangle of sides of lengths a meters, b meters, and c meters; about the side of length a meters.

In Exercises 19 through 22, find for the given lamina each of the following: (a) the moment of inertia about the x axis; (b) the moment of inertia about the y axis; (c) the radius of gyration about the x axis; (d) the polar moment of inertia.

19. The lamina of Exercise 1.

20. The lamina of Exercise 4.

21. The lamina of Exercise 9.

22. The lamina of Exercise 10.

23. A homogeneous lamina of area density ρ slugs/ft² is in the shape of the region bounded by an isosceles triangle having a base of length b ft and an altitude of length h ft. Find the radius of gyration of the lamina about its line of symmetry.

24. A homogeneous lamina of area density ρ slugs/ft² is in the shape of the region bounded by the curve $x = \sqrt{y}$, the x axis, and the line $x = a$, where $a > 0$. Find the moment of inertia of the lamina about the line $x = a$.

25. A lamina is in the shape of the region enclosed by the parabola $y = 2x - x^2$ and the x axis. Find the moment of inertia of the lamina about the line $y = 4$ if the area density varies as its distance from the line $y = 4$. Mass is measured in kilograms and distance is measured in meters.

20.4 THE DOUBLE INTEGRAL IN POLAR COORDINATES

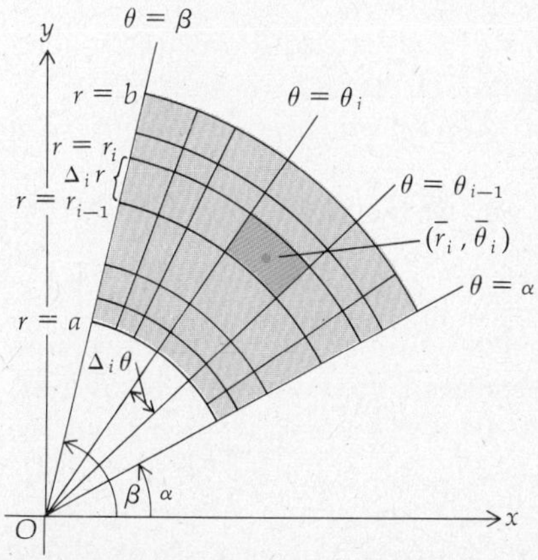

FIGURE 20.4.1

We now show how the double integral of a function on a closed region in the polar coordinate plane can be defined. We begin by considering the simplest kind of region. Let R be the region bounded by the rays $\theta = \alpha$ and $\theta = \beta$ and by the circles $r = a$ and $r = b$. Then let Δ be a *partition* of this region obtained by drawing rays through the pole and circles having centers at the pole. This is shown in Fig. 20.4.1. We obtain a network of subregions that we call "curved" rectangles. The norm $\|\Delta\|$ of the partition is the length of the longest of the diagonals of the "curved" rectangles. Let the number of subregions be n, and let $\Delta_i A$ be the measure of area of the ith "curved" rectangle. Because the area of the ith subregion is the difference of the areas of two circular sectors,

$$\Delta_i A = \tfrac{1}{2} r_i^2(\theta_i - \theta_{i-1}) - \tfrac{1}{2} r_{i-1}^2(\theta_i - \theta_{i-1})$$
$$= \tfrac{1}{2}(r_i - r_{i-1})(r_i + r_{i-1})(\theta_i - \theta_{i-1})$$

Let $\bar{r}_i = \frac{1}{2}(r_i + r_{i-1})$, $\Delta_i r = r_i - r_{i-1}$, and $\Delta_i \theta = \theta_i - \theta_{i-1}$. Then

$$\Delta_i A = \bar{r}_i \, \Delta_i r \, \Delta_i \theta$$

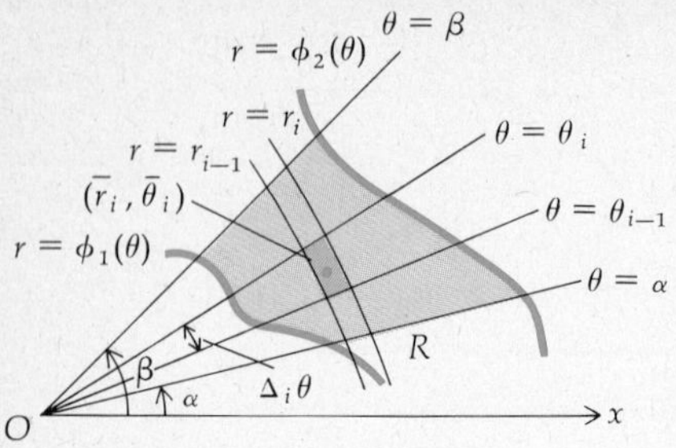

FIGURE 20.4.2

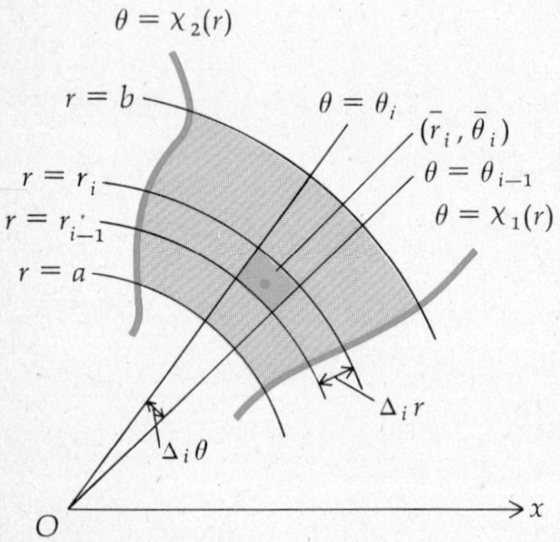

FIGURE 20.4.3

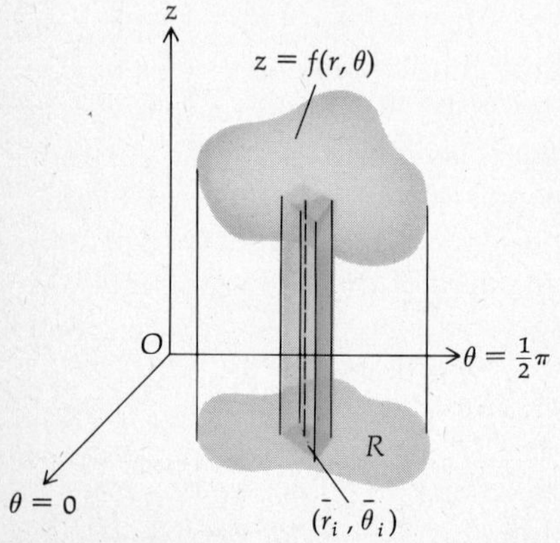

FIGURE 20.4.4

Take the point $(\bar{r}_i, \bar{\theta}_i)$ in the ith subregion, where $\theta_{i-1} \le \bar{\theta}_i \le \theta_i$, and form the sum

$$\sum_{i=1}^{n} f(\bar{r}_i, \bar{\theta}_i)\,\Delta_i A = \sum_{i=1}^{n} f(\bar{r}_i, \bar{\theta}_i)\bar{r}_i\,\Delta_i r\,\Delta_i\theta \tag{1}$$

It can be shown that if f is continuous on the region R, then the limit of the sum in (1), as $\|\Delta\|$ approaches zero, exists and that this limit will be the double integral of f on R. We write either

$$\lim_{\|\Delta\|\to 0} \sum_{i=1}^{n} f(\bar{r}_i, \bar{\theta}_i)\,\Delta_i A = \iint_R f(r, \theta)\,dA \tag{2}$$

or

$$\lim_{\|\Delta\|\to 0} \sum_{i=1}^{n} f(\bar{r}_i, \bar{\theta}_i)\bar{r}_i\,\Delta_i r\,\Delta_i\theta = \iint_R f(r, \theta) r\,dr\,d\theta \tag{3}$$

Observe that in polar coordinates, $dA = r\,dr\,d\theta$.

The double integral can be shown to be equal to an iterated integral having one of two possible forms:

$$\iint_R f(r, \theta)\,dA = \int_\alpha^\beta \int_a^b f(r, \theta) r\,dr\,d\theta = \int_a^b \int_\alpha^\beta f(r, \theta)\,r\,d\theta\,dr \tag{4}$$

We can define the double integral of a continuous function f of two variables on closed regions of the polar coordinate plane other than the one previously considered. For example, consider the region R bounded by $r = \phi_1(\theta)$ and $r = \phi_2(\theta)$, where ϕ_1 and ϕ_2 are smooth functions, and by the lines $\theta = \alpha$ and $\theta = \beta$. See Fig. 20.4.2. In the figure, $\phi_1(\theta) \le \phi_2(\theta)$ for all θ in the closed interval $[\alpha, \beta]$. Then it can be shown that the double integral of f on R exists and equals an iterated integral, and we have

$$\iint_R f(r, \theta)\,dA = \int_\alpha^\beta \int_{\phi_1(\theta)}^{\phi_2(\theta)} f(r, \theta) r\,dr\,d\theta \tag{5}$$

If the region R is bounded by the curves $\theta = \chi_1(r)$ and $\theta = \chi_2(r)$, where χ_1 and χ_2 are smooth functions, and by the circles $r = a$ and $r = b$, as shown in Fig. 20.4.3, where $\chi_1(r) \le \chi_2(r)$ for all r in the closed interval $[a, b]$, then

$$\iint_R f(r, \theta)\,dA = \int_a^b \int_{\chi_1(r)}^{\chi_2(r)} f(r, \theta) r\,d\theta\,dr \tag{6}$$

We can interpret the double integral of a function on a closed region in the polar coordinate plane as the measure of the volume of a solid by using cylindrical coordinates. Figure 20.4.4 shows a solid having as its base a region R in the polar coordinate plane and bounded above by the surface $z = f(r, \theta)$, where f is continuous on R and $f(x, y) \ge 0$ on R. Take a partition of R giving a network of n "curved" rectangles. Construct the

n solids for which the ith solid has as its base the ith "curved" rectangle and as the measure of its altitude $f(\bar{r}_i, \bar{\theta}_i)$, where $(\bar{r}_i, \bar{\theta}_i)$ is in the ith subregion. Figure 20.4.4 shows the ith solid. The measure of the volume of the ith solid is

$$f(\bar{r}_i, \bar{\theta}_i)\ \Delta_i A = f(\bar{r}_i, \bar{\theta}_i)\bar{r}_i\ \Delta_i r\ \Delta_i \theta$$

The sum of the measures of the volumes of the n solids is

$$\sum_{i=1}^{n} f(\bar{r}_i, \bar{\theta}_i)\bar{r}_i\ \Delta_i r\ \Delta_i \theta$$

If V is the measure of the volume of the given solid, then

$$V = \lim_{\|\Delta\| \to 0} \sum_{i=1}^{n} f(\bar{r}_i, \bar{\theta}_i)\bar{r}_i\ \Delta_i r\ \Delta_i \theta = \iint_R f(r, \theta) r\ dr\ d\theta \tag{7}$$

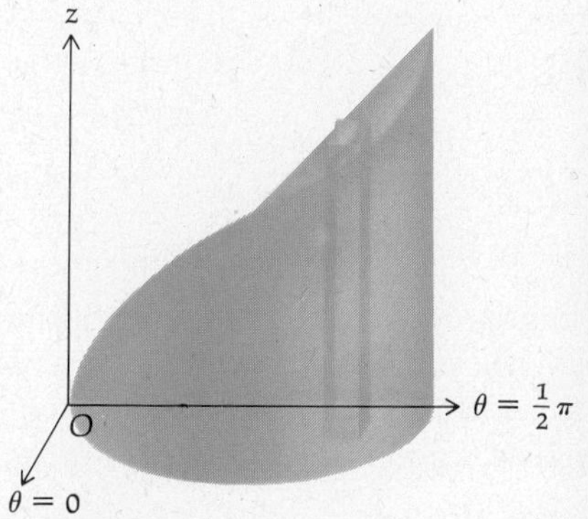

FIGURE 20.4.5

Example 1 Find the volume of the solid in the first octant bounded by the cone $z = r$ and the cylinder $r = 3 \sin \theta$.

Solution The solid and the ith element are shown in Fig. 20.4.5. Using formula (7) with $f(r, \theta) = r$, we have, where V cubic units is the volume of the given solid,

$$\begin{aligned} V &= \lim_{\|\Delta\| \to 0} \sum_{i=1}^{n} \bar{r}_i \cdot \bar{r}_i\ \Delta_i r\ \Delta_i \theta \\ &= \iint_R r^2\ dr\ d\theta \\ &= \int_0^{\pi/2} \int_0^{3 \sin \theta} r^2\ dr\ d\theta \\ &= \int_0^{\pi/2} \Big[\tfrac{1}{3} r^3 \Big]_0^{3 \sin \theta} d\theta \\ &= 9 \int_0^{\pi/2} \sin^3 \theta\ d\theta \\ &= -9 \cos \theta + 3 \cos^3 \theta \Big]_0^{\pi/2} \\ &= 6 \end{aligned}$$

The volume is therefore 6 cubic units.

Example 2 Find the mass of the lamina in the shape of the region inside the semicircle $r = a \cos \theta$, $0 \le \theta \le \frac{1}{2}\pi$, and whose measure of area density at any point is proportional to the measure of its distance from the pole. The mass is measured in kilograms and distance is measured in meters.

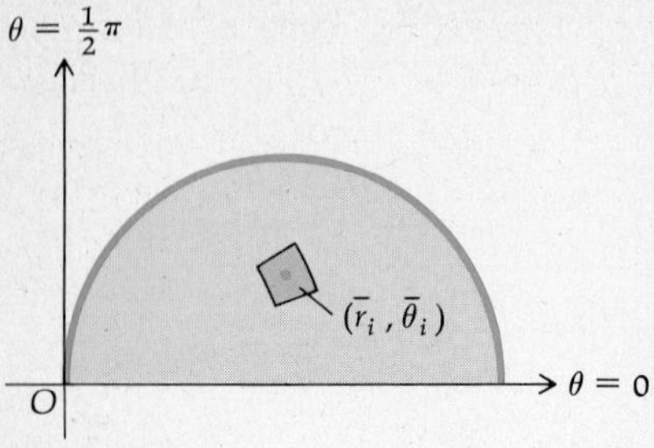

FIGURE 20.4.6

Solution Figure 20.4.6 shows a sketch of the lamina and the ith "curved" rectangle. The area density at the point (r, θ) is kr kg/m², where k is a constant. If M kg is the mass of the lamina, then

$$\begin{aligned} M &= \lim_{\|\Delta\| \to 0} \sum_{i=1}^{n} (k\bar{r}_i)\bar{r}_i \, \Delta_i r \, \Delta_i \theta \\ &= \iint_R kr^2 \, dr \, d\theta \\ &= k \int_0^{\pi/2} \int_0^{a\cos\theta} r^2 \, dr \, d\theta \\ &= \tfrac{1}{3}ka^3 \int_0^{\pi/2} \cos^3\theta \, d\theta \\ &= \tfrac{1}{3}ka^3 \Big[\sin\theta - \tfrac{1}{3}\sin^3\theta \Big]_0^{\pi/2} \\ &= \tfrac{2}{9}ka^3 \end{aligned}$$

Therefore the mass is $\frac{2}{9}ka^3$ kg.

Example 3 Find the center of mass of the lamina in Example 2.

Solution Let the cartesian coordinates of the center of mass of the lamina be $\bar{x}$ and $\bar{y}$, where, as is customary, the x axis is along the polar axis and the y axis is along the $\frac{1}{2}\pi$ axis. Let the cartesian coordinate representation of the point $(\bar{r}_i, \bar{\theta}_i)$ be $(\bar{x}_i, \bar{y}_i)$. Then if M_x kg-m is the moment of mass of the lamina with respect to the x axis,

$$M_x = \lim_{\|\Delta\| \to 0} \sum_{i=1}^{n} \bar{y}_i (k\bar{r}_i)\bar{r}_i \, \Delta_i r \, \Delta_i \theta$$

Replacing $\bar{y}_i$ by $\bar{r}_i \sin\bar{\theta}_i$ we get

$$\begin{aligned} M_x &= \lim_{\|\Delta\| \to 0} \sum_{i=1}^{n} k\bar{r}_i^{\,3} \sin\bar{\theta}_i \, \Delta_i r \, \Delta_i \theta \\ &= \iint_R kr^3 \sin\theta \, dr \, d\theta \\ &= k \int_0^{\pi/2} \int_0^{a\cos\theta} r^3 \sin\theta \, dr \, d\theta \\ &= \tfrac{1}{4}ka^4 \int_0^{\pi/2} \cos^4\theta \sin\theta \, d\theta \\ &= -\tfrac{1}{20}ka^4 \cos^5\theta \Big]_0^{\pi/2} \\ &= \tfrac{1}{20}ka^4 \end{aligned}$$

If M_y kg-m is the moment of mass of the lamina with respect to the y axis, then

$$M_y = \lim_{\|\Delta\| \to 0} \sum_{i=1}^{n} \bar{x}_i (k\bar{r}_i)\bar{r}_i \,\Delta_i r\, \Delta_i \theta$$

Replacing $\bar{x}_i$ by $\bar{r}_i \cos \bar{\theta}_i$ we have

$$\begin{aligned} M_y &= \lim_{\|\Delta\| \to 0} \sum_{i=1}^{n} k\bar{r}_i^{\,3} \cos \bar{\theta}_i \,\Delta_i r\, \Delta_i \theta \\ &= \iint_R kr^3 \cos \theta \, dr\, d\theta \\ &= k \int_0^{\pi/2} \int_0^{a\cos\theta} r^3 \cos\theta \, dr\, d\theta \\ &= \tfrac{1}{4}ka^4 \int_0^{\pi/2} \cos^5 \theta \, d\theta \\ &= \tfrac{1}{4}ka^4 \Big[\sin\theta - \tfrac{2}{3}\sin^3\theta + \tfrac{1}{5}\sin^5\theta \Big]_0^{\pi/2} \\ &= \tfrac{2}{15}ka^4 \end{aligned}$$

Therefore

$$\bar{x} = \frac{M_y}{M} = \frac{\frac{2}{15}ka^4}{\frac{2}{9}ka^3} = \frac{3}{5}a$$

and

$$\bar{y} = \frac{M_x}{M} = \frac{\frac{1}{20}ka^4}{\frac{2}{9}ka^3} = \frac{9}{40}a$$

Hence the center of mass is at the point $(\frac{3}{5}a, \frac{9}{40}a)$.

The area of a region in the polar plane can be found by double integration. The following example illustrates the method.

Example 4 Find by double integration the area of the region enclosed by one leaf of the rose $r = \sin 3\theta$.

Solution Figure 20.4.7 shows the region and the ith "curved" rectangle. If A square units is the area of the region, then

$$\begin{aligned} A &= \lim_{\|\Delta\| \to 0} \sum_{i=1}^{n} \Delta_i A \\ &= \lim_{\|\Delta\| \to 0} \sum_{i=1}^{n} \bar{r}_i \,\Delta_i r\, \Delta_i \theta \end{aligned}$$

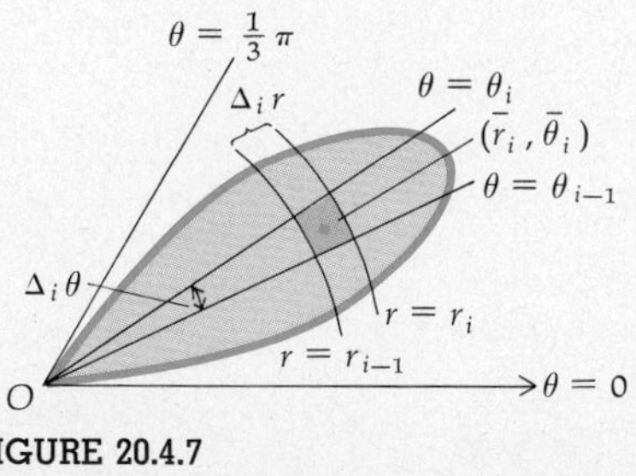

FIGURE 20.4.7

$$= \iint_R r\, dr\, d\theta$$

$$= \int_0^{\pi/3} \int_0^{\sin 3\theta} r\, dr\, d\theta$$

$$= \frac{1}{2} \int_0^{\pi/3} \sin^2 3\theta\, d\theta$$

$$= \tfrac{1}{4}\theta - \tfrac{1}{24} \sin 6\theta \Big]_0^{\pi/3}$$

$$= \tfrac{1}{12}\pi$$

Hence the area is $\frac{1}{12}\pi$ square units.

Sometimes it is easier to evaluate a double integral by using polar coordinates instead of cartesian coordinates. Such a situation is shown in the following example.

Example 5 Evaluate the double integral $\iint_R e^{-(x^2+y^2)}\, dA$ where the region R is in the first quadrant and bounded by the circle $x^2 + y^2 = a^2$ and the coordinate axes.

Solution Because $x^2 + y^2 = r^2$, and $dA = r\, dr\, d\theta$,

$$\iint_R e^{-(x^2+y^2)}\, dA = \iint_R e^{-r^2}\, r\, dr\, d\theta$$

$$= \int_0^{\pi/2} \int_0^a e^{-r^2}\, r\, dr\, d\theta$$

$$= -\frac{1}{2} \int_0^{\pi/2} \left[e^{-r^2} \right]_0^a d\theta$$

$$= -\frac{1}{2} \int_0^{\pi/2} (e^{-a^2} - 1)\, d\theta$$

$$= \tfrac{1}{4}\pi(1 - e^{-a^2})$$

EXERCISES 20.4

In Exercises 1 through 6, use double integrals to find the area of the given region.

1. The region inside the cardioid $r = 2(1 + \sin\theta)$.
2. One leaf of the rose $r = a \cos 2\theta$.
3. The region inside the cardioid $r = a(1 + \cos\theta)$ and outside the circle $r = a$.
4. The region inside the circle $r = 1$ and outside the lemniscate $r^2 = \cos 2\theta$.
5. The region inside the large loop of the limaçon $r = 2 - 4 \sin\theta$ and outside the small loop.
6. The region inside the limaçon $r = 3 - \cos\theta$ and outside the circle $r = 5 \cos\theta$.

In Exercises 7 through 12, find the volume of the given solid.

7. The solid bounded by the ellipsoid $z^2 + 9r^2 = 9$.

8. The solid cut out of the sphere $z^2 + r^2 = 4$ by the cylinder $r = 1$.

9. The solid cut out of the sphere $z^2 + r^2 = 16$ by the cylinder $r = 4 \cos \theta$.

10. The solid above the polar plane bounded by the cone $z = 2r$ and the cylinder $r = 1 - \cos \theta$.

11. The solid bounded by the paraboloid $z = 4 - r^2$, the cylinder $r = 1$, and the polar plane.

12. The solid above the paraboloid $z = r^2$ and below the plane $z = 2r \sin \theta$.

In Exercises 13 through 19, find the mass and center of mass of the given lamina if the area density is as indicated. Mass is measured in kilograms and distance is measured in meters.

13. A lamina in the shape of the region of Exercise 1. The area density varies as the distance from the pole.

14. A lamina in the shape of the region of Exercise 2. The area density varies as the distance from the pole.

15. A lamina in the shape of the region inside the limaçon $r = 2 - \cos \theta$. The area density varies as the distance from the pole.

16. A lamina in the shape of the region bounded by the limaçon $r = 2 + \cos \theta$, $0 \le \theta \le \pi$, and the polar axis. The area density at any point is $k \sin \theta$ kg/m^2.

17. The lamina of Exercise 16. The area density at any point is $kr \sin \theta$ kg/m^2.

18. A lamina in the shape of the region of Exercise 6. The area density varies as the distance from the pole.

19. A lamina in the shape of the region of Exercise 5. The area density varies as the distance from the pole.

In Exercises 20 through 24, find the moment of inertia of the given lamina about the indicated axis or point if the area density is as indicated. Mass is measured in kilograms and distance is measured in meters.

20. A lamina in the shape of the region enclosed by the circle $r = \sin \theta$; about the $\frac{1}{2}\pi$ axis. The area density at any point is k kg/m^2.

21. The lamina of Exercise 20; about the polar axis. The area density at any point is k kg/m^2.

22. A lamina in the shape of the region bounded by the cardioid $r = a(1 - \cos \theta)$; about the pole. The area density at any point is k kg/m^2.

23. A lamina in the shape of the region bounded by the cardioid $r = a(1 + \cos \theta)$ and the circle $r = 2a \cos \theta$; about the pole. The area density at any point is k kg/m^2.

24. A lamina in the shape of the region enclosed by the lemniscate $r^2 = a^2 \cos 2\theta$; about the polar axis. The area density at any point is k kg/m^2.

25. A homogeneous lamina is in the shape of the region enclosed by one loop of the lemniscate $r^2 = \cos 2\theta$. Find the radius of gyration of the lamina about an axis perpendicular to the polar plane at the pole.

26. A lamina is in the shape of the region enclosed by the circle $r = 4$, and the area density varies as the distance from the pole. Find the radius of gyration of the lamina about an axis perpendicular to the polar plane at the pole.

27. Evaluate by polar coordinates the double integral $\iint_R e^{x^2+y^2}\, dA$ where R is the region bounded by the circles $x^2 + y^2 = 1$ and $x^2 + y^2 = 9$.

28. Evaluate by polar coordinates the double integral $\iint_R \frac{x}{\sqrt{x^2 + y^2}}\, dA$ where R is the region in the first quadrant bounded by the circle $x^2 + y^2 = 1$ and the coordinate axes.

20.5 AREA OF A SURFACE

The double integral can be used to determine the area of the portion of the surface $z = f(x, y)$ that lies over a closed region R in the xy plane. To show this we must first define what we mean by the measure of this area and then obtain a formula for computing it. Assume that f and its first partial derivatives are continuous on R and suppose that $f(x, y) > 0$ on R. Let Δ be a partition of R into n rectangular subregions. The ith rectangle has dimensions of measures $\Delta_i x$ and $\Delta_i y$ and an area of measure $\Delta_i A$. Let (ξ_i, γ_i) be any point in the ith rectangle, and at the point $Q(\xi_i, \gamma_i, f(\xi_i, \gamma_i))$ on the surface consider the tangent plane to the surface. Project vertically upward the ith rectangle onto the tangent plane and let $\Delta_i\sigma$ be the measure of the area of this projection. Figure 20.5.1 shows the region R, the portion of the surface above R, the ith rectangular subregion of R, and the projection of the ith rectangle onto the tangent plane to the surface at Q. The number $\Delta_i\sigma$ is an approximation to the measure of the area of the piece of the surface that lies above the ith rectangle. Because there are n such pieces, the summation

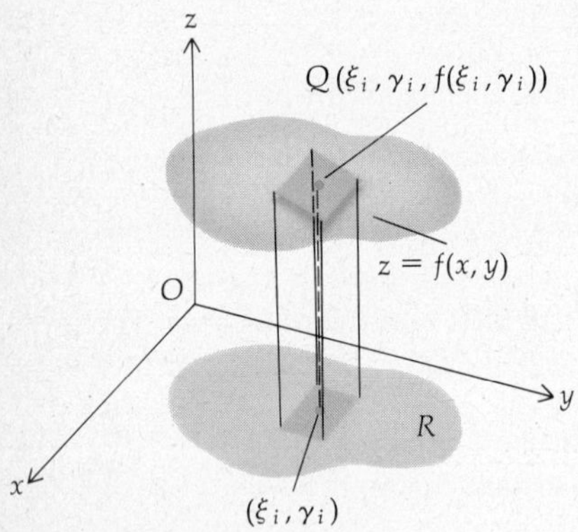

FIGURE 20.5.1

$$\sum_{i=1}^{n} \Delta_i\sigma$$

is an approximation to the measure σ of the area of the portion of the surface that lies above R. This leads to defining σ as follows:

$$\sigma = \lim_{\|\Delta\| \to 0} \sum_{i=1}^{n} \Delta_i\sigma \tag{1}$$

We now need to obtain a formula for computing the limit in Eq. (1). To do this we find a formula for computing $\Delta_i\sigma$ as the measure of the area of a parallelogram. For simplicity in computation take the point (ξ_i, γ_i) in the ith rectangle at the corner (x_{i-1}, y_{i-1}). Let **A** and **B** be vectors having as representations the directed line segments having initial points at Q and forming the two adjacent sides of the parallelogram whose area has measure $\Delta_i\sigma$. See Fig. 20.5.2. Then $\Delta_i\sigma = |\mathbf{A} \times \mathbf{B}|$. Because

$$\mathbf{A} = \Delta_i x\mathbf{i} + f_x(\xi_i, \gamma_i)\,\Delta_i x\mathbf{k}$$

and

$$\mathbf{B} = \Delta_i y\mathbf{j} + f_y(\xi_i, \gamma_i)\,\Delta_i y\mathbf{k}$$

it follows that

$$\mathbf{A} \times \mathbf{B} = \begin{vmatrix} \mathbf{i} & \mathbf{j} & \mathbf{k} \\ \Delta_i x & 0 & f_x(\xi_i, \gamma_i)\,\Delta_i x \\ 0 & \Delta_i y & f_y(\xi_i, \gamma_i)\,\Delta_i y \end{vmatrix}$$

$$= -\Delta_i x\,\Delta_i y f_x(\xi_i, \gamma_i)\mathbf{i} - \Delta_i x\,\Delta_i y f_y(\xi_i, \gamma_i)\mathbf{j} + \Delta_i x\,\Delta_i y\mathbf{k}$$

Therefore

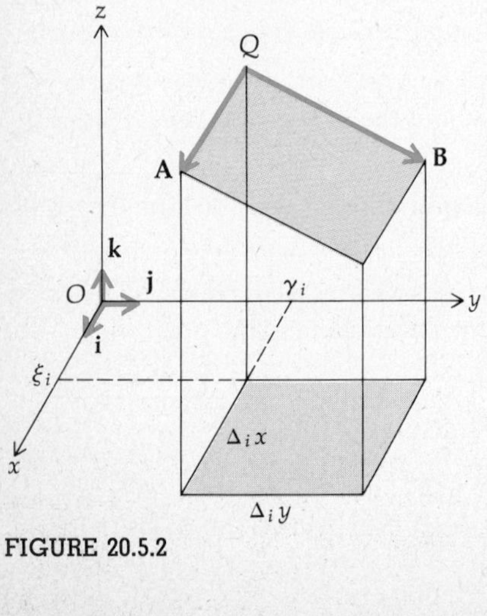

FIGURE 20.5.2

$$\Delta_i\sigma = |\mathbf{A} \times \mathbf{B}| = \sqrt{f_x^{\,2}(\xi_i, \gamma_i) + f_y^{\,2}(\xi_i, \gamma_i) + 1}\;\Delta_i x\,\Delta_i y \tag{2}$$

Substituting from (2) into (1) we get

$$\sigma = \lim_{\|\Delta\| \to 0} \sum_{i=1}^{n} \sqrt{f_x^2(\xi_i, \gamma_i) + f_y^2(\xi_i, \gamma_i) + 1}\, \Delta_i x\, \Delta_i y$$

This limit is a double integral that exists on R because of the continuity of f_x and f_y on R. We have, then, the following theorem.

20.5.1 Theorem Suppose that f and its first partial derivatives are continuous on the closed region R in the xy plane. Then if σ is the measure of the area of the surface $z = f(x, y)$ that lies over R,

$$\sigma = \iint_R \sqrt{f_x^2(x, y) + f_y^2(x, y) + 1}\, dx\, dy \tag{3}$$

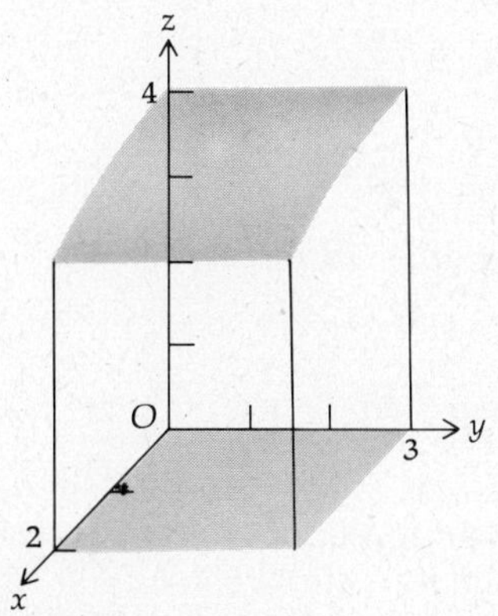

FIGURE 20.5.3

Example 1 Find the area of the surface that is cut from the cylinder $x^2 + z^2 = 16$ by the planes $x = 0$, $x = 2$, $y = 0$, and $y = 3$.

Solution The given surface is shown in Fig. 20.5.3. The region R is the rectangle in the first quadrant of the xy plane bounded by the lines $x = 2$ and $y = 3$. The surface has the equation $x^2 + z^2 = 16$. Solving for z we get $z = \sqrt{16 - x^2}$. Hence $f(x, y) = \sqrt{16 - x^2}$. So if σ is the measure of the area of the surface, then from Theorem 20.5.1,

$$\begin{aligned}
\sigma &= \iint_R \sqrt{f_x^2(x, y) + f_y^2(x, y) + 1}\, dx\, dy \\
&= \int_0^3 \int_0^2 \sqrt{\left(\frac{-x}{\sqrt{16 - x^2}}\right)^2 + 0 + 1}\, dx\, dy \\
&= \int_0^3 \int_0^2 \frac{4}{\sqrt{16 - x^2}}\, dx\, dy \\
&= 4 \int_0^3 \left[\sin^{-1} \tfrac{1}{4}x\right]_0^2 dy \\
&= 4 \int_0^3 \tfrac{1}{6}\pi\, dy \\
&= 2\pi
\end{aligned}$$

The surface area is therefore 2π square units.

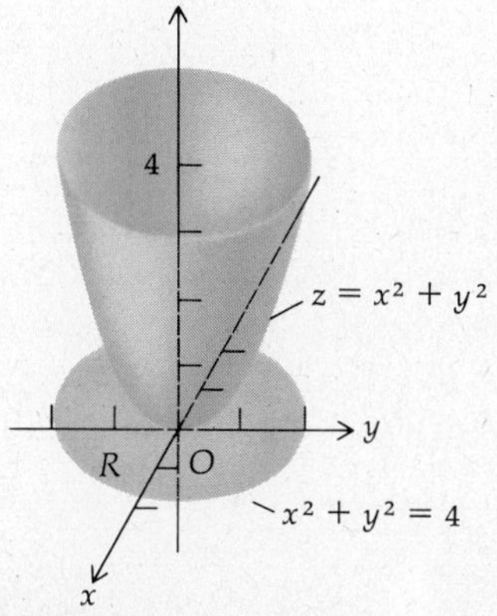

FIGURE 20.5.4

Example 2 Find the area of the paraboloid $z = x^2 + y^2$ below the plane $z = 4$.

Solution Figure 20.5.4 shows the given surface. From the equation of the paraboloid we see that $f(x, y) = x^2 + y^2$. The closed region in the xy plane bounded by the circle $x^2 + y^2 = 4$ is the region R. If σ square units is the required surface area, then from Theorem 20.5.1,

$$\sigma = \iint_R \sqrt{f_x^2(x, y) + f_y^2(x, y) + 1}\, dx\, dy$$

$$= \iint_R \sqrt{4(x^2 + y^2) + 1}\, dx\, dy$$

Because the integrand contains the terms $4(x^2 + y^2)$, the evaluation of the double integral is simplified by using polar coordinates. Then $x^2 + y^2 = r^2$, and $dx\, dy = dA = r\, dr\, d\theta$. Furthermore, the limits for r are from 0 to 2 and the limits for θ are from 0 to 2π. Therefore

$$\sigma = \iint_R \sqrt{4r^2 + 1}\, r\, dr\, d\theta$$

$$= \int_0^{2\pi} \int_0^2 \sqrt{4r^2 + 1}\, r\, dr\, d\theta$$

$$= \int_0^{2\pi} \left[\tfrac{1}{12}(4r^2 + 1)^{3/2} \right]_0^2 d\theta$$

$$= \tfrac{1}{6}\pi(17\sqrt{17} - 1)$$

Hence the area of the paraboloid below the given plane is $\frac{1}{6}\pi(17\sqrt{17} - 1)$ square units.

Example 3 Find the area of the top half of the sphere $x^2 + y^2 + z^2 = a^2$.

Solution The hemisphere is shown in Fig. 20.5.5. Solving the equation of the sphere for z and setting this equal to $f(x, y)$ we get

$$f(x, y) = \sqrt{a^2 - x^2 - y^2}$$

Because $f_x(x, y) = -x/\sqrt{a^2 - x^2 - y^2}$ and $f_y(x, y) = -y/\sqrt{a^2 - x^2 - y^2}$, f_x and f_y are not defined on the circle $x^2 + y^2 = a^2$, which is the boundary of the region R in the xy plane. Furthermore, the double integral obtained from (3) is

$$\iint_R \frac{a}{\sqrt{a^2 - x^2 - y^2}}\, dx\, dy$$

which is improper because the integrand has an infinite discontinuity at each point of the boundary of R. We can take care of this situation by considering the region R' as that bounded by the circle $x^2 + y^2 = b^2$, where $b < a$, and then take the limit as $b \to a^-$. Furthermore, the computation is simplified if the double integral is evaluated by an iterated integral using polar coordinates. Then

$$\sigma = \lim_{b \to a^-} \int_0^b \int_0^{2\pi} \frac{a}{\sqrt{a^2 - r^2}}\, r\, d\theta\, dr$$

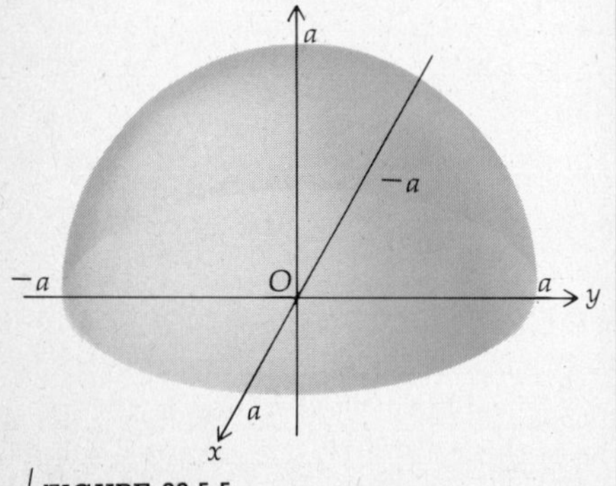

FIGURE 20.5.5

$$= 2\pi a \lim_{b \to a^-} \int_0^b \frac{r}{\sqrt{a^2 - r^2}}\, dr$$
$$= 2\pi a \lim_{b \to a^-} \left[-\sqrt{a^2 - r^2} \right]_0^b$$
$$= 2\pi a \lim_{b \to a^-} [-\sqrt{a^2 - b^2} + a]$$
$$= 2\pi a^2$$

The area of the hemisphere is therefore $2\pi a^2$ square units.

Consider now the curve $y = F(x)$ with $a \le x \le b$, $F(x) > 0$ on $[a, b]$ and F' continuous on $[a, b]$. If this curve is rotated about the x axis, we obtain a surface of revolution. From Sec. 17.7 an equation of this surface is

$$y^2 + z^2 = [F(x)]^2 \tag{4}$$

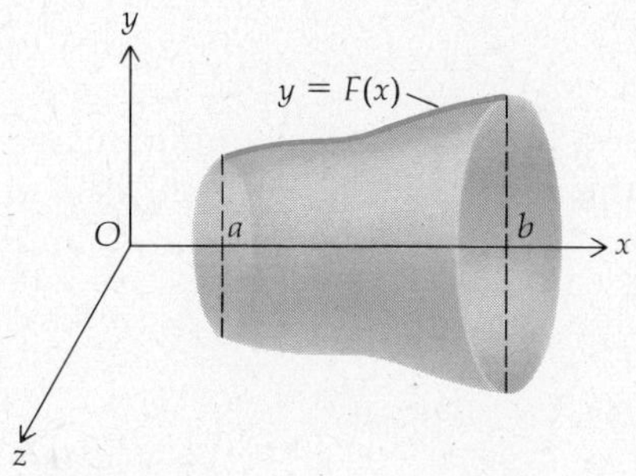

FIGURE 20.5.6

Figure 20.5.6 shows the surface of revolution. In the figure the xy plane is in the plane of the paper; however, we still have a right-handed system. We wish to obtain a formula for finding the measure of the area of this surface of revolution by using Theorem 20.5.1. From properties of symmetry, the measure of the area of the surface above the xz plane and in front of the xy plane is one-fourth of the measure of the area of the entire surface. Solving (4) for z (neglecting the negative square root because $z \ge 0$), we get $f(x, y) = \sqrt{[F(x)]^2 - y^2}$. The region R in the xy plane is that bounded by the x axis, the curve $y = F(x)$, and the lines $x = a$ and $x = b$. Computing the partial derivatives of f we obtain

$$f_x(x, y) = \frac{F(x)F'(x)}{\sqrt{[F(x)]^2 - y^2}} \qquad f_y(x, y) = \frac{-y}{\sqrt{[F(x)]^2 - y^2}}$$

We see that $f_x(x, y)$ and $f_y(x, y)$ do not exist on part of the boundary of R (when $y = -F(x)$ and when $y = F(x)$). The double integral obtained from (3) is

$$\iint_R \sqrt{\frac{[F(x)]^2[F'(x)]^2}{[F(x)]^2 - y^2} + \frac{y^2}{[F(x)]^2 - y^2} + 1}\; dy\, dx$$
$$= \iint_R \frac{F(x)\sqrt{[F'(x)]^2 + 1}}{\sqrt{[F(x)]^2 - y^2}}\, dy\, dx$$

This double integral is improper because the integrand has an infinite discontinuity at each point of the boundary of R where $y = -F(x)$ and $y = F(x)$. Hence we evaluate the double integral by an iterated integral for which the inner integral is improper.

$$\sigma = 4 \int_a^b \left[F(x)\sqrt{[F'(x)]^2 + 1} \int_0^{F(x)} \frac{dy}{\sqrt{[F(x)]^2 - y^2}} \right] dx \tag{5}$$

where

$$\int_0^{F(x)} \frac{dy}{\sqrt{[F(x)]^2 - y^2}} = \lim_{\epsilon \to 0^+} \int_0^{F(x)-\epsilon} \frac{dy}{\sqrt{[F(x)]^2 - y^2}}$$

$$= \lim_{\epsilon \to 0^+} \left[\sin^{-1} \frac{y}{F(x)} \right]_0^{F(x)-\epsilon}$$

$$= \lim_{\epsilon \to 0^+} \sin^{-1} \left(1 - \frac{\epsilon}{F(x)} \right)$$

$$= \tfrac{1}{2}\pi$$

Therefore, from (5),

$$\sigma = 2\pi \int_a^b F(x) \sqrt{[F'(x)]^2 + 1}\, dx$$

We state this result as a theorem, where F is replaced by f.

20.5.2 Theorem Suppose that the function f is positive on $[a, b]$ and f' is continuous on $[a, b]$. If σ is the measure of the area of the surface of revolution obtained by revolving the curve $y = f(x)$, with $a \le x \le b$, about the x axis, then

$$\sigma = 2\pi \int_a^b f(x) \sqrt{[f'(x)]^2 + 1}\, dx$$

Example 4 Find the area of the paraboloid of revolution generated by revolving the top half of the parabola $y^2 = 4px$, with $0 \le x \le h$, about the x axis.

Solution The paraboloid of revolution is shown in Fig. 20.5.7. Solving the equation of the parabola for y ($y \ge 0$) we obtain $y = 2p^{1/2}x^{1/2}$. So if σ square units is the area of the surface, from Theorem 20.5.2, with $f(x) = 2p^{1/2}x^{1/2}$,

$$\sigma = 2\pi \int_0^h 2p^{1/2}x^{1/2} \sqrt{\frac{p}{x} + 1}\, dx$$

$$= 4\pi p^{1/2} \int_0^h \sqrt{p + x}\, dx$$

$$= \tfrac{8}{3}\pi p^{1/2}(p + x)^{3/2} \Big]_0^h$$

$$= \tfrac{8}{3}\pi\left(\sqrt{p(p + h)^3} - p^2\right)$$

The area of the paraboloid of revolution is therefore $\frac{8}{3}\pi(\sqrt{p(p + h)^3} - p^2)$ square units.

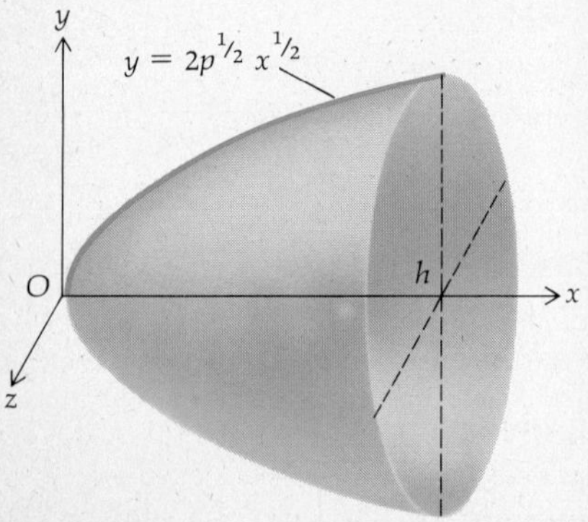

FIGURE 20.5.7

EXERCISES 20.5

1. Find the area of the surface cut from the plane $2x + y + z = 4$ by the planes $x = 0$, $x = 1$, $y = 0$, and $y = 1$.

2. Find the area of the surface cut from the plane $z - 2x - y = 5$ by the planes $x = 0$, $x = 2$, $y = 0$, and $y = 4$.

3. Find the area of the portion of the surface of the plane $36x + 16y + 9z = 144$ that is cut by the coordinate planes.

4. Find the area of the surface that is cut from the plane $z = ax + by$ by the planes $x = 0$, $x = a$, $y = 0$, and $y = b$, where $a > 0$ and $b > 0$.

5. Find the area of the surface in the first octant that is cut from the cylinder $x^2 + y^2 = 9$ by the plane $x = z$.

6. Find the area of the surface that is cut from the cylinder $x^2 + y^2 = 25$ by the planes $x = 0$, $x = 1$, $z = 1$, and $z = 3$.

7. Let R be the triangular region in the xy plane with vertices at $(0, 0, 0)$, $(0, 4, 0)$, and $(2, 4, 0)$. Find the surface area of the part of the graph of $z - 5x - y^2 = 2$ that lies above R.

8. Find the area of the surface in the first octant cut from the cone $x^2 + y^2 = z^2$ by the plane $x + y = 4$.

9. Find the area of the portion of the surface of the sphere $x^2 + y^2 + z^2 = 4x$ that is cut out by one nappe of the cone $y^2 + z^2 = x^2$.

10. Find the area of the portion of the surface of the sphere $x^2 + y^2 + z^2 = 36$ that lies within the cylinder $x^2 + y^2 = 9$.

11. Find the area of the portion of the surface of the sphere $x^2 + y^2 + z^2 = 4z$ that lies within the paraboloid $x^2 + y^2 = 3z$.

12. For the sphere and paraboloid of Exercise 11, find the area of the portion of the surface of the paraboloid that lies within the sphere.

13. The line segment from the origin to the point (a, b) is revolved about the x axis. Find the area of the surface of the cone generated.

14. Derive the formula for the area of the surface of a sphere by revolving a semicircle about its diameter.

15. Find the area of the surface of revolution obtained by revolving the arc of the catenary $y = a \cosh(x/a)$ from $x = 0$ to $x = a$ about the y axis.

16. Find the area of the surface of revolution obtained by revolving the catenary of Exercise 15 about the x axis.

17. The loop of the curve $18y^2 = x(6 - x)^2$ is revolved about the x axis. Find the area of the surface of revolution generated.

18. Find the area of the surface of revolution generated by revolving the arc of the curve $y = \ln x$ from $x = 1$ to $x = 2$ about the y axis.

19. Find the area of the portion of the plane $x = z$ that lies between the planes $y = 0$ and $y = 6$ and within the hyperboloid $9x^2 - 4y^2 + 16z^2 = 144$.

20. Find the area of the surface cut from the hyperbolic paraboloid $y^2 - x^2 = 6z$ by the cylinder $x^2 + y^2 = 36$.

21. Suppose that f and its first partial derivatives are continuous on the closed region R in the xy plane. Show that if σ is the measure of the area of the portion of the surface $z = f(x, y)$ that lies over R, then

$$\sigma = \iint_R \left| \nabla g(x, y, z) \right| dx\, dy$$

where $g(x, y, z) = z - f(x, y)$.

20.6 GREEN'S THEOREM

There is a theorem that expresses a double integral over a plane region R in terms of a line integral around a curve that is a boundary of R. This theorem is named after the English mathematician and physicist George

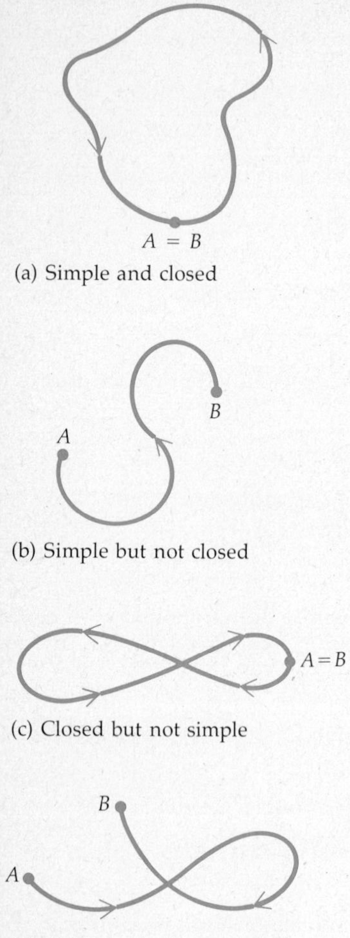

(a) Simple and closed

(b) Simple but not closed

(c) Closed but not simple

(d) Neither simple nor closed

FIGURE 20.6.1

Green (1793–1841), who introduced the theorem in a paper on the applications of mathematics to electricity and magnetism. Before stating the theorem, it is necessary to review and introduce some terminology pertaining to plane curves.

In Sec. 19.6 we stated that if a curve C is defined by the parametric equations

$$x = f(t) \qquad y = g(t) \qquad a \leq t \leq b \tag{1}$$

then C is said to be smooth on the closed interval $[a, b]$ if f' and g' are continuous on $[a, b]$ and $f'(t)$ and $g'(t)$ are not both zero at every point in the open interval (a, b). If for a curve C defined by parametric equations (1) the initial point $A(f(a), g(a))$ and the terminal point $B(f(b), g(b))$ coincide, then curve C is said to be **closed.** A curve C is said to be **simple** if it does not intersect itself between the points A and B, that is, if $(f(t_1), g(t_1)) \neq (f(t_2), g(t_2))$ for all t_1 and t_2 in the open interval (a, b). The circle and the ellipse are examples of smooth simple closed curves. In Fig. 20.6.1(a)–(d) there are further examples of smooth curves that may or may not be simple and closed. In (a) the curve is both simple and closed; in (b) the curve is simple but not closed; in (c) the curve is closed but not simple; and in (d) the curve is neither simple nor closed.

Recall from Sec. 19.6 that a curve C is said to be sectionally smooth on an interval I if I can be partitioned into a finite number of subintervals on which C is smooth. In the statement of Green's theorem we refer to a line integral around a sectionally smooth simple closed curve C that forms the boundary of a region R in the plane, and the direction along C is counterclockwise. In Fig. 20.6.2 such a region R is shown with the required boundary curve C. The line integral around C in the counterclockwise direction is denoted by $\oint_C$.

20.6.1 Theorem
Green's Theorem

Let M and N be functions of two variables x and y such that they have continuous first partial derivatives on an open disk B in R^2. If C is a sectionally smooth simple closed curve lying entirely in B, and if R is the region defined by C, then

$$\oint_C M(x, y)\, dx + N(x, y)\, dy = \iint_R \left(\frac{\partial N}{\partial x} - \frac{\partial M}{\partial y} \right) dA \tag{2}$$

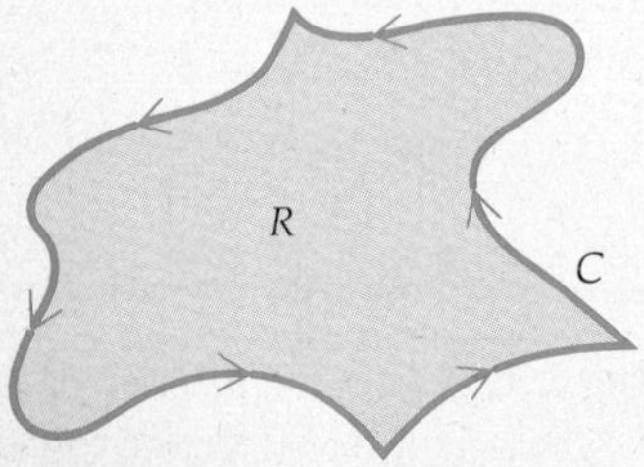

FIGURE 20.6.2

The proof of Green's theorem for all regions bounded by curves that are sectionally smooth, simple, and closed belongs to a course in advanced calculus. However, we shall prove the theorem for a particular kind of region, one for which each horizontal line and each vertical line intersect it in at most two points. The proof follows.

Proof Let R be a region in the xy plane that can be defined by either

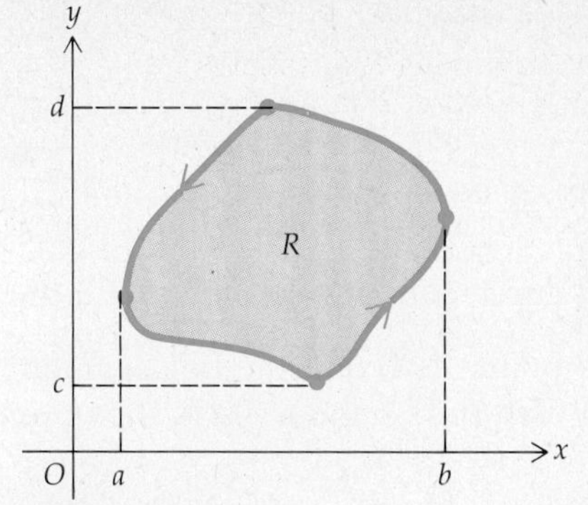

FIGURE 20.6.3

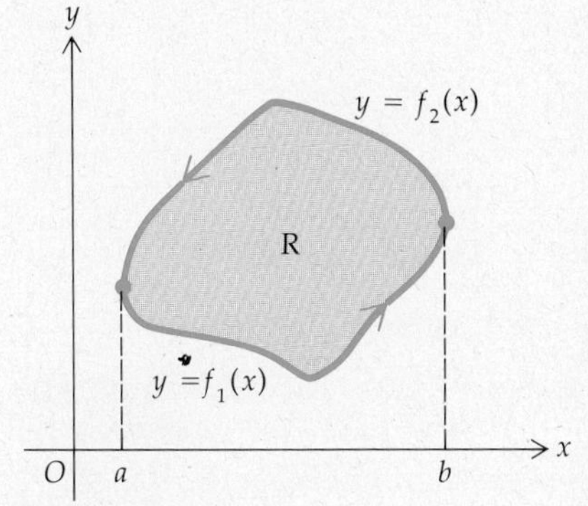

FIGURE 20.6.4

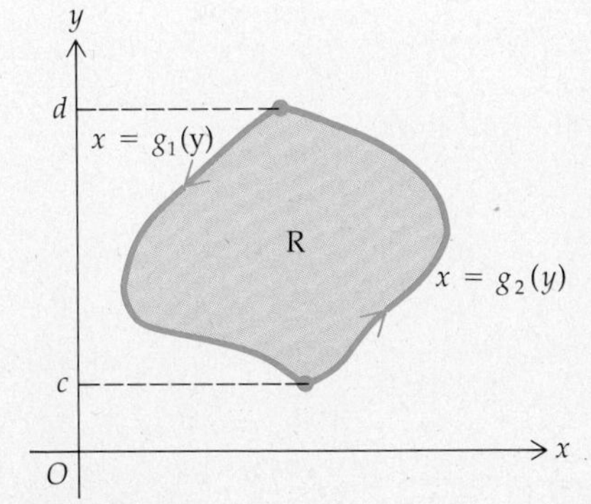

FIGURE 20.6.5

$$R = \{(x, y) | \; a \le x \le b, f_1(x) \le y \le f_2(x)\} \tag{3}$$

or

$$R = \{(x, y) | \; c \le y \le d, g_1(y) \le x \le g_2(y)\} \tag{4}$$

where the functions f_1, f_2, g_1, and g_2 are smooth. Figure 20.6.3 shows such a region R. Region R is regarded as being defined by (3) in Fig. 20.6.4 and by (4) in Fig. 20.6.5.

The proof of (2) will consist of showing that

$$\oint_C M(x, y)\, dx = -\iint_R \frac{\partial M}{\partial y}\, dA \tag{5}$$

and

$$\oint_C N(x, y)\, dy = \iint_R \frac{\partial N}{\partial x}\, dA \tag{6}$$

To prove (5) we treat R as a region defined by (3). Refer to Fig. 20.6.4. Let C_1 be the graph of $y = f_1(x)$ from $x = a$ to $x = b$; that is, C_1 is the lower part of the boundary curve C going from left to right. Let C_2 be the graph of $y = f_2(x)$ from $x = b$ to $x = a$; that is, C_2 is the upper part of the boundary curve C going from right to left. Consider the line integral $\oint_C M(x, y)\, dx$.

$$\begin{aligned}\oint_C M(x, y)\, dx &= \int_{C_1} M(x, y)\, dx + \int_{C_2} M(x, y)\, dx \\ &= \int_a^b M(x, f_1(x))\, dx + \int_b^a M(x, f_2(x))\, dx \\ &= \int_a^b M(x, f_1(x))\, dx - \int_a^b M(x, f_2(x))\, dx \\ &= \int_a^b [M(x, f_1(x)) - M(x, f_2(x))]\, dx \qquad (7)\end{aligned}$$

We now deal with the double integral $\iint_R \frac{\partial M}{\partial y}\, dA$ where R is still considered to be defined by (3). Then

$$\begin{aligned}\iint_R \frac{\partial M}{\partial y}\, dA &= \int_a^b \int_{f_1(x)}^{f_2(x)} \frac{\partial M}{\partial y}\, dy\, dx \\ &= \int_a^b \left(\int_{f_1(x)}^{f_2(x)} \frac{\partial M}{\partial y}\, dy \right) dx \\ &= \int_a^b M(x, y) \Big]_{f_1(x)}^{f_2(x)}\, dx \\ &= \int_a^b [M(x, f_2(x)) - M(x, f_1(x))]\, dx \qquad (8)\end{aligned}$$

By comparing (7) and (8) it follows that (5) holds.

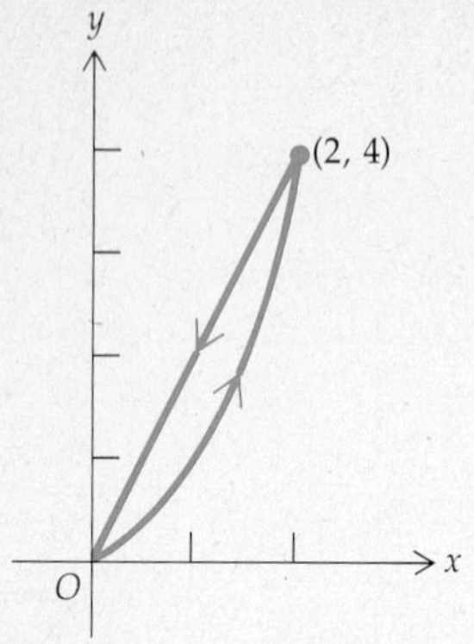

FIGURE 20.6.6

To prove (6), R is regarded as a region defined by (4) as in Fig. 20.6.5. The details of the proof are left as an exercise (see Exercise 39).

By adding corresponding members of Eqs. (5) and (6) we obtain Green's theorem for this region R. ■

● ILLUSTRATION 1 We use Green's theorem to evaluate the line integral $\oint_C y^2\,dx + 4xy\,dy$ where C is the closed curve consisting of the arc of the parabola $y = x^2$ from the origin to the point (2, 4) and the line segment from (2, 4) to the origin. The region R with the boundary C is shown in Fig. 20.6.6. From Green's theorem,

$$\begin{aligned}\oint_C y^2\,dx + 4xy\,dy &= \iint_R \left[\frac{\partial}{\partial x}(4xy) - \frac{\partial}{\partial y}(y^2)\right] dA\\ &= \int_0^2 \int_{x^2}^{2x} (4y - 2y)\,dy\,dx\\ &= \int_0^2 y^2\Big]_{x^2}^{2x} dx\\ &= \int_0^2 (4x^2 - x^4)\,dx\\ &= \tfrac{4}{3}x^3 - \tfrac{1}{5}x^5\Big]_0^2\\ &= \tfrac{64}{15}\end{aligned}$$

To show the advantage of using Green's theorem we evaluate the same line integral by the method of Sec. 19.6. If C_1 is the arc of the parabola $y = x^2$ from (0, 0) to (2, 4) and C_2 is the line segment from (2, 4) to (0, 0), then

$$\oint_C y^2\,dx + 4xy\,dy = \oint_{C_1} y^2\,dx + 4xy\,dy + \oint_{C_2} y^2\,dx + 4xy\,dy$$

Parametric equations for C_1 are

$$x = t \qquad y = t^2 \qquad 0 \le t \le 2$$

Therefore

$$\begin{aligned}\oint_{C_1} y^2\,dx + 4xy\,dy &= \int_0^2 (t^2)^2\,dt + 4(t)(t^2)(2t\,dt)\\ &= \int_0^2 9t^4\,dt\\ &= \tfrac{9}{5}t^5\Big]_0^2\\ &= \tfrac{288}{5}\end{aligned}$$

Arc C_2 can be represented parametrically by

$$x = t \qquad y = 2t \qquad \text{from } t = 2 \text{ to } t = 0$$

Thus

$$\begin{aligned}\oint_{C_2} y^2\,dx + 4xy\,dy &= \int_2^0 (2t)^2\,dt + 4(t)(2t)(2\,dt)\\ &= \int_2^0 20t^2\,dt\\ &= \tfrac{20}{3}t^3\Big]_2^0\\ &= -\tfrac{160}{3}\end{aligned}$$

Hence

$$\begin{aligned}\oint_C y^2\,dx + 4xy\,dy &= \tfrac{288}{5} - \tfrac{160}{3}\\ &= \tfrac{64}{15}\end{aligned}$$

which agrees with the result obtained by using Green's theorem. ●

Example 1 Use Green's theorem to find the total work done in moving an object in the counterclockwise direction once around the circle $x^2 + y^2 = a^2$ if the motion is caused by the force field $\mathbf{F}(x, y) = (\sin x - y)\mathbf{i} + (e^y - x^2)\mathbf{j}$. Assume the arc is measured in meters and the force is measured in newtons.

Solution If W joules is the work done, then

$$W = \oint_C (\sin x - y)\,dx + (e^y - x^2)\,dy$$

where C is the circle $x^2 + y^2 = a^2$. From Green's theorem,

$$\begin{aligned}W &= \iint_R \left[\frac{\partial}{\partial x}(e^y - x^2) - \frac{\partial}{\partial y}(\sin x - y)\right] dA\\ &= \iint_R (-2x + 1)\,dA\end{aligned}$$

We use polar coordinates to evaluate the double integral, with $x = r\cos\theta$ and $dA = r\,dr\,d\theta$. Then

$$\begin{aligned}W &= \int_0^{2\pi}\int_0^a (-2r\cos\theta + 1)r\,dr\,d\theta\\ &= \int_0^{2\pi}\int_0^a (-2r^2\cos\theta + r)\,dr\,d\theta\\ &= \int_0^{2\pi} -\frac{2}{3}r^3\cos\theta + \frac{r^2}{2}\Big]_0^a\,d\theta\end{aligned}$$

$$= \int_0^{2\pi} \left(-\frac{2}{3}a^3 \cos\theta + \frac{a^2}{2}\right) d\theta$$

$$= -\frac{2}{3}a^3 \sin\theta + \frac{a^2}{2}\theta \Big]_0^{2\pi}$$

$$= \pi a^2$$

Therefore the work done is πa^2 joules.

The following theorem, which is a consequence of Green's theorem, gives a useful method for computing the area of a region bounded by a sectionally smooth simple closed curve.

20.6.2 Theorem If R is a region having as its boundary a sectionally smooth simple closed curve C, and A square units is the area of R, then

$$A = \frac{1}{2}\oint_C x\,dy - y\,dx$$

Proof In the statement of Green's theorem, let $M(x, y) = -\frac{1}{2}y$ and $N(x, y) = \frac{1}{2}x$. Then

$$\oint_C -\tfrac{1}{2}y\,dx + \tfrac{1}{2}x\,dy = \iint_R \left[\frac{\partial}{\partial x}(\tfrac{1}{2}x) - \frac{\partial}{\partial y}(-\tfrac{1}{2}y)\right] dA$$

$$= \iint_R (\tfrac{1}{2} + \tfrac{1}{2})\,dA$$

$$= \iint_R dA$$

Because $\iint_R dA$ is the measure of the area of R,

$$\frac{1}{2}\oint_C x\,dy - y\,dx = A$$

■

Example 2 Use Theorem 20.6.2 to find the area of the region enclosed by the ellipse $x^2/a^2 + y^2/b^2 = 1$.

Solution Parametric equations for the ellipse are

$$x = a\cos t \qquad y = b\sin t \qquad 0 \le t \le 2\pi$$

Then $dx = -a\sin t\,dt$ and $dy = b\cos t\,dt$. If C is the ellipse and A square units is the area of the region enclosed by C, then from Theorem 20.6.2,

$$A = \frac{1}{2}\oint_C x\,dy - y\,dx$$

$$= \frac{1}{2}\int_0^{2\pi} [(a\cos t)(b\cos t\,dt) - (b\sin t)(-a\sin t\,dt)]$$

$$= \frac{1}{2}\int_0^{2\pi} ab(\cos^2 t + \sin^2 t)\,dt$$

$$= \tfrac{1}{2}ab\int_0^{2\pi} dt$$

$$= \pi ab$$

Therefore the area is πab square units.

Example 3 Use Green's theorem to evaluate the line integral $\oint_C (x^4 - 3y)\,dx + (2y^3 + 4x)\,dy$ if C is the ellipse $9x^2 + 4y^2 = 36$.

Solution From Green's theorem,

$$\oint_C (x^4 - 3y)\,dx + (2y^3 + 4x)\,dy = \iint_R \left[\frac{\partial}{\partial x}(2y^3 + 4x) - \frac{\partial}{\partial y}(x^4 - 3y)\right] dA$$

$$= \iint_R (4 + 3)\,dA$$

$$= 7\iint_R dA$$

The double integral $\iint_R dA$ is the measure of the area of the region enclosed by the ellipse. The equation of the ellipse is $x^2/4 + y^2/9 = 1$, and so from Example 2 with $a = 2$ and $b = 3$, the measure of the area of the region enclosed by the ellipse is 6π. Therefore

$$\oint_C (x^4 - 3y)\,dx + (2y^3 + 4x)\,dy = 42\pi$$

There are two vector forms of Green's theorem that we shall proceed to obtain. The two scalar quantities given in the following definition are needed for this development.

20.6.3 Definition If $\mathbf{F}(x, y) = M(x, y)\mathbf{i} + N(x, y)\mathbf{j}$, then the **scalar curl** of $\mathbf{F}$ is given by

$$\operatorname{curl} \mathbf{F} = \frac{\partial N}{\partial x} - \frac{\partial M}{\partial y}$$

and the **divergence** of $\mathbf{F}$ is given by

$$\operatorname{div} \mathbf{F} = \frac{\partial M}{\partial x} + \frac{\partial N}{\partial y}$$

Suppose a vector equation of a curve C is

$$\mathbf{R}(s) = x\mathbf{i} + y\mathbf{j}$$

and $x = f(s)$ and $y = g(s)$ where s units is the length of arc measured from a particular point P_0 on C to the point P on C. Then

$$D_s\mathbf{R}(s) = \frac{dx}{ds}\mathbf{i} + \frac{dy}{ds}\mathbf{j}$$

From Theorem 16.8.3, if $\mathbf{T}(s)$ is the unit tangent vector of C at P,

$$\mathbf{T}(s) = \frac{dx}{ds}\mathbf{i} + \frac{dy}{ds}\mathbf{j} \tag{9}$$

With the notation

$$d\mathbf{R} = dx\mathbf{i} + dy\mathbf{j}$$

we write the relationship given by (9) in the form

$$\mathbf{T}(s)\,ds = d\mathbf{R} \tag{10}$$

If

$$\mathbf{F}(x, y) = M(x, y)\mathbf{i} + N(x, y)\mathbf{j}$$

then

$$\mathbf{F}(x, y)\cdot d\mathbf{R} = M(x, y)\,dx + N(x, y)\,dy \tag{11}$$

Substituting from (10) into (11) we have

$$\mathbf{F}(x, y)\cdot\mathbf{T}(s)\,ds = M(x, y)\,dx + N(x, y)\,dy \tag{12}$$

Thus from (12) and Definition 20.6.3, Eq. (2) of Green's theorem can be written

$$\oint_C \mathbf{F}(x, y)\cdot T(s)\,ds = \iint_R \text{curl}\,\mathbf{F}\,dA$$

This vector form of Green's theorem is stated formally as the following theorem named after the Irish mathematician and physicist George Stokes (1819–1903).

20.6.4 Theorem
Stokes' Theorem in the Plane

Let the functions M and N, the curve C, and the region R be as defined in Green's theorem. If $\mathbf{F}(x, y) = M(x, y)\mathbf{i} + N(x, y)\mathbf{j}$ and $\mathbf{T}(s)$ is the unit tangent vector of C at P where s units is the length of arc measured from a particular point P_0 on C to P, then

$$\oint_C \mathbf{F}\cdot\mathbf{T}\,ds = \iint_R \text{curl}\,\mathbf{F}\,dA$$

Example 4 Verify Stokes' theorem in the plane if $\mathbf{F}(x, y) = 2y\mathbf{i} + 5x\mathbf{j}$ and R is the region bounded by the circle $x^2 + y^2 = 1$.

Solution The boundary of R is the unit circle that can be represented parametrically by

$$x = \cos s \qquad y = \sin s \qquad 0 \le s \le 2\pi$$

where s units is the length of arc from the point where $s = 0$ to the point P on C. Then a vector equation of C is

$$\mathbf{R}(s) = \cos s\mathbf{i} + \sin s\mathbf{j}$$

and because $\mathbf{T}(s) = D_s\mathbf{R}(s)$,

$$\mathbf{T}(s) = -\sin s\mathbf{i} + \cos s\mathbf{j}$$

At a point $P(\cos s, \sin s)$ on C, $\mathbf{F}$ is $2 \sin s\mathbf{i} + 5 \cos s\mathbf{j}$. Therefore

$$\begin{aligned}
\oint_C \mathbf{F}\cdot\mathbf{T}\, ds &= \int_0^{2\pi} (2 \sin s\mathbf{i} + 5 \cos s\mathbf{j})\cdot(-\sin s\mathbf{i} + \cos s\mathbf{j})\, ds \\
&= \int_0^{2\pi} (-2 \sin^2 s + 5 \cos^2 s)\, ds \\
&= -2 \int_0^{2\pi} \frac{1 - \cos 2s}{2}\, ds + 5 \int_0^{2\pi} \frac{1 + \cos 2s}{2}\, ds \\
&= -s + \tfrac{1}{2} \sin 2s + \tfrac{5}{2}s + \tfrac{5}{4} \sin 2s \Big]_0^{2\pi} \\
&= \tfrac{3}{2}s + \tfrac{7}{4} \sin 2s \Big]_0^{2\pi} \\
&= 3\pi
\end{aligned}$$

Because

$$\begin{aligned}
\operatorname{curl} \mathbf{F} &= \frac{\partial N}{\partial x} - \frac{\partial M}{\partial y} \\
&= \frac{\partial}{\partial x}(5x) - \frac{\partial}{\partial y}(2y) \\
&= 5 - 2 \\
&= 3
\end{aligned}$$

and $\iint_R dA = \pi$, then

$$\begin{aligned}
\iint_R \operatorname{curl} \mathbf{F}\, dA &= \iint_R 3\, dA \\
&= 3\pi
\end{aligned}$$

Thus Stokes' theorem in the plane is verified for this $\mathbf{F}$ and R.

To obtain the second vector form of Green's theorem consider Eq. (10), which expresses $d\mathbf{R}$ in terms of the unit tangent vector $\mathbf{T}(s)$. Writing the equation with $dx\mathbf{i} + dy\mathbf{j}$ in place of $d\mathbf{R}$ we have

$$dx\mathbf{i} + dy\mathbf{j} = \mathbf{T}(s)\, ds$$

The vector $\mathbf{N}(s)$ defined by the equation

$$dy\mathbf{i} - dx\mathbf{j} = \mathbf{N}(s)\, ds$$

is a unit normal vector of C at P. To verify this fact observe that

$$\begin{aligned}(\mathbf{T}(s)\, ds)\cdot(\mathbf{N}(s)\, ds) &= (dx\mathbf{i} + dy\mathbf{j})\cdot(dy\mathbf{i} - dx\mathbf{j})\\ &= 0\end{aligned}$$

and the magnitudes of $\mathbf{T}(s)\, ds$ and $\mathbf{N}(s)\, ds$ are equal. Because

$$\begin{aligned}\mathbf{F}(x, y)\cdot\mathbf{N}(s)\, ds &= [M(x, y)\mathbf{i} + N(x, y)\mathbf{j}]\cdot(dy\mathbf{i} - dx\mathbf{j})\\ &= M(x, y)\, dy - N(x, y)\, dx\end{aligned}$$

then

$$\oint_C \mathbf{F}(x, y)\cdot\mathbf{N}(s)\, ds = \oint_C -N(x, y)\, dx + M(x, y)\, dy \tag{13}$$

We apply Green's theorem to the line integral on the right side of (13) and we have

$$\begin{aligned}\oint_C \mathbf{F}(x, y)\cdot\mathbf{N}(s)\, ds &= \iint_R \left[\frac{\partial M}{\partial x} - \frac{\partial}{\partial y}(-N)\right] dA\\ &= \iint_R \left(\frac{\partial M}{\partial x} + \frac{\partial N}{\partial y}\right) dA\\ &= \iint_R \operatorname{div} \mathbf{F}\, dA\end{aligned}$$

Therefore we have the following theorem.

20.6.5 Theorem
The Divergence Theorem in the Plane

Let the functions M and N, the curve C, and the region R be as defined in Green's theorem. If $\mathbf{F}(x, y) = M(x, y)\mathbf{i} + N(x, y)\mathbf{j}$ and $\mathbf{N}(s)$ is a unit normal vector of C at P where s units is the length of arc measured from a particular point P_0 on C to P, then

$$\oint_C \mathbf{F}\cdot\mathbf{N}\, ds = \iint_R \operatorname{div} \mathbf{F}\, dA$$

Example 5 Verify the divergence theorem in the plane for $\mathbf{F}$ and R of Example 4.

Solution In Example 4,

$$\mathbf{F}(x, y) = 2y\mathbf{i} + 5x\mathbf{j}$$

and the unit tangent vector of C at P is defined by

$$\mathbf{T}(s) = -\sin s\mathbf{i} + \cos s\mathbf{j}$$

If $\mathbf{N}(s)$ is a unit normal vector of C at P, then

$$\begin{aligned}\mathbf{N}(s) &= \frac{D_s\mathbf{T}(s)}{|D_s\mathbf{T}(s)|}\\ &= -\cos s\mathbf{i} - \sin s\mathbf{j}\end{aligned}$$

Then

$$\begin{aligned}\oint_C \mathbf{F}\cdot\mathbf{N}\,ds &= \int_0^{2\pi} (2\sin s\mathbf{i} + 5\cos s\mathbf{j})\cdot(-\cos s\mathbf{i} - \sin s\mathbf{j})\,ds\\ &= \int_0^{2\pi} (-2\sin s\cos s - 5\sin s\cos s)\,ds\\ &= -7\int_0^{2\pi} \sin s\cos s\,ds\\ &= -\tfrac{7}{2}\sin^2 s\Big]_0^{2\pi}\\ &= 0\end{aligned}$$

Because

$$\begin{aligned}\operatorname{div}\mathbf{F} &= \frac{\partial M}{\partial x} + \frac{\partial N}{\partial y}\\ &= \frac{\partial}{\partial x}(2y) + \frac{\partial}{\partial y}(5x)\\ &= 0\end{aligned}$$

then

$$\iint_R \operatorname{div}\mathbf{F}\,dA = 0$$

We have therefore verified the divergence theorem in the plane for this $\mathbf{F}$ and R.

EXERCISES 20.6

In Exercises 1 through 9, evaluate the line integral by Green's theorem. Then verify the result by the method of Sec. 19.6.

1. $\oint_C 4y\,dx + 3x\,dy$, where C is the square with vertices at (0, 0), (1, 0), (1, 1), and (0, 1).

2. $\oint_C y^2\,dx + x^2\,dy$, where C is the square of Exercise 1.

3. $\oint_C 2xy\,dx - x^2y\,dy$, where C is the triangle with vertices at $(0, 0)$, $(1, 0)$, and $(0, 1)$.

4. The line integral of Exercise 3, where C is the triangle with vertices at $(0, 0)$, $(1, 0)$, and $(1, 1)$.

5. The line integral of Exercise 3, where C is the rectangle with vertices at $(0, 0)$, $(1, 0)$, $(1, 2)$, and $(0, 2)$.

6. $\oint_C (x^2 - y^2)\,dx + 2xy\,dy$, where C is the circle $x^2 + y^2 = 1$.

7. $\oint_C x^2y\,dx - y^2x\,dy$, where C is the circle $x^2 + y^2 = 1$.

8. The line integral of Exercise 6, where C is the closed curve consisting of the arc of $4y = x^3$ from $(0, 0)$ to $(2, 2)$ and the line segment from $(2, 2)$ to $(0, 0)$.

9. The line integral of Exercise 7, where C is the closed curve of Exercise 8.

In Exercises 10 through 22, use Green's theorem to evaluate the line integral.

10. $\oint_C \sqrt{y}\,dx + \sqrt{x}\,dy$, where C is the closed curve determined by the x axis, the line $x = 1$, and the curve $y = x^2$.

11. $\oint_C (x + y)\,dx + xy\,dy$, where C is the closed curve determined by the x axis, the line $x = 2$, and the curve $4y = x^3$.

12. $\oint_C (x^2 + y)\,dx$, where C is the closed curve determined by the x axis and the parabola $y = 4 - x^2$.

13. $\oint_C (-x^2 + x)\,dy$, where C is the closed curve determined by the line $x - 2y = 0$ and the parabola $x = 2y^2$.

14. $\oint_C e^{x+y}\,dx + e^{x+y}\,dy$, where C is the circle $x^2 + y^2 = 4$.

15. $\oint_C \cos y\,dx + \cos x\,dy$, where C is the rectangle with vertices at $(0, 0)$, $(\frac{1}{3}\pi, 0)$, $(\frac{1}{3}\pi, \frac{1}{4}\pi)$, and $(0, \frac{1}{4}\pi)$.

16. $\oint_C x \sin y\,dx - y \cos x\,dy$, where C is the rectangle with vertices at $(0, 0)$, $(\frac{1}{2}\pi, 0)$, $(\frac{1}{2}\pi, \frac{1}{4}\pi)$, and $(0, \frac{1}{4}\pi)$.

17. $\oint_C (\sin^4 x + e^{2x})\,dx + (\cos^3 y - e^y)\,dy$, where C is the curve $x^4 + y^4 = 16$.

18. $\oint_C e^y \cos x\,dx + e^y \sin x\,dy$, where C is the curve $x^6 + y^4 = 10$.

19. $\oint_C \dfrac{x^2y}{x^2 + 1}\,dx - \tan^{-1} x\,dy$, where C is the ellipse $4x^2 + 25y^2 = 100$.

20. $\oint_C \tan y\,dx + x \tan^2 y\,dy$, where C is the ellipse $x^2 + 4y^2 = 1$.

21. $\oint_C (e^x - x^2y)\,dx + 3x^2y\,dy$, where C is the closed curve determined by $y = x^2$ and $x = y^2$.

22. $\oint_C 2xy\,dx + 4x^2y\,dy$, where C is the closed curve determined by the x axis, the y axis, and the arc of the circle $x^2 + y^2 = 4$ in the first quadrant.

In Exercises 23 through 30, use Theorem 20.6.2 to find the area of the given region.

23. The region having as its boundary the quadrilateral with vertices at (0, 0), (4, 0), (3, 2), and (1, 1).

24. The region having as its boundary the triangle with vertices at (0, 0), $(a, 0)$, and $(0, b)$ where $a > 0$ and $b > 0$.

25. The region having as its boundary the circle $x^2 + y^2 = a^2$.

26. The region bounded below by the x axis and above by one arch of the cycloid having parametric equations

$$x = t - \sin t \qquad y = 1 - \cos t \qquad 0 \le t \le 2\pi$$

27. The region bounded by the hypocycloid having parametric equations

$$x = a\cos^3 t \qquad y = a\sin^3 t \qquad a > 0 \qquad 0 \le t \le 2\pi$$

28. The region bounded by the parabola $y = 2x^2$ and the line $y = 8x$.

29. The region bounded by the graphs of $y = x^2$ and $y = \sqrt{x}$.

30. The region bounded by the graphs of $y = x^3$ and $y = \sqrt{x}$.

In Exercises 31 through 34, use Green's theorem to find the total work done in moving an object in the counterclockwise direction once around the given curve C if the motion is caused by the given force field $\mathbf{F}(x, y)$. *Assume the arc is measured in meters and the force is measured in newtons.*

31. C is the ellipse $x^2 + 4y^2 = 16$; $\mathbf{F}(x, y) = (3x + y)\mathbf{i} + (4x - 5y)\mathbf{j}$.

32. C is the circle $x^2 + y^2 = 25$; $\mathbf{F}(x, y) = (e^x + y^2)\mathbf{i} + (x^2y + \cos y)\mathbf{j}$.

33. C is the triangle with vertices at (0, 0), (2, 0), and (0, 2); $\mathbf{F}(x, y) = (e^{x^2} + y^2)\mathbf{i} + (e^{y^2} + x^2)\mathbf{j}$.

34. C consists of the top half of the ellipse $9x^2 + 4y^2 = 36$ and the interval $[-2, 2]$ on the x axis; $\mathbf{F}(x, y) = (xy + y^2)\mathbf{i} + xy\mathbf{j}$.

35. Verify Stokes' theorem in the plane if $\mathbf{F}(x, y) = x^2\mathbf{i} + y^2\mathbf{j}$ and R is the region bounded by the ellipse $4x^2 + 25y^2 = 100$.

36. Verify Stokes' theorem in the plane if $\mathbf{F}(x, y) = 3y\mathbf{i} - 2x\mathbf{j}$ and R is the region bounded by $x^{2/3} + y^{2/3} = 1$.

37. Verify the divergence theorem in the plane for $\mathbf{F}$ and R of Exercise 35.

38. Verify the divergence theorem in the plane for $\mathbf{F}$ and R of Exercise 36.

39. Prove $\oint_C N(x, y)\,dy = \iint_R \frac{\partial N}{\partial x}\,dA$ if R is the region defined by $R = \{(x, y) \mid c \le y \le d, g_1(y) \le x \le g_2(y)\}$ where g_1 and g_2 are smooth.

20.7 THE TRIPLE INTEGRAL

The extension of the double integral to the triple integral is analogous to the extension of the single integral to the double integral. The simplest type of region in R^3 is a rectangular parallelepiped that is bounded by six planes: $x = a_1$, $x = a_2$, $y = b_1$, $y = b_2$, $z = c_1$, and $z = c_2$, with $a_1 < a_2$, $b_1 < b_2$, and $c_1 < c_2$. Let f be a function of three variables and suppose that f is continuous on such a region S. A partition of this region is formed by dividing S into rectangular boxes by drawing planes parallel to the coordinate planes. Denote such a partition by Δ and suppose that n is the number of boxes. Let $\Delta_i V$ be the measure of the volume of the ith box. Choose an arbitrary point (ξ_i, γ_i, μ_i) in the ith box. Form the sum

$$\sum_{i=1}^{n} f(\xi_i, \gamma_i, \mu_i)\,\Delta_i V \tag{1}$$

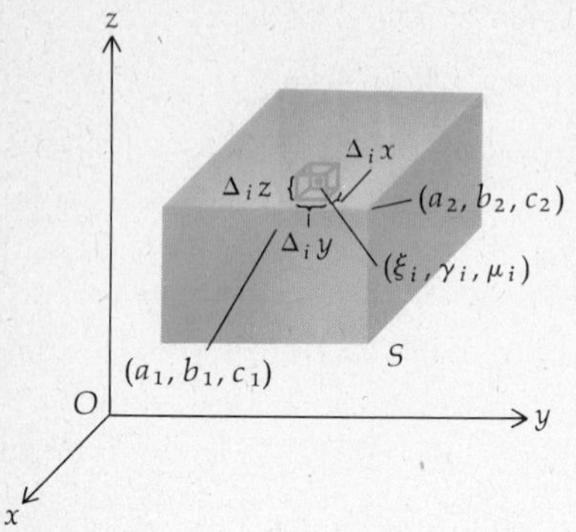

FIGURE 20.7.1

Refer to Fig. 20.7.1, which shows the rectangular parallelepiped together with the ith box. The *norm* $\|\Delta\|$ of the partition is the length of the longest diagonal of the boxes. If the sums of form (1) approach a limit as $\|\Delta\|$ approaches zero for any choices of the points (ξ_i, γ_i, μ_i), then we call this limit the **triple integral** of f on S and write

$$\lim_{\|\Delta\| \to 0} \sum_{i=1}^{n} f(\xi_i, \gamma_i, \mu_i)\, \Delta_i V = \iiint_S f(x, y, z)\, dV$$

A sufficient condition for the triple integral of f on S to exist is that f be continuous on S.

Analogous to a double integral being equal to a twice-iterated integral, the triple integral is equal to a thrice-iterated integral. When S is the rectangular parallelepiped described above, and f is continuous on S, then

$$\iiint_S f(x, y, z)\, dV = \int_{a_1}^{a_2} \int_{b_1}^{b_2} \int_{c_1}^{c_2} f(x, y, z)\, dz\, dy\, dx$$

Example 1 Evaluate the triple integral $\iiint_S xy \sin yz\, dV$ if S is the rectangular parallelepiped bounded by the planes $x = \pi$, $y = \frac{1}{2}\pi$, $z = \frac{1}{3}\pi$, and the coordinate planes.

Solution

$$\begin{aligned}
\iiint_S xy \sin yz\, dV &= \int_0^{\pi} \int_0^{\pi/2} \int_0^{\pi/3} xy \sin yz\, dz\, dy\, dx \\
&= \int_0^{\pi} \int_0^{\pi/2} \Big[-x \cos yz \Big]_0^{\pi/3} dy\, dx \\
&= \int_0^{\pi} \int_0^{\pi/2} x(1 - \cos \tfrac{1}{3}\pi y)\, dy\, dx \\
&= \int_0^{\pi} x\left(y - \frac{3}{\pi} \sin \frac{1}{3}\pi y \right)\Big]_0^{\pi/2} dx \\
&= \int_0^{\pi} x\left(\frac{\pi}{2} - \frac{3}{\pi} \sin \frac{\pi^2}{6} \right) dx \\
&= \frac{x^2}{2}\left(\frac{\pi}{2} - \frac{3}{\pi} \sin \frac{\pi^2}{6} \right)\Big]_0^{\pi} \\
&= \frac{\pi}{4}\left(\pi^2 - 6 \sin \frac{\pi^2}{6} \right)
\end{aligned}$$

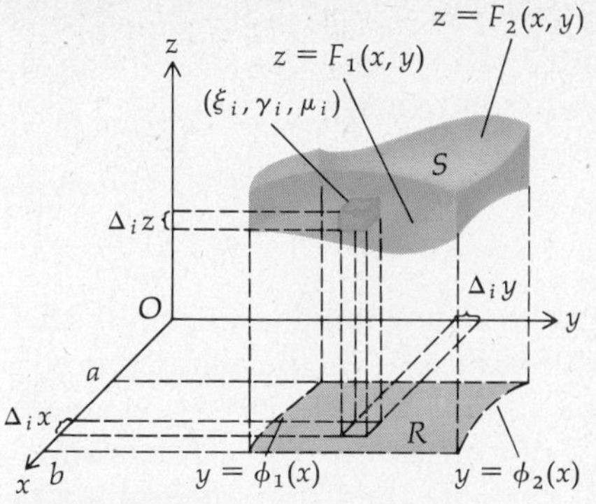

FIGURE 20.7.2

We now discuss how to define the triple integral of a continuous function of three variables on a region in R^3 other than a rectangular parallelepiped. Let S be the closed three-dimensional region that is bounded by the planes $x = a$ and $x = b$, the cylinders $y = \phi_1(x)$ and $y = \phi_2(x)$, and the surfaces $z = F_1(x, y)$ and $z = F_2(x, y)$, where the functions ϕ_1, ϕ_2, F_1, and F_2 are smooth (i.e., they have continuous derivatives or partial derivatives). See Fig. 20.7.2. Construct planes parallel to the coordinate planes, thereby forming a set of rectangular parallelepipeds that completely cover S. The parallelepipeds that are entirely inside S or on the boundary of S form a **partition** Δ of S. Choose some system of numbering so that they are numbered from 1 to n. The norm $\|\Delta\|$ of this partition of S is the length of the longest diagonal of any parallelepiped belonging to the partition. Let the measure of the volume of the ith parallelepiped be $\Delta_i V$. Let f be a function of three variables that is continuous on S and let (ξ_i, γ_i, μ_i) be an arbitrary point in the ith parallelepiped. Form the sum

$$\sum_{i=1}^{n} f(\xi_i, \gamma_i, \mu_i)\, \Delta_i V \tag{2}$$

If the sums of form (2) have a limit as $\|\Delta\|$ approaches zero, and if this limit is independent of the choice of the partitioning planes and the choices of the arbitrary points (ξ_i, γ_i, μ_i) in each parallelepiped, then this limit is called the **triple integral** of f on S, and we write

$$\lim_{\|\Delta\| \to 0} \sum_{i=1}^{n} f(\xi_i, \gamma_i, \mu_i)\, \Delta_i V = \iiint_S f(x, y, z)\, dV \tag{3}$$

It can be proved in advanced calculus that a sufficient condition for the limit in (3) to exist is that f be continuous on S. Furthermore, under the condition imposed on the functions ϕ_1, ϕ_2, F_1, and F_2 that they be smooth, it can also be proved that the triple integral can be evaluated by the iterated integral

$$\int_a^b \int_{\phi_1(x)}^{\phi_2(x)} \int_{F_1(x,y)}^{F_2(x,y)} f(x, y, z)\, dz\, dy\, dx$$

Just as the double integral can be interpreted as the measure of the area of a plane region when $f(x, y) = 1$ on R^1, the triple integral can be interpreted as the measure of the volume of a three-dimensional region. If $f(x, y, z) = 1$ on S, then Eq. (3) becomes

$$\lim_{\|\Delta\| \to 0} \sum_{i=1}^{n} \Delta_i V = \iiint_S dV$$

and the triple integral is the measure of the volume of the region S.

Example 2 Find by triple integration the volume of the solid of Example 3 in Sec. 20.2.

Solution The solid is above the xy plane bounded by the elliptic paraboloid $z = x^2 + 4y^2$ and the cylinder $x^2 + 4y^2 = 4$. It is shown in Fig. 20.2.6. If V cubic units is the volume of the solid, then

$$V = \lim_{\|\Delta\| \to 0} \sum_{i=1}^{n} \Delta_i V = \iiint_S dV \tag{4}$$

where S is the region bounded by the solid. The z limits are from 0 (the value of z on the xy plane) to $x^2 + 4y^2$ (the value of z on the elliptic paraboloid). The y limits for one-fourth of the volume are from 0 (the value of y on the xz plane) to $\frac{1}{2}\sqrt{4 - x^2}$ (the value of y on the cylinder). The x limits for the first octant are from 0 to 2. We evaluate the triple integral in (4) by an iterated integral and obtain

$$V = 4\int_0^2 \int_0^{\sqrt{4-x^2}/2} \int_0^{x^2+4y^2} dz\, dy\, dx$$

$$= 4\int_0^2 \int_0^{\sqrt{4-x^2}/2} (x^2 + 4y^2)\, dy\, dx$$

This is the same twice-iterated integral that we obtained in Example 3 in Sec. 20.2, and so the remainder of the solution is the same.

Example 3 Find the volume of the solid bounded by the cylinder $x^2 + y^2 = 25$, the plane $x + y + z = 8$ and the xy plane.

Solution The solid is shown in Fig. 20.7.3. The z limits for the iterated integral are from 0 to $8 - x - y$ (the value of z on the plane). The y limits are obtained from the boundary region in the xy plane, which is the circle $x^2 + y^2 = 25$. Hence the y limits are from $-\sqrt{25 - x^2}$ to $\sqrt{25 - x^2}$. The x limits are from -5 to 5. If V cubic units is the required volume,

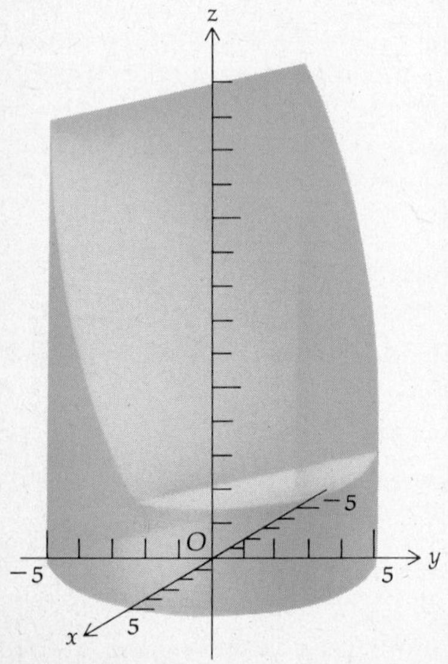

FIGURE 20.7.3

$$V = \lim_{\|\Delta\| \to 0} \sum_{i=1}^{n} \Delta_i V$$

$$= \iiint_S dV$$

$$= \int_{-5}^{5} \int_{-\sqrt{25-x^2}}^{\sqrt{25-x^2}} \int_0^{8-x-y} dz\, dy\, dx$$

$$= \int_{-5}^{5} \int_{-\sqrt{25-x^2}}^{\sqrt{25-x^2}} (8 - x - y)\, dy\, dx$$

$$= \int_{-5}^{5} \left[(8 - x)y - \tfrac{1}{2}y^2 \right]_{-\sqrt{25-x^2}}^{\sqrt{25-x^2}} dx$$

$$= 2\int_{-5}^{5} (8 - x)\sqrt{25 - x^2}\, dx$$

$$= 16\int_{-5}^{5} \sqrt{25 - x^2}\, dx + \int_{-5}^{5} \sqrt{25 - x^2}(-2x)\, dx$$

$$= 16\left(\tfrac{1}{2}x\sqrt{25 - x^2} + \tfrac{25}{2}\sin^{-1}\tfrac{1}{5}x\right) + \tfrac{2}{3}(25 - x^2)^{3/2}\Big]_{-5}^{5}$$

$$= 200\pi$$

The volume is therefore 200π cubic units.

Example 4 Find the mass of the solid above the xy plane bounded by the cone $9x^2 + z^2 = y^2$ and the plane $y = 9$ if the measure of the volume density at any point (x, y, z) in the solid is proportional to the measure of the distance of the point from the xy plane.

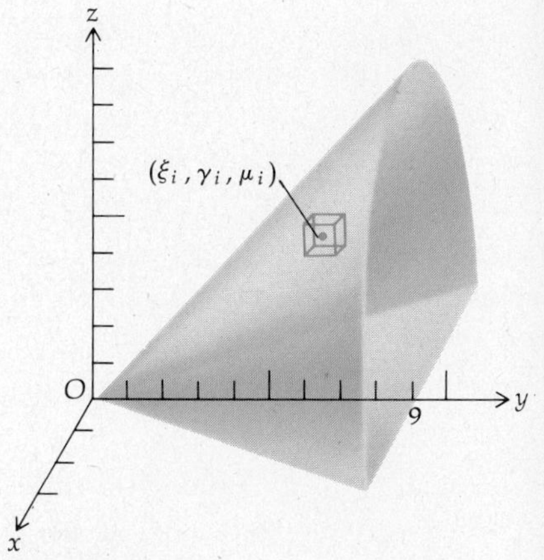

FIGURE 20.7.4

Solution Figure 20.7.4 shows the solid. Let M kg be the mass of the solid, and let the distance be measured in meters. Then the volume density at any point (x, y, z) in the solid is kz kg/m^3, where k is a constant. Then if (ξ_i, γ_i, μ_i) is any point in the ith rectangular parallelepiped of the partition,

$$M = \lim_{\|\Delta\| \to 0} \sum_{i=1}^{n} k\mu_i\, \Delta_i V$$

$$= \iiint_S kz\, dV$$

$$= 2k \int_0^9 \int_0^{y/3} \int_0^{\sqrt{y^2 - 9x^2}} z\, dz\, dx\, dy$$

$$= 2k \int_0^9 \int_0^{y/3} \left[\tfrac{1}{2}z^2\right]_0^{\sqrt{y^2-9x^2}} dx\, dy$$

$$= k \int_0^9 \int_0^{y/3} (y^2 - 9x^2)\, dx\, dy$$

$$= \tfrac{2}{9}k \int_0^9 y^3\, dy$$

$$= \tfrac{729}{2}k$$

The mass is therefore $\frac{729}{2}k$ kg.

EXERCISES 20.7

In Exercises 1 through 4, evaluate the iterated integral.

1. $\displaystyle\int_0^1 \int_0^{1-x} \int_{2y}^{1+y^2} x\, dz\, dy\, dx$ 2. $\displaystyle\int_1^2 \int_y^{y^2} \int_0^{\ln x} ye^z\, dz\, dx\, dy$ 3. $\displaystyle\int_0^{\pi/2} \int_z^{\pi/2} \int_0^{xz} \cos\frac{y}{z}\, dy\, dx\, dz$ 4. $\displaystyle\int_0^2 \int_0^y \int_0^{\sqrt{3}z} \frac{z}{x^2 + z^2}\, dx\, dz\, dy$

In Exercises 5 through 10, evaluate the triple integral.

5. $\displaystyle\iiint_S y\, dV$ if S is the region bounded by the tetrahedron formed by the plane $12x + 20y + 15z = 60$ and the coordinate planes.

6. $\iiint\limits_S (x^2 + z^2)\, dV$ if S is the same region as in Exercise 5.

7. $\iiint\limits_S z\, dV$ if S is the region bounded by the tetrahedron having vertices (0, 0, 0), (1, 1, 0), (1, 0, 0), and (1, 0, 1).

8. $\iiint\limits_S yz\, dV$ if S is the same region as in Exercise 7.

9. $\iiint\limits_S (xz + 3z)\, dV$ if S is the region bounded by the cylinder $x^2 + z^2 = 9$ and the planes $x + y = 3$, $z = 0$, and $y = 0$, above the xy plane.

10. $\iiint\limits_S xyz\, dV$, if S is the region bounded by the cylinders $x^2 + y^2 = 4$ and $x^2 + z^2 = 4$.

In Exercises 11 through 23, use triple integration.

11. Find the volume of the solid in the first octant bounded below by the xy plane, above by the plane $z = y$, and laterally by the cylinder $y^2 = x$ and the plane $x = 1$.

12. Find the volume of the solid in the first octant bounded by the cylinder $x^2 + z^2 = 16$, the plane $x + y = 2$, and the three coordinate planes.

13. Find the volume of the solid in the first octant bounded by the cylinders $x^2 + y^2 = 4$ and $x^2 + 2z = 4$ and the three coordinate planes.

14. Find the volume of the solid bounded by the elliptic cone $4x^2 + 9y^2 - 36z^2 = 0$ and the plane $z = 1$.

15. Find the volume of the solid above the elliptic paraboloid $3x^2 + y^2 = z$ and below the cylinder $x^2 + z = 4$.

16. Find the volume of the solid enclosed by the sphere $x^2 + y^2 + z^2 = a^2$.

17. Find the volume of the solid enclosed by the ellipsoid

$$\frac{x^2}{a^2} + \frac{y^2}{b^2} + \frac{z^2}{c^2} = 1$$

18. Find the volume of the solid bounded by the cylinders $z = 5x^2$ and $z = 3 - x^2$, the plane $y + z = 4$, and the xz plane.

19. Find the mass of the homogeneous solid bounded by the cylinder $z = 4 - x^2$, the plane $y = 5$, and the coordinate planes, if the volume density at any point is ρ kg/m^3.

20. Find the mass of the solid enclosed by the tetrahedron formed by the plane $100x + 25y + 16z = 400$ and the coordinate planes if the volume density varies as the distance from the yz plane. The volume density is measured in kilograms per cubic meter.

21. Find the mass of the solid bounded by the cylinders $x = z^2$ and $y = x^2$, and the planes $x = 1$, $y = 0$, and $z = 0$. The volume density varies as the product of the distances from the three coordinate planes, and it is measured in kilograms per cubic meter.

22. Find the mass of the solid bounded by the surface $z = 4 - 4x^2 - y^2$ and the xy plane. The volume density at any point of the solid is $3z|x|$ kg/m^3.

23. Find the mass of the solid bounded by the surface $z = xy$, and the planes $x = 1$, $y = 1$ and $z = 0$. The volume density at any point of the solid is $3\sqrt{x^2 + y^2}$ kg/m^3.

24. A solid has the shape of a right-circular cylinder of base radius r meters and height h meters. Find the mass of the solid if the volume density varies as the distance from one of the bases. The volume density is measured in kilograms per cubic meter.

20.8 THE TRIPLE INTEGRAL IN CYLINDRICAL AND SPHERICAL COORDINATES

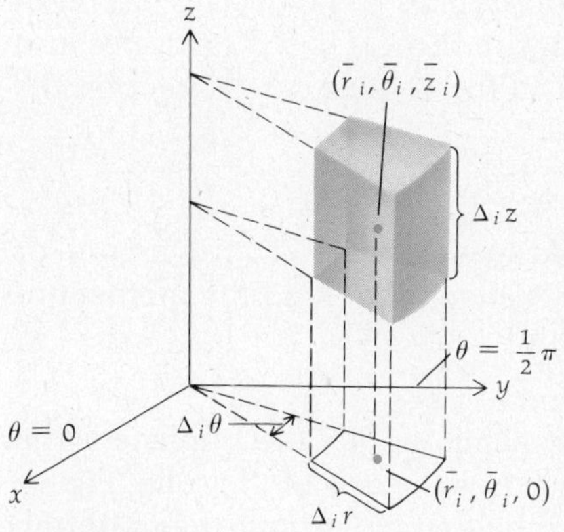

FIGURE 20.8.1

If a region S in R^3 has an axis of symmetry, triple integrals on S are easier to evaluate if cylindrical coordinates are used. If there is symmetry with respect to a point, it is convenient to choose that point as the origin and to use spherical coordinates. In this section we discuss the triple integral in these coordinates and apply them to physical problems.

To define the triple integral in cylindrical coordinates we construct a partition of the region S by drawing planes through the z axis, planes perpendicular to the z axis, and right-circular cylinders having the z axis as axis. A typical subregion is shown in Fig. 20.8.1. The elements of the constructed partition lie entirely in S. We call this partition a **cylindrical partition.** The measure of the length of the longest "diagonal" of any of the subregions is the **norm** of the partition. Let n be the number of subregions of the partition and $\Delta_i V$ be the measure of the volume of the ith subregion. The measure of the area of the base is $\bar{r}_i \, \Delta_i r \, \Delta_i \theta$, where $\bar{r}_i = \frac{1}{2}(r_i + r_{i-1})$. Hence, if $\Delta_i z$ is the measure of the altitude of the ith subregion,

$$\Delta_i V = \bar{r}_i \, \Delta_i r \, \Delta_i \theta \, \Delta_i z$$

Let f be a function of r, θ, and z, and suppose that f is continuous on S. Choose a point $(\bar{r}_i, \bar{\theta}_i, \bar{z}_i)$ in the ith subregion such that $\theta_{i-1} \leq \bar{\theta}_i \leq \theta_i$, and $z_{i-1} \leq \bar{z}_i \leq z_i$. Form the sum

$$\sum_{i=1}^{n} f(\bar{r}_i, \bar{\theta}_i, \bar{z}_i) \, \Delta_i V = \sum_{i=1}^{n} f(\bar{r}_i, \bar{\theta}_i, \bar{z}_i) \bar{r}_i \, \Delta_i r \, \Delta_i \theta \, \Delta_i z \tag{1}$$

As the norm of Δ approaches zero, it can be shown, under suitable conditions on S, that the limit of the sums of form (1) exist. This limit is called the **triple integral in cylindrical coordinates** of the function f on S, and we write

$$\lim_{\|\Delta\| \to 0} \sum_{i=1}^{n} f(\bar{r}_i, \bar{\theta}_i, \bar{z}_i) \, \Delta_i V = \iiint_S f(r, \theta, z) \, dV \tag{2}$$

or

$$\lim_{\|\Delta\| \to 0} \sum_{i=1}^{n} f(\bar{r}_i, \bar{\theta}_i, \bar{z}_i) \bar{r}_i \, \Delta_i r \, \Delta_i \theta \, \Delta_i z = \iiint_S f(r, \theta, z) r \, dr \, d\theta \, dz \tag{3}$$

Note that in cylindrical coordinates, $dV = r \, dr \, d\theta \, dz$. We can evaluate the triple integral in (2) and (3) by an iterated integral. For instance, suppose that the region S in R^3 is bounded by the planes $\theta = \alpha$ and $\theta = \beta$, with $\alpha < \beta$, by the cylinders $r = \lambda_1(\theta)$ and $r = \lambda_2(\theta)$, where λ_1 and λ_2 are smooth on $[\alpha, \beta]$ and $\lambda_1(\theta) \leq \lambda_2(\theta)$ for $\alpha \leq \theta \leq \beta$, and by the surfaces $z = F_1(r, \theta)$ and $z = F_2(r, \theta)$, where F_1 and F_2 are functions of two variables that are smooth on some region R in the polar plane bounded by the curves $r = \lambda_1(\theta)$, $r = \lambda_2(\theta)$, $\theta = \alpha$, and $\theta = \beta$. Furthermore, suppose that $F_1(r, \theta) \leq$

$F_2(r, \theta)$ for every point (r, θ) in R. Then the triple integral can be evaluated by an iterated integral by the formula

$$\iiint_S f(r, \theta, z) r\, dr\, d\theta\, dz = \int_\alpha^\beta \int_{\lambda_1(\theta)}^{\lambda_2(\theta)} \int_{F_1(r,\theta)}^{F_2(r,\theta)} f(r, \theta, z) r\, dz\, dr\, d\theta \tag{4}$$

There are five other iterated integrals that can be used to evaluate the triple integral in (4) because there are six possible permutations of the three variables r, θ, and z.

Triple integrals and cylindrical coordinates are especially useful in finding the moment of inertia of a solid with respect to the z axis because the distance from the z axis to a point in the solid is determined by the coordinate r.

Example 1 A homogeneous solid in the shape of a right-circular cylinder has a radius of 2 m and an altitude of 4 m. Find the moment of inertia of the solid with respect to its axis.

Solution Choose the coordinate planes so that the xy plane is the plane of the base of the solid and the z axis is the axis of the solid. Figure 20.8.2 shows the portion of the solid in the first octant together with the ith subregion of a cylindrical partition. Using cylindrical coordinates and taking the point $(\bar{r}_i, \bar{\theta}_i, \bar{z}_i)$ in the ith subregion with k kg/m³ as the volume density at any point, then if I_z kg-m² is the moment of inertia of the solid with respect to the z axis,

$$I_z = \lim_{\|\Delta\| \to 0} \sum_{i=1}^{n} \bar{r}_i^{\,2} k\, \Delta_i V$$

$$= \iiint_S kr^2\, dV$$

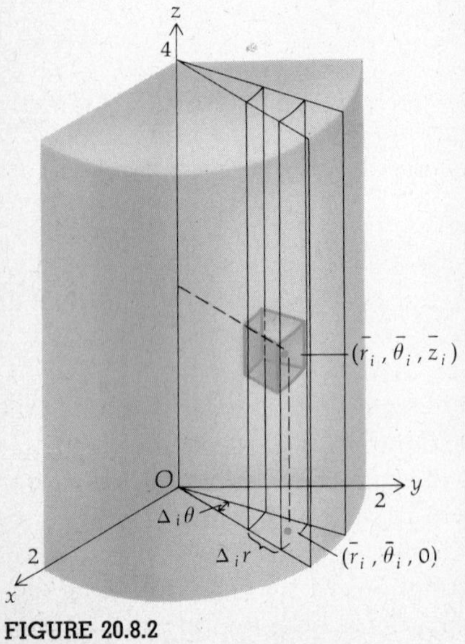

FIGURE 20.8.2

There are six different possible orders of integration. Figure 20.8.2 shows the order $dz\, dr\, d\theta$. Using this order, we have

$$I_z = \iiint_S kr^2\, dz\, r\, dr\, d\theta$$

$$= 4k \int_0^{\pi/2} \int_0^2 \int_0^4 r^3\, dz\, dr\, d\theta$$

In the first integration the blocks are summed from $z = 0$ to $z = 4$; the blocks become a column. In the second integration the columns are summed from $r = 0$ to $r = 2$; the columns become a wedge-shaped slice of the cylinder. In the third integration the wedge-shaped slice is rotated from $\theta = 0$ to $\theta = \frac{1}{2}\pi$; this sweeps the wedge about the entire three-dimensional region in the first octant. We multiply by 4 to obtain the entire volume. Performing the integration we obtain

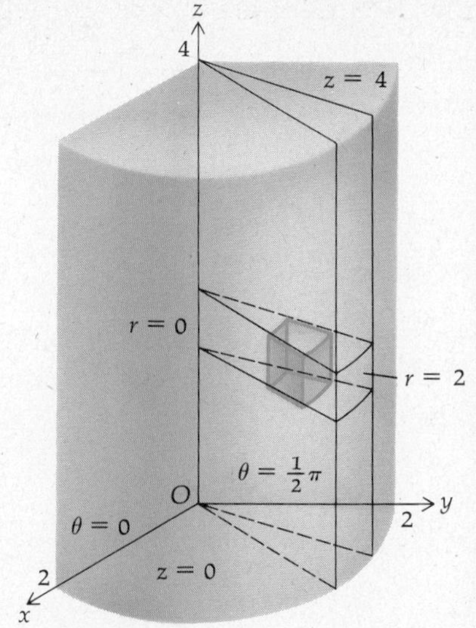

FIGURE 20.8.3

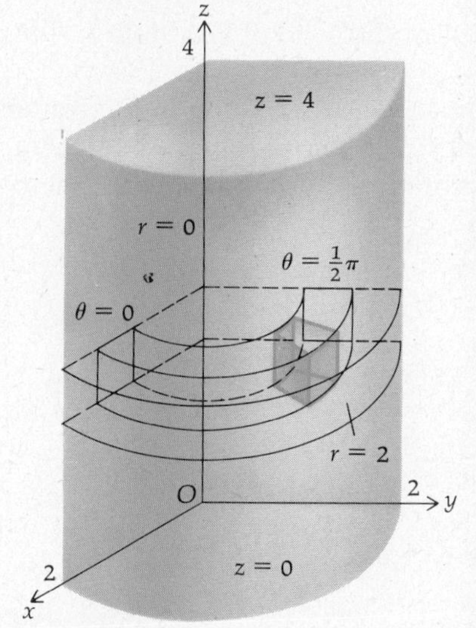

FIGURE 20.8.4

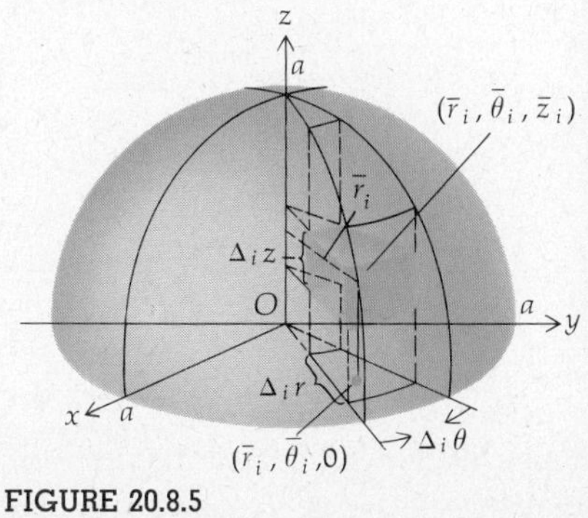

FIGURE 20.8.5

$$I_z = 16k \int_0^{\pi/2} \int_0^2 r^3\, dr\, d\theta$$

$$= 64k \int_0^{\pi/2} d\theta$$

$$= 32k\pi$$

Hence the moment of inertia is $32k\pi$ kg-m^2.

Example 2 Solve Example 1 by taking the order of integration as (a) $dr\, dz\, d\theta$; (b) $d\theta\, dr\, dz$.

Solution

(a) Figure 20.8.3 represents the order $dr\, dz\, d\theta$. It shows the block summed from $r = 0$ to $r = 2$ to give a wedge-shaped sector. We then sum from $z = 0$ to $z = 4$ to give a wedge-shaped slice. The slice is rotated from $\theta = 0$ to $\theta = \frac{1}{2}\pi$ to cover the first octant. Then

$$I_z = 4k \int_0^{\pi/2} \int_0^4 \int_0^2 r^3 dr\, dz\, d\theta$$

$$= 32k\pi$$

(b) Figure 20.8.4 represents the order $d\theta\, dr\, dz$. It shows the blocks summed from $\theta = 0$ to $\theta = \frac{1}{2}\pi$ to give a hollow ring inside the cylinder. These hollow rings are summed from $r = 0$ to $r = 2$ to give a horizontal slice of the cylinder. The horizontal slices are summed from $z = 0$ to $z = 4$. Therefore

$$I_z = 4k \int_0^4 \int_0^2 \int_0^{\pi/2} r^3\, d\theta\, dr\, dz$$

$$= 32k\pi$$

Example 3 Find the mass of a solid hemisphere of radius a meters if the volume density at any point is proportional to the distance of the point from the axis of the solid and is measured in kilograms per cubic meter.

Solution If we choose the coordinate planes so that the origin is at the center of the sphere and the z axis is the axis of the solid, then an equation of the hemispherical surface above the xy plane is $z = \sqrt{a^2 - x^2 - y^2}$. Figure 20.8.5 shows this surface and the solid together with the ith subregion of a cylindrical partition. An equation of the hemisphere in cylindrical coordinates is $z = \sqrt{a^2 - r^2}$. If $(\bar{r}_i, \bar{\theta}_i, \bar{z}_i)$ is a point in the ith subregion, the volume density at this point is $k\bar{r}_i$ kg/m^3, where k is a constant; and if M kg is the mass of the solid, then

$$M = \lim_{\|\Delta\| \to 0} \sum_{i=1}^{n} k\bar{r}_i \Delta_i V$$

$$= \iiint_S kr\, dV$$

$$= k \int_0^{2\pi} \int_0^a \int_0^{\sqrt{a^2-r^2}} r^2\, dz\, dr\, d\theta$$

$$= k \int_0^{2\pi} \int_0^a r^2 \sqrt{a^2 - r^2}\, dr\, d\theta$$

$$= k \int_0^{2\pi} \left[-\tfrac{1}{4} r(a^2 - r^2)^{3/2} + \tfrac{1}{8} a^2 r \sqrt{a^2 - r^2} + \tfrac{1}{8} a^4 \sin^{-1} \frac{r}{a} \right]_0^a d\theta$$

$$= \tfrac{1}{16} k a^4 \pi \int_0^{2\pi} d\theta$$

$$= \tfrac{1}{8} k a^4 \pi^2$$

The mass of the solid hemisphere is therefore $\frac{1}{8}ka^4\pi^2$ kg.

Example 4 Find the center of mass of the solid of Example 3.

Solution Let the cartesian coordinate representation of the center of mass be $(\bar{x}, \bar{y}, \bar{z})$. Because of symmetry, $\bar{x} = \bar{y} = 0$. We need to calculate $\bar{z}$. If M_{xy} kg-m is the moment of mass of the solid with respect to the xy plane, then

$$M_{xy} = \lim_{\|\Delta\| \to 0} \sum_{i=1}^{n} \bar{z}_i (k\bar{r}_i)\, \Delta_i V$$

$$= \iiint_S kzr\, dV$$

$$= k \int_0^{2\pi} \int_0^a \int_0^{\sqrt{a^2-r^2}} zr^2\, dz\, dr\, d\theta$$

$$= \tfrac{1}{2} k \int_0^{2\pi} \int_0^a (a^2 - r^2) r^2\, dr\, d\theta$$

$$= \tfrac{1}{15} k a^5 \int_0^{2\pi} d\theta$$

$$= \tfrac{2}{15} k a^5 \pi$$

Because $M\bar{z} = M_{xy}$, we get

$$\bar{z} = \frac{M_{xy}}{M} = \frac{\frac{2}{15} k a^5 \pi}{\frac{1}{8} k a^4 \pi^2} = \frac{16}{15\pi} a$$

The center of mass is therefore on the axis of the solid at a distance of $16a/15\pi$ meters from the plane of the base.

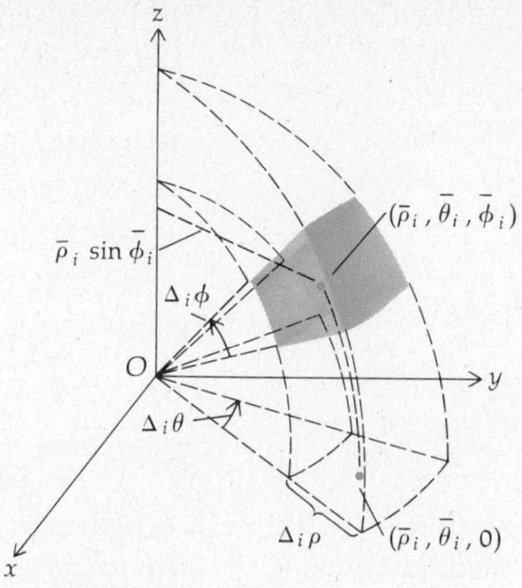

FIGURE 20.8.6

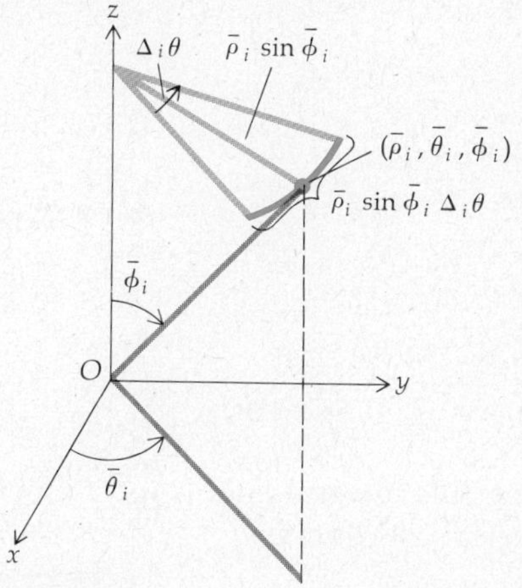

FIGURE 20.8.7

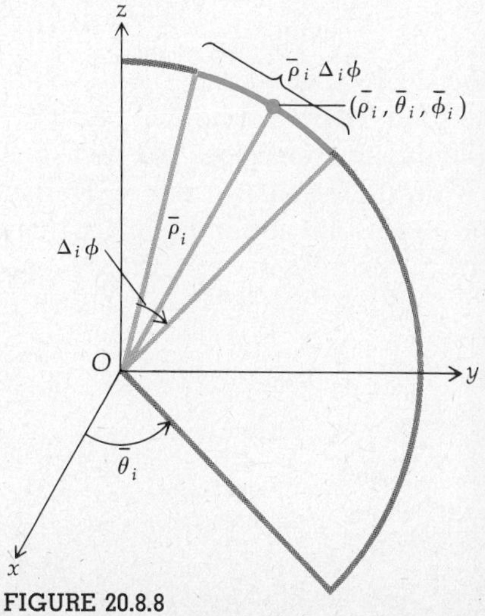

FIGURE 20.8.8

We now proceed to define the triple integral in spherical coordinates. A spherical partition of the three-dimensional region S is formed by planes containing the z axis, spheres with centers at the origin, and circular cones having vertices at the origin and the z axis as the axis. A typical subregion of the partition is shown in Fig. 20.8.6. If $\Delta_i V$ is the measure of the volume of the ith subregion, and $(\bar{\rho}_i, \bar{\theta}_i, \bar{\phi}_i)$ is a point in it, we can get an approximation to $\Delta_i V$ by considering the region as if it were a rectangular parallelepiped and taking the product of the measures of the three dimensions. These measures are $\bar{\rho}_i \sin \bar{\phi}_i \, \Delta_i \theta$, $\bar{\rho}_i \, \Delta_i \phi$, and $\Delta_i \rho$. Figures 20.8.7 and 20.8.8 show how the first two measures are obtained, and Fig. 20.8.6 shows the dimension of measure $\Delta_i \rho$. Hence

$$\Delta_i V = \bar{\rho}_i^{\,2} \sin \bar{\phi}_i \, \Delta_i \rho \, \Delta_i \theta \, \Delta_i \phi$$

The **triple integral in spherical coordinates** of a function f on S is given by

$$\lim_{\|\Delta\| \to 0} \sum_{i=1}^{n} f(\bar{\rho}_i, \bar{\theta}_i, \bar{\phi}_i) \, \Delta_i V = \iiint_S f(\rho, \theta, \phi) \, dV \tag{5}$$

or

$$\lim_{\|\Delta\| \to 0} \sum_{i=1}^{n} f(\bar{\rho}_i, \bar{\theta}_i, \bar{\phi}_i) \bar{\rho}_i^{\,2} \sin \bar{\phi}_i \, \Delta_i \rho \, \Delta_i \theta \, \Delta_i \phi = \iiint_S f(\rho, \theta, \phi) \rho^2 \sin \phi \, d\rho \, d\theta \, d\phi \tag{6}$$

Observe that in spherical coordinates, $dV = \rho^2 \sin \phi \, d\rho \, d\theta \, d\phi$. The triple integrals in (5) or (6) can be evaluated by an iterated integral. Spherical coordinates are especially useful in some problems involving spheres, as shown in the following example.

Example 5 Find the mass of the solid hemisphere of Example 3 if the volume density at any point is proportional to the distance of the point from the center of the base.

Solution If $(\bar{\rho}_i, \bar{\theta}_i, \bar{\phi}_i)$ is a point in the ith subregion of a spherical partition, the volume density at this point is $k\bar{\rho}_i$ kg/m³, where k is a constant. If M kg is the mass of the solid, then

$$\begin{aligned} M &= \lim_{\|\Delta\| \to 0} \sum_{i=1}^{n} k\bar{\rho}_i \, \Delta_i V \\ &= \iiint_S k\rho \, dV \\ &= 4k \int_0^{\pi/2} \int_0^{\pi/2} \int_0^{a} \rho^3 \sin \phi \, d\rho \, d\phi \, d\theta \end{aligned}$$

$$= a^4k \int_0^{\pi/2} \int_0^{\pi/2} \sin \phi \, d\phi \, d\theta$$

$$= a^4k \int_0^{\pi/2} \Big[-\cos \phi \Big]_0^{\pi/2} d\theta$$

$$= a^4k \int_0^{\pi/2} d\theta$$

$$= \tfrac{1}{2}a^4k\pi$$

Hence the mass of the solid hemisphere is $\frac{1}{2}a^4k\pi$ kg.

It is interesting to compare the solution of Example 5, which uses spherical coordinates, with what is entailed when using cartesian coordinates. By the latter method, a partition of S is formed by dividing S into rectangular boxes by drawing planes parallel to the coordinate planes. If (ξ_i, γ_i, μ_i) is any point in the ith subregion, and because $\rho = \sqrt{x^2 + y^2 + z^2}$, then

$$M = \lim_{\|\Delta\| \to 0} \sum_{i=1}^{n} k\sqrt{\xi_i{}^2 + \gamma_i{}^2 + \mu_i{}^2}\, \Delta_i V$$

$$= \iiint_S k\sqrt{x^2 + y^2 + z^2}\, dV$$

$$= 4k \int_0^a \int_0^{\sqrt{a^2 - z^2}} \int_0^{\sqrt{a^2 - y^2 - z^2}} \sqrt{x^2 + y^2 + z^2}\, dx\, dy\, dz$$

The computation involved in evaluating this integral is obviously much more complicated than that using spherical coordinates.

Example 6 A homogeneous solid is bounded above by the sphere $\rho = a$ and below by the cone $\phi = \alpha$, where $0 < \alpha < \frac{1}{2}\pi$. Find the moment of inertia of the solid about the z axis.

Solution The solid is shown in Fig. 20.8.9. Let k kg/m³ be the constant volume density at any point of the solid. Form a spherical partition of the solid and let $(\bar{\rho}_i, \bar{\theta}_i, \bar{\phi}_i)$ be a point in the ith subregion. The measure of the distance of the point $(\bar{\rho}_i, \bar{\theta}_i, \bar{\phi}_i)$ from the z axis is $\bar{\rho}_i \sin \bar{\phi}_i$. Hence if I_z kg-m² is the moment of inertia of the given solid about the z axis, then

$$I_z = \lim_{\|\Delta\| \to 0} \sum_{i=1}^{n} (\bar{\rho}_i \sin \bar{\phi}_i)^2 k\, \Delta_i V$$

$$= \iiint_S k\rho^2 \sin^2 \phi\, dV$$

$$= k \int_0^{\alpha} \int_0^{2\pi} \int_0^{a} (\rho^2 \sin^2 \phi)\rho^2 \sin \phi\, d\rho\, d\theta\, d\phi$$

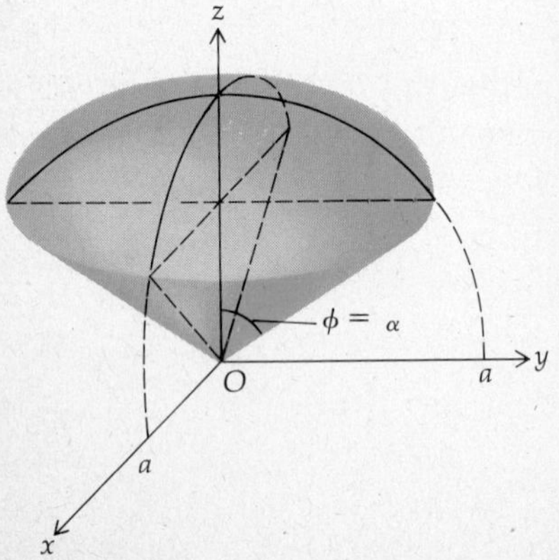

FIGURE 20.8.9

$$= \tfrac{1}{5}ka^5 \int_0^{\alpha} \int_0^{2\pi} \sin^3 \phi \, d\theta \, d\phi$$

$$= \tfrac{2}{5}ka^5\pi \int_0^{\alpha} \sin^3 \phi \, d\phi$$

$$= \tfrac{2}{5}ka^5\pi \Big[-\cos \phi + \tfrac{1}{3} \cos^3 \phi \Big]_0^{\alpha}$$

$$= \tfrac{2}{15}ka^5\pi(\cos^3 \alpha - 3 \cos \alpha + 2)$$

The moment of inertia of the solid about the z axis is therefore

$$\tfrac{2}{15}ka^5\pi(\cos^3 \alpha - 3 \cos \alpha + 2) \text{ kg-m}^2$$

EXERCISES 20.8

In Exercises 1 through 6, evaluate the iterated integral.

1. $\displaystyle\int_0^{\pi/4} \int_0^{a} \int_0^{r\cos\theta} r \sec^3 \theta \, dz \, dr \, d\theta$ **2.** $\displaystyle\int_0^{\pi/4} \int_{2\sin\theta}^{2\cos\theta} \int_0^{r\sin\theta} r^2 \cos \theta \, dz \, dr \, d\theta$ **3.** $\displaystyle\int_0^{\pi} \int_2^{4} \int_0^{1} re^z \, dz \, dr \, d\theta$

4. $\displaystyle\int_0^{2\pi} \int_0^{\pi} \int_0^{2} \rho^3 \sin \phi \, d\rho \, d\phi \, d\theta$ **5.** $\displaystyle\int_0^{\pi/4} \int_0^{2a\cos\phi} \int_0^{2\pi} \rho^2 \sin \phi \, d\theta \, d\rho \, d\phi$ **6.** $\displaystyle\int_{\pi/4}^{\pi/2} \int_{\pi/4}^{\phi} \int_0^{a\csc\theta} \rho^3 \sin^2 \theta \sin \phi \, d\rho \, d\theta \, d\phi$

7. Find the volume of the solid enclosed by the sphere $x^2 + y^2 + z^2 = a^2$ by using (a) cylindrical coordinates and (b) spherical coordinates.

8. If S is the solid in the first octant bounded by the sphere $x^2 + y^2 + z^2 = 16$ and the coordinate planes, evaluate the triple integral $\displaystyle\iiint_S xyz \, dV$ by three methods: (a) using spherical coordinates; (b) using rectangular coordinates; (c) using cylindrical coordinates.

In Exercises 9 through 12, use cylindrical coordinates.

9. Find the mass of the solid bounded by a sphere of radius a meters if the volume density varies as the square of the distance from the center. The volume density is measured in kilograms per cubic meter.

10. Find the mass of the solid in the first octant inside the cylinder $x^2 + y^2 = 4x$ and under the sphere $x^2 + y^2 + z^2 = 16$. The volume density varies as the distance from the xy plane, and it is measured in kilograms per cubic meter.

11. Find the moment of inertia with respect to the z axis of the homogeneous solid bounded by the cylinder $r = 5$, the cone $z = r$, and the xy plane. The volume density at any point is k slugs/ft^3.

12. Find the moment of inertia of the solid bounded by a right-circular cylinder of altitude h meters and radius a meters, with respect to the axis of the cylinder. The volume density varies as the distance from the axis of the cylinder, and it is measured in kilograms per cubic meter.

In Exercises 13 through 16, use spherical coordinates.

13. Find the center of mass of the solid bounded by the hemisphere of Example 5. The volume density is the same as that in Example 5.

14. Find the moment of inertia with respect to the z axis of the homogeneous solid bounded by the sphere $x^2 + y^2 + z^2 = 4$. The volume density at any point is k slugs/ft^3.

15. Find the moment of inertia with respect to the z axis of the homogeneous solid inside the cylinder $x^2 + y^2 - 2x = 0$, below the cone $x^2 + y^2 = z^2$, and above the xy plane. The volume density at any point is k kg/m³.

16. Find the mass of a spherical solid of radius a meters if the volume density at each point is proportional to the distance of the point from the center of the sphere. The volume density is measured in kilograms per cubic meter.

In Exercises 17 through 22, use the coordinate system that you decide is best for the problem.

17. Find the mass of a solid hemisphere of radius 2 m if the volume density varies as the distance from the center of the base and is measured in kilograms per cubic meter.

18. Find the mass of the homogeneous solid inside the paraboloid $3x^2 + 3y^2 = z$ and outside the cone $x^2 + y^2 = z^2$ if the constant volume density is k kg/m³.

19. Find the center of mass of the solid inside the paraboloid $x^2 + y^2 = z$ and outside the cone $x^2 + y^2 = z^2$. The constant volume density is k kg/m³.

20. Find the moment of inertia with respect to the z axis of the homogeneous solid of Exercise 19.

21. Find the moment of inertia about a diameter of the solid between two concentric spheres having radii a ft and $2a$ ft. The volume density varies inversely as the square of the distance from the center, and it is measured in slugs per cubic foot.

22. Find the mass of the solid of Exercise 21. The volume density is the same as that in Exercise 21.

In Exercises 23 through 26, evaluate the iterated integral by using either cylindrical or spherical coordinates.

23. $\int_0^4 \int_0^3 \int_0^{\sqrt{9-x^2}} \sqrt{x^2 + y^2}\, dy\, dx\, dz$ **24.** $\int_0^2 \int_0^{\sqrt{4-y^2}} \int_0^{\sqrt{4-x^2-y^2}} \frac{1}{x^2 + y^2 + z^2}\, dz\, dx\, dy$

25. $\int_0^1 \int_0^{\sqrt{1-y^2}} \int_{\sqrt{x^2+y^2}}^{\sqrt{2-x^2-y^2}} z^2\, dz\, dx\, dy$ **26.** $\int_0^2 \int_0^{\sqrt{1-x^2}} \int_0^{\sqrt{1-x^2-y^2}} \frac{z}{\sqrt{x^2 + y^2}}\, dz\, dy\, dx$

REVIEW EXERCISES FOR CHAPTER 20

In Exercises 1 through 8, evaluate the given iterated integral.

1. $\int_0^1 \int_x^{\sqrt{x}} x^2 y\, dy\, dx$ **2.** $\int_{-2}^2 \int_{-\sqrt{4-y^2}}^{\sqrt{4-y^2}} xy\, dx\, dy$ **3.** $\int_0^{\pi/2} \int_0^{2\sin\theta} r\cos^2\theta\, dr\, d\theta$

4. $\int_0^{\pi} \int_0^{3(1+\cos\theta)} r^2 \sin\theta\, dr\, d\theta$ **5.** $\int_0^1 \int_0^z \int_0^{y+z} e^x e^y e^z\, dx\, dy\, dz$ **6.** $\int_1^2 \int_3^x \int_0^{\sqrt{3}y} \frac{y}{y^2 + z^2}\, dz\, dy\, dx$

7. $\int_0^{\pi/2} \int_{\pi/6}^{\pi/2} \int_0^2 \rho^3 \sin\phi \cos\phi\, d\rho\, d\phi\, d\theta$ **8.** $\int_0^a \int_0^{\pi/2} \int_0^{\sqrt{a^2-z^2}} zre^{-r^2}\, dr\, d\theta\, dz$

In Exercises 9 through 12, evaluate the multiple integral.

9. $\iint_R xy\, dA$; R is the region in the first quadrant bounded by the circle $x^2 + y^2 = 1$ and the coordinate axes.

10. $\iint_R (x + y)\, dA$; R is the region bounded by the curve $y = \cos x$ and the x axis from $x = -\frac{1}{2}\pi$ to $x = \frac{1}{2}\pi$.

11. $\iiint_S z^2\,dV$; S is the region bounded by the cylinders $x^2 + z = 1$ and $y^2 + z = 1$ and the xy plane.

12. $\iiint_S y\cos(x + z)\,dV$; S is the region bounded by the cylinder $x = y^2$ and the planes $x + z = \frac{1}{2}\pi$, $y = 0$, and $z = 0$.

13. Evaluate by polar coordinates the double integral

$$\iint_R \frac{1}{x^2 + y^2}\,dA$$

where R is the region in the first quadrant bounded by the two circles $x^2 + y^2 = 1$ and $x^2 + y^2 = 4$.

14. Evaluate by polar coordinates the iterated integral $\int_0^1 \int_{\sqrt{3}y}^{\sqrt{4-y^2}} \ln(x^2 + y^2)\,dx\,dy$.

In Exercises 15 and 16, evaluate the iterated integral by reversing the order of integration.

15. $\int_0^1 \int_x^1 \sin y^2\,dy\,dx$ **16.** $\int_0^1 \int_0^{\cos^{-1} y} e^{\sin x}\,dx\,dy$

In Exercises 17 and 18, evaluate the line integral by Green's theorem.

17. $\oint_C (3x + 2y)\,dx + (3x + y^2)\,dy$, where C is the ellipse $16x^2 + 9y^2 = 144$.

18. $\oint_c \ln(y + 1)\,dx - \dfrac{xy}{y + 1}\,dy$, where C is the closed curve determined by the curve $\sqrt{x} + \sqrt{y} = 2$ and the intervals $[0, 4]$ on the x and y axes.

In Exercises 19 and 20, evaluate the iterated integral by changing to either cylindrical or spherical coordinates.

19. $\int_0^3 \int_0^{\sqrt{9-x^2}} \int_0^2 \sqrt{x^2 + y^2}\,dz\,dy\,dx$ **20.** $\int_0^2 \int_0^{\sqrt{4-x^2}} \int_0^{\sqrt{4-x^2-y^2}} z\sqrt{4 - x^2 - y^2}\,dz\,dy\,dx$

21. Use double integration to find the area of the region in the first quadrant bounded by the parabolas $x^2 = 4y$ and $x^2 = 8 - 4y$ by two methods: (a) integrate first with respect to x; (b) integrate first with respect to y.

22. Use double integration to find the area of the region in the xy plane bounded by the parabolas $y = 9 - x^2$ and $y = x^2 + 1$.

23. Find the area of the region enclosed by the parabola $y = x^2$ and the line $y = x + 2$ by using Theorem 20.6.2.

24. Use double integration to find the volume of the solid above the xy plane bounded by the cylinder $x^2 + y^2 = 16$ and the plane $z = 2y$ by two methods: (a) integrate first with respect to x; (b) integrate first with respect to y.

25. Find the volume of the solid bounded by the surfaces $x^2 = 4y$, $y^2 = 4x$, and $x^2 = z - y$.

26. Find the mass of the lamina in the shape of the region bounded by the parabola $y = x^2$ and the line $x - y + 2 = 0$ if the area density at any point is x^2y^2 kg/m².

27. Find the area of the surface of the cylinder $x^2 + y^2 = 9$ lying in the first octant and between the planes $x = z$ and $3x = z$.

28. Find the area of the surface of the part of the cylinder $x^2 + y^2 = a^2$ that lies inside the cylinder $y^2 + z^2 = a^2$.

29. Use double integration to find the area of the region inside the circle $r = 1$ and to the right of the parabola $r(1 + \cos\theta) = 1$.

30. Find the mass of the lamina in the shape of the region exterior to the limaçon $r = 3 - \cos\theta$ and interior to the circle $r = 5\cos\theta$ if the area density at any point is $2|\sin\theta|$ kg/m².

31. Find the center of mass of the rectangular lamina bounded by the lines $x = 3$ and $y = 2$ and the coordinate axes if the area density at any point is xy^2 kg/m².

32. Find the center of mass of the lamina in the shape of the region bounded by the parabolas $x^2 = 4 + 4y$ and $x^2 = 4 - 8y$ if the area density at any point is kx^2 kg/m^2.

33. Find the mass of the lamina in the shape of the region bounded by the polar axis and the curve $r = \cos 2\theta$, where $0 \leq \theta \leq \frac{1}{4}\pi$. The area density at any point is $r\theta$ kg/m^2.

34. Find the moment of inertia about the x axis of the lamina in the shape of the region bounded by the circle $x^2 + y^2 = a^2$ if the area density at any point is $k\sqrt{x^2 + y^2}$ kg/m^2.

35. Use Green's theorem to find the total work done in moving an object in the counterclockwise direction once around the ellipse $4x^2 + y^2 = 16$ if the motion is caused by the force field $\mathbf{F}(x, y) = (e^x + 2y)\mathbf{i} + (x^2 - \tan y)\mathbf{j}$. Assume the arc is measured in meters and the force is measured in newtons.

36. Verify Stokes' theorem in the plane if $\mathbf{F}(x, y) = x^2\mathbf{i} - 2xy\mathbf{j}$ and R is the region bounded by the circle $x^2 + y^2 = 9$.

37. Verify the divergence theorem in the plane for $\mathbf{F}$ and R of Exercise 36.

38. Use triple integration to find the volume of the solid in the first octant bounded by the plane $y + z = 8$, the cylinder $y = 2x^2$, the xy plane, and the yz plane.

39. Find the moment of inertia about the x axis of the lamina in the shape of the region bounded by the curve $y = e^x$, the line $x = 2$, and the coordinate axes if the area density at any point is xy kg/m^2.

40. Find the moment of inertia of the lamina of Exercise 39 about the y axis.

41. Find the moment of inertia with respect to the $\frac{1}{2}\pi$ axis of the homogeneous lamina in the shape of the region bounded by the curve $r^2 = 4\cos 2\theta$ if the area density at any point is k kg/m^2.

42. Find the mass of the lamina of Exercise 41.

43. Find the polar moment of inertia and the corresponding radius of gyration of the lamina of Exercise 41.

44. Find the moment of inertia about the y axis of the lamina in the shape of the region bounded by the parabola $y = x - x^2$ and the line $x + y = 0$ if the area density at any point is $(x + y)$ kg/m^2.

45. Find the mass of the solid bounded by the spheres $x^2 + y^2 + z^2 = 4$ and $x^2 + y^2 + z^2 = 9$ if the volume density at any point is $k\sqrt{x^2 + y^2 + z^2}$ kg/m^3.

46. Find the moment of inertia about the z axis of the solid of Exercise 45.

47. The homogeneous solid bounded by the cone $z^2 = 4x^2 + 4y^2$ between the planes $z = 0$ and $z = 4$ has a volume density at any point of k kg/m^3. Find the moment of inertia about the z axis for this solid.

48. Find the center of mass of the solid bounded by the sphere $x^2 + y^2 + z^2 - 6z = 0$ and the cone $x^2 + y^2 = z^2$, and above the cone, if the volume density at any point is kz kg/m^3.

Appendix

TABLE 1 Powers and Roots

n	n^2	$\sqrt{n}$	n^3	$\sqrt[3]{n}$	n	n^2	$\sqrt{n}$	n^3	$\sqrt[3]{n}$
1	1	1.000	1	1.000	51	2,601	7.141	132,651	3.708
2	4	1.414	8	1.260	52	2,704	7.211	140,608	3.732
3	9	1.732	27	1.442	53	2,809	7.280	148,877	3.756
4	16	2.000	64	1.587	54	2,916	7.348	157,464	3.780
5	25	2.236	125	1.710	55	3,025	7.416	166,375	3.803
6	36	2.449	216	1.817	56	3,136	7.483	175,616	3.826
7	49	2.646	343	1.913	57	3,249	7.550	185,193	3.848
8	64	2.828	512	2.000	58	3,364	7.616	195,112	3.871
9	81	3.000	729	2.080	59	3,481	7.681	205,379	3.893
10	100	3.162	1,000	2.154	60	3,600	7.746	216,000	3.915
11	121	3.317	1,331	2.224	61	3,721	7.810	226,981	3.936
12	144	3.464	1,728	2.289	62	3,844	7.874	238,328	3.958
13	169	3.606	2,197	2.351	63	3,969	7.937	250,047	3.979
14	196	3.742	2,744	2.410	64	4,096	8.000	262,144	4.000
15	225	3.873	3,375	2.466	65	4,225	8.062	274,625	4.021
16	256	4.000	4,096	2.520	66	4,356	8.124	287,496	4.041
17	289	4.123	4,913	2.571	67	4,489	8.185	300,763	4.062
18	324	4.243	5,832	2.621	68	4,624	8.246	314,432	4.082
19	361	4.359	6,859	2.668	69	4,761	8.307	328,509	4.102
20	400	4.472	8,000	2.714	70	4,900	8.367	343,000	4.121
21	441	4.583	9,261	2.759	71	5,041	8.426	357,911	4.141
22	484	4.690	10,648	2.802	72	5,184	8.485	373,248	4.160
23	529	4.796	12,167	2.844	73	5,329	8.544	389,017	4.179
24	576	4.899	13,824	2.884	74	5,476	8.602	405,224	4.198
25	625	5.000	15,625	2.924	75	5,625	8.660	421,875	4.217
26	676	5.099	17,576	2.962	76	5,776	8.718	438,976	4.236
27	729	5.196	19,683	3.000	77	5,929	8.775	456,533	4.254
28	784	5.291	21,952	3.037	78	6,084	8.832	474,552	4.273
29	841	5.385	24,389	3.072	79	6,241	8.888	493,039	4.291
30	900	5.477	27,000	3.107	80	6,400	8.944	512,000	4.309
31	961	5.568	29,791	3.141	81	6,561	9.000	531,441	4.327
32	1,024	5.657	32,768	3.175	82	6,724	9.055	551,368	4.344
33	1,089	5.745	35,937	3.208	83	6,889	9.110	571,787	4.362
34	1,156	5.831	39,304	3.240	84	7,056	9.165	592,704	4.380
35	1,225	5.916	42,875	3.271	85	7,225	9.220	614,125	4.397
36	1,296	6.000	46,656	3.302	86	7,396	9.274	636,056	4.414
37	1,369	6.083	50,653	3.332	87	7,569	9.327	658,503	4.431
38	1,444	6.164	54,872	3.362	88	7,744	9.381	681,472	4.448
39	1,521	6.245	59,319	3.391	89	7,921	9.434	704,969	4.465
40	1,600	6.325	64,000	3.420	90	8,100	9.487	729,000	4.481
41	1,681	6.403	68,921	3.448	91	8,281	9.539	753,571	4.498
42	1,764	6.481	74,088	3.476	92	8,464	9.592	778,688	4.514
43	1,849	6.557	79,507	3.503	93	8,649	9.643	804,357	4.531
44	1,936	6.633	85,184	3.530	94	8,836	9.695	830,584	4.547
45	2,025	6.708	91,125	3.557	95	9,025	9.747	857,375	4.563
46	2,116	6.782	97,336	3.583	96	9,216	9.798	884,736	4.579
47	2,209	6.856	103,823	3.609	97	9,409	9.849	912,673	4.595
48	2,304	6.928	110,592	3.634	98	9,604	9.899	941,192	4.610
49	2,401	7.000	117,649	3.659	99	9,801	9.950	970,299	4.626
50	2,500	7.071	125,000	3.684	100	10,000	10.000	1,000,000	4.642

TABLE 2 Natural Logarithms

N	0	1	2	3	4	5	6	7	8	9
1.0	0000	0100	0198	0296	0392	0488	0583	0677	0770	0862
1.1	0953	1044	1133	1222	1310	1398	1484	1570	1655	1740
1.2	1823	1906	1989	2070	2151	2231	2311	2390	2469	2546
1.3	2624	2700	2776	2852	2927	3001	3075	3148	3221	3293
1.4	3365	3436	3507	3577	3646	3716	3784	3853	3920	3988
1.5	4055	4121	4187	4253	4318	4383	4447	4511	4574	4637
1.6	4700	4762	4824	4886	4947	5008	5068	5128	5188	5247
1.7	5306	5365	5423	5481	5539	5596	5653	5710	5766	5822
1.8	5878	5933	5988	6043	6098	6152	6206	6259	6313	6366
1.9	6419	6471	6523	6575	6627	6678	6729	6780	6831	6881
2.0	6931	6981	7031	7080	7129	7178	7227	7275	7324	7372
2.1	7419	7467	7514	7561	7608	7655	7701	7747	7793	7839
2.2	7885	7930	7975	8020	8065	8109	8154	8198	8242	8286
2.3	8329	8372	8416	8459	8502	8544	8587	8629	8671	8713
2.4	8755	8796	8838	8879	8920	8961	9002	9042	9083	9123
2.5	9163	9203	9243	9282	9322	9361	9400	9439	9478	9517
2.6	9555	9594	9632	9670	9708	9746	9783	9821	9858	9895
2.7	9933	9969	*0006	*0043	*0080	*0116	*0152	*0188	*0225	*0260
2.8	1.0296	0332	0367	0403	0438	0473	0508	0543	0578	0613
2.9	0647	0682	0716	0750	0784	0818	0852	0886	0919	0953
3.0	1.0986	1019	1053	1086	1119	1151	1184	1217	1249	1282
3.1	1314	1346	1378	1410	1442	1474	1506	1537	1569	1600
3.2	1632	1663	1694	1725	1756	1787	1817	1848	1878	1909
3.3	1939	1969	2000	2030	2060	2090	2119	2149	2179	2208
3.4	2238	2267	2296	2326	2355	2384	2413	2442	2470	2499
3.5	1.2528	2556	2585	2613	2641	2669	2698	2726	2754	2782
3.6	2809	2837	2865	2892	2920	2947	2975	3002	3029	3056
3.7	3083	3110	3137	3164	3191	3218	3244	3271	3297	3324
3.8	3350	3376	3403	3429	3455	3481	3507	3533	3558	3584
3.9	3610	3635	3661	3686	3712	3737	3762	3788	3813	3838
4.0	1.3863	3888	3913	3938	3962	3987	4012	4036	4061	4085
4.1	4110	4134	4159	4183	4207	4231	4255	4279	4303	4327
4.2	4351	4375	4398	4422	4446	4469	4493	4516	4540	4563
4.3	4586	4609	4633	4656	4679	4702	4725	4748	4770	4793
4.4	4816	4839	4861	4884	4907	4929	4951	4974	4996	5019
4.5	1.5041	5063	5085	5107	5129	5151	5173	5195	5217	5239
4.6	5261	5282	5304	5326	5347	5369	5390	5412	5433	5454
4.7	5476	5497	5518	5539	5560	5581	5602	5623	5644	5665
4.8	5686	5707	5728	5748	5769	5790	5810	5831	5851	5872
4.9	5892	5913	5933	5953	5974	5994	6014	6034	6054	6074
5.0	1.6094	6114	6134	6154	6174	6194	6214	6233	6253	6273
5.1	6292	6312	6332	6351	6371	6390	6409	6429	6448	6467
5.2	6487	6506	6525	6544	6563	6582	6601	6620	6639	6658
5.3	6677	6696	6715	6734	6752	6771	6790	6808	6827	6845
5.4	6864	6882	6901	6919	6938	6956	6974	6993	7011	7029

TABLE 2 (Continued)

N	0	1	2	3	4	5	6	7	8	9
5.5	1.7047	7066	7084	7102	7120	7138	7156	7174	7192	7210
5.6	7228	7246	7263	7281	7299	7317	7334	7352	7370	7387
5.7	7405	7422	7440	7457	7475	7492	7509	7527	7544	7561
5.8	7579	7596	7613	7630	7647	7664	7681	7699	7716	7733
5.9	7750	7766	7783	7800	7817	7834	7851	7867	7884	7901
6.0	1.7918	7934	7951	7967	7984	8001	8017	8034	8050	8066
6.1	8083	8099	8116	8132	8148	8165	8181	8197	8213	8229
6.2	8245	8262	8278	8294	8310	8326	8342	8358	8374	8390
6.3	8405	8421	8437	8453	8469	8485	8500	8516	8532	8547
6.4	8563	8579	8594	8610	8625	8641	8656	8672	8687	8703
6.5	1.8718	8733	8749	8764	8779	8795	8810	8825	8840	8856
6.6	8871	8886	8901	8916	8931	8946	8961	8976	8991	9006
6.7	9021	9036	9051	9066	9081	9095	9110	9125	9140	9155
6.8	9169	9184	9199	9213	9228	9242	9257	9272	9286	9301
6.9	9315	9330	9344	9359	9373	9387	9402	9416	9430	9445
7.0	1.9459	9473	9488	9502	9516	9530	9544	9559	9573	9587
7.1	9601	9615	9629	9643	9657	9671	9685	9699	9713	9727
7.2	9741	9755	9769	9782	9796	9810	9824	9838	9851	9865
7.3	9879	9892	9906	9920	9933	9947	9961	9974	9988	*0001
7.4	2.0015	0028	0042	0055	0069	0082	0096	0109	0122	0136
7.5	2.0149	0162	0176	0189	0202	0215	0229	0242	0255	0268
7.6	0281	0295	0308	0321	0334	0347	0360	0373	0386	0399
7.7	0412	0425	0438	0451	0464	0477	0490	0503	0516	0528
7.8	0541	0554	0567	0580	0592	0605	0618	0630	0643	0656
7.9	0669	0681	0694	0707	0719	0732	0744	0757	0769	0782
8.0	2.0794	0807	0819	0832	0844	0857	0869	0882	0894	0906
8.1	0919	0931	0943	0956	0968	0980	0992	1005	1017	1029
8.2	1041	1054	1066	1078	1090	1102	1114	1126	1138	1150
8.3	1163	1175	1187	1199	1211	1223	1235	1247	1258	1270
8.4	1282	1294	1306	1318	1330	1342	1353	1365	1377	1389
8.5	2.1401	1412	1424	1436	1448	1459	1471	1483	1494	1506
8.6	1518	1529	1541	1552	1564	1576	1587	1599	1610	1622
8.7	1633	1645	1656	1668	1679	1691	1702	1713	1725	1736
8.8	1748	1759	1770	1782	1793	1804	1815	1827	1838	1849
8.9	1861	1872	1883	1894	1905	1917	1928	1939	1950	1961
9.0	2.1972	1983	1994	2006	2017	2028	2039	2050	2061	2072
9.1	2083	2094	2105	2116	2127	2138	2148	2159	2170	2181
9.2	2192	2203	2214	2225	2235	2246	2257	2268	2279	2289
9.3	2300	2311	2322	2332	2343	2354	2364	2375	2386	2396
9.4	2407	2418	2428	2439	2450	2460	2471	2481	2492	2502
9.5	2.2513	2523	2534	2544	2555	2565	2576	2586	2597	2607
9.6	2618	2628	2638	2649	2659	2670	2680	2690	2701	2711
9.7	2721	2732	2742	2752	2762	2773	2783	2793	2803	2814
9.8	2824	2834	2844	2854	2865	2875	2885	2895	2905	2915
9.9	2925	2935	2946	2956	2966	2976	2986	2996	3006	3016

Use ln 10 = 2.30259 to find logarithms of numbers greater than 10 or less than 1. *Example:* $\ln 220 = \ln 2.2 + 2 \ln 10 = 0.7885 + 2(2.30259) = 5.3937$.

TABLE 3 Exponential Functions

x	e^x	$\log_{10}(e^x)$	e^{-x}	x	e^x	$\log_{10}(e^x)$	e^{-x}
0.00	1.0000	0.00000	1.000000	**0.50**	1.6487	0.21715	0.606531
0.01	1.0101	.00434	0.990050	0.51	1.6653	.22149	.600496
0.02	1.0202	.00869	.980199	0.52	1.6820	.22583	.594521
0.03	1.0305	.01303	.970446	0.53	1.6989	.23018	.588605
0.04	1.0408	.01737	.960789	0.54	1.7160	.23452	.582748
0.05	1.0513	0.02171	0.951229	**0.55**	1.7333	0.23886	0.576950
0.06	1.0618	.02606	.941765	0.56	1.7507	.24320	.571209
0.07	1.0725	.03040	.932394	0.57	1.7683	.24755	.565525
0.08	1.0833	.03474	.923116	0.58	1.7860	.25189	.559898
0.09	1.0942	.03909	.913931	0.59	1.8040	.35623	.554327
0.10	1.1052	0.04343	0.904837	**0.60**	1.8221	0.26058	0.548812
0.11	1.1163	.04777	.895834	0.61	1.8404	.26492	.543351
0.12	1.1275	.05212	.886920	0.62	1.8589	.26926	.537944
0.13	1.1388	.05646	.878095	0.63	1.8776	.27361	.532592
0.14	1.1503	.06080	.869358	0.64	1.8965	.27795	.527292
0.15	1.1618	0.06514	0.860708	**0.65**	1.9155	0.28229	0.522046
0.16	1.1735	.06949	.852144	0.66	1.9348	.28663	.516851
0.17	1.1853	.07383	.843665	0.67	1.9542	.29098	.511709
0.18	1.1972	.07817	.835270	0.68	1.9739	.29532	.506617
0.19	1.2092	.08252	.826959	0.69	1.9937	.29966	.501576
0.20	1.2214	0.08686	0.818731	**0.70**	2.0138	0.30401	0.496585
0.21	1.2337	.09120	.810584	0.71	2.0340	.30835	.491644
0.22	1.2461	.09554	.802519	0.72	2.0544	.31269	.486752
0.23	1.2586	.09989	.794534	0.73	2.0751	.31703	.481909
0.24	1.2712	.10423	.786628	0.74	2.0959	.32138	.477114
0.25	1.2840	0.10857	0.778801	**0.75**	2.1170	0.32572	0.472367
0.26	1.2969	.11292	.771052	0.76	2.1383	.33006	.467666
0.27	1.3100	.11726	.763379	0.77	2.1598	.33441	.463013
0.28	1.3231	.12160	.755784	0.78	2.1815	.33875	.458406
0.29	1.3364	.12595	.748264	0.79	2.2034	.34309	.453845
0.30	1.3499	0.13029	0.740818	**0.80**	2.2255	0.34744	0.449329
0.31	1.3634	.13463	.733447	0.81	2.2479	.35178	.444858
0.32	1.3771	.13897	.726149	0.82	2.2705	.35612	.440432
0.33	1.3910	.14332	.718924	0.83	2.2933	.36046	.436049
0.34	1.4049	.14766	.711770	0.84	2.3164	.36481	.431711
0.35	1.4191	0.15200	0.704688	**0.85**	2.3396	0.36915	0.427415
0.36	1.4333	.15635	.697676	0.86	2.3632	.37349	.423162
0.37	1.4477	.16069	.690734	0.87	2.3869	.37784	.418952
0.38	1.4623	.16503	.683861	0.88	2.4109	.38218	.414783
0.39	1.4770	.16937	.677057	0.89	2.4351	.38652	.410656
0.40	1.4918	0.17372	0.670320	**0.90**	2.4596	0.39087	0.406570
0.41	1.5068	.17806	.663650	0.91	2.4843	.39521	.402524
0.42	1.5220	.18240	.657047	0.92	2.5093	.39955	.398519
0.43	1.5373	.18675	.650509	0.93	2.5345	.40389	.394554
0.44	1.5527	.19109	.644036	0.94	2.5600	.40824	.390628
0.45	1.5683	0.19543	0.637628	**0.95**	2.5857	0.41258	0.386741
0.46	1.5841	.19978	.631284	0.96	2.6117	.41692	.382893
0.47	1.6000	.20412	.625002	0.97	2.6379	.42127	.379083
0.48	1.6161	.20846	.618783	0.98	2.6645	.42561	.375311
0.49	1.6323	.21280	.612626	0.99	2.6912	.42995	.371577
0.50	1.6487	0.21715	0.606531	**1.00**	2.7183	0.43429	0.367879

TABLE 3 (Continued)

x	e^x	$\log_{10}(e^x)$	e^{-x}	x	e^x	$\log_{10}(e^x)$	e^{-x}
1.00	2.7183	0.43429	0.367879	**1.50**	4.4817	0.65144	0.223130
1.01	2.7456	.43864	.364219	1.51	4.5267	.65578	.220910
1.02	2.7732	.44298	.360595	1.52	4.5722	.66013	.218712
1.03	2.8011	.44732	.357007	1.53	4.6182	.66447	.216536
1.04	2.8292	.45167	.353455	1.54	4.6646	.66881	.214381
1.05	2.8577	0.45601	0.349938	**1.55**	4.7115	0.67316	0.212248
1.06	2.8864	.46035	.346456	1.56	4.7588	.67750	.210136
1.07	2.9154	.46470	.343009	1.57	4.8066	.68184	.208045
1.08	2.9447	.46904	.339596	1.58	4.8550	.68619	.205975
1.09	2.9743	.47338	.336216	1.59	4.9037	.69053	.203926
1.10	3.0042	0.47772	0.332871	**1.60**	4.9530	0.69487	0.201897
1.11	3.0344	.48207	.329559	1.61	5.0028	.69921	.199888
1.12	3.0649	.48641	.326280	1.62	5.0531	.70356	.197899
1.13	3.0957	.49075	.323033	1.63	5.1039	.70790	.195930
1.14	3.1268	.49510	.319819	1.64	5.1552	.71224	.193980
1.15	3.1582	0.49944	0.316637	**1.65**	5.2070	0.71659	0.192050
1.16	3.1899	.50378	.313486	1.66	5.2593	.72093	.190139
1.17	3.2220	.50812	.310367	1.67	5.3122	.72527	.188247
1.18	3.2544	.51247	.307279	1.68	5.3656	.72961	.186374
1.19	3.2871	.51681	.304221	1.69	5.4195	.73396	.184520
1.20	3.3201	0.52115	0.301194	**1.70**	5.4739	0.73830	0.182684
1.21	3.3535	.52550	.298197	1.71	5.5290	.74264	.180866
1.22	3.3872	.52984	.295230	1.72	5.5845	.74699	.179066
1.23	3.4212	.53418	.292293	1.73	5.6407	.75133	.177284
1.24	3.4556	.53853	.289384	1.74	5.6973	.75567	.175520
1.25	3.4903	0.54287	0.286505	**1.75**	5.7546	0.76002	0.173774
1.26	3.5254	.54721	.283654	1.76	5.8124	.76436	.172045
1.27	3.5609	.55155	.280832	1.77	5.8709	.76870	.170333
1.28	3.5966	.55590	.278037	1.78	5.9299	.77304	.168638
1.29	3.6328	.56024	.275271	1.79	5.9895	.77739	.166960
1.30	3.6693	0.56458	0.272532	**1.80**	6.0496	0.78173	0.165299
1.31	3.7062	.56893	.269820	1.81	6.1104	.78607	.163654
1.32	3.7434	.57327	.267135	1.82	6.1719	.79042	.162026
1.33	3.7810	.57761	.264477	1.83	6.2339	.79476	.160414
1.34	3.8190	.58195	.261846	1.84	6.2965	.79910	.158817
1.35	3.8574	0.58630	0.259240	**1.85**	6.3598	0.80344	0.157237
1.36	3.8962	.59064	.256661	1.86	6.4237	.80779	.155673
1.37	3.9354	.59498	.254107	1.87	6.4483	.81213	.154124
1.38	3.9749	.59933	.251579	1.88	6.5535	.81647	.152590
1.39	4.0149	.60367	.249075	1.89	6.6194	.82082	.151072
1.40	4.0552	0.60801	0.246597	**1.90**	6.6859	0.82516	0.149569
1.41	4.0960	.61236	.244143	1.91	6.7531	.82950	.148080
1.42	4.1371	.61670	.241714	1.92	6.8210	.83385	.146607
1.43	4.1787	.62104	.239309	1.93	6.8895	.83819	.145148
1.44	4.2207	.62538	.236928	1.94	6.9588	.84253	.143704
1.45	4.2631	0.62973	0.234570	**1.95**	7.0287	0.84687	0.142274
1.46	4.3060	.63407	.232236	1.96	7.0993	.85122	.140858
1.47	4.3492	.63841	.229925	1.97	7.1707	.85556	.139457
1.48	4.3929	.64276	.227638	1.98	7.2427	.85990	.138069
1.49	4.4371	.64710	.225373	1.99	7.3155	.86425	.136695
1.50	4.4817	0.65144	0.223130	**2.00**	7.3891	0.86859	0.135335

TABLE 3 (Continued)

x	e^x	$\log_{10}(e^x)$	e^{-x}	x	e^x	$\log_{10}(e^x)$	e^{-x}
2.00	7.3891	0.86859	0.135335	**2.50**	12.182	1.08574	0.082085
2.01	7.4633	.87293	.133989	2.51	12.305	1.09008	.081268
2.02	7.5383	.87727	.132655	2.52	12.429	1.09442	.080460
2.03	7.6141	.88162	.131336	2.53	12.554	1.09877	.079659
2.04	7.6906	.88596	.130029	2.54	12.680	1.10311	.078866
2.05	7.7679	0.89030	0.128735	**2.55**	12.807	1.10745	0.078082
2.06	7.8460	.89465	.127454	2.56	12.936	1.11179	.077305
2.07	7.9248	.89899	.126186	2.57	13.066	1.11614	.076536
2.08	8.0045	.90333	.124930	2.58	13.197	1.12048	.075774
2.09	8.0849	.90756	.123687	2.59	13.330	1.12482	.075020
2.10	8.1662	0.91202	0.122456	**2.60**	13.464	1.12917	0.074274
2.11	8.2482	.91636	.121238	2.61	13.599	1.13351	.073535
2.12	8.3311	.92070	.120032	2.62	13.736	1.13785	.072803
2.13	8.4149	.92505	.118837	2.63	13.874	1.14219	.072078
2.14	8.4994	.92939	.117655	2.64	14.013	1.14654	.071361
2.15	8.5849	0.93373	0.116484	**2.65**	14.154	1.15088	0.070651
2.16	8.6711	.93808	.115325	2.66	14.296	1.15522	.069948
2.17	8.7583	.94242	.114178	2.67	14.440	1.15957	.069252
2.18	8.8463	.94676	.113042	2.68	14.585	1.16391	.068563
2.19	8.9352	.95110	.111917	2.69	14.732	1.16825	.067881
2.20	9.0250	0.95545	0.110803	**2.70**	14.880	1.17260	0.067206
2.21	9.1157	.95979	.109701	2.71	15.029	1.17694	.066537
2.22	9.2073	.96413	.108609	2.72	15.180	1.18128	.065875
2.23	9.2999	.96848	.107528	2.73	15.333	1.18562	.065219
2.24	9.3933	.97282	.106459	2.74	15.487	1.18997	.064570
2.25	9.4877	0.97716	0.105399	**2.75**	15.643	1.19431	0.063928
2.26	9.5831	.98151	.104350	2.76	15.800	1.19865	.063292
2.27	9.6794	.98585	.103312	2.77	15.959	1.20300	.062662
2.28	9.7767	.99019	.102284	2.78	16.119	1.20734	.062039
2.29	9.8749	.99453	.101266	2.79	16.281	1.21168	.061421
2.30	9.9742	0.99888	0.100259	**2.80**	16.445	1.21602	0.060810
2.31	10.074	1.00322	.099261	2.81	16.610	1.22037	.060205
2.32	10.176	1.00756	.098274	2.82	16.777	1.22471	.059606
2.33	10.278	1.01191	.097296	2.83	16.945	1.22905	.059013
2.34	10.381	1.01625	.096328	2.84	17.116	1.23340	.058426
2.35	10.486	1.02059	0.095369	**2.85**	17.288	1.23774	0.057844
2.36	10.591	1.02493	.094420	2.86	17.462	1.24208	.057269
2.37	10.697	1.02928	.093481	2.87	17.637	1.24643	.056699
2.38	10.805	1.03362	.092551	2.88	17.814	1.25077	.056135
2.39	10.913	1.03796	.091630	2.89	17.993	1.25511	.055576
2.40	11.023	1.04231	0.090718	**2.90**	18.174	1.25945	0.055023
2.41	11.134	1.04665	.089815	2.91	18.357	1.26380	.054476
2.42	11.246	1.05099	.088922	2.92	18.541	1.26814	.053934
2.43	11.359	1.05534	.088037	2.93	18.728	1.27248	.053397
2.44	11.473	1.05968	.087161	2.94	18.916	1.27683	.052866
2.45	11.588	1.06402	0.086294	**2.95**	19.106	1.28117	0.052340
2.46	11.705	1.06836	.085435	2.96	19.298	1.28551	.051819
2.47	11.822	1.07271	.084585	2.97	19.492	1.28985	.051303
2.48	11.941	1.07705	.083743	2.98	19.688	1.29420	.050793
2.49	12.061	1.08139	.082910	2.99	19.886	1.29854	.050287
2.50	12.182	1.08574	0.082085	**3.00**	20.086	1.30288	0.049787

TABLE 3 (Continued)

x	e^x	$\log_{10}(e^x)$	e^{-x}	x	e^x	$\log_{10}(e^x)$	e^{-x}
3.00	20.086	1.30288	0.049787	**3.50**	33.115	1.52003	0.030197
3.01	20.287	1.30723	.049292	3.51	33.448	1.52437	.029897
3.02	20.491	1.31157	.048801	3.52	33.784	1.52872	.029599
3.03	20.697	1.31591	.048316	3.53	34.124	1.53306	.029305
3.04	20.905	1.32026	.047835	3.54	34.467	1.53740	.029013
3.05	21.115	1.32460	0.047359	**3.55**	34.813	1.54175	0.028725
3.06	21.328	1.32894	.046888	3.56	35.163	1.54609	.028439
3.07	21.542	1.33328	.046421	3.57	35.517	1.55043	.028156
3.08	21.758	1.33763	.045959	3.58	35.874	1.55477	.027876
3.09	21.977	1.34197	.045502	3.59	36.234	1.55912	.027598
3.10	22.198	1.34631	0.045049	**3.60**	36.598	1.56346	0.027324
3.11	22.421	1.35066	.044601	3.61	36.966	1.56780	.027052
3.12	22.646	1.35500	.044157	3.62	37.338	1.57215	.026783
3.13	22.874	1.35934	.043718	3.63	37.713	1.57649	.026516
3.14	23.104	1.36368	.043283	3.64	38.092	1.58083	.026252
3.15	23.336	1.36803	0.042852	**3.65**	38.475	1.58517	0.025991
3.16	23.571	1.37237	.042426	3.66	38.861	1.58952	.025733
3.17	23.807	1.36671	.042004	3.67	39.252	1.59386	.025476
3.18	24.047	1.38106	.041586	3.68	39.646	1.59820	.025223
3.19	24.288	1.38540	.041172	3.69	40.045	1.60255	.024972
3.20	24.533	1.38974	0.040764	**3.70**	40.447	1.60689	0.024724
3.21	24.779	1.39409	.040357	3.71	40.854	1.61123	.024478
3.22	25.028	1.39843	.039955	3.72	41.264	1.61558	.024234
3.23	25.280	1.40277	.039557	3.73	41.679	1.61992	.023993
3.24	25.534	1.40711	.039164	3.74	42.098	1.62426	.023754
3.25	25.790	1.41146	0.038774	**3.75**	42.521	1.62860	0.023518
3.26	26.050	1.41580	.038388	3.76	42.948	1.63295	.023284
3.27	26.311	1.42014	.038006	3.77	43.380	1.63729	.023052
3.28	26.576	1.42449	.037628	3.78	43.816	1.64163	.022823
3.29	26.843	1.42883	.037254	3.79	44.256	1.64598	.022596
3.30	27.113	1.44317	0.036883	**3.80**	44.701	1.65032	0.022371
3.31	27.385	1.43751	.036516	3.81	45.150	1.65466	.022148
3.32	27.660	1.44186	.036153	3.82	45.604	1.65900	.021928
3.33	27.938	1.44620	.035793	3.83	46.063	1.66335	.021710
3.34	28.219	1.45054	.035437	3.84	46.525	1.66769	.021494
3.35	28.503	1.45489	0.035084	**3.85**	46.993	1.67203	0.021280
3.36	28.789	1.45923	.034735	3.86	47.465	1.67638	.021068
3.37	29.079	1.46357	.034390	3.87	47.942	1.68072	.020858
3.38	29.371	1.46792	.034047	3.88	48.424	1.68506	.020651
3.39	29.666	1.47226	.033709	3.89	48.911	1.68941	.020445
3.40	29.964	1.47660	0.033373	**3.90**	49.402	1.69375	0.020242
3.41	30.265	1.48094	.033041	3.91	49.899	1.69809	.020041
3.42	30.569	1.48529	.032712	3.92	50.400	1.70243	.019840
3.43	30.877	1.48963	.032387	3.93	50.907	1.70678	.019644
3.44	31.187	1.49397	.032065	3.94	51.419	1.71112	.019448
3.45	31.500	1.49832	0.031746	**3.95**	51.935	1.71546	0.019255
3.46	31.817	1.50266	.031430	3.96	52.457	1.71981	.019063
3.47	32.137	1.50700	.031117	3.97	52.985	1.72415	.018873
3.48	32.460	1.51134	.030807	3.98	53.517	1.72849	.018686
3.49	32.786	1.51569	.030501	3.99	54.055	1.73283	.018500
3.50	33.115	1.52003	0.030197	**4.00**	54.598	1.73718	0.018316

TABLE 3 (Continued)

x	e^x	$\log_{10}(e^x)$	e^{-x}	x	e^x	$\log_{10}(e^x)$	e^{-x}
4.00	54.598	1.73718	0.018316	**4.50**	90.017	1.95433	0.011109
4.01	55.147	1.74152	.018133	4.51	90.922	1.95867	.010998
4.02	55.701	1.74586	.017953	4.52	91.836	1.96301	.010889
4.03	56.261	1.75021	.017774	4.53	92.759	1.96735	.010781
4.04	56.826	1.75455	.017597	4.54	93.691	1.97170	.010673
4.05	57.397	1.75889	0.017422	**4.55**	94.632	1.97604	0.010567
4.06	57.974	1.76324	.017249	4.56	95.583	1.98038	.010462
4.07	58.577	1.76758	.017077	4.57	96.544	1.98473	.010358
4.08	59.145	1.77192	.016907	4.58	97.514	1.98907	.010255
4.09	59.740	1.77626	.016739	4.59	98.494	1.99341	.010153
4.10	60.340	1.78061	0.016573	**4.60**	99.484	1.99775	0.010052
4.11	60.947	1.78495	.016408	4.61	100.48	2.00210	.009952
4.12	61.559	1.78929	.016245	4.62	101.49	2.00644	.009853
4.13	62.178	1.79364	.016083	4.63	102.51	2.01078	.009755
4.14	62.803	1.79798	.015923	4.64	103.54	2.01513	.009658
4.15	63.434	1.80232	0.015764	**4.65**	104.58	2.01947	0.009562
4.16	64.072	1.80667	.015608	4.66	105.64	2.02381	.009466
4.17	64.715	1.81101	.015452	4.67	106.70	2.02816	.009372
4.18	65.366	1.81535	.015299	4.68	107.77	2.03250	.009279
4.19	66.023	1.81969	.015146	4.69	108.85	2.03684	.009187
4.20	66.686	1.82404	0.014996	**4.70**	109.95	2.04118	0.009095
4.21	67.357	1.82838	.014846	4.71	111.05	2.04553	.009005
4.22	68.033	1.83272	.014699	4.72	112.17	2.04987	.008915
4.23	68.717	1.83707	.014552	4.73	113.30	2.05421	.008826
4.24	69.408	1.84141	.014408	4.74	114.43	2.05856	.008739
4.25	70.105	1.84575	0.014264	**4.75**	115.58	2.06290	0.008652
4.26	70.810	1.85009	.014122	4.76	116.75	2.06724	.008566
4.27	71.522	1.85444	.013982	4.77	117.92	2.07158	.008480
4.28	72.240	1.85878	.013843	4.78	119.10	2.07593	.008396
4.29	72.966	1.86312	.013705	4.79	120.30	2.08027	.008312
4.30	73.700	1.86747	0.013569	**4.80**	121.51	2.08461	0.008230
4.31	74.440	1.87181	.013434	4.81	122.73	2.08896	.008148
4.32	75.189	1.87615	.013300	4.82	123.97	2.09330	.008067
4.33	75.944	1.88050	.013168	4.83	125.21	2.09764	.007987
4.34	76.708	1.88484	.013037	4.84	126.47	2.10199	.007907
4.35	77.478	1.88918	0.012907	**4.85**	127.74	2.10633	0.007828
4.36	78.257	1.89352	.012778	4.86	129.02	2.11067	.007750
4.37	79.044	1.89787	.012651	4.87	130.32	2.11501	.007673
4.38	79.838	1.90221	.012525	4.88	131.63	2.11936	.007597
4.39	80.640	1.90655	.012401	4.89	132.95	2.12370	.007521
4.40	81.451	1.91090	0.012277	**4.90**	134.29	2.12804	0.007477
4.41	82.269	1.91524	.012155	4.91	135.64	2.13239	.007372
4.42	83.096	1.91958	.012034	4.92	137.00	2.13673	.007299
4.43	83.931	1.92392	.011914	4.93	138.38	2.14107	.007227
4.44	84.775	1.92827	.011796	4.94	139.77	2.14541	.007155
4.45	85.627	1.93261	0.011679	**4.95**	141.17	2.14976	0.007083
4.46	86.488	1.93695	.011562	4.96	142.59	2.15410	.007013
4.47	87.357	1.94130	.011447	4.97	144.03	2.15844	.006943
4.48	88.235	1.94564	.011333	4.98	145.47	2.16279	.006874
4.49	89.121	1.94998	.011221	4.99	146.94	2.16713	.006806
4.50	90.017	1.95433	0.011109	**5.00**	148.41	2.17147	0.006738

TABLE 3 (Continued)

x	e^x	$\log_{10}(e^x)$	e^{-x}	x	e^x	$\log_{10}(e^x)$	e^{-x}
5.00	148.41	2.17147	0.006738	**5.50**	244.69	2.38862	0.0040868
5.01	149.90	2.17582	.006671	5.55	257.24	2.41033	.0038875
5.02	151.41	2.18016	.006605	5.60	270.43	2.43205	.0036979
5.03	152.93	2.18450	.006539	5.65	284.29	2.45376	.0035175
5.04	154.47	2.18884	.006474	5.70	298.87	2.47548	.0033460
5.05	156.02	2.19319	0.006409	**5.75**	314.19	2.49719	0.0031828
5.06	157.59	2.19753	.006346	5.80	330.30	2.51891	.0030276
5.07	159.17	2.20187	.006282	5.85	347.23	2.54062	.0028799
5.08	160.77	2.20622	.006220	5.90	365.04	2.56234	.0027394
5.09	162.39	2.21056	.006158	5.95	383.75	2.58405	.0026058
5.10	164.02	2.21490	0.006097	**6.00**	403.43	2.60577	0.0024788
5.11	165.67	2.21924	.006036	6.05	424.11	2.62748	.0023579
5.12	167.34	2.22359	.005976	6.10	445.86	2.64920	.0022429
5.13	169.02	2.22793	.005917	6.15	468.72	2.67091	.0021335
5.14	170.72	2.23227	.005858	6.20	492.75	2.69263	.0020294
5.15	172.43	2.23662	0.005799	**6.25**	518.01	2.71434	0.0019305
5.16	174.16	2.24096	.005742	6.30	544.57	2.73606	.0018363
5.17	175.91	2.24530	.005685	6.35	572.49	2.75777	.0017467
5.18	177.68	2.24965	.005628	6.40	601.85	2.77948	.0016616
5.19	179.47	2.25399	.005572	6.45	632.70	2.80120	.0015805
5.20	181.27	2.25833	0.005517	**6.50**	665.14	2.82291	0.0015034
5.21	183.09	2.26267	.005462	6.55	699.24	2.84463	.0014301
5.22	184.93	2.26702	.005407	6.60	735.10	2.86634	.0013604
5.23	186.79	2.27136	.005354	6.65	772.78	2.88806	.0012940
5.24	188.67	2.27570	.005300	6.70	812.41	2.90977	.0012309
5.25	190.57	2.28005	0.005248	**6.75**	854.06	2.93149	0.0011709
5.26	192.48	2.28439	.005195	6.80	897.85	2.95320	.0011138
5.27	194.42	2.28873	.005144	6.85	943.88	2.97492	.0010595
5.28	196.37	2.29307	.005092	6.90	992.27	2.99663	.0010078
5.29	198.34	2.29742	.005042	6.95	1043.1	3.01835	.0009586
5.30	200.34	2.30176	0.004992	**7.00**	1096.6	3.04006	0.0009119
5.31	202.35	2.30610	.004942	7.05	1152.9	3.06178	.0008674
5.32	204.38	2.31045	.004893	7.10	1212.0	3.08349	.0008251
5.33	206.44	2.31479	.004844	7.15	1274.1	3.10521	.0007849
5.34	208.51	2.31913	.004796	7.20	1339.4	3.12692	.0007466
5.35	210.61	2.32348	0.004748	**7.25**	1408.1	3.14863	0.0007102
5.36	212.72	2.32782	.004701	7.30	1480.3	3.17035	.0006755
5.37	214.86	2.33216	.004654	7.35	1556.2	3.19206	.0006426
5.38	217.02	2.33650	.004608	7.40	1636.0	3.21378	.0006113
5.39	219.20	2.34085	.004562	7.45	1719.9	3.23549	.0005814
5.40	221.41	2.34519	0.004517	**7.50**	1808.0	3.25721	0.0005531
5.41	223.63	2.34953	.004472	7.55	1900.7	3.27892	.0005261
5.42	225.88	2.35388	.004427	7.60	1998.2	3.30064	.0005005
5.43	228.15	2.35822	.004383	7.65	2100.6	3.32235	.0004760
5.44	230.44	2.36256	.004339	7.70	2208.3	3.34407	.0004528
5.45	232.76	2.36690	0.004296	**7.75**	2321.6	3.36578	0.0004307
5.46	235.10	2.37125	.004254	7.80	2440.6	3.38750	.0004097
5.47	237.46	2.37559	.004211	7.85	2565.7	3.40921	.0003898
5.48	239.85	2.37993	.004169	7.90	2697.3	3.43093	.0003707
5.49	242.26	2.38428	.004128	7.95	2835.6	3.45264	.0003527
5.50	244.69	2.38862	0.004087	**8.00**	2981.0	3.47436	0.0003355

TABLE 3 (Continued)

x	e^x	$\log_{10}(e^x)$	e^{-x}	x	e^x	$\log_{10}(e^x)$	e^{-x}
8.00	2981.0	3.47436	0.0003355	**9.00**	8103.1	3.90865	0.0001234
8.05	3133.8	3.49607	.0003191	9.05	8518.5	3.93037	.0001174
8.10	3294.5	3.51779	.0003035	9.10	8955.3	3.95208	.0001117
8.15	3463.4	3.53950	.0002887	9.15	9414.4	3.97379	.0001062
8.20	3641.0	3.56121	.0002747	9.20	9897.1	3.99551	.0001010
8.25	3827.6	3.58293	0.0002613	**9.25**	10405	4.01722	0.0000961
8.30	4023.9	3.60464	.0002485	9.30	10938	4.03894	.0000914
8.35	4230.2	3.62636	.0002364	9.35	11499	4.06065	.0000870
8.40	4447.1	3.64807	.0002249	9.40	12088	4.08237	.0000827
8.45	4675.1	3.66979	.0002139	9.45	12708	4.10408	.0000787
8.50	4914.8	3.69150	0.0002036	**9.50**	13360	4.12580	0.0000749
8.55	5166.8	3.71322	.0001935	9.55	14045	4.14751	.0000712
8.60	5431.7	3.73493	.0001841	9.60	14765	4.16923	.0000677
8.65	5710.0	3.75665	.0001751	9.65	15522	4.19094	.0000644
8.70	6002.9	3.77836	.0001666	9.70	16318	4.21266	.0000613
8.75	6310.7	3.80008	0.0001585	**9.75**	17154	4.23437	0.0000583
8.80	6634.2	3.82179	.0001507	9.80	18034	4.25609	.0000555
8.85	6974.4	3.84351	.0001434	9.85	18958	4.27780	.0000527
8.90	7332.0	3.86522	.0001364	9.90	19930	4.29952	.0000502
8.95	7707.9	3.88694	.0001297	9.95	20952	4.32123	0.0000477
9.00	8103.1	3.90865	0.0001234	**10.00**	22026	4.34294	0.0000454

TABLE 4 Hyperbolic Functions

x	sinh x	cosh x	tanh x
0	.00000	1.0000	.00000
0.1	.10017	1.0050	.09967
0.2	.20134	1.0201	.19738
0.3	.30452	1.0453	.29131
0.4	.41075	1.0811	.37995
0.5	.52110	1.1276	.46212
0.6	.63665	1.1855	.53705
0.7	.75858	1.2552	.60437
0.8	.88811	1.3374	.66404
0.9	1.0265	1.4331	.71630
1.0	1.1752	1.5431	.76159
1.1	1.3356	1.6685	.80050
1.2	1.5095	1.8107	.83365
1.3	1.6984	1.9709	.86172
1.4	1.9043	2.1509	.88535
1.5	2.1293	2.3524	.90515
1.6	2.3756	2.5775	.92167
1.7	2.6456	2.8283	.93541
1.8	2.9422	3.1075	.94681
1.9	3.2682	3.4177	.95624
2.0	3.6269	3.7622	.96403
2.1	4.0219	4.1443	.97045
2.2	4.4571	4.5679	.97574
2.3	4.9370	5.0372	.98010
2.4	5.4662	5.5569	.98367
2.5	6.0502	6.1323	.98661
2.6	6.6947	6.7690	.98903
2.7	7.4063	7.4735	.99101
2.8	8.1919	8.2527	.99263
2.9	9.0596	9.1146	.99396
3.0	10.018	10.068	.99505
3.1	11.076	11.122	.99595
3.2	12.246	12.287	.99668
3.3	13.538	13.575	.99728
3.4	14.965	14.999	.99777
3.5	16.543	16.573	.99818
3.6	18.285	18.313	.99851
3.7	20.211	20.236	.99878
3.8	22.339	22.362	.99900
3.9	24.691	24.711	.99918
4.0	27.290	27.308	.99933
4.1	30.162	30.178	.99945
4.2	33.336	33.351	.99955
4.3	36.843	36.857	.99963
4.4	40.719	40.732	.99970
4.5	45.003	45.014	.99975
4.6	49.737	49.747	.99980
4.7	54.969	54.978	.99983
4.8	60.751	60.759	.99986
4.9	67.141	67.149	.99989
5.0	74.203	74.210	.99991

TABLE 5 Trigonometric Functions

Degrees	Radians	Sin	Cos	Tan	Cot		
0	0.0000	0.0000	1.0000	0.0000		1.5708	90
1	0.0175	0.0175	0.9998	0.0175	57.290	1.5533	89
2	0.0349	0.0349	0.9994	0.0349	28.636	1.5359	88
3	0.0524	0.0523	0.9986	0.0524	19.081	1.5184	87
4	0.0698	0.0698	0.9976	0.0699	14.301	1.5010	86
5	0.0873	0.0872	0.9962	0.0875	11.430	1.4835	85
6	0.1047	0.1045	0.9945	0.1051	9.5144	1.4661	84
7	0.1222	0.1219	0.9925	0.1228	8.1443	1.4486	83
8	0.1396	0.1392	0.9903	0.1405	7.1154	1.4312	82
9	0.1571	0.1564	0.9877	0.1584	6.3138	1.4137	81
10	0.1745	0.1736	0.9848	0.1763	5.6713	1.3963	80
11	0.1920	0.1908	0.9816	0.1944	5.1446	1.3788	79
12	0.2094	0.2079	0.9781	0.2126	4.7046	1.3614	78
13	0.2269	0.2250	0.9744	0.2309	4.3315	1.3439	77
14	0.2443	0.2419	0.9703	0.2493	4.0108	1.3265	76
15	0.2618	0.2588	0.9659	0.2679	3.7321	1.3090	75
16	0.2793	0.2756	0.9613	0.2867	3.4874	1.2915	74
17	0.2967	0.2924	0.9563	0.3057	3.2709	1.2741	73
18	0.3142	0.3090	0.9511	0.3249	3.0777	1.2566	72
19	0.3316	0.3256	0.9455	0.3443	2.9042	1.2392	71
20	0.3491	0.3420	0.9397	0.3640	2.7475	1.2217	70
21	0.3665	0.3584	0.9336	0.3839	2.6051	1.2043	69
22	0.3840	0.3746	0.9272	0.4040	2.4751	1.1868	68
23	0.4014	0.3907	0.9205	0.4245	2.3559	1.1694	67
24	0.4189	0.4067	0.9135	0.4452	2.2460	1.1519	66
25	0.4363	0.4226	0.9063	0.4663	2.1445	1.1345	65
26	0.4538	0.4384	0.8988	0.4877	2.0503	1.1170	64
27	0.4712	0.4540	0.8910	0.5095	1.9626	1.0996	63
28	0.4887	0.4695	0.8829	0.5317	1.8807	1.0821	62
29	0.5061	0.4848	0.8746	0.5543	1.8040	1.0647	61
30	0.5236	0.5000	0.8660	0.5774	1.7321	1.0472	60
31	0.5411	0.5150	0.8572	0.6009	1.6643	1.0297	59
32	0.5585	0.5299	0.8480	0.6249	1.6003	1.0123	58
33	0.5760	0.5446	0.8387	0.6494	1.5399	0.9948	57
34	0.5934	0.5592	0.8290	0.6745	1.4826	0.9774	56
35	0.6109	0.5736	0.8192	0.7002	1.4281	0.9599	55
36	0.6283	0.5878	0.8090	0.7265	1.3764	0.9425	54
37	0.6458	0.6018	0.7986	0.7536	1.3270	0.9250	53
38	0.6632	0.6157	0.7880	0.7813	1.2799	0.9076	52
39	0.6807	0.6293	0.7771	0.8098	1.2349	0.8901	51
40	0.6981	0.6428	0.7660	0.8391	1.1918	0.8727	50
41	0.7156	0.6561	0.7547	0.8693	1.1504	0.8552	49
42	0.7330	0.6691	0.7431	0.9004	1.1106	0.8378	48
43	0.7505	0.6820	0.7314	0.9325	1.0724	0.8203	47
44	0.7679	0.6947	0.7193	0.9657	1.0355	0.8029	46
45	0.7854	0.7071	0.7071	1.0000	1.0000	0.7854	45
		Cos	Sin	Cot	Tan	Radians	Degrees

TABLE 6 Common Logarithms

N	0	1	2	3	4	5	6	7	8	9
10	0000	0043	0086	0128	0170	0212	0253	0294	0334	0374
11	0414	0453	0492	0531	0569	0607	0645	0682	0719	0755
12	0792	0828	0864	0899	0934	0969	1004	1038	1072	1106
13	1139	1173	1206	1239	1271	1303	1335	1367	1399	1430
14	1461	1492	1523	1553	1584	1614	1644	1673	1703	1732
15	1761	1790	1818	1847	1875	1903	1931	1959	1987	2014
16	2041	2068	2095	2122	2148	2175	2201	2227	2253	2279
17	2304	2330	2355	2380	2405	2430	2455	2480	2504	2529
18	2553	2577	2601	2625	2648	2672	2695	2718	2742	2765
19	2788	2810	2833	2856	2878	2900	2923	2945	2967	2989
20	3010	3032	3054	3075	3096	3118	3139	3160	3181	3201
21	3222	3243	3263	3284	3304	3324	3345	3365	3385	3404
22	3424	3444	3464	3483	3502	3522	3541	3560	3579	3598
23	3617	3636	3655	3674	3692	3711	3729	3747	3766	3784
24	3802	3820	3838	3856	3874	3892	3909	3927	3945	3962
25	3979	3997	4014	4031	4048	4065	4082	4099	4116	4133
26	4150	4166	4183	4200	4216	4232	4249	4265	4281	4298
27	4314	4330	4346	4362	4378	4393	4409	4425	4440	4456
28	4472	4487	4502	4518	4533	4548	4564	4579	4594	4609
29	4624	4639	4654	4669	4683	4698	4713	4728	4742	4757
30	4771	4786	4800	4814	4829	4843	4857	4871	4886	4900
31	4914	4928	4942	4955	4969	4983	4997	5011	5024	5038
32	5051	5065	5079	5092	5105	5119	5132	5145	5159	5172
33	5185	5198	5211	5224	5237	5250	5263	5276	5289	5302
34	5315	5328	5340	5353	5366	5378	5391	5403	5416	5428
35	5441	5453	5465	5478	5490	5502	5514	5527	5539	5551
36	5563	5575	5587	5599	5611	5623	5635	5647	5658	5670
37	5682	5694	5705	5717	5729	5740	5752	5763	5775	5786
38	5798	5809	5821	5832	5843	5855	5866	5877	5888	5899
39	5911	5922	5933	5944	5955	5966	5977	5988	5999	6010
40	6021	6031	6042	6053	6064	6075	6085	6096	6107	6117
41	6128	6138	6149	6160	6170	6180	6191	6201	6212	6222
42	6232	6243	6253	6263	6274	6284	6294	6304	6314	6325
43	6335	6345	6355	6365	6375	6385	6395	6405	6415	6425
44	6435	6444	6454	6464	6474	6484	6493	6503	6513	6522
45	6532	6542	6551	6561	6571	6580	6590	6599	6609	6618
46	6628	6637	6646	6656	6665	6675	6684	6693	6702	6712
47	6721	6730	6739	6749	6758	6767	6776	6785	6794	6803
48	6812	6821	6830	6839	6848	6857	6866	6875	6884	6893
49	6902	6911	6920	6928	6937	6946	6955	6964	6972	6981
50	6990	6998	7007	7016	7024	7033	7042	7050	7059	7067
51	7076	7084	7093	7101	7110	7118	7126	7135	7143	7152
52	7160	7168	7177	7185	7193	7202	7210	7218	7226	7235
53	7243	7251	7259	7267	7275	7284	7292	7300	7308	7316
54	7324	7332	7340	7348	7356	7364	7372	7380	7388	7396

TABLE 6 (Continued)

N	0	1	2	3	4	5	6	7	8	9
55	7404	7412	7419	7427	7435	7443	7451	7459	7466	7474
56	7482	7490	7497	7505	7513	7520	7528	7536	7543	7551
57	7559	7566	7574	7582	7589	7597	7604	7612	7619	7627
58	7634	7642	7649	7657	7664	7672	7679	7686	7694	7701
59	7709	7716	7723	7731	7738	7745	7752	7760	7767	7774
60	7782	7789	7796	7803	7810	7818	7825	7832	7839	7846
61	7853	7860	7868	7875	7882	7889	7896	7903	7910	7917
62	7924	7931	7938	7945	7952	7959	7966	7973	7980	7987
63	7993	8000	8007	8014	8021	8028	8035	8041	8048	8055
64	8062	8069	8075	8082	8089	8096	8102	8109	8116	8122
65	8129	8136	8142	8149	8156	8162	8169	8176	8182	8189
66	8195	8202	8209	8215	8222	8228	8235	8241	8248	8254
67	8261	8267	8274	8280	8287	8293	8299	8306	8312	8319
68	8325	8331	8338	8344	8351	8357	8363	8370	8376	8382
69	8388	8395	8401	8407	8414	8420	8426	8432	8439	8445
70	8451	8457	8463	8470	8476	8482	8488	8494	8500	8506
71	8513	8519	8525	8531	8537	8543	8549	8555	8561	8567
72	8573	8579	8585	8591	8597	8603	8609	8615	8621	8627
73	8633	8639	8645	8651	8657	8663	8669	8675	8681	8686
74	8692	8698	8704	8710	8716	8722	8727	8733	8739	8745
75	8751	8756	8762	8768	8774	8779	8785	8791	8797	8802
76	8808	8814	8820	8825	8831	8837	8842	8848	8854	8859
77	8865	8871	8876	8882	8887	8893	8899	8904	8910	8915
78	8921	8927	8932	8938	8943	8949	8954	8960	8965	8971
79	8976	8982	8987	8993	8998	9004	9009	9015	9020	9025
80	9031	9036	9042	9047	9053	9058	9063	9069	9074	9079
81	9085	9090	9096	9101	9106	9112	9117	9122	9128	9133
82	9138	9143	9149	9154	9159	9165	9170	9175	9180	9186
83	9191	9196	9201	9206	9212	9217	9222	9227	9232	9238
84	9243	9248	9253	9258	9263	9269	9274	9279	9284	9289
85	9294	9299	9304	9309	9315	9320	9325	9330	9335	9340
86	9345	9350	9355	9360	9365	9370	9375	9380	9385	9390
87	9395	9400	9405	9410	9415	9420	9425	9430	9435	9440
88	9445	9450	9455	9460	9465	9469	9474	9479	9484	9489
89	9494	9499	9504	9509	9513	9518	9523	9528	9533	9538
90	9542	9547	9552	9557	9562	9566	9571	9576	9581	9586
91	9590	9595	9600	9605	9609	9614	9619	9624	9628	9633
92	9638	9643	9647	9652	9657	9661	9666	9671	9675	9680
93	9685	9689	9694	9699	9703	9708	9713	9717	9722	9727
94	9731	9736	9741	9745	9750	9754	9759	9763	9768	9773
95	9777	9782	9786	9791	9795	9800	9805	9809	9814	9818
96	9823	9827	9832	9836	9841	9845	9850	9854	9859	9863
97	9868	9872	9877	9881	9886	9890	9894	9899	9903	9908
98	9912	9917	9921	9926	9930	9934	9939	9943	9948	9952
99	9956	9961	9965	9969	9974	9978	9983	9987	9991	9996

TABLE 7 The Greek Alphabet

α	alpha	ι	iota	ρ	rho
β	beta	κ	kappa	σ	sigma
γ	gamma	λ	lambda	τ	tau
δ	delta	μ	mu	υ	upsilon
ϵ	epsilon	ν	nu	ϕ	phi
ζ	zeta	ξ	xi	χ	chi
η	eta	o	omicron	ψ	psi
θ	theta	π	pi	ω	omega

Answers to Odd-Numbered Exercises

EXERCISES 1.1 (Page 13)

1. $(-2, +\infty)$ **3.** $(-\infty, \frac{3}{4}]$ **5.** $[4, 8]$ **7.** $(-\frac{5}{3}, \frac{4}{3}]$ **9.** $(-\infty, -\frac{1}{2}) \cup (0, +\infty)$ **11.** $(-\infty, -1) \cup (\frac{1}{3}, 3)$ **13.** $(-\infty, -2) \cup (2, +\infty)$ **15.** $(-\infty, -5) \cup (3, +\infty)$ **17.** $[-1, \frac{1}{2}]$ **19.** $(-3, \frac{3}{4})$ **21.** $(\frac{3}{2}, \frac{31}{14}] \cup (\frac{7}{3}, +\infty)$ **23.** $\{-\frac{5}{2}, 1\}$ **25.** $\{-\frac{1}{4}, 4\}$ **27.** $\{-\frac{2}{3}, \frac{1}{2}\}$ **29.** $\{\frac{4}{3}, 3\}$ **31.** $[\frac{5}{8}, +\infty)$ **33.** $(-\infty, -2] \cup [5, +\infty)$ **35.** $(-\infty, 1] \cup [4, +\infty)$ **37.** $(-11, 3)$ **39.** $[\frac{2}{3}, 2]$ **41.** $(-\infty, 1) \cup (4, +\infty)$ **43.** $(1, +\infty)$ **45.** $[-\frac{9}{2}, \frac{3}{2}]$ **47.** $(-\infty, \frac{10}{9}) \cup (2, +\infty)$ **49.** $|x| > |a|$ **51.** $|x - 2| > 2$

EXERCISES 1.2 (Page 19)

1. (a) $(1, 2)$; (b) $(-1, -2)$; (c) $(-1, 2)$; (d) $(-2, 1)$ **3.** (a) $(2, -2)$; (b) $(-2, 2)$; (c) $(-2, -2)$; (d) does not apply **5.** (a) $(-1, 3)$; (b) $(1, -3)$; (c) $(1, 3)$; (d) $(-3, -1)$ **43.** (a) $y = 0$; (b) $x = 0$; (c) $xy = 0$ (Sketches of the graphs for Exercises 7 through 41 appear in Figs. EX 1.2-7 through EX 1.2-41.)

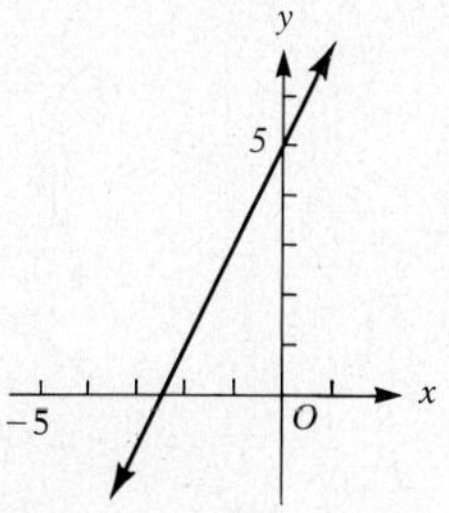

Figure EX 1.2-7

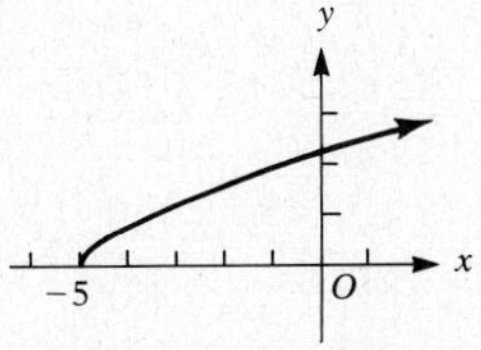

Figure EX 1.2-9

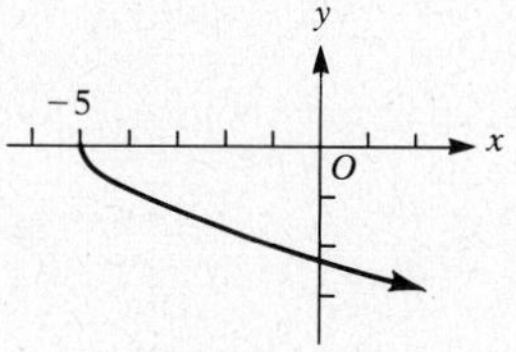

Figure EX 1.2-11

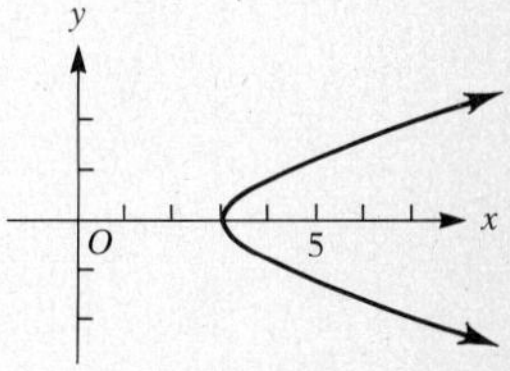

Figure EX 1.2-13

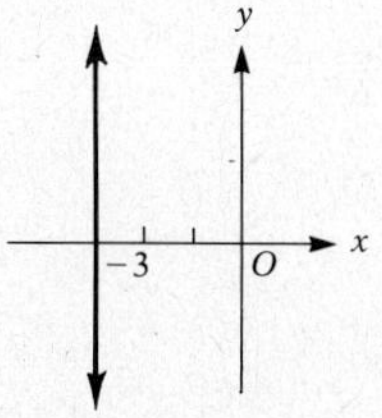

Figure EX 1.2-15

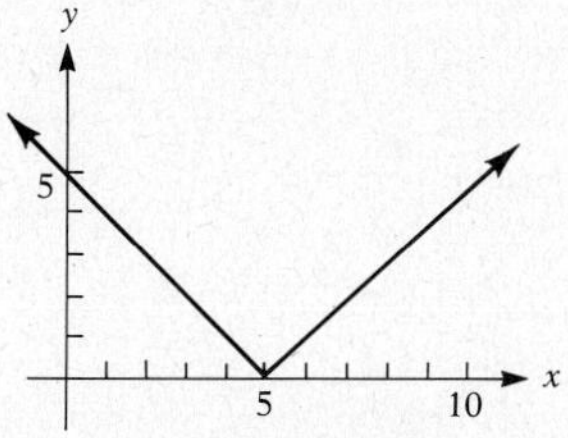

Figure EX 1.2-17

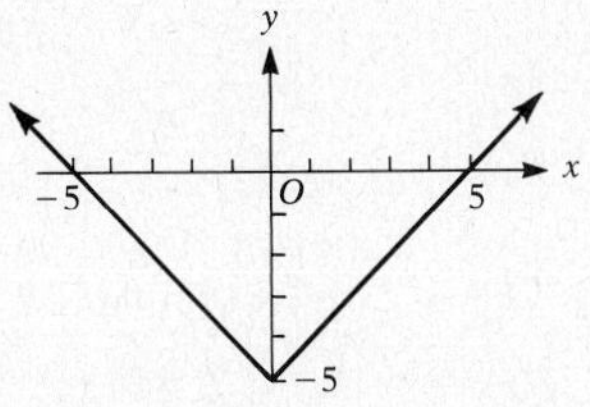

Figure EX 1.2-19

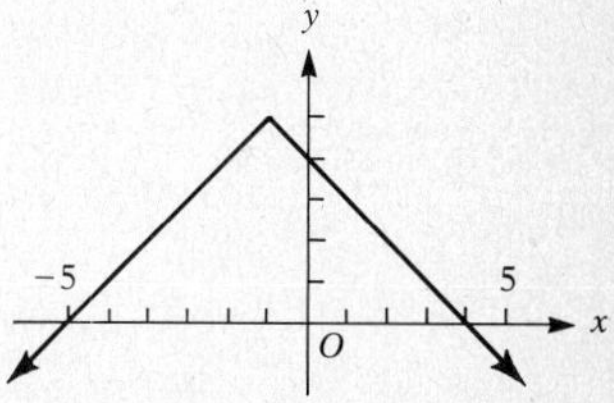

Figure EX 1.2-21

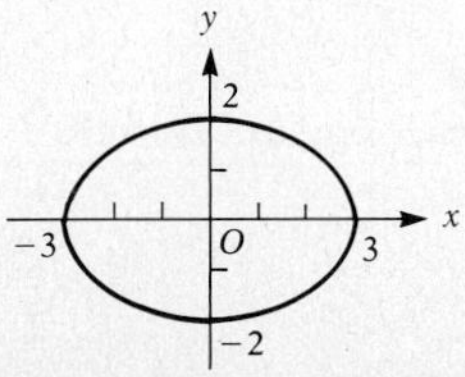

Figure EX 1.2-23

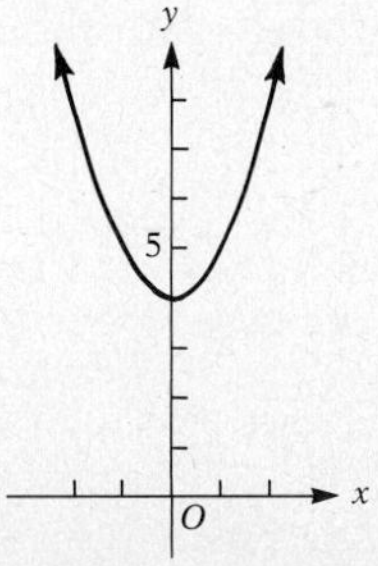

Figure EX 1.2-25

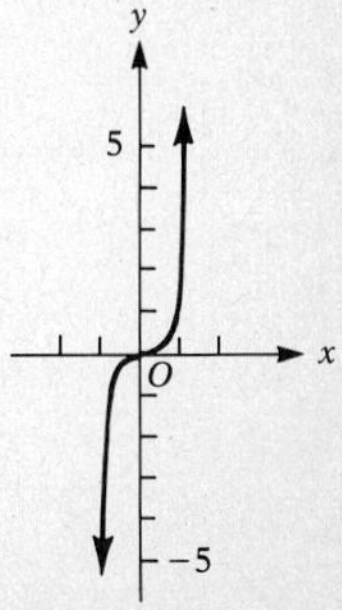

Figure EX 1.2-27

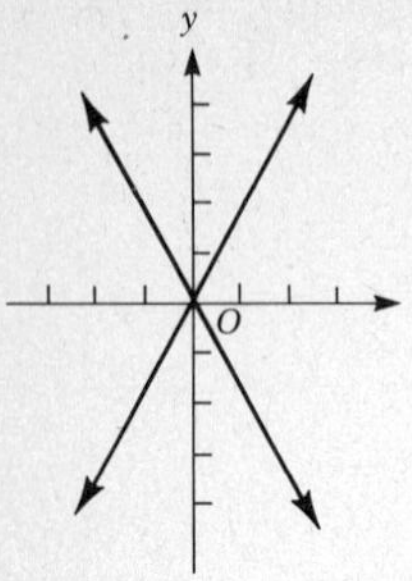

Figure EX 1.2-29

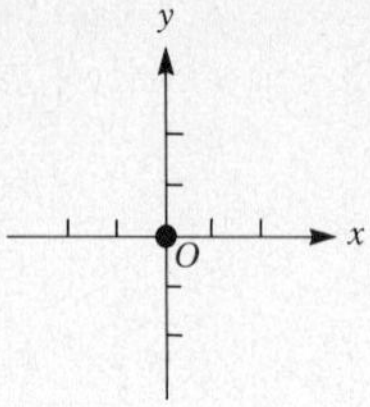

Figure EX 1.2-31

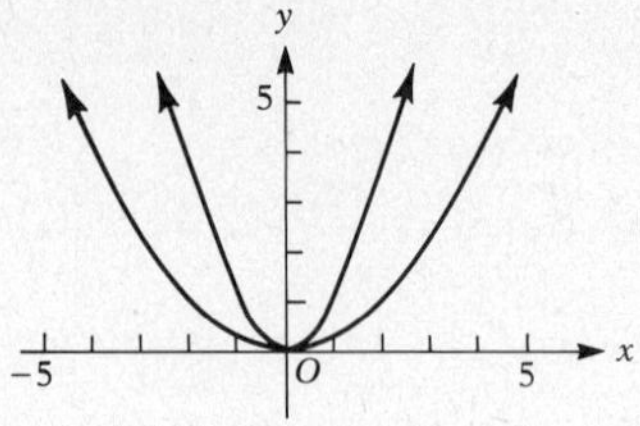

Figure EX 1.2-33

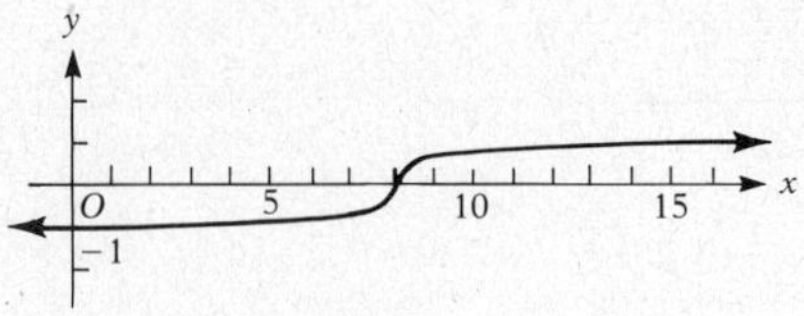

Figure EX 1.2-35

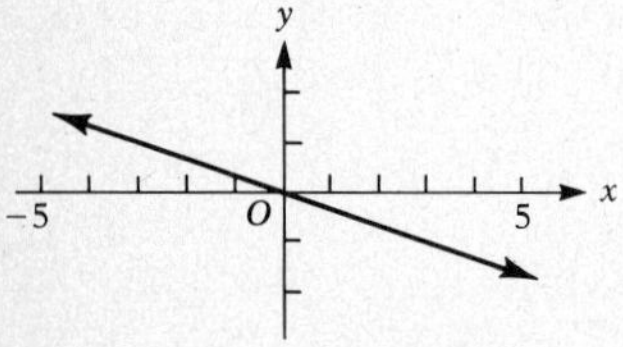

Figure EX 1.2-37(a)

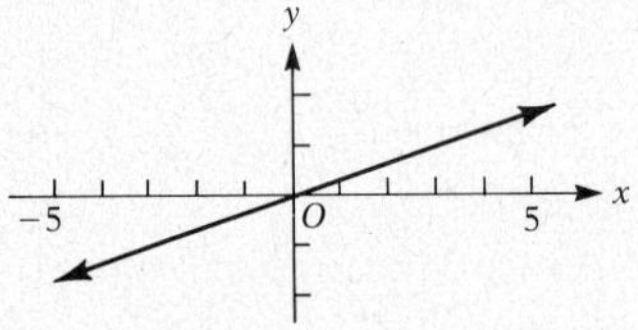

Figure EX 1.2-37(b)

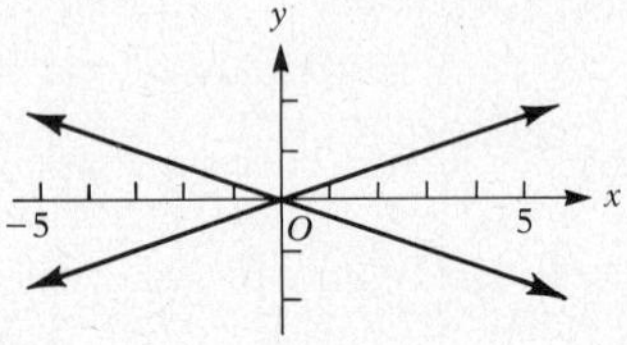

Figure EX 1.2-37(c)

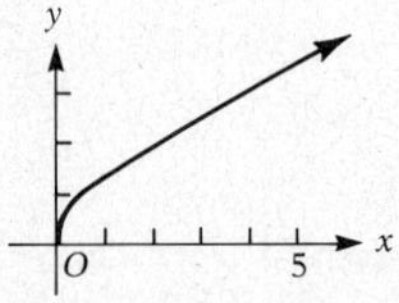

Figure EX 1.2-39(a)

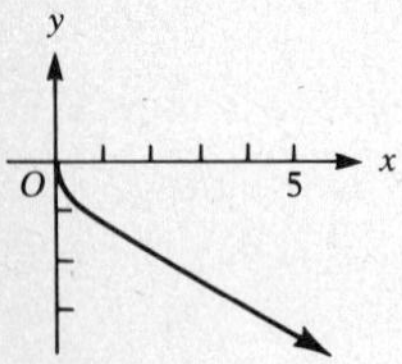

Figure EX 1.2-39(b)

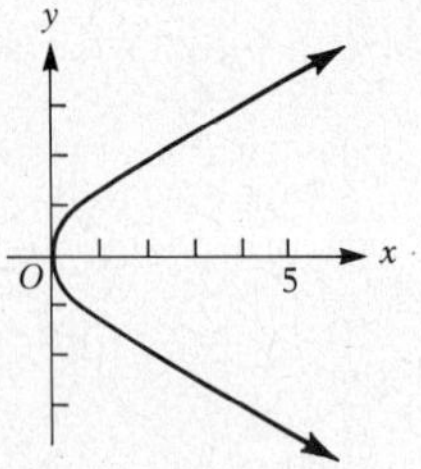

Figure EX 1.2-39(c)

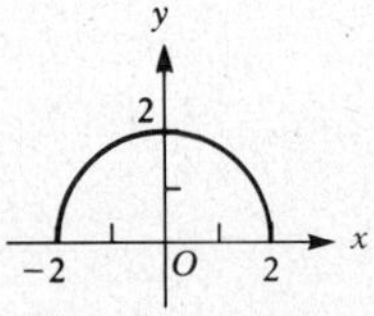

Figure EX 1.2-41(a)

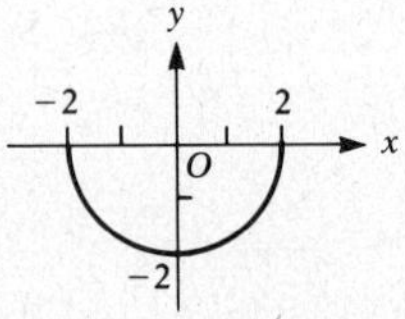

Figure EX 1.2-41(b)

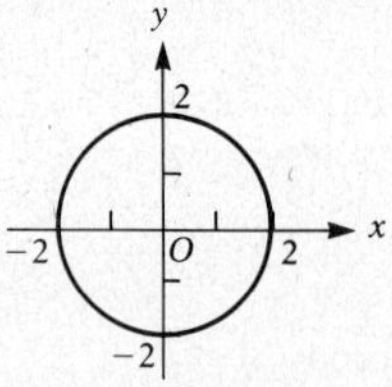

Figure EX 1.2-41(c)

EXERCISES 1.3 (Page 28)

1. $\sqrt{26}$; $\frac{1}{2}\sqrt{89}$; $\frac{1}{2}\sqrt{53}$ **3.** $|\overline{AB}| = \sqrt{41}$, $|\overline{AC}| = \sqrt{41}$, $|\overline{BC}| = \sqrt{82}$, and $|\overline{AB}|^2 + |\overline{AC}|^2 = |\overline{BC}|^2$; $\frac{41}{2}$ **9.** $\sqrt{578}$ **11.** $(-8, 12)$ **13.** $(-2 + \frac{3}{2}\sqrt{3}, \frac{3}{2} + 2\sqrt{3})$ and $(-2 - \frac{3}{2}\sqrt{3}, \frac{3}{2} - 2\sqrt{3})$ **15.** $77x^2 + 90xy + 21y^2 - 122x - 66y - 55 = 0$ **17.** $x^2 + y^2 - 8x + 6y = 0$ **19.** $x^2 + y^2 + 10x + 24y + 160 = 0$ **21.** $x^2 + y^2 - 2x - 4y - 8 = 0$ **23.** $(3, 4)$; 4 **25.** $(0, -\frac{2}{3})$; $\frac{5}{3}$ **27.** circle **29.** the empty set **31.** circle **37.** $(-\frac{9}{4}, \frac{17}{4})$, $(\frac{1}{2}, \frac{11}{2})$, $(\frac{13}{4}, \frac{27}{4})$ **39.** $(\frac{9}{4}, \frac{11}{4})$ **41.** $(-\frac{3}{2}, \frac{7}{2})$ **43.** $(5, -7)$ **45.** $2x + y = 5$

EXERCISES 1.4 (Page 38)

1. -1 **3.** $-\frac{1}{7}$ **5.** $4x - y = 11$ **7.** $4x - 3y + 12 = 0$ **9.** $x + 2y - 4 = 0$ **11.** $y = -7$ **13.** $\sqrt{3}x - y + (2\sqrt{3} - 5) = 0$ **15.** $x + y = 0$ **17.** $\frac{2}{3}$ **19.** 0 **21.** $\frac{x}{\frac{8}{5}} + \frac{y}{8} = 1$ **25.** (a) collinear; (b) not collinear **27.** (a) not collinear; (b) collinear **29.** slopes of two sides are $-\frac{4}{3}$ **31.** (a) $2x + 3y + 7 = 0$; (b) $\sqrt{13}$ **33.** (a) $-\frac{A}{B}$; (b) $-\frac{C}{B}$; (c) $-\frac{C}{A}$; (d) $Bx - Ay = 0$ **35.** $9x - 4y - 11 = 0$; $y = 1$; $9x + 4y - 19 = 0$. They intersect at $(\frac{5}{3}, 1)$. **37.** $2x + 3y - 12 = 0$; $(2 + 2\sqrt{2})x + (3 - 3\sqrt{2})y = 12$; $(2 - 2\sqrt{2})x + (3 + 3\sqrt{2}y) = 12$ **39.** (a) $x = 1$; (b) $y = 1$; (c) $x = 1$; (d) $2x + y - 3 = 0$; (e) $x - 2y + 1 = 0$; (f) $x + y - 2 = 0$ **41.** $x^2 + y^2 + 6x + 10y + 9 = 0$ **43.** $x^2 + y^2 - 4x - 4y - 2 = 0$ **45.** $3x + 4y - 19 = 0$
51. $r^2(m^2 + 1) = b^2$

EXERCISES 1.5 (Page 45)

Sketches of the graphs appear in Figs. EX 1.5-1 through EX 1.5-39. **1.** domain: $(-\infty, +\infty)$; range: $(-\infty, +\infty)$ **3.** domain: $(-\infty, +\infty)$; range: $[-6, +\infty)$ **5.** domain: $[-1, +\infty)$; range: $[0, +\infty)$ **7.** domain: $[\frac{4}{3}, +\infty)$; range: $[0, +\infty)$ **9.** domain: $(-\infty, -2] \cup [2, +\infty)$; range: $[0, +\infty)$ **11.** domain: $(-\infty, +\infty)$; range: $[0, +\infty)$ **13.** domain: $\{x \mid x \neq 1\}$; range: $\{y \mid y \neq -2\}$ **15.** domain: $(-\infty, +\infty)$; range: $\{-2, 2\}$ **17.** domain: $(-\infty, +\infty)$, $R = \{y \mid y \neq 3\}$ **19.** domain: $(-\infty, +\infty)$; range: $[-4, +\infty)$ **21.** domain: $(-\infty, +\infty)$; range: $(-\infty, -2) \cup [1, +\infty)$ **23.** domain: $\{x \mid x \neq -5 \text{ and } x \neq -1\}$; range: $\{y \mid y \neq -7 \text{ and } y \neq -3\}$

25. domain: $(-\infty, -1] \cup [4, +\infty)$; range: $[0, +\infty)$ **27.** domain: $\{x \mid x \neq 2\}$; range: $[0, +\infty)$ **29.** domain: $\{x \mid x \neq -5\}$; range: $[-6, +\infty)$ **31.** domain: $(-\infty, +\infty)$; range: $[1, +\infty)$ **33.** domain: $(-\infty, +\infty)$; range: $[-1, +\infty)$ **35.** domain: $(-\infty, +\infty)$; range: {integers} **37.** domain: $(-\infty, +\infty)$; range: $[0, 1)$ **39.** domain: $\{x \mid x \neq 0\}$; range: $(-\infty, -1] \cup \{0\} \cup (\frac{1}{2}, 1]$

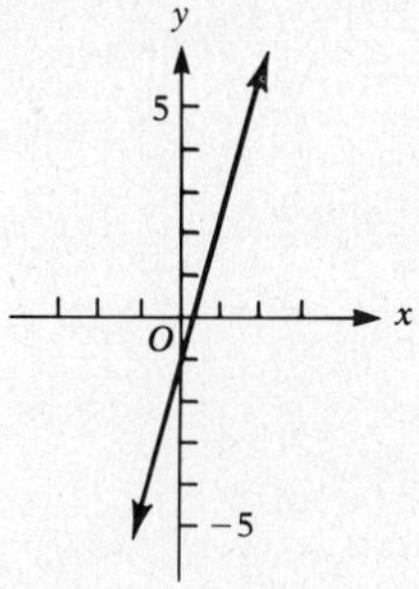

Figure EX 1.5-1

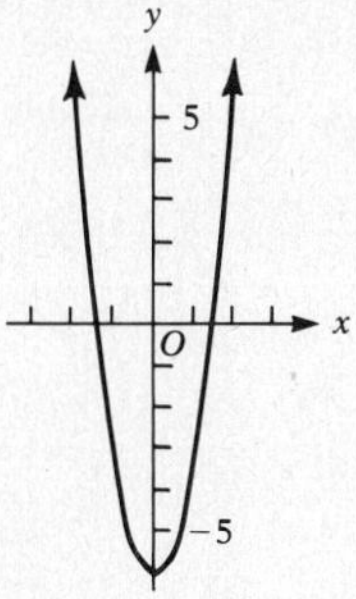

Figure EX 1.5-3

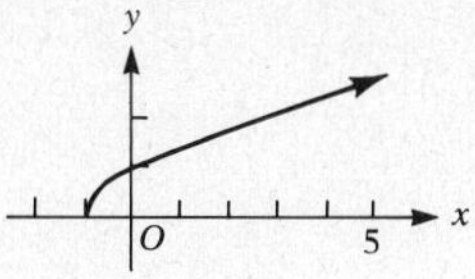

Figure EX 1.5-5

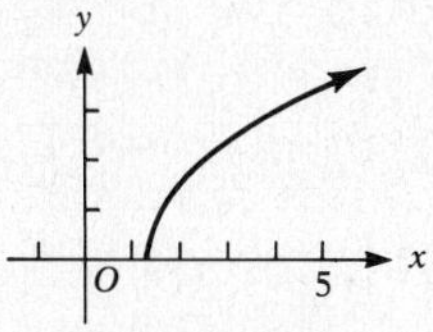

Figure EX 1.5-7

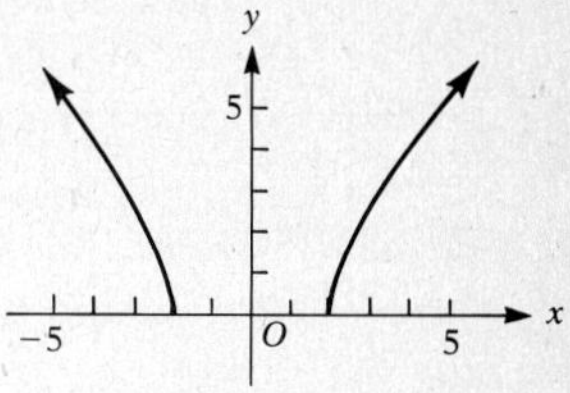

Figure EX 1.5-9

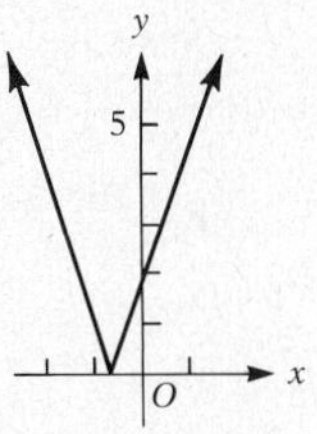

Figure EX 1.5-11

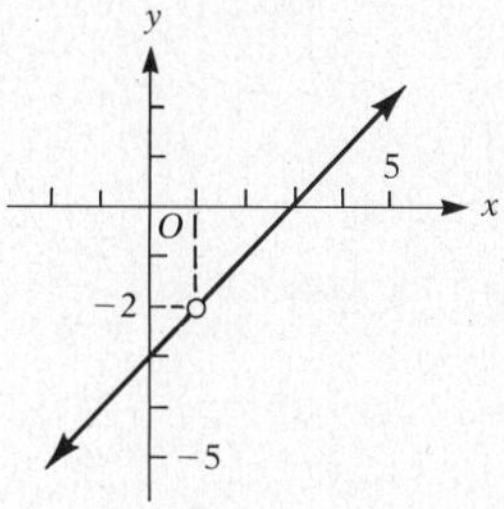

Figure EX 1.5-13

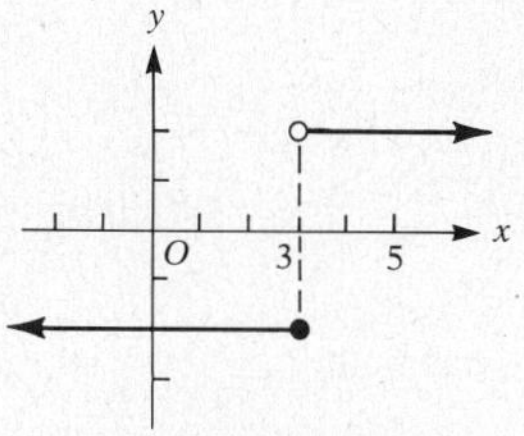

Figure EX 1.5-15

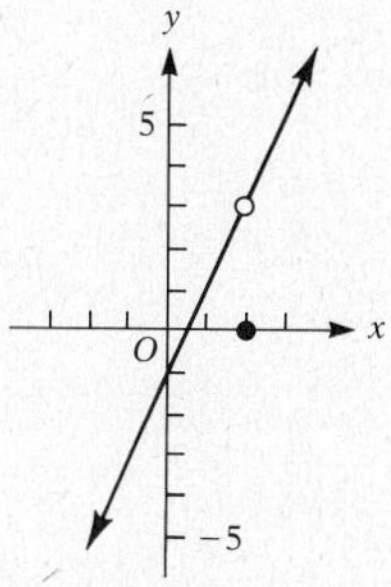

Figure EX 1.5-17

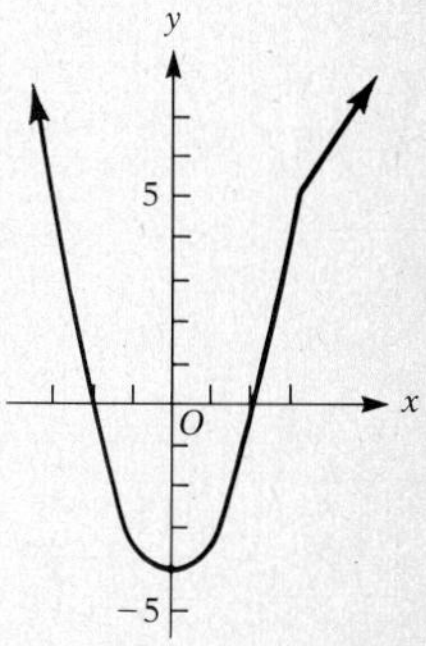

Figure EX 1.5-19

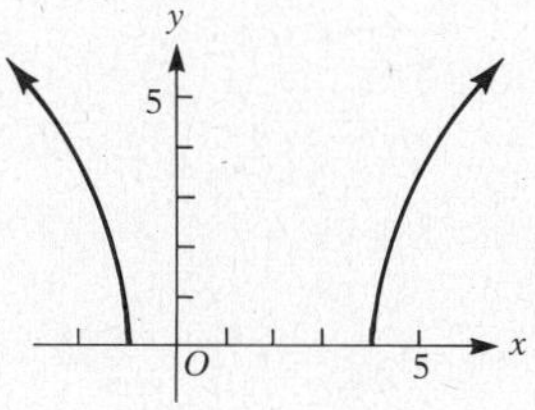

Figure EX 1.5-25

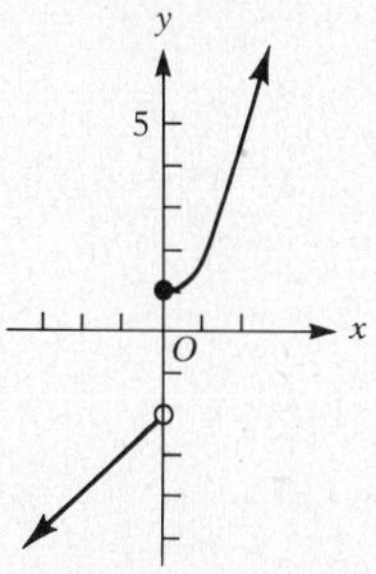

Figure EX 1.5-21

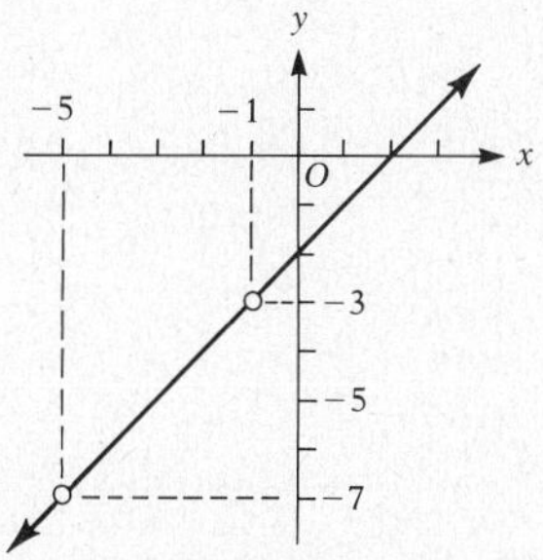

Figure EX 1.5-23

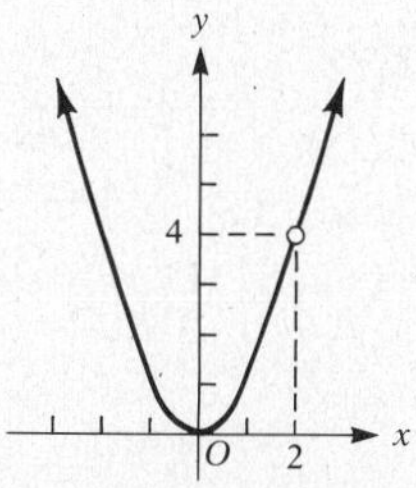

Figure EX 1.5-27

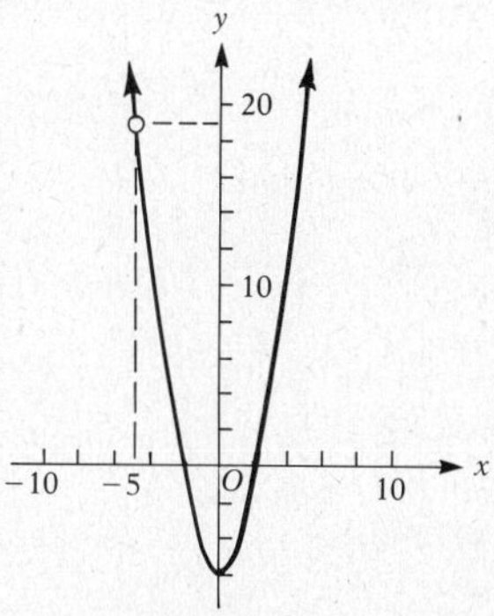

Figure EX 1.5-29

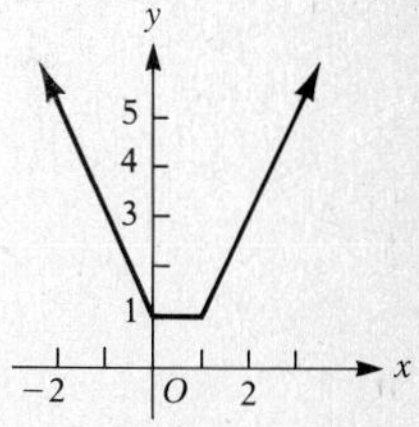

Figure EX 1.5-31

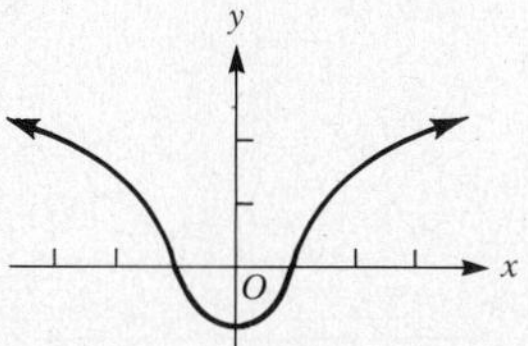

Figure EX 1.5-33

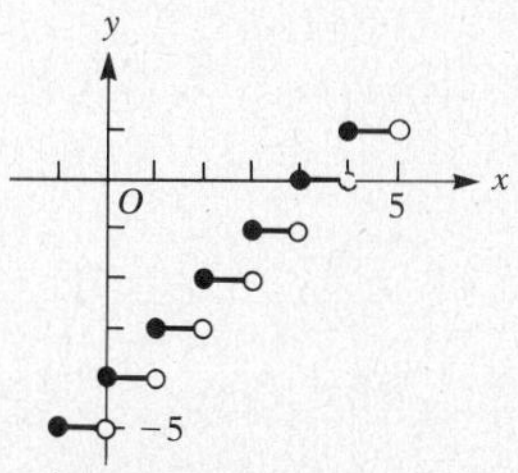

Figure EX 1.5-35

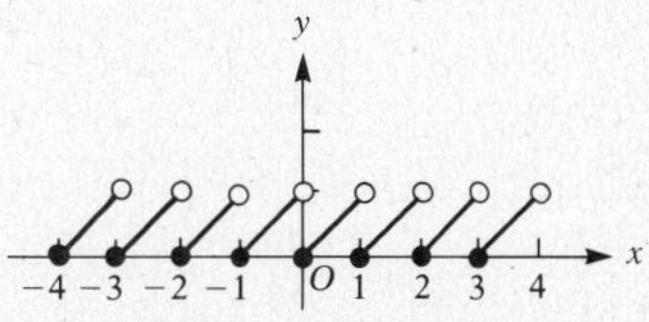

Figure EX 1.5-37

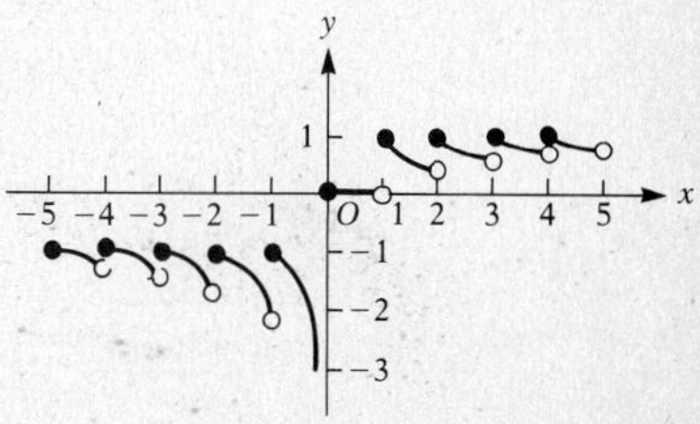

Figure EX 1.5-39

EXERCISE 1.6 (Page 51)

1. (a) 5; (b) -5; (c) -1; (d) $2a + 1$; (e) $2x + 1$; (f) $4x - 1$; (g) $4x - 2$; (h) $2x + 2h - 1$; (i) $2x + 2h - 2$; (j) 2 **3.** (a) -5; (b) -6; (c) -3; (d) 30; (e) $2h^2 + 9h + 4$; (f) $8x^4 + 10x^2 - 3$; (g) $2x^4 - 7x^2$; (h) $2x^2 + (4h + 5)x + (2h^2 + 5h - 3)$; (i) $2x^2 + 5x + (2h^2 + 5h - 6)$; (j) $4x + 2h + 5$ **5.** (a) 1; (b) $\sqrt{11}$; (c) 2; (d) $3\sqrt{7}$; (e) $\sqrt{4x + 9}$; (f) $2/(\sqrt{2x + 2h + 3} + \sqrt{2x + 3})$ **7.** (a) 1; (b) -1; (c) 1; (d) -1; (e) 1 if $x \leq 0$ and -1 if $x > 0$; (f) 1 if $x \geq -1$ and -1 if $x < -1$; (g) 1; (h) -1 if $x \neq 0$ and 1 if $x = 0$ **9.** (a) $x^2 + x - 6$; domain: $(-\infty, +\infty)$; (b) $-x^2 + x - 4$, domain: $(-\infty, +\infty)$; (c) $x^3 - 5x^2 - x + 5$, domain: $(-\infty, +\infty)$; (d) $\dfrac{x - 5}{x^2 - 1}$, domain: $\{x \mid x \neq -1, x \neq 1\}$; (e) $\dfrac{x^2 - 1}{x - 5}$, domain: $\{x \mid x \neq 5\}$; (f) $x^2 - 6$, domain: $(-\infty, +\infty)$; (g) $x^2 - 10x + 24$, domain: $(-\infty, +\infty)$ **11.** (a) $\dfrac{x^2 + 2x - 1}{x^2 - x}$, domain: $\{x \mid x \neq 0, x \neq 1\}$; (b) $\dfrac{x^2 + 1}{x^2 - x}$, domain: $\{x \mid x \neq 0, x \neq 1\}$; (c) $\dfrac{x + 1}{x^2 - x}$, domain: $\{x \mid x \neq 0, x \neq 1\}$; (d) $\dfrac{x^2 + x}{x - 1}$, domain: $\{x \mid x \neq 0, x \neq 1\}$; (e) $\dfrac{x - 1}{x^2 + x}$, domain: $\{x \mid x \neq -1, x \neq 0, x \neq 1\}$; (f) $\dfrac{1 + x}{1 - x}$, domain: $\{x \mid x \neq 0, x \neq 1\}$; (g) $\dfrac{x - 1}{x + 1}$, domain: $\{x \mid x \neq -1, x \neq 1\}$ **13.** (a) $\sqrt{x} + x^2 - 1$, domain: $[0, +\infty)$; (b) $\sqrt{x} - x^2 + 1$, domain: $[0, +\infty)$; (c) $\sqrt{x}(x^2 - 1)$, domain: $[0, +\infty)$; (d) $\dfrac{\sqrt{x}}{x^2 - 1}$, domain: $[0, 1) \cup (1, +\infty)$; (b) $\dfrac{x^2 - 1}{\sqrt{x}}$, domain: $(0, +\infty)$; (f) $\sqrt{x^2 - 1}$, domain: $(-\infty, -1] \cup [1, +\infty)$; (g) $x - 1$, domain: $[0, +\infty)$ **15.** (a) $x^2 + 3x - 1$, domain: $(-\infty, +\infty)$; (b) $x^2 - 3x + 3$, domain: $(-\infty, +\infty)$; (c) $3x^3 - 2x^2 + 3x - 2$, domain: $[-\infty, +\infty)$; (d) $\dfrac{x^2 + 1}{3x - 2}$, domain: $\{x \mid x \neq \frac{2}{3}\}$; (e) $\dfrac{3x - 2}{x^2 + 1}$, domain: $(-\infty, +\infty)$; (f) $9x^2 - 12x + 5$, domain: $(-\infty, +\infty)$; (g) $3x^2 + 1$, domain: $(-\infty, +\infty)$ **17.** (a) $\dfrac{x^2 + 2x - 2}{x^2 - x - 2}$, domain: $\{x \mid x \neq -1, x \neq 2\}$; (b) $\dfrac{-x^2 - 2}{x^2 - x - 2}$, domain: $\{x \mid x \neq -1, x \neq 2\}$; (c) $\dfrac{x}{x^2 - x - 2}$, domain: $\{x \mid x \neq -1, x \neq 2\}$; (d) $\dfrac{x - 2}{x^2 + x}$, domain: $\{x \mid x \neq -1, x \neq 0, x \neq 2\}$; (e) $\dfrac{x^2 + x}{x - 2}$, domain: $\{x \mid x \neq -1, x \neq 2\}$; (f) $\dfrac{x - 2}{2x - 2}$, domain: $\{x \mid x \neq 1, x \neq 2\}$; (g) $-\dfrac{1}{2x + 1}$, domain: $\{x \mid x \neq -1, x \neq -\frac{1}{2}\}$ **19.** (a) $2x^2 - 3$, domain: $(-\infty, +\infty)$; (b) $4x^2 - 12x + 9$, domain: $(-\infty, +\infty)$; (c) $4x - 9$, domain: $(-\infty, +\infty)$ **21.** (a) 0; (b) 4; (c) $4 - 2x$ **23.** (a) even; (b) neither; (c) odd; (d) even; (e) odd; (f) odd; (g) neither; (h) even; (i) even; (j) odd **27.** (a) $-7 + 2x$; (b) $(x^2 + 2) + 0$; (c) $-1 + x^3$; (d) $(x^4 + 3) + (x^3 - x)$; (e) $0 + \dfrac{1}{x}$; (f) $\dfrac{x^2 + 1}{x^2 - 1} + \dfrac{-2x}{x^2 - 1}$; (g) $\frac{1}{2}[2|x| + |x - 1| + |x + 1|] + \frac{1}{2}[|x - 1| - |x + 1|]$ **29.** (a) even; (b) odd; (c) even; (d) even Sketches of the graphs for Exercises 31–45 appear in Figs. EX 1.6-31 through EX 1.6-45. **31.** $u(x) = \begin{cases} 0 & \text{if } x < 0 \\ 1 & \text{if } x \geq 0 \end{cases}$ **33.** $f(x) = \begin{cases} 0 & \text{if } x < 1 \\ 1 & \text{if } x \geq 1 \end{cases}$

35. $g(x) = \begin{cases} 0 & \text{if } x < -1 \\ 1 & \text{if } x = -1 \\ 2 & \text{if } -1 < x < 1 \\ 1 & \text{if } x = 1 \\ 0 & \text{if } x > 1 \end{cases}$ **37.** $h(x) = \begin{cases} 0 & \text{if } x < 0 \\ x & \text{if } x \geq 0 \end{cases}$ **39.** $f(x) = \begin{cases} 0 & \text{if } x < 0 \\ x - 1 & \text{if } x \geq 0 \end{cases}$ **41.** $G(x) = \begin{cases} 0 & \text{if } x < -1 \\ x + 1 & \text{if } -1 \leq x \leq 0 \\ 1 & \text{if } x > 0 \end{cases}$

43. $h(x) = \begin{cases} -1 & \text{if } x \leq 0 \\ 0 & \text{if } x > 0 \end{cases}$ **45.** $g(x) = \begin{cases} -1 & \text{if } x < 0 \\ 0 & \text{if } x = 0 \\ x + 1 & \text{if } x > 0 \end{cases}$ **47.** $(f \circ g)(x) = \begin{cases} 2 & \text{if } x < 0 \\ x & \text{if } 0 \leq x \leq 1 \\ 2 & \text{if } x > 1 \end{cases}$

49. $(f \circ g)(x) = \begin{cases} 1 & \text{if } x < 0 \\ 4x^2 & \text{if } 0 \leq x \leq \frac{1}{2} \\ 0 & \text{if } \frac{1}{2} < x \leq 1 \\ 1 & \text{if } x > 1 \end{cases}$ **51.** $g(x) = x - 3$; $g(x) = 1 - x$

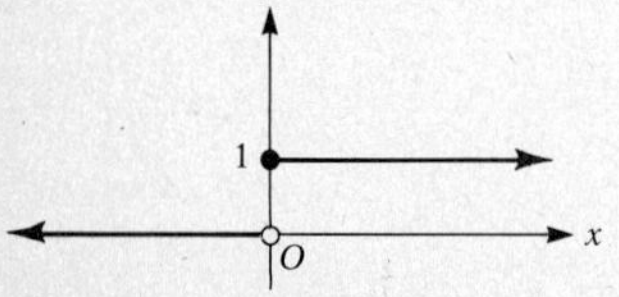

Figure EX 1.6-31

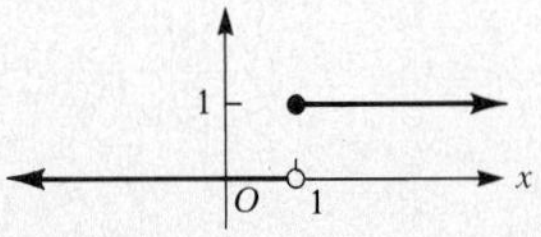

Figure EX 1.6-33

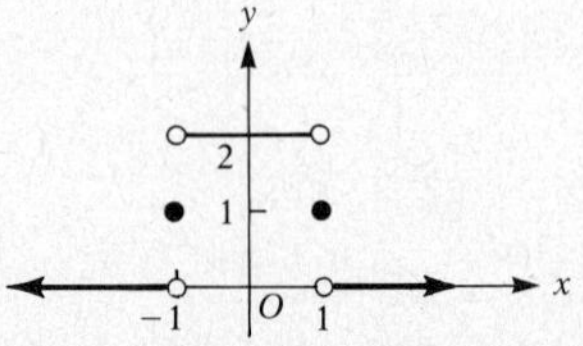

Figure EX 1.6-35

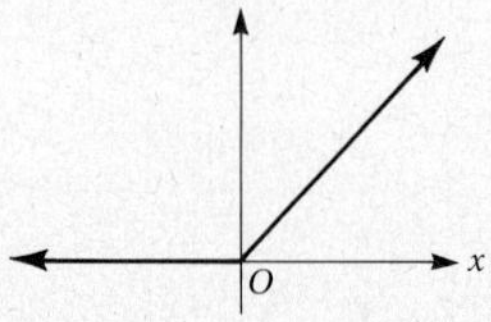

Figure EX 1.6-37

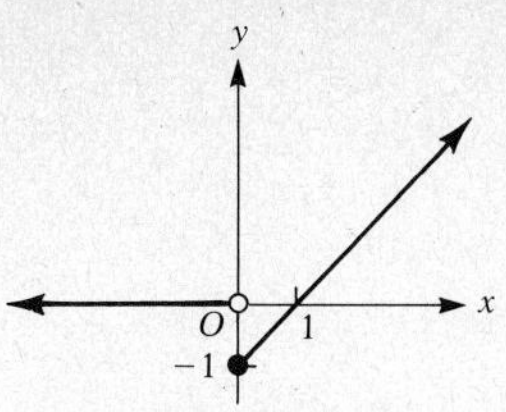

Figure EX 1.6-39

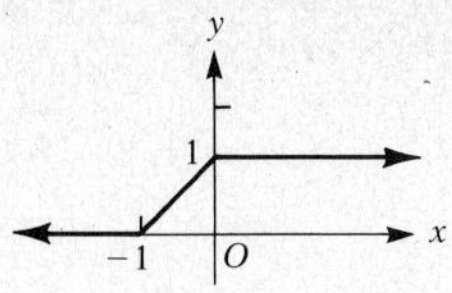

Figure EX 1.6-41

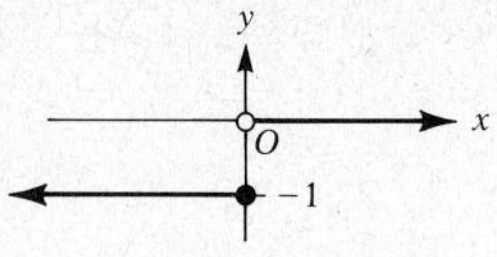

Figure EX 1.6-43

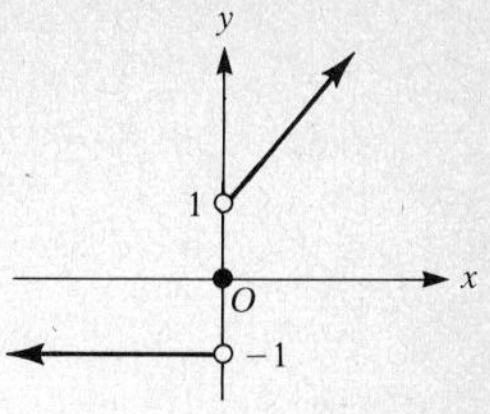

Figure EX 1.6-45

EXERCISES 1.7 (Page 60)

1. (a) $\frac{1}{3}\pi$; (b) $\frac{3}{4}\pi$; (c) $\frac{7}{6}\pi$; (d) $-\frac{5}{6}\pi$; (e) $\frac{1}{9}\pi$; (f) $\frac{5}{2}\pi$; (g) $-\frac{5}{12}\pi$; (h) $\frac{5}{9}\pi$ **3.** (a) 45°; (b) 120°; (c) 330°; (d) −90°; (e) 28° 39′; (f) 540°; (g) −114° 36′; (h) 15° **5.** (a) $\frac{1}{2}$; (b) $\frac{1}{2}\sqrt{2}$; (c) −1; (d) $-\frac{1}{2}$ **7.** (a) $-\frac{1}{2}\sqrt{3}$; (b) $\frac{1}{2}\sqrt{2}$; (c) −1; (d) 0 **9.** (a) $\frac{1}{3}\pi, \frac{2}{3}\pi$; (b) $\frac{3}{4}\pi, \frac{5}{4}\pi$ **11.** (a) $\frac{1}{2}\pi, \frac{3}{2}\pi$; (b) $\frac{7}{6}\pi, \frac{11}{6}\pi$ **13.** (a) $\frac{1}{4}(\sqrt{6} - \sqrt{2})$; (b) $\frac{1}{4}(\sqrt{2} - \sqrt{6})$ **15.** (a) $\frac{1}{2}\sqrt{2 + \sqrt{3}}$; (b) $\frac{1}{2}\sqrt{2 - \sqrt{2}}$ **23.** $3 \sin t - 4 \sin^3 t$ **25.** (a) $\frac{1}{4}(\sqrt{2} + \sqrt{6})$; (b) $-\frac{1}{4}(\sqrt{2} + \sqrt{6})$ **27.** (a) $\frac{1}{4}(\sqrt{6} - \sqrt{2})$; (b) $-\frac{1}{2}\sqrt{2 + \sqrt{2}}$ **29.** (a) $-\cos t$; (b) $-\cos(\frac{1}{2}\pi - t)$; (c) $-\cos t$; (d) $\cos(\frac{1}{2}\pi - t)$ **31.** (a) $\cos(\frac{1}{2}\pi - t)$; (b) $-\cos t$; (c) $-\cos(\frac{1}{2}\pi - t)$; (d) $-\cos t$ **33.** (a) $\frac{1}{2}\pi + 2k\pi$, where k is any integer; (b) $2k\pi$, where k is any integer **35.** (a) $t = \frac{1}{6}\pi + 2k\pi$ and $t = \frac{5}{6}\pi + 2k\pi$, where k is any integer; (b) $\pm\frac{1}{3}\pi + 2k\pi$, where k is any integer

REVIEW EXERCISES FOR CHAPTER 1 (Page 61)

1. $[5, +\infty)$ **3.** $(-\infty, -\frac{1}{3}) \cup (1, +\infty)$ **5.** $(-\frac{3}{2}, 1)$ **7.** $[-\frac{12}{5}, \frac{6}{5}]$ **9.** $\{-3, \frac{9}{2}\}$ **11.** $\{-2, 2\}$ **13.** See Fig. REV 1-13. **15.** See Fig. REV 1-15. **17.** See Fig. REV 1-17. **19.** −4, 14 **21.** $x^2 + y^2 + 6x - 8y + 15 = 0$ **25.** 15 **27.** $(-2, -5)$; $(3, -6)$ **31.** $x^2 + y^2 - 2x - 8y - 3 = 0$ **33.** $(\frac{3}{2}, -1)$; 1 **35.** $5x + 2y - 19 = 0$ **37.** $0 \le k \le 2$ **39.** $7x^2 + 7y^2 + 11x - 19y - 6 = 0$ **41.** $12x + 3y - 2 = 0$ **43.** $x = 0$; $3x + 4y = 0$ **45.** domain: $[-\frac{5}{2}, 0)$, range: $[0, +\infty)$; see Fig. REV 1-45 **47.** domain: $\{x \mid x \ne -\frac{5}{2}, x \ne 2\}$, range: $\{y \mid y \ne \frac{1}{2}, y \ne 5\}$; see Fig. REV 1-47 **49.** $G(x) = \begin{cases} 0 & \text{if } x < -3 \\ x + 3 & \text{if } x \ge -3 \end{cases}$, domain: $(-\infty, +\infty)$, range: $[0, +\infty)$; see Fig. REV 1-49 **51.** domain: $(-\infty, +\infty)$, range: $[-4, +\infty)$; see Fig. REV 1-51 **53.** (a) 35; (b) $3x^4 + x^2 + 5$; (c) $6x + 3h - 1$ **55.** (a) odd; (b) even; (c) neither; (d) odd **57.** (a) $x^2 + 4x - 7$, domain: $(-\infty, +\infty)$; (b) $x^2 - 4x - 1$, domain: $(-\infty, +\infty)$; (c) $4x^3 - 3x^2 - 16x + 12$, domain: $(-\infty, +\infty)$; (d) $\dfrac{x^2 - 4}{4x - 3}$, domain: $\{x \mid x \ne \frac{3}{4}\}$; (e) $\dfrac{4x - 3}{x^2 - 4}$, domain: $\{x \mid x \ne \pm 2\}$; (f) $16x^2 - 24x + 5$, domain: $(-\infty, +\infty)$; (g) $4x^2 - 19$, domain: $(-\infty, +\infty)$ **59.** (a) $\dfrac{x^2 - 2x + 1}{x^2 - 2x - 3}$, domain: $\{x \mid x \ne -1, x \ne 3\}$; (b) $\dfrac{-x^2 + 4x + 1}{x^2 - 2x - 3}$, domain: $\{x \mid x \ne -1, x \ne 3\}$; (c) $\dfrac{x}{x^2 - 2x - 3}$, domain: $\{x \mid x \ne -1, x \ne 3\}$; (d) $\dfrac{x + 1}{x^2 - 3x}$, domain: $\{x \mid x \ne -1, x \ne 0, x \ne 3\}$; (e) $\dfrac{x^2 - 3x}{x + 1}$, domain: $\{x \mid x \ne -1, x \ne 3\}$; (f) $-\dfrac{x + 1}{2x + 3}$, domain: $\{x \mid x \ne -\frac{3}{2}, x \ne -1\}$; (g) $\dfrac{1}{x - 2}$, domain: $\{x \mid x \ne 2, x \ne 3\}$ **61.** (a) $\frac{1}{6}\pi, \frac{5}{6}\pi$; (b) π; (c) $0, \pi$ **63.** (a) $\frac{1}{2}\sqrt{3}$; (b) $-\frac{1}{2}\sqrt{2}$; (c) $-\frac{1}{2}$; (d) $-\frac{1}{2}\sqrt{2 + \sqrt{3}}$

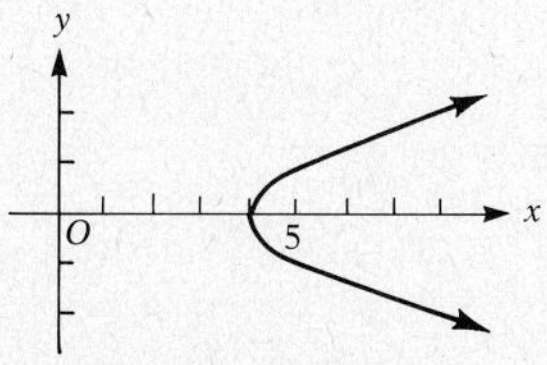

Figure REV1-13

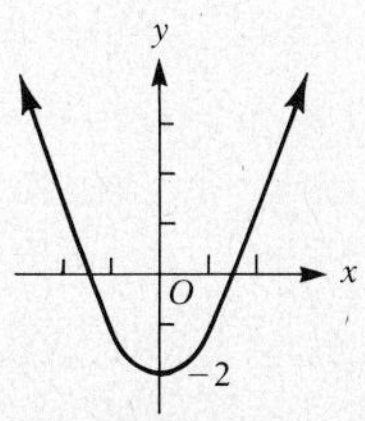

Figure REV1-15

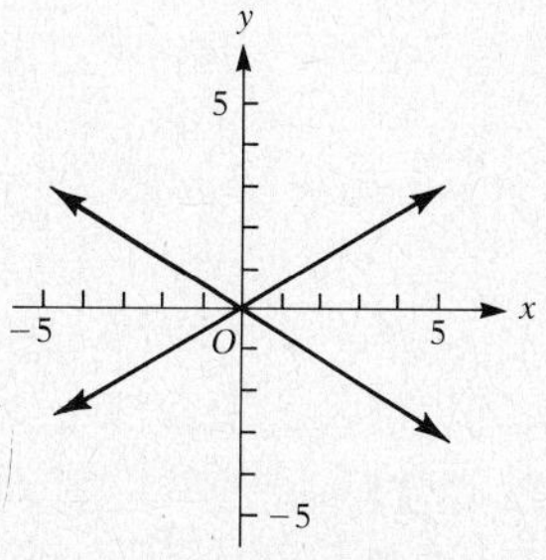

Figure REV1-17

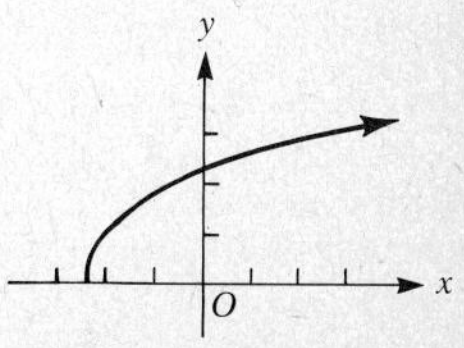

Figure REV1-45

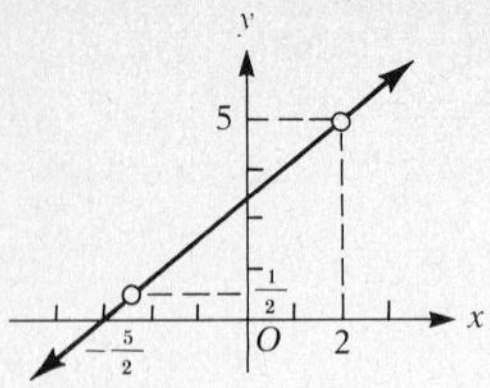

Figure REV1-47

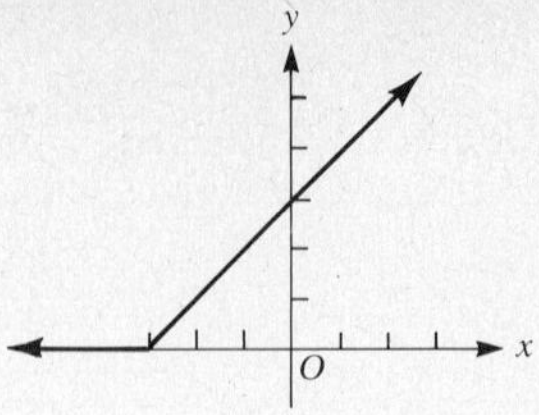

Figure REV1-49

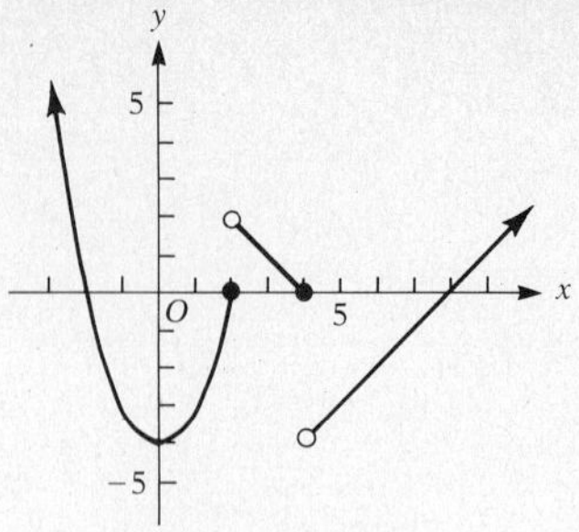

Figure REV1-51

EXERCISES 2.1 (Page 73)

1. 0.005 **3.** 0.005 **5.** $\frac{1}{1400}$ **7.** $\frac{1}{3000}$ **9.** 0.002 **11.** 0.01 **13.** 0.006 **15.** $\frac{1}{2}\epsilon$ **17.** $\frac{1}{5}\epsilon$ **19.** $\frac{1}{3}\epsilon$ **21.** ϵ **23.** $\min(1, \frac{1}{3}\epsilon)$ **25.** $\min(1, \frac{1}{8}\epsilon)$ **27.** $\min(1, \frac{1}{6}\epsilon)$ **29.** $\min(1, 6\epsilon)$ **31.** $\min(1, \frac{1}{2}\epsilon)$ **33.** $\min(\frac{1}{2}, \frac{1}{4}\epsilon)$ **35.** $\min(1, \frac{2}{3}\epsilon)$ **37.** $\min(1, (2\sqrt{2} + 3)\epsilon)$ **39.** $\min(1, 2\epsilon)$ **41.** $\min(1, \frac{1}{2}(9\sqrt{7} + 21)\epsilon)$ **43.** $\delta = \min(a, \epsilon/3a)$ **45.** $\delta = \min(a, \sqrt{a}\epsilon)$ **47.** $\delta = \min(-a, \epsilon/7a^2)$

EXERCISES 2.2 (Page 84)

1. 8 **3.** 7 **5.** 0 **7.** $\frac{1}{2}$ **9.** $-\frac{1}{22}$ **11.** $\frac{1}{4}$ **13.** -6 **15.** $\frac{1}{7}$ **17.** $\frac{16}{7}$ **19.** 12 **21.** $\frac{3}{2}$ **23.** $\frac{2}{3}$ **25.** $\sqrt{\frac{6}{5}}$ **27.** $\frac{1}{6}$ **29.** $\frac{1}{4}\sqrt{2}$ **31.** $\frac{1}{3}$ **33.** -1 **35.** $\frac{11}{17}$ **41.** (a) 0

EXERCISES 2.3 (Page 88)

1. (a) -3; (b) 2; (c) does not exist because $\lim_{x \to 1^+} f(x) \neq \lim_{x \to 1^-} f(x)$ **3.** (a) 8; (b) 0; (c) does not exist because $\lim_{t \to -4^+} f(t) \neq \lim_{t \to -4^-} f(t)$ **5.** (a) 4; (b) 4; (c) 4 **7.** (a) 5; (b) 5; (c) 5 **9.** (a) 0; (b) 0; (c) 0 **11.** (a) 0; (b) 0; (c) 0 **13.** (a) -4; (b) -4; (c) -4 **15.** (a) 1; (b) -1; (c) does not exist because $\lim_{x \to 0^+} f(x) \neq \lim_{x \to 0^-} f(x)$ **17.** (a) 2; (b) 0; (c) 0; (d) -2 **19.** (a) 0; (b) 0; (c) 0 **21.** (a) 0; (b) 0; (c) 0; (d) 0; (e) 0; (f) 0 **23.** (a) -2; (b) 2; (c) does not exist **25.** (a) 2; (b) 1; (c) does not exist **27.** (a) -1; (b) 1; (c) does not exist **29.** -6 **31.** $a = -\frac{3}{2}, b = 1$ **35.** $f(x) \cdot g(x) = \begin{cases} x^4 + 3x^2 & \text{if } x \leq 1 \\ 2x + 2 & \text{if } x > 1 \end{cases}$

EXERCISES 2.4 (Page 99)

1. $+\infty$ **3.** $+\infty$ 5. $-\infty$ **7.** $-\infty$ **9.** $+\infty$ **11.** $+\infty$ **13.** $+\infty$ **15.** $-\infty$ **17.** $-\infty$ **19.** $+\infty$ **21.** $-\infty$ **23.** $+\infty$ **25.** $+\infty$ **27.** $-\infty$ **29.** $-\infty$ **31.** $-\infty$ **33.** (a) $x = 0$; (b) $x = 0$; (c) $x = 0$; (d) $x = 0$ **35.** $x = 4$ **37.** $x = -3$ **39.** $x = -3$ **41.** $x = -5, x = -3$ **43.** $x = \frac{7}{2}, x = -\frac{5}{2}$ **45.** take $\delta = \sqrt{3/N}$

EXERCISES 2.5 (Page 110)

1. 0 **3.** $\frac{2}{5}$ **5.** $-\frac{2}{5}$ **7.** $\frac{7}{3}$ **9.** 0 **11.** $+\infty$ **13.** $\frac{1}{2}$ **15.** $+\infty$ **17.** $-\infty$ **19.** 1 **21.** -1 **23.** 0 **25.** $-\infty$ **27.** 0 **29.** $y = 2; x = 3$ **31.** $y = 1; x = 0$ **33.** $y = 0; x = -2; x = 2$ **35.** $y = 4; x = -3; x = 3$ **37.** $y = 0; x = \frac{2}{3}; x = -\frac{5}{2}$ **39.** $x = -\sqrt{2}; x = \sqrt{2}$ **41.** $y = \frac{2}{3}; x = \frac{4}{3}$ **43.** $y = 1; y = -1$ **45.** $y = -1; y = 1; x = 3$ **47.** take $N = 1 + \dfrac{1}{\epsilon}$ **49.** take $N = (\epsilon - 7)/2$ **51.** f is defined on some open interval $(a, +\infty)$ and for any $M < 0$ there exists an $N > 0$ such that $f(x) < M$ whenever $x > N$. (b) f is defined on some open interval $(-\infty, b)$ and for any $M > 0$ there exists an $N < 0$ such that $f(x) > M$ whenever $x < N$. (c) f is defined on some open interval $(-\infty, b)$ and for any $M < 0$ there exists an $N < 0$ such that $f(x) < M$ whenever $x < N$. **55.** (a) $C(x) = 16x + 32{,}400x^{-1}$; $(0, +\infty)$; (b) $+\infty$; $+\infty$ **57.** (a) $A(x) = 3x + 30 + 48x^{-1}$; $(0, +\infty)$; (b) $+\infty$, $+\infty$

EXERCISES 2.6 (Page 114)

1. -3; $f(-3)$ does not exist **3.** -3; $\lim_{x \to -3} g(x) \neq g(-3)$ **5.** $-3, 2$; $h(-3)$ and $h(2)$ do not exist **7.** 4; $\lim_{x \to 4} f(x)$ does not exist **9.** $-2, 2$; $F(-2)$ and $F(2)$ do not exist **11.** $-2, 2$; $G(-2)$ and $G(2)$ do not exist **13.** 0; $\lim_{x \to 0} f(x)$ does not exist **15.** 2, $\lim_{t \to 2} g(t) \neq g(2)$ **17.** 0; $\lim_{x \to 0} g(x)$ does not exist **19.** 3; $f(3)$ does not exist **21.** all integers; $\lim_{x \to k} [\![x]\!]$ does not exist if k is any integer **23.** 0; $\lim_{x \to 0} \operatorname{sgn} x$ does not exist **25.** 0; $\lim_{x \to 0} g(x)$ does not exist **27.** removable; 4 **29.** essential **31.** removable; 0 **33.** essential **35.** removable; $-\frac{1}{6}$ **37.** removable; $\frac{1}{48}$ **39.** f is continuous everywhere **41.** $\dfrac{1}{3a^2}$ **43.** 50 and 200; $\lim_{x \to 50} C(x)$ and $\lim_{x \to 200} C(x)$ do not exist

EXERCISES 2.7 (Page 124)

1. all real numbers **3.** all real numbers **5.** all real numbers except 3 **7.** all real numbers except -1 and 1 **9.** all real numbers except -2 and 2 **11.** all real numbers except -4, -3, and 1 **13.** $(-\infty, -4) \cup (4, +\infty)$ **15.** all real numbers **21.** $\frac{1}{5}$ **23.** -1 **25.** -2 **27.** 15 **29.** $(f \circ g)(x) = \sqrt{x^3}$; $(0, +\infty)$ **31.** $(f \circ g)(x) = \dfrac{1}{\sqrt{x-2}}$; $(2, +\infty)$ **33.** $(f \circ g)(x) = \dfrac{1}{\sqrt{x}-2}$; $(0, 4) \cup (4, +\infty)$ **35.** $(f \circ g)(x) = \dfrac{\sqrt{x}+1}{\sqrt{x}-1}$; $(0, 1) \cup (1, +\infty)$ **37.** $(-3, 3)$ **39.** $(-\infty, -4) \cup (4, +\infty)$ **41.** $(-5, 2)$ **43.** all real numbers **45.** all real numbers except 0 **47.** $(0, +\infty)$ **49.** all real numbers except -1, 1, and 3 **51.** all real numbers except $\frac{1}{2}n$ where n is any integer **53.** $(-1, 0) \cup (0, 1)$

EXERCISES 2.8 (Page 129)

1. continuous; discontinuous; discontinuous; continuous; discontinuous; continuous **3.** continuous; continuous; discontinuous; continuous; discontinuous; continuous **5.** continuous; continuous; continuous; continuous; discontinuous **7.** continuous; discontinuous; discontinuous; discontinuous; continuous **9.** continuous; continuous; continuous; continuous; discontinuous; discontinuous **11.** continuous; discontinuous; discontinuous; discontinuous **13.** discontinuous; continuous; continuous; continuous; discontinuous; discontinuous; discontinuous; continuous **15.** $(-\infty, -2)$; $(-2, 2)$; $(2, +\infty)$ **17.** $(-\infty, -3]$; $[4, +\infty)$ **19.** $(-\infty, -1)$; $(-1, 1)$; $(1, +\infty)$ **21.** continuous on $(0, 1)$ and all intervals $[k, k+1)$ where k is any integer except zero **29.** 5 **31.** $c = -3$ and $k = 4$ **37.** $c = \frac{1}{2}(1 + \sqrt{5})$ **39.** $c = -4$ **41.** f is discontinuous at -2 **43.** f is discontinuous at 1 **45.** no; f will be continuous on $[a, c]$ if $\lim\limits_{x \to b^-} g(x)$ exists and is equal to $h(b)$ **47.** $\sqrt{10} - 3$

EXERCISES 2.9 (Page 137)

1. 4 **3.** $\frac{9}{7}$ **5.** $\frac{3}{5}$ **7.** $\frac{1}{9}$ **9.** 0 **11.** 0 **13.** 12 **15.** $\frac{1}{2}$ **17.** 0 **19.** $+\infty$ **21.** 0 **23.** -1 **25.** 3 **27.** 0 **29.** -4 **31.** 1 **33.** 0

EXERCISES 2.10 (Page 142)

1. any interval (a, b) where $-2 < a < 3$ and $b > 3$ **3.** any interval (a, b) where $0 \le a < 1$ and $b > 1$ **5.** any interval (a, b) where $a < -3$ and $-3 < b < 3$ **7.** any interval (a, b) where $-4 \le a < 0$ and $b > 0$ **9.** (a) $(-\infty, -1) \cup (0, +\infty)$; (b) (a, b) where $a < k < b \le -1$ or $0 \le a < k < b$ **11.** (a) all numbers in the interval $(-\frac{7}{2}, \frac{3}{2})$; (b) (a, b) where $-\frac{7}{2} \le a < b \le -2$ or $-2 \le a < b \le \frac{3}{2}$

REVIEW EXERCISES FOR CHAPTER 2 (Page 143)

1. 9 **3.** -6 **5.** $\sqrt[3]{2}$ **7.** $-\frac{1}{6}$ **9.** $\frac{1}{2}\epsilon$ **11.** $\min(1, \frac{1}{9}\epsilon)$ **13.** $\min(1, \epsilon)$ **15.** $\min(1, \epsilon)$ **17.** $\min(1, 5\epsilon)$ **19.** $-2, 1$; $f(-2)$ and $f(1)$ do not exist **21.** -2; $\lim\limits_{x \to -2} g(x)$ does not exist **23.** f is continuous on $(-\infty, -\frac{3}{2}]$ and $[\frac{3}{2}, +\infty)$ **25.** g is continuous on $[3, 4)$ **27.** $-\frac{5}{2}$ **29.** 8 **31.** $\frac{1}{3}$ **33.** $-\infty$ **35.** $+\infty$ **37.** $-\infty$ **39.** $-\frac{1}{2}$ **41.** $\frac{5}{2}$ **43.** 0 **45.** $\dfrac{2}{3a^{1/3}}$ **47.** $-\frac{1}{2}$ **49.** (a) yes; (b) no **51.** removable; $\frac{6}{5}$ **53.** essential **55.** $y = \frac{5}{4}$; $x = -\frac{7}{4}$ **57.** $y = \frac{4}{3}$ **59.** $y = 5$; $x = -2$; $x = 2$ **61.** $(f \circ g)(x) = \sqrt{x-1}$; $(1, +\infty)$ **63.** $(f \circ g)(x) = \begin{cases} 1 & \text{if } x < -1 \\ 0 & \text{if } x = -1 \\ -1 & \text{if } -1 < x < 1 \\ 0 & \text{if } x = 1 \\ 1 & \text{if } x > 1 \end{cases}$; all real numbers except -1 and 1 **65.** $c = \sqrt{7}$ **67.** 0 **69.** $a = 10, b = -23$ **71.** take $\delta = \min(1, \frac{1}{3}\epsilon^2)$ **73.** take $N = -\dfrac{3}{\epsilon}$ **75.** (b) any interval (a, b) where $-2 \le a < -\frac{3}{2} < b \le \frac{3}{2}$ **77.** (b) all values of a; (c) all nonintegers **79.** $\operatorname{sgn} x$ **81.** (b) $f(x) = \begin{cases} 1 & \text{if } x \ne 0 \\ 0 & \text{if } x = 0 \end{cases}$

EXERCISES 3.1 (Page 151)

1. $-2x_1$ **3.** $-4x_1 + 4$ **5.** $3x_1^2$ **7.** $3x_1^2 - 3$ **9.** $\dfrac{1}{2\sqrt{4 - x_1}}$ **11.** $3x_1^2 - 8x_1 + 4$ **13.** $8x + y + 9 = 0$; $x - 8y + 58 = 0$ **15.** $6x - y - 16 = 0$; $x + 6y - 52 = 0$ **17.** $2x + 5y - 17 = 0$; $5x - 2y + 30 = 0$ **19.** $2x + 3y - 12 = 0$; $3x - 2y - 5 = 0$ **21.** $4x + y = 0$; $x - 4y = 0$ **23.** $x - 12y + 16 = 0$; $12x + y - 98 = 0$ **25.** $8x - y - 5 = 0$ **27.** $4x - 4y - 1 = 0$ **29.** $2x - y - 2 = 0$ **31.** $(12 - 2\sqrt{30})x - y - 30 + 4\sqrt{30} = 0$; $(12 + 2\sqrt{30})x - y - 30 - 4\sqrt{30} = 0$

EXERCISES 3.2 (Page 156)

1. 7 **3.** 0 **5.** $-4x$ **7.** $8x + 5$ **9.** $-3x^2$ **11.** $\dfrac{1}{2\sqrt{x}}$ **13.** $-\dfrac{1}{(x+1)^2}$ **15.** $-\dfrac{2}{x^3} - 1$ **17.** $-\dfrac{1}{2(x+1)^{3/2}}$ **19.** -6 **21.** $-\frac{1}{216}$ **23.** $\frac{5}{4}$ **25.** 3 **27.** -12 **29.** $-\frac{1}{27}$ **31.** $-\dfrac{8}{x^3} + 3$ **33.** $-\dfrac{7}{2\sqrt{2-7x}}$ **35.** $-\dfrac{2}{3(1-2x)^{2/3}}$ **37.** $g(a)$ **39.** $\dfrac{1}{3(x-1)^{2/3}}$; no **41.** $-\dfrac{1}{(1-3x)^{2/3}}$; $\frac{1}{3}$ **43.** $\dfrac{2a}{x^3}$

EXERCISES 3.3 (Page 163)

1. (b) yes; (c) 1, -1; (d) no **3.** (b) yes; (c) -1, 1; (d) no **5.** (b) yes; (c) 0, 1; (d) no **7.** (b) yes; (c) 0, 0; (d) yes **9.** (b) yes; (c) does not exist, 0; (d) no **11.** (b) yes; (c) 8, 8; (d) yes **13.** (b) yes; (c) neither exists; (d) no **15.** (b) yes; (c) -6, -6; (d) yes **23.** $f'_+(0) = 0$ **25.** (a) $\frac{7}{5}$; (b) no **27.** $a = 8$, $b = -9$ **29.** (a) 0; (b) 1; (c) does not exist **35.** (a) $I(x) = \begin{cases} 15x & \text{if } 0 \le x \le 150 \\ 22.5x - 0.05x^2 & \text{if } 150 < x \le 250 \end{cases}$

EXERCISES 3.4 (Page 172)

1. 7 **3.** $-2 - 2x$ **5.** $3x^2 - 6x + 5$ **7.** $x^7 - 4x^3$ **9.** $t^3 - t$ **11.** $4\pi r^2$ **13.** $2x + 3 - \dfrac{2}{x^3}$ **15.** $16x^3 + \dfrac{1}{x^5}$ **17.** $-\dfrac{6}{x^3} - \dfrac{20}{x^5}$ **19.** $3\sqrt{3}s^2 - 2\sqrt{3}s$ **21.** $70x^6 + 60x^4 - 15x^2 - 6$ **23.** $-18y^2(7 - 3y^3)$ **25.** $10x^4 - 24x^3 + 12x^2 + 2x - 3$ **27.** $-\dfrac{1}{(x-1)^2}$ **29.** $-\dfrac{4(x+1)}{(x-1)^3}$ **31.** $\dfrac{5(1-2x^2)}{(1+2x^2)^2}$ **33.** $\dfrac{48x^2}{(x^3+8)^2}$ **35.** $\dfrac{6(x^2+10x+1)}{(x+5)^2}$ **37.** $12x - y = 20$ **39.** $x + 20y + 96 = 0$ **41.** $2x - y = 3$ **43.** $x + 8y + 2 = 0$; $x + 8y - 2 = 0$ **45.** $28x - y = 99$; $4x - y = 3$ **49.** $2(3x+2)(6x^2 + 2x - 3)$ **51.** $3(2x^2 + x + 1)^2(4x + 1)$

EXERCISES 3.5 (Page 181)

1. $v(t) = 6t$; 18 **3.** $v(t) = -\dfrac{1}{4t^2}$; -1 **5.** $v(t) = 6t^2 - 2t$; 8 **7.** $v(t) = \dfrac{8}{(4+t)^2}$; $\frac{1}{2}$ **9.** $t < -3$, moving to right; $-3 < t < 1$, moving to left; $t > 1$, moving to right; changes direction when $t = -3$ and $t = 1$ **11.** $t < -2$, moving to right; $-2 < t < \frac{1}{2}$, moving to left; $t > \frac{1}{2}$, moving to right; changes direction when $t = -2$ and $t = \frac{1}{2}$ **13.** $t < -3$, moving to left; $-3 < t < 3$, moving to right; $t > 3$, moving to left; changes direction when $t = -3$ and $t = 3$ **15.** (a) -32 ft/sec; (b) -64 ft/sec; (c) 4 sec; (d) -128 ft/sec **17.** 160 cm/sec **19.** (a) $(20t_1 + 24)$ ft/sec; (b) $\frac{6}{5}$ sec **21.** (a) 8.6; (b) 8.3; (c) 8.1; (d) 8 **23.** (a) 104π; (b) 101π; (c) 100.1π; (d) 100π **25.** (a) $-\frac{1}{16}$, $-\frac{1}{4}$; (b) $-\frac{1}{100}$, $-\frac{1}{10}$ **27.** (a) -24 units per dollar; (b) -23 units per dollar **29.** (a) 18,750 liters per minute; (b) 17,500 liters per minute **31.** $D_I R = -\dfrac{E}{I^2}$ **33.** (a) -2.9 degrees per hour; (b) -3 degrees per hour **35.** (a) 920 people per year; (b) 1400 people per year; (c) 6.1%; (d) 6.4% **37.** (a) profitable; (b) not profitable; (c) 90

EXERCISES 3.6 (Page 187)

1. $3\cos x$ **3.** $2(\cos t - t\sin t)$ **5.** $x\cos x$ **7.** $4\cos 2x$ **9.** $-x^2\sin x$ **11.** $-\dfrac{2(z+1)\sin z + 2\cos z}{(z+1)^2}$ **13.** $\dfrac{1}{\cos x - 1}$ **15.** $\dfrac{2\cos y}{(1-\sin y)^2}$ **17.** $(1-\cos x)(x + \cos x) + (1 - \sin x)(x - \sin x)$ **19.** 1 **21.** $-\dfrac{2}{\pi}$ **23.** -3π **25.** 2 **27.** $\sqrt{2}$ **29.** $-\frac{10}{3}$ **35.** $3\sec^3 x + 3\sec x\tan^2 x$ **37.** $\dfrac{-5\csc t\cot t}{(\csc t + 2)^2}$ **39.** $x + y = \pi$ **41.** $x + 2y = 6$ **43.** (a) $\dfrac{\sqrt{2}}{9}W$; (b) $2W$

EXERCISES 3.7 (Page 194)

1. $6(2x+1)^2$ **3.** $8(x+2)(x^2+4x-5)^3$ **5.** $2(2t^4 - 7t^3 + 2t - 1)(8t^3 - 21t^2 + 2)$ **7.** $\dfrac{-4x}{(x^2+4)^3}$ **9.** $-12(\sin 3x + \cos 4x)$ **11.** $-4t\sin 2t^2 - 2\sin 2t$ **13.** $36\cos 4x\cos 3x - 48\sin 4x\sin 3x$ **15.** $6(3u^2+5)^2(3u-1)(12u^2 - 3u + 5)$ **17.** $-2(2x-5)^{-2}(4x+3)^{-3}(12x - 17)$ **19.** $\dfrac{18(y-7)}{(y+2)^3}$ **21.** $\dfrac{-2(14x+3)}{(7x^2+3x-1)^2}$ **23.** $2(r^2+1)^2(2r+5)(8r^2+15r+2)$ **25.** $\dfrac{2z(z^2-5)^2(z^2+22)}{(z^2+4)^3}$ **27.** $\dfrac{4(4x-1)^2(x^2+2)^3(21x^4 - 3x^3 + 49x^2 - 4x + 30)}{(3x^2+5)^3}$ **29.** $6t\sin(6t^2 - 2)$

31. $6(\sin^2 x - x^2)^2(\sin x \cos x - x)$ **33.** $\dfrac{6 \cos 2y(\sin^2 2y + 2)}{(\cos^2 2y + 1)^2}$ **35.** $-12 \cos 3x \sin(\sin 3x)$ **37.** $24x + y + 39 = 0$; $y = 0$; $y = 1$; $y = 0$; $24x - y - 39 = 0$ **39.** $3x - 4y + 5 = 0$ **41.** $6x - 18\sqrt{3}y = 2\pi + 27\sqrt{3}$ **43.** (a) $\dfrac{8t(t^2 - 1)}{(t^2 + 1)^3}$ cm/sec; (b) 0 cm/sec, $\frac{48}{125}$ cm/sec **45.** (a) $(5\pi \cos \pi t - 3\pi \sin \pi t)$ cm/sec; (b) $\sqrt{2}\pi$ cm/sec, 3π cm/sec **47.** (a) $3x^4$; (b) $6x^5$ **55.** 2.7 km/min

EXERCISES 3.8 (Page 198)

1. $x^{-1/2}(2 - \frac{5}{2}x^{-1})$ **3.** $\dfrac{4x}{\sqrt{1 + 4x^2}}$ **5.** $\dfrac{-2}{(5 - 3x)^{1/3}}$ **7.** $\dfrac{y}{(25 - y^2)^{3/2}}$ **9.** $-\dfrac{\sin \sqrt{t}}{\sqrt{t}}$ **11.** $-\dfrac{3 \cos 3x}{2(\sin 3x)^{3/2}}$ **13.** $\dfrac{17}{2(3x + 1)^{3/2}(2x - 5)^{1/2}}$ **15.** $\dfrac{6x^2 - 10x + 1}{3\sqrt[3]{(2x^3 - 5x^2 + x)^2}}$ **17.** $\dfrac{1}{\sqrt{2t}}\left(1 - \dfrac{1}{t}\right)$ **19.** $\frac{1}{4}x(5 - x^2)^{-1/2}(x^3 + 1)^{-3/4}(-7x^3 + 15x - 4)$ **21.** $\dfrac{1}{x^2\sqrt{x^2 - 1}}$ **23.** $\dfrac{\cos t}{(1 - \sin t)^{3/2}(1 + \sin t)^{1/2}}$ **25.** $\dfrac{-y \sin \sqrt{y^2 + 1}}{\sqrt{y^2 + 1}}$ **27.** $\dfrac{x + 5}{6\sqrt{x - 1}\sqrt[3]{(x + 1)^4}}$ **29.** $\dfrac{-1}{4\sqrt{9 + \sqrt{9 - x}}\sqrt{9 - x}}$ **31.** $\dfrac{-\sin 4x}{\sqrt{1 + \cos^2 2x}}$ **33.** $2x + 3y = 6$; $2\sqrt[3]{4}x + 3y = 3\sqrt[3]{4}$; $x = 1$; $2\sqrt[3]{4}x - 3y = \sqrt[3]{4}$; $2x - 3y + 2 = 0$ **35.** $4x - 5y + 9 = 0$ **37.** $x + 4y = 0$ **39.** (a) 0; (b) $\frac{1}{2}$; (c) no value of t **41.** 100 **43.** $\dfrac{2x(x^2 - 4)}{|x^2 - 4|}$ **45.** $3x|x|$ **51.** $\frac{4}{3}(3x + 2)^3(x^2 - 1)^{-1/3}(12x^2 + 2x - 9)$ **53.** $\frac{1}{6}(t^3 - 2t + 1)^{1/2}(t^2 + t + 5)^{-2/3}(31t^4 + 29t^3 + 109t^2 - 18t - 88)$

EXERCISES 3.9 (Pages 203)

1. $-\dfrac{x}{y}$ **3.** $\dfrac{8y - 3x^2}{3y^2 - 8x}$ **5.** $-\dfrac{y^2}{x^2}$ **7.** $-\dfrac{\sqrt{y}}{\sqrt{x}}$ **9.** $\dfrac{x - xy^2}{x^2y - y}$ **11.** $\dfrac{3x^2 - 4xy - 1}{2x^2 + 2}$ **13.** $\dfrac{y^{2/3} + y}{24x^{2/3}y^{5/3} - x}$ **15.** $\dfrac{y + 4\sqrt{xy}}{\sqrt{x} - x}$ **17.** $\dfrac{2 + 5x^2 - 4x^{3/2}y}{2x^{1/2}(x^2 + 3)}$ **19.** $\dfrac{\sin(x - y)}{\sin(x - y) - 1}$ **21.** $\dfrac{y \sin x - \sin y}{x \cos y + \cos x}$ **23.** $\dfrac{3x^2 - 4y}{4x - 3y^2}$ **25.** $\dfrac{3x^2 - y^3}{x^3 - 6xy}$ **27.** $\dfrac{x^3 + 8y^3}{4x^3 - 3x^2y}$ **29.** $2x + y = 4$ **31.** $5x - 7y + 11 = 0$ **33.** -1 **35.** $(1, 0)$ **37.** (a) $f_1(x) = 2\sqrt{x - 2}$, domain: $x \geq 2$; $f_2(x) = -2\sqrt{x - 2}$, domain: $x \geq 2$; (d) $f_1'(x) = (x - 2)^{-1/2}$, domain: $x > 2$; $f_2'(x) = -(x - 2)^{-1/2}$, domain: $x > 2$; (e) $\dfrac{2}{y}$; (f) $x - y - 1 = 0$; $x + y - 1 = 0$ **39.** (a) $f_1(x) = \sqrt{x^2 - 9}$, domain: $|x| \geq 3$; $f_2(x) = -\sqrt{x^2 - 9}$, domain: $|x| \geq 3$; (d) $f_1'(x) = x(x^2 - 9)^{-1/2}$, domain: $|x| > 3$; $f_2'(x) = -x(x^2 - 9)^{-1/2}$, domain: $|x| > 3$; (e) $\dfrac{x}{y}$; (f) $5x + 4y + 9 = 0$; $5x - 4y + 9 = 0$ **41.** (a) $f_1(x) = \sqrt{8 - x^2 + 2x} + 2$, domain: $-2 \leq x \leq 4$; $f_2(x) = -\sqrt{8 - x^2 + 2x} + 2$, domain: $-2 \leq x \leq 4$; (d) $f_1'(x) = (1 - x)(8 - x^2 + 2x)^{-1/2}$, domain: $-2 < x < 4$; $f_2'(x) = (x - 1)(8 - x^2 + 2x)^{-1/2}$, domain: $-2 < x < 4$; (e) $\dfrac{1 - x}{y - 2}$; (f) $y - 5 = 0$, $y + 1 = 0$

EXERCISES 3.10 (Page 209)

1. $f'(x) = 5x^4 - 6x^2 + 1$; $f''(x) = 20x^3 - 12x$ **3.** $g'(s) = 8s^3 - 12s^2 + 7$; $g''(s) = 24s^2 - 24s$ **5.** $f'(x) = \frac{5}{2}x^{3/2} - 5$; $f''(x) = \frac{15}{4}x^{1/2}$ **7.** $f'(x) = x(x^2 + 1)^{-1/2}$; $f''(x) = (x^2 + 1)^{-3/2}$ **9.** $f'(t) = -8t \sin t^2$; $f''(t) = -16t^2 \cos t^2 - 8 \sin t^2$ **11.** $G'(x) = -2x(3 + 2x^2)^{-3/2}$; $G''(x) = (8x^2 - 6)(3 + 2x^2)^{-5/2}$ **13.** $g'(x) = \dfrac{8x}{(x^2 + 4)^2}$; $g''(x) = \dfrac{32 - 24x^2}{(x^2 + 4)^3}$ **15.** $f'(x) = \dfrac{\cos x}{2\sqrt{\sin x + 1}}$; $f''(x) = -\frac{1}{4}\sqrt{\sin x + 1}$ **17.** $24x$ **19.** $1152(2x - 1)^{-5}$ **21.** $\dfrac{18 + 6x}{(1 - x)^5}$ **23.** $-32(\sin 2x + \cos 2x)$ **29.** $-\dfrac{3a^4x^2}{y^7}$ **31.** $-\frac{7}{4}$; 5 **33.** $v = 3t^2 - 18t + 15$; $a = 6t - 18$; when $0 < t < 1$, the particle is at the right of the origin; it is moving to the right, the velocity is decreasing, and the speed is decreasing; when $1 < t < \dfrac{9 - \sqrt{21}}{2}$, the particle is at the right of the origin, it is moving to the left, the velocity is decreasing, and the speed is increasing; when $\dfrac{9 - \sqrt{21}}{2} < t < 3$, the particle is at the left of the origin, it is moving to the left, the velocity is decreasing, and the speed is increasing; when $3 < t < 5$, the particle is at the left of the origin, it is moving to the left, the velocity is increasing, and the speed is decreasing; when $5 < t < \dfrac{9 + \sqrt{21}}{2}$, the particle is at the left of the origin, it is moving to the right, the velocity is increasing, and the speed is increasing; when $\dfrac{9 + \sqrt{21}}{2} < t < +\infty$, the particle is at the right of the origin, it is moving to the right, the velocity is increasing, and the speed is increasing **35.** $\frac{3}{2}$ sec; $\frac{7}{4}$ ft; $-\frac{1}{4}$ ft/sec **37.** $\frac{1}{2}$ sec; $\frac{249}{80}$ ft; $-\frac{11}{8}$ ft/sec **39.** $\frac{3}{2}$ sec; $\frac{4}{3}\sqrt{6}$ ft; $\frac{2}{3}\sqrt{6}$ ft/sec **43.** $f'(x) = 2|x|$, domain: $(-\infty, +\infty)$;

$f''(x) = 2\frac{|x|}{x}$, domain: $(-\infty, 0) \cup (0, +\infty)$ **45.** $f'(x) = 4|x^3|$, domain: $(-\infty, +\infty)$; $f''(x) = 12x|x|$, domain: $(-\infty, +\infty)$ **47.** $f'''(x) = 24|x|$ **49.** $h''(x) = (f'' \circ g)(x)(g'(x))^2 + (f' \circ g)(x)g''(x)$

REVIEW EXERCISES FOR CHAPTER 3 (Page 210)

1. $-\frac{8}{x^3} + \frac{12}{x^5}$ **3.** $60t^4 - 39t^2 - 6t - 4$ **5.** $\frac{-6x^2}{(x^3 - 1)^2}$ **7.** $4(2s^3 - 3s + 7)^3(6s^2 - 3)$ **9.** $\frac{1 - 2x^3}{3x^{2/3}(x^3 + 1)^{4/3}}$ **11.** $x(4x^2 - 13)(x^2 - 1)^{1/2}(x^2 - 4)^{-1/2}$ **13.** $(x + 1)\sin x + x \cos x$ **15.** $-3 \sin 3w \cos(\cos 3w) + 6 \sin 4w$ **17.** $\frac{8x}{3y^2 - 8y}$ **19.** $\frac{1}{\sqrt{x^2 - 1}\,(x - \sqrt{x^2 - 1})}$ **21.** $\frac{\cos x \cos y}{\sin x \sin y}$ **23.** $2x + 3[x^3 + (x^4 + x)^2]^2[3x^2 + 2(x^4 + x)(4x^3 + 1)]$ **25.** (a) $t = 3, t = 8$; (b) when $t = 0$, $v = 24$, and the particle is moving to the right; when $t = 3$, $v = -15$, and the particle is moving to the left; when $t = 8$, $v = 40$, and the particle is moving to the right **27.** $v^2 = 4(5 - s)(s - 3)$; $a = 16 - 4s$ **29.** (a) 32 ft/sec; (b) 256 ft; (c) 7 sec; (d) -128 ft/sec **31.** $5x - 4y - 6 = 0$; $4x + 5y - 13 = 0$ **33.** $12x - 6\sqrt{3}y = 3\sqrt{3} + 8\pi$ **35.** $(-1, 0)$ **37.** $-3(3 - 2x)^{-5/2}$ **39.** $x < -3$ or $x > -1$ **41.** (a) 8.005; (b) 8 **43.** (a) 3312.3; (b) 3212.5 **45.** $\frac{2}{\sqrt{4x - 3}}$ **47.** $-\frac{1}{3}$ **49.** $2(|x + 1| - |x|)\left(\frac{x + 1}{|x + 1|} - \frac{x}{|x|}\right)$ **51.** $\frac{1}{2}$ **53.** (b) continuous at 3; (c) not differentiable at 3 **55.** $a = -\frac{1}{2}$; $b = \frac{3}{2}$ **57.** (b) 0; (c) 0 **63.** $(f \circ g)'(0) = 2$ **65.** $f(x) = |x|$, $g(x) = x^2$ **69.** $P'(r)$

EXERCISES 4.1 (Page 219)

1. -3 **3.** -2 **5.** $-\frac{\sqrt{3}}{2}$ **7.** $-\frac{3}{4}$ **9.** $\frac{9}{5}$ ft/sec **11.** $\frac{1}{2\pi}$ ft/min **13.** $\frac{5}{8\pi}$ m/min **15.** $\frac{25}{3}$ ft/sec **17.** $\frac{6}{25\pi}$ m/min **19.** 1800 lb/ft^2 per min **21.** 128π cm^2/sec **23.** 14 ft/sec **25.** \$1020 per week **27.** 875 units per month **29.** decreasing at the rate of 55 shirts per week **33.** 22 m^3/min **35.** $\frac{1}{194}(3\sqrt{97} + 97)$ ft/sec ≈ 0.65 ft/sec **37.** decreasing at the rate of $\frac{1}{48}$ rad/sec

EXERCISES 4.2 (Page 228)

1. $-5, \frac{1}{3}$ **3.** $-3, -1, 1$ **5.** 0, 2 **7.** $-2, 0, 2$ **9.** no critical numbers **11.** no critical numbers **13.** $\frac{1}{6}k\pi$, where k is any integer **15.** $\frac{1}{8}(2k + 1)\pi$, where k is any integer **17.** abs min: $f(2) = -2$ **19.** no absolute extrema **21.** abs min: $f(-\frac{2}{3}\pi) = -1$; abs max: $f(0) = 2$ **23.** abs min: $f(-3) = 0$ **25.** abs min: $h(5) = 1$ **27.** abs min: $F(4) = 1$ **29.** abs min: $g(0) = 2$ **31.** abs max: $f(5) = 2$ **33.** abs min: $f(2) = 0$ **35.** abs min: $g(-3) = -46$; abs max: $g(-1) = -10$ **37.** abs min: $f(-2) = 0$; abs max: $f(-4) = 144$ **39.** abs min: $f(2) = 0$; abs max: $f(3) = 25$ **41.** abs min: $f(-\frac{1}{2}\pi) = -2$; abs max: $f(\frac{1}{2}\pi) = 2$ **43.** abs min: $f(-1) = -1$; abs max: $f(2) = \frac{1}{2}$ **45.** abs min: $f(1) = -2$; abs max: $f(0) = -\frac{1}{3}$ **47.** abs min: $F(-3) = -13$; abs max: $F(3) = 7$ **49.** abs min: $f(-1) = 0$; abs max: $f(1) = \sqrt[3]{4}$

EXERCISES 4.3 (Page 234)

1. 2500 cm^2 **3.** $\frac{1}{2}$ **5.** $12\sqrt{3}$ sq units **7.** $\frac{5}{3}$ in. **9.** 60 m by 120 m **11.** from A to P to C, where P is 8 km down the beach from B **13.** from A to P to C, where P is 4 km down the river from B **15.** $\frac{1}{4}\pi$ **17.** 225 **19.** 400 **21.** 30 **23.** radius is $3\sqrt{2}$ in., height is $6\sqrt{2}$ in. **25.** A, 7 machines; B, 8 machines **27.** (a) 3.4 units; (b) 9.4 units **29.** (a) radius of circle is $\frac{5}{\pi + 4}$ ft and length of side of square is $\frac{10}{\pi + 4}$ ft; (b) radius of circle is $\frac{5}{\pi}$ ft and there is no square

EXERCISES 4.4 (Page 241)

1. 2 **3.** $\frac{1}{4}\pi$ **5.** $\frac{1}{2}$ **7.** $\frac{8}{27}$ **9.** 0 **11.** (b) (i), (ii), (iii) satisfied; (c) $(\frac{3}{4}, -\frac{9}{8}\sqrt[3]{6})$ **13.** (b) (i) not satisfied **15.** (b) (ii) not satisfied **17.** 4 **19.** $\cos c = \frac{2}{\pi}$; $c \approx 0.8807$ **21.** (i) not satisfied **23.** (ii) not satisfied **29.** $\frac{3}{2}$

EXERCISES 4.5 (Page 247)

1. (a) and (b) $f(2) = -5$, rel min; (c) $[2, +\infty)$; (d) $(-\infty, 2]$ **3.** (a) and (b) $f(-\frac{1}{3}) = \frac{5}{27}$, rel max; $f(1) = -1$, rel min; (c) $(-\infty, -\frac{1}{3}]$, $[1, +\infty)$; (d) $[-\frac{1}{3}, 1]$ **5.** (a) and (b) $f(0) = 2$, rel max; $f(3) = -25$, rel min; (c) $(-\infty, 0]$, $[3, +\infty)$; (d) $[0, 3]$ **7.** (a) and (b) f has a rel max value of 4 if $x = (4k + 1)\pi$, where k is any integer; f has a rel min value of -4 if $x = (4k + 3)\pi$, where k is any integer; (c) $[(4k - 1)\pi, (4k + 1)\pi]$, where k is any integer; (d) $[(4k + 1)\pi, (4k + 3)\pi]$, where k is any integer **9.** (a) and (b) $f(0) = 0$, rel min; $f(1) = \frac{1}{4}$, rel max; $f(2) = 0$, rel min; (c) $[0, 1]$, $[2, +\infty)$; (d) $(-\infty, 0]$, $[1, 2]$ **11.** (a) and (b) $f(-2) = -\frac{1}{15}$, rel max; $f(-1) = -\frac{23}{15}$, rel min; $f(1) = \frac{53}{15}$, rel max; $f(2) = \frac{31}{15}$, rel min; (c) $(-\infty, -2]$, $[-1, 1]$, $[2, +\infty)$; $[-2, -1]$, $[1, 2]$ **13.** (a) and (b) no relative extrema; (c) $(0, +\infty)$; (d) nowhere **15.** (a) and (b) $f(\sqrt[3]{2}) = \frac{3}{2}\sqrt[3]{2}$, rel min; (c) $(-\infty, 0)$, $[\sqrt[3]{2}, +\infty)$; (d) $(0, \sqrt[3]{2}]$ **17.** (a) and (b) $f(\frac{1}{3})$

$= \frac{3456}{3125}$, rel max; $f(1) = 0$, rel min; (c) $(-\infty, \frac{1}{5}]$, $[1, +\infty)$; (d) $[\frac{1}{5}, 1]$ **19.** (a) and (b) $f(2) = 4$, rel max; (c) $(-\infty, 2]$; (d) $[2, 3]$ **21.** (a) and (b) $f(-1) = 2$, rel max; $f(1) = -2$, rel min; (c) $(-\infty, -1]$, $[1, +\infty)$; (d) $[-1, 1]$ **23.** (a) and (b) $f(4) = 2$, rel max; (c) $(-\infty, -2]$; (d) $[4, +\infty)$ **25.** (a) and (b) $f(\frac{1}{8}) = -\frac{1}{4}$, rel min; (c) $[\frac{1}{8}, +\infty)$; (d) $(-\infty, \frac{1}{8}]$ **27.** (a) and (b) $f(-1) = 0$, rel max; $f(1) = -\sqrt[3]{4}$, rel min; (c) $(-\infty, -1]$, $[1, +\infty)$; (d) $[-1, 1]$ **29.** (a) and (b) $f(4) = 9$, rel max; (c) $(-\infty, 4]$; (d) $[4, +\infty)$ **31.** (a) and (b) $f(-2) = 5$, rel max; $f(0) = 1$, rel min; (c) $(-\infty, -2]$, $[0, +\infty)$; (d) $[-2, 0]$ **33.** (a) and (b) $f(-1) = 2$, rel max; $f(0) = 1$, rel min; $f(2) = 5$, rel max; (c) $(-\infty, -1]$, $[0, 2]$; (d) $[-1, 0]$, $[2, +\infty)$ **35.** (a) and (b) $f(-9) = -8$, rel min; $f(-4) = -5$, rel min; $f(2) = -7$, rel min; $f(-7) = -4$, rel max; $f(0) = -3$, rel max; (c) $[-9, -7]$, $[-4, 0]$, $[2, +\infty)$; (d) $(-\infty, -9]$, $[-7, -4]$, $[0, 2]$ **37.** (a) and (b) no relative extrema; (c) $[0, +\infty)$; (d) nowhere **39.** (a) and (b) $f(4) = \frac{1}{4}\sqrt[3]{4}$, rel max; (c) $(-4, 4]$; (d) $(-\infty, -4)$, $[4, +\infty)$ **43.** $a = -3, b = 7$ **45.** $a = -2, b = 9, c = -12, d = 7$

EXERCISES 4.6 (Page 254)

1. $f(\frac{1}{3}) = \frac{2}{3}$, rel min **3.** $f(\frac{3}{2}) = \frac{81}{4}$, rel max; $f(-1) = -11$, rel min **5.** $g(0) = 3$, rel max; $g(2) = \frac{5}{3}$, rel min **7.** $f(4) = 0$, rel min **9.** $h(-\frac{3}{4}) = -\frac{99}{256}$, rel min; $h(0) = 0$, rel max; $h(1) = -\frac{5}{6}$, rel min **11.** F has a rel min value of -1 when $x = \frac{1}{3}(2k + 1)\pi$, where k is any integer; F has a rel max value of 1 when $x = \frac{2}{3}k\pi$, where k is any integer **13.** $f(-\frac{1}{2}) = -\frac{27}{16}$, rel min **15.** $f(1) = 8$, rel min **17.** $h(-2) = -2$, rel min **19.** no relative extrema **21.** $F(27) = 9$, rel max **23.** $h(-1) = -2$, rel min; $h(1) = 2$, rel max **25.** $f(0) = 0$, abs min **27.** no absolute extrema **29.** $g(\frac{1}{4}) = \frac{3}{4}$, abs min **31.** no absolute extrema **33.** $f(-2) = -13$, abs min **35.** $g(-1) = -2$, abs max **37.** $f(0) = 0$, abs min; $f(\sqrt{2}) = \frac{1}{18}\sqrt{3}$, abs max **39.** f has a rel min at $x = \frac{1}{2}\sqrt{2}$ and a rel max at $x = -\frac{1}{2}\sqrt{3}$

EXERCISES 4.7 (Page 259)

1. 45 m × 60 m **3.** depth is one-half the length of the base **5.** $2x - y + 1 = 0$ **7.** $(\frac{9}{5}, \frac{3}{5})$ **9.** $30(\sqrt{10} + 1)$ ft by $44(\sqrt{10} + 1)$ ft **11.** ratio of height of rectangle to radius of semicircle is $\frac{1}{4}(4 + \pi)$ **13.** breadth is a cm and depth is $\sqrt{3}a$ cm **15.** $10\sqrt[3]{10}$ knots **17.** $1\frac{11}{25}$ sec **19.** $\sqrt{2}$ **21.** $2\sqrt{2}$ **23.** $\frac{4}{3}a$

EXERCISES 4.8 (Page 268)

1. concave downward for $x < 0$; concave upward for $x > 0$; (0, 0) pt. of infl. **3.** concave downward for $x < -\frac{1}{2}$; concave upward for $x > -\frac{1}{2}$; $(-\frac{1}{2}, 5)$ pt. of infl. **5.** concave upward everywhere **7.** concave downward for $x < -1$ and $0 < x < 1$; concave upward for $-1 < x < 0$ and $x > 1$; (0, 0) pt. of infl. **9.** concave downward for $-\pi < x < -\frac{3}{4}\pi$, $-\frac{1}{4}\pi < x < \frac{1}{4}\pi$, and $\frac{3}{4}\pi < x < \pi$; concave upward for $-\frac{3}{4}\pi < x < -\frac{1}{4}\pi$, and $\frac{1}{4}\pi < x < \frac{3}{4}\pi$; $(-\frac{3}{4}\pi, 0)$, $(-\frac{1}{4}\pi, 0)$, $(\frac{1}{4}\pi, 0)$, and $(\frac{3}{4}\pi, 0)$ pts. of infl. **11.** concave upward for $x < 2$; concave downward for $x > 2$; (2, 0) pt. of infl. **13.** concave upward for $x < 0$; concave downward for $x > 0$; (0, 0) pt. of infl. **15.** concave upward for $x < -4$; concave downward for $x > -4$ **17.** concave downward for $x < \dfrac{24 - \sqrt{6}}{10}$ and $\dfrac{24 + \sqrt{6}}{10} < x < 3$; concave upward for $\dfrac{24 - \sqrt{6}}{10} < x < \dfrac{24 + \sqrt{6}}{10}$ and $x > 3$; (3, 0) pt. of infl.; pts. of infl. at $x = \dfrac{24 \pm \sqrt{6}}{10}$ **19.** concave upward for $2 < x < 8$; concave downward for $x > 8$; $(8, \frac{4}{3}\sqrt{6})$ pt. of infl. **21.** $a = -1, b = 3$ **23.** $a = 2, b = -6, c = 0, d = 3$ (Sketches of the graphs for Exercises 25–39 appear in Figs. EX 4.8-25 through 4.8-39.) **41.** yes **43.** 10:40 A.M.

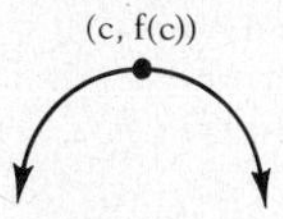

Figure EX 4.8-25

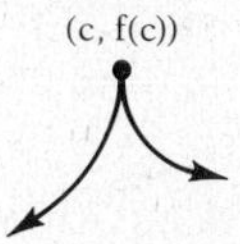

Figure EX 4.8-27

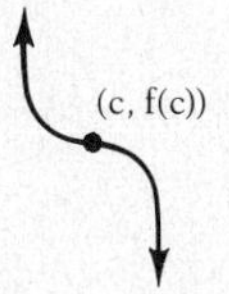

Figure EX 4.8-29

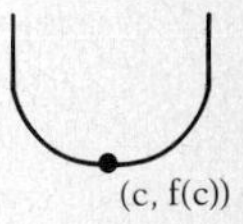

Figure EX 4.8-31

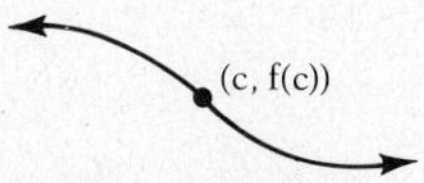

Figure EX 4.8-33

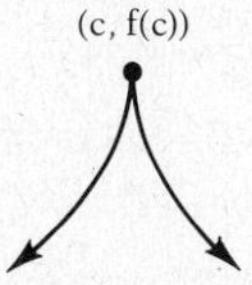

Figure EX 4.8-35

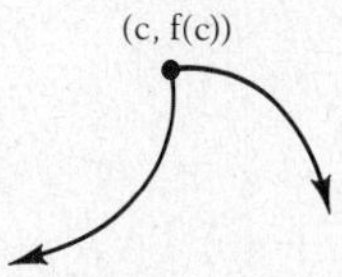

Figure EX 4.8-37

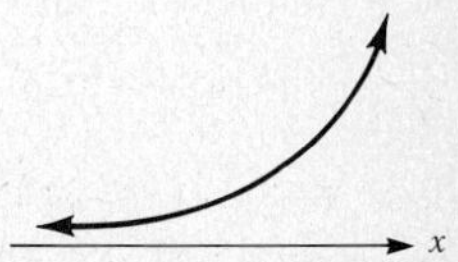

Figure EX 4.8-39

EXERCISES 4.9 (Page 273)

(Sketches of the graphs appear in Figs. EX 4.9-1 through EX 4.9-43)

1. $f(-1) = 5$, rel max; $f(1) = -3$, rel min; (0, 1), pt. of infl.; f increasing on $(-\infty, -1]$ and $[1, +\infty)$; f decreasing on $[-1, 1]$; graph concave downward for $x < 0$; graph concave upward for $x > 0$ **3.** $f(\frac{3}{2}) = -\frac{27}{16}$, rel min; (0, 0), (1, −1), pts. of infl.; f increasing on

$[\frac{3}{2}, +\infty)$; f decreasing on $(-\infty, \frac{3}{2}]$; graph concave upward for $x < 0$ and $x > 1$; graph concave downward for $0 < x < 1$ **5.** $f(-3) = 5$, rel max; $f(-\frac{1}{3}) = -\frac{121}{27}$, rel min; $(-\frac{5}{3}, \frac{7}{27})$, pt. of infl.; f increasing on $(-\infty, -3]$ and $[-\frac{1}{3}, +\infty)$; f decreasing on $[-3, -\frac{1}{3}]$; graph concave downward for $x < -\frac{5}{3}$; graph concave upward for $x > -\frac{5}{3}$ **7.** $f(0) = 1$, rel min; $(\frac{1}{2}, \frac{23}{16})$, $(1, 2)$, pts. of infl.; f decreasing on $(-\infty, 0]$; f increasing on $[0, +\infty)$; graph concave upward for $x < \frac{1}{2}$ and $x > 1$; graph concave downward for $\frac{1}{2} < x < 1$ **9.** $f(-1) = \frac{7}{12}$, rel min; $f(0) = 1$, rel max; $f(2) = -\frac{5}{3}$, rel min; pts. of infl. at $x = \frac{1}{3}(1 \pm \sqrt{7})$; f decreasing on $(-\infty, -1]$ and $[0, 2]$; f increasing on $[-1, 0]$ and $[2, +\infty)$; graph concave upward for $x < \frac{1}{3}(1 - \sqrt{7})$ and $x > \frac{1}{3}(1 + \sqrt{7})$; graph concave downward for $\frac{1}{3}(1 - \sqrt{7}) < x < \frac{1}{3}(1 + \sqrt{7})$ **11.** $f(0) = 0$, rel min; no pts. of infl.; f decreasing on $(-\infty, 0]$; f increasing on $[0, +\infty)$; graph concave upward everywhere **13.** $f(0) = 0$, rel min; no pts. of infl.; f decreasing on $(-\infty, 0]$; f increasing on $[0, +\infty)$; graph concave upward everywhere **15.** no relative extrema; $(0, 0)$ pt. of infl. with horizontal tangent; f increasing on $(-\infty, +\infty)$; graph concave downward on $(-\infty, 0)$; graph concave upward on $(0, +\infty)$ **17.** no relative extrema; $(2, 0)$ pt. of infl.; f decreasing on $(-\infty, +\infty)$; graph concave upward for $x < 2$; graph concave downward for $x > 2$ **19.** $f\left(\frac{4}{5}\right) = \frac{26{,}244}{3{,}125}$, rel max; $f(2) = 0$, rel min; pts. of infl. at $(-1, 0)$ and $x = \frac{1}{10}(8 \pm 3\sqrt{6})$; f increasing on $(-\infty, \frac{4}{5}]$ and $[2, +\infty)$; f decreasing on $[\frac{4}{5}, 2]$; graph concave downward for $x < -1$ and $\frac{1}{10}(8 - 3\sqrt{6}) < x < \frac{1}{10}(8 + 3\sqrt{6})$; graph concave upward for $-1 < x < \frac{1}{10}(8 - 3\sqrt{6})$ and $x > \frac{1}{10}(8 + 3\sqrt{6})$ **21.** $f(-\frac{4}{3}) = \frac{256}{81}$, rel max; $f(0) = 0$, rel min; $(-1, 2)$, pt. of infl.; f increasing on $(-\infty, -\frac{4}{3}]$ and $[0, +\infty)$; f decreasing on $[-\frac{4}{3}, 0]$; graph concave downward for $x < -1$; graph concave upward for $x > -1$ **23.** $f(-\frac{1}{2}\pi) = -3$, rel min; $f(0) = 3$, rel max; $f(\frac{1}{2}\pi) = 0$, rel min; $(-\frac{3}{4}\pi, 0)$, $(-\frac{1}{4}\pi, 0)$, $(\frac{1}{4}\pi, 0)$, $(\frac{3}{4}\pi, 0)$, pts. of infl.; f decreasing on $[-\pi, -\frac{1}{2}\pi]$ and $[0, \frac{1}{2}\pi]$; f increasing on $[-\frac{1}{2}\pi, 0]$ and $[\frac{1}{2}\pi, \pi]$; graph concave downward for $-\pi < x < -\frac{3}{4}\pi$, $-\frac{1}{4}\pi < x < \frac{1}{4}\pi$, and $\frac{3}{4}\pi < x < \pi$; graph concave upward for $-\frac{3}{4}\pi < x < -\frac{1}{4}\pi$ and $\frac{1}{4}\pi < x < \frac{3}{4}\pi$ **25.** no relative extrema; no pts. of infl.; f increasing on $[0, \frac{1}{2}\pi]$ and $[\frac{1}{2}\pi, \pi]$; graph concave downward for $0 < x < \frac{1}{2}\pi$ and $\frac{1}{2}\pi < x < \pi$ **27.** $f(0) = 0$, rel max; $f(2) = 4$, rel min; no pts. of infl.; f increasing on $(-\infty, 0]$ and $[2, +\infty)$; f decreasing on $[0, 1)$ and $(1, 2]$; graph concave downward for $x < 1$; graph concave upward for $x > 1$ **29.** $f(0) = -1$, rel max; no pts. of infl.; f increasing on $(-\infty, -1)$ and $(-1, 0]$; f decreasing on $[0, 1)$ and $(1, +\infty)$; graph concave upward for $x < -1$ and $x > 1$; graph concave downward for $-1 < x < 1$ **31.** $f(-1) = -1$, rel min; $f(1) = 1$, rel max; $(-\sqrt{3}, -\frac{1}{2}\sqrt{3})$ and $(\sqrt{3}, \frac{1}{2}\sqrt{3})$, pts. of infl.; f decreasing in $(-\infty, -1]$ and $[1, +\infty)$; f increasing on $[-1, 1]$; graph concave downward for $x < -\sqrt{3}$ and $0 < x < \sqrt{3}$; graph concave upward for $-\sqrt{3} < x < 0$ and $x > \sqrt{3}$ **33.** $f(0) = 0$, rel min; $f(1) = 1$, rel max; f decreasing on $(-\infty, 0)$ and $[1, +\infty)$; f increasing on $(0, 1]$; graph concave downward for $x < 0$ and $x > 0$
35. $f(1) = -1$, rel min; f decreasing on $(-\infty, 1]$; f increasing on $[1, +\infty)$; graph concave upward for all x **37.** no relative extrema; $(3, 2)$, pt. of infl.; f increasing on $(-\infty, +\infty)$; graph concave downward for $x > 3$; graph concave upward for $x < 3$ **39.** no relative extrema; $(3, 2)$, pt. of infl. with horizontal tangent; f increasing on $(-\infty, +\infty)$; graph concave downward for $x < 3$; graph concave upward for $x > 3$ **41.** $f(0) = 0$, rel min; $f(\frac{16}{5}) = \frac{512}{125}\sqrt{5}$, rel max; pt. of infl. at $x = \frac{1}{15}(48 - 8\sqrt{6})$; f decreasing on $(-\infty, 0]$ and $[\frac{16}{5}, 4]$; f increasing on $[0, \frac{16}{5}]$; graph concave upward for $x < \frac{1}{15}(48 - 8\sqrt{6})$; graph concave downward for $\frac{1}{15}(48 - 8\sqrt{6}) < x < 4$
43. $f(-\frac{2}{3}) = \frac{4}{9}\sqrt{6}$, rel max; f increasing on $(-\infty, -\frac{2}{3}]$; f decreasing on $[-\frac{2}{3}, 0]$; graph concave downward for $x < 0$

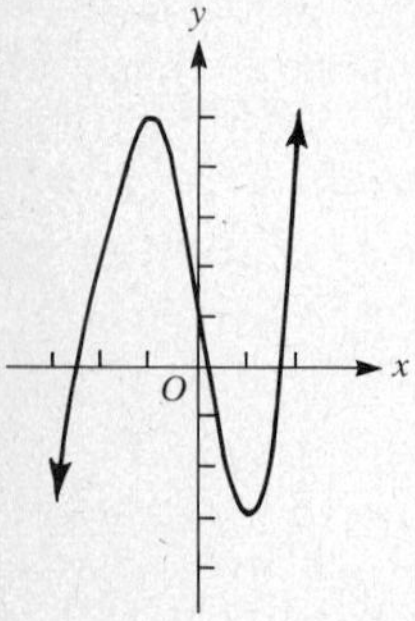

Figure EX 4.9-1

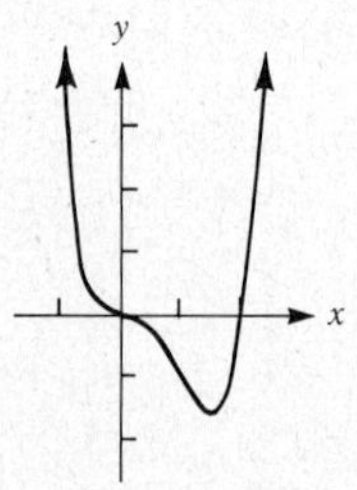

Figure EX 4.9-3

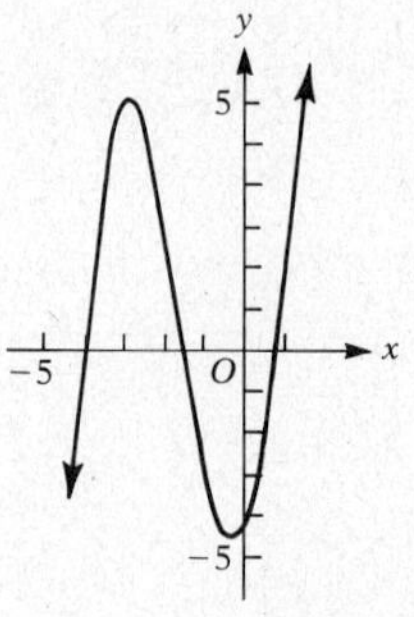

Figure EX 4.9-5

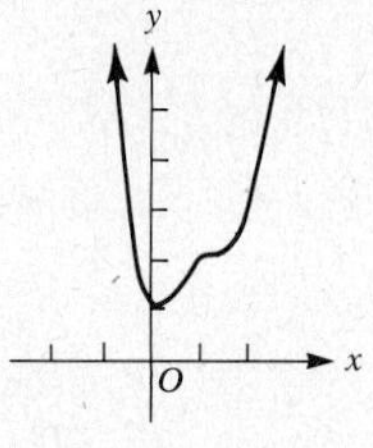

Figure EX 4.9-7

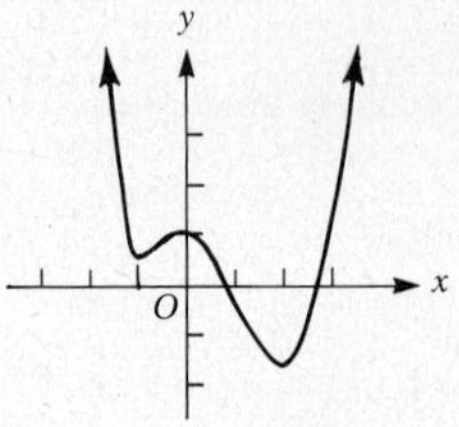

Figure EX 4.9-9

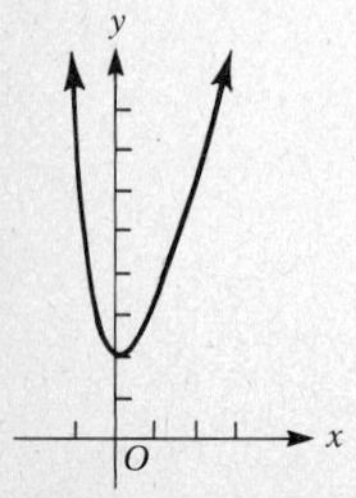

Figure EX 4.9-11

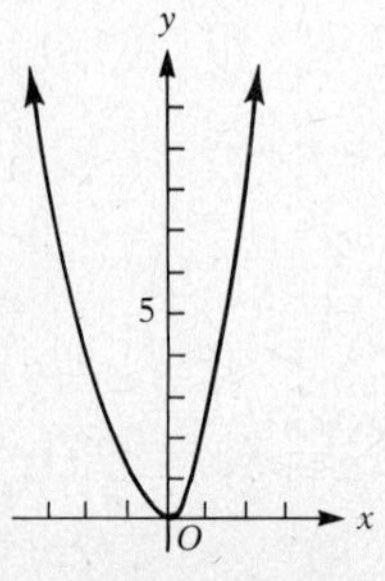

Figure EX 4.9-13

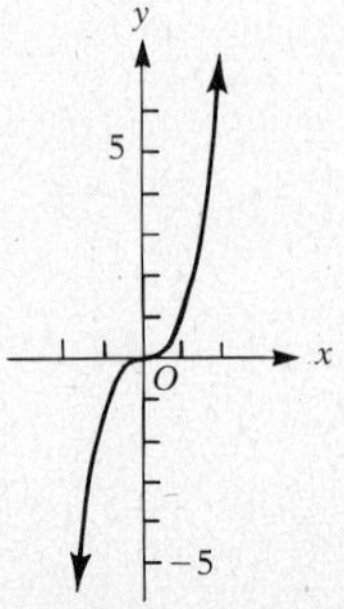

Figure EX 4.9-15

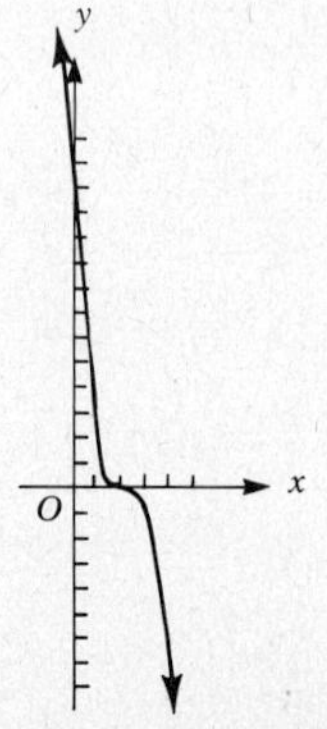

Figure EX 4.9-17

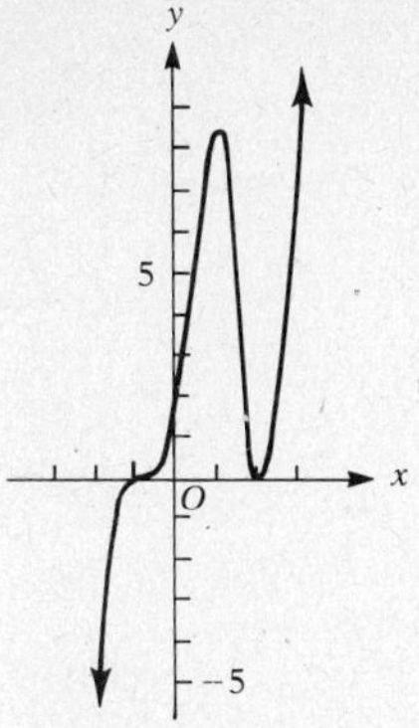

Figure EX 4.9-19

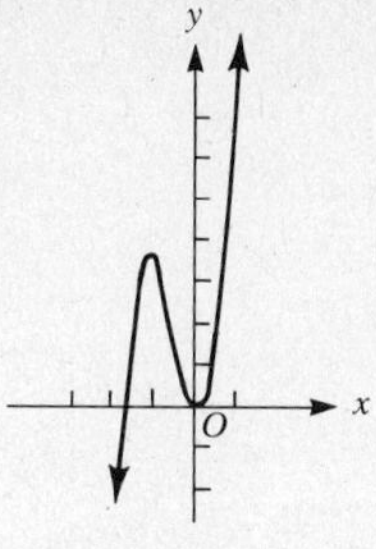

Figure EX 4.9-21

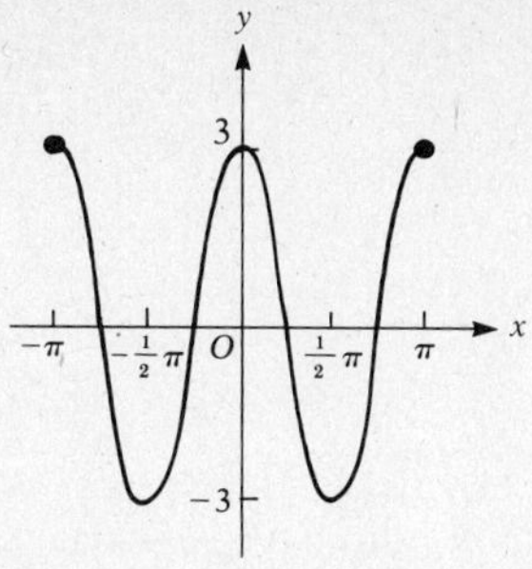

Figure EX 4.9-23

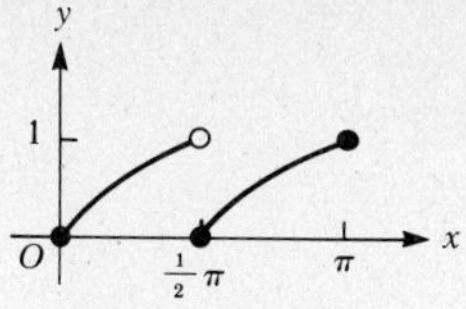

Figure EX 4.9-25

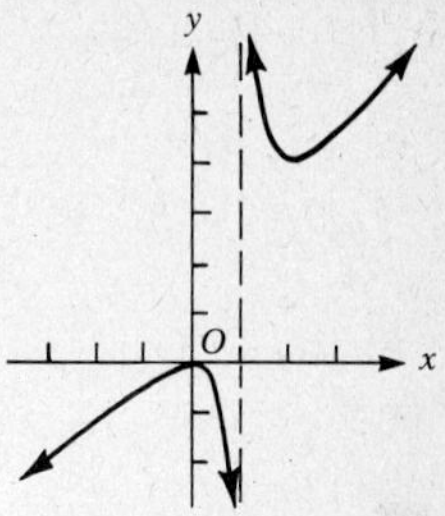

Figure EX 4.9-27

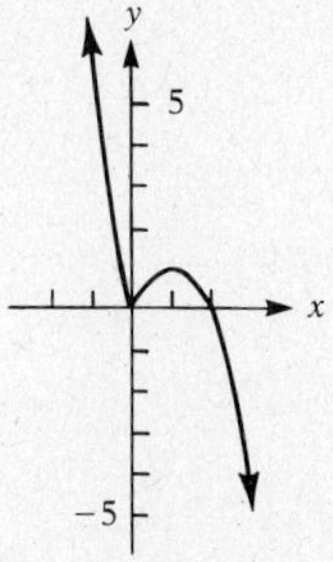

Figure EX 4.9-33

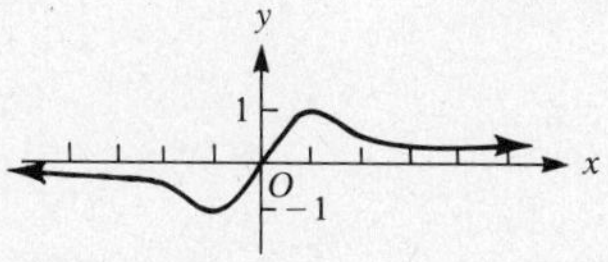

Figure EX 4.9-31

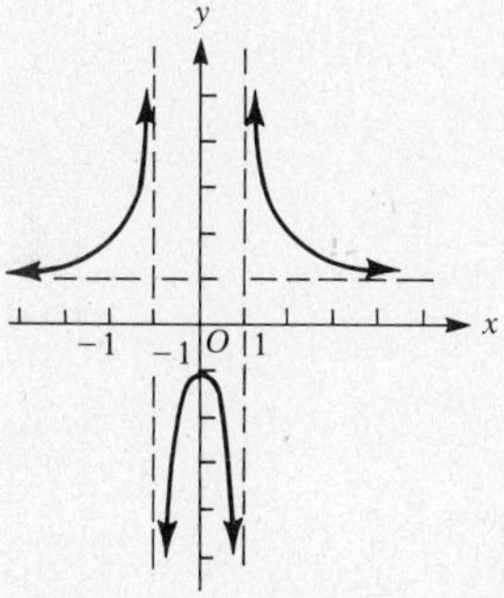

Figure EX 4.9-29

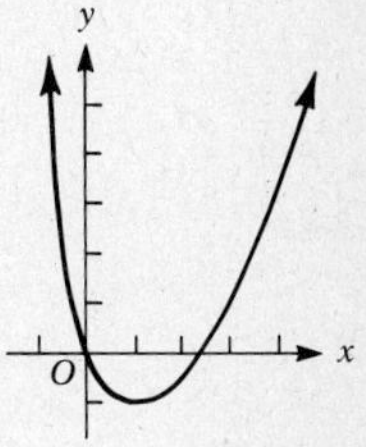

Figure EX 4.9-35

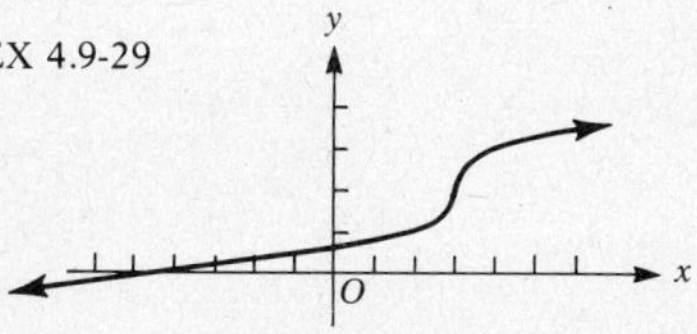

Figure EX 4.9-37

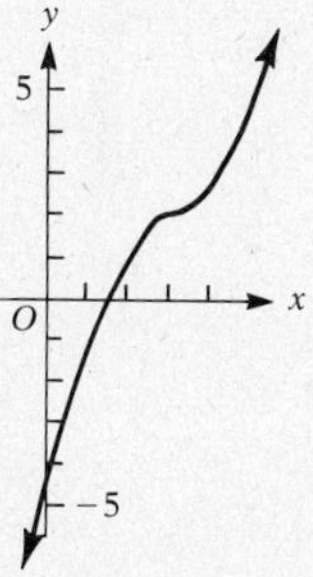

Figure EX 4.9-39

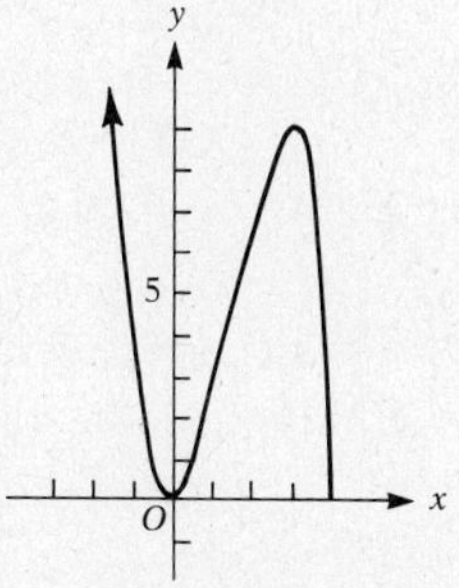

Figure EX 4.9-41

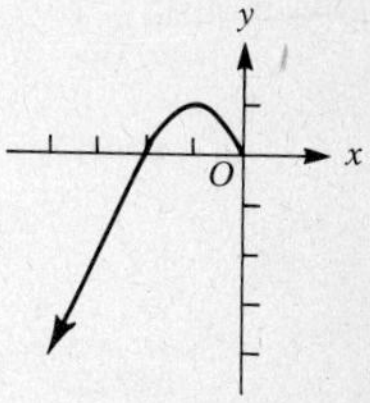

Figure EX 4.9-43

EXERCISES 4.10 (Page 285)

1. (a) $C'(x) = 3 + 2x$; (b) \$83; (c) \$84 **3.** (a) 50 cents per liter; (b) 25 **5.** (a) $Q(x) = x + 4 + \frac{8}{x}$; (b) $C'(x) = 2x + 4$; (c) \$9.66 **7.** (a) $Q(x) = 3x - 6 + \frac{4}{x}$; (b) $C'(x) = 6x - 6$; (c) $[1, +\infty)$; (d) \$0.93 **9.** (a) $[2, +\infty)$; (b) $C'(x) = x^2 - 4x + 5$; (c) decreasing on $[0, 2]$ and increasing on $[2, +\infty)$; (d) graph concave downward for $0 < x < 2$, graph concave upward for $x > 2$, pt. of infl. at $(2, \frac{20}{3})$, infl. tangent: $3x - 3y + 14 = 0$ **11.** (a) $C(x) = 3x + 400$; (b) $Q(x) = 3 + \frac{400}{x}$; (c) $C'(x) = 3$; (e) 953 **13.** (a) $R'(x) = 600 - \frac{3}{20}x^2$; (b) \$540; (c) \$536.95 **15.** (a) $R(x) = 100x\sqrt{36 - x^2}$; (b) $R'(x) = \frac{3600 - 200x^2}{\sqrt{36 - x^2}}$; (c) \$1800 **17.** 80, \$3400 **19.** (a) $P(x) = -2x^2 + 120x - 300$; (c) $R'(x) = 140 - 2x$, $C'(x) = 2x + 20$; (d) \$1500 **21.** (a) $R'(x) = 6 - \frac{1}{5}\sqrt{x - 100} - \frac{x}{10\sqrt{x - 100}}$; $C'(x) = 2$; (b) 200 or 100 **23.** (a) 1875; (b) 62.5 cents; (c) \$503.12 **25.** (a) 1812; (b) 63.8 cents; (c) \$397.30 **27.** (a) \$8.67; (b) \$9.18

REVIEW EXERCISES FOR CHAPTER 4 (Page 287)

1. abs min: $f(-5) = 0$ **3.** abs min: $f(1) = -\frac{1}{2}$; abs max: $f(2) = 64$ **5.** abs min: $f(3) = 0$; abs max: $f(0) = 9$ **7.** abs min: $f(\sqrt{6}) = 0$; abs max: $f(5) = 361$ **9.** abs min: $f(-\frac{1}{6}\pi) = -2$; abs max: $f(\frac{1}{6}\pi) = 2$ **11.** $f(-2) = 0$, rel max; $f(0) = -4$, rel min; $(-1, -2)$, pt. of infl.; f increasing on $(-\infty, -2]$ and $[0, +\infty)$; f decreasing on $[-2, 0]$; graph concave downward for $x < -1$; graph concave upward for $x > -1$ **13.** $f\left(\frac{8}{5}\right) = \frac{839{,}808}{3125}$, rel max; $f(4) = 0$, rel min; pts. of infl. at $x = -2$ and $x = \frac{1}{5}(8 \pm 3\sqrt{6})$; f increasing on $(-\infty, \frac{8}{5}]$ and $[4, +\infty)$; f decreasing on $[\frac{8}{5}, 4]$; graph concave upward for $-2 < x < \frac{1}{5}(8 - 3\sqrt{6})$ and $\frac{1}{5}(8 + 3\sqrt{6}) < x < +\infty$; graph concave downward for $-\infty < x < -2$ and $\frac{1}{5}(8 - 3\sqrt{6}) < x < \frac{1}{5}(8 + 3\sqrt{6})$ **15.** no relative extrema; pt. of infl. at $x = 3$; f increasing on $(-\infty, +\infty)$; graph concave upward for $3 < x < +\infty$; graph concave downward for $-\infty < x < 3$ **17.** $f(0) = 0$, rel max; $f(6) = 12$, rel min; no pts. of infl.; f increasing on $(-\infty, 0]$ and $[6, +\infty)$; f decreasing on $[0, 3)$ and $(3, 6]$; graph concave downward for $x < 3$; graph concave upward for $x > 3$ **19.** $f(-\frac{1}{8}\pi) = -\sqrt{2}$, rel min; $f(\frac{3}{8}\pi) = \sqrt{2}$, rel max; $(\frac{1}{8}\pi, 0)$, pt. of infl.; f decreasing on $[-\frac{3}{8}\pi, -\frac{1}{8}\pi]$ and $[\frac{3}{8}\pi, \frac{5}{8}\pi]$; f increasing on $[-\frac{1}{8}\pi, \frac{3}{8}\pi]$; graph concave upward for $-\frac{3}{8}\pi < x < \frac{1}{8}\pi$; graph concave downward for $\frac{1}{8}\pi < x < \frac{5}{8}\pi$ **21.** 6, 6 **23.** $\frac{1}{2}\sqrt{17}$ **27.** $\frac{3}{2}$ sec; velocity of particle moving horizontally is 1 cm/sec; velocity of particle moving vertically is 3 cm/sec **29.** 10.6 knots **31.** $\dfrac{512}{625\pi}$ in./sec ≈ 0.26 in./sec **33.** $\frac{1}{3}(1 - \sqrt{13})$ **35.** $-\frac{13}{4}$ **41.** $\sqrt{A^2 + B^2}$ **43.** pt. of infl. at $(0, 2)$; graph concave downward for $x < 0$; graph concave upward for $x > 0$ **45.** $a = 2, b = -6, c = 3$ **51.** 12 km from point on bank nearest A **53.** $(h^{2/3} + w^{2/3})^{3/2}$ meters **55.** radius is $\frac{3}{2}r$ cm and altitude is $3h$ cm **57.** 9 m by 18 m **59.** the wire should be cut in half **61.** (a) increasing at a rate of \$10,400 per year; (b) increasing at a rate of \$800 per year **63.** (a) $P(x) = \dfrac{36 - 4x}{x + 3}$; (b) $R(x) = \dfrac{3600x - 400x^2}{x + 3}$; (c) $R'(x) = -\dfrac{400(x + 9)(x - 3)}{(x + 3)^2}$; (d) \$1200 **65.** 1000, \$11

EXERCISES 5.1 (Page 301)

3. (a) $2x\,\Delta x + (\Delta x)^2$; (b) $2x\,\Delta x$; (c) $(\Delta x)^2$ **5.** (a) $(-3 - 4x)\,\Delta x - 2(\Delta x)^2$; (b) $(-3 - 4x)\,\Delta x$; (c) $-2(\Delta x)^2$ **7.** (a) $\sqrt{x + \Delta x} - \sqrt{x}$; (b) $\dfrac{\Delta x}{2\sqrt{x}}$; (c) $-\dfrac{(\sqrt{x} - \sqrt{x + \Delta x})^2}{2\sqrt{x}}$ **9.** (a) $\dfrac{-2\,\Delta x}{(x - 1)(x + \Delta x - 1)}$; (b) $\dfrac{-2\,\Delta x}{(x - 1)^2}$; (c) $\dfrac{2(\Delta x)^2}{(x - 1)^2(x + \Delta x - 1)}$ **11.** (a) 0.0309; (b) 0.03; (c) 0.0009 **13.** (a) $\frac{1}{42} \approx 0.0238$; (b) $\frac{1}{40} = 0.025$; (c) $-\frac{1}{840} \approx -0.0012$ **15.** (a) -0.875; (b) -1.5; (c) 0.625 **17.** $3(3x^2 - 2x + 1)^2(6x - 2)\,dx$ **19.** $\dfrac{(14x^2 + 18x)\,dx}{3(2x + 3)^{2/3}}$ **21.** $\dfrac{(1 - 2\sin x + 2\cos x)\,dx}{(2 - \sin x)^2}$ **23.** $\dfrac{dx}{(x + 1)^{3/2}(x - 1)^{1/2}}$ **25.** $-\dfrac{3x}{4y}$ **27.** $-\dfrac{\sqrt{y}}{\sqrt{x}}$ **29.** 1 **31.** $18t^5 - 56t^3 + 38t$ **33.** $6t^5 - 4t^3 - 2t - 9t(t^2 + 1)^{1/2} + 6t(t^2 + 1)^{-1/2}$ **35.** $\dfrac{8t(16t^4 - 8t^2y + 8t^2 - 2y + 1)}{16t^4 + 8t^2 - y^2 + 1}$ **37.** (a) $dx = 4t\,dt$, $dy = 3\,dt$; (b) $\dfrac{3}{4t}$; (c) $\dfrac{9}{4y}$ **39.** (a) $dx = -2\sin t\,dt$, $dy = 3\cos t\,dt$; (b) $-\dfrac{3\cos t}{2\sin t}$ (c) $-\dfrac{9x}{4y}$ **41.** 6.125 **43.** 3.009 **45.** 0.0993 **47.** -0.849 **49.** (a) 6.75 cm³; (b) 0.3 cm² **51.** $\frac{12}{5}\pi$ m³ **53.** 10 ft³ **55.** 4%

EXERCISES 5.2 (Page 312)

1. $\frac{3}{5}x^5 + C$ **3.** $3t - t^2 + \frac{1}{3}t^3 + C$ **5.** $\frac{8}{5}x^5 + x^4 - 2x^3 - 2x^2 + 5x + C$ **7.** $\frac{2}{5}x^{5/2} + \frac{2}{3}x^{3/2} + C$ **9.** $\frac{2}{5}x^{5/2} - \frac{1}{2}x^2 + C$ **11.** $-\dfrac{1}{x^2} - \dfrac{3}{x} + 5x + C$ **13.** $\frac{2}{5}x^{5/2} + \frac{8}{3}x^{3/2} - 8x^{1/2} + C$ **15.** $\frac{2}{3}x\sqrt{2x} - \sqrt{2x} + C$ **17.** $-\frac{1}{6}(1 - 4y)^{3/2} + C$ **19.** $-\frac{3}{8}(6 - 2x)^{4/3} + C$ **21.** $\frac{1}{3}(x^2 - 9)^{3/2} + C$ **23.** $\frac{2}{9}(x^3 - 1)^{3/2} + C$ **25.** $-\frac{3}{8}(9 - 4x^2)^{5/3} + C$ **27.** $\dfrac{1}{32(1 - 2y^4)^4} + C$ **29.** $\frac{1}{3}\sqrt{3s^2 + 1} + C$ **31.** $\frac{2}{3}(t + 3)^{3/2} - 6(t + 3)^{1/2} + C$ **33.** $-2\left(1 + \dfrac{1}{3x}\right)^{3/2} + C$ **35.** $-\frac{2}{7}(3 - x)^{7/2} + \frac{12}{5}(3 - x)^{5/2} - 6(3 - x)^{3/2} + C$ **37.** $\frac{2}{3}\sqrt{x^3 + 3x^2 + 1} + C$ **39.** $-\dfrac{1}{4(3x^4 + 2x^2 + 1)} + C$ **41.** $\frac{3}{4}(3 - y)^{4/3} - 18(3 - y)^{1/3} + C$ **43.** $\frac{3}{5}(r^{1/3} + 2)^5 + C$ **45.** $\sqrt{x^2 + 4} + \dfrac{4}{\sqrt{x^2 + 4}} + C$ **47.** (a) $2x^4 + 4x^3 + 3x^2 + x + C$; (b) $\frac{1}{8}(2x + 1)^4 + C$ **49.** (a) $\frac{2}{3}x^{3/2} - 2x + 2x^{1/2} + C$; (b) $\frac{2}{3}(\sqrt{x} - 1)^3 + C$ **51.** g is not differentiable on $(-1, 1)$

EXERCISES 5.3 (Page 318)

1. $-\frac{1}{4}\cos 4x + C$ **3.** $8\sin\frac{1}{2}x + C$ **5.** $-3\cos t + \frac{1}{3}\sin 3t + C$ **7.** $\frac{1}{4}(2 + \sin x)^4 + C$ **9.** $-\frac{3}{2}(1 + \cos x)^{4/3} + C$ **11.** $-\frac{1}{3}\sqrt{1 - 2\sin 3x} + C$ **13.** $\cos(\cos x) + C$ **15.** $-\frac{1}{3}\cos^3 t + C$ **17.** $\frac{1}{3}\cos^3 x - \cos x + C$ **19.** $\frac{3}{8}z - \frac{1}{4}\sin 2z + \frac{1}{32}\sin 4z + C$ **21.** $\frac{1}{2}x + \frac{1}{2}\sin x + C$ **23.** $\frac{1}{3}\sin^3 x - \frac{1}{5}\sin^5 x + C$ **25.** $-\frac{1}{3}\cos^3 x + \frac{2}{5}\cos^5 x - \frac{1}{7}\cos^7 x + C$ **27.** $\frac{1}{8}t - \frac{1}{96}\sin 12t + C$ **29.** $\frac{1}{2}\sin^{2/3} 3x - \frac{1}{8}\sin^{8/3} 3x + C$ **31.** $\frac{1}{14}\sin 7x + \frac{1}{2}\sin x + C$ **33.** $-\frac{1}{16}\cos 8y + \frac{1}{4}\cos 2y + C$ **35.** $t - \sin t - \frac{1}{8}\sin 4t + \frac{1}{5}\sin 5t - \frac{1}{12}\sin 6t + C$ **37.** (a) $\sin^2 x + C$; (b) $-\cos^2 x + C$; (c) $-\frac{1}{2}\cos 2x + C$

EXERCISES 5.4 (Page 324)

1. $y = 2x^2 - 5x + C$ **3.** $y = x^3 + x^2 - 7x + C$ **5.** $u = 2v + \frac{3}{v} + C$ **7.** $y = -\frac{3}{2}\cos 2x + C$ **9.** $3x^2y + Cy + 2 = 0$ **11.** $2\sqrt{1 + y^2} = 3x^2 + C$ **13.** $2y - \sin 2y = 4\sin x + C$ **15.** $12y = 5x^4 + 6x^2 + C_1x + C_2$ **17.** $u = \frac{4}{135}(3v + 1)^{5/2} + C_1v + C_2$ **19.** $3y = -6\sin x - \sin^3 x + 3\cos x + C_1x + C_2$ **21.** $3y = x^3 - 3x^2 - 12x + 18$ **23.** $x^2 = 4y^2$ **25.** $4\sin 3x + 6\cos 2y + 7 = 0$ **27.** $12y = x^4 + 6x^3 - 10x + 27$ **29.** $u = 3v^4 + 4v^3 + 2v^2 + 2v$ **31.** $y = x^2 - 3x + 2$ **33.** $3y = -2x^3 + 3x^2 + 2x + 6$ **35.** $12y = -x^4 + 6x^2 - 20x + 27$ **37.** \$325 **39.** 117π m³ **41.** $x^2 + 2y^2 = C$

EXERCISES 5.5 (Page 328)

1. $v = \frac{1}{3}(2t + 4)^{3/2} - \frac{8}{3}$ **3.** $v = 2 + 5t - t^2$, $s = 2t + \frac{5}{2}t^2 - \frac{1}{3}t^3$ **5.** $v = \frac{1}{3}t^3 + t^2 - 4$, $s = \frac{1}{12}t^4 + \frac{1}{3}t^3 - 4t + 1$ **7.** $v = -2\sqrt{2}\sin(2t - \frac{1}{4}\pi)$, $s = \sqrt{2}\cos(2t - \frac{1}{4}\pi)$ **9.** $1600s = v^2 + 1200$ **11.** $5s^2 + 4s = v^2 + 12$ **13.** (a) $\frac{5}{4}$ sec; (b) 20 ft/sec; (c) $\frac{5}{8}$ sec; (d) $\frac{25}{4}$ ft **15.** (a) $\frac{1}{4}\sqrt{555}$ sec; (b) $8\sqrt{555}$ ft/sec **17.** (a) 3.4 sec; (b) 99 ft/sec **19.** (a) $v^2 = -64s + 1600$; (b) 24 ft/sec **21.** 1.62 m/sec² **23.** (a) 3.47 sec; (b) 48.22 m **25.** 20 m/sec or 72 km/hr **27.** (a) 3 m/sec; (b) 9 m/sec; (c) 2 sec

EXERCISES 5.6 (Page 332)

1. (a) $R(x) = 12x - \frac{3}{4}x^2$; (b) $p = 12 - \frac{3}{4}x$ **3.** $C(x) = x^2 + x + 6$ **5.** $C(x) = \frac{6}{5}\sqrt{5x + 4} + \frac{38}{5}$ **7.** $p(x + 5) = 4x + 22$ **9.** $C(x) = 3x^2 - 17x + 47$ **11.** (a) $R(x) = 16x - x^3$; (b) $p = 16 - x^2$ **13.** $C(x) = x^3 + 4x^2 + 4x + 6$ **15.** \$3375 **17.** (a) \$2.50; (b) \$3 **19.** \$5 **21.** \$3.25 million

REVIEW EXERCISES FOR CHAPTER 5 (Page 333)

1. $\frac{1}{2}x^4 - \frac{1}{3}x^3 + 3x + C$ **3.** $2y^2 + 4y^{3/2} + C$ **5.** $-\frac{1}{3}\cos 3t + C$ **7.** $-\frac{1}{3}\cos^3 x + C$ **9.** $-\frac{2}{3x^3} + \frac{5}{x} + C$ **11.** $\frac{5}{9}(2 + 3x^2)^{3/2} + C$ **13.** $\frac{2}{3}x\sqrt{3x} + \frac{2}{5}\sqrt{5x} + C$ **15.** $\frac{1}{2}\sqrt{x^4 + 2x^2} + C$ **17.** $\frac{1}{6}(2s + 3)^{3/2} - \frac{3}{2}(2s + 3)^{1/2} + C$ **19.** $\frac{3}{8}t - \frac{1}{2}\sin t + \frac{1}{16}\sin 2t + C$ **21.** $\frac{1}{16}\cos^8 2x - \frac{1}{12}\cos^6 2x + C$ **23.** $\frac{1}{420}(4x + 3)^{3/2}(30x^2 - 18x + 79) + C$ **25.** $\frac{1}{2}(y^2 - 1)^{-1} = x^{-1} + C$ **27.** $y + y^{-1} = -x + C$ **29.** $y = \frac{1}{15}(2x - 1)^{5/2} + C_1x + C_2$ **31.** (a) -0.16; (b) -0.64 **33.** 0.4125 **35.** 2π in.³ **37.** $x^4 - 24x^2 + 36x + 12y - 1 = 0$ **39.** 0 **41.** $\frac{dy}{dt} = \left(\frac{4y - 24x^2}{81y^2 - 4x}\right)\left(\frac{3t^2 + 2tx}{3x^2 - t^2}\right)$ **43.** (a) $\frac{1}{9}(x^3 + 1)^3 + C$; (b) $\frac{1}{9}x^9 + \frac{1}{3}x^6 + \frac{1}{3}x^3 + C$ **45.** $y = \frac{4}{15}(x + 4)^{5/2} + \frac{1}{3}(6 - 32\sqrt{2})x - \frac{1}{15}(75 - 128\sqrt{2})$ **47.** $v = 3\sin 2t + 3$, $s = -\frac{3}{2}\cos 2t + 3t + \frac{1}{2}(5 - 3\pi)$ **49.** $\frac{25\sqrt{817} - 625}{8}$ sec ≈ 11.2 sec **51.** (a) 1 sec; (b) -80 ft/sec **53.** 4738 units **55.** an allowable error of $\frac{5}{24}$ in. **57.** an error of approximately $\frac{\pi^2}{1610t}$ sec **61.** $C(x) = \frac{8}{3}\sqrt{3x + 1} + \frac{28}{3}$

EXERCISES 6.1 (Page 347)

1. 51 **3.** $\frac{73}{12}$ **5.** $\frac{63}{4}$ **7.** $\frac{7}{12}$ **13.** 10,400 **15.** $10(10^n - 1)$ **17.** $\frac{100}{101}$ **19.** $n^4 - \frac{2}{3}n^3 - 3n^2 - \frac{4}{3}n$ **21.** $\frac{8}{3}$ sq units **23.** 15 sq units **25.** $\frac{5}{3}$ sq units **27.** 9 sq units **29.** $\frac{3}{5}$ sq units **31.** $\frac{17}{4}$ sq units **33.** $\frac{27}{4}$ sq units **35.** $\frac{1}{2}m(b^2 - a^2)$ sq units **37.** $\frac{1}{2}h(b_1 + b_2)$ sq units

EXERCISES 6.2 (Page 356)

1. $\frac{247}{32}$ **3.** $\frac{1469}{1320}$ **5.** 0.835 **7.** $\frac{1}{24}(10 + \sqrt{2} + 3\sqrt{3})\pi$ **9.** 9 **11.** $\frac{8}{3}$ **13.** $\frac{15}{4}$ **15.** 66 **17.** 4 **19.** 20 sq units **21.** 44 sq units **23.** $\frac{5}{3}$ sq units **25.** $\frac{305}{6}$ sq units **27.** $\frac{31}{4}$ sq units **29.** $\int_0^2 x^2\,dx$ **31.** $\int_0^1 \frac{1}{x^2}\,dx$

EXERCISES 6.3 (Page 367)

1. 12 **3.** $4\sqrt{5}$ **5.** -5 **7.** 15 **9.** 0 **11.** -21 **13.** $-\frac{3}{2}$ **15.** $4 + \pi$ **17.** $\frac{33}{2}\pi$ **19.** [24, 56] **21.** [0, 64] **23.** [0, 27] **25.** [0, 576] **27.** $[\frac{1}{12}\pi, \frac{1}{12}\sqrt{3}\pi]$ **29.** [0, 6] **31.** $[-3, \frac{3}{2}]$ **33.** $[-5\pi, 0]$ **35.** $\geq$ **37.** $\leq$

EXERCISES 6.4 (Page 371)

1. $\frac{2}{3}\sqrt{3}$ **3.** $\frac{1}{2}\sqrt[3]{30}$ **5.** $-2 + \sqrt{21}$ **7.** 0 **17.** average value is $\frac{1}{2}$ at $x = \frac{1}{2}$ **19.** $\frac{2}{\pi}$ **21.** $v = 32t$; 32 **23.** π

EXERCISES 6.5 (Page 381)

1. 12 **3.** 36 **5.** $\frac{3}{2}$ **7.** $\frac{3}{16}$ **9.** $\frac{134}{3}$ **11.** -8 **13.** 1 **15.** $\frac{2}{9}(27 - 2\sqrt{2})$ **17.** $2 - \sqrt[3]{2}$ **19.** $\frac{104}{5}$ **21.** $\frac{29}{2}$ **23.** $\frac{2}{3}\sqrt{2}$ **25.** $\frac{256}{15}$ **27.** $\frac{5}{6}$ **29.** $\frac{6215}{12}$ **31.** $\frac{11}{6}$ **33.** $\frac{1}{8}$ **35.** $\frac{47}{480}$ **37.** $\sqrt{4 + x^6}$ **39.** $-\sqrt{\sin x}$ **41.** $\dfrac{2}{3 + x^2}$ **43.** $\frac{32}{3}$ sq units **45.** $\frac{22}{3}$ sq units **47.** $\frac{52}{3}$ sq units **49.** 2 **51.** average value is $\frac{32}{3}$ at $x = \frac{4}{3}(3 - \sqrt{3})$ **53.** $\dfrac{42{,}304}{175}$ **55.** $\dfrac{4}{\pi}$ **59.** $\frac{188}{3}$

EXERCISES 6.6 (Page 392)

1. approx: 4.250; exact: 4 **3.** 0.696 **5.** approx: 0; exact: 0 **7.** 0.880 **9.** 0.248 **11.** 3.689 **13.** $-0.5 \le \epsilon_T \le 0$ **15.** $-0.007 \le \epsilon_T \le -0.001$ **17.** $0 \le \epsilon_T \le 0.072$ **19.** (a) 4; (b) 4 **21.** 0.693 **23.** 0.6045 **25.** 0 **27.** $-0.0005 \le \epsilon_s \le 0$ **29.** 0.237 **31.** 1.568 **33.** 1.402 **35.** 3.090 **37.** (a) 56; (b) 56 **39.** (a) 222; (b) 222

REVIEW EXERCISES FOR CHAPTER 6 (Page 393)

1. 6 **3.** $-\frac{27}{4}$ **5.** 0 **7.** $\frac{2}{3}(\sqrt{14} - \sqrt{2})$ **9.** $\frac{2}{3}\sqrt{3} - 1$ **11.** $\frac{2}{27}(20\sqrt{10} + \frac{64}{5})$ **13.** $\frac{652}{15}$ **15.** $\frac{313}{2}$ **17.** $\frac{1}{2}$ **19.** 12; $\sqrt{3}$ **21.** 4, 100, 656, 560 **27.** $[3\sqrt{5}, 9]$ **29.** $[0, \pi]$ **31.** $-(3x^2 - 4)^{3/2}$ **33.** 18 sq units **35.** $\frac{224}{3}$ sq units **37.** $\frac{1}{3}(40\sqrt{5} - 20)$ sq. units **39.** 0 **41.** $\dfrac{42{,}304}{175}$ **45.** \$1197. **47.** 1.624 **49.** 1.563 **51.** $\sqrt{3} + \frac{1}{6}\pi$ **53.** (a) $\dfrac{1}{r}$; (b) 0

EXERCISES 7.1 (Page 403)

1. $\frac{32}{3}$ sq units **3.** $\frac{32}{3}$ sq units **5.** $\frac{1}{6}$ sq units **7.** $\frac{12}{5}$ sq units **9.** $\frac{9}{2}$ sq units **11.** $\frac{8}{3}\sqrt{2}$ sq units **13.** $\frac{5}{12}$ sq units **15.** $\frac{27}{10}$ sq units **17.** $\frac{64}{3}$ sq units **19.** $\frac{253}{12}$ sq units **21.** $\frac{37}{12}$ sq units **23.** $(\sqrt{2} - 1)$ sq units **25.** $\frac{7}{3}$ sq units **27.** 12 sq units **29.** $\frac{128}{5}$ sq units **31.** 64 sq units **33.** $\left(\dfrac{\pi}{2} - 1\right)$ sq units **35.** $\frac{16}{3}p^2$ sq units **37.** 32 **39.** $-\dfrac{8}{m^4}$ **41.** $\frac{3}{2}K$

EXERCISES 7.2 (Page 409)

1. 64π cu units **3.** $\frac{704}{5}\pi$ cu units **5.** $\frac{384}{7}\pi$ cu units **7.** $\frac{3456}{35}\pi$ cu units **9.** $\frac{256}{15}\pi$ cu units **11.** $\frac{128}{5}\pi$ cu units **13.** $\frac{128}{7}\pi$ cu units **15.** $\frac{4}{3}\pi r^3$ cu units **17.** $\frac{1}{2}\pi^2$ cu units **19.** $\frac{5}{8}\pi^2$ cu units **21.** $\frac{1250}{3}\pi$ cu units **23.** $\frac{64}{5}\pi$ cu units **25.** $\frac{261}{32}\pi$ cu units **27.** 16π cu units **29.** 180π cm³ **31.** $\frac{16}{3}\pi$ cu units **33.** 2

EXERCISES 7.3 (Page 414)

13. $\frac{1}{2}\pi$ cu units **15.** $\frac{3}{10}\pi$ cu units **17.** $\frac{5}{6}\pi$ cu units **19.** $\frac{49}{30}\pi$ cu units **21.** 16π cu units **23.** $\frac{512}{15}\pi$ **25.** $\frac{32}{15}\pi a^3$ cu units **27.** $\frac{32}{15}\pi$ cu units **29.** $\frac{1}{2}(2 + \sqrt{2})\pi$ cu units **31.** $\frac{7}{2}\pi$ cu units **33.** $\frac{224}{3}\pi$ cu units **35.** 24π cu units **37.** $\sqrt[3]{2744}$

EXERCISES 7.4 (Page 418)

1. $32\sqrt{2}$ cu units **3.** $\frac{1372}{3}\sqrt{3}$ cm³ **5.** $\frac{686}{3}$ cm³ **7.** $\frac{8}{3}r^3$ cu units **9.** 14 in.³ **11.** $\frac{128}{3}$ cu units **13.** 1944 cm³ **15.** $\frac{3969}{10}$ cu units **17.** $\frac{2}{3}r^3$ cm³ **19.** $\pi r^2 h$ cu units **21.** $\frac{1}{3}\pi h(r_1^2 + r_2^2 + r_1 r_2)$ cu units

EXERCISES 7.5 (Page 424)

1. $\frac{158}{3}$ ft-lb **3.** $\frac{1076}{15}$ joules **5.** $\frac{1}{2}(3 + \sqrt{393})$ **7.** 180 in.-lb **9.** 8 joules **11.** 1350 ergs **13.** $6562.5w$ ft-lb **15.** $256\pi w$ ft-lb **17.** 100,000 ft-lb **19.** 5500 ft-lb **21.** $1{,}017{,}938\pi$ joules **23.** $9{,}196{,}875\pi$ joules **25.** $\frac{144}{55}w$ sec **27.** $2\sqrt{3}$ ft

EXERCISES 7.6 (Page 431)

1. $\frac{14}{3}$ **3.** $\frac{33}{16}$ **5.** $\frac{1}{27}(97^{3/2} - 125)$ **7.** 12 **9.** $\frac{22}{3}$ **11.** $\frac{9}{8}$ **13.** $\dfrac{8a^3 - (a^2 + 3b^2)^{3/2}}{8(a^2 - b^2)}$ **15.** $2\sqrt{3} - \frac{4}{3}$ **17.** $2\sqrt{2}$ **19.** $\displaystyle\int_0^{\pi} \sqrt{1 + \cos^2 x}\, dx$; $\frac{1}{6}\pi(2\sqrt{6} + \sqrt{2} + 1) \approx 3.829$

REVIEW EXERCISES FOR CHAPTER 7 (Page 432)

1. $\frac{1}{12}$ sq unit **3.** 36 sq units **5.** $\frac{37}{12}$ sq units **7.** 250π cu units **9.** 1024 cu units **11.** $\frac{7}{3}$ sq units **13.** 3π cu units **15.** $\frac{16}{3}\sqrt{3}$ **17.** $\frac{25}{6}\pi$ cu units **19.** $2\sqrt{2}$ sq units **21.** 6000 ergs **23.** 400 ft-lb **25.** 44,145,000 joules **27.** $\frac{2752}{3}w\pi$ ft-lb

29. $\frac{13{,}600}{3} w\pi$ ft-lb **31.** 90 ft³ **33.** 558π cm³ **35.** $\sqrt[7]{2744}$ **37.** $\frac{1}{5320}[(2251)^{3/2} - \frac{1}{8}(10999)^{3/2}]$ **39.** $\frac{1}{4}\pi^2$ cu units

EXERCISES 8.1 (Page 443)

1. one-to-one **3.** not one-to-one **5.** one-to-one **7.** one-to-one **9.** one-to-one **11.** one-to-one **13.** not one-to-one **15.** $f^{-1}(x) = \frac{1}{5}(x + 7)$; domain: $(-\infty, +\infty)$, range: $(-\infty, +\infty)$ **17.** no inverse **19.** $f^{-1}(x) = 4 - \sqrt[3]{x}$; domain: $(-\infty, +\infty)$, range: $(-\infty, +\infty)$ **21.** $h^{-1}(x) = \frac{1}{2}x^2 + 3$; domain: $[0, +\infty)$, range: $[3, +\infty)$ **23.** $F^{-1}(x) = x^3 - 1$; domain: $(-\infty, +\infty)$, range: $(-\infty, +\infty)$ **25.** no inverse **27.** $f^{-1}(x) = \frac{1}{32}x^5$; domain: $(-\infty, +\infty)$, range: $(-\infty, +\infty)$ **29.** $f^{-1}(x) = \dfrac{x+3}{1-x}$; domain: $\{x \mid x \neq 1\}$; range: $\{y \mid y \neq -1\}$ **31.** $g^{-1}(x) = \sqrt{x-5}$; domain: $[5, +\infty)$, range: $[0, +\infty)$ **33.** $f^{-1}(x) = \frac{1}{2}(\sqrt[3]{x} - 1)$; domain: $[0, 8]$, range: $[-\frac{1}{2}, \frac{1}{2}]$ **35.** $F^{-1}(x) = \sqrt{9 - x^2}$; domain: $[0, 3]$, range: $[0, 3]$ **37.** (b) $f^{-1}(x) = \frac{1}{4}(x + 3)$ **39.** (b) $f^{-1}(x) = \sqrt[3]{x-2}$ **41.** (b) $f^{-1}(x) = \dfrac{4x-1}{3-2x}$ **43.** (b) $f_1(x) = x^2 + 4, x \geq 0$; $f_2(x) = x^2 + 4, x \leq 0$ (c) $f_1^{-1}(x) = \sqrt{x-4}$, domain: $[4, +\infty)$; $f_2^{-1}(x) = -\sqrt{x-4}$; domain: $[4, +\infty)$ **45.** (b) $f_1(x) = \sqrt{9-x^2}, 0 \leq x \leq 3$; $f_2(x) = \sqrt{9 - x^2}, -3 \leq x \leq 0$; (c) $f_1^{-1}(x) = \sqrt{9-x^2}$, domain: $[0, 3]$; $f_2^{-1}(x) = -\sqrt{9-x^2}$, domain: $[0, 3]$ **47.** $f^{-1}(x) = \begin{cases} x & \text{if } x < 1 \\ \sqrt{x} & \text{if } 1 \leq x \leq 81 \\ \frac{1}{729}x^2 & \text{if } x > 81 \end{cases}$

EXERCISES 8.2 (Page 452)

11. $\frac{2}{3}$ **13.** $\frac{1}{10}$ **15.** $\frac{1}{12}$ **17.** $\frac{1}{21}$ **19.** -2 **21.** $-\frac{1}{4}$ **23.** $\frac{1}{3}$ **25.** $\dfrac{1}{\sqrt{15}}$ **27.** (b) 6; (c) $\frac{1}{6}$ **29.** (a) $f_1(x) = \sqrt{9-x^2}$; $f_2(x) = -\sqrt{9-x^2}$; (b) neither has an inverse; (c) $D_x y = -\dfrac{x}{y}$, $D_y x = -\dfrac{y}{x}$ **31.** (a) $f(x) = \dfrac{4}{x}$; (b) $f^{-1}(x) = \dfrac{4}{x}$, domain: $\{x \mid x \neq 0\}$; (c) $D_x y = -\dfrac{y}{x}$, $D_y x = -\dfrac{x}{y}$ **33.** (a) $f(x) = \dfrac{2x^2+1}{3x}$; (b) no inverse; (c) $D_x y = \dfrac{4x-3y}{3x}$, $D_y x = \dfrac{3x}{4x-3y}$

EXERCISES 8.3 (Page 462)

1. $\dfrac{5}{4+5x}$ **3.** $\dfrac{5}{8+10x}$ **5.** $\dfrac{6}{3t+1}$ **7.** $\dfrac{6\ln(3t+1)}{3t+1}$ **9.** $-\dfrac{2x}{12-3x^2}$ **11.** $\dfrac{5\cos 5y}{\sin 5y}$ **13.** $-\dfrac{\sin(\ln x)}{x}$ **15.** $-\dfrac{17}{2(2w-5)(3w+1)}$ **17.** $\dfrac{\ln x - 1}{(\ln x)^2}$ **19.** $\dfrac{1-2x-x^2}{3(x+1)(x^2+1)}$ **21.** $\dfrac{1}{2(1+\sqrt{x+1})}$ **23.** $-\dfrac{xy+y}{xy+x}$ **25.** $x + y$ **27.** $\dfrac{4x^2y - xy - 2y}{6xy^2 + x}$ **39.** $2x + 4y + 4\ln 2 - 1 = 0$ **41.** $40x - y = 40$ **43.** $-\frac{1}{2}$ **45.** (a) \$5 per \$1 change in budget; (b) \$688

EXERCISES 8.4 (Page 466)

1. $\dfrac{3x^2}{x^3+1}$ **3.** $-\dfrac{3\sin 3x}{\cos 3x}$ **5.** $2x(x+1)^6(x-1)^2(6x^2-2x-1)$ **7.** $\dfrac{-13x^2-4x+8}{(x-4)^4}$ **9.** $\dfrac{8x^9 - 4x^7 + 15x^2 + 10}{5(x^7+1)^{6/5}}$ **11.** $\dfrac{3(3x+4)}{2(x+1)^{3/2}(x+2)^{3/2}}$ **13.** $\dfrac{x\ln(x^2-1)}{\sqrt{x^2+1}} + \dfrac{2x\sqrt{x^2+1}}{x^2-1}$ **15.** $-\frac{1}{2}\ln|3-2x| + C$ **17.** $\frac{3}{2}\ln(x^2+4) + C$ **19.** $\frac{1}{5}\ln|5x^3-1| + C$ **21.** $\frac{1}{2}\ln|1 + 2\sin t| + C$ **23.** $x^2 + 4\ln|x^2-4| + C$ **25.** $\ln|\ln x| + C$ **27.** $\frac{1}{3}\ln^3 3x + C$ **29.** $\ln|(\ln x)^2 + \ln x| + C$ **31.** $\ln 5$ **33.** $4 + \ln 2$ **35.** $\dfrac{1}{\ln 4}$ **37.** $\ln 4$ sq units ≈ 1.386 sq units **39.** $\frac{1}{4}\ln 5 \approx 0.402$ **41.** $2000\ln 2$ lb/ft² $\approx$ 1386 lb/ft² **43.** $\pi(11 + 8\ln 2)$ cu units ≈ 51.977 cu units

EXERCISES 8.5 (Page 474)

1. $5e^{5x}$ **3.** $-6xe^{-3x^2}$ **5.** $-e^{\cos x}\sin x$ **7.** $e^{2x}\cos e^x + e^x \sin e^x$ **9.** $\dfrac{2e^{2x}(x-1)}{x^3}$ **11.** $\dfrac{4}{(e^x+e^{-x})^2}$ **13.** $2x$ **15.** $x^x(1 + \ln x)$ **17.** $\dfrac{e^x(1-e^y)}{e^y(e^x-1)}$ **19.** $-\dfrac{y^2+2ye^{2x}}{2e^{2x}+3xy}$ **21.** $-\frac{1}{5}e^{2-5x} + C$ **23.** $e^x - e^{-x} + C$ **25.** $\dfrac{1}{6(1-2e^{3x})} + C$ **27.** $e^x - 3\ln(e^x+3) + C$ **29.** e^2 **31.** 2 **33.** $\frac{1}{2}$ **35.** $\frac{1}{2}(e^4 - 1)$ **39.** $f(1) = \dfrac{1}{e}$, rel max; $(2, 2e^{-2})$, pt. of infl.; f increasing on $(-\infty, 1]$; f decreasing on $[1, +\infty)$; graph concave downward for $x < 2$; graph concave upward for $x > 2$ **41.** 2.66 **43.** 2.56

45. $x + 2y = 1 + \ln 2$ **47.** $\frac{1}{2}(e^2 - e^{-2} - 4)$ cu units ≈ 1.627 cu units **49.** -9.17 lb/ft^2 per sec **51.** 0.006 **53.** $(e^3 + \frac{1}{2})$ ft ≈ 20.586 ft **55.** $\frac{1}{2}\pi w(e^{-2} - e^{-8})$ ft-lb

EXERCISES 8.6 (Page 482)

1. $(5 \ln 3)3^{5x}$ **3.** $4^{3t^2}(\ln 4)6t$ **5.** $4^{\sin 2x}(2 \ln 4) \cos 2x$ **7.** $2^{5x}3^{4x^2}(5 \ln 2 + 8x \ln 3)$ **9.** $\dfrac{1}{x^2}\log_{10}\dfrac{e}{x}$ **11.** $\dfrac{\log_a e}{2x\sqrt{\log_a x}}$ **13.** $\dfrac{(\log_{10} e)^2}{(x+1)\log_{10}(x+1)}$ **15.** $-3^{t^2}\sin 3^{t^2}(2t \ln 3)$ **17.** $x^{\sqrt{x}-(1/2)}(1 + \frac{1}{2}\ln x)$ **19.** $z^{\sin z - 1}[z(\ln z)\cos z + \sin z]$ **21.** $(\cos x)^{\sin x - 1}(\cos^2 x \ln(\cos x) - \sin^2 x)$ **23.** $x^{e^x - 1}e^x(x \ln x + 1)$ **25.** $\dfrac{3^{2x}}{2 \ln 3} + C$ **27.** $\dfrac{a^t e^t}{1 + \ln a} + C$ **29.** $\dfrac{10^{x^3}}{3 \ln 10} + C$ **31.** $\dfrac{6^{e^y}}{\ln 6} + C$ **33.** 0.621 **35.** 2.999 **47.** (a) 61 sales per day; (b) 2.26 sales per day **49.** (a) $y = 200 \cdot 2^{t/10}$; (b) \$12,800; (c) \$877 per year **51.** $\left(e - 1 - \dfrac{1}{\ln 2}\right)$ sq units ≈ 0.276 sq units **53.** $\left(\dfrac{4}{\ln 3} - 1\right)\pi$ cu units ≈ 8.297 cu units

EXERCISES 8.7 (Page 490)

1. 68.4 years **3.** \$2734 **5.** $8000\sqrt{2} \approx 11{,}300$ **7.** 43.9 g **9.** 10.6 years **11.** 69.9 **13.** (a) 102 sec; (b) 42.1° **15.** (a) 96%; (b) 66% **17.** 118.7 **19.** (a) \$52.59; (b) 10.52% **21.** 15.9 years **23.** 15,100 years

REVIEW EXERCISES FOR CHAPTER 8 (Page 491)

1. (a) $f^{-1}(x) = \sqrt[3]{x+4}$, domain: $(-\infty, +\infty)$, range: $(-\infty, +\infty)$ **3.** no inverse **5.** $f^{-1}(x) = \dfrac{4}{3-x}$; domain: $\{x \mid x \neq 3\}$; range: $\{y \mid y \neq 0\}$ **7.** (b) $f^{-1}(x) = x^3 - 1$ **9.** $\frac{1}{4}$ **11.** $\frac{1}{12}$ **13.** $\dfrac{8 \ln x}{x}$ **15.** $(-4 \ln 2)2^{\cos 4x}\sin 4x$ **17.** $\dfrac{(4 - x^2)e^{x/(4+x^2)}}{(4 + x^2)^2}$ **19.** $\dfrac{\log_{10} e}{(1 - x^2)\sqrt{\log_{10}\dfrac{1+x}{1-x}}}$ **21.** 0 **23.** $x^{xe^x + e^x - 1}(xe^x \ln^2 x + e^x \ln x + 1)$ **25.** $\frac{3}{2}\ln(1 + e^{2x}) + C$ **27.** $\dfrac{1}{3}\left(e^{3x} + \dfrac{2^{3x}}{\ln 2} + C\right)$ **29.** $\frac{1}{6}\sqrt{1 + e^{6x^2}} + C$ **31.** $\dfrac{2}{3 \ln 2}\sqrt{3 \cdot 2^x + 4} + C$ **33.** $\frac{1}{3}(e^8 - 1)$ **35.** $\frac{3}{2}\ln 2$ **37.** $1 + 5 \ln \frac{3}{4}$ **39.** $-\dfrac{ye^x + e^y + 1}{e^x + xe^y + 1}$ **41.** 5.004 **43.** $v = e^t - e^{-t} + 1$; $s = e^t + e^{-t} + t$ **45.** $\frac{1}{2}\pi(1 - e^{-2b})$; $\frac{1}{2}\pi$ **47.** 8212 years **49.** $g(x) = -e^x$; domain: $(-\infty, +\infty)$ **53.** $3000 \ln \frac{3}{2}$ in.-lb **55.** 8.66 years **57.** $(1, e)$ **59.** $\operatorname{sgn} t(1 - e^{-|t|})$

EXERCISES 9.1 (Page 502)

3. $-4 \csc 4x \cot 4x$ **5.** $2 \tan x \sec^2 x$ **7.** $2x \sec x^2 \tan x^2$ **9.** $-3e^x \csc^2 e^x$ **11.** $2 \tan 2x$ **13.** $-\dfrac{3 \csc^2 3r}{2\sqrt{\cot 3r}}$ **15.** $4 \tan^5 x + 6 \tan^3 x + 2 \tan x$ **17.** $-3t^2 \csc(t^3 + 1)\cot(t^3 + 1)$ **19.** $(-3 \ln 2)2^{\csc 3x}\csc 3x \cot 3x$ **21.** $4 \cot t \csc^2 t$ **23.** $5 \sec 5x$ **25.** $-\cot^2 z \csc z(4 \cot^2 z + 3)$ **27.** $\dfrac{\sin^2 x \cos x + \cos x - 4}{\cos^2 x(\cos x - 4)^2}$ **29.** $(\sin x)^{\tan x}(\sec^2 x \ln(\sin x) + 1)$ **31.** $\dfrac{1}{\sqrt{1 - \cot^2 x}}(\sec^2 x + \sin^2 x - 3\cos^2 x)$ **33.** $\dfrac{\sec^5 x \sqrt[3]{\tan x + 1}}{\sqrt{\csc^3 x - 4}}\left(5 \tan x + \dfrac{\sec^2 x}{3(\tan x + 1)} + \dfrac{3 \csc^3 x \cot x}{2(\csc^3 x - 4)}\right)$ **35.** $\sec^2 x \tan x \sin^2 y \tan y$ **37.** $\dfrac{2 \tan x \sec^2 x + \csc^2(x - y)}{2 \sec^2 y \tan y + \csc^2(x - y)}$ **43.** $y = \sqrt{2}(x + 1 - \frac{1}{4}\pi)$ **45.** $\frac{2000}{9}$ ft/sec **47.** $5\sqrt{5}$ ft

EXERCISES 9.2 (Page 506)

1. (a) $\frac{1}{2}$; (b) $\frac{10}{11}$; (c) $-\frac{7}{4}$; (d) $\frac{7}{31}$; (e) $\frac{7}{26}$ **3.** 27°, 45°, 108° **5.** $3x - y + 7 = 0$; $x + 3y - 11 = 0$ **9.** $-\frac{19}{3} - \frac{1}{3}\sqrt{370}$ **13.** $(\frac{1}{4}(1 + 4n)\pi, \frac{1}{2}\sqrt{2})$, where n is any even integer; $(\frac{1}{4}(1 + 4n)\pi, -\frac{1}{2}\sqrt{2})$, where n is any odd integer; 109° 30′ **15.** $\dfrac{-2 \ln a}{1 - \ln^2 a}$

EXERCISES 9.3 (Page 511)

1. $\frac{1}{5}\tan 5x + C$ **3.** $\frac{1}{2}\ln|\sec 2w| + C$ **5.** $-\frac{1}{3}\csc 3x + C$ **7.** $\frac{1}{10}\ln|\csc 10t - \cot 10t| + C$ **9.** $-\frac{1}{10}\cot 5x^2 + C$ **11.** $\ln|\sin e^t| + C$ **13.** $\frac{1}{6}\ln|\sec 3y^2 + \tan 3y^2| + C$ **15.** $-\frac{1}{2}\csc^2 x - \ln|\sin x| + C$ **17.** $-\frac{1}{6}\cot 6r - r + C$ **19.** $\frac{1}{7}\tan^7 x + \frac{1}{9}\tan^9 x + C$ **21.** $-\frac{1}{9}\cot^3 3x - \frac{1}{15}\cot^5 3x + C$ **23.** $\frac{1}{2}(\tan 2x - \cot 2x) + C$ **25.** $2 \sec w - \tan w + C$ **27.** $\frac{1}{12}\tan^4 3x - \frac{1}{6}\tan^2 3x + \frac{1}{3}\ln|\sec 3x| + C$ **29.** $u - 2\tan\frac{1}{2}u + C$ **31.** $-\frac{1}{3}\csc^3 x + C$ **33.** $\frac{1}{4}\tan^4(\ln x) + \frac{1}{3}\tan^6(\ln x) + \frac{1}{8}\tan^8(\ln x) + C$ **35.** $\frac{1}{8}\ln 2$ **37.** $\frac{56}{15}$ **39.** $\frac{1}{5}$ **49.** $\ln(2 + \sqrt{3})$ **51.** $\frac{4}{3}\pi$ cu units

EXERCISES 9.4 (Page 517)

1. (a) $\frac{1}{4}\pi$; (b) $\frac{1}{4}\pi$; (c) $-\frac{1}{4}\pi$; (d) $\frac{3}{4}\pi$; (e) $\frac{1}{4}\pi$; (f) $-\frac{1}{4}\pi$ **3.** (a) $-\frac{1}{2}\pi$; (b) π; (c) $-\frac{1}{4}\pi$; (d) $\frac{3}{4}\pi$; (e) π; (f) $-\frac{1}{2}\pi$ **5.** (a) $\frac{2}{3}\sqrt{2}$; (b) $\frac{1}{4}\sqrt{2}$; (c) $2\sqrt{2}$; (d) $\frac{3}{4}\sqrt{2}$; (e) 3 **7.** (a) $-\frac{2}{5}\sqrt{5}$; (b) $\frac{1}{5}\sqrt{5}$; (c) $-\frac{1}{2}$; (d) $\sqrt{5}$; (e) $-\frac{1}{2}\sqrt{5}$ **9.** (a) $-\frac{2}{3}$; (b) $\frac{1}{3}\sqrt{5}$; (c) $-\frac{2}{5}\sqrt{5}$; (d) $-\frac{1}{2}\sqrt{5}$; (e) $\frac{3}{5}\sqrt{5}$
11. (a) $\sqrt{3}$; (b) $\frac{1}{7}\sqrt{21}$ **13.** (a) $\frac{1}{2}\sqrt{3}$; (b) $\frac{1}{2}\sqrt{3}$ **15.** (a) $\frac{1}{2}\pi$; (b) $-\frac{1}{4}\pi$; (c) π; (d) $\frac{3}{4}\pi$ **17.** $\frac{119}{169}$ **19.** $\frac{2}{9}(1+\sqrt{10})$
21. $\frac{1}{27}(7\sqrt{5}+8\sqrt{2})$ **23.** $\frac{1}{39}(48-25\sqrt{3})$ **25.** $\frac{1}{15}(4\sqrt{10}+\sqrt{5})$

EXERCISES 9.5 (Page 523)

1. $\dfrac{1}{\sqrt{4-x^2}}$ **3.** $\dfrac{2}{1+4x^2}$ **5.** $-\dfrac{1}{\sqrt{x-x^2}}$ **7.** $\dfrac{1}{|t|\sqrt{25t^2-1}}$ **9.** $-\dfrac{x}{|x|\sqrt{1-x^2}}$ **11.** $\dfrac{4}{4+x^2}$ **13.** $\sin^{-1} 2y + \dfrac{2y}{\sqrt{1-4y^2}}$
15. $\csc^{-1}\dfrac{1}{x} + \dfrac{x}{\sqrt{1-x^2}}$ **17.** $-\dfrac{\cos x}{|\cos x|}$ **19.** $\dfrac{2x}{(1+x^4)\tan^{-1} x^2}$ **21.** $2\sqrt{4-x^2}$ **23.** 0 **25.** $\cot^{-1} x$ **27.** $-e^x - \dfrac{1}{\sqrt{1-x^2}}$
29. $\dfrac{(1+y^2)(3x^2+\sin y)}{1-x\cos y(1+y^2)}$ **31.** $\frac{52}{3}\pi$ km/min **33.** 0.078 rad/sec **35.** 8 ft/sec **37.** 0.547 **39.** $\dfrac{6}{x\sqrt{x^2-64}}$

EXERCISES 9.6 (Page 528)

1. $\frac{1}{2}\sin^{-1} 2x + C$ **3.** $\dfrac{1}{12}\tan^{-1}\dfrac{3x}{4} + C$ **5.** $\dfrac{1}{2}\tan^{-1}\dfrac{x-1}{2} + C$ **7.** $\dfrac{1}{16}\sec^{-1}\left|\dfrac{x}{4}\right| + C$ **9.** $\dfrac{\sqrt{5}}{5}\sin^{-1}\dfrac{\sqrt{10}}{2} + C$ **11.** $\dfrac{1}{6}\sin^{-1}\dfrac{3r^2}{4} + C$ **13.** $\dfrac{1}{\sqrt{7}}\tan^{-1}\dfrac{e^x}{\sqrt{7}} + C$ **15.** $2\tan^{-1}\sqrt{x} + C$ **17.** $\dfrac{2}{\sqrt{7}}\tan^{-1}\dfrac{2x-1}{\sqrt{7}} + C$ **19.** $\cos^{-1}\dfrac{1-x}{4} + C$ **21.** $\cos^{-1}\dfrac{1+x}{2} - \sqrt{3-2x-x^2} + C$ **23.** $\sin^{-1}\dfrac{1+x}{\sqrt{5}} - \sqrt{4-2x-x^2} + C$ **25.** $\frac{1}{2}x^2 + 2x + \frac{5}{4}\ln(2x^2-4x+3) - \frac{1}{2}\sqrt{2}\tan^{-1}\sqrt{2}(x-1) + C$
27. $\frac{1}{4}\pi + \frac{1}{2}\ln 2$ **29.** $\frac{1}{3}\pi$ **31.** $\tan^{-1} e - \frac{1}{4}\pi$ **33.** $\frac{1}{4}\pi$ **35.** $\frac{1}{3}\pi$ sq units **37.** $2\pi r$

EXERCISES 9.7 (Page 535)

17. $\dfrac{4}{5}\operatorname{sech}^2\dfrac{4x+1}{5}$ **19.** $2e^{2y}\cosh e^{2y}$ **21.** $-8\operatorname{sech}^2 4w \tanh 4w$ **23.** e^{2x} **25.** $2\operatorname{csch} 2t$ **27.** $2x\operatorname{sech} x^2$
29. $x^{\sinh x - 1}(x\cosh x \ln x + \sinh x)$ **35.** $2\cosh\sqrt{x} + C$ **37.** $\frac{1}{224}\sinh 28t + \frac{1}{28}\sinh 14t + \frac{3}{8}t + C$ **39.** $\frac{1}{2}\ln^2 \cosh x + C$
41. $e^{\cosh t} + C$ **43.** $\frac{1}{3}\tanh 3x - \frac{1}{9}\tanh^3 3x + C$ **45.** $\frac{4}{5}$ **47.** $\frac{1}{3}\cosh^3 2 - \cosh 2 + \frac{2}{3}$ **49.** $\operatorname{sech}(0) = 1$, rel max; pts of infl at $x = \pm\ln(1+\sqrt{2})$; sech increasing on $(-\infty, 0]$; sech decreasing on $[0, +\infty)$; graph concave upward for $x < -\ln(1+\sqrt{2})$ and $x > \ln(1+\sqrt{2})$; graph concave downward for $-\ln(1+\sqrt{2}) < x < \ln(1+\sqrt{2})$ **51.** $a^2\sinh\dfrac{x_1}{a}$ sq units
53. $v = e^{-ct/2}[(B - \frac{1}{2}cA)\sinh t + (A - \frac{1}{2}cB)\cosh t]$; $a = e^{-ct/2}[(A - cB + \frac{1}{4}c^2A)\sinh t + (B - cA + \frac{1}{4}c^2B)\cosh t]$; $a = K_1 s + K_2 v$, where $K_1 = 1 - \frac{1}{4}c^2$ and $K_2 = -c$

EXERCISES 9.8 (Page 541)

11. $\ln\frac{1}{4}(1+\sqrt{17})$ **13.** $\frac{1}{2}\ln 3$ **15.** $\dfrac{2x}{\sqrt{x^4+1}}$ **17.** $\dfrac{4}{1-16x^2}$ **19.** $-\dfrac{1}{2x+3x^2}$ **21.** $2x\left(\cosh^{-1} x^2 + \dfrac{x^2}{\sqrt{x^4-1}}\right)$
23. $|\sec x|$ **25.** $-\dfrac{\csc x \cot x}{|\cot x|}$ **27.** $\dfrac{6z(\coth^{-1} z^2)^2}{1-z^4}$ **29.** $-e^x \csc e^x$ **31.** $\sinh^{-1} x$

REVIEW EXERCISES FOR CHAPTER 9 (Page 542)

1. $6\tan 3x \sec^2 3x$ **3.** $-\sec x \csc x$ **5.** $\sec^2 x(e^{\tan x} + 2\tan x)$ **7.** $\dfrac{2^x \ln 2}{1+2^{2x}}$ **9.** $6\cosh^2 2x \sinh 2x$ **11.** $5(\log_{10} e)\sec 5x$
13. $2w\sinh^{-1} 2w + \dfrac{2w^2}{\sqrt{4w^2+1}}$ **15.** $\dfrac{\sin x - \cos x - 1}{1+\sin 2x}$ **17.** $-\dfrac{2\cot x(x^2\csc^2 x - x\cot x + \csc^2 x)}{(1+x^2)^2}$ **19.** $\dfrac{1}{w^2+1}$
21. $\dfrac{\operatorname{sech}^2\sqrt{x}}{2\sqrt{x}(1+\tanh^2\sqrt{x})}$ **23.** $-e^x \csc e^x$ **25.** $\dfrac{1}{t}\sin t + (\ln t)\cos t$ **27.** $\dfrac{(\tan x)^{1/x^2}(x\sec^2 x - 2\tan x \ln \tan x)}{x^3 \tan x}$ **29.** 106°, 90°, 112°, 52° **31.** 4 **33.** $\dfrac{y - \sec^2(x+y)}{\sec^2(x+y) - x}$ **35.** $\dfrac{6xy - 9y^4 - x^4y^2}{3x^2 + 18xy^3 + 2x^5y}$ **37.** $\frac{1}{6}\ln|\sec 3x^2| + C$ **39.** $\frac{3}{2}\sin^{-1} x^2 + C$
41. $-\frac{1}{4}\cot^4 e^x - \frac{1}{6}\cot^6 e^x + C$ **43.** $\dfrac{2}{\sqrt{31}}\tan^{-1}\dfrac{4x+3}{\sqrt{31}} + C$ **45.** $\ln|\cosh x| + \frac{1}{2}\operatorname{sech}^2 x + C$ **47.** $2\sec w - \tan w + C$
49. $\dfrac{\sqrt{2}}{4}\sec^{-1}\dfrac{\sqrt{2}e^x}{4} + C$ **51.** $2\tan\sqrt{t} + \frac{2}{3}\tan^3\sqrt{t} + C$ **53.** $\frac{856}{105}$ **55.** $\frac{2}{3}\pi + \sqrt{3} - 2$ **57.** $\tanh 1$ **59.** $\frac{1}{2}\pi$

61. $\ln(2+\sqrt{3})$ **63.** $\frac{1}{2}\pi - 1$ **65.** $3x + 6\sqrt{3}y = 2\pi - 12\sqrt{3}$ **69.** $\pi(\frac{15}{32} - \frac{1}{2}\ln 2)$ **71.** (a) 120 rad/hr; (b) 60 rad/hr
73. decreasing at the rate of 60 rad/hr **75.** amplitude is $\frac{1}{2}\sqrt{2}(1+\sqrt{3})$ **77.** $\frac{1}{10}\pi$ hr; he walks all the way

EXERCISES 10.2 (Page 552)

1. $\frac{1}{3}xe^{3x} - \frac{1}{9}e^{3x} + C$ **3.** $x\sec x - \ln|\sec x + \tan x| + C$ **5.** $x\ln x - x + C$ **7.** $x\ln^2 x - 2x\ln x + 2x + C$
9. $\frac{1}{2}\tan^{-1} x(x^2+1) - \frac{1}{2}x + C$ **11.** $\dfrac{e^x}{x+1} + C$ **13.** $-\cos x\ln(\cos x) + \cos x + C$ **15.** $\frac{1}{2}e^x(\cos x + \sin x) + C$
17. $-x^2\sqrt{1-x^2} - \frac{2}{3}(1-x^2)^{3/2} + C$ **19.** $x^2\cosh x - 2x\sinh x + 2\cosh x + C$ **21.** $\frac{1}{4}\sec^3 x\tan x$
$+ \frac{3}{8}[\sec x\tan x + \ln|\sec x + \tan x|] + C$ **23.** $2\sqrt{z}\cot^{-1}\sqrt{z} + \ln(1+z) + C$ **25.** $\dfrac{36}{\ln 3} - \dfrac{36}{(\ln 3)^2} + \dfrac{16}{(\ln 3)^3}$ **27.** $\frac{9}{16}$
29. $\frac{1}{4}(3e^4+1)$ **31.** $\frac{4}{25}(e^{3\pi/4}+1)$ **33.** $\frac{4}{3}\pi - \sqrt{3}$ **35.** (e^2+1) sq units **37.** $\frac{1}{2}\pi(3e^4+1)$ cu units **39.** $(8 - 24e^{-2})$ sq units **41.** $\frac{1}{4}(1-9e^{-8})w\pi$ ft-lb **43.** $6320 **45.** (b) $\frac{1}{5}\sec^4 x\tan x + \frac{4}{15}\sec^2 x\tan x + \frac{8}{15}\tan x + C$
47. (b) $\frac{1}{5}x^5(\ln x)^2 - \frac{2}{25}x^5\ln x + \frac{2}{125}x^5 + C$

EXERCISES 10.3 (Page 558)

1. $-\dfrac{(4-x^2)^{1/2}}{4x} + C$ **3.** $\dfrac{1}{2}\ln\left|\dfrac{x}{2+\sqrt{x^2+4}}\right| + C$ **5.** $\dfrac{1}{5}\ln\left|\dfrac{5-\sqrt{25-x^2}}{x}\right| + C$ **7.** $\ln|x + \sqrt{x^2-a^2}| + C$ **9.** $\dfrac{1}{4}\tan^{-1}\dfrac{1}{2}x - \dfrac{x}{2(x^2+4)} + C$ **11.** $-\frac{1}{9}x(4x^2-9)^{-1/2} + C$ **13.** $\frac{1}{5}\ln(\sqrt{t^4+25} - 5) - \frac{2}{5}\ln t + C$ **15.** $\ln|x + 2 + \sqrt{4x+x^2}| + C$
17. $\dfrac{x+2}{9\sqrt{5-4x-x^2}} + C$ **19.** $\dfrac{\tan x}{4\sqrt{4-\tan^2 x}} + C$ **21.** $\frac{1}{3}\sqrt{\ln^2 w - 4}(8 + \ln^2 w) + C$ **23.** $-\dfrac{e^t+4}{9\sqrt{e^{2t}+8e^t+7}} + C$
25. $\frac{128}{3} - 24\sqrt{3}$ **27.** $\frac{1}{27}(6 - 2\sqrt{3})$ **29.** $\frac{1}{2}\cos^{-1}\frac{1}{3} - \frac{1}{6}\pi$ **31.** $\frac{625}{16}\pi$ **33.** (a) $\sec^{-1}|\frac{2}{3}x| + C$; (b) $-\frac{5}{2}\sqrt{3-2x^2} + C$
35. $\ln\frac{1}{3}(2\sqrt{5} + \sqrt{10} - \sqrt{2} - 1) + \sqrt{10} - \sqrt{2}$ **37.** $\frac{81}{16}\pi^2$ cu units **39.** $11,201,300

EXERCISES 10.4 (Page 565)

1. $\dfrac{1}{4}\ln\left|\dfrac{x-2}{x+2}\right| + C$ **3.** $\ln|C(x-2)^2(x+2)^3|$ **5.** $\ln\left|\dfrac{C(w+4)^3}{2w-1}\right|$ **7.** $\dfrac{1}{4}\ln\left|\dfrac{Cx^4(2x+1)^3}{2x-1}\right|$ **9.** $\dfrac{1}{9}\ln\left|\dfrac{x+3}{x}\right| - \dfrac{1}{3x} + C$
11. $2\ln\left|\dfrac{x+1}{x}\right| - \dfrac{1}{x} - \dfrac{1}{x+1} + C$ **13.** $\dfrac{3}{x+1} + \ln|x+1| - \dfrac{1}{2}\ln|2x+3| + C$ **15.** $\dfrac{5}{16(z+2)} - \dfrac{7}{16(z-2)} + \dfrac{1}{32}\ln\left|\dfrac{z+2}{z-2}\right| + C$
17. $\dfrac{1}{2}x^2 + 2x - \dfrac{3}{x-1} - \ln|x^2+2x-3| + C$ **19.** $-\ln(3x+2)^{2/3}(x-1)^2 - \dfrac{1}{3(3x+2)} - \dfrac{3}{x-1} + C$ **21.** $4\ln\frac{4}{3} - \frac{3}{2}$
23. $\ln\frac{27}{4} - 2$ **25.** $13\ln 2 - 4\ln 5$ **27.** $\ln\frac{7}{2} - \frac{5}{3}$ **29.** $\ln 4.5$ sq units **31.** $2\pi(2 + 6\ln 3 - 2\ln 2)$ cu units **33.** $\frac{31}{19}$
35. 7.4 lb

EXERCISES 10.5 (Page 570)

1. $\dfrac{1}{2}\ln\left|\dfrac{Cx^2}{2x^2+1}\right|$ **3.** $\dfrac{1}{8}\ln\left|\dfrac{2x-1}{2x+1}\right| - \dfrac{1}{4}\tan^{-1} 2x + C$ **5.** $\frac{1}{10}\ln|(t^2+1)(2t+1)^3| + \frac{2}{5}\tan^{-1} t + C$ **7.** $\ln|x-1| + \tan^{-1} x + C$
9. $\dfrac{1}{2}\ln\left(\dfrac{Cx^2}{x^2+x+1}\right) - \dfrac{1}{\sqrt{3}}\tan^{-1}\left(\dfrac{2x+1}{\sqrt{3}}\right)$ **11.** $\ln\left(\dfrac{Cx^2}{x^2+1}\right) - \dfrac{1}{2}\tan^{-1} x - \dfrac{x}{2(x^2+1)}$ **13.** $\dfrac{5}{2}\ln(z^2-2z+5) - \dfrac{65}{16}\tan^{-1}\left(\dfrac{z-1}{2}\right)$ $+ \dfrac{-47z+15}{8(z^2-2z+5)} + C$ **15.** $\dfrac{5}{162}\ln|9x^2+3x+1| - \dfrac{2}{81}\ln|3x-1| + \dfrac{5}{9\sqrt{3}}\tan^{-1}\left(\dfrac{6x+1}{\sqrt{3}}\right) + C$ **17.** $\dfrac{1}{6}\tan^{-1}\dfrac{2}{3}x + \dfrac{x}{4x^2+9} + C$
19. $\ln|1+\tan x| + \dfrac{2}{\sqrt{3}}\tan^{-1}\left(\dfrac{2\tan x - 1}{\sqrt{3}}\right) + C$ **21.** $6\ln 2$ **23.** $\ln\frac{12}{5} + \frac{3}{2}\ln\frac{20}{13}$ **25.** $\frac{1}{4}\pi$ **27.** $\frac{3}{4}\ln 2 + \frac{5}{8}\pi$ **29.** $\frac{3}{8}\ln\frac{125}{41}$
31. $\dfrac{1}{6}\ln 3 + \dfrac{\sqrt{3}}{18}\pi$ **33.** $\dfrac{3}{50}\ln\dfrac{(t_1+2)^2}{4(t_1^2+1)} - \dfrac{7}{5(t_1+2)} - \dfrac{4}{25}\tan^{-1} t_1 + \dfrac{7}{10}$

EXERCISES 10.6 (Page 575)

1. $\sinh^{-1}\frac{1}{2}x + C = \ln\frac{1}{2}(x + \sqrt{x^2+4}) + C$ **3.** $\frac{1}{2}\cosh^{-1} x^2 + C = \frac{1}{2}\ln(x^2 + \sqrt{x^4-1}) + C$ **5.** $\begin{cases} -\frac{1}{12}\tanh^{-1}(\frac{3}{4}x) + C & \text{if } |x| < \frac{4}{3} \\ -\frac{1}{12}\coth^{-1}(\frac{3}{4}x) + C & \text{if } |x| > \frac{4}{3} \end{cases}$
$= \dfrac{1}{24}\ln\left|\dfrac{4-3x}{4+3x}\right| + C$ **7.** $\sinh^{-1}\dfrac{\sin x}{\sqrt{3}} + C = \ln(\sin x + \sqrt{4 - \cos^2 x}) + C$ **9.** $\dfrac{1}{\sqrt{5}}\cosh^{-1}(\sqrt{5}e^t) + C$

$= \frac{1}{\sqrt{5}} \ln(\sqrt{5}e^t + \sqrt{5e^{2t} - 1}) + C$ **11.** $\left\{\begin{matrix} \frac{1}{\sqrt{6}} \tanh^{-1}\left(\frac{x+2}{\sqrt{6}}\right) + C & \text{if } |x+2| < \sqrt{6} \\ \frac{1}{\sqrt{6}} \coth^{-1}\left(\frac{x+2}{\sqrt{6}}\right) + C & \text{if } |x+2| > \sqrt{6} \end{matrix}\right\} = \frac{1}{2\sqrt{6}} \ln\left|\frac{\sqrt{6}+2+x}{\sqrt{6}-2-x}\right| + C$

13. $\begin{cases} -\tanh^{-1}(x+5) + C & \text{if } |x+5| < 1 \\ -\coth^{-1}(x+5) + C & \text{if } |x+5| > 1 \end{cases} = -\frac{1}{2} \ln\left|\frac{x+6}{x+4}\right| + C$ **15.** $\frac{3}{2} \sinh^{-1}\left(\frac{2 \ln w}{3}\right) + C = \frac{3}{2} \ln(2 \ln w + \sqrt{4 \ln^2 w + 9}) + C$

17. $\ln\left(\frac{5+\sqrt{21}}{3+\sqrt{5}}\right)$ **19.** $\ln 3$ **21.** $\frac{1}{3} \ln \frac{7+\sqrt{40}}{4+\sqrt{7}}$

EXERCISES 10.7 (Page 578)

1. $\frac{6}{\sqrt{15}} \tan^{-1}\left(\frac{1}{\sqrt{15}} \tan \frac{x}{2}\right) + C$ **3.** $\frac{\sqrt{15}}{5} \ln\left|\frac{\sqrt{15} + \tan \frac{1}{2}x}{\sqrt{15} - \tan \frac{1}{2}x}\right| + C$ **5.** $\frac{1}{2} \ln|\tan \frac{1}{2}x| - \frac{1}{4} \tan^2 \frac{1}{2}x + C$ **7.** $\frac{1}{4} \ln\left|\frac{3 \tan \frac{1}{2}x - 9}{3 \tan \frac{1}{2}x - 1}\right| + C$

9. $\frac{1}{5} \ln\left|\frac{3 \tan \frac{1}{2}x - 1}{\tan \frac{1}{2}x + 3}\right| + C$ **11.** $\sqrt{2} \ln\left|\frac{\tan x + \sqrt{2}}{\tan x - \sqrt{2}}\right| + C$ **13.** $\frac{1}{2}x + \frac{\sqrt{3}}{6} \ln\left|\frac{\sqrt{3} - \tan \frac{1}{2}x}{\sqrt{3} + \tan \frac{1}{2}x}\right| + C$ **15.** $-\frac{1}{2} \ln|13 - \tan^2 x| + C$

17. $2 \tan^{-1}(2 + \tan \frac{1}{2}x) + C$ **19.** $\frac{1}{4} \ln 3$ **21.** $2\sqrt{3} \ln(1 + \sqrt{3})$ **23.** $\frac{1}{2}\sqrt{3} \ln(1 + \frac{1}{2}\sqrt{3})$

EXERCISES 10.8 (Page 581)

1. $\frac{2}{3}x^{3/2} - 3x + 18\sqrt{x} - 54 \ln(3 + \sqrt{x}) + C$ **3.** $\ln\left|\frac{\sqrt{1+4x} - 1}{\sqrt{1+4x} + 1}\right| + C$ **5.** $-2\sqrt{1+x} + \sqrt{2} \ln\left|\frac{\sqrt{1+x} + \sqrt{2}}{\sqrt{1+x} - \sqrt{2}}\right| + C$

7. $\frac{3}{2}(x-2)^{2/3} - 3(x-2)^{1/3} + 3 \ln|1 + (x-2)^{1/3}| + C$ **9.** $2\sqrt{2x} + 2\sqrt{x+4} + 4\sqrt{2} \ln\left|\frac{(2\sqrt{2} - \sqrt{x+4})(\sqrt{x} - 2)}{x-4}\right| + C$

11. $\frac{1}{9}\sqrt{1+2x^3}(2x^3 + 7) + C$ **13.** $3 \tan^{-1}\sqrt[6]{x} + \frac{3\sqrt[6]{x}}{1 + \sqrt[3]{x}} + C$ **15.** $\begin{cases} \ln\left|\frac{Cx}{x + 2 + 2\sqrt{1+x+x^2}}\right| & \text{if } x > 0 \\ -\ln\left|\frac{Cx}{x + 2 - 2\sqrt{1+x+x^2}}\right| & \text{if } x < 0 \end{cases}$

17. $\begin{cases} -\sin^{-1}\left(\frac{1-2x}{\sqrt{13}x}\right) + C & \text{if } x > 0 \\ \sin^{-1}\left(\frac{1-2x}{\sqrt{13}x}\right) + C & \text{if } x < 0 \end{cases}$ **19.** $\begin{cases} \frac{\sqrt{5x^2+8x-1}}{x} - 4 \sin^{-1}\left(\frac{1-4x}{\sqrt{21}x}\right) + C & \text{if } x > 0 \\ \frac{\sqrt{5x^2+8x-1}}{x} + 4 \sin^{-1}\left(\frac{1-4x}{\sqrt{21}x}\right) + C & \text{if } x < 0 \end{cases}$ **21.** $4 - 2 \ln 3$ **23.** $\ln \frac{11}{10}$

25. $\frac{1}{9}(54 - 2\sqrt[3]{3})$ **27.** $\frac{3}{2}\pi - \frac{152}{35}$ **29.** $2 \ln|\sqrt{x} - 1| + C$

REVIEW EXERCISES FOR CHAPTER 10 (Page 582)

1. $\frac{1}{8}x - \frac{1}{128} \sin 16x + C$ **3.** $-2\sqrt{4 - e^x} + C$ **5.** $(x+1) \tan^{-1} \sqrt{x} - \sqrt{x} + C$ **7.** $\frac{1}{2}x + \frac{3}{4} \sin \frac{2}{3}x + C$ **9.** $\ln|x - 1| - 2(x-1)^{-1} - (x-1)^{-2} + C$ **11.** $\frac{1}{4} \sin 2x - \frac{1}{8} \sin 4x + C$ **13.** $3 \ln\left|\frac{x^{1/3}}{1 + x^{1/3}}\right| + C$ **15.** $\frac{1}{3} \tan 3x - \frac{1}{3} \cot 3x + \frac{2}{3} \ln|\tan 3x| + C$

17. $2t + \ln \frac{t^2}{(t+2)^{10}} - \frac{15}{t+2} + C$ **19.** $x - \tan^{-1} x + \frac{1}{2} \ln\left|\frac{x-1}{x+1}\right| + C$ **21.** $\frac{1}{16}x - \frac{1}{192} \sin 12x - \frac{1}{144} \sin^3 6x + C$

23. $\sin^{-1}\left(\frac{r+2}{\sqrt{7}}\right) + C$ **25.** $\frac{1}{2}x^2 \sin x^2 + \frac{1}{2} \cos x^2 + C$ **27.** $\frac{2}{17}e^{t/2}(4 \sin 2t + \cos 2t) + C$ **29.** $\frac{1}{4} \tan^{-1}(\frac{1}{2} \sin^2 x) + C$

31. $\begin{cases} \frac{1}{n}\left(-\cos nx + \frac{2}{3}\cos^3 nx - \frac{1}{5}\cos^5 nx\right) + C & \text{if } n \neq 0 \\ C & \text{if } n = 0 \end{cases}$ **33.** $-\frac{1}{4} \csc^3 x \cot x - \frac{3}{8} \csc x \cot x + \frac{3}{8} \ln|\csc x - \cot x| + C$

35. $2 \ln|y - 2| - 8(y-2)^{-1} - \frac{9}{2}(y-2)^{-2} + C$ **37.** $-\tan^{-1}(\cos x) + C$ **39.** $2 \sin^{-1}\left(\frac{t-2}{2}\right) + \frac{1}{2}(t-2)(4t - t^2)^{1/2} + C$

41. $\frac{1}{3} \ln\left|1 - \frac{1}{x^3}\right| + C$ **43.** $\frac{1}{3} \sin^{-1}(\frac{3}{2}e^x) + C$ **45.** $-\frac{1}{15} \cot^5 3x - \frac{1}{9} \cot^3 3x + C$ **47.** $\frac{1}{3}x^3 \sin^{-1} x + \frac{1}{9}(x^2 + 2)\sqrt{1 - x^2} + C$

49. $\tan^{-1}(\cos x) + C$ **51.** $\frac{2}{3} \sec^{-1}|2 \sin 3t| + C$ **53.** $-\frac{(x^2 + a^2)^{3/2}}{3a^2x^3} + C$ **55.** $\sqrt{2t} - \sqrt{1 - 2t} \sin^{-1} \sqrt{2t} + C$

57. $\frac{4}{3}\sqrt{2 + \sqrt{x-1}}(\sqrt{x-1} - 4) + C$ **59.** $\frac{\sqrt{2}}{4} \ln\left|\frac{\tan x - \sqrt{2 \tan x} + 1}{\tan x + \sqrt{2 \tan x} + 1}\right| + \frac{\sqrt{2}}{2} \tan^{-1}(\sqrt{2 \tan x} - 1) + \frac{\sqrt{2}}{2} \tan^{-1}(\sqrt{2 \tan x} + 1) + C$

61. $\begin{cases} \dfrac{x^{n+1}\ln x}{n+1} - \dfrac{x^{n+1}}{(n+1)^2} + C & \text{if } n \neq -1 \\ \frac{1}{2}\ln^2 x + C & \text{if } n = -1 \end{cases}$ **63.** 4 **65.** $\frac{1}{2} + 2\ln\frac{6}{5}$ **67.** $\frac{16}{3} - \frac{8}{3}\sqrt{2}$ **69.** $\frac{4}{3}\sqrt{3} - \frac{1}{2}\pi$ **71.** $\frac{4}{3}$
73. $\frac{1}{2} - \frac{1}{4}\ln 2$ **75.** $a^2(\frac{1}{2}\pi - \frac{9}{8}\sqrt{3})$ **77.** $\frac{1}{2}\ln\frac{9}{2} - \frac{1}{6}\pi$ **79.** 5 **81.** $\frac{1}{6} + \ln\frac{3}{2}$ **83.** $\frac{4}{3}$ **85.** $1 - \frac{1}{2}\ln 3$ **87.** $\frac{1}{24}\pi$
89. $\sqrt{3} - \frac{1}{2}\ln(2 + \sqrt{3})$ **91.** $\frac{1}{5}\ln\frac{3}{2}$ **93.** $\frac{256}{15}$ **97.** $9\sqrt{2} - 3\sqrt{5} + \frac{3}{2}\ln(2\sqrt{10} + 3\sqrt{5} - 4\sqrt{2} - 6)$ **99.** $\frac{1}{8}\pi$ sq units
101. $\pi(e^2\ln 2 + \frac{1}{2}e^2 + \frac{1}{8})$ cu units **103.** (a) $x = 300\left(\dfrac{18^t - 17^t}{3\cdot 18^t - 2\cdot 17^t}\right)$; (b) 35.94 lb

EXERCISES 11.1 (Page 591)

1. 250 lb **3.** 4000 dynes **5.** $\frac{3}{2}$ m/sec^2 **7.** $\frac{8}{3}$ slugs **9.** 4 **11.** 6 **13.** 54 kg; $\frac{11}{3}$ m from one end **15.** 171 slugs; 5.92 in. from one end **17.** 42 g; $\frac{44}{7}$ cm from left end **19.** 31.5 kg; $\frac{18}{7}$ m from the end having the greater density **21.** 16 slugs; $\frac{16}{5}$ ft from one end **23.** 1.2 m from the end having the greater density **25.** $\dfrac{392}{60 + 27\ln 3}$ cm from the left end **27.** $2(1 - e^{-6})$ kg; $\dfrac{e^6 - 7}{e^6 - 1}$ m from one end **29.** $8\ln 2$ g; $\left(\dfrac{15}{4\ln 2} - 1\right)$ cm from one end **31.** $\dfrac{20}{L^2}x^2$ slugs/ft

EXERCISES 11.2 (Page 601)

1. $(2, \frac{1}{3})$ **3.** $\frac{29}{7}$ **5.** $(\frac{2}{3}, 1)$ **7.** $(0, \frac{8}{5})$ **9.** $(0, \frac{12}{5})$ **11.** $(\frac{16}{15}, \frac{64}{21})$ **13.** $(\frac{1}{2}, -\frac{3}{2})$ **15.** $\left(\dfrac{2e^3 + 1}{e^3 - 1}, \dfrac{e^3 + 1}{4}\right)$
17. $\left(\dfrac{20 - 15\cos^{-1}\frac{3}{5}}{5\ln 3 - 4}, \dfrac{26}{225(5\ln 3 - 4)}\right)$ **19.** $\left(\dfrac{8\ln 3 - 8}{2 - \ln 3}, \dfrac{2}{9(2 - \ln 3)}\right)$ **21.** $\frac{5}{3}p$ **23.** $(2, 0)$ **25.** $\frac{4}{3}k\ln 2$ kg **27.** the point on the bisecting radial line whose distance from the center of the circle is $\dfrac{4}{3\pi}$ times the radius **29.** $\frac{1}{3}\pi r^2 h$ cu units **31.** $\frac{1}{2}r^3(\pi + \frac{4}{3})$

EXERCISES 11.3 (Page 609)

1. $(0, \frac{8}{5}, 0)$ **3.** $(\frac{1}{2}, 0, 0)$ **5.** $(\frac{16}{5}, 0, 0)$ **7.** $(2, \frac{10}{7}, 0)$ **9.** $(0, \frac{7}{3}, 0)$ **11.** $(0, \frac{13}{5}, 0)$ **13.** $(\frac{5}{16}, 4, 0)$ **15.** $(p, \frac{5}{8}p, 0)$
17. $\left(\dfrac{\pi^2 + 4}{4\pi}, 0, 0\right)$ **19.** $\left(\dfrac{e^2 - 3}{2e^2 - 2}, 0, 0\right)$ **21.** on axis and $\frac{3}{4}h$ units from vertex **23.** 2 **25.** $(\frac{4}{5}, 0, 0)$ **27.** $(\frac{8}{9}, 0, 0)$

EXERCISES 11.4 (Page 615)

1. 320ρ lb **3.** 64ρ lb **5.** 2.25ρ lb **7.** 941,760 nt **9.** 4,087,500 nt **11.** $(4 - \frac{1}{4}\pi)\rho$ lb **13.** $(\frac{8}{3}\pi + 3\sqrt{3})\rho$ lb
15. $\sqrt[3]{16.31}$ m ≈ 2.54 m **17.** $11{,}250\sqrt{3}\rho$ lb **19.** $250\sqrt{409}\rho$ lb **23.** $\frac{1}{9}\sqrt{3}\rho$ lb **25.** $100{,}000\rho$ ft-lb

REVIEW EXERCISES FOR CHAPTER 11 (Page 616)

1. $\frac{2}{13}$ **3.** $(\frac{3}{2}, 2)$ **5.** $\frac{104}{3}$ slugs; $\frac{298}{65}$ in. from the left end **7.** $\frac{1}{3}k(1 - e^{-9})$ kg; $\dfrac{e^9 - 10}{3(e^9 - 1)}$ m from one end **9.** $(\frac{9}{8}, \frac{18}{5})$
11. $\left(\dfrac{8e^3 + 4}{9e^2 + 9}, \dfrac{10e^3 - 4}{27e^2 + 27}\right)$ **13.** $(\frac{1}{2}\pi - 1, \frac{1}{2})$ **15.** $\left(0, \dfrac{32}{15\pi}\right)$ **17.** $(0, \frac{8}{5})$ **19.** $\frac{256}{3}\pi$ m^3 **21.** $\left(0, \dfrac{e^2 + 1}{8}, 0\right)$ **23.** $(0, 2, 0)$
25. $\left(\dfrac{\pi^2 - 4}{4\pi}, 0, 0\right)$ **27.** $\frac{18}{5}\rho$ lb **29.** 756 lb **31.** 3ρ lb

EXERCISES 12.1 (Page 623)

1. (a) $(-4, \frac{5}{4}\pi)$; (b) $(4, -\frac{7}{4}\pi)$; (c) $(-4, -\frac{3}{4}\pi)$ **3.** (a) $(-2, \frac{3}{2}\pi)$; (b) $(2, -\frac{3}{2}\pi)$; (c) $(-2, -\frac{1}{2}\pi)$ **5.** (a) $(-\sqrt{2}, \frac{3}{4}\pi)$; (b) $(\sqrt{2}, -\frac{1}{4}\pi)$; (c) $(-\sqrt{2}, -\frac{5}{4}\pi)$ **7.** (a) $(-2, \frac{3}{4}\pi)$; (b) $(-2, -\frac{5}{4}\pi)$; (c) $(2, \frac{15}{4}\pi)$ **9.** $(3, \frac{4}{3}\pi)$; $(-3, \frac{1}{3}\pi)$ **11.** $(-4, -\frac{7}{6}\pi)$; $(4, -\frac{1}{6}\pi)$ **13.** $(-2, \frac{3}{4}\pi)$; $(2, \frac{7}{4}\pi)$ **15.** $(2, 2\pi + 6)$; $(-2, 6 - \pi)$ **17.** (a) $(-3, 0)$; (b) $(-1, -1)$; (c) $(2, -2\sqrt{3})$; (d) $(\frac{1}{2}\sqrt{3}, -\frac{1}{2})$ **19.** (a) $(\sqrt{2}, \frac{7}{4}\pi)$; (b) $(2, \frac{5}{6}\pi)$; (c) $(2\sqrt{2}, \frac{1}{4}\pi)$; (d) $(5, \pi)$ **21.** $r = |a|$ **23.** $r = \dfrac{2}{1 - \cos\theta}$ **25.** $r = 6\sin\theta$ **27.** $r^2 = 4\cos 2\theta$ **29.** $r = \dfrac{3a\sin 2\theta}{2(\sin^3\theta + \cos^3\theta)}$
31. $(x^2 + y^2)^2 = 4xy$ **33.** $(x^2 + y^2)^3 = x^2$ **35.** $y = x\tan(x^2 + y^2)$ **37.** $x = -1$ **39.** $4x^2 - 5y^2 - 36y - 36 = 0$

EXERCISES 12.2 (Page 630)

41. $\theta = \frac{2}{3}\pi, \theta = \frac{4}{3}\pi$ **43.** $\theta = \frac{1}{4}\pi, \theta = \frac{3}{4}\pi$ **47.** -1

EXERCISES 12.3 (Page 634)

1. $(\frac{3}{2}, \frac{1}{6}\pi)$; $(\frac{3}{2}, \frac{5}{6}\pi)$ **3.** pole; $(\sqrt{2}, \frac{1}{4}\pi)$ **5.** $(\frac{1}{2}\pi, \frac{1}{8}\pi)$ **7.** $(1, \frac{3}{2}\pi)$; $(1, \frac{1}{2}\pi)$; $(\frac{1}{2}, \frac{4}{3}\pi)$; $(\frac{1}{2}, \frac{2}{3}\pi)$; pole; (0.22, 2.47); (0.22, 3.82)
9. pole; $(\frac{1}{2}\sqrt{2}, \frac{1}{8}(2n + 1)\pi)$, where $n = 0, 1, \ldots, 7$ **11.** pole; $(\sqrt{15}, \cos^{-1}\frac{1}{4})$; $(\sqrt{15}, \pi - \cos^{-1}\frac{1}{4})$ **13.** $(4\sqrt{2}, \frac{1}{4}\pi)$ **15.** $(4, 1)$;

$(-4, -1)$ **17.** $(3, \frac{1}{3}\pi)$; $(3, -\frac{1}{3}\pi)$ **19.** pole; $(\frac{1}{2}\sqrt{3}, \frac{1}{3}\pi)$; $(-\frac{1}{2}\sqrt{3}, -\frac{1}{3}\pi)$ **21.** $(2\sqrt{2}, \frac{1}{4}\pi)$ **23.** pole; $(\frac{1}{2}\sqrt{2}, \frac{1}{6}(2n+1)\pi)$, where $n = 0, 1, 2, 3, 4, 5$

EXERCISES 12.4 (Page 637)

1. $\frac{3}{4}\pi$ **3.** $153° 26'$ **5.** $38° 9'$ **7.** $\frac{5}{6}\pi$ **9.** $\frac{1}{3}\pi$ **11.** $\frac{1}{2}\pi$ **13.** $\frac{1}{3}\pi$ **15.** 0 at pole; $\frac{1}{2}\pi$ at $(1, \pi)$; $\frac{1}{2}\pi$ at $(1, 0)$ **17.** $0°$ at $(0, \frac{1}{2}\pi)$; $79° 6'$ at $(\frac{1}{2}\sqrt{3}, \frac{1}{6}\pi)$; $79° 6'$ at $(-\frac{1}{2}\sqrt{3}, \frac{5}{6}\pi)$

EXERCISES 12.5 (Page 642)

1. $\frac{9}{4}\pi$ sq units **3.** 4π sq units **5.** 4 sq units **7.** $\frac{9}{16}\pi^3$ sq units **9.** $\frac{9}{8}\pi$ sq units **11.** $(\frac{11}{4}\pi - \frac{11}{2}\sin^{-1}\frac{1}{3} - 3\sqrt{2})$ sq units **13.** $(\frac{19}{3}\pi - \frac{11}{2}\sqrt{3})$ sq units **15.** $(\frac{9}{2}\pi - 9)$ sq units **17.** $a^2(2 - \frac{1}{4}\pi)$ sq units **19.** $\frac{1}{2}(\pi + 1)$ sq units **21.** $\frac{1}{2}\pi a^2$ sq units **23.** 4 sq units **25.** $16a^2\pi^3$ sq units

REVIEW EXERCISES FOR CHAPTER 12 (Page 643)

1. $r^2(4\cos^2\theta - 9\sin^2\theta) = 36$ **3.** $r = 9\cos\theta - 8\sin\theta$ **5.** $4x^4 + 8x^2y^2 + 4y^4 + 36x^3 + 36xy^2 - 81y^2 = 0$ **7.** $(x^2 + y^2)^2 = 4x^3 - 12xy^2$ **15.** $3\sqrt{3}$ **17.** $\theta = \frac{1}{2}\pi$ **19.** $a^2(\frac{1}{3}\pi + \frac{1}{2}\sqrt{3})$ sq units **21.** $\left(\dfrac{e^{4k\pi} - 1}{4k}\right)$ sq units **23.** $r = \dfrac{\pi^2}{\sin\theta - \pi\cos\theta}$ **25.** pole **27.** $\frac{1}{6}\pi, \frac{5}{6}\pi, \frac{1}{2}\pi$ **29.** $(16\pi - 24\sqrt{3})$ sq units; $(32\pi + 24\sqrt{3})$ sq units **31.** (a) $4r\sin\theta - r^2\cos^2\theta = 4$; (b) $x^2 = 4y - 4$ **33.** $r^2 - 2r_0 r\cos(\theta - \theta_0) + r_0{}^2 = a^2$

EXERCISES 13.1 (Page 648)

1. $(0, 1)$; $y = -1$; 4 **3.** $(-2, 0)$; $x = 2$; 8 **5.** $(0, -\frac{1}{4})$; $y = \frac{1}{4}$; 1 **7.** $(\frac{9}{8}, 0)$; $x = -\frac{9}{8}$; $\frac{9}{2}$ **9.** $y^2 = 20x$ **11.** $x^2 = -8y$ **13.** $y^2 = 2x$ **15.** $y^2 = -6x$ **17.** $y^2 = 10x$ **19.** $x^2 = -y$ **21.** $k = -\frac{5}{2}$; $x^2 = -10y$ **23.** 16.6 m **27.** $\frac{32}{45}$ in. **29.** $y^2 - 10x - 10y + 20 = 0$ **31.** $x^2 + y^2 + 10y = 0$ **33.** (a) $y = x$; (b) $(\frac{1}{2}, \frac{1}{2})$; (c) $2\sqrt{2}$

EXERCISES 13.2 (Page 654)

1. $x'^2 + y'^2 = 13$ **3.** $y'^2 = 6x'$ **5.** $x'^2 + 4y'^2 = 4$ **7.** $y' = 2x'^3$ **9.** $y' = \ln x'$ **11.** $y' = \cosh x'$ **13.** $x'^2 + 4y'^2 = 16$ **15.** $3x'^2 - 2y'^2 = 6$ **17.** $(-3, \frac{1}{4})$; $(-3, -\frac{3}{4})$; $x = -3$; $y = \frac{5}{4}$ **19.** $(1, -5)$; $(-\frac{1}{2}, -5)$; $y = -5$; $x = \frac{5}{2}$ **21.** $(\frac{2}{3}, 1)$; $(\frac{7}{24}, 1)$; $y = 1$; $x = \frac{25}{24}$ **23.** $y^2 + 20x - 8y - 24 = 0$ **25.** $x^2 + 2x - 8y + 41 = 0$ **27.** $x^2 - 6x - 6y - 3 = 0$; $x^2 - 6x + 6y + 21 = 0$ **29.** $y^2 - 4x - 4y - 12 = 0$ **31.** $y^2 - 2x - 8y + 25 = 0$; $y^2 - 18x - 8y + 169 = 0$ **33.** $\left(-\dfrac{b}{2a}, \dfrac{4ac - b^2}{4a}\right)$ **35.** $x^2 - 2x + 4y + 1 = 0$; $x^2 - 2x - 16y - 79 = 0$; $x^2 - 2x - 4y - 31 = 0$; $x^2 - 2x + 16y + 49 = 0$ **37.** $x'^3 - y'^2 = 0$ **41.** translate the axes so that the new origin is $(\frac{1}{4}\pi, -3)$

EXERCISES 13.3 (Page 659)

1. $3x^2 - 24x + 4y^2 = 0$; ellipse **3.** $x^2 - 3y^2 + 30y - 27 = 0$; hyperbola **5.** $8x^2 - 24x - y^2 + 4y - 4 = 0$; hyperbola **7.** $4x^2 + 32x - 12y + 49 = 0$; parabola **9.** $43x^2 + 30xy + 3y^2 + 4x + 4y - 4 = 0$; hyperbola **11.** $16x^2 + 4xy + 19y^2 - 152x + 116y + 496 = 0$; ellipse **13.** $\frac{3}{5}$ **15.** 1 **17.** 1 **19.** $e = 1$: $y = 0$; $e < 1$: $\dfrac{x^2}{a^2} + \dfrac{y^2}{b^2} = 1$; $e > 1$: $\dfrac{x^2}{a^2} - \dfrac{y^2}{b^2} = 1$

EXERCISES 13.4 (Page 666)

1. (a) parabola; (b) hyperbola; (c) ellipse; (d) circle **3.** (a) 1; (b) parabola; (c) $r\cos\theta = -2$ **5.** (a) $\frac{1}{2}$; (b) ellipse; (c) $r\sin\theta = 5$ **7.** (a) $\frac{2}{3}$; (b) ellipse; (c) $r\cos\theta = -3$ **9.** (a) $\frac{6}{5}$; (b) hyperbola; (c) $2r\sin\theta = -3$ **11.** (a) $\frac{4}{7}$; (b) ellipse; (c) $r\sin\theta = -5$ **13.** (a) $\frac{5}{4}$; (b) hyperbola; (c) $r\cos\theta = 2$ **15.** $r = \dfrac{8}{1 - \sin\theta}$ **17.** $r = \dfrac{36}{3 + 4\cos\theta}$ **19.** $r = \dfrac{3}{2 - \cos\theta}$ **21.** (a) $r = \dfrac{5}{1 - 3\cos\theta}$; (b) $3r\cos\theta = -5$ **23.** $\frac{16}{3}\sqrt{3}\pi$ sq units **29.** (a) $r = \dfrac{40{,}000{,}000}{1 - \cos\theta}$; (b) 20,000,000 miles **31.** $r^2 = \dfrac{a^2(1 - e^2)}{1 - e^2\cos^2\theta}$

EXERCISES 13.5 (Page 673)

1. vertices: $(\pm 3, 0)$; foci: $(\pm\sqrt{5}, 0)$; directrices: $x = \pm\frac{9}{5}\sqrt{5}$; $e = \frac{1}{3}\sqrt{5}$; ends of minor axis: $(0, \pm 2)$ **3.** vertices: $(\pm 3, 0)$; foci: $(\pm\sqrt{3}, 0)$; directrices: $x = \pm 3\sqrt{3}$; $e = \frac{1}{3}\sqrt{3}$; ends of minor axis: $(0, \pm\sqrt{6})$ **5.** vertices: $(\pm 2, 0)$; foci: $(\pm\sqrt{13}, 0)$; directrices: $x = \pm\frac{4}{13}\sqrt{13}$; $e = \frac{1}{2}\sqrt{13}$; $2a = 4$; $2b = 6$ **7.** vertices: $(\pm\frac{1}{3}, 0)$; foci: $(\pm\frac{5}{12}, 0)$; directrices: $x = \pm\frac{4}{15}$; $e = \frac{5}{4}$; $2a = \frac{2}{3}$; $2b = \frac{1}{2}$ **9.** $9x^2 - 4y^2 = 36$ **11.** $16x^2 + 25y^2 = 100$ **13.** $32x^2 - 33y^2 - 380 = 0$ **15.** $2x + 3y - 12 = 0$ **17.** $\sqrt{2}y - kx - 1 = 0$

19. $7x^2 - 4y^2 = 28$ **21.** $\frac{42}{5}$ m **23.** 98π in.^{3} **25.** the right branch of the hyperbola $16x^2 - 9y^2 = 14{,}400$ **27.** $\frac{4}{3}\pi ab^2$ cu units **29.** 2400ρ ft-lb

EXERCISES 13.6 (Page 679)

1. $e = \frac{1}{3}\sqrt{3}$; center: $(2, 3)$; foci: $(2 \pm \sqrt{3}, 3)$; directrices: $x = 2 \pm 3\sqrt{3}$ **3.** $e = \frac{1}{5}\sqrt{10}$; center: $(0, \frac{1}{2})$; foci: $(0, \frac{1}{2} \pm \frac{1}{10}\sqrt{170})$; directrices: $y = \frac{1}{2} \pm \frac{1}{4}\sqrt{170}$ **5.** point-circle $(-\frac{5}{2}, 4)$ **7.** $e = \frac{1}{5}\sqrt{10}$; center: $(1, 0)$; foci: $(\pm\sqrt{2} + 1, 0)$; directrices: $x = 1 \pm \frac{5}{2}\sqrt{2}$ **9.** point-ellipse **11.** $3x^2 + 4y^2 = 300$ **13.** $x^2 + 4y^2 = 4$ **15.** $\dfrac{(x-4)^2}{25} + \dfrac{(y+2)^2}{9} = 1$ **17.** $\dfrac{(y-3)^2}{64} + \dfrac{(x+1)^2}{48} = 1$

21. $\dfrac{(x-1)^2}{64} + \dfrac{(y+5)^2}{39} = 1$ **23.** $(h \pm \sqrt{a^2 - b^2}, k)$ **25.** $18\pi\rho$ lb

EXERCISES 13.7 (Page 687)

1. $e = \sqrt{2}$; center: $(-4, -1)$; foci: $(-4 \pm 6\sqrt{2}, -1)$; directrices: $x = -4 \pm 3\sqrt{2}$; asymptotes: $x - y + 3 = 0, x + y + 5 = 0$
3. $e = \sqrt{3}$; center: $(-3, -1)$; foci: $(-3, -1 \pm \frac{2}{3}\sqrt{6})$; directrices: $y = -1 \pm \frac{2}{9}\sqrt{6}$; asymptotes: $\pm x + \sqrt{2}y + \sqrt{2} \pm 3 = 0$
5. $e = \frac{1}{2}\sqrt{7}$; center: $(-1, 4)$; foci: $(-1, -3), (-1, 11)$; directrices: $y = 0, y = 8$; asymptotes: $\pm 2x + \sqrt{3}y \pm 2 - 4\sqrt{3} = 0$
7. $e = \frac{1}{3}\sqrt{13}$; center: $(1, -2)$; foci: $(1, -2 \pm \sqrt{13})$; directrices: $y = -2 \pm \frac{9}{13}\sqrt{13}$; asymptotes: $3x + 2y + 1 = 0, 3x - 2y - 7 = 0$
9. $25x^2 - 144y^2 = 14{,}400$ **11.** $\dfrac{(y+1)^2}{144} - \dfrac{(x+2)^2}{81} = 1$ **13.** $\dfrac{(x+3)^2}{36} - \dfrac{(y-1)^2}{81} = 1$ **15.** $\dfrac{(y+2)^2}{9} - \dfrac{(x-1)^2}{4} = 1$

17. (a) $\dfrac{(x+2)^2}{1} - \dfrac{(y+1)^2}{4} = 1$; (b) $2x - y + 3 = 0$; $2x + y + 5 = 0$ **21.** $\dfrac{(y-5)^2}{4} - \dfrac{(x-2)^2}{12} = 1$ **23.** $(3, 4)$

EXERCISES 13.8 (Page 692)

1. $\bar{y}^2 - \bar{x}^2 = 8$ **5.** $16\bar{y}^2 - 9\bar{x}^2 = 36$ **7.** $\bar{x}^2 - \bar{y}^2 = 16$ **9.** $9\bar{x}^2 + 4\bar{y}^2 = 36$ **11.** $3\bar{x}'^2 + \bar{y}'^2 = 18$ **13.** $\bar{x}'^2 + 4\bar{y}'^2 = 16$
15. $\sqrt{2}\bar{x}'^2 = \bar{y}'$ **17.** $\bar{x}'^2 - 4\bar{y}'^2 = 16$

REVIEW EXERCISES FOR CHAPTER 13 (Page 693)

1. $x^2 = -12y$; 12 **3.** (a) $(2, 3)$; (b) $(0, 3)$; (c) $y = 3$; (d) $x = 4$ **5.** $x^2 - 3y^2 - 2x - 16y - 19 = 0$; hyperbola **7.** $5x^2 + 9y^2 - 78x - 54y + 297 = 0$; ellipse **9.** $9(x-1)^2 + 5(y-2)^2 = 405$ **11.** $3x^2 - y^2 - 3 = 0$ **13.** $49(x + \frac{53}{7})^2 + 112(y-3)^2 = 576$

15. $r = \dfrac{8}{1 - 3\cos\theta}$ **17.** $r = \dfrac{6}{1 - \sin\theta}$ **19.** (a) $e = \frac{1}{2}$; (b) ellipse; (c) $r\sin\theta = -2$ **21.** (a) $e = \frac{3}{2}$; (b) hyperbola; (c) $3r\cos\theta = 4$ **23.** $e = \frac{1}{2}\sqrt{3}$; center: $(-3, 8)$; foci: $(-3, 8 \pm 2\sqrt{3})$; directrices: $y = 8 \pm \frac{8}{3}\sqrt{3}$ **25.** $e = \sqrt{26}$; center: $(-1, 3)$; foci: $(-1 \pm \sqrt{26}, 3)$; directrices: $x = -1 \pm \frac{1}{26}\sqrt{26}$; asymptotes: $5x + y + 2 = 0, 5x - y + 8 = 0$ **27.** $21\bar{x}'^2 - 49\bar{y}'^2 = 72$

29. $r = \dfrac{1}{1 - \cos\theta}$ **31.** $4\sqrt{3}\pi\, a^2 b$ cu units **33.** $(x-5)^2 = 8(y-1)$ **35.** 9π cm **37.** 600 mi. **39.** $(3 \pm \frac{3}{4}\sqrt{6})$ million miles **41.** the left branch of the hyperbola $16x^2 - 9y^2 = 1{,}440{,}000$ **47.** first quadrant part of a parabola

EXERCISES 14.1 (Page 704)

1. $\frac{4}{3}$ **3.** $\frac{1}{2}\pi$ **5.** $-\frac{11}{4}$ **7.** 1 **9.** 2 **11.** $+\infty$ **13.** $-\frac{1}{8}$ **15.** $\ln\frac{2}{3}$ **17.** $\frac{1}{6}$ **19.** $\frac{1}{3}$ **21.** $\frac{1}{2}$ **23.** 1

25. $\frac{3}{5}$ **27.** 2 **29.** -1 **31.** $\dfrac{Et}{L}$ **33.** $a = -3, b = \frac{9}{2}$

EXERCISE 14.2 (Page 709)

1. 0 **3.** 0 **5.** 1 **7.** 0 **9.** 0 **11.** $\frac{1}{2}$ **13.** 1 **15.** $\frac{1}{2}$ **17.** 1 **19.** e^2 **21.** e^a **23.** 1 **25.** e^2
27. $e^{-1/3}$ **29.** $\frac{1}{2}$ **31.** 0 **33.** 1 **35.** 1 **39.** 2

EXERCISES 14.3 (Page 715)

1. 1 **3.** $-\dfrac{1}{2\ln 5}$ **5.** 1 **7.** divergent **9.** divergent **11.** $\frac{1}{3}\pi$ **13.** 2 **15.** 1 **17.** divergent

19. (a) divergent; (b) 0 **21.** (a) 0 **23.** $\frac{1}{2}\pi$ **25.** $\frac{1}{2}\pi$ **27.** $\dfrac{1000}{0.08 + \ln 2}$ dollars $\approx$ \$1293.41 **31.** $n = \frac{3}{2}$; $\frac{3}{4}\ln\frac{7}{16}$

EXERCISES 14.4 (Page 719)

1. 2 **3.** -4 **5.** $\frac{1}{3}\pi$ **7.** divergent **9.** divergent **11.** divergent **13.** divergent **15.** divergent **17.** 0

19. divergent **21.** 0 **23.** $\frac{1}{3}\pi$ **25.** 0 **27.** $n > -1$; $\dfrac{1}{n+1}$ **29.** $n > -1$; $\dfrac{2}{(n+1)^3}$ **31.** yes; 6π

EXERCISES 14.5 (Page 725)

1. $P_3(x) = -1 - (x-1) - (x-1)^2 - (x-3)^3$; $R_3(x) = \dfrac{(x-1)^4}{(\xi-2)^5}$, ξ between 1 and x **3.** $P_3(x) = 8 + 3(x-4) + \frac{3}{16}(x-4)^2 - \frac{1}{128}(x-4)^3$; $R_3(x) = \dfrac{3(x-4)^4}{128\xi^{5/2}}$, ξ between 4 and x **5.** $P_3(x) = \frac{1}{2} + \frac{1}{2}\sqrt{3}(x - \frac{1}{6}\pi) - \frac{1}{4}(x - \frac{1}{6}\pi)^2 - \frac{1}{12}\sqrt{3}(x - \frac{1}{6}\pi)^3$; $R_3(x) = \frac{1}{24}\sin\xi(x - \frac{1}{6}\pi)^4$, ξ between $\frac{1}{6}\pi$ and x **7.** $P_4(x) = x + \frac{1}{6}x^3$; $R_4(x) = \frac{1}{120}\cosh\xi x^5$, ξ between 0 and x **9.** $P_3(x) = x - 1 - \frac{1}{2}(x-1)^2 + \frac{1}{3}(x-1)^3$; $R_3(x) = -\frac{1}{4}\xi^{-4}(x-1)^4$, ξ between 1 and x **11.** $P_3(x) = -\ln 2 - \sqrt{3}(x - \frac{1}{3}\pi) - 2(x - \frac{1}{3}\pi)^2 - \frac{4}{3}\sqrt{3}(x - \frac{1}{3}\pi)^3$; $R_3(x) = -\frac{1}{12}(2\sec^2\xi\tan^2\xi + \sec^4\xi)(x - \frac{1}{3}\pi)^4$, ξ between $\frac{1}{3}\pi$ and x **13.** $P_3(x) = 1 + \frac{3}{2}x + \frac{3}{8}x^2 - \frac{1}{16}x^3$; $R_3(x) = \frac{3}{128}(1+\xi)^{-5/2}x^4$, ξ between 0 and x **15.** 2.71828 **17.** $|\text{error}| < \dfrac{(0.1)^4}{24} < 0.000005$ **19.** 0.515 **21.** 0.1823 **23.** $\dfrac{55\sqrt{2}}{672}$, $|\text{error}| < \frac{1}{7680}\sqrt{2}$ **27.** $x \approx \dfrac{\pi}{2(1+m)}$ **29.** $2(x-1) + 5(x-1)^2 + 3(x-1)^3 + (x-1)^4$

REVIEW EXERCISES FOR CHAPTER 14 (Page 726)

1. $+\infty$ **3.** $\frac{1}{3}$ **5.** $+\infty$ **7.** 1 **9.** $-\frac{1}{2}$ **11.** e^{12} **13.** 0 **15.** 1 **17.** e **19.** divergent **21.** $\frac{1}{2}$ **23.** divergent **25.** $\dfrac{32}{\ln 2}$ **27.** divergent **29.** $\frac{1}{4}\pi$ **31.** divergent **33.** (b) $f'(0) = \frac{1}{2}$ **35.** (a) no; (b) 0 **37.** (a) divergent; (b) 0 **39.** $P_6(x) = 1 - \frac{1}{2}x^2 + \frac{1}{24}x^4 - \frac{1}{720}x^6$; $R_6(x) = \frac{1}{5040}\sin\xi x^7$, ξ between 0 and x **41.** $P_4(x) = \dfrac{1}{3} - \dfrac{1}{54}(x-9) + \dfrac{1}{648}(x-9)^2 - \dfrac{5}{34{,}992}(x-9)^3 + \dfrac{35}{2{,}519{,}424}(x-9)^4$; $R_4(x) = -\frac{63}{256}\xi^{-11/2}(x-9)^5$, ξ between 9 and x **43.** $+\infty$ **45.** $\frac{64}{3}$ **47.** 1 **49.** \$152,500

EXERCISES 15.1 (Page 736)

1. $\frac{1}{2}$ **3.** divergent **5.** -2 **7.** 0 **9.** 1 **11.** divergent **13.** divergent **15.** $e^{1/3}$ **17.** 1 **19.** 0 **21.** choose $N = 1 + \dfrac{3}{\epsilon}$ **23.** choose $N = \dfrac{6}{\epsilon} - \dfrac{3}{2}$ **25.** choose $N = \dfrac{1}{5}\sqrt{\dfrac{2-5\epsilon}{\epsilon}}$

EXERCISES 15.2 (Page 742)

1. increasing **3.** decreasing **5.** not monotonic **7.** not monotonic **9.** decreasing **11.** increasing after the first two terms **13.** increasing **15.** decreasing **19.** not bounded **31.** $\left\{\dfrac{(-1)^{n+1}}{n}\right\}$

EXERCISES 15.3 (Page 754)

1. $s_n = \dfrac{n}{2n+1}$; $\dfrac{1}{2}$ **3.** $s_n = \dfrac{5n}{3n+1}$; $\dfrac{5}{3}$ **5.** $s_n = \ln\left(\dfrac{1}{n+1}\right)$; divergent **7.** $s_n = \dfrac{5}{2}\left(1 - \dfrac{1}{5^n}\right)$; $\dfrac{5}{2}$ **9.** $\displaystyle\sum_{n=1}^{+\infty} \frac{2}{(3n-2)(3n+1)}$; $\dfrac{2}{3}$ **11.** $\displaystyle\frac{1}{2} - \sum_{n=2}^{+\infty}\frac{1}{2^n}$; 0 **13.** $\displaystyle\sum_{n=1}^{+\infty}\ln\left(\frac{2n+1}{2n-1}\right)$; divergent **15.** divergent **17.** 2 **19.** divergent **21.** 1 **23.** $\dfrac{1}{e-1}$ **25.** divergent **27.** $\frac{3}{2}$ **29.** divergent **31.** $\frac{3}{11}$ **33.** $\frac{137}{111}$ **35.** 8 m **37.** $\displaystyle\sum_{n=1}^{+\infty}\frac{1}{n}$ and $\displaystyle\sum_{n=1}^{+\infty}\frac{1}{n+1}$ are both divergent; $\displaystyle\sum_{n=1}^{+\infty}\frac{1}{n^2+n}$ is convergent

EXERCISES 15.4 (Page 764)

1. convergent **3.** convergent **5.** divergent **7.** convergent **9.** divergent **11.** convergent **13.** divergent **15.** convergent **17.** divergent **19.** convergent **21.** divergent **23.** convergent **25.** convergent

EXERCISES 15.5 (Page 768)

1. divergent **3.** convergent **5.** convergent **7.** convergent **9.** divergent **11.** convergent **13.** convergent **15.** convergent **17.** convergent **19.** convergent **21.** divergent **27.** $\displaystyle 0.7032 < \sum_{m=50}^{100}\frac{1}{m} < 0.7134$

EXERCISES 15.6 (Page 778)

1. convergent **3.** convergent **5.** convergent **7.** divergent **9.** convergent **11.** $|R_4| < \frac{1}{5}$ **13.** $|R_4| < \frac{1}{81}$

15. 0.112 **17.** 0.406 **19.** absolutely convergent **21.** absolutely convergent **23.** divergent **25.** absolutely convergent **27.** absolutely convergent **29.** absolutely convergent **31.** divergent

EXERCISES 15.7 (Page 785)

1. $[-1, 1)$ **3.** $[-1, 1]$ **5.** $[-\frac{1}{2}, \frac{1}{2}]$ **7.** $(-3, 3)$ **9.** $(-\infty, +\infty)$ **11.** 0 **13.** $(-9, 9]$ **15.** $(0, 2]$ **17.** $\left(-\frac{1}{e^2}, \frac{1}{e^2}\right)$ **19.** $[-1, 1]$ **21.** $(-4, 6)$ **23.** $[4, 6)$ **25.** $[-1, 1]$ **27.** $(-e, e)$ **29.** $+\infty$

EXERCISES 15.8 (Page 794)

1. (a) $R = 1$; $[-1, 1]$; (b) $\sum_{n=1}^{+\infty} \frac{x^{n-1}}{n}$; $R = 1$; (c) $[-1, 1)$ **3.** (a) $R = 1$; $[-1, 1)$; (b) $\sum_{n=1}^{+\infty} \sqrt{n}x^{n-1}$; $R = 1$; (c) $(-1, 1)$ **5.** (a) $R = +\infty$; $(-\infty, +\infty)$; (b) $\sum_{n=1}^{+\infty} (-1)^{n-1} \frac{x^{2n-2}}{(2n-2)!}$; $R = +\infty$; (c) $(-\infty, +\infty)$ **7.** (a) $R = \frac{1}{3}$; $(0, \frac{2}{3})$; (b) $\sum_{n=1}^{+\infty} 3n(n+1)(3x-1)^n$; $R = \frac{1}{3}$; (c) $(0, \frac{2}{3})$ **9.** (a) $R = 3$; $[-2, 4)$; (b) $\sum_{n=1}^{+\infty} \frac{(x-1)^{n-1}}{3^n}$; $R = 3$; (c) $(-2, 4)$ **11.** $\frac{1}{2}\sum_{n=2}^{+\infty} n(n-1)x^{n-2}$ **13.** $\sum_{n=0}^{+\infty} (-1)^n(n+1)x^n$ **15.** (a) $\sum_{n=0}^{+\infty} 2^n x^n$ if $|x| < \frac{1}{2}$; (b) $\sum_{n=1}^{+\infty} n2^n x^{n-1}$ if $|x| < \frac{1}{2}$ **17.** (a) $\sum_{n=0}^{+\infty} \frac{x^{2n}}{n!}$; (b) $\sum_{n=0}^{+\infty} \frac{x^{2n+1}}{n!}$ **19.** 0.60653 **21.** (a) $\sum_{n=0}^{+\infty} \frac{x^{2n+1}}{(2n+1)!}$; (b) $\sum_{n=0}^{+\infty} \frac{x^{2n}}{(2n)!}$ **23.** 2 **25.** (a) $\sum_{n=0}^{+\infty} (-1)^n \frac{x^{n+2}}{n!}$ **27.** $\sum_{n=0}^{+\infty} \frac{x^n}{n!}$

EXERCISES 15.9 (Page 802)

1. (a) 0.288; (b) 0.288 **3.** (a) 1.718; (b) 1.718 **5.** 0.485 **7.** 0.747 **9.** 0.041 **11.** 0.450 **13.** 0.248 **15.** 1.318 **17.** $\sum_{n=0}^{+\infty} \frac{x^{2n+1}}{(2n+1)!}$ **19.** 0.2450 **21.** 0.221 **23.** $\sum_{n=0}^{+\infty} \frac{x^{2n+1}}{2n+1}$ **29.** $\sum_{n=0}^{+\infty} \frac{(-1)^n x^{2n+1}}{(2n+1)!}$; $R = +\infty$

EXERCISES 15.10 (Page 812)

7. $\ln 2 + \sum_{n=1}^{+\infty} (-1)^{n-1} \frac{(x-2)^n}{n2^n}$ **9.** $\ln 2 + \sum_{n=1}^{+\infty} (-1)^{n-1} \frac{(x-1)^n}{n2^n}$; $R = 2$ **11.** $2 + \frac{1}{4}(x-4) + 2\sum_{n=2}^{+\infty} (-1)^{n-1} \frac{1 \cdot 3 \cdot 5 \cdot \ldots \cdot (2n-3)(x-4)^n}{2 \cdot 4 \cdot 6 \cdot \ldots \cdot (2n) \cdot 4^n}$; $R = 4$ **13.** $\frac{1}{2} - \frac{1}{2}\sqrt{3}(x - \frac{1}{3}\pi) - \frac{1}{4}(x - \frac{1}{3}\pi)^2 + \frac{1}{12}\sqrt{3}(x - \frac{1}{3}\pi)^3 + \frac{1}{48}(x - \frac{1}{3}\pi)^4 - \cdots$; $R = +\infty$ **15.** $\frac{1}{2}\sum_{n=1}^{+\infty} \frac{(-1)^{n-1}(2x)^{2n}}{(2n)!}$ **17.** (a) $x + \frac{1}{3}x^3 + \frac{2}{15}x^5$; (b) $1 + x^2 + \frac{2}{3}x^4$; (c) $\frac{1}{2}x^2 + \frac{1}{12}x^4 + \frac{1}{45}x^6$ **19.** 0.5299 **21.** 1.97435 **23.** -0.2231 **25.** 2.7182818 **27.** 0.0415 **29.** 0.0048 **31.** 0.2398 **33.** $a_4 = 3$; $a_3 = -5$; $a_2 = 2$; $a_1 = -1$; $a_0 = 6$

EXERCISES 15.11 (Page 817)

1. $1 + \frac{x}{2} + \sum_{n=2}^{+\infty} \frac{(-1)^n(-1) \cdot 1 \cdot 3 \cdot \ldots \cdot (2n-3)}{2^n n!} x^n$; $R = 1$ **3.** $\frac{1}{2} + \frac{1}{2}\sum_{n=1}^{+\infty} \frac{(-1)^n \cdot 1 \cdot 3 \cdot 5 \cdot \ldots \cdot (2n-1)x^n}{8^n n!}$; $R = 4$ **5.** $1 + \sum_{n=1}^{+\infty} \frac{(-1)2 \cdot 5 \cdot 8 \cdot \ldots \cdot (3n-4)}{3^n n!} x^{3n}$; $R = 1$ **7.** $\frac{1}{3} + \frac{1}{3}\sum_{n=1}^{+\infty} \frac{(-1)^n 1 \cdot 3 \cdot 5 \cdot \ldots \cdot (2n-1)x^{4n}}{18^n n!}$; $R = \sqrt{3}$ **9.** $x^2 + \sum_{n=1}^{+\infty} \frac{(-1)^n \cdot 1 \cdot 3 \cdot 5 \cdot \ldots \cdot (2n-1)x^{n+2}}{2^n n!}$; $R = 1$ **11.** 4.899 **13.** 5.010 **15.** 0.198 **17.** $x + \sum_{n=1}^{+\infty} \frac{(-1)^n \cdot 1 \cdot 3 \cdot 5 \cdot \ldots \cdot (2n-1)x^{2n+1}}{2^n n!(2n+1)}$; $R = 1$ **19.** 2.0271 **21.** 0.4970 **23.** 0.5082 **25.** $\frac{1}{p+1}x^{p+1} + \sum_{n=1}^{+\infty} \frac{1 \cdot 3 \cdot 5 \cdot \ldots \cdot (2n-1)x^{2n+p+1}}{2^n n!(2n+p+1)}$; $R = 1$

REVIEW EXERCISES FOR CHAPTER 15 (page 818)

1. $1, \frac{3}{2}, \frac{9}{5}, 2; 3$ **3.** $0, \frac{3}{5}, \frac{4}{5}, \frac{15}{17}; 1$ **5.** 1, 3, 1, 3; no limit **7.** $4, \frac{81}{16}, \frac{4096}{729}, \frac{390{,}625}{65{,}536}; e^2$ **9.** $\frac{3}{16}, \frac{15}{64}, \frac{63}{256}, \frac{255}{1024}; s_n = \frac{4^n - 1}{4^{n+1}}; \frac{1}{4}$
11. convergent; 3 **13.** divergent **15.** convergent; $4 + 2\sqrt{3}$ **17.** convergent; $\frac{1}{6}$ **19.** convergent **21.** divergent
23. convergent **25.** divergent **27.** convergent **29.** divergent **31.** divergent **33.** $\frac{437}{330}$ **35.** absolutely convergent
37. conditionally convergent **39.** divergent **41.** absolutely convergent **43.** $[-1, 1)$ **45.** $[-3, 3]$ **47.** $x = 3$
49. $(-7, 5)$ **51.** $(-1, 3)$ **53.** 0.1974 **55.** 5.0658 **57.** 0.9986 **59.** 1.6094 **61.** 0.0124 **63.** $\sum_{n=0}^{+\infty} \frac{(\ln a)^n x^n}{n!}$; $(-\infty, +\infty)$ **65.** $\frac{3}{4}\sum_{n=1}^{+\infty} \frac{(-1)^{n-1}(3^{2n} - 1)x^{2n+1}}{(2n+1)!}; (-\infty, +\infty)$ **67.** $\sum_{n=1}^{+\infty} \frac{(-1)^n(3x + \pi)^{2n-1}}{(2n-1)!}$ **69.** $\sum_{n=1}^{+\infty} -\frac{(x+1)^n}{n}$

EXERCISES 16.1 (Page 825)

1. 5 **3.** 2 **5.** $\sqrt{11}$ **7.** $\langle 2, -3\rangle$ **9.** $\langle -2, -7\rangle$ **11.** $\langle 5, 6\rangle$ **13.** $(-4, 3)$ **15.** $(12, -5)$ **17.** $\langle -1, 9\rangle$
19. $\langle 1, -5\rangle$ **21.** $\langle -2, 2\rangle$ **23.** $\langle 7, 3\rangle$ **25.** $\langle -9, -4\rangle$ **27.** $\langle \sqrt{2}, \sqrt{3}\rangle$ **29.** (a) $\langle 1, -2\rangle$; (b) $\langle 1, -2\rangle$ **31.** $\langle -2, 7\rangle$
33. $\sqrt{74}$ **35.** $\sqrt{1061}$ **37.** $2\sqrt{26}$ **39.** $h = \frac{1}{4}, k = \frac{1}{2}$

EXERCISES 16.2 (Page 830)

1. $10\mathbf{i} + 15\mathbf{j}$ **3.** $-24\mathbf{i} + 6\mathbf{j}$ **5.** $6\mathbf{i} + 2\mathbf{j}$ **7.** $2\sqrt{10}$ **9.** $\sqrt{13} + \sqrt{17}$ **11.** $-14\mathbf{i} + 21\mathbf{j}$ **13.** $7\sqrt{13}$ **15.** $5\sqrt{13} - 6\sqrt{17}$ **17.** (a) $-28\mathbf{i} + 6\mathbf{j}$; (b) $2\sqrt{205}$ **19.** $\frac{11}{\sqrt{137}}\mathbf{i} + \frac{4}{\sqrt{137}}\mathbf{j}$ **21.** (a) $5(\frac{3}{5}\mathbf{i} - \frac{4}{5}\mathbf{j})$; (b) $\frac{3}{5}\mathbf{i} - \frac{4}{5}\mathbf{j}$
23. (a) $2\sqrt{2}(\cos \frac{1}{4}\pi\mathbf{i} + \sin \frac{1}{4}\pi\mathbf{j})$; (b) $\frac{1}{2}\sqrt{2}\mathbf{i} + \frac{1}{2}\sqrt{2}\mathbf{j}$ **25.** (a) $8(\cos \frac{2}{3}\pi\mathbf{i} + \sin \frac{2}{3}\pi\mathbf{j})$; (b) $-\frac{1}{2}\mathbf{i} + \frac{1}{2}\sqrt{3}\mathbf{j}$ **27.** (a) $16(\cos \pi\mathbf{i} + \sin \pi\mathbf{j})$; (b) $-\mathbf{i}$ **29.** $h = 2, k = 3$

EXERCISES 16.3 (Page 837)

1. 10 **3.** -1 **11.** $\frac{1}{10}\sqrt{2}$ **13.** $-\frac{16}{65}$ **15.** $10, -\frac{2}{5}$ **17.** (a) 0; (b) no k **19.** $\frac{-240 + \sqrt{85{,}683}}{407}$ **21.** $-\frac{4}{17}\sqrt{17}\mathbf{i} + \frac{1}{17}\sqrt{17}\mathbf{j}$; $\frac{4}{17}\sqrt{17}\mathbf{i} - \frac{1}{17}\sqrt{17}\mathbf{j}$ **23.** $\pm\frac{1}{|A|}(a_2\mathbf{i} - a_1\mathbf{j})$ **25.** $-\frac{112}{17}\mathbf{i} + \frac{96}{17}\mathbf{j}$ **27.** $\frac{29}{50}\sqrt{50}$ **29.** (a) 24 ft-lb; (b) $24\sqrt{3}$ ft-lb
31. $(18 - 9\sqrt{3})$ ft-lb ≈ 2.41 ft-lb

EXERCISES 16.4 (Page 845)

1. $(-\infty, 0) \cup (0, 4]$ **3.** $[-1, 1]$ **5.** all real numbers not in $(-4, 3)$ **7.** $\frac{4}{3}t, \frac{4}{9}$ **9.** $\frac{1 + \ln t}{te^t(2 + t)}$; $\frac{(2 + t) - (1 + \ln t)(2 + 4t + t^2)}{t^3e^{2t}(2 + t)^3}$ **11.** $-\frac{b}{a}\cot t; -\frac{b}{a^2}\csc^3 t$ **13.** $(y - 1)^2 = x$ **15.** $25x^2 - 9y^2 = 225$ **17.** $x^2 - y^2 = 1, x \geq 1$
19. $9x^2 + 16y^2 = 144$ **21.** $5\sqrt{3}x + 2y = 20$ **23.** $y = 1; x = -1$ **27.** $\frac{dy}{dx} = 0; \frac{d^2y}{dx^2} = -\frac{1}{4a}; \frac{d^3y}{dx^3} = 0$ **31.** $x^{2/3} + y^{2/3} = a^{2/3}$
35. $3\pi a^2$ sq units

EXERCISES 16.5 (Page 853)

1. $4\mathbf{i} + 4\mathbf{j}$ **3.** $2\mathbf{i}$ **5.** $\mathbf{i}$ **7.** $\mathbf{R}'(t) = 2e^{2t}\mathbf{i} + \frac{1}{t}\mathbf{j}; \mathbf{R}''(t) = 4e^{2t}\mathbf{i} - \frac{1}{t^2}\mathbf{j}$ **9.** $\mathbf{R}'(t) = (1 + t^2)^{-1}\mathbf{i} + 2^t \ln 2\mathbf{j}; \mathbf{R}''(t) = -2t(1 + t^2)^{-2}\mathbf{i} + 2^t(\ln 2)^2\mathbf{j}$ **11.** $\mathbf{R}'(t) = -2t(t^2 + 4)^{-2}\mathbf{i} - \frac{5}{2}(1 - 5t)^{-1/2}\mathbf{j}; \mathbf{R}''(t) = (6t^2 - 8)(t^2 + 4)^{-3}\mathbf{i} - \frac{25}{4}(1 - 5t)^{-3/2}\mathbf{j}$
13. $\mathbf{R}'(t) = 10 \cos 2t\mathbf{i} - 4 \sec 4t \tan 4t\mathbf{j}; \mathbf{R}''(t) = -20 \sin 2t\mathbf{i} + (16 \sec 4t - 32 \sec^3 4t)\mathbf{j}$ **15.** $(2t - 3)(2t^2 - 6t + 5)^{-1/2}$ **17.** $20t$
19. $8e^{4t} - 8e^{-4t}$ **25.** $\ln|\sec t|\mathbf{i} - \ln|t|\mathbf{j} + \mathbf{C}$ **27.** $\frac{1}{3}e^{3t}\mathbf{i} + \ln|t - 1|\mathbf{j} + \mathbf{C}$ **29.** $(t \ln t - t)\mathbf{i} + \frac{1}{3}t^3\mathbf{j} + \mathbf{C}$
31. $\frac{1}{2}(t - \frac{1}{2}\sin 2t - \pi)\mathbf{i} + (t + \frac{1}{2}\sin 2t - \pi)\mathbf{j}$ **33.** $x^2 + y^2 = 1; 0$ **35.** 0

EXERCISES 16.6 (Page 860)

1. $1 + \frac{1}{2}\sqrt{2}\ln(1 + \sqrt{2})$ **3.** $2\sqrt{10} + \sqrt{2}\ln(2 + \sqrt{5})$ **5.** $\frac{2}{27}[(40)^{3/2} - (13)^{3/2}]$ **7.** 120 **9.** $6a$ **11.** 8π
13. $a[\ln \cosh 2 + \ln \cosh 1]$ **15.** 5π **17.** $2\pi a$ **19.** 12 **21.** $\frac{1}{2}\sqrt{5}(e^8 - 1)$ **23.** $\frac{1}{2}a(\theta_1 - \frac{3}{2}\sin \frac{2}{3}\theta_1)$ **25.** 200 m

EXERCISES 16.7 (Page 866)

1. (a) $2t\mathbf{i} + \mathbf{j}$; (b) $2\mathbf{i}$; (c) $\sqrt{37}$; (d) 2 **3.** (a) $-10 \sin 2t\mathbf{i} + 6 \cos 2t\mathbf{j}$; (b) $-20 \cos 2t\mathbf{i} - 12 \sin 2t\mathbf{j}$; (c) 10; (d) 12 **5.** (a) $\mathbf{i} + \tan t\mathbf{j}$;

(b) $\sec^2 t\mathbf{j}$; (c) $\sqrt{2}$; (d) 2 **7.** (a) $\cos t\mathbf{i} + \sec^2 t\mathbf{j}$; (b) $-\sin t\mathbf{i} + 2\sec^2 t \tan t\mathbf{j}$; (c) $\frac{1}{6}\sqrt{91}$; (d) $\frac{1}{18}\sqrt{849}$ **9.** (a) $2\mathbf{i} + 6\mathbf{j}$; (b) $2\mathbf{j}$; (c) $2\sqrt{10}$; (d) 2 **11.** (a) $2\mathbf{i} + 8\mathbf{j}$; (b) $2\mathbf{i} + 16\mathbf{j}$; (c) $2\sqrt{17}$; (d) $2\sqrt{65}$ **13.** (a) $3\mathbf{j}$; (b) $-4\mathbf{i}$; (c) 3; (d) 4 **15.** (a) $\mathbf{i} + \sqrt{3}\mathbf{j}$; (b) $-\sqrt{3}\mathbf{i} + \mathbf{j}$; (c) 2; (d) 2 **17.** $\dfrac{2t-3}{t-1}\mathbf{i} + \dfrac{4-2t-t^2}{2}\mathbf{j}$ **19.** $(e^{-t} + 3t - 1)\mathbf{i} + (\frac{1}{2}e^{2t} + \frac{5}{2})\mathbf{j}$ **21.** (a) $\dfrac{390{,}625}{2}$ ft; (b) $\dfrac{390{,}625}{8}$ ft; (c) $1250\sqrt{2}\mathbf{i} - 1250\sqrt{2}\mathbf{j}$ **23.** $(25 + \sqrt{631})$ sec; $(20{,}000\sqrt{3} + 800\sqrt{1893})$ ft **25.** 283 ft/sec **27.** $40° 8'$

EXERCISES 16.8 (Page 871)

1. $\mathbf{T}(t) = \dfrac{t^2-1}{t^2+1}\mathbf{i} + \dfrac{2t}{t^2+1}\mathbf{j}$; $\mathbf{N}(t) = \dfrac{2t}{t^2+1}\mathbf{i} + \dfrac{1-t^2}{t^2+1}\mathbf{j}$ **3.** $\mathbf{T}(t) = \dfrac{e^{2t}}{\sqrt{e^{4t}+1}}\mathbf{i} - \dfrac{1}{\sqrt{e^{4t}+1}}\mathbf{j}$; $\mathbf{N}(t) = \dfrac{1}{\sqrt{e^{4t}+1}}\mathbf{i} + \dfrac{e^{2t}}{\sqrt{e^{4t}+1}}\mathbf{j}$
5. $\mathbf{T}(t) = -\sin t\mathbf{i} + \cos t\mathbf{j}$; $\mathbf{N}(t) = -\cos t\mathbf{i} - \sin t\mathbf{j}$ **7.** $\mathbf{T}(t) = -\sin kt\mathbf{i} + \cos kt\mathbf{j}$; $\mathbf{N}(t) = -\cos kt\mathbf{i} - \sin kt\mathbf{j}$
9. $\mathbf{T}(t) = -(1 + \cot^4 t)^{-1/2}\mathbf{i} + (1 + \tan^4 t)^{-1/2}\mathbf{j}$; $\mathbf{N}(t) = \dfrac{-\cos^2 t}{\sqrt{\sin^4 t + \cos^4 t}}\mathbf{i} - \dfrac{\sin^2 t}{\sqrt{\sin^4 t + \cos^4 t}}\mathbf{j}$ **11.** $\mathbf{T}(t) = \dfrac{2t}{\sqrt{4t^2 + e^{2t}}}\mathbf{i} + \dfrac{e^t}{\sqrt{4t^2 + e^{2t}}}\mathbf{j}$; $\mathbf{N}(t) = \dfrac{e^t(1-t)}{|t-1|\sqrt{4t^2+e^{2t}}}\mathbf{i} + \dfrac{2t(t-1)}{|t-1|\sqrt{4t^2+e^{2t}}}\mathbf{j}$ **13.** $\mathbf{T}(t) = \sin t\mathbf{i} - \cos t\mathbf{j}$; $\mathbf{N}(t) = \cos t\mathbf{i} + \sin t\mathbf{j}$ **15.** $\frac{27}{185}\sqrt{37}$
19. $x = 2 + \cos s, y = 3 + \sin s$ **21.** $x = \frac{1}{9}[(27s+8)^{2/3} - 4], y = \frac{1}{27}[(27s+8)^{2/3} - 4]^{3/2}$ **23.** $x = a\left(\dfrac{3a-2s}{3a}\right)^{3/2}$, $y = a\left(\dfrac{2s}{3a}\right)^{3/2}$ **25.** $s = \sqrt{2}(e^t - 1)$

EXERCISES 16.9 (Page 878)

1. $\frac{1}{8}\sqrt{2}$; $4\sqrt{2}$ **3.** $\frac{1}{4}\sqrt{2}$; $2\sqrt{2}$ **5.** $\dfrac{\sqrt{2}\,|1-t^2|^3}{(1+6t^2+t^4)^{3/2}}$; $\sqrt{2}$; $\frac{1}{2}\sqrt{2}$ **7.** $\frac{1}{2}$; 2 **9.** $\frac{1}{4}\sqrt{2}$; $2\sqrt{2}$ **11.** $\frac{4}{49}\sqrt{7}$; $\frac{7}{4}\sqrt{7}$ **13.** $\frac{2}{289}\sqrt{17}$; $\frac{17}{2}\sqrt{17}$ **15.** $\dfrac{(2-x^2)^{3/2}}{|x|}$ **17.** $\frac{1}{576}(16x^2 + 81y^2)^{3/2}$ **19.** $\dfrac{2(x+y)^{3/2}}{a^{1/2}}$ **21.** $4|a \sin \frac{1}{2}t|$ **25.** $(3, 9)$ **27.** $(-3, -1)$
29. $x^2 + y^2 + 4x - 6y + 5 = 0$ **31.** $\frac{23}{98}\sqrt{7}$; $\frac{14}{23}\sqrt{7}$ **33.** $\dfrac{1}{16|a|}$; $16|a|$ **37.** 2; $\frac{1}{2}$; $(0, -\frac{1}{2})$ **39.** $\left(3x + 2p, -\dfrac{y^3}{4p^2}\right)$
41. $\left(\dfrac{a^2-b^2}{a}\cos^3 t, \dfrac{b^2-a^2}{b}\sin^3 t\right)$

EXERCISES 16.10 (Page 882)

1. $\mathbf{V}(t) = (3t^2 - 3)\mathbf{i} + 6t\mathbf{j}$; $\mathbf{A}(t) = 6t\mathbf{i} + 6\mathbf{j}$; $A_T(t) = 6t$; $A_N(t) = 6$ **3.** $\mathbf{V}(t) = t\cos t\mathbf{i} + t\sin t\mathbf{j}$; $\mathbf{A}(t) = (\cos t - t\sin t)\mathbf{i} + (\sin t + t\cos t)\mathbf{j}$; $A_T(t) = 1$; $A_N(t) = t$ **5.** $\mathbf{V}(\frac{1}{4}\pi) = \frac{3}{4}\sqrt{2}\mathbf{i} - \frac{3}{4}\sqrt{2}\mathbf{j}$; $\mathbf{A}(\frac{1}{4}\pi) = \frac{3}{4}\sqrt{2}\mathbf{i} + \frac{3}{4}\sqrt{2}\mathbf{j}$; $A_T(\frac{1}{4}\pi) = 0$; $A_N(\frac{1}{4}\pi) = \frac{3}{2}$ **7.** $\mathbf{V}(t) = 2\mathbf{i} + 2t\mathbf{j}$; $\mathbf{A}(t) = 2\mathbf{j}$; $\mathbf{T}(t) = \dfrac{1}{\sqrt{1+t^2}}\mathbf{i} + \dfrac{t}{\sqrt{1+t^2}}\mathbf{j}$; $\mathbf{N}(t) = \dfrac{-t}{\sqrt{1+t^2}}\mathbf{i} + \dfrac{1}{\sqrt{1+t^2}}\mathbf{j}$; $|\mathbf{V}(t)| = 2\sqrt{1+t^2}$; $A_T(t) = \dfrac{2t}{\sqrt{1+t^2}}$; $A_N(t) = \dfrac{2}{\sqrt{1+t^2}}$; $K(t) = \dfrac{1}{2(1+t^2)^{3/2}}$; $\mathbf{V}(2) = 2\mathbf{i} + 4\mathbf{j}$; $\mathbf{T}(2) = \dfrac{1}{\sqrt{5}}\mathbf{i} + \dfrac{2}{\sqrt{5}}\mathbf{j}$; $\mathbf{A}(2) = 2\mathbf{j}$; $\mathbf{N}(2) = \dfrac{-2}{\sqrt{5}}\mathbf{i} + \dfrac{1}{\sqrt{5}}\mathbf{j}$; $|\mathbf{V}(2)| = 2\sqrt{5}$; $A_T(2) = \dfrac{4}{\sqrt{5}}$; $A_N(2) = \dfrac{2}{\sqrt{5}}$; $K(2) = \dfrac{1}{10\sqrt{5}}$ **9.** $\mathbf{V}(t) = -15\sin 3t\mathbf{i} + 15\cos 3t\mathbf{j}$; $\mathbf{A}(t) = -45\cos 3t\mathbf{i} - 45\sin 3t\mathbf{j}$; $\mathbf{T}(t) = -\sin 3t\mathbf{i} + \cos 3t\mathbf{j}$; $\mathbf{N}(t) = -\cos 3t\mathbf{i} - \sin 3t\mathbf{j}$; $|\mathbf{V}(t)| = 15$; $A_T(t) = 0$; $A_N(t) = 45$; $K(t) = \frac{1}{5}$; $\mathbf{V}(\frac{1}{3}\pi) = -15\mathbf{j}$; $\mathbf{A}(\frac{1}{3}\pi) = 45\mathbf{i}$; $\mathbf{T}(\frac{1}{3}\pi) = -\mathbf{j}$; $\mathbf{N}(\frac{1}{3}\pi) = \mathbf{i}$; $|\mathbf{V}(\frac{1}{3}\pi)| = 15$
11. $\mathbf{V}(t) = e^t\mathbf{i} - e^{-t}\mathbf{j}$; $\mathbf{A}(t) = e^t\mathbf{i} + e^{-t}\mathbf{j}$; $\mathbf{T}(t) = \dfrac{e^{2t}}{\sqrt{e^{4t}+1}}\mathbf{i} - \dfrac{1}{\sqrt{e^{4t}+1}}\mathbf{j}$; $\mathbf{N}(t) = \dfrac{1}{\sqrt{e^{4t}+1}}\mathbf{i} + \dfrac{e^{2t}}{\sqrt{e^{4t}+1}}\mathbf{j}$; $|\mathbf{V}(t)| = \dfrac{\sqrt{e^{4t}+1}}{e^t}$; $A_T(t) = \dfrac{e^{4t}-1}{e^t\sqrt{e^{4t}+1}}$; $A_N(t) = \dfrac{2e^t}{\sqrt{e^{4t}+1}}$; $K(t) = \dfrac{2e^{3t}}{(e^{4t}+1)^{3/2}}$; $\mathbf{V}(0) = \mathbf{i} - \mathbf{j}$; $\mathbf{A}(0) = \mathbf{i} + \mathbf{j}$; $\mathbf{T}(0) = \dfrac{1}{\sqrt{2}}\mathbf{i} - \dfrac{1}{\sqrt{2}}\mathbf{j}$; $\mathbf{N}(0) = \dfrac{1}{\sqrt{2}}\mathbf{i} + \dfrac{1}{\sqrt{2}}\mathbf{j}$; $|\mathbf{V}(0)| = \sqrt{2}$; $A_T(0) = 0$; $A_N(0) = \sqrt{2}$; $K(0) = \dfrac{1}{\sqrt{2}}$ **13.** (a) $\mathbf{i} + 4\mathbf{j}$; (b) $\mathbf{i} + 8\mathbf{j}$; (c) $8\mathbf{j}$; (d) $\frac{64}{65}\sqrt{65}$; (e) $\frac{8}{65}\sqrt{65}$ **15.** $\mathbf{R} = 2\mathbf{i} + 4\mathbf{j}$; if k = constant speed, $\mathbf{V} = \dfrac{k}{\sqrt{2}}\mathbf{i} + \dfrac{k}{\sqrt{2}}\mathbf{j}$; $\mathbf{A} = \dfrac{k^2}{16}\mathbf{i} - \dfrac{k^2}{16}\mathbf{j}$; $\mathbf{T} = \dfrac{1}{\sqrt{2}}\mathbf{i} + \dfrac{1}{\sqrt{2}}\mathbf{j}$; $\mathbf{N} = \dfrac{1}{\sqrt{2}}\mathbf{i} - \dfrac{1}{\sqrt{2}}\mathbf{j}$; $A_T = 0$; $A_N = \dfrac{k^2\sqrt{2}}{16}$

REVIEW EXERCISES FOR CHAPTER 16 (Page 883)

1. $-25\mathbf{i} + 63\mathbf{j}$ **3.** $\sqrt{4594} \approx 67.8$ **5.** $15\sqrt{2} - 14\sqrt{13} \approx 71.7$ **7.** 92 **9.** $\dfrac{9}{\sqrt{106}}\mathbf{i} - \dfrac{5}{\sqrt{106}}\mathbf{j}$ **11.** $h = -\frac{1}{2}$; $k = \frac{1}{2}$
13. $\dfrac{-38}{\sqrt{50}}$ **15.** (a) all real numbers in $[0, +\infty)$ except $t = 1$; (b) $\frac{1}{2}\mathbf{i} + \frac{1}{2}\mathbf{j}$; (c) $\dfrac{-1}{(t+1)^2}\mathbf{i} + \dfrac{2t^{1/2} - t - 1}{2t^{1/2}(t-1)^2}\mathbf{j}$ **17.** $\dfrac{dy}{dx} = \dfrac{1}{6t}$;

$\frac{d^2y}{dx^2} = -\frac{1}{108t^3}$ **19.** $x = 12; y = 16; y = -16$ **21.** $-\frac{4(t^{12} + t^{10}) + 144(t^2 - 1)^3}{t^9(t^2 - 1)^3}$ **23.** $\frac{3}{2}\sqrt{37} + \frac{1}{4}\ln(6 + \sqrt{37})$ **29.** $|at|$

31. $\frac{2}{75}; \frac{75}{2}$ **33.** $\frac{dy}{dx} = -\tan t; \frac{d^2y}{dx^2} = \frac{1}{3a}\sec^4 t \csc t$ **35.** $A_T = \frac{-8 + 4t}{\sqrt{25 - 16t + 4t^2}}; A_N = \frac{6}{\sqrt{25 - 16t + 4t^2}}$ **37.** $\mathbf{T}(t) = \frac{e^t - e^{-t}}{e^t + e^{-t}}\mathbf{i}$
$+ \frac{2}{e^t + e^{-t}}\mathbf{j}; \mathbf{N}(t) = \frac{2}{e^t + e^{-t}}\mathbf{i} + \frac{e^{-t} + e^t}{e^t + e^{-t}}\mathbf{j}$ **39.** $x = 2(3s + 17\sqrt{17})^{2/3} - 34; y = \frac{1}{3}\{(3s + 17\sqrt{17})^{2/3} - 16\}^{3/2}$

41. $h = \frac{1}{64}(v_0 \sin \alpha)^2$ **47.** $\mathbf{V}(t) = 2 \sinh 2t\mathbf{i} + 2 \cosh 2t\mathbf{j}; \mathbf{A}(t) = 4 \cosh 2t\mathbf{i} + 4 \sinh 2t\mathbf{j}; |\mathbf{V}(t)| = 2\sqrt{\cosh 4t}; A_T(t) = \frac{4 \sinh 4t}{\sqrt{\cosh 4t}};$
$A_N(t) = \frac{4}{\sqrt{\cosh 4t}}$

EXERCISES 17.1 (Page 891)

1. (b) (7, 2, 0), (0, 0, 3), (0, 2, 0), (0, 2, 3), (7, 0, 3), (7, 0, 0); (c) $\sqrt{62}$ **3.** (b) (2, 1, 2), (−1, 3, 2), (−1, 1, 5), (2, 3, 2), (−1, 3, 5), (2, 1, 5); (c) $\sqrt{22}$ **5.** (b) (3, −1, 0), (3, 3, 0), (1, 3, 0), (1, 3, 5), (1, −1, 5), (3, −1, 5); (c) $3\sqrt{5}$ **7.** (a) 3; (b) $(2, 5, \frac{5}{2})$ **9.** (a) $\frac{13}{2}$; (b) $(\frac{5}{4}, -1, 2)$ **11.** (a) $7\sqrt{2}$; (b) $(-1, \frac{9}{2}, -\frac{1}{2})$ **13.** $(\pm 4\sqrt{6}, 4, 2)$ **17.** (a) $|\overline{AB}| = 9\sqrt{2}; |\overline{AC}| = 2\sqrt{62}; |\overline{BC}| = \sqrt{62}$; (b) midpoint of AB: $(\frac{1}{2}, 1, \frac{3}{2})$; midpoint of AC: (−1, 2, 5); midpoint of BC: $(-\frac{5}{2}, 8, \frac{7}{2})$ **21.** sphere with center at (4, −2, −1) and $r = 5$ **23.** the point (0, 0, 3) **25.** the empty set **27.** $x^2 + (y - 1)^2 + (z + 4)^2 = 9$

EXERCISES 17.2 (Page 898)

1. $\langle 21, -13, -2 \rangle$ **3.** $\langle -25, -26, 5 \rangle$ **5.** $7\sqrt{59} - 5\sqrt{41}$ **7.** $\sqrt{1326}$ **9.** $\langle -19, -16, -1 \rangle$
11. $\langle -6\sqrt{91}, -8\sqrt{91}, -2\sqrt{91} \rangle$ **13.** $a = b = 0$ **21.** $\frac{4}{\sqrt{89}}; \frac{3}{\sqrt{89}}; \frac{8}{\sqrt{89}}$ **23.** $-\frac{6}{\sqrt{86}}; -\frac{1}{\sqrt{86}}; -\frac{7}{\sqrt{86}}$ **25.** $(\frac{13}{3}, 0, -\frac{4}{3})$
27. $(\frac{17}{3}, \frac{4}{3}, -6)$ **29.** $7(-\frac{6}{7}\mathbf{i} + \frac{2}{7}\mathbf{j} + \frac{3}{7}\mathbf{k})$ **31.** $\sqrt{14}\left(-\frac{2}{\sqrt{14}}\mathbf{i} + \frac{1}{\sqrt{14}}\mathbf{j} - \frac{3}{\sqrt{14}}\mathbf{k}\right)$ **33.** $\langle \frac{1}{9}, \frac{8}{9}, \frac{4}{9} \rangle$ **35.** $\left\langle -\frac{1}{\sqrt{3}}, \frac{1}{\sqrt{3}}, \frac{1}{\sqrt{3}} \right\rangle$
37. $\cos^{-1}\frac{1}{\sqrt{3}}$ or $\cos^{-1}\left(-\frac{1}{\sqrt{3}}\right)$ **39.** (c) $r = 8, s = -7, t = 5$

EXERCISES 17.3 (Page 902)

1. −44 **3.** −468 **5.** −31 **7.** $\langle -84, 198, 124 \rangle$ **9.** $-\frac{13}{54}\sqrt{6}$ **11.** (a) −3; (b) $\langle 2, 1, -2 \rangle$ **13.** (a) $-\frac{13}{9}\sqrt{6}$; (b) $\langle -\frac{26}{27}, -\frac{91}{27}, \frac{13}{27} \rangle$ **21.** $\frac{1}{11}\sqrt{4422}$ **25.** $\frac{7}{2}\sqrt{3}$ sq units **27.** $-\frac{43}{30}\sqrt{6}$ **29.** 25 ft-lb **31.** $\frac{5}{3}\sqrt{6}$ ft-lb **33.** $\frac{2}{5}$

EXERCISES 17.4 (Page 909)

1. $x + 2y - 3z + 1 = 0$ **3.** $y - z + 3 = 0$ **5.** $x - 3y - 4z - 3 = 0$ **7.** $3x + 2y + 6z = 23$ **9.** $\langle \frac{2}{3}, -\frac{1}{3}, \frac{2}{3} \rangle$; $\langle -\frac{2}{3}, \frac{1}{3}, -\frac{2}{3} \rangle$ **11.** $\langle \frac{4}{13}, \frac{3}{13}, -\frac{12}{13} \rangle; \langle -\frac{4}{13}, -\frac{3}{13}, \frac{12}{13} \rangle$ **13.** $\left\langle \frac{3}{\sqrt{13}}, 0, \frac{2}{\sqrt{13}} \right\rangle; \left\langle -\frac{3}{\sqrt{13}}, 0, -\frac{2}{\sqrt{13}} \right\rangle$ **15.** $5x - 3y + 7z + 14 = 0$
17. $2x - y - z + 1 = 0$ **19.** $4y - 3z - 1 = 0$ and $z = 1$ **21.** $\frac{8}{21}$ **23.** $-\frac{16}{45}$ **25.** $\frac{16}{15}\sqrt{6}$ **27.** $\frac{3}{2}$

EXERCISES 17.5 (Page 914)

1. $x = 1 + 4t, y = 2 - 3t, z = 1; \frac{x - 1}{4} = \frac{y - 2}{-3}, z = 1$ **3.** $x = 13t, y = -12t, z = -8t; \frac{x}{13} = \frac{y}{-12} = \frac{z}{-8}$ **5.** $x = -2 + t$, $y = -8t, z = 3 + \frac{13}{2}t; \frac{x + 2}{2} = \frac{y}{-16} = \frac{z - 3}{13}$ **7.** $x = 4 + t, y = -5 + 3t, z = 20 - 6t; \frac{x - 4}{1} = \frac{y + 5}{3} = \frac{z - 20}{-6}$ **9.** $\frac{x - \frac{1}{7}}{2} = \frac{y}{7}$ $= \frac{z - \frac{10}{7}}{13}$ **13.** $8x - y - 66 = 0; 13x - 5z - 102 = 0; 13y - 40z + 42 = 0$ **15.** $4x + y + 3 = 0; 3x - z + 4 = 0; 3y + 4z - 7 = 0$
17. $\frac{5}{18}\sqrt{6}$ **19.** $4x + 7y - 3z + 7 = 0$ **21.** $4x + 2y - 3z + 5 = 0$ **23.** $(\frac{5}{3}, -\frac{17}{6}, \frac{5}{12})$ **25.** $\frac{x - 3}{1} = \frac{y - 6}{2} = \frac{z - 4}{1}$ **27.** $\frac{2}{5}\sqrt{70}$

EXERCISES 17.6 (Page 924)

1. $\langle 7, 13, -11 \rangle$ **3.** −490 **11.** $\langle 9, -1, -23 \rangle$ **19.** $\frac{2}{3}\sqrt{2}$ **21.** $\sqrt{89}$ sq units **23.** $9\sqrt{29}$ sq units **25.** $5x - 2y + 7z = 0$ **29.** $\frac{1}{\sqrt{3}}(\mathbf{i} + \mathbf{j} - \mathbf{k})$ **31.** $\frac{1}{\sqrt{6}}\mathbf{i} + \frac{2}{\sqrt{6}}\mathbf{j} + \frac{1}{\sqrt{6}}\mathbf{k}$ **33.** 20 cu units **35.** $\frac{38}{3\sqrt{78}}$

EXERCISES 17.7 (Page 930)

13. $x^2 + z^2 = 4y$ **15.** $x^2 + 4y^2 + 4z^2 = 16$ **17.** $y^2 + z^2 = \sin^2 x$ **19.** $x^2 - z^2 = 4$; z axis **21.** $z = \sqrt{|y|}$; y axis
23. $y^2 = 9x^2$; y axis

EXERCISES 17.8 (Page 936)

1. elliptic hyperboloid of one sheet **3.** hyperbolic paraboloid **5.** hyperbolic cylinder **7.** ellipsoid **9.** elliptic hyperboloid of one sheet **11.** elliptic cone **13.** elliptic paraboloid **15.** hyperbolic paraboloid **17.** elliptic hyperboloid of two sheets **19.** (a) $1 < |k| < \sqrt{2}$; (b) $|k| < 1$ **21.** vertex: $(1, -\frac{1}{3}, 0)$; focus: $(1, 0, 0)$ **25.** 8π cu units **27.** $\dfrac{abh^2}{2c}\pi$ cu units

EXERCISES 17.9 (Page 943)

1. $\dfrac{1}{\sqrt{4t^2+5}}(\mathbf{i} - 2t\mathbf{j} - 2\mathbf{k})$ **3.** $\mathbf{T}(t) = \frac{1}{3}\sqrt{3}[(\cos t - \sin t)\mathbf{i} + (\cos t + \sin t)\mathbf{j} + \mathbf{k}]$

5. $\dfrac{1}{\sqrt{4t^2+29}}[2(\cos t - t\sin t)\mathbf{i} + 5\mathbf{j} + 2(\sin t + t\cos t)\mathbf{k}]$ **7.** $\sqrt{21} + \frac{3}{2} + \frac{5}{4}\ln(4 + \sqrt{21})$ **9.** $\sqrt{3}(e^3 - 1)$ **11.** 13 **15.** $\mathbf{R}(t) = t\mathbf{i} + e^t\mathbf{j} + te^t\mathbf{k}$ **17.** $-\frac{3}{5}$ **19.** $\mathbf{T}(1) = \frac{1}{14}\sqrt{14}\mathbf{i} + \frac{1}{7}\sqrt{14}\mathbf{j} + \frac{3}{14}\sqrt{14}\mathbf{k}$; $\mathbf{N}(1) = -\frac{11}{266}\sqrt{266}\mathbf{i} - \frac{4}{133}\sqrt{266}\mathbf{j} + \frac{9}{266}\sqrt{266}\mathbf{k}$; $\mathbf{B}(1) = \dfrac{\sqrt{19}}{19}(3\mathbf{i} - 3\mathbf{j} + \mathbf{k})$; $K(1) = \frac{1}{98}\sqrt{266}$ **21.** $\mathbf{T}(-1) = \frac{1}{3}(\mathbf{i} + 2\mathbf{j} - 2\mathbf{k})$; $\mathbf{N}(-1) = \frac{2}{15}\sqrt{5}\mathbf{i} - \frac{1}{3}\sqrt{5}\mathbf{j} - \frac{4}{15}\sqrt{5}\mathbf{k}$; $\mathbf{B}(-1) = -\frac{2}{5}\sqrt{5}\mathbf{i} - \frac{1}{5}\sqrt{5}\mathbf{k}$; $K(-1) = \frac{2}{27}\sqrt{5}$ **23.** $\mathbf{T}(0) = \frac{1}{3}\sqrt{3}(\mathbf{i} + \mathbf{j} + \mathbf{k})$; $\mathbf{N}(0) = -\frac{1}{2}\sqrt{2}(\mathbf{i} - \mathbf{j})$; $\mathbf{B}(0) = -\frac{1}{6}\sqrt{6}(\mathbf{i} + \mathbf{j} - 2\mathbf{k})$; $K(0) = \frac{1}{3}\sqrt{2}$ **25.** $\mathbf{V}(\frac{1}{2}\pi) = -a\mathbf{i} + \mathbf{k}$; $\mathbf{A}(\frac{1}{2}\pi) = -a\mathbf{j}$; $|\mathbf{V}(\frac{1}{2}\pi)| = \sqrt{a^2+1}$ **27.** $\mathbf{V}(1) = 2e^2\mathbf{i} - 2e^{-2}\mathbf{j} + 3e^2\mathbf{k}$; $\mathbf{A}(1) = 4e^2\mathbf{i} + 4e^{-2}\mathbf{j} + 8e^2\mathbf{k}$; $|\mathbf{V}(1)| = \sqrt{13e^4 + 4e^{-4}}$ **31.** 2

EXERCISES 17.10 (Page 949)

1. (a) $(0, 3, 5)$; (b) $(-\frac{7}{2}, \frac{7}{2}\sqrt{3}, -4)$; (c) $(\cos 1, \sin 1, 1)$ **3.** (a) $(\sqrt{6}, \sqrt{2}, 2\sqrt{2})$; (b) $(0, 2\sqrt{3}, 2)$; (c) $(\frac{1}{2}\sqrt{3}, \frac{3}{2}, -\sqrt{3})$ **5.** (a) $(2, \frac{2}{3}\pi, -2\sqrt{3})$; (b) $(0, \frac{3}{4}\pi, -\sqrt{2})$; (c) $(\sqrt{6}, \frac{1}{3}\pi, \sqrt{6})$ **7.** ellipsoid; $r^2 + 4z^2 = 16$ **9.** elliptic paraboloid; $r^2 = 3z$ **11.** elliptic cone; $r^2\cos 2\theta = 3z^2$ **13.** sphere; $\rho^2 - 9\rho\cos\phi = 0$ **15.** right circular cylinder; $\rho^2\sin^2\phi = 9$ **17.** sphere; $\rho = 8\sin\phi\cos\theta$ **19.** (a) right circular cylinder; $x^2 + y^2 = 16$; (b) plane through z axis; $y = x$ **21.** $x^2 - y^2 = z^3$ **23.** (a) sphere; $x^2 + y^2 + z^2 = 81$; (b) plane through z axis; $x = y$; (c) cone with vertex at origin; $z = \sqrt{x^2+y^2}$ **25.** right circular cylinder; $x^2 + y^2 = 36$ **27.** $x\sqrt{x^2+y^2+z^2} = 2y$ **31.** (b) $2\pi\sqrt{a^2+1}$

REVIEW EXERCISES FOR CHAPTER 17 (Page 950)

3. the x axis **5.** the circle in the xz plane with center at the origin and radius 2 **7.** the plane perpendicular to the xy plane and intersecting the xy plane in the line $y = x$ **9.** the solid of revolution generated by revolving $y^2 = 9z$ about the z axis **11.** the solid of revolution generated by revolving $y = x$ about the x axis **13.** $\mathbf{i} + 26\mathbf{j} - 16\mathbf{k}$ **15.** -3 **17.** $7\sqrt{1270}$ **19.** $\langle 60, -40, 80\rangle$ **21.** 16 **23.** 295 **25.** $(-3, \sqrt{167}, 1), (-3, -\sqrt{167}, 1)$ **27.** $(x+2)^2 + (y+1)^2 + (z-3)^2 = 17$ **29.** $z^2 = e^{4y}$ or $x^2 = e^{4y}$; the y axis **31.** 3 **33.** (a) $\cos\alpha = -\dfrac{7}{\sqrt{78}}$, $\cos\beta = -\dfrac{5}{\sqrt{78}}$, $\cos\gamma = -\dfrac{2}{\sqrt{78}}$; (b) $-\dfrac{7}{\sqrt{78}}\mathbf{i} - \dfrac{5}{\sqrt{78}}\mathbf{j} - \dfrac{2}{\sqrt{78}}\mathbf{k}$ **35.** (a) $-\frac{4}{5}\sqrt{3}$, (b) $-\frac{28}{25}\mathbf{i} + \frac{4}{25}\mathbf{j} - \frac{4}{5}\mathbf{k}$ **37.** $x - 6y - 10z + 23 = 0$ **39.** $\frac{47}{16}\sqrt{2}$ **41.** 3 **43.** $\frac{1}{3}\sqrt{3}$ **45.** $\dfrac{x}{4} = \dfrac{y}{-3} = \dfrac{z}{1}$; $x = 4t$, $y = -3t$, $z = t$ **49.** $\frac{54}{25}\pi$ sq units **51.** 24 cu units **53.** $\frac{25}{4}[\sqrt{2} + \ln(\sqrt{2}+1)]$ **55.** $\mathbf{V}(\frac{1}{2}\pi) = -\frac{1}{2}\pi\mathbf{i} + \mathbf{j} + \mathbf{k}$; $\mathbf{A}(\frac{1}{2}\pi) = -2\mathbf{i} - \frac{1}{2}\pi\mathbf{j}$; $|\mathbf{V}(\frac{1}{2}\pi)| = \frac{1}{2}\sqrt{8+\pi^2}$ **57.** $\mathbf{T}(\frac{1}{2}\pi) = -\dfrac{3}{\sqrt{13}}\mathbf{i} - \dfrac{2}{\sqrt{13}}\mathbf{j}$; $\mathbf{N}(\frac{1}{2}\pi) = -\mathbf{k}$; $\mathbf{B}(\frac{1}{2}\pi) = \dfrac{2}{\sqrt{13}}\mathbf{i} - \dfrac{3}{\sqrt{13}}\mathbf{j}$ **59.** $(\frac{3}{2}\sqrt{3}, \pi, \frac{3}{2})$ **61.** (a) $z = r^2(1 + \sin 2\theta) + 1$; (b) $r^2(25\cos^2\theta + 4\sin^2\theta) = 100$

EXERCISES 18.1 (Page 960)

1. (a) $-\frac{1}{7}$; (b) $\dfrac{x^2+y^2}{x^2-y^2}$; (c) $\dfrac{x^2+2xy+y^2}{x^2-2xy+y^2}$; (d) 0; (e) the set of all points (x, y) in R^2 except those on the line $x = y$; (f) $(-\infty, +\infty)$ **3.** domain: set of all points (x, y) in R^2 interior to and on the circumference of the circle $x^2 + y^2 = 25$; range: $[0, 5]$ **5.** domain: set of all points (x, y) in R^2 interior to and on the circumference of the circle $x^2 + y^2 = 25$ except those on the line $x = 0$; range: $(-\infty, +\infty)$ **7.** domain: set of all points (x, y) in R^2 interior to and on the circumference of the circle $x^2 + y^2 = 25$ as well as all points on the y axis; range: $[0, 5]$ **9.** domain: set of all points (x, y) in R^2 interior to the circle $x^2 + y^2 = 25$ and all points on the y axis except $(0, 5)$ and $(0, -5)$; range: $(-\infty, +\infty)$ **11.** domain: set of all points (x, y) in R^2 except those on the line $x = y$; range: $(-\infty, +\infty)$ **13.** domain: set of all points (x, y) in R^2 except those on the x axis; range: $(-\infty, +\infty)$ **15.** domain: set of all points (x, y) in R^2 for which $xy > 1$; range: $(-\infty, +\infty)$ **17.** domain: set of all points (x, y, z) in R^3 on and above the plane $z = 2$; range: $(-\infty, +\infty)$ **19.** domain: set of all points (x, y, z) in R^3 for which $|x| \leq 1$ and $|y| \leq 1$; range: $(-\pi, 2\pi)$ **21.** domain: set of all points (x, y) in R^2; range: $[0, +\infty)$ **23.** domain: set of all points (x, y) in R^2; range: $(-\infty, 16]$ **25.** domain: set of all points (x, y) in R^2 for which $x + y^2 \leq 10$; range: $[0, +\infty)$ **33.** $h(x, y) = \sin^{-1}\sqrt{1 - x^2 - y^2}h$; domain: set of all points (x, y) in R^2 interior to and on the circle $x^2 + y^2 = 1$ **35.** (a) 2; (b) 6; (c) $\sqrt{x} - y^2$; (d) $|x - y|$; (e) $|x - y|$

EXERCISES 18.2 (Page 972)

1. 0 **3.** 12 **5.** 2 **7.** $\frac{1}{2}$ **9.** $\delta = \frac{1}{7}\epsilon$ **11.** $\delta = \frac{1}{5}\epsilon$ **13.** $\delta = \min(1, \frac{1}{6}\epsilon)$ **15.** $\delta = \min(1, \frac{1}{8}\epsilon)$ **23.** 0 **25.** 0 **27.** $\frac{1}{4}\pi$ **29.** $\frac{1}{2}$ **31.** limit exists and is 0; take $\delta = \sqrt{\epsilon}$ **33.** limit does not exist **35.** limit exists and is 0; take $\delta = \frac{1}{2}\epsilon$ **39.** $\frac{1}{2}$ **41.** $\frac{1}{2}$

EXERCISES 18.3 (Page 978)

1. continuous at every point (x, y) in R^2 that is not on the line $y = 1$ **3.** continuous at every point (x, y) in R^2 that is not on the y axis **5.** continuous at every point (x, y) in R^2 that is not on the line $y = 2x$ **7.** continuous at every point (x, y) in R^2 that is interior to the circle $x^2 + y^2 = 25$ **9.** continuous at every point in R^2 **11.** continuous at every point $(x, y) \neq (0, 0)$ in R^2 **13.** continuous at every point in R^2 **17.** all points (x, y) in R^2 that are interior to the circle $x^2 + y^2 = 16$ **19.** all points (x, y) in R^2 that are exterior to the ellipse $4x^2 + 9y^2 = 36$ **21.** all points (x, y) in R^2 for which $|xy| \geq 1$ **23.** all points (x, y) in R^2 in the first and third quadrants for which $|x + y| \leq 1$ **25.** all points in R^2 **27.** removable; $f(0, 0) = 0$ **29.** essential **31.** removable; $f(0, 0) = 0$ **35.** continuous at every point (x, y, z) in R^3 that is exterior to the sphere $x^2 + y^2 + z^2 = 1$ **37.** continuous at all points in R^3

EXERCISES 18.4 (Page 986)

1. 6 **3.** $3x - 2y$ **5.** $\dfrac{x}{\sqrt{x^2 + y^2}}$ **7.** $x^2 - 6xy + 2z$ **9.** $xy + yt + zt$ **11.** 4 **13.** $\dfrac{x}{\sqrt{x^2 + y^2}}$ **15.** $-2 \sin 3\theta \sin 2\phi$ **17.** $\dfrac{e^{y/x}}{xy}\left(y \ln \dfrac{x^2}{y} - x\right)$ **19.** $\dfrac{-z}{(x^2 + y^2 + z^2)^{3/2}}$ **21.** $4xy + \dfrac{1}{z}$ **23.** $xze^{xyz} + \dfrac{3xz^2}{z^4 + 9x^2y^2}$ **25.** (a) -1; (b) 12 **27.** $-\ln \sin x$; $\ln \sin y$ **31.** (a) 1; (b) 1 **33.** (a) -2; (b) 0 **35.** 4 **37.** -4 deg/cm; -8 deg/cm **39.** (a) $\dfrac{100}{i^2}\left[\dfrac{9i + 1}{(1 + i)^9} - 1\right]$; (b) $\dfrac{1}{0.0036}\left[\dfrac{1.54}{(1.06)^9} - 1\right] \approx -24.4$; (c) $\dfrac{5000 \ln 1.06}{3(1.06)^t}$; (d) $-\dfrac{5000 \ln 1.06}{3(1.06)^8} \approx -61$

EXERCISES 18.5 (Page 999)

1. (a) $3(\Delta x)^2 + 2(\Delta x)(\Delta y) - (\Delta y)^2 + 14\,\Delta x - 6\,\Delta y$; (b) 0.5411; (c) $14\,\Delta x - 6\,\Delta y$; (d) 0.54 **3.** (a) $(2 + \Delta x)(-4 + \Delta y)e^{(2+\Delta x)(-4+\Delta y)} + 8e^{-8}$; (b) -0.0026; (c) $28e^{-8}\,\Delta x - 14e^{-8}\,\Delta y$; (d) -0.0019 **5.** (a) $\Delta x + 4\,\Delta y + (\Delta x)(\Delta y) + \ln(1 + \Delta y) + \ln(5 + \Delta z) - \ln 5$; (b) 0.2141; (c) $\Delta x + 5\Delta y + \frac{1}{5}\Delta z$; (d) 0.214 **7.** $(12x^2 - y^2)\,dx + (3 - 2xy)\,dy$ **9.** $(\cos y - y \cos x)\,dx + (-x \sin y - \sin x)\,dy$ **11.** $\dfrac{2x\,dx + 2y\,dy + 2z\,dz}{x^2 + y^2 + z^2}$ **13.** $\tan^{-1} z\,dx - \dfrac{2y}{z}\,dy + \left(\dfrac{x}{1 + z^2} + \dfrac{y^2}{z^2}\right)dz$ **15.** (a) $2(x_0y_0 - y_0)\,\Delta x + (x_0^2 - 2x_0)\,\Delta y + (y_0\,\Delta x + \Delta x\,\Delta y)\,\Delta x + 2(x_0\,\Delta x - \Delta x)\,\Delta y$; (b) $\epsilon_1 = y_0\,\Delta x + \Delta x\,\Delta y$; $\epsilon_2 = 2(x_0\,\Delta x - \Delta x)$ **17.** (a) $\dfrac{2x_0y_0\,\Delta x + y_0(\Delta x)^2 - x_0^2\,\Delta y}{y_0^2 + y_0\,\Delta y}$; (b) $\epsilon_1 = \dfrac{y_0^2\,\Delta x}{y_0^3 + y_0^2\,\Delta y}$; $\epsilon_2 = \dfrac{x_0^2\,\Delta y - 2x_0y_0\,\Delta x}{y_0^3 + y_0^2\,\Delta y}$ **25.** $\xi_1 = \frac{1}{3}\sqrt{156}$ **27.** $\xi_2 = \sqrt{5} - 1$ **31.** (a) $(y_0 - z_0)\,\Delta x + x_0\,\Delta y + (2z_0 - x_0)\,\Delta z - (\Delta z)(\Delta x) + (\Delta x)(\Delta y) + (\Delta z)(\Delta z)$; (b) $\epsilon_1 = -\Delta z$, $\epsilon_2 = \Delta x$, $\epsilon_3 = \Delta z$ **35.** 7.36 m³ **37.** 0.14 cm; 1.4% **39.** $\frac{13}{1600}$; 0.325% **41.** \$2400 **43.**

$$D_1 f(x, y) = \begin{cases} 2x \sin \dfrac{1}{\sqrt{x^2 + y^2}} - \dfrac{x}{\sqrt{x^2 + y^2}} \cos \dfrac{1}{\sqrt{x^2 + y^2}} & \text{if } (x, y) \neq (0, 0) \\ 0 & \text{if } xy = (0, 0) \end{cases};$$

$$D_2 f(x, y) = \begin{cases} 2y \sin \dfrac{1}{\sqrt{x^2 + y^2}} - \dfrac{y}{\sqrt{x^2 + y^2}} \cos \dfrac{1}{\sqrt{x^2 + y^2}} & \text{if } (x, y) \neq (0, 0) \\ 0 & \text{if } (x, y) = (0, 0) \end{cases}$$

EXERCISES 18.6 (Page 1007)

1. $\dfrac{\partial u}{\partial r}$: (a) $6x - 2y$; (b) $16r - 10s$; $\dfrac{\partial u}{\partial s}$: (a) $-2x - 4y$; (b) $-10r - 6s$ **3.** $\dfrac{\partial u}{\partial r}$: (a) $13x - 2y + 5$; (b) $24r - 41s + 5$; $\dfrac{\partial u}{\partial s}$: (a) $-17x - 7y - 10$; (b) $-41r + 44s - 10$ **5.** $\dfrac{\partial u}{\partial r}$: (a) $\dfrac{2e^{y/x}}{x^2}(2x \sin t - y \cos t)$; (b) 0; $\dfrac{\partial u}{\partial t}$: (a) $\dfrac{2re^{y/x}}{x^2}(y \sin t + 2x \cos t)$; (b) $2e^{2\tan t} \sec^2 t$ **7.** $\dfrac{\partial u}{\partial r} = 2r(2x + y) + 3x$; $\dfrac{\partial u}{\partial s} = 2s(2x + y) - 2x$ **9.** $\dfrac{\partial u}{\partial r} = \dfrac{6re^s + s \cos rs}{\sqrt{1 - (3x + y)^2}}$; $\dfrac{\partial u}{\partial s} = \dfrac{3r^2e^s + r \cos rs}{\sqrt{1 - (3x + y)^2}}$ **11.** $\dfrac{\partial u}{\partial r} = \dfrac{6s}{x^2} \sinh \dfrac{y}{x}(xe^r - ry)$; $\dfrac{\partial u}{\partial s} = \dfrac{3}{x^2} \sinh \dfrac{y}{x}(2xe^r - yr^2) = 0$ **13.** $\dfrac{\partial u}{\partial r} = 2x \sin \phi \cos \theta + 2y \sin \phi \sin \theta + 2z \cos \phi$; $\dfrac{\partial u}{\partial \phi} = 2xr \cos \phi \cos \theta + 2yr \cos \phi \sin \theta - 2zr \sin \phi$; $\dfrac{\partial u}{\partial \theta} = -2xr \sin \phi \sin \theta + 2yr \sin \phi \cos \theta$ **15.** (a) $e^x(\cos t - y \sin t) + e^y(x \cos t - \sin t)$;

(b) $e^{\cos t}(\cos t - \sin^2 t) + e^{\sin t}(\cos^2 t - \sin t)$ **17.** (a) $\dfrac{x \sec^2 t - y \sin t + z \cos t}{\sqrt{x^2 + y^2 + z^2}}$; (b) $\tan t \sec t$ **19.** $\dfrac{txe^t - y}{t(x^2 + y^2)}$

21. $\dfrac{x + y + 2t + ty - tx}{t(y + t)^2}$ **23.** $\dfrac{\partial z}{\partial x} = \dfrac{3y - 6x - 4z}{2z + 4x}$; $\dfrac{\partial z}{\partial y} = \dfrac{3x - 2y}{2z + 4x}$ **25.** $\dfrac{\partial z}{\partial x} = -\dfrac{z}{x}$; $\dfrac{\partial z}{\partial y} = \dfrac{xyz + 1}{3xy \tan 3xz - xy^2}$

31. $\dfrac{\partial u}{\partial v} = \cos w \sinh v \dfrac{\partial u}{\partial x} + \sin w \cosh v \dfrac{\partial u}{\partial y}$; $\dfrac{\partial u}{\partial w} = -\sin w \cosh v \dfrac{\partial u}{\partial x} + \cos w \sinh v \dfrac{\partial u}{\partial y}$ **33.** decreasing at a rate of $\frac{8}{61}$ rad/min **35.** increasing at a rate of 3840π cm³/min **37.** increasing at a rate of 16 cm³/min

EXERCISES 18.7 (Page 1015)

1. (a) $\dfrac{2}{y} - \dfrac{6y}{x^4}$; (b) $\dfrac{2x^2}{y^3}$ **3.** (a) $4e^{2x} \sin y$; (b) $-e^{2x} \sin y$ **5.** (a) $2 \tan^{-1} \dfrac{y}{x} - \dfrac{2xy}{x^2 + y^2}$; (b) $2 \tan^{-1} \dfrac{y}{x} + \dfrac{2xy}{x^2 + y^2}$ **7.** (a) $3y \cosh x$; (b) $4x \sinh y$ **9.** (a) $e^x \cos y - \dfrac{2x \ln y}{(1 + x^2)^2}$; (b) $-e^x \cos y - \dfrac{\tan^{-1} x}{y^2}$ **11.** (a) $12x + 20y$; (b) $12x + 20y$ **13.** (a) 0; (b) e^y

15. (a) $-2e^z \sin e^z$; (b) $-2we^z(\sin e^z + e^z \cos e^z)$ **17.** (a) $\dfrac{-320rst}{(r^2 + 4s^2 - 5t^2)^3}$; (b) $\dfrac{16r(5t^2 + 12s^2 - r^2)}{(r^2 + 4s^2 - 5t^2)^3}$ **25.** neither exist

27. $f_{12}(0, 0) = -1$; $f_{21}(0, 0) = 1$ **31.** $6se^{r-s}(2 + r) - 8e^{-2s}$ **33.** $10 \cos^2 \theta + 8$ **35.** $-10r \sin 2\theta$

REVIEW EXERCISES FOR CHAPTER 18 (Page 1017)

1. $\dfrac{2x}{3y^2}$; $\dfrac{y - 2x^2}{3y^3}$; $-\dfrac{4x}{3y^3}$ **3.** $t^2 \cos st^2 + te^s$; $2st \cos st^2 + e^s$; $2t(\cos st^2 - st^2 \sin st^2) + e^s$ **5.** $\dfrac{1}{uw^2}$; $-\dfrac{2}{uw^3}$; $\dfrac{2}{u^2w^3}$ **7.** (a) $\dfrac{\partial u}{\partial t} = \dfrac{6y(x + y)}{x^2 + y^2} + 3 \ln(x^2 + y^2)$; $\dfrac{\partial u}{\partial s} = \dfrac{4y(x - y)}{x^2 + y^2} - 2 \ln(x^2 + y^2)$; (b) $\dfrac{\partial u}{\partial t} = (3t - 2s)\dfrac{18t}{4s^2 + 9t^2} + 3 \ln(8s^2 + 18t^2)$; $\dfrac{\partial u}{\partial s} = (3t - 2s)\dfrac{8s}{4s^2 + 9t^2} - 2 \ln(8s^2 + 18t^2)$ **9.** (a) $18xyse^{3rs} + 6yse^{3rs} + 9x^2r^2s^2 + 6xr^2s^2 - 9zr^2s^2$; (b) $6r^3s^3e^{3rs}(3e^{3rs} + 1) + 3r^2s^2(3e^{6rs} + 2e^{3rs} - 3 \ln 4)$

11. (a) $3x \cos t - 4(y + 2x) \sin t$; (b) $12 \cos 2t - 16 \sin 2t$; $\left.\dfrac{du}{dt}\right]_{t=\pi/4} = -16$ **13.** (a) $3(-1 + \Delta x)(3 + \Delta y)^2 - 5(-1 + \Delta x)(2 + \Delta z)^2 - 2(-1 + \Delta x)(3 + \Delta y)(2 + \Delta z) - 5$; (b) -0.47; (c) $-5\,\Delta x - 14\,\Delta y + 26\,\Delta z$; (d) -0.48 **15.** domain: set of all points (x, y) in R^2 such that $|x| \ge |y|$; range: $[0, \frac{1}{2}\pi)$ **17.** domain: all points (x, y) in R^2 interior to the ellipse $4x^2 + y^2 = 36$ and all points (x, y) in R^2 on the y axis; range: $(-\infty, +\infty)$ **19.** domain: all points (x, y, z) in R^3 such that $z > 0$; range: $(-\infty, +\infty)$ **21.** domain: all points (x, y, z) in R^3 except those for which $y = \pm z$; range: $(-\infty, +\infty)$ **23.** (a) -2; (b) $\frac{5}{3}$; (c) $\frac{2}{3}y(x^2 + 1)$; (d) $\dfrac{4x^2 + 9y^2}{4x^2}$ **25.** $\frac{1}{6}\pi$

27. $\delta = \min(1, \frac{1}{35}\epsilon)$ **29.** limit exists and is 0; take $\delta = \sqrt[4]{\epsilon}$ **31.** limit does not exist **33.** continuous at all points (x, y) in R^2 not on the lines $x = \pm 2y$ **35.** continuous at all points (x, y) in R^2 except $(x, y) = (2n + 1, 2m + 1)$, where n and m are any integers **37.** continuous at all points (x, y) in R^2 **39.** (a) 12; (b) -3 **43.** 39¢ **45.** -3200π cm³/min **47.** $-\frac{201}{1280}$ nt/cm² per min

EXERCISES 19.1 (Page 1028)

1. $2\sqrt{2}x + 5\sqrt{2}y$ **3.** $3x + \sqrt{2}y + 4z$ **5.** $\dfrac{17}{13(x - y)^2}$ **7.** $(8x - 3y)\mathbf{i} + (2y - 3x)\mathbf{j}$ **9.** $\dfrac{x}{x^2 + y^2}\mathbf{i} + \dfrac{y}{x^2 + y^2}\mathbf{j}$

11. $\dfrac{y + z}{(x + z)^2}\mathbf{i} - \dfrac{1}{x + z}\mathbf{j} - \dfrac{x - y}{(x + z)^2}\mathbf{k}$ **13.** $e^{-2y} \sec z(\mathbf{i} - 2x\mathbf{j} + x \tan z\mathbf{k})$ **15.** 6 **17.** -42

19. -2 **21.** $-3e^{\pi/4} \cos \frac{1}{12}\pi$ **23.** (a) $\langle -4, -4\rangle$; (b) $-2 - 2\sqrt{3}$ **25.** $\langle -12, 2, 14\rangle$; (b) $\frac{6}{7}$ **29.** $\frac{3}{20}\pi + \frac{2}{5}$; $\frac{1}{4}\sqrt{\pi^2 + 4}$

31. $-\frac{29}{11}$; $\sqrt{21}$ **33.** $\theta = \tan^{-1} \dfrac{3}{3\pi + 1}$ **35.** (a) $6\sqrt{2}$ degrees per meter; (b) direction of the vector $\dfrac{1}{\sqrt{2}}\mathbf{i} + \dfrac{1}{\sqrt{2}}\mathbf{j}$ **37.** (a) -1; (b) direction of the vector $-\mathbf{j}$ and magnitude 2

EXERCISES 19.2 (Page 1033)

1. $2x - 2y + 3z = 17$; $\dfrac{x - 2}{2} = \dfrac{y + 2}{-2} = \dfrac{z - 3}{3}$ **3.** $4x + 8y + 3z + 22 = 0$; $\dfrac{x + 2}{4} = \dfrac{y + 4}{8} = \dfrac{z - 6}{3}$ **5.** $ex - y = 0$; $\dfrac{x - 1}{-e} = \dfrac{y - e}{1}, z = 0$ **7.** $x - y - 3 = 0$; $\dfrac{x - 6}{1} = \dfrac{y - 3}{-1}, z = 3$ **9.** $x + 2y + 2z - 8 = 0$; $\dfrac{x - 4}{1} = \dfrac{y - 1}{2} = \dfrac{z - 1}{2}$

11. $3x - 2y - 6z + 84 = 0$; $\dfrac{x + 8}{-3} = \dfrac{y - 27}{2} = \dfrac{z - 1}{6}$ **13.** $\dfrac{x - 2}{4} = \dfrac{y + 2}{-1} = \dfrac{z}{20}$ **15.** $x = 4, y = 16$ **17.** $\dfrac{x}{1 - 8\pi} = \dfrac{y - 2}{-2\pi} = \dfrac{z - 1}{-1}$ **19.** surfaces are tangent

EXERCISES 19.3 (Page 1045)

1. $f(4, -\frac{1}{2}) = -\frac{133}{4}$, rel min **3.** $f(-\frac{1}{4}, 16) = -12$, rel max **5.** no relative extrema **7.** no relative extrema **9.** $f(1, 1) = -5$, rel min; $f(-1, -3) = 31$, rel max **11.** $f(\frac{1}{3}\pi, \frac{1}{3}\pi) = \frac{3}{2}\sqrt{3}$, rel max; $f(\frac{5}{3}\pi, \frac{5}{3}\pi) = -\frac{3}{2}\sqrt{3}$, rel min **13.** 8, 8, 8 **15.** $l:w:h = 1:1:\frac{1}{2}$ **17.** $\frac{16}{3}\sqrt{3}$ cu units **19.** $(\frac{41}{14}, -\frac{5}{7}, \frac{33}{14})$ **21.** $\left(0, \dfrac{1}{\sqrt{17}}, -\dfrac{4}{\sqrt{17}}\right)$ and $\left(0, -\dfrac{1}{\sqrt{17}}, \dfrac{4}{\sqrt{17}}\right)$; 1 **23.** length of the base is $2\frac{2}{3}$ ft; width of the base is 2 ft; depth is 3 ft **25.** 3 machine-hours and 9 person-hours **27.** \$2000

EXERCISES 19.4 (Page 1052)

1. (0, 0) and (0, 4) **3.** $(\frac{6}{7}, -\frac{4}{7}, \frac{2}{7})$ **5.** $f(0, 3) = f(\pm\frac{1}{2}\sqrt{35}, \frac{1}{2}) = 3$, rel max; $f(0, -3) = -3$, rel min **7.** $f(-\frac{2}{3}\sqrt{3}, -\frac{1}{3}\sqrt{6}, -\frac{1}{3}\sqrt{3}) = f(-\frac{2}{3}\sqrt{3}, \frac{1}{3}\sqrt{6}, \frac{1}{3}\sqrt{3}) = f(\frac{2}{3}\sqrt{3}, -\frac{1}{3}\sqrt{6}, \frac{1}{3}\sqrt{3}) = f(\frac{2}{3}\sqrt{3}, \frac{1}{3}\sqrt{6}, -\frac{1}{3}\sqrt{3}) = -\frac{2}{9}\sqrt{6}$, rel min; $f(\frac{2}{3}\sqrt{3}, \frac{1}{3}\sqrt{6}, \frac{1}{3}\sqrt{3}) = f(-\frac{2}{3}\sqrt{3}, -\frac{1}{3}\sqrt{6}, \frac{1}{3}\sqrt{3}) = f(-\frac{2}{3}\sqrt{3}, \frac{1}{3}\sqrt{6}, -\frac{1}{3}\sqrt{3}) = f(\frac{2}{3}\sqrt{3}, -\frac{1}{3}\sqrt{6}, -\frac{1}{3}\sqrt{3}) = \frac{2}{9}\sqrt{6}$, rel max **9.** 3 **11.** $3\sqrt{3}$ **13.** (a) 12; (b) 4; (c) 1 **15.** $\frac{4}{13}\sqrt{26}$ **17.** 2, 6 **19.** $\frac{37}{13}$ **21.** $\frac{7}{32}$ **33.** hottest at $(\pm\frac{1}{2}\sqrt{3}, -\frac{1}{2})$; coldest at $(0, \frac{1}{2})$

EXERCISES 19.5 (Page 1058)

1. $f(x, y) = 2x^2 - \frac{3}{2}y^2 + C$ **3.** $f(x, y) = 3x^2 - 5xy + 2y^3 + C$ **5.** $f(x, y) = 2x^3y^2 - 7x^2y + 3x - 8y + C$ **7.** not a gradient **9.** $f(x, y) = \dfrac{2x^2 - 2y^2 - x}{2xy^2} + C$ **11.** $f(x, y) = x^2 \cos y - x + C$ **13.** $f(x, y) = x^2 \sec 2y + C$ **15.** $f(x, y) = xe^y - x^2y + C$ **17.** $f(x, y, z) = 2xy - 5xz + 8yz + C$ **19.** $f(x, y, z) = xe^y \sin z + C$ **21.** $f(x, y, z) = 2x^2y + 3xyz - 5yz^2 - 2x + z + C$ **23.** $f(x, y, z) = xz \tan y + C$ **25.** $f(x, y, z) = e^{x+z} + e^y \ln z - e^x \ln y + C$

EXERCISES 19.6 (Page 1067)

1. $\frac{16}{3}$ **3.** $\frac{4}{3}$ **5.** $-\frac{5}{12}$ **7.** $\frac{1477}{2}$ **9.** 1 **11.** $\pi - \frac{8}{3}$ **13.** 8π **15.** $\frac{11}{6}$ **17.** $\frac{35}{2}$ **19.** $\pi(a^2 + 2a)$ **21.** 8 **23.** $\frac{4}{3}$ joules **25.** $20\frac{3}{4}$ joules **27.** $27\frac{3}{4}$ joules **29.** $(\frac{1}{16}\pi a^4 + a^2)$ joules **31.** 3 joules **33.** $(e^2 + e^4 + e^8 - 3)$ joules **35.** $2\frac{1}{2}$ joules

EXERCISES 19.7 (Page 1075)

1. $\phi(x, y) = xy + C$ **3.** $\phi(x, y) = e^x \sin y + C$ **5.** $\phi(x, y) = x^2y^2 - xy^3 + 2y + C$ **7.** $\phi(x, y, z) = \frac{1}{3}x^3 + \frac{1}{2}z^2 - xy + 3yz + C$ **9.** $\phi(x, y, z) = x^2 \cos y - yz^2 - 3x + 2z + C$ **11.** 2 **13.** e^2 **15.** -4 **17.** 15 **19.** -14 **21.** $\frac{13}{2}$ **23.** 4 **25.** 0 **27.** 3 **29.** 4

REVIEW EXERCISES FOR CHAPTER 19 (Page 1077)

1. 14 **3.** $-\frac{5}{2}\sqrt{2}$ **5.** $-\frac{8}{3}$ **7.** (a) $\frac{1}{2}\mathbf{i} + \frac{1}{2}\mathbf{j}$; (b) $\frac{1}{4}(1 + \sqrt{3})$ **9.** (a) $12\mathbf{i} + 3\mathbf{j} - 4\mathbf{k}$; (b) $\frac{13}{7}\sqrt{14}$ **11.** $f(x, y) = e^{x^2} \ln y + C$ **13.** $f(x, y, z) = \dfrac{y}{x + z} - \dfrac{1}{x} - \dfrac{2}{z} + C$ **15.** $4x + 2y + z - 12 = 0$; $\dfrac{x - 2}{4} = \dfrac{y - 1}{2} = \dfrac{z - 2}{1}$ **17.** $\dfrac{x - 1}{17} = \dfrac{y + 2}{20} = \dfrac{z - 11}{-4}$ **19.** (a) the direction of the vector $-12\mathbf{i} - 150\mathbf{j}$; (b) ascending; (c) the direction of either of the vectors: $\dfrac{25}{\sqrt{629}}\mathbf{i} - \dfrac{2}{\sqrt{629}}\mathbf{j}$ or $-\dfrac{25}{\sqrt{629}}\mathbf{i} + \dfrac{2}{\sqrt{629}}\mathbf{j}$ **21.** $f(-1, -1) = 1$, rel max **23.** $f(\pm\sqrt{5}, 0) = 10$, rel min **25.** $f(9, 11, 15) = -362$, rel max **27.** $\frac{7}{6}\sqrt{6}$ **29.** $\frac{44}{3} - 3\pi$ **31.** $\frac{27}{10}$ **33.** $\frac{1568}{15}$ joules **35.** $\frac{23}{6}$ joules **37.** $9e^2 - 1$ **39.** 13 **41.** $2\sqrt{3}$ by $\frac{2}{3}\sqrt{3}$ by $2\sqrt{3}$ **43.** (a) $-\dfrac{3\sqrt{3} + 2}{11}$ degrees per cm; (b) $\frac{2}{11}\sqrt{13}$ degrees per cm in the direction of the vector $-\dfrac{3}{\sqrt{13}}\mathbf{i} - \dfrac{2}{\sqrt{13}}\mathbf{j}$ **45.** $f(\frac{1}{3}\sqrt{3}, \frac{1}{3}\sqrt{6}, \frac{1}{3}\sqrt{6}) = \frac{2}{9}\sqrt{3}$, rel max; $f(-\frac{1}{3}\sqrt{3}, -\frac{1}{3}\sqrt{6}, -\frac{1}{3}\sqrt{6}) = -\frac{2}{9}\sqrt{3}$, rel min **47.** $\phi(x, y, z) = y^2e^{3z} + z^2 \tan x + C$ **49.** square base and a depth that is one-half the length of a side of the base **51.** $\frac{1}{2}\pi$

EXERCISES 20.1 (Page 1086)

1. 45 **3.** 50 **5.** 1368 **7.** 704 **9.** $\frac{203}{4}$ **11.** 1376 **13.** 68.6 **15.** 38.2 **17.** [0, 24] **19.** $[1, e]$ **21.** $[0, \frac{3}{2}\sqrt{3}\pi^2]$

EXERCISES 20.2 (Page 1093)

1. 42 **3.** $\frac{98}{3}$ **5.** $-\frac{49}{5}$ **7.** $\frac{1}{3}$ **9.** $\frac{1}{3}$ **11.** 45 **13.** $\frac{152}{3}$ **15.** $\frac{3}{2}\pi$ **17.** $\frac{864}{4}$ **19.** $\frac{512}{3}$ cu units **21.** $\frac{16}{3}$ cu units **23.** $\dfrac{15\pi - 32}{120}$ cu units **25.** $\frac{1}{12}$ sq units **27.** 72 sq units **29.** $\dfrac{c}{ab}\displaystyle\int_{-a}^{a}\int_{-(b/a)\sqrt{a^2-x^2}}^{(b/a)\sqrt{a^2-x^2}} \sqrt{a^2b^2 - b^2x^2 - a^2y^2}\, dy\, dx$ **31.** $\frac{337}{30}$ cu units

33. (b) $\frac{2}{3}a^3$; (c) $\frac{2}{3}\int_0^a \int_0^{\sqrt{a^2-x^2}} (2x + y)\, dx\, dy$ **35.** 0

EXERCISES 20.3 (Page 1100)

1. 12 kg; $(2, \frac{3}{2})$ **3.** $\frac{27}{2}$ kg; $(\frac{6}{5}, \frac{9}{5})$ **5.** $\frac{176}{15}k$ kg; $(\frac{35}{22}, \frac{102}{77})$ **7.** $\frac{2}{3}ka^3$ kg; $(\frac{3}{32}a(2 + \pi), \frac{3}{32}a(2 + \pi))$ **9.** $\frac{1}{4}k\pi$ kg; $\left(\frac{\pi}{2}, \frac{16}{9\pi}\right)$ **11.** $\frac{4}{3}$ kg; $(\frac{6}{5}, \frac{6}{5})$ **13.** 9ρ kg-m^2 **15.** $\frac{1}{2}\pi\rho a^4$ kg-m^2 **17.** $\frac{96}{35}\rho$ kg-m^2 **19.** (a) $\frac{144}{5}$ kg-m^2; (b) 54 kg-m^2; (c) $\frac{2}{5}\sqrt{15}$ m; (d) $\frac{414}{5}$ kg-m^2 **21.** (a) $\frac{3}{32}\pi k$ kg-m^2; (b) $\frac{1}{24}\pi(2\pi^2 - 3)k$ kg-m^2; (c) $\frac{1}{4}\sqrt{6}$ m; (d) $(\frac{1}{12}\pi^3 - \frac{1}{32}\pi)k$ kg-m^2 **23.** $\frac{1}{12}b\sqrt{6}$ ft **25.** $\frac{19{,}904}{315}k$ kg-m^2

EXERCISES 20.4 (Page 1106)

1. 6π sq units **3.** $\frac{1}{4}a^2(8 + \pi)$ sq units **5.** $(4\pi + 12\sqrt{3})$ sq units **7.** 4π cu units **9.** $\frac{128}{9}(3\pi - 4)$ cu units **11.** $\frac{7}{2}\pi$ cu units **13.** $\frac{40}{3}\pi k$ kg; $(0, \frac{21}{10})$ **15.** $\frac{22}{3}k\pi$ kg; $(-\frac{57}{44}, 0)$ **17.** $\frac{20}{3}k$ kg; $(\frac{23}{25}, \frac{531}{1280}\pi)$ **19.** $\frac{112}{3}\pi k$ kg; $(0, -\frac{24}{7})$ **21.** $\frac{5}{64}k\pi$ kg-m^2 **23.** $\frac{11}{16}k\pi a^4$ kg-m^2 **25.** $\frac{1}{4}\sqrt{2\pi}$ m **27.** $\pi e(e^8 - 1)$

EXERCISES 20.5 (Page 1113)

1. $\sqrt{6}$ sq units **3.** $2\sqrt{1633}$ sq units **5.** 9 sq units **7.** $\frac{1}{12}(135\sqrt{10} - 13\sqrt{26})$ sq units **9.** 8π sq units **11.** 12π sq units **13.** $\pi b\sqrt{a^2 + b^2}$ sq units **15.** $2\pi a^2(1 - e^{-1})$ sq units **17.** 12π sq units **19.** $\frac{72}{5}[2 + \sqrt{2}\ln(1 + \sqrt{2})]$ sq units

EXERCISES 20.6 (Page 1123)

1. -1 **3.** $-\frac{5}{12}$ **5.** -4 **7.** $-\frac{1}{2}\pi$ **9.** $-\frac{32}{15}$ **11.** $-\frac{3}{7}$ **13.** $\frac{1}{5}$ **15.** $\frac{1}{24}(5 - 4\sqrt{2})\pi$ **17.** 0 **19.** -10π **21.** $\frac{41}{70}$ **23.** $\frac{9}{2}$ sq units **25.** πa^2 sq units **27.** $\frac{3}{8}\pi a^2$ sq units **29.** $\frac{1}{3}$ sq units **31.** 24π joules **33.** 0

EXERCISES 20.7 (Page 1129)

1. $\frac{1}{10}$ **3.** $\frac{1}{2}\pi - 1$ **5.** $\frac{15}{2}$ **7.** $\frac{1}{24}$ **9.** $\frac{648}{5}$ **11.** $\frac{1}{4}$ cu units **13.** $\frac{3}{2}\pi$ cu units **15.** 4π cu units **17.** $\frac{4}{3}\pi abc$ cu units **19.** $\frac{80}{3}\rho$ kg **21.** $\frac{1}{28}k$ kg **23.** $\frac{2}{5}(2\sqrt{2} - 1)$ kg

EXERCISES 20.8 (Page 1137)

1. $\frac{1}{3}a^3$ **3.** $6\pi(e - 1)$ **5.** πa^3 **7.** $\frac{4}{3}\pi a^3$ **9.** $\frac{4}{5}a^5\pi k$ kg **11.** $1250\pi k$ slug-ft^2 **13.** $(0, 0, \frac{2}{5}a)$ **15.** $\frac{512}{75}k$ kg-m^2 **17.** $8k\pi$ kg **19.** $(0, 0, \frac{1}{2})$ **21.** $\frac{56}{9}\pi a^3 k$ slug-ft^2 **23.** 18π **25.** $\frac{1}{15}\pi(2\sqrt{2} - 1)$

REVIEW EXERCISES FOR CHAPTER 20 (Page 1138)

1. $\frac{1}{40}$ **3.** $\frac{1}{8}\pi$ **5.** $\frac{1}{8}e^4 - \frac{3}{4}e^2 + e - \frac{3}{8}$ **7.** $\frac{3}{4}\pi$ **9.** $\frac{1}{8}$ **11.** $\frac{1}{3}$ **13.** $\frac{1}{2}\pi\ln 2$ **15.** $\frac{1}{2}(1 - \cos 1)$ **17.** 12π **19.** 9π **21.** $\frac{8}{3}$ sq units **23.** $\frac{9}{2}$ sq units **25.** $\frac{1104}{35}$ cu units **27.** 18 sq units **29.** $(\frac{1}{2}\pi - \frac{2}{3})$ sq units **31.** $(2, \frac{3}{2})$ **33.** $\frac{1}{108}(3\pi - 7)$ kg **35.** $\frac{16}{3}(8 - 3\pi)$ joules **39.** $\frac{1}{64}(7e^8 + 1)$ kg-m^2 **41.** $k(\pi + \frac{8}{3})$ kg-m^2 **43.** $2k\pi$ kg-m^2; $\frac{1}{2}\sqrt{2\pi}$ m **45.** $65k\pi$ kg **47.** $\frac{32}{5}k\pi$ kg-m^2

Index

82 83 84 9 8 7 6 5 4 3 2

44. $\displaystyle\int \frac{u^2\,du}{\sqrt{a^2-u^2}} = -\frac{u}{2}\sqrt{a^2-u^2} + \frac{a^2}{2}\sin^{-1}\frac{u}{a} + C$

45. $\displaystyle\int \frac{du}{u\sqrt{a^2-u^2}} = -\frac{1}{a}\ln\left|\frac{a+\sqrt{a^2-u^2}}{u}\right| + C$

$\displaystyle = -\frac{1}{a}\cosh^{-1}\frac{a}{u} + C$

46. $\displaystyle\int \frac{du}{u^2\sqrt{a^2-u^2}} = -\frac{\sqrt{a^2-u^2}}{a^2u} + C$

47. $\displaystyle\int (a^2-u^2)^{3/2}\,du = -\frac{u}{8}(2u^2-5a^2)\sqrt{a^2-u^2} + \frac{3a^4}{8}\sin^{-1}\frac{u}{a} + C$

48. $\displaystyle\int \frac{du}{(a^2-u^2)^{3/2}} = \frac{u}{a^2\sqrt{a^2-u^2}} + C$

Forms Containing $2au - u^2$

49. $\displaystyle\int \sqrt{2au-u^2}\,du = \frac{u-a}{2}\sqrt{2au-u^2} + \frac{a^2}{2}\cos^{-1}\left(1-\frac{u}{a}\right) + C$

50. $\displaystyle\int u\sqrt{2au-u^2}\,du = \frac{2u^2-au-3a^2}{6}\sqrt{2au-u^2} + \frac{a^3}{2}\cos^{-1}\left(1-\frac{u}{a}\right) + C$

51. $\displaystyle\int \frac{\sqrt{2au-u^2}\,du}{u} = \sqrt{2au-u^2} + a\cos^{-1}\left(1-\frac{u}{a}\right) + C$

52. $\displaystyle\int \frac{\sqrt{2au-u^2}\,du}{u^2} = -\frac{2\sqrt{2au-u^2}}{u} - \cos^{-1}\left(1-\frac{u}{a}\right) + C$

53. $\displaystyle\int \frac{du}{\sqrt{2au-u^2}} = \cos^{-1}\left(1-\frac{u}{a}\right) + C$

54. $\displaystyle\int \frac{u\,du}{\sqrt{2au-u^2}} = -\sqrt{2au-u^2} + a\cos^{-1}\left(1-\frac{u}{a}\right) + C$

55. $\displaystyle\int \frac{u^2\,du}{\sqrt{2au-u^2}} = -\frac{(u+3a)}{2}\sqrt{2au-u^2} + \frac{3a^2}{2}\cos^{-1}\left(1-\frac{u}{a}\right) + C$

56. $\displaystyle\int \frac{du}{u\sqrt{2au-u^2}} = -\frac{\sqrt{2au-u^2}}{au} + C$

57. $\displaystyle\int \frac{du}{(2au-u^2)^{3/2}} = \frac{u-a}{a^2\sqrt{2au-u^2}} + C$

58. $\displaystyle\int \frac{u\,du}{(2au-u^2)^{3/2}} = \frac{u}{a\sqrt{2au-u^2}} + C$

Forms Containing Trigonometric Functions

59. $\displaystyle\int \sin u\,du = -\cos u + C$

60. $\displaystyle\int \cos u\,du = \sin u + C$

61. $\displaystyle\int \tan u\,du = \ln|\sec u| + C$

62. $\displaystyle\int \cot u\,du = \ln|\sin u| + C$

63. $\displaystyle\int \sec u\,du = \ln|\sec u + \tan u| + C = \ln|\tan(\tfrac{1}{4}\pi + \tfrac{1}{2}u)| + C$

64. $\displaystyle\int \csc u\,du = \ln|\csc u - \cot u| + C = \ln|\tan \tfrac{1}{2}u| + C$

65. $\displaystyle\int \sec^2 u\,du = \tan u + C$

66. $\displaystyle\int \csc^2 u\,du = -\cot u + C$

67. $\displaystyle\int \sec u \tan u\,du = \sec u + C$

68. $\displaystyle\int \csc u \cot u\,du = -\csc u + C$

69. $\displaystyle\int \sin^2 u\,du = \tfrac{1}{2}u - \tfrac{1}{4}\sin 2u + C$

70. $\displaystyle\int \cos^2 u\,du = \tfrac{1}{2}u + \tfrac{1}{4}\sin 2u + C$

71. $\displaystyle\int \tan^2 u\,du = \tan u - u + C$

72. $\displaystyle\int \cot^2 u\,du = -\cot u - u + C$

73. $\displaystyle\int \sin^n u\,du = -\frac{1}{n}\sin^{n-1} u\cos u + \frac{n-1}{n}\int \sin^{n-2} u\,du$

74. $\displaystyle\int \cos^n u\,du = \frac{1}{n}\cos^{n-1} u\sin u + \frac{n-1}{n}\int \cos^{n-2} u\,du$

75. $\displaystyle\int \tan^n u\,du = \frac{1}{n-1}\tan^{n-1} u - \int \tan^{n-2} u\,du$

76. $\displaystyle\int \cot^n u\,du = -\frac{1}{n-1}\cot^{n-1} u - \int \cot^{n-2} u\,du$

77. $\displaystyle\int \sec^n u\,du = \frac{1}{n-1}\sec^{n-2} u\tan u + \frac{n-2}{n-1}\int \sec^{n-2} u\,du$

78. $\displaystyle\int \csc^n u\,du = -\frac{1}{n-1}\csc^{n-2} u\cot u + \frac{n-2}{n-1}\int \csc^{n-2} u\,du$

79. $\displaystyle\int \sin mu \sin nu\,du = -\frac{\sin(m+n)u}{2(m+n)} + \frac{\sin(m-n)u}{2(m-n)} + C$

80. $\displaystyle\int \cos mu \cos nu\,du = \frac{\sin(m+n)u}{2(m+n)} + \frac{\sin(m-n)u}{2(m-n)} + C$

81. $\displaystyle\int \sin mu \cos nu\,du = -\frac{\cos(m+n)u}{2(m+n)} - \frac{\cos(m-n)u}{2(m-n)} + C$

82. $\displaystyle\int u \sin u\,du = \sin u - u\cos u + C$

83. $\displaystyle\int u \cos u\,du = \cos u + u\sin u + C$

84. $\displaystyle\int u^2 \sin u\,du = 2u\sin u + (2-u^2)\cos u + C$